2014 International Power Electronics Conference

(IPEC-Hiroshima 2014 ECCE-ASIA)

Hiroshima, Japan
18-21 May 2014

Pages 1-842

IEEE Catalog Number: CFP14CPB-POD
ISBN: 978-1-4799-2706-7

Copyright © 2014 by the Institute of Electrical and Electronic Engineers, Inc
All Rights Reserved

Copyright and Reprint Permissions: Abstracting is permitted with credit to the source. Libraries are permitted to photocopy beyond the limit of U.S. copyright law for private use of patrons those articles in this volume that carry a code at the bottom of the first page, provided the per-copy fee indicated in the code is paid through Copyright Clearance Center, 222 Rosewood Drive, Danvers, MA 01923.

For other copying, reprint or republication permission, write to IEEE Copyrights Manager, IEEE Service Center, 445 Hoes Lane, Piscataway, NJ 08854. All rights reserved.

***This publication is a representation of what appears in the IEEE Digital Libraries. Some format issues inherent in the e-media version may also appear in this print version.

IEEE Catalog Number: CFP14CPB-POD
ISBN 13:   978-1-4799-2706-7

Additional Copies of This Publication Are Available From:

Curran Associates, Inc
57 Morehouse Lane
Red Hook, NY 12571 USA
Phone: (845) 758-0400
Fax: (845) 758-2633
E-mail: curran@proceedings.com
Web: www.proceedings.com

2014 International Power Electronics Conference (IPEC-Hiroshima 2014 ECCE-ASIA)

Hiroshima, Japan
18-21 May 2014

IEEE Catalog Number: CFP14CPB-POD
ISBN: 978-1-47992-706-7

TABLE OF CONTENTS

A NOVEL CONTROL SCHEME FOR THREE-LEVEL FULL-BRIDGE CONVERTER ACHIEVING LOW THD OUTPUT VOLTAGE 66
Liu, Jilong ; Xiao, Fei ; Chen, Wei ; Yang, Guorun

PARALLEL CONNECTED THREE PHASE INVERTERS BASED ON MODULAR DESIGN AND DISTRIBUTED CONTROL 72
Xiao, Fei ; Chen, Wei ; Liu, Jilong ; Wang, Hengli

EFFICIENCY INVESTIGATIONS OF A 3KW T-TYPE INVERTER FOR SWITCHING FREQUENCIES UP TO 100 KHZ 78
Anthon, Alexander ; Zhang, Zhe ; Andersen, Michael A.E. ; Franke, Toke

MINIATURIZATION OF THE BOOST-UP TYPE ACTIVE BUFFER CIRCUIT IN A SINGLE-PHASE INVERTER 84
Watanabe, Hiroki ; Koiwa, Kazuhiro ; Itoh, Jun-ichi ; Ohnuma, Yoshiya ; Miyawaki, Satoshi

TESTING FACILITY USING LARGE CAPACITY INVERTER 92
Ishimaru, Yusuke ; Adachi, Mitsuo ; Tsukakoshi, Masahiko ; Nakamura, Ritaka ; Masuda, Hiroyuki ; Ogashi, Yoshihiro ; Tsuboi, Yuichi

PERFORMANCE EVALUATION UNDER THE ACTUAL OPERATING CONDITION OF A LARGE CAPACITY VSI INVERTER FOR STEEL MILL APPLICATIONS 97
Mamun, Mostafa ; Yoshizawa, Daisuke ; Mukunoki, Makoto

A SOFT-SWITCHING SINGLE-PHASE UNIFIED POWER QUALITY CONDITIONER 105
Jiang, Maoh-Chin ; Chang, Kai-Chi ; Lu, Kao-Yi ; Shih, Bing-Jyun ; Liu, Tai-Chun

NOVEL THREE-PHASE PWM AC-AC CONVERTERS SOLVING COMMUTATION PROBLEM 110
Khan, Ashraf Ali ; Shin, Hyunhak ; Cha, Honnyong ; Kim, Heung-Geun

EXPERIMENTAL INVESTIGATION OF NORMALLY-ON TYPE BIDIRECTIONAL SWITCH FOR INDIRECT MATRIX CONVERTERS 117
Sung, Kyungmin ; Iijima, Ryuji ; Nishizawa, Shinichi ; Norigoe, Isami ; Ohashi, Hiromichi

VISUALIZATION OF PWM WAVEFORMS OF OUTPUT VOLTAGE AND INPUT CURRENT FOR A DIRECT MATRIX CONVERTER 123
Asai, Inami ; Takeshita, Takaharu

SPACE VECTOR MODULATION BASED ON VIRTUAL INDIRECT CONTROL FOR HIGH FREQUENCY AC-LINKED MATRIX CONVERTER 130
Inoue, Keita ; Shioda, Masashi ; Katade, Motohumi ; Goto, Akira ; Morishita, Shin ; Itoh, Junichi ; Koiwa, Kazuhiro

A FUNDAMENTAL VERIFICATION OF A SINGLE-PHASE TO THREE-PHASE MATRIX CONVERTER WITH A PDM CONTROL BASED ON SPACE VECTOR MODULATION 138
Nakata, Yuki ; Itoh, Jun-ichi

STEADY STATE CHARACTERISTICS OF THE BOOST-TYPE MATRIX CONVERTER FOR STAND-ALONE POWER SOURCE 146
Nagano, Y. ; Yamamura, N. ; Ishida, M. ; Hirokado, K.

DESIGN PROCEDURE FOR OUTPUT CURRENT CONTROL AND DAMPING CONTROL OF MATRIX CONVERTER 152
Takahashi, Hiroki ; Itoh, Jun-ichi

A NOVEL LCL FILTER PARAMETER DESIGN METHOD BASING ON RESONANT FREQUENCY OPTIMIZATION OF THREE-LEVEL NPC GRID CONNECTED INVERTER 160
Li, Ning ; Wang, Yue ; Niu, Ruigen ; Guo, Wei ; Lei, Wanjun ; Wang, Zhao'An

DESIGN AND ANALYSIS OF ISOLATED BI-DIRECTIONAL DC/DC CONVERTER USING QUASI-RESONANT ZVS 166
Noh, Yong-Su ; Won, Chung-Yuen ; Oh, Min-Seok ; Jeon, Jin-Yong ; Jung, Yong-Chae

AN ACTIVE-CLAMPING ZVS FLYBACK CONVERTER WITH INTEGRATED TRANSFORMER 172
Lin, Jing-Yuan ; Lo, Yu-Kang ; Chiu, Huang-Jen ; Wang, Chao-Fu ; Lin, Chien-Yu

PFM AND PWM HYBRID CONTROLLED LLC CONVERTER 177
Yamamoto, Junichi ; Zaitsu, Toshiyuki ; Abe, Seiya ; Ninomiya, Tamotsu

DISCUSSIONS ON VARIOUS VOLTAGE EQUALIZERS FOR EDLCS USING CW CIRCUIT 183
Khant, Hlaing Kyi Pyar ; Matsui, Keiju ; Hasegawa, Masaru ; Yasubayashi, Mikio ; Umeno, Masayoshi ; Ooishi, Eiji

ISOLATION SYSTEM WITH WIRELESS POWER TRANSFER FOR MULTIPLE GATE DRIVER SUPPLIES OF A MEDIUM VOLTAGE INVERTER 191
Kusaka, Keisuke ; Orikawa, Koji ; Itoh, Jun-ichi ; Morita, Kazunori ; Hirao, Kuniaki

STUDY AND IMPLEMENTATION OF A 15-W POWER AMPLIFIER FOR PIEZOELECTRIC ACTUATOR 199
Lo, Yu-Kang ; Chiu, Huang-Jen ; Liu, Yu-Chen ; Lin, Chung-Yi ; Cheng, Shih-Jen ; Yang, CS

ISOLATED VOLTAGE-BOOSTING CONVERTER 204
Hwu, K.I. ; Jiang, W.Z. ; Shieh, Jenn-Jong

HIGH VOLTAGE CONVERSION RATIO CASCADE BOOST CONVERTER WITH DC SNUBBER 208
Lee, Yuang-Shung ; Yu, Ling-Chia ; Chou, Tzu-Han

DESIGN-ORIENTED ANALYSIS OF RESONANCE DAMPING AND HARMONIC COMPENSATION FOR LCL-FILTERED VOLTAGE SOURCE CONVERTERS 216
Wang, Xiongfei ; Blaabjerg, Frede ; Loh, Poh Chiang

STATE-SPACE AVERAGE MODELING OF BIDIRECTIONAL DC-DC CONVERTER FOR BATTERY CHARGER USING LCLC FILTER..224
Moon, Sang-Ho ; Jou, Sung-Tak ; Lee, Kyo-Beum

A NEW SVPWM STRATEGY FOR INPUT SWITCHED MULTILEVEL CONVERTER..230
Xiong, Li ; Prasanna, U.R. ; Bilal, Akin ; Rajashekara, Kaushik

ESD RELIABILITY INFLUENCE OF A 60 V POWER LDMOS BY THE FOD-BASED (& DOTTED-OD) DRAIN..236
Chen, Shen-Li ; Lee, Min-Hua

ENHANCED TRANSVERSE-FLUX MOTOR WITH TORUS COILS..240
Tanaka, Junya ; Sakai, Kazuto

THE INFLUENCE OF MAGNETIC PROPERTIES OF PERMANENT MAGNET ON THE PERFORMANCE OF IPMSM FOR AUTOMOTIVE APPLICATION..246
Yoshioka, S. ; Morimoto, S. ; Sanada, M. ; Inoue, Y.

CHARACTERISTICS OF INTERIOR PERMANENT MAGNET SYNCHRONOUS MOTOR WITH IMPERFECT MAGNETS..252
Shinagawa, Syuhei ; Ishikawa, Takeo ; Kurita, Nobuyuki

STUDY OF STATOR STRUCTURE TO IMPROVE RELUCTANCE TORQUE FOR IPMSM WITH CONCENTRATED WINDING..258
Morikawa, R. ; Sanada, M. ; Morimoto, S. ; Inoue, Y.

DEVELOPMENT AND VERIFICATION OF ENERGY-ACCURATE SIMULATION MODELS FOR PERMANENT MAGNET SYNCHRONOUS MOTORS IN AUTOMATION SYSTEMS..264
Blank, Frederic ; Roth-Stielow, Jorg

COMPARISON OF THE RESISTANCE- AND INDUCTANCE-BASED SALIENCY OF A PMSM DUE TO A SHORT-CIRCUITED ROTOR WINDING..270
Graus, Johannes ; Rambetius, Alexander ; Hahn, Ingo

DESIGN AND OPTIMIZATION OF HIGH-SPEED SWITCHED RELUCTANCE MOTOR USING SOFT MAGNETIC COMPOSITE MATERIAL..278
Gaing, Zwe-Lee ; Kuo, Kuan-Yi ; Hu, Jia-Sheng ; Hsieh, Min-Fu ; Tsai, Ming-Hsiao

INFLUENCE OF PULSE WIDTH MODULATION (PWM) ON THE IRON LOSSES OF ELECTRICAL STEEL......283
Boehm, Andreas ; Hahn, Ingo

INVESTIGATION ON IRON LOSS CHARACTERISTICS IN STAR-CONNECTION AND DELTA-CONNECTION UNDER THREE PHASE PWM INVERTER EXCITATION..289
Odawara, Shunya ; Fujisaki, Keisuke ; Fukuhara, Shuhei

OPTIMIZATION ON ARRANGEMENT OF PERMANENT MAGNETS FOR MAGNETIC LEVITATION SYSTEM FOR THIN STEEL PLATE (FUNDAMENTAL CONSIDERATION ON LEVITATION PROBABILITY)..294
Ishii, Hirotaka ; Hasegawa, Shinya ; Narita, Takayoshi ; Oshinoya, Yasuo

EFFECT OF A MAGNETIC FIELD FROM THE HORIZONTAL DIRECTION ON A MAGNETICALLY LEVITATED STEEL PLATE (FUNDAMENTAL CONSIDERATIONS ON THE SHAPE ANALYSIS OF ULTRATHIN STEEL PLATE)..299
Kurihara, Takeshi ; Hasegawa, Shinya ; Narita, Takayoshi ; Oshinoya, Yasuo

NOVEL MAGNETIC STRUCTURE OF INTEGRATED DIFFERENTIAL-MODE AND COMMON-MODE INDUCTORS TO SUPPRESS DC SATURATION..304
Umetani, Kazuhiro ; Tera, Takahiro ; Shirakawa, Kazuhiro

A NOVEL CONTROL METHOD IN FLUX-WEAKENING REGION FOR EFFICIENT OPERATION OF INTERIOR PERMANENT MAGNET SYNCHRONOUS MOTOR..312
Ueda, K. ; Morimoto, S. ; Inoue, Y. ; Sanada, M.

IMPLEMENTATION OF THE MTPA AND MTPV CONTROL WITH ONLINE PARAMETER IDENTIFICATION FOR A HIGH SPEED IPMSM USED AS TRACTION DRIVE..318
Nguyen, Quoc Khanh ; Petrich, Matthias ; Roth-Stielow, Jorg

CORRECTION OF REFERENCE FLUX FOR MTPA CONTROL IN DIRECT TORQUE CONTROLLED INTERIOR PERMANENT MAGNET SYNCHRONOUS MOTOR DRIVES..324
Shinohara, Atsushi ; Inoue, Yukinori ; Morimoto, Shigeo ; Sanada, Masayuki

VOLTAGE REGULATION AND MAXIMUM OUTPUT POWER TRACKING OF A 4.5KW PERMANENT-MAGNET SYNCHRONOUS GENERATOR..330
Chang, Yuan-Chih ; Chang, Hsiu-Feng ; Dai, Wei-Fu ; Wu, Chun-Wei

A NOVEL FLUX-WEAKENING CONTROL METHOD BASED ON SINGLE CURRENT REGULATOR FOR PERMANENT MAGNET SYNCHRONOUS MOTOR..335
Fang, Xiaocun ; Hu, Taiyuan ; Lin, Fei ; Yang, Zhongping

PREDICTIVE CURRENT CONTROL METHOD IN INDUCTION MOTOR SPEED SENSORLESS DRIVE..........341
Wei, Sun ; Yong, Yu ; Dianguo, Xu ; Jin, Xu ; Li, Ding

REAL-TIME IMPLEMENTATION OF AN ONLINE MODEL PREDICTIVE CONTROL FOR IPMSM USING PARALLEL COMPUTING ON FPGA..346
Leuer, Michael ; Bocker, Joachim

AN INTEGRAL SLIDING-MODE CONTROLLER FOR ENERGY EFFICIENCY IMPROVEMENT IN AC POWER SOURCE SUPPLIED AC MACHINE DRIVES..351
Shieh, Hsin-Jang ; Chen, Ying-Zuo

PERFORMANCE IMPROVEMENT OF ULTRA-HIGH-SPEED PMSM DRIVE SYSTEM BASED ON DTC BY USING SIC INVERTER..356
Togashi, Ryo ; Inoue, Yukinori ; Morimoto, Shigeo ; Sanada, Masayuki

MATHEMATICAL MODEL FOR HIGH-EFFICIENCY CONTROL OF PERMANENT-MAGNET SYNCHRONOUS MOTOR IN STATOR FLUX LINKAGE SYNCHRONOUS FRAME..363
Inoue, Tatsuki ; Inoue, Yukinori ; Morimoto, Shigeo ; Sanada, Masayuki

WIDE-SPEED-RANGE OPERATION OF DTC-BASED PMSM DRIVE SYSTEM USING MTPF CONTROL..........................370
Inoue, Yukinori ; Ichiya, Takahiro ; Morimoto, Shigeo ; Sanada, Masayuki

AN INDUSTRIAL LOW-VOLTAGE INVERTER FOR PRM CONTROL 376
Nakamura, M. ; Oka, T. ; Oishi, K.

OPTIMAL PULSE PATTERN DETERMINATION BASED ON PULSE HARMONIC MODULATION................................383
Furukawa, Kimihisa ; Ajima, Toshiyuki ; Miyazaki, Hideki

METHOD FOR AUTO-TUNING OF CURRENT AND SPEED CONTROLLER IN IPMSM DRIVE SYSTEM BASED ON PARAMETER IDENTIFICATION 390
Tadokoro, D. ; Morimoto, S. ; Inoue, Y. ; Sanada, M.

COMPARATIVE STUDY OF PWM STRATEGIES FOR THREE-PHASE OPEN-END WINDING INDUCTION MOTOR DRIVES 395
Zhu, B. ; Prasanna, U.R. ; Rajashekara, K. ; Kubo, H.

10MW,3.3MWH ENERGY STORAGE SYSTEM CONSISTING OF 4000 FLYWHEELS CONTROLLED BY ICT NETWORK FOR SHORT CYCLE POWER FLUCTUATION COMPENSATION 403
Kato, Koji ; Ishigma, Satoru ; Nakajima, Yoichiro ; Arai, Haruki ; Ueda, Tetsuya ; Iwata, Tetsuki ; Ito, Yoichi ; Sugao, Kazumi

VERSATILE POWER TRANSFER STRATEGIES OF PV-BATTERY HYBRID SYSTEM FOR RESIDENTIAL USE WITH ENERGY MANAGEMENT SYSTEM..409
Choi, Seong-Chon ; Sin, Min-ho ; Kim, Dong-Rak ; Won, Chung-Yuen ; Jung, Yong-Chae

HIGH-EFFICIENCY AND COST-MINIMIZATION METHOD OF ENERGY STORAGE SYSTEM WITH MULTI STORAGE DEVICES FOR GRID CONNECTION...415
Haga, Hitoshi ; Shimao, Toshihiro ; Kondo, Seiji ; Kato, Koji ; Itoh, Youichi ; Arimatsu, Kenji ; Matsuda, Katsuhiro

BIDIRECTIONAL DC-DC CONVERTER WITH MULTIPLE SWITCHED-CAPACITOR CELLS 421
Lee, Yuang-Shung ; Huang, Hsin-Wei ; Chou, Tzu-Han

SWITCHED-CAPACITOR CHARGE EQUALIZATION CIRCUIT FOR SERIES-CONNECTED BATTERIES 429
Hsieh, Yao-Ching ; Cai, Zheng-Xiu ; Wu, Wen-Zhe

PERFORMANCE ANALYSIS OF UNITL-H6 INVERTER WITH SIC MOSFETS 433
Barater, Davide ; Buticchi, Giampaolo ; Concari, Carlo ; Franceschini, Giovanni ; Gurpinar, Emre ; De, Dipankar ; Castellazzi, Alberto

MAXIMUM POWER POINT TRACKING OF GRID-TIED PHOTOVOLTAIC POWER SYSTEMS 440
Lee, Ya-Ting ; Chiu, Chian-Song ; Chiu, Tse-Wei

A NEW VOLTAGE TYPE MAGNETICALLY COUPLED T-SOURCE INVERTER..446
Tran, Q.V. ; Low, K.S.

A HIGH EFFICIENCY HYBRID 7-LEVEL INVERTER WITH SINGLE DC SOURCE 452
Yanhong, Zhang ; Kazuya, Ogura ; Oi, Kazunobu

OPTIMAL IDLING CONTROL STRATEGY FOR THREE-PORT FULL-BRIDGE CONVERTER 458
Jiang, Yongjie ; Liu, Fuxin ; Ruan, Xinbo ; Wang, Lipeng

FILTER DESIGN FOR THREE-LEVEL GRID-CONNECTED INVERTER WITH LOW SWITCHING FREQUENCY 465
Ren, Kangle ; Zhang, Xing ; Wang, Fusheng ; Tu, Yunwu ; Wang, Lingxiang ; Deng, Lirong

A NOVEL EFFICIENT T TYPE THREE LEVEL NEUTRAL-POINT-CLAMPED INVERTER FOR RENEWABLE ENERGY SYSTEM 470
Wu, Wenlong ; Wang, Fei ; Wang, Yong

A NOVEL NEUTRAL POINT VOLTAGE AUTOMATIC BALANCING CARRIER-BASED MODULATION STRATEGY OF THREE-LEVEL NPC CONVERTER 475
Li, Ning ; Wang, Yue ; Niu, Ruigen ; Guo, Wei ; Lei, Wanjun ; Wang, Zhao'An

A HIGH VOLTAGE GAIN SWITCHED-COUPLED-INDUCTOR QUASI-Z-SOURCE INVERTER................................480
Ahmed, Furqan ; Cha, Honnyong ; Kim, Su-Han ; Kim, Heung-Geun

A NOVEL CONTROL STRATEGY TO SUPPRESS DC CURRENT INJECTION TO THE GRID FOR THREE-PHASE PV INVERTER 485
Zhang, Tao ; He, Guofeng ; Chen, Min ; Xu, Dehong

CLC FILTER DESIGN OF A FLYBACK-INVERTER FOR PHOTOVOLTAIC SYSTEMS 493
Shin, Yesl ; Lee, June-Hee ; Lee, June-Seok ; Lee, Kyo-Beum

THREE-PHASE INVERTER TOPOLOGIES FOR GRID-CONNECTED PHOTOVOLTAIC SYSTEMS................................498
Ozkan, Ziya ; Hava, Ahmet M.

A THREE-PORT TOPOLOGY COMPARISON FOR A LOW POWER STAND-ALONE PHOTOVOLTAIC SYSTEM..506
Mira, Maria C. ; Knott, Arnold ; Andersen, Michael A.E.

EFFECT OF CONVENTIONAL GRID-VOLTAGE FEEDFORWARD ON THE OUTPUT IMPEDANCE OF A THREE-PHASE PHOTOVOLTAIC INVERTER..514
Messo, T. ; Jokipii, J. ; Suntio, T.

POWER AMPLIFIER SUITABLE FOR PHOTOVOLTAIC CELL BOOSTER 522
Kohama, Teruhiko ; Sogawa, Yuki ; Tsuji, Satoshi

REALIZATION STUDY OF INTERLEAVED PV MICROINVERTER BY QUADRATURE-PHASE-SHIFT SPWM CONTROL 526
Hsieh, Hung-I ; Hsieh, Guan-Cyun ; Hou, Jiaxin

CURRENT SENSORLESS MPPT METHOD FOR A PV FLYBACK MICROINVERTERS USING A DUAL-MODE 532

Lee, June-Hee ; Lee, June-Seok ; Lee, Kyo-Beum

A NOVEL METHOD OF SUPPRESSING INRUSH CURRENTS OF SQUIRREL-CAGE INDUCTION MACHINE USING MATRIX CONVERTER IN WIND POWER GENERATION SYSTEMS 538

Yamada, Hiroaki ; Hanamoto, Tsuyoshi

NONLINEAR PITCH CONTROL DESIGN FOR LOAD REDUCTION ON WIND TURBINES 543

Xiao, Shuai ; Yang, Geng ; Geng, Hua

DEVICE LOADING OF MODULAR MULTILEVEL CONVERTER MMC IN WIND POWER APPLICATION 548

Popova, L. ; Pyrhonen, J. ; Ma, K. ; Blaabjerg, F.

A NOVEL OPTIMAL DESIGN OF DFIG CROWBAR RESISTOR DURING GRID FAULTS 555

Hu, Sheng ; Zou, XuDong ; Kang, Yong

DC-VOLTAGE REGULATION OF A FIVE LEVELS NEUTRAL POINT CLAMPED CASCADED CONVERTER FOR WIND ENERGY CONVERSION SYSTEM 560

Merahi, Farid ; Mekhilef, Saad ; Berkouk, El Madjid

A REACTIVE POWER SHARING METHOD BASED ON VIRTUAL CAPACITOR IN ISLANDING MICROGRID 567

Xu, Haizhen ; Zhang, Xing ; Liu, Fang ; Shi, Rongliang ; Yu, Changzhou ; Zhao, Wei ; Yu, Yong ; Cao, Wei

STORAGE CAPACITY PERFORMANCE FOR HYBRID PV/DIESEL SYSTEM IN SABAH MALAYSIA 573

Hidayat, Nabil M ; Kari, Mat Nasir ; Mohd Arif, Mohd Johari

NEW TECHNIQUES FOR MEASURING ISLANDED MICROGRID IMPEDANCE CHARACTERISTICS BASED ON CURRENT INJECTION 577

Hou, Lixiang ; Liu, Baoquan ; Shi, Hongtao ; Yi, Hao ; Zhuo, Fang

A GENERAL FRAMEWORK TO DESIGN OPERATION MODES OF DC MICROGRIDS WITHOUT COMMUNICATION LINKS 582

Pan, Miao ; Shen, Na ; Yang, Geng ; Morita, Kazunori ; Ogura, Kazuya ; Wu, Weiyang

IMPLEMENTATION DESIGN OF THE CONVERTER-BASED GALVANIC ISOLATION FOR LOW VOLTAGE DC DISTRIBUTION 587

Mattsson, A. ; Vaisanen, V. ; Nuutinen, P. ; Kaipia, T. ; Lana, A. ; Peltoniemi, P. ; Silventoinen, P. ; Partanen, J.

PEAK DETECTION METHOD USING TWO-DELTA OPERATION FOR SINGLE VOLTAGE SAG 595

Lee, Woo-Cheol ; Lee, Taeck-Kie

LINE LOSS MINIMIZATION IN RADIAL DISTRIBUTION SYSTEM USING MULTIPLE STATCOMS AND STATIC CAPACITORS 601

Miyazaki, Kensuke ; Takeshita, Takaharu

A NOVEL CONTROL METHOD FOR INDIVIDUAL DC VOLTAGE BALANCING IN H-BRIDGE CASCADED STATCOM 609

Xu, Rong ; Yu, Yong ; Yang, Rongfeng ; Qu, Lizhi ; Sun, Wei ; Xu, Dianguo

RESEARCH ON THE CONTROL STRATEGY OF STATCOM BASED ON MODULAR MULTILEVEL CONVERTER 614

Zhang, Wei ; Gao, Qiang ; Su, Bonan ; Jin, Miaoxin ; Xu, Dianguo ; Liu, Jianyu

FAULT DIAGNOSIS IN LARGE FORMAT LIFEPO4 ESS APPLICATION THROUGH DWT-BASED MRA 619

Kim, Jonghoon

COMPARISON OF DIFFERENT IGBT BASED DESIGNS OF POWER ELECTRONIC TRANSFORMER 624

Wang, Xinyu ; Ouyang, Shaodi ; Liu, Jinjun ; Meng, Fei ; Javed, Riffat

SEMI-ADAPTIVE HARMONIC CONTROL FOR POWER BALANCING DEVICE FOR AC TRACTION 629

Akagi, Masataka ; Tsuruta, Hironori ; Oso, Hiroshi

RESEARCH OF EFFICIENT MAIN POWER EQUIPMENT USING SIC POWER DEVICE 634

Shinbo, Mitsuo ; Sonoda, Hideki ; Ishida, Takahito ; Abiko, Hiroshi ; Shibanuma, Kenichi ; Chiba, Yoshinori

A HIGH PERFORMANCE CONTROL STRATEGY FOR THREE-LEVEL NPC EMU CONVERTERS 640

Song Kejian ; Wu Mingli ; Wang Hui ; Agelidis, Vassilios Georgios

A DESIGN OF INRUSH CURRENT IDENTIFICATION SYSTEM FOR HIGH-SPEED TRAIN'S TRACTION TRANSFORMER 647

Yu, Weikai ; Liu, Xiankai ; Zhang, Yuzhuo ; Cao, Yuan ; Ma, Weigang ; Hei, Xinhong ; Huang, Zhenhui ; Jiang, Dawang

CURRENT SOURCE INVERTER BASED CASCADED SOLID STATE TRANSFORMER FOR AC TO DC POWER CONVERSION 651

Roy, Sudhin ; De, Ankan ; Bhattacharya, Subhashish

EVALUATION OF HIGH VOLTAGE 15 KV SIC IGBT AND 10 KV SIC MOSFET FOR ZVS AND ZCS HIGH POWER DC -DC CONVERTERS 656

Moballegh, Shiva ; Madhusoodhanan, Sachin ; Bhattacharya, Subhashish

THE DIRECT YAW-MOMENT CONTROL TO FOLLOW THE NEUTRAL STEERING PATH REGARDLESS OF VELOCITY 664

Jang, Young-Jin ; Nam, Kwang-Hee

NEXT-GENERATION IGBT MODULE STRUCTURE FOR HYBRID VEHICLE WITH HIGH COOLING PERFORMANCE AND HIGH TEMPERATURE OPERATION 671

Morozumi, Akira ; Gohara, Hiromichi ; Momose, Fumihiko ; Saito, Takashi ; Nishimura, Yoshitaka ; Mochizuki, Eiji ; Takahashi, Yoshikazu

INTEGRATION OF PLUG-IN ELECTRIC VEHICLES IN POWER SYSTEMS USING CHARGING MODE SWITCHING 677

Wen-Tai Li ; Wen, Chao-Kai ; Chen, Jung-Chieh ; Teng, Jen-Hao ; Ting, Pangan

A NOVEL COMPENSATION METHOD FOR A MOTOR PHASE CURRENT SENSOR OFFSET ERROR VARIED DURING A VSI-MOTOR DRIVE 682

Tamura, Hiroshi ; Noto, Yasuo ; Ajima, Toshiyuki ; Itoh, Jun-ichi

INVESTIGATION OF CALCULATION METHOD OF LOSSES IN PWM INVERTER WITH VOLTAGE BOOSTER USING BOTH DC LINK VOLTAGE CONTROL AND FLUX WEAKENING CONTROL 689

Imakiire, Akihiro ; Hikita, Masayuki ; Yamamoto, Kichiro ; Yonemori, Ryo

DYNAMIC AND STEADY-STATE BEHAVIOR OF A PARALLELING THREE-PHASE AC-TO-DC CONVERTER WITH REDUCED DC BUS CAPACITOR 694

Kamnarn, Uthen ; Kanthaphayao, Yutthana ; Chunkag, Viboon

REACTIVE POWER LOSS OPTIMIZATION METHOD FOR BI-DIRECTIONAL ISOLATED DC-DC CONVERTERS 702

Wen, Huiqing

POWER SUPPLY FOR A WIRELESS SENSOR NETWORK: AIRLINER FLIGHT TEST CASE STUDY 707

Durand Estebe, P. ; Boitier, V. ; Bafleur, M. ; Dilhac, J-M. ; Berhouet, S.

A CONFIGURABLE THREE-PHASED INVERTER FOR TEACHING POWER ELECTRONICS 712

Kern, Ansgar

A BACHELOR-STUDENT PROJECT: BUCK-BOOST OPERATION OF AN INTEGRATED H-BRIDGE FOR VARIABLE-SPEED ENERGY STORAGE SYSTEMS USING MEASUREMENT COILS IN THE STATOR OF A DC-MACHINE 718

De Belie, Frederik ; Darba, Araz ; Melkebeek, Jan

DEVELOPMENT OF A WEB-BASED REMOTE EXPERIMENT SYSTEM FOR ELECTRICAL MACHINERY LEARNERS 724

Ishibashi, Makoto ; Fukumoto, Hisao ; Furukawa, Tatsuya ; Itoh, Hideaki ; Ohchi, Masashi

DEVELOPMENT OF POWER MEASUREMENT SYSTEM IN SIMULATED MICRO GRID SYSTEM FOR EDUCATION 730

Hira, Yuki ; Furukawa, Tatsuya ; Yakabe, Seichiro ; Fukumoto, Hisao ; Itoh, Hideaki ; Ohchi, Masashi

POWER ELECTRONIC TECHNOLOGIES FOR FLEXIBLE DC DISTRIBUTION GRIDS 736

De Doncker, Rik W.

2.5KV, 200KW BI-DIRECTIONAL ISOLATED DC/DC CONVERTER FOR MEDIUM-VOLTAGE APPLICATIONS 744

Matsuoka, Yuji ; Wada, Keiji ; Nakahara, Mizuki ; Takao, Kazuto ; Kyungmin Sung ; Ohashi, Hiromichi ; Nishizawa, Shinichi

POWER-LOSS BREAKDOWN OF A 750-V, 100-KW, 20-KHZ BIDIRECTIONAL ISOLATED DC-DC CONVERTER USING SIC-MOSFET/SBD DUAL MODULES 750

Akagi, Hirofumi ; Yamagishi, Tatsuya ; Tan, Nadia M.L. ; Kinouchi, Shin-ichi ; Miyazaki, Yuji ; Koyama, Masato

DESIGN CONSIDERATIONS OF A 15KV SIC IGBT ENABLED HIGH-FREQUENCY ISOLATED DC-DC CONVERTER 758

Tripathi, Awneesh ; Mainali, Krishna ; Patel, Dhaval ; Kadavelugu, Arun ; Hazra, Samir ; Bhattacharya, Subhashish ; Hatua, Kamalesh

COMMON-MODE CURRENTS IN MULTI-CELL SOLID-STATE TRANSFORMERS 766

Huber, Jonas E. ; Kolar, Johann W.

SINGLE-STAGE RECONFIGURABLE DC/DC CONVERTER FOR WIDE INPUT VOLTAGE RANGE OPERATION IN HEVS 774

Zeljkovic, Sandra ; Reiter, Tomas ; Gerling, Dieter

A TWO STAGE DC/DC CONVERTER WITH WIDE INPUT RANGE FOR EV 782

Peng Wen ; Changsheng Hu ; Haitao Yang ; Longlong Zhang ; Cheng Deng ; Yashun Li ; Dehong Xu

INTERMEDIATE AND LIGHT LOAD EFFICIENCY IMPROVEMENT OF A HIGH-POWER DENSITY BIDIRECTIONAL DC-DC CONVERTER IN HYBRID ELECTRIC VEHICLES WITH MR FLUID GAP INDUCTOR 790

Ahmed, Furqan ; Su-Han Kim ; Cha, Honnyong ; Kim, Dong-Hun ; Heung-Geun Kim

REGENERATIVE CONTROL OF BI-DIRECTIONAL DC-DC CONVERTER CONTROLLING VARIABLE DC-LINK FOR FCEV 796

Il-Kuen Won ; An-Yeol Ko ; Do-Yun Kim ; Chung-Yuen Won ; Young-Ryul Kim

LARGE DRIVING RANGE INCREASE OF SERIES CHOPPER BASED POWER TRAIN USING MOTOR TEST BENCH 801

Hosoyamada, Yu ; Takeda, Masashi ; Motoi, Naoki ; Kawamura, Atsuo

THE POWER ELECTRONICS PROGRAM AT BEIJING JIAOTONG UNIVERSITY 807

Fei Lin ; Zhongping Yang ; Zheng, T.Q.

EFFORTS FOR POWER ELECTRONICS EDUCATION IN A START-UP COMPANY 811

Hattori, Fumiya ; Imaoka, Jun ; Ishitobi, Manabu ; Nagai, Shinichiroh ; Yamamoto, Masayoshi

EDUCATION FOR THE ENGINEERS OF TRACTION POWER SUPPLY DIVISION IN EAST JAPAN RAILWAY COMPANY 817

Takino, Toshiaki ; Iwakami, Tetsuro

SUCCESSFUL ONLINE EDUCATION - GECKOCIRCUITS AS OPEN-SOURCE SIMULATION PLATFORM 821

Musing, Andreas ; Kolar, Johann W.

AN ELECTRIC VEHICLE PROJECT FOR ECO-RUN RACE 829

Yamagata, Shinichi ; Oda, Yoshinori ; Tanai, Masanobu ; Sung, Kyungmin

MULTI-LOOP CONTROLLER DESIGN FOR DIODE-ASSISTED BUCK-BOOST VOLTAGE SOURCE INVERTER 835

Yan Zhang ; Jinjun Liu ; Xiaolong Ma ; Junjie Feng

VOLUME 2

REAL-TIME SIMULATION OF WIND TURBINE CONVERTER-GRID SYSTEMS 843
Shah, Shahil ; Vieto, Ignacio ; Nian Heng ; Sun, Jian

TECHNOLOGIES FOR MITIGATING FLUCTUATION CAUSED BY RENEWABLE ENERGY SOURCES 850
Katoh, Shuji ; Ohara, Shinya ; Itoh, Tomomichi

RELIABILITY-ORIENTED ENERGY STORAGE SIZING IN WIND POWER SYSTEMS 857
Zian Qin ; Liserre, Marco ; Blaabjerg, Frede ; Poh Chiang Loh

A MULTI-LEVEL VIRTUAL CONDUCTOR AS A BACKBONE OF A DC POWER ROUTING SYSTEM 863
Ramadan, Husam A. ; Imamura, Yasutaka ; Kawachi, Konosuke ; Yang, Sihun ; Shoyama, Masahito

SEMI-NUMERICAL METHOD FOR LOSS-CALCULATION IN FOIL-WINDINGS EXPOSED TO AN AIR-GAP FIELD 868
Leuenberger, D. ; Biela, J.

LOSS REDUCTION OF LAMINATED CORE INDUCTOR USED IN ON-BOARD CHARGER FOR EVS 876
Tera, Takahiro ; Taki, Hiroshi ; Shimizu, Toshihisa

FEASIBLE EVALUATIONS OF COUPLED MULTILAYER CHIP INDUCTOR FOR POL CONVERTER 883
Imaoka, Jun ; Kimura, Shota ; Itoh, Yuki ; Yamamoto, Masayoshi ; Suzuki, Michiaki ; Kawano, Kenji

OPTIMAL INDUCTOR DESIGN FOR 3-PHASE VOLTAGE-SOURCE PWM CONVERTERS CONSIDERING DIFFERENT MAGNETIC MATERIALS AND A WIDE SWITCHING FREQUENCY RANGE 891
Burkart, Ralph M. ; Uemura, Hirofumi ; Kolar, Johann W.

COMPARATIVE ANALYSIS OF INDUCTOR CONCEPTS FOR HIGH PEAK LOAD LOW DUTY CYCLE OPERATION 899
Leibl, Michael ; Kolar, Johann W.

INITIAL POSITION ESTIMATION FOR IPMSMS USING COMB FILTERS AND EFFECTS ON VARIOUS INJECTED SIGNAL FREQUENCIES 907
Suzuki, Toshiki ; Tomita, Mutuwo ; Hasegawa, Masaru ; Doki, Shinji

ADAPTIVE SIGNAL INJECTION METHOD COMBINED WITH EEMF BASED POSITION SENSORLESS CONTROL OF IPMSM DRIVES 914
Ohnuma, Takumi ; Makaino, Yuki ; Saitoh, Ryoh

STUDY OF LOW SPEED SENSORLESS DRIVES FOR SPMSM BY CONTROLLING ELLIPTICAL INDUCTANCE 919
Maekawa, Sari ; Hinata, Toshifumi ; Suzuki, Nobuyuki ; Kubota, Hisao

SUPPRESSION OF INJECTION VOLTAGE DISTURBANCE FOR HIGH FREQUENCY SQUARE-WAVE INJECTION SENSORLESS DRIVE WITH REGULATION OF INDUCED HIGH FREQUENCY CURRENT RIPPLE 925
Dongouk Kim ; Yong-Cheol Kwon ; Seung-Ki Sul ; Jang-Hwan Kim ; Rae-Sung Yu

APPLICATION TREND OF SALIENCY-BASED SENSORLESS DRIVES 933
Yamazaki, Akira ; Ide, Kozo

SWITCHING-LEVEL SIMULATION MODEL OF MMC-BASED BACK-TO-BACK CONVERTER FOR HVDC APPLICATION 937
Byung Moon Han ; Jong kyou Jeong

POWER-CELL SWITCHING-CYCLE CAPACITOR VOLTAGE CONTROL FOR THE MODULAR MULTILEVEL CONVERTERS 944
Wang, Jun ; Burgos, Rolando ; Boroyevich, Dushan ; Bo Wen

A COMPARISON OF MODULAR MULTILEVEL ENERGY CONVERSION PROCESSES: DC/AC VERSUS DC/DC 951
Kish, Gregory J. ; Lehn, Peter W.

A NOVEL TOPOLOGY OF WIND POWER PLANT SUITABLE FOR DC POWER TRANSMISSION SYSTEMS 959
Nishikata, Shoji ; Tatsuta, Fujio ; Suzuki, Katsumi

AN IMPEDANCE-BASED APPROACH TO HVDC SYSTEM STABILITY ANALYSIS AND CONTROL DEVELOPMENT 967
Liu, Hanchao ; Shah, Shahil ; Sun, Jian

TOPOLOGY EVALUATION OF SLOTLESS BEARINGLESS MOTORS WITH TOROIDAL WINDINGS 975
Steinert, Daniel ; Nussbaumer, Thomas ; Kolar, Johann W.

WINDING ARRANGEMENT IN SINGLE-DRIVE BEARINGLESS MOTOR WITH RADIAL GAP 982
Sugimoto, Hiroya ; Tanaka, Seiyu ; Chiba, Akira ; Rahman, M.A.

DEVELOPMENT OF A ONE-AXIS ACTIVELY REGULATED BEARINGLESS MOTOR WITH A REPULSIVE TYPE PASSIVE MAGNETIC BEARING 988
Asama, Junichi ; Watanabe, Daisuke ; Oiwa, Takaaki ; Chiba, Akira

CONTROL CHARACTERISTICS OF 8/10 AND 12/14 BEARINGLESS SWITCHED RELUCTANCE MOTOR 994
Zhenyao Xu ; Dong-Hee Lee ; Jin-Woo Ahn

BASIC CHARACTERISTIC OF A TWO-UNIT OUTER ROTOR TYPE BEARINGLESS MOTOR WITH CONSEQUENT POLE PERMANENT MAGNET STRUCTURE 1000
Takemoto, Masatsugu

VOLTAGE RIPPLE ELIMINATION IN INDUCTOR-LESS AC-TO-AC CONVERTERS FOR MULTI-POLE PERMANENT MAGNET SYNCHRONOUS GENERATORS 1006

Tanaka, Koutaro ; Fujita, Hideaki

A NEW SVM METHOD TO REDUCE COMMON-MODE VOLTAGE IN DIRECT MATRIX CONVERTER 1013

Huu-Nhan Nguyen ; Hong-Hee Lee

EXPERIMENTAL VERIFICATION OF HIGH FREQUENCY LINK DC-AC CONVERTER USING PULSE DENSITY MODULATION AT SECONDARY MATRIX CONVERTER 1021

Itoh, Jun-ichi ; Oshima, Ryo ; Takahashi, Hiroki

LOSS ANALYSIS AND DESIGN METHOD FOR HIGH EFFICIENCY MATRIX CONVERTER 1028

Koiwa, Kazuhiro ; Goh Teck Chiang ; Itoh, Jun-ichi

CAPACITOR CLAMPED MULTI-LEVEL MATRIX CONVERTER 1036

Raju, Siddharth ; Mohan, Ned

EUROPEAN TRENDS AND TECHNOLOGIES IN TRACTION 1043

Drofenik, Uwe ; Canales, Francisco

CO-PHASE POWER SUPPLY SYSTEM FOR HSR 1050

Qunzhan Li ; Wei Liu ; Zeliang Shu ; Shaofeng Xie ; Fulin Zhou

THE APPLICATION OF ELECTRONIC FREQUENCY CONVERTER TO THE SHINKANSEN RAILYARD POWER SUPPLY 1054

Shimizu, Toshimasa ; Kunomura, Ken ; Kai, Masahiko ; Onishi, Mitsuru ; Masuzawa, Hiroshi ; Miyajima, Hiroki ; Otsuki, Midori ; Tsuruma, Yoshinori

APPLICATION EXAMPLES OF ENERGY SAVING MEASURES IN JAPANESE DC FEEDING SYSTEM 1062

Suzuki, Takashi ; Hayashiya, Hitoshi ; Yamanoi, Takashi ; Kawahara, Keiji

LITHIUM ION BATTERY APPLICATION IN TRACTION POWER SUPPLY SYSTEM 1068

Teshima, Masato ; Takahashi, Hirotaka

INTEGRATED ISOLATION AND VOLTAGE BALANCING LINK OF 3-PHASE 3-LEVEL PWM RECTIFIER AND INVERTER SYSTEMS 1073

Boillat, David O. ; Kolar, Johann W.

VOLTAGE STEP-UP CONVERTER BASED ON MULTISTAGE STACKED BOOST ARCHITECTURE (MSBA) 1081

Rufer, Alfred ; Barrade, Philippe ; Steinke, Gina

COMPARISON OF CASCADED MULTILEVEL CONVERTER TOPOLOGIES FOR AC/AC CONVERSION 1087

Ilves, Kalle ; Bessegato, Luca ; Norrga, Staffan

EVALUATION OF ISOLATED THREE-PHASE AC-DC CONVERTER USING MODULAR MULTILEVEL CONVERTER TOPOLOGY 1095

Nakanishi, Toshiki ; Itoh, Jun-ichi

SELF-DECOUPLED DUAL PICK-UP COILS WITH LARGE LATERAL TOLERANCE FOR ROADWAY POWERED ELECTRIC VEHICLES 1103

Choi, Su Y. ; Lee, Sung W. ; Lee, Eun S. ; Jeong, Seog Y. ; Gu, Beom W. ; Rim, Chun T.

CONTACTLESS POWER TRANSFER SYSTEM SUITABLE FOR LOW VOLTAGE AND LARGE CURRENT CHARGING FOR EDLCS 1109

Kudo, Takahiro ; Toi, Takahiro ; Kaneko, Yasuyoshi ; Abe, Shigeru

EXCITATION SYSTEM BY CONTACTLESS POWER TRANSFER SYSTEM WITH THE PRIMARY SERIES CAPACITOR METHOD 1115

Nozawa, Ryosuke ; Kobayashi, Ryota ; Tanifuji, Hikaru ; Kaneko, Yasuyoshi ; Abe, Shigeru

DESIGN OF FERRITE CORES OF INDUCTIVE POWER COLLECTION COILS FOR MOVING VEHICLES 1122

Shimode, Daisuke ; Murai, Toshiaki ; Sawada, Tadashi

TORQUE/CURRENT RATIO IMPROVEMENT AND VIBRATION REDUCTION OF SWITCHED RELUCTANCE MOTORS USING MULTI-STAGE STRUCTURE 1128

Matsui, Ryota ; Nakao, Noriya ; Akatsu, Kan

IMPROVEMENT OF EFFICIENCY BY STEPPED-SKEWING ROTOR FOR SWITCHED RELUCTANCE MOTORS 1135

Sugiura, Makoto ; Ishihara, Yuji ; Ishikawa, Hiroki ; Naitoh, Haruo

A SINGLE PHASE SRM DRIVEN BY COMMERCIAL AC POWER SUPPLY 1141

Aiso, Kohei ; Nakao, Noriya ; Akatsu, Kan

FAST ANALYTICAL MODEL OF SWITCHED RELUCTANCE MACHINE 1148

Smaka, Senad ; Masic, Semsudin ; Cosovic, Mirsad

DETAILED ANALYSIS AND A GENERAL DESIGN PROCEDURE OF DAMPED LCL FILTERS IN THREE PHASE VOLTAGE SOURCE CONVERTERS 1155

Baoquan Liu ; Shaohui Zhong ; Yixin Zhu ; Hao Yi ; Fang Zhuo

70 KHZ, 15 KW SILICON-CARBIDE MOSFET INVERTER FOR INDUSTRIAL INDUCTION HEATING SYSTEMS 1160

Komeda, Shohei ; Tsuboi, Yoshiki ; Fujita, Hideaki

A STUDY ON EFFICIENCY IMPROVEMENT OF HIGH-FREQUENCY CURRENT OUTPUT INVERTER BASED ON IMMITTANCE CONVERSION ELEMENT 1166

Suzuki, Shun ; Shimizu, Toshihisa

HIGH-SPEED SWITCHING METHOD OF MOSFET USING VOLTAGE BOOST AUXILIARY CIRCUIT FED BY GATE DRIVE POWER SUPPLY 1173

Noguchi, Toshihiko ; Murata, Munehiro

OPERATING STRATEGY FOR BI-DIRECTIONAL LLC RESONANT CONVERTER WITH SEAMLESS OPERATION 1179

Abe, Seiya ; Yamamoto, Junichi ; Zaitsu, Toshiyuki ; Ninomiya, Tamotsu

NEGATIVE SEQUENCE CURRENT INJECTION CONTROL ALGORITHM COMPENSATING FOR UNBALANCED PCC VOLTAGE IN MEDIUM VOLTAGE PMSG WIND TURBINES 1185

Jayoon Kang ; Daesu Han ; Suh, Yongsug ; Byoungchang Jung ; Jeongjoong Kim ; Jonghyung Park ; Youngjoon Choi

OPTIMIZATION OF AN OFF-GRID HYBRID SYSTEM FOR SUPPLYING OFFSHORE PLATFORMS IN ARCTIC CLIMATES 1193

Kalogera, Maria ; Bauer, Pavol

ACTIVE DAMPING CONTROL OF LLCL FILTERS FOR THREE-LEVEL T-TYPE GRID CONVERTERS 1201

Alemi, Payam ; Lee, Dong-Choon

DEVELOPING A NEW TOPOLOGY FOR THE DC-DC CONVERTER USED IN FUEL CELL-ELECTRIC DOUBLE LAYER CAPACITOR HYBRID POWER SOURCE SYSTEM FOR MOBILE DEVICES 1207

Tosaka, Shuhei ; Yamanaka, Tatsuya ; Katayama, Noboru ; Hayase, Masanori ; Dowaki, Kiyoshi ; Kogoshi, Sumio

MULTIPLE OUTPUT CHARGER BASED ON PHASE SHIFT FULL BRIDGE CONVERTER WITH NOVEL TIME DIVISION MULTIPLE CONTROL TECHNIQUE 1214

Van-Long Tran ; Woojin Choi

DC-BREAKER FOR A MULTI-MEGAWATT BATTERY ENERGY STORAGE SYSTEM 1220

Demetriades, Georgios D. ; Hermansson, Willy ; Svensson, Jan R ; Papastergiou, Konstantinos ; Larsson, Tomas

ENERGY MANAGEMENT METHOD USING THE IIR FILTER FOR PEMFC-SUPERCAPACITOR HYBRID POWER SOURCE 1227

Yamanaka, Tatsuya ; Katayama, Noboru ; Tosaka, Shuhei ; Kogoshi, Sumio

ADVANCED TORQUE AND CURRENT CONTROL TECHNIQUES FOR PMSMS WITH A REAL-TIME SIMULATOR INSTALLED BEHAVIOR MOTOR MODEL 1234

Tanabe, Ryo ; Akatsu, Kan

COMPENSATION OF THE CURRENT MEASUREMENT ERROR WITH PERIODIC DISTURBANCE OBSERVER FOR MOTOR DRIVE 1242

Yamaguchi, Takashi ; Tadano, Yugo ; Hoshi, Nobukazu

RAPID AND STABLE SPEED CONTROL OF SPMSM BASED ON CURRENT DIFFERENTIAL SIGNAL 1247

Kitajima, Jun ; Ohishi, Kiyoshi

PARALLEL CONNECTED MULTIPLE DRIVE SYSTEM USING SMALL AUXILIARY INVERTER FOR NUMBERS OF PMSM 1253

Nagano, Tsuyoshi ; Itoh, Jun-chi

A TRANSFORMER INRUSH REDUCTION TECHNIQUE FOR LOW-VOLTAGE RIDE-THROUGH OPERATION OF RENEWABLE CONVERTERS 1261

Hsin-Chih Chen ; Ping-Heng Wu ; Cheng, Po-Tai

A CELL CAPACITOR ENERGY BALANCING CONTROL OF MODULAR MULTILEVEL CONVERTER CONSIDERING THE UNBALANCED AC GRID CONDITIONS 1268

Jung, Jae-Jung ; Shenghui Cui ; Kim, Sungmin ; Sul, Seung-Ki

FAULT CURRENT LIMITATION USING THYRISTOR BASED DEVICES 1276

Komatsu, Wilson ; Giaretta, Antonio Ricardo ; de Miranda, Rubens Domingos ; Jardini, Jose Antonio ; Casolari, Ronaldo Pedro ; Vasquez-Arnez, Ricardo Leon ; Hojo, Toshiaki ; Carvalho, Eden Luiz ; Maezono, Paulo Koiti

DC-DC BOOST CONVERTER BASED MSHE-PWM CASCADED MULTILEVEL INVERTER CONTROL FOR STATCOM SYSTEMS 1283

Law, Kah Haw ; Dahidah, Mohamed S.A.

NOVEL PRINCIPLE FOR FLUX SENSING IN THE APPLICATION OF A DC + AC CURRENT SENSOR 1291

Schrittwieser, L. ; Mauerer, M. ; Bortis, D. ; Ortiz, G. ; Kolar, J.W.

UTILIZING VOLTAGE MEASUREMENT OF FET SWITCH FOR MPPT OF DC ENERGY SOURCE 1299

Kimura, Noriyuki ; Niijima, Koji ; Morizane, Toshimitsu ; Omori, Hideki

HIGH FREQUENCY TRANSFORMER BASED ON A COUPLED INDUCTOR TOPOLOGY WITH DIELECTRIC ISOLATION 1303

Amanci, Adrian Z. ; Dawson, Francis P. ; Ruda, Harry E.

CONCEPT AND EXPERIMENTAL EVALUATION OF A NOVEL DC- 100MHZ WIRELESS OSCILLOSCOPE 1309

Lobsiger, Yanick ; Ortiz, Gabriel ; Bortis, Dominik ; Kolar, Johann W.

INTRODUCTION AND EFFECTIVENESS OF STATCOM TO THE INDEPENDENT POWER SYSTEM OF JR EAST 1317

Omi, Masataro ; Kotegawa, Ryo ; Ando, Masato ; Masui, Takeshi ; Horita, Yasuhisa

THE ANALYSIS OF TIME-VARYING RESONANCES IN THE POWER SUPPLY LINE OF HIGH SPEED TRAINS 1322

Chu, Xi ; Lin, Fei ; Yang, Zhongping

FUZZY FEED-FORWARD CHARGE/DISCHARGE CONTROL OF STATIONARY ENERGY STORAGE SYSTEMS FOR DC ELECTRIC RAILWAYS 1328

Kikuchi, Takuya ; Taga, Hironori ; Takagi, Ryo

TRAIN GROUP CONTROL FOR ENERGY-SAVING DC-ELECTRIC RAILWAY OPERATION 1334

Watanabe, Shoichiro ; Koseki, Takafumi

TRANSFORMER-LESS UNIFIED POWER FLOW CONTROLLER USING THE CASCADE MULTILEVEL INVERTER 1342

Fang Zheng ; Shao Zhang ; Shuitao Yang ; Gunasekaran, Deepak ; Karki, Ujjwal

A NEW POWER FLOW CONTROLLER USING SIX MULTILEVEL CASCADED CONVERTERS FOR DISTRIBUTION SYSTEMS..1350

Tsuruta, Ryoji ; Hosaka, Tatsuya ; Fujita, Hideaki

A PROPOSAL OF MODULAR MULTILEVEL CONVERTER APPLYING THREE WINDING TRANSFORMER..1357

Tamada, Shunsuke ; Nakazawa, Yosuke ; Irokawa, Shoichi

BACK-TO-BACK SYSTEM FOR FIVE-LEVEL CONVERTER WITH COMMON FLYING CAPACITORS 1365

Hasegawa, Isamu ; Urushibata, Shota ; Kondo, Takeshi ; Hirao, Kuniaki ; Kodama, Takashi ; Hui Zhang

HARMONIC MODELING OF A VEHICLE TRACTION CIRCUIT TOWARDS THE DC BUS..........................1373

Haghbin, Saeid ; Karvonen, Andreas ; Thiringer, Torbjorn

AC/DC CONVERTER BASED ON INSTANTANEOUS POWER BALANCE CONTROL FOR REDUCING DC-LINK CAPACITANCE 1379

Tokumasu, Akira ; Taki, Hiroshi ; Shirakawa, Kazuhiro ; Wada, Keiji

MODULAR CONVERTER ARCHITECTURE FOR MEDIUM VOLTAGE ULTRA FAST EV CHARGING STATIONS: DUAL HALF-BRIDGE-BASED ISOLATION STAGE 1386

Vasiladiotis, Michail ; Bahrani, Behrooz ; Burger, Niklaus ; Rufer, Alfred

NEW INTERLEAVED CURRENT-FED RESONANT CONVERTER WITH SIGNIFICANTLY REDUCED HIGH CURRENT OUTPUT FILTER FOR EV AND HEV APPLICATION..1394

Moon, Dongok ; Park, Junsung ; Choi, Sewan

15 PHASE INDUCTION MOTOR DRIVE WITH 1:3:5 SPEED RATIOS USING POLE PHASE MODULATION...................1400

Umesh B S ; Sivakumar K

MATHEMATICAL MODEL OF NOVEL WOUND-FIELD SYNCHRONOUS MOTOR SELF-EXCITED BY SPACE HARMONICS..1405

Aoyama, Masahiro ; Noguchi, Toshihiko

DUAL PURPOSE NO VOLTAGE WINDING DESIGN FOR THE BEARINGLESS AC HOMOPOLAR AND CONSEQUENT POLE MOTORS 1412

Severson, Eric ; Nilssen, Robert ; Undeland, Tore ; Mohan, Ned

HARVESTING ENERGY FROM SHIP ROLLING USING AN ECCENTRIC DISK REVOLVING IN A HULA-HOOP MOTION 1420

Yu-Jen Wang

LOAD-INDEPENDENT CURRENT OUTPUT OF INDUCTIVE POWER TRANSFER CONVERTERS WITH OPTIMIZED EFFICIENCY 1425

Zhang, Wei ; Wong, Siu-Chung ; Tse, Chi K. ; Chen, Qianhong

VOLTAGE CONTROL OF INDUCTIVE CONTACTLESS POWER TRANSFER SYSTEM WITH COAXIAL CORELESS TRANSFORMER FOR DC POWER DISTRIBUTION..1430

Miiura, Yushi ; Ojika, Satoshi ; Ise, Tomofumi

CONTACTLESS HIGH POWER TRANSFORMER TECHNOLOGIES FOR RAILWAY VEHICLES...................1438

Kondo, Keiichiro ; Yamamoto, Kohei ; Kitazawa, Satochi

TWO-SWITCH VOLTAGE EQUALIZER BASED ON HALF-BRIDGE CONVERTER WITH MULTI-STACKED CURRENT DOUBLERS FOR SERIES-CONNECTED BATTERIES 1444

Uno, Masatoshi ; Kukita, Akio

OPTIMAL ENERGY STORAGE SYSTEM PLANNING FOR MICROGRIDS WITH CONTRACT CAPACITY CONSTRAINT 1452

Shu-Hung Liao ; Jen-Hao Teng ; Yung-Ching Huang ; Dong-Jing Lee

OPTIMAL ZERO SEQUENCE INJECTION IN MULTILEVEL CASCADED H-BRIDGE CONVERTER UNDER UNBALANCED PHOTOVOLTAIC POWER GENERATION 1458

Yu, Yifan ; Konstantinou, Georgios ; Hredzak, Branislav ; Agelidis, Vassilios G.

SIMPLE METHOD FOR MEASURING OUTPUT IMPEDANCE OF A THREE-PHASE INVERTER IN DQ-DOMAIN 1466

Jokipii, Juha ; Messo, Tuomas ; Suntio, Teuvo

ANALYSIS AND DESIGN OF POWER MANAGEMENT SCHEME FOR AN ON-BOARD SOLAR ENERGY STORAGE SYSTEM 1471

Jiang, W. ; Yu, F.Y. ; Lin, Z.Y. ; Wu, G.F. ; Chen, H. ; Hashimoto, S

LVRT CONTROL STRATEGY OF CSC-DPMSG-WGS UNDER UNBALANCED GRID FAULTS....................1476

Meiqin Mao ; Yong Ding ; Shiting Weng ; Liuchen Chang

A NEW CURRENT CONTROL DROOP STRATEGY FOR VSI-BASED ISLANDED MICROGRIDS 1482

Shoeiby, B. ; Davoodnezhad, R. ; Holmes, D.G. ; McGrath, B.P.

POWER EXCHANGE USING PFC FOR MICRO GRID..1490

Sakai, Tomoyasu ; Takeda, Takashi ; Yukita, Kazuto ; Goto, Yasuyuki ; Ichiyanagi, Katsuhiro ; Morita, Hiroshi

DETERMINATION OF ROTOR TEMPERATURE FOR AN INTERIOR PERMANENT MAGNET SYNCHRONOUS MACHINE USING A PRECISE FLUX OBSERVER..1501

Specht, Andreas ; Wallscheid, Oliver ; Bocker, Joachim

MONITORING CRITICAL TEMPERATURES IN PERMANENT MAGNET SYNCHRONOUS MOTORS USING LOW-ORDER THERMAL MODELS 1508

Huber, Tobias ; Peters, Wilhelm ; Bocker, Joachim

ROBUST CURRENT CONTROL INSENSITIVE TO GAIN DEVIATION AND OFFSET OF INVERTER DC-LINK CURRENT SENSOR FOR SPMSM...1516

Matsuura, Kei ; Ando, Itaru ; Ohishi, Kiyoshi ; Matsuhashi, Masataka

AUTO-TUNING METHOD OF INDUCTANCES FOR PERMANENT MAGNET SYNCHRONOUS MOTORS 1522

Nomura, Naofumi ; Higuchi, Shinichi

AN IMPEDANCE-BASED STABILITY ANALYSIS METHOD FOR PARALLELED VOLTAGE SOURCE CONVERTERS 1529

Wang, Xiongfei ; Blaabjerg, Frede ; Loh, Poh Chiang

DYNAMIC CHARACTERISTICS AND STABILITY COMPARISONS BETWEEN VIRTUAL SYNCHRONOUS GENERATOR AND DROOP CONTROL IN INVERTER-BASED DISTRIBUTED GENERATORS 1536

Jia Liu ; Miura, Yushi ; Ise, Toshifumi

EMBEDDED LIMITATIONS AND PROTECTIONS FOR DROOP-BASED CONTROL SCHEMES WITH CASCADED LOOPS IN THE SYNCHRONOUS REFERENCE FRAME 1544

D'Arco, Salvatore ; Guidi, Giuseppe ; Suul, Jon Are

VIRTUAL SYNCHRONOUS GENERATOR CONTROL WITH DOUBLE DECOUPLED SYNCHRONOUS REFERENCE FRAME FOR SINGLE-PHASE INVERTER 1552

Hirase, Yuko ; Noro, Osamu ; Yoshimura, Eiji ; Nakagawa, Hidehiko ; Sakimoto, Kenichi ; Shindo, Yuji

CONTACTLESS DC CONNECTOR BASED ON GAN LLC CONVERTER FOR NEXT GENERATION DATA CENTERS 1560

Hayashi, Yusuke ; Toyoda, Hajime ; Ise, Toshifumi ; Matsumoto, Akira

ANALYSIS OF MIS-INTERRUPTION OF SEMICONDUCTOR BREAKER IN DC POWER FEEDING SYSTEM 1567

Murai, Kensuke ; Kanai, Yasuyuki ; Asakimori, Koki ; Babasaki, Tadatoshi

A RELIABLE ELECTRONIC CHOKE WITH NO NEED OF GAIN ADJUSTMENT FOR WIRE COMMUNICATION SYSTEM 1575

Katsuki, Akihiko ; Nakamura, Tatsuya ; Mizuki, Tatsuya ; Shibahara, Kohei ; Abe, Tomohiko ; Ikeda, Tomohiko ; Maeyama, Shigetaka

DESIGN OF NEW CONTROL STRATEGIES FOR A FOUR-LEG THREE-PHASE INVERTER TO ELIMINATE THE NEUTRAL CURRENT UNDER UNBALANCED LOADS 1580

Zhao-Qin Guo ; Panda, Sanjib Kumar ; Prasanna, I.V.

RESEARCH TRENDS OF MODULAR MULTILEVEL CASCADE INVERTER (MMCI-DSCC)-BASED MEDIUM-VOLTAGE MOTOR DRIVES IN A LOW-SPEED RANGE 1586

Okazaki, Yuhei ; Matsui, Hitoshi ; Hagiwara, Makoto ; Akagi, Hirofumi

AN INPUT SWITCHED MULTILEVEL INVERTER FOR OPEN-END WINDING INDUCTION MOTOR DRIVE 1594

Zhu, B. ; Jia, Y. ; Prasanna, U.R. ; Rajashekara, K. ; Kubo, H.

VARIABLE CARRIER FREQUENCY MIXED PWM TECHNIQUE BASED ON CURRENT RIPPLE PREDICTION FOR REDUCED SWITCHING LOSS 1601

Kubo, Hajime ; Yamamoto, Yasuhiro

SLIDING MODE PWM FOR EFFECTIVE CURRENT CONTROL IN SWITCHED RELUCTANCE MACHINE DRIVES 1606

Manolas, Iakovos ; Papafotiou, Georgios ; Manias, Stefanos N.

EXPERIMENTAL VERIFICATION OF AN EMC FILTER USED FOR PWM INVERTER WITH WIDE BAND-GAP DEVICES 1613

Itoh, Jun-ichi ; Araki, Takahiro ; Orikawa, Koji

PACKAGING FOR SIC POWER DEVICE 1621

Funaki, Tsuyoshi

SOLID STATE TRANSFORMER AND MV GRID TIE APPLICATIONS ENABLED BY 15 KV SIC IGBTS AND 10 KV SIC MOSFETS BASED MULTILEVEL CONVERTERS 1626

Madhusoodhanan, Sachin ; Tripathi, Awneesh ; Patel, Dhaval ; Mainali, Krishna ; Kadavelugu, Arun ; Hazra, Samir ; Bhattacharya, Subhashish ; Hatua, Kamalesh

VOLUME 3

GENERALIZED MODULAR MULTILEVEL CONVERTER AND MODULATION 1634

Hui Liu ; Loh, Poh Chiang ; Blaabjerg, Frede

AVERAGE POWER CONTROL OF DC BUS VOLTAGES OF CASCADED H-BRIDGE MULTILEVEL CONVERTERS 1639

Lee, Chia-Tse ; Chen, Hsin-Chih ; Ching-Wei Wang ; Ching-Hsiang Yang ; Cheng, Po-Tai

ANALYSIS AND COMPARISON OF HIGH POWER SEMICONDUCTOR DEVICE LOSSES IN 5MW PMSG MV WIND TURBINES 1646

Kihyun Lee ; Kyungsub Jung ; Seunghoo Song ; Suh, Yongsug ; Changwoo Kim ; Hyoyol Yoo ; Sunsoon Park

APPLICATION OF MODULAR MATRIX CONVERTER TO WIND TURBINE GENERATOR 1654

Inomata, Kentaro ; Hara, Hidenori ; Morimoto, Shinya ; Fujii, Junji ; Takeda, Kotaro ; Yamamoto, Eiji

FREE MOTION MECHANICAL POWER FACTOR; COMPARISON BETWEEN ROBOTS IN DIFFERENT STRUCTURE AND COORDINATE 1660

Mizoguchi, Takahiro ; Nozaki, Takahiro ; Ohnishi, Kouhei

ANALYSIS OF SETTLING BEHAVIOR AND DESIGN OF CASCADED PRECISE POSITIONING CONTROL IN PRESENCE OF NONLINEAR FRICTION 1665

Ruderman, Michael ; Iwasaki, Makoto

FIELD AND BENCH TEST EVALUATION OF RANGE EXTENSION CONTROL SYSTEM FOR ELECTRIC VEHICLES BASED ON FRONT AND REAR DRIVING-BRAKING FORCE DISTRIBUTIONS 1671

Fujimoto, Hiroshi ; Harada, Shingo ; Goto, Yuichi ; Kawano, Daisuke ; Sato, Koji ; Matsuo, Yusuke

VIBRATION SUPPRESSION OF INTEGRATED RESONANT AND TIME DELAY SYSTEM BY REFLECTED WAVE REJECTION 1679

Saito, Eiichi ; Oboe, Roberto ; Katsura, Seiichiro

THRUST CHARACTERISTICS IMPROVEMENT OF A CIRCULAR SHAFT MOTOR FOR DIRECT-DRIVE APPLICATIONS 1685

Omura, Mototsugu ; Shimono, Tomoyuki ; Fujimoto, Yasutaka

DESIGN OF A BEARINGLESS FLUX-SWITCHING SLICE MOTOR 1691

Gruber, Wolfgang ; Radman, Karlo ; Schob, Reto.T.

PROPOSAL OF A PERMANENT MAGNET HYBRID TYPE AXIAL MAGNETICALLY LEVITATED MOTOR 1697

Kurita, Nobuyuki ; Ishikawa, Takeo ; Takada, Hiromu ; Suzuki, Genri

COMPARISON OF HIGH SPEED BEARINGLESS DRIVE TOPOLOGIES WITH COMBINED WINDINGS 1701

Mitterhofer, Hubert ; Mrak, Branimir ; Gruber, Wolfgang

HIGH-SPEED MAGNETICALLY LEVITATED REACTION WHEEL DEMONSTRATOR 1707

Zwyssig, Christof ; Baumgartner, Thomas ; Kolar, Johann W.

STABILIZED SUSPENSION CONTROL CONSIDERING ARMATURE REACTION IN A D-Q AXIS CURRENT CONTROL BEARINGLESS MOTOR 1715

Ooshima, Masahide ; Kumakura, Yoshito

ANALYSIS AND DESIGN OF A HIGH-FREQUENCY ISOLATED DUAL-TANK LCL RESONANT AC-DC CONVERTER 1721

Du, Yimian ; Bhat, Ashoka K.S.

VERIFICATION OF LLC RESONANT CONVERTER APPLIED A CURRENT-BALANCING HIGH-FREQUENCY TRANSFORMER WITH MULTI-OUTPUT WINDINGS 1728

Araki, Jun ; Shinozaki, Ikki ; Funato, Hirohito ; Ogasawara, Satoshi ; Murakami, Daichi ; Hirota, Yukitsugu ; Mihara, Teruyoshi ; Mouri, Masayuki ; Okazaki, Fumihiro

LIGHT-LOAD EFFICIENCY IMPROVEMENT STRATEGY FOR LLC RESONANT CONVERTER UTILIZING A STEP-GAP TRANSFORMER 1734

Huang, Wen-Nan ; Lee, Shiu-Hui ; Chen, Ching-Guo

A NOVEL ACCURATE PRIMARY SIDE CONTROL (PSC) METHOD FOR HALF-BRIDGE (HB) LLC CONVERTER 1738

Jae-Bum Lee ; Kim, Chong-Eun ; Jae-Hyun Kim ; Cheol-O Yeon ; Young-Do Kim ; Moon, Gun-Woo

A SIMPLE CONTROL SCHEME FOR IMPROVING LIGHT-LOAD EFFICIENCY IN A FULL-BRIDGE LLC RESONANT CONVERTER 1743

Kim, Jae-Hyun ; Kim, Chong-Eun ; Lee, Jae-Bum ; Young-Do Kim ; Han-Shin Youn ; Moon, Gun-Woo

POWER CONDITIONER FOR STABILIZING POWER DISTURBANCE CAUSED OF WIND TURBINE GENERATOR SYSTEM 1748

Saga, Yasunao ; Fujii, Kansuke ; Yoda, Kazuyuki

A FRONT-TO-FRONT (FTF) SYSTEM CONSISTING OF MULTIPLE MODULAR MULTILEVEL CASCADE CONVERTERS FOR OFFSHORE WIND FARMS 1761

Sasongko, Firman ; Hagiwara, Makoto ; Akagi, Hirofumi

MODELLING, DESIGN AND CONTROL OF GRID CONNECTED CONVERTER FOR HIGH ALTITUDE WIND POWER APPLICATION 1775

Adhikari, Jeevan ; Rathore, Akshay K. ; Panda, S K

PRACTICAL STUDY OF A HIGH STEP-DOWN CONVERTER 1781

Jinno, Masahito ; Su, Hong-Wei ; Tsai, Jiung-Lin ; Matsuo, Hirofumi

GENERALIZED MODELING AND OPTIMIZATION OF A BIDIRECTIONAL DUAL ACTIVE BRIDGE DC-DC CONVERTER INCLUDING FREQUENCY VARIATION 1788

Jauch, Felix ; Biela, Jurgen

BALANCED DISCHARGING OF POWER BANK WITH BUCK-BOOST BATTERY POWER MODULES 1796

Moo, Chin-Sien ; Wu, Tsung-Hsi ; Hou, Chih-Hao ; Hsieh, Yao-Ching

Y-SOURCE IMPEDANCE-NETWORK-BASED ISOLATED BOOST DC/DC CONVERTER 1801

Siwakoti, Yam P. ; Town, Graham E. ; Loh, Poh Chiang ; Blaabjerg, Frede

MULTI-PHASE DC-DC CONVERTER WITH RIPPLE-LESS OPERATION FOR THERMO-ELECTRIC GENERATOR 1806

Kimura, Noriyuki ; Niijima, Koji ; Morizane, Toshimitsu ; Omori, Hideki

POSITION SENSORLESS START-UP METHOD OF SURFACE PERMANENT MAGNET SYNCHRONOUS MOTOR USING NONLINEAR ROTOR POSITION OBSERVER 1811

Hanamoto, Tsuyoshi ; Yamada, Hiroaki ; Okuyama, Yoshihiro

SENSORLESS CONTROL OF PMSM FOR THE WHOLE SPEED RANGE USING TWO-DEGREE-OF-FREEDOM CURRENT CONTROL AND HF TEST CURRENT INJECTION FOR LOW SPEED RANGE 1816

Seilmeier, Markus ; Piepenbreier, Bernhard

ELLIPSE-TRAJECTORY-ORIENTED VECTOR CONTROL FOR ENERGY EFFICIENT/WIDE-SPEED-RANGE DRIVES OF SENSORLESS PMSM 1824

Shinnaka, Shinji ; Amano, Yuki

DEVELOPMENT OF POSITION SENSORLESS CONTROL FOR PERMANENT-MAGNET SYNCHRONOUS GENERATOR DRIVE 1832

Chang, Yuan-Chih ; Lin, Chia-Yu ; Dai, Wei-Fu ; Wu, Chun-Wei

CONTROL OF A 750KW PERMANENT MAGNET SYNCHRONOUS MOTOR 1837

Liping Zheng ; Dong Le

REGIONAL SMART GRID OF ISLAND IN CHINA WITH MULTIFOLD RENEWABLE ENERGY 1842
Xu Cai ; Zheng Li

STABILIZING SMALL ISLAND POWER SYSTEM WITH RENEWABLES BY USE OF POWER CONDITIONING SYSTEMS - JAPANESE ISLAND SYSTEM CASE - 1849
Baba, Jumpei

POWER ELECTRONICS SOLUTIONS APPLIED TO A VARIETY OF DEMONSTRATIVE MICROGRID PROJECTS 1855
Ueda, Yoshinobu

MOVING TOWARDS THE SMART GRID: THE NORWEGIAN CASE 1861
Fosso, Olav B. ; Molinas, Marta ; Sand, Kjell ; Coldevin, Grete H.

POWER ELECTRONICS TECHNOLOGY IN SMART GRID PROJECTS -APPLICATIONS AND EXPERIENCES- 1868
Kobayashi, Takenori

EV AND HEV MOTOR DEVELOPMENT IN TOSHIBA 1874
Arata, Masanori ; Kurihara, Yoshihiro ; Misu, Daisuke ; Matsubara, Masakatsu

MOTOR STATOR WITH THICK RECTANGULAR WIRE LAP WINDING FOR HEVS 1880
Ishigami, Takashi ; Tanaka, Yuichiro ; Homma, Hiroshi

COMPARISON STUDY OF VARIOUS MOTORS FOR EVS AND THE POTENTIALITY OF A FERRITE MAGNET MOTOR 1886
Matsuhashi, Daiki ; Matsuo, Keisuke ; Okitsu, Takashi ; Ashikaga, Tadashi ; Mizuno, Takayuki

OPTIMAL FIELD EXCITATION CONTROL OF A CLAW POLE MOTOR FOR HYBRID ELECTRIC VEHICLE 1892
Azuma, M. ; Hazeyama, M. ; Morita, M. ; Kuroda, Y. ; Daikoku, A. ; Inoue, M.

A WIDE SPEED RANGE HIGH EFFICIENCY EV DRIVE SYSTEM USING WINDING CHANGEOVER TECHNIQUE AND SIC DEVICES 1898
Takatsuka, Yushi ; Hara, Hidenori ; Yamada, Kenji ; Maemura, Akihiko ; Kume, Tsuneo

PERFORMANCE COMPARISON OF A GAN GIT AND A SI IGBT FOR HIGH-SPEED DRIVE APPLICATIONS 1904
Tuysuz, Arda ; Bosshard, Roman ; Kolar, Johann W.

WIDE-BAND GAP DEVICES IN PV SYSTEMS - OPPORTUNITIES AND CHALLENGES 1912
Sintamarean, C. ; Eni, E. ; Blaabjerg, F. ; Teodorescu, R. ; Wang, H.

POWER ELECTRONICS EQUIPMENTS APPLYING NOVEL SIC POWER SEMICONDUCTOR MODULES 1920
Mino, Kazuaki ; Yamada, Ryuji ; Kimura, Hiroshi ; Matsumoto, Yasushi

EMI PREDICTION METHOD FOR SIC INVERTER BY THE MODELING OF STRUCTURE AND THE ACCURATE MODEL OF POWER DEVICE 1929
Maekawa, Sari ; Tsuda, Junichi ; Kuzumaki, Atsuhiko ; Matsumoto, Shuhei ; Mochikawa, Hiroshi ; Kubota, Hisao

SYSTEM INTEGRATION OF GAN TECHNOLOGY 1935
Ferreira, J.A. ; Popovic, J. ; van Wyk, J.D. ; Pansier, F.

POWER LOSSES OF MULTILEVEL CONVERTERS IN TERMS OF THE NUMBER OF THE OUTPUT VOLTAGE LEVELS 1943
Kashihara, Yugo ; Itoh, Jum-ichi

A LARGE CAPACITY 3-LEVEL IEGT INVERTER 1950
Yoshizawa, Daisuke ; Mukunoki, Makoto ; Omote, Kenichiro ; Hayashi, Makoto ; Isida, Takashi

VIBRATION SUPPRESSING CONTROL METHOD OF ANGULAR TRANSMISSION ERROR OF CYCLOID GEAR FOR INDUSTRIAL ROBOTS 1956
Yoshioka, Takashi ; Hirano, Yosei ; Ohishi, Kiyoshi ; Miyazaki, Toshimasa ; Yokokura, Yuki

AN ADVANCED POSITION CONTROL OF OVERHEAD CRANE BY SWAY SUPPRESSION METHOD EMULATING NATURAL DAMPING 1962
Kurabayashi, Toshiyuki ; Yang Chuan ; Murakami, Toshiyuki

A ROBOTIC CANE FOR WALKING ASSISTANCE 1968
Shimizu, Kyohei ; Smadi, Issam ; Fujimoto, Yasutaka

HAND POSITION ESTIMATION IN BINOCULAR VISUAL SPACE USING LINEAR APPROXIMATION OF KINEMATICS 1974
Komada, Satoshi ; Turpin, Santiago ; Hashimoto, Kento ; Yashiro, Daisuke ; Hirai, Junji

CONTACT STATE RECOGNITION BASED ON HAPTIC SIGNAL PROCESSING FOR ROBOTIC TOOL USE 1978
Matsuzaki, Ryohei ; Okuma, Jun ; Sakaino, Sho ; Tsuji, Toshiaki

RECENT TECHNICAL TRENDS IN MAGNETIC MATERIALS 1984
Wajima, Kiyoshi ; Toda, Hiroaki ; Kosaka, Takashi ; Marukawa, Yasuhiro ; Ishihara, Chio

MULTI-DOMAIN CO-SIMULATION WITH NUMERICALLY IDENTIFIED PMSM INTERWORKING AT HILS FOR ELECTRIC PROPULSION 1990
Park, Gyeong-Jae ; Jung, Hochang ; Kim, Yong-Jae ; Jung, Sang-Yong

RECENT TECHNICAL TRENDS IN PMSM 1997
Morimoto, Shigeo ; Asano, Yoshinari ; Kosaka, Takashi ; Enomoto, Yuji

RECENT TECHNICAL TRENDS IN SRM AND FSM 2004
Kano, Yoshiaki

RECENT TECHNICAL TRENDS IN VARIABLE FLUX MOTORS 2011
Toba, Akio ; Daikoku, Akihiro ; Nishiyama, Noriyoshi ; Yoshikawa, Yuichi ; Kawazoe, Yosuke

A GENERAL DISCRETE TIME MODEL TO EVALUATE ACTIVE DAMPING OF GRID CONVERTERS WITH LCL FILTERS 2019
Parker, S.G. ; McGrath, B.P. ; Holmes, D.G.

ANALYSIS AND REDUCTION OF POWER LOSSES IN PV CONVERTERS FOR GRID CONNECTION TO LOW-VOLTAGE THREE-PHASE THREE-WIRE SYSTEMS 2027
Amma, Ryosuke ; Fujita, Hideaki

DESIGN OF GRID CONNECTED PWM CONVERTERS CONSIDERING TOPOLOGY AND PWM METHODS FOR LOW-VOLTAGE RENEWABLE ENERGY APPLICATIONS 2034
Kantar, Emre ; Hava, Ahmet M.

PERFORMANCE OF DEAD TIME COMPENSATION METHODS IN THREE-PHASE GRID-CONNECTION CONVERTERS 2042
Mannen, Tomoyuki ; Fujita, Hideaki

D-S DIGITAL CONTROL FOR THREE-PHASE BI-DIRECTIONAL INVERTERS 2050
Wu, T.-F. ; Chang, C.-H. ; Lin, L.-C.

EXPECTATIONS OF NEXT-GENERATION POWER DEVICES FOR HOME AND CONSUMER APPLIANCES 2058
Kanouda, Akihiko ; Shoji, Hiroyuki ; Shimada, Takae ; Okubo, Toshikazu

APPLICATION TREND AND FORESIGHT OF SIC POWER DEVICES TO AIR CONDITIONERS 2064
Kamikura, Mamoru ; Murata, Yuichiro ; Kutsuki, Tomohiro ; Saito, Katsuhiko

RECENT TECHNICAL TRENDS AND FUTURE PROSPECTS OF IGBTS AND POWER MOSFETS 2068
Ogura, Tsuneo

RECENT DEVELOPMENT AND FUTURE PROSPECTS OF POWER SIC DEVICES 2074
Nakamura, T. ; Nakano, Y. ; Aketa, M. ; Hanada, T.

RECENT ADVANCES AND FUTURE PROSPECTS ON GAN-BASED POWER DEVICES 2075
Ueda, Tetsuzo

SCALING AND BALANCING OF MULTI-CELL CONVERTERS 2079
Kasper, Matthias ; Bortis, Dominik ; Kolar, Johann W.

HYBRID MODULATED UNIVERSAL SOFT-SWITCHING CURRENT-FED DC/DC CONVERTER FOR WIDE VOLTAGE REGULATION FOR PV/FUEL CELLS/BATTERY APPLICATIONS 2087
Moorthy, Radha Sree Krishna ; Rathore, Akshay Kumar

HIGH EFFICIENCY POWER CONVERTERS FOR BATTERY ENERGY STORAGE SYSTEMS 2095
Kawakami, Noriko ; Iijima, Yukihia ; Li, Haiqing ; Ota, Satoru

IMPLEMENTATION OF BRIDGELESS CUK POWER FACTOR CORRECTOR WITH POSITIVE OUTPUT VOLTAGE 2100
Yang, Hong-Tzer ; Chiang, Hsin-Wei

A NOVEL SYNCHRONOUS RECTIFIER METHOD FOR A LLC RESONANT CONVERTER WITH VOLTAGE-DOUBLER RECTIFIER 2108
Murata, Koji ; Kurokawa, Fujio

LATEST DEVELOPMENTS IN INCREASING THE POWER DENSITY OF TRACTION DRIVES 2113
Bakran, Mark-M. ; Marz, Andreas ; Laska, Bernd ; Krafft, Eberhard ; Korner, Olaf ; Nagel, Andreas

CATENARY AND STORAGE BATTERY HYBRID SYSTEM FOR ELECTRIC RAILCAR SERIES EV-E301 2120
Kono, Y. ; Shiraki, N. ; Yokoyama, H. ; Furuta, R.

TECHNOLOGY FOR ENERGY-SAVING RAILWAY OPERATION THROUGH POWER-LIMITING BRAKES—A CASE STUDY AT AN URBAN RAILWAY 2126
Koseki, Takafumi ; Watanabe, Shoichiro ; Hamazaki, Yasuhiro ; Kondo, Keiichiro ; Hasegawa, Tomonori ; Mizuma, Takeshi

AN OVERVIEW ON BRAKING ENERGY REGENERATION TECHNOLOGIES IN CHINESE URBAN RAILWAY TRANSPORTATION 2133
Yang, Zhongping ; Xia, Huan ; Wang, Bin ; Lin, Fei

TRACTION INVERTER THAT APPLIES COMPACT 3.3 KV / 1200 A SIC HYBRID MODULE 2140
Ishikawa, Katsumi ; Yukutake, Seigo ; Kono, Yasuhiko ; Ogawa, Kazutoshi ; Kameshiro, Norifumi

POWER ELECTRONIC-BASED PROTECTION FOR DIRECT-CURRENT POWER DISTRIBUTION IN MICRO-GRIDS 2145
Tseng, K.J. ; Luo, Guomin

A CONCEPT OF HIGH POWER DC/DC CONVERTER WITH DOUBLE LOW POWER OUTPUTS 2152
Hojo, Masahide ; Nishioka, Tomoya ; Yamanaka, Kenji

PERFORMANCE EVALUATION FOR GRID IMPEDANCE BASED ISLANDING DETECTION METHOD 2156
Liu, Ning ; Aljankawey, A.S. ; Diduch, C.P. ; Chang, L. ; Mao, Meiqin ; Yazdkhasti, Pegah ; Su, Jianhui

IDENTIFYING NATURAL DEGRADATION/AGING IN POWER MOSFETS IN A LIVE GRID-TIED PV INVERTER USING SPREAD SPECTRUM TIME DOMAIN REFLECTOMETRY 2161
Li, Qian ; Khan, Faisal H.

CONTROL METHOD FOR INDUCTIVE POWER TRANSFER WITH HIGH PARTIAL-LOAD EFFICIENCY AND RESONANCE TRACKING 2167
Bosshard, R. ; Kolar, J.W. ; Wunsch, B.

STANDARD MODELS FOR SMART GRID SIMULATIONS 2175
Noda, Taku ; Nagashima, Tomohiro ; Sekisue, Takayuki ; Kabasawa, Yuichiro ; Kato, Shinji ; Sekiba, Yoichi ; Tokuda, Hirokazu ; Kounoto, Masaaki

MODEL DEVELOPMENT FOR MOTOR DRIVE SYSTEM SIMULATIONS 2183
Ishikawa, Hiroki ; Abe, Takashi ; Kato, Toshiji ; Kubota, Yutaka ; Shimomura, Junichi ; Kohno, Yusuke ; Ikeda, Masahiro ; Umeda, Nobuhiro ; Kimura, Noriyuki ; Shigematsu, Koichi ; Inoue, Yukinori

PRACTICAL SIMULATION EXAMPLES OF AUTOMOTIVE AND POWER SUPPLY SYSTEMS............2189
Abe, Takashi ; Fukushima, Kentaro ; Sekisue, Takayuki ; Shigematsu, Koichi ; Ichihara, Junichi ; Kato, Toshiji ; Ishikawa, Hiroki ; Kouno, Yusuke ; Konoto, Masaaki ; Saito, Ryoji ; Nishida, Yasuyuki

ADMITTANCE MATRICES OF VOLTAGE SOURCE CONVERTERS FOR DISTRIBUTED GENERATORS......2195
Lian, K.L. ; Huang, T.D.

FPGA-BASED SIMULATION OF POWER ELECTRONICS USING ITERATIVE METHODS............2202
Zhang, Huiguo ; Sun, Jian

GALLIUM ARSENIDE IC TECHNOLOGY FOR POWER SUPPLIES ON CHIP............2208
Pala, Vipindas ; Peng, Han ; Hella, Mona ; Chow, T.Paul

SILICON ON NANOCRYSTALLINE AND MICROCRYSTALLINE DIAMOND STACKING STRUCTURE FOR POWER SUPPLY ON CHIP............2212
Yamada, Takatoshi ; Hasegawa, Masataka

A NOVEL LOAD REGULATION TECHNIQUE FOR POWER-SOC WITH PARALLEL CONNECTED POLS............2216
Abe, Seiya ; Matsumoto, Satoshi ; Hidaka, Akira ; Rikitake, Jungo ; Ninomiya, Tamotsu

MATRIX-POL ARCHITECTURE FOR INTEGRATED POWER SUPPLY............2222
Ishizuka, Yoichi ; Shibahara, Ryota ; Ninomiya, Tamotsu ; Tanaka, Kiminori ; Abe, Seiya

ON-CHIP BUCK CONVERTER WITH SPIRAL FERRITE INDUCTOR AND REDUCING IR DROP IN 3D STACKED INTEGRATION............2228
Fuketa, Hiroshi ; Shinozuka, Yasuhiro ; Ishida, Koichi ; Takamiya, Makoto ; Sakurai, Takayasu

DCM ANALYSIS OF A SINGLE SIC SWITCH BASED ZVZCS TAPPED BOOST CONVERTER............2232
Choi, Bo H. ; Lee, Eun S. ; Kim, Ji H. ; Rim, Chun T.

EFFECT OF INPUT AND OUTPUT TERMINAL SOURCES ON DYNAMIC BEHAVIOR OF SWITCHED-MODE CONVERTERS............2240
Suntio, T. ; Viinamaki, J. ; Jokipii, J. ; Messo, T. ; Sitbon, M. ; Kuperman, A.

A FULLY SOFT-SWITCHED MULTIPHASE DC-DC CONVERTER WITH REDUCED SWITCH COUNT FOR HIGH POWER APPLICATION............2247
Kim, Minjae ; Yang, Daeki ; Choi, Sewan

A STATIC CHARACTERISTIC ANALYSIS OF PROPOSED BI-DIRECTIONAL DUAL ACTIVE BRIDGE DC-DC CONVERTER............2252
Nagata, Shun ; Takasaki, Mika ; Furukawa, Yutaka ; Hirose, Toshiro ; Ishizuka, Yoichi

HYBRID BATTERY CHARGING SYSTEM COMBINING OBC WITH LDC FOR ELECTRIC VEHICLES............2260
Kim, Seonghye ; Kang, Feel-soon

TRANSIENT BEHAVIOR OF THE DUAL ACTIVE BRIDGE CONVERTER IN HIGH EFFICIENT ENERGY CONVERSION SYSTEM............2266
Aoyama, Kohei ; Motoi, Naoki ; Tsuruta, Yukinori ; Kawamura, Atsuo

STATE-OF-CHARGE ESTIMATION FOR LITHIUM-ION BATTERY PACK USING RECONSTRUCTED OPEN-CIRCUIT-VOLTAGE CURVE............2272
Chun, Chang Yoon ; Seo, Gab-Su ; Yoon, Sung Hyun ; Cho, Bo-Hyung

SYSTEM DESIGN OF ELECTRIC ASSISTED BICYCLE USING EDLCS AND WIRELESS CHARGER............2277
Itoh, Jun-ichi ; Noguchi, Kenji ; Orikawa, Koji

STUDY ON LOW-LOSS GATE DRIVE CIRCUIT FOR HIGH EFFICIENCY SERVER POWER SUPPLY USING NORMALLY-OFF SIC-JFET............2285
Katoh, Kaoru ; Ishikawa, Katsumi ; Hatanaka, Ayumu ; Ogawa, Kazutoshi ; Akiyama, Satoru ; Ogawa, Takashi ; Yokoyama, Natsuki ; Maru, Naoki ; Takahashi, Osamu ; Nishisu, Koji

A SHORT CIRCUIT PROTECTION METHOD BASED ON A GATE CHARGE CHARACTERISTIC............2290
Horiguchi, Takeshi ; Kinouchi, Shin-ichi ; Nakayama, Yasushi ; Oi, Takeshi ; Urushibata, Hiroaki ; Okamoto, Shoji ; Tominaga, Shinji ; Akagi, Hirofumi

HIGHLY RELIABLE 1200-V P-TYPE MOSFET FOR LEVEL-SHIFT CIRCUIT USED IN DRIVER IC............2297
Sakurai, Naoki ; Hakutou, Takuma ; Yura, Masashi

A NEW LEVEL UP SHIFTER FOR HVICS WITH HIGH NOISE TOLERANCE............2302
Akahane, Masashi ; Jonishi, Akihiro ; Yamaji, Masaharu ; Kanno, Hiroshi ; Tanaka, Takahide ; Nishio, Haruhiko ; Sumida, Hitoshi

OUTPUT RIPPLE MINIMIZATION OF SINGLE-STAGE POWER FACTOR CORRECTED BI-DIRECTIONAL BUCK AC/DC CONVERTER............2310
Veerasamy, Balaji ; Kitagawa, Wataru ; Takeshita, Takaharu

THREE-PHASE ISOLATED FULL-BRIDGE BOOST PFC WITH FLYBACK PASSIVE AUXILIARY CONVERTER............2318
Meng, Tao ; Yu, Shuai ; Ben, Hongqi ; Wei, Guo ; Sun, Shaohua

CONTROL AND EXPERIMENT OF A MODULAR PUSH-PULL PWM CONVERTER FOR A BATTERY ENERGY STORAGE SYSTEM............2323
Hagiwara, Makoto ; Akagi, Hirofumi

ACTIVE FRONT-END TOPOLOGY FOR 5 LEVEL MEDIUM VOLTAGE DRIVE SYSTEM WITH ISOLATED DC BUS............2330
Oka, Toshiaki ; Kusunoki, Hironobu ; Tsukakoshi, Masahiko ; Kleinecke, John ; Daskalos, Mike

A DUAL ACTIVE BRIDGE DC-DC CONVERTER WITH OPTIMAL DC-LINK VOLTAGE SCALING AND FLYBACK MODE FOR ENHANCED LOW-POWER OPERATION IN HYBRID PV/STORAGE SYSTEMS............2336
Poshtkouhi, Shahab ; Trescases, Olivier

NOVEL MODULAR MULTIPLE-INPUT BIDIRECTIONAL DC-DC POWER CONVERTER (MIPC)............2343
Hintz, Andrew ; Prasanna, Udupi.R. ; Rajashekara, Kaushik

SINGLE-SWITCH PWM CONVERTER INTEGRATING VOLTAGE EQUALIZER FOR PHOTOVOLTAIC MODULES UNDER PARTIAL SHADING..........2351

Uno, Masatoshi ; Kukita, Akio

NEW DC RAIL SIDE SOFT-SWITCHING PWM DC-DC CONVERTER WITH VOLTAGE DOUBLER RECTIFIER FOR PV GENERATION INTERFACE..........2359

Sayed, Khairy ; Kwon, Soon-Kurl ; Nishida, Katsumi ; Nakaoka, Mutsuo

MODELING METHOD OF STRAY MAGNETIC COUPLINGS IN AN EMC FILTER FOR A SIC SOLAR INVERTER..........2366

Masuzawa, Takashi ; Hoene, Eckart ; Hoffmann, Stefan ; Lang, Klaus-Dieter

DC BUS VOLTAGE EMI MITIGATION IN THREE-PHASE ACTIVE RECTIFIERS USING A VIRTUAL NEUTRAL FILTER..........2372

Parker, S.G. ; Segaran, D.S. ; Holmes, D.G. ; McGrath, B.P.

EFFECTS OF TRANSFORMER STRUCTURES ON THE NOISE BALANCING AND CANCELLATION MECHANISMS OF SWITCHING POWER CONVERTERS..........2380

Hsieh, Hung-I ; Shih, Sheng-Fang

A NOVEL TECHNIQUE FOR REDUCING LEAKAGE CURRENT BY APPLICATION OF ZERO-SEQUENCE VOLTAGE..........2385

Ayano, Hideki ; Murakami, Kouhei ; Matsui, Yoshihiro

AC-CHOPPERS USING INSTANTANEOUS VOLTAGE CONTROL TECHNIQUE TO SOLVE VOLTAGE SAG PROBLEMS..........2392

Khomfoi, Surin

VOLTAGE REGULATION IN DISTRIBUTION SYSTEM USING THE COMBINED DVR..........2400

Nakamura, Sota ; Aoki, Mutsumi ; Ukai, Hiroyuki

NONLINEAR CONTROL OF THREE-PHASE FOUR-WIRE DYNAMIC VOLTAGE RESTORERS FOR DISTRIBUTION SYSTEM..........2406

Jeong, Seon-Yeong ; Nguyen, Thanh Hai ; Lee, Dong-Choon ; Kim, Jang-Mok

VOLUME 4

DISTURBANCE CALCULATION BASED ON SPACE VECTOR DOT PRODUCT: APPLICATIONS TO COMPENSATORS..........2413

de Carvalho, Kelly Caroline Mingorancia ; Ama, Naji Rajai Nasri ; Komatsu, Wilson ; Martinz, Fernando Ortiz ; Figueredo, Ricardo Souza ; Matakas, Lourenco

PROPOSAL OF 6TH RADIAL FORCE CONTROL BASED ON FLUX LINKAGE..........2421

Kanematsu, Masato ; Miyajima, Takayuki ; Fujimoto, Hiroshi ; Hori, Yoichi ; Enomoto, Toshio ; Kondou, Masahiko ; Komiya, Hiroshi ; Yoshimoto, Kantaro ; Miyakawa, Takayuki

AIR GAP CONTROL OF MULTI-PHASE TRANSVERSE FLUX PERMANENT MAGNET LINEAR SYNCHRONOUS MOTOR BY USING INDEPENDENT VECTOR CONTROL..........2427

Hwang, Seon-Hwan ; Bang, Deok-Je ; Kim, Ji-Won

MODIFIED DIRECT INSTANTANEOUS TORQUE CONTROL OF SWITCHED RELUCTANCE MOTOR WITH HIGH TORQUE PER AMPERE AND REDUCED SOURCE CURRENT RIPPLE..........2433

Suryadevara, Rohit ; Fernandes, B.G.

CONTROL OF WOUND FIELD SYNCHRONOUS MOTOR INTEGRATED WITH ZSI..........2438

Tajima, G. ; Kosaka, T. ; Matsui, N. ; Tonogi, K. ; Minoshima, N. ; Yoshida, T.

A NOVEL IPMSM MODEL FOR ROBUST POSITION SENSORLESS CONTROL TO MAGNETIC SATURATION..........2445

Matsumoto, Atsushi ; Hasegawa, Masaru ; Doki, Shinji

MOTOR DRIVE SYSTEM USING NONLINEAR MATHEMATICAL MODEL FOR PERMANENT MAGNET SYNCHRONOUS MOTORS..........2451

Iwaji, Yoshitaka ; Nakatsugawa, Junnosuke ; Sakai, Toshifumi ; Aoyagi, Shigehisa ; Nagura, Hirokazu

SENSORLESS-ORIENTED DESIGN OF IPMSM..........2457

Kano, Yoshiaki

NOISE REDUCTION METHOD BY INJECTED FREQUENCY CONTROL FOR POSITION SENSORLESS CONTROL OF PERMANENT MAGNET SYNCHRONOUS MOTOR..........2465

Taniguchi, Shun ; Yasui, Kazuya ; Yuki, Kazuaki

FORCE SENSORLESS BILATERAL CONTROL USING A DYNAMICAL ASYMMETRIC COMPENSATOR..........2470

Hama, Ryota ; Imai, Jun ; Takahashi, Akiko ; Funabiki, Shigeyuki

DESIGN OF M-IPD CONTROLLER OF MULTI-INERTIA SYSTEM USING DIFFERENTIAL EVOLUTION..........2476

Ikeda, Hidehiro ; Tsuyoshi, Hanamoto

A GUIDE TO DESIGN DISTURBANCE OBSERVER BASED MOTION CONTROL SYSTEMS..........2483

Sariyildiz, Emre ; Ohnishi, Kouhei

IDENTIFICATION OF TWO-MASS MECHANICAL SYSTEMS USING TORQUE EXCITATION: DESIGN AND EXPERIMENTAL EVALUATION..........2489

Saarakkala, Seppo E. ; Hinkkanen, Marko

INDUCTOR LOSS CALCULATION OF COUPLED INDUCTORS FOR HIGH POWER DENSITY BOOST CONVERTER..........2497

Itoh, Yuki ; Kimura, Shota ; Imaoka, Jun ; Yamamoto, Masayoshi

1.2KW DUAL-ACTIVE BRIDGE CONVERTER USING SIC POWER MOSFETS AND PLANAR MAGNETICS..........2503

De, D. ; Castellazzi, A. ; Lamantia, A.

ANALYSIS OF HYSTERESIS AND EDDY-CURRENT LOSSES FOR A MEDIUM-FREQUENCY TRANSFORMER IN AN ISOLATED DC-DC CONVERTER 2511

Nakahara, Mizuki ; Wada, Keiji

EXPERIMENTAL VERIFICATION OF CAPACITIVE POWER TRANSFER USING ONE PULSE SWITCHING ACTIVE CAPACITOR FOR PRACTICAL USE 2517

Kitabayashi, Tatsuaki ; Funato, Hirohito ; Kobayashi, Hiroya ; Yamaichi, Katsuya

A SINGLE-STAGE HIGH-PF DRIVER FOR SUPPLYING A T8-TYPE LED LAMP 2523

Cheng, Chun-An ; Chang, Chien-Hsuan ; Cheng, Hung-Liang ; Chung, Tsung-Yuan

ELIMINATION OF ELECTROLYTIC CAPACITOR IN AC-DC SYSTEM OF LED DRIVER 2529

Mustapa, Rijalul Fahmi ; Hidayat, Nabil M ; Tukiman, Rahayu

A NOVEL BRIDGELESS BOOST HALF-BRIDGE ZVS-PWM SINGLE-STAGE UTILITY FREQUENCY AC-HIGH FREQUENCY AC RESONANT CONVERTER FOR DOMESTIC INDUCTION HEATERS 2533

Mishima, Tomoakzu ; Nakagawa, Yuki ; Nakaoka, Mutsuo

APPLICATION OF VIRTUAL VALIDATION SYSTEM FOR INVERTER HEAT PUMP SYSTEM 2541

Kanamori, Masaki ; Noda, Koji ; Endo, Takahisa ; Suzuki, Nobuyuki

TEST SETUP FOR ACCELERATED TEST OF HIGH POWER IGBT MODULES WITH ONLINE MONITORING OF VCE AND VF VOLTAGE DURING CONVERTER OPERATION 2547

de Vega, Angel Ruiz ; Ghimire, Pramod ; Pedersen, Kristian Bonderup ; Trintis, Ionut ; Beczckowski, Szymon ; Munk-Nielsen, Stig ; Rannestad, Bjorn ; Thogersen, Paul

DESIGN OF HIGH-SPEED IGBT-BASED SWITCHING MODULES FOR PULSED POWER APPLICATIONS 2554

Kluge, Andreas ; Goehler, Lutz ; Gueldner, Henry ; Trompa, Thomas ; Mory, David ; Segsa, Karl-Heinz

COMPARATIVE SUITABILITY EVALUATION OF REVERSE-BLOCKING IGBTS FOR CURRENT-SOURCE BASED CONVERTER 2562

De, Ankan ; Roy, Sudhin ; Bhattacharya, Subhashish

NEW REVERSE-CONDUCTING IGBT (1200V) WITH REVOLUTIONARY COMPACT PACKAGE 2569

Takahashi, K. ; Yoshida, S. ; Noguchi, S. ; Kuribayashi, H. ; Nashida, N. ; Kobayashi, Y. ; Kobayashi, H. ; Mochizuki, K. ; Ikeda, Y. ; Ikawa, O.

AN IMPROVED MODULATED CARRIER CONTROL OF SINGLE-PHASE CCM BOOST PFC CONVERTER 2575

Kim, Hyejin ; Cho, Bo-Hyung ; Choi, Hangseok

MODIFIED INTERLEAVED CURRENT SENSORLESS CONTROL FOR THREE-LEVEL BOOST PFC CONVERTER WITH ASYMMETRIC LOADS 2580

Chen, Hung-Chi ; Liao, Jhen-Yu

A NOVEL CRITICAL-CONDUCTION-MODE BRIDGELESS INTERLEAVED BOOST PFC RECTIFIER 2587

Cao, Guoen ; Kim, Hee-Jun

ANALYSIS AND DESIGN OF A PUSH-PULL SINGLE-STAGE FLYBACK POWER FACTOR CORRECTOR 2593

Lo, Yu-Kang ; Chiu, Huang-Jen ; Liu, Yu-Chen ; Lin, Chung-Yi ; Cheng, Shih-Jen ; Yang, CS

LINEAR OVER-MODULATION STRATEGY FOR CURRENT CONTROL IN PHOTOVOLTAIC INVERTER 2598

Park, Yongsoon ; Sul, Seung-Ki ; Hong, Ki-Nam

DESIGN OF DECENTRALIZED VOLTAGE CONTROL FOR PV INVERTERS TO MITIGATE VOLTAGE RISE IN DISTRIBUTION POWER SYSTEM WITHOUT COMMUNICATION 2606

Lee, Tzung-Lin ; Yang, Shih-Sian ; Hu, Shang-Hung

STABILITY ANALYSIS AND ACTIVE DAMPING FOR LLCL-FILTER BASED GRID-CONNECTED INVERTERS 2610

Huang, Min ; Blaabjerg, Frede ; Loh, Poh Chiang ; Wu, Weimin

INTEGRATED COMMON AND DIFFERENTIAL MODE FILTER APPLIED TO A SINGLE-PHASE TRANSFORMERLESS PV MICROINVERTER WITH LOW LEAKAGE CURRENT 2618

Figueredo, Ricardo Souza ; de Carvalho, Kelly Caroline Mingorancia ; Matakas, Lourenco

DESIGN AND INTEGRATION OF INTERPHASE INDUCTORS FOR INTERLEAVED THREE PHASE VOLTAGE-SOURCE-INVERTERS IN DC-FED MOTOR DRIVE SYSTEMS 2626

Zhang, Xuning ; Boroyevich, Dushan ; Burgos, Rolando

A NOVEL TRANSFORMER MODEL USING MAGNETIC CIRCUIT 2632

Nakamurame, Fuminori ; Ise, Toshifumi

HARDWARE-IN-THE-LOOP SIMULATION OF A MACHINE MODEL WITH REAL-TIME ANIMATION 2638

Xiaojie Zhuang ; Hibino, Shinya ; Harakawa, Masaya ; Terabe, Ryosuke ; Ozaki, Takayuki ; Nagano, Tetsuaki

DEVELOPMENT OF REAL TIME DIGITAL SIMULATOR FOR SELF-COMMUTATED SVC TO SUPPRESS VOLTAGE FLICKER 2644

Terao, Yutaka ; Shishida, Yasuhiro ; Tsuruma, Yoshinori ; Ishizuka, Tomotsugu ; Aoyama, Fumio ; Yoshino, Teruo ; Kato, Yutaka ; Belanger, Jean

OPERATIONAL ASPECTS AND POWER ARCHITECTURE DESIGN FOR A MICROGRID TO INCREASE THE USE OF RENEWABLE ENERGY IN WIRELESS COMMUNICATION NETWORKS 2649

Kwasinski, Alexis ; Kwasinski, Andres

P+ MULTIPLE RESONANT CONTROL FOR OUTPUT VOLTAGE REGULATION OF MICROGRID WITH UNBALANCED AND NONLINEAR LOADS 2656

Kyungbae Lim ; Jaeho Choi ; Juyoung Jang ; Junghum Lee ; Jaesig Kim

130MVA-STATCOM FOR TRANSIENT STABILITY IMPROVEMENT 2663

Imanishi, Takao ; Nagatomo, Yoshinobu ; Iwasaki, Shinya ; Masaki, Kenji ; Fujii, Toshiyuki ; Ieda, Jun

IMPROVED DROOP CONTROLLER FOR MICROGRID INVERTER CONSIDERING THE LINE IMPEDANCE MISMATCHING 2668

Du Yan ; Liuchen Chang ; Meiqin Mao ; Jianhui Su ; Ning Liu

SUPPRESSION CONTROL METHOD FOR IRON LOSS OF MATRIX MOTOR UNDER FLUX WEAKENING UTILIZING INDIVIDUAL WINDING CURRENT CONTROL 2673

Hijikata, Hiroki ; Akatsu, Kan ; Miyama, Yoshihiro ; Arita, Hideaki ; Daikoku, Akihiro

PERFORMANCE ANALYSIS OF A NEW CONCENTRATEDWINDING INTERIOR PERMANENT MAGNET SYNCHRONOUS MACHINE UNDER FIELD ORIENTED CONTROL 2679

Nguyen, D. ; Dutta, R. ; Fletcher, J. ; Rahman, F. ; Lovatt, Howard

ONLINE PARTICLE SWARM OPTIMIZATION FOR SENSORLESS IPMSM DRIVES CONSIDERING PARAMETER VARIATION 2686

Song, Z.Q. ; Xiao, D. ; Rahman, M.F.

A DTC-PWM CONTROL SCHEME OF PMSM BASED ON 12-SECTORS DIVISION AND SPEED INFORMATION 2693

Yunchang Kwak ; Jin-Woo Ahn ; Dong-Hee Lee

CONTROL OF POWER FLOW BETWEEN THE WIND GENERATOR AND NETWORK 2700

Stumpf, Peter ; Nagy, Istvan ; Vajk, Istvan

ADVANCES IN NANOGRID TECHNOLOGY AND ITS INTEGRATION INTO RURAL ELECTRIFICATION IN INDIA 2707

Mishra, Santanu ; Ray, Olive

STUDY AND IMPLEMENTATION OF SEVEN-LEVEL INVERTER USING COUPLED INDUCTOR AND SWITCHED-CAPACITOR 2714

Yi-Chun Lin ; Jiann-Fuh Chen ; Wen-Chien Hsu ; Sheng-Kai Kao

CASCADED MULTILEVEL CONVERTER BASED BIDIRECTIONAL INDUCTIVE POWER TRANSFER (BIPT) SYSTEM 2722

Bac Xuan Nguyen ; Vilathgamuwa, D.M. ; Foo, Gilbert ; Ong, Andrew ; Sampath, Prasad K. ; Madawala, Udaya K.

UNDERSAMPLING CONTROL OF A BIDIRECTIONAL CASCADED BUCK+BOOST DC-DC CONVERTER 2729

Rosekeit, Martin ; Joebges, Philipp ; Lelie, Markus ; Sauer, Dirk Uwe ; De Doncker, Rik W.

SUB-MICROSECOND RESPONSE DIGITAL CONTROLLER FOR POL 2737

Nonaka, Hirotaka ; Ishizuka, Yoichi ; Mii, Kenji ; Takenami, Fumiaki ; Kanemoto, Daisuke

GAIN CONTROLLED HIGH EFFICIENCY POWER FACTOR CORRECTION CIRCUIT 2745

Yonezawa, Yu ; Nakao, Hiroshi ; Sasaki, Tomotake ; Matsui, Yoshinobu ; Nakashima, Yoshiyasu ; Kaneko, Junji ; Shimamori, Hiroshi ; Yoshino, Yukio ; Hisato, Hosoyama ; Atsushi, Manabe ; Motizuki, Shun ; Yamashita, Shigeharu

DESIGN OF QUASI-RESONANT FLYBACK CONVERTER CONTROL IC WITH DCM AND CCM OPERATION 2750

Kai-Hui Chen ; Tsorng-Juu Liang

LOAD TRANSIENT RESPONSE IMPROVEMENT BASED ON PID CONTROL 2754

Yau, Y.T. ; Hwu, K.I.

AN ACTIVE-CLAMPING FORWARD CONVERTER WITH NON-LINEAR STEP-DOWN CONVERSION 2758

Jing-Yuan Lin ; Yu-Kang Lo ; Huang-Jen Chiu ; Chao-Fu Wang ; Chien-Yu Lin

SWITCHING LOSS MINIMIZATION OF 3-PHASE INTERLEAVED BIDIRECTIONAL DC-DC CONVERTER 2763

Eui-Cheol Nho ; Jae-Hun Jung ; Hak-Soo Kim ; In-Dong Kim ; Heung-Geun Kim ; Tae-Won Chun

MODIFIED THREE-PHASE THREE-LEVEL DC-DC CONVERTER -ADOPTING ASYMMETRICAL DUTY CYCLE CONTROL 2768

Yue Chen ; Xuling Chen ; Liu, Fuxin ; Ruan, Xinbo

DEADBEAT CONTROL OF POWER LEVELING UNIT WITH BIDIRECTIONAL BUCK/BOOST DC/DC CONVERTER 2775

Hamasaki, Shin-ichi ; Mukai, Ryosuke ; Yano, Yoshihiro ; Tsuji, Mineo

DESIGN OF OPTIMIZED ON-OFF CONTROL TO IMPROVE EFFICIENCY OF PARALLELED CONVERTER SYSTEM 2781

Kohama, Teruhiko ; Sogawa, Yuki ; Tsuji, Satoshi

EFFICIENCY IMPROVEMENTS IN A SINGLE ACTIVE BRIDGE MODULAR DC-DC CONVERTER WITH SNUBBER CAPACITANCE OPTIMISATION 2787

Ting, Yeh ; de Haan, Sjoerd ; Ferreira, Jan A.

A WIRELESS POWER TRANSFER SYSTEM OPTIMIZED FOR HIGH EFFICIENCY AND HIGH POWER APPLICATIONS 2794

Bani Shamseh, Mohammad ; Kawamura, Atsuo ; Yuzurihara, Itsuo ; Takayanagi, Atsushi

NON-ITERATIVE LCL FILTER DESIGN FOR THREE-PHASE TWO-LEVEL VOLTAGE-SOURCE PWM CONVERTERS 2802

Byung-Geuk Cho ; Seung-Ki Sul

DSP-BASED INTERLEAVED BUCK POWER FACTOR CORRECTOR 2810

Yu-Chen Liu ; Tsan Chen ; Po-Jung Tseng ; Yu-Kang Lo ; Huang-Jen Chiu

THE AVERAGE MODEL OF A THREE-PHASE THREE-STAGE POWER ELECTRONIC TRANSFORMER 2815

Shaodi Ouyang ; Liu, Jinjun ; Wang, Xinyu ; Wang, Xiaojian ; Fei Meng ; Riffat, Javid

A MULTI-CARRIER PWM FOR AC-DC-AC CONVERTER WITHOUT DC LINK ELECTROLYTIC CAPACITOR 2821

Chung-Chuan Hou ; Hsin-Ping Su

A DECOUPLING OFFSET-BASED PWM CONTROL FOR A MULTILEVEL INVERTER UNDER DC VOLTAGE UNBALANCE 2826

Nho Van Nguyen ; Tam Khanh Tu Nguyen ; Lee, Hong-Hee

?-? PARETO OPTIMIZATION OF 3-PHASE 3-LEVEL T-TYPE AC-DC-AC CONVERTER COMPRISING SI AND SIC HYBRID POWER STAGE 2834

Uemura, Hirofumi ; Krismer, Florian ; Okuma, Yasuhiro ; Kolar, Johann W.

PRACTICAL INVESTIGATION OF THE GATE BIAS EFFECT ON THE REVERSE RECOVERY BEHAVIOR OF THE BODY DIODE IN POWER MOSFETS 2842

Lindberg-Poulsen, Kristian ; Petersen, Lars Press ; Ouyang, Ziwei ; Andersen, Michael A.E.

AN ONLINE VCE MEASUREMENT AND TEMPERATURE ESTIMATION METHOD FOR HIGH POWER IGBT MODULE IN NORMAL PWM OPERATION 2850

Ghimire, Pramod ; de Vega, Angel Ruiz ; Beczkowski, Szymon ; Munk-Nielsen, Stig ; Rannested, Bjorn ; Thogersen, Paul Bach

EVALUATION ON IRON LOSS CHARACTERISTICS IN SERIES CONNECTION AND PARALLEL CONNECTION OF LOADS WITH INVERTER EXCITATION 2856

Odawara, Shunya ; Fujisaki, Keisuke

LOSS AND THERMAL MODEL FOR POWER SEMICONDUCTORS INCLUDING DEVICE RATING INFORMATION 2862

Ma, K. ; Bahman, A.S. ; Beczkowski, S.M. ; Blaabjerg, F.

IMPROVING RELIABILITY OF IGBT SURFACE ELECTRODE FOR 200 C OPERATION 2870

Nishimura, Tomohiro ; Ikeda, Yoshinari ; Hokazono, Hiroaki ; Mochizuki, Eiji ; Takahashi, Yoshikazu

INFLUENCE OF CARRIER FREQUENCY ON IRON LOSS TAKING ACCOUNT OF DEAD TIME EFFECT 2874

Kogi, Ryosuke ; Odawara, Shunya ; Fujisaki, Keisuke

DECREASE OF SIC-BJT DRIVER LOSSES BY ONE-STEP COMMUTATION 2881

Barth, Henry ; Hofmann, Wilfried

POWER PROFILE BASED SELECTION AND OPERATION OPTIMIZATION OF PARALLEL-CONNECTED POWER CONVERTER COMBINATIONS 2887

Vogt, T. ; Peters, A. ; Frohleke, N. ; Bocker, J. ; Kempen, S.

A NOVEL POWER LOSS CALCULATION METHOD FOR IGBTS IN POWER CONVERTERS VIA CHAOTIC SPWM CONTROL 2893

Boyu Wang ; Li, Hong ; Xiaojie You ; Trillion Zheng

LOSS ANALYSIS AND SOFT-SWITCHING CHARACTERISTICS OF FLYBACK-FORWARD HIGH GAIN DC/DC CONVERTER WITH GAN FET 2899

Zhang Yajing ; Zheng, Trillion Q. ; Li Yan

INSULATED METAL SUBSTRATE FOR POWER MODULES USING ANODIC OXIDE FILM OF ALUMINUM 2904

Tokuyama, Takeshi ; Kusukawa, Jumpei ; Nakatsu, Kinya

A FAST-TRANSIENT-RESPONSE BUCK CONVERTER WITH SPLIT-TYPE III COMPENSATION AND CHARGE-PUMP CIRCUIT TECHNIQUE 2910

Chen, Jiann-Jong ; Wei-Ting Hsu ; Jih-Hua Yu ; Hwang, Yuh-Shyan ; Cheng-Chieh Yu

ADVANTAGES OF LOW PARASITIC INDUCTANCE PACKAGES OF POWER MOSFET FOR SERVER POWER APPLICATIONS 2914

Wonsuk Choi ; Dongkook Son ; Dongwook Kim

MODULAR INTEGRATION OF A MATRIX CONVERTER 2920

Solomon, Adane Kassa ; Skuriat, Robert ; Castellazzi, Alberto ; Wheeler, Pat

A MODULAR NANOSECOND PULSE GENERATION SYSTEM FOR PLASMA-ASSISTED IGNITION 2926

Peng Gao ; Fletcher, John ; O'Byrne, Sean

DEVELOPMENT OF A SINGLE SWITCH CELL FOR MODULAR NANOSECOND PULSE GENERATION SYSTEMS 2932

Peng Gao ; Fletcher, John ; O'Byrne, Sean

ADVANTAGE OF SUPER JUNCTION MOSFET FOR POWER SUPPLY APPLICATION 2939

Tabira, K. ; Watanabe, S. ; Shimatou, T. ; Watashima, T. ; Takenoiri, S.

STUDY ON AN ACCURATE CALCULATION OF THE CONDUCTED EMI NOISE OF THE POWER CONVERTERS 2944

Omata, Shinpei ; Shimizu, Toshihisa

AN EXACT DISCRETE-TIME MODEL CONSIDERING DEAD-TIME NONLINEARITY FOR AN H-BRIDGE GRID-CONNECTED INVERTER 2950

Xie, Ruiliang ; Hao, Xiang ; Yang, Xu ; Chen, Wenjie ; Huang, Lang ; Chao Wang

THEORETICAL ANALYSIS OF THE DUALITY PRINCIPLE APPLIED TO INTERLEAVED TOPOLOGIES 2954

Caris, M.L.A. ; Huisman, H. ; Duarte, J.L.

A NEW IMPEDANCE MEASUREMENT METHOD BASED ON HIGH FREQUENCY COMPENSATION 2960

Yue, Xiaolong ; Zhuo, Fang ; Hao Yi

NUMERICAL AND EXPERIMENTAL INVESTIGATION OF PARASITIC EDGE CAPACITANCE FOR PHOTOVOLTAIC PANEL 2967

Wenjie Chen ; Xiaomei Song ; Hao Huang ; Xu Yang

VEHICLE INTERIOR NOISE CONTROL OF ULTRA-COMPACT ELECTRIC VEHICLE (FUNDAMENTAL CONSIDERATION USING RECTANGULAR ENCLOSURE) 2972

Kato, Taro ; Kato, Hideaki ; Oshinoya, Yasuo ; Suzuki, Ryosuke ; Hasegawa, Shinya

CONSIDERATION FOR THE PROPAGATION PATH OF CONDUCTIVE NOISE IN AIR CONDITIONERS 2977

Tokiwa, Tsuyoshi ; Kanamori, Masaki ; Endo, Takahisa ; Iida, Mikiya ; Ogasawara, Satoshi ; Yizhanyi Tang

IRON LOSS EVALUATION OF IRON POWDER CORE SUITABLE FOR INDUCTOR USED IN POWER CONVERTERS 2983

Mori, Tomohiro ; Igarashi, Kazunori ; Kanagawa, Kinji ; Yamashita, Nobuyuki ; Shimizu, Toshihisa ; Bizen, Yosio

OPTIMIZED TUNING METHOD OF STATIONARY FRAME PROPORTIONAL RESONANT CURRENT CONTROLLERS 2988

Martinz, Fernando Ortiz ; de Carvalho, Kelly Caroline Mingorancia ; Ama, Naji Rajai Nasri ; Komatsu, Wilson ; Matakas, Lourenco

INSTANTANEOUS POWER THEORY APPLIED TO POWER CONDITIONING UNDER DISTORTED MAINS VOLTAGES: A MATLAB/SIMULINK APPROACH 2996

Nicolae, Petre-Marian ; Popa, Lucian-Dinut ; Nicolae, Marian-Stefan ; Nicolae, Ileana-Diana

THE RESEARCH ON RELIABILITY AND REAL-TIME OF THE SCHEME OF PROCESS LAYER GOOSE NETWORK IN SMART SUBSTATION BASED ON ARTIFICIAL COBWEB TOPOLOGY STRUCTURE 3002

Liu, Xiaosheng ; Zhu, Honglin ; Xu, Dianguo ; Li, Yanxiang

EFFICIENCY IMPROVEMENT OF A SELF-START TYPE PERMANENT MAGNET SYNCHRONOUS MOTOR 3007

Saikusa, H. ; Arikawa, S. ; Higuchi, T. ; Yokoi, Y. ; Abe, T.

CONSIDERATION OF OPTIMAL NUMBER OF POLES AND FREQUENCY FOR HIGH-EFFICIENCY PERMANENT MAGNET MOTOR 3012

Misu, Daisuke ; Matsushita, Makoto ; Takeuchi, Katsutoku ; Oishi, Koji ; Kawamura, Mitsuhiro

BASIC STUDY ON THE SUITABLE STRUCTURE OF A PERMANENT MAGNET SYNCHRONOUS MOTOR WITH A POWDER MAGNETIC CORE 3018

Hashimoto, Shizuka ; Sanada, Masayuki ; Morimoto, Shigeo ; Inoue, Yukinori

CHARACTERISTICS OF A HALF-WAVE RECTIFIED BRUSHLESS SYNCHRONOUS GENERATOR 3024

Hirakawa, Yuki ; Higuchi, Tsuyoshi ; Yokoi, Yuichi ; Abe, Takashi

MODELING OF WOUND ROTOR SYNCHRONOUS MACHINES CONSIDERING HARMONICS, GEOMETRIC SALIENCIES AND SATURATION INDUCED SALIENCIES 3029

Rambetius, Alexander ; Luthardt, Sven ; Piepenbreier, Bernhard

DESIGN AND COMPARISON OF HIGH FREQUENCY TRANSFORMERS USING FOIL AND ROUND WINDINGS 3037

Iyer, Kartik V ; Robbins, William P ; Mohan, Ned

A METHOD TO CALCULATE THE PERFORMANCE OF LINEAR INDUCTION MOTORS USING SIMPLE TWO-PHASE MODEL 3044

Hirahara, Hideaki ; Yamamoto, Shu ; Ara, Takahiro ; Shimizu, Toshihisa

AN ESP DOWNHOLE PARAMETERS MONITORING SYSTEM BASED ON CURRENT LOOP TRANSMISSION METHOD 3050

Jin Miaoxin ; Zhang Wei ; Gao Qiang ; Xu Dianguo

BENDING MAGNETIC LEVITATION CONTROL FOR THIN STEEL PLATE (EXPERIMENTAL CONSIDERATION USING SLIDING MODE CONTROL) 3055

Yonezawa, Hikaru ; Narita, Takayoshi ; Oshinoya, Yasuo ; Marumori, Hiroki ; Hasegawa, Shinya

TRANSFORMER WINDING LOSSES WITH ROUND CONDUCTORS FOR DUTY-CYCLE REGULATED SQUARE WAVES 3061

Iyer, Kartik V ; Robbins, William P ; Basu, Kaushik ; Mohan, Ned

SIMULATION OF RESIN MOLDED TYPE SENSOR IN POLE SWITCH FOR POWER DELIBERY SYSTEMS 3067

Furukawa, Tatsuya ; Muta, Shoichiro ; Fukumoto, Hisao ; Itoh, Hideaki ; Ohchi, Masashi

ROBUST STARTUP CONTROL OF SENSORLESS PMSM DRIVES WITH SELF-COMMISSIONING 3072

Lin, Chiao-Chien ; Tzou, Ying-Yu

POSITION SENSORLESS CONTROL OF PMSM WITH A LOW-FREQUENCY SIGNAL INJECTION 3079

Nimura, Tomohiro ; Doki, Shinji ; Fujitsuna, Masami

A COMPARISON OF DIFFERENT SENSORLESS POSITION ACQUISITION METHODS AT LOW SPEEDS FOR A PERMANENT MAGNET SYNCHRONOUS MACHINE IN VEHICLE APPLICATIONS 3085

Lehmann, Oliver ; Zehelein, Matthias ; Schuster, Johannes ; Roth-Stielow, Jorg

STABILITY COMPARISON OF IPMSM SENSORLESS VECTOR CONTROL SYSTEMS USING EXTENDED EMF 3093

Tsuji, Mineo ; Mizusaki, Hiroshi ; Hamasaki, Sin-ichi

INDUCTION MACHINE BASED FLYWHEEL SPEED ESTIMATION AT STAND-BY MODE 3099

Liu, Rongqiang ; Xu, David

SYMMETRICAL SIGNALING SYSTEM FOR SENSOR-LESS SRM DRIVE 3106

Yamamoto, Kenji ; Takahashi, Hisashi ; Ushiro, Nobumasa ; Shirasawa, Koki

DIGITAL INTEGRATORS FOR CONDITION MONITORING: A DC AND MULTITONE SIGNAL ANALYSIS 3111

Peretti, L.

AUDIBLE NOISE REDUCTION METHOD IN IPMSM POSITION SENSORLESS CONTROL BASED ON HIGH-FREQUENCY CURRENT INJECTION 3119

Tauchi, Yuki ; Kubota, Hisao

A NOVEL DESIGN FOR INDUCTION MOTOR FLUX ESTIMATION USING IMPULSIVE OBSERVER 3124

Peng Wang ; Yan Li ; Jianwen Zhang ; Xu Cai ; Zhengzhi Han

LOAD TORQUE AND INERTIA SIMULATION BASED ON DOUBLE-STATOR PERMANENT-MAGNET SYNCHRONOUS MOTOR 3129

Zhe Wang ; Mingyan Wang ; Ben Guo ; Chai Feng

INDEPENDENT SPEED AND POSITION CONTROL OF TWO PERMANENT MAGNET SYNCHRONOUS MOTORS FED BY A FOUR-LEG INVERTER 3134

Kubo, Yuji ; Moroi, Takayuki ; Kouki, Matsuse ; Kubota, Hisao ; Rajashekara, Kaushik

MINIMIZATION OF STATOR CURRENTS FOR MONO INVERTER DUAL PARALLEL PMSM DRIVE SYSTEM 3140
Yongjae Lee ; Ha, Jung-Ik

PERFORMANCE COMPARISON OF INVERTER AND DRIVE CONFIGURATIONS WITH OPEN-END AND STAR-CONNECTED WINDINGS 3145
Neubert, Markus ; Koschik, Stefan ; De Doncker, Rik W.

INPUT CURRENT HARMONICS REDUCTION CONTROL FOR ELECTROLYTIC CAPACITOR LESS INVERTER BASED IPMSM DRIVE SYSTEM 3153
Abe, Kodai ; Ohishi, Kiyoshi ; Haga, Hitoshi

NONCONTACT GUIDE SYSTEM FOR TRAVELING ELASTIC STEEL PLATES (THEORETICAL STUDY ON THE SHAPE OF TRAVELING STEEL PLATE) 3159
Sakaba, Kouichi ; Hasegawa, Shinya ; Narita, Takayoshi ; Oshinoya, Yasuo

ACTIVE SEAT SUSPENSION FOR ULTRA-COMPACT VEHICLE (FUNDAMENTAL CONSIDERATION ON ELECTROMYOGRAM WHEN FALL FROM THE BUMP) 3162
Mashino, Masahiro ; Sunaga, Keita ; Hasegawa, Shinya ; Ishida, Masaki ; Kato, Hideaki ; Oshinoya, Yasuo

ADAPTIVE CURRENT TRACKING OF THREE-PHASE ACTIVE POWER FILTER USING BACKSTEPPING CONTROL 3168
Yunmei Fang ; Juntao Fei ; Shixi Hou ; Weili Dai

FAST IDENTIFICATION OF RESONANCE CHARACTERISTIC FOR 2-MASS SYSTEM WITH ELASTIC LOAD 3174
Ming Yang ; Liang Hao ; Dianguo Xu

AUTONOMOUS NAVIGATION SYSTEM BASED ON COLLISION DANGER-DEGREE FOR UNMANNED GROUND VEHICLE 3179
Yasuno, Takashi ; Tanaka, Daiki ; Kuwahara, Akinobu

A HIGH-PERFORMANCE BIDIRECTIONAL DC-DC CONVERTER FOR DC MICRO-GRID SYSTEM APPLICATION 3185
Shu-Wei Kuo ; Yu-Kang Lo ; Huang-Jen Chiu ; Shih-Jen Cheng ; Chung-Yi Lin ; Yang, CS

VOLUME 5

IMPROVEMENT IN EFFICIENCY OF LED LIGHTING SYSTEM 3190
Hwu, K.I. ; Jiang, W.Z. ; Jenn-Jong Shieh

COMPARISON AND EVALUATION OF VIBRATION-BASED PIEZOELECTRIC POWER GENERATORS 3194
Basari, Amat A. ; Awaji, Sosuke ; Hashimoto, Seiji ; Kasai, Makoto ; Suto, Kenji ; Kumagai, Shunji ; Kasai, Makoto ; Suto, Kenji ; Wei Jiang ; Shuren Wang

BATTERY SELECTION FOR HYBRID ENERGY SYSTEMS AND THERMAL MANAGEMENT IN ARCTIC CLIMATES 3200
Kalogera, Maria ; Bauer, Pavol

100KW PV PCS WITH NATURAL CONVECTION COOLING FOR OUTDOOR INSTALLATION 3207
Jin, Yasuhiro ; Matsuoka, Kazumasa ; Takahashi, Takehiro ; Takahashi, Nobuhiro

A NEW PLL BASED ON FAST POSITIVE AND NEGATIVE SEQUENCE DECOMPOSITION ALGORITHM WITH MATRIX OPERATION UNDER DISTORTED GRID CONDITIONS 3213
Shaohua Sun ; Hongqi Ben ; Tao Meng ; Jinyong Zhang

PERFORMANCE IMPROVEMENT OF PHOTOVOLTAIC POWER GENERATION SYSTEMS USING ON-OFF CONTROL METHODS 3218
Kenji, Matsumoto ; Nomura, Shinichi

LOW VOLTAGE PV POWER INTEGRATION INTO MEDIUM VOLTAGE GRID USING HIGH VOLTAGE SIC DEVICES 3225
Chattopadhyay, Ritwik ; Bhattacharya, Subhashish ; Foureaux, Nicole C. ; Silva, Sidelmo M. ; Braz Cardoso, F. ; de Paula, Helder ; Pires, Igor A. ; Cortizio, Porfirio C. ; Moraes, Lenin ; de S.Brito, Jose A.

A NOVEL GLOBAL MAXIMUM POWER POINT TRACKING METHOD FOR PHOTOVOLTAIC GENERATION SYSTEM OPERATING UNDER PARTIALLY SHADED CONDITION 3233
Jing-Hsiao Chen ; Yu-Shan Cheng ; Shun-Chung Wang ; Huang, Jia-Wei ; Liu, Yi-Hua

AN APPLICATION OF Z-SOURCE CONVERTER TO BATTERIES CHARGE WITH A PHOTOVOLTAIC SYSTEM 3239
Razik, H. ; Zitouni, Y. ; Maret, C.

PCS WITH SCANNING-TYPE MPPT CONTROL FOR INDUSTRIAL GRID-CONNECTED PV POWER GENERATION SYSTEM 3244
Itako, Kazutaka

FEASIBLE METHOD OF CALCULATING LEAKAGE REACTANCE OF 9-WINDING TRANSFORMER FOR HIGH-VOLTAGE INVERTER SYSTEM 3249
Fukumoto, Hisao ; Furukawa, Tatsuya ; Itoh, Hideaki ; Ohchi, Masashi

HIGH POWER HVDC-DC CONVERTERS FOR THE INTERCONNECTION OF HVDC LINES WITH DIFFERENT LINE TOPOLOGIES 3255
Schon, Andre ; Bakran, Mark-M.

CHARACTERIZATION OF A CURRENT SHUNT AND AN INDUCTIVE VOLTAGE DIVIDER FOR PMU CALIBRATION 3263
Kon, Saytaro ; Yamada, Tatsuji

DISTRIBUTED SERIES/HYBRID-SHUNT COMPENSATION FOR HARMONIC MITIGATION IN COMMERCIAL FACILITIES 3270

Diniz, Rogerio Azevedo ; Pires, Igor A. ; Franca, Gleisson J. ; Cardoso, Braz J.

ROBUST CONTROL DESIGN FOR THE VOLTAGE TRACKING LOOP OF A DVR.................3278

Ferrari, Bruno Augusto ; Ama, Naji Rajai Nasri ; de Carvalho, Kelly Caroline Mingorancia ; Martinz, Fernando Ortiz ; Matakas, Lourenco

MULTI-PORT SOLID STATE TRANSFORMER FOR INTER-GRID POWER FLOW CONTROL 3286

Roy, Sudhin ; De, Ankan ; Bhattacharya, Subhashish

REACTIVE POWER CONTROL STRATEGY BASED ON DC CAPACITOR VOLTAGE CONTROL FOR ACTIVE LOAD BALANCER IN THREE-PHASE FOUR-WIRE DISTRIBUTION SYSTEMS 3292

Tint Soe Win ; Hisada, Yoshihiro ; Tanaka, Toshihiko ; Hiraki, Eiji ; Okamoto, Masayuki ; Lee, Seong Ryong

VOLTAGE SAG RIDE-THROUGH PERFORMANCE OF VIRTUAL SYNCHRONOUS GENERATOR.................3298

Alipoor, Jaber ; Miura, Yushi ; Ise, Toshifumi

CONTROL OF DISTRIBUTED GENERATION SYSTEMS UNDER UNBALANCED VOLTAGE CONDITIONS 3306

Kabiri, R. ; Holmes, D.G. ; McGrath, B.P.

STABILITY ANALYSIS OF GRID-CONNECTED INVERTERS WITH LCL-FILTER BASED ON HARMONIC BALANCE AND FLOQUET THEORY...3314

Jing Bian ; Hong Li ; Zheng, Trillion Q.

COMPARATIVE EVALUATION OF PASSIVE DAMPING TOPOLOGIES FOR PARALLEL GRID-CONNECTED CONVERTERS WITH LCL FILTERS 3320

Beres, Remus ; Wang, Xiongfei ; Blaabjerg, Frede ; Bak, Claus Leth ; Liserre, Marco

STUDY AND IMPLEMENTATION OF A SEPIC LED DRIVER WITH ADJUSTABLE OUTPUT VOLTAGE 3328

Po-Jung Tseng ; Yu-Chen Liu ; Yu-Kang Lo ; Chiu, Huang-Jen ; Yun-Chu Chiu

AN INTERLEAVED SINGLE-STAGE LLC RESONANT CONVERTER USED FOR MULTI-CHANNEL LED DRIVING 3333

Chang, Chien-Hsuan ; Cheng, Chun-An ; Jinno, Masahito ; Cheng, Hung-Liang

A NOVEL TYPE OF WIRELESS V2H SYSTEM WITH BIDIRECTIONAL RESONANT SINGLE-ENDED INVERTER 3341

Fukuoka, Hiroki ; Iga, Yuichi ; Omori, Hideki ; Morizane, Tosimitsu ; Kimura, Noriyuki ; Nakaoka, Mutuo

DESIGN AND IMPLEMENTATION OF AN INTERLEAVED BCM BOOST PFC CONTROL IC 3346

Kuan-Hsien Chou ; Tsorng-Juu Liang ; Kai-Hui Chen ; Ji-Shiang Lee

LOW CAPACITIVE INDUCTORS FOR FAST SWITCHING DEVICES IN ACTIVE POWER FACTOR CORRECTION APPLICATIONS 3352

Hernandez, Juan C. ; Petersen, Lars P. ; Andersen, Michael A.E.

TEMPERATURE-ROBUST LC3 LED DRIVER WITH LOW THD, HIGH EFFICIENCY, AND LONG LIFE.........3358

Lee, Eun S. ; Choi, Bo H. ; Cheon, Jun P. ; Kim, Bong C. ; Rim, Chun T.

OPTIMIZING REPULSIVE LORENTZ FORCES FOR A LEVITATING INDUCTION COOKER.................3365

Zingerli, Claudius M. ; Nussbaumer, Thomas ; Kolar, Johann W.

DESIGN OF A MODULAR RESONANT CONVERTER FOR 25KV-8A DC POWER SUPPLY OF RF CAVITIES 3371

Siemaszko, Daniel ; Pittet, Serge ; Aguglia, Davide ; de Mallac, Louis

A NOVEL TRANSFORMER-LESS INTERLEAVED FOUR-PHASE HIGH STEP-DOWN DC CONVERTER WITH LOW SWITCH VOLTAGE STRESS 3379

Ching-Tasi Pan ; Chen-Feng Chuang ; Chia-Chi Chu ; Hao-Chien Cheng

EFFICIENCY IMPROVEMENT OF POWER SUPPLY WITH TRANSIENT CURRENT CIRCUIT USING DIGITAL CONTROL 3386

Takashita, Haruomi ; Shoyama, Masahito ; Yonezawa, Yu ; Nakashima, Yoshiyasu

ULTRA HIGH STEP-DOWN CONVERTER 3392

Yau, Y.T. ; Hwu, K.I.

DIGITAL CONTROL OF PWM INVERTER USING ULTRA HIGH SPEED NETWORK FOR FEEDBACK SIGNALS WITH COMMUNICATION DISTURBANCE OBSERVER BASED ON ROCKET I/O PROTOCOL 3397

Saito, Ryo ; Tsuchida, Kazuo ; Yokoyama, Tomoki

100 KHZ DC CHOPPER DIGITALLY GATE CONTROLLED WITH PARTIAL TURN- OFF SWITCHING USING SIC-MOSFET AND FPGA 3403

Tsuruta, Yukinori ; Kawamura, Atsuo

VARIABLE CARRIER DEADBEAT CONTROL WITH DIGITAL HYSTERESIS METHOD USING SOC-FPGA FOR UTILITY INTERACTIVE INVERTER 3410

Ohashi, Shunsuke ; Yoshida, Morito ; Yokoyama, Tomoki

A SPACE VECTOR MODULATION STRATEGY FOR THREE-LEVEL OPERATION BASED ON DUAL TWO-LEVEL VOLTAGE SOURCE INVERTERS 3417

Kumsuwan, Yuttana ; Srirattanawichaikul, Watcharin

INVESTIGATION ON THE PARALLEL OPERATION OF ALL-GAN POWER MODULE AND THERMAL PERFORMANCE EVALUATION 3425

Cheng, Stone ; Po-Chien Chou

FULL SILICON CARBIDE BOOST CHOPPER MODULE FOR HIGH FREQUENCY AND HIGH TEMPERATURE OPERATION 3432

Pettersson, Sami ; Kicin, Slavo ; Holm, Toni ; Bianda, Enea ; Canales, Francisco

DEVELOPMENT OF ULTRAHIGH VOLTAGE SIC POWER DEVICES 3440

Fukuda, Kenji ; Okamoto, Dai ; Harada, Shinsuke ; Tanaka, Yasunori ; Yonezawa, Yoshiyuki ; Deguchi, Tadayoshi ; Katakami, Shuji ; Ishimori, Hitoshi ; Takasu, Shinji ; Arai, Manabu ; Takenaka, Kensuke ; Fujisawa, Hiroyuki ; Takei, Manabu ; Matsumoto, Kazushi ; Ohse, Naoyuki ; Ryo, Mina ; Ota, Chiharu ; Takao, Kazuto ; Mizukami, Makoto ; Kato, Tomohisa ; Izumi, Toru ; Hayashi, Toshihiko ; Nakayama, Koji ; Asano, Katsunori ; Okumura, Hajime ; Kimoto, Tsunenobu

HIGH SWITCHING PERFORMANCE OF 1.7KV, 50A SIC POWER MOSFET OVER SI IGBT FOR ADVANCED POWER CONVERSION APPLICATIONS 3447

Hazra, Samir ; De, Ankan ; Bhattacharya, Subhashish ; Lin Cheng ; Palmour, John ; Schupbach, Marcelo ; Hull, Brett ; Allen, Scott

CONTROL METHOD FOR FIVE LEVEL CONVERTER WITH COMMON FLYING CAPACITORS TO AVOID VOLTAGE LEVEL SKIP 3455

Wei Yan ; Hui Zhang ; Ogura, Kazuya ; Urushibata, Shota

LOW-COMPLEXITY ANALYTICAL APPROXIMATIONS OF SWITCHING FREQUENCY HARMONICS OF 3-PHASE N-LEVEL VOLTAGE-SOURCE PWM CONVERTERS 3460

Burkart, Ralph M. ; Kolar, Johann W.

DYNAMIC VOLTAGE BALANCING ALGORITHM FOR MODULAR MULTILEVEL CONVERTER WITH THREE-LEVEL FLYING CAPACITOR SUBMODULES 3468

Dekka, Apparao ; Wu, Bin ; Zargari, Navid R.

MODULAR MEDIUM VOLTAGE DRIVE FOR DEMANDING APPLICATIONS 3476

Dujic, Drazen ; Wahlstroem, Jonas ; Marrero Sosa, Juan Alberto ; Fritz, Dominik

ASYMMETRICAL FAULT RIDE-THROUGH OF THREE-PHASE PV SYSTEMS USING FOUR-WIRE DC-AC CONVERTERS 3482

Iyer, Shivkumar ; Bin Wu ; Yunwei Li ; Singh, B.N.

OPERATION MODE ANALYSIS FOR SOLVING THE PARTIAL SHADOW IN A NOVEL PV POWER GENERATION SYSTEM 3489

Qi Zhang ; Xiangdong Sun ; Yanru Zhong ; Lie Guo ; Matsui, Mikihiko

ANALYSIS OF PARTIAL POWER PROCESSING DISTRIBUTED MPPT FOR A PV POWERED ELECTRIC AIRCRAFT 3496

Marzouk, Ahmad Diab ; Fournier-Bidoz, Sebastien ; Yablecki, Jessica ; McLean, Kenneth ; Trescases, Olivier

IMPACTS OF RECTIFIER CIRCUIT LOADS ON ISLANDING DETECTION OF PHOTOVOLTAIC SYSTEMS 3503

Yoshida, Yoshiaki ; Suzuki, Hirokazu

INDUCTION MOTOR MADE OF SMC 3509

Morimoto, Masayuki ; Inamori, Mamiko

ESTIMATION AND COMPARISON OF THE WINDAGE LOSS OF A 60 KW SWITCHED RELUCTANCE MOTOR FOR HYBRID ELECTRIC VEHICLES 3513

Kiyota, Kyohei ; Kakishima, Takeo ; Chiba, Akira

DEVELOPMENT OF HIGH-POWER PMASYNRM USING FERRITE MAGNETS FOR REDUCING RARE-EARTH MATERIAL USE 3519

Sanada, Masayuki ; Morimoto, Shigeo ; Inoue, Yukinori

CONSIDERATION OF 10KW IN-WHEEL TYPE AXIAL-GAP MOTOR USING FERRITE PERMANENT MAGNETS 3525

Sone, Kodai ; Takemoto, Masatsugu ; Ogasawara, Satoshi ; Takezaki, Kenichi ; Hino, Wataru

POWER CONTROL METHOD FOR MULTI-PARALLEL DC DISTRIBUTION SYSTEM THROUGH THE EQUIVALENT CIRCUIT MODEL 3532

Seok-Jin Hong ; Soo-Cheol Shin ; Hee-Jun Lee ; Chung-Yuen Won ; Taeck-Kie Lee

A COMMUNICATION-LESS DISTRIBUTED VOLTAGE CONTROL STRATEGY FOR A MULTI-BUS AC ISLANDED MICROGRID 3538

Wang, Yanbo ; Yongdong Tan ; Chen, Zhe ; Wang, Xiongfei ; Tian, Yanjun

AN ENHANCED LOAD POWER SHARING STRATEGY FOR LOW-VOLTAGE MICROGRIDS BASED ON INVERSE-DROOP CONTROL METHOD 3546

Yixin Zhu ; Fang Zhuo ; Baoquan Liu ; Hao Yi

ADDING VIRTUAL RESISTANCE IN SOURCE SIDE CONVERTERS FOR STABILIZATION OF CASCADED CONNECTED TWO STAGE CONVERTER SYSTEMS WITH CONSTANT POWER LOADS IN DC MICROGRIDS 3553

Mingfei Wu ; Lu, Dylan D.C.

EXPANSION OF OPERATING RANGE AND IMPROVEMENT OF TORQUE RESPONSE OF PMSM DRIVE BY USING MODEL PREDICTIVE CONTROL 3557

N/A

NONLINEAR MODEL PREDICTIVE TORQUE CONTROL OF A LOAD COMMUTATED INVERTER AND SYNCHRONOUS MACHINE 3563

Almer, Stefan ; Besselmann, Thomas ; Ferreau, Joachim

MODEL PREDICTIVE CURRENT CONTROL FOR PMSM CONSIDERING NUMBER OF SWITCHING OPERATIONS 3568

Zanma, Tadanao ; Yasumura, Yuji ; Liu, KangZhi

PREDICTIVE INDIRECT MATRIX CONVERTER FED TORQUE RIPPLE MINIMIZATION WITH WEIGHTING FACTOR OPTIMIZATION 3574

Uddin, Muslem ; Mekhilef, Saad ; Rivera, Marco ; Rodriguez, Jose

HIGH-POWER DENSITY HYBRID CONVERTER TOPOLOGIES FOR LOW-POWER DC-DC SMPS 3582

Radic, Aleksandar ; Ahssanuzzaman, S.M. ; Mahdavikhah, Behzad ; Prodic, Aleksandar

COUPLED INDUCTOR BASED CURRENT-FED SWITCHED INVERTER FOR LOW VOLTAGE RENEWABLE INTERFACE 3587

Nag, Soumya Shubhra ; Mishra, Santanu Kumar

A SEMI-ISOLATED MULTI-INPUT CONVERTER FOR HYBRID PV/WIND POWER CHARGER SYSTEM 3592

Cheng-Wei Chen ; Kun-Hung Chen ; Chen, Yaow-Ming

HFL PV MICRO-INVERTER WITH FRONT-END CURRENT-FED CONVERTER AND HALF-WAVE CYCLOCONVERTER 3598

Nayanasiri, D.R. ; Vilathgamuwa, D.M. ; Maskell, D.L.

COMPREHENSIVE STUDY ABOUT STABILITY ISSUES OF MULTI-MODULE DISTRIBUTED SYSTEM 3604

Liu, Fangcheng ; Liu, Jinjun ; Zhang, Haodong ; Xue, Danhong ; Dou, Qinyun

CHARACTERISTICS STUDY OF NEURAL NETWORK AIDED DIGITAL CONTROL FOR DC-DC CONVERTER 3611

Maruta, Hidenori ; Motomura, Masashi ; Kurokawa, Fujio

ZERO CURRENT SWITCHING CURRENT-FED PARALLEL RESONANT PUSH-PULL (CFPRPP) CONVERTER 3616

Moorthy, Radha Sree Krishna ; Rathore, Akshay Kumar

CHARACTERISTICS OF TRANSMISSION CARRIER IN A NEW WIRE COMMUNICATION SYSTEM BY THE USE OF HIGH-RIPPLE DC-DC CONVERTER 3624

Katsuki, Akihiko ; Mizuki, Tatsuya ; Shibahara, Kohei ; Morita, Kosuke ; Masutomo, Kazufumi ; Maeyama, Shigetaka

5MHZ PWM-CONTROLLED CURRENT-MODE RESONANT DC-DC CONVERTER USING GAN-FETS 3630

Hariya, Akinori ; Yanagi, Hiroshige ; Ishizuka, Yoichi ; Matsuura, Ken ; Tomioka, Satoshi ; Ninomiya, Tamotsu

DESIGN AND PERFORMANCE EVALUATION OF DIGITAL CONTROL FOR LLC SERIES RESONANT DC-TO-DC CONVERTERS 3638

Pidaparthy, Syam Kumar ; Choi, Byungcho ; Jang, Jinhaeng

EXPERIMENTAL VERIFICATION OF NOISELESS SAMPLING FOR BUCK CHOPPER CIRCUIT WITH CURRENT CONTROL 3646

Takeuchi, Shun ; Wada, Keiji

CONTROL CHARACTERISTICS IMPROVEMENT OF FULL-BRIDGE DC-DC CONVERTER WITH SNUBBER CAPACITOR 3652

Domoto, Kazuhide ; Ishizuka, Yoichi ; Abe, Seiya ; Ninomiya, Tamotsu

DCM CONTROL METHOD OF BOOST CONVERTER BASED ON CONVENTIONAL CCM CONTROL 3659

Le Hoai Nam ; Orikawa, Koji ; Itoh, Jun-ichi

TECHNICAL ASSESSMENT OF LOAD COMMUTATION SWITCH IN HYBRID HVDC BREAKER 3667

Hassanpoor, Arman ; Hafner, Jurgen ; Jacobson, Bjorn

CONTROL OF HEXAGONAL MODULAR MULTILEVEL CONVERTER FOR 3-PHASE BTB SYSTEM 3674

Hamasaki, Shin-ichi ; Okamura, Kazuki ; Tsubakidani, Takashi ; Tsuji, Mineo

A SYNTHESIZED CAPACITORS VOLTAGE CONTROL FOR MODULAR MULTILEVEL CONVERTER IN HVDC APPLICATION 3680

Rongfeng Yang ; Shunke Sui ; Binbin Li ; Wei Wang ; Dianguo Xu

OPERATING PHASE AND FREQUENCY SELECTION OF LOW FREQUENCY AC TRANSMISSION SYSTEM USING CYCLOCONVERTERS 3687

Achara, Pichetjamroen ; Ise, Toshifumi

FAST ACTING DC CIRCUIT BREAKER FOR HVDC TRANSMISSION LINE BASED ON DC/DC CHOPPER 3695

Liangyi Tang ; Bin Wu ; Yaramasu, Venkata ; Weirong Chen ; Athab, Hussain S.

1700V SI-IGBT AND SIC-SBD HYBRID MODULE FOR AC690V INVERTER SYSTEM 3702

Haining Wang ; Ikawa, O. ; Miyashita, S. ; Nishimura, T. ; Igarashi, S.

SWITCHING SIMULATION OF SIC HIGH-POWER MODULE WITH LOW PARASITIC INDUCTANCE 3707

Yamamoto, Takashi ; Hasegawa, Kohei ; Ishida, Masaaki ; Takao, Kazuto

SWITCHING PERFORMANCE OF PARALLEL-CONNECTED POWER MODULES WITH SIC MOSFETS 3712

Colmenares, Juan ; Peftitsis, Dimosthenis ; Nee, Hans-Peter ; Rabkowski, Jacek

BUILT-IN RELIABILITY DESIGN OF A HIGH-FREQUENCY SIC MOSFET POWER MODULE 3718

Jianfeng Li ; Gurpinar, Emre ; Lopez-Arevalo, Saul ; Castellazzi, Alberto ; Mills, Liam

EXPERIMENTAL SWITCHING FREQUENCY LIMITS OF 15 KV SIC N-IGBT MODULE 3726

Kadavelugu, Arun ; Bhattacharya, Subhashish ; Ryu, Sei-Hyung ; Van Brunt, Edward ; Grider, Dave ; Leslie, Scott

SELECTION OF SUITABLE CARRIER-BASED PWM METHOD FOR MODULAR MULTILEVEL CONVERTER 3734

Ciftci, Baris ; Erturk, Feyzullah ; Hava, Ahmet M.

CONTROL AND EXPERIMENT OF A 380-V, 15-KW MOTOR DRIVE USING MODULAR MULTILEVEL CASCADE CONVERTER BASED ON TRIPLE-STAR BRIDGE CELLS (MMCC-TSBC) 3742

Kawamura, Wataru ; Hagiwara, Makoto ; Akagi, Hirofumi

A POWER ELECTRONIC TRANSFORMER WITH SINUSOIDAL VOLTAGES AND CURRENTS USING MODULAR MULTILEVEL CONVERTER 3750

Sahoo, Ashish Kumar ; Mohan, Ned

VARYING AND UNEQUAL CARRIER FREQUENCY PWM TECHNIQUES FOR MODULAR MULTILEVEL CONVERTERS 3758

Konstantinou, Georgios ; Darus, Rosheila ; Pou, Josep ; Ceballos, Salvador ; Agelidis, Vassilios G.

COMPARISON OF PHASE-SHIFTED AND LEVEL-SHIFTED PWM IN THE MODULAR MULTILEVEL CONVERTER 3764

Darus, Rosheila ; Konstantinou, Georgios ; Pou, Josep ; Ceballos, Salvador ; Agelidis, Vassilios G.

A SINGLE-PHASE POWER CONDITIONER WITH A BUCK-BOOST-TYPE POWER DECOUPLING CIRCUIT 3771
Yamaguchi, Shota ; Shimizu, Toshihisa

A NOVEL ASYMMETRICAL FLC-BASED MPPT TECHNIQUE FOR PHOTOVOLTAIC GENERATION SYSTEM 3778
Yi-Hsun Chiu ; Yu-Shan Cheng ; Yi-Hua Liu ; Shun-Chung Wang ; Zong-Zhen Yang

A NOVEL CURRENT LINK DISTRIBUTED MPPT PV SYSTEM - OVERALL SYSTEM PROTOTYPING AND EVALUATION 3784
Mikihiko ; Toru ; Akira ; Xiang-Dong Sun ; Byung-Gyu Yu

POWER FLOW CONTROL AND MPPT PARAMETER SELECTION FOR RESIDENTIAL GRID-CONNECTED PV SYSTEMS WITH BATTERY STORAGE 3789
Chokchai, Chuenwattanapraniti

A MAXIMUM POWER POINT TRACKING METHOD WITH RIPPLE CURRENT ORIENTATION 3796
Moo, Chin-Sien ; Wu, Gwo-Bin

OUTPUT CHARACTERISTICS OF A SURFACE PERMANENT MAGNET-TYPE VERNIER MOTOR - COMPARISON OF TEST RESULTS AND CALCULATION 3801
Kataoka, Yasuhiro ; Takayama, Masakazu ; Anazawa, Yoshihisa ; Matsushima, Yoshitarou

TOPOLOGY OPTIMIZATION FOR SKEW OF SPMSM BY USING MULTI-STEP PARALLEL GA 3809
Kitagawa, Wataru ; Takeshita, Takaharu

LOSS MINIMIZATION DESIGN USING MAGNETIC EQUIVALENT CIRCUIT FOR A PERMANENT MAGNET SYNCHRONOUS MOTOR 3815
Sato, Daisuke ; Itoh, Jun-ichi

THE PROPOSAL OF A NEW MOTOR WHICH HAS A HIGH WINDING FACTOR AND A HIGH SLOT FILL FACTOR 3823
Makita, Shinji ; Ito, Yasuhide ; Aoyama, Tomohiro ; Doki, Shinji

VARIABLE LEAKAGE FLUX INTERIOR PERMANENT MAGNET SYNCHRONOUS MACHINE FOR IMPROVING EFFICIENCY ON DUTY CYCLE 3828
Minowa, Masanao ; Hijikata, Hiroki ; Akatsu, Kan ; Kato, Takashi

HISTORY AND TRENDS OF CONVERTER TECHNOLOGY FOR DC AND AC TRANSMISSION IN JAPAN 3834
Yoshino, Teruo

ACCURATE OUTPUT POWER CONTROL OF CONVERTERS FOR MICROGRIDS BASED ON LOCAL MEASUREMENT AND UNIFIED CONTROL 3842
Meiqin Mao ; Zheng Dong ; Yong Ding ; Liuchen Chang

IMPEDANCE-BASED ANALYSIS OF ACTIVE FREQUENCY DRIFT ISLANDING DETECTION METHOD FOR GRID-TIED INVERTER SYSTEM 3850
Wen, Bo ; Boroyevich, Dushan ; Burgos, Rolando ; Shen, Zhiyu ; Mattavelli, Paolo

DEVELOPMENT OF 200-MVAR CLASS THYRISTOR SWITCHED CAPACITOR SUPPORTING FAULT RIDE-THROUGH 3857
Ohtake, Asuka ; Fei Zhang ; Fujimoto, Takafumi ; Nakayama, Naoyuki

DETAILED ANALYSIS AND DESIGN OF A THREE-PHASE PHASE-MODULAR ISOLATED MATRIX-TYPE PFC RECTIFIER 3864
Cortes, Patricio ; Fassler, Lukas ; Bortis, Dominik ; Kolar, Johann W. ; Silva, Marcelo

AN ENERGY SAVING DRIVE METHOD OF AN INDUCTION MOTOR WITH THE SUPPRESSION OF SUDDEN ACCELERATION AND DECELERATION 3872
Asano, Yuji ; Inoue, Kaoru ; Kotera, Keito ; Kato, Toshiji

FIELD ORIENTED CONTROL OF SENSORLESS LINEAR INDUCTION MOTOR USING MATRIX CONVERTER 3877
Sayed, Mahmoud A. ; Mohamed, Essam Ebaid ; Mohamed, Tarek Hassan ; Takeshita, Takaharu

A STATOR-EQUATION-BASED REDUCED-ORDER OBSERVER FOR POSITION-SENSORLESS VECTOR CONTROL SYSTEM OF DOUBLY-FED INDUCTION MACHINES 3885
Smiththisomboon, Somrat ; Suwankawin, Surapong

INPUT CURRENT RIPPLE ANALYSIS OF INVERTER FED DUAL THREE-PHASE AC MOTORS 3893
Dahono, Pekik Argo ; Satria, Andri

OFFLINE EXTRACTION OF INDUCTION MACHINE PARAMETERS FOR CONTROL STRATEGY SYNTHESIS 3898
Koschik, Stefan ; Bauer, Florian ; De Doncker, R.W.

HIGH CURRENT PLANAR TRANSFORMER FOR VERY HIGH EFFICIENCY ISOLATED BOOST DC-DC CONVERTERS 3905
Pittini, Riccardo ; Zhe Zhang ; Andersen, Michael A.E.

HIGH VOLTAGE-GAIN INTERLEAVED BOOST DC-DC CONVERTER DISCARDED ELECTROLYTIC CAPACITOR 3913
Nha, Quang Trong ; Huang-Jen Chiu ; Yu-Kang Lo ; Pham Phu Hieu

PARALLEL BI-DIRECTIONAL DC-DC CONVERTER FOR ENERGY STORAGE SYSTEM 3920
Ouchi, Takayuki ; Kanoda, Akihiko ; Takahashi, Naoya

CHARGING SCENARIO OF SERIAL BATTERY POWER MODULES WITH BUCK-BOOST CONVERTERS 3928
Jhen-Yu Jian ; Chu-Shen Chang ; Moo, Chin-Sien ; Hau-Chen Yen

COMPARATIVE THERMAL PERFORMANCE EVALUATION OF SIC MOSFETS AND SI MOSFET FOR 1.2 KW 300 KHZ DC-DC BOOST CONVERTER AS A SOLAR PV PRE-REGULATOR 3933
Taekyun Kim ; Minsoo Jang ; Agelidis, Vassilios G.

TOLERANCE ANALYSIS OF A CONSTANT-ON TIME CURRENT-MODE VOLTAGE REGULATOR WITH ADAPTIVE VOLTAGE POSITION FEATURE 3938

Chih Wei Chen ; Dan Chen ; Shin Shiung Wang

FPGA-BASED DIGITAL-CONTROLLED POWER CONVERTER DESIGNED WITH UNIVERSAL INPUT MEETING 80 PLUS PLATINUM EFFICIENCY CODE AND STANDBY POWER CODE FOR SEVER POWER APPLICATIONS 3942

Lai, Yen-Shin ; Ho, Kung-Min

STATIC AND DYNAMIC ANALYSES OF DIGITAL PEAK CURRENT MODE DC-DC CONVERTER 3950

Kajiwara, Kazuhiro ; Kurokawa, Fujio ; Shibata, Yuichiro

EXTENDED DISCRETE CONTROL OF CLASS E AMPLIFIER IN ORDER TO ACHIEVE NOMINAL OPERATION 3955

Suetsugu, Tadashi ; Xiuqin Wei ; Kuga, Shotaro

ADAPTIVE POWER EFFICIENCY CONTROL BY COMPUTER POWER CONSUMPTION PREDICTION USING PERFORMANCE COUNTERS 3959

Kawaguchi, Shinichi ; Yachi, Toshiaki

Author Index

The 2014 International Power Electronics Conference - ECCE ASIA -

IPEC - Hiroshima 2014

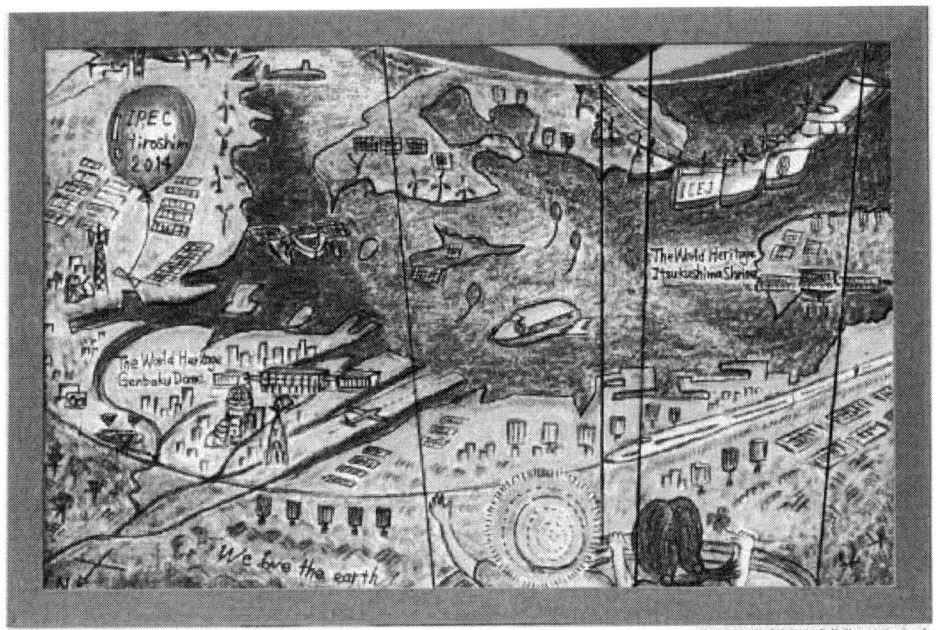

Illustrated by Mikihiko Matsui

May 18th-21st, 2014
International Conference Center Hiroshima
-Power Electronics for Peaceful World-
Sponsored by IEEJ Industry Applications Society

ECCE Asia Cooperation
  IEEJ Industry Applications Society (IEEJ IAS)
  The Korean Institute of Power Electronics (KIPE)
  China Electrotechnical Society (CES)

Technical Co-Sponsorship
  IEEE Power Electronics Society (IEEE-PELS)
  IEEE Industry Applications Society (IEEE-IAS)

In Cooperation with
  The European Power Electronics and Drives Association
  (EPE)

Welcome Message

On behalf of the Organizing, Steering and Technical Program Committees, we sincerely welcome you to the 2014 International Power Electronics Conference -ECCE Asia-, which we simply call IPEC-Hiroshima 2014. The first IPEC was held in 1983, and since then, in 1990, 1995, 2000, and 2005, IPECs have been held in different locations, sponsored by IEEJ. Since 2010, IPEC has had the additional function of hosting the power electronics conference series, which is held in the three-country rotation of China, Japan and Korea, as one of the ECCE Asia conferences through Technical Co-sponsorship with the IEEE Power Electronics Society and the IEEE Industry Applications Society.

The seventh International Power Electronics Conference, IPEC-Hiroshima 2014 -ECCE Asia-, is held from May 18 to May 21, 2014, in Hiroshima, Japan. The conference venue is the International Conference Center Hiroshima, which is located in Hiroshima Peace Memorial Park. Power Electronics has been providing numerous new technologies in the fields of electric energy conversion and motor drive systems for more than 40 years. In recent years, global energy and environmental issues have become more serious, and power electronics is expected to play a key role in solving the problems. IPEC-Hiroshima 2014 will provide a unique opportunity for researchers, engineers, and academics from all over the world to present and exchange the latest information on power electronics, motor drives, and related subjects.

The Technical Program Committee received about 788 digests submitted from 33 countries and areas. Finally, 465 papers and 168 Organized Session papers were accepted for presentation. We would like to thank the committee members and reviewers for their support of the IPEC.

IPEC-Hiroshima 2014 will start with an Industrial Seminar in the afternoon of May 18, followed by a welcome reception. Prior to the Industrial Seminar, a Ph.D. Candidate Meeting will be held in conjunction with the IPEC Steering Committee and the IEEE PELS Young Professionals Membership Committee in the morning of May 18. The Plenary Session will be held in the morning of May 19. At that time, winners of the Isao Takahashi Award and the Prize Paper Award will be announced, followed by the plenary presentations. The Regular Sessions consist of both Oral and Poster presentations, and they are open from May 19 to 21. The Organized Session will present invited papers. A banquet is scheduled for the evening of May 20. Finally, I would like to express special thanks to the many companies that supported the conference as sponsors or exhibitors. I would also like to thank the Japan Society for the Promotion of Science (Grant-in Aid for Scientific Research) and the Hiroshima Convention & Visitors Bureau for their financial support.

Hiroshima, the conference venue, is the capital city of Hiroshima Prefecture and situated in the western region of Japan's main island. Hiroshima is well known for its world heritage site, Miyajima Itsukushima Shrine, which has a history dating back 1400 years. Everyone will enjoy the beautiful landscape of the shrine built on the waterfront of the Inland Sea. Visitors can also savor oysters and other delicious seafood, as well as the original thin pancake cuisine known as "okonomiyaki." We are sure that all participants will have a memorable and enjoyable time in Hiroshima. If you need any help during the conference, please let us know or notify any member of the Steering Committee. We would be very happy to serve you and help you enjoy the conference.

Organizing Committee Chairperson
Atsuo Kawamura

Steering Committee Chairperson
Toshihisa Shimizu

Technical Program Committee Chairperson
Takaharu Takeshita

Atsuo Kawamura
Chairperson
Organizing Committee

Toshihisa Shimizu
Chairperson
Steering Committee

Takaharu Takeshita
Chairperson
Technical Program Committee

Gap in pagination due to formatting issues.

Pages 1-65

The 2014 International Power Electronics Conference

A Novel Control Scheme for Three-Level Full-Bridge Converter Achieving Low THD Output Voltage

Jilong Liu[1], Fei Xiao[2], Wei Chen[2], Guorun Yang[2]

(1) School of Electrical Engineering, Xi'an Jiaotong University
Xi'an, Shaanxi, CHINA 710049
(2) National Key Laboratory for Vessel Integrated Power System Technology
Naval University of Engineering, Wuhan, CHINA 430033
Email: 66976@163.com, xfeyninger@gmail.com, tsenwe@163.com, woyaodang-jiangjun@163.com

Abstract—A novel control scheme called symmetrical chop plus phase-shift (SCPS) control, allowing six switches of the three-level full-bridge converter to achieve zero voltage switching (ZVS) and reducing the total harmonic distortion (THD) of the transformer primary side voltage is proposed in this paper. The traditional control schemes for three-level full-bridge converter to achieve ZVS ignored the impact of the high THD of transformer primary side voltage. The high THD will cause large power loss in transformer and is not conducive to delivery energy for transformer. In the SCPS control, the output voltage of full bridge is symmetrical in half of switching period and the THD is reduced. At the same time, two switches have not achieved zero voltage on. This paper presents switching process and conditions of this control scheme. THD of the output voltage is analyzed and compared with CPS control. The performance of the proposed control scheme was verified on a prototype.

Keywords—three-level converter; full-bridge converter; symmetrical chop plus phase-shift; total harmonic distortion; soft switching

I. INTRODUCTION

The three-level full-bridge dc/dc converter has drawn much interest in high power and high input voltage applications. There are two capacitors in series at the input side of three-level converter. Center tap of the two capacitors provide neutral voltage for the converter. Each phase leg of the three-level converter has two pairs of switching devices in series. The center of each device pair is clamped to the neutral through clamping diodes. In this kind of converters, voltage stress of main switches is half of input voltage by using the neutral point clamped (NPC) technique [1, 2].

The switching loss in three-level converters is very high because they are always used in high input voltage and high current applications. Soft switching technique should better be used in three-level converters to decrease power loss and expand its application range [3]. At the same time, the THD of transformer primary side voltage should be kept low because voltage and current harmonics will result in higher power loss and excessive temperature rise which will eventually diminish the useful life of transformers [4].

In [5], three-level half-bridge converter can achieve soft switching is proposed. Soft switching three-level converters

were classified into two kinds: zero voltage switching (ZVS) and zero voltage zero current switching (ZVZCS). In [6], a three-level converter designed with ZVS techniques, is proposed and is used in ship electric power distribution systems.

A ZVS PWM hybrid three-level full-bridge converter, which has a three-level leg and a two-level leg, is proposed in [7]. The input filter and output filter will be significantly reduced. But the switches of two-level leg sustain the input voltage and those of three-level leg sustain half of the input voltage. Using different kind of switches will increase the cost of the converter.

Three-level full-bridge PWM converter that achieves ZVS was proposed in [8]. This converter, different from the hybrid three-level converter, has only one kind of switch. A control scheme called chop plus phase-shift (CPS) is used for this converter. Two main switches are operating at chopping mode and other switches are operating at phase-shift mode.

Another control scheme, similar to CPS control, for three-level full-bridge PWM converter is proposed in [9]. In this control scheme, all switches are working at phase-shift mode. It is called double phase-shift (DPS) control and it improves efficiency by reducing losses of switches' body diodes without adding additional components compared with CPS control.

An improved three-level full-bridge dc/dc converter for a wind turbine in a dc grid by inserting a passive filter into the dc/dc converter is proposed in [10]. The passive filter can effectively reduce the voltage stress of the medium frequency transformer [10]. In [11], a new ZVS PWM three-level full-bridge dc/dc converter, which is composed of a TL FB converter and a TL half-bridge (HB) converter is proposed. The converter can achieve ZVS for all power switches in wide load range and even at light load condition.

However, converter in [11] is relatively complex and the transformer is especially difficult to make. Converter in [10] doesn't achieve soft switching. The output voltage of three-level full-bridge converter using CPS control in [8] or DPS control in [9] has high voltage and current harmonics which will cause high power loss in the transformer.

A novel control scheme called symmetrical chop plus phase-shift (SCPS) control is proposed in this paper. This

978-1-4799-2706-7/14 $31.00 © 2014 IEEE

control scheme allows six switches of the converter to achieve ZVS and reduces the THD of the transformer primary side voltage. Operation principle and zero voltage switching are analyzed. THD of the transformer primary side voltage is calculated. Simulation results of SCPS control scheme are given in this paper.

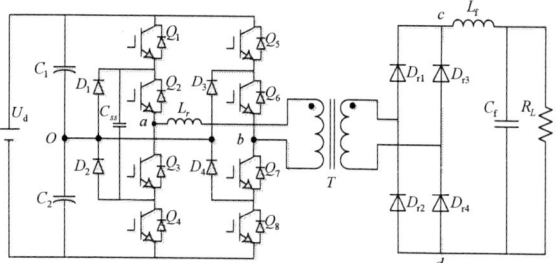

Fig.1 Three-level full-bridge converter

II. OPERATION PRINCIPLE

A NPC three-level full-bridge converter is shown in Figure 1. The converter has two legs, eight switches (IGBT) Q_1-Q_8, four clamping diodes D_1-D_4. Each of the IGBTs has an intrinsic capacitor. These capacitors are named C_1-C_8 and they are treated as resonant capacitors to achieve ZVS. The resonant inductor is L_r and it includes leakage inductor of the transformer. Full-bridge rectifier includes four diodes D_{r1}-D_{r4}. The inductor L_f and the capacitor C_f constitute the low-pass filter.

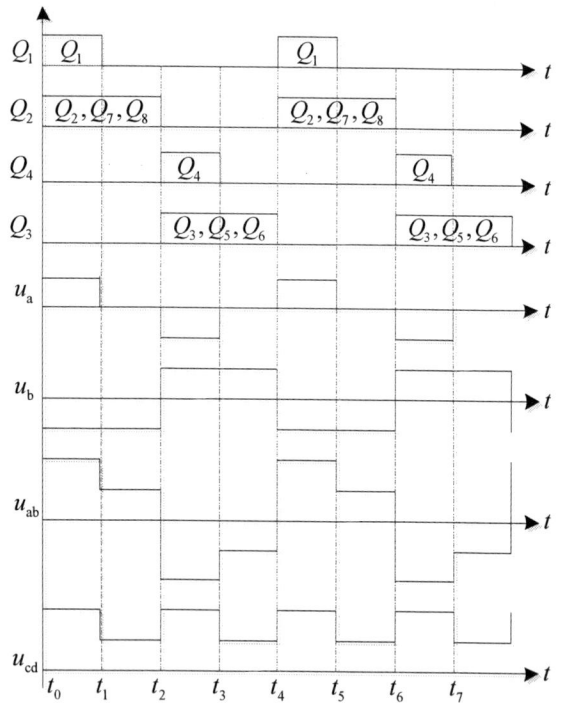

Fig.2 Waveform of three-level mode in CPS control scheme.

Fig 2 shows the key waveforms in CPS control scheme for three-level full-bridge converter. In CPS control scheme, as

shown in Fig 2, Q_2 and Q_3 are switched out of phase. Q_5 and Q_6 are switched in phase. Q_7 and Q_8 are switched in phase. Q_5&Q_6 and Q_7&Q_8 are switched out of phase. In CPS control, Q_1 and Q_2 are turned on at the same time. Q_4 and Q_3 are turned on at the same time. In the CPS control, transformer primary side voltage is not symmetrical in half switching period and it is neither odd function nor even function. In the CPS control, zero voltage switching can be achieved, but the THD of transformer voltage is high.

So the key problem of CPS control is that the THD of transformer voltage is high. The key reason is the position of chopping switches Q_1 and Q_4 signals. If the position of chopping switches Q_1 and Q_4 signals are shifted right, transformer primary side voltage will become even function. The THD of transformer voltage will be decreased and the performance of the transformer will be improved.

Fig 3 shows the key waveforms in SCPS control scheme for three-level full-bridge converter. In SCPS control scheme, as shown in Fig 3, Q_2 and Q_3 are switched out of phase. Q_5 and Q_6 are switched in phase. Q_7 and Q_8 are switched in phase. Q_5&Q_6 and Q_7&Q_8 are switched out of phase. The left leg is operating like a three-level half-bridge converter while the right leg is operating like a traditional two-level leg in phase-shift full-bridge converter.

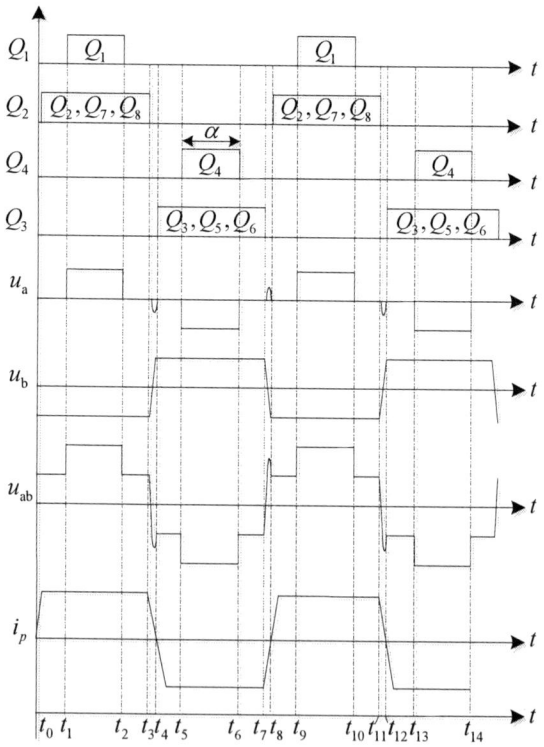

Fig.3 Waveform of three-level mode in SCPS control scheme.

In SCPS control scheme, Q_1 and Q_4 are operating at chopping mode while other six switches are operating at phase-shift mode. Therefore Q_1 and Q_4 are called chopping switches. Q_2 and Q_3 are switching leading to Q_5&Q_6 and Q_7&Q_8. Therefore Q_2 and Q_3 are called leading switches,

Q_5&Q_6 and Q_7&Q_8 are called lagging switches. The chop angle of Q_1 and Q_4 is called α and phase-shift angle of other six switches is called θ. When θ is zero, Q_2 and Q_7&Q_8 are switching in phase, the converter is operating at three-level mode as shown in Fig 3. Two-level mode is relatively simple and it is described in [8].

$$\begin{cases} \alpha > 0, \theta = 0 & \text{(three-level)} \\ \alpha = 0, \theta > 0 & \text{(two-level)} \end{cases} \tag{1}$$

By using SCPS control, output voltage of left leg u_a and output voltage of right leg u_b are out of phase. Left and right sides of u_{ab} is symmetrical in half of switching period. This is a key difference between SCPS control scheme and other non-symmetrical control schemes. When the converter is operating at three-level mode, the main switches Q_1 and Q_4 are working at hard switching mode. When the converter is operating at two-level mode, the main switches Q_1 and Q_4 are not working.

Compared with the CPS and DPS control schemes, SCPS control scheme can reduce THD of transformer primary voltage. The transformer primary voltage is odd function to time while those in CPS and DPS control are neither odd function nor even function.

III. Zero Voltage Switching Conditions

The key waveforms of three-level full-bridge converter under SCPS control scheme is shown in Fig 3. All the eight switching modes in three-level mode are analyzed below.

1) Mode 1 [t_1-t_2]: Prior to t_1, Q_2 and Q_7&Q_8 are conducting, D_{r1} and D_{r4} are on, D_{r2} and D_{r3} are off. At t_1, Q_1 is turned on, intrinsic capacitor of Q_1 (C_1) starts discharge and intrinsic capacitor of Q_4 (C_4) starts charge. During this mode, transformer primary side current can be seen as a constant value. At t_1, Q_1 doesn't achieve zero voltage on.

2) Mode 2 [t_2-t_3]: Prior to t_2, Q_1, Q_2 and Q_7&Q_8 are conducting. At t_2, Q_1 is turned off, intrinsic capacitor of Q_1 (C_1) starts charge and intrinsic capacitor of Q_4 (C_4) starts discharge. Because of the capacitor C_1, Q_1 achieves zero voltage off. During this mode, transformer primary side current can be also seen as a constant value.

3) Mode 3 [t_3-t_4]: At t_3, Q_2 and Q_7&Q_8 are turned off. Because of the capacitor C_2, C_7 and C_8, Q_2 and Q_7&Q_8 achieve zero voltage off. Current in the resonant inductor L_r provides energy to charge C_2, C_7 and C_8, to discharge C_3, C_5 and C_6. The output current is I_o and ratio of primary and secondary windings of transformer is K. The capacitor values of all the intrinsic capacitor are C_{int}. During mode 3, Q_2 and Q_7&Q_8 are turned off, and transformer primary side voltage is zero. The four rectifier diodes will be on during this mode. The transformer primary side current decreases resonantly, as shown in equation (2).

$$i_p(t) = \frac{I_o}{K} \cos \omega_r (t - t_3) \tag{2}$$

Where,

$$\omega_r = \frac{1}{\sqrt{6 C_{int} L_r}} \tag{3}$$

Before the transformer primary side current decreases to zero, charge and discharge during mode 3 should be over. C_2, C_7 and C_8 will be charged to $U_d/2$, C_3, C_5 and C_6 will be discharged to zero. The resonant inductor should have enough energy to complete the charge and discharge operation.

$$\frac{1}{2} L_r \left(\frac{I_o}{K}\right)^2 \geq \frac{3}{4} C_{int} U_d^2 \tag{4}$$

The minimum load current to keep safe ZVS is

$$I_{omin} = K U_d \sqrt{\frac{3 C_{int}}{2 L_r}} \tag{5}$$

Mode 3 interval is the dead zone of the switches. The resonant period is T_r and the discharge process will be over at $1/4\ T_r$. So the dead zone time should be longer than $1/4$ of the resonant period T_r.

$$t_4 - t_3 \geq \frac{1}{2} \pi \sqrt{6 C_{int} L_r} \tag{6}$$

If the resonant inductor has more energy after the charge and discharge process is completed, diodes of Q_3, Q_4 and Q_5&Q_6 will be on. This will cause a voltage spike on transformer primary side voltage.

4) Mode 4 [t_4-t_5]: At t_4, Q_3, Q_5 and Q_6 are turned on at zero voltage condition. The transformer primary side current will change its direction, D_{r2} and D_{r3} are on, D_{r1} and D_{r4} are off. The power source starts to supply energy to the load.

5) Mode 5 [t_5-t_6]: Prior to t_5, Q_3 and Q_5&Q_6 are conducting, D_{r2} and D_{r3} are on, D_{r1} and D_{r4} are off. At t_5, Q_4 is turned on, intrinsic capacitor of Q_4 (C_4) starts discharge and intrinsic capacitor of Q_1 (C_1) starts charge. During this mode, transformer primary side current can be seen as a constant value. At t_4, Q_4 doesn't achieve zero voltage on.

6) Mode 6 [t_6-t_7]: Prior to t_6, Q_3, Q_4 and Q_5&Q_6 are conducting. At t_6, Q_4 is turned off, intrinsic capacitor of Q_4 (C_4) starts charge and intrinsic capacitor of Q_1 (C_1) starts discharge. Because of the capacitor C_4, Q_4 achieves zero voltage off.

7) Mode 7 [t_7-t_8]: At t_7, Q_3 and Q_5&Q_6 are turned off. Because of the capacitor C_3, C_5 and C_6, Q_3 and Q_5&Q_6 achieve zero voltage off. Current in the resonant inductor L_r provides energy to charge C_3, C_5 and C_6, to discharge C_2, C_7 and C_8. Mode 7 is similar to mode 3. Transformer primary side current decreases resonant to zero. C_3, C_5 and C_6 will be charged to $U_d/2$, C_2, C_7 and C_8 will be discharged to zero. If formula (4) is satisfied, the resonant inductor will have enough energy to complete the charge and discharge operation. Also, the dead zone time should satisfy the formula (7). During this mode, transformer primary side voltage is zero and the four rectifier diodes will be on.

$$t_8 - t_7 \geq \frac{1}{2} \pi \sqrt{6 C_{int} L_r} \tag{7}$$

If the resonant inductor has more energy after the charge and discharge process is completed, diodes of Q_1, Q_2 and Q_7&Q_8 will be on. This will cause a voltage spike on transformer primary side voltage.

8) Mode 8 [t_8-t_9]: At t_8, Q_2, Q_7 and Q_8 are turned on at zero voltage condition. The transformer primary side current will change its direction, D_{r1} and D_{r4} are on, D_{r2} and D_{r3} are off. The power source starts to supply energy to the load.

IV. THD OF OUTPUT VOLTAGE

In the CPS or DPS control, transformer primary side voltage is not symmetrical in half switching period and it is neither odd function nor even function. Fig 4 shows the waveform of one full switching period in CPS or DPS control while Fig 5 shows the waveform of one full switching period in SCPS control.

Switching frequency is selected as the base frequency and both of two waveforms can be done Fourier analysis. THD of the two waveforms can be calculated. THD of the waveform in CPS or DPS control is shown in (13) and THD of the waveform in SCPS control is shown in (14).

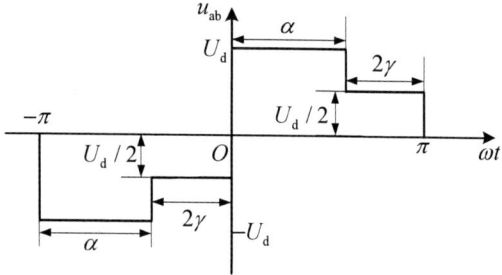

Fig.4 Waveform of transformer primary side voltage with CPS or DPS control

The switching period is defined as T. So the angle frequency of the converter is

$$\omega = 2\pi / T \qquad (8)$$

The RMS value of transformer primary side voltage in Fig 4 is shown in (9).

$$U_{rms1} = \sqrt{\frac{1}{\pi}\int_0^\pi u_{ab}{}^2(\omega t)d(\omega t)}$$
$$= U_d\sqrt{\frac{\pi + 3\alpha}{4\pi}} \qquad (9)$$

The Fourier analysis of waveform in Fig 4 is shown in (10)-(12).

$$f_1(t) = \sum_{n=1}^{\infty}(a_{n1}\cos n\omega t + b_{n1}\sin n\omega t) \qquad (10)$$

$$a_{n1} = \frac{1}{\pi}\int_{-\pi}^{\pi} u_{ab}(\omega t)\cos n\omega t \cdot d(\omega t)(n=1,2,3\text{L })$$
$$= \frac{U_d}{2n\pi}\left[\sin(n\pi - n\alpha) - \sin n\pi + \sin n\alpha\right] \qquad (11)$$

$$b_{n1} = \frac{1}{\pi}\int_{-\pi}^{\pi} u_{ab}(\omega t)\sin n\omega t \cdot d(\omega t)(n=1,2,3\text{L })$$
$$= \frac{U_d}{2n\pi}\left[\cos(na - n\pi) - 3\cos n\pi + 3 - \cos n\alpha\right] \qquad (12)$$

The THD of waveform in Fig 4 can be got from (9)-(12) and it is shown in (13). So the THD is function of the chopping angle α.

$$THD_1 = \sqrt{\frac{\pi^2 + 3\alpha\pi - 20 + 12\cos\alpha}{20 - 12\cos\alpha}} \qquad (13)$$

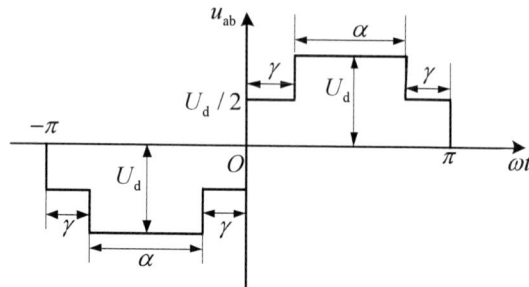

Fig.5 Waveform of transformer primary side voltage with SCPS control

$$U_{rms2} = U_d\sqrt{\frac{\pi + 3\alpha}{4\pi}} \qquad (14)$$

The RMS value of transformer primary side voltage in Fig 4 is shown in (14). The waveform got by SCPS control is even function and there is no DC component in this waveform. So the Fourier analysis is shown in (15).

$$f_2(t) = \sum_{n=1}^{\infty}\left[\frac{4U_d}{\pi}\sin^2(\frac{n\pi}{4} + \frac{n\alpha}{4})\cdot\sin n\omega t)\right] \qquad (15)$$

The THD of waveform in Fig 5 can be got from (9) (14) (15) and it is shown in (16). The THD is function of the chopping angle α.

$$THD_2 = \sqrt{\frac{\pi(\pi + 3\alpha)}{32\sin^4(\frac{\pi + \alpha}{4})} - 1} \qquad (16)$$

Per-unit fundamental RMS voltage of transformer primary side under the two control methods is shown in Fig 6. When the chopping angle is zero or π, the RMS values under the two methods are same. When $0 < \alpha < \pi$, the RMS value under SCPS control is larger than that under CPS control.

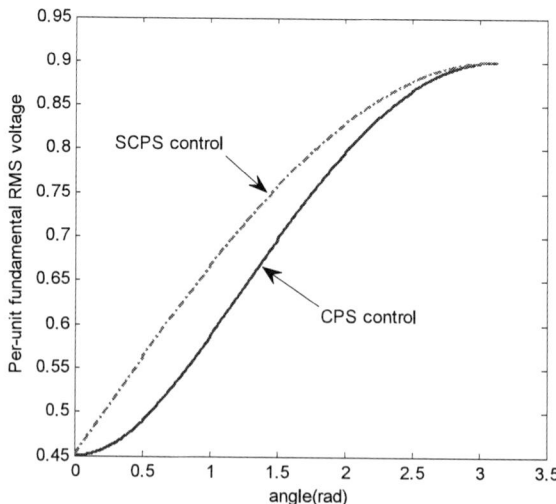

Fig.6 Per-unit fundamental RMS voltage of transformer primary side under the two control methods

The curves of the THD are shown in Fig 7. It can be seen that THD of the waveform in CPS or DPS control is much larger than that in SCPS control, especially when the overlap angle is small. THD of the waveform in CPS or DPS control can be even up to 0.73.

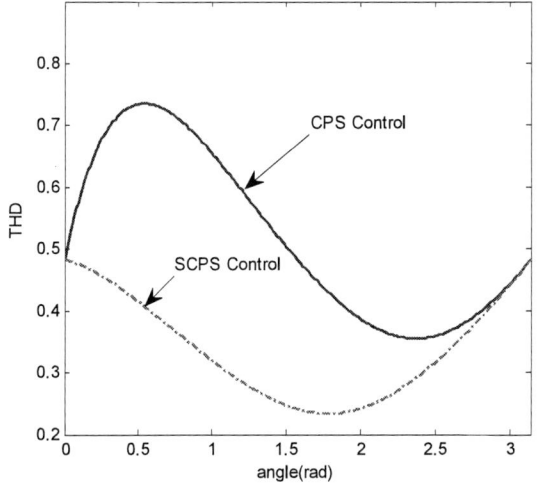

Fig.7 Theoretical THD of transformer primary side voltage under the two control schemes

V. SIMULATION AND EXPERIMENT RESULTS

Simulation results to verify the SCPS control scheme is shown in Fig 8. The input voltage is 1000V and output voltage is 350V. Fig 8 shows the key waveforms of SCPS control, including output voltage and current of the full-bridge. The four waveforms are respectively output voltage of left leg, output voltage of right leg, transformer primary side voltage and transformer primary side input current. It can be seen that the simulation results are same with the analyzed waveforms shown in Fig 3.

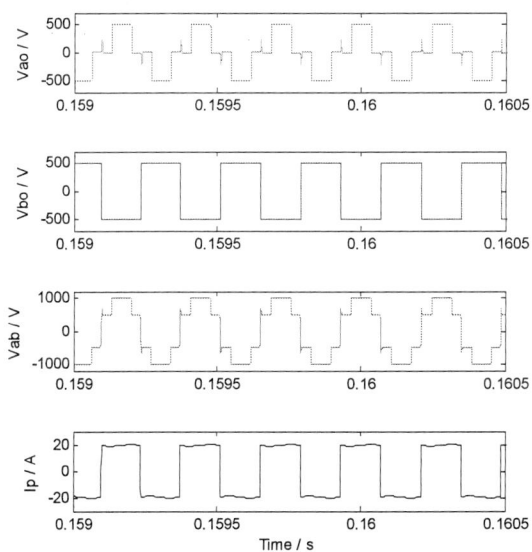

Fig.8 Simulation waveforms of SCPS control in three-level mode.

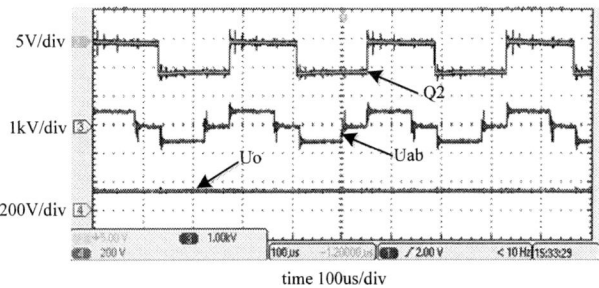

Fig.9 Control method waveforms of TL FB converter SCPS control in two-level mode (they are drive signal of Q2, transformer primary side voltage and DC output voltage).

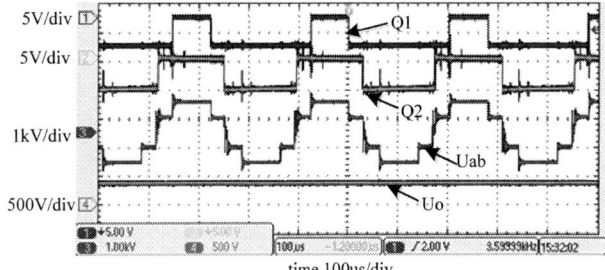

Fig.10 Control method waveforms of TL FB converter SCPS control in three-level mode (they are drive signal of Q1, drive signal of Q2, transformer primary side voltage and DC output voltage).

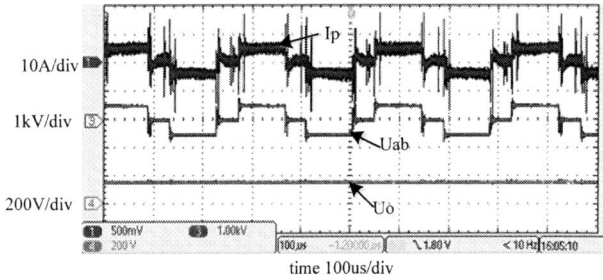

Fig.11 Transformer primary side current waveforms of TL FB converter SCPS control in two-level mode (they are transformer primary side current, transformer primary side voltage and DC output voltage).

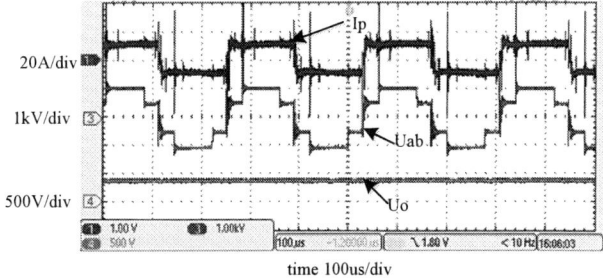

Fig.12 Transformer primary side current waveforms of TL FB converter SCPS control in three-level mode (they are transformer primary side current, transformer primary side voltage and DC output voltage).

Fig 9 shows the experiment waveforms of SCPS control method in two-level mode. This mode is used when the input voltage is very high or a low output voltage is needed. In this mode, Q_1 and Q_4 will be never turned on. All the six working switches can achieve ZVS. The two-level waveforms in SCPS control scheme are same with those in CPS control scheme. Fig 10 shows the experiment waveforms of SCPS control method in three-level mode. This mode is the key difference with CPS control. In this mode, main switches Q_1 and Q_4 start working. Main switches Q_1 and Q_4 give up achieving ZVS to keep the THD of transformer primary side voltage low.

Fig 11 shows the transformer primary side current in two-level mode. Fig 12 shows the transformer primary side current in three-level mode. Transformer voltage and current in fig 12 are same with those in fig 8 which are got by simulation.

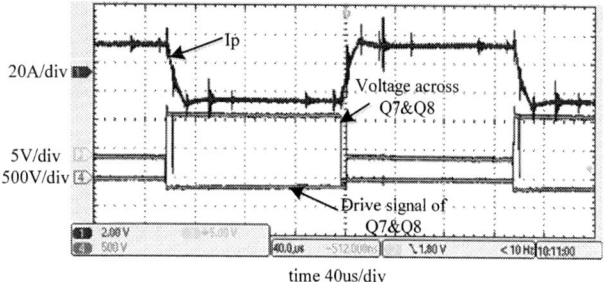

time 40us/div

Fig.13 Zero voltage switching waveforms of switches Q_7 and Q_8

Fig 13 shows the ZVS waveforms of switches Q_7 and Q_8. As described in chapter II, Q_7 and Q_8 are switched in phase. In fig 13, channel 4 is the voltage across Q_7 and Q_8. Channel 2 is the drive signal of Q_7 and Q_8. Before high drive signal comes, the voltage across Q_7 and Q_8 turns to zero.

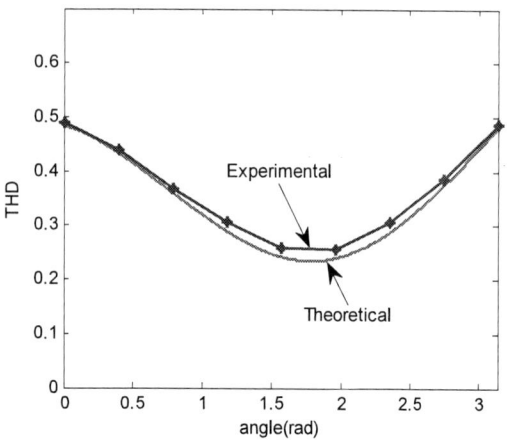

Fig.14 Theoretical THD and experimental THD of transformer primary side voltage under SCPS control scheme

Fig 14 shows theoretical THD and experimental THD of transformer primary side voltage under SCPS control scheme. The experimental value is a little higher than theoretical value. This is influenced by the voltage spikes as shown in the simulation waveforms (Fig 8) and experiment waveforms (Fig12). When the chopping angle is π, there are no voltage spikes any more. At this time, theoretical THD value and experimental THD value are same.

VI. CONCLUSIONS

A novel symmetrical chop plus phase-shift control scheme for three-level full-bridge converter is proposed in this paper. With this control scheme, voltage stress of all main switches is half of the input voltage and output filter inductor is reduced. Six main switches of the converter can achieve zero voltage switching and two main switches can't achieve zero voltage on. The resonant inductor provides energy to allow the switches to achieve ZVS. Compared with the CPS or DPS control scheme, THD of transformer primary side voltage in this control scheme is highly reduced. Simulation results and experiment results are given and the analysis of this control scheme is verified.

ACKNOWLEDGMENT

This project is supported by "Program for New Century Excellent Talents in University" (NCET-11-0871) and supported by "National Natural Science Foundation of China" (51177170).

REFERENCES

[1] Jih-Sheng Lai, Fang Zheng Peng, "Multilevel Converters-A New Breed of Power Converters," *IEEE Trans. On Industrial Applications*, vol. 32, no. 3, pp. 509-517, 1996.

[2] Yungtaek Jang, Milan M. Jovanovic, "A New Three-Level Soft-Switched Converter," *IEEE Trans. on Power Electronics*, vol. 20, no. 1, pp. 75-81, 2005.

[3] Yilei Gu, Zhengyu Lu, Lijun Hang, "Three-Level LLC Series Resonant DC/DC Converter," *IEEE Trans. on Power Electronics*, vol. 20, no. 4, pp. 781-789, 2005.

[4] M. Shareghi, B.T. Phung, M.S. Naderi, "Effects of Current and Voltage Harmonics on Distribution Transformer Losses," *IEEE International Conference on Condition Monitoring and Diagnosis*, pp. 633-636, 2005.

[5] Xinbo Ruan, Linquan Zhou, "Soft-Switching PWM Three-Level Converters," *IEEE Trans. on Power Electronics*, vol. 16, no. 5, pp.612-622, 2001.

[6] Byeong-Mun Song, Robert McDowell, "A Three-Level DC-DC Converter with Wide-Input Voltage Operations for Ship-Electric-Power-Distribution Systems" *IEEE Trans. on Plasma Science*, vol. 32, no. 5, pp. 1856-1863, 2004.

[7] Xinbo Ruan, Zhiying Chen, Wu Chen, "Zero-Voltage-Switching PWM Hybrid Full-Bridge Three-Level Converter," *IEEE Trans. on Power Electronics*, vol. 20, no. 2, pp.395-404, 2005.

[8] Zhiliang Zhang, Xinbo Ruan, "Zero-Voltage-Switching PWM Full-Bridge Three-Level Converter," *The 4th International Power Electronics and Motion Control Conference*, vol. 3, pp. 1085-1090, 2004.

[9] Zhiliang Zhang, Xinbo Ruan, "A Novel Double Phase-Shift Control Scheme for Full-Bridge Three-Level Converter," *Annual Applied Power Electronics Conference and Exposition*, vol. 2, pp. 1240-1245, 2005.

[10] Fujin Deng, Zhe Chen, "Control of Improved Full-Bridge Three-Level DCDC Converter for Wind Turbines in a DC Grid," *IEEE Trans. on Power Electronics*, vol. 28, no. 1, pp.314-324, 2013.

[11] Yong Shi, Xu Yang, "Zero-Voltage Switching PWM Three-Level Full-Bridge DC–DC Converter With Wide ZVS Load Range," *IEEE Trans. on Power Electronics*, vol. 28, no. 10, pp.4511-4524, 2013.

The 2014 International Power Electronics Conference

Parallel Connected Three Phase Inverters Based on Modular Design and Distributed Control

Fei Xiao[1], Wei Chen[1], Jilong Liu[2], Hengli Wang[1]

(1) National Key Laboratory for Vessel Integrated Power System Technology
Naval University of Engineering, Wuhan, CHINA 430033
(2) School of Electrical Engineering, Xi'an Jiaotong University
Xi'an, Shaanxi, CHINA 710049
Email: xfeyninger@gmail.com, tsenwe@163.com, 66976@163.com, wanghengli1984@126.com

Abstract—This paper investigates the characteristics of parallel connected three phase inverters based on modular design and distributed control. Every inverter module has a local hardware manager and the whole system has a central application manager. The hardware managers and application manager are connected together by the optical fiber to build a fiber ring net structure. The fiber ring net can achieve high communication rate and synchronous drive signals in different local controllers. Application manager will decide the number of operating modules and every inverter module can run or stop dynamically. The system structure and system model vary with the number of operating modules. The controller in application manager should be designed based on every possible model. In other words, the controller should adapt to the worst situation. In this converter, current sharing performance is good with no current sharing method used. A prototype is made to verify the design and analysis.

Keywords—three phase inverter; parallel connected; modular design; distributed control; variable structure

I. INTRODUCTION

Parallel connected inverters can be used in low voltage high current applications to avoid using extra expensive large current switches. The benefits of connecting the inverters in parallel include flexibility in planning, economy in design effort, easy redundancy operation, high efficiency and capacity for expansion [1]. Furthermore, if proper control scheme is used in parallel connected inverters, the low frequency isolate transformer can be omitted. So parallel connected inverters have been hot research topic in recent year.

The most important issue in parallel connected inverters is how to make the inverters share the load current equally. When inverters are connected directly, circulating currents will automatically be generated through other converters due to the asynchronous switching operations and system parameter differences. This will result in unbalance load current between inverters. This will also increase the total harmonic distortion (THD) of the output current [2].

There are mainly three control modes in parallel inverters: central mode control, master and slave mode control and wireless mode control. Additional current sharing control scheme should be added to balance the load current and avoid

circulating current. A number of methods have been proposed to decrease circulating current [1-2] and achieve balanced load current [3-7].

In [1], detailed analysis of the zero-sequence circulating current is provided and a novel dual-modulator compensation technique is proposed to eliminate the zero-sequence circulating current. Model of the zero-sequence circulating current in parallel multi-phase converters is proposed in [2]. An independent controller can be used to eliminate the zero-sequence circulating current. And it is adaptable to any N number parallel M phase converters. Model of the zero-sequence circulating current for parallel three phase boost rectifiers with different load sharing is proposed in [3].

A multi-inverter system with instantaneous average-current-sharing scheme is presented in [4]. A current sharing bus is used and stability of the current-sharing controller is discussed. In [5], the output impedance of the inverters operating in parallel with master slave current sharing scheme is modeled and a stability criterion was proposed to predict the stability of parallel inverters.

A wireless control technique derived from the droop method is proposed in [6] and the output impedance of the inverters is investigated. In [7], the power-sharing control loops are based on the P/Q droop method, novel control loops to achieve both stable output impedance and proper power balance are proposed.

However, a master inverter without zero-sequence control is needed in [1] [2] and it should be chosen carefully and appropriate protections should be incorporated into the system. A current sharing bus is needed in [4] and it is not easy to achieve modular design. With power-sharing control loops based on the P/Q droop method, the system stability margin and output characteristics is affected.

Fiber ring net for distributed control is introduced in [8-10], and it is an important strategy to achieve modularization and standardization of power electronics devices.

In this paper, parallel connected three phase inverters based on modular design and distributed control is researched. The controller has two levels: central controller and local controllers. Central controller and local controllers build a fiber ring net, as shown in [8-10]. Central controller completes

978-1-4799-2706-7/14 $31.00 © 2014 IEEE

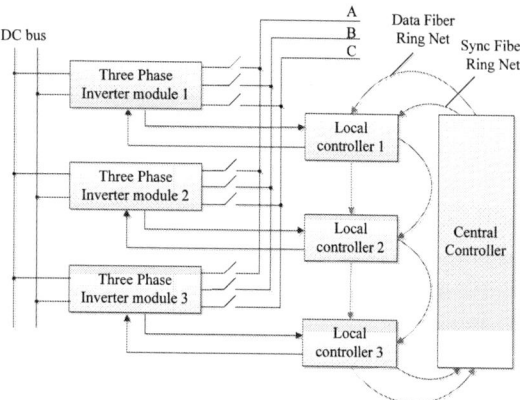

Fig 1. Structure of parallel inverters based on fiber ring net.

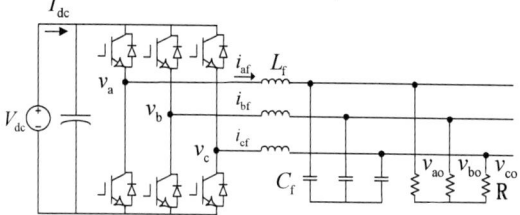

Fig 2. Topology of the inverter module with load

main control algorithm, while local controllers complete the sampling, driving and protection. Data exchange is completed by the fiber ring net. It is easy to achieve redundancy because every inverter module can run or stop dynamically, decided by the central controller. The System structure is simple and reliable. It can avoid zero-sequence circulating current inherently and achieve good load current sharing.

II. SYSTEM STRUCTURE AND FIBER RING NET

A. System Structure

Structure of parallel inverters, without load connected, based on fiber ring net is shown in Fig 1. The system has three inverter modules which are input parallel connected and output parallel connected (IPOP). Central controller completes main control algorithm, while local controllers complete the sampling, driving and protection.

The fiber ring net is also shown in Fig 1. The fiber ring net has two rings: data fiber ring and synchronization fiber ring. Data fiber ring is used for data exchange between central controller and local controllers. The data for exchange include voltage, current, temperature, duty cycle and control command. Synchronization fiber ring is used to achieve synchronous driving signals between local controllers.

Communication rate of the data fiber ring can be 100Mbps and synchronization accuracy of sync fiber ring can be 5ns-10ns. Such high communication rate will satisfy the data exchange need in one switching cycle in parallel connected inverters.

First of all, topology of the three phase inverter module will be discussed. Details of the inverter modules with resistance load connected are shown in Fig 2. The three phase output

voltage is $\mathbf{V}_o=[v_{ao},\ v_{bo},\ v_{co}]^T$. The three phase inductor current is $\mathbf{I}_L=[i_{af},\ i_{bf},\ i_{cf}]^T$. The three phase duty cycle is $\mathbf{D}=[d_a,\ d_b,\ d_c]^T$. The state space equation of three phase inverter is shown in (1).

$$\begin{cases} \dfrac{d}{dt}\mathbf{I}_L = \dfrac{1}{L_f}\mathbf{D}V_{dc} - \dfrac{1}{L_f}\mathbf{V}_o \\[2mm] \dfrac{d}{dt}\mathbf{V}_o = \dfrac{1}{C_f}\mathbf{I}_L - \dfrac{1}{RC_f}\mathbf{V}_o \\[2mm] i_{dc} = \mathbf{D}^T\mathbf{I}_L \end{cases} \tag{1}$$

Transform equation (1) into DQ coordinate and do small signal linearization. AC small signal model of three-phase inverter module in the DQ coordinate is shown in Fig 3. There are coupled items between DQ axes cause by the coordinate transformation. After decoupling, the DQ axes can be controlled in D axis and Q axis respectively.

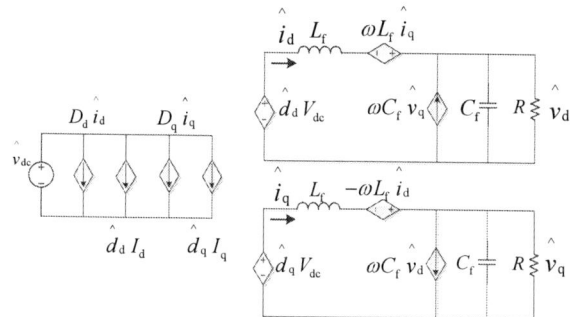

Fig 3. AC small signal model of three-phase inverter module in the DQ coordinate

$$\begin{cases} G_{id} = \dfrac{\hat{i}_d(s)}{\hat{d}_d(s)} = \dfrac{V_{dc}(sRC_f+1)}{s^2L_fRC_f+sL_f+R} \\[3mm] G_{iq} = \dfrac{\hat{i}_q(s)}{\hat{d}_q(s)} = \dfrac{V_{dc}(sRC_f+1)}{s^2L_fRC_f+sL_f+R} \end{cases} \tag{2}$$

$$\begin{cases} G_{ud} = \dfrac{\hat{v}_d(s)}{\hat{i}_d(s)} = \dfrac{R}{sRC_f+1} \\[3mm] G_{uq} = \dfrac{\hat{v}_q(s)}{\hat{i}_q(s)} = \dfrac{R}{sRC_f+1} \end{cases} \tag{3}$$

Dual control loops are used for the inverter: inner current control loop and outer voltage control loop. After decoupling, the transfer functions of current open loop and voltage open loop can be got for the equivalent AC small signal circuit in Fig 3, as shown in (2) (3). This is the equivalent AC small signal model of one inverter module. Because only one central controller completes the control algorithm of the whole parallel connected system, model of the parallel connected inverters should be discussed and built.

B. Switchable Fiber Ring Net

Fig 4 shows the structure of the slave node in switchable fiber ring net[10]. There is a PECL-to-LVDS chip in the slave node for converting the receiving PECL type signal to LVDS type signal. The converted LVDS signal is sent to the FPGA's LVDS port. This signal can be the data source sent to the next node directly. The output PECL signal of CY7C9689A is also converted to LVDS signal which is sent to FPGA's LVDS port. One of the two signals will be selected as the data source sent to the next node. The logic controller in the FPGA will determine which LVDS signal is converted to PECL type signal and transmitted to the next node through the optical fiber.

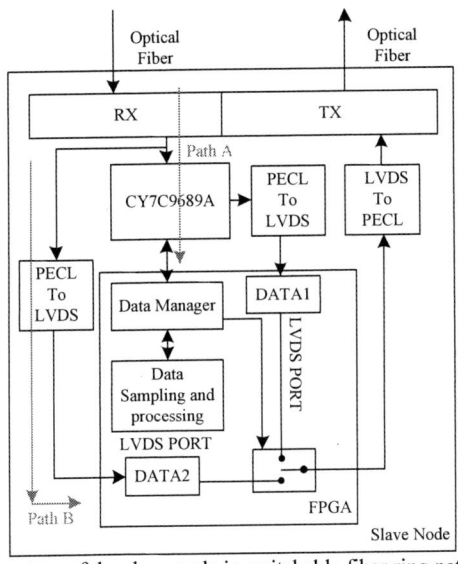

Fig 4. Structure of the slave node in switchable fiber ring net

The slave node has two operating states: normal state and transmitting state. In a normal state, the LVDS type signal received from the previous node is selected as the data source to be transmitted to the next node. Therefore the coming character (DATA2) is received by this node and transmitted to the next node approximately at the same time. In a transmitting state, the data source transmitted to the next node is changed to DATA1, as shown in Fig 4. Using this topology, it is possible to shorten the communication time and allow more time for the master to complete the function of computing and control in a period of switching time. More details of switchable fiber ring net can be found in [10].

III. MODEL OF PARALLEL INVERTERS

In the parallel connected three phase inverters, feedback variables of inner current control loop are sum of three phase inductor currents of all the inverter modules. The variable used for decoupling is also sum of three phase inductor currents. Feedback variables of outer voltage control loop are the voltages of load. The central controller collects all these needed data from the local controllers by the fiber ring net. At the same time, central controller sends command to every local controller through the fiber ring net. The sync fiber ring can achieve synchronous action between local controllers and the driving signals can be synchronous. Switches in different modules can be driven with almost same signals controlled by

the central controller. This is a key element to achieve current sharing and avoid zero-sequence circulating current.

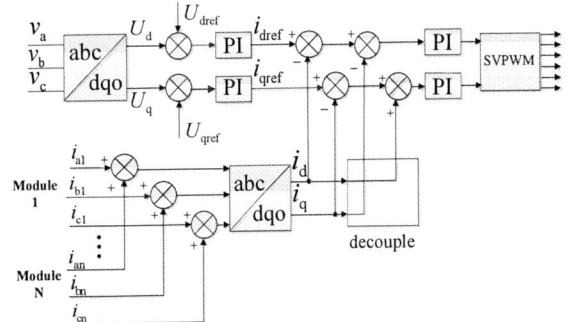

Fig 5 Control block diagram of parallel connected inverters

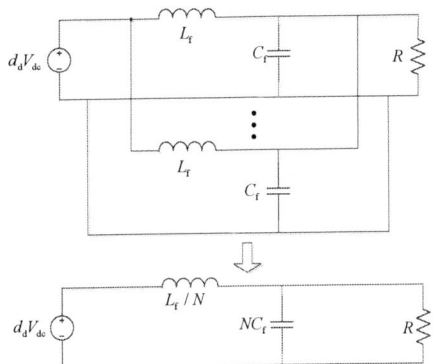

Fig 6. Model of parallel connected inverters in D axis

The control block diagram in central controller of parallel connected inverters is shown in Fig 5. Sum of the inductor currents is regarded as feedback variable of the inner loop. Input voltages and output voltages are same between different inverter modules. The parallel inverters are regarded as one system. The equivalent model of parallel connected inverters in D axis is shown in Fig 6 and that in Q axis is same with Fig 6.

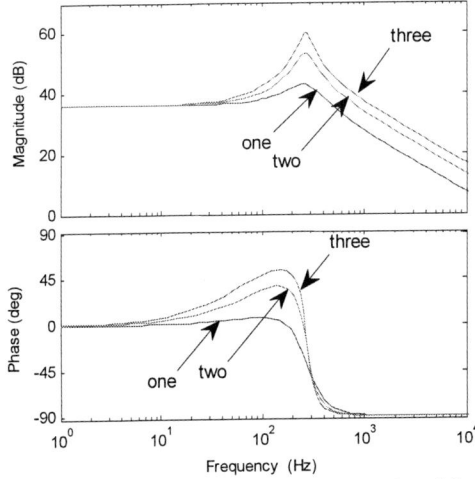

Fig. 7 Inner loop Bode plot with different amount of modules

978-1-4799-2706-7/14 $31.00 © 2014 IEEE

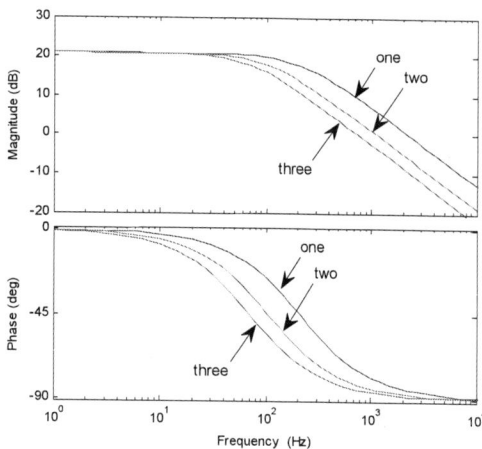

Fig. 8 Outer loop Bode plot with different amount of modules

So the system structure and system model vary with the amount of operating modules which are at run mode. Central controller will decide the amount of operating modules based on the load and every inverter module can run or stop dynamically. The inverter modules at stop mode can disconnect from the system by blocking the driving signals and turn off the contactor. The system is actually a variable structure system. The inner loop Bode plot with different amount of modules is shown in Fig 7 and outer loop Bode plot with different amount of modules is shown in Fig 8. Quality factor of the inverter is shown in (4). Resonant frequency of the inverter is shown in (5). It can be seen that resonant frequency doesn't change with the amount of parallel inverters. However, the Quality factor of N parallel inverters is N times the single inverter module got from (4) and Fig 6.

$$Q = R\sqrt{\frac{C_f}{L_f}} \qquad (4)$$

$$f_r = \frac{1}{2\pi}\frac{1}{\sqrt{L_f C_f}} \qquad (5)$$

The controller of the system should change with the amount of modules which are at run mode. Several PI parameters can be stored in the controller and the controller will decide which parameter is used to get better dynamic property.

IV. LOAD CURRENT SHARING

In the parallel connected inverters, input voltages and output voltages are same. The current sharing of each inverter module is only influenced by the filter inductor. For example, currents of phase A satisfy the equation (6). Phase of the three currents is exactly same and amplitude is inverse proportional to the value of inductor as shown in (7).

$$\dot{D}_a V_{dc} - \dot{V}_{ao} = \dot{I}_{a1}\omega L_{f1} = \dot{I}_{a2}\omega L_{f2} = \dot{I}_{a3}\omega L_{f3} \qquad (6)$$

$$\frac{I_i}{I_j} = \frac{L_j}{L_i} \qquad (1 \le i, j \le 3) \qquad (7)$$

The circulating current of the parallel connected inverters is determined by the difference of the output currents. From (7),

we can know that circulating current mainly depends on the distributed parameter of the filter inductors when the input voltage and load are same.

V. EXPERIMENT RESULTS

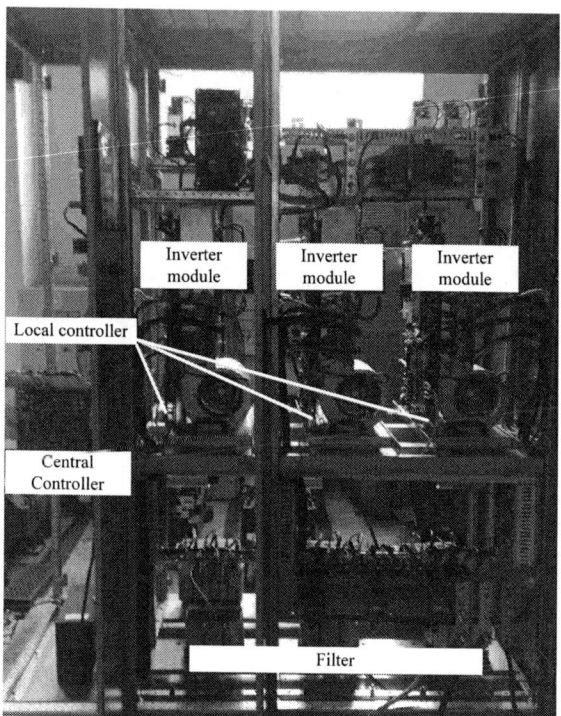

Fig. 9 The prototype of three-phase inverter system based on modular design and distributed control

A 12kW prototype which is constituted of three inverter modules is made. The device is shown in Fig 9. The prototype is based on modular design and distributed control technique using high speed fiber ring net. Power of every inverter module is about 4kW with resistance load connected. The filters of all the three modules are same. Total three phase output currents of the three phase inverter are shown in Fig 10. The phase output currents of the three modules are shown in Fig 11.

It can be seen that without any current sharing controller used, current sharing performance of the system is good. Both phase and amplitude of currents in phase A are same. This means that every module provides same active and reactive power to the load.

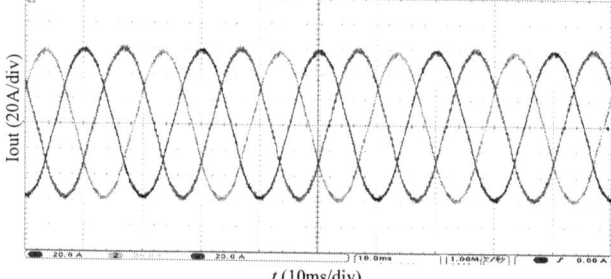

Fig. 10. Total output currents of the three phase inverter.

The 2014 International Power Electronics Conference

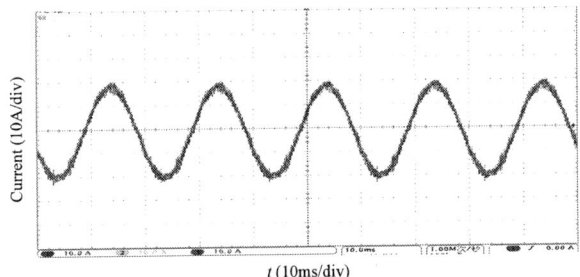

Fig. 11 The phase A output currents of the three modules.

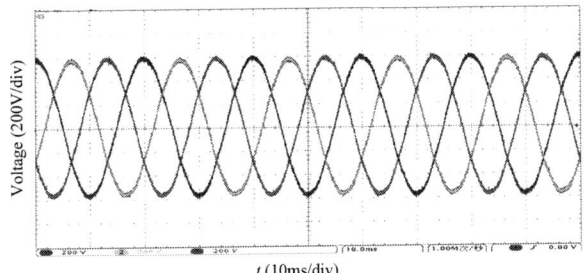

Fig. 12 Output three phase line voltage of the three phase inverter.

TABLE I
HARMONIC CONTENT OF THE OUTPUT VOLTAGE

harmonic order	3	5	7	9
Amplitude (V)	1.8	1.2	1.3	0.2
Percent (%)	0.8	0.5	0.6	0.1

TABLE II
HARMONIC CONTENT OF THE OUTPUT CURRENT

harmonic order	3	5	7	9
Amplitude (A)	0.01	0.02	0.06	0.01
Percent (%)	0.1	0.2	0.6	0.1

Harmonic content of the output three phase line voltages is shown in Table I and Harmonic content of the output three phase line currents is shown in Table II. The output voltages and currents have very low THD and they are respectively 1.1% and 0.9% measured through Power quality analyzer FLUKE 43B.

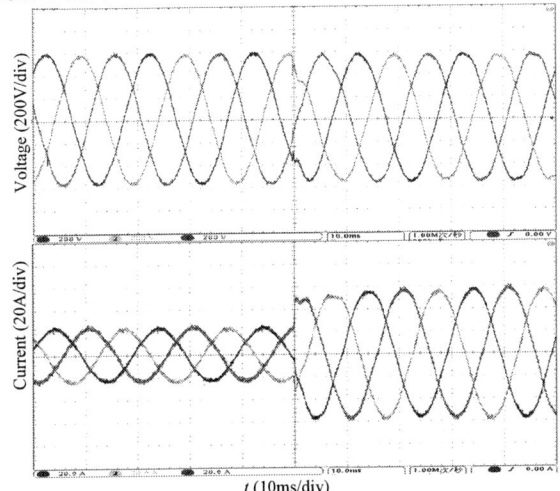

Fig. 13 Three phase line voltages and currents under step load addition of the three phase inverter

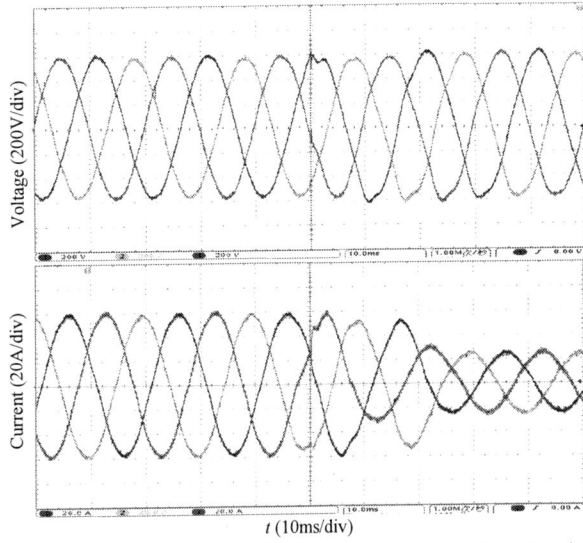

Fig. 14 Three phase line voltages and currents under step load reduction of the three phase inverter

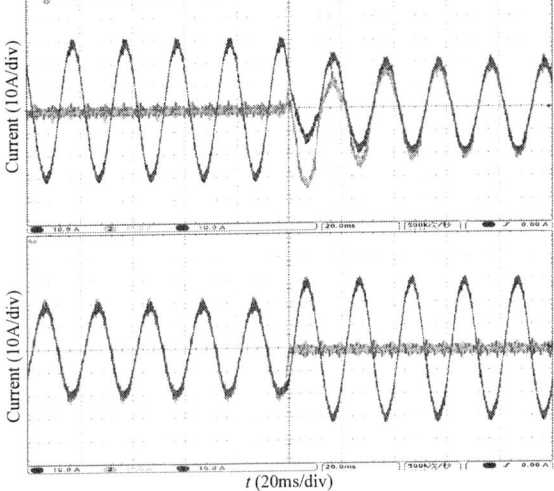

Fig. 15 Dynamic waveforms when amount of operating modules which are at run mode changes suddenly, two modules to three modules and three modules to two modules.

Fig 13 shows the dynamic waveforms under step load addition. Three phase output line voltages have light distortion and three phase output currents change from light load condition to heavy load condition suddenly. Fig 14 shows the dynamic waveforms under step load reduction. Three phase output line voltages have light distortion and three phase output currents change from heavy load condition to light load condition suddenly. From fig 13 and fig 14, it can be seen that the three phase inverter system based on modular design and distributed control has good dynamic performance.

Fig 15 shows the load sharing performance under dynamic operation. In the above picture, the amount of operating modules which are at run mode changes from two to three. It takes about one line period achieving load current sharing. In the second picture of fig 15, the amount of operating modules which are at run mode changes from three to two. When the module 3 stops running, the current in module 3 transfers to module 1 and 2 immediately. So, central controller can decide the amount of operating modules based on the load. The three phase inverter based on modular design and distributed control will achieve high reliability and flexibility.

VI. CONCLUSIONS

Structure of parallel connected three phase inverters based on modular design and distributed control is proposed in this paper. In the system, every inverter module has a local hardware manager and the whole system has a central application manager. They exchange sampling data, duty cycle and command by the high speed fiber ring net. The fiber ring net also achieves synchronous driving signals between different local controllers. AC small signal model of three-phase inverter module is given. When the system has a number of inverter modules parallel connected, the AC small signal model of the inverter system is changing. The controller in application manager is designed based on every possible model. Current sharing of the inverter modules is only affected by the filter inductor. By using synchronous driving signals, zero-sequence circulating current path is avoided. A prototype is made and the experiment results verify the design and analysis in this paper.

ACKNOWLEDGMENT

This project is supported by "Program for New Century Excellent Talents in University" (NCET-11-0871) and supported by "National Natural Science Foundation of China" (51177170).

REFERENCES

[1] Tsung-Po Chen, "Dual-modulator compensation technique for parallel inverters using space-vector modulation," *IEEE Trans. on Industry Electronics*, vol. 56, no. 8, pp. 3004-3012, 2009.

[2] Zhihong Ye, Boroyevich D, Lee F.C, "Modeling and control of zero-sequence current in parallel multi-phase converter," *Power Electronics Specialists Conference*, vol. 2, pp. 680-685, 2000.

[3] Ching-Tsai Pan, Hsinchu, Yi-Hung Liao, "Modeling and Control of Circulating Currents for Parallel Three-Phase Boost Rectifiers With Different Load Sharing," *IEEE Trans. on Industry Electronics*, vol. 55, no. 7, pp. 2776-2785, 2008.

[4] Xiao Sun, Lee, Y.-S, Dehong Xu, "Modeling, Analysis, and Implementation of Parallel Multi-Inverter Systems With Instantaneous Average-Current-Sharing Scheme," *IEEE Trans. on Power Electronics*, vol. 18, no. 3, pp. 844-856, 2003.

[5] Zeng Liu, Jinjun Liu, Yalin Zhao, Weihan Bao, "Output Impedance Modeling and Stability Criterion for Parallel Inverters with Master-slave Sharing Scheme in AC Distributed Power System," *Applied Power Electronics Conference and Exposition 2012*, pp. 1907-1913, 2012.

[6] Guerrero J.M, Garcia de Vicuna L, Miret J, "Output Impedance Performance for Parallel Operation of UPS Inverters Using Wireless and Average Current-Sharing Controllers," *Power Electronics Specialists Conference 2004*, vol. 4, pp. 2482-2488, 2004.

[7] Guerrero J.M, Garcia De Vicuna L, Matas J, "Output Impedance Design of Parallel-Connected UPS Inverters With Wireless Load-Sharing Control," *IEEE Trans. on Industry Electronics*, vol. 52, no. 4, pp. 1126-1135, 2005.

[8] Francis G, Burgos R, Wang F, "A Universal Controller for Distributed Control of Power Electronics Conversion Systems," *Computers in Power Electronics, 2006*, pp. 8-14, 2006.

[9] Sun Chi, Zhang Cheng; Ai Sheng, "Topology and Protocol of Practical High-speed Fiber Ring Net for Large-capacity Power Electronic Systems," *Proceedings of the CSEE*, vol. 32, no. 15, pp. 63-73, 2012.

[10] Jilong LIU, Fei XIAO, Xu YANG. "Study on Synchronization Methods in Switchable Fiber Ring Net for Distributed Control," *IECON2013，39th Annual Conference of the IEEE Industrial Electronics Society*, pp. 3780-3785, 2013.

Efficiency Investigations of a 3 kW T-Type Inverter for Switching Frequencies up to 100 kHz

Alexander Anthon, Zhe Zhang, Michael A.E. Andersen
Dept. of Electrical Engineering
Technical University of Denmark
Kgs. Lyngby, Denmark
jant@elektro.dtu.dk

Dr.-Ing. Toke Franke
Danfoss Silicon Power
Flensburg, Germany

Abstract—**This paper deals with a 3 kW multilevel inverter used for PV applications. A comparison has been made based on simulations using IGBTs and SiC MOSFETs to see how much efficiency can be gained when SiC diodes are used. A prototype with the same IGBTs and SiC MOSFETs has been built but using regular soft-recovery Si diodes instead of SiC diodes. Efficiencies and switching transitions for different switching frequencies up to 100 kHz have been measured. Thermal investigations of both IGBTs and SiC MOSFETs have been conducted to analyze the feasibility of increased switching frequencies. When SiC MOSFETs are used in combination with Si diodes, switching frequencies could be doubled achieving the same efficiencies than the IGBT converter.**

Keywords—SiC MOSFET, IGBT, multilevel inverter, reverse recovery current

I. INTRODUCTION

Photovoltaic (PV) systems have become more and more attractive in recent years. Especially residential PV inverter systems gained much attraction. Due to the low efficiency of the PV panels themselves, much attention must be paid in the design of the PV inverter which leads to a strong demand for low cost and high efficiency power converters. Two-level inverters have the advantage of having a lower cost factor due to the smaller amount of components, being simple in structure and control but suffer from a strong switching frequency and power depending efficiency as well as a relatively large output filter [1]. Multilevel topologies such as the Neutral-Point-Clamped (NPC) inverter have, on the other hand, efficiencies which are less depending on the switching frequency and they give a good compromise between system complexity, cost and efficiency [2]-[3]. Among the three-level inverter topologies, the T-Type inverter (also called Conergy [4] or BSNPC [5]) shows a higher efficiency than the NPC counterpart for low to medium switching frequencies [3]. Furthermore, the efficiency of the T-Type inverter can be improved by using Silicon Carbide (SiC) switching devices in order to reduce switching losses by increased switching transitions and hence increase the overall efficiency. Previous work has shown that SiC switching devices such as normally-on/off SiC JFETs, SiC BJTs and SiC MOSFETs show superior switching performance in various applications over their

silicon counterparts, [6]-[7]. An all SiC MOSFET T-Type inverter has been introduced in [8] achieving efficiencies over 98 %. A major aspect when using fast switching SiC devices is to equip the converter with SiC diodes instead of Si diodes in order to keep the switching losses low; otherwise the reverse recovery current caused by a high di/dt will increase the switching losses again and hence dampen the efficiency improvements. The feasibility of using SiC MOSFETs in the T-Type converter is investigated on a practical approach in this paper. Two 3 kW T-Type inverters equipped with 1200 V IGBTs and 1200 V SiC MOSFETs are compared for different power levels and switching frequencies. In Section II the topology including its modulation and current commutation is explained. Simulations of the topology have been carried out in Section III, in which expected efficiencies are obtained and a breakdown loss analysis is conducted. Practical results and efficiency measurements of a 3 kW prototype are introduced in section Section IV. Efficiency investigations for increased switching frequencies are investigated in Section V.

II. THE T-TYPE INVERTER

The T-Type inverter is a derivation from the NPC inverter. One phase leg comprises of four switching devices and four diodes as shown in Fig. 1. The output voltage of the inverter has three states with reference to the midpoint M, i.e. $+0.5V_{DC}$, 0 and $-0.5V_{DC}$. It is a commonly used topology in three-phase PV inverters in the medium power range and rather low switching frequencies of up to 16 kHz. Switches S_1 and S_3 including their free-wheeling diodes D_3 and D_4 require a breakdown voltage of at least the full DC link voltage V_{DC} whereas switches S_3, S_4 and the diodes D_1 and D_2 require a breakdown voltage of at least half the DC link voltage. In PV inverter systems, the DC link voltage can usually increase up to 1000 V, so S_1, S_2, D_3 and D_4 are 1200 V and S_3, S_4, D_1 and D_2 are 600 V devices to have a margin for overvoltages. A sinusoidal output voltage can be obtained by having switches S_1 and S_2 operated at a chosen switching frequency whereas switches S_3 and S_4 operate at grid frequency as shown in Fig. 1. The T-Type topology benefits from having lower conduction losses than its NPC counterpart because only

The 2014 International Power Electronics Conference

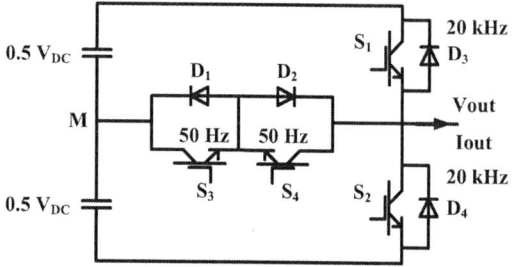

(a) Schematic of a single phase T-Type inverter with midpoint connection

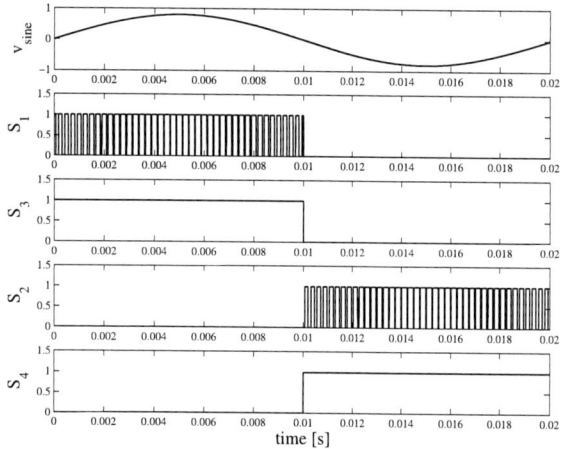

(b) Output voltage creation of the T-Type inverter using sinusoidal PWM

Fig. 1. Schematic of a single phase T-Type inverter and sinusoidal PWM scheme

TABLE I. SPECIFICATIONS

Symbol	Meaning	Value
L	Output filter inductance	3 mH
V_{DC}	DC link voltage	800 V
V_{out}	Filtered output voltage, RMS	230 V
P_{out}	Output power	250 W to 3000 W

one switch conducts current at the same time. The current commutations for a resistive load are shown in Fig. 2.

III. SIMULATION RESULTS WITH SiC DIODES

The simulations were done in PLECS and the semiconductor parameters were taken from their datasheets. The specifications for the inverter are shown in Table I. Switches S_1 and S_2 are chosen to be IGBTs due to their higher breakdown capabilities compared to Si MOSFETs. Their SiC counterpart will be a 1200 V SiC MOSFET C2M0080120D from Cree. Switches S_3 and S_4 are chosen to be IGBTs in both configurations due to their low switching frequency requirements. The diodes D_1 and D_2 are SiC diodes to show possible achievable efficiencies when no reverse recovery is taken into account. Table II shows the semiconductors used in the simulations. The Si converter comprises of 1200 V IGBTs and the SiC converter comprises of 1200 V SiC MOSFETs.

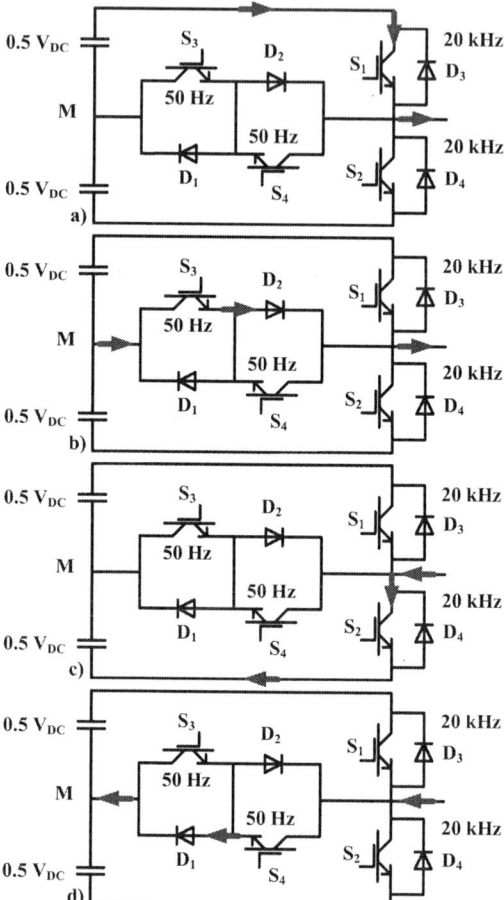

Fig. 2. Current paths in the T-Type inverter. a) Positive output voltage b) Zero output voltage c) Negative output voltage d) Zero output voltage

TABLE II. SEMICONDUCTORS

Version	D_1 and D_2	S_1 and S_2	S_3 and S_4
Si Converter	C4D20120A	IKW15N120T2	IKP15N60T
SiC Converter	C4D20120A	C2M0080120D	IKP15N60T

The simulation results of the T-Type inverter for 16 kHz and 30 kHz are shown in Fig. 3. At a switching frequency of 16 kHz, a maximum efficiency of 97.9 % is achieved when IGBTs are used and 98.6 % when SiC MOSFETs are used. A larger efficiency difference between the IGBT version and SiC MOSFET version can be obtained if the switching frequency is increased to 30 kHz. Then a maximum efficiency of 97 % with IGBTs and 98.2 % with SiC MOSFETs are achieved. Although the specifications do not exactly match with [8], the results are close to what has been presented in previous work so that the simulations can be considered a proper representation of what to expect. A breakdown loss analysis has been conducted to show the loss distribution of the converter system. Apart from the semiconductors, losses in the filter inductor as well as the DC link capacitors have been included. The results are shown in Fig. 3. It can be seen that due to the modulation applied, switching losses mainly occur in the 1200 V switches.

978-1-4799-2706-7/14 $31.00 © 2014 IEEE

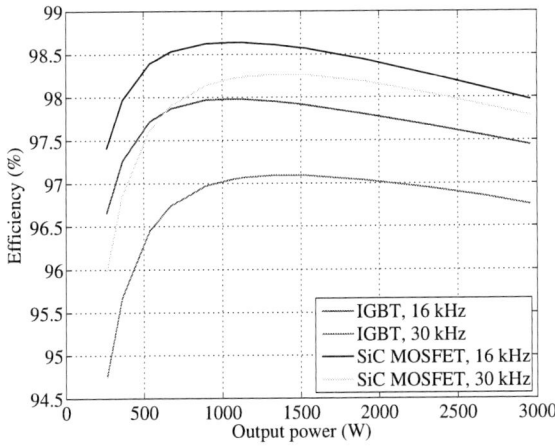

(a) Simulation results of T-Type inverter using 1200 V IGBTs with $R_g = 2.2\,\Omega$ and 1200 V SiC MOSFETs with $R_g = 5\,\Omega$

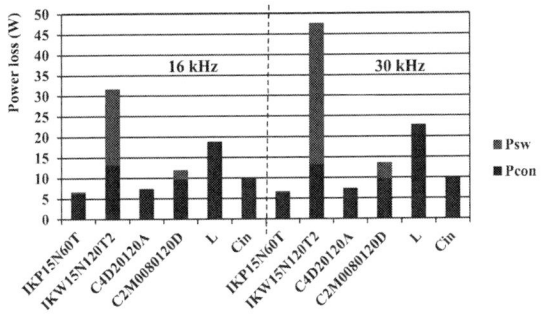

(b) Breakdown analysis of loss distribution in the T-Type inverter at a full power of 3 kW

Fig. 3. Simulation results of T-Type inverter using 1200 V IGBTs and 1200 V SiC MOSFETs

Fig. 4. Prototype of a 3 kW T-Type inverter. The dimensions of the printed circuit board are 8.5 cm by 7 cm

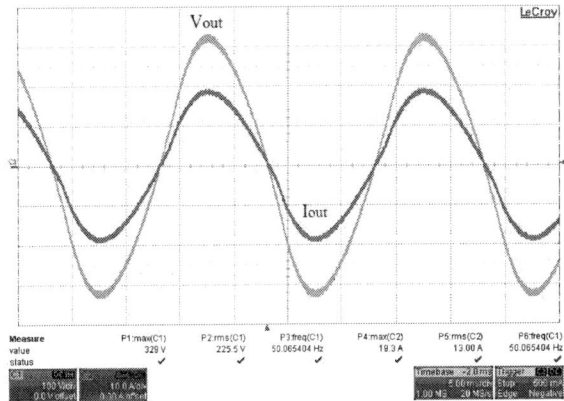

Fig. 5. Filtered output waveforms at an output power of 3 kW and a switching frequency of 16 kHz

Hence the switching frequency is a limiting factor for the efficiency of the T-Type inverter. However, switching losses can be reduced by using SiC switching devices. The effect of the fast switching capabilities of SiC devices becomes more important when a higher power density is targeted because switching losses in regular IGBTs become dominant degrading overall efficiency. Based on the simulations, switching and conduction losses in the 1200 V IGBT are relatively balanced at a switching frequency of 16 kHz whereas switching losses of the SiC MOSFETs are still smaller than the conduction losses at a switching frequency of 30 kHz. Both the size of the filter inductor and the DC link capacitors were kept constant, though a redesign of these could have reduced losses at increased switching frequencies. However, a main requirement to the simulated efficiencies is that the diodes D_1 and D_2 do not show any reverse recovery current.

IV. EXPERIMENTAL RESULTS

To see how the T-Type inverter performs with IGBTs and SiC MOSFETs, a prototype has been built which is shown in Fig. 4. For both the 1200 V IGBTs and SiC MOSFETs, a TO-247 package was used having the same pinning and hence the same printed circuit board (PCB) and layout could be used for a fair comparison. For layout optimization, S_3 and D_1 are packed in one TO-220 package and so are S_4 and D_2. Hence the whole converter could be built with four discrete devices. Only the gate drivers (Dr1 - Dr4) for the IGBTs and SiC MOSFETs were adjusted to stay within their absolute maximum ratings for the Gate-Source voltage. The IGBTs were switched on and off with a Gate-Source voltage of ± 15 V whereas the SiC MOSFETs were switched on with a Gate-Source voltage of 19 V and switched off with a Gate-Source voltage of -5 V. For further comparisons to the simulations, the prototype is equipped with soft-recovery Si diodes instead of SiC diodes. At full power, the filtered output voltage and current are shown in Fig. 5.

A N4L PPA5500 power analyzer was used for efficiency measurements. A first comparison is made with an IGBT version having a gate resistance of 2.2 Ω and a SiC MOSFET version having a gate resistance of 5 Ω. The results are shown in Fig. 6.

It can be seen in Fig. 6 that the efficiency could be improved when a SiC MOSFET with a gate resistance of 5 Ω is implemented. However, efficiency improvements are larger as the switching frequency is increased. At 16 kHz, a maximum efficiency improvement of 0.3 % is achieved. Increasing switching frequency to 30 kHz leads to a maximum efficiency improvement of 0.8 %. It can furthermore be seen that the SiC MOSFET inverter has similar efficiencies at 30 kHz than the IGBT inverter at 16 kHz. The switching frequency for the SiC converter is therefore increased to 60 kHz and plotted in Fig. 6. It can be seen that the SiC converter at 60 kHz has similar efficiencies than the IGBT converter at 30 kHz which yields to the conclusion that the switching frequency can be doubled when SiC MOSFETs are implemented without degrading the efficiency. The case temperatures of the IGBTs and SiC MOSFETs were measured to get a comparison of the power dissipation in such devices. The operating conditions are at full power, i.e. 3 kW and 20 kHz for the IGBT and 30 kHz for the SiC MOSFET. The case temperatures were measured with an infrared camera and the results are shown in Fig. 7.

It is seen in Fig. 7 that even though the switching frequency is increased, the case temperature for the SiC MOSFET is around 10 °C lower. The thermal resistance of the 1200 V IGBT is given in the datasheet to be 0.63 K/W and the thermal resistance for the SiC MOSFET is given to be 0.60 K/W. Hence the junction temperature of the SiC MOSFET is around 10 °C lower as well. That the case temperature of the 600 V IGBT is higher than the case temperature of the SiC MOSFET can be explained by the fact that a regular TO-220 package for the IGBT was used. In that package, the IGBT comes along with a Si soft recovery free-wheeling diode. These free-wheeling diodes for the two 600 V IGBTs are used to be D_1 and D_2. As a consequence, the TO-220 package withstands the power dissipation for both the

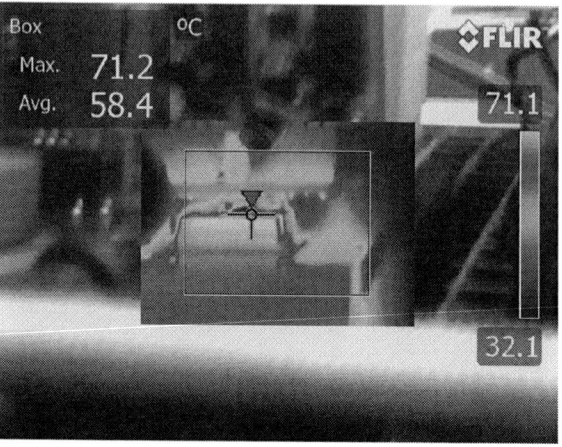

(a) Case temperature of IGBT at 20 kHz and 3 kW

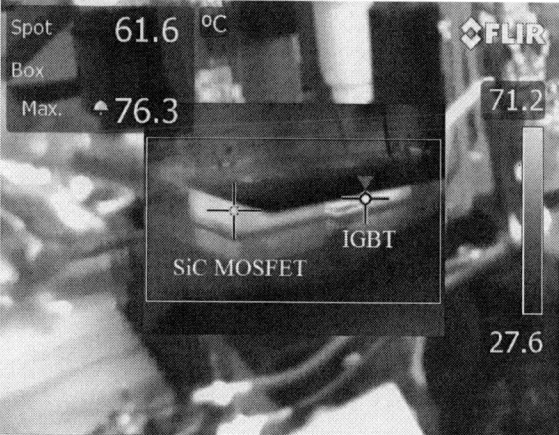

(b) Case temperatures of SiC MOSFET and 600 V IGBT+Diode at 30 kHz and 3 kW

Fig. 7.   Temperature measurements of 1200 V switching devices

IGBT and the free-wheeling diode. A switching transition for both turn on and turn off of the SiC MOSFET has been captured. The gate resistance is kept to be 5 Ω, output power is 900 W and switching frequency is 16 kHz. The current was measured with a Rogowski coil having a 20 MHz bandwidth limitation. The Drain-Source voltage was measured with a high voltage probe with a 400 MHz bandwidth limitation and the Gate-Source voltage was measured with a voltage probe having a 500 MHz bandwidth limitation. Furthermore, the time delay of 24 ns of the Rogowski coil was compensated in the measurements and the attenuation for the current measurement was set such that 2 V/div equals to 2 A/div. The transitions are shown in Fig. 8. It can be seen in Fig. 8a that the SiC MOSFET switches 400 V within 30 ns resulting in a dv/dt of more than 13 kV/µs. A maximum dv/dt was measured to be 25 kV/µs. The current rises 2 A within 4 ns resulting in a di/dt of 500 A/µs. The peak current is measured to be 10 A. During the turn off transition as shown in Fig. 8b, the maximum dv/dt is measured to be 20 kV/µs. The maximum di/dt is 400 A/µs. For comparison, the IGBT switched 400 V within 120 ns resulting in a dv/dt of 3 kV/µs.

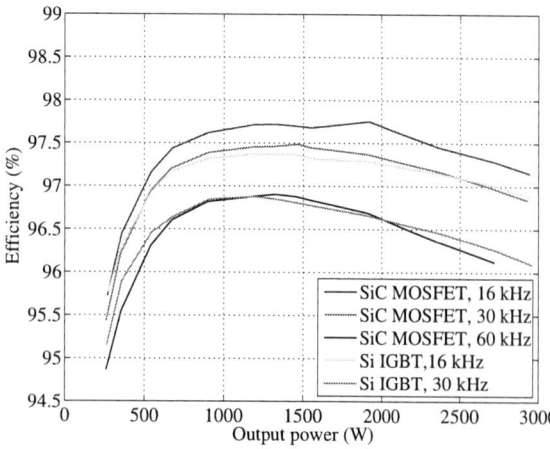

Fig. 6.   Measured efficiencies of IGBT and SiC T-Type inverter at different switching frequencies

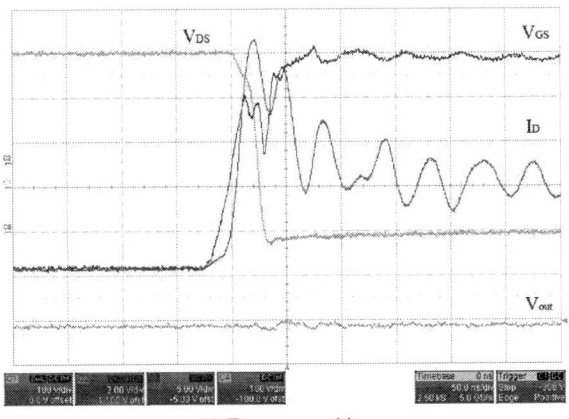

(a) Turn on transition

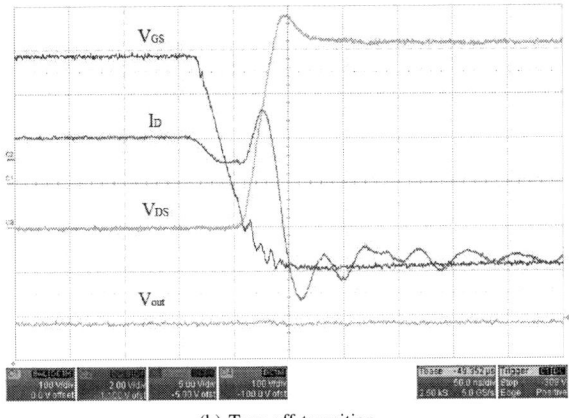

(b) Turn off transition

Fig. 8. Turn on and turn off transition of SiC MOSFET

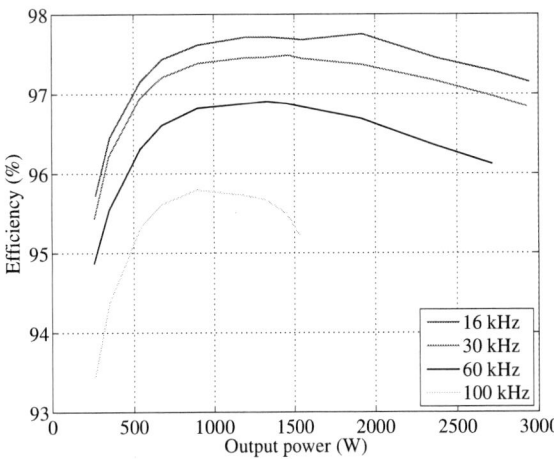

Fig. 9. Measured efficiencies of SiC T-Type inverter for switching frequencies up to 100 kHz

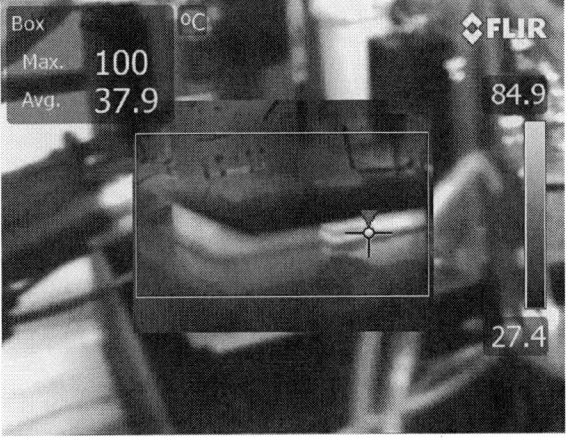

Fig. 10. Case temperature of 600 V devices in the TO-220 package at a switching frequency of 60 kHz and an output power of 2.7 kW

V. EFFICIENCY INVESTIGATIONS FOR INCREASED SWITCHING FREQUENCIES

It is seen that the efficiencies could be improved when SiC MOSFETs are implemented and the switching frequency could be doubled achieving the same efficiencies when IGBTs are used. It is therefore of interest to furthermore increase the switching frequency and to see how it affects the efficiency. As a last operating point, the switching frequency is increased to 100 kHz. The efficiency curves for the SiC converter at different switching frequencies are shown in Fig. 9.

It can be seen that the overall efficiency dramatically drops as the switching frequency increases up to 100 kHz. Also, the maximum efficiency point is shifted down to a lower power operating point compared to lower switching frequencies. The measurements were limited to a maximum power of 1.6 kW as the case temperature of the TO-220 packages became too high and hence the risk of a thermal damage was increased. However, the case temperature of the SiC MOSFETs were still below 80 °C at an output power of 1.6 kW. So the limiting factor are the 600 V devices in the TO-220 package. A thermal picture of the TO-220 package at an operating point of 60 kHz and 2.7 kW was taken to verify the limiting factor at increased switching frequencies. The result is shown in Fig. 10.

The case temperature of the TO-220 package is measured to be 100 °C and is much higher than the case temperature of the SiC MOSFET, as it can be seen in the scale on the right hand side of Fig. 10.

VI. CONCLUSIONS

In this paper the feasibility of SiC switching devices on a 3 kW T-Type inverter topology for PV applications has been investigated. Simulations with regular IGBTs and SiC MOSFETs have been carried out including a breakdown loss analysis to investigate the loss contribution on the overall efficiency. It is shown that efficiency improvements can be achieved when SiC MOSFETs are equipped in combination with SiC diodes. A prototype has been built using the same IGBTs and SiC MOSFETs but regular Si diodes instead of SiC diodes. Efficiency measurements have been done to see how much the reverse recovery current of the Si diodes will affect the overall efficiency. Using Si diodes instead of SiC diodes, efficiency improvements could be achieved but not as much as it could be in the simulations with SiC diodes. However, switching frequency could be doubled achieving the similar efficiency curves when IGBTs are used. Switching frequencies were increased up to 100 kHz to see how much efficiency drop one might expect. The limiting factor at increased switching frequencies are the 600 V devices in a TO-220 package. Using external SiC diodes in combination with 600 V IGBTs could furthermore improve efficiencies and enable higher switching frequencies.

REFERENCES

[1] R. Teichmann and S. Bernet, "A comparison of three-level converters versus two-level converters for low-voltage drives, traction, and utility applications," *IEEE Trans. Ind. Appl.*, vol. 41, no. 3, pp. 855–865, 2005.

[2] P. Alemi and D.-C. Lee, "Power loss comparison in two- and three-level PWM converters," in *Power Electronics and ECCE Asia (ICPE & ECCE), 2011 IEEE 8th International Conference on*, 2011, pp. 1452–1457.

[3] M. Schweizer, I. Lizama, T. Friedli, and J. W. Kolar, "Comparison of the chip area usage of 2-level and 3-level voltage source converter topologies," in *IECON 2010 - 36th Annual Conference on IEEE Industrial Electronics Society*, 2010, pp. 391–396.

[4] L. Ma, T. Kerekes, R. Teodorescu, J. Xinmin, D. Floricau, and M. Liserre, "The high efficiency transformer-less PV inverter topologies derived from npc topology," in *Power Electronics and Applications, 2009. EPE '09. 13th European Conference on*, 2009, pp. 1–10.

[5] J. Pinne, A. Gruber, K. Rigbers, E. Sawadski, and T. Napierala, "Optimization and comparison of two three-phase inverter topologies using analytic behavioural and loss models," in *Energy Conversion Congress and Exposition (ECCE), 2012 IEEE*, 2012, pp. 4396–4403.

[6] M. Oestling, R. Ghandi, and C.-M. Zetterling, "SiC power devices — present status, applications and future perspective," in *Power Semiconductor Devices and ICs (ISPSD), 2011 IEEE 23rd International Symposium on*, 2011, pp. 10–15.

[7] Y. Gao, Q. Huang, S. Krishnaswami, J. Richmond, and A. K. Agarwal, "Comparison of static and switching characteristics of 1200v 4H-SiC BJT and 1200v si-IGBT," in *Industry Applications Conference, 2006. 41st IAS Annual Meeting. Conference Record of the 2006 IEEE*, vol. 1, 2006, pp. 325–329.

[8] D. De, A. Castellazzi, A. Solomon, A. Trentin, M. Minami, and T. Hikihara, "An all SiC MOSFET high performance PV converter cell," in *Power Electronics and Applications (EPE), 2013 15th European Conference on*, 2013, pp. 1–10.

Miniaturization of the Boost-up type Active Buffer Circuit in a Single-phase Inverter

Hiroki Watanabe, Kazuhiro Koiwa, Jun-ichi Itoh
Department of Electrical Engineering
Nagaoka University of Technology
Nagaoka, Niigata, Japan
hwatanabe@stn.nagaokaut.ac.jp

Yoshiya Ohnuma, Satoshi Miyawaki
Nagaoka Power Electronics Co.Ltd
Nagaoka, Niigata, Japan
ohnuma@npe.co.jp

Abstract— This paper discusses miniaturization of a single-phase grid connected inverter, which has power decoupling function for PV. The power ripple of twice grid frequency is compensated in the proposed circuit with the small capacitor (50 μF) at 200 W. However, the buffer inductor in the proposed circuit becomes large because the buffer inductor current is fluctuated at twice grid frequency by power ripple compensation, and the switching ripple is large at the peak current. In this paper, in order to reduce the size of the buffer inductor, the buffer inductance is optimized in terms of the volume and efficiency according to the switching frequency. As an experimental result, it is confirmed that the input current ripple is reduced by 90 % in the proposed circuit, and the output current THD is 3.5 %. On the other hand, the buffer inductance is reduced by 73 % at 64 kHz of the switching frequency. In addition, the volume of the proposed circuit can be reduced by 37% in compared to the conventional circuit. Finally, from the evaluation of the power density using the Pareto front curve, the proposed circuit can design the high power density in compared to the conventional circuit.

Keywords— *grid connected inverter; power ripple compensation; loss analysis; power density*

I. INTRODUCTION

The dramatically cost down of photovoltaic (PV) has urged the uses of solar energy becoming popular in the interest of energy saving. In general, the power conversion system employs a grid-connected inverter with the boost-up chopper to connect the PV modules to a single-phase grid system.

The grid-connected inverter for the PV applications can be categorized into two types. The first type uses a large power capacity inverter to connect multiple numbers of PV modules in a series connection. The second type employs a small power capacity inverter, which is also known as the "micro-inverter", and then it is connected to each of the single PV cell [1]-[4]. The "micro-inverter" offers better features because of the following reasons; (i) Maximum Power Point Tracking control (MPPT) can be easily applied in each of the individual PV cell. (ii) Optimal design for control is simple. (iii) Capacity of the PV system can be easily modified due to the simple structure.

However, micro-inverters have problems regarding

cost and size, since the micro-inverter need a lot of numbers for PV system in comparison with the first type system which only employs a large power capacity single-phase inverter.

Although the lifetime of a PV module is namely 25 years, the lifetime of the electrolytic capacitor is typically 1000-7000 hours at 105 degree Celsius operating temperature. As a result, periodically maintenance is required for the conventional micro-inverter. Besides, the volume of the electrolytic capacitor dominates the total volume of the micro-inverter. Moreover, the initial charging for the electrolytic capacitor is needed.

In order to remove the large electrolytic capacitor, power decoupling methods between DC side and AC side known as the DC link active filters have been studied such as [5]-[9].

These topologies technically can reduce the capacitance value of the DC link capacitor. However, extra devices and passive components are required. Consequently, the efficiency is degraded. Additionally, the cost is higher because of the additional circuit.

The authors have proposed a circuit topology, which comprises the single-phase inverter and the active buffer based on a boost-chopper between DC side and AC side [10]. The active buffer circuit is employed to achieve boost-up function for PV voltage and to decouple the power fluctuation with a small capacitor simultaneously. The proposed circuit has the following features; (i) A resonant DC/DC converter is applied with the zero current switching (ZCS), (ii) the miniaturization and cost reduction are achieved because general 6 in 1 modules can be used in proposed circuit. However, the buffer inductor current in the proposed circuit should be controlled to twice frequency of the power grid in order to compensate the power fluctuation. As a result, the volume of the boost-up inductor becomes large in comparison to a general boost chopper.

In this paper, the volume of the buffer inductor in the proposed circuit is evaluated by experiment. In order to reduce the size of the buffer inductor, the buffer inductance is optimized according to the switching frequency. Although, the switching loss in the active buffer is increased, the copper loss of the inductor is reduced at the same ripple current.

The remainder of this paper is organized as follows; first, the constitution of the proposed circuit is shown. After that, the principle of the power decoupling control strategy is described. In addition, the designing the buffer capacitor and buffer inductor for miniaturization is described. Finally, the power density of the proposed circuit is evaluated using the Pareto front curve.

II. CIRCUIT CONFIGURATION

Figure 1 shows the circuit configuration of a conventional micro-inverter for PV. The DC voltage of the single PV cell is relatively low, generally ranging from 25 V to 50 V. On the other hand, the power grid voltage is typically from 110 V to 240 V. For the reason, the boost-up chopper is necessary for the grid-connected inverter. L_{dc} and C_{dc} are the components of an input filter to remove the switching ripple in the DC link

The power ripple which has twice of the power grid frequency occurs in the input DC current. Because of the large current ripple, the effeciency of the PV is decreased. In order to suppress the power ripple, the large electolytic capacitor is connected to the DC link part. The electrolytic capacitor requires large space and an initial charge circuit. Thus, regular maintenance is required. As a result, the conventional circuit becomes bulky and high cost.

Figure 2 (a) shows the circuit configuration of the isolated resonance DC/DC converter. This circuit boost the input voltage v_{in} tenfold, and the zero current switching (ZCS) control is applied to reduce the switching loss.

Figure 2 (b) shows the circuit configuration of the active buffer and inverter circuit. The active circuit consists of a single-phase inverter, a boost-up chopper and a small capacitor C_5 to absorb the power ripple. In comparison with the conventional circuit, the advantages of the propsed circuit are smaller size and longer lifetime because the small film or ceramic capacitor is used instead of an electrolytic capacitor. In addition, the number of the components in the proposed circuit is not increased from the conventional micro-inverter as shown in Figuer 1.

III. POWER DECOUPLING AND CONTROL STRATEGY

Figure 3 shows the principle of the power decoupling between the DC and AC sides. When both the input voltage and current waveforms are sinusoidal, the instantaneous output power pout is expressed as

$$p_{out} = \frac{V_{out} I_{out}}{2}(1 - \cos 2\omega t) \qquad (1)$$

where, V_{out} is the peak voltage, I_{out} is the peak current, and the angular frequency of the output voltage. From (1), the power ripple, that contains twice frequency of the power grid, appears at the DC link.

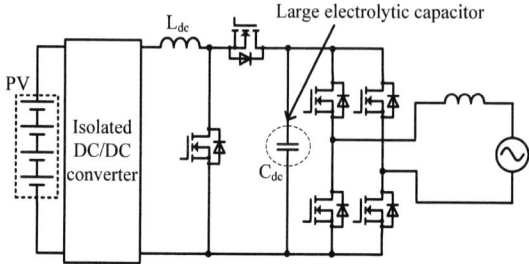

Fig.1 Conventional circuit for PV micro inverter.

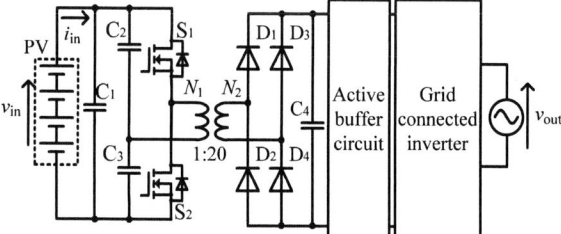

(a) Isolated resonance DC/DC converter.

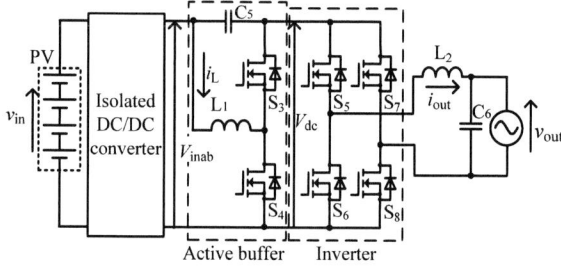

(b) Active buffer, inverter circuit.

Fig.2 Circuit configuration of the proposed circuit.

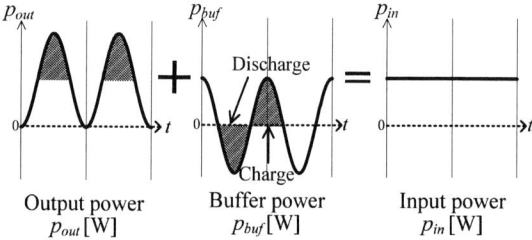

Fig.3 Compensation principle of the power ripple with active buffer.

In order to absorb the power ripple, the instantaneous power p_{buf} of an active buffer, should be controlled by

$$p_{buf} = \frac{1}{2}V_{out} I_{out} \cos 2\omega t \qquad (2)$$

where, the polarity of the p_{buf} is defined as positive when the active buffer discharges. It is noted that the mean power of the active buffer is zero because the active buffer does not generate the power.

Figure 4 shows the control block diagram for the proposed system. It is noted that automatic voltage regulator (AVR) is implemented to control the voltage of the buffer capacitor C_b. However, high speed response is required in the AVR because the control reference signal frequency is twice of the single-phase power grid frequency. In order to solve this problem, controlling the capacitor voltage which is implemented on the rotational d-q frame, has been discussed in Refs. [11]. In this method, the AVR does not require a high speed response. However, the control structure is complicated owing to using the rotational frame. The automatic current control (ACR) is applied to the active buffer circuit. In addition, in order to control the output voltage, the AVR is applied to the inverter.

In the paper, the current reference i_{amp}^* which intendeds to compensate the power ripple is added into the buffer inductor current reference of an ACR. Thus, the AVR which has high speed response is not required in the proposed circuit. The instantaneous buffer inductor current reference i_L^* is expressed by (3)

$$i_L^* = i_{amp}^* + i_{in}^* = i_c^* + i_{in}^* \tag{3}$$

where, i_{amp}^* is the power ripple compensation current reference, and i_{in}^* is the DC component in the buffer inductor current.

The power ripple compensation current reference i_{amp}^* is calculated by output power pout and buffer capacitor voltage v_c Thus, the power ripple compensation current is expressed by (4)

$$i_{amp}^* = i_c^* = \frac{p_{out}}{v_c}\cos(2\omega t) \tag{4}$$

On the other hand, the DC component in the buffer inductor current i_{in}^* is expressed by (5)

$$i_{in}^* = \frac{p_{in}}{v_{in}} \tag{5}$$

where, p_{in} is the input power, and v_{in} is the transformer secondary voltage. Then, the current reference i_{amp}^* and the DC current i_{in}^* are calculated by (4) and (5).

Relationship among the DC link voltage v_{inv},

transformer secondary voltage v_{rec} and buffer capacitor voltage v_c can be expressed by (6).

$$v_{inv} = v_{rec} + v_c \tag{6}$$

It is note that, the inverter input voltage v_{inv} is fluctuated owing to the power ripple from single phase grid. In addition, maximum power point tracking (MPPT) is implemented in order to keep the output power at the maximum point. In order to control the DC link voltage v_{inv}, the reference value of the AVR is higher than the grid voltage. However, the buffer capacitor voltage v_c fluctuates owing to twice frequency of the single-phase grid. As a result, the distortion in the output current occurs. Additionally, in order to solve this problem, a band eliminate filter (BEF) is applied in the detection value of the DC link voltage Further the AVR controls the average value of the DC link voltage. In order to connect the single phase grid, it is necessary that the phase angle of the inverter output current corresponds to the grid voltage using the PLL.

Figure 5 shows the maximum power point tracking (MPPT), known as the hill climbing method. The MPPT is controlled in four modes as illustrated as mode1, mode2. In mode1, the input current is increased to observe the maximum point of input power. In mode2, the present value is compared with the last value, if the present value of the input power is lower than 20% of the previous maximum power point, then the input current is increased in order to maintain the point of maximum power. As a result, the input power is controlled by (7)

$$P_{th} \leq P_{in} \leq P_{max} \tag{7}$$

IV. DESIGN METHOD OF THE COMPONENT

A. Buffer capacitor C_5

The buffer capacitor C_5 is expressed from P_{out} which is the electric storage energy to compensate the power ripple, by (8)

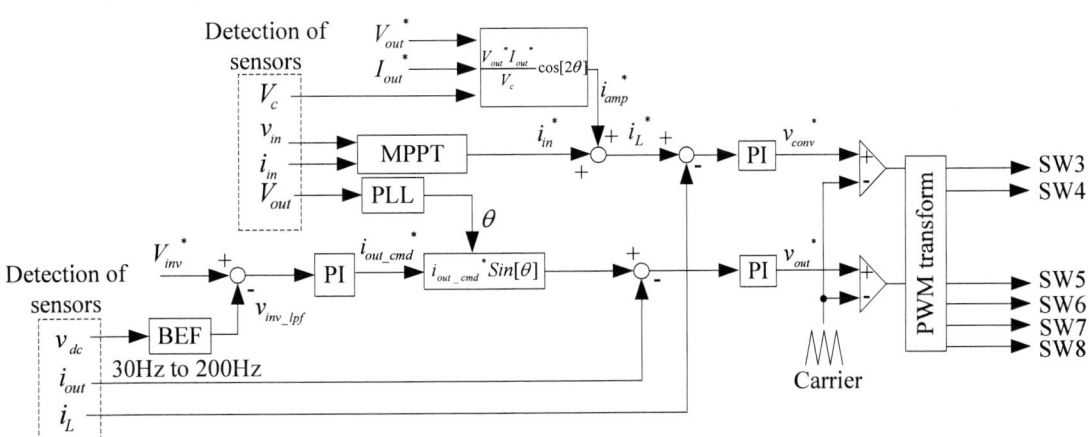

Fig.4 Control block diagram for the proposed circuit.

$$C_5 = \frac{2P_{out}}{\omega V_c^2 r_c} \tag{8}$$

where, r_c is the ratio of the voltage ripple that is expressed by (9)

$$r_c = \frac{\Delta V_c / 2}{V_c} \tag{9}$$

where, ΔV is the pulsatile voltage. In order to reduce the C_5, ΔV is fluctuated widely. In other words, the peak voltage increases. Thus, C_5 is needed to design less than the blocking voltage.

Figure 6 shows the relationship between the ΔV and the capacitance of C_5. According to Figure 6, the C_5 is reduced by 50 µF at 63.6 V. In this paper, C_5 is used the ceramic capacitor what blocking voltage is 200 V.

B. Buffer inductor L_1

The buffer inductance L_1 is designed from ripple current of buffer inductor current i_L. L_1 is obtained by (10).

$$L_1 \geq \frac{V_{in}}{2r_l I_{in} f_{sw}} \frac{V_{out} - V_{in}}{V_{out}} \tag{10}$$

where, r_c is the ratio of the voltage ripple that is expressed by (11)

$$r_l = \frac{\Delta I_L / 2}{I_L} \tag{11}$$

From (10), Relationship between the buffer inductance L_f and switching frequency f_{sw} is inverse variation. Thus, L_f can reduce by increasing the switching frequency.

Figure 7 shows the relationship between buffer inductance and carrier frequency. As shown in fig.7, the inductance can be reduced by 73% when the switching frequency is 64 kHz.

V. EXPERIMENTAL RESULTS

A. Fundamental operation

In order to demonstrate the validity of the proposed circuit, the proposed circuit is demonstrated by using a 200-W prototype. In is noted that the proposed circuit is connected to 200 V grid. Table 1 lists the experimental parameters.

Figure 8(a) shows the experimental results without the power decoupling control. It is noted that, the active buffer is operated as the boost-up-chopper, that the power ripple is not compensated. According to Figure 8(a), the input current ripple is fluctuated at approximately 6 A (peak-to-peak).

Figure 8(b) shows the operation waveforms that the power decoupling control is applied. As a result, the input

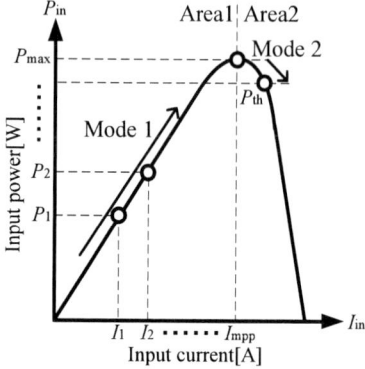

Fig.5. Behavior of the operation point in MPPT.

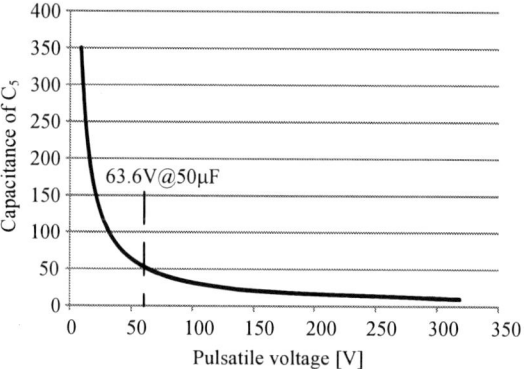

Fig.6 Relationship between the pulsatile voltage and capacitance of C_5.

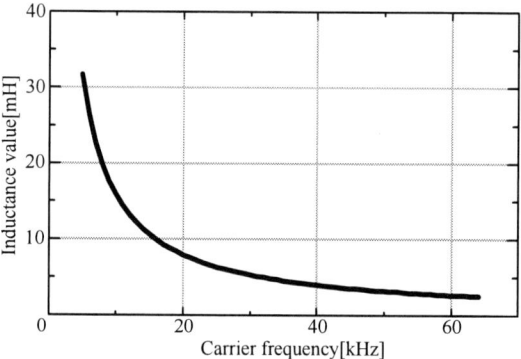

Fig.7 Relationship between buffer inductance and carrier frequency.

Table1. Experimental condition.

Rated Power P_{out}		200W
Input voltage V_{in}		36V
Grid voltage V_{out}		200V
Grid frequency f		50Hz
Switching frequency f_{sw}	DC/DC converter	150kHz
	Active buffer , Inverter	64kHz
Response angular frequency	ACR(active buffer)	4000rad/s
	ACR(Inverter)	4000rad/s
	AVR	50rad/s

current fluctuation is reduced to less than 0.5 A (peak-to-peak). In addition, the output power factor is 0.99 at the maximum point. Additionally, the output current waveform is sinusoidal waveform.

Figure 9(a) shows the buffer capacitor voltage and buffer inductor current waveforms without the power decoupling control. According to Figure 9(a), the buffer inductor current is not fluctuated twice frequency of a single phase grid.

Figure 9(b) shows the buffer capacitor voltage and buffer inductor current waveforms with the power decoupling control. According to figure 9(b), the buffer capacitor voltage V_c is fluctuated at approximately 65.2 V (peak to peak) due to the power decoupling control. The results illustrated that the buffer inductor current is fluctuating because the active buffer circuit is decoupling the power fluctuation.

Figure 10 shows the result of harmonic analysis. As the result, the second order harmonic component is reduced by 87.7% compared to that without the power decoupling control. However, the second order harmonic component is not suppressed by 12.3%. This is because; (i) the compensation value is not enough at 200 W, and (ii) the phase of buffer power p_{buf} is not agreed to the single-phase power grid.

Figure 11 shows the buffer inductor current ripple characteristics for the switching frequency. According to Figure 11, the experimental results almost agree with the design value. Thus, the validity of the design is confirmed. However, the experimental results are lower than the design value. This is because the inductance value is large than the design value. Thus, the experimental value conforms to theoretical value by using the design value.

Figure 12(a) show the efficiency and power factor of the proposed system in subjecting to the output power. According to Figure 12(a), the maximum efficiency of the DC/DC converter is 96.2%, and the maximum efficiency of the inverter which includes the active buffer is 95.5%. However, when the output power is 50W, the efficiency becomes low owing to low power factor.

Figure 12(b) shows the output current THD and input current ripple of the proposed system in subjecting to the output power. According to Figure 12(b), the input current ripple is 12.3% when the output power is 150 W. In addition, the output current THD of the inverter is less than 5% when the output power is more than 100 W.

Figure 13 shows the experimental results of the MPPT. In this paper, the MPPT is demonstrated in three conditions. When the maximum input power is changed at 100 W, 150 W, and 200 W, the maximum power is tracked by MPPT control. In order to simulate the output power of PV, the resistor is connected to the DC power supply in series. The input current is increases when the output power is increased. As the

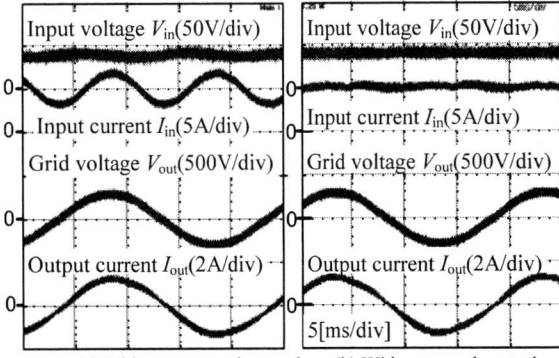

(a) Without proposed control   (b) With proposed control
Fig. 8 Experimental results of the power decoupling control.

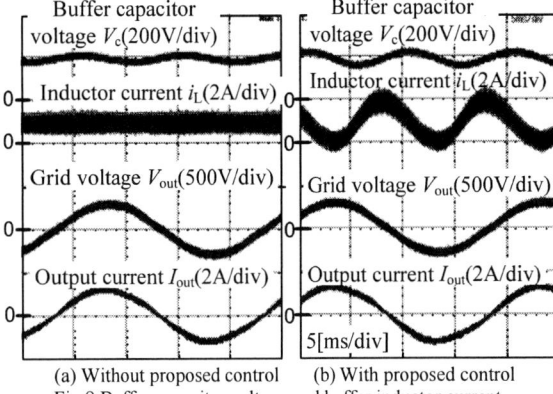

(a) Without proposed control   (b) With proposed control
Fig.9 Buffer capacitor voltage and buffer inductor current.

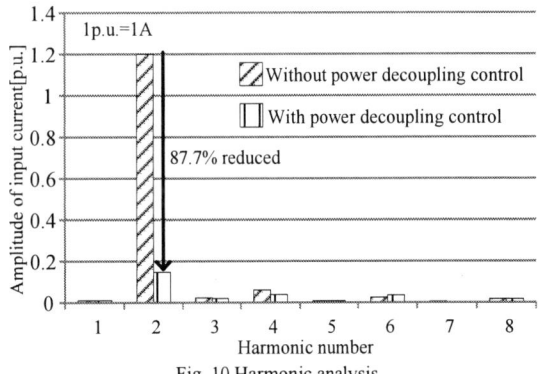

Fig. 10 Harmonic analysis.

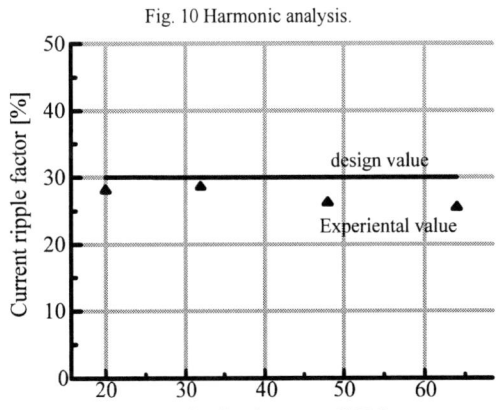

Fig.11 Buffer inductor current ripple factor.

results, the voltage drop of the resistor is increased. In other words, the input voltage of the converter decreases. For this reason, the maximum power point can be obtained. However, when the maximum input power is 200W, the fluctuation of the power differs in the previous condition, this cause that the input current limit is set to 8 A. In order to solve this problem the input current limit becomes higher. On the other hand, the fluctuation of the input power is large. This is because the permissible fluctuation range of the input power is wide. Thus, in order to decrease the fluctuation of the input power, it is necessary to expend the permissible fluctuation range of the input power.

B. Loss analysis

Figure 14 shows the loss analysis results. According to Figure 14, the no load loss is large compared to other loss. This is because the stray capacitance of the switching device is large. In order to reduce the no load loss, It is necessary to optimize the switching device in terms of converter capacity. Furthermore, the volume of the buffer inductor can be reduced the volume. However, the switching loss increases owing to the switching frequency. In order to reduce the switching loss, the active buffer applies the zero voltage switching (ZVS).

Figure 15 shows the relationship between the carrier frequency and efficiency. According to Figure 15, the efficiency of the active buffer and inverter is decreased when the switching frequency increased. This is because the no load loss and switching loss increase. According to figure 3 and figure 15, relationship between the high efficiency and miniaturization of the volume of the input inductor has a trade-off. Thus, the efficiency is evaluated with the power density [12].

C. High power density design

Figure 16 (a) shows the volume comparative results of the conventional circuit and proposed circuit. It is noted that, the volume of the heat sink is calculated using CSPI (Cooling System Performance Index) [13]. The volume of the heat sink $Vol_{cooling}$ is expressed by

$$Vol_{cooling} = \frac{1}{R_{th} \times CSPI} = \frac{P_{loss}}{(T_j - T_a) \times CSPI} \quad (12)$$

where, R_{th} is the thermal resistance of the cooling system, T_j is is the junction temperature of the switching device, T_a is the ambient temperature, and P_{loss} is the power loss of the switching device. Table 2 lists the parameters to calculate the $vol_{cooling}$.

In this figure, the volume of the proposed system is compared to the conventional circuit without the isolated DC/DC converter. It is noted that the smoothing capacitor C_{dc} is designed from the ripple current and ripple voltage. The maximum ripple current is expressed by

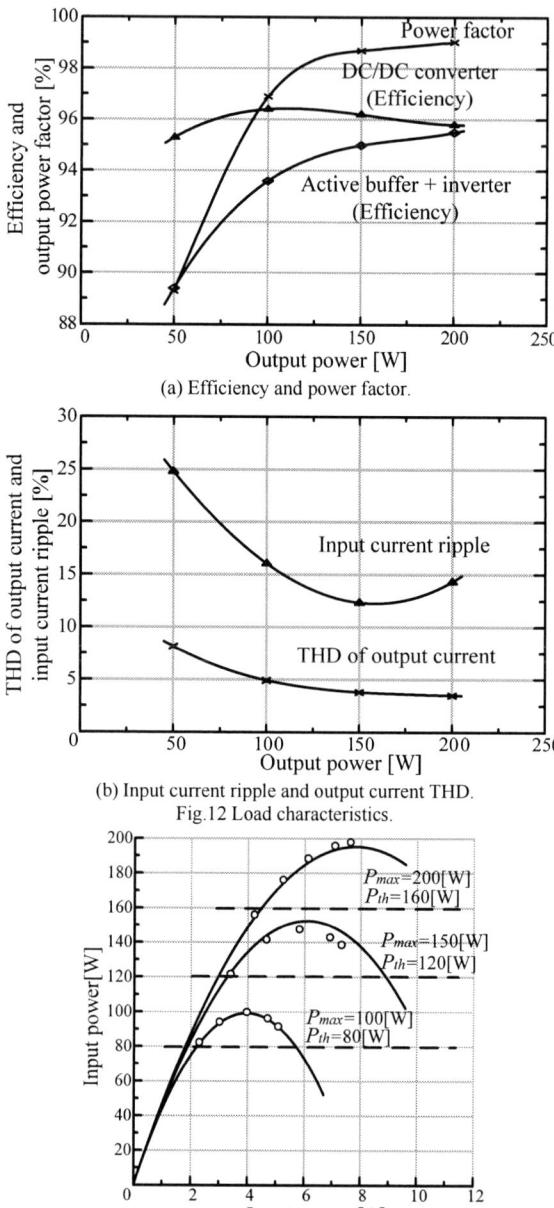

(a) Efficiency and power factor.

(b) Input current ripple and output current THD.

Fig.12 Load characteristics.

Fig.13. Experimental results of MPPT.

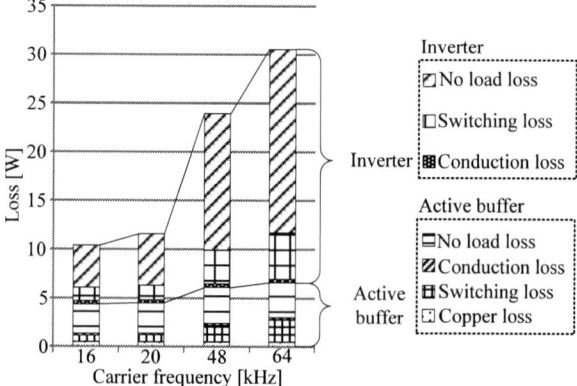

Fig.14 Converter loss analysis results.

$$i_{ripple} = \frac{V_{max} I_{max}}{V_{dc}} \cos 2\omega t \qquad (13)$$

where, V_{max} is maximum grid voltage, and I_{max} is maximum output current of inverter, V_{dc} is DC link voltage. Capacitor current i_c is expressed by

$$i_c = C_{dc} \frac{dv_c}{dt} \qquad (14)$$

from (14), C_{dc} is expressed by

$$C_{dc} = \frac{\Delta i_c}{\Delta v_c} \Delta t \qquad (15)$$

where, Δt is reciprocal number of the twice grid frequency component. Δv_c and Δi_c equal to the ripple voltage and ripple current. They are needed to design less than 10% of maximum value. From (13) and (15), C_{dc} need more than 500µF.

According to figure 16 (a), the volume of proposed circuit is reduced by 37%. This is because the smoothing capacitor is not needed in the proposed circuit, and the small capacitor can be applied C_5. Thus, the proposed circuit is smaller than the conventional circuit.

Figure 16 (b) shows the volume comparison when the switching frequency increases to 64 kHz. According to figure 16 (b), the volume of the proposed circuit at 64 kHz switching frequency is reduced by 61% in comparison that the switching frequency is 16 kHz. This is because the volume of the buffer inductor can be reduced by increasing the switching frequency. However, the volume of the heat sink increases owing to the switching loss. Thus, relationship between the high switching frequency and miniaturization of the volume of the buffer inductor has a trade-off. In addition, in order to reduce the volume of the heat sink. It is necessary to improve the efficiency.

Figure 17 shows the Pareto front curve of the conventional circuit and proposed circuit at the range of switching frequency from 20 kHz to 300 kHz. The inductor volume is evaluated by the Area product [14] According to Figure 17, the power density of the proposed circuit is high more than the conventional circuit. The constitution of the proposed circuit is equal to the conventional circuit nearly. However, the proposed circuit no need the smoothing capacitor on DC link, and buffer capacitor is small. Thus, the proposed circuit can design the high power density more than the conventional circuit. Experimental verification by high power density design is the future work.

VI. CONCLUSION

This paper discussed the miniaturization of the boost-up type active buffer in a single-phase grid connected inverter. The proposed circuit was experimented by using

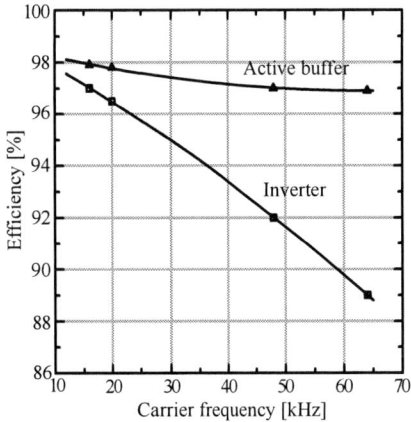

Fig.15 Relationship between the carrier frequency and efficiency.

Table.1. parameters to calculate the *volume*.

T_j	120℃
T_a	25℃
CSPI	3
Switching device S3-S6 Volume	SCH2080KE (Rohm) 1.67cm^3
Buffer capacitor C5 Volume	EVS20329S2G306MS09 (Murata manufacturing Co.Ltd) 2.56cm^3
Smoothing capacitor C_{dc} Volume	NU series (Nichicon) 167cm^3
Buffer inductor L$_1$ Volume	33cm^3

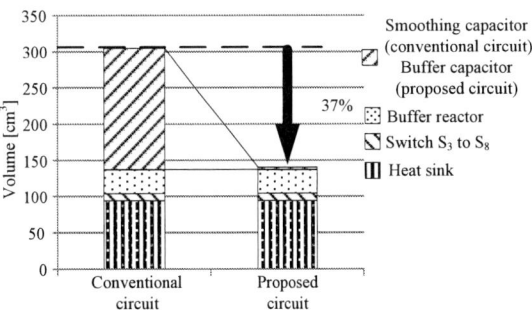

(a) Volume estimates of the proposed circuit.

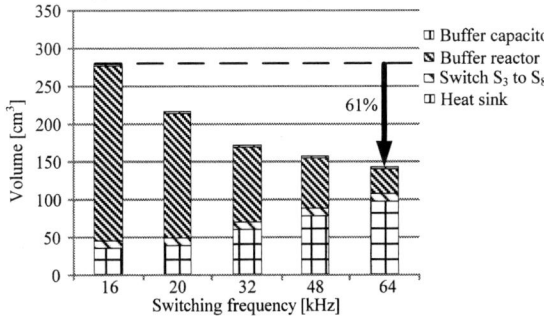

(b) Volume comparison when the switching frequency increase to 64kHz.

Fig.16 Volume estimates for the proposed circuit.

a 200W prototype. From the volume estimate, the volume of proposed circuit is reduced by 37% in compared to the conventional circuit. This is because the proposed circuit has no the small capacitor can be used as the buffer capacitor and C_5, can apply the small capacitor. From the experimental results, the output THD is 3.5% at 200 W output power, and the input current fluctuation is reduced by 90% owing to the power decoupling control. From the evaluation of the power density using the Pareto front curve, the power density of the proposed circuit is higher than that of the conventional circuit. Thus, high power density for the proposed circuit can be designed.

In the future work, the efficiency will be increased.

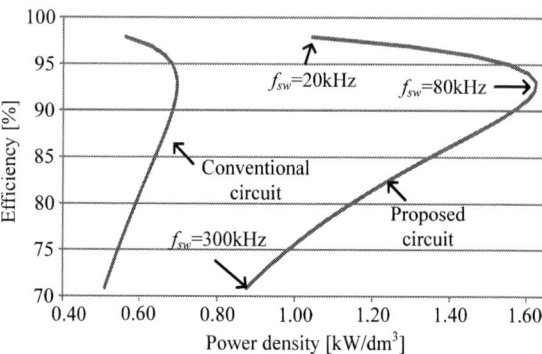

Fig.17 Pareto front curve of the conventional circuit and proposed circuit.

REFERENCES

[1] S. B. Kjaer, JK Pedersen: "A review of single-phase grid-connected inverters for photovoltaic modules", IEEE Trans., Vol. 41, No. 5, pp. 1292-1306 (2005).

[2] H. Hu, S. Harb, N. Kutkut, I. Batarseh, Z. J. Shen: "Power Decoupling Techniques for Micro-inverters in PV Systems-a Review", ECCE2010, Vol. 32826, No. 2, pp. 3235-3240 (2010)s", IEEE Trans., Vol. 41, No. 5, pp. 1292-1306 (2005).

[3] T. Shimizu, K. Wada, N. Nakamura: "Flyback-Type Single-Phase Utility Interactive Inverter With Power Pulsation Decoupling on the DC Input for an AC Photovoltaic Module System", IEEE Trans., Vol. 21, No. 5, pp. 1264-1272 (2006)

[4] D. Cao, S. Jiang, X. Yu: "Low-Cost Semi-Z-source Inverter for Single-Phase Photovoltaic Systems", IEEE Trans., Vol. 26, No. 12, pp. 3514-35-23 (2011)

[5] F.Shinjo, K.Wada, T.Shimizu:"A Single-Phase Grid-Connected Inverter with a Power Decoupling Function " PESC 2007, pp.1245-1249, (2007)

[6] Kuo-Hen Chao, Po-Tai Cheng: "Power decoupling methods for single-phase three-poles AC/DC converters", ECCE 2009, pp3742-3747,(2009)

[7] F. Schimpf, L. Norum: "Effective Use of Film Capacitors in Single-Phase PV-inverters by Active Power Decoupling", IECON 2010, Vol. , No. , pp. 2784-2789 (2010)

[8] Y. Ohnuma, J. Itoh: "Comparison of Boost Chopper and Active Buffer as Single to Three Phase Converter", IEEE ECCE2011, Vol. , No. , pp. 515-521 (2011)

[9] T. Shimizu, S. Suzuki: "A single-phase Grid-Connected Inverter with Power Decoupling Function", IPEC 2010, Vol. , No. , pp. 2918-2923 (2010)

[10] J. Itoh, H.Watanabe, K.Koiwa, Y. Ohnuma: "Experimental verification of single-phase inverter with power decoupling function using boost chopper", EPE '13-ECCE Europe, the 15th European Conference on Power Electronics and Applications, Vol. , No. , pp. (2013)

[11] FANG Xiong, WANG Yue, LI Ming, LIU Jinjun: "A novel frequency-adaptive PLL for single-phase grid-connected converters", ECCE2010, Vol. , No. , pp. 414-419 (2010)

[12] Y. Kashihara, J. Itoh: "Parformance Evaluation among Four types of Five-level Topologies using Pareto Front Curves", IEEE Energy Conversion Congress and Exposition, Vol. , No. , pp. 1296-1303 (2013)

[13] J. Itoh, T. Araki: "Volume Evaluation of a PWM Inverter with Wide Band-Gap Devices for Motor Drive System", 5th IEEE Annual International Energy Conversion Congress and Exhibition, Vol. , No. 4-4-2, pp. 372-378 (2013)

[14] Wm. T. McLyman' Transformer and Inductor Design Handbook' CRC Press, 2004

Testing Facility Using Large Capacity Inverter

Yusuke Ishimaru, Mitsuo Adachi,
Masahiko Tsukakoshi, Ritaka Nakamura,
Hiroyuki Masuda, Yoshihiro Ogashi,
Yuichi Tsuboi
Toshiba Mitsubishi-Electric Industrial Systems Corporation
Tokyo, Japan

Abstract— **It is an ideal to perform a combined test of an inverter and a motor for a smooth start-up at the site. There is a method to carry out a full load test without driving a machine. It is called "Back-to-back test". It is possible to obtain 25MW output of testing facility in our company. In this paper, it is introduced the performance evaluation of the test results such as, torque ripple analysis of the proposed system.**

Keywords— *VSI, Back-to-back test, 5-level inverter, Fixed Pulse PWM, Torsional vibration, Air-gap torque, Shaft torque*

I. INTRODUCTION

The standard performance test of a Voltage Source Inverter (VSI) is performed by connecting a reactor between a rectifier and an inverter as a load which resembles the rated load by applying required voltage and current. This method has advantages which can reduce an equipment size and power consumption. However, it cannot check the operating characteristics acquired by combining with a motor. Moreover, it cannot measure torque ripple of the shaft. There is a method called "Back-to-back test" which combines 2 sets of VSIs and a motor for the solution. At present, there is no other method to perform the full load test which combines a VSI and a motor without using the actual load.

The largest domestic testing facility is developed which can perform the actual load testing up to 25MW. In this paper, it is introduced the outline of testing facility and the results obtained for torque ripple evaluation.

II. CHARACTERISTIC OF TESTING FACILITY

There are following characteristics in the testing facility of back-to-back test.

(1) The full load test of the VSI and a motor is possible.

(2) The load test of 25 MW is possible with the testing facility.

(3) The test facility only needs to supply power for the test circuit losses, so a test can be carried out for a VSI system exceeding the power supply capacity.

(4) Fixed pulse mode which reduces low order torque ripple is used. So, the torque which measures performance will not be affected [1].

III. CONFIGURATION OF TESTING FACILITY

Fig.1 shows the test circuit configuration. Table I shows the equipment rating of testing facility, and Table II shows the equipment rating of testing inverter. Power supply voltage to the facility is 3.3kV, and stepped up to 11kV by a transformer TR-A. It supplies power to transformers TR-B1 and TR-B2. TR-B1 is the transformer which supplies power to the testing facility that functions as a load. The voltage is stepped down from 11kV to 3.15kV for the inverter of the testing facility. The inverter of the testing facility is used for regenerative operation at this time. TR-B2 is a transformer which supplies power to the testing inverter and motor where performance characteristics were measured. Fig. 2 shows the external view of the indoor installed equipment. Fig. 3 shows the external view of the outdoor installed equipment.

TABLE I
RATING OF TESTING FACILITY

Name	Rating
Transformer (TR-A)	1Ry: 3.30kV-60Hz-6MVA 2Ry: 11.00kV-60Hz-6MVA
Transformer (TR-B1)	1Ry: 3x(11.00kV-60Hz-10MVA) 2Ry: 3x(3.15kV-10MVA)
Transformer (TR-B2)	1Ry: 3x(11.00kV-60Hz-10MVA) (-10deg / 0deg / +10deg) 2Ry: 3x(2.10kV-2x5MVA)
Regenerative inverter	Input: 24MVA-3.95kV-3.51kA Output: 30MVA-7.20kV-2.40kA

TABLE II
RATING OF TESTING INVERTER

Name	Rating
Testing inverter	Input: 22MVA-6x1.76kV-1.20kA Output: 20MVA-6.00kV-1.93kA

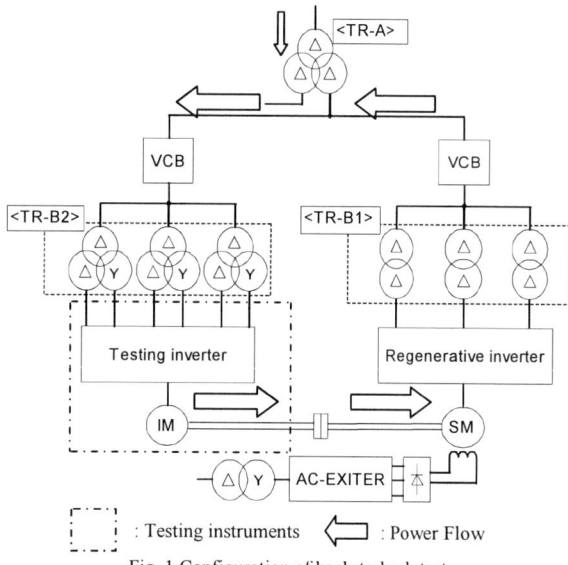

Fig. 1 Configuration of back-to-back test

Fig. 2 Indoor installed equipment

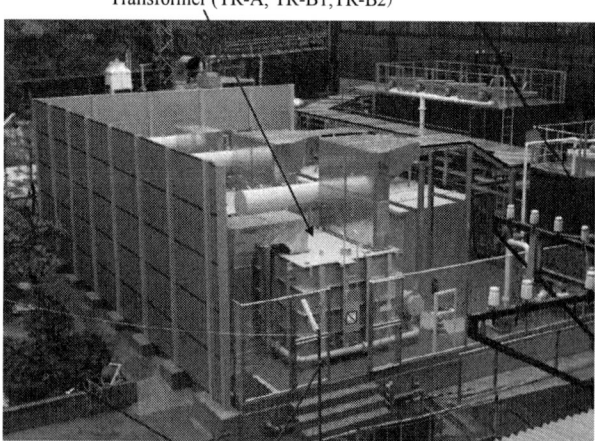

Fig. 3 Outdoor installed equipment

IV. CHARACTERISTIC OF A REGENERATIVE INVERTER

A. Configulation of the main circuit

Fig. 4 shows the main circuit configuration of a regenerative inverter. It has the composition per one leg that it is connected the turn-off device to the arm of P and N side in two series and is connected to the clamp diodes. This leg is configured from a full bridge circuit which is a single phase connection. It is configured with a 3-phase inverter by the star wire connection of one of the two lines on the output side. The output of five levels is possible for an output phase voltage, and the output of nine levels is possible for a line-to-line voltage by this composition [2] [3] [4].

The power semiconductor device, GCT thyristor of 6 kV - 6 kA rating is adopted as a turn-off device. The operation of motoring and regenerating at 25 MW is possible by combining with this 5-level inverter composition.

Since the DC circuit of each phase is independent, it has 3 sets of rectifiers. 3-level PWM rectifier can perform four quadrant operation of an electric motor. Moreover, since the PWM rectifier can control the circuit with a power-factor of 1 by switching, it is possible to configure a small power supply system.

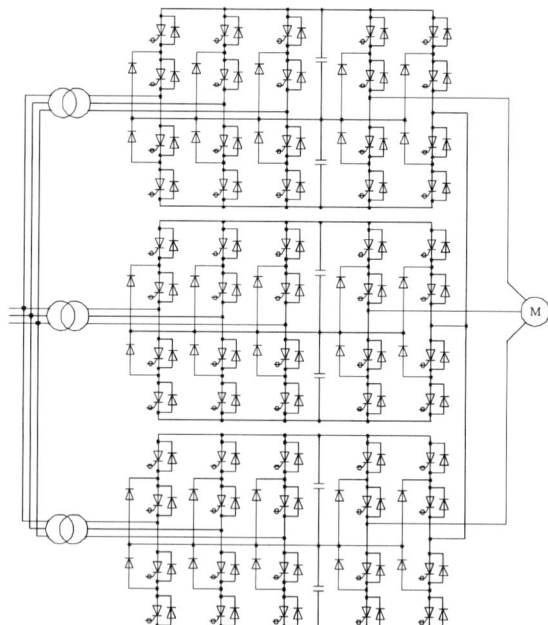

Fig. 4 Main circuit configuration of a regenerative inverter

B. Fixed Pulse PWM mode

When a three-phase power is supplied to the motor by the PWM VSI, frequency other than the fundamental frequency is included in the output voltage and the output current. There is a possibility that the torque ripple causes resonance to the natural frequency of the machine system such as the motor couplings, and it may damage itself. The resonance frequency of the machine system is between 10Hz and 100Hz. Since the switching of a VSI is performed at a higher frequency from 400 to 600Hz or more, it is very rare that this causes the vibration problem directly. There are two causes of producing the torque ripple of low frequency by the VSI. First, the vibration of the machine system is amplified by the control of the inverter. Second, the vibration elements of a low frequency will be synthesized the frequencies of the inverter control. For example, the vibration elements of a low frequency might be created from the output frequency, the switching frequency, and the control sampling, etc.

This VSI has adopted the fixed pulse pattern system, in order to aim at low order harmonics reduction [5] [6]. Fig. 5 shows an example of a fixed pulse pattern for legs of the U-phase of the 5-level inverter. The patterns were prepared to eliminate low frequency components. The fixed N-pulses pattern is called with the number of the pulses of the half cycle of the output phase voltage for every leg. The equation (1), (2) of the fixed 3-pulses pattern is shown. In case of the fixed 3-pulse pattern, it can eliminate the harmonics of 5th, 7th, 11th and 13th.

Moreover, the low order harmonics of the area which causes resonance with motor coupling are reduced by increasing the number of pulses as speed decreases.

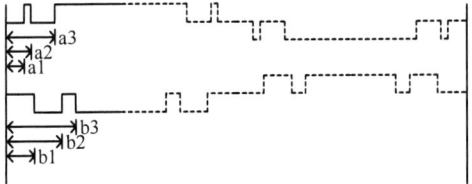

Fig. 5 Example of the fixed pulse pattern of the 5-level inverter

$$V_n = 4 \times \frac{2}{n\pi}\{(cosna1 - cosna2 + cosna3) + (cosna1 - cosna2 + cosna3)\}$$
(1)

$$V_1 = m, V_5 = 0, V_7 = 0, V_{11} = 0, V_{13} = 0$$
(2)

m: modulation factor

V. METHOD OF BACK-TO-BACK TEST

The testing inverter and motor for performance measurement is used for powering operation, and the inverter in test facility is used for regenerative operation as a load. This configuration circulates power between the powering inverter and regenerative inverter. Therefore, the facility only supplies power for the test circuit loss which allows the testing of inverter and motor that are much larger than the facility power supply.

VI. EVALUATION OF TORQUE RIPPLE BY LOAD TEST RESULT

The main method of measuring the torque is to use a strain gauge. However, the method of using the strain gauge is difficult in case of fitting to a high speed motor. There is a method of calculating a shaft torque more easily than the method of measuring with a strain gauge. The air gap torque is calculated from a voltage and a current waveform, and shaft torque is calculated from the resonance magnification of the shaft calculated beforehand. Recently, although it is becoming possible to predict torque ripple by improvement in simulation technology, evaluation with an actual measurement value and an estimate value is not yet enough. At this time, the actual measurement value and the estimated value are evaluated in an actual loading test, and it introduces that the good results were obtained.

A. Shaft Torque Measurement

Fig. 6 shows the measurement point of shaft torque of the test system. Torsional distortion of high speed flexible joint is measured by using strain gauge and telemeter, and thus shaft torque is calculated.

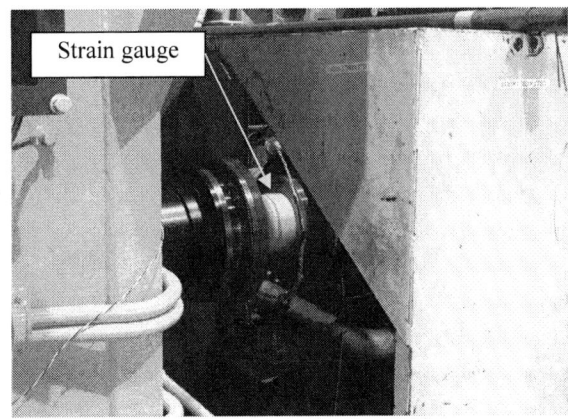

Fig. 6 Shaft Torque Measurement Point

B. Torsional Vibration Analysis of the Motor

Torsional vibration analysis is performed to survey the torque ripple by the shaft, especially to analyze the strength of the elasticity joint. Fig. 7 shows an analysis typical model for this test system. The analytical model is divided into 25 mass points between a motor and a generator.

The vibration analysis is performed by MATLAB with this model. It is obtained 30.0Hz for the lowest torsional vibration in which a motor and a generator are set in reverse phase vibration mode. In addition, torsional natural frequency is found also for 30.0Hz and 657.7Hz with in 657.7Hz of operation frequency. Fig. 8 shows the results of torsional inherence mode for 30.0Hz, 306.9Hz, 360.9Hz and 657.7Hz. The results show the distribution of the amplitude of vibration. It is obtained for the low frequencies that the normalized modal vector increases gradually with the increase of number of inertia.

Fig. 9 shows the relationship between frequency and torque amplitude at the coupling when the motor torque. It shows that a resonance point is at 30 Hz from Fig. 9.

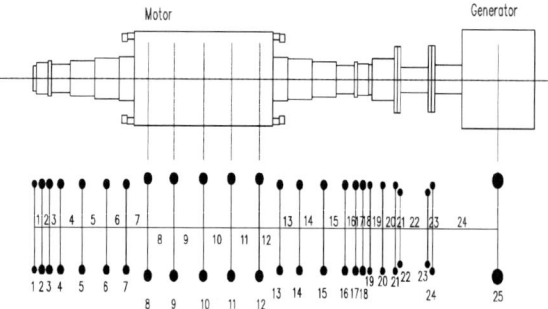

Fig. 7 Torsional Vibration Analysis Typical Model

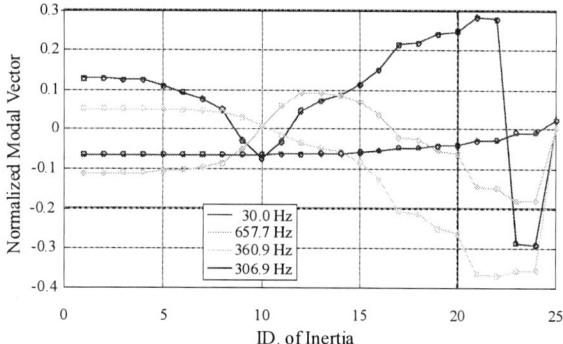

Fig. 8 Torsional Inherence Mode

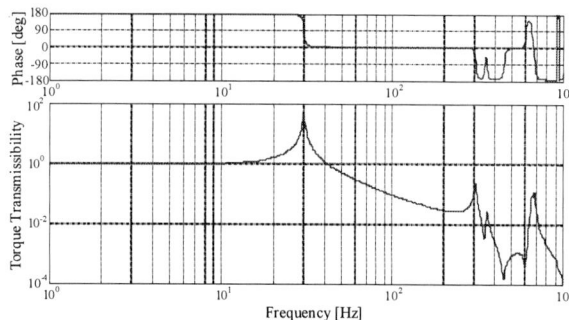

Fig. 9 Relationship between frequency and torque amplitude at the coupling when the motor torque

C. Calculating method of Air-gap Torque ripple

Air gap torque was calculated with stator voltage, current, and resistance as follows. [7]

(1) Measuring the stator resistance of a motor: R

(2) Rotating the motor at constant speed and record the stator voltage: V and the stator current: I for a certain period.

(3) Calculating Eq. (3) with the data recorded.

$$\frac{d\Phi}{dt} = V - RI \qquad (3)$$

(4) Executing Fast Fourier Transform analysis (FFT) about $d\Phi/dt$ calculated.

(5) Calculating the flux linkage: Φ by integrating the FFT result of $d\Phi/dt$ with respect to the each frequency components.

$$\Phi = \int (V - RI)dt \qquad (4)$$

(6) Calculating the torque by Eq.(5)

$$\tau = P_m(\Phi \times I) \qquad (5)$$

Φ : Flux linkage
V : Stator voltage
I : Stator current
R : Stator resistance
τ : Motor torque
P_m : Number of pole pairs

D. Evaluation of Torque ripple

It is carried out the test with conditions shown in Table III. The output of U-phase voltage and current waveforms of testing inverter is shown in Fig. 10. The air gap torques of motor and generator are calculated from the voltage and current by the method of section C. The shaft torque is calculated by the air-gap torque of the motor and generator and the transfer function of their mechanical system respectively. The damping coefficient was set to 0.01 of a general value. Fig.11 shows the test result and the calculated value of the shaft torque. Blue points show the calculated torque and the red points show the torque measured with the strain gauge. It is confirmed that the same size torque ripple as the almost same frequency component appeared from the comparative result. Therefore, it is confirmed that the calculated value and the actual measurement value were well in agreement from this result.

Moreover, it is confirmed by the results of Fig. 11 that the shaft torque value is smaller as 2.3% for the resonance point. This shows that the low order torque ripple is reduced by applying a fixed pulse pattern system.

TABLE III
TESTING CONDITION OF VSI

Output voltage	6.0kV
Output current	1.9kA
Output capacity	20MVA

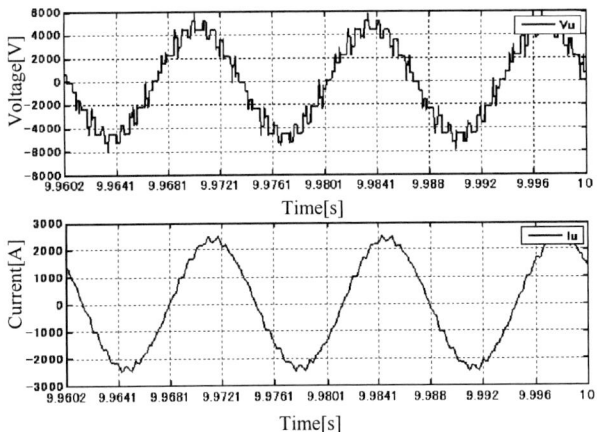

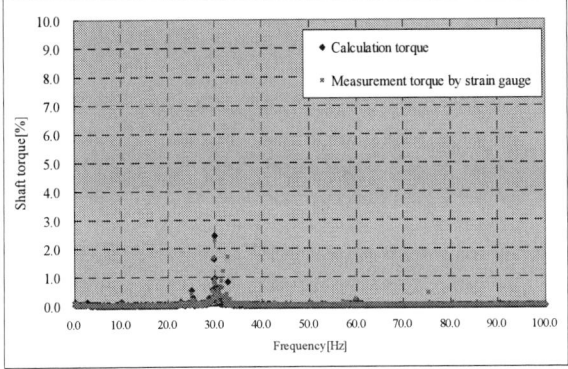

Fig.10 The output U-phase voltage and current waveform (Testing inverter)

Fig.11 Test result and calculation value of shaft torque

VII. CONCLUSION

The torque ripple measurement which cannot be measured only with the VSI test was obtained by this test using the back-to-back test facility. For a motor, it was possible to obtain data such as temperature and loss at conditions similar to the actual operation at site. The configuration only requires power supply equivalent to circuit and motor loss, which allows testing of much larger capacity inverter and motor than the facility power supply. Moreover, if this testing facility is more systematized, it is possible to realize the minimization of time and cost.

REFERENCES

[1] M. Tsukakoshi, M. A. Mamun, K. Hashimura, H. Hosoda, S. C. Peak, "Introduction of a Large Scale High Efficiency 5-level IEGT Inverter for Oil and Gas Industry", *IEEE Energy Conversion Congress and Exposition (ECCE 2010)*, pp.4313-4320, 2010.

[2] M. Yamamoto, K. Satoh, T. Nakagawa, A. Kawakami, " GCT (gate commutated turn-off) thyristor and gate drive circuit ", Power Electronics Specialists Conference, 1998. PESC 98 Record, Vol.2, pp. 1711 – 1715, 17-22 May 1998

[3] A. Nabae, I. Takahashi, and H. Agaki, "A new neutral-point-clamped PWM inverter", IEEE Trans. IA.17, 518-523 (1981).

[4] D. Yoshizawa, K. Takao, M. Mukunoki, Y. Shimomura, "The large Capacity 5 level GCT Inverter for OIL & GAS plant

application," *The Institute of Electrical Engineers of Japan, Industry Application Society*, 2008 Conference Proceedings, pp.I359-I362, Aug. 2008.(In Japanese)

[5] M. Tsukakoshi, M. A. Mamun, K. Hashimura and H. Hosoda, "Performance evaluation of a large capacity VSD system for oil and gas industry," IEEE Energy Conversion Congress and Exposition (ECCE 2009), San Jose, CA,USA, pp. 3485-3492, Sep. 2009.

[6] M. Tsukakoshi, M. A. Mamun, K. Hashimura and H. Hosoda, J. Sakaguchi, L. Ben-Brahim, "Novel Torque Ripple Minimization Control for 25MW Variable Speed Drive System Fed by Multilevel Voltage Source Inverter", Proc. of 39th Turbomachinary Symposium, Houston, TX, USA, pp. 193-200, Oct. 2010.

[7] T. Kojima, H. Kometani, M. Kawamura, Y. Tsuboi, R. Nakamura, H. Masuda, Y. Ohgashi "Experimental Verification of Torque Ripple Calculation for Induction and Synchronous Motor", *Electric Power and Energy Conversion Systems*, 2013.

Performance Evaluation under the Actual Operating Condition of a Large Capacity VSI Inverter for Steel Mill Applications

Mostafa Mamun, Daisuke Yoshizawa and Makoto Mukunoki

Drive Systems Development Section, Power Electronics Systems Division,
Toshiba Mitsubishi-Electric Industrial Systems Corporation, TMEIC
1, Toshiba-cho, Fuchu-shi,Tokyo 183-8511, Japan.
{MOSTAFA.mamun, YOSHIZAWA.daisuke, MUKUNOKI.makoto} @tmeic.co.jp

Abstract— Large capacity motor drive equipment, especially those with low harmonics and high efficiency, are in demand in the industrial field. In general Voltage Source Inverters (VSI) have low harmonics and high efficiency and have been widely applied in many steel plants. However, the cycloconverter drives which were applied in 1980's are still operating in many steel plants. The cycloconverter has low power factor and it is necessary to include harmonic filters to reduce the input power harmonics, making the whole equipment larger. Compared to the conventional cycloconverter, IGBT/IEGT VSI has the benefits of high input power factor and better speed control accuracy, including motor and load response. Moreover, the drive footprint is significantly smaller. Also, some existing cycloconverter drives have forcing voltages around 3600V. Until now there was no drive equipment that could directly replace all previous AC drives in both capacity and voltage. The rated capacity of the latest individual drive module is 9MVA with a maximum output of 3650V. With a power density of 1.3MVA/m³ this compact structural configuration offers space-saving advantages for new and upgraded installations.

Keywords— 3-level inverter; Voltage Source Inverters (VSI); drive equipment; steel mill application.

I. Introduction

Many large capacity motor drive equipments have been used in steel mills for several years. Especially low harmonics with high-efficiency motor drives have a great demand in the industrial field. Generally, a VSI (Voltage Source Inverter) has low harmonics and high-efficiency. So, it has spread widely and is applied in the steel plants. However, the cycloconverter drives which have been applied in 1980's are still operating in the steel plants. The cycloconverter has low power factor and it is necessary to mount the harmonic filter to reduce the input power harmonics which makes the whole equipment larger in size. Comparing to the conventional cycloconverter, IGBT/IEGT VSI has the benefits of high input power factor, better speed control accuracy with speed response and load response. Moreover, it is possible to obtain the footprint downsizing. So, the renewal of the currently installed drive equipments is increasing now. Although the

cycloconverter drives are applied to the AC motor drives including 3300V rating synchronous motor, there was no drive equipment line-up of the required large capacity for updating until now. Thus, it is introduced the development of a large capacity 3-level IEGT inverter in which the output capacity is increased based on the conventional drive equipments is applicable to 3300V rating synchronous motor.

The rated output capacity of this inverter is 9MVA and the maximum output voltage is 3650V. It is possible to connect maximum 4 banks in parallel. It is also applicable to set up a common converter by connecting several numbers of inverters with the common converter. The power density 1.3MVA/m³ of this inverter has been confirmed as its compact structural configuration with space-saving. In this paper, it is also introduced the features of the IEGT inverter with the results of a full power test obtained by the 3-level IEGT inverter which confirms 9MVA rated output capacity.

II. Specification and Characteristics of the IEGT

A. Specification of the IEGT

A newly developed IEGT (Injection Enhanced Gate Transistor), 4500V/4200A rating is applied to develop this inverter equipment. Fig. 1 shows the general overview and internal configuration of the IEGT. The device is configured with 42 pieces of individual IEGT chips installed inside the press pack. The press pack packaging has been used for many years on thyristor based devices. It is designed to make good contact for all the IEGT chips and therefore maximize the package rating. One advantage of the IEGT is that the gate driver requires a voltage signal with a simple circuit with relatively low power. It is possible to obtain higher capacity by connecting internal IEGT chips in parallel and has good performance results using as a main drive in the steel plant [1].

In general, if the IGBT is made with high withstand voltage, it is generated a few carriers around gate and emitter which produces comparatively high on state voltage. However, it is realized a lower on state voltage even it is made with high withstand voltage by improving the gate structure in IEGT. This was made possible by improving the individual chip characteristics to include a lower on state voltage and faster switching [2].

Fig. 1. General overview and internal configuration of an IEGT (4500V/4200A rating).

Comparing to the conventional IEGT, it is possible to apply higher DC voltage. And it shows lower switching losses so that it can be applied in large capacity inverters. The new type IEGT (ST2100GXH26A) of Toshiba is applied in this development and the main characteristic of this IEGT is shown in Table I.

TABLE I
CHARACTERISTICS OF THE IEGT (ST2100GXH26A)

Items	Ratings	Conditions
Continous applied DC voltage	3000V	Failure rate: 100FIT
Collector-emitter voltage	4500V	VGE=-15V, Tj=125°C
Maximum collector cutoff current	4200A	Vcp=4500V, Vcc=3000V, VGE=±15V, Tj□125°C, non-repetition
Collector-emitter saturation voltage	3.0V (typical)	Ic=2100A,VGE=15V, Tj=125°C
Turn-on switching loss	20J (maximum)	Vcc=3000V, Ic=2100A, Tj=125 °C, VGE=±15V

B. Characteristics of the IEGT

Fig. 2 shows an example of cutoff characteristics of the IEGT and Fig. 3 shows the turn-on waveform of the IEGT. It is confirmed by the results that it is possible to cutoff for more than 3000V DC with this IEGT.

On state voltage of this IEGT is lower and it shows about 3.0V (typical value). Comparing to the conventional IEGT, it is possible to apply higher DC voltage (3000V). And it shows lower switching losses so that it can be applied in large capacity inverters.

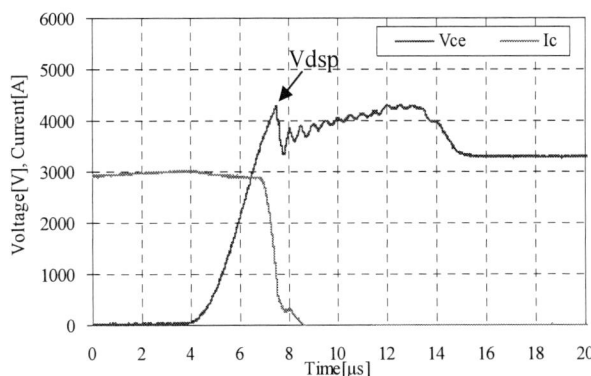

Fig. 2. Cut-off characteristics of the IEGT.

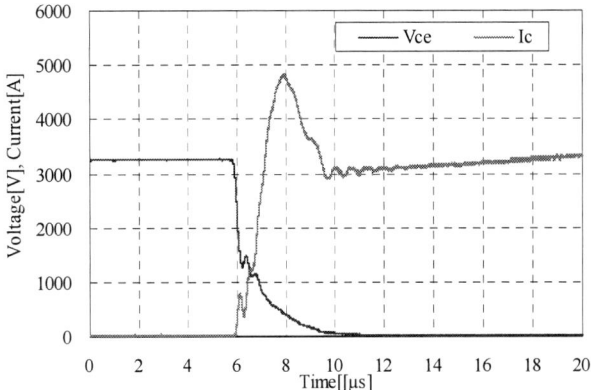

Fig. 3. Turn-on waveform of the IEGT

In order to reduce the wiring inductance, it is designed to obtain a compact structure of the IEGT stacks and a block of clamp snubber capacitors for low inductance structure so that the maximum surge voltage (Vdsp) can be controlled to 4200V.

By adding a super saturation reactor, turn-on di/dt of the IEGT becomes a smooth curve and it is possible to reduce the loss of IEGT about 40%.

III. Large Capacity VSI Inverter

A 3-level inverter is developed by using the new type IEGT (ST2100GXH26A). In a 3-level inverter, which is commonly used for higher power VSI, the output voltage and current waveforms are improved due to the greater number of voltage levels. The efficiency of 3-level inverters at full load is also higher than that of 2-level inverters, which means better energy handling of the system. A better efficiency at rated power also means a smaller heat sink and better reliability [3]. The efficiency of 3-level inverters at small power is also improved.

The 2014 International Power Electronics Conference

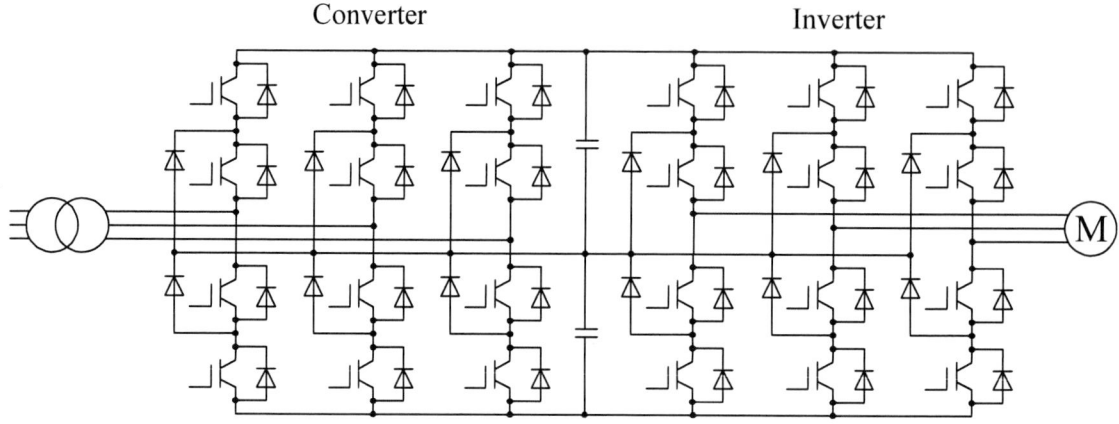

Fig. 4. Configuration of a 3-level IEGT inverter

A. System Configuration

Fig. 4 shows the configuration of a 3-level IEGT VSI. Note that in IEC terminology "converter" refers to the entire circuit of converting constant frequency and voltage to variable frequency and voltage. In this paper the common usage term "converter" applies to the process of converting AC to DC and "inverter" refers to the inversion of DC to AC. VSI refers to either conversion process.

In the 3-level power circuit commonly used for higher power VSI, the output voltage and current waveforms are improved due to the increased number of voltage levels over the 2-level type. Both inverter and converter are identically configured using IEGT power devices, diodes and capacitors in single phase assemblies as shown in Fig. 5. The rated capacity of the unit discussed is 9MVA with the ability to connect 4 banks in parallel for a rating of 36MVA. Because the converter and inverter power bridges are modular designs, they can also be arranged in a common converter configuration by connecting several inverters to the DC bus established by one or more converters.

The control regulators are contained in a separate panel that can be remotely mounted, minimizing arc flash exposure by minimizing the amount of time personnel are in the power room. The control panel interface signals to the main circuit panels are a small number of fiber optic cables for noise immunity.

B. Main Circuit Configuration

The 9MVA rated output capacity was possible by the availability of new IEGT power devices incorporated into a new physical package. The previous 3-level IEGT converter-inverter has been available since 1999 in a physically larger 8MVA configuration. The main switching devices consist of 4500V/4200A IEGTs with free-wheeling diodes, neutral point clamping diodes and components for overvoltage control. The six identical per-phase circuits of Fig. 4 are shown in Fig. 5 as an individual phase assembly. Compared to the previous

circuit implementation [4], it reduces the number of parts by applying 2-in-1 type clamping snubber capacitors in the main circuit configuration. Smaller crimping type diodes are also used for the bypass diodes. Moreover, the loss of a device is reduced by introducing a super saturation reactor as usual.

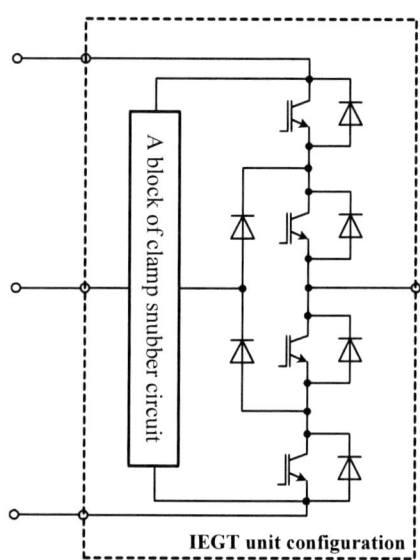

Fig. 5. Main circuit configuration (one phase of a 3-level inverter)

C. Main Circuit Panel Configuration

Fig. 6 shows the power unit of IEGT. One unit consists of an IEGT stack, a diode stack and a block of clamp snubber circuit. At the time of device failure, the whole unit is exchanged so that it is possible to shortening the MTTR (Mean Time To Repair) of the equipment. Moreover, fully oil less configuration is realized as a whole equipment by using resin mold capacitors for the smoothing capacitors and the

978-1-4799-2706-7/14 $31.00 © 2014 IEEE

clamp snubber capacitors. There is also no concern of an oil leakage compared with the conventional oil capacitors, and the compact design confirms the reduction in size and weight. Furthermore, since resin itself is an insulator, it is possible to realize with a high density package.

Fig. 6. IEGT power unit

Aluminum (Al) is used in the water cooled heat sink instead of copper (Cu) for cost saving and weight reduction [5]. Although aluminum is inferior to copper in thermal conductivity, heat thermal resistance has been reduced by design of the shape and internal flow channel of the heat sink. The characteristic of the heat sink is shown in Fig. 7, showing that the developed aluminum heat sink has lower thermal resistance compared to the conventional copper heat sink. The development of an improved aluminum heat sink allowed for the 9MVA output in a more compact 3-level IEGT power unit. Compared to the conventional copper heat sink, the aluminum heat sink provides for a IEGT power unit weight reduction of approximately 20%.

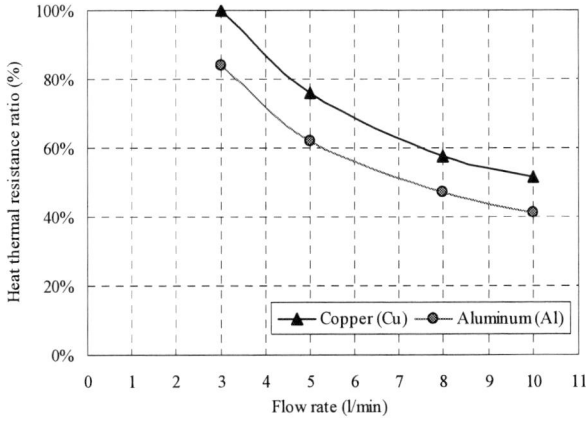

Fig. 7. Characteristics of the heat sink

D. Overview of the Equipment and Its Specification

Fig. 8 is a picture of the 3-level 9MVA IEGT converter-inverter. The dimensions in meters are 2.0W x 2.3H x 1.5D which is 1.3MVA/m^3 for the 9MVA package. Of course the converter-inverter package has a rating of 9MVA in (converter) and 9MVA out (inverter) so each power section can be considered to have twice the power density or 2.6 MVA/m^3. The control panel (0.8W x 2.3H x 0.75D) and pure water cooling unit are not shown. The previous generation measured 2.4W x 2.3H x1.65D and was available in 8MVA and 10MVA ratings. The power density for the earlier 10MVA design is nearly 1.1 MVA/m^3 so there is nearly a 20% improvement in power density and over 30% improvement in floor space requirements.

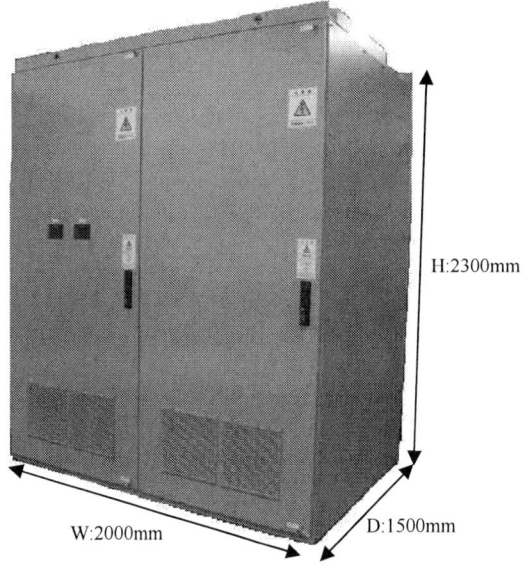

(a) Outside View

(b) Front side view

Fig. 8. Picture of the 3-level 9MVA IEGT converter and inverter

The equipment specification is shown in Table II. Compared to the previous 3-level IEGT VSI [4], the rated output voltage was increased to 3650Vrms from 3500Vrms. Along with this the rated DC voltage was also increased to 2600Vdc from 2430Vdc. The maximum output frequency was increased from 60Hz to 75Hz. Application of the new type IEGT power device and previously mentioned design improvements made it possible to increase the power density and performance.

TABLE II
SPECIFICATION OF THE NEWLY DEVELOPED 3-LEVEL IEGT INVERTER

Items	Specification
Main circuit configuration	3-level method
Rated capacity	9MVA
AC output voltage	3650Vrms
AC output current	1430Arms (overload 150%-1 minute)
Output frequency	0 to 75 Hz
Cooling method	De-ionized (pure) water
Efficiency	99%
Dimension (m)	2.0W x 2.3H x 1.5H
Power density	1.3 MVA/m^3
Weight of power unit	3800kg (Fig. 8-b)

Note that the AC output current rating includes an overload of 150% for one minute. This is equivalent to 2145Arms or 13.5MVA when compared to AC-AC converters with no overload rating included.

E. Fixed Pulse Pattern (Five-pulse) PWM Control

The 3-level IEGT converter uses the same equipment as the inverter and both are operated in PWM mode. The PWM VSI has an advantage over cycloconverters in that they do not generate sub-harmonics of the converter input current that vary with output frequency; that is, motor speed. However the conventional PWM firing method cannot sufficiently eliminate lower order harmonics such as the 5th, 7th, 11th, 13th, 17th and so on to meet IEEE 519 and various other local standards. In this section an optimization method of reducing converter current harmonics in the PWM mode is presented.

Fig. 9 shows the typical gate pulse and phase voltage waveforms of a PWM converter in Fixed Pulse Pattern (FPP) mode where each switching device turns on five times per half cycle of the voltage reference. In this mode, the switching frequency is equal to the case of three-level triangular PWM where carrier frequency is 540Hz for a source frequency of 60Hz. The pulse pattern is determined by five switching angles α, β, γ, δ and ε. These can be used as design parameters to optimize source current harmonics. The nth order harmonic of output voltage $F(n)$ is given as:

$$F(n)=\frac{4V_{dc}}{n\pi}\{\cos(n\alpha)-\cos(n\beta)+\cos(n\gamma)-\cos(n\delta)+\cos(n\varepsilon)\} \quad (1)$$

$$\text{for } n=1, \ 6k\pm1 \ (k=1,2,3,\cdots)$$

and Vdc represents the DC voltage of each smoothing capacitor.

In the FPP mode the voltage amplitude of the fundamental component is fixed. Currents in the power source are controlled only by changing the phase of output voltages.

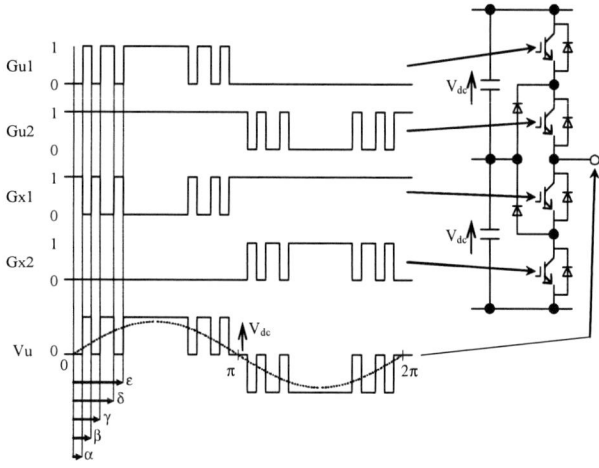

Fig. 9. Fixed Pulse Pattern (Five-pulse) PWM mode

Fig. 10(a) shows an example of waveforms of the conventional PWM and Fig. 10(b) shows the FPP PWM mode. Shown are the input current, the diode current, the IEGT current and the output voltage. In FPP lower harmonics of the input current is observed due to reducing switching times. At the same time switching losses and efficiency are improved. The carrier frequency of triangular PWM is 540Hz and the number of fixed pulse is five. Both have equivalent switching frequencies.

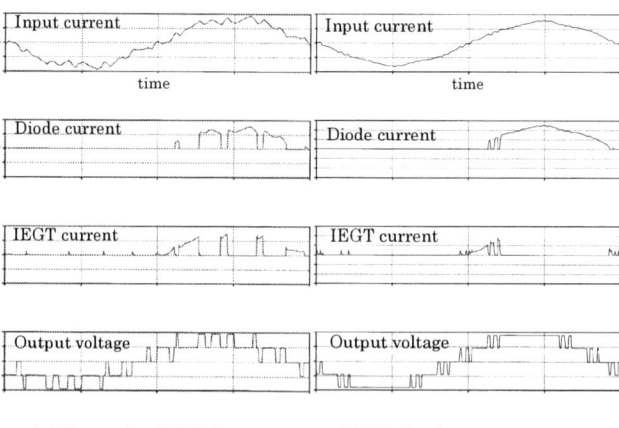

(a) Conventional PWM　　　(b) Fixed pulse pattern (FPP) PWM

Fig. 10 PWM waveforms

978-1-4799-2706-7/14 $31.00 © 2014 IEEE

IV. Test Results and Performance Evaluation

For the development of the proposed 3-level IEGT inverter, single phase runback test is performed to check the voltage distribution in the equipment. And the temperature of main devices, capacitors and fuses are also checked and confirmed the satisfactory results according to the design and specification of those particular components.

A. Runback Test Verification

Older thyristor drive equipment was commonly tested separately with low voltage and rated current and also with rated voltage and small current. However, this practice is not sufficient for inverters using turn-on/turn-off devices such as IGBT, GCT or IEGT. The current and voltage stresses of these devices depend on the output voltage, current, frequency and power factor. It is more appropriate to test them according to the actual operating conditions. A special *Runback Test* system has been developed for this purpose. The *Runback Test* is done for all of the IEGT stacks before shipment. The inverter and converter are controlled as if they are in the actual field operation. The output of the inverter, frequency, voltage, current and power factor can be controlled. This means that all kinds of operational modes, including low frequency large current tests, can be performed at the factory [6].

In the Runback Test circuit most of the test power is circulated and only a small power supply is required to make up the losses. This means very large capacity equipment can be tested at the factory. Fig. 11 shows the configuration of the test circuit.

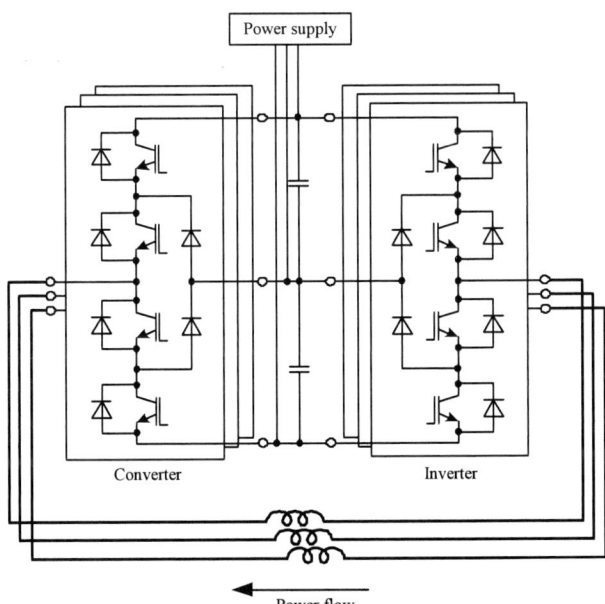

Fig. 11. Configuration of the runback test circuit

During the development of the new 3-level IEGT inverter single phase runback tests were performed to check the voltage distribution within the equipment. The temperature of main devices, capacitors and fuses were also checked and confirmed to be satisfactory according to the design and specification of those particular components.

First, rated voltage and current tests were carried out as a verification of the development. In this test power was circulated in the converter and the inverter by applying the same voltage and current as real operational conditions. Only the component losses needed to be supplied from the test system power supply. The test condition is shown below.

- Inverter output current: 1430Arms (100%), overload 150%-1 minute
- Inverter output voltage: 3650Vrms

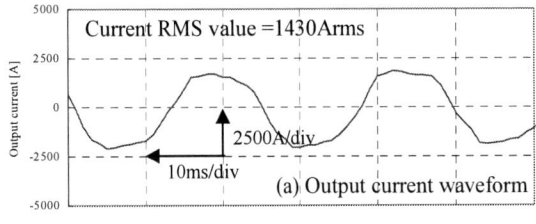

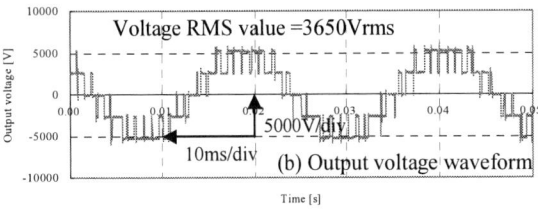

Fig. 12. Runback test results: (a) U-phase output current and (b) line-to-line output voltage waveform

Fig. 12 shows the U-phase output current and line-to-line output voltage waveform of the inverter. It is confirmed from this runback test that the output voltage and current are obtained at rated operation. The test is also carried out for 150%-1 minute (under overload condition) and it is found a satisfactory result.

Moreover, the equipment efficiency is confirmed 99% at the time of rated operation obtained from the test results shown in Fig. 13.

The system efficiency is calculated for the main circuit block. Here, the main circuit block includes converter and inverter. Fig. 13 shows the proposed 3-level IEGT inverter efficiency for the load changes from 50% to 150%. This is the actual test result obtained by the run back test in the factory. For both inverter and converter, water cooling loss and the common air cooling loss are considered to calculate the efficiency of the main circuit block. In the inverter and converter, water cooling loss mainly includes the combined losses of IEGT devices, fly wheel diodes, clamp diodes, snubber diodes, snubber resistors and discharge resistors. And air cooling loss includes the combined losses of voltage divider resistors, snubber capacitors, main fuses, cores, bus bars, smoothing capacitors, gate and control power supplies.

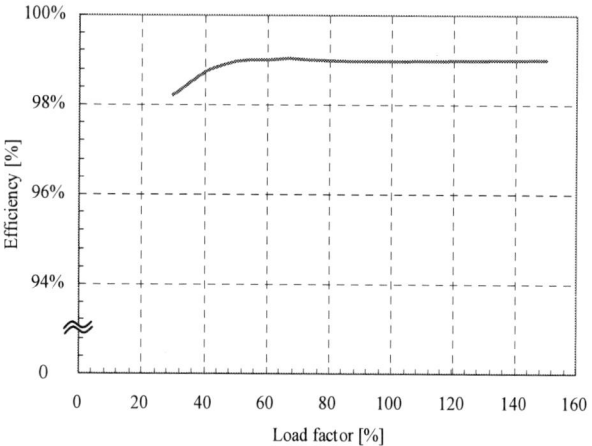

Fig. 13. Equipment efficiency of the newly developed 3-level IEGT inverter

$$Efficiency\,(\%) = \frac{Rated\ output \times Load\ factor[\%]}{Rated\ output \times Load\ factor[\%] + Total\ loss} \times 100 \quad (2)$$

Finally, the efficiency of the 3-level IEGT inverter main circuit block is obtained by (2). It shows that the efficiency is about 99% even the load changes from 50% to 150%.

B. Field Test Result

Fig. 14 shows the field test result of the inverter output current at overload 122% operation. It shows the inverter output current 1750Arms (1430Arms at 100% rated output current). In general, it is possible to obtain clean output current when the steps of the output voltage go higher in number. Here, the inverter output current is near to sinusoidal curve obtained in this study.

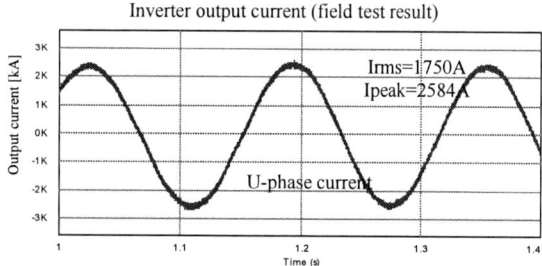

Fig. 14. Field test result: Inverter output current (U-phase) at 122% overload operation

V.   Conclusions

Due to the increasing demand, it is expected to expand the production efficiency of the steel mill plants. However, in order to keep the operation availability sufficiently high, the system reliability has significant importance. For such large plants, the capacity of the drive equipment is also required to be increased to megawatt or tens of megawatt range with high reliability and high efficiency.

In this paper, the recent technical trend in the field of the large drive equipment based on VSI is built with the large capacity semiconductor devices. The large capacity semiconductor and its application technology help to realize simple inverters with small number of parts. This type of drive equipment is expected to be reliable and the field experience showed the expectation.

A 9MVA rating 3-level IEGT inverter is developed to apply in the steel plant. With the realization of downsizing and light weight by reconfiguring the power unit structure and material, the newly developed IEGT is also able to successfully use for this equipment. The improvement in cooling efficiency has realized 9MVA. Moreover, it corresponds to the common converter system with the availability of connecting maximum 4 banks in parallel. The application of this 3-level IEGT inverter for a new installation and a renewal of existing drive equipment in the steel plant are expected for both domestic and overseas in near future.

978-1-4799-2706-7/14 $31.00 © 2014 IEEE

References

[1] S. Kodama and T. Ishida, "Development of drive systems in metal rolling lines", *2010 Annual Conference of IEE of Japan, Industry Applications Society*, Tokyo, Japan, 1-S5-1, pp. 29-34, August 2010.

[2] H. Hosoda, S. Kodama and R. Tessendorf, "Large PWM inverters for rolling mills", *AIST Iron & Steel Technology (Journal Article)*, PA, USA, PR-PM0108-4, January 2008.

[3] M. Tsukakoshi, M. A. Mamun, K. Hashimura, H. Hosoda and S. C. Peak, "Introduction of a large scale high efficiency 5-level IEGT inverter for oil and gas industry," *IEEE Energy Conversion Congress and Exposition (ECCE 2009)*, San Jose, CA,USA, pp. 4313-4320, September 2010.

[4] K. Ichikawa, M. Tsukakoshi and R. Nakajima, "Higher efficiency three-level inverter employing IEGTs", *Applied Power Electronics Conference and Exposition (APEC2004)*, USA, vol. 3, pp. 1663-1668, February 2004.

[5] R. Nakajima, M. Mukunoki and K. Omote, "Application of an aluminum material to the large IEGT stack", *The Institute of Electrical Engineers of Japan (IEEJ)*, Hiroshima, Japan, session 4-154, pp. 265, March 2012.

[6] H. Hosoda, M. A. Mamun and T. Yoshino, "Trends in MW-rated VSI technology and reliability for adjustable speed drives", *The Applied Power Electronics Conference and Exposition (APEC 2010)*, Palm Springs, CA ,USA, pp. 1261-1265, February 2010.

A Soft-Switching Single-Phase Unified Power Quality Conditioner

Maoh-Chin Jiang, *Member, IEEE*, Kai-Chi Chang
Department of Electrical Engineering
National Ilan University
Yilan, Taiwan, R.O.C

Kao-Yi Lu, Bing-Jyun Shih, Tai-Chun Liu
Department of Electrical Engineering
National Ilan University
Yilan, Taiwan, R.O.C

Abstract—A soft-switching single-phase unified power quality conditioner (UPQC) using simple resonant units is proposed in this paper. In the proposed UPQC, all main switches operate at zero-voltage-switching (ZVS) turn-on and all auxiliary switches operate at zero-current-switching (ZCS) turn-off. The proposed UPQC consists of a soft-switching series active power filter and a soft-switching shunt active power filter. The soft-switching series active power filter can compensate the ac source voltage fluctuation. It can be used to regulate both the over-voltage and under-voltage situations. On the other hand, the soft-switching shunt active power filter can compensate the current harmonics and improve the power factor. Finally, some simulation results are presented for verification.

Keywords—soft-switch, unified power quality conditioner, zero-voltage-switching (ZVS), zero-current-switching (ZCS).

I. INTRODUCTION

Recently, power quality has become an interesting issue because several precise devices are widely used in our life. All of these precise devices are sensitive to the power quality delivered to them. Also, poor power quality may result in equipment malfunction or even get damaged. Moreover, power interruption, harmonics and voltage unstable are the major cause of poor power quality.

In order to conform to the related standards and satisfy the industrial requirement, it is necessary to improve power quality. There are two aspects can be improved in power quality. One is to compensate current-based distortions such as current harmonics and reactive power. The other is to compensate voltage-based distortions such as voltage flickers, voltage sags and voltage swells. In recent years, active power filters (APFs) become one of the best tools for harmonic elimination, voltage regulation and reactive power compensation. The APFs are classified as series APF, shunt APF and hybrid APF [1]. The unified power quality conditioner (UPQC), a type of hybrid APF, is formed with a series APF and a shunt APF. The series APF suppresses voltage-based distortions and maintains a stable load voltage. At the same time, shunt APF compensates reactive current and eliminates current-based distortions to achieve a unity power factor [2]. Hence, UPQC is a multifunction device which can solve both voltage-based and current-based distortions [3]-[5].

The configuration of a common UPQC is shown in Fig. 1. It is an association with a series and a shunt APF connected to a common dc link. The series APF compensates the voltage-based distortions of source voltage so that the load voltage is sinusoidal and balanced. The shunt APF eliminates the source current harmonic components so that the source current is sinusoidal and distortion free. It is apparent that UPQC not only compensates load voltage but eliminates source current harmonics to improve power quality. The traditional UPQC is usually operated at hard-switching [5]-[8]. This hard-switching technique with lowly switching frequency may result in lowly compensated performance of UPQC. The compensated performance is increased by increasing switching frequency. However, when operating at high switching frequency, there are some incoming problems such as higher switching stresses, more switching losses and serious electromagnetic interferences (EMI). As a result, soft-switching technique is proposed to solve these problems and several studies aimed at soft-switching technique are proposed in recent years.

A soft-switching single-phase UPQC based upon the concept of the auxiliary resonant commutated pole inverter (ARCPI) is presented in this paper [9]. Rather than the traditional ARCPI, only one dc capacitor is needed to construct the common dc-link and two resonant tanks are used in the proposed UPQC. Also, both series and shunt APFs need only one resonant tank to achieve that all power switches can operate at ZVS turn-on. Furthermore, the reduction in the number of resonant tanks not only simplifies the control circuit but also reduces the cost. Finally, some simulation results are presented to verify the feasibility of the proposed soft-switching single-phase UPQC.

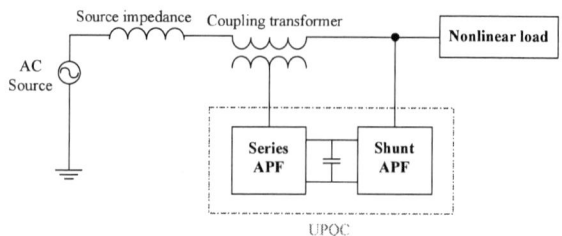

Fig. 1. System configuration of an UPQC.

II. PROPOSED SOFT-SWITCHING SINGLE-PHASE UPQC

A. System Configuration

Fig. 2 shows the system configuration of the proposed soft-switching UPQC which consists of a soft-switching shunt APF and a soft-switching series APF. The soft-switching shunt APF eliminates the current harmonics and compensates the reactive current to achieve a unity power factor. The soft-switching series APF is used to compensate voltage-based distortions such as voltage harmonics, voltage sags and voltage swells. Thus, the load voltage maintains a stable sinusoidal voltage.

The soft-switching principle is using a suitable conducting time to create resonant condition. The resonant inductors and the parasitic capacitors of main switches form a resonant tank so that the drain voltage drops to zero before main switches are turned on. As a result, all main switches operate at zero-voltage-switching (ZVS) turn-on. Moreover, switching losses and switching stresses on main switches can be significantly reduced by using ZVS. Also, the resonant inductor current drops to zero before the auxiliary switches are turned off. Hence, all auxiliary switches operate at zero-current-switching (ZCS) turn-off. In order to minimize switching frequency of auxiliary switches and reduce conducting losses, a nature resonant control strategy is added in the proposed UPQC. Because the resonant current only passes through the auxiliary circuit, the voltage stress and current stress will not increase on the main switches. Finally, it can effectively improve the EMI, efficiency and power density by using the soft-switching technique.

B. Operation Principle

The operation principle of the proposed UPQC will be analyzed in this section. Because the operation modes of the shunt APF are similar to the series APF, only the shunt APF will be clarified in this part for simplifying the analysis. In order to simplify the analysis, the following assumptions are made during one switching cycle: (i) all components and devices are ideal; (ii) the dc capacitor C_{dc} is large enough that the dc-link voltage V_{dc} can be assumed to be constant and ripple free; (iii) the input boost inductor L_c is large enough that the compensating current i_c can be assumed to be constant, i.e., $i_c=I_c$; (iv) all values of resonant capacitors are identical, i.e., $C_{r1}=C_{r2}=C_{r3}=C_{r4}=C_r$, where C_r is the sum of the parasitic capacitor of MOSFET. Based on these assumptions, circuit operations in one switching cycle can be divided into six modes. The six dynamic equivalent circuits of the proposed shunt APF are shown in Fig. 3.

Mode 1 ($t < t_0$): Fig. 3(a) shows the initial mode of the shunt APF. Assume that the line voltage is in the positive half-cycle, switches S_1 and S_2 are conducting the compensating current I_c, and that the initial conditions are the resonant voltages $v_{cr1}(t) = v_{cr2}(t) = 0$ and the resonant voltages $v_{cr3}(t) = v_{cr4}(t) = V_{dc}$. The auxiliary switches S_5 and S_6 are turned off and the resonant current $i_{Lr}(t) = 0$.

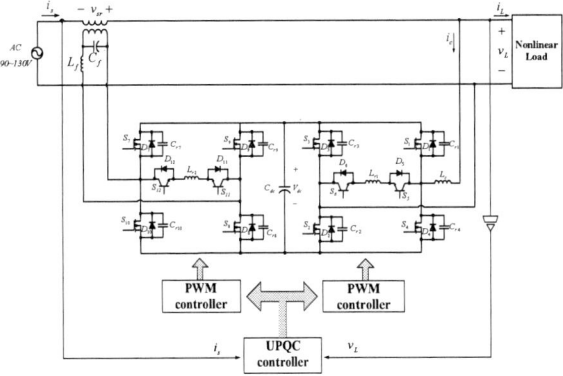

Fig. 2. System configuration of the proposed UPQC.

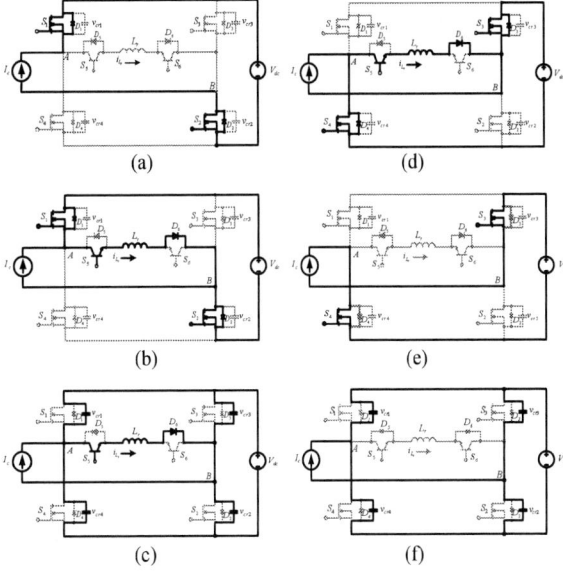

Fig. 3. Operation modes of the proposed soft-switching shunt APF.

Mode 2 ($t_0 \le t \le t_1$): As shown in Fig. 3(b), this mode begins when the auxiliary switch S_5 is turned on with ZCS at $t = t_0$. The resonant inductor L_r charges linearly from dc-link voltage V_{dc}. The resonant inductor current $i_{Lr}(t)$ is increased linearly. This mode ends when the resonant current rises to $I_{c,max}$.

Mode 3 ($t_1 \le t \le t_2$): As shown in Fig. 3(c), this mode begins when switches S_1 and S_2 are turned off at $t = t_1$. Then resonant inductor L_r and the resonant capacitors C_{r1}, C_{r2}, C_{r3} and C_{r4} form a resonant tank. At first, the resonant current $i_{Lr}(t)$ increases and then decreases when it reaches its peak value. This mode is finished when the resonant voltages $v_{cr3}(t)$ and $v_{cr4}(t)$ drop to zero and the resonant voltages $v_{cr1}(t)$ and $v_{cr2}(t)$ rise to V_{dc}.

Mode 4 ($t_2 \le t \le t_3$): As shown in Fig. 3(d), since the resonant capacitor voltages $v_{cr3}(t)$ and $v_{cr4}(t)$ equal zero, the body diodes D_3 and D_4 start to conduct the resonant current. Switches S_3 and S_4 are turned on during this interval with ZVS commutation. The energy stored in the resonant inductor L_r is delivered back to the dc-link voltage. The resonant current i_{Lr} (t) is decreased linearly. This mode ends when the resonant current i_{Lr} (t) drops to zero.

Mode 5 ($t_3 \leq t \leq t_4$): As shown in Fig. 3(e), since the resonant current equals zero, the auxiliary switch S_5 is turned off with ZCS at $t = t_3$. Switches S_3 and S_4 are conducting the compensating current. The resonant voltages v_{cr3} $(t) = v_{cr4}$ $(t) = 0$ and the resonant voltages v_{cr1} $(t) = v_{cr2}$ $(t) = V_{dc}$. Also, the auxiliary switches S_5 and S_6 are turned off and the resonant current i_{Lr} $(t) = 0$.

Mode 6 ($t_4 \leq t \leq t_5$): As shown in Fig. 3(f), this mode begins when the switches S_3 and S_4 are turned off at $t = t_4$. When all main switches are turned off, system transfers to blanking time. In order to maintain the compensating current continuous, the resonant capacitors start to charge or discharge. This mode is finished when the resonant voltages v_{cr1} (t) and v_{cr2} (t) drop to zero and the resonant voltages v_{cr3} (t) and v_{cr4} (t) rise to V_{dc}.

After mode 6, since the resonant capacitor voltages v_{cr1} (t) and v_{cr2} (t) equal zero, the body diodes D_1 and D_2 start to conduct the compensating current at $t = t_5$. Switches S_1 and S_2 are turned on during this interval with ZVS commutation. Then the circuit operation is returned to Mode 1. From above analysis, one can find that all main switches operate at ZVS turn-on, and all auxiliary switches operate at ZCS turn-off. Moreover, the resonant phenomenon only occurs in a short blanking time, it will not increase the voltage and current stress on main switches.

III. SIMULATION RESULTS

A simulation configuration was constructed to demonstrate the feasibility and performance of the proposed soft-switching UPQC. Figures 4 and 5 show the commutation phenomenon for the main switches (S_1, S_2, S_3 and S_4) and the auxiliary switches (S_5 and S_6) of the proposed shunt APF, respectively. Figure 4 shows that drain voltage decreased to zero before main switches are turned on. Hence, all main switches operate at ZVS turn-on. Also, Figure 5 shows that the resonant currents have been decreased to zero before auxiliary switches are turned off. Figure 5 demonstrates that ZCS is achieved at a constant frequency for the auxiliary switches. As a result, the system switching losses can be significantly reduced. Figures 6 and 7 show the simulation results at different resonant modes. Figs. 6(a) and 7(a) show the auxiliary resonant mode which can ensure that all main switches operating at ZVS but it will increase the conducting loss. Figs. 6(b) and 7(b) show the proposed natural resonant mode that can reduce the conducting losses of auxiliary switches. When the compensating current i_c or the output inductor current i_{sr} are between ± 1 A, all main switches operate at auxiliary resonant mode. Natural resonant technique can reduce the switching frequency of auxiliary switches. Finally, the

whole efficiency of the proposed UPQC that uses a natural resonant technique is 4% higher than the conventional auxiliary resonant method. Fig. 8 shows the steady-state operation results of the input voltage, v_s, input current, i_s, load voltage, v_L, and load current, i_L, under a 400W nonlinear load operating at different input voltage. As shown in Fig. 8, one can find that the load voltage remains a stable voltage with lowly THD under different input voltage. Moreover, the input current is sinusoidal and in phase with the input voltage, so that it can achieve a unity power factor. As shown in Fig. 8(a), the THD of the input current is 3.58% and the THD of the load voltage is 2.05%. Also, when the input voltage is over-voltage, the THD of the input current is 3.72 and the THD of the load voltage is 2.55% as shown in Fig. 8(b). After all, the power factor is about 0.995 to demonstrate that the proposed UPQC provides a good power factor. Fig. 9 shows the transient responses under a step load change from 200W to 400W operating at a different voltage level. As shown in Fig. 9, it is obvious that the input current changes immediately to the response the load changes. Moreover, the load voltage still maintains a stable and sinusoidal waveform, no matter what the load changes. Consequently, one can find that the proposed UPQC has good steady-state and transient performance.

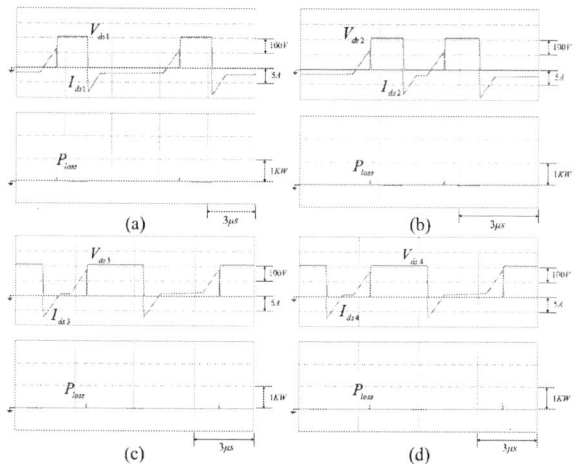

Fig. 4. The commutation phenomenon for main switches of the proposed shunt APF.

IV. CONCLUSIONS

This paper proposes a soft-switching single-phase unified power quality conditioner, which consists of a soft-switching shunt APF and a soft-switching series APF. In the proposed UPQC, all main switches operate at ZVS turn-on and all auxiliary operate at ZCS turn-off by using simple resonant units. The whole efficiency is 4% higher than the conventional resonant control by using the proposed natural resonant control. The THD of source current is 45%, the THD of load voltage is 7.52% and the power factor is 0.77 before compensating. After using the proposed UPQC, the THD of source current is significantly suppressed to 3.58%, the THD of load voltage is 2.55% and the power factor is about 0.995.

Clearly, the proposed UPQC is good at eliminating the current harmonics, compensating voltage distortions and improving system power factor. Moreover, in the transient period, the source current remains a sinusoidal waveform in phase with source voltage and the load voltage still maintain 110V with stable and sinusoidal waveforms. It is evident that the proposed UPQC has good steady-state and transient performances.

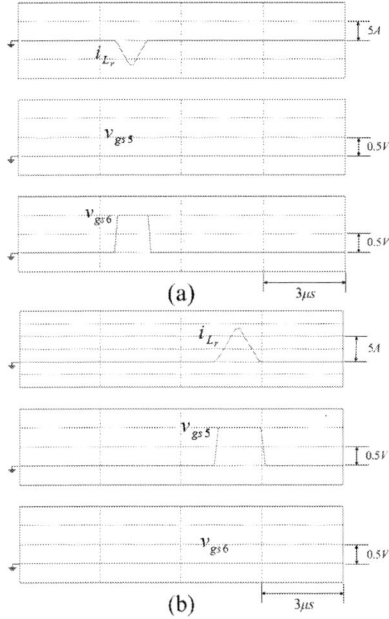

Fig. 5. The commutation phenomenon for auxiliary switches of the proposed shunt APF.

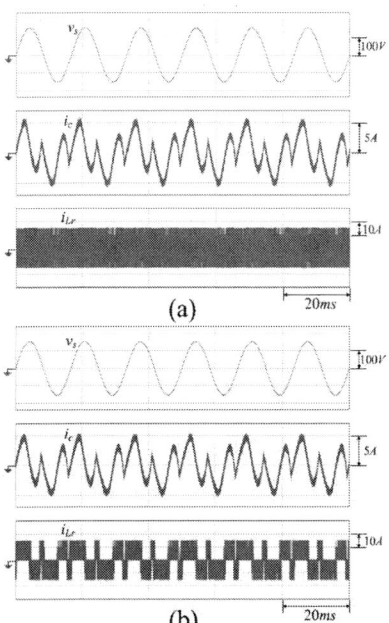

Fig. 6. Simulation results at different resonant modes of the proposed shunt APF (a) auxiliary resonant mode (b) natural resonant mode.

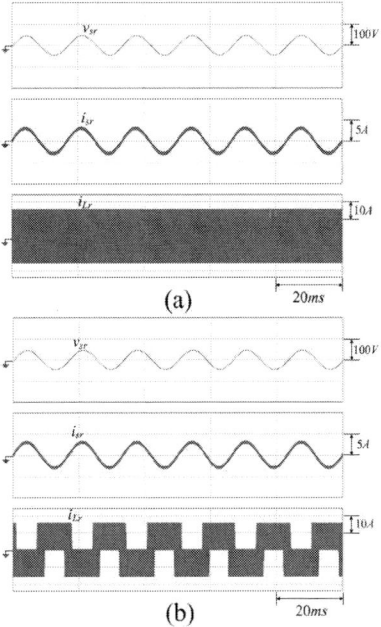

Fig. 7. Simulation results at different resonant modes of the proposed series APF (a) auxiliary resonant mode (b) natural resonant mode.

Acknowledgment

The authors would like to acknowledge the financial support of the National Science Council of Taiwan, R. O. C. through grant number NSC 102-2221-E-197-008.

References

[1] H. Fujita and H. Akagi, "The Unified Power Quality Conditioner: The integration of series and shunt-active filters," IEEE Trans. Power Electronics, vol. 13, no. 2, pp. 315-322, Mar. 1998.

[2] V. Khadkikar and A. Chandra, "A New Control Philosophy for a Unified Power Quality Conditioner (UPQC) to Coordinate Load-Reactive Power Demand Between Shunt and Series Inverter," IEEE Trans. Power Deli., vol. 23, no.4, pp. 2522-2534, Oct. 2008.

[3] F. Mehdi and A. Saeed, "Online Wavelet Transform-Based Control Strategy for UPQC Control System," *IEEE Trans. Power Delivery*, vol. 22, no. 1, pp. 481-491, Jan. 2007.

[4] H. R. Mohammadi, A. Y. Varjani and H. Mokhtari, "Multiconverter Unified Power-Quality Conditioning System: MC-UPQC," *IEEE Trans. Power Deli.*, vol. 24, no. 3, pp. 1679-1686, July 2009.

[5] Y. Y. Kolhatkar and S. P. Das, "Experimental Investigation of a Single-Phase UPQC With Minimum VA Loading," *IEEE Trans. Power Deli.* vol. 22, no. 1, pp. 373-380, Jan. 2007.

[6] V. Khadkikar, A. Chandra, A. O. Barry and T. D. Nguyen, "Analysis of Power Flow in UPQC during Voltage Sag and Swell Conditions for Selection of Device Ratings," in *CCECE 6th Canadian Conf. on Electrical and Computer Engineering*, pp. 867-872, May 2006.

[7] A. Kazemi, M. Sarlak and M. Barkhordary, " An adaptive noise canceling method for single-phase unified power quality conditioner," in 1*st IEEE Conf. on Ind. Electronics and Applicat.*, pp. 1-6, May 2006.

[8] J. M. Correa, S. Chakraborty, M. G. Simoes and F. A. Farret, "A single phase high frequency AC microgrid with an unified power quality conditioner," in 38th *IAS Annual Meeting Record of the Ind. Applicat. Conf.*, vol. 2, pp. 956-962, Oct. 2003.

[9] R. W. De Doncker and J. P. Lyons, "The auxiliary resonant commutated pole converter," in *IAS Annual Meeting Record of the Ind. Applicat.* Conf., pp. 1228-1235, Oct. 1990.

[10] C. C. Chan, K. T. Chau and J. Yao, "Soft-switching vector control for resonant snubber based inverters," in 23*rd Int. Conf. on Ind.*

Electronics, Control and Instrumentation, vol. 2, pp. 453-458, Nov. 1997.

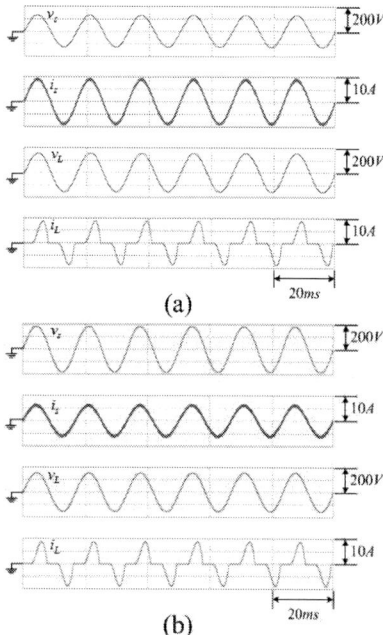

(a)

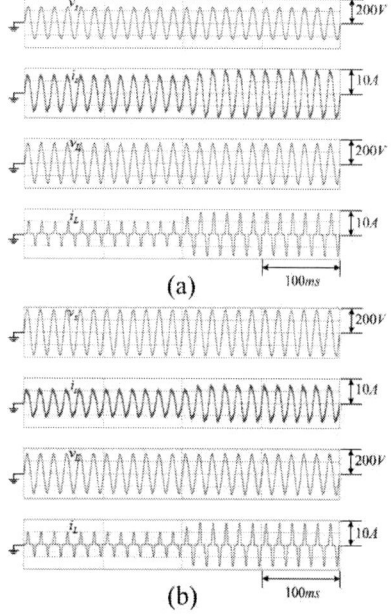

(b)

Fig. 8. Steady-state operation results under a 400W nonlinear load operating at different input voltage (a) under-voltage (b) over-voltage.

Fig. 9. Transient responses under a step load change from 200W to 400W operating at different input voltage (a) under-voltage (b) over-voltage.

The 2014 International Power Electronics Conference

Novel Three-Phase PWM AC-AC Converters Solving Commutation Problem

Ashraf Ali Khan, Hyunhak Shin, Honnyong Cha,
School of Energy Engineering
Kyungpook National University
Daegu, Korea
08beeashrafa@seecs.edu.pk

Heung-Geun Kim
School of Electrical Engineering
Kyungpook National University
Daegu, Korea

Abstract— This paper proposes a novel three-phase PWM ac-ac converters, presents detailed circuit operation and operating principle. The proposed converters employs the basic switching cell structure and coupled inductor to solve commutation problem without sensing input voltage polarity. The commutation scheme allows that the proposed converters can be short and open circuited without generating high switch voltage and current surge moreover the proposed ac-ac converters can be operated with highly distorted input voltage. Theoretical and practical analysis results are given using boost converter as an example and can be extended easily to buck converter. To prove the performance and effectiveness of proposed converters a 500W prototype of proposed converter is built and tested.

Keywords— *Commutation, coupled inductor, switching cell, three phase ac-ac converter.*

I. INTRODUCTION

PWM rectifier followed by a PWM voltage source inverter with dc link is the preferred converter in many industrial applications for ac-ac power conversion that requires variable frequency and voltage, however for the applications that demands only voltage regulation the direct PWM ac-ac converter is a practical choice in terms of volume and cost. Fig.1 shows the direct PWM ac-ac converter is a practical choice to achieve lower cost and smaller volume, recently a Z-source PWM ac-ac converter as shown in Fig. has been introduced to have both buck and boost functions [1], [2], inspection reveals that the conventional converters and Z-source PWM ac-ac converter inherits the commutation problem. The commutation problem does not occur in ideal PWM signals. However, due to limited switching speed of switches and delay time, there practically exists an instant of overlap or dead time among the switches. During the overlap-time the input voltage will be short-circuited and switches are damaged permanently, similarly in case of dead time no current path is available for inductor current and causes destruction of switches because of over voltage.

A natural trend to address the problem of commutation and to provide smooth current transition has been to adopt converters or topologies with bulky RC snubbers

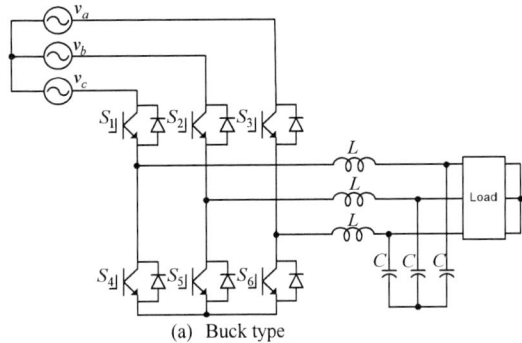
(a) Buck type

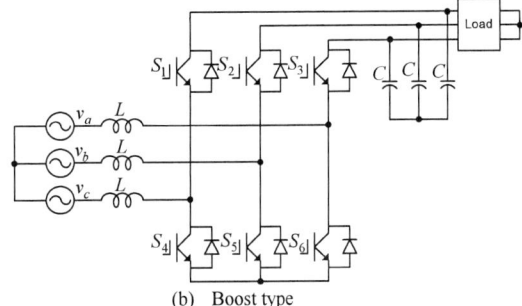
(b) Boost type

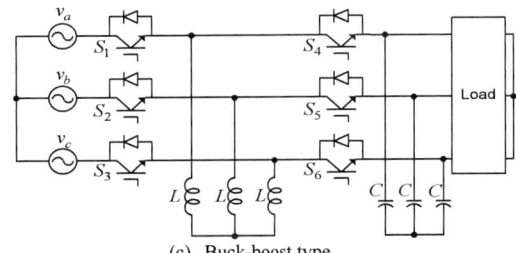
(c) Buck-boost type

Fig. 1. Conventional three-phase direct PWM ac-ac converters

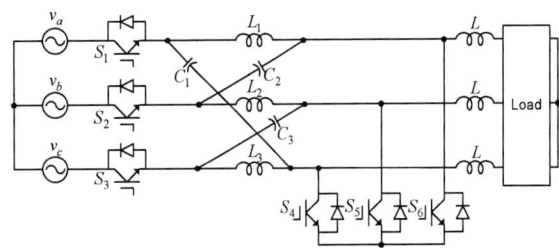

Fig. 2. Three-phase direct Z-source PWM ac-ac converter

978-1-4799-2706-7/14 $31.00 © 2014 IEEE

[2], [3] to protect switches. Several approaches of soft commutation have been introduced and employed successfully in the research so far but most of them sense input voltage polarity and unpractical to use with distorted input voltage or current around zero crossing.

In an effort to address this issue, a converter that also minimize the size requirement of input and output inductors significantly in boost and buck respectively have been introduced. In this paper, novel three-phase PWM ac-ac converters without having commutation problem are proposed. The proposed converters employs the basic switching cell structure and coupled inductor [7] as shown in Fig.3 to solve commutation problem without sensing input voltage polarity, with this structure the proposed converters can be short and open-circuit without damaging switching devices. As the proposed converters do not need to sense the voltage polarity for commutation Therefore, the proposed converters don't have commutation problem with highly distorted input voltage.

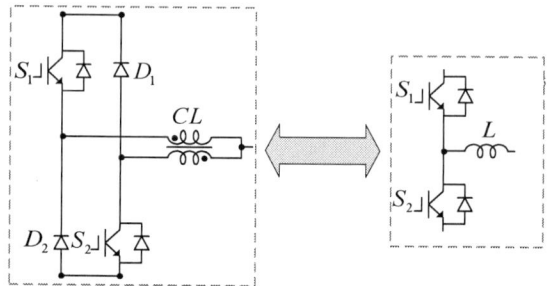

Fig. 3. Switching cell with coupled inductor

II. PROPOSED THREE-PHASE AC-AC CONVERTERS AND THEIR OPERATION

Unlike the conventional three phase ac-ac converters the proposed converters shown in Fig. 4 employs the basic switching cell structure and coupled inductor shown in Fig. 3, [4~6].

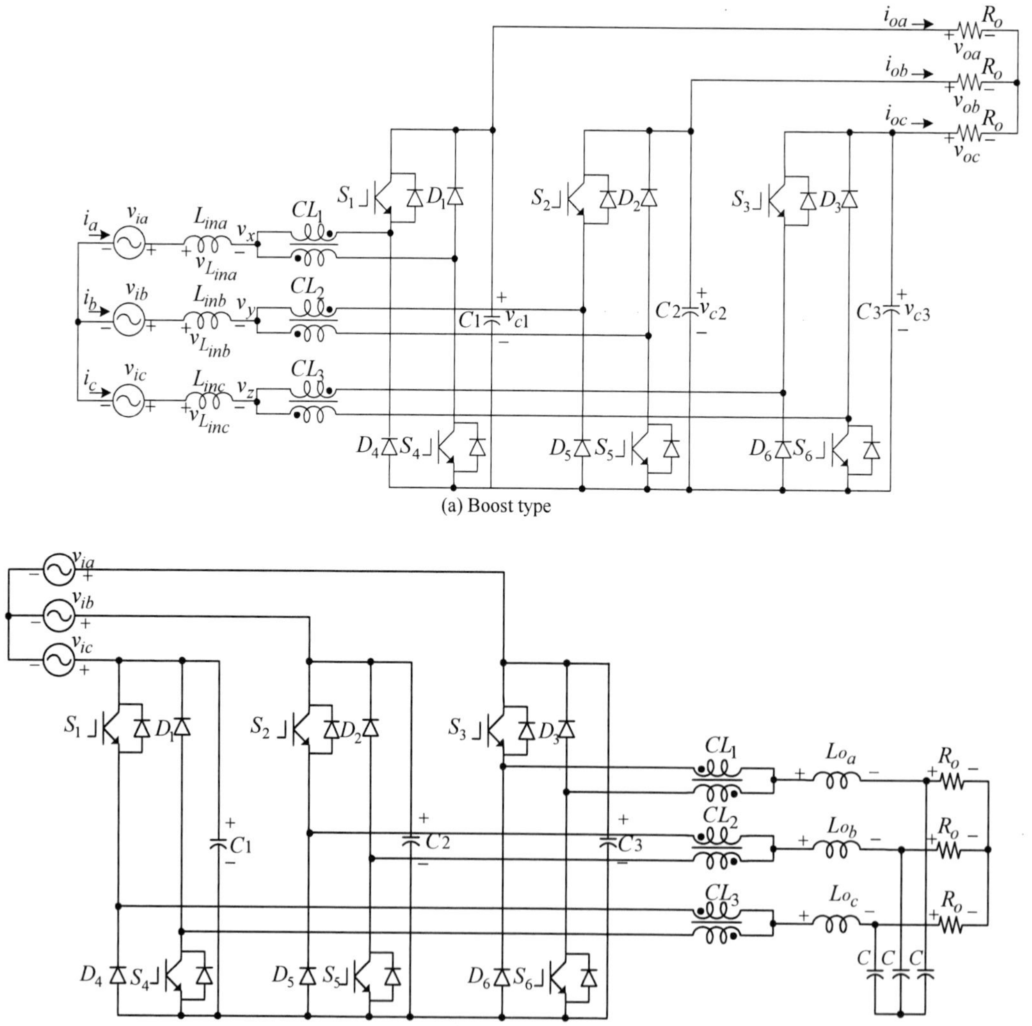

(a) Boost type

(b) Buck type
Fig. 4. Proposed three phase ac-ac PWM converters

978-1-4799-2706-7/14 $31.00 © 2014 IEEE 111

Three coupled inductors (CL_1, CL_2, CL_3) are inserted between the switch arms (S_1, S_4), (S_2, S_5) and (S_3, S_6). These coupled inductors serve to limit the currents when all switches are turned-on like a short circuit. The three capacitors C_1, C_2 and C_3 are added to provide a current path when (S_1, S_4), (S_2, S_5) and (S_3, S_6) are turned-off. An addition to this, these capacitors serve as the input filter capacitors for the buck-type and output filter capacitors for the boost-type converters. These capacitors also suppress the switch voltage overshoot caused by the circuit stray inductance works as a snubber capacitor.

In this paper, the boost type three phase ac-ac converter is considered and the buck type can be analyzed similarly. To generate PWM gate signals for proposed converter, two carrier signals (v_{ca1} and v_{ca2}) 180° out of phase shown in Fig. 5 are compared with the reference voltage (v_{ref}). As shown in Fig. 7, the proposed converters has four continuous conduction modes and three unipolar voltage levels (0, v_{cx}, $v_{cx}/2$) for each phase leg, where x corresponds to 1,2,3 for phase leg a, b and c respectively. The converter has two distinct current components, input inductor current and common mode current as shown in Fig. 6. The average of the winding currents in each leg is used to calculate common mode currents (icm_a, icm_b, icm_c) and differences of winding currents results input inductors currents (i_a, i_b, i_c) shown in Fig. 6 corresponding to phase leg a, b and c respectively, regardless of modes operations following relationships always apply.

$$icm_k = \frac{i_{Lk1} + i_{Lk2}}{2} \tag{1}$$

$$i_k = i_{Lk2} - i_{Lk1} \tag{2}$$

$$i_{Lk1} = icm_k - \frac{i_k}{2} \tag{3}$$

$$i_{Lk2} = icm_k + \frac{i_k}{2} \tag{4}$$

where $k = a, b, c$ for phase leg a, b and c respectively. All the coupled inductors are assumed to be identical in analysis with coupling coefficient close to unity.

The all upper switches (S_1, S_2, S_3) and lower switches (S_4, S_5, S_6) are turned-on in mode 1 and all switches are turned-off in mode 3. In mode 1, currents are limited by the coupled inductors and there exists currents freewheeling paths in mode 2 and 3. All the switches in the circuit can be short-and open-circuited without generating high switch surge voltage and current. Therefore, the commutation problem can be perfectly solved with the proposed converters. Moreover, a little mismatch in gate signal of S_1 and S_4 (or S_2 and S_5, S_3 and S_6) can also be solved with the proposed structure. Proposed converters can be operated even when the input voltage is highly distorted. Therefore proposed

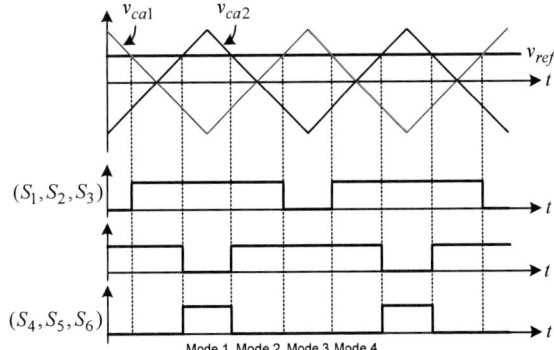
Fig. 5. Gate signal generation of the proposed ac-ac converters

converters do not need to sense the input voltage polarity as it has no commutation problem.

In analysis only half portions of input voltages are considered, input and circulating currents are shown for phase a, other phases follows the same analysis.

A. Mode 1 $[0 \sim DTs]$

In this energy storage mode all the switches S_1-S_6 are turned on, when switches are turned on input energy is stored in the input inductors. The voltages and currents are related as follows

$$\begin{cases} v_{xy} = \dfrac{v_{c1}}{2} - \dfrac{v_{c2}}{2} = \dfrac{v_{ab}}{2} \\[2mm] v_{yz} = \dfrac{v_{c2}}{2} - \dfrac{v_{c3}}{2} = \dfrac{v_{bc}}{2} \\[2mm] v_{zx} = \dfrac{v_{c3}}{2} - \dfrac{v_{c1}}{2} = \dfrac{v_{ca}}{2} \end{cases} \tag{5}$$

$$\begin{cases} v_{Linab} = v_{iab} - v_{xy} = v_{iab} - \dfrac{v_{ab}}{2} \\[2mm] v_{Linbc} = v_{ibc} - v_{yz} = v_{ibc} - \dfrac{v_{bc}}{2} \\[2mm] v_{Linca} = v_{ica} - v_{zx} = v_{ica} - \dfrac{v_{ca}}{2} \end{cases} \tag{6}$$

$$\begin{cases} v_{Linab} = v_{Lina} - v_{Linb} \\ v_{Linbc} = v_{Linb} - v_{Linc} \\ v_{Linca} = v_{Linc} - v_{Lina} \end{cases} \tag{7}$$

using (6) and (7) voltages across input inductors are obtained as shown in (9) following (8).

$$v_{Lina} = v_{ia} - \frac{v_{oa}}{2} + v_{Linb} - v_{ib} + \frac{v_{ob}}{2} \tag{8}$$

$$v_{Link} = v_{ik} - \frac{v_{ok}}{2} \tag{9}$$

where $k = a, b, c$ for phase leg a, b and c respectively.

$$\frac{d}{dt}\begin{bmatrix} icm_a \\ icm_b \\ icm_c \end{bmatrix} = \frac{1}{4Ls}\begin{bmatrix} v_{c1} \\ v_{c2} \\ v_{c3} \end{bmatrix} \tag{10}$$

978-1-4799-2706-7/14 $31.00 © 2014 IEEE

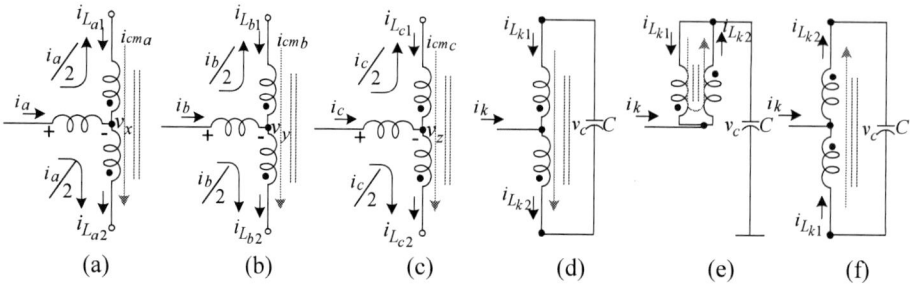

Fig. 6. Input currents division, circulation of common mode currents and switching states for proposed converters. (a), (b) and (c) Division of input and circulating currents in three phase ac-ac design. (d) All switches on. (e) P-cells switches on. (f) All switches off.

$$\frac{d}{dt}\begin{bmatrix} i_a \\ i_b \\ i_c \end{bmatrix} = \frac{1}{L}\begin{bmatrix} v_{L_{ina}} \\ v_{L_{inb}} \\ v_{L_{inc}} \end{bmatrix} \qquad (11)$$

Ls is the self inductance of coupled inductor and L is the input inductor inductance.

B. Mode 2 $[DTs \sim 0.5Ts]$

All P-cells switches are turned on while all N-cells are turned off. The diodes (D_1, D_6) are turned on due to freewheeling and input energy is transferred to load. The voltage and current relationships are given as follows

$$v_x = v_{c1}, v_y = v_{c2}, v_z = v_{c3} \qquad (12)$$

$$\begin{cases} v_{xy} = v_{c1} - v_{c2} = v_{ab} \\ v_{yz} = v_{c2} - v_{c3} = v_{bc} \\ v_{zx} = v_{c3} - v_{c1} = v_{ca} \end{cases} \qquad (13)$$

$$\begin{cases} v_{L_{inab}} = v_{iab} - v_{xy} = v_{iab} - v_{ab} \\ v_{L_{inbc}} = v_{ibc} - v_{yz} = v_{ibc} - v_{bc} \\ v_{L_{inca}} = v_{ica} - v_{zx} = v_{ica} - v_{ca} \end{cases} \qquad (14)$$

$$v_{L_{ink}} = v_{ik} - v_{ok} \qquad (15)$$

$$\frac{dicm_k}{dt} = 0 \qquad (16)$$

where $k = a,b,c$ for phase leg a,b and c respectively

C. Mode 3 $[0.5Ts \sim (0.5+D)Ts]$

In this mode all switches are turned off and all diodes are turned on, similar to mode 3 input energy is again stored in input inductors, the voltages v_x , v_y , v_z , v_{xy} , v_{yz} , v_{zx} , $v_{L_{ina}}$, $v_{L_{inb}}$, $v_{L_{inc}}$ all follows the same relationships as described in mode 1. The common mode currents are as follows

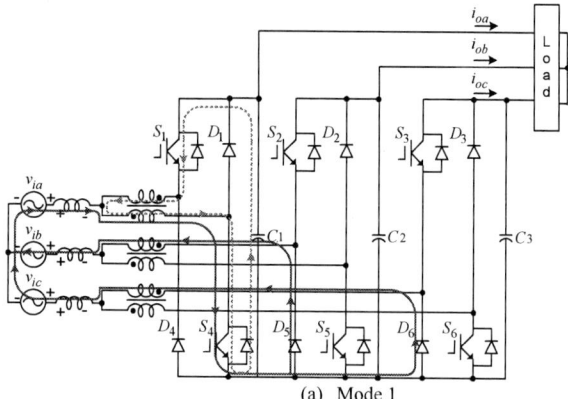

(a)　Mode 1

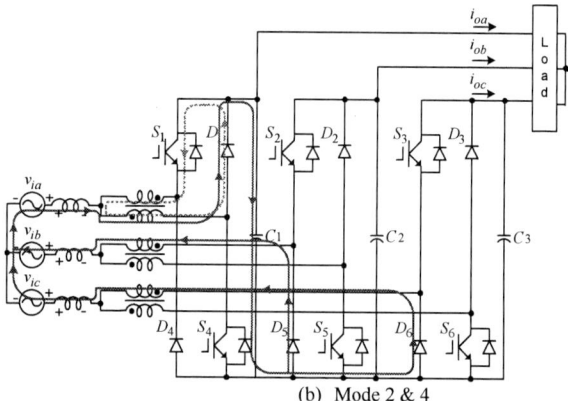

(b)　Mode 2 & 4

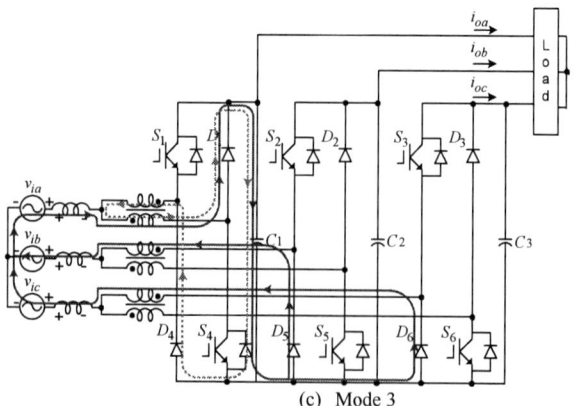

(c)　Mode 3

Fig. 7. Continuous conduction modes of proposed converter

978-1-4799-2706-7/14 $31.00 © 2014 IEEE　　113

$$\frac{d}{dt}\begin{bmatrix} icm_a \\ icm_b \\ icm_c \end{bmatrix} = -\frac{1}{4Ls}\begin{bmatrix} v_{c1} \\ v_{c2} \\ v_{c3} \end{bmatrix} \qquad (17)$$

D. Mode 4 $\left[(0.5+D)Ts \sim Ts\right]$

This mode is same as mode 2
Gain of proposed converter is same as conventional converter and is derived by applying flux balance condition to input inductors

$$D\left(v_{iab} - \frac{v_{ab}}{2}\right) = \left(D - \frac{1}{2}\right)(v_{iab} - v_{ab})$$

$$\frac{v_{ab}}{v_{iab}} = \frac{1}{(1-D)} \qquad (18)$$

D is the duty ratio of proposed three phase ac-ac converter defined in Fig. 8. Theoretical key operational waveforms of converters are shown in Fig. 8 for $D < 0.5$ and can be analyzed for other half portion of duty.

III. THEORETICAL ANALYSIS OF CURRENT RIPPLES

In this section, mathematical analysis of current ripples of the input inductors and coupled inductors is developed. The input inductors in the proposed converter experiences twice the switching frequency so the size of the input inductors can be reduced considerably, the coupled inductors winding currents can be calculated from circulating currents and input inductors currents.

The current ripples of the coupled inductors are calculated using (10) and are at its maximum when $D = 0.5$. The winding current ripples are as

$$\Delta\begin{bmatrix} icm_a \\ icm_b \\ icm_c \end{bmatrix} = \frac{DTs}{4Ls}\begin{bmatrix} v_{c1} \\ v_{c2} \\ v_{c3} \end{bmatrix} \qquad (19)$$

Ts is the time period of proposed converter.

$$\begin{cases} v_{ab} = v_{c1} - v_{c2} = v_{oa} - v_{ob} \\ v_{bc} = v_{c2} - v_{c3} = v_{ob} - v_{oc} \\ v_{ca} = v_{c3} - v_{c1} = v_{oc} - v_{oa} \end{cases} \qquad (20)$$

$$\begin{cases} v_{c1} = v_{oa} + (v_{c2} - v_{ob}) \\ v_{c2} = v_{ob} + (v_{c3} - v_{oc}) \\ v_{c3} = v_{oc} + (v_{c1} - v_{oa}) \end{cases} \qquad (21)$$

The relation between output phase and capacitor voltage is derived as follows

$$\begin{cases} v_{c1} - v_{oa} = v_{oa\max} = v_{ob\max} = v_{oc\max} \\ v_{c2} - v_{ob} = v_{oa\max} = v_{ob\max} = v_{oc\max} \\ v_{c3} - v_{oc} = v_{oa\max} = v_{ob\max} = v_{oc\max} \end{cases} \qquad (22)$$

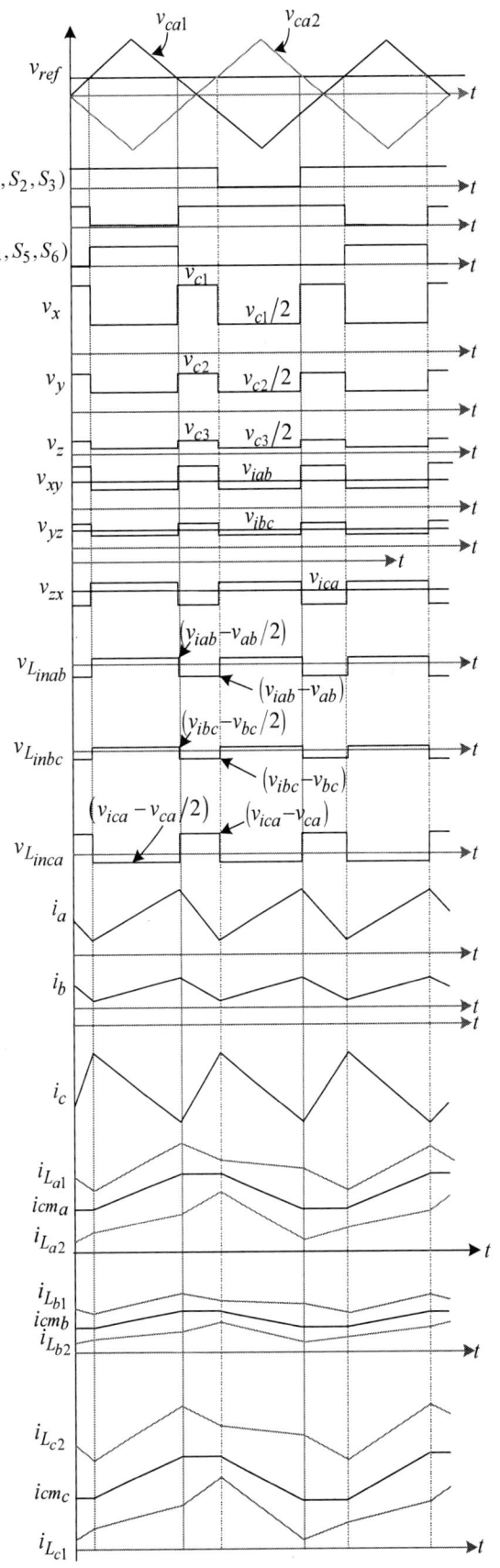

Fig. 8. Theoretical operational waveforms of proposed boost converter.

Where v_{oa}, v_{ob}, v_{oc} are phase voltages and v_{ab}, v_{bc} and v_{ca} are line-line voltages .

$$\begin{cases} v_{c1\max} = 2v_{oa\max} \\ v_{c2\max} = 2v_{ob\max} \\ v_{c3\max} = 2v_{oc\max} \end{cases} \qquad (23)$$

from (19)-(23), the maximum current ripples in (CL_1, CL_2, CL_3) can be obtained and coupled inductors are designed accordingly.

Referring to equations (11) and (18) the current ripples in the input inductors are calculated as shown

$$\Delta i_k = \frac{(0.5 - D)D}{f_s Lin_k} v_{ok} \qquad (24)$$

$k = a, b, c$ for phase leg a, b, c respectively and f_s is switching frequency of proposed converter, maximum current ripples in input inductors occurs when output voltages are maximum and $D = 0.25$, and the D at which maximum current ripples in input inductors occurs is obtained by applying derivative test on (24).

IV. EXPERIMENTAL RESULTS

A boost type 500 W prototype ac-ac converter is built and tested. Table I shows the electrical specifications of the proposed converter.

TABLE I
ELECTRICAL SPECIFICATIONS

Input voltage range	110~220 Vrms/60 Hz
Output voltage	220 Vrms/60 Hz
Output	500 W
Fsw	20 kHz
IGBT	SKM75GB063D
Cin, C1, C2, C3	5 uF
Coupled inductor	240 uH

Intentionally less than 1 us time delay is generated between the switch gate signals to prove the robustness of the proposed converter. Fig. 9 Shows the ideal gate signals and mismatched gate signal. Fig. 10(a) shows the experimental waveforms with the mismatched gate signals shown in Fig. 9(b). Fig. 10(b) shows the zoom-in waveforms of 10(a). Fig. 10(c) shows current waveforms of the input, output, and coupled inductors.

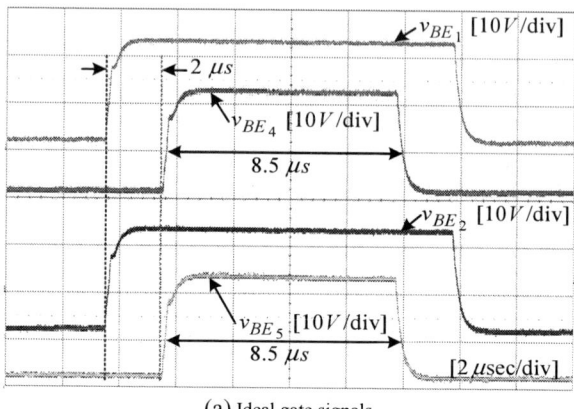

(a) Ideal gate signals

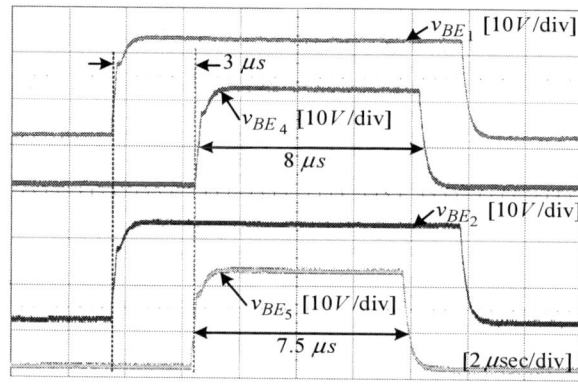

(b) Mismatched gate signal
Fig. 9. Gate signals

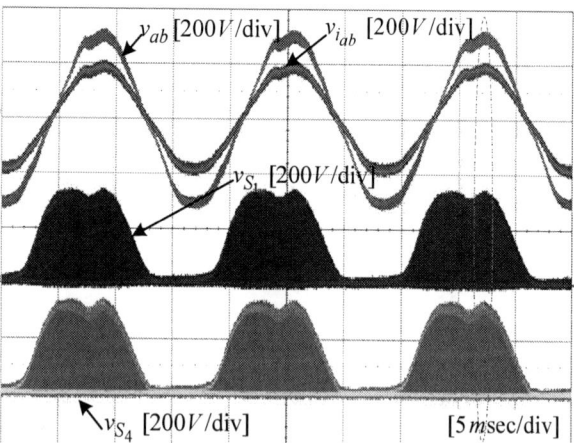

10. (a) Input, output, and switch voltages

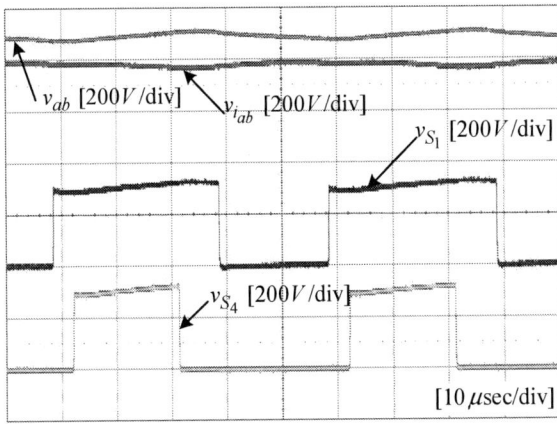

(b) Zoom-in waveforms of 10. (a)

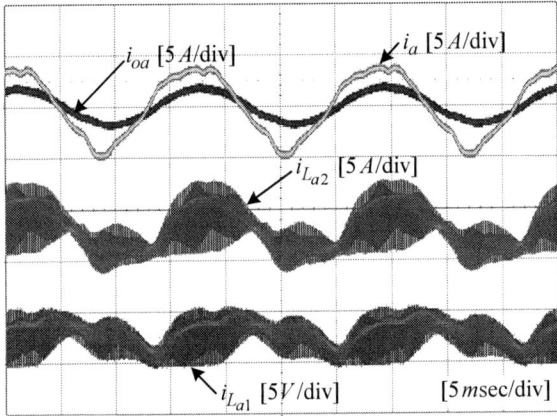

(c) Input, output and coupled inductor current

Fig. 10. Experimental waveforms with slight time differences in the gate signals

V. CONCLUSIONS

In this paper, three phase ac-ac converters solving commutation problems were described. The proposed converters has four continuous conduction modes and three unipolar voltage levels. The proposed converters can be short and open circuited without generating high voltage and current surge. Unlike the conventional ac-ac converters the proposed converters do not need to sense input voltage polarity therefore proposed converters can be operated even with highly distorted input. The size requirements of input inductors can be reduced considerably with the proposed converters.

ACKNOWLEDGMENT

This research was supported by Basic Science Research Program through the National Research Foundation of Korea (NRF) funded by Ministry of Education, Sciences and Technology (2011-0029721).

REFERENCES

[1] P. Fang Zheng, C. Lihua, and Z. Fan, "Simple topologies of PWM AC-AC converters," *Power Electronics Letters, IEEE,* vol. 1, no. 1, pp. 10-13, Mar. 2003.

[2] B. H. Kwon, B. D. Min, and J. H. Kim, "Novel topologies of AC choppers," *Electric Power* Applications, *IEEE Proceedings -,* vol. 143, no. 4, pp. 323-330, July 1996.

[3] T. Yu, X. Shaojun, and Z. Chaohua, "Z-Source AC-AC Converters Solving Commutation Problem," *Power Electronics, IEEE Transactions on,* vol. 22, no. 6, pp. 2146-2154, Nov. 2007.

[4] L. M. Tolbert, P. Fang Zheng, F. H. Khan *et al.,* "Switching cells and their implications for power electronic circuits," Proc. pp. 773-779.

[5] J. Salmon, A. Knight, J. Ewanchuk *et al.,* "Multi-level single phase boost rectifiers using coupled inductors," Proc. pp. 3156-3163.

[6] F. H. Khan, L. M. Tolbert, and F. Z. Peng, "Deriving New Topologies of DC-DC Converters Featuring Basic Switching Cells," Proc. pp. 328-3

[7] Hyunhak Shin, Honnyong Cha, Heung-Geun Kim, "A Novel Single-Phase PWM AC-AC Converters without having Commutation Problem," in *Energy Conversion Congress and Exposition (ECCE),* 2013 , pp. 2355-2362.

[8] C. Chapelsky, J. Salmon, and A. M. Knight, "High-Quality Single Phase Power Conversion by Reconsidering the Magnetic Components in the Output Stage-Building a Better Half Bridge," *Industry Applications, IEEE Transactions on,* vol. 45, no. 6, pp.2048-2055, Nov./Dec. 2009.

[9] L. Chen and F. Zheng Peng, "Dead-time elimination for voltage source inverters," *IEEE Trans. Power Electron.,* vol. 23, no. 2, pp. 574–580, Mar. 2008.

[10] B. A. Welchko, S. E. Schulz, and S. Hiti, "Effects and compensation of dead-time and minimum pulse-width limitations in two-level PWM voltage source inverters," in *Conf. Rec. IEEE IAS Annu. Meeting,* 2006, pp. 889–896.

[11] J.Salmon, J.Ewanchuk, and A.M.Knight, "PWM Inverters Usisng Split-Wound Coupled Inductors," Industry Applications, IEEE Transactions on, vol. 45, no. 6, pp. 2001-2009, 2009

[12] A. Knight, J. Ewanchuk, and J. Salmon, "Coupled three-phase inductors for interleaved inverter switching," *IEEE Trans. Magn.,* vol. 44, no. 11, pp. 4119–4122, Nov. 2008.

Experimental Investigation of Normally-On Type Bidirectional Switch for Indirect Matrix Converters

Kyungmin Sung and Ryuji Iijima
Depart. of Electrical and Electronic System Engineering
Ibaraki National College of Technology
Hitachinaka, Ibaraki, Japan

Shinichi Nishizawa, Isami Norigoe, and
Hiromichi Ohashi
National Institute of Advanced Industrial
Science and Technology
Tsukuba, Ibaraki, Japan

Abstract— In this paper, the novel normally-on type bidirectional switch, in which is comprised of SiC-JFET, SiC-SBD, and Si-IGBT is proposed for one of a protection method of the Matrix Converter (MC). When the MC becomes gate block situation, a diode clamp circuit or auxiliary circuits keeps inductive load current loop in a conventional MC utilized a bidirectional switch. We focus that the normally-on type SiC-JFET becomes turn-on state, when zero gate bias voltage and a SiC-devices have a good tolerance capability for short current than silicon devices.

These characteristic of normally-on type SiC-JFET is used to replace diode clamp circuit in MC driver system. The experimentation based on indirect MC induction motor driver system was carried out. The experimental result of IM driver shows that the proposed bidirectional switch can overcome a generated inductive load current by IM. Finally, in order to design heat sink, the power loss of each devices of proposed switch was estimated by experimental results.

Keywords— *Matrix Converters, Bidirectional switch, SiC-JFET, Induction Motor.*

I. INTRODUCTION

When an emergency situation as power supply become cut off suddenly or gate block, a conventional Matrix Converters (MC) must have a diode clamp circuit or an auxiliary bypass circuit using bidirectional switch for insure freewheeling load current loop to prevent surge voltage from opening circuit in inductive load. Many protection methods for MC under fault condition in which is a power device as MOSFET and IGBTs become open or short circuit suddenly are proposed.[1-8] Normally, the diode clamp circuit consists of a three-phase diode rectifier and a capacitor with resistors is the simplest and reliable method for protection opening circuit situation of MC. [1,2]

High speed fuse and circuit breaker are effective for protect power devices from over current by short circuit in the main circuit of MC or load side. Also, high speed control and gate block signal interruption by microprocessor can achieve a good protection ability in general commercial inverter drive system recently. As well as in the MC drive application, these over current protection methods are available. However, in case of

protecting of surge voltage by opening load current is completely distinct between MC and inverter system. MC cannot interrupt gate block signal without any auxiliary circuit as voltage clamp circuit because it has not any passive elements. Therefore, A circuit of maintain load current loop is necessarily for unexpected situation as a gate block. The other protection method also had been reported in instead of the diode clamp circuit.[3-8] The common method in these reports has a bidirectional switch like a thyristor and an auxiliary current pass beside main circuit on condition that the power for gate driver and controller have to be supplied. Because the voltage source inverter has freewheeling diode and a capacitor of DC link, the load current loop is not a problem when the sudden gate block or a halting of main power devices. But, adopting diode clamp circuit has many disadvantages in terms of a converter volume, lifetime, and maintenance in application field.

On the other side, the normally-on type SiC-JFET (Silicon Carbide – Junction Field Effect Transistor) and SiC-SIT (Static Induction Transistor) is reported its good switching performance and tolerance against a short current. [9] The normally-on means that these devices become turn-on state under near zero gate voltage or zero gate bias voltage.

In this paper, we propose a novel protection method to keep inductive load current loop using a normally-on type bidirectional switch. The proposed method does not require any additional circuits for protection of opening circuit. Only, two switching devices are changed by normally-on type SiC-JFET instead of normally-off device. When the gate block arises, the SiC-JFET and diode maintains a one side current loop as a freewheeling diode in voltage source inverter. In this time, the switching combination of MC becomes zero voltage vector state. Three output phases are connected to one input phase. The load current is dissipated by internal resistance between Induction Motor (IM) and the main circuit of MC.

The proposed method is verified by experimental results of IM drive system. In addition, the power loss of devices in which consisted of proposed switch in MC for design cooling system is calculated. The final purpose of

The 2014 International Power Electronics Conference

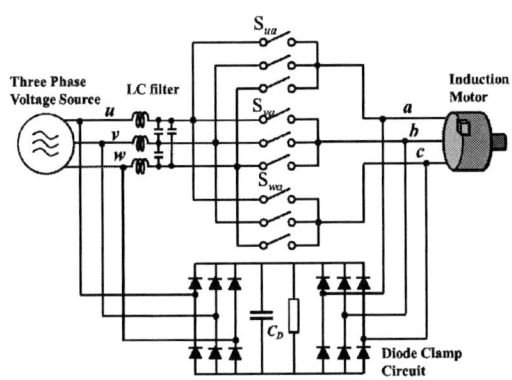

Fig. 1. Direct matrix converter.

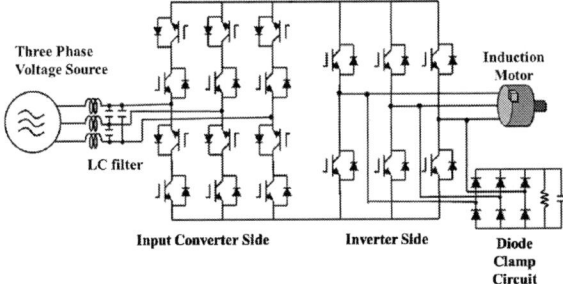

Fig. 2. Indirect matrix converter.

this study is to investigate the proposed normally-on type switch can replace the diode clamp circuit from the field of MC drive system.

II.   PROPOSED METHOD

A. Matrix Converters

The MC has two circuit topologies. One is direct conversion and the other is indirect conversion. Fig. 1 and 2 shows two type MC. One is direct MC, the other is indirect MC. The main circuit of direct MC is composed of nine bidirectional switch, in which used two switching device and two diode as depicted in Fig. 3 (a).

On the other hand, the indirect MC is composed of three-phase input bidirectional PWM converter and three-phase inverter. There is no capacitor of DC link, but DC link part exists in this circuit. The output inverter circuit is same a conventional inverter system. The converter side of indirect MC utilize bidirectional switch like direct MC. Also, each bidirectional switch is consisted of connected two normally-off type IGBTs or MOSFETs and two diodes. The reason of the surge voltage problem by open current, both MC have to need diode clamp circuit. About the diode clamp circuit and capacitor value had been reported in [1,2]

B. Proposed bidirectional Switch

The proposed switch is comprised of one IGBT with freewheeling diode and one normally-on type SiC-JFET with SiC-SBD. The Fig. 3(b) shows the proposed

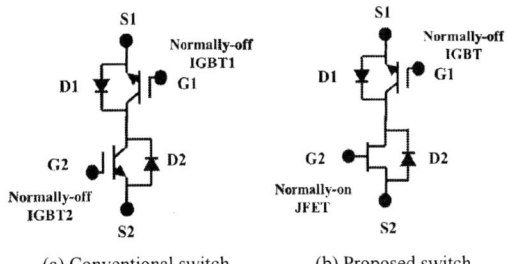

(a) Conventional switch   (b) Proposed switch
Fig. 3. Bidirectional switches.

TABLE I
CONDUCTING MODE OF NORMALLY-ON TYPE
BIDIRECTIONAL SWITCH

G1	G2	Voltage difference	Current flow	
			Normally-off type AC switch	Normally-on type AC switch
L	H	S1>S2	S1-D1-IGBT2 -S2	S1-D1-JFET -S2
L	H	S1<S2	S2-D2-IGBT1 Blocking	S2-D2-IGBT Blocking
H	H	S1>S2	Bidirectional conduction	
H	H	S1<S2		
L	L	S1>S2	Bidirectional blocking	
L	L	S1<S2		
H	L	S1>S2	S1-D1-IGBT2 Blocking	S1-D1-JFET Blocking
H	L	S1<S2	S2-D2-IGBT1 -S1	S2-D2-IGBT -S1
0	0	S1>S2	Bidirectional Blocking	S1-D1-JFET -S2
0	0	S1<S2	Bidirectional Blocking	S2-D2-IGBT Blocking

bidirectional switch and Table. I shows a conducting mode by gate voltage level.

G1 and G2 mean that gate bias state of switching devices. L is off gate bias state and H is on state respectively. In case of both G1 and G2 becomes H level, the current flow is determined by bias voltage between S1 and S2. The zero in Table 1 means zero gate voltage. It is not floating voltage. Normally, normally-off type power switching devices as IGBTs and a MOSFETs become turn-off state, when the gate-emitter voltage of IGBT is under gate threshold voltage. A gate driver circuit output zero or negative bias voltage for turn-off.

On the other hand, when the gate-source voltage of SiC-JFET is near to zero or zero level, the SiC-JFET becomes a turn-on state. So that, the proposed switch conduct only one-way through from the parallel diode of IGBT to the SiC-JFET when the G1 and G2 become zero.

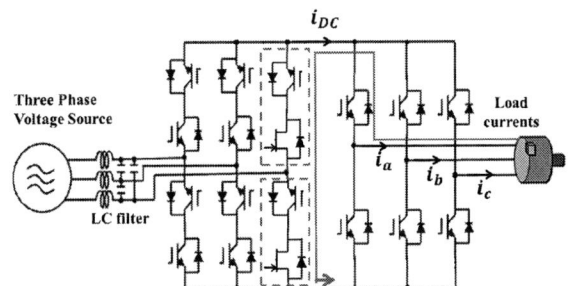

Fig. 4. Bidirectional switches.

978-1-4799-2706-7/14 $31.00 © 2014 IEEE   118

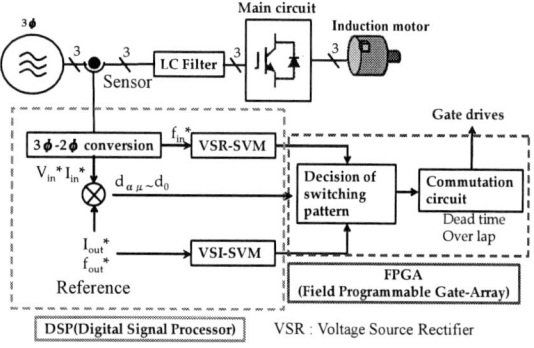

Fig. 5. Bidirectional switches.

Fig. 6. Photo of indirect MC.

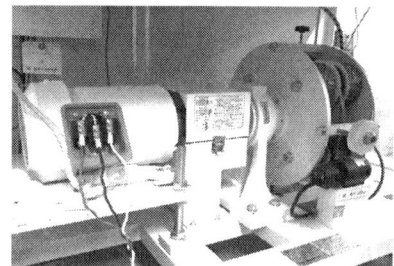

Fig. 7. Induction motor and eddy-current braking system.

TABLE II
SPECIFICATION OF POWER DEVICES

Si-IGBT	V_{CE}:1200V
	I_C:20A
SiC-JFET SemiSouth SJDP120R085	V_{DS}:1200V
	I_D:27A
SiC-SBD Infineon IDV06S60C	V_{DC}:600V
	I_F:6A
Switching frequency	12kHz

C. *Proposed Indirect MC*

In this paper, we firstly confirm the validity of the proposed switch to apply indirect MC. But, because these two conversion method have same conversion principle, then the proposed method can apply both MC. The proposed method is that only one leg of three-phase of input converter side has two normally-on type proposed bidirectional switches, which is consisted by one IGBT with diode and one normally-on SiC-JFET with diode. The output load side is three-phase inverter circuit. The Fig. 4 depicts conducting of load current when the gate block arises.

In this situation, the load motor becomes generation mode. The three-phase current by electromotive force of motor circulate through the one leg of normally-on type switches of converter side, freewheeling diode of inverter and a winding of IM.

The internal resistance of diodes, SiC-JFETs, and motor is considered not zero value. Therefore, the generated electric power of motor is consumed in these internal resistances as heat of joule energy. The enough cooling ability for withstand the joule energy must be taken. Also, according to [9], the short-circuit capability of a SiC-devices has over three times tolerance against short-current than a Si-devices. It means that the proposed AC switch using SiC-SBD and SiC-JFET with effective heat sink can overcome the generated load current of IM.

III. HARDWARE IMPLEMENTATION

In this paper, the Space Vector Modulation (SVM) is used for control method of proposed indirect MC. [12] Commonly, SVM method is implemented easily by a Digital Signal Processor (DSP) and Field-programmable Gate Array (FPGA). The Fig. 5 shows the configuration of the implemented control circuit in this paper. Firstly, the DSP receives the input of a three-phase voltage and a current signal data from each sensor. Next, these signal are converted a 3-phase to a 2-phase and decides the space vector of input voltage, next send to the FPGA. The FPGA decides switching patterns from the received data of the DSP and calculates an overlap and generates a dead time sequentially. Finally, commutation signals of each switch are outputted a gate-driving circuit.

The photo of prototype indirect MC is shown in Fig. 6. The input converter and output inverter circuit is implemented by one printed circuit board. An electrolytic capacitor for DC link is cancelled. Power devices specification of the proposed indirect MC are listed in Table II.

In order to confirm the validity of the proposed bidirectional switch, we carry out experiment by 1.5kW rating, 4-pole three-phase IM with indirect MC. A 6.3kg weighted copper disk is connected to shaft of IM as a load. The inertia moment of this disk is approximately 0.7kgm^2. Because of using this heavy disk, if the indirect MC be stopped operating at a certain moment, the IM will be rotated for a while. Also, the eddy-current brake system is used to deliver rotating torque to IM. These IM and braking system is shown in Fig. 7.

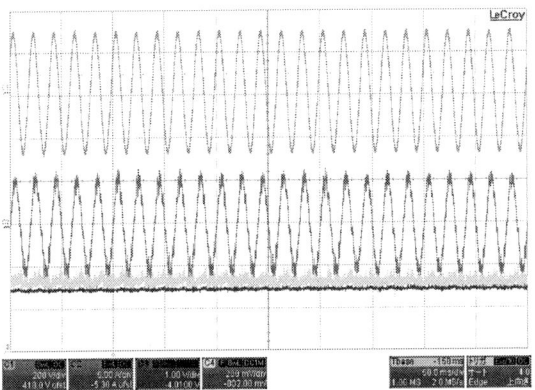

(a) Input voltage, input current, speed and torque signal waveforms.

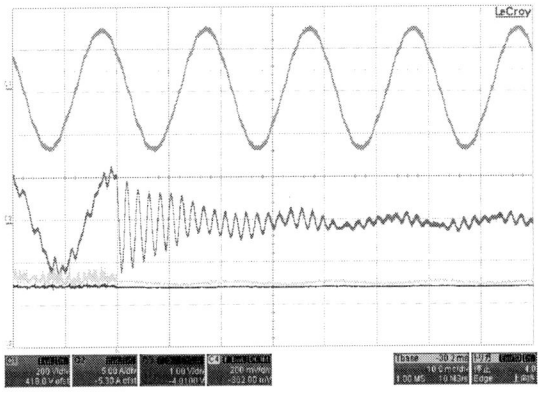

(a) Input voltage, input current, speed and torque signal waveforms.

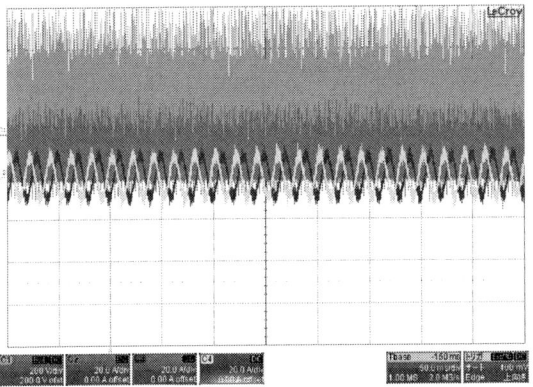

(b) DC link voltage, DC link current, and load currents of inverter side.
Fig. 8. Waveforms of input converter and output inverter side in normal operation state.

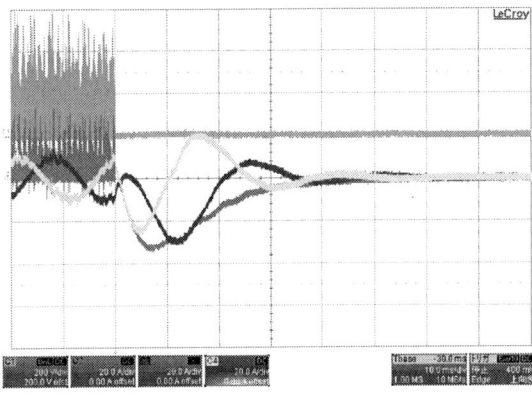

(b) DC link voltage, DC link current, and load currents of inverter side.
Fig. 9. Waveforms of input converter and output inverter side when the gate block arises.

IV.   EXPERIMENTAL RESULTS

In experimentation, firstly, the proposed bidirectional switch is comprised of two IGBT modules that is a 1200V and 40A rating. One is operated normally-off type and the other is operated by normally-on in instead of normally-on type SiC-JFET because of its low current conduct rating.   The normally-on type IGBT become turn-on state when the gate block arises.

The waveforms of input line-to-line voltage, input current, rotating speed signal, and torque signal in normal driver condition are shown in Fig. 8 (a). The input line-to-line voltage is between the terminals of condenser of input LC filter. Also, the load side waveforms of DC link voltage, current, and output two phase currents are shown in Fig. 8 (b). In this case, supply input voltage is 200Vrms and the MC convert from 50Hz input to 50Hz output. The load condition is that output torque is 3.05Nm and rotating speed is 1405rpm. Output power of IM is approximately 500Watt.

When the gate block is generated in suddenly, each waveforms of input converter side and output inverter side is shown in Fig. 9 (a) and (b) simultaneously

In this case, gate signals of all switching devices become zero voltage at a certain moment. It means that the three phase current of IM circulate freewheeling diode of inverter and DC link through the proposed switches and to return to the IM. Three phase output currents are changed opposite direction at this moment. A sum of output currents is equal to DC link current, in which is the conduct current of the leg composed of the proposed switches. And then, the DC link current is changed from positive to negative direction at gate block time. As this result, the proposed switch can conduct an inductive load current at gate block.

Fig. 10 shows peak current waveforms of DC link in each load torque. In the zero load torque, the maximum peak current is generated in Fig.10 (a). These peak currents are decreased in proportion to weight of load torque. The value of peak current by weight of load torque and each rms terminal voltage of IM are shown in Fig. 11. According to these results, the peak current value is decreased in proportion to the weight of load torque and the terminal voltage. It means that when the lightest load condition, the generated current by IM becomes the maximum value.

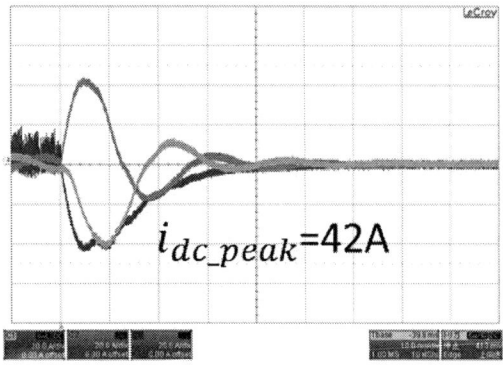

(a) When the load torque is approximately zero value.

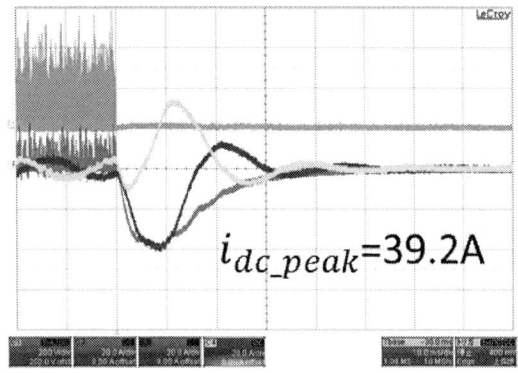

(b) When the load torque is 1.0Nm.

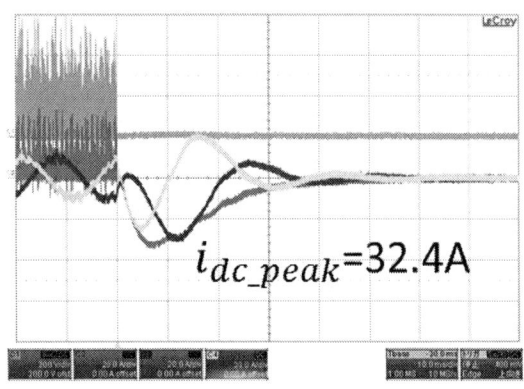

(c) When the load torque is 3.05Nm.
Fig. 10. Waveforms of load current and DC link current when the gate block arises.

In Fig. 12, the conducting time of circulated current by the weight of load torque and the terminal voltage of IM. These conducting times are form when the load current is changed to opposite direction to when the DC link current becomes same value which is before gate block. These conducting times show almost same period in any load conditions and terminal voltages.

Finally, we calculate the joule energy by DC link current that is a sum of load current by measured waveforms and conducting time of devices. Table III shows the calculated results. In this case, a loss of the diode and SiC-JFET is calculated by the measured

current value of the Si-IGBT and on-resistance value of a datasheet.

V. CONCLUSION

The novel normally-on type bidirectional switch which consisted of SiC-JFET and with Si-IGBT has been presented to eliminate the diode clamp circuit or other protection circuit in indirect MC. The experimentation of IM driver system used proposed indirect MC was carried out to confirm the validity of the proposed protection method. According to the experimental result, the proposed switch shows that it can replace diode clamp circuit from MC. Also, the power loss of the proposed switch is investigated on deferent load conditions and terminal voltages.

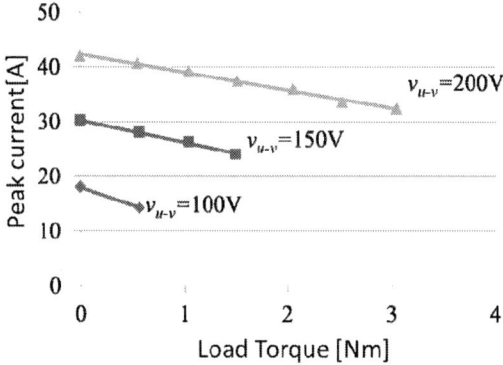

Fig. 11. Peak current by weight of load torque and terminal voltages of IM.

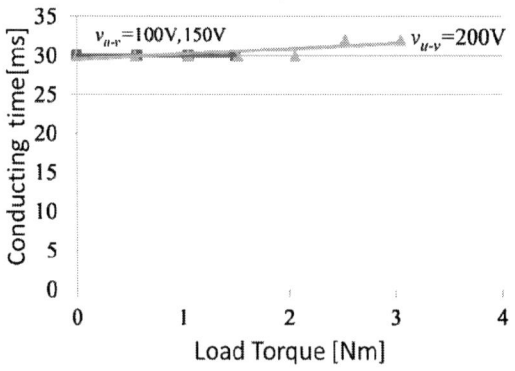

Fig. 12. Conducting time by weight of load torque and terminal voltages of IM.

TABLE III
POWER LOSS OF EACH DEVICE

	Normally-off type AC switch Si-IGBT+Si-IGBT		Normally-on type AC switch Si-IGBT+SiC-JFET	
	W[J]	P_{max}[W]	W[J]	P_{max}[W]
IGBT V_{CE}=1.75[V] 125°C	1.1	73.5		
Diode V_F=2.2[V] 125°C	1.4	92.4	1.4	92.4
SiC-JFET R_{DS}=0.16[O] 125°C			4.2	282.2

REFERENCES

[1] Jochen Mahlein, Manfred Bruckmann, and Michael Braun, "Passive Protection Strategy for a Drive System With a Matrix Converter and an Induction Machine", *IEEE Trans. Ind Electron.* vol. 49, no. 2, pp. 297-303, Apr. 2002.

[2] C. Klimpner and F. Blaabjerg, "Experimental Evaluation of Ride-through Capabilities for a Matrix Converter under Short Power Interruptions," *IEEE Trans. Ind Electron.* vol. 49, no. 2, pp. 315-324, Apr. 2002.

[3] Sudarat Khwan-on, Liliana de Lillo, Lee Empringham, and Pat Wheeler, " Fault-Tolerant Matrix Converter Motor Drives With Fault Detection of Open Switch Faults", *IEEE Trans. Ind Electron.* vol.59, no. 1, pp. 257-268, Jan. 2012.

[4] H. J. Cha and Prasad N. Enjeti, "A New Ride-Through Approach for Matrix Converter Fed Adjustable Speed Drives", Industry Applications Conference, 2002. 37th IAS Annual Meeting. vol. 4, pp.2555-2560.

[5] B. W. Augdahl, H. L. Hess, and B. K. Johnson, "Output Protection Strategies for Matrix Converters in Distributed Generation Applications", *Industry Applications Conference, 2006. 41st IAS Annual Meeting*, vol.4, pp.2082-2089

[6] Christian Klumpner, "An Indirect Matrix Converter with a Cost Effective Protection and Control", *Power Electronics and Applications, 2005 European Conference*, pages:11, 2005.

[7] Lina Wang, Fuyuan Xu, Kai Sun, and Lipei Huang, "A Novel Safe Shutdown Strategy for Matrix Converter Even Under Fault Condition", *APEC* 2005. Twentieth Annual IEEE, vol:3, pp.1786-1790

[8] Jon Andreu, Jose Miguel De Diege, Inigo Martinez de Alwgria, Inigo Kortabarria, Jose Luis Martin, and Salvador Ceballos, "New Protection Circuit for High-Speed Switching and Start-Up of a Practical Matrix Converer", *IEEE Trans. Ind Electron.* vol. 55, no. 8, pp. 3100-3114, Aug. 2008.

[9] K. Yano, Y. Tanaka, T. Yatuno, and K. Arai, "Short-Circuit Capability of SiC-Buried-Gate Static Induction Transistors: Basic Mechanism and Impacts of Channel Width on Short-Circuit Performance," *IEEE Trans. On Electron Devices*, vol. 57, no. 4, pp.919-927, Apr. 2010.

[10] Jun-Ichi Itoh and Ken-Ichi Nagayoshi, "A New Bidirectional Switch With Regenerative Snubber to Realize a Simple Series Connection for Matrix Converters", *IEEE Trans. PE*, vol. 24, no. 3, Mar. 2009, pp.822-829

[11] M. Jussila, M. Eskola, and H. Tuusa, "Analysis of non-idealities in direct and indirect matrix converters", *Power Electronics and Applications, 2005 European Conference*, pages:10, 2005.

[12] L. Huber and D. Borojevic, "Space vector modulated three-phase to three-phase matrix converter with input power factor correction," *IEEE Trans. Ind. Appl.*, vol.31, no.6, pp1234-1246, Nov./Dec. 1995.

[13] J. Kim and S. Sul, "New control scheme for AC-DC-AC converter without DC link Electrolytic capacitor", *Power Electronics Specialists conf.*1993, pp.300-306.

[14] A. Ecklebe, A. Lindemann, and S. Schulz, " Bidirectional Switch Commutation for a Matrix Converter Supplying a Series Resonant Load", *IEEE Trans. PE*, vol. 24, no. 5, pp.1173-1181, May 2009.

[15] Iman Lorzadeh, Ebrahim Farjah, and Omid Lorzadeh, "Fault-Tolerant Matrix Converter Topologies and Switching Function Algorithms for AC Motor Drives with Delta Connection Windings", *Power Electronics Electrical Drives, Automation and Motion(SPEEDAM)*, pp.1651-1657, 2010

Visualization of PWM Waveforms of Output Voltage and Input Current for a Direct Matrix Converter

Inami Asai, and Takaharu Takeshita

Nagoya Institute of Technology, Nagoya, Japan

Email: 23517501@stn.nitech.ac.jp, take@nitech.ac.jp

Abstract—The explanations of the PWM strategies of a direct matrix converter are classified into "Direct AC/AC converter", "Virtual AC/DC/AC Conversion" and "Instantaneous Space Vector Diagram". Since the PWM strategies of matrix converters are complex, it may be difficult to understand the same PWM strategies under the different explanations. This paper presents the visualization of the PWM waveforms of the output voltage and the input current during the input and output voltage periods. From the visualized waveforms of the output voltage and the input current, the difference and evaluation among several PWM strategies can be easily realized. For the two PWM strategies which are derived from the fundamental equations of the duty cycles of a matrix converter, the visualized waveforms of the output voltage and the input current are derived. The proposed visualization can be extended to the transient state under the change of the output voltage reference. The effectiveness of the visualization of PWM strategy has been verified by comparison between the visualized and experimental waveforms.

Keywords—*matrix converter, PWM strategy, visualization, duty cycles*

I. INTRODUCTION

A matrix converter can directly convert from three-phase AC voltage to three-phase AC voltage with arbitrary amplitude and frequency. In comparison with a rectifier-inverter system, the matrix converter can achieve high efficiency, high power density, and high reliability. The output voltages and input currents of the matrix converter can be simultaneously controlled by nine bi-directional switches, Because 27 switching patterns exist, various PWM strategies have been proposed. The explanations of the PWM strategy for the matrix converter is classified into "Direct AC/AC Conversion [1]-[5]", "Virtual AC/DC/AC Conversion [6]", and "Instantaneous Space Vector Diagram [7]". But, the explanations are usually based on the duration of the control period when the state variables are treated as constant values. Therefore, it is difficult to evaluate and compare among the different PWM strategies based on the visualized waveforms of the output voltage and the input current.

As a quantitative evaluation of the PWM waveforms during one control period for the matrix converter, the instantaneous effective value of the input current and the output voltage has been proposed [8]. Also, the visible evaluation using the space vector diagram for

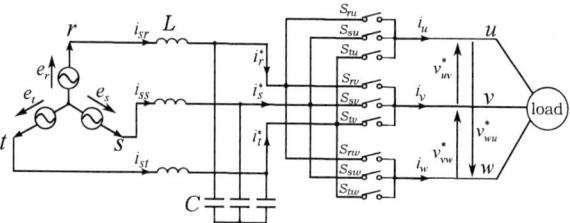

Fig. 1. Matrix converter

several PWM strategies has been proposed [9]. However, the visualization of the waveforms during the input and output voltage periods cannot be realized, because these evaluations are based on the duration of the control period.

This paper presents the visualization of the PWM waveforms of the output voltage and the input current during the input and output voltage periods. The fundamental relations of the duty cycles are derived from the input current references and the output voltage references. Because the fundamental relations are treated as continuous time-varying values, the PWM strategy can be easily extended to the input and output voltage periods. In this paper, the visualization for two PWM strategies for reducing number of commutations [2], [3] under low carrier frequency is realized. The output voltage and input current waveforms during the input and output voltage periods for two PWM strategies are compared. It is possible to visually evaluate the harmonics characteristics of the output voltage and input current. The proposed visualization can be extended to the transient state under the change of the output voltage reference. For the practical use, the relations of the behaviors between the visible waveforms under the low carrier frequency and the experimental waveforms under actual carrier frequency are clarified.

II. VISUALIZATION OF PWM WAVEFORMS

A. Model of Matrix Converter

Fig.1 shows a configuration of a three-phase to three-phase matrix converter. The matrix converter has nine bi-directional switches S_{ru}-S_{tw}. The input side is connected to a three-phase power supply with phase voltages e_r, e_s, and e_t through an LC filter, and the output side is

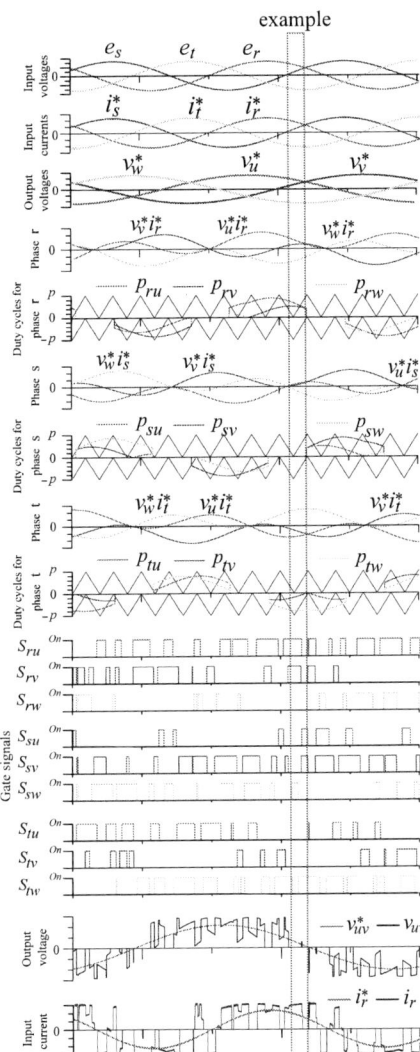

Fig. 2. Visualized PWM strategy under unity input power factor (Method 1)

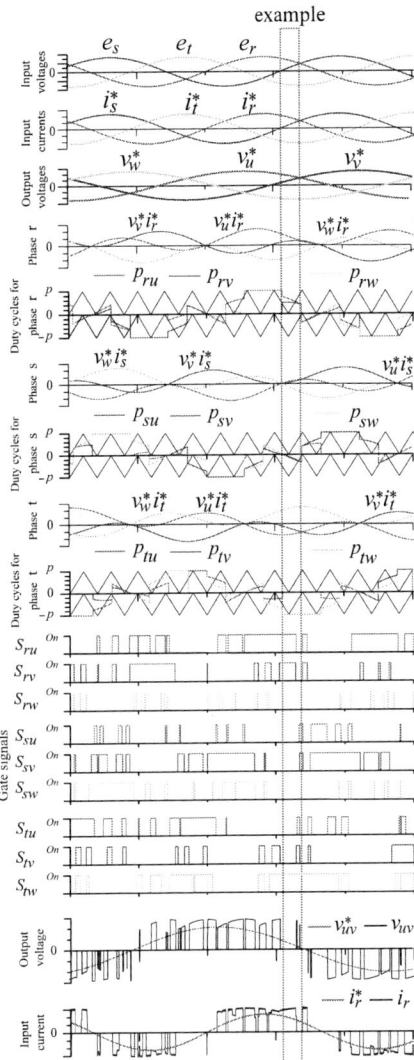

Fig. 3. Visualized PWM strategy under unity input power factor (Method 2)

connected to a three-phase load. All duty cycles of the bidirectional switches are decided to simultaneously realize the output line voltages v_{uv}, v_{vw}, v_{wu} and input currents i_r, i_s, i_t corresponding to the reference values.

B. Visualization in Steady State

Fig.2 and Fig.3 show the proposed visualization of the PWM waveforms of the output voltage and the input current during the input and output voltage periods. These two PWM strategies realize the output voltage and input current references by only four commutations during the control period. The effective value of the output voltage reference is 0.75 times of the effective value of the input voltage and the input power factor is unity. In Method 1 in Fig.2, the output voltage waveform v_{uv} consists of the input voltage pulses with the small voltage difference from the output voltage reference. In Method 2 in Fig.3, one of the output phase switches is conducted during the whole control period, and the switches in the other

two phases are under PWM control. These two PWM strategies realize the output voltage and input current references by only four commutations during the control period.

In order to generate the duty cycles, the instantaneous powers p_{ru}-p_{tw} through the switches, that are obtained by the product of the input current references and the output voltage references, are calculated according to the input phase angle. The switches without instantaneous power in Fig.2 and Fig.3 are "the disconnected switch" or "the connection determined by the other switches on same output phase". The gate signals of switching devices S_{ru}-S_{tw} are obtained from the comparison between the triangular carriers and the instantaneous powers p_{ru}-p_{tw}.

Since the behavior waveforms of the output voltage, the input current and the comparisons between the triangular carriers and the instantaneous powers are drawn in one figure, the features and the harmonic characteristics for each control method are visually understood. For example, the following is visually understood. The output voltage waveform in Method 1 has short duration of zero voltage and consists of the voltage pulses with the small voltage difference from the output voltage reference compared with the output voltage waveform in Method 2.

In the part of example in Fig.2 and Fig.3, the relations of the output voltages $v_u^* \geq v_v^* \geq v_w^*$ and the input current references $i_r^* \geq i_s^* \geq 0 \geq i_t^*$ are obtained. The duty cycles in the example are obtained in (1) and (2), respectively.

$$
\left.
\begin{aligned}
d_{ru} &= \frac{v_{uw}^*}{p}i_r^*, \quad d_{rv} = \frac{v_{vw}^*}{p}i_r^* \\
d_{rw} &= 0, \quad -d_{tu} = 0 \\
-d_{tv} &= \frac{v_{uv}^*}{p}i_t^*, \quad -d_{tw} = \frac{v_{uw}^*}{p}i_t^* \\
d_{su} &= 1 - d_{ru} - d_{tu} \\
d_{sv} &= 1 - d_{rv} - d_{tv} \\
d_{sw} &= 1 - d_{rw} - d_{tw}
\end{aligned}
\right\}
\tag{1}
$$

$$
\left.
\begin{aligned}
d_{ru} &= \frac{v_{uw}^*}{p}i_r^*, \quad d_{rv} = \frac{v_{vw}^*}{p}i_r^* \\
d_{rw} &= 0, \quad -d_{su} = \frac{v_{wu}^*}{p}i_s^* \\
-d_{sv} &= \frac{v_{uv}^*}{p}i_s^*, \quad -d_{sw} = 0 \\
d_{tu} &= 1 - d_{ru} - d_{su} \\
d_{tv} &= 1 - d_{rv} - d_{sv} \\
d_{tw} &= 1
\end{aligned}
\right\}
\tag{2}
$$

where, p is the instantaneous input power as follows

$$
p = e_r i_r^* + e_s i_s^* + e_t i_t^*
\tag{3}
$$

The derivation of the duty cycles are explained on the next section.

C. Derivation of Duty Cycles

1) Definition of voltage and current: The input phase voltages e_r, e_s, and e_t are expressed by (4) using the effective value of input line voltage E and the angular frequency ω.

$$
\begin{bmatrix} e_r \\ e_s \\ e_t \end{bmatrix} = \sqrt{\frac{2}{3}} E \begin{bmatrix} \cos\theta \\ \cos(\theta - 2\pi/3) \\ \cos(\theta + 2\pi/3) \end{bmatrix}
\tag{4}
$$

$$
\theta = \omega t
\tag{5}
$$

The input current references i_r^*, i_s^*, and i_t^* (* indicates a reference) are expressed by (6) using the effective value I^* and the power factor angle φ^* of the input current reference.

$$
\begin{bmatrix} i_r^* \\ i_s^* \\ i_t^* \end{bmatrix} = \sqrt{2}I^* \begin{bmatrix} \cos(\theta + \varphi^*) \\ \cos(\theta + \varphi^* - 2\pi/3) \\ \cos(\theta + \varphi^* + 2\pi/3) \end{bmatrix}
\tag{6}
$$

In Fig.2 and Fig.3, in order to be unity power factor, the input power factor angle reference $\varphi^* = 0$. And, because the input current effective value I^* is decided by the balance between instantaneous input and output

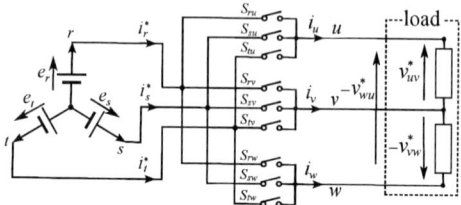

Fig. 4. Analytical model of matrix converter

power, it is possible to give $I^*=1$ for deriving the duty cycles.

The output voltage reference v_u^*, v_v^* and v_w^* are expressed by using the effective value of output line voltage V_L^*, the output phase θ_L^* and the output angular frequency ω_L^*.

$$
\begin{bmatrix} v_u^* \\ v_v^* \\ v_w^* \end{bmatrix} = \sqrt{\frac{2}{3}}V_L^* \begin{bmatrix} \cos\theta_L^* \\ \cos(\theta_L^* - 2\pi/3) \\ \cos(\theta_L^* + 2\pi/3) \end{bmatrix}
\tag{7}
$$

$$
\theta_L^* = \omega_L^* t
\tag{8}
$$

2) Relations among duty cycles: Fig.4 shows the analytical model without the input LC filter, for deriving the duty cycles during control period. Assuming that the control period T_s is short enough compared to the time constant of the circuit, the input voltages are treated as constant values during the control period. The duty cycles have the following relations between the continuity of the load current and prevention of a short circuit in the input line voltage.

$$
\left.
\begin{aligned}
d_{ru} + d_{su} + d_{tu} &= 1 \\
d_{rv} + d_{sv} + d_{tv} &= 1 \\
d_{rw} + d_{sw} + d_{tw} &= 1
\end{aligned}
\right\}
\tag{9}
$$

The input current reference i_r^*, i_s^*, and i_t^* are obtained by the product of the duty cycles and the output currents.

$$
\left.
\begin{aligned}
i_r^* &= d_{ru}i_u + d_{rv}i_v + d_{rw}i_w = (d_{ru}-d_{rv})i_u - (d_{rv}-d_{rw})i_w \\
i_s^* &= d_{su}i_u + d_{sv}i_v + d_{sw}i_w = (d_{su}-d_{sv})i_u - (d_{sv}-d_{sw})i_w \\
i_t^* &= d_{tu}i_u + d_{tv}i_v + d_{tw}i_w = (d_{tu}-d_{tv})i_u - (d_{tv}-d_{tw})i_w
\end{aligned}
\right\}
\tag{10}
$$

When the switching losses are ignored, the instantaneous input power p is equal to the instantaneous output power p_{out}. Then, the following equations are obtained.

$$
p = e_r i_r^* + e_s i_s^* + e_t i_t^* = \sqrt{3}EI^*\cos\varphi^*
\tag{11}
$$

$$
p = p_o = v_u^* i_u + v_v^* i_v + v_w^* i_w = v_{uv}^* i_u - v_{vw}^* i_w
\tag{12}
$$

The duty cycles must satisfy (10) and (12) for the arbitrary input currents i_r^*, i_s^* and i_t^* in (11). By the relationship of (12), i_r^*, i_s^* and i_t^* are expressed as follow:

$$
\left.
\begin{aligned}
i_r^* &= \frac{i_r^*}{p}\{i_v(v_v^* - v_u^*) + i_w(v_w^* - v_u^*)\} \\
i_s^* &= \frac{i_s^*}{p}\{i_v(v_v^* - v_u^*) + i_w(v_w^* - v_u^*)\} \\
i_t^* &= \frac{i_t^*}{p}\{i_v(v_v^* - v_u^*) + i_w(v_w^* - v_u^*)\}
\end{aligned}
\right\}
\tag{13}
$$

The duty cycles are obtained from the equivalence between (10) and (13). The following 6 equations are obtained by comparing with (10) and (12) about each coefficient of i_r^*, i_s^* and i_t^*.

$$d_{ru} - d_{rv} = \frac{v_{uv}^*}{p} i_r^* = \frac{v_u^* i_r^* - v_v^* i_r^*}{p} \quad (14)$$

$$d_{rv} - d_{rw} = \frac{v_{vw}^*}{p} i_r^* = \frac{v_v^* i_r^* - v_w^* i_r^*}{p} \quad (15)$$

$$d_{su} - d_{sv} = \frac{v_{uv}^*}{p} i_s^* = \frac{v_u^* i_s^* - v_v^* i_s^*}{p} \quad (16)$$

$$d_{sv} - d_{sw} = \frac{v_{vw}^*}{p} i_s^* = \frac{v_v^* i_s^* - v_w^* i_s^*}{p} \quad (17)$$

$$d_{tu} - d_{tv} = \frac{v_{uv}^*}{p} i_t^* = \frac{v_u^* i_t^* - v_v^* i_t^*}{p} \quad (18)$$

$$d_{tv} - d_{tw} = \frac{v_{vw}^*}{p} i_t^* = \frac{v_v^* i_t^* - v_w^* i_t^*}{p} \quad (19)$$

(14) $\sim$ (19) are the general relations of the duty cycles for a three-phase to three-phase matrix converter, because these equations are derived by the definitional identity of reference values (10).

In (14) $\sim$ (19), when the two duty cycles are determined, the rest seven duty cycles are uniquely determined. In both PWM strategies of Fig.2 and Fig.3, the number of commutations are reduced by giving zero duty cycle to the two switches. For example, (16) is obtained by (14), (18) and the first and second lines of (9) as follows;

$$
\begin{aligned}
d_{su} - d_{sv} &= -(d_{ru} - d_{rv}) - (d_{tu} - d_{tv}) \\
&= -\frac{v_{uv}^*}{p} i_r^* - \frac{v_{uv}^*}{p} i_t^* = \frac{v_{uv}^*}{p} i_s^* \quad (20)
\end{aligned}
$$

Similarly, (17) is obtained by (15), (19) and the second and third lines of (9) as follows;

$$
\begin{aligned}
d_{sv} - d_{sw} &= -(d_{rv} - d_{rw}) - (d_{tv} - d_{tw}) \\
&= -\frac{v_{vw}^*}{p} i_r^* - \frac{v_{vw}^*}{p} i_t^* = \frac{v_{vw}^*}{p} i_s^* \quad (21)
\end{aligned}
$$

The selection methods of the zero duty switches are different in each PWM strategy. In the PWM strategy of Fig.2, the switches with a large deference between the input voltage and the output voltage (S_{rw} and S_{tu} on the example part of Fig.2) are selected as zero duty. In contrast, in the PWM strategy of Fig.3, the switches that has the largest instantaneous power (S_{ru} or S_{tw} on the example part of Fig.3) is kept connecting during a control period. Therefore, the two switches on the same output phase are given zero duty cycle. However, every duty cycle must be greater than or equal to zero, so the zero duty cycles are determined by positive and negative sign of the input currents and the output line voltages. According to these selection methods, (1) and (2) are obtained from (14) $\sim$ (19).

Concerning (14) $\sim$ (21), the equation of duty cycles are generally expressed by the instantaneous powers at both denominator and numerator as follows;

$$d_{lm} = \frac{p_{lm}}{p}, \quad (l = r, s, t, \quad m = u, v, w) \quad (22)$$

D. Duty Cycles of Method 1

In this section, the process for deriving the duty cycles and generating the PWM pattern of Method 1 [2] in Fig.2 are explained.

In the part of example in Fig.2, the range of the input phase θ and the output phase θ_L are $0 \leq \theta \leq \pi/3$ and $0 \leq \theta_L \leq \pi/3$. In these ranges, the relation of the input voltages and the output voltages are $e_r \geq e_s \geq e_t$ and $v_u^* \geq v_v^* \geq v_w^*$. The relation of the input currents $i_r^* \geq i_s^* \geq 0 \geq i_t^*$, because the input power factor $\varphi^* = 0$. Therefore, following relations "$i_r^* \geq 0$, $v_{uv}^* \geq 0$, $v_{vw}^* \geq 0$" are obtained. The following inequality is obtained, because the right side of (14) and (15) are both positive.

$$d_{ru} > d_{rv} > d_{rw} \geq 0 \quad (23)$$

From (23), the zero duty cycle must be d_{rw} on the duty cycles of r phase, so the following equations are obtained by (14), (15) and (22).

$$d_{ru} = \frac{p_{ru}}{p} = \frac{v_{uw}^*}{p} i_r^* = \frac{v_u^* i_r^* - v_w^* i_r^*}{p} \quad (24)$$

$$d_{rv} = \frac{p_{rv}}{p} = \frac{v_{vw}^*}{p} i_r^* = \frac{v_v^* i_r^* - v_w^* i_r^*}{p} \quad (25)$$

$$d_{rw} = \frac{p_{rw}}{p} = \frac{v_{ww}^*}{p} i_r^* = \frac{v_w^* i_r^* - v_w^* i_r^*}{p} = 0 \quad (26)$$

Similarly, concerning the relations "$i_t^* \leq 0$, $v_{uv}^* \geq 0$, $v_{vw}^* \geq 0$", the following inequality is obtained, because the right side of (18) and (19) are both negative.

$$-d_{tw} < -d_{tv} < -d_{tu} \leq 0 \quad (27)$$

From (27), the zero duty cycle must be d_{tu} on the duty cycles of t phase, so the following equations are obtained by (18), (19) and (22).

$$-d_{tu} = \frac{-p_{tu}}{p} = \frac{v_{uu}^*}{p} i_t^* = \frac{v_u^* i_t^* - v_u^* i_t^*}{p} = 0 \quad (28)$$

$$-d_{tv} = \frac{-p_{tv}}{p} = \frac{v_{uv}^*}{p} i_t^* = \frac{v_u^* i_t^* - v_v^* i_t^*}{p} \quad (29)$$

$$-d_{tw} = \frac{-p_{tw}}{p} = \frac{v_{uw}^*}{p} i_t^* = \frac{v_u^* i_t^* - v_w^* i_t^*}{p} \quad (30)$$

At this point, the duty cycles of r phase and t phase are decided by (24) $\sim$ (26), (28) $\sim$ (30). The duty cycles of s phase are obtained as the rest of duration by using (9).

$$\left.\begin{aligned}
d_{su} &= 1 - d_{ru} - d_{tu} \\
d_{sv} &= 1 - d_{rv} - d_{tv} \\
d_{sw} &= 1 - d_{rw} - d_{tw}
\end{aligned}\right\} \quad (31)$$

Fig.5 shows the example part of Fig.2. In Fig.2, the powers of r phase p_{ru}, p_{rv}, p_{rw} and the powers of t phase $-p_{tu}$, $-p_{tv}$, $-p_{tw}$ are drawn on the same graph. Concerning r phase, when p_{ru}, p_{rv}, p_{rw} are larger than the triangular carrier that is the range between zero and the instantaneous power p, each duty signal of S_{ru}, S_{rv}, S_{rw} are turn-on state. Concerning t phase, when $-p_{tu}$, $-p_{tv}$, $-p_{tw}$ are less than the triangular carrier that is the range between zero and the instantaneous power $-p$, each duty signal of S_{tu}, S_{tv}, S_{tw} are turn-on state. According to (1), the duty signal of s phase turn on when the both

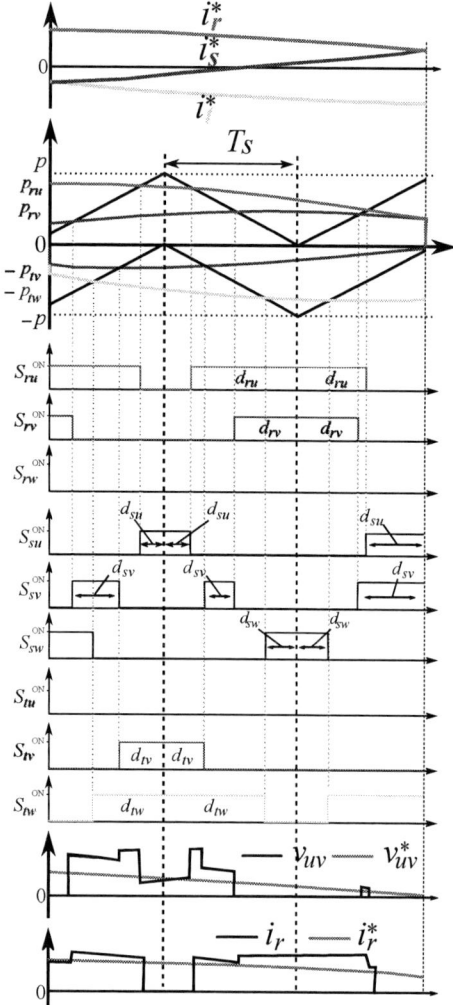

Fig. 5. Generation of duty cycles

signal of r phase and t phase are turn-off state on each output phase. In

E. Duty Cycles of Method 2

this section, the process for deriving the duty cycles and generating the PWM patterns of Method 2 [3] in Fig.3 are explained.

Just as Method 1, the range of explanation is the part of example in Fig.3. Therefore, the relation of the input voltages and the output voltages are $e_r \geq e_s \geq e_t$ and $v_u^* \geq v_v^* \geq v_w^*$. The relation of the input currents $i_r^* \geq i_s^* \geq 0 \geq i_t^*$, because Method 2 is the the input power factor $\varphi^* = 0$. Additionally, Method 2 has the feature that one of the output phase switches is conducted during the whole control period. Therefore, the two duty cycles are determined as zero, the six switches of the rest output phase can be derived. The switch that keep on conducting are decided by the sign of i_s^*, because all of duty cycles must not be less than zero.

In case of $i_s^* \geq 0$, the following relations $i_r^* \geq 0$, $i_s^* \geq 0$, $i_t^* \leq 0$, $v_{uv}^* \geq 0$, $v_{vw}^* \geq 0$ are obtained. By this

relation, the following inequality are obtained, because the right side of (14)~(17) are positive and the right side of (18), (19) are negative.

$$1 \geq d_{ru} > d_{rv} > d_{rw} \geq 0 \tag{32}$$
$$1 \geq d_{su} > d_{sv} > d_{sw} \geq 0 \tag{33}$$
$$1 \geq d_{tw} > d_{tv} > d_{tu} \geq 0 \tag{34}$$

In the each output phase, one switch is kept conducting during one control period, then, the other two switches are not conducted according to the prevention of a short circuit in the input line voltage in (9). Only w phase meet this condition in (32)~(34), the duty cycles of w phase are decided as follows;

$$d_{rw} = 0, \quad d_{sw} = 0, \quad d_{tw} = 1 \tag{35}$$

The following equations are obtained by substituting (35) into (14)~(19).

$$d_{ru} = \frac{p_{ru}}{p} = \frac{v_{uw}^*}{p} i_r^* = \frac{v_u^* i_r^* - v_w^* i_r^*}{p} \tag{36}$$

$$d_{rv} = \frac{p_{rv}}{p} = \frac{v_{vw}^*}{p} i_r^* = \frac{v_v^* i_r^* - v_w^* i_r^*}{p} \tag{37}$$

$$d_{rw} = \frac{p_{rw}}{p} = \frac{v_{ww}^*}{p} i_r^* = \frac{v_w^* i_r^* - v_w^* i_r^*}{p} = 0 \tag{38}$$

$$-d_{su} = \frac{-p_{su}}{p} = \frac{v_{wu}^*}{p} i_r^* = \frac{v_w^* i_r^* - v_u^* i_r^*}{p} \tag{39}$$

$$-d_{sv} = \frac{-p_{sv}}{p} = \frac{v_{wv}^*}{p} i_r^* = \frac{v_w^* i_r^* - v_v^* i_r^*}{p} \tag{40}$$

$$-d_{sw} = \frac{-p_{sw}}{p} = \frac{v_{ww}^*}{p} i_r^* = \frac{v_w^* i_r^* - v_w^* i_r^*}{p} = 0 \tag{41}$$

$$d_{tu} = 1 - d_{ru} - d_{su} \tag{42}$$
$$d_{tv} = 1 - d_{rv} - d_{sv} \tag{43}$$
$$d_{tw} = 1 \tag{44}$$

In the case of $i_s^* \leq 0$, the following relations $i_r^* \geq 0$, $i_s^* \leq 0$, $i_t^* \leq 0$, $v_{uv}^* \geq 0$, $v_{vw}^* \geq 0$ are obtained. The following inequality are obtained by substituting this relations into the right side of (14)~(19).

$$1 \geq d_{ru} > d_{rv} > d_{rw} \geq 0 \tag{45}$$
$$1 \geq d_{sw} > d_{sv} > d_{su} \geq 0 \tag{46}$$
$$1 \geq d_{tw} > d_{tv} > d_{tu} \geq 0 \tag{47}$$

The switch S_{ru} that keep on conducting are decided by (45)~(47) and (9), the duty cycles of u phase are decided as follows;

$$d_{ru} = 1, \quad d_{su} = 0, \quad d_{tu} = 0 \tag{48}$$

The following equations are obtained by substituting (48) into (14)~(19).

$$\left.\begin{array}{l}
d_{ru} = 1, \quad d_{rv} = 1 - d_{sv} - d_{tv} \\
d_{rw} = 1 - d_{sw} - d_{tw} \\
d_{su} = 0, \quad d_{sv} = \dfrac{v_{vu}^*}{p} i_s^*, \quad d_{sw} = \dfrac{v_{wu}^*}{p} i_s^* \\
-d_{tu} = 0, \quad -d_{tv} = \dfrac{v_{uv}^*}{p} i_t^*, \quad -d_{tw} = \dfrac{v_{uw}^*}{p} i_t^*
\end{array}\right\} \tag{49}$$

978-1-4799-2706-7/14 $31.00 © 2014 IEEE

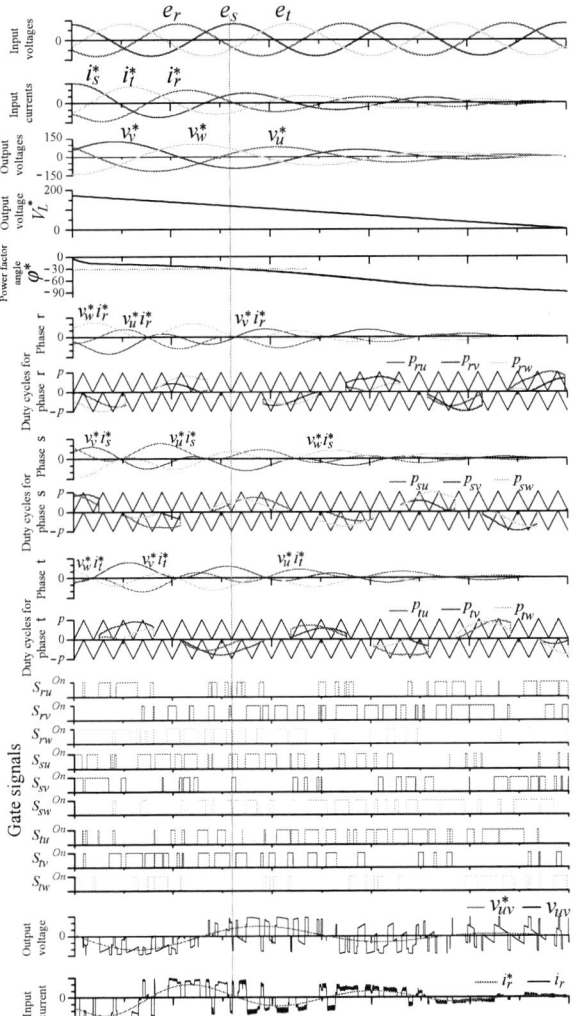

Fig. 6. Visualized PWM strategy in transient state

F. Visualization in Transient State

Fig.6 shows the behaviors of PWM waveforms of Method 1 in transient state when the output voltage reference V_L^* is decreasing from 0.86 times of the input voltage (maximum voltage) to zero. The proposed visualization method can be applied to transient state. The output power is decreasing and the input power factor angle is lagging to keep unity source power factor. The dotted line shows the point of the input power factor angle $\varphi^* = -\pi/6$. In $\varphi^* > -\pi/6$, the output voltage waveform v_{uv} consists of the voltage pulse of the same sign with the reference. However, in $\varphi^* < -\pi/6$, the output voltage waveform v_{uv} includes the voltage pulse of the opposite sign with the reference. The feature of the PWM strategy including the transient state is clarified by the visualization.

III. EXPERIMENTAL RESULTS

A. Experimental System Configuration

Fig.7 shows the experimental system configuration. TableI lists the specifications for the experiment. The

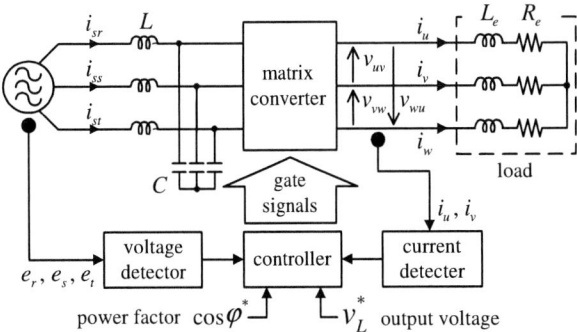

Fig. 7. Experimental system configuration

TABLE I. SPECIFICATIONS FOR EXPERIMENTS

Source voltage E, ω	200 V, $2\pi \times 60$ rad/s
Input filter L, C	1.13 mH, 13.2 μF
Load R_e, L_e	40 Ω, 8 mH
Output voltage reference V_L^*, ω_L^*	0~173V, $2\pi \times 40$ rad/s
Carrier frequency $f_s (= 1/2T_s)$	10 kHz

voltage source is 200 V and 60 Hz, and the load is an inductive load. The controller is a DSP (Digital Signal Processer, TI, TMS320C6713). The visualization of PWM strategy has been explained with the carrier frequency of 500 Hz in Chapter II. On the experimental system for practical use, the carrier frequency is changed to 10 kHz. The behavior of the visible PWM strategy of Method 1 is verified on the experimental system.

B. Experimental Waveforms

Fig.8 shows the experimental waveforms when the output voltage reference value V_L^* is decreasing from 173 V to 0 V. As similarly in Fig.6, the output voltage waveform v_{uv} consists of the voltage pulses of the same sign with the reference during the input power factor angle of $\varphi^* > -\pi/6$ (red dotted line). The output voltage waveform v_{uv} include the voltage pulses of the opposite sign with the reference during $\varphi^* < -\pi/6$. The input power factor is controlled to keep the unity source power factor. The phase of the source current i_{sr} is coincident with the phase of the source voltage e_r when the input power factor angle $\varphi^* < -\pi/3$ at the blue line in Fig.8. When the output voltage reference V_L^* is further reduced, the input current i_r is decreasing. Finally, the input current i_r is zero and the source current i_{sr} flows the leading current by the LC filter.

From the experimental results, the relation between the behavior of the visible PWM strategy and the behavior of the actual system with higher switching frequency has been verified.

IV. CONCLUSIONS

In this paper, the duty cycles expressed by the product of the input current reference and the output voltage reference are derived. Because the duty cycles are treated as continuous time-varying values, the PWM strategy

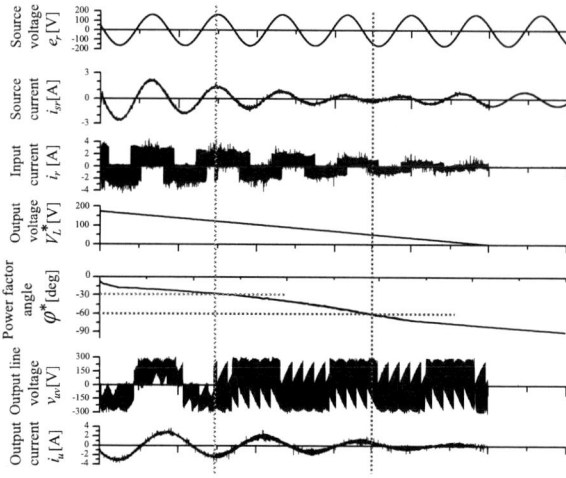

Fig. 8. Experimental waveforms

[9] J. Haruna and J. Itoh: "Method for Visualizing Switching Patterns for a Matrix Converter Using Instantaneous Space Vector Diagrams", *IEEJ Trans. IA*, Vol.131, No.2, pp.144-150 (2011) (in Japanese)

can be easily extended to the input and output voltage periods. The visualization for two PWM strategies under low carrier frequency is realized. The visualization can be realized in not only steady state, but also transient state. The evaluation for the PWM strategy can be realized by using the visualized waveforms of the input currents and the output voltages. The practical use of the visualized PWM strategy has been verified by the experimental results under the carrier frequency of 10 kHz.

This work is partial supported by " Grant - in - Aid for scientific Research (B) #24360107 "

REFERENCES

[1] M. Venturini: "A new sine wave in sine wave out, conversion technique which eliminates reactive elements", *Proc. POWER-CON 7*, pp.E3_1-E3_15, (1980)

[2] T. Takeshita and H. Shimada: "Matrix Converter Control Using Direct AC/AC Conversion Approach to Reduce Output Voltage Harmonics", *IEEJ Trans. IA*, Vol.126, No.6, pp.778-787 (2006) (in Japanese)

[3] A. Ishiguro, T. Furuhashi, M.Ishida, S. Okuma: "Output Voltage Control Method for PWM-Controlled Cycloconverters Using Instantaneous Values of Input Line to Line Voltages", *IEEJ Trans. IA*, Vol.111, No.3, pp.201-207 (1991) (in Japanese)

[4] S. Ishikawa and T. Takeshita: "Input Power Factor Control of Three-Phase to Three-Phase Matrix Converters", *IEEJ Trans. IA*, Vol.129, No.3, pp.258-266 (2009) (in Japanese)

[5] I. Asai and T. Takeshita: "Duty Cycles for Three-Phase of Three-Phase Matrix Converters", *IEEJ Trans. IA*, Vol.131, No.9, pp.1173-1174 (2011) (in Japanese)

[6] A. Odaka, I. Sato, H. Ohguchi, Y. Tamai, H. Mine and J. Itoh: "A PAM Control Method for Matrix Converter Based on Virtual AC/DC/AC Conversion Method", *IEEJ Trans. IA*, Vol.126, No.9, pp.1185-1192 (2006) (in Japanese)

[7] Y. Tadano, S. Hamada, S. Urushibata, M. Nomura, Y. Sato and M. Ishida: "A Space Vector Modulation Scheme for Matrix Converter that Gives Top Priority to the Improvement of the Output Control Performance", *IEEJ Trans. IA*, Vol.128, No.5, pp.631-641 (2008) (in Japanese)

[8] T. Takeshita, S. Ishikawa and Y. Andou: "Instantaneous Effective Values Theory and Its Application to Output Voltage Harmonics Suppression of Matrix Converters", *IEEJ Trans. IA*, Vol.130, No.12, pp.1290-1297 (2010) (in Japanese)

Space Vector Modulation based on Virtual Indirect Control for High frequency AC-linked Matrix Converter

Keita Inoue, Masashi shioda, Motohumi Katade,
Akira Goto, Shin Morishita
San-Eisha ,Ltd.
5-2-1, Ebara, Shinagawa, Tokyo, Japan
shioda-masashi@san-eisha.co.jp

Junichi Itoh, Kazuhiro Koiwa
Department of Electrical engineering
Nagaoka University of Technology, NUT
Niigata, Japan
newkoiwa@stn.nagaokaut.ac.jp

Abstract— **This paper proposes a space vector modulation of a high frequency AC-inked matrix converter which is applied to an electric transformer system. The one of problem of a commercial frequency transformer is bulky and heavy. In order to resolve this problem, we propose to use a high frequency transformer driven by three-phase to single phase matrix converter. In this paper, the virtual indirect control method with a space vector modulation is applied to the primary and secondary side matrix converter. Particularly, the control of the secondary side matrix converter should be considered because the input is rectangle which has multiple levels of voltage. The fundamental operation of the proposed system is demonstrated by experiment and simulation with 1-kW prototype. Besides, the volume and weight of the proposed system are compared with that of the commercial frequency transformer system at 50 kVA. As the result, it is confirmed that the volume and weight of the proposed system are reduced to 15% and 5%, respectively in comparison with that of the transformer system.**

Keywords— *matrix converter, transformer, space vector modulation, virtual indirect control*

I. INTRODUCTION

Recently, distributed power supplies which uses renewable energy sources, such as a photovoltaic and a fuel cell, and the ac power supply such as a wind turbine, which can install near the demand area, are attracted. In these systems, commercial frequency transformers are used from the view point of isolation and protection. However, the transformer at the commercial frequency is large. For example, the volume and the weight of the transformer at 50 kVA are $2.2 \times 10^2 \mathrm{dm}^3$, and 315kg respectively. It is very bulky and heavy.

In order to reduce the volume and weight of the transformer, the high frequency transformer systems which comprises a PWM rectifier, an isolated DC-DC converter and a PWM inverter, have been discussed. In this method, the size of the transformer is reduced. However, a large electrolytic capacitor is necessary in the

DC link in order to smooth the DC link voltage. Besides, the lifetime of the system depends on the electrolytic capacitor. In addition, the power loss is increased owing to the four times conversion from the input to the output. In order to resolve these problems, a matrix converter, the direct AC-AC converters have been studied [1]-[3].

Further, the high frequency conversion systems which use the three-phase to single-phase matrix converter, have been proposed [1]-[4]. In this system, the electrolytic capacitor is not necessary owing to the absence of the DC link in the matrix converter. Additionally, the efficiency can be improved because the conversion stage in the system is reduced to two times in comparison with the conventional system. However, according to the modulation method in Ref. [5] and [6], the number of switching during one carrier period is not optimized because of the use of the carrier comparison modulation method. On the other hand, a space vector modulation based on a virtual indirect control method has been introduced to the control strategy of the primary matrix converter [7] as the isolated AC to DC converter system. Then, the features of this control method are as follows: (i) the number of switching is lesser than that of Ref. [5] and [6], and (ii) simple control strategy. However, the control performance of the space vector modulation based on the virtual indirect control is not applied to a bipolar direction power flow system. The asynchronous of the secondary control to the primary control is problematic by general PWM control because the secondary input voltage is reactangle which has multiple levels of voltage. In addition, the evaluation of the volume reduction has not been reported in pastworks in comparison between the high frequency ac link system and the commercial frequency transformer.

In this paper, the space vector modulation based on virtual indirect control is applied to primary and secondary side matrix converter. The control method of the secondary matrix converter is considered. Furthermore, the volume and weight of the proposed

978-1-4799-2706-7/14 $31.00 © 2014 IEEE   130

system are compared with a conventional system using a commercial frequency transformer.

The remainder of this paper is organized as follows. First, the circuit structure and control strategies are described. Next, the proposed system is compared with the conventional system from the view of the volume and weight. Finally, the fundamental operation of the proposed system is demonstrated in simulation and experiment.

II. CIRCUIT STRUCTURE

Fig. 1 shows the conventional circuit with a high frequency transformer. This circuit comprises a 3-phase PWM rectifier, DC-DC converter and a 3-phase PWM inverter. The power loss of this system is larger owing to the four times conversion from the input to the output. Additionally, this circuit has a large electrolytic capacitor. Thus, it is difficult to miniaturize system.

Fig. 2 shows a proposed circuit using three-phase to single phase matrix converter. The proposed system can achieve high efficiency because the power conversion stages are reduced in comparison with the conventional system. Furthermore, the volume and weight of the system are reduced owing to the absence of the electrolytic capacitor.

III. EVALUATION OF VOLUME

A. Volume of Transformer

In order to validate the proposed system, the volume and weight are calculated. For example, the volume of transformer is expressed by (1).

$$Vol_{trans} = K_{vol}\left(\frac{2L_1 I_1^2}{K_u J B_m}\right)^{\frac{3}{4}} \tag{1}$$

Where K_{vol} is the form factor of the core, L_1 is the magnetizing inductance of primary side, I_1 is the magnetizing current, K_u is the occupancy rate, J is the current density, B_m is the saturation magnetic flux density. Table.4 shows the several values of the 50 kVA conventional systems. Moreover, Vol_{com} is calculated by (1) as the volume of the commercial transformer. The magnetizing inductance of primary side L_1 is expressed by (2) in case of commercial frequency. On the other hand, the magnetizing inductance of the high frequency transformer is expressed by (3).

$$L = \frac{V_{max}}{4 f I_{max}} \tag{2}$$

$$L = \frac{V_{max}}{2\pi f I_{max}} \tag{3}$$

Where V_{max} is the max voltage, f is the frequency, I_{max} is the max current.

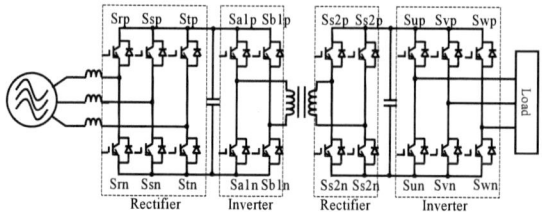

Fig. 1 Conventional circuit which consists of a PWM rectifier, DC-DC converter and a PWM inverter

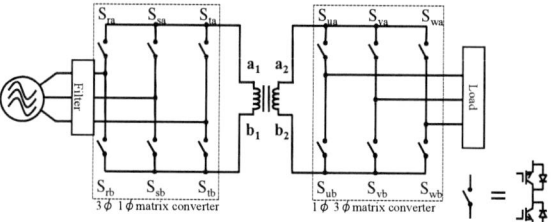

Fig. 2 Proposed circuit using three-phase to single phase matrix converter

Table.1 The values for calculating volume

	Commercial frency	High frequency
Form factor of the core K_{vol}	23.0	23.0
Inductance of primary side L_1	4.4×10^{-3}	3.50×10^{-5}
Magnetizing current I_1	272 A	272 A
Occupancy rate K_u	0.5	0.50
Current density J	4.00 A/mm^2	4.00 A/mm^2
saturation magnetic flux density B_m	5.00×10^{-1}	1.23
Max votage V_{max}	283 V	283 V
Frequency f	50 Hz	10 kHz
Max current I_{max}	204 A	204 Akake
Volume	1.83×10^{2} dm^3	12.0 dm^3

Above all, the transformer of commercial frequency is 1.83×10^{2} dm^3, the volume of high frequency transformer is 12.0 dm^3. On the other hand, in case of the proposed system, the values are shown in Table 3. Therefore, the high frequency transformer of the proposed system is 12.0 dm^3.

B. Heat-sink

About the total volume of the proposed system, it consists of the heat-sink, input filter, switching devices and the high frequency transformer. The volume of matrix converter is discussed in[14][15]. The volume of these components is described in [8][9]. First, the volume of the heat-sink Vol_{fin} is calculated by

$$Vol_{fin} = \frac{1}{CSPI \times R_{th}} \tag{4}$$

where *CSPI* is the cooling system performance index. The volume of a heat-sink is described by [9][10]. In this paper, *CSPI* is set to 5. In addition, the thermal resistance of the heat-sink R_{th} is expressed by

$$R_{th} = \frac{T_j - T_a}{P_{loss}} \tag{5}$$

where T_j is the junction temperature of the switching devices, T_a is the ambient temperature, and P_{loss} is the power loss of the converter. It is assumed that the efficiency of the converter is 95%.

C. Input LC filter

Next, the capacitor volume of the input filter Vol_c is expressed by (6) from the view point of the energy density. How to calculate the volume from the point of the energy density is disscued in [11], the volume of the converter is disscued in [14][15].

$$Vol_c = \frac{1}{8.8 k_e k_z^2} C V_{in}^2 \tag{6}$$

It is noted that the k_e is the dielectric coefficient, k_z is the coefficient of the voltage proof, C is the capacitance of the capacitor, and V_{in} is the maximum value of the input voltage. Finally, the volume of the inductor Vol_L is indicated by (7) from the view of the area product [9].

$$Vol_L = K_{vol} \left[\frac{LI^2}{K_u B_m J} \right]^{\frac{3}{4}} \tag{7}$$

It is noted that K_{vol} depends on the configuration of the inductor core, L is the inductance of the inductor, I is the input current. In this case, the volume of IGBTs is 1.2×10^{-1} dm^3. Additionally, the volume of the high frequency transformer is 12.0 dm^3 by (1). Thus, the volume of the proposed system is approximately 2.6×10^1 dm^3. On the other hand, the weight of the proposed system is 12.6 kg. This value of weight is calculated by the actual devices which we select (the heat-sink: 0.3 kg, switching devices: 0.45×24 kg) and the high frequency transformer. The weight of this transformer is simply calculated by the ratio of the frequency. It is calculated to be 1.6kg. Therefore, the total weight is 12.6 kg.

D. The total volume comparison

Fig. 3 shows the volume of the proposed system in comparison with the conventional system with 50 kVA. Moreover, Fig. 4 shows the breakdown of the parts. According to the result, the volume of the proposed system is 85% lesser than in comparison with that of the conventional system.

Fig. 5 shows the weight of the proposed system compared with the conventional system with 50 kVA. The weight of the proposed system is lesser than 4% in comparison with that of the conventional system. Based

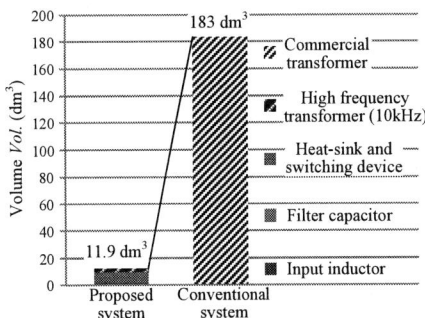

Fig. 3 The comparison of the volume of the proposed system and conventional system

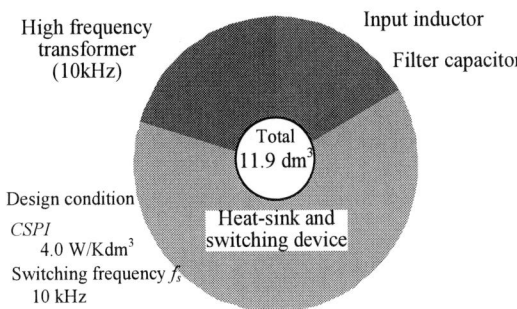

Fig.4 The ratio of the parts of the proposed system

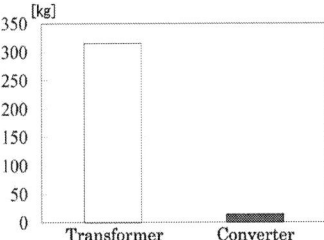

Fig.5 The weight of the converter system are compared with the transformer system at 50 kVA

on these result, it is confirmed that the proposed system is smaller and lighter than the conventional system

IV. CONTROL STRATEGY

Fig.6 shows the control strategy of the proposed system. First, the system gets the both input and output command. i_r, i_s, i_t is the input current command, v_u, v_v, v_w is the output voltage command. In case of primary side, single leg modulation is applied while space vector modulation is applied. The duty command T_{11}, T_{21} is calculated. Next, switching pulses are generated by career comparison. On the other hand, in case of secondary side, space vector modulation is applied while the duty values should be made appropriate for the synchronous to primary side. Last, the synchronized duty values are translated to the switching pulses of the secondary side.

A. Virtual indirect control

The control strategy of the matrix converter is complicated. In order to control the matrix converter simply, virtual indirect control is applied. Virtual indirect control is disscused in [12], [13]. The relationship between the input phase voltage ${}^t[v_r, v_s, v_t]$ and the transformer voltage ${}^t[v_a, v_b]$ is expressed by

$$\begin{bmatrix} v_a \\ v_b \end{bmatrix} = \begin{bmatrix} S_{ra} & S_{sa} & S_{ta} \\ S_{rb} & S_{sb} & S_{tb} \end{bmatrix} \begin{bmatrix} v_r \\ v_s \\ v_t \end{bmatrix} \tag{8}$$

where S_{xy} is the switching function of the switch S_{xy}. In the matrix converter, the relationship between the output and input voltage can be expressed in (8), where S_{xy} represents the switching intervals for the matrix converter. In addition, the switching intervals for Fig. 1 are described by (9), where the left block is the full-bridge inverter and the right block is the PWM rectifier.

$$\begin{bmatrix} v_a \\ v_b \end{bmatrix} = \begin{bmatrix} S_{ap} & S_{an} \\ S_{bp} & S_{bn} \end{bmatrix} \begin{bmatrix} S_{rp} & S_{sp} & S_{tp} \\ S_{rn} & S_{sn} & S_{tn} \end{bmatrix} \begin{bmatrix} v_r \\ v_s \\ v_t \end{bmatrix} \tag{9}$$

Based on (9), the switching units in (8) is expressed as (10), that is, the switching units are the multiplication between the left block and the right

$$\begin{bmatrix} S_{ra} & S_{sa} & S_{ta} \\ S_{rb} & S_{sb} & S_{tb} \end{bmatrix} = \begin{bmatrix} S_{a1p} & S_{a1n} \\ S_{b1p} & S_{b1n} \end{bmatrix} \begin{bmatrix} S_{rp} & S_{sp} & S_{tp} \\ S_{rn} & S_{sn} & S_{tn} \end{bmatrix} \tag{10}$$

The switching function of the direct converter can be obtained by a synthesizing of these switching functions. Similarly the switching function in the secondary matrix converter can be expressed by (11).

$$\begin{bmatrix} S_{ua} & S_{va} & S_{wa} \\ S_{ub} & S_{vb} & S_{wb} \end{bmatrix} = \begin{bmatrix} S_{a2p} & S_{a2n} \\ S_{b2p} & S_{b2n} \end{bmatrix} \begin{bmatrix} S_{up} & S_{vp} & S_{wp} \\ S_{un} & S_{vn} & S_{wn} \end{bmatrix} \tag{11}$$

B. Space vector modulation

The space vector modulation is applied to the virtual indirect controlled high frequency AC-linked matrix converter. It is noted that the duality of power conversion circuit is used.

Fig. 7 shows the vector diagram and the switching patterns for the PWM rectifier. It is noted that the fundamental vectors and the switching patterns are

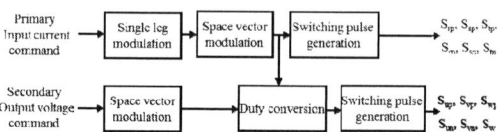

Fig.6 The block diaglam of proposed system

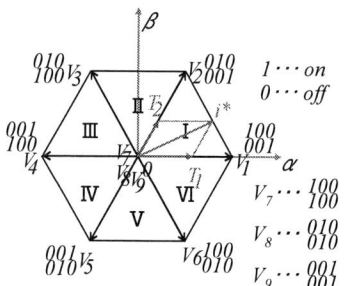

Fig.1 Vector diagram of PWM rectifier

determined from the area of the input current command. Furthermore, the switching duty T_{11}, T_{21}, and T_{z1} are expressed by (12), (13), and (14)

$$T_1 = \frac{1}{|A|} \begin{vmatrix} v_\alpha & V_{2\alpha} \\ v_\beta & V_{2\beta} \end{vmatrix} \tag{12}$$

$$T_2 = \frac{1}{|A|} \begin{vmatrix} V_{1\alpha} & v_\alpha \\ V_{1\beta} & v_\beta \end{vmatrix} \tag{13}$$

$$T_z = 1 - (T_1 - T_2) \tag{14}$$

where v_α and v_β, are current command vector of α and β element, and $V_{1\alpha}$, $V_{1\beta}$, $v_{2\alpha}$ and $v_{2\beta}$ are fundamental vectors of the α and β element.

The transformer voltage is generated by the virtual inverter. In addition, the virtual rectifier is used in order to rectify the output voltage of the transformer. It is noted that the switching functions of the virtual inverter at the primary matrix converter and the virtual rectifier at the secondary matrix converter are expressed by (15).

$$\begin{bmatrix} S_{a1p} & S_{a1n} \\ S_{b1p} & S_{b1n} \end{bmatrix} = \begin{bmatrix} S_{a2p} & S_{a2n} \\ S_{b2p} & S_{b2n} \end{bmatrix} = \begin{bmatrix} 0 & 1 \\ 1 & 0 \end{bmatrix}, \begin{bmatrix} 0 & 1 \\ 1 & 0 \end{bmatrix} \tag{15}$$

The virtual inverter at the primary matrix converter and the virtual rectifier at the secondary matrix converter are alternately switched at the switching frequency by using (15).The switching pattern of the primary side matrix converter and the secondary side matrix converter are generated by substituting (15) to (10), (11).

978-1-4799-2706-7/14 $31.00 © 2014 IEEE

C. Control method of the secondary matrix converter

Fig. 8 shows the input voltage waveform of the secondary side matrix converter. It is note that T_{11} is the on-duty of the voltage V_{11}, T_{21} is the duty of V_2. The input voltage of the secondary side matrix converter is the square wave of V_1 and V_2. Therefore, general PWM control cannot be applied . In order to resolve this problem, the duty command of the secondary side matrix converter should be compensated by the duty command of the primary side matrix converter. Accordingly, when the input voltage is two levels, the general PWM control can be applied.

Fig. 9 shows the duty compensation method for the secondary matrix converter. Thus, T_a, T_b, T_c, T_d, T_e which can divide to the several levels of output voltage of the secondary side matrix converter, are expressed by (16).

$$
\begin{aligned}
T_a &= T_{11} \times T_{12} \\
T_b &= T_{11} \times T_{22} \\
T_c &= T_{z2} \\
T_d &= T_{21} \times T_{22} \\
T_e &= T_{21} \times T_{12}
\end{aligned}
\tag{16}
$$

It is noted that T_{12} is the duty of the secondary output V_1, T_{22} is is the duty of the secondary output V_2, and T_{z2} is is the duty of the secondary output zero voltage. Additionally, d_a, d_b, d_c, d_d are expressed by (17).

$$
\begin{aligned}
d_a &= T_e + T_b + T_c + T_d \\
d_b &= T_e + T_b + T_c \\
d_c &= T_e + T_b \\
d_d &= T_e
\end{aligned}
\tag{17}
$$

Finally, the switching pulses of the secondary circuit are generated by comparing these duties with the carrier. In this method, the average current becomes constant during the each control cycle. Moreover, secondary circuit control is synchronized with the primary circuit control.

V. SIMULATION AND EXPERIMENTAL RESULT

In order to confirm the validity of the proposed system, the proposed system is demonstrated with a 1-kVA prototype in simulation and experiment. Table.3 lists the simulation and experimental conditions. As the simulation verification, the 50Hz/50Hz conversion is

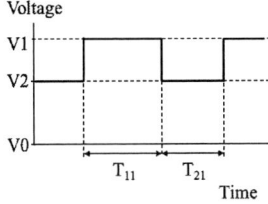

Fig.4 The input voltage of the virtual inverter of the secondary circuit

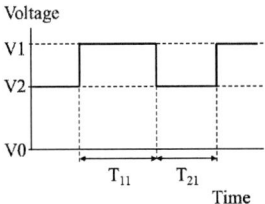

Fig.2 The input voltage of the virtual inverter of the secondary circuit

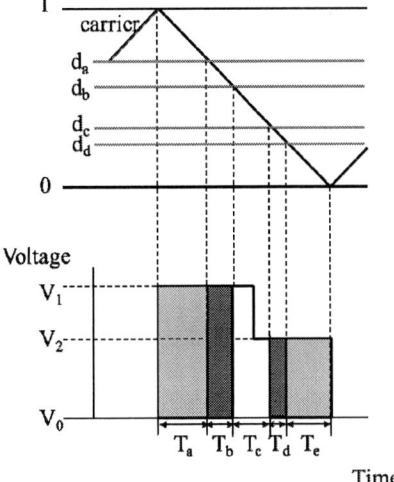

Fig.3 The diagram of the carrier comparison modulation processing to control the secondary circuit.

Table.2 the simulation and experimental conditions

	simulation	experiment
Input line to line voltage	200[V]	100[V]
Input frequency	50[Hz]	50[Hz]
Primary output Frequency	10[kHz]	10[kHz]
Output frequency	30,50,60[Hz]	50[Hz]
Carrier frequency	10[kHz]	10[kHz]
Input inductor	2[mH]	2[mH]
Turn ratio of transformer	1:1.5	1:1.5
The load in experiment	48[Ω],5[mH]	48[Ω]
The commutation-time in experiment	0μsec	3μsec

demonstrated. Moreover, frequency conversion vilification (50Hz/30Hz, 50Hz/60Hz) is demonstrated. Finaly, the experiment result of 50Hz/50Hz conversion is indicated. In this experiment, 4 step voltage commutation is applied to the primary control, 4 step current commutation is applied to the secondary control [16].

A. Simulation results

Fig. 10 shows the simulation result of the proposed system. As shown in Fig. 10, the total harmonic distortion (THD) of the input current is 2 %. Moreover, the unity power factor is obtained. On the other hand, the output voltage and current are in phase, the unity power factor is obtained. In addition, the output voltage is sinusoidal and 140 V. THD of the output voltage and current are 1%.

From these results, it is confirmed that the proposed system is suggested to be operated. Further, 240 V line is provided in the proposed system in simulation.

Fig. 11 shows the secondary input voltage and the output line voltage. The input voltage of the secondary matrix converter is rectangular waveform. The cycle of the primary output voltage is 100 μsec. The polarity of the wave is replaced every 50 μsec cycles. Therefore, the high frequency transformer is correctly operated. On the other hand, the 3 phase line output voltage is organized. This feature is that the PWM control is applied to several levels of voltage.

Fig. 12 and Fig. 13 shows 50Hz/30Hz conversion and 50Hz/60Hz conversion simulation result. It is observed that the Input and output waveform are sinusoidal. Moreover the power factor of input and output is unity. Even in the conditions of the frequency conversion, there are very few distortion, the values of THD is about 2%. This results indicates that the proposed control method can be applied to frequency conversion.

B. Experimental results

Fig. 14 shows the experimental result. As shown in Fig. 14, the input current waveform is distorted(THD 20%). This is caused by the output voltage which is poor sinusoidal. This poor sinusoidal is caused by low precision of the switching. This control method requires fast switching speed the same to 30 kHz or more, therefore, IGBTs is less suitable. Furthermore, the synchronous missing is occurred by the secondary circuit

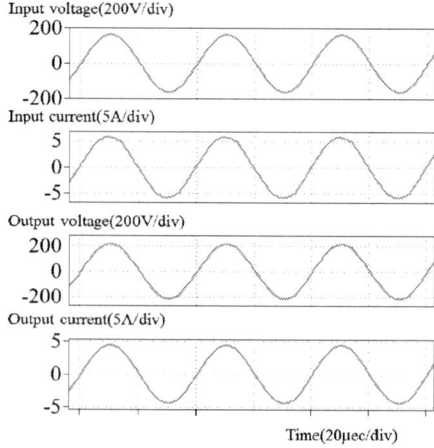

Fig.5 Simulation result

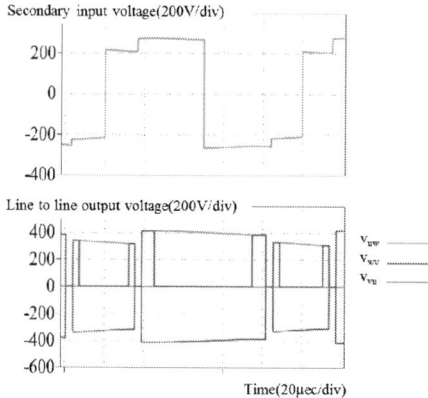

Fig.11 The secondary input voltage and the output line voltage

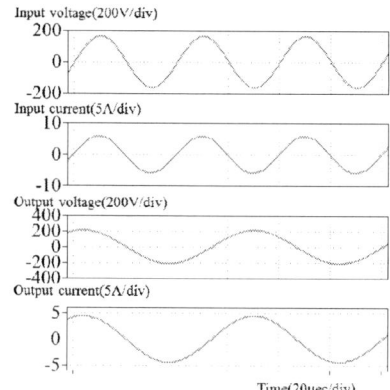

Fig.12 Simulation result

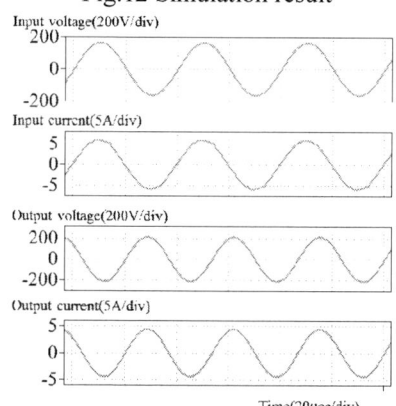

Fig.13 Simulation result

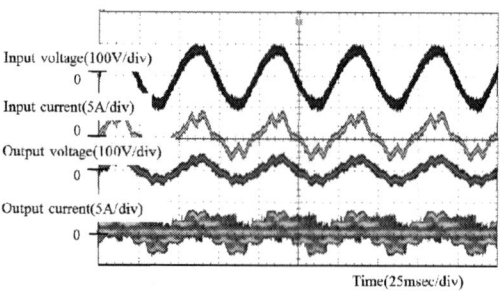

Fig.14 experimental result

with primary circuit which depends on the accuracy of the duty values.

The total efficiency is 86 %. The breakdown of this is the primary circuit efficiency(96%), high frequency transformer efficiency(97%), secondary circuit efficiency (93%). In this case, the secondary loss of the power is the higher than the primary. It is thought because the secondary actual switching frequency is three times as high as the primary switching frequency.

Fig. 15 shows the secondary high frequency link voltage and output line voltage. It is shown that the Secondary high frequency voltage is consists of the two levels voltage vector V_1, V_2. That is thought that the primary converter is successflly operated. Next, the line voltage is divided from the high frequency voltage. These several block of the voltage V_1, V_2 is controled by the ratio of the secondary duty T_{12}, T_{22}, T_{z2}. However, the unexpected zero voltage is observed. It is pointed by arrow A. This phenomenon is occured when the high frequency voltage polality is replaced. In order to resolve this problem, the configuration of the secondary zero vector have to be considered.

VI. CONCLUSION

This paper discussed a high frequency ac-linked matrix converter for a transformer system by applying virtual indirect control and space vector modulation.

The validity of the proposed system was demonstrated in simulation and experiment. In this simulation, the output current THD is 2.0%. The output voltage of the converter is sinusoidal.

Finally, the volume and weight of the proposed system were calculated. As the result, the volume and weight of the proposed system are lesser than 15% and 4%, respectively in comparison with that of the transformer system. Thus, the proposed system could be smaller and lighter than the transformer system.

However, in the experiment of the prototype, It is becomed apparent that there are some problems. It is difficult to sinchronize the secondary circuit control to the primary. In order to synchronize, the fast switching speed, higher precision of the primary side duty which depends on the detection accuracy. That is more, it is essential for the proposed control system to improve the configuration of the voltage vector.

On the other hand, it is expected the application of PDM (pulse density modulation) [17]-[20]. Applying PDM possibly cause the accuracy improvement of the modulation. Further, the improvement in efficiency is occurred by applying PDM bacause of the zero voltage switching. Moreover, next generation semiconductor device assists the realization in this method.

In future works, the accuracy improvement of the secondary control is fundamental. In order to realize this, we consider applying the higher switching speed, or appllying other control method like PDM in addition.

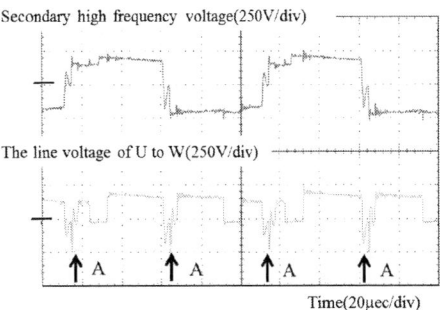

Fig.15 Secondary high frequency link voltage and line voltage

REFERENCES

[1] Bo Wen, X. Zhang, Q. Wang, R. Burgos, P. Mattavelli, D. Boroyevich:"Comparison of Three-Phase AC-AC Matrix Converter and Voltage DC-Link Back-to-Back Converter Systems", IEEE Trans., Vol. 59, No. 12, pp. 4487-4510 (2012)

[2] P. W. Wheeler, J Rodoriguez, J. C. Clare, L. Empringham: "Matrix Converters: A technology Review" IEEE Transactions, Vo 49, No. 2, pp274-288, 2002.

[3] K. Inagaki, T. Furuhashi, A.Ishiguro, M. Ishida, S. Okuma, Y. Uchikawa: "A Waveform Control Method of AC to DC Converters with High-Frequency Links" in Japanese

[4] Garcia-Gil, R.; Espi, J.M., Dede, E.J.; and Sanchis-Kilders, E. "A bidirectional and isolated three-phase rectifier with soft-switching operation", IEEE Trans. On Industrial Electronics, Vol. 52, Issue. 3, pp.

[5] J.Itoh,T.Iida,A.Odaka:"Realization of High Efficiency AC link Converter System based on AC/AC Direct Conbersion Techniques with RB-IGBT", The 32th Annual Conference of the IEEE Industrial Electronics Society(2006)

[6] D. Matsumura, J.Itoh, S.Kondo:"A loss reduction method of high frequency AC link converter based on direct type power converter"The Institute of Electrical Engineers of Japan(2005)

[7] Y.Ohnuma, J.Itoh"Space Vector Modulation for a Single Phase to Three Phase Converter Using an Actibe Buffer", IPEC (2010)

[8] Thomas Friedli, Johann W. Kolar, Jose Rodriguez, Ptrick W. Wheeler: "Comparative Evaluation of Three-Phase AC/AC Matrix Converter and Voltage DC-Link Back-to-Back Converter Systems", IEEE Trans., Vol. 59, No. 12 pp 4487-4510 (2012)

[9] Y.Kashihara, J.Itoh: "Performance Evaluation among Four types of Five-level Topologies using Pareto Front Curves"

[10] U.DROFENIK, G. LAIMER, J. W. KOLAR: "Theoretical Converter Power Density Limits for Forced Convection Cooling", Proceedings of the International PCIM Europe Conference, Vol. , No. , pp. 608-609(2005)

[11] U. Badstuebner, J. Miniboeck, J. W. Kolar: "Experimental verification of the efficiency/power-density (η-ρ) Pareto Front of single-phase double-boost and TCM PFC rectifier systems", APEC, pp. 1050-1057 (2013)

[12] J. Itoh, I. Sato, H. Ohguchi, K. Sato, A. Odaka, N. Eguchi: "A Control Method for the Matrix Converter Based on Virtual AC/DC/AC Conversion Using Carrier Comparison Method", IEEJ Trans. D, Vol. 124, No. 5, pp. 457-463 (2004)

[13] J. Itoh, H. Kodachi, A. Odaka, I. Sato, H. Ohguchi, H. Umida: "A High Performance Control Method for the

Matrix Converter Based on PWM generation of Virtual AC/DC/AC Conversion", JIASC IEEJ, Vol. , No. , pp. I-303-I-308 (2004)

[14] Bo Wen, X. Zhang, Q. Wang, R. Burgos, P. Mattavelli, D. Boroyevich: "Comparison of Three-Phase AC-AC Matrix Converter and Voltage DC-Link Back-to-Back Converter Topologies Based on EMI Filter", ECCE US, Vol. , No. , pp. 2698-2706 (2013)

[15] R. Lai, F. Wang, R. Burgos, Y. Pei, D. Boroyevich, B. Wang, T. A. Lipo, V. D. Immanuel, K. J. Karimi: "A Systematic Topology Evaluation Methodology for High-Density Three-Phase PWM AC-AC Converters", IEEE Trans., Vol. 23, No. 6, pp. 2665-2680 (2008)

[16] K. Kato, J. Itoh: "Development of a Novel Commutation Method which Drastically Suppresses Commutation Failure of a Matrix Converter", IEEJ Trans. D, Vol. 127, No. 8, pp. 829-836 (2007)

[17] Y. Nakata, J. Itoh: "An Experimental Verifications and Analysis of a Single-phase to Three-phase Matrix Converter using PDM Control Method for High-freuqnecy Application", IEEE 9th PEDS, No. 383, (2011)

[18] Y. L. Feng, Y. Konishi, M. Nakaoka, "Current-fed soft-switching inverter with PDM-PWM control scheme for ozone generation tube drive", IEEJ Transactions on Industry Applications, Vol.120, No.10 pp.1239-1240, October 2000 (in Japanese)

[19] P.K.Sood and T.A.Lipo : "Power Conversion Distribution System using a High-Frequency AC Link," IEEE Trans. on IA, Vol.IA-24, No.2, pp.228-300 (1988)

[20] Hisayuki Sugimura, Bishwajit Saha, H. Omori, Hyun-Woo Lee, M. Nakaoka, "Single reverse blocking switch type pulse density modulation controlled ZVS inverter with boost transformer for dielectric barrier discharge lamp dimmer", IEEE 5th International Power Electronics and Motion Control Conference, 2006, vol. 2, pp. 1 - 5, August 2006

A Fundamental Verification of a Single-phase to Three-phase Matrix Converter with a PDM Control based on Space Vector Modulation

Yuki Nakata and Jun-ichi Itoh
Department of Electrical, Electronics and Information Engineering
Nagaoka University of Technology
Nagaoka, Niigata, Japan
nakata@nagaokaut.ac.jp, itoh@stn.nagokaut.ac.jp

Abstract—**This paper discusses pulse density modulation (PDM) control methods for a single-phase to three-phase matrix converter for high-frequency applications such as the grid interface converter for a wireless power transfer system and high-frequency transformer link system. The input frequency for this converter is assumed as several hundred kHz, and at the same time outputting a commercial frequency, i.e. 50 Hz or 60 Hz. The proposed circuit achieves high efficiency by zero voltage switching from a PDM control method. In the PDM control using delta-sigma conversion, the output waveform has inverse voltage pulse and clamp phenomena on the output voltage waveform. Hence, an improvement method of the output waveform based on SVM is proposed. In this paper, the experimental results using the prototype circuit of indirect matrix converter (IMC) and conventional matrix converter (CMC) with the two PDM control methods are compared. As a result, the THD of the output voltage with delta-sigma conversion and PDM pattern method based on SVM are 5.96% and 2.15% respectively in the experiment using IMC. Furthermore, the maximum efficiency has been improved from 93.4% to 97.3% by applying the proposed PDM method. From the results, the validity of PDM control based on SVM has been confirmed. In addition, a prototype of direct type circuit (CMC) has been built and tested. From the results, the switching at zero voltage and clear sinusoidal output waveform has been confirmed too. As a result, the output voltage THD, which is controlled by the proposed PDM based on SVM are 3.25%.**

Keywords— delta-sigma conversion, PDM control, space vector modulation, wireless power transfer

I. INTRODUCTION

In recent years, wireless power transfer systems have been actively researched [1-4]. In the wireless power transfer system, the frequency of the generated voltage at the receiving coil is from tens of kHz to several MHz, which is identical to the power source frequency. Accordingly, in order to connect this system to a load, an interface converter which converts the received power into a controlled output power is required. The characteristic of this interface converter must have a high input frequency (several hundred kHz) and a low output frequency (50 Hz or 60 Hz) which is suitable for commercial power grid. That is, an AC-to-AC converter

is generally clarified as the interface converter for this system. In general, back-to-back (BTB) system, which is constructed by a PWM rectifier, a smoothing capacitor and a PWM inverter, is used as this AC-to-AC converter. However, it is difficult to use the PWM rectifier in the input side due to high-frequency input.

On the other hand, matrix converters (MC) have been attracted attentions as the AC interface converter for the wireless system, because it delivers advantages in terms of a size reduction and an energy saving owning to high efficiency [5-7]. However, the implementation of the MC in the high-frequency power source has not been reported.

There are two type topologies for the matrix converter, so called indirect matrix converter (IMC) and conventional matrix converter (CMC). The IMC consists of a diode rectifier and three-phase inverter, and the configuration and the control of the circuit are simpler than the CMC circuit. However, the IMC has low efficiency because of the twice power conversion. Furthermore, the IMC circuit is a unidirectional power flow converter because the diode rectifier is used in the input side. On the other hand, the CMC consists of bidirectional switches. The CMC has high efficiency because of the only once power conversion. Additionally, this circuit is a bidirectional power flow converter because the bidirectional switches are used. However, control of this circuit is more complicated in comparison than that of the IMC because the input voltage polarity of the circuit alternates.

The authors have previously proposed a Pulse density modulation (PDM) [8-11] control method for the high-frequency power source [12-13]. The proposed PDM control method is applied to the proposed circuit by using the half cycle of the input voltage as a pulse for PDM control. Therefore the proposed circuit achieves the reduction of the switching loss by the switching at zero input voltage. The PDM signals can be obtained from delta-sigma conversion. However, in the delta-sigma conversion, the inverse voltage pulses and clamp phenomena occur in the output voltage waveform of the IMC. When the phase angle between the output voltage

978-1-4799-2706-7/14 $31.00 © 2014 IEEE

and current becomes larger instantaneously, the DC link current flow backward to the power source and DC link voltage is clamp to snubber capacitor voltage because the DC link current cannot flow toward the power source. Therefore the clamp phenomenon of the output voltage occurs. Additionally, in CMC circuit, the clam phenomenon does not occur because the power regeneration is allowed in the CMC.

In this paper, the PDM control based on space vector modulation (SVM) is proposed in order to solve these problems and improve the quality of output waveform. The proposed method generates switching patterns based on SVM in order to minimize the phase angle between output voltage and current. Therefore, there are no clamp phenomena on the output voltage waveform of the IMC. This paper compares and investigates the two PDM control methods, which are based on the delta-sigma conversion and SVM are compared and investigated from the experimental results using the prototype circuit of IMC and CMC. Firstly, the configurations of the proposed circuit are introduced. Secondly, the PDM control strategy using the delta-sigma conversion and the proposed PDM method based on the SVM are discussed. Next, the validity of the proposed method is confirmed in the experimental results using the prototype circuit of IMC. Finally, the proposed PDM control method is applied to a prototype circuit of CMC, and the basic operation of the proposed PDM control in direct type circuit is confirmed.

II. CIRCUIT CONFIGURATION

A. System Configuration

Fig. 1 shows a configuration of the wireless power transfer system. The receiving coil voltage is a high-frequency AC, which has same as the frequency of the source. In order to connect to the commercial grid, a single phase AC to three phase AC converter is required as an interface converter.

The interface converter inputs several-hundred-kHz sinusoidal waveform, and outputs the low-frequency waveform such as 50 Hz and 60 Hz. Hence, PDM control which uses a half cycle of the input voltage as a pulse can be applied to the interface converter.

Therefore, the use of the matrix converter as an interface converter for wireless power transfer system is proposed. The matrix converter has high frequency by applying proposed PDM control. The detail of PDM control is explained in next section.

B. Indirect Single-phase to Three-phase Matrix Converter

Fig. 2 shows circuit configuration of the proposed single-phase to three-phase IMC. This circuit is constructed from a diode rectifier as an input interface and a three-phase inverter for the rear side. Since this converter does not required electrolytic capacitors in the DC link, the lifetime is longer and the size is more compact than the conventional system, which is

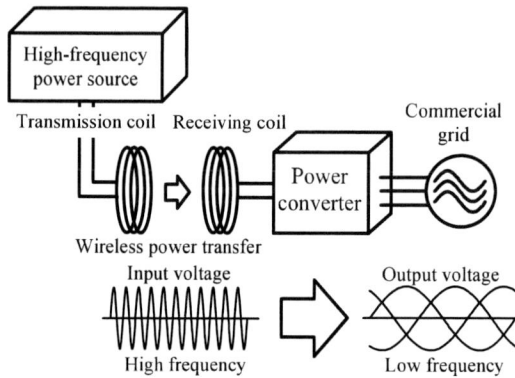

Fig. 1. Configuration of the wireless power transfer system.

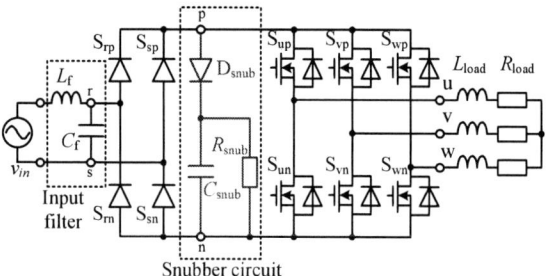

Fig. 2. Single-phase to three-phase indirect matrix converter.

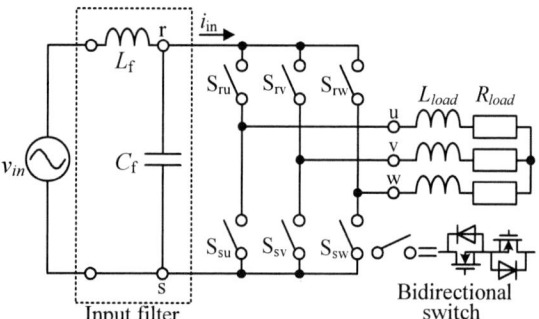

Fig. 3. Single-phase to three-phase conventional matrix converter.

constructed by a PWM rectifier, electrolytic capacitors and a three-phase PWM inverter.

The IMC has low efficiency because the power conversion number of the IMC is twice. Furthermore, the IMC circuit is a unidirectional power flow converter because the diode rectifier is used in the input side. However, the control of the circuit are simpler than the CMC circuit because the inverter is controlled only.

Note that, in the experimental circuit, a snubber circuit is connected to the DC link as a protection circuit. It is constructed by a diode, a small capacitor and a resistor.

The prototype of this circuit configuration is used in the first experiment, in order to compare two PDM methods, which is explained in chapter III and confirm the validity of the proposed method explained in section III-C.

C. Single-phase to Three-phase Matrix Converter

Fig. 3 shows the circuit configuration of the proposed single-phase to three-phase CMC. This circuit is

978-1-4799-2706-7/14 $31.00 © 2014 IEEE

constructed by six bidirectional switches. Since this converter does not require electrolytic capacitors in the DC link, the lifetime is longer and the size is more compact than that of the conventional system, which is constructed by a PWM rectifier, electrolytic capacitors and a three-phase PWM inverter. Further, the efficiency of a CMC is higher than a conventional system because the conversion number is reduced in comparison with the conventional system. However, control of this circuit is more complicated in comparison than that of the IMC because the input voltage polarity of the circuit alternates.

Note that, in the experimental circuit, a snubber circuit is connected to the output side of the converter as a protection circuit. It is constructed by a diode, a small capacitor and a resistor.

The prototype of this circuit configuration is used in the second experiment, in order to confirm the basic operation and validity of proposed PDM method explained in section III-C in CMC.

D. Design of the Input Filter

The impedance matching is important on the circuit for the high-frequency application. In this system, the input filter is connected to the input side as an impedance matching circuit, which can match the impedance of the voltage source and the circuit. The design method, which matches the impedance Z_{in} to 50 Ω is explained in this section because the general high-frequency power sources have 50 Ω of matching impedance.

Fig. 4 shows the configuration of input filter. It is constructed from a reactor L_f and a capacitor C_f. The impedance of the load connected to the filter is expressed by resister R_{load} as shown in Fig. 4. The value of L_f and C_f are decided in order that the real part of the synthetic impedance equals to 50 Ω and the imaginary part equals to 0 Ω to achieve the 50 Ω of matching impedance. L_f and C_f are calculated by (1) and (2), where, $\omega=2\pi f$ is the input voltage angular frequency. L_f and C_f are 80 μH and 16 nF respectively, when the frequency of f is 100 kHz, and R_{load} is 100 Ω. These values are used in the experiment.

$$L_f = \frac{C_f R_{load}^2}{1+(\omega C_f R_{load})^2} \qquad (1)$$

$$C_f = \frac{1}{\omega R_{load}} \sqrt{\frac{R_{load}}{50}-1} \qquad (2)$$

III. CONTROL STRATEGYS

A. Concept of PDM Control

A PDM control method is applied to the proposed system in order to reduce the switching loss of the converter. PDM controls the density and the plus/minus of the constant-width pulse, and then these pulse signals are used as the output unit.

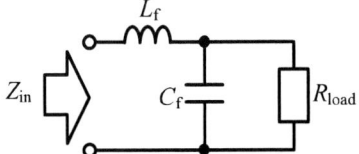

Fig. 4. Configuration of the input filter.

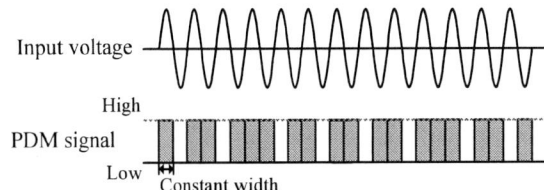
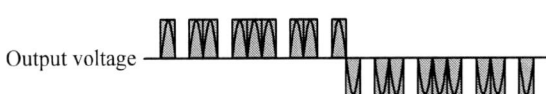

Fig. 5. PDM control waveform of proposed circuit.

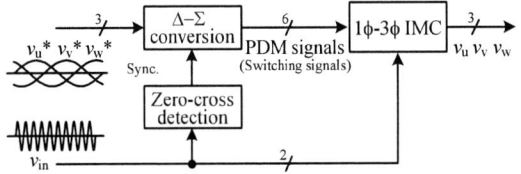

(a) Control block diagram using delta-sigma conversion (overall).

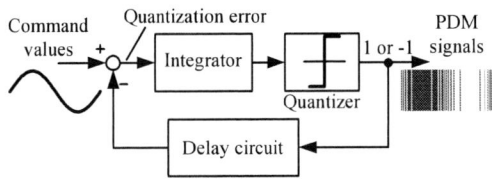

(b) Block diagram of the delta-sigma conversion.

Fig. 6. PDM control block diagram using delta-sigma conversion.

Fig. 5 shows a frame format of PDM control waveform for the single-phase to three-phase CMC. Assuming that the single-phase to three-phase CMC is connecting to a wireless power transfer system, it is receiving high frequency sinusoidal voltage as an input. Therefore the PDM control can be applied to the proposed circuit by using the half cycle of the input voltage as a pulse for PDM control as shown in Fig. 5.

B. PDM Control using Delta-sigma Conversion

Fig. 6 shows the PDM control block diagram using delta-sigma conversion. The PDM signals used for the switching can be obtained by applying the delta-sigma conversion to each phase command values (v_u^*, v_v^*, v_w^*). In general, delta-sigma conversion is used for analog-digital conversion. These PDM signals are used to turn on/off the inverter arm of IMC shown in Fig. 2.

Additionally, the zero cross points exist with respects to the frequency since the input voltage is a sinusoidal waveform. The turn-on and off of the each switching devices are implemented at every zero cross points of the input voltage in order to achieve the switching at zero voltage. The loss from the switching devices at the converter can reduce drastically because the switching loss can be decreased nearly to zero by the implementation of the switching at zero voltage.

However, there are problems in this method, due to the inverse voltage pulses and clamp phenomena that are constant voltage areas occur shown as Fig. 8. Delta-sigma conversion generates the inverse voltage pulses in order to cancel the quantization error shown in Fig. 5(b). In the clamped parts, switching loss increases because the switching at zero voltage cannot be achieved. The clamp phenomenon occurs because when the phase angle between the output voltage and current becomes larger than 30 degrees, the DC link current flows backward to the power source in the inverter side. As a result, the DC link voltage is equaled to the snubber capacitor voltage.

In order to resolve this problem, the PDM based on SVM is proposed to apply for the generator of PDM patterns. The detail of this method is explained in the next section.

C. PDM Control based on Space Vector Modulation

Fig. 7(a) shows the PDM signal generation block based on the SVM. The selected vector signals generated from the SVM is the input to the D flip flop (D-FF). The plus/minus detection signal from the input voltage is used to detect the zero cross points of input voltage, and this signal is an input to the CLK of the D-FF. The output of the D-FF, "Q" is synchronized at the edge of plus/minus signal, which is the zero cross point of the input voltage. After that, the switching signals which correspond to the vectors in Fig. 7(b), are generated from output signals of the D-FF using pattern table in Fig. 7(c). These signals are Switching signals for IMC shown in Fig. 2.

Furthermore, in order to apply this method to CMC shown in Fig. 3, the switching signals are exclusive NORed with the polarity of the input voltage, because it is necessary to exchange the switching signals of the upper arm and the lower arm depending on the polarity of the input voltage.

Therefore, the switching patterns generated by SVM are quantized by the pulse whose width is half cycle of the input voltage and the switching timing is synchronized with the zero cross point of the input voltage. As a result, this proposed method based on SVM can achieve switching at zero voltage.

IV. EXPERIMENTAL RESULTS USING INDIRECT MATRIX CONVERTER AND CONVENTIONAL MATRIX CONVERTER

In this chapter, a prototype circuit of IMC and CMC are built and tested in order to demonstrate the validity of the proposed PDM methods.

First, the single-phase to three-phase IMC was tested in the experiment with the two PDM control methods in section IV-A and IV-B. A high-frequency power source, where the matching impedance is 50 Ω is used as the input of the circuit in order to confirm the basic principle of the two PDM control methods. Table I shows the

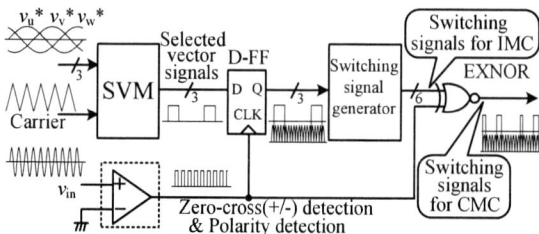

(a) PDM signals generation block diagram based on SVM.

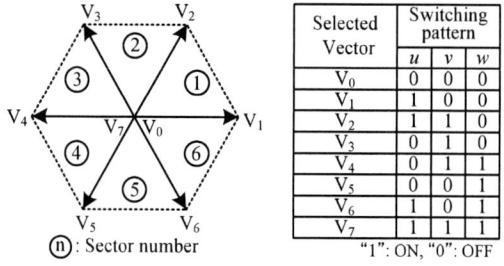

Selected Vector	Switching pattern		
	u	v	w
V_0	0	0	0
V_1	1	0	0
V_2	1	1	0
V_3	0	1	0
V_4	0	1	1
V_5	0	0	1
V_6	1	0	1
V_7	1	1	1

(n): Sector number   "1": ON, "0": OFF

(b) Space vector diagram.   (c) Switching pattern table.

Fig. 7. Control block diagram of the PDM control based on SVM

Table I. Experimental Parameters for IMC.

Parameter	Value
Input voltage	200 V
Input frequency	100 kHz
Output line-to-line voltage	100 V
Output frequency	50 Hz
Carrier frequency (in proposed PDM control based on SVM)	5 kHz
Load R_{load}	100 Ω
Load L_{load}	10 mH

Table II. Experimental Conditions for CMC.

Parameter	Value
Input voltage	70.7 V
Input frequency	100 kHz
Output line-to-line voltage	40 V
Output frequency	50 Hz
Carrier frequency	5 kHz
Load R_{load}	16 Ω
Load L_{load}	10 mH

experimental conditions.

Secondly, the single-phase to three-phase CMC was tested in the experiment with the proposed PDM control methods, which is based on SVM. A high-frequency power source, where the matching impedance is 0 Ω is used as the input of the circuit in this experiment. Table II shows the experimental conditions in order to demonstrate the validity of the proposed PDM methods in CMC in section IV-C.

A. Experimental Results using IMC with PDM Control using Delta-sigma Conversion

Fig. 8(a) shows the operation waveforms of the proposed circuit with delta-sigma conversion. From the result, the output voltage and current are 50-Hz sinusoidal waveforms. The PDM control operation using delta-sigma conversion can be verified and confirmed at the experimental results. Similarly to the simulation results, certain voltage pulses are inversed due to the failure on the pattern conversion.

Fig. 8(b) shows the extended view of the interval "A" in Fig. 8(a). The results confirm that the switching at zero voltage is approximately performed at the zero cross point of the input voltage waveform. However, the switching has a 1 μs delay due to the following reasons; (i) detection of the zero cross points is slow and (ii) the dead time of the inverter. The delay of the zero cross point detection is approximately 0.5 μs, and dead time of the inverter is set to 0.5 μs. The delay can be improved by modifying the zero cross point detection circuit. Thus, the dead time period can be improved by studying and evaluating the device parameters. Additionally, although the switching has a short delay, the surge voltage is lower than that of the hard switching method. However, due to the clamp phenomena, as covered in the section III-B, switching loss increases because switching at zero voltage cannot be achieved. Therefore, the circuit has low efficiency in this experiment.

In addition, Fig. 9 shows the harmonics analysis on the output voltage and input current. From (a), output voltage does not include low-order harmonic components, and the integral-multiple harmonic nearly closed to 200 kHz. The frequency is DC link voltage fluctuating frequency is included in the high-order harmonic components. The output voltage THD of 5.96% is obtained. Moreover, (b) shows that the input current has the integral-multiple harmonic nearly closed to 100 kHz, and therefore input current THD is 79.7%.

B. Experimental Results using IMC with PDM Control based on Space Vector Modulation

Fig. 10(a) shows the operation waveforms of the proposed circuit in the experiment with the PDM control based on SVM. The result confirms that 50 Hz sinusoidal waveforms are obtained on the output voltage and current. The PDM control operation based on SVM can be verified and confirmed from the experimental results. Notice there are no inverse voltage pulses in this case.

Fig. 10(b) shows the extended view of the interval "B" in Fig. 10(a). As a result, the output line-line voltage v_{uv} confirms that the switching of the inverter is happened at approximately each zero cross point of the 100 kHz sinusoidal input waveform. However, the switching has a 1 μs delay due to the same reasons as mentioned in the explanation of Fig. 8.

Fig. 11 shows the harmonics analysis on the output voltage from experimental result with the PDM control based on SVM. From the result, the low-order harmonic components closely to 50 Hz are not included in the output voltage, and the output voltage THD of 2.15% can be obtained. Therefore, the improvement of the output voltage waveform is confirmed. In this case, output voltage includes of the integral-multiple harmonic nearly to the carrier frequency of 5 kHz, which is the switching frequency. Additionally, the integral-multiple harmonic nearly to the 200 kHz is included in the output voltage

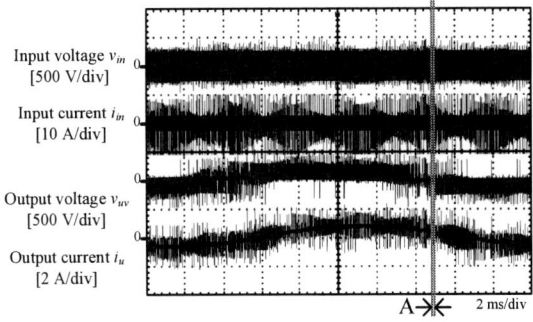

(a) Input and output operation waveforms.

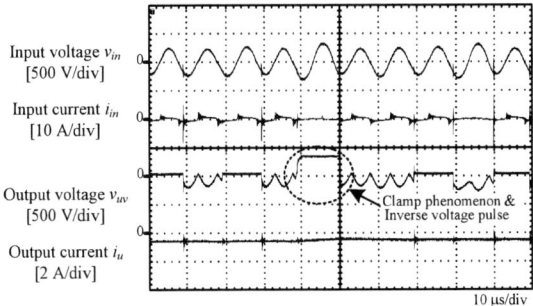

(b) Extended view of each operation waveform.

Fig. 8. Operation waveforms of the proposed circuit in the experiment with PDM using delta-sigma conversion in IMC.

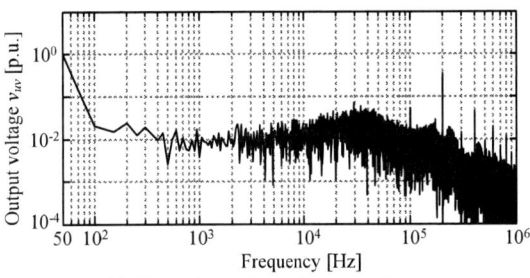

(a) Harmonics analysis of output voltage.

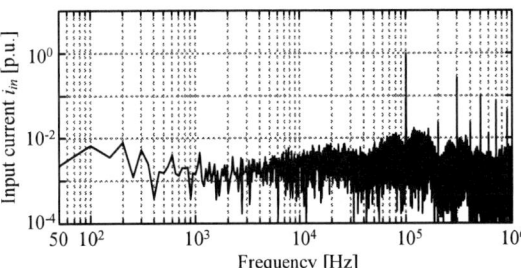

(b) Harmonics analysis of input current.

Fig. 9. Harmonics analysis of output voltage and input current with PDM control using delta-sigma conversion in IMC.

because of the same reason as case of Fig. 9. Furthermore, input current includes the integral-multiple harmonic nearly closed to 100 kHz and input current THD is 55.5%.

From these results, the validity of the proposed PDM control for waveform improvement is confirmed. Therefore, in next chapter, the proposed PDM control method is applied to CMC (direct type circuit) in order to check the basic operation of the PDM control in a direct

978-1-4799-2706-7/14 $31.00 © 2014 IEEE

type circuit.

C. Experimental Results using CMC with PDM Control based on Space Vector Modulation at low voltage condition

Fig. 12(a) shows the operation waveforms of the proposed circuit in the experiment with PDM control based on SVM. The result confirms that 50 Hz sinusoidal waveforms are obtained on the output voltage and current. The PDM control operation based on SVM can be verified and confirmed at the experimental results. However, the output waveform has a surge voltage because of the switching delay due to the zero cross point detection circuit.

Fig. 12(b) shows the extended view of the interval "C" in Fig. 12(a). As a result, the output line-to-line voltage v_{uv} confirms that the switching of the CMC is happened at approximately each zero cross point of the 100 kHz sinusoidal input waveform. Therefore, basic operation of the proposed PDM control based on SVM is confirmed.

Fig. 13(a) shows the harmonics analysis on the output voltage from experimental result with PDM control based on SVM. From the result, the low-order harmonic components of 50 Hz are not included in the output voltage, and the output voltage THD of 3.25% can be obtained. In addition, the output voltage includes the integral-multiple harmonic nearly to the carrier frequency of 5 kHz, which is the control cycle. Additionally, the integral-multiple harmonic nearly to the 200 kHz is included in the output voltage in the high-order harmonic components because the output voltage has the fluctuation, which is twice of the input frequency.

Furthermore, Fig. 13(b) shows the harmonics analysis on the input current. From the result, the input current includes the integral-multiple harmonic nearly closed to 100 kHz and the input current THD is 48.5%, which is large value. However, the harmonics can be eliminated by using small input filter with small capacitor because the cutoff frequency can be very high frequency.

Therefore, the basic operation of the proposed PDM control method in direct type circuit is confirmed from these results.

V. EFFICIENCY CHARACTERRISTICS AND LOSS ANALYSIS

Fig. 14 shows the efficiency characteristics of the indirect type circuit (IMC) with PDM control using delta-sigma conversion and PDM control based on SVM. The input and output voltage conditions are same as shown in table 1, and the output power is controlled by changing the load value. From the result, the efficiency of the circuit is improved as output power increases in both the PDM control methods. Furthermore, the efficiency for the PDM control based on SVM is higher than that for the PDM control using delta-sigma conversion at all measurement regions. This is because the switching loss decreases due to resolution of the clamp phenomenon and reduction of the switching frequency applying the PDM

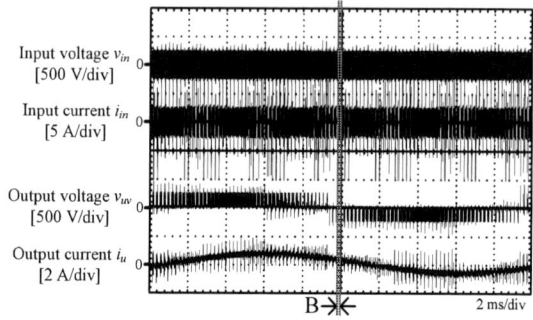

(a) Input and output operation waveforms.

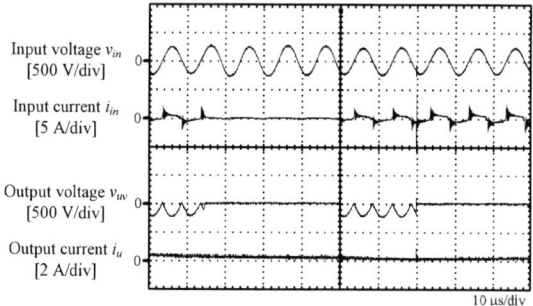

(b) Extended view of each operation waveform.

Fig. 10. Operation waveforms of the proposed circuit in the experiment based on SVM in IMC.

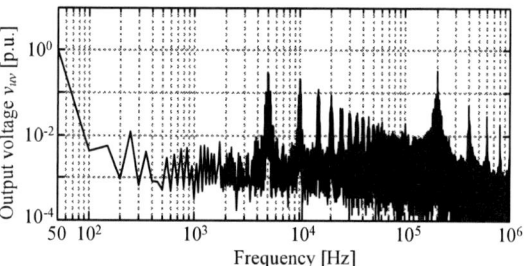

(a) Harmonics analysis of output voltage.

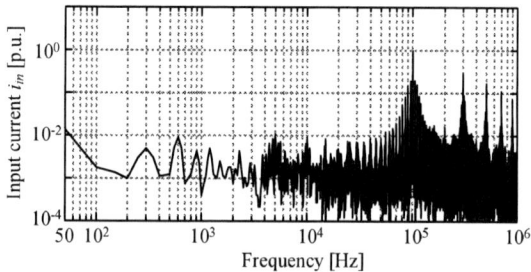

(b) Harmonics analysis of input current.

Fig. 11. Harmonics analysis of output voltage and input current with PDM control based on SVM in IMC.

control based on SVM. As a result, the validity of PDM control based on SVM is confirmed about efficiency improvement.

In addition, the maximum efficiency with PDM control using delta-sigma conversion and with PDM control based on SVM are 93.4% and 97.3% respectively.

Fig. 15 shows the efficiency characteristics of the direct type circuit (CMC) with PDM control based on SVM. The input and output voltage conditions are same as shown in table II, and the output power is controlled by changing the load value as the case of IMC. From the result, the maximum efficiency is 91.5% at load of 35 W. This value is lower than that of IMC. The cause of low efficiency in the CMC circuit is investigated from the result of the loss analysis.

Fig. 16 shows the result of the loss analysis. From the results, it is confirmed that the cause of low efficiency in the CMC circuit is the conduction loss of the switching device is larger than that of IMC. In this experiment, the current value, which flows through switching devices is larger than case of IMC to output same power because the input and output voltage of the circuit is lower than the case of IMC. It is a limitation of the voltage source, which is used in this experiment. In addition, the bidirectional switches have high on-resistance because the bidirectional switches used in CMC construct of two unidirectional switching devices connected in anti-series.

Therefore, the efficiency of IMC improves by making an experiment at high voltage condition and choosing the switching devices, which have lower on-resistance.

It is noted that the recovery losses of the freewheeling diodes (FWD) and diode rectifier are not included in this simulation. Thus the value of the total loss is lower than actual value.

VI. CONCLUSIONS

In this paper, the implementation of PDM control methods in a single-phase to three-phase matrix converter for high frequency application is discussed and evaluated by the experiments. In the PDM control using delta-sigma conversion, the output waveform has inverse voltage pulse and clamp phenomena on the output voltage waveform. Hence, an improvement method of the output waveform based on SVM is proposed. In SVM, the phase error between the voltage and current are dramatically reduced.

At first a prototype of indirect type circuit has been built and tested. From the results, the switching at zero voltage and clear sinusoidal output waveform have been confirmed. On the other hand, it is confirmed that the PDM control based on SVM can resolve the problems with clamp phenomenon. As a result, the THD of the output voltage with delta-sigma conversion and PDM pattern method based on SVM are 5.96% and 2.15% respectively. Furthermore, the maximum efficiency has been improved from 93.4% to 97.3% by applying the proposed PDM method. From the results, the validity of PDM control based on SVM has been confirmed.

Secondly, a prototype of direct type circuit was built and tested. From the results, the switching at zero voltage and clear sinusoidal output waveform have been confirmed too. As a result, the output voltage THD and input current THD, which is controlled by the proposed PDM based on SVM are 3.25% and 48.5%, respectively.

In addition, the AC-to-AC converter for the high-frequency application such as the proposed circuit is necessary in other case. For example, the MC for high-

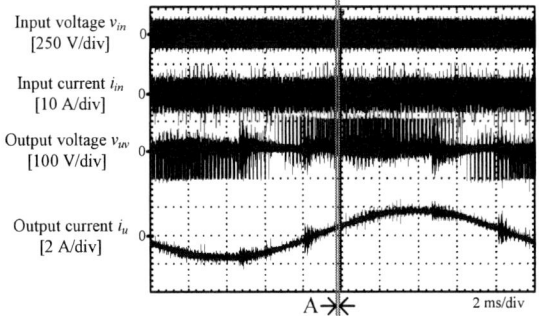
(a) Input and output operation waveforms.

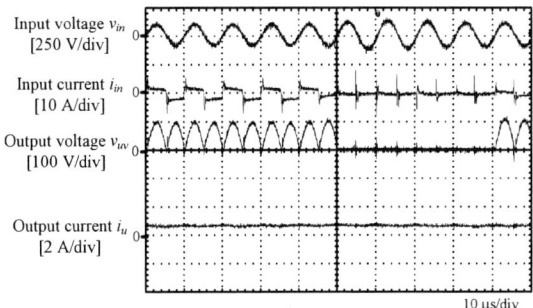
(b) Extended each operation waveform.
Fig. 12. Operation waveforms of the proposed circuit in the experiment based on SVM in CMC.

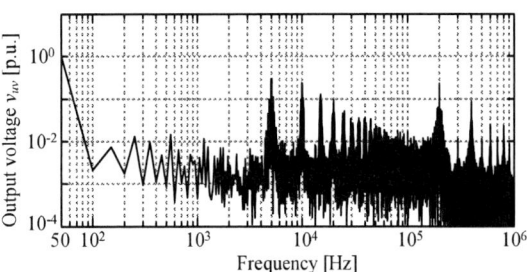
(a) Harmonics analysis of output voltage.

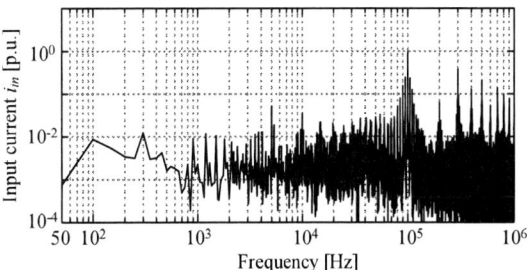
(b) Harmonics analysis of input current.
Fig. 13. Harmonics analysis of output voltage and input current with PDM control based on SVM in CMC.

frequency application can be applied to trans-linked converter, whose secondary side is connected to commercial grid. For downsizing the converter, the transformer for high-frequency link size is a major bottleneck [14]. Therefore, the size of the transformer can be small by higher frequency converter using high-speed switching devices such as SiC and GaN.

In future work, efficiency characteristics of the proposed circuit will be measured, and the zero cross

978-1-4799-2706-7/14 $31.00 © 2014 IEEE   144

point detection circuit will be improved to reduce the switching loss and the switching surge.

REFERENCES

[1] Takehiro IMURA, Yoichi HORI : "Wireless power transfer using electromagnetic resonant coupling", The Journal of The Institute of Electrical Engineers of Japan, vol.129, No.7 pp.414-417, July 2009 (in Japanese)

[2] Keisuke Kusaka and Jun-ichi Itoh, "Experimental verification of rectifiers with SiC/GaN for wireless power transfer using a magnetic resonance coupling", The 9th IEEE International Conference on Power Electronics and Drive Systems, pp. 1094-1099, December 2011

[3] Keisuke Kusaka, Satoshi Miyawaki and Jun-ichi Itoh : "A experimental evaluation of a SiC schottky barrier rectifier with a magnetic resonant coupling for contactless power transfer as a power supply", 2010 Annual Conference of IEEJ, Industry Applications Society, No.1-41, 2010 (in Japanese)

[4] A.Kurs, A. Karalis, R. Moffatt, J. D. Joannopoulos, P. Fisher and M. Soljačić, "Wireless power transfer via strongly coupled magnetic resonances", Science, vol.317, pp.83-86, June 2007

[5] Patrick W. Wheeler, José Rodríguez, Jon C. Clare, Lee Empringham and Alejandro Weinstein, "Matrix converters : A technology review", IEEE Transactions on Industry Electronics, Vol. 49, No. 2, pp.274-288, April 2002

[6] Yugo Tadano, Shizunori Hamada, Shota Urushibata, Masakatsu Nomura, Yukihiko Sato and Muneaki Ishida, "A space vector modulation scheme for matrix converter that gives top priority to the improvement of the output control performance", IEEJ Transactions on Industry Applications, Vol.128, No.5, pp.631-641, May 2008 (in Japanese)

[7] Jun-ichi Itoh, Ikuya Sato, Hideki Ohguchi, Kazuhisa Sato, Akihiro Odaka and Naoya Eguchi, "A control method for the matrix converter based on virtual AC/DC/AC conversion using carrier comparison method", IEEJ Transactions on Industry Applications, vol.124, No.5, pp.457-463, May 2004 (in Japanese)

[8] Y. L. Feng, Y. Konishi, M. Nakaoka, "Current-fed soft-switching inverter with PDM-PWM control scheme for ozone generation tube drive", IEEJ Transactions on Industry Applications, Vol.120, No.10 pp.1239-1240, October 2000 (in Japanese)

[9] P.K.Sood and T.A.Lipo : "Power Conversion Distribution System using a High-Frequency AC Link," IEEE Trans. on IA, Vol.IA-24, No.2, pp.228-300 (1988)

[10] Hisayuki Sugimura, Bishwajit Saha, H. Omori, Hyun-Woo Lee, M. Nakaoka, "Single reverse blocking switch type pulse density modulation controlled ZVS inverter with boost transformer for dielectric barrier discharge lamp dimmer", IEEE 5th International Power Electronics and Motion Control Conference, 2006, vol. 2, pp. 1 - 5, August 2006

[11] Abdelhalim Sandali, Ahmed Cheriti and Pierre Sicard, "Design considerations for PDM Ac/ac converter implementation", Applied Power Electronics Conference 2007, pp. 1678 - 1683, February 2007

[12] Yuki Nakata and Jun-ichi Itoh, "An experimental verification and analysis of a single-phase to three-phase matrix converter using PDM control method for high-frequency applications", The 9th IEEE International Conference on Power Electronics and Drive Systems, No. 383, pp.1084-1089, December 2011

[13] Yuki Nakata and Jun-ichi Itoh, "Control methods of an indirect-type Single-phase to three-phase matrix converter for high-frequency applications", Workshop on Semiconductor Power Converter, SPC-12-029, January 2012 (in Japanese)

[14] Masayoshi Yamamoto, Hiroyuki Horii, "Trans-linked single phase interleaved PFC converter", IEEJ Transactions on Industry Applications, Vol.130, No.6, pp.828-829, June 2010 (in Japanese)

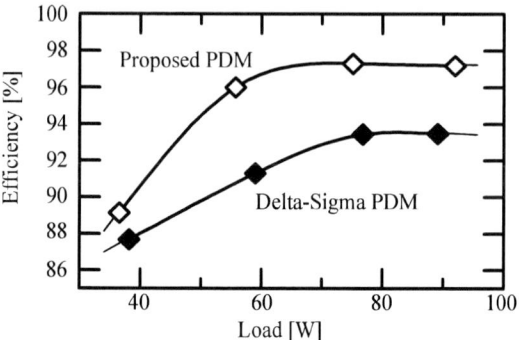

Fig. 14. Characteristics of the proposed circuit's efficiency in the experiment using IMC.

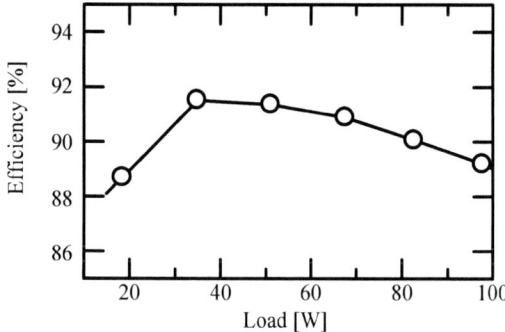

Fig. 15. Characteristics of the proposed circuit's efficiency in the experiment using CMC.

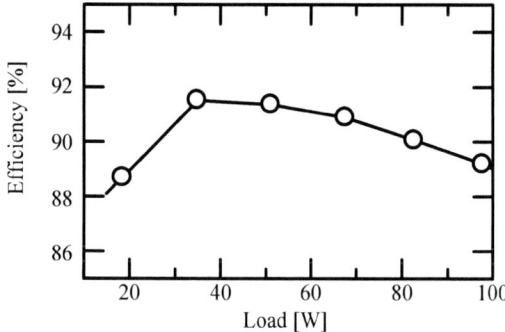

Fig. 16. Loss analysis of the IMC and the CMC with 100-W load.

The 2014 International Power Electronics Conference

Steady State Characteristics of the Boost-Type Matrix Converter for Stand-Alone Power Source

Y.Nagano, N.Yamamura, M.Ishida, K. Hirokado

Mie-university, 1577, Kurimamachiya,Tsu,Mie,Japan. 514-8507

Abstract— This paper describes steady state characteristics of the boost-type matrix converter for three-phase four wire system. The converter is intended for use in a stand-alone power source with constant voltage and frequency. The conventional method however is not able to control the output-voltage in the case of low power factor load. To solve the problem, we propose a new control method of the matrix converter by reviewing the conventional control method. To apply this system for stand-alone power source, we analyze steady state characteristics, and formulate relationship among circuit parameters, the output voltage and output power.

Keywords— *matrix conver, boost-type, three-phase four wire, stand-alone source*

I. INTRODUCTION

A boost-type matrix converter is the matrix converter having a function of voltage boost. The merit of the boost-type matrix converter is to expand the input voltage range of use. The boost-type matrix converter is intended for use in a stand-alone power source with constant output voltage and frequency, even in the case that the voltage and frequency of the input power generator are variable.

Thus, it is necessary to establish output voltage for various loads including no power-factor load. However, the conventional control method is not able to control the output voltage in the case of low power-factor load [1]. To cope with this problem, the new control method was proposed [2].

However, the new control method has not been analyzed to establish the design method for the given output voltage and circuit parameters. So, as preparation for it, this paper presents analytical results of the steady state characteristics of the boost-type matrix converter.

II. THE BOOST-TYPE MATRIX CONVERTER

A. Circuit Configuration

Fig. 1 shows the circuit configuration of the boost-type matrix converter for the three-phase four wire system. The main circuit is composed of three balanced voltage sources, input L_{in} filters, nine bidirectional switches S_{a1}~S_{c3}, output filters C_{out}, and three-phase loads.

Relationship between the output currents i_u, i_v, i_w averaged in every control period T_S and the input currents i_a, i_b, i_c of the boost-type matrix converter is given as follows:

$$\begin{bmatrix} \bar{i_u} \\ \bar{i_v} \\ \bar{i_w} \end{bmatrix} = \begin{bmatrix} a_1 & b_1 & c_1 \\ a_2 & b_2 & c_2 \\ a_3 & b_3 & c_3 \end{bmatrix} \begin{bmatrix} i_a \\ i_b \\ i_c \end{bmatrix} \tag{1}$$

Relationship between the input voltages v_a, v_b, v_c averaged in every control period T_S and the output voltages v_u, v_v, v_w of the boost-type matrix converter is given as follows:

$$\begin{bmatrix} \bar{v_a} \\ \bar{v_b} \\ \bar{v_c} \end{bmatrix} = \begin{bmatrix} a_1 & a_2 & a_3 \\ b_1 & b_2 & b_3 \\ c_1 & c_2 & c_3 \end{bmatrix} \begin{bmatrix} v_u \\ v_v \\ v_w \end{bmatrix} \tag{2}$$

Here, we define control functions a_1~c_3 as duty factors in every control period T_S corresponding to the switches S_{a1}~S_{c3}, respectively.

B. Control Functions

Two type of the control functions, method I and method II, are derived by using the principle of the coordinate transformation. Spatial vectors rotating with angular speed of ω_S in fixed coordinate space $(\alpha_S$-$\beta_S)$ are transformed to spatial vectors rotating with angular speed of ω_L viewed from the coordinate space $(\alpha_L$-$\beta_L)$ rotating with angular speed of ω_S+ω_L or ω_S-ω_L.

The control functions of the method I are given as follows:

$$\begin{bmatrix} a_1 & b_1 & c_1 \\ a_2 & b_2 & c_2 \\ a_3 & b_3 & c_3 \end{bmatrix} = \begin{bmatrix} f_1 & f_2 & f_3 \\ f_2 & f_3 & f_1 \\ f_3 & f_1 & f_2 \end{bmatrix} \tag{3}$$

$$\begin{bmatrix} f_1 \\ f_2 \\ f_3 \end{bmatrix} = A \begin{bmatrix} \cos(\theta_S + \theta_L + \varphi) \\ \cos(\theta_S - 2\pi/3 + \theta_L + \varphi) \\ \cos(\theta_S + 2\pi/3 + \theta_L + \varphi) \end{bmatrix} + \begin{bmatrix} 1/3 \\ 1/3 \\ 1/3 \end{bmatrix} \tag{4}$$

The control functions of the method II are given as follows:

$$\begin{bmatrix} a_1 & b_1 & c_1 \\ a_2 & b_2 & c_2 \\ a_3 & b_3 & c_3 \end{bmatrix} = \begin{bmatrix} g_1 & g_2 & g_3 \\ g_3 & g_1 & g_2 \\ g_2 & g_3 & g_1 \end{bmatrix} \tag{5}$$

$$\begin{bmatrix} g_1 \\ g_2 \\ g_3 \end{bmatrix} = A \begin{bmatrix} \cos(\theta_S - \theta_L - \varphi) \\ \cos(\theta_S - 2\pi/3 - \theta_L - \varphi) \\ \cos(\theta_S + 2\pi/3 - \theta_L - \varphi) \end{bmatrix} + \begin{bmatrix} 1/3 \\ 1/3 \\ 1/3 \end{bmatrix} \tag{6}$$

Here, A is amplitude modulation factor. θ_S and ω_S are the phase angle and its angular speed of the source

978-1-4799-2706-7/14 $31.00 © 2014 IEEE

voltage, respectively. θ_L and ω_L are the phase angle and its angular speed reference angle of the output voltage of the converter, respectively. φ is the phase command of the output voltage, which is defined as the initial phase of the output current vector $\dot{I}_L$ referred the input current vector $\dot{I}_S$ and also that of the input voltage vector $\dot{V}_S$ referred to the output voltage vector $\dot{V}_L$ when $\theta_L = \theta_S = 0$.

C. Relationship between the Input and the Output voltages and currets of the Boost-Type Matrix Converter

The source voltages (spatial vector $\dot{V}_{S0}$) are given as follows:

$$\begin{bmatrix} v_{a0} \\ v_{b0} \\ v_{c0} \end{bmatrix} = V_{S0} \begin{bmatrix} \cos\theta_S \\ \cos(\theta_S - 2\pi/3) \\ \cos(\theta_S + 2\pi/3) \end{bmatrix} \quad (7)$$

Here, V_{S0} is the amplitude of the source voltages. And, the input currents (spatial vector $\dot{I}_S$) are given as follows:

$$\begin{bmatrix} i_a \\ i_b \\ i_c \end{bmatrix} = I_S \begin{bmatrix} \cos(\theta_S - \varphi_S) \\ \cos(\theta_S - 2\pi/3 - \varphi_S) \\ \cos(\theta_S + 2\pi/3 - \varphi_S) \end{bmatrix} \quad (8)$$

Here, I_S is the amplitude of the input currents. The average values of the output currents (spatial vector $\dot{I}_L$) controlled by method I are obtained from (1),(3),(4) and (8) as follows:

$$\begin{bmatrix} \bar{i}_u \\ \bar{i}_v \\ \bar{i}_w \end{bmatrix} = \frac{3}{2} A I_S \begin{bmatrix} \cos(\theta_L + \varphi_S + \varphi) \\ \cos(\theta_L - 2\pi/3 + \varphi_S + \varphi) \\ \cos(\theta_L + 2\pi/3 + \varphi_S + \varphi) \end{bmatrix} \quad (9)$$

And, the average values of the output currents (spatial vector $\dot{I}_L$) controlled by method II are obtained from (1),(3),(4) and (8) as follow as follows:

$$\begin{bmatrix} \bar{i}_u \\ \bar{i}_v \\ \bar{i}_w \end{bmatrix} = \frac{3}{2} A I_S \begin{bmatrix} \cos(\theta_L - \varphi_S + \varphi) \\ \cos(\theta_L - 2\pi/3 - \varphi_S + \varphi) \\ \cos(\theta_L + 2\pi/3 - \varphi_S + \varphi) \end{bmatrix} \quad (10)$$

The output voltages (spatial vector $\dot{V}_L$) are established by output currents flowing into output filter C_{out} and the three-phase loads as follows:

$$\begin{bmatrix} v_u \\ v_v \\ v_w \end{bmatrix} = V_L \begin{bmatrix} \cos\theta_L \\ \cos(\theta_L - 2\pi/3) \\ \cos(\theta_L + 2\pi/3) \end{bmatrix} \quad (11)$$

Here, we introduce output-side d-q coordinates synchronously rotating with the output voltage vector $\dot{V}_L$. And moreover, the current phase command φ is set as follows for the method I and method II, respectively, in order to coincide the output vector $\dot{V}_L$ with the d axis.

Method I... $\varphi = -\varphi_S - \varphi_L$ (12)

Method II... $\varphi = \varphi_S - \varphi_L$ (13)

Then, in the case of the method I, the average values of the input voltages (spatial vector $\dot{V}_S$) are obtained from (2),(3),(4),(11) and (12) as follows:

$$\begin{bmatrix} \bar{v}_a \\ \bar{v}_b \\ \bar{v}_c \end{bmatrix} = \frac{3}{2} A V_L \begin{bmatrix} \cos(\theta_S - \varphi_S - \varphi_L) \\ \cos(\theta_S - 2\pi/3 - \varphi_S - \varphi_L) \\ \cos(\theta_S + 2\pi/3 - \varphi_S - \varphi_L) \end{bmatrix} \quad (14)$$

In the case of the method II, the average values of the input voltages (spatial vector $\dot{V}_S$) are obtained from (2),(5),(6),(11) and (13) as follows:

$$\begin{bmatrix} \bar{v}_a \\ \bar{v}_b \\ \bar{v}_c \end{bmatrix} = \frac{3}{2} A V_L \begin{bmatrix} \cos(\theta_S - \varphi_S + \varphi_L) \\ \cos(\theta_S - 2\pi/3 - \varphi_S + \varphi_L) \\ \cos(\theta_S + 2\pi/3 - \varphi_S + \varphi_L) \end{bmatrix} \quad (15)$$

Here, we also define input side d-q coordinates synchronously rotating with the source voltage vector $\dot{V}_{S0}$ where the d-axis coincides with $\dot{V}_{S0}$

Fig. 2 shows relationships of the spatial vectors on dq coordinates in the case of the method I. The inverted load factor angle φ_L appears in the phase of the input currents with respect to that of the input voltages. Fig. 3 shows relationships of the spatial vectors on dq coordinates in the case of the method II. The load factor angle φ_L appears in the phase of the input currents with respect to that of the input voltages.

Relationships between amplitudes of the input voltages V_S, the output voltages V_L, the output currents I_L and the input currents I_S are obtained from (8),(9),(10),(11),(14) and (15) as follows:

$$V_S = (3A/2) \cdot V_L, \quad I_L = (3A/2) \cdot I_S \quad (16)$$

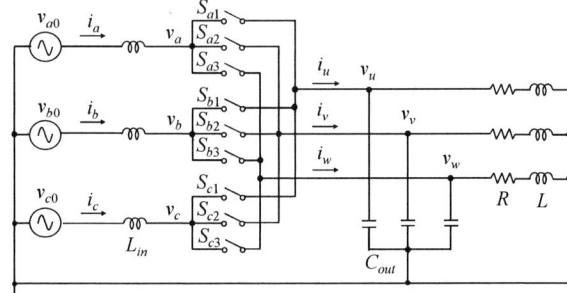

Fig. 1. Configuration of Boost-type matrix converter

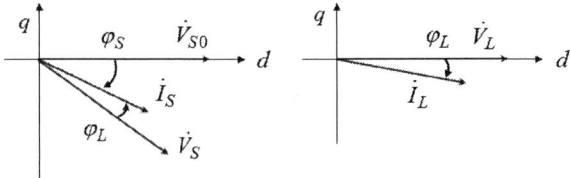

(a) input   (b)output

Fig. 2. Spatial vector in case of control method I

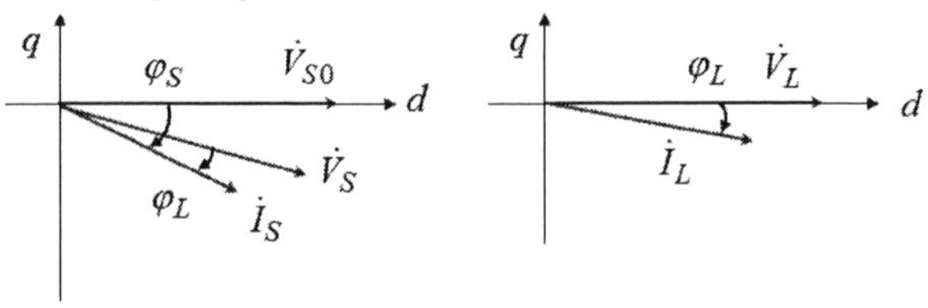

(a) input   (b)output

Fig. 3. Spatial vector in case of control method II

III. PROPOSE THE ANALYTICAL MODEL OF THE BOOST-TYPE MATRIX CONVERTER IN STEADY STATE

A. Proposal of the Steady-State Analytical Model

We propose a steady-state analytical model of the boost-type matrix converter. The bidirectional switches and three phase balanced voltage source are assumed to be ideal. The three-phase loads are supposed to be balanced.

Therefore, we analyze the steady-state characteristics of the boost-type matrix converter using one-phase equivalent circuit as shown in Fig. 4. In Fig. 4, the connected load is expressed as an admittance $\dot{Y}_{L0}$, and $\dot{Y}_L$ is a total admittance denoted as $Y_L \angle \varphi_L$ including the output filter. The matrix converter (MC) is regarded as a transformer. Note that the frequencies on the input and the output sides of MC are different with each other. Fig. 5 shows the equivalent circuit referred to the primary side. In Fig. 5, $\dot{Y}'_L$ dontes referred $\dot{Y}_L$ to the primary side and obtained from Fig. 2, Fig. 3 and (16) as follows:

Method I

$$\dot{Y}'_L = \frac{\dot{I}_S}{\dot{V}_S} = \frac{I_S}{V_S} \angle -\varphi_L = \left(\frac{2}{3A}\right)^2 \cdot Y_L \angle -\varphi_L \qquad (17)$$

Method II

$$\dot{Y}'_L = \frac{\dot{I}_S}{\dot{V}_S} = \frac{I_S}{V_S} \angle +\varphi_L = \left(\frac{2}{3A}\right)^2 \cdot Y_L \angle +\varphi_L \qquad (18)$$

The angle of the Y'_L is inverse of the angle of the Y_L in the case of the method I. But, in the case of the method II, the angle of Y'_L is the same angle as Y_L. The input voltages of the MC for the method I or method II is obtained from Fig. 5 and (17) or (18) as follows:

Method I

$$\dot{V}_S = \frac{\dot{V}_{S0}}{1 + j\omega_S L_{in}(2/3A)^2 Y_L \angle -\varphi_L} \qquad (19)$$

Method II

$$\dot{V}_S = \frac{\dot{V}_{S0}}{1 + j\omega_S L_{in}(2/3A)^2 Y_L \angle +\varphi_L} \qquad (20)$$

The output voltage V_L is obtained from (16) and (19) or (20) as follows:

Method I

$$\dot{V}_L = \frac{(3A/2)V_{S0}}{\sqrt{\{\omega_S L_{in} G\}^2 + \{(3A/2)^2 + \omega_S L_{in} B\}^2}} \qquad (21)$$

Method II

$$\dot{V}_L = \frac{(3A/2)V_{S0}}{\sqrt{\{\omega_S L_{in} G\}^2 + \{(3A/2)^2 - \omega_S L_{in} B\}^2}} \qquad (22)$$

B. Validity of the Analytical model

The validity of the equations (21) and (22) is verified by comparing with circuit simulation results. Fig .6 shows the simulation circuit of the system. In the circuit simulation, the bidirectional switches and the three phases balanced voltage source are assumed to be ideal. The three-phase loads are supposed to be balanced. Table

I shows the simulation parameters of the system. Three cases (a),(b) and (c) of the load parameters shown in Table II are simulated. Fig. 7 shows the simulation result of the output voltage for the method I in the case of (c), no-load. Fig. 8 shows FFT result of the voltage waveform in Fig.7. The amplitude of the output voltage is 228.32Vrms. And, the value obtained from equation (21) is 230.46Vrms. Table III shows the comparison results of the circuit simulation and the values obtained from the equations (21) and (22) in the case of the connected loads is (a),(b) and (c) as shown in Table II, respectively.

Approximation errors exist between the simulation results and analytical ones as shown in Table III (for method I) and Table IV (for method II). The reason of the errors is considered that there exist some ripple components in the output voltages and the input currents while no ripple components are assumed in the analysis of the input voltage V_S and the output voltage V_L. From Table III and Table IV, it is found that approximation errors are small, and it can be ignored. Therefore, the steady state characteristics is approximately expressed by the equation (21),(22) and Fig. 5.

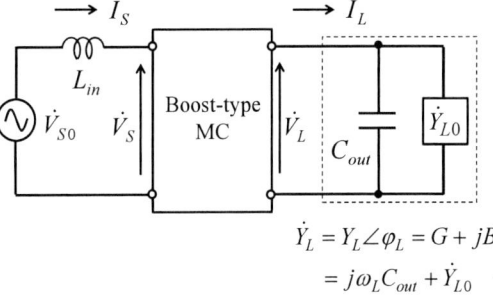

$$\dot{Y}_L = Y_L \angle \varphi_L = G + jB$$
$$= j\omega_L C_{out} + \dot{Y}_{L0}$$

Fig. 4. Single phase equivalent circuit of the boost-type matrix converter.

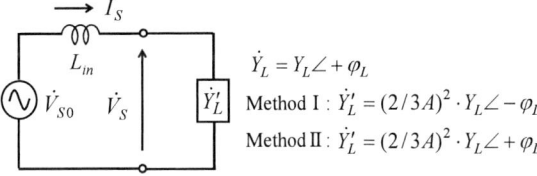

Fig. 5. Primary-side equivalent circuit of Fig. 4.

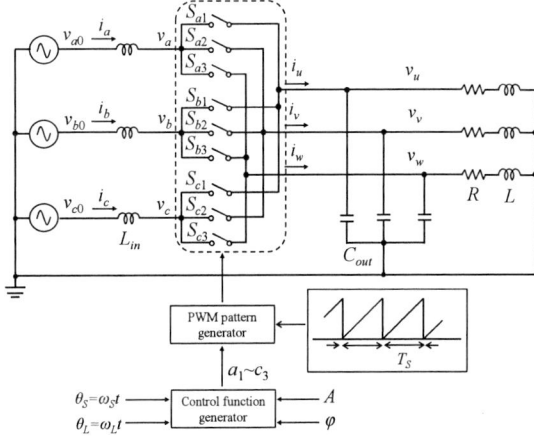

Fig. 6. Control system configuration of MC for sumulation.

TABLE I
CIRCUIT PARAMETERS

Symbol	Meaning	Value
V_{S0}	Source voltage	100Vrms
ω_S	Input angular frequency	30Hz
L_{in}	Input filter reactor	0.01H
ω_L	Output angular frequency	60Hz
C_{out}	Output filter capacitor	10μF
A	Amplitude modulation factor	0.25
φ	Phase command	0deg

TABLE II
CONNECTED LOAD

Case	Connected Load	Value
(a)	R	100Ω
(b)	R, L	100Ω,0.1H
(c)	no-load	

TABLE III
THE OUTPUT VOLTAGE (METHOD I)

Case	CIRCUIT SIMULATION	ANALYTICAL MODEL	ERROR
(a)	228.32Vrms	230.01 Vrms	0.74%
(b)	237.80Vrms	239.47 Vrms	0.70%
(c)	230.46 Vrms	231.56 Vrms	0.47%

TABLE IV
THE OUTPUT VOLTAGE (METHOD II)

Case	CIRCUIT SIMULATION	ANALYTICAL MODEL	ERROR
(a)	309.25 Vrms	310.47 Vrms	0.40%
(b)	297.78 Vrms	296.19 Vrms	0.54%
(c)	316.82 Vrms	314.32 Vrms	0.78%

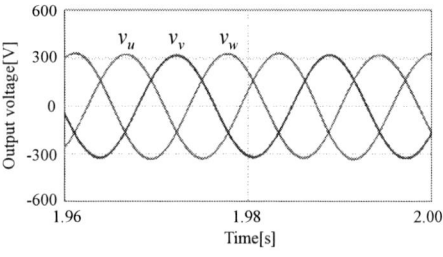

Fig. 7 Simulation result of the output voltages v_u, v_v, v_w.

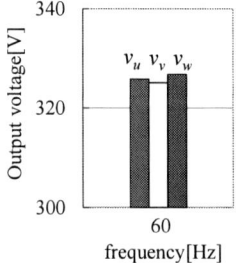

Fig. 8 FFT result of the output voltages in Fig. 7.

IV. THE LOAD AREA OF GENERATIONG THE OUTPUT VOLTAGE

The load condition (G, B) for the given output voltage V_L is obtained from (21) and (22) for various values of A as follows:

Method I

$$G^2 + \left\{ B + \omega_L C_{out} + \frac{(3A/2)^2}{\omega_S L_{in}} \right\}^2 = \frac{(3A/2)\cdot(V_{S0}/V_L)}{\omega_S L_{in}} \quad (23)$$

Method II

$$G^2 + \left\{ B + \omega_L C_{out} - \frac{(3A/2)^2}{\omega_S L_{in}} \right\}^2 = \frac{(3A/2)\cdot(V_{S0}/V_L)}{\omega_S L_{in}} \quad (24)$$

Moreover, the region of A is obtained from (3) to (6) considering $0 \le a_1 \sim c_3 \le 1$ and as a result $0 \le A \le 1/3$.

Therefore, the load area for the given output voltage is obtained from (23) and (24) as follows:

Method I

$$B \le -\omega_S L_{in} \left(\frac{V_L}{V_{S0}} \right)^2 G^2 + \frac{1}{4\omega_S L_{in}} \left(\frac{V_{S0}}{V_L} \right)^2 \quad (25)$$

$$G^2 + \left(B + \frac{1}{4\omega_S L_{in}} \right)^2 \le \left(\frac{1}{2\omega_S L_{in}} \cdot \frac{V_{S0}}{V_L} \right)^2$$
$$※B \le -\frac{1}{4\omega_S L_{in}} \left\{ 1 - 2 \left(\frac{V_{S0}}{V_L} \right)^2 \right\} \quad (26)$$

Method II

$$B \ge \omega_S L_{in} \left(\frac{V_L}{V_{S0}} \right)^2 G^2 - \frac{1}{4\omega_S L_{in}} \left(\frac{V_{S0}}{V_L} \right)^2 \quad (27)$$

$$G^2 + \left(B - \frac{1}{4\omega_S L_{in}} \right)^2 \le \left(\frac{1}{2\omega_S L_{in}} \cdot \frac{V_{S0}}{V_L} \right)^2$$
$$※B > \frac{1}{4\omega_S L_{in}} \left\{ 1 - 2 \left(\frac{V_{S0}}{V_L} \right)^2 \right\} \quad (28)$$

Here, the load area to generate the desired output voltage is given by fulfilling (25) and (26) for the control method I. And it is given by fulfilling (27) and (28) for the control method II. In the case of the parameters of the boost-type matrix converter as listed in Table I, the load areas are calculated as shown in Fig. 4.1.

The load areas for the method I and method II have symmetrical relations to the line of $B= -j\omega_L C_{out}$. The reason of this is that in the case of the method I, the inverted output-side power factor angle φ_L appears in the phase of the input currents with respect to that of the input voltages, and in the case of the method II, the same φ_L appears there. Therefore, the method I is suitable for inductive loads and method II is suitable for captive loads as shown Fig. 4.1.

The steady state operations in the case of the method II are simulated. Fig. 4.2 shows enlarged view of Fig. 4.1 (b). Figs. 4.3, 4.4, 4.5 and 4.6 show the input space vectors at the load conditions corresponding to the marked points ×, ▲, ■ in Fig. 4.2.

In the case of the mark × (capacitive load), the output susceptance becomes large due to the output filter susceptance. The reactance negative by this susceptance and the reactance of the input filter reactor are canceled each other. So, the input reactance becomes small viewing from the power source, resulting in large input current as shown in Fig. 4.3.

In the case of the mark ▲ (inductive load), the output susceptance becomes small due to the output filter susceptance, but still negative. So, the input reactance becomes large viewing form power source, resulting in small current as shown in Fig. 4.4.

There exist two values of the amplitude modulation factor A in any load point below the load power circle in the case of $A=1/3$. In the case of the mark ■ (resistance load), for example, the output load $\dot{Y}_{L0}$ is pure conductance that is pure resistance load. So the output power factor is a little leading, due to the output filter capacitor that result in the leading power factor on the input side of the MC. In this case $A=0.178$ and $A=0.050$ are satisfied with the condition of (24). Fig. 4.5 shows the input current and voltage vectors obtained from analytical equation (21). Fig. 4.8 shows the simulation results of the input current and source voltage waveforms in the case of $A=0.178$. Fig. 4.6 shows the input current and voltage vectors obtained from analytical equation (22). Fig. 4.9 shows the simulation results of the input current and source voltage waveforms in the case of $A=0.050$ in this case, the input current is very large. The figures show the input current is small in case $A=0.178$. So, it is desired to select the larger value of the amplitude modulation factor A between the two solutions of (22) for A.

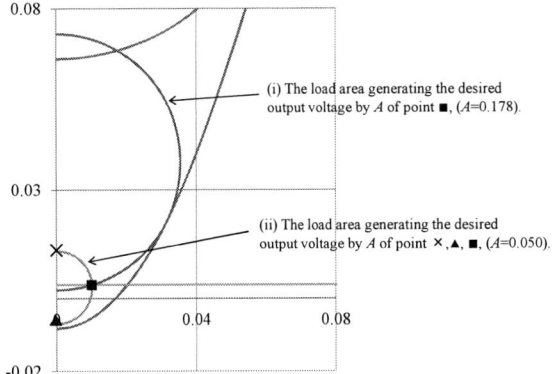

Fig. 4.2 enlarged view of Fig.4.1(b).

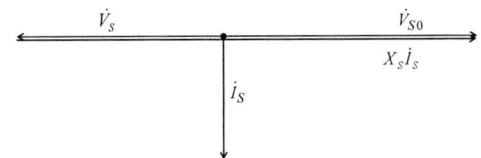

Fig. 4.3 Input vectors of mark ×.

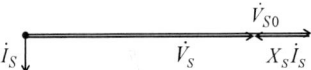

Fig. 4.4 Input vectors of mark ▲.

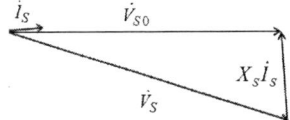

Fig. 4.5 Input vectors of mark ■ ($A=0.178$).

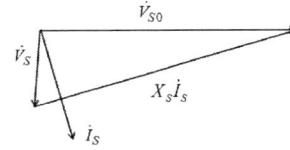

Fig. 4.6 Input vectors of sign ■ ($A=0.050$).

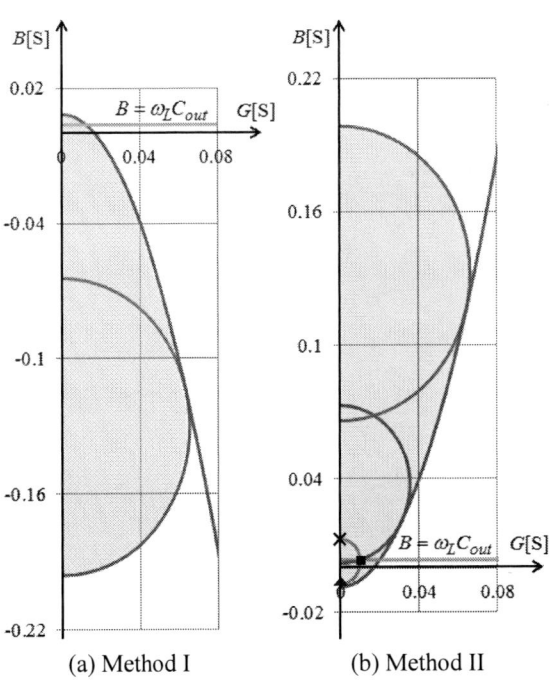

Fig. 4.1 the load areas Y_{L0}

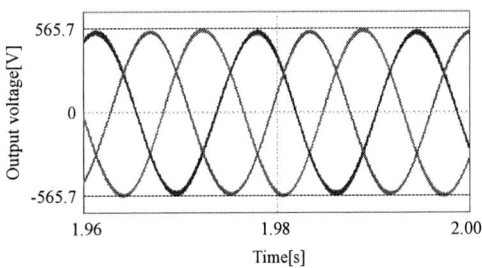

Fig. 4.7 The simulation result of the output voltages v_u, v_v, v_w marked ■ in Fig.4.1($A=0.178$).

978-1-4799-2706-7/14 $31.00 © 2014 IEEE   150

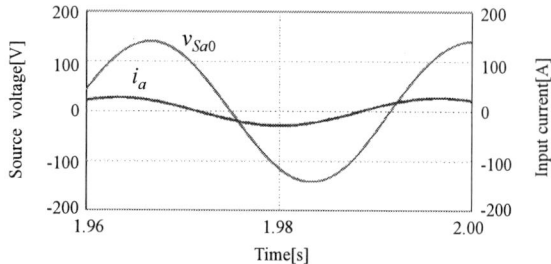

Fig. 4.8 Simulation results of the source voltage v_{Sa0} and the input current i_a marked ■ in Fig.4.1 (A=0.178).

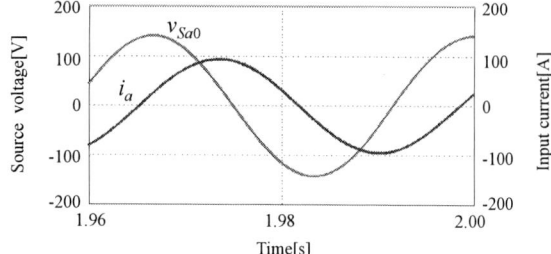

Fig. 4.9 Simulation results of the source voltage v_{Sa0} and the input current i_a marked ■ in Fig.4.1 (A=0.050).

V. CONCLUSIONS

In this paper, we analyzed the steady state characteristics of the boost-type matrix converter (MC). For two types of the proposed control methods of MC method I and method II, we showed the relationships among the output voltage, the output power and the circuit parameters (input source voltage input filter reactance, output filter capacitors).

The input/output modulation factor of the MC, and figures of input/output current and voltage vectors are shown by obtained from the analytical results.

We also showed the simulation results concerning the circuit behavior, which verify the analytical results. So, the usefulness of the analytical results were conformed by the simulation results.

The problem of the proposed MC is that the input power factor is not able to control flexibly. So, it is necessary to investigate and establish the input power factor control method.

REFERENCES

[1] T. Maegima, N. Yamamura, M. Ishida "The control method for the boost-type matrix converter," *Joint Technical Meeting on Electron Devices and Semiconductor Power Converter*, EDD-10-084,SPC-10-111,2010

[2] A. Inoue, N. Yamamura, M. Ishida "Improvement of the control method for the boost-type matrix converter," *Joint Technical Meeting on Electron Devices and Semiconductor Power Converter*, EDD-10-085,SPC-10-142,2010

Design Procedure for Output Current Control and Damping Control of Matrix Converter

Hiroki Takahashi and Jun-ichi Itoh
Dept. of Energy and Environmental Science
Nagaoka University of Technology
Nagaoka, Niigata, Japan
thiroki@stn.nagaokaut.ac.jp and itoh@vos.nagaokaut.ac.jp

Abstract— This paper presents a design procedure for an output current control and an output damping control of a matrix converter to suppress LC filter resonance and improve transient current response. With a conventional design method which gives preference to stability, the output damping control causes a large output current overshoot and a control bandwidth of the output current is affected. Therefore, in order to obtain a desired transient response keeping a stable operation, this paper describes a modified control block diagram which a reference filter is added in to suppress the output current overshoot and a detailed design procedure using Bode-diagrams and a flowchart. From the experimental results, the output damping control designed with the proposed method suppresses the filter resonance, which results in stable operation. In addition, the proposed design method reduces the output current overshoot of 60% and an error between the desired and the obtained control bandwidth in comparison with the conventional method in experiments.

Keywords— *Matrix converter, Resonance, Damping control*

I. INTRODUCTION

Recently, a matrix converter has attracted a lot of attentions as an interface converter of a wind turbine system and an elevator because this converter has no DC energy buffer such as bulky electrolytic capacitors [1]-[6]. A matrix converter promises to achieve higher efficiency, smaller size and longer life-time compared to a conventional BTB (back to back) system which consists of a PWM rectifier and a PWM inverter.

A matrix converter requires LC filters in the grid side to eliminate harmonic current due to switching operation. However, a matrix converter has a problem that a filter LC resonance is excited by a grid voltage fluctuation and an input current fluctuation. In general, the filter resonance is suppressed by using damping resistors connected in parallel with the filter inductors [7]. However, when a transformer is connected to the input side of the matrix converter such as a wind turbine system and an elevator which requires isolation between a load and the grid, a leakage inductance of the transformer is used as the filter inductors and the damping resistors are not inserted. Therefore, a suppression method for the filter resonance is needed.

In past works, some papers about stability analysis of a matrix converter to prevent the filter resonance have been presented [8]-[11]. These papers investigate the stability taking account into a detection of the input voltage with a filter [8], digital control [9]-[10] and matrix converter losses [10]. In addition, an approach to the stability based on admittance of a matrix converter has reported [11]. These investigations clarify stable limitations of a matrix converter operation with each method and suppress the filter resonance. However, the stability of a matrix converter can be improved more from the point of view of a current control of a matrix converter.

As control strategies to suppress the filter resonance, damping controls have been proposed [12]-[14]. The damping controls are separated into two types. The first one is a damping control combined with the input current control of the matrix converter [12]-[13] and another one is combined with the output current control [14]. Then, the output damping control has an advantage when the matrix converter is applied to a wind farm and an elevator which requires a field oriented control for accurate speed control. The advantage is that a required feedback control is on the output stage only while the matrix converter with the input damping control needs feedback loops on both of the input and the output sides.

The authors have already proposed the output damping control and its parameter design method [15]. The presented design method is based on a gain margin of the matrix converter and suppresses the filter resonance. However, the damping control designed with the conventional method based on the gain margin generates a large overshoot of the output current. Furthermore, a detailed design procedure to earn an intended control bandwidth of the output current has not been presented.

This paper proposes a modified control block diagram and a design procedure for the output current control and the output damping control to suppress the output current overshoot and to obtain an intended control bandwidth. The output current overshoot is reduced by an added reference filter for pole-zero cancellation and designed with an approximate model of the matrix converter. Moreover, a desired control bandwidth of the output current and a stable operation are achieved by a proposed design flowchart based on Bode-diagrams. The proposed reference filter and the design procedure result in fine output performance compared with the conventional method. This paper is organized as follows; firstly, an integrated block model and an approximate model of the matrix converter to design the output damping control and the output current control including the reference filter are described; secondly, the design flowchart for the controls is presented; finally, the stability and the transient response are evaluated in simulation and experiment.

II. SYSTEM CONFIGURATION

Fig. 1 shows a system block diagram of the matrix converter with the output damping control. For simplicity, the input transformer is assumed as an ideal transformer and a leakage inductance L_f, and the ideal transformer is omitted. For the same reason, a load motor is replaced with an R-L load. In addition, an output current control with a PI controller is used because a wind turbine system and an elevator require a field oriented control for accurate speed control. The output damping control which has a damping gain K_d and a damping HPF (high pass filter) of a time constant T_{hpf} is combined with the output current control feedback and suppresses the filter resonance. However, the output damping control generates a large current overshoot in a transient state because a closed-loop transfer function of the output current control has a zero due to the damping control. Hence, a reference filter $F(s)$ is added for a pole-zero cancellation. In order to derive an appropriate equation of $F(s)$ and obtain an intended control bandwidth and a stable operation, this paper deals with Bode-diagrams derived from an integrated block model which is composed of a main circuit and a control diagram. Moreover, a design procedure for the output damping control and the output current control is proposed.

III. DESIGN MODEL

A. Modeling of Matrix Converter based on Space Vector

In order to derive the integrated block model of a matrix converter, an output voltage and an input current equations are required. Then, these equations of a matrix converter based on the duty-cycle space vector have been presented [16]. In this paper, the integrated block model of the matrix converter which is composed of the main circuit and the control diagram is derived by using the duty-cycle space vector.

Fig. 2 shows the integrated block model of the matrix converter. Variables in bold font represent their space vectors. For example, an input voltage vector v_{in} is defined as following.

$$v_{in} = \frac{2}{3}\left(v_r + v_s e^{j\frac{2}{3}\pi} + v_t e^{-j\frac{2}{3}\pi}\right) \quad (1)$$

where, v_r is the input R-phase voltage, v_s is the input S-phase voltage and v_t is the input T-phase voltage.

In order to connect the input and the output circuits of the matrix converter, the duty-cycle space vector m_d and

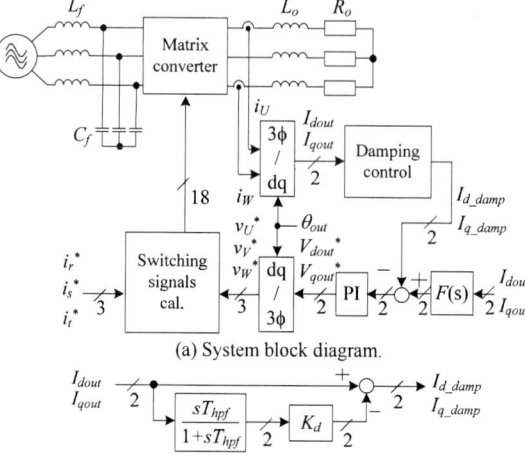

(a) System block diagram.

(b) Output damping control.

Fig. 1. Matrix converter with the output damping control to suppress the filter resonance and the proposed reference filter $F(s)$ to reduce the output current overshoot.

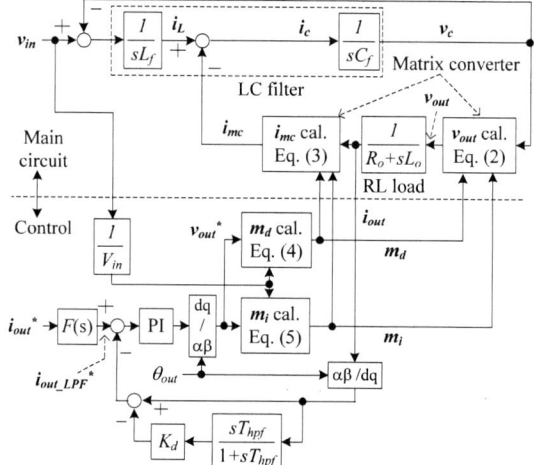

Fig. 2. Integrated block model of the matrix converter which is composed of the main circuit and the control diagram. This block diagram is based on the duty-cycle space vector.

m_i are introduced. Then, an output voltage vector v_{out} and an input current vector i_{mc} are yielded by (2) and (3).

$$v_{out} = \frac{3}{2}\overline{m_i}v_c + \frac{3}{2}m_d\overline{v_c} \quad (2)$$

$$i_{mc} = \frac{3}{2}m_i i_{out} + \frac{3}{2}m_d\overline{i_{out}} \quad (3)$$

where, v_c is a filter capacitor voltage vector and i_{out} is an output current vector. Note that a bar placed over a variable indicates complex conjugate. In contrast, m_d and m_i are represented by follows.

$$m_d = \frac{v_{in}}{3V_i}v_{out}^* \quad (4)$$

$$m_i = \frac{v_{in}}{3V_i}\overline{v_{out}^*} \quad (5)$$

where, $v_{out}{}^*$ is an output voltage reference vector and V_i is amplitude of the input voltage.

However, (2) and (3) are non-linear equations because time constants of fluctuation of v_c, m_d, m_i and i_{out} are close to each other. Therefore, a linear approximation method around a steady operating point is applied to these non-linear parts in order to draw Bode-diagrams. Equations (2) to (5) are separated into steady and differential components, and the differential components are expressed as (6) to (9).

$$\Delta v_{out} = \frac{3}{2}\left(\overline{\Delta m_i} v_{cs} + \overline{m_{is}} \Delta v_c\right) + \frac{3}{2}\left(\Delta m_d \overline{v_{cs}} + m_{ds} \overline{\Delta v_c}\right) \quad (6)$$

$$\Delta i_{mc} = \frac{3}{2}\left(\Delta m_i i_{outs} + m_{is} \Delta i_{out}\right) + \frac{3}{2}\left(\Delta m_d \overline{i_{outs}} + m_{ds} \overline{\Delta i_{out}}\right) \quad (7)$$

$$\Delta m_d = \frac{v_{ins}}{3V_i} \Delta v_{out}{}^* \quad (8)$$

$$\Delta m_i = \frac{v_{ins}}{3V_i} \overline{\Delta v_{out}{}^*} \quad (9)$$

where, suffix "s" represents the steady component based on its fundamental frequency while "Δ" means the differential component in a transient state. It should be noted that Δv_{in} is not considered because the output current response is evaluated in this paper.

Fig. 3 shows a linearized block model of the matrix converter regarding the differential components. Frequency characteristics of the output current control and the stability are obtained by using Bode-diagram. Note that analyses should be implemented with DC mode in which an input and an output angles are fixed because a rotating frequency of the steady vector does not equal to a rotating frequency of the differential vector.

B. Verification of the linearized model with simulation

Table 1 and Table 2 present the main circuit and the control parameters of the circuit model as illustrated in Fig. 1 and the linearized model as shown in Fig. 3. This section shows a simulation result without the output current control and the output damping control in order to evaluate a verification of the linearized model. Note that the inductor and capacitor parameters are normalized based on 50 Hz. In addition, a carrier frequency is increased to 100 kHz in this section only.

Fig. 4 shows an indicial response of the circuit model and the linearized model with an open-loop control in a simulation. It should be noticed that all waveforms are applied with a 1 kHz cut-off frequency LPF (low pass filter) in order to observe these average waveform without switching ripples. From Fig. 4, it is confirmed that the linearized model waveforms correspond to the circuit model result. Furthermore, error between the circuit model and the linearized model waveforms is less

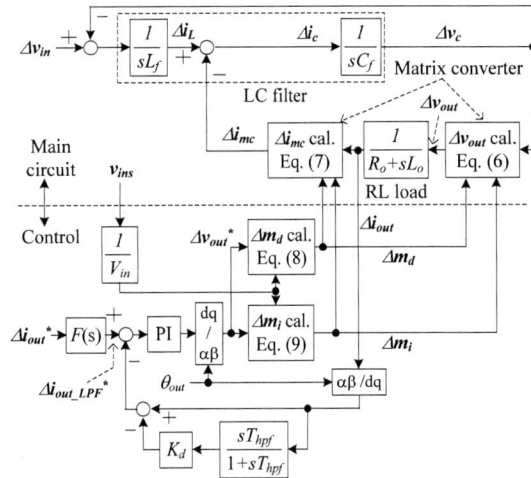

Fig. 3. Linearized block model of the matrix converter focusing on the differential components. Stability and a control bandwidth of the output current are evaluated by analyzing this block model.

TABLE 1. CIRCUIT PARAMETERS OF THE MATRIX CONVERTER

Input line voltage	200 V	Rated output voltage	173 V
Rated power	3 kW	Carrier frequency	10 kHz
Input filter L (L_f)	10.0 mH (23.6%)	Load resistance (R_o)	12.7 Ω (127%)
Input filter C (C_f)	4.55 μF (1.91%)	Load inductance (L_o)	6.27 mH (19.7%)
Input voltage angle	105 deg.	Output voltage angle	15 deg.

TABLE 2. CONTROL PARAMETERS FOR OPEN-LOOP CONTROL

Open-loop control	d-axis voltage command (steady state)	0.5 p.u.
	d-axis voltage command (step input)	0.01 p.u.
	q-axis voltage command (steady state)	0 p.u.
	q-axis voltage command (step input)	0 p.u.

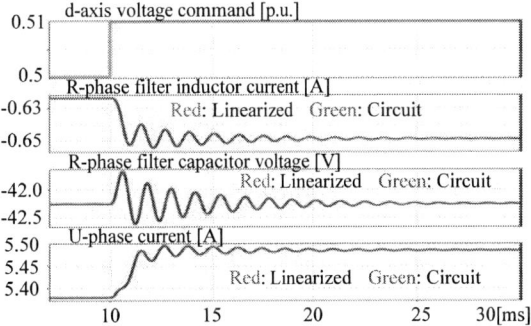

Fig. 4. Indicial response of the circuit model and the linearized model with an open-loop control in a simulation. It is confirmed that the linearized model result corresponds to the circuit model result.

than 1% in a steady state. Thus, the linearized model is valid for modeling of the matrix converter. In consequence, the output current control including the proposed reference filter and the output damping control are designed with Bode-diagrams derived from the linearized model in the next chapter.

IV. PROPOSED DESIGN PROCEDURE OF CONTROL PARAMETERS

A. Approximate Model to Design Proposed Reference Filter and Control Parameters

In order to design the output current control with the proposed reference filter and the output damping control, Bode-diagrams of the linearized block model in Fig. 3 should be clarified. However, the linearized block model is a fifth order transfer function, which is caused by the input LC filter, the load inductor, the PI controller and the output damping control, and the transfer function is very complicated. Therefore, an approximate model based on a third order transfer function is used to design these controls.

Fig. 5 shows the approximate block model between $\Delta i_{out\_LPF}^{*}$ and Δi_{out} in Fig. 3 without the output damping control. The approximate model is based on the gain curve of the linearized block model in Fig. 3 in order to obtain the proposed reference filter configuration easily. It should be noted that the output damping control is not applied when a forward transfer function in the approximate model is calculated. The approximate model is a second order transfer function which has a damping factor ζ and a natural angular frequency ω_n. The transfer function in Fig. 5 approximates a resonance point characteristic of the linearized block model. Therefore, ζ and ω_n are expressed by (10) and (11).

$$\zeta = \sqrt{\frac{1}{2} - \frac{1}{2}\sqrt{1 - \frac{1}{M_p^{2}}}} \qquad (10)$$

$$\omega_n = \omega_p \left(1 - 1/M_p^{2}\right)^{\frac{1}{4}} \qquad (11)$$

where, M_p is a peak gain of the resonance point and ω_p is a resonance angular frequency which are measured with the gain characteristic of the linearized block model of the matrix converter in Fig. 3.

Table 3 shows parameters of the PI controller. These parameters should be designed with a flowchart as presented in the next section. However, in order to clarify the validity of the approximate block model in Fig. 5, the PI parameters are set to the linearized model in Fig. 3.

Fig. 6 shows a gain characteristic of the closed-loop transfer function between $\Delta i_{out\_LPF}^{*}$ and Δi_{out} without the output damping control. From Fig. 6, the resonance peak gain and the resonance frequency of the approximate model designed with (10) and (11) corresponds to the linearized block model of the matrix converter. In addition, the gain characteristic of the approximate model is similar to the linearized model. In a frequency region higher than 1 kHz, gain error between the linearized model and the approximate model can be neglected because the region is not available for the current control due to detection delay and PWM delay of the real system if the carrier frequency is set to 10 kHz as shown in Table 1. Thus, the approximate model in Fig. 6 is valid for the

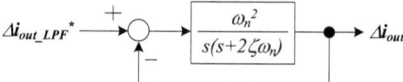

Fig. 5. Approximate block model between $\Delta i_{out\_LPF}^{*}$ and Δi_{out} without the output damping control to simplify the design procedure. The transfer function of this block diagram becomes the standard form of the second order system.

TABLE 3. PARAMETERS OF THE PI CONTROLLER

PI controller	Current command (step input)	0.01 p.u.
	Current command (steady state)	0.4 p.u.
	ACR natural frequency	650 Hz

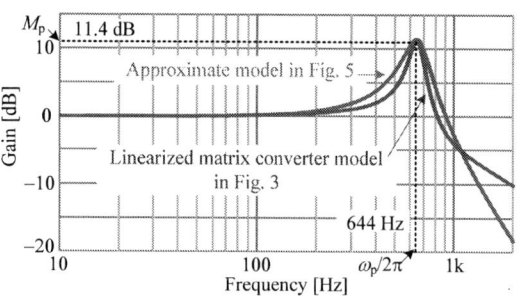

Fig. 6. Gain characteristic of the closed-loop transfer function between $\Delta i_{out\_LPF}^{*}$ and Δi_{out} without the output damping control. The gain curve of the approximate model is similar to the linearized model.

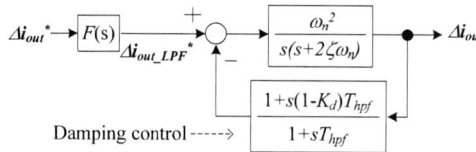

Fig. 7. Approximate block model between Δi_{out}^{*} and Δi_{out} taking the output damping control into account. The proposed reference filter $F(s)$ is derived from this block model.

modeling of the linearized block model of the matrix converter without the output damping control.

Fig. 7 shows the approximate block model between Δi_{out}^{*} and Δi_{out} taking the output damping control into account. First, a transfer function of the proposed reference filter $F(s)$ to suppress the output current overshoot is derived. A closed-loop transfer function between $\Delta i_{out\_LPF}^{*}$ and Δi_{out} is presented by (12).

$$\frac{\Delta i_{out}}{\Delta i_{out\_LPF}^{*}} = \frac{\dfrac{\omega_n^{2}}{T_{hpf}}\left(1 + sT_{hpf}\right)}{s^{3} + \left(\dfrac{1}{T_{hpf}} + 2\zeta\omega_n\right)s^{2} + \omega_n\left(\dfrac{2\zeta}{T_{hpf}} + \omega_n(1 - K_d)\right)s + \dfrac{\omega_n^{2}}{T_{hpf}}} \qquad (12)$$

It is obvious that the output damping control generates a large current overshoot because the closed-loop transfer function between $\Delta i_{out\_LPF}^{*}$ and Δi_{out} has a zero depending on T_{hpf}. In order to suppress the output current overshoot, the zero in (12) should be cancelled. Hence, $F(s)$ is defined as (13).

$$F(s) = \frac{1}{1 + sT_{hpf}} \qquad (13)$$

Hence, the proposed reference filter is designed easily when the T_{hpf} is decided in the proposed design procedure.

B. Design Flowchart

In order to design the output current control and the output damping control, this section proposes a design flowchart.

Fig. 8 shows a proposed design flowchart for the output current control and the output damping control to ensure a desired control bandwidth and the stability. Input parameters into the flowchart are determined by specifications and outputs of the flowchart are the proportional gain K_p, the integral time T_i of the PI controller, K_d and T_{hpf} of the output damping control.

1) Step 1: Calculation of K_p and T_i of PI Controller

First, K_p and T_i should be calculated in order to draw a Bode-diagram as shown in Fig. 6. Then, K_p and T_i are designed by (14) and (15), supposing that a plant of the controller is an R-L model.

$$K_p = \omega_{c\_design} L_o \qquad (14)$$

$$T_i = \frac{L_o}{R_o} \qquad (15)$$

where, ω_{c_design} is a desired cut-off angular frequency of the output current control. It should be noted that the design concept of (14) and (15) is based on the first order transfer function of the output current control loop neglecting the input LC filter and the output damping control. However, the actual cut-off angular frequency is changed by the resonance characteristic and the output damping control.

2) Step 2: Simulation to Measure M_p and ω_p

In order to calculate parameters of the approximate model for a simple design process, a Bode-diagram between $\Delta i_{out\_LPF}^*$ and Δi_{out} of the linearized matrix converter block model such as Fig. 6 is drawn by a simulator. In this paper, Piece-wise Linear Electrical Circuit Simulation (PLECS) is used as a simulator.

3) Step 3: Calculation of ζ and ω_n

In order to obtain the approximate model of the matrix converter, ζ and ω_n are calculated by (10) and (11).

4) Step 4: Decision of K_d and T_{hpf}

By using (12) and (13), a Bode-diagram between Δi_{out}^* and Δi_{out} of the approximate block model in Fig. 7 is illustrated. Then, K_d and T_{hpf} is decided with the Bode-diagram to earn an intended control bandwidth of the output current.

Fig. 9 shows gain characteristics between Δi_{out}^* and Δi_{out} of the approximate block model when the K_d and f_{hpf} are changed respectively. Note that f_{hpf} is a cut-off frequency of a HPF of the output damping control. Fig. 9 (a) shows a gain characteristic when K_d is changed and f_{hpf} is a constant of 50 Hz, and (b) shows a result when f_{hpf}

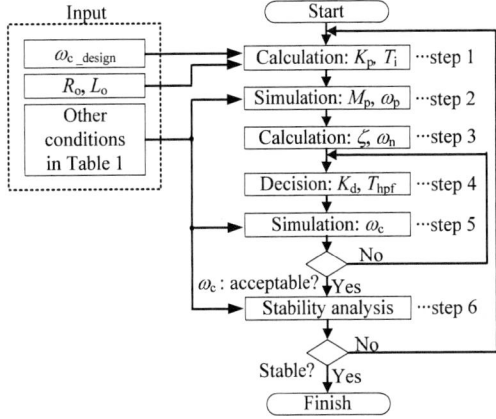

Fig. 8. Proposed design flowchart for the output current control and the output damping control.

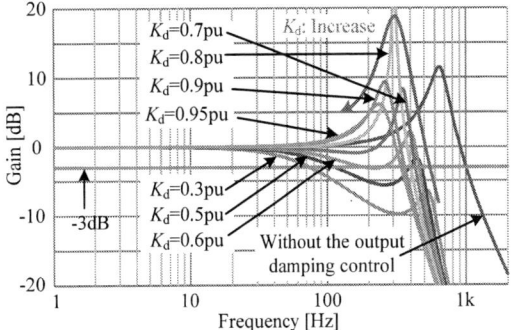

(a) K_d is changed (f_{hpf} is a constant of 50 Hz).

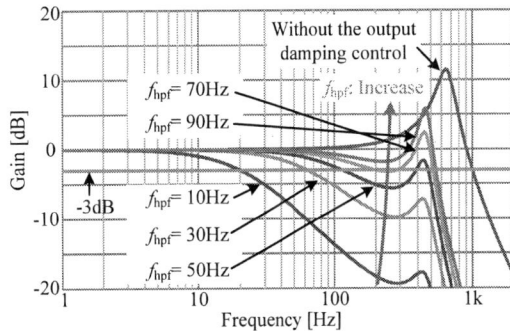

(b) f_{hpf} is changed (K_d is a constant of 0.5 p.u.).

Fig. 9. Gain characteristics between Δi_{out}^* and Δi_{out} of the approximate block model when the K_d and f_{hpf} are changed respectively.

is changed and K_d is a constant of 0.5 p.u.. From Fig. 9 (a), the cut-off frequency of the output current control increases as K_d is increased from 0.3 p.u. to 0.6 p.u.. However, the cut-off frequency is decreased in the region where K_d is over 0.6 p.u.. This is affected the characteristic around the resonance point. On the other hand, it is confirmed that the cut-off frequency increases as the f_{hpf} is increased from Fig. 9 (b). Therefore, an intended control bandwidth of the output current is yielded by adjusting the damping parameters K_d and f_{hpf}.

5) Simulation to Confirm Cut-off Frequency of Output Current Control

A Bode-diagram between Δi_{out}^* and Δi_{out} of the linearized block model in Fig. 3 with the decided damping parameters is drawn and the control bandwidth of the output current ω_c is evaluated. If the obtained control bandwidth is not acceptable, the damping parameters should be redesigned by going back to Step 4.

6) Stability Analysis

Finally, the stability of the system is analyzed with a Nyquist-diagram. If the obtained stability does not satisfy the specification of stability, the PI parameters and the damping parameters which influence the frequency characteristic of the open-loop transfer function of the output current control should be adjusted again. For example, ω_{c_design} is reduced or T_{hpf} is increased from the original value in order to increase the stability and K_d is redesigned to earn the intended control bandwidth.

V. COMPARISON RESULTS BETWEEN THE PROPOSED AND CONVENTIONAL METHODS IN SIMULATION

Table 4 presents the designed parameters of the output damping control with the proposed design flowchart in Fig. 8 and the conventional method. It should be noted that the conventional method is based on the desired gain margin [15]. In the conventional method, the PI parameters are designed with (14) and (15) in common with the proposed procedure. In addition, the damping parameters are calculated by the following equations.

$$\left(g_a + g_m\right) + 20\log_{10}\left(1 - K_d\right) = 0 \quad (16)$$

$$T_{hpf} = \frac{5}{2\pi\left(1 - K_d\right)f_{cp}} \quad (17)$$

where, f_{cp} is a phase-crossover frequency of a gain curve of the output current control without the output damping control, g_a is a gain at f_{cp} without the damping control and g_m is a desired gain margin which is set to the same as the designed result with the proposed method for comparison.

Fig. 10 shows gain characteristics between $\Delta i_{out}{}^{*}$ and Δi_{out} with the proposed and the conventional methods. The gain curve of the approximate model meets -3 dB at 650 Hz as shown in Table 3 by the proposed design flowchart. In addition, the proposed method mitigates error between the desired and the obtained cut-off frequencies by 1/4 in comparison with the conventional method. This is because the conventional method gives preference to the stability design over adjusting the control bandwidth. Moreover, the gain curve with the conventional method increases in the region around 100 Hz because of the zero generated by the output damping control and causes a large output current overshoot.

Fig. 11 shows the Nyquist-diagram of the linearized block model of the matrix converter between $\Delta i_{out\_LPF}{}^{*}$ and Δi_{out} with the proposed and the conventional procedures. As a result of the proposed procedure, the Nyquist-diagram of the proposed method has a gain margin of 3.85 dB and the system is stable, which is equivalent to suppression of the filter resonance. It should be noticed that the experimental setup has more stability because the loss of the matrix converter behaves as a

TABLE 4. DESIGNED PARAMETERS OF THE OUTPUT DAMPING CONTROL.

Proposed method	Damping gain (K_d)	0.60 p.u.
	Damping HPF time constant (T_{hpf})	0.64 ms
Conventional method	Damping gain (K_d)	0.56 p.u.
	Damping HPF time constant (T_{hpf})	3.1 ms

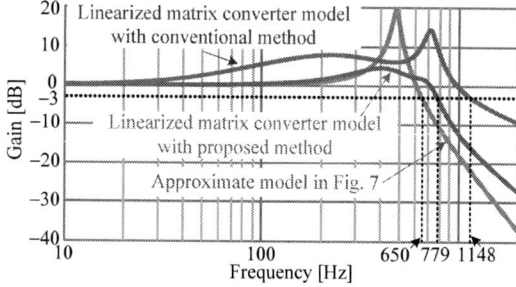

Fig. 10. Gain characteristics of the closed-loop transfer function between $\Delta i_{out}{}^{*}$ and Δi_{out} with the proposed and the conventional methods. The proposed method mitigates the error between the desired and the yielded cut-off frequencies by 1/4 compared to the conventional method.

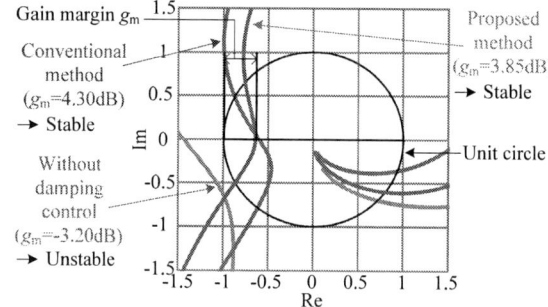

Fig. 11. Nyquist-diagram of the linearized block model of the matrix converter between $\Delta i_{out\_LPF}{}^{*}$ and Δi_{out} with the proposed and the conventional design procedures. The proposed design procedure yields a gain margin of 3.85 dB and the system is stable.

damping resistor [10]. On the other hand, the conventional method also stabilizes the system with the gain margin of 4.30 dB. Then, error of the gain margin between the intended value of 3.85 dB and the actual is caused by the break point approximation of a Bode-diagram for deriving (17) [15]. Therefore, it is confirmed that the proposed procedure yields the stability and the desired control bandwidth of the output current.

Fig. 12 shows the output d-axis current waveform in simulations with the linearized block model in Fig. 3. From Fig. 12 (a), the system without the damping control becomes unstable right after the step input because the system gain margin is -3.20 dB as shown in Fig. 11. However, the output damping control designed with the proposed and the conventional methods stabilizes the system. Then, the proposed method reduces the d-axis current overshoot by 67% because of the proposed reference filter $F(s)$ in comparison with the conventional method. Hence, it is confirmed that the proposed method achieves less overshoot of the output current than the conventional method in the simulations.

VI. EXPERIMENTAL RESULTS

This chapter evaluates the stability and the transient response of a matrix converter prototype as shown in

Fig.1. The prototype parameters are almost consistent with the parameters shown from Table 1 to Table 4. However, the input voltage is 50 Hz while the output voltage is 30 Hz. In addition, an output d-axis current reference is separated into a steady component of 0.4 p.u. and a step component of 0.05 p.u.. A modulation strategy of the prototype employs the method as introduced in [3].

Fig. 13 shows the input and output waveforms of the matrix converter obtained by experiments. Fig. 13 (a) shows a result without the output damping control and (b) shows a result with the output damping control designed with the proposed method. In Fig. 13 (a), the filter resonance is excited, and the input current and the output current have resonant distortions. It is noted that the experimental result in Fig. 13 (a) shows a stability limit although the simulation result in Fig. 12 becomes unstable. This is because loss of a matrix converter behaves as a damping resistor as mentioned in the previous chapter. The input current THD (total harmonic distortion) is 52.9% and the output current THD is 15.7% from Fig. 13 (a). In contrast, the filter resonance is suppressed and the system is stabilized by the output damping control designed with the proposed method. As a result, the input and output current THDs are reduced to 10.3% and 4.43%, respectively. A main cause of remaining distortion in the input current is generated by commutation failure because the input voltage detection for the commutation is implemented at a power supply not the filter capacitor voltage in the experiment. The input current will be improved by applying a hybrid commutation method [17]. Thus, the designed output damping control mitigates the resonance distortions in the input current by 81%.

Fig. 14 shows the dq-axis output current in a transient response with the damping control designed with the conventional and the proposed methods. It should be noted that the experimental results after this paragraph are obtained in the condition of a fixed output current angle in common with simulations. The conventional and the proposed methods stabilize the matrix converter and the output current converges toward the reference. However, the damping control designed with the conventional method generates a d-axis current overshoot of 188% due to the zero in the closed-loop transfer function. On the other hand, the proposed method reduces the d-axis current overshoot to 76% because of the pole-zero cancellation owing to the proposed reference filter $F(s)$. Hence, the proposed method reduces the output current overshoot in transient response by 60%.

Fig. 15 shows gain curve characteristics of the output current control designed with the proposed and the conventional methods in simulations and experiments. It should be noted that the intended control bandwidth is 650 Hz and the sinusoidal waveform as the d-axis current reference is used in the experiment. The proposed method mitigates the error between the desired and the obtained cut-off frequencies in comparison with the conventional method since the cut-off frequency of the conventional

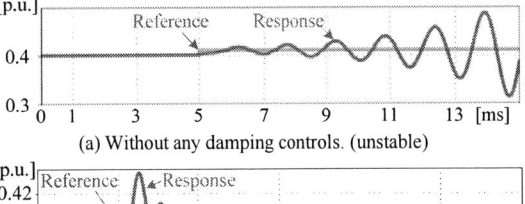

(a) Without any damping controls. (unstable)

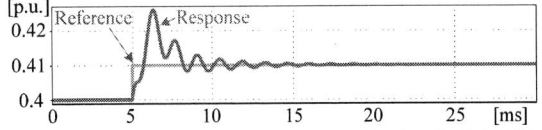

(b) Damping control designed with the conventional method. (stable)
(d-axis current overshoot: 159%)

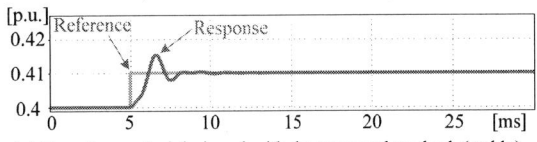

(c) Damping control designed with the proposed method. (stable)
(d-axis current overshoot: 52.6%)

Fig. 12. Output d-axis current waveform in simulation. The d-axis current overshoot is suppressed by 67 % owing to the proposed method.

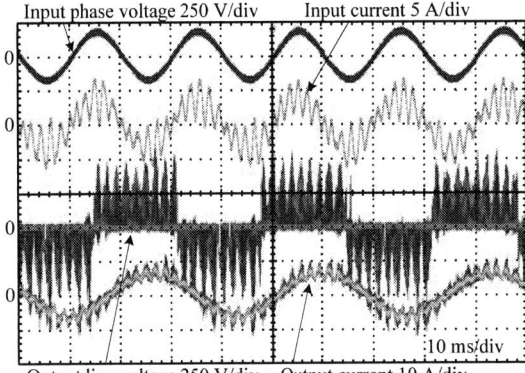

(a) Without the output damping control which results in the input current THD of 52.9%.

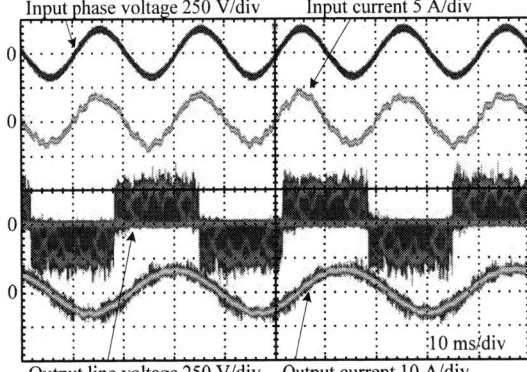

(b) With the output damping control designed with the proposed method which results in the input current THD of 10.3%.

Fig. 13. Input and output waveforms of the matrix converter as illustrated in Fig. 1 in experiments. The output damping control designed with the proposed method suppresses the filter resonance and reduces the input current THD by 81%.

method exceeds 1 kHz because of the damping control. In addition, the proposed method reduces gain in lower frequency band around 100 Hz compared to the conventional method because of the proposed reference filter $F(s)$. As a result, the proposed method reduces the output current overshoot as shown in Fig. 14. It should be noted that the error between the experimental and the simulation results in Fig. 15 is caused by digital control and a full consideration about the difference will be reported in the future. Therefore, it is confirmed that the proposed design method reduces the output current overshoot and the error of the control bandwidth of the output current.

VII. CONCLUSION

This paper presents a design procedure for the output current control and the output damping control of a matrix converter to suppress the LC filter resonance and improve a transient response. With the conventional design method, the output damping control causes a large output current overshoot and a control bandwidth of the output current is affected. In the proposed design procedure, a reference filter to suppress an output current overshoot and a design flow chart to obtain a desired control bandwidth of the output current and the stability are proposed. From experimental results, the output damping control designed with the proposed method suppresses the filter resonance, which results in a stable operation. In addition, the proposed design method reduces the output current overshoot of 60% and error between the desired and the obtained control bandwidth in comparison with the conventional method which is based on a gain margin. Therefore, the validity of the proposed design procedure is confirmed.

REFERENCES

[1] P. W. Wheeler, J. Rodriguez, J. C. Clare, L. Empringham: "Matrix Converters: A Technology Review", IEEE Trans. Ind. Electron., Vol. 49, No. 2, pp. 274-288 (2002)

[2] T. Friedli, J. W. Kolar: "Milestones in Matrix Converter Research", IEEJ Journal I. A., Vol. 1, No. 1, pp. 2-14 (2012)

[3] J. Itoh, I. Sato, A. Odaka, H. Ohguchi, H. Kodachi, N. Eguchi: "A Novel Approach to Practical Matrix Converter Motor Drive System With Reverse Blocking IGBT", IEEE Trans. Power Electron., Vol. 20, No. 6, pp. 1356-1363 (2005)

[4] C. Klumpner, F. Blaabjerg, I. Boldea, P. Nielsen: "New Modulation Method for Matrix Converters", IEEE Trans. Ind. Appl., Vol. 42, No. 3, pp. 797-806 (2006)

[5] F. Blaabjerg, D. Casadei, C. Klumpner, M. Matteini: "Comparison of Two Current Modulation Strategies for Matrix Converters Under Unbalanced Input Voltage Conditions", IEEE Trans. Ind. Electron., Vol. 49, No. 2, pp. 289-296 (2002)

[6] M. Rivera, J. Rodriguez, J. Espinoza, T. Friedli, J. W. Kolar, A. Wilson, C. A. Rojas: "Imposed Sinusoidal Source and Load Currents for an Indirect Matrix Converter", IEEE Trans. Ind. Electron., Vol. 59, No. 9, pp. 3427-3435 (2012)

[7] D. Casadei, G. Serra, A. Tani, L. Zarri: "Stability Analysis of Electric Drives Fed by Matrix Converters", Proc. of the 2002 IEEE-ISIE, Vol. 4, No. 8-11, pp. 1108-1113 (2002)

[8] F. Liu, C. Klumpner, F. Blaabjerg: "A Robust Method to Improve Stability in Matrix Converters ", Proc. 35th PESC 2004, pp. 3560-3566 (2004)

[9] C. A. J. Ruse, J. C. Clare, C. Klumpner: "Numerical Approach for Guaranteeing Stable Design of Practical Matrix Converter Drive Systems", Proc. 32nd IECON 2006, pp. 2630-2635 (2006)

[10] D. Casadei, G. Serra, A. Tani, A. Trentin, L. Zarri: "Theoretical and Experimental Investigation of the Stability of Matrix Converters", IEEE Trans. Ind. Electron., Vol. 52, No. 5, pp. 1409-1419 (2005)

[11] Y. Sun, M. Su, X. Li, H. Wang, W. Gui: "A General Constructive Approach to Matrix Converter Stabilization", IEEE Trans. Power Electron., Vol. 28, No. 1, pp. 418-431 (2013)

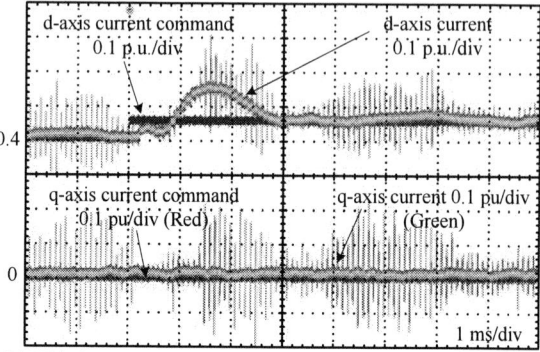

(a) With the conventional method. The d-axis current overshoot is 188%.

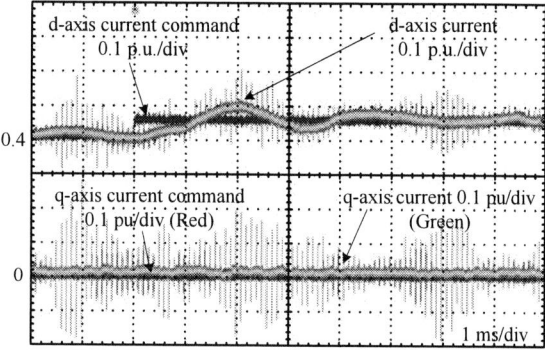

(b) With the proposed method. The d-axis current overshoot is 76%.
Fig. 14. Output dq-axis current response with the damping control designed with the conventional and the proposed methods. The proposed method reduces the current overshoot in transient response by 60%.

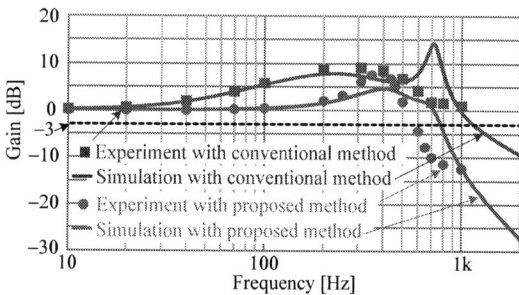

Fig. 15. Gain characteristics of the closed-loop transfer function between Δi_{out}^* and Δi_{out} with the proposed and the conventional methods in simulation and experiment.

[12] M. Rivera, C. Rojas, J. Rodriguez, P. W. Wheeler, B. Wu, J. Espinoza: "Predictive Current Control With Input Filter Resonance Mitigation for a Direct Matrix Converter", IEEE Trans. Power Electron., Vol. 26, No. 10, pp. 2794-2803 (2011)

[13] T. Nunokawa, T. Takeshita:"Resonance Suppression Control in Complex Frame for Three-Phase to Three-Phase Matrix Converters", EPE2007 (2007)

[14] J. Haruna, J. Itoh: "Control Strategy for a Matrix Converter with a Generator and a Motor", Proc. 26th IEEE APEC, pp. 1782-1789 (2011)

[15] H. Takahashi, J. Itoh: "Stability Analysis of Damping Control to Suppress Filter Resonance in Multi-modular Matrix Converter", IEEE Energy Conversion Congress and Expo 2013, pp. 448-455 (2013)

[16] D. Casadei, G. Serra, A. Tani, L. Zarri: "Matrix Converter Modulation Strategies: Anew General Approach Based on Space-Vector Representation of the Switch State", IEEE Trans. Ind. Electron., Vol. 49, No. 2, pp. 370-381 (2002)

[17] K.Kato, J. Itoh: "Improvement of Input Current Waveforms for a Matrix Converter Using a Novel Hybrid Commutation Method", Proc. PCC 2007, pp. 763-768 (2007)

The 2014 International Power Electronics Conference

A Novel LCL Filter Parameter Design Method Basing on Resonant Frequency Optimization of Three-level NPC Grid Connected Inverter

LI Ning, WANG Yue, NIU Ruigen, GUO Wei, LEI Wanjun, WANG Zhao'an

School of Electrical Engineering
Xi'an Jiaotong University
Xi'an, China
ningli@stu.xjtu.edu.cn

Abstract—**For high power three-level NPC grid connected inverter, which has the characteristics of low switching frequency and complex harmonic distribution of output current waveform, it deduces the total output filter inductor expression and researches the relational expression between inductance ratio, damping ratio and resonant frequency, harmonic attenuation ratio, as well as damping power. Based on this, an optimized resonant frequency LCL filter design scheme which can also reduce the damping power loss of three-level NPC grid connected inverter is proposed. Through the comprehensive theoretical analysis, the proposed scheme calculates the range of key parameters of the LCL output filter, meanwhile, it ensures the filtering effect as well as reduces the system cost. By paralleling a small damping inductor with the damping resistor, it reduces the damping loss and increases the efficiency of the system. Feasibility and effectiveness of the optimized scheme are validated by simulation and experimental results.**

Keywords—*NPC grid connedted inverter, optimized resonant frequency, LCL filter, damping inductor, damping resistor.*

I. INTRODUCTION

Three-level NPC converter is one of the most widely used multi-level converters which are often used in various high power low switching frequency applications[1]. Science the decrease of switching frequency and increase of output power, the traditional L output filter cannot satisfy the demand of current THD characteristic, system dynamic performance and cost. As a result, LCL filter which has a better output current THD performance is more and more used in three-level NPC grid connected inverter[2][3]. However, the LCL filter has a zero impedance resonance point which will enlarge the current harmonic at the resonant frequency. In order to get a better filter results, the resonance must be eliminated through hardware of software methods[4]-[19]. The software LCL filter resonance harmonic suppression (active damping method) methods modify the control algorithm to eliminate the resonance phenomenon, and it will increase the feedback signal numbers and decrease the robustness and reliability of the system[7]-[11]. The hardware LCL filter resonance harmonic suppression (passive damping method) methods use damping resistor

in series with the capacitor to suppress the harmonic[12][13]. The hardware methods are more stable and reliable; as a result, it is widely used in many industry applications.

LCL filter parameter design is an important part of grid connected inverter design. So far, there are many research results about LCL parameter design while none of them is an accepted design criteria. A more authoritative design method is put forward in [6], but this method is aimed at traditional two level gird inverter, and in the design process, the key parameter need to be adjusted two or three times in order to get a satisfactory results. In some papers, scholars present a LCL filter parameter design method basing on three dimensional figures; nevertheless the method is complicated in practical applications[20][21].

In this paper, it firstly analyzes the influence of the key parameters of LCL filter to system performance in three-level NPC grid connected inverter, and then it proposed a novel LCL filter parameter design method basing on resonant frequency optimization. On the one hand the new method ensures the filtering effect, and on the other hand it reduces the loss of damping resistor as far as possible. Finally, the superiority of the new method is verified by simulation and experiment results.

II. S DOMAIN ANALYSIS OF LCL FILTER IN THREE-LEVEL NPC GRID CONNECTED INVERTER

Fig.1 shows the diagram of three-level grid-connected NPC converter, the LCL filter with a damping resistor is used to reduce the harmonic current injection into the grid. Fig.2 shows the equivalent model of the system in fundamental frequency and high frequency.

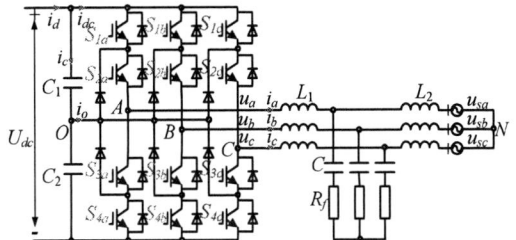

Fig. 1. Three-level grid-connected NPC inverter with a LCL filter

978-1-4799-2706-7/14 $31.00 © 2014 IEEE

160

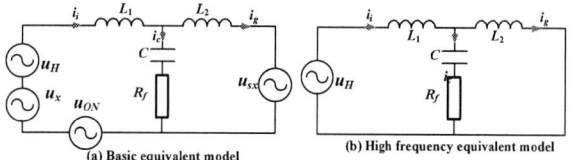

Fig. 2. Equivalent model of the LCL filter

$$\begin{cases} u_{ON} = -\dfrac{1}{3}\sum u_x & (x=a,b,c,\ S_x=1\ or\ 0\ or\ -1) \\ u_x = S_x \cdot U_{dc}/2 \end{cases} \quad (1)$$

In fig.2, u_x and u_H stands for the output fundamental and high frequency component of NPC inverter, and u_{ON} stands for the common-mode voltage, as shown in equation (1). In complex frequency domain (s domain), the transfer function of LCL filter is shown in equation (2). Equation (3) shows the undamped oscillation angular frequency of the system.

$$\begin{cases} \dfrac{I_g(s)}{U_H(s)} = \dfrac{R_f Cs+1}{L_1 L_2 Cs^3 + (L_1+L_2)R_f Cs^2 + (L_1+L_2)s} \\[2mm] \dfrac{I_i(s)}{U_H(s)} = \dfrac{CL_2 s^2 + R_f Cs+1}{L_1 L_2 Cs^3 + (L_1+L_2)R_f Cs^2 + (L_1+L_2)s} \\[2mm] \dfrac{I_c(s)}{U_H(s)} = \dfrac{CL_2 s^2}{L_1 L_2 Cs^3 + (L_1+L_2)R_f Cs^2 + (L_1+L_2)s} \\[2mm] \dfrac{I_g(s)}{I_i(s)} = \dfrac{R_f Cs+1}{L_2 Cs^2 + R_f Cs+1} \end{cases} \quad (2)$$

$$\omega_{res} = \sqrt{\dfrac{L_1+L_2}{L_1 L_2 C}} \quad (3)$$

III. RESONANT FREQUENCY OPTIMIZATION DESIGN METHOD OF THE LCL FILTER IN NPC THREE-LEVEL GRID CONNECTED INVERTER

A. Design of the capacitance and total inductance

In this paper, the first step of LCL filter design is the capacitance design, its design formula is the same with [6], as shown in equation (4).

$$C \le \dfrac{Q_C}{\omega_0 U_{line}^{\,2}} = \dfrac{bP_{rated}}{\omega_0 U_{line}^{\,2}} \quad (4)$$

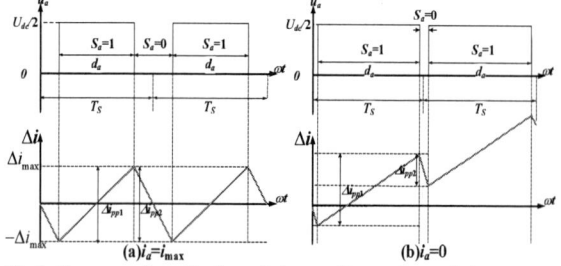

Fig.3 Corresponding situation of phase voltage pulse and phase current ripple (phase a)

The second step of LCL filter design is the total inductance L_T ($L_T = L_1 + L_2$) design. In this paper, the grid

current ripple requirement determines the lower limit of L_T, and the current dynamic demand determines the upper limit of L_T. The max current ripple appears near the maximum of phase current, as shown in fig.3(a), and the most stringent dynamic demand appears at the zero point of phase current, as shown in fig.3(b). Equation (5) and (6) has shown the current ripple relation at the two moments. After some analysis, the total inductance show satisfies equation (7).

$$\begin{cases} \Delta i_{pp1} = \dfrac{U_{dc}}{6} \cdot \dfrac{(2-S_{b1}-S_{c1})-E_m}{L_T} \cdot d_a T_S \\[2mm] \Delta i_{pp2} = \dfrac{U_{dc}}{6} \cdot \dfrac{(-S_{b0}-S_{c0})-E_m}{L_T} \cdot (1-d_a)T_S \end{cases} \quad (5)$$

$$\begin{cases} \Delta i_{pp1} = \dfrac{U_{dc}}{6} \cdot \dfrac{(2-S_{b1}-S_{c1})}{L_T} \cdot d_a T_S \\[2mm] \Delta i_{pp2} = \dfrac{U_{dc}}{6} \cdot \dfrac{(-S_{b0}-S_{c0})}{L_T} \cdot (1-d_a)T_S \end{cases} \quad (6)$$

$$\dfrac{2U_{dc}^{\,2}+3U_{dc}E_m-9E_m^{\,2}}{18 I_{rippleM} U_{dc}} T_S \le L_T \le \dfrac{U_{dc}}{6 I_m \omega} \quad (7)$$

B. The other key parameters effect to the system performance

In section A, the capacitance and total inductance is determined, in this section, the other key parameters effect to the system performance is researched.
In this paper, k and ξ are defined as the inductance ratio and the damping ratio, as shown in equation (8).

$$\begin{cases} L_1 = kL_T \\[2mm] \xi = \dfrac{\omega_{res} R_f C}{2} \end{cases} \quad (8)$$

Using equation (8), equation (2) can be further expressed as equation (9) and the undamped oscillation frequency is shown in equation (10).

$$\begin{cases} \dfrac{I_g(s)}{U_H(s)} = \dfrac{1}{L_T s} \dfrac{(2\xi\omega_{res}s+\omega_{res}^{\,2})}{s^2+2\xi\omega_{res}s+\omega_{res}^{\,2}} \\[2mm] \dfrac{I_i(s)}{U_H(s)} = \dfrac{1}{L_T s} \dfrac{(s^2/k+2\xi\omega_{res}s+\omega_{res}^{\,2})}{s^2+2\xi\omega_{res}s+\omega_{res}^{\,2}} \\[2mm] \dfrac{I_c(s)}{U_H(s)} = \dfrac{1}{L_T s} \dfrac{s^2/k}{s^2+2\xi\omega_{res}s+\omega_{res}^{\,2}} \\[2mm] \dfrac{I_g(s)}{I_i(s)} = \dfrac{2\xi\omega_{res}s+\omega_{res}^{\,2}}{s^2/k+2\xi\omega_{res}s+\omega_{res}^{\,2}} \end{cases} \quad (9)$$

$$f_{res} = \dfrac{1}{2\pi} \cdot \sqrt{\dfrac{1}{k(1-k)L_T C}} \quad (10)$$

The 2014 International Power Electronics Conference

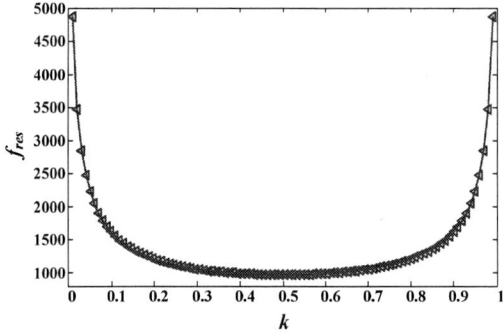

Fig.4 Diagram between undamped oscillation frequency f_{res} and inductance ratio k

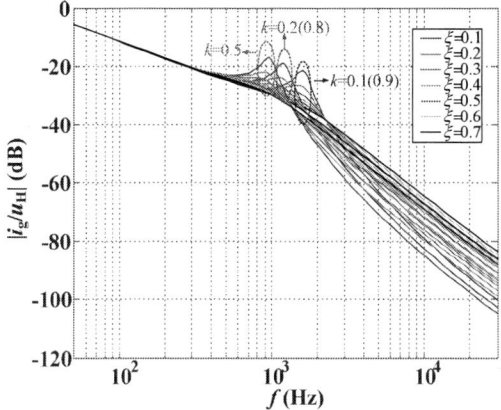

Fig.5 Diagram between amplitude ratio $|i_g/u_H|$, inductance ratio k and damping ratio ξ

If the total inductance L_T equals 6mH, the total capacitance C equals 18uF, the system switching frequency f_s equals 3kHz, the diagram between undamped oscillation frequency f_{res} and inductance ratio k is shown in fig.4 and the diagram between amplitude ratio $|i_g/u_H|$, inductance ratio k and damping ratio ξ is shown in fig.5.

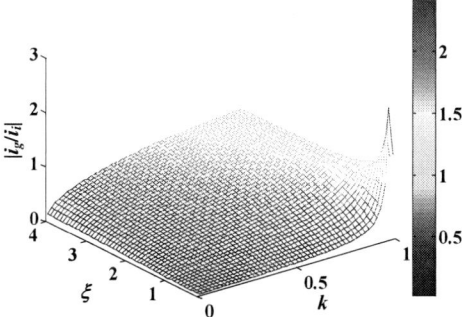

Fig.6 Diagram between amplitude ratio $|i_g/i_i|$, inductance ratio k and damping ratio ξ

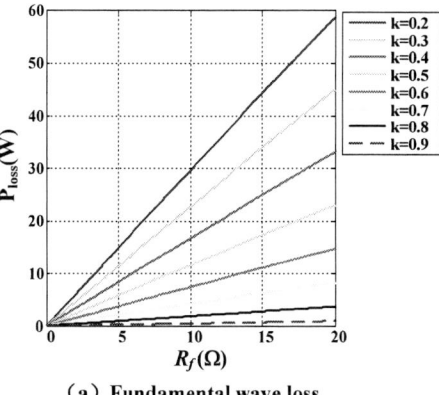

(a) Fundamental wave loss

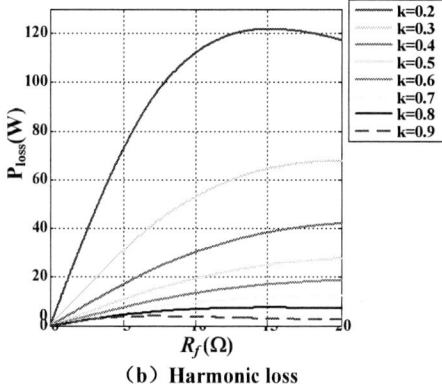

(b) Harmonic loss

Fig.7 Relation between power loss of R_f, inductance ratio k and resistance of R_f

Fig.6 shows the diagram between amplitude ratio $|i_g/i_i|$, inductance ratio k and damping ratio ξ and fig.7 shows the relation between damping resistor power loss, inductance ratio k and resistance of R_f. From all of the 4 figures, tab.1 can be got.

Tab.1 The influence of k and ξ to system performance

	k	ξ
f_{res}	$k=0.5$, $f_{res}=f_{res\min}$, k→0(1), $f_{res}\nearrow$	---
$|i_g/u_H|$	$k=0.5$ $f_{res}=f_{res\min}$; k→0(1), $f_{res}\nearrow$	$\xi\searrow$, Resonance peak$\nearrow$, $\xi\nearrow$, Resonance peak$\searrow$,
$|i_g/i_i|$	$k\nearrow$, $|i_g/i_i|\nearrow$ (main)	$\xi\nearrow$, $|i_g/i_i|\nearrow$ (secondary)
P_{loss}	$k\nearrow$, $P_{loss}\searrow$	$\xi\nearrow$, P_{loss} first $\nearrow$ then $\searrow$

From tab.1, it is clear that there are no optimum parameters which will make all of the system performance have the best effect. All of the LCL filter design methods do trade-off in order to satisfy the most important performance metrics.

C. The resonant frequency optimization design method

In this paper, the LCL filter is used in NPC three-level inverter which has the characteristic of very low switching frequency so the resonant frequency is the most important performance metrics. In order to get the best resonant frequency optimization characteristic, the two inductors should satisfy the following equation:

$$L_1 = L_2 = 0.5L_T \tag{11}$$

The damping resistor design is the core part of the novel LCL design method. In this paper, it uses the attenuation effect at undamped resonant frequency to design the lower value of R_f, and it uses the current harmonic attenuation ratio at switching frequency to design the upper value of R_f. At undamped resonant angle frequency ω_{res} and switching angle frequency ω_s the transfer function is shown in equation (12) and (13).

$$\frac{|i_g|}{|u_H|} = \frac{\sqrt{4\xi^2+1}}{2\xi L_T \omega_{res}} = \frac{\sqrt{1+1/4\xi^2}}{L_T \omega_{res}} \tag{12}$$

$$\frac{|i_g|}{|i_i|} = \frac{\sqrt{\omega_{res}^4 + 4\xi^2 \omega_{res}^2 \omega_s^2}}{\sqrt{(\omega_{res}^2 - \omega_s^2/k)^2 + 4\xi^2 \omega_{res}^2 \omega_s^2}} \tag{13}$$

If the max harmonic current attenuation ratio at ω_{res} is κ and the max harmonic current attenuation ratio ω_s at is γ. After some derivation, the damping resistor should satisfy equation (14).

$$\frac{1}{\sqrt{16\kappa^2 - 4C/L_T}} \le R_f \le \frac{1}{4}\sqrt{\frac{(49\gamma^2-1)L_T}{(1-\gamma^2)C}} \tag{14}$$

D. Power losse reduction method of the damping resistor R_f

The power loss on damping resistor contains the fundamental wave loss and harmonic loss, as shown in equation (15). The fundamental current I_{cx} and harmonic current I_{cHx} flow through R_f are shown in equation (16).

$$P_{loss} \approx 3(I_{cHa}^2 + I_{ca}^2)R_f \tag{15}$$

$$\begin{cases} I_{cx} = \dfrac{\omega_1^2/k}{L_T \omega_1 \sqrt{(\omega_{res}^2 - \omega_1^2)^2 + 4\xi^2 \omega_{res}^2 \omega_1^2}} U_x \\[4mm] I_{cHx} = \dfrac{\omega_s^2/k}{L_T \omega_s \sqrt{(\omega_{res}^2 - \omega_s^2)^2 + 4\xi^2 \omega_{res}^2 \omega_s^2}} U_{Hx} \end{cases} \tag{16}$$

From tab.1 and the above analysis, the drawback of k=0.5 is the power loss on R_f is larger than the other LCL design method. The novel LCL filter design method uses an improvement in hardware in order to reduce the power loss on R_f.

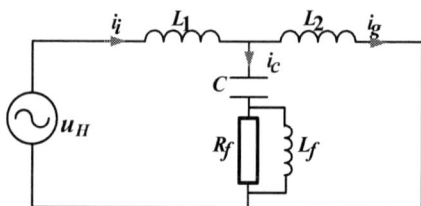

Fig.8 Equivalent model of LCL filter with inductor Lf

$$\begin{cases} \dfrac{I_g(s)}{U_H(s)} = \dfrac{R_f L_f Cs^2 + L_f s + R_f}{\left(\begin{array}{c} L_1 L_2 L_f Cs^4 + [L_1L_2 + (L_1+L_2)L_f]R_f Cs^3 \\ + (L_1+L_2)L_f s^2 + (L_1+L_2)R_f s \end{array}\right)} \\[6mm] \dfrac{I_i(s)}{U_H(s)} = \dfrac{CL_2 s^2 + R_f Cs + 1}{L_1 L_2 Cs^3 + (L_1+L_2)R_f Cs^2 + (L_1+L_2)s} \\[6mm] \dfrac{I_c(s)}{U_H(s)} = \dfrac{CL_2 L_f s^2 + CL_2 R_f s}{\left(\begin{array}{c} L_1 L_2 L_f Cs^3 + [L_1L_2 + (L_1+L_2)L_f]R_f Cs^2 \\ + (L_1+L_2)L_f s + (L_1+L_2)R_f \end{array}\right)} \\[6mm] \dfrac{I_g(s)}{I_i(s)} = \dfrac{R_f L_f Cs^2 + L_f s + R_f}{L_2 L_f Cs^3 + (L_2+L_f)R_f Cs^2 + L_f s + R_f} \end{cases} \tag{17}$$

$$\alpha = \omega_s L_f / R_f \tag{18}$$

Fig.8 shows the finally used LCL filter configuration of this paper, with the damping inductor L_f, the power loss on R_f is significantly reduced[22]. Equation (17) shows the system transfer function with the damping inductor L_f. If α is defined as the impedance ratio which is shown in equation (18), the amplitude relation between grid current i_g and inverter voltage harmonic u_H can be shown as fig.9. And the damping resistor loss ratio with and without damping inductor can be shown as fig.10. From the two figures, the addition of L_f can significant reduces the power loss on R_f while slightly increases the transfer function gain at ω_{res} if α is designed meticulously.

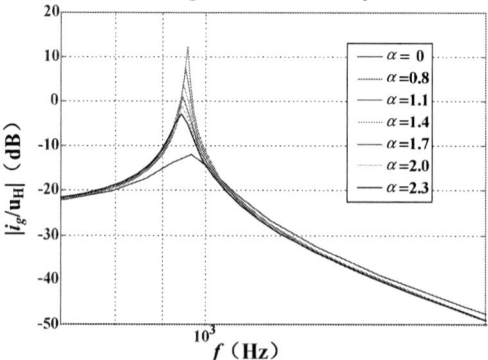

Fig.9 Bode diagram of the LCL filter with damping inductor Lf

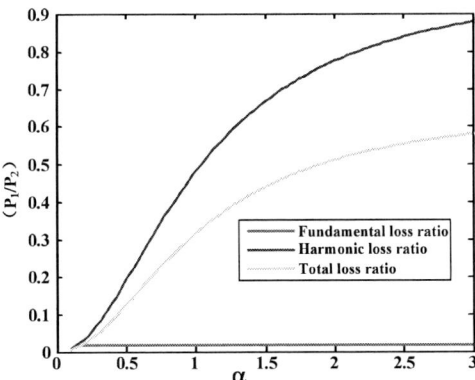

Fig.10 LCL filter damping resistor loss ratio with and without damping inductor

IV. SIMULATION AND EXPERIMENT VERIFICATION

In order to verify the analysis in this paper, simulation and experiment platform is constructed. The rated power is 10kW, the DC voltage U_{dc} equals 750V, and the two DC capacitor $C_1(C_2)$ equals 2000uF, U_{line}=380V, f_0=50Hz. b=10%, $I_{rippleM}$=20%I_{rated}.

Using the design method proposed in this paper, the key parameters of LCL filter should satisfy the following equation.

$$\begin{cases} 5.497mH \le L_T \le 18.5178mH \\ 6.1441uF \le C \le 22.0436uF \\ 0.35396\Omega \le R_f \le 3.9478\Omega \\ 0.018778mH \le L_f \le 0.20944mH \end{cases} \quad (19)$$

In this paper, 6 group parameters are compared in order to verify the superiority of the proposed method. Equation (20) gives the detail parameters, among them, (a) is the design results using the novel method proposed in this paper, (b)-(f) are the contrast groups. Fig.11 and fig.12 show the simulation and experiment results of the 6 group LCL filter parameters, it is clear that the parameter (a) has the best system performance.

$$\begin{cases} (a): L_1 = L_2 = 3mH, C = 18uF, R_f = 1\Omega, L_f = 0.08mH \\ (b): L_1 = L_2 = 3mH, C = 18uF, R_f = 1\Omega, L_f = 0 \\ (c): L_1 = 5mH, L_2 = 1mH, C = 18uF, R_f = 1\Omega, L_f = 0 \\ (d): L_1 = 5.5mH, L_2 = 0.5mH, C = 18uF, R_f = 1\Omega, L_f = 0 \\ (e): L_1 = L_2 = 3mH, C = 18uF, R_f = 0, L_f = 0 \\ (f): L_1 = 5mH, L_2 = 1mH, C = 18uF, R_f = 0, L_f = 0 \end{cases} \quad (20)$$

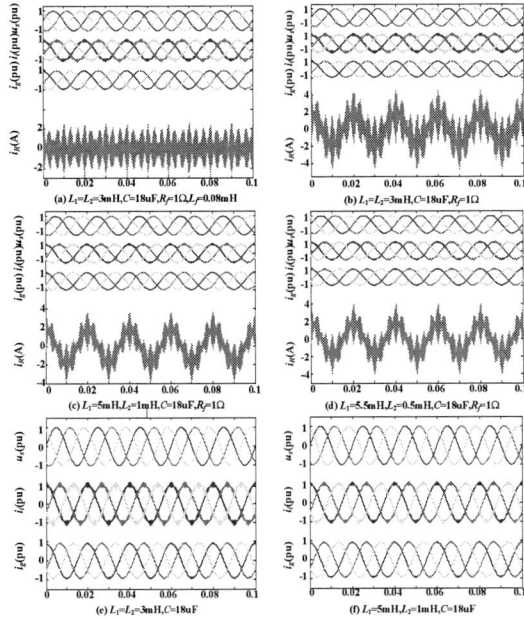

Fig.11 Simulation results of LCL filter under different key parameters

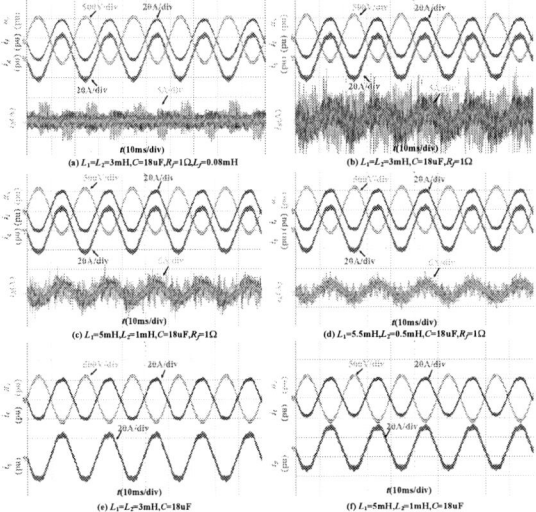

Fig.12 Experiment results of LCL filter under different key parameters

V. CONCLUSION

This work proposed a novel LCL filter parameter design method used in NPC three-level grid connected inverter which bases on the resonant frequency optimization in order to get the best filter effect. In order to reduce the power loss on damping resistor, a small inductor is parallel with R_f. This paper deduces all the parameter range in order to get more accurate design result.

REFERENCES

[1] Nabae A, Takahashi I, Akagi H. A New Neutral Point Clamped PWM Inverter [J]. *IEEE Transaction on Industry Applications*, 1981, 17(5): 518-523.

[2] Blaabjerg F, Teodorescu R, Liserre M. Overview of control and grid synchronization for distributed power generation systems[J]. *IEEE Transactions on Industrial Electronics*, 2006,53(5):1398−1409

[3] Parikshith P, John V. Filter optimization for grid interactive Voltage source inverters[J]. *IEEE Transactions on Industrial Electronics*, 2010, 57(12) : 4106−4114.

[4] Tang Yi, Loh P C, Wang Peng, et al. Exploring inherent damping characteristic of LCL-filters for three-phase grid-connected voltage source inverters[J]. *IEEE Trans. on Power Electronics*, 2012, 27(3) : 1433-1443.

[5] Jalili K, Bernet S. Design of LCL filters of active-front-end two-level voltage-source converters[J]. *IEEE Transactions on Industrial Electronics*, 2009, 56 (5):1674-1689.

[6] Marco Liserr, Frede Blaabjerg, Steffan Hansen. Design and control of an LCL-filter-based three phase active rectifier[J]. *IEEE Transactions on Industry Applications*, 2005,41(5):1281-1290.

[7] Xu Jinming, Xie Shaojun, Xiao Huafeng. Research on control mechanism of active damping for LCL filters[J]. *Proceedings of the CSEE*, 2012, 32(9): 27-33(in Chinese).

[8] CHEN Guozhu, ZHAO Wenqiang. Virtual Resistor Control Strategy of Parallel Active Power Filter with LCL Filter[J]. *High Voltage Engineering*, 2010, 36(7) : 1827-1832(in Chinese).

[9] YIN Jingyuan, JIN Xinmin, WU Xuezhi,et al.Active Damping Control Strategy for LCL Filter Based on Band Pass Filter [J].*Power System Technology*, 2013, 37(8): 2376-2382(in Chinese).

[10] LIU Jilong, MA Weiming, XIAO Fei, et al. An active damping control strategy and design method for LCL filter[J]. *Electric Machines And Control*, 2013, 17(5): 22-27(in Chinese).

[11] Zhang Xueguang, Liu Yicheng, Wang Rui, et al. A Novel Active Damping Control Strategy for PWM Converter with LCL Filter[J]. *Transactions Of China Electrotechnical Society*, 2011, 26(10) : 188-192(in Chinese).

[12] Wang Yaoqiang, Wu Fengjiang, Sun Li, et al. Optimized design of LCL filter for minimal damping power loss[J]. *Proceedings of the CSEE*, 2010, 30(27): 90-95(in Chinese)

[13] LEI Yi, ZHAO Zhengming, LU Sizhao. Hybrid control of active and passive damping for grid connected PV inverter with filter[J]. *Electric Power Automation Equipment*, 2012, 32(11): 23-27(in Chinese).

[14] Shen G Q, Xu D H, Cao L P, et al. An improved control strategy for grid-connected voltage source inverters with a LCL filter [J]. *IEEE Transactions on Power Electronics*, 2008, 23(4): 1899-1906.

[15] Dannehl J, Fuchs F W, Hansen S, et al. PI state space current control of grid-connected PWM converters with LCL filters [J]. *IEEE Trans. on Power Electronics*, 2010, 25(9): 2320-2330

[16] Bao Chenlei, Ruan Xinbo, Wang Xuehua, et al. Design of grid-connected inverters with LCL filter based on PI regulator and capacitor current feedback active damping[J]. *Proceedings of the CSEE*, 2012, 32(25): 133-142 (in Chinese).

[17] Wang Yingjie, Wu Xiaojie, Dai Peng, et al. PWM rectifier with LCL filter based on parameter indentification and a new active damping [J]. *Proceedings of the CSEE*, 2012, 32(15): 31-39(in Chinese).

[18] Wang Yaoqiang, Wu Fengjiang, Sun Li, et al. Control strategy for grid-connected inverter with an LCL output filter[J]. *Proceedings of the CSEE*, 2011, 31(12): 34-39(in Chinese).

[19] Qiu Zhiling, Yang Enxing, Kong Jie, et al. Current loop control approach for LCL-based shunt active power filter[J]. *Proceedings of the CSEE*, 2009, 29(18): 15-20(in Chinese).

[20] Karshenas, H R, and Saghafi H. Basic criteria in designing LCL filters for grid connected converters[C].// *IEEE International Symposium on Industrial Electronics*, New York, USA, 2006.

[21] Fei, L, Xiaoming Z, Yan Z, et al. Design and research on parameter of LCL filter in three-phase grid-connected inverter.//

IEEE International Power Electronics and Motion Control Conference, Wuhan, China, 2009.

[22] Wang, T C Y, Zhihong Y, Gautam S, et al. Output Filter Design for a Grid-interconnected Three-phase Inverter[C].// *Power Electronics Specialist Conference*, Acapulco, Mexico, 2003.

Design and Analysis of Isolated Bi-directional DC/DC Converter using Quasi-Resonant ZVS

Yong-Su Noh[1], Chung-Yuen Won[*]
Dept. of Information and Communication Engineering
Sungkyunkwan University
Suwon, Korea
nys66@skku.edu

Min-Seok Oh[2]
R&D Center
INTECH-FA
Yongin, Korea

Jin-Yong Jeon[3]
Materials R&D Center
Samsung Electronics
Yongin, Korea

Yong-Chae Jung[4]
Dept. of Electronic Engineering
Namseoul University
Cheonan, Korea

Abstract— **Conventional dual half-bridge converter has been previously shown to have advantages in low voltage and high current input applications. This characteristic is suited to photovoltaic application or battery application likes energy storage system (ESS) or uninterruptible power supply (UPS). However, conventional dual half-bridge converter cannot charge the battery because the power flow is one-way. Also, when switches are turned off, high voltage spikes occur by parasitic inductance. To solve this problem, a DC/DC converter which can operate in bi-direction and eliminated voltage spikes are proposed in this paper. Also proposed converter is operated under zero voltage switching (ZVS) condition. In this paper, operation principles and design method of proposed converter are described. These are verified by PSIM simulation and experimental results.**

Keywords— *Dual half bridge converter, bi-directional DC/DC converter, current-fed DC/DC converter, Quasi-resonant ZVS.*

I. INTRODUCTION

Recently, there is a growing need to renewable energy likes photovoltaic (PV), wind power for environmental pollution. And energy storage system (ESS) is also received worldwide interest because increasing of distributed power system and importance of grid stability [1]-[3]. In case of topology for ESS, it should be considered efficiency and low input current ripple for stability of battery. And enough voltage gain to satisfy operation for grid connection is needed. In general, isolated current source type DC/DC converter is used for ESS system for high voltage step-up conversion. Since dual half bridge converter has boost converter characteristic, it is suited to PV or ESS application [4]-[6]. Also it has other advantages which are low-voltage, high-current input characteristic, property of high step up capability, small input current ripple by using two inductors, galvanic isolation and low turn-ratio. However, conventional dual half bridge converter cannot charge the battery because the power flow is one-way. Also, when switches are turned off, high voltage spikes occur by parasitic inductance. To solve this problem, a DC/DC converter which can operate in bi-direction and eliminated voltage spikes are proposed in this paper.

In this paper, isolated bi-directional DC/DC converter using quasi-resonant for zero voltage switching (ZVS) is proposed. Since proposed circuit is operated as a resonant

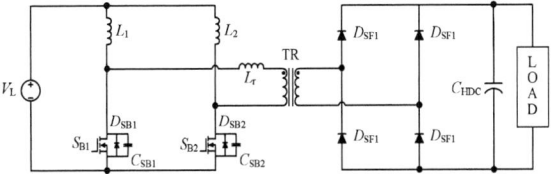

Fig. 1. The dual half-bridge converter [4]-[6].

converter, output voltage should be controlled by switching frequency. The proposed converter consists of dual half-bridge converter as low-voltage stage, full-bridge converter as high-voltage stage, and resonance components. The dual half-bridge converter and full-bridge converter are operated under ZVS condition in both operation modes using resonance components. Since all switches of proposed converter are operated under ZVS condition, switching loss will be reduced and system efficiency will be increased.

In this paper, the operation modes of proposed converter are analyzed. And design method of main inductor and resonance components is introduced. The validity of proposed converter is verified by PSIM simulation and experimental results.

II. OPERATION PRINCIPLES AND MODES

The proposed bi-directional DC/DC converter using quasi-resonance for ZVS method is shown as Fig. 2. The low-voltage stage (i.e. Left side of transformer) consists of dual half bridge converter and resonance components which are resonance inductor L_r, resonance capacitor C_{SBx}. And high-voltage stage (i.e. Right side of transformer) consists of full-bridge converter to rectify bipolar voltage of transformer secondary side and resonance capacitor C_{SFx}. By using resonance components (i.e. L_r, C_{SBx}), the bi-directional converter low-voltage stage can be achieved

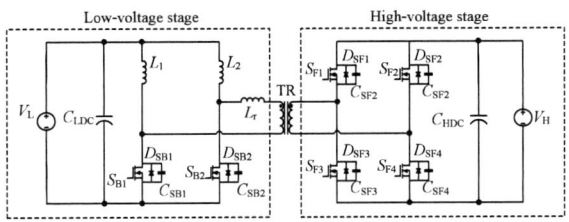

Fig. 2. The proposed bidirectional DC/DC converter.

ZVS condition. Also full-bridge converter can be operated under ZVS condition by resonance components C_{SFx} and reflected inductance L_r.

Analyses of proposed converter operations are performed as below. To analyze the operation characteristic of proposed converter simply, some assumptions are needed.

1) Low-voltage stage inductors (L_x), resonance capacitors (C_{SBx}) and high-voltage stage capacitance value ($C_{SFx} \sim C_{SF4}$) have same inductance and capacitance values in each components, respectively.

2) Input and output DC voltages are constant and ripple-free in steady state.

3) Parasitic resistances of switches, power line and transformer are neglected.

A. Boost Operation

When proposed converter is operated as boost mode, switch S_{B1} and S_{B2} of low-voltage stage converter can be achieved ZVS condition by the resonance inductor L_r and resonance capacitor C_{SBx}. Also high-voltage stage switches $S_{F1} \sim S_{F4}$ can be operated under ZVS condition because the resonance current of low-voltage stage is transferred to secondary side by transformer. The key waveforms of proposed converter in boost mode are shown as Fig. 3. Q_{SBx} and Q_{SFx} are gate signal of switch S_{Bx} and S_{Fx}, i_{Lx} and i_{Lr} are current flowing on inductor L_x and L_r respectively. As shown in Fig. 3, currents of low-voltage stage inductor L_1 and L_2 are not affected by duty cycle of switch S_{F1} and S_{F2}. This means that voltage gain of proposed converter does not affected switch duty cycle such as a conventional boost converter. The voltage gain of proposed converter is mainly affected by resonant components L_r, switching frequency f_{sw} and one of resonance capacitor C_{SBx}. Analysis of the proposed converter boost mode is described as below.

Mode 1 *($t_0 \le t < t_1$):* At t_0, one of gate signals (i.e. Q_{SB1} in Fig. 3) is turned-off and the other still turn-on condition. As shown in Fig. 3, resonance capacitor will be charged and discharged by resonance current i_{Lr}. This mode ends when voltage across the switch S_{B1} (i.e. V_{SB1}) is zero. Voltage of switch V_{SB1} can be calculated as:

$$V_{SB1}(t) = V_{SBx\_m}\sin(\omega_r t + \alpha) + V_{SBx\_C} \tag{1}$$

$$V_{SBx\_m} = \sqrt{\left(\frac{i_{CSBx}(0)}{C_{SBx}\omega_r}\right)^2 + V_{SBx\_C}^{\,2}} \tag{2}$$

$$V_{SBx\_C} = \frac{V_L + mV_{T1}}{1 + m} \tag{3}$$

where V_{SBx_C} is offset voltage of S_{Bx} and V_{SBx_m} is maximum value of voltage across S_{Bx} exclude V_{SBx_C}. And V_{T1} is transformer primary side voltage, m represents inductance ratio which means $m = L_x / L_r$ and resonance angular frequency ω_r is:

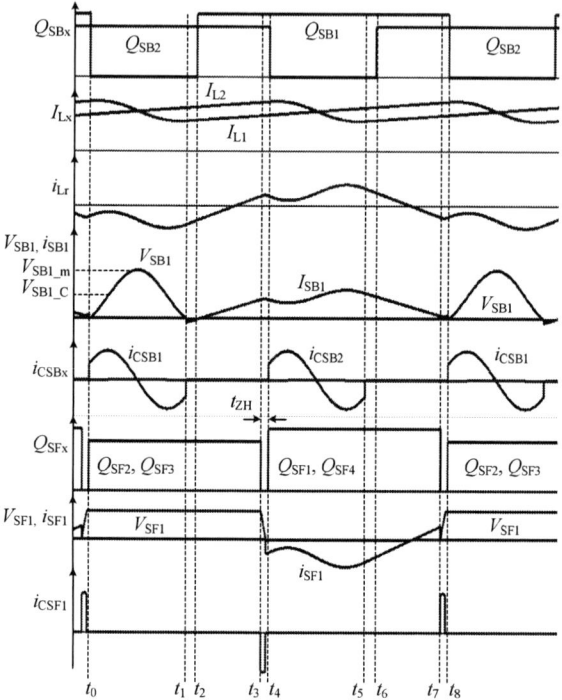

Fig. 3. The key waveforms of proposed converter in boost operation.

$$\omega_r = \sqrt{\frac{L_r + L_x}{L_x L_r C_r}} = \sqrt{1 + \frac{1}{m}}\;\omega_{r,LC} \tag{4}$$

where $\omega_{r,LC}$ represents the resonance angular frequency between L_r and C_{SBx}. If the inductance ratio m is larger than 10, we can approximate as $V_{SBx\_C} \approx V_{T1}$ and $\omega_r \approx \omega_{r,LC}$. Using (1), current flowing resonance inductor L_r (i.e. i_{Lr}) can be calculated as (5).

$$i_{Lr}(t) = \frac{V_{SBx\_m}}{\omega_r L_r}\cos(\omega_r t + \alpha) + i_{Lr}(0) \tag{5}$$

where $i_{Lr}(0)$ is initial current value of L_r in this mode.

Mode 2 *($t_1 \le t < t_2$):* In this mode, the current flowing resonant capacitor i_{CSB1} is flowing through the switch anti-parallel diode D_{SB1} because the D_{SB1} is changed forward bias at t_1. So, voltage of switch S_{B1} is remained as zero during mode 2. Since the voltage of switch S_{B1} will be increased when i_{CSB1} reaches to zero (i.e. End of mode 2, $t = t_2$), switch S_{B1} must be turned on before end of mode 2 to satisfy ZVS condition. In this mode, i_{Lr} and i_{SB1} are calculated as (7),

$$i_{Lr}(t) = \frac{V_{T1}}{L_r}t + i_{Lr}(t_1) \tag{6}$$

$$i_{SB1}(t) = \left(\frac{V_{T1}}{L_r} + \frac{V_L}{L_x}\right)t + i_{SB1}(t_1^+) \tag{7}$$

where $i_{SB1}(t_1^+) = i_{Lr}(t_1) + i_{L1}(t_1)$.

The 2014 International Power Electronics Conference

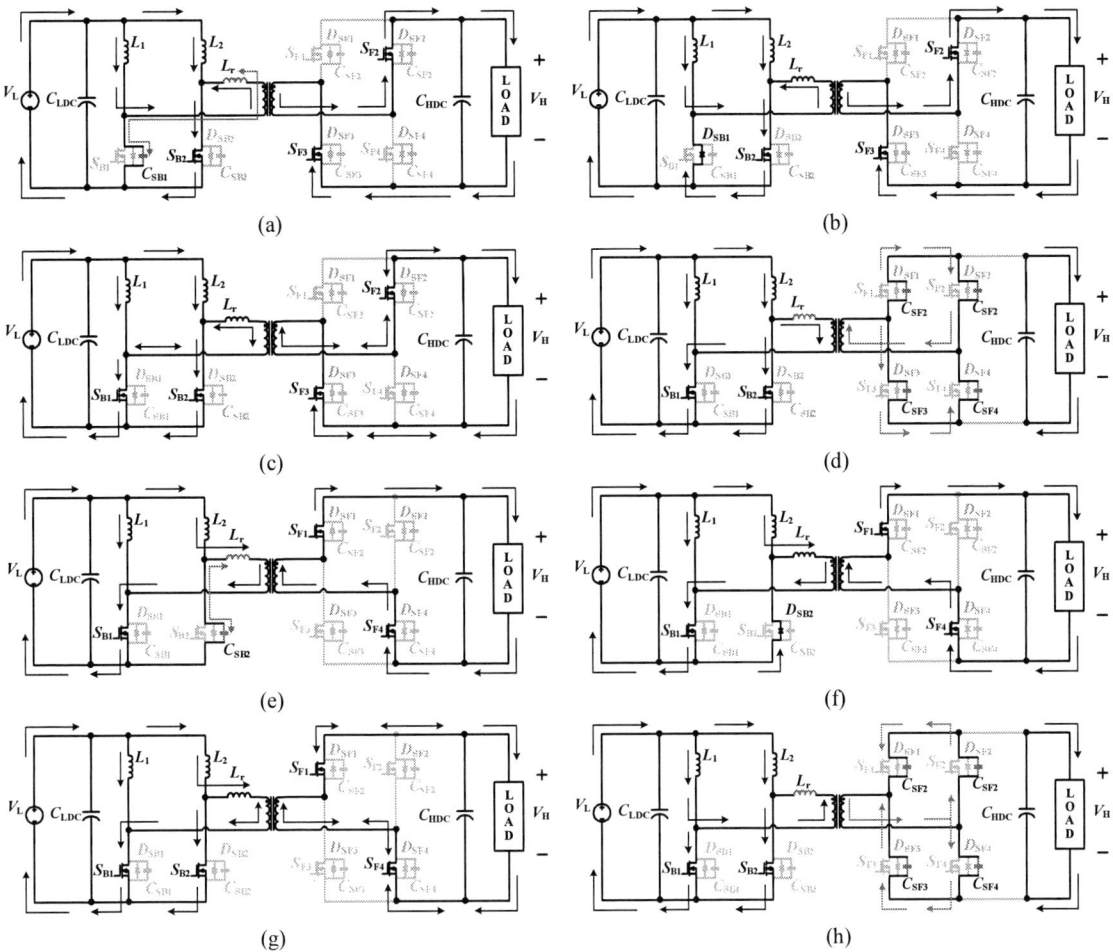

Fig. 4. The operation modes of the proposed bi-directional converter in boost operation (a) Mode 1, (b) Mode 2, (c) Mode 3, (d) Mode 4, (e) Mode 5, (f) Mode 6, (g) Mode 7 and (h) Mode 8.

Mode 3 *($t_2 \leq t < t_3$):* At t_2, S_{Bx} is turned-on under ZVS condition. In mode 3, current flowing through anti-parallel diode of switch D_{SB1} in mode 2 flows switch S_{Bx} and increased linearly. From t_2, both switches of low-voltage stage are turn-on. The current of resonance inductor and switch are:

$$i_{Lr}(t) = \frac{V_{T1}}{L_r}t + i_{Lr}(t_2) \tag{8}$$

$$i_{SB1}(t) = \left(\frac{V_{T1}}{L_r} + \frac{V_L}{L_x}\right)t + i_{SB1}(t_2) \tag{9}$$

Mode 4 *($t_3 \leq t < t_4$):* To transfer power from low-voltage stage to high-voltage stage, turn-on time of high-voltage stage switches are synchronized with turn-off time of low-voltage stage switches. In this mode, enough dead time (i.e. t_{ZH}) between high-voltage switches should be considered for ZVS condition. During the dead time of high-voltage stage switches, total energy in output capacitor of switch S_{Fx} should be discharged by current

which transfer from low-voltage stage i_{Lr}. By this reason, dead time t_{ZH} of high-voltage stage should be satisfied below condition:

$$t_{ZH} > \frac{2nC_{SFx}V_H}{i_{Lr}(t_3)} \tag{10}$$

Mode 5-8 *($t_4 \leq t < t_8$):* Since low-voltage stage consists of two legs which has same switch and inductor, operation mode of other leg is similarly operated as mode 1-4. Even though C_{SB1} does not participate in resonance at mode 5-8, currents i_{Lr} and I_{SB1} do not operate linearly by resonance of other leg (i.e. L_r and C_{SB2} in Fig. 3). The current i_{Lr} in mode 5 can be calculated as (11).

$$i_{Lr}(t) = -\frac{V_{SBx\_m}}{\omega_r L_r}\cos(\omega_r t + \alpha) + i_{Lr}(t_4) \tag{11}$$

The current i_{Lr} in mode 6 and 7 is calculated as,

$$i_{Lr}(t) = -\frac{V_{T1}}{L_r}t + i_{Lr}(t_5) \tag{12}$$

978-1-4799-2706-7/14 $31.00 © 2014 IEEE   168

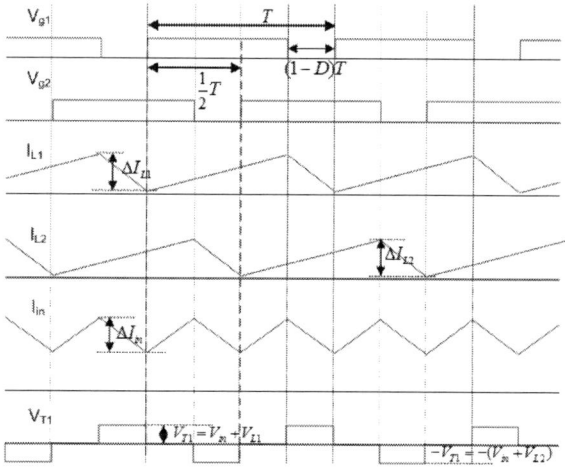

Fig. 5. Switching signals and current waveforms of proposed converter.

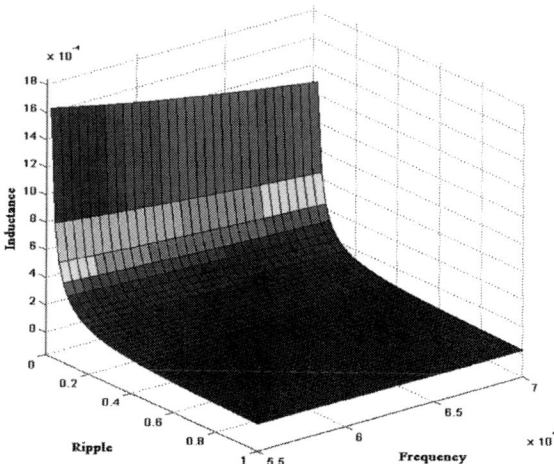

Fig. 6. Main inductance selection.

B. Buck Operation

When the proposed converter is operated as buck mode, high-voltage stage switches $S_{F1} \sim S_{F4}$ are operated under ZVS condition. To make ZVS condition, dead time between high-voltage stage switches is needed likes for boost operation. And low-voltage switches are operated under ZVS condition by resonance components L_r and C_{SBx}. In buck mode operation, low-voltage stage switches are operated same as boost mode.

III. DESIGN OF PROPOSED CONVERTER

The proposed bi-directional DC/DC converter is composed of dual half-bridge converter and the full-bridge converter. In low-voltage stage, the dual half bridge converter has two main inductors and switches, and resonant inductor and capacitors. In high-voltage stage, the full-bridge converter is composed of four switches and resonant capacitors. In this paper, design of main inductor and resonance components to reduce input current ripple and satisfy ZVS conditions. Table I is system basic parameters to design in this paper.

A. Main Inductor

The proposed converter main inductor L_x accumulates energy in magnetic fields when two switches of low-voltage stage are turned-on. And one of switches is turned-off, accumulated energy in main inductor L_x can transfer to high-voltage stage through transformer. To

TABLE I
BASIC PARAMETERS OF PROPOSED CONVERTER

Parameter	Symbol	Value	Unit
Low voltage	V_L	30	V
High voltage	V_H	200	V
Power rate	P	250	W
Switching frequency	f_{sw}	55 ~ 70	kHz

satisfy these operations, duty ratio of the switches in low-voltage stage should be controlled over 0.5.

Current flow of low-voltage stage is divided into two inductors. By this reason, current ripple of low voltage stage is smaller than the current of using conventional one-inductor topologies.

Voltage of transformer can be described as,

$$V_{T1} = L_1 \times \Delta I_{LV} \left\{ \frac{1}{\left(-2D^2 + 3D - 1 \right)T} \right\} \quad (13)$$

where ΔI_{LV} is low-voltage stage current ripple. The low-voltage current ripple can be calculated as,

$$\Delta I_{Lx} = r I_{Lx,ave} \quad (14)$$

where r is ripple ratio (i.e. $0.05 < r < 0.3$) and $I_{Lx,ave}$ is average value of current flowing L_x. Using (13) and (14), main inductor can be calculated as (15),

$$L_x = \frac{V_L}{2 \times \left(\Delta I_{LV} \right) \times f_{sw}} \quad (15)$$

Since large main inductor current affects transformer power loss or battery characteristic, the input current ripple ratio is less than 10% in general. However, small input current ripple can bring large inductor size. So, main inductance should be designed considering inductance, size, and current ripple. In this paper, main inductor is selected considering switching frequency and current ripple. Figure 6 shows inductance curve according to the switching frequency and ripple ratio. As shown in Fig. 6, main inductance is designed as 480µH by selecting current ripple ratio under 3% and maximum switching frequency is 70kHz.

B. Resonance Components

As shown in Table I, switching frequency is varied between 55kHz and 70kHz to control output voltage. To avoid interference by switching frequency, resonance frequency should be selected higher than switching

The 2014 International Power Electronics Conference

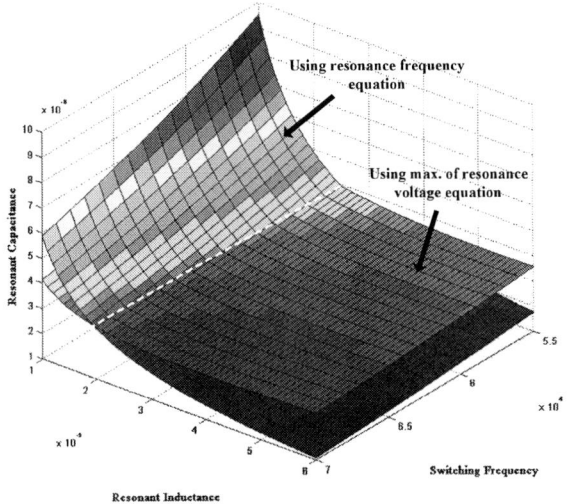

Fig. 7. Resonance components selection curve.

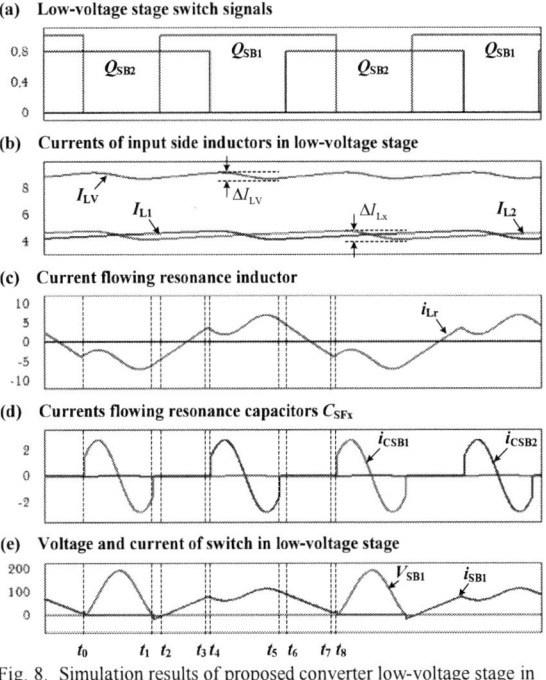

Fig. 8. Simulation results of proposed converter low-voltage stage in boost operation (a) gate signals, (b) currents of input side, (c) current flowing resonance inductor, (d) current flowing resonance capacitor C_{SBx} and (e) voltage and current of S_{B1}.

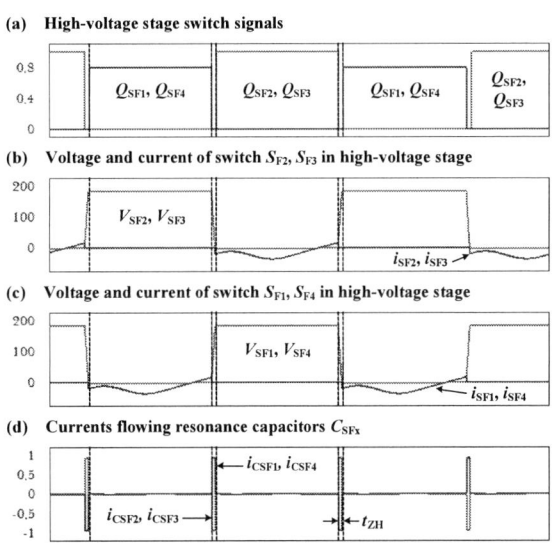

Fig. 9. Simulation results of high-voltage stage in boost operation (a) gate signals, (b) voltage and current of S_{F2}, S_{F3}, (c) voltage and current of S_{F1}, S_{F4}, (d) current flowing resonance capacitor C_{SFx}.

frequency. However, if the resonance frequency is too high, switches of proposed converter cannot satisfy ZVS condition. By these reasons, resonance frequency is selected as three times than maximum switching frequency in this paper.

And resonance voltage is affected by resonance components as (2). By this reason, resonance components are calculated after selecting maximum resonance voltage. Using (2) and (4), resonance capacitance curves can be drawn as Fig. 7. Resonance components should be selected To satisfy both conditions. In this paper, resonant inductance and capacitance are selected as 15μH and 38nF, respectively.

IV. SIMULATION RESULTS

In order to verify proposed converter operation, PSIM simulation is performed. The system basic parameters are given in Table I and designed inductors and capacitor are given in Table II.

Figure 8 is simulation result of low-voltage stage in boost operation. As shown in Fig. 8(a), duty ratio of gate signals are higher than 0.5. And low-voltage resonance is occurred during $t_0 \sim t_1$ and $t_4 \sim t_5$ as Fig. 8(c). Before t_2, switch S_{B1} should be turn on to satisfy ZVS condition.

Figure 8(e) shows voltage and current of switch in low-voltage stage operated under ZVS condition.

Figure 9 is simulation results of high-voltage stage. Switches of high-voltage stage are operated as almost half duty. As shown in Fig. 9(b) and (c), all switches of high-voltage stage are operated under ZVS condition. And enough dead time for charge or discharge capacitor C_{SFx} is need to make ZVS condition in high-voltage stage as Fig. 9(d).

TABLE II
SYSTEM PARAMETERS OF PROPOSED CONVERTER

Parameter	Symbol	Value	Unit
Main inductance	L_x	480	μH
Resonance inductance	L_r	15	μH
Resonance capacitance in low-voltage stage	C_{SBx}	38	nF
Resonance capacitance in high-voltage stage	C_{SFx}	1	nF
Transformer turns ratio	n	2	

978-1-4799-2706-7/14 $31.00 © 2014 IEEE

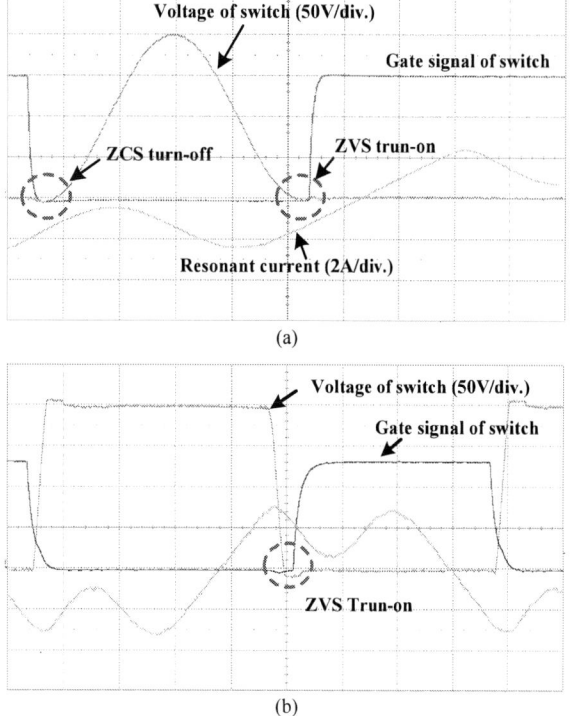

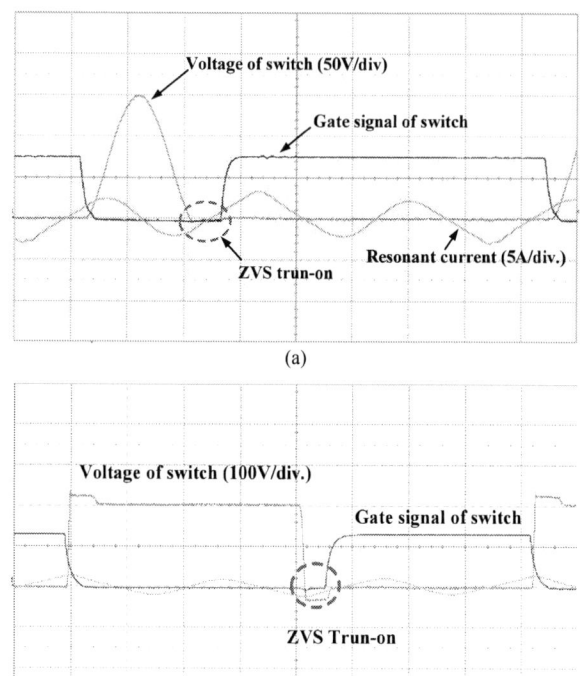

(a)

(a)

(b)

(b)

Fig. 10. Switch voltages and gate signals of low-voltage stage and (b) high-voltage stage of proposed converter in boost operation.

Fig. 11. Switch voltages and gate signals of low-voltage stage and (b) high-voltage stage of proposed converter in buck operation.

V. EXPERIMENTAL RESULTS

The proposed DC/DC converter has been examined. Rated power of prototype is 250W, and nominal voltage of V_L and V_H are controlled as 30V and 200V, respectively. To achieve output voltage regulating and ZVS condition, switching frequency and duty control are adopted. In this system, the DSP TMS320F28035 is used as the controller.

The circuit waveforms in steady state are shown in Fig. 10 and Fig. 11. Low-voltage stage switch is turned-on at zero voltage in both modes. To satisfy ZVS condition, switch duty ratio should be controlled to turn on after voltage of switch is zero voltage. Since all switches of proposed converter are operated in ZVS condition, system efficiency will be increased. In this prototype experiment, system efficiency reaches 93% in boost mode and 92% in buck mode at rated power.

VI. CONCLUSIONS

In this paper, the isolated bi-directional DC/DC converter for battery application is proposed and theoretical analysis and design method are described. Proposed converter can operate as boost operation and buck operation. All switches of proposed DC/DC converter are operated under ZVS condition. To satisfy ZVS condition widely operation range, frequency and switch duty control method was applied. PSIM simulations and experiment are performed to verify proposed method.

ACKNOWLEDGMENT

This work was supported by Samsung Electronics Co. Ltd..(No. IO131217-01110-01).

REFERENCES

[1] B. Y. Choi, Y. S. Noh, Y. H. Ji, B. K. Lee and C. Y. Won, "Battery-integrated power optimizer for PV-battery hybrid power generation system", *IEEE Conf. of Vehicle Power and Propulsion Conference (VPPC)*, pp. 1343-1348, 2012,

[2] B. Espinar and D. Mayer, "The role of energy storage for mini-grid stabilization", *IEA, Report IEA-PVPS*, 2011

[3] M. S. Moghaddam and A. Hajizadeh, "Control of hybrid PV/Fuel cell/Battery power system", *in Proc. Power Electronics, Drives and Energy System (PEDES) Conference*, pp. 1-7, 2010.

[4] P. J. Wolfs, "A current-sourced DC-DC converter derived via the duality principle from the half-bridge converter", *IEEE Trans. On Industrial Electronics*, vol. 40, no. 1, pp. 139-144, 1993.

[5] P. J. Wolfs and Q. Li, "An analysis of a resonant half bridge dual converter operating in continuous and discontinuous modes", *IEEE Power Electronics Specialist Conference (PESC)*, vol. 3, pp. 1313-1318, 2002.

[6] B. Yuan, X. Yang, X. Zeng, J. Duan, J. Zhai and D. Li, "Analysis and design of a high step-up current-fed multiresonant DC-DC converter with low circulating energy and zero current switching for all active switches", *IEEE Trans. On Industrial Electronics*, vol. 59, no. 2, pp. 964-978, 2012.

An Active-Clamping ZVS Flyback Converter with Integrated Transformer

Jing-Yuan Lin
Dept. of Electrical Engineering

National Taitung College

Taitung City, Taiwan R.O.C
jylin@ntc.edu.tw

Yu-Kang Lo
Dept. of Electronic Engineering

National Taiwan University of
Science and Technology
Taipei City, Taiwan R.O.C
yklo@mail.ntust.edu.tw

Huang-Jen Chiu
Dept. of Electronic Engineering

National Taiwan University of
Science and Technology
Taipei City, Taiwan R.O.C
hjchiu@mail.ntust.edu.tw

Chao-Fu Wang
Dept. of Electronic
Engineering
National Taiwan University of
Science and Technology
Taipei City, Taiwan R.O.C
D9902204@mail.ntust.edu.tw

Chien-Yu Lin
Dept. of Electronic
Engineering
National Taiwan University of
Science and Technology
Taipei City, Taiwan R.O.C
D9902209@mail.ntust.edu.tw

Abstract - **This paper propose an active-clamping flyback converter using a integrated transformer. The proposed converter is composed of two active-clamp flyback converter. The presented converter can equally share the total load current between two secondaries so that the rectifier diode conduction loss can be decreased. Otherwise, the main switch of any converter as auxiliary switch for the other one, so that only two switches are required and both can achieve zero-voltage-switching (ZVS) operation. The transformers of converter are integrated into one magnetic core, therefore, the volume and copper loss of transformer can be reduced. Detailed analysis and design of this integrated magnetic active-clamping flyback converter are described. Experimental results are recorded for a prototype converter with an AC input voltage ranging from 85 to 135 V, an output voltage of 24 V and an output current of 5 A, operating at a switching frequency of 100 kHz.**

I. Introduction

The flyback converters are widely using in the switching mode power supply design to provide regulated output voltages for low-power applications. The transformer in the flyback converter is used to achieve both electric isolation and energy storage. The switch in the flyback converter is conventionally operated at hard switching. Therefore high voltage and current spikes suffered from the transformer leakage inductance are imposed on the switch [1-4]. A new two-transformer active-clamping ZVS flyback converter [5], which is mainly composed of two active-clamping flyback converters, by utilizing two separate transformers, the large DC flux can be equally shared between the two primaries to decrease the transformer core loss and copper loss. However, it still requires two independent transformers, increasing the circuit size, cost and complexity. According to above statements, an active-clamping flyback converter using an integrated transformer is proposed in this paper, the advantage of this proposed topology is described in reference [5]. It is noteworthy that the proposed circuit combines two separated magnetizing inductance in a single core. A single EE core is used for all magnetic components, including two transformers. The primary and secondary windings of transformer are wounded on outer legs of EE core, all three legs are gapped. By this integrated magnetic structure, the flux is cancelled in the center leg [6, 7], thus, it not only reduces more core loss but also decreases the volume of magnetic component. Detailed analysis of the circuit operations, steady-state behaviors, and DC gain of the proposed converter are given in the

following sections. Design procedures are formulated to facilitate the design process. Theoretical discussions are validated with experimental results on a prototype converter that delivers a 24-V/5-A output from an AC input ranging from 85 to 135 V at a maximum efficiency of 93 %.

II. Circuit description and principle operation of proposed converter

A. Circuit Description

Fig. 1 shows the schematic of the proposed converter using an integrated transformer, which is derived from two active-clamping flyback converter in a single core. Fig. 2 shows the structure of magnetic component for this proposed circuit.

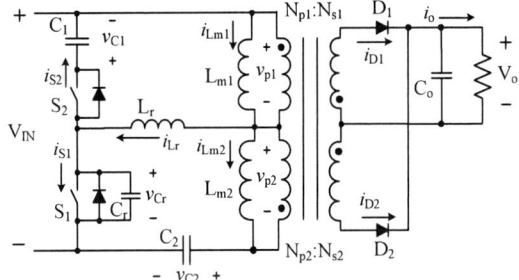

Fig. 1 schematic of the proposed active-clamping converter

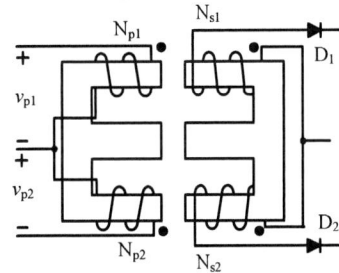

Fig. 2 the structure of magnetic component for the proposed circuit

To analyze the proposed active-clamping flyback converter, the following assumptions are made.
- The clamping capacitor C_1 and C_2 are larger than resonant capacitor C_r, moreover, the steady-state voltage of clamping capacitor V_{C1} and V_{C2} almost equal to constant.
- The turns ratio of the transformer windings is n = N_{p1}/N_{s1}, also N_{p1} = N_{p2} and N_{s1} = N_{s2}, The

magnetizing inductances of the two turns are equal $L_{m1} = L_{m2} = L_m$.

- The conduction times of S_1 and S_2 are DT_s and $(1-D)T_s$, respectively, where D is the duty cycle of Q_1, and T_s is the switching period.
- The energy stored in the resonant inductor is greater than the energy stored in the resonant capacitor to achieve ZVS operation for the active switches.
- The resonant inductance L_r is much smaller than the magnetizing inductance L_m.

From the above assumptions, the voltages of the clamping capacitors C_1 and C_2 in the steady state can be calculated. In addition, the transfer ratio of V_O to V_{IN} can also be obtained. When S_1 is turned on, the voltage across the N_{p1} (or the magnetizing inductor L_{m1}) approximates the input voltage V_I. On the other hand, when S_2 turns on, the voltage across the L_{m1} is about $-V_{C1}$. From the flux balance of L_{m1} under the steady state, V_{C1} can be determined as

$$V_{C1} = \frac{D}{1-D} V_{IN} \tag{1}$$

Similarly, when Q_1 is turned on, the voltage across the L_{m2} is V_{C2}. When Q_1 turns off, the voltage across L_{m2} is about $(V_I + V_{C1} - V_{C2})$. Then from (1) and the inductor flux balance, the steady-state value of V_{C2} can be found to be

$$V_{C2} = V_{IN} \tag{2}$$

Also from the flux balance of two primary turns, the voltage gain of the proposed active-clamping flyback can be obtained as

$$\frac{V_O}{V_{IN}} = \frac{1}{n} \frac{D}{1-D} \tag{3}$$

B. Integrated Magnetic Analysis

Fig. 3 shows the transformer in Fig. 1 is replaced by an integrated magnetic component. To ease analysis and discussion, the following analyze is under assumption that the leakage flux is negligible. From Fig. 3, that shows the flux path in core, Φ_1 and Φ_2 are the flux of two outer legs, Φ_c is the flux of the center leg, Φ_c can be obtained by subtracting the flux of two outer legs.

$$\Phi_c = |\Phi_1 - \Phi_2| \tag{4}$$

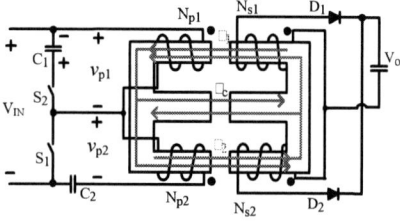

Fig. 3 the proposed circuit using an integrated magnetic

The steady state operation can simplify to two stages [6], the equivalent magnetic reluctance circuit of each stage as shown in Fig. 4, where R_o and R_c is the reluctance of each outer legs and center leg.

1. Stage 1

Stage 1 is the time interval of S_1 turned on and S_2 turned off. It shows in Fig. 4(a). The voltage applied on the v_{p1} and v_{p2} are positive, it induces the voltage on secondary is negative so that the output diodes are blocked. During this time, both the flux Φ_1 and Φ_2 are increase. The magnetic reluctance circuit for this stage shown in fig. 4(c) gives the following equations.

$$N_{p1}i_{p1} = \Phi_1 R_o + \Phi_c R_c$$
$$N_{p2}i_{p2} = \Phi_2 R_o - \Phi_c R_c \tag{5}$$

2. Stage 2

Stage 2 is the time interval of S_2 turned on and S_1 turned off. It shows in Fig. 4(b). The voltage applied on the v_{p1} and v_{p2} are negative, it induces the voltage on secondaries is positive, thus force output diode D_1 and D_2 conduct, secondary current i_{D1} and i_{D2} share the output current i_o. During this time, both the fluxes Φ_1 and Φ_2 are decrease. The magnetic reluctance circuit for this stage shown in fig. 4(c) gives the following equations.

$$N_{p1}i_{p1} + N_{s1}i_{D1} = \Phi_1 R_o + \Phi_c R_c$$
$$N_{p2}i_{p2} + N_{s2}i_{D2} = \Phi_2 R_o - \Phi_c R_c \tag{6}$$

According to Faraday's law, we have

$$\Delta\Phi_1 = \Delta\Phi_2 = \frac{V_{IN} \cdot D}{N_p \cdot f_{sw}} \tag{7}$$

During stage 1, the current of switch $S_1 (i_{S1})$ is the sum of two primary winding current i_{p1} and i_{p2}, thus the ripple current of switch S_1 is Δi_{S1} can be obtained as :

$$N_p \cdot (\Delta i_{p1} + \Delta i_{p2}) = N_p \cdot \Delta i_{S1} = R_o \cdot (\Delta\Phi_1 + \Delta\Phi_2) \tag{8}$$

Substituting (7) in (8) yields:

$$\Delta i_{S1} = \frac{2 \cdot R_o \cdot V_{IN} \cdot D}{N_p^2 \cdot Ae_o \cdot f_{sw}} \tag{9}$$

Where Ae_o is the cross-sectional areas of two outer legs. Comparing to original single transformer active-clamping flyback topology at same power specification, each primary turn share the power storage in transformer, it decrease the power stress of magnetic component. Following same analyze, the ripple current of output Δi_o also can be obtained. During stage 2, because of the average value of $(i_{p1} + i_{p2})$ equal to zero, output ripple current Δi_o can be derived as:

$$\Delta i_{D1} + \Delta i_{D2} = \Delta i_o = \frac{2 \cdot R_o \cdot V_o \cdot (1-D)}{N_s^2 \cdot Ae_o \cdot f_{sw}} \tag{10}$$

It can be know in (10), output current is the sum of two secondary winding current i_{D1} and i_{D2}, thus, the current stress and conduction loss of each output rectifier can be decreased.

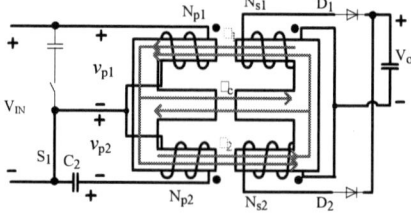

(a) Flux path in core of stage 1

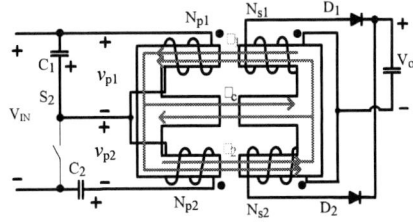

(b) Flux path in core of stage 2

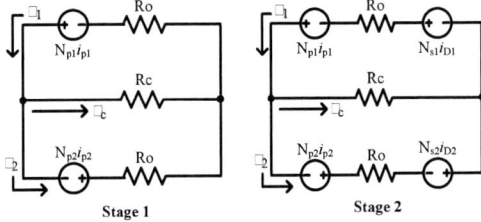

(c) Magnetic reluctance circuit of each stages

Fig. 4 steady state operation of proposed circuit

The flux rate of the two outer legs are equal during each stage, thus, the flux of the center leg is almost zero, the proposed integrated magnetic has a lower core loss in center leg. An important factor is the DC flux, it can be calculated with aid of the circuit as shown in Fig. 4. The fluxes Φ_1, Φ_2 and Φ_c in core can be determined as:

$$\Phi_1 = \frac{1}{(1+k) \cdot R_o} \left[N_p i_{p1} + N_s i_{D1} + k \cdot (N_p i_{p2} + N_s i_{D2}) \right] \quad (11)$$

$$\Phi_2 = \frac{1}{(1+k) \cdot R_o} \left[N_p i_{p2} + N_s i_{D2} + k \cdot (N_p i_{p1} + N_s i_{D1}) \right] \quad (12)$$

$$\Phi_c = \frac{(1-k)}{(1+k) \cdot R_o} \left[N_p i_{p1} + N_s i_{D1} - N_p i_{p2} - N_s i_{D2} \right] \quad (13)$$

Where

$$k = \frac{R_c}{R_o + R_c} \quad (14)$$

The average value of each parameter is

$$I_{p1} + I_{p2} = 0, \quad I_{p1} - I_{p2} = \frac{I_o \cdot N_s}{N_p \cdot (1-D)}, \quad I_{D1} = I_{D2} = \frac{I_o}{2} \quad (15)$$

From above equation, the ripple current of switch S_1 and output rectifiers can be reduce by reducing the reluctance of outer leg R_o. Furthermore, k equals unit when R_o becomes smaller than R_c, the DC flux of each leg is derived as:

$$\Phi_{1,DC} = \frac{N_s \cdot I_o}{(1+k) \cdot R_o} \quad (16)$$

$$\Phi_{2,DC} = \frac{N_s \cdot I_o}{(1+k) \cdot R_o} \quad (17)$$

$$\Phi_{c,DC} = \frac{(1-k) \cdot N_s \cdot I_o}{(1+k) \cdot R_o \cdot (1-D)} \quad (18)$$

The peak of flux is sum of DC value and half of AC value.

$$\Phi_{1,peak} = \frac{L_s \cdot I_o}{N_s} + \frac{1}{2} \cdot \frac{Vo \cdot (1-D)}{N_s \cdot fsw} \quad (19)$$

$$\Phi_{2,peak} = \frac{L_s \cdot I_o}{N_s} + \frac{1}{2} \cdot \frac{Vo \cdot (1-D)}{N_s \cdot fsw} \quad (20)$$

$$\Phi_{c,peak} = \frac{(1-k)}{(1-D)} \cdot \frac{L_s \cdot I_o}{N_s} \quad (21)$$

Where

$$L_s = \frac{N_s^2}{(1+k) \cdot R_o} \quad (22)$$

To calculate the turn ratio and duty cycle range under the specifications, choose the maximum AC flux density and DC flux density according to magnetic material, solving the turn of N_s and N_p, designing the air-gap of center leg and calculate peak flux density from (19) to (21). To avoid the saturation of magnetic core, the following equation should be satisfied.

$$B_1 \leq B_{1,peak} = \frac{L_s \cdot I_o}{Ae_o \cdot N_s} + \frac{1}{2} \cdot \frac{Vo \cdot (1-D)}{Ae_o \cdot N_s \cdot fsw} \quad (23)$$

$$B_2 \leq B_{2,peak} = \frac{L_s \cdot I_o}{Ae_o \cdot N_s} + \frac{1}{2} \cdot \frac{Vo \cdot (1-D)}{Ae_o \cdot N_s \cdot fsw}$$

$$B_c \leq B_{c,peak} = \frac{(1-k)}{(1-D)} \cdot \frac{L_s \cdot I_o}{Ae \cdot N_s} \quad (24)$$

Where Ae is the cross area of center leg.

III. EXPERIMENTAL RESULTS

In order to verify the theoretical analysis, a 150-W prototype converter is built and tested in the laboratory. The implementation of the proposed active-clamping flyback converter with integrated transformer is shown in Fig. 5. The experimental results are obtained with the following parameters.

Input AC voltage range: 85 ~ 135 V_{rms}
Output voltage: $V_O = 24V$
Rated output current: $I_O = 5A$
Switching frequency: $f_s = 100$ kHz
Maximum duty cycle: $D_{max} = 0.45$
Minimum duty cycle: $D_{min} = 0.15$
Turns ratio: $n = N_p/N_s = 40/8$
Clamping capacitances: $C_1 = 0.33 \; \mu F$, $C_2 = 200 \; \mu F$
Output capacitance: $C_O = 560 \; \mu F$
Conversion efficiency: $\eta > 0.9$

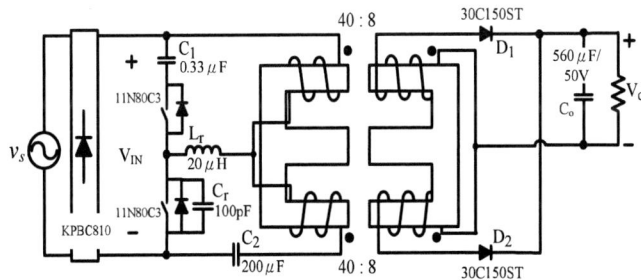

Fig. 5 The implementation of the proposed active-clamping flyback converter.

An EE-42 core is used for the integrated transformer. The magnetizing inductance of each primary is form basic in doctor energy storage equation to calculate, therefore converter to use the magnetizing inductances 500uH. The clamp capacitance C1 and C2 are calculated from reference [5], therefore clamp capacitance C1 to use standard value 0.33uF, C2 to use 200uF for avoid ripple voltage. To ensure the ZVS operation, a 50-ns dead time, which is about fourth of the resonant period is inserted between the gate signals of S_1 and S_2. The output capacitance, C_{OSS}, of a 11N80C3 MOSFET is about 50 pF at a 200-V drain-to-source voltage. Therefore, the equivalent resonant capacitance C_r including the output capacitances of S_1 and S_2 and the parasitic capacitance across the transformer primary winding is about 100 pF. The resonant inductance L_r is about $20\,\mu$H. The Co is selected 560 uF. Fig. 6 illustrates the experimental results of the gate signal v_{GS1}, the resonant inductor current i_{Lr}, and secondary diode currents i_{D1} and i_{D2} at 20% and 100% output load condition. As S_1 turns off, the energy stored in the transformer primaries is transferred to the output load and the transformer secondary diodes conduct. Also in Fig. 8 it can be seen that the two rectifier diodes equally share the load current to decrease the diode conduction losses. Fig. 9 illustrates the measured results of gate signals v_{GS1}, resonant inductor current i_{Lr}, switching current of i_{S1} and i_{S2}. Fig. 10 shows the measured efficiencies of the proposed active-clamping flyback converter with integrated transformer at a 110-V_{rms} input voltage for different output powers. The average efficiency when the system is active is above 90 %. At the rated full load, the conversion efficiency above 92 %.

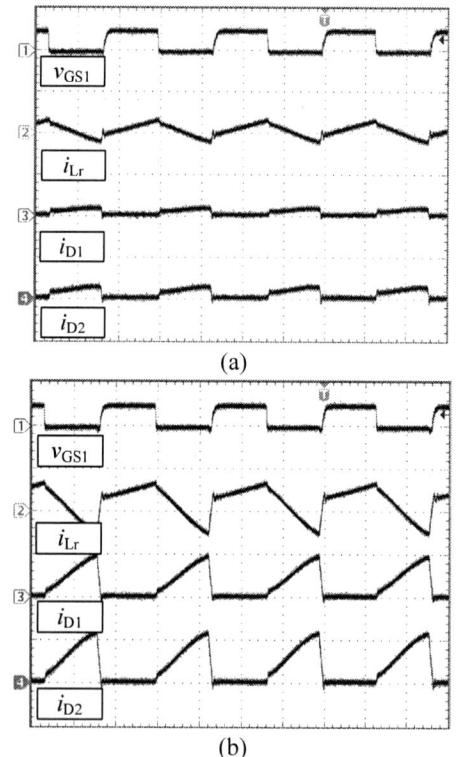

(a)

(b)

Fig. 8 Waveforms of v_{GS1}, i_{Lr}, i_{D1} and i_{D2} at a load current of (a) 20%, and (b) 100%. (v_{GS1}: 25 V/div., i_{Lr}: 5 A/div., i_{D1}/ i_{D2}: 10 A/div., Time: 4μs/div.)

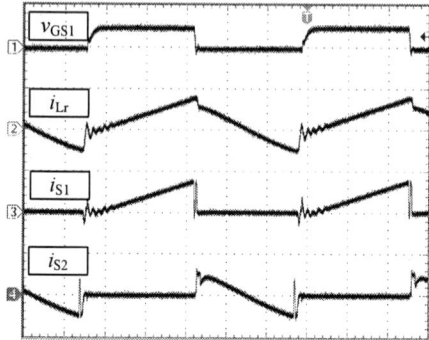

Fig. 11 Waveforms of v_{GS1}, i_{Lr}, i_{S1} and i_{S2} at 20% load current. (v_{GS1}: 25 V/div., i_{Lr}: 2 A/div., i_{S1}: 2 A/div., i_{S2}: 2 A/div., Time: 2μs/div.)

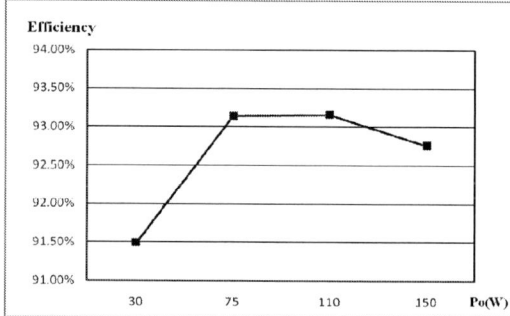

Fig. 12 Efficiencies of the proposed integrated-transformer active-clamping flyback converter at different output powers.

IV. CONCLUSION

This paper presents a active-clamping flyback converter with integrated transformer to facilitate a low-profile implementation while gaining a high conversion efficiency via ZVS of the active switches. Two switches are respectively the auxiliary cells for achieving ZVS operations for each other. The load current is equally shared between the two secondary rectifier diodes to reduce the conduction loss. An integrated magnetic component not only decrease the size of transformer, but also improve the loss and current stress because the flux rate of center leg of EE core is reduced. The experimental results on a 150-W prototype are recorded to verify the theoretical analysis. The average active-mode efficiency is above 90%. The proposed active-clamping flyback converter is suitable for the low-to-medium power applications with larger load currents.

REFERENCES

[1] C. S. Leu, G. Hua, F. C. Lee, and C. Zhou, "Analysis and design of RCD clamp forward converter," in Proc. 7th High Frequency Power Conversion Conf., 1992, pp. 198–208.

[2] C. T Coi, C. K. Li, and S.K. Kok, "Control of an active clamp discontinuous conduction mode flyback converter," in Proc. IEEE-PEDS, vol. 2, 1999, pp 1120-1123.

[3] R. Watson, F. C. Lee, and G. C. Hua, "Utilization of an active-clamp circuit to achieve soft switching in flyback

converters," IEEE Trans. Power Electron., vol. 11, no. 1, pp. 162-169, 1996.

[4] P. Aiou, A. Bakkali, I. Barbero, J. A. Cobos, M. Rascon, "A low power topology derived from flyback with active clamp based on a very simple transformer," in proc. 21rd Ann.Appl. power Electron. Conf., 2006, pp, 6.

[5] Y. K. Lo, J. Y. Lin,. "Active-Clamping ZVS Flyback Converter Employing Two Transformers," IEEE Trans. Power Electron., vol. 22, NO. 6, 2007.

[6] Qianhong Chen, Yang Feng, Linquan Zhou, Jian Wang, Xinbo Ruan,. "An improved Active Clamp Forward Converter with Integrated Magnetic," IEEE Power Electronics Specialists Conference, 2007. PESC 2007.

[7] Peng Xu, Qiaoqiao Wu, Pit-Leong Wong, Lee, F.C. "A novel integrated current doubler rectifier," IEEE Applied Power Electronics Conference and Exposition, 2000. APEC 2000.

PFM and PWM Hybrid controlled LLC converter

Junichi Yamamoto
Texas Instruments Japan Ltd.
Tokyo, Japan
j_yamamoto1@ti.com

Seiya Abe
International Centre for the Study of East Asian
Development (ICSEAD)
Kitakyushu, Japan
abe@icsead.or.jp

Toshiyuki Zaitsu
Texas Instruments Japan Ltd.
Tokyo, Japan
t_zaitsu@ti.com

Tamotsu Ninomiya
International Centre for the Study of East Asian
Development (ICSEAD)
Kitakyushu, Japan
t_ninomiya@icsead.or.jp

Abstract—**This paper proposes the new LLC converter with unique control method. Generally, LLC resonant converter, which is controlled by PFM, has the drawback of (1) difficulty of PWM controllability, (2) difficulty of constant current limitation, and (3) narrow input voltage range limitation. These difficulties come from the resonant operation. In order to overcome those drawbacks, a hybrid control of PFM and PWM is employed for LLC. The resonant frequency is set to lower frequency to generate triangle current waveform, and the output voltage can be controlled by PWM and PFM through adjustment of resonant inductor energy. The experimental results for 48Vin, 16Vo, 80W board achieves 90% efficiency and validated this method is useful practically.**

Keywords— LLC converter,Half-bridge converter ,PWM, PFM

I. INTRODUCTION

It is well known that LLC resonant converter has been used widely in DTV (Digital TV) power supply for a long time because of high efficiency, low noise and low profile. Recently, the LLC converter is expected for new application such server, battery charger and industrial application. Because LLC operation can achieve soft switching with simple 2 switches structure as half-bridge. Usually PSFB (Phase Sifted Full Bridge) is widely used for those application in order to operate in ZVS (Zero-Voltage-Switching) which leads to high efficiency and low noise. However, PSFB is very complicated because it is 4 switches structure and has driving circuit for each switch. In a high power such 3kW or higher, PSFB should be a good choice. However, in a medium power range less than a few kilo watts, PSFB has too much components which leads to higher cost. This is why the LLC resonant converter is interested in this power range.

In a server, battery charger, and industrial application, the power supplies are required to have a constant current control and cover wide input voltage. However, the conventional LLC resonant converter is not able to have the constant current characteristic neither cover wide input voltage because of the resonant operation.

In order to overcome those drawbacks, a hybrid control of PFM (Pulse Frequency Modulation) and PWM (Pulse Width Modulation) is proposed in this paper. The proposed LLC converter has Full-wave voltage doubler rectification is employed at secondary side. The resonant frequency of the proposed LLC converter is set to lower than conventional LLC converter to generate triangle current waveform. The output voltage is decided by the stored energy of the resonant inductance, and then the output voltage can be controlled by PWM and PFM through adjustment of resonant inductor energy.

Therefore, an appropriative combination of PFM and PWM can realize the constant current control and reduce the effective current which has benefit at light load.

In this paper, the conventional LLC converter is summarized in section II, the proposed LLC converter is discussed with unique topology and analysis of the operation in each stage in section III, and then experimental result is reported in section IV.

II. CONVENTIONAL LLC RESONANT CONVERTER

Figure 1 (a) shows the conventional LLC resonant converter [1-6]. It is controlled by PFM due to its resonant frequency characteristic as shown in Fig.1 (b). The electrical power is transferred to secondary side by resonant fashion. So, the conventional LLC resonant converter can achieve ZCS (Zero-Current-Switching) turn-off at secondary side diode (D1, D2), and ZVS

(Zero-Voltage- switching) turn on at primary MOSFET (Q1, Q2) in a particular frequency range as shown in Fig.1(c).

Those behavior makes LLC resonant converter achieve high efficiency, and low noise. However, the conventional LLC converter has the drawbacks as follows[1-3];

(1) Difficulty of PWM controllability
(2) Difficulty of constant output current limit for battery load
(3) Narrow input voltage range limitation.

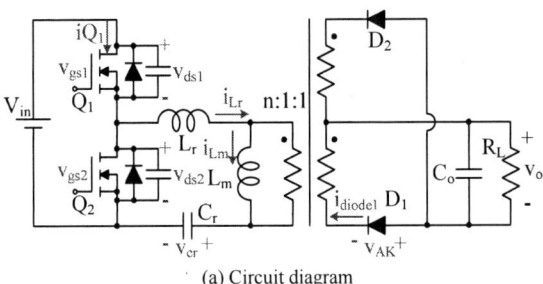

(a) Circuit diagram

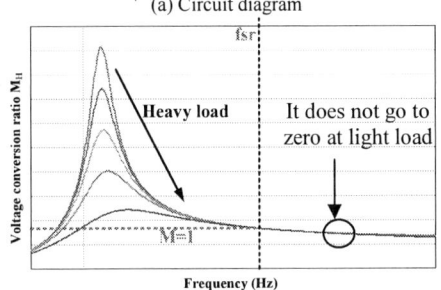

(b) Frequency characteristic curve of output voltage

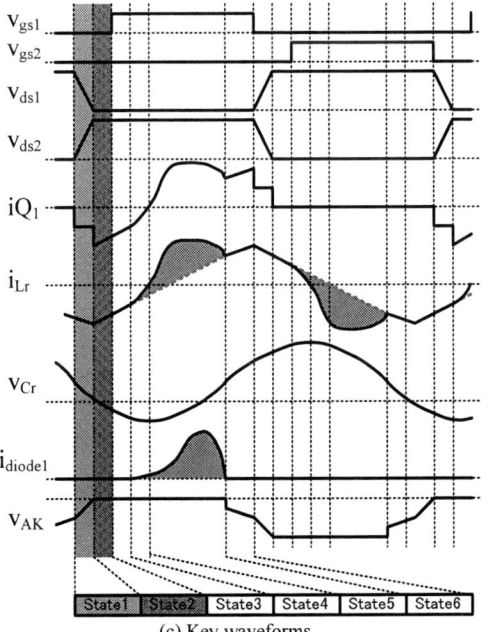

(c) Key waveforms
Fig. 1. Conventional LLC resonant converter.

The difficulty of PWM controllability is explained below. The current goes through the LLC power stage is approximately sinusoidal waveform.

Only the fundamental frequency component of pulse waveform can be transferred to secondary side. However, the amplitude of fundamental frequency component of the pulse is $\left(\frac{4}{\pi}\right)\sin(D\pi)$ by Fourier series expansion, where D is duty of pulse wave. So, $\left(\frac{4}{\pi}\right)\sin(D\pi)$ does not have good linearity versus D. This is the reason why resonant converter can't be controlled by PWM [7].

The difficulty of constant output current limit is explained below.

As shown in Fig.1(b), the output voltage never go down to zero by PFM in a certain light load. Meanwhile conventional PWM control can easily implement the constant current limit at any current level. Hence, the conventional LLC resonant converter has to have hiccup current protection which never allowed to battery application. Also, it can only operate in narrow input voltage range limitation because of the limitation of frequency variation range as shown in Fig.1 (b) .

III. PROPOSED HYBRID LLC CONVERTER

In order to overcome mentioned above drawbacks, an unique idea is introduced to both LLC topology and control scheme. At the secondary side, the full-wave voltage doubler rectification topology is employed as shown in Fig. 2. The auxiliary capacitor Ca is set to a large value, which makes the resonant frequency lower, to generate a triangle current waveform. It is not like conventional LLC resonant converter which generates resonant current waveform.

A. Maximum Duty Ratio

Figure 3 and 4 show the key waveforms and the operation mode of each state of the proposed LLC converter with Duty=0.46, respectively. The operation of the proposed hybrid LLC converter can be divided into 4 states in a half switching cycle.

State 1 - This state begins when MOSFET Q2 turns off. During this period, the current iQ1 flowing through MOSFET Q1 is negative, and it discharges MOSFET Q1 parasitic capacitor to ensure ZVS operation. This state will be end when the voltage Vds1 reaches zero (and the voltage of MOSFET Vds2 reaches Vin).

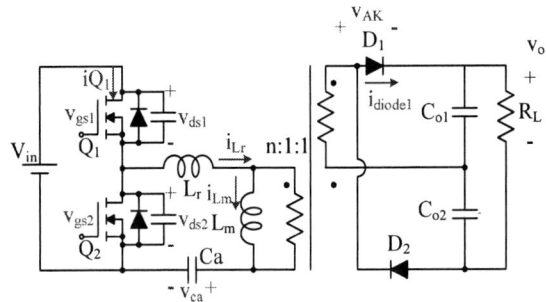

Fig. 2. Proposed Hybrid controlled LLC converter.

The 2014 International Power Electronics Conference

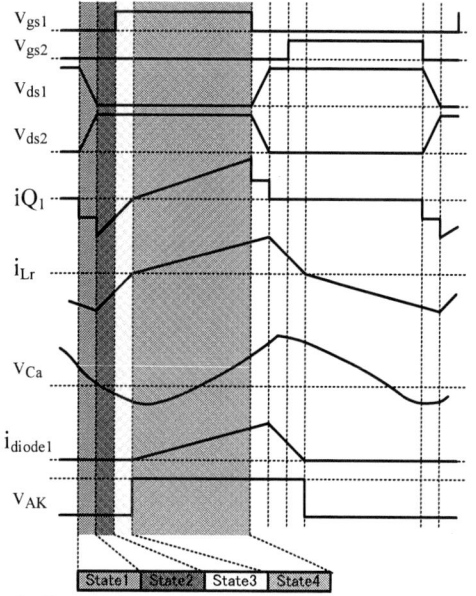

Fig. 3. Key waveforms of proposed Hybrid LLC(D=0.46:Max).

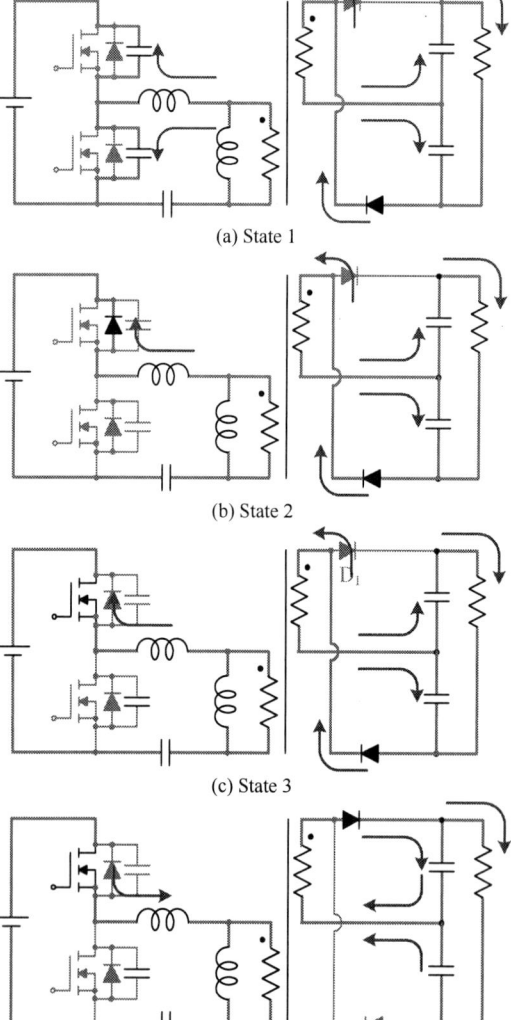

(a) State 1

(b) State 2

(c) State 3

(d) State 4

Fig. 4. Operating state(D=0.46:Max).

State 2 - The current iQ1 flows through the body diode of MOSFET Q1, and then the body diode connects to Vin. This state will be end when the MOSFET Q1 turns on.

State 3 – This state starts when MOSFET Q1 turns on under the zero Vds1 voltage. The drain current of Q1 flows from source to drain because of negative current. The energy stored in the inductor Lr, Lm and the auxiliary capacitor Ca is fed back to the input source Vin, discharging the capacitor Ca. This state will be end when the D1 turns on.

State 4 –The inductor current iLr becomes positive, and then secondary side diode D1 turns on. The power is transferred from primary side to the load R_L. Secondary side ripple current flows through Co1 and Co2. This state will be end when Q1 turns off.

B. Small Duty Ratio

Figure 5 and 6 show the key waveforms and the operation mode of each state of proposed hybrid controlled LLC converter with Duty=0.25, respectively. The operation of the proposed hybrid LLC converter can be divided into 5 states in a half switching cycle.

State 1 - This state begins when MOSFET Q_2 turns off. Vds1 and Vds2 are freely resonant.

State 2 – This state starts when MOSFET Q_1 turns on. The current iLr ramps up from zero, and then secondary side diode D1 turns on. The power is transferred from primary side to the load R_L. Secondary side ripple current flows through Co1 and Co2. This state will be end when Q1 turns off.

State 3 –The current iLr discharge rapidly the MOSFET Q2 parasitic capacitance. This state will be end when Vds2 reaches zero.

State 4 –The body diode of Q2 turns on. Secondary side diode current i_{diode} continue to flow. This state will be end when i_{diode} reaches zero.

State 5 –The inductor current iLr becomes negative, so secondary side diode D2 turns on. And Vds1 and Vds2 are freely resonating through the resonant inductor Lr, the parasitic capacitor of MOSFET Q1 and Q2.

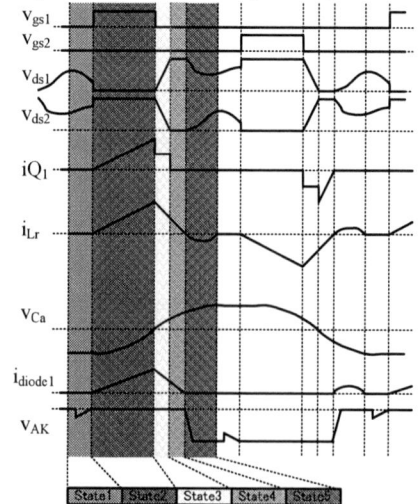

Fig. 5. Key waveforms of proposed Hybrid LLC(D=0.25).

IV. EXPERIMENTAL WORKS

In order to verify the operation characteristics of proposed hybrid LLC converter, the prototype circuit has been implemented. The circuit parameters and specifications are shown in Table 1.

Figure 7 and 8 show experimental waveforms when Duty=0.46 and Duty=0.25, respectively. The operations of this converter have been explained in the previous section.

Table 1 The circuit parameters and specifications

Sings	Parameters	Value
Vin	Input voltage	48V
n	Turns ratio of transformer	12:6
Lr	Series resonant inductance	5.2uH
Cr	Series resonant capacitance	10uF
fsr	Series resonant frequency	22kHz
fsw	Switching frequency	100kHz

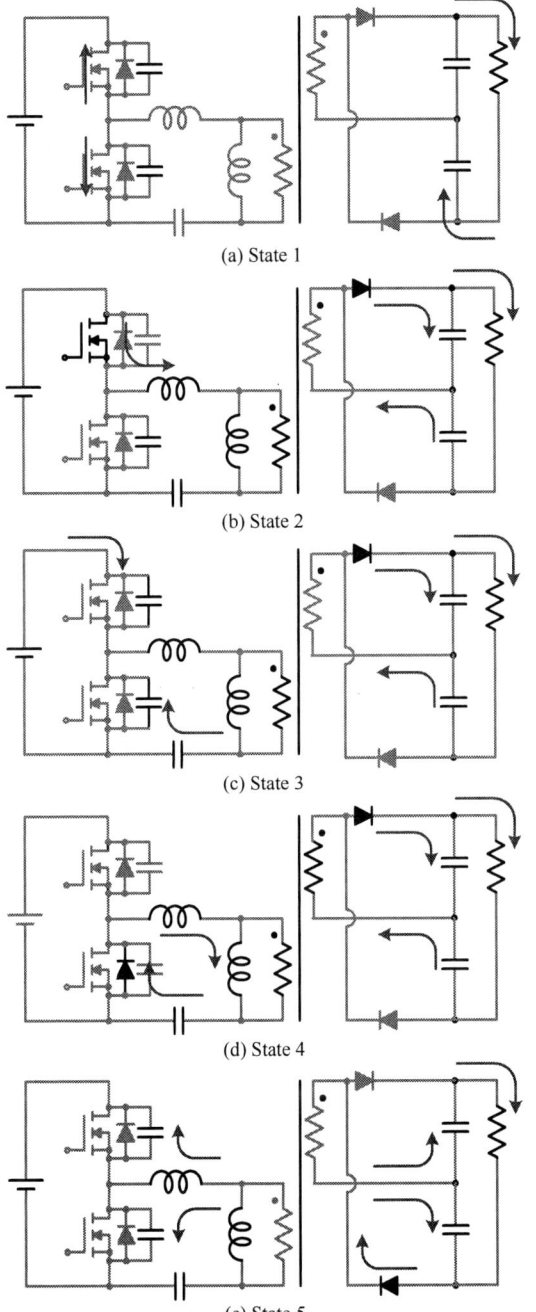

(a) State 1

(b) State 2

(c) State 3

(d) State 4

(e) State 5

Fig. 6. Operating state(D=0.25).

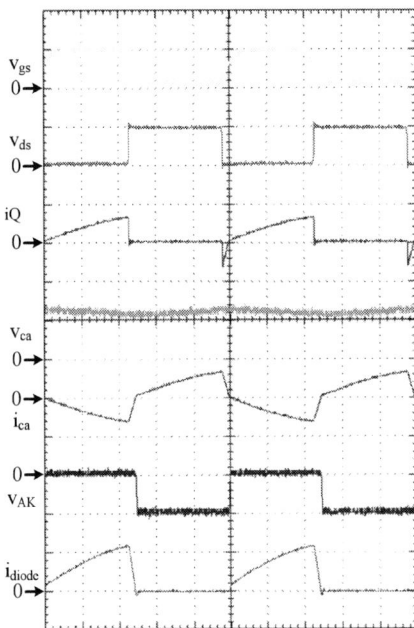

Fig. 7. Experimental waveforms (D=0.46)

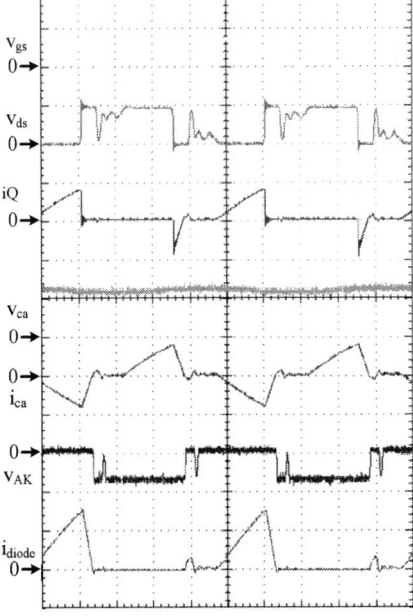

Fig. 8. Experimental waveforms (D=0.25)

From the experimental key waveforms, non-resonant operation has been verified, and the current flowing through the axially capacitor Ca increases and decreases linearly. Furthermore, the ZVS operation of primary side MOSFET has been achieved at turn-on. Moreover, the secondary side diode current decreases linearly at turn-off and the current reaches zero smoothly, because di/dt is limited by Lr. Therefore, there is much smaller reverse recovery loss than the conventional LLC resonant converter.

Figure 9 shows the experimental results of frequency characteristics. As shown in these figures, the output voltage is decreasing with higher frequency in both case (D=Max and D=0.3).

Figure 10 shows the experimental results of control characteristics against duty ratio with various loads. In the heavy load, when switching frequency fs is higher such 180 kHz, the linearity of duty ratio is not good as shown in Fig. 10 (a). Therefore, PFM is more useful in a heavy load. Meanwhile, in the light load, the linearity of duty ratio is good as shown in Fig. 10 (c), and PWM is more useful in the light load.

These control strategies are implemented to this hybrid LLC converter to establish the constant voltage (CV) and constant current (CC) characteristic. The control sequence for CV-CC control is shown as follows;

The output voltage is controlled by PWM in light load case, and the duty ratio becomes larger toward to heavy load. When the duty ratio is saturated, the output voltage control scheme changes to PFM. In this case the switching frequency starts maximum value. The switching frequency becomes lower toward to heavy load. The constant current limitation starts at required load.

The switching frequency becomes higher in order to keep constant current. When the switching frequency is saturated, the control scheme changes to PWM.

Figure 11 shows the experimental result of over current protection control. As shown in Fig. 11, the constant current limitation starts at Iout=3.0A, and desirable constant current limit characteristic have been achieved.

Figure 12 shows the efficiency characteristics. The efficiency is achieved over 85% and the maximum is around 90%. Moreover, the efficiency characteristics do not depend heavily on load current.

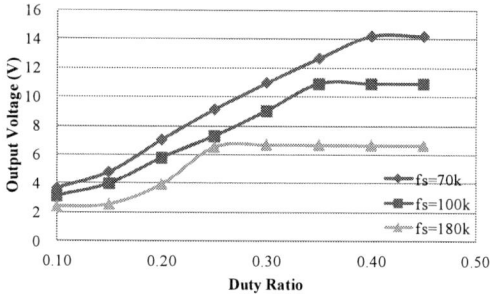

(a) Ro=2.4 Ohm

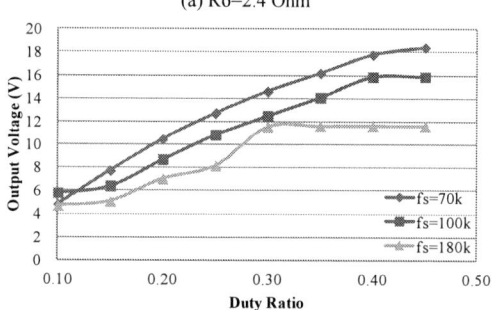

(b) Ro=5 Ohm

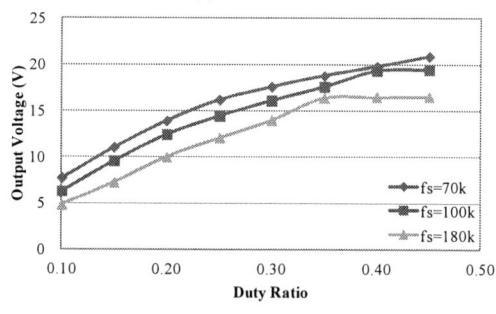

(c) Ro=10 Ohm

Fig. 10. Output voltage vs Duty

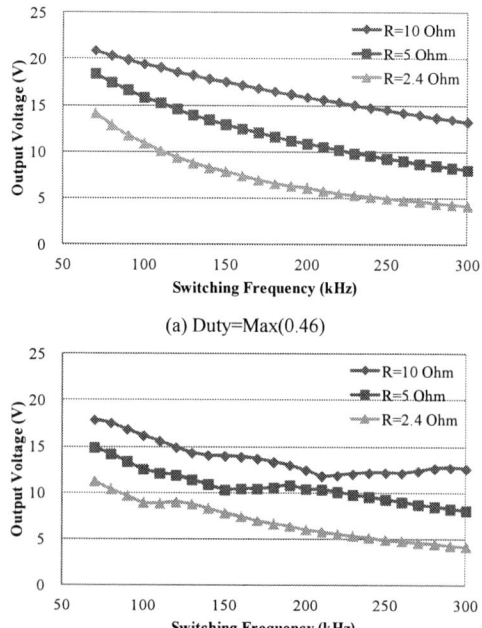

(a) Duty=Max(0.46)

(b) Duty= 0.30

Fig. 9. Output voltage vs switching frequency

Fig. 11. Over current protection control (OCP point = 3.0A)

978-1-4799-2706-7/14 $31.00 © 2014 IEEE

V. CONCLUSIONS

In this paper, the hybrid control with PFM and PWM for LLC converter was proposed. The full-wave voltage doubler rectification topology at secondary side was employed. The auxiliary capacitor value in LLC element is set to a large value to generate triangle waveform current. Also, The output voltage is decided by the stored energy of the resonant inductance, and then the output voltage can be controlled by PWM and PFM through adjustment of resonant inductor energy.

The series resonant frequency is set to far below than switching frequency which still keeps PFM controllability as well as PWM controllability. The key waveforms and operation mode of each state was explained. The prototype board was implemented to verify this hybrid controlled LLC converter. The experimental results at 48Vin, 16Vo, 80W board achieved 90% efficiency and validated this method is useful practically.

REFERENCES

[1] B. Yang, F. C. Lee, A. J. Zhang, G. Huang, "LLC Resonant Converter for front end DC/DC conversion," in IEEE-APEC 2002, pp. 1108-1112.

[2] J. H. Jung, H. S. Kim, J. H. Kim, M. H. Ryu, J. W. Baek,"High Efficiency Bidirectional LLC Resonant Converter for 380V DC Power Distribution System Using Digital Control Scheme," APEC'12, pp. 532-538, 2012.

[3] Ashoka K. S. Bhat, "A Generalized Steady-State Analysis of Resonant Converters Using Two-Port Model anf Fourier-Sries Approach," IEEE Trans. on P. E., vol. 13, No. 1, pp. 142-151, 1998.

[4] J. F. Lazar and R. Martinelli, "Steady-State Analysis of the LLC Series Resonant Converter," APEC'01, pp. 728-735, 2001.

[5] T. Liu, Z. Zhou, A. Xiong, J. Zeng, J, Ying,"A Novel Precise Design Method for LLC Series Resonant Converter," INTELEC'06, pp. 1-6, 2006.

[6] V. Vorperian, S. Cuk, "A Complete DC Analysis of The Series Resonant Converter," PESC'82, pp.85-100, 1982.

[7] T. Zaitsu, et al, "PWM-controlled Current-Mode Resonant Converter Using an Active-Clamp Technique," IEEE PESC'96, pp. 89-93, 1996.

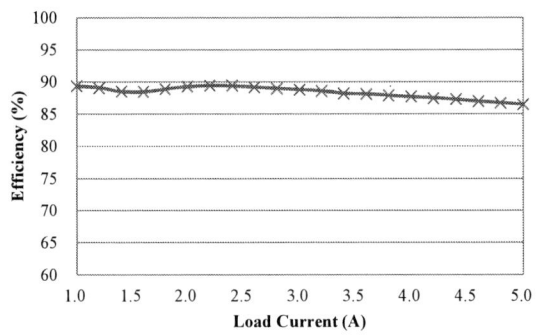

Fig. 12. Efficiency with PWM and PFM control (Vo=16V)

Discussions on Various Voltage Equalizers for EDLCs using CW circuit

Hlaing Kyi Pyar Khant, Keiju Matsui,
Masaru Hasegawa, Mikio Yasubayashi
Masayoshi Umeno
Chubu University, Kasugai, Japan

Eiji Ooishi
Minna-Denryoku, Inc.
Setagaya-Monozukuri-Gakkou
Setagaya, Tokyo, Japan

Abstract – **EDLCs offer high energy density and long life span, so various applications may be anticipated in the realm of energy storage devices, such as those used in electric vehicles or electric power stabilization in power systems, etc. However, since the voltage limit is low of the devices, it is necessary to connect them in series or parallel. In addition, it is required that they be used in the region of their critical voltage limit or capacity limit. In order to apply them efficiently, the devices should be used with balanced voltage. In this paper, some novel voltage balancers are presented, employing a CW (Cockcroft-Walton) circuit as a basic construction. Characteristics of proposed circuits are analyzed and improved especially about circuit construction strategies.**

Keywords – Supercapacitor; EDLC; Equalizer; Cockcroft -Walton circuit; Buck-boost chopper

I. Introduction

Various energy storage devices have been examined and reported. Among them, EDLCs (Electric Double Layer Capacitors) , which are often called EDLCs, can offer high energy storage performance in terms of surge power, efficiency, cold temperature operation and large number of energy cycles [1].

For these reasons, in power compensating equipments for voltage fluctuations or instantaneous voltage drop in the power systems, EDLCs are expected to be applied as energy storage equipments. Additionally, in various vehicles, such as electric cars and trains, these applications have just been introduced. In such EDLCs, however, as voltage limit of devices is low, it is necessary to connect them in series or parallel configurations, and to use them in the vicinity of their voltage limit. Consequently, in order to use efficiently, these devices must be used in well balanced manner. Amongst various voltage balancing techniques, a balancing method using resistors can be applied as the most fundamental, simple and effective solution [2]. When considering the power losses, however, such methods have restricted application in practice. Another method, using Zener diodes, has been discussed and evaluated [1, 3]. In considering the energy consumption of such Zener diodes, the power capacity of the system may be limited. Methods employing chopper circuits have also been proposed [3, 4], but the number of switching devices and their accompanied control circuitry is increased, leading to the high cost of such systems. Other original strategies have been presented and components [5]. Though the inverter circuit is complicated and charging operations are needed, such

methods are suited to the required increased capacities. The most orthodox method is thought to be the forward converter method, using transformers, which accompany each EDLC, and charge and discharge through their primary and secondary windings [6,7]. Similarly to solution [3], their controls may be complicated by many devices like transformers. Although such devices are necessary, however, their size is very small. Thus, this technique is expected to be widely used in extensive applications like the electric vehicles [8,9].

Considering the various types of EDLCs voltage balancers, perfect or even adequate solutions have not obtained. However, in the future, various other methods will be studied and proposed. In the light of the above research into voltage balancers for EDLCs, we had initially studied reference [10,11], and derived novel methods which was examined and discussed in [10-12]. An alternative approach to voltage balancing was presented, employing a Cockcroft-Walton circuit (CW circuit), which was invented long ago [13] for high dc voltage generation and employs numbers of capacitors and diodes. By means of this CW circuit, EDLCs having different capacitances were made to provide identical voltage. Voltage balancing is achieved with a high frequency power supply or buck-boost chopper. Their analyzed results and the mechanism are present and discuss[7]. In this paper, these results are applied to balance the small voltages of EDLCs. Another splendid equalizer is also proposed [15,16], in which the cells are controlled by means of a string of reference capacitors and double groups of many switches. The balancing operation is a little analogous to the CW circuit, so its principle is interesting. The circuit and its operation, however, is a little complicated, many switches should be provided and yet the operational principle is entirely different. Under such background, in March, 2008, a novel voltage equalizer using CW circuit had been proposed [14] firstly in the world by one of the authors. Various further versions and their results are to be presented and discussed including experiment.

II. Fundamental Circuit Configuration

A. Operation Analysis in High Frequency Power supply

Fig.1 shows the proposed circuit for the balanced charging of EDLCs, using CW (Cockcroft-Walton) circuit. C_1^* to C_5^*, on the left hand side, indicate, for example, electrolytic capacitors, which have relatively uniform values and can be obtained at low cost. e is the high frequency power supply. The purpose is not to supply the output power to these, but to supply relatively reduced power in order to balance the EDLC voltage on the right hand side. By means of adjusting the voltage or frequency, voltage equalizing conditions can be controlled. Inductor L is connected to suppress the inrush current into lower impedance of capacitor. By operating e, EDLCs. can be held effectively in voltage equalization between device.

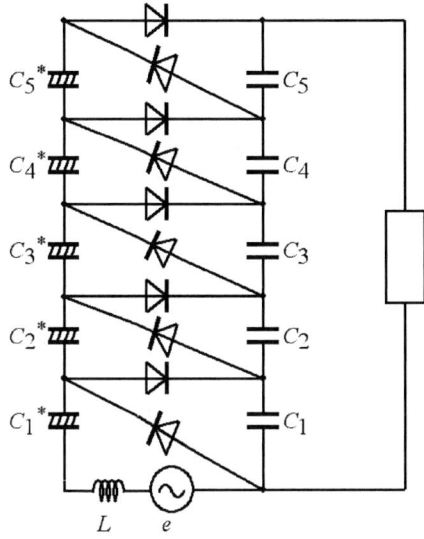

Fig.1. Basic voltage equalizer with CW circuit.

Let us discuss, for example, charging C_1 on the first stage. C_1^* is charged in the first half cycle to $V_{c1}^* = E$, where E is average value over half cycle. After that, in the second half cycle C_1 is charged by the power supply e and C_1^* voltage. As a result, C_1 is charged according to the following equation

$$V_{C1} = 2\frac{C_1^*}{C_1}Ef + V_{O1} \qquad (1)$$

where f is frequency of ac power supply, and V_{o1} is the initial voltage of C_1.

Above equations are for the first stage capacitor ones. Also in a case of upper stage, analogous equations can be developed. In such a way, the operation is moved upwards. Thus,

$$V_{C2}^* = V_{C1} \qquad (2)$$

As C_2^* is much smaller than C_1, the voltage drop of C_1 on the way of operation can be ignored and obtained this equality with a small error. On the next step from C_2^* to C_2 in the upper stage, the analogous equation compared to above ones can be obtained as follows,

$$V_{C2} = 2\frac{C_2^*}{C_2}V_{C2}^* + V_{O2} \qquad (3)$$

As general equations,

$$V_{Cn} = 2\frac{C_n^*}{C_n}V_{Cn}^* + V_{On} \qquad (4)$$

In the power transmission from large capacitance C_n to reduced capacitance C_{n+1}^*, the voltage drop of the capacitor on the side of supply may not be considered. From such reasons, the following equality can be obtained.

$$V_{C(n+1)}^* = V_{Cn} \qquad (5)$$

In a case of power transmission from C_n^* to C_n, the capacitor voltage is fairly reduced during the power transmission. For example, V_{C1}^* in (3) or V_{C2}^* in (3) is reduced on the way of power transmission. In comparing between V_{C1} in (1) that is not much reduced in voltage and V_{C2} in (3) or V_{Cn} in (4), the voltage increasing degree ΔV_{Cn} are different each other, that is

$$\Delta V_{C1} \geq \Delta V_{C2} \qquad (6)$$

In the upper stage compared to the first stage, the equality like (7) and (2) are established and the incremental component per cycle is almost the same, that is

$$\Delta V_{Cn} = \Delta V_{C(n+1)} \qquad (7)$$

This equation demonstrates that charging time from V_{C1}^* to V_{C1} is faster than the others on the upper stage.

On the other hand, the values of C_n^* and C_n are equal in the usual CW circuit, where, for example, in the steady state, V_{C1} equals $2E$. In the proposed method, by the second half cycle, the first stage capacitor C_1 is charged, due to the first and second operations. Consequently, by means of selecting the operating frequency of the high frequency power supply as $f = C_1/C_1^*$, V_{C1} becomes $2E$, being the value in the usual CW circuit. In the following stages, similar charging equations can be derived. In such a way, the whole voltage can be controlled by frequency control of the high frequency power supply. One reason for an unbalanced voltage on a EDLC is due to deteriorated performance by aging. In such a case, the corresponding EDLC voltage is somewhat increased. The charging of normal EDLCs is prevented by that the deteriorated one. In the proposed method, however, additional charge on EDLC bringing some increased voltage can be drawn towards the following stage. In such a way, the excess electric charge can be transmitted towards upper stages in succession. The influence due to increased voltage on the deteriorated capacitor can be reduced towards equilibrium. In another EDLC with reduced voltage, deficiency in electric charge can be supplied in an analogous operation, as explained above.

III. BUCK-BOOST CHOPPER METHOD

For another example of circuit configuration, we have proposed a modified circuit, which is constructed using buck-boost chopper. Their chopping frequency is high. By switching on and off, a high frequency voltage is generated across the set up inductance, which plays the role of high frequency voltage supply, as above just described ac power supply. In an analogous manner, each voltage on the EDLCs can be controlled uniformly.

The 2014 International Power Electronics Conference

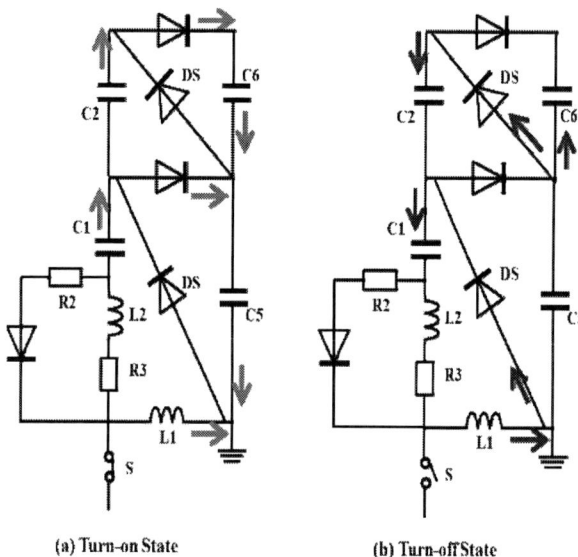

(a) Turn-on State   (b) Turn-off State

Fig.2. CW equalizer with buck-boost chopper.

Fig.2 shows operation circuit of another equalizer with buck boost chopper at turning-on (a) and turning-off (b). At turning-on, the charges of left side capacitors are transmitted toward the right side ones. On the other side, at turning-off, the stored energy in the inductor L_1 is discharging C_1 through D_s. At the same time, excess C_5 charge is discharging toward C_2, just upper side of C_1. By means of such repetitive operations, each capacitor voltage is equilibrated between devices. In the figure, when the switch S is turned on, a dc power is supplied by external circuit, for example, dc voltage across C_5 and C_6. L_2 and R_3 are operate to suppress the inrush current at turned-on. Bypass diode is used to suppress the surge voltage at the turned-off.

Fig.3 shows various operational waveforms when the equalizing characteristics among EDLCs can be obtained by chopper excitation method in Fig.2. Because of stable state after reaching to equalization, each current is fairly reduced. Be means of such repetitive operations, capacitor voltage is equilibrated between every capacitor.

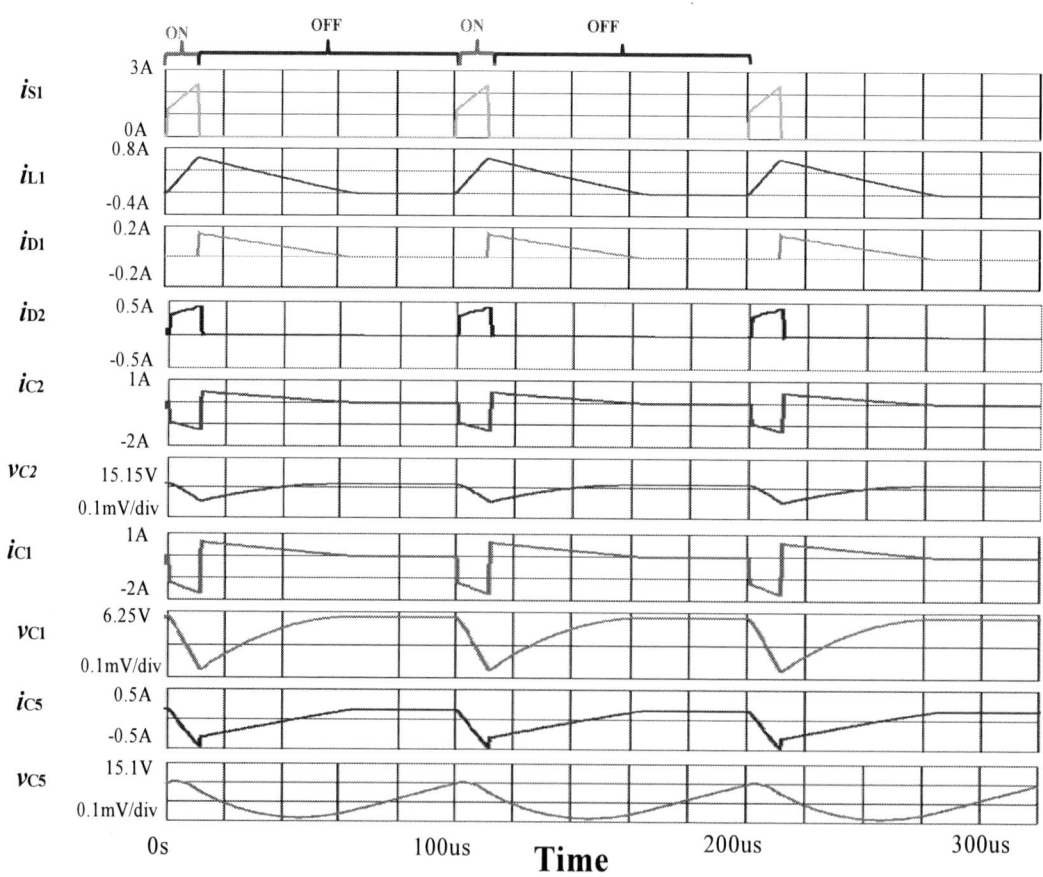

Fig.3. Various operating waveforms.

978-1-4799-2706-7/14 $31.00 © 2014 IEEE   185

IV. CONSTRUCTION WITH EDLCs ONLY

A. Discussion of the arrangement of capacitors

In the previously mentioned EDLCs, the voltage is fairly large because of like module circuit constructed in series and parallel connection of EDLC cells. In practice, however, the limiting voltage is very small, of the order of 2.7V or 3.7V, etc. Consequently, it would be of concern that, as there are many diodes in a series connection, such a small EDLC voltage can not exceed the forward voltage drop of such diodes. And also, as can be seen above results because of reduced electrolytic capacitor, it would take long time to charge or discharge, and yet internal impedance of electrolytic capacitor is fairly large. In order to resolve these concerns, the electrolytic capacitors are replaced by EDLCs for reference capacitor. At the same time, to increase the storage capacity of the above proposed circuit, some discussions are performed.

In order to confirm this, some circuit constructions are presented and discussed.

The equilization theory can be explained as following section.

B. First version in totally constructed by EDLC

For further development, some voltage equalizers are presented and discussed where every capacitor is constructed by EDLC only as described. As a result, the total capacitance and usage ability can be increased.

Fig.3(a) is a fundamental voltage equalizer using ac power supply, e as an exciter. Both C_1 to C_3 on the left hand side and C_4 to C_6 on the right hand side are entirely EDLCs to charge and discharge with voltage equalization function. L_1 to L_3 are for rush current suppression, where L_2 is for charging and L_3 is for discharging C_1 to C_3 with respect to external power supply. After C_1 is charged by ac power supply e, the following discharge current due to reversed voltage of e is prevented by diode D_1. At this time, the electric charge of C_1 is transmitted to C_4 through L_3-C_1-C_4, and can be obtained an equality, $V_{C1} = V_{C4}$

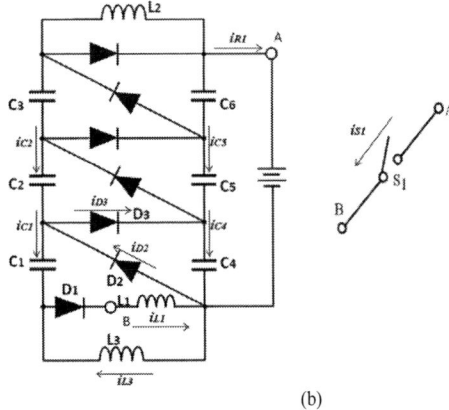

(b)

Fig.4. Further development for equalizers with EDLC only.
ac voltage exciter(a) and chopper exciter (b).

Fig.4(b) shows another voltage equalizer with chopper exciter having EDLC only. A dc voltage is applied across the terminal A and B. As fundamental buck-boost chopper, L_1 is stored at turned-on and the stored energy in delivered to C_1 at turned-off. The excess charge of C_1 is delivered to C_4 through L_3 at turned-on. Thus, C_1 and C_4 voltage is identical. At the second stage, as each capacitor can be operated like the first stage operation. In such way, subsequent operation at the upper side, an analog operation can be repeated like fundamental circuit. Thus, C_1 to C_6 can be charged in equalization. This circuit merit is that as the whole capacitors are employed as storage devices, the device utilization factor can be much improved. As the whole capacitors can play the role of storage devices having identical voltage, there is no need to make a particular specification as compared to the basic CW equalizer having different value of electrolytic capacitors and voltage. This circuit remarkable advantage is that both capacitors in the column can be given by the desired energy storage devices. The total capacity can be increased because of EDLC only.

V. DOUBLE SWITCHES METHOD WITH EDLC ONLY

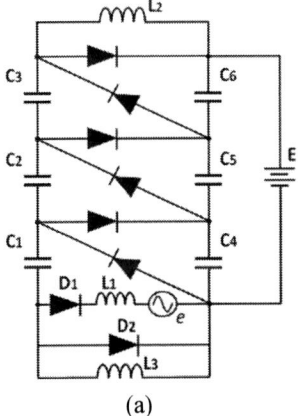

(a)

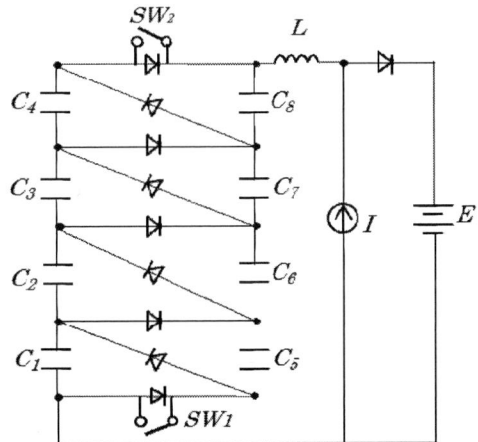

Fig.5. Double switches equalizers.

978-1-4799-2706-7/14 $31.00 © 2014 IEEE

Fig.5 shows another version of voltage equalizer using modified CW circuit having double switches. As comparison to above mentioned buck-boost chopper or others with EDLC only, their main switches are removed, the inductors of chopper is replaced by diodes and double switches S_1, S_2 are equipped, which are turned on and off periodically. By means of this, simple equalizer can be obtained.

A. Equalizing operation

a) <u>circuit operation at turned-off.</u> The charging current is flowing through double paths root 1 and root 2 in Fig.6. C_1 or C_8 shares each current in parallel path whose voltage becomes about double compared to others.

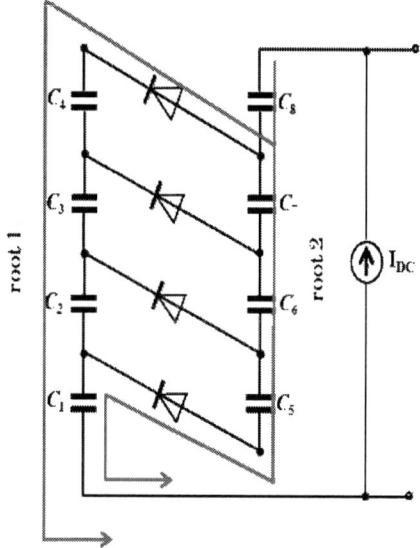

Fig.6. The circuit operation at switch turned-off.

b) <u>Circuit operation at turned-on.</u> Due to turned-off operation, for example, as C_8 voltage is higher than C_4 voltage, excess charge is delivered from C_8 to C_4 through root 3 as shown in Fig.7. Also in the lower stage, the excess charge is delivered from C_1 to C_5 through root 4 in the similar reason.

B. Voltage equalizing operation for EDLC deterioration

In Fig.8, for a case of deterioration of C_5, for instance of aging, an excess charge of C_5 is delivered to C_2 through root 5, because the closed loop due to root 2 is constituted. As usual, the charging current through root 2 is larger than trough root 5, so the excess charge on C_5 can be delivered toward C_2 capacitor for voltage equilibration .

In Fig.9, for a case of deterioration of C_2 , as root 4 is constructed, the excess charge of C_2 is delivered to C_6.

Under such consideration, the excess charge of left hand side capacitor is delivered toward that opposite side capacitor at the right hand side capacitor is delivered toward just upper side capacitor of the opposite side. The above switch S_1 might be unnecessary. For turned-off period, however, the excess charge of C_8 due to root 1 and root 2 must be delivered to C_4.

Thus, both switches are necessary for the transmittal operation of excess charge from the right hand side capacitor to just lower side capacitor in the opposite side of the left.

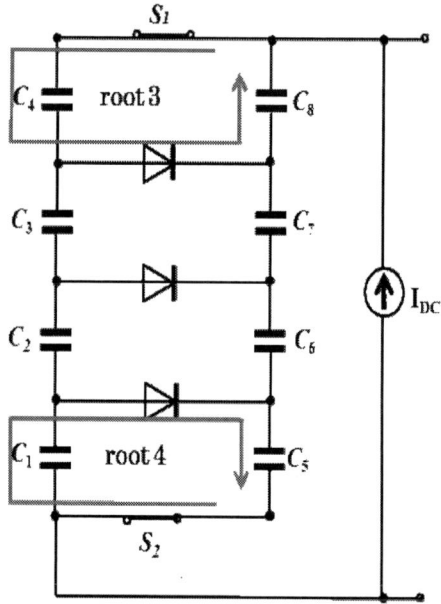

Fig.7. The circuit operation at switch turned-on.

A.

Fig.8. Voltage equalizing operation for C_5 deteriorated.

hand side. For above description, simultaneous operation of both switches was verified and discussed. The simultaneous operation of both ones is not always necessary. Even though with alternating switching, operation can be performed successfully, where one switch is open and other is closed, or vice versa. Even though the frequency is different each other, the voltage equalization can be performed.

The 2014 International Power Electronics Conference

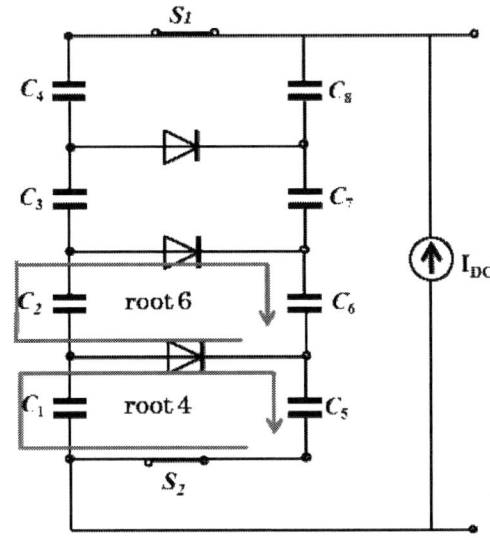

Fig.9. Voltage equalizing operation for C_2 deteriorated.

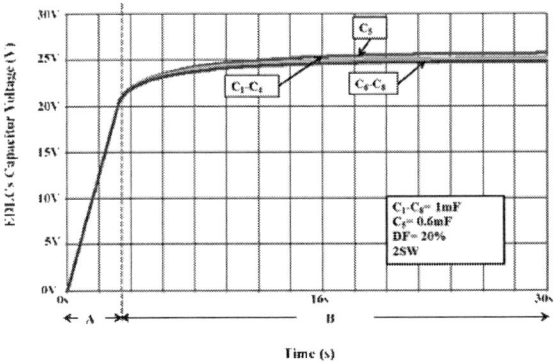

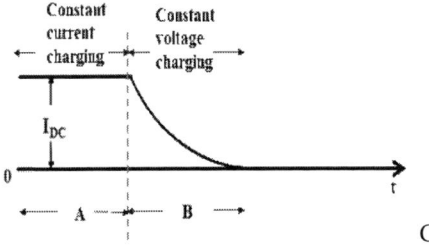

Fig.10. Equalizing characteristic with double switches.

In Fig.10, region A is called constant current charging and B is constant voltage one. This figure shows equalization characteristic, where C_1 to C_8 = 1mF except C_5 = 0.6mF reduced by deterioration, and duty ratio = 20%, I_{DC} = 0.1A, E = 100V. For the first rising region shown by A, the equalizing characteristic can be kept satisfactorily as shown, but for the region B shows discrepancies in the curves, which becomes slightly out of desired value of 25V. The reason can be explained as follows; in the region A, compensating effect due to switching is more significant. Meanwhile, in the region B, such effect due to switching is a little worse. In order to improve the equalization characteristic, the duty factor is

increased to 30%, where such result is shown in Fig.11. In the figure, by means of merely turned-on and off, satisfactory characteristic can be obtained, where every EDLC voltage is completely converging toward the desired value of 25V.

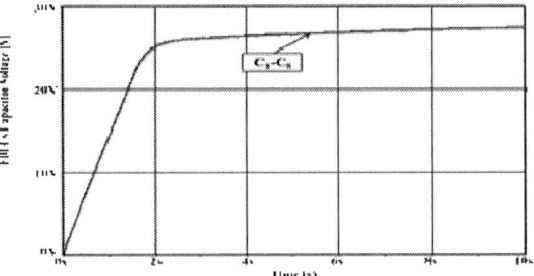

Fig.11. Satisfactory equalizing characteristic.

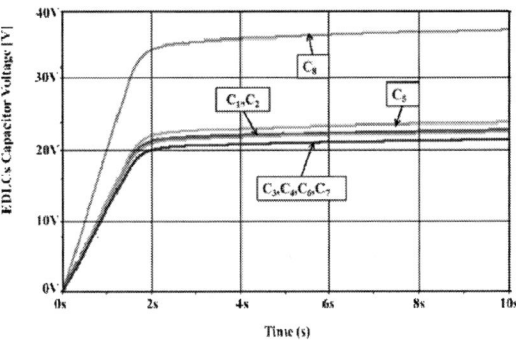

Fig.12. Unsuccessful equalizing characteristic without S1

On the other hand, when one switch S_1 is removed from the equalizer, equalizing characteristic is lost, whose characteristic is shown in Fig.12. C_8 voltage is increased about double, because current is flowing in parallel path through C_8. For C_1 voltage, however, by attached switch voltage is somewhat compensated.

C. Periodical charging and discharging characteristic

In this section, the charging and discharging characteristics are to be verified and discussed. In Fig.13, at turned-on for switch S_A, the circuit is in charging condition, and at turned-on of S_B, the the circuit is discharging one. Both operation is performed by usual current source. Equalizing characteristic is confirmed in both period.

D. Charging and discharging characteristics

Fig.14 shows charging and discharging characteristics where each duration is charged over periodically. Except C_5= 0.6mF, every capacitance 1mF, duty ratio is 20%. During turned-on, every curve is rising as almost straight line, where it can be seen that equalization characteristic can be achieved. On the other hand, during turned-off, for the deteriorated capacitor C_5 and C_1 in parallel connection, discharging rate is increased, so their voltage discharging ratio becomes higher. Equalization characteristic is deteriorated significantly.

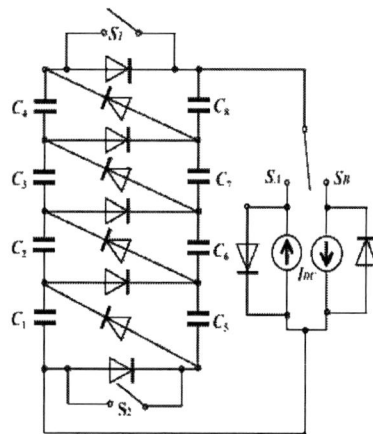

Fig.13. Periodical charging and discharging circuit.

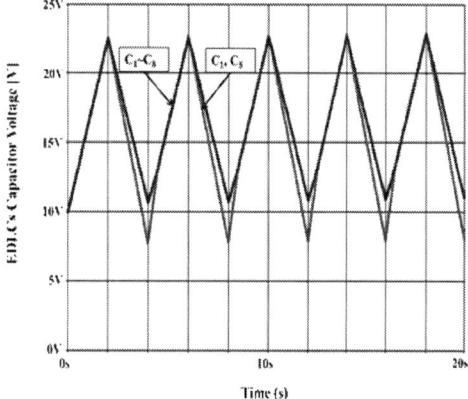

Fig.14. Periodical charge and discharge characteristics.

VI. EXPERIMENT AND ITS DISCUSSIONS

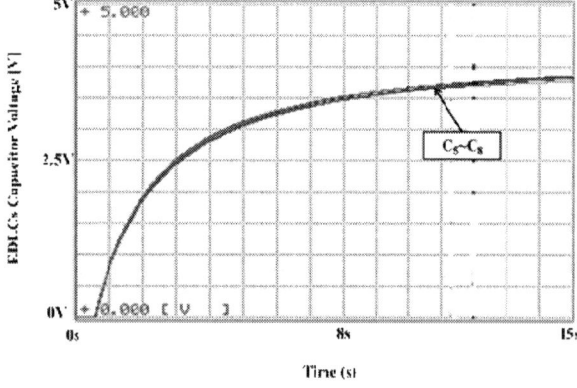

Fig.15. Equalizing characteristic at experiment.

Fig.15 shows voltage equalizing characteristic at trial experiment. The term "trial" means that instead of EDLC, electrolytic capacitors are used because sometimes unexpected excess charge was generated in a process of various experiments. Because of application of higher voltage against EDLC having lower voltage limit, electrolytic capacitor was chosen as experimental capacitor. External power supply is 12V dc supply having series resistance of 100 ohm which behaves like a current source. Each capacitance is 6.8 mF except assumed deteriorated C_5 of 3.4mF. Switching frequency is 60Hz in which both switch operate simultaneously. Every curve lies on single line. It can be said that equalization characteristic can be performed successfully.

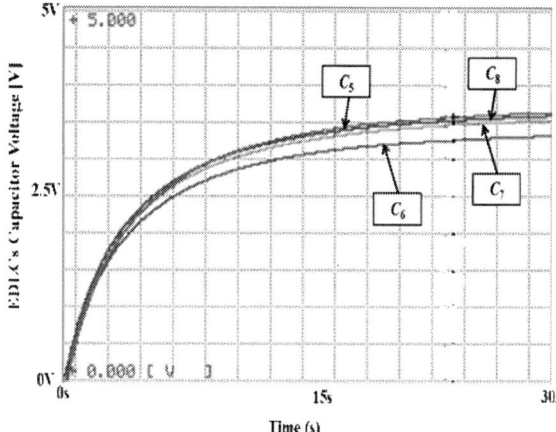

Fig.16. Non-equalizing characteristic without switching.

Fig.16 shows non-equalizing characteristic where double switches are always turned-on. Due to deteriorated C5, equarization could not be obtained. Experimental parameters are similar to above experiment. As shown, deterioration brings capacitance lower, so its voltage is higher as usual.

VII. CONCLUSIONS

A novel voltage equalizer using a CW circuit with chopper has been proposed and discussed. It is easy to apply, because of its simple and concise construction. The purpose of this system is not to obtain boosted power, but to obtain balanced voltages. Since such delivered power is not large, the high frequency power supply or dc chopper is small in size. A system that could deliver a high power may be possible, but possibly, with a much reduced efficiency. In such a system, the proposed modified construction may be effective

The proposed circuit using a CW approach is compared with the forward converter method [9-10], which is called the conventional method in this paper and that application might be most orthodox way at this time.

i) The conventional method needs a transformer, even though the size is small because of high frequency used.

Transformer and its design are complicated to every EDLC installation. The proposed method does not need particular design, and can be used as general-purpose.

ii) The proposed method needs a corresponding number of complementary capacitors. However, it may be possible that such capacitors can be replaced by EDLC, in which case, the system storage capacity could be significntly increased.

iii) The proposed method has a little deteriorated characteristic, in terms of circuit response, because the command is gradually delivered from the bottom to the top side. The conventional method is advantageous in relation to response. However, the aging time lapse is long, so quick response may not be needed for voltage balancer of EDLCs.

iv) For a case of using EDLC only having double switches, circuit becomes in much simple construction and in practical use, but equalizing operation can be obtained only at charging period. During discharging period such equalizing characteristic could not obtained unfortunately. In this period, however, such feature could be accepted because a fatal over voltage would not be generated

ACKNOWLEDGEMENT

This research is partly supported by a grant of the NEDO (New Energy and Industry Technology Development Organization.

REFERENCES

[1] Michio Okamura : "Electric Double Layer Capacitors and its Energy Storage Systems", 3rd Edition, Nikkan Kogyo Shinbun-sha, 2005

[2] Akitoshi Minemura, Masahiro Yashiro, Yasuyoshi Kaneko, Shigeru Abe : "Equalization of the Voltages Using Passive Resistors for Electric Double Layer Capacitors", The 2007 National Convention Record of IEE Japan, no.4-018, 2007

[3] Philippe Barrade, Serge Pittet, and Alfer Rufer, "Energy storage system using a series connection of EDLCs, with an active   device for equalizing the voltages" IPEC-Tokyo-2000, pp.1555-1560, 2000

[4] Alfer Rufer, and Pilippe Barrade, "A EDLC-Based Energy-Storage System for Elevators With Soft Commutated Interface", IEEE Transaction On Industry Applications, VOL.38, NO.5, SEPTEMBER/OCTOBER 2002

[5] Takatsugu Kishi, Toshihisa Shimizu, "A Study of Voltage Balancer for Electric Double Layer Capacitors", Technical Meeting on Semicondector Power Converter SPC-04-37, 2004

[6] Kazuya Mori, Akio Hasebe, Kiko Tsuruga, Takahiko Itoh, Sumiko Seki, "Voltage Balancer for Electric Double Layer Capacitors", The 2001 National Convention Record of IEE Japan, no.4-207, 2001

[7] Eiji Sakai, Koosuke Harada, S.Muta, Kiyomi Yamasaki, "Swiching Converters using Double-Layer Capacitors as Power Backup", The 19th International Telecommunication Energy Conference, Proceedings of IEEE-Intelec 1997, pp.611-616, Oct.,1997

[8] Nasser H. Kutkur, Deepak M. Divan and Donald W. Novotny, "Charge Equalization for Series Connected Battery Strings", IEEE Transaction on Industry Applications, vol.31, no.3, pp.562-568, May/June, 1995

[9] H.Sakamoto,K.Murata,E.Sakai,,K.Nishijima,K.Harada,S.Taniguchi, K.Yamasaki,G.Akiyoshi, "Voltage Balanced Charging of Series Connected Battery Cells", The 20th International Telecommunication Energy Conference, Proceedings of IEEE-Intelec 1998, pp.311-315, Oct.,1998

[10] Keiju Matsui, Hiroto Shimada, Masaru Hasegawa, "Novel Voltage Balancer for an Electric Double Layer Capacitor by using Forward Converter ",The 4th International Telecommunication Energy Special Conference, Vienna Austria, Proceedings of IEEE-telescon 2009, II.3-1,pp.1-6, May, 2009

[11] Kimihiro Nishijima, Hiroshi Sakamoto, Koosuke Harada, "Voltage Equalizing System for Series Connected Battery Cells", IEICE Trans. On B,vol.J84-B,no.9, pp.1701-1708, Sep. 2001

[12] Jonathan W. Kimball, B. T. Kuhn and P. T. Krein, "Increased Performance of Battery Packs by Active Equalization," IEEE Vehicle Power and Propulsion Conference, pp. 323–327, Sep. 2007.

[13] J. D. Cockcroft, E. T. S. Walton : "Further development on the method of obtaining high velocity positive ions", Proc. Royal Society London, UK, 1932.

[14] Keiju Matsui, Isamu Yamamoto, Masaru Hasegawa, Hiroto Shimada," A Novel Voltage Balancer for EDLCs Using Cockcroft-Walton Circuit", 2008 National Convention Record IEE Japan vol.4, 4-138, pp.230-231, March 2008

[15] Nasser H. Kutkur, Herman L.N.Wiegman, Deepak M. Divan and Donald W. Novotny, "Design Considerations for Charge Equalization of an Electric Vehicle Battery System", IEEE Transaction on Industry Applications, vol.38, no.5,pp.28-35, Sep/Oct. 1999

[16] K. Matsui, T. Suzuki, H. Shimada, M. Hasegawa, K.Ando, "Further Development on Voltage Balancer for EDLCs Employing Cockcroft-Walton Circuit",Proceedings of IEEE-2009, pp.PES-4.1-6

Isolation System with Wireless Power Transfer for Multiple Gate Driver Supplies of a Medium Voltage Inverter

Keisuke Kusaka, Koji Orikawa and Jun-ichi Itoh
Dept. of Energy and Environmental
Nagaoka University of Technology
Nagaoka, Niigata, Japan
kusaka@stn.nagaokaut.ac.jp

Kazunori Morita and Kuniaki Hirao
Research & Development Group
Meidensha Corporation
Numazu, Japan

Abstract— In this paper, a multiple wireless power transfer system for multiple gate driver supplies of a medium voltage inverter is developed. The proposed isolation system achieves a galvanic isolation with an air-gap of 50 mm using a wireless power transfer with magnetic resonance coupling. It easily respects the standard of galvanic isolation, which is established by International electrotechnical commission (IEC). Moreover, the power is supplied from one transmitting board to six gate drivers without a solid magnetic core. In this paper, the isolation system is developed and tested. It is clarified that the isolation system transmits power of not less than 300 mW to each gate drivers beyond an air-gap. However, sum of the output power of the each receiving board are limited up to approximately 3.5 W because of a voltage drop in the equivalent series resistances of the transmission coils.

Keywords— *Galvanic isolation, Medium voltage inverter, Wireless power transfer, Magnetic Resonance Coupling*

I. INTRODUCTION

In recent years, system voltage of a three-phase medium voltage inverter for a motor drive system is rising to 3.3 kV and 6.6 kV. In the medium voltage inverter, galvanic isolations are required at gate driver supplies. The safety standards, which are established by IEC [1], have to be satisfied for safety. The safety standards require a minimum clearance of 14 mm and a creepage distance of 81 mm when a system voltage of the inverter is 6.6 kV, a comparative tracking index (CTI) is $100 \leq CTI < 400$ and a pollution degree is two [1].

In general, isolation transformers with solid magnetic cores are used for a galvanic isolation. However, it causes an increase in a cost because it is typically custom-built. Moreover, the isolation transformer is huge in order to obtain a high isolation voltage. For example, a typical dimension of the isolation transformer, which have an isolation voltage of 20 kV_{rms} for 10 s, are 200 mm × 200 mm × 200 mm at a weight of approximately 5.5 kg [2]. These transformers are placed at each gate driver supplies.

In order to achieve a cost reduction and a downsizing of the

isolation system, a single-chip DC-isolated gate drive IC has been demonstrated [3-5]. It supplies power using a microwave from a bottom layer of a sapphire substrate to a top layer. In this method, galvanic isolation is achieved by the sapphire substrate. It can downsize an isolation system vastly. However, it does not satisfy the safety standards because both of an isolation distance and a creepage distance are not enough.

Meanwhile, J. W. Kolar et al. proposed the isolation system with using a printed circuit board (PCB) [6]. It achieves a galvanic isolation with a coreless transformer. In this system, one transmitting side transmits power to a receiving side one-by-one (1×1). Thus, the systems are also required at each gate driver supply. Therefore the reduction of the volume is limited.

In this paper, a galvanic isolation system with a multiple wireless power transfer system with magnetic resonance coupling for a medium voltage inverter is developed. The isolation system transmits power from one transmitting board to six receiving boards (1×6) beyond an air-gap of 50 mm. Because all of the isolation system is constructed by the PCBs. The isolation system contributes a cost reduction and a downsizing of the isolation system. Besides, the insulation with an air-gap of 50 mm easily respects IEC standard of a clearance and a creepage distance, when the system voltage of the inverter is 6.6 kV. Moreover, the air-gap of 50 mm reduces a common-mode current, which is induced by high dv/dt of the medium voltage inverter, to a small-signal circuit.

II. PROPOSED ISOLATION SYSTEM

A. System Configuration

Fig. 1 shows the system configuration of the developed galvanic isolation system for a medium voltage inverter. The isolation system consists of the transmitting board and six receiving boards. The power consumption in the gate drive units (GDUs) is supplied through the isolation system from the low-voltage power supply of 48 V in a medium voltage inverter.

978-1-4799-2706-7/14 $31.00 © 2014 IEEE

Incidentally, magnetic resonance coupling achieves a wireless power transfer with a resonance with a high-quality factor Q of the transmission coils in a middle-range transmission distance [7-12]. In this system, 2.18 MHz is used as the transmission frequency because a high-frequency transmission is required in order to downsize the transmission coils. Then the isolation system does not require a transformer with a magnetic core. In the conventional system, the transformer with a magnetic core prevents an isolation system from a cost reduction and a downsizing. In contrast, the isolation system is constructed only by the PCBs in this system. The PCBs can be manufactured easily in comparison with the transformer.

Fig. 2 shows the positional relationship of the transmitting board and the receiving boards. The maximum size of the system is constrained up to 300 mm × 150 mm × 150 mm for a reason of the space limitations of the medium voltage inverter. The receiving boards are placed at top and bottom of the transmitting board. Each distance between the receiving boards and the transmitting board is kept at not less than 50 mm for the galvanic isolation. It contributes the high isolation voltage and a low common-mode current through leakage capacitances. It is enough to fulfill the safety standards of IEC [1] when the operating voltage of the medium voltage inverter is 6.6 kV.

The transmitting boards consist of a high-frequency inverter, a series resonance capacitor and a transmitting coil for a wireless power transfer system. The inverter is operated by a square-wave operation with an output frequency of 2.18 MHz because high-frequency operation is necessary in order to downsize the transmission coils on the PCB. On the other hand, the receiving boards consist of a receiving coil, series resonance capacitor and a diode bridge rectifier.

Fig. 3 shows the photograph of the prototype of the isolation system. Each board is placed according to the Fig. 2. Table I indicates the specifications of transmitting coil and receiving coils.

B. Rated Power of Gate Driver Supplies

The isolation system transmits the power consumption of the gate drivers. In this subsection, required power of the each gate driver is calculated.

Fig. 4 shows the five-level diode-clamped multilevel inverter as a medium voltage inverter, which has a rated output voltage of 6.6 kV and a rated output power of 1 MVA [13]. Each switching device is a string of three 1.7-kV IGBTs connected in series. The power consumption of a gate resistance P_G of an IGBT is calculated by (1) where f_c is a switching frequency, Q_g is total gate charge and V_{GE} are gate-emitter voltage of the IGBT.

$$P_G = f_c \left(\left| +Q_g \right| + \left| -Q_g \right| \right) \left(\left| +V_{GE} \right| + \left| -V_{GE} \right| \right) \qquad (1)$$

From eq. (1), the power consumption is calculated as about

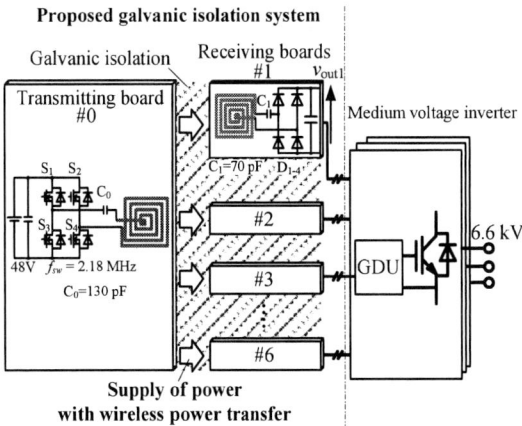

Fig. 1. System configuration of the developed isolation system. Power is transmitted from the transmitting board to the six receiving boards by a wireless power transfer with magnetic resonance coupling.

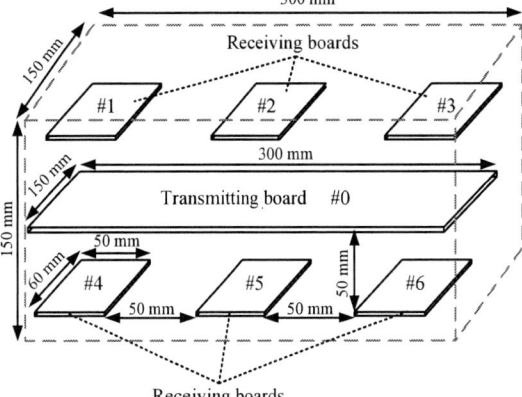

Fig. 2. Placements of each boards of the proposed isolation system. The receiving boards are placed up and down to the transmitting board. Each board are placed keeping an air-gap of 50 mm.

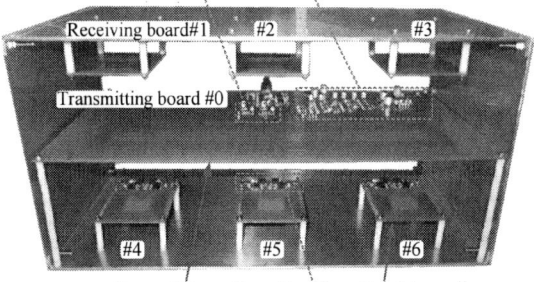

Fig. 3. Photograph of proposed isolation system.

60 mW where a switching frequency of the medium voltage inverter is 1 kHz, total gate charges $\pm Q_g$ are ± 1000 nC and gate-emitter voltage is ± 15 V. Note that the values, which is used in this calculation, are typical value of an IGBT (V_{CE} = 1700V, I_C = 150 A). Considering a power loss in a gate driver circuit, power of at least 120 mW is required per one receiving board as the output power of the isolation system.

978-1-4799-2706-7/14 $31.00 © 2014 IEEE

III. FREQUENCY CHARACTERISTICS WITH ELECTROMAGNETIC ANALYSIS

In this section, the part of the wireless power transfer in the isolation system is analyzed with the electromagnetic analysis with Agilent advanced design system (ADS). In ADS, a 3-D model is analyzed by the momentum method.

Table I shows the specifications of the coils for the analysis. The isolation system transmits power with the wireless power transfer with the series resonance capacitors. The windings of the coils are made up on the surfaces of the PCBs. In the system, series resonance capacitor in the transmitting side and series resonance capacitors in the receiving side, which is called as S/S resonance, are used. The resonance capacitors on the transmitting board and the receiving boards are 130 pF and 70 pF, respectively. The wireless power transfer with S/S resonance type has an advantage compared to other method such as a series resonance and parallel resonance (S/P), parallel resonance and parallel resonances (P/P) from the standpoint of the variation of a coupling coefficient and a load. The load characteristics do not affect the resonance frequency in S/S resonance. It is suitable characteristic for the isolation system because the loads have different value in each GDU in the isolation system. Also, the coupling coefficient does not affect the resonance frequency in the S/S resonance. The coupling coefficients vary among the receiving boards. It means that, the isolation system with S/S resonance can be operated at a constant operating frequency without a complex control.

Fig. 5 presents the definition of S-parameters. The S-parameter is one of the methods to express the characteristics of a multi-terminal circuit. The relationship between a square root of input power and a square root of output power is expressed by the matrix shown as eq. (2) where a_1 is a square root of travelling power, b_1 is a square root of reflected power in the primary side, a_2 is a square root of travelling power and b_2 is a square root of reflected power in the secondary side. In this paper, the S-parameters are used to analyze the characteristics of the isolation system.

$$\begin{pmatrix} a_1 \\ b_1 \end{pmatrix} = \begin{pmatrix} S(1,1) & S(1,2) \\ S(2,1) & S(2,2) \end{pmatrix} \begin{pmatrix} a_2 \\ b_2 \end{pmatrix} \quad (2)$$

Fig. 6 shows the definition of S-parameters in the system. In this consideration, a S-parameter $S(n,0)$ is especially focused. The S-parameter $S(n,0)$ is the ratio of a square root of power from the transmitting board to the receiving boards, where n is the number of the receiving boards ($n = 1, 2, \cdots, 6$). In the isolation system, the power flow is limited to one-way; from the transmitting board to the receiving board. Thus, the transmission from the receiving boards (#n) to the transmitting board (#0) can be ignored.

Fig. 7 shows the S-parameters $S(n,0)$ of the system. The S-parameters $S(2,0)$ and $S(5,0)$ reach to -9 dB at the resonance frequency. It means that the each ratio of the output power of

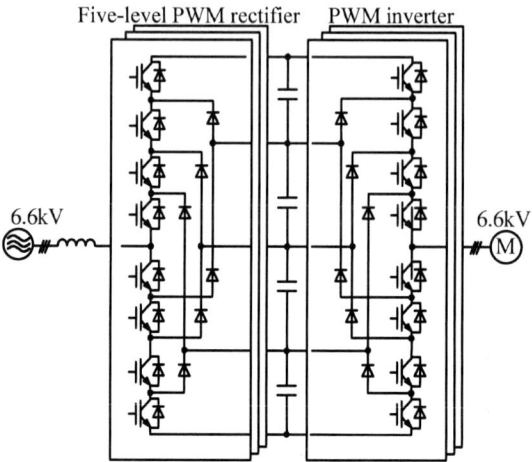

Fig. 4. Assumed 6.6 kV, 1 MVA five-level diode-clamped medium voltage invertere.

Table I. Specification of coils.

(a) Transmitting coil.

Items	Values		Remarks
Number of turn	12	turn	Short type
Width	250	mm	
Depth	40	mm	
Line width	1	mm	
Gap between windings	0.7	mm	
Thickness of PCB	1.6	mm	FR-4
Film thickness of copper	70	μm	

(b) Receiving coils.

Items	Values		Remarks
Number of turns	40	turn	Short type
Outer diameter	44	mm	
Inner diameter	22	mm	
Gap between windings	0.4	mm	
Line width	0.2	mm	
Thickness of PCB	1.6	mm	FR-4

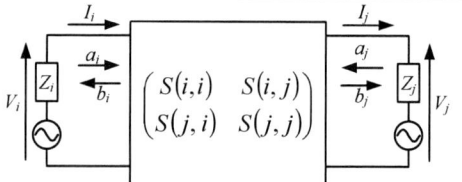

Fig. 5. Definition of S-parameters.

the receiving boards (#2 and #5) to input power of the transmitting board are 13.2%. In contrast, the S-parameters $S(1,0)$, $S(3,0)$, $S(4,0)$ and $S(6,0)$ cannot reach to -10 dB. Thus, the each ratio of the output power of the receiving boards (#1, #3, #4 and #6) to input power calculated as 7.4 %. The non-uniform transmitted power is caused by the difference in the coupling coefficient among the boards. The receiving boards, which are placed in center (#2 and #5) are coupled to the transmitting board (#0) strongly compared to other boards.

Fig. 8 shows the effect between the abutting receiving

The 2014 International Power Electronics Conference

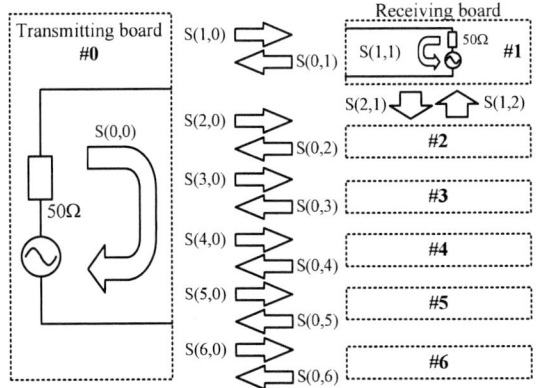

Fig. 6. S-parameters in the isolation system. All of the boards are numbered from #0 to #6. All of the analyses with S-parameters in this paper are held with a characteristic impedance of 50 Ω.

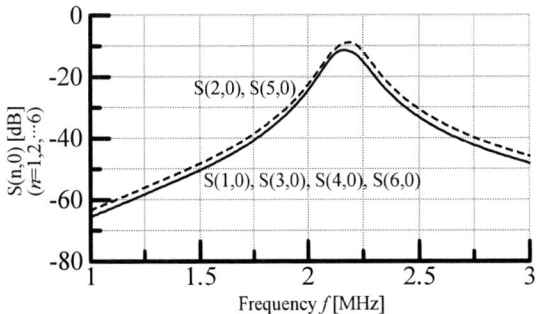

Fig. 7. Positional dependence of the receiving boards on the S-parameters. The S-parameters between the transmitting board #0 and the receiving board #2, 5 are larger than the S-parameter between the receiving board #0 and #1, 3, 4, 6.

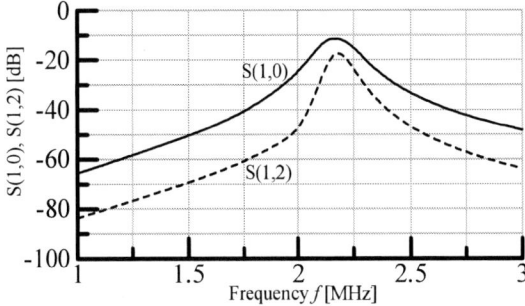

Fig. 8. Effect between the abutting receiving board. The transmission between the receiving boards such as S(1,2) is smaller than the transmission from #0 to #1.

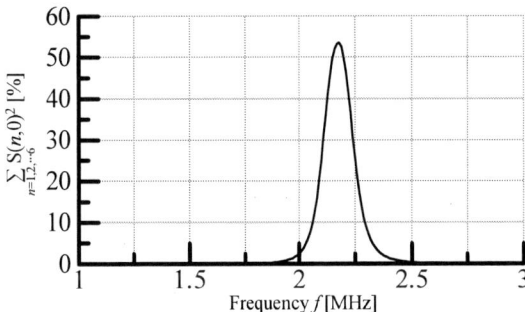

Fig. 9. Transmission efficiency of the isolation system. The efficiency is analyzed from the 3-D model with ADS.

boards. The transmission between the receiving boards is significantly small than the transmission between the transmitting board (#0) and the receiving boards (#1-6). For example, the transmission from the transmitting board (#0) to the receiving board (#1) S(1,0) reaches to -11.4 dB. On the other hand, the transmission from the receiving board (#2) to the receiving board (#1) S(1,2) is a -17.3 dB. Thus, the coupling between the receiving boards can be ignored in the system.

Fig. 9 shows the simulated transmission efficiency. The simulation is held without the converters such as an inverter and a rectifier. The transmission efficiency expresses the ratio of sum of the received power on the receiving boards to the transmitted power from the transmitting board. Thus, the transmission efficiency η_T is calculated by (3). It should be noted that the output impedance of the power supply and the load impedance, which are used for an analysis, are 50 Ω.

$$\eta_T = \sum_{n=1,2,\cdots 6} S(n,0)^2 \tag{3}$$

At a resonance frequency of 2.18 MHz, the total transmission efficiency increases drastically because magnetic resonance coupling achieves the efficient wireless power transfer with a high quality factor Q. The maximum transmission efficiency reaches to 53.6% at 2.18 MHz.

In the isolation system, low transmission efficiency is accepted because the power loss in the proposed isolation system is extremely low compared to the power loss of a medium voltage inverter, typically. Thus, the power loss of the isolation system can be ignored.

IV. TIME-DOMAIN ANALYSIS

A. Equivalent Circuit

Fig. 10 presents the equivalent circuit of the wireless power transfer [14-15] where r_{0-6} are the equivalent series resistances of the windings and C_{0-6} are the series resonance capacitors. The equivalent circuit of the multiple wireless power transfer

is obtained as transformers with multiple windings where k is the coupling coefficient among the each winding, which is expressed as (4). Note that, the subscript indicates the number of the transmitting board.

$$\mathbf{k} = \begin{pmatrix} 0 & k_{01} & k_{02} & \cdots & k_{06} \\ k_{10} & 0 & k_{12} & \cdots & k_{16} \\ k_{20} & k_{21} & 0 & \cdots & k_{26} \\ \vdots & \vdots & \vdots & \ddots & \vdots \\ k_{60} & k_{61} & k_{62} & \cdots & 0 \end{pmatrix} \tag{4}$$

If the self-inductance of the each winding L is expressed as (5), the leakage inductance L_{le} and magnetizing inductance L_m

are provided as (6) and (7), respectively. It should be noted that, k_{ij} is equals to k_{ji} $(i, j = 0, 1, \cdots, 6)$.

$$\mathbf{L} = \begin{pmatrix} L_0 & L_1 & L_2 & L_3 & L_4 & L_5 & L_6 \end{pmatrix} \tag{5}$$

$$\mathbf{L_{le}} = \mathbf{L} - L_0 \begin{pmatrix} 1 & 1 & 1 & 1 & 1 & 1 & 1 \end{pmatrix} \mathbf{k} \tag{6}$$

$$\mathbf{L}_m = \mathbf{L}\mathbf{k} \tag{7}$$

The parameters; the self-inductance, the series resonance capacitors and the coupling coefficients are introduced by the analysis results with ADS at the resonance frequency of 2.18 MHz. Table II shows the derived parameters. In Fig. 10, the coupling coefficients among the receiving boards are ignored because the effect of these coupling is significantly small. It means that the receiving boards are only magnetically coupled to the transmitting board.

Fig. 11 shows the comparison results of the frequency characteristics of the F-parameters. The F-parameters are compared between the equivalent circuit and the 3-D model with ADS in frequency characteristics. ADS indicates the circuit characteristic as S-parameters, typically. Thus, the analysis results are converted to the F-parameters according to eq. (8).

$$\begin{pmatrix} \dot{V}_i \\ \dot{I}_i \end{pmatrix} = \begin{pmatrix} A & B \\ C & D \end{pmatrix} \begin{pmatrix} \dot{V}_j \\ \dot{I}_j \end{pmatrix}$$

$$= \frac{1}{2S_{ji}} \begin{pmatrix} (1+S_{ii})(1-S_{jj}) + S_{ji}S_{ij} & \{(1+S_{ii})(1+S_{jj}) - S_{ji}S_{ij}\}\dot{Z}_j \\ \{(1-S_{ii})(1-S_{jj}) - S_{ji}S_{ij}\}\frac{1}{\dot{Z}_i} & \{(1-S_{ii})(1+S_{jj}) + S_{ji}S_{ij}\}\frac{\dot{Z}_j}{\dot{Z}_i} \end{pmatrix} \tag{8}$$

The equivalent circuit model shows the good agreement with the 3-D model. The error is caused by the unconsidered parameters; parasitic capacitances between the layers of the PCB, parasitic capacitances between the windings and a dielectric loss of the PCB. These parameters cannot be derived even by the simulation. Furthermore, the error occur owing to the non-linearly characteristics of an impedance of the transmission coils because the impedance of the coils are not proportional to a frequency in the high-frequency region. However, the difference does not cause a fatal error on the time-domain analysis. Thus, the operation model is discussed using the equivalent circuit model in the next subsection.

B. Operation modes

Fig. 12 shows the equivalent circuit with converters for the analysis of the operation modes of the system. The series resonance capacitors are designed according to eq. (9) where ω_{sw} is $2\pi f_{sw}$. The resonance capacitance C_0 on the transmitting board is selected in order to obtain a resonance with the self-inductance of the transmitting coil. Similarly, the resonance capacitances on the receiving boards are selected to resonate with the each self-inductance L of the receiving boards.

$$\mathbf{C} = \frac{1}{\omega_{sw}^2 \mathbf{L}} \tag{9}$$

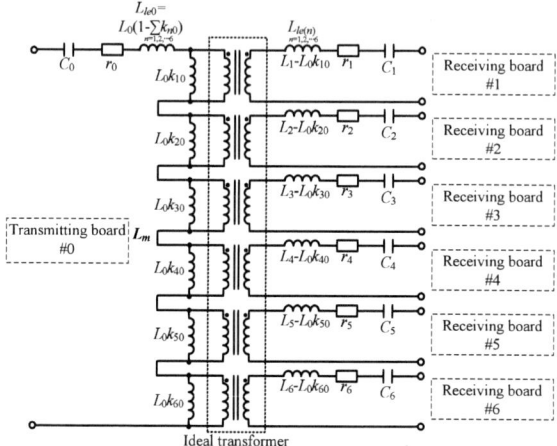

Fig. 10. Simplified equivalent circuit of the proposed isolation system.

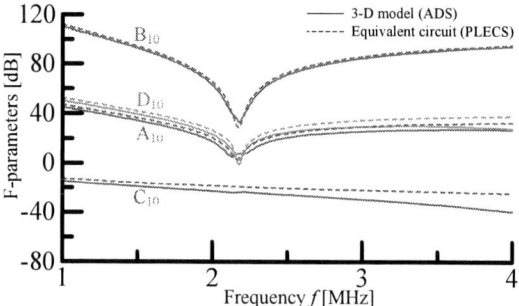

Fig. 11. Comparison results of frequency characteristics between the 3-D model with ADS and the equivalent circuit mode.

Table II. Derived parameter value from the 3-D analysis.

Items		Symbol	Values
Equivalent series resistances (ESRs)	Board #0	r_0	12.9 Ω
	Board #1	r_1	11.7 Ω
	Board #2	r_2	11.6 Ω
	Board #3	r_3	11.7 Ω
	Board #4	r_4	11.7 Ω
	Board #5	r_5	11.6 Ω
	Board #6	r_6	11.7 Ω
Self-inductances	Board #0	L_0	41.8 μH
	Board #1	L_1	76.5 μH
	Board #2	L_2	75.9 μH
	Board #3	L_3	76.5 μH
	Board #4	L_4	76.5 μH
	Board #5	L_5	75.9 μH
	Board #6	L_6	76.5 μH
Coupling coefficients		k_{01}	0.016
		k_{02}	0.026
		k_{03}	0.025
		k_{04}	0.016
		k_{05}	0.026
		k_{06}	0.025

Fig. 13 shows the vector diagrams of the isolation system. Owing to the resonance based on eq. (9), the inverter voltage and the current are in phase. The terminal voltage of the resonance capacitor and the inductance, become extremely high with opposite directions. Focusing on the receiving side, the resonance capacitor C_1 and the leakage inductance L_{le1} resonate.

Fig. 14 illustrates the simplified operation waveforms of the isolation system.

1) Mode I

The inverter current i_{inv} starts to flow through the MOSFETs and the coils because the MOSFETs S_1 and S_4 are turn-on. Owing to the resonance, the input current becomes a sinusoidal and it is in phase to the inverter output voltage v_{inv} during a period.

2) Mode II

The inverter current i_{inv} commutates to the diodes D_2 and D_4 because the all of MOSFETs turn-off during the dead-time T_d. If the dead-time is significantly short compared to the switching period, the MOSFETs achieve a zero current turn-off owing to the current resonance. After the half of the switching period, the inverter current crosses zero and commutate to the diodes D_1 and D_3. The inverter output voltage is decided by the polarity of the inverter output current.

3) Mode III

The MOSFETs S_2 and S_4 turn-on. The inverter current commutates to the MOSFETs S_2 and S_4 from the diodes. Also, the inverter current is sinusoidal owing to the resonances.

4) Mode IV

The MOSFETs S_2 and S_4 have turned-off at the start of this mode. Owing to the current resonance, the MOSFETs turn-off around a zero current. It contributes a reduction of the switching loss. The inverter currents commutate to the diodes D_1 and D_4. The directions of the inverter current change to a positive from a negative.

V. EXPERIMENTAL RESULTS

A. Fundamental Verifications

Fig. 15 shows the operation waveforms with the prototype, which is shown in Fig. 3. Fig. 15 (a) is the waveforms with no-load condition. Fig. 15 (b) is the loaded waveforms with a resistance load of 107 Ω where the input power of the isolation system is 31.3 W. In this experiment, resistances are used instead of the gate drivers for simplicity.

From the experimental waveforms, it is demonstrated that the wireless power transfer with an air-gap of 50 mm is achieved. An air-gap of 50 mm behaves as an isolation distance. Therefore, it is clear that the isolation system satisfies the safety standards, which is established by IEC [1], when a system voltage of the medium voltage inverter is 6.6 kV.

When the load is not connected, the rectifier input voltage v_{in1} is a sinusoidal because the diodes in the receiving boards do not turn-on. In contrast, the rectifier input voltage becomes

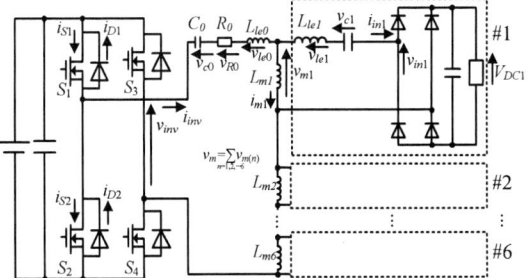

Fig. 12. Equivalent circuit with converters. The ideal transformers are omitted for simplicity.

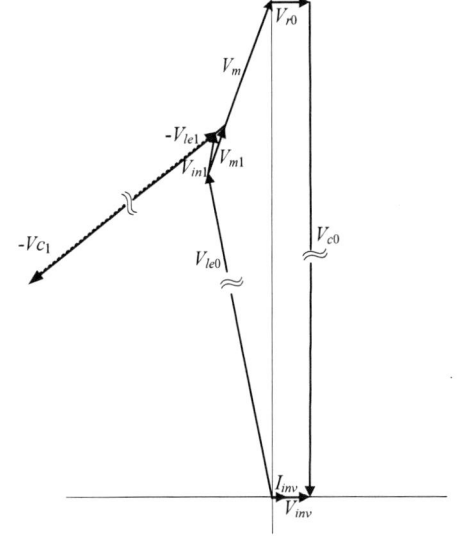

Fig. 13. Vector diagram of the wireless power transfer. The series resonance capacitor C_0 resonates with the sum of leakage inductance L_{le0} and the magnetizing inductance L_m. The self-inductance L_2 and the series resonance capacitor C_1 resonate.

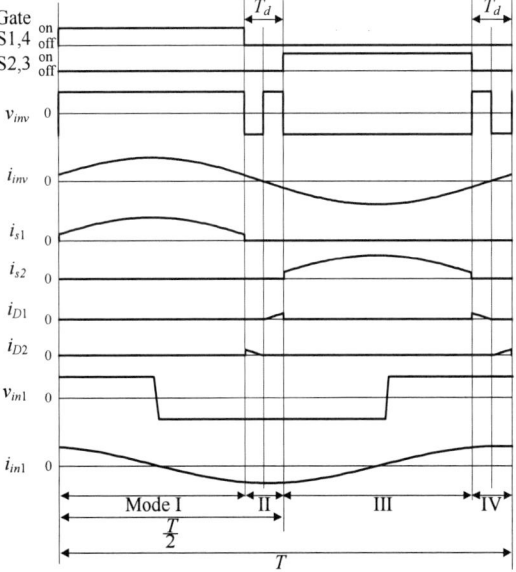

Fig. 14. Simplified operation waveforms. The MOSFETs achieve a zero-current turn-off because of the current resonance.

a trapezoidal wave when the load is connected. This is caused by the sinusoidal-rectifier input current i_{in1}.

As an output voltage, a DC voltage V_{DC1} is obtained despite a load value. However, the output voltage decreases when the load is connected. It is caused by the voltage drop on the equivalent series resistances of the transmission coils r_{0-6}. Incidentally, the others receiving boards simultaneously output the DC voltage. In this paper, the waveforms are omitted due to the page limitations.

Fig. 16 shows the output power of each receiving board. The output power varies among the position of the receiving boards. The difference between the output power is caused by the accuracy of the chassis and differences of the coupling coefficients. In the experimental setup, the output power of the receiving boards (#1, 2, 3) are smaller than that of (#4, 5, 6), respectively because the difference in the transmission distance occur between the upper boards and lower boards. Furthermore, the output power of the receiving boards #2 and #5 are larger than that of adjacent boards, respectively. Thus, the output power of #5 has maximum output power. The minimum output power reaches to 320 mW. It is confirmed that the isolation system supplies the power, which is required in the gate driver supplies.

Fig. 17 shows the total output power P_{out_total} characteristics. The total output power is calculated as (10).

$$P_{out\_total} = \sum_{n=1,2,\cdots,6} P_{out}(n) \qquad (10)$$

The output power increases as a load resistance decreases in the interval of the load resistance from 125 Ω to 750 Ω. However, the output power stops to increase when the small resistance such as 107 Ω is connected because the voltage drop on the equivalent series resistances of the transmission coils increases owing to the increment of load current.

B. Operation with Gate Drivers

In this subsection, the prototype with the gate drivers is tested. The gate drivers, which are operated at a switching frequency of 1 kHz, a gate-emitter voltage of ±15V, are connected as a load. Note that DC/DC converters as voltage regulators are connected at the front end of the gate drivers. The DC/DC converter output voltages of ±15 V. Moreover, the photocouplers (Toshiba, TLP250) are used in order to drive the IGBTs in spite of the deficient isolation distance because an isolation of the PWM signal is not a main topic of this paper.

Fig. 18 shows the operation waveforms where the capacitors of 33 nF is connected instead of the IGBTs. Even if the gate drivers are connected as a load, the operation system achieves the wireless power transfer beyond an air-gap of 50 mm.

From the experimental results, it is confirmed that the proposed isolation system can be used as an isolation system for the medium voltage inverter.

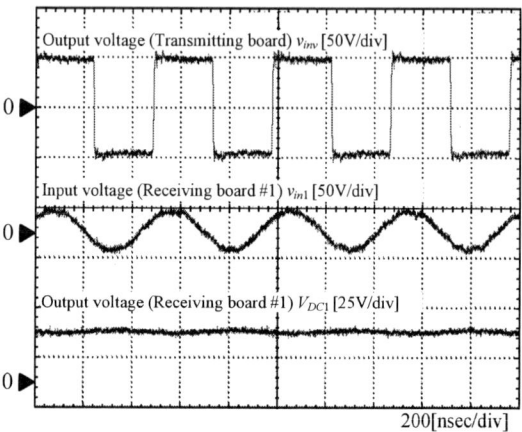

(a)   No-load.

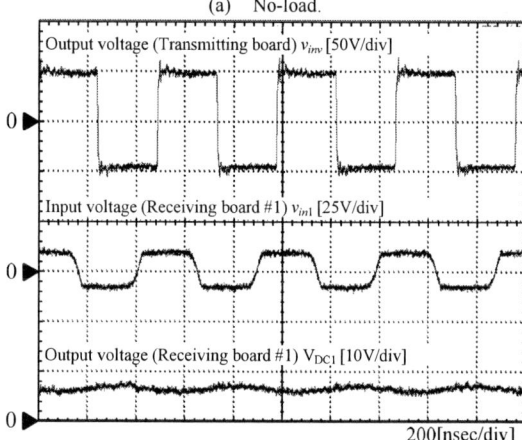

(b)   Loaded.

Fig. 15. Operation waveforms. The load resistance of 107 Ω is connected as a load instead of a gate driver supplies.

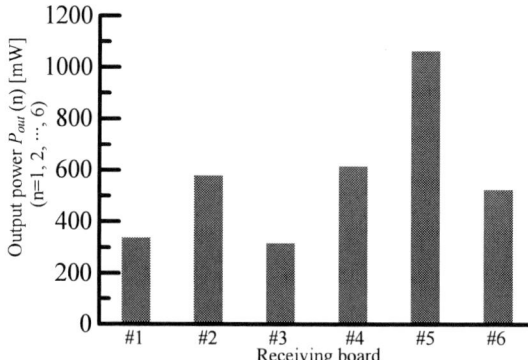

Fig. 16. Operation waveforms. The output power varies among the position of receiving boards. The difference of output power is caused by the difference in the accuracy of the chassis.

VI.   Conclusions

In this paper, the isolation system for gate drivers of a medium voltage inverter with the system voltage of 6.6 kV is reported. The isolation system transmits the power to the six gate drivers beyond the air-gap of 50 mm. The isolation distance satisfies the safety standards in IEC [1]. The isolation system consists of the only seven PCBs. It contributes a cost and weight reduction because it does not need a magnetic core. The experimental results demonstrate that the isolation system supplies the power of not less than 320 mW to the each receiving boards. It is enough power to drive the IGBT in the medium voltage inverter.

REFERENCES

[1] International Electrotechnical Commission (IEC): "Adjustable speed electrical power drive systems – Part 5-1: safety requirements – Electrical, thermal and energy", *IEC 61800-5-1* (2007)

[2] Christoph Marxgut, jurgen Biela, Johann W. Kolar, Reto Steiner, Peter K. Steimer: "DC-DC Converter for Gate Power Supplies with an Optimal Air Transformer", *Applied Power Electronics Conference and Exposition 2010*, pp. 1865-1870 (2010)

[3] S. Nagai, N. Negoro, T. Fukuda, N. Otsuka, H. Sakai, T. Ueda, et al.: "A DC-isolated gate drive IC with drive-by-microwave technology for power switching devices", *International Solid-State Circuits Conference 2012*, pp. 404-406 (2012)

[4] S. Nagai, T. Fukuda, N. Otsuka, D. Ueda, N. Negoro, H. Sakai, et al.: "A one-chip isolated gate driver with an electromagnetic resonant coupler using a SPDT switch", *24th IEEE International Symposium on Power Semiconductor Devices and ICs 2012*, pp. 73-76 (2012)

[5] S. Nagai, N. Negoro, T. Fukuda, H. Sakai, T. Ueda, T. Tanaka, et al.: "Drive-by-Microwave technologies for isolated direct gate drivers", *IEEE Microwave Workshop Series on Innovative Wireless Power Transmission*: Technologies, Systems, and Applications 2012, Vol. , No. , pp. 267-270 (2012)

[6] R. Steiner, P. K. Steimer, F. Krismer, J. W. Kolar: "Contactless Energy transmission for an Isolated 100W Gate Driver Supply of a Medium Voltage Converter", *35th Annual Conference of the IEEE Industrial Electronics Society*, pp. 302-307 (2009)

[7] S. Lee, R. D. Lorenz: "Development and Validation of Model for 95%-Efficiency 200-W Wireless Power Transfer Over a 30-cm Air-gap", *IEEE Trans. On Industry Applications*, Vol. 47, No. 6, pp. 2495-2504 (2011)

[8] T. Imura, Y. Hori: "Maximizing Air Gap and Efficiency of Magnetic Resonant Coupling for Wireless Power Transfer Using Equivalent Circuit and Neumann Formula", *IEEE Trans. On Industrial Electronics*, Vol. 58, No. 10, pp. 4746-4752 (2011)

[9] E. Waffenschmidt, T. Staring: "Limitation of inductive power transfer for consumer applications", *European Conference on Power Electronics and Applications*, pp. 1-10 (2009)

[10] S. Cheon, Y. Kim, S. Kang, M. L. Lee, J. Lee, T. Zyung: "Circuit-Model-Based Analysis of a Wireless Energy-Transfer System via Coupled Magnetic Resonances", *IEEE Trans. On Industrial Electronics*, Vol. 58, No. 7, pp. 2906-2914 (2011)

[11] A. P. Sample, D. A. Meyer, J. R. Smith: "Analysis, Experimental results, and Range Adaptation of Magnetically Coupled Resonators for Wireless Power Transfer", *IEEE Trans. On Industrial Electronics*, Vol. 58, No. 2, pp. 544-554 (2011)

[12] S. Lee, R. D. Lorenz: "A Design Methodology for Multi-kW, Large Airgap, MHz Frequency, Wireless Power Transfer Systems", *IEEE ECCE 2011*, pp. 3503-3510 (2011)

[13] N. Hatti, Y. Kondo, H. Akagi: "Five-Level Diode-Clamped PWM Converters Connected Back-to-Back for Motor Drives", *IEEE Trans. On Industry Applications*, Vol. 44, No. 4, pp. 1268-1276 (2008)

[14] D. Ahn, S. Hong: "A Study on Magnetic Field Repeater in Wireless Power Transfer", *IEEE Trans. On Industrial Electronics*, Vol. 60, No. 1, pp. 360-371 (2013)

[15] T. Imura: "Equivalent Circuit for Repeater Antenna for Wireless Power Transfer via Magnetic Resonant Coupling Considering Signed Coupling", *6th IEEE Conf. On Industrial Electronics and Applications 2011*, pp. 1501-1506 (2011)

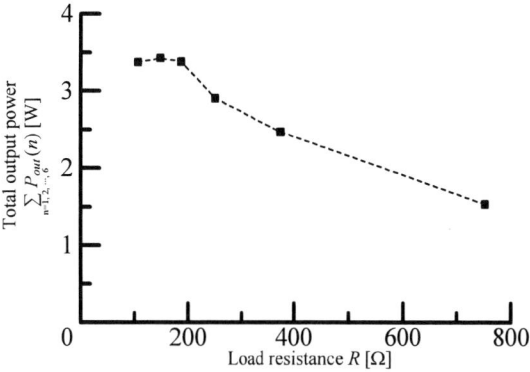

Fig. 17. Output power vs. load resistance. Owing to the voltage drop of the equivalent series resistance of the coils, the total output power is constrained at approximately 3.5 W.

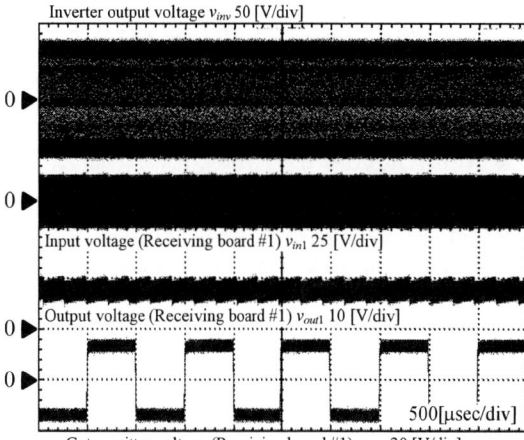

(a) Gate-emitter voltage with proposed isolation system.

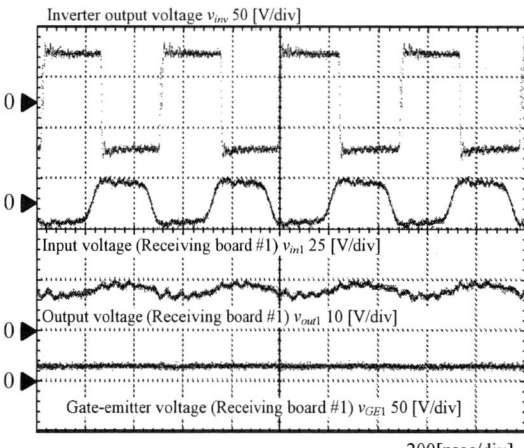

(b) Extended operation waveforms of (a).

Fig. 18. Operation waveforms of the proposed isolation system with the gate drivers.

Study and Implementation of a 15-W Power Amplifier for Piezoelectric Actuator

Yu-Kang Lo, Huang-Jen Chiu, and Yu-Chen Liu
Department of Electronic Engineering
National Taiwan University of Science and Technology
Taipei City, Taiwan, ROC
Email: yklo@mail.ntust.edu.tw

Chung-Yi Lin, Shih-Jen Cheng, and CS Yang
Flextronics Power
New Taipei City, Taiwan, ROC
Email: keyboard.lin@flextronics.com

Abstract-This paper implements a 15-W power amplifier for piezoelectric actuators which is mainly composed of a flyback converter and a power operational amplifier. The flyback converter offers a variable DC output voltage to the power operational amplifier, which outputs an amplified sinusoidal wave with a voltage gain of 20 V/V and a DC bias of 100 V to drive the piezoelectric actuator. The output voltage of the flyback converter which varies with the amplitude of the input signal can reduce the power losses, and the whole power conversion efficiency of the amplifier can thus be promoted up to 40%. From the experimental results, the implemented prototype possesses some features, such as a nearly constant output-to-input voltage gain, a high slew rate, a high input impedance, a low output impedance, and low output voltage ripple.

KEYWORDS: Piezoelectric actuator power amplifier, flyback converter, and power operational amplifier.

I. INTRODUCTION

Piezoelectric actuators are with many features of microminiaturized application, non-electromagnetic interference, higher conversion efficiency, and non-flammable. The utilizations of piezoelectric actuators are widely in fields of not only information technology and robotics but also biomedical engineering and energy engineering, such as industrial precision positioning, ultrasound, and camera module, etc [1]-[5].

There are various kinds of power amplifier for driving the piezoelectric actuators. Class A/B power amplifier is widely adopted in the output stage of operational amplifier due to its advantage of combining the high linearity and low distortion features of class A power amplifier with the higher conversion efficiency of class B power amplifier. Among the operational amplifiers, power operational amplifier, the one of operational amplifiers being able to output high voltage and high current, is suitable for driving resistive load, inductive load, and especially the capacitive load like piezoelectric actuator since utilizing the class A/B power amplifier as the output stage.

In this paper, a power operational amplifier is used to output an amplified sinusoidal wave with a constant voltage gain and a DC bias to drive the piezoelectric actuator, of which the supply voltage is offered by a flyback converter. In addition, the larger voltage difference between the supply voltage and the output

voltage across the internal class A/B operation output stage of the power operational amplifier results in larger power loss. An approach that making the supply voltage offered by the flyback converter automatically track with the peak value of the sinusoidal input signal is proposed to resolve this issue. The overall conversion efficiency of the proposed power amplifier can thus be improved. In the following sections, the detailed analysis and experimental results on a prototype will be given.

II. ANALYSIS OF THE PROPOSED POWER AMPLIFIER

Figure 1 shows the block diagram of the proposed power amplifier. It is mainly composed of a flyback converter and a power operational amplifier. The universal line-in voltage is fed into the flyback converter through the EMI filters and rectifier with low-line voltage doubler. The input voltage range of the flyback converter can thus be reduced due to the adoption of the rectifier with low-line voltage doubler. The flyback converter offers a variable DC output voltage, V_{cc}, to the power operational amplifier, which outputs an amplified sinusoidal wave with a voltage gain of 20 V/V and a DC bias of 100 V to drive the piezoelectric actuator.

The sinusoidal input signal v_{in} is first fed into the instrumentation amplifier to avoid the load effect and the common-mode noise issue. The instrumentation amplifier then faithfully outputs the sinusoidal vin to the following non-inverting summer which adds the DC bias voltage $5V_{dc}$ with v_{in}. Finally, the output of the non-inverting summer, $v_{in}+5V_{dc}$, is amplified by the power operational amplifier with a non-inverting gain of 20 V/V to drive the piezoelectric actuator. At the same time, the instrumentation amplifier also transmits v_{in} to the reference voltage generator which generates the reference voltage V_{ref} tracking with the peak value of v_{in} to the flyback converter. Therefore, the output voltage of the flyback converter can vary with the amplitude of input signal, which improves the overall conversion efficiency of the proposed power amplifier.

978-1-4799-2706-7/14 $31.00 © 2014 IEEE

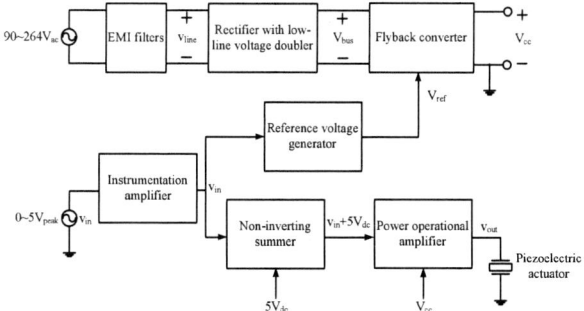

Fig. 1. The block diagram of the proposed power amplifier.

Some key sub-circuits of the proposed power amplifier are analyzed as follows.

Power operational amplifier

Power operational amplifier with the supply voltage usually can be up to few hundred volts is an operational amplifier featuring the capabilities of high output voltage and/or high output current. The internal output stage biased for class A/B operation results in excellent linearity. Referring to Figure 1, the power operational amplifier here is used to amplify the front-end output v_{in} $+5V_{dc}$ with a non-inverting gain of 20 V/V to drive the piezoelectric actuator. The schematics of power operational amplifier are shown in Figure 2 of which the supply voltage V_{cc} generated from the flyback converter.

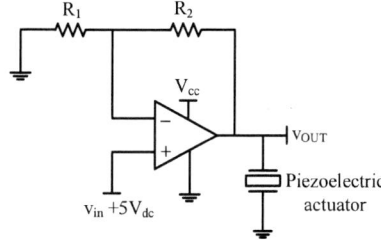

Fig. 2. The schematics of power operational amplifier.

Reference voltage generator

In Figure 2, the voltage difference between the supply voltage V_{cc} and the output voltage v_{OUT} across the internal class A/B operation output stage of the power operational amplifier results in large power loss, which deteriorates the whole conversion efficiency. An approach that the reference voltage of the flyback converter, V_{ref}, tracking with the peak value of the sinusoidal input signal v_{in} can make V_{cc} follow with v_{in} is proposed to resolve this issue. The voltage difference between V_{cc} and v_{OUT} is kept at about 20 V here.

As shown in Figure 3, the reference voltage generation process mainly composes of two procedures. Firstly, v_{OUT_ins}, the output of the instrumental amplifier faithfully duplicates the sinusoidal input signal v_{in}, and the peak value of v_{in} is calibrated through the peak detector composed of A_1, D_1, C_1, and R_1. Since V_f, the forward voltage drop across D_1, would result in a little deviation of the peak value calibration, the following non-inverting summer composed of A_2, R_2, R_3, R_4, and R_5 as well as an extra compensation voltage equaling to V_f is cascaded to ideally output V_a, the peak value of V_{in}. Finally, in order

to realize the relationship of keeping the voltage difference between V_{cc} and v_{OUT} in 20 V shown in Table 1, the rest components in Figure 3 is needed to generate the required V_{ref}.

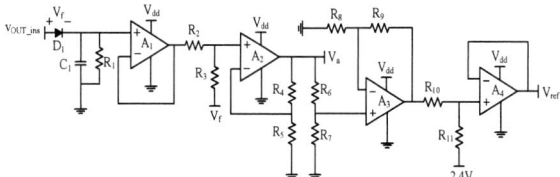

Fig. 3. The schematics of reference voltage generator.

TABLE 1
THE RELATIONSHIP OF THE KEY PARAMETERS.

v_{in} (V)	0	1	2	3	4	5
v_{OUT} (V)	100	120	140	160	180	200
V_{cc} (V)	120	140	160	180	200	220
V_{ref} (V)	1.2	1.4	1.6	1.8	2.0	2.2

III. EXPERIMENTAL RESULTS

A prototype is built to conduct the experiments. Some important circuit specifications and components of the flyback converter are listed in Table 2. In addition, the specifications of the power amplifier are also listed in Table 3.

TABLE 2
THE SPECIFICATIONS AND COMPONENTS OF THE FLYBACK CONVERTER.

Line-in voltage	$90 \sim 264$ V_{rms}
Input DC voltage	$V_{bus} = 250 \sim 370$ V_{dc}
Output DC voltage	$V_{cc} = 120 \sim 220$ V_{dc}
Maximum output power	$P_{o\_max} = 40$ W
Maximum duty cycle	$D_{max} = 35\%$
Switching frequency	$f_{sw} = 50$ kHz
Inductance	$L_m = 1.6$ mH
Primary-side winding turns	$N_p = 44$ turns
Secondary-side winding turns	$N_s = 31$ turns
Magnetic core	TDK PQ26/20 PC40

TABLE 3
THE SPECIFICATIONS OF THE POWER AMPLIFIER.

Input signal	$v_{in} = 0 \sim 5$ V_{peak} sinusoidal wave
Output signal	$v_{OUT} = 0 \sim 200$ V_{pp} sinusoidal wave
Output DC bias voltage	100 V
Output power	15 W
Voltage gain	20 V/V
Output slew rate	400 V/ms
Input impedance	33 kΩ
Output impedance	$< 10 \ \Omega$

TL494, a well-known PWM controller, is adopted for controlling the flyback converter. Figures 4(a) and 4(b) show the measured waveforms of gate-to-source voltage and drain-to-source voltage across the switch of the flyback converter under the same output DC voltage V_{cc}, maximum output power but different line-in voltage. It can be seen apparently that the designed flyback converter actually operates under DCM to adapt the output voltage in the wide range.

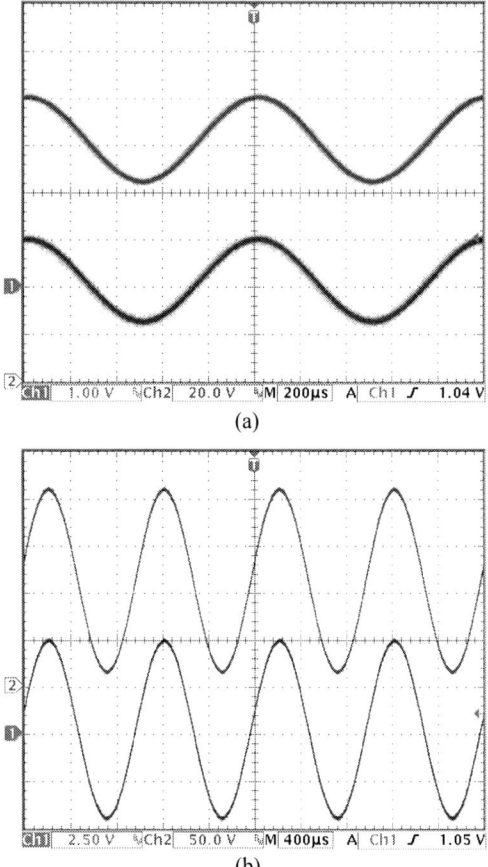

(a)

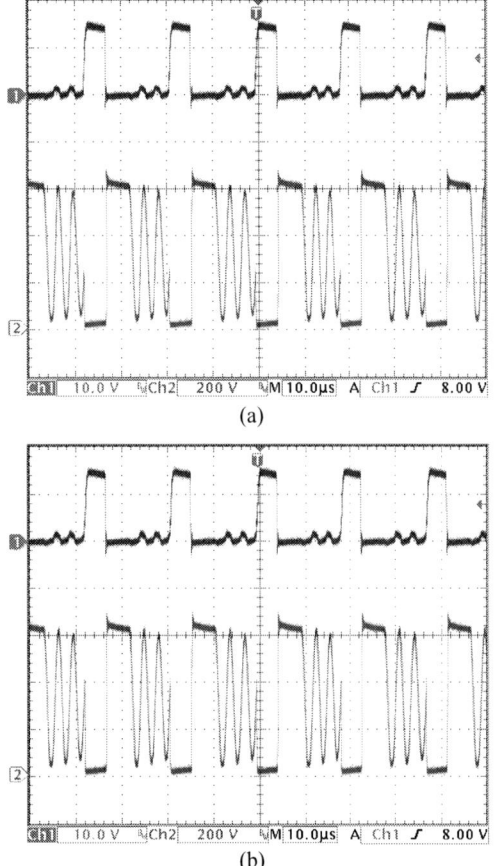

(a)

(b)

Fig. 5. Measured sinusoidal waveforms of v_{in} (Ch1) and v_{OUT} (Ch2) under (a) 110 V_{rms} line-in voltage, V_{cc} = 120 V and v_{in} = 1 V_{peak} (b) 110 V_{rms} line-in voltage, V_{cc} = 220 V and v_{in} = 5 V_{peak}.

(b)

Fig. 4. Measured maximum-output-power waveforms of v_{GS} (Ch1) and v_{DS} (Ch2) under (a) 110 V_{rms} line-in voltage and V_{cc} = 220 V (b) 220 V_{rms} line-in voltage and V_{cc} = 220 V.

Figures 5(a) and 5(b) show the measured sinusoidal waveforms of different input signal v_{in} and output signal v_{OUT} under 110 V_{rms} line-in voltage. It can be seen obviously that the designed power amplifier can generate almost non-distorted sinusoidal wave outputs with a voltage gain of 20 V/V and a DC bias of 100 V for driving the piezoelectric actuator.

Figures 6 shows the measured slew rate waveforms of vin and v_{OUT} to see the output slew rate. The measured output slew rate results are over the specification listed in Table 3 very much.

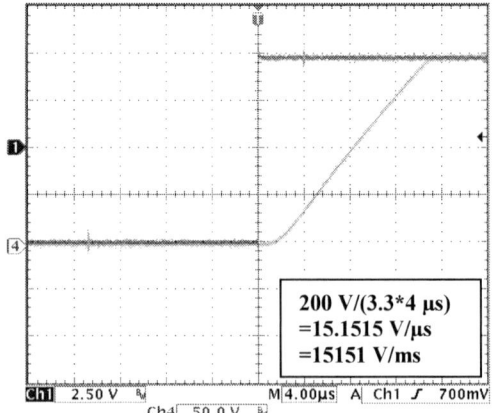

Fig. 6. Measured slew rate waveforms of v_{in} (Ch1) and v_{OUT} (Ch4) under v_{in} = 5 V_{peak} and V_{cc} = 220 V.

Figure 7 shows the expected and actual voltage gain curves of the power amplifier. The actual voltage gain curve is quiet linear and closed to the expected one. The feasibility of the constant voltage gain specification listed in Table 3 is implemented nearly. Figure 8 shows the expected and actual output voltage curves of the flyback converter. Although the measured actual output voltage

978-1-4799-2706-7/14 $31.00 © 2014 IEEE

results of the flyback converter have slight deviations from the expected ones, the goal of keeping about 20 V voltage difference between V_{cc} and v_{OUT} is achieved to reduce the power loss of the internal class A/B operation output stage of the power operational amplifier. Figures 9 and 10 show the overall efficiency curves at $110V_{rms}$ and $220V_{rms}$ line-in voltages, respectively. It can be seen that much efficiency improves both at $110V_{rms}$ and $220V_{rms}$, up to 40% increase at zero input signal v_{in}, if making V_{cc} offered by the flyback converter track with v_{in} and keep about 20 V voltage difference between V_{cc} and v_{OUT}.

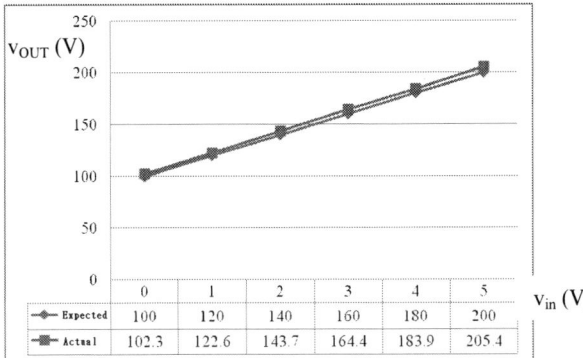

Fig. 7. Voltage gain curves of the power amplifier.

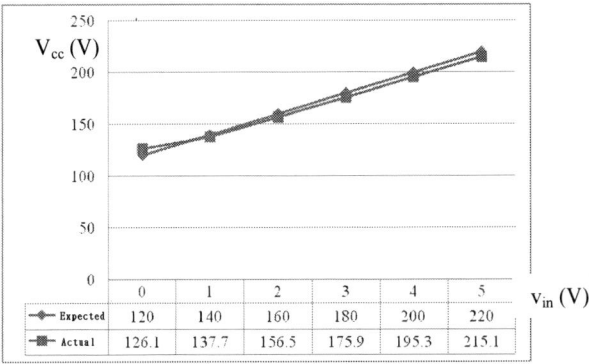

Fig. 8. Output voltage curves of the flyback converter.

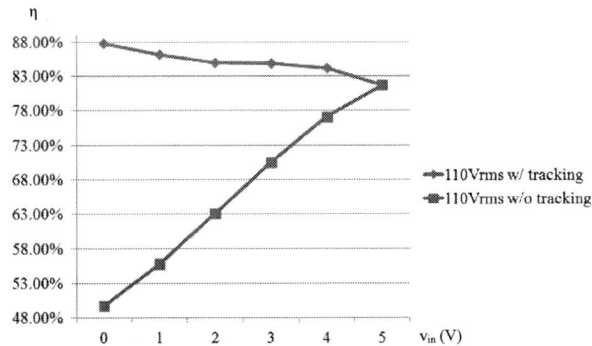

Fig. 9. Overall efficiency curves of the proposed power amplifier at $110V_{rms}$ line-in voltage.

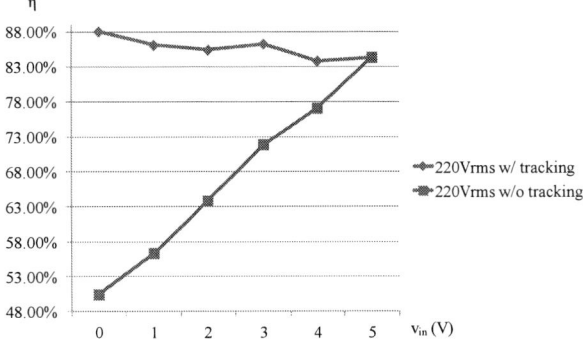

Fig. 10. Overall efficiency curves of the proposed power amplifier at $220V_{rms}$ line-in voltage.

Table 4 shows the input impedance and output impedance of the power amplifier. The measured results meet the specifications listed in Table 3 very much, especially the ultra high input impedance. This is due to the adoption of the instrumentation amplifier at the input shown in Figure 4.

TABLE 4

MEASURED INPUT IMPEDANCE AND OUTPUT IMPEDANCE OF THE POWER AMPLIFIER.

Input impedance	Output impedance
99.99 MΩ	0.257 Ω

IV. CONCLUSIONS

In this paper, the detailed sub-circuits of the proposed power amplifier are analyzed. The ways how to output an amplified sinusoidal wave with a voltage gain of 20 V/V and a DC bias of 100 V to drive the piezoelectric actuator as well as make the supply voltage of power operational amplifier track with the peak value of the sinusoidal input signal are shown. From the experimental results, besides the almost non-distorted sinusoidal wave outputs, the implemented prototype possesses features of a nearly constant output-to-input voltage gain, a high slew rate, a high input impedance, a low output impedance, and low output voltage ripple (about 40 mV$_{pp}$). In addition, the output voltage of the flyback converter which varies with the amplitude of the input signal can reduce the power losses, and the whole power conversion efficiency of the amplifier can thus be promoted up to 40%.

REFERENCES

[1] K. Uchino, "Piezoelectric actuators 2006, expansion from IT/robotics to ecological/energy applications," *International Journal of Electroceram*, vol. 20, pp. 301-311, 2008.

[2] C. Park, S. Cha, Y. Lee, O. Kwon, D. Park, K. Kwon, and J. Lee, "A highly accurate piezoelectric actuator driver IC for auto-focus in camera module of mobile phone," *Proceedings of IEEE ISCAS 2010*, pp. 1284-1287, 2010.

[3] J.G.G. Stephan and B. Maundy, "A novel circuit element and its application in signal amplification," *International Journal of Circuit Theory and Applications*, vol. 36, pp. 219-231, 2008.

[4] M. Porfiri, F. dell'Isola, and M.F.M. Frattale, "Circuit analog of a beam and its application to multimodal vibration damping, using piezoelectric transducers," *International Journal of Circuit Theory and Applications*, vol. 32, pp. 167-198, 2004.

[5] G. Gnad and R. Kasper, "A power drive control for piezoelectric actuators," *Proceedings of IEEE International Symposium on Industrial Electronics 2004*, vol. 2, pp. 963-967, 2004.

Isolated Voltage-Boosting Converter

K. I. Hwu[1], *Member, IEEE*, W. Z. Jiang[2], *Student Member, IEEE*, and Jenn-Jong Shieh[3]
Department of Electrical Engineering, National Taipei University of Technology, Taipei, Taiwan[1, 2]
Department of Electrical and Electronic Engineering, Ta Hwa University of Science and Technology, Hsinchu, Taiwan[3]
E-mail[1]: eaglehwu@ntut.edu.tw, E-mail[2]: newjerusalem333@gmail.com, E-mail[3]: eesjj@tust.edu.tw

Abstract—In this paper, a single switch isolated voltage-boosting converter is presented, which is derived from the traditional flyback converter and charge pump concept. The proposed converter possesses an output inductor, so the output current is non-pulsating. Moreover, there are several advantages of the proposed converter over the traditional flyback converter, such as higher voltage conversion ratio with only additional four passive elements, and smaller voltage and current ripples, except the same switch voltage stress. Finally, some experimental results are provided to verify the effectiveness of the proposed converter.

Keywords—Charge pump, flyback, isolated, voltage-boosting, voltage conversion ratio.

I. INTRODUCTION

In recent years, the popularity of renewable energy systems has been increasing because of their low pollution levels. These renewable energy systems include solar cells, fuel cells, electrical vehicles, etc. In such applications, to stabilize and regulate the output voltages which the renewable energy systems produce, in general the traditional boost converter and the buck-boost converter [1] are widely used. However, even though the converters mentioned above are simple in structures, it is quite difficult to achieve higher output voltages due to the limitations of the parasitic resistances. To obtain a higher output voltage, many non-isolated voltage-boosting converters using different voltage-boosting techniques are presented over the past few decades. These voltage-boosting techniques involve switched-inductor [2], [3], coupled inductor with turns ratio [4]-[9], switched-capacitor [10], auxiliary transformer with turns ratios [11], [12], etc.

In [3] and [9], the output terminal is floating, leading to application complexity. In [2] and [4]-[10], these converters contain too many components, thereby making the converters relatively complicated. In [1]-[12], the output currents are pulsating, therefore causing the output voltage ripples to tend to be high, which are not suitable for the electrical devices that need low voltage ripples.

Based on the mentioned above, a voltage-boosting converter is presented, named KY converter [13], which is famous for its continuous output current and simple structure. However, the voltage conversion ratio of the KY converter is not high enough. For more research, a series of converters with high voltage conversion ratios derived from the KY converter [14]-[20] has been gradually engineered. Although these converters have higher voltage conversion ratios than that of the KY converter, they all belong to the non-isolated converters. In some applications, an isolated converter is preferred to meet the safety standards owing to its galvanic isolation. Therefore, the traditional flyback converter is quite appealing in industrial applications because of its low component count, simple

structure and low cost. Nonetheless, it suffers from several limitations like pulsating output current and low voltage conversion ratio, which are not desirable for some loads that need low output current ripples and high voltages.

Accordingly, a novel isolated voltage-boosting converter is presented herein, which possesses a coupled inductor with a turns ratio for both isolation between the input and the output and enhancement of the output voltage, an output inductor for reduction of the output current ripple, two capacitors for further upgrade of the output voltage. Furthermore, another problem in designing a voltage-boosting converter is the voltage stress across the switch. Without the voltage spikes considered, the voltage stress of the proposed isolated converter is the same as that of the traditional flyback converter. As a result, compared with the traditional flyback converter except the same switch voltage stress, there are two additional advantages of the proposed converter: higher voltage conversion ratio with only additional four passive elements required, and smaller voltage and current ripples. Moreover, to protect the power switch, the voltage spike should be suppressed to an acceptable level. There are various techniques to accomplish the goal. The common method used in industrial applications is the RCD snubber, which is adopted in the proposed converter.

In this paper, a detailed description along with some experimental results is given to provide the effectiveness of the proposed converter.

II. OVERALL SYSTEM CONFIGURATION

Fig. 1 shows the proposed converter, which contains one MOSFET switch S_1, one coupled inductor composed of the primary winding with N_p turns and the secondary winding with N_s turns, two charge pump capacitors C_1 and C_2, two diodes D_1 and D_2, one output inductor L_o, one output capacitor C_o. In addition, the input voltage is denoted by V_i, the output voltage is signified by V_o, and the output resistor is represented by R_o.

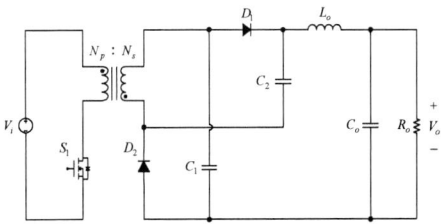

Fig. 1. Proposed isolated voltage-boosting converter.

III. BASIC OPERATING PRINCIPLES

For analysis convenience, there are some assumptions to be made as follows.

(1) The coupled inductor is modeled as an ideal transformer except that one magnetizing inductor L_m is connected in parallel with the primary winding.
(2) The MOSFET switch and the diodes are viewed as ideal components.
(3) The values of all the capacitors are large enough such that the voltages across them are kept constant at some values.
(4) The capacitances of C_1 and C_2 are equal.
(5) The magnitude of the switching ripple is negligible. Therefore, the small ripple approximation will be adopted in the following analysis.
(6) The turns ratio n of the coupled inductor is defined as N_s/N_p.
(7) The magnetizing inductor L_m operates in the CCM, and the output inductor L_o operates in the positive current region (PCR).

The following analysis contains the explanation of the power flow path for each mode as well as the corresponding equations and voltage conversion ratio. Due to the last assumption mentioned above, there are two operating modes in the proposed converter. In addition, the input current is denoted by i_i, the current in the N_p winding is signified by i_{Np}, the current in the N_s winding is represented by i_{Ns}, the current in L_m is denoted by i_{Lm}, the current in L_o is indicated by i_{Lo}, the current in R_o is signified by I_o, and the voltage on L_o is described by v_{Lo}. On the other hand, the voltage across L_m or the voltage across the N_p winding is signified by v_{Np}, the voltage across the N_s winding is represented by v_{Ns}, the voltage on C_1 is indicated by V_{C1}, the voltage on C_2 is denoted by V_{C2}, and the voltage on L_o is described by v_{Lo}. Furthermore, the key waveforms for the proposed converter operating in this case are shown in Fig. 2.

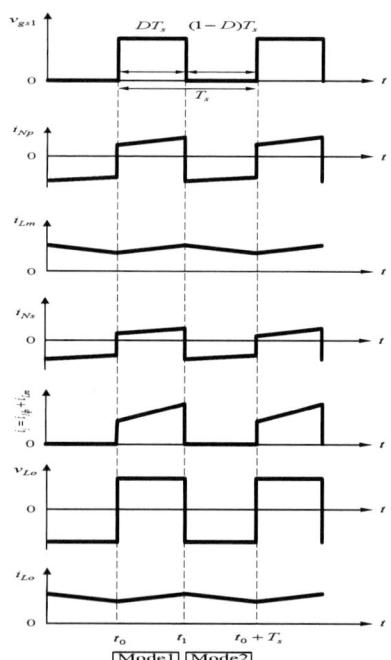

Fig. 2. Key waveforms for L_m operating in the CCM and L_o operating in the PCR.

Mode 1 [$t_0 \le t \le t_1$]

During this interval, as shown in Fig. 3, S_1 is turned on. Therefore, the input voltage V_i is imposed on N_p, thus causing L_m to be magnetized and the voltage across N_s to be induced, equal to $V_i \times N_s / N_p$. In the meantime, D_1 and D_2 become reverse-biased, and the voltage across L_o, v_{Lo}, is a positive value, equal to $v_{Ns} + V_{C1} + V_{C2} - V_o$, thus making L_o magnetized. Also, the induced voltage $V_i \times N_s / N_p$ across N_s, together with the voltages across C_1 and C_2, provides the energy to the load. Hence, the corresponding equations are shown below:

$$v_{Np} = V_i \tag{1}$$

$$v_{Lo} = v_{Ns} + V_{C1} + V_{C2} - V_o = V_i \times \frac{N_s}{N_p} + V_{C1} + V_{C2} - V_o \tag{2}$$

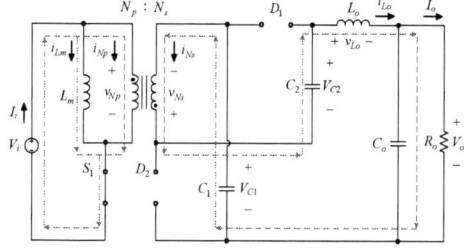

Fig. 3. Power flow in mode 1.

Mode 2 [$t_1 \le t \le t_0 + T_s$]

During this interval, as shown in Fig. 4, S_1 is turned off. S_1 is turned off and D_1 and D_2 become forward-biased. Therefore, the voltage $-V_{C1} \times N_p / N_s$ is imposed on winding N_p, thereby causing the magnetizing inductor L_m to be demagnetized. The voltage across L_o is a negative value, equal to $V_{C1} - V_o$ or $V_{C2} - V_o$, thus causing L_o to be demagnetized. Accordingly, the energy stored in L_m via N_s, together with the energy stored in L_o, provides the energy to C_1, C_2 and the load. Hence, the corresponding equations are shown below:

$$v_{Np} = -V_{C1} \times \frac{N_p}{N_s} \tag{3}$$

$$v_{Lo} = V_{C1} - V_o \quad \text{or} \quad v_{Lo} = V_{C2} - V_o \tag{4}$$

By applying the voltage-second balance principle to L_m over one switching period, the following equation can be obtained:

$$V_i \times D + (-V_{C1} \times \frac{N_p}{N_s}) \times (1-D) = 0 \tag{5}$$

Also, by rearranging the above equation, the voltages across C_1 and C_2 can be obtained to be

$$V_{C1} = V_{C2} = \frac{N_s}{N_p} \times \frac{D}{1-D} \times V_i \tag{6}$$

Similarly, by applying the voltage-second balance principle to L_o over one switching period, the following equation can be obtained:

$$(V_i \times N_s / N_p + V_{C1} + V_{C2} - V_o) \times D + (V_{C1} - V_o) \times (1 - D) = 0 \quad (7)$$

Next, based on (6) and (7), the corresponding voltage conversion ratio can be expressed to be

$$\frac{V_o}{V_i} = 2 \times \frac{N_s}{N_p} \times \frac{D}{1 - D} \quad (8)$$

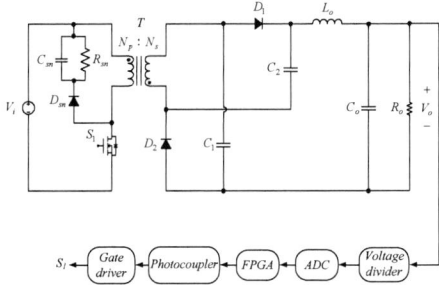

Fig. 4. Power flow in mode 2.

IV. CONTROL METHOD APPLIED WITH DESIGN CONSIDERATIONS

Fig. 5 shows the overall system block diagram. The feedback control loop contains one voltage divider, one digital-to-analog converter (ADC), one FPGA, which is the control kernel, one photocoupler and one gate driver. The system specifications and used component names of the proposed converter are shown in Tables I and II, respectively.

Fig. 5. Proposed overall system block diagram.

Table I System specifications of the proposed converter

System parameters	Specs.
Input voltage (V_i)	12V
Rated output voltage (V_o)	72V
Rated output current ($I_{o,rated}$)/power ($P_{o,rated}$)	0.833A/60W
Minimum output current ($I_{o,min}$)/power ($P_{o,min}$)	0.1A/7.2W
Switching frequency (f_s)	100kHz

Table II Components used in the proposed converter

Components	Specifications
MOSFET switches S_1	V_{DS}=100V, I_D=120A,
Diode D_1, D_2	V_R=120V, $I_{F(AV)}$=20A,
Charge pump capacitor C_1, C_2	150μF/50V
Output capacitor C_o	440μF/100V
Coupled inductor T	N_p:N_s=1:3, L_m=48.6μH, L_{l1}=0.39μH
Output inductor L_o	L_o=565μH
FPGA	EP1C3T100
Gate driver	TC4420

Photocoupler	TLP250
ADC	ADC7476
RCD snubber	R_{sn}=250Ω/5W, C_{sn}=0.8μF

V. EXPERIMENTAL RESULTS

Figs. 6 to 9 show the measured waveforms at rated load. Furthermore, from Fig. 8, the voltages across C_1 and C_2 are approximately equal, and the heavier the load is, the slightly higher the voltage across C_1 and C_2. This is because the voltages across C_1 and C_2 are functions of the duty cycle D, which is varied with the load so as to maintain the output voltage at the constant value. Moreover, Fig. 9 shows the waveforms of output voltage and current in the steady state at rated load.

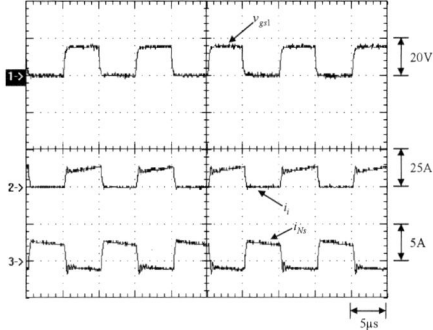

Fig. 6. Waveforms at rated load: (1) v_{gs1}; (2) i_i; (3) i_{Ns}.

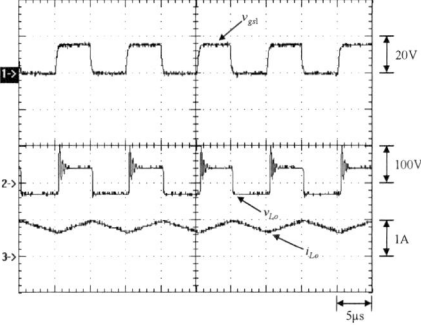

Fig. 7. Waveforms at rated load: (1) v_{gs1}; (2) v_{Lo}; (3) i_{Lo}.

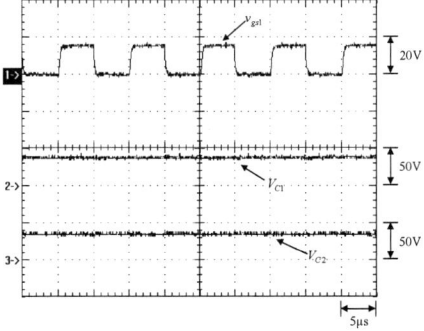

Fig. 8. Waveforms at rated load: (1) v_{gs1}; (2) V_{C1}; (3) V_{C2}.

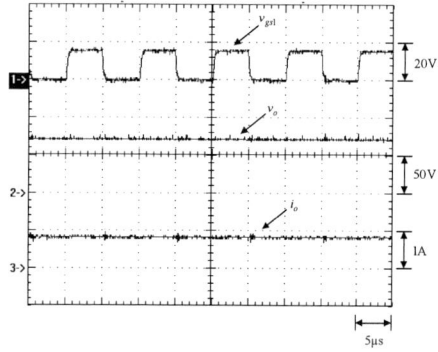

Fig. 9. Waveforms at rated load in steady state: (1) v_{gs1}; (2) v_o; (3) i_o.

VI. CONCLUSION

A novel isolated voltage-boosting converter is presented herein. By combining the coupled inductor with the turns ratio, and the switched capacitor, the corresponding voltage conversion ratio is higher than that of the traditional flyback converter. Furthermore, the proposed converter has one output inductor such that the output current is non-pulsating and hence the output voltage is of small ripple. Moreover, the structure of the proposed converter is quite simple and very suitable for industrial applications.

REFERENCES

[1] R. W. Erickson and D. Maksimovic, *Fundamentals of Power Electronics*, 2nd ed., Norwell: KLuwer Academic Publishers, 2001.

[2] B. R. Lin, F. Y. Hsieh and J. J. Chen, "Analysis and implementation of a bidirectional converter with high converter ratio," *IEEE ICIT'08*, pp.1-6, 2008.

[3] B. Axelrod, Y. Berkovich and A. Ioinovici, "Switched-capacitor/switched-inductor structures for getting transformerless hybrid DC-DC PWM converters," *IEEE Transactions on Circuits and System I: Regular Paper*, vol. 55, no. 2, pp. 687-696, 2008.

[4] K. B. Park, G. W. Moon and M. J. Youn, "Nonisolated high step-up stacked converter based on boost-integrated isolated converter," *IEEE Transactions on Power Electronics*, vol. 26 no. 2, pp. 577-587, 2011.

[5] Y. Deng, Q. Rong, Y. Zhao, J. Shi and X. He, " Single switch high step-up converters with built-in transformer voltage multiplier cell," *IEEE Transactions on Power Electronics*, vol. 27 no. 8, pp. 3557-3567, 2012.

[6] Wuhua Li, Weichen Li, Xiangning He, D. Xu and Bin Wu, "General derivation law of nonisolated high-step-up interleaved converters with built-in transformer," *IEEE Transactions on Industrial Electronics*, vol. 59, no. 3, pp. 1650-1661, 2012.

[7] C. M. Lai, C. T. Pan and M. C. Cheng, "High-efficiency modular high step-up interleaved boost converter for dc-microgrid applications," *IEEE Transactions on Industrial Electronics*, vol. 48 no. 1, pp.161-171, 2012.

[8] R. J. Wai and R. Y. Duan, "Hgih step-up converter with coupled-inductor," *IEEE Transactions on Power Electronics*, vol. 20 no. 5, pp.1025-1035, 2005.

[9] R. J. Wai and R. Y. Duan, "High-efficiency power conversion for low power fuel cell generation system," *IEEE Transactions on Power Electronics*, vol. 20 no. 4, pp.847-856, 2005.

[10] F. L. Luo, "Analysis of super-lift luo-converters with capacitor voltage drop," *IEEE ICIEA ' 08*, pp. 417-422, 2008.

[11] K. B. Park, G. W. Moon and M. J. Youn, "High step-up boost converter integrated with voltage-doubler," *IEEE ECCE'10*, pp. 810-816, 2010.

[12] C. T. Pan and C. M. Lai, "A high-efficiency high step-up converter with low switch voltage stress for fuel-cell system applications," *IEEE Transactions on Industrial Electronics*, vol. 57, no. 6, pp. 1998-2006, 2010.

[13] K. I. Hwu and Y. T. Yau, "KY converter and its derivatives," *IEEE Transactions on Power Electronics*, vol. 57, no. 6, pp. 128-137, 2009.

[14] K. I. Hwu, Y. T. Yau and Y. H. Chen "A novel voltage-boosting converter with passive voltage clamping," *IEEE ICSET'08*, pp.351-354, 2008.

[15] K. I. Hwu and Y. T. Yau, "Two types of KY buck-boost converter," *IEEE Transactions on Industrial Electronics*, vol. 56, no. 8, pp. 2970-2980, 2009.

[16] K. I. Hwu and Y. T. Yau, "A KY boost converter," *IEEE Transactions on Power Electronics*, vol. 25, no. 11, pp. 2699-2703, 2010.

[17] K. I. Hwu and Y. T. Yau, "Inductor-coupled KY boost converter," *IET Electronics Lettlers*, vol. 46, no. 24, pp. 1624-1626, 2010.

[18] K. I. Hwu, K. W. Huang and W. C. Tu, "Step-up converter combining KY and buck-boost converters," *IET Electronics Letters*, vol. 47. no. 12, pp. 722-724, 2011.

[19] K. I. Hwu and W. C. Tu, "Voltage-boosting converters with hybrid energy pumping," *IEEE Transactions on Power Electronics*, vol. 5, no. 2, pp. 185-195, 2012.

[20] K. I. Hwu and Y. T. Yau, "High step-up converter based on charge pump and boost converter," *IEEE Transactions on Power Electronics*, vol. 27, no. 5, pp. 2484-2494, 2012.

High Voltage Conversion Ratio Cascade Boost Converter with DC Snubber

Yuang-Shung Lee[1]

Department of Electrical Engineering
Fu Jen Catholic University
New Taipei City, Taiwan
002877@mail.fju.edu.tw[1]

Ling-Chia Yu[2], Tzu-Han Chou[3]

Department of Electrical Engineering
Fu Jen Catholic University
New Taipei City, Taiwan
401216236@mail.fju.edu.tw[2],401216250@mail.fju.edu.tw[3]

Abstract— This paper proposed a high voltage conversion ratio cascade boost converter topology composed two stages of converters. First stage boosts the input source voltage twice larger by using the double-mode switched-capacitor cells converter (SCC). Second stage provides a high voltage conversion ratio boost converter with a couple inductor and diode-capacitor (DC) snubber. The converter topology analysis and design are described in detail. Results of simulation and experimental measurements are carried out to verify the performances of the proposed high voltage conversion ratio cascade boost converter under different operation conduction modes.

Keywords— Cascade boost converter, coupled inductor, DC snubber, high voltage gain.

I. INTRODUCTION

Today, more industrial applications require DC-DC converters to provide large conversion ratios with a good efficiency, such as battery powered devices, uninterruptible power supplies (UPS), photovoltaic (PV) systems, and some energy harvesting systems.

Several researches have proposed high step-up DC-DC converters[1-7]. A boost converter with coupled inductors and a buck–boost type of active-clamp circuit has a high step-up voltage ratio proposed in [1], without a DC snubber at secondary side of transformer, the voltage ratio in [1] is always smaller than the voltage gain of proposed paper. A high step-up converter was presented in [2], where the converter was suitable only for low-current applications. In some application as high currents needed, the voltage drops across the capacitor is too large and creating a large charging current through the charge pump capacitor. A dc/dc converter with a buck-boost type of active clamp circuit in parallel which is ZVS of power switches and ZCS of diodes is proposed in [3]. A high step up boost converter with couple inductor and charge pump is investigated [5]. The conversion ratio of the converter in above two paper are lower than the voltage gain of the proposed converter. A family of high step up interleaved converters with winding cross coupled inductors designed to reboost the output voltage and reduce the switch voltage stress in the converter was shown in [8-13].

This paper presents a converter that has high gain because of a DC snubber on the secondary side of the coupled inductor. This converter needs only one duty cycle and a complementary duty cycle for the switches,

making it easier to control and obtain higher gain than the converter proposed in [9]. If another inductor series is added with the capacitor C_{1b}, the resonant frequency of the LC becomes higher than the switching frequency of the main switch, and the first stage can be designed to operate at ZCS [9]. Fig. 1 shows the fundamental scheme of the proposed converter without synchronous rectifiers. The proposed converter efficiency can also be improved with synchronous rectifiers replacing diodes D_{P2} and D_{N2} for reduced diode forward voltage drops and power losses as shown in Fig. 1.

The topologies description and the operating principles for the proposed high gain converter are elucidated in Section II . Section III shows the converter circuit analysis and modes derivation. Section IV demonstrates that the results of simulations corresponds to completely the experiments under continuous and discontinuous conduction modes (CCM and DCM). Section V shows the conclusion.

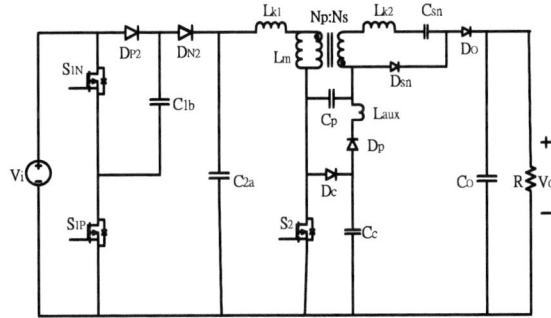

Fig. 1. The proposed converter with DC snubber circuit

II. TOPOLOGIES DESCRIPTION

Fig. 1 shows a switched-capacitor converter cascaded with a coupled inductor converter with DC snubber for high voltage conversion ratio boost scheme. The first stage converter can double boost the input source voltage using SCC. The second stage conversion scheme is a high gain coupled inductor boost converter with DC snubber. The operating duty cycle for the three switches in the proposed circuit, give switches S_{1P}, S_2 the same duty cycle and S_{1N} has complementary duty cycle, or give switch S_{1N}, S_2 the same duty cycle and S_{1P} the complementary duty cycle. The proposed converter in any of the above conditions would produce the same conversion ratio. In this paper, we choose switches S_{1N}, S_2 to have the same duty cycle with S_{1P} having the complementary duty cycle for research purposes. The typical waveforms of the proposed converter are shown in Fig. 2. Figs. 3 (a)-(e) show the equivalent circuits for the proposed converter under different operating intervals.

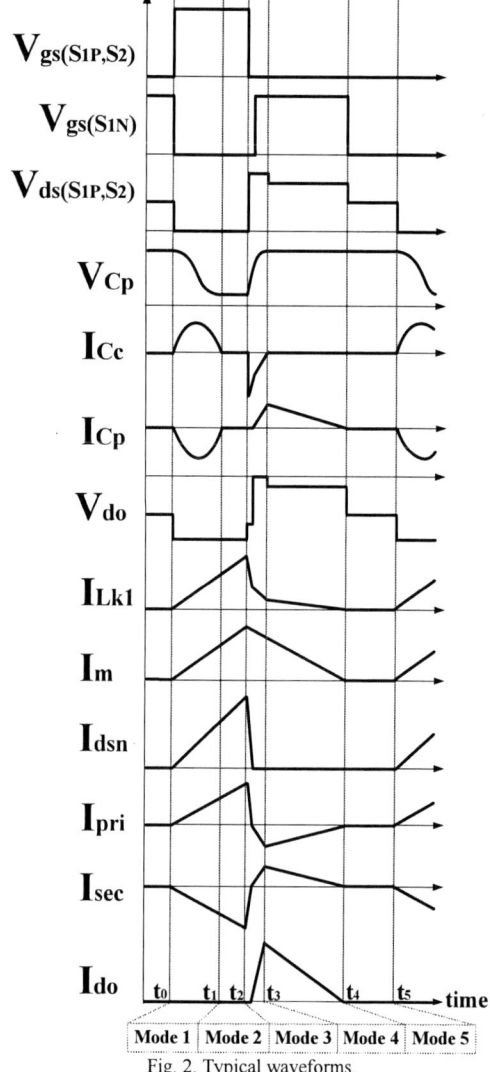

Fig. 2. Typical waveforms.

The proposed converter topological modes and working principles are described as follows:

1) Mode 1 ($t_0 \sim t_1$):

In this mode, switch S_{1N} is turned off. Switches S_{1P} and S_2 are turned on as shown in Fig. 3 (a). The capacitor C_{1b} is charged by the source voltage V_i. The magnetizing inductor current i_m increases gradually in an approximately linear way. Flowing through the diode D_p, capacitor C_c is charged to the half time of this interval, capacitor C_c is then discharged to zero at the end of this mode. Capacitor C_p takes the opposite action with C_c. Capacitor C_{sn} of the DC snubber at the secondary side transformer is then discharged through the diode D_{sn}.

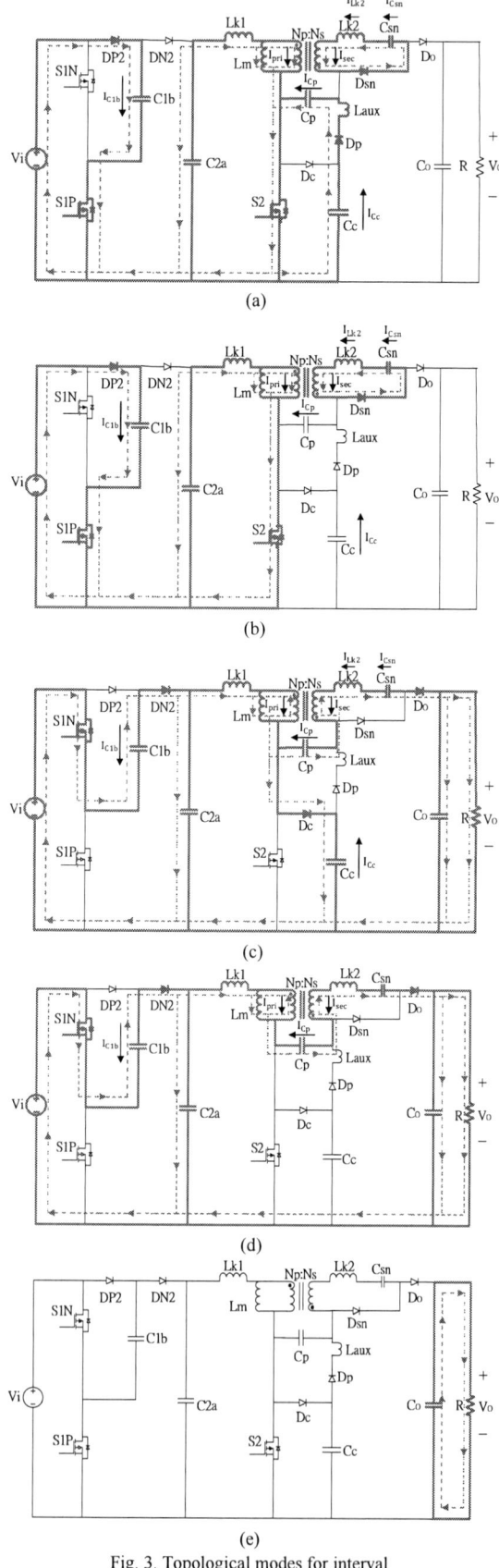

Fig. 3. Topological modes for interval
(a) $t_0 \sim t_1$ (b) $t_1 \sim t_2$ (c) $t_2 \sim t_3$ (d) $t_3 \sim t_4$ (e) $t_4 \sim t_5$

2) Mode 2 ($t_1 \sim t_2$)

Switches S_{1P} and S_2 are still turned on since mode 1, and switch S_{1N} are also turned off in this interval as shown as Fig. 3 (b). Magnetizing inductor L_m and the primary inductor of the transformer are charged linearly. The energy of capacitors C_c and C_p are both discharged to zero in mode 1, and thus there is no current flowing through C_c and C_p. The secondary transformer maintains the same operation as mode 1.

3) Mode 3 ($t_2 \sim t_3$):

At $t = t_2$, switch S_{1P} and main switch S_2 are both turned off. Switch S_{1N} is turned on instead as shown in Fig. 3 (c). Within a very small interval at the beginning of this mode, the magnetizing inductor L_m is started to discharge. The primary current of the transformer does not change direction immediately. The operation of all of the elements on the secondary side do not be changed in this small interval. The proposed circuit operation is ignored because the time of this extraordinary interval is very small. The behavior of the circuit in the remaining time of this mode is taken as the main operation of this stage. Since the magnetizing inductor current i_m decreases linearly, capacitors C_c and C_p are charged, and the secondary side current charges C_{sn} and C_o through the output diode D_o.

4) Mode 4 ($t_3 \sim t_4$):

This mode begins when the switching capacitor C_c is charged to zero, and C_p starts to discharge as shown in Fig. 3 (d). There is no current through C_c. Because C_p is discharged, the current from the output diode D_o also decreases.

If the proposed circuit is operated at CCM, then only four modes (mode 1- mode 4) exist. Moreover, the converter is operated at DCM, when $t = t_4$, the circuit would transfer to the transient state as the description in mode 5.

5) Mode 5 ($t_4 \sim t_5$):

At t4, switches S_{1P}, S_{1N}, S_2 are all turned off. Output capacitor C_o is discharged its remaining energy to the output. There is no current in any inductor elements in the circuit as shown in Fig. 3 (e). This mode ends at t_5, and returns to the initial condition for the next period of mode 1.

III.   ANALYSIS AND MODE DERIVATION

A.   Continuous conduction mode

We designed switches S_{1P} and S_2 to operate at same duty cycle (D), with switch S_{1N} operates at the complementary duty cycle (1-D). We separated the proposed converter into two stages. Switches S_{1P} and S_{1N}, diodes D_{p2} and D_{n2}, capacitor C_{1b}, comprise the first stage . The remaining circuit elements comprise the second stage.

In the first stage, when switch S_{1P} turns on, C_{1b} is charged by the input source, and $V_{C_{1b}}$ equals V_i. Switch S_{1N} then turns on and switch S_{1P} turns off. The output

energy of the first stage is charged to C_{2a} as the input source V_s for the second stage, which is equal to $2V_i$, ($V_s = V_{C2a} = 2V_i$).

Referring to Figs. 3 (a) and (b), in the second stage, in mode 1 and mode 2, the main switch S_2 is turned on, and the magnetizing inductor voltage L_m (V_{L_m}) equals V_s. In mode 3 and mode 4, V_{Lm} is equal to $V_S - V_{Cc}$. According to the voltage-second balance principle :

$$V_S \cdot D \cdot T_S + (V_S - V_{Cc}) \cdot (1 - D) \cdot T_S = 0 \tag{1}$$

The coupled inductor turn ratio is defined as :

$$n = \frac{N_s}{N_p} \tag{2}$$

When the proposed circuit operates at modes 3 and 4,

$$V_{Cc} + V_{Cp} - n(V_S - V_{Cc}) + nV_S - V_O = 0 \tag{3}$$

$$V_{Cc} = \frac{V_O}{(n+2)} \tag{4}$$

The voltage-second balance equations (1) via (4) become

$$V_S \cdot D \cdot T_S + \left(V_S - \frac{V_O}{(n+2)}\right) \cdot (1 - D) \cdot T_S = 0 \tag{5}$$

The voltage conversion ratio in the second stage can then be derived as:

$$M_{stage2} = \frac{V_O}{V_S} = \frac{(n+2)}{(1-D)} \tag{6}$$

Finally, the gain operating at continuous conduction mode (CCM) can be integrated as :

$$M = \frac{V_O}{V_i} = \frac{V_O}{V_S} \cdot \frac{V_S}{V_i} = 2 \cdot M_{stage2} = \frac{2 \cdot (n+2)}{(1-D)} \tag{7}$$

The corresponding three-dimensional gain plotted under different coupled inductor turn ratio and duty cycles D is shown in Fig. 4. The operation point chosen to verify the performance of the proposed converter is M (2.7,0.3). The maximum gain is 47 at D = 0.8. The minimum gain is 4 when D = 0 and n = 0.

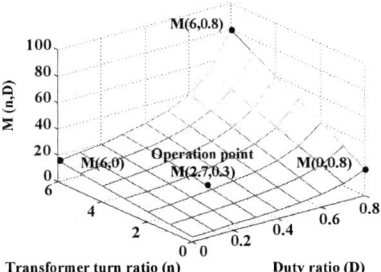

Fig. 4. 3D plot for the voltage conversion ratio

B.   Discontinuous conduction mode

Assume the duty cycle of mode 1 is d_1, the duty cycle of mode 2 is d_2, the duty cycle of mode 3 is d_3, the duty cycle of mode 4 is d_4, and duty cycle of mode 5 is d_5. Because the main switch S_2 is turned on when the converter is operating at mode 1 and mode 2, the sum of these two duty cycle modes equals the duty cycle of the main switch S_2, which means $d_1 + d_2 = D$. In modes 3 and 4, switches S_{1P} and S_2 are turned off, and switch S_{1N} is turned on instead. Assume the operated duty cycle of S_{1N} is D_L. We can know $d_3 + d_4 = D_L$. The last mode5 is

978-1-4799-2706-7/14 $31.00 © 2014 IEEE   210

operating when no switch is turned on, $d_5 = 1 - D - D_L$.

As discussed, in modes 1 and 2, the main switch S_2 is turned on (D), and the voltage on the magnetizing inductor L_m (V_{Lm}) is equal to V_s. In modes 3 and 4, V_{Lm} is equal to $V_S - V_{Cc}$. Using the voltage-second balance principle:

$$V_S D T_S + (V_S - V_{Cc}) D_L T_S + 0 \cdot (1 - D - D_L) T_S = 0 \tag{8}$$

To determine D_L, the ratio between d_3 and d_4 is required. Referring to Fig. 5, using the area of the triangle with $I_{m(MAX)}$ and d_3 equals the area of the triangle $I_{Do(MAX)}$ and $(d_3 + d_4)$, and obtains the ratio between d_4 and d_3:

$$I_{m(MAX)} \cdot d_3 = I_{Do(MAX)} \cdot (d_3 + d_4) \tag{9}$$

$$\frac{I_{Do(MAX)}}{\frac{I_{m(MAX)}}{n+1}} = \frac{d_4}{d_3 + d_4} \tag{10}$$

$$\frac{I_{m(MAX)}}{n+1} \cdot d_4 = I_{m(MAX)} \cdot d_3 \tag{11}$$

$$d_4 = (n+1) \cdot d_3 \tag{12}$$

$$d_4/d_3 = (n+1) \tag{13}$$

According to equations $I_{Do(avg)} = I_{O(avg)}$, $I_{m(MAX)} = (1/L_m) \cdot V_S \cdot D \cdot T_S$, and $\tau_m = L_m/R \cdot T_S$. We can get D_L:

$$D_L = 2 \cdot (n+2) \cdot M_{2stage} \cdot \tau_m \cdot (1/D) \tag{14}$$

Taking the results into (8), we can get:

$$2\tau_m M_{2stage}^2 - 2(n+2)\tau_m M_{2stage} - D^2 = 0 \tag{15}$$

The voltage conversion ratio of the whole converter operating at DCM can be obtained as:

$$M_{DCM} = (n+2) + \sqrt{(n+2)^2 + (2 \cdot D^2/\tau_m)} \tag{16}$$

Using the CCM and the DCM, the boundary condition curve of the proposed converter for the specified load condition is shown in Fig. 6.

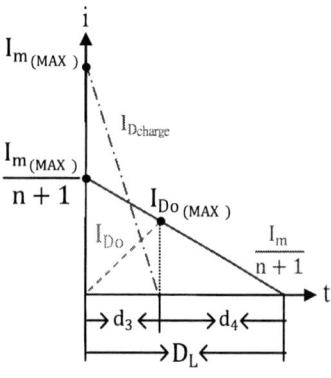

Fig. 5. The zoom in partial wave from modes 3 and 4

Since the voltage conversion ratio of CCM and DCM are both known, let (7) equal to (16):

$$\frac{2 \cdot (n+2)}{(1-D)} = (n+2) + \sqrt{(n+2)^2 + \frac{2 \cdot D^2}{\tau_m}} \tag{17}$$

$$\frac{(2+n) \cdot (1+D)}{(1-D)} = \sqrt{(n+2)^2 + \frac{2 \cdot D^2}{\tau_m}} \tag{18}$$

$$\frac{(2+n)^2 \cdot (1+D)^2}{(1-D)^2} = (n+2)^2 + \frac{2 \cdot D^2}{\tau_m} \tag{19}$$

$$\frac{(n+2)^2 \cdot 4 \cdot D}{(1-D)^2} = \frac{2 \cdot D^2}{\tau_m} \tag{20}$$

We can get the boundary condition in (21)

$$\tau_m = \frac{(1-D)^2 \cdot D}{2 \cdot (n+2)^2} \tag{21}$$

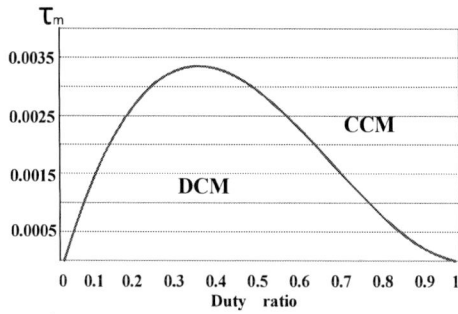

Fig. 6. Boundary condition of proposed converter

The proposed converter can achieve a much higher voltage conversion ratio from the selected multiple mode switched-capacitor and multiple charge pump cells in the conversion scheme. Fig. 7 shows the proposed high gain boost converter topologies with the m-mode switched-capacitor cells and single charge pump capacitor, the voltage conversion ratio is expressed in (22).

$$M_{SC}(D) = m \cdot (n+2)/(1-D) \tag{22}$$

Where m is the number of SCC.

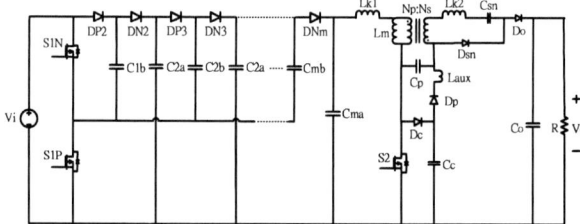

Fig. 7. The proposed converter with m-mode SCC

The extended proposed high gain boost converter topologies with the double-mode switched-capacitor cells and the Ncp multiple charge-pump capacitor cells is shown in Fig. 8, the voltage conversion ratio can be expressed as (23).

$$M_{CP}(D) = 2 \cdot (n + N_{CP} + 1)/(1-D) \tag{23}$$

Where Ncp is the number of charge pump capacitor cells. Fig. 9 (a) shows the voltage conversion ratio with different m and D for the proposed converter with m-mode SCC compared with the different high gain boost conversion schemes. The voltage conversion ratio of the proposed cascade boost converter is the highest compared with the high efficiency converter in [1], high step-up converter with coupled-inductor in [4], high step-up converter based on charge pump and boost converter in [5], and high gain boost converter with quasi-resonant switched capacitor and coupled inductor in [9].

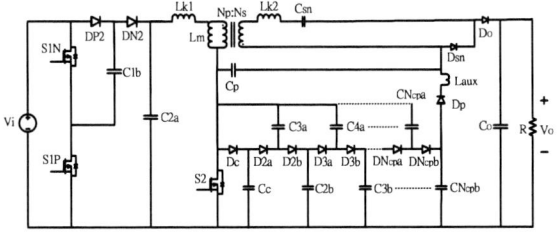

Fig. 8. The proposed converter with double-mode SCC and Ncp charge-pump capacitor cells

Fig. 9 (b) shows the voltage conversion ratio with different Ncp and D for the proposed converter with Ncp charge-pump capacitor cells under n = 2.7 and m = 1 in CCM operation.

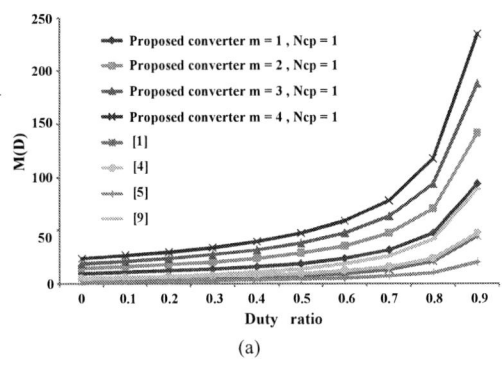

(a)

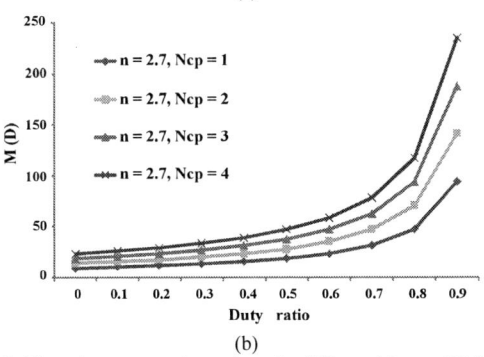

(b)

Fig. 9. The voltage conversion ratio under different (a) m and D for the proposed converter with m-mode SCC compared with different schemes (b) Ncp and D for the proposed converter with Ncp charge-pump capacitor cells

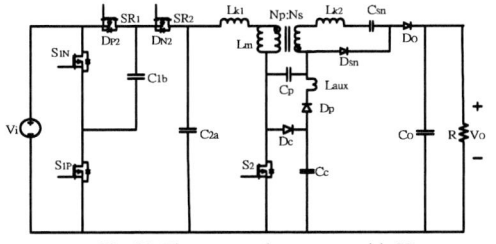

Fig. 10. The proposed converter with SR

Fig. 10 shows the proposed converter with synchronous rectification (SR), for achieving higher

efficiency, replacing diode D_{P2} and D_{N2} into MOSFETs, SR_1 and SR_2 to reduce the power losses of the converter. The efficiency of proposed converter with SR is higher than the converter without SR, the experimental results are shown in next section.

IV.   SIMULATION & EXPERIMENT

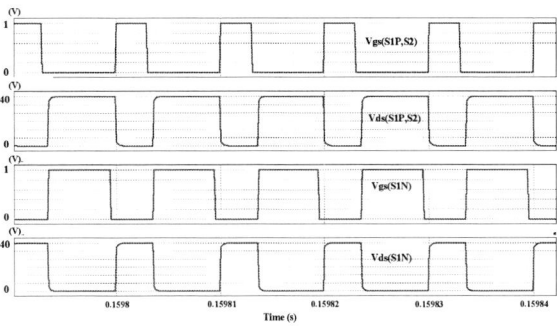

(a)

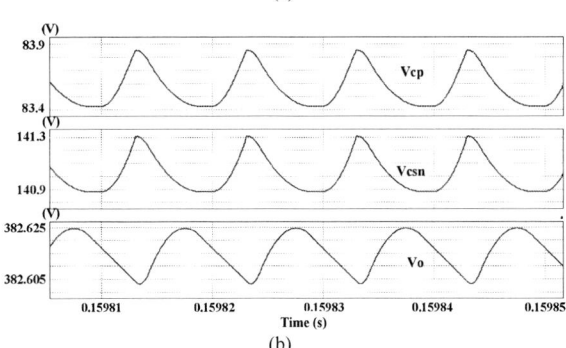

(b)

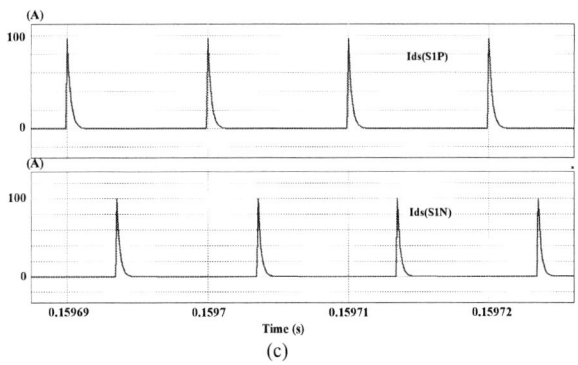

(c)

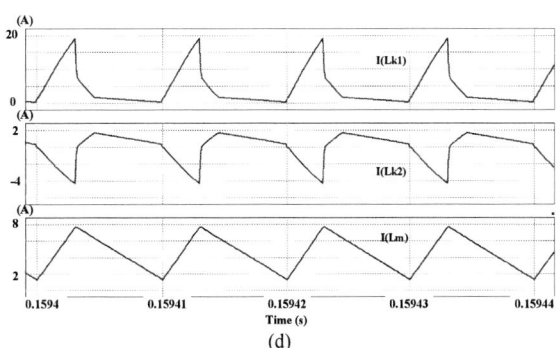

(d)

Fig. 11. Results of Simulation for the corresponding waveforms in CCM (a) Vgs(S1P,S2,S1N), Vds(S1P,S2,S1N) (b) Vcp, Vcsn, Vo (c) Ids(S1P), Ids(S1N) (d) ILk1, ILk2 , ILm

978-1-4799-2706-7/14 $31.00 © 2014 IEEE   212

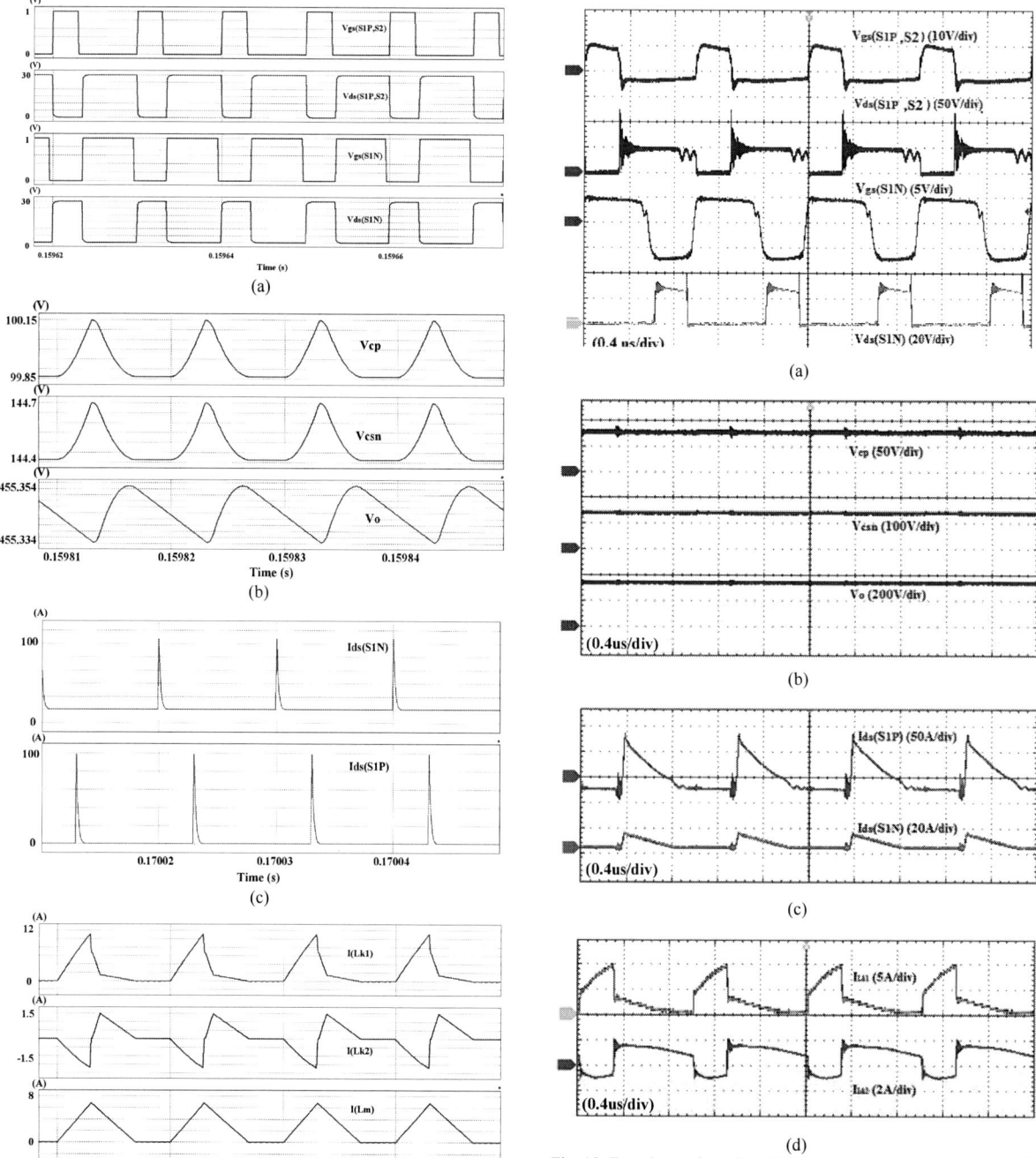

Fig. 12. Results of Simulation for the corresponding waveforms in DCM (a) Vgs(S1P,S2,S1N), Vds(S1P,S2,S1N) (b) Vcp, Vcsn, Vo (c) Ids(S1P), Ids(S1N) (d) ILk1, ILk2, ILm

Fig. 13. Experimental results of the corresponding waveforms in CCM (a) Vgs(S2), Vgs(S1P), Vds(S1P) (b) Vcp, Vcsn, Vo (c) Ids(S1P), Ids(S1N) (d) Ilk1, Ilk2

A prototype circuit was implemented to verify and validate the performance of the proposed high voltage gain converter with DC snubber circuit. The results of simulation and experiments were carried out for the proposed converter with the conversion ratio on the designed converter parameters : m = 2, Ncp = 1, and n = 2.7, which was shown as in Fig. 1. The duty cycles $D_{1P} = D_2 = 0.25$, $D_{1N} = 0.75$, were selected for the switches, S_{1P}, S_2, and S_{1N}, respectively.

The prototype converter had the following specifications: 30 (V) input voltage, 120 W output power (CCM) and 100 W output power (DCM), the duty cycle is 0.25, the switching frequency is selected at 100 kHz. The first stage switches S_{1N} and S_{1P} are IRF3410, the main switch S_2 is IRFB4227. Coupled inductor uses ETD-39 core, primary leakage inductor L_{k1} is 0.4 µH, secondary leakage inductor L_{k2} is 3.1 µH, magnetizing inductor L_m is 2.4 µH. The coupled inductor primary winding N_P is 10 turns, secondary winding Ns is 27 turns, The SCC capacitors are selected C_{1a} = 22 µF, C_{1b} = 10 µF, charge pump capacitors are the $C_c = C_p$ = 10 µF, capacitor in DC snubber C_{sn} = 10 μF, output capacitor C_o = 68 μF. The small auxiliary inductor L_{aux} = 734 nH, diode D_{P2}, D_{N2} and D_{sn} are SR4060, charge pump diode D_{cc}, D_{cp} and output diode D_o are STPS10H100CT. The system employs DSP dsPIC30F2020 to drive the converter main switches S_2, S_{1N}, and S_{1P}.

The results of simulation for $V_{ds(S1p)}$, V_{Cp}, V_{Csn}, V_o, $I_{ds(S1n)}$, $I_{ds(S1p)}$, I_{Lk1}, I_{Lk2}, and I_{Lm} of the proposed converter are in CCM shown in Fig. 11. The magnetizing current in Lm can be seen to operate at CCM shown in Fig. 13 (d). The results of simulation of proposed converter in DCM shown in Fig. 12. Fig. 13 (b) shows the magnetizing current operated at DCM. The simulation output voltage in CCM is 382 (V), in DCM is 454 (V).

Fig. 13 shows the experimental results of the corresponding waveforms in the same converter in CCM. The experimental results of the corresponding waveforms in DCM are shown in Fig. 14 . The experimental output voltage in CCM is 364(V), in DCM is 448(V). Due to the coupled coefficient kc and ESR voltage drop in the converter loop, the experimental output voltage is smaller than the simulation output voltage.

Because the magnetizing inductor current cannot be measured directly, we saved the physical measured data from the primary and secondary transformer current, then multiplied the secondary current with the transformer turn ratio, and added this to secondary current for depicted the magnetizting current of the couple inductor. We obtained the magnetizing inductor current as shown in Figs. 15 (a) and (b) in CCM and DCM, respectively. The primary side current spikes can be absorbed using an active snubber circuit installed on the lower voltage side of the coupled inductor [10].

The efficiencies of the proposed converter with and without SR under various loads are shown in Fig. 16. The maximum efficiency of the proposed converter without SR is 90% and the average efficiency is over 85% under the measured prototype. The maximum efficiency of the proposed converter with SR is over 91% and the average efficiency is over 85%. The hardware implemented photograph of the proposed converter with SR is shown in Fig. 17.

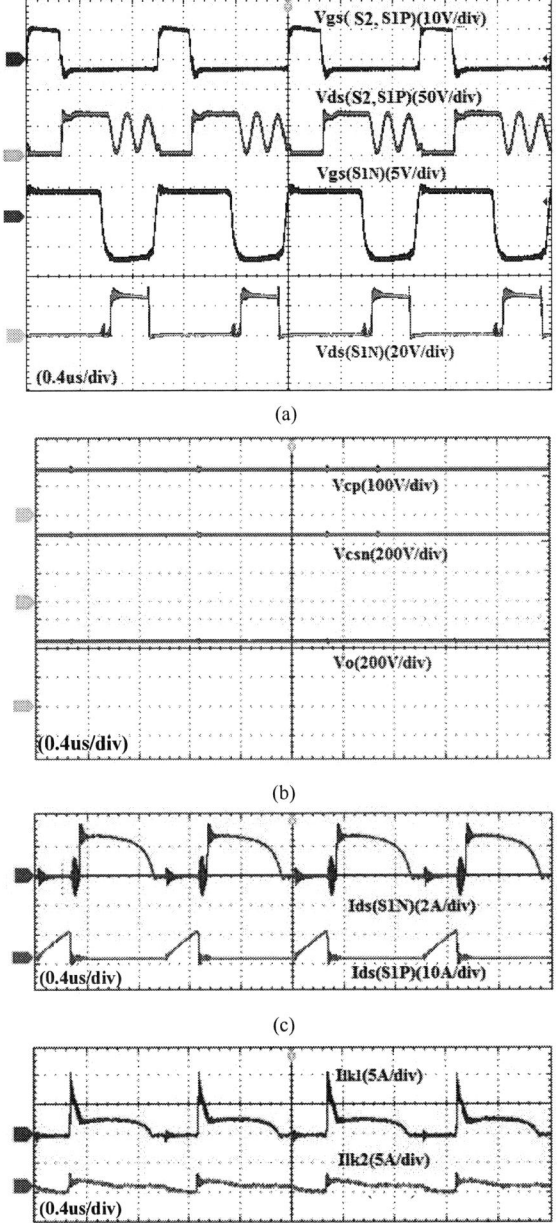

Fig. 14. Experimental results of the corresponding waveforms in DCM
(a) Vgs(S2), Vgs(S1P), Vds(S1P) (b) Vcp, Vcsn, Vo
(c) Ids(S1P), Ids(S1N) (d) Ilk1, Ilk2

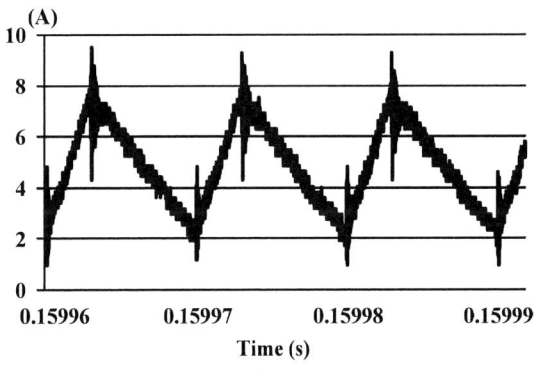

(a)

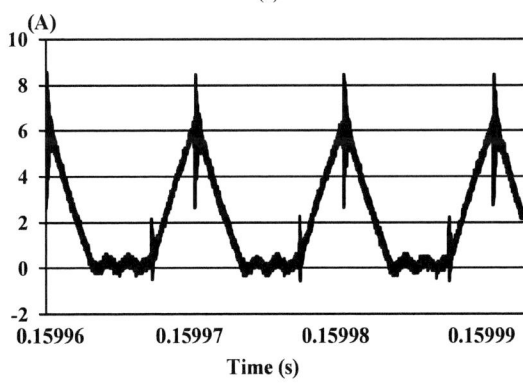

(b)

Fig. 15. Experimental results of the magnetizing current (a) Output power 200W (CCM) (b) Output power 88W (DCM)

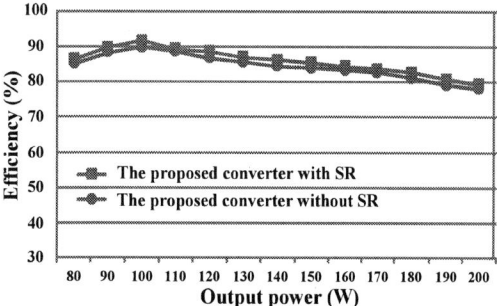

Fig. 16. Experimental results of converter efficiency

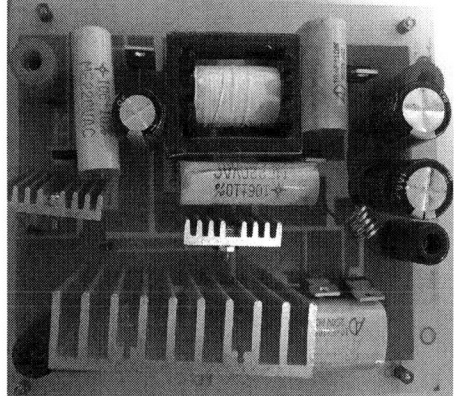

Fig. 17. PCB board of the proposed converter with SR

V. CONCLUSIONS

A switched-capacitor converter cascaded a coupled inductor converter with a DC snubber for high voltage conversion ratio boost converter was developed to operate in different conduction modes. The highest gain of the proposed converter in practical implementation was nearly 20. It has a high conversion ratio because of the DC snubber circuit and cascaded conversion structure. Only one duty cycle with a complementary duty cycle are needed to control the three switches. The proposed converter also can easily to obtain higher voltage conversion ratio by extend m-mode SCC and/or multiple charge-pump capacitor cells. And the proposed converter with SR can obtain a better efficiency. To achieve higher efficiency, ZCS turn-off of diodes and ZVS turn-on of power switches will be researched in the future study.

REFERENCES

[1] W. Yu, C. Hutchens, J. S. Lai, J. Zhang, G. Lisi, A. Djabbari, G. Smith, and T. Hegarty, "High efficiency converter with charge pump and coupled inductor for wide input photovoltaic AC module applications," *IEEE Energy Conversion Congress and Exposition (ECCE)*, pp. 3895-3900, 2009.

[2] Y. P. Ko, Y. S. Lee, and W. H. Chao, "Analysis, design and implementation of fuzzy logic controlled quasi-resonant zero-current switching switched-capacitor bidirectional converter," *IET Power Electronics*, vol. 6, no. 4, July 2011, pp.683-692.

[3] B. R. Lin and J. Y. Dong, "Analysis and implementation of an active clamping zero-voltage turn-on switching/zero-current turn-off switching converter," *IET Power Electronics*, vol.3, no. 3, pp.429-437, 2010.

[4] R. J. Wai and R. Y. Duan, "High step-up converter with coupled inductor," *IEEE Trans. on Power Electronics*, vol. 20, no.5, pp.1025-1035, 2005.

[5] K. I. Hwu and Y. T. Yau, "High step-up converter based on charge pump and boost converter," *International Power Electronics Conference (IPEC)*, pp.1038-1041, 2010.

[6] W. Li and X. He, "A family of interleaved dc/dc converter deduced from a basic cell with winding-cross-coupled inductor (WCCIs) for high step-up or step-down conversions," *IEEE Trans. on Power Electronics*, vol. 22, no. 4, pp. 1499-1507, 2008.

[7] W. Li and X. He, "Review of nonisolated high-step-up dc/dc converters in photovoltaic grid-connected applications," *IEEE Trans. on Industrial Electronics*, vol. 58, no. 4, pp. 1239-1250, 2011.

[8] N. Vazquez, L. Estrada, C. Hernandez, and E. Rodriguez, "The tapped-inductor boost converter," *IEEE International Symposium on Industrial Electronics (ISIE)*, pp.538-543, 2007.

[9] Y. S. Lee, S. S. Huang, and W. Y. Yang, "High gain boost converter with quasi-resonant switched capacitor and coupled inductor," *IEEE 7th International Power Electronics and Motion Control Conference - ECCE Asia*, pp.1817-1821, 2012.

[10] Y. S. Lee, S. S. Huang, and W. Y. Yang, "High conversion ratio converter with ZVS snubber circuit for PV charger application," *The International Conference of Power Electronics and Drive Systems (PEDS)*, pp.924-930, 2011.

[11] Y. P. B. Yeung, K. W. E. Cheng, D. Sutanto, and S. L. Ho, "Zero-current switching switched-capacitor quasi-resonant step-down converter," *IEE Proc. Electr. Power Appl.*, vol. 149, pp. 111-121, 2002

[12] O. Abutbul, A. Gherlitz, Y. Berkovich, and A. Ioinovici,"Step-up switching-mode converter with high voltage gain using a switched-capacitor circuit," *IEEE Trans. on Circuit and Systems-I: Fundamental Theory and Applications*, vol. 50, no. 8, pp.1098-1102, 2003.

[13] Y. S. Lee and Y. P. Ko, "Switched-capacitor bidirectional converter performance comparison with and without quasi-resonant zero-current switching," *IET Power Electronics*, vol. 3, iss. 2, pp. 269-278, 2010.

[14] R. J. Wai, C. Y. Lin, R. Y. Duan, and Y. R. Chang, "High-efficiency dc-dc converter with high voltage gain and reduced switch stress," *IEEE Trans. on Industrial Electronics*, vol. 54, no.1, pp. 354-364, 2007.

[15] K. C. Tseng and T. J. Liang, "Novel high-efficiency step-up converter," *Proc. Inst. Elect. Eng.*, vol. 151, pp. 182–190, 2004.

[16] Q. Zhao and F. C. Lee, "High-efficiency, high step-up dc–dc converters," *IEEE Trans. Power Electron.*, vol. 18, no. 1, pp. 65–73, Jan. 2003.

[17] B. Axelrod, Y. Berkovich, and A. Ioinovici, "Switched coupled-inductor cell for dc-dc converter swith very large conversion ratio," *IEEE IECON*, pp. 2366–2371, 2006.

[18] A. A. Fardoun and E. H. Ismail, "Ultra step-up dc–dc converter with reduced switch stress," *IEEE Trans. Ind. Appl.*, vol. 46, no. 5, pp. 2025–2034, Sep./Oct. 2010.

[19] K. K. Law, K. W. E. Cheng, and Y. P. Byeung, "Design and analysis of switched-capacitor-based step-up resonant converters, " *IEEE Trans. Circuits Syst.*, vol. 52, no. 5, pp. 943–948, 2005

[20] H. S. H. Chung and W. L. Cheung, "Generalized structure of bi-directional switched-capacitor dc/dc converter," *IEEE Trans. Circuits Syst.-I*, vol. 50, no. 6, pp. 743–753, 2003.

Design-Oriented Analysis of Resonance Damping and Harmonic Compensation for *LCL*-Filtered Voltage Source Converters

Xiongfei Wang, Frede Blaabjerg, and Poh Chiang Loh
Department of Energy Technology, Aalborg University, Aalborg, Denmark
xwa@et.aau.dk, fbl@et.aau.dk, pcl@et.aau.dk

Abstract— **This paper addresses the interaction between harmonic resonant controllers and active damping of *LCL* resonance in voltage source converters. A virtual series R-C damper in parallel with the filter capacitor is proposed with the capacitor current feedback loop. The phase lag resulting from the digital computation and modulation delays is thus compensated by the virtual capacitor. The frequency region that allows using resonant controllers is then identified with the virtual damper. It reveals that the upper frequency limit of harmonic resonant controllers can be increased above the gain crossover frequency defined by the proportional gain of current controller. This is of particular interest for high-performance active harmonic filtering applications and low-pulse-ratio converters. Case studies in experiments validate the theoretical analysis.**

Keywords— Harmonic compensation, resonant controller, active damping, LCL resonance

I. Introduction

LCL filters have widely been found in grid-connected voltage source converters, owing to their better switching ripple attenuation and smaller volumes compared to *L* filters [1]. But on the other hand, the *LCL* resonance characteristic challenges the current control of converters and worsens the sensitivity to grid voltage harmonics [2]. These issues may be exacerbated in paralleled converters with *LCL* filters such as in renewable power plants and microgrids, in which multiple resonance frequencies may be introduced challenging the system stability [3], [4]. Furthermore, with the increased penetration of nonlinear load, the harmonic distortion in the power grids is getting worse. The control of current harmonics is important for grid-connected converters to meet the grid codes and to furnish the active harmonic filtering functions [5].

The design of resonant current controllers for voltage source converters with *L* filters has been well developed, where the harmonics up to the Nyquist frequency can be controlled [6]-[8]. However, for *LCL*-filtered converters, the use of resonant controllers is challenged by the *LCL* resonance, and particularly for the harmonics close to the resonance frequency of the filter. Traditionally, the resonant controllers are merely employed below the gain crossover frequency of the current control loop which is

This work was supported by European Research Council (ERC) under the European Union's Seventh Framework Program (FP/2007-2013)/ERC Grant Agreement n. [321149-Harmony].

defined by the proportional gain [4], [9], [10]. Thus, the stability of resonant controllers can be decoupled from the effect of *LCL* resonance, whereas the frequency range of harmonic compensation is limited compared to the converters with *L* filters. This drawback may be more obvious for low-pulse-ratio converters in which the ratio of the switching frequency to the fundamental frequency is low [11], [12]. The variation of grid impedance is another challenge in this case, which may shift the gain crossover frequency in a wide range and may reduce the Phase Margin (PM) of the resonant controllers near the gain crossover frequency resulting in unstable harmonic compensation [10]. It is, therefore, of particular interest to broaden the frequency region for resonant controllers by means of *LCL* resonance damping techniques.

Extensive research has gone into the damping of *LCL* filter resonance for voltage source converters, which can be attained by either passive dampers or active damping control concept [13]-[15]. Among these approaches, the capacitor-current-based active damping control is found easy-to-use and effective [2], [15]-[20]. In this method, the proportional gain of the capacitor current feedback loop is basically equivalent to a virtual resistor in parallel with the filter capacitor [17]. But considering the effect of the digital computation and Pulse Width Modulation (PWM) delays, the virtual resistor turns into a virtual impedance. The imaginary part of this virtual impedance changes the actual *LCL* resonance frequency, while the real part may be negative depending on the ratio of *LCL* resonance frequency to the control frequency [18]. Such a negative virtual resistance brings the open-loop Right Half-Plane (RHP) poles into the current control loop resulting in a non-minimum phase behavior [19].

Several solutions have recently been reported to avoid the non-minimum phase system induced by the capacitor-current-based active damping. A direct way is to reduce the computation delay by shifting the sampling instant of the capacitor current [19]. It is easy to implement but is susceptible to the aliasing of capacitor current. Another attractive way is to predict the capacitor current by means of either the observer [20], or the discrete-time derivative controller [21]. A good robustness of resonance damping against the grid impedance variation is obtained in these works, yet few of them addresses the interaction of the active damping with resonant controllers and, particularly when the current harmonics close to the *LCL* resonance frequency need to be controlled.

This paper presents a design-oriented analysis for the interaction of the capacitor-current-based active damping and the harmonic resonant controllers. A virtual series R-C damper in parallel with the filter capacitor is proposed by means of a first-order high-pass filter in the capacitor current feedback loop. Thus, the effect of computation and PWM delays in the active damping can be mitigated by the virtual capacitor. Then, based on the virtual R-C damper, the frequency region that allows using harmonic resonant controllers is investigated. The effect of active damping control parameters on the stability of resonant controllers is analyzed. It shows that with a robust design of active damping, the resonant controller can be applied for the harmonics above the gain crossover frequency defined by the proportional gain of current control loop. The design constraint of the virtual series R-C damper is thus formulated. Case studies in experiments validate the theoretical analysis.

II. REVIEW OF CAPACITOR-CURRENT-BASED ACTIVE DAMPING AND HARMONIC RESONANT CONTROLLERS

A. System Description

Fig. 1 illustrates a three-phase grid-connected voltage source converter with the *LCL* filter. The DC-link voltage is assumed to be constant, and the bandwidth of the grid synchronization loop is designed to be lower than the grid fundamental frequency in order not to induce instabilities [22]. The grid current (i_2) is controlled synchronous with the Point of Common Coupling (PCC) voltage. The main parameters of the converter are listed in Table I, which are used in this study.

Fig. 2 depicts the per-phase diagram of the grid current control, since the three-phase systems without the neutral wire can be transformed into two decoupled single-phase systems in the stationary $\alpha\beta$-frame [23]. $G_c(s)$ denotes the proportional plus multi-resonant current controllers. $G_d(s)$ represents the effect of the digital computation and PWM delays in the continuous form, which are given by

$$G_c(s) = K_p + \frac{K_{i1}s}{s^2 + \omega_1^2} + \sum_{h=5,7,...n} K_{ih} \frac{s\cos(\theta_h) - h\omega_1\sin(\theta_h)}{s^2 + h^2\omega_1^2} \tag{1}$$

$$G_d(s) = e^{-1.5T_s s} \tag{2}$$

where ω_1 is the fundamental frequency of grid, T_s is the sampling period, and θ_h is the phase leading angle.

B. Capacitor-Current-Based Active Damping

The capacitor-current-based active damping is attained by a proportional feedback of the filter capacitor current, as depicted in Fig. 2. Thus, a double-loop current control structure is formed with the grid current feedback loop. The instability of the inner active damping loop results in a non-minimum phase behavior of the outer grid current control loop, which consequently imposes the constraint on the design of the current controller [16].

Fig. 3 illustrates the physical insight of the inner active

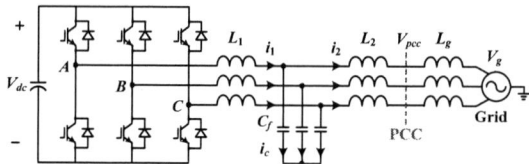

Fig. 1. Three-phase grid-connected voltage source converter with the *LCL* filter.

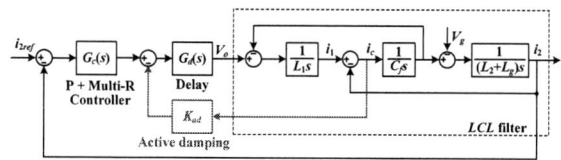

Fig. 2. Per-phase diagram of the grid current control with the capacitor-current-based active damping.

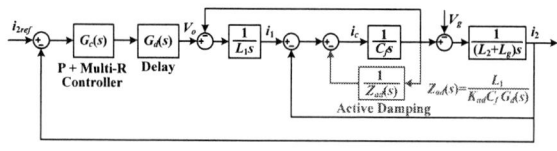

(a)

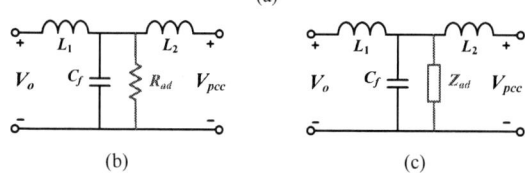

(b)   (c)

Fig. 3. Equivalent transformation of the capacitor-current-based active damping loop. (a) Block diagram representation. (b) Equivalent circuit of the *LCL* filter without the delay effect. (c) Equivalent circuit of the *LCL* filter with the delay effect.

TABLE I
MAIN PARAMETERS OF GRID-CONNECTED CONVERTER

Symbol	Meaning	Value
V_g	Grid voltage	400 V
f_1	Grid frequency	50 Hz
f_{sw}	Switching frequency	10 kHz
f_s	Sampling frequency	10 kHz
T_s	Sampling period	100 μs
V_{dc}	DC-link voltage	800 V
L_1	Filter inductor 1	3.6 mH
L_2	Filter inductor 2	1 mH
C_f	Filter capacitor	4.7 μF

damping loop by means of block diagram transformation. It is seen that the virtual resistor in parallel with the filter capacitor is basically synthesized by the proportional gain (K_{ad}), which is shown in Fig. 3 (b). However, considering the effect of computation and PWM delays, the virtual resistor turns into the virtual impedance, as shown in Fig. 3(c), which is given by [19]

$$Z_{ad}(j\omega) = \frac{L_1}{K_{ad}C_f}\left[\cos(1.5\omega T_s) + j\sin(1.5\omega T_s)\right] \tag{3}$$

Consequently, the actual *LCL* resonance frequency is changed by the imaginary part of virtual impedance. The real part of virtual impedance becomes negative when the

LCL resonance frequency is between the one-sixth of the control frequency ($\omega_s/6$) and the Nyquist frequency ($\omega_s/2$). The negative resistance brings the open-loop RHP poles into the outer grid current control loop.

It is worth noting that the frequency $\omega_s/6$ has been identified as the critical frequency in grid current control loop and the active damping is only needed for the *LCL* resonance frequency higher than $\omega_s/6$ [17]. However, the wide variation of grid impedance in weak grids shifts the *LCL* resonance frequency across $\omega_s/6$, which may result in unstable grid current control loop, when the resonance frequency is close to $\omega_s/6$.

C. Harmonic Resonant Controllers

The implementation of resonant current controllers has been well documented in [6]-[8]. The current harmonics up to the Nyquist frequency can be compensated with a proper discretization method and phase compensation.

Fig. 4 (a) depicts the basic structure of the resonant current controller based on two integrators. It consists of the forward and backward Euler integrators. The transfer function can be derived as

$$G_{Rh}(z) = K_{ih}T_s \frac{z^{-1}\left[\cos(\theta_h) - h\omega_1 T_s \sin(\theta_h)\right] - z^{-2}\cos(\theta_h)}{1 - 2z^{-1}(1 - h^2\omega_1^2 T_s^2/2) + z^{-2}} \tag{4}$$

This form of discretization is computationally efficient but with low accuracy resulting from the displacement of resonant poles and zeros. The term $(1 - h^2\omega_1^2 T_s^2/2)$ in the denominator of (4) is a second-order Taylor series of the trigonometric term $\cos(h\omega_1 T_s)$ required for the accurate placement of resonant poles [7]. The discrepancy leads to the frequency errors of resonant peaks, which are further exacerbated with the increase of T_s and $h\omega_1$ [8]. Further, compared to the resonant controller with accurate phase compensation, which is given by

$$G_{Rh}(z) = K_{ih}T_s \frac{\cos(\theta_h) - z^{-1}\cos(\theta_h - h\omega_1 T_s)}{1 - 2z^{-1}\cos(h\omega_1 T_s) + z^{-2}} \tag{5}$$

a correction of zeros is needed for the resonant controller with two integrators to obtain good performance.

Fig. 4 (b) shows the diagram of the resonant controller with the improved accuracy [8], which is adopted in this work for harmonic compensation. The sixth-order Taylor series of $\cos(h\omega_1 T_s)$ is used for the correction of resonant poles, and the phase lead input is modified to reduce the inaccuracies of zeros.

The improved resonant controller works well in the *L*-filtered converters, as the phase lead θ_h is readily derived. However, for the *LCL*-filtered converters, the calculation of phase lead is complicated by the phase change around the *LCL* resonance frequency. Also, the non-minimum phase behavior induced by the inner active damping loop limits the frequency range for stable resonant controllers. Hence, it is needed to identify the upper frequency limit of using harmonic resonant controllers in the *LCL*-filtered converters.

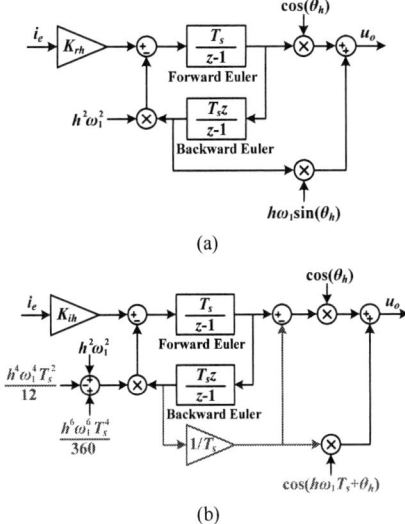

(a)

(b)

Fig. 4. Block diagrams of the resonant controller with two integrators. (a) Basic structure. (b) Improved structure.

III. PROPOSED VIRTUAL SERIES R-C DAMPER

A. Basic Principle

Fig. 5 shows the block diagram of the proposed active damping control scheme. Unlike the proportional gain in Fig. 2, the first-order high-pass filter is introduced in the capacitor current feedback loop.

First, in order to see the physical property of the high-pass filter, the effect of computation and PWM delays is neglected. Fig. 6 shows the equivalent circuit of the *LCL* filter with the high-pass filter. It is seen that the virtual series R-C damper in parallel with the filter capacitor is synthesized, which can be given by

$$R_{rc} = \frac{L_1}{K_{rc}C_f}, \quad C_{rc} = \frac{K_{rc}C_f}{L_1\omega_{rc}}, \quad \omega_{rc} = \frac{1}{R_{rc}C_{rc}} \tag{6}$$

where K_{rc} and ω_{rc} are the gain and cut-off frequency of the high-pass filter, respectively.

Then, considering the delay effect, the virtual damper is changed as the following virtual impedance

$$Z_{rc}(j\omega) = \frac{L_1}{K_{rc}C_f}(1 - j\frac{\omega_{rc}}{\omega})\left[\cos(1.5\omega T_s) + j\sin(1.5\omega T_s)\right] \tag{7}$$

Thus, the frequency at which the real part of the virtual impedance becomes negative (ω_{nr}) can be derived, which is plotted in Fig. 7 as a function of the cut-off frequency of the high-pass filter. It can be seen that at the zero cut-off frequency, $\omega_{nr} = \omega_s/6$. As the increase of the filter cut-off frequency, ω_{nr} is increased.

B. Comparative Analysis in Frequency-Domain

To illustrate the effect of the proposed active damping approach, a comparative analysis with the conventional capacitor-current-based active damping for the different grid inductances is presented in the following. Table II summarizes the main controller parameters.

Fig. 8 shows the frequency response of the open-loop

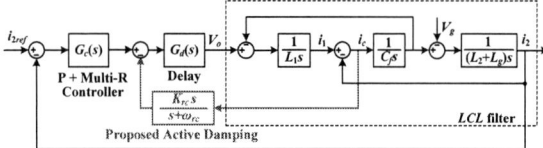

Fig. 5. Block diagram of the proposed active damping control scheme.

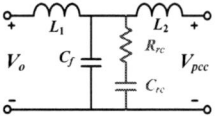

Fig. 6. Equivalent circuit of the *LCL* filter with the high-pass filter.

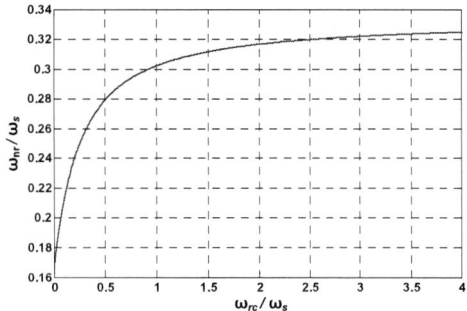

Fig. 7. Relationship between the frequency where the virtual resistance becomes negative and the cut-off frequency of high-pass filter.

TABLE II
MAIN CONTROLLER PARAMETERS

Symbol	Meaning	Value
K_p	Proportional gain	20
K_{i1}	Fundamental frequency resonant controller gain	1000
K_{ih}	Harmonic resonant controllers gains	1000
K_{rc}	High-pass filter gain	15
ω_{rc}	Cut-off frequency of high-pass filter	$0.2\omega_s$ rad/s

gain of the grid current control loop with the proportional current controller only. The proportional gain is designed based on the *LCL* filter inductors (L_1+L_2) with the PM of 45° [16]. It can be seen that the *LCL* resonance frequency is reduced as the increase of grid inductance, and the current control loop turns unstable when the resonance frequency is below $\omega_s/6$.

Fig. 9 shows the frequency response of the open-loop gain of grid current control loop with the conventional capacitor-current-based active damping control. The non-minimum phase behavior induced by the negative virtual resistance is observed. Instead of the resonance damping, the resonant peaks are shifted by the virtual impedance. Based on the Nyquist stability criterion, it is found that the system are stable for the cases of L_g=0 mH and L_g=9 mH, and are marginally stable for the case of L_g=4.5 mH. This is because the resonance frequency in this case is closer to the frequency for negative virtual resistance ω_{nr}.

Fig. 10 shows the frequency response of the open-loop gain of grid current control loop with the proposed virtual series R-C damper. The non-minimum phase behavior is mitigated by the virtual capacitor. The resonant peaks are

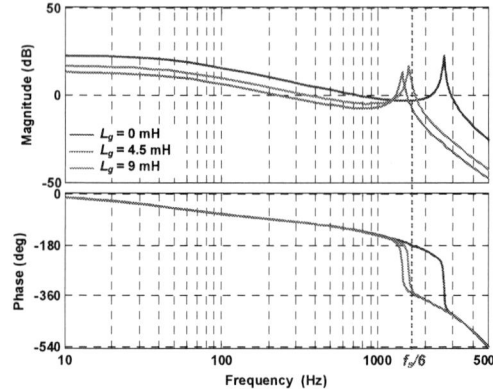

Fig. 8. Frequency response of the open-loop gain of the grid current control with the proportional current controller only.

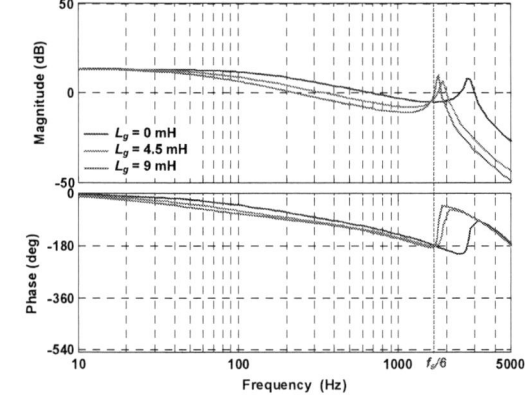

Fig. 9. Frequency response of the open-loop gain of the grid current control with the conventional capacitor-current-based active damping.

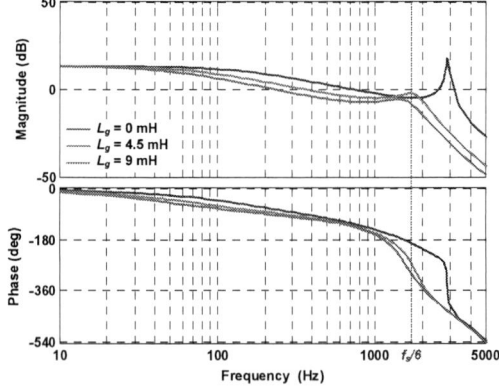

Fig. 10. Frequency response of the open-loop gain of the grid current control with the proposed virtual series R-C damper.

well dampened for the cases of L_g=4.5 mH and L_g=9 mH, while the resonant peak is shifted in the case of L_g=0 mH. This is because the frequency ω_{nr} is shifted closer to the resonance frequency in the case of L_g=0 mH.

IV. DESIGN-ORIENTED ANALYSIS

A. Design of Virtual Series R-C Damper

In order to achieve a robust design of the virtual series R-C damper against the grid impedance variation, a root locus analysis in the *z*-domain is presented as follows:

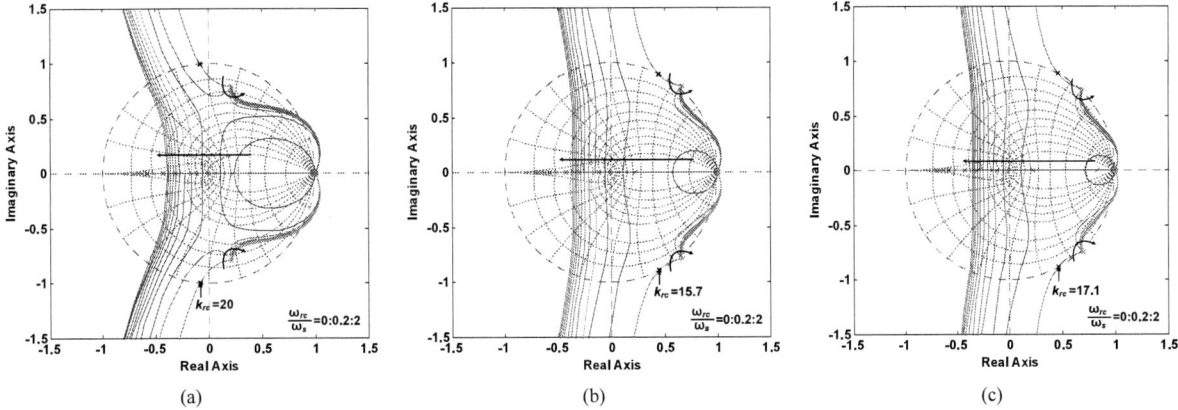

Fig. 11. Root loci of the outer grid current control loop with the different grid inductance (L_g) in the discrete-time z-domain. (a) L_g=0 mH, f_{res}=2.6 kHz. (b) L_g=4.5 mH, f_{res}=1.57 kHz. (c) L_g=9 mH, f_{res}=1.42 kHz.

The Zero-Order-Hold (ZOH) transform is used for the transfer function from the inverter output voltage to filter capacitor current, and the impulse invariant transform is applied for the transfer function of filter capacitor current to grid current [16], which are given by

$$G_{vc}(s) = \left.\frac{i_c}{V_o}\right| = \frac{1}{L_1}\frac{s}{\left(s^2 + \omega_{res}^2\right)}, \quad G_{cg}(s) = \frac{i_g}{i_c} = \frac{1}{L_{to}C_f s^2}$$

(8)

$$G_{vc}(z) = \frac{\sin(\omega_{res}T_s)}{\omega_{res}L_1}\frac{(z-1)}{\left[z^2 - 2z\cos(\omega_{res}T_s)+1\right]}$$

(9)

$$G_{cg}(z) = \frac{T_s z}{\left(L_2 + L_g\right)C_f\left(z-1\right)^2}$$

(10)

where ω_{res} is the LCL resonance frequency. The high-pass filter is discretized by the Tustin transform.

Fig. 11 depicts the effect of the high-pass filter gain K_{rc} on the closed-loop poles trajectory of the grid current control loop, and the different cut-off frequencies of the high-pass filter are swept with different grid inductances.

Fig. 11 (a) shows that the poles track well inside the unit circle for L_g=0 mH, owing to the inherent damping effect of the computation and PWM delays. In contrast, in Figs. 11 (b) and (c), the poles are initially outside the unit circle and track back inside with the increase of K_{rc}. This is because the resonance frequencies in these cases are lower than $\omega_s/6$. The stable region of K_{rc} is reduced as the resonance frequency becomes closer to the frequency for the virtual resistance becoming negative (ω_{nr}). Also, Figs. 11 (b) and (c) show that the cases with resonance frequency close to ω_{nr} has less damping.

In all three cases, the increase of the cut-off frequency ω_{rc} forces the poles to track inside the unit circle, and the stable region of K_{rc} is broadened. However, the change of root loci becomes small for high ω_{rc}. This agrees with the analysis in Fig. 7 where the ω_{nr} has few change for high ω_{rc}. Hence, to design the virtual series R-C damper, the frequency ω_{rc} can first be determined from Fig. 7 in order

that the frequency ω_{nr} is well separated from the range of LCL resonance frequency. Then the root locus analysis in Fig. 11 can be used to find the high-pass filter gain.

B. Frequency Region of Harmonic Control

From Fig. 8, it can be observed that the gain crossover frequency is significantly reduced as the increase of grid inductance. The frequency region of the harmonic control is limited if resonant controllers are only applied below the gain crossover frequency. Consequently, the resonant controllers with proper phase compensation are needed to broaden the frequency region of harmonic control.

Basically, the LCL resonance frequency imposes the upper frequency limit of resonant controllers, due to the sharp phase transition at the LCL resonance frequency. The phase leading angle of the resonant controllers below the LCL resonance frequency can be approximated as

$$\theta_h = \frac{\pi}{2} + 1.5h\omega_1 T_s$$

(11)

However, from Fig. 10, it is seen that the virtual series R-C damper brings in phase change for the frequencies near the LCL resonance. Using (11) may cause the inaccurate phase compensation of resonant controllers.

Fig. 12 depicts the effect of the high-pass filter on the phase change around the LCL resonance frequency in the case of L_g=9 mH. Fig. 12 (a) depicts the effect of K_{rc} for the case of ω_{rc}=0.2ω_s. It is seen that the phase lag caused by the high-pass filter is increased with K_{rc}. Fig. 12 (b) depicts the case of ω_{rc}=3ω_s. Compared to Fig. 12 (a), it is clear that the increase of ω_{rc} requires an increase of K_{rc} in order to obtain the same level of damping. The phase lag is further increased for the same level of damping.

Fig. 13 depicts such phase lag effect on the stability of harmonic resonant controller. Two high-pass filters with the same level of damping but the different ω_{rc} and K_{rc} are compared. The phase leading angles for the resonant controllers are calculated by (11). It can be seen that the phase lag resulting from the increase of ω_{rc} reduces the PM of the high-order harmonic resonant controllers (23rd) and even result in instability (25th). Hence, apart from the presence of the negative virtual resistance, the stability of

978-1-4799-2706-7/14 $31.00 © 2014 IEEE

The 2014 International Power Electronics Conference

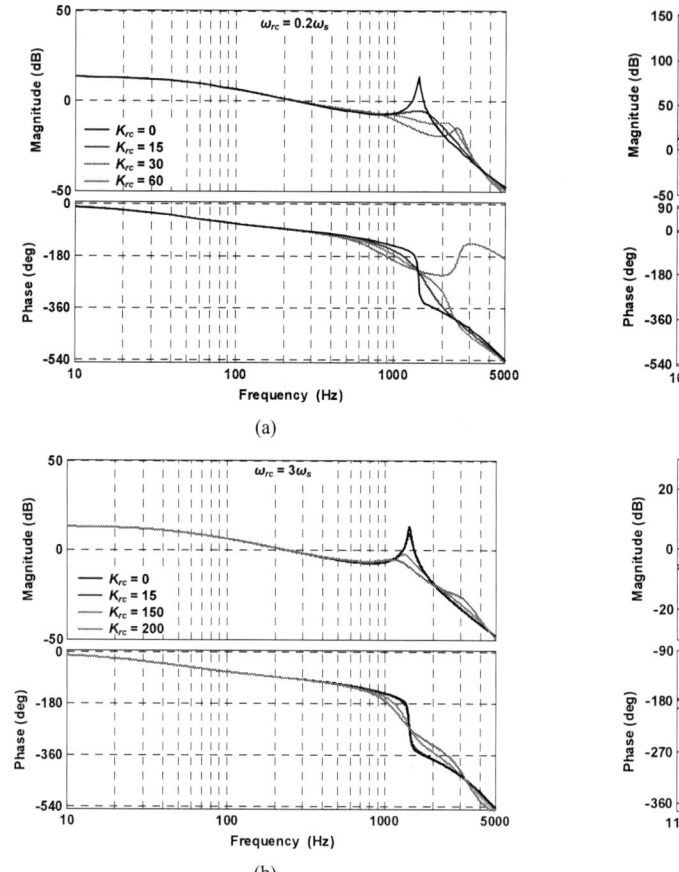

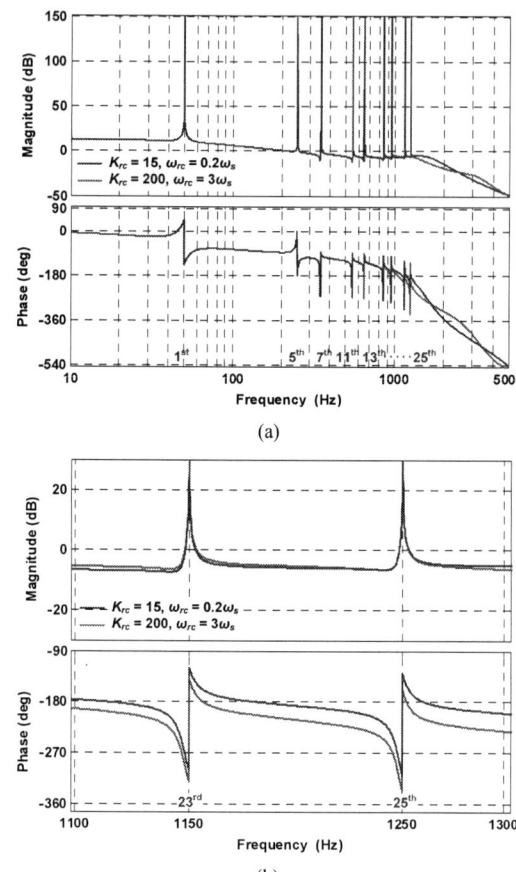

Fig. 12. Effect of high-pass filters on the phase change close to the *LCL* resonance frequency. (a) Low cut-off frequency (ω_{rc}=0.2ω_s). (b) High cut-off frequency (ω_{rc}=3ω_s).

Fig. 13. Comparison between the effect of phase lag resulting from the different high-pass filter design. (a) Full view. (b) Zoomed out for the 23rd and 25th harmonic resonant controllers.

resonant controllers for the harmonics near the resonance frequency imposes the upper limit on the value of the cut-off frequency of high-pass filter.

V. EXPERIMENTAL RESULTS

To validate the theoretical analysis, the experimental case studies are carried out based on the parameters listed in Table I and II. The California Instruments MX-series AC power source is used for grid emulation. The Danfoss frequency converter powered by the constant DC voltage source is controlled as the grid-connected voltage source converter. The control system is implemented in DS1006 dSPACE system, where the DS5101 is used for the PWM generation in synchronous with the Analog/Digital (A/D) sampling circuit.

A. Active Damping of LCL resonance

Fig. 14 shows the measured A-phase PCC voltage and grid current without using the active damping control. It can be seen that the system becomes unstable in the cases with LCL resonance frequencies below ω_s/6. This agrees well with the frequency-domain analysis in Fig. 8.

Figs. 15 and 16 compare the conventional capacitor-current-based active damping and the proposed virtual series R-C damper. The step response of grid current is tested. The measured results with the conventional active

damping method is shown in Fig. 15. The current control keeps stable in the cases of L_g=0 mH and L_g=9 mH, but it is unstable for L_g=4.5 mH. Furthermore, the transient oscillation in Fig. 5 (c) indicates the less damping in the case of L_g=9 mH. Fig. 16 shows the measured results with the proposed scheme. The system keeps stable in all the cases. The transient performance of the step response is also improved with more damping. These tests confirm both the frequency-domain analysis in Figs. 9 and 10, and the z-domain root locus analysis in Fig. 11.

B. Harmonic Compensation

To evaluate the effect of the proposed active damping on harmonic resonant controllers, the square waveform is generated from the AC power source and the grid current is controlled to be sinusoidal. Two different designs of the high-pass filters with the parameters given in Fig. 13 are compared by the harmonic spectra of grid current.

Fig. 17 shows the measured PCC voltage and grid current with low cut-off frequency of the high-pass filter, ω_{rc}=0.2ω_s. Fig. 17 (a) show the test results in the case of L_g=0 mH, in which up to the 43rd harmonics can be controlled by the resonant controllers with the phase leading angle derived by (1). However, as the increase of L_g, the decrease of the *LCL* resonance frequency reduces the frequency region of the harmonic compensation. Only

978-1-4799-2706-7/14 $31.00 © 2014 IEEE

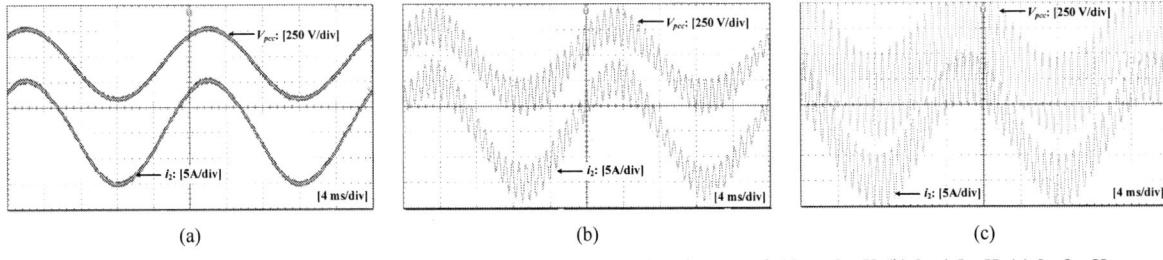

Fig. 14. Measured A-phase PCC voltage and grid current without using the active damping control. (a) L_g=0 mH. (b) L_g=4.5 mH. (c) L_g=9 mH.

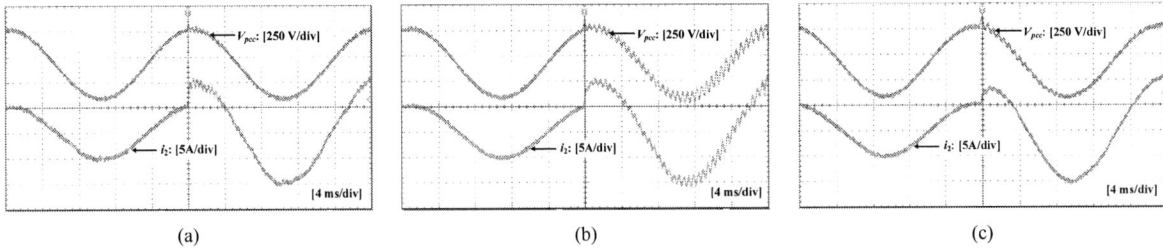

Fig. 15. Measured A-phase PCC voltage and grid current with the conventional active damping control. (a) L_g=0 mH. (b) L_g=4.5 mH. (c) L_g=9 mH.

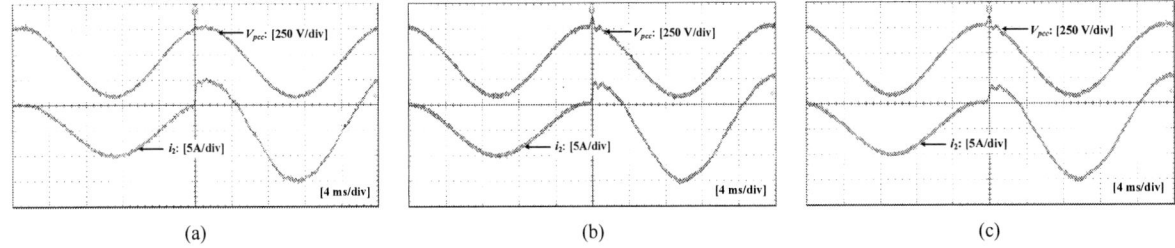

Fig. 16. Measured A-phase PCC voltage and grid current with the proposed active damping control. (a) L_g=0 mH. (b) L_g=4.5 mH. (c) L_g=9 mH.

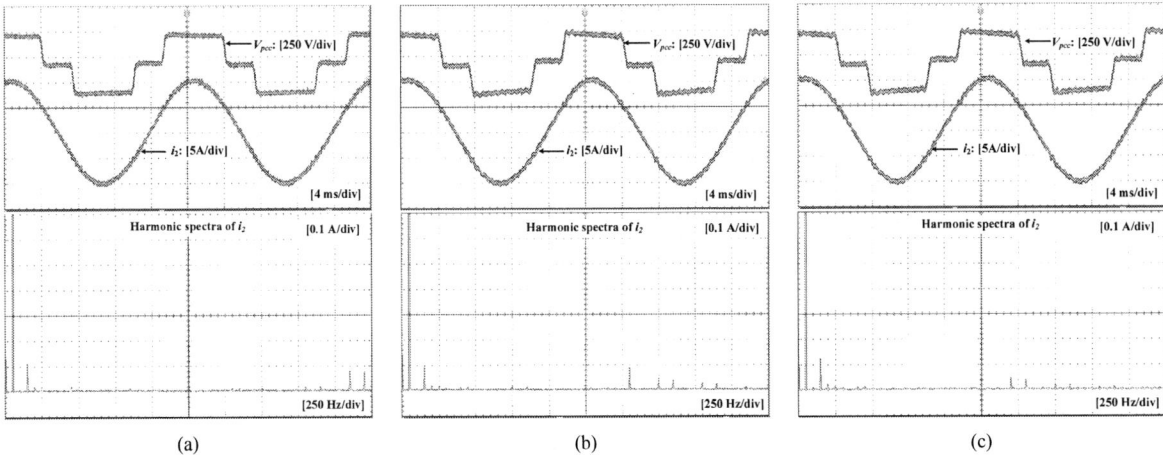

Fig. 17. Measured A-phase PCC voltage and grid current with the low cut-off frequency of high-pass filter (ω_{rc}=0.2ω_s) for harmonic compensation. (a) L_g=0 mH. (b) L_g=4.5 mH. (c) L_g=9 mH.

the harmonics up to 29th and 25th are controlled in Fig. 17 (b) and (c), respectively.

Fig. 18 shows the measured waveforms with the high cut-off frequency of the high-pass filter, ω_{rc}=3ω_s. The phase lag effect resulting from the increase of the cut-off frequency of high-pass filter can be clearly observed in Figs. 18 (b) and (c). Compared to Figs. 17 (b) and (c), it is seen that the 29th harmonics in Fig. 18 (b) and the 25th harmonics in Fig. 18 (c) are not compensated. The results in Fig. 17 (c) and Fig. 18 (c) confirm the analysis result

in Fig. 13. In contrast, the small discrepancy between Fig. 17 (a) and Fig. 18 (a) indicates that the increase of grid inductance exacerbates the phase lag effect caused by the high-pass filter.

VI. CONCLUSIONS

This paper has discussed the stability of the capacitor-current-based active damping and resonant controllers for current harmonics control. A virtual series R-C damper in parallel with the filter capacitor has been proposed, which

978-1-4799-2706-7/14 $31.00 © 2014 IEEE

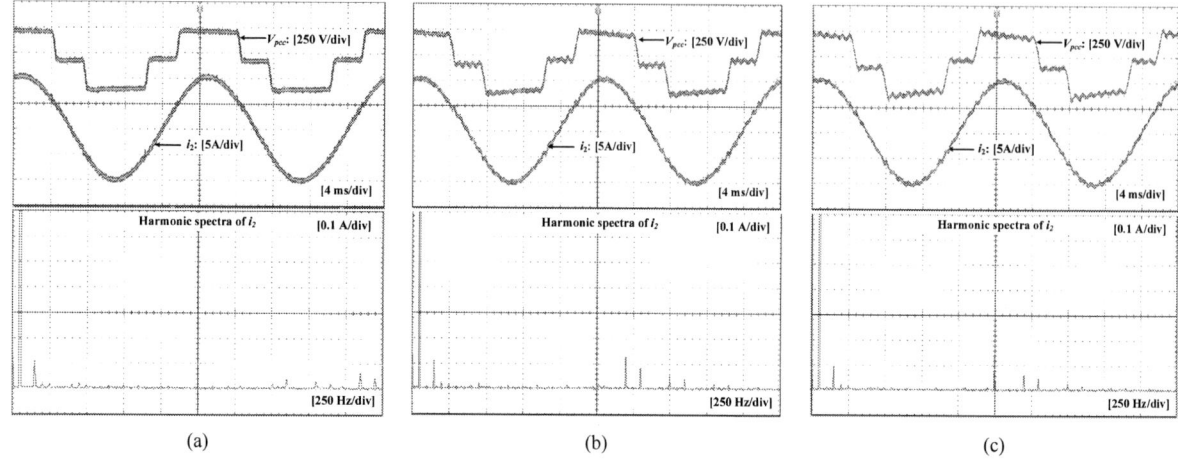

Fig. 18. Measured A-phase PCC voltage and grid current with the high cut-off frequency of high-pass filter ($\omega_{rc}=3\omega_s$) for harmonic compensation. (a) L_g=0 mH. (b) L_g=4.5 mH. (c) L_g=9 mH.

is realized by means of a first-order high-pass filter in the capacitor current feedback loop. The instability effect of computation and PWM delays in conventional capacitor current based active damping is addressed. A design-oriented analysis of the proposed active damping concept has been presented considering the interaction with the high-order harmonic resonant controllers under the wide variation of grid impedance. The design constraints of the high-pass filter have been analyzed by the z-domain root locus analysis and in the continuous frequency-domain. The experimental case studies validated the theoretical analysis and the performance of the proposed approach.

REFERENCES

[1] M. Liserre, F. Blaabjerg, and S. Hansen, "Design and control of an *LCL*-filter-based three-phase active rectifiers," *IEEE Trans. Ind. Appl.*, vol. 41, no. 5, pp. 1281-1291, Sept./Oct. 2005.

[2] E. Twining and D. G. Holmes, "Grid current regulation of a three-phase voltage source inverter with an LCL input filter," *IEEE Trans. Power Electron.*, vol. 18, no. 3, pp. 888-895, May 2003.

[3] X. Wang, F. Blaabjerg, and W. Wu, "Modeling and analysis of harmonic stability in an AC power-electronics-based power system," *IEEE Trans. Power Electron.*, in press, Feb. 2014.

[4] X. Wang. F. Blaabjerg, M. Liserre, Z. Chen, J. He, and Y. W. Li, "An active damper for stabilizing power-electronics-based AC systems," *IEEE Trans. Power Electron.*, vol. 29, no. 7, pp. 3318-3329, Jul. 2014.

[5] X. Wang, F. Blaabjerg, and Z. Chen, "Synthesis of variable harmonic impedance in inverter-interfaced distributed generation unit for harmonic damping throughout a distribution network," *IEEE Trans. Ind. Appl.*, vol. 48, no. 4, pp. 1407-1417, Jul./Aug. 2012.

[6] A. G. Yepes, F. Freijedo, J. Gandoy, O. Lopez, J. Malvar, and P. Comesana, "Effects of discretization methods on the performance of resonant controllers," *IEEE Trans. Power Electron.*, vol. 25, no. 7, pp. 1692-1712, Jul. 2010.

[7] A. G. Yepes, F. D. Freijedo, O. Lopez, and J. D. Gandoy, "Analysis and design of resonant current controllers for voltage-source converters by means of Nyquist diagrams and sensitive function," *IEEE Trans. Ind. Electron.*, vol. 58, no. 11, pp. 5231-5250, Nov. 2011.

[8] A. G. Yepes, F. Freijedo, O. Lopez, and J. Gandoy, "High-performance digital resonant controllers implemented with two integrators," *IEEE Trans. Power Electron.*, vol. 26, no. 2, pp. 563-576, Feb. 2011.

[9] F. Wang, J. L. Duarte, M. A. M. Hendrix, and P. F. Ribeiro, "Modeling and analysis of grid harmonic distortion impact of aggregated DG inverters," *IEEE Trans. Power Electron.*, vol. 26, no. 3, pp. 786-797, Mar. 2011.

[10] M. Liserre, R. Teodorescu, and F. Blaabjerg, "Stability of photovoltaic and wind turbine grid-connected inverters for a large set of grid impedance values," *IEEE Trans. Power Electron.*, vol. 21, no. 1, pp. 263-272, Jan. 2006.

[11] Z. Li, Y. Li, P. Wang, H. Zhu, C. Liu, and F. Gao, "Single-loop digital control of high-power 400-Hz ground power unit for airplanes," *IEEE Trans. Ind. Electron.*, vol. 57, no. 2, pp. 532-543, Feb. 2010.

[12] R. Venturini, P. Mattavelli, P. Zanchetta, M. Sumner, "Adaptive selective compensation for variable frequency active power filters in more electrical aircraft," *IEEE Trans. Aero. Electron. Syst.*, vol. 48, no. 2, pp. 1319-1328, Apr. 2012.

[13] R. N. Beres, X. Wang, F. Blaabjerg, C. L. Bak, and M. Liserre, "A review of passive filters for grid-connected voltage source converters," in *Proc. IEEE APEC* 2014, in press.

[14] J. Dannehl, M. Liserre, and F. W. Fuchs, "Filter-based active damping of voltage source converters with *LCL* filter," *IEEE Trans. Ind. Electron.*, vol. 58, no. 8, pp. 3623-3633, Aug. 2011.

[15] J. Dannehl, F. W. Fuchs, S. Hansen, and P. B. Thogersen, "Investigation of active damping approaches for PI-based current control of grid-connected pulse width modulation converters with LCL filters," *IEEE Trans. Ind. Appl.*, vol. 46, no. 4, pp. 1509-1517, Jul./Aug. 2010.

[16] S. Parker, B. P. McGrath, and D. G. Holmes, "Region of active damping control for LCL filters," *IEEE Trans. Ind. Appl.*, vol. 50, no. 1, pp. 424-432, Jan./Feb. 2014.

[17] Y. Lei, Z. Zhao, F. He, S. Lu, and L. Yin, "An improved virtual resistance damping method for grid-connected inverters with LCL filters," in *Proc. IEEE ECCE* 2011, pp. 3816-3822.

[18] C. Bao, X. Ruan, X. Wang, W. Li, D. Pan, and K. Weng, "Step-by-step controller design for LCL-type grid-connected inverter with capacitor-current-feedback active-damping," *IEEE Trans. Power Electron.*, vol. 29, no. 3, pp. 1239-1253, Mar. 2014.

[19] D. Pan, X. Ruan, C. Bao, W. Li, and X. Wang, "Capacitor-current-feedback active damping with reduced computation delay for improving robustness of *LCL*-type grid-connected inverter," *IEEE Trans. Power Electron.*, vol. 29, no. 7, pp. 3414-3427, Jul. 2014.

[20] V. Miskovic, V. Blasko, T. Jahns, A. Smith, C. Romenesko, "Observer based active damping of LCL resonance in grid connected voltage source converters" in *Proc. IEEE ECCE* 2013, pp. 4850-4856.

[21] M. Wagner, T. Barth, C. Ditmanson, R. Alvarez, and S. Bernet, "Discrete-time optimal active damping of LCL resonance in grid connected converters by proportional capacitor current feedback," in *Proc. IEEE ECCE* 2013, pp. 721-727.

[22] L. Harnefors, M. Bongiorno, and S. Lundberg, "Input-admittance calculation and shaping for controlled voltage-source converters," *IEEE Trans. Ind. Electron.*, vol. 54, no. 6, pp. 3323-3334, Dec. 2007.

[23] S. Buso and P. Mattavelli, *Digital Control in Power Electronics*, San Francisco, CA: Morgan & Claypool Publ., 2006.

State-Space Average Modeling of Bidirectional DC-DC Converter for Battery Charger Using LCLC Filter

Sang-Ho Moon
Department of Electrical and
Computer Engineering.
Ajou University, Suwon, Korea
E-mail: skythefly3@ajou.ac.kr

Sung-Tak Jou
Department of Electrical and
Computer Engineering.
Ajou University, Suwon, Korea
E-mail: jst@ajou.ac.kr

Kyo-Beum Lee
Department of Electrical and
Computer Engineering.
Ajou University, Suwon, Korea
E-mail: kyl@ajou.ac.kr

Abstract— In this paper, a controller design is introduced for a interleaved dc-dc converter with an LCLC-filter. In order to design an accurate controller of the bidirectional converter, a new model is proposed. First, it introduces the bidirectional converter small-signal model using a state-space averaging method and derives the required transfer functions. Second, transfer functions of the bidirectional converter are derived by using the state-space averaging method. Next, the frequency characteristics of the converter are discussed. Finally, the simulation results are analyzed to verify the proposed state-space model of the bidirectional converter

Keywords— *Bidirectional, DC-DC converter, Battery charger, LCLC filter, State-space averaging*

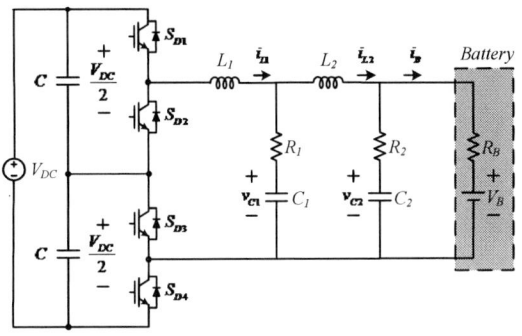

Fig. 1. The bidirectional DC-DC converter system

I. INTRODUCTION

A bidirectional DC-DC converter system has been widely used in hybrid electric vehicle and fuel cell applications [1]. A battery charge-discharge system, which is used in the bidirectional dc-dc converter and battery. Directions of the charge and the discharge current are changed from time to time to perform the power conversion. Therefore, the dynamic analysis and control design of the DC-DC converter is very important to increase the performance of the converter system [2].

In the proposed paper, a control model using the state-space averaging method is designed. The state-space averaging method makes use of this description to derive the small-signal averaged equation of pulse-width modulation (PWM) switching converters. The advantage of the state space average method is that the results can be generalized. In other words, given the state equation of the original form, small-signal average model can be obtained [3-4].

To satisfy the required current ripple of the battery (5 mA) and response time (10 ms), this paper presents an LCLC filter design and a dual current proportional–integral (PI)-based control. The dual current controller has faster response characteristics than the single-loop control.

This paper presents the performance of a bidirectional DC–DC converter using an LCLC filter for high-voltage low-current battery applications.

II. CIRCUIT TOPOLOGY

A. Interleaved model

The bidirectional DC–DC converter is composed of a high DC-link input voltage and low output voltage of the battery. The converter allows the battery to charge–discharge in two directions to deliver power and to provide a faster response to the output. A multi-level DC–DC converter with a small voltage-unbalanced condition is used in this study. The battery charge–discharge system uses the interleaved method to reduce the ripple of the input current between the battery and the converter and to increase the efficiency. In addition, the capacity and size of the converter are reduced [5]. The proposed bidirectional DC–DC converter system is shown in Fig. 1. The output side consists of the LCLC filter and the battery. The LCLC filter is composed of outer inductor L_1, outer capacitor C_1, inner inductor L_2, and inner capacitor C_2. The internal resistance of the battery and that of the capacitors must consider the current ripple that flows through the inner capacitor to reduce the current ripple that flows through the battery. $RC1$, $RC2$, and RB are the internal resistances of the outer capacitor, inner capacitor, and battery, respectively. The internal resistance of the battery changes according to the state of charge–discharge, but it is modeled as an average value. The bidirectional DC– DC converter operation is shown in Fig. 2.

978-1-4799-2706-7/14 $31.00 © 2014 IEEE

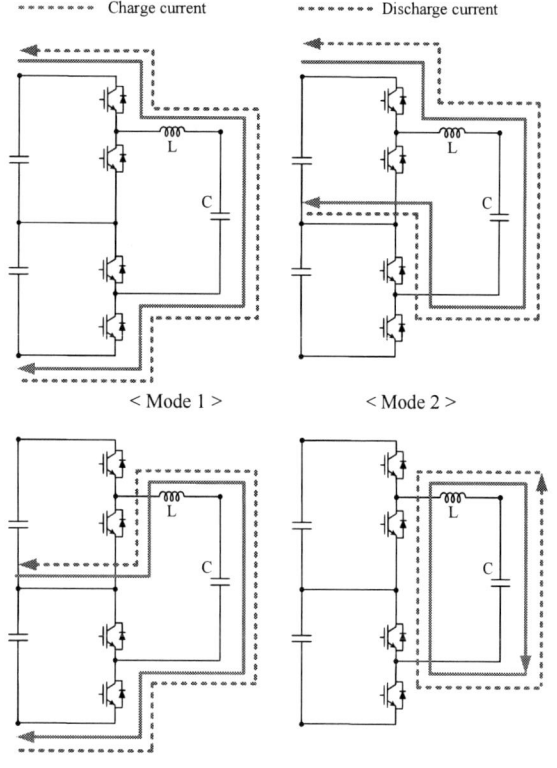

<Mode 1 >   < Mode 2 >

<Mode 3 >   < Mode 4 >
Fig. 2. Converter operating mode.

TABLE I
LCLC-FILTER PARAMETER

Symbol	Parameters	Value
L_1	Outer inductor	6.6 [mH]
L_2	Inner inductor	120 [μH]
C_1	Outer capacitor	2.17 [mF]
C_2	Inner capacitor	470 [μF]

B. LCLC-filter design

The filter characteristics are related to the internal resistance of the output capacitor. If the size of the output capacitor of the filter is increased, the internal resistance of the output capacitor decreases, and the filter can reduce more ripples. However, this results in slow response characteristics of the system. If the size of the output capacitor is small, the internal resistance of the capacitor is larger than that of the battery, and the current ripple flows through the battery. For this reason, the LC filter suffers from the limitations of its filter design with a small ripple. In this study, by adding another LC to the LC filter, an LCLC filter is obtained to reduce the size of the filter. The filter design based on the Bessel filter has a maximum flat delay response for control stability. It also has a structure with a fast response and smaller size.

The current ripples at the outer inductor side and the battery side are designed to be 2 and 5 mA, respectively. The result of the LCLC-filter design is listed in Table I.

III. PROPOSED CONTROLLER DESIGN

A. State-space averaging method

In the linear network, the derivative of state variables can be expressed as the linear combination of the independent inputs of the system and state variables themselves. In the LCLC-filter, there are four state variables: two inductor voltages and two capacitor currents. Four equations are applied to the state space averaging method: two node equations and two loop equations. To model the current flowing through the battery, the system requires the battery side current. For this reason, the current equation of the battery side is added as output vector.

When the switch is OFF, inductor currents, capacitor voltages and output battery current are

When the switching state is OFF, the currents and voltages are expressed as

$$v_{C1} = -L_1 \frac{di_{L1}}{dt} - C_1 R_1 \frac{dv_{C1}}{dt} \tag{1}$$

$$i_{L1} = i_{L2} + C_1 \frac{dv_{C1}}{dt} \tag{2}$$

$$v_{C2} = v_{C1} + C_1 R_1 \frac{dv_{C1}}{dt} - L_2 \frac{di_{L2}}{dt} - C_2 R_2 \frac{dv_{C2}}{dt} \tag{3}$$

$$i_{L2} = \frac{v_{C2} + C_2 R_2 \frac{dv_{C2}}{dt} - v_B}{R_B} + C_2 \frac{dv_{C2}}{dt} \tag{4}$$

$$i_B = \frac{v_{C2} + R_2 C_2 \frac{dv_{C2}}{dt} - v_B}{R_B}. \tag{5}$$

These equations can be written in the following state-space form when the switching state is OFF:

$$\frac{d}{dt} x = A_1 x + B_1 u \tag{6}$$

$$y = C_1 x + E_1 u. \tag{7}$$

When the switching state is ON, the currents and voltages are

$$v_{C1} = V_{DC} - L_1 \frac{di_{L1}}{dt} - C_1 R_1 \frac{dv_{C1}}{dt} \tag{8}$$

$$i_{L1} = i_{L2} + C_1 \frac{dv_{C1}}{dt} \tag{9}$$

$$v_{C2} = v_{C1} + C_1 R_1 \frac{dv_{C1}}{dt} - L_2 \frac{di_{L2}}{dt} - C_2 R_2 \frac{dv_{C2}}{dt} \tag{10}$$

$$i_{L2} = \frac{v_{C2} + C_2 R_2 \frac{dv_{C2}}{dt} - v_B}{R_B} + C_2 \frac{dv_{C2}}{dt} \tag{11}$$

978-1-4799-2706-7/14 $31.00 © 2014 IEEE

$$i_B = \frac{v_{C2} + R_2 C_2 \dfrac{dv_{C2}}{dt} - v_B}{R_B}. \tag{12}$$

These equations can be written in the following state-space form when the switching state is ON:

$$\frac{d}{dt}x = A_2 x + B_2 u \tag{13}$$

$$y = C_2 x + E_2 u. \tag{14}$$

Averaging each state by using the duty ratio as a weighting factor. When the switch is OFF, OFF duty ratio d_1 is multiplied to (6) and (7); when the switch is ON, ON duty ratio d_2 is multiplied to (13) and (14). (15) and (16) can be obtained by adding these two values.

$$\frac{d}{dt}x = (A_1 d_1 + A_2 d_2)x + (B_1 d_1 + B_2 d_2)u \tag{15}$$

$$y = (C_1 d_1 + C_2 d_2)x + (E_1 d_1 + E_2 d_2)u \tag{16}$$

To derive the DC model, the DC terms are set to zero. The state-space averaged model that describes the converter in equilibrium is expressed as

$$0 = AX + BU \tag{17}$$

$$Y = CX + EU \tag{18}$$

where the averaged matrices are expressed as (19) (shown in the Appendix).

Equations (17) and (18) can be solved to find the equilibrium state and output vectors, which are expressed as (20) and (21) (shown in the Appendix). The DC model yields the DC information (steady-state behavior) and can be used for the loss estimation.

To construct a small-signal AC model at a quiescent operating point, each vector is composed of the equilibrium DC components and AC variations.

$$\begin{aligned} x &= X + \hat{x}, \quad y = Y + \hat{y}, \quad u = U + \hat{u}, \\ d_1 &= D_1 + \hat{d}_1, \quad d_2 = D_2 + \hat{d}_2. \end{aligned} \tag{22}$$

Here, X, Y, U, D_1, and D_2 are the equilibrium DC components in the quiescent operating point, and hats above the variables are the small AC variations. By assuming that the small AC variations are much smaller than the DC components, the nonlinear converter equations can be linearized. The substitution of (15) and (16) into (22) yields the following:

$$\begin{aligned} \frac{d}{dt}(X + \hat{x}) &= (A_1 d_1 + A_2 d_2)(X + \hat{x}) \\ &+ (B_1 d_1 + B_2 d_2)(U + \hat{u}) \end{aligned} \tag{23}$$

$$\begin{aligned} (Y + \hat{y}) &= (C_1 d_1 + C_2 d_2)(X + \hat{x}) \\ &+ (E_1 d_1 + E_2 d_2)(U + \hat{u}). \end{aligned} \tag{24}$$

To obtain the AC model (small-signal) equation, we neglect the nonlinear product terms. The AC model equations are expressed as (25) and (26).

$$\frac{d}{dt}\hat{x} = A\hat{x} + B\hat{u} + F\hat{d} \tag{25}$$

$$\hat{y} = C\hat{x} + E\hat{u} + G\hat{d} \tag{26}$$

where

$$F = (A_1 - A_2)X + (B_1 - B_2)U \tag{27}$$

$$G = (C_1 - C_2)X + (E_1 - E_2)U. \tag{28}$$

Transform the AC state-space model into frequency domains (Laplace transform, s-domain). The hat above the variables will be neglected for simplicity. The small signal of a variable is denoted using lower case letter. The dynamic equations are as follow:

$$\frac{d}{dt}x = Ax + Bu + Fd \tag{29}$$

$$y = Cx + Eu + Gd. \tag{30}$$

The Laplace transform of (29) is expressed as

$$sx(s) = Ax(s) + Bu(s) + Fd(s). \tag{31}$$

Simplifying (31) yields

$$\begin{aligned} x(s) &= (sI - A)^{-1}Bu(s) + (sI - A)^{-1}Fd(s) \\ &= H_u(s)u(s) + H_d(s)d(s). \end{aligned} \tag{32}$$

From (32), we can derive the following:

$$H_u(s) = \frac{x(s)}{u(s)} = (sI - A)^{-1}B \tag{33}$$

$$H_d(s) = \frac{x(s)}{d(s)} = (sI - A)^{-1}F. \tag{34}$$

Supposing that $H_u(s)$ does not change, $H_d(s)$ is the converter control-to-inductor voltages, capacitor currents transfer function.

The Laplace transform of (30) is expressed as

$$y(s) = Cx(s) + Eu(s) + Gd(s). \tag{35}$$

The substitution of (32) into (35) yields the following:

$$\begin{aligned} y(s) &= [C(sI - A)^{-1}B + E]u(s) \\ &+ [C(sI - A)^{-1}F + G]d(s) \end{aligned} \tag{36}$$

$$= G_u(s)u(s) + G_d(s)d(s). \tag{37}$$

From (36) and (37), we can derive the following:

$$G_u(s) = \frac{y(s)}{u(s)} = C(sI - A)^{-1}B + E \tag{38}$$

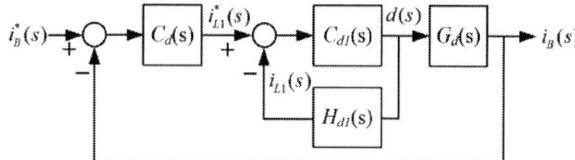

Fig. 3. Block diagram of dual current controller.

TABLE II
CIRCUIT PARAMETERS

Symbol	Parameters	Value
V_{DC}	Input DC voltage	1650 [V]
V_B	Output battery voltage	330 [V]
f_{SW}	Switching frequency	7.8 [kHz]
L_1	Outer inductor	6.6 [mH]
L_2	Inner inductor	120 [μH]
C_1	Outer capacitor	2.17 [mF]
C_2	Inner capacitor	470 [μF]
R_{C1}	ESR of C_1	61.1 [mΩ]
R_{C2}	ESR of C_2	282 [mΩ]
R_B	ESR of the battery	324 [mΩ]

$$G_d(s) = \frac{y(s)}{d(s)} = C(sI - A)^{-1}F + G. \qquad (39)$$

Supposing that the $G_u(s)$ does not change, the $G_d(s)$ is the converter control-to-output current transfer function. Using (34) and (39), the transfer function of the desired inputs and outputs, can be derived.

In this paper, double loop current controller composed inner current controller and outer current controller is presented. The transfer function of outer current controller H_{d1} is the first row of the matrix H_d because the outer current controller controls current flowing through the outer inductor. The transfer function of outer current controller is H_{d1} and expressed as (40). The transfer function of inner current controller is G_d and expressed as (41).

B. Dual current controller

The controller is designed as a dual current controller, which can operate stably on the basis of the calculated transfer function using the state-average equation. The block diagram of the DC–DC converter system is shown in Fig. 5. Using the state-space averaging method, a stable dual current controller can be designed. The block diagram of the converter system is composed of the battery current (i_B), duty cycle (d), outer inductor current (i_{L1}), transfer function of the control-to-battery current ($G_d(s)$), transfer function of the control-to-outer inductor current ($H_{d1}(s)$), PI controller of the outer current ($C_d(s)$), and PI controller of the inner current ($C_{d1}(s)$).

The received reference battery current passes through the PI controller. The inner current controller receives the reference current from the outer current controller as an input. Then, the inner current controller determines the duty. The duty passes through the transfer function, and the system finally outputs the battery current.

The proposed current controller uses a dual current controller in the battery charge–discharge system. Using the dual current controller increases the complexity of the controller design but yields fast dynamic characteristics

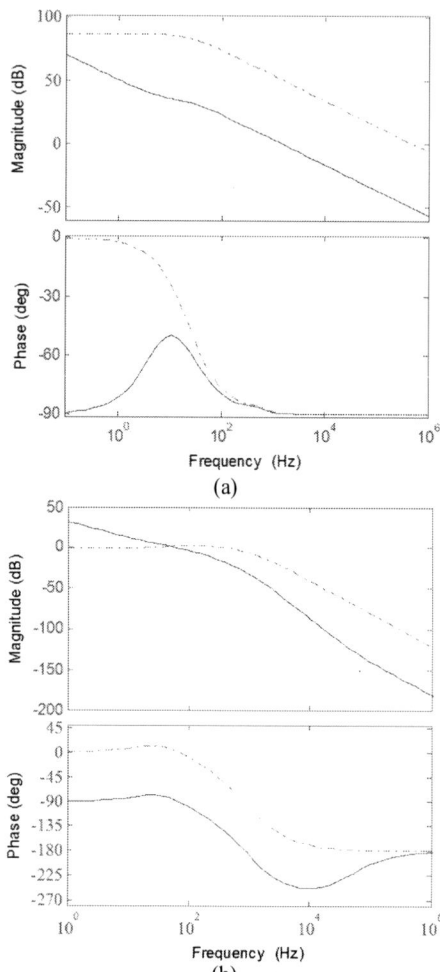

Fig. 4. Frequency characteristic of the system. (a) Bode plot of inner current controller. (b) Bode plot of outer current controller.

owing to the large bandwidth. The dual current controller process is described as follows: initially, the current control loop, which is an inner loop, is designed. While the stability of the controller is maintained, the battery current controller is designed as the outer loop to realize the required performance.

C. Frequency characteristics

The frequency characteristics for the open-loop transfer function obtained in (34) and (39) are analyzed. The parameters of the system are listed in Table II. Fig. 4(a) shows the frequency characteristic of the inner current controller. The dashed–dotted line shows the bode plot of $H_{d1}(s)$. As shown in Fig. 4(a), the $H_{d1}(s)$ transfer function characteristic is stable, but the cross frequency is 800 MHz, which is too high to be affected by the switching ripple. For this reason, by adding PI controller $C_i(s)$, the system can obtain the required characteristics. The solid line shows the bode plot of $H_{d1}(s)$ multiplied by $C_{d1}(s)$. The system design of the phase margin is 90°, and the crossover frequency of 3 kHz does not affect the switching frequency.

Fig. 4(b) shows the frequency characteristics of the outer current controller. The dashed–dotted line shows the bode plot of $G_d(s)$. As shown in this figure, the $G_d(s)$ transfer function characteristic is stable, but the cross frequency is almost 0 Hz, which is too small for fast response characteristics. Therefore, by adding PI controller $C_d(s)$, the system can obtain the required characteristics. The solid line shows the bode plot of $G_d(s)$ multiplied by $C_d(s)$. The system design of the phase margin is 78°, and the crossover frequency of 90 Hz does not affect the inner current controller.

Thus, to obtain a stable operation, the system must have not only a sufficient phase margin but also a gain margin in the region so that it cannot be affected by other control frequencies.

IV. SIMULATION RESULTS

The response time must be under 10 ms. In addition, the current ripple of the outer inductor must not be more than 2 A, and the current ripple of the battery should not be more than 5 mA.

Fig. 5(a) shows the simulation results when the PSIM software is used when the current reference changes from -10 to 10 A, and Fig. 5(b) shows the simulation result when the current reference changes from 10 to -10 A. The red line shows the battery current; the time required to reach from the transient state to the steady state is 8 ms, which satisfies the required transient response characteristic (10 ms). Fig. 6 shows the current ripple of the outer inductor and the battery of the DC–DC converter. The upper line shows the current ripple of the outer inductor. The lower line shows current ripple of the battery of the DC–DC converter. We can see that the peak-to-peak value of the inductor current ripple is within 2 A during the DC–DC converter switching. And the current ripple of the battery is within 5 mA, as designed.

V. CONCLUSION

This paper has presented the design of a 15-kW battery charge–discharge system controller for nonlinear dynamic system. This system can develop a structure with a fast response and smaller size using an LCLC filter. The system model includes the parasitic elements such as the capacitor internal resistance, which improves the accuracy of the controller design.

The simulation results show that the system using the frequency characteristics is stable and satisfies the required dynamic characteristics and current ripples. The effectiveness of the designed controller and the LCLC filter was verified by the simulation results.

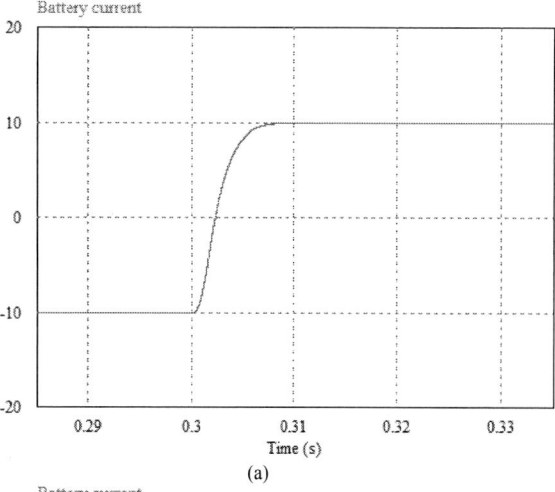

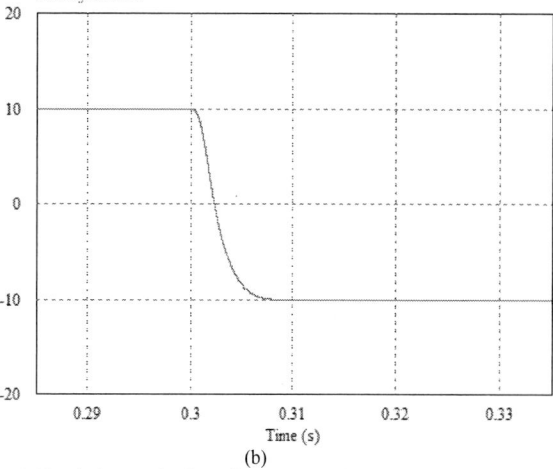

Fig. 5. Simulation result of transient response depending on (a) current change from -10 to 10 A and (b) current change from 10 to -10 A.

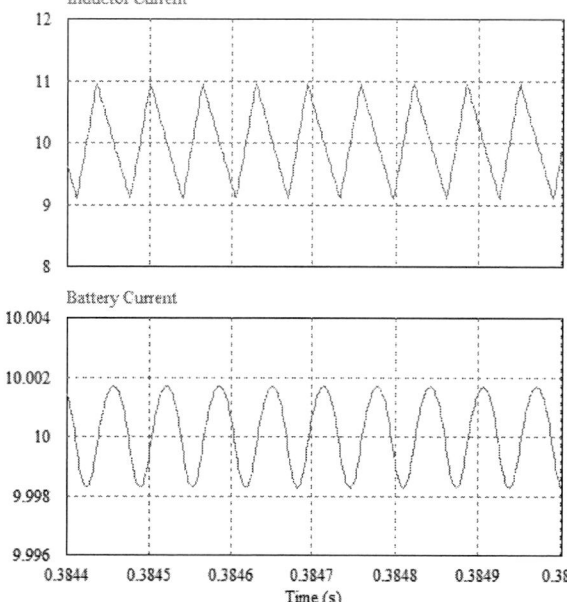

Fig. 6. Simulation results of current ripple of inductor and battery.

REFERENCES

[1] X. Li, W. Zhang, H. Li, R. Xie and D. Xu "Design and control of bi-directional DC/DC converter for 30kW fuel cell power system," in Proc. 8th Int. Conf. Power Electron. ECCE Asia (ICPE & ECCE), pp. 1024 – 1030, 2011

[2] S. Inoue and H. Akagi "A Bidirectional DC-DC Converter for an Energy Storage System With Galvanic Isolation" IEEE Trans. On Power Electronics, vol. 22, no. 6, pp 2299 – 2306, 2007.

[3] S. Yang, K. Goto, Y. Imamura and M. Shoyama "Dynamic Characteristics Model of Bi-directional DC-DC Converter using

State-Space Averaging Method," in Proc. 34th Int. Telecommunications Energy Conference (INTELEC), pp. 1-5, 2012.

[4] G. Feng, E. Meyer and Y.F. Liu "A New Digital Control Algorithm to Achieve Optimal Dynamic Performance in DC-to-DC Converters," *IEEE Trans. Power Electronics*, Vol. 22, No. 4, pp. 1489-1498, Jul. 2007.

[5] A. C. Schittler, D. Pappis, C. Rech and A. Campos "Generalized state-space model for the interleaved buck converter," in Proc. Power Electronics Conference (COBEP), pp. 451 – 457, 2011.

APPENDIX

$$A = \begin{bmatrix} -\dfrac{R_{C1}}{L_1} & \dfrac{R_{C1}}{L_1} & -\dfrac{1}{L_1} & 0 \\ \dfrac{R_{C1}(R_{C2}+R_B)}{(R_{C2}+R_B)L_2} & -\dfrac{R_{C1}R_{C2}+R_{C2}R_B+R_BR_{C1}}{(R_{C2}+R_B)L_2} & \dfrac{1}{L_2} & -\dfrac{R_B}{L_2(R_{C2}+R_B)} \\ \dfrac{1}{C_1} & -\dfrac{1}{C_1} & 0 & 0 \\ 0 & \dfrac{R_B}{C_2(R_{C2}+R_B)} & 0 & -\dfrac{1}{C_2(R_{C2}+R_B)} \end{bmatrix}, \quad B = \begin{bmatrix} \dfrac{D}{2L_1} & 0 \\ 0 & -\dfrac{R_{C2}}{(R_{C2}+R_B)L_2} \\ 0 & 0 \\ 0 & \dfrac{1}{(R_{C2}+R_B)C_2} \end{bmatrix} \tag{19}$$

$$C = \begin{bmatrix} 0 & \dfrac{R_{C2}}{R_B+R_{C2}} & 0 & \dfrac{1}{R_B+R_{C2}} \end{bmatrix}, \quad E = \begin{bmatrix} 0 & -\dfrac{1}{R_B+R_{C2}} \end{bmatrix}$$

$$X = -A^{-1}BU = \begin{bmatrix} \dfrac{DV_{DC}}{R_B} - \left(\dfrac{R_{C2}}{R_B(R_B+R_{C2})} + \dfrac{1}{R_B+R_{C2}} \right)V_B \\ \dfrac{DV_{DC}}{R_B} - \left(\dfrac{R_{C2}}{R_B(R_B+R_{C2})} + \dfrac{1}{R_B+R_{C2}} \right)V_B \\ DV_{DC} \\ DV_{DC} \end{bmatrix} \tag{20}$$

$$Y = CX + EU = \begin{bmatrix} \dfrac{\left(\dfrac{R_{C2}L_1}{(R_B+R_{C2})R_B} + \dfrac{L_1}{R_B+R_{C2}} \right)DV_{DC}}{L_1} - \left(\dfrac{\left(\dfrac{R_{C2}L_2}{(R_B+R_{C2})R_B} + \dfrac{L_2}{R_B+R_{C2}} \right)R_{C2}}{(R_B+R_{C2})L_2} + \dfrac{1}{R_B+R_{C2}} \right)V_B \end{bmatrix} \tag{21}$$

$$H_{d1} = \frac{(L_2C_1C_2(R_{C2}+R_B))s^3 + (C_1C_2(R_{C1}R_{C2}+R_{C1}R_B+R_{C2}R_B)+C_1L_2)s^2 + (C_2(R_{C2}+R_B)+C_1(R_{C1}+R_B))s+1)V_{DC}}{\begin{pmatrix} (L_1L_2C_1C_2(R_{C2}+R_B))s^4 + (L_1L_2C_1+L_1C_1C_2(R_{C1}R_{C2}+R_{C2}R_B+R_{C1}R_B)+L_2C_1C_2R_{C1}(R_{C2}+R_B))s^3 + (R_{C1}C_1L_2 \\ +L_2C_2(R_{C2}+R_B)+L_1C_2(R_{C2}+R_B)+L_1C_1(R_{C1}+R_B)+R_{C1}R_{C2}R_BC_1C_2)s^2 + (R_{C2}R_BC_2+R_{C1}R_BC_1+L_1+L_2)s+R_B \end{pmatrix}} \tag{40}$$

$$G_d = \frac{(R_{C1}R_{C2}C_1C_2(R_{C2}+R_B)s^2 + (R_{C2}C_2(R_2+R_B)+R_{C1}C_1(R_{C2}+R_B))s+R_{C2}+R_B)V_{DC}}{(R_2+R_B)\begin{pmatrix} ((L_1L_2C_1C_2(R_{C2}+R_B))s^4 + (L_1L_2C_1+R_{C1}R_{C2}C_1C_2(L_1+L_2)+R_BC_1C_2(R_{C1}L_1+R_{C1}L_2+R_{C2}L_1))s^3 \\ +(R_{C1}L_2C_1+R_BL_2C_2+R_{C2}L_2C_2+R_BL_1C_2+R_{C2}L_1C_2+R_{C1}L_1C_1+L_1C_1R_B+R_{C1}R_{C2}R_BC_1C_2)s^2 \\ +(R_B(R_{C2}C_2+R_{C1}C_1)+L_1+L_2)s+R_B \end{pmatrix}} \tag{41}$$

978-1-4799-2706-7/14 $31.00 © 2014 IEEE

A New SVPWM Strategy for Input Switched Multilevel Converter

Li Xiong, U. R. Prasanna, Akin Bilal, Kaushik Rajashekara
Power Electronics and Drives Laboratory http://www.utdallas.edu/pedl/
The University of Texas at Dallas, Richardson, TX, 75080
Email: {li.xiong, prasanna.ur, bilal.akin, k.raja}@utdallas.edu

Abstract- This paper proposes a new space vector pulse width modulation (SVPWM) strategy for a hybrid input switched multilevel converter. The converter topology is a cascaded connection of input switched and H-bridge converters, which can provide 2N+1 voltage levels for each phase. The proposed SVPWM strategy treats the multilevel converter as a two-level converter by introducing an offset vector. With generated switching pattern, it is possible to obtain minimum number of switching-state transitions and switch the H-bridge close to line frequency; thus, the switching losses can be significantly reduced. The proposed SVPWM strategy has been realized on a seven-level converter to demonstrate the principle of operation. The advantages of the proposed scheme and architecture are verified by both simulation and experimental results.

I INTRODUCTION

The multilevel converter topologies in general can generate high-quality voltage waveforms where power semiconductor switches are operated at a very low frequency [1]. Due to the low switching frequency, the power can be efficiently transferred with better power quality and low electromagnetic interference (EMI) noise. Additionally, multilevel converters feature several dc links, allowing independent voltage control and MPP tracking in each string. There are three traditional multilevel converter topologies; the diode clamped (DCC), flying capacitor clamped (FCC), and cascaded H-bridge (CHB) converters. Among them, the three-level DCC converter is widely accepted in industries for different applications. However, it has several disadvantages at higher voltage levels such as higher cost due to more number of power devices, deviation in dc link capacitor voltage and so on. These issues are also valid for flying capacitor based multilevel converters. Though cascade H-bridge converters do not have dc link capacitor voltage drifting issue, multiple isolated DC sources, which are usually provided by phase shifted transformers, are needed. The phase shifted transformers are bulky and expensive, which impairs the power density and cost-effectiveness of the converters. To address the issues associated with the traditional multilevel converters mentioned above, a number of multilevel converters with hybrid configurations such as input switched converters, are proposed in [2]-[5].

Several space vector modulation strategies are proposed to modulate the output voltage of multilevel converters in the literature [6]-[11]. In [7], a SVPWM modulation strategy has been proposed for three-level converter. The idea is to divide the complete space vector into six two-level sectors. However, a systematic switching sequence has not been presented and the computation is intensive due to the rotating d-q plane. .. In this paper, a new SVPWM control strategy for an input switched multilevel converter topology, which treats the multilevel converter as a two-level converter, is proposed to modulate the output voltages, allowing simple implementation. The proposed technique relies on introducing an offset vector close to the reference vector, shifting the origin to this offset vector, and operating in a two-level hexagon sector. A simple yet effective algorithm is proposed to generate the offset vector with minimum amount of computation. The procedure of synthesizing the reference voltage vector is a simple three-step process: a) locating the offset vector, b) determining switching pattern, and c) calculating dwelling time of each nearest vector (NV). The proposed SVPWM technique is analyzed on a seven-level input switched converter configuration. With the proposed SVPWM strategy, the switching frequency of the input switched converter can be significantly decreased, and the H-bridge can be switched at close to line frequency.

This paper is organized as follows. In section II, operating principles and switching states of the hybrid input switched multilevel converter, are introduced. The new SVPWM strategy is described in Section III. The simulation and experimental results are provided in Section IV to illustrate the effectiveness of the proposed SVPWM strategy. Finally, the conclusions derived from the study are presented in section V.

II POWER CIRCUIT OPERATION PRINCIPLES

The seven-level input switched multilevel converter consists of two stages as shown in Fig. 1. The first stage consists of three power arms, which are named as input switched power converter in the rest . this paper. Each power arm includes two devices in back-to-back configuration. The dc link consists of three capacitors $C1$, $C2$, and $C3$, which divide the link voltage equally to $V_{dc}/3$. The two power switches in each arm are gated on/off in a complementary pattern. The on/off state of each power switch is determined by the direction of load current and the voltage level of the first stage. When S_1 is turned-on and other switches are off, the output voltage of the first stage is V_{dc}, the current flows through S_1 and D_4. Similarly, other output voltages can be achieved by different switching states. The converter has twenty four switching states which can generate seven different voltage levels V_{dc}, $2V_{dc}/3$, $V_{dc}/3$, 0, $-V_{dc}$, $-2V_{dc}/3$, and $-V_{dc}/3$. The switching states are combinations of the six switching states provided by the input switched power converter and the four switching states of the H-bridge.

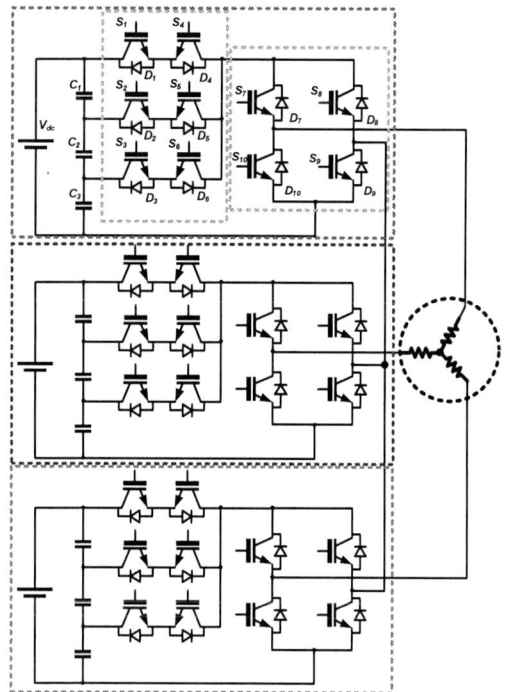

Fig. 1 Power circuit diagram of the three-phase input switched seven level converter configuration

III MODULATION STRATEGY DESCRIPTION

For a multilevel converter, the reference voltage vector can be synthesized by three nearest vectors as defined in [12]. The relationship can be described as

$$T_s \vec{V}_{ref} = d_1 T_s \vec{V}_1 + d_2 T_s \vec{V}_2 + d_3 T_s \vec{V}_3 \quad (1)$$

Where $\vec{V}_{ref}$ is the reference voltage vector; $\vec{V}_1, \vec{V}_2$, and $\vec{V}_3$ are the three NVs; T_s is the switching time period of the carrier signal; $d_1 T_s$, $d_2 T_s$, and $d_3 T_s$ are the dwelling time for the three NVs; $d_i = \dfrac{T_i}{T_s} (i=1,2,3)$. For each phase, the output voltage can be expressed as in Eq. (2).

$$V_i = (S_i - 3) \times \frac{V_{dc}}{3} (i = a, b, c) \quad (2)$$

S_i can be any of the seven states from the set {0, 1, 2, 3, 4, 5, 6}, and can be used to represent the voltage level of each phase. Then (S_a, S_b, S_c) can be utilized to signify a voltage vector. For example, $(S_a, S_b, S_c) = (3, 4, 5)$ means that the phase A voltage is 0, phase B voltage is $V_{dc}/3$, and phase C voltage is $2V_{dc}/3$. The resultant space vector diagram of the seven-level converter is shown in Fig. 2. By introducing an offset vector, the reference voltage vector can be rewritten as

$$\vec{V}_{ref} = d_1 \vec{V}_1 + d_2 \vec{V}_2 + d_3 \vec{V}_3$$
$$= d_1(\vec{V}_{offset} + \vec{V}_1') + d_2(\vec{V}_{offset} + \vec{V}_2') + d_3(\vec{V}_{offset} + \vec{V}_3') \quad (3)$$
$$= \vec{V}_{offset} + d_1 \vec{V}_1' + d_2 \vec{V}_2' + d_3 \vec{V}_3'$$

Eq. (3) can also be rephrased by Eq. (4) as

$$\vec{V}_{ref} = \vec{V}_{offset} + \vec{V}_{refx} \quad (4)$$

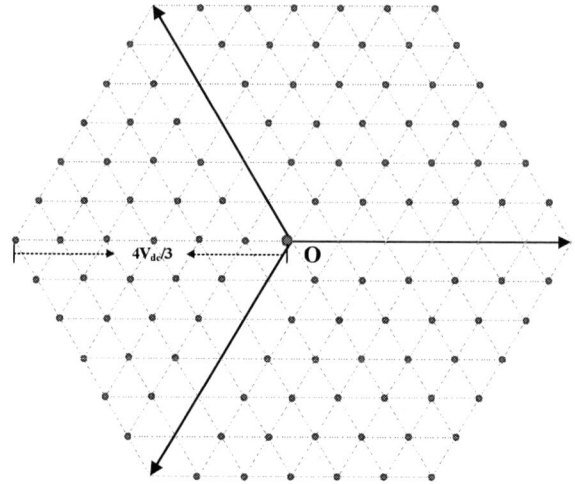

Fig. 2 Space vector diagram of the seven level converter

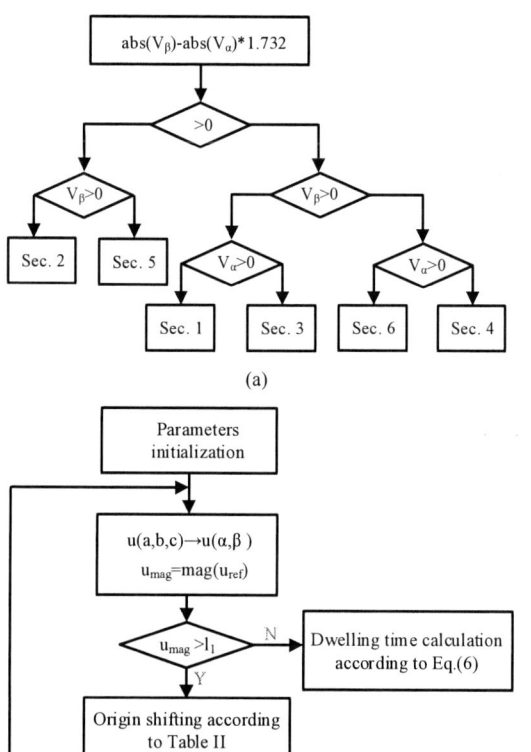

Fig. 3. Flowcharts for; (a) sector number determination, (b) Offset vector location.

For a given reference vector, different combinations of offset and remaining reference vectors can be found. The combination which contains the smallest remaining reference vector is preferred because the remaining reference vector can be synthesized by the vectors in the sub vector space in a way similar to that for two level inverters with minimum switching-state transitions. The resultant sub vector space is illustrated in Fig. 4. $\vec{V}_1$, $\vec{V}_2$, and $\vec{V}_3$ are the three NVs as shown with dashed black arrow line; the blue solid arrow line indicates the reference voltage vector $\vec{V}_{ref}$; the offset vector $\vec{V}_{offset}$ and remaining reference vector $\vec{V}_{refx}$ are expressed with red solid arrow line; the purple hexagon displays the sub vector space which will be used to synthesize the remaining reference vector; $\vec{V}_1'$, $\vec{V}_2'$, and $\vec{V}_3'$ are the three vectors in the sub vector space. Here, three NVs are selected such that the resultant remaining reference vector becomes the smallest. The offset vector in Fig. 4(a) is $\vec{V}_1$. There are two-coordinate systems, where the origin of the initial coordinate system, O, is shifted by an offset vector, where the offset vector becomes the origin of the new coordinate system, O'. The reference vector $\vec{V}_{ref}$ in the coordinate system with O as the origin is equivalent to the remaining reference vector $\vec{V}_{refx}$ in the new coordinate system with O' as the origin. The remaining reference vector can be calculated once the offset vector is determined, which can be synthesized in sub vector space, as shown in Fig. 4(b), in a similar way to that for two level inverters.

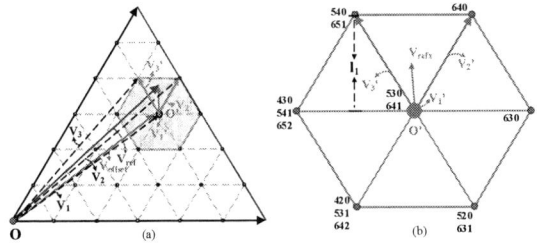

Fig. 4. Sub vector space synthesis with offset vector; (a) space vector diagram of sector one, (b) sub sector used for synthesizing the remaining reference vector.

The procedure of synthesizing the reference voltage vector is a three-step process. Firstly, the position of the offset vector needs to be located, then the switching sequence should be determined, finally, the dwelling time of the NVs are calculated.

A. Locating Offset Vector

The process to search for the combination with smallest remaining reference vector aims to enclose the remaining reference vector by the smallest hexagon in the space vector diagram by shifting the coordinate origin. At each point, the origin of the coordinate can be moved along six directions, i.e. along A, B, and C, and as well as the opposite directions of A, B, and C. The direction to choose is determined by the relative position relationship between the reference vector and the coordinate system. There are a number of paths to shift the origin of the coordinate system closer to the destined point, while the one with minimum number of steps is preferred due to the reduced time and computation cost. The sector number (SN) which the remaining reference vector

locates in current coordinate system determines the direction of movement of next step. This relationship is summarized in the Table I, and the flowchart of the implementation is given in Fig. 3(b).

TABLE I SECTOR NUMBER, DIRECTION, AND POSITION UPDATE

SN	Direction	Position Update	
1	A	$u_\alpha' = u_\alpha - \dfrac{1}{3\sqrt{3}}$	$u_\beta' = u_\beta$
2	-C	$u_\alpha' = u_\alpha - \dfrac{1}{6\sqrt{3}}$	$u_\beta' = u_\beta - \dfrac{1}{6}$
3	B	$u_\alpha' = u_\alpha + \dfrac{1}{6\sqrt{3}}$	$u_\beta' = u_\beta - \dfrac{1}{6}$
4	-A	$u_\alpha' = u_\alpha + \dfrac{1}{3\sqrt{3}}$	$u_\beta' = u_\beta$
5	C	$u_\alpha' = u_\alpha + \dfrac{1}{6\sqrt{3}}$	$u_\beta' = u_\beta + \dfrac{1}{6}$
6	-B	$u_\alpha' = u_\alpha - \dfrac{1}{6\sqrt{3}}$	$u_\beta' = u_\beta + \dfrac{1}{6}$

After each step, the position information of the remaining reference vector is updated accordingly, as shown in Table I. The voltages in Table I are normalized by the maximum available reference voltage. The origin of the coordinate system continues to be shifted till the magnitude of the remaining reference vector is smaller than l_1 which is shown in Fig. 4(b). The flowchart of the searching process is shown in Fig. 3(b). An example is given in Fig. 5 to illustrate the searching procedure.

B. Switching sequence determination

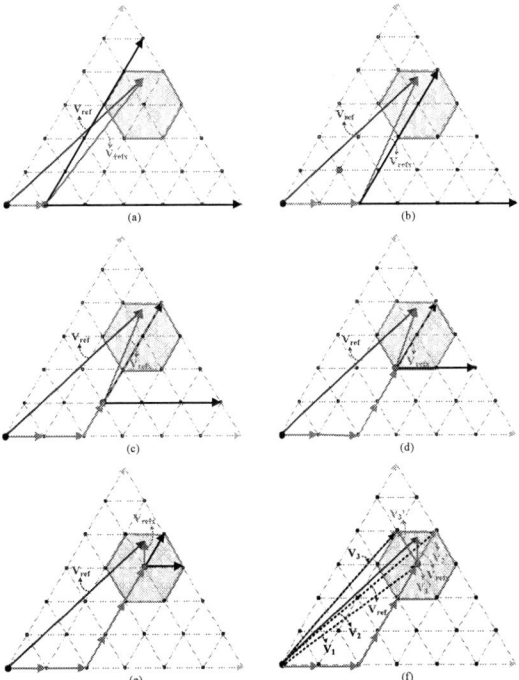

Fig. 5. Process of offset vector determination.

The switching sequence can be determined according to Table II. Each switching sequence is comprised of four space vectors, i. e. $\vec{V}_1$, $\vec{V}_2$, $\vec{V}_3$ and $\vec{V}_4$, $\vec{V}_1$ and $\vec{V}_4$ are the offset vectors determined by the offset vector locating process as described in Section III.A. There are six switching patterns available to be chosen. The switch pattern is determined by the sector number, where the remaining reference vector locates in the sub vector space. For the case shown in Fig. 4(b), the remaining reference vector locates in Sector 2. Then switching pattern for Sector 2 from Table II should be adopted for the switching sequence generation. The switching sequence is $(S_a, S_b, S_c) \rightarrow (S_a+1, S_b, S_c) \rightarrow (S_a+1, S_b, S_c+1) \rightarrow (S_a+1, S_b+1, S_c+1)$. The switching sequence in other sectors can be selected accordingly.

TABLE II SECTOR NUMBER AND SWITCHING SEQUENCE

SN	V_1	V_2	V_3	V_4
1	(S_a, S_b, S_c)	(S_a+1, S_b, S_c)	(S_a+1, S_b+1, S_c)	(S_a+1, S_b+1, S_c+1)
2	(S_a, S_b, S_c)	(S_a, S_b+1, S_c)	(S_a+1, S_b+1, S_c)	(S_a+1, S_b+1, S_c+1)
3	(S_a, S_b, S_c)	(S_a, S_b+1, S_c)	(S_a, S_b+1, S_c+1)	(S_a+1, S_b+1, S_c+1)
4	(S_a, S_b, S_c)	(S_a, S_b, S_c+1)	(S_a, S_b+1, S_c+1)	(S_a+1, S_b+1, S_c+1)
5	(S_a, S_b, S_c)	(S_a, S_b, S_c+1)	(S_a+1, S_b, S_c+1)	(S_a+1, S_b+1, S_c+1)
6	(S_a, S_b, S_c)	(S_a+1, S_b, S_c)	(S_a+1, S_b, S_c+1)	(S_a+1, S_b+1, S_c+1)

C. Dwelling time calculation

Using Eq. (3), the remaining voltage reference can be expressed as in Eq. (5)

$$\vec{V}_{refx} = d_1 \vec{V}_1' + d_2 \vec{V}_2' + d_3 \vec{V}_3' \qquad (5)$$

The dwelling times can be calculated as

$$\begin{cases} d_2 T_s = 6 \times [V_\alpha \sin(\frac{n}{3}\pi) - V_\beta \cos(\frac{n}{3}\pi)] T_s \\ d_3 T_s = -6 \times [V_\alpha \sin(\frac{n-1}{3}\pi) - V_\beta \cos(\frac{n-1}{3}\pi)] T_s \\ d_1 T_s = (1 - d_1 - d_2) T_s \end{cases} \qquad (6)$$

Where, V_α and V_β are the projections of the remaining reference vector on the alpha-beta plane, and n denotes the sector number within the sub vector space. Two duty cycle generation methods for switching vectors are shown in Fig. 6.

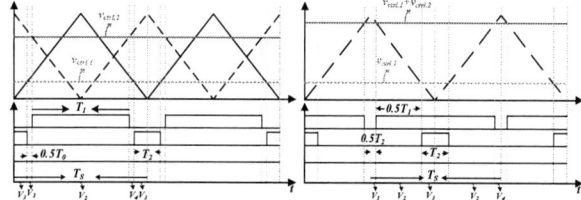

Fig. 6. Duty cycle generation of switching vectors; (a) Using two phase shifted triangle carriers, (b) Using a single triangle carrier.

IV SIMULATION AND EXPERIMENTAL RESULTS

The simulation and experiments are conducted to verify the effectiveness and viability of the seven-level input switched converter modulated with the proposed SVPWM strategy. The parameters used in the simulations are listed in Table III.

TABLE III SIMULATION PARAMETERS

DC link voltage	300V
Switching frequency	5kHz
DC link capacitor	2.2mF(each)
R	57.7Ω
L	23mH

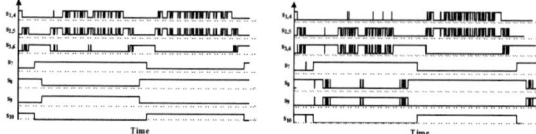

Fig. 7. Gating pulse patterns when; (a) modulation index = 0.8, (b) modulation index = 0.5.

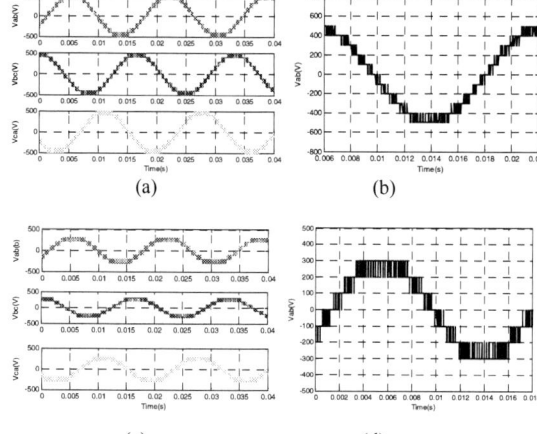

Fig. 8 (a) (b) Simulation results of three phase line voltages when modulation index = 0.8; (c) (d) Simulation results of three phase line voltages when modulation index = 0.5

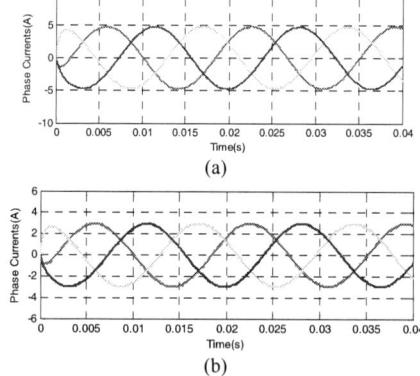

Fig. 9 (a) Simulation results of three phase phase currents when modulation index = 0.8; (b)Simulation results of phase currents when modulation index = 0.5

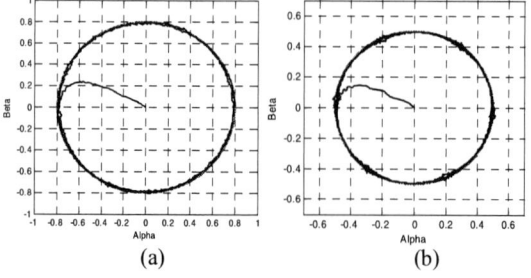

Fig. 10 (a) Simulation results of three phase currents in polar plane when modulation index = 0.8; (b) Simulation results of three phase currents in polar plane when modulation index = 0.5

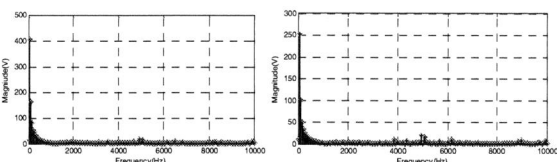

Fig. 11 (a) Frequency spectrum analysis of line voltages when modulation index = 0.8, THD = 12.56%; (b) Frequency spectrum analysis of line voltages when modulation index = 0.5, THD = 18.29%

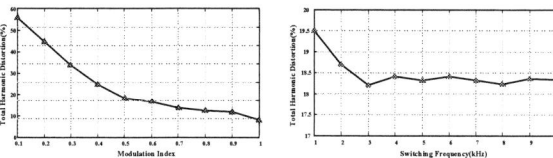

Fig. 12. THD plot as a function of; (a) modulation index when switching frequency is 5 kHz; (b) switching frequency when modulation index is 0.5.

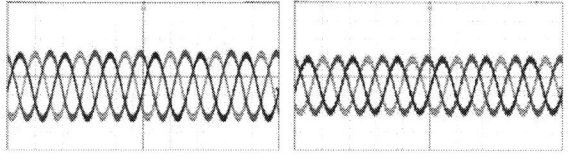

Fig. 13 (a) (b) Experimenal results of three phase line voltages when modulation index = 0.8; (c) (d) Simulation results of three phase line voltages when modulation index = 0.5 (100V/div, 10ms/div)

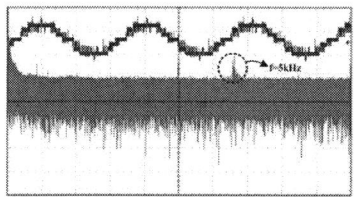

Fig. 14. Experimental results of line currents; (a) modulation index = 0.8 (2A/div, 10ms/div); (b) modulation index = 0.5 (2A/div, 10ms/div).

Fig. 15 Frequency spectrum of line voltage when modulation index = 0.5.

The corresponding results are presented in Fig. 7 to Fig. 15. The switching gating pulses when the modulation index is 0.8 and 0.5 are shown in Fig. 7(a) and Fig. 7(b) respectively, from which, it can be clearly seen that the switch pairs S_1 and S_4, S_2 and S_5, and S_3 and S_6 are not gated on at the same time. As mentioned before, the power switches in the H-bridge stage almost switch at line frequency.

The waveforms of the three phase line voltages, phase voltages, and line currents when the modulation index is equal to 0.8 and 0.5 are displayed in Fig. 8 and Fig. 9 respectively. The line currents waveforms in alpha-beta plane are shown in Fig. 10. In comparison of the results under both modulation indexes, it can be observed that

more voltage levels are available when the modulation index is higher, which provides lower voltage THD and better power quality. To support this, the frequency spectrum analysis of the line voltages for both cases is shown in Fig. 11. The line voltage THD is 12.56% when modulation index is 0.8, while the THD is 18.29% when the modulation index is 0.5. The harmonics around the switching frequency, 5 kHz, can be observed from the figure. The relationship of the voltage THD and the modulation index for the given switching frequency of 5 kHz is plotted in Fig. 12(a). As it can be seen, the voltage THD decreases as the modulation index increases. It is worth mentioning that inverters usually operate with high modulation index to maximize the utilization of dc link capacitors. The relationship of voltage THD and switching frequency for the given modulation index of 0.5, is shown in Fig. 12(b). From the figure, it can be seen that the voltage THD does not significantly decrease as switching frequency reaches to a specific level. This relationship provides insightful information and guidance on the determination of switching frequency, addressing the trade-off between voltage THD and power loss. The experimental results are presented in Fig. 13 to Fig. 15. As seen from the results, the experimental results are consistent with the simulation results.

V CONCLUSION

This paper presents a SVPWM strategy for input switched multi-level converters. The proposed SVPWM strategy treats the multilevel converter as a two-level converter by introducing an offset vector for output voltage modulation. The offset vector and systematic switching sequence determination procedures are provided within the framework of the paper. The simulation and experimental results are presented to verify the validity of the proposed strategy.

REFERENCES

[1] S. Kouro, M. Malinowski, K. Gopakumar, J. Pou, L. G. Franquelo, B. Wu, J. Rodriguez, M. A. Pérez, and J. I. Leon, "Recent Advances and Industrial Applications of Multilevel Converters," *IEEE Trans. Ind. Electron.*, vol. 57, no. 8, pp. 2553-2580, Aug. 2010.

[2] Y. Zhang, G. P. Adam, T. C. Lim, S. J. Finney, and B. W. Williams, "Hybrid Multilevel Converter: Capacitor Voltage Balancing Limits and its Extension," *IEEE Trans. Ind. Informat.*, vol. 9, no. 4, pp. 2063-2073, Nov. 2013.

[3] S. Debnath and M. Saeedifard, "A New Hybrid Modular Multilevel Converter for Grid Connection of Large Wind Turbines," *IEEE Trans. Sustain. Energy*, vol. 4, no. 4, pp. 1053-1064, Oct. 2013.

[4] S. Pulikanti, G. Konstantinou, and V. G. Agelidis, "Hybrid Seven-Level Cascaded Active-Neutral-Point-Clamped Based Multilevel Converter under SHE-PWM," *IEEE Trans. Ind. Electron.*, vol. 60, no. 11, pp. 4794-4804, Nov. 2013.

[5] J. Ebrahimi, E. Babaei, and G. B. Gharehpetian, "A New Multilevel Converter Topology With Reduced Number of Power Electronic Components," *IEEE Trans. Ind. Electron.*, vol. 59, no. 2, pp. 655-667, Nov. 2011.

[6] Y. Deng, K. H. Teo, and R. G. Harley, "A Fast and Generalized Space Vector Modulation Scheme for Multilevel Inverters." Applied Power Electronics Conference and Exposition (APEC), 2013 Twenty-Eighth Annual IEEE. IEEE, 2013..

[7] J. H. Seo, C. Ho Choi, and D. S. Hyun, "A New Simplified Space-Vector PWM Method for Three-Level Inverters," *IEEE Trans. Power Electron.*, vol. 16, no. 4, pp. 545-550, Jul. 2001.

[8] M. Saeedifard, P. M. Barbosa, P. K. Steimer, "Operation and Control of a Hybrid Seven-Level Converter," *IEEE Trans. Power Electron.*, vol. 27, no. 2, pp. 652-660, Feb. 2012.

[9] J. I. Leon, S. Vazquez, J. A. Sanchez, R. Portillo, L. G. Franquelo, J. M. Carrasco, and E. Dominguez, "Conventional Space-Vector Modulation Techniques Versus the Single-Phase Modulator for Multilevel Converters," *IEEE Trans. Ind. Electron.*, vol. 57, no. 7, pp. 2473-2482, Jul. 2010.

[10] Ó. López, J. Álvarez, J. Doval-Gandoy, F. D. Freijedo, "Multilevel Multiphase Space Vector PWM Algorithm," *IEEE Trans. Ind. Electron.*, vol. 55, no. 5, pp. 1933-1942, May 2008.

[11] L. G. Franquelo, J. Nápoles, R. C. P. Guisado, J. I. León, and M. A. Aguirre, "A Flexible Selective Harmonic Mitigation Technique to Meet Grid Codes in Three-Level PWM Converters," *IEEE Trans. Ind. Electron.*, vol. 54, no. 6, pp. 3022-3029, Dec. 2007.

[12] Rodríguez, J., Correa, P., & Moran, L. (2001). A vector control technique for medium voltage multilevel inverters. In Applied Power Electronics Conference and Exposition, 2001. APEC 2001. Sixteenth Annual IEEE (Vol. 1, pp. 173-178). IEEE.

The 2014 International Power Electronics Conference

ESD Reliability Influence of a 60 V Power LDMOS by the FOD-Based (& Dotted-OD) Drain

Shen-Li Chen
Department of Electronic Engineering
National United University
MiaoLi City, Taiwan
jackchen@nuu.edu.tw

Min-Hua Lee
Department of Electronic Engineering
National United University
MiaoLi City, Taiwan
secret9610@hotmail.com

Abstract — **An LDMOS (lateral-diffused MOS) is often used to as the ESD device in a high-voltage circuit for its low on-resistance benefit. But, it has several serious disadvantages, including the V_h value is not high enough and the device in a multi-finger structure can't completely turn on which resulting in the ESD capability per unit length is very low. So, the non-uniform turned-on phenomenon is seriously impacted the robustness of ESD reliability. Therefore, this paper is based on the drain FOD structure of an nLDMOS, and which will change the OD structure for contacts located in the drain-side. The OD structure will renew as some dotted-ODs layout. Experimental results show that the dotted-OD layout has a higher ESD capability than the FOD structure, and the layout type of dotted-OD will affect the ESD capability of an HV component, where a uniformly distributed type of dotted-OD will have a highest I_{t2} value, the I_{t2} value is increased about 12% as compared with the traditional LDMOS. The V_h value will increase with the contacts number increasing within the dotted-OD, therefore, this structure can also effectively improve the latch-up (LU) immunity.**

Keywords— Electrostatic discharge (ESD), Field oxide device (FOD), n-channel lateral-diffused MOS (nLDMOS), Transmission-line pulse (TLP).

I. INTRODUCTION

Recently, the high-voltage (HV) LDMOS devices are widely used in communication modules, electronic power-switch components, power management ICs, automotive electronics, LED illuminations and LCD drivers [1]-[3]. The HV devices have become more and more important. However, the operating voltage of an HV device is very high which indeed need a better reliability [4]-[5]. An LDMOS is still a surface dominated MOSFET structure and most of the LDMOS on-state conduction flows along the silicon surface and high electric field around silicon surface accelerated the impact ionization. Nevertheless, LDMOS transistors are inherently frail with respect to

ESD stress and enhancing their ESD robustness has been an on-going target.

II. LAYOUT DESIGN OF DUTs

In this paper, adding the FOD structure in the drain-side of an HV nLDMOS is regarded as basic architecture. Then, changing the contact layout of oxide definition (OD) area in the drain-side makes the OD layout renew as the dotted-OD layout as shown in Figs 1~4. As a result, the ESD clamp ability of these new nLDMOS DUTs will be investigated. All test DUTs for this work are fabricated by a TSMC 0.25 μm HV 60 V process. A multi-finger structure of GGnMOS used in this work, the channel length (L) of each finger nLDMOS is kept to be 2 μm, channel width (W_f) of each finger nLDMOS is 100 μm, and total channel width (W_{tot}) is kept constancy, 600 μm. And, all the stripe number (M) is 6.

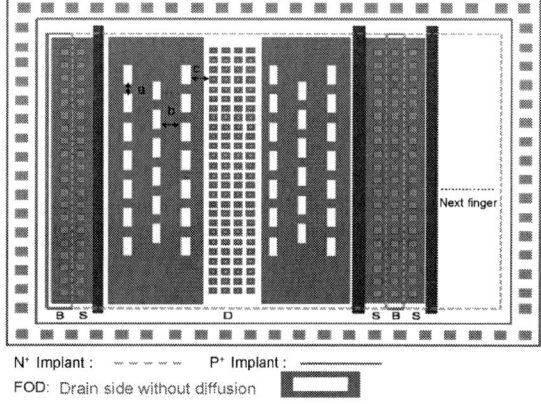

Fig. 1. Schematic layout of an HV nLDMOS with adding FOD & dotted-OD structures in the drain-side, where the dotted-OD is uniform distributed and contained one contact.

978-1-4799-2706-7/14 $31.00 © 2014 IEEE

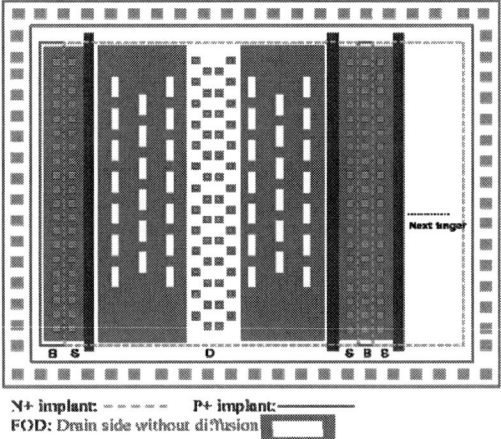

N+ implant: - - - - P+ implant:——
FOD: Drain side without diffusion [▭]

Fig. 2. Schematic layout of an HV nLDMOS with adding FOD & dot-OD structures in the drain-side, where the dot-OD is discrete distributed and contained one contact.

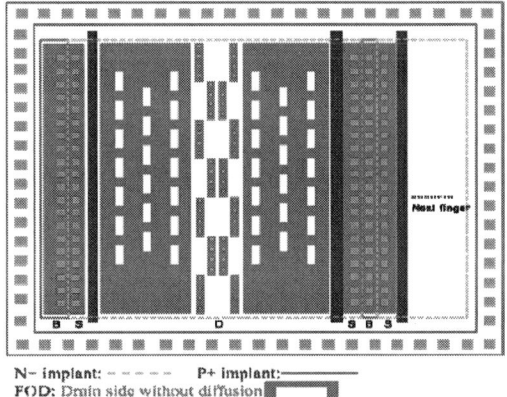

N- implant: - - - - P+ implant:——
FOD: Drain side without diffusion [▭]

Fig. 3. Schematic layout of an HV nLDMOS with adding FOD & dotted-OD structures in the drain-side, where the dotted-OD is gap distributed and contained four contacts.

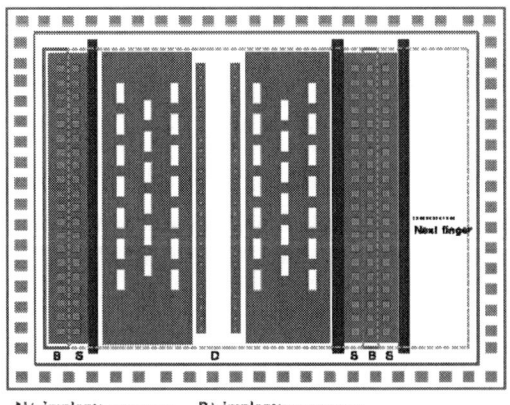

N+ implant: - - - - P+ implant:——
FOD: Drain side without diffusion [▭]

Fig. 4. Schematic layout of an HV nLDMOS with adding FOD & dotted-OD structures in the drain-side, where the dotted-OD is continuous distributed and contained 92 contacts.

Here, the goal of this paper is to adjust the layout manner and location of drain contacts which means adjusting the series resistance and achieving the uniform conducting. Therefore, some layout parameters of an FOD were fixed (column # of FOD is fourteen; FOD length/width= 3 μm× 0.6 μm; a= 0.6 μm, b= 0.6 μm, c= 0.54 μm). The parameter "a" is the spacing between one FOD block and next one FOD block in the same column of FODs; parameter "b" is the distance between the nearest parallel FOD columns; and parameter "c" is the distance between the nearby FOD edges to the drain-contact edge. Then, the dotted-OD contacts number within the central area changed from the minimum one contact (shown in Fig. 2), and gradually increased the contacts number within the OD to 2, 4 (shown in Fig. 3), 23, 46, and eventually increased to the entire 92 (shown in Fig. 4). Meanwhile, as the contacts number increased within the dot OD, the dotted-OD layout will be gradually become the stripe-OD layout and the form of the non-OD layout will distribute in the central zone of drain-side. Moreover, one type of the uniformly distributed dotted-OD layout (shown in Fig. 1) designed to investigate the influence of dotted-OD on ESD protection robustness, while the dotted-OD has only one contact.

The layout structure we need that is completely symmetric. Both minimum distances should be suited between a contact to a dotted-OD and a drain-side contact to OD on the vertical direction while channel width is 100 μm. According to the aforementioned description, the maximum number of contacts in a column is 92 and that is the reason why the maximum number of contacts of dotted-OD is 92. And, the 92 number could be divisible by the contacts which surrounded by dotted-OD, i.e., 1, 2, 4, 23, 46, and 92 contacts. Therefore, these six kinds of contact variation used in this work. In this way, there are two structures taken as for reference DUTs. One is a traditional nLDMOS without adding FOD & dotted-OD, and the other is an nLDMOS only with adding FODs.

III. TESTING EQUIPMENT SYSTEM

A transmission-line pulse (TLP) system for experimental testing is controlled by the LabVIEW software, which managed the electronic instrument of subsystem such as the ESD pulse generator, the high-frequency digital oscilloscope and the digital power electric meter instruments, in order to achieve the automatic measurement. This machine can provide a continuous step-high square wave to a DUT, and shortly raise time of the continuous square wave can also simulate transient noise of an ESD. This HBM-like system has used the short square wave with 100ns pulse widths and 10ns rising/falling times to evaluate the voltage and current response of a DUT.

978-1-4799-2706-7/14 $31.00 © 2014 IEEE

IV. MEASUREMENT RESULTS AND DISCUSSION

With adding FOD and dotted-OD on drain-side will be discussed as following: one is the traditional nLDMOS not with any FOD and dotted-OD structures and compared with the only FOD adding or only dotted-OD adding, by changing the layout manner of dotted-OD to study the influence on ESD immunity. Second is changing the number of contacts on dotted-OD structures.

The test results of snapback I-V curves of part one are shown in Fig. 5. As only adding the FOD structure in an nLDMOS structure, the resistance of FOD structure is not large enough to cause the I_{t2} value of nLDMOS increased. However, when an nLDMOS device is added the dotted-OD structure, the corresponding I_{t2} value will increase significantly. Then, it shows that a dotted-OD structure is good for ESD current dissipation than an FOD structure, it can make the component with a high ESD current capability. The only adding dotted-OD structure not only has a higher I_{t2} value but also has a higher V_h value than a traditional nLDMOS, so that the dotted-OD structure is worth further developing. However, when a dotted-OD combined with an FOD structure, its I_{t2} value will increase unremarkable, particularly in the discrete layout types have a worse I_{t2} values. The reason for the resistance of an FOD is not large enough to apply at high voltage components, so when adding an FOD structure on a dotted-OD area will weaken the ESD capability. The number of contacts on drain-side of a discrete-type layout is less than of a uniform-type layout, which will result in a current crowding effect, and make the I_{t2} value of components decreased. However, the most important feature of adding an FOD structure is the V_h value obviously increased, it is good for latch-up immunity in the high voltage environment. And, the layout-type distribution of snapback key parameters of V_{t1}, V_h and I_{t2} are shown in Figs 6~7.

From Fig. 8, the second part of contact number modulation in FOD & dotted-OD structures, it can be found that an FOD structure combined with a dotted-OD structure in the drain-side can result in the V_h and I_{t2} values increased. As a stripe OD contains a more contact numbers, it will have a higher V_h value. When the contact number of an OD area is 46, it has the maximum value of V_h. The last DUT of dotted-OD type (dotOD92) is with a different layout of the former groups, it is a long stripe OD (shown in Fig. 4). This layout-type makes the drain-side resistance will decrease slightly, which is resulted in the V_h value followed too. In addition, the layout-type distribution of V_{t1} and V_h are shown in Fig. 9. In FOD & dotted-OD structures, the distribution aspect of I_{t2} will behave as a bell shape shown in Fig. 10. In the experimental dotOD2 and dotOD4 situations, these DUTs

had a higher I_{t2} value. Therefore, a proper design of experimental parameters will be good for component reliability.

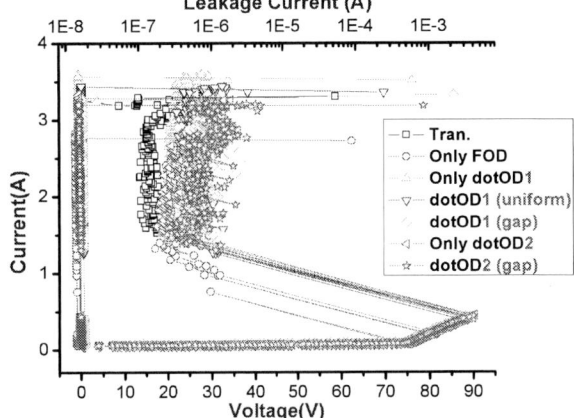

Fig. 5. Snapback I-V curves & leakage currents of HV nLDMOS DUTs as adding FOD & dotted-OD structures in the drain-side.

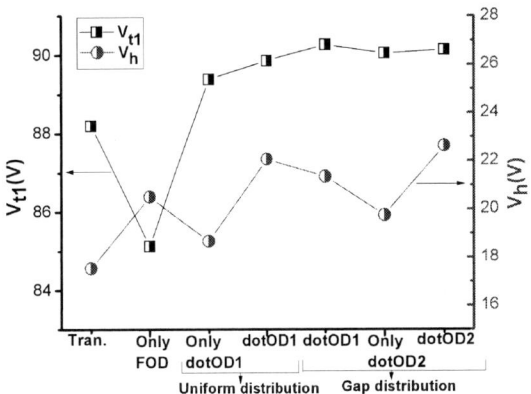

Fig. 6. V_{t1} and V_h distributions by different drain-side engineering.

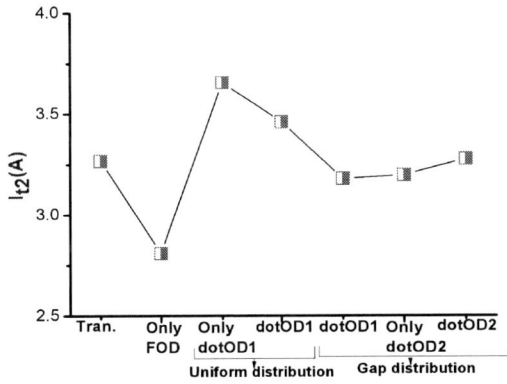

Fig. 7. I_{t2} distributions by different drain-side engineering.

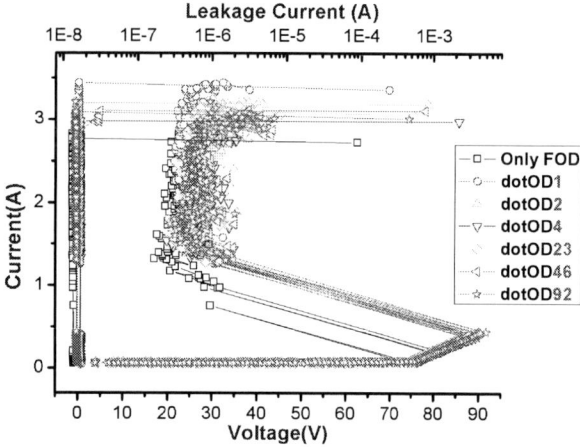

Fig. 8. Snapback I-V curves & leakage currents of HV nLDMOS DUTs as adding FOD & dotted-OD structures in the drain-side.

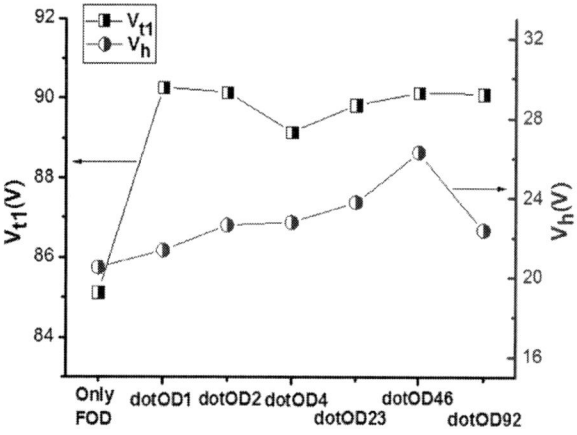

Fig. 9. V_{t1} and V_h distributions by different drain-side engineering.

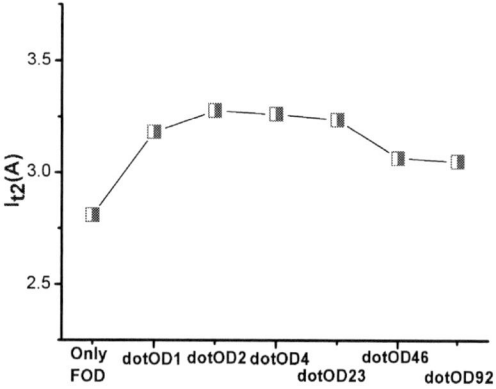

Fig. 10. I_{t2} distributions by different drain-side engineering.

V. CONCLUSION

This paper is focused on two crucial subject, one is the comparison of an traditional structure, FOD structure, and dotted-OD structures in the drain-side; the other is FOD structures combined with a dotted-OD structure modulation. After systematic tests, it is found that in high-voltage nLDMOS components, adding an FOD structure in the drain side will cause the I_{t2} value of a DUT decreased, but it will with a higher LU immunity. And, dotted-OD structures in the drain side can improve the I_{t2} value of a DUT; meanwhile, the V_h value will not be sacrificed. So that these two structures combined together should be haven a high ESD capability and high LU immunity. However, although the combination of these two structures have a high V_h value, but the results of I_{t2} compared with a traditional nLDMOS DUT is not increasing so high. Nevertheless, experimental results showed that the I_{t2} value of a uniform-layout type is better than a gap-layout type. In other words, the dotted-OD structure and uniform-layout type should be adopted in the further application.

ACKNOWLEDGMENT

In this work, authors would like to thank the National Chip Implementation Center in Taiwan for providing the process information and fabrication platform.

REFERENCES

[1] Yongqiang Zhou, Wanrong Zhang, Lixin Wang, Hongyun Xie, Chunbao Ding, "A 230 watts RF LDMOS high power amplifier for WCDMA application," *IEEE 13th International Conference on Communication Technology*, 2011, pp. 298-301.

[2] Yong-Keon Choi, Il-Yong Park, Hee-Sung Oh, Wook Lee, Nam-Joo Kim, Kwang-Dong Yoo, " Implementation of low Vgs (1.8V) 12V RF-LDMOS for high-frequency DC-DC converter applications," *24th International Symposium on Power Semiconductor Devices and ICs (ISPSD)*, 2012, pp. 125-128.

[3] A.D. Sagneri, D.I. Anderson, D.J. Perreault, " Optimization of Integrated Transistors for Very High Frequency DC–DC Converters," *IEEE Transactions on Power Electronics*, vol.28, issue:7, pp. 3614-3626, 2013.

[4] A.A. Salman, F. Farbiz, A. Appaswamy, H. Kunz, G. Boselli, M. Dissegna, "Engineering optimal high current characteristics of high voltage DENMOS," *IEEE International Reliability Physics Symposium*, 2012, pp. 3E.1.1-3E.1.6.

[5] Qinsong Qian, Qinsong Qian, Siyang Liu, Weijun Wan, Weifeng Sun, "Reliability Concern on Extended E-SOA of SOI Power Devices With P-Sink Structures," *IEEE Transactions on Device and Materials Reliability*, vol.13, issue:1, pp. 161-166, 2013.

Enhanced Transverse-Flux Motor with Torus Coils

Junya Tanaka
Graduate School of Engineering, Open Information Systems
Toyo University
Kawagoe-shi, Saitama, Japan

Kazuto Sakai
Department of Electrical, Electronic and Computer Engineering
Toyo University
Kawagoe-shi, Saitama, Japan
k_sakai@toyo.jp

Abstract—Two of the most important characteristics of electric vehicle motors are low cost and high reliability. Armature windings are difficult to manufacture with automated techniques and are not very reliable. Thus, to obtain a motor with low cost and high reliability, we have investigated a transverse-flux motor (TFM) containing a torus coil. In this study, we propose an enhanced TFM and discuss its principles and basic characteristics. The rotor has circumferential permanent magnets, and the stator has torus coils and axial 9-shaped cores. The stator core is arranged alternately in the circumferential direction, and has a radial and axial air gap. Our analysis confirms that the enhanced TFM produces a high torque per motor volume of 4.74 kN·m/m³.

Keywords—high-efficiency permanent magnet motor, transverse flux, torus

I. INTRODUCTION

Permanent magnet (PM) motors are used in electric vehicles, trains, and wind generators because of their high efficiency and high power density. For these applications, it is important that the motor be low-cost and highly reliable. However, the armature windings of these motors make automated manufacture more difficult, and automated winding reduces reliability. A transverse-flux motor (TFM) with torus coils was previously investigated for a direct-drive linear motor and a wind generator [1–4]. In this study, we propose a novel PM motor, enhanced with torus coils, to be used as a low-cost, high-quality motor, and investigate its principles and basic characteristics.

II. PRINCIPLES OF THE TFM

Figure 1 shows the basic configuration of an enhanced TFM. Figure 2 shows the A–R cross-section of the basic configuration of the enhanced TFM. The stator consists of torus coils and 9-shaped cores. The 9-shaped cores surround the center of a torus coil and are arranged in the poloidal direction. Each 9-shaped core is arranged in the circumferential direction and in the reverse side of the adjacent core each other, as shown in Figure 1. The permanent N- and S-pole magnets are arranged in the rotor core in the circumferential direction.

Next, we describe the flux flows produced by the armature current and PMs. Figure 3 shows the magnetic circuit in the enhanced TFM. The armature current in each 9-shaped stator core produces a reverse flow of the flux in the axial and radial air gap to the adjacent core. Normally, a TFM has one stator core per pole pair [1, 2], and the pole width of one core occupies the electrical angle between 120° and 180° within a cycle. None of the sections in any core occupy a greater extent in the circumferential direction because the stator core displaced by an electrical angle of 180° produces a minus torque. However, the enhanced TFM has two alternately arranged 9-shaped cores per pole pair. This configuration of the enhanced TFM allows the flux to flow in the PM in each air gap in the reverse direction, as shown in Figure 3.

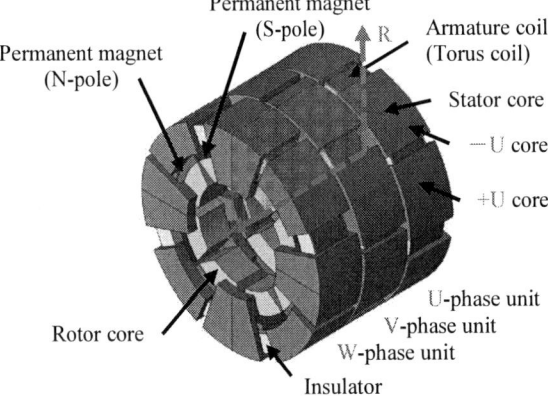

Fig. 1. The configuration of the enhanced TFM.

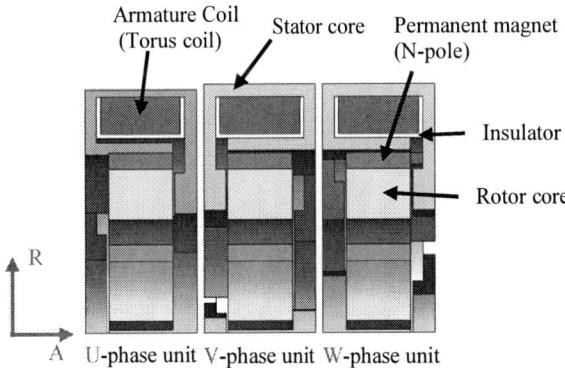

Fig. 2. The configuration of an enhanced TFM (the A-R cross-section).

978-1-4799-2706-7/14 $31.00 © 2014 IEEE

The 9-shaped cores are composed of an electromagnetic steel or dust core, and the three units are phase-shifted by 120°, as shown in Figure 4. When configuring the 9-shaped cores using electromagnetic steel, it is configured by several divided cores axially laminated. The −U core is arranged at 165° instead of at a 180° displacement from the +U core, as shown in Figure 5. This configuration is designed to reduce torque ripple. The pole arc of each stator core occupies an electrical angle of 148°. Therefore, the stator of the enhanced TFM is 82% occupied within a cycle in the circumferential direction by the stator core.

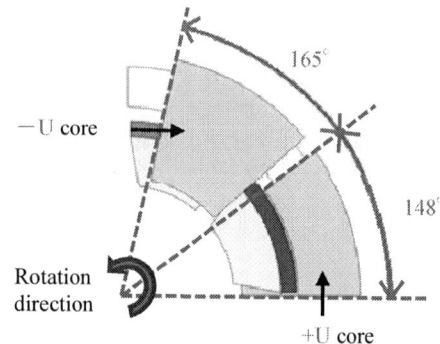

Fig. 5. Arrangement of the stator core.

III. ANALYTICAL MODEL

We performed a three-dimensional (3D) electromagnetic analysis to verify the feasibility of the driving force for the enhanced TFM. Figure 6 shows the analytical model of the enhanced TFM designed for the 3D finite-element method (FEM). The analytical 3D model is a partial model of a one-pole pair. Table 1 gives the motor specifications of the analytical model. We analyzed the TFM model using the 3D-FEM software JMAG.

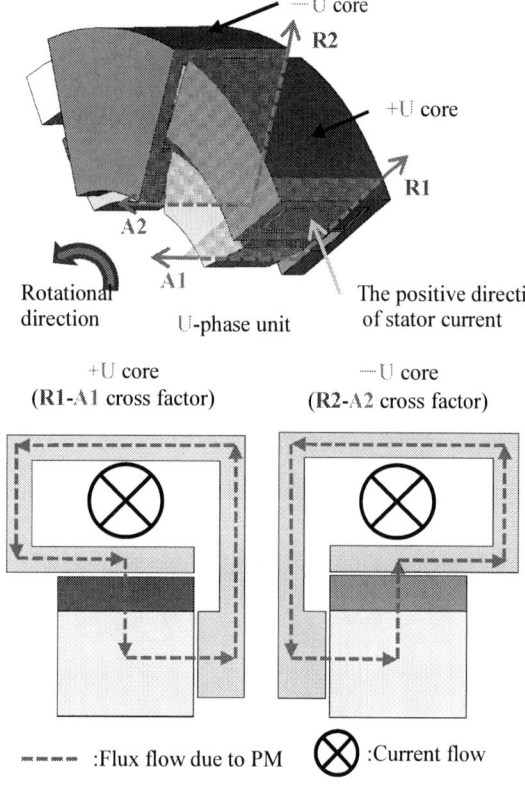

Fig. 3. Magnetic circuit of an enhanced TFM.

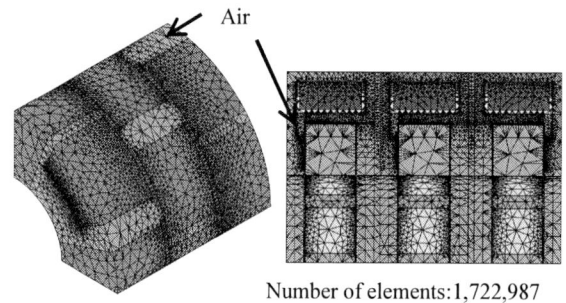

Number of elements:1,722,987

Fig. 6. A 3D-FEM model of the enhanced TFM.

TABLE I
SPECIFICATIONS FOR THE ENHANCED TFM MODEL

Phase/Pole	3 Phase / 8 Pole
Diameter of stator	120 mm
Number of turns	157/phase
Frequency	200 Hz
Rotational speed	3000 rpm
Rated current	4 Arms
Current density	4.21 A/mm^2
Air gap length	0.5 mm
Coil resistance	0.94 Ω/phase
Core material	JFE_steel_35JN300
Permanent magnet	NEOMAX-42AH
Insulator	Non-magnetic material

IV. RESULTS OF ANALYSIS

We clarified the basic no-load and load characteristics using a FEM magnetic field analysis.

A. No-Load Characteristics

Torque is produced by interaction between the rotating field and the alternate magnetic flux in the air gap produced by the PMs. We verified that the 9-shaped cores

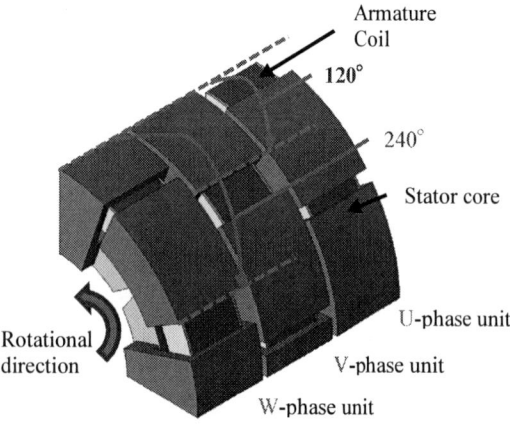

Fig. 4. Arrangement of the phase unit.

978-1-4799-2706-7/14 $31.00 © 2014 IEEE

and PMs generate an alternating magnetic flux distribution in the air gap. Figure 7 shows the distribution of the magnetic flux density in three units for no load. The transverse alternating fields interact with the current flowing in the circumferential direction. Figure 8 shows the 8-pole wave form of the induced voltage for no load. This indicates that the enhanced TFM can form a multi-pole from three torus coils. Figure 9 shows each harmonic component of the induced voltage. Because the third harmonic is 7.5% and the fifth harmonic is 6.9%, a few harmonics will lead to a small torque ripple and slight core loss.

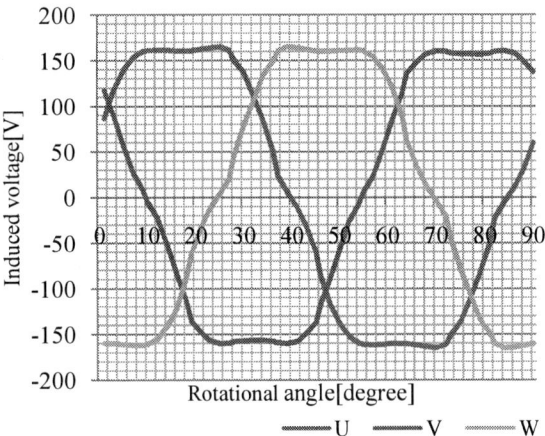

Fig. 8. The wave form of the induced voltage in the enhanced TFM.

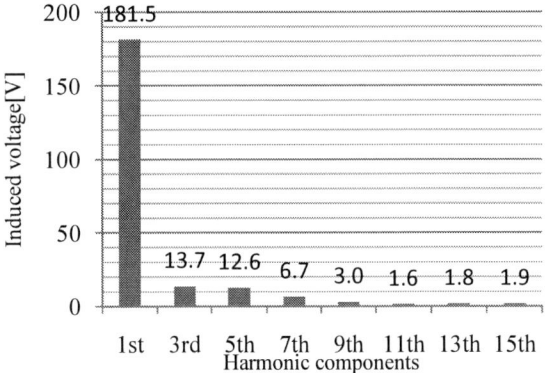

Fig. 9. Harmonic components of the induced voltage in the enhanced TFM.

Magnetic flux density [T]

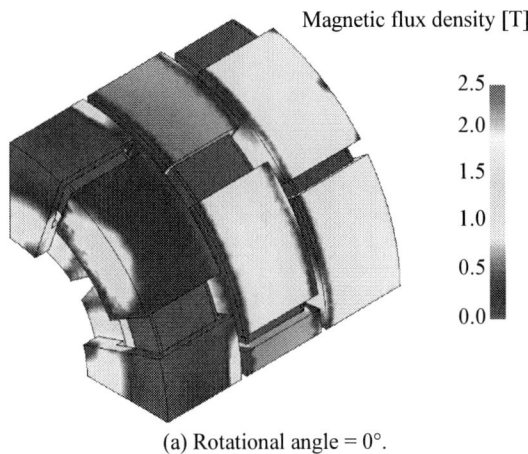

(a) Rotational angle = 0°.

Magnetic flux density [T]

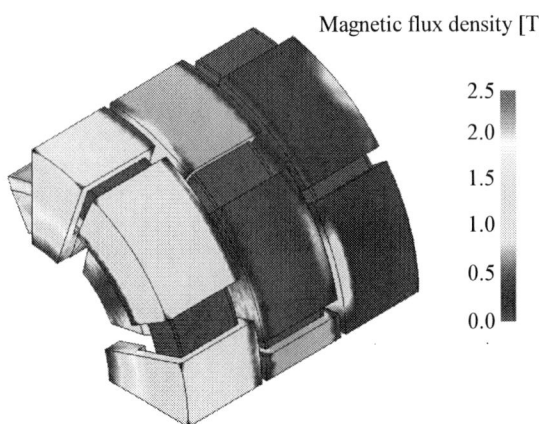

(b) Rotational angle = 90°

Fig. 7. Distribution of the magnetic flux density of the enhanced TFM for no-load.

B. Load Characteristics

Figure 10 shows the on-load distribution of the magnetic flux density of the three units at the rated current. The magnetic flux distribution shifts from an adjoining unit of a different phase and changes periodically.

C. Torque Characteristics

We now discuss the torque characteristics during rotation. Based on magnetic-field analysis Figure 11 shows the torque wave form for the rated load and no-load. Because the average torque per motor volume is 4,74 N·m/m³, the motor can produce a high torque. At about 18.4% of maximum torque, a slight torque ripple is observed.

D. Loss Characteristics

Figure 12 shows the iron loss of the enhanced TFM for the rated load and no-load cases at varied rotational speeds from 3,000 to 15,000 rpm, based on magnetic-field analysis. When the motor is rotating for the rated load case, the iron loss is 50.1 W. Copper loss of the motor is 45 W for the rated load case, per our calculation.

From these results, we obtain motor efficiency of about 94%.

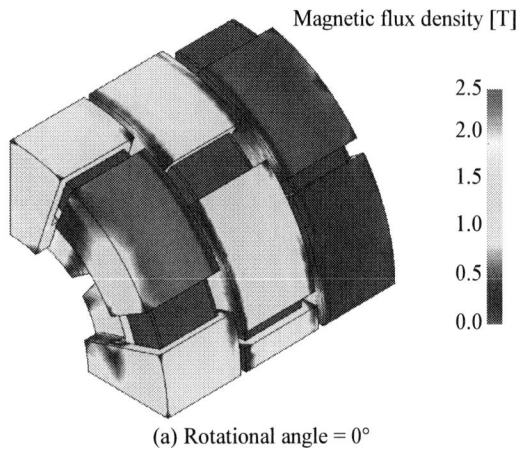

Magnetic flux density [T]

2.5
2.0
1.5
1.0
0.5
0.0

(a) Rotational angle = 0°

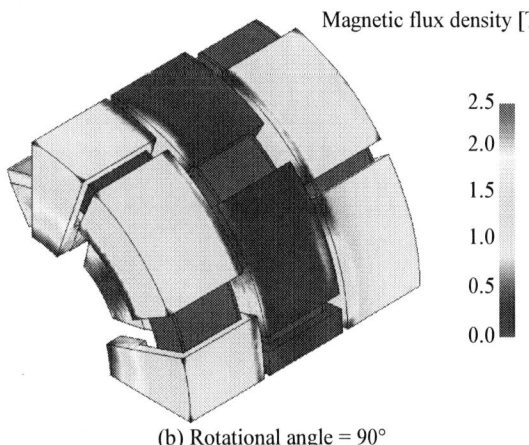

Magnetic flux density [T]

2.5
2.0
1.5
1.0
0.5
0.0

(b) Rotational angle = 90°

Fig. 10. Distribution of the magnetic flux density of the enhanced TFM at the rated load.

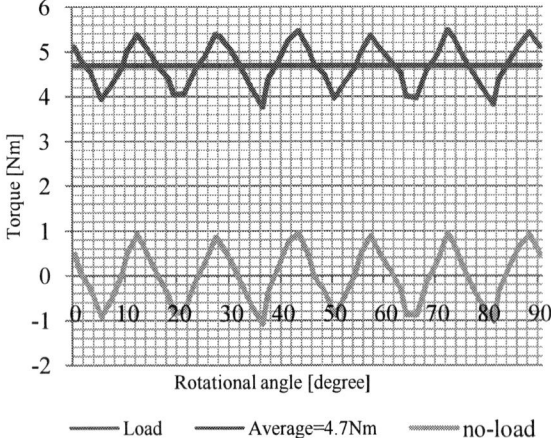

Fig. 11. Torque characteristics of the enhanced TFM.

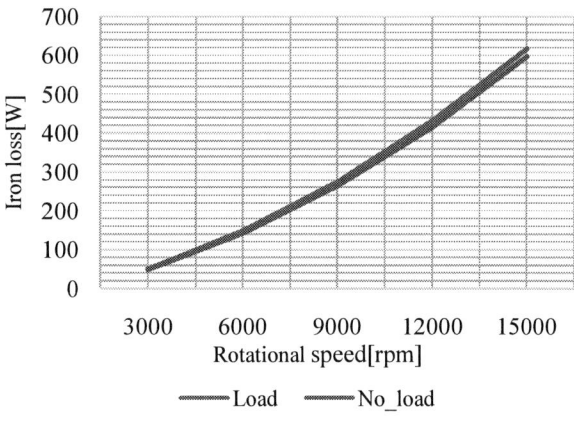

Fig. 12. Iron loss in the enhanced TFM.

E. Reduction of Torque Ripple

We alternated the arrangement of the 9-shaped cores to reduce torque ripple in the enhanced TFM. The original arrangement, shown in Figure 5, is hereafter called arrangement A. Figure 13 shows two core arrangements to reduce torque ripple. Figure 13(a) and Figure 13(b) are hereafter called arrangement B and C, respectively. Figure 14 shows the harmonic components in the torque for the rated load of each arrangement. Figure 15 shows torque waveforms for the rated load in each arrangement. Figure 16 shows the average torque of these waveforms. Harmonic components in the case of C show radical torque ripple in the enhanced TFM. The results indicate that the larger torque ripple components are sixth and twelfth. This arrangement has large torque ripple about 49.5%. So the case of core arrangement A shows arrangement to reduce twelfth component of torque ripple in the enhanced TFM. And the case of B shows arrangement to reduce sixth and twelfth component of torque ripple in the enhanced TFM. As a result, core arrangement A and B has slight torque ripple about 18.4% and 8.7 % of that in arrangement C. But average torques of core arrangement A and B decreases by about 3.4% and 6.8% of that in arrangement C. From these results, we learn that alternation of the stator cores arrangement is an acceptable way to reduce torque ripple.

The 2014 International Power Electronics Conference

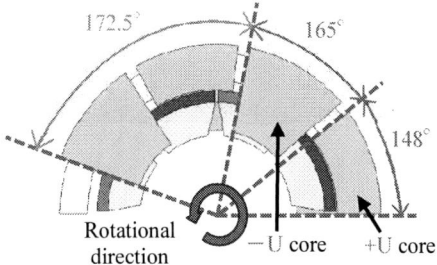

Fig. 13(a). Core arrangement B.

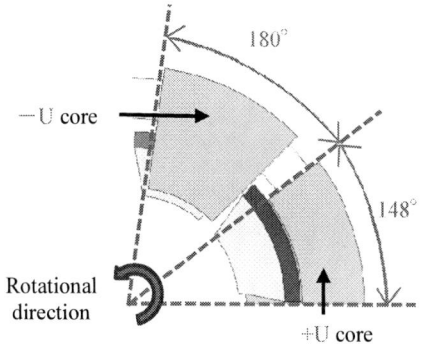

Fig. 13(b). Core arrangement C.

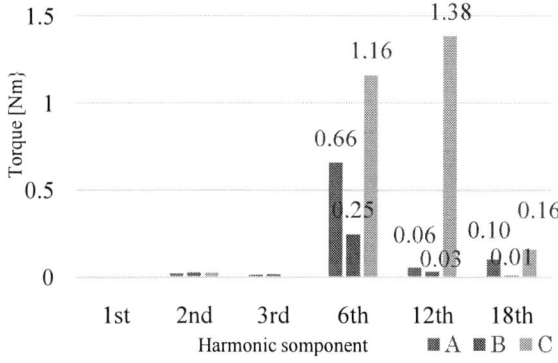

Fig. 14. Harmonic components of the rated load torque in the each core arrangement.

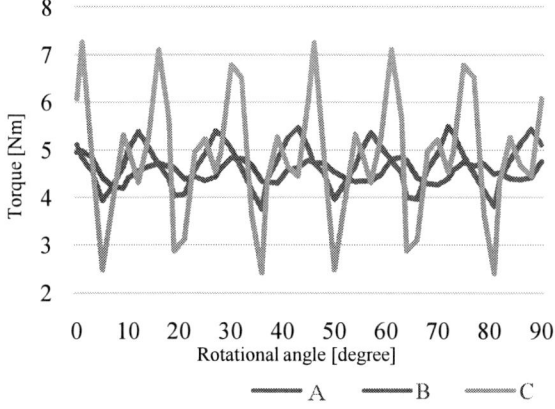

Fig. 15. Torque waveform in each core arrangement.

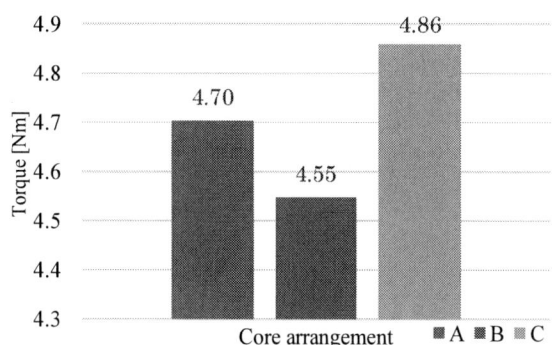

Fig. 16. Average rated load torque in each core arrangement.

F. Comparison of Torque Characteristics of Enhanced TFM and Conventional Motor

We compared enhanced TFM with conventional Interior Permanent magnet motor (IPMSM). We performed electromagnetic analysis to verify the feasibility of the drive force for the conventional IPMSM. Table 2 shows specifications of IPMSM for comparison with enhanced TFM. For comparison with enhanced TFM on a like-for-like basis in terms of machine specifications, we configured the same machine volume and the same value ampere-turn as IPMSM. Figure 17 shows average torque of the enhanced TFM and IPMSM at rated torque. The results indicate that the enhanced TFM has a higher torque per motor volume than the conventional motor.

TABLE II
SPECIFICATIONS OF CONVENTIONAL IPMSM.

Phase/Pole	3 Phase/8 Pole
Diameter of stator	120 mm
Number of turns	157/phase
Frequency	200 Hz
Rotational speed	3000 rpm
Rated current	4 Arms
Current density	2.60 A/mm^2
Air gap length	0.5 mm
Coil resistance	0.29 Ω/phase
Core material	JFE_steel_35JN300
Permanent magnet	NEOMAX-42AH
Insulator	Non-magnetic material

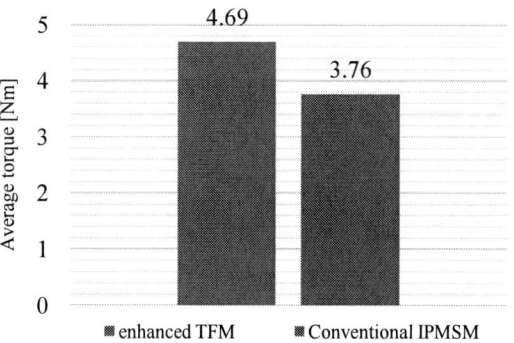

Fig. 17. Average torque in the enhanced TFM and conventional IPMSM at rated torque.

978-1-4799-2706-7/14 $31.00 © 2014 IEEE   244

V. CONCLUSIONS

In this study, we proposed a PM motor (enhanced TFM) that uses only three torus coils to form a multi-pole. The results indicate that the enhanced TFM produces a high torque per motor volume. Furthermore, the stator with the torus coils has no coil-end. Therefore, this motor can reduce copper loss by decreasing the resistance of armature windings.

ACKNOWLEDGMENT

This work was supported by the Japan Society for the Promotion of Science KAKENHI Grant Number 24560346.

REFERENCES

[1] H. Yokose, K. Sakai, and S. Kuramochi, "Principle and basic characteristics on variable-magnetic-force transverse flux motor," in *Proc. ICEMS2012*, paper no.DS3G1-8, pp.1–4.

[2] W. M. Arshad, T. Bäckström, and C. Sadaragani, "Analytical design and analysis procedure for a transverse flux machine," *IEEE International Electric Machines and Drives Conf.*, 20, pp.17–2001.

[3] H.-J. Kim, J. Nakatsugawa, K. Sakai, and H. Shibata, "High acceleration linear motor, 'Tunnel Actuator'," *J. Magn. Soc. Jpn.*,no. 29, pp.199–204, 2005.

[4] Dipl.-Ing. Peter Seibold, Dipl.-Ing. Frieder Schuller, Prof. Dr.-Ing. Nejila Parspour, Dipl.-Ing. Manuel Gärtner, "Transverse Flux Machine in a diesel-electric drive system" ATZ of highway, pp.68-76Edition: 2012-02.

The Influence of Magnetic Properties of Permanent Magnet on the Performance of IPMSM for Automotive Application

S. Yoshioka, S. Morimoto, M. Sanada, and Y. Inoue

Graduate School of Engineering
Osaka Prefecture University
1-1 Gakuen-cho, Naka-ku, Sakai, Osaka, Japan
E-mail: st106045@edu.osakafu-u.ac.jp

Abstract—**High power density is required for motors intended for automotive applications. Therefore, interior permanent magnet synchronous motors (IPMSMs) are commonly used in hybrid electric vehicles and electric vehicles. It is known that the characteristics of the magnetic materials are a significant factor with respect to the performance of the IPMSM, and the magnet is particularly important. In addition, currently, the high-performance magnets that have a high residual magnetic flux density are being developed, and technics for applying these magnets to motors are required.**

In this paper, three motors having different magnets with different residual magnetic flux densities are investigated for use in automotive applications. As a result, by using a strong magnet, the total torque and output power are increased, and the high-efficiency operating region is widened and shifted to the low-speed, high-torque range. Moreover, the maximum efficiencies at frequently used operating points are compared. In the city-driving, the efficiency is approximately the same. However, the efficiency in highway driving decreases when the strong magnet is used.

Keywords— *interior permanent magnet synchronous motor (IPMSM), automotive application, high-performance magnet, efficiency map*

I. INTRODUCTION

Hybrid electric vehicles (HEVs) and electric vehicles (EVs) have been considered as low-emission vehicles that will provide energy savings and help to realize a low-carbon society. In order to realize these goals, highly efficient motors are also required. Motors for traction drive applications have numerous requirements, such as high power density because of limited mounting space, maintenance-free, and high efficiency in order to improve fuel efficiency [1], [2]. In particular, the efficiency under light load is important because such an operating region is most commonly used [3].

In order to satisfy these requirements, an interior permanent magnet synchronous motor (IPMSM) is adopted for these vehicle driving applications [4] - [6]. The performance of the IPMSM is considered to be affected by magnetic materials, especially by the permanent magnet. In addition, high-performance

This work was partly supported by "Future Pioneering Projects, Development of magnetic material technology for high-efficiency motors" in Japan.

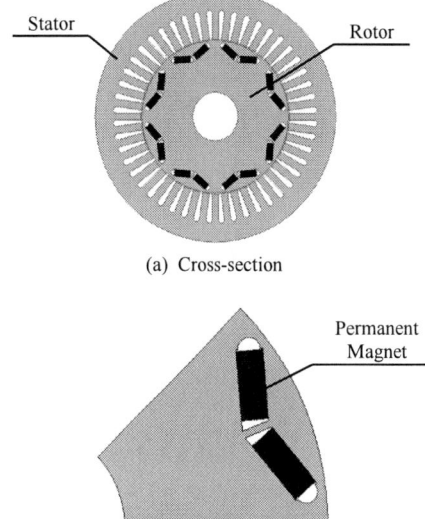

(a) Cross-section

(b) Rotor structure (1 pole)

Fig. 1. Analysis model.

magnets, which have high residual magnetic flux density and are unlikely to be demagnetized, are being developed.

In this paper, three IPMSMs having different magnets with different residual magnetic flux densities are examined in order to determine the influence of the magnetic properties of permanent magnets on the performance of the IPMSM in automotive applications. This paper is intended to provide a guideline for designing the optimum rotor structure for a high-performance magnet, which will be developed in the near future.

II. ANALYSIS MODEL

The analysis model is shown in Fig. 1. This model is an eight-pole machine and has a distributed winding stator with 48 slots. This model is designed based on the traction motor of the current HEV [7], [8]. The arrangement of permanent magnets is V-shaped and has one layer, and the volume of the permanent magnet is 100 cm^3. The motor specifications are shown in Table I, and the magnet specifications are shown in Table II.

TABLE I
MOTOR SPECIFICATIONS

Item (Unit)	Value
Stator diameter (mm)	264
Rotor diameter (mm)	160.4
Shaft diameter (mm)	51
Air gap length (mm)	0.75
Thickness of steel (mm)	50
Armature winding (turns/phase)	88
Winding resistance (Ω)	0.096

TABLE II
MAGNET SPECIFICATIONS (60 °C)

Item (Unit)	Type A	Type B	Type C
PM material code	NMX-34GH	NMX-45GH	NMX-S54
Remnant magnetic flux density (T)	1.12	1.28	1.42
Coercive force (kA/m)	847.3	971.6	571.0

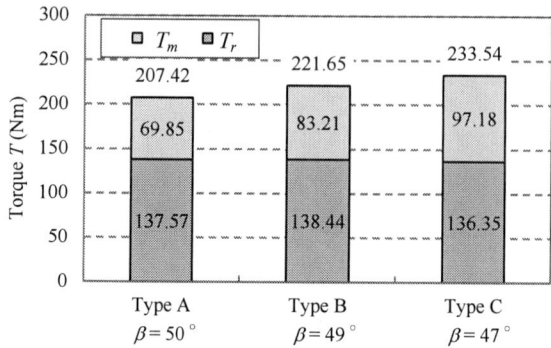

Fig. 2. Maximum torque characteristics (134 A).

TABLE III
PARAMETERS AT MAXIMUM TORQUE (134 A)

Item (Unit)	Type A	Type B	Type C
Ψ_a (Wb)	0.117	0.137	0.153
L_d (mH)	0.96	0.92	0.88
L_q (mH)	2.25	2.22	2.15
Ψ_{dmin} (Wb)	-0.157	-0.130	-0.107

Three IPMSMs using different magnets with different residual magnetic flux densities are examined. Permanent magnets with material codes of NMX-34GH, NMX-45GH, and NMX-54GH are used for Types A, B, and C, respectively. The magnets NMX-45GH (Type B) and NMX-54GH (Type C) are not suitable for automotive applications because of their irreversible demagnetization at high operating temperatures. However, in this study, these magnets are substituted for the high-performance magnets, based on the assumption that a strong magnet, for which irreversible demagnetization would be difficult, is developed. In this study, the temperature of the magnets is set to 60°C, and the performances of three types of IPMSM are analyzed using a 2-D finite element (FE) method.

III. ANALYSIS RESULTS

A. Torque characteristics

The total torque T, magnet torque T_m, and reluctance torque T_r under the maximum torque per ampere (MTPA) condition are shown in Fig. 2. The current phase angle β in the MTPA condition is also shown in Fig. 2. Here, β is the leading angle of the current vector from the q-axis. The current is set to 134 A, which corresponds to a current density of 17.6 A/mm². The typical parameters of the analysis models at maximum torque are listed in Table III. The maximum torques of Type A, B, and C motors are 207.4 Nm at $\beta = 50°$, 221.7 Nm at $\beta = 49°$, and 233.5 Nm at $\beta = 47°$, respectively. As shown in Fig. 3 and Table III, the PM flux linkage Ψ_a increases in proportion to the residual magnetic flux density, and accordingly the magnet torque also increases. The d- and q-axis inductances L_d and L_q, respectively, are similar for all of the motors because their structures are the same. However, when using a stronger magnet, L_d and L_q

become slightly smaller by about the same amount because of the magnetic saturation. Therefore, L_q - L_d becomes approximately constant. As a result, the reluctance torque is approximately the same for all of the motors.

The parameter Ψ_{dmin} in Table III is the minimum d-axis flux linkage, which is defined as following equation:

$$\Psi_{dmin} = \Psi_a - L_d I_{am} \qquad (1)$$

where I_{am} is a maximum current ampere. Ψ_{dmin} is an important parameter for determining the characteristics of the high-speed constant-power operation [9]. In the case of $\Psi_{dmin} < 0$, the maximum power decreases as Ψ_{dmin} decreases, and the operating speed is not limited. The condition $\Psi_{dmin} = 0$ can achieve optimum constant-power operation, the speed range of which becomes infinity, while maintaining the highest possible output power. As shown in the Table III, Ψ_{dmin} is negative in all motors. For a stronger magnet, Ψ_a increases, and Ψ_{dmin} approaches 0.

B. Speed-versus-power curve and efficiency map

The speed-versus-power curve is shown in Fig. 3, where the maximum current ampere I_{am} is 134 A and the maximum terminal voltage V_{am} is set to 507 V because the DC bus voltage V_{DC} is 650 V. As shown in Table III, using the stronger magnet, Ψ_a increases and the output power also increases. In the high-speed region, all of the motors are controlled by the flux-weakening control. In the high-speed flux-weakening operating region, the power decreases as the speed increases because Ψ_{dmin} of all of the models is negative.

The comparison of copper loss and iron loss for each motor is shown in Fig. 4. For each motor, the contour plots of copper loss for 500 W and 3000 W are shown in Fig. 5 (a), and the contour plots of iron loss for 100 W

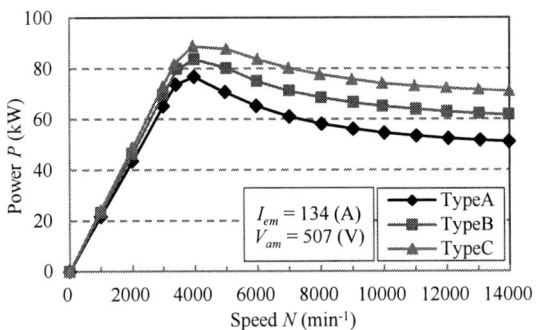

Fig. 3. Power-versus-speed curve.

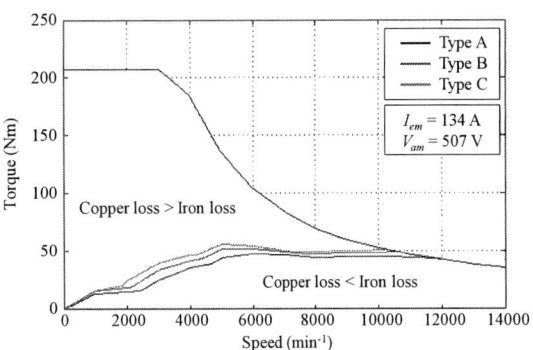

Fig. 4. Main loss.

and 500 W are shown in Fig. 5 (b). The contour plots of efficiency for 95 % and 97.5 % for each motor are shown in Fig. 5 (c). The copper loss and iron loss are considered in calculating the efficiency. The copper loss is caused by the winding resistance, and the iron loss, which consists of the hysteresis loss and eddy current loss, is caused by the flux variation in the steel core. The copper loss, iron loss, and efficiency are calculated using the following equations:

$$W_c = 3R_a I_e^2 \tag{2}$$

$$W_h = \sum_{e=1}^{nelem} \left\{ \sum_{k=1}^{n} a(B_k) \times f_k \right\} \times V_e \tag{3}$$

$$W_e = \sum_{e=1}^{nelem} \left\{ \sum_{k=1}^{n} b(B_k, f_k) \times f_k^2 \right\} \times V_e \tag{4}$$

$$W_i = W_h + W_e \tag{5}$$

$$\eta = \frac{\omega T - W_c}{\omega T + W_i} \tag{6}$$

where, W_c is the copper loss, R_a is the winding resistance, I_e is the root mean square value of the phase current, W_i is the iron loss, W_h is the hysteresis loss, W_e is the eddy current loss, nelem is the number of the FEM analysis model, e is the element number, f_k is the harmonic order frequency, k is the harmonic order, V_e is the volume of the e-th element, n is the maximum harmonic order, B_k is the amplitude of the flux density harmonic k in the radial and tangential directions of the e-th element, $a(B_k)$ and $b(B_k, f_k)$ are evaluated based on the loss curve of the material data, η is the efficiency, and ω is the mechanical angular velocity.

Fig. 4 shows the majority loss for the entire operating region. In the low-speed, heavy-load region, the copper loss is large. On the other hand, the iron loss is large in the high-speed region.

Fig. 5 (a) shows that when using a strong magnet, the copper loss is smaller in the low-speed region, especially in the heavy-load region. This is because the model using the strong magnet is able to obtain high torque and has low copper loss. On the other hand, the copper loss is high in the high-speed region. In terms of iron loss from Fig. 5 (b), there is little change in the low-speed region. In the high-speed region, the iron loss increases for the

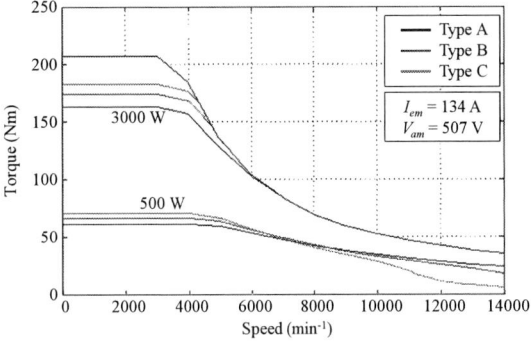

(a) Copper loss map.

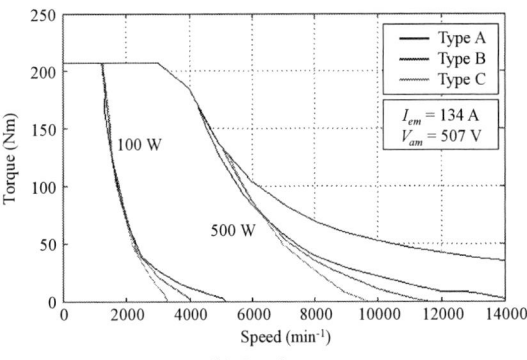

(b) Iron loss map.

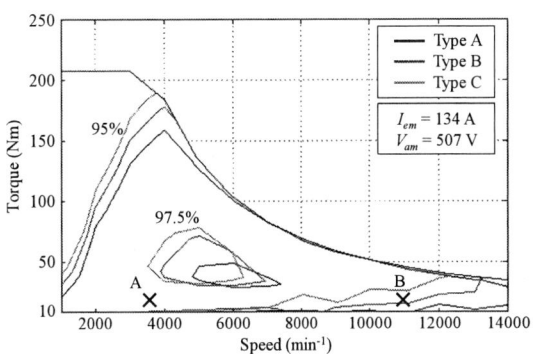

(c) Efficiency map.

Fig. 5. Losses and efficiency.

978-1-4799-2706-7/14 $31.00 © 2014 IEEE

TABLE IV
COMPARISON OF TYPICAL PROPERTIES OF ANALYSIS MODEL
(A) CITY-DRIVING EVALUATION POINT (POINT A)

Item (Unit)	Type A	Type B	Type C
Phase current I_e (A)	17.5	15.6	14.2
Phase of current β (°)	25	22	20
Flux linkage Ψ_o (Wb)	0.149	0.159	0.171
Induced electromotive force (V)	270	289	310
Copper loss W_c (W)	88.20	70.09	58.07
Iron loss W_i (W)	107.61	121.36	136.84
Total loss (W)	195.81	191.45	194.91
Maximum efficiency η (%)	97.36	97.42	97.36

(B) HIGHWAY-DRIVING EVALUATION POINT (POINT B)

Item (Unit)	Type A	Type B	Type C
Phase current I_e (A)	27.8	32.5	38.8
Phase of current β (°)	68.4	74.5	78.7
Flux linkage Ψ_o (Wb)	0.089	0.089	0.089
Induced electromotive force (V)	504	504	507
Copper loss W_c (W)	222.58	304.20	433.57
Iron loss W_i (W)	585.47	733.64	910.71
Total loss (W)	808.05	1037.84	1344.28
Maximum efficiency η (%)	96.52	95.57	94.28

stronger magnet. As shown in Fig. 5 (c), all of the motors exhibit high efficiency of greater than 95 % over most of the operating region. The Type A motor realizes high efficiency in the high-speed, light-load region, and the Type C motor realizes high efficiency in the low-speed, heavy-load region. As shown in Fig. 4, the ratio of copper loss to the total loss is high in the low-speed, heavy-load region. Hence, the Type C motor with a strong magnet realizes high efficiency in the low-speed, heavy-load region.

By using the stronger magnet, the high-efficiency region with an efficiency that exceeds 97.5 % is extended and shifted to the low-speed, heavy-load region. When using the stronger magnet, the region with an efficiency that exceeds 95 % expands, but the efficiency decreases in the high-speed region. Reduction in efficiency in the high-speed region is a problem.

C. Performance evaluation at frequent driving points

In the traction motor, the ability to achieve high torque over a wide speed range is required. However, another important aspect in this application is that the frequent operating points are concentrated under light-load conditions, such as normal city driving and highway driving [3]. Therefore, two evaluation points for such frequent driving are assumed: point A (20 Nm, 3500 min⁻¹) is the city-driving evaluation point, and point B (20 Nm, 11000 min⁻¹) is the highway-driving evaluation point. Then, the maximum efficiencies at these points are listed in Table IV.

At the city-driving evaluation point (point A), the maximum efficiency is approximately the same for all of the motors. This is because the copper loss decreases and the iron loss increases with use of the strong magnet. Thus, as a result, there is little change in the total loss. With respect to the copper loss, the current required to obtain 20 Nm is reduced because the magnet torque is increased by using the stronger magnet. On the other

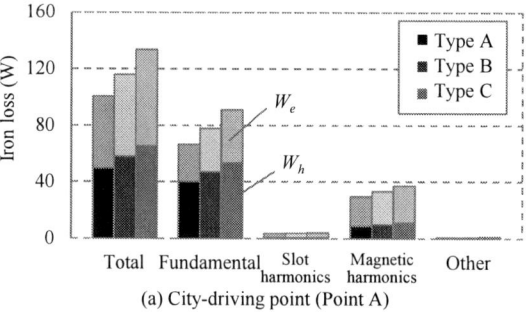

(a) City-driving point (Point A)

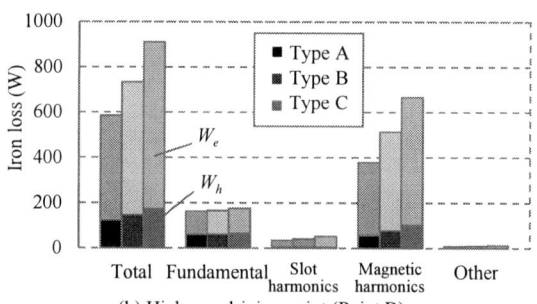

(b) Highway-driving point (Point B).
Fig. 6. Separated iron losses due to these origins.

hand, for the iron loss, the flux linkage Ψ_o increases as shown in Table IV, and thus the iron loss also increases.

At the highway-driving evaluation point (point B), the maximum efficiency of the Type A motor is the highest. In the IPMSM using the stronger magnet, the PM flux linkage Ψ_a increases as shown in Table III, and the terminal voltage reaches the maximum voltage at lower speed. Moreover, the copper loss increases because the current required for the flux-weakening becomes large at high speed.

Fig. 6 shows the separated iron losses due to these origins [10] - [13]. The origins consist of the fundamental component (first-order component), the slot harmonics

978-1-4799-2706-7/14 $31.00 © 2014 IEEE   249

The 2014 International Power Electronics Conference

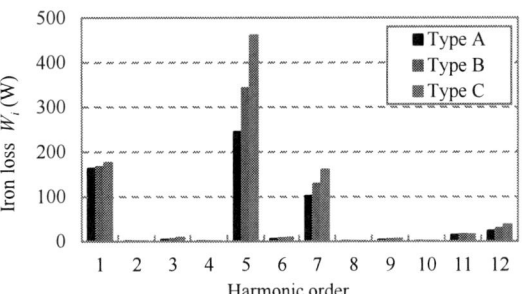

(a) Total

(b) 1st-order component

(c) 12th-order component

Fig. 7. Iron loss density distribution of Type A (Point A)

generated with rotation (12th-order, 24th-order, etc), the magnetic harmonic components (odd-order components excluding the fundamental components; third-order, fifth-order, seventh-order, etc), and the other. The orders of slot harmonics are the 12-th order slot harmonics due to the number of stator teeth per pole pair because the analysis model is an eight-pole machine with 48 slots. As the magnetic harmonics, the influence of harmonics generated by the rotor magnet and the stator winding is considered. The iron loss density distributions are shown in Fig. 7. In order to show the distribution of each cause, Fig. 7 shows (a) total iron loss, (b) first-order component iron loss and (c) 12th-order component iron loss.

At the city-driving evaluation point, the majority of the iron loss is caused by the fundamental component. The induced terminal voltages generated by the PM flux do not reach V_{am}. Therefore, in this operating region, flux-weakening control is not used. Therefore, the fundamental magnetic flux is increased by using a strong magnet, and the iron loss is increased. In addition, the magnetic harmonics is also increased by using a strong magnet. From Fig. 7, the majority of the iron loss occurs in the stator. In terms of each component, the iron loss caused by slot harmonics is generated only on the rotor surface, and the iron loss caused by magnetic harmonics is generated only on the stator teeth.

At the highway-driving evaluation point, the majority of the iron loss is caused by magnetic harmonics. In the high-speed region, the terminal voltage reaches the maximum voltage V_{am}, and the motor is under the flux-weakening control. Therefore, the iron loss due to the fundamental component is approximately the same regardless of the strength of the magnets. The harmonic components of the iron loss for each model are shown in Fig. 8, which indicates that using a strong magnet increases the iron loss in each order harmonic, especially

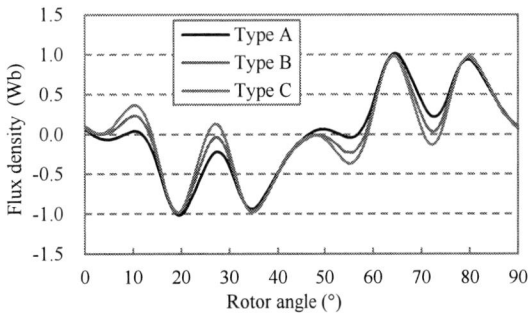

Fig. 8. Iron loss harmonic components (Point B).

Fig. 9. The variation of flux density at the center of the teeth in the radial direction (Point B).

in the fifth-order harmonic. As shown in Fig. 7, the greatest portion of the iron loss appears to be generated in the stator teeth, the variation of the flux density in the radial direction at the center of the teeth is shown in Fig. 9, and the harmonic components given by the Fourier transformation are shown in Fig. 10. Figs. 9 and 10 show the reason for the increase in the fifth-order iron loss. The fifth-order component of the flux density at the teeth increased significantly. By using stronger magnets, since

978-1-4799-2706-7/14 $31.00 © 2014 IEEE   250

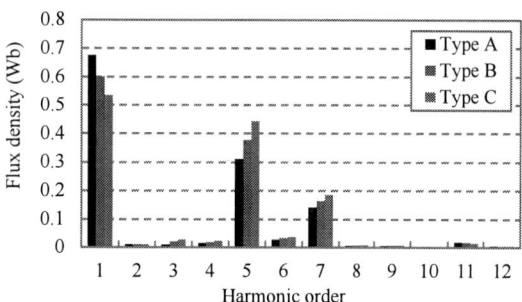

Fig. 10. Flux density harmonic components at the center of the teeth in the radial direction (Point B).

the current required for the flux-weakening control is increased and the current phase angle β increase, the variation of the magnetic flux density is complicated and harmonic components increase. This is why the iron loss increases despite the slight change in the first-order iron loss. Thus, the efficiency of the Type A motor with a weak magnet becomes highest at operating point B.

When using the stronger magnet, the iron loss caused by the harmonic components increases in the high-speed region. Since the harmonic components are related to the PM flux and the stator winding, taking into account the magnet arrangements and flux barriers will reduce the iron loss to some extent. Therefore, permanent magnets should be placed so as not to contain the harmonic components and thereby to reduce iron loss.

IV. CONCLUSIONS

In this paper, we investigated the influence of the magnetic properties of the permanent magnet on the performance of the IPMSM for automotive applications. Based on the analysis results, the effects of using a strong magnet, which has a high residual magnetic flux density, are summarized as follows.

The total torque increases because the PM flux and the magnet torque increase, and the output power also increased because Ψ_{dmin} is larger and approaches 0. From the loss maps, in the low-speed region, the copper loss is smaller and the iron loss is approximately the same when using the stronger magnet. In the high-speed region, the copper loss and iron loss also increase. The efficiency map indicates that the region in which the efficiency exceeds 97.5% is larger and is shifted to the low-speed, high-torque region. From the evaluation point, the maximum efficiency is approximately the same at the city-driving evaluation point. However, maximum efficiency is lower at the highway-driving evaluation point. For the iron loss at the evaluation point, the fundamental component is larger at the city-driving evaluation point, and the harmonic component is larger because of the flux-weakening control at the highway-driving evaluation point. The iron loss is a major loss in the high-speed region and should be considered in deciding the magnet arrangement in order to reduce the harmonic component when using a strong magnet.

REFERENCES

[1] I. Boldea, L. Tutelea, C. I. Pitic, "PM-Assisted Reluctance Synchronous Motor/Generator (PM-RSM) for Mild Hybrid Vehicles: Electromagnetic Design", *IEEE Transaction Industrial Applications*, Vol. 40, No. 2, pp. 492-498 (2004)

[2] K. Sakai, K. Hagiwara, Y. Hirano, "High-Power and High-Efficiency Permanent-Magnet Reluctance Motor for Hybrid Electric Vehicles", Toshiba review, Vol. 60, No. 11, pp. 41-44 (2005)

[3] M. Kamiya, "Development of Traction Drive Motors for the Toyota Hybrid System", *IEEJ Transaction Industrial Applications*, Vol. 126, No. 4, pp. 473-479 (2006)

[4] G. Pellegrino, A. Vagati, B. Boazzo, P. Guglielmi, "Comparison of Induction and PM Synchronous Motor Drives for EV Application Including Design Examples", *IEEE Transaction Industrial Applications*, Vol. 48, No. 6, pp. 2322-2332 (2012)

[5] D. G. Dorrell, Min-Fu Hsieh and A. M. Knight, "Alternative Rotor Designs for High Performance Brushless Permanent Magnet Machines for Hybrid Electric Vehicles", *IEEE Transactions on Magnetics*, Vol. 48, No. 2, pp. 835-838 (2012)

[6] G. PellegriNo, A. Vagati, P. Guglielmi, B. Boazzo, "Performance Comparison Between Surface-Mounted and Interior PM Motor Drives for Electric Vehicle Application", *IEEE transactions on industrial electronics*, Vol. 59, No. 2, pp. 803-811 (2012)

[7] M. Olszewski, "Evaluation of the 2010 Toyota Prius Hybrid Electric Drive System", *Oak Ridge Nat. Lab.*, U. S. Dept. Energy (2011)

[8] K. Kiyota, A. Chiba, "Design and Analysis of a 60kW Switched Reluctance Motor for Hybrid Electric Vehicles", *Proc. of the 2011 Annual Conference of Industry Applications Society*, IEEE of Japan, No. 3-87, pp. 401-406 (2011) (in Japanese)

[9] S. Morimoto, "Trend of Permanent Magnet Synchronous Machines", *IEEJ Transaction*, 2, pp. 101-108 (2007)

[10] Seok-Hee Han, T. M. Jahns, Z. Q. Zhu, "Design Tradeoffs Between Stator Core Loss and Torque Ripple in IPM Machines" *IEEE Transactions on Industry Applications*, Vol. 46, No. 1, pp. 187-195 (2010)

[11] K. Yamazaki, H. Ishigami, "Rotor-Shape Optimization of Interior-Permanent-Magnet Motors to Reduce Harmonic Iron Losses", *IEEE Transactions on Industrial Electronics*, Vol. 57, No. 1, pp. 61-69 (2010)

[12] K. Yamazaki and Y. Seto, "Iron Loss Analysis of Interior Permanent Magnet Synchronous Motors − Variation of Main Loss Factors Due to Driving Condition", *IEEE Transactions on Industry Applications*, Vol. 42, No. 4, pp. 1045-1052 (2006)

[13] S. Jang-Ho, W. Dong-Kyun, C. Tae-Kyung and J. Hyun-Kyo, "A Study on Loss Characteristics of IPMSM for FCEV Considering the Rotating Field", *IEEE Transactions on Magnetics*, Vol. 46, No. 8, pp. 3213-3216 (2010)

Characteristics of Interior Permanent Magnet Synchronous Motor with Imperfect Magnets

Syuhei Shinagawa Takeo Ishikawa Nobuyuki Kurita
Division of Electronics and Informatics
Gunma University
Kiryu, Japan
t13801444@gunma-u.ac.jp

Abstract— This paper investigates the effect of demagnetization of the permanent magnets (PM) on the performance of interior permanent magnet motors. In order to remove the individual difference of the motors, several shapes of PMs are prepared, and the experiments are carried out with the same stator. This paper takes into account two types control strategy, V/f constant control and vector control. It is clarified that the stator current and the vibration acceleration are useful for the fault detection of demagnetized PMs under controlled by V/f constant. Under the vector control, the stator voltage and the vibration acceleration are useful for the detection of demagnetized PMs. Moreover, it is found that the fundamental component of stator voltage of the motor with magnets demagnetized in the axial direction is larger than that with magnets demagnetized in the radial direction.

Keywords— Demagnetization, failure diagnosis, permanent magnet synchronous motor

I. INTRODUCTION

Permanent magnet synchronous motors (PMSMs) are widely used in home appliances and industrial applications, such as air conditioners and hybrid vehicles and so on, because of their high efficiency and high-power density. As these motors require a high driving reliability, the detection of failure of the demagnetization of PMs is becoming important. PMs of the PMSM can be demagnetized by temperature, large stator currents, large short-circuit currents produced by inverter or stator faults, and the aging of the magnet itself. Moreover, since the PM material is often magnetized after it is built in the devices, there is a possibility that the PM is not in the state of complete magnetization.

There have been several research studies for detecting the demagnetization of PM. Rajagopalan et al. made a magnet defect by chipping off a part of the magnet and measured stator current [1]. Urresty, Romeral, and Ortega et al. have made a lot of contributions [2]–[8]. They analyzed the stator current of PM motor by several methods, where one of three pairs of poles was demagnetized by 50 %, namely, the amount of 8.3 % of PM was decreased. So, in these past studies, the demagnetization of PMs is relatively large. In order to deal with a less demagnetized situation, we have

manufactured two interior PM motors, where the thickness of one of four PMs is reduced by 10 % and 20 % in order to imitate the demagnetization of PM, namely, the amount of 2.5 % and 5 % of the PM is decreased [9], [10]. When dealing with the less demagnetized situation, slight difference of motors except for PMs could affect the experimental results.

In order to remove the individual difference of the motors, we prepare several shapes of PMs. And then we imitate the demagnetization or imperfect situation of PMs by inserting these PMs. This paper deals with two types of PMs to imitate the demagnetization. One is a situation that the PM of one pole is reduced in the radial direction. The other is a situation that the PM of one pole is reduced in the axial direction. The volume of one of four PMs is reduced by 10%, 20 % and 30 %, namely, the amount of 2.5 %, 5 % and 7.5 % of the PM is decreased. This paper clarifies the effect of the demagnetization of PMs on the performance of PMSMs under controlled by both a V/f constant and vector strategy.

II. PMSM WITH PERMANENT MAGNET DEFECT

Fig. 1 shows the rotor configuration of the experimental PMSM. This motor is a 1.5-kW, 3000-min⁻¹, 4.8-N·m, 5.6-A, four-pole machine. One of four poles is demagnetized. Fig. 2 shows the demagnetization in radial direction, we call this situation radial demagnetization hereafter. One pole is composed of four PMs, and the thickness of four PMs which compose one pole is reduced by 10 %, 20 % and 30 % in order to imitate the demagnetization or imperfect magnetization. Fig. 3 shows the demagnetization in axial direction, we call this situation axial demagnetization hereafter. Axial length of two of four PMs is reduced by 20 %, 40 % and 60 %. Therefore, the PM volume of one pole for axial demagnetization is same as that for radial demagnetization. Nonmagnetic materials are inserted into the reduction region in order to remove eccentricity. Fig. 4 shows a photograph of the rotor of the experimental PMSM. When performing experiments with several motors, slight difference can affect the motor performance especially if the amount of demagnetization is little. In order to avoid this problem, we insert the PMs to the same rotor.

978-1-4799-2706-7/14 $31.00 © 2014 IEEE

Fig. 5 shows a block diagram for the PMSM controlled by V/f constant strategy. The s-function makes the reference of three stator voltages proportional to the inverter frequency f. This block diagram was made on DS1102 DSP board, and a block DS1102 PWM sends a duty cycle to three-phase inverter. Fig. 6 shows a typical block diagram for the vector-controlled PMSM system. In this diagram, two control loops are used: one is the inner loop to regulate the stator currents by detecting the rotor angle, and the other is the outer loop to control the motor speed. Fig. 7 shows the experimental setup. The system is composed of an inverter, the IPMSM coupled with an encoder to defect the rotor position, a torque meter, and a hysteresis brake. Physical variables to be detected are stator voltage, stator current, the angular velocity, the developed torque, and the vibration. A piezoelectric accelerometer PV08A is attached on the top of the stator shown in Fig. 7 and is connected to a charge amplifier UV-06, which is connected to a signal analyzer SA-01A4 and them to a PC where a software for SA-01A4 is installed.

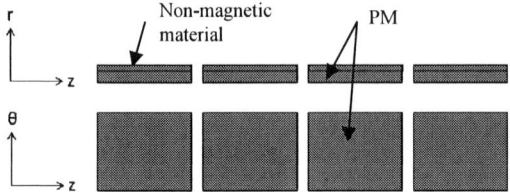

Fig. 2. PMs for radial demagnetization

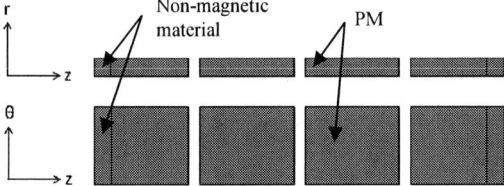

Fig. 3. PMs for axial demagnetization

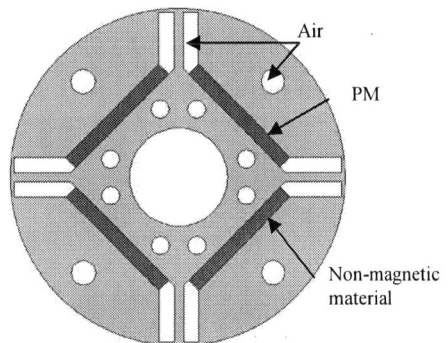

Fig. 1. Rotor configuration

Fig. 4. Rotor of the experimental PMSM.

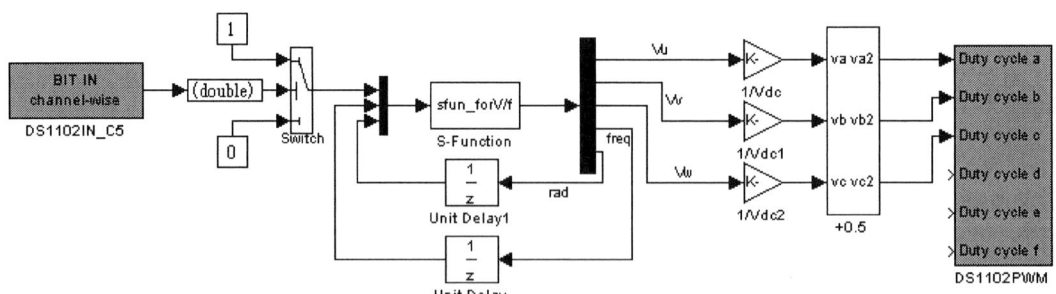

Fig. 5. Block diagram for PMSM controlled by V/f strategy.

978-1-4799-2706-7/14 $31.00 © 2014 IEEE

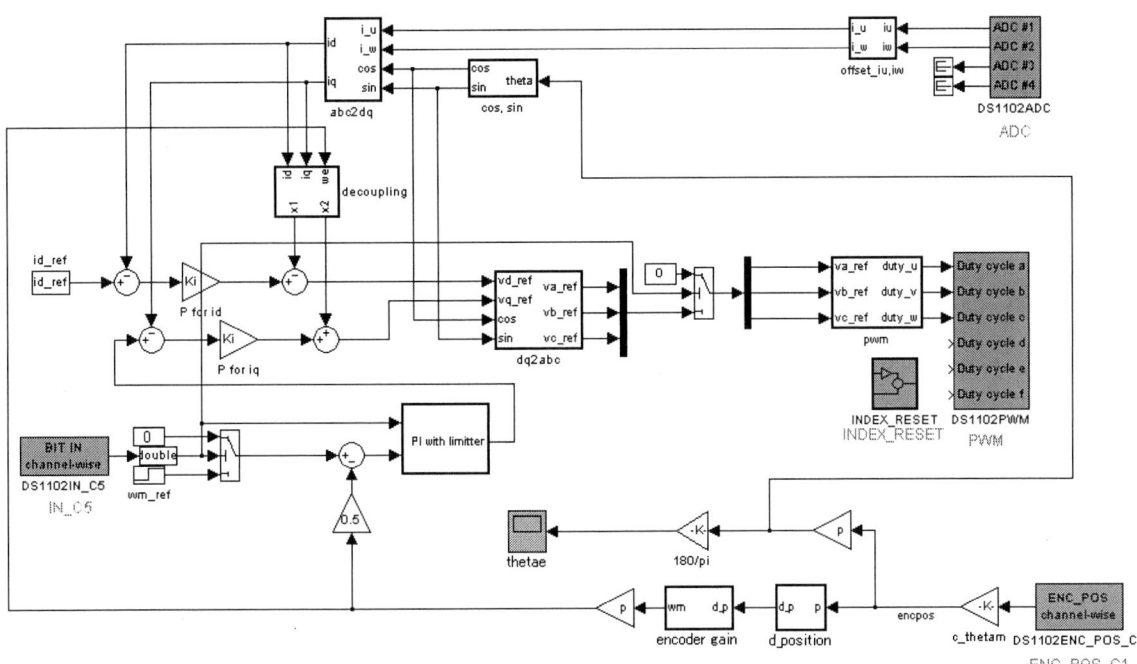

Fig. 6. Block diagram for a brushless dc motor controlled by the vector strategy.

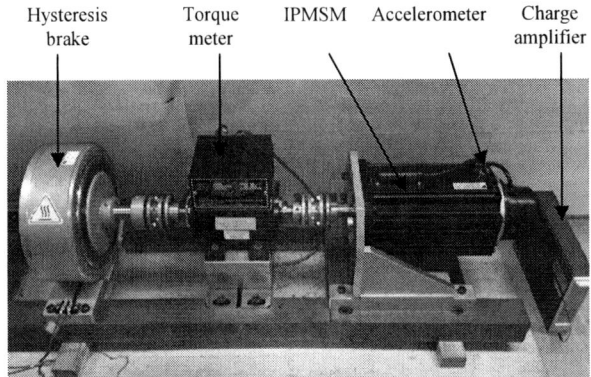

Fig. 7. Experimental setup.

III. MEASUREMENT RESULTS UNDER V/F CONTROL

The speed of IPMSM is controlled to a constant value by the V/f strategy. Load of 0 %, 10 %, 20 % and 30 % of rated torque is applied by the hysteresis brake. The measurement was carried out three times. Fig. 8 shows the fundamental component of the stator current of the motor with the radial demagnetization. It is found that there is a difference when the load torque is small. The fundamental component of the stator current is of the order of the healthy motor the motor with the PM reduce by 10 %, 20 %, 30 %. Fig. 9 shows the fundamental component of the stator current of the motor with the axial demagnetization. The characteristics are approximately the same as that of radial demagnetization. As a result, we can distinguish the demagnetization

using the fundamental stator current especially in a small torque range. Fig. 10 shows the fundamental component of the stator voltage of the motor with the radial demagnetization, and Fig. 11 for the axial demagnetization. The stator voltage, that is, the output voltage of the inverter decreases, when the output torque becomes large. This is due to not enough capacity of input capacitor and voltage drop of transistors in inverter circuit. It is found that it is difficult to distinguish the demagnetization using the fundamental stator voltage in this control strategy.

Fig. 12 shows the 11th component of vibration of the motor with the radial demagnetization, that is 550 Hz. It is found that there is a difference when the load torque is large. Fig. 13 shows the 11th component of vibration of the motor with the axial demagnetization. The characteristics are approximately the same as that of radial demagnetization.

IV. MEASUREMENT RESULTS UNDER VECTOR CONTROL

This chapter discusses the motor performance under a closed loop control. The speed of IPMSM is controlled to a constant value by the vector strategy. Load of 0 %, 10 %, 20 %, 30 % and 40 % of rated torque is applied by the hysteresis brake. The measurement was carried out three times. Fig. 14 shows the fundamental component of the stator current of the motor with the radial demagnetization. The fundamental stator current is approximately proportional to the load torque. Fig. 15 shows the fundamental component of the stator current of

The 2014 International Power Electronics Conference

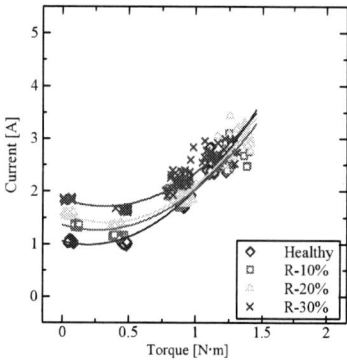

Fig. 8. Fundamental component of the stator current of the motor with the radial demagnetization.

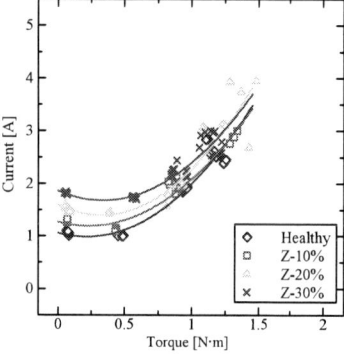

Fig. 9. Fundamental component of the stator current of the motor with the axial demagnetization.

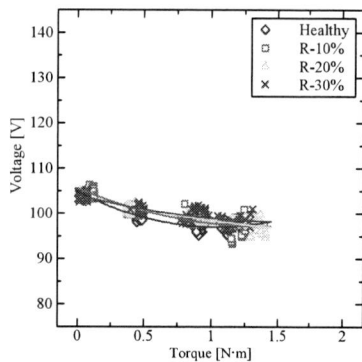

Fig. 10. Fundamental component of the stator voltage of the motor with the radial demagnetization.

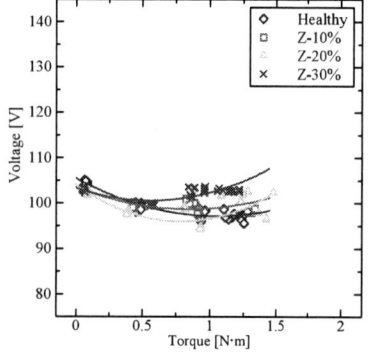

Fig. 11. Fundamental component of the stator voltage of the motor with the axial demagnetization.

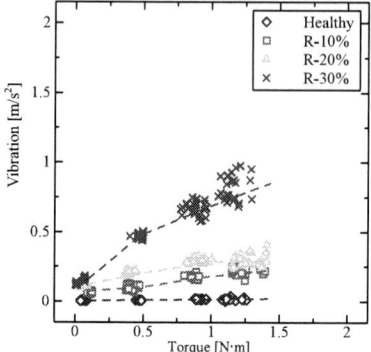

Fig. 12. 11th component of the vibration of the motor with the radial demagnetization.

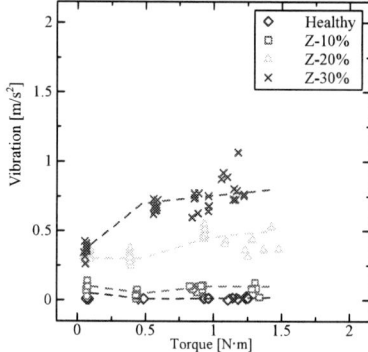

Fig. 13. 11th component of the vibration of the motor with the axial demagnetization.

the motor with the axial demagnetization. The characteristics are approximately the same as that of radial demagnetization. It is difficult to distinguish the demagnetization using the fundamental stator current in this control strategy.

Fig. 16 shows the fundamental component of the stator voltage of the motor with the radial demagnetization. We can find a difference in the stator voltage between the motor with the demagnetized and the healthy machine. The fundamental component of stator voltage is of the order of the motor with the PM reduced by 30 %, 20 % and 10 %, and the healthy motor. Fig. 17 shows the fundamental component of the stator voltage of the motor with the axial demagnetization. We can find a difference in the stator voltages between the motor with the demagnetized magnet and the healthy machine. The order of the fundamental component of stator voltage of the motor with the axial magnetization is the same as that of radial demagnetization. It is found that the fundamental component of stator voltage of the motor with PMs demagnetized in the axial direction is larger than that with PMs demagnetized in the radial direction.

Next, we investigate the harmonic components of the stator voltage. Fig. 18 shows the Fourier analysis of the stator voltage of the motor with the radial demagnetization at no load condition. Although peaks are at 2, 5, 7, 10, 11, 13 and 14th harmonic components of supply frequency, difference due to demagnetization is observed at the 10, 11, 13 and 14th components. Among there, Fig. 19 shows the 11th component of stator voltage

978-1-4799-2706-7/14 $31.00 © 2014 IEEE

255

of the motor with the radial demagnetization. It is found that there is a difference when the load torque is small. We also investigate the harmonic components of the stator voltage of the motor with axial demagnetization. The characteristics are approximately the same as that of radial demagnetization.

Fig. 20 shows the Fourier analysis of vibration of the motor with the radial demagnetization when the load torque is 40 % of rated load condition. Peaks are at 0, 4, 5, 8, 11, 12, 13, 23, 24 and 25th components of supply frequency. Difference due to demagnetization is observed at the 11, 23 and 25th components. Among there, Fig. 21 shows the 11th component of vibration of the motor with the radial demagnetization. It is found that there is a difference when the load torque is large. Fig. 22 shows the Fourier analysis of vibration of the motor with the axial demagnetization when the load torque is 40 % of rated load condition. Peaks are observed at the same components as those of the motor with the radial demagnetization. Difference due to demagnetization is observed at the same components of the motor with the radial demagnetization component namely, 11, 23 and 25th components. Fig. 23 shows the 11th component of vibration of the motor with the axial demagnetization. It is found that there is a difference when the load torque is large. The characteristics are approximately the same as that of radial demagnetization.

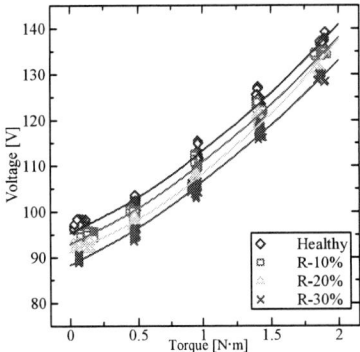

Fig. 16. Fundamental component of the stator voltage of the motor with the radial demagnetization.

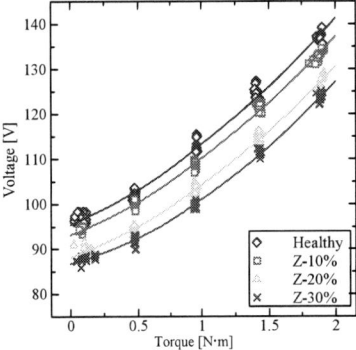

Fig. 17. Fundamental component of the stator voltage of the motor with the axial demagnetization.

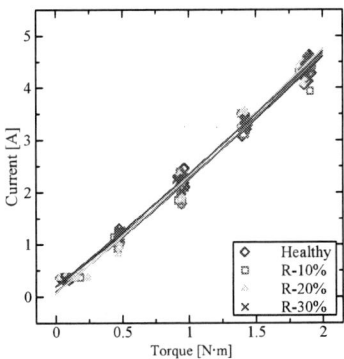

Fig. 14. Fundamental component of the stator current of the motor with the radial demagnetization.

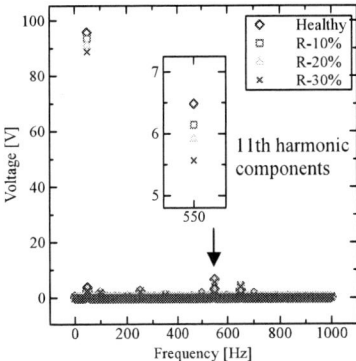

Fig. 18. Fourier analysis of the stator voltage of the motor with the radial demagnetization.

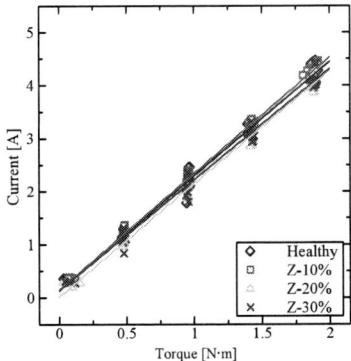

Fig. 15. Fundamental component of the stator current of the motor with the axial demagnetization.

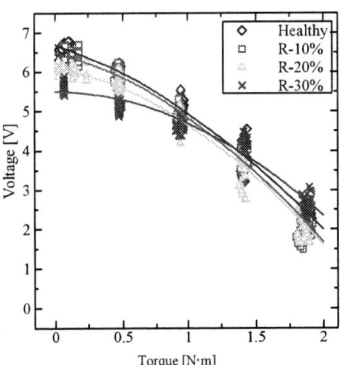

Fig. 19. 11th component of the stator voltage of the motor with the radial demagnetization.

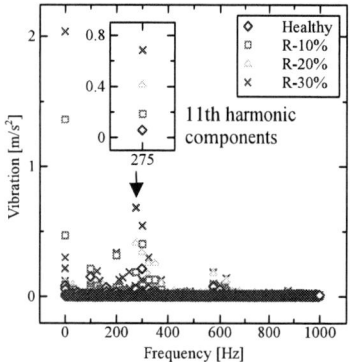

Fig. 20. Fourier analysis of the vibration of the motor with the radial demagnetization.

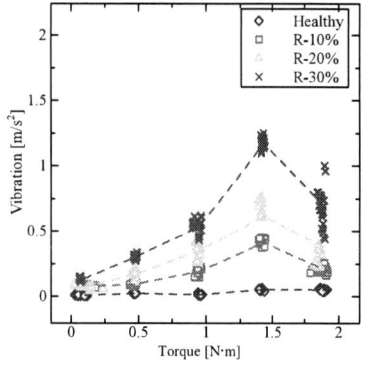

Fig. 21. 11th component of the vibration of the motor with the radial demagnetization.

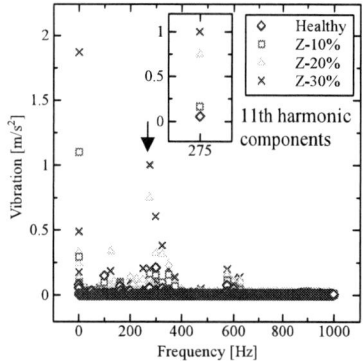

Fig. 22. Fourier analysis of the vibration of the motor with the axial demagnetization.

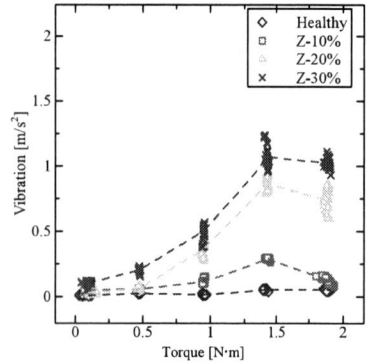

Fig. 23. 11th component of the vibration of the motor with the axial demagnetization.

V. CONCLUSIONS

In the paper, we have investigated the effect of less demagnetized situations of the PMs in the IPMSM. We have prepared several shapes of PMs in order to remove the slight difference of the motors, and have dealt with two types of PMs to imitate the demagnetization. This paper has clarified that the stator current and the vibration acceleration are useful for the fault detection of demagnetized PM of the PMSM under V/f control. It has also clarified that the stator voltage and the vibration acceleration are useful for the fault detection of demagnetized PM under vector control. Moreover, it has been found that the fundamental component of stator voltage of the motor with PMs demagnetized in the axial direction is larger than that with PMs demagnetized in the radial direction.

REFERENCES

[1] S. Rajagopalan, R. W. Le, T. G. Habetler, and R. G. Harley, "Diagnosis of potential rotor faults in brushless DC machines," in Proc. Inst. Elect. Eng. Conf Publ., 2004, vol. 2, pp. 668–673.

[2] J. Rosero, L. Romeral, J. Cusido, and J. A. Ortega, "Fault detection by means of wavelet transform in a PMSM under demagnetization," in Proc. IEEE 33rd Annu. Conf. Ind. Electron. Soc., 2007, pp. 1149–1154.

[3] A. G. Espinosa, J. A. Rosero, J. Cusido, L. Romeral, and J. A. Ortega, "Fault detection by means of Hilbert Huang transform of the stator current in a PMSM with demagnetization," IEEE Trans. Energy Convers., vol. 25, no. 2, pp. 312–318, Jun. 2010.

[4] J. R.R. Ruiz, J.A.Rosero, A. G. Espinosa, and L. Romeral, "Detection of demagnetization faults in permanent-magnet synchronous motors under nonstationary conditions," IEEE Trans. Magn., vol. 45, no. 7, pp. 2961–2969, Jul. 2009.

[5] M.D. Prieto,A.G.Espinosa, J. R. Ruiz, J.C.Urresty, and J. A. Ortega,"Feature extraction of demagnetization faults in permanent-magnet synchronous motors based on box-counting fractal dimension," IEEE Trans. Ind. Electron., vol. 58, no. 5, pp. 1594–1605, May 2011.

[6] J. Urresty, J. Riba, H. Saavedra, and J. Romeral, "Analysis of demagnetization faults in surface-mounted permanent magnet synchronous motors with symmetric windings," in Proc. IEEE Int. Symp. Diagnost. Elect. Mach., Power Electron. Drives, Sep. 2011, pp. 240–245.

[7] J. Urresty, J. R. Riba, M. Delgado, and L. Romeral, "Detection of demagnetization faults in surface-mounted permanent magnet synchronous motors by means of the zero-sequence voltage component," IEEE Trans. Energy Convers., vol. 27, no. 1, pp. 42–51, Mar. 2012.

[8] J. Urresty, R. J. Riba, and L. Romeral, "A back-emf based method to detect magnet failures in PMSMs," IEEE Trans. Magn., vol. 49, no. 1, pt. 3, Jan. 2013.

[9] T. Ishikawa, Y. Seki, N. Kurita, T. Matsuura, "Failure diagnosis of brushless DC motor with rotor magnet defect," 2011 IEEE International Electric Machines & Drives Conference (IEMDC), pp.1552-1556, May 2011.

[10] T. Ishikawa, Y. Seki, and N. Kurita, "Analysis for Fault Detection of Vector-Controlled Permanent Magnet Synchronous Motor With Permanent Magnet Defect, " IEEE Transaction on Magnetics, vol. 49, no. 5, pp. 2331-2334, MAY 2013

Study of Stator Structure to Improve Reluctance Torque for IPMSM with Concentrated Winding

R. Morikawa, M. Sanada, S. Morimoto and Y. Inoue

Graduate School of Engineering
Osaka Prefecture University
1-1 Gakuen-cho, Naka-ku, Osaka, Japan
st106043@edu.osakafu-u.ac.jp

Abstract – **Interior permanent magnet synchronous motors (IPMSMs) that employ rare-earth magnets are high-performance motors and are used in many applications. However, it is desirable to reduce the number of rare-earth magnets used in IPMSMs because of the high cost. If a non-rare earth magnet is used in IPMSMs with concentrated windings, the motor generates reluctance torque principally; however, this reluctance torque is significantly decreased. Therefore, it is necessary to improve the reluctance torque in concentrated-winding IPMSMs. In this study, we aim to improve the reluctance torque for IPMSMs with concentrated windings by investigating the stator structure. The influence of the length and thickness of the pole horn on reluctance torque is examined and suitable lengths and thicknesses of a pole horn are clarified by finite element analysis.**

Index Terms–**IPMSM, concentrated winding, stator, pole horn, reluctance torque.**

I. INTRODUCTION

Interior permanent magnet synchronous motors (IPMSMs) are widely used because of their high power and high efficiency [1], [2]. Rare-earth magnets are used in IPMSMs; however, it is desired to reduce the number of rare-earth magnets because of the high cost and unstable supply. A motor employing a rare-earth magnet generates magnet torque principally, whereas a motor using non-rare earth magnet generates reluctance torque principally [3] because the magnetic coercive force is deteriorated, and as a result the magnet torque is low. Tow principal winding methods are used in the manufacture of IPMSMs, one is concentrated winding and the other is distributed winding [4]. Although IPMSMs with concentrated windings have lower torque because their reluctance torque is less than that of IPMSMs with distributed windings [5], it is desirable to use a concentrated-winding stator because a concentrated winding reduces copper loss and is easy to produce compared to a distributed-winding stator [6]. However, if a concentrated-winding stator is used for IPMSMs with non-rare earth magnets, the total torque decrease because non-rare earth IPMSMs generate reluctance torque principally, and reluctance torque is relatively small. Therefore, it is necessary to improve the reluctance torque of non-rare earth IPMSMs with concentrated windings.

In this study, we aim to improve the reluctance torque of IPMSMs with concentrated windings by investigating the structure of the stator. The influence of the length and thickness of the pole horn on reluctance torque is examined for IPMSMs with concentrated windings.

II. INFLUENCE OF POLE HORN ON TORQUE IN RARE-EARTH MAGNET MODEL

A. Analysis model

Fig. 1 shows the analysis model used in our experiment and Table I lists its specifications. In this study, the influence of the pole horn length and pole horn thickness on reluctance torque is examined. Pole horn length is expressed by pole horn angle θ, which is set to 0° when there is no pole horn, as shown in Fig. 1. The input current I_e is set to 3 A. In this model, the number of stator poles is 6 and the teeth width is 15 mm, therefore the maximum value of θ is 15°. Because the number of turns per phase is 232 regardless of the shape of the pole horn, the maximum value of pole horn thickness δ is set to 3 mm considering the negative influence of δ on the winding space. Thus the examination is conducted under the conditions that the pole horn thickness is $\delta = 1$ mm, 2 mm, and 3 mm. The basic torque equation for IPMSMs is shown in Eq. (1).

$$T = P_n \Psi_a I_a \cos \beta + \frac{P_n}{2}(L_q - L_d)I_a{}^2 \sin 2\beta \qquad (1)$$

where P_n is the number of pole pairs, Ψ_a is the armature flux linkage resulting from the permanent magnets on the d-q coordinate, I_a is the input current expressed on the d-q coordinate, L_d is the d-axis inductance, L_q is the q-axis inductance, and β is the leading phase angle of I_a from the

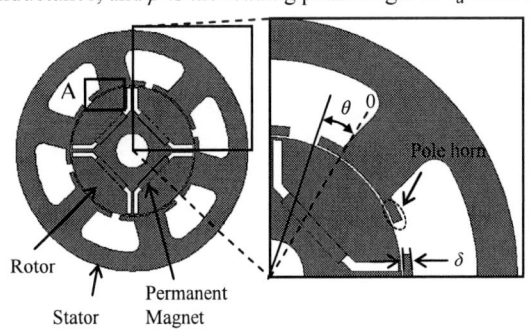

Fig. 1. Analysis model of IPMSMs with concentrated windings.

TABLE I
SPECIFICATIONS OF ANALYSIS MODEL

Item [unit]	Value
Number of poles	4
Stator outer diameter (mm)	112
Stator inner diameter (mm)	61
Rotor diameter (mm)	60
Air gap length (mm)	0.5
Shaft diameter (mm)	16
Back yoke width (mm)	9
Teeth width (mm)	15
Stack length (mm)	40
Number of turns (turn/phase)	232
Coercive force of PM (kA/m)	915
Steel material	50H470

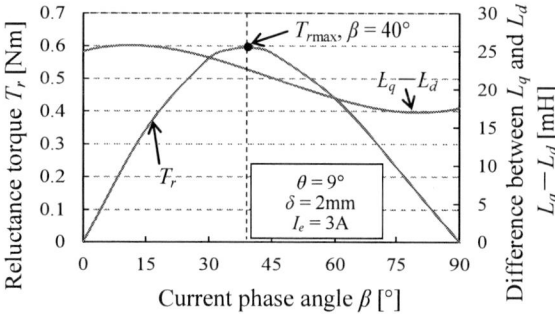

Fig. 2 Influence of β on reluctance torque.

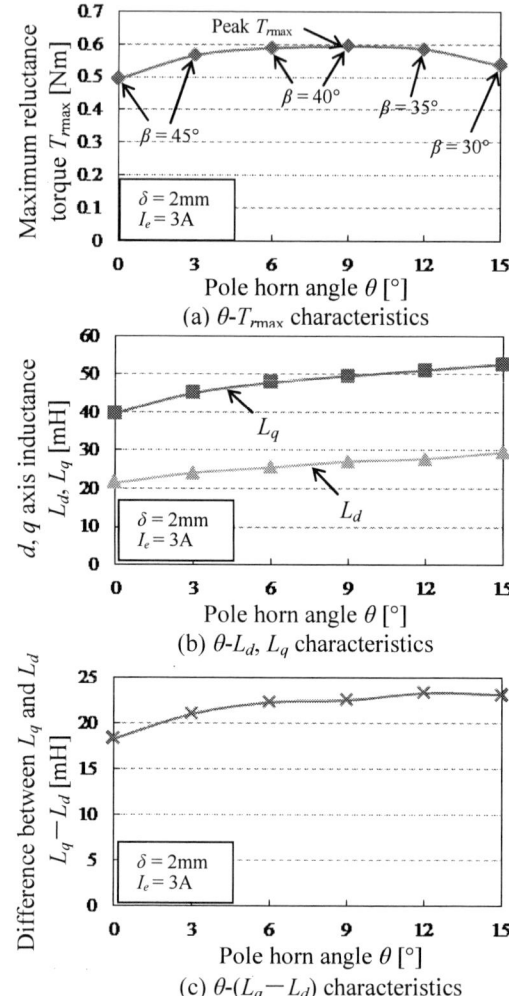

(a) θ-$T_{r\max}$ characteristics

(b) θ-L_d, L_q characteristics

(c) θ-(L_q-L_d) characteristics

Fig. 3. Influence of θ on $T_{r\max}$ and motor parameters.

q-axis.

In Eq. (1), the first term in the right-hand side is the magnet torque component T_m, which depends on the magnetic flux, and the second term in the right-hand side is the reluctance torque component T_r, which depends on β and the difference between L_q and L_d ($L_q - L_d$). Fig. 2 shows the influence of β on reluctance torque under the condition that $\theta = 9°$, $\delta = 2$ mm, and $I_e = 3$ A.

As shown in Fig. 2, the value of $L_q - L_d$ changes by β and the reluctance torque becomes maximum ($T_{r\max}$) at the suitable value of β. Therefore, it is necessary to set β so that it increases the reluctance torque.

B. Influence of pole horn angle θ

Fig. 3 shows the influence of pole horn angle θ on maximum reluctance torque ($T_{r\max}$), L_d, L_q, and $L_q - L_d$ under the condition that $\delta = 2$ mm and $I_e = 3$ A. Fig. 4 shows the inductance characteristics against β. As shown in Fig. 3 (a), $T_{r\max}$ has a peak against θ. Because Eq. (1) shows that reluctance torque is influenced by $L_q - L_d$, it is necessary to consider the inductance characteristics against θ, as shown in Figs. 3 (b), (c). As shown in Fig. 3 (b), L_d and L_q increase with the increase in θ. Moreover, Fig. 3 (c) shows that $L_q - L_d$ increases because the increase of L_q is larger than that of L_d. $T_{r\max}$ has a peak because $L_q - L_d$ has a peak at $\theta = 12°$. However, the value

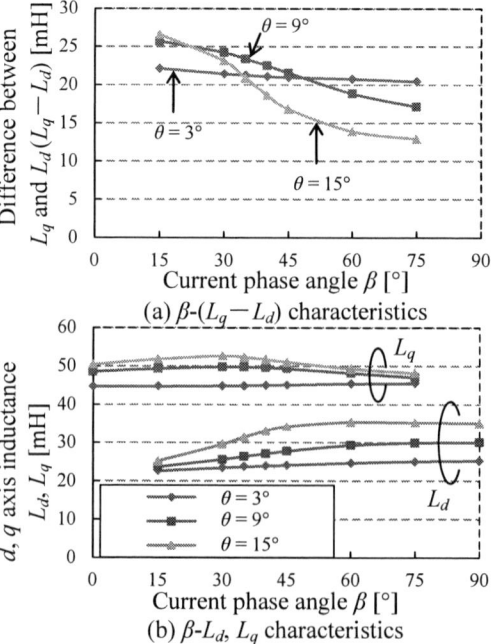

(a) β-(L_q-L_d) characteristics

(b) β-L_d, L_q characteristics

Fig. 4. Influence of β on inductance characteristics.

of θ where T_{rmax} and $L_q - L_d$ become a peak is different because β of peak T_{rmax} decreases with the increase in θ.

As shown in Fig. 4 (a), the $L_q - L_d$ characteristic is tilted, and $L_q - L_d$ decreases with the increase in β. Therefore β of peak T_{rmax} becomes small because the value of $L_q - L_d$ is large in the small region of β. The reason for this gradient is assumed to be that the d-axis current increases with the increase in β and this weakens the magnetic flux. Thus, as shown in Fig. 4 (b), the magnetic saturation is eased and L_d increases. Because the more θ increases, the more flux flows through the pole horn, the influence of increasing L_d by saturation easing becomes large at a large value of θ. Therefore, the gradient characteristic of $L_q - L_d$ becomes large for a large value of θ. Thus as θ increases, the β of peak T_{rmax} decreases. Accordingly, T_{rmax} has a peak with respect to θ and as a result T_{rmax} becomes maximum at $\theta = 9°$ because of the good combination of β and $L_q - L_d$ under the condition that $\delta = 2$ mm and $I_e = 3$ A.

C. Influence of pole horn thickness δ

Fig. 5 (a) shows the influence of pole horn thickness δ on T_{rmax} and $L_q - L_d$ under the condition that $I_e = 3$ A and $\theta = 9°$. Fig. 5 (b) shows the influence of pole horn thickness δ on a suitable θ. As shown in Fig. 5 (a), T_{rmax} and $L_q - L_d$ increase with an increase in δ, because the iron core of the pole horn becomes thick with the increase in δ and the reluctance becomes small. Therefore, the thicker δ becomes, the larger T_{rmax} becomes.

As shown in Fig. 5 (b), θ of peak T_{rmax} is deferent only when $\delta = 1$ mm, and T_{rmax} become maximum at $\theta = 6°$. Fig. 6 shows the magnetic flux density distribution near pole horn area "A" in Fig. 1. In Fig. 6, the area of high magnetic flux density ($B \geq 1.8 T$) is enclosed in a white dashed line. Fig. 7 shows inductance characteristics against θ under the conditions that $I_e = 3$ A and $\delta = 1$ mm, 3 mm. As shown in Fig. 6, the influence of magnetic saturation becomes strong with the decrease in δ. This is the reason that the increase in L_q at $\delta = 1$ mm is suppressed in comparison with that at $\delta = 3$ mm and $L_q - L_d$ has a peak at $\theta = 6°$. These are the reasons that T_{rmax} become maximum at $\theta = 6°$ only when δ is set to 1 mm.

D. Influence of pole horn angle θ and pole horn thickness δ

In the preceding section, the influence of θ on T_{rmax} and that of δ are examined independently. In this section, the influence of θ and δ is examined at the same time. Fig. 8 shows the influence of θ and δ on maximum reluctance torque and maximum total torque T under the condition that $I_e = 3$ A. As a result, reluctance torque becomes maximum at $\theta = 9°$ and $\delta = 3$ mm. However, as shown in Fig. 8 (b), total torque becomes maximum at $\theta = 12°$ and $\delta = 3$ mm because a longer pole horn length corresponds to a greater torque. However, at $\theta = 15°$, the flux is short at the edge of the pole horn; so total torque becomes maximum at $\theta = 12°$. Therefore, the model of peak T_{rmax} and the model of peak T are different.

In these analysis models, the space factor is not considered and the number of turns per phase is 232 in

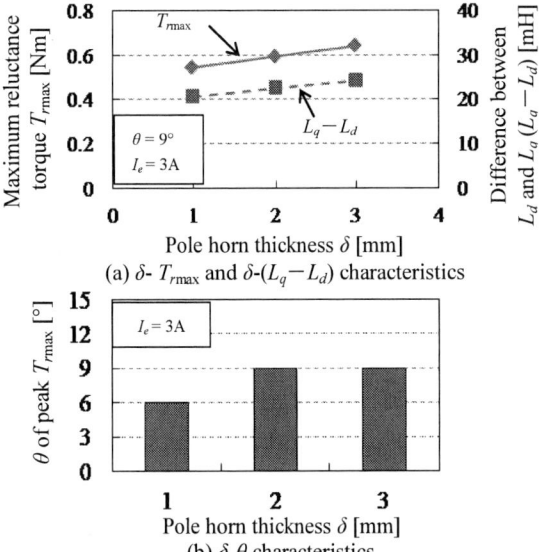

(a) δ- T_{rmax} and δ-($L_q - L_d$) characteristics

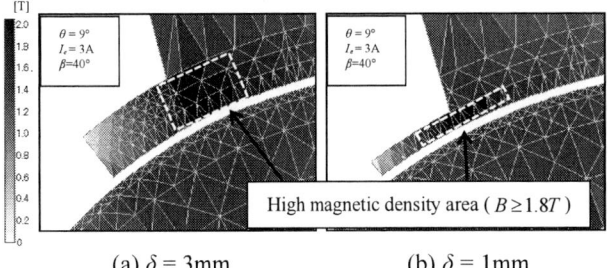

(b) δ-θ characteristics

Fig. 5. Influence of δ on T_{rmax} and inductance characteristics.

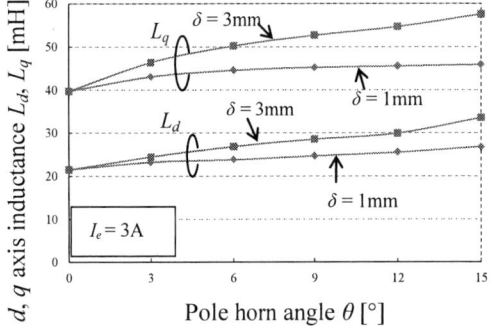

(a) $\delta = 3$mm   (b) $\delta = 1$mm

Fig. 6. Magnetic flux density distribution near the pole horn.

Fig. 7. Influence of θ on inductance characteristics.

order to clarify the influence of the pole horn shape. However, the space factor is important from the viewpoint of practicality. Therefore, it is necessary to examine the analysis model under the condition that the space factor is maintained at 50 %.

Fig. 9 shows the influence of θ and δ on maximum reluctance torque and maximum total torque under the condition that the space factor is kept constant at 50% by changing the number of turns per phase. To maintain the space factor, the number of turns per phase is changed to 236, 226, and 214 at $\delta = 1$ mm, 2 mm, and 3 mm, respectively. Fig. 9 (a) shows that reluctance torque decreases at $\delta = 3$ mm because the number of turns per phase decreases to 214. Conversely, reluctance torque increases at $\delta = 1$ mm because the number of turns per phase is increased to 236; however, reluctance torque is influenced by the saturation of the pole horn. Therefore, reluctance torque becomes maximum at $\delta = 2$ mm, $\theta = 9°$.

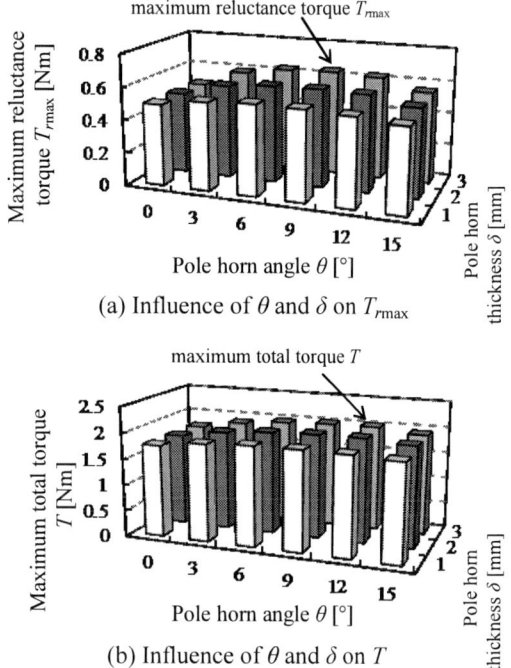

(a) Influence of θ and δ on $T_{r\text{max}}$

(b) Influence of θ and δ on T

Fig. 8. Influence of θ and δ on maximum reluctance torque and maximum total torque ($I_e = 3$ A).

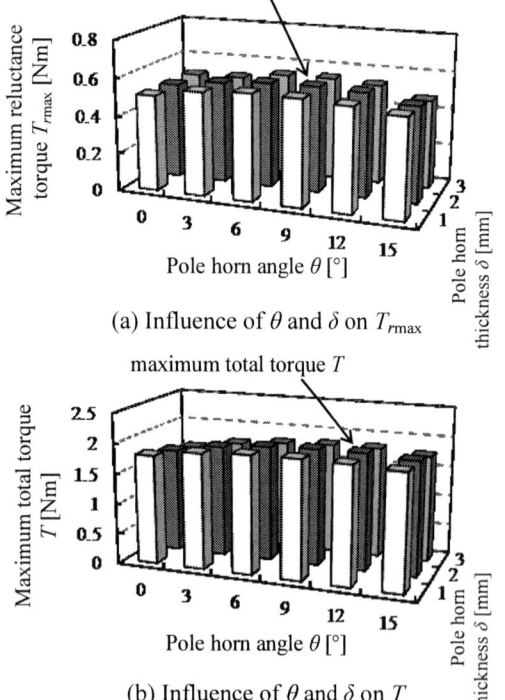

(a) Influence of θ and δ on $T_{r\text{max}}$

(b) Influence of θ and δ on T

Fig. 9. Influence of θ and δ on maximum reluctance torque and maximum total torque under the condition that the space factor is maintained at 50% ($I_e = 3$ A).

Total torque, which is shown in Fig. 9 (b), is also subject to the influence of the number of turns per phase and total torque becomes maximum at $\delta = 2$ mm, $\theta = 12°$.

From these results, it is shown that the influence of the number of turns on reluctance torque is larger than the influence of pole horn thickness.

III. INFLUENCE OF POLE HORN ON TORQUE IN FERRITE MAGNET MODEL

A. Analysis model

The influence of the pole horn on torque and reluctance torque is shown in the preceding section by examination of a rare-earth magnet model. In this section, the rotor shape and magnet used in the analysis model are changed in order to examine whether the influence of pole horn shape depends on the rotor shape. Permanent magnets are used to meet the prerequisite that rare-earth magnets are not used.

Fig. 10 shows the other analysis model in which a ferrite magnet with a coercive force of 300 kA/m is used. This model has a double-layer rotor [3] to enhance reluctance torque. The other specifications of this model are the same as the rare-earth magnet model, and the number of turns per phase is 232 in order to examine these models under the same conditions.

B. Influence of pole horn angle θ

Fig. 11 shows the influence of pole horn angle θ on maximum reluctance torque ($T_{r\text{max}}$), L_d, L_q, and $L_q - L_d$, under the condition that $\delta = 2$ mm and $I_e = 3$ A. As shown in Fig. 11, this tendency is almost the same as in Fig. 3.

L_d of the ferrite magnet model becomes low because the reluctance of the d-axis becomes large due to the double-layer structure. The flux barrier of the rotor is formed in accordance with the q-axis and as a result of suppressing the decrease in of L_q. Therefore, the value of L_q becomes smaller than that of L_d, and thus $L_q - L_d$ in the ferrite magnet model becomes larger than that in the rare-earth magnet model.

The result shows that the influence of $L_q - L_d$ on reluctance torque is larger in the ferrite magnet model; therefore, maximum reluctance torque is almost the same at $\theta = 9°$ and $\theta = 12°$ compared with Fig. 3 (a).

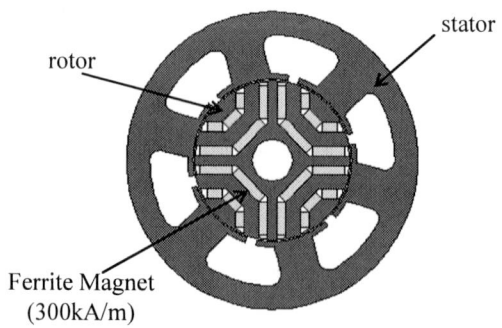

Fig. 10. Analysis model with ferrite magnets.

C. Influence of pole horn thickness δ

Fig. 12 shows the influence of pole horn thickness δ on $T_{r\max}$ and $L_q - L_d$ under the same conditions as in Fig. 5 (a), where $I_e = 3$ A and $\theta = 9°$. As shown in Fig. 12, this tendency is the same as in Fig. 5 (a), which shows that $T_{r\max}$ and $L_q - L_d$ increase as δ increases. Only the values of $T_{r\max}$ and $L_q - L_d$ become large, as described in Section III-B. Therefore, it is shown that the influence of pole horn thickness does not depend on the rotor shape and magnets as discussed in Sections II-C and III-C.

D. Influence of pole horn angle θ and pole horn thickness δ

In this section, we examine the influence of θ and δ on maximum reluctance torque and maximum total torque as shown in Section II-D. Fig. 13 shows the ferrite magnet

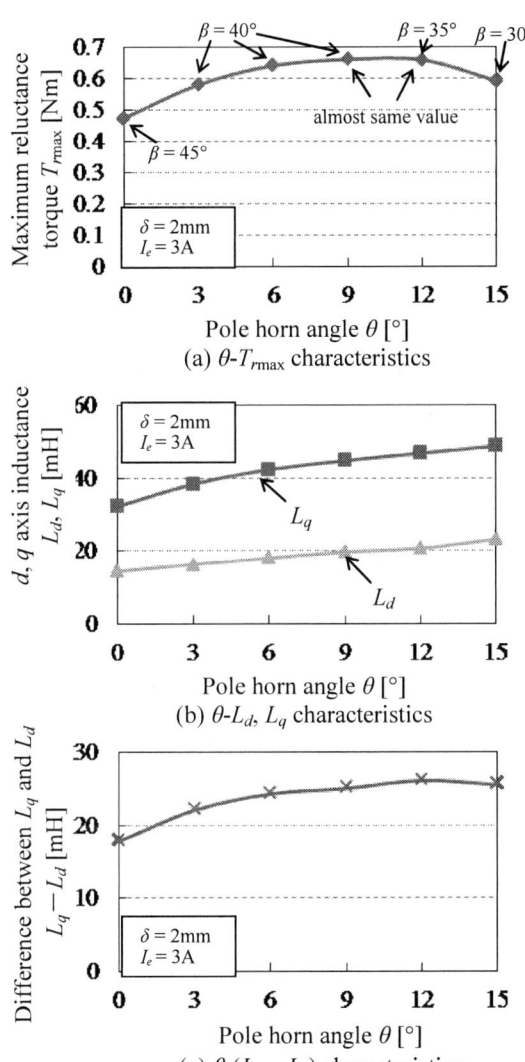

(a) θ-$T_{r\max}$ characteristics

(b) θ-L_d, L_q characteristics

(c) θ-($L_q - L_d$) characteristics

Fig. 11. Influence of θ on $T_{r\max}$ and motor parameters.

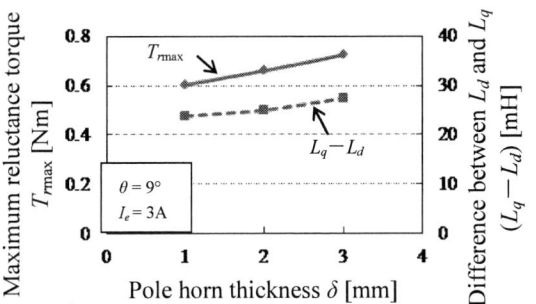

Fig. 12. Influence of δ on $T_{r\max}$ and inductance characteristics.

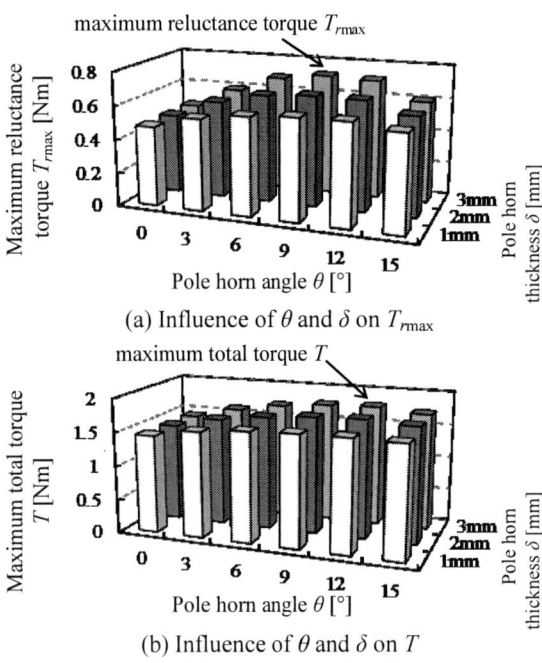

(a) Influence of θ and δ on $T_{r\max}$

(b) Influence of θ and δ on T

Fig. 13. Influence of θ and δ on maximum reluctance torque and maximum total torque under the condition that the space factor is not maintained at 50% ($I_e = 3$ A).

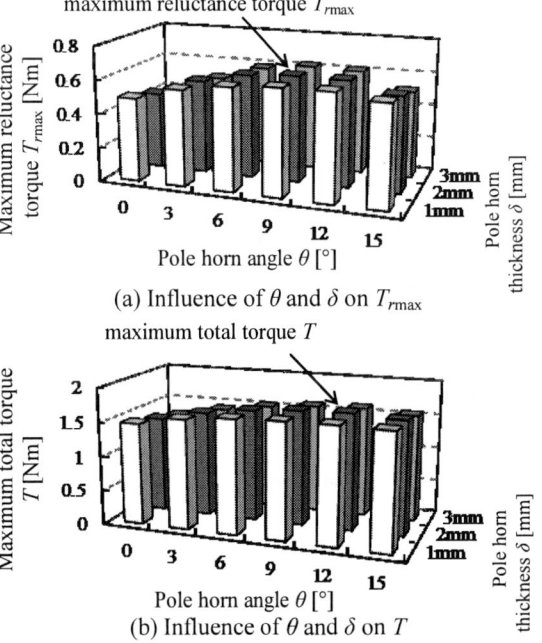

(a) Influence of θ and δ on $T_{r\max}$

(b) Influence of θ and δ on T

Fig. 14. Influence of θ and δ on maximum reluctance torque and maximum total torque under the condition that the space factor is maintained at 50% ($I_e = 3$ A).

978-1-4799-2706-7/14 $31.00 © 2014 IEEE

model under the condition that the number of turns per phase is 232. Fig. 14 shows the same model under the condition that the space factor is maintained at 50%.

Figs. 13 and 14 indicate that total torque becomes maximum at $\theta = 12°$ and reluctance torque becomes maximum at $\theta = 9°$, which is the same as in Figs. 8 and 9. Likewise, as in Figs. 8 and 9, total torque and reluctance torque increase with an increase in δ under the condition that the number of turns per phase is 232; however, these torques become maximum $\delta = 2$ mm under the condition that the space factor is maintained at 50%.

IV. CONCLUSION

In this study, we aimed to improve the reluctance torque for IPMSMs with concentrated windings by investigating the stator structure. To do this, we examined the influence of pole horn thickness and length on reluctance torque and total torque.

In experiments using a rare-earth model and a ferrite magnet model under the condition that the space factor is maintained at 50%, reluctance torque became maximum at $\theta = 9°$, $\delta = 2$ mm and total torque became maximum at $\theta = 12°$, $\delta = 2$ mm. However, when the number of turns per phase was 232 and the space factor was not considered, the value of δ changed to 3 mm because the number of turns increased to 232 at $\delta = 3$ mm from 226 at $\delta = 2$ mm.

The results of the experiments show that maximum total torque is achieved by making the pole horn as long as possible without allowing flux leakage at the edge of the pole horn, and making the pole horn thin for increasing the number of turns per phase but without oversaturation at the pole horn. In this study, total torque became maximum at $\theta = 12°$, $\delta = 2$ mm.

In the future, we plan to investigate other motor design parameters such as back yoke thickness and teeth width.

REFERENCES

[1] M. Kamiya, "Development of Traction Drive Motors for the Toyota Hybrid System," *IEEJ Transactions on Industry Applications,* vol. 126, no. 4, pp. 473-479, 2006.

[2] C. S. Jin, D. S. Jung, K. C. Kim, Y. D. Chun, H. W. Lee, J. Lee, "A Study on Improvement Magnetic Torque Characteristics of IPMSM for Direct Drive Washing Machine," *IEEE Transactions on Magnetics,* vol. 45, no. 6, pp. 2811-2814, 2009.

[3] S. Morimoto, M. Sanada, and Y. Takeda, "Performance of PM-Assisted Synchronous Reluctance Motor for High-Efficiency and Wide Constant-Power Operation," *IEEE Transactions Industry Applications,* vol. 37, no. 5, pp. 1234-1240, 2001.

[4] J. Wang, Z. P. Xia, D. Howe, and S. A. Long, "Comparative study of 3-phase permanent magnet brushless machines with concentrated, distributed and modular windings," *Proc. of, PEMD* 2006. pp. 489-493, 2006

[5] J. Okamoto, M. Sanada, S. Morimoto, Y. Inoue, "Influence of Magnet and Flux Barrier Arrangement for IPMSM with Concentrated Winding," *Proc. of IEEE PEDS* 2013, pp. 741-745, 2013.

[6] J. M. Park, S. I. Kim, J. P. Hong, Jung Ho Lee, "Rotor Design on Torque Ripple Reduction for a Synchronous Reluctance Motor With Concentrated Winding Using Response Surface Methodology," *IEEE Transactions on magnetics,* vol. 42, no. 10, pp. 3479-3481, 2006.

[7] S. Morimoto, "Trend of Permanent Magnet Synchronous Machines," *IEEJ Transactions on electrical and electronic engineering,* vol. 2, pp. 101-108, 2007.

Development and verification of energy-accurate simulation models for permanent magnet synchronous motors in automation systems

Frederic Blank, Jörg Roth-Stielow
Institute of Power Electronics and Electrical Drives
University of Stuttgart
Stuttgart, Germany
blank@ilea.uni-stuttgart.de

Abstract— In order to evaluate the energy efficiency of electrical drive systems for automation, simulations are made. For this purpose energy-accurate models are needed, which include all significant losses. The development and verification of an optimized simulation model for a permanent magnet synchronous motor with a nominal power of 1 kW is presented in this paper, using measurements in different operating points. It is not necessary to know the design data of the motor or finite-element-models for this. Influences are analyzed and the model is verified using measurements and statistical methods. With the help of the presented method, simulation models can be developed to compare the energy efficiency of different motors application dependent and therefore save energy.

Keywords—energy efficiency, losses, permanent magnet synchronous motor, simulation model

I. INTRODUCTION

Energy savings in automation systems often aim at increasing the energy efficiency. Through the use of efficient components large energy savings can be achieved. This has been confirmed in various studies, especially for drives in stationary applications. In these applications, the potential of savings is calculated using a given efficiency of the component in a given operating point. This paper's focus is on drive systems of small power ($\leq 5\,\text{kW}$) which are common in automation systems and often operate in a non-steady-mode. They accelerate and decelerate in short cycles. Even if the absolute power and therefore the saving potential of a single drive is small, they appear in large numbers in production sites and operate many hours a day and so the saving potential is added up. Permanent magnet synchronous motors (PMSM) are common in these drive systems. Therefore investigations and measurements on the losses were made on different PMSM.

II. ENERGY-ACCURATE MODEL

In this paper, an energy-accurate model is developed for an industry-standard PMSM. The motor parameters are listed in Table I. The motor is operated with a standard voltage source inverter using pulse-with-modulation and a field orientated current control. The

model has to describe all significant losses which are introduced in section III and the parameters are estimated in section IV. With the developed formulas and parameters the losses are calculated and compared to two more common models. The new model is verified using measurements (section V). As the accuracy of the measurements limits the accuracy of the model, the issue of accuracy is discussed in section VI.

TABLE I
MOTOR-PARAMETERS (FESTO EMMS-AS-70-M)

Symbol	Meaning	Value
T_N	Nominal torque	2.29 Nm
n_N	Nominal rotational speed	4100 rpm
I_N	Nominal current	2.6 A
R_S	Winding resistance (20°C)	3.31 Ω

In order to estimate the power losses of the motor, it is measured on a motor test bench with a load motor, a high-precision power analyzer and a torque-meter. The electrical and mechanical powers are measured in 430 static operating points. And the power loss P_L is calculated, see Fig. 1. In the upper half (1st quadrant) of the speed-torque-plane the PMSM is in motor mode and in the lower half (2nd quadrant) in generator mode. The other two quadrants are symmetrical.

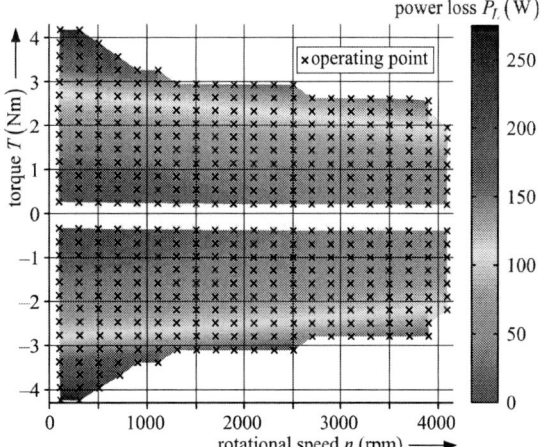

Fig. 1. Measured power loss in the speed-torque-plane.

With this data, the efficiency of the motor is calculated as shown in Fig. 2. The efficiency varies significantly with the operating point. The maximum is accomplished near the nominal operating point at the right upper end.

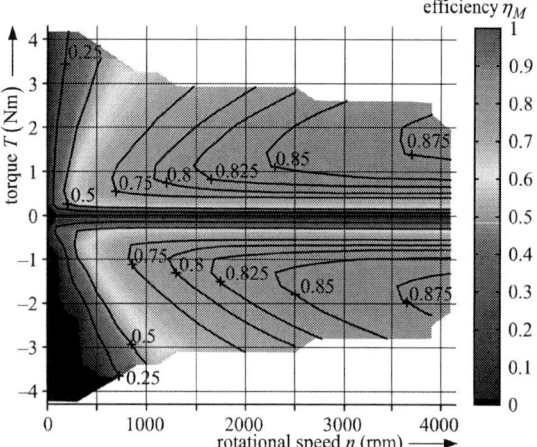

Fig. 2. Measured efficiency in the speed-torque-plane.

When the motor operates in start-stop-mode it passes through different operation points in the speed-torque-plane. The efficiency cannot be evaluated using the nominal efficiency. As an alternative approach, the losses of the PMSM are accurately investigated. With the knowledge of the losses in quantity and dependencies, an energy-accurate simulation model is developed [1]. It is derived and verified using the measurements at the different operating points. The energy consumption of the motor in non-steady-operation can then be estimated by simulations with this model.

A simplified sectional drawing of the PMSM is depicted in Fig. 3. The motor contains a holding brake which is an option for servo motors. It is required for safety reasons when the motor drives an axis for vertical movements. It must be powered for release and closes instantly after shutdown or electricity failure.

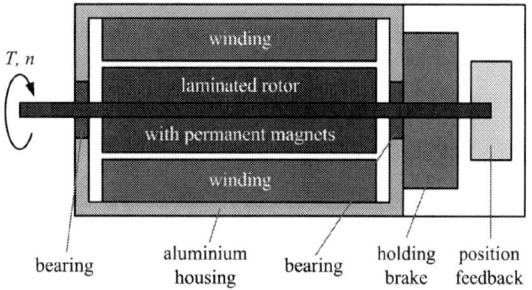

Fig. 3. Simplified sectional drawing of the motor.

III. MOTOR LOSSES

The common field-orientated equivalent circuit used for the development of control systems, is shown in Fig. 4. It only contains the copper losses, described by R_S. Although other losses can often be neglected for the control, they appear in electrical machines and have a significant influence on the efficiency [2]. In this section, the different losses are described briefly.

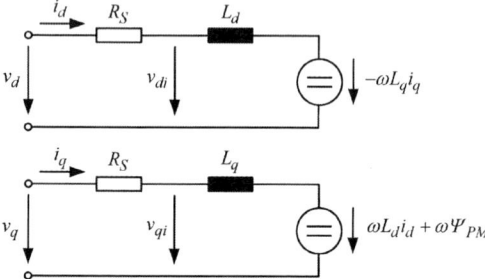

Fig. 4. Field-oriented equivalent circuit of a PMSM.

A. Copper losses

The current through the windings is approximately proportional to the inner torque of the motor and causes losses. As the resistance of copper varies with its temperature, the temperature has a significant influence on the copper losses $P_{L,Cu}$.

B. Frictional losses

The frictional losses $P_{L,F}$ are mechanical losses of ball bearings and air in the gap. They can be described by a speed-dependent torque $T_F(n)$.

C. Iron losses

As the stator consists of iron and the magnetic flux through the stator changes with the electrical frequency of the rotor, eddy-current- and hysteresis-losses appear in the stator. They vary with the intensity of the magnetic flux and the electrical frequency which is proportional to the mechanical speed. As the iron losses $P_{L,Fe}$ and the frictional losses both depend on the speed of the motor, it is difficult to determine them independently.

D. Holding brake

The optional holding brake must be powered with a constant voltage for being released and closes automatically without power supply. The power loss $P_{L,H}$ is almost constant, with a little variation in the temperature. The power loss of the holding brake is neglected in the following loss-analysis because it can be added up independently.

IV. ESTIMATION OF PARAMETERS

The dependencies and parameters of the different losses are determined by measurements.

A. Copper losses

The resistance of the windings is measured at a temperature of 20°C. Due to the resistance changes approximately linearly with the temperature the resistance at different winding temperatures is calculated with eq. (1). By using the motor currents, the copper losses $P_{L,Cu}$ are calculated with eq. (2).

$$R(\vartheta) = R_{20} \cdot \left(1 + \alpha_{20°C} \cdot \left(\vartheta - \vartheta_{20°C}\right)\right) \tag{1}$$

with $\alpha_{20°C} = 3.93 \cdot 10^{-3} \, \text{K}^{-1}$

$$P_{V,Cu} = \tfrac{3}{2} \cdot \left(i_q^2 + i_d^2\right) \cdot R(\vartheta) \tag{2}$$

To determine the temperature of the windings, a simplified thermal model of the motor is developed, based on [3]. With this model, the temperatures of the motor windings and the brake are determined. The simplified thermal model is shown in Fig. 5. The time constant of the thermal model is much greater (typically 1 to 2 h) than the positioning cycle time in non-steady-operation (typ. 2 to 20 s). Therefore, only the mean power losses and the thermal resistances R_{th} are relevant for the temperatures in the steady state. In the thermal model, there are two sources of losses, since they influence the motor at different locations. $P_{L,W}$ is the sum of the copper, frictional and iron losses. $P_{L,B}$ is the power of the holding brake. The thermal resistances R_{th} describe the heat transfer between the components (B=brake, W=windings, H=housing) and the ambience (A).

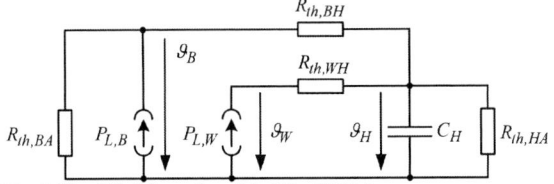

Fig. 5. Simplified thermal model of the PMSM.

The parameters of the thermal model were determined using a dc-current to heat up the motor similar to [6]. As the temperature of the windings could not be measured directly, it is calculated with the actual resistance of the windings using eq. (1). With the thermal model, the temperature of the winding is estimated and used to determine the resistance R_S in simulations. The resulting copper losses $P_{L,Cu}$ are depicted in Fig. 6.

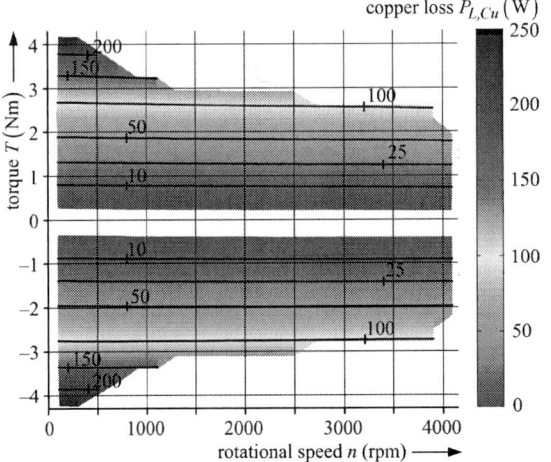

Fig. 6. Calculated copper losses $P_{L,Cu}$ in the speed-torque-plane.

B. Frictional losses

The frictional losses $P_{L,F}$ are determined by propelling the PMSM with the load motor on the motor test bench and measuring the torque at different speeds with a torque-meter. The result is displayed in Fig. 7. For inclusion in the simulation model, the measured torque is approximated by a second-order polynomial with the parameters in eq. (3). However, this torque cannot be assigned to the frictional losses only. The magnetic flux

of the permanent magnets causes iron losses, even if the PMSM is propelled externally. The frictional losses $P_{L,F}$ are calculated in eq. (4) using $T_F(n)$. In Fig. 8, the calculated frictional-losses are depicted.

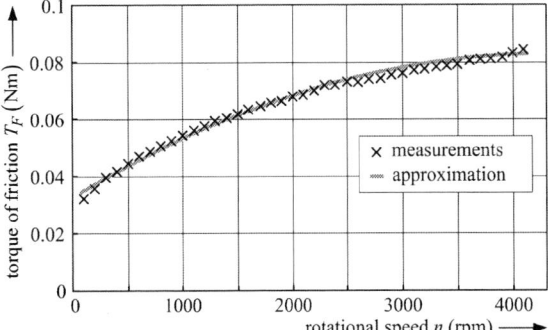

Fig. 7. Measured and approximated torque of friction as function of rotational speed.

$$T_F(n) = t_2 \cdot n^2 + t_1 \cdot n + t_0$$

$$\text{with} \quad t_2 = -9.41 \cdot 10^{-9} \; Nm s^2$$

$$t_1 = 1.35 \cdot 10^{-6} \; Nm s \tag{3}$$

$$t_0 = 33.3 \cdot 10^{-3} \; Nm$$

$$P_{L,F} = 2\pi \cdot n \cdot T_F(n) \tag{4}$$

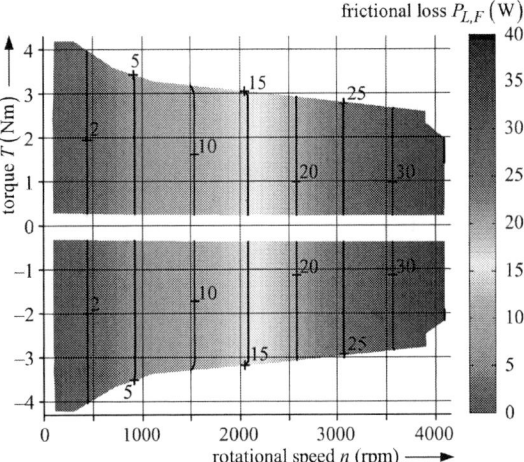

Fig. 8. Calculated frictional losses $P_{L,F}$ in the speed-torque-plane.

C. Iron losses and other losses

As mentioned above, the reproduction of the frictional losses includes an unknown part of the iron losses. They both depend on the speed and are impossible to decompose without modifying the motor. The energy-accurate model needs to reproduce the sum of losses with the correct dependencies for all operating points. To achieve this, the remaining power of losses $P_{L,rest}$ are calculated for every measured operating point (5).

$$P_{L,rest} = P_L - P_{L,Cu} - P_{L,F} \tag{5}$$

1) Iron loss model with R_{Fe}

A first approach to model the iron losses is depicted in Fig. 9. The iron losses are approximated by the resistance R_{Fe} in the equivalent circuit. This approach is common for motors powered with line frequency. The iron losses is calculated using eq. (6).

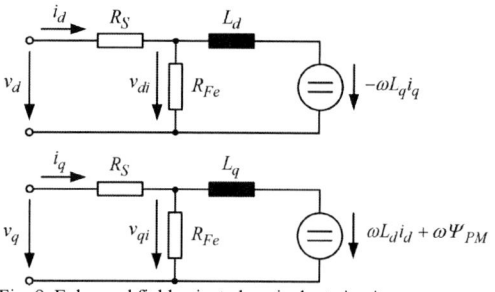

Fig. 9. Enhanced field-oriented equivalent circuit.

$$P_{L,Fe} = \frac{\left(v_{qi}^2 + v_{di}^2\right)}{R_{Fe}} \tag{6}$$

The parameter R_{Fe} must be determined, so that the equation $P_{L,Fe} = P_{L,rest}$ is fulfilled in the best possible way for all operating points. For this purpose, a numerical optimization is applied using the least-square-method. This gives an ordinary least squares solution to the linear system described by the equations (7) by using the measurements of all N operating points. The result is shown in section V.

$$A \cdot x = b$$

$$A = \left[\left(v_{qi,1}^2 + v_{di,1}^2\right) \quad \cdots \quad \left(v_{qi,N}^2 + v_{di,N}^2\right) \right]^T \tag{7}$$

$$x = 1/R_{Fe} \; ; \; b = \left[P_{V,residual,1} \quad \cdots \quad P_{V,residual,N} \right]^T$$

2) Iron loss model with k_h and k_e

As the first approach has significant deviations, a second approach is examined. For the dependencies of hysteresis- and eddy-current-losses, approach (8) was derived [4] with the two parameters k_h and k_e for the two loss mechanisms. v_{qi} and v_{di} are the inner voltages in the motor model, see Fig. 4.

$$P_{L,Fe} = \frac{\left(v_{qi}^2 + v_{di}^2\right)}{\omega_{el}} \cdot k_h + \left(v_{qi}^2 + v_{di}^2\right) \cdot k_e \tag{8}$$

The two parameters k_h and k_e must be determined, so that the equation $P_{L,Fe} = P_{L,rest}$ is fulfilled. This is done similar to the first approach using the linear system described by the equations (9). The calculated iron losses $P_{L,Fe}$ with eq. (8) are depicted in Fig. 10. The losses increase with higher speeds and loads.

$$A \cdot x = b$$

$$A = \begin{bmatrix} \left(v_{qi,1}^2 + v_{di,1}^2\right)\big/\omega_{el,1} & \left(v_{qi,1}^2 + v_{di,1}^2\right) \\ \vdots & \vdots \\ \left(v_{qi,N}^2 + v_{di,N}^2\right)\big/\omega_{el,N} & \left(v_{qi,N}^2 + v_{di,N}^2\right) \end{bmatrix} \tag{9}$$

$$x = \begin{bmatrix} k_h & k_e \end{bmatrix}^T \; ; \; b = \begin{bmatrix} P_{V,residual,1} \cdots P_{V,residual,N} \end{bmatrix}^T$$

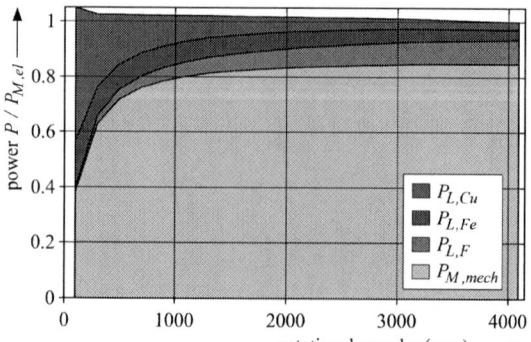

Fig. 10. Calculated iron losses $P_{L,Fe}$ in the speed-torque-plane.

D. Calculation and decomposition of losses

With the presented dependencies and determined parameters, the different power losses are calculated for every operating point. Fig. 11 depicts the decomposition for a load-torque of 0.82 Nm and Fig. 12 for a load-torque of 2.6 Nm. The variables are listed in Table II. All powers are normalized to the electrical power $P_{M,el}$ of the machine and therefore $P_{M,mech} / P_{M,el}$ corresponds to the efficiency of the motor. $P_{M,el}$ and $P_{M,mech}$ are measured and the power losses are calculated. The difference between the sum (blue) and 1 is due to the limited inaccuracy of the model. This is more significant at low torque and speed, where the total power is small. The copper-losses dominate at lower speeds and especially at higher torques, whereas the friction and iron losses dominate at high speeds and low torques.

Fig. 11. Decomposition of powers for $T \approx 0.82$ Nm.

The 2014 International Power Electronics Conference

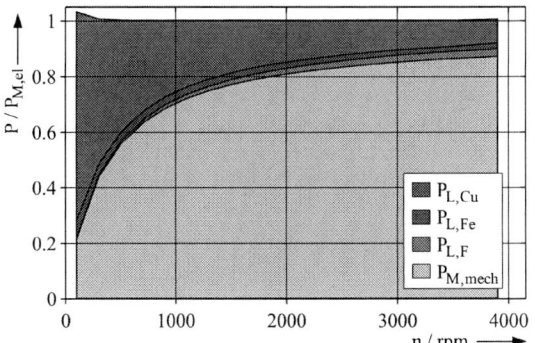

Fig. 12. Decomposition of powers for $T \approx 2.6$ Nm.

TABLE II
MEASURED AND CALCULATED POWERS

Symbol	Meaning
$P_{M,el}$	Measured electrical power of motor
$P_{M,mech}$	Measured mechanical power of motor
$P_{L,Cu}$	Calculated copper power loss
$P_{L,Fe}$	Calculated iron power loss
$P_{L,F}$	Calculated frictional power loss

V. MODEL VERIFICATION

The energy-accuracy of the model has to be verified in all possible operating points. The losses are calculated using the presented dependencies and parameters in all measured operating points. These calculated power losses are then compared to the measured power loss P_L in the different operating points and the deviation P_{Error} is calculated in eq. (10).

$$P_{Error} = P_L - P_{L,Cu} - P_{L,F} - P_{L,Fe} \qquad (10)$$

In a perfect model, P_{Error} would equal to zero for all points. The accomplished deviation is depicted in Fig. 13. The nominal operation area $-T_N \leq T \leq T_N$ has a deviation smaller than 5 W which is 0.5% of the nominal power. Lower conformity is achieved at high torque because the winding temperatures were rising very fast during the measurements and therefore the deviations results from the simplified thermal model.

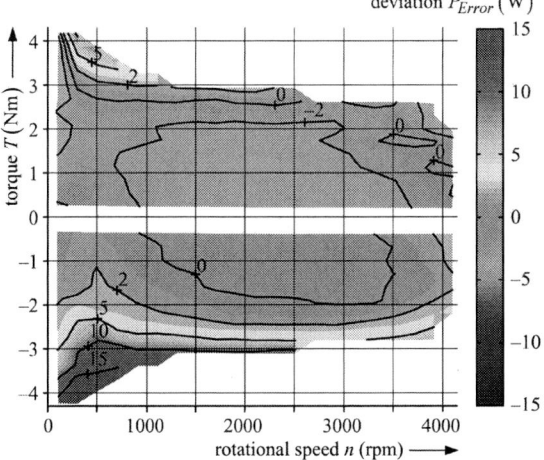

Fig. 13. Deviation P_{Error} in speed-torque-plane.

The relative frequency of the deviations is shown as histogram in Fig. 14, in order to evaluate the accuracy of the model. As the investigated points are equally spread over the possible operating area, this gives an overview over the precision of the model.

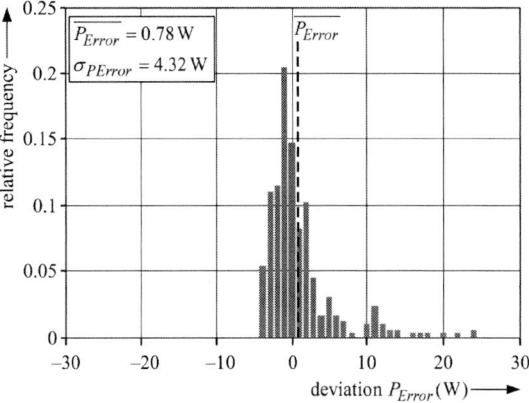

Fig. 14. Frequency distribution of deviations between calculated and measured losses in all operating points.

To evaluate and validate the accuracy of the model, this frequency distribution is characterized by two statistical variables: the mean value $\overline{P}_{Error}$ (11) and standard deviation $\sigma_{P_{Error}}$ (12) for the power P_{Error}.

$$\overline{P}_{Error} = \frac{1}{N} \sum_{i=1}^{N} P_{Error,i} \qquad (11)$$

$$\sigma_{P_{Error}} = \sqrt{\frac{1}{N-1} \sum_{i=1}^{N} \left(P_{Error,i} - \overline{P}_{Error} \right)^2} \qquad (12)$$

The losses were calculated for two alternative models in order to compare the presented model with other possibilities: A simple model with a constant stator resistance and friction torque and without iron losses and a second model with the iron losses approximated by the equivalent resistance R_{Fe}. The resulting mean errors and standard deviations are listed in Table III. The developed model has the smallest mean error and standard deviation. This indicates that this model includes the significant losses with the correct quantity and dependencies.

TABLE III
COMPARISON OF DIFFERENT MOTOR MODELS

Model	Mean error $\overline{P}_{Error}$	Standard deviation $\sigma_{P_{Error}}$
Simple model with R_S = const., T_F = const. and no iron losses	22.42 W	11.57 W
Model with simple iron losses with R_S = const., T_F = const. and R_{Fe} for the iron losses	6.90 W	9.83 W
Developed model with all presented details	0.78 W	4.32 W

VI. LIMIT OF ACCURACY

As the development of the model is based on direct measurements of the losses on the motor-test-bench, the model can only be as exact as these measurements are. Because of this, it is very important to examine and minimize the uncertainties in measurement. Every component in the measurement-system has an influence on the overall accuracy. The power-loss is calculated in eq. (13) with the measured electrical and mechanical powers. The electrical power is measured with a power-analyzer (ZES Zimmer LMG500 and PSU200) and the mechanical power with a torque-meter (Lorenz Messtechnik DR-2208 30/3Nm).

$$P_L = \left| P_{M,el} - P_{M,mech} \right| = \left| UI\lambda - 2\pi Tn \right| \qquad (13)$$

Even though the used devices are highly accurate and offsets are adjusted, the uncertainties of every device still limit the accuracy of P_L. The combined uncertainty in the measurements is analyzed according to [6] and [7]. The uncertainty for a single measurement is calculated using eq. (14). The sensitivity factors c_{i,P_L} describe the influence of the uncertainties $u(x_k)$ on the power loss. The factors are calculated with eq. (15) for the quantities x_k. Every measuring device has a certain uncertainty depending on the measured value (m.v.) and the full scale value (f.s.). All factors are listed in Table IV.

$$\Delta P = \sqrt{\sum_i c_{i,P_L}^2 \cdot u^2(x_i)} \qquad (14)$$

$$c_i = \frac{\delta P_L}{\delta x_i} \qquad (15)$$

TABLE IV
FACTORS FOR ACCURACY ANALYSIS

quantity x_i	sensitivity factor c_{i,P_L}	standard uncertainty u_i
T	$\frac{\partial P_L}{\partial T} = 2\pi n$	$\left(T_{f.s.} \cdot 2 + T_{m.v.} \cdot 0.5 \right) \cdot 10^{-3}$
n	$\frac{\partial P_L}{\partial n} = 2\pi T$	$\left(n_{f.s.} \cdot 0 + n_{m.v.} \cdot 0.1 \right) \cdot 10^{-3}$
I	$\frac{\partial P_L}{\partial I} = U \cdot \lambda$	$\left(I_{f.s.} \cdot 0.05 + I_{m.v.} \cdot 0.15 \right) \cdot 10^{-3}$
$P_{M,el}$	$\frac{\partial P_L}{\partial P_{M,el}} = 1$	$\left(P_{f.s.} \cdot 0.3 + P_{m.v.} \cdot 0.28 \right) \cdot 10^{-3}$

The combined uncertainty is calculated for every operating point because the measured value and the full scale value vary in the measured 430 points. Afterwards, they are combined as rectangular distribution to a frequency distribution of uncertainties over all points. The resulting histogram is depicted in Fig. 15. The distribution is characterized by a standard deviation $\sigma_{\Delta P}$. In this case the uncertainties in measurement are factor 2.4 smaller than the standard deviation of the developed model. This has to be checked because the model can only be as accurate as the measurement is. As a matter of principle, the maximum achievable accuracy of the model will always be limited by the measurement.

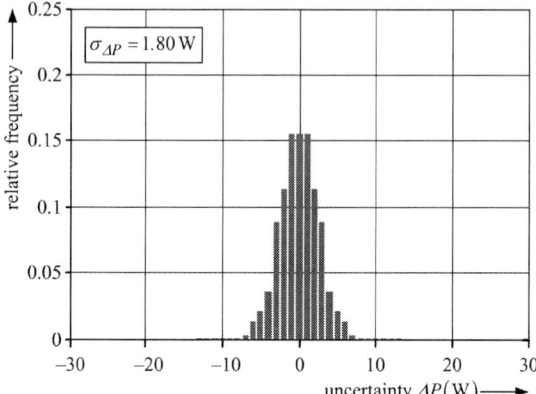

Fig. 15. Frequency distribution of combined uncertainties in measurement

VII. CONCLUSION

All significant losses of the PMSM for non-steady-operations are analyzed in order to develop an energy-accurate model. With the dependencies found and the parameters measured the losses in the motor are calculated and the loss model is developed. It is verified by measurements using statistical parameters. The model has a standard deviation $\sigma = 4.32\,\text{W}$ for all operating points. This is small regarding the nominal power of $1\,\text{kW}$ and compared to alternative models. The maximum reachable accuracy is limited by uncertainties in the measurement, which are needed for verification of the model.

REFERENCES

[1] F. Blank and J. Roth-Stielow, "Energieexakte Simulationsmodelle zur Bewertung der Energieeffizienz" *Tagungsband VDE Kongress 2011*, Würzburg, 2011.

[2] N. Urasaki, T. Senjyu et al., "An accurate modeling for permanent magnet synchronous motor drives", *APEC 2000 Applied Power Electronics Conference and Exposition*, vol.1, pp.387-392, 2000.

[3] A. Boglietti, A. Cavagnino et al., "A simplified thermal model for variable speed self cooled industrial induction motor", *Industry Applications Conference 37th IAS Annual Meeting*, vol. 2, pp. 723-730, 2002.

[4] A. Boglietti, A. Vallan, "Measurement of Housing Thermal Resistances in Industrial Motors", *Instrumentation and Measurement Technology Conference, IMTC 2006, Proceedings of the IEEE*, pp. 1321-1325, 2006.

[5] V. Navrapescu, M. Popescu et al., "Modelling of iron losses in salient pole permanent magnet synchronous motors", *ICPE '07, 7th International Conference on Power Electronics*, pp.352- 357, 2007.

[6] K.-D. Sommer, B. Siebert, "Praxisgerechtes Bestimmen der Messunsicherheit nach GUM", *Technisches Messen 71*, Oldenbourg Verlag, 2004.

[7] *JCGM 100:2008 – Evaluation of measurement data — Guide to the expression of uncertainty in measurement*, Joint Committee for Guides in Metrology, 2008.

Comparison of the Resistance- and Inductance-Based Saliency of a PMSM due to a Short-Circuited Rotor Winding

Johannes Graus, Alexander Rambetius and Ingo Hahn, *Senior Member, IEEE*

Chair of Electrical Drives and Machines, University Erlangen-Nuremberg
Cauerstrasse 9, 91058 Erlangen, Germany
johannes.graus@fau.de

Abstract— Sensorless position estimation based on signal-injection methods always requires a resistance- or inductance-based anisotropy. Therefore, a short-circuited electrical winding can be attached on the rotor. As will be derived in this paper, this method leads to both a resistance- and inductance-based saliency. By an experimental test setup, both saliencies are analyzed and compared towards their accuracy and frequency-dependence. Thereby, for the resistance-based saliency a characteristic behavior is found that can be explained by the interaction between the rotor winding and the eddy current effect.

Keywords— Eddy currents, interior permanent magnet (IPM) motor, ringed pole, sensorless control.

I. INTRODUCTION

An essential drawback of permanent magnet synchronous machines (PMSM) is the need of a position encoder to be operated in position, velocity or torque control. For small machines these encoders can account for up to 35 % [1] of the machine costs. Another disadvantage of a mechanical sensor is its sensitivity against environmental influences. In order to overcome these issues, many approaches have been published in the last years, which focus on the estimation of the rotor position by evaluating the electrical behavior of the machine. Thus, the mechanical encoder can be omitted in many cases.

Considering high speed, the position information is contained in the back-EMF. However, at low speed or standstill active position detection methods must be applied to estimate the rotor angle based on the position-dependent electromagnetic properties of the machine. Therefore, the methods proposed in [2] and [3] use the position-dependency of the hysteretic and eddy current losses that influence the high frequency resistance. Another technique, which is used in [4], [5] and [6], exploits the magnetic saliency of the rotor. Therefore, the inductances in the direct and quadrature axis must differ in all operating points. The major problem that arises in practice is the influence of saturation which causes the inductances' load- and position-dependency [7], [8]. In consequence, the magnetic anisotropy may vanish at certain operating points, so that the position information is lost. Another saturation induced issue is the effect of

cross-coupling which directly leads to an error in the estimated position [9]. To overcome these problems, the influence of the machine design on the differential inductances was investigated either by magnetic equivalent circuit models [10] or finite element simulations [11]. The results show that the feasibility for sensorless operation mainly depends on the given rotor configuration but can be slightly improved by design optimizations.

A different approach to obtain an electromagnetic saliency is the use of a short-circuited winding on the rotor [12]. As a main advantage, it almost exclusively interacts with the HF-test-signal which enables a nearly independent design of the HF-behavior of the machine. It was also found that the sensorless performance of a so called ringed-pole surface-mounted permanent magnet (SPM) synchronous machine, which has a short-circuited ring around each magnet, is much less sensitive towards magnetic saturation than machines that provide only a geometric saliency [13]. However, in [13] and [14] the concept of a short-circuited winding is only applied to SPM machines since they do not provide a significant geometric saliency. In contrast, interior permanent magnet (IPM) machines have a magnetic saliency and hence, are feasible for sensorless operation without an additional rotor winding. However, the IPM motor as well as all other rotor configurations investigated in [1] show a current limit at which the difference of the differential inductances vanishes due to saturation. This suggests that beside the IPM motor other rotor types could also benefit from a short-circuited rotor winding.

As a first step towards such a new rotor configuration, the influence of a short-circuited rotor winding is investigated fundamentally in this paper. For that purpose, a two-phase synchronous machine was developed to measure the frequency- and position-dependent influence of a short-circuited rotor winding on the resulting HF-resistance and -inductance of the stator winding.

While in [13], [14] and [15] only the inductance-based saliency is considered for sensorless control, in this paper the resistance-based saliency due to a short-circuited rotor winding, as suggested in [16], is also discussed and compared with regard to the estimation error.

978-1-4799-2706-7/14 $31.00 © 2014 IEEE

Furthermore, the impact of the test signal frequency on the accuracy of the position estimation is examined.

II. SHORT-CIRCUITED WINDING

To derive the basic influence of a short-circuited rotor coil on the impedance of the stator winding, the model in Fig. 1 is used. It consists of two magnetically coupled coils with the ohmic resistance $R_{S/R}$, the self-inductance $L_{S/R}$ and the mutual-inductance M_{SR}.

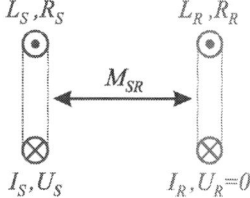

Fig. 1. Magnetically coupled stator and rotor coil to show the principle of a short-circuited winding.

Using the complex phasor notation for ac steady state analysis, the two voltage equations can be expressed as

$$U_S = R_S I_S + j\omega L_S I_S + j\omega M_{SR} I_R \tag{1}$$

$$U_R = R_R I_R + j\omega L_R I_R + j\omega M_{SR} I_R = 0. \tag{2}$$

Since the rotor coil is short-circuited, its voltage equation equals zero. Extracting I_R from (2) and substituting in (1), the effective impedance of the stator coil Z_S results:

$$Z_S = \frac{U_S}{I_S} = \underbrace{\left(R_S + \frac{\omega^2 R_S M_{SR}^2}{R_R^2 + \omega^2 L_R^2}\right)}_{R_{S,eff}} + j\omega \underbrace{\left(L_S - \frac{\omega^2 L_R M_{SR}^2}{R_R^2 + \omega^2 L_R^2}\right)}_{L_{S,eff}} \tag{3}$$

By separating (3) into its real and imaginary part, an effective resistance $R_{S,eff}$ and an effective inductance $L_{S,eff}$ of the stator coil can be defined. It becomes obvious that the short-circuited winding leads to an increase of the effective resistance $R_{S,eff}$ and to a drop of the effective inductance $L_{S,eff}$. For small frequencies, the rotor coil does not affect the stator coil

$$Z_{S,\omega\to 0} = R_S + j\omega L_S , \tag{4}$$

while for very high frequencies ($\omega \gg R_S/L_S$) the effective resistance and inductance become constant:

$$Z_{S,\omega\to\infty} = \underbrace{\left(R_S + R_R \frac{M_{SR}^2}{L_R^2}\right)}_{R_{S,eff,\infty}} + j\omega \underbrace{\left(L_S - L_R \frac{M_{SR}^2}{L_R^2}\right)}_{L_{S,eff,\infty}} \tag{5}$$

The fraction in (5) is a measure for the coupling of both coils and should be as high as possible. At first sight, (5) suggests that a high resistance R_R results in a high effective resistance $R_{S,eff}$. However, this goes along with a shift of the edge frequency of the denominator in (3) to higher frequencies.

III. CARRIER-SIGNAL INJECTION

In the following, the derived theory will be transferred to the machine model to establish a link between the short-circuited rotor winding and its influence on the sensorless performance.

Considering standstill, the voltage equations of a synchronous machine in the rotor-fixed reference frame can be described as

$$\begin{bmatrix} U_d \\ U_q \end{bmatrix} = \begin{bmatrix} Z_{dd} & Z_{dq} \\ Z_{dq} & Z_{qq} \end{bmatrix} \begin{bmatrix} I_d \\ I_q \end{bmatrix} \text{ with } Z_{xy} = R_{xy} + sL_{xy}. \tag{6}$$

In (6), Z_{dd} stands for the HF-impedance of the direct axis, which consists of the resistance R_{dd} and the inductance L_{dd} multiplied by the Laplace operator s. The same holds for the impedance of the quadrature axis Z_{qq}. If x and y are unequal, the mutual quantities result, which can be non-zero if the machine is not symmetric or if saturation occurs. However, in the following derivation the machine is assumed as linear so that all resistances and inductances are constant.

One way to extract the position information from the impedance matrix in (6) is the injection of an alternating test current in the d-axis $I_{d,HF}$ of an arbitrary synchronous reference frame, which is misaligned with the real rotor-fixed reference frame by the error angle φ_{err}, while the q-current $I_{q,HF}$ is controlled to zero. These currents are then transformed in the real reference frame, evaluated, and transformed back into the estimated reference frame, as derived in [1]. These transformations yield the voltage response in the q-axis $U_{q,HF}$ and the corresponding transfer function

$$\frac{U_{q,HF}}{I_{d,HF}} = Z_{diff} \sin(2\varphi_{err}) + Z_{dq} \cos(2\varphi_{err}) \tag{7}$$

with

$$Z_{diff} = \frac{1}{2}\left(Z_{qq} - Z_{dd}\right). \tag{8}$$

Neglecting Z_{dq}, equation (7) shows that the voltage response in the q-axis of the estimated reference frame is directly proportional to the sine of twice the error angle φ_{err}. Thus, it can be used as a feedback for a position control loop, which controls this voltage and thereby the error angle to zero, so the estimated reference frame is correctly aligned with the actual rotor position. Therefore, a difference in the self-impedances is required according to (8). However, in practice the mutual-impedance Z_{dq} cannot be neglected which leads to a non-zero voltage response even if the rotor position is correctly estimated so that φ_{err} is zero. In this case, the influence of the mutual-inductance must be compensated by the control algorithm to minimize the estimation error.

Hereinafter, it is assumed that the machine without a short-circuited rotor winding is symmetrical, so that the self-impedances Z_{dd} and Z_{qq} are equal and the cross-coupling impedance Z_{dq} vanishes. Hence, the impedance matrix in (6) does not contain position information and the voltage response in (7) yields zero. Under this assumption, in the following discussion any sensorless capabilities can be attributed to the short-circuited rotor winding.

A. Resistance-Based Saliency Tracking

If the resistance-based saliency of the machine is used for the position estimation, only the real part of the transfer function (7) is evaluated. Assuming $s = j\omega$ in (6) yields

$$Re\left\{\frac{U_{q,HF}}{I_{d,HF}}\right\} = R_{diff}\,sin(2\varphi_{err}) + R_{dq}\,cos(2\varphi_{err}) \quad (9)$$

with

$$R_{diff} = \frac{1}{2}(R_{qq} - R_{dd}). \quad (10)$$

With regard to the evaluation of the measurement results, it is useful to calculate the position error for the case that the voltage response in (9) is controlled to zero without compensating the mutual-resistance. Setting (9) equal to zero and solving for φ_{err} leads to the uncompensated estimation error ε_R:

$$\varepsilon_R = \frac{1}{2}arctan\left(-\frac{R_{dq}}{R_{diff}}\right) = \frac{1}{2}arctan\left(-\frac{2\,R_{dq}}{R_{qq}-R_{dd}}\right) \quad (11)$$

According to (11), a non-zero difference of the self-resistances is a prerequisite for a small error angle. Furthermore, the mutual-resistance leads directly to a position error. However, this error can be minimized if the difference R_{diff} is high enough.

B. Inductance-Based Saliency Tracking

By using the imaginary part of the transfer function (7) instead, the inductance-based saliency can be tracked:

$$Im\left\{\frac{U_{q,HF}}{I_{d,HF}}\right\} = L_{diff}\,sin(2\varphi_{err}) + L_{dq}\,cos(2\varphi_{err}) \quad (12)$$

with

$$L_{diff} = \frac{1}{2}(L_{qq} - L_{dd}). \quad (13)$$

Similar to (11), an uncompensated estimation error ε_L can be expressed:

$$\varepsilon_L = \frac{1}{2}arctan\left(-\frac{L_{dq}}{L_{diff}}\right) = \frac{1}{2}arctan\left(-\frac{2\,L_{dq}}{L_{qq}-L_{dd}}\right) \quad (14)$$

It becomes obvious that a non-zero difference of the self-inductances and a small mutual-inductance is the prerequisite for accurate position estimation.

C. Influence of the Short-Circuited Winding

Up to now, the influence of the short-circuited winding was not included in the equations for the carrier-signal injection. Therefore, the short-circuited rotor winding is now assumed to be located in the direct-axis of the rotor, so that it only interacts with the voltage equation of the d-axis in the first line of (6) without affecting the q-axis. Under these assumptions, the theory of the two-coil model in Fig. 1 can be applied. Assuming a sufficiently high frequency, (5) can be adopted to describe the self-impedance of the d-axes including the influence of the rotor winding:

$$Z_{dd,\omega\rightarrow\infty} = \underbrace{\left(R_{dd} + R_r\,\frac{M_{dr}^2}{L_r^2}\right)}_{R_{dd,eff,\infty}} + j\omega\,\underbrace{\left(L_{dd} - L_r\,\frac{M_{dr}^2}{L_r^2}\right)}_{L_{dd,eff,\infty}} \quad (15)$$

In (15), the index r is used for the quantities of the short-circuited rotor winding. Since the self-impedances of the d- and q-axis were initially assumed as equal, the resistive and inductive difference (10) and (13) yields

$$R_{diff} = -\frac{R_r M_{dr}^2}{2L_r^2} \quad \text{and} \quad L_{diff} = \frac{L_r M_{dr}^2}{2L_r^2}. \quad (16)$$

According to (16), a high mutual-inductance M_{dr} is decisive for a high saliency. Therefore, the coil sides of the rotor winding should be located on the rotor surface.

Under the given assumptions Z_{dq} and hence, the error angle (11) and (14) are zero. However, this is not necessarily the case when the experimental results are considered.

IV. EXPERIMENTAL TEST SETUP

To investigate the influence of a short-circuited winding on the rotor, a two-phase synchronous machine was designed and built. According to Fig. 2, the stator has four teeth and a concentrated winding. The coils of each phase, which are located opposite to each other, are pairwisely connected in series. To investigate different rotor winding configurations in future, the rotor has 24 slots, which are evenly distributed around the circumference. Since the rotor has neither a geometrical saliency nor permanent magnets its mean torque is zero. Fig. 2 refers to a rotor angle of zero degrees for which the d-axis of the rotor coincides with the middle axes of coil A1. The main machine parameters are listed in Table I.

For the experiments, which are presented in this paper, a coil width of 105 degrees was chosen for the rotor winding. This value results from preceding simulations, based on an analytical machine model, in which the coil parameters were optimized.

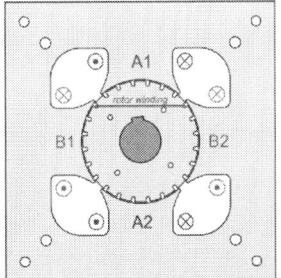

Fig. 2. Cross-section of the two-phase synchronous machine: scheme (left) and experimental machine (right).

TABLE I
MAIN DATA OF THE EXPERIMENTAL MACHINE

Parameter	Value
Stator outer dimensions	230 mm x 230 mm
Stator inner diameter	102 mm
Stack length	100 mm
Number of turns per stator coil	40
Wire diameter of the stator coil	1 mm
Rotor outer diameter	100 mm
Cross-section of the rotor coil	3 mm x 1.5 mm

978-1-4799-2706-7/14 $31.00 © 2014 IEEE

A. Measurement Procedure

The measurements of the coil and phase impedances as well as the voltage transfer functions between two coils or phases were carried out using a *Bode 100 Vector Network Analyzer*. In order to obtain the impedance matrix in (6), the machine parameters are identified in the stator-fixed reference frame, first. Considering standstill, the voltage equations of a synchronous machine in the stator-fixed reference frame can be described as

$$\begin{bmatrix} U_\alpha \\ U_\beta \end{bmatrix} = \begin{bmatrix} Z_{\alpha\alpha} & Z_{\alpha\beta} \\ Z_{\beta\alpha} & Z_{\beta\beta} \end{bmatrix} \begin{bmatrix} I_\alpha \\ I_\beta \end{bmatrix}. \tag{17}$$

For the case that phase β is open-circuited, i.e. $I_\beta = 0$, (17) can be simplified to

$$\begin{bmatrix} U_\alpha \\ U_\beta \end{bmatrix} = \begin{bmatrix} Z_{\alpha\alpha} \\ Z_{\beta\alpha} \end{bmatrix} I_\alpha. \tag{18}$$

In consequence, the self-impedance $Z_{\alpha\alpha}$ can be identified by an impedance measurement of phase α:

$$Z_{\alpha\alpha} = \frac{U_\alpha}{I_\alpha} \tag{19}$$

With (19), the remaining mutual-impedance $Z_{\beta\alpha}$ can be calculated from the voltage transfer function between both phases:

$$\frac{U_\beta}{U_\alpha} = \frac{Z_{\beta\alpha}}{Z_{\alpha\alpha}} \tag{20}$$

In an analogous manner, $Z_{\beta\beta}$ and $Z_{\alpha\beta}$ can be identified. Afterwards, the impedance matrix in (17) is transformed into the rotor-fixed reference frame.

All measurements refer to no-load condition, since the influence of saturation is not subject of this paper.

B. Impedance Measurement

For the position depicted in Fig. 2, the impedance of coil A1 in Fig. 3 results. Above the edge frequency of about 40 Hz the inductance drops to a nearly constant value according to (5). In a similar way, the resistance increases and remains nearly constant up to approximately 500 Hz.

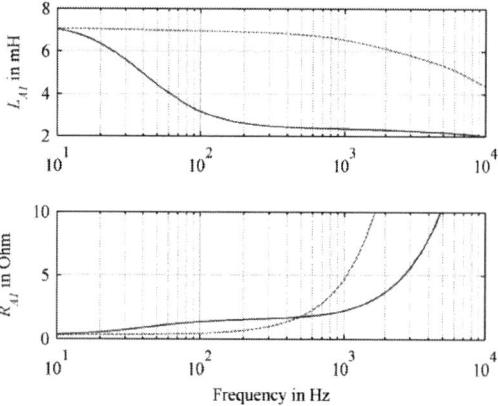

Fig. 3. Inductance (top) and resistance (bottom) of coil A1 for the position depicted in Fig. 2 with (−) and without (--) the short-circuited rotor winding.

However, above this frequency the rotor winding leads to a lower resistance compared to the case without the rotor winding. This phenomenon cannot be explained by the model in Fig. 1, which predicts constant values for frequencies much higher than the corner frequency. Nevertheless, this additional frequency range features a considerable resistance-based saliency and could be useful with regard to sensorless position estimation. Therefore, an extended model is presented in the next chapter that can explain the results in Fig. 3.

V. THREE COIL MODEL

When the high frequency behavior of electrical machines is discussed, eddy currents play a major role. Although in a real machine eddy currents are spread over the entire magnetically active parts, it appears likely that their influence on the stator impedance can be represented by a single short-circuited winding. Due to the lamination, the resistance of such an equivalent winding must be much higher than that of the real short-circuited rotor winding, of course. According to this theory, an additional coil is added to the existing model from Fig. 1 which leads to the three coil model in Fig. 4. Thereby, the subscript 'S' denotes the stator coil, while 'R' and 'E' mark the coils, which represent the rotor winding and the equivalent winding to account for the eddy currents, respectively.

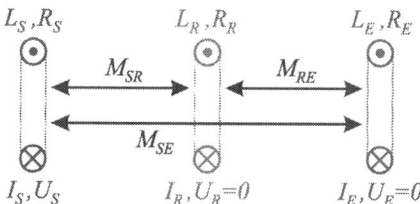

Fig. 4. Three magnetically coupled coils to model the influence of a short-circuited rotor winding including eddy currents.

Considering that only the voltage in the stator coil is non-zero, the voltage equations can be written as follows:

$$\begin{bmatrix} U_S \\ 0 \\ 0 \end{bmatrix} = \begin{bmatrix} R_S + sL_S & sM_{SR} & sM_{SE} \\ sM_{SR} & R_R + sL_R & sM_{RE} \\ sM_{SE} & sM_{RE} & R_E + sL_E \end{bmatrix} \begin{bmatrix} I_S \\ I_R \\ I_E \end{bmatrix} \tag{21}$$

This equation system can then be solved for the effective impedance of the stator coil:

$$Z_S = \frac{U_S}{I_S} = R_{S,eff} + j\omega L_{S,eff} \tag{22}$$

Due to the lack of space, the analytical solution is not included in this paper. However, the influence of the additional coil will be discussed by simulation results. Therefore, the parameters of the impedance matrix in (21) were adjusted manually, so that the effective resistance and inductance in (22) approximately match the measurements in Fig. 3. The resulting parameters are listed in Table II. The resistance and self-inductance of the stator and rotor coils were measured with an *LCR*-meter. Considering that a self-inductance is proportional to the square of the number of turns and that a mutual-inductance is proportional to the number of turns of both coupled coils, the mutual-inductance between the stator and rotor coil M_{RS} can be approximated by the expression in Table II. Thereby, the coefficient can be explained by the fact that both coils are not perfectly coupled. For the third coil, which represents the eddy currents, best agreement could be reached when it was parameterized just like the rotor coil, except for its resistance. It was chosen hundred times higher what can be attributed to the lamination of the electrical steel sheet.

For these parameters, the three coil model from Fig. 4 yields the solid lines in Fig. 5. A comparison with Fig. 3 shows that the measurement results can be accurately reproduced by the presented three coil model. Furthermore, it can explain why a short-circuited rotor winding leads to both a higher and a lower effective resistance compared to the case without a rotor winding. With regard to sensorless position detection, this is an interesting aspect since it basically offers the opportunity to actively influence the sign of the resistive saliency of the machine according to (10).

For the case that only the stator and rotor coil are taken into account, the effective impedance of the stator coil can be calculated by (3), which leads to the dotted results in Fig. 5. Obviously, the two coil model that was also used in [15], is only suitable for the prediction of the effective inductance within a limited frequency range.

TABLE II
PARAMETERS OF THE THREE COIL MODEL IN FIG. 4 USED FOR FIG. 5

Parameter	Value	Description
R_S	300 mΩ	*measured*
L_S	7.3 mH	*measured*
R_R	1.3 mΩ	*measured*
L_R	4.4 µH	*measured*
R_E	130 mΩ	$\approx 100 \cdot R_R$
L_E	4.4 µH	$\approx L_R$
M_{SR}	146 µH	$\approx 0.8 \cdot L_S$ / *(stator number of turns)*
M_{SE}	146 µH	$\approx M_{SR}$
M_{RE}	3 µH	$\approx 0.66 \cdot L_S$ / *(stator number of turns)²*

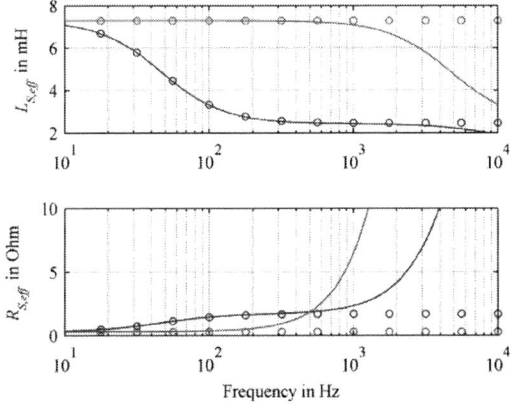

Fig. 5. Effective inductance (top) and resistance (bottom) of the stator coil of the three coil model in Fig. 4 with (−) and without (--) the short-circuited rotor winding (o/o: eddy currents neglected).

VI. ROTOR ANGLE DEPENDENCY

For sensorless position estimation, a saliency must exist in all rotor positions. Therefore, the impedances of both phases were measured in different positions and transformed into the rotor-fixed reference frame. The results for the inductances are shown in Fig. 6. For low frequencies, L_{dd} and L_{qq} are approximately equal despite of the position-dependent variations due to manufacturing inaccuracies. Additionally, the stator slotting leads to a 4th harmonic in the results. Above the corner frequency of the short-circuited rotor winding of about 40 Hz L_{dd} drops while L_{qq} remains nearly constant which leads to a high inductive saliency L_{diff} over a wide frequency range. In order to evaluate the feasibility for sensorless position estimation, the mutual-inductance L_{dq}, which is obviously non-zero, must also be taken into account. This is done in the following chapter.

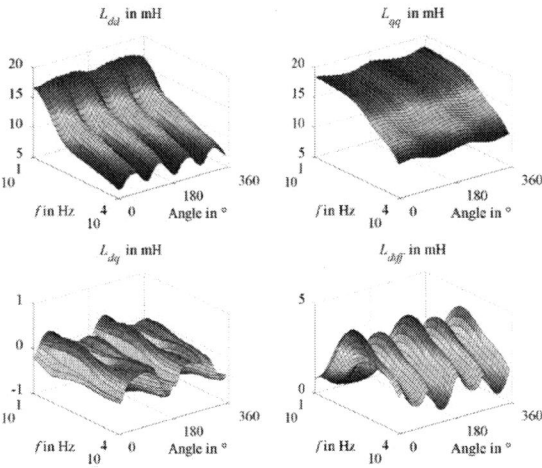

Fig. 6. Position- and frequency-dependence of the self- and mutual-inductances and the resulting saliency with short-circuited rotor winding.

For the effective resistances, the two frequency ranges mentioned in the previous chapter are discussed separately. For the lower frequency range between 10 Hz and 500 Hz the resistances in Fig. 7 result. For the sake of clarity, the frequency axis is reversed compared to Fig. 6. In good agreement with the results from the three coil model, the effective resistance in the d-axis R_{dd} increases while R_{qq} stays nearly constant. According to (10), this leads to a negative difference R_{diff}. However, while approaching 500 Hz, this saliency vanishes. The mutual-resistance R_{dq} increases with the frequency.

For the higher frequency range between 1 kHz and 10 kHz a similar behavior can be observed in Fig. 8. Compared to Fig. 7, the main difference is the sign of the resistive saliency R_{diff}. This is caused by the lower slope of R_{dd} due to the short-circuited rotor winding in combination with the influence of the eddy currents. Hence, the resistive saliency increases with frequency. However, this does not necessarily yield a small position error, since the ripple of the mutual-resistance R_{dq} also increases with the frequency. Therefore, the error angles from (11) and (14) are used in the next section to evaluate and compare the sensorless performance.

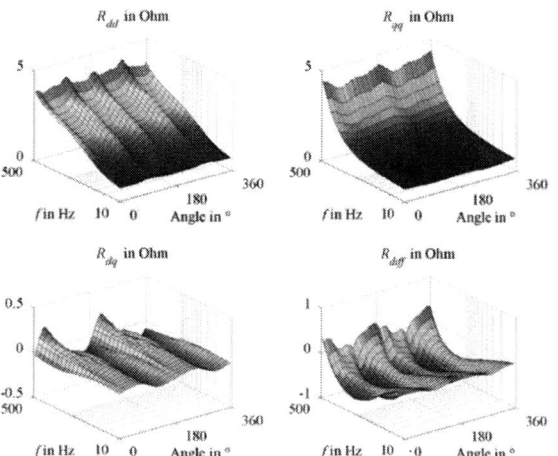

Fig. 7. Position- and frequency-dependence of the self- and mutual-resistances and the resulting saliency with short-circuited rotor winding between 10 Hz and 500 Hz.

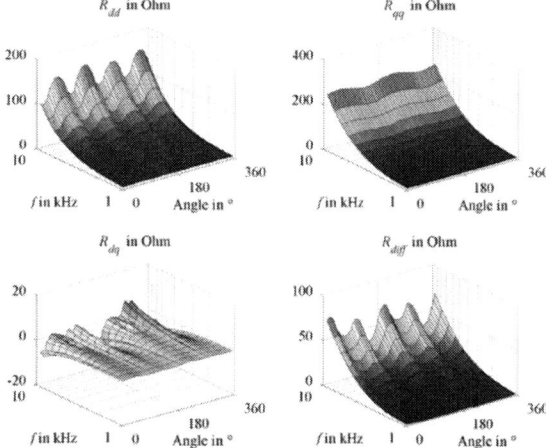

Fig. 8. Position- and frequency-dependence of the self- and mutual-resistances and the resulting saliency with short-circuited rotor winding between 1 kHz and 10 kHz.

VII. COMPARISON OF R- AND L-BASED SALIENCY TRACKING

In order to evaluate the accuracy of the position estimation, the position error was calculated according to (11) from the measurement results for each position and frequency. Considering the maximum absolute value of the position error over one revolution $\varepsilon_{L,max}$ as worst case, the frequency-dependence of the position error in Fig. 9 results. Over a wide frequency range, the accuracy is relatively high and nearly independent from the test signal frequency. In this range, the mutual-inductance is low in relation to the inductive saliency. In Fig. 9, $L_{diff,min}$ refers to the minimal absolute value of (13) over one revolution which is considered as worst case with regard to sensorless position estimation. By contrast, for the mutual-inductance the maximum absolute value $L_{dq,max}$ is used.

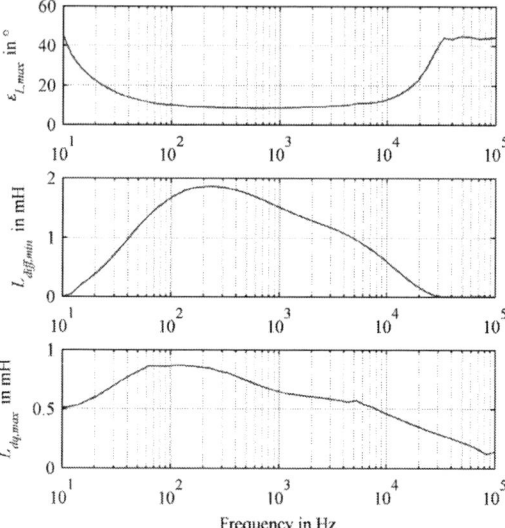

Fig. 9. Maximum absolute value of the error angle, minimum absolute value of the inductive saliency and maximum absolute value of the mutual-inductance over one revolution due to the rotor winding.

In an analogous manner, the worst case quantities for the resistance-based saliency were calculated. As discussed in the previous sections, for the resistance two characteristic frequency ranges occur, which differ considerably in magnitude. Therefore, the results in Fig. 10 are depicted separately with independent scaling. In both ranges, the maximum position error $\varepsilon_{R,max}$ has nearly the same minimal value. Consequently, the ratio between the maximum absolute value of the mutual-resistance $R_{dq,max}$ and the minimum resistive saliency $R_{diff,min}$ must be approximately equal in both frequency ranges. Nevertheless, the resistive saliency in the high frequency range is about 250 times higher compared to the lower range.

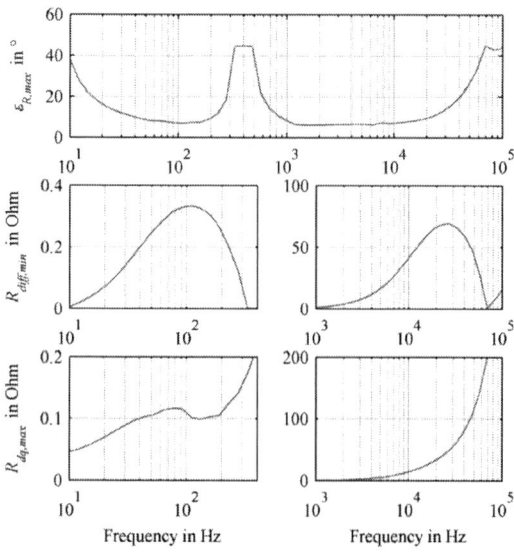

Fig. 10. Maximum absolute value of the error angle, minimum absolute value of the resistive saliency and maximum absolute value of the mutual-resistance over one revolution due to the rotor winding.

For a better comparison of the resistance- and inductance-based saliency tracking, Fig. 11 shows both maximum error angles together in one diagram. It becomes obvious that for the presented machine the resistance-based saliency tracking yields the smallest position error of about 8 degrees. However, with a test signal frequency of approximately 400 Hz the position estimation is impossible. This is not the case for the inductance-based saliency, which can be tracked for any frequency between 200 Hz and 20 kHz, although the accuracy is slightly worse. Finally, the relative saliencies are discussed, which are defined by

$$L_{diff,rel} = \frac{(L_{qq}-L_{dd})}{L_{qq}} \qquad (23)$$

for the inductances and

$$R_{diff,rel} = \frac{(R_{qq}-R_{dd})}{R_{qq}} \qquad (24)$$

for the resistances. Again, considering the minimum of the absolute value over one revolution as worst case, $L_{diff,rel,min}$ and $R_{diff,rel,min}$ result, which are depicted in Fig. 11. The relative resistive saliency is about twice as high as the inductive one.

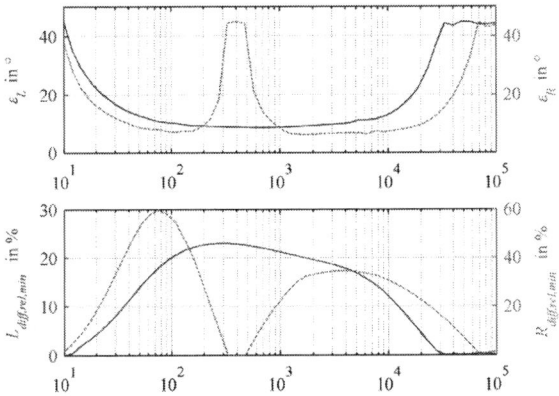

Fig. 11. Maximum absolute value of the error angle and minimum absolute value of the relative saliency over one revolution for the inductance- (−) and resistance-based (--) saliency tracking.

VIII. CONCLUSION AND OUTLOOK

Due to the short-circuited rotor winding, the presented machine features both an inductance- and resistance-based saliency which provide good capability for sensorless position estimation. It could also be shown by the suggested three coil model that the interaction with the eddy currents can be useful for a resistive saliency. However, many aspects were not yet considered and hence, will be subject to future research. First, saturation was neglected. However, saturation will lead to a lower coupling M_{dr} in (16) and in consequence, reduce both saliencies. Second, the short-circuited winding increases the total losses of the machine [17]. Therefore, one major task is the optimization of the rotor winding to find a good compromise between sensorless performance and efficiency. As mentioned in the introduction, a short-circuited rotor winding could also be used to extend the

current limit of machine types other than surface-mounted permanent magnet synchronous machines. Therefore, an interior permanent magnet rotor with an integrated rotor winding will be investigated in future work.

ACKNOWLEDGMENT

This work was supported by the Deutsche Forschungs-gemeinschaft under grant Ha 6081/1-1.

REFERENCES

[1] I. Hahn, "Differential magnetic anisotropy - prerequisite for rotor position detection of PM-synchronous machines with signal injection methods," in *First Symposium on Sensorless Control for Electrical Drives (SLED),2010*, pp. 40–49.

[2] Shih-Chin Yang and R. D. Lorenz, "Surface Permanent Magnet Synchronous Machine Position Estimation at Low Speed Using Eddy-Current-Reflected Asymmetric Resistance," *Power Electronics, IEEE Transactions on*, vol. 27, no. 5, pp. 2595–2604, 2012.

[3] P. Garcia, F. Briz, D. Reigosa, C. Blanco, and J. M. Guerrero, "On the use of high frequency inductance vs. high frequency resistance for sensorless control of AC machines," *Sensorless Control for Electrical Drives (SLED), 2011 Symposium on*, pp. 90–95, 2011.

[4] D. Raca, P. Garcia, D. D. Reigosa, F. Briz, and R. D. Lorenz, "Carrier-Signal Selection for Sensorless Control of PM Synchronous Machines at Zero and Very Low Speeds," *Industry Applications, IEEE Transactions on*, vol. 46, no. 1, pp. 167–178, 2010.

[5] M. Linke, R. Kennel, and J. Holtz, "Sensorless speed and position control of synchronous machines using alternating carrier injection," in *Electric Machines and Drives Conference, 2003. IEMDC'03. IEEE International*, 2003, pp. 1211–1217 vol.2.

[6] J. Reill, B. Piepenbreier, and I. Hahn, "Utilisation of magnetic saliency for sensorless-control of permanent-magnet synchronous motors," in *Power Electronics Electrical Drives Automation and Motion (SPEEDAM), 2010 International Symposium on*, 2010, pp. 1484–1489.

[7] M. Seilmeier, "Modelling of electrically excited synchronous machine (EESM) considering nonlinear material characteristics and multiple saliencies," in *Power Electronics and Applications (EPE 2011), Proceedings of the 2011-14th European Conference on*, 2011, pp. 1–10.

[8] N. Bianchi, E. Fornasiero, and S. Bolognani, "Effect of stator and rotor saturation on sensorless rotor position detection," in *Energy Conversion Congress and Exposition (ECCE), 2011 IEEE*, 2011, pp. 1528–1535.

[9] P. Guglielmi, M. Pastorelli, and A. Vagati, "Cross saturation effects in IPM motors and related impact on zero-speed sensorless control," in *Industry Applications Conference, 2005. 40th IAS Annual Meeting. Conference Record of the 2005*, 2005, pp. 2546–2552, vol. 4.

[10] J. Graus, A. Boehm, I. Hahn, "Influence of Stator Saturation on the Differential Inductances of PM-Synchronous Machines with Concentrated Winding," *Sensorless Control for Electrical Drives (SLED), 2012 IEEE Symposium on*, 2012.

[11] N. Bianchi, S. Bolognani, "Influence of Rotor Geometry of an IPM Motor on Sensorless Control Feasibility," *Industry Applications, IEEE Transactions on*, vol. 43, no. 1, pp. 87–96, jan.-feb. 2007.

[12] T. A. Nondahl, C. Ray, P. B. Schmidt, "A permanent magnet rotor containing an electrical winding to improve detection of motor angular position," in *Industry Applications Conference, 1998. 33rd IAS Annual Meeting*, vol.1, pp. 359–363, 1998.

[13] A. Faggion, E. Fornasiero, N. Bianchi, S. Bolognani, "Sensorless Capability of Fractional-Slot Surface-Mounted PM Motors," *Industry Applications, IEEE Transactions on*, vol. 49, no. 3, pp. 1325–1332, may-june 2013.

[14] M. Morandin, S. Bolognani, A. Faggion, "Outer-rotor ringed-pole SPM starter-alternator suited for sensorless drives," *Sensorless Control for Electrical Drives (SLED), 2011 Symposium on*, pp. 96–101, 2011.

[15] A. Faggion, N. Bianchi, and S. Bolognani, "Ringed-Pole Permanent-Magnet Synchronous Motor for Position Sensorless Drives," *Industry Applications, IEEE Transactions on*, vol. 47, no. 4, pp. 1759–1766, 2011.

[16] Shih-Chin Yang and R. D. Lorenz, "Comparison of resistance-based and inductance-based self-sensing controls for surface permanent-magnet machines using high-frequency signal injection," *Industry Applications, IEEE Transactions on*, vol. 48, no. 3, pp. 977–986, 2012.

[17] D. Mingardi, E. Fornasiero, N. Bianchi, S. Bolognani, and A. Faggion, "Ring losses evaluation in ringed pole PM motors," *Sensorless Control for Electrical Drives and Predictive Control of Electrical Drives and Power Electronics (SLED/PRECEDE), 2013 IEEE International Symposium on*, pp. 1–8, 2013.

Design and Optimization of High-Speed Switched Reluctance Motor Using Soft Magnetic Composite Material

Zwe-Lee Gaing
Dep. of Electrical Engineering
Kao Yuan University
Kaohsiung City 821, Taiwan

Kuan-Yi Kuo,   Jia-Sheng Hu
Dep. of Greenergy Technology
National University of Tainan
Tainan City 700, Taiwan

Min-Fu Hsieh,   Ming-Hsiao Tsai
Electric Motor Technology Research Center
National Cheng Kung University
Tainan City 701, Taiwan

Abstract— In this paper, an optimal design of a high-speed switched reluctance motor (SRM) with lower torque ripple using the Taguchi method for household appliances has been presented. To enhance the efficiency and reduce the manufacture processes of the proposed high-speed double-salient SRM, the soft magnetic composite (SMC) material is adopted. Moreover, to decrease noise and vibration of the double-salient SRM, torque ripple must be reduced; thus, the best SRM geometry should be found. To find the best geometric parameters, the Taguchi method and the finite element method (FEM) has been employed to achieve the target in this paper. The experimental results have shown that the proposed SRM can achieve the design goal for lower torque ripple and higher efficiency which is suitable for application to household electric power blenders and food processors.

Keywords— switched reluctance motor, soft magnetic composite, Taguchi method, finite element method.

I. INTRODUCTION

Reluctance motors possess the characteristics of high power density, simple structure, robust construction, low manufacturing cost, etc., making it a very popular subject of research. Reluctance motors feature the advantages of high fault tolerance, safety, and reliability for application in high speed home appliances, like food processors and blenders [1][2].

Soft magnetic composites (SMCs) are high purity metal powder particles surrounded by an ultra-thin layer of inorganic insulating film, as shown in Fig. 1. Utilizing the powder metallurgy technology, SMCs are easy to be die-casted into any shape, for example, iron cores of motors, which are designed with 3D technology, as in depicted Fig. 2. Traditional motor rotors made with silicon steel sheets are manufactured by stacking and pressing, which are limited to 2D structures. On the contrary, die-casting 3D technology allows SMC to a reduction in manufacturing processes and materials consumption [3][4].

On the other hand, SMCs have the magnificent electromagnetic property of reducing eddy currents of the motor core, which are induced by the changes of the external high-frequency magnetic flux, to decrease the eddy losses. Figure 3 shows a 6-slot 4-pole(6/4) SRM motor, of which the cores are made with SMC Somaloy 500 1P and 50CS600 silicon steel sheets made by the China Steel Cooperation. Motor performance is shown in Table I. As seen in Fig. 3(a), at speeds above 15,000rpm, the iron loss of Somaloy 500 is lower than 50CS600. Furthermore, Fig. 3(b) shows both the eddy current losses and the hysteresis losses produced by the

50CS600 lamination and the Somaloy 500 material, respectively. Therefore, the SMC materials are suitable to applications containing low pole numbers and high-speed operation, which can efficiently reduce iron loss and enhance performance.

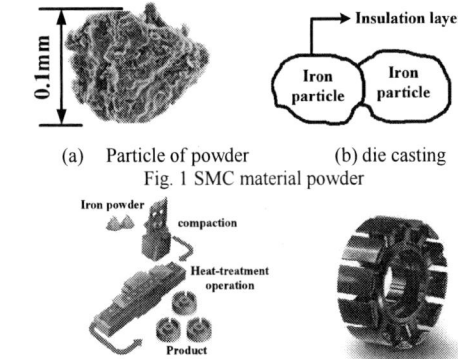

(a)   Particle of powder   (b) die casting
Fig. 1 SMC material powder

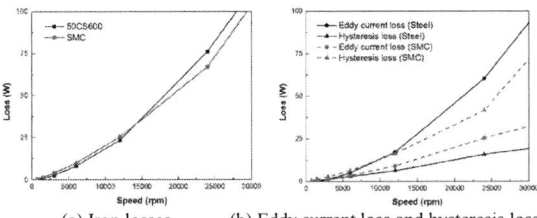

Fig. 2 SMC die-casting forming

(a) Iron losses   (b) Eddy current loss and hysteresis loss
Fig. 3 Comparisons of iron cores made of different materials

However, regardless of core materials, the double salient pole geometry and drive mode of reluctance motors produce obvious torque ripples, which are the main factors that cause vibration and noise. In order to reduce torque ripples, in many previous studies, special current wave control modes or complicated control algorithms have been adopted in the drivers, which increase the difficulties in SRM control as well as the costs of driver production and maintenance [5][6].

TABLE I
COMPARISON OF PERFORMANCES

Core materials Performances	50CS600	Somaloy 500
Torque (Nm)	0.30	0.31
T_{max} (Nm)	0.60	0.60
T_{min} (Nm)	0.10	0.10
T_{ripple} coefficient	1.61	1.61
Current (A)	7.08	7.65
Power (W)	713.83	713.83
Efficiency (%)	77.8	78.7

In order to effectively reduce the production cost and torque ripples of high-speed SRMs, this paper proposes the utilization of SMC as core materials and takes the four reducing torque ripple strategies into consideration simultaneously when designing the motor structure, which are (I) changing the excitation trigger angle of the stator's windings, (II) notched rotor, (III) uneven air-gap width, and (IV) changing the pole arc angle of the rotor, so that the output performances of the motor can meet the design requirements [7].

In this paper, a 6S4P(6/4), 100V, rated speed 22000rpm and rated output power 700W SRM is taken as the study case, which is adopted as the power source of the domestic food processor shown in Fig. 4; the layout of the driver circuit is dispatched as Fig. 4(e), and the geometric parameters are listed in Table II.

TABLE II
GEOMETRIC PARAMETERS OF 6/4 SRM

Pole		4
Slot		6
Voltage (V)		100
Winding	Turn	37
	Single-phase resistance (Ω)	0.23
Rated speed (rpm)		22,000
Rated Power/Torque(W/Nm)		680/
Rotor size (Radius)		22.5mm×15.6mm×45mm
Stator size (Radius)		40mm×32mm×45mm
Air gap width (mm)		0.5
Material of iron core		Somaloy 500

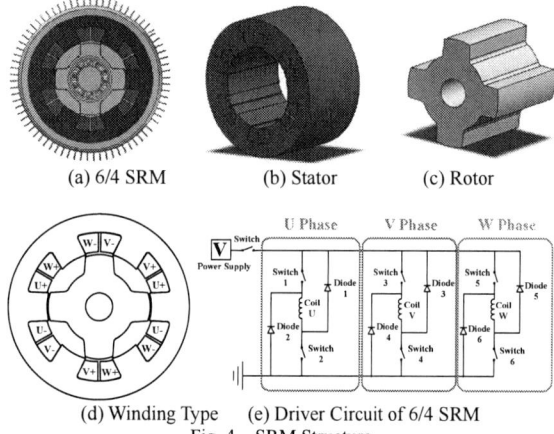

(a) 6/4 SRM   (b) Stator   (c) Rotor

(d) Winding Type   (e) Driver Circuit of 6/4 SRM
Fig. 4   SRM Structure

II. MATHEMATICAL MODE OF RELUCTANCE MACHINE

A. Torque Equations of Reluctance Machine

The magnetic lines of a magnetic circuit will stretch out along the path of least reluctance and form a closed loop. Hence, if the magnetic sheets of the stator and rotor are not in parallel alignment, they will be influenced by the magnetic field effect and shift towards the magnetic path of least reluctance, thus producing a distracting force, i.e., reluctance force. Reluctance motors use such reluctance force to induce the rotation of the rotor to generate torque.

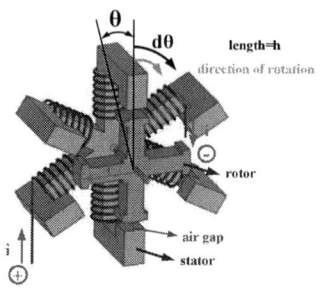

Fig. 5 Simplified structure of a 4-pole reluctance machine.

Figure 5 shows a simplified structure of a 4-pole reluctance machine. Given the structure of a linear system and via electromechanical energy conversion, the coenergy $W_f^{'}$ of the system is

$$W_f^{'} = \frac{1}{2} L(\theta) \cdot i^2 \qquad (1)$$

with a torque T_f of

$$T_f = \frac{\partial W_f^{'}}{\partial \theta} = \frac{1}{2} i^2 \frac{\partial L(\theta)}{\partial \theta} . \qquad (2)$$

According to the symmetry characteristics of the magnetic circuit, the inductance L generated in a pole-pair can be expressed as [5]

$$L(\theta) = \frac{N^2}{\Re(\theta)} = \frac{N^2 \mu_o (r + 0.5g)\theta}{2g} . \qquad (3)$$

Hence, the coenergy $W_f^{'}$ of the 4-pole reluctance machine is [9]

$$W_f^{'} = 2\mu_o g H_{ag}^2 h(r + 0.5g)\theta \qquad (4)$$

According to (1) and (2), the torque T_f generated by the system can be derived as

$$T_f = \frac{\partial W_f^{'}}{\partial \theta} = \frac{2 \cdot g B_{ag}^2 h(r + 0.5g)}{\mu_o} \qquad (5)$$

By (5), the reluctance machine takes advantages of the double salient poles structure and unaligned arc angles of the stator and rotor, i.e., $\theta \neq 0$; thus, the reluctance force and torque are produced. However, the torque of the reluctance machine is related to structural geometric parameters, such as machine length h, air gap width g, rotor radius r, air-gap flux density B_{ag}, etc. Therefore, how to select suitable geometric parameters for designing a high-performance SRM is a serious issue.

B. Torque Ripple

Torque ripple is produced due to the alignment effect of the magnetic pole teeth of the motor stator and rotor, and the discontinuous input current occurred when switching the driver. Figure 6 presents the torque ripples of the reluctance motor, which causes vibration and noise in motors.

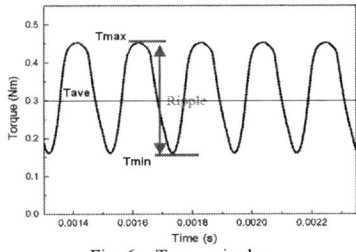

Fig. 6   Torque ripples.

In order to measure torque ripples, in this paper, the coefficient of torque ripples are defined as

$$T_{ripple} = \frac{T_{\max} - T_{\min}}{T_{ave}} \qquad (6)$$

where T_{ripple} denotes to the coefficient of torque ripples, T_{max} and T_{min} represent to the maximal and minimal values of torque, respectively, and T_{ave} is the average value of the difference between T_{max} and T_{min} [5].

III. TORQUE RIPPLE REDUCTION STRATEGIES

Referring to [5][7], when the strategies are applied to reduce torque ripples, the output torque or efficiency of SRM may drop accordingly. In order to effectively reduce torque ripples, four effective strategies are proposed, which are:

A. Changing excitation trigger angle of the stator's windings

In order to understand the impact of the excitation trigger angle on SRM performances, in terms of output torque and torque ripples, the different excitation trigger angles are analyzed, as shown in Fig. 7. The step angle θ_{step} of a three-phase 6/4 SRM is 30 degrees.

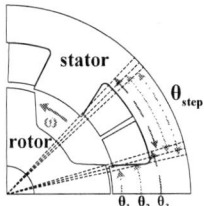

Fig. 7   Rotor position of advanced excitation

B. Notched rotor

Due to the high-density powder molding technology and high-temperature sintering are employed, the mechanical strength of powder metallurgy material is sufficient. Hence, the approach of forming notches in the rotor of the 6/4 SRM, as shown in Fig. 8, is adopted. For realizing the influences of notching angle to motor output performance and torque ripples, different notching angles are evaluated.

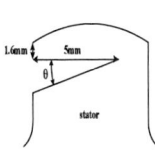 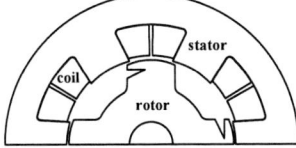

Fig. 8   Notching angle of rotor.

C. Uneven air-gap width

On original design, the air-gap width between the stator and rotor of a 6/4 SRM is set to be 0.5mm. Uneven air-gap width geometry is employed for reducing the torque ripple in this paper. The endeavor herein is to adjust the rotor salient pole center from 4mm to 6mm upwards, while maintaining the stator in the same position, as shown in Fig. 9.

D. Changing the pole arc angle of the rotor

This strategy attempts to change the pole arc angle between the stator and rotor to understand its influences on motor performance. The original rotor pole arc angle of a 6/4 SRM is set to be 32 degrees. In this paper, many different pole arc angles are examed, as illustrated in Fig. 10, and the output performance is evaluated.

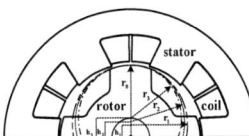 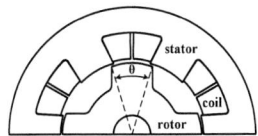

Fig. 9 Uneven air-gap width   Fig. 10   Pole arc angle of rotor.

IV. SRM OPTIMIZATION BY USING TAGUCHI METHOD

A. Control factor selection and orthogonal array establishment

The Taguchi method is one of the best robust control design methods, which adopts the least experiments to determine the best process parameter combination and eliminate the variations of the quality characteristics. This design method of optimization was developed by Dr. Genichi Taguchi, who extended experiments with orthogonal arrays and quantified the impact of control factor settings on output performance with a ratio of signal to noise (S/N)[8].

In order to optimize the performance of SRM, the Taguchi method is applied to optimize the quality objective of SRM. In this paper, four geometric parameters are selected to be the control factors, i.e. excitation trigger angle, notching angle of rotor, uneven air-gap width, and pole arc angle of rotor. The configuration of the control factors and levels are shown in Table III, and the corresponding $L^9(3^4)$ orthogonal array is depicted in Table IV.

TABLE III
CONFIGURATION OF CONTROL FACTORS AND LEVELS

Factor \ Level	1	2	3
A. Excitation trigger angle	12°	13°	14°
B. Notching angle	25°	26°	27°
C. Uneven air-gap width(mm)	4	5	6
D. Pole arc angle of rotor	32°	33°	34°

B. Signal-to-noise (S/N) ratio and responding table

The quality target of this research is to achieve low torque ripple; therefore, the-lower-the-better (LB) approach of the Taguchi Method is employed for analysis, and the signal-to-noise (S/N) ratio is given by

$$S/N = -10 \cdot log(MSD) = -10 \cdot log\left[\frac{1}{n}\sum_{i=1}^{n} y_i^2\right] \quad (dB) \quad (7)$$

where MSD denotes to mean square deviation, y_i and n represent the experimental values and experiment times, respectively. The experimental results and computed S/N ratios are shown in Table IV.

TABLE IV.
$L^9(3^4)$ ORTHOGONAL TABLE AND S/N RATIOS

No.	A	B	C	D	S/N
1	12	25.	4	32	-3.16
2	12	26	5	33	-2.28
3	12	27	6	34	-1.30
4	13	25	5	34	-0.97
5	13	26	6	32	-1.37
6	13	27	4	33	-1.88
7	14	25	6	33	-0.33
8	14	26	4	34	-0.68
9	14	27	5	32	-1.36

According to the S/N ratios in Table IV, the responding table of control factors is obtained as Table V and Fig. 12; herewith, the best combination is gathered, i.e. $A_3B_2C_3D_3$.

TABLE V
RESPONDING TABLE OF CONTROL FACTORS

Factor / Level	A	B	C	D
1	-2.24	-1.49	-1.91	-1.96
2	-1.41	-1.44	-1.54	-1.50
3	-0.79	-1.51	-1.00	-0.98
Max-Min	**1.45**	**0.07**	**0.91**	**0.98**

C. Analysis of variance (ANOVA)

In order to understand the contribution of each control factor to quality characteristics, the analysis of variance (ANOVA) is employed [8], and the results are listed in Table VI, in which Factor A (excitation trigger angle) makes the greatest contribution, followed by Factor D (pole arc angle of the rotor); Factor B (notched rotor) is last.

TABLE VI
ANOVA IN TORQUE RIPPLE

Factor	SS	DOF	Var.	F	PC(%)
A	3.1758	2	1.5879	-7.2079	53.99
B	0.0078	2	0.0039	-0.0177	0.13
C	1.2566	2	0.6283	-2.8520	21.36
D	1.4424	2	0.7212	-3.2737	24.52
e	-0.4406	2	-	-	-
e^T	-0.4406	[2]	-2E-01	2.5007	-
T(Sum)	5.442	10	-	-	100.0

D. Verifying experiment

The optimal parameter combination ($A_3B_2C_3D_3$) is obtained with the application of the Taguchi Method, with which the SRM is optimized by an excitation trigger angle of $14°$(A_3), a notching rotor angle of $26°$(B_2), 6mm uneven air-gap(C_3), and a pole arc angle of $34°$(D_3), as shown in Fig. 11.

Figures 12–14 illlustrate the simulation verification, which verifies that the optimal parameter combination $A_3B_2C_3D_3$ obtained by the Taguchi method is capable of achieving the quality objective of low torque ripples, i.e., lower T_{ripple} coeffient, referring to Table VII.

V. CONCLUSION

SMC materials are appliable of die-casted forming, which possesses the mechanical characteristics of reducing the production cost and the consuption of materials and are further equipped with the magnetic

properties of lower eddy current and eddy loss under a high-frequency magnetic flux enviroment. In this paper, SMCs are applied as the iron core materials of a high-speed SRM. Meanwhile, for reducing torque ripples, noises, and vibrations of the proposed SRM are adopted as control factors; the Taguchi method integrated with FEM analysis is emplyed to determine the best parameter combination for optimizing SRM to achieve the quality target of low torque ripple. The experimental results have shown that the proposed SRM not only satisfying the motor's specifications but also the T_{ripple} is reduced by 45.3% which is suitable for application to high-speed household appliances.

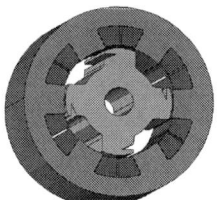

Fig. 11 Proposed SRM model using SMC material

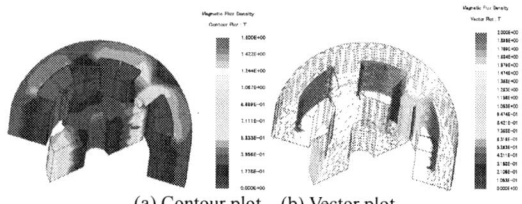

(a) Contour plot   (b) Vector plot
Fig. 12 Magnetic flux density.

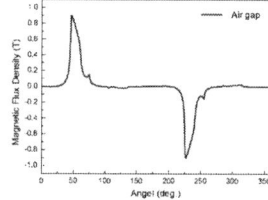

Fig. 13 Flux density of air-gap   Fig. 14   Verifying experiment
(U-phase excitation)

TABLE VII
COMPARISONS BETWEEN BEFORE AND AFTER OPTIMIZATION.

Performances	Before	After
Speed (rpm)	22,000	22,000
Average Torque (Nm)	0.3248	0.2957
Output Power (W)	747.91	680.90
T_{max} (Nm)	0.6917	0.4130
T_{min} (Nm)	0.1047	0.1202
$T_{max}-T_{min}$ (Nm)	0.5870	0.2928
T_{ripple} coefficient	1.81	0.99

VI. ACKNOWLEGMENT

The authors gratefully acknowledge financial supports from the National Science Council of R.O.C. under contract NSC 101-2221-E-244-011 and the Ministry of Economic Affairs of R.O.C. under contract MOEA 100-EC-17-A-05-S1-192.

REFERENCES

[1] M. Krishnamurthy *et al.*, "Making the case for applications of

switched reluctance motor technology in automotive products," *IEEE Trans. on Power Electron.*, vol. 21, no. 3, pp. 659–675, May 2006.

[2] K. M. Rahman and S. E. Schulz, "Design of high-efficiency and high-torque-density switched reluctance motor for vehicle propul- sion," *IEEE Trans. Ind. Appl.*, vol. 38, no. 6, pp. 1500–1507, Nov./Dec.2002.

[3] A. G. Jack, P. G. Dickinson, D. Stephenson, J. S. Burdess, N. Fawcett, and J. T. Evans, "Permanent-Magnet Machines with Powdered Iron Cores and Prepressed Windings," *IEEE Transaction on Industry Appli.*, vol. 36, no. 4, pp.1077-1085, 2000.

[4] M. Persson, G. Nord, L. O. Pennander, G. J. Atkinson, and A. G. Jack, "Development of Somaloy Components for a BLDC motor in a Scroll Compressor Application," *PM2006 Conference Proceeding*, September 24-28, 2006.

[5] T. J. E. Miller, "Switched Reluctance Motor and Their Control," Oxford Press, New York, 1993.

[6] S. Lei and W. Jianhua, "Switched reluctance motor vibration prediction: From low frequency to high frequency," in *IEEE Int. Electric Ma- chines and Drives Conf.*, pp. 971–978, May 3-6, 2009.

[7] G. Li, J. Ojeda, S. Hlioui, E. Hoang, M. Lecrivain, and M. Gabsi, "Modification in Rotor Pole Geometry of Mutually Coupled Switched Reluctance Machine for Torque Ripple Mitigating," *IEEE Trans. on Magnetics,* vol. 48, no. 6, pp. 2025–2034, June 2012.

[8] S. Brisset, F. Gillon, S. Vivier and P. Brochet, "Optimization with Experimental Design: An Approach Using Taguchi's Methodology and Finite Element Simulations," *IEEE Trans. on Magnetics*, Vol.37, No.5, pp.3530-3533, Sept. 2001.

Influence of Pulse Width Modulation (PWM) on the Iron Losses of Electrical Steel

Andreas Boehm

Chair of Electrical Drives and Machines
University of Erlangen-Nuremberg
Erlangen, Germany
andreas.boehm@fau.de

Ingo Hahn, *Senior Member IEEE*

Chair of Electrical Drives and Machines
University of Erlangen-Nuremberg
Erlangen, Germany
ingo.hahn@ieee.org

Abstract-This paper deals with the measurement of the magnetic properties of electrical steel under the influence of pulse width modulation (PWM) in comparison to an ideal sinusoidal waveform. At first the theory and practice of measuring the magnetic properties with the ring core method are described. Next the measurement setup and results are depicted. The results show the well-known fact that a sinusoidal voltage has lower losses than pulse width modulated voltage. The relative additional losses of PWM increase with the fundamental frequency. Furthermore a high PWM frequency shows fewer losses than a low PWM frequency. Finally a voltage created with maximal modulation index and minimal DC voltage leads to a decrease of the losses in comparison to the same effective voltage created by a lower modulation index and higher DC voltage.

Index Terms- Electrical steel, inverter supply, iron losses, modulation index, pulse width modulation.

I. INTRODUCTION

Nowadays inverter fed three phase electrical machines are widely spread throughout industry and daily life. For synchronous machines, especially permanent magnet synchronous machines (PMSM), which have become quite popular in the last three decades, an inverter supply is needed. Although induction machines can be connected directly to the grid, an inverter is needed if the machines run in a speed controlled application. The commonly used control methods are either the open loop V/f control or the closed loop field oriented control.

Besides the ohmic losses in electrical machine's windings the iron losses (hysteresis and eddy current losses) represent a large proportion of the total losses. Usually manufacturers of electrical steel specify the magnetic properties for single frequencies without any harmonic content in the flux density as it is required in the Epstein frame measurement standard [1]. In order to take the influence of inverter supply into account measurements with a PWM voltage supply are carried out with the ring core method. Previous publications have mostly used Epstein frames to determine the influence of the PWM [2], [3]. In [4] the ring core method was used to determine the iron losses for a fundamental frequency of 50 Hz. This paper will also analyze the influence of different fundamental frequencies.

II. DETERMINATION OF THE MAGNETIC PROPERTIES

To obtain the magnetic properties the magnetization curve is determined by the so called ring core method. Therefore a ring specimen of electrical steel sheet is produced and wound with two coils each, a primary and a secondary coil. The primary coil is fed by a sinusoidal voltage or current source. In the Epstein frame measurement standard [1] the primary voltage is controlled so that a sinusoidal voltage at the secondary coil is reached. For this paper the primary coil is fed by a pulse width modulated sinusoidal voltage and for the purpose of comparison by an ideal sinusoidal voltage. The current I_p on the primary side and the voltage V_S on the open-circuit secondary side are measured. To gain the magnetic field strength H in a ring specimen the magneto-motive force Θ is calculated as follows:

$$\oint \vec{H}\,\vec{ds} = NI_p = \Theta \tag{1}$$

with N being the number of turns of the primary coil. Solving (1) for H leads to:

$$H(r_e) = \frac{NI_p}{2\pi r_e} \tag{2}$$

The effective radius r_e of the ring specimens can be calculated with the two core constants C_1 and C_2 which are given in DIN EN 60205 [5]:

$$C_1 = \frac{2\pi}{h_e \ln\left(\frac{r_1}{r_2}\right)} \tag{3}$$

$$C_2 = \frac{2\pi\left(\frac{1}{r_2} - \frac{1}{r_1}\right)}{h_e^2 \ln^3\left(\frac{r_1}{r_2}\right)} \tag{4}$$

with h_e being the height of the ring, r_1 the outer radius and r_2 the inner radius. The effective radius r_e of the core can then be determined by:

$$r_e = \frac{C_1^2}{2\pi C_2} = \frac{\ln\left(\frac{r_1}{r_2}\right)}{\frac{1}{r_2} - \frac{1}{r_1}} \tag{5}$$

Since the voltage V_S of the secondary coil is the time derivation of the flux linkage Ψ, the flux ϕ can be determined by the following equation:

$$\int_{t_0}^{t} V_S \, dt = \Psi(t) = N\phi(t) \tag{6}$$

The flux density can be calculated as:

$$B(t) = \frac{\phi(t)}{A_e} \tag{7}$$

The effective cross section area A_e is also given in [5]:

$$A_e = \frac{C_1}{C_2} = h_e \frac{\ln^2\left(\frac{r_1}{r_2}\right)}{\frac{1}{r_2} - \frac{1}{r_1}} \tag{8}$$

Having calculated B and H the magnetization curve or B(H)-loop can be created as can be seen in the following figure.

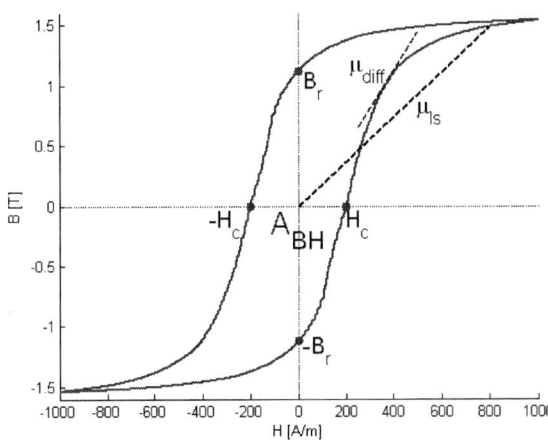

Fig. 1. Ordinary B(H)-loop with coercive field strength, remanence flux density, differential and large signal permeability and the occupied BH area.

The B(H)-loop contains all the information of the magnetic properties of the material such as the coercive field strength H_C, the remanence flux density B_r, the differential and large signal permeabilities μ_{diff} and μ_{ls} respectively and the area A_{BH} of the B(H)-loop which is equal to the iron losses of one cycle per effective volume:

$$A_{BH} = \frac{W_{Fe}}{V_e} = \int_{H_{min}}^{H_{max}} (B_u(H) - B_l(H)) \, dH \tag{9}$$

With B_u and B_l being the values of the flux density of the upper and lower branch of the loop respectively. The effective volume V_e is given by:

$$V_e = A_e 2\pi r_e \tag{10}$$

The specific iron losses p_{Fe} (in [W/kg]) can be determined to:

$$p_{Fe} = f \frac{A_{BH}}{\rho_{Fe}} \tag{11}$$

with f being the fundamental frequency of the voltage source on the primary side and ρ_{Fe} being the specific mass density of the considered iron core material.

In [6] it is shown that the losses can also be determined with the measured primary current and secondary voltage:

$$p_{Fe} = \frac{1}{\rho_{Fe} V_e T} \int_{t=0}^{T} I_p(t) V_s(t) \, dt \tag{12}$$

The next figure shows the primary voltages (V_p) and primary currents (I_p) for an ideal sinusoidal and a pulse width modulated primary voltage.

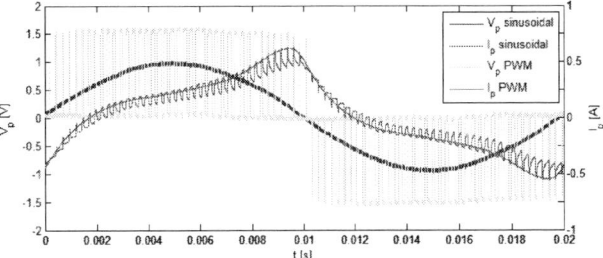

Fig. 2. Primary voltage and current measured on a ring core specimen for an ideal sinusoidal and a pulse width modulated primary voltage.

The sinusoidal primary voltage effects a smooth primary current, while the PWM voltage, which jumps between zero and the DC-link voltage leads to a ripple in the primary current. The resulting B(H)-curves can be seen in the following figure.

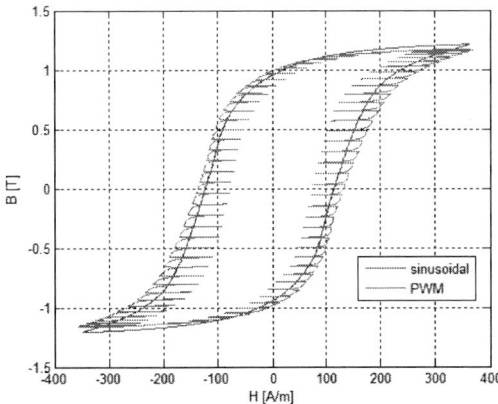

Fig. 3. B(H)-curves for sinusoidal and PWM supply.

While the sinusoidal feeding of the specimen results in a smooth B(H)-curve, the current ripple in the primary PWM current (Fig. 2, green) effects the drops of the field strength in the B(H)-loop.

III. Measurement Setup

To perform the measurements the following setup of Fig. 4 and Fig. 5 is used. The primary coil of the specimen is supplied by a power amplifier that gets the signal from a function generator which is controlled by Matlab. For the calculation of the B(H)-loops the current of the primary coil and the voltage on the secondary coil are measured with a 12-bit oscilloscope. For the purpose of verification the primary voltage is also recorded. The oscilloscope is controlled and read-out by the same Matlab file that generates the signals for the function generator.

For feeding the specimen with a PWM voltage the arbitrary mode of the function generator is used. The PWM waveform for one fundamental period is calculated in Matlab with up to 16000 data points since this is the limit of the function generator.

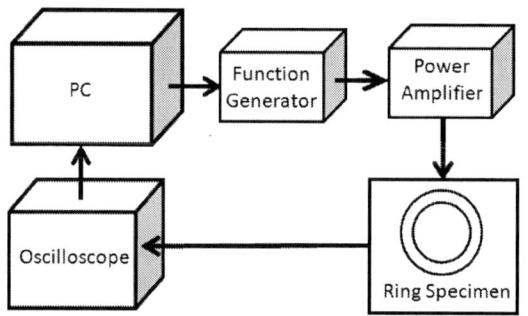

Fig. 4. Measurement setup: ring specimen; oscilloscope for measuring the current of the primary coil and the voltage of the secondary coil; function generator for controlling the power amplifier that supplies the specimen; PC for controlling the function generator and for reading out the oscilloscope.

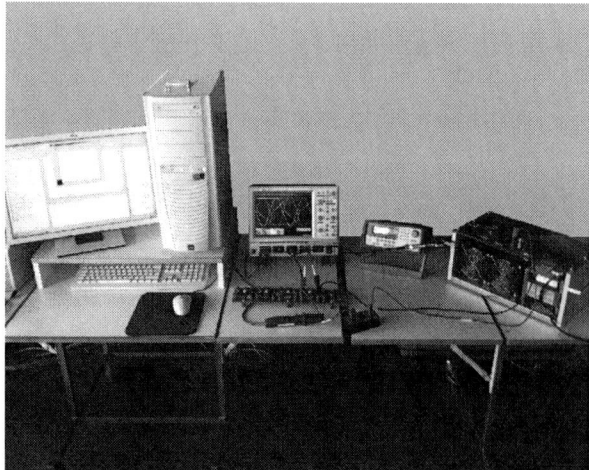

Fig. 5. Measurement setup from left to right: PC, oscilloscope, function generator and power amplifier; in front: ring core specimens, clamp meter and fuse box.

IV. Measurement Results

For the measurements three ring core specimens made of different materials are chosen (cf. TABLE 1). The measurements are carried out for different fundamental and PWM carrier frequencies. The following table shows the specifications of the three ring core samples.

TABLE 1: SPECIFICATIONS OF THE THREE RING CORE SAMPLES

M800-50A	
layers / height	10 / 5 mm
inner diameter	45 mm
outer diameter	55 mm
primary winding	95 turns
secondary winding	95 turns

M270-35A	
layers / height	15 / 5.25 mm
inner diameter	45 mm
outer diameter	55 mm
primary winding	200 turns
secondary winding	100 turns

3P Somaloy 700 HR	
layers / height	1 / 5 mm
inner diameter	45 mm
outer diameter	55 mm
primary winding	100 turns
secondary winding	100 turns

TABLE 2: DC-LINK VOLTAGES

	M800-50A	M270-35A	Somaloy
50 Hz	1.6 V	3.3 V	2.5 V
100 Hz	3.2 V	6.6 V	4 V
200 Hz	6.4 V	13.2 V	6.25 V
400 Hz	12.8 V	26.4 V	11 V

M800-50A is chosen due to its high quality commutation curve (cf. Fig. 6) and M270-35A due to its low losses (cf. Fig. 9). The soft magnetic composite 3P Somaloy 700HR is used for comparison reasons in order to determine the origin of the additional losses in a PWM-fed specimen. The DC-link voltage of the two electrical steels is rising linear with frequency (cf. TABLE 2) since its impedance shows a very inductive behavior. In contrast to that, the resistive part of the Somaloy's impedance cannot be neglected and consequently the DC-link voltage cannot be raised linear with the frequency.

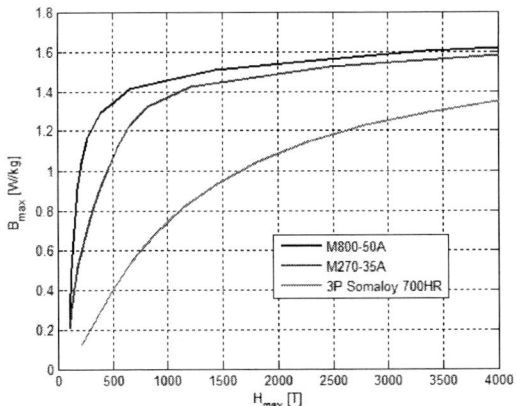

Fig. 6: Commutation curves of the three considered materials.

The first measurement is carried out at 50 Hz for the M800-50A specimen for different PWM frequencies. The following figure shows the results for the iron losses.

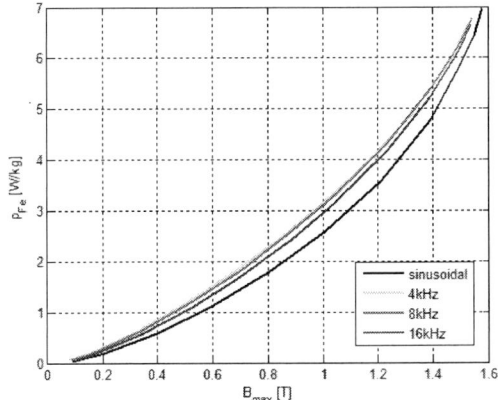

Fig. 7. Iron losses for M800-50A at 50 Hz for a sinusoidal voltage and for voltages of different PWM carrier frequencies.

In order to point out the influence of the PWM the next figure shows the corresponding relative additional losses based on the losses for the ideal sinusoidal waveform.

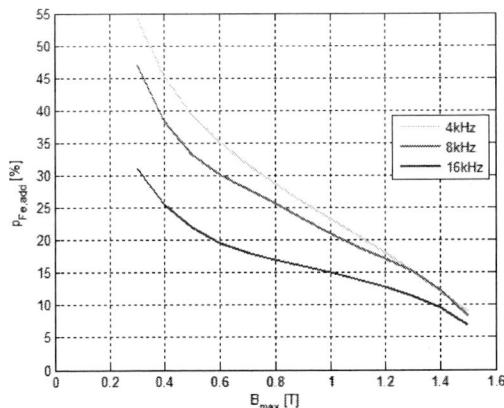

Fig. 8. Relative additional losses at 50Hz fundamental frequency for different PWM carrier frequencies.

It can be seen that the additional losses decrease with higher PWM carrier frequency. This can be explained by

the lower current ripple. For high flux densities, i.e. high modulation indices, the additional losses drop to about 5-10% of the ideal sinusoidal voltage. However it has to be mentioned that a high PWM carrier frequency will also increase the switching losses of the inverter and thus one has to find a minimum between iron and switching losses.

The next two figures show the measurements for the M270-35A. The losses at 1.5 T are higher than the guaranteed losses of 2.7 W/kg since the material was thermally deteriorated during the laser cutting process. The results show a similar behavior as the M800-50A.

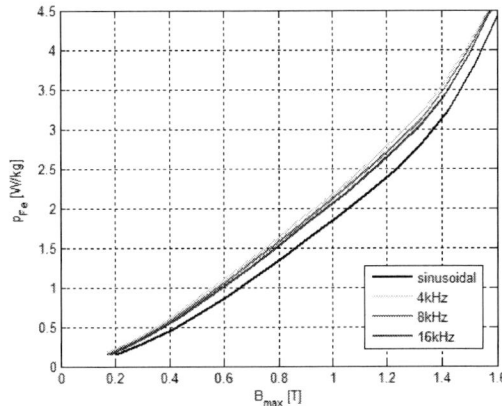

Fig. 9: Iron losses for M270-35A for a sinusoidal voltage and for voltages of different PWM frequencies.

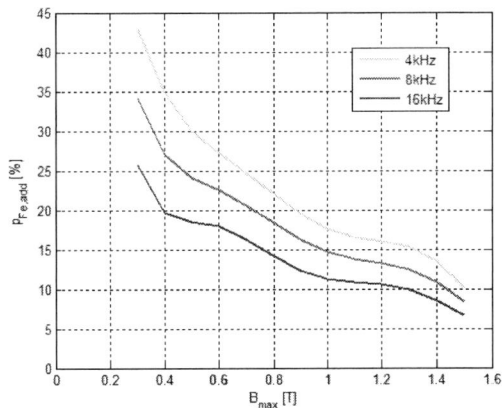

Fig. 10: Relative additional losses for M270-35A at 50Hz fundamental frequency for different PWM carrier frequencies.

The third material, the soft magnetic composite 3P Somaloy 700HR has a very high specific resistance of about 600 μΩm [7] (electrical steel has a typical specific resistance of 0.48μΩm [8]). Due to the high resistance of the SMC, the eddy current losses are very low. The next figure shows that the PWM modulation has hardly an influence on the iron losses compared to a sinusoidal fed specimen. Nevertheless the total iron losses for the SMC are higher than those of the electrical steels at a frequency of 50 Hz. However for high fundamental frequencies SMCs have lower losses than electrical steel [9].

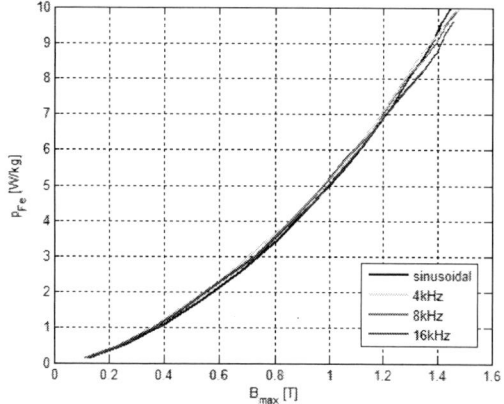

Fig. 11: Iron losses for 3P Somaloy 700HR at 50 Hz for a sinusoidal voltage and for voltages of different PWM carrier frequencies.

The next figure compares the additional losses of the three materials for a fundamental frequency of 400 Hz and a PWM frequency of 8 kHz. The SMC shows only slightly increased losses in comparison to a sinusoidal voltage, the two electrical steels show a massive increase of the losses, especially for low modulation indices. The reason for this behaviour are additional eddy currents induced by the larger amount of harmonics in the PWM modulated voltage. The increase of the losses due to hysteresis effects caused by the PWM is quite low since the magnetic domains will not switch the polarity as long as the amplitudes of the harmonics stay under a certain limit.

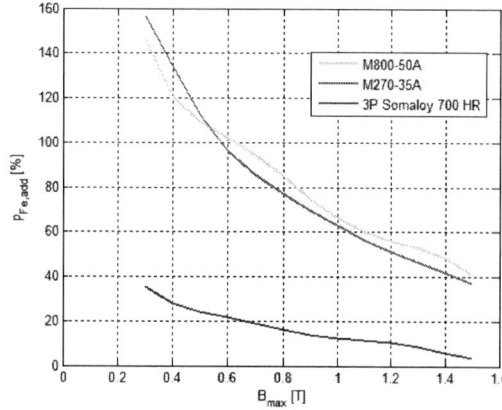

Fig. 12: Comparison of the relative additional losses for the three materials at a fundamental frequency of 400 Hz and a PWM frequency of 8 kHz.

The next figure shows the iron losses of M800-50A for different frequencies for ideal sinusoidal voltages.

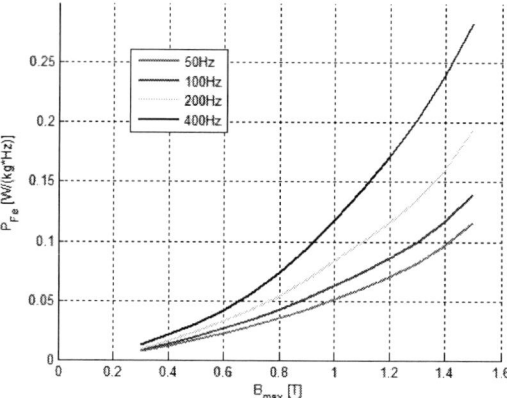

Fig. 13: Iron losses of M800-50A for different frequencies for sinusoidal voltages.

The next figure shows the additional losses if the voltages of Fig. 13 are not ideal sinusoidal but created with a PWM frequency of 4 kHz.

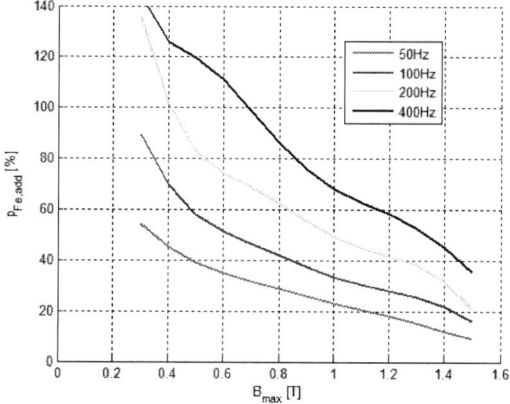

Fig. 14. Influence of fundamental frequency on the iron losses at a PWM frequency of 4 kHz.

The measurement results show that the relative additional losses increase with increasing fundamental frequency, since the ratio of PWM carrier frequency to fundamental frequency decreases and consequently the harmonics are rising. For example at 400 Hz a fundamental period is only created by 10 PWM periods.

For the last measurements the test procedure is changed. The different values of the primary voltage are not adjusted by variation of the modulation index, but by changing the DC-link voltage. Therefore the modulation index is set to the maximum and consequently the DC voltage can be reduced to a minimum.

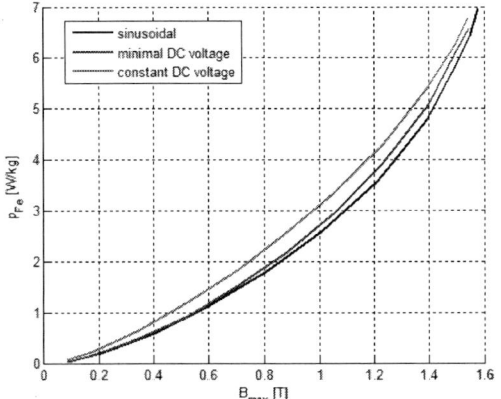

Fig. 15. Iron losses for a sinusoidal primary voltage, a PWM with minimal DC voltage and maximal modulation index and a PWM with constant DC voltage and variable modulation index for a fundamental frequency of 50 Hz and a PWM carrier frequency of 8 kHz.

In the following figure it can be seen that the corresponding additional losses for a DC voltage variation is about constant and only about 5 % higher than for an ideal sinusoidal voltage.

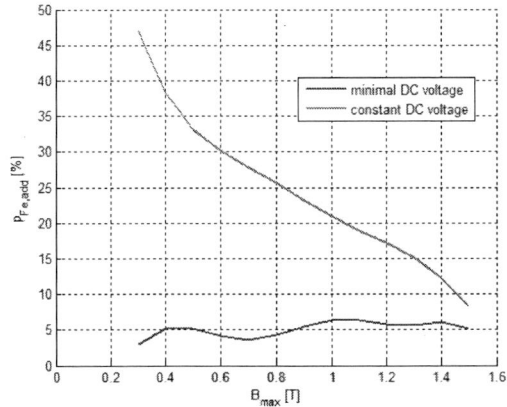

Fig. 16. Relative additional iron losses for a PWM voltage with minimal DC voltage at maximal modulation index and a PWM with constant DC voltage at variable modulation index.

V. CONCLUSION AND OUTLOOK

The presented measurement results have shown that PWM voltages increase the iron losses of electrical steel, especially at low modulation index operation. The increase is mostly based on higher eddy current losses since hardly conductive soft magnetic composites show only a slight rise of the total losses. For normal, grid operated inverters, the DC voltage of three phase inverters is about constant at 1.35 times the effective grid voltage and consequently a low modulation index cannot be avoided in low effective voltage machine operation. However, especially for battery powered applications,

e.g. electric vehicles, a reduction of the DC voltage could improve the efficiency. The reduction of the DC voltage can be achieved by switching to a parallel connection of the batteries at low effective voltage motor operation.

For future research measurements will be carried out for further materials and waveforms, e.g. the influence of a block commutation of BLDC-motors will be investigated.

REFERENCES

[1] "Magnetic materials – Part 2: Methods of the magnetic properties of electrical steel strip and sheet by means of an Epstein frame" (IEC 60404-2:1996 + A1:2008); German version EN 60404-2:1998 + A1:2008.

[2] Boglietti, A.; Cavagnino, A.; Mthombeni, T.L.; Pillay, P., "Comparison of lamination iron losses supplied by PWM voltages: US and European experiences," *IEEE International Conference on Electric Machines and Drives 2005*, vol., no., pp.1431,1436, 15-15 May 2005.

[3] J. Chen; W. Ma; D. Wang; X. Yu; Y. Guo, "Development of automatic iron loss measurement system of magnetic material with PWM excitation," *2011 International Conference on Materials for Renewable Energy & Environment (ICMREE)*, , vol.1, no., pp.1013,1017, 20-22 May 2011.

[4] Kawabe, M.; Nomiyama, T.; Shiozaki, A.; Kaihara, H.; Takahashi, N.; Nakano, M., "Behavior of Minor Loop and Iron Loss Under Constant Voltage Type PWM Inverter Excitation", *IEEE Transactions on Magnetics*, vol.48, no.11, pp.3458,3461, Nov. 2012.

[5] "Calculation of the effective parameters of magnetic piece parts" (IEC 60205:2006 + A1:2009); German version EN 60205:2006 + A1:2009.

[6] F. Fiorillo, "Measurement and Characterization of Magnetic Materials", Elsevier, 2004.

[7] Höganäs; (2011, March); Somaloy® Technology for Electric Motors. Retrieved 02 28, 2014, from http://www.hoganas.com/en/Segments/Somaloy-Technology/downloads/

[8] EMT; Typische Werkstoffeigenschaften nach Angaben eines Herstellers; Retrieved 28 Feb 2014, from http://www.elektrobleche.de.

[9] A. Boehm, I. Hahn, "Comparison of Soft Magnetic Composites (SMCs) and electrical steel", EDPC 2012, Nuremberg, Germany.

Investigation on Iron Loss Characteristics in Star-Connection and Delta-Connection under Three Phase PWM Inverter Excitation

Shunya Odawara
(Toyota Technological Institute)
Department of engineering
2-12-1 Hisaskata, Tempaku
Nagoya, Aichi, Japan
sodawara@toyota-ti.ac.jp

Keisuke Fujisaki
(Toyota Technological Institute)
Department of engineering
2-12-1 Hisaskata, Tempaku
Nagoya, Aichi, Japan
fujisaki@toyota-ti.ac.jp

Shuhei Fukuhara
(Toyota Technological Institute)
Department of engineering
2-12-1 Hisaskata, Tempaku
Nagoya, Aichi, Japan
sd09079@toyota-ti.ac.jp

Abstract— An iron loss characteristic with three ring specimens connected in a star type and in a delta type on three phase inverter excitation is investigated. It is indicated that iron loss obtained from star-connection is about 14 % larger than that obtained from delta-connection. This difference is caused by potential of neutral point having the third harmonics on star-connection. Therefore, although the zero-phase current may flow, the drive of motor by delta-connection should be performed to reduce the iron loss. In star-connection, since a neutral point exists, a potential on neutral point sways by three phase inverter excitation. As a result, BH curve is expanded by this fluctuation.

Keywords— *Iron loss, three phase inverter, star-connection, delta-connection.*

I. INTRODUCTION

Evaluation of iron loss characteristics becomes important as high-efficiency of electrical machines is promoted. Many electrical machines are composed of magnetic materials such as electrical steel sheet. The iron loss of electrical steel sheet is generally evaluated by sinusoidal excitation without time-harmonics due to standard [1]-[2]. However, actual electrical machine is driven by inverter to control various velocities easily, so understanding of iron loss characteristics by using inverter excitation is necessary for high-efficiency of electrical machines.

In our previous research, the iron loss characteristics by using single phase Pulse-Width-Modulation (PWM) inverter have been revealed [3]-[6]. However, most electrical machines such as especially electric motor are driven by three phase excitation. Therefore, by using three phase PWM inverter, the influence of three phase PWM inverter excitation on iron loss characteristics is investigated in this research.

As specimens for evaluation, three ring specimens composed of electrical steel sheet are connected in a star and in a delta of most common three phase connections, and these rings are excited by three phase PWM inverter. In some conventional researches, the evaluations of iron loss by using three phase inverter are carried out [7]-[8],

but investigation which paid attention to difference of connecting method is not performed until now. Thus the investigation of the iron loss characteristic by the difference between the star-connection and the delta-connection is performed.

II. EVALUATION METHOD

A. Three Phase PWM Inverter

In this research, the commercial three phase inverter (MWINV-9R122A) produced by Myway Plus Corporation is used. Fig. 1 shows the circuit diagram of three phase inverter. The fundamental frequency f_o, the carrier frequency f_c and the modulation factor m are set to 50 Hz, 10 kHz and 0.6, respectively. The applied DC voltage V_{dc} in Fig. 1 is decided so that maximum flux density B_{max} in ring specimen becomes 1 T. These conditions of inverter excitation are shown in Table I.

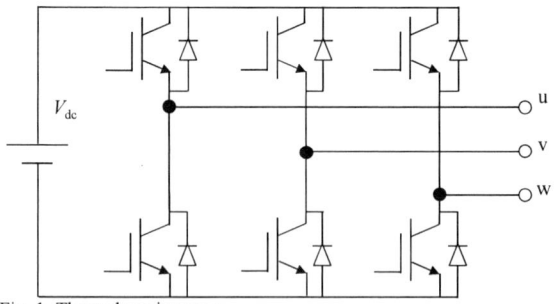

Fig. 1. Three phase inverter.

TABLE I. EXCITING CONDITIONS.

Fundamental frequency	50 Hz
Carrier frequency	10 kHz
Modulation factor	0.6
Magnetic flux density in ring	1 T

B. Connection Method of Three Rings

The ring specimen for evaluation of iron loss is composed of laminated electrical steel sheet (35H300). The specifications of this ring specimen are shown in Table II and three rings are made so that the specifications on each ring become the same as much as possible.

To carry out the three phase inverter excitation, these three rings are connected in star type and in delta type as shown in Fig. 2 since these connection method are most general in three phase excitation. As a feature of star-connection in Fig. 2(a), the neutral point exists. On the other hand, as a feature of delta-connection in Fig. 2(b), the zero-phase current circulating in circuit when the circuit is unbalanced exists. Usually, on motor driving system, the star-connection is more commonly used.

TABLE II. SPECIFICATIONS OF RING SPECIMEN.

Material	35H300
Outside diameter	127 mm
Inside diameter	102 mm
Height	7 mm
Primary coil winding number	254 turn
Secondary coil winding number	254 turn

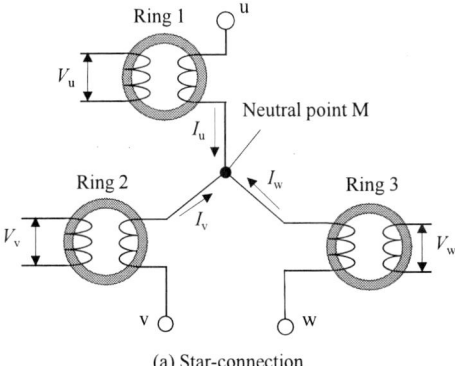

(a) Star-connection

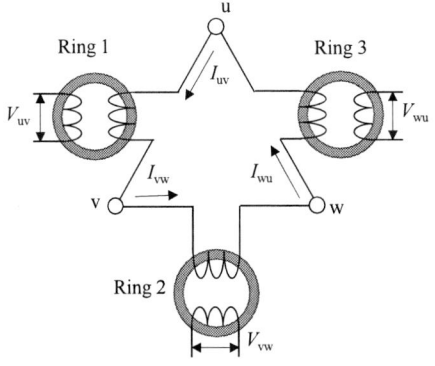

(b) Delta-connection

Fig. 2. Star-connection and delta-connection.

C. Calculation of Iron Loss

In order to calculate iron loss, the primary current I and the secondary voltage V on each ring are obtained by using AD converter. Magnetic field intensity H and magnetic flux density B in ring are calculated by following equations (1), (2). The iron loss W is obtained from integration of H and B as following equation (3).

$$H_i = \frac{N_1 I_i}{L} \ \ [\text{A/m}] \tag{1}$$

$$B_i = \frac{1}{N_2 S} \int V_i dt \ [\text{T}] \tag{2}$$

$$W_i = \frac{f_o}{\rho} \int H_i dB_i \ [\text{W/kg}] \tag{3}$$

where N_1, N_2, L, S and ρ in equation are the primary windings, the secondary windings, the magnetic path length, the cross section area and the density of electrical steel sheet, respectively. The subscript (i) indicates each ring number.

In measurement, since the maximum flux density B_{max} does not become just 1 T, the normalization of iron loss on just $B_{max} = 1$ T is performed by using following equation (4) because iron loss is proportional to the square of magnetic flux density.

$$W_n = W_i \times 1^2 / B_{max}^2 \ [\text{W/kg}] \tag{4}$$

where W_n and B_{max} are the normalized loss and the measured maximum magnetic flux density, respectively.

III. EXPERIMENTAL RESULTS

The iron losses obtained from star-connection and delta-connection are shown in Table III. In addition, the iron losses on single ring by using u-v phase of three phase PWM inverter excitation are also shown for reference in Table III. First, it is obtained as theoretical that the applied DC voltage V_{dc} of star-connection is the square root of three times that of delta-connection.

In Table III, although the iron losses of single ring and delta-connection are almost the same, the iron losses of star-connection are about 14 % larger than that of delta-connection. Fig. 3 shows the BH curves obtained from star-connection and delta-connection on ring 1. When both BH curves are compared with, BH curve of star-connection is expanded in direction of H than BH curve of delta-connection. Iron loss is calculated by equation (3) and this equation means that the iron loss is proportional to internal area of BH curve. Therefore, in star-connection with large internal area, iron losses become larger than delta-connection.

In next chapter, the cause by which the BH curve is expanded on star-connection is discussed.

TABLE III. IRON LOSSES ON EACH CONDITION.

Ring no.	Star-connection		Delta-connection		Single	
	V_{dc} [V]	W_n [W/kg]	V_{dc} [V]	W_n [W/kg]	V_{dc} [V]	W_n [W/kg]
1		1.80		1.57	19.4	1.51
2	32.5	1.73	18.8	1.54	19.3	1.50
3		1.76		1.54	19.4	1.55

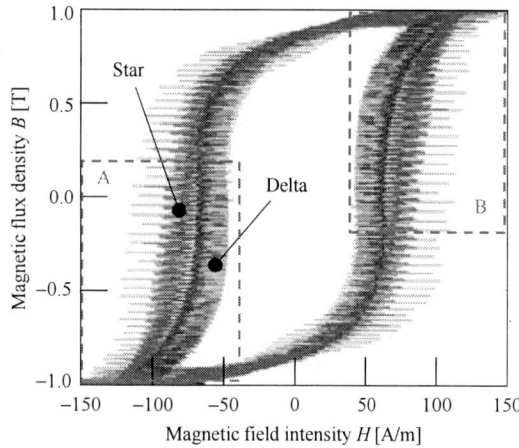

Fig. 3. BH curves of ring 1 obtained from star-connection and delta-connection.

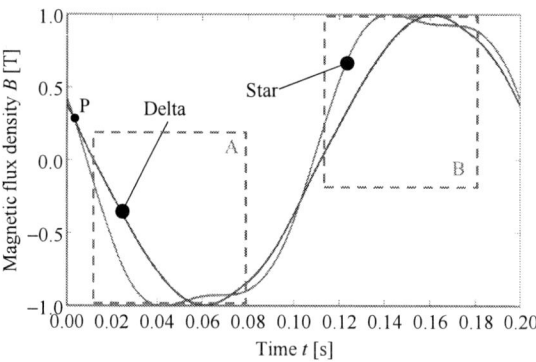

Fig. 6. Waveforms of magnetic flux density B.

IV. DISCUSSION

First, waveforms of magnetic flux density B and magnetic field intensity H used for making BH curves are shown in Figs. 6 and 7, respectively. In Fig. 6, although magnetic flux density of delta-connection becomes sinusoidal waveform, that of star-connection is distorted. In Fig. 7, the waveforms of magnetic field intensity obtained from star-connection and delta-connection differ from each other. These factors cause the difference of BH curve. In particular, when point P in which B and H of delta-connection and star-connection are the same becomes basis, B of star-connection becomes the minimum early than that of delta-connection. As a result, since H of star-connection becomes the minimum earlier than that of delta-connection, H of star-connection begins to change a little early. This phenomenon causes the expanding in direction of H.

Next, reason which the distortion of magnetic flux density of star-connection is caused is verified. Figs. 8 and 9 show the secondary voltage waveform and the primary current waveform. Especially, voltage of star-connection has third harmonic. Magnetic flux density B is calculated by using secondary voltage V as equation (2), so this third harmonics cause the distortion of B of star-connection. The current waveforms of star-connection and delta-connection also differ due to difference of voltage waveform.

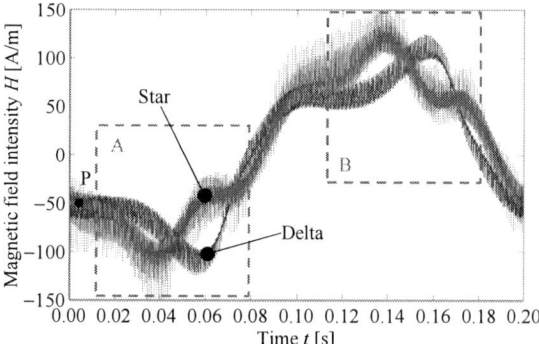

Fig. 7. Waveforms of magnetic field intensity H.

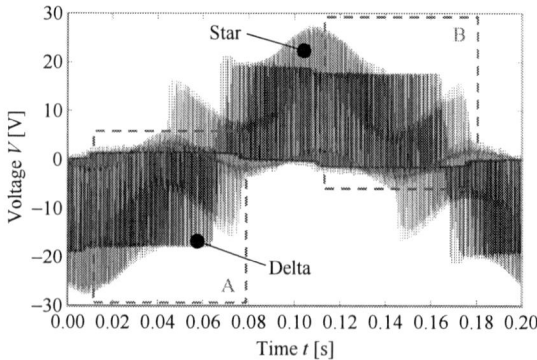

Fig. 8. Waveforms of secondary voltage V.

The 2014 International Power Electronics Conference

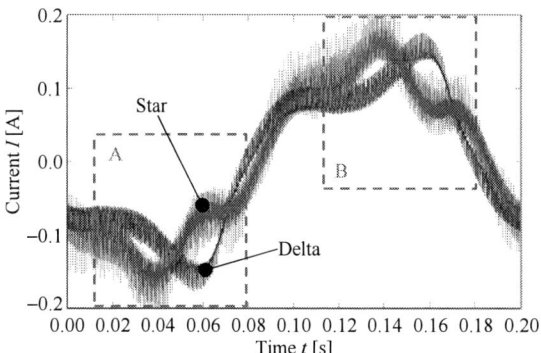

Fig. 9. Waveforms of primary current I.

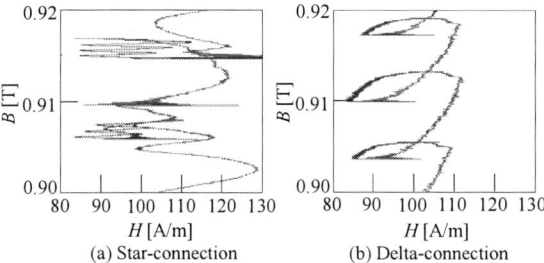

(a) Star-connection

(b) Delta-connection

Fig. 10. Enlargement of minor loops near maximum flux density.

We have described the relationship between iron loss and minor loop of BH curve [3]-[4], so the enlargement of minor loop near maximum flux density is shown in Fig. 10. However, the minor loop obtained from star-connection is not clearly, and it seems that the iron loss increase by expanding in direction of H on star-connection is dominant. Therefore, in this paper, only influence of waveform distortion is discussed.

V. THEORETICAL VERIFICATION

Third harmonics on voltage waveform of star-connection is caused by potential on neutral point. In this chapter, by theoretical verification using a simple rectangular wave, the cause which potential on neutral point sways is explained.

Fig. 11 shows the three phase inverter and the three loads connected in star type. It is assumed that impedances Z of each load are the same. As a switching signal, on-signal or off-signal is applied in each switching device S_i as shown in Fig. 12. The bar (-) means negation. At the time of (i) in Fig. 12, the electrical circuit becomes Fig. 13(a). In this case, the voltage V_{mo} between neutral point M and ground O becomes as following equation (5).

$$V_{mo} = -\frac{E}{2} + \frac{Z_2}{\dfrac{Z_1 Z_3}{Z_1 + Z_3} + Z_2} E = \frac{E}{6} \quad (\because Z_1 = Z_2 = Z_3) \quad (5)$$

On the other hand, at the time of (ii) in Fig. 12, the electrical circuit becomes Fig. 13(b) and the voltage V_{mo} becomes as following equation (6).

$$V_{mo} = -\frac{E}{2} + \frac{\dfrac{Z_2 Z_3}{Z_2 + Z_3}}{Z_1 + \dfrac{Z_2 Z_3}{Z_2 + Z_3}} E = -\frac{E}{6} \quad (\because Z_1 = Z_2 = Z_3) \quad (6)$$

Similarly, the waveform calculated to from (i) to (vi) is shown in Fig. 12. The waveform of V_{mo} has three times period in one period of switching waveform.

As a result, the V_{um} becomes as shown in Fig. 12. This V_{um} corresponds to voltage of star-connection in Fig. 8. In Fig. 12, it is confirmed that the V_{um} does not sway by the third harmonics of neutral point potential theoretically. However, actually, impedances of three ring specimens are not the same because the characteristics differ in each ring as shown in Table III, so the electrical circuit must become unbalanced. By this unbalanced circuit, it is presumed that V_{mo} is distorted and V_{um} obtained from V_u and V_{mo} has just third harmonics resulting from this unbalance.

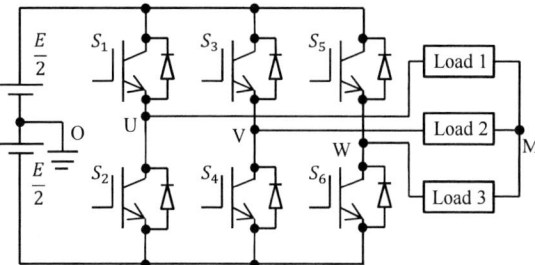

Fig. 11. Three phase star-connection for verification.

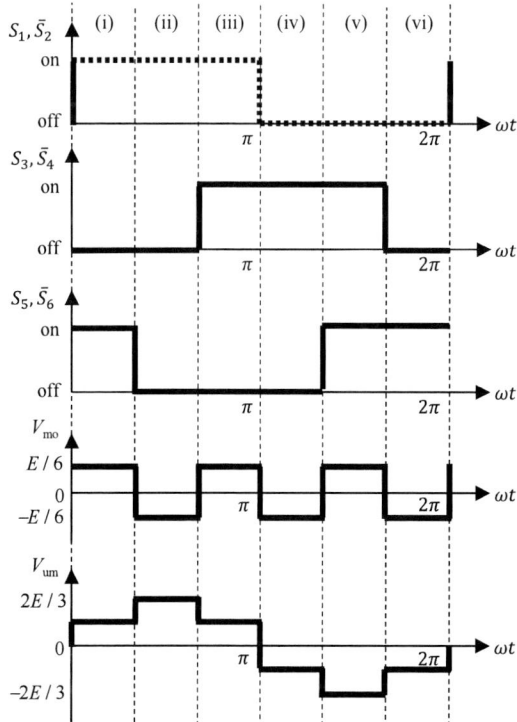

Fig. 12. Waveform obtained from rectangular wave.

978-1-4799-2706-7/14 $31.00 © 2014 IEEE   292

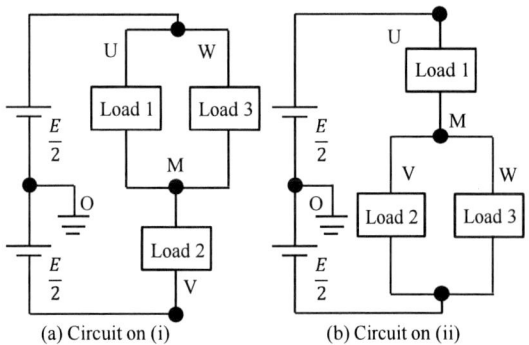

(a) Circuit on (i)   (b) Circuit on (ii)

Fig. 13. Electrical circuit diagram.

VI. CONCLUTION

In this research, the iron loss characteristics in star-connection and in delta-connection are evaluated by three phase PWM inverter excitation. The iron loss obtained from star-connection is about 14 % larger than the iron loss obtained from delta-connection. This difference is caused by expanding BH curve of star-connection due to potential of neutral point having the third harmonics on star-connection. Therefore, although the zero-phase current may flow, the driving by delta-connection should be carried out in order to reduce the iron loss.

REFERENCES

[1] International Electrotechnical Commission, 60404-3, Second edition, 1992.

[2] Japanese Industrial Standard, C2556, 1996.

[3] K. Fujisaki, and S. Liu : "Magnetic Hysteresis Curve Influenced by Power-Semiconductor Characteristics in PWM Inverter", *Journal of Applied Physics*, 115, 17A321, 2014.

[4] D. Kayamori, and K. Fujisaki: "Influence of Power Semiconductor On-Voltage on Iron Loss of Inverter-fed", *IEEE Power Electronics and Drive Systems (PEDS)*, 9034, 2013.

[5] K. Fujisaki, R. Yamada, and T. Kusakabe : "Difference in Iron Loss and Magnetic Characteristics for Magnetic Excitation by PWM Inverter and Linear Amplifier", *IEEJ transactions on industry applications* 133(1), 69-76, 2013 (in Japanese).

[6] K. Fujisaki, M. Sakai, S. Takeda: "Motor Loss Increment of Induction Motor Driven by PWM Inverter in Comparison with Inverter Circuit Loss," *ICEMS2012 Sapporo (15th International Conference on Electrical Machines and Systems)*, Sapporo, LS3B-3, October 2012.

[7] H. Matsumori, T. Shimizu, K. Takano, and H. Ishii: "Comparison between single phase and three phase of PWM inverters in Iron Loss measurement," *Power Electronics and Motion Control Conference (EPE/PEMC), 2012 15th International*, DS1d.3-1 - DS1d.3-8, 2012.

[8] S. Tadsuan, and C. Tangsiriworakul: "Design and comparison of iron losses mathematical model with single phase and three phase PWMinverter supply," *Industrial Technology, 2008. ICIT 2008. IEEE International Conference on*, pp. 1-6, 2008.

Optimization on Arrangement of Permanent Magnets for Magnetic Levitation System for Thin Steel Plate (Fundamental Consideration on Levitation Probability)

Hirotaka Ishii
Tokai University
Undergraduated Dep. of Prime Mover Eng.
4-1-1, Kitakaname,Hiratsuka-shi,Kanagawa,Japan

Takayoshi Narita
Tokai University
Graduate Student, Course of Science & Tech.
4-1-1, Kitakaname,Hiratsuka-shi,Kanagawa,Japan

Shinya Hasegawa
Tokai University
Dep. of Prime Mover Eng.
4-1-1, Kitakaname,Hiratsuka-shi,Kanagawa,Japan

Yasuo Oshinoya
Tokai University
Dep. of Prime Mover Eng.
4-1-1, Kitakaname,Hiratsuka-shi,Kanagawa,Japan
ossy@keyaki.cc.u-tokai.ac.jp

Abstract— **For thin steel plates, which are used in many industrial products including those of the automobile industry, we have proposed a magnetic levitation control system and confirmed its realization by means of a digital control experiment. However, the use of a limited number of electromagnets cannot suppress static deflection and high-order-mode elastic vibration, which are characteristics of a flexible magnetic material. To solve this problem, we have proposed a hybrid levitation control system for thin steel plate using the magnetic force generated by permanent magnets, which have no operational costs. In this study, we attempt to determine the optimal gap, placement and number of permanent magnets to reduce the deflection of a thin steel plate under the generated magnetic field. To verify the usefulness of optimal placement of the permanent magnets, experiments concerning vibration were performed on a magnetically levitated thin steel plate**

Keywords— Steel plate, Permanent magnet, FDM, Genetic algorithm, Optimal placement

I. INTRODUCTION

Recently, demand in improving of quality for thin steel plates used in various industrial products is increasing. However, thin steel plates are generally conveyed while in contact with rollers in the conveyance process. This leads to deterioration in the quality of the plate surface, such as flaws and peeling of the plated layer. To solve these problems, research on noncontact conveyance using magnetic levitation technology has been actively carried out [1]-[2]. The researchers in our laboratory have been involved in examining noncontact magnetic levitation control of the conveyance of a rectangular thin steel plate.[3] For stable levitation, we proposed a hybrid magnetic levitation control system for a thin steel plate using the magnetic force of permanent

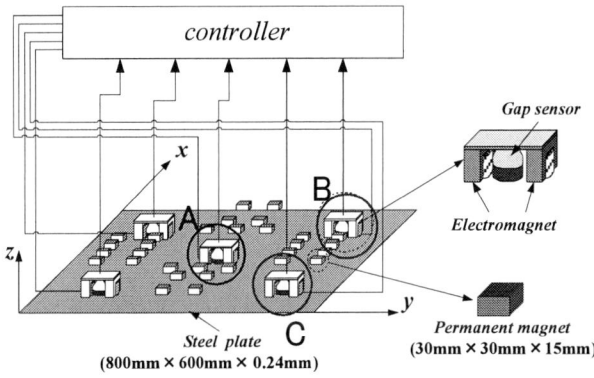

Fig. 1. Electromagnetic levitation control system for a steel plate with permanent magnets.

magnets in an almost uniform distribution above the steel plate at the areas where no electromagnets are placed, as shown in Fig. 1[4]. In the previous studies, fundamental research on the optimal conditions in terms of the number and position of permanent magnets, was carried out, in which the attractive force of the permanent magnet applied to the rectangular thin steel plate was simplified and modeled. Furthermore, the optimal position and uniform distribution of the permanent magnets were experimentally examined to verify the effectiveness of the permanent magnets [5]. In this study, the gap, number and optimal position of permanent magnets for suppressing the displacement of a magnetically levitated thin steel plate were investigated using thin steel plates (thickness: 0.24mm). We performed magnetic levitation experiment for the steel plates of different thicknesses, and the usefulness of permanent magnets was examined.

978-1-4799-2706-7/14 $31.00 © 2014 IEEE

II. Optimization of Gap, Placement and Number of Permanent Magnets

A. Analysis model

The rectangular galvanized steel plate [SS400, 800 mm (L) × 600 mm (W)] shown in Fig. 1 was used. In this study, the displacement of the rectangular thin steel plate subjected to gravity and the attractive force of the permanent magnets is calculated. The equations for the static displacement of the rectangular thin steel plate are expressed as

$$DV^4 z = f - \rho h g \qquad (1)$$

$$D = \frac{Eh^3}{12(1-v^2)}, \ \nabla^4 = \frac{\partial^4}{\partial x^4} + 2\frac{\partial^4}{\partial x^2 \partial y^2} + \frac{\partial^4}{\partial y^4}$$

where E is the Young's modulus of the thin steel plate [N/m^2], h is the plate thickness [m], v is the Poisson ratio, x and y are the coordinates in the width and longitudinal directions [m], respectively, z is the vertical displacement of the plate [m], f is the dynamic magnetic force applied to the plate from the vertical direction by the permanent magnets [N/m^2], ρ is the plate density [kg/m^3], and g is the acceleration due to gravity [m/s^2].

Using eq. (1), the displacement of the thin steel plate is calculated by the finite difference method (FDM). We calculated as the steel plate is simply supported at the position of the electromagnets. When the distribution of the attractive force is rigorously taken into consideration, as shown in Fig. 2, considerable calculation time is expected for the analysis of the attractive force applied to the rectangular thin steel plate by the permanent magnets. Therefore, analysis is carried out using a model in which the attractive force is assumed to be concentrated at one point, as shown in Fig. 3. The size of FDM mesh is 10mm ×10mm.

B. Search method of optimization

In this study, the distance between the surface of each permanent magnet and the thin steel plate (hereafter, gap) is set in the range of 40-80mm, considering that the permanent magnets are installed in the electromagnetic levitation control system shown in Fig. 1. The gap is changed by increments of 5 mm. Here, the distance of the permanent magnets above the thin steel plate was the same for all magnets.

The optimal number and position of the permanent magnets were sought for this range of the gap at the same time. It is practically impossible to search for the optimal position of the permanent magnets by experiment because the number of combinations of search patterns is huge. Therefore, the optimal position is sought using a genetic algorithm, the effectiveness of which was previously confirmed when searching for the optimal position of permanent magnets using a free-free beam as an optimization algorithm. Fig. 4 shows a flowchart of the genetic algorithm. First, the number and initial

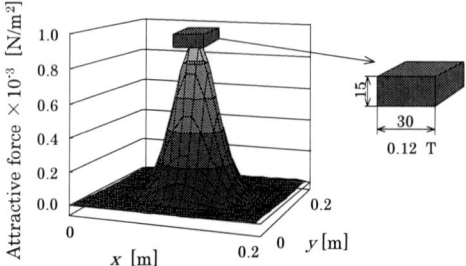

Fig. 2. Distribution of attractive force of permanent magnet.

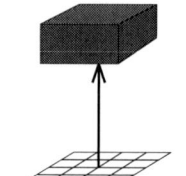

Fig. 3. Model of attractive force.

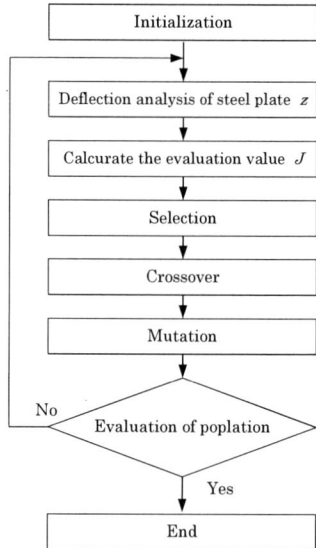

Fig. 4. Flow chart of genetic algorithm.

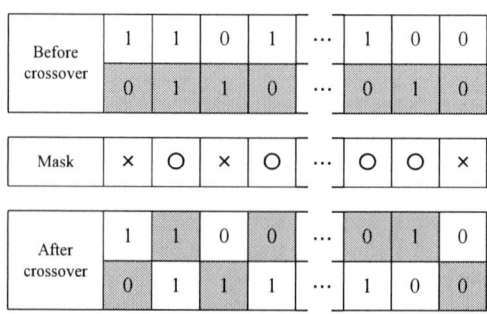

Fig. 5. Schematic illustration of uniform crossover.

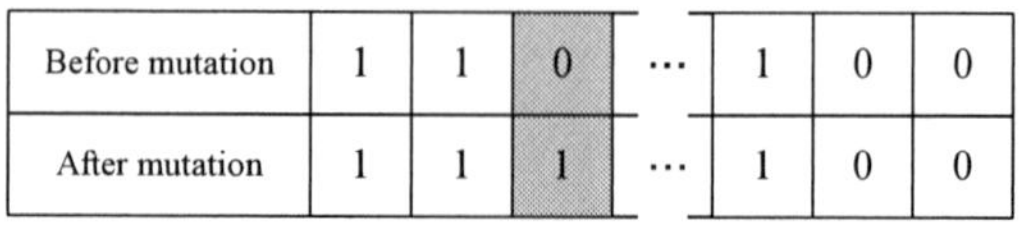

Fig. 6. Schematic illustration of mutation.

position of permanent magnets are randomly determined (referred to as initialization in Fig. 4; number of initial groups: 32). Then the shape of the thin steel plate subjected to gravity and the attractive force of the permanent magnets is determined (referred to as "deflection analysis of steel plate") and the evaluation function is calculated ("calculation of the evaluation value"). Using the evaluation function, candidates for the optimal position are selected ("selection"; the elite preservation rule by which the top 50% are selected unconditionally is applied). Furthermore, from among the candidates, new positions are generated with a certain probability to obtain an optimal solution ("crossover" is 90% in uniform crossover, as shown in Fig. 5).

However, because there is a possibility that a group including similar position patterns gives a local solution ,the position patterns are dispersed with a probability of 5% to maintain the diversity of position patterns ("mutation" as shown in Fig. 6).

The computation ends when the final value of the evaluation function does not change for over 200 generations. The number of permanent magnets is determined such that the total attractive force of the permanent magnet is no more than 50% of the total weight of the thin steel plate, considering the controllability the magnetically levitated thin steel plate.

In addition, an evaluation value is calculated to compare the shape of the thin steel plate for different positions of permanent magnets. The purpose of this study is to suppress the displacement of the thin steel plate. Accordingly, the evaluation value J used to search for the optimal position of permanent magnets is set as

$$J = \frac{J_Z}{J_{Z0}} \times w_Z + \frac{J_D}{J_{D0}} \times w_D \qquad (2)$$

$$J_Z = \frac{\sum_{i=1}^{N} |z_i|}{N}, \quad J_D = |z_{\max}|, \quad w_Z + w_D = 1$$

where J_z is the evaluation function of the mean deflection, J_D is the evaluation function of the maximum deflection, $z_{\max}$ is the maximum deflection of the thin steel plate and z_i is the displacement at each analysis point on the thin steel plate. N is the total number of analysis points.

As reference evaluation values, J_{Z0} is the mean deflection and J_{D0} is the maximum deflection of the thin steel plate when no permanent magnets are used, and each term in eq. (2) is nondimensionalized. w_Z and w_D are the weight coefficients for the mean deflection and maximum deflection of the thin steel plate, respectively. $w_Z = 0.5$ and $w_D = 0.5$, for which deflection is expected to be most greatly suppressed from the results of simulations, were adopted. The values of J are obtained at each gap, and we assume the gap of minimum value of J the optimal gap. The values of J are obtained at each gap (40-80mm,5mm interval), and we assume the gap of minimum value of J the optimal gap.

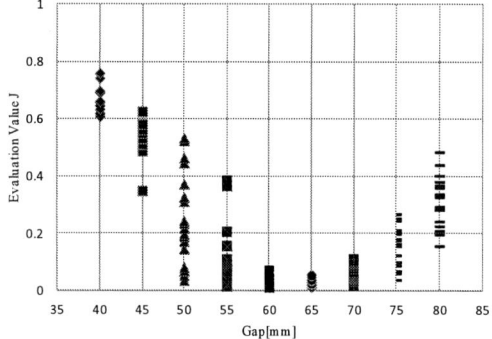

Fig. 7. Evaluation value of theoretical result.

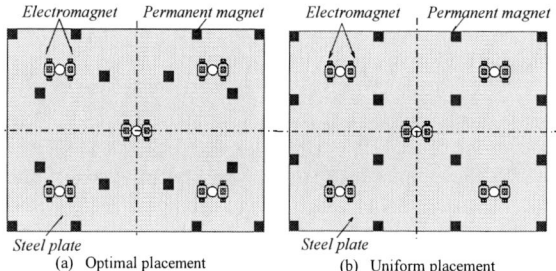

(a) Optimal placement   (b) Uniform placement

Fig. 8. Placement of permanent magnets

(Number of permanent magnets = 16).

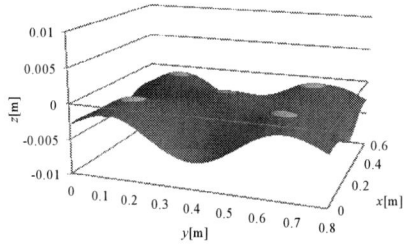

(a) Without permanent magnet

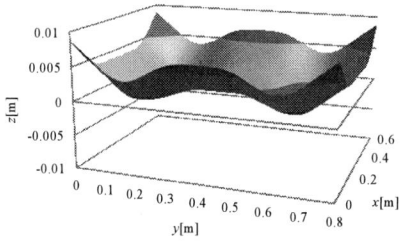

(b) Uniform placement

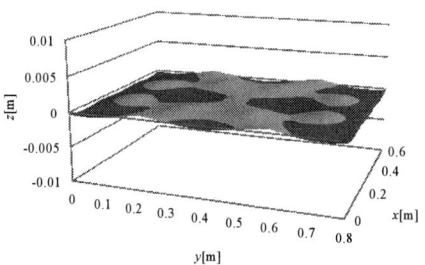

(c) Optimal placement

Fig. 9. Shape of thin steel plate

(Plate thickness = 0.24mm).

C. Results of optimization

To search for the optimal conditions in terms of the gap, number and position of permanent magnets to minimize the displacement of the thin steel plate (thickness: 0.24mm) was used. Figure 7 shows the relationship between J and the gap obtained in the search for the optimal position. The displacement can be maximally suppressed at a gap of 60 mm for the plate thicknesses of 0.24 mm and the optimal numbers of permanent magnet for the plate thickness of 0.24 mm is 16.

To experimentally examine the relationship between the gap, number and position of permanent magnets and J for the plate, magnetic levitation experiments were performed, adopting the optimal position of permanent magnets that induces the minimum J obtained in the analysis. Fig. 8 (a) show the optimal position of permanent magnets for the plate thicknesses of 0.24 mm for which $J = 7.29 \times 10^{-3}$. Fig. 8 (b) shows the case in which the permanent magnets are uniformly distributed. Fig. 9 (a) show the shape of the thin steel plate without permanent magnet. Fig. 9 (b) show the shape of the thin steel plate corresponding to the positions shown in Fig. 8(b). Fig. 9 (c) shows the shape of the thin steel plate corresponding to the positions shown in Fig. 8 (a). The number of permanent magnets is 16 in Fig. 8 (a) and 8 (b). The experimental results in Fig. 9 confirmed that the optimal placement can suppress the displacement.

III. APPLICATION OF OPTIMAL POSITION TO ELECTROMAGNETIC LEVITATION CONTROL SYSTEM

To verify the effectiveness of the search for the optimal position of permanent magnets, magnetic levitation experiments using the rectangular thin steel plates each placement of permanent magnets were carried out by placing the permanent magnets in the experimental setup shown in Fig. 1. Each gap was set to the optimal distance. We compared the following two cases: (1) when permanent magnets are almost uniformly distributed above the steel plate at the positions where no electromagnets are placed, (2) when permanent magnets are optimally placed as shown in Fig. 8(a). Sinusoidal wave with a natural frequency of 8.65 Hz, corresponding to the 1st elastic modes, was applied to the y-axis direction of the thin steel plate as a disturbance to induce elastic vibration in the plate. Fig. 10 show the waveforms, measured by gap sensor attached the electromagnets as shown in Fig.1, point A, B and C, (a) with almost uniformly distributed permanent magnets, (b)with optimally placed permanent magnets. In the experiment, standard deviation of the waveform measured at point B and C is suppressed by approximately 30% when permanent magnets are optimally placed and by approximately 35% when permanent magnets are optimally placed. As we compare standard deviation in order to show the utility of optimal placement, optimal placement is 35% smaller than uniform placement. As a result, the sinusoidal wave due to disturbance can be

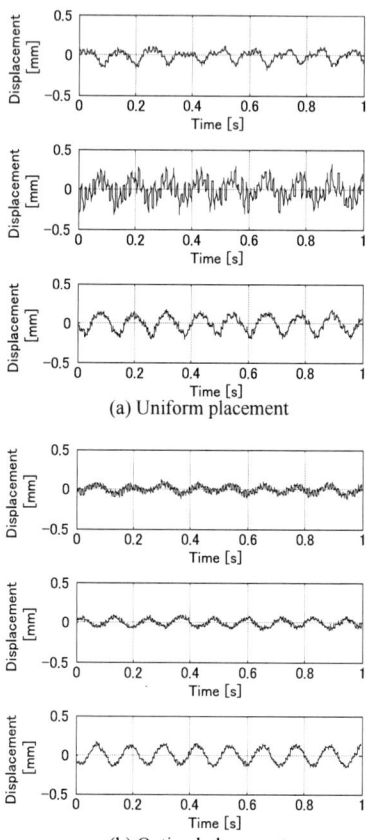

(a) Uniform placement

(b) Optimal placement

Fig. 10. Experimental results (Plate thickness=0.24mm).

suppressed by optimally positioning the permanent magnets, although the number of permanent magnets used was the same for the placements. Therefore, the usefulness of the search for the optimal position of the permanent magnets for a given plate thickness performed in this study was confirmed.

IV. CONCLUSIONS

As optimization on arrangement of permanent magnets for magnetic levitation system, the gap, number and position of permanent magnets were experimentally examined for optimal position. In addition, the levitation stability of the thin steel plate upon the application of a disturbance was confirmed to be improved in the levitation experiments using optimally placed permanent magnets to suppress the deflection of the levitated plate. The optimal position of the permanent magnets obtained in this study was confirmed to be valid for magnetically levitated thin steel plate. As for our future work, we want to clarify the relationship of the number of the permanent magnet and the plate thickness.

REFERENCES

[1] Y. Maruyama, T. Mizuno, M. Takasaki, and Y. Ishino,. "Development of 3-DOF Active Control Type Magnetically Suspended Gyro (Mechanical Systems)" *Transactions of the JSME Series C*, vol. 76, pp. 1445-1451, 2010, (in Japanese).

[2] H. Aburano, H. Miyazaki, T. Ohji, K. Amei, and M. Sakui, "Motion Tests on a Levitated Body Using a Magnetic Levitation System for Three-Dimensional Motion with Partial Zero Power

Control" *J. Magn. Soc. Jpn.*, vol. 35, pp. 123-127, 2011, (in Japanese).

[3] J. Clerk Maxwell, "A treatise on electricity and magnetism," *IEEE Trans. on Industry Applications*, vol. 589, no. 2, pp. 68-73, 2010.

[4] Y. Oshinoya, S. Kobayashi, and K. Tanno "Optimal Electromagnetic Levitation Control of a Thin Rectangular Steel Plate with Two Opposite Edges Reinforced by a Beam" *Transactions of the JSME Series C*, vol. 62, pp. 3067-3073, 1996, (in Japanese).

[5] Y. Oshinoya, and K. Ishibashi, "Development of Electromagnetic Levitation Control Device for a Rectangular Sheet Steel" *Transactions of the JSME Series C*, pp. 2855-2862, 2001, (in Japanese).

[6] S. Hasegawa, Y. Oshinoya, and K. Ishibashi "Consideration on Elastic Vibration Control of a Magnetically Levitated Thin Steel Plate Using Sliding Mode Control (Mechanical System)" *Transactions of the JSME Series C*, vol. 74, pp. 823-832, 2008, (in Japanese).

The 2014 International Power Electronics Conference

Effect of a Magnetic Field from the Horizontal Direction on a Magnetically Levitated Steel Plate (Fundamental Considerations on the Shape Analysis of Ultrathin Steel Plate)

Takeshi Kurihara
Tokai University
Undergraduate, Dep. of Prime Mover Eng.
4-1-1, Kitakaname,Hiratsuka-shi,Kanagawa,Japan

Takayoshi Narita
Tokai University
Graduate Student, Course of Science & Tech.
4-1-1, Kitakaname,Hiratsuka-shi,Kanagawa,Japan

Shinya Hasegawa
Tokai University
Department of Prime Mover Eng.
4-1-1, Kitakaname,Hiratsuka-shi,Kanagawa,Japan

Yasuo Oshinoya
Tokai University
Department of Prime Mover Eng.
4-1-1, Kitakaname,Hiratsuka-shi,Kanagawa,Japan
ossy@keyaki.cc.u-tokai.ac.jp

Abstract— **Thin steel plates are widely used in various industrial products. But there are problems where deterioration of surface quality and metal plating occurred by transporting. As a solution to these problems, a noncontact transport of steel plates using electromagnetic force has been proposed. However, there is a risk that side slipping or the dropping of the plate may occur owing to inertial force because the levitation control system does not provide a restraining for the direction of travel. Therefore we have proposed the addition of electromagnetic actuators to control the horizontal motion of levitated steel plate. In this study, we calculated the shape of steel plate with horizontal attractive force using Finite Difference Method (FDM). It was confirmed that the vertical displacement of the steel plate can be effectively suppressed by using magnetic field in the horizontal direction by performed simulation.**

Keywords— Electromagnetic levitation control, Ultrathin steel plate, Simulation, FDM.

I. INTRODUCTION

Thin steel plates are widely used in various industrial products. But there are problems where deterioration of surface quality and metal plating occurred by transporting [1]. As a solution to these problems, a noncontact transport of steel plates using electromagnetic force has been proposed [2,3]. In particular, for the transport of a cut piece of a steel plate by electromagnetic levitation, the problems of side slipping and dropping must be resolved.

Previously, our research group examined a 0.3-mm-thick rectangular steel plate as a sample, and confirmed that noncontact transport is possible when applying magnetic fields to the magnetically levitated steel plate in the horizontal direction [4,5].

Moreover, we performed a similar examination for an ultrathin steel plate with a thickness of 0.18 mm, and demonstrated that the noncontact transport of ultrathin steel plates, which have been difficult to levitate using the conventional system, is also possible [6,7]. However, the effect of applying magnetic fields from the horizontal direction on a magnetically levitated less than 0.18 mm ultrathin steel plate has not yet been examined.

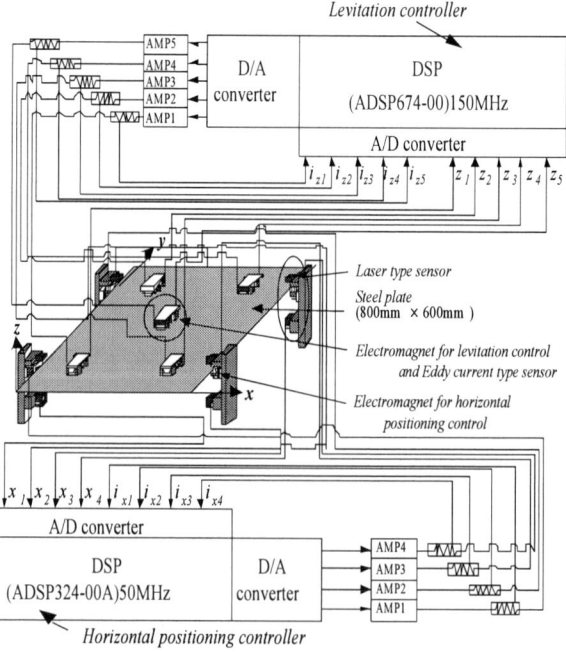

Fig. 1. Electromagnetic levitation system with horizontal positioning controller.

978-1-4799-2706-7/14 $31.00 © 2014 IEEE

In this study, we considered about the possibility for levitated ultrathin steel plates. Furthermore, we examined how the effect of applying magnetic field from horizontal direction affected the possibility of this experiment. First, we performed an electromagnetic field analysis on 0.09 mm, 0.12 mm, and 0.15 mm steel plate, and calculated the distribution of the attractive force induced by electromagnets. Using the Finite Difference Method (FDM), we obtained the shape of the steel plate when the force applied by the electromagnets acted in the horizontal direction. Finally, we reviewed the effect of a horizontal attractive force that applied to levitated ultrathin steel plate on its deflection.

II. ELECTROMAGNETIC LEVITATION SYSTEM

A. *Outline of System*

Figure 1 is as schematic illustration of the electromagnetic levitation control system with a horizontal positioning controller. A zinc-coated steel plate (length 800mm, width 600mm, material SS400 steel) is levitated and positioned in a noncontact mode by attractive forces of electromagnets which are controlled by feedback signals from gap sensors.

The vertical displacement of the plate and the electromagnet are acquired using five eddy-current- type displacement sensors and five resistors connected in series to the magnetic circuit, respectively.

These values are converted from analog to digital in the sampling interval of 1.0 ms, to allow them to be processed in a Digital Signal Processor. The five velocities of the plate are detected by differentiating the signals from the displacement sensors in the computer.

The control forces (voltage) are calculated in the DSP and are fed to the five electromagnets through the D/A converter and five power amplifiers.

The horizontal displacement of the plate and the electromagnet are acquired using four laser beam displacement sensors and four resistors connected in series to the magnetic circuit, respectively. The four velocities of the plate are detected by differentiating the signals from the displacement sensors in the computer. The control forces are calculated on a computer using these twelve values (sampling interval is 1.0 ms). These sensors and electromagnets are installed in the frame.

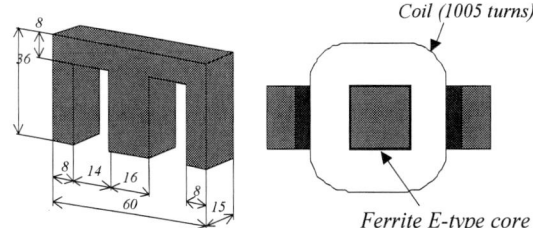

Fig. 3. Configuration of electromagnet.

B. *Electromagnetic Levitation Control System*

A single degree of freedom model in which the design of the control system is simplified through consideration of centralized control at the location of the electromagnet is considered. The steel plate is imaginarily divided into five parts, and each part is modeled as a single-degree-of-freedom electromagnetic levitation model, as shown in Figure 2. Consequently, the displacement, the velocity and the coil current detected at one electromagnet position are used for the control of that electromagnet. In this study, eddy current sensors are used to detect vertical displacements of the steel plate. In an equilibrium levitation state, magnetic forces are determined so as to balance with gravity. The equation of small vertical motion around the equilibrium state of the steel plate subjected to magnetic forces is expressed as

$$m_z \ddot{z} = 2f_z \tag{1}$$

where $m_z = m/5$ [kg], m: the mass of magnetic levitation steel plate [kg], z: vertical displacement [m], and $f_z(t)$: dynamic magnetic force [N].

Figure 3 shows a schematic illustration of the electromagnet. The number of turns of electromagnet coil is 1005 (diameter of wire is 0.5mm), and the sectional area passing the magnetic flux of the E - type core, which was made from ferrite, is 225mm². Characteristics of the electromagnet are estimated on the basis of the following assumptions: The permeability of the core is infinity.

The eddy current inside the core is neglected, and the inductance of the electromagnetic coil is expressed as a sum of the component inversely proportional to the gap between the steel plate and magnet and the component of leakage inductance. If deviation around the static equilibrium state is very small, the characteristic equations of the electromagnet are linearized as

$$f_z = \frac{2F_z}{Z_0} z + \frac{2F_z}{I_z} i_z \tag{2}$$

$$\frac{d}{dt} i_z = -\frac{L_{eff}}{L_z} \cdot \frac{I_z}{Z_0{}^2} \dot{z} - \frac{R_z}{2L_z} i_z + \frac{1}{2L_z} v_z \tag{3}$$

$$L_z = \frac{L_{eff}}{Z_0} + L_{lea} \tag{4}$$

Using the state vector, the equations (1) ~ (4) are written as the following state equations:

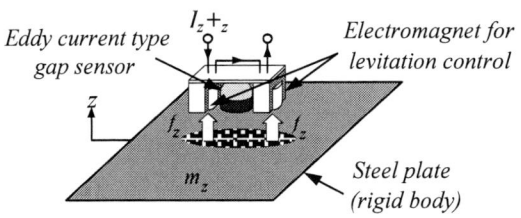

Fig. 2. Theoretical model of levitation control of the steel plate.

$$\dot{z} = A_z z + B_z v_z \tag{5}$$

$$z = \begin{bmatrix} z & \dot{z} & i_z \end{bmatrix}^{\mathrm{T}} \tag{6}$$

where F_z: magnetic force of the coupled magnets in the equilibrium state [N], Z_0: gap between the steel plate and electromagnet in the equilibrium state [m], I_z : current of the coupled magnets in the equilibrium state [A], i_z: dynamic current of the coupled magnets [A], L_z: inductance of one magnet coil in the equilibrium state [H], R_z: resistance of the coupled magnet coils [Ω], v_z: dynamic voltage of the coupled magnets [V], L_{eff}/Z_0: effective inductance of the one magnet coil [H], and L_{lea}: leakage inductance of the one magnet coil [H].

C. Horizontal Positioning Control System

Horizontal positioning control was carried out using the electromagnets installed at two edges of the steel plate. The banded laser beam sensor, keeping a 5mm clearance between the edge of the steel plate and each electromagnet surface, measured the horizontal displacement.

For the basic examination, the horizontal motion of the steel plate was modeled to have single degree of freedom, as shown in Figure 4.

Therefore, the same attractive forces were generated from two electromagnets placed at one side of the steel plate. The equation of small horizontal motion around the equilibrium state of the steel plate subjected to the same static magnetic forces from the electromagnets at two edges is expressed as

$$m\ddot{x} = f_1 - f_2 = f_x \tag{7}$$

$$f_x = \frac{4F_x}{X_0} x + \frac{4F_x}{I_x} i_x \tag{8}$$

$$\frac{d}{dt} i_x = -\frac{L_{xeff}}{L_x} \cdot \frac{I_x}{X_0^{\,2}} \dot{x} - \frac{R_x}{2L_x} i_x + \frac{1}{2L_x} v_x \tag{9}$$

$$L_x = \frac{L_{xeff}}{X_0} + L_{xlea} \tag{10}$$

Using the state vector, equations (7) ~ (10) are written as the following state equations:

$$\dot{x} = A_x\, x + B_x\, v_x \tag{11}$$

$$x = \begin{bmatrix} x & \dot{x} & i_x \end{bmatrix}^{\mathrm{T}} \tag{12}$$

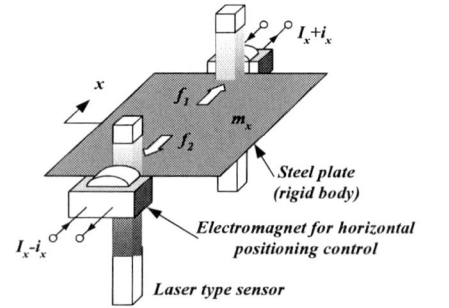

Fig. 4. Theoretical model of horizontal positioning control of the steel plate.

where F_x: magnetic force of the coupled magnets in the equilibrium state [N], X_0: gap between steel plate and electromagnet in the equilibrium state [m], I_x : current of the coupled magnets in the equilibrium state [A], i_x: dynamic current of the coupled magnets [A], L_x: inductance of the one magnet coil in the equilibrium state [H], R_x: resistance of the coupled magnet coils [Ω], v_x: dynamic voltage of the coupled magnets [V], L_{xeff}/X_0: effective inductance of the one magnet coil [H], and L_{xlea}: leakage inductance of the one magnet coil [H].

III. SHAPE ANALYSIS OF STEEL PLATE

Table 1 shows the specifications of the system. The distributions of the vertical attractive force of the electromagnets and the horizontal tension were calculated using JMAG software, which is used to analyze electromagnetic fields, to examine the effect of horizontal magnetic fields on a levitated steel plate.

The shape of the thin steel plate while being levitated by the above-mentioned attractive force is determined below. The equations for the static displacement of the rectangular thin steel plate are expressed as

$$D\nabla^4 z = f_{hz} + f_{hx}\frac{\partial^2}{\partial x^2} z - \rho h g \tag{13}$$

$$D = \frac{Eh^3}{12(1-v^2)},\ \nabla^4 = \frac{\partial^4}{\partial x^4} + 2\frac{\partial^4}{\partial x^2 \partial y^2} + \frac{\partial^4}{\partial y^4} \tag{14}$$

where E is the Young's modulus of the thin steel plate [N/m^2], h is the plate thickness [m], v is the Poisson ratio, x and y are the coordinates in the width and longitudinal directions [m], respectively, f_{hz} is the dynamic magnetic force applied to the plate from the vertical direction by the electromagnets [N/m^2], f_{hx} is the dynamic magnetic force applied to the plate from the horizontal direction by the electromagnets [N/m], ρ is the plate density [kg/m^3], and g is the acceleration due to gravity [m/s^2].

Using eq. (13) and (14), the displacement of the thin steel plate is calculated by the FDM.

We postulated that the steel plate stand still by positioning control from electromagnets. Then, we considered the steel plate was supported at the position of the electromagnets for levitation control and applied the attractive force from electromagnets for horizontal positioning control, regularly.

TABLE I
SYMBOLS AND VALUES

Symbol	Value
E	217GPa
v	0.30
ρ	7.27×10^3kg/m^3
g	9.81m/s^2
h	0.09, 0.12, 0.15×10^{-3}m
Z_0, X_0	5.00×10^{-3}m
R_z	21.0Ω
L_z	1.41×10^{-1}H
L_{EFF}	2.55×10^{-4}Hm
L_{LEA}	9.00×10^{-2}H
L_X	7.05×10^{-2}H
L_{XEFF}	1.78×10^{-4}Hm
L_{XLEA}	4.50×10^{-2}H

The 2014 International Power Electronics Conference

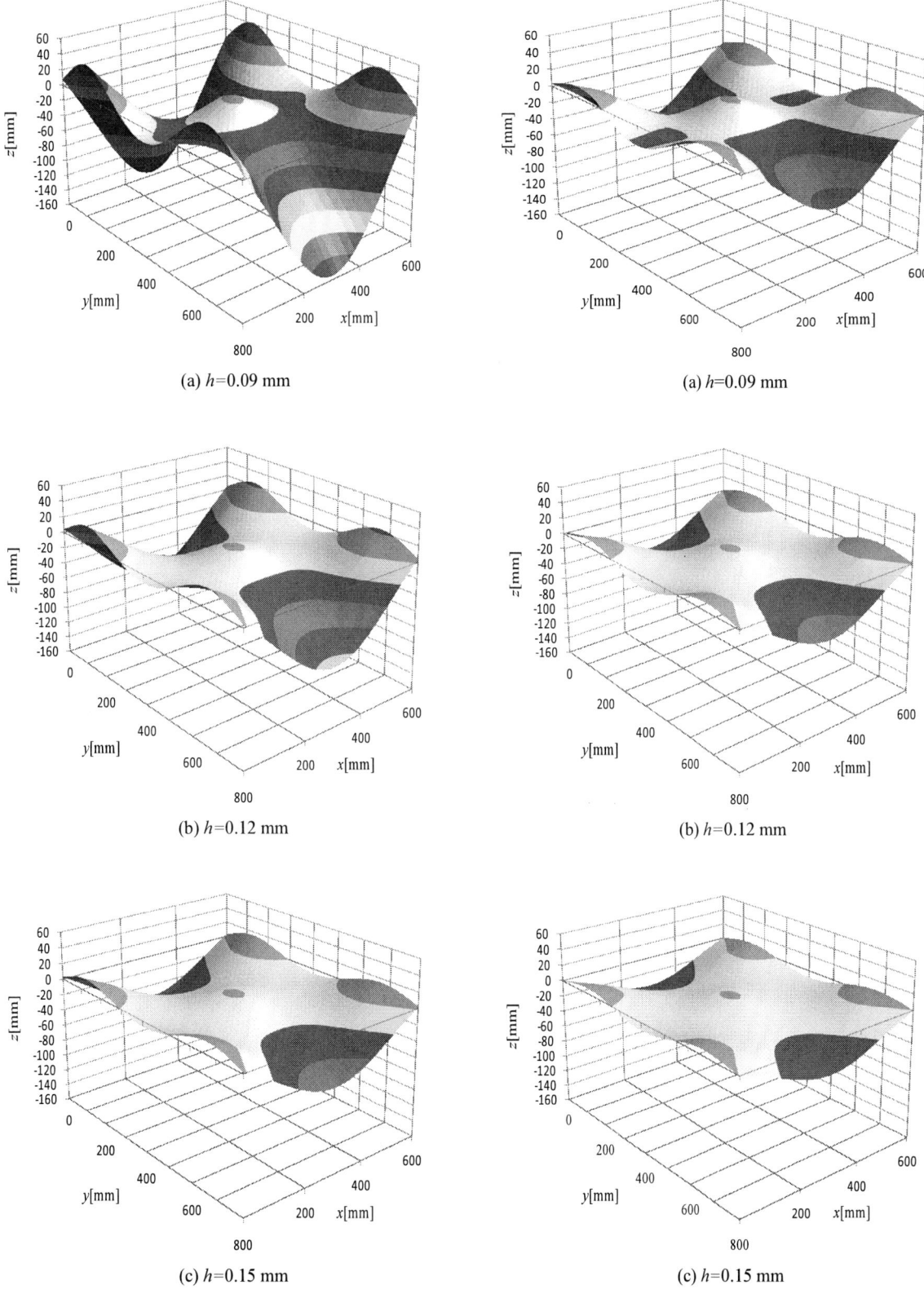

(a) h=0.09 mm

(b) h=0.12 mm

(c) h=0.15 mm

Fig. 5. Shape of thin steel plate as bird's eye view.

(without horizontal positioning control)

(a) h=0.09 mm

(b) h=0.12 mm

(c) h=0.15 mm

Fig. 6. Shape of thin steel plate as bird's eye view.

(with horizontal position control: I_x=0.5A)

978-1-4799-2706-7/14 $31.00 © 2014 IEEE   302

Figure 5 (a), (b), and(c) show analytical results of ultrathin steel plate without horizontal positioning control, revealing the regions of the steel plate that were severely bent. Figure 6 (a), (b), and(c) are results of ultrathin steel plate applying horizontal positioning control. According to these figures, we were able to confirm that the deflection of steel plate increase as the thickness of the steel plate decrease. Furthermore, the deflection of steel plates respectively restrained when magnetic fields from the horizontal direction applied. Thus, it was confirmed that the horizontally applied magnetic fields contribute to raising the edges of the thin steel plate weighed down by gravity, making it almost flat.

IV. CONCLUSION

In this study, we examined the effect of applying magnetic fields in the horizontal direction on rectangular ultrathin steel plates with thickness of 0.09 mm, 0.12 mm, and 0.15 mm which were levitated in the simulation. From the simulation results, it was revealed that the deflection of the magnetically levitated steel plate was suppressed and that the edges of the steel plate weighed down by gravity were raised, making the steel plate almost flat, which have effect on prevent problems, such as side slipping and the dropping of the plate. In the future, we will carry out experiment practically and compare the shape of steel plate by using simulation with the aim of realizing the stable control of steel plates during levitation.

REFERENCES

[1] Y. Oshinoya and T. Shimogo, "Electro-Magnetic Levitation Control of a Travelling Elastic Plate," *Proc. of Int.Conf. on Advanced Mechatronics*, pp. 845-850, 1998.

[2] Y. Oshinoya, S.Kobayashi and K. Tanno, "Optimal Electromagnetic Levitation Control of a Thin Rectangular Steel Plate with Two Opposite Edges Reinforced by a Beam," *Trans. of Japan Society of Mechanical Engineers*, vol. 62, no. 600-C, pp. 127-133, 1998, (in Japanese).

[3] H. Abunoya, H. Miyazaki, T. Ohji, K. Amei and M. Sakui, "Motion Tests on a Levitated Body Using a Magnetic Levitation System for Three-Dimensional Motion with Partial Zero Power Control," *Journal of the Magnetics Society of Japan*, vol. 35, no. 2, pp. 123-127, 2011, (in Japanese).

[4] Y. Oshinoya, "Study on Electromagnetic Levitation Conveyance of a Steel Plate with Free Edges by Using Electromagnets and Permanent Magnets," *Journal of The Japan Society Applied Electromagnetics and Mechanics*, vol. 6, no. 4, pp. 364-371, 1998, (in Japanese).

[5] S. Hasegawa, T. Obata, Y. Oshinoya and K. Ishibashi, "Study on Noncontact Support and Transportation of Rectangular Thin Steel Plate," *Proceedings of the School of Engineering the School of Information Technology and Electronics Tokai University, Series E*, vol. 27, pp. 1-12, 2002.

[6] S. Akiyoshi, T. Saito, T. Iwata, U. Matrika, Y. Oshinoya, K. Ishibashi and H. Kasuya, "Fundamental Study on Horizontal Noncontact Positioning Control for a Magnetically Levitated Ultra-thin Steel Plate," *The International Conference on Electrical Engineering (ICEE) 2008 OKINAWA*, CD-ROM P-068, 2008.

[7] T. Narita, Y. Oshinoya, and S. Hasegawa, "Study on Horizontal Noncontact Positioning Control for a Magnetically Levitated Thin Steel Plate (Experimental Considerations on Elastic Vibration Control under Transport)," *Journal of Council on Electrical Engineering*, vol. 1, no. 3, pp. 292-297, 2011.

Novel Magnetic Structure of Integrated Differential-Mode and Common-Mode Inductors to Suppress DC Saturation

Kazuhiro Umetani, Takahiro Tera, Kazuhiro Shirakawa

R&D Center
DENSO CORPORATION
1-1 Showa-cho, Kariya, Aichi 448-8661, Japan
KAZUHIRO_Z_UMETANI@denso.co.jp

Abstract— **Integrating differential-mode (DM) and common-mode (CM) inductors onto a single core has been expected to miniaturize EMI filters. However, the previously-reported magnetic structure often suffers from saturation by DC current, hindering miniaturization by integration. This problem seems exacerbated by the fact that this conventional structure tends to induce large DC flux because its equivalent number of turns for the DM inductance is restricted to only half of the total number of turns. This paper addresses the problem by proposing a novel structure that assigns more turns to the DM inductance to suppress the DC flux more effectively. This paper confirmed theoretically and experimentally that the proposed structure is equivalent to series-connected DM and CM inductors. Additionally, analytical estimation revealed that the proposed structure reduced core volume by 41% compared to the conventional structure under the same amount of copper. These results suggest effectiveness of the proposed structure for miniaturizing EMI filters.**

Keywords— Integrated magnetic component, EMI filter, Differential-mode inductor, Common-mode inductor.

I. INTRODUCTION

Recently, high power density is intensely required for switching converters. Because magnetic devices for EMI filters are voluminous components in the converters, a number of techniques have been proposed to miniaturize differential-mode (DM) and common-mode (CM) inductors [1]–[11].

One of promising approaches is to integrate a DM inductor and a CM inductor into a device. Particularly, highly integrated structures are proposed based on planar magnetic cores [1]–[3]. These structures are beneficial in possibilities of further implementing capacitors by inserting dielectric layer between layers of planar windings. However, these structures seem to suffer from excessive copper loss in high power applications because planar core generally needs long wire length for the windings. The same benefit and problem also tend to occur in the structures in which conductive foils are used as windings [4] because the foils tend to have large DC resistance. Therefore, high power applications still seem to need integration techniques based on bulk core with

windings of thick wire.

Techniques of this type have also been reported by a number of studies. These techniques seem classifiable into the following two major categories. One is structural integration, which implements DM and CM inductors on separate magnetic cores partly sharing the windings [5]–[8]. Techniques of this category are beneficial in reducing the dead space because the cores can be closely placed by sharing the windings. The other category is magnetic integration, which allows sharing not only the windings but also the core between the DM and CM inductors [9]–[11]. Magnetic integration possibly enables further miniaturizing because the total core volume can also be reduced by sharing magnetic paths.

On the other hand, magnetic integration has a drawback that the CM inductance, as well as the DM inductance, may saturate by magnetic saturation of the shared magnetic paths, which can be caused by the DC component in DM current. This possibly hinders reducing the core volume because the cross-section of the magnetic paths may be designed to expand in order to meet requirement for the saturation property not only of the DM inductance but also of the CM inductance.

A remedy probably lies in increasing the equivalent number of turns N_{DM} that links with the flux induced by the DM current. According to the analogy to the basic inductor with a single magnetic path [12], we can define N_{DM} as the ratio of the total flux linkage to the flux, when only DM current is applied. Hence, we obtain the relation (1), if we assume constant DM inductance L_{DM}.

$$N_{DM} \equiv \frac{L_{DM} I_{DM}}{\phi_{DM}},$$

$$\therefore \phi_{DM} = \frac{L_{DM} I_{DM}}{N_{DM}}, \tag{1}$$

where ϕ_{DM} is the flux induced by the DM current I_{DM}.

Accordingly, we can express the DC flux ϕ_{DC} induced by the DC component I_{DC} of I_{DM} as in (2).

$$\phi_{DC} = \frac{L_{DM} I_{DC}}{N_{DM}}. \tag{2}$$

The 2014 International Power Electronics Conference

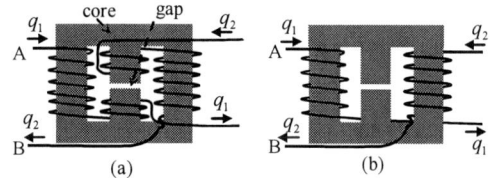

Fig. 1. Proposed magnetic structure (a) and conventional magnetic structure (b).

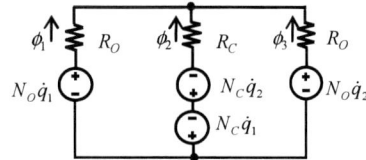

Fig. 2. Magnetic circuit model of the proposed structure.

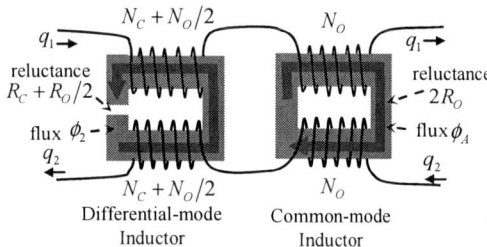

Fig. 3. Equivalent circuit of the proposed structure.

Because L_{DM} and the maximum value of I_{DC} are generally specified as requirement, increasing N_{DM} is indispensable to suppressing ϕ_{DC} in order to avoid the DC saturation without expanding the cross-section of the flux path. However, as shown in this paper, N_{DM} is restricted to only half of the total number of turns in the conventional magnetic structure proposed in the prior techniques [9]–[11].

To address the problem, this paper proposes a novel magnetic structure. In the proposed structure, more winding turns can be assigned to N_{DM} than the conventional structure. This alleviates the magnetic saturation. As a result, further reduction in the core volume is probably expected, if magnetic saturation is a determining factor in designing the cross-section area of the magnetic paths as is often the cases when large L_{DM} or large I_{DC} is specified.

This paper discusses the proposed structure in the following four sections. Section 2 discusses the operating principle of the proposed structure theoretically. Then, Section 3 confirms it experimentally. Section 3 also confirms that the discrete DM and CM inductors are miniaturized by integrating them into the proposed structure. Section 4 analytically estimates improvement

of core reduction effect by suppressing the DC flux using the proposed structure. In the estimation, core volume is compared between the proposed and conventional structures under the same total amount of copper and under the same specification in which magnetic saturation dominantly determines the cross-section area of magnetic paths. Finally, Section 5 presents the conclusions.

II.   PROPOSED MAGNETIC STRUCTURE

Figure 1(a) illustrates the proposed magnetic structure. The structure has a core with three legs. The center leg has a gap and two windings with the same number of turns. The windings on the center leg are both wound so that DM current induce the same direction of flux. And each of the outer legs has a winding connected in series with a winding on the center leg. The windings on the outer legs have the same number of turns and are wound so that DM current induce the flux in the outer leg in the direction that reinforces the flux in the center leg.

Meanwhile, the conventional magnetic structure proposed in [9]–[11] is magnetically equivalent to Fig. 1(b). It differs from the proposed structure in the windings on the center leg.

Electrical functions of the proposed structure are equivalent to series-connected DM and CM inductors, as well as the conventional structure. We can show the fact utilizing the Lagrangian modeling [13].

As discussed in [13], the Lagrangian modeling is useful to transform an integrated magnetic component into an equivalent circuit of basic transformers and inductors. In the method, we first translate the physical magnetic structure into Lagrangian, which is directly configurable from their electric and magnetic network. Then, we apply point transformation [14] to the Lagrangian, obtaining another Lagrangian that belongs to a circuit of basic transformers and inductors, each of which consists of a single closed magnetic path. Finally, we again translate the resultant Lagrangian to obtain the equivalent circuit.

Now, we apply this method to the proposed structure. The magnetic circuit of the proposed structure can be expressed as Fig. 2. N_C and N_O are the number of turns of the windings on the center leg and the outer legs, respectively. And R_C is the reluctance of the center leg. The outer legs are designed to have the same reluctance R_O according to the designing concept of the proposed structure. We denote the electric charge that flows through the winding A and B as q_1 and q_2, respectively. Then, translating Fig. 2 yields Lagrangian L as in (3).

$$
\begin{aligned}
L = & \; N_O \dot{q}_1 \phi_1 + N_O \dot{q}_2 \phi_3 - N_C \dot{q}_1 \phi_2 - N_C \dot{q}_2 \phi_2 \\
& - R_O {\phi_1}^2 / 2 - R_C {\phi_2}^2 / 2 - R_O {\phi_3}^2 / 2 \\
& + \lambda(\phi_1 + \phi_2 + \phi_3),
\end{aligned} \tag{3}
$$

where λ is a Lagrangian multiplier; and ϕ_{1-3} are the fluxes of the left outer leg, the center leg, and the right outer leg, respectively. A dot over a variable indicates its

TABLE I
REQUIREMENT SPECIFICATIONS AND EVALUATION RESULTS OF THE PROTOTYPES

	Requirement	Proposed Structure	Discrete Inductors
DM Inductance[a]	60μH	61μH	60μH
DM Saturation Property[b]	−25%	−5.8%	−25%
CM Inductance[a]	440μH	476μH	447μH
CM Saturation Property[b]	−25%	−20%	——
DC Resistance	16.5mΩ	16.1mΩ	16.4mΩ

a) Inductance is specified when DC current of 16A is applied in DM current.
b) Inductance decrease ratio between DC current of 22.5A and 0A applied in DM current.

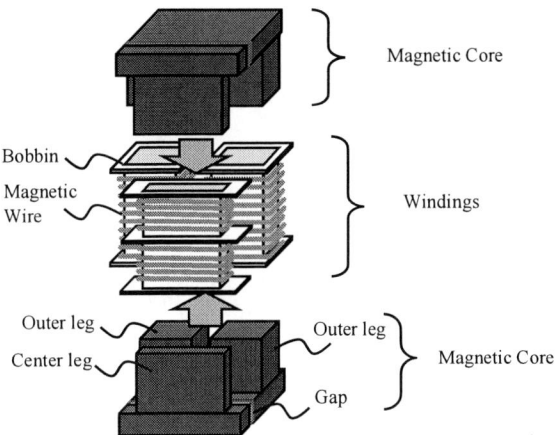

Fig. 4. Physical structure of the prototype of the proposed structure.

time derivative.

The Lagrangian multiplier can be eliminated by substituting $\phi_3 = -\phi_1 - \phi_2$ into (3). Then, we have (4).

$$L = N_O\dot{q}_1\phi_1 - N_O\dot{q}_2(\phi_1 + \phi_2) - N_C\dot{q}_1\phi_2 - N_C\dot{q}_2\phi_2 \\ - R_O\phi_1{}^2/2 - R_C\phi_2{}^2/2 - R_O(\phi_1 + \phi_2)^2/2. \quad (4)$$

Next, we perform a point transformation on the result. The purpose of this transformation is to convert the magnetic energy terms in (4), i.e. the fifth, sixth, and seventh terms, into a diagonal form of the flux. Then, the resultant Lagrangian corresponds to a circuit of magnetic components each made of a single magnetic path. Introducing a flux ϕ_A defined as $\phi_A = \phi_1 + \phi_2/2$ to eliminate ϕ_1, we obtain (5).

$$L = N_O(\dot{q}_1 - \dot{q}_2)\phi_A - (N_C + N_O/2)(\dot{q}_1 + \dot{q}_2)\phi_2 \\ - R_O\phi_A{}^2 - (R_C + R_O/2)\phi_2{}^2/2. \quad (5)$$

Equation (5) can be translated into a series connection of DM and CM inductors as illustrated in Fig. 3. The flux ϕ_2 constitutes a DM inductor that has two windings with the number of turns $N_C + N_O/2$, whereas ϕ_A constitutes a CM inductor that has two windings with the number of turns N_O. Note that $N_C = 0$ corresponds to the conventional structure. Because $N_C = 0$ in Fig. 3 gives the equivalent circuit for the conventional structure, the number of turns

on its equivalent DM inductor equals to only half of the total number of turns on the core structure. On the other hand, the proposed structure increases the number of turns on the DM inductor by $2N_C$ by adding two windings with the number of turns N_C to the center leg. And it still keeps the number of turns for the CM inductor unchanged.

Now, we examine whether the proposed structure enables its DM inductor to have greater number of turns than the conventional structure. We compare the proposed and conventional structures under the same total wire length and the same core dimension. Note that half of the DC flux in the DM inductor flows in the outer legs, because the DC flux is the DC component of ϕ_2 and the fluxes of the outer legs contains half of ϕ_2 according to $\phi_1 = \phi_A - \phi_2/2$ and $\phi_3 = -\phi_A - \phi_2/2$. Therefore, the cross-section area A_O of the outer leg should be designed at least greater than half of the cross-section area A_C of the center leg. Hence, we have (6).

$$A_C \leq 2A_O. \quad (6)$$

If we assume that the aspect ratio of the center leg is the same as that of the outer leg, we obtain the following relation (7) between the perimeter l_C of the center leg and the perimeter l_O of the outer leg using the fact that the perimeter is proportional to the square root of the cross-section area.

$$l_C \leq \sqrt{2}l_O, \\ \therefore l_C < 2l_O. \quad (7)$$

As we have found in Fig. 3, one turn on the center leg is equivalent in the DM inductor to two turns on the outer leg. Therefore, the proposed structure can implement the DM inductor with the same number of turns by smaller total wire length than the conventional structure. In other words, the proposed structure can implement the DM inductor with a greater number of turns by the same total wire length.

This indicates that the proposed structure is beneficial in suppressing the DC flux induced by the DC component in the DM current. Because the DC flux flows in both the center and outer legs, the DC flux increases not only R_C but also R_O. As a result, the DC flux can saturate not only the DM inductance but also the CM inductance. Hence, the proposed structure is beneficial in suppressing saturation of both the DM and CM inductance, thus avoiding expanding both the center and outer legs to ensure necessary tolerance to saturation.

On the other hand, the proposed structure has a drawback that its CM inductor has a smaller number of turns than the conventional structure under the same wire length. This indicates that the proposed structure requires less reluctance for R_O in order to keep the same CM inductance as the conventional structure. If reducing R_O inevitably requires for expanding the cross-section of the outer legs, the proposed structure does not lead to effective reduction in the core volume. However, it seems effective in otherwise conditions, for example when

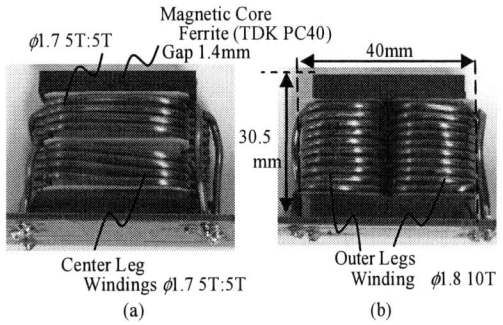

Fig. 5. Photographs of the prototype of the proposed structure. (a) The front side. (b) The rear side.

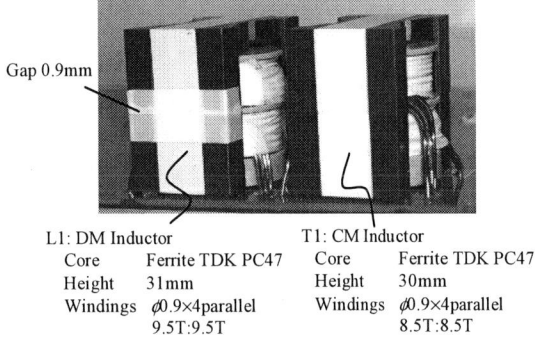

L1: DM Inductor
Core   Ferrite TDK PC47
Height   31mm
Windings ϕ0.9×4parallel
    9.5T:9.5T

T1: CM Inductor
Core   Ferrite TDK PC47
Height   30mm
Windings ϕ0.9×4parallel
    8.5T:8.5T

Fig. 6. Photograph of the prototype of the conventional structure.

tolerance to the DC flux determines the cross-section area rather than requirement for R_O. We present a case study of estimating reduction of the core volume under this condition in Section 4.

III. EXPERIMENT

This section experimentally confirms the electrical functions of the proposed structure and its miniaturizing effect on discrete DM and CM inductors.

A. Prototypes

We developed two prototypes that implement the DM and CM inductance under the same requirement specification presented in Table 1. One is the proposed structure. And the other is series-connected discrete DM and CM inductors. This specification was designed for a part of an input filter of a PFC converter, whose maximum input current was set at 16Arms. Hence, we specified the DM and CM inductance at 16A, which is near the average input current. And we specified the saturation inductance at the maximum instantaneous input current, i.e. 22.5A.

Both prototypes were made on ferrite cores with similar permeability and saturation flux density. We designed these prototypes to have the same vertical

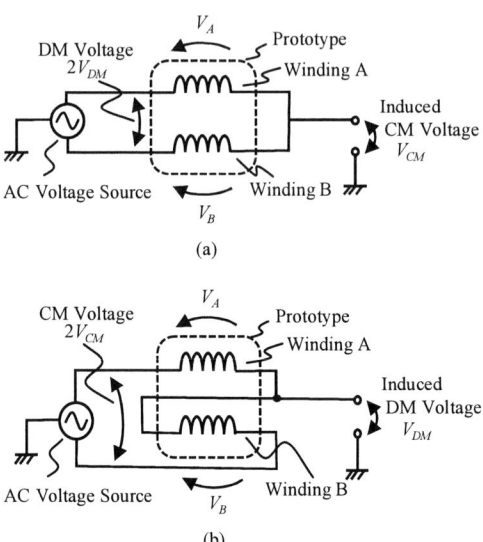

Fig. 7. Evalution circuits of the conversion ratios between DM and CM noise. (a) Evaluation of CM noise response by exciting DM noise. (b) Evaluation of DM noise response by exciting CM noise.

dimension and the same average height so that the horizontal dimension reflects the volume.

Figure 4 illustrates the physical structure of the prototype of the proposed structure. In the prototype, we placed the flattened center leg in the front side and the outer legs in the rear side. This disposition is beneficial in enhancing the CM inductance by minimizing flux path length through the two outer legs. Additionally, for easy assembly, we equipped two gaps to implement R_C (the gap on the center leg in Fig. 1) at the top and bottom beams near the center leg, respectively. We equipped no gap on the outer legs. The photographs of the prototype are presented in Fig. 5.

The cross-section area of the center leg was designed so that the maximum instantaneous input current induces the saturation flux density there. Meanwhile, we designed the cross-section area of the outer leg 1.19 times as great as that of the center leg. Because the DC flux in the outer leg is half of that in the center leg, the DC flux density in the outer leg does not exceed 42% of the saturation flux density. Owing to this, saturation of the CM inductance was expected to be suppressed below −25%.

On the other hand, the prototype of the discrete inductors was made on two basic EE cores, as shown in Fig. 6. We designed wire of their windings to have the similar cross-section area as the prototype of the proposed structure.

B. Functional Equivalence Between the Prototypes

We confirmed that the proposed structure was functionally equivalent to series-connected discrete inductors by evaluating the conversion ratios between DM and CM noise, i.e. CM voltage response to DM noise

The 2014 International Power Electronics Conference

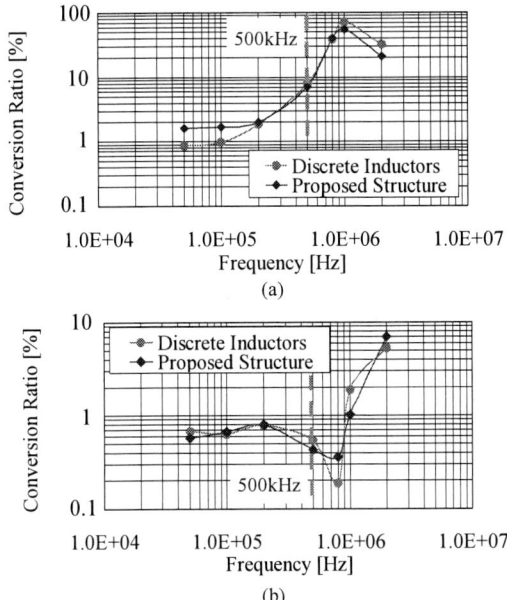

(a)

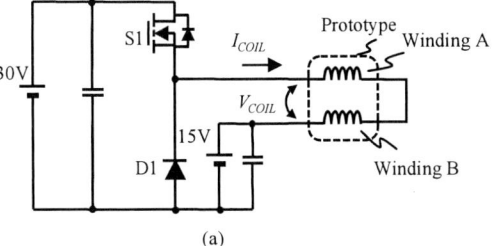

(a)

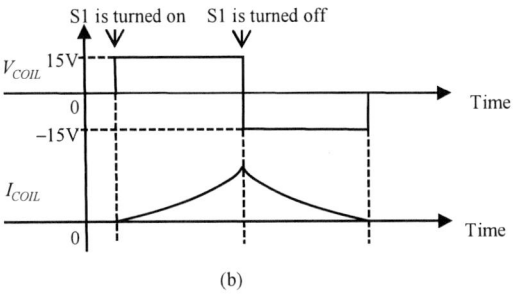

(b)

Fig. 9. Method to evaluate the saturation property of the DM inductance L_{DM}. (a) The evaluation circuit. (b) The voltage and current waveforms in the evaluation process.

(b)

Fig. 8. Measured conversion ratios. (a) The ratio of CM noise response to DM noise excitation. (b) The ratio of DM noise response to CM noise excitation.

excitation and DM voltage response to CM noise excitation. The conversion ratios vanish in series-connected ideal DM and CM inductors. Hence, we expected that the prototype of the proposed structure shows as small conversion ratios as those of the prototype of the discrete inductors.

Evaluation circuit is presented in Fig. 7. We connected the windings A and B in series as shown in Fig. 7(a) and Fig. 7(b). Then, we applied AC voltage signal with the amplitude of ±5Vpeak to the series-connected windings.

Now, we express the voltage induced in each winding using the DM voltage V_{DM} and the CM voltage V_{CM}. Then, we obtain (8), if we denote the induced voltage in the winding A and B as V_A and V_B, respectively.

$$\begin{cases} V_A = V_{CM} + V_{DM}, \\ V_B = V_{CM} - V_{DM}. \end{cases} \quad (8)$$

Note that the AC signal voltage equals to $V_A - V_B$, i.e. $2V_{DM}$, in Fig. 7(a) and to $V_A + V_B$, i.e. $2V_{CM}$, in Fig. 7(b). Hence, the AC signal is a DM voltage source that excites DM noise current in Fig. 7(a) and a CM voltage source that excites CM noise current in Fig. 7(b).

We connected the midpoint between the terminals of the AC signal to the ground. Then, we measured the voltage potential of the connecting point of the winding A and B. The measured voltage represents to the CM voltage response V_{CM} in Fig 7(a) and the DM voltage response V_{DM} in Fig. 7(b). We obtained the conversion ratios by normalizing the amplitude of the measured voltage by half of the amplitude of the AC signal voltage. Then, the normalized voltage in Fig. 7(a) corresponds to the ratio of CM noise response to DM noise excitation,

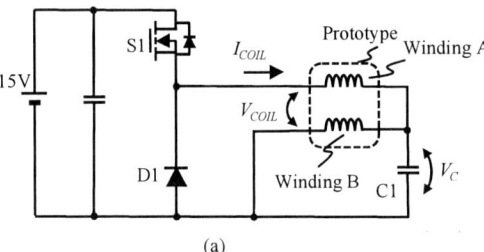

(a)

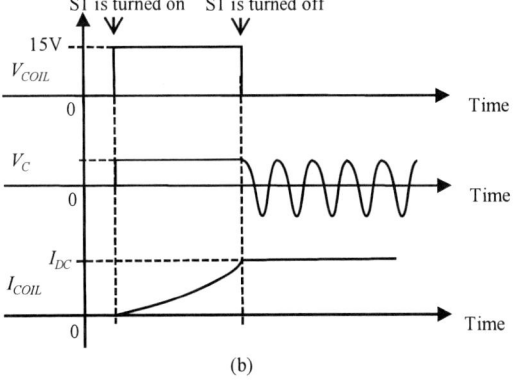

(b)

Fig. 10. Method to evaluate the saturation property of the CM inductance L_{CM}. (a) The evaluation circuit. (b) The voltage and current waveforms in the evaluation process.

and that in Fig. 7(b) corresponds to the ratio of DM noise response to CM noise excitation.

We examined the conversion ratios in the frequency range below 500kHz, because the dimensional resonance and the magnetic resonance may degrade soft-magnetic

978-1-4799-2706-7/14 $31.00 © 2014 IEEE   308

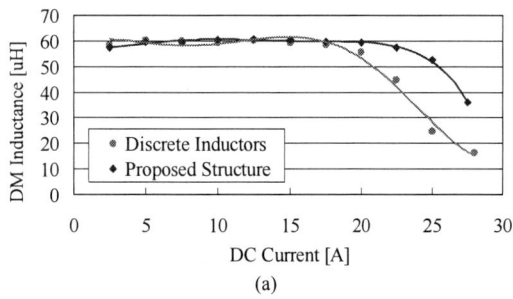

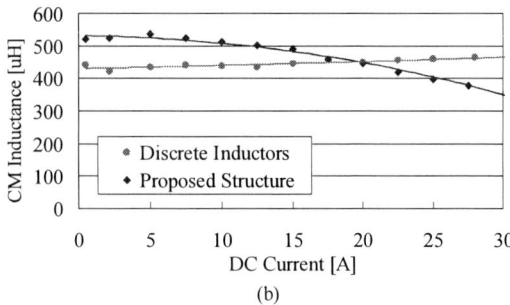

Fig. 11. Measured saturation property. (a) The DM inductance L_{DM}. (b) The CM inductance L_{CM}.

property of the ferrite core above the frequency. Figure 8 shows the results. The ratios of the proposed structure were found approximately as small as those of the discrete inductors. Both of the prototypes showed the ratio of CM noise response smaller than 7% and the ratio of DM noise response smaller than 1% below 500kHz. Hence, the results supported that the two prototypes are approximately equivalent each to the other in their electrical functions.

C. Saturation Property of the DM and CM inductance

Next, we confirmed that the two prototypes have similar filtering capability by evaluating the saturation property of the DM and CM inductance.

Figure 9(a) illustrates the evaluation circuit of the saturation property of the DM inductance. The windings A and B were connected in series in the similar manner as in Fig. 7(a). Therefore, DM voltage was applied to the magnetic device during the on-state of the switch S1. We held S1 in the on-state until the induced DC current sufficiently saturated the magnetic device. At the same time, we measured the applied voltage V_{COIL} and the current I_{COIL}. The current I_{COIL} increased monotonically during the on-state of S1 as illustrated in Fig. 9(b). Hence, we obtained the DM inductance L_{DM} as the differential inductance [12] defined by (9).

$$L_{DM} = \frac{V_{COIL}}{dI_{COIL}/dt} \qquad (9)$$

where t is the time. The saturation property was obtained by determining L_{DM} as the function of I_{COIL}.

The method to evaluate the saturation property of the

CM inductance is slightly more complicated than the method for the DM inductance. Figure 10(a) illustrates the evaluation circuit. In this experiment, we connected the capacitor C1 with capacitance of 1nF between the ground and the connecting point of the windings A and B. Then, we held the switch S1 in the on-state until DC current increased to the predetermined level I_{DC} as illustrated in Fig. 10(b). After the turn-off of S1, the DC current circulated through the diode D1. The circulating DC current maintained itself for a while, because no DM voltage was applied to the magnetic device.

At the same time, an LC oscillation occurred between the capacitor C1 and the magnetic device under test. This oscillation was excited at the turn-off of S1, because the voltage V_C of the connecting point of the windings was approximately half of the applied voltage V_{COIL} at the turn-off of S1 and then V_C was going to settle finally to zero as the oscillation was dissipated. As a result, the voltage and current waveforms can be illustrated as Fig. 10(b). Note that the voltage V_C equals to the CM voltage V_{CM} when the DC current circulates through D1. Therefore, this oscillation corresponds to the LC oscillation between C1 and the CM inductance L_{CM} of the magnetic device. Hence, we obtained L_{CM} according to (10).

$$L_{CM} = \frac{1}{C_1 \omega_{OSC}^2} \qquad (10)$$

where C_1 is the capacitance of C1 and ω_{OSC} is the angular frequency of the oscillation. We obtained the saturation property by determining L_{CM} at various levels of I_{DC}.

The measurement results of the saturation property of L_{DM} and L_{CM} are presented in Fig. 11. Both of the prototypes were found to have similar DM inductance. And the proposed structure showed slightly better saturation property of the DM inductance. As for the CM inductance, only the proposed structure showed saturation. Nonetheless, it always showed the CM inductance within 92%–118% of that of the discrete inductors. To summarize, the prototype of the proposed structure was found to have similar or slightly better filtering capability compared to the prototype of the discrete inductors.

D. Comparison of the Volume

Finally, we compared the volume between the prototypes. The result is shown in Fig. 12. The proposed structure decreased the total volume including the dead space by 31%. This reduction was contributed not only by eliminating dead space but also by reducing the core. According to comparison of core volume, the proposed structure was found to have reduced the core by 17%. Consequently, miniaturizing by the proposed structure was successfully confirmed.

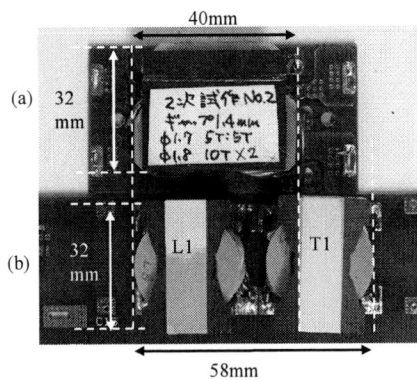

Fig. 12. Comparison of the volume between the prototypes. (a) The proposed structure. (b) The discrete inductors.

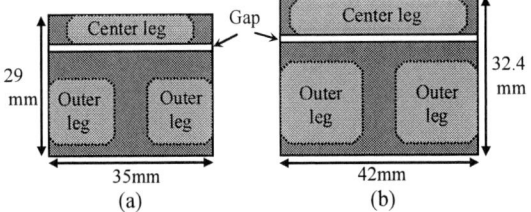

Fig. 13. Top view of the core in the prototype of the proposed structure (a) and the estimated core of the conventional structure (b).

IV. IMPROVEMENT OF CORE REDUCTION EFFECT

This section evaluates improvement of core reduction effect by suppressing the DC flux using the proposed structure in comparison with the conventional structure shown in Fig. 1(b). We briefly estimates the core dimension of the conventional structure, when the same specification in Table I is applied and the same physical core structure as Fig. 4 is employed. We determine its core dimension by modifying the core dimension of the prototype of the proposed structure discussed in the previous section. Then, we compare volume of the estimated core with that of the proposed structure.

When estimating the conventional structure, we keep the total amount of copper the same as the prototype of the proposed structure. Meanwhile, we expand the cross-section of magnetic paths to meet the requirement for the saturation property and adjust R_C and R_O to meet the requirement for the inductance. For convenience, we keep the aspect ratio of the cross-section of the center and outer legs the same as that in the prototype of the proposed structure, when we modify the cross-section. And we assume to adjust permeability of the core material without changing the saturation property, when we adjust R_O. Furthermore, we assume that reluctance of air gap mainly contributes R_C and we adjust the gap length to obtain appropriate value for R_C.

In the first step, we directly apply the conventional structure to the prototype of the proposed structure without changing the magnetic core. In order to keep both the total DC resistance and the total copper volume the same as the prototype of the proposed structure, we keep the total wire length unchanged. As a result, we set the number of turns N_{O_temp} of the outer leg windings of the conventional structure at 16.

Next, we modify the magnetic core. We assume to enlarge the cross-section area of the outer leg by a factor α. Then, the number of turns N_{O_mod} after this modification is also changed according to (11), because the perimeter of the cross-section is expanded by $\sqrt{\alpha}$.

$$N_{O\_mod} = N_{O\_temp}/\sqrt{\alpha} = 16/\sqrt{\alpha}. \quad (11)$$

Now, we consider the DC flux, which is induced by the DC component in DM current, to estimate necessary value for α to implement the required saturation property of the CM inductance. We denote the maximum DC flux after this modification as ϕ_{2_mod}, and that in the prototype of the proposed structure as ϕ_{2_org}. Because we keep the DM inductance L_{DM} unchanged and we require the same maximum DC current, we have the relation (12) according to (2) and Fig. 3.

$$\frac{\phi_{2\_mod}}{\phi_{2\_org}} = \frac{2N_{C\_org} + N_{O\_org}}{N_{O\_mod}} = \frac{20}{N_{O\_mod}}, \quad (12)$$

where N_{C_org} and N_{O_org} are the numbers of turns of the windings on the center and outer legs of the prototype of the proposed structure, respectively.

On the other hand, the maximum DC flux density in the outer leg must be unchanged in order to maintain the same saturation property for the CM inductance. Hence, we require the relation (13).

$$\alpha = \frac{\phi_{2\_mod}}{\phi_{2\_org}}. \quad (13)$$

Equations (11), (12), and (13) determine α and N_{O_mod}.

$$\alpha = 1.56, \quad N_{O\_mod} \approx 13. \quad (14)$$

We also need to expand the cross-section of the center leg by α, because the maximum DC flux density in the center leg must also be kept the same as that of the prototype of the proposed structure in order to maintain the saturation property unchanged for the DM inductance. According to the same reason, we also expand the cross-section of the top and bottom beams by α. Meanwhile, we keep the height of the legs unchanged.

Finally, we obtain the core dimension as presented in Fig. 13. The total core volume for the conventional structure is estimated as $3.5 \times 10^4 \text{mm}^3$, whereas that for the proposed structure is $2.0 \times 10^4 \text{mm}^3$. Consequently, the core volume reduction ratio of the proposed structure to the conventional structure is estimated as -41%. Consequently, the result supports improvement of core reduction effect by the proposed structure.

V. CONCLUSION

Magnetic integration is an attractive technique to

miniaturize EMI filters. Prior works have reported EMI filters that applied the technique to integrate a DM inductor and a CM inductor. However, the magnetic structure proposed in these works tends to suffer from wrong saturation property of the DM and CM inductance, because equivalent number of turns for the DM inductance is restricted only to half of total number of turns and is probably insufficient to suppress DC flux induction in many cases. This possibly entails expanding cross-section of magnetic path to ensure necessary saturation property, thus hindering miniaturization by magnetic integration.

This paper proposed a novel structure that allows assigning more turns to the DM inductance. We confirmed that the proposed structure is equivalent to series-connected DM and CM inductors both theoretically and experimentally. And we further confirmed experimentally that the proposed structure miniaturized discrete DM and CM inductors.

An analytical estimation was carried out to evaluate improvement of core reduction effect by the proposed structure in comparison with the conventional structure. The result revealed that the proposed structure reduced the core volume by 41% compared to the conventional structure under the same amount of copper and under the same specification, in which saturation by the DC flux is a determining factor of the cross-section area of magnetic paths.

These results support effectiveness of the proposed structure for miniaturizing EMI filters.

REFERENCES

[1] R. Chen, J. D. van Wyk, S. Wang, and W. G. Odentaal, "Technologies and characteristics of the integrated EMI filters for switch mode power supplies," in *Proc. IEEE Power Electron. Specialist Conf. (PESC)*, Aachen, Germany, 2004, vol. 4, pp. 4873-4880.

[2] R. Chen, J. D. van Wyk, S. Wang, and W. G. Odentaal, "Improving the characteristics of integrated EMI filters by embedded conductive layers," *IEEE Trans. Power Electron.*, vol. 20, no. 3, pp. 611-619, May 2005.

[3] J. Biela, A. Wirthmueller, R. Waespe, M. L. Heldwein, K. Raggl, and J. W. Kolar, "Passive and active hybrid integrated EMI filters," *IEEE Trans. Power Electron.*, vol. 24, no. 5, pp. 1340-1349, May 2009.

[4] X. Wu, D. Xu, Z. Wen, Y. Okuma, and K. Mino, "Design, modeling, and improvement of integrated EMI filter with flexible multilayer foils," *IEEE Trans. Power Electron.*, vol. 26, no. 5, pp. 1344-1354, May 2011.

[5] P. Boonma, V. Tarateeraseth, and W. Khan-ngern, "A new technique of integrated EMI inductor using optimizing inductor-volume approach," in *Proc. Int. Power Electron. Conf. (IPEC)*, Niigata, Japan, 2005.

[6] L. Nan and Y. Yugang, "A common mode and differential mode integrated EMI filter," in *Proc. IEEE Int. Power Electron. and Motion Control Conf. (IPEMC)*, Shanghia, China, 2006, vol. 1, pp. 1-5.

[7] R. Lai, Y. Maillet, F. Wang, S. Wang, R. Burgos, and D. Boroyevich, "An integrated EMI choke for differential-mode and common-mode noise suppression," *IEEE Trans. Power Electron.*, vol. 25, no. 3, pp. 539-544, Mar. 2010.

[8] W. Tan, C. Cuellar, X. Margueron, and N. Idir, "A common-mode choke using troid-EQ mixed structure," *IEEE Trans. Power Electron.*, vol. 28, no. 1, pp. 31-35, Jan. 2013.

[9] A. K. Upadhyay, "Integrated common mode and differential mode inductor device," U.S. Patent 5 313 176, May 17, 1994.

[10] T. P. Gilmore and G. Ray, "Method of configuring common mode/ differential mode choke," U.S. Patent 6 768 408 B2, Jul. 27, 2004.

[11] F. Luo, D. Boroyevich, P. Mattevelli, K. Ngo, D. Gilham, and N. Gazel, "An integrated common mode and differential mode choke for EMI suppression using magnetic epoxy mixture," in *Proc. IEEE Appl. Power Electron. Conf. Expo.*, 2011, pp. 1715-1720.

[12] A. V. den Bossche, and V. C. Valchev, "Fundamentals of magnetic theory," in *Inductors and transformers for power electronics*, Florida: CRC Press, 2005, pp. 19-23.

[13] K. Umetani, "A Generalized Method for Lagrangian Modeling of Power Conversion Circuit with Integrated Magnetic Components," *IEEJ Trans. Elect. and Electron. Eng.*, vol. 7, issue S1, pp. S146-S152, Nov. 2012.

[14] L. D. Landau, and E. M. Lifshitz, "Canonical transformations," in *Mechanics (Course of Theoretical Physics 1)*, Oxford, U. K.: Butterworth-Heinemann, 1976, pp.143-146.

A Novel Control Method in Flux-weakening Region for Efficient Operation of Interior Permanent Magnet Synchronous Motor

K. Ueda, S. Morimoto, Y. Inoue, and M. Sanada

Graduate School of Engineering
Osaka Prefecture University
1-1 Gakuen-cho, Naka-ku, Sakai, Osaka, Japan
E-mail : mx104008@edu.osakafu-u.ac.jp

Abstract— **This paper proposes a novel control method for an interior permanent magnet synchronous motor in the flux-weakening region. The proposed method consists of a maximum torque per flux (MTPF) control, and an efficient flux-weakening control under a light load. In order to simplify the MTPF control, a method for approximating the MTPF curve is proposed. In the flux-weakening control, more efficient control is applied by minimizing the armature current under a light load. In the proposed efficient flux-weakening control, the current command generation method, in which the armature current is adopted for the operation along the constant- voltage ellipse, is proposed. The validity of the proposed control method is verified based on simulation and experimental results.**

Keywords— IPMSM, Flux-weakening control, High efficiency, MTPF control

I. INTRODUCTION

The interior permanent magnet synchronous motor (IPMSM) is capable of variable speed operation over a wide speed region compared to a surface permanent magnet synchronous motor (SPMSM)[1]. This operation is made possible by the maximum output control, in which the current vector is operated at the optimal operating condition [2]. The IPMSM control method in the high-speed region is dependent on $\Psi_{dmin} (=\Psi_a - L_d I_{am})$ [3], where Ψ_a is the stator flux linkage due to the permanent magnet, L_d is the d-axis inductance, and I_{am} is the armature current limit. If Ψ_{dmin} is negative, maximum torque can be obtained in the high-speed flux-weakening region by applying maximum torque per flux (MTPF) control.

In recent years, the development of the IPMSM, which uses a smaller amount of, or even no, expensive rare-earth magnet, has progressed. In such PMSMs, Ψ_{dmin} becomes negative, and thus MTPF control must be applied at high speeds. The applicability of MTPF control has spread with the use of low-cost magnets which have a weak magnetic force [4]. However, the method of generating the current command value in the MTPF region is complex [5].

The purpose of this study is to develop a novel control method in the flux-weakening region for high-efficiency operation [6]. A straight-line approximation of the MTPF curve is used in order to simplify the control algorithm of the MTPF. In order to achieve high-efficiency operation in the flux-weakening region, the current vector of the IPMSM is operated along a constant-voltage ellipse under the high-speed, light load condition. The validity of the proposed control method is verified based on simulations and experiment.

II. MAXIMUM OUTPUT CONTROL

This section examines the current vector control method used to obtain the maximum output under the limiting voltage and current conditions given as follows:

$$V_o = \omega \sqrt{\left(\Psi_a + L_d i_d\right)^2 + \left(L_q i_q\right)^2} \le V_{om} \tag{1}$$

$$I_a = \sqrt{i_d^2 + i_q^2} \le I_a \tag{2}$$

where V_o is the induced voltage, ω is the electrical rotor angular velocity, i_d and i_q are the armature currents in the d-q frame, L_q is the q-axis inductance, V_{am} is the armature voltage limit, and I_a is the armature current.

Fig. 1 shows the characteristic curves on a current vector plane at 2,000 r/min, where the machine parameters of the tested IPMSM listed in Table I are used. The maximum torque per ampere (MTPA) curve and the MTPF curve indicate the current vector trajectory when MTPA control and MTPF control are applied. In Fig. 1, the constant-current limit circle indicates the armature current limit. The constant voltage ellipse indicates the corresponding voltage limits at a rotor speed of 2,000 r/min. Fig. 2 shows a close-up of the area inside the broken-line rectangle in Fig. 1. Point A is the intersection of the MTPF curve and the constant voltage ellipse. The maximum q-axis current on the constant- voltage ellipse is maximum at Point B. Points C and D are located at the intersection of the constant-torque curve and the characteristic curves.

A. Maximum Torque per Ampere Control

Maximum torque per ampere (MTPA) control is performed at the intersection of the current limit circle and the MTPA curve. This control method is applied in the speed region in which the induced voltage V_o does not exceed the limiting voltage V_{om}. In this case, point I is the operating point generating maximum torque by MTPA control considering the current limit. In this control region, the armature current I_a is equal to the current limit I_{am}, and the induced voltage V_o is less than the voltage limit V_{om}.

B. Flux-weakening Control for Maximizing Torque

The induced voltage increases with increasing rotor speed. When the motor speed reaches the base speed, the induced voltage becomes equal to the voltage limit value. Hence, the induced voltage is controlled to the voltage limit V_{om} by flux-weakening (FW) control above the base speed. In this case, the current vector is controlled along the current limit circle indicated by arrow II. In this control region, the voltage and current are equal to the limiting values.

C. Maximum Torque per Flux Control

In the case of operation along the current limit circle beyond the base speed, the operating limit occurs at the rotor speed at which the operating point is located at the intersection of the d-axis and the current limiting circle. If $\Psi_{dmin} (= \Psi_a - L_d I_{am}) < 0$, the operation of MTPF control can be applied to an IPMSM. Hence, the operating region of the IPMSM can be expanded by operating along the MTPF curve, as indicated by arrow III in Fig. 1. However, the algorithm to determine the operating point of MTPF control is complex.

Hence, this paper proposes a method for generating the current command value using a straight-line approximation of the MTPF curve. A straight-line approximation of the MTPF curve is determined to be an accurate approximation of the linear region of the MTPF curve using the following equation:

$$ i_q = -0.226 i_d - 0.172 \quad (3) $$

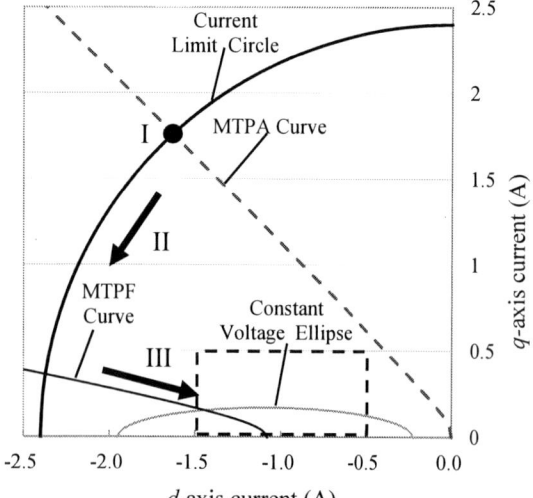

Fig. 1. Characteristic curves on the current vector plane (2,000 r/min).

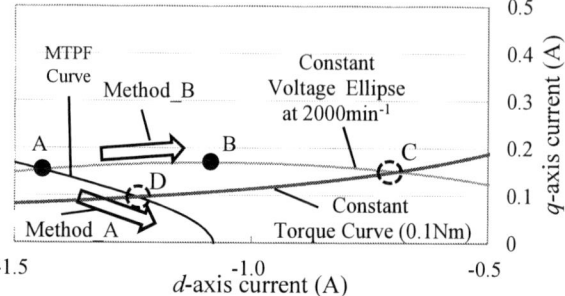

Fig. 2. Close-up of the area indicated by the broken line in Fig. 1.

Table 1. Parameters of the IPMSM and the controller

Item [Unit]	Value
Number of pole pairs	2
Armature flux linkage Ψ_a [Wb]	0.078
d-axis inductance L_d [mH]	72.2
q-axis inductance L_q [mH]	368
Current limit I_{am} [A]	2.4
Voltage limit V_{am} [V]	91.85
Base speed [r/min]	200

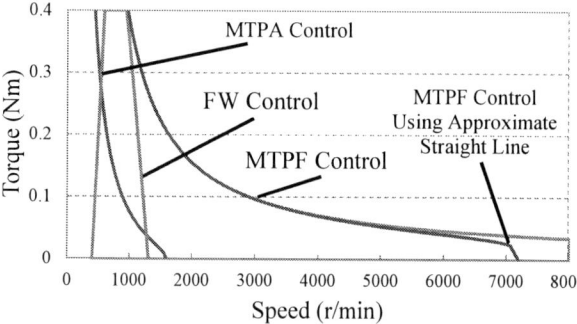

Fig. 3. Torque applying the maximum output control.

Fig. 3 shows the torque trajectory obtained by applying each control method. This figure indicates that operation over a wide speed region is possible by appropriately switching the maximum output control. In addition, the approximation error in the actual operating region is minimized by accurate approximation of the linear region. Therefore MTPF control is possible by applying the proposed method using the straight-line approximation below 5,000 r/min.

III. CONTROL METHOD IN THE FLUX-WEAKENING REGION

A. Current Vector Generation Method in the Flux-weakening Region

In a high-speed region in which the MTPF control should be applied to obtain the maximum output, we discuss the current vector generation method under a light-load.

In the first method, the operation of the current vector along the MTPF approximation inside the constant-voltage ellipse is considered [7]. This method is referred to herein as Method_A. Hence, the operating point is present inside a constant-voltage ellipse, and the induced voltage V_o is less than the limit voltage V_{om}. For example, point C in Fig. 2 corresponds to the operating point at a load torque of 0.1 Nm. In the second method, the operation of current vector along the constant-voltage ellipse is considered [8]. This method is referred to herein as Method_B. In this method, the current vector is operated at the intersection of the constant-voltage ellipse and the constant-torque curve. Point D in Fig. 2 corresponds to the operating point at a load torque of 0.1 Nm. Method_B appears to be more efficient compared to Method_A because the copper loss decreases with decreasing current [9]. Fig. 4 shows the armature current versus torque characteristics under the control of Method_A and Method_B. This figure shows that Method_B can reduce the armature current compared to Method_A. Therefore, Method_B is applied as the current vector generation method in the proposed IPMSM control system.

B. Generation of Current Commands

Fig. 5 (a) shows the general method for generating the *d*- and *q*-axis current commands. In this method, *q*-axis current command i_q* is determined by the speed error $\Delta\omega$. Then, the *d*-axis current command i_d* is determined using the current vector control algorithm. In this conventional method, the current vector moves from point A to point B in Fig. 2 as the load torque decreases. In this case, the *q*-axis current must increase. Fig. 5(b) shows the characteristic of the *q*-axis current variation. Points A, B, and C correspond to operating points A, B, and C in Fig. 2. The MTPF control is switched to operation along the constant-voltage ellipse at point A. Point C is the operating point at a load torque of 0.1 Nm. The *q*-axis current increases with decreasing torque using the conventional method. Therefore, stable operation cannot be achieved by the conventional method. Thus, as shown in Fig. 6(a), in the proposed method, an armature current command I_a* is first generated from $\Delta\omega$, and the current phase angle β* corresponding to the current command I_a* is then determined using (4). Based on I_a* and β*, the *d*- and *q*-axis current commands are determined using (5) and (6). Fig. 6(b) shows the characteristic of the armature current variation. In the operation along the constant-voltage ellipse, the output torque increases with increasing armature current. However, if the armature current is greater than the current at the intersection point of the constant-voltage ellipse and the MTPF curve, the output torque decreases with increasing armature current. Thus, the limitation of the armature current is determined by the intersection point of the constant-voltage ellipse and the MTPF curve.

$$\beta^* = \sin^{-1}\frac{L_d\Psi_a + \sqrt{(L_d\Psi_a)^2 - (L_d^2 - L_q^2)(\Psi_a^2 + (L_q I_a^*)^2 - \left(\frac{V_{om}}{\omega}\right)^2)}}{(L_d^2 - L_q^2)I_a^*} \quad (4)$$

$$i_q^* = I_a^*\cos\beta^* \quad (5)$$

$$i_d^* = -I_a^*\sin\beta^* \quad (6)$$

IV. ANALYSIS RESULTS

The analysis was performed using the parameters listed in Table I. The straight-line approximation of the

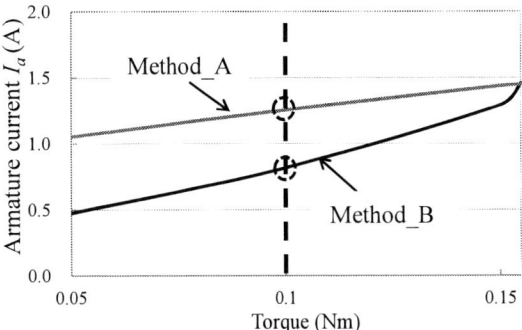

Fig. 4. Comparison of the armature currents corresponding to Method_A and Method_B.

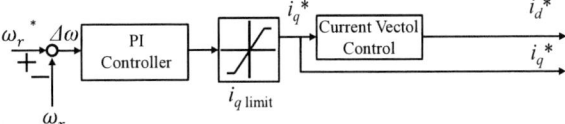

(a) Generation of current commands based on *q*-axis current command .

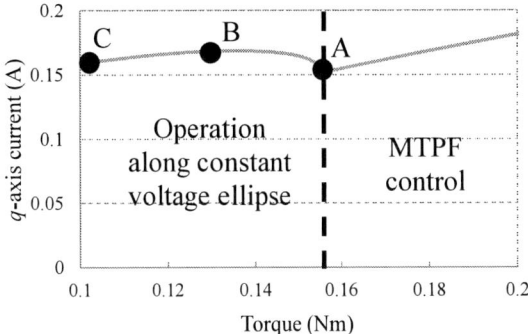

(b) *q*-axis current vs. torque characteristics (2,000 r/min).
Fig. 5. Operation along the constant voltage ellipse (Conventional method).

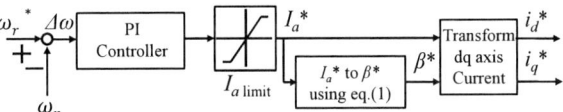

(a) Generation of current commands based on armature current command.

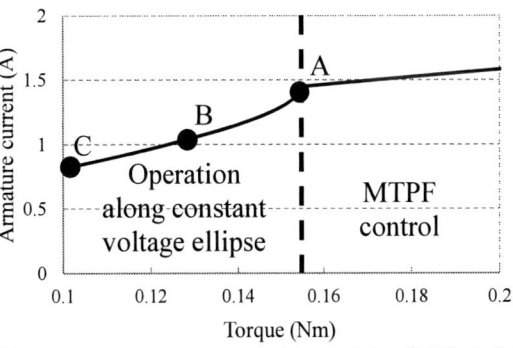

(b) Armature current vs. torque characteristics (2,000 r/min).
Fig. 6. Operation along constant voltage ellipse (Proposed method).

MTPF curve is determined to be an accurate approximation of the linear region of the MTPF curve. The analysis results show the acceleration characteristics, where rotor speed changed from 100 to 2,000 r/min under a light-load of 0.1 Nm. Fig. 7 shows the responses of the rotor speed, output torque and armature current controlled by Method_A and Method_B. The rotor speed and output torque of Method_A and Method_B are approximately the same. However, the armature current of Method_B is less than that of Method_A after the speed reaches the commanded speed. Fig. 8 shows the current vector trajectory under the operation shown in Fig. 7. Fig. 8(a) shows that the operation along the straight-line approximation of the MTPF curve is possible using the Method_A. Fig. 8(b) shows that operation along the constant-voltage ellipse is possible using the Method_B.

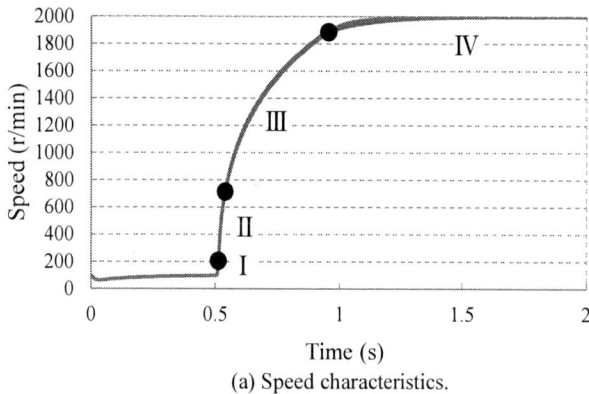

(a) Speed characteristics.

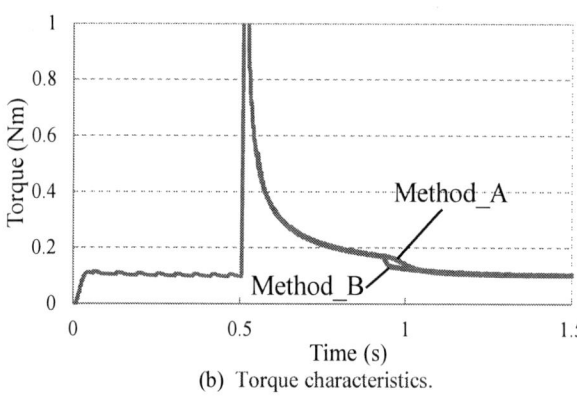

(b) Torque characteristics.

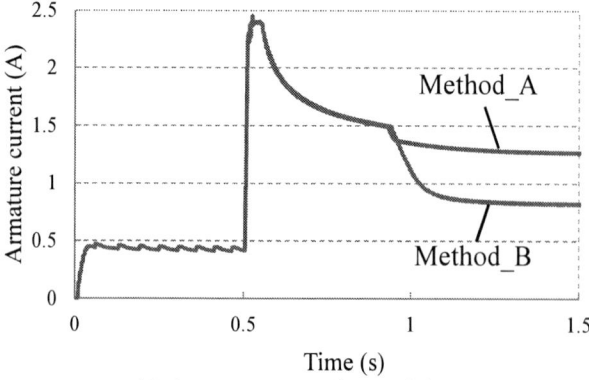

(c) Armature current characteristics.
Fig. 7. Characteristics of Method_A and Method_B.

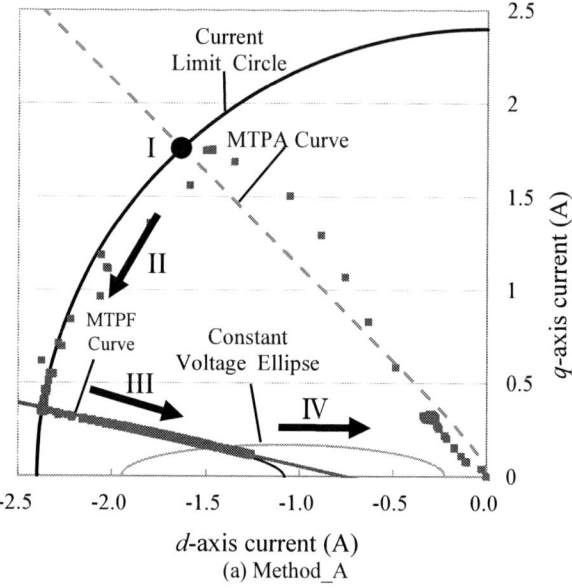

(a) Method_A

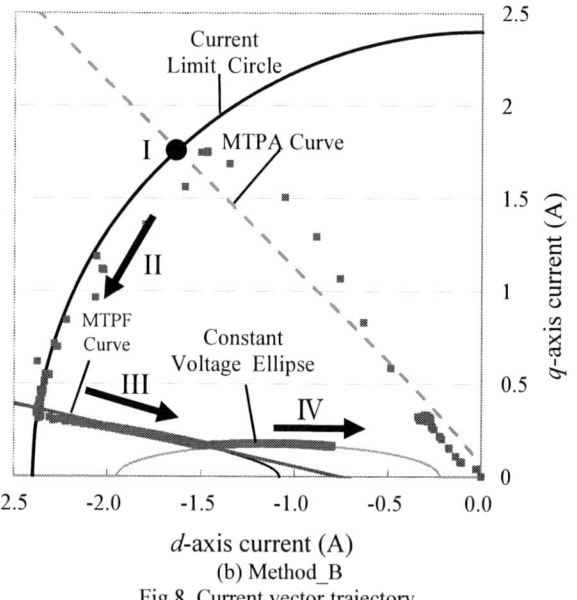

(b) Method_B
Fig.8. Current vector trajectory.

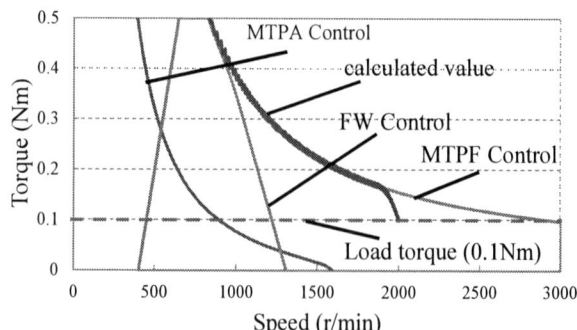

Fig. 9. Torque corresponding to the rotor speed.

The 2014 International Power Electronics Conference

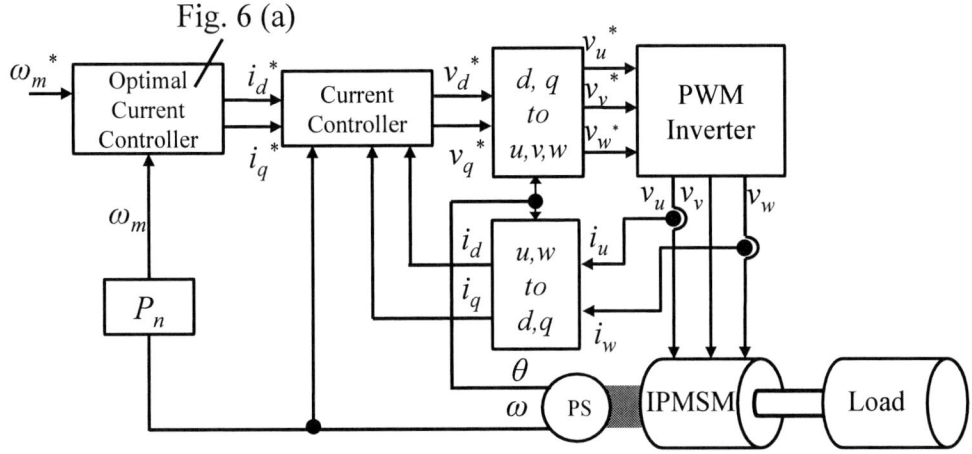

Fig. 6 (a)

Fig. 10. IPMSM drive system based on current vector control.

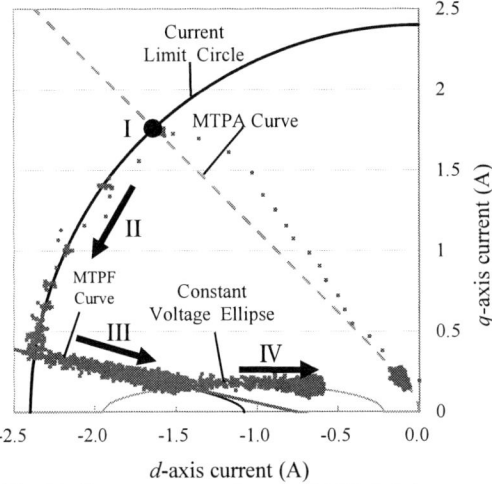

Fig. 11. Current vector trajectory (2,000 r/min).

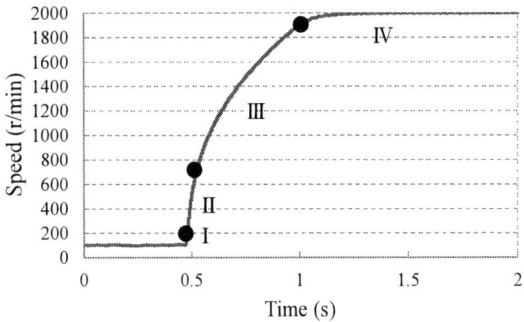

Fig. 12. Speed characteristics.

Fig. 9 shows the calculated torque under the operation conditions shown in Fig. 7. Fig. 9 also shows the theoretical value for each control method. The output torque was not observed to differ between the calculated value and the theoretical value in the region which is used in actual operation. Therefore, the effect of the approximation error is minimized.

Based on the analysis results, the following results were obtained. The MTPF control is possible using the straight-line approximation of the MTPF curve. Operation using Method_B is more efficient compared to that using Method_A under a light load.

V. EXPERIMENTAL RESULT

The experiment was carried out under same conditions as the analysis. The purpose of the experiment was to realize efficient operation along the constant-voltage ellipse and MTPF control in the high-speed region. Fig. 10 shows the IPMSM drive system used in the experiment. The experimental results reveal the acceleration characteristics, where the rotor speed

changed from 100 to 2,000 r/min under a light-load of 0.1 Nm. Fig. 11 shows the current vector trajectory. The three control methods are confirmed to switch properly with increasing rotor speed. Fig. 12 shows that desirable speed characteristics are realized. The results of the experiment and analysis were confirmed to be similar.

VI. CONCLUTIONS

A novel control method for use in the flux-weakening region for operation at optimal torque point has been proposed. The simulation and experimental results confirmed the possibility of expanding the operation region by applying the simplified control method using approximate MTPF control algorithm. In addition, efficient operation is possible using Method_B. Thus, wide, efficient operation is possible by combining the proposed methods.

References

[1] S. Morimoto, M. Sanada, and Y. Takeda, "Performance of PM-Assisted Synchronous Reluctance Motor for High-Efficiency and Wide Constant-Power Operation," *IEEE Transactions on Industry Applications*, vol.37, no.5, pp.1234-1240, 2001.

[2] M. Michael, B. Joachim, "Optimum Control for Interior Permanent Magnet Synchronous Motors (IPMSM) in Constant Torque and Flux-weakening Range," *IEEE Power Electronics and Motion Control Conference*, pp.282-286, 2006.

[3] H. Liu, Z. Q. Zhu, "Flux-weakening Control of Nonsalient Pole PMSM Having Large Winding Inductance, Accounting for Resistive Voltage Drop and Inverter Nonlinearities, *IEEE*

Transactions on Power Electronics, vol. 27, no. 2, pp. 942-952 2012.

[4] T. Kato, N, Limsuwan, "Rare earth reduction using a novel variable magnetomotive force, flux intensified IPM machine," *IEEE Energy Conversion Congress and Exposition*, pp. 4346 - 4353,15-20, 2012.

[5] P. Vaclavek, and P. Blaha, "Interior Permanent Magnet Synchronous Machine High Speed Operation using Field Weakening Control Strategy," *12th WSEAS International Conference on SYSTEMS*, Heraklion, Greece, pp. 581-586, 2008.

[6] S. Chi, L. Xu, "Efficiency-Optimized Flux-weakening Control of PMSM Incorporating Speed Regulation," *IEEE Power Electronics Specialists Conference*, pp. 1627-1633, 2007.

[7] G. Pellegrino, "Direct-Flux Vector Control of IPM Motor Drives in the Maximum Torque Per Voltage Speed Range," *IEEE Transactions on Industrial Electronics*, vol. 59, no. 10, 2012.

[8] G. G. López, S. Gunawan, and E. Walters "Optimum Torque Control of Permanent-Magnet AC Machines in the Field-Weakened Region," *IEEE Transactions on Industry Applications*, vol. 41, no.4, pp.1020-1028, 2005.

[9] J. Chen, K. Chin, "Minimum Copper Loss Flux-weakening Control of Surface Mounted Permanent Magnet Synchronous Motors," *IEEE Transactions on Power Electronics*, vol.18, no.4, pp. 929-936, 2003.

Implementation of the MTPA and MTPV control with online parameter identification for a high speed IPMSM used as traction drive

Quoc Khanh Nguyen, Matthias Petrich, Jörg Roth-Stielow
Institute of Power Electronics and Electrical Drives
University of Stuttgart
Stuttgart, Germany
nguyen@ilea.uni-stuttgart.de

Abstract—In this paper, the problem of how to implement the MTPA/MTPV control for an energy efficient operation of a high speed Interior Permanent Magnet Synchronous Motor (IPMSM) used as traction drive is considered. This control method depends on the inductances L_d, L_q, the flux linkage Ψ_{PM} and the stator resistance R_s which might vary during operation. The parameter variation causes miscalculation of the set point currents I_d and I_q for the inner current control system and thus a wrong torque will be set. Consequently the IPMSM will not be operating in the optimal operation point which yields to a reduction of the total energy efficiency and the performance. As a consequence, this paper proposes the implementation of the the Recursive Least Square Estimation (RLS) for a high speed and high performance IPMSM. With this online identification method the variable parameters are estimated and adapted to the MTPA and MTPV control strategy.

Keywords—MTPA/MTPV control, online parameter identification, synchronous motor

I. Introduction

Due to the high power density the IPMSMs are widely used as traction drives for electrical vehicles (EV). In order to achieve high compactness and to abstain from a multi-stage gearbox an electrical traction drive usually has a wide speed range (up to 12,000 rpm) to cover the high driving speed of the EV. In case of IPMSM the high speed can be achieved by operating in the field weakening region. To realize an operation with optimal energy efficiency and high performance the field oriented current control (FOC) with maximal torque per ampere (MTPA) in base speed region and maximal torque per voltage (MTPV) in the field weakening region must be adapted. However, the calculation of this control strategy depends on the parameters of the machine which might vary unpredictably during the operation. In the case of an IPMSM designed for high power and high speed this parameter variation is especially distinct.

To solve this problem the parameters must be adaptively tuned during the operation for the correct calculation of the MTPA and MTPV control strategy. The estimation of the parameters can either be realized offline or online [1]. In the offline method the parameters are measured on a test bench and the measurement results are stored in look up tables. The parameters of IPMSMs produced in line production can vary due to the complex manufacturing process. This fact is the disadvantage of the offline method.

On the contrary to the offline measurement, the online parameter identification considers the parameter variation of line produced IPMSMs. In [1] and [2] the recursive least square (RLS) method is proposed for the identification of the parameters of a IPMSM. The key advantages of this method in comparison with others are simple implementation and constant computing time [1], [3]. In [1] and [2] this online identification method is extensively investigated and implemented for low speed and low power IPMSMs. In this paper we investigate how it can be extended to apply on a high speed and high power IPMSM used as traction drive for EVs.

II. Model and Control of the IPMSM

The IPMSM can be modeled in the rotor reference frame with

$$U_d = R_s \cdot I_d + L_d \cdot \frac{d}{dt} I_d - \Omega \cdot L_q \cdot I_q \tag{1}$$

$$U_q = R_s \cdot I_q + L_q \cdot \frac{d}{dt} I_q + \Omega \cdot L_d \cdot I_d + \Omega \cdot \Psi_{PM} \tag{2}$$

$$T_e = \frac{3}{2} \cdot zp \cdot I_q \cdot \left((L_d - L_q) \cdot I_d + \Psi_{PM} \right) \tag{3}$$

where U_d and U_q are the d- and q-axis voltages, I_d and I_q are the d- and q-axis currents, L_d and L_q are the d- and q-axis inductances. R_s represents the stator resistance, Ψ_{PM} is the flux linkage of the permanent magnet, T_e is the electromagnetic torque, zp is the number of the pole pair and Ω is the electrical angular speed of the machine. Fig. 1 shows the structure of the control system wherein the stator currents I_d and I_q are controlled in an inner loop (FO Current Controller). The set point currents I_d^{Set} and I_q^{Set} are calculated with the MTPA/MTPV control strategy in the outer loop. The input of this control strategy is the set point torque given by the driver via the throttle pedal of the EV. The MTPA strategy is adapted in the base speed range of the machine. In this range the current is limited. The most energy efficient set point value of the current I_d for a given torque T_e can be estimated by solving the equation (4) numerically,

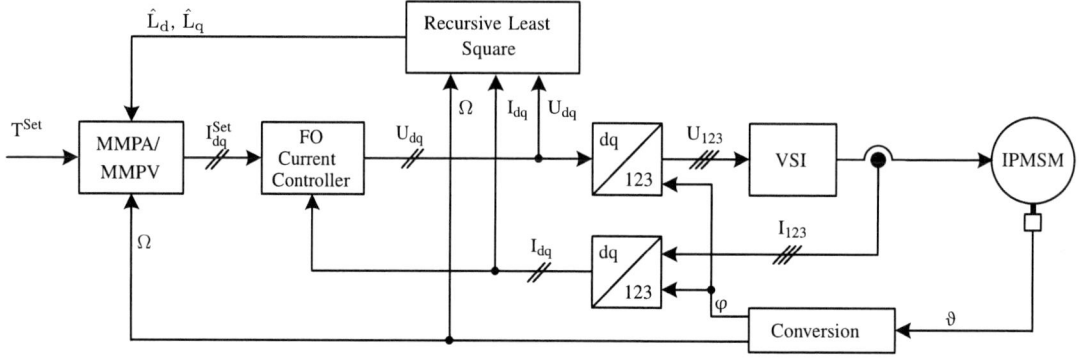

Fig. 1. Structure of the control system

compare [4].

$$A^2 I_d^4 + 3ABI_d^3 + 3B^2 I_d^2 + \frac{B^3}{A} I_d = T_e^2 \quad (4)$$

with

$$A = \frac{3}{2} \cdot zp \cdot (L_d - L_q) \quad (5)$$

$$B = \frac{3}{2} \cdot zp \cdot \Psi_{PM} \quad (6)$$

The set point value of the current I_q can then be estimated with

$$I_q = \sqrt{\frac{I_d \cdot \Psi_{PM}}{L_d - L_q} + I_d^2} \quad (7)$$

In the flux weakening the voltage constraint has to be considered with increasing speed in addition to the current constraint. The MTPV strategy is adapted and the set point of the current I_d can be calculated by numerically solving the equation

$$2 \cdot L_d \cdot I_d^2 + C \cdot I_d + D = 0 \quad (8)$$

with

$$C = 3 \cdot L_d \cdot \Psi_{PM} + \frac{L_d^2}{L_d - L_q} \cdot \Psi_{PM} \quad (9)$$

and

$$D = \Psi_{PM}^2 \cdot \left(\frac{L_d}{L_d - L_q} + 1 \right) - \left(\frac{U_{max}}{\Omega} \right)^2 \quad (10)$$

The set point value of the current I_q can then be estimated with

$$I_q = \sqrt{\frac{U_{max}^2}{\Omega^2 \cdot L_q^2} - \left(\frac{L_d \cdot I_d + \Psi_{PM}}{L_q^2} \right)^2} \quad (11)$$

The equations (4) - (11) shows that the effectiveness of the MTPA and MTPV strategy depends on the accuracy of the machine parameters R_s, L_d, L_q and Ψ_{PM}.

TABLE I.   MACHINE PARAMETERS

Symbol	Meaning	Value
U_{DC}	DC Link Voltage	350 V
I_N / I_{max}	Rated/ Peak Current	200 A/480 A
n_N	Rated Speed	4,000 rpm
n_{max}	Maximum Speed	12,000 rpm
P_N / P_{max}	Rated/ Peak Power	56 kW/123 kW
T_N / T_{max}	Rated/ Peak Torque	135 Nm/290 Nm
$R_s^{23°C}$	Stator Resistance measured at 23° C	13 mΩ
L_{dN}	d-Axis Stator Inductance	90 μH
L_{qN}	d-Axis Stator Inductance	410 μH
Ψ_{PM}	Rotor Flux Linkage	5.8 mVs
zp	Number of Pole Pair	4

III.   PARAMETER VARIATION

A. Temperature-sensitive parameters

The data and the parameters of the machine investigated in this work are summarized in Table I. Herein the stator resistance is given for a temperature of 20 °C. This parameter is generally temperature-sensitive and can be described by

$$R_s(\vartheta) = R_s^{23°C} \cdot [1 + \alpha \cdot (\vartheta - 23°C)] \quad (12)$$

with $\alpha = 3.9 \times 10^{-3} 1/K$ as the temperature coefficient of the resistance for copper. At the maximum operating temperature $\vartheta = 100°C$ of the windings the stator resistance R_s can reach 16.9 mΩ. This means that the stator resistance has a maximum deviation of 3.9 mΩ. For the calculation of the MTPA and MTPV control strategy the resistive voltage drop is relevant. With the peak current of 480 A the deviation of the voltage drop on the stator resistance will be approximately 2 V. In comparison to the motor voltage this voltage deviation is very small and can be neglected. Hence the parameter R_s is assumed to be constant in this work.

The second temperature-sensitive parameter is the flux linkage Ψ_{PM} induced by the permanent magnet. The investigated machine is equipped with neodymium magnets. The flux temperature coefficient of this magnet material is about -0.1 %/K. This means in worst case of 60 °C operating temperature of the rotor, Ψ_{PM} decreases by 3.7 %. In comparison with the derivation caused by the

The 2014 International Power Electronics Conference

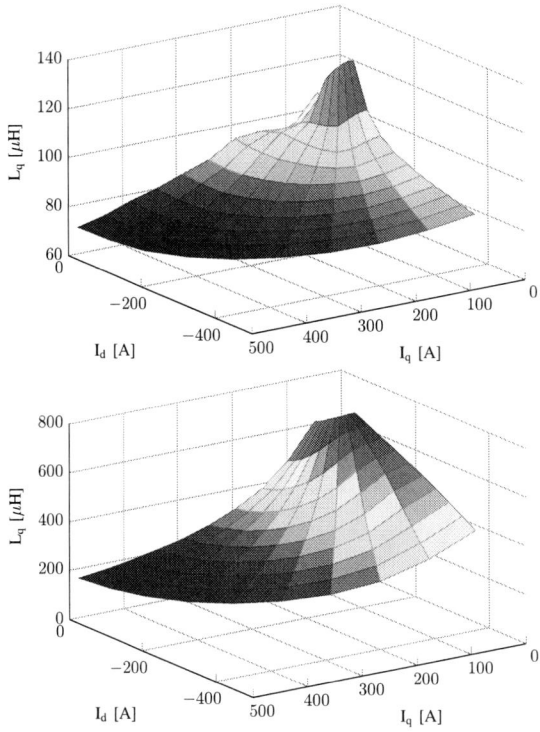

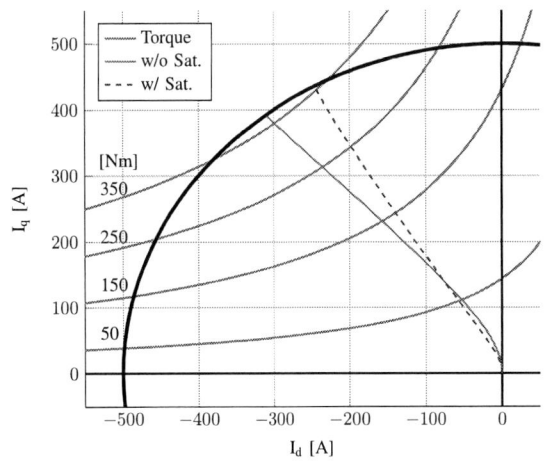

Fig. 3. Comparison of the current locus simulated with and without saturation at speed 2000 rpm

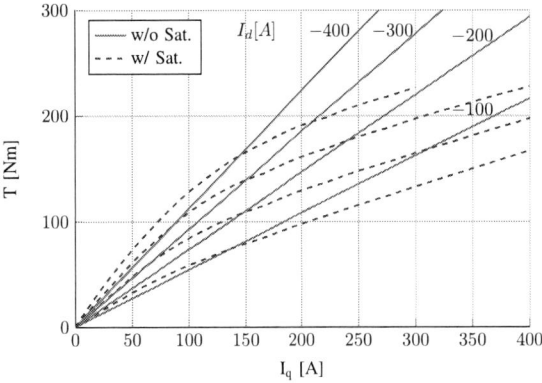

Fig. 4. Comparison of motor torque depending on current simulated with and without saturation

Fig. 2. Saturated inductances L_d and L_q depending on current

flux saturation described in the following the derivation of Ψ_{PM} is small and therefore can also be neglected.

B. Flux saturation

In general the stator flux linkage of the IPMSM can be described by

$$\Psi_d = L_d(I_d, I_q) \cdot I_d + \Psi_{PM} \tag{13}$$

$$\Psi_q = L_q(I_d, I_q) \cdot I_q \tag{14}$$

The functions $L_d(I_d, I_q)$ and $L_q(I_d, I_q)$ also contain the cross coupling inductances L_{dq} and L_{qd}. In case of saturation the flux linkages change nonlinear with the current. Hence the inductances L_d and L_q which are defined by

$$L_d = \frac{\partial \Psi_d}{\partial I_d}\Big|_{I_q=const.} \tag{15}$$

$$L_q = \frac{\partial \Psi_q}{\partial I_q}\Big|_{I_d=const.} \tag{16}$$

are not constant and the variation of these parameters have to be involved in the calculation of the MTPA and the MTPV control. In Fig. 2 the offline measured inductances L_d and L_q of the machine used in this paper are plotted depending on the stator currents I_d and I_q. Herein the inductance L_d is reduced by about 50 % and L_q even by 70 % in comparison to the no-load and therefore no current condition.

The effect of the flux saturation is shown in Fig. 3 and Fig. 4. In Fig. 3 the optimal current locus calculated with the MTPA control is simulated on two conditions: flux

saturation considered (dashed blue) and no flux saturation considered (red). Due to the saturation effect both current loci differ, especially at the high toque range. Fig. 4 shows the motor torque as function of the stator currents I_d and I_q. If the flux saturation is not considered (red) the motor torque changes linearly with the current I_q when I_d is constant. On the saturated condition (dashed blue) the motor torque is not linear with the current I_q. In comparison with the linear case it even reduces. Thus the flux saturation causes the miscalculation of the currents if it is not involved in the MTPA/MTPV control and this leads to an inefficient operation of the IPMSM.

IV. Online Parameter Identification

As described above the variations of the parameters R_s and Ψ_{PM} are small, so they are neglected. Therefore only the flux saturation is taken into account in the calculation of the MTPA/MTPV control strategy. In [1] and [2] the recursive least square method was proposed for such calculation. For every computing step k the output vector $\underline{y}$ and the feedback matrix $\underline{X}$ are given from measurements. The estimated parameter vector $\hat{\underline{\Theta}}$

978-1-4799-2706-7/14 $31.00 © 2014 IEEE   320

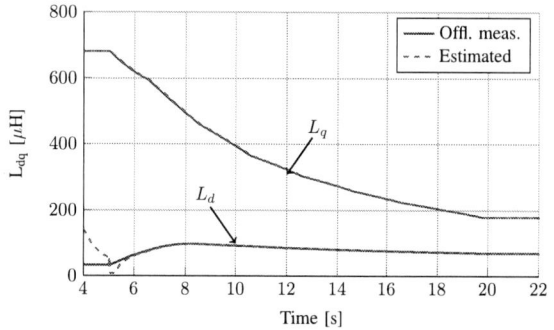

Fig. 5. Comparison of measured and estimated inductances

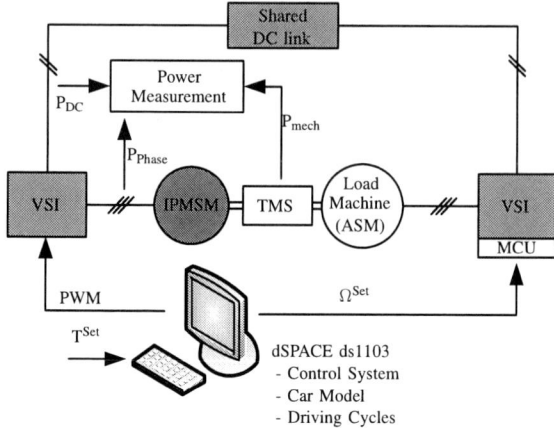

Fig. 6. Overview of the test bench

can then be online calculated with

$$\hat{\underline{\Theta}}[k] = \hat{\underline{\Theta}}[k-1] + \underline{K}[k] \cdot \underline{\varepsilon}[k] \qquad (17)$$

herein the estimation error is given by

$$\underline{\varepsilon}[k] = \underline{y}[k] - \underline{\underline{X}}[k] \cdot \hat{\underline{\Theta}}[k-1] \qquad (18)$$

The correction gain matrix $\underline{\underline{K}}$ is defined by

$$\underline{\underline{K}}[k] = \frac{\underline{\underline{P}}[k-1] \cdot \underline{\underline{X}}^T[k]}{\lambda \cdot \underline{\underline{I}} + \underline{\underline{X}}[k] \cdot \underline{\underline{P}}[k-1] \cdot \underline{\underline{X}}^T[k]} \qquad (19)$$

with

$$\underline{\underline{P}}[k] = \left(\lambda \cdot \underline{\underline{I}}\right)^{-1} \cdot \left[\underline{\underline{P}}[k-1] - \underline{\underline{K}}[k] \cdot \underline{\underline{X}}[k] \cdot \underline{\underline{P}}[k-1]\right] \qquad (20)$$

The parameter $\lambda \in [0,1]$ is the forgetting factor which weights the measurement from the time step $[k-j]$ with λ^j [3].

Applied to the IPMSM the output vector $\underline{y}$, the feedback matrix $\underline{\underline{X}}$ and the parameter vector $\underline{\Theta}$ are given with

$$\underbrace{\begin{bmatrix} U_d - R_s \cdot I_d \\ U_q - R_s \cdot I_q - \Omega \cdot \Psi_{PM} \end{bmatrix}}_{\underline{y}} = \underbrace{\begin{bmatrix} -\Omega \cdot I_q & 0 \\ 0 & \Omega \cdot I_d \end{bmatrix}}_{\underline{\underline{X}}} \cdot \underbrace{\begin{bmatrix} L_q \\ L_d \end{bmatrix}}_{\underline{\Theta}} \qquad (21)$$

V. SIMULATION RESULTS

In this section, simulation results of the control system and the RLS identification using the parameters of the machine investigated in this paper are reported. The results of the RLS identification are shown in Fig. 5 wherein the online identified inductances L_d and L_q are compared with the offline measured inductances. In this simulation the speed of the IPMSM is set to 2,000 rpm and the torque ramps up to the maximum value starting at the time 5 s. Due to the increase of the currents with the torque the machine becomes saturated and the inductances L_d and L_q vary strongly. As shown in the Fig. 5 the both online identified inductances converge toward the offline measured values.

VI. EXPERIMENTAL RESULTS

In order to verify the results, the RLS identification was implemented on the real machine and tested on the test bench. Fig. 6 shows the overview of the test bench. Herein the control system with the MTPA/MTPV control strategy and the RLS identification were implemented on a dSPACE ds1103 rapid prototyping system. The speed control of the asynchronous load machine is realized on a micro-controller unit (MCU). With a torque measuring shaft (TMS) the actual torque was measured.

In order to verify the result concerning the energy efficiency, a car model and a driving cycle were implemented on the ds1103 system. With these additional implementation the set point values for the speed Ω^{Set} and the torque T^{Set} are calculated and given to the control system of the IPMSM and the load machine on the test bench. This structure (Fig. 9) allows to test the IPMSM's performance in compliance on the condition of a real car. For the energy investigation the electrical input and the mechanical output energy are introduced:

$$E_{el} = \int_{T_C} P_{el} dt \qquad (22)$$

$$E_{mech} = \int_{T_C} P_{mech} dt. \qquad (23)$$

Herein the electrical power P_{el} and the mechanical power P_{mech} are measured with the power analyzer (Fig. 6). The integration time T_C is the duration of the used driving cycle. For comparison the energy efficiency is defined:

$$\varepsilon = \frac{E_{mech}}{E_{el}} \qquad (24)$$

In comparison with the power efficiency η, this parameter takes into account the dynamic behaviors and the start-stop-operation of the traction drive and it is therefore more appropriate for this application. The disadvantage is its dependence on the used driving cycle [5]. In the first measurement the effectiveness and the robustness of the RLS identification are shown in Fig. 7. Herein the speed

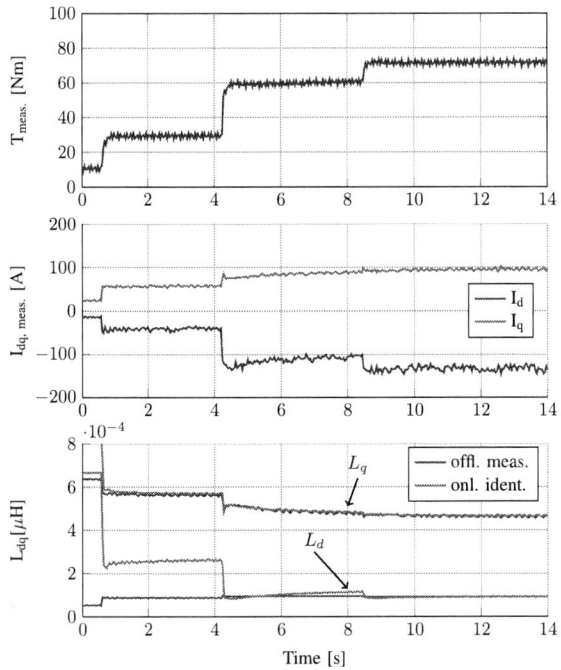

Fig. 7. Result of the online-identification for various torque steps at 4,000 rpm

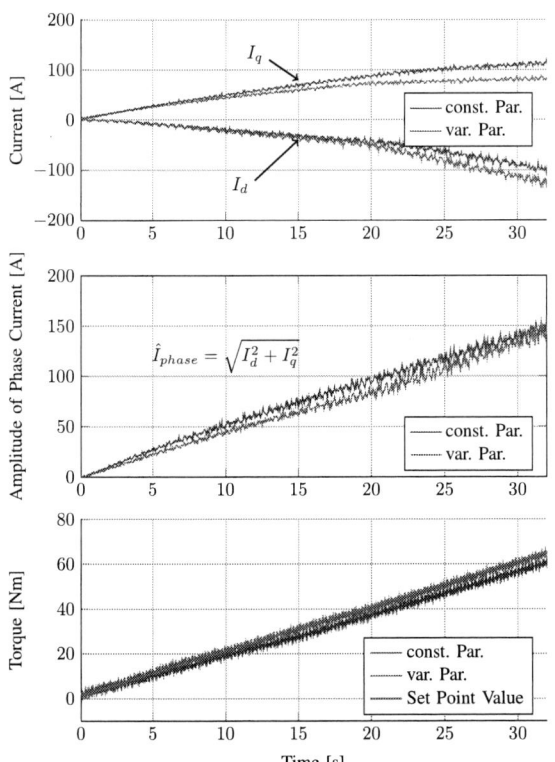

Fig. 8. Comparison of measured torque when operated with constant and variable parameters L_d and L_q

of the IPMSM is set to 4,000 rpm and the torque steps from 10 Nm to 30 Nm, from 30 Nm to 60 Nm and from 60 Nm to 70 Nm (top diagram). The resulting currents I_d and I_q are also plotted (middle diagram). It is clearly that the MTPA control sets the current I_d to the values of about -30 A at 30 Nm and -120 A at 70 Nm. Due to the relation between the torque T_e and the currents I_d, I_q from eq. (3), this control method aims to use the reluctance torque effectively with the negative values of I_d. It should be noticed that the IPMSM is operated in the base speed region in this measurement and only the MTPA control is active.

The bottom diagram of Fig. 7 compares the offline measured and the online identified inductances L_d and L_q. In this measurement it discovered that L_d and L_q can be estimated accurately if the absolute value of the I_d and I_q, respectively, is higher than 50 A. If the absolute value of I_d or I_q is smaller than 50 A, L_d or L_q will be estimated incorrectly. The cause for this inaccuracy lies in the calculation of the RLS algorithm where I_d and I_q occurs in the denominator of a fraction. So every inaccuracy (e.g. measuring inaccuracy) in the small current range leads to the wrong value of the estimated inductance. To overcome this drawback the rated constant values of L_d and L_q from table I are chosen in the calculation of the MTPA/MTPV control for the small current range. For the current range above 50 A the controller switchs over to the result of the RLS identification and the estimated L_d and L_q are online adapted. This does not have a negative effect on the result of the MTPA/MTPV control because in the small current range the IPMSM is not saturated and therefore the influence on the energy efficiency is negligible.

The second measurement is shown in Fig. 8 where the torque is ramped up from 0 Nm to 70 Nm at the constant speed of 4,000 rpm. The resulting currents I_d and I_q are plotted in the top diagram for the two cases: the parameters L_d, L_q are assumed as constant (blue) and are online variably adapted with the estimated values (green) in the MTPA/MTPV control. In the middle diagram the amplitude of the phase current $\hat{I}_{phase}$ is shown and in the bottom diagram the set value of the torque (red) and the measured torque are plotted, also for the aforementioned two cases. It is obvious that in the first case the actual torque deviates from the set point value (blue vs. red) and in the second case the measured torque converges toward the set point value correctly (green vs. red). Furthermore with variable parameters L_d and L_q, the combination of currents I_d and I_q is chosen more optimally and hence the amplitude of the phase current $\hat{I}_{phase}$ is smaller in most cases.

In the third measurement the car model was used and parametrized for the electric smart (smart ed). As driving cycle for the measurement the low and the medium speed parts from the *Worldwide harmonized Light vehicles Test Procedures* (WLTP, Class 3) are chosen. The speed diagram of this driving cycle is plotted in Fig. 10 (top). In the middle diagram the motor torque which is calculated with the car model and set by the motor control is shown. In the

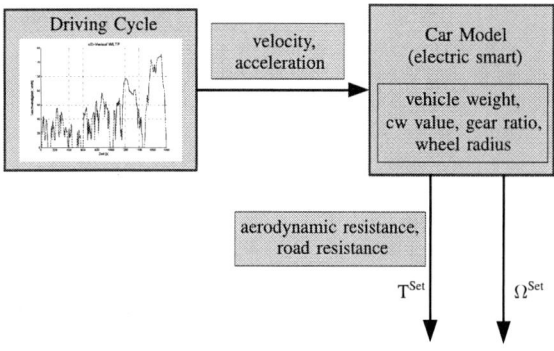

Fig. 9.   Generating of Ω^{Set} and T^{Set} from a driving cycle and a car model

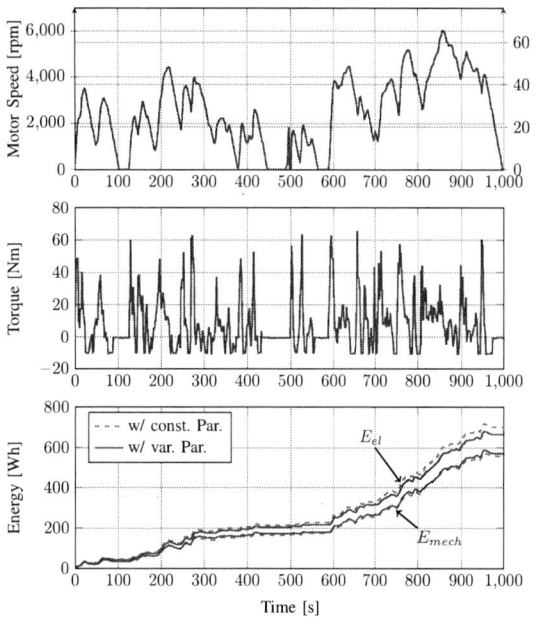

Fig. 10.   Energy Measurement

TABLE II.   ENERGY AND ENERGY EFFICIENCY

	Oper. w/ const. Parameters	Oper. w/ var. Parameters
E_{el}	700 Wh	665 Wh
E_{mech}	556 Wh	569 Wh
ε	79 %	85 %

For the online estimation of L_d and L_q the recursive least square method has been implemented and experimental measurement results showed that these two parameters can be accurately estimated online if the IPMSM is under load. With the implemented MTPA/MTPV in combination with the RLS method the desired torque is correctly calculated and set. Furthermore the proposed control improves the energy consumption of the IPMSM. Measurements have shown that using the correct values for the inductances L_d and L_q, under the chosen test condition, leads to an improvement of the efficiency ε by around 6 %.

REFERENCES

[1] S. J. Underwood and I. Husain, "Online parameter estimation and adaptive control of permanent-magnet synchronous machines." *IEEE Transactions on Industrial Electronics*, vol. 57, no. 7, pp. 2435–2443, 2010.

[2] N. Y. Senjyu T., "Parameter identification for interior permanent-magnet synchronous motor," *Proceeding of International Conference on Electrical Machines and Systems*, 2007.

[3] D. Schröder, *Intelligente Verfahren: Identifikation Und Regelung Nichtlinearer Systeme.* Springer Verlag Heidelberg Berlin, 2010.

[4] D. Schröder, *Elektrische Antriebe - Regelung von Antriebssystemen.* Springer Verlag Heidelberg Berlin, 2009.

[5] F. Blank and J. Roth-Stielow, "Bewertungsmethode für die energieeffizienz eines elektrischen antriebssystems," *SPS/IPC/Drives 2010 Tagungsband*, pp. 321–329, 2010.

bottom diagram of Fig. 10 the electrical input energy E_{el} and the mechanical output energy E_{mech} of the IPMSM are plotted for the two operating conditions: Parameters L_d and L_q are set constant (dashed, red) and parameters L_d and L_q changes adaptively (blue). From this diagram it is cognizable that the operation with variable L_d and L_q improves the energy consumption of the IPMSM. This result is obvious by comparing the absolute values of the end energies end the efficiencies in table II.

VII.   CONCLUSION

To achieve the high energy efficiency of IPMSMs used as traction drives for EVs the MTPA/MTPV control should be applied in combination with the field oriented current control. In this paper this control strategy has been implemented for such a high speed IPMSM. It has been shown that the magnetic saturation of the machine plays an important role and so the variation of the inductances L_d and L_q have been considered in the control strategy.

Correction of Reference Flux for MTPA Control in Direct Torque Controlled Interior Permanent Magnet Synchronous Motor Drives

Atsushi Shinohara, Yukinori Inoue, Shigeo Morimoto, Masayuki Sanada

Graduate School of Engineering
Osaka Prefecture University
Sakai, Japan
ss106030@edu.osakafu-u.ac.jp

Abstract—This paper proposes a maximum torque per ampere control strategy for direct torque controlled interior permanent magnet synchronous motor drives. The feature of the proposed method is the ability to search the optimal current directly and successively. Therefore, variations of the parameters can be considered. The simulation and experimental results validate the proposed method.

Keywords—permanent magnet synchronous motors, direct torque control, maximum torque per ampere control, local search algorithm.

I. Introduction

Recently, permanent magnet synchronous motors (PMSMs) have been applied in many applications due to their many attractive characteristics, such as their high power and efficiency and their lack of need for maintenance. Consequently, many researchers have investigated control methods for PMSM drives for high-performance control. Generally, *d*- and *q*-axis current control is applied for high-performance motor drives. Many control laws for high-performance drives are based on a mathematical model in the *d-q* rotating frame that is synchronized with the rotor.

A maximum torque per ampere (MTPA) control method is proposed as one of the current vector control methods for high-performance drives [1],[2]. The purpose of MTPA control is to minimize copper loss. MTPA control is applied in the low-speed range because the back electromotive force is small, and the voltage limitation can be neglected. Ideally, the MTPA condition is independent of speed [1]. In the case of interior PMSMs (IPMSMs), MTPA control allows the exploitation of both magnet torque and reluctance torque under a constant current amplitude. However, *d*- and *q*-axis inductances which are required for MTPA control vary due to magnetic saturation. In current-control-based IPMSM drives, some strategies can obtain the MTPA condition without knowledge of the motor parameters in advance. These methods are based on parameter estimation during operation [3],[4].

Direct torque control (DTC) was first proposed for induction motor drives [5]. DTC can be applied to any AC motor [6]–[8], including PMSMs. When compared to current control, DTC possesses advantages such as less parameter dependency, no requirement for rotor position, and fast torque response. DTC controls the stator flux linkage vector and requires two references: torque and stator flux linkage magnitude. DTC has no current control loops, and the current is not regulated directly [8]. Hence, it is difficult to control the current in DTC drives.

In DTC systems, the reference stator flux linkage for MTPA control is obtained using a look-up table (LUT) because the relationship between the torque and stator flux linkage is more complex than *d*- and *q*-axis currents, which are applied as the references in current-control-based motor drives. The relationship is influenced by the *d*- and *q*-axis inductances and the magnet flux linkage. Therefore, the components of LUT for MTPA control have dependency on the motor parameters, and therefore, it cannot allow mismatches of the parameters. Moreover, the estimation error of the stator flux linkage affects the optimal reference stator flux linkage value for MTPA control. The error of the stator flux linkage almost depends on the rotational speed of the stator flux linkage vector [9],[10]. Hence, the error of the stator flux linkage depends on the rotor speed, and the optimal reference stator flux linkage value for MTPA control also has speed dependency.

This paper proposes an MTPA control strategy in direct torque controlled IPMSM drive systems. The proposed method is based on the sequential variation of the actual current to an optimal point by correcting the reference stator flux linkage with a local search algorithm in order to minimize the current. It can adapt the reference stator flux linkage to inductance variation and stator flux linkage estimation error. The simulation results presented in this paper show that the proposed method can obtain the MTPA condition under inductance variation and flux linkage estimation error. The experimental results validate the utility of the proposed method.

II. Direct Torque Control System

The equations of the IPMSM in the stator reference frame are given by (1) and (2),

$$\begin{bmatrix} v_\alpha \\ v_\beta \end{bmatrix} = R_a \begin{bmatrix} i_\alpha \\ i_\beta \end{bmatrix} + \frac{d}{dt}\begin{bmatrix} \psi_\alpha \\ \psi_\beta \end{bmatrix} \qquad (1)$$

$$T_e = P_n(\psi_\alpha i_\beta - \psi_\beta i_\alpha) \qquad (2)$$

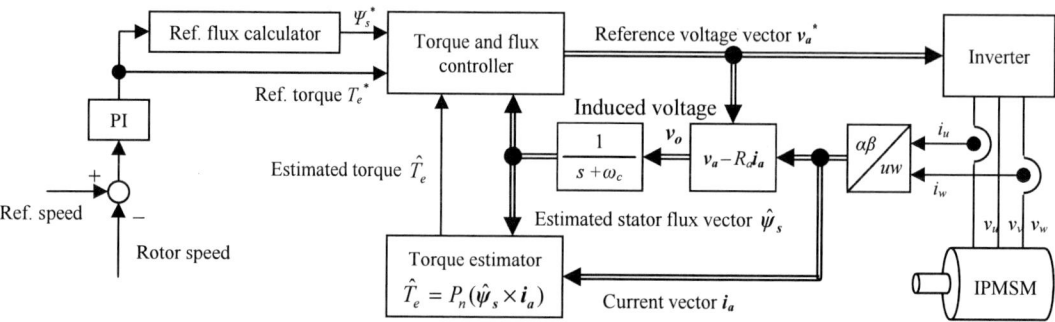

Fig. 1. Direct torque controlled IPMSM drive system.

where v_α and v_β are the α- and β-axis components of the voltage vector $\boldsymbol{v_a}$, i_α and i_β are the α- and β-axis components of the armature current vector $\boldsymbol{i_a}$, ψ_α and ψ_β are the α- and β-axis components of the stator flux linkage vector $\boldsymbol{\psi_s}$, respectively, R_a is the armature resistance, P_n is the number of pole pairs, and T_e is the electromagnetic torque.

Fig. 1 shows a block diagram of the direct torque controlled IPMSM drive system. In DTC, the stator flux linkage vector is estimated with an integrator based on (1). However, pure integrators have problems, such as the saturation of the integrator and the initial conditions [9],[10]. Thus, in this paper, a low-pass filter is applied as the integrator, as shown in Fig. 1. As is well known, this causes flux estimation error.

III. MTPA CONTROL IN DTC

A. MTPA Control Based on IPMSM Model

The equations for the IPMSM in the d-q reference frame are given by (3) and (4),

$$T_e = P_n\{\Psi_a + (L_d - L_q)i_d\}i_q \tag{3}$$

$$\Psi_s^2 = (\Psi_a + L_d i_d)^2 + (L_q i_q)^2 \tag{4}$$

where i_d and i_q are the d- and q-axis components of the armature current vector, L_d and L_q are the d- and q-axis inductances, and Ψ_a and Ψ_s are the magnet and stator flux linkages, respectively.

The IPMSM can use reluctance torque, which depends on both the d- and q-axis currents, because of magnetic saliency. Therefore, the armature current vector can be optimally controlled in order to produce maximum torque for a constant current. From (3), the relationship between i_d and i_q under MTPA control is derived as follows [1]:

$$i_d = \frac{\Psi_a}{2(L_q - L_d)} - \sqrt{\frac{\Psi_a^2}{4(L_q - L_d)^2} + i_q^2} \tag{5}$$

In (5), i_d is negative if $L_q > L_d$.

Rewriting (5), i_d and i_q have the following relationship:

$$i_q^2 = i_d\left(i_d - \frac{\Psi_a}{L_S}\right) \tag{6}$$

where $L_S = L_q - L_d$, which is the difference between the q- and d-axis inductances.

B. Relationship Between Torque and Flux

In a speed control system, the reference torque is calculated from the difference between the reference speed and the rotor speed. Thus, it is required to calculate the reference stator flux linkage from the reference torque. However, the relationship shown in (5) under the MTPA condition is unavailable directly in DTC because DTC requires a relationship between torque and flux. Therefore, (5) is modified in order to be applied to DTC as follows: (7) can be derived from (3) and (6), and $\Psi_R = L_S i_d$ is treated as a variable.

$$\left(\frac{L_S T_e}{P_n}\right)^2 = \Psi_R(\Psi_R + \Psi_a)^3 \tag{7}$$

Equation (7) is a quartic equation for Ψ_R. The algebraic solutions of a quartic equation exist. Additionally, Ψ_R should be a real number and cannot be negative (i.e., $\Psi_R \geq 0$) based on (5). Thus, the solution Ψ_R of (7) is decided uniquely as follows:

$$\Psi_R = \frac{\Psi_a}{4}\left\{(1 + K)\left(1 + \sqrt{\frac{2}{K} - 1}\right) - 1\right\} \tag{8}$$

where

$$K = \sqrt{\frac{1}{2}\left\{\left(\sqrt{3Z^2 + 1} + 1\right)^{\frac{1}{3}} - \left(\sqrt{3Z^2 + 1} - 1\right)^{\frac{1}{3}}\right\}^3} \tag{9}$$

$$Z = \frac{16}{9}\frac{L_S T_e}{P_n \Psi_a^2} \tag{10}$$

On the other hand, using (4), (6), and Ψ_R, the reference flux for MTPA control $\Psi_{s\text{-}MTPA}$ is given as follows:

$$\Psi_{s\text{-}MTPA}^2 = \frac{L_q^2 + L_d^2}{L_S^2}\Psi_R^2 + \left(1 + \frac{L_d^2}{L_S^2}\right)\Psi_a \Psi_R + \Psi_a^2 \tag{11}$$

$\Psi_{s\text{-}MTPA}$ can be calculated as follows. First, Ψ_R is calculated from the reference torque T_e^* using (8)–(10). Next, $\Psi_{s\text{-}MTPA}$ is calculated from Ψ_R using (11). Then, $\Psi_{s\text{-}MTPA}$ is applied to the reference flux Ψ_s^*.

C. Validation of Reference Flux for MTPA Control

In the MTPA condition, the current is minimized under constant torque. Thus, to validate the MTPA condition, the reference flux minimizing current must be searched under constant torque. In this paper, the correction rate ε is added to $\Psi_{s\text{-}MTPA}$ in order to vary the

978-1-4799-2706-7/14 $31.00 © 2014 IEEE

reference flux for searching the MTPA control points, as shown in (12).

$$\Psi_s^* = (1+\varepsilon)\Psi_{s-MTPA} \tag{12}$$

The MTPA condition can be obtained with Ψ_{s-MTPA} when the current is minimized at $\varepsilon=0$.

The reference flux for MTPA control Ψ_{s-MTPA} described in the preceding section is based on (5), where the motor parameters are assumed to be constant. However, the inductance varies because of magnetic saturation. This causes MTPA operating point variation. Additionally, the operating point also varies because of flux estimation error that results from an imperfect integrator in the DTC systems. For example, the operating points satisfied with MTPA are shown in Fig. 2. The armature current is evaluated with respect to ε under constant load, and the line in Fig. 2 is the trajectory of the minimum current with respect to a constant load, which is regarded as the MTPA operating point.

Four models are evaluated in this section. Table I shows the parameters. Model 2 is the case with constant parameters and a pure integrator in the flux estimator. Model 1 is the case considering q-axis inductance variation due to magnetic saturation. In this study, the q-axis inductance of the IPMSM is modeled as a function of current, as shown in (13).

$$L_q = 24.3 - 0.7 \mid i_q \mid \quad [\text{mH}] \tag{13}$$

On the other hand, a constant q-axis inductance is used for the MTPA control in DTC.

Models 3 and 4 are the cases with an imperfect integrator in the flux estimator. The other parameters are listed in Table II.

From comparing Models 1 and 2, the parameter is mismatched due to the magnetic saturation results in the MTPA operating point variation. On the other hand, to compare Models 2, 3, and 4, the flux estimation error also causes MTPA operating point variation. Moreover, comparing Models 3 and 4, where the difference is only rotor speed, the MTPA operating point is different. It shows that the MTPA operating point also depends on the rotor speed. Because of these problems, deciding the operating point analytically is very difficult. Therefore, the MTPA condition is not always obtained with Ψ_{s-MTPA}. In the next section, a correction method is proposed that makes the reference flux obtain the MTPA condition.

D. Correction of Reference Flux for MTPA Control

The characteristics of the armature current I_a versus the correction rate ε with (12) under constant torque has the minimum point shown in Fig. 3. If ε minimizing I_a can be obtained in some way, MTPA control can be obtained by correcting Ψ_{s-MTPA} with the adjustment of ε.

In this paper, (14) is used for the adjustment of ε.

$$\varepsilon[n+1] = \varepsilon[n] - \Delta\varepsilon \operatorname{sgn}\left(\frac{I_a[n]-I_a[n-1]}{\varepsilon[n]-\varepsilon[n-1]}\right) \tag{14}$$

This equation means that ε is changed in steps of $\Delta\varepsilon$ depending on the sign of the derivative of I_a on the I_a-ε plane under constant torque. The behavior of (14) is similar to a local search algorithm. For example, in Fig. 3,

the operating points of $N[0]$ and $N[1]$ are known. In this case, $I_a[n]$ and $\varepsilon[n]$ are the discretized armature current I_a and correction rate ε, respectively. First, $\varepsilon[2]$ is decided as the value $\varepsilon[1]$ decreased by $\Delta\varepsilon$ because the derivative of the line between $N[0]$ and $N[1]$ is positive. Therefore, the operating point is changed to point $N[2]$. Next, $\varepsilon[3]$ is decided by $N[1]$ and $N[2]$ in the same way as $\varepsilon[2]$. The minimum I_a can be obtained successively to repeat this algorithm.

However, when $I_a[n]=I_a[n-1]$, the method cannot be applied because $\varepsilon[n+1]$ becomes the same as $\varepsilon[n]$, and $\varepsilon[n+2]$ cannot be defined due to division by zero. Additionally, this method is sensitive to current ripple. To avoid these issues, (15) is applied when $I_a[n]-I_a[n-1] \leq \Delta I_{min}$.

$$\varepsilon[n+1] = \varepsilon[n] - \Delta\varepsilon \operatorname{sgn}\left(\varepsilon[n]-\varepsilon[n-1]\right) \tag{15}$$

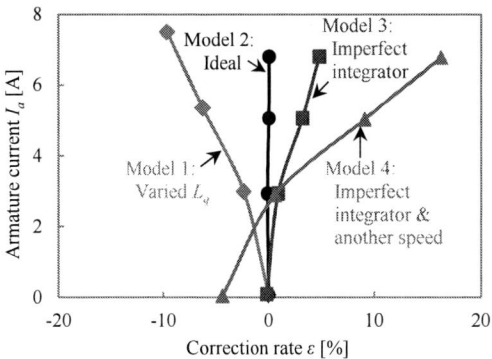

Fig. 2. MTPA operating points for the cases in Table I.

TABLE I
CONDITIONS FOR VALIDATING MTPA CONTROL POINTS

Parameter characteristics	Model 1	Model 2	Model 3	Model 4
q-axis inductance L_q in IPMSM	(13)	24.3 mH		
Cut-off frequency of flux estimator ω_c	0 (Pure integrator)	20 rad/s		
Rotor speed [min⁻¹]	1000			300
Other Parameters	In Table II			

TABLE II
CONSTANT PARAMETERS FOR MTPA CONTROL

Parameter characteristics	Value
Number of pole pairs P_n	2
Armature resistance R_a	0.824 Ω
Magnet flux linkage Ψ_a	78.5 mWb
d-axis inductance L_d	9.67 mH
q-axis inductance L_q in DTC	24.3 mH
Limitation of armature current I_{am}	8.66 A
Control period of DTC	100 μs

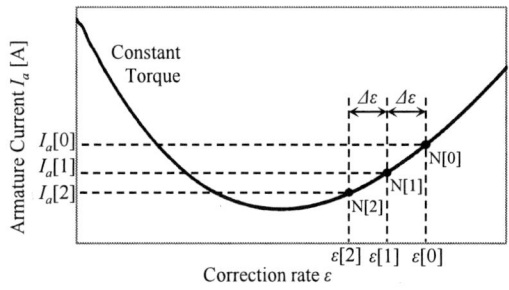

Fig. 3. Behavior of ε with (14) on $I_a-\varepsilon$ plane.

Noting that the magnitude of $(\varepsilon[n+1] - \varepsilon[n])$ equals $\varDelta\varepsilon$, ε shuttles between $\varepsilon[n]$ and $\varepsilon[n-1]$ by this equation. In the correction method, ε is never constant because (14) or (15) are always applied, and both (14) and (15) vary ε. The feature makes it easy to deal with the load variations. The proposed system for calculating the reference flux for MTPA control is shown in Fig. 4.

IV. SIMULATION RESULTS

In this section, the proposed method is validated by simulation. Fig. 4 is a block diagram of the proposed reference flux calculator.

Tables II and III show the parameters of the IPMSM and DTC in the simulation. The q-axis inductance of the IPMSM is given by (13), whereas a constant L_q is used in the DTC system. In this paper, the cut-off frequency of the flux estimator is 20 rad/s, which is sufficiently smaller than the operating frequency.

First, the proposed method is evaluated in the transient state. The reference speed is 1000 min^{-1}. Fig. 5 shows the characteristics of the rotor speed, torque, current, and ε. A load of 1.5 Nm is applied at 1 s. As shown in Fig. 2, ε satisfied with the MTPA condition depends on the armature current. In Fig. 5, ε converges to a certain value, the armature current also converges to the optimal current, and then, the rotor speed converges to 1000 min^{-1}, which is the reference speed.

Fig. 5(c) shows the trajectory of the operating point. At first, the correction rate ε stays near zero because ε is determined by (15) under no load. Point A is the operating point before load is applied. When a load torque is applied, the motor torque and current are stepped, and the operating point is moved to point B. Obviously, point B is not the MTPA point, and the operating point is then moved to point C by the proposed method because of the derivation of the constant torque curve at point B. If the motor torque is kept at 1.57 Nm, the operating point fits the constant torque curve of 1.57 Nm like in Fig. 3; however, the motor torque converges to 1.5 Nm as shown in Fig. 5(a), and the operating point is also moved to the constant torque curve of 1.5 Nm. As the result, there seems to be no negative effects under speed control, and the proposed method can be applied to the transient state.

Next, the proposed method is evaluated in the steady state. Table IV shows the armature current versus constant load torque at a rotor speed of 1000 min^{-1}. Table IV also shows the error rates of the current for the two methods with respect to the optimal current. From the comparison of the proposed method and the MTPA calculation without correction ($\varepsilon=0$), the proposed method can reduce the current under constant load. Moreover, the armature current with the proposed

method is almost the same as the optimal current. Therefore, the proposed method can achieve the MTPA condition.

TABLE III
CONDITIONS IN SIMULATIONS AND EXPERIMENTS

Parameter characteristics	Simulation	Experiment
q-axis inductance L_q in IPMSM	(11)	(11)
Cut-off frequency of flux estimator ω_c	20 rad/s	20 rad/s
Control period of (14)	5 ms	30 ms
Step of error $\varDelta\varepsilon$	0.01 %	1 %
Minimum value of difference of current to apply (14) $\varDelta I_{min}$	0.433 mA (0.005% of I_{am})	0 A
Other Parameters	In Table II	In Table II

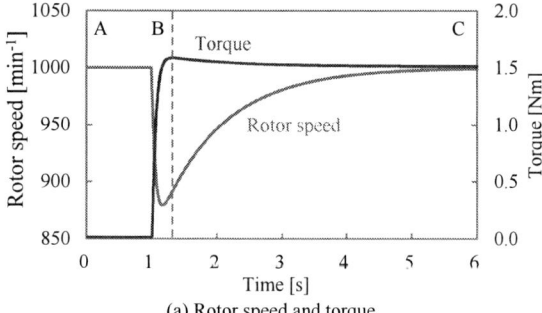

(a) Rotor speed and torque.

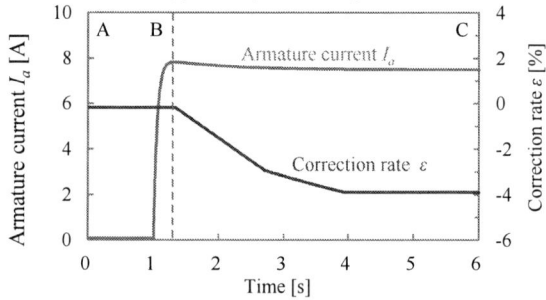

(b) Armature current I_a and correction rate ε.

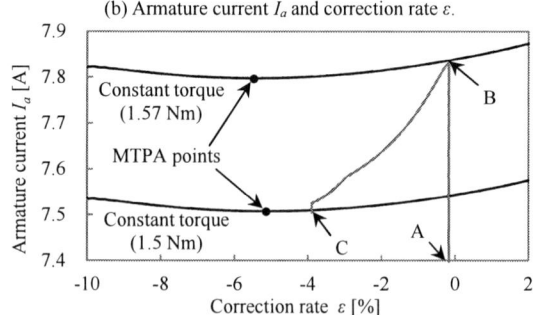

(c) Trajectory of control point by (14) and (15).
Fig. 5. Characteristics of transient state in simulation
(reference speed is 1000 min^{-1}, 1.5 Nm load).

TABLE IV
SIMULATION RESULTS WITH MTPA CONTROL

Load [Nm]	Armature current I_a [A]			Error of armature current [%]	
	Without correction ($\varepsilon = 0$)	Proposed	Optimal	$\varepsilon = 0$	Proposed
0.0	0.0876	0.0867	0.0867	1.02	0.00
0.5	2.9892	2.9875	2.9875	0.06	0.00
1.0	5.3760	5.3652	5.3645	0.21	0.01
1.5	7.5392	7.5054	7.5018	0.50	0.05

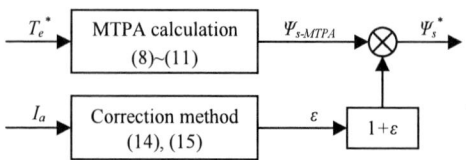

Fig. 4. Block diagram of proposed method.

In addition, the effect of the d-axis inductance mismatch is validated. The reference speed and load torque are the same as the previous evaluation, and the d-axis inductance L_d used in DTC is set to 14.505 mH, which is considered to be a 50% error of L_d. Fig. 6 shows the characteristics of the current and the correction rate ε. In the case of ε=0, as shown in Fig. 6(a), the armature current is increased by mismatches in the inductance. The armature current is larger than 7.5 A, which is the optimal value for the MTPA condition. On the other hand, when the proposed method is applied, the current is kept near 7.5 A, as shown in Fig. 6(b), which is the optimal value under a 1.5 Nm load. In Fig. 6(b), because L_d and $\Psi_{s\text{-}MTPA}$ are different from the case in Fig. 5, the convergent value for ε is different from the value in Fig. 5(b). As the result, the proposed method can realize MTPA control under inductance mismatches.

V. EXPERIMENTAL RESULTS

In this section, the proposed method for MTPA control is validated by experiment. The same simulation system is applied in the experiment. The DC link voltage is 150 V, and an insulated gate bipolar transistor inverter is applied. The DTC scheme is similar to the method reported in [7], where three-phase reference voltages are compared to the PWM carrier signal. The carrier frequency is 20 kHz. A hysteresis break is applied as the load. The other parameters of the IPMSM and DTC are the same as in Tables II and III. These results are obtained under a constant load of 1.5 Nm and a constant speed of 1000 min^{-1}. The experimental DTC system includes a speed sensor, which is used for rotor speed control.

In the experiment, the characteristics of the armature current I_a and the electrical angle of the rotor position θ in the case of ε=0 are shown in Fig. 7. The armature current I_a has a large ripple, which is synchronized to the rotor position. The peak-to-peak amplitude of the current ripple reaches 1.5 A, and the expected behavior of (14) may not be obtained. Therefore, the average current during the rotation period of the electrical angle is applied as follows:

$$I_m = \frac{1}{2\pi}\int_0^{2\pi} I_a d\theta \qquad (15)$$

In actual implementation, the armature current I_a is accumulated, and the cumulative current over some interval and the cumulative current divided by the interval is treated as I_m when θ exceeds 2π. Although (15) uses the knowledge of the rotor position θ, θ can be replaced by the integration of the rotor speed because θ is used only for determining the periodicity, and the periodicity is independent of the error of the initial value of θ.

Fig. 7 also shows the characteristics of I_m. Compared to I_a, the peak-to-peak amplitude of the ripple of I_m is reduced to about 0.3 A. As a result, I_m is a better value to use as the current in (14) than I_a. Moreover, because the application of I_m is based on considering the current ripple, ΔI_{min} is set to zero. When I_m is replaced with I_a in (14), the calculation periods of (14) and (15) T_{sc} can be calculated as follows:

$$T_{sc} = \frac{60}{P_n N} \qquad (16)$$

where N is the rotor speed, whose unit is min^{-1}. In this case, T_{sc} is 30 ms because the rotor speed is 1000 min^{-1}, and the number of pole pairs P_n is two. T_{sc} is also equal to the period of the current ripple.

First, in order to search for the value of ε that minimizes the armature current, the current is verified under various value of ε. Fig. 8 shows the armature current characteristics with respect to the value of ε. The smallest current is obtained near 0%.

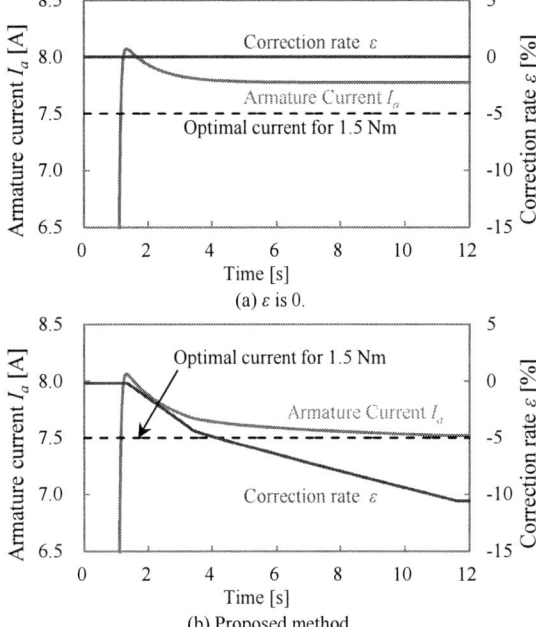

(a) ε is 0.

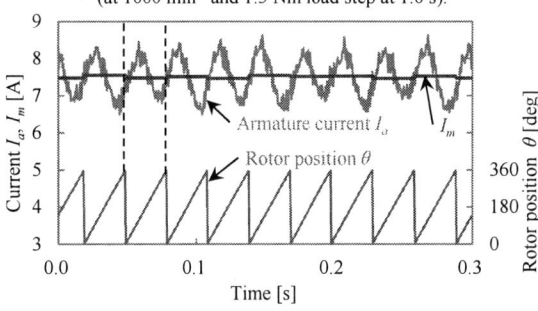

(b) Proposed method.

Fig. 6. Characteristics of current under 50% error of L_d in DTC (at 1000 min^{-1} and 1.5 Nm load step at 1.0 s).

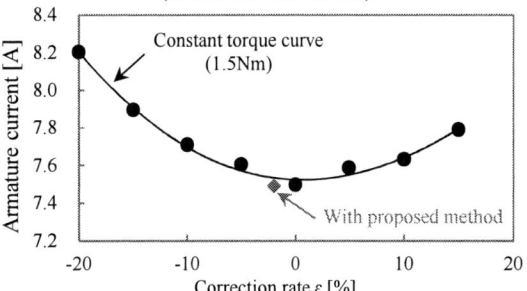

Fig. 7. Characteristics of armature current and rotor position with ε=0. (at 1000 min^{-1} and 1.5 Nm)

Fig. 8. Characteristics of current versus ε in experiment (at 1000 min^{-1} and 1.5 Nm).

Next, the proposed method is applied, and the correction characteristic is verified. Here, the initial values $\varepsilon[0]$ and $\varepsilon[1]$ are set to -19% and -18%, respectively, which are obviously different from the MTPA operating point. Fig. 9 shows the experimental results for the proposed method. Compared to Fig. 8, the current with the proposed method becomes smaller, and ε converges near 0%, where the ε minimizing current exists. As shown in Fig. 9, ε is varied every 30 ms, which is the updating cycle of I_m. At point D, the converged value of ε is varied. This is caused by the varied I_m and ε, as noted before. However, this issue does not have a major effect on the MTPA condition like I_m does. Moreover, Fig. 10 shows the convergence performance of the proposed method. In Fig. 10, the proposed method is applied at 0.5 s. Before the method is applied, I_m exceeds 8 A and the MTPA condition cannot be easily obtained. When the method is applied, ε increases, and I_m is reduced. Furthermore, this condition is also plotted in Fig. 7. The obtained I_m and ε with the proposed method fit in a constant torque curve and ε is close to 0%, which is the MTPA operating point. Therefore, the proposed method can follow the MTPA condition.

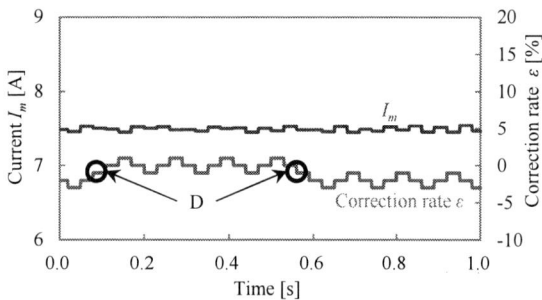

Fig. 9. Characteristics of current and ε with proposed method in experiment. (at 1000 min^{-1} and 1.5 Nm)

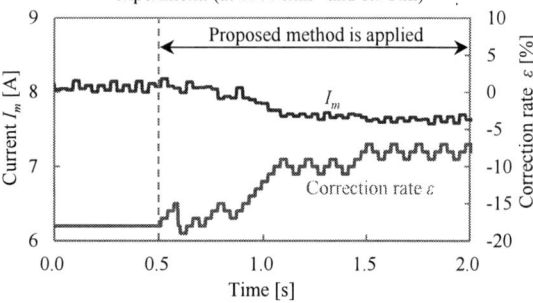

Fig. 10. Characteristics of converging current and ε in experiment. (at 1000 min^{-1} and 1.5 Nm)

VI. CONCLUSION

This paper proposed a correction method for the reference flux in the MTPA condition. The features of the proposed method are using the current as information and reducing the current sequentially with a local search algorithm to vary the reference flux directly. The simulation and experimental results confirmed that the inductance variation and flux estimation error do not affect the performance of the proposed method.

REFERENCES

[1] S. Morimoto, M. Sanada, and Y. Takeda, "Wide-speed operation of interior permanent magnet synchronous motors with high performance current regulator," IEEE Transactions on Industry Applications, vol. 30, no. 4, pp. 920-926, Jul./Aug. 1994.

[2] T. M. Jahns, B. K. Gerald, and W. N. Thomas, "Interior permanent magnet synchronous motors for adjustable speed drives," IEEE Transactions on Industry Applications, vol. IA-22, no. 4, pp. 738-747, Jul./Aug. 1986.

[3] Y. Inoue, Y. Kawaguchi, S. Morimoto, M. Sanada, "Performance improvement of sensorless IPMSM drives in a low-speed-region using online parameter identificaton," IEEE Transactions on Industry Applications, vol. 47, no. 2, pp. 798-804, Mar./Apr. 2011.

[4] P. Niazi, A. Toliyat, A. Goodarzi, "Robust maximum torque per ampere (MTPA) control of PM-Assisted SynRM for traction applications," IEEE Transactions on Vehicular Technology, vol. 56, no. 4, pp. 1538-1545, Jul. 2007.

[5] I. Takahashi and T. Noguchi, "A new quick-response and high-efficiency control strategy of an induction motor," IEEE Transactions on Industry Applications, vol. IA-22, no 5, pp. 820-827, Sep. 1986.

[6] M. F. Rahman, L. Zhong, K. W. Lim, "A direct-torque-controlled interior permanent magnet synchronous motor drive incorporating field weakening," IEEE Transactions on Industry Applications, vol. 34, no. 6, pp. 1246-1253, Nov./Dec. 1998.

[7] L.Tang, L. Zhong, M. F. Rahman, Hu, Y., "A novel direct torque controlled interior permanent magnet synchronous machine drive with low ripple in flux and torque and fixed switching frequency," IEEE Transactions on Power Electronics, vol. 19, no. 2, pp. 346-356, Mar. 2004.

[8] G. S. Buja, M. P. Kazmierkowski, "Direct Torque Control of PWM Inverter-Fed AC Motors - A Survey," IEEE Transactions on Industrial Electronics, vol. 51, no. 4, pp. 744-757, Aug. 2004.

[9] Jin Hu and Bin Wu, "New integration algorithms for estimating motor flux over a wide speed range," IEEE Transactions on Power Electronics, vol. 13, no. 5, pp. 969-977, Sep. 1998.

[10] Nik Rumzi, Nik Idris, and Abdul Halim Mohamed Yatim, "An improved stator flux estimation in steady-state operation for direct torque control of induction machines," IEEE Transactions on Industry Applications, vol. 38, no. 1, pp. 110-116, Jan./Feb. 2002.

Voltage Regulation and Maximum Output Power Tracking of a 4.5kW Permanent-Magnet Synchronous Generator

Yuan-Chih Chang, Hsiu-Feng Chang, Wei-Fu Dai and Chun-Wei Wu

Elegant Power Application Research Center (EPARC), Department of Electrical Engineering
National Chung Cheng University
Min-Hsiung, Chia-Yi 62102, Taiwan, ROC
Email:ycchang@ccu.edu.tw

Abstract-Wind energy is one of the renewable energy been developing in recent years. Wind turbines are generally connected to the utility grid as the generation resources. In local application, the wind generator can be implemented in the DC or AC micro-grid. In the DC micro-grid, the output AC/DC converter of wind generator should accomplish the AC/DC power conversion and maximum power tracking. This paper develops a 4.5 kW permanent-magnet generator drive to convert AC output of wind generator to DC micro-grid. The generator is driven by a permanent-magnet synchronous motor to simulate the wind turbine. The DC-link voltage, line voltages and line currents of the generator are measured to regulate the output voltage and power. The Division-Summation (D-Σ) current control scheme is adopted in this paper. The rotor reference frame transformation is unnecessary in this algorithm, which reduces the execution time of the control unit. Constant output power control, DC-link voltage regulation and maximum power tracking are achieved. The DC-link voltage is regulated by the one line-cycle regulation scheme. The DC-link voltage of the PMSG can be regulated in one electrical period under output power variation. The maximum output power tracking is achieved via perturb and observe method. The current control, voltage regulation and maximum power tracking are verified by experimental results.

Keywords— Voltage regulation, maximum output power tracking, permanent magnet synchronous generator, division-summation current control.

I. INTRODUCTION

The generator implemented in the wind generation system [1-2] have four main types: (i) fixed-speed system: the squirrel-cage induction generator (SCIG) is adopted; (ii) limited variable-speed system: the wound-rotor induction generator (WRIG) is implemented; (iii) variable-speed system with partially-rated power converter: the doubly-fed induction generator (DFIG) is applied; (iv) variable-speed system with fully-rated power converter: the permanent-magnet synchronous generator (PMSG) is performed to accomplish the reactive power compensation. The conventional circuit configurations [3] of the PMSG-based wind generation system can be classified into diode front-end type and back-to-back type. A novel circuit topology called Z-source converter [4] is proposed to achieve maximum power tracking and grid connecting simultaneously. The generating performance such as total switching device power (TSDP) and total harmonic distortions (THD) is improved.

The maximum power point tracking (MPPT) [5] is the key technology in wind generation system. The common MPPT algorithms can be categorized to three types: tip speed ratio control, perturb and observe (P&O) control and optimum relationship-based control. These algorithms can be combined [6] to achieve MPPT without anemometer and system preknowledge. Fast and accurate response to wind speed variation is obtained. In connecting to grid, the low-voltage ride-through (LVRT) capability of the wind energy conversion system is important. A novel algorithm [7] applying feedback linearization is proposed to accomplish this scheme. In addition, the grid synchronization of grid-connected wind power system is significant. The harmonic and reactive power compensation capability [8] is adopted to achieve MPPT and grid-connection.

In this paper, a 4.5kW permanent magnet synchronous generator is developed to simulate the wind energy conversion system applied in the DC micro-grid. The system configuration and parameter estimation is introduced. A D-Σ current control algorithm is proposed to tracking winding current command. This method does not need the rotor reference frame transformation (abc to dq), therefore the execution time of the control unit can be shorten. Since the inverter of PMSG in DC micro-grid system should accomplish voltage regulation and maximum power tracking in the meantime, the one-line-cycle regulation scheme and P&O method are implemented. The DC-link voltage can be regulated in one electrical cycle under output power variation. The response time of the proposed algorithm is quite faster than the conventional methods. The voltage regulation and maximum power tracking are verified by some experimental results.

978-1-4799-2706-7/14 $31.00 © 2014 IEEE

The 2014 International Power Electronics Conference

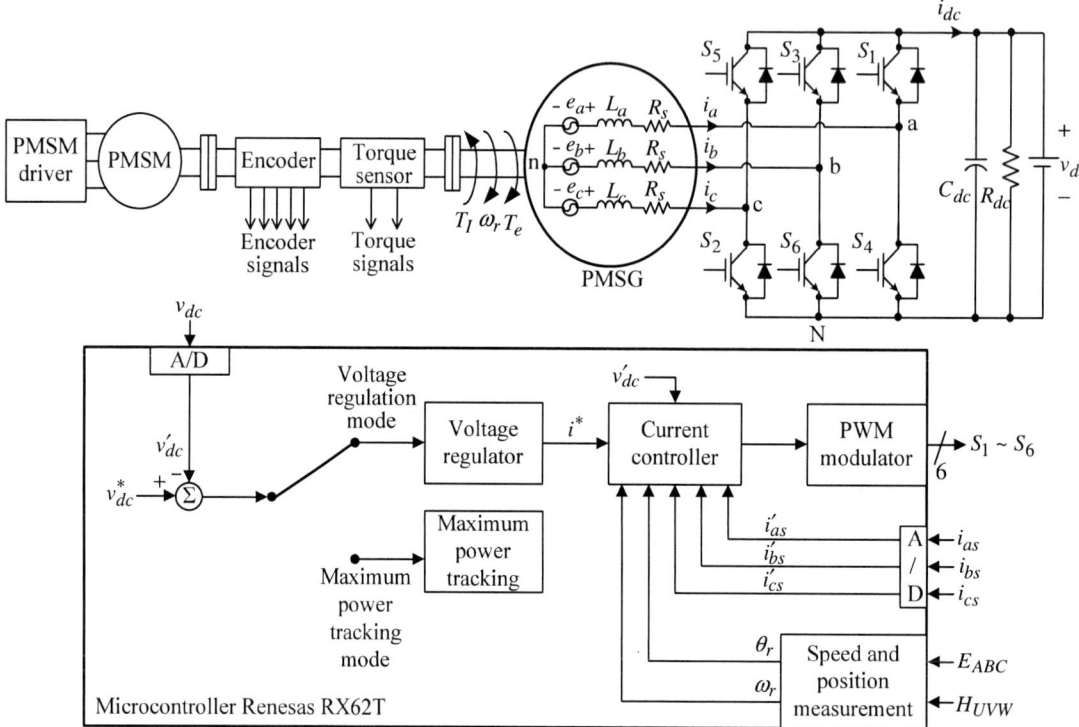

Fig. 1 System configuration of the developed permanent magnet synchronous generator.

II. SYSTEM CONFIGURATION

The system configuration of the developed permanent-magnet synchronous generator (PMSG) is shown in Fig. 1. The generator is driven by a permanent-magnet synchronous motor (PMSM). The speed variation of wind energy system is controlled via the PMSM driver. Generator winding currents and DC-link voltage are sensed and filtered using the auxiliary circuits. The position and speed information are obtained by the assembled optical encoder with 1024 pulses/rev resolution. Two operating modes including voltage regulation and maximum power tracking are implemented in this paper. The current command is determined by the outer loop and then utilized in the D-Σ current control scheme. The pulse-width modulation (PWM) signals are generated by the microcontroller Renesas RX62T.

The circuit parameters of PMSG are first estimated to design the current control scheme. In the single-phase equivalent circuit of sinusoidal wave PMSG, the back-EMF e_{as} can be expressed as the function of speed ω_r and rotor position θ_r. The back-EMF constant k_e can be measured by driving PMSG at no-load condition. The back-EMF constant is estimated as $k_e = 99.79$ V/krpm.

In the estimation of winding resistance and inductance for Y-connected PMSM without neutral line, the PMSM is excited by ac source in two windings. The d-axis and q-axis inductances are found as $L_d = 1.896\,\text{mH}$ and $L_q = 2.131\,\text{mH}$. The winding resistance is obtained as $R_s = 0.48\,\Omega$.

III. DIVISION-SUMMATION (D-Σ) CURRENT CONTROL

The division-summation current control is derived from the space-vector. The time sections of winding currents and the space-vector diagram are shown in Fig. 2(a). The switching mechanism is determined by the reference voltage $\mathbf{v}_{ref}$ and eight basic vectors $\mathbf{v}_1 - \mathbf{v}_8$. Each vector represents one switching state, eg. $\mathbf{v}_1(100)$ means that the lower switch of phase-a is turn-on and the lower switches of phase-b and phase-c are turn-off. For instance, the switching sequence in sector I is $\mathbf{v}_8 \rightarrow \mathbf{v}_1 \rightarrow \mathbf{v}_2 \rightarrow \mathbf{v}_7 \rightarrow \mathbf{v}_2 \rightarrow \mathbf{v}_1 \rightarrow \mathbf{v}_8$. The switching state of the conventional space-vector modulation and the two-phase modulation are shown in Fig. 2(b).

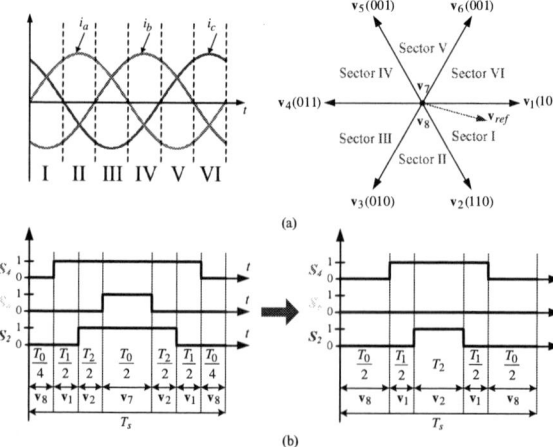

Fig. 2 The proposed D-Σ current control scheme: (a) Time sections of the winding current and the space-vector diagram; (b) the space-vector modulation and two-phase modulation.

A. Division (D)

In the switching sequence shown in Fig. 2(b), there are three time intervals (T_0, T_1, T_2) representing the space vectors ($\mathbf{v}_8, \mathbf{v}_1, \mathbf{v}_2$). Since the equivalent circuits of these states can be found, the voltage equations of time intervals T_0, T_1, T_2 are derived as:

$$T_0 : \begin{bmatrix} v_{dc} \\ -v_{dc} \end{bmatrix} = \begin{bmatrix} L_a & -L_b \\ L_c & L_b + L_c \end{bmatrix} \begin{bmatrix} \Delta i_{a0} \\ \Delta i_{b0} \end{bmatrix} \frac{1}{T_0} + \begin{bmatrix} e_{ab} \\ e_{bc} \end{bmatrix} \tag{1}$$

$$T_1 : \begin{bmatrix} 0 \\ -v_{dc} \end{bmatrix} = \begin{bmatrix} L_a & -L_b \\ L_c & L_b + L_c \end{bmatrix} \begin{bmatrix} \Delta i_{a1} \\ \Delta i_{b1} \end{bmatrix} \frac{1}{T_1} + \begin{bmatrix} e_{ab} \\ e_{bc} \end{bmatrix} \tag{2}$$

$$T_2 : \begin{bmatrix} 0 \\ 0 \end{bmatrix} = \begin{bmatrix} L_a & -L_b \\ L_c & L_b + L_c \end{bmatrix} \begin{bmatrix} \Delta i_{a2} \\ \Delta i_{b2} \end{bmatrix} \frac{1}{T_2} + \begin{bmatrix} e_{ab} \\ e_{bc} \end{bmatrix} \tag{3}$$

where L_x represents the inductance of phase-x, Δi_{xn} is the current variation of phase-x in time interval T_n, e_{xy} is the line-to-line back-EMF of phase-xy, v_{dc} is the DC-link voltage. The digital control will use the current command variation and feedback current to derive the control law.

B. Summation (Σ)

The current variations in (1)-(3) are difficult to found in the digital control because the winding current is sensed only once in one switching cycle. Therefore, the control laws of time intervals T_0, T_1, T_2 cannot be directly derived from (1)-(3). To solve this problem, the voltage equations are rewritten as:

$$T_0 : \begin{bmatrix} \Delta i_{a0} \\ \Delta i_{b0} \end{bmatrix} = \frac{T_0}{L_T^2} \begin{bmatrix} L_b + L_c & L_b \\ -L_c & L_a \end{bmatrix} \begin{bmatrix} v_{dc} - e_{ab} \\ -v_{dc} - e_{bc} \end{bmatrix} \tag{4}$$

$$T_1 : \begin{bmatrix} \Delta i_{a1} \\ \Delta i_{b1} \end{bmatrix} = \frac{T_1}{L_T^2} \begin{bmatrix} L_b + L_c & L_b \\ -L_c & L_a \end{bmatrix} \begin{bmatrix} -e_{ab} \\ -v_{dc} - e_{bc} \end{bmatrix} \tag{5}$$

$$T_2 : \begin{bmatrix} \Delta i_{a2} \\ \Delta i_{b2} \end{bmatrix} = \frac{T_2}{L_T^2} \begin{bmatrix} L_b + L_c & L_b \\ -L_c & L_a \end{bmatrix} \begin{bmatrix} -e_{ab} \\ -e_{bc} \end{bmatrix} \tag{6}$$

where $L_T^2 = L_a L_b + L_b L_c + L_c L_a$.

The total current variation is obtained via summarizing these three equations:

$$\begin{bmatrix} \Delta i_a \\ \Delta i_b \end{bmatrix} = \frac{T}{L_T^2} \begin{bmatrix} L_b + L_c & L_b \\ -L_c & L_a \end{bmatrix} \begin{bmatrix} -e_{ab} \\ -e_{bc} \end{bmatrix}$$
$$+ \frac{1}{L_T^2} \begin{bmatrix} L_b + L_c & L_b \\ -L_c & L_a \end{bmatrix} \begin{bmatrix} -v_{dc} & 0 \\ v_{dc} & v_{dc} \end{bmatrix} \begin{bmatrix} T_0 \\ T_1 \end{bmatrix} \tag{7}$$

where $\Delta i_x = \Delta i_{x0} + \Delta i_{x1} + \Delta i_{x2}$.

The time intervals T_0, T_1 can be found to be:

$$\begin{bmatrix} T_0 \\ T_1 \end{bmatrix} = \frac{1}{-v_{dc}} \begin{bmatrix} -L_a \Delta i_a + L_b \Delta i_b - e_{ab}T \\ L_a \Delta i_a + L_c \Delta i_b + (e_{ab} + e_{bc})T \end{bmatrix} \tag{8}$$

The duty ratios in section I can be obtained:

$$\begin{bmatrix} D_{S4} \\ D_{S6} \\ D_{S2} \end{bmatrix} = \begin{bmatrix} \dfrac{(L_a + L_b)\Delta i_a + L_b \Delta i_c}{v_{dc}T} \\ 0 \\ \dfrac{L_b \Delta i_a + (L_b + L_c)\Delta i_c}{v_{dc}T} \end{bmatrix} + \begin{bmatrix} 1 - \dfrac{e_{ab}}{v_{dc}} \\ 0 \\ 1 + \dfrac{e_{bc}}{v_{dc}} \end{bmatrix} \tag{9}$$

where $D_{S4} = (T_1 + T_2)/T$, $D_{S2} = T_2/T$.

The duty ratios of the other sections can be derived via the same procedure.

IV. VOLTAGE REGULATION AND MAXIMUM POWER TRACKING

A. Voltage regulation

The DC-link voltage can be regulated by adjusting the current command of PMSG. The voltage regulation algorithm adopted in this paper is the one-line-cycle regulation. By assuming that the power losses of inverter are neglected, the power conversion between PMSG and DC-link is:

$$P_{PMSG} = \sqrt{3} \times E_{as} \times I_{as} = V_{dc} \times I_{dc} \tag{10}$$

where P_{PMSG} is the generated power of PMSG, E_{as} is the rms value of phase back-EMF, I_{as} is the rms value of phase current, (V_{dc}, I_{dc}) are the average value of DC-link voltage and current.

The desired phase current of PMSG is:

$$I_{as} = \frac{V_{dc} \times I_{dc}}{\sqrt{3} \times E_{as}} = k \times I_{dc} \tag{11}$$

When the load variation occurs, the DC-link voltage in digital control will increase or decrease from $V_{dc}(n-1)$ to $V_{dc}(n)$. The DC-link current should be compensated as:

$$\Delta I_{dc} = C_{dc} \frac{V_{dc}(n-1) - V_{dc}(n)}{T_l}$$
$$= f_l \times C_{dc} \times (V_{dc}(n-1) - V_{dc}(n)) \tag{12}$$

where ΔI_{dc} is compensation value of the DC-link current, $f_l = 1/T_l$ is the electrical frequency of PMSG, C_{dc} is the DC-link capacitance.

The phase current $I_{as}(n)$ can be found from ΔI_{dc}:

$$I_{as}(n) = I_{as}(n-1) + k \times \frac{C_{dc}(V_{dc}(n-1) - V_{dc}(n))}{T_l} \tag{13}$$

Therefore, the DC-link voltage variation caused by power unbalance can be compensated by adjusting the winding current command. In the experimental results, the voltage reference value is set as 380V.

B. Maximum power tracking

The maximum power tracking algorithm implemented in this study is the perturb and observe (P&O) method. The current command is disturbed by a small value in one electrical frequency cycle. Then the output power is calculated and compared with the last value. If the power is increasing, the current command is increased in the next cycle. If the power is decreasing, the command is decreased from the original value in the next cycle.

Due to the lack of wind turbine, the maximum power tracking is verified by the circuit configuration shown in Fig. 3. The PMSG windings are connected with three-phase balance resistors R_M. The single-phase equivalent circuit is also shown in Fig.3. According to the maximum power transferring theorem, if the equivalent resistance of the inverter is equal to the arranged resistance R_M, the maximum power transferring is achieved. Therefore, by setting different value of R_M under different speeds, the maximum power tracking can be validated.

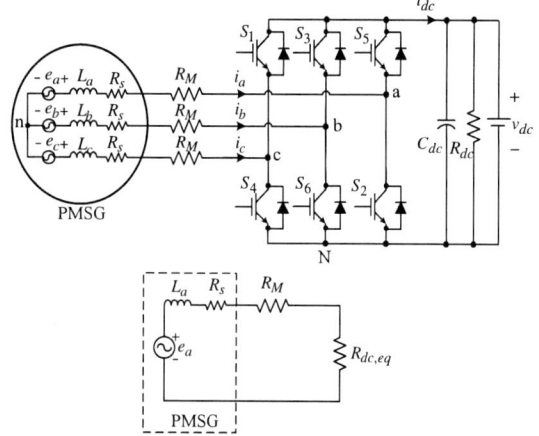

Fig. 3 Circuit configuration to verify the maximum power tracking via maximum power transferring theorem.

V. EXPERIMENTAL RESULTS

First, the effectiveness of the proposed D-Σ current control is verified. Fig. 4 shows the current command and winding current at ω_r = 900rpm, i_a^* = 10A. The experimental results show that the winding current tracks the current command very well.

Fig. 5(a) shows the voltage regulation under output power variation at ω_r = 1800rpm. The winding currents are simultaneously adjusted to achieve voltage regulation. Fig. 5(b) is the voltage regulation under output power variation at ω_r = 1200rpm. It can be found that the DC-link voltage drops due to the output power variation, then the current command is increased to balance the power and therefore regulate DC-link voltage at 380V. Fig. 5(c) shows the voltage regulation via conventional voltage controller. It can be found that the response time of the proposed voltage regulation scheme is quite faster than the conventional controller.

In the maximum power tracking verification, R_M = 5Ω and R_M = 10Ω are chosen at ω_r = 600rpm and ω_r = 1800rpm, respectively. The measured winding current waveforms are shown in Fig. 6(a) and Fig. 6(b). The maximum power tracking accuracy is 99.9% and 99.8%.

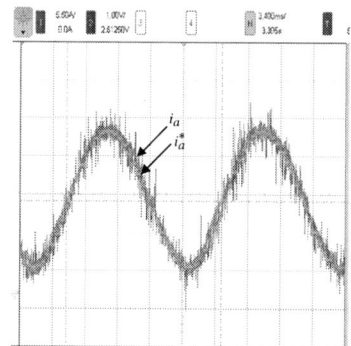

Fig. 4 Current command and winding current at ω_r =900 rpm, i_a^* =10A.

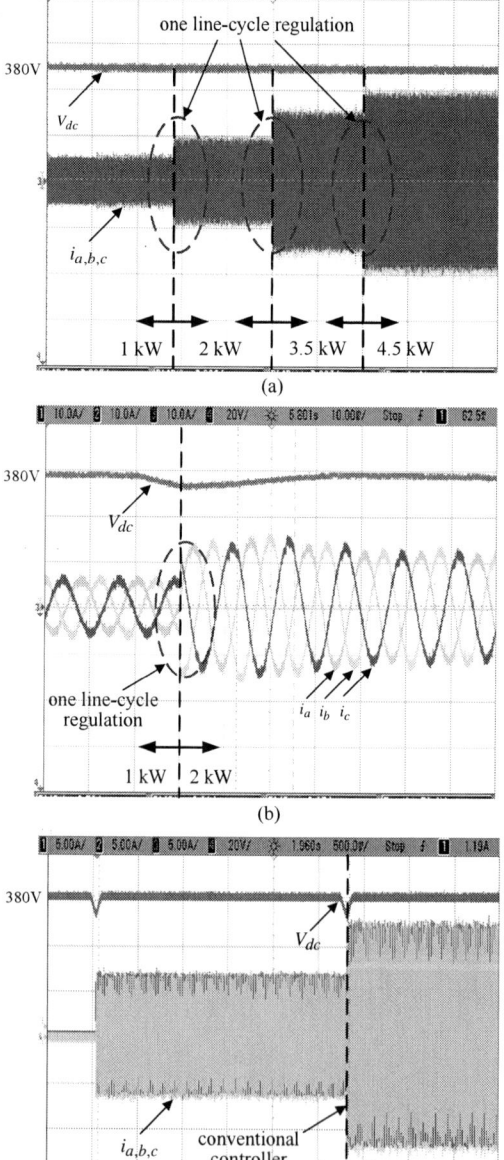

Fig. 5 Measured winding currents and DC-link voltage under output power variation: (a) 1800rpm; (b) 1200rpm; (c) 1200rpm with conventional voltage controller.

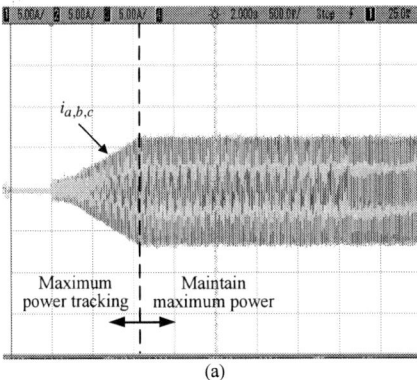

(a)

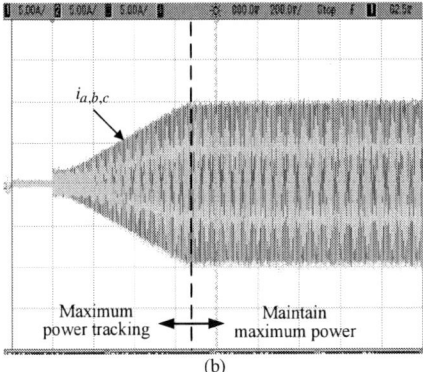

(b)

Fig. 6 Measured winding currents at maximum power tracking mode: (a) $R_M = 5\Omega$, $\omega_r = 600$rpm; (b) $R_M = 10\Omega$, $\omega_r = 1800$rpm.

VI. CONCLUSIONS

A 4.5kW permanent-magnet synchronous generator with DC-link voltage regulation and maximum power tracking is developed. The D-Σ current control is proposed to fulfill the winding current tracking. The DC-link voltage is well regulated by the implemented one-line-cycle regulation under output power variation at different speed. The maximum power tracking of the inverter is verified via the maximum power transferring theorem. The maximum tracking accuracy is above 99.8%.

ACKNOWLEDGMENT

This research was supported by the National Science Council, Taiwan, ROC, under the Grant of NSC 102-2221-E-194-029.

REFERENCES

[1] J. M. Carrasco, L. G. Franquelo, J. T. Bialasiewicz, E. Galvan, R. C. P. Guisado, A. M. Prats, J. I. Leon, and N. Moreno-Alfonso, "Power electronic systems for the grid integration of renewable energy sources: A survey," IEEE Trans. Ind. Electron., vol. 53, no. 4, pp. 1002–1016, 2006.

[2] Z. Chen, J. M. Guerrero, and F. Blaabjerg, "A review of the state of the art of power electronics for wind turbines," IEEE Trans. Power Electron., vol. 24, no. 8, pp. 1859–1875, 2009.

[3] S. Zhang, K. J. Tseng, D. M. Vilathgamuwa, T. D. Nguyen and X. Y. Wang, "Design of a robust grid interface system for PMSG-Based wind turbine generators," IEEE Trans. Ind. Electron., vol. 58, no. 1, pp. 316-328, 2011.

[4] S. M. Dehghan, M. Mohamadian and A. Y. Varjani, "A new variable-speed wind energy conversion system using permanent-magnet synchronous generator and Z-source inverter," IEEE Trans. Energy Convers., vol. 24, no. 3, pp. 714-724, 2009.

[5] Q. Wang and L. C. Chang, "An intelligent maximum power extraction algorithm for inverter-based variable speed wind turbine systems," IEEE Trans. Power Electron., vol. 19, no. 5, pp. 1242–1249, 2004.

[6] Y. Xia, K. H. Ahmed and B. W. Williams, "A new maximum power point tracking technique for permanent magnet synchronous generator based wind energy conversion system," IEEE Trans. Power Electron., vol. 26, no. 12, pp. 3609–3620, 2011.

[7] K. H. Kim, Y. C. Jeung and D. C. Lee, "LVRT scheme of PMSG wind power systems based on feedback linearization," IEEE Trans. Power Electron., vol. 27, no. 5, pp. 2376–2384, 2012.

[8] M. Singh, V. Khadkikar and A. Chandra, "Grid-synchronization with harmonics and reactive power compensation capability of a permanent magnet synchronous generator-based variable speed wind energy conversion system," IET Power Electron., vol. 4, no. 1, pp. 122-130, 2011.

A Novel Flux-Weakening Control Method Based on Single Current Regulator for Permanent Magnet Synchronous Motor

Xiaocun Fang, Taiyuan Hu, Fei Lin,and Zhongping Yang
School of Electrical Engineering
Beijing Jiaotong University
Beijing, China

Abstract--Traditional flux-weakening control algorithms for PMSM (permanent magnet synchronous motor) apply two independent current regulators. The two current regulators conflict with each other in high speed operation region, leading to worse system performance. Based on the d-q current cross-coupling effect, deep flux-weakening control of PMSM can be achieved by using single current regulator. Q-axis voltage given scheme is a key issue of this method, which influences PMSM's voltage utilization ratio, efficiency and load capability. This paper presents a new q-axis voltage given method. It can adjust q-axis voltage to motor operating condition. This method does not depend on motor parameters or look-up table, is easy to achieve. This method is verified by the results of simulation and experiment.

Keywords—PMSM, Flux-weakening Control, Current Cross-coupling, Single Current Regulator, Q-axis Voltage

I. INTRODUCTION

Permanent Magnet Synchronous Motor (PMSM) is widely applied in various fields because of its simple structure, high efficiency and high power density. Flux-weakening control is desired for a wider speed range operation. But it is not easy to realize deep field flux-weakening operation.

Traditional flux-weakening control algorithms for PMSM apply two independent current regulators. The d-axis current i_d and q-axis current i_q are controlled respectively. References [1-3] calculate current commands from the machine equations, so the method is parameter sensitive and is rarely used. References [4-6] get current commands from a table based on experimental results, so the method requires large experimental work and is short of transportability. References [7-8] get current commands by limiting and regulating the voltage commands, the method is useable but still not competent for deep field flux-weakening. In deep flux-weakening operation, the q-axis current and d-axis current couple deeply and the system will be unstable [9-11].

To achieve higher speed range, references [9-11] proposed SCR (single current regulator) control. The method utilizes the coupling between q-axis and d-axis

currents. And it has only one d-axis current regulator and gives q-axis voltage directly. If the given q-axis voltage is fixed, the PMSM's voltage utilization ratio, efficient and load capacity will decline. In order to improve the motor efficiency and load capacity, reference [9] changes the q-axis voltage command by a linear function of q-axis voltage and current. The method is parameter sensitive and the efficient and load capacity is not improved enough to the expected results. References [10-11] give the q-axis voltage command by looking up a table based on experimental results. The constant power range of a 50kW experiment PMSM was expanded to 7 times of constant torque range. But the method requires large experimental work and is short of transportability.

Based on the idea of a single current regulator, the references [12-13] propose a flux-weakening control method named SCR-VAC (single current regulator and voltage angle control). Compared to SCR-FQV (single current regulator and fixed q-axis voltage), SCR-VAC can improve PMSM's voltage utilization ratio, efficiency and load capability. But the d-axis control can be only used for motion and the q-axis control can be only used for generation. SCR-VAC is not sufficiently practical. Reference [xx] proposes another VAC flux-weakening control without current regulator. The method can control motor working at the optimal point. But it need machine equations to correct the flux-weakening operating point, if not the PMSM may be operation unsteadily.

The flux-weakening control method with single current regulator and given q-axis voltage has the advantages of simple structure, fast dynamic response, and good robustness of parameters etc. The problem is how to adjust q-axis voltage to motor operating condition. This paper proposes a new method to optimize q-axis voltage. The method does not rely on motor parameters and does not look up table. It makes full use of the motor DC voltage and makes the motor operate at the optimal operating point. The method is verified by the results of simulation and experiment in this paper.

II. ANALYSIS OF FLUX-WEAKENING CONTROL

The voltage equations of PMSM in the synchronous rotation coordinate system at steady state can be

978-1-4799-2706-7/14 $31.00 © 2014 IEEE

expressed as follows:

$$\begin{cases} u_d = R_s i_d - \omega_r L_q i_q \\ u_q = R_s i_q + \omega_r (L_d i_d + \psi_f) \end{cases} \quad (1)$$

where

u_d, u_q is d- and q-axis components of stator terminal voltage,

i_d, i_q is d- and q-axis components of stator current,

L_d, L_q is d- and q-axis components of armature winding self-inductance,

ω_r is the electrical rotor angle velocity,

R_s is the stator phase resistance,

ψ_f is the flux linkage of the permanent magnet.

The torque is obtained as follows:

$$T_e = \frac{3}{2} P_n [\psi_f i_q + (L_d - L_q) i_d i_q] \quad (2)$$

where P_n is the number of poles.

As the limited of the inverter output voltage and current, voltage and current of PMSM satisfy the following relations:

$$u_d^2 + u_q^2 \le u_{s\max}^2 \quad (3)$$

$$i_d^2 + i_q^2 \le i_{s\max}^2 \quad (4)$$

where u_{smax} is the maximum amplitude of voltage vector, i_{smax} is the maximum amplitude of current vector.

When the motor speed is higher than the rated speed, the stator resistant can be neglected. The formula (3) can be expressed as follows:

$$(L_q i_q)^2 + (L_d i_d + \psi_f)^2 \le \left(\frac{u_{s\max}}{\omega_r}\right)^2 \quad (5)$$

Equation (2), (4) and (5) in i_d-i_q plane are shown in figure 1.

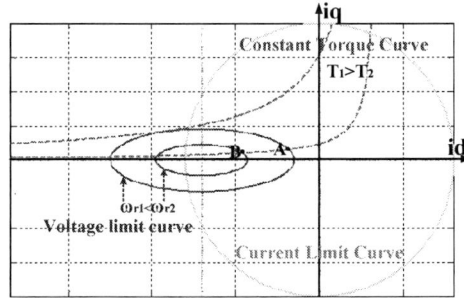

Fig. 1. Steady-state stator current vector trajectory.

The flux-weakening operation area is the coincidence region of current limit circle and voltage limit ellipse. With the rise of motor speed, voltage limit ellipse narrow and the flux-weakening operation area narrow. When the motor is in steady state, state current vector trajectory should satisfy the motor constant torque curve formula (2). When the motor stable operating point is the intersection point of constant torque curve and voltage limit ellipse, input voltage is fully utilized and the stable operating point is the optimal operating point. In figure 1, when the motor load torque is T_2 and speed is ω_l, the

motor optimal stable operating point is the intersection point A of constant torque curve and voltage limit ellipse. With speed rise to ω_2, the motor optimal stable operating point is the intersection point B of constant torque curve and voltage limit ellipse. When speed is ω_l, the maximum output torque of motor is T_l, which is determined by the intersection point of voltage ellipse and current circle.

III. ANALYSIS OF FLUX-WEAKENING CONTROL BASED ON SINGLE CURRENT REGULATOR AND GIVEN Q-AXIS VOLTAGE

The second equation in equation (1) can be rewritten as:

$$i_q = -\frac{\omega_r L_d}{R_s} i_d + \frac{u_q - \omega_r \psi_f}{R_s} \quad (6)$$

Due to u_q is limited by DC bus voltage when the motor operate in flux-weakening operation area at a certain speed, i_d and i_q cannot be controlled independently any more as the two variables of equation (6). When u_q is fixed and speed is certain, equation (6) is a straight line in i_d-i_q plane. The higher the speed is, the greater the slope of the line determined by equation (6) is, and the stronger the coupling between i_d and i_q is.

Using coupling relationship between i_d and i_q, flux-weakening control strategy based on SCR-FQV is built. The block diagram of flux-weakening control based on SCR-FQV is show in figure 2.

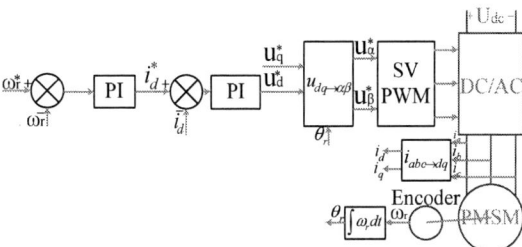

Fig. 2. Flux-weakening control based on SCR-FQV.

The speed regulator generates a current command including both torque and demagnetizing current information. Current vector trajectory of the i_d-i_q coupling line determined by equation (6) is show in figure 3.

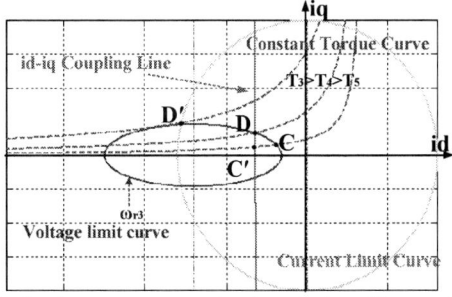

Fig. 3. Steady-state stator current vector trajectory of flux-weakening control based on SCR-VQV.

With output torque T_5, the motor stable operating point is point C', which is the intersection point of the constant torque curve and the i_d-i_q coupling line. Point C' is in the voltage limit ellipse determined by ω_{r1}. The current vector amplitude determined by point C′ is larger than the current vector amplitude determined by point C. Motor efficiency declines.

With the rise of load torque, the maximum output torque of motor is T_4, which is determined by the intersection point D of the voltage limit ellipse and the i_d-i_q coupling line. This torque is lower than the maximum output torque of motor T_5. The motor output maximum torque declines.

To solve this problem, this paper proposed a new given method of q-axis voltage, which called flux-weakening control method based on SCR-VQV (single current regulator and variable q-axis voltage).

As previously described, the intersection point of constant torque curve and voltage limit ellipse is the optimal operating stable point. So the q-axis voltage command is given as follow to control the motor working on the voltage limit ellipse:

When the speed is positive

$$u_q^* = \sqrt{u_{s\max}^2 - u_d^{*2}} \tag{7}$$

When the speed is negative

$$u_q^* = -\sqrt{u_{s\max}^2 - u_d^{*2}} \tag{8}$$

The reverse situation is similar to the forward, this paper only discuss the forward situation.

The diagram of flux-weakening control based on SCR-VQV is show in figure 4.

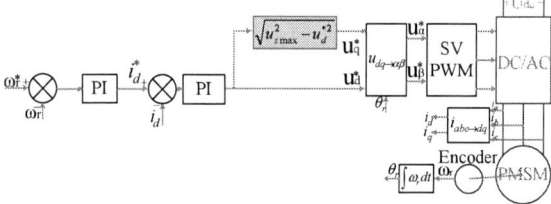

Fig. 4. Flux-weakening control based on SVR-VQV.

The motor stable operating point meets equation (2), (3) and (6). So the operating point is the intersection point of the constant torque curve and the voltage limit ellipse, such as the point F in figure 5. The efficient of the motor is improved.

According to equation (7), $u_q^* \in [0, u_{s\max}]$. With the change of q-axis voltage, the line determined by q-axis voltage equation is moved parallel in i_d-i_q plane. It is shown in figure 5.

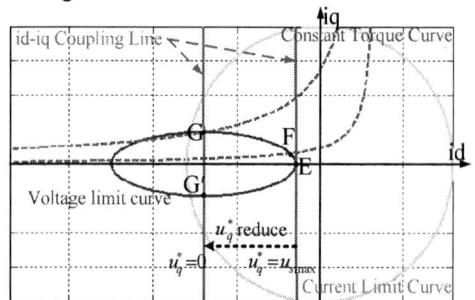

Fig. 5. Steady-state stator current vector trajectory of flux-weakening control based on SCR-VQV.

When the q-axis voltage is the maximum value, point E is the intersection point of the i_d-i_q coupling line and the voltage limit ellipse. When the q-axis voltage is the minimum value, point G and G′ is the intersection point of i_d-i_q coupling line and the voltage limit ellipse. The motor stable operating point is on the arc GFEG′ of the voltage limit ellipse. The motor output maximum torque is improved.

According to steady-state performance analysis, the PMSM efficiency and load capacity can be improved controlled by SCR-VQV compared to SCR-FQV flux-weakening method.

When speed or load torque command is increased, the dynamically adjusting process of the motor controlled by SCR-VQV flux-weakening method is shown in Figure 6.

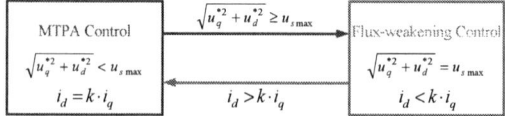

Fig. 6 Dynamically adjusting process of PMSM controlled by SCR-VQV flux-weakening method

When speed or load torque command is increased, the i_d command decreases, then u_d command decreases. According to equation (7), the decrease of u_d command reduces the u_q command. According to PMSM model, i_d and i_q can be calculated by:

$$\begin{cases} i_d = \dfrac{1}{L_d} \int (u_d - R_s i_d + \omega_r L_q i_q) dt \\ i_q = \dfrac{1}{L_q} \int (u_q - R_s i_q - \omega_r L_d i_d - \omega_r \psi_f) dt \end{cases} \tag{9}$$

According to equation (9), the decrease of u_d and u_q command reduces i_d and increases i_q. Then the output torque increases. The motor reaches a new equilibrium.

IV. SIMULATION AND EXPERIMENT RESULTS

In order to verify the above analysis, this paper sets up a simulation model of drive system and an experimental platform.

MTPA (maximum torque per ampere) control method is applied when the motor speed is low or the load is light. The switching between MTPA and flux-weakening control is shown in figure 7.

Fig. 7. The switching rules between MTPA and flux-weakening control based on single current regulator.

The equation $i_d = k \cdot i_q$ is the linearization of MTPA curve. Where k≤0. When the PMSM is salient, k=0.

The simulation model parameters are shown in table I.

TABLE I
PARAMETERS OF SIMULATION MODEL

Item	Value
Rated DC bus voltage	300V
Rated phase current	56.6A
D-axis induction	2.5mH
Q-axis induction	7.5mH
Flux linkage	0.175Wb

The DC bus voltage is fixed at 300V. SVPWM modulation is applied. And constant amplitude principle is observed in coordinate transformation. So the maximum amplitude of stator voltage vector is $300/\sqrt{3} \approx 172$ V.

The motor speed command is 1500 rpm at 0 sec and increase to 3000 rpm at 1 sec. After 2 sec, the motor load torque increases 5 Nm per sec. The motor speed, output torque, current and voltage waveforms are shown in figure 8, figure 9 and figure 10. The left side waveforms belong to SCR-VQV flux-weakening control. The right side waveforms belong to SCR-FQV flux-weakening control.

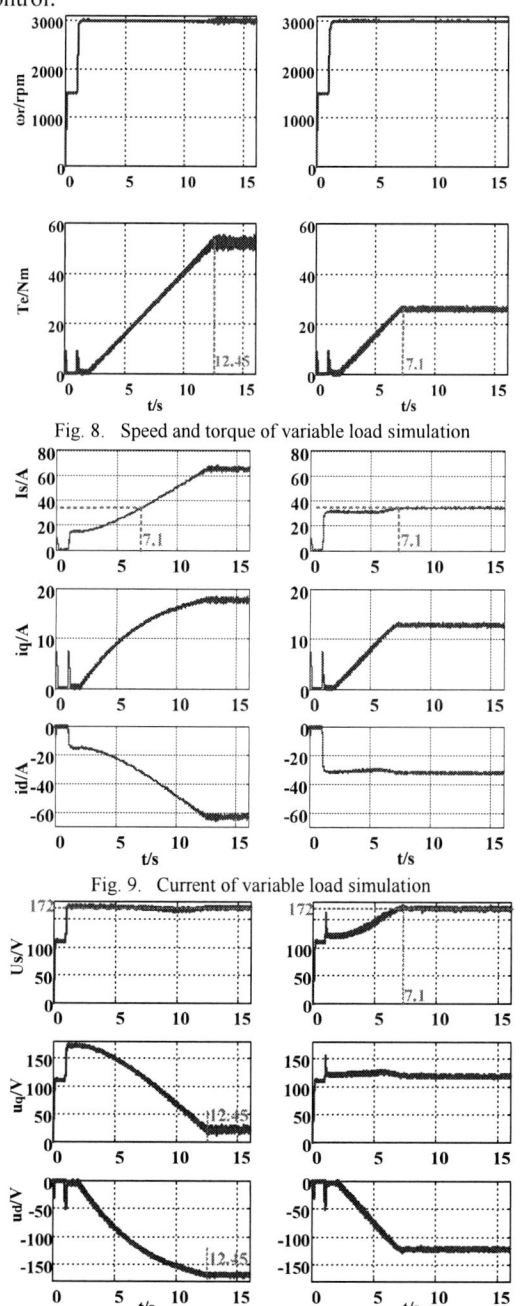

Fig. 8.   Speed and torque of variable load simulation

Fig. 9.   Current of variable load simulation

Fig. 10   Voltage of variable load simulation

In figure 8, the motor output torque can increase to 53

Nm with variable q-axis voltage. The motor output torque just can increase to 25.5 Nm with fixed q-axis voltage. The maximum motor output torque is improved with variable q-axis voltage.

In figure 9, with the same speed and output torque of motor, the current vector amplitude with variable q-axis voltage is smaller than the current vector amplitude with fixed q-axis voltage. The motor efficient is improved with variable q-axis voltage.

In figure 10, the voltage vector amplitude is always the maximum value with variable q-axis voltage. The voltage vector amplitude is smaller than the maximum value with fixed q-axis voltage. The motor stable operating points are on the voltage limit ellipse with variable q-axis voltage. The motor stable operating points are in the voltage limit ellipse with fixed q-axis voltage.

According to the simulation results, the flux-weakening control with variable q-axis voltage can fully utilize the DC voltage and make the motor working at the optimal operating point. The efficiency and load capacity of motor are improved obviously.

The d-axis current and q-axis current in i_d-i_q plane are shown in figure 11.

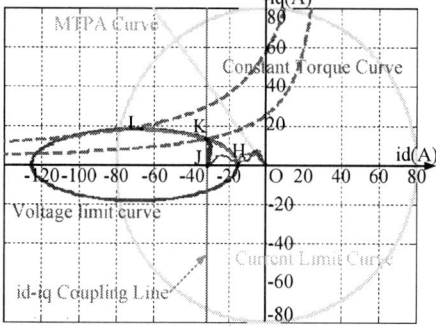

Fig. 11.   Trajectory of stator current vector of variable load simulation.

The trajectory of stator current vector of flux-weakening control based on SCR-VQV is the red curve OHL which is on the current limit ellipse. The trajectory of stator current vector of flux-weakening control based on SCR-FQV is the blue curve OJK which is in the voltage limit ellipse and on the i_d-i_q coupling line.

The PMSM in braking operation condition is simulated. The simulation result is shown in figure 12.

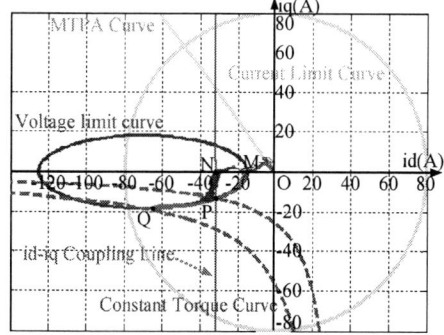

Fig. 12.   Trajectory of stator current vector of variable load simulation in braking condition.

When is the motor speed reach 3000 rpm, the reverse load torque increases 5 Nm per sec. The maximum motor reverse output torque under SCR-VQV flux-weakening control is 50 Nm. The maximum motor reverse output torque under SCR-FQV flux-weakening control is 30 Nm. The trajectory of stator current vector of flux-weakening control based on SCR-VQV is the red curve OMQ which is on the current limit ellipse. The trajectory of stator current vector of flux-weakening control based on SCR-FQV is the blue curve ONP which is in the voltage limit ellipse and on the i_d-i_q coupling line. Simulation result in braking condition is similar to simulation result in traction condition.

The simulation result is consistent with the theoretical result entirely.

The experimental parameters are shown in table II.

TABLE II
PARAMETERS OF EXPERIMENT

Item	Value
Rated DC bus voltage	560V
Rated phase current	12A
Rated speed	1500rpm
Rated torque	35Nm

Constant total power principle is observed in coordinate transformation. So the maximum amplitude of stator voltage vector is $560/\sqrt{2} \approx 380$ V. The maximum value of the current vector amplitude is $12 \times \sqrt{3} \approx 20.8$ A, and limited to 20A in the control program.

In order to verify the dynamic performance of PMSM flux-weakening control based on SCR-VQV, speed step response experiments have made. The motor load is 6Nm and constant. The motor accelerates to 2100rpm and get into flux-weakening area. At 2 sec, the speed command step to 2300rmp. At 3.5 sec, the speed command step back to 2100rmp and keep constant. The motor experimental speed and current waveforms are shown in figure 13.

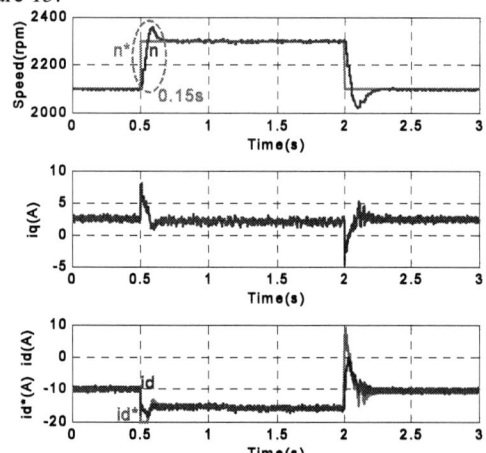

Fig. 13 Waveforms of dynamic regulation of experiment based on SCR-VQV

Dynamically adjusting time of the motor speed is 0.15s. The actual d-axis current coincides with the d-axis current command well. The PMSM has a good dynamic performance under flux-weakening control based on SCR-VQV.

In order to verify flux-weakening control based on SCR-VQV can improve the PMSM's voltage utilization ratio, efficiency and load capability compared to flux-weakening control based on SCR-FQV. Control modes

switching experiments have made. The motor accelerates to 2200rpm under flux-weakening control based on SCR-FQV. At 2 sec, the control mode switches to flux-weakening control based on SCR-VQV. At 3.5 sec, the control mode switches back to flux-weakening control based on SCR-FQV. The motor experimental speed, voltage and current waveforms are shown in figure 14, 15 and 16.

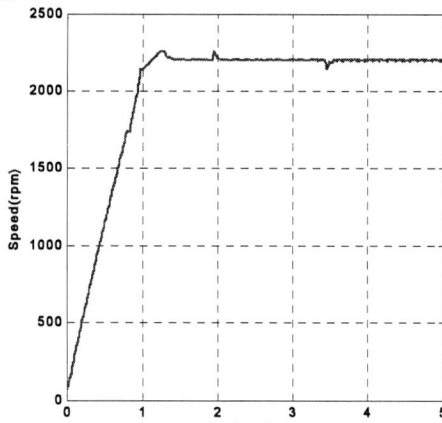

Fig. 14 Speed of control modes switching experiment

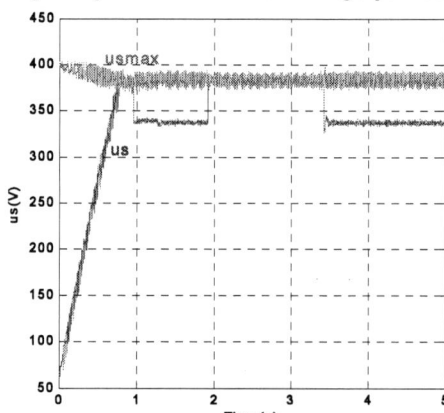

Fig. 15 Voltage of control modes switching experiment

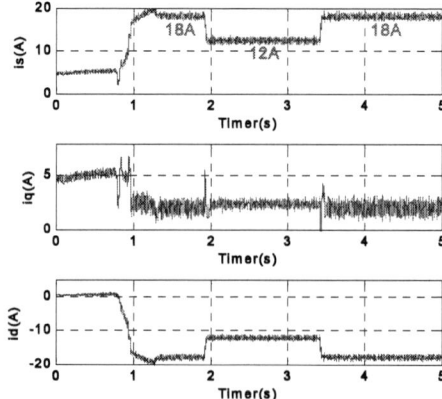

Fig. 16 Current of control modes switching experiment

According to figure 15, the voltage vector amplitude under flux-weakening control based on SCR-FQV is always smaller than the maximum value, but the voltage vector amplitude under flux-weakening control based on SCR-VQV is always the maximum value.

According to figure 16, the current vector amplitude

under flux-weakening control based on SCR-FQV is bigger than the one under flux-weakening control based on SCR-FQV, which are 18A to 12A.

The results of control modes switching experiment are consistent with the theoretical and simulation results well. The stable operating point of PMSM under flux-weakening control based on SCR-VQV is on the voltage limit ellipse. The efficiency and load capability is optimized compared to flux-weakening control based on SCR-FQV.

V. CONCLUSIONS

This paper presents a new method of given q-axis voltage, named flux-weakening control based on SCR-VQV. The new flux-weakening control strategy can fully utilize the DC voltage and make the motor working at the optimal operating point on the basis of maintaining the original advantages. The efficiency and load capacity of motor are improved.

The flux-weakening control based on SCR-VQV will be used to control 300kW traction PMSM for metro train next. The performance under the background of low switching frequency and high power will be researched.

REFERENCES

[1] S. Morimoto, M. Sanada, and Y. Takeda, "Wide-speed operation of interior permanent magnet synchronous motors with high-performance current regulator," IEEE Transactions on Industry Applications, vol. 30, no. 4, pp. 920-926, 1994.

[2] Z. X. Fu, "Pseudo constant power times speed operation in the field weakening region of IPM synchronous machines," Industry Applications Conference, 2003.

[3] P. Ching-Tsai, and S. M. Sue, "A linear maximum torque per ampere control for IPMSM drives over full-speed range," IEEE Transaction on Energy Conversion, vol. 20, no. 2, pp. 359-366, 2005.

[4] B. Bon-Ho, N. Patel, S. Schulz et al. , "New field weakening technique for high saliency interior permanent magnet motor," Industry Applications Conference, 2003. Conference Record of the 38th IAS Annual Meeting. pp. 898-905 vol.2, 2003.

[5] R. U. Lenke, R. W. De Doncker, K. Mu-Shin et al. , "Field Weakening Control of Interior Permanent Magnet Machine using Improved Current Interpolation Technique," Power Electronics Specialists Conference, 2006. PESC '06. 37th IEEE, pp. 1-5, 2006.

[6] J. Ottosson, and M. Alakula, "A compact field weakening controller implementation," International Symposium on Power Electronics, Electrical Drives, Automation and Motion, 2006. SPEEDAM 2006. pp. 696-700, 2006.

[7] K. Jang-Mok, and S. Seung-Ki, "Speed control of interior permanent magnet synchronous motor drive for the flux weakening operation," IEEE Transactions on Industry Applications, vol. 33, no. 1, pp. 43-48, 1997.

[8] Paicu C, Tutelea L, Andreescu Gl, Blaabjerg F, Lascu C, Boldea L. Wide speed range sensorless control of PM-RSM via "active flux model" [C]. IEEE Energy Conversion Congress and Exposition, San Joe, CA, United states, 2009.

[9] Song C, Xu L, Zhang Z. Efficiency-Optimized Flux-Weakening Control of PMSM Incorporating Speed Regulation[C]. Power Electronics Specialists Conference, Orlando, FL, United states, 2007.

[10] Xu L, Zhang Y, Guven M. A New Method to Optimize Q-axis Voltage for Deep Flux Weakening Control of IPM Machines Based on Single Current Regulator[C]. International Conference on Machines and Systems, Wuhan, China, 2008.

[11] Zhang Y, Xu L, Guven M, Song C. Experimental Verification of Deep Flux-Weakening Operation of A 50 kW IPM Machine by Using Single Current Regulator. IEEE Trans. on Industry Applications, 2011, 47(1): 128-133.

[12] Zhu L, Xue S, Wen X, Li Y, Kong L. A New Deep Field-Weakening Strategy of IPM Machines Based on Single Current Regulator and Voltage Angle Control[C]. Energy Conversion Congress and Exposition, Atlanta, GA, United states, 2010.

[13] Zhu L, Wen X, Feng Z, Kong, Zhang B. Deep field-weakening control of PMSMs for both motion and generation operation[C]. International Conference on Eletrical Machines and Systems, Beijing, China. 2011.

[14] Miyajima T, Fujimoto H, Fujitsuna M. A Precise Model Based Design of Voltage Phase Controller for IPMSM [J]. IEEE Trans on Power Electronics, 2013, 28(12): 5655-5664.

Predictive Current Control Method in Induction Motor Speed Sensorless Drive

Sun Wei, Yu Yong, Xu Dianguo, Xu Jin, Ding Li
School of Electrical and Electronic Engineering
Harbin Institute of Technology
Harbin, China

***Abstract*-The sensorless vector control method in induction motor drives is utilized more and more popular in industry applications. The traditional current loop controller is PI regulator but its bandwidth is limited by switching frequency. In this paper a predictive current controller is proposed to solve the limitation of bandwidth in sensorless vector control system. The stability condition of proposed method is analyzed. Meanwhile the sensitivity of predictive system to motor parameters is analyzed and a robust predictive control method is proposed. To compared with bandwidth of PI control system, bandwidth of predictive is tested. The experiments results demonstrate that the bandwidth of proposed predictive system is higher than traditional PI regulator and the feasibility of proposed method is also proved.**

I. INTRODUCTION

The speed sensorless vector control method is widely applied in induction applications because it can reduce dependency on speed sensor that can not be utilized in execrable situation [1]. Furthermore stator current can be decoupled to flux and torque current (named as d-axis current and q-axis current) separately by vector control method and system dynamic response is improved [2].

In traditional current loop the decoupling flux and torque current are controlled by PI regulator [3]. PI regulator is simple and can achieve fine control effectiveness. However, the bandwidth of current loop controlled by PI regulator, is limited by switching frequency and digital delay in digital control system. Consequently the bandwidth of current loop can not achieve very high because of restriction of stability margin that is related to digital delay and switching frequency. Meanwhile the faster system dynamic response is required in recent induction applications, so it is necessary to focus on the predictive control method which is reported to have better properties than PI regulator [4-8].

Although predictive current control has more efficient effect, some problem still exists in predictive control system. In [9], the predictive method is reported to be sensitive to changes of motor parameters, and the inaccuracies can lead to be unstable when motor operates long time. To solve this problem, a predictive current system that is insensitive to motor parameters, is proposed in [10].

In this paper, the predictive current control method with robust coefficients is proposed. The restriction of improvement for current bandwidth in PI regulator is ana-

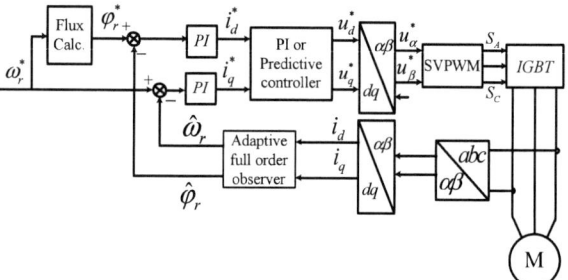

(a) Sensorless vector control block diagram.

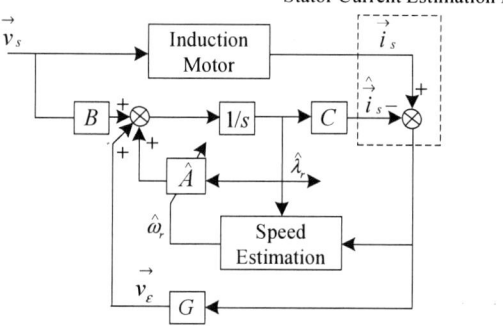

(b) Structure figure of adaptive full order observer
Figure. 1. Structure figure of control system.

lyzed. For predictive current controller, the sensitivity to motor parameters is also analyzed and two robust coefficients are introduced to system to reduce sensitivity to motor parameters. To get accurate sensitivity to motor parameters, the relationship between robust coefficient and tolerance for change of motor parameters, is presented.

II. STATE OBSERVER

A. Adaptive Full Order Observer

In the stationary reference the dynamic model of an induction motor can be expressed as follows:

$$p\vec{\lambda}_s = A_{11}\vec{\lambda}_s + A_{12}\vec{\lambda}_r + I\vec{v}_s \quad (1)$$

$$p\vec{\lambda}_r = A_{21}\vec{\lambda}_s + A_{22}\vec{\lambda}_r \quad (2)$$

$$\vec{i}_s = C_1\vec{\lambda}_s + C_2\vec{\lambda}_r \quad (3)$$

where

$$\vec{\lambda}_s = \begin{bmatrix} \lambda_{s\alpha} & \lambda_{s\beta} \end{bmatrix}^T, \vec{\lambda}_r = \begin{bmatrix} \lambda_{r\alpha} & \lambda_{r\beta} \end{bmatrix}^T, \vec{v}_s = \begin{bmatrix} v_{s\alpha} & v_{s\beta} \end{bmatrix}^T,$$

$$\vec{i}_s = \begin{bmatrix} i_{s\alpha} & i_{s\beta} \end{bmatrix}^T, A_{11} = -a_{11}I, A_{12} = a_{12}I, A_{21} = a_{21}I,$$

978-1-4799-2706-7/14 $31.00 © 2014 IEEE

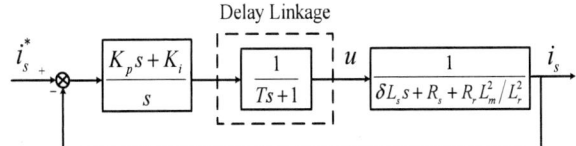

Fig. 2 Block diagram of current loop with PI regulator

$$A_{22} = -a_{22}I + \omega_r J, \quad C_1 = c_{11}I, \quad C_2 = -c_{12}I,$$

$$I = \begin{bmatrix} 1 & 0 \\ 0 & 1 \end{bmatrix}, \quad J = \begin{bmatrix} 0 & -1 \\ 1 & 0 \end{bmatrix},$$

$$a_{11} = \frac{R_s}{\delta L_s}, \quad a_{12} = \frac{R_s L_m}{\delta L_s L_r}, \quad a_{21} = \frac{R_r L_m}{\delta L_s L_r}, \quad a_{22} = \frac{R_r}{\delta L_r},$$

$$c_{11} = \frac{1}{\delta L_s}, \quad c_{12} = \frac{L_m}{\delta L_s L_r}, \quad \delta = 1 - \frac{L_m^2}{L_s L_r}$$

The adaptive full-order observer shown in Fig.1 (a) is can be constructed using (1), (2) and (3) as follows:

$$p\vec{\hat{\lambda}}_s = A_{11}\vec{\hat{\lambda}}_s + A_{12}\vec{\hat{\lambda}}_r + I\vec{v}_s + G(\vec{i}_s - \vec{\hat{i}}_s) \qquad (4)$$

$$p\vec{\hat{\lambda}}_r = A_{21}\vec{\hat{\lambda}}_s + \hat{A}_{22}\vec{\hat{\lambda}}_r \qquad (5)$$

$$\vec{\hat{i}}_s = C_1\vec{\hat{\lambda}}_s + C_2\vec{\hat{\lambda}}_r \qquad (6)$$

where $G = \begin{bmatrix} g_1 & -g_2 & g_3 & -g_4 \\ g_2 & g_1 & g_4 & g_3 \end{bmatrix}^T$

The rotor speed is estimated using Lyapunov's theorem and transformed to synchronously rotating reference:

$$\hat{\omega}_r = k_p(i_{sq} - \hat{i}_{sq})\hat{\lambda}_{rd} + k_i \int (i_{sq} - \hat{i}_{sq})\hat{\lambda}_{rd} dt \qquad (7)$$

The symbol "^" signifies the estimated values and variables. The $\hat{A}_{22}$ is the matrix of coefficient in which ω_r is replaced by estimated speed $\hat{\omega}_r$.

B. Traditional PI Controller in Current Loop

Based on induction motor equations, the block diagram of current loop in sensorless vector control system is shown in Fig. 2. The open loop transfer function of current loop is:

$$G_{op}(s) = \frac{K_p(s + K_i/K_p)}{\delta L_s s(Ts+1)(s + \frac{R_s + R_r L_m^2/L_r^2}{\delta L_s})} \qquad (8)$$

To offset great inertia linkage in (8), the coefficients in PI controller are described by:

$$\frac{K_i}{K_p} = \frac{R_s + R_r L_m^2/L_r^2}{\delta L_s} \qquad (9)$$

Considering (8) and (9), the open loop transfer function can be simplified as followed:

$$G_{op}(s) = \frac{K_p}{\delta L_s s(Ts+1)} \qquad (10)$$

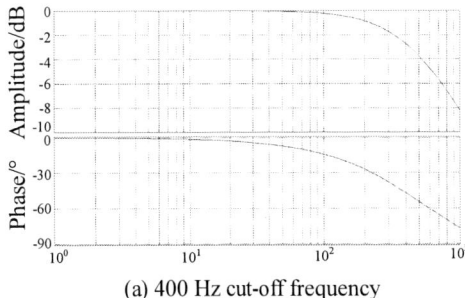

(a) 400 Hz cut-off frequency

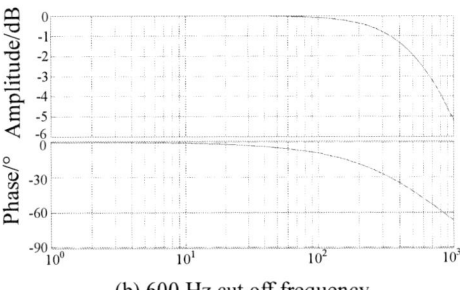

(b) 600 Hz cut-off frequency

Fig. 3 Bode diagrams with different cut-off frequency

According to (10) and analyzed in signal frequency domain, the cut-off frequency of current loop can be obtained:

$$\omega_b = \frac{K_p}{\delta L_s} \qquad (11)$$

Considering the system stability, cut-off frequency must meet the condition:

$$\omega_b < \frac{1}{T} \qquad (12)$$

On the basis of (12), the system stability margin γ can be derived as followed:

$$\gamma = 90° - \arctan \omega_b T \qquad (13)$$

Based on (9) and (11), the proportional coefficient K_p and integral coefficient K_i in PI controller can be obtained be setting cut-off frequency.

The Bode diagram of current loop transfer function is shown in Fig. 3. In Fig. 3 (a), cut-off frequency is set at 400Hz, and in Fig.3 (b) cut-off frequency is 600Hz. As Fig. 3 showed, the higher cut-off frequency is set, the higher bandwidth of current loop is. However in order to ensure system stability and make better dynamic characteristics, the highest cut-off frequency is 500Hz as usual. As a result the bandwidth of current loop is less than 500Hz.

III. PREDICTIVE CURRENT CONTROL METHOD

With an ideal condition, reference voltage u*(k) should be utilized in period (k) to (k+1) as shown in Fig. 4. However, due to the hysteresis characteristic in digital control system, u*(k) always delay behind period (k) to (k+1).

978-1-4799-2706-7/14 $31.00 © 2014 IEEE

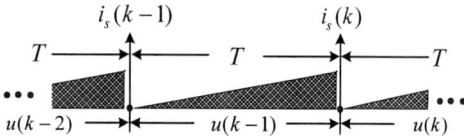

Fig. 4 Notation of switching period in predictive control system

As it is analyzed in last section, the hysteresis in controller restricts improvement of current loop bandwidth. Nevertheless in predictive current control system, reference voltage u*(k) can be obtained in period (k-1) to (k). So the hysteresis problem can be resolved.

For small control period T and rotor flux orientation, the section of rotor flux in rotating q-axis λ_{rq} is zero, and $\lambda_{rd} = L_m i_{sd}$. The mathematical dynamic equations of induction motor in rotating reference are possible to convert to discrete form:

$$
\begin{cases}
v_{sd}(k)=P_1 i_{sd}(k)-P_2 i_{sd}(k+1) \\
\qquad +P_4 \omega_e i_{sq}(k) \\
v_{sq}(k)=P_6 i_{sq}(k)-P_2 i_{sq}(k+1) \\
\qquad -P_4 \omega_e i_{sd}(k)-P_5 \omega_r i_{sd}(k)
\end{cases}
\tag{14}
$$

where

$$
P_1=\frac{\delta L_s}{T}-R_s, P_2=\frac{\delta L_s}{T}, P_3=\frac{T}{\delta L_s}, P_4=\delta L_s, P_5=\frac{L_m^2}{L_r}
$$

$$
P_6=\frac{\delta L_s}{T}-R_s-\frac{L_m^2 R_r}{L_r^2}
$$

In (14) the variables of $v_{sd}(k)$, $v_{sq}(k)$, $i_{sd}(k+1)$ and $i_{sq}(k+1)$ are separately set as $u^*_{sd}(k)$, $u^*_{sq}(k)$. $i^*_{sd}(k+1)$ and $i^*_{sq}(k+1)$. And the symbol "*" signifies reference value. Consequently the (14) can be converted to

$$
\begin{cases}
u^*_d(k)=P_1 i_{sd}(k)-P_2 i^*_{sd}(k+1) \\
\qquad +P_4 \omega_e i_{sq}(k) \\
u^*_q(k)=P_6 i_{sq}(k)-P_2 i^*_{sq}(k+1) \\
\qquad -P_4 \omega_e i_{sd}(k)-P_5 \omega_r i_{sd}(k)
\end{cases}
\tag{15}
$$

Considering (15) in period (k) to (k+1), $i^*_{sd}(k+1)$ and $i^*_{sq}(k+1)$ are the reference current that obtained from output of speed PI regulator. The value of $i_{sd}(k)$ and $i_{sq}(k)$ is sampling current. Consequently the predictive current controller can be derived from (15). The reference voltage u*$_{sd}$(k) and u*$_{sq}$(k) will be utilized in period (k+1) to (k+2) to make the sampling current equals to reference current..

However when motor parameters are not accurate enough, the predictive current system may be unstable when motor is operating. So a robust coefficient will be introduced to make system insensitive to motor parameters.

Firstly considering motor mathematical equations, the sensitivity of predictive current system to motor parameters is analyzed as followed:

Taking stator current in rotating d-axis as a example, the interference in (14) and (15) is neglected

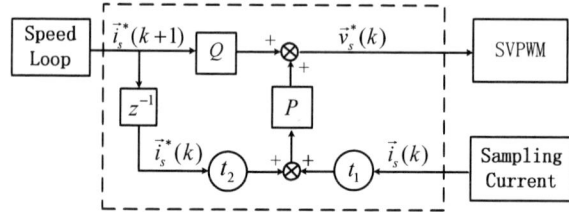

Fig. 5 Structure of predictive current controller

$$
\begin{cases}
v_{sd}(k)=P'_6 i_{sd}(k)-P'_2 i_{sd}(k+1) \\
u^*_d(k)=P_6 i_{sd}(k)-P_2 i^*_{sd}(k+1)
\end{cases}
\tag{16}
$$

If real motor parameters is different from model motor parameters, and $v_{sd}(k)$ is equal to $u^*_d(k)$ in (16), a new equation is derived as followed:

$$
H' i_{sd}(k)-H' i_{sd}(k+1)=H i_{sd}(k)-H i^*_{sd}(k+1) \tag{17}
$$

where

P'_2, P'_6, H' : real motor parameters;

P_2, P_6, H : model motor parameters

Considering period (k) to (k+1), the discrete transfer function of predictive system is derived from (17):

$$
G(z)=\frac{i_{sd}(z)}{i^*_{sd}(z)}=\frac{Hz}{H'z+H-H'} \tag{18}
$$

The characteristic equation of (18) is:

$$
H'z+H-H'=0 \tag{19}
$$

The criteria of robust stability in continuous-time system can not be utilized for predictive current control system which is a discrete system. Therefore with the purpose of using criteria of robust stability in continuous-time system, it is necessary to convert (19) to w-domain with $z = (w+1)/(w-1)$, and the pole of transfer function in w-domain is

$$
w=1-\frac{2H'}{H} \tag{20}
$$

In order to make system stable, the real part of pole as shown in (20) must be negative. Hence derived from (20), the predictive current system is unstable when H′ <0.5H.

Because predictive current control system is sensitivity to motor parameters, two coefficients are introduced to the control system. The sampling current in (15) can be described by

$$
i_{sd}(k)=t_1 i_{sd}(k)+t_2 i^*_{sd}(k), \quad i_{sq}(k)=t_1 i_{sq}(k)+t_2 i^*_{sq}(k)
$$

where t$_1$>0, t$_2$>0, t$_1$+t$_2$=1.And the structure of predictive current controller is shown in Fig. 5. Predictive reference voltage is

$$
\vec{v}_s(k)=P\vec{i}_s(k)+Q\vec{i}^*_s(k+1) \tag{21}
$$

Where

$$
P=\begin{bmatrix} P_1 & P_4\omega_e \\ -P_4\omega_e-P_5\omega_r & P_6 \end{bmatrix}, Q=\begin{bmatrix} P_2 & 0 \\ 0 & P_2 \end{bmatrix}
$$

With coefficient t_1 and t_2, the discrete transfer function of predictive system is

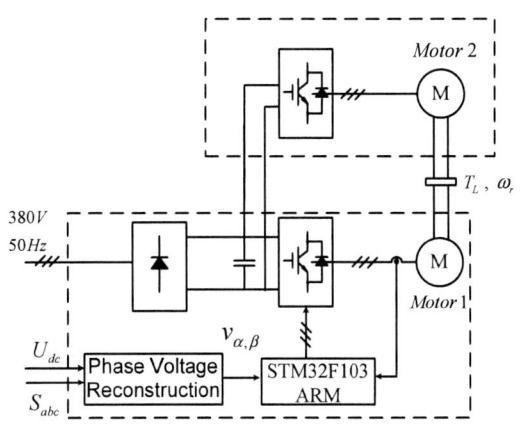

Fig. 6 Structure of experimental platform

$$G'(z) = \frac{i_{sd}(z)}{i_{sd}^*(z)} = \frac{Hz + H't_2 - Ht_2}{H'z + Ht_1 - H't_1} \quad (22)$$

The characteristic equations in w-domain is

$$(H' + Ht_1 - H't_1)w - (Ht_1 - H't_1 - H') = 0 \quad (23)$$

The pole of transfer function in w-domain is

$$w = \frac{Ht_1 - H't_1 - H'}{Ht_1 - H't_1 + H'} \quad (24)$$

Based on the criteria of robust stability in continuous-time system, predictive current control system is stable when

$$t_1(1 - \frac{H'}{H}) - \frac{H'}{H} < 0 \quad (25)$$

From (25) the relationship between coefficient t_1 and error of motor parameters is obtained as shown in Table I. In Table I, it is obvious that the smaller t_1 is set, the stronger the tolerance of motor parameters error is. And robustness of predictive current control system becomes strong.

TABLE I
RELATIONSHIP BETWEEN t_1 AND ERROR OF MOTOR PARAMETERS

t_1	H'/H
1.0	2
0.9	2.1
0.8	2.2
0.7	2.4
0.6	2.6
0.5	3
0.4	3.5
0.3	4.3
0.2	6
0.1	11

IV. EXPERIMENTAL RESULTS

The proposed predictive current method is verified using a laboratory platform built around a STM32F103 ARM and switching frequency is 6 KHz. The experimental platform adopting 7.5 kW induction motor is shown in Fig.6. Motor 1 is the tested motor and motor 2 provide load for motor 1. The tested motor parameters are given in Table □.

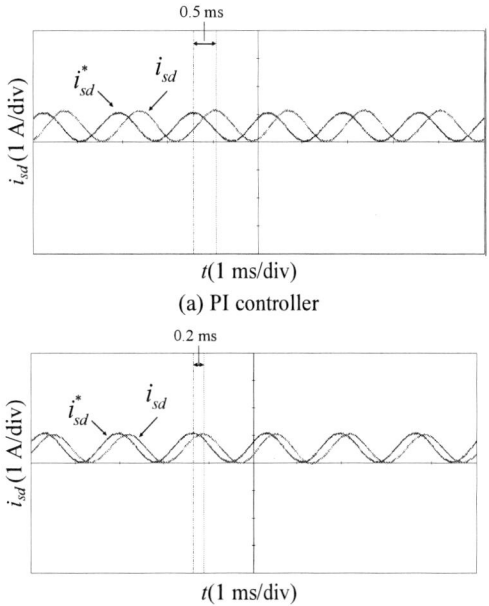

(a) PI controller

(b) Predictive current controller

Fig. 7 Current loop bandwidth test with injected 600Hz sine signal

The experiment of current loop bandwidth is shown in Fig.7. The cut-off frequency of traditional PI controller is 600Hz, which is the highest cut-off frequency that PI controller can reach. Meanwhile in predictive current controller, robust coefficient t_1 is 0.85. In Fig. 7 (a), A 600Hz sine signal is injected to the d-axis reference current. The phase angle of sampling current delay behind reference current about 108°. However, in predictive control system the phase angle of sampling current delay behind reference current about 43°.

In Fig. 8, the experiments of motor speed reverse with 50% of rated load torque at 0.2-p.u.speed is presented, The flux current component of stator current is kept constant. The estimated speed is also correct with little oscillations. And the system operates stably.

In Fig. 9, the motor operates at 0.2-p.u. speed with increasing load torque which is from 0% to 100% rated load torque. As Fig. 9 (a) shown, the flux current component and torque current component is kept stably. In Fig. 9 (b), the estimated speed is kept constant except small oscillation.

TABLE II
PARAMETERS OF TESTED INDUCTION MOTOR

Symbol	Parameters	Values
P_e	Rated Power	7.5kW
V_e	Rated Voltage	380V
I_e	Rated Current	15.6A
f_e	Rated Frequency	50Hz
n_e	Rated Speed	1440rpm
P_n	Pole Pair	2
R_s	Stator Resistance	0.567Ω
R_r	Rotor Resistance	0.441Ω
L_m	Magnetizing Inductance	110.1mH
L_s	Stator Self-inductance	114.1mH
L_r	Rotor Self-inductance	114.1mH

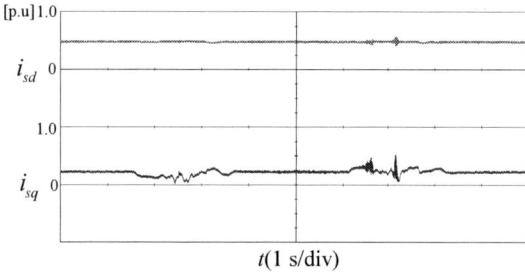

(a) d-axis and q-axis component of stator current

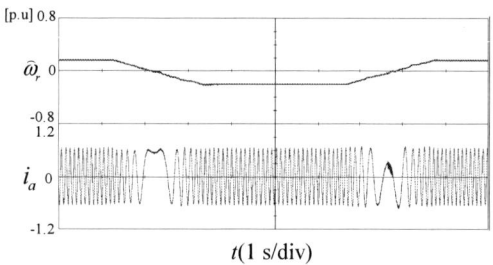

(b) Estimated speed and stator current

Fig. 8 System response of speed reverse with 50% of rated torque at 0.2-p.u. speed

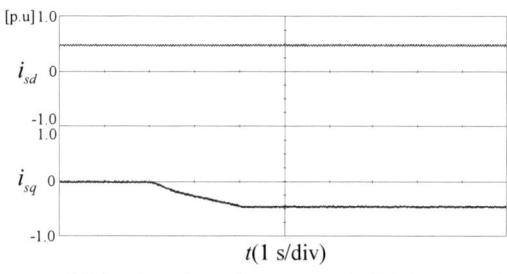

(a) d-axis and q-axis component of stator current

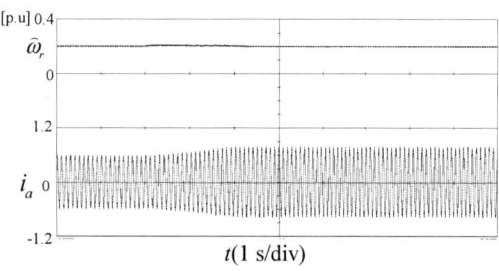

(b) Estimated speed and stator current

Fig. 9 System response with load torque increased from 0% to 100% of rated torque at 0.2-p.u. speed

V. CONCLUSION

The sensorless vector control method in induction motor drives is utilized more and more popular in industry applications. The traditional current loop controller is PI controller but its bandwidth is limited by switching frequency. In this paper a predictive current controller is proposed to solve the limitation of bandwidth in sensorless vector control system. The stability condition of proposed method is analyzed.

Bandwidth of predictive and PI controller is tested. Through experiments it is proved that bandwidth of proposed predictive system is higher than traditional PI controller.

REFERENCES

[1] J. Holtz, "Sensorless control of induction motor," *Proc. IEEE*, vol. 90, no. 8, pp. 1358-1394, 2002.
[2] A.V. Ravi Teja, C. Chakraborty, "A New Model Reference Adaptive Controller for Four Quadrant Vector Controlled Induction Motor Drives," *IEEE Trans. on Industrial Electronics*, vol. 59, no. 10, pp. 3757-3767, 2012.
[3] T. Orlowska-Kowalska, M. Dybkowski, "Stator-Current-Based MRAS Estimator for a Wide Range Speed-Sensorless Induction Motor Drive," *IEEE Trans. on Industrial Electronics*, vol. 57, no. 4, pp. 1296-1038, 2010.
[4] S. Mariethoz, A. Domahidi, M. Morari, "High-Bandwidth Explicit Model Predictive Control of Electrical Drives," *IEEE Trans. on Industry Applications*, vol. 48, no. 6, pp. 1980-1992, 2012.
[5] M. Cirrincione, M. Pucci, "An MRAS-based sensorless high-performance induction motor drive with a predictive adaptive model," *IEEE Trans. on Industrial Electronics*, vol. 52, no. 2, pp. 532-551, 2005.
[6] L. Ben-Brahim, A. Kawamura, "Digital current regulation of field-oriented controlled induction motor based on predictive flux observer," *IEEE Trans. on Industry Applications*, vol. 27, no. 5, pp. 956-961, 1991.
[7] J. Rodriguez, J. Pontt, "Predictive Current Control of a Voltage Source Inverter," *IEEE Trans. on Industrial Electronics*, vol. 54, no. 1, pp. 495-503, 2007.
[8] P. Cortes, J. Rodriguez, "Delay Compensation in Model Predictive Current Control of a Three-Phase Inverter," *IEEE Trans. on Industrial Electronics*, vol. 59, no. 2, pp. 1323-1325, 2012.
[9] J. Moreno, J. Huerta, "A Robust Predictive Current Control for Three-Phase Grid-Connected Inverters." *IEEE Trans. on Industrial Electronics*, vol. 56, no. 6, pp. 1993-2004, 2009.
[10] J. Guzinski, H. Abu-Rub, "Speed Sensorless Induction Motor Drive With Predictive Current Controller," *IEEE Trans. on Industrial Electronics*, vol. 60, no. 2, pp. 699-709, 2013.

Real-Time Implementation of an Online Model Predictive Control for IPMSM Using Parallel Computing on FPGA

Michael Leuer, Joachim Böcker
Power Electronics and Electrical Drives
Paderborn University
D-33095 Paderborn, Germany
leuer@lea.upb.de, boecker@lea.upb.de

Abstract—**Model Predictive Control (MPC) offers a variety of advantages compared to conventional control methods. The problem with MPC is the high computational cost and the associated long control cycle time. This makes MPC unattractive for processes with small time constants, as in permanent magnet synchronous motors with interior magnets (IPMSM) for electric vehicles. In this paper a Model Predictive Control method for nonlinear systems with inherent output saturation is presented. This approach offers real-time capability for online MPC even for processes with time constants in the millisecond range. This becomes feasible by the possibility of parallel computation, as provided by a FPGA (Field Programmable Gate Array). This can overcome the drawbacks of the high computational effort. After the functional principle of this real-time MPC approach is presented, the resulting performance is shown by simulation results and the real-time capability is verified by test results of an IPMSM for automotive applications on a testbench.**

Keywords—*IPMSM Drive Control, Model Predictive Control, Real-Time Implementation*

I. INTRODUCTION

Interior permanent magnet synchronous motors (IPMSM) are favoured for traction drives due to their high power and torque densities and high efficiency [1]. To fully exploit the performance of an electric drive, adequate control is essential. Conventional control for IPMSM is in general based on PI-controls, such as introduced in [2], [3]. For the PI-control of an IPMSM a decoupling of the current components is frequently applied. However, the decoupling can only be guaranteed in the non-saturated case. Moreover a trade-off between short rise time and minimum overshoot has to be made to avoid excessive overshoot even in worst-case operating points because the PI-controller parameters are calculated only once during the design process. Thus the control dynamics are only satisfactory in linear operating points, which are not influenced by saturation due to actuator control limits. A suitable solution is to use reconfigurable controls [4]. However, it can be problematic to switch between the different control algorithms because of transients. That is why the principle of the Model Predictive Control (MPC) is more effective. MPCs are based on the solution of a dynamic optimization problem on a moving horizon. A system model is used to predict the state variable

trajectory over a specified time horizon. Thereby the actuating variables can be scheduled so that a faster and better damped transient response is achieved in comparison to conventional concepts. Control and state variable constraints are directly taken into account in the optimization process. Hence, even further optimization goals such as the minimization of the cupper losses can be considered in the optimization process [5]. Despite many advantages of MPC approaches, enormous demands on processing power for high-dynamic control systems, as they are common in the electric drive technology, have to be faced. However, there are new MPC approaches which enable a reduction of the online computation time. Here, the explicit MPC [6] has to be mentioned. The optimization is performed offline in the controller design. By doing so, the online computational complexity of the MPC is drastically reduced. Such explicit MPC for IPMSM drives have been presented in [7], [8] and [9]. As a fast online method the sub-optimal MPC, as presented in [10] is worth mentioning. This method reduces the computational complexity by calculating only a certain number of iterations for the optimization in each sampling step. The established MPC calculates the optimal actuating variable from an infinite number of possibilities with an optimization algorithm. The Vector-MPC [11] only considers the adjustable voltage vectors of the switching states to solve the optimization problem best. In this MPC approach the online minimization problem is solved with a resource-efficient optimization algorithm. The basic functioning as well as the simulation results have already been shown in [11], so this contribution focusses on the real-time capability of this Model Predictive Control.

II. VECTOR-MPC APPROACH

In conventional MPC the optimization algorithm calculates an optimal control sequence with respect to a defined quality function and given limitations. As a result, a mean voltage value is calculated as the actuating variable. These mean values can be generated from adjustable voltage vectors by using a pulse width modulation (PWM).

In the method presented here, the optimization problem is addressed differently. With the inverter (see Figure 1) 2^3 switch states can be connected, so 8 voltage vectors can be

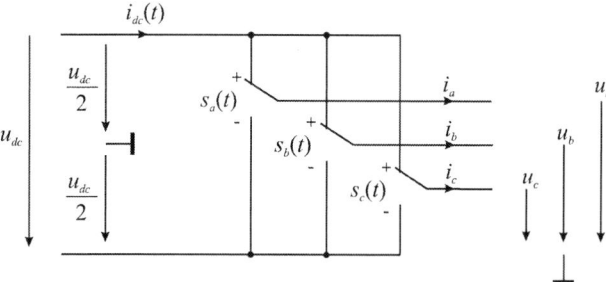

Fig. 1. Block diagram of the inverter

realized. These voltage vectors are shown in Figure 2. Since the voltage vectors $\underline{v}_0$ and $\underline{v}_7$ are the same, only 7 different voltage vectors are possible. The following proposal, called Vector-MPC, uses an optimization algorithm to compute the optimal voltage vector, which can be implemented with high computational efficiency. Here, it is investigated which of the 8 voltage vectors minimizes the quality function best. The basic procedure has been introduced in [9] as an offline MPC. While

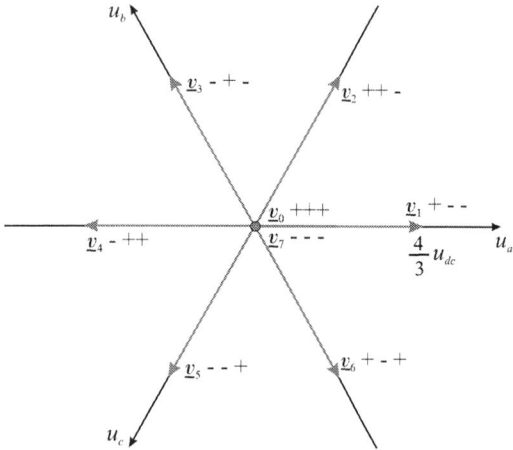

Fig. 2. Chooseable voltage vectors of the inverter

in conventional MPC all actuating variables must be taken into consideration, the Vector-MPC only considers 8 realizable voltage vectors. By doing so, the computational cost can be significantly reduced, so that a reduction of the controller cycle time is possible. The fact that switching is only possible to the discrete sampling periods leads to high current ripple. To lower it to an acceptable level, further reduction of the control cycle time is required.

A. Working Principle of the Real-time Online Vector-MPC

In order to find the optimal voltage vector for the next output step, the eight possible voltage vectors are applied to the plant-model contained in the MPC and each model response is calculated. In this way, the resulting quality function for each voltage vector can be calculated. As output variable, the voltage vector is chosen which fulfills the smallest value of the quality function. This procedure allows saving

the computation and time consuming optimization algorithm. Furthermore, this algorithm is suitable for the implementation on a FPGA, as the model calculations and the calculation of the quality functionals associated with the voltage vectors can be performed in parallel. Figure 3 represents the basic work principle of the Vector-MPC for a prediction horizon of $n_s = 1$ step. A schematic flowchart of this MPC-approach is shown in [11]. A control cycle time of $10\,\mu s$ can be realized easily by splitting the control algorithm into parts processed be either the dSPACE DS1006 quad-core processor board or the dSPACE DS5203 FPGA board (see Figure 4). As advantages of this Vector-MPC approach the possibility of parallelization (see Figure 3), the low computational complexity, as well as the optimal use of the available DC-link voltage can be mentioned. While the consideration of the output limit in a conventional MPC optimization algorithm may require a high computational effort, the output limit is implied by the fact, that the Vector-MPC directly sets the full voltage vector. Unlike other control algorithms, the whole spanned hexagon from the eight possible voltage vectors (see Figure 2) can be chosen as the working area. Moreover, no further processing of the actuating variables, generated by the controller, by PWM or similar components is necessary because this approach directly generates the gate signals for the inverter.

B. Vector-MPC Implementation

The actual control algorithm is performed in a parallel manner on the FPGA board to achieve a controller cycle time of $10\,\mu s$. The calculation of the state model and the update of the non-linear plant model, is executed on the processor board with a cycle time of $100\,\mu s$. The resulting error during this 10 times longer cycle time is very small, as the state matrices are changing only slowly. The data exchange between the two control boards is done using the peripheral high-speed bus (PHS). While splitting up the control algorithm, it

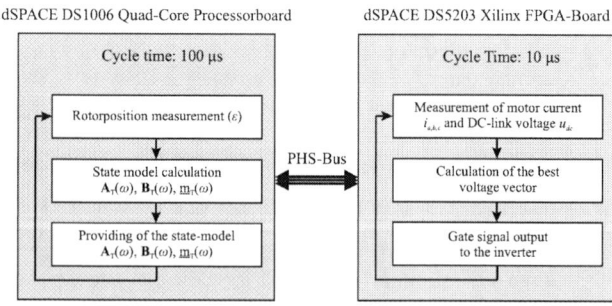

Fig. 4. Distribution of the control algorithm on the processor and FPGA board

is necessary to pay particular attention to the consideration of the different cycle times. Also the output control signals have to be taken into consideration because of the limited switching frequency of the converter. If the switching of the inverter is allowed in every cycle of the Vector-MPC, a very good controller performance will be achieved. On the other hand, the inverter can be thermally destroyed due to the high switching losses. Therefore, the switching frequency was

978-1-4799-2706-7/14 $31.00 © 2014 IEEE   347

The 2014 International Power Electronics Conference

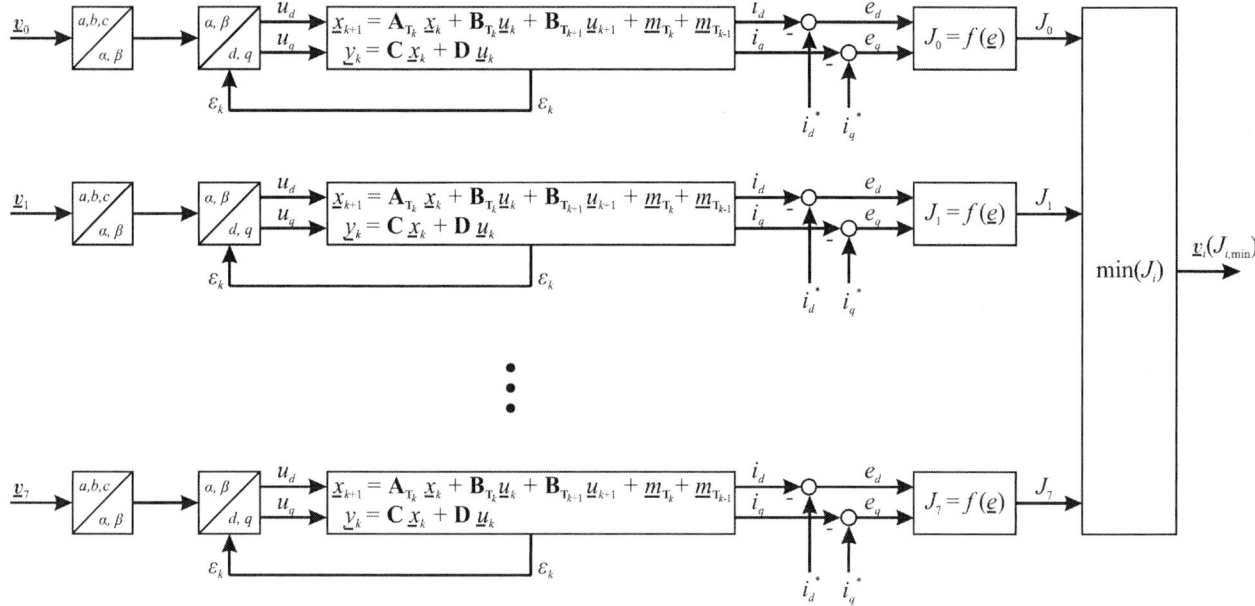

Fig. 3. Basic work principle of the Vector-MPC

limited to $10\,\text{kHz}$. So the Vector-MPC and the PI control with PWM will have the same maximum switching period for the following comparisons.

Following successful implementation of the Vector-MPC scheme using MATLAB Simulink and Xilinx System Generator, the performance of the Vector-MPC method was shown by simulations and the real-time capability was demonstrated by successful implementation on a testbench.

C. Quality Function Design

After implementation of the parameterization, the MPC has to be configured. In the Model Predictive Control this is very easy because the MPC is based on the optimization of a quality function. Therefore it is only necessary to define a suitable quality function with appropriate weighting matrices. For the presented Vector-MPC the following quality function is selected:

$$J = \sum_{i=0}^{n_s} \underline{e}_i^{\text{T}} \mathbf{Q} \underline{e}_i + \Delta \underline{u}_i^{\text{T}} \mathbf{R} \Delta \underline{u}_i + \Delta \underline{s}_i^{\text{T}} \mathbf{S} \underline{s}_i + F^{\Delta f_i} \longrightarrow \min \tag{1}$$

with

$\underline{e} = \underline{r}_k - \underline{x}_k$ ... control error after execution of the respective voltage vector;

$\Delta \underline{u} = \underline{u}_k - \underline{u}_{k+1}$ ... gradient of the respective voltage vector in (d, q)-coordinates;

$\Delta \underline{s}$ ... number of necessary switching procedures for the respective voltage vector;

Δf ... number of switching procedures during the last five sampling periods that are higher than one;

$\mathbf{Q}, \mathbf{R}, \mathbf{S}$ ... weighting matrix of the control error, the gradient and the switching procedures;

F ... weighting parameter of the switching frequency.

By strong weighting of the control error, a possible quick adjustment of the set value is reached, while a strong weighing of the voltage vector gradient results a slowly changing actuating variable sequence. By weighing the switching operations, the switching frequency can be affected. That way, an optimal trade-off between switching losses and dynamics of the control has to be found. The controller design, and the configuration of the MPC can be represented as Pareto optimization.

III. SIMULATION RESULTS

To verify the operation of the Vector-MPC technique, a simulative comparison of the MPC and the PI controller was performed. The PI-controller was designed according to the magnitude optimum method. The simulation results are shown in Figure 5. It can be seen that the MPC approach is superior to the PI control with regard to the control performance. The MPC excels in short rise time and low overshoot. The weighting of the switching frequency and the gradient of the voltage vectors make it possible to reduce the switching frequency in stationary operation compared to the PI control, without increasing the current ripple too much. To investigate the dynamics of the Vector-MPC with respect to its small-signal behaviour, the simulation results for a small step response are shown in Figure 6. Comparing Figure 5 with Figure 6, we see the benefits of the non-linear control algorithm over the linear PI-controller. While the step response of the linear PI control in non-saturated case has the same shape for both the small signal behaviour, as well as the large-signal behaviour it can be seen that the MPC always seeks the quickest way to compensate the control error. A further reduction of the current ripple would be possible with an even smaller controller step-size. Since the calculation time on the FPGA is less than $1\,\mu\text{s}$

978-1-4799-2706-7/14 $31.00 © 2014 IEEE   348

The 2014 International Power Electronics Conference

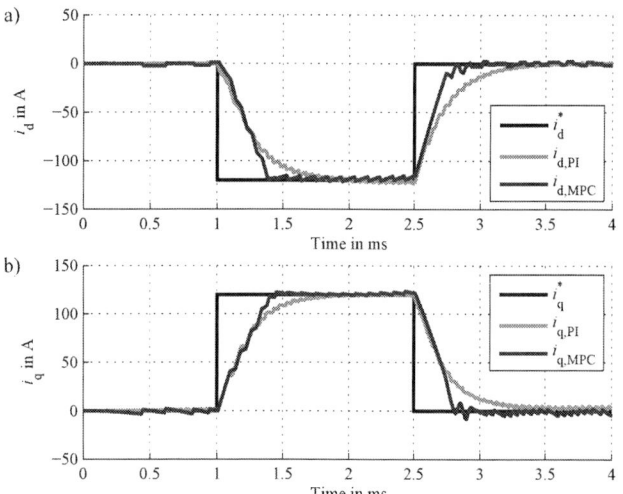

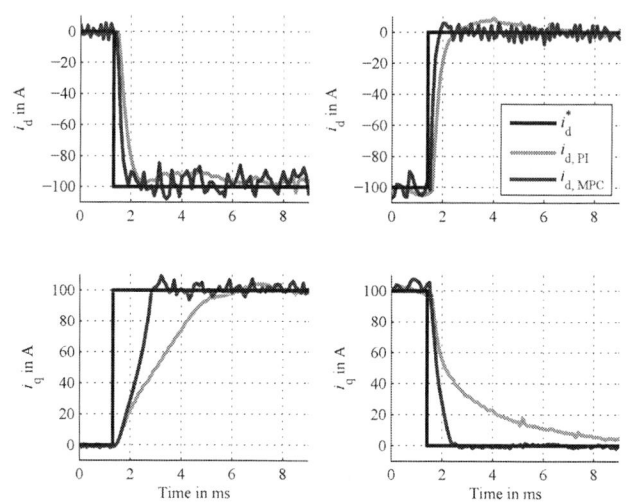

Fig. 5. Simulation of large-signal step response of the Vector-MPC (blue curve) compared to the PI control (green curve); a) Motor current in d-direction; b) Motor current in q-direction

Fig. 7. Measurement of the large-signal step response of the Vector-MPC (blue curve) compared to the PI control (green curve)

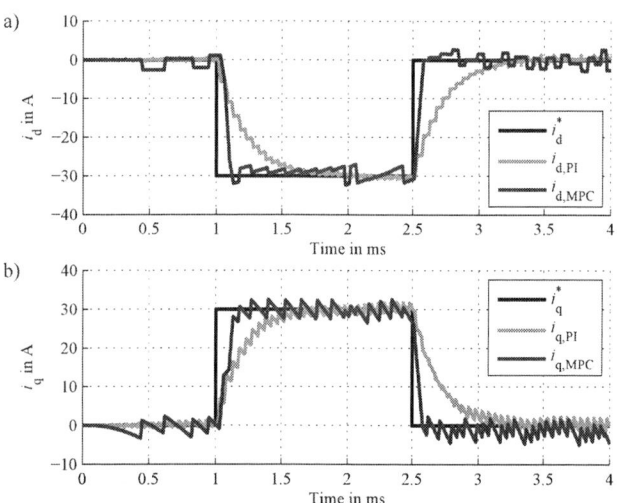

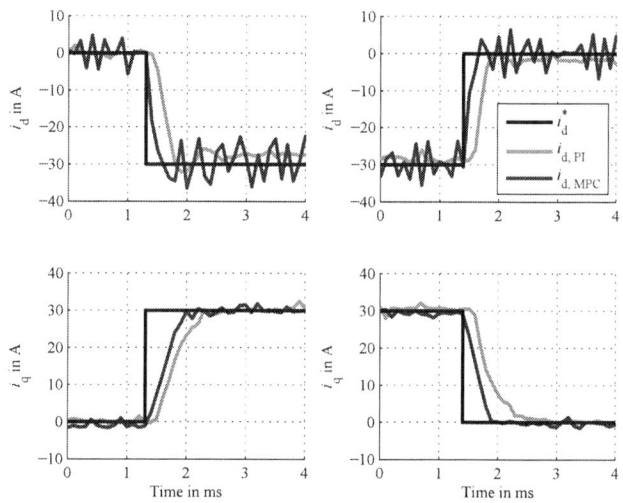

Fig. 6. Simulation of small-signal step response of the Vector-MPC (blue curve) compared to the PI control (green curve); a) Motor current in d-direction; b) Motor current in q-direction

Fig. 8. Measurement of the small-signal step response of the Vector-MPC (blue curve) compared to the PI control (green curve)

the controller cycle time could be reduced by the factor 10. The calculation time for the processor-board is less than $40\,\mu s$. So there is also the possibility to reduce this controller cycle time, too.

IV. MEASUREMENT RESULTS

To verify the real-time capability of the Vektor-MPC approach, the control-algorithm was tested at a testbench. The controlled IPMSM was a motor manufactured by Brusa for the use in automotive applications. The measurement results for the large-signal step response are shown in Figure 7. Figure 8 shows the step response of the small-signal behavior. As can

be seen from the measurement results from Figure 7 and 8, the implementations of the PI control as well as the Vector-MPC at the testbench are able to control the motor currents quite well. This shows that the real-time implementation of the Vector-MPC is possible, even for such highly dynamic processes as an IPMSM. In addition it can be seen that the MPC realizes, as expected from the simulation results, shorter rise times compared to the PI control. As can be seen from Figure 7 (top right), also the overshoot is lower in the MPC-controlled loop. However, the optained current ripple is significantly higher in the Vector-MPC compared to the PI control. This effect is particularly noticeable at the small-signal behavior (Figure 8). This effect was already seen in Figure 6 in the simulation. The larger current ripple of the Vector-MPC at the

978-1-4799-2706-7/14 $31.00 © 2014 IEEE

test bench measurements here is probably due to the poorer response to disturbances of the MPC. To minimize this current ripple further work is still needed. The reason, that the current ripple in d-direction is higher than in q-direction is due to the characteristic of the IPMSM ($L_d < L_q$).

V. CONCLUSIONS

It was shown that a real-time online MPC can be implemented using this Vector-MPC. This method does not rely on a complex and thus computationally expensive optimization algorithm and therefore turns out to be suitable for controlling processes with small time constants, such as IPMSM. Moreover the step response demonstrated that the dynamics of the drive can be improved by the use of Model Predictive Control. However, additional work is required for a further reduction of the current ripple. With the help of measurements it was shown that real-time implementation of the Vector-MPC is possible.

ACKNOWLEDGMENT

This work was devoloped as part of the DFG-programm "BO 2535/8-1" and was funded by the DFG (Deutsche Forschungsgemeinschaft).

REFERENCES

[1] T. Finken and M. Felden and K. Hameyer, *Comparison and design of different electrical machine types regarding their applicability in hybrid electrical vehicles.* ICEM2008, Vilamoura, Portugal, 2008.

[2] W. Peters and T. Huber and J. Böcker, *Control Realization for an Interior Permanent Magnet Synchronous Motor (IPMSM) in Automotive Drive Trains.* PCIM2011, Nürnberg, Deutschland, 2011.

[3] C. Mademlis and I. Kioskeridis and N. Margaris, *Optimal Efficiency Control Strategy for Interior Permanent-Magnetsynchronous Motor Drives.* IEEE Transactions on Energy Conversion, Vol. 19, No. 4, 2004.

[4] S. Mathapati and J. Böcker, *Dynamically Reconfigurable Control Structure for Induction Motor Drives on FPGA Control Platform.* EPE Journal, Vol. 20-1, 2010.

[5] M. Leuer and A. Rüting and J. Böcker, *Efficiency-Optimized Model Predictive Torque Control for IPMSM.* Energycon2014, Dubrovnik, Croatia, 2014.

[6] A. Bemporad and M. Morani and V. Dua and E. Pistikopoulos, *The explicit solution of Model Predictive Control via multiparametric Quadratic Programming.* Proceedings of the American Control Conference, 2000.

[7] S. Bolognani, R. Kennel, S. Kuehl, G. Paccagnella, *Speed and Current Model Predictive Control of an IPM Synchronous Motor Drive.* IEMDC2011, Niagara Falls, Canada, 2011.

[8] A. Horch, *Angewandte Regelung und Optimierung in der Prozessindustrie.*, ABB, Germany, 2010.

[9] H. Kobayashi and H. Kitagawa and S. Doki and S. Okuma, *Realization of a Fast Current Control System of PMSM based on Model Predictive Control.* IECON2008, Orlando (FL), USA, 2008.

[10] K. Graichen and M. Egretzberger and A. Kugi, *A Suboptimal Approach to Real-time Model Predictive Control of Nonlinear Systems.* at - Automatisierungstechnik, 2010.

[11] M. Leuer and J. Böcker, *Fast Online Model Predictive Control of IPMSM Using Parallel Computing on FPGA.* IEMDC2013, Chicago (IL), USA, 2013.

An Integral Sliding-Mode Controller for Energy Efficiency Improvement in AC Power Source Supplied AC Machine Drives

Hsin-Jang Shieh
Department of Electrical Engineering
National Dong Hwa University
Hualien 97401, Taiwan
E-mail: hjshieh@mail.ndhu.edu.tw

Ying-Zuo Chen
Department of Electrical Engineering
National Dong Hwa University
Hualien 97401, Taiwan
E-mail: d9923002@ems.ndhu.edu.tw

Abstract—**This paper develops an integration sliding-mode controller (ISMC) for a full-bridge single-ended-primary-inductance converter (SEPIC) supplied to an inverter-fed AC motor drive. Configuration of the developed SEPIC-based system is that a full diode bridge connected AC-to-DC SEPIC converter with power factor correction (PFC) is built for stably supplying DC voltage to the inverter. Under this configuration, the ISMC is applied for voltage control of the SEPIC converter with PFC of the AC-side line. With the PFC, the energy efficiency measured from the AC power source to the inverter-fed AC drive can substantially be improved. To perform the proposed ISMC, a Simulink model consisting of the AC source, diode-bridge rectifier, SEPIC, voltage-type inverter, and induction motor drive was built in a computer. Simulation results from the built Simulink model are illustrated to demonstrate the validity of the proposed ISMC for applications.**

Keywords—*Energy efficiency, Motor drives, Power factor correction (PFC), Single Ended Primary Inductance Converter (SEPIC), Integration sliding mode control (ISMC).*

I. INTRODUCTION

In recent years, the issue on energy-efficient electric drives has been paid much attention in many applications. In general, there is low energy efficiency in the AC power source when supplying to AC motor drives. This is because the power factor (PF) in the AC-side line is often deteriorated due to complicated switching of the inverter with inductive AC machines. The deteriorated PF usually results in low energy efficiency such as to reduce the power efficiency of the inverter-fed AC machine drives [1], [2]. Among AC machine drives, there has been shown that induction motor drives have many advantages for practical applications. Therefore, many studies on energy savings of the induction drive systems have been published in many publications.

To reduce the energy loss of the AC power source to induction motor drives, the PFC techniques for AC line are usually presented. This is because the PFC often can make the non-resistive load for AC power source become the nearly pure resistive load [3] so that the power factor close to unity can be obtained. In order to achieve an improved PF on the induction drive systems, the sliding-mode control (SMC) with PFC function is developed in this paper, since there are many advantages of the SMC, such as improved transient response and robustness to disturbances [4]-[7].

For the SEPIC converter, single-phase AC voltage source is fed and the pulse-width modulation (PWM) inverter-driven induction motor is loaded. Therefore, to increase the energy efficiency of the drive system, the PFC applied to the AC-to-DC converter is a popular and effective scheme [1], [8], [9]. Because of the benefits of the SEPIC converter, such as continuous input and output currents, small output filter, and wide range in output voltage [10]-[12], the DC-to-DC SEPIC topology is adopted as a part of the AC-to-DC converter and the PFC technique is implemented in the converter.

In this paper, the main objective is to design the integral siding-mode controller (ISMC) and apply the ISMC to the SEPIC converter. With use of the ISMC, the PF of the AC-side line can be regulated nearly to unity and the DC voltage which supplies to the inverter-fed induction motor drive can stably and robustly be controlled. At last, simulations through Smulink model of the AC source-powered induction drive system are performed. Simulation results are illustrated to show the validity of the proposed control system in the paper.

II. SEPIC-CASCADED PWM INVERTER FOR AC MOTOR DRIVE

The system setup of the pulse-width modulation (PWM) inverter-fed motor drive with SEPIC cascaded is drawn in Fig. 1, in which a full-bridge diode rectifier, a SEPIC converter, and a voltage-type PWM inverter are configured. In this configuration, the SEPIC converter shown in Fig. 2 is employed to step up and down the voltage so that a suitable voltage can be supplied to the inverter. Moreover, for improving the efficiency of the AC power source supply to the PWM inverter, the PFC scheme on the SEPIC converter is also carried out.

978-1-4799-2706-7/14 $31.00 © 2014 IEEE

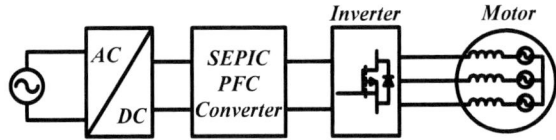

Fig. 1. Block diagram of the SEPIC-cascaded PWM inverter system for AC motor drives.

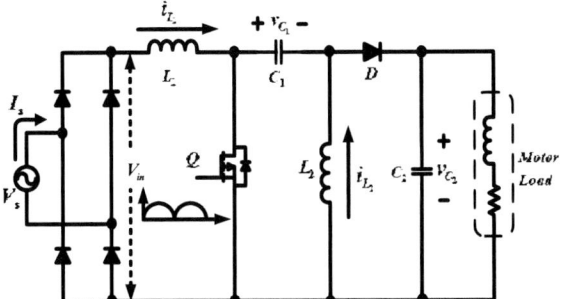

Fig. 2. Circuit diagram of the SEPIC-based converter

In order to obtain an improved efficiency of the power source by the PFC scheme, a controller that can effectively regulate the AC voltage and current to be in phase is developed. To develop the controller, we derive the dynamic model of the SEPIC converter as shown in Fig. 2. In Fig. 2, a power switch Q with duty cycle d, a diode D, two inductors L_1 and L_2, two capacitors C_1 and C_2 and a resistor-inductor load are given.

Without loss of generality, the state-space averaging method is used for deriving the dynamic model for SEPIC converter. Under the continue conduction mode (CCM), the state-space averaged model of the SEPIC converter can be derived as follows:

$$L_1 \frac{di_{L_1}}{dt} = v_{in} - (1-d)(v_{c_1} + v_{c_2}) \tag{1}$$

$$L_2 \frac{di_{L_2}}{dt} = d \cdot v_{c_1} - (1-d)v_{c_2} \tag{2}$$

$$C_1 \frac{dv_{c_1}}{dt} = (1-d)i_{L_1} - d \cdot i_{L_2} \tag{3}$$

$$C_2 \frac{dv_{c_2}}{dt} = (1-d)(i_{L_1} + i_{L_2}) - \frac{v_{c_2}}{R_o} \tag{4}$$

where d denotes the duty cycle of switch Q, R_O is an equivalent load made by the AC machine.

It has shown that when the input voltage and current are in phase, the PF is equal to unity, the maximum power factor that the AC power source can be. In this condition of PF = 1, the following equations can be obtained.

$$v_{in} = V_s\sqrt{2} \cdot |\sin \omega t| \tag{5}$$

$$i_{L_1} = I_s\sqrt{2} \cdot |\sin \omega t| \tag{6}$$

where V_s and I_s denote the RMS values of the AC source voltage and current, respectively. v_{in} is the voltage obtained from the full-bridges diodes rectifier, and ω is the frequency of the AC source.

III. DESIGN OF THE PFC SEPIC CONVERTER CONTROLLER

Controller design for the power converter can usually be conducted in two steps: a) current control in inner loop and b) voltage control in outer loop. The detailed controller configuration is shown in Fig. 3.

a) Inner-Loop Current Control

In order to obtain an improved PF, the sliding-mode current controller is developed and the output of the controller is transformed to a typical PWM control signal. In the controller development, the SEPIC converter is operated in continuous conduction mode (CCM). To develop the current controller, only (1) is taken into account. Therefore, we define the state variables in the following for the controller design

$$e_1 = i_{L_1} - i_{ref} \tag{7}$$

$$e_2 = \int (i_{L_1} - i_{ref}) \, dt \tag{8}$$

where i_{ref} is the reference inductor current.

For the sliding-mode control, a stable sliding surface is given as follows:

$$S = \alpha e_2 + e_1 \tag{9}$$

where α represents a positive coefficient. Taking the time derivative of (9) yields

$$\dot{S} = \alpha \dot{e}_2 + \dot{e}_1 \tag{10}$$

Let the following equalities be:

$$\dot{i_{L_1}} = \frac{v_{in}}{L_1} - \frac{(1-d) \cdot (v_{c_1} + v_{c_2})}{L_1} + \delta \ , \ \delta = \frac{\Delta v_{in}}{L_1} \tag{11}$$

where δ denotes the uncertainties introduced by v_{in}. Substituting (7) and (8) into (10) gives

$$\dot{S} = \alpha \dot{e}_2 + \dot{e}_1$$
$$= \alpha \left(\frac{v_{in}}{L_1} - \frac{v_{c_1} + v_{c_2}}{L_1} + \frac{v_{c_1} + v_{c_2}}{L_1} \cdot d \right) + i_{L_1} - i_{ref} + \delta \tag{12}$$

From the sliding surface, a sliding-mode controller that satisfies the hitting condition and stabilize the sliding surface is given as follows.

$$d = \frac{L_1}{\alpha(v_c + v_o)} \left[-\alpha \frac{v_{in}}{L_1} + \alpha \frac{v_c + v_o}{L_1} - i_{L_1} + i_{ref} - (\xi + \beta)sign(s) \right] \tag{13}$$

where $sign(\cdot)$ is a sign function assigned as

978-1-4799-2706-7/14 $31.00 © 2014 IEEE

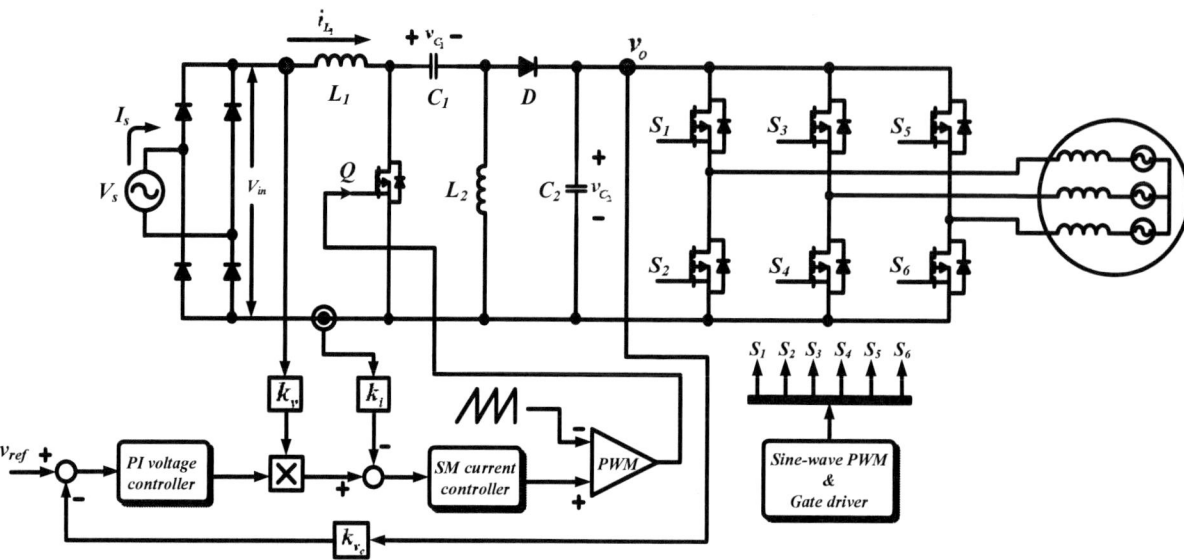

Fig. 3. Control system of the SEPIC-based PFC converter

$$sign(s) = \begin{cases} +1 & ,s > 0 \\ -1 & ,s < 0 \end{cases}$$

ξ is designed as an upper bound on the perturbation term, $|\delta| \leq \xi$, and $\beta > 0$ is a suitable control gain,

In order to obtain the existence of the sliding mode for the control system, that is, [4]:

$$\lim_{t \to 0} S\dot{S} < 0 \tag{14}$$

Furthermore, the Lyapunov function for stability of the sliding manifold $S = 0$ is given by

$$V = \frac{1}{2}S^2 \tag{15}$$

Taking the time derivative of (15) yields:

$$
\begin{aligned}
\dot{V} &= S\dot{S} \\
&= S\left[\alpha\left(\frac{v_{in}}{L_1} - \frac{v_{c_1} + v_{c_2}}{L_1} + \frac{v_{c_1} + v_{c_2}}{L_1} \cdot d\right) + i_{L_1} \right.\\
&\quad \left. - i_{ref} + \delta\right]
\end{aligned}
\tag{16}
$$

Substituting (13) into (16) gives

$$
\begin{aligned}
S\dot{S} &= -(\xi + \beta)sign(s) \cdot S + \delta \cdot S \\
&= -\beta|S| - \xi|S| + \delta \cdot S \\
&\leq -\beta|S| - \xi|S|\left(1 - \frac{\delta \cdot s}{\xi|S|}\right) \leq 0
\end{aligned}
\tag{17}
$$

From (17) and Lyapunov stability theorem, it shows that $\dot{S} \to 0$ as $t \to \infty$ and $S \to 0$ as $t \to \infty$. Also, this implies that the asymptotic stability of the proposed controller is guaranteed. Therefore, using the controller of (13), the stability of the current control can be achieved.

b) Outer-Loop Voltage Control

In general, a simple voltage controller that can stably regulate the output voltage of the SEPIC converter is of a proportional-integral (PI) controller. This is because the PI controller is capable of reducing the tracking error. Hence, the DC voltage error e_v is defined as $e_v = v_{ref} - v_o$ and then the PI controller can be expressed by

$$u_v = K_p e_v + K_i \int e_v \, dt \tag{18}$$

where K_p and K_i are the proportional and integral gains, respectively. With use of the PI voltage controller, the output voltage of the SEPIC converter can be stably regulated to the reference voltage command.

IV. SIMULINK RESULTS

To demonstrate the validity of the developed SEPIC-base PFC motor drive with the proposed controller, a single-phase voltage source powered SEPIC converter was built by Simulink model. In the Simulink model, the proposed controller was implemented for simulations. Simulation results are illustrated to confirm the validity of the proposed controller. All the values of the parameters taken in simulation work are listed as follows.

Input voltage (V_s)	= 110 V
Mains frequency (f_o)	= 60 H_Z
Switching frequency (f_s)	= 50k H_Z
DC inductor (L_1)	= 1 mH
DC inductor (L_2)	= 10 mH
DC capacitor (C_1)	= 10 mF
DC capacitor (C_2)	= 10 mF
Inverter switching frequency	= 1k H_Z
Induction motor rated power	= 1 HP
Induction motor rated voltage	= 130 V
Induction motor rated frequency	= 60 H_Z
Induction motor pole	= 4 pole

The 2014 International Power Electronics Conference

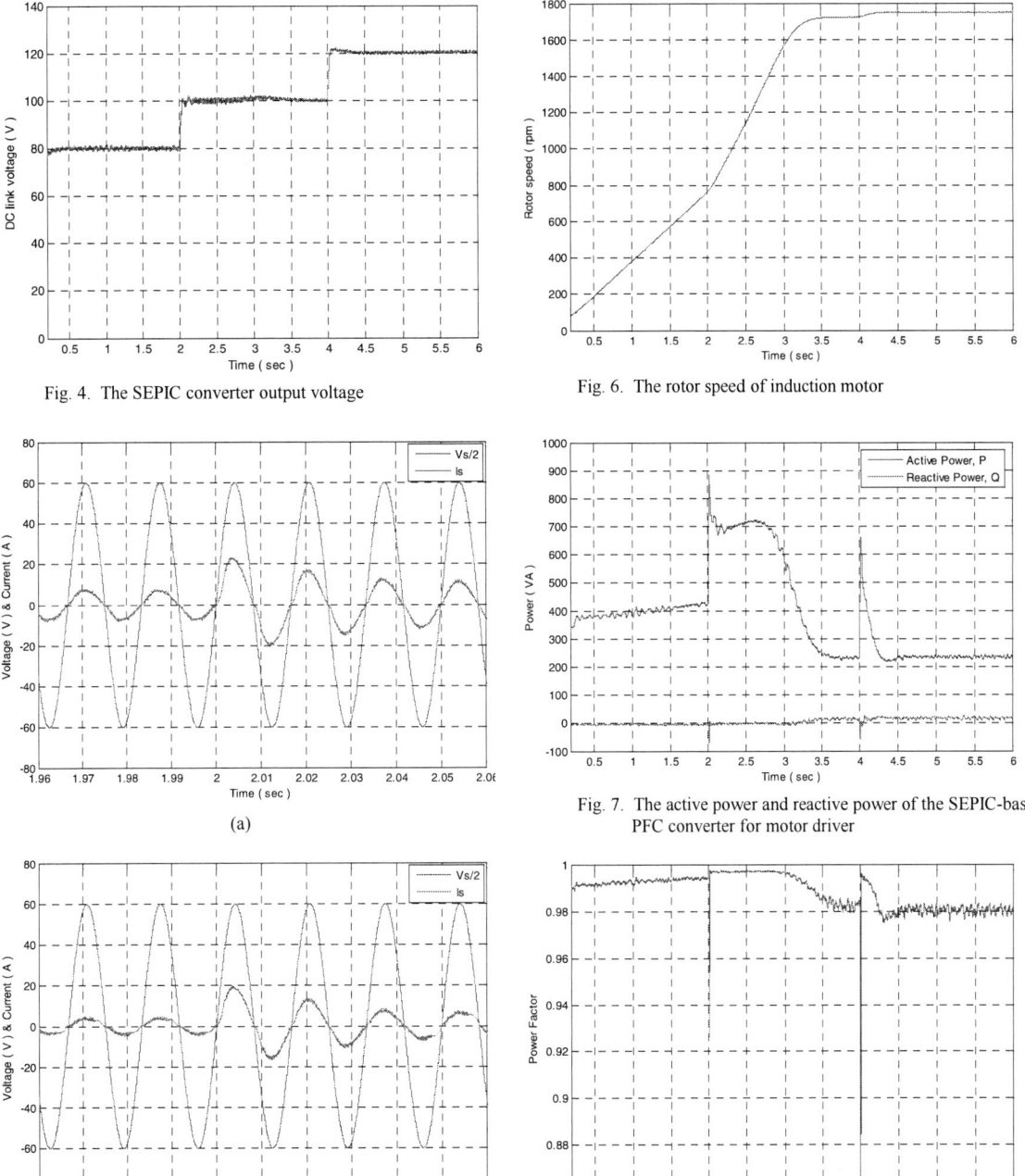

Fig. 4. The SEPIC converter output voltage

(a)

(b)

Fig. 5. The AC side voltage and the current change to (a) 100V at 2 sec (b) 120V at 4 sec of the SEPIC-base PFC converter.

Fig. 6. The rotor speed of induction motor

Fig. 7. The active power and reactive power of the SEPIC-base PFC converter for motor driver

Fig. 8. The power factor of the SEPIC-base PFC converter for motor driver.

In the simulation, the reference voltage command is assumed to be changeable. The initial command value of the voltage was set to 80V. Then, the command was changed to 100V at 2 s and changed to 120V at 4 s. Fig. 4 shows the output voltage response of the SEPIC converter with the proposed controller. Fig. 5 shows the waveforms of the AC-side voltage (V_s) and current (I_s), in which the magnitude scale in y-axis is a half of the real measured voltage. From view of Fig. 5, it can be found that the AC current and the AC voltage are in phase under the sliding-

mode PFC control. Also, from Fig. 5 one can see that even when the step change in reference command of the output voltage, the AC-side current is still sinusoidal and almost has the same phase as the voltage. From these results, it shows that the power factor close to unity is achieved.

Fig. 6 shows the corresponding rotor speed response of induction motor under the case given in Figs. 4 and 5. From Fig. 6 it can be observed that the DC link voltage affects the speed time-response.

978-1-4799-2706-7/14 $31.00 © 2014 IEEE

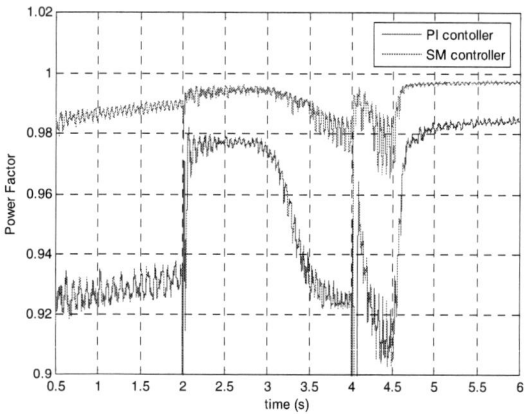

Fig. 9. Dynamic response of power factor regulation by using PI power factor controller and ISMC power factor controller, respectively.

To show the improvement of the power factor, the active (P) and reactive (Q) power time-response which is obtained under the operating condition of Figs. 4-6 are shown in Fig. 7. From view of Fig. 7, the reactive power is substantially reduced closely to zero even when the induction motor drive is running. In this case, the power factor of the AC-side is kept at 0.98 and above, which is shown in Fig. 8. Therefore, from these results, it confirms that the proposed PFC-SEPIC converter is valid for the variable-frequency AC machine drive system.

For comparison convenience, simulation result with the conventional PI power-factor controller and with the proposed ISMC is illustrated in Fig. 9. In Fig. 9, step changes in AC drive load from 1 to 10 Nm at 2 s, and from 10 to 1 Nm at 4 s are respectively given. From view of Fig. 9, it can be found that the power factor improvement by the proposed ISMC is better than that by the PI controller. Consequently, the effectiveness of the proposed controller in the SEPIC converter is confirmed by the simulation work.

V. CONCLUSIONS

A SEPIC-based PFC converter for motor driver is designed and then simulation by using Matlab/Simulink to validation of designs. According to simulation results, the implementation of the PI voltage control and the sliding mode current control has been found to offer excellent time response of DC link voltage performance and acceptable power quality. The PFC SEPIC converter has ensured near unity PF in a wide range of the speed and the input ac voltage. By using fixed frequency and SEPIC (buck/boost) PFC topology is used as this is the best option for applications having rated DC voltage higher than single phase supply RMS voltage. Therefore, the technique cans widespread application to other motor drives.

ACKNOWLEDGMENT

The authors would like to greatly thank for the financial support for this work by the National Science Council of Taiwan under grant NSC101-2628-E-259-002.

REFERENCES

[1] B. K. Bose, *Modern Power Electronics and AC Drives.* Prentice Hall PTR, 2002.

[2] A. Consoli, M. cacciato, A. Testa and F. Gennaro, "Single chip integration for motor drive converters with power factor capability," *IEEE Trans. Power Electronics*, Vol. 19, no. 6, pp. 1372-1379, 2004.

[3] M. K. Kazimierczuk and D. Czarkowski, *Resonant Power Converters 2/E.* Wiley, 2011.

[4] U. Itkis, *Control System of Variable Structure.* New York: Wiley, 1976.

[5] A. J. Koshkouei, K. J. Burnham and A. S. I. Zinober, "Dynamic sliding mode control design," *IEE Proc. Control Theory and Applications*, Vol. 152, no. 4, pp.392-396, 2005.

[6] Y. Shtessel, S. Baev and H. Biglari , "Unity Power Factor Control in Three-Phase AC/DC Boost Converter Using Sliding Modes," *IEEE Trans. Industrial Electronics*, Vol. 55, no. 11, pp. 3874-3882, 2008.

[7] G. Chu, S. C. Tan, C. K. Tse and S. C. Wong, "General control for boost PFC converter from a sliding mode viewpoint," in *IEEE Power Electronics Specialists Conf.*, pp. 4452-4456, Jun. 2008.

[8] J. Y. Chai and C. M. Liaw, "Development of a Switched-Reluctance Motor Drive With PFC Front End," *IEEE Trans. Energy Conversion*, Vol. 24, no. 1, pp. 30-42, 2009.

[9] B. Singh and S. Singh, "Single-phase power factor controller topologies for permanent magnet brushless DC motor drives," *IET Power Electronics*, Vol. 3, no. 2, pp. 147-175, 2010.

[10] H. Macbahi, J. Xu, A. Cheriti, and V. Rajagopalan, "A soft-switched SEPIC based AC-DC converter with unity power factor and sinusoidal input current," *IEEE Proc. INTELEC*, pp. 663–668, Oct. 1998.

[11] L. H. Gonzalez, Z. B. Brito and D. S. V. Estrada, "Analysis, simulation and physical implementation: Modified-SEPIC DC/DC converter," *IEEE Proc. ICEEE.*, pp. 389–392, Sept. 2005.

[12] H. Y. Kanaan and K. Al-Haddad, "Small-signal averaged model and carrier-based linear control of a SEPIC-type Power Factor Correction circuit," in *30th IEEE International Telecommunications Energy Conf.*, pp. 1-6, Sept. 2008.

[13] L. S. Yang and T. J. Liang, "Analysis and design of a novel single-phase PFC AC-DC step-up/down converter," *IEEE Proc.26th APEC*, pp. 1470-1474, Mar. 2011.

Performance Improvement of Ultra-High-Speed PMSM Drive System based on DTC by using SiC Inverter

Ryo Togashi, Yukinori Inoue, Shigeo Morimoto, and Masayuki Sanada

Graduate School of Engineering
Osaka Prefecture University
Sakai, Japan
sv116046@edu.osakafu-u.ac.jp

Abstract— This paper examines the control performance of a permanent magnet synchronous motor drive system based on direct torque control (DTC) using a SiC-MOSFET inverter (SiC inverter). The dead-time of the inverter affects the performance of the motor control. In high-speed drives, the dead-time effect increases because the control period becomes shorter. By using a SiC inverter for which the switching speed is faster than that of the Si-IGBT inverter, the dead-time can be shortened, and performance improvement of the control is expected. This paper also proposes a sensorless starting method using DTC. It is difficult to start the motor in a drive system without a position sensor. In the conventional ultra-high-speed drive using DTC, the motor is started by open-loop control. Experiments reveal the effectiveness of using a SiC inverter and shortening the dead-time. In addition, the experimental results reveal the applicability of the proposed starting method.

Keywords— *Direct torque control (DTC), permanent-magnet synchronous motor (PMSM), ultra-high-speed drive, SiC inverter*

I. INTRODUCTION

Due to advances in power electronics and high-performance microprocessors and the development of low-iron-loss magnetic materials, permanent magnet synchronous motors (PMSMs) have recently been driven at ultra-high speeds. Ultra-high-speed PMSMs are used in a number of applications, including compressors, superchargers, and air blowers [1]-[3].

Current control or direct torque control (DTC) is used for high-efficiency driving with wide ranges and variable speeds. Current control often works on the d-q frame based on the rotor magnet flux, whereas DTC is operated in a stationary reference frame based on the stator winding. DTC was first proposed for induction machine drives but can also be applied to PMSM drives [4]. In high-speed drives, since the fundamental frequency becomes high, a short sampling period is required for stable control, and so the speed sensor becomes unreliable. DTC is therefore suitable for ultra-high-speed PMSM drives because of its simple structure and sensorless feature.

The dead-time is required in order to avoid cross-conduction current of switching devices of the inverter used for the motor drive, but the dead-time causes problems such as output current distortion and voltage error [5]. In particular, in high-speed drives, the voltage error, which can be expressed in the form of the dead-time divided by the control period, increases, thus degrading the control performance.

The sensorless drive system for an ultra-high-speed PMSM based on DTC has been proposed in [6]. This paper examines the control performance of the ultra-high-speed PMSM drive system based on DTC by shortening the dead-time using the SiC inverter. A performance improvement is expected by shortening the dead-time, because the voltage error is proportional to the dead-time.

This paper also proposes a sensorless starting method using DTC. Start the PMSM in a position sensorless drive system is difficult. In the conventional ultra-high-speed drive based on DTC, the motor is started by open-loop control [6]. The proposed method achieves sensorless starting by changing the reference flux depending on the rotor speed.

In the proposed system, the reference flux vector calculator (RFVC) DTC [7] is applied for ultra-high speeds (rated speed: 42,000 min^{-1}). The experimental results demonstrate the effectiveness of the SiC inverter. In addition, the proposed sensorless starting method is experimentally demonstrated to be valid.

II. DIRECT-TORQUE-CONTROL-BASED MOTOR DRIVE SYSTEM

A. Definition of Coordinate Axes

DTC operates in the α-β frame, which is a stationary reference frame. Fig. 1 shows the relationship between the d-q and α-β frames. The stator flux linkage position θ_s is defined as the angle between the stator flux linkage vector Ψ_s and the α-axis, and δ is the torque angle.

978-1-4799-2706-7/14 $31.00 © 2014 IEEE

The 2014 International Power Electronics Conference

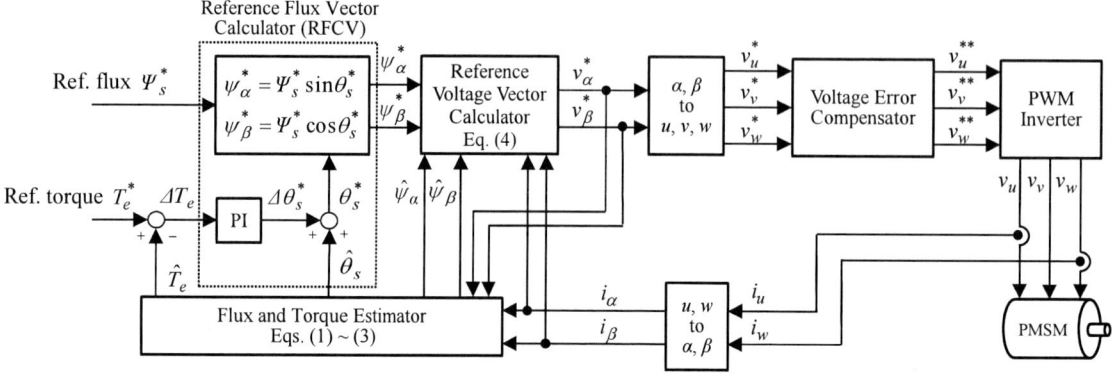

Fig. 2. Motor drive system based on RFVC DTC.

B. Reference Flux Vector Calculator DTC

Fig. 2 shows a block diagram of the PMSM drive system using RFVC DTC. The stator flux linkage is estimated in the α-β frame. Therefore, the DTC system is suitable for drives without position sensors. The estimated stator flux linkage ($\hat{\psi}_\alpha$, $\hat{\psi}_\beta$), the estimated torque ($\hat{T}_e$), the estimated position of the stator flux linkage ($\hat{\theta}_s$), and the reference voltages (v_α^*, v_β^*) are calculated as follows:

$$\begin{cases} \hat{\psi}_\alpha = \int (v_\alpha^* - R_a i_\alpha)dt \\ \hat{\psi}_\beta = \int (v_\beta^* - R_a i_\beta)dt \end{cases} \tag{1}$$

$$\hat{T}_e = P_n(\hat{\psi}_\alpha i_\beta - \hat{\psi}_\beta i_\alpha) \tag{2}$$

$$\hat{\theta}_s = \tan^{-1} \frac{\hat{\psi}_\alpha}{\hat{\psi}_\beta} \tag{3}$$

$$\begin{cases} v_\alpha^* = \dfrac{\psi_\alpha^* - \hat{\psi}_\alpha}{t_s} + R_a i_\alpha \\ v_\beta^* = \dfrac{\psi_\beta^* - \hat{\psi}_\beta}{t_s} + R_a i_\beta \end{cases} \tag{4}$$

where R_a is the armature resistance, i_α and i_β are the α- and β-axis components of the armature currents, P_n is the number of pole pairs, $\hat{\theta}_s$ is the estimated position of the stator flux linkage in the α-β frame, and t_s is the sampling period.

The dashed line shown in Fig. 2 indicates the RFVC block, which calculates the reference flux vector using a proportional and integral (PI) controller.

C. Voltage Error due to the Inverter

In this study, considering the voltage error of the inverter, voltage error compensation is applied to this system. The voltage error of the dead-time is assumed to be much larger than the voltage drop of the switching device due to the short sampling period. The relationship between the reference

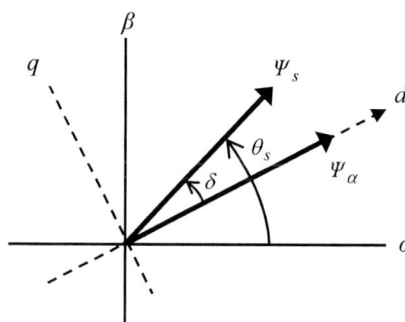

Fig. 1. Definition of coordinate axes.

voltages and the inverter output voltages is expressed as follows [8]:

$$\begin{bmatrix} v_u \\ v_v \\ v_w \end{bmatrix} = \begin{bmatrix} v_u^{**} \\ v_v^{**} \\ v_w^{**} \end{bmatrix} - \Delta V \begin{bmatrix} \text{sgn}(i_u) \\ \text{sgn}(i_v) \\ \text{sgn}(i_w) \end{bmatrix} \tag{5}$$

$$\Delta V = \frac{t_d}{t_s} V_{DC} \tag{6}$$

where v_u, v_v, and v_w represent the three-phase inverter output voltages, v_u^{**}, v_v^{**}, and v_w^{**} represent the compensated three-phase reference voltages, ΔV is the voltage error due to the dead time of the inverter, i_u, i_v, and i_w are the three-phase currents, sgn() is the sign function, t_d is the dead-time of the inverter, and V_{DC} is the DC link voltage. In this paper, t_s is equal to the inverse of the PWM carrier frequency.

D. Speed Sensorless Drive

Fig. 3 shows the composition of the speed estimator and the reference flux calculator. The estimated rotor speed ($\hat{\omega}_m$) is calculated by taking the difference of the estimated position of the stator flux linkage ($\hat{\theta}_s$) and using a low-pass filter (LPF)

978-1-4799-2706-7/14 $31.00 © 2014 IEEE   357

because the rotor velocity corresponds to the velocity of the stator flux-linkage vector in a steady state. A PI controller is used for speed control, and the relationship between the torque and the flux is obtained using maximum torque per ampere (MTPA) control for a high-efficiency drive.

III. SENSORLESS STARTING METHOD

The induced voltage V_o and the stator flux linkage Ψ_s are related as follows:

$$V_o = \omega \Psi_s \tag{7}$$

where ω is the electrical angular velocity.

In the low-speed region, the induced voltage becomes small. Based on the reference voltage, a sufficient voltage cannot be provided to the motor because the modulation factor of the inverter is close to zero. In addition, the voltage error in the inverter cannot be ignored. The stator flux linkage of the DTC is estimated by integrating the voltage, as shown in (1), so that the stator flux linkage estimation is not performed correctly in the low-speed region.

In this paper, the reference flux is calculated in inverse proportion to the rotor speed in order to ensure that the armature voltage is sufficiently high. In order to apply MTPA control when the rotor speed becomes sufficiently high, the reference flux is calculated as follows:

$$\Psi_s^* = \frac{K_1}{\hat{\omega}_m} + \Psi_{s-MTPA}^* \tag{8}$$

where K_1 is the constant of proportionality, and Ψ_{s-MTPA}^* is the reference flux given by MTPA control.

Fig. 4 shows the reference fluxes of the conventional method and the proposed method. Fig. 5 shows the composition of the reference flux calculator of the proposed method.

Substituting Ψ_s^* in (8) for Ψ_s in (7), the induced voltage V_o is expressed as follows:

$$\begin{aligned} V_o &= \omega \left(\frac{K_1}{\hat{\omega}_m} + \Psi_{s-MTPA}^* \right) \\ &= P_n K_1 + P_n \hat{\omega}_m \Psi_{s-MTPA}^* \end{aligned} \tag{9}$$

Fig. 6 shows the voltage characteristic of (9). The induced voltage becomes $P_n K_1$ when $\hat{\omega}_m = 0$ (the motor is standstill) using the proposed method.

IV. EXPERIMENTAL RESULTS

A. Experimental Setup

Table I lists the parameters of the PMSM drive system tested in this paper. Using the motor for a vacuum cleaner, the load torque is proportional to the square of the rotor speed. All of the controls are processed through a digital signal processor. The sampling period is 62.5 μs, which is equal to the inverse of the PWM carrier frequency (16 kHz). An SCH2080KE semiconductor (ROHM Corp., Ltd.) is used for the switching

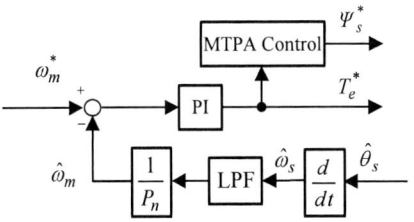

Fig. 3. Speed estimator and reference calculator.

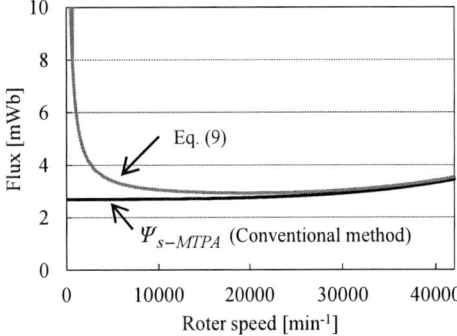

Fig. 4. Reference flux.

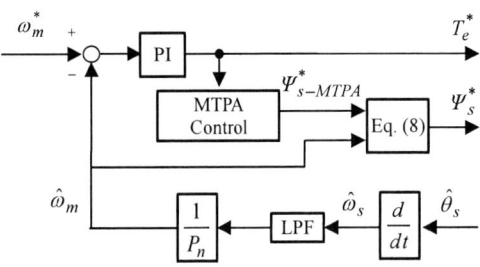

Fig. 5. Proposed reference calculator.

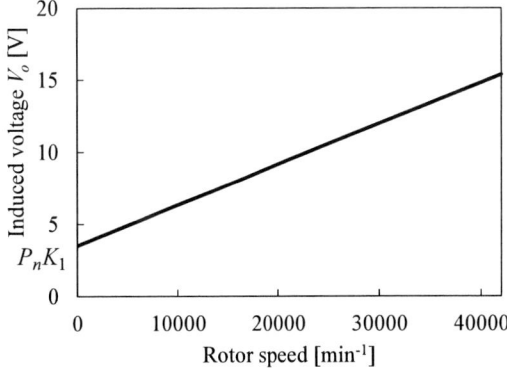

Fig. 6. Voltage characteristic of (9).

The 2014 International Power Electronics Conference

TABLE I
PARAMETERS OF PMSM DRIVE SYSTEM

Rated speed	42000 min^{-1}
Rated current	16 A
Armature resistance R_a	0.05 Ω
Magnet flux Ψ_a	2.7 mWb
d-axis inductance L_d	0.07 mH
q-axis inductance L_q	0.1 mH
Number of pole pairs P_n	1
Inertia moment J_m	$5.0 \times 10^{-6}\,\mathrm{kg \cdot m^2}$
Load torque T_{load}	$3.7 \times 10^{-9} \times \omega_m^2$ [Nm]
DC link voltage V_{DC}	40 V

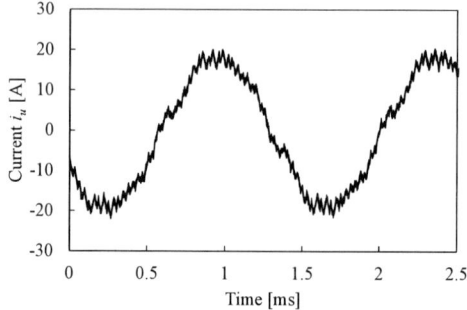

(a) Si-IGBT inverter ($t_d = 3.5\ \mu s$)

devices of the SiC-MOSFET inverter, and a PS21767 semiconductor (Mitsubishi Electric Corp.) is used as a Si-IGBT inverter.

B. Performance Comparion by Changing the Inverter and the Deadtime

Driving the ultra-high-speed PMSM at the rated speed of 42,000 min^{-1}, we compared the control performance by changing the type of switching device and the dead-time. The dead-time of the Si-IGBT inverter is selected to be 3.5 μs, considering the limitation of the MWINV-5R022 inverter unit (Myway Plus Corp.). Similarly, the dead-time for the SiC-MOSFET inverter MWINV-1044-SiC (Myway Plus Corp.) is chosen to be 0.5 μs.

Fig. 7 shows the u-phase current as measured using a current probe. The current waveform is controlled sinusoidally, and the PMSM is found to be operated at 42,000 min^{-1} because a single period of the u-phase current is approximately 1.4 ms. The current ripples appear at the zero crossing point because voltage error compensation is applied using the sign of the current. Table II shows the results of frequency analysis of the u-phase current. The ratios shown in Table II are the harmonic components divided by the fundamental component. Comparing the Si-IGBT inverter and the SiC-MOSFET inverter when the dead-time is set to 3.5 μs, the harmonic components are not so different. However, the amplitude of the fundamental component of the SiC-MOSFET inverter is smaller than that of the Si-IGBT inverter. This is because MTPA control performs better as a result of the reduction of the flux estimation error in DTC due to the decrease in the voltage error in the inverter, as shown in Fig. 8. The voltage error is calculated from the deference between the reference armature voltage and the line voltage.

In the case of the SiC-MOSFET inverter, the amplitude of the fundamental component and the ratio of the harmonic component in the u-phase current are reduced by setting the dead-time to 0.5 μs, as shown in Table II. The reason for this is that the voltage error due to the dead-time is reduced and the control is better performed by setting a small dead-time.

Fig. 9 shows the torque waveform. The estimated torque is calculated by (2). When the dead-time is 0.5 μs, torque ripples are reduced and so the average of the estimated torque is reduced. The reason for this is that the current ripples are

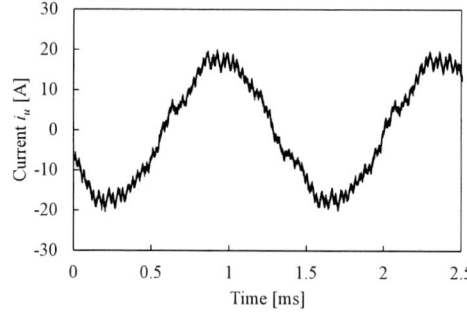

(b) SiC-MOSFET inverter ($t_d = 3.5\ \mu s$)

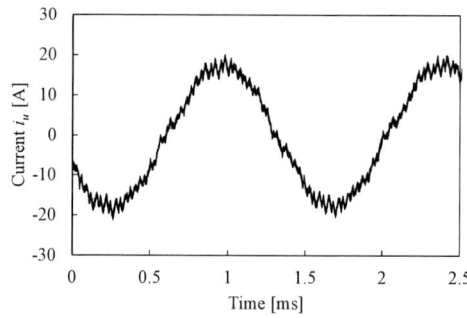

(c) SiC-MOSFET inverter ($t_d = 0.5\ \mu s$)

Fig. 7. Waveform of u-phase current.

TABLE II
FREQUENCY ANALYSIS OF THE U-PHASE CURRENT

Switching device (Dead-time t_d)	u-phase current				
	fundamental component	harmonic component			
		5th	7th	11th	13th
Si-IGBT (3.5 μs)	18.40 A	1.00 A (5.4%)	0.92 A (5.0%)	0.08 A (0.4%)	0.09 A (0.5%)
SiC-MOSFET (3.5 μs)	17.63 A	0.83 A (4.7%)	0.89 A (5.4%)	0.12 A (0.7%)	0.15 A (0.9%)
SiC-MOSFET (0.5 μs)	17.42 A	0.44 A (2.5%)	0.69 A (4.0%)	0.10 A (0.6%)	0.04 A (0.3%)

978-1-4799-2706-7/14 $31.00 © 2014 IEEE

The 2014 International Power Electronics Conference

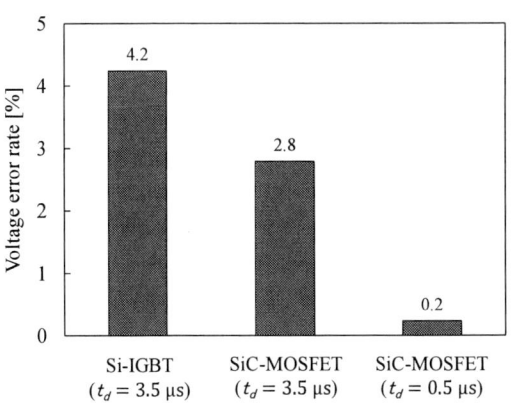

Fig. 8. Voltage error

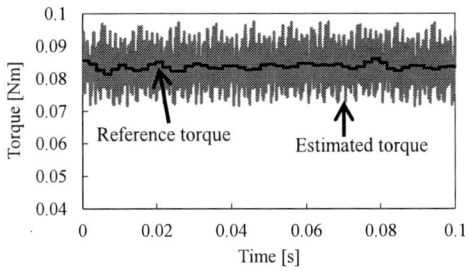

(a) Si-IGBT inverter (t_d = 3.5 μs).

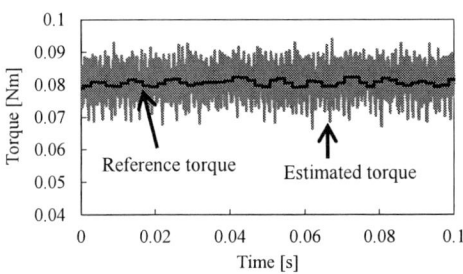

(b) SiC-MOSFET inverter (t_d = 3.5 μs).

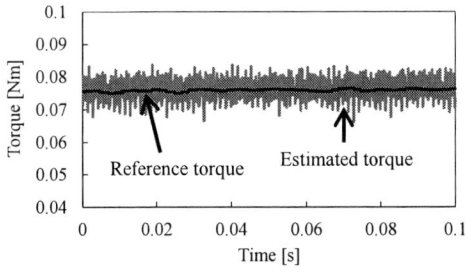

(c) SiC-MOSFET inverter (t_d = 0.5 μs).

Fig. 9. Torque waveform.

reduced, as shown in Table II.

Fig. 10 shows the flux waveform. There is an offset between the estimated flux and the reference flux, which originates from the effect of using an LPF for the voltage integration in the flux estimator and the discretization error.

The flux ripples are reduced due to torque ripple reduction when the dead-time is 0.5 μs. In this system, the estimated rotor speed $\hat{\omega}_m$ is calculated based on the rotor position $\hat{\theta}_s$, which is calculated based on the estimated stator flux $\hat{\Psi}_s$. Therefore, the ripples of the estimated speed are also reduced, as shown in Fig. 11.

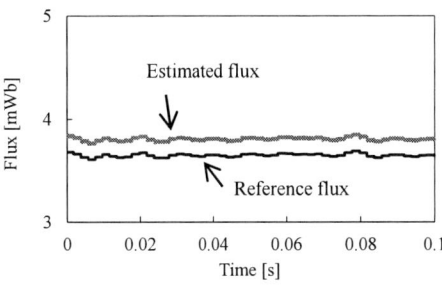

(a) Si-IGBT inverter (t_d = 3.5 μs).

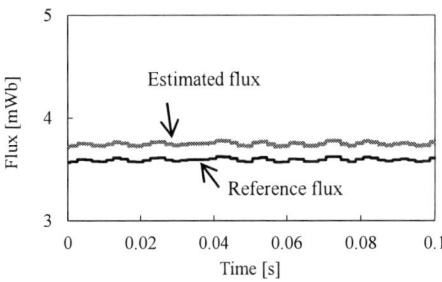

(b) SiC-MOSFET inverter (t_d = 3.5 μs).

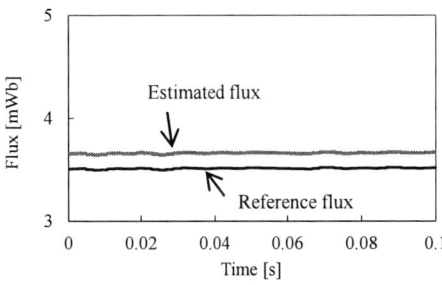

(c) SiC-MOSFET inverter (t_d = 0.5 μs).

Fig. 10. Flux waveform.

The 2014 International Power Electronics Conference

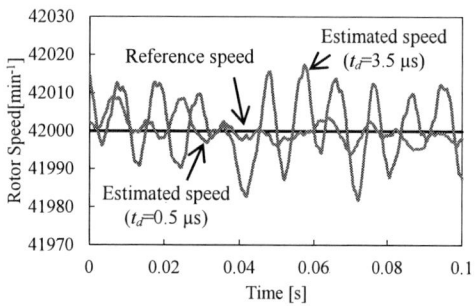

Fig. 11. Speed waveform (SiC-MOSFET inverter).

C. Performance of the Sensorless Starting Method

In this paper, K_1 in (8) is set to 3.5, and the upper limit of Ψ_s^* is set to 18 mWb in order to prevent the reference flux from diverging.

Fig. 12 shows the starting characteristics. In this experiment, the acceleration rate is limited in order to avoid step out. The motor is started at 0.5 s. The PMSM can be started up from standstill because the amplitude of the u-phase voltage is maintained approximately constant, as shown in Fig. 12(d), and thus a sufficient voltage is provided to the motor in the low-speed region. In addition, the reference flux is reduced according to the rotor speed and approaches the reference flux given by MTPA control.

Fig. 13 shows the acceleration characteristics of the proposed method when the rotor speed is increased from standstill to 42,000 min^{-1}. The acceleration rate is limited in the low-speed region in order to prevent the rotor from step out. The proposed method allows smooth acceleration without switching the control method.

V. CONCLUSIONS

This paper examined the control performance in the case of changing the dead-time for an ultra-high-speed PMSM drive system using a SiC inverter. The experiment revealed the effectiveness of using the SiC inverter. The control performance of the motor drive system is improved by using the SiC inverter because the voltage error is reduced due to the small voltage drop of the switching devices of the SiC inverter. In addition, the control performance is improved by setting a small dead-time because the voltage error due to the dead-time is reduced. A sensorless starting method using DTC was also proposed, in which the reference flux changes according to the rotor speed in order to keep the voltage approximately constant in a low-speed region. Using the proposed method, providing the reference flux to keep the armature voltage constant, sensorless starting is demonstrated experimentally to be possible, and the rotor speed can be increased smoothly without switching the control method.

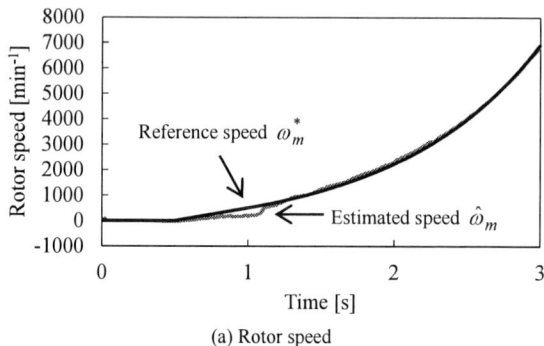

(a) Rotor speed

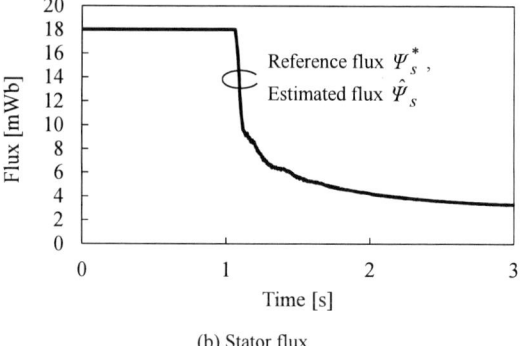

(b) Stator flux

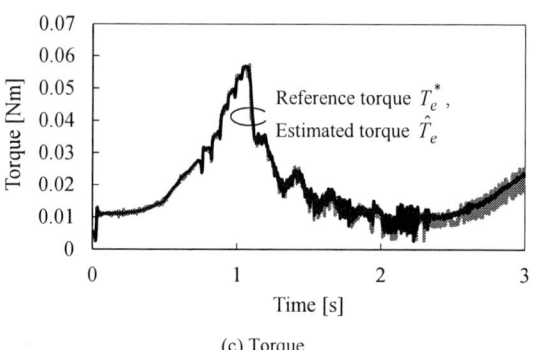

(c) Torque

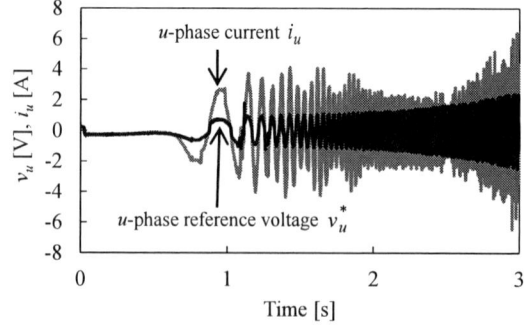

(d) u-phase reference voltage, u-phase current

Fig. 12. Characteristics of startup.

978-1-4799-2706-7/14 $31.00 © 2014 IEEE

The 2014 International Power Electronics Conference

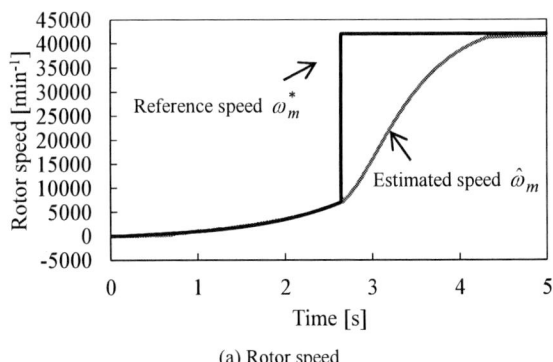

(a) Rotor speed

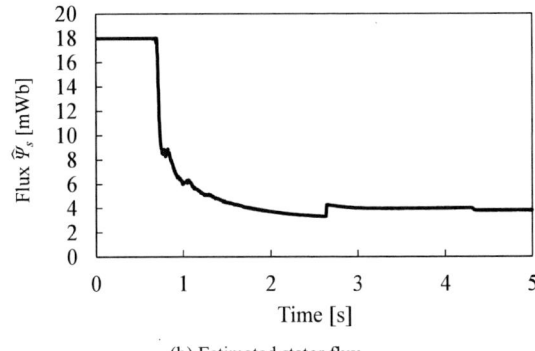

(b) Estimated stator flux

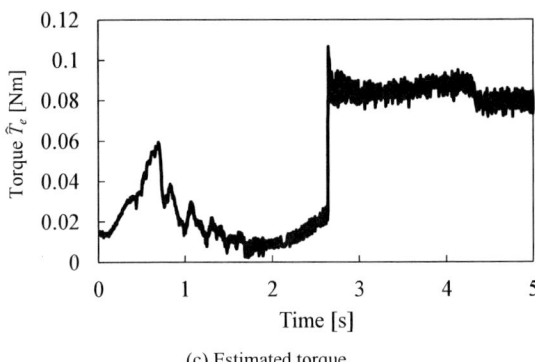

(c) Estimated torque

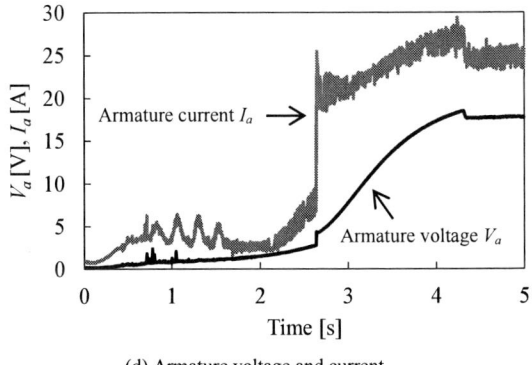

(d) Armature voltage and current

Fig. 13. Acceleration characteristics for the proposed method.

REFERENCES

[1] B. Bae, S. Sul, and J. Kwon, and J. Byeon "Implementation of Sensorless Vector Control for Super-High-Speed PMSM of Turbo-Compressor," *IEEE Trans. on Industry Applications*, vol. 39, no. 3, pp. 811-818, 2003.

[2] T. Noguchi and T. Wada, "1.5-kW 150,000-r/min Ultra High-Speed PM Motor Fed by 12-V Power Supply for Automotive Supercharger," *Proceeding of the 13th European Conference on Power Electronics and Applications (EPE2009)*, 2009.

[3] D. K. Hong, B. C. Woo, J. Y. Lee, and D. H. Koo, "Ultra high speed motor supported by air foil bearings for air blower cooling fuel cells," *IEEE Trans. Magnetics*, vol. 48, no. 2, pp. 871–874, 2012.

[4] I. Takahashi, T. Noguchi, "A New Quick-Response and High-Efficiency Control Strategy of an Induction Motor", *IEEE Trans. on Industry Applications*, vol. 22, no. 5, pp. 820-827, 1986.

[5] L. Chen and F. Z. Peng, "Dead-time elimination for voltage source inverters," *IEEE Trans. Power Electron.*, vol. 23, no. 2, pp. 574–580, Mar. 2008.

[6] J. Yoshimoto, Y. Inoue, S. Morimoto, M. Sanada, "Ultra-High-Speed PMSM Sensorless Drive Using Direct Torque Control," *Proceedings of the 15th International Conference on Electrical Machines and Systems (ICEMS 2012)*, DS4G1-5, 2012.

[7] L.Tang, L. Zhong, M. Rahman, and Y. Hu, "A Novel Direct Torque Control for Interior Permanent-Magnet Synchronous Machine Drive With Low Ripple in Torque and Flux-A Speed-Sensorless Approach," *IEEE Trans. on Industry Applications*, vol. 39, no. 6, pp. 1748-1756, 2003.

[8] S. Morimoto, M. Sanada, Y. Takeda, "Mechanical Sensorless Drives of IPMSM with Online Parameter Identification," *IEEE Trans. on Industry Applications*, vol.42, no. 5, pp. 1241-1248, 2006.

Mathematical Model for High-Efficiency Control of Permanent-Magnet Synchronous Motor in Stator Flux Linkage Synchronous Frame

Tatsuki Inoue, Yukinori Inoue, Shigeo Morimoto, and Masayuki Sanada

Graduate School of Engineering
Osaka Prefecture University
Sakai, Japan
st106002@edu.osakafu-u.ac.jp

Abstract—A mathematical model is proposed for the high-efficiency control law of a permanent-magnet synchronous motor (PMSM) in the stator flux linkage synchronous frame (M-T frame). The mathematical model takes into consideration magnetic saturation and cross-magnetization effects. In taking into consideration magnetic saturation effects, most of the mathematical models in the *d-q* frame are complex. The M-T frame has several advantages, such as the simplicity of the flux-weakening control law and its applicability to various motors. In this paper, the control characteristics of PMSM in the M-T frame are analyzed, and a mathematical model for estimating the high-efficiency control trajectory in the M-T frame is proposed. In addition, a method of simply determining the parameters for the mathematical model is proposed. The validity of the proposed model and method is confirmed by comparing the analytical and experimental results.

Keywords—Permanent-magnet synchronous motor, M-T frame, maximum torque per ampere, magnetic saturation, mathematical model, direct torque control

I. INTRODUCTION

Permanent-magnet synchronous motor (PMSM) drives are used extensively as high-efficiency motors in industrial applications and household electrical appliances. Interior permanent-magnet synchronous motor (IPMSM) drives, which possess high-efficiency and wide-speed-range capabilities, are used in various applications. However, when the PMSM is required of increasing torque or miniaturization, magnetic saturation and cross-magnetization effects that occur in the PMSM are significant.

Thus far, in PMSM control laws, the mathematical models in the rotor synchronous frame (*d-q* frame) are commonly used. However, when magnetic saturation occurs, the *d*- and *q*-axis inductances vary, and high-efficiency control is difficult. Therefore, a novel mathematical model, which can take into consideration the magnetic saturation effect, is required for high-performance drives [1]-[4].

Several studies have examined extended models in the *d-q* frame in order to take into consideration magnetic saturation and cross-magnetization effects [4]-[6]. However, most of the mathematical models used in these studies are complex and are difficult to apply to a PMSM drive system in practice.

This paper focuses on the stator flux linkage synchronous frame (M-T frame) and examines a high-efficiency control law that can take into consideration the saturation effect in the M-T frame. Since mathematical models in the M-T frame are commonly nonlinear models, the control laws of the M-T frame were not seriously examined. However, the M-T frame has several advantages. Flux-weakening control in the M-T frame is affected by the motor parameters to a lesser extent [7], and the M-T frame can be applied to various motors [8].

The remainder of this paper is organized as follows. In Section II, the characteristic of the maximum torque per ampere (MTPA) in the M-T frame is analyzed. The MTPA control law minimizes copper loss. Therefore, the MTPA control law is a high-efficiency control law. In addition, a mathematical model of the MTPA condition is proposed, and the curve drawn by the proposed model is compared to analysis results. The proposed model takes into account magnetic saturation. In Section III, a method for determining the parameters of the proposed model is discussed. In Section IV, the validity of the proposed model and the proposed method is verified based on experimental results for PMSM and synchronous reluctance motor (SynRM).

II. MTPA CURVE IN M-T FRAME

A. Control Characteristics in M-T Frame

Fig. 1 shows the vector diagram in the M-T frame. Here, ψ_a is the stator flux linkage vector due to permanent magnet, ψ_s is the stator flux linkage vector, i_a is the armature current vector, and i_d, i_q, i_M, and i_T are the *d*-, *q*-, M-, and T-axis components, respectively, of i_a. The *d-q* frame is based on the direction of the permanent-magnet stator flux vector ψ_a, whereas the M-T frame is based on the direction of the stator flux linkage vector ψ_s.

Fig. 2 shows the MTPA, flux-weakening, and constant torque trajectories in the i_q-i_d and Ψ_s-i_T planes, where Ψ_s is the stator flux linkage. The analyzed curves are calculated based on the parameters of the Type-A motor shown in Table I. The Type-A motor is an IPMSM, and

The 2014 International Power Electronics Conference

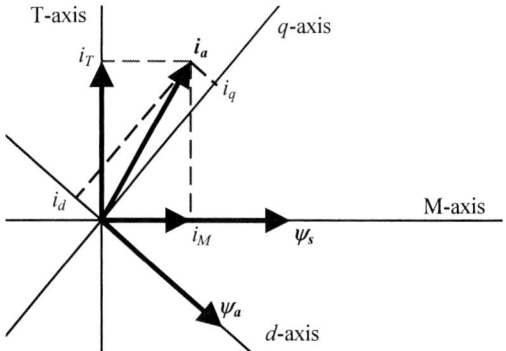

Fig. 1. Vector diagram in the M-T frame.

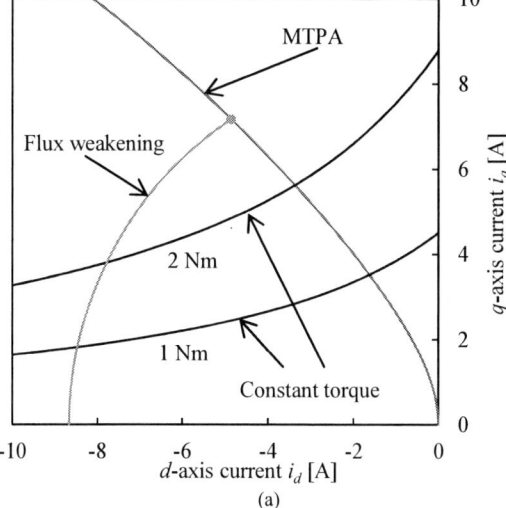

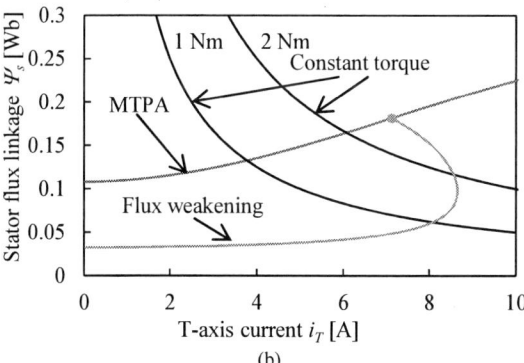

(b)

Fig. 2. Control trajectories. (a) i_q-i_d plane. (b) Ψ_s - i_T plane.

the Type-B motor is a SynRM, the permanent-magnet flux Ψ_a of which is zero, that uses only reluctance torque. The saturation effect is assumed to have been taken into consideration in only the axis having the greater inductance, and the inductance decreases linearly as the armature current increases. In the d- and q-axis current control method, the mathematical model of the MTPA curve shown in Fig. 2(a) is reported in [9]. However, an LUT or linear approximate equation is required for the reference flux calculation because the relationship between the torque T_e and the stator flux Ψ_s cannot be modeled as a mathematical model. A maximum power (flux-weakening) control law in the M-T frame, which depends only on the armature resistance was previously proposed in [7]. Therefore, in the flux-weakening control region, the PMSM can be concisely control. However, a simple MTPA control law in the M-T frame has not yet been proposed. In this section, an approximate mathematical model of the MTPA curve is proposed.

B. Mathmatical Model of MTPA Curve

Fig. 3 shows the characteristics of the MTPA curve for various magnet fluxes, as shown in Table II. The parameters applied to drawing MTPA curves are based on the Type-A motor. In Table II, the magnet fluxes Ψ_a of the Type-AF1 and Type-AF2 motors are 50% and 25% of that of the Type-A motor. Fig. 3 shows that Ψ_s corresponds to Ψ_a at $i_T = 0$. In addition, as the magnet flux decreases, the gradient of the curve increases.

Fig. 4 shows the characteristics of the MTPA curve for various q-axis inductances, as shown in Table III. In Table III, the q-axis inductances L_q of the Type-AL1 and

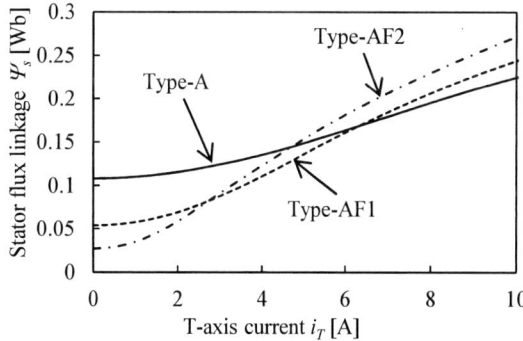

Fig. 3. Characteristics of the MTPA curves for various permanent-magnet fluxes.

TABLE I
MACHINE PARAMETERS USED IN ANALYSIS AND EXPERIMENT

Symbol	Meaning	Value	
		Type-A	Type-B
P_n	No. of pole pairs	2	2
Ψ_a	Permanent-magnet flux	0.108 Wb	0 Wb
L_d	d-axis inductance	8.7 mH	$399.8 - 44.3\lvert i_d\rvert$ mH
L_q	q-axis inductance	$28.3 - 0.657\lvert i_q\rvert$ mH	55 mH
R_a	Armature resistance	0.64 Ω	11.45 Ω
I_{am}	Limiting current	8.66 A	2.2 A

TABLE II
MACHINE PARAMETERS FOR VARIOUS PERMANENT-MAGNET FLUXES

Symbol	Meaning	Value	
		Type-AF1	Type-AF2
P_n	No. of pole pairs	2	2
Ψ_a	Permanent-magnet flux	0.054 Wb	0.027 Wb
L_d	d-axis inductance	8.7 mH	8.7 mH
L_q	q-axis inductance	$28.3 - 0.657\lvert i_q\rvert$ mH	$28.3 - 0.657\lvert i_q\rvert$ mH
R_a	Armature resistance	0.64 Ω	0.64 Ω
I_{am}	Limiting current	8.66 A	8.66 A

Type-AL2 motors are 125% and 75% of that of the Type-A motor. Based on Fig. 4, as the q-axis inductance increases, the gradient of the curve increases. However, in the large-current region, the variation of the gradient is small.

According to the above discussions, the MTPA curve in the Ψ_s - i_T plane is modeled as follows:

$$\Psi_s = \frac{2}{\pi}\left(L_T - b_T\, i_T\right)i_T \tan^{-1}\left(\frac{L_k}{\Psi_a}i_T\right) + \Psi_a \qquad (1)$$

where L_T, L_k, and b_T are the parameters of (1). Here, L_T denotes an inductance, L_k affects the shape of the curve, and b_T is a coefficient of i_T as determined by magnetic saturation. In addition, the arctangent in (1) expresses the shape of the MTPA curve, as shown in Fig. 4.

In the case of a SynRM, since Ψ_a is equal to zero in (1), the mathematical model can be written as

$$\Psi_s = \left(L_T - b_T\, i_T\right)i_T. \qquad (2)$$

Based on (2), in the case of a SynRM, L_k can be ignored.

C. Comparison with Analysis Results

Fig. 5 shows the analytical MTPA curve and the curve of (1). The parameters for the Type-A, Type-AF1, and Type-AF2 motors shown in Tables I and II are used for the evaluation. The parameters determined using (1) are as follows:

Type-A:
$L_T = 44.7$ mH, $L_k = 8.5$ mH, and $b_T = 1.7$ mH/A
Type-AF1:
$L_T = 46.8$ mH, $L_k = 8.5$ mH, and $b_T = 1.7$ mH/A
Type-AF2:
$L_T = 47.2$ mH, $L_k = 8.5$ mH, and $b_T = 1.7$ mH/A.

According to Fig. 5, the proposed model can be used to draw the MTPA curve of each motor.

Fig. 6 shows the analytical MTPA curve and the curve given by (1). The parameters of the Type-AL1 and Type-AL2 motors listed in Table III are applied. The parameters determined by (1) are as follows:

Type-AL1:
$L_T = 65.0$ mH, $L_k = 8.5$ mH, and $b_T = 2.1$ mH/A
Type-AL2:
$L_T = 25.0$ mH, $L_k = 8.5$ mH, and $b_T = 1.2$ mH/A.

According to Fig. 6, the proposed model can also draw the MTPA curves of motors that have various inductances. Therefore, the efficacy of the model is confirmed.

Fig. 7 shows the MTPA curve of the Type-B motor and the curve given by (2). The parameters determined using (2) are as follows:

Type-B: $L_T = 465.5$ mH and $b_T = 73.6$ mH/A.

According to Fig. 7, efficacy of the proposed model in the SynRM is also confirmed.

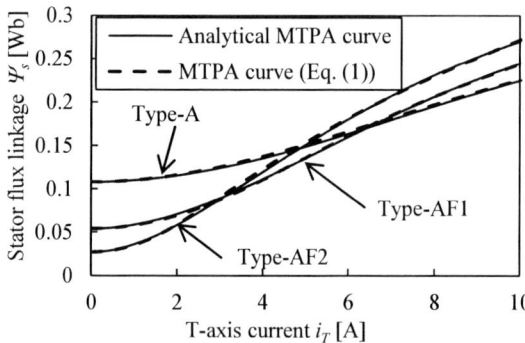

Fig. 5. Comparisons of the MTPA curve and the mathematical model (1) (Type-A, Type-AF1, and Type-AF2 motors).

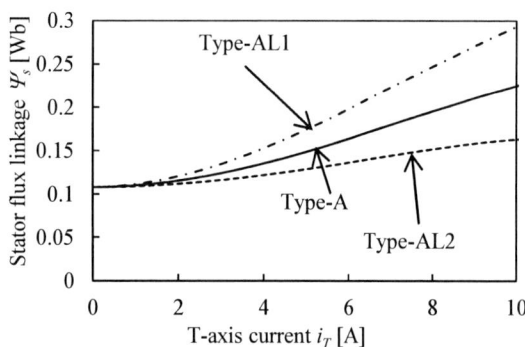

Fig. 4. Characteristics of the MTPA curves for various q-axis inductances.

TABLE III
MACHINE PARAMETERS FOR VARIOUS INDUCTANCES

Symbol	Meaning	Value	
		Type-AL1	Type-AL2
P_n	No. of pole pairs	2	2
Ψ_a	Permanent-magnet flux	0.108 Wb	0.108 Wb
L_d	d-axis inductance	8.7 mH	8.7 mH
L_q	q-axis inductance	$35.4 - 0.657\lvert i_q\rvert$ mH	$21.3 - 0.657\,\lvert i_q\rvert$mH
R_a	Armature resistance	0.64 Ω	0.64 Ω
I_{am}	Limiting current	8.66 A	8.66 A

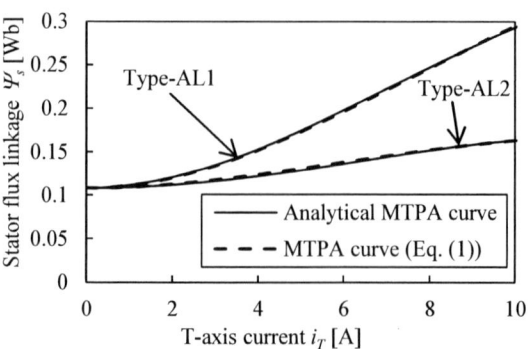

Fig. 6. Comparisons of the MTPA curve and the mathematical model (1) (Type-AL1 and Type-AL2 motors).

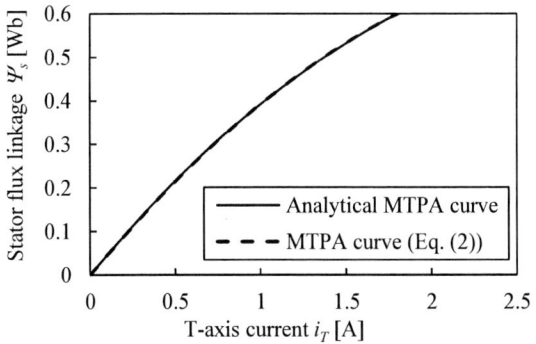

Fig. 7. Comparison of the MTPA curve of the Type-B motor and the mathematical model given by (2).

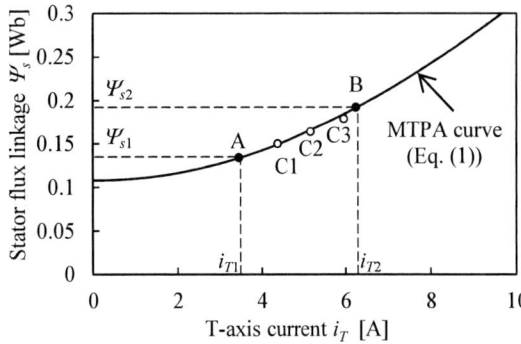

Fig. 8. Explanation of the proposed method.

III. METHOD OF SIMPLY DETERMINING PARAMETERS OF THE PROPOSED MODEL

A mathematical model of the MTPA curve in the Ψ_s - i_T plane has been proposed. However, since the proposed model has three variable parameters (L_T, b_T, and L_k), determination of the parameters is troublesome. In this section, a method of simply determining the parameters of the proposed model is described.

A. Method for PMSM

Fig. 8 shows an explanation of the method for simply determining the parameters of the proposed model. Two operating points, A and B, on the approximate MTPA curve are measured, as shown in Fig. 8, when (3) is obtained by substituting the values for two operating points in (1), as follows:

$$
\begin{bmatrix}
\left(\Psi_{s1}-\Psi_a\right)\Big/\left(\dfrac{2}{\pi}\tan^{-1}\left(\dfrac{L_k}{\Psi_a}i_{T1}\right)\right) \\[2mm]
\left(\Psi_{s2}-\Psi_a\right)\Big/\left(\dfrac{2}{\pi}\tan^{-1}\left(\dfrac{L_k}{\Psi_a}i_{T2}\right)\right)
\end{bmatrix}
=\begin{bmatrix}-i_{T1}^{2} & i_{T1} \\ -i_{T2}^{2} & i_{T2}\end{bmatrix}\begin{bmatrix}b_T \\ L_T\end{bmatrix}
\tag{3}
$$

where i_{T1} and i_{T2} are the T-axis currents at points A and B, and Ψ_{s1} and Ψ_{s2} are the flux linkages at points A and B.

When the value of L_k is given, (3) becomes a solvable linear equation in terms of L_T and b_T. From (3), L_T and b_T are calculated as follows:

$$
\begin{bmatrix}L_T \\ b_T\end{bmatrix}=
\begin{bmatrix}\dfrac{i_{T1}i_{T2}}{i_{T2}-i_{T1}}\left(Z_1-Z_2\right) \\[3mm]
\dfrac{i_{T1}Z_1-i_{T2}Z_2}{i_{T2}-i_{T1}}\end{bmatrix}
$$

$$
\begin{bmatrix}Z_1 \\ Z_2\end{bmatrix}=
\begin{bmatrix}
\left(\Psi_{s1}-\Psi_a\right)\Big/\left(\dfrac{2}{\pi}i_{T1}^{2}\tan^{-1}\left(\dfrac{L_k}{\Psi_a}i_{T1}\right)\right) \\[3mm]
\left(\Psi_{s2}-\Psi_a\right)\Big/\left(\dfrac{2}{\pi}i_{T2}^{2}\tan^{-1}\left(\dfrac{L_k}{\Psi_a}i_{T2}\right)\right)
\end{bmatrix}.
\tag{4}
$$

Therefore, the parameters are determined from L_k. The MTPA curve, which is drawn by the proposed method for determining the parameters, passes through points A and B, as shown in Fig. 8.

The method of determining L_k is used of numerical analysis, for example, searching the best L_k which minimizes the root mean square (RMS) of the errors between the curve of (1) and measured points C1, C2, and C3 as shown in Fig. 8. However, since the range of L_k is obscure, the range analyzed by the numerical analysis is wide. Therefore, L_{k_max}, which is the maximum value of L_k, is defined.

Here, L_k is included in the arctangent in (1), and the value of the arctangent converged to $\pi/2$ as the independent variable increases. Therefore, L_{k_max} should be a sufficiently large value as the arctangent converges to $\pi/2$ at small T-axis current. In this paper, L_{k_max} is decided as follows:

$$
L_{k\_max}=\frac{10\Psi_a}{I_{am}}\tan\left(\frac{2\pi}{5}\right)
\tag{5}
$$

where I_{am} is the limiting armature current. L_{k_max} is determined such that the value of the arctangent corresponds to 80% of $\pi/2$ at i_T of $I_{am}/10$.

The process of the proposed method for determining the parameters is summarized as follows:

1) Two points (for example, points A and B in Fig. 8) are measured based on the experimental results.
2) At least one point (for example, point C1, C2, or C3 in Fig. 8) is measured based on the experimental results.
3) L_{k_max} is calculated using (5).
4) L_T and b_T are calculated within the range of $0 < L_k < L_{k\_max}$.
5) L_T, L_k, and b_T are determined by minimizing the RMS.

The RMS of the flux errors between the MTPA curve given by (1) and the measured points ε is defined as follows:

$$
\varepsilon=\sqrt{\frac{\displaystyle\sum_{j=1}^{n}\left(\Psi_{s\_MTPAj}-\Psi_{sj}\right)^2}{n}}
\tag{6}
$$

where n is the number of measured points, i_{Tj} and Ψ_{sj} are the T-axis current and the stator flux linkage of the j'th measured point, respectively, and Ψ_{s_MTPAj} is the stator flux linkage, which is obtained by substituting i_{Tj} for (1).

B. Method for SynRM

In the case of a SynRM, since the mathematical model of the MTPA curve is (2), (4) is written as follows:

$$
\begin{bmatrix} L_T \\ b_T \end{bmatrix} = \begin{bmatrix} \dfrac{i_{T1} i_{T2}}{i_{T2} - i_{T1}} (Z_1 - Z_2) \\ \dfrac{i_{T1} Z_1 - i_{T2} Z_2}{i_{T2} - i_{T1}} \end{bmatrix}
$$

$$
\begin{bmatrix} Z_1 \\ Z_2 \end{bmatrix} = \begin{bmatrix} \Psi_{s1} / i_{T1}^2 \\ \Psi_{s2} / i_{T2}^2 \end{bmatrix} . \tag{7}
$$

Therefore, L_T and b_T are calculated by (7), and L_k has no influence on parameter calculation.

IV. EXPERIMENTAL RESULTS

The Type-A and Type-B motors shown in Table I were considered in this experiment. Fig. 9 shows the experimental system. Here, N and N_{ref} are the measured speed and the reference speed, respectively, T_{ref} is the reference torque of a load, i_u and i_w are the measured values of the u- and w-phase currents, respectively, and v_u^*, v_v^*, and v_w^* are the reference values of the u-, v-, and w-phase voltages, respectively. The T-axis current and stator flux linkage for minimizing the armature current for each torque were measured, and parameters L_T, L_k, and b_T were obtained using the proposed method. The loads in the experimental systems of the Type-A and Type-B motors were a servo motor and powder brake, respectively. Direct torque control (DTC) is used for the motor drive.

The stator flux linkages Ψ_α and Ψ_β in the α-β frame are obtained by the following equations [10]:

$$
\Psi_\alpha = \int (v_\alpha - R_a\, i_\alpha)dt + \Psi_{\alpha 0} \tag{8}
$$

$$
\Psi_\beta = \int (v_\beta - R_a\, i_\beta)dt + \Psi_{\beta 0} \tag{9}
$$

where v_α and v_β are the α- and β-axis armature voltages, respectively, i_α and i_β are the α- and β-axis armature currents, respectively, $\Psi_{\alpha 0}$ and $\Psi_{\beta 0}$ are the α- and β-axis initial flux linkages, respectively, and R_a is the armature resistance.

From (8) and (9), the angle θ_s formed by the α-axis and the stator flux linkage vector, the T-axis current i_T, and the stator flux linkage Ψ_s are obtained by

$$
\theta_s = \tan^{-1}\left(\frac{\Psi_\beta}{\Psi_\alpha}\right) \tag{10}
$$

$$
i_T = -i_\alpha \sin\theta_s + i_\beta \cos\theta_s \tag{11}
$$

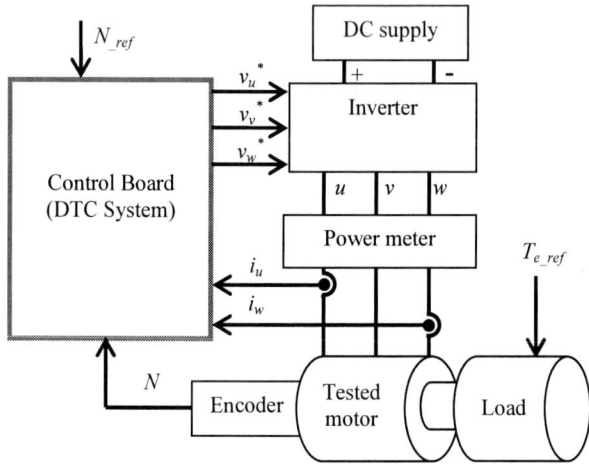

Fig. 9. Experimental system.

$$
\Psi_s = \sqrt{\Psi_\alpha^2 + \Psi_\beta^2} . \tag{12}
$$

In this experiment, the voltages required for (8) and (9) were estimated from the reference voltages to an inverter, and integrals were substituted for first-order lag filters. In addition, the voltage errors, which occurred in the inverter, were compensated. The operating conditions of the Type-A and Type-B motors were at 1,000 min^{-1} and 500 min^{-1}, respectively.

Fig. 10 shows the experimental results for the Type-A motor. Two measured points used in determining the parameters are points A and B shown in Fig. 10(a). Here, L_{k_max} is 384 mH. Fig. 10(a) shows the MTPA curve of (1) and the measured points. Fig. 10(b) shows the RMS of the flux errors ε. Fig. 10(c) shows the parameters L_T and b_T. The best value of L_k is the value that minimizes error ε in the range of $0 < L_k < L_{k\_max}$, as shown in Fig. 10(b). As shown in Fig. 10(b), the minimum ε is 5.185 mWb. As shown in Fig. 10(c), the parameters are determined to be as follows:

$L_k = 17$ mH, $L_T = 18.9$ mH, and $b_T = -1.31$ mH/A.

Points A and B in Fig. 10(a) were confirmed to lie on the MTPA curve of the mathematical model given by (1). In addition, the MTPA curve of (1) was confirmed to be in better agreement with the experimental results. Therefore, the proposed model is a valid model for an actual machine, and the proposed method can effectively determine the parameters of (1).

However, note that the reliability of the MTPA curve is influenced by the two measured points used for the proposed method. For example, Fig. 11 shows the MTPA curve in the case of selecting points B and C used in the parameter determination, as shown in Fig. 11(a). The parameters are determined to be as follows:

$L_k = 0.01$ mH, $L_T = -23.3$ H, and $b_T = -9.57$ H/A.

As shown in Fig. 11(b), the minimum ε is 16.53 mWb, which is larger than that in the case of selecting points A and B. The two points should be selected precisely, and a mutual distance should be maintained.

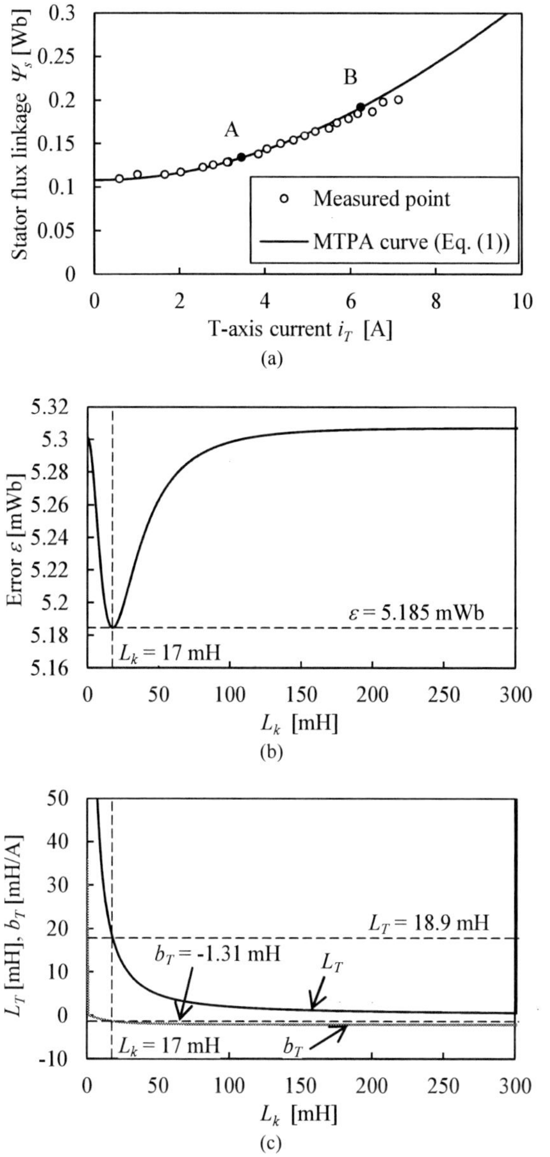

Fig. 10. Experimental results for the Type-A motor. (a) Comparison of the MTPA curve given by (1) and the measured points. (b) RMS of the flux errors. (c) L_T and b_T.

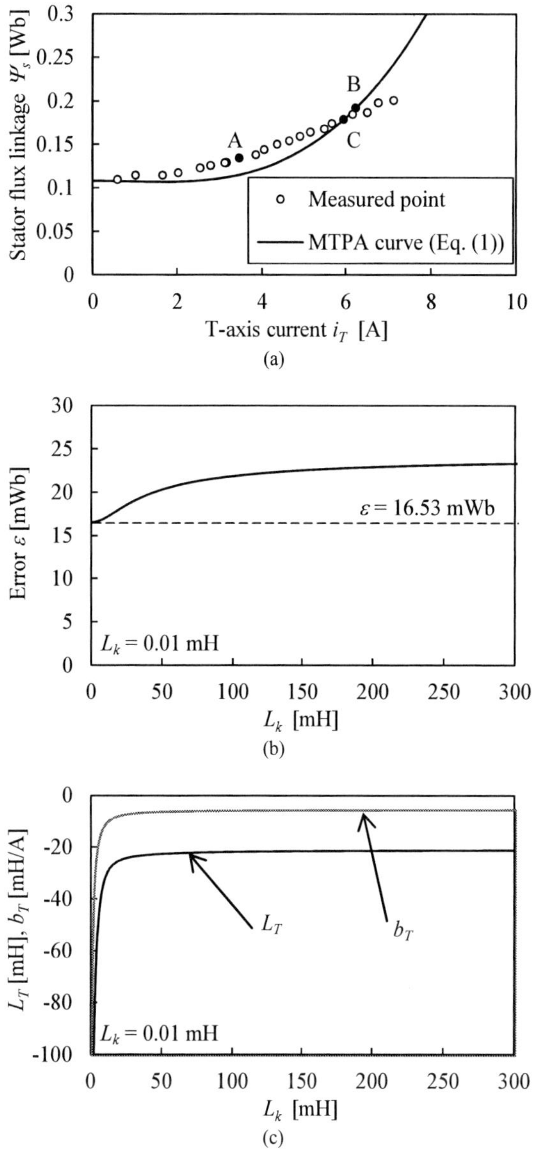

Fig. 11. Experimental results in the case of selecting points B and C. (a) Comparison of the MTPA curve given by (1) and the measured points. (b) RMS of the errors. (c) L_T and b_T.

Fig. 12 shows experimental results for the Type-B motor. Since the Type-B motor is a SynRM, L_T and b_T are simply decided by (7). Two measured points applied to the proposed method are points A and B, as shown in Fig. 12. The parameters are determined to be as follows:

$L_T = 464$ mH and $b_T = 51.6$ mH/A.

Fig. 12 confirms that the MTPA curve of the mathematical model given by (2), which uses the parameters determined by (7), is in better agreement with the experimental results.

V. CONCLUSIONS

In this paper, the characteristics of the MTPA control law taking into consideration magnetic saturation effect

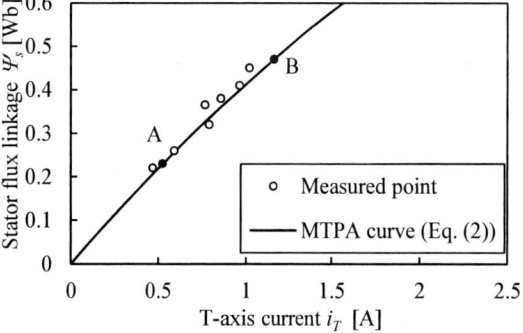

Fig. 12. Experimental results for the Type-B motor.

are analyzed in the M-T frame, and a mathematical model for drawing the MTPA curve is proposed. In addition, a method for simply determining the parameters used in the mathematical model is proposed. Based on the analysis and experimental results, the proposed model can be expressed for the MTPA curve taking into consideration the magnetic saturation effect in the M-T frame, and the proposed method was confirmed to be appropriate for determining the parameters for the mathematical model.

REFERENCES

[1] N. Bianchi, E. Fornasiero, and S. Bolognani, "Effect of Stator and Rotor Saturation on Sensorless Rotor Position Detection," *IEEE Transactions on Industry Applications*, vol. 49, no. 3, pp. 1333-1342, 2013.

[2] E. Levi, and V. A. Levi, "Impact of Dynamic Cross-Saturation on Accuracy of Saturated Synchronous Machine Models," *IEEE Transactions on Energy Conversion*, vol. 15, no. 2, pp. 224-230, 2000.

[3] T. Frenzke, "Impacts of Cross-Saturation on Sensorless Control of Surface Permanent Magnet Synchronous Motors," *EPE'05 11th European Conference on Power Electronics and Applications*, CD-ROM, 2005.

[4] B. Štumberger, G. Štumberger, D. Dolinar, A. Hamler, and M. Trlep, "Evaluation of Saturation and Cross-Magnetization Effects in Interior Permanent-Magnet Synchronous Motor," *IEEE Transactions on Industry Applications*, vol. 39, no. 5, pp. 1264-1271, 2003.

[5] K. J. Meessen, P. Thelin, J. Soulard, and E. A. Lomonova, "Inductance Calculations of Permanent-Magnet Synchronous Machines Including Flux Change and Self- and Cross-Saturations," *IEEE Transactions on Magnetics*, vol. 44, no. 10, pp. 2324-2331, 2008.

[6] S. Lee, Y. Jeong, Y. Kim, and S. Jung, "Novel Analysis and Design Methodology of Interior Permanent-Magnet Synchronous Motor Using Newly Adopted Synthetic Flux Linkage," *IEEE Transactions on Industrial Electronics*, vol. 58, no. 9, pp. 3806-3814, 2011.

[7] Y. Inoue, S. Morimoto, and M. Sanada, "Comparative Study of PMSM Drive Systems Based on Current Control and Direct Torque Control in Flux-Weakening Control Region," *IEEE Transactions on Industry Applications*, vol. 48, no. 6, pp. 2382-2389, 2011.

[8] G. Pellegrino, R. I. Bojoi, and P. Guglielmi, "Unified Direc-Flux Vector Control for AC Motor Drives," *IEEE Transactions on Industry Applications*, vol. 47, no. 5, pp. 2093-2102, 2011.

[9] S. Morimoto, M. Sanada, and Y. Takeda, "Wide-Speed Operation of Interior Permanent Magnet Synchronous Motors with High-Performance Current Regulator," *IEEE Transactions on Industry Applications*, vol. 30, no. 4, pp. 920-926, 1994.

[10] M. F. Rahman, L. Zhong, and K. W. Lim, "A Direct Torque-Controlled Interior Permanent Magnet Synchronous Motor Drive Incorporating Field Weakening," *IEEE Transactions on Industry Applications*, vol. 34, no. 6, pp. 1246-1253, 1998.

Wide-Speed-Range Operation of DTC-Based PMSM Drive System Using MTPF Control

Yukinori Inoue, Takahiro Ichiya, Shigeo Morimoto, Masayuki Sanada

Graduate School of Engineering
Osaka Prefecture University
Sakai, Japan
inoue@eis.osakafu-u.ac.jp

Abstract— **This paper proposes a control method of permanent-magnet synchronous motor (PMSM) drive system based on direct torque control (DTC) in high speed region. In high speed region, when the magnet flux of the PMSM is relatively weak, maximum torque per flux (MTPF) control is used for wide-speed-range operation. In the DTC-based motor drive system, the reference torque and the reference flux are required for optimal vector control. The proposed method is based on torque limitation using the maximum torque angle. Simulation and experimental results show the effectiveness of the proposed method. The proposed PMSM drive system with DTC has advantage including simplicity of calculation.**

Keywords— *Direct torque control, maximum torque per flux, permanent magnet synchronous motor, wide speed range operation.*

I. INTRODUCTION

Permanent-magnet synchronous motors (PMSMs) are used in many applications, including electric vehicles, air conditioners and industrial servo drives. The PMSM can achieve high-performance and high-efficiency drives. Generally, PMSM drives used for wide-speed-range operation are controlled by appropriate ways. The most of PMSM drives have been controlled in the *d-q* frame, which is synchronized with the rotor. The maximum torque per ampere (MTPA) control is adopted in low speed region and flux-weakening (FW) control is applied as the motor accelerates and the armature voltage reaches its limiting voltage which corresponds to the maximum available voltage of the inverter. Moreover, if the motor drives are required to expand the operating range, the maximum torque per flux (MTPF) control is applied [1]–[9]. In the *d*- and *q*-axis current control, two variables (e.g., the armature current I_a and the current phase β) are controlled depending on operating condition [1]. A control method by one variable (e.g., utilization of only the *d*-axis current or the current phase) is also proposed in [2] and [3]. On the other hand, methods applied to direct torque control (DTC) have been proposed. In [4], the reference torque and flux are determined by an approximate equation. In [5], the MTPF control is based on torque angle control, and an estimation method of torque angle is proposed. In [6]–[8], a proportional and integral (PI) controller is used for torque angle control,

and the torque angle is calculated from the stator flux-linkage vector and the rotor position. A method for synchronous reluctance motors is also proposed in [9]. Consequently, it is complicated to calculate reference values.

This paper proposes a method of the MTPF control using maximum torque angle determined from the stator flux linkage. In the proposed method, only one variable, that is reference torque, is used for the MTPF control. The reference flux is separately calculated by the FW control law. The torque angle estimation is unnecessary for the MTPF control. The proposed method is suitable for the DTC-based PMSM drive system. In addition, influence of parameter variation on the control characteristic is discussed. Through MATLAB/Simulink simulation and experiment, the effectiveness of the proposed control technique is confirmed.

II. CONTROL METHOD FOR WIDE-SPEED-RANGE OPERATION

A. Torque Limiting and Flux-Weakening Control

The controls of the PMSM are restricted by the limitations of current and voltage. Table I shows the optimal controls with respect to operating condition. The DTC requires the reference torque and the reference flux for high-efficiency and wide-speed-range operation. In the proposed system, the control laws are based on the mathematical model in a rotating M–T reference frame which is synchronized with the stator flux-linkage vector. Fig. 1 shows a vector diagram and the definition of coordinate axes. For current limitation, the torque is

TABLE I SUMMARY OF CONTROL METHODS

Speed range	Low		High
Control method	MTPA	FW	MTPF
Voltage	$V_a < V_{am}$	$V_a = V_{am}$	
Current	$I_a \leq I_{am}$		$I_a < I_{am}$
Torque limiting	$T_{lim} = P_n \Psi_s \sqrt{I_{am}^2 - i_M^2}$		Eqs. (3)-(6)
Reference flux	MTPA Look up Table	$\Psi_{s-FW} = 1/\omega \left\{ -R_a i_T + \sqrt{V_{am}^2 - (R_a i_M)^2} \right\}$	

controlled so as not to exceed the T_{lim} as the following equation [10].

$$T_{lim} = P_n \Psi_s \sqrt{I_{am}^2 - i_M^2} \qquad (1)$$

where P_n is the number of pole pairs, Ψ_s is the amplitude of the stator flux linkage, I_{am} is the limiting armature current, and i_M is the M-axis current.

In low speed region, the MTPA control is applied, and the reference flux is calculated by a look-up table. As the rotor speed increases, the armature voltage V_a reaches the limiting voltage V_{am}. Considering the voltage limitation, the FW control is applied and the reference flux is given by

$$\Psi_{s-FW} = \frac{1}{\omega}\left\{-R_a i_T + \sqrt{V_{am}^2 - (R_a i_M)^2}\right\} \qquad (2)$$

where ω is the electrical rotor angular velocity, R_a is the armature resistance, and i_T is the T-axis current.

B. MTPF Control Based on Maximum Torque Angle

The torque limiting of (1) can be used in the operating range using the FW control. In higher speed, however, the torque limiting based on the MTPF is used for expansion of operating speed range.

The electromagnetic torque of the PMSM is determined by the stator flux linkage Ψ_s and the torque angle δ. The maximum torque T_{em-dm} is calculated from the maximum torque angle δ_m, as the following equation [11].

$$T_{em-dm} = \frac{P_n \Psi_s}{2 L_d L_q}\left\{2\Psi_a L_q \sin\delta_m + \Psi_a (L_d - L_q)\sin 2\delta_m\right\} \qquad (3)$$

$$\delta_m = \cos^{-1}\left[\frac{1}{4}\left\{a/\Psi_s - \sqrt{(a/\Psi_s)^2 + 8}\right\}\right] \qquad (4)$$

$$a = \frac{L_q}{L_q - L_d}\Psi_a \qquad (5)$$

where Ψ_a is the magnet flux linkage, L_d and L_q are the d- and q-axis inductances.

Fig. 2 shows the relationship between the torque and the torque angle for rotor speeds of 1000 min⁻¹, 2000 min⁻¹ and 3000 min⁻¹. The operating point for the MTPF is the maximum value of each torque curve, which is given by (3)-(5). In the DTC, however, if the torque angle δ exceeds the maximum torque angle (e.g., δ_{m1}, δ_{m2} and δ_{m3} shown in Fig. 2), the torque control becomes unstable. Hence, δ should be below δ_m for both steady and transient states. In the proposed method, the reference torque is given by

$$T_{e-MTPF} = k \cdot T_{em-dm} \qquad (6)$$

where k is a constant ($0 < k \leq 1$).

C. DTC-based PMSM Drive System

Fig. 3 shows a block diagram of the proposed system based on the direct torque controlled PMSM drives. The DTC estimates the stator flux-linkage vector. The estimation is based on voltage integration. The estimated vector of the stator flux linkage is given by

$$\hat{\psi}_s = \int (v_a - R_a i_a)dt \qquad (7)$$

where v_a is the armature voltage vector, and i_a is the

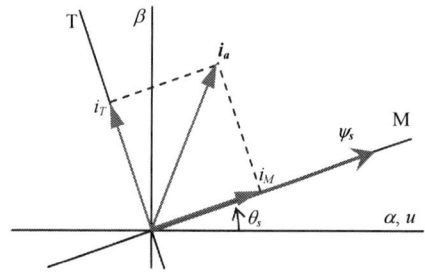

Fig. 1. Vector diagram of stator flux and armature current.

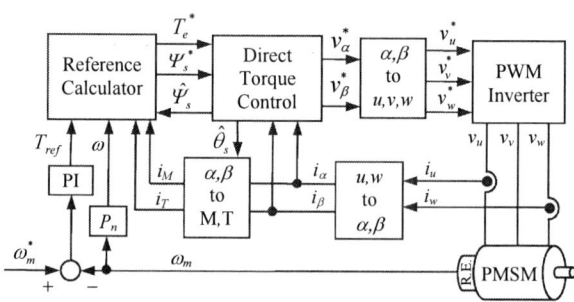

Fig. 3. Direct torque control-based PMSM drive system.

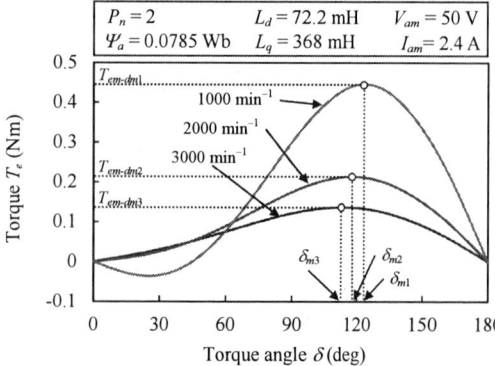

Fig. 2. Torque versus torque angle characteristics.

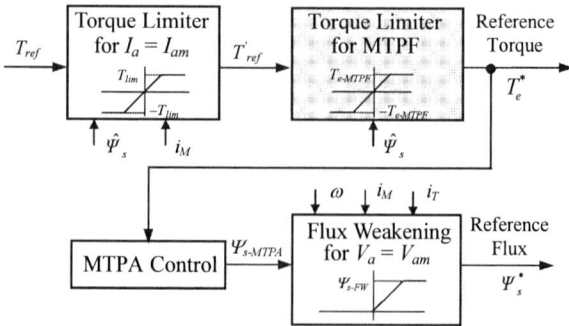

Fig. 4. Calculator of reference torque and flux.

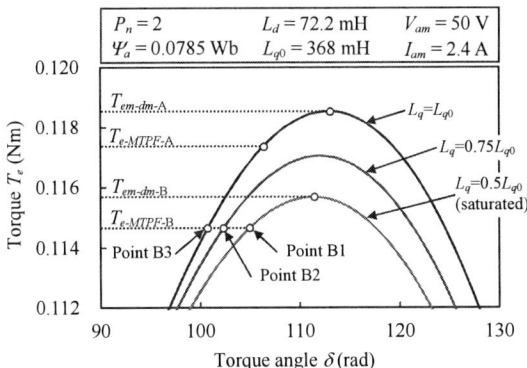

| $P_n = 2$ | $L_d = 72.2$ mH | $V_{am} = 50$ V |
| $\Psi_a = 0.0785$ Wb | $L_{q0} = 368$ mH | $I_{am} = 2.4$ A |

Fig. 5. Effect of coefficient k and inductance variation.
($\Psi_s = 0.0487$ Wb, k=0.991, 3500 min^{-1})

TABLE II SYSTEM PARAMETERS

Number of pole pairs P_n	2
d-axis inductance L_d	72.2 mH
q-axis inductance L_q	368 mH
Permanent magnet flux Ψ_a	0.0785 Wb
Armature resistance R_a	12.04 Ω
Limiting armature current I_{am}	2.4 A
Limiting armature voltage V_{am}	50 V
Base speed	195 min^{-1}
Rated torque	1.97 Nm
DC link voltage of inverter	83 V
Control period	100 μs

armature current vector.

Fig. 4 shows the reference calculator adopting MTPF control. The reference torque and the reference flux are calculated according to the operating condition, as shown in Table I. In the proposed method, the stator flux linkage Ψ_s replaces $\hat{\Psi}_s$, which is the estimated stator flux linkage.

D. Influence of Parameter Variation

In the proposed MTPF control based on (3)-(6), the inductance variation affects the operating characteristic in high speed region. Fig. 5 shows a torque characteristic with respect to the torque angle. $T_{em\text{-}dm\text{-}A}$ is the maximum torque under the nominal stator flux, which is determined from the nominal value of the q-axis inductance L_{q0}. $T_{em\text{-}dm\text{-}B}$ is the maximum torque under the saturated stator flux, which is determined from half of the nominal q-axis inductance. Since the reference torque is given by (6), the reference values $T_{e\text{-}MTPF\text{-}A}$ and $T_{e\text{-}MTPF\text{-}B}$ are calculated from $T_{em\text{-}dm\text{-}A}$ and $T_{em\text{-}dm\text{-}B}$, respectively. Generally, the inductance decreases due to magnetic saturation as the armature current increases. When the reference torque is $T_{e\text{-}MTPF\text{-}A}$, there is no intersection point of $T_{e\text{-}MTPF\text{-}A}$ with the torque curve under the saturated stator flux. Hence, the DTC cannot operate at the torque of $T_{e\text{-}MTPF\text{-}A}$. If the reference torque is given by $T_{e\text{-}MTPF\text{-}B}$, there are some intersection such as points B2 and B3. In this case, the DTC operate at the torque of $T_{e\text{-}MTPF\text{-}B}$ because the actual torque can follow the reference torque. Hence, stable control can be achieved.

Consequently, in the proposed system, the inductance under the rated current is used for the maximum torque calculation of (3).

III. SIMULATION RESULTS

A. Parameters of Tested System

The parameters of the tested PMSM drive system are shown in Table II. In the tested system, k of (6) is 0.991. In this paper, the reference flux vector calculation DTC [12], [13] is applied as a DTC, but the proposed method can be applied to any type of the DTC.

B. Acceleration Characteristic

Fig. 6 shows the acceleration characteristics, where the reference speed is changed from 0 to 4000 min^{-1} under no load condition. The armature voltage V_a is kept to the limiting voltage V_{am} and the armature current I_a is reduced from the limiting current I_{am}, because of MTPF condition. From the result of the torque angle, the torque angle δ doesn't exceed the maximum torque angle δ_m in both steady and transient states, although difference between δ_m and δ appears. In the proposed method, since the coefficient k is used for torque limitation based on the MTPF control law, the torque angle can be indirectly maintained within the maximum torque angle.

Therefore, in high speed region, it is confirmed that the proposed method of the MTPF control is effective for wide-speed-range operation and torque angle limitation.

C. Considerations in determining factor k

The coefficient k in (6) affects the operating characteristic under the MTPF control. In actual system, k should be less than unity because this condition is required to operate at torque angles not in excess of the maximum torque angle. The model errors such as control delay and parameter variation result in runaway torque angle and unstable control. For example, the flux estimation error comes from the discretization error and the resistance variation because the flux estimation is based on (7).

Fig. 7 shows influence of coefficient k on the MTPF control, where the rotor speed is changed from 0 to 4000 min^{-1}. From Fig. 7(a), the acceleration time increases as the coefficient k decrease, and thus, the acceleration time under a coefficient k of 0.7 is 0.25 seconds slower than that under k of 0.991. This is because the torque angle is restricted by smaller value, as shown in Fig. 7(b). The proposed system cannot generate enough torque for high-speed operation. Fig. 7(c) shows comparison of the torque trajectories. From this result, however, the operating region under k of 0.7 is approximately equal to that under k of 0.991 and theoretical operating region. Therefore, large k is preferred, but the influence of coefficient k on the operating region is negligible.

978-1-4799-2706-7/14 $31.00 © 2014 IEEE

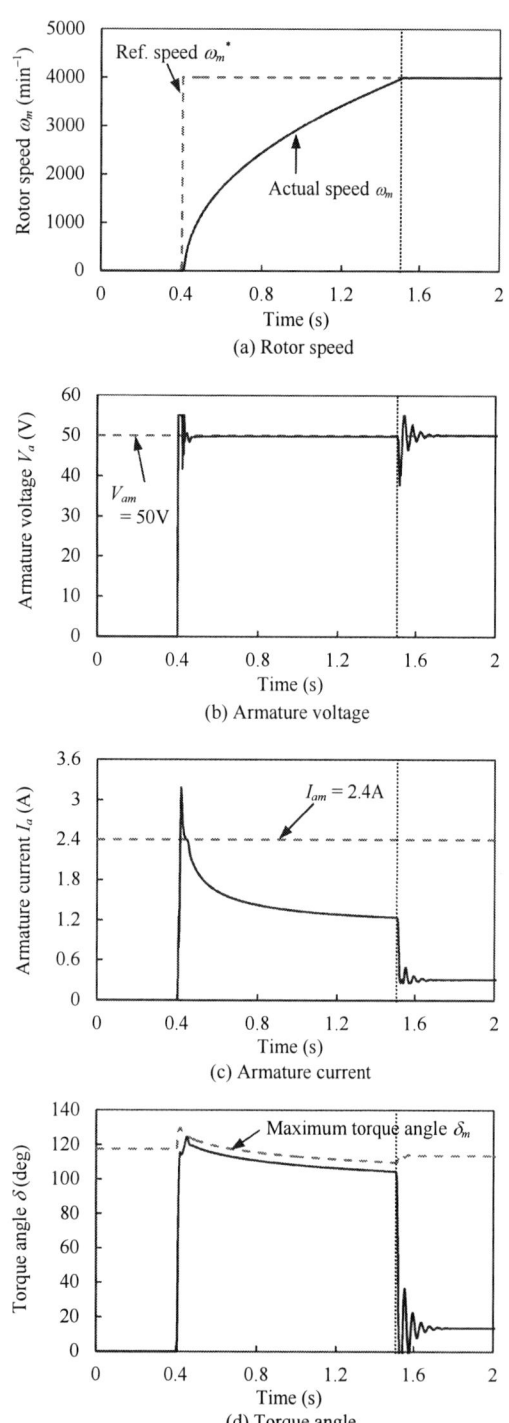

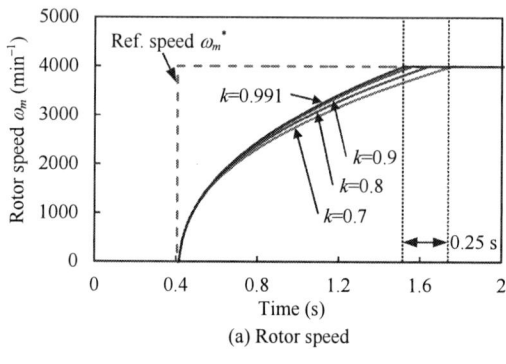

(a) Rotor speed

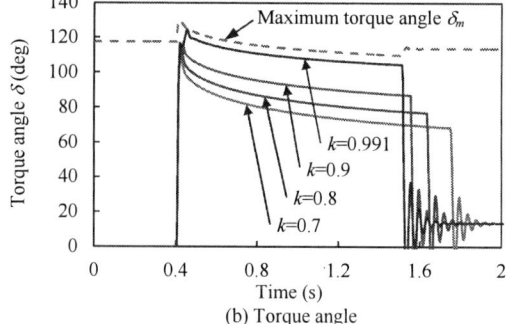

(b) Torque angle

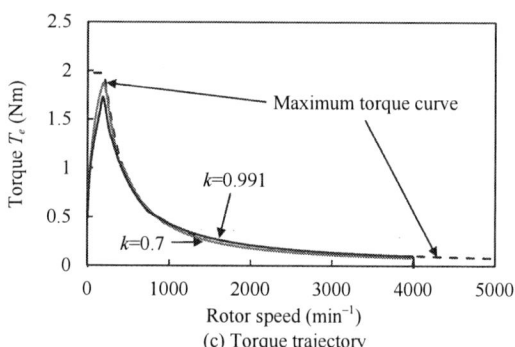

(c) Torque trajectory

Fig. 7. Influence of coefficient k on MTPF control.
(0 min⁻¹ to 4000 min⁻¹, no load).

Fig. 6. Simulation result of MTPF control.
(0 min⁻¹ to 4000 min⁻¹, no load, $k = 0.991$).

IV. EXPERIMENTAL RESULTS

A. Experimental Setups

The effectiveness of the proposed method is also evaluated by using the experimental results. All of the controls are processed through a digital signal processor (Texas Instruments TMS320C6713). An insulated gate bipolar transistor module (Mitsubishi Elec., PM30RSF060) is used for the inverter. The carrier frequency of the pulse width modulation is 10 kHz. The rotor speed is detected by an incremental encoder attached to the tested motor. Flux estimation is based on a first-order low-pass filter in order to avoid divergence of the estimated flux from DC offset caused by the current transducer and initial error of the estimated value.

B. Steady State Characteristics

Fig. 8 shows the trajectory of maximum power operation. The theoretical trajectories of the MTPF, MTPA and FW controls are also shown in this result. From Fig. 8, it is confirmed that the control method transition from the MTPA to the FW at the operating point P_1. The control method is changed to the MTPF at

The 2014 International Power Electronics Conference

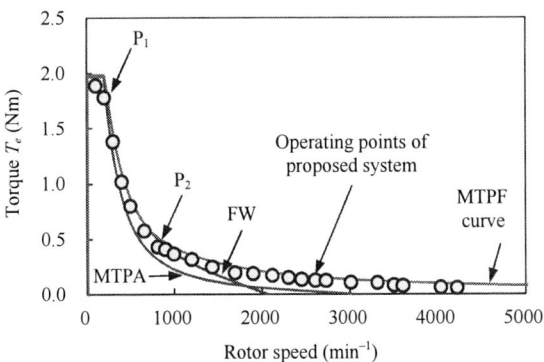

Fig. 8. Experimental result of maximum power operation.
$$(V_a = V_{am}, \ I_a \leq I_{am})$$

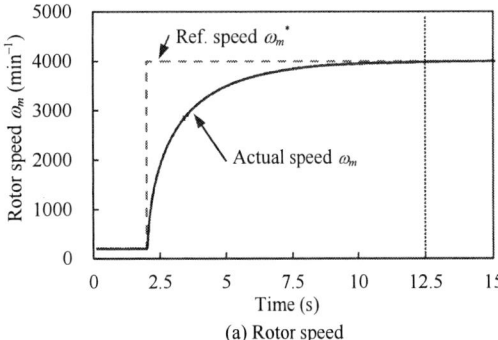

(a) Rotor speed

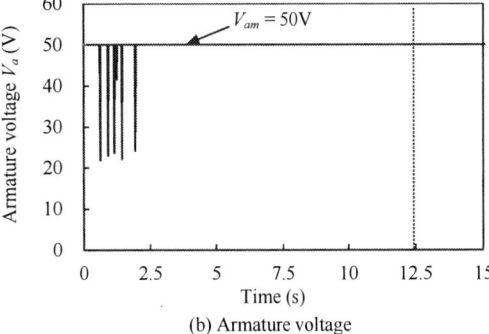

(b) Armature voltage

the operating point P_2. Finally, the rotor speed exceeds 4000 min^{-1}. When only the FW control is applied, speeds more than 2000 min^{-1} are unavailable, as shown in Fig. 8. The MTPF control is performed properly in high speed region.

C. Transient State Characteristics

Fig. 9 shows the experimental result of acceleration characteristics, where the reference speed is changed from 200 min^{-1} to 4000 min^{-1} under no-load condition. From the beginning, the FW control law of (2) is applied, and thus the armature voltage is maintained at its limiting value because the initial rotor speed is same as the base speed of the tested motor. In Fig. 9(c), the armature current I_a decreases due to the MTPF control. The rotor speed is maintained at the reference speed 4000 min^{-1} after 12.5 s. Fig. 10 shows the reference values of stator flux and torque. The control modes are changed smoothly.

V. CONCLUSIONS

This paper proposed a method of the MTPF control based on the maximum torque angle in high speed region. Both simulation and experimental results was provided to evaluate the control characteristics of the proposed method. The proposed method is effective for stable operation of the DTC-based PMSM drive system and wide-speed-range operation.

REFERENCES

[1] S. Morimoto, Y. Takeda, K. Hatanaka, Y. Tong, and T. Hirasa, "Expansion of operating limits for permanent magnet motor by current vector control considering inverter capacity", *IEEE Trans. Industry Applications*, IA-26, no. 5, pp. 866-871, 1990.

[2] Y. Zhang, L. Xu, M. K. Guven, S. Chi, and M. S. Illindala, "Experimental verification of deep flux-weakening operation of a 50 kW IPM machine by using single current regulator," *Proceedings of the 1st IEEE Energy Congress Conference and Exposition (ECCE 2009)*, pp. 3647-3652, 2009.

[3] Y. Wang, X. Wen, and F. Zhao, "A proposed control strategy of PMSM for deep field-weakening and square-wave mode," *Proceedings of the 15th International Conference on Electrical Machines and Systems (ICEMS 2012)*, 2012.

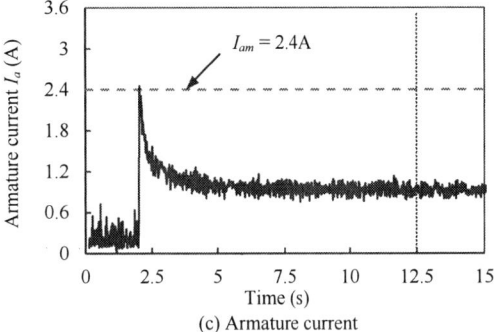

(c) Armature current

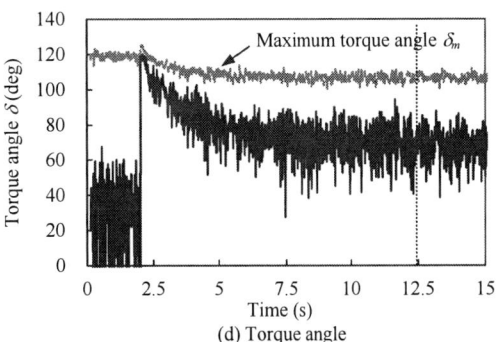

(d) Torque angle

Fig. 9. Experimental result of acceleration characteristic.
(200 min^{-1} to 4000 min^{-1}, no load, $k = 0.991$).

[4] J. Faiz, and S. H. M.-Zonoozi, "A novel technique for estimation and control of stator flux of a salient-pole PMSM in DTC method based on MTPF," *IEEE Trans. Industrial Electronics*, vol. 50, no. 2, pp. 262-271, 2003.

[5] J. Luukko, O. Pyrhönen, M. Niemela, and J. Pyrhönen, "Limitation of the load angle in a direct-torque-controlled synchronous machine drive," *IEEE Trans. Industrial Electronics*, vol. 51, no. 4, pp. 793-798, 2004.

978-1-4799-2706-7/14 $31.00 © 2014 IEEE   374

The 2014 International Power Electronics Conference

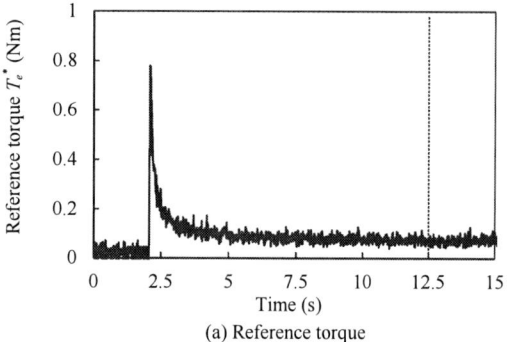

(a) Reference torque

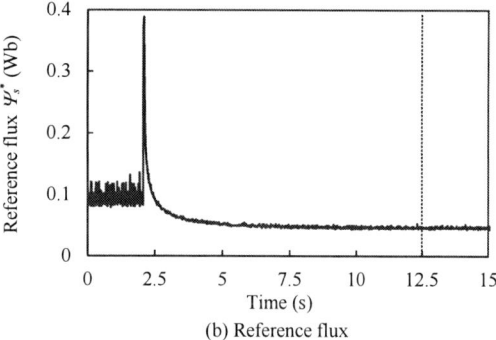

(b) Reference flux

Fig. 10. Experimental result of proposed reference calculator.
(200 min^{-1} to 4000 min^{-1}, no load, $k = 0.991$).

[6] G. Pellegrino, E. Armand, and P. Guglielmi, "Direct flux field-oriented control of IPM drives with variable DC link in the field-weakening region," *IEEE Trans. Industry Applications*, vol. 45, no. 5, pp. 1619-1627, 2009.

[7] G. Pellegrino, R. I. Bojoi, and P. Guglielmi, "Unified direct-flux vector control for AC motor drives," *IEEE Trans. Industry Applications*, vol. 47, no. 5, pp. 2093-2102, 2011.

[8] G. Pellegrino, E. Armando, and P. Guglielmi, "Direct-flux vector control of IPM motor drives in the maximum torque per voltage speed range," *IEEE Trans. Industrial Electronics*, vol. 59, no. 10, pp. 3780-3788, 2012.

[9] Y. Inoue, S. Morimoto, and M. Sanada, "A novel control scheme for maximum power operation of synchronous reluctance motors including maximum torque per flux control," *IEEE Trans. Industry Applications*, vol. 47, no. 1, pp. 115-121, 2011.

[10] Y. Inoue, S. Morimoto, and M. Sanada, "Comparative study of PMSM drive systems based on current control and direct torque control in flux-weakening control region," *IEEE Trans. Industry Applications*, vol. 48, no. 6, pp. 2382-2389, 2012.

[11] M. F. Rahman, L. Zhonge, and K. W. Lim, "A direct torque-controlled interior permanent magnet synchronous motor drive incorporating field weakening", *IEEE Trans. Industry Applications*, vol. 34, no. 6, pp. 1246-1253, 1998.

[12] L. Tang, L. Zhong, M. F. Rahman, and Y. Hu, "A novel direct torque controlled interior permanent magnet synchronous machine drive with low ripple in flux and torque and fixed switching frequency," *IEEE Trans. Power Electronics*, vol. 19, no. 2, pp. 346–354, 2004.

[13] Y. Inoue, S. Morimoto, and M. Sanada, "Examination and linearization of torque control system for direct torque controlled IPMSM," *IEEE Trans. Industry Applications*, vol. 46, no. 1, pp. 159-166, 2010.

An Industrial Low-voltage Inverter for PRM Control

M. Nakamura*, T. Oka*, and K. Oishi**

Toshiba Mitsubishi-electric Industrial Systems Corporation
* 1 Toshiba-cho, Fuchu-shi, Tokyo, 183-8511, JAPAN
** 6-14 Maruo-cho, Nagasaki-shi, Nagasaki, 852-8004, JAPAN

Abstract— The PRM which uses reluctance torque positively and reduced the amount of magnets relatively, and the industrial low-voltage inverter to which the PRM control function is applied has been developed. Though exact motor parameters may not be known, the inverter has a control method which on-site PMM control commissioning can be performed easily for such the case. The combination test of the PRM and the inverter was carried out, and each performance was checked.

Keywords— PRM, Vector control, PMM, Inverter

I. INTRODUCTION

In the industrial low-voltage motor & drive system for steel and paper industries, the induction motors (IM) are widely applied. Moreover, in the motor & drive system for the industry, the needs for the permanent magnet synchronous motor (PMM) are growing from the demand of energy saving and high-efficiency in recent years.

On the other hand, the manufacturing cost of the PMM is getting higher due to price increase of the rare earth required for a permanent magnet. From the viewpoint of a cost-cutting, the PMM which reduced the amount of the permanent magnet used is demanded. One of the solution is the PRM[1] (Permanent magnet Reluctance Motor) which is combined the permanent magnet motor and the reluctance motor.

In the rolling line of steel, or the paper machine line of paper manufacture, the motor & drive system consisting of many low-voltage small power motors and industrial low-voltage inverters are applied. Most of such motors are the induction motor (IM) formerly, but the motor & drive system replaced with the PRM can improve the total system efficiency.

This time, the PRM control function was added to the industrial low-voltage inverter TMdrive™-10e2, and also the low-voltage PRM TM21-FP was developed, thus the configurations and the combination test results are presented.

II. OVERVIEW OF THE PRM

The PRM uses less permanent magnet than the conventional permanent magnet motor. Fig. 1 shows a basic rotor section of the traditional IPM and the PRM.

Reluctance torque of the PRM is generated higher and an amount of permanent magnet usage is relatively lower.

The composition of the PRM is a union of the permanent magnet motor and the reluctance motor, and generated reluctance torque is increased and the amount of the permanent magnet used is reduced relatively.

Since the amount of the permanent magnet usage of the PRM is decreased, the induced voltage which occurs on a permanent magnet becomes lower than the conventional permanent magnet motor. This can widen a range which can drop motor voltage without negating magnetic flux. Therefore this can widely expand a range of constant voltage control in higher rotation speed, thus the PRM is suitable for the use with the wide adjustable-speed range of a constant power range.

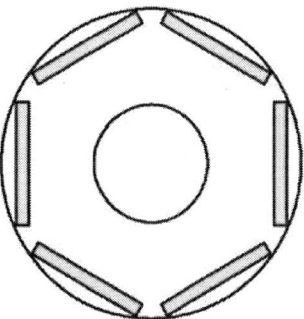

(a) The traditional IPM

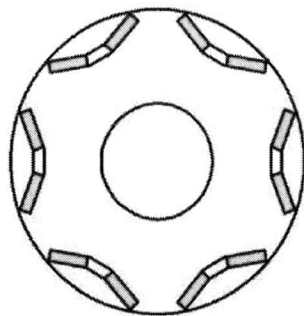

(b) The PRM

Fig.1 Basic rotor section

The 2014 International Power Electronics Conference

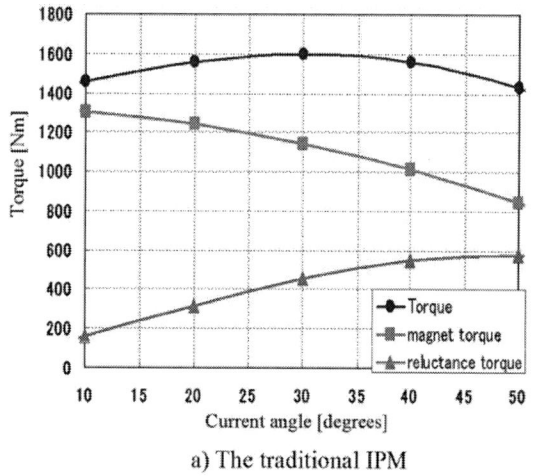

a) The traditional IPM

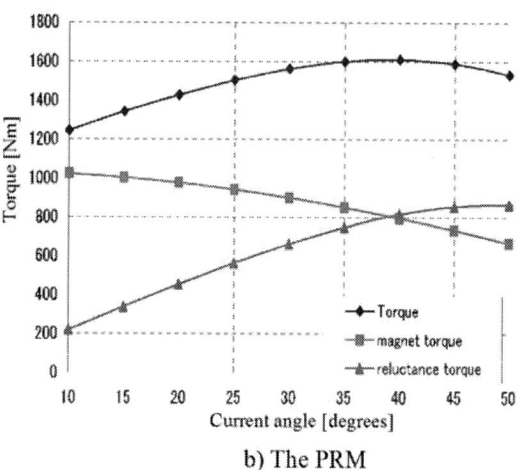

b) The PRM

Fig. 2 Current angle vs. Torque curve

About the PRM, a permanent magnet is arranged in a V-shape in a rotor core so that the reluctance torque increases, and flux barriers are also applied to make a magnetic salient of the rotor large. Moreover, centrifugal force is distributed by making size of a permanent magnet smaller and distributing arrangement. This has contributed to improvement in high-speed of a rotor, and rotor diameter expansion.

Fig. 2 shows current angle vs. torque characteristics of a conventional permanent magnet motor and the PRM. Generated torque is a sum of magnetic torque and reluctance torque. The reluctance torque which accounts for the total torque becomes approximately 50% in the PRM, while the conventional type permanent magnet motor was approximately 25%. Moreover the amount of the magnet used was reduced 40% compared to a conventional permanent magnet motor. Thus, instead of reducing the amount of the permanent magnet used, torque equivalent to the permanent magnet motor of a conventional type can be generated by generating reluctance torque greatly.

III.   THE PRM CONTROL MODEL

The d-axis is defined in the direction of a magnetic pole of the PRM, and the q-axis is defined in the rectangular direction. The PRM voltage equation in the d-q axis coordinate system which rotates with a rotor is shown as;

$$\begin{bmatrix} v_d \\ v_q \end{bmatrix} = \begin{bmatrix} R_a + pL_d & -\omega L_q \\ \omega L_d & R_a + pL_q \end{bmatrix} \begin{bmatrix} i_d \\ i_q \end{bmatrix} + \begin{bmatrix} 0 \\ \omega \Psi_a \end{bmatrix} \quad (1)$$

where,
v_d, v_q : the d-axis and the q-axis armature voltages
i_d, i_q : the d-axis and the q-axis armature currents
L_d : the d-axis inductance

L_q : the q-axis inductance
Ψ_a : armature interlinkage flux of permanent magnet
ω : rotating speed of a rotor
p : differential operator

Shown in equation (1), v_q is determined by induced voltage proportional to ω, voltage drop caused by i_q, and an armature reaction caused by i_d.

Armature interlinkage flux is,

$$\begin{bmatrix} \Psi_{0d} \\ \Psi_{0q} \end{bmatrix} = \begin{bmatrix} L_d & 0 \\ 0 & L_q \end{bmatrix} \begin{bmatrix} i_d \\ i_q \end{bmatrix} + \begin{bmatrix} \Psi_a \\ 0 \end{bmatrix} \quad (2)$$

Torque T which is generated by the PRM, which the number of poles is P, is given by the cross product of current vector $\boldsymbol{i_a}$ and armature interlinkage flux $\boldsymbol{\Psi_0}$.

$$\boldsymbol{T} = \frac{P}{2} \boldsymbol{\Psi_0} \times \boldsymbol{i_a} + \frac{P}{2}(L_d - L_q)\boldsymbol{i_d} \times \boldsymbol{i_q} \quad (3)$$

When the current vector is described with an absolute value i_a and current angle β (an angle between the q-axis and the current vector), the torque T is described as shown;

$$T = \frac{P}{2} \left\{ \Psi_a I_a \cos \beta + \frac{1}{2}(L_q - L_d)I_a^2 \sin 2\beta \right\}$$

$$= T_m + T_r \quad (4)$$

The first term of right-hand side in equation (4) shows magnetic torque generated by the permanent magnet, and the second term shows reluctance torque generated by counter-saliency. Since the PRM has counter-saliency $L_d < L_q$, the minus d-axis current flowing and reluctance torque contributes to output torque. Moreover, it acts so that the d-axis armature reaction magnetic flux may weaken the magnetic flux of a permanent magnet by the minus d-axis current.

978-1-4799-2706-7/14 $31.00 © 2014 IEEE

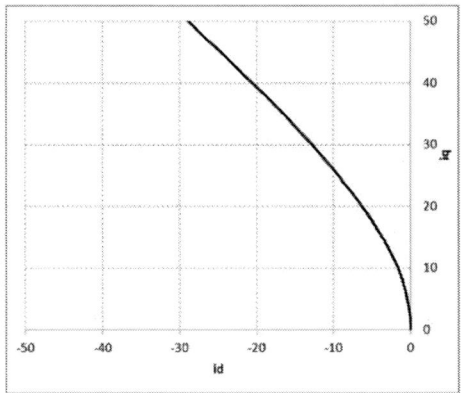

Fig. 3 Maximum ratio of Torque/Current curve

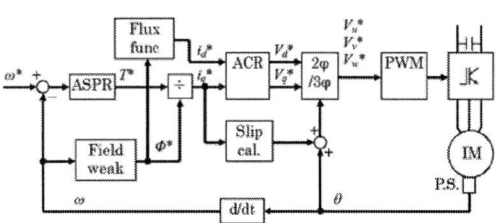

Fig. 4 IM vector control block diagram

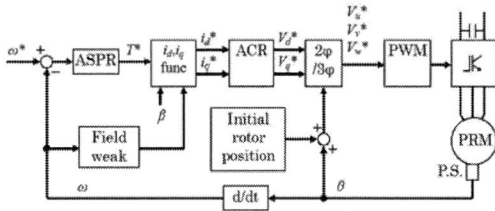

Fig. 5 PRM vector control block diagram

Equation (4) shows that a generated torque changes with β also the same current. Generally there are some d-q axis current determination methods in the vector control of the PMM[2]. One of these, there is a method of choosing β which makes the ratio of torque and current the maximum[3]. Thereby, the PRM can generate torque efficiently. This condition is given by the following formula as relation between i_d and i_q.

$$i_d = \frac{\Psi_a}{2(L_q - L_d)} - \sqrt{\frac{\Psi_a^2}{4(L_q - L_d)^2} + i_q^2} \qquad (1)$$

When this relation is plotted, it will become a curve like Fig. 3.

IV. THE INVERTER FOR THE PRM CONTROL

The industrial low-voltage inverter was developed to drive the IM. Fig. 4 shows the overview of the IM vector control block diagram of the inverter. The q-axis current reference (i_q*) is generated from the torque reference generated by the speed control (ASPR), the d-axis and the q-axis current control is performed with also the d-axis current reference (i_d*) considered field weakening, and the modulation ratio reference to PWM circuit is generated.

On the other hand, in the vector control of the PRM, it is necessary to pass current with the optimal current angle so that the motor can generate torque efficiently. Generally there are some methods of the d-axis current and the q-axis current determination in the vector control of the PMM, the PRM vector control of the inverter uses the d-axis and the q-axis current determination method based on a control to make the maximum ratio of torque and current. The control method is the method which determines i_d and i_q so that the motor torque can become

the maximum with the same current value ($=\sqrt{i_d^2 + i_q^2}$).

In addition, to perform field weakening of the PRM, control which passes negative i_d and negates flux of the magnet is also performed. Fig. 5 shows the overview of the PRM vector control block diagram of the inverter. Meanwhile, this control can be applied not only to the PRM but also to the IPM.

A resolver is used in order to detect the magnetic pole position of the PRM. Although the resolver could detect absolute angle of rotation, the magnetic pole angle and the resolver detection angle have shifted only by attaching a resolver to the PRM. Therefore, an offset addition which rectifies an error of a resolver detection angle and a magnetic pole angle (= control phase angle) is performed. The offset can be easily adjusted with the method of pulling the rotor of the PRM with passing direct-current to the PRM.

It is necessary to perform on-site commissioning of the motor which the parameters are unknown in the industries where the inverters are applied. Therefore, straight line approximation with the current angle is used instead of the curve that the maximum ratio of torque and current, and simplification of the on-site commissioning is attained.

V. TEST RESULTS OF COMBINATION WITH THE PRM AND THE INVERTER

A. Test facilities

A test which combined the PRM and the inverter was carried out. Table I and Table II show major data of the PRM and the inverter used for the test, respectively.

PRM1 is a medium-speed motor of the base speed 1500min^{-1} (electric frequency of 75 Hz), and for an industry use like a fan / a pump drive etc. PRM2 are the low speed and high torque motor of the base speed 270min^{-1} (electric frequency of 13.5 Hz), and they are designed for the auxiliary machinery motors of a steel rolling line. The inverter is the low-voltage small capacity of 150kVA and 460Vac output, and is designed for the general industries including a steel rolling line.

Table I PRMs for the test

	PRM1	PRM2
Output power [kW]	110	40
Induced voltage at 1500min^{-1} [V]	261.1	251.4
Rated voltage [V]	400	400
Rated current [A]	204	85
Rotor speed [min^{-1}]	1500/1800	270/460
Number of poles	6	6
Rated torque [Nm]	700	1414
Winding resistance [Ω]	0.0079	0.0642
L_q [mH]	2.63	32.6
L_d [mH]	0.89	11.2
Salient ratio	2.96	2.91

Table II Inverter for the test

	INV for PRM1	INV for PRM2
Frame size	150	
Rated DC voltage (equipment) [V]	680	
Rated DC voltage (usage) [V]	620	
Rated output voltage (equipment) [V]	460	
Rated output voltage (usage) [V]	420	
Rated current (equipment) [A]	204	204
Rated current (usage) [A]	204	85

Fig. 6 and Fig. 7 show appearances of the PRM and the inverter used for the test respectively. The IM of the comparable output power was coupled for a load of the PRM. Another inverter sharing DC power source drove the IM, Back-To-Back operation of the PRM and the IM was performed.

(a) PRM1

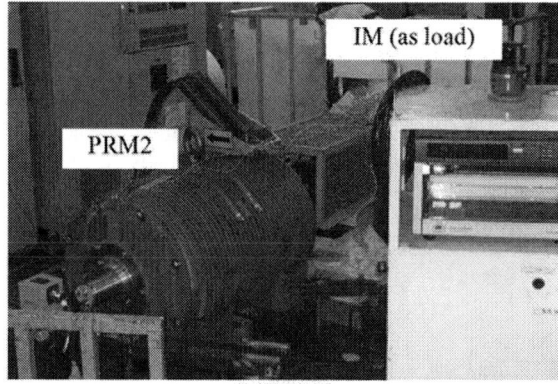

(b) PRM2

Fig. 6 Appearance of the tested PRMs

Fig. 7 Appearance of the Inverter

B. Test results

Fig. 8 shows the characteristics of current angle vs. motor current at the base speed of each motor (PRM1: 1500 min⁻¹, PRM2: 270 min⁻¹). PRM1, in case of 100% load and 120% load, current became the lowest around 40 degrees of current angle. In case of 50% load, current became the lowest around 30 degrees of current angle. However, in comparison with current around 40 degrees, the difference was small. From this point, torque / current ratio can be approximately made into the minimum by fixing a current angle by 30 degrees to 40 degrees.

PRM2, in the case of 100% of load, current became the minimum by current angle 30 degrees to 40 degrees. In the case of 200% of load, a current angle was not able to be made smaller than 50 degrees by restrictions of motor voltage. From this point, the control which fixes a current angle by 30 degrees to 40 degrees when motor voltage is below voltage restrictions, and changes a current angle from this when motor voltage exceeds the

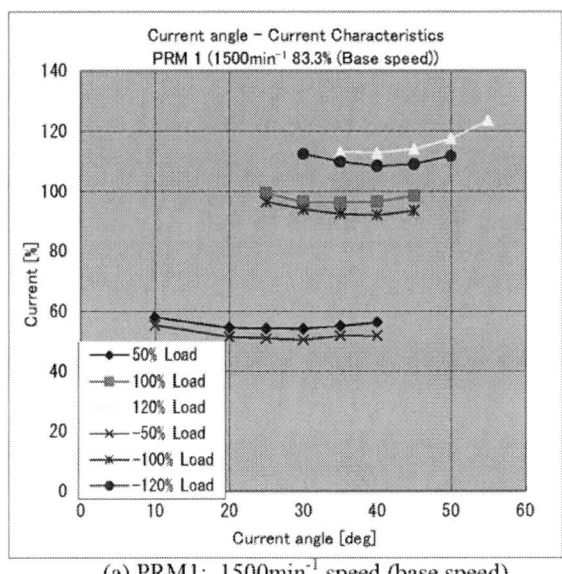

(a) PRM1: 1500min⁻¹ speed (base speed)

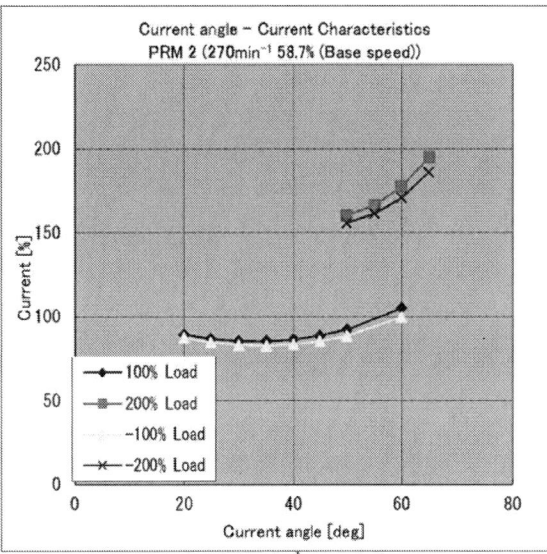

(b) PRM2: 270min⁻¹ speed (base speed)

Fig. 8 Experimental result of current angle vs. current

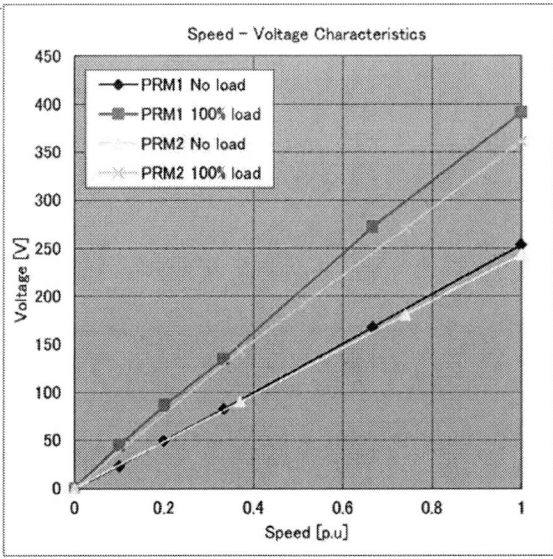

Fig. 9 Experimental result of speed vs. voltage

restriction, and controls a voltage surge, is carried out.

Fig. 9 shows the motor voltage characteristic against the motor speed at the time of no-load and 100% load. 1 p.u is the base speed here. The current angle at the time of 100% load was set to 40 degrees.

At the base speed, although the motor voltage reached near 400V as motor rating voltage at the time of 100% load, no-load voltage is 250V and falls to 63%.

From this, even if it was the speed range exceeding the base speed, it was shown by reducing output torque that it can operate on the voltage below rated voltage, without performing magnetic flux weakening control.

Fig. 10 shows the characteristic of the motor voltage against a current angle.

It turns out that the d-axis current becomes large in the minus direction, the action which negates the magnetic flux of a rotor magnet works, and motor voltage falls, so that a current angle becomes large.

However, if a current angle becomes large too much, motor current will increase as shown in Fig. 8 and the efficiency will decrease. Moreover, it becomes easy to receive restrictions of the current limiting of a motor or an inverter.

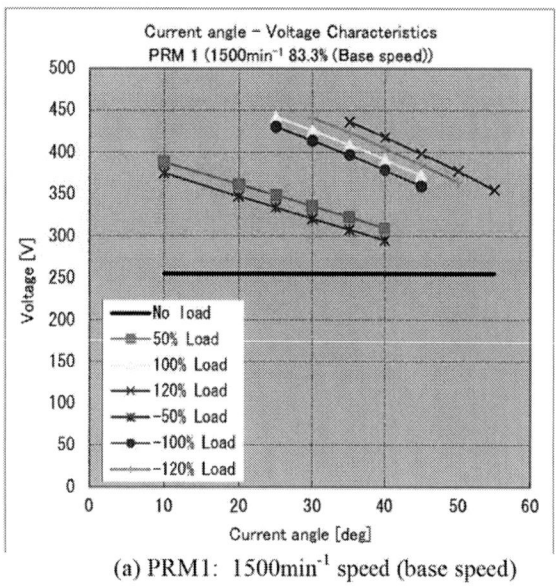

(a) PRM1: 1500min⁻¹ speed (base speed)

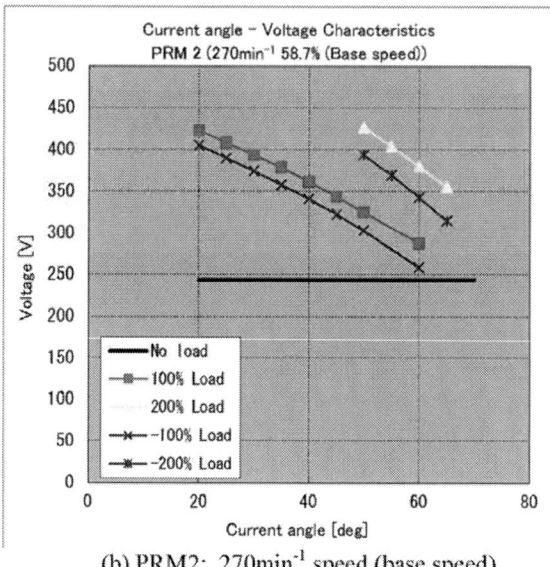

(b) PRM2: 270min⁻¹ speed (base speed)

Fig. 10 Experimental result of current angle vs. voltage

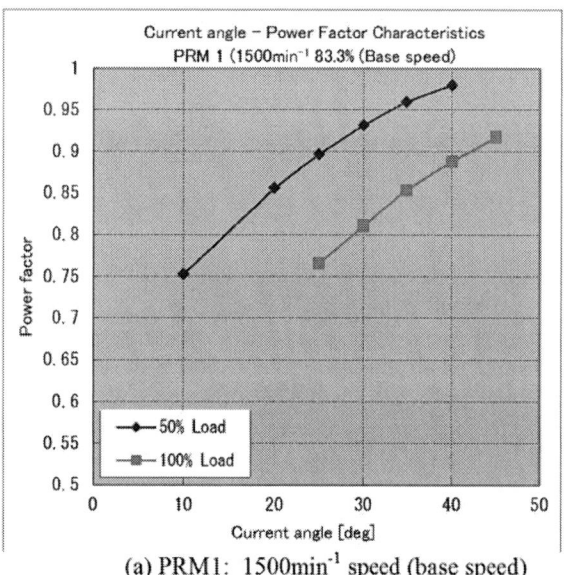

(a) PRM1: 1500min⁻¹ speed (base speed)

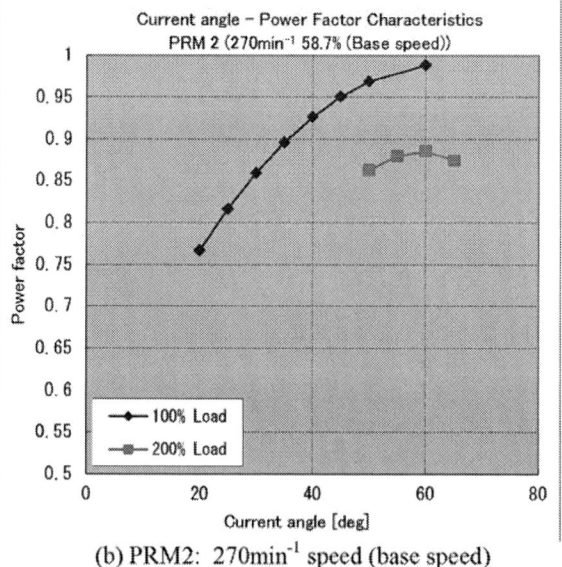

(b) PRM2: 270min⁻¹ speed (base speed)

Fig. 11 Experimental result of current angle vs. power factor

Fig. 11 shows the power factor characteristics against a current angle.

For the current angle between 30 degrees and 40 degrees, although a power-factor exceeds 0.9 at the time of 50% of load, it is 0.8 to 0.9 at the time of 100% of load in PRM1. Moreover, it has a power-factor of 0.85 to 0.92 at the time of 100% load in PRM2.

For setting a power-factor to 1.0, it is necessary to make a current angle larger than 40 degrees. That is, since it leads to flow more current, in the PRM, it turns out that power-factor 1 control is not necessarily the advantageous control method.

Fig. 12 shows the characteristics of acceleration and deceleration operation of PRM1 with a constant load. In this test, it is restricted so that motor voltage might not become more than the voltage at 1500min⁻¹.

978-1-4799-2706-7/14 $31.00 © 2014 IEEE 381

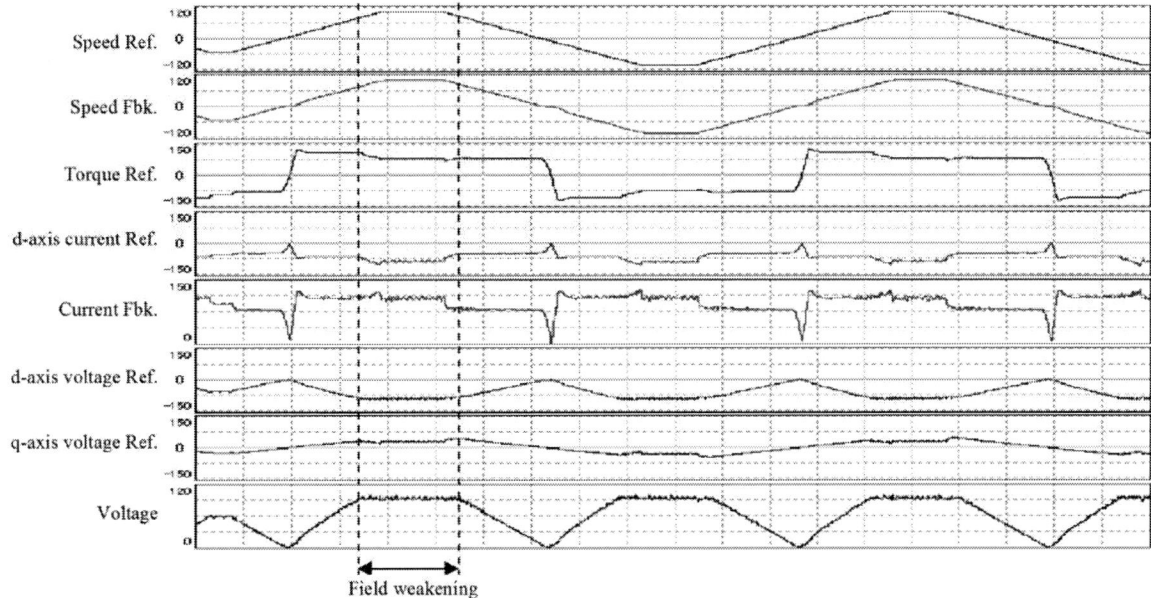

Fig. 12 Experimental result of acceleration / deceleration characteristics

Motor voltage is rising in proportion to the speed from zero speed to 1500min^{-1}. At the speed beyond 1500min^{-1}, by magnetic flux weakening control, the d-axis current was made to increase and the rise of voltage is controlled. As a result, even if the speed rises, it turns out that it is controlled so that the motor voltage becomes fixed.

VI. CONCLUSION

The load test of the PRM which uses reluctance torque positively and reduced the amount of magnets relatively and the industrial low-voltage inverter to which the PRM control function is applied has been carried out, and each performance has been checked.

In addition, though the exact motor parameters may not be known, the inverter has a control method which the on-site PMM control commissioning can be performed easily for such a case.

By the trend of recent years orienting energy saving and the high efficiency, the opportunity which applies the PMM to the industrial plant is growing. The PRM has the feature which has less amount of the magnet used, and the range of voltage fixed control is wide. The PRM and the inverter are one of the solutions of this market.

REFERENCES

[1] K. Sakai, M. Arata, T. Tajima, "High-Efficiency and High-Performance Motor for Energy Saving in Systems", Toshiba review Vol. 55 No.9, 2000 (in Japanese)

[2] S. Morimoto, Y. Takeda, T. Hirasa, "Current Phase Control Methods for Permanent Magnet Synchronous Motors", IEEE Trans. on Power Electronics, PE-5, No.2, pp. 133-139 (1990)

[3] S. Morimoto, Y. Takeda, K. Hatanaka, Y. Tong, T. Hirasa, "Design and Control System of Inverter-driven Permanent Magnet Synchronous Motors for High Torque Operation", IEEE Trans. on Industry Applications, IA-29, No.6, pp. 1150-1155 (1993)

The 2014 International Power Electronics Conference

Optimal Pulse Pattern Determination Based on Pulse Harmonic Modulation

Kimihisa Furukawa*, Toshiyuki Ajima**

Power Electronics Systems Division
Hitachi, Ltd., Hitachi Research Laboratory
Hitachi City Ibaraki, Japan
* kimihisa.furukawa.jz@hitachi.com
** toshiyuki.ajima.fz@hitachi.com

Hideki Miyazaki

Sawa Works
Hitachi Automotive Systems, Ltd.
Hitachinaka city Ibaraki, Japan
hideki.miyazaki.bz@hitachi.com

Abstract— **The design of optimized modulation schemes at low pulse ratios, e.g. Selective Harmonic Elimination (SHE), has been widely studied. However, most previous work has focused on half-wave symmetry pulse patterns. For variable speed and variable torque applications, non-half-wave symmetry SHE modulators depending on the power factor angle still achieve highly efficient inverters in many power factors. Unfortunately, few papers have studied inverter loss on non-half-wave symmetry SHE modulators. Thus, this paper proposes our original SHE method: Pulse Harmonic Modulation (PHM). Its mechanisms are explained in detail. Additionally, different pulse patterns of PHM are quantitatively compared. On the basis of this comparison, several design criteria are proposed for modulators at low pulse ratios.**

Keywords— PWM, Inverter, IGBT, Loss

I. INTRODUCTION

The design of high-power density inverters has been widely discussed for some time [1]. Such inverters typically use IGBT devices. For such inverter designs, the power capability and the power quality are always challenging. Unfortunately, semiconductor devices generate considerable losses at each switching event. To avoid an unnecessary power de-rating, the switching frequencies for such inverters are typically extremely low, e.g. five pulses per 180 deg fundamental [2] – [6]. The current ripple or the power quality (THD) of the inverter is a major design challenge. At such low pulse ratios, it is highly attractive to explore optimal modulation schemes to enhance the power capability without dramatically sacrificing the power quality of the inverter.

Numerous works on the modulation schemes have been presented [1] – [6]. However, most previous work has focused on center aligned pulse patterns. For variable speed and variable torque applications, non-center aligned SHE modulators depending on the current phase still achieve highly efficient inverters in a wide range of power factors. Unfortunately, few papers have covered this topic. It is hard to find literature describing systematic studies on inverter loss of non-center carrier SHE modulators. Thus, this paper proposes our original SHE method: Pulse Harmonic Modulation (PHM) [7].

This paper provides an insight into the impact of modulation schemes on the power capability of high-power inverters when the inverter operates at extremely low pulse ratios. This paper proposes novel modulation schemes, focusing on non-half-wave symmetry schemes using selectable zero vector, which is two degrees of freedom at V0 and V7. The characteristics of six types of pulse patterns are explained and quantitatively compared. On the basis of this comparison, several optimization criteria are summarized.

The study is based on a two-level inverter that uses IGBT devices. The load is a PM motor. The performance of different modulators is simulated in a time domain based on MATLAB. The simulator was developed explicitly for the performance design of motor drive systems. The simulation speed and precision of this tool have been proven.

II. PRINCIPLE OF PULSE HARMONIC MODULATION

The concept is to superimpose more two-pulse sets so that they cancel out each other's specified harmonics. Here, the unit pulse string is $f_0(\theta)$, the period of the string is 2π rad., amplitude of positive and negative pulses are normalized to 1, and pulse width is δ rad. To eliminate cosine harmonics and even-order harmonics, $f_0(\theta)$ is a half-wave symmetry and an odd-function symmetry. Therefore, Fourier series expansion of $f_0(\theta)$ is made up of the sum of the sine components with odd-order harmonics (third, fifth, seventh, etc.). The harmonic amplitude of kth order harmonics, $F_0(k)$, is given in (1), derived from the definition of the Fourier coefficient.

$$F_0(k) = \frac{1}{\pi} \int_0^{2\pi} f(\theta) \sin k\theta \, d\theta = \frac{4}{k\pi} \cos k\left(\frac{\delta}{2} - \frac{\pi}{2}\right) \quad (1)$$

Consider the cancellation of the third harmonic by the superposition shown in Fig. 1. The pulse string A is a shifted waveform of unit pulse $f_0(\theta)$ by $-\pi/6$ rad., and the pulse string B is a shifted waveform of unit pulse $f_0(\theta)$ by $+\pi/6$ rad. The phase relationship of the third harmonic component between A and B is an opposite phase because the phase difference between A and B is $\pi/3$ rad. Superimposing pulse strings A and B will lead

978-1-4799-2706-7/14 $31.00 © 2014 IEEE

to the cancellation of the third order harmonic. This phenomenon is represented as Fourier coefficient $F_{AB}(k)$ in eq. (2).

$$F_{AB}(k) = \frac{1}{\pi}\int_0^{2\pi} f\left(\theta - \frac{1}{3}\cdot\frac{\pi}{2}\right)\sin k\theta\, d\theta$$
$$+\frac{1}{\pi}\int_0^{2\pi} f\left(\theta + \frac{1}{3}\cdot\frac{\pi}{2}\right)\sin k\theta\, d\theta$$
$$= 2\cos k\left(\frac{1}{3}\cdot\frac{\pi}{2}\right)\cdot F_0(k) \tag{2}$$

Similarly, Fig. 2 illustrates the cancellation of the third and fifth harmonics. Both pulse strings C and D are shifted waveforms indicated in Fig. 2; therefore, there is no third harmonic in C or D. The fifth harmonic cancellation is accomplished by superimposing C and D, where the phase difference between C and D is $\pi/5$ rad. As a consequence, the combined pulse of C and D includes neither third nor fifth harmonics.

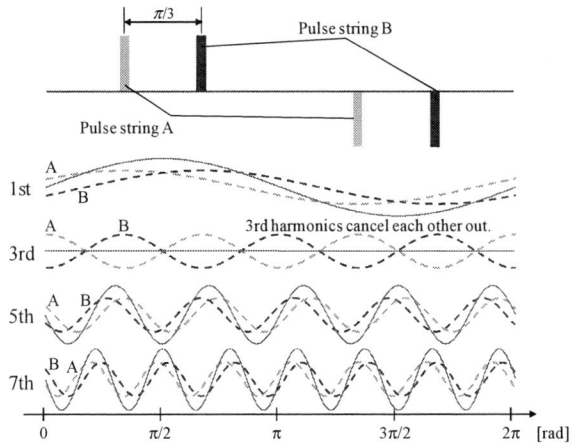

Fig. 1: Third harmonic cancellation.

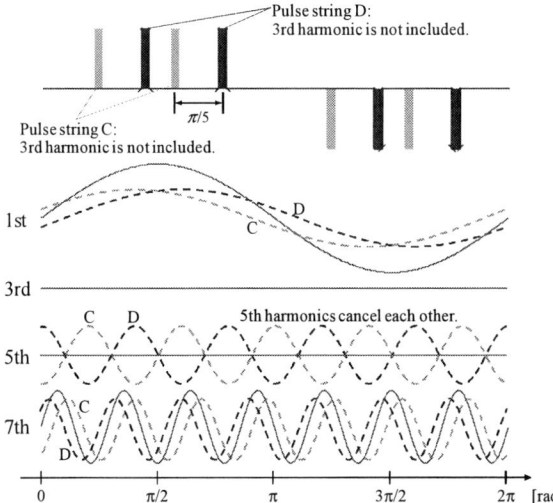

Fig. 2: Third and fifth harmonics cancellation.

Therefore, Fourier coefficient for a three-level voltage is made up of a product of the cosine components with cancellation orders of harmonics. The problems of (2) and (3) can generalize down to M-pieces of a harmonics cancellation problem in the following form:

$$F(k) = \left(2^M \prod_{j=1}^{M}\cos k\varphi_j\right)\cdot F_0(k) \tag{3}$$

where φ_j is represented using cancellation orders of n_j as

$$\varphi_j = \frac{l_j}{n_j}\cdot\frac{\pi}{2}, \quad 1\le l_j < n_j \text{ and } \{l_j, n_j\}\in odd.$$

For example, one of the cancellation harmonics is fifth order, $n_j = 5$, and the selectable value of l_j is 1 or 3. Therefore, φ_j is represented as $\varphi_j \in \{\pi/10, 3\pi/10\}$. The harmonic content $F(k)$ given in (3) is expanded from cosine product to cosine sum represented as

$$F(k) = \frac{4}{k\pi}\sum_{\{e_j\}\in S}\cos k(e_1\varphi_1 + \cdots + e_M\varphi_M)\cdot\cos k\left(\frac{\delta}{2} - \frac{\pi}{2}\right)$$
$$= \frac{4}{k\pi}\sum_{\{e_j\}\in S}\left(\cos k\left(\left(\frac{\pi}{2} - \frac{\delta}{2}\right) - (e_1\varphi_1 + \cdots + e_M\varphi_M)\right)\right.$$
$$\left. -\cos k\left(\left(\frac{\pi}{2} + \frac{\delta}{2}\right) - (e_1\varphi_1 + \cdots + e_M\varphi_M)\right)\right) \tag{4}$$

where $S = \{1, -1\}^M$, i.e., S is the direct product of $\{1, -1\}$. For example,

$$S = \{(1, 1), (-1, 1), (1, -1), (-1, -1)\} \text{ at } M = 2,$$

and $(e_1, e_2)\in S$.

Now, from past research [1]-[3], the magnitude of the harmonic content $F(k)$ can be formulized in another form using the switching angle as

$$F(k) = \frac{4}{k\pi}\sum_{i=1}^{N}(\cos k\alpha_{2i-1} - \cos k\alpha_{2i}) \tag{5}$$

where α_{2i-1} and α_{2i} are the switching angles (i is a positive integer) and N is the number of pulses, and these parameters are defined between 0 rad. and $\pi/2$ rad. The switching angles α_{2i-1} and α_{2i} represent a rising edge and a trailing edge, respectively. Then, compare (2) and (3) in the condition of $N = 2^M$. Both their cosine terms are equivalent, and the switching angles are represented as

$$\{\alpha_{2i-1}\} = \left\{\left\|\left(\frac{\pi}{2} - \frac{\delta}{2}\right) - |e_1\varphi_1 + \cdots + e_M\varphi_M|\right\|\right\},$$
$$\{\alpha_{2i}\} = \left\{\left\|\left(\frac{\pi}{2} + \frac{\delta}{2}\right) - |e_1\varphi_1 + \cdots + e_M\varphi_M|\right\|\right\} \tag{6}$$

where $1\le i < 2^M$, i is the positive integer, and $\{e_i\}\in S$.

Here, the modulation index is equivalent to fundamental component $F(1)$ controlled by pulse width 'δ.' From (1) and (2), δ is represented as

$$\delta = 2\arcsin\left(\frac{\pi\sqrt{3}\cdot F(1)}{2^{M+3}}\frac{1}{\prod_{j=1}^{M}\cos k\varphi_j}\right). \quad (7)$$

In a two-level inverter, each leg voltage is two levels, but each line voltage indicates three levels; a two-level pulse pattern can be derived from a three-level pulse pattern on the basis of space-vector representation. The space-vector of the two-level inverter consists of eight voltage vectors, i.e., $S0$–$S7$. The selectable zero vectors, V0 and V7, are indicated in Table I. These vectors are normalized by DC voltage, and the relationship between the line voltage pattern and the leg voltage pattern is clarified. In vectors of $S1$–$S6$, voltage states between leg voltages and line voltages correspond one-to-one, i.e., leg voltages can be uniquely specified from line voltages. On the other hand, vectors of $S0$ and $S7$ represent different voltage states in leg voltages, but they indicate a zero voltage state in line voltages. Therefore, a selectable switch state has two degrees of freedom at S0 and S7.

The process to derive a two-level switching pattern eliminating the fifth and seventh harmonics is indicated below. Fig. 3 shows a specific example to cancel third, fifth, and seventh harmonics. Here, parameters of the calculation conditions are $M=3$, $l_1/n_1 = 1/3$, $l_2/n_2 = 1/5$, $l_3/n_3 = 3/7$. Fig. 4 shows a block diagram of an applied control scheme.

TABLE I
Relationship of line voltage versus leg voltage.

Mode	Input: Line Voltage			Output: Leg Voltage for Controlling Inverter		
	Vuv	Vvw	Vwu	Vu	Vv	Vw
V1	+1	0	-1	1	0	0
V2	-1	+1	0	0	1	0
V3	0	+1	-1	1	1	0
V4	0	-1	+1	0	0	1
V5	+1	-1	0	1	0	1
V6	-1	0	+1	0	1	1
V7	0	0	0	1	1	1
V0				0	0	0

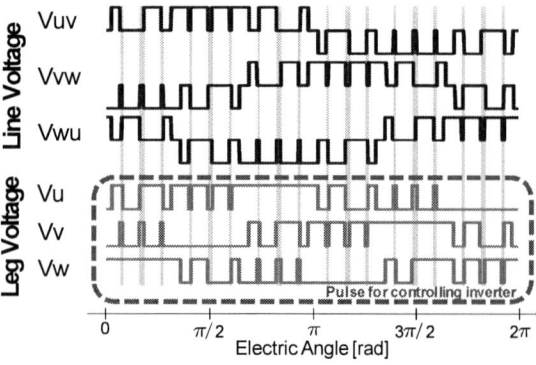

Fig. 3: Two-level switching angle derivation.

III. VARIOUS PULSE PATTERNS BASED ON ZERO VECTOR SELECTION

This section describes the pulse pattern selection to minimize inverter loss applied for a three-phase two-level inverter.

At the derivation of two-level leg voltage, this freedom is usable, e.g., the switching frequency of the power device can be minimized, and switching phases are selectable at a relatively low current. By using this method, a two-level pulse pattern is created. By choosing the zero vector in accordance with the maximum current, a PWM pulse pattern is made that has little switching at the time of the maximum current, as shown in Fig. 5. In the powering operation, the power factor is lagging, and in regenerative operation, current peak is leading.

To minimize inverter loss, especially switching loss of power devices, power losses depend on phase current condition. It is effective to reduce switching numbers on current peak. Fig. 6 indicates one of the pulse patterns of phase U and corresponding zero vector angles between 0 and π rad. The pulse patterns are in the condition of $M=3$, $l_1/n_1 = 1/3$, $l_2/n_2 = 1/5$, $l_3/n_3 = 3/7$. Modulation indexes of pulse patterns are 0.2, 0.6, and 1.0. Zero vector is derived from the relationship of the three-phase switching edge in Fig. 4. Pulse width and zero vector angles vary linearly depending on the modulation index. Zero vector angles are divided in 18 sectors between 0 and 2π rad. (Fig. 6 indicates partial angle range.) Therefore, all combinations of pulse patterns are $2^{18} = 262144$ sets, though usable combinations are limited in accordance with the next conditions.

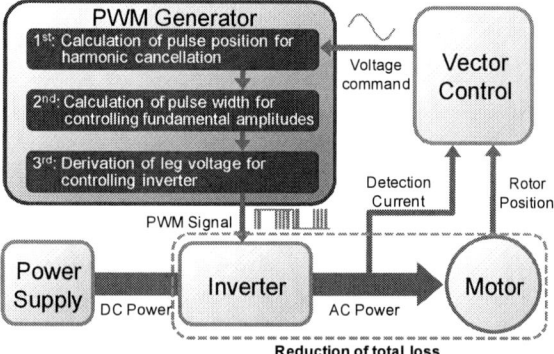

Fig. 4: Block diagram of proposed motor drive control.

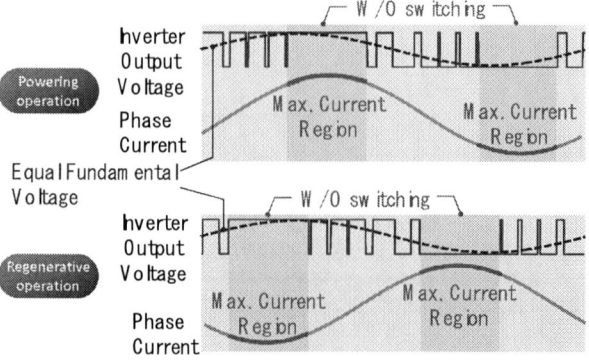

Fig. 5: Concept of pulse patterns selection to minimize

The conditions are to prevent power loss variability at each switching device in a three-phase inverter and to minimize power loss. The power loss variability can be avoided by two restrictions. One is $3\pi/2$ rad. repetitive symmetry zero vector patterns selection. This is effective at preventing the loss unbalancing among three phases. The other is π rad. symmetry zero vector patterns selection as an odd function. This is effective at preventing the loss unbalancing between high and low side arms. As a result of this restriction, zero vector angles are categorized as a, b, c, $\bar{a}$, $\bar{b}$, $\bar{c}$ in Fig. 6. Either V0 or V7 is selected for each of the six parameters. a and $\bar{a}$ are selected for different zero vectors, i.e., the combination of $(a,\bar{a})$ is (V0, V7) or (V7, V0). $(b,\bar{b})$ and $(c,\bar{c})$ are selected in the same manner. If the combination of (a,b,c) is selected, $(\bar{a},\bar{b},\bar{c})$ is automatically decided. Therefore combinations of $(a,b,c,\bar{a},\bar{b},\bar{c})$ are narrowed down to $2^3 = 8$ sets.

Finally, power loss is minimized like in a discontinuous PWM, i.e., the arrangement of the same type of zero vectors is gathered in adjacent areas when possible, and two types of zero vectors are not arranged alternately. Eventually, combinations of zero vectors are six sets as listed in Table II. For the results, Fig. 7 shows

six sets of PHM leg voltage patterns at phase U under the condition of fifth and seventh harmonics elimination at modulation factor MF=1.0. These leg voltage patterns have a unique line voltage pattern, and different zero vectors are selected. Fig. 8 shows the relationship of leg voltage patterns and the modulation factor at phase U under the condition of fifth and seventh harmonics elimination. The pulse patterns in Fig. 7 (A)-(F) correspond to Fig. 8 (A)-(F) of MF=1.0. In Fig. 8, white areas indicate high level voltage regions, and daubed areas are low level voltage regions. Differences in each leg voltage depend on the selection of zero vectors. Fig. 6 shows the zero vector regions in Fig. 8 (A)-(F), and daubed areas indicate zero vector region selectable from V0 and V7.

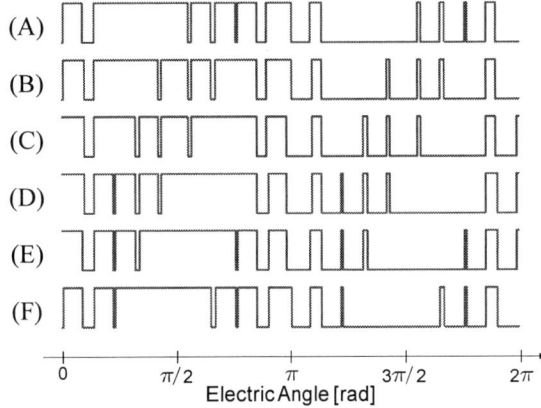

Fig. 7: Switching angle of 5th and 7th elimination.

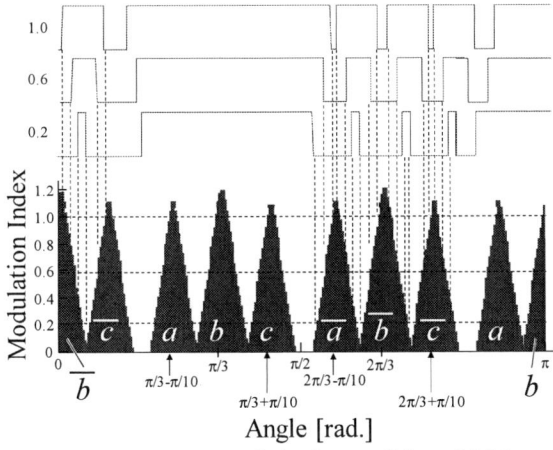

Fig. 6: Zero vector angle in the condition of M=3, $l_1/n_1 = 1/3$, $l_2/n_2 = 1/5$, $l_3/n_3 = 3/7$.

TABLE II
Zero vector combination of pulse pattern.

	a	b	c	$\bar{a}$	$\bar{b}$	$\bar{c}$
A	V7	V7	V7	V0	V0	V0
B	V7	V7	V0	V0	V0	V7
C	V7	V0	V0	V0	V7	V7
D	V0	V0	V0	V7	V7	V7
E	V0	V0	V7	V7	V7	V0
F	V0	V7	V7	V7	V0	V0

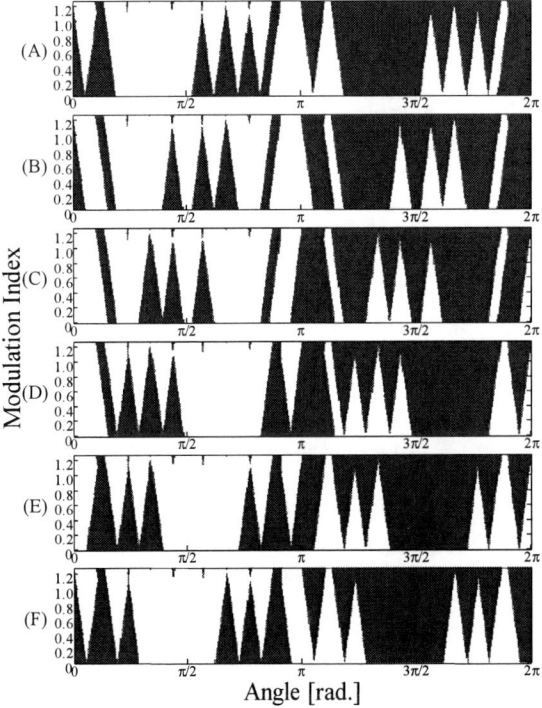

Fig. 8: Modulation factor dependency of switching angle.

IV. POWER LOSS CALCULATION

IGBTs and the diodes of a three-phase inverter generate power losses. The power losses are attributed to switching losses and conduction losses. The switching losses consist of turn-on and turn-off losses of the IGBTs and recovery losses of the diodes. The conduction losses depend on both the IGBT and the diodes.

During the turn-on transient, turn-on loss of the IGBT E_{on} is represented as

$$E_{on} = \int_{t_0}^{t_1} v_{ce}(t) \cdot i_c(t) dt. \tag{8}$$

where t_0 and t_1 represent turn-on transient duration, v_{ce} is corrector-emitter voltage of IGBT, i_{ce} is IGBT current. Generally, E_{on} is approximately proportional to phase current and DC voltage. Turn-off loss of the IGBT E_{off} and recovery loss of the diodes E_{rr} are defined similarly.

Conduction loss (dissipation power) of the IGBT P_{cond} is represented as

$$P_{cond} = v_{ce(on)}(t) \cdot i_c(t) dt. \tag{9}$$

where $v_{ce(on)}$ is the on-state corrector-emitter voltage of IGBT. Generally, $v_{ce(on)}$ is a function of phase current and not relevant to DC voltage. Forward voltage v_{fd} and conduction loss of the diode P_{fd} are defined similarly.

In this paper, the switching losses of IGBTs and diodes are simulated using a look-up-table function. The arguments of the function are phase current and DC voltage, and the output of the function is switching loss as E_{on}, E_{off}, and E_{rr}. Losses are derived at every switching instant. The conduction losses of IGBT and diode are simulated using the look-up-table function. The arguments of the function are phase current, and the output of the function is on state corrector-emitter voltage $v_{ce(on)}$ and forward voltage v_{fd}.

In the simulation, AC phase current is 173 Arms, DC link voltage is 300V, electric frequency is 800 Hz, and power factor angle and modulation factor are changed. Here, the power factor angle is between a leg voltage and phase current, and the positive power factor angle is defined by a current delayed state with reference to leg voltage angle. Simulation is executed at the six pulse patterns indicated in Fig. 8. Switching loss and conduction loss are calculated at $0 \leq \varphi < 2\pi$ and $0 \leq MF < 1$.

Fig. 9 indicates an example of a simulated wave form at $\varphi = 0$ rad., MF=1.0, and pulse pattern (A): (a) is voltage and current of phase U, (b) is switching loss, and (c) is conduction loss. Here, (c) and (d) are characteristics of high side devices (IGBT and diode) of phase U. The results indicated that switching loss is generated at the switching edge. The phase current influences switching loss and conduction loss. The comparisons of loss characteristics among different conditions are evaluated by averaged losses per 2π rad.

Fig. 10 shows power loss characteristics of phase U including the six pulse patterns of (A)-(F) indicated

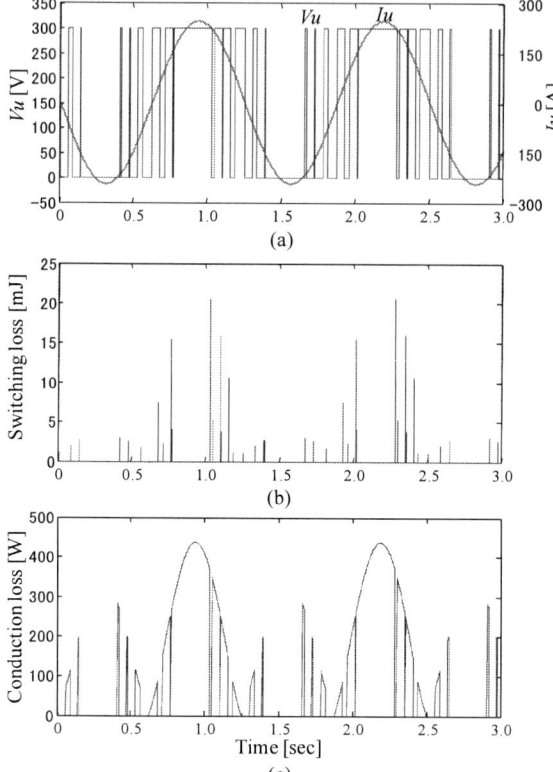

Fig. 9: Switching angle.

in Fig. 8, where modulation factor is MF=0.5: (a) is switching loss, (b) is conduction loss, and (c) is total power loss. In Fig 10 (a), the pattern of minimum switching loss varies depending on power losses. Around $\varphi = -\pi/4$ rad. is in pattern (A) and maximum switching loss is realized at pattern (D). This is explained as follows. The power factor angle is $\varphi = -\pi/4$ rad., positive peak of phase current is at $\theta = \pi/4$ rad, and negative peak is at $\theta = 5\pi/4$ rad. There is no switching edge in pattern (A) around $\theta = \pi/4$ rad. and $5\pi/4$ rad. Therefore, the switching loss of pattern (A) is minimized relative to other patterns. On the other hand, the switching edge of pattern (D) is concentrated around $\theta = \pi/4$ rad. and $5\pi/4$ rad. Therefore, the switching loss of pattern (D) is maximized. Similarly, the power factor angle is $\varphi = 3\pi/4$ rad., negative peak of phase current is at $\theta = \pi/4$ rad., and positive peak is at $\theta = 5\pi/4$ rad. The switching loss is the same as $\varphi = -\pi/4$ rad., i.e., the switching loss of pattern (A) is minimized and that of pattern (D) is maximized.

For the same reason, the switching loss of pattern (D) is minimized and that of pattern (A) is maximized at $\varphi = \pi/4$ rad. and $\varphi = -3\pi/4$ rad.

The conduction losses in Fig. 10 (b) show marginal differences depending on the pulse patterns. There are two reasons for the difference. One is the distribution ratio of diode conduction loss and IGBT conduction loss. The other is non-linearity of conductive losses (diode and IGBT) characteristics.

The 2014 International Power Electronics Conference

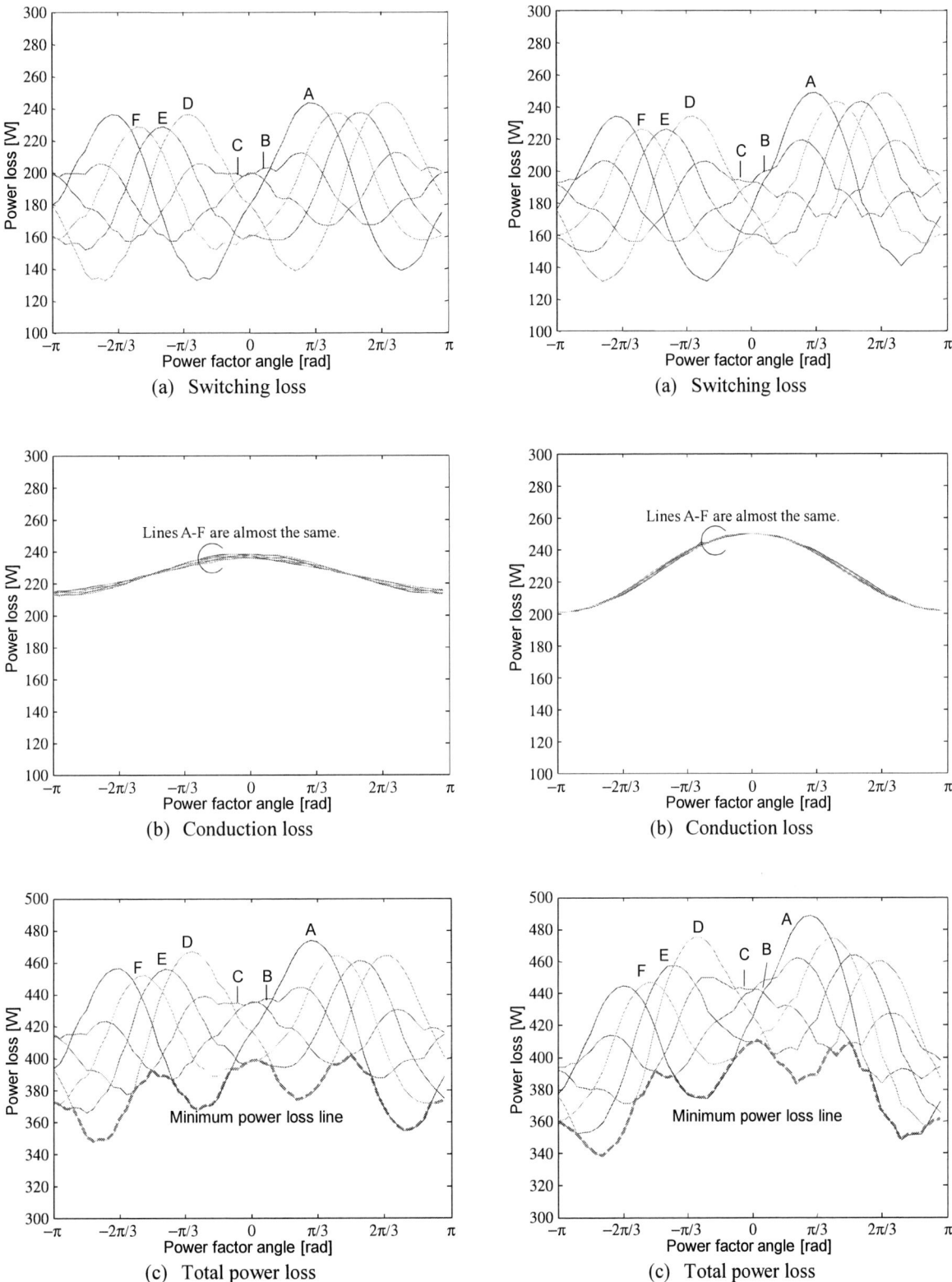

Fig. 10: Pulse patterns dependence on loss at MF=0.5.

Fig. 11: Pulse patterns dependence on Loss at MF=1.0.

978-1-4799-2706-7/14 $31.00 © 2014 IEEE

388

The total power loss in Fig. 10 (c) is the summation of switching loss and conduction loss. The minimum power loss characteristics are indicated by the dashed line. The total loss characteristics on pulse pattern dependency are mainly affected by switching loss characteristics shown in (a), i.e., the pulse pattern of the minimal switching loss corresponds to that of the minimal total power loss at an arbitrary power factor angle.

Fig. 11 indicates the power losses at MF=1.0. The switching loss order of pulse patterns at MF=1.0 is the same as that at MF=0.5 in the arbitrary power factor. By the same token, the total power loss order of pulse patterns at MF=1.0 is the same as that at MF=0.5 in the arbitrary power factor. The conduction loss at MF=1.0 does not vary depending on pulse patterns.

Fig. 12 shows three-dimensional relationships among total power loss, modulation factor, and power factor angle: (a) shows the loss driven at pulse pattern A as an example of single pulse pattern driving, and (b) shows the loss driven at selected pulse pattern as to power factor angle to minimize power losses.

CONCLUSIONS

This paper introduced a new harmonic elimination PWM strategy, 'PHM': a modulation method that can be designed independently of both motor current harmonics and loss minimization of power device. This paper shows patterns of fifth and seventh harmonics eliminated by PHM. The line voltage pattern is unique, though the number of leg voltage patterns is six depending on the zero vectors selection.

By using these leg voltage patterns, power losses of a inverter are simulated. In the calculation, AC phase current is 173 Arms, DC link voltage is 300V, electric frequency is 800 Hz, and power factor angle is changed.

The results represent power losses per phase, consisting of high side and low side IGBTs and diodes. The power losses vary depending on the power factor angle. This characteristic is attributed to switching losses of devices. The power losses can be minimized by choosing the most suitable pulse in accordance with the power factor.

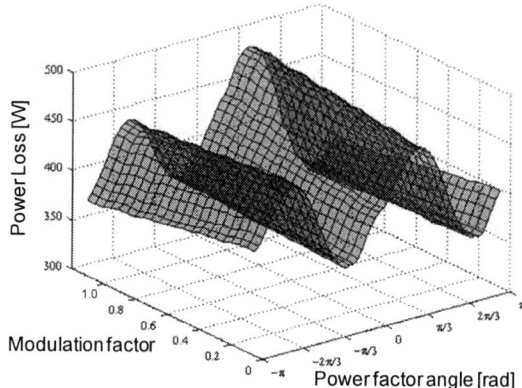

(a) Pulse pattern "A"

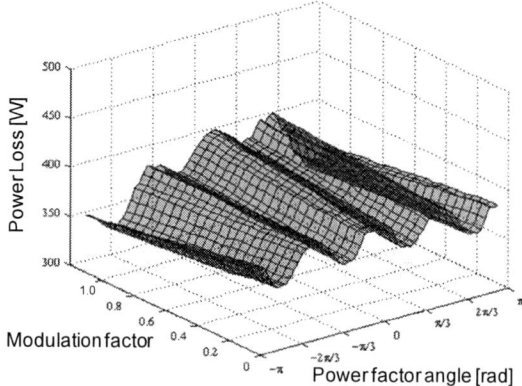

(b) Pulse pattern of minimum power loss

Fig. 12: Power loss characteristic depending on modulation factor and power factor angle

REFERENCES

[1] Holtz Joachim, Oikonomou Nikolaos, "Estimation of the Fundamental Current in Low-Switching-Frequency High Dynamic Medium-Voltage Drives," *IEEE Trans. on Industry Applications*, vol. 44, no. 5, 2008.

[2] Holtz Joachim, "Pulsewidth Modulation for Electronic Power Conversion," *Proc. of the IEEE*, vol. 82, No. 8, 1994.

[3] Patel Hasmukh S., Hoft Richard G., "Generalized Techniques of Harmonic Elimination and Voltage Control in Thyristor Inverters : Part I–Harmonic Elimination," *IEEE Trans. on Industry Applications*, vol. IA-9, no. 3, 1973.

[4] Patel Hasmukh S., Hoft Richard G., "Generalized Techniques of Harmonic Elimination and Voltage Control in Thyristor Inverters : Part II – Voltage Control Techniques," *IEEE Trans. on Industry Applications*, vol. IA-10, no. 5, 1974.

[5] Zach Franz C., Ertl Hans, "Efficiency optimal control for AC drives with PWM inverters," *IEEE Trans. on Industry Applications*, vol. IA-21, no. 4, 1985.

[6] Buja Giuseppe S., "Optimum Output Waveforms in PWM Inverters," *IEEE Trans. on Industry Applications*, vol. IA-16, no. 6, 1980.

[7] Furukawa Kimihisa, Miyazaki Hideki, "Solution for SHE-PWM: Non-iterative Computation Algorithm Based on Trigonometric Harmonic Cancellation Rule," EPE 2013 ECCE Europe 15th European Conference on Power Electronics and Applications, 2013.

Method for Auto-tuning of Current and Speed Controller in IPMSM Drive System Based on Parameter Identification

D. Tadokoro, S. Morimoto, Y. Inoue, M. Sanada

Graduate School of Engineering
Osaka Prefecture University
1-1 Gakuen-Cho, Naka-ku, Sakai, Osaka, Japan
E-mail: st106021@edu.osakafu-u.ac.jp

Abstract—**In motor drive systems, the most suitable gains change according to the operating state. As such, in order to achieve a high-performance control, the gains of the current and speed controller must be tuned according to the operating state. In this study, we propose an auto-tuning method for a current and speed controller based on the parameter identification. First, the resistance, including the on-resistance of the IGBT, and the *d*- and *q*-axis inductances are identified at standstill. Second, after the motor starts up, the magnet flux linkage, the inertia moment, and the viscous friction coefficient are also identified using the parameters identified at standstill. Finally, the controller gains are determined by substituting the identified values into the derived relationship between control gains and motor parameters. The effectiveness of the proposed method is verified experimentally.**

Keywords— interior permanent magnet synchronous motor (IPMSM), identification, auto-tuning, control system.

I. INTRODUCTION

The realization of high performance control and highly efficient operation by vector control of an interior permanent magnet synchronous motor (IPMSM) requires the motor parameters to be determined exactly and the appropriate control based on the motor model must be performed [1]-[3]. However, since the motor parameters continue to change during driving, the appropriate parameters should be obtained according to the driving state [4]-[6]. In this study, parameters are identified based on reference voltage, detected current, and so on, according to the driving condition. Moreover, the relationship between the control gains and the motor parameters required to achieve appropriate drive is established. The gains of the current and speed controller are optimized using the identified values. The effectiveness of the proposed method is verified through several experimental results.

II. CONTROL SYSTEM

A. Mathematical Model of the IPMSM

In the voltage source PWM inverter that is generally used in motor drives, the error between the armature voltage that is applied to the inverter and the reference voltage is generated by the voltage drop due to the switching devices and the dead time of the drive circuit. Fig. 1 shows the proposed IPMSM drive system. In this study, we investigate the auto-tuning IPMSM drive based on the voltage equation given by (1) considering the voltage error. The *d*- and *q*-axis voltage drops caused by the deadtime and the switching devices are given by (2) [7]. This equation represents the area surrounded by the broken line in Fig. 1.

$$\begin{bmatrix} v_{d\_ref} \\ v_{q\_ref} \end{bmatrix} = \begin{bmatrix} R+pL_d & -\omega L_q \\ \omega L_d & R+pL_q \end{bmatrix}\begin{bmatrix} i_d \\ i_q \end{bmatrix} + \begin{bmatrix} 0 \\ \omega\Psi_a \end{bmatrix} + \begin{bmatrix} \Delta v_d \\ \Delta v_q \end{bmatrix} \quad (1)$$

$$\begin{bmatrix} \Delta v_d \\ \Delta v_q \end{bmatrix} = \sqrt{\frac{2}{3}}\Delta V \begin{bmatrix} \cos\theta & \cos\left(\theta-\frac{2\pi}{3}\right) & \cos\left(\theta+\frac{2\pi}{3}\right) \\ -\sin\theta & -\sin\left(\theta-\frac{2\pi}{3}\right) & -\sin\left(\theta+\frac{2\pi}{3}\right) \end{bmatrix}\begin{bmatrix} \text{sgn}(i_u) \\ \text{sgn}(i_v) \\ \text{sgn}(i_w) \end{bmatrix} \equiv \Delta V \begin{bmatrix} D_d \\ D_q \end{bmatrix} \quad (2)$$

In the above equations, v_{d_ref} and v_{q_ref} are the *d*- and *q*-axis reference voltages, i_d and i_q are the *d*- and *q*-axis armature currents, R is the resistance, L_d and L_q are the *d*- and *q*-axis inductances, Ψ_a is the stator flux linkage by the rotor magnet, and D_d and D_q are state variables that are dependent on the rotor position and the phase currents, respectively.

B. Online Parameter Identification Algorithm

The recursive least squares (RLS) method is used in the identification algorithm [7]-[8]. The general formula of the algorithm is represented as follows:

$$\Theta(k) = \Theta(k-1) + \frac{P(k-1)Z(k)}{\lambda + Z^T(k)P(k-1)Z(k)}\left(Y(k) - Z^T(k)\Theta(k-1)\right) \quad (3)$$

$$P(k) = \frac{1}{\lambda}\left(P(k-1) - \frac{P(k-1)Z(k)Z^T(k)P(k-1)}{\lambda + Z^T(k)P(k-1)Z(k)}\right) \quad (4)$$

where Θ is the parameter coefficient vector, Θ is the estimated parameter coefficient vector, P is the covariance matrix, Y is the output, Z is the input vector, and λ is the forgetting factor, where $\lambda < 1$.

C. Identification of Electrical Parameters

The motor parameters are identified using the voltage equation for the IPMSM given by (1). By discretizing (1) by sampling time T_{se}, the parameter identification models of the IPMSM considering the voltage drop by the inverter are as follows:

$$v_{d\_ref}(k-1) = Ri_d(k-1) - L_q\omega(k-1)i_q(k-1)$$
$$+ \frac{L_d}{T_{se}}\{i_d(k) - i_d(k-1)\} + \Delta VD_d(k-1) \quad (5)$$

$$v_{q\_ref}(k-1) = Ri_q(k-1) + L_d\omega(k-1)i_d(k-1)$$
$$+ \frac{L_q}{T_{se}}\{i_q(k) - i_q(k-1)\} + \psi_a\omega(k-1) + \Delta VD_q(k-1) \quad (6)$$

The identification models for the IPMSM at standstill and in the steady state with regard to the current are obtained by substituting $\omega = 0$, $i_d(k) = i_d(k\text{-}1)$, and $i_q(k) = i_q(k\text{-}1)$ into (5) and (6) as follows:

$$v_{d\_ref}(k) - \Delta V \cdot D_d(k) = R \cdot i_d(k) \quad (7)$$

$$v_{q\_ref}(k) - \Delta V \cdot D_q(k) = R \cdot i_q(k) \quad (8)$$

Based on the preceding model, the parameter R can be identified. For example, the d-axis identification model based on (7) is obtained as follows:

$$Y(k) = v_{d\_ref}(k) - \Delta V \cdot D_d(k) \quad (9)$$

$$Z(k) = i_d(k) \quad (10)$$

$$\Theta(k) = \hat{R}(k) \quad (11)$$

After the resistance R has been identified, the identification model for identifying L_d and L_q at standstill and at the transient state of the current is given as follows based on (5) and (6):

$$i_d(k) - i_d(k-1) = \frac{T_{se}}{L_d}\{v_{d\_ref}(k) - \hat{R}i_d(k) - \Delta V \cdot D_d(k)\} \quad (12)$$

$$i_q(k) - i_q(k-1) = \frac{T_{se}}{L_q}\{v_{q\_ref}(k) - \hat{R}i_q(k) - \Delta V \cdot D_q(k)\} \quad (13)$$

Only the magnet flux linkage Ψ_a cannot be identified at standstill. As such, Ψ_a is identified at the operating state. When R, L_d, and L_q are identified, the identification model for identifying Ψ_a at the operating state based on

(6) is as follows:

$$Y(k) = \frac{\hat{L}_q}{T_s}\{i_q(k+1) - i_q(k)\} + \hat{L}_d\omega(k)i_d(k) - v_{q\_ref}(k) + \hat{R}i_q(k) \quad (14)$$

$$Z(k) = \omega(k) \quad (15)$$

$$\Theta(k) = \hat{\Psi}_a(k) \quad (16)$$

D. Identification of Mechanical Parameters

The identification of the mechanical system parameters is based on the following motion equation in the same way as the identification of the electrical system parameter:

$$T = J_m\frac{d\omega_m}{dt} + D_r\omega_m + T_L \quad (17)$$

where ω_m is the mechanical angular velocity, and T_L is the disturbance load torque. The torque is calculated from the following formula based on the identified parameters $\hat{\Psi}_a, \hat{L}_d$ and $\hat{L}_q$

$$T = P_n\{\hat{\Psi}_a i_q + (\hat{L}_d - \hat{L}_q)i_d i_q\} \quad (18)$$

By discretizing by forward difference, the following equation was derived from (17). In this case, the sampling period T_{sm} is not the same as that of the current control system:

$$J_m\frac{\omega_m(k) - \omega_m(k-1)}{T_{sm}} + T_L(k) = T - D_r\omega_m(k-1) \quad (19)$$

For example, the input and output in the identification model for J_m and T_L are obtained as follows:

$$Y(k) = T - D_r\omega_m(k-1) \quad (20)$$

$$Z(k) = \left[\frac{\omega_m(k) - \omega_m(k-1)}{T_{sm}} \quad 1\right]^T \quad (21)$$

$$\Theta(k) = \left[\hat{J}_m \quad \hat{T}_L\right]^T \quad (22)$$

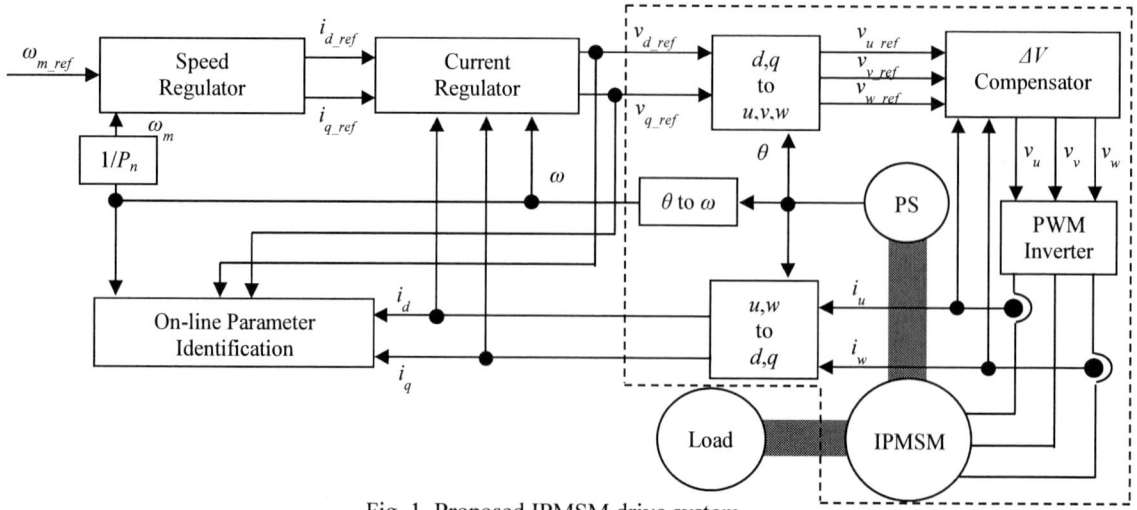

Fig. 1. Proposed IPMSM drive system.

III. Relations Between Parameters And Control Gains

A block diagram of the current control system considering the decoupling compensation is shown in Fig. 2. The transfer function is derived from Fig. 2 as follows:

$$G(s) = \frac{i}{i^*} = \frac{K_p\left(s + \frac{K_i}{K_p}\right)}{sL\left(s + \frac{R}{L}\right) + K_p\left(s + \frac{K_i}{K_p}\right)} \quad (23)$$

If K_i/K_p is set to R/L, the transfer function given by (23) becomes a simple first-order delay system. By determining the time constant τ of the first-order system, the proportional gain K_p and the integral gain K_i can be determined as follows:

$$K_p = \frac{L}{\tau}, \quad K_i = K_p \frac{R}{L} \quad (24)$$

The current control gains for each axis current controller can be obtained by exchanging L for L_d and L_q in (24), and the speed control gains can be determined in the same manner. The details of the current and speed regulator are shown in Fig. 3.

IV. Auto-tuning Method

This paper proposes an auto-tuning method for current and speed controller gains according to the operating state. Furthermore, an auto-tuning method using parameter identification is proposed. Therefore, the proposed method must identify all of the motor parameters, such as the resistance R, the d- and q-axis inductances L_d and L_q, the inertia moment J_m, and the viscous friction coefficient D_r in the system shown in Fig. 1. First, the parameters, such as R, L_d, and L_q, are identified at standstill. Second, the magnet flux linkage Ψ_a, which cannot be identified at standstill, is identified at the operating state. When R, L_d, and L_q are identified, the auto-tuning gains of the current regulator can be determined using (24). After operating using auto-tuning gains of the current regulators, J_m and D_r can be identified at the operating state. Finally, the auto-tuning gains of the speed regulator are determined by substituting the identified values of J_m and D_r into (24) according to the operating state.

V. Experimental Results

Table I lists the measured motor parameters for the IPMSM used in this experiment. The maximum torque per ampere control is applied in this experiment. In identification, the sampling time T_{se} was set to 100 μs and the sampling period T_{sm} was set to 5 ms. Furthermore, all identifications were started from 0.2 s. Fig. 4 shows the identification results for R at standstill. At this time, a square wave current (amplitude: 1.0 A, period: 0.02 s) is injected to the d-axis current reference

in order to identify R. Fig. 5 shows the identification results for L_d at the operating state. At this time, a square wave current (amplitude: 1.5 A, period: 0.02 s) is injected to the d-axis current reference at the steady state of 1,000 min^{-1} in order to identify L_d. Fig. 6 shows the identification results for L_q. For similar conditions, a square wave current (amplitude: 1.0 A, period: 0.02 s) is injected to the q-axis current reference. Fig. 7 shows the responses of the q-axis current before and after auto-tuning. The initial values of the gains before auto-tuning have been determined by substituting the initial value of the parameters before identification into (24). The results of this experiment revealed that the error in identifying R is 4.62%, the error in identifying L_d is -7.48%, and the error in identifying L_q is -3.24%. The convergence time is short and all parameters except for L_q converge by the first current injection. Furthermore, the effectiveness of

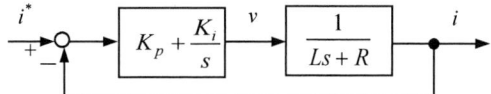

Fig. 2. Block diagram of the current control system.

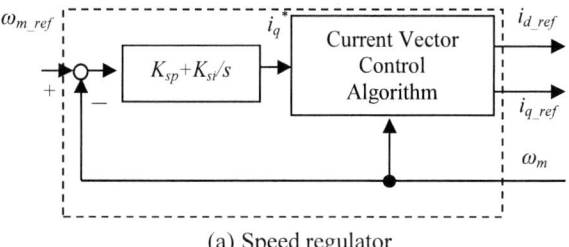

(a) Speed regulator

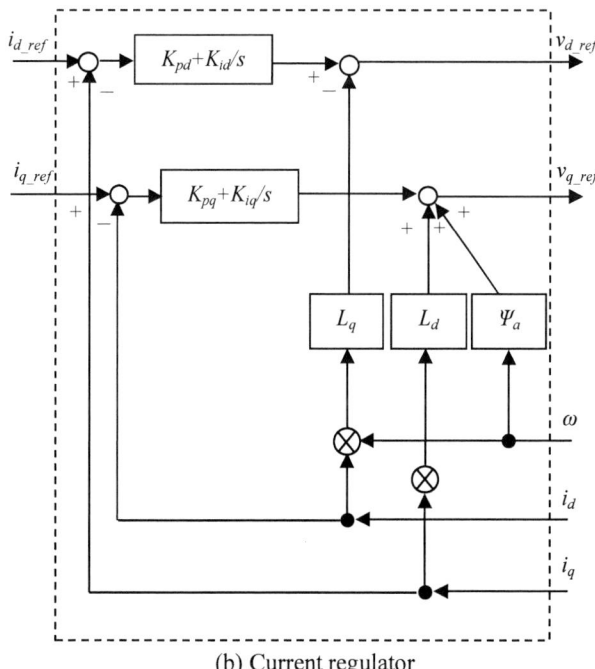

(b) Current regulator

Fig.3. Detail of the block diagram of the regulators.

the proposed auto-tuning method with respect to the response of the current without overshoot has been verified by Fig. 7.

0 Fig. 8 shows the identification results for Ψ_a with an error of 2.72%. This identified value of Ψ_a is used to identify the mechanical parameters. Fig. 9 shows the identification results for J_m with an error of 4.89%, and Fig. 10 shows the identification results for D_r with an error of -8.34%. In the identification for J_m, the speed reference changes between 1000min^{-1} and 1500min^{-1} stepwise. Fig. 11 shows the responses of the speed before and after auto-tuning. The effectiveness of the proposed auto-tuning method with respect to the response of the speed without overshoot is verified by Fig. 11.

VI. CONCLUSION

This paper proposed a method for auto-tuning the current and speed controller through the use of parameter identification. In the parameter identification, all of the motor parameters in the IPMSM drive system are identified with an error of ±8%. Moreover, the current and speed controller is optimized by the proposed auto-tuning method. As a result, the responses of the current and speed without overshoot have been validated experimentally.

TABLE I
PARAMETERS OF THE IPMSM

Symbol	Meaning	Value
P_n	Number of pole pairs	2
L_d	d-axis inductance	4.95mH
L_q	q-axis inductance	10.42mH
Ψ_a	Magnet flux linkage	0.11Wb
R	Resistance	0.325Ω
J_m	Inertia moment	1.84×10^{-3}kgm^2
D_r	Viscous friction coefficient	2.39×10^{-4} Nm/(rad/s)
V_{DC}	DC link voltage	280V
ΔV	Compensation voltage for deadtime	14V
f_{car}	PWM carrier frequency	10kHz
T_D	Deadtime	5µs

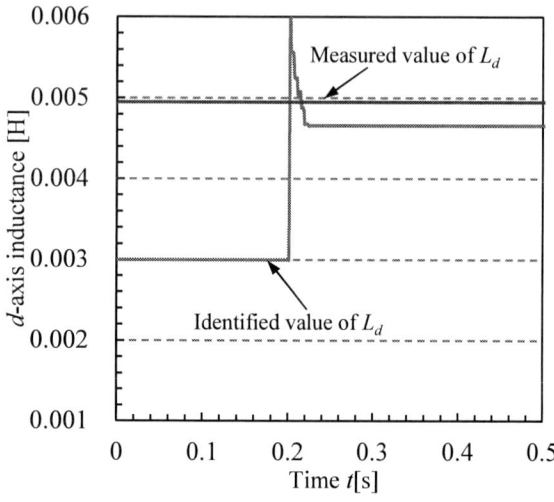

Fig. 5. Response of the identified value of L_d.

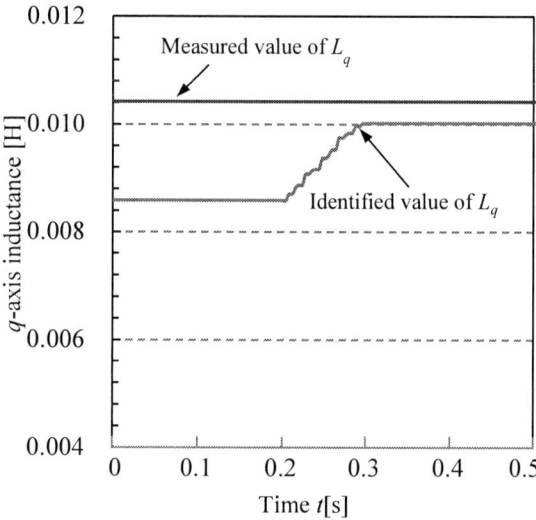

Fig. 6. Response of the identified value of L_q.

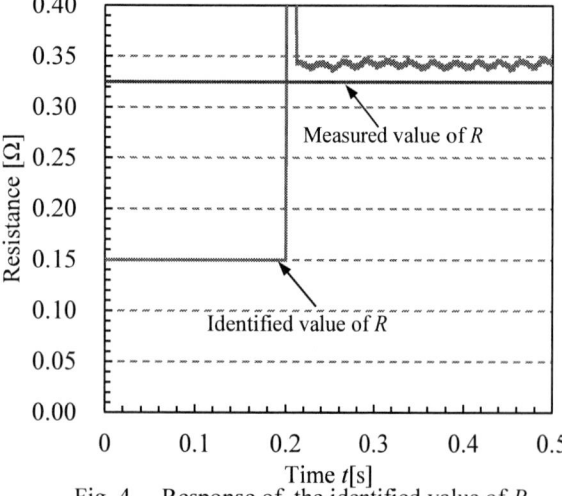

Fig. 4. Response of the identified value of R.

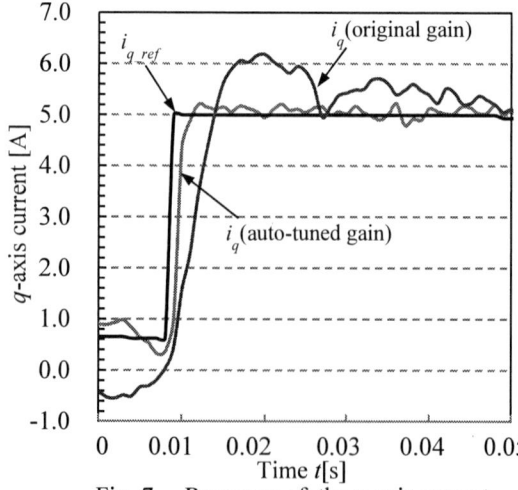

Fig. 7. Response of the q-axis current.

978-1-4799-2706-7/14 $31.00 © 2014 IEEE

The 2014 International Power Electronics Conference

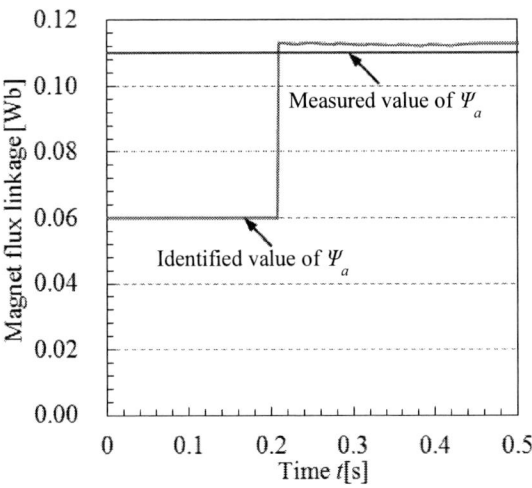

Fig. 8. Response of the identified value of Ψ_a.

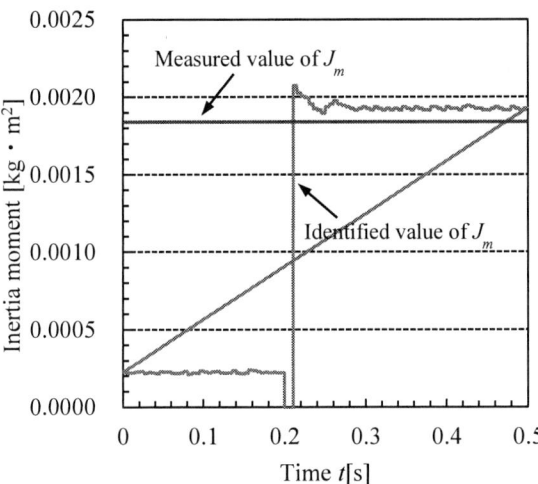

Fig. 9. Response of the identified value of J_m.

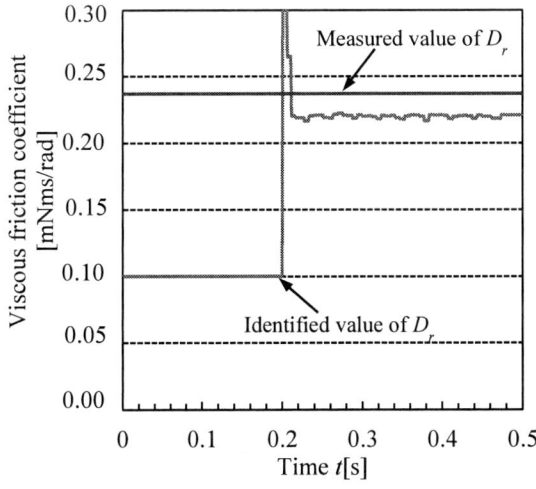

Fig. 10. Response of the identified value of D_r.

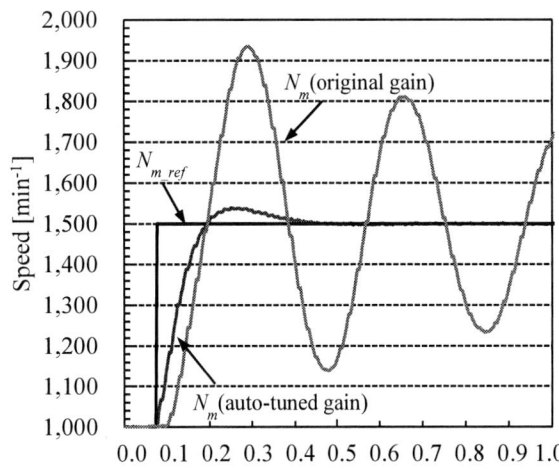

Fig. 11. Response of the speed.

References

[1] T. M. Jahns, G. B. Kliman, and T.W.Newmann, "Interior permanent magnet synchronous motors for adjustable-speed drives," *IEEE Trans. Ind. Appl.* , vol. IA-22, no. 4 , pp. 738-747, Jul. /Aug. 1986

[2] S. Morimoto, Y. Takeda, T. Hirasa, "Current phase control methods for permanent magnet synchronous motors, " *IEEE Trans. Power Electron.* , vol. 5, no. 2, pp. 133-139, Apr. 1990.

[3] S. Morimoto, M. Sanada, and Y. Takeda, "Wide speed operation of interior permanent magnet synchronous motors with high performance current regulator, " IEEE Trans. Ind. Appl. , vol. 30, no. 4, pp. 920-926, Jul. / Aug. 1994.

[4] S. Ichikawa, A. Iwata, M. Tomita, S. Doki, "Sensorless control of synchronous reluctance motors using an online parameter identification method taking into account magnetic saturation," *PESC2004*, vol.5, pp. 3311-3316, 2004.

[5] B. N. Mobarakeh, F. M. Tabar, and F. M. Sargos, "Online identification of PMSM electrical parameters based on decoupling control," *in Proc. IEEE IAS Annu. Meeting*, pp. 266-273, 2002.

[6] S. Ichikawa, M. Tomita, S. Doki, S. Okuma, "Sensorless Control of Permanent Magnet Synchronous Motors Using Online Parameter Identification Based on System Identification Theory," *IEEE Trans. on Industrial Electronics*, vol.53, no. 2, pp. 363-372, 2006.

[7] S. Morimoto, M. Sanada, Y. Takeda, "Mechanical Sensorless Drives of IPMSM With Online Parameter Identification," *IEEE Transactions on Industry Applications,* vol. 42, no. 5, pp. 1241-1248, Sep/Oct, 2006.

[8] S. Morimoto, A. Shinmei, M. Sanada, and Y. Takeda, "Parameter Identification Of PMSM Drive System Including Inverter," *in Proc. ICEM*, pp. 1-4, 2004.

978-1-4799-2706-7/14 $31.00 © 2014 IEEE

The 2014 International Power Electronics Conference

Comparative Study of PWM Strategies for Three-Phase Open-end Winding Induction Motor Drives

B. Zhu[1], U. R. Prasanna[1], K. Rajashekara[1] and H. Kubo[2]

[1]University of Texas at Dallas
Richardson, Texas, USA

[2]Meidensha Corporation
Numazu-shi, Shizuoka, Japan

Abstract – **Open-end winding machine configuration provides better fault tolerance and flexibility for control compared with a conventional three-phase neutral connected electric motor. In this paper, three types of pulse width modulation (PWM) schemes - sinusoidal PWM, space vector PWM, and symmetrical sinusoidal PWM for open-end winding induction motor drives are being examined. The three PWM schemes are simulated in MATLAB/Simulink environment and their harmonic spectra are investigated for a 5.5 kW open-end winding induction motor. The current ripple under each PWM method is analyzed. Finally, the optimal switching strategy is discussed based on harmonic content, switching losses, and current ripple.**

Keywords – *current ripple, open-end winding induction motor, PWM strategies, zero sequence cancellation*

I. INTRODUCTION

Open-end winding electric machine is also known as split phase machine or separately fed machine, obtained by removing the neutral point of the stator winding of a conventional electric machine. These machines are being examined for various applications such as propulsion motor in electric and hybrid vehicles, fault tolerant generator in aircraft systems, ship propulsion, and industrial drive systems. The advantages of this type of machine configuration are [1-2]:

(a) No pole switching constraint
(b) In case of open or short circuit in one of the phases, the magneto motive force (MMF) can be maintained by using the remaining two phases. This feature can be extended to multi-phase machines.
(c) Independent control of switches in each phase, which brings independent control of the stator current.
(d) Reduced power rating of individual switch: Feeding power from both ends of the winding by DC source, the switches in each phase bear half the power compared to that in neutral-connected configuration.
(e) Reduction in switching loss and total harmonic distortion (THD).

The open-end winding three-phase induction motor controlled from two two-level three-phase inverters is shown in Fig. 1. Instead of calling it as an open-end winding machine, it is also referred to as separately-fed machine as shown in Fig. 2 [1-2]. The strategy in these papers is to overcome the pole switching constraint of three-phase bridge inverters, where the switching states are limited to eight including two zero switching states.

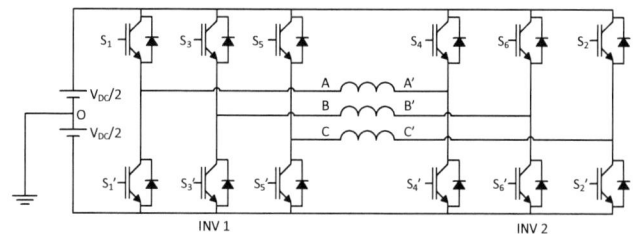

Fig. 1 Open-end Winding Configuration with Two Three-phase Bridge Inverters. [5]

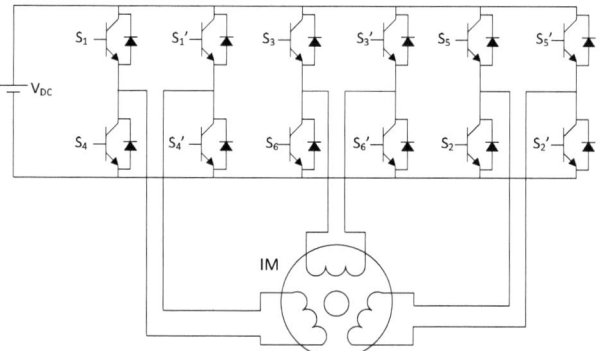

Fig. 2 Induction Motor Fed by Three Independent Single-phase Bridge Inverters [1-2]

Many researchers have investigated the modulation strategies to achieve flexibility of control, reduction of harmonics, and elimination of common mode voltage in open-end winding machines [3-12]. Depending on topology of inverters or number of DC sources being used for the open-end winding machines, various PWM strategies are proposed in the literature.

With a configuration of three single-phase H-bridges, a PWM scheme with controllable grid power factor and higher phase voltage is proposed in [3]. When the topology is configured as dual two-level three-phase inverters, a high power three-phase synchronous motor with open windings is presented in [4]. Using the same power circuits, a space vector PWM strategy using the redundancy in switching states is presented in [5]; a PWM strategy with improved DC-bus utilization is also illustrated in [6].

Based on the number of isolated DC links, two possible topologies are presented in literature. Topologies shown in Fig. 1 and Fig. 2 use only one DC-link and it can be represented either as three single-phase H-bridges or as two three-phase inverters. The second topology is shown in Fig. 3, which uses two isolated DC sources to avoid the zero-sequence currents [5, 8]. A scheme using two isolated DC sources feeding from each ends with zero common mode voltage is illustrated in [8]. With only one DC source, a PWM scheme with common mode elimination is presented in [9].

As isolated power sources are limited in many applications, this paper focuses on the configuration with only one DC voltage source for comparing different PWM strategies for control of the inverters feeding the open-end winding machine.

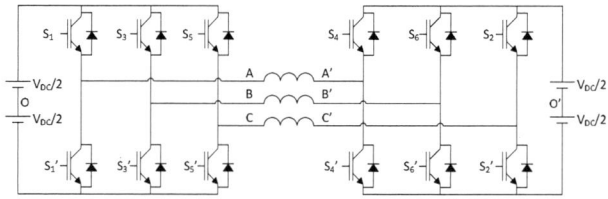

Fig. 3 Open-end Winding Configuration with Two Isolated DC-links

In this paper, implementation of three PWM strategies is discussed for controlling the open-end winding machines: sinusoidal PWM (SPWM), space vector PWM (SVPWM) and symmetrical sinusoidal PWM (SSPWM). A bipolar SPWM method using the circuit configuration in Fig. 2 is analyzed. And an SVPWM scheme using dual three-phase inverters configuration in Fig. 1 is investigated to eliminate the common mode current. Two implementation methods for SSPWM are also presented. The voltage harmonic spectra under different carrier frequencies and for various modulation indices are simulated using Matlab/Simulink. An open-end winding induction motor model is built in Simulink, and current ripples of the stator and rotor windings during steady state are investigated. A comparison of the harmonics, fundamental voltage, the switching losses, and current ripples under these PWM strategies are presented.

II. PWM SCHEMES FOR OPEN-END WINDING MOTOR DRIVES

A. SPWM

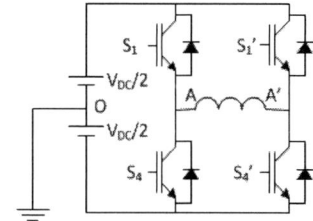

Fig. 4 One Phase of Open-end Winding Motor Drive

Fig. 4 shows one phase of the open-end winding of induction motor driven by an H-bridge converter. For bipolar SPWM, two sine waves which are phase shifted by θ are compared with a triangular carrier wave. The output voltage for phase-A, $V_{AA'}$ can be expressed as follows:

$$V_{AA'} = \frac{V_{DC}}{2}[\sin(wt) - \sin(wt - \theta)] \qquad (1)$$

The output voltages for phase-B and phase-C can be respectively obtained by shifting phase-A voltage by 120° and 240°. (1) can be also expressed as:

$$V_{AA'} = V_{DC} \sin\left(\frac{\theta}{2}\right) \cos\left(wt - \frac{\theta}{2}\right) \qquad (2)$$

From (2), it is shown that the phase shift θ between the two references has influence on the amplitude of the output phase voltage. With θ varying from 0° to 360°, the amplitude of output phase voltage varies between 0 and V_{DC}. During the process, when $\theta = 180°$, the phase voltage reaches its maximum amplitude of V_{DC}, while with phase shift other than 180°, the magnitude of output phase voltage is less than V_{DC}.

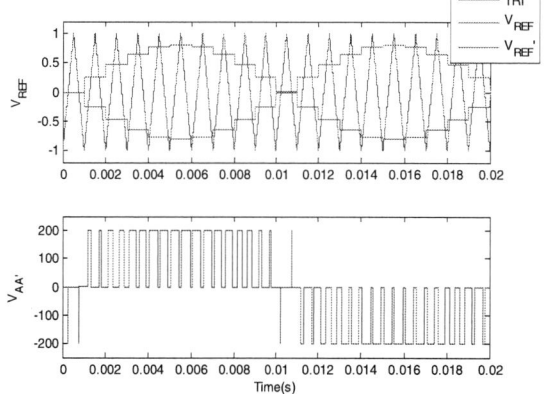

Fig. 5 Modulation Waveform for SPWM for One Phase

Fig. 5 shows the modulation process of SPWM for one period, the two references V_{REF} and V_{REF}' have a phase difference $\theta = 180°$. The switching rule of the gates (S_1, S_4, S_1' and S_4') is listed in Table 1.

TABLE 1
GATING RULE FOR SWITCHES OF H-BRIDGE UNDER SPWM

Gate	Gating Signal
S_1	$V_{REF} > TRI$
S_4	$\overline{V_{REF} > TRI}$
S_1'	$V_{REF}' > TRI$
S_4'	$\overline{V_{REF}' > TRI}$

Depending on the switching states, either positive DC bus voltage or negative DC bus voltage is applied to the motor winding. For the other two phases, the sinusoidal wave is shifted by -2π/3 and -4π/3. The maximum peak phase voltage V_{AA}' in linear modulation range that can be obtained from this scheme is V_{DC}.

B. SVPWM

To implement the space vector PWM, the three single-phase H-bridges are reconfigured as two three-phase inverters (INV 1 and INV 2) with a common DC link as shown in Fig. 1. The phase voltage of each winding can have three values ($+V_{dc}$, 0, $-V_{dc}$), which forms a three-level configuration. And each two-level inverter has eight switching states, as shown in Fig. 6; hence a total number of 64 switching combinations can be distributed in the 24 possible sectors in a three-level structure [5]. Among 64 switching combinations, there are only 19 non-duplicated voltage vectors, which are represented as A to S in Fig. 6.

The zero sequence voltage for open-end winding machine is defined as:

$$V_0 = \frac{1}{3}(V_{AA'} + V_{BB'} + V_{CC'}) \qquad (3)$$

Where $V_{AA'}$, $V_{BB'}$ and $V_{CC'}$ are the voltages across the three-phase windings. Due to the open-end winding configuration, some switching combinations may generate zero sequence voltage, which results in common mode current. For instance, when the switching state is 1-2', INV-1 is in the state 100, and INV-2 is in the state 110. Hence $V_{AA'} = 0$, $V_{BB'} = -V_{DC}$, $V_{CC'} = 0$, and the zero sequence voltage V_0 is $-V_{DC}/3$. Similarly, the zero sequence voltage generated under each switching combination is listed in Table 2.

TABLE 2
ZERO SEQUENCE VOLTAGE OF EACH SWITCHING COMBINATION

Zero Sequence	Switching combinations
V_{DC}	8-7'
$2V_{DC}/3$	2-7', 4-7', 6-7', 8-2', 8-4', 8-6'
$V_{DC}/3$	1-7', 2-1', 2-3', 2-5', 3-7', 4-1', 4-3', 4-5', 5-7', 6-1', 6-3', 6-5', 8-1', 8-3', 8-5'
0	1-1', 2-2', 3-3', 4-4', 5-5', 6-6', 7-7', 8-8', 1-3', 1-5', 2-4', 2-6', 3-1', 3-5', 4-2', 4-6', 5-1', 5-3', 6-2', 6-4'
$-V_{DC}/3$	1-2', 1-4', 1-6', 2-8', 3-2', 3-4', 3-6', 4-8', 5-2', 5-4', 5-6', 6-8', 7-1', 7-3', 7-5'
$-2V_{DC}/3$	1-8', 3-8', 5-8', 7-2', 7-4', 7-6',
$-V_{DC}$	7-8'

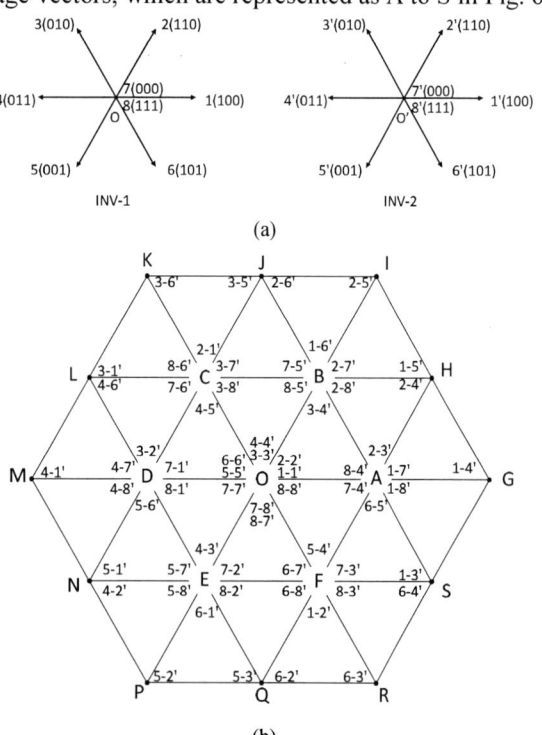

Fig. 6 (a) Switching States of Two Individual Three-phase Inverters (b) Space Vector Combinations of the Dual Three-phase Inverters Fed Open-end Winding Induction Motor

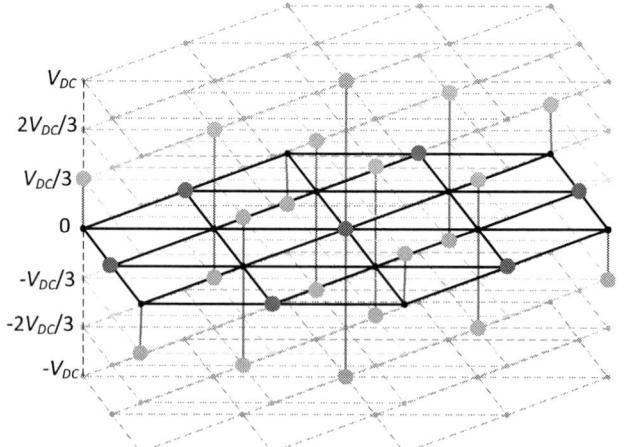

Fig.7 Zero Sequence Voltage of Each Switching Combination

To eliminate the zero sequence voltage, voltage vectors of the two inverters must be selected carefully so that in each switch interval, $V_{AA'} + V_{BB'} + V_{CC'} = 0$. One way to eliminate the zero sequence voltage is to use only voltage vectors that do not contain zero sequence components. As shown in Fig. 7, the three-level hexagon is mapped on a 3-

D space, in which the three-level hexagon is located on the x-y plane, and the magnitude of zero sequence voltage is represented along z axis. From Fig. 7, it is shown that only seven voltage vectors (marked by red dots), which includes 20 switching combinations, do not contain zero sequence voltage. By using these seven vectors in SVPWM, no zero sequence voltage is generated; hence the zero sequence current is eliminated.

From the observation of switching combination given in Table 2, it is clear that in the switching combinations which do not result any zero sequence, the phase difference between the two inverters is 120°. As the frequency of common mode voltage is three times or multiple of three times of fundamental frequency, the phase difference between the third order terms of the two inverter legs is 360°. Note that the voltage across the windings is the difference between the pole voltages of the two inverter legs of the H-bridges, the third order term is eliminated due to the subtraction of two phase-shifted voltage references. The maximum phase voltage $V_{AA'}$ in linear range that can be obtained from this scheme is V_{DC}, when the modulation index is 1.15.

C. SSPWM

Two PWM strategies called symmetrical sinusoidal PWM techniques are proposed in [1-2] to reduce the harmonics and switching losses in the separately fed three-phase induction machine. In both the schemes, N pulses in a half cycle are placed at the center of each time period C_i that are located at equally divided time space, where i varies from 1 to N. The width of each pulse P_i is obtained in two ways. In one scheme, the pulse width is proportional to the sine of the angle corresponding to the midpoint of the interval. In another scheme, it is proportional to the incremental area under the sine wave within the limits of the interval. The different methods for calculating the pulse widths are illustrated below.

1) SSPWM – Scheme 1

Consider a sinusoidal wave, whose half cycle is divided into N equal intervals. The center point of each pulse C_i can be calculated by:

$$C_i = (2i - 1) \times \frac{\pi}{2N} , i = 1, 2, \cdots, N \quad (4)$$

The pulse width of the ith pulse can be obtained by:

$$P_i = T_S W \sin(C_i) , i = 1, 2, \cdots, N \quad (5)$$

Where W is the width modulation index, ranging from 0 to 1 for linear modulation, and greater than 1 for over modulation region.

2) SSPWM – Scheme 2

The positions of pulse center points can be obtained using the same method as in Scheme 1. To calculate the width of each pulse P_i, the half sine wave is first divided into N intervals with same width, and the pulse width P_i is proportional to the area of the ith interval. Thus, the pulse width P_i is given by:

$$P_i = T_S W \cdot \int_{(i-1)\frac{\pi}{N}}^{(i)\frac{\pi}{N}} \sin(wt) \, dt , i = 1, 2 \dots N \quad (6)$$

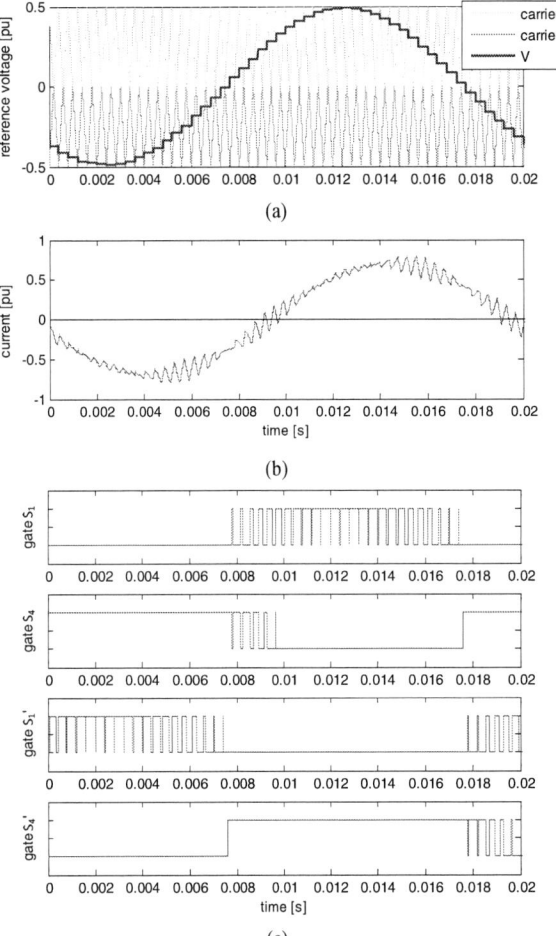

Fig. 8 (a) Reference Wave and Carriers (b) Load Current (c) Gating Signals of the H-bridge

Fig. 8 demonstrates the method to implement SSPWM. Two triangular carriers are shown in Fig. 8 (a), the top one has a maximum value $V_{DC}/2$ and minimum value 0, and the bottom one has a minimum value -$V_{DC}/2$ and maximum value 0. The value of reference wave is zero order hold in each triangular wave period, and it is equal to the pulse width calculated in (5) or (6). Fig. 8 (b) shows the load current. Based on the sign of the reference wave and the load current, the gating rule for the 4 switches of the H-bridge is listed in Table 3, the corresponding gating signals are shown in Fig. 8 (c).

978-1-4799-2706-7/14 $31.00 © 2014 IEEE   398

TABLE 3
GATING RULE FOR SWITCHES OF H-BRIDGE UNDER SSPWM

	V > 0 I > 0	V > 0 I < 0	V < 0 I > 0	V < 0 I < 0
S_1	PWM	PWM	OFF	OFF
S_4	OFF	$\overline{\text{PWM}}$	ON	ON
S_1'	OFF	OFF	PWM	PWM
S_4'	ON	ON	$\overline{\text{PWM}}$	OFF

Where 'PWM' means the corresponding switch switches at carrier frequency to provide PWM signal, '$\overline{\text{PWM}}$' means the device switches complementary to the one in the 'PWM' state. 'ON' means that the switch is conducting throughout the interval and 'OFF' means that the switch is off during the interval.

III. HARMONIC ANALYSIS

To analyze the harmonics of different PWM schemes, specifications and parameters are the same for each PWM strategy so that the results can be compared. The parameters for analyzing the harmonics are listed in Table 4.

TABLE 4
PARAMETERS FOR HARMONIC ANALYSIS

Parameters	Value
V_{DC}	200V
f	60Hz
Fs	960/1440/1680
m_f	16/24/28
M	0.1-1.2

Where m_f indicates the frequency modulation index, for SPWM, $m_f = Fs/f = N$, which means there are N triangles in one cycle. For SPWM and SVPWM, there are N pulses in half cycle; for SSPWM, there will be N/2 pulses in half cycle. And M represents the modulation index.

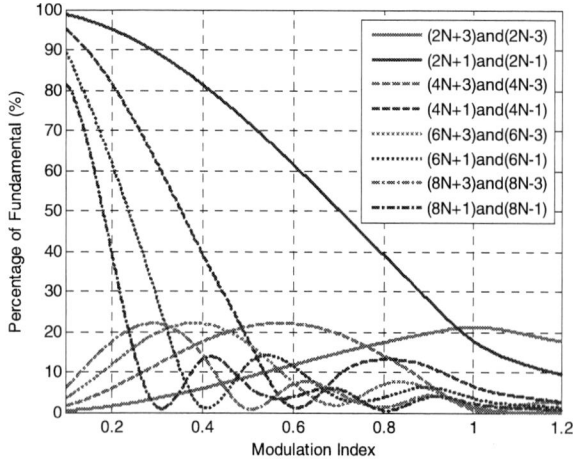

Fig. 9 Harmonics vs. Modulation Index in SPWM.

Fig. 9 illustrates the relationship between modulation index and the predominant harmonics. The predominant harmonics are $(2N\pm1)^{th}$, $(4N\pm1)^{th}$, $(6N\pm1)^{th}$ and $(8N\pm1)^{th}$, and the first harmonic with predominant value is $(2N-1)^{th}$. When modulation index is low, the value of harmonics of order $(2N\pm3)^{th}$ is small, but as modulation index increases, the value of $(2N\pm3)^{th}$ harmonics will be not negligible.

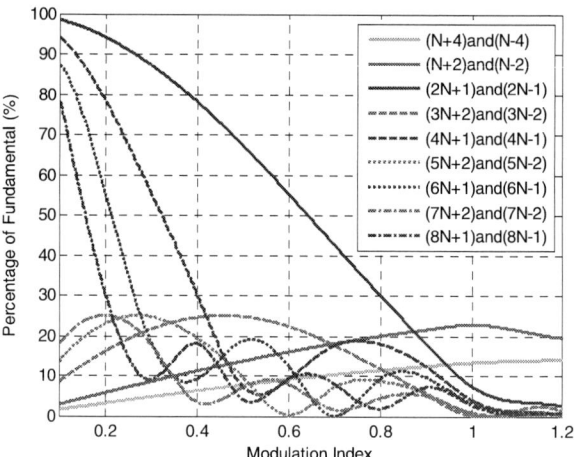

Fig. 10 Harmonics vs. Modulation Index in SVPWM.

In Fig. 10, it is shown that when modulation index is less than 0.2, $(2N\pm1)^{th}$, $(4N\pm1)^{th}$, $(6N\pm1)^{th}$ and $(8N\pm1)^{th}$ orders of harmonics dominate. As modulation index increases, the $(4N\pm1)^{th}$, $(6N\pm1)^{th}$ and $(8N\pm1)^{th}$ harmonics drop quickly, but the $(2N\pm1)^{th}$ still has a strong influence. The $(N\pm2)^{th}$ harmonic increases as a function of modulation index increases, and when modulation index is approaching to 1, it is the dominant harmonic. The first harmonic with non-negligible value seen is $(N-4)^{th}$.

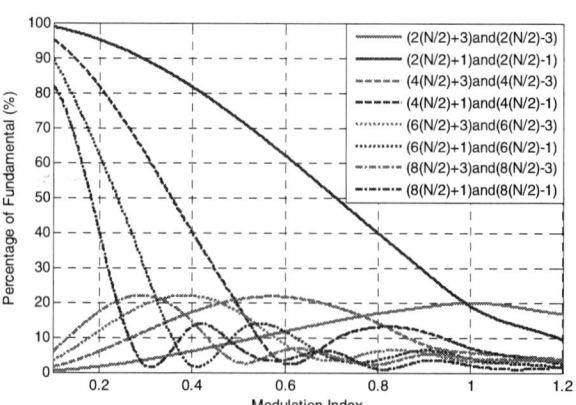

Fig. 11 Harmonics vs. Modulation Index in SSPWM Scheme 1.

The 2014 International Power Electronics Conference

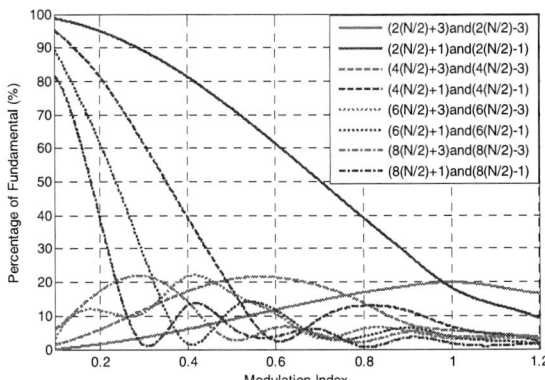

Fig. 12 Harmonics vs. Modulation Index in SSPWM Scheme 2.

Fig. 11 and Fig. 12 depict the relationship between harmonic amplitude and modulation index. It can be noticed that only minor difference exists between two schemes of SSPWM. The harmonics vs. modulation index plot shows that SSPWM shares certain similarities with SPWM in the harmonic spectra. The predominant harmonics in SSPWM are $(2(N/2)\pm1)^{th}$, $(4(N/2)\pm1)^{th}$, $(6(N/2)\pm1)^{th}$ and $(8(N/2)\pm1)^{th}$, where N/2 denotes the number of pulses in half cycle and $(2(N/2)\pm3)^{th}$ has a stronger influence as modulation index increases. The first dominant harmonic expected to meet in the harmonic spectra is the $(2(N/2)-1)^{th}$.

It is noticed that when m_f increases, i.e. when N increases, the harmonic with lowest order and higher amplitude shifts to a higher frequency. And this can be interpreted in two aspects. On one hand, when investigating certain orders of harmonics, increasing m_f does not lower the amplitude of the harmonics, it only shifts the harmonics to a higher frequency band. On the other hand, when looking into a certain frequency band, increasing m_f shifts the harmonics to a higher frequency band, while a better frequency response can be obtained in the lower frequency band.

The relationship between fundamental voltages and modulation index can be seen in Fig. 13 (a). The linear region of modulation index is 0 to 1 for SPWM and it is 0 to 1.15 for SVPWM. In all the three modulation schemes, when modulation index is at the maximum value in the linear region (1 for SPWM/SSPWM, 1.15 for SVPWM), fundamental voltage obtained is $\frac{V_{dc}}{\sqrt{2}} = 0.707 V_{dc}$, as shown in the horizontal line in Fig. 13 (a). However, when looking into the same modulation index value, 0.8 for example, the vertical line in Fig. 13 (a) indicates that SSPWM and SPWM maintain a higher fundamental voltage than SVPWM. The fundamental voltage of SSPWM and SPWM is $\frac{2}{\sqrt{3}} = 1.15$ times of the SVPWM fundamental.

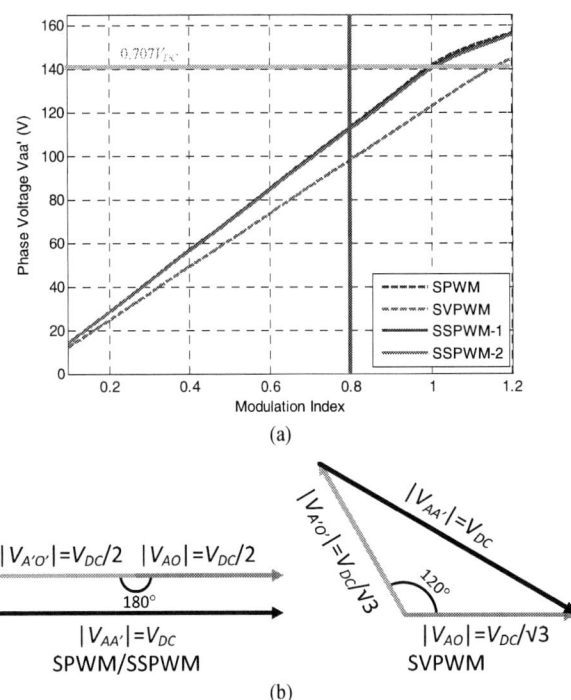

(a)

(b)

Fig. 13 (a) Fundamental Voltages Varying with Respect to Modulation Index (b) Synthesis of Phase Voltage in SPWM, SSPWM and SVPWM

Unlike in the neutral connected motor case, in which SVPWM can have 1.15 times larger fundamental voltage and linear region than SPWM, SVPWM loses this advantage due to the phase difference between the two inverters. As shown in Fig. 13 (b), when modulation index is 1, the amplitude of pole voltages in SPWM/SSPWM is $V_{DC}/2$. As the phase difference across the two legs of the H-bridge is 180°, the voltage across the phase winding is $V_{DC}/2 - (-V_{DC}/2) = V_{DC}$. For SVPWM case, when modulation index is 1.15, the amplitude of pole voltage is $V_{DC}/\sqrt{3}$. As the phase difference of the two three-phase inverter is 120°, the voltage across the phase winding is still V_{DC}, as shown in Fig. 13 (b).

The total harmonic distortion (THD) can be calculated using (7):

$$THD = \frac{V_H}{V_F} \qquad (7)$$

Where $V_H = \sqrt{V_2^2 + V_3^2 + V_4^2 + \cdots + V_n^2}$. V_n is the RMS value of the nth harmonic, and V_F is the fundamental voltage.

The relationship between THD and modulation index is shown in Fig. 14. It can be seen that when modulation index is low (less than 0.2), the THDs of SPWM and SVPWM are lower than that of SSPWM. But as modulation index increases, THD of SSPWM drops faster than SPWM and SVPWM. When modulation index is

978-1-4799-2706-7/14 $31.00 © 2014 IEEE   400

between 0.6 and 1.0, which is the working region for most of the motors, the difference in THD between SSPWM and SPWM reduces to less than 5%. As switching frequency increases, the first dominant harmonic is shifted to a higher frequency band, and the lower frequency band showing minor difference among the three PWM methods.

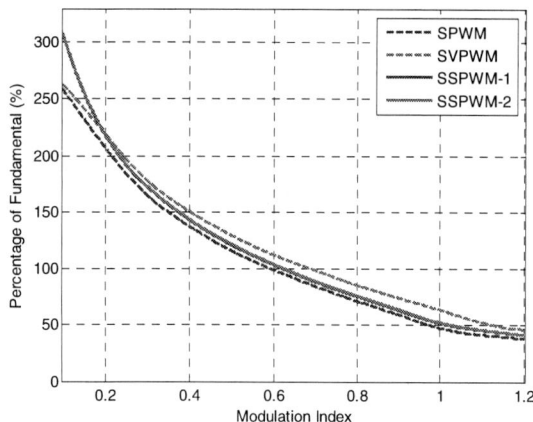

Fig. 14 THD vs. Modulation Index

IV. STEADY STATE PERFORMANCE FOR OPEN-END WINDING INDUCTION MOTOR

In order to compare the steady state performances of the PWM schemes, a 5.5 kW, 4 pole open-end winding induction motor is modeled in Simulink the carrier frequency is chosen to be 2.5 kHz. Field orientation control (FOC) method is used to regulate the current. The measured current is transformed to synchronous frame, in which the d axis is along the direction of the flux.

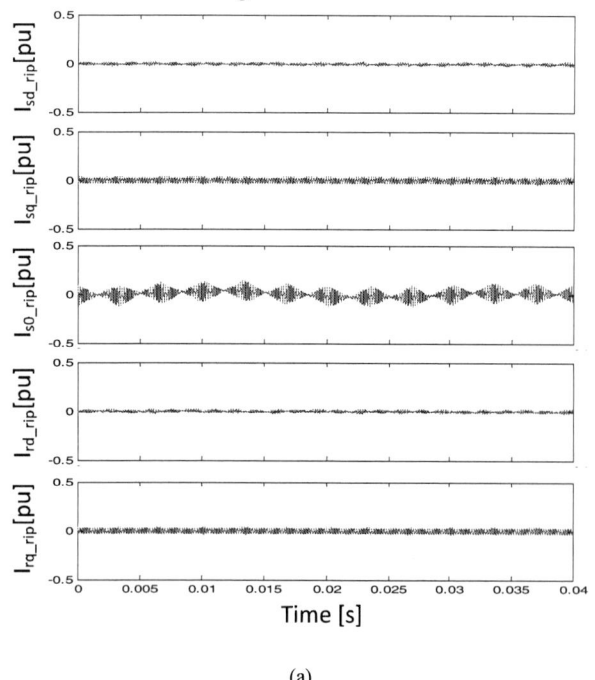

(a)

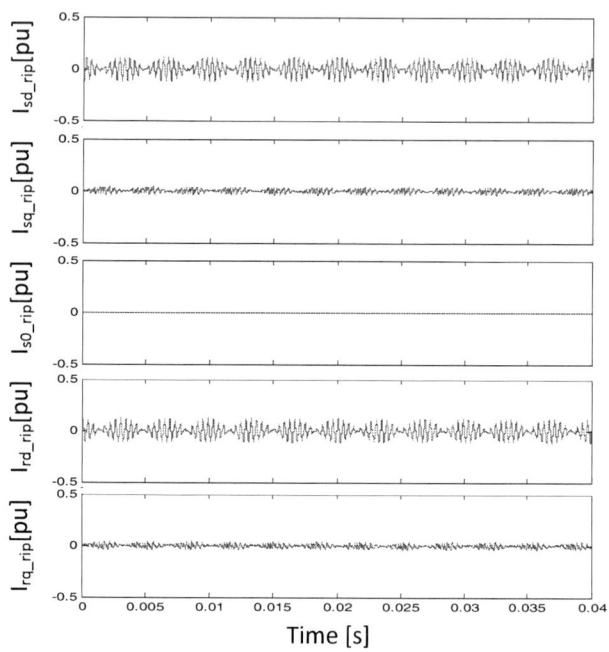

(b)

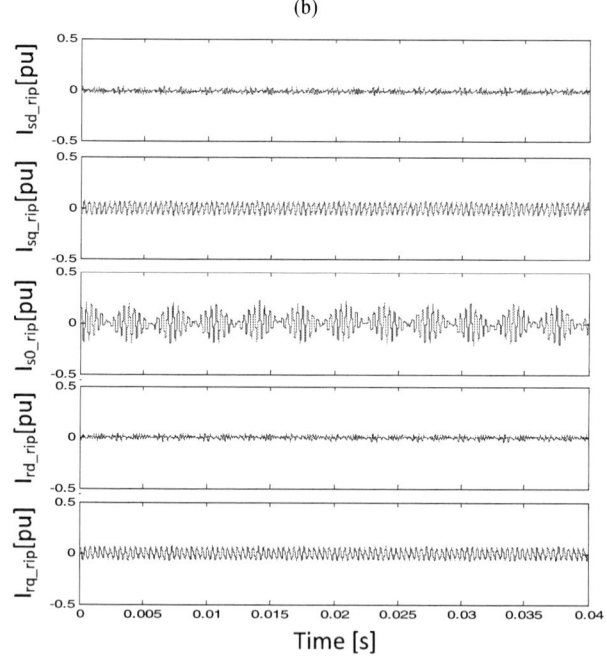

(c)

Fig. 15 (a) Current Ripple under SPWM (b) Current Ripple under SVPWM (c) Current Ripple under SSPWM

Fig. 15 shows the current ripple in both stator winding and rotor winding. The current ripple is defined as:

$$I_{x\_rip} = I_{x\_reference} - I_{x_{measured}} \quad (5)$$

x refers to rd, rq, sd, sq, $s0$, which represent the rotor d axis component, rotor q axis component, stator d axis component, stator q axis component and stator zero sequence component.

From Fig. 15 (a) and (c), it can be observed that during the steady state, SPWM and SSPWM both present large zero sequence current, because both of the schemes utilize the switching combinations which can result in zero sequence voltage. Besides zero sequence current, the ripple magnitude of SPWM in d axis and q axis current is less than that of SSPWM, which means SPWM provides a better approximation about the reference wave. On the other hand, SVPWM generates no zero sequence, as shown in Fig. 15 (b); however, large d axis current ripple is observed in both stator winding and rotor windings.

Table 5 gives a summary of the steady state performances of the three PWM schemes for open-end winding induction motor. It is shown that among the three PWM schemes, SPWM presents the lowest voltage and current total harmonic distortion (THD); SVPWM guarantees absence of zero sequence currents; and SSPWM offers a lowest switching frequency per device, which is defined as switching counts of all devices divided by device numbers. As switching loss is proportional to switching frequency, large switching loss reduction can be expected in SSPWM strategy.

TABLE 5

PERFORMANCES OF SPWM, SVPWM AND SSPWM FOR AN 5.5 KW OPEN-END WINDING INDUCTION MOTOR

	SPWM	SVPWM	SSPWM
Switching Freq/device	2.5kHz	2.5kHz	0.78kHz
V THD (%)	54.7	55.1	55.4
I THD (%)	5.5	6.2	11.0
$I_{dS\ ripple}$ [pu]	0.048	0.240	0.086
$I_{qs\ ripple}$ [pu]	0.080	0.087	0.150
$I_{0S\ ripple}$ [pu]	0.260	0	0.420
$I_{dr\ ripple}$ [pu]	0.046	0.230	0.083
$I_{qr\ ripple}$ [pu]	0.077	0.084	0.150

V. CONCLUSION

In this paper, three PWM strategies, i.e. SPWM, SVPWM with common mode current elimination, and SSPWM are presented. The harmonic spectra of the three PWM schemes are analyzed and compared for different carrier frequencies and modulation indices. A test has been performed for an open-end winding induction motor to compare the steady state performances of the three PWM schemes. The results show that SPWM presents the lowest total harmonic distortion (THD) among the three schemes, SVPWM does not generate zero sequence currents, and SSPWM provides lowest switching loss. Based on different requirement of specific motor drives, corresponding PWM strategy can be chosen to eliminate zero sequence current, to minimize THD, or to obtain the lowest switching losses.

REFERENCES

[1] K.S.Rajashekara and J.Vithayathil, "Microcomputer Based Symme-trical Sinusoidal Pulse Width Modulated Inverter", *IEEE IECI Proceedings*, San Francisco, USA, PP. 34-38, November 1981.

[2] K.S. Rajashekara and V. Rajagopalan, "Simulation of SSPWM Inverter-Fed Induction Motor", *International Symposium on Modeling and Simulation of Electrical Machines, Converters and Power Systems*, August 24-25, 1987, Quebec City, Canada.

[3] Gupta, R.K.; Somani, A.; Mohapatra, K.K.; Mohan, N., "Space vector PWM for a direct matrix converter based open-end winding ac drives with enhanced capabilities," Applied Power Electronics Conference and Exposition (APEC), 2010 Twenty-Fifth Annual IEEE , vol., no., pp.901,908, 21-25 Feb. 2010

[4] H. Stemmler and P. Guggenbach, "Configurations of high-power voltage source inverter drives," inProc. Eur. Power Elec. Appl. Conf., Brighton, U.K., 1993, vol. 5, pp. 7–14.

[5] Shivakumar, E.G. ; Gopakumar, K. ; Sinha, S.K. ; Pittet, A. ; Ranganathan, V.T. "Space Vector PWM Control of Dual Inverter Fed Open-end Winding Induction Motor Drive", *APEC 2001*, Anaheim, pp 399-405.

[6] Somani, A. ; Gupta, R.K. ; Mohapatra, K.K. ; Basu, Kaushik ; Mohan, N. "Modulation strategies for direct-link drive for open-end winding AC machines", *Electric Machines and Drives Conference*, 2009. *IEMDC '09*. IEEE International, Pp 1863-1868

[7] Yang Wang ; Panda, D. ; Lipo, T.A. ; Di Pan, "Open-Winding Power Conversion Systems Fed by Half-Controlled Converters". *IEEE Transactions on Power Electronics*, Volume:28, Issue: 5, 2013, Pp 2427-2436

[8] Baiju, M. R.; Mohapatra, K.K.; Kanchan, R. S.; Gopakumar, K., "A dual two-level inverter scheme with common mode voltage elimination for an induction motor drive," *Power Electronics, IEEE Transactions on* , vol.19, no.3, pp.794-805, May 2004

[9] Somasekhar, V.T.; Gopakumar, K.; Baiju, M. R., "Dual two-level inverter scheme for an open-end winding induction motor drive with a single DC power supply and improved DC bus utilisation," *Electric Power Applications, IEE Proceedings -* , vol.151, no.2, pp.230,238, Mar 2004

[10] K. Gopakumar, V. T. Ranganathan, and S. R. Bhat, "Split-Phase Induction Motor Operation from PWM Voltage Source Inverter, ' IEEE Transactions on Industry Applications, Vol. 29, NO. 5, September/October 1993, pp. 927-932

[11] Levi, E.; Bodo, N.; Dordevic, O.; Jones, M., "Recent advances in power electronic converter control for multiphase drive systems," *Electrical Machines Design Control and Diagnosis (WEMDCD)*, 2013 IEEE Workshop on , vol., no., pp.158,167, 11-12 March 2013

[12] T.A. Lipo, "New Open Winding Machine Concepts", Course Notes for ECE 511, Department of Electrical and Computer Engineering, University of Wisconsin-Madison, 2013

The 2014 International Power Electronics Conference

10MW,3.3MWh Energy Storage System consisting of 4000 Flywheels controlled by ICT network for Short Cycle Power Fluctuation Compensation

Koji Kato[*1], Satoru Ishigma[*1] Yoichiro Nakajima[*1], Haruki Arai[*1], Tetsuya Ueda[*1],
Tetsuki Iwata[*1], Yoichi Ito[*1], Kazumi Sugao[*2]

[*1]Sanken Electric CO., LTD., [*2]Belltec
[*1]Kawagoe city, Japan, [*2]Chiba city, Japan
[*1]k.kato@sanken-ele.co.jp

Abstract— This paper proposes a 10MW 3.3MJ Energy Storage System consisting of 4000 Flywheels controlled by ICT network. The flywheel has a lot of advantages such as an environmental friendly, long life time, easy manufacturing and etc. The proposed system consists of the many small capacity flywheels and the high speed communication network. Therefore proposed system has a high flexibility from small capacity to large capacity. The 90% efficiency of charge and discharge for the flywheels are confirmed by experimental results. Moreover 1kWh energy capacity and 200W loss of rated flywheel rotation speed are obtained.

Keywords— *The authors shall provide up to 4 keywords or phrases (in alphabetical order and separated by commas) to help identify the major topics of the paper.*

I. Introduction

The renewable energy such as PV and wind power is introduced into the power grid by earthquake disaster reconstructions in Japan. There is a ministry plan that the renewable energy will be introduced to 50 % of the consumed power until 2050 in Japan. Renewable energy systems, especially a wind turbine and a photovoltaic cell, generated a power fluctuation due to the meteorological conditions. Therefore problem of renewable energy into power grid is frequency and voltage fluctuation of power grid [1]-[9]. To avoid frequency fluctuation of power grid, the energy storage system is located in the power grid.

There are a lot of energy storage devices such as lead acid battery (Lab), lithium ion battery (Lib), electric double layer capacitor (EDLC), Flywheel (FW) and etc. To compensate the short cycle frequency fluctuation, the energy storage device repeats charge and discharge. The Lab and Lib has a problem of short lifetime. In particular, the lifetime depends on an ambient and the number of charge and discharge cycle. On the other hands, the EDLC and FW don't depend on the number of charge and discharge cycle. Therefore the EDLC and FW are suitable for the compensation of short cycle frequency

fluctuation. However the lifetime of EDLC depends on the ambient same to the chemical battery. As a result, the energy storage devices with chemical materials such as Lab, Lib and EDLC will be high cost to replace the energy storage devices.

On the other hand, the flywheel has some advantages as follows.

· Environmental friendly
· Low maintenance cost
· Long lifetime due to no chemical structure
· High rate charge and discharge characteristic

Therefore FW is better than the EDLC for energy storage device of the compensation of short cycle frequency fluctuation. However there are some problems for the high capacity of the FW. The huge capacity FW is realized by the custom-made FW without flexibility. Therefore it is difficult to use the general-purpose application.

There is some discussion of the large energy capacity approach such an ultra high speed rotation of the FW [9]. The kinetic energy which is stored in the FW is proportional to the square of the rotational speed. The magnetic bearings have been studied in order to achieve high energy density It can lead to high cost and complexity of the control system of the FW. Because the magnetic bearing FW needs a big protection cases, vacuum cases, vacuum pumps and magnetic bearing controllers.

This paper proposes a 10MW 3.3MJ Energy Storage System consisting of 4000 FW to compensate the short cycle frequency fluctuation. To achieve the large capacity FW energy storage system, the small capacity FW is connected to power grid in parallel. Each FW are controlled by using the high speed communication network. The proposed system is emulated a large capacity FW by the high speed communication network. Therefore the FW design method of the low cost and high efficiency is important.

978-1-4799-2706-7/14 $31.00 © 2014 IEEE

In this paper, the proposed FW uses a pivot bearing which is realized low loss and low cost. In addition, the proposed FW rotating speed is around 6000 rpm. Therefore it is unnecessary to use the expensive protection case. The 5 type FW energy storage devices (FWED) prototype are produced experimentally.

The fundamental operation of the proposed system has confirmed by experimental results of 2nd prototype FWED. The 90% efficiency of charge and discharge for the flywheels are confirmed by experimental results. Moreover 1kWh energy capacity and 200W loss of rated flywheel rotation speed are obtained. The validity of the proposed system is confirmed by 100kVA FW energy storage system.

II. Flywheel Energy Storage System

A. FW Energy Storage Devices

To compensate the short cycle fluctuation of the power grid frequency, the governor-free operation and the load frequency control are used for the load fluctuation of power grid. The purpose of proposed system is to develop the power fluctuation compensation system using FW energy storage system.

Figure 1 shows the comparison with the energy storage devices for the power density and the energy density. The energy density (Wh/kg) of the lead-acid batteries is the best of them. The power density (W/kg) of the electrolytic capacitors is the best of them. The FW and EDLC are intermediate positions in them.

Table 1 shows the comparison with the power storage devices for the cycle life and the storage capacity. FW and EDLC, which has long cycle life, are suitable to compensate the short cycle frequency fluctuation. However, EDLC is weak against heat and cold. It is difficult to use the EDLC in the cold area and warm area. On the other hands, FW doesn't degrade with temperature variation and charge-discharge cycle. Therefore the FW is the best of them.

B. Proposed Flywheel Energy Storage System

Figure 2 shows the energy storage system using FW. The conventional FW energy storage system[4] as shown in Fig.3(a) consists of the large size FW. The large size FW has economies of scale. However, the large size FW lacks versatility. Moreover it is difficult to manufacture the large size FW.

Figure 3(b) shows the proposed system. The proposed system consists of the small size FW in parallel and the high speed communication network to realize the virtual large size FW. The proposed system has advantage of the versatility and easy manufacturing.

Table 2 shows the specifications of the proposed system. The proposed system needs 10MW and 20 minute energy storage capacity to compensate the short

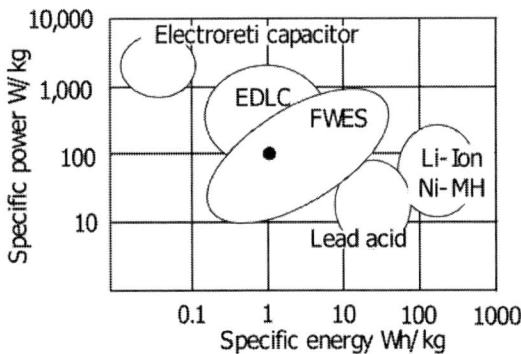

Fig. 1. Comparison with power storage devices for power density and energy density.

Table 1 comparison with the cycle life and the storage capacity.

	LaB	LiB	FW	EDLC
Energy capacity [Wh/kg]	50	100	1	1
Cycle life	1500	3000	300000	100000

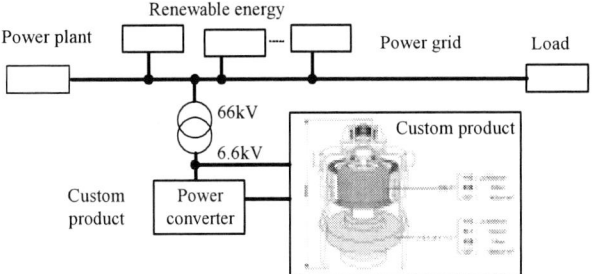

Flywheel energy storage device 20MVA 200MJ

(a) Conventional system

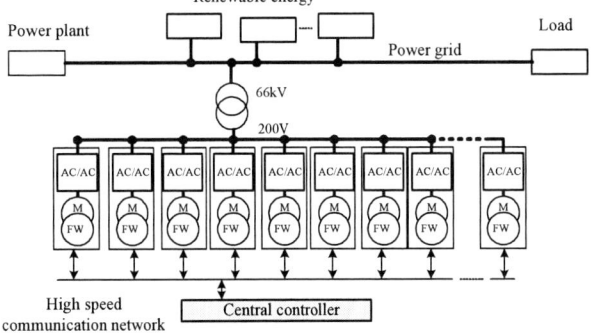

(b) Proposed system

Fig.2 Flywheel energy storage system

Table 2 Specification of proposed system

key technique		specification
FWES	output power	2.5kW (minimum)
	capacity	3MJ
	size	Less than 1m
	rotation speed	Less than 10000rpm
	Efficiency	80%
High speed communication network	Number of FW	4000 unit
	speed	12.5 usec/unit

cycle power fluctuation. Therefore the system is composed by 4000unit FWs at 3MJ and the high speed network to control each FW. The minimum output power of FW is 2.5kW and 80% efficiency of charge and discharge including the power converter. Moreover FW size is lower than 1 meter, FW rotation speed is lower than 10000 rpm to achieve the easy manufacturing and low cost. The communication speed of the communication network is lower than 12.5 μsec per unit to provide the control command to each FW. Therefore Ether-CAT [6] is used for the proposed system.

C. Communication Network

Figure 3 shows the communication network configuration of the proposed system. The proposed system consists of the 2 type network and FWES controller. The high speed network must be sent the active power commands to all FW within 50m seconds to compensate the short cycle power fluctuation. Therefore the high speed network uses the "Ether-CAT" protocol[6]. The low speed network is used to monitor the FWSS conditions such as FWED temperature, rotation speed, alarm and etc. The FWES controller calculates the active power command of each FWSS depending on the power levering command of power grid controller.

III. Flywheel Energy Storage Devices

There is widely discussion of the magnetic levitation bearing for the high power density of FW. The kinetic energy which is stored in the FW is proportional to the square of the rotational speed. The magnetic levitation bearings have been studied in order to achieve high energy density. It can lead to high cost and complexity of the control system of the FW. Because the magnetic bearing FW needs a big protection cases, vacuum cases, vacuum pumps and magnetic bearing controllers.

The purpose of this study is to develop the low cost FW. The proposed FW is unapplied to the magnetic

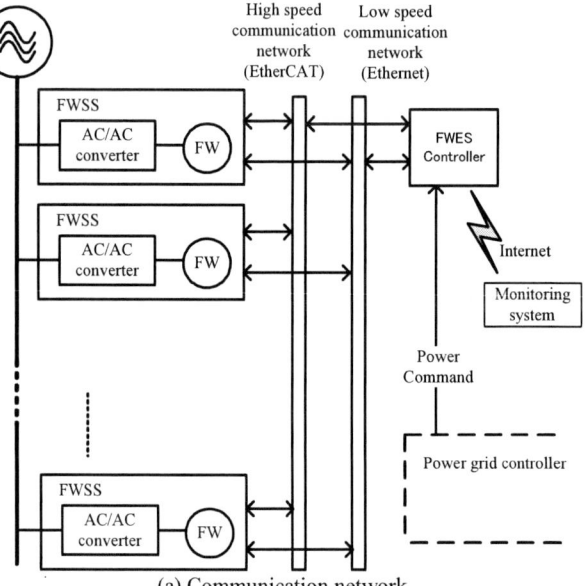

(a) Communication network

(b) FWES controller

Fig.3 Communication network configuration

Table 3 Specification of FW prototype

	Conventional model	1st prototype	2nd prototype
Outline			
Energy capacity	1MJ	3MJ	3.6MJ
Rotation speed	6000rpm	6000rpm	6000rpm
Bearing	Pivot	Pivot	Pivot
Motor	Custom IM	Custom IM	Custom IM
Size	φ518 x t143	φ518 x t372	φ694 x t156
Weight	130kg	470kg	386kg

levitation bearing and the ultra high speed rotating speed FW. Therefore proposed FW is applied to the pivot bearing.

A. Specification of Prototype FW

Table 3 shows the specification of the prototype FW energy storage devices. There are 3 type FWs such as 1MJ FW, 3MJ FW and 3.6MJ FW. The FW weight is increased to improve the energy storage capacity. The diameter of the prototype FW rotor is less than 1m to use the generic metalworking machines. The rotation speed of the prototype FW is less than 10000rpm to use the general bearing. Under those conditions, the energy storage capacity of the prototype FW is realized more than 3MJ.

B. Pivot Bearing

Figure 4 shows the pivot bearing diagram. The pivot bearing consists of the conical axial end and the concave bearing. There is a spiral groove on the conical axial end. The concave bearing is suffused with the oil. The pivot bearing generates the levitation pressure owing to the spiral groove on the conical axial end catches up the oil when the conical axial end is rotated at a high speed. Therefore low loss is achieved to levitate the pivot bearing.

C. Free Run Characteristic

Figure 5 shows the outline of the FW energy storage device and free run characteristic. The FW is continued to rotate by charging energy. However rotation speed of the FW is slowly decreased by energy losses which are windage losses, bearing losses and etc.

In this result, the 2nd prototype FW can realize the high energy capacity.

D. FWED Control method

Figure 6 shows the control block diagram of the low-loss control for the inverter. The flux command □d* is calculated from eq. (1). Therefore the copper loss of the motor is decreased by minimum exciting current.

$$I = \frac{T_m^*}{T_{mn}} \cdot \frac{I_n}{\sqrt{2}} \qquad (1),$$

where, Tmn is nominal value of torque, In is nominal value of motor current.

Figure 7 shows the experimental results of the low-loss control method. In the light load, motor efficiency is improved about 10% by low-loss control method. The validity of low-loss control method is confirmed from experimental results.

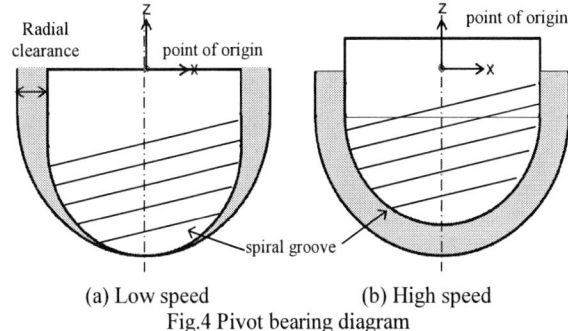

(a) Low speed   (b) High speed
Fig.4 Pivot bearing diagram

(a) Outline of FW energy storage device

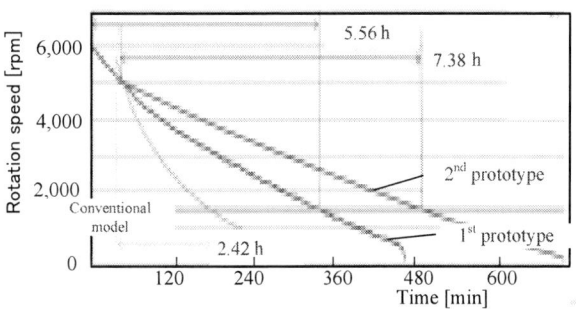

(b)Free run characteristic
Fig. 5. Prototype FW energy storage devices

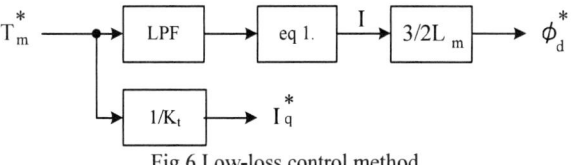

Fig.6 Low-loss control method

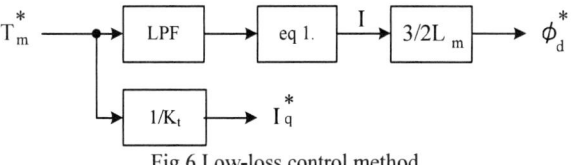

Fig.7 Experimental results of low-loss control method

978-1-4799-2706-7/14 $31.00 © 2014 IEEE

IV. Experimental Results

A. Test facility

The proposed system consists of the 4000 FWED and communication network. Therefore each FWEDs characteristic is important. This chapter shows the experimental results of 2nd prototype FWED.

Figure 8 shows the test facilities of the proposed FWES. The experimental system consists of the 5 type FWED prototype of 30kVA, the inverter board of 37kVA and converter of 100kVA.

B. Charge and Discharge Characteristic

Figure 9 shows the charge and discharge results of the 2nd prototype. In this result, the FW accelerates or accelerates within a specified period of time. Table 4 shows the charge and discharge efficiency. The charge energy of 3.45MJ and discharge energy of 2.98MJ are obtained. The efficiency of charge and discharge is 86.5%. The 1kWh energy capacity is confirmed by experimental results.

Figure 10 shows the charge and discharge characteristics of 2nd prototype FWED. In this result, desired characteristics of charge and discharge for the 2nd prototype FWED are obtained.

C. Charge and Discharge Efficiency

Figure 11 shows the charge and discharge efficiency under the vacuum and helium-filled condition of FW. In this result, maximum efficiency is over 90%. The efficiency of vacuum condition is higher than helium-filled condition for the long acceleration and deceleration time. Because the windage loss of vacuum condition is lower than helium-filled condition. On the other hand, the almost same efficiency are obtained in the vacuum and helium-filled condition when the short acceleration and deceleration time. The short acceleration and deceleration time mean to input or output high power from FW. Therefore the FW has strong advantage of the high rate

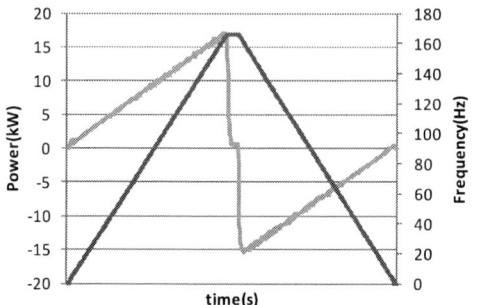

Fig.9 charge and discharge results

Table 4 Charge and Discharge efficiency

Charge energy [MJ]	3.45
Discharge energy [MJ]	2.98
Efficiency [%]	86.5

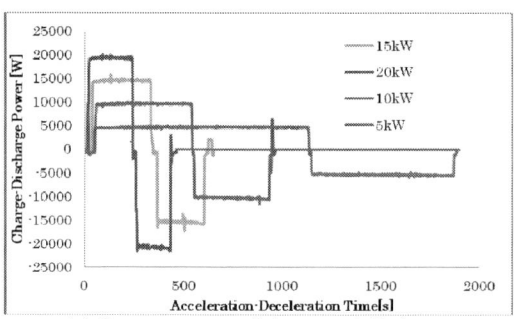

(a) Constant power of charge and discharge

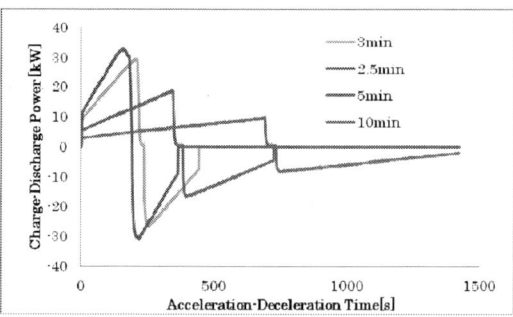

(b) Variable power of charge and discharge
Fig.10 Charge and discharge characteristics

(a) Prototype FWEDs

(b) Test inverter and converter

Fig.8 Test facility

charge and discharge. It is noted that the FW has no limitation of the rate of charge and discharge. On the other hand, general purpose lithium-ion battery has the rate of charge and discharge of 2C.

D. Loss Analysis

Figure 12 shows the loss analysis results under the vacuum and helium-filled condition of FW. The loss of FW is increased by high rotation speed. The windage loss is dominant. Therefore the loss is drastically decreased by vacuum for the FW. In this result, 200W losses of the rated rotation speed are obtained.

v. Conclusions

This paper introduces the FW energy storage system to compensate the short cycle frequency fluctuation. The system achieves the large energy capacity of FW with a high speed communication technology.
The 90% efficiency of charge and discharge for the flywheels are confirmed by experimental results. Moreover 1kWh energy capacity and 200W loss of rated flywheel rotation speed are obtained. The validity of the proposed system is confirmed by 100kVA FW energy storage system.

References

[1] O. Wasynczuk, D.T. Man, and J.P. Sullivan: "Dynamic Behavior of a Class of Wind Turbine Generators During Random Fluctuations" IEEE Transactions on Power Apparatus Sysr., Vol. PAS-100, No. 6, pp2837-2854, 1981

[2] J. Baba, et al:" Combined power supply method for micro grid by use of several type of distributed power generation systems" Proc. of EPE 2005, CD-ROM, Dresden, Germany, 2005

[3] T.Tanabe, S.Suzuki, Y.Ueda, T.Ito, S.Numata, E.Shimoda, T.Funabashi, and R.Yokoyama: "Control Performance Verification of Power System Stabilizer with an EDLC in Islanded Microgrid" IEEJ Trans. PE, Vol.129, No.1, pp.139-147, 209(in Japanese).

[4] H. Kameno, A. Kubo and R. Takahata, "Basic Design of 1kWh Class Compact Flywheel Energy Storage System", Koyo Engssineering Journal No.163, March 2003

[5] Y. Ito and S. Ishiguma, "Uninterruptible Power Supply with Function of Absorbing Regenerative Energy ", Conf. Rec. of IPEC-Sapporo, IEEJ (2010)

[6] T.Matukawa, M.Kanke, R.Shimada, et al:" A 215MVA Flywheel Moter-generator with 4GJ Discharge Energy for JT-60 Troidal Field Coil Power Supply System " IEEE Trans. on Energy Conv. EC-2,262, 1987.

[7] http://www.cat.com/power-generation/ups

[8] http://www.activepower.com/CompanyInformation/

[9] M. Murakami, et al.:"Large Levitation Force due to Flux Pinning in YBaCuO Superconductors Fabricated by MPMG Process" Jpn.J.Appl. Phys, 29,p.L 1991

[10] EtherCAT "http://www.ethercat.org/"

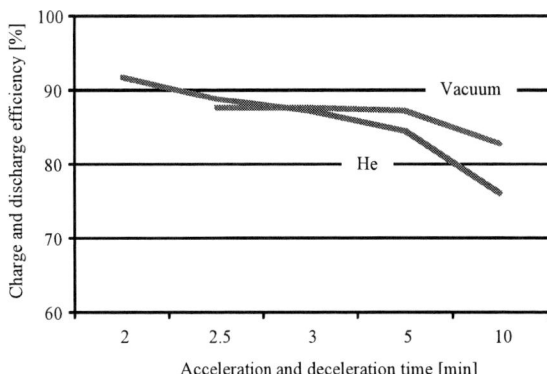

Fig. 11. Charge and discharge efficiency

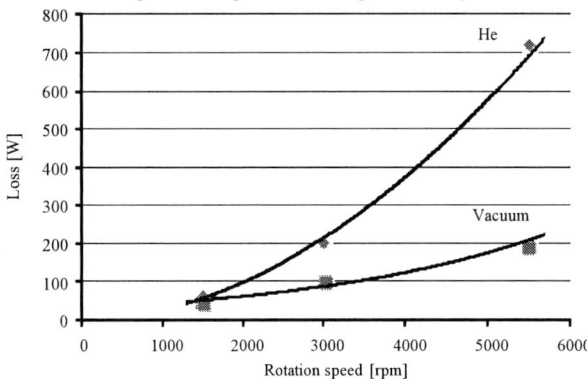

Fig.12 Loss analysis results

The 2014 International Power Electronics Conference

Versatile Power Transfer Strategies of PV-Battery Hybrid System for Residential Use with Energy Management System

Seong-Chon Choi[1], Min-ho Sin[2], Dong-Rak Kim[3],
Chung-Yuen Won[*]
Dept. of Information and Communication Engineering
Sungkyunkwan University
Suwon, Republic of Korea
chon8787@skku.edu

Yong-Chae Jung[4]
Dept. of Electronic Engineering
Namseoul University
Suwon, Republic of Korea

Abstract— **This paper presents versatile power transfer strategies (PTS) of the grid-connected hybrid system based on energy management system (EMS). The hybrid system is composed of power conditioning system (PCS) for photovoltaic (PV), energy storage system (ESS) and battery management system (BMS). To effectively operate the grid-connected PV-battery hybrid system, it is necessary to operate the system with command of EMS considering various conditions through communication between EMS and hybrid system. The proposed PTS can be applied to EMS, taking into account the output fluctuation of renewable energy, load demand, state of the grid, peak load period and state of charge (SOC) for battery to effectively operate the grid-connected hybrid system. The high reliability and economic efficiency of system is achieved by applying the proposed PTS to hybrid system through EMS. The experimental results were presented to verify the proposed operation.**

Keywords— *Energy management system, PV-Battery hybrid system, Power transfer strategie .*

I. INTRODUCTION

Recently, the renewable energy is widely used in countries all over the world due to the growing energy demand and rapid depletion of fossil-fuel reserves. Among these resources, the PV energy is most suitable for household application compared with other resources because of low cost of installation and maintenance. However, The PV energy has disadvantage that the output power of PV depends on the weather conditions and cell temperature. The output of PV falls dramatically when a small part of PV module or PV array is shaded though it is in the sunlight. Therefore, there is a limit to supply stable power to the load and grid. So, the hybrid system including the batteries has been researched in order to overcome these inherent drawbacks[1][2].

Most applications of hybrid-system are for stand-alone to supply local load, such as island and mountainous territory where energy is not enough[3]. However, in household applications, a grid-connected systems, contrary to the stand-alone systems, has several advantages. It supplies power from grid to local load when it generates insufficient power from PV and

battery[4]. Therefore, the grid-connected system is relatively stable and effective to share load as compared with stand-alone type and has the economic efficiency since it can sell power generated by PV when the power rates are respectively expensive. For this reasons, the installation of the grid-connected system are constantly increased. To effectively operate the grid-connected hybrid system, it is necessary to have versatile power transfer with respect to operation of the system. Many power control algorithms have been reported in the literature[5]~[7]. However, most of them are to balance the load and regulate the output power.

The Fig. 1 presents the overall configuration of PV-battery hybrid system with EMS for residential use with EMS. This paper proposes power transfer strategies of PV-battery hybrid system, unlike conventional systems, taking into account the output fluctuation of renewable energy, the load demand, the state of grid and the SOC of the battery in order to apply the EMS and optimize household applications, including peak load period that has expensive power rates. The proposed PTS is to coordinates the four modes so that all requirements are satisfied. The first mode is that PV power generation and battery dis/charge modes are combined, having three kinds of sub-mode. The second mode is battery dis/charge mode and has two sub-modes in detail. The third mode is emergency battery charge mode and operates when the SOC for battery falls to the predefined limit. The fourth mode is uninterrupted power supply (UPS) mode.

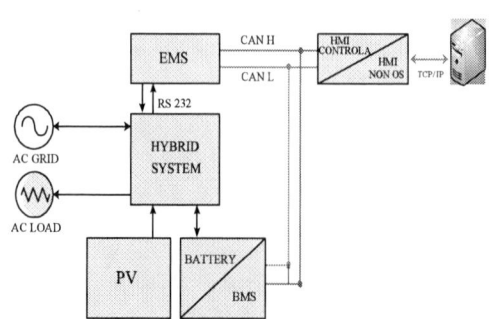

Fig. 1. The block diagram of hybrid system with EMS.

978-1-4799-2706-7/14 $31.00 © 2014 IEEE   409

Section II explains the overall configuration of hybrid-system in this paper. In Section III, the control methods of the system are discussed. Section IV introduces the proposed PTS for hybrid system with EMS. Section V verifies the PTS with experimental results which demonstrate efficient and economical operation of the hybrid system with EMS for residential use.

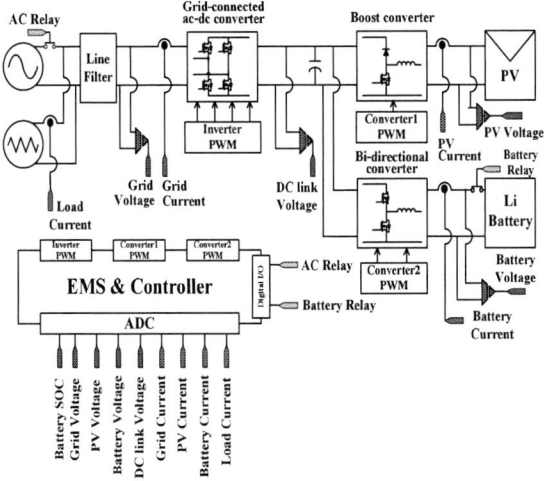

Fig. 2. The overall configuration of grid-connected hybrid system with EMS.

II. CONFIGURATION OF HYBRID SYSTEM WITH EMS

As shown in Fig. 2, the hybrid system consists of grid-connected single-phase ac-dc converter, a boost converter for PV generation and a bi-directional dc-dc converter for battery dis/charge. The single-phase ac-dc PWM converter has single-phase input 220V/60Hz, dc-link output 380V and switching frequency 15kHz. In this system, both output terminal of PV and input terminal of the bi-directional dc-dc converter which charges and discharges battery are used as common dc-bus.

III. CONTROL METHOD OF HYBRID SYSTEM

Fig. 3 shows operation mode and control strategy of the grid-connected hybrid system based on EMS. The EMS determines the system operation by considering the state of power system, amount of PV power generation, load demand and the battery SOC. In addition, it transfers operation modes and reference values to hybrid system. The hybrid system has major four modes. The power generated from PV is used to fill the load demand and transferred to the grid when the surplus power is generated in Mode 1. Also, effective energy management can be achieved by using sub-mode(a)~(c). The Mode 2 is for the operation of bi-directional converter connected to the battery. Similarly, economic efficiency and reliability of the system can be improved by sub-mode(a)~(b), considering SOC of battery and peak load period that has expensive power rates. The emergency battery charge mode secures the minimum capacity of battery in preparation for UPS mode. In UPS mode, energy is independently transferred to the load when grid fault is occurred.

The grid connected ac-dc converter performs phase locked loop by detecting and calculating grid voltage. Also, it carries out power factor control and regulates dc-link voltage for grid connection except UPS mode. The bi-directional converter carries out constant current control with dis/charge current reference transferred through EMS. The boost converter performs maximum power point tracking (MPPT) using perturb and observe (P&O) algorithm when PV power is generated. In UPS mode, the bi-directional converter regulates the dc-link voltage as 380V and the grid connected ac-dc converter performs the output voltage control for 220V/60Hz by operating as an inverter.

IV. PROPOED VERSATILE POWER TRANSFER STRATEGIES

To select the mentioned four modes, it is necessary that output variation of PV, requirements of the load, state of the grid and SOC of battery are considered. Fig. 4 shows

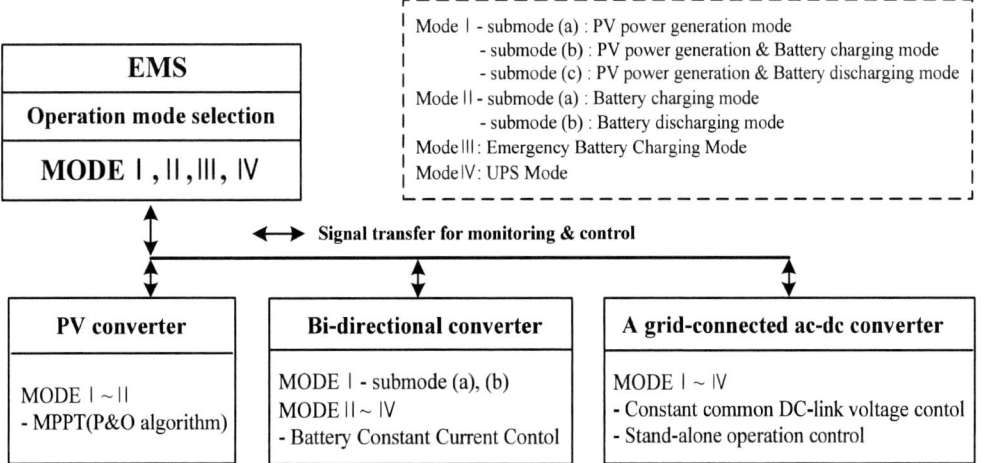

Fig. 3. The operation mode and control strategy of the grid-connected hybrid system with EMS.

proposed PTS considering requirements of the load, output fluctuation of PV and SOC of battery. According to SOC of battery can be separated into three areas. Area 1 indicates the state that the battery is fully charged. Area2 indicates SOC that the battery can be charged or discharged freely. Area3 is the minimum SOC of battery saved for UPS mode. Fig. 5 shows the proposed PTS algorithm diagram of hybrid system with EMS

1) Mode 1-submode (a) $[P_{pv} > P_{Load}, Area3]$: PV power is directly transferred to grid and load. In this mode, amount of PV power generation is larger than power which the load requires because of the high isolation. The generated power transferring to grid can be expressed by

$$P_{grid} = P_{pv} \qquad (1)$$

2) Mode 1-submode (b) $[P_{pv} > P_{Load}, Area2]$: The hybrid system charges the battery with surplus power from PV except load demand. The PV power should be stored in the battery because it is more effective to transfer power in peak load period. The generated power transferred to the grid can be expressed by

$$P_{grid} = P_{pv} - P_{batt} \qquad (2)$$

At this time, the EMS transfers operation mode instructions and calculated battery discharge current reference value. The battery discharge current reference can be obtained by

$$I_{batt(ref)} = \frac{P_{batt}}{V_{batt}} = \frac{P_{pv} - P_{load}}{V_{batt}} \qquad (3)$$

3) Mode 1-submode (c) $[P_{pv} < P_{Load}, Area2, Area3]$: The hybrid system transfers power of PV and battery to the load and gird. In this mode, PV power is smaller than

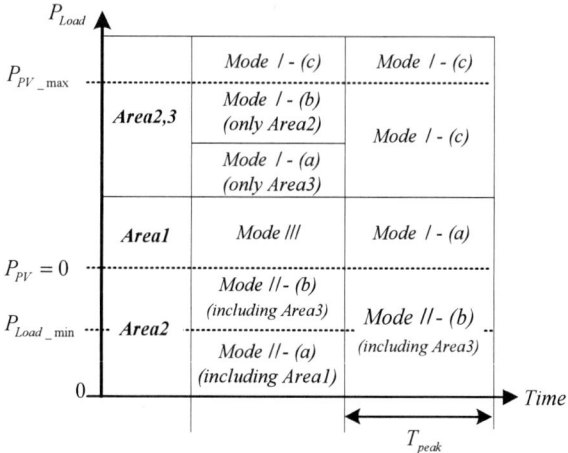

Fig . 4. The proposed PTS for hybrid system with EMS.

the power which the load requires. So, insufficient power should be compensated from the battery. The power transferred to grid can be given in

$$P_{grid} = P_{PV} + P_{batt} \qquad (4)$$

This is operated according to the reference from EMS during T_{peak} even though condition '$P_{pv} < P_{load}$' is not satisfied. The battery discharge current reference can be obtained by

$$I_{batt(ref)} = -\left(\frac{P_{batt}}{V_{batt}}\right) = \frac{\left[\left(P_{pv} + P_{peak}\right) - P_{load}\right]}{V_{batt}} \qquad (5)$$

Where P_{peak} is the power which is transferred to the grid additionally when the power rates are expensive during peak load period.

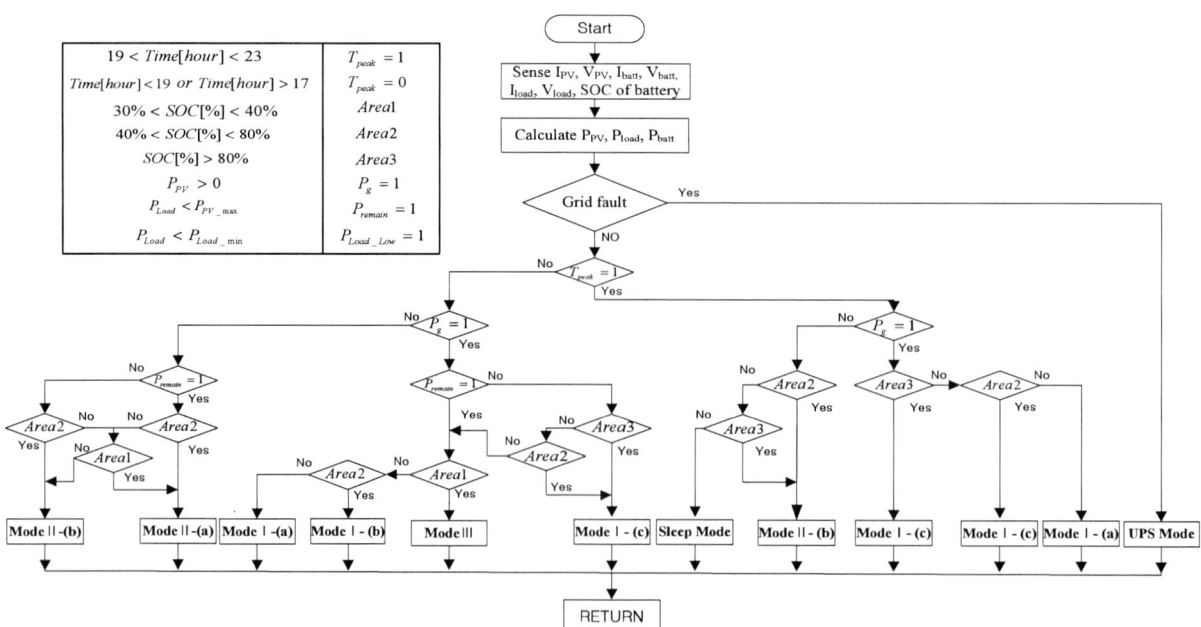

Fig. 5. The proposed PTS algorithm diagram of hybrid system with EMS.

5) Mode 2-submode (a) [$P_{pv} = 0$, $P_{Load} < P_{Load\_min}$, *Area1*, *Area2*]: In this mode, the battery is charged by the grid when load demand is smaller than minimum load power P_{Load_min}. Also, this mode is operated if it is not the time of peak load period. The bi-directional converter charges the battery with rated constant current. The power transferred to grid can be expressed by

$$P_{grid} = -P_{batt} \qquad (8)$$

4) Mode 2-submode (b) [$P_{pv} = 0$, *Area2*, *Area3*] : The hybrid system transfers the battery power to the load without PV power. In addition, in the event of peak load, it generates energy by using the battery power to the grid because the power rates in this time are expensive. The power transferred to grid can be expressed by

$$P_{grid} = P_{batt} \qquad (6)$$

In this mode, the battery discharge current reference can be obtained by

$$I_{batt(ref)} = -\left(\frac{P_{batt}}{V_{batt}}\right) = -\frac{\left(P_{peak} + P_{load}\right)}{V_{batt}} \qquad (7)$$

6) Mode 3 [$P_{pv} > 0$, *Area1*, *Area2*] : In this mode, the battery is charged by the grid when the SOC of battery falls until the predefined limit.

7) Mode 4 [*Area1*, *Area2*, *Area3*] : This mode is uninterrupted power supply mode. In UPS mode, energy is independently transferred to the load when grid fault is occurred.

V. EXPERIMENT

To verify the proposed power flow strategy, the PV-battery hybrid system was built up and the experiments were performed. The EMS represents PV power, grid power, battery power, load power and battery SOC obtained by BMS. Fig. 6 shows the experimental set of hybrid system. The controller of the hybrid system uses digital signal processor (Model : TMS320F28335). PV simulator is used to generate PV power. The battery bank includes four battery modules and one module is composed of 14-cells of battery in series. And, the capacity of battery is 40Ah. The parameters of experimental set are shown in Table I.

TABLE I
EXPERIMENTAL PARAMETERS OF HYBRID SYSTEM

Parameter	Value	Unit
Rated power	3	kW
Grid voltage	220	V
Switching frequency	15	kHz
DC-link voltage	400	V
DC-link Capacitor,	1700	μF
Battery bank Terminal voltage	207.2	V
Battery dis/charge current [Max]	15	A

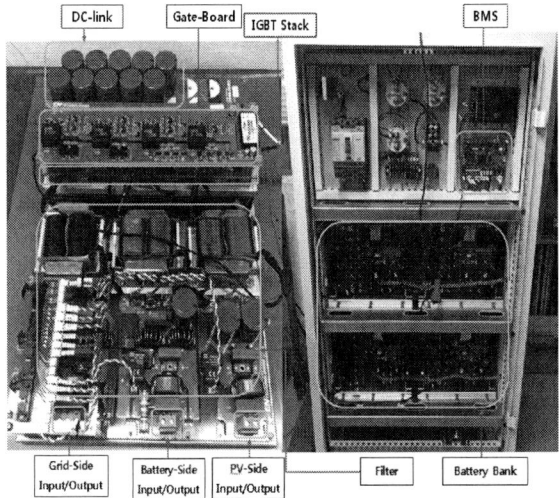

Fig. 6. The experimental set of the hybrid system with BMS.

The system operation of Mode 1-submode (a)~(c) is confirmed by Fig. 7~9. The phase of grid current is opposite to that of grid voltage because the PV power and battery power is transferred into grid shown in Fig. 8~10. Fig. 7 shows that PV power generated by PV simulator of about 3kW is injected into grid. As shown in Fig. 8, PV power of about 3kW is transferred to grid and battery when PV power is larger than load power. The battery has to be charged in order to sell the power in peak load period with relatively expensive power rates. The each power that is delivered into grid and battery is 1.5kW. Fig. 9 shows that the insufficient power is compensated by discharging battery since PV generation power of about 1.5kW is less than load power.

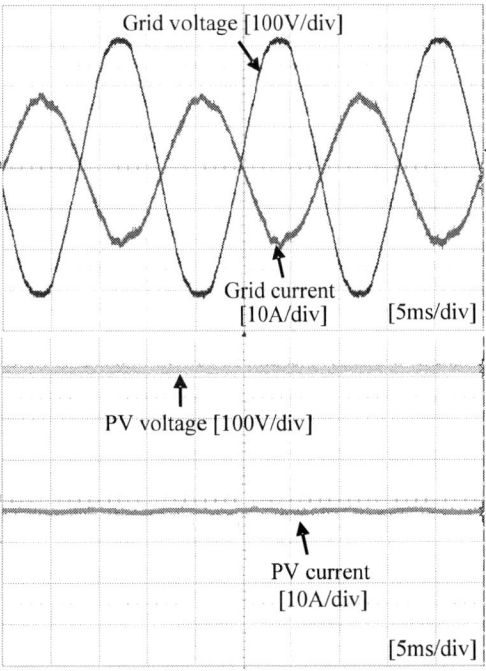

Fig. 7. The experimental results of Mode 1-submode (a).

978-1-4799-2706-7/14 $31.00 © 2014 IEEE

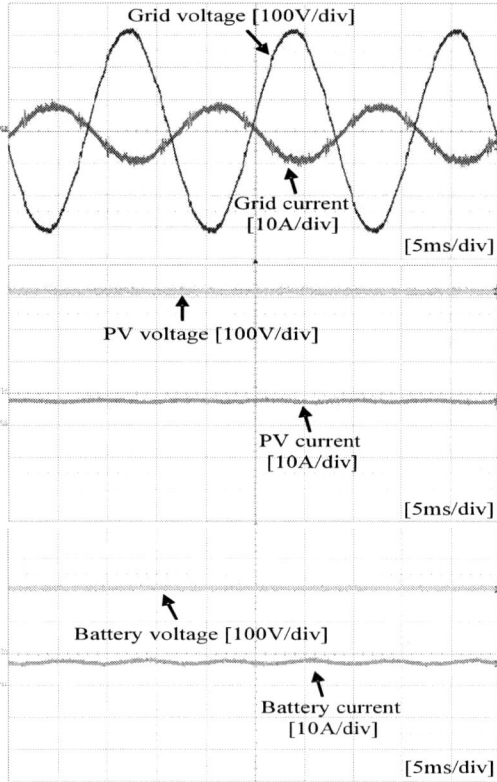

Fig. 8. The experimental results of Mode 1-submode (b).

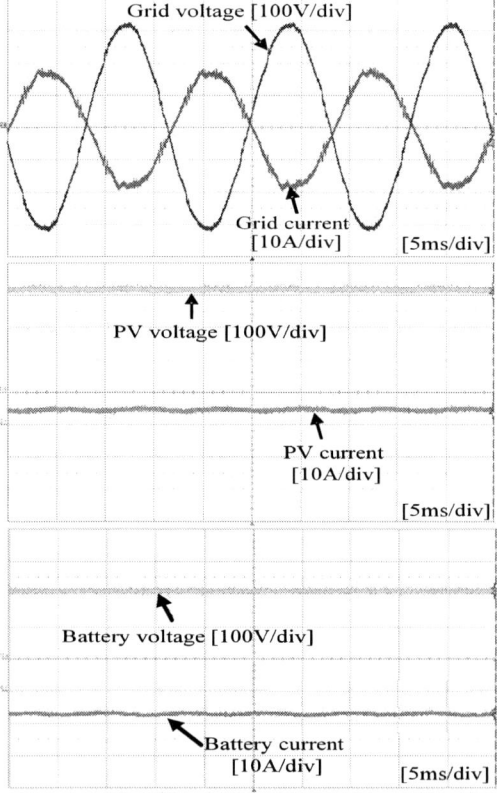

Fig. 9. Experimental results of Mode 1-submode (c).

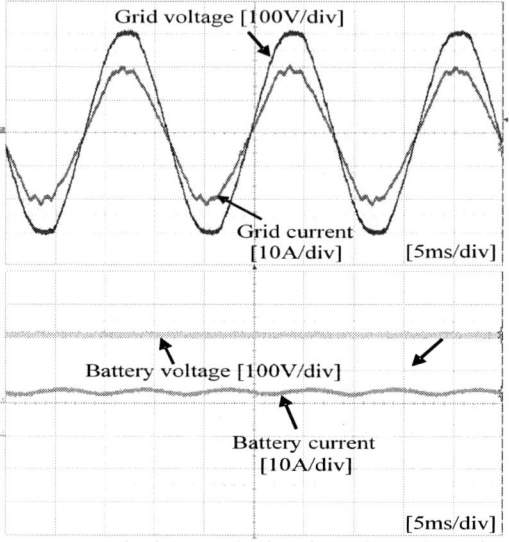

Fig. 10. The experimental results of Mode 2- submode (a).

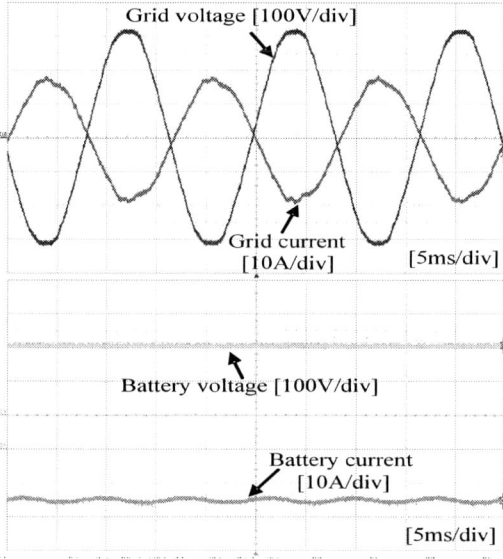

Fig. 11. The experimental results of Mode 2-submode (b).

The system operation of Mode 2 - submode (a)~(b) is confirmed by Fig. 10~11. Also, they show that battery dis/charging current are constant. Fig. 10 shows that the battery is charged with power 3kW at Mode 2 - sub mode (a). In this mode, since PV power is not generated, The energy is supplied from grid if the demand of load is less than predefined minimum load power and the time is not peak load period. Fig. 11 shows that battery of about 3kW is delivered to grid and load at Mode 2-submode (b).

Fig. 12 presents experimental results that present PV generation power, load demand power and dis/charge power per day when the proposed PTS is applied to hybrid system with EMS. It is assumed that peak load period is for T_{peak} = 17~19 hour and minimum load power

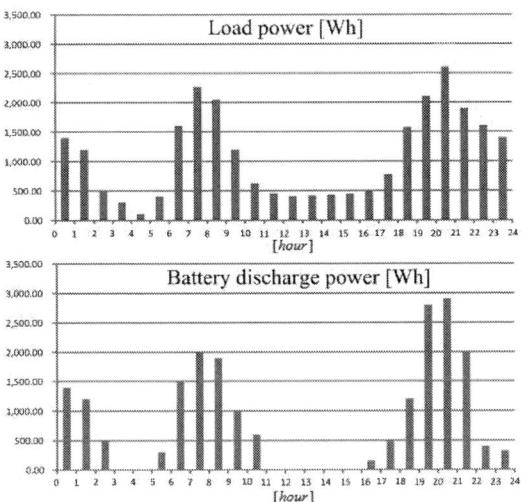
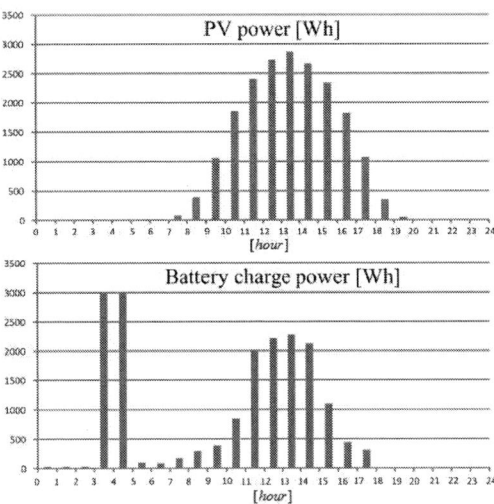

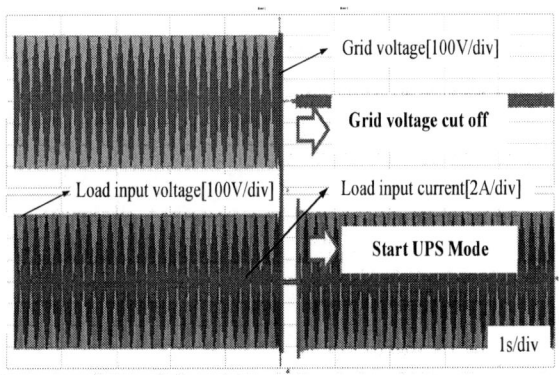

Fig. 12. Experimental results of applying to the proposed PTS.

is at P_{Load_min} = 300W. Fig. 13 shows waveforms of hybrid system when operation of system is changed to UPS mode. If a grid failure occurs, the grid and load are separated. And then, the power corresponding to P_{Load} = 300W is transferred to the load from battery within 300 ms.

Fig. 13. Experimental results of UPS mode.

VI. CONCLUSION

This paper proposes a power transfer strategies of grid-connected hybrid system which takes into account the output fluctuation of renewable energy, the load demand, the state of grid, peak load period and the SOC of the battery. The power flow strategy is optimized for the household load demand. Also, economical effect is considered by charging energy generated by PV to the battery during the daytime and generating the power when power rates are expensive with peak load period, Also, the stability of the system is improved by considering the battery SOC while the minimum battery capacity is guaranteed as readiness for the unexpected UPS mode. The proposed PTS were verified to implement the experiment.

ACKNOWLEDGMENT

This work was supported by Samsung Electronics Co. Ltd.(No. 1000871)

REFERENCES

[1] S. K. Kim, "Dynamic modeling and control of a grid-connected hybrid generation system with versatile power transfer," *IEEE Trans. on Industry. Electronics*, vol.55, no. 4, pp. 1677-1688, Apr. 2008.

[2] H. Kim, "Reconfigurable Solar Converter: A Single-Stage Power Conversion PV-Battery System," *IEEE Trans. on Power Electronics*, vol. 28, no. 8, pp. 3788-3797, Aug. 2013.

[3] B. Belvedere, "A microcontroller-based power management system for standalone microgrids with hybrid power supply," *IEEE Trans. on Sustainable Energy*, vol. 3, no. 3, pp. 422-431, Jul 2012.

[4] D.H. Jang and S. K. Han, "High power density and low cost photovoltaic power conditioning system with energy storage system," in *Proc. IPEC Conf*, 2010, pp. 235-240.

[5] D. H Jang and S. K. Han, "Low Cost High Power Density Photovoltaic Power Conditioning System with an Energy Storage System," *Journal of Power Electronics*, vol.12, no.3, pp. 487-494, 2012.

[6] H. Kim, "Reconfigurable Solar Converter: A Single-Stage Power Conversion PV-Battery System," *IEEE Trans. on Power Electronics*, vol. 28, no. 8, pp. 3788-3797, Aug. 2013.

[7] L. N. Khanh, "Power-Management Strategies for a Grid-Connected PV-FC Hybrid System," *IEEE Trans, on Power Delivery*, vol.25, no.3, pp. 1874-1882, Jul. 2010.

High-efficiency and Cost-minimization Method of Energy Storage System with Multi Storage Devices for Grid Connection

Hitoshi Haga[*1], Toshihiro Shimao[*1], Seiji Kondo[*1], Koji Kato[*2], Youichi Itoh[*2],
Kenji Arimatsu[*3], and Katsuhiro Matsuda[*3]

[*1]Nagaoka University of Technology, [*2]Sanken Electric Co., Ltd, [*3]Tohoku Electric Power Co., Inc.

[*1]Nagaoka-city, Japan, [*2]Kawagoe-city, Japan, [*3]Sendai-city, Japan

hagah@vos.nagaokaut.ac.jp, tshimao@stn.nagaokaut.ac.jp, kondo@vos.nagaokaut.ac.jp,
k.kato@sanken-ele.co.jp, y.ito@sanken-ele.co.jp, arimatsu.kenji.wt@tohoku-epco.co.jp

Abstract— an energy storage system is integrated into a power-grid to compensate for the power fluctuation from renewable energy systems. This paper proposes a high-efficiency and low-cost energy storage system for the power-grid. The proposed system combines multi-energy storage devices such as a LaB and an EDLC for reducing the cost incurred by the system for the complete operation period. This paper also proposes a control method to improve the converter efficiency. The proposed method divides the amplitude of the compensation power reference into the LaB and the EDLC. A 10-kW prototype system has been designed and implemented to validate the proposed architecture and system performance. Experimental results demonstrate that the proposed control method effectively compensated for the power fluctuation from the renewable energy system. Simulation analysis confirms that the proposed system reduces the cost and losses in comparison with the conventional system.

Keywords— *renewable energy , power grid, multi storage devices, high efficiency, cost minimization.*

I. INTRODUCTION

With the growing global focus on environment and the energy resources, renewable energy systems are increasingly being integrated into power grids [1]-[5]. A primary issue in most renewable energy systems such as solar photovoltaic (PV) is inherent to its nature. Therefore, these systems require energy storage capabilities to compensate for the output power fluctuation at all times [6]-[9]. The compensation systems generally interface with the grid through power converters and energy storage devices. The energy storage devices such as lead acid batteries (LaBs) or electric double-layer capacitors (EDLCs) can be used to provide fast frequency regulation, and load following. However, these energy storage devices and power converters are bulky and expensive. In addition, the power loss in power converters is also quite large. The LaB is lifetime-limiting factor of the system. The energy storage system which combines two types of energy

devices is effective for low-cost and small size of the system. In [10], the compensation power is divided at frequency. High frequency component of the compensation power is supplied by EDLC, and low frequency component of the compensation power is supplied by LaB. The system is effective for small size of the system. However, it is necessary for the system to improve efficiency.

This paper proposes a high-efficiency and cost-minimization method for grid-power fluctuation compensation system. The proposed system combines a LaB and an EDLC. The proposed control method divides the amplitude of the compensation power reference, and the divided references provides to the LaB and the EDLC respectively. The proposed method improves the converter efficiency and reduces the cost required by the system for the complete operation period (20 years).

First, this paper discusses the principle of the proposed method. Next, experimental results demonstrate that the proposed control method effectively compensated for the power fluctuation from the PV unit. The experimental results confirm that the proposed method improves the converter efficiency than that of conventional method. A

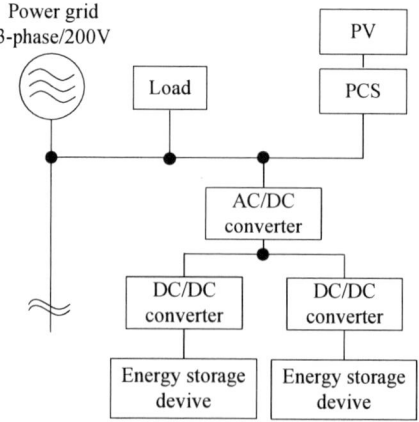

Fig.1. Configuration of the power grid.

Simulation analysis confirms that the proposed system reduces the cost in comparison with the conventional system.

II. PRINCIPLE OF ENERGY STORAGE SYSTEM WITH MULTI STORAGE DEVICES

Fig.1 shows the system configuration. A renewable energy system such as PV and the energy storage system are connected to the three-phase power-grid. The energy storage system is integrated into a power-grid to compensate for the power fluctuation from the renewable energy systems. The energy storage system consists of an AC/DC converter, a bi-directional DC/DC converter, and an energy storage device. This paper assumes two types of energy devices.

Fig.2 and 3 show a principle of compensation method using two types of energy devices. Fig.2 shows the conventional method (frequency division allotment method: FDAM). In the conventional system, the compensation power reference p^{ref} is divided at frequency.

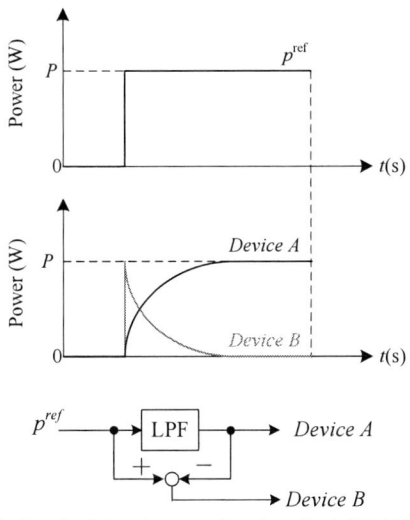

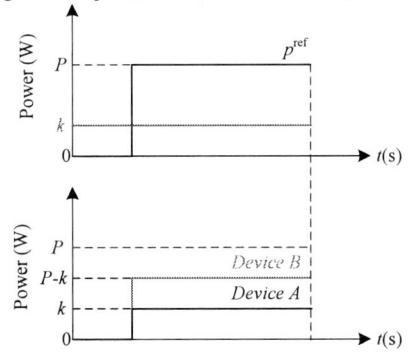

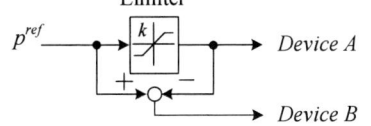

Fig.2. Principle of conventional method (FDAM).

Fig.3. Principle of proposed method (ADAM).

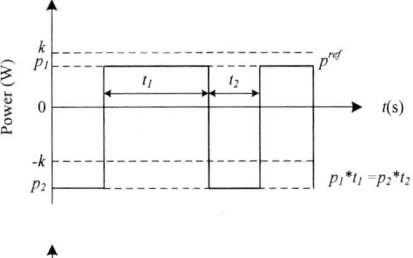

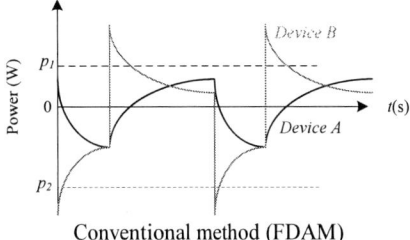

Conventional method (FDAM)

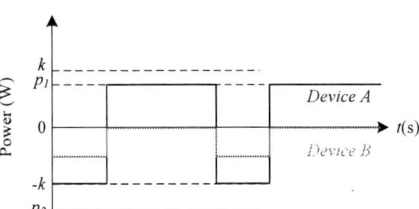

Proposed method (ADAM)

Fig.4. Example of the compensation power reference.

Fig.2 also shows control block diagram for the conventional method. p^{ref} is divided by using low-pass filter. The low frequency component of p^{ref} is supplies to the device A, and high frequency component of p^{ref} is supplied to the device B. In the conventional system, each power converter needs electricity capacity of P [W].

Fig.3 shows the proposed method (amplitude division allotment method: ADAM). In the proposed system, the compensation power reference p^{ref} is divided at amplitude k (W). Fig.3 also shows control block diagram for the proposed method. p^{ref} is divided by using the limiter. The reference which is lower than the limiter level is supplied to the device A. The reference which is higher than the limiter level is supplies to the device B. In the proposed system, the power converter for the A needs electricity capacity of k [W]. The converter for the B needs electricity capacity of $(P-k)$ [W]. The proposed system reduces the rated power of the power converter.

Fig.4 shows an example of the compensation power reference p^{ref}. In Fig.4, the positive polarity shows a charge mode, and the negative polarity shows a discharge mode. In the conventional method, the offset component isn't included in both devices. However, the proposed method includes the offset component in both devices. This means that the proposed method needs management of state of charge (SOC) more.

III. SYSTEM CONFIGURATION

A. Conventional control method (FDAM)

Fig.5 (a) shows a system configuration of the

conventional compensation method using multi-energy storage devices [2]. Fig.5 (b) shows a control block diagram of the conventional method. The LaB and the EDLC are used for the energy storage devices. The compensation power reference is divided between the LaB and EDLC by a filter. The LaB takes the low-frequency component of the compensation power reference, and the EDLC takes the high-frequency component.

B. Proposed control method (ADAM)

Fig.6 (a) shows the principle of the proposed compensation method using multi-energy storage devices.

Fig.6 (b) shows a control block diagram of the proposed method. LaB and EDLC are used for the energy storage devices. The proposed method divides the amplitude of the compensation power reference between LaB and EDLC using the amplitude division allotment method (ADAM). The compensation power reference is divided between LaB and EDLC by an amplitude limiter. The LaB takes the compensation power reference that is smaller than the limiter, while the EDLC takes the higher compensation power reference. Therefore, the DC/DC converter for EDLC works on the condition that the reference is higher than the limiter. The converter loss of the proposed ADAM can be lower than that of the conventional control method. Furthermore, the proposed

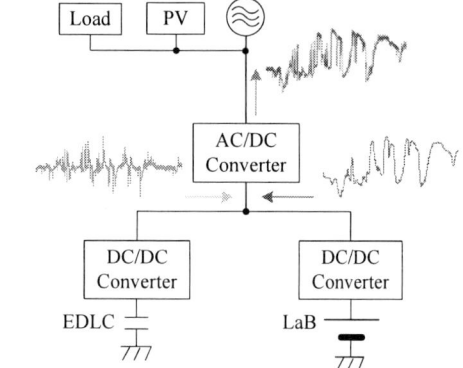

(a) principle of compensation method

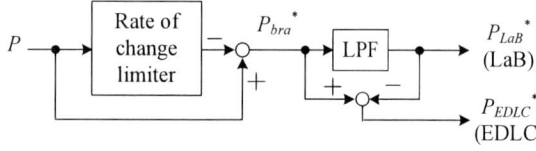

(b) control block diagram
Fig.5. Conventional method.

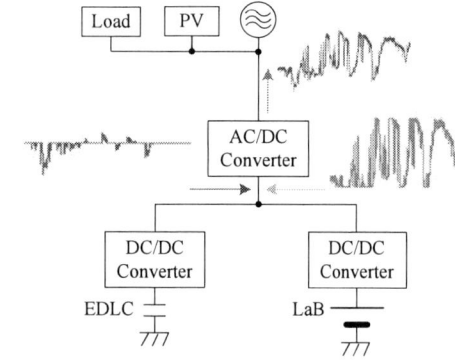

(a) principle of compensation method

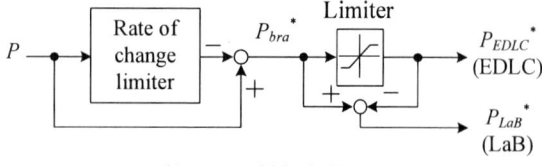

(b) control block diagram
Fig.6. Proposed method.

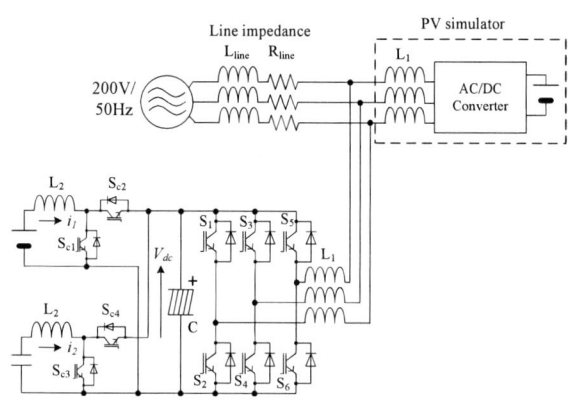

Fig.7. Simulation system.

Table 1. System parameters.

PV rated power	10 kW
Line impedance R_{line}	0.16 Ω
Line impedance L_{line}	0.512mH
AC inductance L_1	2mH
DC link capacitance C	2200uF
Chopper inductance L_2	5.37mH

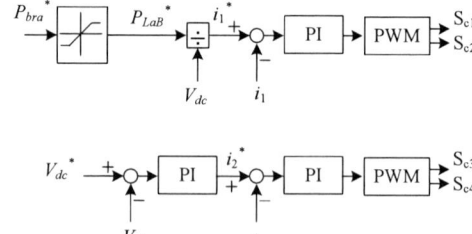

Fig.8. Proposed control block diagram of DC/DC converter.

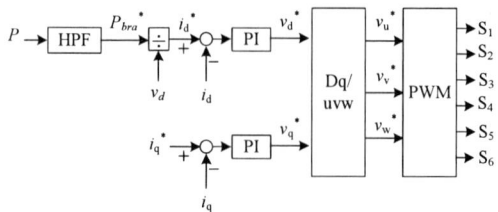

Fig.9. Proposed control block diagram of AC/DC converter.

method reduces the charge/discharge time of the LaB.

IV. SIMULATION AND EXPERIMENTAL RESULTS

A. Proposed control block diagram and Simulation results

First, the present study confirms the validity of the proposed ADAM through simulation.

Fig.7 shows the simulation system. Table 1 shows the system parameters. The PV unit is simulated by the PV simulator in Fig.7, which consists of a three-phase voltage source inverter (VSI) and a voltage source. The output power pattern P is provided from the measurement of the real PV output power. The energy storage system consists of an AC/DC converter and a bi-directional DC/DC converter. The DC/DC converters are connected to the DC-link in parallel.

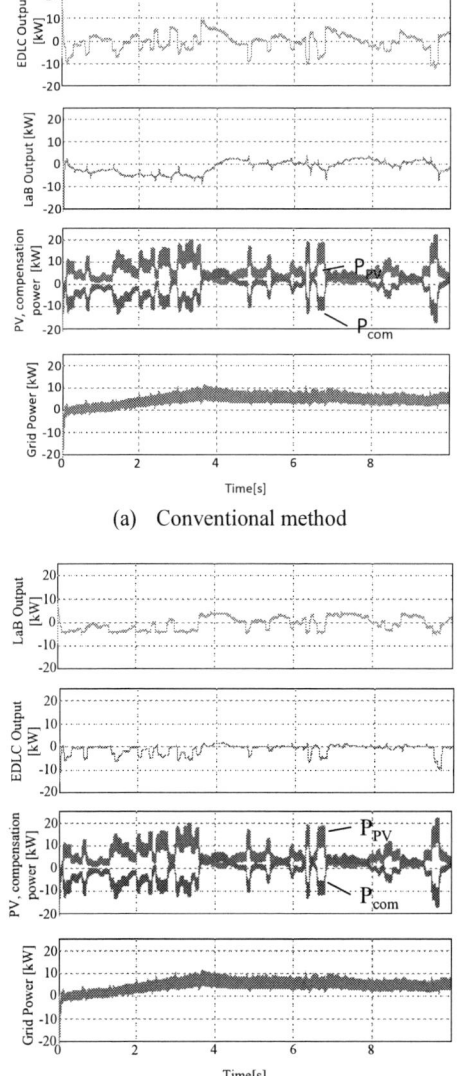

(a) Conventional method

(b) Proposed method
Fig.10. Simulation result.

Fig. 8 shows the proposed ADAM control block diagrams of both the LaB and EDLC energy storage systems. The LaB energy storage system controls the output current i_1 of the DC/DC converter. The current reference i_1^* is obtained by DC-link voltage V_{dc} and the output power reference P_{LaB}^*. The amplitude limiter output of P_{bra}^* is P_{LaB}^*. P_{bra}^* is the compensation power reference, which is obtained by the grid-power and rate of change limiter (limit value is set to 10[W/s]). The EDLC energy storage system controls the both of DC-link voltage V_{dc} and its output current i_2.

Fig.9 shows the control block diagram of the AC/DC converter. The AC/DC converter regulates the inverter output power to compensate for the PV output power fluctuation.

Fig.10 shows the simulation result. Fig.10 (a) shows the result of the conventional method. Fig.10 (b) shows the result of the proposed ADAM. The proposed method compensates for power fluctuation from the PV unit. In the conventional method, the maximum output power of the EDLC is higher than 10 kW. The LaB and EDLC energy storage systems always work to compensate for the power fluctuations. However, the proposed method restricts the maximum output power of the LaB to 4kW (limit value). There is the period when the EDLC does not work to compensate for power fluctuations.

B. Experimental result

Next, this paper confirms the effectiveness of the proposed method through an experiment.

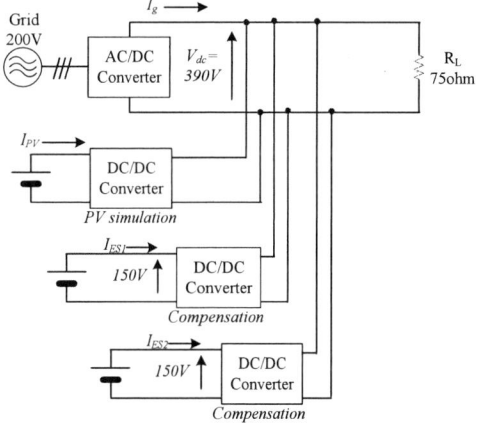

Fig.11. Experimental system.

AC/DC converter   DC/DC converter
Fig.12. Developed power converter.

Fig.11 shows the experimental system. In the experimental system, a PV simulator is connected to the DC-link in the AC/DC converter. The PV simulator consists of a bi-directional DC/DC converter and a DC voltage source. The bi-directional DC/DC converter for the LaB and the EDLC energy storage systems refers to a switched capacitor converter (SCC) in this paper. Fig.12 shows the developed power converter.

Fig.13 shows the experimental result of the conventional method (FDAM). The time constant of LPF in Fig.13 (a) is 10[min]. The time constant of LPF in Fig.13 (b) is 1[min].

Fig.14 shows the experimental result of the proposed method (ADAM). The limit value in Fig.14 (a) was set to 300[W]. The limit value in Fig.14 (b) was set to 150[W]. The LaB takes the compensation power reference which is less than the limiter value. The EDLC takes the compensation power reference which is greater than the limiter value. The proposed method compensates for power fluctuation from the PV unit.

C. Comparisons of the converter losses

Fig.15 shows an experimental result of loss comparison. The result is measured by using Fig.13 and 14. The proposed ADAM control method reduces the loss by 0.3 kWh in comparison with the conventional control

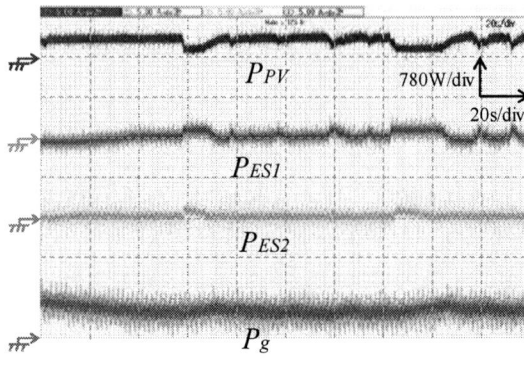

(a) 300W

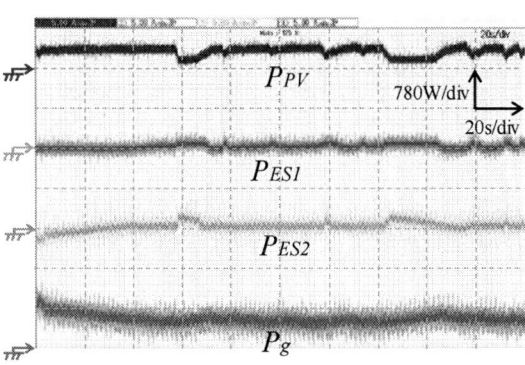

(b) 150W

Fig.14. Experimental result of the proposed method (ADAM)

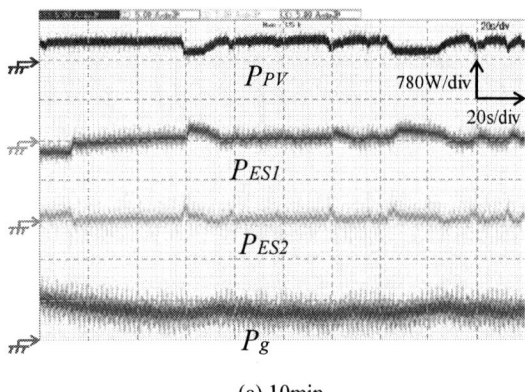

(a) 10min

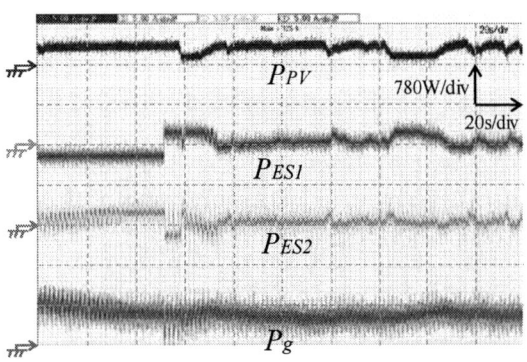

(b) 1min

Fig.13. Experimental result of the conventional method (FDAM)

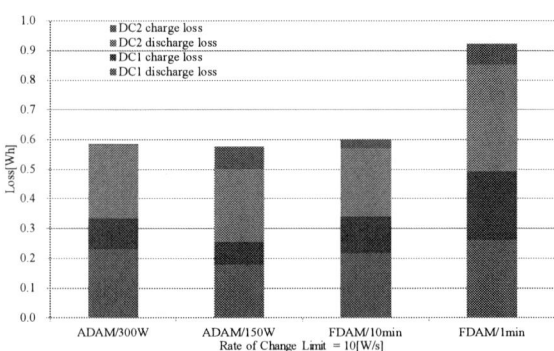

Fig.15. Converter loss comparison.

method.

Fig.16 shows a simulation result of loss comparison. These losses were calculated using data from the PV output power for one month of measurement.

The proposed ADAM control method reduced the converter loss better than the conventional control method because the proposed method controls its output current by using an amplitude limiter. In addition, the low-frequency compensation interfered the high-frequency compensation in the conventional method. Hence, the output compensation power of the conventional method increased more than that of the proposed method. The proposed ADAM control method reduces the loss by 2 kWh in comparison with the

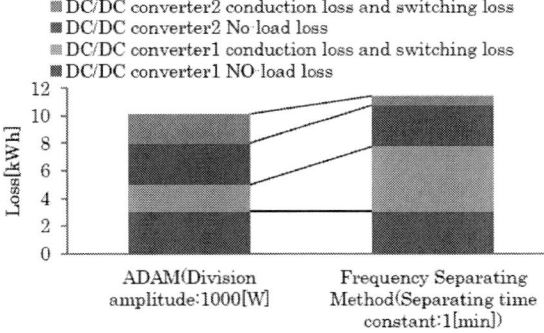

■ DC/DC converter2 conduction loss and switching loss
■ DC/DC converter2 No load loss
■ DC/DC converter1 conduction loss and switching loss
■ DC/DC converter1 NO load loss

Fig.16. Converter loss comparison.

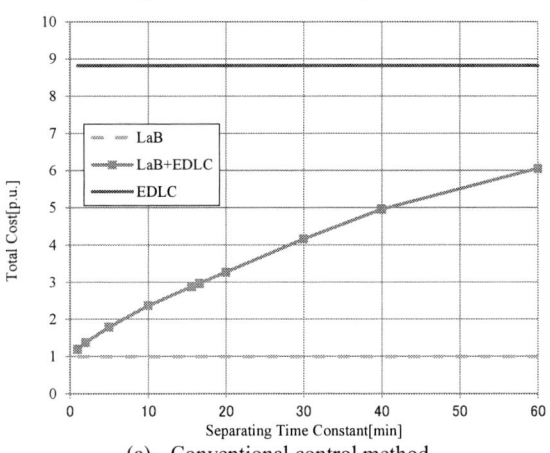

(a) Conventional control method

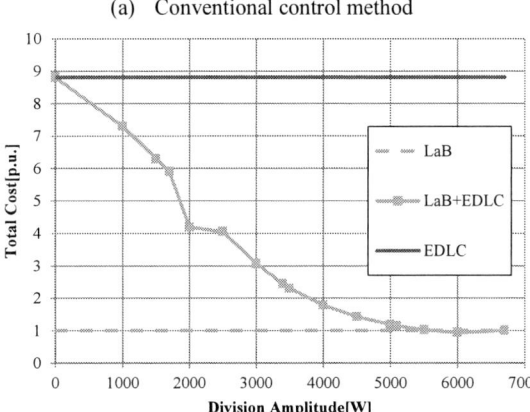

(b) Proposed ADAM

Fig.17. Characteristics of DC/DC converter efficiency.

Table 2. Parameters of LaB and EDLC.

	LaB	EDLC
Cost per kWh [yen/kWh]	50	6300
Cycle limitation [year]	0.521	6300
Rated current [A]	0.1C	Free

conventional control method.

D. Comparisons of the cost

Fig.17 shows the cost calculation results. Table 2 shows the parameters of the energy storage devices that this study used for the cost calculation. These parameters are based on a marketing research result. Fig.17 shows

the total cost considering the replacement time based on the life of the energy storage device. This paper assumes that the operational life of the energy storage system is 20 years. In addition, the assumed fluctuation power pattern from the PV unit is shown in Fig.10.

Fig.17 (a) shows the cost calculation results of the conventional method. The total cost changes for different frequencies. The minimum total cost is shown in the result. Fig.17 (b) shows the cost calculation results of the proposed method. The total cost changes for different limit values. The minimum total cost is also shown in the result. The proposed ADAM has a lower total cost than the conventional method.

V. CONCLUSIONS

This paper proposes a high-efficiency, cost-minimization method of an energy storage system for a grid-power fluctuation compensation system. The effectiveness of the proposed method was proven experimentally in comparison with the conventional method. The proposed control method reduces the loss by 0.3 kWh in comparison with the conventional control method. In addition, the proposed ADAM has a lower total cost than the conventional method.

REFERENCES

[1] F. Blaabjerg, R. Teodorescu, M. Liserre, and A. V. Timbus, "Overview of Control and Grid Synchronization for Distributed Power Generation Systems," *IEEE Trans. On Industrial Electronics*, vol. 53, no. 5, pp. 1398-1409, 2006.

[2] E. Koutroulis and F. Blaabjerg, "Design Optimization of Transformerless Grid-Connected PV Inverters Including Reliability," *IEEE Trans. On Power Electronics*, vol.28, no.1, pp.325-335, 2013.

[3] J. V. Appen, T. Stetz, M. Braun, and A. Schmiegel, "Local Voltage Control Strategies for PV Storage Systems in Distribution Grids," *IEEE Trans. On Smart Grid*, vol. 5, no. 2, pp.1002-1009, 2014.

[4] P. M. Carvalho, P. F. Correia, and L. A. Ferreira, "Distributed reactive power generation control for voltage rise mitigation in dist. networks," *IEEE Trans. Power Syst.*, vol. 23, pp. 766–772, 2008.

[5] R. Tonkoski, L. A. Lopes, and T. H. El-Fouly, "Coordinated active power curtailment of grid connected PV inverters for overvoltage prevention," *IEEE Trans. Sustain. Energy*, vol. 2, pp. 139–147, 2011.

[6] M. J. E. Alam, K. M. Muttaqi, and D. Sutanto, "Mitigation of Rooftop Solar PV Impacts and Evening Peak Support by Managing Available Capacity of Distributed Energy Storage Systems," *IEEE Trans. On Power Systems*, vol.28, no.4, pp.3874-3884, 2013.

[7] A. Canova, L. Giaccone, F. Spertino, and M. Tartaglia, "Electrical impact of photovoltaic plant in distributed network," *IEEE Trans. Ind. Appl.*, vol. 45, pp. 341–347, 2009.

[8] R. A. Walling, R. Saint, R. C. Dugan, J. Burke, and L. A. Kojovic, "Summary of distributed resources impact on power delivery systems," *IEEE Trans. Power Del.*, vol. 23, pp. 1636–1644, 2008.

[9] M. Takagi, Y. Iwafune, K. Yamaji, H. Yamamoto, K. Okano, R. Hiwatari, and T. Ikeya, "Economic Value of PV Energy Storage Using Batteries of Battery-Switch Stations," *IEEE Trans. On Sustainable Energy*, vol.4, no.1, pp.164-173, 2013.

[10] S. Suzuki, S. Maeshima, K. Morino, E. Shimoda, H. Sugihara, and T. Esaki, "Site Test in Micro-Grid into which a large amount of PV power generation system are introduced," *The 2008 Annual Conference of Power and Energy Society, IEEJ*, no. 166, pp. 11-12, 2008.

The 2014 International Power Electronics Conference

Bidirectional DC-DC Converter with Multiple Switched-Capacitor Cells

Yuang-Shung Lee[1]

Department of Electrical Engineering
Fu Jen Catholic University
New Taipei City, Taiwan
002877@mail.fju.edu.tw[1]

Hsin-Wei Huang[2], Tzu-Han Chou[3]

Department of Electrical Engineering
Fu Jen Catholic University
New Taipei City, Taiwan
david.sumida@gmail.com[2], 401216250@mail.fju.edu.tw[3]

Abstract-A new bidirectional DC-DC converter with multiple switched-capacitor (SC) cells is presented in this paper. This device provides voltage conversion ratios from double-mode versus half-mode to n-modeversus 1/n -mode by switching the solid-state relay(SSR) into various numbers of switched-capacitor cells and power MOSFET switches. In addition, the output inductor proposed to construct the voltage conversion ratio of the converter can be adjusted by controlling the switch duty ratios. The converter working principle is described in the detailedconverter topology analysis. Simulation and experimental results are carraied out verify and validate the proposed converter performance in different operation modes.

Keywords- Bidirectional converter, switch capacitor converter, soft switching, synchronous rectification.

I. INTRODUCTION

With renewable system development, energy storage systems research has become critical issue in recent years. The bidirectional DC-DC converter along with energy storage has become a major option for many power conversion systems, such as hybrid vehicles, solar energy storage systems and so on[1-3]. Clean energy resources such as photovoltaic arrays and wind turbines have recently been exploited for developing renewable electrical power generation systems. The bidirectional DC-DC converter is usually used to transfer the solar energy into a capacitive energy source. A switched-capacitor based bidirectional DC-DC converter providing step-down / step-up voltage capability is proposed in [4-7]. Battery equalization schemes where the stronger energy of series connected battery system is transferred into the weaker energy subsystem using a bidirectional power flow control scheme were discussed in [8-11].There are many practical applications such as vehicles, renewable power systems and DC backup energy systems requiring a high voltage transfer ratio bidirectional DC-DC converters. In this respect, multiple switched-capacitor cell technology designs are employed to increase the voltage conversion ratio and power rating of the bidirectional power converters for renewable energy storage applications. The multiple SC bidirectional converters offer a tempting solution for the applications mentioned above when different voltage levels have to be matched [12-16]. In addition, soft switching techniques such as zero-voltage switching

(ZVS) and Zero-current switching (ZCS) are designed to increase the efficiency by decreasing the switching losses [17-18].The proposed bidirectional DC-DC converter includes five switches, five capacitors and an inductor designed to operate in different conduction modes. Simulation and experimental results verify the effectiveness of the proposed converter with different mode switched-capacitor sunder wide output voltage range, high voltage gain and higher efficiency due to the ZVS soft-switching and synchronous rectification designs

II. CONVERTER TOPOLOGY ANALYSIS

A. Continuous conduction mode

Figure 1 shows the proposed new switch-capacitor bidirectional DC-DC converter circuit. Different from the traditional switched-capacitor DC-DC converter, the inductor is connected to the output stage of the converter in order to control the voltage conversion ratio using the switching duty ratio. The low pass filter is composed of L and C_s for filtered the high frequency switching noises of the converter. Switches Q_{N1} and Q_{P1} can control the forward power flow from the source voltage to the output voltage. Switches Q_{N2}, $Q_{P2}$and Q_{P3} can control the reverse power flow from the output voltage to the input source voltage. The Q_{N2}, D_{N2}, Q_{P2}, D_{P2}, Q_{P3}, and D_{P3} are designed into a synchronous rectifiers in order to reduce the conduction losses for improved the converter efficiency.

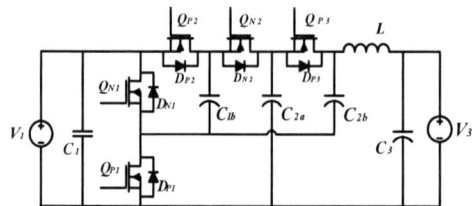

Fig. 1Switch-capacitor bidirectional DC-DC converter

Figure 2 shows the typical waveforms of the proposed converter. Figures 3(a)-(d) show the equivalent circuit for the proposed switch-capacitor DC-DC converter with forward power flow control. Switched-capacitors $C_{1b} = C_{2b}$, and $C_{2a}$is large enough to store energy to maintain constant voltage. The operating principle in continuous inductor current conduction mode(CCM) is described in detail as follows:

978-1-4799-2706-7/14 $31.00 © 2014 IEEE   421

Stage 1[$t_0 \leqq t < t_1$]

Switch Q_{P1}, Q_{P2} and Q_{P3} are turned on and Q_{N1}, Q_{N2} are turned off, as shown in Fig.3(a). The C_{1b} is charged by the input voltage V_1 and C_{2b} is charged by the C_{2a}. The state equations in this stage can be expressed as:

$$L\frac{di_{L1}}{dt} = v_{c2b} - v_3 \cong 2V_1 - V_3 \tag{1}$$

$$v_{c2b} = v_{c2a} \cong 2V_1 \tag{2}$$

$$v_{c1b} = V_1 \tag{3}$$

Stage 2 [$t_1 \leqq t < t_2$]

All switches are turned off as shown in Fig 3(b). In this transient stage, the diode D_{P1}, D_{P2}, and D_{P3}still turn on because of the inductive effect. Therefore, the switches Q_{P1}, Q_{P2} and Q_{P3}canbe designed to turn off at ZVS in this stage.

Stage 3 [$t_2 < t < t_3$]

Switch Q_{N1} and$Q_{N2}$is turned on and switches Q_{P1}, Q_{P2}, Q_{P3} are turned off as shown in Fig.3(c). In this interval the input voltage V_1 is series connected with the switch-capacitor C_{1b}, C_{2b} and charged to output inductor L. The state equations in this stage can be described as:

$$L\frac{di_{L1}}{dt} = V_1 + v_{c2b} - v_3 \cong 3V_1 - V_3 \tag{4}$$

$$v_{c2a} = V_1 + v_{c1b} \cong 2V_1 \tag{5}$$

Stage 4[$t_3 < t < t_4$]

All switches are turned off in this transient stage as shown in Fig 3(d). Diode D_{P1}, D_{P2}, and D_{P3} are turned on by the inductive effect. This stage ends at t_4 when switches Q_{P1}, Q_{P2} and $Q_{P3}$are turned on at ZVS, and returns to the initial condition for the next period.

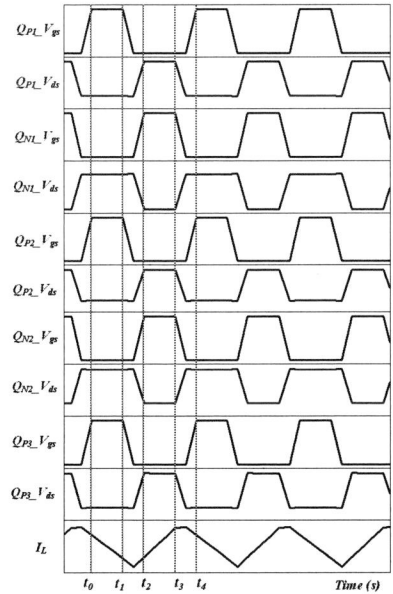

Fig. 2 Typical waveforms under forward power flow

According to the voltage-seconds balance relationship on the output inductor, we can derive the forward voltage conversion ratio as:

$$\langle v_L \rangle = \frac{1}{T_s}\left[\int_0^{DT_s}(3V_1 - V_3)\,dt + \int_0^{DT_s}(2V_1 - V_3)\,dt\right] \tag{6}$$

$$M_F = \frac{V_3}{V_1} = 2 + D \tag{7}$$

Where D is the Q_{N1} and Q_{P1} duty ratio, which can beranged in $0 \leqq D \leqq 1$. And it is called second-mode switched-capacitor bidirectional converter.

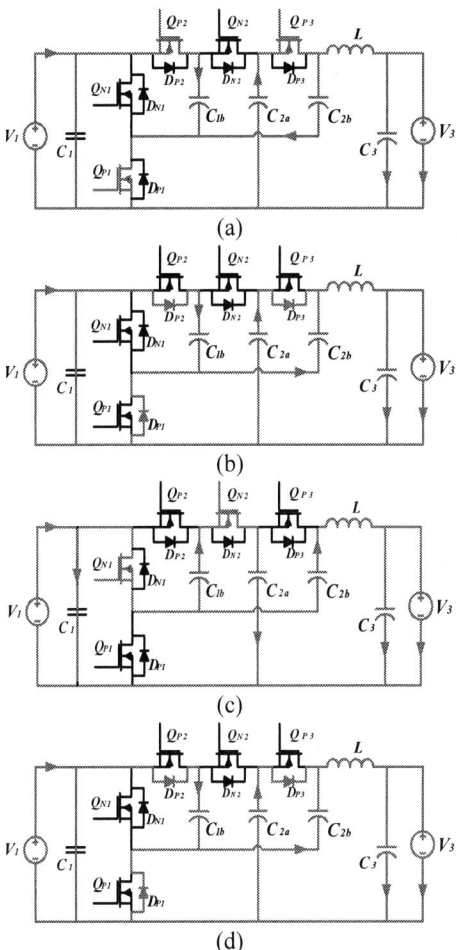

Fig. 3Topological modes for different intervals under forward power flow (a)$t_0 \sim t_1$(b)$t_1 \sim t_2$ (c) $t_2 \sim t_3$ (d) $t_3 \sim t_4$

In addition, it can also operate at reverse power flow control. Figure 4 shows the typical waveforms under reverse power flow of the proposed converter.Figs.5 (a)-(d) shows the equivalent circuit for the proposed switch-capacitor DC-DC converter with reverse power flow control. The analysis for the equivalent circuit is the same as that for the forward power flow control. The operating principle is described in detail as follows:

Stage 1[$t_0 \leqq t < t_1$]

The switches Q_{P1}, Q_{P2} and $Q_{P3}$are turned on, switches Q_{N1}, and Q_{N2} are turned off as shown in Fig. 5(a). The state equations in this stage can be depicted as:

$$L\frac{di_{L1}}{dt} = V_3 - v_{c2b} - \cong V_3 - 2V_1 \tag{8}$$

$$v_{c2b} = v_{c2a} \cong 2V_1 \tag{9}$$

$$v_{c1b} = V_1 \qquad (10)$$

Stage 2 [$t_1 \leqq t < t_2$]

Switches Q_{N1} and Q_{N2} are still turned off and switches Q_{P1}, Q_{P2} and Q_{P3} are turning off in this transient stage as shown in Fig. 5(b). At the same time, diodes D_{N1}, and D_{N2}, are turning on due to the inductive effect. It makes switches Q_{N1} and Q_{N2} can be operated at ZVS turn on.

Stage 3 [$t_2 \leqq t < t_3$]

During this interval, switches Q_{N1} and Q_{N2} are turned on and switches Q_{P1}, Q_{P2} and Q_{P3} are still turn off as shown in Fig. 5(c). The state equations in this stage can be expressed as:

$$L\frac{di_{L1}}{dt} = V_1 - v_{c2b} + V_3 \cong V_3 - V_1 \qquad (11)$$

$$v_{c2a} = V_1 + v_{c1b} \cong 2V_1 \qquad (12)$$

Stage 4 [$DT_s \leqq t < T_s$]

Diodes D_{N1} and D_{N2} are turning on when the switches Q_{N1} and Q_{N2} are turning off and switches Q_{P1}, Q_{P2} and Q_{P3} still turned off in this transient stage as shown in Fig. 5(d). This stage ends at t_4 and the switches Q_{N1} and Q_{N2} are turned on at ZVS, and returns to the initial condition for the next period.

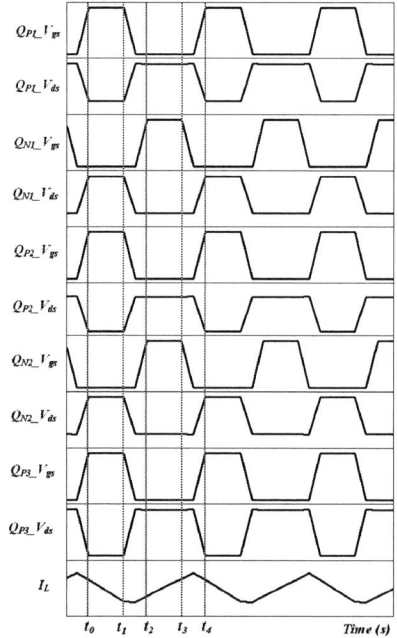

Fig.4 Typical waveforms under reverse power flow

According to the voltage-seconds balance relationship on the inductor, the reverse voltage conversion ratio can be derived as shown in the reciprocal expression of M_F as:

$$M_R = \frac{V_1}{V_3} = \frac{1}{2+D} \qquad (13)$$

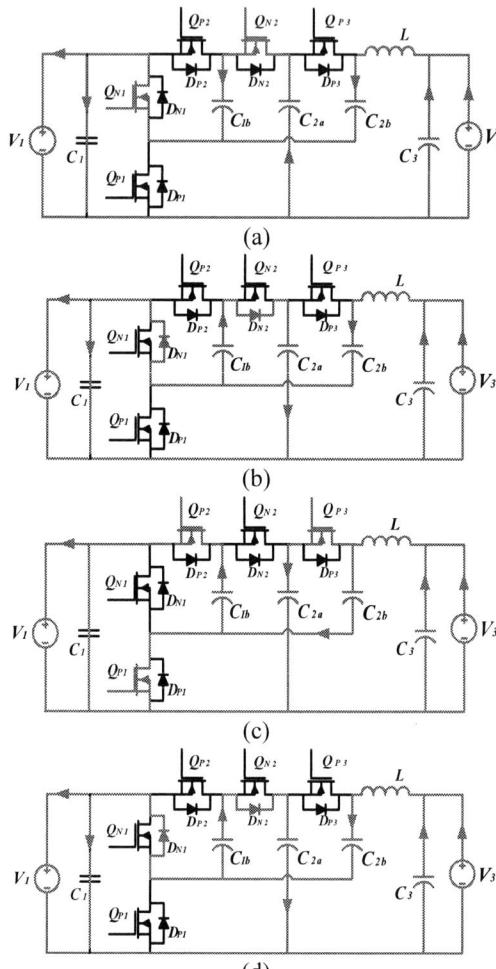

(a)

(b)

(c)

(d)

Fig. 5 Topological modes for different intervals under reverse power flow (a)$t_0 \sim t_1$ (b)$t_1 \sim t_2$ (c) $t_2 \sim t_3$ (d) $t_3 \sim t_4$

B. Discontinuous conduction mode

The inductance design and output ripple current calculation are described for the forward mode in this subsection. The forward conversion equivalent schematic can be simplified as shown in Figs.6 (a)-(c). According to the inductor voltage and current timing diagram shown in Fig. 7, the maximum inductor voltage V_L^{+max} and the minimum inductor voltage V_L^{-max} can be expressed as (14) and (15):

$$V_L^{+max} = L\frac{di_L}{dt} = 3V_1 - V_3 \qquad (14)$$

$$V_L^{-max} = L\frac{di_L}{dt} = 2V_1 - V_3 \qquad (15)$$

The maximum inductor voltage I_L^{+max} can be calculated as (16):

$$i_L^{+max} = \frac{1}{L}(1 - D)V_1DT_s \qquad (16)$$

From equation (16) we can derive the output ripple current in boundary conduction mode as:

$$\Delta i_L = i_L^{+max} = \frac{1}{L}(1 - D)V_1DT_s$$

$$(17)$$

Fig. 7 shows the inductor voltage and current timing diagram in discontinuous conduction mode (DCM) and boundary conduction mode (BCM). While in DCM, the inductor current will be equal to 0 in the negative half-cycle period. According to the vol-second balance, the voltage conversion ratio in discontinuous mode can be derived as (18):

$$M_{DCM} = \frac{V_3}{V_1} = \frac{2D_L + 3D}{D + D_L} \tag{18}$$

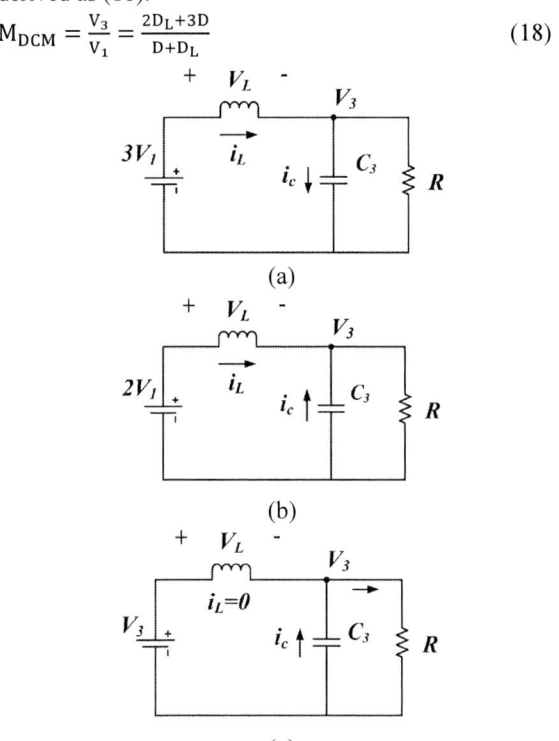

Fig. 6 Simplified cycle schematic under forward power flow
(a) $0 \leqq t < DT_s$ (b) $DT_s \leqq t < (D+D_L)T_s$ (c) $(D+D_L) T_s \leqq t < T_s$

Where D_L is the duty between the peak inductor current and first decreased to zero states. The boundary conduction mode inductor current will equal exactly 0 in the negative half cycle. At the same time, the output current is equal to the steady state inductor current. Therefore, we can know that the current ripple is equal to twice the output current, and we can use this design to meet the inductance value needs. The steady state inductor current ripple for boundary conduction is $2I_o$, and the output current and voltage are $I_o = V_3/R$ and $V_3 = (2+D)V_1$. Substituting these values into equation (18) will obtain the normalized critical time constant in BCM:

$$\tau_{crit} = \frac{L}{RT_s} = D(1-D)/(2+D) \tag{19}$$

And the voltage conversion ratio can be derived as:

$$M_{DCM} = \frac{V_3}{V_1} = \frac{3D}{\tau_{crit}+D} \tag{20}$$

The verified inductor current ripple boundary curve with respect to the duty ratio between the discontinuous conduction mode and continuous conduction mode for the specified parameters of the proposed converter is shown in Fig. 8.

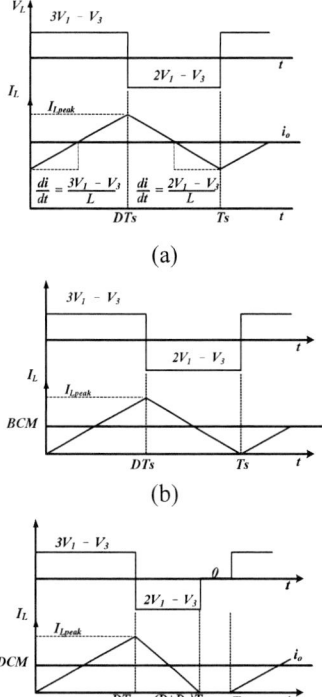

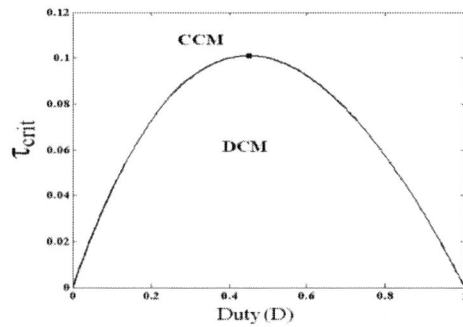

Fig. 7 Inductor voltage and current timing diagram in(a) CCM (b)BCM (c) DCM

Fig. 8 Ripple current with respect to duty ratio in boundary conduction mode

III. EXTENDED PROPOSED CONVERTER

The proposed switch-capacitor bidirectional DC-DC converter can be extended into the design shown in Fig. 9 by increasing the number of switched-capacitor cells, where the SSR S.W. is the selected switch using the solid-state relay (SSR). The converter can adjust different voltage conversion ratio when switching the manual switch. The voltage conversion ratios can be expanded from 1+D(first-mode, $V_5 = (1 + D)V_1$) to 5+D (fifth-mode, $V_5 = (5 + D)V_1$), and their reciprocal function under the reverse power flow control schemes. In addition, the multi-phase implementation in the proposed conversion structure to reach higher output power and reduce current ripple. This design uses a staggered approach that provides the same control switching frequency with interleaved technology. Figure 10 shows the proposed multi-phase switch-capacitor bidirectional DC-DC converter with second-mode switched capacitor cell.

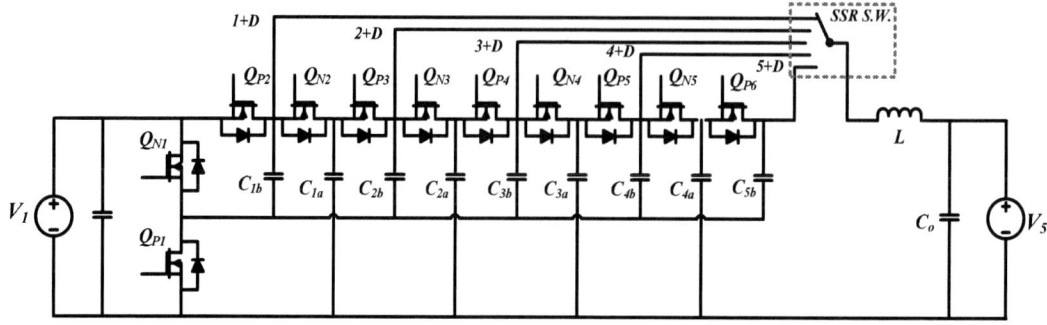

Fig. 9Extended multiple switch-capacitor cells bidirectional DC-DC converter with SSR switches

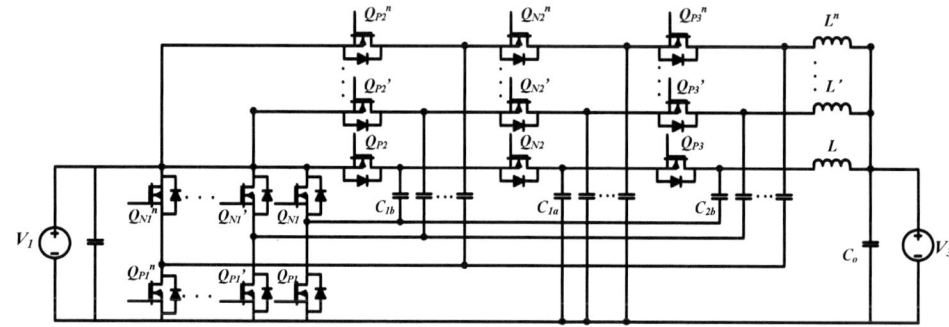

Fig. 10Extended multi-phase switch-capacitor cells bidirectional DC-DC converter

IV. SIMULATION & EXPERIMENT

Simulations and experiments were carried out on the second-mode and fifth-mode switch-capacitor bidirectional DC-DC converter with and without synchronous rectification to verify the performance of proposed converters. The implemented prototype circuit parameters are listed as follows: 40 (V) input voltage, 98.7 (V) output voltage, 100-250 W output power and 155 kHz switching frequency. Switched-capacitors C_{1b}=C_{2b}=C_{3b}=10μF. C_{2a} =100 μF. The output capacitor is selected100 μF. The output inductor is 200μH. MOSFET switches are IRF4310, Schottky diode are SBR4060PT. In addition, the extended circuit parameters are listed as follows: 55 (V) input voltage, 299.75 (V) output voltage, 100-300 W output power and 155 kHz switching frequency. Switched-capacitors C_{1b}= C_{2b}= C_{3b} =C_{4b} =C_{5b} = 10μF. C_{2a}=C_{3a}=C_{4a}=C_{5a}=100μF.The output capacitor is chosen100 μF. The input inductor is 10 μH, output inductor is 200μH. MOSFET switches are IRF3710S, Schottky diode are SBR4060PT. Figs. 11 (a)-(b) show the simulation results for the proposed converter under the forward and reverse power flow, respectively. In forward power flow, we used only the MOSFET Q_{N1} and Q_{P1}, where the duty ratios D_{QN1} = D_{QP1} = 0.45.In the reverse power flow, the employed MOSFET were Q_{N2}, Q_{P2} and Q_{P3}, where the duty ratios D_{QN2} = D_{QP2} =D_{QP3} = 0.45. The simulation output voltage is V_1 = 40(V) and V_3 =98.7 (V) under forward and reverse power flow control schemes, respectively. Figs. 12(a)-(b) show the simulation results for the fifth-mode converter. The simulation output voltage is V_1 = 55(V) and V_5 =299.8 (V). While the switch-capacitor cells increase in fifth-

mode, the efficiency will decrease at the same time. For the reason, the MOSFET driver with synchronous rectification to maintain efficiency becomes necessary. It can be seen the ZVS soft-switching performance on the main switches.

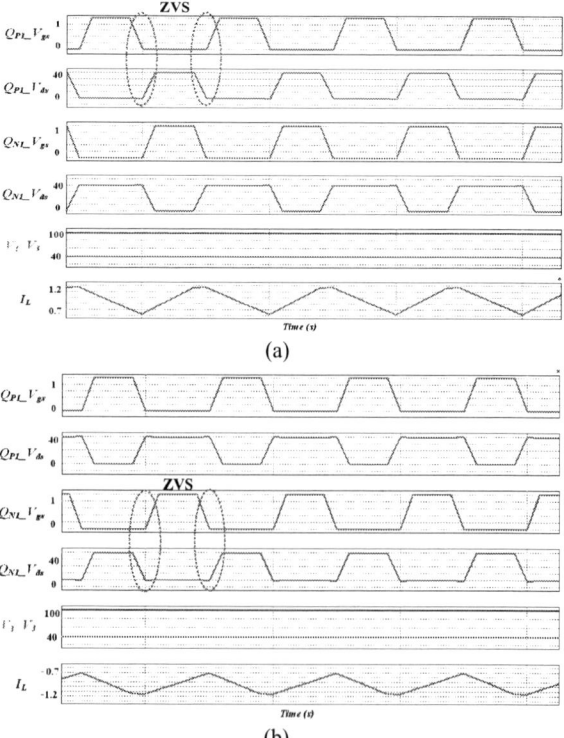

Fig. 11 Simulation waveforms (a) forward (b) reverse power flow

The 2014 International Power Electronics Conference

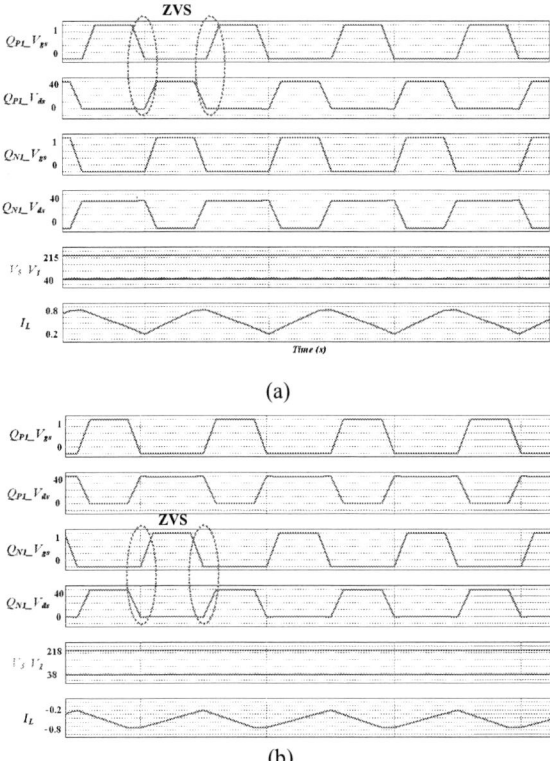

(a)

(b)

Fig. 12 Simulation waveforms under extended converter (a) forward (b) reverse power flow

Fig.13 shows the corresponding experimental results under forward power flow without and with synchronous rectification. In Fig. 13(a), we only drive the MOSFET Q_{P1} and Q_{N1} in forward power flow. Current just flow through the body diode in others MOSFET. The measured input and corresponding output voltages are $V_1 = 30$ (V) $V_3 = 68.7$ (V) and $V_1 = 30$ (V) $V_3 = 70.8$ (V)shown in Figs. 13 (a) and (b), respectively. It can be seen that the transient spike during switching in Fig.13 (a) is higher than Fig. 13 (b), it cause the output voltage and efficiency lower. Fig. 14shows the corresponding experimental results under reverse power flow. The inductor current is reversed. Different from forward power flow, the MOSFET we drive Q_{P2}, Q_{N2} and Q_{P3} in Fig. 14 (a). The measured input and corresponding output voltages are $V_3= 70$ (V)$V_1= 28.8$ (V) and $V_3= 70$ (V)$V_3 = 25.6$ (V) shown in Figs. 14 (a) and (b),respectively. The main switches are verified to operate at ZVS soft-switching.

Figs. 15(a)-(c) show the corresponding experimental results under forward power flow of the proposed converter operated at the different output power for 60W (DCM),86W (BCM) and 150W (CCM), respectively. At the same time, the measured input and corresponding output voltages are $V_1 = 55$ (V)$V_3 = 308.7$ (V)while the output power is 60 W operated at discontinuous conduction mode, the output is higher than the theoretical output voltage due to the DCM effect; $V_1 = 55$ (V)$V_3 = 293.8$ (V)while the output power is 86 W operated at the boundary conduction mode; and $V_1 = 55$ (V)$V_3 = 276.8$ (V) while the output power is

150Woperated at continuous conduction mode. These values are smaller than the specified output voltage due to the ESR voltage drop in the converter loop. Figs. 16 (a)-(b) show the converter efficiencies for different load conditions under forward and reverse power flow. The blue line and red line are the efficiencies of the converter with and without synchronous rectification, respectively. In the forward power flow, the maximum efficiency exceeds 95.8 %, and the average efficiency is about 92.8 %. The maximum efficiency in reverse power flow control scheme is 95.5% lower than that is in the forward power flow scheme due to the increased power losses of the power semiconductor switches.

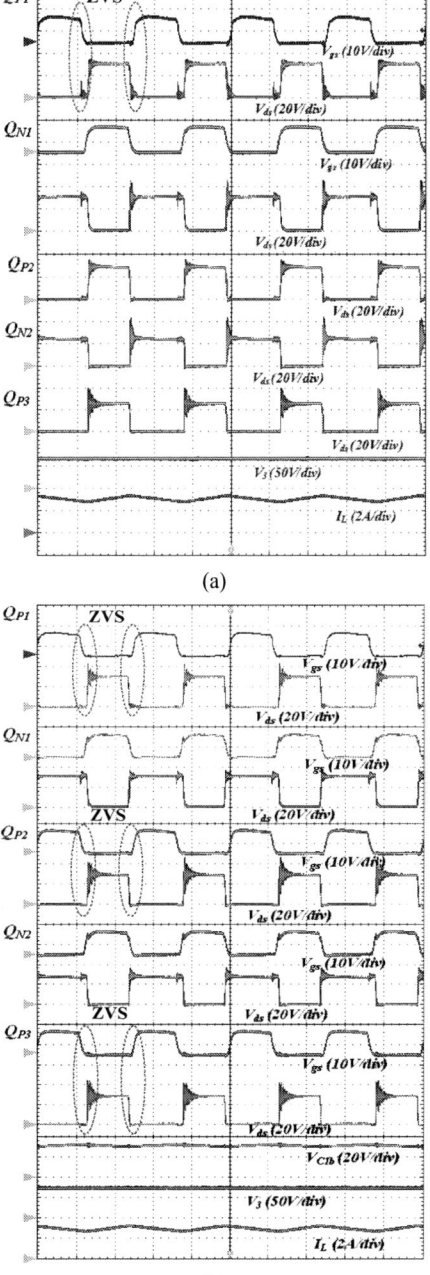

(a)

(b)

Fig. 13 Experimental results for forward converter (a) without (b) with synchronous rectification

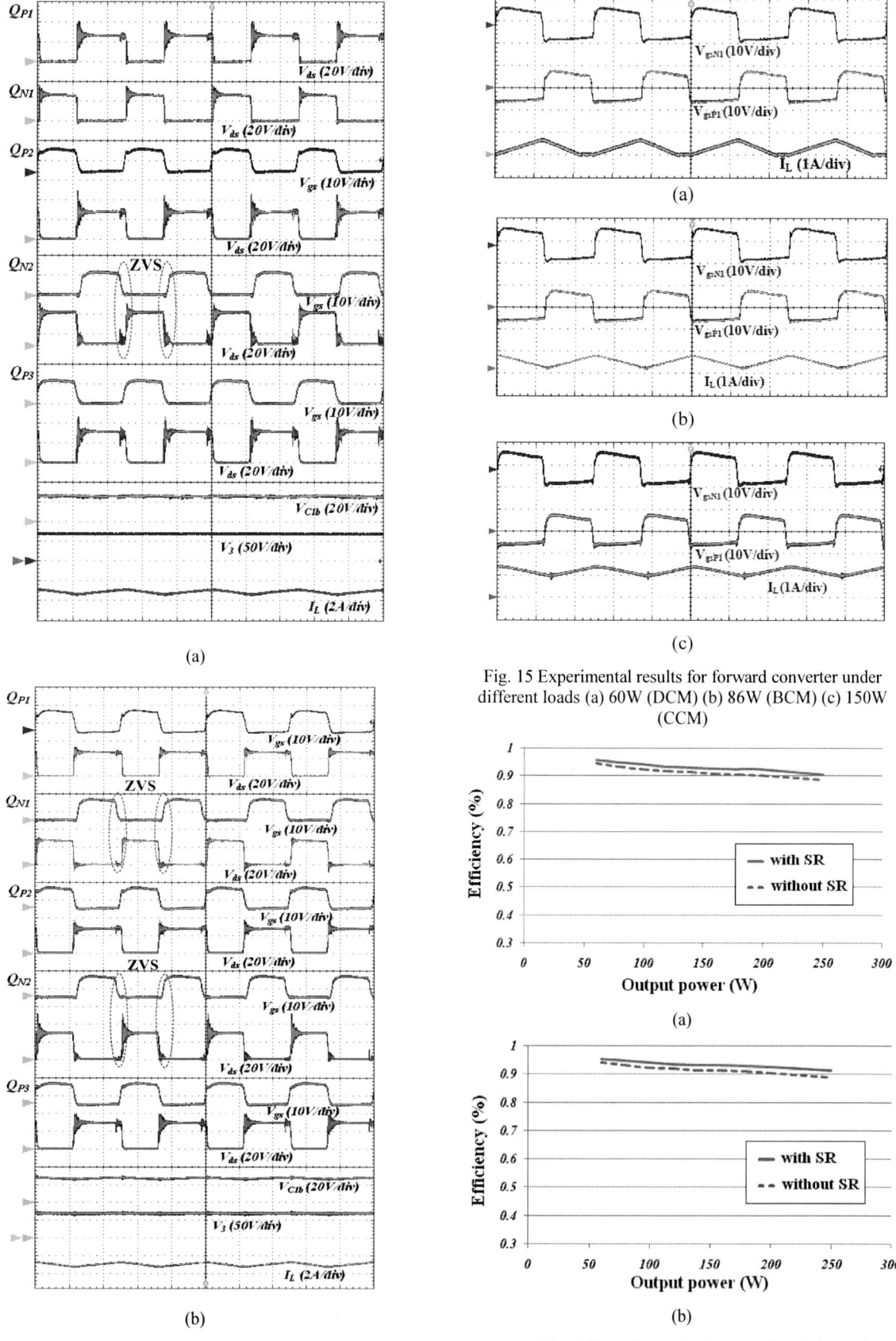

Fig. 15 Experimental results for forward converter under different loads (a) 60W (DCM) (b) 86W (BCM) (c) 150W (CCM)

(a)

(b)

Fig. 14 Experimental results for reverse converter (a) without (b)with synchronous rectification

Fig. 16Experimental efficiency for (a) forward (b)reverse schemes

V. CONCLUSION

This paper presents a new bidirectional DC-DC converter with switched-capacitor cells and the extended proposed converter. The advantage of the proposed converter is high efficiency and high voltage conversion. This converter design can be easily extended to a bidirectional DC-DC converter using multiple switched-capacitor cells under an adjustable voltage conversion scheme. Moreover, the converter efficiency can be increased by using the synchronous rectification and ZVS soft switching designed in the main switch. The proposed bidirectional converter can be applied in battery charging/discharging applications and battery management system.

REFERENCES

[1] Y. S. Lee, Y. P. Ko, M. W. Cheng, and L. J. Liu, "Multiphase zero-current switching bidirectional converters and battery energy storage application," IEEE Trans. on Power Electronics, vol. 28, no. 8, pp.3806-3815, 2013.

[2] W. Yu, H. Qian and J. S. Lai, "Design of high-efficiency bidirectional DC-DC converter and high-precision efficiency measurement," IEEE Trans. on Power Electronics, vol. 25, no. 3, pp.650-658, 2010.

[3] W. Yu, and J. S. Lai, "Ultra high efficiency bidirectional DC-DC converter with multi-frequency pulse width modulation," IEEE APEC, pp1079-1084, 2008.

[4] Y. S. Lee, Y. Y. Chiu, and M. W. Cheng, "Inverting ZCS switched-capacitor bi-directional converter," IEEE Power Electronic Specialists Conference, pp.1-6, 2006

[5] Y. S. Lee and Y. P. Ko, "Switched-capacitor bidirectional converter performance comparison with and without quasi-resonant zero-current switching," IET Power Electronics, vol. 3, no. 2, pp. 269–278, Mar. 2010.

[6] P. Jose and N. Mohan, "A novel bidirectional DC-DC converter with ZVS and interleaving for dual voltage systems in automobiles," IEEE IAS, 2002, pp.131-1314

[7] Y. S. Lee, Y. P. Ko, and C. A. Chi, "A novel QRZCS switched-capacitor bidirectional converter," IEEE International Conference on Power Electronics and Drive System, pp.151-156, 2007

[8] Y. S. Lee and G. T. Cheng, "Quasi-resonant zero currents witching bidirectional converter for battery equalization application, "IEEE Trans. on Power Electronics, vol.21, no. 5, pp. 1213-1224, 2006.

[9] H. S. H. Chung and W. L. Cheung, "Generalized structure of bidirectional switched-capacitor DC/DC converter," IEEE Trans. Circuits and Systems-I: Fundamental Theory and Applications, vol. 50, no. 6, pp.743-753, 2003.

[10] H. S. H. Chung and W. C. Chow," Development of a switched-capacitor DC/DC converter with bidirectional power flow," IEEE Trans. Circuits and Systems-I: Fundamental Theory and Applications, vol. 47, no. 9, pp.1383-1389, 2000.

[11] H. Funato, T. Ishikawa and K. Kamiyama," Transient response of three phase variable inductance realized by variable active-passive reactance(VAPAR),"IEEE.APEC, vol. 2, pp. 1281-1286, 2001.

[12] Z, J. Lai, R. Kim, and W. Yu, "High-power density design of a soft-switching high-power bidirectional dc-dc converter, " IEEE Trans. Power Electron., vol. 22, no. 4,pp. 1145-1153, Jul. 2007.

[13] H. L. Chan, K. W. E. Cheng and D. Sutanto," Phase-shift controlled DC-DC converter with bidirectional power flow," IEE Proc.-Electr. Power Appl., vol. 148, pp. 193-201, 2001.

[14] W. Yu, H. Qian and J.S. Lai," Design of High-Efficiency Bidirectional DC-DC Converter and High-Precision Efficiency Measurement," IEEE Trans. Power Electron, pp. 650-658, 2010.

[15] P. L. Wong, Peng Xu, P. Yang, and F. C. Lee," Performance improvements of interleaving VRMs with coupling inductors, " IEEE Trans. Power Electronics, vol. 16, no. 4, pp. 499-507, 2007.

[16] J. Li, A. Stratakos, A. Schultz and C. R. Sullivan, "Using coupled inductors to enhance transient performance of multi-phase buck converters," IEEE APEC04,vol. 2, pp. 1289-1293.

[17] R. Y. Duan, and J. D. Lee. "Soft switching bidirectional DC-DC converter with coupled inductor," IEEE Industrial Electronics and Applications (ICIEA), pp.2797-2802, 2011.

[18] H. Tao, A Kotsopoulos, J. L. Duarte, and M. A. M Hendrix, "Transformer-coupled multiport ZVS bidirectional DC–DC converter with wide input range," IEEE Trans. on Power Electronics, vol. 23, no. 2,pp. 771-781, 2008.

Switched-Capacitor Charge Equalization Circuit for Series-Connected Batteries

Yao-Ching Hsieh   Zheng-Xiu Cai   Wen-Zhe Wu

Dept. of Electrical Engineering
National Dong Hwa University
Hualien, Taiwan
ychsieh@mail.ndhu.edu.tw

Abstract- In this paper, a novel switched-capacitor charge-equalization circuit is proposed. Electric charge is transferred to a capacitor from two series-connected cells; whereas charge is returned back one of the cells by the charged capacitor. In this way, a larger equalization current can be achieved, and therefore effectively shorten the equalization time. In order to enhance the energy efficiency, voltage-sensing and comparison circuits are added. By comparing the voltages of the adjacent cells, switches are differentially operated to allow selective energy transfer. Thereafter, meaningless energy transfers are reduced and the equalization circuit can achieve higher efficiency.

Keywords: series-connected batteries, switched-capacitor, charge-equalization circuit

I. INTRODUCTION

Secondary batteries are widely applied in modern appliances. Apart from 3C gadgets, high-power and high-voltage applications such as electric vehicles require series-connected batteries to serve as power sources [1]. Due to the divergences of batteries intrinsic natures and operation environments, even though the in/out current of each battery is just the same, the *SOC* (state-of-charge) of the batteries differs. The discrepancies between battery *SOCs* might get worse and worse after cycles of operations. This will greatly shorten the cycle life of the battery string. In order to avoid the premature cycle loss of battery strings or packs, some techniques should be exerted on the battery string to avert the diversities. This is just what charge-equalization circuit will do [2].

There are many papers presented on charge-equalization circuits [3-9]. These circuits make use of converters, capacitors, or transformers to fulfill the purpose of current diversion. By adequate control schemes, the electric charge is transferred from charge-rich batteries to the poor ones. This paper proposes a new type of switched-capacitor charge-equalization circuit for series-connected batteries. Unlike other switched-capacitor circuits, this proposed circuit can charge the capacitor by doubled-number of cells; whereas the capacitor recharges the cells individually. This is accomplished by rearrange the active switch arrays and revised gating signal generating kernel. Furthermore, to achieve better operation efficiency, comparator circuits are applied to produce prejudiced

charge for the cells. By this scheme, useless charge transferences can be avoided.

II. CONVENTIONAL SWITCHED-CAPACITOR CHARGE-EQUALIZATION CIRCUIT

A four series battery string is taken for example in this paper. There are 8 active switches and 3 capacitors. The circuit topology is shown in Fig.1. The active switches are activated by simple PWM gating signals. Since the operations are similar for each capacitor, two adjacent batteries are taken as an example here. Assuming battery voltages v_{B1} is greater than v_{B2}, the gating signals and ideal capacitor voltage and current waveforms are illustrated in Fig. 2. The operation of this circuit can be divided to two modes as follows:

(i)   Mode i (t_0-t_1)

At instant t_0, switches S_{1a} and S_{2a} are turned on; the topology is shown in Fig. 3(a). Battery B_1 is discharged to capacitor C_1 and elevates capacitor voltage v_{C1}. As the switches are turned off, this mode terminates and the circuit idles.

(ii) Mode ii (t_2-t_3)

Switches S_{1b} and S_{2b} are turned on at t_2 as shown in Fig. 3(b). During this mode, i_{C1} is negative, *i.e.* C_1 releases its energy to battery B_2. When the switches are turned off, the circuit idles again and then returns back to Model i.

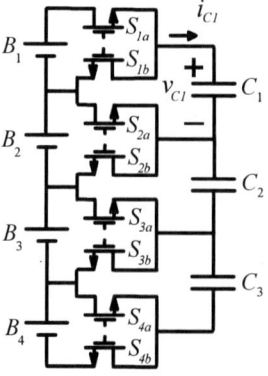

Fig. 1 Switched-capacitor charge-equalization circuit

978-1-4799-2706-7/14 $31.00 © 2014 IEEE

The 2014 International Power Electronics Conference

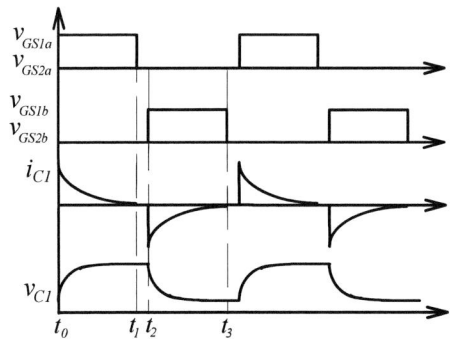

Fig. 2 Expected waveforms of the switched-capacitor circuit

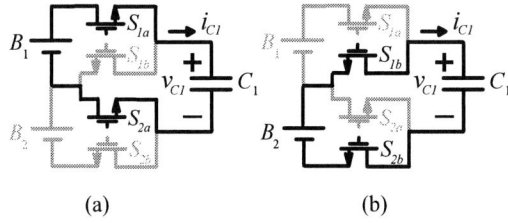

(a)　　　　　　　　　(b)

Fig. 3 Operation modes

Repeating the above operations, the excessive charges will be redistributed to the other batteries and achieves charge equalization. However, since the redistribution current is proportional to the battery voltage difference; whereas close to equalization, the smaller the voltage difference, and the lower current. This ends up with excessively long equalization time.

III. PROPOSED CHARGE-EQUALIZATION CIRCUIT

A novel switched-capacitor circuit is proposed to fulfill the charge-equalization purpose. There are 12 active switches and 3 capacitors in this 4-cell battery string as shown in Fig. 4. Besides, voltage comparator circuits are included to pick out the higher voltage of the two adjacent cells. Although more circuit elements and more complicated control effort are required; the equalization efficiency is improved. Because the operations of all the switched-capacitor modules are similar, the B_1, B_2, and C_1, module is used to explain the operation modes of this circuit. Without losing generality, voltage of B_1 is assumed to be higher than that of B_2 here. The theoretical waveforms are illustrated in Fig. 5.

(i) Mode I (t_0-t_1)

At t_0, switches S_{1a} and S_{1d} are turned on; the conduction loop is illustrated in Fig. 6(a). In this mode, batteries B_1 and B_2 are series-connected to charge capacitor C_1. Whereas the two switches are simultaneously turned off, this mode terminates and all the currents stop flowing. Therefore, the next mode is in quiescent status and is omitted.

(ii) Mode III (t_2-t_3)

Switches S_{1a} and S_{1c} are turned on at t_2. The conduction path is depicted in Fig. 6(b). Because the

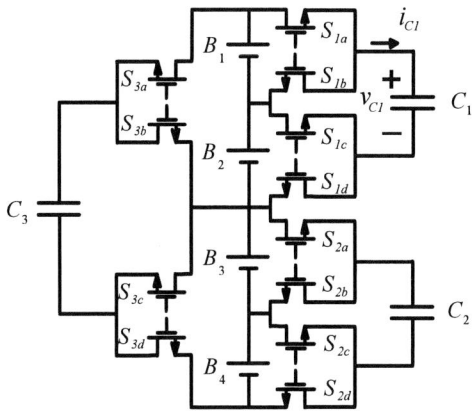

Fig. 4 A novel switched-capacitor charge-equalization circuit

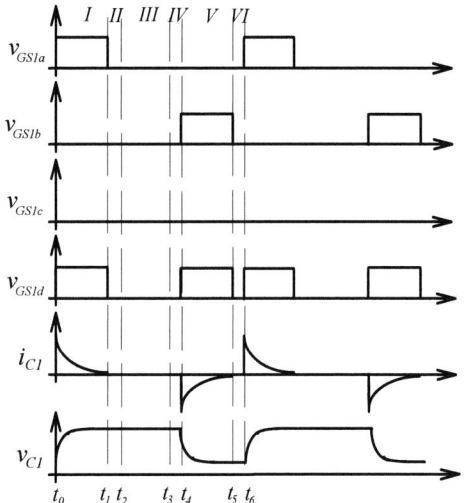

Fig. 5 Theoretical waveforms for the proposed circuit

voltage on capacitor C_1 is higher than V_{B1}, C_1 will transfer its charge to B_1. Again, when the two switches are off, the circuit goes into a quiescent mode.

(iii) Mode V (t_4-t_5)

In this mode, switches S_{1b} and S_{1d} are turned on as shown in Fig. 6(c). Similarly, C_1 has higher voltage than battery B_2; and therefore the charge will be transferred from C_1 to B_2. This mode also finishes when both the switches are off. And after the following quiescent mode, the circuit operation returns back to mode I and initiates a fresh new cycle.

Although there are three operation modes, except the quiescent modes, II, IV, VI; modes III and V are exclusive. That is, if $V_{B1} < V_{B2}$, only mode III is activated; otherwise, mode V is the survived mode. For example, here $V_{B1} > V_{B2}$ is assumed, mode III is omitted, as Fig. 5 depicted. By this selectively operating the switches, the useless, and more likely harmful, energy transfer can be avoided; contrarily, energy is transferred to the proper cell. The energy transference efficiency can be promoted.

978-1-4799-2706-7/14 $31.00 © 2014 IEEE　　　　430

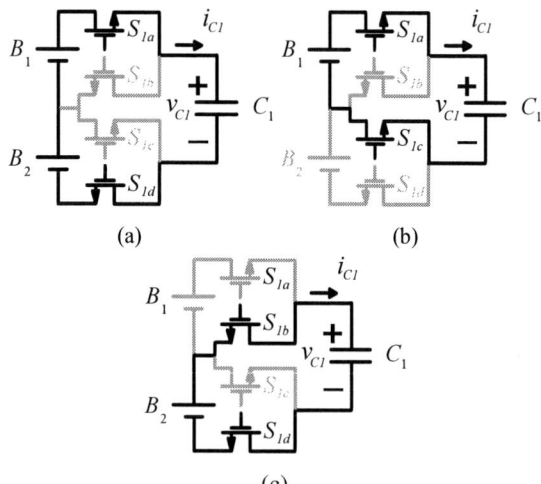

(a)   (b)

(c)

Fig. 6 Operation modes for the proposed circuit

In the 4-cell example of this proposed circuit shown in Fig. 4, on the right are the equalization circuits to balance the adjacent cells; on the left is the one that equalizes the upper and lower 2-cell combinations. Except the two-cell combinations, the operation is the same with the right-hand part.

IV. EXPERIMENTAL RESULTS

A Conventional and the proposed prototype circuits for four cells are built for experiments. Figs. 7, 8 are the experimental waveforms associated with the capacitors C_1 and C_2 of the conventional switched-capacitor circuit. In both the two figures, when S_{1a} and S_{2a} turn on, the cells will charge the capacitors. While S_{1b} and S_{2b} turn on, the capacitors release their charge to the other battery cells. The capacitor voltages can be observed to be rise and fall slightly accordingly. Fig. 9 is the charge equalization result. Before the equalization process, the open circuit cell voltages are $V_{B1} = 4.22V$, $V_{B2} = 4.076V$, $V_{B3} = 4.117V$, $V_{B4} = 4.232V$. As expected, the cell voltages are pulled closer as the switched-capacitor circuit operated. However, it took more than 10 hours for the difference to shrink to less than 40mV.

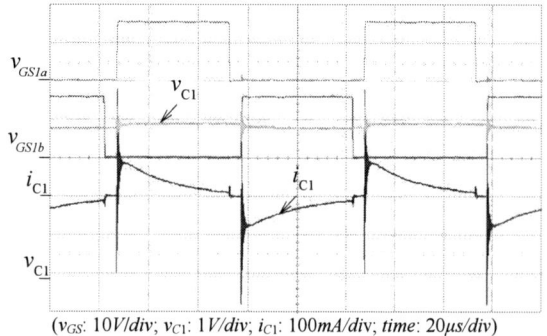

(*v$_{GS}$*: 10*V/div*; *v$_{C1}$*: 1*V/div*; *i$_{C1}$*: 100*mA/div*; *time*: 20*μs/div*)

Fig. 7 Waveforms of conventional switched-capacitor circuit I

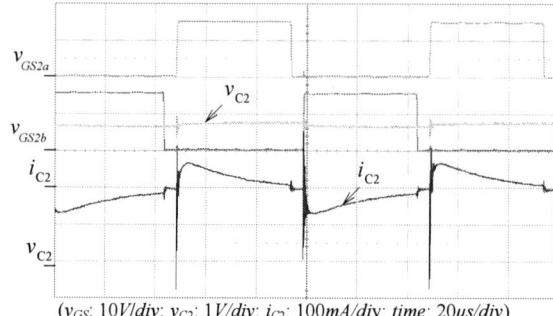

(*v$_{GS}$*: 10*V/div*; *v$_{C2}$*: 1*V/div*; *i$_{C2}$*: 100*mA/div*; *time*: 20*μs/div*)

Fig. 8 Waveforms of conventional switched-capacitor circuit II

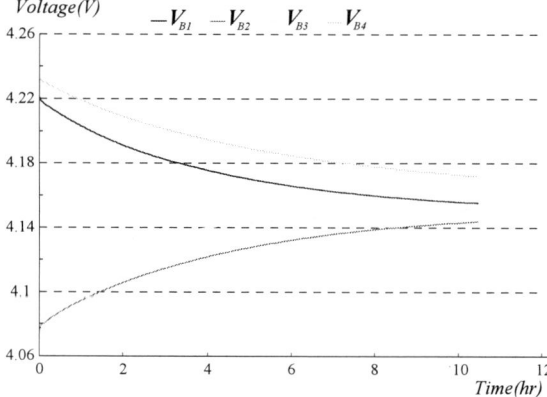

Fig. 9 Charge equalization of conventional switched-capacitor circuit

Figs. 10-13 are the experimental waveforms of the proposed switched-capacitor circuit. Fig.10 illustrates the PWM gating signals of switches associated with capacitor C_1. The associated operation modes are marked. In the results of Figs. 11-13, battery voltages $V_{B1} > V_{B2}$, $V_{B3} > V_{B4}$, and $V_{B1}+V_{B2} > V_{B3}+V_{B4}$ are assumed. In the figures, the capacitors are charged in mode I, and are discharged in mode V; while mode III is not activated.

Fig. 14 records the charge equalization process of the proposed circuit. The open-circuit voltages before equalization are $V_{B1} = 4.258V$, $V_{B2} = 4.013V$,

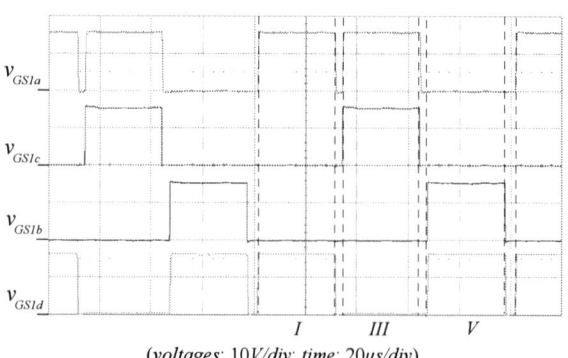

(*voltages*: 10*V/div*; *time*: 20*μs/div*)

Fig. 10 Waveforms of the proposed circuit I

978-1-4799-2706-7/14 $31.00 © 2014 IEEE   431

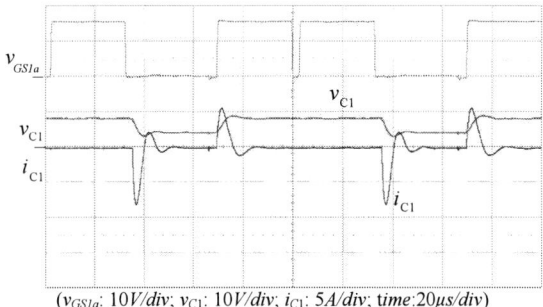

(v_{GS1a}: 10V/div; v_{C1}: 10V/div; i_{C1}: 5A/div; time:20μs/div)

Fig. 11 Waveforms of the proposed circuit II

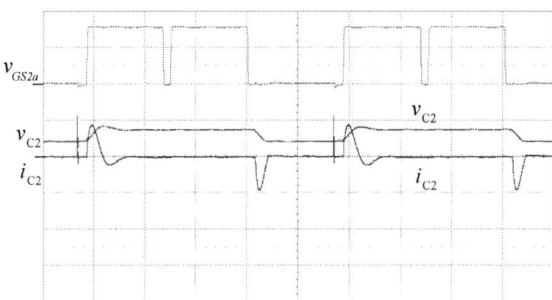

(v_{GS2a}: 10V/div; v_{C2}: 10V/div; i_{C2}: 5A/div; time:20μs/div)

Fig. 12 Waveforms of the proposed circuit III

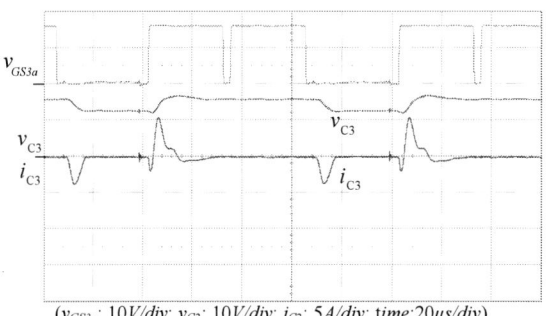

(v_{GS3a}: 10V/div; v_{C3}: 10V/div; i_{C3}: 5A/div; time:20μs/div)

Fig. 13 Waveforms of the proposed circuit IV

V_{B3} =4.051V, and V_{B4} = 3.842V. This result clearly represents that the deviation of the cell voltages is squeezed to less than 10mV within two hours. This result is much better than that of the conventional circuit, and this is owing to the much larger equalization current.

V. CONCLUSION

A novel switched-capacitor charge equalization circuit is proposed in this paper. The capacitor is charged by two series-connected groups of cells, and the charge is then transferred to either group which has the lower

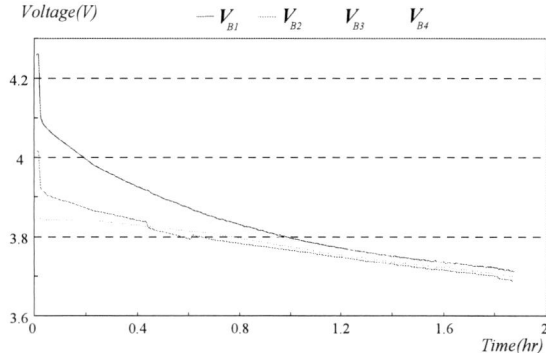

Fig. 14 Charge equalization of the proposed circuit

significantly boosted, especially when the cell *SOC*. By this way, the equalization current can be voltage difference is very small. The charge equalization can be achieved within much shorter time. And the effectiveness of this circuit topology is verified by experimental results.

REFERENCES

[1] M. Yilmaz, P.T. Krein, "Review of Battery Charger Topologies, Charging Power Levels, and Infrastructure for Plug-In Electric and Hybrid Vehicles," *IEEE Trans. on Power Electronics*, vol. 28, no. 5, pp. 2151-2169, 2013.

[2] W. G. Hurley, Y. S. Wong and W. H. Wölfle, "Self-Equalization of Cell Voltages to Prolong the Life of VRLA Batteries in Standby Applications," *IEEE Trans. on Industrial Electronics*, vol. 56, no. 6, pp.2115-2120, 2009.

[3] N. H. Kutkut, "A Modular Non-dissipative Current Diverter for EV Battery Charge Equalization," *13th Applied Power Electronics Conference*, pp.686-690, 1998.

[4] C. Pascual and P. T. Krein, "Switched capacitor system for automatic series battery equalization," in *Proc. 12th IEEE Applied Power Electronics Conference*, pp. 848-854, 1997.

[5] C. S. Moo, Y. C. Hsieh, I S. Tsai, and J. C. Cheng, "Dynamic Charge Equalisation for Series-Connected Batteries," *IEE Proceedings Electric Power Application*, vol. 150, no. 5, pp.501-505, 2003.

[6] R. Lu, C. Zhu, L. Tian, and Q. Wang, "Super-Capacitor Stacks Management System with Dynamic Equalization Techniques," *IEEE Trans. on Magnetics*, vol. 43, no.1, pp. 254-258, 2007.

[7] Y. S. Lee, and M. W. Cheng, "Intelligent Control Battery Equalization for Series Connected Lithium-Ion Battery Strings," *IEEE Trans. on Industrial Electronics*, vol. 52, no. 5, pp. 1297-1307, 2005.

[8] M. Tang and T. Stuart, "Selective buck-boost equalizer for series battery packs," *IEEE Trans. on Aerospace and Electronic Systems*, vol. 36, pp. 201-211, 2000.

[9] N. H. Kutkut, H. L. N. Wiegman, D. M. Divan, and D. W. Novotny, "Design considerations for charge equalization of an electric vehicle battery system," *IEEE Trans. Industry Applications*, vol. 35, no. 1, pp. 28–35, 1999.

The 2014 International Power Electronics Conference

Performance Analysis of UniTL-H6 Inverter with SiC MOSFETs

Davide Barater, Giampaolo Buticchi,
Carlo Concari, Giovanni Franceschini
Department of Information Engineering
University of Parma
Parma, Italy
davide.barater@nemo.unipr.it

Emre Gurpinar, Dipankar De, Alberto Castellazzi
Power Electronics, Machines and Control Group
The University of Nottingham
Nottingham, UK
emre.gurpinar@nottingham.ac.uk

Abstract—**A transformerless single-phase PV inverter has been tested with latest generation 650V/ 20A SiC MOSFETs from ROHM. Test results show that the inverter can be operated with high efficiency at high switching frequencies due to high switching and conduction performance of the devices. Results also show that reduction in output filter size, reduced harmonic distortion and increase in power density are enabled by SiC MOSFETs for PV inverters without compromising on the efficiency of the system.**

Keywords—Renewable energy, SiC MOSFET, Single phase inverters, Transformerless PV inverter

I. INTRODUCTION

Renewable energy exploitation has been in the spotlight of scientific research in the recent years. In this framework, photovoltaic energy harvesting has been one of the most studied topics, and researchers are still active to increase the performance of the power electronics that compose a photovoltaic system. These systems can be either stand-alone, where the energy is stored in batteries and the power converters supply local loads, or grid-connected, where no energy storage is present, and the maximum available power of the photovoltaic field is transferred to the grid. The absence of expensive and bulky energy storage elements is the greatest advantage of this latter kind of systems. The first design of grid-connected inverters featured a line-frequency transformer for the coupling to the mains. In recent converters, high-frequency transformer coupling or transformerless inverters are preferred, due to higher system efficiency and lower converter cost.

The main problem that arises with transformerless topologies is due to the photovoltaic panels' parasitic capacitance between the panel and the earth connection [1], [2], that causes ground leakage current to flow into the grid. This effect is extremely detrimental for the power quality and can cause the disconnection of the inverter due to the residual current device; Fig. 1 shows the path of the ground leakage current.

The main cause of the ground leakage current is the high-frequency harmonics of the output common-mode voltage of the converter. In a conventional full-bridge

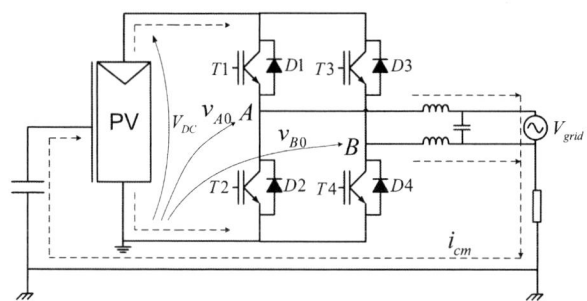

Fig. 1. Ground leakage current path

converter with unipolar modulation and symmetrically split filter inductor at the output, the common mode voltage of the converter will vary from full DC link to half DC link and 0 at the switching switching frequency of the inverter [3]. For this reason, specific topologies were developed to limit the ground leakage current in transformerless full-bridge inverters with unipolar modulation. These inverters usually feature additional devices that detach the converter from the grid during the freewheeling phases and keep the common voltage constant. Fig. 2 shows the most widespread approaches to solve the ground leakage current issue: DC or AC decoupling. DC or AC decoupling switches are used to disconnect the grid from the DC supply during freewheeling phase, i.e., when zero voltage is supplied. The converter with DC decoupling is called H5 (e.g. used in SMA converters) [4]. The DC link switch and the bottom switches in full-bridge are switching at high frequency and top switches in full-bridge are switching at grid frequency. The converter witch AC decoupling is called HERIC (e.g. used in Sunways converters) [5]. In HERIC, the switches in full-bridge are switching at high frequency and AC decoupling switches are switching at grid frequency. Several analysis of these topologies are reported in literature [6]–[9] and all of these topologies were tested with standard Si devices, typically with 600V breakdown voltage capability.

The aim of this paper is to evaluate the performance of H6 inverter with UniTL PWM strategy, which is proposed in [10], and 650V SiC MOSFETs. In Section II, details of

978-1-4799-2706-7/14 $31.00 © 2014 IEEE   433

H6 inverter and UniTL PWM strategy will be explained.

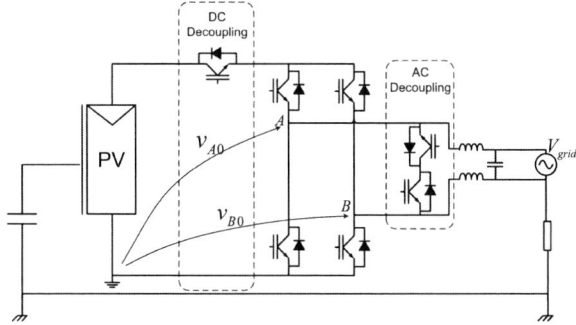

Fig. 2. H-Bridge converter with DC or AC decoupling

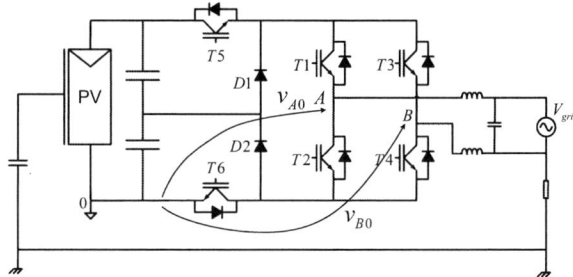

Fig. 3. The H6 topology

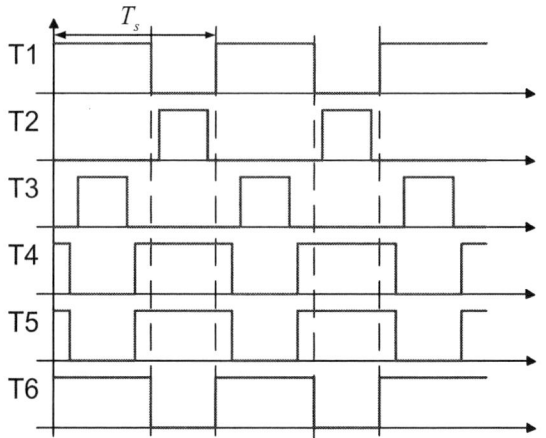

Fig. 4. UniTL PWM signals

In Section III, the test setup and experiment conditions will be presented. In Section IV, detailed performance analysis of the converter with STGW35HF60WD Si IGBTs and SCT2120AF SiC MOSFETs at different load, switching and temperature conditions will be presented.

II. H6 TOPOLOGY WITH UNITL PWM STRATEGY

The H6 architecture was first proposed in [11] and is shown in Fig. 3. In addition to the H5 topology, another switch (T6) in the lower rail of the DC Link is present. D_1 and D_2 diodes in H6 are optional devices that do not

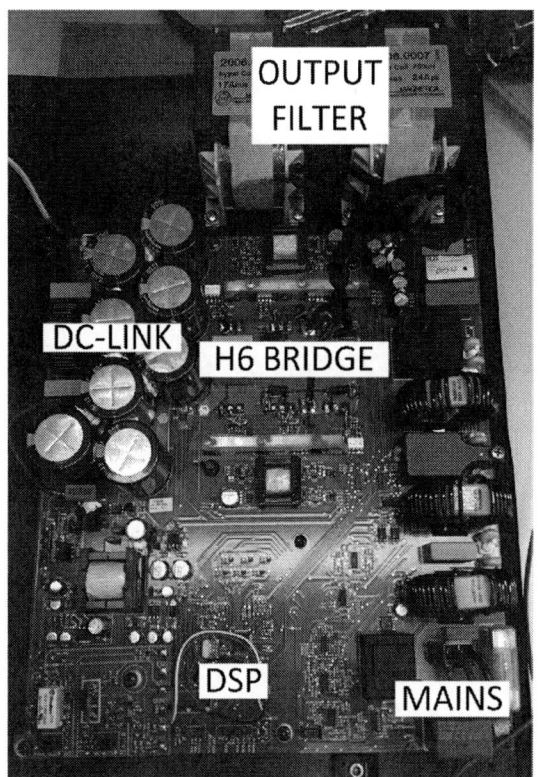

Fig. 5. Prototype converter

conduct current but ensure fixing common-mode voltage to $V_{DC}/2$ in case of an asymmetric switching behaviour in the full-bridge. In [12] an alternative modulation strategy named UniTL was proposed for driving the same topology. The main advantage of this strategy respect to the one proposed in [11] is that the effective output current ripple is at twice the devices' switching frequency, enabling a downsize of the output filter inductor. The PWM signals are shown in Fig. 4.

As can be see in 4 , during a switching period Ts, the H-bridge legs are driven by different duty cycles (T1 and T3): this means that the freewheeling will happen in both the upper and lower parts of the H-bridge, thus effectively doubling the output voltage frequency. A dead time exists between the commutations of the complementary pairs (T1-T2 and T3-T4). The DC decoupling transistors are switched off at the beginning of each freewheeling phase. In particular T5 is switched off when the output current freewheels in the upper part of the H-bridge, when the current freewheels in the lower part, T6 device is to be switched off. As a matter of fact, a lead-lag time between the commutations of the DC decoupling and H-bridge transistors can be adopted to reduce the common mode voltage [10]. Although four devices are conducting in active states, the H6 topology with UniTL PWM scheme promises higher efficiency in comparison to full-bridge with unipolar and bipolar modulations due to lower switching losses. A theoretical loss analysis of these inverter is presented in [12].

978-1-4799-2706-7/14 $31.00 © 2014 IEEE

The 2014 International Power Electronics Conference

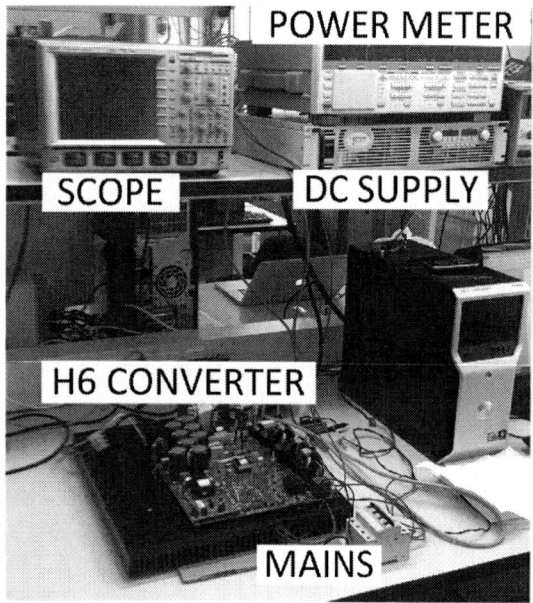

Fig. 6. Test bed

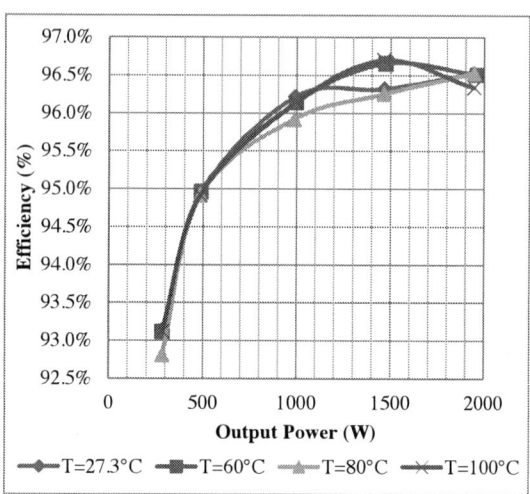

Fig. 7. Efficiency of converter with Si IGBTs at 10kHz switching frequency for different device case temperatures

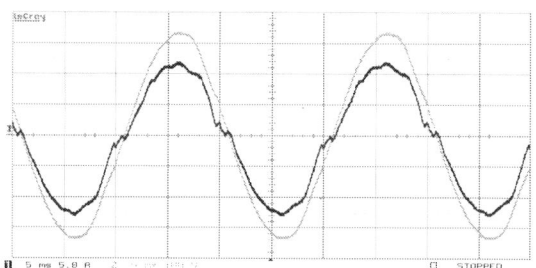

Fig. 8. Grid voltage and current waveforms with Si IGBTs at 10kHz switching frequency (CH1: Grid current 5A/div, CH2: Grid voltage 100V/div)

III. TEST SETUP

A commercial converter (Fig. 5) embedding the H6 bridge with UniTL PWM modulation is used to evaluate the performance of the 650V SiC Devices in comparison to 600V Si IGBTs. This converter features a Freescale MC56F8323 DSP for grid connected control of the converter. The output LC filter is composed of two inductors Lf = 750μH and a Cf = 5μF capacitor. The experimental setup is shown in Fig. 6, corresponding to the schematic of Fig. 3, where the PV panels are substituted with a DC Power supply. The isolated gate drives provide 0-18V gate signal for SiC devices and 0-15V for Si IGBTs. Input DC link voltage of the converter is fixed to 400V and output grid voltage is 230V$_{rms}$ with 50Hz grid frequency. In order to decouple the device case temperature from the load, a full controllable heat sink is used where the temperature of the heat sink can be controlled independently from the load by cooling fans and heating resistors.

IV. EXPERIMENTAL RESULTS

The converter is tested at different load, switching frequency and temperature conditions. In this section, firstly, results with Si IGBT devices will be presented. (Section IV-A). After that, result with SiC MOSFETs will be discussed (Section IV-B) and finally, the performance comparison of Si IGBT and SiC MOSFET in H6 converter will be evaluated (Section IV-C).

A. H6 with 600V Si IGBTs

The converter is rated at 2kW output power and with Si IGBTs, it is tested from 300W to 2kW at 10, 20 and 40kHz switching frequencies. During the tests at different switching frequencies, the case temperature

of the IGBTs is also varied from 27.3°C to 100°C in order to observe the performance of devices under different temperature conditions. The dead-time between T1-T2 and T3-T4 complementary switches must be set to 1μs at all switching frequencies in order to prevent short circuits at DC link due to the tail currents of IGBTs during turn-off. The efficiency curves and grid waveforms with IGBTs at 10kHz switching frequency for different device case temperatures are given in Fig. 7 and Fig. 8 respectively. Efficiency curves in Fig. 7 show that maximum efficiency of the converter is below 97% under all load conditions.

The efficiency waveforms and grid waveforms at 20kHz switching frequency are given in Fig. 9 and Fig. 10 respectively. The efficiency waveforms clearly show that the device temperature has a significant effect on performance of IGBTs and at half load condition, the efficiency difference between 27.3°C and 100°C case temperatures is approximately 0.5%. In comparison to Fig. 7, there is 0.3% drop in the peak efficiency of the converter. From Fig 10, the distortion in the grid current due to large dead-time can be clearly seen.

978-1-4799-2706-7/14 $31.00 © 2014 IEEE

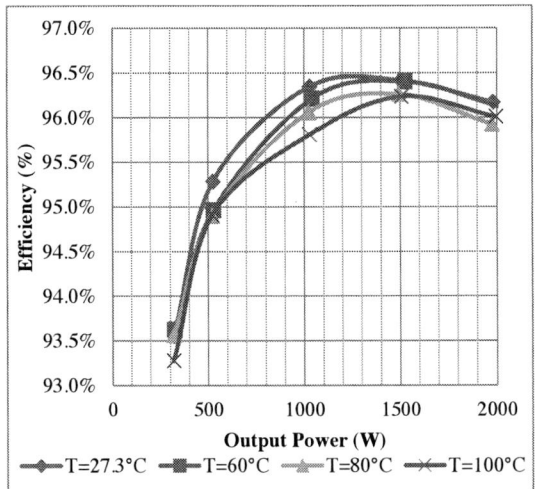

Fig. 9. Efficiency of converter with Si IGBTs at 20kHz switching frequency for different device case temperatures

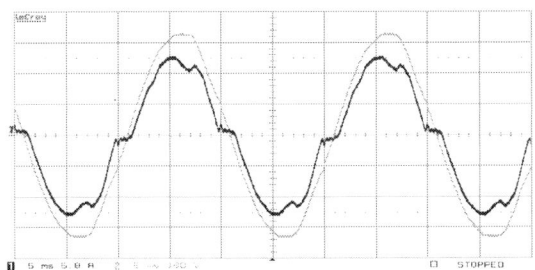

Fig. 10. Grid voltage and current waveforms with Si IGBTs at 20kHz switching frequency (CH1: Grid current 5A/div, CH2: Grid voltage 100V/div)

For final tests of the converter with Si IGBTs, the switching frequency is increased to 40kHz and efficiency waveforms are presented in Fig. 11. The efficiency waveforms show that operating the converter at 40kHz switching frequency causes 1.3% and 1% peak efficiency drop in comparison to operating at 10 and 20kHz switching frequencies respectively. In addition to this, similar to Fig. 7 and Fig. 9, Fig. 11 also shows that increase in device case temperature causes up to 0.5% drop in overall efficiency.

B. H6 with 650V SiC MOSFETs

The H6 converter is tested with SiC MOSFETs in similar conditions to the test with Si IGBTs in order to see the impact of replacing Si devices with SiC devices. The only difference is the dead-time reduction from 1μ to 250ns due to high switching performance of SiC MOSFET in comparison to Si IGBT. The results at 10kHz switching frequency are presented in Fig. 12 and Fig. 13. Efficiency waveforms in Fig. 12 shows that the converter reaches 97.2% peak efficiency at half load. Moreover, the variations in device case temperature variation do not cause significant loss of performance over wide temperature and load range. The grid current waveform in Fig. 13 shows that the reduction in dead-

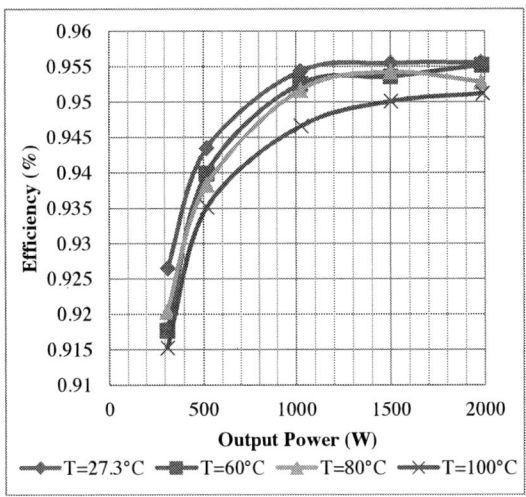

Fig. 11. Efficiency of converter Si IGBTs at 40kHz switching frequency for different device case temperatures

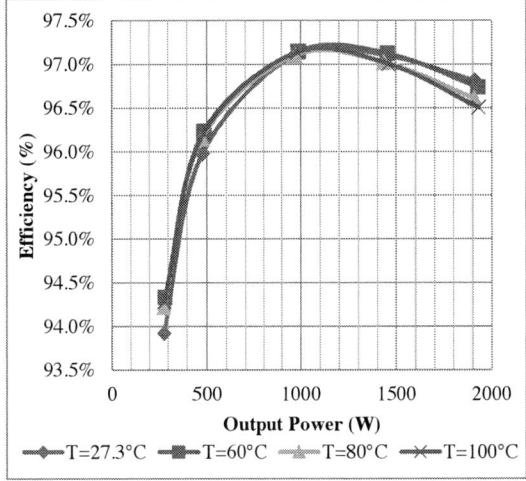

Fig. 12. Efficiency of converter with SiC MOSFETs at 10kHz switching frequency for different device case temperatures

time from 1μs to 250ns reduces the harmonic distortion injected to the grid.

Efficiency and grid voltage and current waveforms at 20kHz switching frequency are given in Fig. 14 and Fig. 15 respectively. In comparison to Fig 12, although the switching frequency of the converter is doubled, the efficiency of the converter with different device case temperatures does not change due to high switching performance of the devices and the grid waveforms in Fig. 15 accompany these results with low harmonic distortion in grid current.

The final results of the converter with SiC devices is at 40kHz switching frequency. The efficiency waveforms are presented in Fig. 16 and grid waveforms are presented in Fig. 17. Similar to the performance at 10 and 20kHz switching frequencies, the efficiency of the converter is not affected from the increase in device case temperature and there is 0.4% efficiency drop in comparison to

The 2014 International Power Electronics Conference

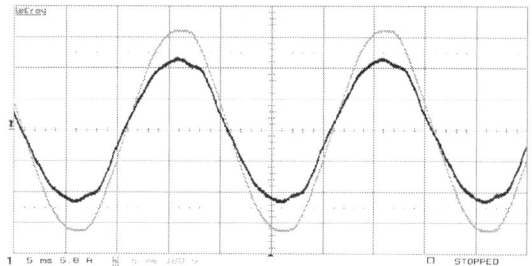

Fig. 13.   Grid voltage and current waveforms with SiC MOSFETs at 10kHz switching frequency (CH1: Grid current 5A/div, CH2: Grid voltage 100V/div)

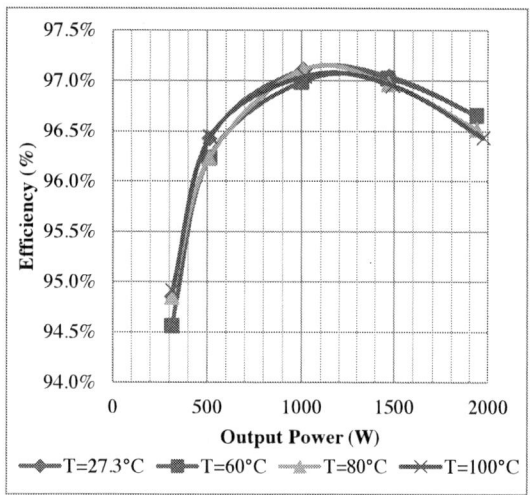

Fig. 14.   Efficiency of converter with SiC MOSFETs at 20kHz switching frequency for different device case temperatures

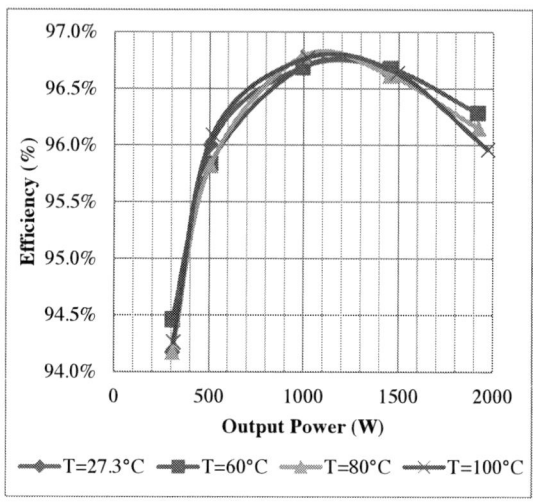

Fig. 16.   Efficiency of converter with SiC MOSFETs at 40kHz switching frequency for different device case temperatures

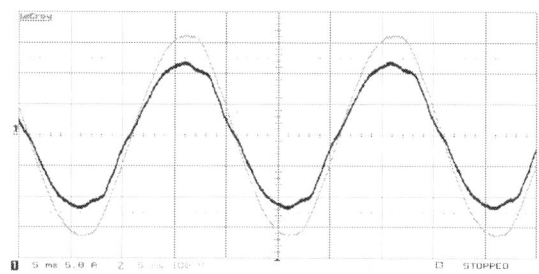

Fig. 17.   Grid voltage and current waveforms with SiC MOSFETs at 40kHz switching frequency (CH1: Grid current 5A/div, CH2: Grid voltage 100V/div)

operation at 10kHz switching frequency. In similar to the waveforms in Fig. 13 and Fig. 15, the grid current waveform in Fig. 17 is not affected by the dead-time commutation.

C. Comparative Analysis of H6 with Si and SiC Devices

The efficiency comparison of the converter with Si and SiC devices at 20kHz switching frequency for low and high device case temperature values respectively present Fig. 18 and Fig. 19. Fig. 18 and 19 show that

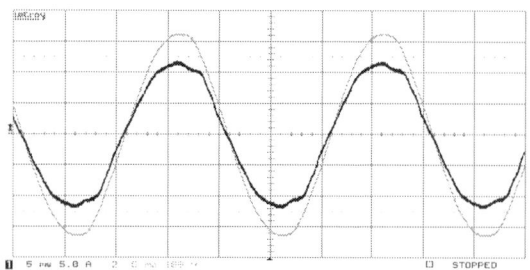

Fig. 15.   Grid voltage and current waveforms with SiC MOSFETs at 20kHz switching frequency (CH1: Grid current 5A/div, CH2: Grid voltage 100V/div)

replacing Si devices with SiC devices brings from 0.5% to 1% efficiency increase depending on the output power level. The comparison of two devices at 40kHz switching frequency for 80°C and 100°C case temperature values is given in Fig. 20. The comparison in Fig. 20 clearly shows the robust performance of SiC devices at high temperature operation and when the case temperature is increased to 100°C, the efficiency difference goes up to 2.3% at 1kW output power. In addition to that, converter operates above 96% efficiency in a wide load range with SiC devices at 40kHz switching frequency and high case temperature values. In comparison to SiC, the efficiency of the converter with Si IGBTs is below 95% at 100°C case temperature. These results clearly show that it is possible to reduce the size of passive components in the converter and the size of the heat sink without compromising the performance of the converter replacing Si IGBTs with SiC MOSFETs of comparable voltage rating.

V.   CONCLUSIONS

In this paper, performance analysis of UniTL-H6 grid connected converter with 600V Si IGBTs and 650V SiC MOSFETs is discussed. The converter is tested with both devices at different load, switching frequency and device

978-1-4799-2706-7/14 $31.00 © 2014 IEEE   437

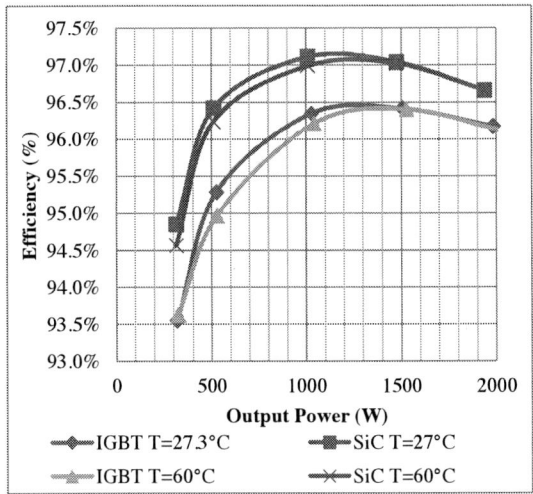

Fig. 18. Efficiency comparison of the converter with SiC MOSFETs and Si IGBTs at 20kHz switching frequency for different device case temperatures

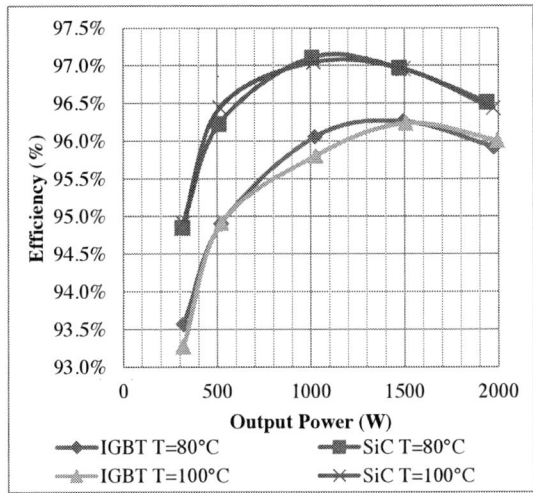

Fig. 19. Efficiency comparison of the converter with SiC MOSFETs and Si IGBTs at 20kHz switching frequency for different device case temperatures

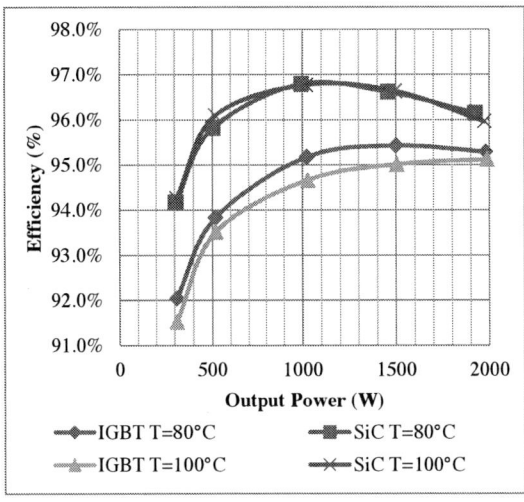

Fig. 20. Efficiency comparison of the converter with SiC MOSFETs and Si IGBTs at 40kHz switching frequency for different device case temperatures

case temperatures in order to evaluate the performance of the devices under different conditions. The results show that it is possible to increase the efficiency of the converter up to 1% by just replacing the Si devices with SiC devices. Another option possible option is to reduce the size of passive components by increasing the switching frequency by four times and reducing the size of heat sink by increasing the operating temperature of SiC devices without compromising the overall efficiency of the system. In addition to that, high switching performance SiC MOSFETs allow shorter dead time between complementary switches that leads to less harmonic distortion at the grid current at high switching frequencies. The comparison of Si IGBTs and SiC MOSFETs clearly show that SiC devices are promising devices for transformerless PV inverters and they will be used more widely in renewable technologies as the semiconductor

companies broaden their SiC product range. The current research in photovoltaic systems is currently focused on increasing performance of the system and maximising the energy transfer from PV panel to grid in order to reduce payback time of system capital cost. In this case, up to 1% efficiency increase by just replacing semiconductor devices or maintaining the efficiency of the system while reducing size of passive and mechanical components will be attractive options for PV inverter manufacturers. To summarise, here are the key benefits of using SiC MOSFETs:

• SiC MOSFETs do not require separate anti-parallel diode (reduced component counts)

• SiC MOSFETs improves efficiency and at same switching frequency or enables higher switching frequencies for same efficiency

• SiC MOSFET performance only minimally affected by increase of switching frequency and device case temperature; and provided very stable performance that leads to high reliability

• SiC MOSFETs' faster switching times enable "clearer" output waveforms by reducing dead-time time between complementary switches.

REFERENCES

[1] H. Xiao and S. Xie, "Leakage Current Analytical Model and Application in Single-Phase Transformerless Photovoltaic Grid-Connected Inverter," *IEEE Transactions on Electromagnetic Compatibility*, vol. 52, no. 4, pp. 902–913, Nov. 2010.

[2] O. Lopez, F. Freijedo, A. Yepes, P. Fernandez-Comesaa, J. Malvar, R. Teodorescu, and J. Doval-Gandoy, "Eliminating Ground Current in a Transformerless Photovoltaic Application," *IEEE Transactions on Energy Conversion*, vol. 25, no. 1, pp. 140–147, Mar. 2010.

[3] E. Gubía, P. Sanchis, A. Ursúa, J. López, and L. Marroyo, "Ground currents in single-phase transformerless photovoltaic systems," *Progress in Photovoltaics: Research and Applications*, vol. 15, no. 7, pp. 629–650, Nov. 2007.

[4] M. Victor, F. Greizer, S. Bremicker, and U. Hübler, "Method of converting a direct current voltage from a source of direct current voltage, more specifically from a photovoltaic source of direct current voltage, into a alternating current voltage," Aug. 12 2008, uS Patent 7,411,802.

[5] H. Schmidt, C. Siedle, and J. Ketterer, "Dc/ac converter to convert direct electric voltage into alternating voltage or into alternating current," May 16 2006, uS Patent 7,046,534.

[6] B. Gu, J. Dominic, J.-s. Lai, C.-L. Chen, T. LaBella, and B. Chen, "High Reliability and Efficiency Single-Phase Transformerless Inverter for Grid-Connected Photovoltaic Systems," *IEEE Transactions on Power Electronics*, vol. 28, no. 5, pp. 2235–2245, May 2013.

[7] L. Zhang, K. Sun, Y. Xing, and M. Xing, "H6 Transformerless Full-Bridge PV Grid-Tied Inverters," *IEEE Transactions on Power Electronics*, vol. 29, no. 3, pp. 1229–1238, Mar. 2014.

[8] S. V. Araujo, P. Zacharias, and R. Mallwitz, "Highly Efficient Single-Phase Transformerless Inverters for Grid-Connected Photovoltaic Systems," *IEEE Transactions on Industrial Electronics*, vol. 57, no. 9, pp. 3118–3128, Sep. 2010.

[9] T. Kerekes, R. Teodorescu, P. Rodriguez, G. Vazquez, and E. Aldabas, "A New High-Efficiency Single-Phase Transformerless PV Inverter Topology," *IEEE Transactions on Industrial Electronics*, vol. 58, no. 1, pp. 184–191, Jan. 2011.

[10] G. Buticchi, D. Barater, E. Lorenzani, and G. Franceschini, "Digital Control of Actual Grid-Connected Converters for Ground Leakage Current Reduction in PV Transformerless Systems," *IEEE Transactions on Industrial Informatics*, vol. 8, no. 3, pp. 563–572, Aug. 2012.

[11] R. Gonzalez, J. Lopez, P. Sanchis, and L. Marroyo, "Transformerless Inverter for Single-Phase Photovoltaic Systems," *IEEE Transactions on Power Electronics*, vol. 22, no. 2, pp. 693–697, Mar. 2007.

[12] D. Barater, G. Buticchi, A. S. Crinto, G. Franceschini, and E. Lorenzani, "Unipolar PWM Strategy for Transformerless PV Grid-Connected Converters," *IEEE Transactions on Energy Conversion*, vol. 27, no. 4, pp. 835–843, Dec. 2012.

The 2014 International Power Electronics Conference

Maximum Power Point Tracking of Grid-tied Photovoltaic Power Systems

Ya-Ting Lee
Dept. of Beauty Science
National Taichung University of Science and Technology
Taichung 40401, Taiwan.

Chian-Song Chiu* and Tse-Wei Chiu
Dept. of Electrical Engineering
Chung-Yuan Christian University
Chungli 32023, Taiwan.
acs.chiu@gmail.com

Abstract—This paper presents high performance maximum power point tracking (MPPT) of a grid-tied PV power system via terminal sliding mode control (TSMC). Based on the proposed TSMC method, the output current of the grid-connected inverter is regulated to the maximum power point in finite time. Moreover, the inverter current is synchronized to the grid voltage, such that the current harmonic is reduced. In addition, the finite-time tracking performance of the TSMC improves the effectiveness of the maximum power point searching algorithm, so that the maximum power is correctly captured from the PV system in finite time. Also, the robustness to uncertainties and disturbances is guaranteed by the TSMC controller. No singular problem exists as similar as traditional TSMC control methods. The simulations and experiments have shown the expected performance.

Keywords—Solar power system; maximum power point tracking; terminal sliding mode control.

I. INTRODUCTION

The grid-tied photovoltaic (PV) power system is one typical PV power generation system which extracts the solar power from the PV array to the utility grid. Since the grid-tied PV system plays a distributed power system to the utility grid, more efficiency is provided to the power system. To capture more solar energy, many maximum power point tracking (MPPT) methods are developed including perturb and observe method [1], [2], incremental conductance method [3], fuzzy logic method [4], [5], neural network method [6], etc. Meanwhile, current control of the grid-connected inverter is used for tracking the reference maximum power current (MPC) and synchronizing the utility grid voltage. Although many relative works are proposed in literature (e.g. [7]-[10]), the control performance depends on the inverter current controller design, such as power quality depending current harmonic, power efficiency depending the tracking error of the MPC. Therefore, it is an important problem to provide fast convergence rate and high accuracy for the MPC tracking of the grid-tied inverter.

Sliding mode control [11] has powerful ability for the control of uncertain systems; therefore, the controlled system with sliding mode exhibits robustness properties with respect to both internal parameter uncertainties and external disturbances. Although the convergence rate of

SMC is arbitrarily fast via higher gain, it provides only asymptotically stable and infinite time convergence. From the pioneering work of [12], terminal sliding mode control (TSMC) achieves finite-time convergence stability. Many application examples of TSMC have been considered to provide fast and finite time control performance as well as high precision [13], [14]. However, due to the use of fractional power functions as the sliding hyperplane, there exist intrinsic singular problems [15], [16], [17]. Therefore, all of the above motivate us to improve the TSMC method for the MPPT of PV power systems. In light of this, if the TMSC can be applied to the PV voltage tracking, the MPV based MPPT will provide better control performance.

In this paper, a terminal sliding mode control (TSMC) based MPPT scheme is introduced for a grid-connected PV power generation system. First, the MPC reference is obtained from the incremental conductance algorithm [3]. Second, the TSMC controller is proposed for the MPC tracking of the single-phase DC/AC full-bridge inverter. The proposed TSMC assures finite-time convergence of the current tracking, such that the current harmonic and power factor (or voltage synchronization) are improved under accurate MPC tracking. Moreover, since the maximum power point of the PV system is accurately followed, the efficiency of the PV power conversion is kept on the best value. Compared with the traditional MPPT methods, faster response and more effective MPPT is achieved. Meanwhile, the stability has high robustness which is lacked in traditional MPPT methods [7]-[10]. In addition, the singularity problem of traditional TSMC [12] is removed in this paper.

This paper is organized as follows. First, the grid-tied PV power generation system is given in Section II. Section III proposes the TSMC based MPPT controller for the PV system, while the strict stability analysis is performed with finite-time convergence. In Section IV, numerical simulations illustrate control performance. Section V shows the experiment results. Finally, some conclusions are made in Section VI.

II. CONFIGURATION OF PHOTOVOLTAIC POWER GENERATION SYSTEM

In this study, the configuration of a grid-connected PV power generation system is depicted in Fig. 1. Here, the

---

*Corresponding author

978-1-4799-2706-7/14 $31.00 © 2014 IEEE

DC/DC boost converter is used to provide a stable voltage for the inverter, while the full-bridge inverter is connected to the single-phase utility grid. To capture the maximum power from the PV array, the PV power is regulated by controlling the grid-connected inverter. Referring [9] and [10], the electric characteristic of the PV array can be described by the current equation:

$$i_{pv} = N_p i_{ph} - N_p i_{rs}(e^{qV_{pv}/N_s A K_o \mathrm{T}} - 1)$$

where V_{pv} and i_{pv} denote the output voltage and current of the PV array, respectively; N_p and N_s are the number of the parallel and series cells, respectively; T is the cell temperature; the electronic charge $q = 1.6 \times 10^{-19}$ C; Boltzmann's constant $K_o = 1.3805 \times 10^{-23}$ J/°K; the ideal P-N junction characteristic factor $A = 1 \sim 5$; i_{ph} is the light-generated current depending on the solar insolation and temperature; and i_{rs} denotes the reverse saturation current. Beside, i_{rs} and i_{ph} are functional of solar insolation and cell temperature in the following form:

$$i_{rs} = I_{rr}(\frac{\mathrm{T}}{\mathrm{T}_r})^3 e^{qE_{gp}(1/\mathrm{T}_r - 1/\mathrm{T})/(K\phi)}$$
$$i_{ph} = (I_{sc} + \mathrm{K}_I(\mathrm{T} - \mathrm{T}_r))\lambda/100$$

where I_{rr} is the reverse saturation current at the reference temperature T_r; $E_{gp} = 1.1ev$ is the band-gap energy of the semiconductor making up the cell; I_{sc} is the short-circuit cell current at reference temperature and insolation; K_I $(A/°\mathrm{K})$ is the short-circuit current temperature coefficient; and λ is the insolation in mW/cm^2. The PV power can be obtained below:

$$\begin{aligned} P_{pv} &= i_{pv} V_{pv} \\ &= N_p i_{pv} V_{pv} - N_p i_{rs} V_{pv}(e^{qV_{pv}/N_s A K_o \mathrm{T}} - 1) \end{aligned} \quad (1)$$

To show the electric feature, we depict the power-voltage characteristic diagram of a PV array as shown in Fig. 2. Obviously, the maximum power changes along various insolation and cell temperature. Furthermore, there exists a unique operational PV voltage V_{pv} such that the output PV power is maximized. When the power slope $dP_{pv}/dv_{pv} = 0$, the system provides the maximum power. However, the maximum power point is hard solved from $\frac{dP_{pv}}{dv_{pv}} = 0$ due to the high nonlinearity. Accordingly, the maximum power generation cannot be easily achieved by traditional methods. In other words, the inverter should be controlled to regulate the PV voltage to the optimal operational point.

To this end, the dynamics of the power inverter is considered as follows:

$$\dot{i}_{L2}(t) = \frac{1}{L_2}(V_{dc}d(t) - i_{L2}(t)r_{L2} - v_g(t)) \quad (2)$$

where V_{dc} is the power inverter input voltage which is the boosted PV voltage from the boost converter; v_g is the grid voltage; i_{L2} is the grid-connected current; L_2 is grid-connected filter inductor; r_{L2} is the internal resistance on the L_2; and $d(t)$ is the control duty input of the inverter.

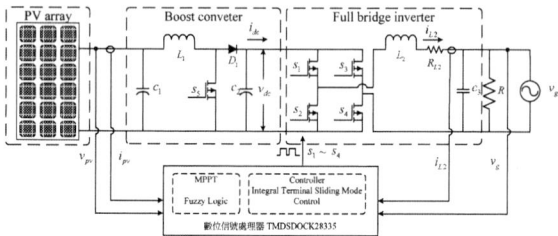

Fig. 1. The control configuration of the grid-tied PV power system.

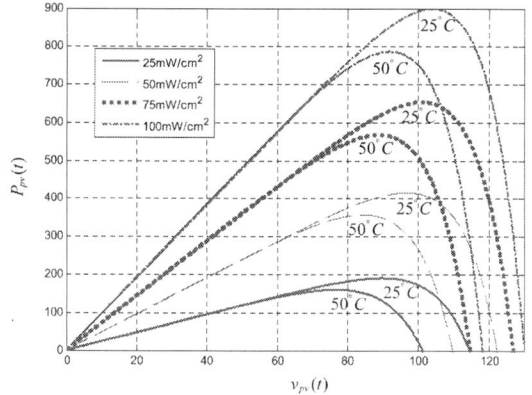

Fig. 2. The electrical characteristics of the PV array.

III. TERMINAL SLIDING MODE MPPT CONTROLLER

To achieve the MPPT under the changing atmosphere, the grid-tied PV system will be controlled to the maximum power point with the maximum power current of the inverter. Meanwhile, the output current of the inverter is synchronized with the grid voltage for a unit power factor. To this end, i_{pv} and V_{pv} are firstly measured to search the reference MPC i_{L2}^*. Then, the TSMC controller is applied to track the reference current i_{L2}^* for MPPT of the grid-tied PV power system.

A. MPC Searching Algorithm

To achieve the maximum power operation, we use an incremental conductance method to search the MPC $i_{L2}^*(t) = i^* \sin(\omega t)$ for the grid-connected inverter, where $\sin(\omega t)$ is the normalized signal of the grid voltage and i^* is the current amplitude determined later. According to the electric power equation (1), the power slope dP_{pv}/dV_{pv} can be expressed as

$$\frac{dP_{pv}}{dV_{pv}} = i_{pv} + V_{pv}\frac{di_{pv}}{dV_{pv}}$$

When the power slope $dP_{pv}/dV_{pv} = 0$, i.e., $\frac{di_{pv}}{dV_{pv}} = -\frac{i_{pv}}{V_{pv}}$, the PV system operates at the maximum power generation. Thus, the tuning law of the MPC is given by the following rules:

$$\begin{cases} i^*(k) = i^*(k-1) + \Delta i^*, & \text{for } \frac{di_{pv}}{dV_{pv}} > -\frac{i_{pv}}{V_{pv}} \\ i^*(k) = i^*(k-1) - \Delta i^*, & \text{for } \frac{di_{pv}}{dV_{pv}} < -\frac{i_{pv}}{V_{pv}} \\ i^*(k) = i^*(k-1), & \text{for } \frac{di_{pv}}{dV_{pv}} = -\frac{i_{pv}}{V_{pv}} \end{cases} \quad (3)$$

where $i^*(k)$ is the magnitude of the reference MPC $i^*_{L2}(t)$ at k-th step; Δi^* is the update parameter which can be adjusted based on the experiment settings. When $\frac{di_{pv}}{dV_{pv}} = -\frac{i_{pv}}{V_{pv}}$, the power slope $dP_{pv}/dV_{pv} = 0$ for the maximum power point generation. Many researches (e.g. [3], [10]) have pointed out that correct maximum power point searching requires the exact tracking of the reference MPC. Thus, it is important to construct an optimal current tracking controller for the grid-connected inverter.

B. Terminal Sliding Mode Controller

To achieve the maximum power tracking, the terminal sliding mode controller is proposed to let the inverter current $i_{L2}(t)$ track the reference MPC $i^*_{L2}(t)$. Once V_{pv} always follows V_{pvd}, the PV power system will move to the maximum power point along the incremental conductance adjusting. First, let $x_1(t) = i_{L2}(t)$, $x_{1d}(t) = i^*_{L2}(t)$ and define the current tracking error $e(t) = x_1(t) - x_{1d}(t)$. Define a terminal sliding mode function $e_R(t)$ as

$$e_R(t) = e(t) + \alpha \int_0^t e^{q/p}(\tau)\, d\tau \quad (4)$$

where $\alpha > 0$; $e^{q/p}(t) = |e(t)|^{q/p} sign(e(t))$; p and q are odd integers satisfying $0 < q < p < 2q$. Let $\dot{e}_\alpha(t) = e^{q/p}(t)$ (i.e., $e_\alpha(t) = \int_0^t e^{q/p}(\tau)\, d\tau$ is an integral error). If $e_R(t) = e(t) + \alpha e_\alpha(t)$ is always kept at zero, then the error equation (4) renders to

$$\dot{e}_\alpha(t) = -\alpha^{q/p} e_\alpha^{q/p}(t). \quad (5)$$

To show the convergence, let us consider a Lyapunov function candidate $V = \frac{1}{2}e_\alpha^2$, the time derivative of V along the error dynamics (5) satisfies

$$\dot{V} = -\alpha^{q/p} e_\alpha^{(q+p)/p}(t)$$

where $e_\alpha^{\frac{p+q}{q}} > 0$ for all $e_\alpha \neq 0$ and even number $p + q$. The error term e_α converges to zero asymptotically. As a result, from solving the above differential equation, the error $e_\alpha(t)$ converges to zero in finite time

$$T_\alpha = \frac{|e_\alpha(0)|^{1-\frac{q}{p}}}{\alpha^{q/p}(1-\frac{q}{p})}$$

The convergence time of $e_\alpha(t)$ and $e(t)$ can be adjusted from choosing proper parameters α, p, q.

To enhance the control performance, we further introduce a double-integral terminal sliding mode function $s(t)$ is defined as

$$s(t) = e_R(t) + \beta e_I(t) \quad (6)$$

where $\beta > 0$, $e_I(t)$ is the error integral from

$$\begin{aligned} \dot{e}_I(t) &= sign(e_R(t)) \\ &= sign(e(t) + \alpha \int_0^t e^{q/p}(\tau)\, d\tau). \end{aligned}$$

Also, $s(t) = 0$ implies the finite-time convergence of the current tracking error because $\dot{e}_I(t) = -sign(e_I(t))$ when $s(t) = 0$. To show the convergence, let us consider a Lyapunov function candidate $V_e = \frac{1}{2}e_I^2$, the time derivative of V_s along the error dynamics $s(t) = 0$ satisfies

$$\dot{V}_e = -|e_I(t)|$$

The error term $e_I(t)$ converges to zero in finite time. Thus, the terminal sliding mode function $s(t)$ should be driven to zero for control objective.

Next, based on the definition of the sliding mode function, the time derivative of the double-integral terminal sliding function $s(t)$ simply arrives with

$$\begin{aligned} \dot{s}(t) &= \dot{e}(t) + \alpha e^{q/p}(t) + \beta \dot{e}_I(t) \\ &= f(x(t)) + g(x(t))d(t) \\ &\quad - \dot{x}_{1d}(t) + \alpha e^{q/p}(t) + \beta \dot{e}_I(t) \end{aligned} \quad (7)$$

where $f(x) = -\frac{1}{L_2}(x_1 r_{L2} + v_g)$, and $g(x) = \frac{V_{dc}}{L_2} \neq 0$ for all $x(t)$. To drive $s(t)$ to zero, the control input $d(t)$ is designed as follows:

$$d(t) = \frac{-1}{g(x)}[f(x) - \dot{x}_{1d} + \alpha e^{q/p} + \beta \dot{e}_I + k_1 s + k_2 sign(s)] \quad (8)$$

with $k_1, k_2 > 0$. By applying the above control law, the closed-loop system is obtained below:

$$\dot{s} = -k_1 s - k_2 sign(s) \quad (9)$$

Therefore, according to Lyapunov stability analysis, the closed-loop control system is ensured with finite-time convergence for the MPC tracking. In other words, the MPC searching algorithm can find the correct maximum power point and the grid-connected inverter achieves the maximum power efficiency.

Theorem 1: Consider the PV grid-connected power system (2) using the TSMC controller (8) with $k_1, k_2 > 0$. The closed-loop control system is ensured with finite-time convergence for the MPC tracking of the inverter output current. □

Furthermore, if considering system uncertainty, the TSMC still assures the stability of the current tracking once the control gain is redesigned. Assume the system with uncertainties in the form:

$$\begin{aligned} f(x) &= \hat{f}(x) + \Delta f(x) \\ g(x) &= \hat{g}(x) + \Delta g(x) \\ \dot{x}_{1d} &= \hat{h}(x) + \Delta h(x) \end{aligned} \quad (10)$$

where $\hat{f}$, $\hat{g}$, and $\hat{h}$ are known and measurable; Δf, Δg, and Δh are the uncertain parts arising from system uncertainties and measurement errors. Meanwhile, the uncertainties satisfy the following boundary conditions:

$$\varpi^{-1} \leq \hat{g}/g \leq \varpi$$
$$|\Delta f| < F$$
$$|\Delta h| < H$$

where $\hat{g}$, $g \neq 0$ and $\varpi \geq 1$; F and H are known upper bounds of the uncertain dynamics of Δf and Δh, respectively. It is worthwhile to note that $\hat{h}$ is the nominal part of $\dot{x}_{1d}$, which is obtained from the difference approximation of $\dot{x}_{1d}$. In other words, the term Δh presents the derivative approximation error. Furthermore, the measurement errors and current ripples on i_{L2} can be included in the uncertainties Δf, Δg, and Δh. This means that the robustness satisfies generality. From the terminal sliding function (6), its derivative with uncertainties is written as

$$
\begin{aligned}
\dot{s}(t) &= \hat{f}(x(t)) + \Delta f(x(t)) + g(x(t))d(t) \\
&\quad - \dot{x}_{1d}(t) + \alpha e^{q/p}(t) + \beta \dot{e}_I(t)
\end{aligned}
$$

The control law is modified as follows

$$
\begin{aligned}
d(t) &= \frac{-1}{\hat{g}}[\hat{f} - \dot{x}_{1d} + \alpha e^{q/p} + \beta \dot{e}_I \\
&\quad + k_1 s + k_2 sign(s)]
\end{aligned} \tag{11}
$$

where $\hat{f}$ and $\hat{g}$ are the nominal part of f and g, respectively. The closed-loop system becomes

$$
\begin{aligned}
\dot{s} &= \Delta f(x) - g(x)\hat{g}^{-1}(x)(k_1 s + k_2 sign(s)) \\
&\quad + (g(x) - \hat{g}(x))\frac{-1}{\hat{g}(x)}[\hat{f} - \dot{x}_{1d} \\
&\quad + \alpha e^{q/p} + \beta \dot{e}_I]
\end{aligned} \tag{12}
$$

Theorem 2: Consider the uncertain PV grid-connected power system (2) using the TSMC controller (11) with k_1, k_2 satisfying following conditions:

$$
\begin{aligned}
k_1 &> 0 \\
k_2 &> \varpi|\Delta f| + \\
&\quad + \varpi(\varpi - 1)\left|\hat{f} + \dot{x}_{1d} + \alpha|e|^{q/p} + \beta\right|
\end{aligned} \tag{13}
$$

The closed-loop control system is ensured with finite-time convergence for the MPC tracking of the inverter output current. $\square$

Proof. To prove the stability, consider the Lyapunov function

$$V_s = \frac{1}{2}s^2$$

and taking the time derivative of V_s along (12). The following is obtained,

$$
\begin{aligned}
\dot{V}_s &= s\Delta f(x) - sg(x)\hat{g}^{-1}(x)(k_1 s + k_2 sign(s)) \\
&\quad + s(g(x) - \hat{g}(x))\frac{-1}{\hat{g}(x)}[\hat{f} - \dot{x}_{1d} + \alpha e^{q/p} + \beta \dot{e}_I] \\
&\leq -k_1\varpi^{-1}s^2 - k_2\varpi^{-1}|s| + |\Delta f||s| \\
&\quad (\varpi - 1)\left|\hat{f} - \dot{x}_{1d} + \alpha e^{q/p} + \beta \dot{e}_I\right||s|
\end{aligned}
$$

If the gain conditions (13) are satisfied, then

$$\dot{V}_s \leq -\eta|s|.$$

The terminal sliding mode function is assured to be asymptotically stable and with finite convergence time. As for the finite time convergence of the tracking error $e(t)$, it is similar to the proof of Theorem 1. Thus, the terminal sliding mode control of the MPC current tracking is achieved with robustness for the uncertain grid-connected inverter. In other words, the MPPT is guaranteed for the PV power system.

Table 1. Parameters of PV module SP75 (36×1 cells)

Parameters	Value
Maximum power	$75W$
Rated voltage	$17V$
Rated current	$4.4A$
Short circuit current	$4.8A$
Open circuit voltage	$21.7V$
Short-circuit current temp. coefficient	$2.06mA/°\text{K}$

IV. NUMERICAL SIMULATIONS

For the verification of the MPP control scheme, a grid-connected PV power system is established with a boost converter, a full bridge inverter, and a $900W$ PV array. The PV array is composed of 6×2 PV panels, where each single PV panel has the electrical characteristic parameters stated in Table 1. From Fig. 1, the boost converter regulates the inverter input voltage to $180V$ by using a PI controller, such that the DC voltage V_{dc} is almost fixed. The parameters of the grid-connected inverter are chosen as $L_2 = 1.82\ mH$, $C_3 = 4\mu F$, $v_g = 110V/60Hz$. To carry out the TSMC based MPPT control, the incremental conductance method stated in Section III is used to determine the reference MPC i_{L2}^* with the incremental value $\triangle i^* = 0.0005$ for each step. The frequency of the searching algorithm is set to 10KHz, which will enhance the speed of MPPT. On the other hand, the TSMC controller parameters are set to $k_1 = 1000$, $k_2 = 10$, $\alpha = 800$, $\beta = 10$, $p = 13$, and $q = 7$. In the following, two scenarios including fixed insolation and varying insolation are simulated to verify the proposed scheme.

First, consider the PV array operated in a fixed insolation $100mW/cm^2$ and cell temperature 25°C. By using the TSMC based MPPT controller (11), simulation results are obtained as shown in Figs. 3 and 4. The PV power system is quickly regulated to the maximum power point. Next, consider a varying atmosphere at 25°C and a piecewise-constant insolation. With the above settings, the PV MPPT control is performed and renders to the control results given in Fig. 5. Figure 6 also shows the comparison of current tracking errors for TSMC and PI controller (which is well tuned). Obviously, the traditional controller renders to residual current error which reduces the PV power efficiency. Therefore, the TSMC based MPPT provides fast and stable maximum power control of grid-tied PV power systems.

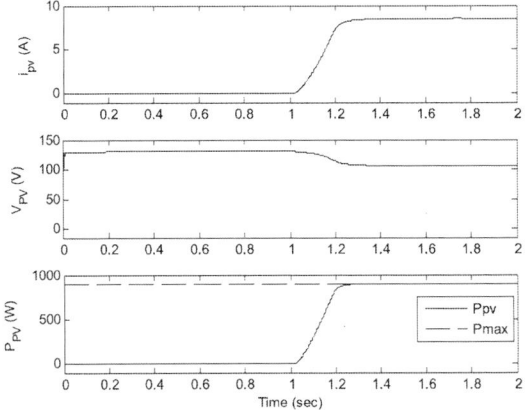

Fig. 3. Control response of the PV output current, voltage, and power.

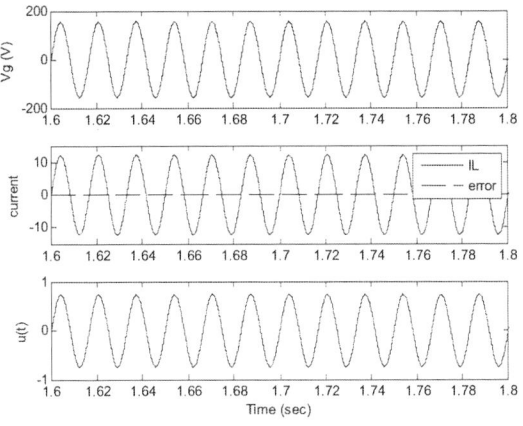

Fig. 4. (a) The grid voltage; (b) inverter current and control error (dashed line); (c) control input.

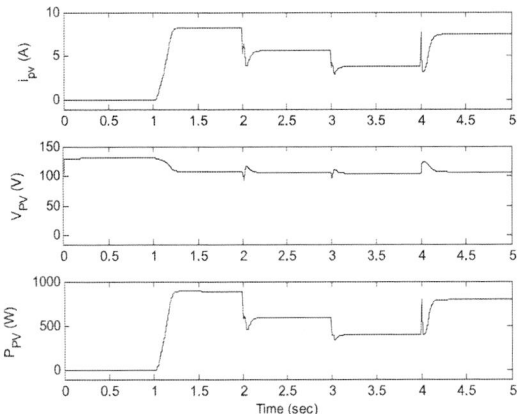

Fig. 5. Control response of the PV output current, voltage, and power for varying insolation.

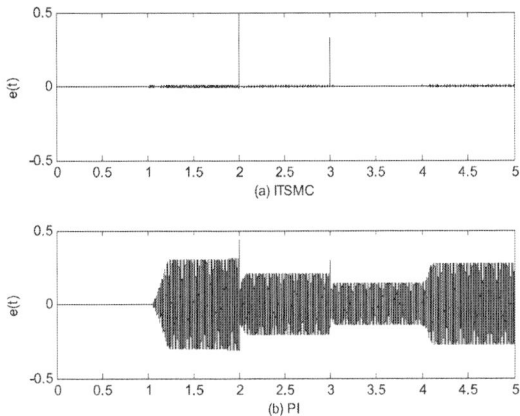

Fig. 6. Error comparison for (a) TSMC; (b) PI control.

V. EXPERIMENTAL RESULTS

Based on the same setting as the numerical simulation, some experiments are performed to verify the validity of the proposed method. For simplification, a programmable power supply (Chroma 62150H-600s) is used to emulate the 900W PV array, which parameters are given in Table 1. Note that the characteristic of the PV array (1) is programmed for different solar insolation and cell temperature. After the boost converter and DC/AC inverter circuit are constructed, a DSP module TMS320F28335 is applied to realize the MPC searching algorithm and TSMC current controller. The control duty input $d(t)$ is directly transformed to switching PWM signals for the IGBT full bridge of the inverter. The switching frequency of the inverter is set to 38KHz. In addition, the TSMC is also compared with the traditional PI controller

$$u(t) = K_p e(t) + K_I \int_0^t e(\tau) d\tau$$

with controller gains $K_p = 2$ and $K_I = 800$. Here the TSMC control parameters are set as same as the simulation examples. First, considering a fixed insolation,

the TSMC controller results in the inverter voltage and output current as show in Fig. 7, where the current is synchronized with the grid voltage. In comparison, Table 2 shows the percentage of different harmonic orders of the current for the TSMC and PI controllers. Obviously, the current harmonic is reduced by the TSMC with finite-time convergence. Furthermore, the power response of three MPPT methods are shown in Fig. 8, where the MPPT controllers act at 3.5 sec. A traditional sliding mode control (SMC) with $p = q = 1$ is also applied to the system. From the results, the proposed TSMC based MPPT has better performance for the grid-tied PV power generation system.

Table 2. Comparison of current harmonic.

Harmonic order	TSMC controller	PI controller
1th~10th	1.6%	2.3%
11th~15th	1.0%	1.5%
16th~22th	1.2%	1.6%

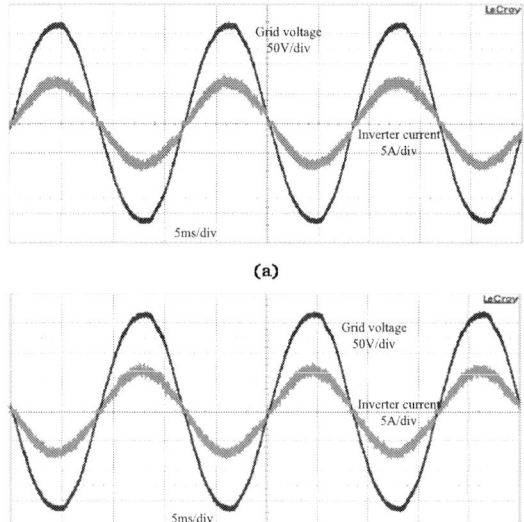

(a)

(b)

Fig. 7.   Experimental inverter voltage and output current for (a)PI control; (b) TSMC.

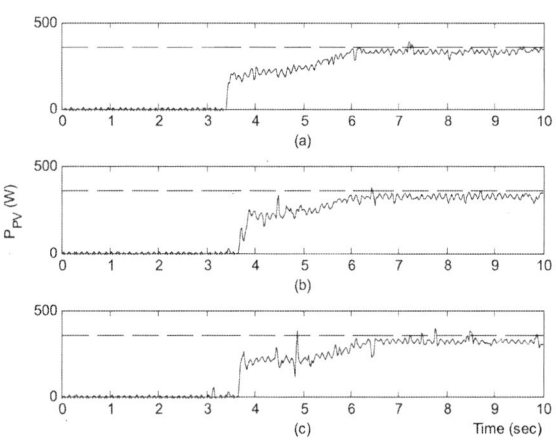

Fig. 8.   Power response of MPPT for (a) TSMC; (b) SMC; (c) PI control.

VI.   CONCLUSIONS

In this paper, the terminal sliding mode control has been introduced for improving the MPPT of the grid-tied PV power generation system. Even if considering uncertainty and changing atmosphere, the grid-tied inverter is controlled with finite-time MPC tracking performance. Due to the faster MPC tracking, the PV system is effectively and quickly regulated to the maximum power point. Compared with the traditional MPPT methods, the PV power extracted by TSMC is higher and more stable. Moreover, by more accurate current tracking, the current harmonic is reduced. Therefore, the TSMC-based MPPT control method assures better tracking performance and higher efficiency for the grid-tied PV power system.

ACKNOWLEDGEMENT

This work was supported by the National Science Council, R.O.C., under Grants NSC-102-2221-E-033-053.

REFERENCES

[1]   Z. Salameh and D. Taylor, "Step-up maximum power point tracker for photovoltaic arrays," *Solar Energy*, vol. 44, no. 1, pp. 57-61, 1990.

[2]   D. Petreus, T. Patarau, S. Daraban, D. Moga, B. Morley, "A novel maximum power point tracker based on analog and digital control loops," *Solar Energy*, vol. 85, pp. 588-600, 2011.

[3]   K. H. Hussein, I. Muta, T. Hoshino, and M. Osakada, "Maximum photovoltaic power tracking: an algorithm for rapidly changing atmospheric condition," *IEE Proc.-Gener. Trans. Distrib.*, vol. 142, no. 1, pp. 59-64, 1995.

[4]   I. H. Altasa and A. M. Sharaf, "A novel maximum power fuzzy logic controller for photovoltaic solar energy systems," *Renewable Energy*, vol. 33, pp. 388-399, 2008.

[5]   S. Lalouni, D. Rekioua, T. Rekioua, and E. Matagne, "Fuzzy logic control of stand-alone photovoltaic system with battery storage," *Journal of Power Sources*, vol. 193, pp. 899-907, 2009.

[6]   M. Taherbaneh and K. Faez, "Maximum power point estimation for photovoltaic systems using neural networks," *Proc. IEEE Int. Conf. Control Autom.*, pp. 1614-1619, 2007.

[7]   L. Wu, Z. Zhao, J. Liu, "A single-stage three-phase grid-connected photovoltaic system with modified MPPT method and reactive power compensation," *IEEE Trans. Energy Conversion*, vol. 22, no. 4, pp. 881-886, 2007.

[8]   L. Hassaine, E. Olias, J. Quintero, A. Barrado, "Digital control based on the shifting phase for grid connected photovoltaic inverter," *Applied Power Electronics Conference and Exposition*, pp. 945-961, 2008.

[9]   I. S. Kim, "Robust maximum power point tracker using sliding mode controller for the three-phase grid-connected photovoltaic system," *Solar Energy*, vol. 81, pp. 405-414, 2007.

[10]   V. Salas, E. Olias, A. Barrado, and A. Lazaro, "Review of the maximum power point tracking algorithms for stand-alone photovoltaic systems," *Solar Energy Materials & Solar Cells*, vol. 90, pp. 1555-1578, 2009.

[11]   J.-J. E. Slotine and W. Li, *Applied Nonlinear Control*. Englewood Cliffs, New Jersey: Prentice-Hall Inc., 1991.

[12]   S. T. Venkataraman, and S. Gulati, "Control of nonlinear systems using terminal sliding modes," *ASME Journal of Dynamic Systems, Measurement and Control*, vol. 115, pp. 554-560, 1993.

[13]   J. Min and Z. Xu, "Backstepping control for a class of uncertain systems based on non-singular terminal sliding mode," *Int. Conf. on Industrial Mechatronics and Automation*, pp. 169-172, 2009.

[14]   E. C. Chang, T. J. Liang, J.-F. Chen, F.-J. Chang, "Real-time implementation of grey fuzzy terminal sliding mode control for PWM DC-AC converters," *IET Power Electronics*, vol. 1, no. 2, pp. 235-244, 2008.

[15]   V. Parra-Vega and G. Hirzinger, "Chattering-free sliding mode control for a class of nonlinear mechanical systems," *International Journal of Robust and Nonlinear Control*, vol. 11, no. 12, pp. 1161-1178, 2004.

[16]   O. Barambones and V. Etxebarria, "Energy-based approach to sliding composite adaptive control for rigid robots with finite error convergence time," *International Journal of Control*, vol. 75, no. 5, pp. 352-359, 2002.

[17]   Z. Man and X. H. Yu, "Terminal sliding mode control of MIMO linear systems," *IEEE Trans. on Circuits and Systems I: Fundamental Theory and Applications*, vol. 44, no. 11, pp. 1065-1070, .1997

A New Voltage Type Magnetically Coupled T-source Inverter

Q.V. Tran, K.S. Low, *Senior Member*, IEEE

School of Electrical & Electronic Engineering, Nanyang Technological University

qvtran@ntu.edu.sg, k.s.low@ieee.org

Abstract-This paper proposes a new impedance-source inverter named as magnetically coupled T-source inverters (MTSI). The proposed circuit employs a three-winding coupled transformer and a capacitor formed in T-shape to produce a higher voltage gain in comparison with the traditional Z-source inverter (ZSI), Trans-Z-source inverter (trans-ZSI) and Γ-source inverters (ΓSI). While maintaining the main features of the previously presented impedance-source network, the new network exhibits an unique characteristic, that is its voltage gain can be raised not only by decreasing the first turn ratio α_{mT}, but also by increasing the second turn ratio β_{mT} of the coupled transformer. Simulation results are conducted to validate the analysis.

I. INTRODUCTION

Rapid development of renewable energy, distributed generation and electric vehicles demands for more challenging converters to be innovated. Traditional voltage source inverter (VSI) or current-source inverter (CSI) has similar limitations and drawbacks. Particularly, the VSI can only be used as a voltage step down inverter and the CSI as a current-buck inverter. Moreover, the dead time is required to protect switching device from short circuit state. Adding step-up dc-dc converter before the inverters can overcome the aforementioned problem. However, this results in a higher cost, lower reliability and no single-stage power conversion.

In recent years, the Z-source inverter and quasi-ZSI [1], [2] were proposed as potential inverters to replace the traditional one. It advantageously utilizes an impedance network and shoot-through of the inverter bridge to boost voltage in VSIs or open circuit to boost current in CSIs. Omission of the dead time and the use of an impedance network also help to increase the immunity of the inverter to electromagnetic interference (EMI) [3]. On the other hand, the voltage-fed Z-source/quasi-Z-source inverters cannot have bidirectional operation unless the diode is replaced with a bidirectional conducting, unidirectional blocking switch [4].

Research trend of the impedance-source inverters has been focused on some prospective applications such as motor drive, grid-connected photovoltaic, hybrid electric vehicle etc. [5]-[8], pulse width modulation techniques for facilitating control of the inverter [9]-[11]. The maximum voltage gain using maximum boost control [9] by turning all the zero states in the traditional VSIs to shoot-through zero states. However, it causes low-frequency ripples in the ac-side fundamental frequency. The achieved voltage gain by constant boost control [10] is slightly less than the previous one but it can remove the aforementioned low–frequency ripple and thus reduce the

size of inductor and capacitor in Z-source network. On the other hand, the concept has been applied to dc-dc converters [12], ac-dc rectifiers [13] and modified to ac-ac converters [14], [15].

Recently, some modifications of impedance network have been proposed to increase the voltage gain in [16]-[21]. These techniques included the switched-component techniques [16]-[17], cascading techniques [18], and magnetically coupled techniques [19]-[21]. Among them, the magnetically coupled techniques are more attractive because it can produce a high gain with the support of coupled transformers or inductors with minimum component counts. T-source inverter and Trans-Z-source inverter can boost the dc voltage by increasing the turn ratio of coupled transformer, whereas Γ-source inverter can obtain the same result via inversed way. It utilizes a smaller turn ratio of coupled transformer to obtain a higher voltage gain. However, the voltage gain of Γ-source inverter is very sensitive with the turn ratio at high voltage gain demanding applications such as EV/HEV motor drive and fuel cell power conversion as well.

In this paper, the magnetically coupled T-source inverter inherits both outstanding features of Trans-Z-source inverter and Γ-source inverter to produce a higher voltage gain by employing three-winding coupled transformer and one capacitor connected in T-shape. Its gain is raised not only by lowering the first transformer turns ratio α_{mT} but also by increasing the second turn ratio β_{mT}. This is a very unique feature that is not available in other impedance source circuits. The special feature and operating performance of the proposed inverter have been validated by theoretical analysis and simulation results.

II. TRANS-Z-SOURCE AND Γ-SOURCE INVERTER

The traditional Z-source inverter is shown in Fig. 1(a). According to [1], the capacitor voltage V_c, the peak dc-link voltage $\hat{v}_i$ and the peak phase output voltage $\hat{v}_{ac}$ of the traditional Z-source inverter can be boosted by increasing the shoot-though ratio D_{st} that are expressed in the following equations.

$$V_c = \frac{1-D_{st}}{1-2D_{st}} V_{dc}$$

$$\hat{v}_i = \frac{1}{1-2D_{st}} V_{dc} \qquad (1)$$

$$\hat{v}_{ac} = \frac{1}{1-2D_{st}} 0.5 M_a V_{dc}$$

where V_{dc}, D_{st}, M_a are the input dc voltage, shoot-though ratio and modulation index of the traditional Z-source inverter, respectively.

The 2014 International Power Electronics Conference

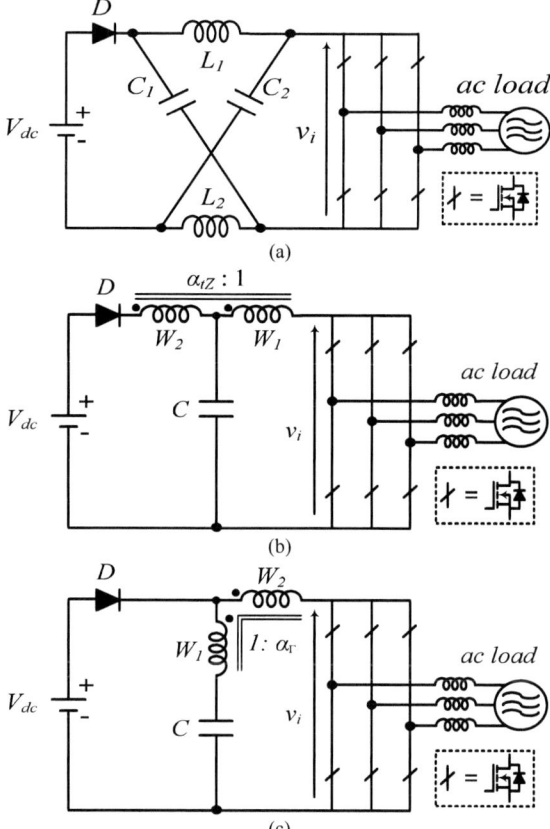

(a)

(b)

(c)

Fig. 1. Magnetically coupled impedance source inverter: (a) Traditional Z-source inverter, (b) Trans-Z-source inverter, (c) Γ-source inverter.

With a larger shoot-through ratio for high gain application, a lower modulation index can cause a higher voltage stress across the switching devices and lead to more distortion of the inverter output voltage. To overcome the drawbacks, the Trans-Z-source inverter and Γ-source inverter shown in Fig. 1(b) and Fig. 1(c), respectively, have been proposed recently in [20], [21]. With a coupled transformer, a high boost gain can be achieved easier than that of the traditional Z-source inverters. Voltage expressions of the Tran-Z-source inverter can be shown as below.

$$V_c = \frac{1 - D_{tZ}}{1 - (1 + \alpha_{tZ})D_{tZ}} V_{dc}$$

$$\hat{v}_i = \frac{1}{1 - (1 + \alpha_{tZ})D_{tZ}} V_{dc} \qquad (2)$$

$$\hat{v}_{ac} = \frac{1}{1 - (1 + \alpha_{tZ})D_{tZ}} 0.5 M_{tZ} V_{dc}$$

where α_{tZ} is the turn ratio of the coupled transformer.

The limit of shoot-through ratio D_{tZ} can be appropriately determined as $0 \leq D_{tZ} < \frac{1}{(1 + \alpha_{tZ})}$ by setting the denominator of (2) to be greater than zero. The range of corresponding modulation index M_{tZ} can therefore be stated as $0 \leq M_{tZ} \leq 1.15(1 - D_{tZ})$. Obviously, the voltage gain of Trans-Z-source inverter can be raised by increasing $\alpha_{tZ} > 1$, which cannot be attained by the traditional Z-source inverter. This result in a smaller

shoot-through ratio, a higher modulation index and a lower total harmonic distortion of the inverter output voltage at the same voltage gain compared with that of the traditional one. It is concluded that the Trans-Z-source has been capable of producing a high voltage gain but is stressed by high instantaneous current $\hat{\imath}_{st}$ during shoot-through state [20]. This pulse current is proportional to transformer turn ratio α_{tZ} to stress on switching device and transformer winding W_l as well. On the other hand, a high transformer turn ratio is also a demerit of the Trans-Z-source inverter since it causes a bulky and high cost inverter.

$$\hat{\imath}_{st} = (1 + \alpha_{tZ})I_{dc} \qquad (3)$$

where I_{dc} presents the dc input current draw from the source.

The Γ-source inverter has been suggested in [21] to improve the performance of the Trans-Z-source inverter. The voltage relationship of the Γ-source inverter can be described as

$$V_c = \frac{1 - D_\Gamma}{1 - \left(1 + \frac{1}{\alpha_\Gamma - 1}\right)D_\Gamma} V_{dc}$$

$$\hat{v}_i = \frac{1}{1 - \left(1 + \frac{1}{\alpha_\Gamma - 1}\right)D_\Gamma} V_{dc} \qquad (4)$$

$$\hat{v}_{ac} = \frac{1}{1 - \left(1 + \frac{1}{\alpha_\Gamma - 1}\right)D_\Gamma} 0.5 M_\Gamma V_{dc}$$

where α_Γ is the turn ratio of coupled transformer.

From (4), it can be observed that the Γ-source inverter can produce a high voltage gain by lowering transformer turn ratio instead of increasing as the Trans-Z-source inverter. By setting the common denominator of the above equations to be greater than zero, the maximum shoot-though ratio can be read as $D_\Gamma = (\alpha_\Gamma - 1)/\alpha_\Gamma$ and the maximum modulation index is also limited at $M_\Gamma = 1.15(1 - D_\Gamma)$. Using smaller turn ratio can generate a higher voltage gain to be a significant feature of the Γ-source inverter which has not been mentioned before. Consequently, its coupled transformer is more compact and lower cost than the Trans-Z-source inverter. However, the voltage gain in terms of the turn ratio would be not a linear relationship and so sensitive at low turn ratio area. A critical small change of the turn ratio can cause a dramatically change of the voltage gain which results in difficulty of transformer design.

III. PROPOSED MAGNETICALLY COUPLED T-SOURCE INVERTER

To overcome the constrains of Γ-source and Trans-Z-source inverter, a magnetically coupled T-source inverter shown in Fig. 2, is proposed in this paper. The proposed circuit employs a three-winding coupled transformer and a capacitor formed in T-shape to produce a high voltage gain. The transformer includes of one primary winding W_1 and two secondary windings W_2 and W_3. The voltage gain of new magnetically coupled T-source inverter which is definitely higher than that of both mentioned inverters, can be raised by employing a three-winding

978-1-4799-2706-7/14 $31.00 © 2014 IEEE

transformer and a capacitor connected in T-shape. The higher voltage gain can be obtained not only by means of reducing the first turn ratio α_{mT}, but also by increasing the second turn ratio β_{mT}, where α_{mT}, β_{mT} is turn ratio of secondary windings W_2, W_3 with respect to primary winding W_1, respectively. This unique feature is a concept that has not been mentioned before. It can broaden promising application of the new inverter in the low input voltage area, such as photovoltaic system or HEV/EV and so on. Some simulations are carried to validate the analysis and performance of the magnetically coupled T-source inverter to be shown next section.

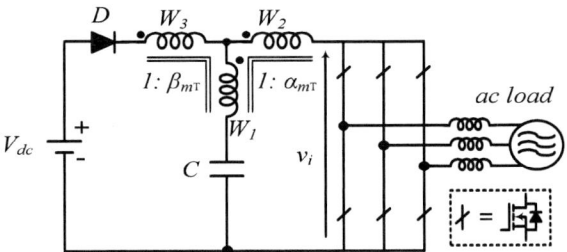

Fig. 2. Proposed magnetically coupled T-source inverter

Similar to the traditional Z-source inverter, the magnetically coupled T-source inverter has an extra shoot-through zero state and non-shoot-though state that consists of six active states and two traditional zero states. The shoot-through zero state can be achieved by short-circuiting both the upper and lower switching devices of any phase leg of the inverter. It contributes to the unique buck–boost feature which has never gotten from the traditional one. For analyzing operation of the magnetically coupled T-source inverter, the overall equivalent circuit viewed from the inverter dc side can be drawn in Fig. 3(a).

From Fig. 3(a), voltage relationships among primary and secondary windings are always obtained as

$$\frac{v_{w2}}{v_{w1}} = \frac{n_{w2}}{n_{w1}} = \alpha_{mT} \tag{5}$$

$$\frac{v_{w3}}{v_{w1}} = \frac{n_{w3}}{n_{w1}} = \beta_{mT} \tag{6}$$

where n_{w1}, n_{w2}, n_{w3} is number of turn of winding W_1, W_2, W_3, respectively. Subscript mT represents the magnetically coupled T-source in short.

In the shoot-through zero state for a time interval of $(D_{mT}T)$, the diode D is blocked, where D_{mT} is shoot-through ratio over one switching period T. Equivalent circuit of the inverter is hence shown in Fig. 3(b). The voltages across primary winding W_1 and the inverter peak dc-link voltage $\hat{v}_i$ can be expressed as

$$v_{w1} = v_{w2} - V_c = \alpha_{mT}v_{w1} - V_c \tag{7}$$

$$v_{w1} = \frac{V_c}{\alpha_{mT}-1} \tag{8}$$

$$\hat{v}_i = 0 \tag{9}$$

In any of the non-shoot-through states for an interval of $(1-D_{mT})T$, the inverter bridge can be modeled as an equivalent current source i_i as shown in Fig. 3(c). Apply Kirchhoff's voltage loop to this state, the voltage relationship can be described in (10).

$$V_{dc} - V_c = v_{w1} + v_{w3} = v_{w1} + \beta_{mT}v_{w1} \tag{10}$$

The voltages across primary winding W_1 and the inverter peak dc-link voltage $\hat{v}_i$ can be derived in (11) and (12).

$$v_{w1} = \frac{V_{dc}-V_c}{\beta_{mT}+1} \tag{11}$$

$$\hat{v}_i = V_{dc} - v_{w2} - v_{w3}$$

$$= V_{dc} - \left(\frac{\alpha_{mT}+\beta_{mT}}{\beta_{mT}+1}\right)(V_{dc} - V_c) \tag{12}$$

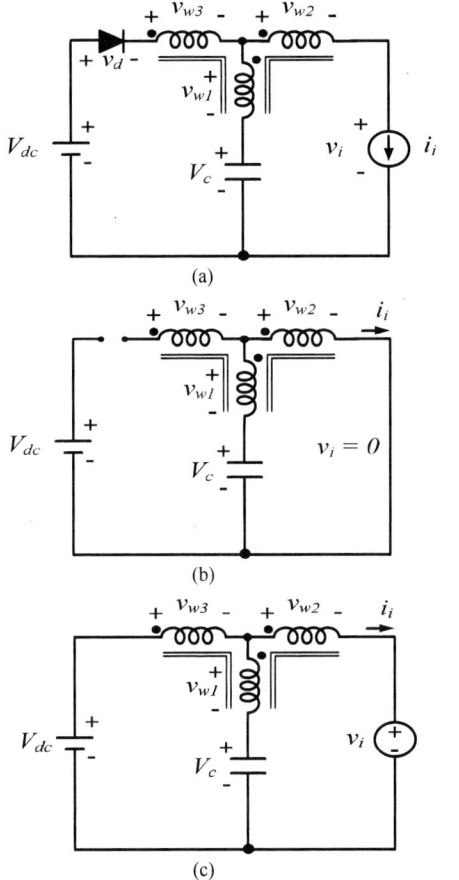

(a)

(b)

(c)

Fig. 3. Proposed magnetically coupled T-source inverter equivalent circuit: (a) MTSI view from dc source; (b) MTSI in shoot-though state; (c) MTSI in non-shoot-through state.

The average voltage of primary winding W_1 must be zero over one switching period in the steady state. Thus, summing (8) and (11) with their respective duty cycles yield

$$\bar{v}_{w1} = (D_{mT}T)\frac{V_c}{\alpha_{mT}-1} + (1-D_{mT})T\frac{V_{dc}-V_c}{\beta_{mT}+1} = 0 \tag{13}$$

The capacitor voltage can be calculated by solving (13)

978-1-4799-2706-7/14 $31.00 © 2014 IEEE   448

$$V_c = \frac{1 - D_{mT}}{1 - \left(1 + \frac{\beta_{mT}+1}{\alpha_{mT}-1}\right)D_{mT}} V_{dc} \qquad (14)$$

From (12) and (14), the peak dc-link voltage of the inverter bridge can be derived as

$$\hat{v}_i = \frac{1}{1 - \left(1 + \frac{\beta_{mT}+1}{\alpha_{mT}-1}\right)D_{mT}} V_{dc} = B_{mT} V_{dc} \qquad (15)$$

where the boost factor is $B_{mT} = \frac{1}{1 - \left(1 + \frac{\beta_{mT}+1}{\alpha_{mT}-1}\right)D_{mT}}$ (16)

Substitute (14) into (12), the peak value of the ac phase voltage from the inverter output can be derived as

$$\hat{v}_{ac} = \frac{1}{1 - \left(1 + \frac{\beta_{mT}+1}{\alpha_{mT}-1}\right)D_{mT}} 0.5 M_{mT} V_{dc} \qquad (17)$$

From (16) and (17), it can be realized that both of the boost factor B_{mT} and input-to-output voltage gain $\{ \hat{v}_{ac}/V_{dc} \}$ of the proposed inverter can be raised apparently not only by lowering its the first turn ratio α_{mT}, but also by increasing the second turn ratio β_{mT}. This is an important characteristics of the proposed magnetically coupled T-source inverter.

IV. MAGNETICALLY COUPLED T-SOURCE INVERTER VERSUS Γ-SOURCE AND TRANS-Z-SOURCE INVERTER

The magnetically coupled T-source inverter can be considered as a general form of the Γ-source inverter since it becomes the Γ-source by setting $\beta_{mT} = 0$. Comparing $\hat{v}_{ac}$ in (17) with (4) reveals that the voltage gain of the proposed magnetically coupled T-source inverter is always higher than the Γ-source inverter for any β_{mT} greater than zero for a given modulation index and shoot-though ratio. Hence, its shoot-though ratio range becomes narrower and modulation index is wider. Its shoot-through ratio range can be determined as $0 \leq D_{mT} < \frac{1}{\left(1 + \frac{\beta_{mT}+1}{\alpha_{mT}-1}\right)}$ and its maximum modulation index therefore is $M_{mT} = 1.15(1 - D_{mT})$.

Fig. 4. shows a boost factor comparison of the proposed inverter with that of the Γ-source inverter and the Trans-Z-source inverter as well. Aim to be easy to comparison, the shoot-through ratio and the first turn ratio of those inverter are set equal as $D_{mT} = D_{tZ} = D_\Gamma = 0.2$ and $\alpha_{mT} = 1.5$, respectively. Whereas the second turn ratio β_{mT} of the proposed inverter is tuned from zero to 0.6. At the given condition, the boost factor value of the Trans-Z-source and the Γ-source inverter is respectively $B_{tZ} = 2$ and $B_\Gamma = 2.5$. The one of the magnetically coupled T-source inverter is impressively boosted to $B_{mT} = 3.57$ and 6.25 in case of $\beta_{mT} = 0.3$ and 0.6, respectively.

Table I shows the comparison of the governing equations for the Trans-Z-source, Γ-source and magnetically coupled T-source inverter. The capacitor voltage, dc-link voltage, peak ac output voltage and reversed diode voltage are concisely shown here for comparison. The range of the shoot-through ratio and the modulation index and turn ratio trend are also mentioned to reveal the outstanding feature of each the impedance-source inverter.

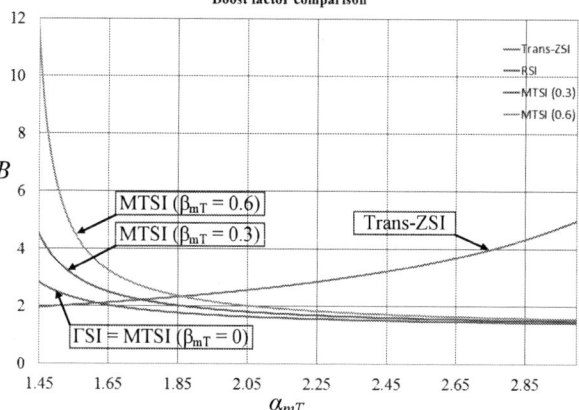

Fig. 4. Boost factor comparison of the magnetically coupled T-source, Γ-source and Trans-Z-source inverter.

V. SIMULATION RESULTS

Simulation study has been conducted to verify the unique gain tuning feature of the magnetically coupled T-source inverter. All simulating parameters are listed in Table II. Modified space vector modulation scheme in [6] is used to adjust the shoot-though ratio. Modulation index and shoot-through ratio are fixed at 0.9 and 0.15 for the following simulations, respectively. The input dc voltage is set as 100 V. A second order low pass filter with a cutoff frequency of 1.125kHz is utilized before a three-phase 20Ω/phase resistive load.

TABLE II.
PARAMETERS USED FOR SIMULATIONS

Parameter/Description	Value
Input voltage V_{dc}	100 V
T-source capacitor C	500 µF
Primary winding turn W_1	n_{W1} = 70 turns
Secondary winding turn W_2	Case 1 (α_{mT} = 1.8): n_{W2} = 126 turns Case 3 (α_{mT} = 1.5): n_{W2} = 105 turns
Secondary winding turn W_3	Case 1 (β_{mT} = 0.7): n_{W2} = 49 turns Case 2 (β_{mT} = 1.0): n_{W2} = 70 turns
Inductive ac filter L_f	2.5 mH/phase
Capacitive ac filter C_f	8 µF/phase
Three-phase Y-connected load R_L	20 Ω/phase
Carrier frequency	5 kHz
Modulating frequency	50 Hz
Shoot-through ratio	0.15
Modulation index	0.9

The first two simulations are fulfilled to proof the effect of increasing turn ratio β_{mT} on raise of the voltage gain. The turn ratio β_{mT} is increased from 0.7 to 1.0 in simulation case 1 and 2, respectively, while keeping the turn ratio α_{mT} at 1.8. Fig. 5 shows simulation results in case 1 where transient and stead state response of the mentioned voltages of the system are shown in Fig. 5(a) and Fig. 5(b), respectively. The capacitor voltage is obviously boosted to 159 V from the dc input voltage of 100V in case 1. Simultaneously, dc-link voltage should be chopping and has a peak of 187 V which approximately reaches to theoretical value. Corresponding ac output voltage from inverter bridge v_{ac} is pure sinusoidal and its peak value is 84.4 V that is almost the same as computing value of 84.7 V. In addition, the shoot-though i_{ST} and input dc current i_{dc} over one switching period used modified space vector modulation scheme suggested in [6], are apparently shown in Fig. 5(c), where the shoot-through state is evenly distributed on three phases leg of the inverter bridge.

By increasing turn ratio β_{mT} to 1.0 in simulation case 2, the simulation results shown in Fig. 6(a) reveals that the peak dc-link voltage $\hat{v}_i$ is raised correspondently from 187 V to 207 V. The inverter peak ac output voltage $\hat{v}_{ac}$ is hence boosted to 92.5 V (or 113 line RMS) in this case. The simulation results have been justified that the voltage gain B_{mT} of the proposed magnetically coupled T-source inverter can be enhanced by increasing the turn ratio β_{mT}.

Fig. 6(b) shows simulation results in case 3, where the turn ratio α_{mT} is reduced from 1.8 to 1.5 while fixing β_{mT} at value of 0.7. This simulation aims to verify that lowering α_{mT} can be also impacted to the raise of the inverter voltage gain. Since reducing turn ratio α_{mT}, the corresponding peak dc-link voltage $\hat{v}_i$ and the capacitor voltage V_c are stepped up to 286 V and 244 V, respectively. The peak ac output voltage $\hat{v}_{ac}$ is therefore impressively boosted to a value of 127 V. Theory and simulation value of the capacitor voltage V_c, peak dc-link voltage $\hat{v}_i$ and peak ac output voltage $\hat{v}_{ac}$ as well of the inverter are summarized concisely case by case in table III. The simulation value of the ac output voltage is dropped a little bit with respected to that of theoretical value would be caused by drop voltage on filter inductor. The matching between simulation results and theoretical analysis consistently proofs the proposed analysis.

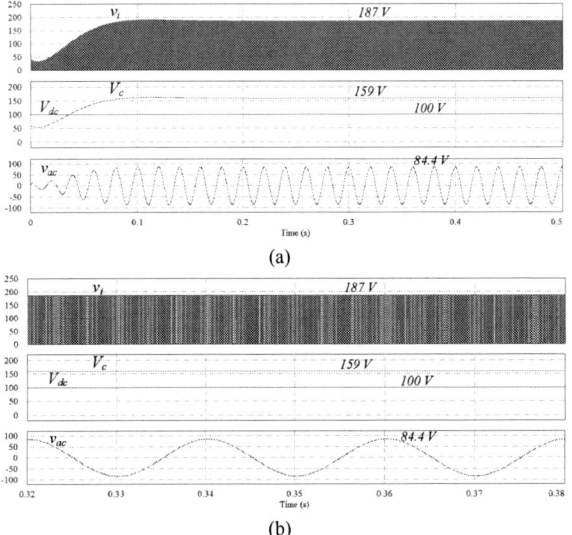

(a)

(b)

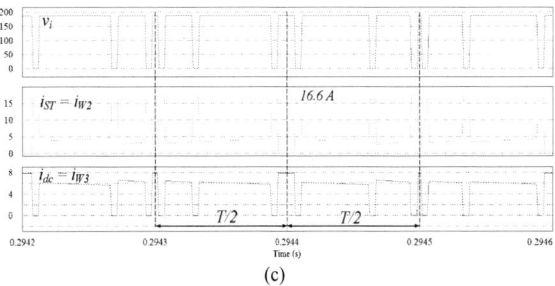

(c)

Fig. 5. Voltage and current response in simulation case 1: (a) Transient response; (b) Steady state response, (c) Current response at steady state.

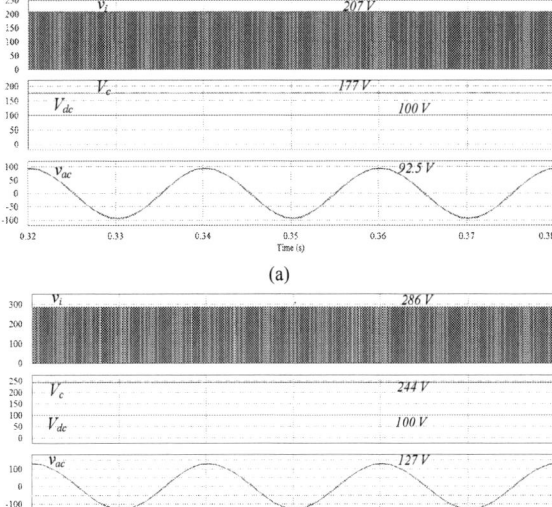

(a)

(b)

Fig. 6. Voltage response in simulation case 2 & 3: (a) Voltage response at steady state in case 2, (b) Voltage response at steady state in case 3.

VI. CONCLUSION

This paper proposes a new magnetically coupled T-source inverter with high boost capability by employing a three-winding coupled transformer and one capacitor formed in T-shape as an impedance network. Its voltage gain can be raised not only by reducing the first turn ratio α_{mT} but also by increasing the second turn ratio β_{mT}. This is a unique feature of the proposed inverter which has been reported. Simulation results have validated the theoretical analysis.

REFERENCES

[1] F. Z. Peng, "Z-source inverter," IEEE Trans. Ind. Appl., vol. 39, no. 2, pp. 504-510, Mar/Apr. 2003.

[2] J. Anderson and F. Z. Peng, "Four quasi-Z-source inverters," in Proc. IEEE Power Electron. Spec. Conf., (PESC'08), pp. 2743-2749.

[3] M. Shen, A. Joseph, J. Wang, F. Z. Peng, and D. J. Adams, "Comparison of traditional inverters and Z-source inverter for fuel cell vehicles," IEEE Trans. Power Electron., vol. 22, no. 4, pp. 1453-1463, Jul. 2007.

TABLE I.

COMPARISON OF THE MAGNETICALLY COUPLED T-SOURCE INVERTER WITH TRANS-Z-SOURCE AND Γ-SOURCE INVERTER

$$\left(K_\Gamma = \frac{1}{\alpha_\Gamma - 1}; \quad K_{mT} = \frac{\beta_{mT}+1}{\alpha_{mT}-1}\right)$$

Feature	Trans-Z-source inverter	Γ-source inverter	Magnetically coupled T-source inverter
Dc-link voltage $\hat{v}_i$	$\dfrac{1}{1-(1+\alpha_{tz})D_{tz}}V_{dc}$	$\dfrac{1}{1-\left(1+\frac{1}{\alpha_\Gamma-1}\right)D_\Gamma}V_{dc}$	$\dfrac{1}{1-\left(1+\frac{\beta_{mT}+1}{\alpha_{mT}-1}\right)D_{mT}}V_{dc}$
Capacitor voltage V_c	$\dfrac{1-D_{tz}}{1-(1+\alpha_{tz})D_{tz}}V_{dc}$	$\dfrac{1-D_\Gamma}{1-\left(1+\frac{1}{\alpha_\Gamma-1}\right)D_\Gamma}V_{dc}$	$\dfrac{1-D_{mT}}{1-\left(1+\frac{\beta_{mT}+1}{\alpha_{mT}-1}\right)D_{mT}}V_{dc}$
Peak output voltage $\hat{v}_{ac}$	$\dfrac{1}{1-(1+\alpha_{tz})D_{tz}}0.5M_{tz}V_{dc}$	$\dfrac{1}{1-\left(1+\frac{1}{\alpha_\Gamma-1}\right)D_\Gamma}0.5M_\Gamma V_{dc}$	$\dfrac{1}{1-\left(1+\frac{\beta_{mT}+1}{\alpha_{mT}-1}\right)D_{mT}}0.5M_{mT}V_{dc}$
Diode voltage v_d	$\dfrac{\alpha_{tz}}{1-(1+\alpha_{tz})D_{tz}}V_{dc}$	$\dfrac{K_\Gamma}{1-(1+K_\Gamma)D_\Gamma}V_{dc}$	$\dfrac{K_{mT}}{1-(1+K_{mT})D_{mT}}V_{dc}$
Turn ratio trend for increasing gain	Increasing : $\alpha_{tz} \geq 1$	Decreasing: $1 \leq \alpha_\Gamma \leq 2$	Decreasing: $1 \leq \alpha_{mT} \leq 2$ Increasing: $\beta_{mT} \geq 0$
Range of shoot-through duty ratio	$0 \leq D_{tz} < \dfrac{1}{1+\alpha_{tz}}$	$0 \leq D_\Gamma < \dfrac{1}{1+\frac{1}{\alpha_\Gamma-1}}$	$0 \leq D_\Gamma < \dfrac{1}{1+\frac{\beta_{mT}+1}{\alpha_{mT}-1}}$
Range of modulation index	$0 \leq M_{tz} \leq 1.15(1-D_{tz})$	$0 \leq M_\Gamma \leq 1.15(1-D_\Gamma)$	$0 \leq M_{mT} \leq 1.15(1-D_{mT})$

TABLE III.

SUMMARY OF THE VOLTAGE OF THE MAGNETICALLY COUPLED T-SOURCE INVERTER IN THEORETICAL ANALYSIS AND SIMULATIONS

Parameter	Case 1: $\alpha_{mT}=1.8, \beta_{mT}=0.7$		Case 2: $\alpha_{mT}=1.8, \beta_{mT}=1.0$		Case 3: $\alpha_{mT}=1.5, \beta_{mT}=0.7$	
	Theoretical value	Simulation value	Theoretical value	Simulation value	Theoretical value	Simulation value
Capacitor voltage V_c	160 V	159 V	179 V	177 V	250	244 V
Peak dc-link voltage $\hat{v}_i$	188 V	187 V	210 V	207 V	294	286 V
Peak ac output voltage $\hat{v}_{ac}$	84.7 V	84.4 V	94.7 V	92.5 V	132.3 V	127 V

[4] H. Xu, F. Z. Peng, L. Chen, and X. Wen, "Analysis and design of Bidirectional Z-source inverter for electrical vehicles," inProc. 23rd IEEE Annu. Appl. Power Electron. Conf. Expo., (APEC'08), pp. 1252–1257.

[5] F. Z. Peng, A. Joseph, J. Wang, M. Shen, L. Chen, Z. Pan, E. OrtizRivera, and Y. Huang, "Z-source inverter for motor drives," IEEE Trans. Power Electron., vol. 20, no. 4, pp. 857-863, Jul. 2005.

[6] Q.V Tran, T.W. Chun, J.RAhn, and H.H Lee,"Algorithms for Controlling Both the DC Boost and AC Output Voltage of Z-Source Inverter,"IEEE Trans. Ind. Electron., vol. 54, no. 5, pp. 2745-2750, Oct. 2007.

[7] M. Hanif, M. Basu, and K. Gaughan, "Understanding the operation of a Z-source inverter for photovoltaic application with a design example," IET Power Electron., vol. 4, no. 3, pp. 278-287, Mar. 2011.

[8] F. Z. Peng, M. Shen, and K. Holland, "Application of Z-source inverter for traction drive of fuel cell - battery hybrid electric vehicles," IEEE Trans. Power Electron., vol. 22, no. 3, pp. 1054-1061, May 2007.

[9] F. Z. Peng, M. Shen, and Z. Qian, "Maximum boost control of the Z-source inverter," IEEE Trans. Power Electron., vol. 20, no. 4, pp. 833–838, Jul. 2005.

[10] M. Shen, J. Wang, A. Joseph, F. Z. Peng, L. M. Tolbert, and D. J. Adams, "Constant boost control of the Z-source inverter to minimize current ripple and voltage stress,"IEEE Trans. Ind. Appl., vol. 42, no. 3, pp. 770–778, May/Jun. 2006.

[11] P. C. Loh, D. M. Vilathgamuwa, Y. S. Lai, G. T. Chua, and Y. W. Li, "Pulse-width modulation of Z-source inverters," IEEE Trans. Power Electron., vol. 20, no. 6, pp. 1346-1355, Nov. 2005.

[12] X. Fang, "A novel Z-source dc-dc converter," in Proc. IEEE-ICIT'08, 2008, pp. 1-4.in Proc. ICEM'10, 2010, pp. 1-6.

[13] X. Ding, Z. Qian, Y. Xie, and Z. Lu, "Three phase Z-source rectifier," in Proc. IEEE-PESC'05, 2005, pp. 494-500.

[14] M. von Zimmermann, S. Labusch, and B. Piepenbreier, "Bidirectional ac-ac Z-source inverter with active rectifier and feed-forward control," in Proc. IEEE-ECCE'10, 2010, pp. 3180-3186.

[15] X. P. Fang, Z. Qian, and F. Z. Peng, "Single-phase Z-source PWM ac-ac converters," IEEE Power Electron. Letters, vol. 3, no. 4, pp. 121-124, Dec. 2005.

[16] M. Zhu, K. Yu, and F. L. Luo, "Switched inductor Z-source inverter," IEEE Trans. Power Electron., vol. 25, no. 8, pp. 2150-2158, Aug. 2010.

[17] M. K. Nguyen, Y. C. Lim, and G. B. Cho, "Switched-inductor quasi Z-source inverter," IEEE Trans. Power Electron., vol. 26, no. 11, pp. 3183-3191, Nov. 2011.

[18] D. Li, F. Gao, P. C. Loh, M. Zhu, and F. Blaabjerg, "Hybrid-source impedance networks: Layouts and generalized cascading concepts," IEEE Trans. Power Electron., vol. 26, no. 7, pp. 2028-2040, Jul. 2011.

[19] R. Strzelecki, M. Adamowicz, N. Strzelecka, and W. Bury, "New type T-source inverter," in Proc. CPE'09, May 2009, pp. 191-195.

[20] W. Qian, F. Z. Peng, and H. Cha, "Trans-Z-source inverters," IEEE Trans. Power Electron., vol. 26, no. 12, pp. 3453-3463, Dec. 2011.

[21] C. Loh, D. Li, and F. Blaabjerg, "Γ-Z-Source Inverters," IEEE Trans. Power Electron., vol. 28, no. 11, pp. 4880-4884, Nov. 2013.

The 2014 International Power Electronics Conference

A High Efficiency Hybrid 7-level Inverter with Single DC Source

Zhang Yanhong, Ogura Kazuya
R&D Division
Meiden Singapore Pte Ltd
Singapore
yanhong.zhang@meidensg.com.sg
ogura.k@meidensg.com.sg

Kazunobu Oi
R&D Group
Meidensha Corporation
Japan
ooi-ka@mb.meidensha.co.jp

Abstract-- **A high efficiency 3-phase hybrid 7-level inverter is presented in this paper. Compared with traditional hybrid 7-level converters, the proposed topology and its T-type cell have lower power loss and simple configuration. It can be easily implemented and applied in renewable energy areas. In the paper, power loss of proposed T-type cell is analyzed, by which conduction loss of T-type cell is proved to be lower than commonly used H-bridge. Therefore, it can be considered to replace H-bridge as a higher efficiency approach in many multi-level topologies. Due to single DC source proposed, DC voltage control scheme for auxiliary inverters is also discussed and presented. A comparison study is carried out by MATLAB + PLECS simulation platform in a grid-connected system.**

Keywords— High efficiency, Hybrid multilevel inverter, T-type cell, Single DC source

I. Introduction

In renewable energy applications, reducing power losses, thus increasing system efficiency is a hot and very important topic which not only researchers but also manufacturers and customers pay much attention to. As studied, among many kinds of approaches for achieving high efficiency, an increased interest has been given to hybrid topologies which integrate more than one topology in a single converter. The main attractive point of hybrid multi-level converter topologies is to guarantee that high power cells work in low switching frequency and only one cell works in high frequency [1][2], which provides the possibility of reducing power losses of switching devices, especially in high power and high voltage applications. Furthermore, the Selective Harmonic Elimination (SHE) PWM strategy enhances the performance of hybrid multi-level topology by reducing harmonics and increasing efficiency [3]. However, hybrid multi-level converter has a drawback that several isolated DC sources are needed for its cascaded auxiliary converters [8], as shown in Fig 1 as an example of 5-level topology, which limits its applications.

Recently, a hybrid multi-level topology with single DC source was proposed in several papers [1][3][6][9]. An example of hybrid 5-level inverter with single DC source is illustrated in Fig. 2, in which DC bus of cascaded auxiliary inverters is equipped with capacitors, and only one DC source is applied to main inverter. Such kinds of topologies greatly simplify the circuit structure and reduce converter volumes. However, on the other hand,

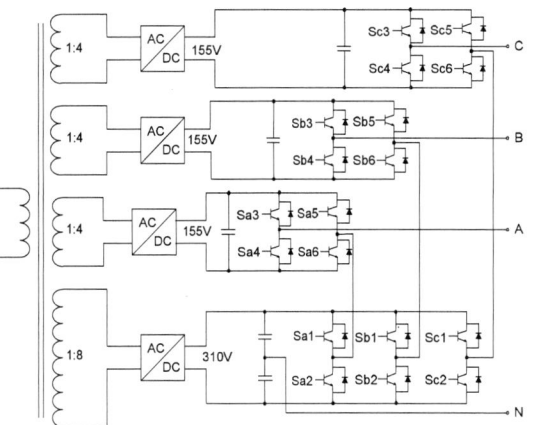

Fig. 1. An example of hybrid inverter with isolated DC sources [8]

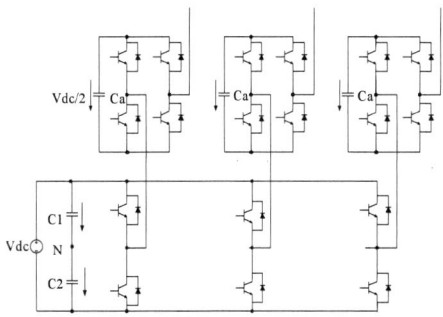

Fig. 2. An example of hybrid inverter with single DC sources [1]

capacitor voltage control approach of auxiliary converters is another challenge.

Besides, as presented in Fig. 1 and 2, most of the existing multilevel topologies use traditional H-bridge or half bridge as a basic cell. Although H-bridge or half-bridge has some advantages, the T-type as a cascaded cell proposed in this paper has a unique feature of lower power loss, which allows the entire topology to reach the goal of reducing power loss further and thus gaining high efficiency.

II. Proposed Topology

The proposed hybrid 7-level topology is presented in Fig. 3, which is composed of a 3-phase 3-level T-type topology as a main inverter, and 3 single-phase 3-level T-type cells (auxiliary inverters) to form a 7-level topology.

978-1-4799-2706-7/14 $31.00 © 2014 IEEE   452

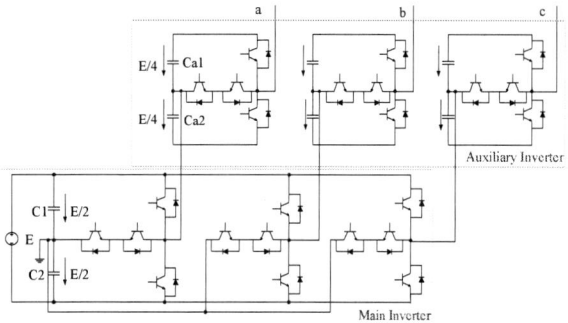

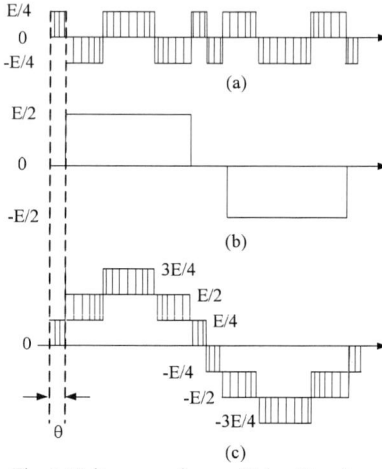

Fig. 3. Proposed 7-level hybrid topology with single DC source

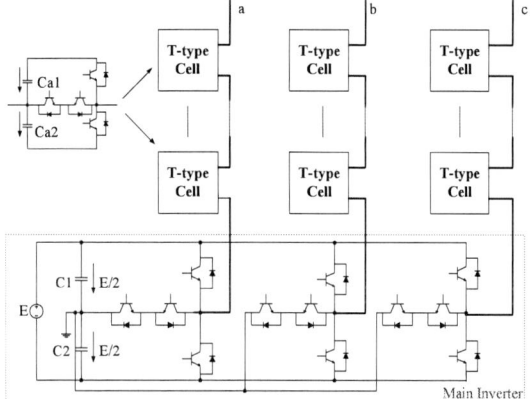

Fig. 4. More levels hybrid topology

Fig. 5. Voltage waveforms of 7-level topology:
(a). Output of auxiliary inverter
(b). Output of main inverter
(c). Integrated output voltage, Van

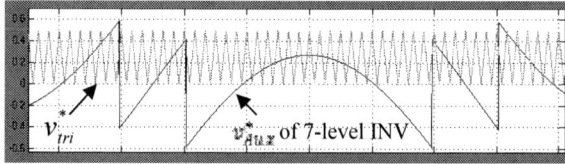

Fig. 6. Carrier-based PWM for auxiliary inverter

In the proposal, only a single DC source is required to supply the inverter. Two capacitors are series connected as DC bus of auxiliary inverters.

Same as H-bridge, if basic T-type cells are extended, it has potential to be configured as a more level topology, as illustrated in Fig.4. However, the paper focuses on the discussion of 7-level topology.

III. Operation Principle and DC Voltage Control

Based on the rule of hybrid multilevel inverter that the main inverter with high power operates at the fundamental frequency, while auxiliary inverters with lower power works at higher switching frequency, the main 3-phase 3-level inverter in proposed 7-level inverter is only required to generate the voltage with fundamental frequency, as shown in the middle waveform of Fig.5 (b). If DC source has voltage value E, it will output 3 voltage levels: E/2, 0, -E/2.

To form the designed output voltage, the reference voltage of auxiliary inverter v^*_{AUX} should be the result of the designed reference voltage v^*_{ref} minus normalized output voltage of main inverter, v^*_M, e.g. $v^*_{AUX} = v^*_{ref} - v^*_m$. Using traditional carrier-based PWM method, the voltage reference v^*_{AUX} compared with a triangle carrier to generate PWM commands for auxiliary inverter as illustrated in Fig.6. Since DC voltage of each capacitor in auxiliary T-type cell is controlled to be E/4, the auxiliary inverters

output 3-level voltages as: E/4, 0 and –E/4 with high switching frequency, as illustrated in the top waveform in Fig.5 (a). The bottom waveform in Fig. 5 (c) is the integration of outputs of main and auxiliary inverters, e.g. Van, which is therefore to be 7-level voltage.

Here, in Fig. 5, θ defined as an angle between voltage zero crossing point to that of main inverter switching to E/2 or –E/2, is determined by SHE as π/10 (ideal value) to eliminate 5th harmonic [10].

To control DC voltage of auxiliary inverters, two kinds of PI controllers, PI_1 and PI_2, are applied. DC voltage is feedback, compared with reference and added through PI_1 to the path of auxiliary inverter control. Meanwhile, through PI_2, it is further controlled by adjusting angle θ of main inverter. The whole system control diagram is presented in Fig. 7.

IV. POWER LOSS ANALYSIS OF PROPOSED T-TYPE CELL

Basically, the conduction loss of switching devices can be expressed by formula (1):

$$P_{con} = \frac{1}{2\pi} \int_0^{2\pi} v * i * d(\omega t) \qquad (1)$$

If the device switches, its current will be pulsed waveform and the conduction loss will become:

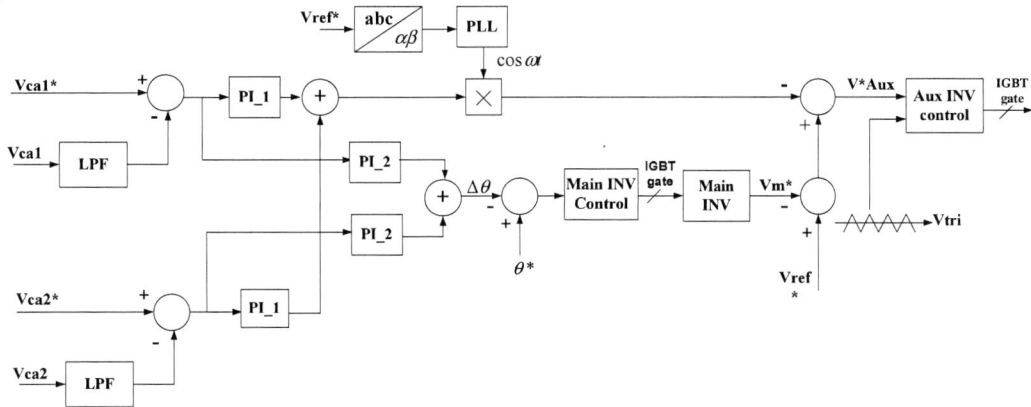

Fig. 7. Control diagram

$$P_{con} = \frac{1}{2\pi} \int\limits_{0}^{2\pi} v * i * D * d(\omega t) \tag{2}$$

where, D is PWM duty cycle and less than 1; v is on-voltage of device which can be expressed as:

$$v = v_{ce} = v_{ceo} + r_{ce} i_c , \qquad \text{for IGBT, and:}$$

$$v = v_d = v_{DO} + r_d i_d , \qquad \text{for free-wheel diode.}$$

Switching loss is caused by overlap of voltage and current of switches during its switching on and off. The turn-on and turn-off energy losses are therefore defined as:

$$W_{on} = \int\limits_{0}^{ton} v * i * dt \tag{3}$$

and,

$$W_{off} = \int\limits_{0}^{toff} v * i * dt \tag{4}$$

t_{on} and t_{off} are overlap time of switching on and off. Then the switching power loss is presented by the product of switching energy and switching frequency f_{sw}:

$$P_{sw} = (W_{on} + W_{off}) * f_{sw} \tag{5}$$

To present the lower power loss characteristics of proposed T-cell, the power loss of traditional H-bridge cell that is commonly used in cascaded converter topologies is discussed first, from which the power loss of proposed T-type cell is compared:

A. Power Loss Features of H-bridge with 3-level Voltage PWM Stratagems

Since different PWM stratagem results in the different power loss, to be a fair comparison, the H-bridge working in 3-level voltage mode as T-type cell is considered. In this case, there are three common PWM stratagems, which are called as unipolar, hybrid-I and hybrid-II. Unipolar PWM stratagem is implemented by: 1) PWM command of device pair S1and S2, which is connected in positive terminal, is a result of reference voltage v_{ref} compared with triangle carrier v_{tri} (device labelled in Fig 12 (a)); while 2) pair of S3 and S4 uses the result of $-v_{ref}$ compared with v_{tri} [12]. Fig. 8 illustrates the

unipolar mechanism and switching command for devices. As the result, inverter output voltage V_{INV} will become 3-level voltage: $+Vc/2$, 0 and $-Vc/2$, as presented in Fig. 9.

With unipolar PWM stratagem, all switching devices S1~S4, switch at the same switching frequency. They have same pattern of current and voltage waveform as shown in Fig. 11 (a), thus result in the same power loss.

However, the concept of hybrid PWM stratagem is allowing some switches to work in lower frequency while others in higher switching frequency to reduce the switching loss. Hybrid-I is defined as a device pair in one leg such as S3 and S4, works in fundamental frequency; while another pair of S1 and S2 operates in high PWM switching frequency. The PWM is still generated by v_{ref} compared with v_{tri}, however, v_{tri} has 0~1 peak to peak value instead of -1~1 in normal triangle carrier. The current and voltage waveform of S1 and S4 in this PWM technic is given by Fig. 11 (b). Another pair of S2 and S3 has same waveform but 180° phase shift.

Certainly, switching loss of S3 and S4 is much lower than S1 and S2. As for conduction loss, if considering IGBT part only, conduction loss of pair of S3 and S4 is higher than S1 and S2 because current of S3 and S4 is continuous waveform, while the one of S1 and S2 is pulsed, e.g. conduction loss expressed in formula (1) > (2) if current envelop is same. But S1 and S2 have diode conduction loss, while S3 and S4 do not have. Thus it is hard to say whose overall conduction loss is lower. It actually depends on loss characteristics of device, PWM duty cycle, etc. However, "saving" in the switching loss is greater than the "wasting" in the conduction loss, especially in high switching applications. The total loss of S3 and S4 is still much lower than S1 and S2, which is proved by simulation result listed in Table I.

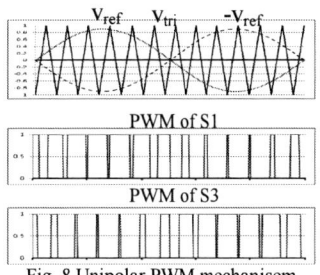

Fig. 8 Unipolar PWM mechanisem

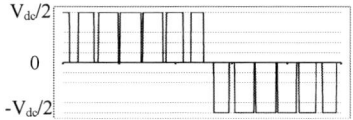

Fig. 9. 3-level output voltage of H-bridge inverter

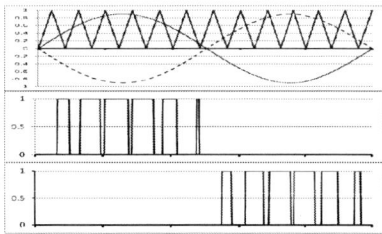

Fig. 10. PWM mechanism of Hybrid-II

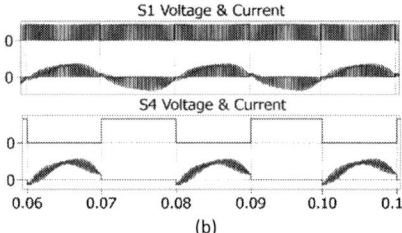

(a)

(b)

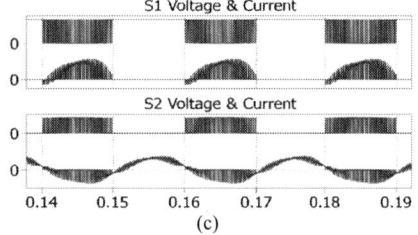

(c)

Fig. 11. Voltage and current waveform of switching device:
(a). Using unipolar PWM scheme
(b). Using hybrid-I PWM scheme
(c). Using hybrid-II PWM scheme

TABLE I
POWER LOSS DISTRIBUTION IN EACH DEVICE OF H-BRIDGE

PWM stratagem	Unipolar	Hybrid-I	Hybrid-II
S1- switching loss	116.6W	117.7W	79W
S2- switching loss	116.6W	117.7W	36.3W
S3- switching loss	115W	0.97W	80W
S4- switching loss	115W	0.97W	36.7W
S1- conduction loss	72.3W	71.7W	51.7W
S2- conduction loss	72.3W	71.7W	97.7
S3- conduction loss	72.5W	75.3W	54.7W
S4- conduction loss	72.6W	75.3W	95W
Total	753W	531.3W	531.3W

Unlike hybrid-I, hybrid-II PWM stratagem can distribute the low switching frequency period to each switching device in order to balance their burden.

The implementation of hybrid-II is same as the method of unipolar PWM generation, but using triangle carrier with 0 ~ 1 peak to peak value. The device current and voltage waveform of hybrid-II are given by Fig. 11 (c).

Under the same condition, it can be observed from Fig. 11 (b) and (c) that the total power loss in hybrid-I and hybrid-II is almost same.

A simulation study of these PWM stratagems, with 3-level output voltage, is carried out in this paper. The power loss result is listed in Table I. The studied H-bridge comes with parameters of 60KVA/415V, PF=0.95, f_{SW}=4KHz and switching device of SKM600GB126D (IGBT).

Thus, based on above discussion and simulation, the conclusion is that: if H-bridge worked in 3-level output voltage mode, unipolar PWM stratagem will generate higher power loss, while hybrid-I and hybrid-II have same and lower power loss. From the loss distribution point of view, hybrid-II is better than hybrid-I. Therefore, H-bridge with hybrid-II PWM stratagem is selected to do power loss comparison with proposed T-type cell.

B. Power Loss Comparison between H-bridge and Proposed T-type Cell

The switching devices in H-bridge and T-type topologies are labeled in Fig. 12. If both topologies are applied as auxiliary inverter in proposed 7-level hybrid inverter, the voltage and current of device pair S1 and S2 will present the waveforms given by Fig. 13, in which an interesting phenomenal can be seen that current of S1 is same in both H-bridge and T-type cell; while voltage is same in some periods, but not in other periods. However, if looking at those different voltage periods carefully, it can be found that the corresponding current in these periods is zero. Zero current means zero power loss. Therefore, it implies that the power loss of device S1 only appears in those periods with same voltage and current, thus, resulting in same power loss of S1 for H-bridge and T-type topology, according to the power loss definitions, expressed in (1), (2), (3), (4) and (5).

If looking at the voltage and current waveforms of device in H-bridge and T-type cell, same voltage but different current are presented. In those periods of different current, such as $\varphi 1 \sim \varphi 2$, S2 in T-type cell conducts a pulsed current, whose conduction loss is expressed by (2); while in H-bridge, a continuous current waveform appears. Once such continuous current periods exist, conduction loss will become:

$$P_{con} = \frac{1}{2\pi}\left[\int_{0}^{\varphi 1} v*i*D*d(\omega t) + \int_{\varphi 1}^{\varphi 2} v*i*d(\omega t) + ...\right]$$
(6)

which is greater than (2) for the same envelope of current waveform. Therefore, conduction loss of S2 is higher in H-bridge than that in T-type cell.

On the other hand, it is also noted that although S2 in T-type conducts pulsed current, in those periods of different current, its voltage is zero, which implies S2 not switching at all. Therefore, switching loss of S2 is zero in those periods for both topologies.

Summarised from above analysis, device pair of S1 and S2 in T-type cell has lower conduction loss and same switching loss, compared with H-bridge. The conclusion is also valid for another pair of device S3 and S4.

978-1-4799-2706-7/14 $31.00 © 2014 IEEE

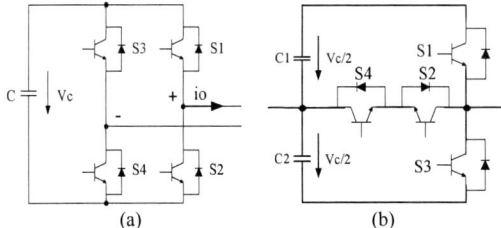

Fig. 12. Topology of H-bridge cell (a) and T-type cell (b)

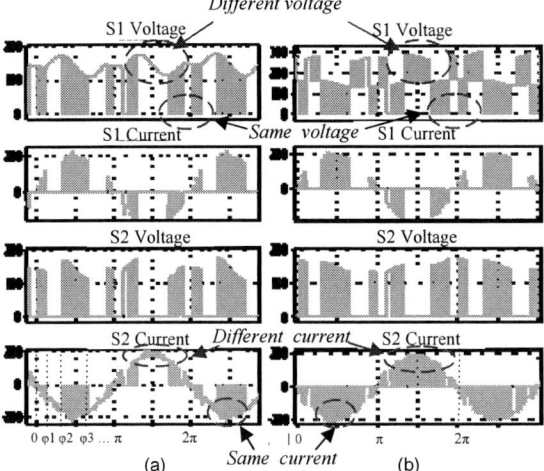

Fig. 13. Voltage and current waveforms of S1 and S2 in (a) H-bridge, (b) T-type cell

V. Simulation Verification

Proposed topology is verified in MATLAB+PLECS simulation platform in a 3-phase grid-connected system. The simulation conditions are listed in Table II.

Simulation result of inverter output voltage, current as well as DC voltage is presented in Fig. 14. Output voltage (line to line) has 13 levels and THD of output current is 4.6%. DC bus voltage of axuliary inverter is controlled as 141V, which has voltage ripple 32V in this case. The voltage ripple is in fact related to load current, capacitor value, system frequency, modulation index, etc. which are not discussed in this paper

The power losses of switching devices in T-type cell and H-bridge are listed in Table III. The result verifies that T-type cell has almost same switching loss, but lower conduction loss as analyzed in section IV. In the case of 415V/100KVA system, 10% power loss reduction is achieved.

To further check the effect of power loss reduction, a comparison study between proposed hybrid 7-level topology and an existing hybrid 7-level topology, as shown in Fig. 15, is also conducted.

TABLE II
SYSTEM PARAMETERS AND CONDITIONS

System		Inverter	
V	415V	fsw	4.5KHz
S	100KVA	Vdc	282V*2
PF	1	DC Cap	4mF
f	50Hz	AC filter L	3.5%
		AC filter C	6%

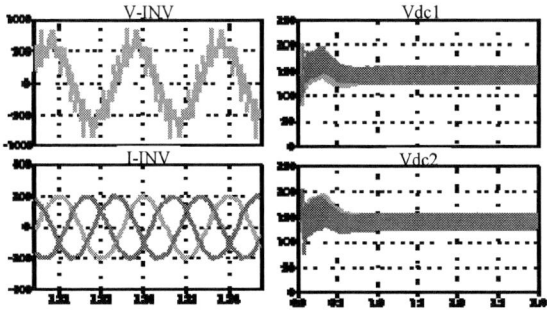

Fig. 14. Voltage and current waveforms

TABLE III
POWER LOSS OF S1 AND S2 IN H-BRIDGE AND T-TYPE CELL
(ONE PHASE, USING IGBT: SKM400GB066D)

Symbol	Meaning	T-type	H-bridge
$P_{CON}\text{-}S1$	S1 Conduction Loss	13W	13W
$P_{SWT}\text{-}S1$	S1 Switching Loss	24W	23W
$P_{CON}\text{-}S2$	S2 Conduction Loss	117W	130W
$P_{SWT}\text{-}S2$	S2 Switching Loss	22W	21W

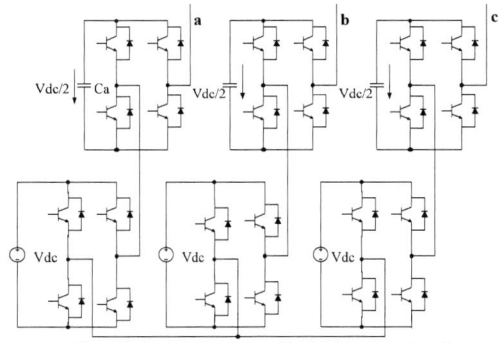

Fig. 15. An existing 7-level hybrid inverter topology

TABLE IV
POWER LOSS OF PROPOSED 7-LEVEL VS EXISTING 7-LEVEL

Symbol	Meaning	Proposed	Existing
$P_{CON}\text{-}Main$	Conduction Loss of Main	515.6W	930.5W
$P_{SWT}\text{-}Main$	Switching Loss of Main	10.8W	5W
$P_{CON}\text{-}Aux$	Conduction Loss of Aux	780.6W	857W
$P_{SWT}\text{-}Aux$	Switching Loss of Aux	278W	261.4W
P_{Total}	Total Loss	1585W	2054W

NOTE: IGBT MODULE, SKM600GB126D, 600A/1200V USED FOR MAIN INV OF PROPOSED TOPOLOGY, WHILE SKM400GB066D, 400A/600V FOR OTHERS

Based on same conditions listed in Table II, the simulation results of power loss in both topologies are given in Table III. In this case, around 23% power loss is reduced due to lower conduction loss of proposed topology.

VI. Conclusion

A high efficiency hybrid 7-level inverter is proposed and analyzed. Due to single DC source used, DC capacitor voltage of auxiliary inverters is effectively controlled by two kinds of PI controllers through main and auxiliary inverters. Proposed T-type cell has been proved with a unique feature of lower conduction loss, which makes the proposed hybrid 7-level topology achieve 23% power loss reduction in 415V/100KVA system, compared with the existing hybrid 7-level topology. Simulation results from a grid-connected system verify that the proposed 7-level inverter can be easily applied in renewable energy area or used as a general propose inverter, especially in high voltage and high power applications to meet the high efficiency requirement.

However, because of the drawback of T-type topology, such as double voltage stress for S1 and S3 and one more DC capacitor needed, a tradeoff should be considered between efficiency and cost, etc.

References

[1] Haiwen Liu, Leon M. Tolbert, Burak Ozpineci, Zhong Du, "Hybrid Multilevel Inverter with Single DC Source", *the 51st IEEE International Midwest Symposium on Circuits and Systems*, 2008

[2] Diorge A. B. Zambra, Cassiano Rech, and José Renes Pinheiro, "Comparison of Neutral-Point-Clamped, Symmetrical, and Hybrid Asymmetrical Multilevel Inverters", *IEEE Trans. on Industrial Electronics*, vol. 57, no. 7, July 2010

[3] Lavanya Gundugula, Anand Kakarapalli, "Control Of An Hybrid Multilevel Converter With Floating DC-Links For The Improvement Of Current Waveform", *IJERA ISSN*, 2248-9622 Vol. 2, Issue 5, September-October 2012, pp.1772-1780

[4] Cassiano Rech, and José Renes Pinheiro, "Hybrid Multilevel Converters: Unified Analysis and Design Considerations", *IEEE Transactions on Industrial Electronics*, vol. 54, no. 2, April 2007

[5] P. Thongprasri, "A 5-Level Three-Phase Cascaded Hybrid Multilevel Inverter", *International Journal of Computer and Electrical Engineering*, vol. 3, no. 6, December 2011

[6] F. Khoucha, A. Ales, A. Khoudiri, K. Marouani, M.E.H. Benbouzid and A. Kheloui, "A 7-Level Single DC Source Cascaded H-Bridge Multilevel Inverters Control Using Hybrid Modulation", *XIX International Conference on Electrical Machines - ICEM* 2010, Rome

[7] Richard Lund, Madhav D. Manjrekar, Peter Steimer, "Control Strategies for a Hybrid Seven-level Inverter", *EPE* 1999

[8] Surin Khomfoi, Nattapat Praisuwanna, Leon M. Tolbert, "A Hybrid Cascaded Multilevel Inverter Application for Renewable Energy Resources Including a Reconfiguration Technique", 978-1-4244-5287-3/10/$26.00 ©2010 IEEE

[9] Jingsheng Liao, Keith Corzine, and Mehdi Ferdowsi, "A New Control Method for Single-DC-Source Cascaded H-Bridge Multilevel Converters Using Phase-Shift Modulation", 978-1-4244-1874-9/08 ©2008 IEEE

[10] A.D. Rajapakse, A.M. Gole, and P.L. Wilson, "Approximate Loss Formulae for Estimation of IGBT Switching Losses through EMTP-type Simulation", *International Conference on Power Systems Transients (IPST'05)*, June 19-23, 2005, paper no. IPST05-184

[11] Zambra, D.A.B, Rech,C., Goncalves, F.A.S., Pinheiron, J.R., "Power Losses Analysis and Cooling System Design of Three Topologies of Multilevel Inverters", *IEEE PESC 2008*, ISSN 0275-9306, pages: 4290-4295

[12] Ned Mohan, Tore M. Undeland, William P. Robbins, "Power Electronics: Converters, Applications, and Design", John Wiley & Sons, Inc 1995

The 2014 International Power Electronics Conference

Optimal Idling Control Strategy for Three-Port Full-Bridge Converter

Yongjie Jiang*, Fuxin Liu*, Xinbo Ruan*, Lipeng Wang**

*: Jiangsu Key Laboratory of New Energy Generation and Power Conversion

Nanjing University of Aeronautics and Astronautics

Nanjing, China

**: CSR Zhuzhou Electric Locomotive Research

Institute CO., LTD

Zhuzhou, China

Abstract—**Multi-port converters (MPCs) are the prospective alternative in the hybrid power system, which can simplify the system structure and reduce the cost. In some operation conditions, one port may quit from the system, leaving this idling port at zero power. Under the exclusive operation, different control strategies can be used. Based on the analysis of the soft-switching ranges and the conduction losses, this paper proposed an optimal idling control strategy that combines the phase-shift control with the PWM control. In order to ensure the conduction losses be minimum, the duty cycle of the idling port is regulated in real time referring to the voltage of the port and the total output power. As an additional benefit, the converter can operate under a wide soft-switching range. Finally, a 1kW prototype converter has been built to verify the theoretical analysis.**

Keywords— Three-port full-bridge converter, phase-shift control, phase-shift plus PWM control, Idling control.

I. INTRODUCTION

The energy crisis and environmental pollution have stimulated extensive research of the electric vehicles and hybrid renewable power systems[1-3]. Instead of using several individual dc/dc converters, a multi-port converter (MPC) can be used to simplify the structure and reduce the overall cost. With the advantages of symmetric structure, simple control strategy and soft-switching characteristics, multi-port full-bridge converters are the prospective alternative in the high power applications[4, 5]. Under some operation conditions, one port may quit from the system, meaning that the converter leaves the idling port at zero power[6, 7]. Though keeping out of the drive signal for the MOSFETs in the idling port, the power will continue to go through the body diodes of MOSFETs. As a result, an active control strategy should be introduced to solve this problem. A phase-shift plus duty cycle control was proposed for the three-port full-bridge converter in [4] to achieve zero power at one specific port. With the duty cycle control, the soft-switching can be realized under a wide range within the voltage variation. However, only the phase-shift control is used to control the power flow to ensure the

circulating power to be minimum, and the function of the duty cycle was not analyzed in detailed. Based on the fundamental harmonic analysis, the constraints between the phase angle and the voltage value make it possible to use a dual voltage close loop to control the idling port[8]. However, the control strategy will make it difficult for the idling port to realize soft-switching. Furthermore, if the idling port voltage is limited by connecting to a voltage source, the control strategy will be disabled.

In this paper, two control strategies, i.e. phase-shift control and phase-shift plus PWM control were adopted and compared to obtain zero power at one port. Based on the constraints between the phase angles and the duty cycle of the idling port, an optimal idling control strategy was proposed to ensure the conduction losses to be minimum within the input voltage and power range. Meanwhile, it can realize soft-switching for switches under a wide voltage range. A 1kW prototype converter has been built to prove the theoretical analysis.

II. IDLING CONTROL STRATEGY

The topology of a three-port full-bridge converter is shown in Fig. 1, in which, V_1, V_2' and V_3' represent for the dc voltages of port 1, port 2 and port 3 respectively. i'_{L1}, i'_{L2} and i'_{L3} are the currents flowing through the series inductance L_1, L'_2 and L'_3.

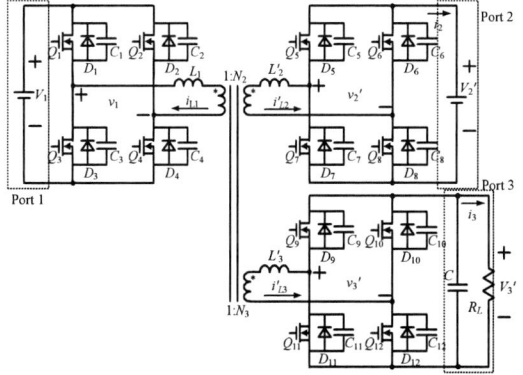

Fig. 1. Three-port full-bridge converter

The Δ-type equivalent circuit of a three-winding transformer is adopted for the convenience to analyze the power flow among each port, as shown in Fig. 2. Assuming that the power of port 2 is zero, the following control strategies can be adopted. If the duty cycles of all ports are equal to one and the shift-phases between each

---

Sponsored by National Natural Science Foundation of China(50807024), the Power Electronics Science and Education Development Program of Delta Environmental & Educational Foundation and Jiangsu province university outstanding science and technology innovation team project.

978-1-4799-2706-7/14 $31.00 © 2014 IEEE   458

port are regulated to control the power, the converter will operate under phase-shift control, and v_1, v_2 and v_3 are both square waves. If the PWM control is introduced at port 2 here, v_2 will represent a quasi square wave, and the converter operates under the phase-shift plus PWM control.

The relationship between the parameters in Fig. 1 and the ones in Fig. 2 is

$$
\begin{cases}
L_{\Delta 12} = L_1 + L_2' / N_2^2 + L_1 L_2' N_3^2 / (L_3' N_2^2) \\
L_{\Delta 23} = L_2' / N_2^2 + L_3' / N_3^2 + L_2' L_3' / (N_2^2 N_3^2 L_1) \\
L_{\Delta 31} = L_3 / N_3^2 + L_1 + L_3 L_1 N_2^2 / (L_2 N_3^2)
\end{cases} \quad (1)
$$

Before analysis, the following assumptions are given: 1) all the switches and diodes are ideal, 2) all the inductances and capacitors are ideal, 3) $C_1=C_2=\cdots=C_{12}$, 4) $L_{\Delta 12}=L_{\Delta 13}=L_{\Delta 23}=L_\Delta$.

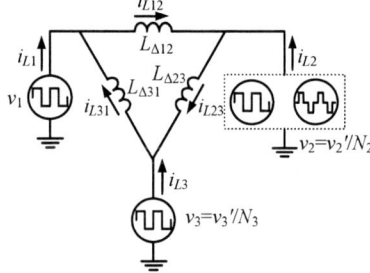

Fig. 2. Δ-type equivalent circuit

A. Phase-shift control

The key waveforms of the equivalent circuit under phase-shift control are shown in Fig. 3, where φ_{12} and φ_{13} are defined as positive when the phase of v_1 is leading to that of v_2 and v_3 at the rising edge. When the instantaneous current values are calculated by the energy conservation and voltage-second balance within one switching period, the power transfer expressions can be obtained as follows,

$$
P_{12} = \frac{V_1 V_2 \varphi_{12}(\pi - \varphi_{12})}{2\pi^2 L_{\Delta 12} f_s} \quad (2)
$$

$$
P_{13} = \frac{V_1 V_3 \varphi_{13}(\pi - \varphi_{13})}{2\pi^2 L_{\Delta 31} f_s} \quad (3)
$$

$$
P_{23} = \frac{V_2 V_3 (\varphi_{13} - \varphi_{12})(\pi - \varphi_{13} + \varphi_{12})}{2\pi^2 L_{\Delta 23} f_s} \quad (4)
$$

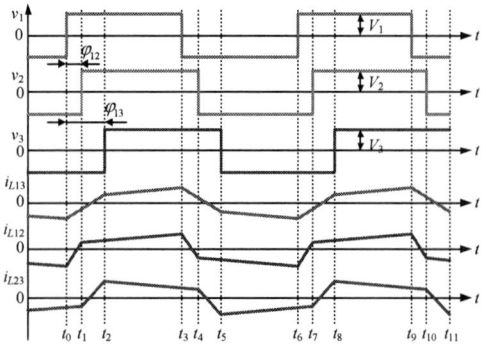

Fig. 3. Key waveforms under phase-shift control

where P_{12}, P_{13} and P_{23} are the power transferred from v_1 to v_2, v_1 to v_3 and v_2 to v_3, respectively. If the power of port 2 is zero, i.e., $P_{12}=P_{23}$, the following constraint can be obtained from (2) and (4),

$$
\varphi_{13} = \frac{-\sqrt{\pi^2 d_{13}^2 + 4 d_{13} \varphi_{12}^2 - 4\pi d_{13} \varphi_{12}}}{2 d_{13}} + \frac{\pi + 2\varphi_{12}}{2} \quad (5)
$$

where $d_{12}=V_2/V_1$ and $d_{13}=V_3/V_1$.

B. Phase-shift plus PWM control

In order to widen the soft-switching range, PWM control can be introduced at port 2 together with the phase-shift control. According to the relationship between the shift-phase and D, the following operation modes can be obtained.

1. Operation mode 1 ($\varphi_{13} \geq \varphi_{12}$, $\varphi_{12}+D\pi \geq \pi$)

The key waveforms of operation mode 1 are given in Fig. 4, from which D can be expressed by

$$
D = 2(t_3 - t_1) / T_s \quad (6)
$$

The power transfer expressions can be achieved with the same analysis method under the phase-shift control.

$$
P_{12} = \frac{\pi D^2 - \pi D + 2D\varphi_{12}}{4\pi L_{\Delta 12} f_s} V_1 V_2 \quad (7)
$$

$$
P_{23} = \frac{\pi^2 D^2 + 2\pi D\varphi_{12} - 2\pi D\varphi_{13} - \pi^2 D}{4\pi^2 L_{\Delta 23} f_s} V_2 V_3 +
$$
$$
\frac{2(\varphi_{12} - \varphi_{13})^2}{4\pi^2 L_{\Delta 23} f_s} V_2 V_3 \quad (8)
$$

$$
P_{13} = \frac{\varphi_{13}(\pi - \varphi_{13})}{2\pi^2 L_{\Delta 31} f_s} V_1 V_3 \quad (9)
$$

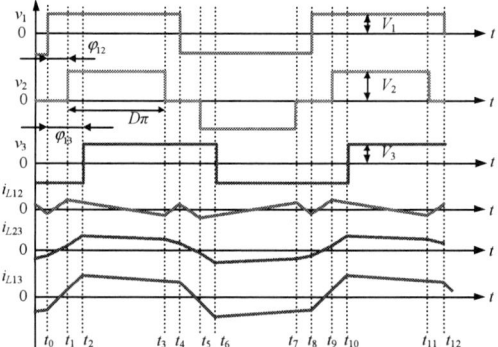

Fig. 4. Key waveforms of operation mode 1

As the power of port 2 is zero, the constraint between different variables can be given by

$$
\varphi_{13} = \varphi_{12} + \frac{\pi D}{2} - \frac{\sqrt{\pi d_{13} D[4\varphi_{12} - 2\pi(1-D-d_{13}) - \pi D d_{13}]}}{2 d_{13}} \quad (10)
$$

According to (10), the relationship between φ_{12} and φ_{13} is depicted in Fig. 5 when D=0.4, D=0.6 and D=0.8. Note that the shadowed area denotes the operation area. It is obvious that the constraint curves are out of the operation area, so the idling control strategy cannot be achieved in operation mode 1.

2. Operation mode 2 ($\varphi_{13} \leq \varphi_{12}$, $\varphi_{12}+D\pi \leq \pi$)

The key waveforms of operation mode 2 are shown in Fig. 6. The power transfer expressions of P_{12} and P_{13} are the same with the ones in operation mode 1.

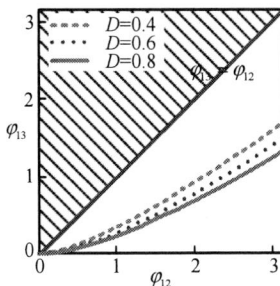

Fig. 5. Operation area and constraint curves of operation mode 1

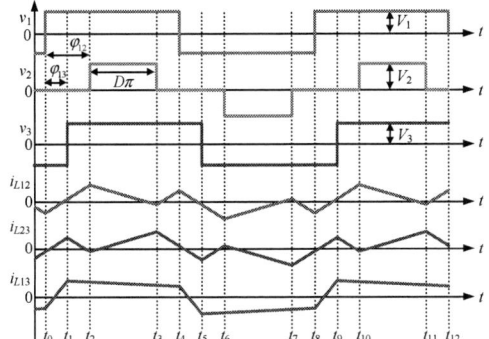

Fig. 6. Key waveforms of operation mode 2

The expression of P_{23} can be represented as

$$P_{23} = -\frac{\pi D^2 - \pi D + 2D\varphi_{12} - 2D\varphi_{13}}{4\pi L_{\Delta 23}f_s}V_2V_3 \quad (11)$$

When $P_{12}=P_{23}$, it follows from (7) and (11) that

$$\varphi_{13} = \frac{2\varphi_{12} - \pi d_{13} - \pi + 2d_{13}\varphi_{12} + \pi D + \pi D d_{13}}{2d_{13}} \quad (12)$$

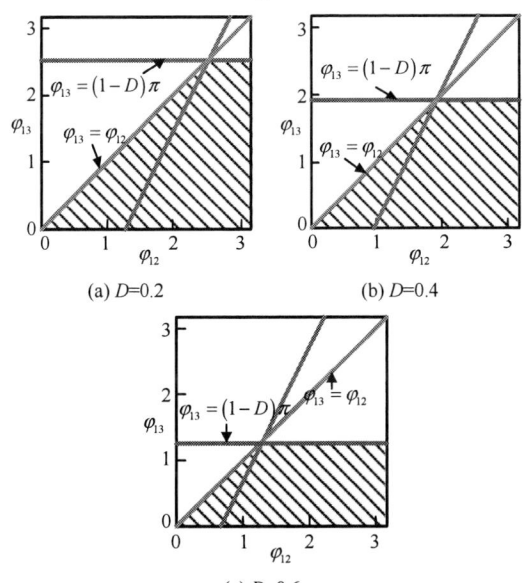

Fig. 7. Operation area and constraint curves of operation mode 2

Fig. 7 shows the operation area and the constraint curves of (12). As illustrated, parts of the constraint curves are in the shadowed areas. As a result, it is possible to adopt phase-shift plus PWM control to realize idling control strategy in operation mode 2.

The other operation modes can be analyzed by the same method mentioned above, and the final conclusion is that only operation condition 2 can achieve zero power transmission at port 2.

III. COMPARISON BETWEEN THE CONTROL STRATEGIES

Switching losses and conduction losses are the dominant factors that affect the efficiency of converter. In order to obtain a better performance, soft-switching ranges and rms value of current are analyzed under the phase-shift control and phase-shift plus PWM control.

A. Analysis of the soft-switching ranges and Rms values of currents

Under phase-shift control, the zero-voltage-switching (ZVS) conditions of all switches are $i_{L1}(t_0)<0$, $i_{L2}(t_1)<0$ and $i_{L3}(t_2)<0$ according to the waveforms in Fig. 3, where, $i_{L1}(t)=i_{L12}(t)+i_{L13}(t)$, $i_{L2}(t)=i_{L23}(t)-i_{L12}(t)$ and $i_{L3}(t)= -i_{L13}(t)-i_{L23}(t)$. When $d_{13}=1$, the soft-switching range is maximum and given by

$$d_{12} \le \frac{\pi + 4\varphi_{12}}{\pi - 2\varphi_{12}} \quad (13)$$

$$d_{12} \ge \frac{\pi - 2\varphi_{12}}{\pi} \quad (14)$$

Eq. (13) represents the soft-switching condition of both port 1 and port 3, and Eq. (14) represents the soft-switching condition of port 2.

Similarly, the soft-switching conditions under phase-shift plus PWM control are

$$\varphi_{12} \ge (\pi + \pi D d_{12} - 2\pi D)/4 \quad (15)$$

$$d_{12} \ge 1 \quad (16)$$

where, Eq. (15) represents the soft-switching condition of both port 1 and port 3, and Eq. (16) represents the one of port 2.

The conduction losses can be expressed as

$$P_{lossi} = 2I^2_{rmsi}R_{ds(on)} \quad (17)$$

$$I^2_{rmsi} = I^2_{1\_rmsi} + I^2_{2\_rmsi} + I^2_{3\_rmsi} \quad (18)$$

where $i=1$ and $i=2$ represent the losses under phase-shift control and phase-shift plus PWM control, respectively. I_{1_rmsi}, I_{2_rmsi} and I_{3_rmsi} are the rms value of current flowing through the series inductances in three ports, and $R_{ds(on)}$ is the conduction resistance of MOSFETs.

B. Comparisons between two control strategies

The specifications of the converter are given as follows: V_1=120V; V_2=135V~270V; V_3=240V; P_1=0~1000W; P_2=0W; P_3=0~1000W; f_s=50kHz. L_Δ is designed as 68.6μH and 56.25μH under phase-shift control and phase-shift plus PWM control, respectively.

1. Comparison on soft-switching ranges

Under phase-shift control, the output power of port 3 is

$$P_3 = \frac{V_1V_2\varphi_{12}(\pi - \varphi_{12})}{2\pi^2 L_{\Delta 12}f_s} + \frac{V_1V_3\varphi_{13}(\pi - \varphi_{13})}{2\pi^2 L_{\Delta 13}f_s} \quad (19)$$

According to (13), (14) and (19), Fig. 8 shows the

relationship between d_{12} and φ_{12} at different P_3. The shadowed area represents the soft-switching range. From Fig. 8, it can be known that the switches are more difficult to realize ZVS when φ_{12} decreases, and when P_3 is small, the range of d_{12} to realize ZVS will be much wider.

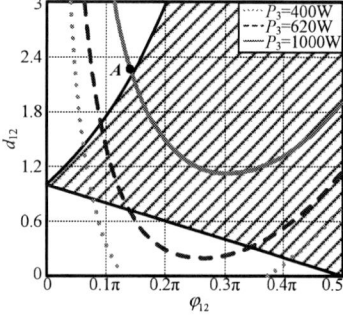

Fig. 8. Soft-switching ranges and power curves under phase-shift control

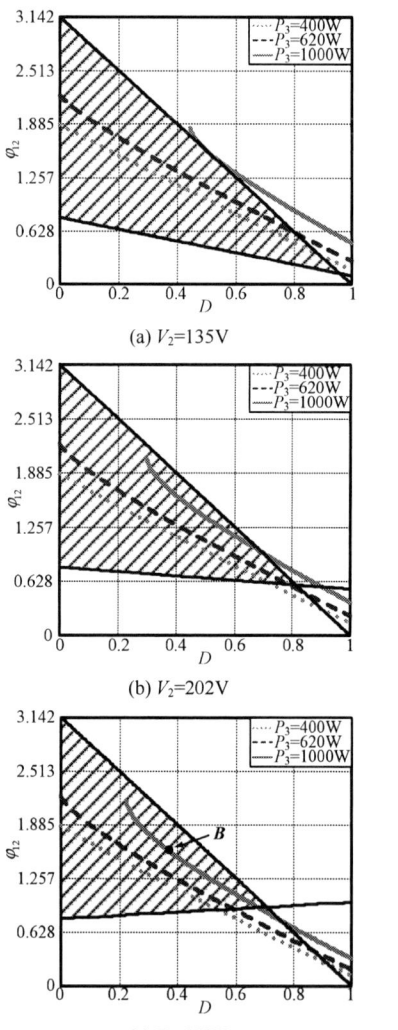

(a) V_2=135V

(b) V_2=202V

(c) V_2=270V

Fig. 9. Soft-switching ranges and power curves under phase-shift plus PWM control

According to (9), (11), (15) and $P_3=P_{23}+P_{13}$, soft-switching ranges and power curves under phase-shift plus PWM control are given in Fig. 9. As depicted, when

V_2 varies from 135V to 270V and P_3 varies from zero to the full load, the converter can always realize ZVS by choosing a proper D.

As a conclusion, wider soft-switching ranges can be achieved under phase-shift plus PWM control compared with the phase-shift control.

2. Comparison on the rms value of currents

According to (5), (18) and $P_3=P_{13}+P_{23}$, Fig. 10 shows the relationship between I^2_{rms1} and P_3 at different V_2 under phase-shift control. As shown, I^2_{rms1} decreases firstly and then rises when V_2 increases under full load condition.

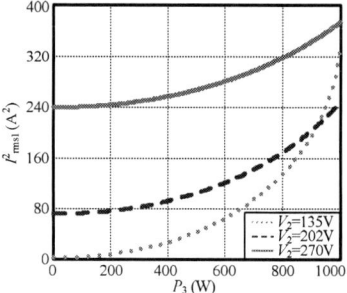

Fig. 10. Curves of the I^2_{rms1} versus P_3 under phase-shift control

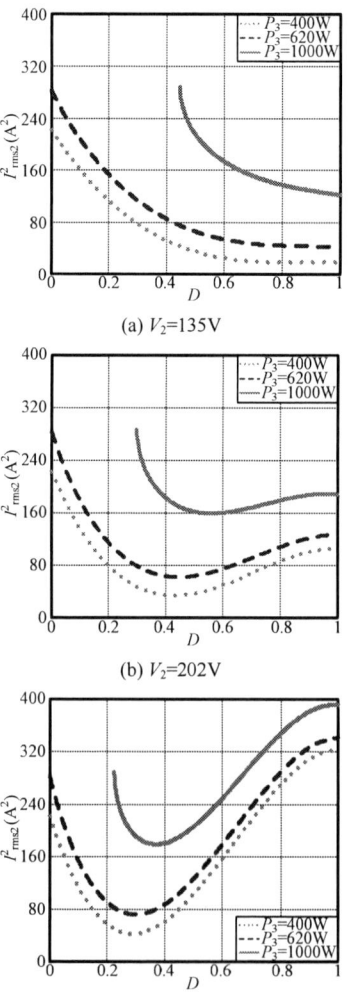

(a) V_2=135V

(b) V_2=202V

(c) V_2=270V

Fig. 11. Curves of I^2_{rms2} versus P_3 under phase-shift plus PWM control

With the same method, the curves of I^2_{rms2} versus P_3 at different V_2 under phase-shift plus PWM control are shown in Fig. 11. It can be found that I^2_{rms2} can be as small as possible by choosing a proper D, and I^2_{rms2} is always smaller than I^2_{rms1} under the same operation condition.

IV. OPTIMAL IDLING CONTROL STRATEGY

From the above analysis, it can be known that the converter can increase the efficiency by choosing a proper D to ensure I^2_{rms2} to be minimum under phase-shift plus PWM control. Combining with the operation ranges and soft-switching ranges in Fig. 9, one can extract a number of points of D, P_3 and V_2 in Fig. 11 which make I^2_{rms2} be minimum under the corresponding operation condition. Quadric curve is used to describe the points so that an approximate relationship between D and P_3 is derived

$$D(P_3) = a(V_2)P_3^2 + b(V_2)P_3 + c(V_2) \quad (20)$$

where

$$a(V_2) = 1.1 \cdot 10^{-11} \cdot V_2^2 - 6.0 \cdot 10^{-9} \cdot V_2 + 9.4 \cdot 10^{-7} \quad (21)$$

$$b(V_2) = 5.8 \cdot 10^{-10} \cdot V_2^2 - 8.7 \cdot 10^{-8} \cdot V_2 - 3.4 \cdot 10^{-5} \quad (22)$$

$$c(V_2) = 2.667 \cdot 10^{-5} \cdot V_2^2 - 0.015 \cdot V_2 + 2.287 \quad (23)$$

From (12) and (20), the relationship between I^2_{rms2} and P_3 is shown in Fig. 12. As show, when V_2 increases at full load, I^2_{rms2} increases firstly and then decreases.

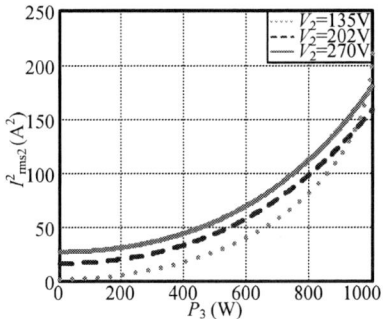

Fig. 12. Curves of I^2_{rms2} versus P_3 under phase-shift plus PWM control

k is defined as the ratio of I^2_{rms2} and I^2_{rms1} to make a clear comparison, which is illustrated in Fig. 13. From Fig. 13, it can be known that when V_2 is small, I^2_{rms2} is close to I^2_{rms1}; when V_2 is large, I^2_{rms2} is apparently smaller than I^2_{rms1} especially at light load.

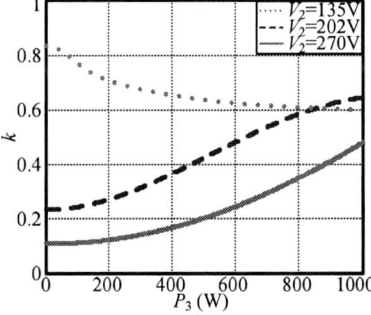

Fig. 13. Curves of k versus P_3

Based on Eq. (20), an optimal control scheme was proposed, as shown in Fig. 14. Three degrees of freedom,

i.e. D, φ_{12} and φ_{13} are available for the system control. The voltage error, $v_{3\_ref} - v_3'$ and the current error, $i_{2\_ref} - i_2$ pass through a PI controller individually and the corresponding outputs are interpreted as φ_{12} and φ_{13}. If the measured V_2' and the product of V_3' and i_3 are available, a proper D_1 in real time will be calculated from (20). D_2 is the critical duty cycle to achieve zero power transmission. By comparing D_1 and D_2, the smaller one is chosen as the optimal duty cycle. From the above analysis, the proposed idling control strategy can both realize soft-switching in a wide operation range and ensure the conduction losses to be minimum.

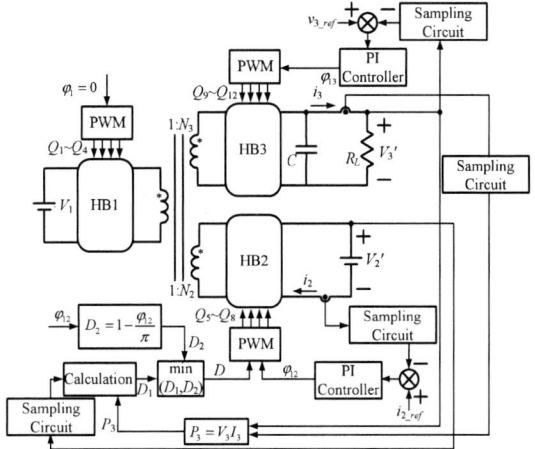

Fig. 14. System scheme under the optimal idling control

V. EXPERIMENTAL RESULTS

A 1kW prototype was built to verify the theory and its parameters are given in section III.

Fig. 15 shows the steady-state waveforms at V_2=270V and P_3=1000W under optimal idling control. As shown in Fig. 15(b), the average value of P_2 is zero, which means the proposed control strategy can achieve zero power at port 2.

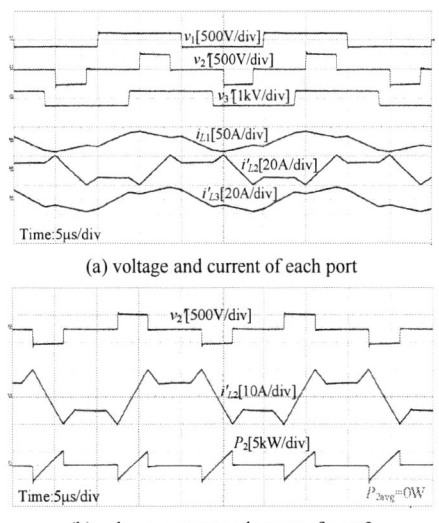

(a) voltage and current of each port

(b) voltage, current and power of port 2

Fig. 15. Steady-state waveforms under optimal idling control

Take port 2 at V_2=270V and P_3=1kW for example. The operation points to be proved are referred to point A in

978-1-4799-2706-7/14 $31.00 © 2014 IEEE

Fig. 8 and point *B* in Fig. 9(c). Fig. 16 shows the gate drive signal v_{GS}, the voltage across the drain and source v_{DS}, and drain current i_D of Q_{12}. From Fig. 16(a), it can be seen that the switch suffers hard-switching and a current spike occurs in i_D. However, under the same operation point with optimal idling control, before the switch turns on, the current flows through the body diode, and v_{DS} is clamped at zero, thus the switch achieves zero-voltage turn-on, as shown in Fig. 16(b).

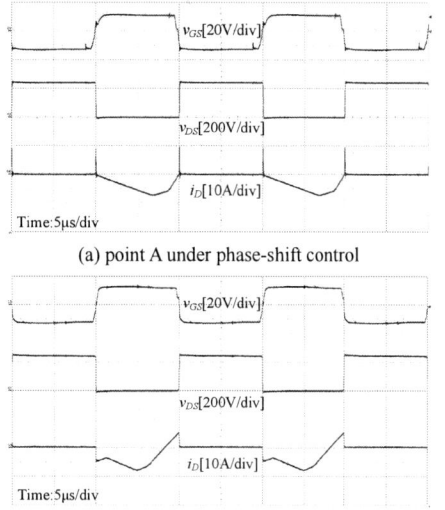

(a) point A under phase-shift control

(b) point B under optimal idling control

Fig. 16. Waveforms of Q_{12} under different control strategy

The measured I^2_{rms1} and I^2_{rms2} are shown in Figs. 17(a) and (b). The tendencies are almost the same with the ones shown in Fig. 10 and Fig. 12 respectively. From Fig. 17(c), it can be seen that I^2_{rms2} is always smaller than I^2_{rms1}.

Fig. 18 shows the efficiencies with different V_2 and P_3

under the two control strategies. As depicted, the efficiencies under optimal idling control are always higher than the ones under phase-shift control. Especially, when V_2 and P_3 are both small, the efficiencies under optimal idling control are much higher than that under phase-shift control for soft-switching realization.

VI. CONCLUSIONS

In this paper, an optimal idling control strategy is proposed for three-port full-bridge converter, based on the phase-shift plus PWM control strategy, which focuses on the minimum conducting losses under different operation conditions. The advantages are as follows: soft-switching can be realized in the whole operation range; the conduction losses are ensured to be minimum so that the efficiency is higher than that under phase-shift control. Experimental results from a 1kW prototype converter are presented to validate the theoretical analysis.

ACKNOWLEDGEMENT

Supported by National Natural Science Foundation of China(50807024), the Power Electronics Science and Education Development Program of Delta Environmental & Educational Foundation and Jiangsu province university outstanding science and technology innovation team project.

REFERENCES

[1] A. Emadi, S. S. Williamson, A. Khaligh, "Power electronics intensive solutions for advanced electric, hybrid electric, and fuel cell vehicular power systems," *IEEE Transactions on Power Electronics*, vol. 21, no. 3, pp. 567-577, 2006.

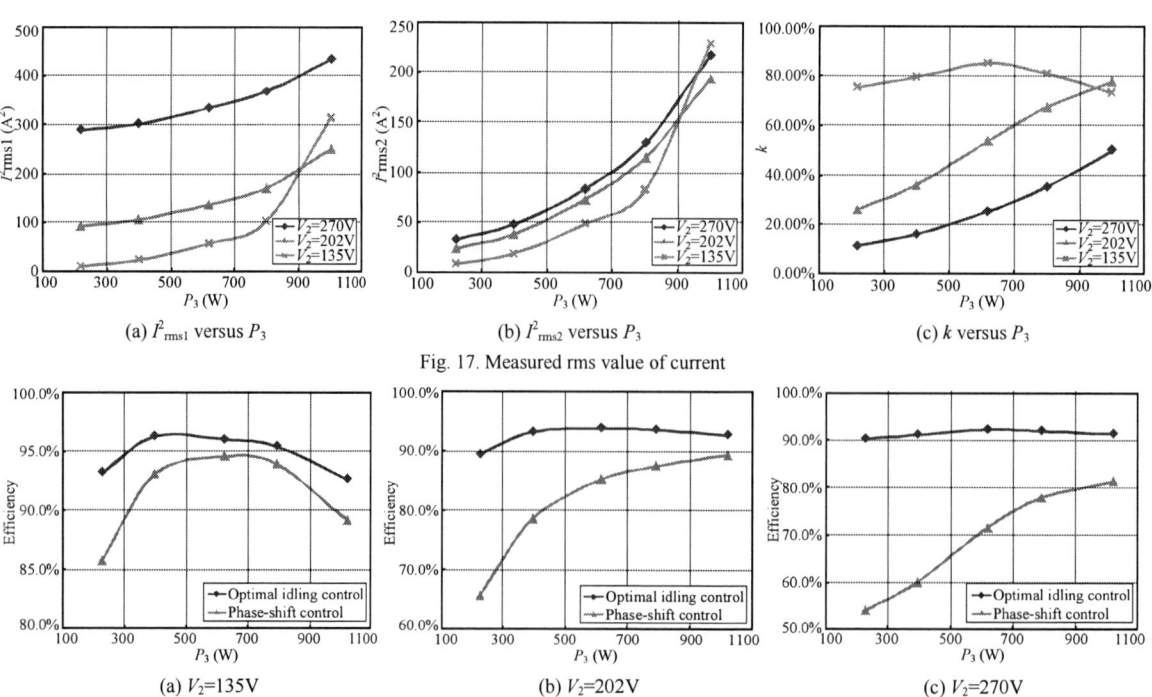

(a) I^2_{rms1} versus P_3 (b) I^2_{rms2} versus P_3 (c) k versus P_3

Fig. 17. Measured rms value of current

(a) V_2=135V   (b) V_2=202V   (c) V_2=270V

Fig. 18. Efficiency with different P_3 at V_2=135V, V_2=202V and V_2=270V

[2] L. D. Jorge, H. Marcel, G. S. Marcelo, "Three-port bidirectional converter for hybrid fuel cell systems," *IEEE Trans. on Power Electronics*, vol. 22, no. 2, pp. 480-487, 2007.

[3] W. Li, J. Xiao, Y. Zhao, "PWM plus phase angle shift(PPAS) control scheme for combined multiport DC/DC converters," *IEEE Transactions on Power Electronics*, vol. 27, no. 3, pp. 1479-1489, 2012.

[4] C. Zhao, S. D. Round, J. W. Kolar, "An isolated three-port bidirectional DC-DC converter with decoupled power flow management," *IEEE Trans. on Power Electronics*, vol. 23, no. 5, pp. 2443-2453, 2008.

[5] H. Tao, J. L. Duarte, M. A. M. Hendrix, "Three-port triple-half-bridge bidirectional converter with zero-voltage switching," *IEEE Trans. on Power Electronics*, vol. 23, no. 2, pp. 782-792, 2008.

[6] Z. Qian, A. R. Osama, A. A. Hussam, et al, "Modeling and control of three-port DC/DC converter interface for satellite applications," *IEEE Transactions on Power Electronics*, vol. 25, no. 3, pp. 637-649, 2010.

[7] K. Grummi, M. Ferdowsi, "Double-input DC-DC power electronic converters for electric-drive vehicles—topology exploration and synthesis using a single-pole-triple-throw switch," *IEEE Transactions on Industrial Electronics*, vol. 57, no. 2, pp. 617-623, 2010.

[8] S. Y. Kim, H. S. Song, K. Nam, "Idling port isolation control of three-port bidirectional converter for EVs," *IEEE Trans. on Power Electronics*, vol. 27, no. 5, pp. 2495-2506, 2012.

Filter Design for Three-Level Grid-connected Inverter with Low Switching Frequency

Kangle Ren, Xing Zhang, Fusheng Wang
School of Electrical Engineering and Automation
Hefei University of Technology
Hefei, China
renkangle6@163.com, honglf@ustc.edu.cn

Yunwu Tu, Lingxiang Wang, Lirong Deng
Sungrow Power Supply Co., Ltd.
Hefei, China
tuyw@sungrowpower.com

Abstract— Aiming at the problems of poor dynamic control performance and difficult filter design for three-level grid-connected inverter with low switching frequency, a three-level discontinuous PWM strategy and its carrier-based implementation method are introduced. Considering the strategy's output harmonic characteristics, an LLCL filter is adopted. Then, aiming at the resonance problem of LLCL filter, two simple passive damping schemes are comparatively analyzed from the aspects of damping effect, filtering performance and damping losses. The conclusion shows that the series resistor structure has better overall performance. Afterwards, the design principle of the filter parameter is proposed based on the bridge-side current ripple analysis and dominant harmonic component attenuation. Correctness and feasibility of the conclusions are validated by experimental results.

Keywords— three-level inverter; discontinuous PWM modulation; LLCL filter; passive damping; parameter design.

I. INTRODUCTION

Due to the low stress across the semiconductors, little harmonic content of the output voltage and high efficiency, the three-level inverter is adopted in new energy power generation, especially in the medium-voltage high-power applications [1-2]. Under these circumstances, the average switching frequency of the semiconductor device is typically less than 1 KHz in order to reduce the switching losses [3-4]. However, low carrier frequency makes the output harmonic component of inverter densely distributed in the low frequency region. In order to filter these low frequency harmonics so as to meet the grid-connected standards [5], a large inductor is needed. Meanwhile, the selection of the filter's resonant frequency will be very difficult. What's more, the low carrier frequency reduces the system stability and control performance.

The carrier frequency of discontinuous pulse width modulation (DPWM) can be increased to 1.5 times of the continuous pulse width modulation (CPWM) with the same device switching losses [6-7]. So the carrier band moves to the higher frequency region and the harmonic performance of the inverter is improved, particularly in the case of the high modulation index. Therefore, adopting the DPWM strategy in the low switching frequency applications can not only improve the system control performance, but also reduce the cost and volume

of the filter. Aiming at the operating characteristics of grid-connected inverter, this paper analyzes the advantage of switching losses reduction and common-mode voltage suppression of DPWMA strategy from the view of vector synthesis, and gives its carrier-based implementation method. Then the well output harmonic performance of DPWMA which is conducive to filter harmonic was analyzed based on the comparison of the output harmonic spectrum between SVPWM and DPWMA.

Because the filter design is closely related to the modulation strategy, this paper introduces LLCL filter as the grid-side filter structure considering the DPWMA harmonic characteristics. The LLCL filter can increase the filtering performance at specific frequency region to reduce the size of the grid-side filter inductor and improve grid-side current [8]. Aiming at the resonance problem, two simple passive damping schemes are comparatively analyzed from the aspects of damping effect, filtering performance and damping losses. Afterwards, the design principle of the filter parameter was proposed based on bridge-side current ripple analysis and the dominant harmonic component attenuation.

Finally, this paper designs a 30KW prototype as an example. Correctness and feasibility of these conclusions proposed in this paper and the design of parameters are validated by experimental results.

II. DPWMA FOR THREE-LEVEL INVERTER

A. Vector synthesis principles and the carrier-based implementation method

The traditional three-level space vector synthesis obeys the nearest three vector principle, and the switching sequence is seven-segment. In order to reduce the switch frequency of the devices, the switch sequence reduces to five-segment in DPWM strategies. Namely, one redundant switch state of the small vector must be abandoned, Where DPWMA strategy only select the switch state with small common-mode voltage. Its clamping region is shown in Fig.1, where yellow, red and blue represent A, B, and C phase clamping region, 1, 0 and -1 represent the clamping level. For example, when the reference voltage vector is located in yellow 1XX area, only 100 state is selected, phase A is clamped at 1 level.

Fig.1 shows that the modulation waves of DPWMA are clamped at the peak region when modulation index is high and clamped at the zero-crossing region when modulation index is low. For the grid-connected inverter, it usually operates in the state of high modulation index and unity power factor, while run in the state of low modulation index and zero power factor for LVRT condition. This means that the clamping regions of DPWMA strategy are almost located at the peak of the current for grid-connected inverter so that reduce switching losses as much as possible. In addition, this strategy can reduce the inverter output common-mode voltage greatly because of the choice of small common-mode voltage switch state. Thus, DPWMA has greater advantages than other DPWM strategies for three-level grid-connected inverter.

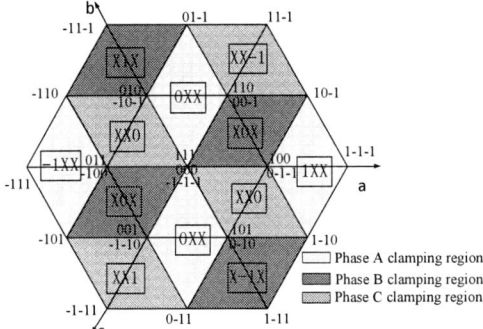

Fig.1. Clamping region of DPWMA strategy

In order to benefit DSP digital implementation, the carrier-based implementation method of DPWMA is researched. From Fig.1, the expression of zero-sequence voltage component can be finally summarized as

$$|V_{0add}| = \min[\min(V_{hx}), \min(V_{xl})] \qquad (1)$$

Where V_{hx} and V_{xl} are the distance from three-phase modulation waves $V_x(x=a, b, c)$ to the corresponding boundary 1, 0 and -1. Namely, when $V_x \geqslant 0$, $V_{hx}=1-V_x$, $V_{xl}=V_x-0$; when $V_x<0$, $V_{hx}=0-V_x$, $V_{xl}=V_x-(-1)$. And when $|V_{0add}|=\min(V_{hx})$, the sign of V_{0add} is positive; when $|V_{0add}|=\min(V_{xl})$, the sign of V_{0add} is negative.

B. Output harmonic performance

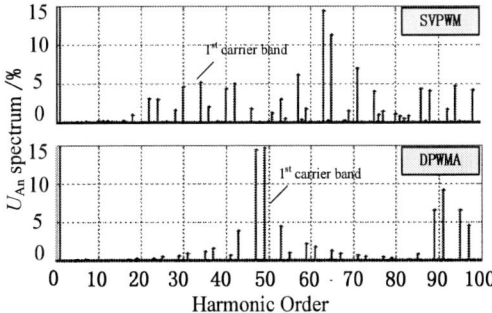

Fig.2. The U_{An} spectrum of the SVPWM and DPWMA.

Fig.2 is the U_{An} spectrum analysis of the SVPWM and DPWMA, where n is the grid neutral point; modulation index is 1.03; the carrier frequency (f_c) of SVPWM and DPWMA are 1.6 KHz and 2.4 KHz respectively, which is to ensure that they have the same device switching frequency.

From Fig.2, the following conclusions can be got.

1) The first carrier band of DPWMA is farther to the fundamental, so the low frequency harmonic components are little, which is conducive to harmonics filtering and resonant frequency selected.

2) Compared to SVPWM, the side-band harmonics of carrier frequency are more concentrated, which benefit to filter the dominant harmonics around the first carrier frequency when adopts LLCL filter.

Taken together, adopting DPWMA strategy can not only improve the system control performance, but also benefit the design of filter. Meanwhile, it reduces the output common-mode voltage and the switching losses.

III. FILTER DESIGN

Considering the concentrated side-band harmonics characteristic of DPWMA, this paper introduces LLCL filter as the grid-side filter structure which is shown in Fig.3, which add an additional inductance L_f in filter capacitor branch to increase the filtering performance at specific frequency, utilizing the L_fC series resonance. Usually, the series resonance frequency is located at f_c.

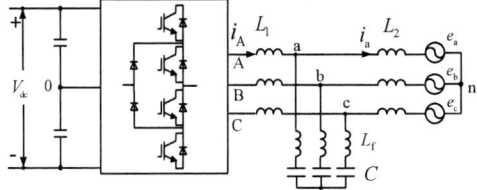

Fig.3. Three-level grid-connected inverter with LLCL filter.

A. Comparison of LLCL filter simple damping schemes

In order to suppress the resonance of LLCL filter, two simple passive damping schemes are shown in Fig.4, which are named as L+R structure and L//R structure, where Rd is damping resistor.

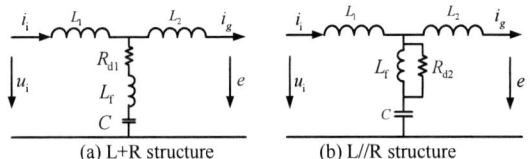

(a) L+R structure   (b) L//R structure

Fig.4. Two simple passive damping schemes of LLCL Filter.

The transfer function of the grid-side current to bridge-side voltage for L+R structure and L//R structure are

$$G_{i_g/U_i}(s)\Big|_{L+R} = \frac{s^2L_fC + sR_{d1}C + 1}{s^3a + s^2bR_{d1}C + sb} \qquad (2)$$

$$G_{i_g/U_i}(s)\Big|_{L//R} = \frac{s^2R_{d2}L_fC + sL_f + R_{d2}}{s^4L_1L_2L_fC + s^3aR_{d2} + s^2bL_f + sbR_{d2}} \qquad (3)$$

Where $a=[L_1L_2+(L_1+L_2)L_f]C$, $b=L_1+L_2$. Usually, the R_d in LCL filter is usually selected as one third of capacitive reactance at resonant frequency [9]. Namely, the damping coefficient ζ is equal to 1/6. According to this principle, the value of R_{d1} can be written

$$R_{d1} = \frac{1}{3}\sqrt{\frac{L_1L_2/(L_1+L_2)+L_f}{C}} \qquad (4)$$

As can be seen from Fig.4 (b), the filter performance

of L//R structure is between no damped LCL filter and no damped LLCL filter with R_{d2} increasing from 0 to infinite. Fig.5 shows the $G_{ig/Ui}(s)$ bode diagrams which including the L+R structure with $R_{d1}=0.7\Omega$ ($\zeta=1/6$), L//R structure with $R_{d2}=0.7\Omega$ and 1.4Ω, LCL structure with $R_d=0.7\Omega$ and LLCL filter undamped, where $L_1=0.9$mH, $L_2=0.3$mH, $C=65\mu$F, $L_f=70\mu$H.

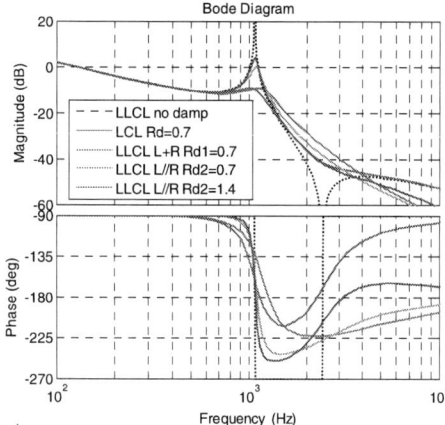

Fig.5. The $G_{ig/Ui}(s)$ bode diagrams of several filter structure.

It can be seen in Fig.5 that although the harmonic attenuation rate around f_c of the two damping schemes are worse than the undamped LLCL filter, they are better than the traditional LCL structure with corresponding damping. If the harmonic attenuation rate around f_c of the two damping schemes are same ($R_{d1}=0.7$, $R_{d2}=1.4$), the resonance peak of L//R structure is much higher than L+R structure. If reducing R_{d2} ($R_{d2}=0.7$), the harmonic attenuation rate becomes poor and the resonance peak is still high. Thus, L//R structure is difficult to balance system stability and filtering requirements. However, L+R structure can guarantee better harmonic attenuation rate around f_c when ensure the damping effect.

The transfer function of the damping resistor current to bridge-side output voltage for LCL filter with series resistor and two LLCL filter passive damping schemes can be easily written From Fig.4, and the their bode diagrams can be got as Fig.6 shown.

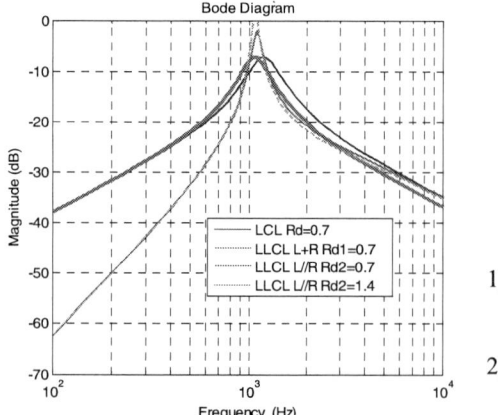

Fig.6. The $G_{iRd/Ui}(s)$ bode diagrams of several filter structure.

It can be seen in Fig.6 that the low frequency component of damping resistor current for L//R structure

is the smallest because of the bypass of L_f, which can effectively reduce the damping resistor losses. However, the high frequency component of L+R structure is small, which makes it also has certain advantages for damping losses when compared to the conventional LCL filter.

Described above, although the damping resistor losses of L//R structure are the least, its filter performance and system stability are difficult to balance. L+R structure has well damping effect, and can guarantee a certain harmonic attenuation rate so that all the current harmonics meet the standard. Meanwhile, its damping losses are smaller than conventional LCL filter structure. Therefore, L+R structure has the best overall performance.

B. Analysis of bridge-side current ripple

From simulation result it can be got that when the modulation index (m) of DPWMA is high, the maximum of current ripple peak-peak value (i_{rpm}) appears in the peak of the modulation wave, where the reference voltage vector locates at a axis in Fig.1. At this point, only vectors 100, 1-1-1 are adopted.

The L_1 voltage of phase a can be expressed as

$$\Delta u_L - u_{a0} - u_{n0} - e_a = [2s_a - (s_b \mid s_c)]V_{dc} / 6 - e_a \quad (5)$$

By (5), the Δu_L of the two vectors are ($1/3 V_{dc}-E_m$), ($2/3 V_{dc}-E_m$), where E_m is the grid phase voltage peak. Define m=2 E_m/V_{dc}. From volt-second balance theory, the valid time T_1, T_2 of the two vectors are

$$T_1 = (2-1.5m)T_s, \quad T_2 = (1.5m-1)T_s \quad (6)$$

Then, the switch sequence and bridge-side current ripple can be shown in Fig.7.

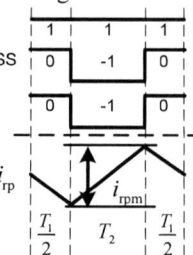

Fig.7. The switch sequence and bridge-side current ripple.

From Fig.7, the current ripple peak-peak value is

$$i_{rpm} = (\frac{2}{3} - \frac{m}{2})(1.5m-1)V_{dc}T_s / L_1 \quad (7)$$

Assuming the expected proportion of i_{rpm} to I_m is K, than L_1 is

$$L_1 \geq (\frac{2}{3} - \frac{m}{2})(1.5m-1)V_{dc}T_s / (KI_m) \quad (8)$$

C. Parameter Design Steps

1) Selecting K to identify the lower limit of L_1 by (8);

2) Selecting the initial value of $C=0.05C_b$. Considering increasing C to reduce inductor size and cost in high-power application, the upper limit of C is set to $0.1C_b$.

3) Calculating L_f by (9);

$$L_f = \frac{1}{(2\pi f_c)^2 C_f} \quad (9)$$

4) Calculating L_2 by (10) to meet the grid standards;

$$I_{g,h,\%}\Big|_{h>35} = \beta_h Z_b \left|G_{i_g/U_i}(jh\omega_b)\right| \le 0.3\% \qquad (10)$$

Where $\beta_h=U_{i,h}/U_{i,1}$, $U_{i,h}(h=1.2\ldots)$ are the harmonics of U_{An}, $G_{ig/Ui}(s)$ is shown by (3), $R_{d1}=0$;

5) Judging (11) whether to set up? If it is not workable, let $C = C + \triangle C$, and skip to step 2);

$$\begin{cases} L_1+L_2 \le 0.1C_b \\ 10f_b < f_{res,LLCL} < 0.5f_s \end{cases} \qquad (11)$$

6) Calculating R_{d1} by (4), skip to step 2) to correct L_2;

IV. EXPERIMENTAL VALIDATION

A. Filter design example

In order to validate the design scheme in this paper, a 30 KW three-level prototype is adopted as a design example. The rated grid line-to-line voltage RMS is 380 V. The DC voltage is 600 V and the carrier frequency is 2.4 KHz.

If K is 35%, L_1 is 0.9mH calculated by (8). Because m is about 1.03 in this example, the dominant harmonics of U_{an} can be got from Fig.2. They are $\beta_{47}(14.5\%)$, $\beta_{49}(14.7\%)$ and $\beta_{91}(9.1\%)$.

According to the parameter design steps and considering parameter errors in practice, the parameters obtained finally are shown in case 1 of Table I.

TABLE I
VALUES OF FILTERS IN EXPERIMENTS

Parameters	Case 1 LLCL (L_f+R_d)	2 LCL	3 LLCL ($L_f//R_d$)	4 LLCL ($L_f//R_d$)
L_1/mH	0.9	0.9	0.9	0.9
L_2/mH	0.4	0.4	0.4	0.4
C_f /μF	65	65	65	65
L_f/μH	70	—	70	70
R_d/Ω	0.7	0.7	0.7	1.4

B. Experimental validation and analysis

In order to validate the correctness and feasibility of the design, a 30KW three-phase three-level experimental system using the parameters in Table I is constructed. The modulation strategy adopts DPWMA.

Fig.8 to Fig.11 are the waveforms of bridge-side current i_A, damping resistor current i_{Rd}, grid-side current i_a and its FFT analysis in case 1 to 4 respectively.

It can be seen from Fig.8 that the i_{rpm} is 23A, which is about 35.3% of the peak value of bridge-side current. It is consistent with the theoretical design. Meanwhile, the damping effect is very well and the most dominant harmonic current content is only 0.3%, which meets the standard [5]. As shown in Fig.9, the i_{rpm} in Case 2 is 26A which is larger than Case 1. The switch frequency harmonic content is about 1%, which is much larger than the Case1 and can't meet the standard.

From Fig.10 it can be seen that although the harmonics around f_{res} can be acceptable, the switch frequency harmonic content is 0.67% that can't meet the standard. However, in Fig.11 the switch frequency harmonics content decrease to 0.35% by increasing R_{d2}, but the harmonics amplitude around f_{res} increase. It reduces system stability.

TABLE Ⅱ is the THD of grid-side current and the damping loss in the four cases. From the table it can be seen that the THD of Case 1 is smallest and that of Case 2 is the largest. While the damping loss in Case 3 and 4 is lower. Although the damping loss in Case 1 is larger than Case 3 or 4, it's lower than Case 2.

In all, the experimental results are consistent with theoretical analysis.

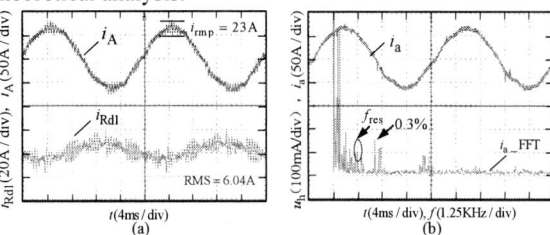

Fig.8. (a) Experimental waveforms of bridge-side current i_A, damping resistor current i_{Rd}, (b) grid-side current i_a and FFT analysis in Case 1.

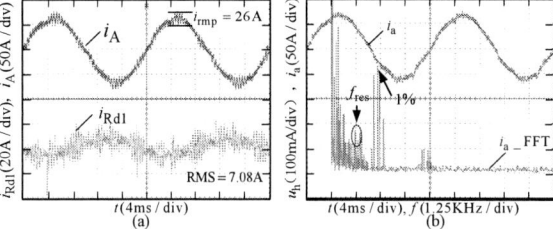

Fig.9. (a) Experimental waveforms of bridge-side current i_A, damping resistor current i_{Rd}, (b) grid-side current i_a and FFT analysis in Case 2.

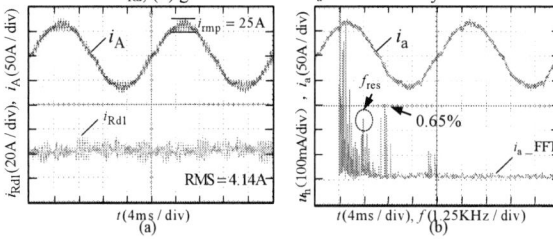

Fig.10. (a) Experimental waveforms of bridge-side current i_A, damping resistor current i_{Rd}, (b) grid-side current i_a and FFT analysis in Case 3.

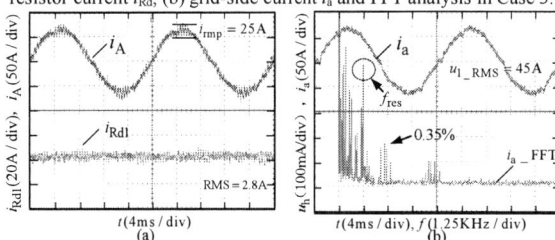

Fig.11. (a) Experimental waveforms of bridge-side current i_A, damping resistor current i_{Rd}, (b) grid-side current i_a and FFT analysis in Case 4.

TABLE Ⅱ
THD% OF GRID-SIDE CURRENT AND POWER LOSS OF DAMPING RESISTOR IN DIFFERENT CASES

THD_i & P_{Rd} \ Case	1	2	3	4
THD/%	3.4	5.8	3.9	3.67
P_{Rd}/w	25.5	35	12	10.9

V. CONCLUSION

This paper introduces the three-level DPWMA strategy which has the advantages of low switching losses, low common-mode voltage and well output harmonic performance, and gives its carrier-based implementation method. Then, considering the harmonic characteristics, the LLCL filter is introduced. Aiming at the resonance problem, two simple passive damping schemes are comparatively analyzed. Thereinto, L//R structure has the least damping loss, but L+R structure can ensure both damping effect and harmonic attenuation. Meanwhile, its damping loss is superior to conventional LCL structure. Finally, the design principles of the filter parameter are proposed based on bridge-side current ripple analysis and the dominant harmonic attenuation. Feasibility and correctness of the conclusions and the design of parameters are validated by experimental results.

REFERENCES

[1] Haitham Abu-Rub, Joachim Holtz, Jose Rodriguez and Ge Baoming, "Medium-voltage multilevel converters-state of the art, challenges, and requirements in industrial applications," *IEEE Transactions on Industrial Electronics.*, vol. 57, no.8, pp. 2581-2596, August. 2010.

[2] A. A. Rockhill, Marco Liserre, Remus Teodorescu and Pedro Rodriguez, "Grid-filter design for a multimegawatt medium-voltage voltage-source inverter" *IEEE Transactions on Industrial Electronics*, vol. 58, no.4, pp. 1205-1217, April. 2011.

[3] T. Brückner and D. G. Holmes, "Optimal pulse-width modulation for three-level inverters," *IEEE Trans. Power Electron*, vol. 20, no. 1, pp. 82-89, Jan. 2005.

[4] Rathore A, Holtz J, Boller T, "Synchronous optimal pulsewidth modulation for low switching frequency control of medium voltage multi-level inverters" *IEEE Transactions on Industrial Electronics*, vol. 57, no.7, pp. 2374-2381, 2010.

[5] IEEE Recommended Practices and Requirements for Harmonic Control in Electrical Power Systems, IEEE Std. 519-1992, May 10, 1992.

[6] Lucian Asiminoaei, Pedro Rodríguez and Frede Blaabjerg, "Application of discontinuous PWM modulation in active power filters" *IEEE Transactions on Power Electronics*, vol. 23, no.4, pp. 1692-1706, 2008.

[7] Luca Dalessandro, Simon D. Round, Uwe Drofenik and Johann W. Kolar, "Discontinuous space-vector modulation for three-level PWM rectifiers," *IEEE Transactions on Power Electronics*, vol. 23, no.2, pp. 530-542, 2008.

[8] Weimin Wu, Yuanbin He, and Frede Blaabjerg, "An LLCL power filter for single-phase grid-tied inverter," *IEEE Transactions on Power Electronics*, vol. 27, no.2, pp. 782-789, 2012.

[9] Marco Liserre, Frede Blaabjerg and Steffan Hansen, "Design and control of an LCL-filter-based three-phase active rectifier," *IEEE Transactions on Industry Applications.*, vol. 41, no.5, pp. 1281-1291, Sept./Oct. 2005.

A Novel Efficient T Type Three Level Neutral-Point-Clamped Inverter for Renewable Energy System

Wenlong Wu, Fei Wang, Yong Wang

Department of Electrical Engineering, Shanghai Jiaotong University

Shanghai, China, 200240

Abstract-**The I type 3L-NPC (Three-level Neutral-Point-Clamped) converter is the most popular multilevel converter used in the renewable energy system for its low switching loss, reduced devices voltage stress. However, it suffers from problems of high conduction losses. The T type 3L-NPC is researched in the literatures to overcome the drawbacks of I type inverter. But, the high side and low side power devices of T type 3L-NPC need to block the whole dc link voltage. The switching frequency is significantly limited due to the high switching losses caused by the high voltage rated power devices. In this paper, based on the drawbacks of both the I and T type 3L-NPC, a novel efficient T type 3L-NPC for renewable energy system is proposed to achieve both low conduction losses and switching losses with high switching frequency. The proposed topology and control strategy are successfully applied in the solar inverter system.**

I. INTRODUCTION

Three level converters are nowadays widely adopted in low voltage renewable energy systems for its low switching loss, reduced devices voltage stress. The basic idea is that the dc link voltage can be split between different capacitors, which can provide intermediate voltage levels between the reference potential and the dc link voltage.

Ref. [1-2] research on the conventional I type 3L-NPC inverter applied in the renewable energy system. The clamping diodes in [1-2] are replaced by the IGBTs to reach active clamping in [3] to improve the semiconductor loss distribution. Ref. [4] proposes a novel high efficiency stacked neutral point clamped 3L-NPC (3L-SNPC) inverter. Ref. [5] proposes an active clamped 3L-SNPC (3L-ASNPC) to improve the 3L-SNPC's efficiency. Moreover, in [6], the 3L-SNPC's structure and control strategy are improved to reduce the power loss, particularly in the low power range when applied in the solar system.

To summarize, the foregoing solutions are all based on or derived from the fundamental 3L-NPC leg. Therefore, they have the common or similar features as described in the following.

This project is supported by National Nature Science Foundation of China through Project 51177100.

Fig. 1 shows the long current paths through two IGBTs or two diodes with switch status as 1100 or 0011.

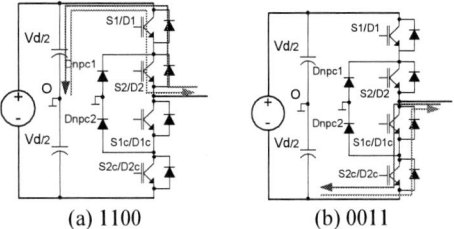

(a) 1100 (b) 0011

Fig. 1 Long current paths of conventional 3L-NPC leg

First of all, the long current paths imply the high conduction loss. Second, higher stray inductance due to the long current paths results in higher power loss and turn-off overvoltage.

It should be mentioned that the foregoing discussed drawbacks of I type 3L-NPC seem more prominent in renewable energy system because they are usually operating in much lower power range than the rated power.

Fig. 2 reveals in the medium and low power range, the on state voltage over load current of IGBT is much higher than the high power situation.

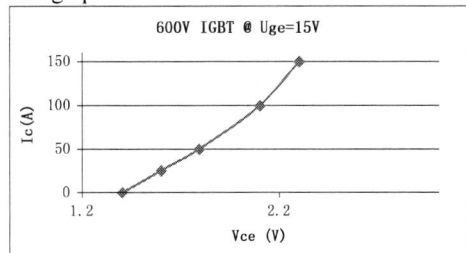

Fig. 2 The IGBT conduction voltage over current

Based on the forgoing recognitions, T type 3L-NPC is being researched increasingly to improve the low voltage renewable energy system efficiency. One leg of T type 3L-NPC is shown in Fig. 3. Ref. [7] evaluates the power loss and control scheme of T type 3L-NPC applied in the low voltage renewable energy system. Ref. [8] applies T type 3L-NPC in the solar system to avoid the high conduction loss.

978-1-4799-2706-7/14 $31.00 © 2014 IEEE   470

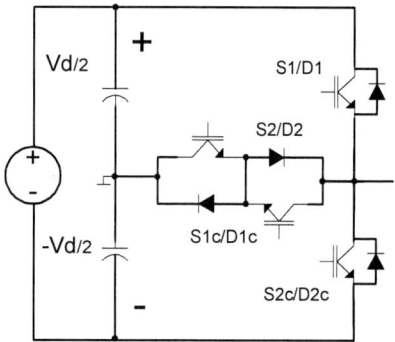

Fig. 3 One leg of the T type 3L-NPC

In the T type 3L-NPC, the long current paths as shown in Fig. 1 are shortened. However, compared with the I type 3L-NPC, the high and low side of T 3L-NPC needs 1200V IGBT which is twice of the I type 3L-NPC because it has to block the whole dc link voltage [7]. The high operation voltage causes high switching loss. Therefore, it is commonly believed that the T type has advantage over I type within a specific low switching frequency range.

To find out the switching frequency limitation for a typical medium power T type solar inverter, the semiconductors power loss calculation and comparison between 5kW T type and I type 3L-NPC leg have been done in this paper. The power devices are selected as shown in Tab. 1 and the power factor is supposed to be unity.

Tab. 1 Power devices selection

I 3L-NPC	T 3L-NPC
Dnpc1=Dnpc2=FFP30S60S	No
S1=S2c=FGH60N60	S1=S2= HGTG30N120CN
S1c=S2= FGH60N60	S1c=S2c= FGH60N60

The PWM strategy is shown in Fig. 4.

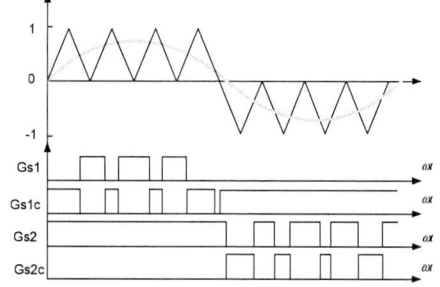

Fig. 4 The PWM strategy for both T and I 3L-NPC

The calculation results are shown in Fig. 5, which shows the T type 3L-NPC only has advantage over I type 3L-NPC within about 12 kHz.

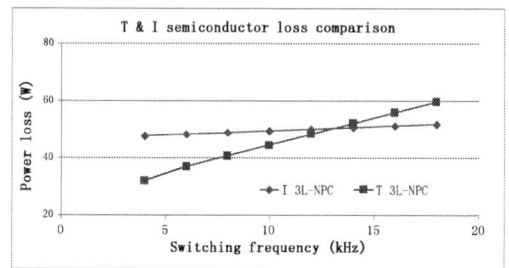

Fig. 5 The power loss comparisons between T and I 3L-NPC

II. PROPOSED NOVEL T 3L-NPC INVERTER AND ITS PWM STRATEGY

Based on the foregoing conclusions, this paper proposes a novel high efficiency T type 3L-NPC inverter for low voltage renewable energy system as shown in Fig. 6. In the proposed scheme, based on the T type 3L-NPC, two additional 600V Coolmosfets S1-aux & S2c-aux are connected between the bus and the middle points of the bidirectional switch to cut down the 1200V IGBTs S1 & S2c switching loss. Moreover, the middle bidirectional switch is also formed by the 600V Coolmosfets to reduce the conduction loss in very low power range.

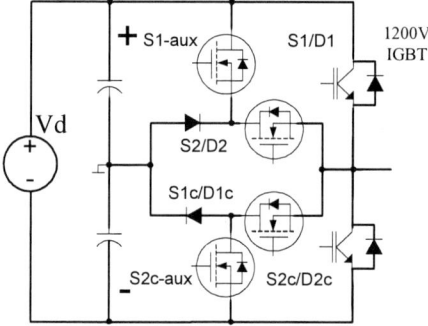

Fig. 6 The proposed novel T type 3L-NPC

The new topology can combine the advantages of both T and I type 3L-NPC. In Fig. 6, the 1200V IGBT's switching loss is moved to the aditional high and low side Coolmosfets by a new PWM strategy as described in the following. Therefore, the uneven power loss distribution is alleviated. The high efficiency and the high switching frequency contradiction in T type 3L-NPC is solved.

In the new PWM strategy, there are three modes proposed to adapt to low, medium and heavy load as shown in Fig. 7 to Fig. 9. The PWM strategy details are investigated by the positive half cycle in the following.

978-1-4799-2706-7/14 $31.00 © 2014 IEEE   471

The 2014 International Power Electronics Conference

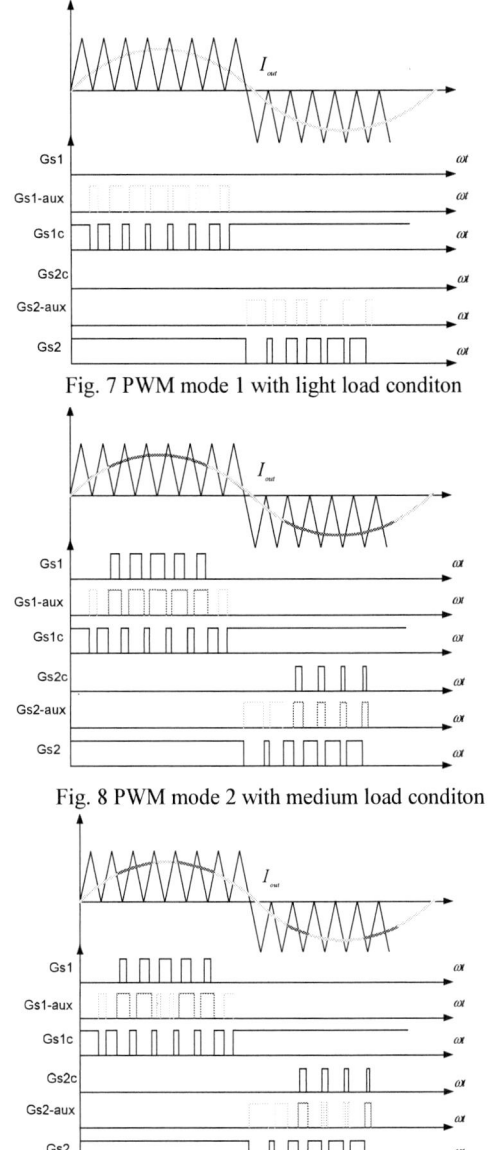

Fig. 7 PWM mode 1 with light load conditon

Fig. 8 PWM mode 2 with medium load conditon

Fig. 9 PWM mode 3 with heavy load conditon

With light load as shown in Fig. 7, IGBT S1 is disabled in posive half cycle. According to the test results and calculation, even S1 is enabled, the current is all flowing from S1-aux & S2 for the low total on state resistance of Coolmosfet. Therefore, it is quite reasonable to disable S1 drive to save the drive power and the switching power. The same analysis can be extended to the negative half cycle.

In Fig. 8 with medium load condition, PWM is divided into two sections shown by blue and red. The PWM strategy is the same with low power conditon in blue section. However, in the red section with relativley larger current, the on state voltage of Coolmosfets increases, causing higher conduction loss. Hence, S1 is enabled to form a paralleled hybrid power device structure to conduct the load current.

With heavy load, the load current is divided into three sections by instant current amplitude responding to three PWM modes as shown by blue, red and green lines in Fig. 9. The blue and red section's PWM strategies are the same with Fig. 8. However, in the high current green section, Coolmosfet S1-aux is only switched on and off shortly to move the switching loss of S1.

With the proposed PWM strategy, the topological state is shown in Fig. 10.

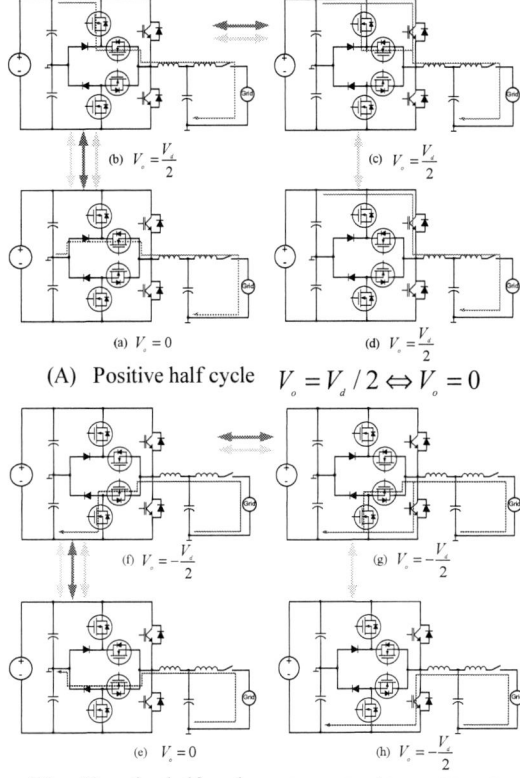

(A)   Positive half cycle $V_o = V_d / 2 \Leftrightarrow V_o = 0$

(B)   Negative half cycle $V_o = -V_d / 2 \Leftrightarrow V_o = 0$

Fig. 10 The topological states with the new PWM strategy

In positive half cycle as shown in (A) of Fig. 10, states (a) and (b) are the two states with light load, the current is only commutating between S1-aux and D2. Though, state (b) shows the same long conduction pass as the conventional I type 3L-NPC, it actually has relatively lower conduction loss due to the lower on state resistance CoolMosfets.

The medium load condition has three states as shown by (a), (b) and (c) with (a) & (c) as the main states, (b) is a short transition state between them. That is to say, Coolmosfet S1-aux is shortly first switched on and shortly later switched off. Hence, the switching loss of 1200V IGBT S1 is moved to the 650V Coolmosfet S1-aux which has much better dynamic performance.

If the current steps into the green region in Fig. 10, which means the load current will mainly conducted by the IGBT S1 when $V_o = V_d / 2$, state (c) is moved to (d) after a short duration as (b) is. Therefore, the current commutation goes through states (a)~(d) with (a),(d) as the main states.

The same analysis can be extended to the negative half cycle as shown from (e)~(h) in Fig. 10.

978-1-4799-2706-7/14 $31.00 © 2014 IEEE   472

It should be mentioned that, currently, the calculation is done offline to control the boundaries, while the online calculation is also possible.

With the novel topology and PWM strategy, the power loss comparisons are also conducted, the results are shown in Fig. 11. It shows the proposed T type 3L-NPC has almost the same power loss trend as the I type 3L-NPC when the switching frequency increases. Moreover, the low power range power loss is particularly better.

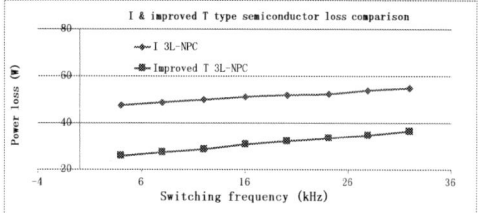

Fig. 11 The semiconductor power loss comparison between I and improved T type 3L-NPC (S1-aux=S2c-aux= IPB60R099C6)

III. EXPERIMENTAL RESULTS

A single phase 3L-NPC solar inverter prototype is built up to verify the proposed novel T type 3L-NPC. The system is based on FPGA XC3AN50TQ144 digital control board. The L value is 0.6mH and C is 4.7uF. The power devices are selected as Tab. 1 with additional Coolmosfet as IPB60R099C6 from Infineon.

Fig. 12~Fig. 17 show the drive signals and currents of S1-aux and S1 in positive half cycle.

Fig. 12 shows IGBT S1 is disabled and the current is all flowing from the Coolmosfet S1-aux with current instant amplitude lower than about 10A. Which is the same with the blue section shown in Fig. 7 - Fig. 9.

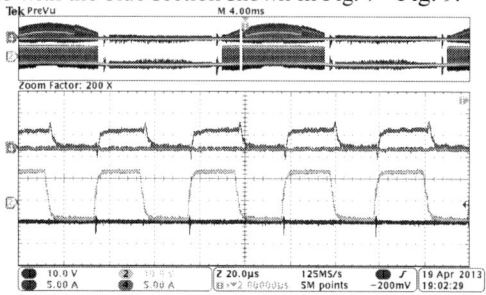

Fig. 12 Current paths in Mode 1 (up to down: S1-aux current(pink), S1 current(green), S1-aux drive(blue), S1 drive(dark blue))

Figure. 13 shows the mode cummutation from mode 1 to mode 2 when load current rises.

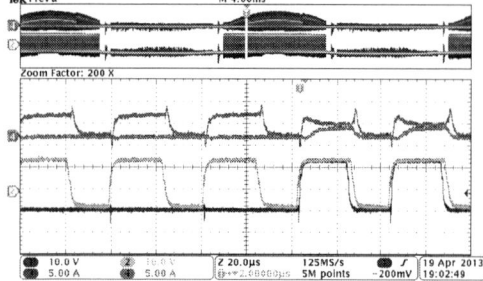

Fig. 13 Transition between mode 1 and mode 2 (up to down: S1-aux current(pink), S1 current(green), S1-aux drive(blue), S1 drive(dark blue))

In PWM mode 2 as shown in Fig. 14, the hybrid parallel structure of Coolmosfet and IGBT conducts current together.

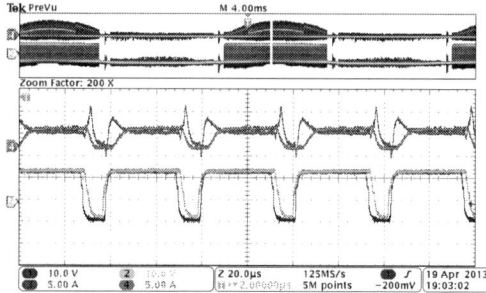

Fig. 14 Current paths in mode 2 (up to down: S1-aux current(pink), S1 current(green), S1-aux drive(blue), S1 drive(dark blue))

Fig.15 shows the commutation between mode 2 and mode 3.

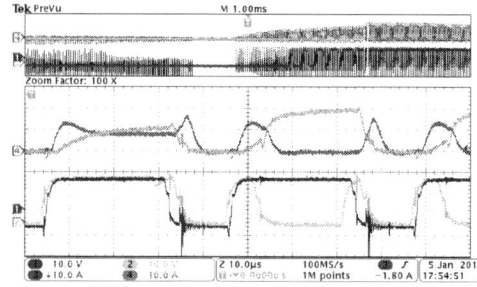

Fig. 15 Transition between Mode 2 and Mode 3 (up to down: S1-aux current(pink), S1 current(green), S1-aux drive(blue), S1 drive(dark blue))

However when the total current is larger than about 20A as shown in Fig. 16, PWM mode changes to the green section described in Fig. 9 where S1-aux is only shortly earlier turned on and shortly later turned off than the IGBT S1.

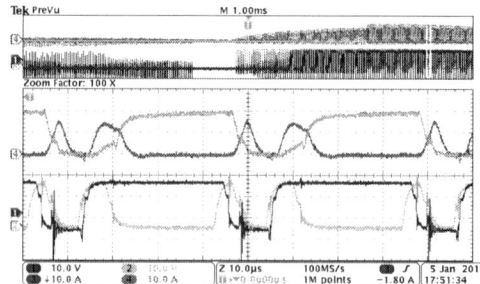

Fig. 16 Current paths in mode 3 (up to down: S1-aux current(pink), S1 current(green), S1-aux drive(blue), S1 drive(dark blue))

The output phase voltage and current is shown in Fig. 17.

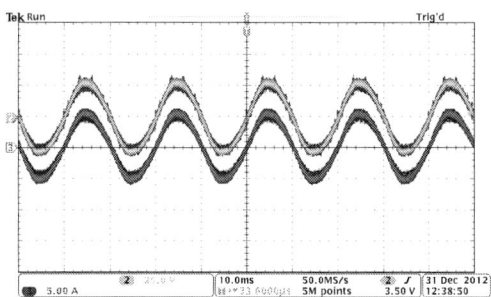

Fig. 17 The output voltage and current

IV. CONCLUSION

In this paper, the advantages and disadvantages of the conventional I and T type 3L-NPC are investigated and summed up. A novel T type 3L-NPC inverter is proposed to combine the advantages of both I and T type 3L-NPC. A new PWM strategy is also proposed to apply in the new topology. The predicted high efficiency is proved by a 5 kW prototype.

REFERENCES

[1] Y. Wang, Q. Gao, X. Cai, "Mixed PWM for Dead-Time Eliminination and Compensation in a Grid-Tied Inverter ," *IEEE Trans. on Industrial Electronics*, Vol. 58, No. 10, pp. 4797-4803, 2011.

[2] O.S. Senturk, L. Helle, S. M. Nielsen, P. Rodriguez, R. Teodorescu, "Power Capability Investigation Based on Electrothermal Models of Press-Pack IGBT Three-Level NPC and ANPC VSCs for Multimegawatt Wind Turbines," *IEEE Trans. on Power Electronics*, Vol. 27 , No. 7, pp. 3195 – 3206, 2012.

[3] J. Li, J.J. Liu, D. Boroyevich, P. Mattavelli, X. Y. Xue, "Three-level Active Neutral-Point-Clamped Zero-Current-Transition Converter for Sustainable Energy Systems," *IEEE Trans. on Power Electronics*, Vol. 26 , No. 12, pp. 3680 – 3693,2011.

[4] D. Floricau, G. Gateau, M. Dumitrescu, R. Teodorescu, "A new stacked NPC converter: 3L-topology and control," *Power Electronics and Applications, 2007 European Conference on*, pp. 1 – 10, 2007.

[5] D. Floricau, G. Gateau, A. Leredde, "New Active Stacked NPC Multilevel Converter: Operation and Features," *IEEE Trans. on Industrial Electronics*, Vol. 57, No. 7, pp. 2272-2278, 2010.

[6] Y. Wang, F. Wang, "Novel Three-Phase Three-Level-Stacked Neutral Point Clamped Grid-Tied Solar Inverter With a Split Phase Controller," *IEEE Trans. on Power Electronics*, Vol. 28, No. 6, pp. 2856 – 2866, 2013.

[7] M. Schweizer, J.W. Kolar, "Design and Implementation of a Highly Efficient Three-Level T-Type Converter for Low-Voltage Applications," *IEEE Trans. on Power Electronics*, Vol. 28, No. 2, pp. 899 – 907, 2013.

[8] Y. Park, S. K. Sul, C.H. Lim, W.C. Kim, S.H. Lee, "Asymmetric Control of DC-Link Voltages for Separate MPPTs in Three-Level Inverters," *IEEE Trans. on Power Electronics*, Vol. 28 , No. 6, pp. 2760 – 2769, 2013.

The 2014 International Power Electronics Conference

A Novel Neutral Point Voltage Automatic Balancing Carrier-Based Modulation Strategy of Three-Level NPC Converter

LI Ning, WANG Yue, NIU Ruigen, GUO Wei, LEI Wanjun, WANG Zhao'an

School of Electrical Engineering
Xi'an Jiaotong University
Xi'an, China
ningli@stu.xjtu.edu.cn

Abstract—**A novel carrier-based modulation strategy of NPC three-level converter which achieves neutral point voltage balance under the whole modulation index and power factor is proposed in this paper. It divides the traditional single modulation wave of each phase in SPWM strategy into two modulation waves, which are compared respectively with the carrier to decide the output switch state of the converter. It analyzes the internal relations of the modulation waves between the new strategy and traditional SPWM strategy, and the 64 solutions of the modulation waves of the new strategy are deduced. On the basis of analysis and derivation, the new strategy is simple and easy to be implemented in digital system. Through the analysis of the DC voltage utilization ratio, the device switching losses and the output phase voltage THD characteristics, the new strategy has the disadvantage of larger switch losses and higher harmonic content. Finally, the analysis of this paper is verified by experiment.**

Keywords— *carrier-based modulation strategy, NPC three-level converter, neutral point voltage balance, whole modulation index and power factor, 64 solutions of the modualtion waves.*

I.   INTRODUCTION

Multilevel converters are commonly used in high power high voltage applications, such as motor drives system, rail transit system and wind generation system. Meticulous and serious work has been done on the topologies, the modulation strategies and the control schemes of multilevel converters over the years. However, further research of multilevel converters is still needed considering the increase of practical problems [1]-[3].

Fig. 1. Scheme of three-level NPC voltage source inverter (VSI)

The neutral point clamped (NPC) inverter is one of the most widely used multilevel converters, as shown in Fig.1. In the above topology, the neutral-point (NP) voltage should be kept at $U_{dc}/2$ for the regular operation of the system. However, the voltage fluctuation always exists in the NP since the NP current i_o is not zero during the practical operation of the system. The voltage fluctuation, if not controlled, would cause the oscillation and unbalance of the two DC capacitor voltages. It is the significant drawback of NPC converter, which has attracted a great deal of research interests. Normally, there are two kinds of methods to achieve the NP voltage control. The first one is achieved by adding an additional hardware or software NP voltage control circuit, and the second is achieved by designing a DC voltage balancing modulation strategy. However, these two methods just focus on the control of the averaged NP voltage, while a low-frequency NP voltage oscillation may appear under some operating conditions, which is seldom considered. And if the NP voltage oscillation is not eliminated, the output voltage will contain low-frequency disturbance [4]-[7].

In this paper, it firstly analyzes the drawback of traditional SPWM strategy used in three-level NPC converter. And then it proposes a novel NP voltage automatic balancing carrier-based modulation strategy of three-level NPC converter. The new strategy applies two modulation waves for each phase [8]-[14], and the NP voltage balance in the full range of modulation index and power factor is realized by the rational selection of amplitude and phase of the two modulation waves. Furthermore, it deduces all the solutions of the modulation waves of the new strategy and analyzes the DC voltage utilization ratio, switching devices losses and output phase voltage THD characteristics of the new strategy. Finally the new strategy is verified by simulation results.

II.   DRAWBACK OF TRADITIONAL SPWM STRATEGY OF NPC THREE-LEVEL CONVERTER

From fig.1, the relation between the instantaneous NP current i_o and the duty ratio of the first switch device (d_{x1})

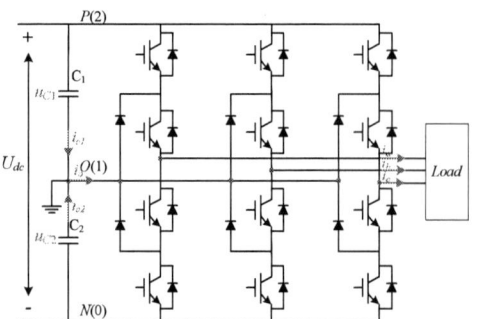

978-1-4799-2706-7/14 $31.00 © 2014 IEEE

475

and the second switch device (d_{x2}) of phase x ($x=a, b, c$) is shown in equation (1).

$$i_o = \sum (d_{x2} - d_{x1}) \cdot i_x (x=a,b,c) \tag{1}$$

where i_x is the load current of phase x. The three phase reference voltage (U_{ra}, U_{rb}, U_{rc}, in P.U.) is shown in equation (2).

$$\begin{cases} U_{ra} = M\sin(\omega t) \\ U_{rb} = M\sin(\omega t - 2\pi/3) \\ U_{rc} = M\sin(\omega t + 2\pi/3) \end{cases} \tag{2}$$

where M is modulation index, ω is the angular frequency of fundamental wave. If symbol I_m represents the peak value of phase current and φ represents the power-factor angle, the three-phase currents of the converter can be given by equation (3).

$$\begin{cases} i_a = I_m \sin(\omega t - \varphi) \\ i_b = I_m \sin(\omega t - 2\pi/3 - \varphi) \\ i_c = I_m \sin(\omega t + 2\pi/3 - \varphi) \end{cases} \tag{3}$$

The duty ratio d_{a1} and d_{a2} are shown in the following equation:

$$\begin{cases} d_{a1} = \begin{cases} M\sin(\omega t), (\omega t \in [0,\pi]) \\ 0, (\omega t \in [\pi, 2\pi]) \end{cases} \\ d_{a2} = \begin{cases} 1, (\omega t \in [0,\pi]) \\ 1 + M\sin(\omega t), (\omega t \in [\pi, 2\pi]) \end{cases} \end{cases} \tag{4}$$

Substitute equation (4) into equation (1) and use it into the other two phases, the NP current i_o can be expressed as equation (5).

$$i_o = \sum [1 - abs(U_{rx})] \cdot i_x (x=a,b,c) \tag{5}$$

Substitute equation (2) and (3) into equation (5), and the following equation can be got[15].

$$i_o = \frac{MI_m}{2}[2\cos(2\omega t - \varphi + (n-1) \cdot \pi/3) - (-1)^n \cos\varphi], \tag{6}$$
$$(\omega t \in [\frac{n\pi}{3}, \frac{(n+1) \cdot \pi}{3}]), n = 0...5$$

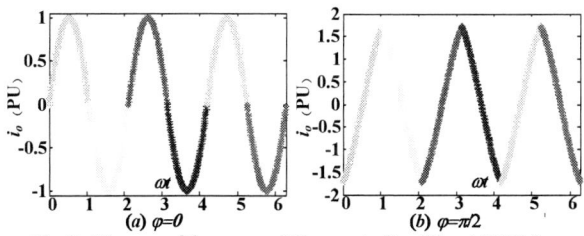

Fig. 2. Diagram of the average NP current of traditional SPWM strategy (P.U. value)

Fig.2 shows the average neutral current of traditional SPWM strategy. It is clear that the average NP current has a low-frequency fluctuation (3ω) which causes the fluctuation of the DC capacitor voltage at the same frequency. In order to balance the DC capacitor voltage, an extra DC capacitor voltage balancing scheme should be added into the control system, if SPWM strategy is used. The zero voltage injection method with SPWM strategy is the most normally used DC capacitor voltage balancing scheme. However, this method cannot eliminate the NP voltage fluctuation within the full range of modulation index and power factor[10].

III. THE NOVEL NP VOLTAGE AUTOMATIC BALANCING CARRIER-BASED MODULATION STRATEGY

In this section, it proposes a novel NP voltage automatic balancing carrier-based modulation strategy under the whole modulation index and power factor. The research idea of the new strategy comes from multiple reference modulation waves for each phase in carrier-based modulation strategy.

A. Relation between the new strategy and SPWM strategy

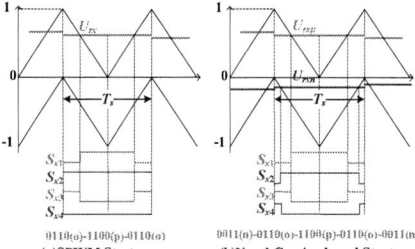

(a)SPWM Strategy   **(b)Novel Carrier-based Strategy**

Fig.3. Relation between the new carrier modulation strategy and traditional SPWM strategy

Fig.3 shows the diagram of relation between the new strategy and traditional SPWM strategy. There are two modulation waves U_{rxp} and U_{rxn} in the new strategy for each phase, and the relation between the modulation waves of the two strategies is shown in equation (7).

$$\begin{cases} U_{rx} = U_{rxp} + U_{rxn} \\ 0 \le U_{rxp} \le 1, -1 \le U_{rxn} \le 0 \ (x=a,b,c) \\ U_{rxp\max} = -U_{rxn\min} \end{cases} \tag{7}$$

where $U_{rxp\max}$ is the maximum value of U_{rxp} and $U_{rxn\min}$ is the minimum value of U_{rxn}. Substitute equation (7) into equation (5), the NP current can be expressed as follows.

$$i_o = \sum (1 + U_{rxn} - U_{rxp}) \cdot i_x (x=a,b,c) \tag{8}$$

In order to eliminate the NP voltage fluctuation, the average value of i_o should be zero in each switching period. Since the three-phase current do not contain zero-sequence component, the following equation is satisfied.

$$i_a + i_b + i_c = 0 \tag{9}$$

Considering the difference of the load conditions and power factor of circuits, a simple and useful solution of $i_o=0$ is

$$1 + U_{ran} - U_{rap} = 1 + U_{rbn} - U_{rbp} = 1 + U_{rcn} - U_{rcp} = k \tag{10}$$

Regarding the relationships (10) and (7), the two modulation waves of each phase in the new strategy are obtained:

$$\begin{cases} U_{rxp} = (U_{rx}+1-k)/2 \\ U_{rxn} = (U_{rx}-1+k)/2 \end{cases} \quad (11)$$

B. Solutions of the modulation waves in the new strategy

From the above analysis, the expressions of the modulation waves are got in equation (11). Since there is only one variable k in the equation, the only problem to solve the modulation-wave expressions is to solve the variable k. In the new modulation strategy, U_{rxp} and U_{rxn} can be set to their extreme values in order to decrease the system switching losses. In such cases, the solutions of k are:

$$\begin{cases} k_{xp(1)} = U_x - 1 \\ k_{xp(0)} = U_x + 1 \\ k_{xn(0)} = -U_x + 1 \\ k_{xn(-1)} = -U_x - 1 \end{cases} \quad (x=a,b,c) \quad (12)$$

where $k_{xy}(i)$ denotes the solution of k when the value of modulation waves U_{xy} ($x=a$, b, c, $y=p$, n) is j (j=1, 0, -1).

Totally, there are 12 solutions of k as shown in equation (12) and any three of them cannot be equal. Any two of the 12 solutions can be equal only in a single point; as a result, the value of U_{rxp} and U_{rxn} can be their extreme value in only 1/6 of one switch cycle in order to keep the symmetry.

If k equals $k_{ap(0)}$, the values of U_{rxp} and U_{rxn} should satisfy the equation (7). The range of ωt which satisfy equation (7) is:

$$\omega t \in [4\pi/3, 5\pi/3] \quad (13)$$

Using the same analysis method in other cases, the solutions of k and the corresponding ranges of ωt can be got in equation (14). Here, U_{rmax} and U_{rxnmin} are the maximum and minimum value of the modulation wave U_{rx} in traditional SPWM strategy.

$$k = \begin{cases} \pm(1+U_{rmin}), \omega t \in [\dfrac{i-1}{3}\pi, \dfrac{i}{3}\pi], i=1,3,5 \\ \pm(1-U_{rmax}), \omega t \in [\dfrac{i-1}{3}\pi, \dfrac{i}{3}\pi], i=2,4,6 \end{cases} \quad (14)$$

The solutions of modulation waves of the new strategy can be got by substituting equation (14) into equation (11). Since the parameter k has two solutions in each 1/6 fundamental frequency, totally, there are 64 solutions of the modulation waves of the new strategy. If U_{rxp} and U_{rxn} are continuous, they should satisfy equation (15).

$$U_{rxy}(\dfrac{i\pi}{3})_- = U_{rxy}(\dfrac{i\pi}{3})_+, (i=0...5) \quad (15)$$

Only two solutions of U_{rxp} and U_{rxn} satisfy equation (15), they are shown in equation (16) and (17). Fig.4 shows the continuous situation of U_{rxp} and U_{rxn}. In other cases, U_{rxp} and U_{rxn} are discontinuous, Fig.5 shows the two discontinuous situation of U_{rxp} and U_{rxn} when they satisfy equation (18) and (19).

$$\begin{bmatrix} U_{rxp} \\ U_{rxn} \end{bmatrix} = \begin{cases} \begin{bmatrix} (U_{rx}-U_{min})/2 \\ (U_{rx}+U_{min})/2 \end{bmatrix}, \omega t \in [\dfrac{i-1}{3}\pi, \dfrac{i}{3}\pi], i=1,3,5 \\ \begin{bmatrix} (U_{rx}+U_{max})/2 \\ (U_{rx}-U_{max})/2 \end{bmatrix}, \omega t \in [\dfrac{i-1}{3}\pi, \dfrac{i}{3}\pi], i=2,4,6 \end{cases} \quad (16)$$

$$\begin{bmatrix} U_{rxp} \\ U_{rxn} \end{bmatrix} = \begin{cases} \begin{bmatrix} (U_{rx}+2+U_{min})/2 \\ (U_{rx}-2-U_{min})/2 \end{bmatrix}, \omega t \in [\dfrac{i-1}{3}\pi, \dfrac{i}{3}\pi], i=1,3,5 \\ \begin{bmatrix} (U_{rx}+2-U_{max})/2 \\ (U_{rx}-2+U_{max})/2 \end{bmatrix}, \omega t \in [\dfrac{i-1}{3}\pi, \dfrac{i}{3}\pi], i=2,4,6 \end{cases} \quad (17)$$

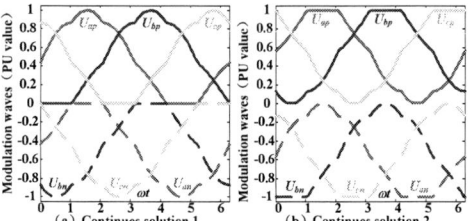

Fig. 4. Diagram of continuous modulation waves of new carrier modulation strategy

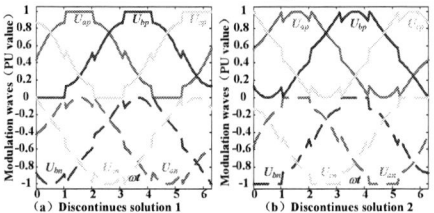

(a) Discontinues solution 1   (b) Discontinues solution 2

Fig. 5. Diagram of discontinuous modulation waves of new carrier modulation strategy

$$\begin{bmatrix} U_{rxp} \\ U_{rxn} \end{bmatrix} = \begin{cases} \begin{bmatrix} (U_{rx}-U_{min})/2 \\ (U_{rx}+U_{min})/2 \end{bmatrix}, \omega t \in [\dfrac{i-1}{3}\pi, \dfrac{i}{3}\pi], i=1,3,5 \\ \begin{bmatrix} (U_{rx}+2-U_{max})/2 \\ (U_{rx}-2+U_{max})/2 \end{bmatrix}, \omega t \in [\dfrac{i-1}{3}\pi, \dfrac{i}{3}\pi], i=2,4,6 \end{cases} \quad (18)$$

$$\begin{bmatrix} U_{rxp} \\ U_{rxn} \end{bmatrix} = \begin{cases} \begin{bmatrix} (U_{rx}+2+U_{min})/2 \\ (U_{rx}-2-U_{min})/2 \end{bmatrix}, \omega t \in [\dfrac{i-1}{3}\pi, \dfrac{i}{3}\pi], i=1,3,5 \\ \begin{bmatrix} (U_{rx}+U_{max})/2 \\ (U_{rx}-U_{max})/2 \end{bmatrix}, \omega t \in [\dfrac{i-1}{3}\pi, \dfrac{i}{3}\pi], i=2,4,6 \end{cases} \quad (19)$$

C. Analysis on the DC voltage utilization ratio and switch device losses

DC voltage utilization ratio is a very important evaluation index of PWM strategy, and it is defined by the ratio of amplitude of output line voltage and total DC voltage. The novel strategy proposed in this paper has the same DC voltage utilization ratio with traditional SPWM strategy. The switch device losses are another important evaluation index of PWM strategy. Comparing with the SPWM strategy, the novel strategy has 5/3 times switch losses of the traditional three-level SPWM strategy.

D. Output THD performance analysis of the new strategy

Using double Fourier analysis and the Jacobi-Anger expansions which are shown in equation (20) to (22), the output waveform THD performance of the new strategy can be solved[16][17].

$$F(\omega_c t, \omega_s t) = \frac{A_{00}}{2} + \sum_{n=1}^{\infty}\{A_{0n}\cos(n\omega_s t) + B_{0n}\sin(n\omega_s t)\}$$

$$+\sum_{m=1}^{\infty}\{A_{m0}\cos(m\omega_c t) + B_{m0}\sin(m\omega_c t)\} \qquad (20)$$

$$+\sum_{m=1}^{\infty}\sum_{n=\pm1}^{\pm\infty}\{A_{mn}\cos(m\omega_c t + n\omega_s t) + B_{mn}\sin(m\omega_c t + n\omega_s t)\}$$

$$C_{mn} = A_{mn} + jB_{mn} = \frac{1}{2\pi^2}\int_{-\pi}^{\pi}\int_{-\pi}^{\pi}F(\omega_c t, \omega_s t)e^{j(m\omega_c t + n\omega_s t)}d\omega_c t d\omega_s t$$

$$(21)$$

$$\begin{cases}\cos(\xi\cos\theta) = J_0(\xi) + 2\sum_{l=1}^{\infty}\cos(l\pi)J_{2l}(\xi)\cos(2l\theta) \\[2mm] \sin(\xi\cos\theta) = 2\sum_{l=0}^{\infty}\cos(l\pi)J_{2l+1}(\xi)\cos[(2l+1)\theta]\end{cases} \qquad (22)$$

The output phase voltage waveform harmonic expressions of the new strategy are shown in equation (23), (24) and (25). Fig.6 shows the THD characteristics of the new carrier modulation strategy in different modulation index.

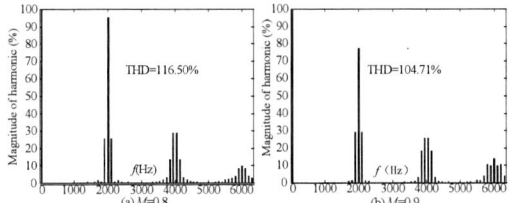

Fig.6. THD characteristics of the new carrier modulation strategy

$$\begin{cases}B_{01} = \dfrac{MU_{dc}}{2} \\[3mm] A_{m0} = \dfrac{4U_{dc}}{m\pi^2}\left|\sin(\dfrac{m\pi}{2})\right|\sum_{l=0}^{\infty}\dfrac{\cos(l\pi)}{2l+1}\cdot X \\[3mm] X = J_{2l+1}(\dfrac{mM\pi}{2})[1-\cos(\dfrac{(2l+1)\pi}{3})] \\[3mm] \quad + J_{2l+1}(\dfrac{\sqrt{3}mM\pi}{2})\cos(\dfrac{(2l+1)\pi}{6}) \\[3mm] \quad + J_{2l+1}(mM\pi)\cos(\dfrac{(2l+1)\pi}{3})\end{cases} \qquad (23)$$

$$C_{mn} = \frac{[1+(-1)^{m+n+1}]U_{dc}}{m\pi^2}\sum_{l=0}^{\infty}\cos(l\pi)\cdot(X+Y+Z),(n=2l+1)$$

$$X = \frac{(-1)^n J_{2l+1}(\dfrac{mM\pi}{2})}{3(2l+1)}\cdot A$$

$$A = [\pi(2l+1)\cos(\frac{2\pi\cdot(2l+1)}{3}) + 3\sin(\frac{2\pi\cdot(2l+1)}{3})] \qquad (24)$$

$$Y = \frac{J_{2l+1}(mM\pi)}{6(2l+1)}[(2l+1)\pi + 3\sin(\frac{2\pi(2l+1)}{3})]$$

$$Z = \frac{\pi}{3}J_{2l+1}(\frac{\sqrt{3}mM\pi}{2})[\cos(\pi\cdot(2l+1)/6)]$$

$$C_{mn} = A\cdot\sum_{l=0}^{\infty}\frac{\cos(l\pi)}{(2l+1)^2 - n^2}(X+Y+Z),(n\neq 2l+1)$$

$$A = \frac{[1+(-1)^{m+n+1}]U_{dc}}{m\pi^2}$$

$$X = J_{2l+1}(\frac{mM\pi}{2})\begin{bmatrix}(2l+1)[(-1)^n e^{j2n\pi/3} + e^{-j2n\pi/3}] \\[1mm] -[1+(-1)^n]\cdot(2l+1)\cos(\dfrac{(2l+1)\pi}{3}) \\[1mm] +jn\cdot[1-(-1)^n]\sin(\dfrac{\pi\cdot(2l+1)}{3})\end{bmatrix}$$

$$Y = -J_{2l+1}(mM\pi)\left[jn\cdot\sin(\frac{\pi\cdot(2l+1)}{3})(e^{jn\pi/3} - e^{j2n\pi/3}) - B\right]$$

$$B = (2l+1)\cdot\cos(\frac{\pi\cdot(2l+1)}{3})(e^{jn\pi/3} + e^{j2n\pi/3})$$

$$Z = J_{2l+1}(\frac{\sqrt{3}mM\pi}{2})[2je^{jn\pi/2}\cdot\sin(\frac{n\pi}{6})\cdot C + D]$$

$$C = (2l+1)\cdot\cos(\frac{(2l+1)\cdot\pi}{2}) - jn\cdot\sin(\frac{(2l+1)\cdot\pi}{2})$$

$$D = [1+(-1)^n](2l+1)\cdot\cos(\frac{(2l+1)\cdot\pi}{6})$$

$$\quad + [(-1)^n - 1]jn\cdot\sin(\frac{(2l+1)\cdot\pi}{6})$$

$$(25)$$

IV. EXPERIMENT VERIFICATION

The proposed new PWM strategy has been verified by experiment results. For the main circuit, 400V DC voltage source supplies resistive three-phase load (R=15Ω) through an output LCL filter (L_1=L_2=3mH, C=17uF) with star-connection. The DC capacitors are constructed using C_1=C_2=1000uF and the switching frequency is 2 kHz.

Fig.7(a) and Fig.7(b) show the results obtained by applying standard SPWM strategy. The variables shown are a line-to-line voltage U_{cb}, the output phase voltage U_{ao}, the voltage on the DC capacitor U_{c1}, and the output load current i_c. It is clear that the significant fundamental low-frequency (the fundamental frequency is 50 Hz in this paper) voltage oscillations on the DC capacitor exists. Fig.7 (c) to Fig.7 (f) show the same results when the proposed strategy is applied. In this case, the voltage on the DC capacitor does not contain any low-frequency oscillation but only high-frequency ripples which are related to the switching frequency.

Fig.8 shows the experiment results of high modulation index and low power factor (U_{dc}=200V, M=0.8, inductive three-phases load, L=20mH). From fig.8, it is clear that the novel strategy does not contain any low frequency oscillation at high modulation index and low power factor.

The 2014 International Power Electronics Conference

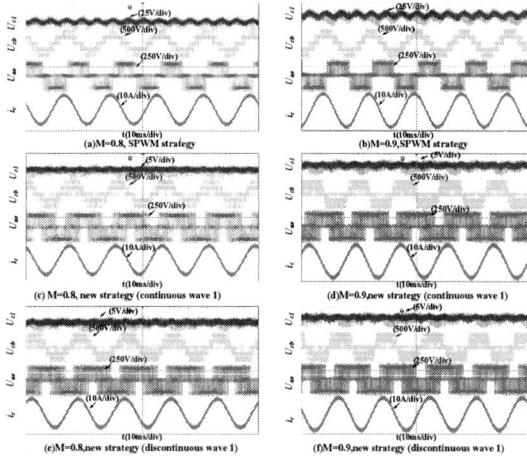

Fig.7 Experiment results comparison between the new carrier modulation strategy and SPWM strategy

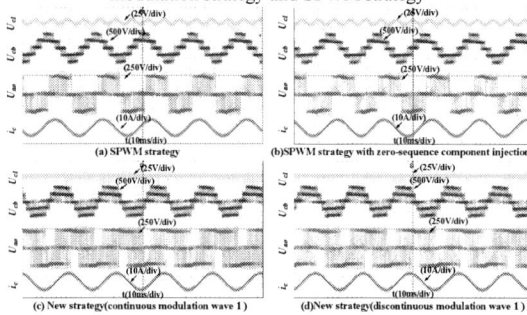

Fig.8 Experiment results under high modulation index and low power factor

V. CONCLUSION

The paper starts by analyzing the drawbacks of traditional PWM strategies used in three-level NPC VSI and then proposes a novel PWM strategy with completely elimination of low frequency oscillation in the DC capacitor voltage under any modulation index and power factor. The operation principle and solving process of the proposed strategy is described in details. Furthermore, the analysis of the new strategy is placed on output phase voltage harmonic characteristic, device switching losses and DC voltage utilization rate. Finally, experiment results are given to verify the theoretical analysis.

REFERENCES

[1] Nabae A, Takahashi I, Akagi H. A New Neutral Point Clamped PWM Inverter [J]. *IEEE Transaction on Industry Applications*, 1981, 17(5): 518-523.

[2] CHEN Bing, XIE Yunxiang, SONG Jingxian. Three-phase Three-wire Three-level APF Based on Vector Mode One-cycle Control During $30° \sim 390°$ [J]. *Automation of Electric Power Systems*, 2007, 31(21): 76-81

[3] ZENG Xiangjun, ZHANG Hongtao, LI Ying,et al. Low Voltage Ride Through Control of Large Direct-drive Wind Turbine with Multiphase PMSG and Hybrid Three-level Full-size Converters[J]. *Automation of Electric Power Systems*, 2012, 36(11): 23-29

[4] JING Wei, TAN Guojun1, YE Zongbin. Comparison of Two-level and Three-level Converters in Permanent Magnet Direct-drive Wind Power Generation System[J]. *Automation of Electric Power Systems*, 2011, 35(6): 92-97

[5] MA Fujun, LUO An, XIONG Qiaopo, et al. A Simplified Three-level Power Conditioner for High-speed Railway System[J]. *Automation of Electric Power Systems*, 2012, 36(16): 108-114

[6] WANG Chong, XING Yan, FANG Yu, et al. Three-level Grid Connected Inverter for Renewable Generation in Ideal Power Grid Condition[J]. *Automation of Electric Power Systems*, 2010, 34(3): 58-62

[7] YAO Wenxi, HU Haibing, XU Haijie, et al. A New PWM Scheme for Multi-level Inverter Based on the Modulation Waves of Three-level SVM[J]. *Automation of Electric Power Systems*, 2007, 31(7): 50-55

[8] Pou J, Zaragoza J, Ceballos S, et al. A Carrier-Based PWM Strategy With Zero-Sequence Voltage Injection for a Three-Level Neutral-Point-Clamped Converter[J]. *IEEE Transactions on Power Electronics*, 2012, 27(2): 642–651.

[9] Chenchen Wang, Yongdong Li. Analysis and Calculation of Zero-Sequence Voltage Considering Neutral-Point Potential Balancing in Three-Level NPC Converters[J]. *IEEE Transactions on Industrial Electronics*, 2010, 57(7): 2262–2271.

[10] Celanovic N, Boroyevich D. A comprehensive study of neutral-point voltage balancing problem in three-level neutral-point-clamped voltage source PWM inverters[J]. *IEEE Transactions on Power Electronics*, 2000, 15(2): 242–249.

[11] Pou J, Zaragoza J, Rodríguez P, et al. Fast-processing modulation strategy for the neutral-point-clamped converter with total elimination of the low-frequency volt-age oscillations in the neutral point[J]. *IEEE Transactions on Industrial Electronics*, 2007, 54(4): 2288–2294.

[12] Tian Kai, Wang Mingyan. Three-level Double-group Double-modulation-wave Carrier-based PWM Method[J]. *Proceedings of the CSEE*, 2010, 30(30): 55-61.

[13] Tian Kai, Wang Ming-yan, Liu Song-bin. Novel Double Modulation Wave Three-level Carrier-based Pulse Width Modulation Method[J]. *Proceedings of the CSEE*, 2009, 29(33): 54-59.

[14] Fu Jiacai, Guo Songlin, Shen Xianqing. Dual-Modulation Waves Based Single-Phase Three-Level Grid-Connected Inverter and Current Improvement Control[J].*Transactions of China Electrotechnical Society*, 2012,27（3）: 159-163.

[15] WANG Liqiao. Modulated Model Analysis of Pulse Width Modulation Technique for Sinusoidal Inverters[J]. *Automation of Electric Power Systems*, 2008, 32(17): 45-49

[16] HOLMES D G, THOMAS A L. Pulse width modulation for power converters: principles and practice[M]. Hoboken, NJ: Wiley-IEEE Press, 2003: 623-633.

[17] Ning Li, Yue Wang, Wulong Cong, et al. Comparative study of four kinds of multicarrier PWM strategies used in NPC three-level converters[C].*//IEEE Energy Conversion Congress and Exposition Asia*, Melbourne, Australia, 2013:183-189

A High Voltage Gain Switched-Coupled-Inductor Quasi-Z-Source Inverter

Furqan Ahmed, Honnyong Cha and Su-Han Kim
School of Energy Engineering
Kyungpook National University
Daegu, Korea
furqanhmd164@gmail.com

Heung-Geun Kim
Department of Electrical Engineering
Daegu, Korea

Abstract— **Impedance-source (Z-source) inverters are widely used to overcome the drawbacks of traditional voltage-source inverters (VSIs) such as no boost ability, waveform distortion because of dead time to avoid shoot-through, and less immunity to EMI noise. Shortcomings associated with the Z-source inverters like discontinuous input current, high voltage stress on capacitors and no common ground point between the dc-source and inverter bridge has been addressed with the development of the quasi-Z-source (qZSI) inverters. This paper proposes design of an improved high voltage gain quasi-Z-source inverter by adopting a combination of switched-capacitor and switched-coupled-inductor (SCL). Use of the coupled-inductor to charge switched-capacitor can effectively enhances the voltage gain and reduces the turns ratio of the coupled-inductor. The proposed SCL-qZSI due to its higher voltage gain, utilizes smaller shoot-through duty ratio and consequently larger modulation index, thus having lower voltage stress on switching devices as compared to the traditional ZSIs and qZSIs. A detailed theoretical analysis of the proposed converter and its comparison with QZSI is given, and followed by simulation and experimental results.**

Keywords— *High voltage gain, switch voltage stress, switched coupled-inductor, Z-source inverter (ZSI).*

I. INTRODUCTION

Traditional voltage-source inverters (VSIs) have some limitations such as: 1) the output line-to-line ac voltage is always smaller than the dc-link voltage which results in only buck mode operation of inverter for dc-ac power conversion. Therefore, in the applications like photovoltaic (PV) modules where input dc voltage is low, an additional boost converter is needed to get desired ac output voltage. 2) Dead time is required to avoid shoot-through of upper and lower switching devices of each phase leg, which results in waveform distortion. To overcome aforementioned problems, impedance-source inverters (ZSIs) have been proposed in [1] which utilize the shoot-through of the inverter bridge to boost the input dc voltage. ZSIs do not require dead-time which results in less distortion of the output waveform. Also, voltage boost and inverter operation is performed in single power conversion stage which leads to high efficiency.

However, there are disadvantages associated with ZSIs such as high voltage stress on switching devices and capacitors and discontinuous input current. In [2], quasi-z-source inverter (qZSI) is proposed which has benefits

of continuous input current, low stress on capacitors and common ground-point between dc voltage source and the inverter-bridge.

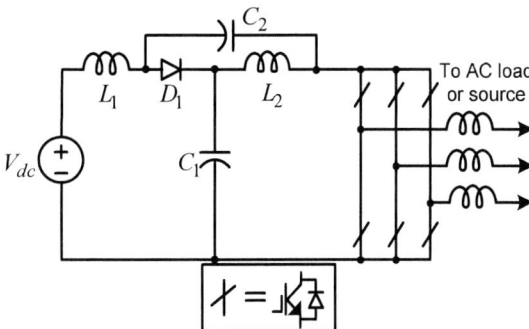

Fig. 1. Traditional qZSI inverter.

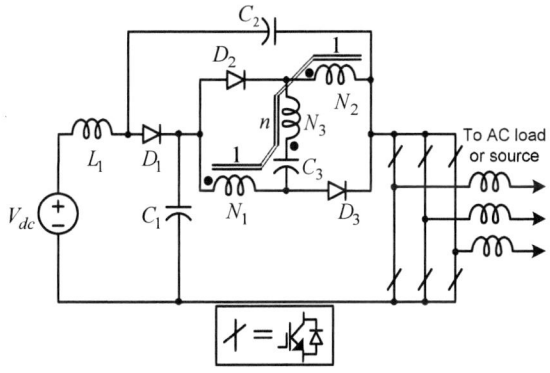

Fig. 2. Proposed SCL-qZSI inverter.

The obtainable voltage gain of both the ZSI and qZSI is limited because for higher voltage gains, the higher shoot-through duty ratio and, consequently, the smaller modulation index has to be used which results in high voltage stress across switching devices and decreased quality of the waveform. Pulse width modulation techniques such as maximum boost control and the constant boost control have been proposed to get maximum possible voltage gain with reduced voltage stress on devices [3]-[4]. In [5]-[8], modified Z-source inverter topologies are proposed to increase voltage boost ability of the inverter. However, either the boost ability is limited or is at the cost of significantly increased number of components and inverter volume. This paper extends the previous work and proposes an improved SCL-qZSI

978-1-4799-2706-7/14 $31.00 © 2014 IEEE

with much higher voltage boost ability compared to the traditional ZSI and qZSI. Circuit configuration of the proposed SCL-qZSI is similar to classic qZSI with the difference being that the inductor L_2 is replaced with switched-coupled-inductor, as shown in Fig. 2. This switched-coupled-inductor is obtained by replacing the two inductors in switched-inductor as proposed in [6], with a coupled-inductor having three separate windings N_1, N_2 and N_3. The switched-capacitor C_3 obtains energy from capacitor C_1 and secondary inductor winding N_3, which enhances the boost ability. The proposed inverter topology due to its higher voltage gain as compared to ZSI and qZSI, requires smaller shoot-through duty ratio to obtain same voltage boost as VSI and qZSI. This feature of the proposed SCL-qZSI helps to reduce voltage stress across the inverter-bridge by utilizing higher modulation index. Therefore, the proposed topology is a best candidates for the applications where input voltage is low, like photovoltaic systems and fuel cells etc.

II. CIRCUIT OPERATION AND DERIVATION OF VOLTAGE GAIN

Fig. 2 shows the circuit topology of the proposed SCL-qZSI which consists of three diodes (D_1, D_2 and D_3), three capacitors (C_1, C_2, and C_3), inductor L_3 and a coupled inductor with three windings (N_1, N_2, and N_3). Windings N_1 and N_2 have same number of turns ($N_1 = N_2$) and turns ratio of winding N_3 to N_1 or N_2 is n ($n = N_3 / N_1 = N_3 / N_2$). There are two operating states namely shoot-through zero and non-shoot-through of the proposed SCL-qZSI, irrespective of the PWM scheme used for control.

During the shoot-through zero state as shown in Fig. 3, diode D_1 is off while diodes D_2 and D_3 are on. Capacitors C_1 and C_2 discharge while capacitor C_3 charges by C_1 through winding N_3. Windings N_1 and N_2 are in parallel with capacitor C_1. Applying KVL yields following equations,

$$\begin{cases} V_{L1} = V_{dc} + V_{C2} \\ V_{N1} = V_{N2} = V_{C1}, V_{N3} = nV_{N1} \\ V_{C3} = V_{N1} + V_{N3} = (n+1)V_{N1} = (n+1)V_{C1} \end{cases} \quad (1)$$

Similarly, in the non-shoot-through states as shown in Fig. 4, diode D_1 is on while diodes D_2 and D_3 are off. Capacitors C_1 and C_2 charges while capacitor C_3 discharges. Again applying KVL gives,

$$\begin{cases} V_{L1} = V_{PN} - V_{dc} - V_{C2} \\ V_{N1} + V_{N2} + V_{N3} = V_{C3} - V_{C2} \\ V_{N1} + V_{N2} + V_{N3} = -V_{PN} + V_{C1} + V_{C3} \end{cases} \quad (2)$$

Due to magnetic coupling $V_{N3} = nV_{N1} = nV_{N2}$ and because of the fact that capacitor acts as a voltage source

and its voltage remains nearly unchanged during a switching cycle, $V_{C3} = (n+1)V_{C1}$.

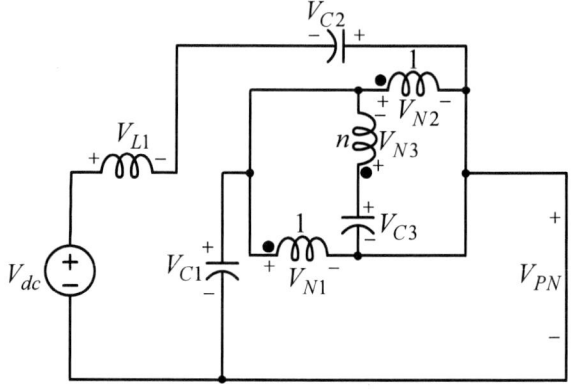

Fig. 3. Equivalent circuit under shoot-through states.

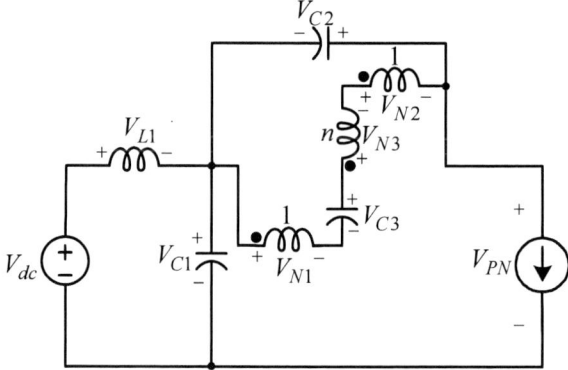

Fig. 4. Equivalent circuit under non-shoot-through states.

Therefore, (2) can be revised as

$$V_{N1} = \frac{V_{C3} - V_{C2}}{2+n} \quad (3)$$

$$V_{N1} = -\frac{V_{PN}}{n+2} + \frac{V_{C3}}{n+1} \quad (4)$$

If $D_{sh}T$ is shoot-through interval and $(1-D_{sh})T$ is non-shoot-through time interval, applying volt-second balance condition on inductor L_1 from (1) and (2), gives

$$(V_{dc} + V_{C2})D_{sh}T = -(V_{PN} - V_{dc} - V_{C2})(1-D_{sh})T \quad (5)$$

Rearranging (5), we get

$$V_{PN} = \frac{1}{1-D_{sh}}(V_{dc} + V_{C2}) \quad (6)$$

Similarly, volt-second balance condition on winding N_1 from (1) and (3), gives

$$\frac{V_{C3}}{n+1}D_{sh}T = -\frac{V_{C3} - V_{C2}}{n+2}(1-D_{sh})T \quad (7)$$

Equation (7) can be revised as,

$$V_{C3} = \frac{(1+n-(1+n)D_{sh})}{n+1+D_{sh}} V_{C2} \qquad (8)$$

Again, volt-second balance condition on winding N_1 or N_2 from (1) and (4), gives

$$\frac{V_{C3}}{n+1} D_{sh} T = -(-\frac{V_{PN}}{n+2} + \frac{V_{C3}}{n+1})(1-D_{sh})T \qquad (9)$$

By solving (9), we get

$$V_{C3} = \frac{(n+1)(1-D_{sh})}{n+2} V_{PN} \qquad (10)$$

Substituting (8) into (10) yields

$$V_{C2} = \frac{n+1+D_{sh}}{n+2} V_{PN} \qquad (11)$$

Substituting (11) into (6) and solving, we get

$$V_{PN} = \frac{n+2}{(1-(3+n)D_{sh})} V_{dc} \qquad (12)$$

Therefore, the boost factor of the proposed SCL-qZSI is given by

$$B = \frac{V_{PN}}{V_{dc}} = \frac{n+2}{(1-(n+3)D_{sh})} \qquad (13)$$

When turns ratio n is 1, boost factor becomes

$$B = \frac{V_{PN}}{V_{dc}} = \frac{3}{(1-4D_{sh})} \qquad (14)$$

Voltage boost factor of the traditional ZSI or qZSI is given by

$$B_{qZSI} = \frac{V_{PN}}{V_{dc}} = \frac{1}{(1-2D_{sh})} \qquad (15)$$

From (14) and (15), it is concluded that the proposed SCL-qZSI has much higher boost ability than the qZSI even with turns ratio of 1, as shown in Fig. 5. The voltage gain G described in [1] can be expressed as

$$G = MB = \frac{\hat{V}_{ac}}{V_{dc}/2} \qquad (16)$$

Where $\hat{V}_{ac}$ is the output peak phase voltage and M is the modulation index. For simple boost control, relation between shoot-through duty ratio and modulation index is $D_{sh} = 1 - M$. Therefore, voltage gain G in terms of

modulation index M for the two inverters can be defined as,

$$G = \frac{3M}{(4M-3)} \qquad (17)$$

$$G_{qZSI} = \frac{M}{(2M-1)} \qquad (18)$$

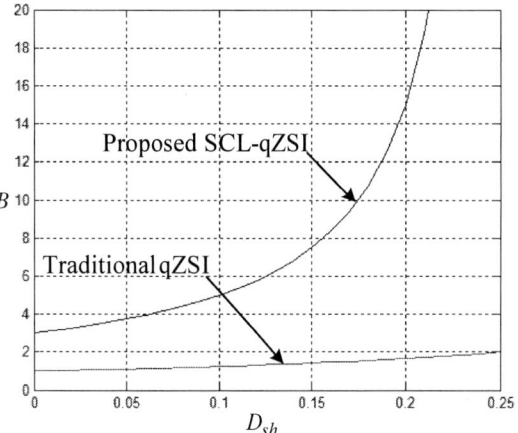

Fig. 5. Comparison of the boost ability for the proposed SCL-qZSI and traditional qZSI.

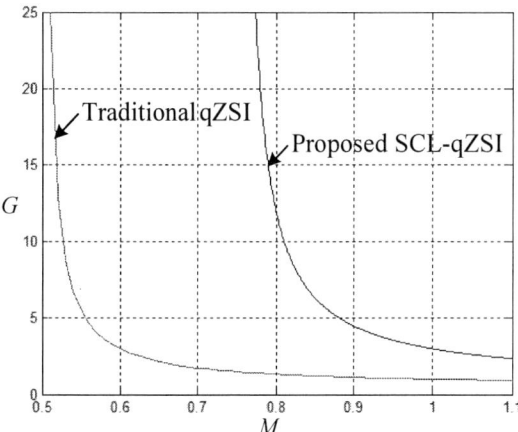

Fig. 6. Comparison of the voltage gain for the proposed SCL-qZSI and traditional qZSI using simple boost control.

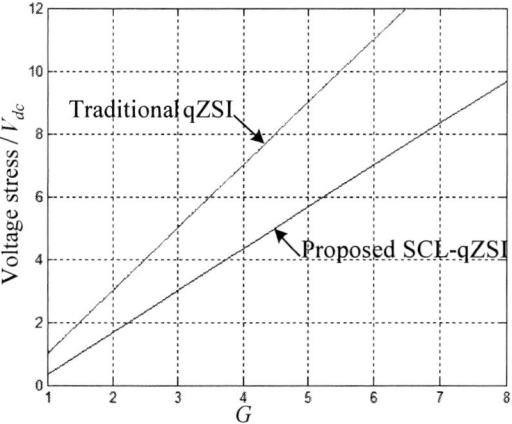

Fig. 7. Switch voltage stress versus voltage gain for the proposed SCL-qZSI and traditional qZSI using simple boost control.

Due to higher boost ability of the proposed SCL-qZSI, compared to ZSI or qZSI, it needs smaller shoot-through duty ratio D_{sh} and consequently, higher modulation index to achieve same voltage gain G as traditional qZSI, which is also shown in Fig. 6. When simple boost control is applied, the voltage stress of the switching devices as expressed in [3] can be revised as,

$$V_{sw,qZSI} = B_{qZSI}\, V_{dc} = 2G - 1 \qquad (19)$$

$$V_{sw} = B V_{dc} = \frac{4G - 3}{3} \qquad (20)$$

Use of the higher modulation index decreases the voltage stress across the switching devices of the proposed SCL-qZSI, as shown in Fig. 7. The voltage stresses across the capacitors and diodes for the proposed topology are given,

$$V_{C1} = \frac{(1-D)}{(1-4D)} V_{dc} \qquad (21)$$

$$V_{C2} = \frac{2+D}{(1-4D)} V_{dc} \qquad (22)$$

$$V_{C3} = \frac{2(1-D)}{(1-4D)} V_{dc} \qquad (23)$$

$$V_{D1} = V_{PN} = \frac{3}{(1-4D_{sh})} \qquad (24)$$

$$V_{D2,D3} = \frac{(6+2D)}{9} V_{PN} \qquad (25)$$

III. SIMULATION RESULTS

Design parameters for the proposed and traditional qZSI are given in Table. I. Fig. 8 and Fig. 9 show the comparison of dc link voltage, output line-to-line voltage and current for the two inverters under the same operational conditions. For the same circuit parameters values and operating conditions, the dc link voltage for the SCL-qZSI is about 400 V, much higher than the 106 V for the traditional qZSI. The SCL-qZSI inverter is able to get rms line-to-line voltage of 220 V, with a small duty ratio of .1. The traditional qZSI inverter on the other hand, can get a very small boost in dc link voltage with shoot-through duty ratio of .1 and, therefore, is able to get rms line-to-line voltage of only 60 V which is far less from the desired output of the 220 V.

Since, the proposed topology is using only .1 shoot-through duty ratio, it enables it to make use of a higher modulation index of .9 which benefits in two aspects, decreased voltage stress on the switching device compared to the traditional qZSI inverter and better quality of the output waveforms. To obtain the desired output line-to-line voltage of 220 V, the traditional qZSI inverter requires a very higher shoot-through duty ratio of larger than .4 and, consequently, it cannot be able to use a

modulation index of more than .6 resulting in degraded output waveforms and very higher voltage stress on active devices. Therefore, the applications like fuel cells and photovoltaic systems where input voltage is low, the proposed inverter is best to use.

TABLE. I CIRCUIT PARAMETERS AND OPERATION CONDITIONS.

Coupled-inductor parameters	• L_m = 38 uH • Turns ratio $n = 1$
qZSI and SCL-qZSI network parameters	• L = 300 uH • C = 100 uF
Operational conditions	• V_{dc} = 86 V f_{sw} = 10 kHz • D_{sh} = 0.1 M = 0.9 • R_{load} = 15 Ω

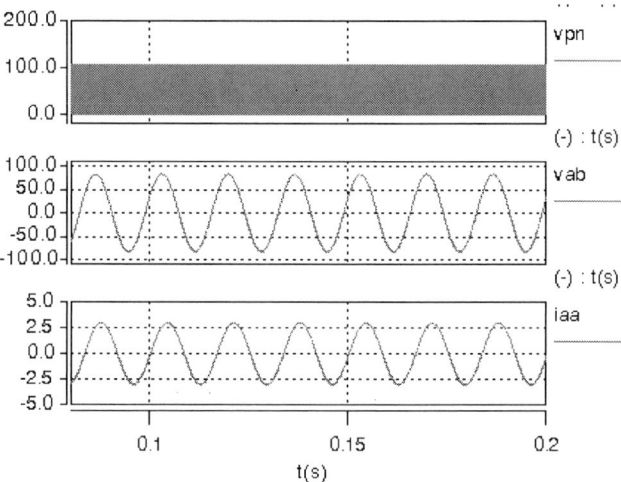

Fig. 8. Simulation waveforms of dc link and output line-to-line voltages, and output phase current for traditional qZSI.

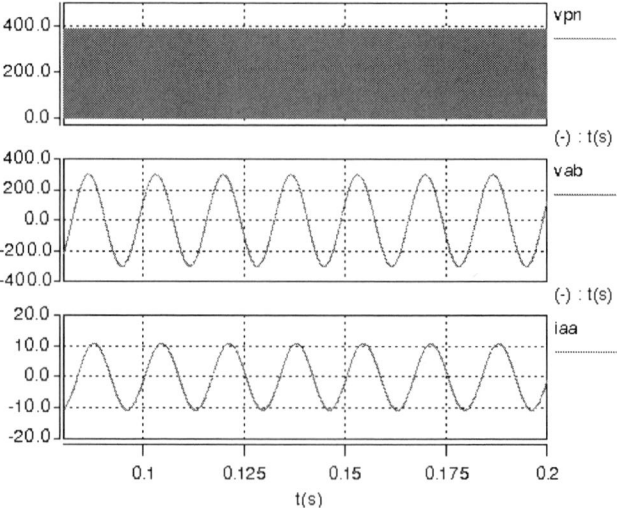

Fig. 9. Simulation waveforms of dc link and output line-to-line voltages, and output phase current for proposed SCL-qZSI.

The 2014 International Power Electronics Conference

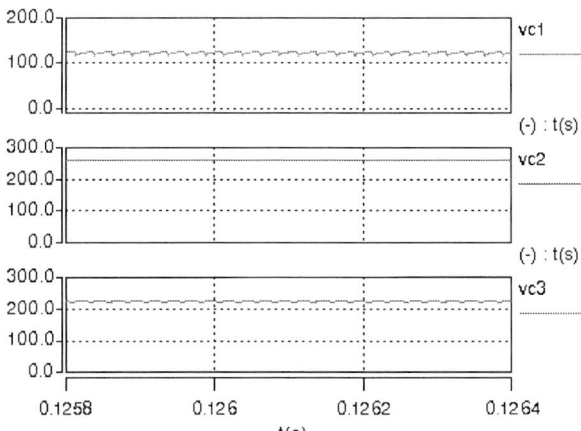

Fig. 10. Simulation waveforms of the capacitors voltages for the proposed SCL-qZSI inverter.

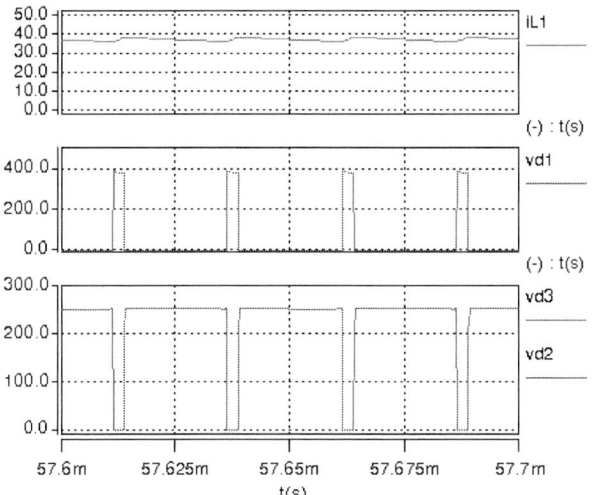

Fig. 11. Simulation waveforms of the input current and diodes voltage Stresses for proposed SCL-qZSI.

IV. CONCLUSIONS

In this paper, an improved high voltage gain switched-coupled-inductor qZSI inverter has been proposed. The proposed inverter successfully combined switched-capacitor and three windings switched-coupled-inductor. Charging of switched-capacitor through coupled-inductor significantly enhanced boost-ability of the proposed inverter. Compared to the traditional qZSI inverter, the proposed inverter requires much smaller shoot-through duty ratio to obtain the same voltage boost. Therefore, the proposed inverter can make use of higher modulation index resulting in better quality of the output waveforms and lower voltage stress on the switching devices. The theoretical analysis followed by simulations has been presented to validate advantages of the proposed topology over the traditional ZSI and qZSI.

ACKNOWLEDGMENT

This research was supported by Basic Science Research Program through the National Research Foundation of Korea(NRF) funded by the Ministry of Education, Science and Technology (NRF-2012R1A1A1044058)

REFERENCES

[1] F. Z. Peng, "Z-source inverter," IEEE Trans. Ind. Appl., vol. 39, no. 2, pp. 504-510, Mar./Apr. 2003.

[2] J. Anderson and F. Z. Peng, "Four-quasi-Z-source inverters," in Proc. IEEE Power Electron. Spec. Conf., (PESC'08), pp. 2743–2749.

[3] F. Z. Peng, M. Shen and Z. Qian, "Maximum boost control of the Z-source inverter," IEEE Trans. Power Electron., vol. 20, no. 4, pp. 833–838, Jul. 2005.

[4] M. Shen, J. Wang, A. Joseph, F. Z. Peng, L. M. Tolbert, and D. J. Adms, "Constant boost control of the Z-source inverter to minimize the current ripple and voltage stress," IEEE Trans. Ind. Applicat., vol. 42, no. 3, pp. 770-778, May./Jun. 2006.

[5] W. Qian, F. Z. Peng and H. Cha, "Trans-Z-source inverters," IEEE Trans. Power Electron., vol. 26, no.12, pp. 3453–3463, Dec. 2011.

[6] M. Zhu, K. Yu, and F. L. Luo, "Switched-inductor Z-source inverter," IEEE Trans. Power Electron., vol. 25, no. 8, pp. 2150–2158, Aug. 2010.

[7] M.-K. Nguyen, Y.-C. Lim, and G.-B. Cho, "Switched-inductor quasi-Z-source inverter," IEEE Trans. Power Electron., vol. 26, no. 11, pp. 3181–3191, Nov. 2011.

[8] C. J. Gajanayake, F. L. Luo, H. B. Gooi, P. L. So, and L. K. Siow, "Extended-boost Z-source inverters," IEEE Trans. Power Electron., vol. 25, no. 10, pp. 2642–2652, Oct. 2010.

[9] K. I. Hwu, C. F. Chuang and W. C. Tu, "High voltage-boosting converters based on bootstrap capacitors and boost inductors," IEEE Trans. Ind. Electron., vol. 60, no. 6, pp. 2178–2193, Jun. 2013.

A Novel Control Strategy to Suppress DC Current Injection to the Grid for Three-Phase PV Inverter

Tao zhang, Guofeng He, Min Chen, Dehong Xu

College of Electrical Engineering

Zhejiang University

Hangzhou, China

xdh@cee.zju.edu.cn

Abstract—**The limitation of DC current injection to grid is critical for transformerless grid-connected PV inverter. In this paper a novel control strategy for three phase PV inverter is proposed in order to suppress DC current injection to the grid. It is based on idea of accurately sensing the DC offset voltage of three-phase PV inverter output. Since each one of the DC components in the three-phase inverter output can be eliminated, DC injection current to the grid can be effectively suppressed. The control scheme is verified on a 20kW three phase PV inverter.**

Keywords—DC current injection; Three-phase PV inverter; DC offset voltage; DC suppression loop; Park transformation

I. INTRODUCTION

For the advantages of higher efficiency, lower weight and lower cost, PV inverter without isolation transformer become more attractive in grid-connected photovoltaic system [1][2]. However, DC current component would inject to the grid for transformerless PV inverter[3], which might cause the transformers saturation, overheating, corrosion of underground equipments, and malfunction of protective equipment in the power system [4][5]. Consequently, standards and regulations have been formulated in order to limit PV inverter DC current injection to the grid [6]-[9]. For example, according to IEEE 929-2000 standard in USA, the maximum DC current permitted without transformer should not exceed 0.5 percent of the rated current. While the United Kingdom formulates the rule that the maximum DC current permitted without transformer should not exceed 5 mA.

At present, the methods to suppress DC injection for single phase PV inverter have been investigated [1], [4]-[5], [10]-[15]. However, the methods to restrict DC injection to the grid in three phase PV inverter have not been reported.

In this paper, a novel control strategy is proposed to suppress DC current injection for three-phase PV inverter. It is based on idea of

accurately sensing the DC offset voltage of three-phase inverter output. Since the DC components in the three-phase inverter output voltage can be eliminated, DC current injection to the grid can be effectively suppressed. Finally, the proposed novel control scheme is verified by the experiment.

II. THE NOVEL DC CONTROL STRATEGY

The control diagram of three phase PV inverter without output isolation-transformer is shown in Fig. 1. Where V_{dc} is DC bus voltage, u_{gA}, u_{gB} and u_{gC} are three phase grid voltage respectively, PLL is phase locked loop to realize the synchronization, i_A, i_B and i_C are three phase grid-connected current respectively. In addition, $i_d{}^*$ and $i_q{}^*$ are the references of inner current loop in dq coordinate frame.

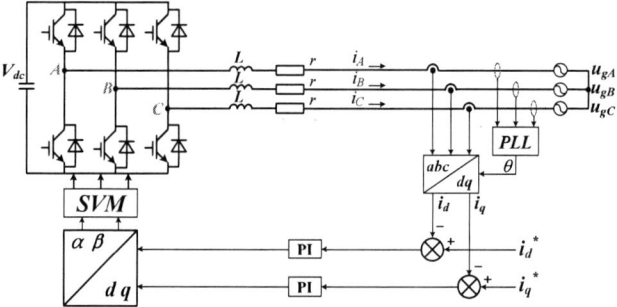

Fig. 1 The original scheme diagram of three phase PV grid-connected inverter

The disparity of power modules, asymmetry of driving pulses, detection error of current can cause DC offset at the inverter output voltage for three-phase PV inverter, which will cause DC current flow into the grid. Therefore, the DC injection may violate the grid connection standards and can not be neglected [15].

In order to effectively suppress DC current injection to the grid, a novel control strategy for a three phase PV inverter without the isolation transformer is shown in Fig. 2. Compared with Fig.1, an extra DC offset voltage suppression loop is added to the original control scheme, which is composed of circuit of differential amplifiers and LPF, DC loop PI controllers and transform matrix T_m.

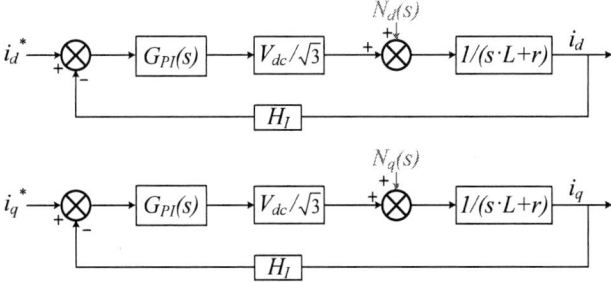

Fig.3 The original control model in *dq* coordinate frame

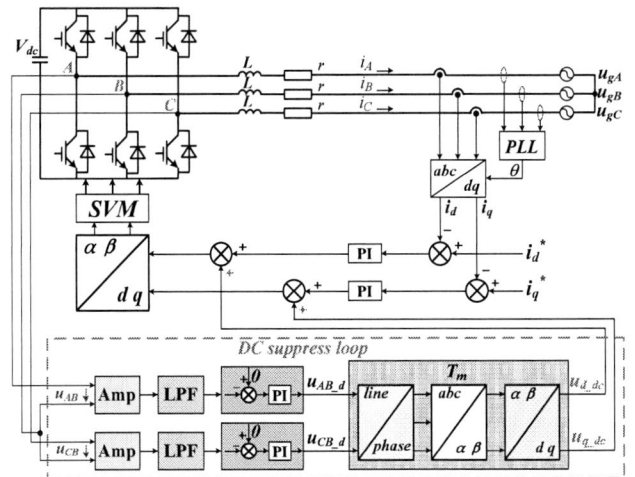

Fig. 2 The novel scheme diagram of three phase PV grid-connected inverter

In Fig.2, differential amplifier of two sampling channels samples the DC offset of the inverter output line voltage u_{AB} and u_{CB} respectively. The LPF is to filter out the fundamental and higher frequency components of sampling signal and remain DC component only. Then the DC component is adjusted by DC loop PI controllers, and the u_{AB_d} and u_{CB_d} from *abc* coordinate frame which is the output of DC loop PI controllers are transformed into *d-q* coordinate frame by transformation matrix T_m. Specially, the transformation process is that the u_{AB_d} and u_{CB_d} are transformed into three phase voltages by matrix $T_{line/phase}$ from line voltage to phase voltage firstly, then the transformation matrix $T_{abc/\alpha\beta}$ from *abc* coordinate frame to $\alpha\beta$ coordinate is used, finally u_{d_dc} and u_{q_dc} which is DC components in *d-q* coordinate frame are achieved by the function of transformation matrix $T_{\alpha\beta/dq}$ from $\alpha\beta$ coordinate frame to *dq* coordinate, hence, the outputs of DC suppression loop , u_{d_dc} and u_{q_dc} are fed back to *dq* axis of current loop.

The advantage of novel control strategy is that the differential amplifier with low offsets, noise and cost and high CMRR (Common-Mode Rejection Ratio) is used to sample DC offset voltage between the bridge-leg middle points, which avoids the zero-drift resulted from Hall-effect sensors. The output voltage u_{AB} and u_{CB} include all information about DC offset of the inverter. DC suppression loop is introduced to suppress inverter output disturbances. Therefore the DC current injected to the grid can be well suppressed.

III.   ANALYSIS OF DISTURBANCE SUPPRESSING EFFECT

A.   The control model of the proposed strategy

Fig.3 shows the original control block of three phase PV grid-connected inverter in *dq* coordinate frame which is derived from Fig.1.

In Fig.3, $G_{PI}(s)=K_p+K_i/s$, K_p and K_i are the parameters of PI controllers of current loop respectively, L is the output filter inductance, r is the equivalent resistance of output filter inductance, and H_I is the feedback coefficient of current loop. From Fig.3, the transfer function from disturbance source $N_d(s)$ to $I_d(s)$, and $N_q(s)$ to $I_q(s)$ with the original control scheme can be expressed as:

$$\frac{I_d(s)}{N_d(s)}=\frac{s}{s^2 \cdot L + s \cdot [r + H_I(V_{dc}/\sqrt{3})K_p] + H_I(V_{dc}/\sqrt{3})K_i} \quad (1)$$

$$\frac{I_q(s)}{N_q(s)}=\frac{s}{s^2 \cdot L + s \cdot [r + H_I(V_{dc}/\sqrt{3})K_p] + H_I(V_{dc}/\sqrt{3})K_i} \quad (2)$$

Fig.4 shows the proposed novel control block in *dq* coordinate frame which is derived from Fig.2.

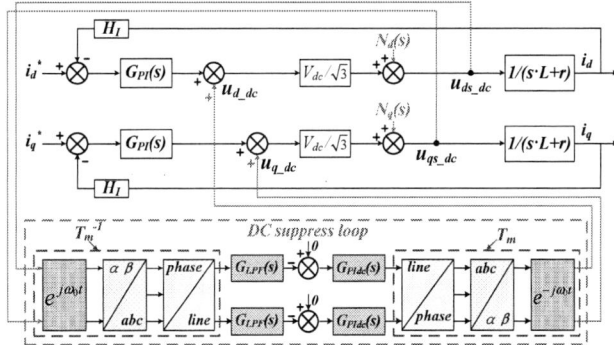

Fig.4 The proposed novel control model in *dq* coordinate frame

Compared with Fig.3, an extra DC suppression loop is added to the original control model. u_{ds_dc} and u_{qs_dc} are two sampling voltage in *dq* coordinate frame respectively, T_m^{-1} is inverse matrix of transformation matrix T_m shown in Fig.2. $G_{LPF}(s)$ is transfer function of differential amplifiers and LPF shown in Fig.2, $G_{PIdc}(s)$ is the transfer function of DC loop PI controller shown in Fig.2, and the $G_{LPF}(s)$ and $G_{PIdc}(s)$ can be given by

$$G_{LPF}(s)=\frac{13}{(4.4s+1)(0.176s+1)} \quad (3)$$

978-1-4799-2706-7/14 $31.00 © 2014 IEEE

$$G_{PIdc}(s) = K_{p\_dc} + \frac{K_{i\_dc}}{s} \tag{4}$$

where K_{p_dc} and K_{i_dc} are the parameters of DC loop PI controller. Fig.5 shows the simplified control model of proposed novel control scheme derived from Fig.4.

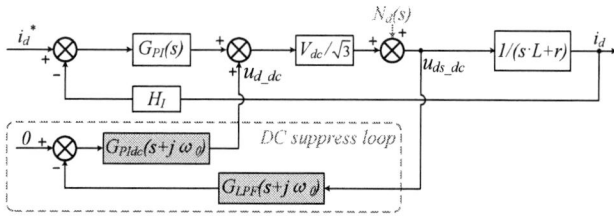

(a). Control model in d axis

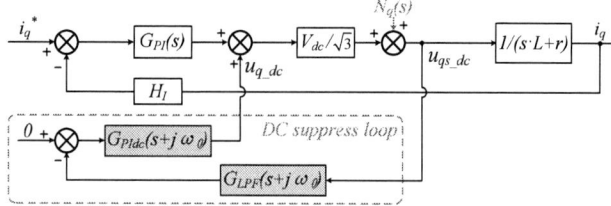

(b). Control model in q axis

Fig.5 Simplified control model block diagram of novel control scheme

In Fig.5, $G_{PIdc}(s+j\omega_0)$ is the PI controller transfer function of DC suppression loop in dq coordinate frame, $G_{LPF}(s+j\omega_0)$ is the LPF transfer function of DC suppression loop in dq coordinate frame, and ω_0 is the angular frequency of fundamental grid voltage. From Fig.5, the transfer function from disturbance source $N_d(s)$ to $I_d(s)$, and $N_q(s)$ to $I_q(s)$ with the novel control scheme can be derived as follows:

$$\frac{I_d(s)}{N_d(s)} =$$
$$\frac{\sqrt{3}}{\sqrt{3}(sL+r)+[G_{PI}(s)V_{dc}H_1+(sL+r)V_{dc}G_{LPF}(s+j\omega_0)G_{PIdc}(s+j\omega_0)]} \tag{5}$$

$$\frac{I_q(s)}{N_q(s)} =$$
$$\frac{\sqrt{3}}{\sqrt{3}(sL+r)+[G_{PI}(s)V_{dc}H_1+(sL+r)V_{dc}G_{LPF}(s+j\omega_0)G_{PIdc}(s+j\omega_0)]} \tag{6}$$

B. *The Analysis of Detecting Error with original control scheme*

In order to analyze the detecting error caused by ADC (analog to digital converter) resolution and accuracy of LEM, the detecting diagram of Fig.2 is shown in Fig6. where, $\triangle I_{g1A}$, $\triangle I_{g1B}$ and $\triangle I_{g1C}$ are each phase grid DC current detecting error caused by ADC resolution, $\triangle U_{AB}$ and $\triangle U_{CB}$ are DC component detecting error of inverter output line voltage caused by ADC resolution, $\triangle I_{g2A}$, $\triangle I_{g2B}$ and $\triangle I_{g2C}$ are detecting error caused by each phase LEM accuracy,

respectively. As the analysis of detecting error in phase A has the same calculation in phase B and phase C, so this paper just take phase A for example to introduce.

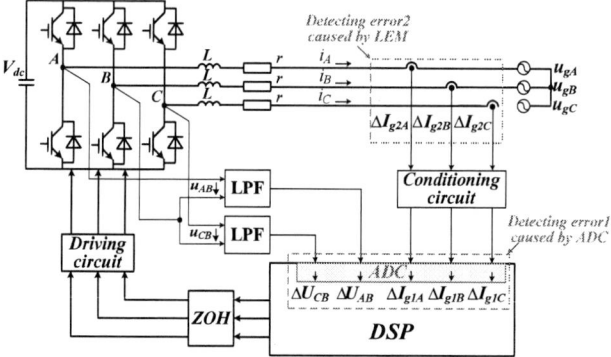

Fig.6 The schematic diagram of detecting error analysis

From Fig.6, the total detecting error $\triangle I_{gA}$ of phase A without DC suppression loop can be calculated as:

$$\Delta I_{gA} = \Delta I_{g1A} + \Delta I_{g2A} \tag{7}$$

If N bits of ADC is adopted, the detecting error of grid DC current $\triangle I_{g1A}$ caused by the ADC resolution can be given by

$$\Delta I_{g1A} = \frac{I_{p\_p}}{2^N} \tag{8}$$

where, I_{p_p} is the rated peak to peak value of the grid current. In addition, the grid DC current detecting error $\triangle I_{g2A}$ caused by the accuracy of LEM is given by

$$\Delta I_{g2A} = e\% \times I_{max\_lem} \tag{9}$$

where, $e\%$ is accuracy of the LEM, I_{max_lem} is the maximum detecting value of the LEM. Therefore, with (8) and (9), $\triangle I_{gA}$ can be calculated as:

$$\Delta I_{gA} = \frac{I_{p\_p}}{2^N} + e\% \times I_{max\_lem} \tag{10}$$

C. *The Analysis of Detecting Error with proposed control scheme*

Fig.7 shows the DC equivalent circuit of the main circuit when only considering the DC component for three phase PV inverter derived from Fig.1.

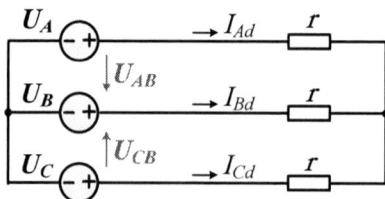

Fig.7 Equivalent model of DC current injection for PV grid inverter

In Fig.7, U_A, U_B, U_C are DC component of inverter output phase voltage, U_{AB}, U_{CB} are DC component of inverter output line voltage, r is equivalent resistance of the output filter inductor, I_{Ad}, I_{Bd} and I_{Cd} is DC injection current respectively. From Fig.7, the relationship between DC injection current and DC component of inverter output line voltage can be given as

$$
\begin{cases}
I_{Ad} = \dfrac{2U_{AB} - U_{CB}}{3 \cdot r} \\[2mm]
I_{Bd} = \dfrac{-U_{AB} - U_{CB}}{3 \cdot r} \\[2mm]
I_{Cd} = \dfrac{-U_{AB} + 2U_{CB}}{3 \cdot r}
\end{cases}
\tag{11}
$$

The detecting error of ΔU_{AB} and ΔU_{CB} caused by the ADC resolution can be given by

$$
\begin{cases}
\Delta U_{AB} = \dfrac{U_{dc\_max}}{2^N} \\[2mm]
\Delta U_{CB} = \dfrac{U_{dc\_max}}{2^N}
\end{cases}
\tag{12}
$$

where, U_{dc_max} is maximum DC offset voltage, combined equation (11) and (12), the grid DC current detecting error for each phase can be derived as:

$$
\begin{cases}
\Delta I_{Ad} = \dfrac{U_{dc\_max}}{2^N \cdot 3 \cdot r} \\[2mm]
\Delta I_{Bd} = \dfrac{-2 \cdot U_{dc\_max}}{2^N \cdot 3 \cdot r} \\[2mm]
\Delta I_{Cd} = \dfrac{U_{dc\_max}}{2^N \cdot 3 \cdot r}
\end{cases}
\tag{13}
$$

Furthermore, the maximum grid DC current detecting error $\Delta I_g'$ with the novel control strategy can be given by

$$
\Delta I_g' = |\Delta I_{Bd}| = \frac{2 \cdot U_{dc\_max}}{2^N \times 3r}
\tag{14}
$$

D.   Comparison of DC Current Suppression Effect

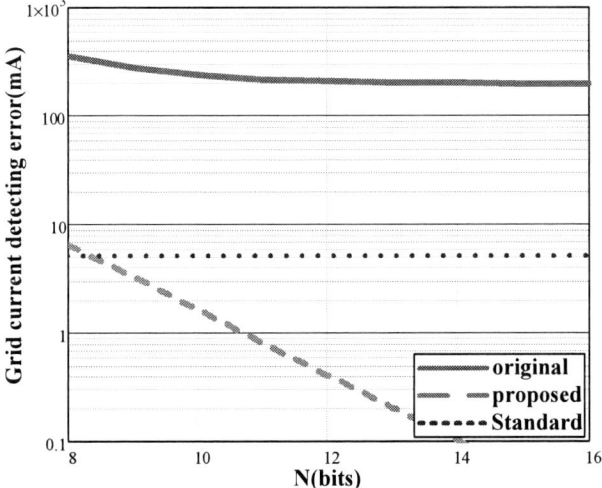

Fig.8. Relationship between the ADC bits and the grid DC current detecting error with the novel control scheme

Considering a three phase PV grid inverter system: the rated power P_e=20 kW, the grid voltage U_g=230V (RMS), the peak to peak value of rated grid current I_{p_p}=41A, the LEM accuracy e% is 0.2%, the maximum detecting value of the LEM I_{max_lem} is 100A, U_{dc_max} is 0.5V, r is 0.3ohm. By substituting the above parameters into (10) and (14), the relationship between the ADC bit and the total grid DC current detecting error with the novel control scheme is shown in Fig.8. The solid line is the relationship between the ADC bits and the grid DC current detecting error with the original control scheme. The dash line is the relationship between the ADC bits and the grid DC current detecting error with the novel control scheme. The dotted line represents the DC current limited standard of United Kingdom [9]. It can be seen that the grid DC current detecting error is lower than the standard value with the novel control scheme if the DSP has 12 bits ADC.

In addition, from equation (1) and (5), using parameters as shown in Table 1, the DC current suppression effect can be plotted in Fig.9. Because DC component can be transformed into negative sequence component by Park transformation in d-q coordinate frame, so it can be seen from Fig.9 that DC current suppression effect with the novel control scheme on the frequency of -50Hz is better than the original control scheme, which indicates that the DC component of inverter grid current can be well suppressed.

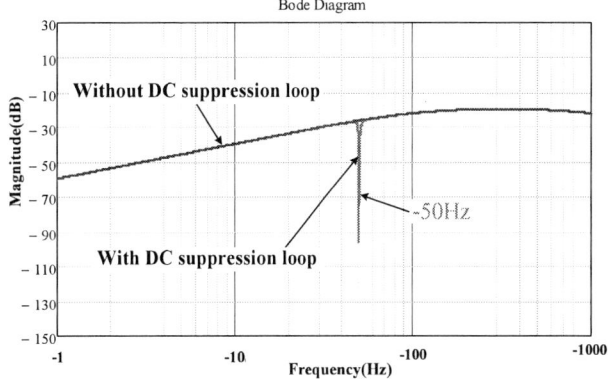

Fig.9. DC current suppression effect

IV.   EXPERIMENT RESULTS

An experimental platform composed of a 20kW three phase PV grid-connected inverter is constructed to verify the proposed novel control strategy. Fig 2 shows the configuration of experimental system, the control scheme is implemented with a TMS320F2808 DSP, and all system parameters are shown in Table 1.

TABLE I PARAMETERS OF THREE PHASE GRID-CONNECTION INVERTER

Parameter	Symbol	Value
Rated power	P_e	20kW
Grid voltage(RMS)	u_{gA}, u_{gB}, u_{gC}	230V
Grid frequency	f	50Hz
DC bus capacitor	C	3400µF
DC bus voltage	Vdc	720V
Switching frequency	f_s	16kHz
Output filter inductance	L	1.5mH
Equivalent resistor of output filter inductance	r	0.3Ω
Feedback gain coefficient of current loop	H_I	0.016
Current loop proportional coefficient	K_p	1.43
Current loop integral coefficient	K_i	903
DC suppression loop proportional coefficient	$K_{p\,dc}$	0.626
DC suppression loop integral coefficient	$K_{i\,dc}$	0.079

Fig.10 shows the DC suppression effect at 15% rated power. Where, channel 1 show the waveform of i_A which is phase A grid-connected current, channel 2 show the waveform of I_{Ad} which is phase A DC injection current, channel 3 show the waveform of I_{Bd} which is phase B DC injection current, and channel 4 show the waveform of I_{Cd} which is phase C DC injection current. In Fig.10(a), DC current injection to the grid is obvious without DC suppression loop at first, but when the DC suppression loop is switched on at the time point t=5s, it can be seen that the DC current injection begin to decrease and is greatly suppressed in steady state finally. Fig.10(b) shows the waveforms of zoom1 without DC suppression loop. It can be seen that the DC current injection to the grid by three phase grid inverter is respectively -333mA, -40mA and 340mA. Fig.10(c) shows the waveforms of zoom2 with DC suppression loop. It can be seen that the DC current injected to the grid can be well suppressed in steady state.

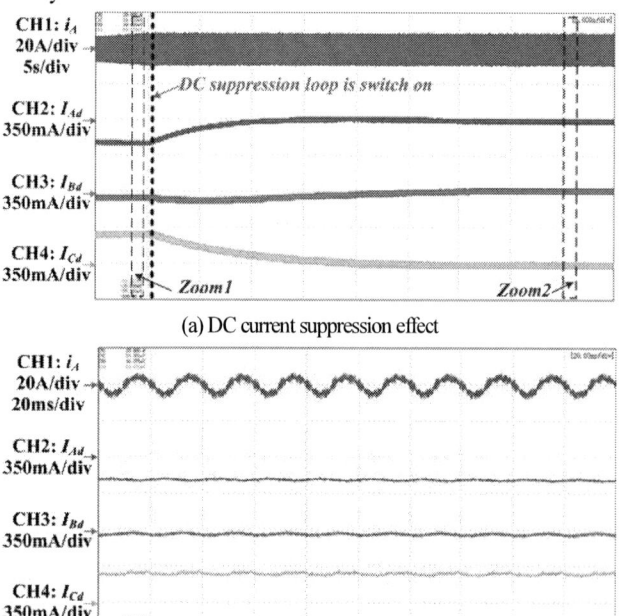

(a) DC current suppression effect

(b) Zoom1 of DC current suppression effect

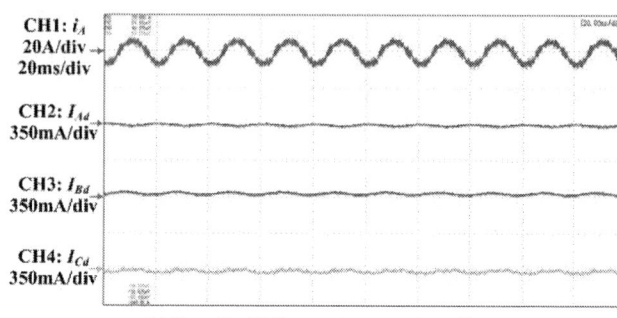

(c) Zoom2 of DC current suppression effect

Fig.10. The waveform of DC current suppression effect (15% rated power)

Fig.11 shows the DC suppression effect at 35% rated power, where, the mean of symbol from channel 1 to channel 4 is same as Fig.10 described. In Fig.11(a), it can be seen that the DC current injection to the grid is obvious without DC suppression loop at first, but when the DC suppression loop is switched on, the DC current injection is greatly suppressed finally. Fig.11(b) shows the waveforms of zoom1 without DC suppression loop. It can be seen that the DC current injection to the grid by three phase grid inverter is respectively -320mA, -30mA and 338mA. Fig.11(c) shows the waveforms of zoom2 with DC suppression loop. It can be seen that the DC current injected to the grid can be well suppressed in steady state.

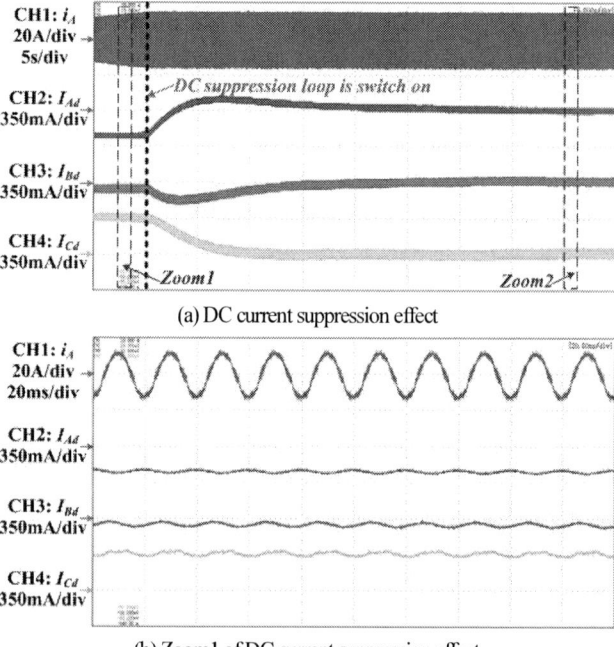

(a) DC current suppression effect

(b) Zoom1 of DC current suppression effect

The 2014 International Power Electronics Conference

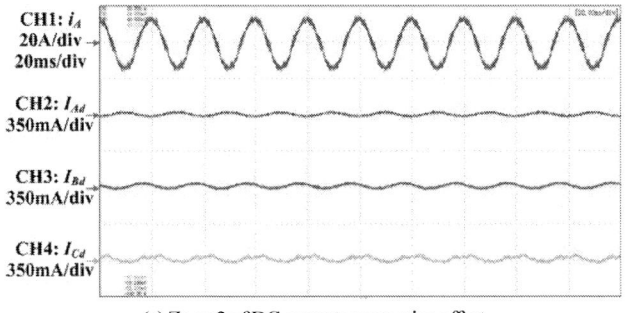

(c) Zoom2 of DC current suppression effect

Fig.11. The waveform of DC current suppression effect (35% rated power)

Fig.12 shows the DC suppression effect at 50% rated power, where, the mean of symbol from channel 1 to channel 4 is same as Fig.10 described. In Fig.12(a), it can be seen that the DC current injection to the grid is obvious without DC suppression loop at first, but when the DC suppression loop is switched on, the DC current injection is greatly suppressed finally. Fig.12(b) shows the waveforms of zoom1 without DC suppression loop. It can be seen that the DC current injection to the grid by three phase grid inverter is respectively -325mA, -33mA and 342mA. Fig.12(c) shows the waveforms of zoom2 with DC suppression loop. It can be seen that the DC current injected to the grid can be well suppressed in steady state.

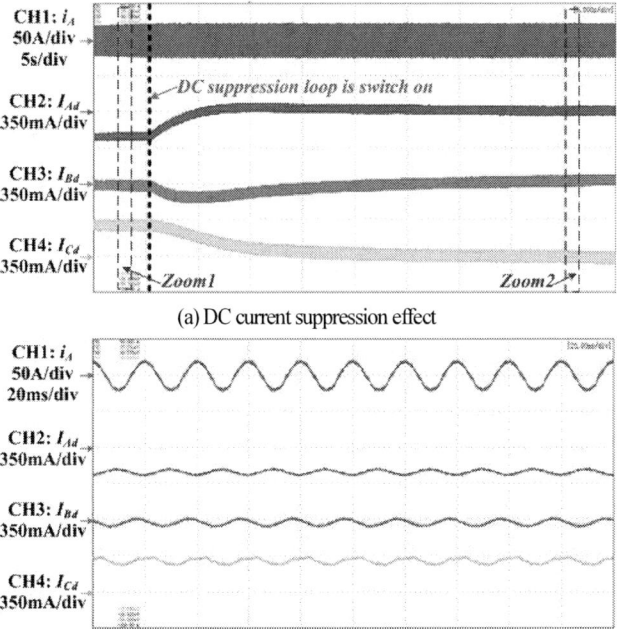

(a) DC current suppression effect

(b) Zoom1 of DC current suppression effect

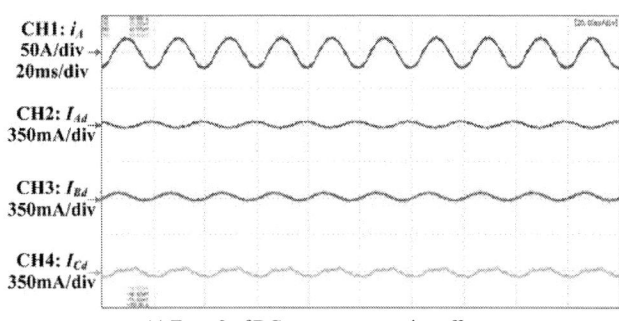

(c) Zoom2 of DC current suppression effect

Fig.12. The waveform of DC current suppression effect (50% rated power)

Fig.13 shows the DC suppression effect at 75% rated power, where, the mean of symbol from channel 1 to channel 4 is same as Fig.10 described. In Fig.13(a), it can be seen that the DC current injection to the grid is obvious without DC suppression loop at first, but when the DC suppression loop is switched on, the DC current injection is greatly suppressed finally. Fig.13(b) shows the waveforms of zoom1 without DC suppression loop. It can be seen that the DC current injection to the grid by three phase grid inverter is respectively -321mA, -28mA and 340mA. Fig.13(c) shows the waveforms of zoom2 with DC suppression loop. It can be seen that the DC current injected to the grid can be well suppressed in steady state.

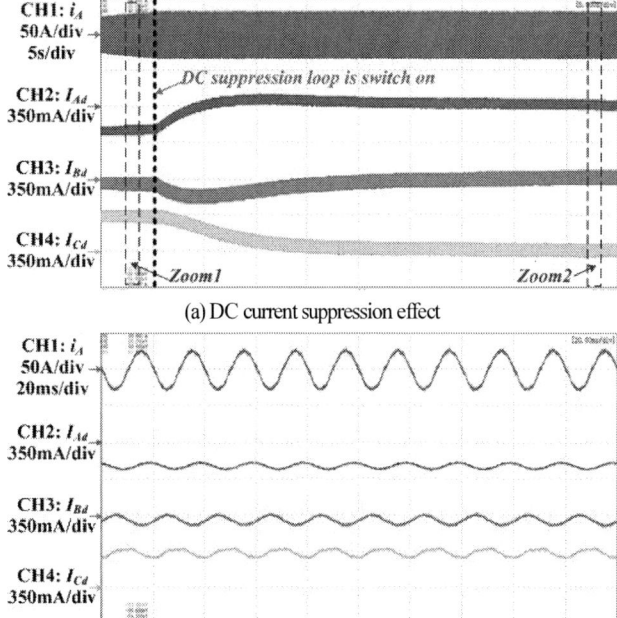

(a) DC current suppression effect

(b) Zoom1 of DC current suppression effect

978-1-4799-2706-7/14 $31.00 © 2014 IEEE

The 2014 International Power Electronics Conference

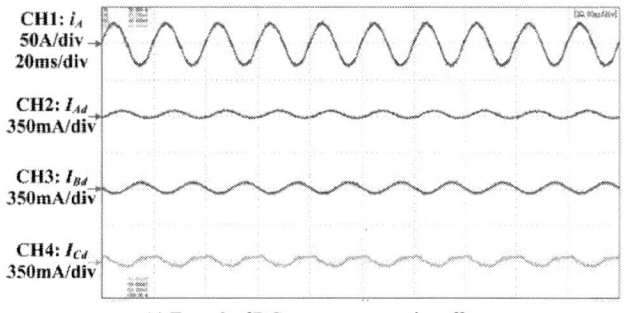

(c) Zoom2 of DC current suppression effect

Fig.13. The waveform of DC current suppression effect (75% rated power)

Fig.14 shows the DC suppression effect at 100% rated power, where, the mean of symbol from channel 1 to channel 4 is same as Fig.10 described. In Fig.14(a), it can be seen that the DC current injection to the grid is obvious without DC suppression loop at first, but when the DC suppression loop is switched on, the DC current injection is greatly suppressed finally. Fig.14(b) shows the waveforms of zoom1 without DC suppression loop. It can be seen that the DC current injection to the grid by three phase grid inverter is respectively -320mA, -31mA and 340mA. Fig.14(c) shows the waveforms of zoom2 with DC suppression loop. It can be seen that the DC current injected to the grid can be well suppressed in steady state.

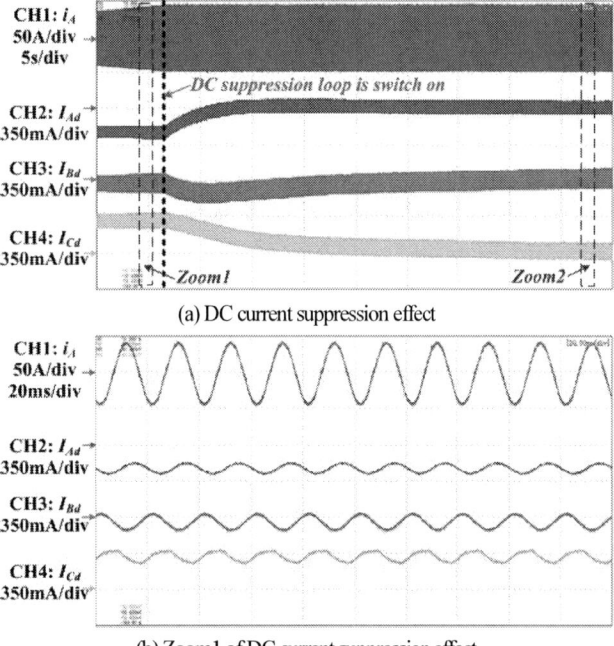

(a) DC current suppression effect

(b) Zoom1 of DC current suppression effect

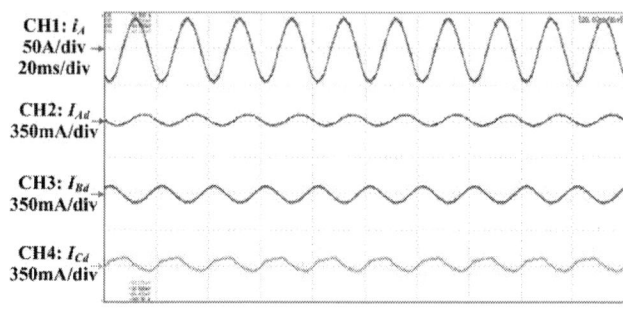

(c) Zoom2 of DC current suppression effect

Fig.14. The waveform of DC current suppression effect (100% rated power)

Fig.15 shows the three phase DC suppression effect with DC suppression loop in different inverter output power. Fig.15(a) shows phase A DC suppression effect, it can be seen that the absolute value of DC current injection is suppressed below 5mA when the inverter output power is changed from 15% rated power to the rated power. Fig.15(b) and Fig.15(c) shows DC suppression effect of phase B and phase C respectively, it can be seen that the absolute value of DC current injection is also suppressed below 5mA from 15% rated power to the rated power. So it can be conclude that the proposed novel DC suppression control strategy can meet the requirement of IEEE DC injection standard.

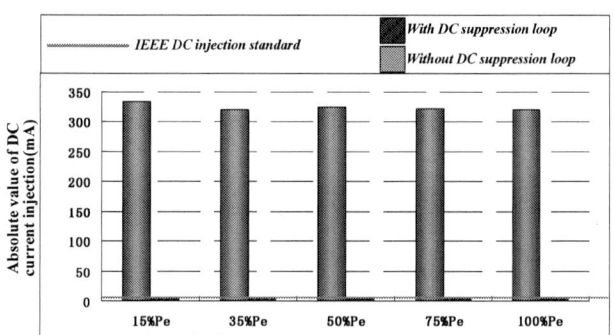

(a) Phase A DC suppression effect

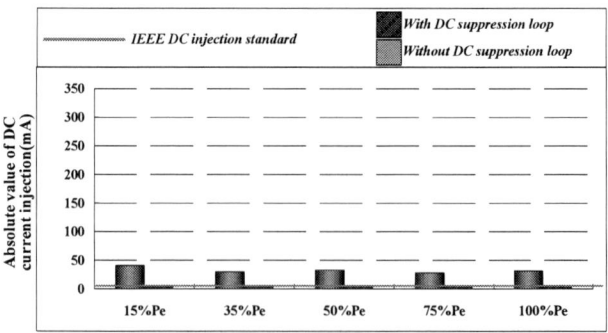

(b) Phase B DC suppression effect

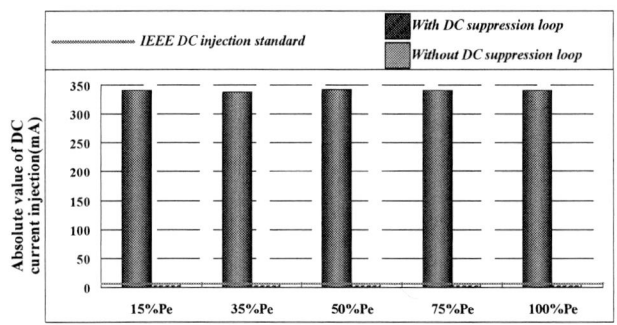

(c) Phase C DC suppression effect

Fig.15. DC suppression effect with different inverter output power

V. CONCLUSION

A novel control strategy to eliminate DC current injection to the grid for three phase PV inverter without the isolation transformer is proposed in this paper. Theoretical analysis is given to show effectiveness of suppressing injecting DC offset of grid inverter. Finally, experiment results verify that the novel control strategy can effectively suppress DC injection current to the grid for three phase PV grid-connected inverter, which is well meet the IEEE standard.

Acknowledgment

This work is supported by the National High Technology Research and Development Program of China 863 Program (2012AA053602, 2012AA053603), the National Natural Science Foundation of China (51277163), the Specialized Research Fund for the Doctoral Program of Higher Education of China (20120101130010), and Zhejiang Key Science and Technology Innovation Group Program (2010R50021).

REFERENCES

[1] R. Gonzalez, E. Gubia, J. Lopez, and L. Marroyo, "Transformerless Single-Phase Multilevel-Based Photovoltaic Inverter", *IEEE Trans. Ind. Electronics*, vol. 55, no. 7, pp. 2694-2702, Jul. 2008.

[2] T. Kerekes, R. Teodorescu, P. Rodriguez, G. Vazquez, and E. Aldabas, "A new high-efficiency single-phase transformerless PV inverter topology," *IEEE Trans. Ind. Electronics*, vol. 58, no. 1, pp. 184-191, Jan. 2011.

[3] D. G Infield, P. Onions, A. D. Simmons, and G A. Smith, "Power Quality From Multiple Grid-Connected Single-Phase Inverters," *IEEE Trans. Ind. Electron.*, vol. 19, no. 4, pp. 1983-1989, Oct. 2004.

[4] S. B. Kjaer, J. K. Pederson, and F. Blaabjerg, "A review of single-phase grid-connected inverters for photovoltaic modules," IEEE Trans. Ind. Electron., vol.

41, no. 5, pp. 1292-1306, Sep. 2005.

[5] W. M. Blewitt, D. J. Atkinson, J. Kelly, and R.A. Lakin, "Approach to low-cost prevention of DC injection in transformerless grid connected inverters," IET Power Electron., vol. 3, no. 1, pp. 111-119, Jan. 2010.

[6] V. salas, E. Olías, M. Alonso, F. Chenlo, and A. Barrado, "DC injection into the network from PV grid inverters," IEEE Photovoltaic Energy Conversion Conference, pp. 2371-2374, Mar. 2006.

[7] AS 4777.2, Grid connection of energy system via inverters Part 2: inverter requirements Australia. 2002.

[8] IEEE 929-2000, IEEE Recommended Practice for Utility interface of photovoltaic (PV) Systems, 3 April, 2000.

[9] ER G83/1, "Recommendations for the connection of small-scale embedded generators (up to 16 A per phase) in parallel with public low-voltage distribution networks". Engineering Recommendation, United Kingdom, September 2003.

[10] R. González, J. López, P. Sanchis, and L. Marroyo, "Transformerless Inverter for Single-Phase Photovoltaic Systems," IEEE Trans. Power Electron., vol. 22, no. 2, pp. 693-697, Mar. 2007.

[11] M. Armstrong, D. J. Atkinson, C. M. Johnson, and T. D. Abeyasekera, "Auto-Calibrating DC Link Sensing Technique for Transformerless, Grid Connected, H-Bridge Inverter Systems," IEEE Trans. Power Electron., vol. 21, no. 5, pp. 1385-1393, Sep. 2006.

[12] L. Bowtell and A. Ahfock, "Direct current offset controller for transformerless single-phase photovoltaic grid-connected inverters," IET Renew. Power Gener., vol. 4, no. 5, pp. 428-437, Sep. 2010.

[13] G Buticchi, E. Lorenzani, and G Franceschini, "A DC Offset Current Compensation Strategy in Transformerless Grid-connected Power Converters," IEEE Trans. Power Delivery., vol. 26, no. 4, pp. 2743-2751, Oct. 2011.

[14] T. L. Ahfock and L. Bowtell, "DC offset elimination in a single-phase grid-connected photovoltaic system," AUPEC, Melbourne, Australia, Dec. 10-13, 2006.

[15] G He M. Chen, N. Su, R. Xie and D. Xu, "A Novel Control Strategy to Suppress DC Current Injection of Single-phase PV Inverter to the Grid, " in Proc. IEEE International Symposium on Power Electronics for Distributed Generation Systems (PEDG), Rogers, Arkansas, USA, July. 8-11, 2013.

CLC Filter Design of a Flyback-Inverter for Photovoltaic Systems

Yesl Shin, June-Hee Lee, June-Seok Lee, and Kyo-Beum Lee
Department of Electrical and Computer Engineering.
Ajou University, Suwon, Korea
E-mail: yeslshin@ajou.ac.kr, ljh20609@ajou.ac.kr, junpb@ajou.ac.kr, kyl@ajou.ac.kr

Abstract— **This paper proposes a CLC filter design of a flyback-inverter for photovoltaic systems. This applied system is small power system; therefore, the capacity and size of the entire system are critical factor. The proposed CLC filter consists of a conventional CL filter and an additional capacitor. The CLC filter, which guarantees the same performance about the current ripple reduction as that of the CL filter, can reduce the size of capacitor and inductor, comparing with components of CL filter. It can help the applied flyback-inverter to be smaller size and cost less. In this paper, the CL filter with the passive damping method is analyzed. Next, the one passive damping method with low power consumption is applied the effect of this method is analyzed. In order to verify the performance of the proposed CLC filter and the validity of passive damping method, a simulation is conducted.**

Keywords— *CL filter, CLC filter, current source inverter (CSI), interleaved flyback inverter, low pass filter, micro-inverter (MIC), passive damping*

I. INTRODUCTION

Performance of photovoltaic system is sensitive of obstacles to even sunshine on solar panels. The obstacles such as trees, dusts and clouds shade on panels. When the shadows cover all over the panel as well as when they cover partially, they cause an critical effect on output efficiency of inverter due to dramatic drop of output current. To get over the mentioned instability of entire power inverting system, micro-inverter (MIC) is developed which can generate power from each panel and sum up later. As each panel has its own inverter, the inverter needs converter part for high boosting ratio since the panel has much lower voltage compared to grid voltage. Transformer with high turn ratio is easily came up for the solution but it leads to high switching loss driven from high voltage and current spikes. This is because that the higher turn ratio a transformer has, the larger parasitic capacitance and leakage inductance it contains which provokes switching loss. For boosting panel's voltage with less power loss, flyback inverter is selected in this paper. Flyback inverter can make high voltage with low transformer turn ratio so it is an adequate low-input-high-output inverter topology and has high efficiency. Still this topology has leakage inductance and the ripple is included on output current [1]-[4].

For reducing current ripple, output filter is a critical composition of the inverter to improve output power

quality and lesson THD (Total Harmonic Distortion). This is because the output current of the system contains high frequency harmonic provoked by PWM switching. For VSI, the simplest output line filter has an inductor only but for better performance of a filter, it can be configured with combinations of capacitors and inductors. Conventionally suggested output filters for VSI vary from simple inductive filter to LC, LCL filter. In comparison with such filters, for CSI, studies of various kinds of output filter is needed beyond CL filter which is commonly used for CSI's output filter [5]-[6]. At resonance frequency of filter configuration, the filter has large amplification and it can cause instability of filter performance. To solve this problem and improve the performance of output filter, damping methods is necessarily applied [7]-[11].

In this paper, a design of CLC output filter for interleaved flyback CSI is proposed. Comparing with CL output filter for CSI, components in filter circuitry show the better performance with smaller size and capacity. For reliability of output filter, passive damping methods applied. According to the location of resistive damping to capacitors or inductors, four damping methods are analyzed by building up transfer functions and plotting bode diagrams. Simulation results prove the validity of the proposed design method of the CLC filter for interleaved flyback CSI.

II. OUTPUT FILTER FOR CSI

Output current contains high frequency harmonics made by switching frequency. To reduce such harmonics output low-pass filter is required. In addition to simple inductive line filter, CL filter has been mostly researched as a line filter for CSI. Furthermore, with an additional

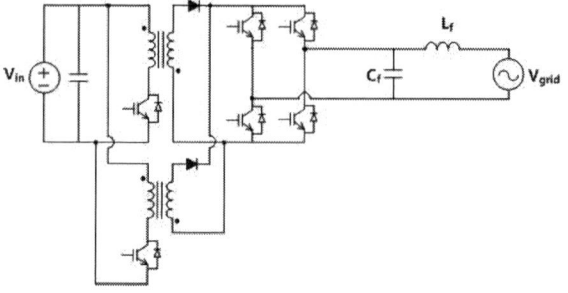

Fig. 1. Interleaved flyback current source inverter with CL filter

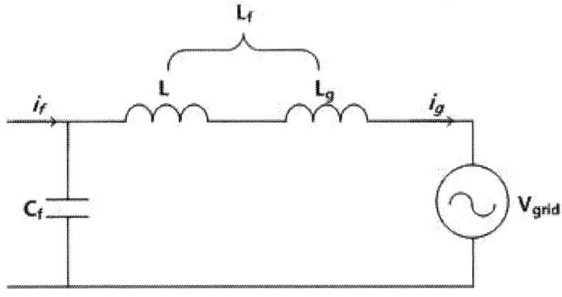

Fig. 2. CL filter circuit.

component to CL filter, CLC filter is suggested to attenuate high frequency harmonics. CLC filter has better performance with components of smaller capacity compared to CL filter.

A. CL Filter

A CL filter is used to deliver the power by coupling the CSI to the voltage source grid. To attenuate switching harmonics, the CL filter is designed as a low-pass filter. The circuit of the interleaved flyback CSI with the capacitive-inductive (CL) filter is shown in Fig. 1. Since the output impedance exists, for the set-up of the transfer function and designing parameters, inductive reactance is considered with the intentionally injected inductor in the filter configuration as shown in Fig. 2.

The performance of the output filter is determined by the location of the resonant frequency of the filter. Hence, the design point is putting the certain harmonic frequency, desired to be reduced, behind the resonant frequency of the filter. To meet the THD criteria in the coupling process, appropriate capacitance and inductance should be determined for the filter's resonant frequency. In this paper for the CL filter, the total output-side inductance amount is included in Lf value like Fig. 2. The transfer function for CL filter is expressed as

$$G_{CL} = \frac{i_g}{i_f} = \frac{1}{s^2 C_f L_f + 1}.$$

(1)

where ig and if are the input and output currents of the filter and Cf and Lf represent capacitance and inductance of the filter respectively. According to the control theory of the second-order system, the damping ratio is 0 and damped natural frequency equals to

$$\omega_d = \omega_n = \frac{1}{\sqrt{C_f L_f}}.$$

(2)

The mentioned CL filter has no damping and its damped natural frequency is the same as its undamped natural frequency and resonant frequency.

The filter's resonant frequency is generated due to passive elements in the filter circuit such as an inductor and a capacitor. It also can be referred as the filter's cutoff frequency. The choice of the proper resonant frequency is substantially related to the PWM switching

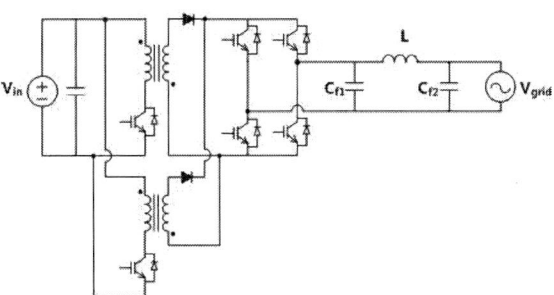

Fig. 3. Interleaved flyback inverter with CLC filter

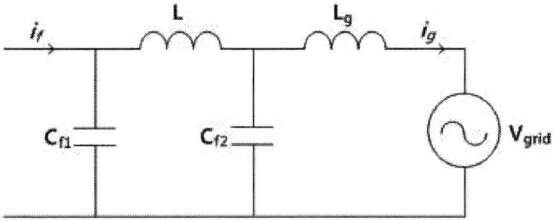

Fig. 4. CLC filter circuit

frequency. Because PWM harmonics are intended to be reduced by the filter and, for low-pass output filter, the resonant frequency of the filter should be set lower than the certain frequency desired to be removed, the resonant frequency should be chosen lower than the PWM switching frequency. The filter's resonant frequency depends on the components in the filter since it is determined by (2). In other words, the resonant frequency is determined by the combination of capacitor and inductor. Generally the larger capacity the components have, the bigger their sizes are. The size difference by the capacity looks more evidently on inductor than capacitor. In this paper, entire system size is one of the most important factors for micro-inverter; therefore the design sets the inductor value first and then chooses capacitor to keep resonant frequency desired constant. If the resonant frequency is not low enough, the switching loss resulted from PWM is increased. But if the resonant frequency is too low, it means the filter requires large inductance and capacitance while causing higher loss in the circuit.

B. CLC Filter

By adding a capacitance component to CL filter, CLC filter can be configured. Additional capacitor is located between intentionally added inductor and grid-side inductor as shown in Fig. 3. Added capacitor is for filtering PWM switching harmonics effectively and at the same time, it helps each component's capacity and size to be reduced. Improved performance of the proposed filter design can be proved through the equations as below

$$G_{CLC} = \frac{i_g}{i_f} = \frac{1}{s^4 C_{f1} C_{f2} L L_g + s^2 (C_{f1} L + C_{f1} L_g + C_{f2} L_g) + 1} . (3)$$

Each parameter indicates the components of the proposed filter circuit in Fig. 4. CLC filter is fourth-order system and its poles are two pairs of complex numbers. Four poles are represented as following.

$$p = \pm j \sqrt{\frac{b \pm c}{2a}}$$
$$a = C_1 C_2 L L_g$$
$$b = C_1 L + C_1 L_g + C_2 L_g$$
$$c = C_1^2 L^2 + 2C_1^2 L L_g - 2a + C_1^2 L_g^2 + 2C_1 C_2 L_g^2 + C_2^2 L_g^2 \qquad (4)$$

Compared to CL filter, CLC filter has two resonant frequencies because one more capacitor is supplemented. However, each resonant frequency is not separately affected by front CL and back CL, additional C and grid-side inductive impedance. Both two resonant frequencies are related to all components' values in the filter as shown in (4). Therefore determining filter's cut-off frequencies are not simple as much as that of CL filter and adequate combinations of each components must be considered.

When choose the resonant frequency for the performance of filter, two kinds of frequency are concerned which are switching frequency and control frequency like for CL filter case. For the output filter, because it has roll-off operation period after each resonant frequencies, both kinds of frequency are set after filter's resonant frequencies. In the case of CL filter, since it has only one resonant frequency, switching and control period harmonics damp with same roll-off rate which is only -20dB/dec. When CLC filter is used, as it has two resonant frequencies, it can make higher frequency from two must-concerned frequencies to attenuate quicker than CL filter. For example, when the control period component set in the first roll-off operation period and the PWM frequency after 2nd resonant frequency, according to its roll-off rate after all resonant frequencies, PWM harmonics' attenuation occurs rapidly with -40dB/dec. Certain frequency component can be selected to be in the periods which are divided by resonant frequencies with arranging resonant frequency of the filter. Design condition for CLC filter is expressed in

$$\omega_{c1} < \omega_{ctrl} < \omega_{c2} < \omega_{PWM} . \qquad (5)$$

Generally in the system, control period is longer than PWM switching period. With CLC output filter, the frequency component which has shorter period is set to be in 2nd roll-off period. In Fig.5, Bode plots of CL filter and CLC filter are shown and it can be proved that CLC filter's damping and roll-off rate are quicker for PWM harmonics. 62.8 kHz and 6280 kHz indicate control period frequency and switching frequency respectively. At each point, magnitude of CLC filter's gain is smaller than CL filter's magnitudes. Resonant points exist one in CL filter's Bode plot and two in CLC filter's Bode plot. At each point, the filter has much larger gain than any other frequency and their frequency can be obtained by equations of transfer functions. One more designing

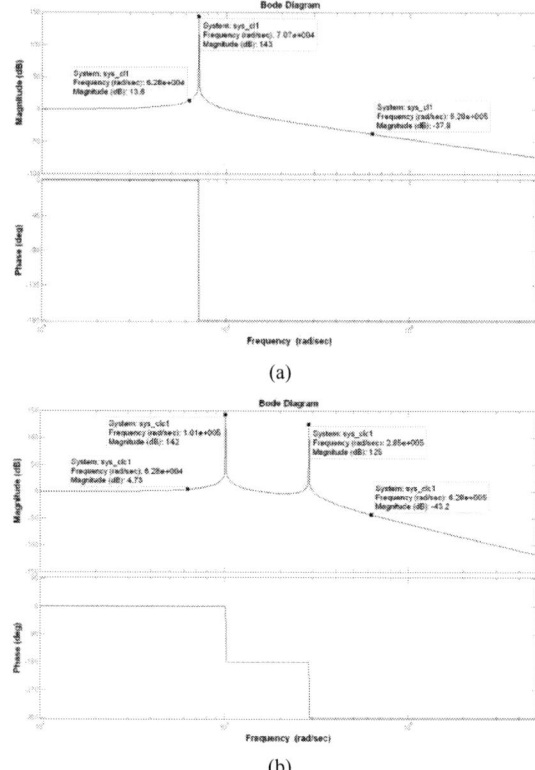

(a)

(b)

Fig. 5. Bode plots of (a) CL filter and (b) CLC filter

feature to take account of is about grid-side inductance. As grid-side inductance is derived from output load, the value of it doesn't set to constant value. Hence, in the design, it must be virtual value and for practical use of the filter, 2nd capacitor's value has to have margin of its capacity.

In this paper, the switching frequency is selected as 50 kHz. By the way the proposed system is operated by 2-phase interleaved method, PWM harmonics of this system appear around 100 kHz which is twice higher value of the PWM switching frequency. According to the system's operation characteristic, the designing criterions should be carefully considered.

III. OUTPUT FILTER WITH PASSIVE DAMPING FOR CSI

In CLC output filter circuit, the resonant frequency is generated due to passive elements in the filter circuit such as inductor and capacitor. At the resonant frequency, the filter has an infinite gain as shown in Fig. 5. There are two ways to avoid the unexpected performance of the filter at resonant frequency to set parameter combinations not to match any operational period with resonant period or to add damping to the filter circuit. First method to avoid resonant frequency's characteristic is not ultimate solution and it still has possibility to get infinite gain with any unpredicted input or disturbance.

To make more reliable filter configuration, output filter with passive damping is suggested. Every possible locations of resistive damping in CL filter are shown in Fig. 6. Those are classified into 4 ways in accordance

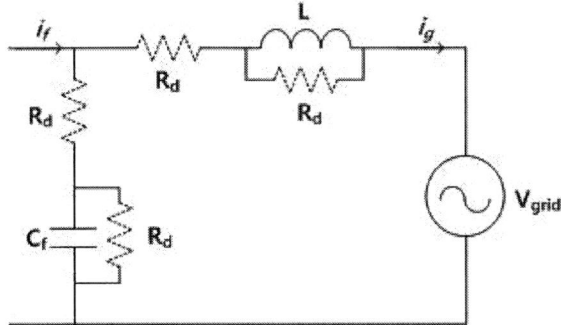

Fig. 6. CL filter circuit with resistive damping.

with whether the resistive damping locates near capacitors or inductors in parallel or series connection.

For CLC filter, there's only difference that one capacitor is added before grid-side. Possible damping positions of additional capacitor are also in parallel or series so CLC filter damping methods are 4 methods, too. As damping methods for CLC filter are same as CL filter, each damping methods' characteristics are researched with those of CL filter.

A. CL Filter with Damping Resistor Parallel with Inductor

One of four damping methods has a damping resistor in parallel with inductor of the filter. It has first-order roll-off rate after the resonant frequency and its transfer function is denoted as below

$$G_{CL\_damped} = \frac{i_g}{i_f} = \frac{sL + R_D}{s^2 C_f L R_D + sL + R_D}. \quad (6)$$

Comparing with other damping types in CL filter configuration, it has much smaller power loss in the resistive damping because the fundamental voltage applied to inductor is expected comparatively small.

B. CLC Filter with Damping Resistor Parallel with Inductor

As damping methods for CLC filter are same as ones for CL filter, this damping method's characteristics are researched with those of CL filter. If this analysis is applied to CLC filter, similar effects can be expected to the CL filter with resistive damping in parallel with inductor. In the configuration of passive damped CLC filter, inductance of grid-side is separated from filter inductor. This characteristic also can be denoted in transfer function presented in

$$G_{CLC\_damped} = \frac{i_g}{i_f} = \frac{sL + R_D}{s^4 A + s^3 B + s^3 C + sD + R_D}$$

$$A = C_{f1} C_{f2} L L_g R_D, B = C_{f2} L L_g$$

$$C = C_{f1} L R_D + C_{f2} L_g R_D + L L_g, D = L + L_g R_D \quad (7)$$

where A, B, C, and D are polynomial's coefficients. The designed CLC filter's performance characteristic is

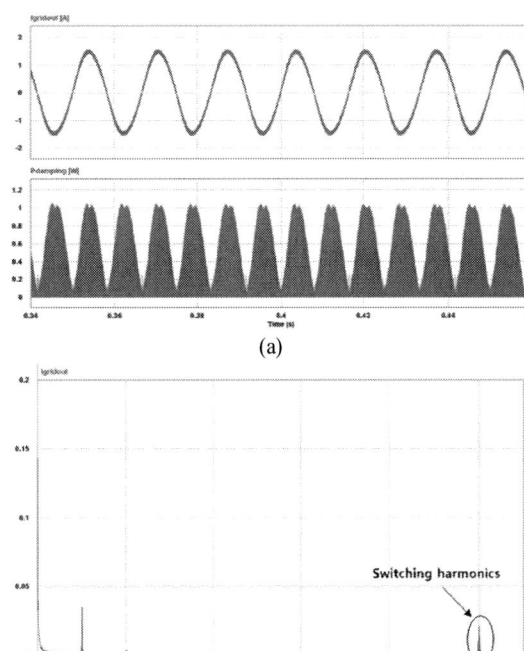

(a)

(b)

Fig. 7. Performance of the inverter with CL filter

(a) waveforms of the output current and power at resistive damping

(b) FFT of the output current

approved by Bode plot of MATLAB simulation, too. The resonant points, the certain frequency points with extremely high gain, are removed by adding resistive damping so, stability of the output filter is guaranteed.

IV. SIMULATIONS

To confirm the validity of the designed CLC filter compared to CL filter, simulation has been conducted using PSIM. The simulation model has been set up under the conditions in Table I.

TABLE I.   SIMULATION PARAMETERS

Parameter	Value
Grid voltage (RMS)	220 V
Fundamental frequency	60 Hz
Switching frequency	50 kHz
Cf of CL filter	0.2 μF
L of CL filter	0.6 mH
Cf1 of CLC filter	0.1 μF
Cf2 of CLC filter	0.1 μF
L of CLC filter	0.15mH

In accordance with IEEE Std 519-1992, the required output current THD is determined to 5% or less. Designed CL and CLC filter's parameters are set as Table I. To approve the performance improvement of output filter from CL filter to CLC filter, two kinds of filters' simulations are conducted. Both models have damping

The 2014 International Power Electronics Conference

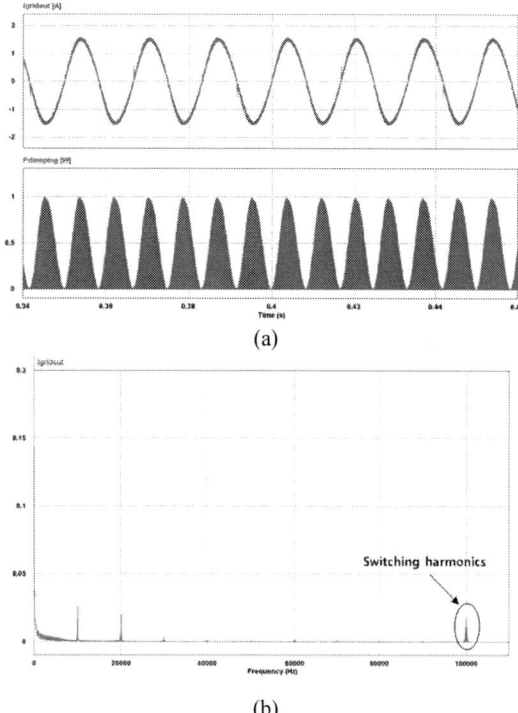

(a)

(b)

Fig. 8. Performance of the inverter with CLC filter

(a) waveforms of the output current and power at resistive damping

(b) FFT of the output current

resistor each and it is implemented by PSIM. Selection of resistive damping depends on dissipated power at the resistive damping and in this paper, the simulation conducted to have 1W or less power loss which occurs at damping. Power loss at the resistive damping can be arranged according to the resistor value and the value has a more effect on current amount across itself than that on filter performance characteristic.

In this simulation, damping resistors are chosen to 200 ohm in CL filter and 6 kilo-ohm in CLC filter. Grid-side inductance is assumed as 0.4mH.

Fig. 7 and Fig. 8 show the waveforms and FFT of output current when CL filter and CLC filter are applied respectively. The figures approve that CLC filter can have the same or better performance than CL filter with much smaller inductance in the output filter. From FFT of output current, Fig. 8(b), it is verified that switching harmonics are reduced with smaller inductance and additional capacitors which have smaller values compared to CL filter's capacitor value as well.

V. CONCLUSION

This paper proposed a design method of CLC filter for 250W photovoltaic interleaved flyback CSI system. The designed CLC filter has been configured with smaller valued and sized components than CL filter's components. With adding the smaller component, output filter performance can be gained as same as CL filter's or can be improved from CL filter's. For output filter's stability of operation, passive damping method is applied.

For more reliable validity of passive damping method, several types of damping methods will be analyzed. The paper also presented the performance of CL filter and CLC filter when damping resistors are applied throughout simulations.

REFERENCES

[1] J. Lee and K. Lee, "Modulation Technique to Reduce Leakage Current in Transformerless Photovoltaic Systems using a Three-Level Inverter," *in IEEE 2013 Applied Power Electronics Conference*, 2013, pp. 700-705.

[2] H. Jeong, G. Kim, and K. Lee, "Second-Order Harmonic Reduction Technique for Photovoltaic Power Conditioning Systems Using a Proportional-Resonant Controller," *Energies*, vol. 6, no. 1, Jan. 2013, pp. 79-96.

[3] H. Hu, Q. Zhang, X. Fang, Z. J. Shen, and I. Batarseh, "A single stage micro-inverter based on a three-port flyback with power decoupling capability," *in IEEE 2011 Energy Conversion Congress and Exposition,* 2011, pp. 1411-1416.

[4] M. Pisano, F. Bizzarri, A. Brambilla, G. Gruosso, and G. Storti Gajani, "Micro-inverter for solar power generation," in *Power Electronics, Electrical Drives, Automation and Motion 2012 International Symposium*, pp. 109 –113, June 2012.

[5] H. Jeong, K. Lee, S. Choi, and W. Choi, "Performance Improvement of LCL-Filter-Based Grid-Connected Inverters Using PQR Power Transformation," *IEEE Trans. Power Electronics*, vol. 25, no. 5, pp. 1320–1330, May 2010.

[6] H. Jeong, D. Yoon, and K. Lee, "Design of an LCL-Filter for Three-Parallel Operation of Power Converters in Wind Turbines," *Jounal of Power Electronics*, vol. 13, no. 3, pp. 437-446, May 2013.

[7] A. S. Morsy, S. Ahmed, P. Enjeti, and A. Massoud, "An active damping technique for a current source inverter employing a virtual negative inductance," in *IEEE Applied Power Electronics Conference*, pp. 63-67, 2010.

[8] Y. W. Li, "Control and resonance damping of voltage-source and current source converters with LC filters," *IEEE Transactions on Industry Applications*, vol. 56, no. 5, pp. 1511–1521, May 2009.

[9] K. Ahmed, S. Finney, and B. Williams, "Passive filter design for three phase inverter interfacing in distributed generation," in *IEEE Compatibility in Power Electronics 2007*, pp. 1–9, June 2007.

[10] Weimin Wu, Yuanbin He, Tianhao Tang, and Frede Blaabjerg. "A New Design Method for the Passive Damped LCL and LLCL Filter-Based Single-Phase Grid-Tied Inverter." *IEEE Trans. Ind. Electron.*, vol. 60, no. 10, pp. 4339-4350, 2013

[11] Channegowda, Parikshith, and Vinod John. "Filter optimization for grid interactive voltage source inverters." *IEEE Trans. Ind. Electron.*, vol. 57, no. 12, pp. 4106-4114, 2010

The 2014 International Power Electronics Conference

Three-Phase Inverter Topologies for Grid-Connected Photovoltaic Systems

Ziya Özkan
Student Member, IEEE
Middle East Technical University
Dept. of Electrical and Electronics Engineering
Ankara, 06800, Turkey
ozziya@metu.edu.tr

Ahmet M. Hava
Member, IEEE
Middle East Technical University
Dept. of Electrical and Electronics Engineering
Ankara, 06800, Turkey
hava@metu.edu.tr

Abstract—**In this paper, the energy conversion efficiency (ECE) and cost characteristics of three-phase photovoltaic (PV) inverters (3P-PVIs) are studied comprehensively based on the operating principles of topologies and with respect to various performance indicators such as semiconductor device count and utilization factors, semiconductor efficiency (SE), dc-bus capacitor and filter inductor voltage/current ripple factors. The evaluation of topologies are based on datasheet parameters and analytical results are provided for 50 kW, 100 kW, and 200 kW power ratings. Such an evaluation of topologies is beneficial to engineers to wisely choose the appropriate topology for both transformerless and transformer based PV systems and to design the inverter accordingly.**

Keywords—Grid-connected, Photovoltaic inverter, Three-phase inverters, Energy conversion efficiency

I. INTRODUCTION

In commercial grid-connected photovoltaic (PV) systems, cost and energy conversion efficiency (ECE) of PV inverters are the two major drivers to determine the payback period of the overall system. According to grid-connection, single-phase PV inverters (1P-PVIs) and three-phase PV inverters (3P-PVIs) are available in the market. 1P-PVIs are available in a power range of 1-5 kW feeding low-voltage distribution systems; whereas the 3P-PVIs are utilized in a power range larger than 10 kW [1] reaching MW levels. In terms of dc-bus capacitor size, 3P-PVIs require fewer amounts of dc-bus capacitors per watt due to symmetrical injection of power causing no second harmonic power oscillations as in the case of 1P-PVI systems. Therefore, to improve inverter lifetime to match the lifetime of PV modules, the dc-bus can be embodied as film capacitors instead of electrolytic capacitors which are life-limiting in 1P-PVIs [1], [2]. Moreover, cost of 3P-PVIs per watts is lower than 1P-PVIs due to fewer components used.

As government incentives in Germany and other pioneering countries favored residential PV installations first, with emphasis on ECE, the evolution of 1P-PVIs took place early and large number of energy efficient topologies have been invented and put to use. Their performances were evaluated, compared, and designs reported well. On the other hand, 3P-PVI systems are only recently evolving and as the application requires high ECE, low cost, and long life, suitable topologies have been under development and several put to use. As a comprehensive work is absent on the ECE and cost characteristics of these topologies, this paper focuses on these issues. Moreover, implementation and test of high power circuits are laborious and require high power equipment, so this study provides a shortcut of evaluation of topologies. Favorable topologies are reviewed, their semiconductor utilization (thus semiconductor cost), dc-bus voltage and filter inductor current ripple characteristics (thus capacitor and magnetics cost), and ECE characteristics are evaluated and compared. Thus, an overall comparison guides the design engineer towards the choice of suitable topology for the intended PV application in a wide power range (10kW-1MW).

II. GENERAL ASPECTS OF 3P-PVIS

3P-PVIs are mostly voltage-source topologies as these topologies have higher ECE and lower cost as compared to current-source topologies [3], and these inverters are configured with transformer or transformerless (Fig.1).

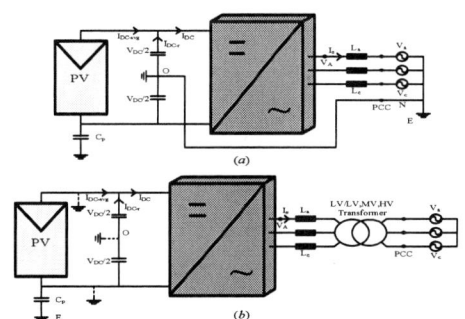

Fig.1 Grid connection of (*a*) transformerless and (*b*) transformer based 3P-PVI topologies.

Transformerless 3P-PVIs are available in power ratings 10 kW to several tens of kW, and they are generally connected to low voltage distribution systems directly. Due to leakage current issues, the midpoint of the inverter dc-bus is grounded. Zero sequence voltage injection for better dc-bus voltage utilization and lower switching losses in such a grounded system is avoided, as

978-1-4799-2706-7/14 $31.00 © 2014 IEEE   498

there is path for zero sequence currents as shown in Fig. 1(a). When transformerless inverters with small power range (several tens of kW) are utilized to establish a large scale PV system, they exhibit better maximum power point tracking characteristics due to the distributed tracking nature. Moreover, due to lack of a line-frequency transformer, transformerless inverters/PV systems exhibit better ECE characteristics as compared to their transformer based counterparts.

The inclusion of a transformer between the power system point of common coupling (PCC) and the inverter is necessary in some countries due to grid connection standards, in some PV module types due to some certain voltage gradient requirement with appropriate grounding, and in solar farm type large-scale PV installations due to the absence of a distribution transformer (or the insufficiency of its power rating) [4]. In transformer-based 3P-PVIs, the transformer secondary winding connection can be made to low-voltage, medium-voltage or high-voltage systems. Due to core and copper losses, the transformer may consume as high as 1-2 % of ECE. Furthermore, the inclusion of the transformer increases the cost, cooling, and real-estate requirements. Despite these drawbacks, the transformer allows the connection to high voltage systems by voltage ratio changing, eases the grounding for various PV module technologies (Fig. 1(b)), reduces capacitive leakage current problems, and allows zero sequence voltage injection to increase the dc-bus voltage utilization and to reduce switching losses of the inverter. Furthermore, transformers with multi-windings allow easy inverter paralleling, thus power rating can be increased with modular inverter units.

III. 3P-PVI TOPOLOGIES

In PV systems, the power is converted from dc to ac at high current rather than high voltage. Thus, power converter topologies with low total series voltage drop are favorable. Although numerous topologies have been reported, of these the following five have found applications, which are two, three, and five-level voltage-source converter topologies. The common sinusoidal pulse width modulation (SPWM) waveforms and output voltages for one phase (V_{AN}) of each of these topologies for dc-bus midpoint grounded (transformerless) case are illustrated in Fig. 2. As the figure demonstrates, the higher the level, the closer to sinusoidal the converter ac output voltage. The foregoing investigation and analyses are based on such dc-bus midpoint grounded systems.

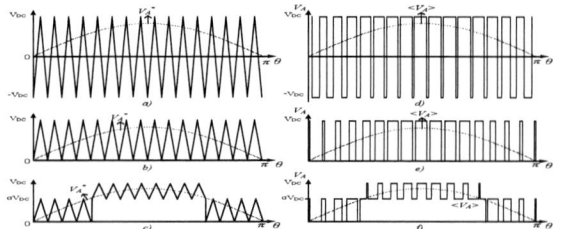

Fig. 2 Common modulation waveforms and output voltages of two, three, and five-level 3P-PVIs.

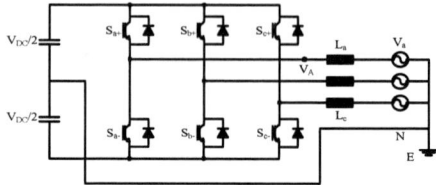

Fig. 3 Grid-connected 2L inverter topology.

Two-level 3P-PVI (2L-3P, Fig. 3): With minimum switch count and lowest conduction losses, this conventional topology is the most established and widely used one. But due to its high switching losses and output voltage/current ripple, its ECE is inferior.

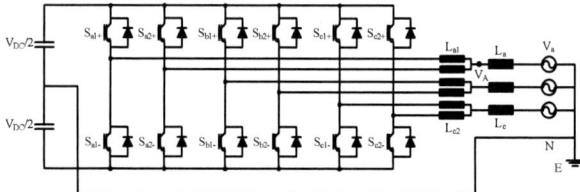

Fig. 4 Grid-connected 2L-SS inverter topology.

The Soft-Switching 3P-PVI (2L-SS, Fig. 4, [5]): The diode reverse recovery based IGBT turn-on speed limitation (thus high E_{on}) and reverse recovery losses (E_{rr}) of the two-level inverter could be avoided with the addition of coupled-inductor and additional inverter leg per-phase. Thus, the switching losses could be reduced, and ECE improved. But the cost and complexity are the handicaps of the topology.

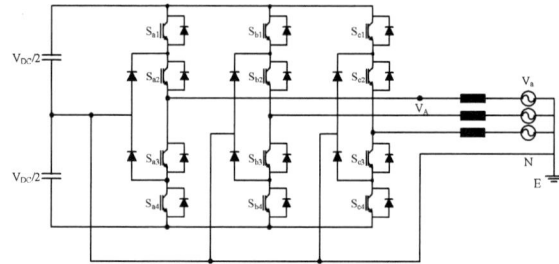

Fig. 5 Grid-connected 3L-NPC inverter topology.

Three-level Neutral Point Clamped 3P-PVI (3L-NPC, Fig. 5, [6]): The three-level NPC inverter provides reduced switching losses, and with lower blocking voltage rating IGBTs, the conduction loss increase can be dominated by switching loss reduction, thus overall enhanced ECE. But, cost and complexity issues exist.

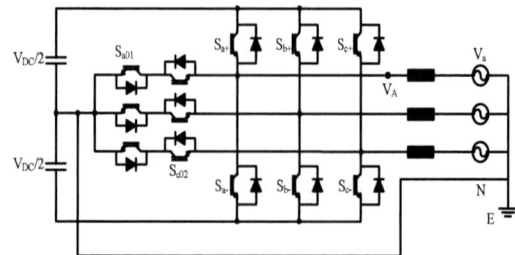

Fig. 6 Grid-connected 3L-T-type inverter topology.

Three-level T-type 3P-PVI (3L-T-type, Fig. 6, [6]): Having the same operational characteristics as the NPC inverter, the difference is in further reduction of conduction losses by combining different types of devices as the clamping and main devices. Thus, lower conduction losses, and higher ECE.

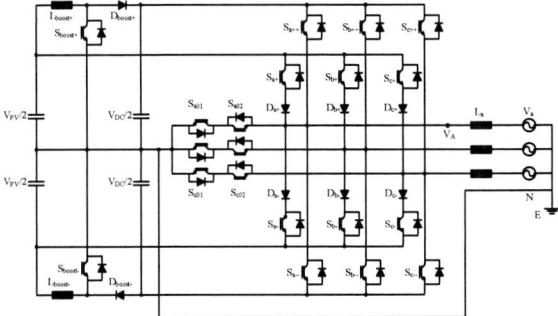

Fig. 7 Grid-connected 5L inverter topology.

Five-level Differential Boost 3P-PVI (5L, Fig. 7, [7]): Allowing two different dc-bus voltage levels and interfacing them with boost converters, the topology provides closer to sinusoidal output voltage pulses and decreases the demand for inductor filter and switching speed. However, the cost, complexity, and lack of reactive power capability are the handicaps of the topology.

IV. CHARACTERISTICS OF 3P-PVI TOPOLOGIES

ECE and cost of 3P-PVIs are the two components determining the payback period of a grid-connected PV system. Relevantly, this section describes/evaluates the performance indices related to cost and ECE of 3P-PVIs. The power semiconductors, the magnetics, and the dc-bus capacitors are the prime elements that affect these characteristics. Thus, focus is placed on these elements in separate sections.

A. Number of Active and Total Switches

In realization of 3P-PVIs, Insulated Gate Bipolar Transistors (IGBTs) are used as active switches, due to their current handling capability. The cost of IGBTs is proportional to their blocking voltage and current capacity. Besides, the cost of diodes becomes considerable as the power rating increases. For active switches, gate drive circuitry becomes more complicated and costly. Therefore, with the current and voltage ratings, the complexity and cost of the converter is proportional to the number of active and total number of switches.

B. Device Utilization Factor

Device utilization factor (C_{DU}) is defined as the ratio of output power of a converter to the total semiconductor kVA of active switches as given in (1). Here, $I_{k\text{-}peak}$ is the peak current at rated power and $V_{k\text{-}block}$ is the blocking voltage of kth switch. C_{DU} is an indicator of effectiveness

of utilization of the semiconductor current and voltage ratings and thus the cost of switches in the converter per watts. In Table III, C_{DU} of each 3P-PVI topology is normalized to the value of the conventional 2L-3P topology, so that a relative comparison of topologies is straightforward.

$$C_{DU} = \frac{P_{rated}}{\sum_k I_{k\text{-}peak} V_{k\text{-}block}} \tag{1}$$

C. Number of Semiconductors on Current Path

In PV applications, as the dc-bus voltage cannot be increased more than 1 kV, the power is increased by increasing the line current. As a result, any voltage drop on the current path decreases ECE significantly. In the 3P-PVIs investigated, the line current is continuous (nearly sinusoidal) and the semiconductors on line current cause a voltage drop on the current path. Therefore, the number of semiconductors on the current path is important (the less, the better) when selecting a topology for high power applications. In 2L-3P, 2L-SS, and NPC topologies number of total semiconductor on the current path are relatively apparent and they are given in Table III. However, in the case of T-type and 5L topologies, the number of semiconductors on the current path is dependent on the modulation index (M, in (2)) and the boost ratio (σ, for the 5L topology in (3)) as in (4) and (5) respectively.

$$M = \frac{V_{s\text{-}1\text{-}p}}{\dfrac{V_{DC}}{2}} \tag{2}$$

$$\sigma = \frac{V_{PV}}{V_{DC}} \tag{3}$$

$$\Theta_{T\text{-}type} = 6 \cdot (1 - M/\pi) \tag{4}$$

$$\Theta_{5L} = 3 \cdot \left[\frac{2 - \sigma - \dfrac{2M}{\pi}\cos(\sin^{-1}(\sigma/M)) - \dfrac{2\sigma}{\pi}\sin^{-1}(\sigma/M)}{1 - \sigma} \right] \tag{5}$$

D. Semiconductor Efficiency (SE)

Semiconductor losses constitute the majority of losses and assess a stringent part of the power converter design especially at high power ratings. Thus, SE of 3P-PVIs are evaluated in this study. Semiconductor losses are mainly comprised of conduction and switching losses. These losses can be calculated based on IGBT/diode datasheet parameters so that a comparative evaluation of 3P-PVIs can be made. In [8] and [9], instantaneous expression of IGBT/diode conduction losses and switching event based expression/approximation of switching losses are given in regards to the switching constraints (such as turn on/off voltage current, gate drive resistor etc.). Further, in many application notes and papers, the evaluation methods of these losses are available [10], [11]. In the evaluation of SE of a power converter, the losses related to each switch are evaluated by relating switch current duty cycle function (CDCF) and the switching constraints to datasheet parameters. The losses of all switches are summed up for a fundamental grid period and they are averaged, so evaluation of SE becomes straightforward. Evaluation of switching constraints of a switch (blocking voltage, on-state current) with respect to grid angle (θ) is also straightforward (blocking voltage of a switch is generally constant, and the on-state current generally varies sinusoidally for half grid period). However, CDCF of a switch should be made clearer as it is not very common in the literature.

The CDCF of a switch is the ratio of the duration of the current passing through the switch during the on state to the switching period. The CDCF of a switch is closely associated to the output voltage duty cycle function ($M\sin\theta$ for $0<\theta<\pi$), and can be derived from the simple volt-seconds balance. The CDCF (or designated as $d_i(\theta)$ in formulas) of a switch is helpful both determining the conduction and switching losses of the switch. For example, for the kth switching period, the conduction losses of an IGBT can be approximated as in (6) (where ω is the angular frequency, T_s is switching period, i_C is IGBT collector current, V_{CE-ON} is the constant component, and R_{CE-ON} is the junction temperature dependent resistive component of on-state voltage drop of IGBT). Moreover, switching losses of a switch can be evaluated by making use of the $d_i(\theta)$ as in (7). At a switching period, if the $d_i(\theta)$ is zero or unity, the switch has no switching losses in that period. Therefore, a function "f" of $d_i(\theta)$ is defined as in (8) and is introduced as a multiplier on the switching losses as in (7).

$$\mathbf{P_{I\text{-}C\text{-}T_s}}[k] = d_i(k\omega T_s) \cdot \left[i_C(k\omega T_s) \cdot V_{CE-ON} + i_C^2(k\omega T_s) \cdot R_{CE-ON}(T_j) \right] \quad (6)$$

$$\mathbf{P_{I\text{-}S\text{-}T_s}}[k] = f(d_i(k\omega T_s)) \cdot \frac{1}{T_s} \left[\begin{array}{l} E_{ON}(I_C(k\omega T_s), V_{block}(k\omega T_s)) \\ + E_{OFF}(I_C(k\omega T_s), V_{block}(k\omega T_s)) \end{array} \right] \quad (7)$$

$$f(d_i(k\omega T_s)) = \text{sgn}(d_i(k\omega T_s)) \cdot \text{sgn}(1 - d_i(k\omega T_s)) \quad (8)$$

In (9)-(15) the CDCF for representative switches of two, three, and five-level 3P-PVIs are given for SPWM. These CDCFs are plotted in Fig. 8 for M=0.9 and σ=0.5 (for the 5L topology) for a half grid period (as the topologies are symmetric half grid period evaluation of losses are sufficient for SE calculation).

$$d_{i-2L-3P_{Sa+}}(\theta) = \frac{1}{2}(1 + M\sin\theta), \quad 0 < \theta < \pi \quad (9)$$

$$d_{i-2L-3P_{Sa-/d}}(\theta) = \frac{1}{2}(1 - M\sin\theta), \quad 0 < \theta < \pi \quad (10)$$

$$d_{i-T-type_{Sa+}}(\theta) = M\sin\theta, \quad 0 < \theta < \pi \quad (11)$$

$$d_{i-T-type_{Sa02}}(\theta) = 1 - M\sin\theta, \quad 0 < \theta < \pi \quad (12)$$

$$d_{i-5L_{Sa++}}(\theta) = \begin{cases} \dfrac{M\sin\theta - \sigma}{1 - \sigma}, & \sigma < M\sin\theta, 0 < \theta < \pi \\ 0, & M\sin\theta < \sigma, 0 < \theta < \pi \end{cases} \quad (13)$$

$$d_{i-5L_{Sa+}}(\theta) = \begin{cases} \dfrac{1 - M\sin\theta}{1 - \sigma}, & \sigma < M\sin\theta, 0 < \theta < \pi \\ \dfrac{M}{\sigma}\sin\theta, & M\sin\theta < \sigma, 0 < \theta < \pi \end{cases} \quad (14)$$

$$d_{i-5L_{Sa02}}(\theta) = \begin{cases} 0, & \sigma < M\sin\theta, 0 < \theta < \pi \\ 1 - \dfrac{M}{\sigma}\sin\theta, & M\sin\theta < \sigma, 0 < \theta < \pi \end{cases} \quad (15)$$

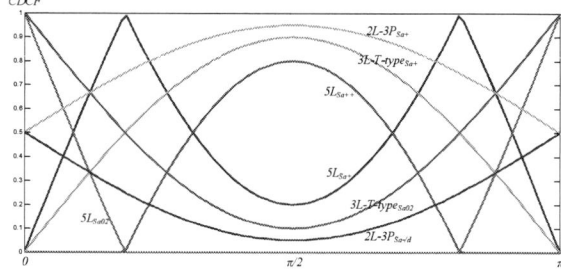

Fig. 8 CDCF of representative switches of representative topologies with different output voltage levels. (M=0.9, σ=0.5 for the 5L topology).

For comparison of topologies in terms of their SE, three sets of semiconductors with appropriate ratings are selected and their datasheets are utilized in 3P-PVI topologies for power ratings of 50 kW, 100 kW, and 200 kW (Table I). Switching frequency (f_s) is adjusted depending on the power rating (as in Table II) and kept same for all topologies at same power at a dc-bus voltage of 750V and M=0.9. As an example, in Fig. 9, the calculated switching and conduction losses of the switches S1, S4, and D3 of the T-type topology are plotted with respect to loading (for 100 kW design, pf=1, M=0.9). Additionally, approximated filter inductor losses (which will be discussed in next section) are plotted in this figure.

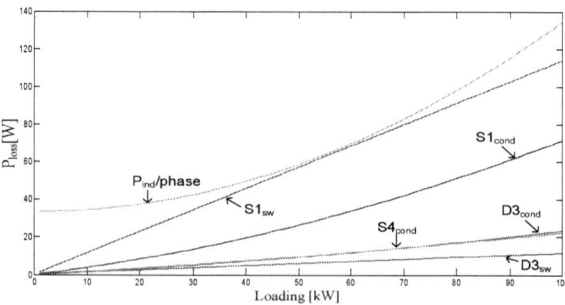

Fig. 9 The T-type topology S1, S4, and D3 loss distribution with respect to loading (100 kW design).

Table I. Semiconductors utilized for SE and ECE calculation

	50 kW	100 kW	200 kW
600 V IGBT	SEMiX202GB066HDs (274 A)	F3I300R12PT4_B26, T2 (300 A)	2MBI600U2E-060 (600A)
1200 V IGBT	SEMiX151GB12E4s (232 A)	F3I300R12PT4_B26, T1 (300 A)	2MBI600VN-120-50 (600 A)
600 V DIODE	SEMiX202GB066HDs (291 A)	F3I300R12PT4_B26, D2 (300 A)	2MBI600U2E-060 (600 A)
1200 V DIODE	SEMiX151GB12E4s (189 A)	F3I300R12PT4_B26, D1 (300 A)	2MBI600VN-120-50 (600 A)

The SE curves of the 3P-PVI topologies are provided at unity power factor with respect to loading in Fig. 10 (a)-(c). Among 3P-PVI topologies, the 5L topology exhibits the highest SE characteristics due to its lowest switching losses of semiconductors switching at less voltage as compared to other topologies. Three-level topologies and the 2L-SS topology follow the 5L topology due to their lower switching losses. Three-level topologies are advantageous due to their semiconductors switching at low voltage, and the 2L-SS topology is advantageous due its soft switching characteristics. Among three-level topologies, the T-type topology exhibits better SE curve than the NPC topology in 100 kW and 200 kW as it has only one semiconductor at active states. Although the 2L-3P topology has very little number of semiconductors on the current path, it switches at full dc-bus voltage therefore it has significant switching losses. Although SE is a good indicator of overall ECE and cost of 3P-PVIs, the inclusion of filter inductor losses and cost are necessary to have an ultimate decision.

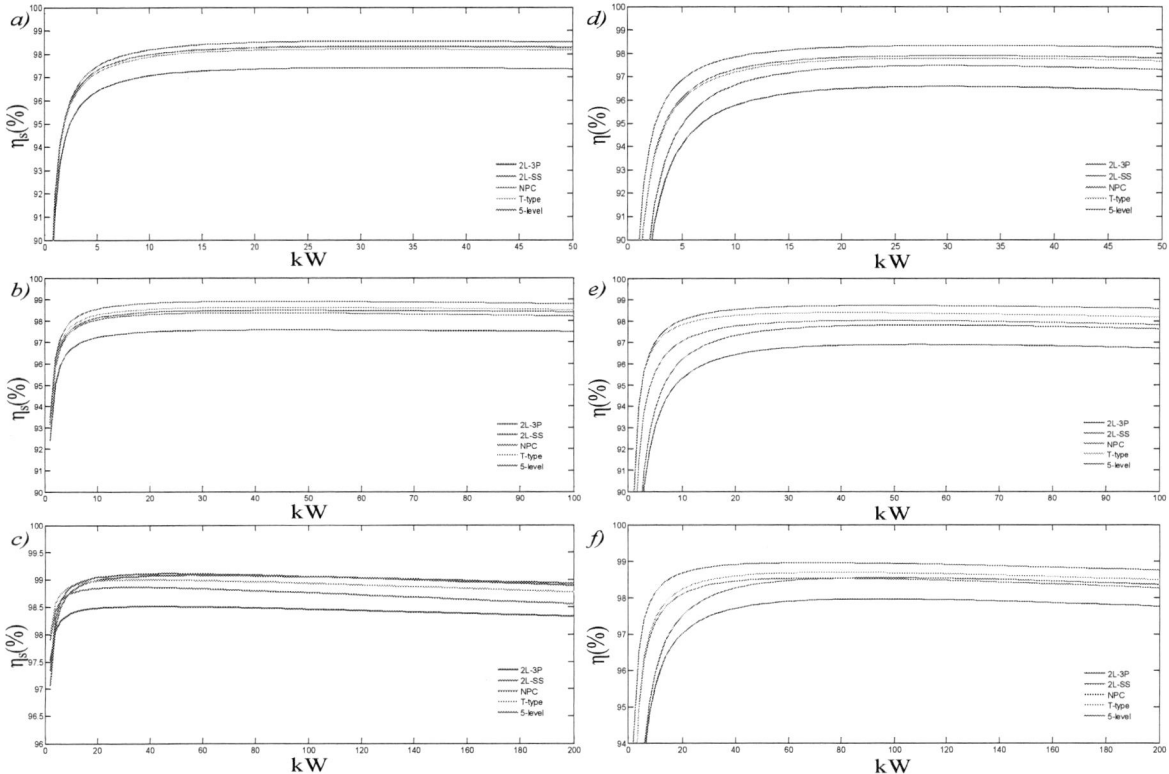

Fig. 10 SEs (a)-(c) and ECEs (d)-(f) of 3P-PVI topologies for 50 kW, 100 kW, and 200 kW.

E. Filter Inductor Size and Losses (FIL)

In addition to semiconductor losses, another major component of losses in 3P-PVIs is the FIL. In three-phase systems LCL filters are utilized so that the PWM current ripple can be suppressed with less inductance. With such a filter, FIL are dominant in the inverter side inductor. In an inverter filter inductor, the losses have two components: core and copper losses. Copper losses are dependent on the inductor current whereas core losses are tightly bonded to filter inductor current ripple (due to dependence of peak to peak magnetic field density, ΔB, thus minor loops, on the ripple current) and f_s as hysteresis losses and eddy current losses. These losses are closely associated with the filter inductor current ripple. Thus, in this work, the filter inductor current ripple characteristics of 3P-PVIs are investigated through ripple factors which are defined as in (16)-(18) for two, three, and five-level 3P-PVIs respectively (for dc-bus midpoint grounded case). These topology specific factors are multiplied by $V_{DC}T_s/2L_F$ to obtain peak to peak current ripple (L_F is the filter inductance). The spatial distributions of these ripple factors are shown in Fig. 11 for two, three and five-level 3P-PVIs. In the figure, it is noticeable that the peak values of the ripple factors are nearly halved at each level increase, so that the peak value of the ripple is nearly halved for the same V_{DC}, T_s, and L_F. Therefore, the required L_F can be halved for the same peak ripple current with higher level inverter. Moreover, as FIL are also a function of the RMS of ripple current (copper losses), the RMS value of the ripple factors are given in Fig. 9 with respect to M. Accordingly, the peak and the RMS values for M=0.9 case are tabulated in Table III.

$$\Gamma_{2L}(\theta) = \frac{1 - M^2 \sin^2(\theta)}{2} \qquad (16)$$

$$\Gamma_{3L}(\theta) = M\sin(\theta) - M^2\sin^2(\theta) \tag{17}$$

$$\Gamma_{5L}(\theta) = \begin{cases} M\sin(\theta) - M^2\sin^2(\theta) & ,M\sin(\theta) < \sigma \\ \dfrac{(1 - M\sin(\theta))\cdot(M\sin(\theta) - \sigma)}{1 - \sigma} & ,\sigma < M\sin(\theta) \end{cases} \tag{18}$$

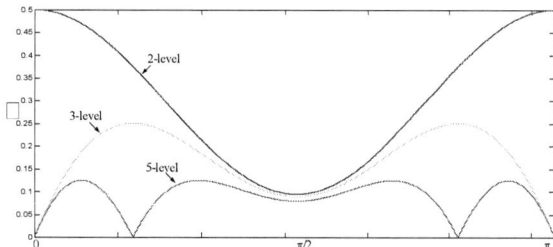

Fig. 11 Ripple factor values for two, three, and five-level inverters for half grid period for M=0.9 (σ=0.5 for the 5L topology).

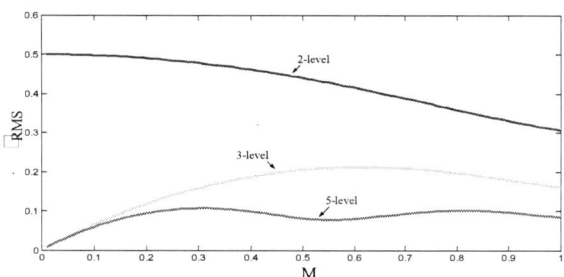

Fig. 12 RMS values of ripple factors (σ=0.5 for the 5L topology).

In order to have a better estimate of ECE, filter inductor losses are approximated as a two coefficient polynomial of line current RMS value (19) and the contribution is merged with the contribution of semiconductor losses where the overall scenario is disclosed for 50 kW, 100 kW, and 200 kW as in Fig. 10 (d)-(f). These coefficients are varied in a way that the filter inductor losses are halved at each level increase as the peak ripple factor is halved (meaning the inductance requirement and associated losses are halved) in the same power level. Full-load filter losses are assumed to be 1%, 0.8%, and 0.6% of full-load power in two-level inverters in 50 kW, 100 kW, and 200 kW systems respectively. In no-load, these losses are assumed to be one quarter of full-load filter losses. Associated polynomial values and switching frequencies are listed in Table II.

$$P_{ind} = k_1 + k_2 I^2_{s-1} \tag{19}$$

Table II. Polynomial coefficients and switching frequencies of topologies with respect to power and number of levels

P	**50 kW**	**100 kW**	**200 kW**
f_s	12.5 kHz	10 kHz	7.5 kHz
2L k_1/k_2	125/0.0328	200/0.0131	300/0.005
3L k_1/k_2	62.5/0.0164	100/0.0066	150/0.0025
5L k_1/k_2	31.25/0.0082	50/0.0033	75/0.00125

F. DC Bus Capacitor Size and Lifetime

The dc-bus capacitors in 3P-PVIs should be kept cool for long lifetime and sizing of the dc-link capacitors should be made accordingly. The can temperature (T_c) is roughly calculated as in (20) where T_a is the ambient temperature, R_{ESR} is the equivalent series resistance of the dc-link capacitor, and $R_{th,c-a}$ is the can to ambient thermal impedance. I_{DC-r} is the ripple current flowing through the dc-link capacitor (shown in Fig. 1) which can be calculated as in (21).

$$T_c = T_a + I^2_{DC-r} \cdot R_{ESR} \cdot R_{th,c-a} \tag{20}$$

$$I_{DC-r} = \sqrt{I^2_{DC} - I^2_{DC-avg}} \tag{21}$$

A figure of merit (called K_{DC}) for evaluating the dc-bus characteristics of 3P-PVIs is the ratio of squares of RMS value of dc-bus current ripple characteristics to the output phase current RMS value (22). Regarding this definition, smaller K_{DC} implies smaller dc-bus capacitor requirement at a given dc-bus voltage, less losses, low temperature and thus longer capacitor life.

$$K_{DC} = I^2_{DC-r} / I^2_{s-1} \tag{22}$$

K_{DC} characteristics of 3P-PVIs investigated in this study are shaped by the output voltage level number, i.e. topologies having the same number of output voltage level have same K_{DC} value for the same operating conditions (such as the modulation index M or the phase angle φ between voltage and current). In the investigation, the duty cycles of the vectors are determined such that the midpoints of the topologies are connected to the neutral wire of the three-phase utility grid (transformerless connection). Hence, no zero-sequence voltage is added to reference phase voltages and same carrier triangle waveform is assumed/used for each phase (SPWM).

In [12] and [13], the values for the two and three-level voltage-source inverters are found as in (23) for the case the star point of the converters is isolated. However, it should be noted that the voltage vector duty cycles are same for the same M and φ, thus the same K_{DC} is obtained and verified by simulations. In Fig. 13, the value of K_{DC} is plotted for two and three-level inverters with respect to M and $cos\varphi$. As seen from the figure, M=0.61 and $cos\varphi$=1 (corresponding to maximum K_{DC} value) is a good design point for the dc-bus capacitors for these 3P-PVIs.

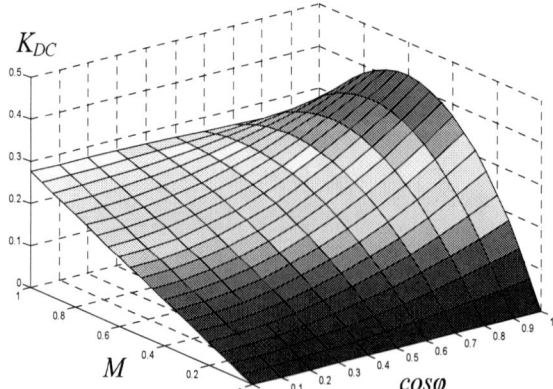

Fig. 13 K_{DC} (M, $cos\varphi$) of two and three-level inverters (SPWM).

$$K_{DC-2/3L} = 2M\left\{\frac{\sqrt{3}}{4\pi} + \left(\frac{\sqrt{3}}{\pi} - \frac{9}{16}M\right)\cos^2(\varphi)\right\} \quad (23)$$

In the 5L 3P-PVI topology, there are two dc-buses, thus the K_{DC} of the topology is characterized as the sum of the upper and lower dc buses as in (24) to reflect the overall inverter side RMS current ripple on the evaluation of the topology. This relation is shown in Fig. 14. However, it should be noted that the boost stage introduces partitioned current so that the higher voltage dc-bus K_{DC} value ($K_{DC-5L-H}$) is obtained as in (25) for an uncorrelated boost stage output current case. Regarding the case, it is natural to expect to have some limitation on the boost ratio (σ) at least due to the capacitor sizing issues. For other topologies, there may be also boost structures affecting the capacitor sizing but these issues are far from the focus of the paper. In Fig. 15, K_{DC} values for two, three, and five-level topologies are illustrated with respect to M ($\sigma=0.5$ for the 5L topology). In regard to the figure, the capacitor size of the 5L topology is nearly the twice of that the ones of two and three level topologies (due to peak K_{DC} values). In Table III, the peak K_{DC} values (K_{DC-p}) of topologies, which are insightful in determining the dc-bus capacitor size of topologies are listed.

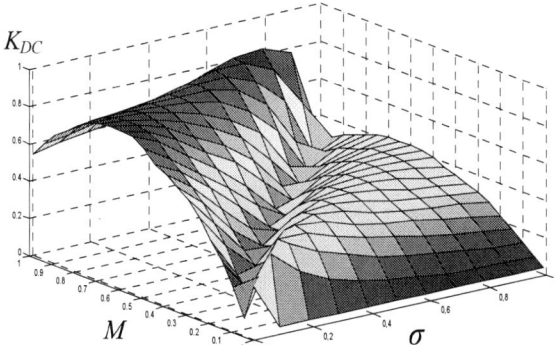

Fig. 14 K_{DC} of the 5L topology with respect to M and σ ($cos\varphi=1$).

$$K_{DC-5L} = K_{DC-5L-H} + K_{DC-5L-L} \quad (24)$$

$$K_{DC-5L-H-U}(M,\sigma) = \frac{I_{DC-H-avg}^2}{I_{s-1}^2\sigma} + K_{DC-5L-H}(M,\sigma) \quad (25)$$

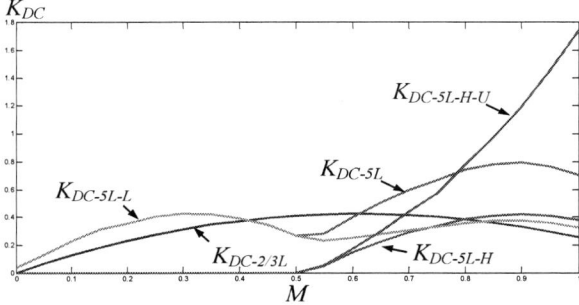

Fig. 15 Ripple factor values for two, three, and five-level inverters for half grid period for M=0.9 ($cos\varphi=1$, $\sigma=0.5$ for the 5L topology).

V. COMPARISON AND EVALUATION

In Table III, numerical values of investigated performance indicators of 3P-PVIs are provided. In terms of number of total and active switches, 2L-3P yields the minimum complexity and this topology yields the minimum number of switches on the current path meaning minimal conduction losses. However, due to full dc-bus switching voltage (V_{DC}), the SE characteristics of this topology deteriorate as its euro efficiency weighted SE ($\eta_{S,EU}$) is the minimum among other topologies. Unlike 2L-3P, in the 2L-SS topology the SE is as high as three-level 3P-PVIs as the switching losses are reduced due to soft turn-on of IGBTs. However, as the switch kVA is doubled and the C_{DU} is halved in the topology, the cost of semiconductors per watt is high. Moreover, FILs in both 2L-3P and 2L-SS are high (thus ECEs of topologies rapidly degrade). Also the filter size is large and the cost is high. Therefore, these topologies are not favorable as 3P-PVIs.

In the 5L topology, switching losses are minimal without any considerable difference in the number of switches on the current path, thus this topology exhibits highest SE among 3P-PVIs. Moreover, FIL in this topology and the required inductance are also minimal, therefore the topology exhibits highest ECE among the 3P-PVIs investigated. However, C_{DU} of the topology is less than a half of C_{DU} of 2L-3P, NPC, and T-type topologies. Therefore, semiconductor cost of the topology is the highest per watt. Furthermore, dc-bus capacitor requirement of the topology is nearly the twice of other topologies, therefore the overall cost of the topology is further increased. Reactive power provision incapability and high complexity of the power circuit are the additional handicaps of the topology as a 3P-PVI.

In three-level topologies, the switching losses are lower as compared to two-level topologies due to halved switching voltage of switches ($V_{DC}/2$) without any increase in the device utilization factor, so this attitude makes three-level topologies favorable in terms of their semiconductor cost. Moreover, dc-bus capacitor requirement of these topologies are low. Due to these cost advantages over the 5L topology, and the ECE advantage over two-level topologies, three-level topologies outperform other 3P-PVI topologies. As shown in Table III, the T-type and the NPC topologies exhibit little difference in terms of SE (as the output voltage levels of these topologies are the same, they have the same FIL). In 50 kW design, the NPC topology has slightly higher SE than the T-type topology, however as the power rating is increased, the T-type topology has higher SE. This is due to the fact that the conduction losses become dominant over switching losses as the power (thus semiconductor current) is increased and due to that, the T-type topology has less number of semiconductors on the current path.

Based on the SE and ECE evaluations of the 50-200kW power ratings, it is observed that the increasing power rating yields higher ECE in general. The ordering of the SE and ECE values for the different topologies remains the same, and the differences remain similar for all the power ratings considered. The SE and ECE based performance indices can be complemented with energy sale price based costs to obtain the conclusive return on investment (ROI) figures to evaluate the topologies and obtain the most feasible solution.

VI. CONCLUSION

In this paper the 3P-PVIs are thoroughly investigated in terms of their SE, overall ECE, cost, and performance characteristics. SE calculations of topologies are performed based on the datasheets for 50 kW, 100 kW, and 200 kW rating PV systems. In the calculation semiconductor datasheets are utilized with a simple analytical CDCF of each switch. With the inclusion of filter inductor losses, merging with semiconductor efficiency, enhanced accuracy of PV system efficiency is obtained. Based on the comparisons for 50-200kW power ratings, the most feasible topologies have been found as the three-level NPC and T-type topologies. The choice between these two depends on the cost and performance characteristics of NPC and T-type modules available in the market. The studies of 50-200kW rating converters can be directly applicable to MW scale PV systems as such systems are generally formed by integration of 100-200kW modules directly paralleling or via a multi-winding transformer.

Table III. Evaluation of 3P-PVIs with respect to performance criteria (M=1 σ=0.5)

TOPOLOGY	2L	NPC	T-Type	2L-SS	5L
Total Switches (PF≠1)	12	30	24	24	-
Total Switches (PF=1)	6	18	18	24	28
Active Switches	6	12	12	12	20
Switches on the Current Path	3	6	4.1	3	4.7
Device Utilization Factor (normalized)	1	1	1	0.5	0.46
Phase to Neutral Output Voltage Level	2	3	3	2	5
Reactive Power Capability	+	+	+	+	-
Γ_p	0.5	0.25	0.25	0.5	0.125
Γ_{RMS-p}	0.5	0.212	0.212	0.5	0.106
K_{DC-p} ($K_{DC-5L-H-p}/K_{DC-5L-L-p}$ for 5L)	0.42	0.42	0.42	0.42	0.44/0.42
$\eta_{S,EU}$ (50kW 3P-PVI) (M=0.9)	97.19	98.11	97.99	98.09	98.32
$\eta_{S,EU}$ (100kW 3P-PVI) (M=0.9)	97.48	98.27	98.53	98.4	98.81
$\eta_{S,EU}$ (200kW 3P-PVI) (M=0.9)	98.44	98.74	98.92	99.02	99.03
η_{EU} (50kW 3P-PVI) (M=0.9)	96.07	97.53	97.42	96.95	98.03
η_{EU} (100kW 3P-PVI) (M=0.9)	96.58	97.80	98.06	97.47	98.57
η_{EU} (200kW 3P-PVI) (M=0.9)	97.75	98.39	98.56	98.32	98.85

REFERENCES

[1] T. Bülo, B. Sahan, C. Nöding, and P. Zacharias "Comparison of three-phase inverter topologies for grid-connected photovoltaic systems," 22nd European Photovoltaic Solar Energy Conference and Exhibition, 3 - 7 September 2007, Milan, Italy.

[2] T. Zhao, V. Bhavaraju, and P. Nirantare "Commercial Scale Transformerless Solar Inverter Technology Review," IEEE-ECCE 2011 Conf., pp 1636-1642, September 2012, Raleigh, North Carolina, USA.

[3] M. Mohr, and F.W. Fuchs, "Comparison of three phase current source inverters and voltage source inverters linked with DC to DC boost converters for fuel cell generation systems," European Conference on Power Electronics and Applications EPE, pp. 1-10, Sept. 2005, Dresden, Germany.

[4] M. Kazanbaş, C. Nöding, H. Can, T. Kleeb, and P. Zacharias, "A new single phase transformerless Photovoltaic inverter topology with coupled inductor," PEMD, University of Bristol, 2012.

[5] B. L. Hesterman, M Ilic, A. B. Malinin, K. N. C. Siddabattula, "Soft switching interleaved power converter," Patent US 6 979 980 B1, Aug. 24, 2004.

[6] A. Nabae, I. Takahashi, H. Akagi, "A New Neutral-Point-Clamped PWM Inverter," IEEE Transactions on Industry Applications, vol.IA-17, no.5, pp.518,523, Sept. 1981.

[7] J. Hantschel "Wechselrichterschaltung für erweiterten Eingangsspannungsbereich," Patent Application DE 10 2006 010 694 A1, March 8, 2006.

[8] Z. Özkan and A. M. Hava, "Energy Conversion Efficiency of Single-Phase Transformerless PV Inverters," ELECO 2013, 8th International Conference on Electrical and Electronics Engineering, 28-30 November 2013, Bursa, Turkey.

[9] Z. Özkan, "Leakage Current and Energy Efficiency Analyses of Single Phase Grid Connected Multi-kVA Transformerless Photovoltaic Inverters," M.Sc. Thesis, Middle East Technical University, Dept. of Electrical and Electronics Eng., Ankara, February 2012.

[10] D. Graovac and M. Pürschel, "IGBT Power Losses Calculation Using the Data-Sheet Parameters," Infineon Application Note, V 1.1, January 2009.

[11] B. Backlund, R. Schnell, U. Schlapbach, R. Fischer, E. Tsyplakov, "Applying IGBTs," ABB Application Note, May 2012.

[12] J. W. Kolar, S. D. Round, "Analytical calculation of the RMS current stress on the DC-link capacitor of voltage-PWM converter systems," Electric Power Applications, Proc. IEE-, vol.153, no.4, pp.535-543, July 2006.

[13] S. Floten and T. S. Haug, "Modulation Methods for Neutral-Point-Clamped Three-Level Inverter," M.Sc. thesis, Norwegian University of Science and Technology, Trondheim, June 2010.

The 2014 International Power Electronics Conference

A Three-Port Topology Comparison for a Low Power Stand-Alone Photovoltaic System

Maria C. Mira, Arnold Knott, Michael A. E. Andersen

Dept. Electrical Engineering
Technical University of Denmark
Oersteds Plads, 349. Kongens Lyngby, Denmark

mmial@elektro.dtu.dk   akn@elektro.dtu.dk   ma@elektro.dtu.dk

Abstract—**Three-port converter (TPC) topologies for renewable energy systems aim to provide higher efficiency and power density than conventional cascaded structures. This work proposes an analytical comparison of different TPC topologies for a photovoltaic LED lamp stand-alone system. A comparison using component stress factor (CSF) is performed, which gives a quantitative measure of the performance of the converter. The candidate topologies are compared to each other according to a defined LED lighting strategy and a solar irradiation profile.**

Keywords— Photovoltaic, three port converter, stand-alone, component stress factor.

I. INTRODUCTION

Due to fossil fuel reserves depletion together with climate concerns, in the last decades renewable energies have become an important part of energy production. Switched-mode power supplies (SMPS) play an important role in the integration of renewable energies due to the requirement of high efficiency conversion.

Solar energy is one of the fastest growing renewable energy sources mainly because sunlight is the most abundant source of energy and is unlimited, clean and free. The major advantage of renewable energy sources, like solar or wind, is the transformation of energy with zero carbon dioxide (CO_2) emissions. However, the main drawback is that the energy source is intermittent in nature since it strongly depends on the weather conditions. In order to overcome this limitation, multiple energy sources are combined to provide a constant power source.

Multi-input converters (MIC) topologies address the issue of interconnecting several energy sources with a single power converter [1]. The common characteristic of MIC is the shared output stage, so the number of components is reduced and the power density is increased. Some MIC make use of magnetic coupling (multiple transformer windings) and present solutions based on flux additivity [2], [3], phase-shifted operation [4], distributed transformers [5] or a four quadrant magnetic structure [6] which allows the two input power stages to deliver power simultaneously. Feasible and unfeasible non-isolated MIC topologies are presented in [7] and a systematic approach to synthesize MIC based on the concept of pulsating voltage source cell (PVSC) and pulsating current source cell (PCSC) is proposed in [8] and [9].

When a storage element is included in a MIC, a stand-alone system is formed. A stand-alone structure is very interesting for powering up systems at remote locations, where cabling is challenging and expensive. It is also of interest in urban areas where not only the cost of cabling but also digging and construction is extremely costly.

The aim of this work is to analyze TPC topologies suitable for low power stand-alone photovoltaic systems and to compare them to conventional cascaded converters. In this type of application, a bidirectional port with a storage element is necessary. The battery will store the excess energy in light or no load conditions and provide energy in case of no input power.

The conventional approach to implement multi-input power systems is to interconnect the elements using different cascaded power converters [10] [11], or parallel [12] as shown in Fig. 1. However, the main disadvantage is the low efficiency of the system due to the power processed in multiple conversion stages. The proposed solution is to implement a multiport converter, which can interface with renewable energies, storage elements and loads simultaneously, as the TPC shown in Fig. 2. Authors claim that the advantage of this method is higher efficiency due to reduced conversion stages and compact packaging [13], [14], [15], [16].

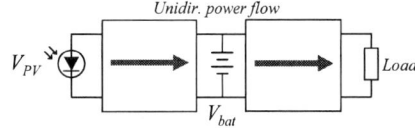

a) Cascade connection

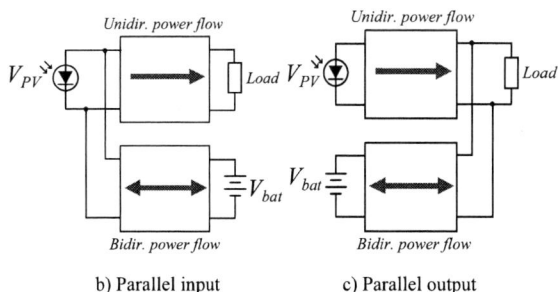

b) Parallel input   c) Parallel output

Fig. 1 Conventional integration method of a multi-input port stand-alone photovoltaic system

978-1-4799-2706-7/14 $31.00 © 2014 IEEE   506

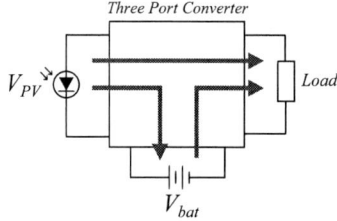

Fig. 2 Three-port stand-alone photovoltaic system

Three different power flows can take place in the system depending on the solar energy and load consumption. The converter will operate in dual input (DI) mode when the load power is higher than the available PV power, so the battery has to deliver the extra required energy. The converter will operate in dual output (DO) mode when the PV power is higher than the load power and the battery has to store the excess energy. During night time, the system will operate in single input single output (SISO) mode delivering power from the battery to the LEDs.

II. SYSTEM SPECIFICATIONS

The stand-alone system under analysis is a photovoltaic LED lamp system with a lithium battery for energy storage. The photovoltaic panel is composed of two paralleled connected monocristaline panels with the specifications shown in Table I. Fig. 3 shows the I-V curve of the PV panel. The maximum power point (MPP) is located at the knee of the curve given by the points V_{mp} and I_{mp}. Equation (1) describes the output current of a photovoltaic cell. The integrated battery is a LiFePO$_4$ (lithium iron phosphate) battery GWL Power WINA with a nominal voltage of $3.2\,V$ and a capacity of $15\,Ah$. The LED lamp is composed of eight series-connected Cree XLamp XP-E with a maximum current of $1\,A$ and a forward voltage drop of $[2.6 - 3.3]\,V$. Fig 4 shows the LED lamp I-V curve extracted from the component datasheet.

$$I = I_{ph} - I_o \left(e^{\frac{V+I \cdot R_s}{V_t}} - 1 \right) \qquad (1)$$

I_{ph}: photogenerated current (A)

I_o: diode's dark saturation current (A)

R_s: series resistance (Ω)

V_t: cell's equivalent thermal voltage. $V_t = AkT/q$ (V)

A: diode ideality factor

k: Boltzmann's constant ($1.380 \cdot 10^{-23}$JK^{-1})

q: charge of an electron ($1.602 \cdot 10^{-19}C$)

T: p-n junction temperature (K)

TABLE I
PV PANEL SPECIFICATIONS @ STC

P_{mp}	8.32 W
V_{mp}	5.4 V
I_{mp}	1.54 A
V_{oc}	6.48 V
I_{sc}	1.65 A

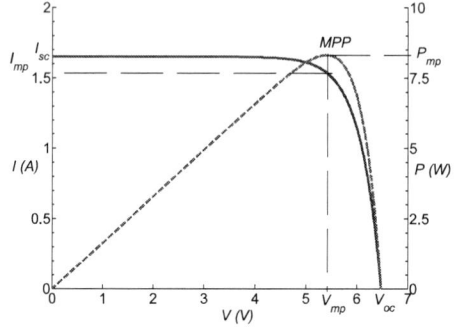

Fig. 3 I-V and P-V characteristic curve of the photovoltaic panel, continuous blue and dotted green line, respectively

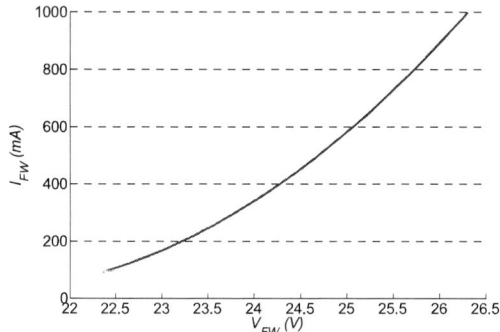

Fig. 4 LED lamp I-V curve (8 series connected LED Cree XLamp XP-E)

III. TOPOLOGY SELECTION

TPC topologies are based on reduced energy processing and shared components to achieve a high efficiency and power density. However, it is often required to add extra switches to provide controllability and/or diodes to configure the power flow path. Therefore, TPC topologies usually need a high number of semiconductors [13], [16], [17].

The conventional cascaded structure chosen as the reference design is a buck and a tapped boost converter connected in series as shown in Fig. 5. In order to drive the LED port, a high step-up ratio is needed from both inputs – battery and PV source. The tapped boost converter provides the required step-up ratio with low amount of components. This topology is selected because it achieves high transformation ratio and makes good use of the coupled inductors, since both windings are active during the discharge. Fig. 6 shows the operating waveforms of the tapped boost converter. The dc voltage transfer functions are obtained from the inductor volt-second balance as shown in (2) for the buck converter and (3) and (4) for to the tapped boost topology.

$$V_{bat} = V_{PV} D_1 \qquad (2)$$

$$V_{bat} \cdot D_2 T + \left(\frac{V_{bat} - V_{LED}}{n + 1} \right) \cdot (1 - D_2) T = 0 \qquad (3)$$

$$V_{LED} = V_{bat} \left(1 + \frac{(n + 1) \cdot D_2}{(1 - D_2)} \right) \qquad (4)$$

The 2014 International Power Electronics Conference

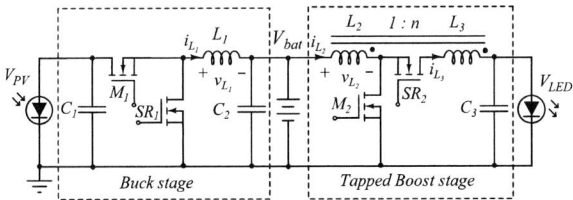

Fig. 5 Conventional cascaded solution. Buck and tapped boost converters series-connected.

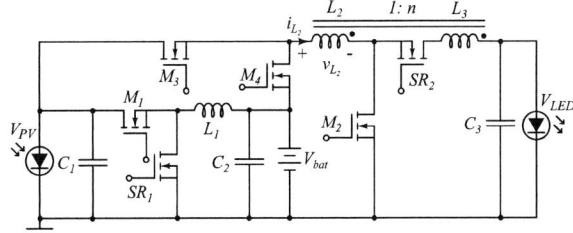

Fig. 7 Non-isolated TPC buck-tapped boost schematic

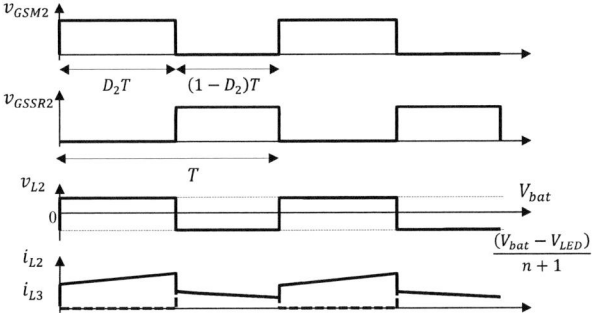

Fig. 6 Tapped boost converter operating waveforms. From top to bottom: Gate source voltage of switches M_2 and SR_2, inductor L_2 voltage and inductor L_2 and L_3 current, continuous and dotted line, respectively

The TPC structures considered have been selected according to the number of components. In order to avoid voltage drops in the power flow path, diodes have been replaced with MOSFETs. However, this solution makes it difficult to drive these components since they are usually not referenced to ground. Moreover, in some topologies these MOSFETs need to be active during the whole period. This complicates the drive circuitry even more since gate drive transformers cannot be used.

Two TPC topologies are analyzed. The first is a non-isolated TPC (NI-TPC). NI-TPC topologies can be derived from dual input converters (DIC) or dual output converters (DOC) as proposed in [16]. The key idea of deriving TPC topologies based on DIC or DOC is to combine power flows with similar characteristics. In a DIC, both the source and the battery can deliver power to the load. For the given specifications, both power flows need to boost the voltage and therefore, can be combined. Hence, a DIC is formed by combining two power flows with similar characteristics. An additional power flow to interconnect the main source and the battery is needed. A buck converter is added to cover the third power flow and therefore a TPC is derived from a DIC topology [16], [14].

The NI-TPC synthesized by combining DI and SISO converters is shown in Fig. 7. This structure is similar to the cascaded conventional buck and tapped boost solution found in SISO mode. In DO mode, it performs like two separate converters. Switch M_3 is always active and switch M_4 is always off. Therefore, there are two direct paths: from PV-LED and PV-battery with a tapped boost and a buck converter respectively.

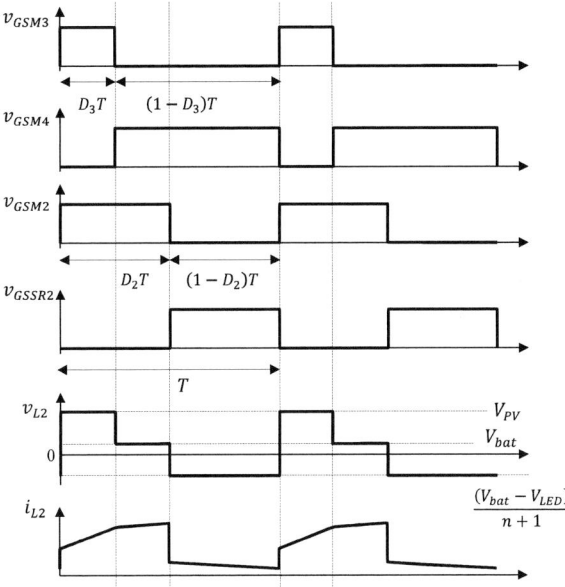

Fig. 8 Non-isolated TPC buck and tapped boost operating waveforms in DI mode. From top to bottom: Gate source voltage of switches M_3, M_4, and M_2, SR_2, inductor L_2 voltage and inductor L_2 current

In DI mode, both the PV panel and the battery are providing energy to the load. The NI-TPC behaves as a single tapped boost converter. The inductor charge time is adjusted with switches M_3 and M_4, which control the power flow from each of the power sources. The NI-TPC operating waveforms in DI mode are shown in Fig. 8. The dc voltage transfer function in DI mode can be calculated by applying the volt-second balance on the inductor L_2 as shown in (5) and (6).

$$V_{PV}D_3 + V_{bat}(D_2 - D_3) + \frac{V_{bat} - V_{LED}}{n + 1}(1 - D_2) = 0 \qquad (5)$$

$$V_{LED} = \frac{(n + 1)V_{PV}D_3 + V_{bat}\big(n(D_2 - D_3) + (1 - D_3)\big)}{(1 - D_2)} \qquad (6)$$

The second topology considered is a half-bridge three-port converter (HB-TPC) [15] [18] as shown in Fig. 9. The full-bridge version (FB-TPC) has also been proposed in literature [19]. However, in this application the HB-TPC is selected because it has a lower number of components. The half-bridge transformer allows for an isolated output port while providing step up/down transformation. In this topology, the magnetizing inductance of the transformer is used as the inductor of the buck converter linking the PV panel with the battery. The HB-TPC operating waveforms in DI mode are shown in Fig. 10.

978-1-4799-2706-7/14 $31.00 © 2014 IEEE   508

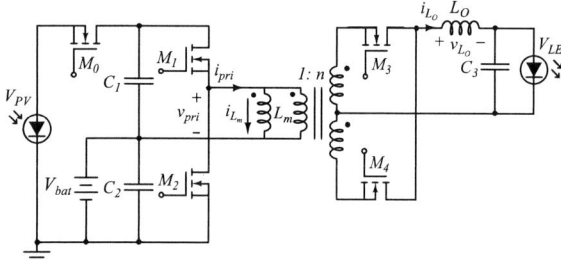

Fig. 9 Half-bridge three-port converter HB-TPC

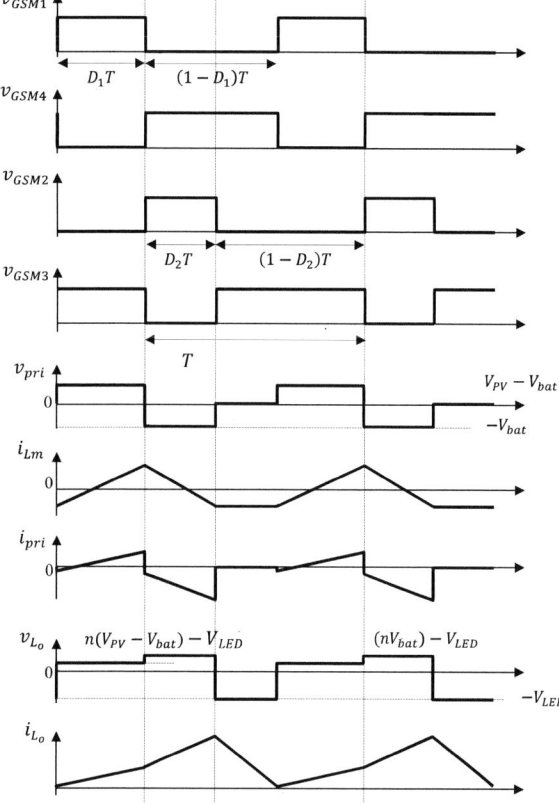

Fig. 10 HB-TPC converter operating waveforms in DI mode. From top to bottom: Gate source voltage of switches M_1, M_4, M_2 and M_3, transformer primary voltage v_{pri}, magnetizing inductor current i_{Lm}, transformer primary current i_{pri}, output inductor voltage v_{L_o} and output inductor current i_{L_o}

Switch M_0 is used to prevent reverse current from flowing to the PV panel and it is active when there is power flow from the PV panel. Switches M_1 and M_4 are driven in complementary mode, like M_2 and M_3. During M_1 conduction time, switch M_3 is also active and the energy from the PV panel is transferred to both the battery and the output port. When switch M_2 is active, rectifier M_3 is also active and the energy from the battery is moved to the output. When the rectifier switches M_3 and M_4 are on, the primary winding is clamped to zero and both the magnetizing and the output inductor are freewheeling.

Two PWM signals are necessary to control the power flow from both the inputs – PV panel and battery – to the

load. The dc voltage transfer functions to regulate the power flow form the PV to the battery and from the battery to the LED are obtained by applying the volt-second balance on the transformer magnetizing inductor and output inductor as presented in (7) and (8).

$$V_{bat} = \frac{V_{PV}D_1}{D_1 + D_2} \tag{7}$$

$$V_{LED} = n[D_1(V_{PV} - V_{bat}) + D_2 V_{bat}] = 2nD_2 V_{bat} \tag{8}$$

In DO mode, the converter operating waveforms are the same as in DI mode except that the average value of the primary and the magnetizing inductance current is positive.

In SISO battery-to-load mode, the converter operates in a forward flyback mode. Switches M_1 and M_2 are driven in complementary mode and M_3 and M_4 are the synchronous rectifiers. When M_2 and M_4 are active, the converter operates in a forward mode. When M_1 and M_3 are on, the converter operates in a flyback mode in order to demagnetize the primary winding. The dc voltage transfer function for this operating mode is shown in (9).

$$V_{LED} = V_{bat} \cdot 2n(1 - D) \tag{9}$$

The main disadvantage of this topology is that under low power conditions at the LED port, there will be a long period where the magnetizing current (or buck inductor current) will be circulating through the transformer secondary windings and synchronous rectifiers, reducing the efficiency of the converter significantly. Moreover, in order to operate in SISO mode from the PV panel to the battery, an extra switch in the return path of the transformer's secondary windings needs to be added so the primary side can be operated as a buck converter.

From the three topologies presented, the conventional cascade solution has the lowest number of semiconductors, followed by the HB-TPC. The topology with highest number of semiconductors is the NI-TPC.

IV. COMPONENT STRESS FACTOR ANALYSIS

To perform a topology comparison, a component stress factor (CSF) is performed [20]. This analysis provides an estimation of the converter stress distribution in the different components. The motivation for using CSF is that it gives a quantitative measure of converter performance. A similar approach called component load factor (CLF) was introduced in [21] and used by [22] to compare single-stage and two-stage PFC solutions. The difference between CLF and CSF is how the individual and total component factors are calculated. In the CSF analysis, in order to make a fair comparison, the same amount of silicon, magnetic winding area and capacitor volume for the components in all the candidate topologies is required. The distribution of resources between the different elements is done according to the specific weighting factor assigned to each component.

978-1-4799-2706-7/14 $31.00 © 2014 IEEE

Three different CSFs are calculated: semiconductor component stress factor (SCSF), winding component stress factor (WCSF) and capacitor component stress factor (CCSF). The stress in a component will depend on the maximum applied voltage and the maximum current.

CSF gives an estimation of the converter's efficiency because it can be correlated to the conduction losses. In a MOSFET the conduction losses are given by the rms current and the MOSFET's on resistance. SCSF is calculated according to the MOSFET's squared maximum voltage and the rms current as shown in (10). Moreover, it is scaled to the processed power, which normalizes the CSF and makes it a dimensionless quantity. WCSF is calculated according to (11). The maximum voltage is defined by (12), which is calculated as the average voltage applied to the inductor winding over a period. CCSF is calculated using the peak voltage of the capacitor and the rms current as shown in (13).

$$SCSF_i = \frac{\sum_j W_j}{W_i} \cdot \frac{V_{DS}^2 \cdot I_{rms}^2}{P^2} \qquad (10)$$

$$WCSF_i = \frac{\sum_j W_j}{W_i} \cdot \frac{V_{max}^2 \cdot I_{rms}^2}{P^2} \qquad (11)$$

$$V_{max} = \sum_i D_i \cdot |V_i| \qquad (12)$$

$$CCSF_i = \frac{\sum_j W_j}{W_i} \cdot \frac{V_{pk}^2 \cdot I_{rms}^2}{P^2} \qquad (13)$$

From (10) to (13) $\frac{\sum_j W_j}{W_i}$ is the weighting factor, which is applied to every component. $\sum_j W_j$ is the sum of the individual weights for the components of the same type in the circuit and W_i is the weight assigned to component i. The weighting factor represents the amount of resources given to a component. In the first CSF iteration, the same component weight is usually given for all the components. Once the components with higher CSF are identified, the weighting factor can be adjusted by giving higher amount of resources to that specific component in order to achieve lower CSF.

After calculating a CSF for each component, the sum of component factors of the same type is performed as shown in (14), (15) and (16).

$$SCSF = \sum_{Semiconductors} SCSF_i \qquad (14)$$

$$WCSF = \sum_{Windings} WCSF_i \qquad (15)$$

$$CCSF = \sum_{Capacitors} CCSF_i \qquad (16)$$

The CSF gives a numerical comparison between components within power converter topologies. This quantitative measure is related to the performance of the converter since component stress can be translated into cost, size and efficiency.

To simplify the calculations, it is assumed that the power losses are neglected ($P_{in} = P_{out}$). It is also assumed that inductors are infinitely large, meaning that the current ripple is null.

A MATLAB program is developed in order to calculate the CSF of the selected topologies for the different operating modes: DI, DO, SISO from the PV panel to the battery and SISO from the battery to the LED port. The calculated rms currents and voltages have been verified with closed loop control simulations in a SPICE-based software.

The CSF is calculated with the same component weight for all the components inside each of the topologies. The resources to be assigned are selected as one unit ($\sum_j W_j = 1$) and distributed equally between the components of the same type.

In order to observe how the stress on the semiconductors moves depending on the power flow, the PV panel is swept from minimum to maximum power. The output power in the LED port is kept at a constant $2.5\,W$. The selected number of turns of the coupled inductors for the cascade solution and the NI-TPC is $n = 5$, which is the value that gives the lower SCSF in SISO mode. The transformer turn ratio in the HB-TPC is chosen as $1\!:\!11$, which is the minimum value required to operate in SISO mode from the battery to the LED. Fig. 11 shows the semiconductor CSF of each of the candidate topologies operating in DI and DO mode. From 0 to 2.5 W the converters operate in DI mode. In this mode, the LED power is higher than the available PV power, so the battery is providing the extra energy. When the PV power is higher than the LED power, the converters operate in DO mode and the energy surplus is stored in the battery. It should be noticed that the CSF is normalized to the processed power. Hence, operating in DI mode the power processed is the output power. On the other hand, when operating in DO mode the total power processed corresponds to the converter input power. It can be observed that for the cascaded solution and the NI-TPC, the higher stress appears on the tapped boost components. Specifically the synchronous rectifier SR_2 has to withstand a large voltage stress. Comparing the cascaded solution to the NI-TPC, it can be seen that the NI-TPC SCSF is degraded due to the increased number of MOSFETs and the fact that the turn ratio of the coupled inductors must be selected according to the stress on both PV-to-LED and battery-to-LED power flows.

When the cascaded converters operate in DI mode, it can be seen that the CSF increases as the power increases since the power must be processed twice. However, even when the power is processed twice, the SCSF is smaller than in the NI-TPC solution. It is interesting to note that

The 2014 International Power Electronics Conference

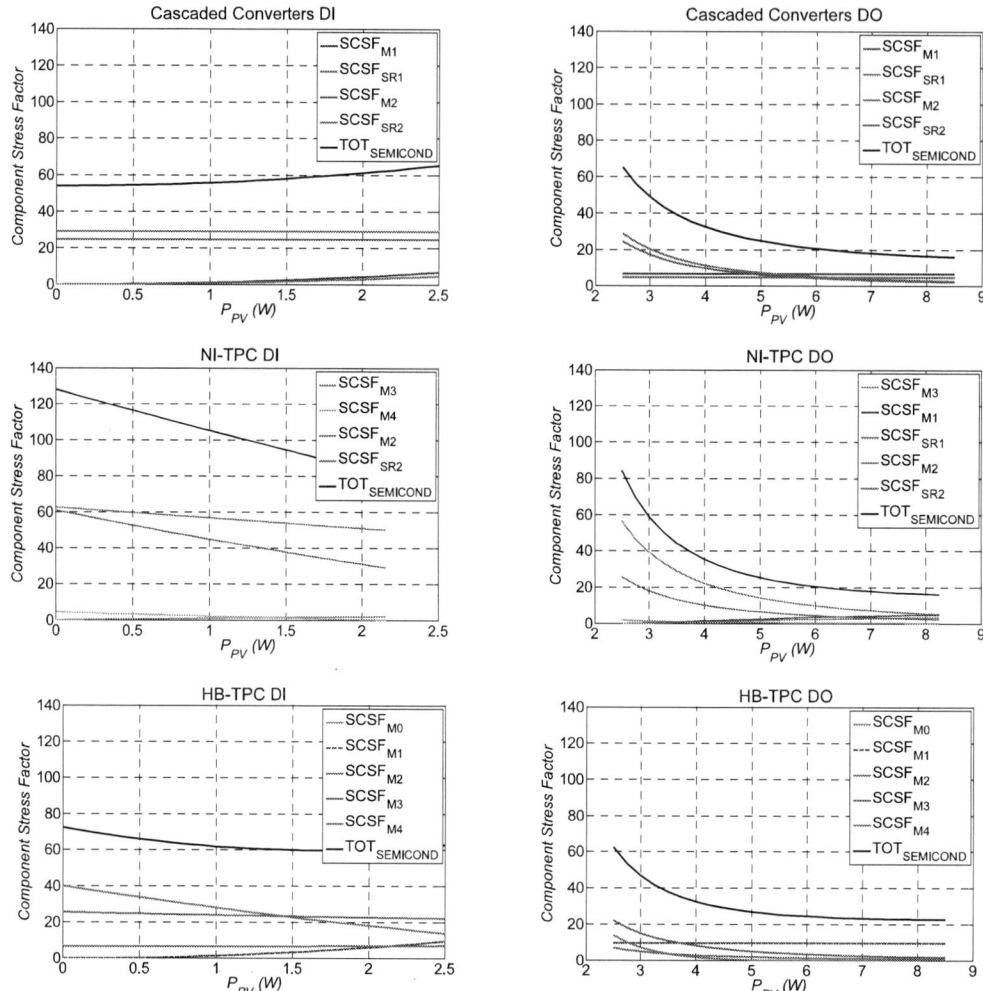

Fig. 11 Semiconductors CSF of the converters operating in DI mode (left column) and DO mode (right column)

for the NI-TPC operating in DI mode, the PV panel can never deliver the full power directly to the LED since the discharge of the inductor is always done through the battery power flow. In the HB-TPC structure, the higher stress in DI mode appears in the switch M_2 and synchronous rectifier M_3. This stress is reduced as the power flow from the panel to the LED increases and moves to the switch M_1 and synchronous rectifier M_4. For all the topologies operating in DO mode, the SCSF decreases as the power flow from the panel to the battery increases since all this power flow is processed by the buck structure.

Depending on the converter operating conditions, LED lighting strategy and the amount of available solar irradiance, the power flows in the system will vary together with the operating voltages, duty cycles and consequently, current and voltage stresses. In such a system, it is important to perform a comparison between the considered topologies for all the possible conditions. This can be done by analyzing the CSF based on real field solar irradiance measurements that will provide a pattern of the PV panel operating voltage

and available power. An irradiance measurement performed over a day [23] is taken as a reference. From the irradiance value, the maximum available power can be obtained using (1). Two different LED lighting patterns are defined in order to compare the CSF analysis for different applications. Fig. 12 shows the maximum available power from the PV panel together with two LED illumination patterns for a period of 24 hours.

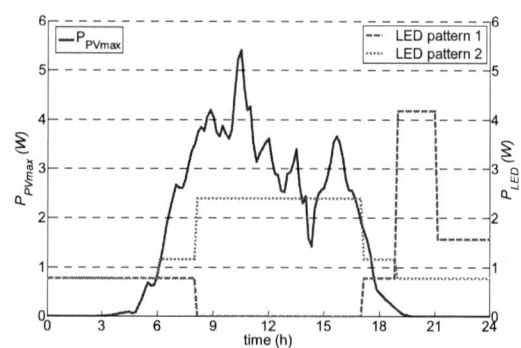

Fig. 12 Maximum PV panel power (blue), LED pattern 1 (green) and LED pattern 2 (red) for a 24 hours period

978-1-4799-2706-7/14 $31.00 © 2014 IEEE   511

The two illumination patterns are defined in order to compare the selected topologies under different operating conditions. The LED pattern 1 (green line) can be used for an application where no light is necessary during the day, like in a garden lighting system. The LED pattern 2 (red line) is defined for an application where the LED power is mostly used during the day, like in a signaling light. In the LED pattern 1, the converter works mostly in SISO PV-to-battery during the day and in SISO battery-to-LED during the night, except for short periods where it operates in DI and DO mode. In the LED pattern 2 the converters never operate in SISO PV-to-battery and work in DI and DO modes.

The CSF at each operating condition is weighted in respect to the maximum processed power and averaged over the whole operating period in order to extract a CSF value that directly relates to the amount of energy lost during the evaluated time. Table II and Table III show the calculated CSF for the cascaded solution and the analyzed TPC topologies for the two lighting pattern. The turn ratio of the tapped boost converter in both the cascaded and the NI-TPC solution is selected to obtain the lowest total CSF. The transformation ratio of the HB-TPC is the minimum value in order to operate in SISO battery-to-LED mode, which also gives the lowest SCSF.

According to the results, it can be seen that the cascaded solution provides the lowest CSF, even when taking into account that the PV power is processed twice. This is due to the fact that the TPC topologies need extra switching elements to control the power flows, which has a negative impact on the amount of available resources for the design of the converter. Further analysis can be carried out by adjusting the different component weights to find an optimal solution for each of the analyzed topologies.

When an irradiance measurement performed over a year (Fig. 13) is used, a different CSF weighting factor can be calculated depending on whether the converter has to be optimized for minimum annual energy loss or for maximum efficiency under low irradiance conditions.

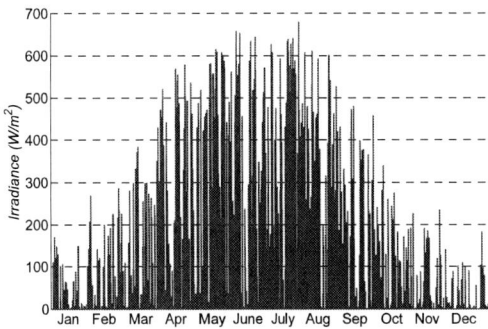

Fig. 13. Annual solar irradiation pattern

V. CONCLUSION

This paper presents a comparison of TPC topologies for a low power PV-battery-LED stand-alone system. The comparison is performed using CSF analysis based on a solar irradiance pattern measurement and a defined LED lighting strategy. The proposed method allows performing a direct comparison of different topologies for a specific application. Two different suitable TPC topologies are compared to a conventional cascaded solution. This is done to evaluate whether having a third power flow improves the system efficiency by reducing the amount of times the power is processed. The results of the CSF analysis employed shows that the cascaded structure provides the best results under the considered lightning patterns. Observing the results of the SCSF, for the first LED lighting pattern considered, the cascaded topology presents a improvement of 35.9% over the NI-TPC topology and a 40.16% over the HB-TPC topology. For the second LED pattern considered, the improvement of the cascaded topology is lower: 15.7% and 25.7% over the NI-TPC and HB-TPC, respectively.

REFERENCES

[1] Hirofumi Matsuo, Wenzhong Lin, Fujio Kurokawa, Tetsuro Shigemizu, and Nobuya Watanabe, "Characteristics of the Multiple-Input DC DC Converter," *IEEE Transactions on Industrial Electronics*, vol. 51, no. 3, pp. 625-631, June 2004.

[2] Yaow-Ming Chen, Yuan-Chuan Liu, and Feng-Yu Wu, "Multi-Input DC/DC Converter Based on the Multiwinding Transformer for Renewable Energy Applications," *IEEE Transactions on Industry Applications*, vol. 38, no. 4, pp. 1096-1104, August 2002.

[3] Yaow-Ming Chen, Yuan-Chuan Liu, and Feng-Yu Wu, "Multiinput Converter with Power Factor Corrention, Maximum Power Point Tracking, and Ripple-Free Input Currents," *IEEE Transactions on Power Electronics*, vol. 19, no. 3, pp. 631-639, May 2004.

[4] H. Matsuo, K. Kobayashi, Y. Sekine, M. Asano, and Lin Wenzhong, "Novel Solar-Cell Power Supply System Using a Multiple-Input DC-DC Converter," *IEEE Transactions on Industrial Electronics*, vol. 53, no. 1, pp. 281-286, February 2006.

[5] Zhe Zhang, Ole C. Thomsen, Michael A.E. Andersen, and H. R. Nielsen, "Dual-Input Isolated Full-Bridge Boost dc-dc Converter based on the Distributed Transformers," *ITE Power Electronics*, vol. 5, no. 7, pp. 1074-1083, August 2012.

[6] Ziwei Ouyang, Zhe Zhang, Michael A. E. Andersen, and Ole C. Thomsen, "Four Quadants Integrated Transformers for Dual-Input Isolated DC-DC Converters," *IEEE Transactions on Power Electronics*, vol. 27, no. 6, pp. 2697-2702, June 2012.

TABLE II
CSF – LED PATTERN 1

Topology	SCSF	WCSF	CCSF
Cascaded $(n = 5)$	10.92	1.76	0.98
NI − TPC $(n = 5)$	17.05	1.64	0.84
HB − TPC $(n = 11)$	18.25	3.51	1.17

TABLE III
CSF – LED PATTERN 2

Topology	SCSF	WCSF	CCSF
Cascaded $(n = 5)$	14.11	2.27	1.66
NI − TPC $(n = 3)$	16.75	1.55	0.82
HB − TPC $(n = 11)$	18.99	2.63	0.80

[7] Alexis Kwansinski, "Identification of Feasible Topologies for Multiple-Input DC-DC Converters," *IEEE Transactions on Power Electronics*, vol. 24, no. 3, pp. 856-861, March 2009.

[8] Yuan-Chuan Liu and Yaow-Ming Chen, "A Systematic Approach to Synthesizing Multi-Input DC-DC Converters," *IEEE Transactions on Power Electronics*, vol. 24, no. 1, pp. 2626-2632, January 2007.

[9] Yan Li, Xinbo Ruan, Dongsheng Yang, Fuxin Liu, and C.K. Tse, "Synthesis of Multiple-Input DC/DC Converters," *IEEE Transactions on Power Electronics*, vol. 25, no. 9, pp. 2372-2385, September 2010.

[10] Nicola Femia, Mario Fortunato, and Massimo Vitelli, "Light-to-Light: PV-Fed LED Lighting Systems," *IEEE transactions on Power Electronics*, vol. 28, no. 8, pp. 4063 - 4073, August 2013.

[11] Jose A. Barros Vieira and Alexandre Manuel Mota, "Implementation of a Stand-Alone Photovoltaic Lighting System with MPPT Battery Charging and LED current control," in *IEEE International Conference on Control Applications - CCA*, Yokohama - Japan, 2010, pp. 185-190.

[12] S. Malo and R Grino, "Design, Construction, and Control of a Stand-Alone Energy-Conditioning System for PEM-Type Fuel Cells," *IEEE Transactions on Power Electronics*, vol. 25, no. 10, pp. 2496-2506 , 2010.

[13] Zihu Zhou, Hongfei Wu, and Yan Xing, "A Non-Isolated Three-Port Converter for Stand-Alone Renewable Power System," in *Industrial Electronics Society IECON 2012*, 2012, pp. 3352-3357.

[14] Hongfei Wu, Yan Xing, Yanbing Xia, and Kai Sun, "A Family of Non-Isolated Three-Port Converters for Stand-Alone Renewable Power System," in *IECON 2011*, Montreal, Canada, 2011, pp. 1030-1035.

[15] Hongfei Wu et al., "A Three-Port Half-Bridge Converter with Synchronous Rectification for Renewable Energy Application," in *IEEE Energy Conversion Congress and Exposition*, 2011.

[16] Hongfei Wu, Kai Sun, Shun Ding, and Yan Xing, "Topology Derivation of NonIsolated Three-Port DC-DC Converters From DIC and DOC," *IEEE Transaction on Powe Electronics*, vol. 28, no. 7, pp. 3297-3307, July 2013.

[17] N. Vazquez, C. M. Sanchez, C. Hernandez, E. Vazquez, and R. Lesso, "A Three Port Converter for Renewable Energy Applications," in *IEEE International Symp. on Industrial Electronics*, 2011, pp. 1735 - 1740.

[18] Hongfei Wu et al., "A Family of Three-Port Half-Bridge Converters for a Stand-Alone Renewable Power System," *IEEE Transactions on Power Electronics*, vol. 26, no. 9, pp. 2697-2706, 2011.

[19] Hongfei Wu, Kai Sun, Runruo Chen, Haibing Hu, and Yan Xing, "Full-Bridge Three-Port Converters With Wide Input Voltage Range for Renewable Power Systems," *IEEE Transactions on Power Electronics*, vol. 27, no. 9, pp. 3965-3974, September 2012.

[20] Ernie Wittenbreder, "Topology Selection by the Numbers," *Technical Witts Inc.*, pp. 32-43, March 2006.

[21] Bruce Carsten, "Converter Component Load Factors; A Performance Limitation of Various Topologies," in *PCI*, Munich, 1988.

[22] Lars Petersen and Michael Andersen, "Two-Stage Power Factor Corrected Power Supplies: the Low Component-Stress Approach," in *Applied Power Electronics Conference and Exposition (APEC)*, 2002.

[23] http://www.soda-is.com.

Effect of Conventional Grid-Voltage Feedforward on the Output Impedance of a Three-Phase Photovoltaic Inverter

T. Messo, J. Jokipii and T. Suntio

Department of Electrical Engineering
Tampere University of Technology, Tampere, Finland
tuomas.messo@tut.fi, juha.jokipii@tut.fi, teuvo.suntio@tut.fi

Abstract—In this paper the small-signal model of a three-phase photovoltaic inverter is upgraded to include the grid-voltage feedforward. The feedforward is shown to increase d and q-components of the inverter output impedance which makes the inverter more insensitive to impedance-based interactions and explains the reported improvement in the output current harmonics. However, using a low-pass filter in the feedforward path is shown to deteoriate the effectiveness of the feedforward already at the frequencies one decade below the cut-off frequency of the low-pass filter. The effect of the feedforward is verified by extracting frequency responses from a switching model and from a prototype inverter operating at reduced voltages.

Keywords—Photovoltaic, grid-voltage feedforward, three-phase inverter, output impedance.

I. INTRODUCTION

Three-phase inverters based on the conventional two-level voltage-source inverter are currently widely utilized in grid-connected photovoltaic and wind-turbine power systems. The increase in the amount of grid-connected renewable energy sources has led to tighter standards regarding power quality and, in particular, harmonic currents generated by the inverters. The harmonic currents can be generated internally due to high-frequency modulation of the switches and externally due to distorted grid voltages [1].

The harmonic currents originating from distorted grid voltages can be mitigated by using a feedforward from the AC-side variables. The feedforward is usually added to the reference values of the duty ratios of the inverter switches. The feedforward can be taken solely from the grid voltages [2], [3], [4]. Alternatively, a combination of AC-side voltages and currents can be used [5] or a feedforward from the current of the output filter capacitor can be used [6].

Different studies have come out to the same conclusion that feedforward from AC-side variables yields better transient response [7] and attenuates high-frequency harmonic currents [6], [8], [9]. The authors in [10] reported the increase in the positive and negative sequence impedances when the grid-voltage feedforward was used.

This paper further develops the existing small-signal model by including the effect of the conventional grid-voltage feedforward. The model shows explicitly that the

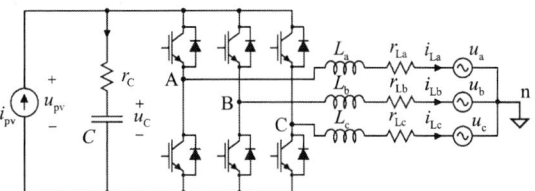

Fig. 1: Power stage of a three-phase PV inverter.

grid-voltage feedforward increases the magnitude of the d and q-components of the inverter output impedance. The model also allows the ideal feedforward gain to be found and the instability originating from the impedance-based interactions to be predicted correctly. Moreover, a low-pass filter in the feedforward path is shown to deteoriate the effect of the feedforward already at the frequencies one decade before the cut-off frequency of the filter. The model is verified by extracting the corresponding frequency responses from a switching model. Experimental measurements from a low-power prototype operating at reduced voltages are used to compare the output impedance with and without the grid-voltage feedforward.

The rest of the paper is organized as follows: The small-signal model of the inverter is derived in Section II. The frequency responses extracted from the switching model and the simulation results are presented in Section III. Experimental evidence from the prototype inverter is presented in Section IV. The most important outcomes of the research are summarized in Section V.

II. SMALL-SIGNAL MODEL OF A THREE-PHASE PV INVERTER

A. Open-loop transfer functions

The power stage of a two-level three-phase photovoltaic inverter is as shown in Fig. 1. The linearized state-space of the inverter can be developed in the synchronous reference frame and is as given in (1) and (2). The vectors **x**, **u**, and **y** contain the state, input and output variables, respectively, as given in (3). The resistance R includes the parasitic resistance of a single switch and the resistance of the output inductor.

$$\frac{d\mathbf{x}}{dt} = \overbrace{\begin{bmatrix} -\left(\frac{R}{L} + \frac{3r_C}{2L}D_d^2\right) & \omega_s\left(1 - \frac{r_C I_{pv}}{U_{pv}}\right) & \frac{D_d}{L} \\ -\omega_s\left(1 + \frac{r_C I_{pv}}{U_{pv}}\right) & -\left(\frac{R}{L} + \frac{3r_C}{2L}D_q^2\right) & \frac{D_q}{L} \\ -\frac{3D_d}{2C} & -\frac{3D_q}{2C} & 0 \end{bmatrix}}^{A} \mathbf{x}$$

$$+ \overbrace{\begin{bmatrix} \frac{r_C D_d}{L} & -\frac{1}{L} & 0 & \frac{U_{pv}-r_C I_{pv}}{L} & 0 \\ \frac{r_C D_q}{L} & 0 & -\frac{1}{L} & -\frac{r_C D_q I_{pv}}{D_d L} & \frac{U_{pv}}{L} \\ \frac{1}{C} & 0 & 0 & -\frac{I_{pv}}{D_d C} & 0 \end{bmatrix}}^{B} \mathbf{u} \tag{1}$$

$$\mathbf{y} = \overbrace{\begin{bmatrix} -\frac{3D_d r_C}{2} & -\frac{3D_q r_C}{2} & 1 \\ 1 & 0 & 0 \\ 0 & 1 & 0 \end{bmatrix}}^{C} \mathbf{x}$$

$$+ \overbrace{\begin{bmatrix} r_C & 0 & 0 & \frac{r_C I_{pv}}{D_d} & 0 \\ 0 & 0 & 0 & 0 & 0 \\ 0 & 0 & 0 & 0 & 0 \end{bmatrix}}^{D} \mathbf{u} \tag{2}$$

$$\mathbf{x} = \begin{bmatrix} \hat{i}_{Ld} & \hat{i}_{Lq} & \hat{u}_C \end{bmatrix}^T$$
$$\mathbf{u} = \begin{bmatrix} \hat{i}_{pv} & \hat{u}_{od} & \hat{u}_{oq} & \hat{d}_d & \hat{d}_q \end{bmatrix}^T \tag{3}$$
$$\mathbf{y} = \begin{bmatrix} \hat{u}_{pv} & \hat{i}_{od} & \hat{i}_{oq} \end{bmatrix}^T$$

The steady-state operating point can be solved from the average model of the inverter presented in [11] and is as given in (4) – (8).

$$D_d = \frac{U_{od} + \sqrt{(-U_{od})^2 + \frac{8}{3}R U_{pv} I_{pv}}}{2U_{pv}} \tag{4}$$

$$D_q = \frac{2\omega_s L I_{pv}}{3D_d U_{pv}} \tag{5}$$

$$I_{Ld} = \frac{2I_{pv}}{3D_d} \tag{6}$$

$$I_{Lq} = 0 \tag{7}$$

$$U_C = \frac{U_{od}}{D_d} + \frac{2RI_{pv}}{3D_d^2} = U_{pv} \tag{8}$$

Transfer functions of the inverter can be solved in the frequency-domain by utilizing (9) and by using a matrix manipulation software such as MATLAB Symbolic Toolbox. The matrix $\mathbf{G}$ in (9) contains the open-loop transfer functions as shown in (10).

$$\mathbf{Y}(s) = \left[\mathbf{C}(s\mathbf{I} - \mathbf{A})^{-1}\mathbf{B} + \mathbf{D}\right]\mathbf{U}(s) = \mathbf{G}\mathbf{U}(s) \tag{9}$$

$$\mathbf{G} = \begin{bmatrix} Z_{in} & T_{oi\text{-}d} & T_{oi\text{-}q} & G_{ci\text{-}d} & G_{ci\text{-}q} \\ G_{io\text{-}d} & -Y_{o\text{-}d} & -Y_{o\text{-}qd} & G_{co\text{-}d} & G_{co\text{-}qd} \\ G_{io\text{-}q} & -Y_{o\text{-}dq} & -Y_{o\text{-}q} & G_{co\text{-}dq} & G_{co\text{-}q} \end{bmatrix} \tag{10}$$

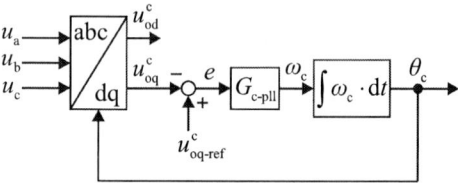

Fig. 2: Control block diagram of a conventional PLL.

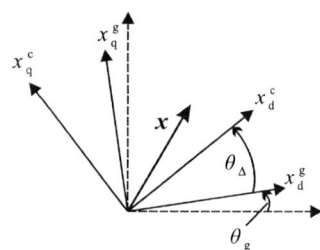

Fig. 3: Synchronous reference frames.

B. The effect of the PLL

The control block diagram of the PLL is as shown in Fig. 2. The error signal e is zero at steady-state and, therefore, the sensed phase angle θ_c equals the grid phase angle.

Fig. 3 shows two phase-shifted synchronous reference frames. The one denoted with superscript 'g' is tied to the three-phase variables of the grid and the one denoted with superscript 'c' is tied to the measured three-phase variables inside the control system. θ_Δ is the angle difference between the reference frames. The angle of the grid reference frame θ_g is selected as zero and, therefore, the space-vector $\mathbf{x}$ can be given in the control system reference frame as in (11).

$$\mathbf{x}^c = x_d^c + \mathrm{j}x_q^c = \mathbf{x}^g\, e^{-\mathrm{j}\theta_\Delta} = \left(x_d^g + \mathrm{j}x_q^g\right) e^{-\mathrm{j}\theta_\Delta} \tag{11}$$

The d and q-components in the control system reference frame can be given as a function of the d and q-components in the grid reference frame by utilizing Euler's formula according to (12) and (13), respectively.

$$x_d^c = x_d^g \cos\theta_\Delta + x_q^g \sin\theta_\Delta \tag{12}$$

$$x_q^c = x_q^g \cos\theta_\Delta - x_d^g \sin\theta_\Delta \tag{13}$$

The relation between the coordinate systems can be approximated as in (14) and (15) because the angle θ_Δ is usually very small at steady-state.

$$x_d^c \approx x_d^g + x_q^g \theta_\Delta \tag{14}$$

$$x_q^c \approx x_q^g - x_d^g \theta_\Delta \tag{15}$$

The relation in (14) and (15) can be linearized by developing the first-order partial derivatives in respect to

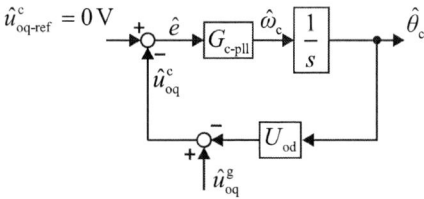

Fig. 4: Linearized control block diagram of the PLL.

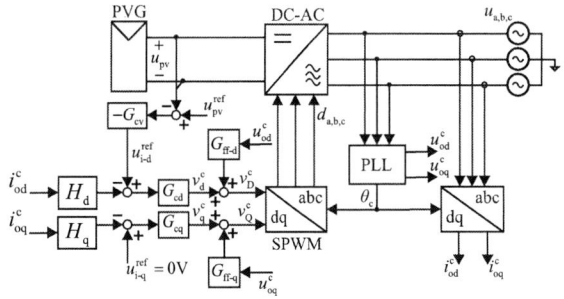

Fig. 5: Control system of the inverter.

each variable and can be written as in (16) and (17) where Θ_Δ is the angle difference at steady-state and X_d and X_q are the steady-state values of the d and q-components.

$$\hat{x}_d^c = \hat{x}_d^g + \Theta_\Delta \hat{x}_q^g + X_q \hat{\theta}_\Delta \tag{16}$$

$$\hat{x}_q^c = \hat{x}_q^g - \Theta_\Delta \hat{x}_d^g - X_d \hat{\theta}_\Delta \tag{17}$$

Output voltages, currents and duty ratios in the control system reference frame can be given as a function of the variables in the grid reference frame as in (18)–(23).

$$\hat{u}_{od}^c = \hat{u}_{od}^g + \Theta_\Delta \hat{u}_{oq}^g + U_{oq} \hat{\theta}_\Delta \tag{18}$$

$$\hat{i}_{od}^c = \hat{i}_{od}^g + \Theta_\Delta \hat{i}_{oq}^g + I_{oq} \hat{\theta}_\Delta \tag{19}$$

$$\hat{d}_d^c = \hat{d}_d^g + \Theta_\Delta \hat{d}_q^g + D_q \hat{\theta}_\Delta \tag{20}$$

$$\hat{u}_{oq}^c = \hat{u}_{oq}^g - \Theta_\Delta \hat{u}_{od}^g - U_{od} \hat{\theta}_\Delta \tag{21}$$

$$\hat{i}_{oq}^c = \hat{i}_{oq}^g - \Theta_\Delta \hat{i}_{od}^g - I_{od} \hat{\theta}_\Delta \tag{22}$$

$$\hat{d}_q^c = \hat{d}_q^g - \Theta_\Delta \hat{d}_d^g - D_d \hat{\theta}_\Delta \tag{23}$$

The steady-state angle between the coordinate systems Θ_Δ, q-component of the grid voltage U_{oq} and the q-component of the output current I_{oq} are zero. In addition, the q-component of the duty ratio D_q is assumed to be small enough to be neglected in the analysis. The steady-state values of U_{od}, I_{od} and D_d can be solved by using (4) – (8). The small-signal angle $\hat{\theta}_\Delta$ remains as the only unknown variable.

The linearized control block diagram of the PLL is as shown in Fig. 4. It is assumed that the small-signal angle $\hat{\theta}_\Delta$ equals the small-signal angle $\hat{\theta}_c$, i.e., the grid frequency is constant. The small-signal angle can be solved from the control block diagram in Fig. 4 and is as given in (24) where the control loop gain of the PLL L_{PLL} is as given in (25).

$$\hat{\theta}_c = \hat{\theta}_\Delta = \frac{1}{U_{od}} \frac{L_{PLL}}{1 + L_{PLL}} \hat{u}_{oq}^g \tag{24}$$

$$L_{PLL} = -\frac{G_{c\text{-pll}} U_{od}}{s} \tag{25}$$

Finally, the relation between the small-signal variables in the two coordinate systems can be presented as given in (26)–(31).

$$\hat{u}_{od}^c = \hat{u}_{od}^g \tag{26}$$

$$\hat{i}_{od}^c = \hat{i}_{od}^g \tag{27}$$

$$\hat{d}_d^c = \hat{d}_d^g \tag{28}$$

$$\hat{u}_{oq}^c = \left(1 - \frac{L_{PLL}}{1 + L_{PLL}}\right) \hat{u}_{oq}^g \tag{29}$$

$$\hat{i}_{oq}^c = \hat{i}_{oq}^g - \frac{I_{od}}{U_{od}} \frac{L_{PLL}}{1 + L_{PLL}} \hat{u}_{oq}^g \tag{30}$$

$$\hat{d}_q^c = \hat{d}_q^g - \frac{D_d}{U_{od}} \frac{L_{PLL}}{1 + L_{PLL}} \hat{u}_{oq}^g \tag{31}$$

The PLL affects the sensed small-signal q-components of currents and voltages and the small-signal q-component of the duty ratio which is fed to the inverter switches. The detailed derivation of the model can be found in [12].

C. Closed-loop transfer functions with grid-voltage feedforward

The control system of the inverter is as shown in Fig. 5 where the sensed variables are denoted by the superscript 'c'. The inverter is controlled using the cascaded control scheme where the DC-link voltage controller G_{cv} gives a reference for the output current controller G_{cd}. The d and q-components of the grid voltages u_{od}^c and u_{oq}^c are sensed, fed through the feedforward gains $G_{ff\text{-}d}$ and $G_{ff\text{-}q}$ and added in the outputs of the corresponding current controllers. The q-component of the output current is regulated at zero by the current controller G_{cq}.

The small-signal voltages that are fed to the SPWM v_D^c and v_Q^c can be given as in (32) if the feedforward gains are assumed to be linear which is usually the case.

$$\begin{aligned} \hat{v}_D^c &= \hat{v}_d^c + G_{ff\text{-}d} \hat{u}_{od}^c \\ \hat{v}_Q^c &= \hat{v}_q^c + G_{ff\text{-}q} \hat{u}_{oq}^c \end{aligned} \tag{32}$$

The grid-voltage feedforward can be included in the control block diagrams which represent the output dynamics related to the d-components and the input dynamics by taking into account (32). The control block diagrams are shown in Figs. 6 and 7 when the output current control loops and the grid-voltage feedforward loops are closed where H_d is the current sensing gain, G_{PWM} is the modulator gain. All other transfer functions are solved

The 2014 International Power Electronics Conference

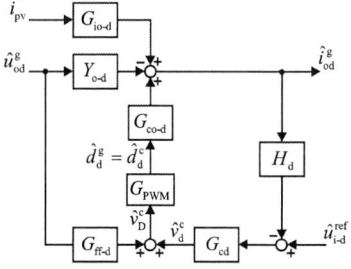

Fig. 6: Output dynamics related to d-components with output current control.

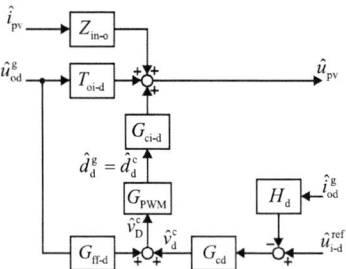

Fig. 7: Input dynamics with output current control.

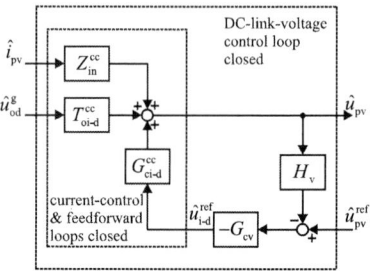

Fig. 8: Input dynamics with DC-link voltage control.

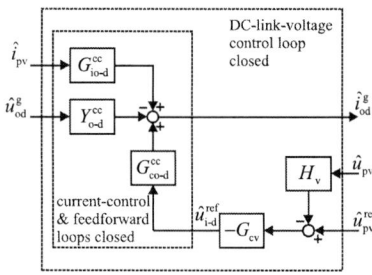

Fig. 9: Output dynamics related to d-components with DC-link voltage control.

numerically using (9). Moreover, it is assumed that the input dynamics depend mainly on the d-components and the cross-coupling transfer functions are neglected.

The transfer functions with output current control and grid-voltage feedforward are as given in (33)–(38) where the superscript 'cc' is used to denote that the output current control loop is closed.

$$G_{\text{io-d}}^{\text{cc}} = \frac{G_{\text{io-d}}}{(1 + L_{\text{d}})} \qquad (33)$$

$$Y_{\text{o-d}}^{\text{cc}} = \frac{Y_{\text{o-d}} - G_{\text{co-d}}G_{\text{PWM}}G_{\text{ff-d}}}{(1 + L_{\text{d}})} \qquad (34)$$

$$G_{\text{co-d}}^{\text{cc}} = \frac{1}{H_{\text{d}}}\frac{L_{\text{d}}}{(1 + L_{\text{d}})} \qquad (35)$$

$$Z_{\text{in}}^{\text{cc}} = \frac{Z_{\text{in}}}{(1 + L_{\text{d}})} + \frac{L_{\text{d}}}{(1 + L_{\text{d}})}Z_{\text{in-d-}\infty} \qquad (36)$$

$$T_{\text{oi-d}}^{\text{cc}} = \frac{T_{\text{oi-d}} + G_{\text{ci-d}}G_{\text{PWM}}G_{\text{ff-d}}}{1 + L_{\text{d}}} + \frac{L_{\text{d}}}{1 + L_{\text{d}}}T_{\text{oi-d-}\infty} \qquad (37)$$

$$G_{\text{ci-d}}^{\text{cc}} = \frac{1}{H_{\text{d}}}\frac{G_{\text{ci-d}}}{G_{\text{co-d}}}\frac{L_{\text{d}}}{(1 + L_{\text{d}})}, \qquad (38)$$

where

$$L_{\text{d}} = G_{\text{co-d}}G_{\text{PWM}}G_{\text{cd}}H_{\text{d}}, \qquad (39)$$

$$Z_{\text{in-d-}\infty} = Z_{\text{in}} - \frac{G_{\text{ci-d}}G_{\text{io-d}}}{G_{\text{co-d}}}, \qquad (40)$$

$$T_{\text{oi-d-}\infty} = T_{\text{oi-d}} + \frac{G_{\text{ci-d}}Y_{\text{o-d}}}{G_{\text{co-d}}}. \qquad (41)$$

The effect of the DC-link voltage control is added in the transfer functions next. The control block diagrams that describe the input dynamics and the output dynamics related to d-components are shown in Figs. 8 and 9 when the DC-link voltage control loop is closed where H_{v} is the voltage sensing gain. The closed-loop transfer functions of the inverter are as given in (42)–(50) where the superscript 'CL' denotes that all the control loops are closed.

$$Z_{\text{in}}^{\text{CL}} = \frac{Z_{\text{in}} + L_{\text{d}}Z_{\text{in-}\infty}}{(1 + L_{\text{dc}})(1 + L_{\text{d}})} \qquad (42)$$

$$T_{\text{oi-d}}^{\text{CL}} = \frac{\hat{u}_{\text{pv}}}{\hat{u}_{\text{od}}^{\text{g}}} = \frac{T_{\text{oi-d}} + G_{\text{ci-d}}G_{\text{PWM}}G_{\text{ff-d}} + L_{\text{d}}T_{\text{oi-d-}\infty}}{(1 + L_{\text{dc}})(1 + L_{\text{d}})} \qquad (43)$$

$$G_{\text{ri}}^{\text{CL}} = \frac{\hat{u}_{\text{pv}}}{\hat{u}_{\text{pv}}^{\text{ref}}} = \frac{1}{H_{\text{v}}}\frac{L_{\text{dc}}}{(1 + L_{\text{dc}})} \qquad (44)$$

$$G_{\text{io-d}}^{\text{CL}} = \frac{\hat{i}_{\text{od}}^{\text{g}}}{\hat{i}_{\text{pv}}} = \frac{G_{\text{io-d}}}{(1 + L_{\text{dc}})(1 + L_{\text{d}})} + \frac{L_{\text{dc}}}{(1 + L_{\text{dc}})}G_{\text{io-d-}\infty} \qquad (45)$$

$$Y_{\text{o-d}}^{\text{CL}} = \frac{Y_{\text{o-d}} - G_{\text{co-d}}G_{\text{PWM}}G_{\text{ff-d}}}{(1 + L_{\text{dc}})(1 + L_{\text{d}})} + \frac{L_{\text{dc}}}{(1 + L_{\text{dc}})}Y_{\text{o-d-}\infty} \qquad (46)$$

$$G_{\text{ro-d}}^{\text{CL}} = \frac{\hat{i}_{\text{od}}^{\text{g}}}{\hat{u}_{\text{pv}}^{\text{ref}}} = \frac{1}{H_{\text{v}}}\frac{G_{\text{co-d}}}{G_{\text{ci-d}}}\frac{L_{\text{dc}}}{(1 + L_{\text{dc}})} \qquad (47)$$

The DC-link voltage control loop gain L_{dc} and the special parameters $G_{\text{io-d-}\infty}$ and $Y_{\text{o-d-}\infty}$ are not affected by the feedforward and can be given as in (48) – (50).

$$L_{\text{dc}} = -G_{\text{ci-d}}^{\text{cc}}G_{\text{cv}}H_{\text{v}} = -\frac{H_{\text{v}}}{H_{\text{d}}}\frac{G_{\text{ci-d}}}{G_{\text{co-d}}}\frac{L_{\text{d}}}{(1 + L_{\text{d}})}G_{\text{cv}} \qquad (48)$$

978-1-4799-2706-7/14 $31.00 © 2014 IEEE   517

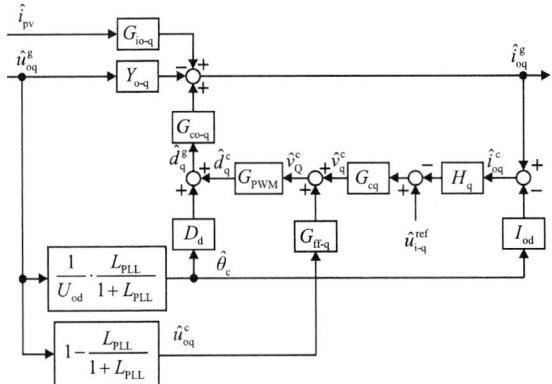

Fig. 10: Output dynamics related to q-components.

$$G_{\text{io-d-}\infty} = G_{\text{io-d}} - \frac{G_{\text{co-d}}}{G_{\text{ci-d}}} Z_{\text{in}} \qquad (49)$$

$$Y_{\text{o-d-}\infty} = Y_{\text{o-d}} + \frac{G_{\text{co-d}}}{G_{\text{ci-d}}} T_{\text{oi-d}} \qquad (50)$$

The PLL affects the sensed q-components of the output current i_{oq}^{c} and voltage u_{oq}^{c} and has to be taken into account by using (26) – (31). The q-component of the reference voltage which is fed to the SPWM can be given as in (51) when the effect of the PLL is taken into account.

$$\hat{v}_{\text{Q}}^{\text{c}} = \hat{v}_{\text{q}}^{\text{c}} + G_{\text{ff-q}} \left(1 - \frac{L_{\text{PLL}}}{1 + L_{\text{PLL}}} \right) \hat{u}_{\text{oq}}^{\text{g}} \qquad (51)$$

The control block diagram that describes the output dynamics related to the q-components is as shown in Fig. 10. The control block diagram can be derived by modifying the one presented in [12] by taking into account (51). The closed-loop transfer functions related to the q-components can be solved from the control block diagram and are as shown in (52)–(55).

$$Y_{\text{o-q}}^{\text{CL}} = \frac{1}{1 + L_{\text{q}}} \left[Y_{\text{o-q}} - G_{\text{co-q}} G_{\text{PWM}} G_{\text{ff-q}} \left(1 - \frac{L_{\text{pll}}}{(1 + L_{\text{pll}})} \right) \right]$$

$$- \frac{L_{\text{q}}}{(1 + L_{\text{q}})} \frac{L_{\text{pll}}}{(1 + L_{\text{pll}})} \frac{I_{\text{od}}}{U_{\text{od}}} - \frac{G_{\text{co-q}}}{(1 + L_{\text{q}})} \frac{L_{\text{pll}}}{(1 + L_{\text{pll}})} \frac{D_{\text{d}}}{U_{\text{od}}} \qquad (52)$$

$$G_{\text{io-q}}^{\text{CL}} = \frac{\hat{i}_{\text{oq}}^{\text{g}}}{\hat{i}_{\text{pv}}} = \frac{G_{\text{io-q}}}{(1 + L_{\text{q}})} \qquad (53)$$

$$G_{\text{ro-q}}^{\text{CL}} = \frac{\hat{i}_{\text{oq}}^{\text{g}}}{\hat{u}_{\text{i-q}}^{\text{ref}}} = \frac{1}{H_{\text{q}}} \frac{L_{\text{q}}}{(1 + L_{\text{q}})} \qquad (54)$$

The current control loop gain L_{q} in (55) and the loop gain of the PLL L_{pll} in (25) are not affected by the feedforward.

$$L_{\text{q}} = G_{\text{co-q}} G_{\text{PWM}} G_{\text{cq}} H_{\text{q}} \qquad (55)$$

TABLE I: Parameters of the simulation model.

U_{MPP}	699.3 V	$U_{\text{grid-rms}}$	230 V	$L_{\text{(a,b,c)}}$	2.46 mH
I_{MPP}	14.66 A	ω_{grid}	$2\pi \cdot 50$ Hz	$r_{\text{L(a,b,c)}}$	25.8 mΩ
r_{pv}	$U_{\text{MPP}}/I_{\text{MPP}}$	C	2.6 mF	r_{sw}	20 mΩ
P_{MPP}	10.25 kW	r_{C}	10 mΩ	f_{s}	10 kHz
L_{f}	475 μH	C_{f}	25 μF	R_{d}	1.46 Ω

D. Ideal Feedforward Gains

The d-component of the output admittance (46) $Y_{\text{o-d}}^{\text{CL}}$ is a sum of two terms. The second term originates from the use of DC-link voltage control and is not affected by the feedforward. However, the first term can be made equal to zero if the feedforward gain $G_{\text{ff-d}}$ is selected as in (56). In other words, a perfectly tuned feedforward cancels out the effect of the open-loop admittance term $Y_{\text{o-d}}$ in (46).

$$G_{\text{ff-d}}^{\text{ideal}} = \frac{Y_{\text{o-d}}}{G_{\text{co-d}} G_{\text{PWM}}} \qquad (56)$$

The two terms with negative signs on the second row of the output admittance q-component $Y_{\text{o-q}}^{\text{CL}}$ in (52) are caused by the PLL and cannot be manipulated using feedforward. However, the admittance terms inside the brackets can be made equal to zero when the feedforward gain $G_{\text{ff-q}}$ is selected as in (57). However, the feedforward does not affect the admittance at frequencies below the crossover frequency of the PLL, i.e., at frequencies where the term $(1 - L_{\text{PLL}}/(1 + L_{\text{PLL}}))$ is close to zero.

$$G_{\text{ff-q}}^{\text{ideal}} = \frac{Y_{\text{o-q}}}{G_{\text{co-q}} G_{\text{PWM}}} \qquad (57)$$

After carefully analysing the ideal feedforward gains in (56) and (57), it was found out that both gains can be approximated as in (58) which is actually the same value that is often used in the literature.

$$G_{\text{ff-d}}^{\text{ideal}} \approx G_{\text{ff-q}}^{\text{ideal}} \approx \frac{1}{U_{\text{dc}}} \qquad (58)$$

It has to be noted that the DC-link voltage in (58) is the steady-state value of the DC-link voltage, not the actual measured DC-link voltage. Including the measured DC-link voltage in the feedforward gain makes the gain nonlinear and adds more loops in the control block diagrams and, thus, changes the dynamics of the inverter. The topic will be addressed in future publications.

III. SIMULATION RESULTS

A switching model of the inverter was implemented using the SimPowerSystems toolbox in MATLAB Simulink. The model was supplied from a voltage-controlled current source which was tuned to replicate the dynamic resistance and the IU-curve of a real PV generator. The reference of the DC-link voltage was set equal to MPP-voltage to exclude the effect of the MPPT-algorithm. The parameters of the simulation model are summarized in Table I.

The 2014 International Power Electronics Conference

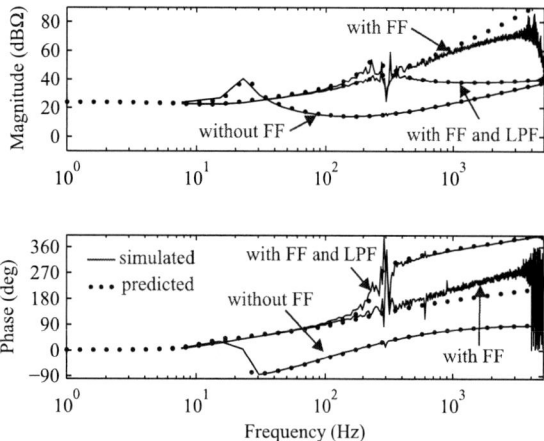

Fig. 11: D-component of the output impedance.

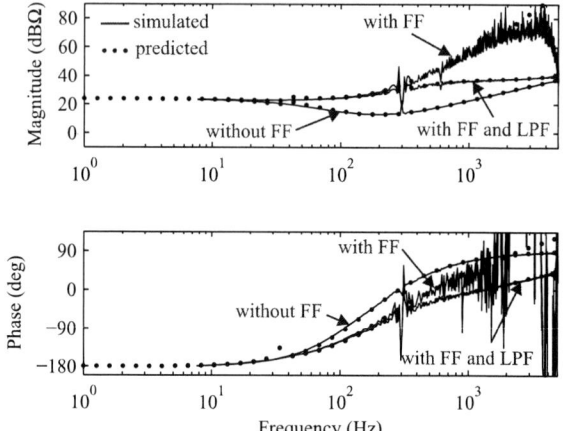

Fig. 12: Q-component of the output impedance.

Figs. 11 and 12 show the simulated and predicted d and q-components of the inverter output impedance in three cases: with and without the grid-voltage feedforward (FF) and when the feedforward includes a low-pass filter (LPF) with a cut-off frequency at 5 kHz. The impedances were extracted using frequency response techniques introduced in [13].

The resonant peak is completely removed from the d-component of the output impedance when the grid-voltage feedforward with ideal gain is used as can be seen in Fig. 11. The magnitude of the impedance is significantly larger at the frequencies higher than the crossover frequency of the DC-link voltage control which was tuned equal to 30 Hz. However, the low-pass filter starts deterorating the effect of the feedforward by decreasing the impedance already at the frequencies one decade lower than the cut-off frequency because the phase of the feedforward gain differs from the ideal gain (58). The q-component of the output impedance experiences similar effect as can be seen in Fig. 12. The deviation in the predicted and simulated impedances in Fig. 11 at high frequencies is propably caused by the fact that the cross-coupling transfer functions were neglected in the analysis.

An inverter with grid-voltage feedforward clearly has larger output impedance at the harmonic frequencies than an inverter which does not utilize feedforward. An output-current controlled inverter should have as large output impedance as possible to be insensitive to the impedance-based interactions originating from the AC-side. The effect of the feedforward was further investigated by connecting the switching model to a weak inductive grid using a three-phase CL-filter with the parameters as given in Table I. The grid inductance L_g was assumed to be 9 mH.

The stability of the system can be verified by examining the impedance ratios at the interface illustrated by the dashed line in Fig. 13. The stability should be evaluated by plotting the ratios of the impedance d and

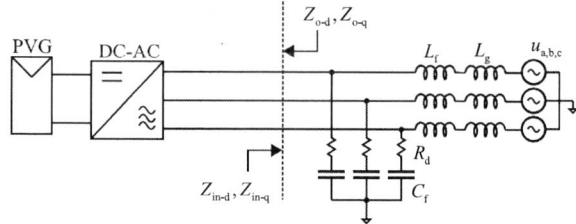

Fig. 13: PV inverter connected to a weak grid using a CL-filter.

q-components, $Z_{in\text{-}d}/Z_{o\text{-}d}$ and $Z_{in\text{-}q}/Z_{o\text{-}q}$, in the complex plane [14]. The system is stable if the impedance ratios satisfy the Nyquist stability criterion. It is important to note that the impedance ratios which are used in the stability analysis of an output-current controlled inverter, such as the PV inverter, have to be the opposite compared to an output-voltage controlled inverter [15]. The input impedance of the CL-filter was derived in the synchronous reference frame as discussed in [12]. The effect of the weak grid can be included in the small-signal model by adding the value of the grid inductance L_g to the value of the filter inductance L_f since they are connected in series in Fig. 13.

Fig. 14 shows the ratio of the impedance d-components $Z_{in\text{-}d}/Z_{o\text{-}d}$ on the complex plane when the grid inductance is equal to 9 mH. The stability of the inverter was examined in two cases. In the first case, the feedforward was not used and, in the second case, the feedforward with the low-pass filter at half the switching frequency was used. The contour does not encircle the critical (-1,0) – point and, thus, the inverter is stable in terms of the d-components regardless of the use of feedforward.

Fig. 15 shows the ratio of the impedance q-components $Z_{in\text{-}q}/Z_{o\text{-}q}$ on the complex plane. The contour

The 2014 International Power Electronics Conference

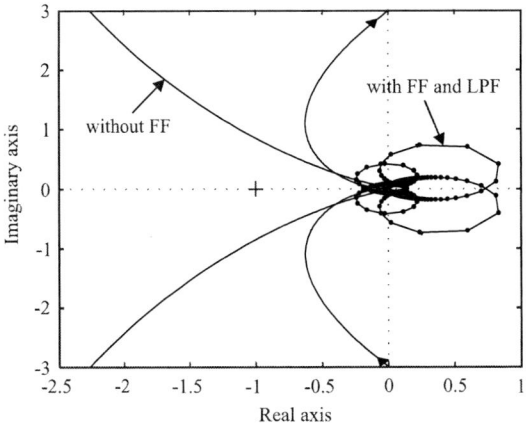

Fig. 14: Ratio of the impedance d-components with and without feedforward.

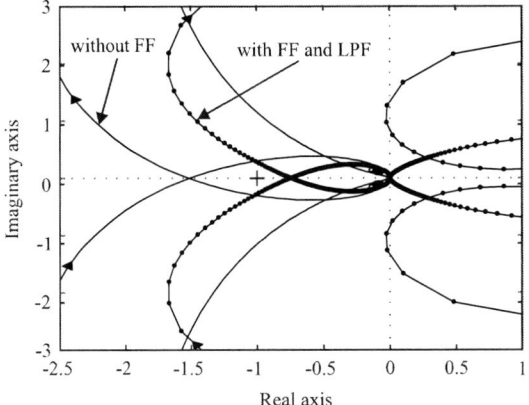

Fig. 15: Ratio of the impedance q-components with and without feedforward.

encircles the (-1,0) – point when the feedforward is deactivated which means that the inverter will be unstable. However, the inverter will be stable when the grid-voltage feedforward is used because the output impedance of the inverter is larger and the contour does not encircle the (-1,0) – point.

The simulated grid voltages and currents are shown in Fig. 16 when the feedforward is deactivated at 0.1 s. The grid current becomes unstable as can be predicted based on Fig. 15. The use of grid-voltage feedforward clearly makes the inverter more insensitive to impedance-based interactions.

IV. EXPERIMENTAL EVIDENCE

A low-power prototype inverter operating at reduced voltages was used to further validate the theoretical findings. The parameters of the prototype are given in Table II. The prototype was supplied from a solar array simulator model E4360 manufactured by Agilent and loaded by a three-phase voltage source model SW5250A

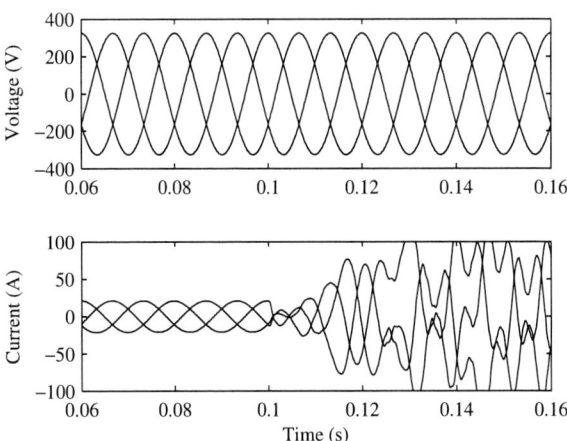

Fig. 16: Grid current becomes unstable due to impedance-based interactions when feedforward is deactivated.

TABLE II: Parameters of the experimental three-phase inverter.

U_{MPP}	30 V	C	300 μF	$L_{a,b,c}$	73 μH
$U_{a,b,c}$	7.07 V_{rms}	r_C	85 mΩ	$r_{L(a,b,c)}$	15 mΩ
I_{MPP}	1 A	ω_s	2·π·50 Hz	f_s	75 kHz

manufactured by Elgar. Three resistors of 1.2 ohms were connected in parallel with the voltages source to dissipate the generated power since the voltage source cannot sink power.

The d and q-components of the inverter output impedance were measured in the synchronous reference frame by making a small-signal perturbation in the voltage references of the three-phase voltage source and feeding either the ratio u_{od}^g/i_{od}^g or u_{oq}^g/i_{oq}^g to a frequency response analyzer manufactured by Venable Instruments. The measured d-components of the output impedance are shown in Fig. 17 with and without the grid-voltage feedforward. The resonant peak is shifted to higher frequencies but it is not removed. However, the magnitude of the impedance is clearly higher at the harmonic frequencies when the feedforward is active.

The measured q-components of the output impedance are shown in Fig. 18 with and without the grid-voltage feedforward. The magnitude has increased 10-20 dB at frequencies higher than the crossover of the PLL which was set at 10 Hz. The low-frequency phase of the impedance is close to -180 degrees which is caused by the PLL and cannot be manipulated by using the conventional grid-voltage feedforward.

V. CONCLUSIONS

Three-phase PV inverters often utilize feedforward from the grid voltages. The feedforward is implemented by adding the sensed d and q-components of the grid

978-1-4799-2706-7/14 $31.00 © 2014 IEEE   520

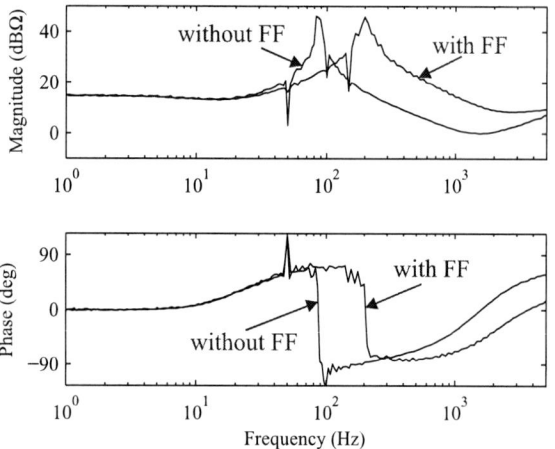

Fig. 17: D-component of the experimental inverter's output impedance with and without feedforward.

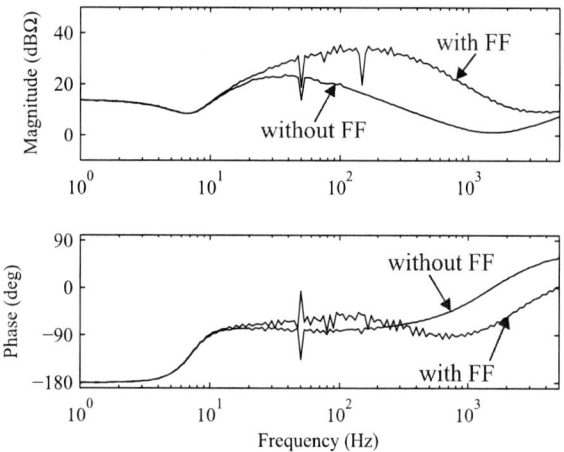

Fig. 18: Q-component of the experimental inverter's output impedance with and without feedforward.

voltages to the outputs of the corresponding current controllers. The effect of the feedforward on the inverter output impedance is explained explicitly based on the developed small-signal model. The model is verified by extracting the output impedances from a switching model by using frequency response techniques. The model predicts that the magnitudes of d and q-components of the inverter output impedance can be increased by tuning the feedforward gain properly. The ideal feedforward gain is derived to be the inverse of DC-link voltage which is also a generally used rule-of-thumb in the literature. Moreover, a low-pass filter in the voltage sensors is shown to deteoriate the effect of the feedforward already one decade before the cut-off frequency of the low-pass filter. However, the grid-voltage feedforward makes the inverter more insensitive to impedance-based interactions

due to the increased output impedance. The effect of the feedforward is verified by extracting the impedances from a switching model and from a low-power prototype operating at reduced voltages.

REFERENCES

[1] J. Enslin and P. Heskes, "Harmonic interaction between a large number of distributed power inverters and the distribution network", *IEEE Transactions on Power Electronics*, vol. 19, no. 6, pp. 1586-1593, 2004.

[2] S.-Y. Park, C.-L. Chen, J.-S. Lai and S.-R. Moon, "Admittance compensation in current loop control for a grid-tie LCL fuel cell inverter", *IEEE Transactions on Power Electronics*, vol. 23, no. 4, pp. 1716-1723, 2008.

[3] L. Harnefors, M. Bongiorno, S. Lundberg, "Input-admittance calculation and shaping for controlled voltage-source converters", *Transactions on Idustrial Electronics*, vol. 54, no. 6, pp. 3323-3334, 2007

[4] A. Timbus, M. Liserre, R. Teodorescu, P. Rodriguez and F. Blaabjerg, "Evaluation of current controllers for distributed power generation systems", *IEEE Transactions on Power Electronics*, vol. 24, no. 3, pp. 654-664, 2009.

[5] X. Wang, X. Ruan, S. Liu and C. K. Tse, "Full feedforward of grid voltage for grid-connected inverter with LCL filter to suppress current distortion due to grid voltage harmonics", *IEEE Transactions on Power Electronics*, vol. 25, no. 12, pp. 3119-3127, 2010.

[6] T. Abeyasekera, C. M. Johnson, D. J. Atkinson and M. Armstrong, "Suppression of line voltage related distortion in current controlled grid connected inverters", *IEEE Transactions on Power Electronics*, vol. 20, no. 6, pp. 1393-1401, 2005.

[7] D. G. Holmes, T. A. Lipo, B. P. McGrath and W. Y. Kong, "Optimized design of stationary frame three phase AC current regulators", *IEEE Transactions on Power Electronics*, vol. 24, no. 11, pp. 2417-2426, 2009.

[8] W. Li, D. Pan and X. Wang, "Full-feedforward schemes of grid voltages for a three-phase LCL-type grid-connected inverter", *IEEE Transactions on Industrial Electronics*, vol. 60, no. 6, pp. 2237-2250, 2013.

[9] Q. Zheng and L. Chang, "Improved current controller based on SVPWM for three-phase grid-connected voltage source inverters", IEEE 36th Power Electronics Specialists Conference, pp. 2912-2917, 2005.

[10] M. Cespedes and J. Sun, "Impedance shaping of three-phase grid-parallel voltage-source converters", Applied Power Electronics Conference and Exposition, pp. 754-760, 2012.

[11] T. Messo, J. Jokipii, J. Puukko, T. Suntio, "Determining the value of DC-link capacitance to ensure stable operation of a three-phase photovoltaic inverter", *IEEE Transactions on Power Electronics*, vol. 29, no. 2, pp. 665-673, 2014.

[12] T. Messo, J. Jokipii and T. Suntio, "Modeling the grid synchronization induced negative-resistor-like behavior in the output impedance of a three-phase photovoltaic inverter", The IEEE 4th International Symposium on Power Electronics for Distributed Generation Systems, pp. 1-7, 2013.

[13] T. Roinila and M. Vilkko and T. Suntio, "Frequency-response measurement of switched-mode power supplies in the presence of nonlinear distortions", *IEEE Transactions on Power Electronics*, vol. 25, no. 8, pp. 2179-2189, 2010.

[14] B. Wen, D. Boroyevich, P. Mattavelli, Z. Shen and R. Burgos, "Experimental verification of the Generalized Nyquist stability criterion for balanced three-phase AC systems in the presence of constant power loads", IEEE Energy Conversion Congress and Exposition, pp. 39263933, 2012.

[15] J. Sun, "Impedance-Based Stability Criterion for Grid-Connected Inverters", *IEEE Transactions on Power Electronics*, vol. 26, no. 11, pp. 3075-3078, 2011.

Power Amplifier Suitable for Photovoltaic Cell Booster

Teruhiko Kohama, Yuki Sogawa, and Satoshi Tsuji
Dept. of Electrical Engineering
Fukuoka University
Fukuoka 814-0180, JAPAN

Abstract— **Power amplifier suitable for booster of small-size nonlinear power source such as photovoltaic (PV) cells is presented. The amplifier is developed to simulate large size of nonlinear power source and is utilized to examine various PV power systems. The amplifier is simple and compact to carry. Principle of PV cell booster with the amplifier is described. Prototype of the amplifier utilized as PV cell booster is built and tested to confirm the effectiveness of the amplifier.**

Keywords— *flyback converter, nonlinear power source, PV booster, photovoltaic cell, power amplifier*

I. INTRODUCTION

Photovoltaic (PV) power generation has been receiving considerable attention as one of the promising alternative energy without emitting CO_2. In order to develop equipment such as inverters and converters suitable for PV power generation system, the equipment must be tested by using actual size of solar panel. However, the panel occupies large area of roof or field and is great expense. To solve the problem photovoltaic power simulator has been developed [1-2]. Normally, the PV power simulator is realized by a computer and a power conversion circuit. It simulates large power PV panel by using data derived from PV equivalent circuit model or stored data by measuring V-I characteristic of actual PV cell. However, immediate change of V-I characteristic due to weather cannot be reflected in the simulator output in real time.

This paper describes a power amplifier which boosts voltage and current in small-size PV cell. The amplifier simulates large size of PV panel and is used for development of various PV power systems. The equipment is small and light enough to carry. The amplifier has following features:

1. Simple configuration of circuits with a conventional flyback converter and a simple control circuit.
2. Immediate changes of voltage and current in small-size PV cell reflect in the output of the amplifier. The amplifier simulates as if it were an actual large power PV panel.
3. Various types of PV cell such as crystalline Si cell, thin-film cell, and organic cell are used to boost their power by just replacing the small-size cell in the amplifier with another type.

4. Easy to carry since the amplifier is compact and light.
5. The amplifier becomes portable equipment by using battery instead of commercial source. This feature allows us to test the PV power system in the place where no power source exists such as mountain and sea.
6. The amplifier can be used not only for PV cell booster but also for other nonlinear V-I power sources.

Principle of new power amplifier is described in section II. Detailed circuit configuration for PV cell booster is shown in section III. A prototype of the booster is build and tested in section IV to confirm the effectiveness of the power amplifier.

II. PRINCIPLE OF POWER AMPLIFIER

Basic principle of new power amplifier suitable for PV cell booster is shown in Fig.1. The booster consists of a voltage amplifier with high input-impedance, a current attenuator, and current controlled current source. Voltage V_{PV} of small-size PV cell is applied to input terminal of the booster. V_{PV} is amplified by the voltage amplifier with gain of K_v. Therefore output voltage V_L reaches K_vV_{PV} and is applied to load. Typical V-I characteristic of PV cell is shown in Figure 2(a) where voltage V_{PV} is nonlinear function of current I_{PV}. It has a maximum power point P_{max} at $V_{PV}=V_{max}$ and $I_{PV}=I_{max}$.

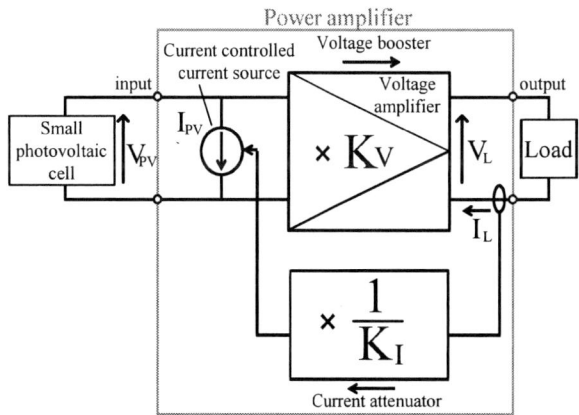

Figure 1. Principle of developed power amplifier.

978-1-4799-2706-7/14 $31.00 © 2014 IEEE   522

To simulate the V-I curve, load current I_L is sensed and fed back to the current controlled current source(CCCS) I_{PV} through the current attenuator to control the cell current. The current feedback enables the output to trace nonlinear V-I characteristic of PV cell. Relationships between I_{PV} and I_L is given by

$$K_I = I_L / I_{PV}. \tag{1}$$

Ideal characteristic of the power amplifier is shown in Fig.2(b) where voltage V_{PV} and current I_{PV} are amplified by factors K_V and K_I, respectively. As a result, maximum output power of the amplifier is $K_V K_I$ times larger than that of small-size PV cell.

III. CIRCUIT CONFIGURATION OF PV CELL BOOSTER

Figure 3 shows circuit configuration of PV cell booster with the amplifier. The voltage amplifier in Fig.1 is realized by a DC-DC converter. A flyback converter shown in Fig.4 is used to cover wide range of V_L. By introducing steady state averaging method[3] ,V_L is given by following equation.

$$\frac{V_L}{V_{in}} = \frac{D}{N(1-D)} \tag{2}$$

where,
V_{in}: DC voltage source derived from AC commercial power source or battery
D: duty ratio of switch Q
N: winding ratio of transformer.

It is noted that PV cell's voltage V_{PV} is only used for voltage reference in PWM control circuit to gain voltage V_L which is proportional to V_{PV}. Load power is delivered from not the PV cell but the DC voltage source V_{in} through DC-DC converter. Considering that V_L' in Fig.3 is assumed V_{PV} due to voltage feedback control, following equation is derived.

$$V_L' = V_{PV} = \frac{R_2}{R_1+R_2} V_L \tag{3}$$

From Eq.(3) voltage gain K_V in Fig.1 is expressed by

$$K_V = \frac{V_L}{V_{PV}} = 1 + \frac{R_1}{R_2} \tag{4}$$

In order to follow a nonlinear V-I characteristic of the small PV cell, a current source is connected in parallel to the small PV cell as a small load whose current is proportional to the load current I_L. The current I_L is sensed with a low resistance R_s and converted to voltage V_{IL}. The current in the PV cell is controlled by a voltage controlled current source(VCCS) I_{PV} with the signal V_{IL}. Detailed circuit for VCCS is shown in Fig.5.

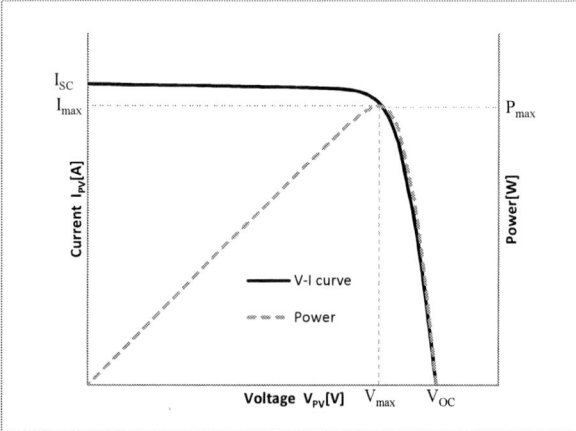

(a) Characteristics of small photovoltaic cell.

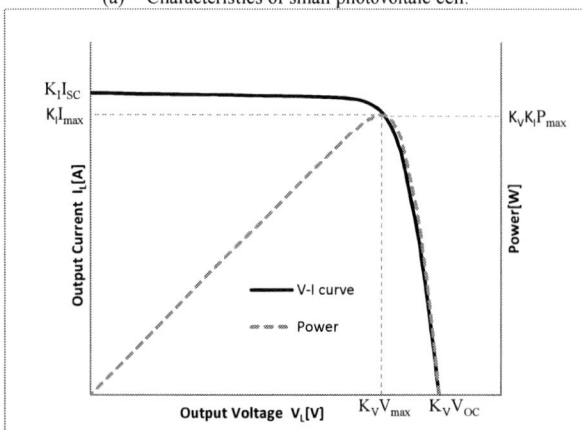

(b) Output characteristics of power amplifier with PV cell.
Figure 2. Input and output V-I characteristics of power amplifier with photovoltaic cell

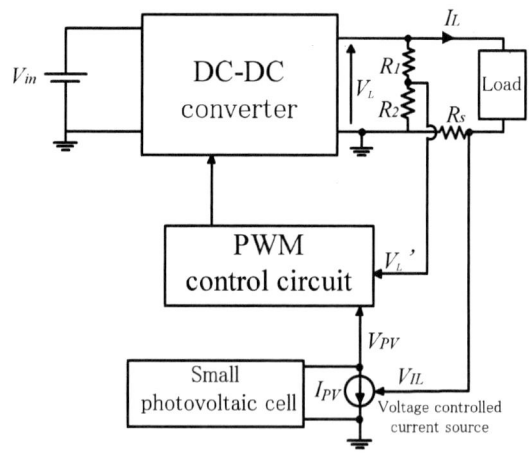

Figure 3. Circuit configuration of PV cell booster.

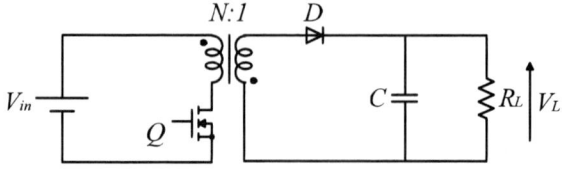

Figure 4. Flyback converter.

978-1-4799-2706-7/14 $31.00 © 2014 IEEE

It consists of a non-inverting amplifier, a differential amplifier, and a feedback loop. V_{IL} is amplified by a factor of $(1+R_f/R_i)$ through the non-inverting amplifier. From Fig.5 output voltage V_2 of the differential amplifier is given by

$$V_2 = -(V_{IL}' - V_{PV}) . \qquad (5)$$

Therefore, current I_{PV} through resistor R_{ref} is given by

$$I_{PV} = \frac{V_{PV} - V_2}{R_{ref}} . \qquad (6)$$

By substituting Eq.(5) and $V_{IL}'=(1+R_f/R_i)V_{IL}$ into (6), I_{PV} is expressed as

$$I_{PV} = \frac{V_{IL}'}{R_{ref}} = \frac{1+R_f/R_i}{R_{ref}} . \qquad (7)$$

Figure 6 shows experimental V-I characteristic of VCCS in Fig.5, where current source I_{PV} is proportional to control voltage V_{IL}.

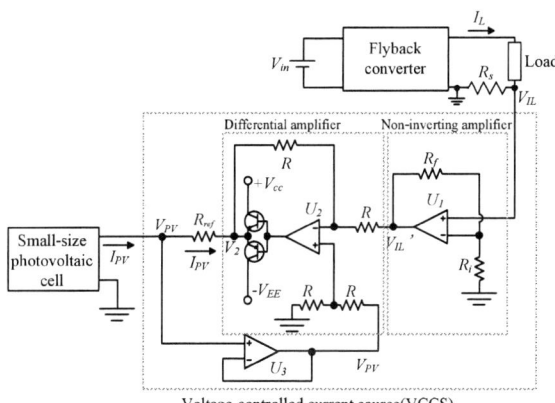

Figure 5. Voltage-controlled current source.

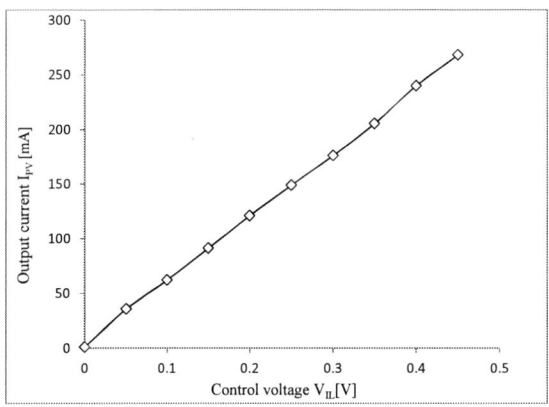

Figure 6 V-I characteristic of VCCS in Fig.5.

IV. EXPERIMENTAL RESULTS

Prototype of a new PV cell booster shown in Fig.7 has built and tested to confirm the effectiveness of the power amplifier. Circuit parameters are as follows:
V_{in}=40[V], C=1320[μF], N=23/15, R_1=24[kΩ], R_2=1[kΩ], R_s=0.22[Ω], f_{sw}=100[kHz].

Figure 8 and Table 1 show characteristics of small-size PV cell used in the booster. In Fig.8, open circuit voltage V_{oc} is 2.2[V] and short circuit current I_{sc} is 0.200[A]. The maximum power P_{max} is 0.324[W] at V_{PV}=1.791[V] and I_{PV}=0.181[A]. The booster's gains K_V and K_I in Fig.1 are set to 25, 8.5 respectively in prototype. Therefore the booster is expected to boost P_{max} in Fig.8 up to $K_V K_I$=25 x 8.5=212.5 times.

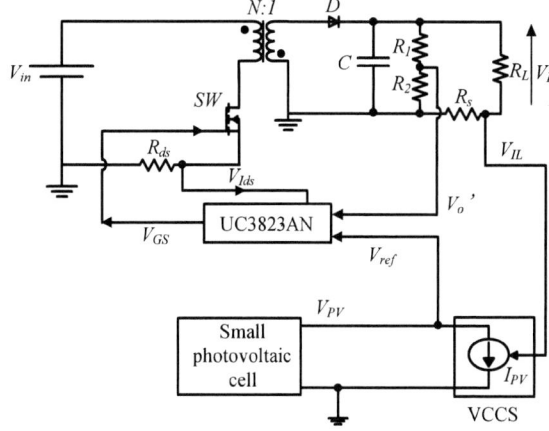

Figure 7. Experimental circuit.

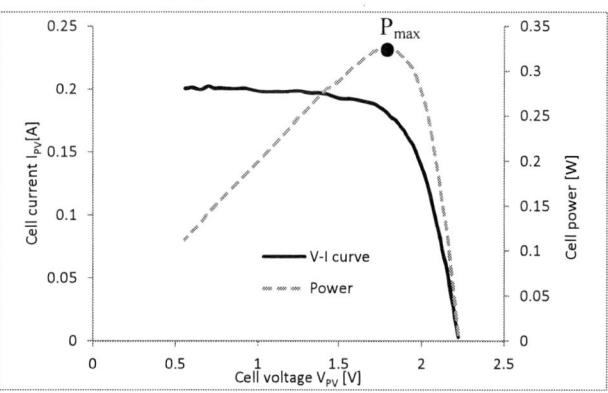

Figure 8. V-I and power curve of small-size PV cell.

TABLE 1
CHARACTERISTICS OF SMALL-SIZE PV CELL

	V_{PV}[V]	I_{PV}[A]	P_{PV}[W]
Light load	2.223	0.003	0.007
Maximum power	1.791	0.181	0.324
Heavy load	0.560	0.200	0.112

The 2014 International Power Electronics Conference

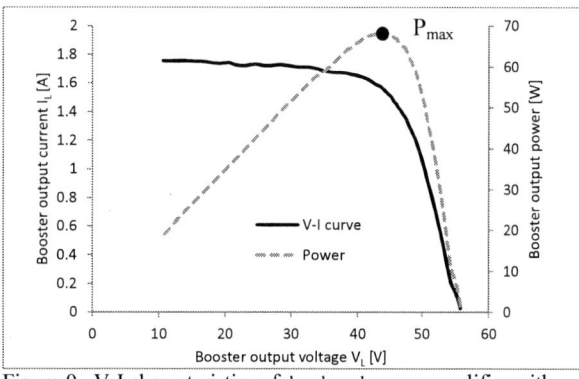

Figure 9. V-I characteristics of developed power amplifier with small PV cell in Fig.8.

TABLE 2
OUTPUT CHARACTERISTICS OF PV CELL BOOSTER

	$V_L[V]$	$I_L[A]$	$P_L[W]$
Light load	55.78	0.028	1.562
Maximum Power	43.46	1.574	68.41
Heavy load	10.60	1.757	18.62

TABLE 3
GAINS OF PV CELL BOOSTER

	Voltage gain K_v	Current gain K_I	Power gain K_vK_I
Light load	25.1	9.3	223.1
Maximum Power	24.2	8.7	212.5
Heavy load	18.9	8.8	166.3

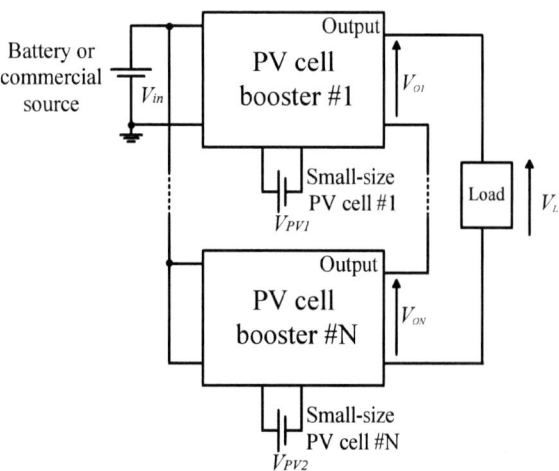

Figure 10. Series connection of PV cell boosters.

Figure 9 and Table 2 show V-I curve of the booster as PV cell in Fig.9 is used. From Tables 1 and 2, measured gains K_V and K_I are shown in Table 3.

From Table 3, gains K_V and K_I at light load and maximum power point are almost the same as specified gains of 25 and 8.5, respectively. At heavy load, measured K_V is decreased to 18.9 compared with specified gain of 25. This is due to the increase in output impedance of flyback converter. The error can be reduced by improving voltage feedback control circuit of the flyback converter.

Output terminals of the several boosters can be connected in series to generate higher voltage since each booster has isolation property. Figure 10 shows the series connection of the boosters. High voltage is applied to the load by the sum of all boosters' output voltages. While, for the input, only one DC source V_{in} is required. This configuration simulates a large PV power source that consists of several panels with different V-I characteristics due to different illumination, temperature, or aged deterioration.

V. CONCLUSIONS

Power amplifier for PV cell booster is developed to test the equipment for various PV power generation systems. The power amplifier is simple and is easily realized by a flyback converter and a simple controller including a controlled current source. In the amplifier, nonlinear V-I characteristics of PV cell immediately reflects in the output. This amplifier can be used not only for PV cell booster but also for any other nonlinear power source.

REFERENCES

[1] S.Gonzalez1, S.Kuszmaul,D.Deuel, R.Lucca,"PV Array Simulator Development and Validation," 35th IEEE Photovoltaic Specialists Conference (PVSC), 2010

[2] Agilent E4360 Modular Solar Array Simulators Datasheet, Agilent Technologies,2011

[3] R.D.Middlebrook and S.Cuk,"A general unified approach to modelling switching converter power stages," IEEE Power Electronics Specialists Conference, pp.18-34,1976

[4] T.Yamabe and T.Kohama,"Development of Photovoltaic Cell Booster," 2011 Annual Conference of I.E.E. of Japan, Industry Applications Society, No.Y-45,2011.

[5] T.Kohama, T.Yamabe, and S.Tsuji,"Development of Photovoltaic Cell Booster Using Flyback Converter", Proceedings of The International Conference on Renewable Energy Research and Applications, 2012.

The 2014 International Power Electronics Conference

Realization Study of Interleaved PV Microinverter by Quadrature-Phase-Shift SPWM Control

Hung-I Hsieh
Department of Electrical Engineering
National Chiayi University
Chiayi, Taiwan
E-mail: hihsieh@mail.ncyu.edu.tw

Guan-Cyun Hsieh and Jiaxin Hou
Department of Electrical Engineering
Chung Yung Christian University
Chung-Li, Taiwan
gchsieh@dec.ee.cycu.edu.tw

Abstract-A PV microinverter implemented using interleaved flyback converter (IFC) is investigated. Double-frequency distortion is found in grid current because of using sawtooth time-based SPWM. A forward compensation circuit that eliminates the distortion at primary side is explored that simplifies the output filter design. Maximum power transfer between PV module, IFC, and grid end is modeled. An elaborate experiment is demonstrated to verify the analysis result and performance predictions.

Keywords—PV microinverter, interleaved flyback converter, SPWM, MPPT

I. INTRODUCTION

To overcome shadow effect on PV array, an AC microinverter suited to a single PV module has been widely developed [1-2]. The major topology of the microinverter is widely implemented by the flyback converter due to its simple structure and low cost. Two-stage unfolding cycloconversion [1-3] for current-source inversion (CSI) and unfolding H-bridge for voltage-source (VS)-to-CSI commutation [1], [4] are often used in microinverter. In order to increase the utilization of PV module and reduce ripple current, an interleaved approach is widely used in PV energy pump [6], [11]. To improve the stability of MPPT behavior, researchers pay more attention to the power decoupling circuit in microinverter study lately [5-8]. In this paper, an interleaved flyback converter (IFC) is used to achieve sinusoidal-CSI to grid through a two-stage unfolding cycloconversion as shown in Fig. 1. Two accurate quadrature-phase-shift triangular references for interleave operation is simply realized by a phase-shift zero-voltage-switching (PS-ZVS) PWM controller. In order to eliminate a double-frequency distortion of the grid current, a forward compensation circuit that consists of an auxiliary winding N_3, a resonant capacitor C_1, and a power switch Q_{1a}, using resonant approach, is presented. This approach not only can eliminate the second harmonic current but also can circulate the resonant current back to an bulk capacitor C_B, saving energy. For simplifying the analysis, a half-circuit of IFC is model to analyze the behaviors of PV energy pump and SPWM modulation. The IFC is operated in discontinuous-

conduction mode (DCM) so as to fit the interleave operation. Maximum power transfer between PV module, IFC, and grid is modeled as well. A power-increment-aided incremental-conductance maximum power point tracking (PI-INC MPPT) with constant-frequency variable-duty (CFVD) control is employed to ensure the PV energy pump accurate and always at MPP on the PV characteristics [11]. Finally, a demonstrate example of 250-W microinverter is explored and experimented to verify the analysis result and performance prediction.

II. INTERLEAVED FRYBACK INVERTER WITH TWO-STAGE UNFOLDING CYCLOCONVERTER

A typical interleaved flyback converter (IFC) used as CSI is explored in Fig. 1(a), in which the two flyback converters are modulated by two quadrature-phase-shift SPWMs at primary side. The two SPWM currents are merged at node M before a CL filter to grid through a two-stage unfolding cycloconverter at secondary side. The waveforms of the modulation and cycloconversion for interleave flyback operation are predicted in Fig. 1(b). For ease of analysis, a simplified circuit model is shown in Fig. 2(a) and the waveforms of interleave energy pump from PV module with SPWM with respect to the gate drives are shown in Fig. 2(b).

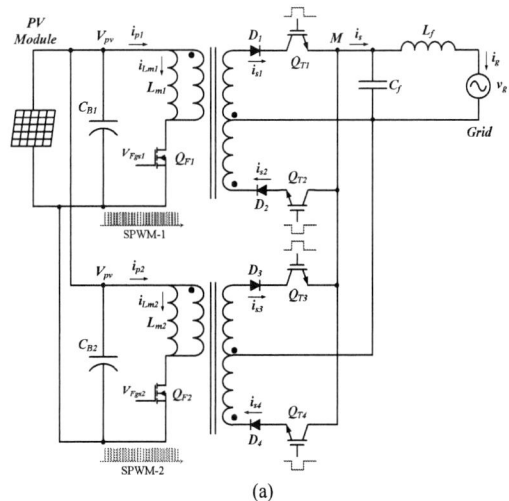

(a)

978-1-4799-2706-7/14 $31.00 © 2014 IEEE

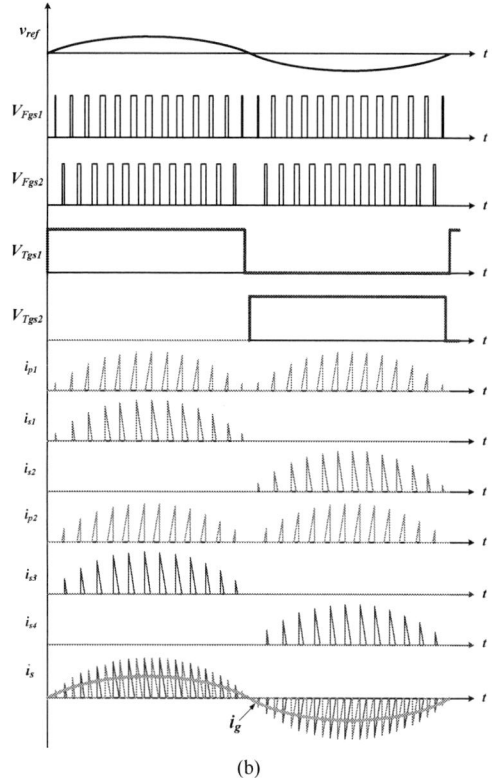

(b)

Fig. 1 (a) Circuit configuration of an interleaved microinverter using flyback converter and (b) the predicted waveforms of state dynamics through an unfolding cycloconversion

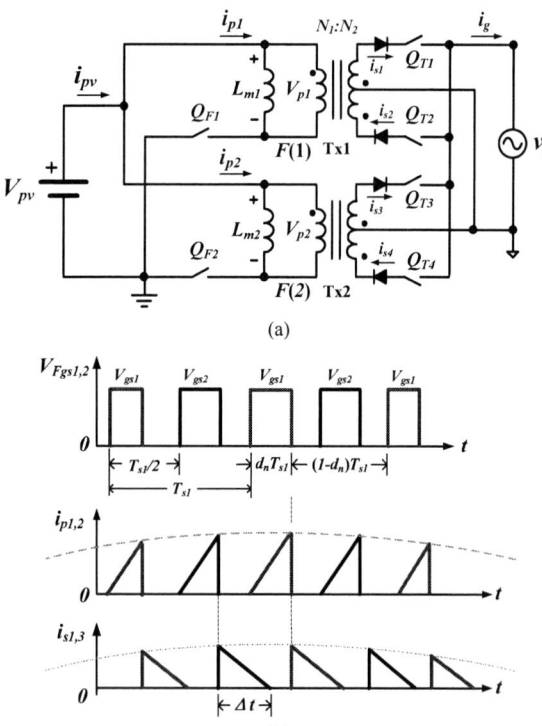

Fig. 2 (a) Simplified model of an interleaved flyback converter, and (b) dynamic waveforms of the flyback conversion in half-interleaved microinverter

In Fig. 2(a), only flyback F(1) is considered for ease of analysis, in which all components for these two flyback converters are assumed identical, such as $L_{m1}=L_{m2}=L_m$. The primary current of Q_{F1} in on period will be

$$i_{p1} = \frac{V_{pv}}{L_{m1}}t \qquad (1)$$

The peak primary current will occur at maximum $\hat{t}_{on}$ in SPWM; that is,

$$\hat{i}_{p1} = \frac{V_{pv}}{L_{m1}}\hat{t}_{on} \qquad (2)$$

and the maximum duty ratio in SPWM can be defined by

$$\hat{d} = \frac{\hat{t}_{on}}{T_s} \qquad (3)$$

We then have the primary current with SPWM given by

$$\begin{aligned} i_{p1}(t) &= \frac{V_{pv}d(t)T_s}{L_{m1}} \\ &= \frac{V_{pv}T_s\hat{d}}{L_{m1}}\sin \omega t \end{aligned} \qquad (4)$$

where $d(t) = \hat{d}\sin \omega t$. For sawtooth time-base reference, the amplitude modulation ratio m_a can be defined by [11]

$$m_a = \frac{\hat{V}_c}{\hat{V}_{tri}} \qquad (5)$$

where the peak control voltage $\hat{V}_c$ is equivalent to the maximum duty ratio $\hat{d}$ occurs at $\omega t = \pi/2$. Since

$$d = \frac{t_{on}}{T_s} = \frac{V_c}{\hat{V}_{tri}} \qquad (6)$$

and

$$\hat{d} = m_a \qquad (7)$$

We then have

$$i_{p1}(t) = \frac{V_{pv}T_s m_a}{L_{m1}}\sin \omega t \qquad (8)$$

and

$$d(t) = m_a \sin \omega t \qquad (9)$$

When the switch Q_{F1} is in off state, the induced voltage from magnetizing current at primary is equal to the flyback voltage from grid utility; that is,

$$\frac{di_{L_m}}{dt} = -\frac{V_g}{L_{m1}}\frac{N_1}{N_2} \qquad (10)$$

where V_g is the segment voltage of the instantaneous $v_g(t)$ at each switching period T_s while the maximum peak $\hat{V}_g$ will present at $\omega t = \pi/2$. In steady-state and during switching period T_s, we can approximately give

$$\frac{\Delta i_{Lm1}}{\Delta t} = \frac{i_{Lm1}}{(1-d)T_s} \approx -\frac{V_g}{L_{m1}}\frac{N_1}{N_2} \qquad (11)$$

and the segment current of i_{Lm1} will be

$$I_{Lm1} \approx -\frac{V_g(1-d)T_s}{L_{m1}}\frac{N_1}{N_2} \qquad (12)$$

During the off-period of Q_{F1}, the current i_{s1} at secondary side with grid voltage considered will be

$$\begin{aligned} i_{s1}(t) &\approx \frac{(1-\hat{d}\sin \omega t)T_s}{L_{s1}}\hat{V}_g \sin \omega t \\ &= \frac{\hat{V}_g \sin \omega t}{L_{s1}} - a(t) \end{aligned} \qquad (13)$$

where

$$a(t) = \frac{\hat{V}_g \hat{d} T_s}{L_{s1}} (\sin \omega t)^2 = \frac{\hat{V}_g \hat{d} T_s}{2L_{s1}} (1 - \cos 2\omega t) \qquad (14)$$

and L_{s1} is the inductance looking into the secondary side by

$$L_{s1} = L_{m1} \left(\frac{N_2}{N_1} \right)^2 \qquad (15)$$

Remarkably, a sine-square term in (14) results a constant and a double-frequency items as in (15), which will distort the grid current, using sawtooth time base for SPWM.

III. FORWARD COMPENSATION IN IFC FOR DISTORTION ELIMINATION

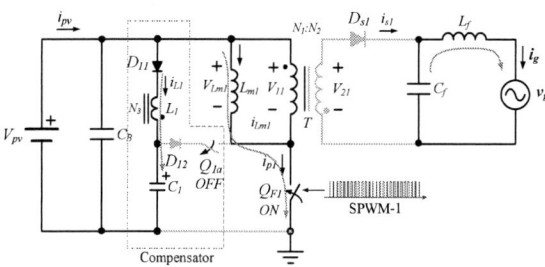

(a)

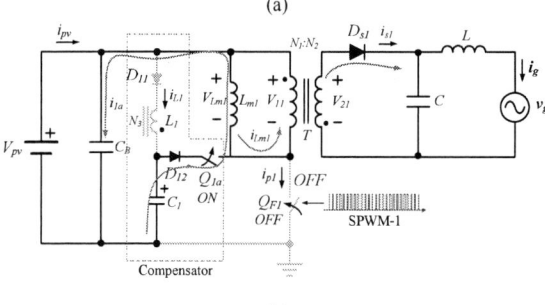

(b)

Fig. 3 Proposed forward compensator for eliminating sin-square term in half interleaved flyback inverter, (a) main switch Q_{1a} in on state and (b) Q_{1a} in off state.

As shown in Fig. 3, the magnetizing current i_{Lm1} at primary, with the reflection of the current i_{s1} in (14) through ampere-turns conserved $i_{Lm1}N_1 = i_{s1}N_2$, can be given by

$$i_{Lm_1}(t) \approx \frac{N_2}{N_1} \left(\frac{\hat{V}_g \sin \omega t}{L_{s1}} - \frac{\hat{V}_g \hat{d} T_s}{L_{s1}} \sin^2 \omega t \right)$$
$$= \frac{N_1}{N_2} \frac{\hat{V}_g \sin \omega t}{L_{m1}} - a'(t) \qquad (16)$$

where

$$a'(t) = \frac{N_1}{N_2} \frac{\hat{V}_g \hat{d} T_s}{L_{m1}} \sin^2 \omega t$$
$$= \frac{N_1}{N_2} \frac{\hat{V}_g \hat{d} T_s}{2L_{m1}} (1 - \cos 2\omega t) \qquad (17)$$

For eliminating the mentioned sin-square distortion in (14), a forward compensation circuit is presented in Fig. 3. An auxiliary winding N_3 in series with a diode D_{11} is built to share part of i_{Lm1}, equal to the amount of $a'(t)$ in (16), to charge C_1 in the on state of Q_{F1}, as shown in Fig. 3(a).

Before Q_{F1} turns on, the voltage on the capacitor C_1 will be V_{pv} through D_{11}. In the off-state of Q_{F1}, the energy stored in C_1 will release to achieve C_1-L_{m1} resonance to cancel the sin-square term through the D_{12}-Q_{12}-L_{m1} loop and eventually goes back to C_B, as Fig. 3(b) shows. The forward compensation result is projected in Fig. 4, in which the resonant current i_{l2} actually reforms the spike portion of the secondary current i_{s1} and eliminates the mentioned double-frequency distortion.

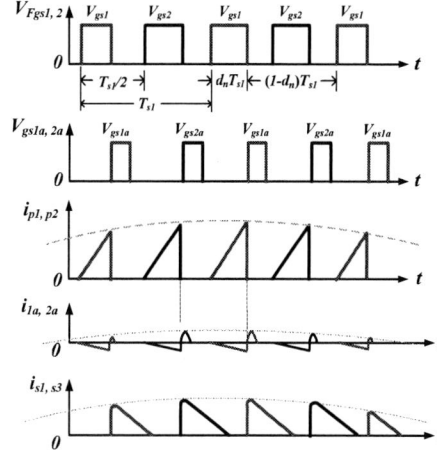

Fig. 4 The projected dynamic waveforms of the secondary current to grid after forward compensation

When Q_{1a} turns off and Q_{F1} turns on, the part current of i_{Lm1} charging to C_1 will be

$$i_3 = \frac{N_1}{N_3} i_{Lm1} \qquad (18)$$

The power in C_1 will be

$$P_{C1} = \frac{1}{2T_s} \left(\frac{N_1}{N_3} i_{Lm1} \right)^2$$
$$= v_{C1} C_1 \frac{dv_{C1}}{dt} \qquad (19)$$

When Q_{1a} turns on and Q_{F1} turns off, the voltage across C_1 will be

$$v_{C1}(t) = \frac{N_1}{N_2} \hat{V}_g \sin \omega t + V_{pv} \qquad (20)$$

We desire to let $i_{C1}(t) = a'(t)$ so as to eliminate the mentioned distortion at grid end. From (17) and (20), the capacitance C_1 will be maximum at $\cos \omega t = -1$, and we can have

$$C_1 = \frac{\hat{d} T_s}{2L_{m1}} \cdot (1 - \hat{d}) T_s \qquad (21)$$

and the auxiliary winding N_3 can also be solved from (19) and (20). The resonant frequency can be given by

$$f_r = \frac{1}{2\pi \sqrt{C_1 L_{m1}}} \qquad (22)$$

IV. MAXIMUM POWER POINT TRACKING

In Fig. 2(a), the peak primary current $\hat{i}_{p1}$ in Q_{F1} on state is equal to the peak $\hat{i}_{pv1}$ drawn from the PV module. This behavior is alternately presented in the two flyback converters, and the peak PV current $\hat{i}_{pv1}$ can be given by

$$\hat{i}_{pv1} = \frac{V_{pv}}{L_{m1}} t_{on} \tag{23}$$

The instantaneous PV current before through the bulk capacitor C_B can then be expressed from (4) and (23) as follows

$$i_{pv}(t) = \frac{V_{pv}\hat{d}}{L_{m1}f_s}\sin\omega t \tag{24}$$

where f_s is switching frequency and ω is the grid frequency. The total average PV current I_{pv} drawn from PV panel with interleave operation can then be approximately given from (24) by

$$I_{pv} = \frac{4V_{pv}\hat{d}}{\pi L_{m1}f_s} \tag{25}$$

If power efficiency η is considered and the loss neglected, the relation of output power P_o (at grid end) and PV power P_{pv} under grid will be

$$\eta I_{pv}V_{pv} = I_g V_g \tag{26}$$

Where V_g and I_g are the rms of the grid voltage and current, respectively; $P_o = I_g V_g$ and $P_{pv} = I_{pv}V_{pv}$. We then have the output current I_g at node M after current synthesis, from (25) and (26); that is,

$$I_g = \frac{4\eta V_{pv}^2 \hat{d}}{\pi V_g L_{m1}f_s} \tag{27}$$
$$= I_S$$

where I_S is the current at node M. The output power to grid will be

$$P_o = \frac{4\eta V_{pv}^2 \hat{d}}{\pi L_{m1}f_s} \tag{28}$$

and then the grid current $i_g(t)$ can be estimated from (27) by

$$i_g(t) = \frac{2\eta V_{pv}^2 \hat{d}}{V_g L_{m1}f_s}\sin\omega t \tag{29}$$
$$= i_s(t)$$

From (29), we then have control-to-output transfer function given by

$$\frac{i_g(t)}{d(t)} = \frac{2\eta V_{pv}^2}{V_g L_{m1}f_s} \tag{30}$$

The grid current $i_g(t)$ generated from the PV module through the IFC under various solar insolation must satisfy the theory of maximum power transfer [9]. Fortunately, the impedance of the PV array is inherently capacitive because of the diffusion and transition capacitances at high frequency; thus, it can conjugate matching the inductive impedance of the IFC. Restated, under maximum power transfer, the impedance Z_{IFC} of IFC should be conjugated to the impedance Z_{pv} of the PV array:

$$Z_{IFC} = \overline{Z}_{pv} \tag{31}$$

where

$$Z_{IFC} = j2\pi f_s L_{m1} \tag{32}$$

and

$$Z_{pv} = \frac{1}{j2\pi f_s C_{pv}} \tag{33}$$

where C_{pv} is the internal capacitance of the PV module. If the internal resistances between the PV array and the IFC are neglected for convenience, then from Eq. (31), at maximum power transfer,

$$L_{m1}C_{pv} = \left(\frac{1}{2\pi f_s}\right)^2 \tag{34}$$

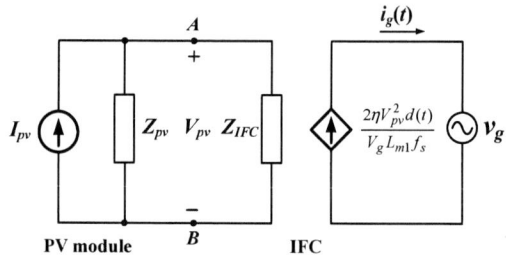

Fig. 5 Maximum power transfer between PV module and grid through IFC

The maximum power transfer between PV module and grid through IFC is modeled in Fig. 5, in which the grid current $i_g(t)$ is linear proportional to the duty variation under constant switching frequency in IFC.

V. DESIGN AND EXPERIMENTS

A. Interleaved sawtooth time-base generation

A 250-W 60Hz PV microinverter drawing energy from two 130-W PV panels connected in series, Kyocera KC230T, is designed and implemented to validate the proposed performance estimation. Quadrature-phase-shift triangular waveforms for SPWM generation are simply achieved by a ZVS-PWM IC UC3875 that can provide an accurate phase shift for interleave operation between the two flyback converter as shown in Fig. 6.

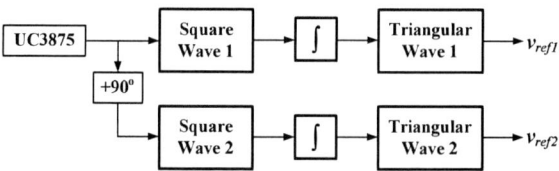

Fig. 6 Quadrature-phase-shift for generating interleave sawtooth time-based signals

Since this approach to grid connection is for a current source inverter (CSI), the two interleaved flyback converters (IFCs) should be modulated by SPWM at primary side before delivering energy to the secondary side. Furthermore, a big ripple fluctuation may occur at the PV module output to interfere the MPPT operation. In words, a power decoupling circuit is necessary to improve the power fluctuation, but not considered in this study [5-7]. We herein use a big bulk capacitor instead of the power decoupling circuit for ease of analysis.

B. Component Calculation

The PV module employed will produce 34.3V and 7.2A at solar insolation 1 kW/m². From the specification, we estimate the parameters as follows. The turns-ratio of the transformer T is $N_1{:}N_2{:}N_3 = 1{:}3{:}2.7$, in which the auxiliary winding N_3 is given by (17) and (19). The magnetizing inductance is 15.6 μH and the resonant capacitance C1 is estimated from (21) as 28 μF (we use 33 μF). The resonant frequency for energy release to eliminate distortion is 7.02 kHz. In order to make MPPT stable, a bulk capacitance 10,000 μF/50V instead of the power decoupling circuit is used. The CF output filter consists of an AC capacitor C_f= 2.2 μF and L_f= 3 mH, of which the response is simulated in Fig. 11. A zero-voltage-switching PWM UC3875 is adopted to provide a quadrature-phase-shift for generating quadrature SPWMs as the two interleave drives. The power switch used is IRFP460 and the power diode is SFA1608G.

C. Maximum Power Point Tracking

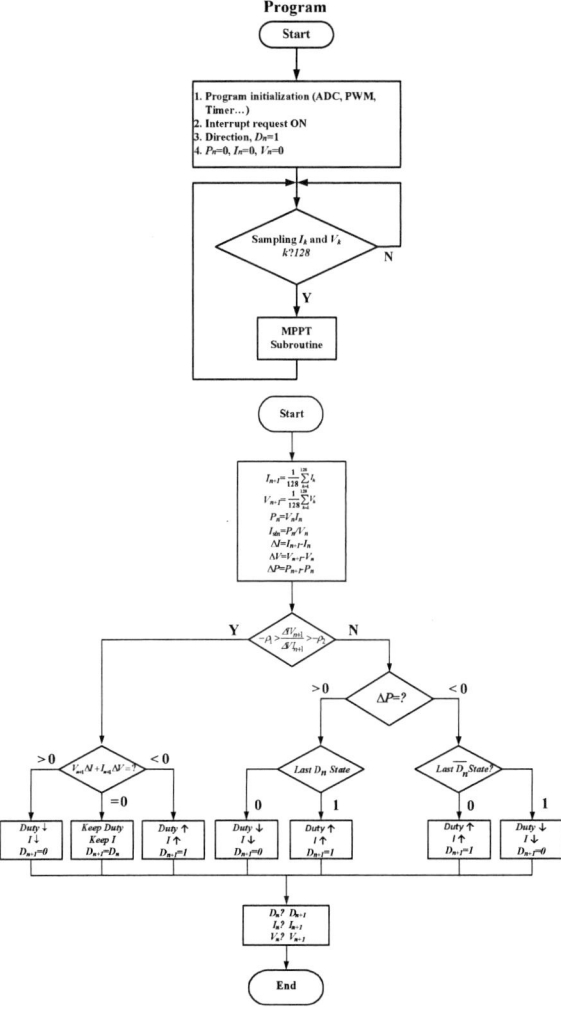

Fig. 7 Tracking flow chart of the PI-INC MPPT using constant-frequency variable-duty (CFVD) control [11]

In this example, the PI-INC) MPPT using CFVD control is adopted [10-11]. The tracking flow chart is described in Fig. 7, in which the program of the PI-INC MPPT is executed by Microchip dsPIC33FJ06GS202. The dynamic tracking of PI-INC MPPT is designed to

suit for maximum power transfer under different solar insolation.

D. Cntrol Strategy

The system control of this microinverter is outlined in Fig. 8, in which PI-INC MPPT will guide the IFC drawing energy from PV module always at MPP without intermittence. The forward compensator adhered on the IFC is programed in the Microchip.

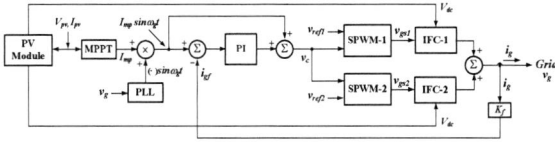

Fig. 8 System control configuration of the interleaved PV microinverter

E. Experiments

The primary and secondary currents of the IFC under solar insolation at 300 W/m^2 and 1kW/m^2 are respectively measured in Fig. 9, in which the waveforms are exactly the same as those predicted in Fig. 1(a). The corresponding grid currents with respective to the grid voltage through CF filter are shown in Fig. 10, respectively. The frequency responses of the CF filter for different solar insolation are simulated in Fig. 11, in which the least phase margin is 36° at 1 kW/m^2 ensuring system stable under all solar insolation.

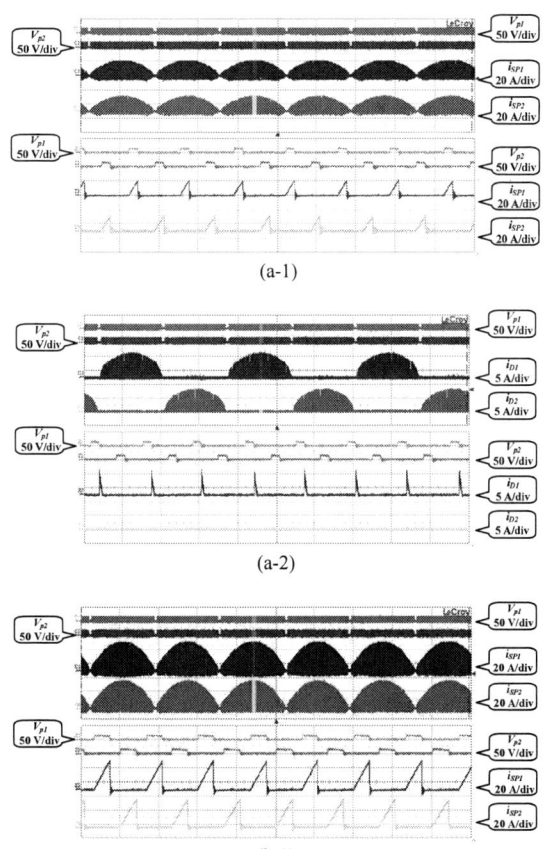

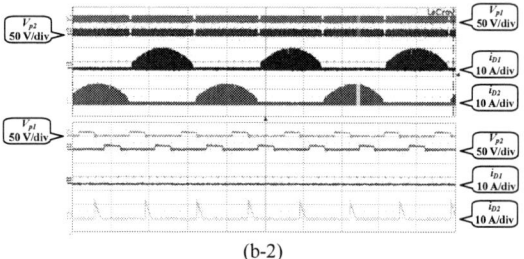

(b-2)

Fig. 9 The interleaved currents and the grid current without using forward compensation: (a-1) and (a-2) at solar insolation 300 W/m² with control duty $d = 0.15$, and (b-1) and (b-2) at 1 kW/m² with $d=0.27$, respectively; all measurements are before low-frequency cyclo-conversion, including amplified waveforms at constant switching frequency $f_s = 15$ kHz; Hor.: 50 μs/div.

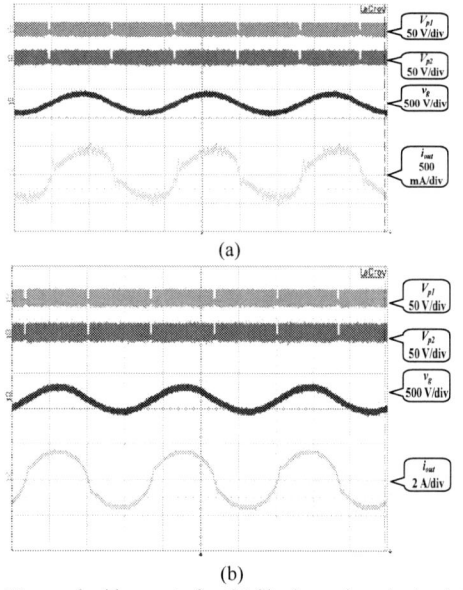

(a)

(b)

Fig. 10 Measured grid current after CF filtering under solar insolation (a) 300 W/m² with control duty $d = 0.15$ and (b) 1kW/m² with $d=0.27$, respectively; switching frequency 15 kHz.

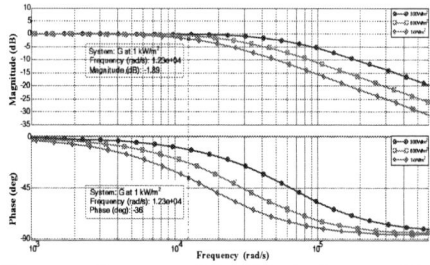

Fig. 11 Frequency Responses of the PV microinverter with CF filter under different solar insolation

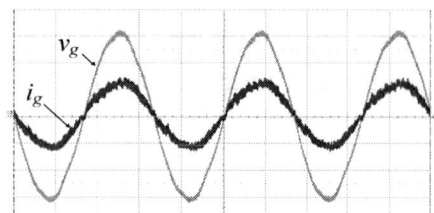

Fig. 12 The grid current after using forward compensation circuit measured at 1 kW/m². Ver: 50V/div. for v_g and 2A/div. for i_g; Hor: 5ms/div.

The grid current after forward compensation is measured as shown in Fig. 12, in which the distorted waveforms in Fig. 10 are removed.

VI. CONCLUSIONS

This paper presents an interleaved PV microinverter using two-stage unfolding commutation to grid connection, in which a forward compensation effectively removes the double-frequency distortion resulting from using sawtooth time-based SPWM. The accurate 90°-phase-shift SPWM simply keeps the IFC operating in DCM and ensures the synthesized current at node M without aliasing. In addition, the PI-INC MPPT using CFVD control can successfully guide the PV microinverter pumping PV energy without intermittence against different solar insolation.

ACKNOWLEDGMENT

The authors would like to thank the National Science Council for the research support of projects: NSC 102-2221-E-415-005 and NSC 101-2221-E-033-051-MY2. Besides, we also thank Mr. Huang Dagang for his assistance in some experiments.

REFERENCES

[1] S. B. Kjaer, J. K. Pedersen, and F. Blaabjerg, "A Review of Single-Phase Grid-Connected Inverters for Photovoltaic Modules," *IEEE Trans. Ind. Appl.*, Vol. 41, No. 5,. pp. 1292-1306, 2005.

[2] T. Shimizu, K. Wada, N. Nakamura, "Flyback-Type Single-Phase Utility Interactive Inverter with Power Pulsation Decoupling on the DC Input for an AC Photovoltaic Module System," *IEEE Trans. Power Electron.*, Vol. 21, Issue 5, pp. 1264-1272, 2006.

[3] F. Zhang and C. Gong, "A New Control Strategy of Single-Stage Flyback Inverter," *IEEE Trans. Ind. Electron.*, Vol. 56, No. 8, pp. 3196-3173, 2009.

[4] Y.H Ji, D.Y. Jung, D.Y. Jung, T.W. Lee, and C.Y. Won, "A current shaping method for PV-AC module DCM-flyback inverter under CCM operation," *Proc. IEEE Power Electron. Motion Control Conf.-ECCE-Asia*, pp. 2598-2605, 2011.

[5] H. Hu, S. Harb, N. H. Kutkut, J. Z. Shen, I. Batarseh, "A Single-Stage Micro-Inverter Without Using Eletrolytic capacitors," *IEEE Trans. Power Electron.*, Vol. 28, Issue 6, pp. 2677-2687, 2013.

[6] Z. Zhang, X.F. He, and Y.F. Liu "An Optimal Control Method for Photovoltaic Grid-Tied-Interleaved Flyback Microinverters to Achieve High Efficiency in Wide Load Range, *IEEE Trans. Power Electron.*,Vol. 28, No. 11, pp. 5074-5087, 2013.

[7] Y. H. Kim, J. G. Kim, Y. H. Ji, C. Y. Won, T. W. Lee, "Flyback Inverter using Voltage Sensorless MPPT for AC Module Systems," *Int'l. Power Electron. Conf. (IPEC)*, pp. 948-953, 2010.

[8] S. Zengin, F. Deveci, M. Boztepe, "Decoupling Capacitor Selection in DCM Flyback PV Microinverters Considering Harmonic Distortion," *IEEE Trans. Power Electron.*, Vol. 28, Issue 2, pp. 816-825, 2013.

[9] C. A. Desoer and E, S. Kuh, *Basic circuit theory*, New York: McGraw-Hill, pp. 322-324, 1969.

[10] K. H. Hussein, I. Muta, T. Hoshino, and M. Osakada, "Maximum photovoltaic power tracking: an algorithm for rapidly changing atmospheric," *IEE Proc. Gener. Transm. Distrib*, vol. 142, no.1, pp. 59-64. January 1995.

[11] G. C. Hsieh, H. I. Hsieh, C. Y. Tsai, and C. H. Wang, "Photovoltaic Power-Increment-Aided Incremental-Conductance MPPT with Two-Phased Tracking" *IEEE Trans. Power Electron.*, Vol. 28, No. 6, pp. 2895 - 2911, 2013.

Current sensorless MPPT Method for a PV Flyback Microinverters Using a Dual-mode

June-Hee Lee
Department of Electrical and
Computer Engineering.
Ajou University, Suwon, Korea
E-mail: ljh20609@ajou.ac.kr

June-Seok Lee
Department of Electrical and
Computer Engineering.
Ajou University, Suwon, Korea
E-mail: junpb@ajou.ac.kr

Kyo-Beum Lee
Department of Electrical and
Computer Engineering.
Ajou University, Suwon, Korea
E-mail: kyl@ajou.ac.kr

Abstract— This paper proposes a current sensorless maximum power point tracking (MPPT) control strategy for a dual-mode photovoltaic (PV) module-type interleaved flyback inverter (ILFI). A flyback inverter, which is one of many transformer topologies, is generally used because of its simplicity and low cost. The topology used in this paper is the interleaved topology; therefore, the secondary ripple current of the transformer can be reduced. The flyback inverter is typically operated in three modes: continuous conduction mode (CCM), discontinuous conduction mode (DCM) and boundary conduction mode (BCM). This paper uses a dual-mode control that includes both the DCM and BCM operations. In addition, the simulations are conducted to verify the performance and effectiveness of the proposed current sensorless MPPT control strategy.

Keywords— *photovoltaic (PV), interleaved flyback inverter (ILFI), current sensorless MPPT, discontinuous conduction mode (DCM), boundary conduction mode (BCM), dual mode*

I. INTRODUCTION

Recently, the development of alternative energy and renewable energy has increased because of the indiscriminate use of fossil fuels and increasing environmental pollution. Photovoltaic (PV) technology converts sunlight into energy through the photoelectric effect, which generates electricity by using a PV cell. Hence, the energy source is clean and unlimited and generates necessary amount of energy. Moreover, PV technology is easy to maintain, can be automated, and has a long life cycle (over 20 years). Therefore, many researches have been conducted on PV generation [1].

A PV power converter system (PV-PCS) uses numerous types of methods to connect the PV cell modules and the PCS: central inverter, multi-string inverter, string inverter, and module-integrated converter (MIC) [2]-[4]. An MIC, which is a PV-PCS attached to a PV cell, is easy to install because the dc line is not required. Moreover, it provides maximum efficiency at the maximum power point (MPP) and can reduce the price of the inverter because it can be produced on a large scale. Therefore, many researches on the MIC have been reported recently [5].

Currently, many studies on the MIC control methods have two reference methods: the discontinuous

conduction mode (DCM) and boundary conduction mode (BCM). The dominant losses of the MIC at light loads include the driving and turn-off losses of the switch and the transformer core loss whereas the dominant losses of the MIC at heavy loads include the conduction loss of the switch and diode and the core and copper losses of the transformer. Therefore, the system loss can be considerably reduced if the system operates in the DCM at a light load and in the BCM at a heavy load. The proposed method is a dual-mode control method [6]-[7].

The PV-PCS has a huge impact on temperature and solar radiation, and the available time is limited. Moreover, the PV array has nonlinear power versus voltage characteristics; therefore, the maximum power tracking technology at the PV cell is an essential part of the PV system. The P&O MPPT control algorithm is used to modify the operating voltage or current of the PV array until the MPP has been reached. For example, if the power output of a cell increases when the voltage applied across a cell is increased, the system increases the operating voltage until the power output begins to decrease. Once this happens, the voltage decreases to return to the MPP. This process continues until the MPP is reached. Therefore, the P&O MPPT control method is generally used because of its simplicity [8].

The conventional MPPT control method for MIC requires voltage and current sensors on the output side of the PV. After obtaining the data from the sensors, the system is controlled using the P&O MPPT control method. In this case, the system cost increases because of the bulky size and expensive price of the sensor [9]. Therefore, a current sensorless MPPT control method for the interleaved flyback inverter (ILFI) with a dual-mode control and only one voltage sensor for MPPT control, is proposed in this study.

The dual-mode control method for the ILFI is discussed in Section II. Section III describes the proposed current sensorless MPPT control scheme. In sections IV, the simulations conducted to prove the feasibility of the proposed current sensorless MPPT control method are presented.

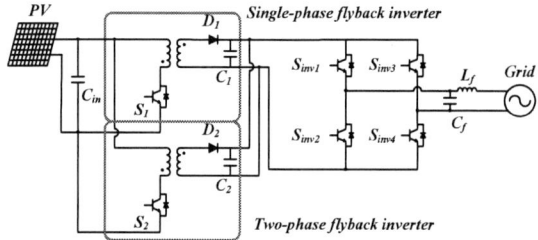

Fig. 1. Grid-connected ILFI.

II. DUAL-MODE CONTROL FOR CURRENT SENSORLESS MPPT OF ILFI

A grid-connected ILFI with a PV is shown in Fig. 1. The system consists of two sets of flyback inverters (with primary-side switches S1 and S2, transformers TR1 and TR2, secondary-side diodes D1 and D2 and secondary-side capacitor C1 and C2), a H-bridge inverter, and an output C-L filter. The ILFI splits the current and transfers the powers, which are shifted by 180°. Therefore, the ILFI is a more comfortable method for reducing the ripple current when compared with the single-stage flyback inverter. Furthermore, by using the interleaved operation, the loss of each switch can be reduced [7]. The H-bridge inverter is built with four switches (Sinv1–Sinv4). When switches S inv1 and Sinv4 are on (Sinv2 and Sinv3 are off), a positive voltage will be applied across the grid. When switches Sinv1 and Sinv4 are off (Sinv2 and Sinv3 are on), the voltage is reversed.

Figs. 2 and 3 show the DCM and BCM operations. As shown in Fig. 2, the transformer current increases linearly during the main switch-on time and decreases linearly during the main switch-off time. During the operation, the dropping of current to zero before the end of the main switch-off time is called the DCM operation. Generally, the DCM operation is used to control ILFIs because it is very simple. In addition, this mode can reduce the frequency-dependent, turn-off, and transformer core losses at light loads [10]. As shown in Fig. 3, the current increases once the main switch-off time ends; this is called the BCM operation. As a result, the switching frequency always varies as soon as the transformer current drops to zero. Therefore, the BCM operation is more complicated than the DCM operation. However, when the load is heavy, the BCM operation is more suitable than the DCM operation because the transformer loss and size can be reduced [10]. The system proposed in this paper operates in the dual mode by taking advantage of each mode. Here, the dual mode indicates that the single-phase DCM operation is applied at a flyback inverter during the light load interval, and the interleaved BCM operation is applied at two flyback inverters during the heavy load interval. Fig. 4 shows the dual-mode operation.

A. Single-stage DCM operation

If the output power is lower than the reference power,

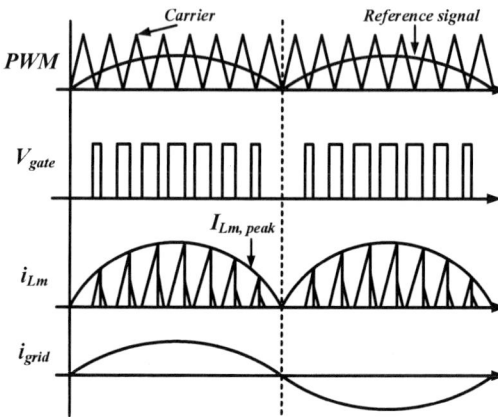

Fig. 2. DCM operation.

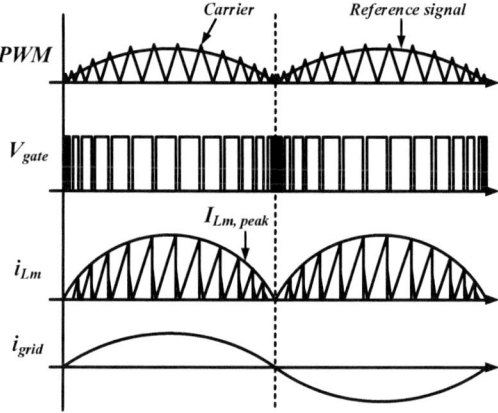

Fig. 3. BCM operation.

the single-phase DCM operation is applied. Therefore, the switching frequency is fixed, but the duty ratio varies. To calculate the duty ratio, the output voltage, which is a grid voltage (V_o), of the flyback inverter is determined as

$$V_o = n \cdot \frac{D}{1-D} \cdot V_{pv} \qquad (1)$$

where is the turn ratio of the transformer. Here, V_{pv} is the input voltage, and D is the duty ratio.

The maximum duty ratio for DCM operation is given by

$$D_{max} = \frac{V_{o,peak}}{nV_{pv} + V_{o,peak}} \qquad (2)$$

Here, $V_{o,peak}$ denotes the peak value of the output voltage. The duty ratio for the DCM operation is always less than D_{max} and can be given by

$$D < \frac{V_{o,peak}}{nV_{pv} + V_{o,peak}} \qquad (3)$$

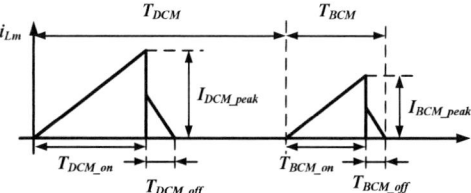

Fig. 5. Operation principle of the DCM and BCM operations.

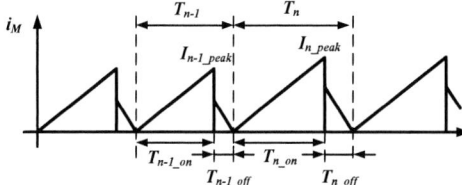

Fig. 6. Operation principle of the BCM operation.

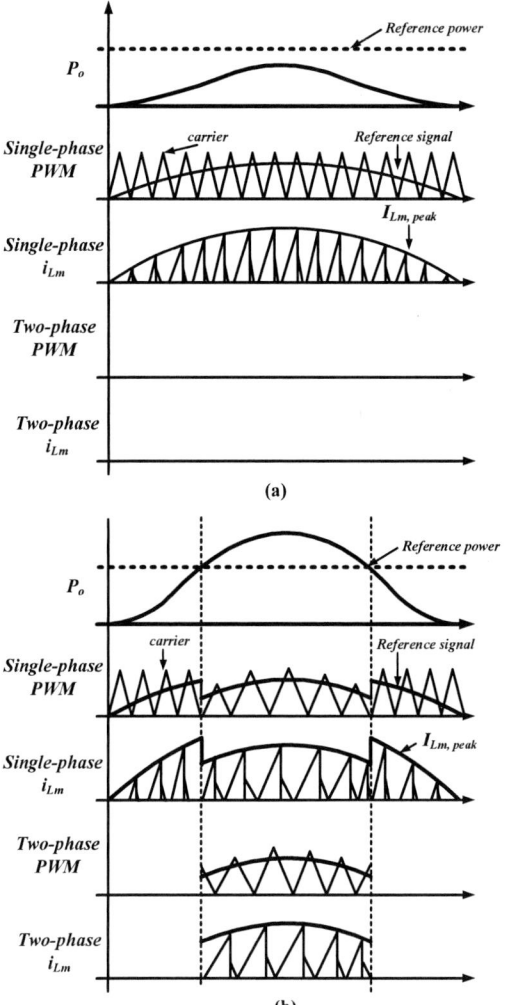

Fig. 4. Operation of the ILFI in the (a) DCM and (b) dual mode.

In DCM operation, the duty ratio is varied while the switching frequency is constant.

B. Interleaved BCM operation

If the output power is greater than the reference power, the interleaved BCM operation is applied at two flyback inverters. The BCM operation must change the duty ratio, and the switching frequency at the secondary current of the transformer drops to zero.

First, the peak value of the primary current ($I_{Lm, peak}$) of the flyback inverter is given as

$$I_{Lm, peak} = \frac{V_i}{L_m} t_{on} = \frac{V_i}{L_m} DT \tag{4}$$

where t_{on} is the switch-on time, and T is the switching period. The primary current (i_{Lm}) has a sinusoidal waveform and is expressed as

$$i_{Lm}(t) = I_{Lm, peak} \sin \omega t, \ 0 \leq \omega t \leq 2\pi \tag{5}$$

The duty ratio has to be modified and can be given as

$$d(t) = D_{\max} \sin \omega t, \ 0 \leq \omega t \leq 2\pi \tag{6}$$

where D_{max} is the duty ratio value of the switching cycle at $\omega t = \pi/2$. In the grid-connected system, the grid frequency is applied to the reference signal.

By combining (1), (4), and (6), the switching time (T) for the BCM operation is calculated as

$$T = \frac{L_m}{V_i \cdot d(t)} I_{Lm, peak} \tag{7}$$

From (7), the variable switching time for the BCM operation can be calculated. In the BCM operation, the square root of $I_{Lm,peak}$ is taken because the interleaved BCM operation is applied.

Fig. 5 shows the transformer current during one switching cycle in the DCM and BCM operations. $I_{Lm, DCM, peak}$ and $T_{DCM, on}$ denote the transformer current and switch-on time, respectively, in the DCM operation. $I_{Lm, BCM, peak}$ and T_{BCM}, denote the transformer current and switch-on time, respectively, in the BCM operation. Equations (8) and (9) represent the average values of primary current during the DCM and BCM operations, respectively, for one switching cycle.

$$I_{DCM, avg} = \frac{I_{Lm, DCM, peak} \cdot T_{DCM, on}}{2T_{DCM}} = \frac{I_{Lm, DCM, peak} \cdot D_{DCM}}{2} \tag{8}$$

$$I_{BCM, avg} = \frac{I_{Lm, BCM, peak} \cdot T_{BCM, on}}{2T_{BCM}} = \frac{I_{Lm, BCM, peak} \cdot D_{BCM}}{2} \tag{9}$$

For the same output power between the DCM and BCM operations, the primary average currents during switching cycle of each mode are the same. Hence, by

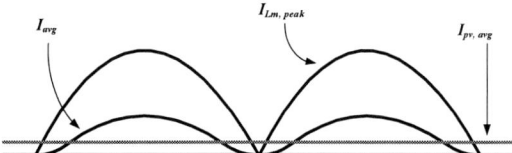

Fig. 7. Average current per half-grid frequency for MPPT control

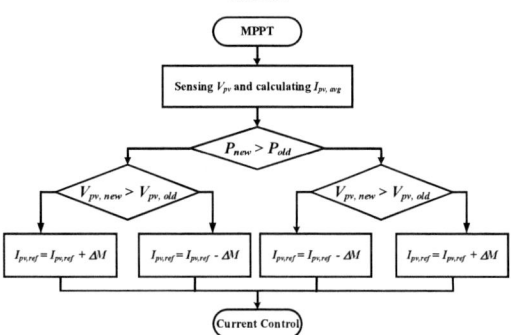

Fig. 8. Flowchart of the current sensorless MPPT control algorithm.

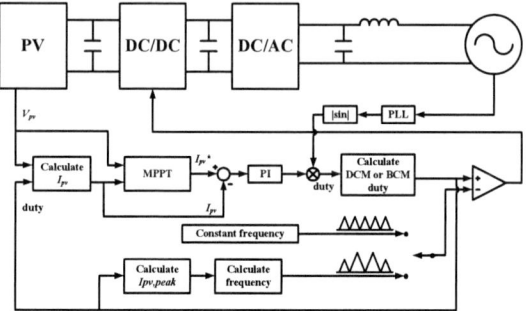

Fig. 9. Block diagram of the proposed method

reduce cost and system size. In this paper, the proposed method calculates the PV current from the average value of the current. Therefore, the maximum power is calculated with the PV voltage and calculated current.

For the MPPT control, initially, the PV output current is calculated using the switching frequency and duty ratio, as given in (8) and (9). The average value of the primary current ($I_{pv,avg}$) for half of the grid period is calculated as

$$ I_{pv,avg} = \frac{\int_0^{T_{half of grid}} I_{avg} dt}{T_{half of grid}} \tag{12} $$

where Thalf of grid denotes half of the grid period. Here, $I_{pv,avg}$ is the dc value in (12), as shown in Fig. 7.

Fig. 8 shows the operation of the current sensorless MPPT control algorithm. First, the PV output power (P_{pv}), given by (13), is calculated from the product of the measured PV voltage and calculated PV current as

$$ P_{pv} = V_{pv} \cdot I_{pv,avg} \tag{13} $$

After the microcontroller unit (MCU) has memorized the previous PV output power (P_{old}) and PV voltage ($V_{pv,old}$), the reference current ($I_{pv,ref}$) is calculated by comparing with the present PV output power (P_{new}) and PV voltage ($V_{pv,new}$) for MPPT control.

Fig. 9 shows block diagram of the proposed dual-mode current sensorless MPPT control method for the ILFI.

IV. SIMULATIONS

In this paper, the simulations are performed using the PSIM tool. The PV module-type ILFI with a dual mode

combining (4), (8), and (9), the BCM operation duty ratio (D_{BCM}) is expressed as

$$ D_{BCM} = D_{DCM} \cdot \sqrt{\frac{T_{DCM}}{T_{BCM}}} \tag{10} $$

As shown in Fig. 6, in the BCM operation, the new duty ratio is used in the same way as in (8)–(10). Therefore, the system can operate in dual modes using (1), (7), and (10).

For using the dual mode, the power (Pout) is calculated instantaneously, and the reference power (P_{ref}) decides the operation mode. From V_{pv} and the primary average current (I_{avg}) of (8) and (9), P_{out} is given as

$$ P_{out} = V_{pv} \cdot I_{avg} \tag{11} $$

If P_{out} is less than P_{ref}, namely, a light load, the single-phase DCM operation is applied at a flyback inverter. On the contrary, if Pout is greater than P_{ref}, mainly, a heavy load, the interleaved BCM operation is applied at two flyback inverters.

III. CURRENT SENSORLESS MPPT CONTROL FOR PV MODULE-TYPE DUAL-MODE INTERLEAVED FLYBACK INVERTER

The P&O MPPT algorithm is generally used because of its ease of implementation. In the conventional P&O MPPT algorithm, the PV output power, which is given by the product of the PV voltage and PV current, must be calculated. Usually, the current sensor is expensive and bulky; hence, it has an effect on the cost of the whole system. Therefore, an alternative method is required to

TABLE I. PARAMETERS FOR SIMULATIONS

Vgrid (rms)	110 Vac (rms)	P_o	110 W
Cin	10.1 mF	Transformer turn ratio	1:2
Cf	0.22 μF	Lm	13.99 μH
Lf	1 mH	Lp,lk	0.01 μH
Rdamping	500 Ω	Frequency (DCM)	50 KHz

The 2014 International Power Electronics Conference

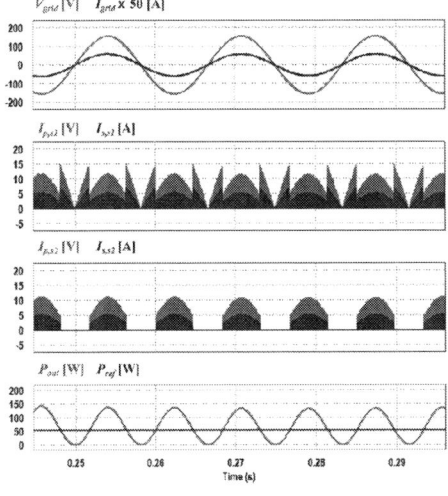

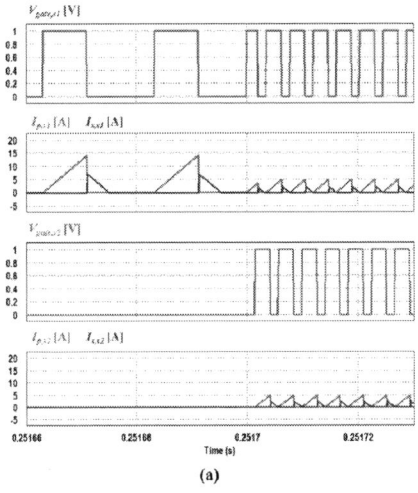

Fig. 10. Waveforms of the dual-mode ILFI.

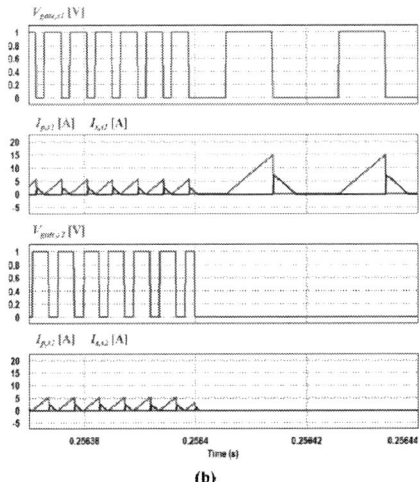

Fig. 11. Closed waveforms of the dual-mode ILFI.

topology, shown in Fig. 1, is used in the simulation. The simulation parameters are listed in Table I.

Fig. 10 shows the operation of the ILFI in with a dual mode. V_{grid} and I_{grid} denote the grid voltage and grid

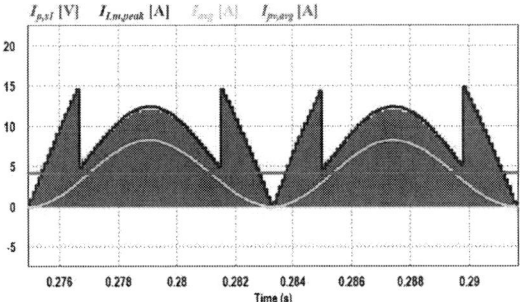

Fig. 12. Key waveform of the dual-mode ILFI.

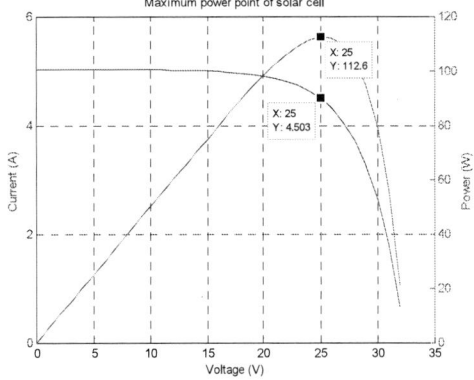

Fig. 13. MATLAB PV-cell module simulation results of the proposed MPPT control scheme.

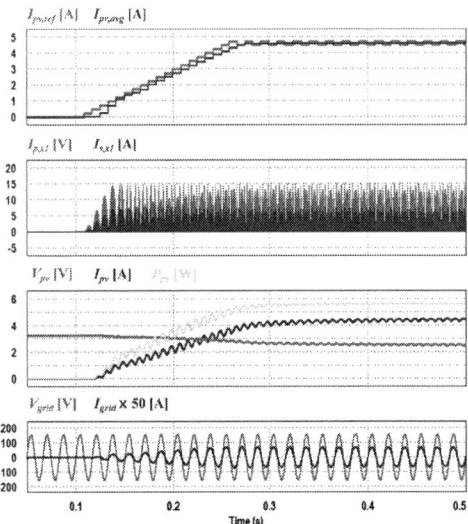

Fig. 14. Proposed MPPT control in the dual-mode ILFI.

current, respectively. $I_{p,s1}$ is the primary current of the single-phase transformer, and $I_{s,s1}$ is its secondary current. $I_{p,s2}$ and $I_{s,s2}$ denote the primary and secondary currents of the two-phase transformer, respectively. P_{out} is the output power calculated from (11), and P_{ref} is the boundary value that separates the operation modes. Here, P_{ref} is set to be 55 W. Therefore, the single-phase DCM operation is applied below P_{ref} whereas the two-phase interleaved

978-1-4799-2706-7/14 $31.00 © 2014 IEEE   536

BCM operation is applied above P_{ref}, as shown in Fig. 10. Once the system enters the two-phase interleaved BCM operation, the frequency and duty ratio are changed by using (7) and (10).

Fig. 11 shows the closed waveforms of the ILFI with a dual mode when the operation mode is changed. Here, $V_{gate,s1}$ and $V_{gate,s2}$ are the switching signals for S1 and S2, respectively. Fig. 11(a) shows that the operation mode is changed from the single-phase DCM operation to the interleaved BCM operation. If the mode enters the interleaved BCM operation, $I_{Lm,peak}$ and the frequency for BCM operation are calculated. Consequently, the new gate signal is sent to the switches. On the contrary, as shown in Fig. 11(b), when the mode changes from the interleaved BCM operation to the single-phase DCM operation, the two-phase flyback inverter switch (S2) turns off.

Fig. 12 shows the key waveform of the system operation. $I_{Lm,peak}$ is the calculated value from (5). $I_{DCM(BCM),avg}$ is the calculated average current, and $I_{pv,avg}$ is the calculated average current during half-grid frequency. From (4)–(9), the calculated values are closely linked. Now, $I_{pv,avg}$ is the average current of the whole system. $I_{Lm,peak}$ is used to calculate $I_{DCM(BCM),avg}$, and $I_{DCM(BCM),avg}$ is used to obtain $I_{pv,avg}$. Further, $I_{pv,avg}$ is the dc value that is applied to the current sensorless MPPT control.

Fig.13 shows the performance of the MATLAB PV-cell module simulation with the proposed MPPT control. In this simulation, the PV-cell module is modeled with the DLL block. Hence, the MPP voltage is 25 V, and the MPP current is 4.503 A. Fig. 14 shows the proposed MPPT control. The proposed MPPT control duty ratio is calculated with I_{ref}, which is generated by the MPPT control, and $I_{pv,avg}$ from (12). As shown in Fig. 14, it is confirmed that $I_{pv,avg}$ follows I_{ref}, and I_{grid} increases when the system power increases. Moreover, the phase of grid voltage is the same as that of the grid current.

V.   CONCLUSION

This paper presents a current sensorless MPPT control strategy for the dual-mode PV module-type ILFI. The conventional MPPT control scheme for the MIC requires voltage and current sensors at the PV output, which increases the system cost because of the large size and high price of the sensor. Although the proposed control scheme is more complicated, the system size is smaller and cheaper than the conventional method.

The simulations results show similar performance thus validating the proposed current sensorless MPPT control method for ILFI.

REFERENCES

[1]   June-Seok Lee and Kyo Beum Lee, "Variable DC-Link Voltage Algorithm with a Wide Range of MPPT for a Two-String PV System," Energies, vol. 6, no. 1, pp. 58-78, Jan. 2013.

[2]   C. Meza, J. J. Negroni, D. Biel, and F. Guinjoan, "Energy-Balance Modeling and Discrete Control for Single-Phase Grid-Connected PV Central Inverters," IEEE Transactions on Industrial Electronics, vol. 55, no. 7, pp. 2734-2743, July 2008

[3]   H. J. Bergveld, D. B¨uthker, C. Castello, T. Doorn, A. d. Jong, R. Otten, and K. Waal, " Module-Level DC/DC Conversion for Photovoltaic Systems: The Delta-Conversion Concept," IEEE Transactions on Power Electronics, vol. 28, no. 4, pp. 2005-2013, April 2013.

[4]   S. A. Khajehoddin, A. Bakhshai, and P. Jain, "A Novel Topology and Control Strategy for Maximum Power Point Trackers and Multi-String Grid-Connected PV Inverters," in IEEE proceeding of APEC, pp. 173-178, 2008.

[5]   A. C. Nanakos, E. C. Tatakis, and Nick P. Papanikolaou, "A Weighted-Efficiency-Oriented Design Methodology of Flyback Inverter for AC Photovoltaic Modules," IEEE Transactions on Power Electronics, vol. 27, no. 7, pp. 3221-3233, July 2012.

[6]   A. C. Kyritsis, E. C. Tatakis and N. P. Papanikolaou, "Optimum Design of the Current-Source Flyback Inverter for Decentralized Grid-Connected Photovoltaic Systems," IEEE Transactions on Energy Conversion, vol. 23, no. 1, pp. 281-293, March 2007.

[7]   Y. H. Ji, D. Y. Jung, J. H. Kim, C. Y. Won and D. S. Oh, "Dual Mode Switching Strategy of Flyback Inverter for Photovoltaic AC Modules," in IEEE proceeding of IPEC, pp. 2924-2929, 2010.

[8]   Y. H. Kim, Y. H. Ji, J. G. Kim, Y. C. Jung, and C. Y. Won, " A New Control Strategy for Improving Weighted Efficiency in Photovoltaic AC Module-Type Interleaved Flyback Inverters," IEEE Transactions on Power Electronics, vol. 28, no. 6, pp. 2688-2699, June 2013.

[9]   N. Kasa, T. Iida and L. Chen, "Flyback Inverter Controlled by Sensorless Current MPPT for Photovoltaic Power System," IEEE Transactions on Industrial Electronics, vol. 52, no. 4, pp. 1145-1152, August 2005.

[10]   Y. Zhang, X. F. He, Z. Zhang and Y. F. Lium "A Hybrid Control Method for Photovoltaic Grid-Connected Interleaved Flyback Micro-Inverter to Achieve High Efficiency in Wide Load Range," in IEEE proceeding of APEC, pp. 751-756, 2013

A Novel Method of Suppressing Inrush Currents of Squirrel-cage Induction Machine using Matrix Converter in Wind Power Generation Systems

Hiroaki Yamada
Graduate School of Life Science and
Systems Engineering
Kyushu Institute of Technology
2-4, Hibikino, Wakamatsu, Kitakyushu, Fukuoka, Japan
hiroaki-ymd@life.kyutech.ac.jp

Tsuyoshi Hanamoto
Graduate School of Life Science and
Systems Engineering
Kyushu Institute of Technology
2-4, Hibikino, Wakamatsu, Kitakyushu, Fukuoka, Japan
hanamoto@life.kyutech.ac.jp

Abstract—**This paper proposes a novel method of suppressing inrush currents of squirrel-cage induction machines using matrix converter in wind power generation systems. In the proposed method, the series-connected side of the matrix converter behaves as a resistor for the induction machine current. Digital computer simulation is implemented to confirm the validity and practicability of the proposed method using PSCAD/EMTDC. Simulation results demonstrate that the proposed inrush current suppressor can suppress the inrush current perfectly.**

Keywords—Inrush current, matrix converter, squirrel-cage induction machine, wind power generation system

I. INTRODUCTION

A number of wind turbines have been constructed and are being used for commercial uses, and their number is increasing yearly. The induction machines (IMs) used with the wind turbines are cumbersome when connecting and disconnecting because wind energy, a natural energy source, is not constant. IMs affect the power quality during the process of connecting the wind turbine to the grid [1] . This process leads to a voltage drop, sometimes over current and a voltage flicker, which are caused by the inrush current of the IM. Soft starters, which consists of anti-parallel thyristors, are widely used for suppressing the inrush current in the large-capacity wind turbines [2]. However it generates harmonic currents and injects them to the grid because the connecting period with the soft starter is less than 5seconds in large-capacity wind turbine systems. Even if each wind turbine injects a small amount of harmonic currents to the grid, a large amount of harmonics is injected by multiple wind turbine systems. These harmonic currents result in harmonic contamination on the grid in wind farm areas. Thuringer et al. have proposed the inrush current suppressor with resistors [3]. The inrush current suppressor suppresses the inrush current by series-connected resistor. No inrush current occurs by using this suppressor. However, the energy loss consumption is very large for suppressing inrush current. To avoid the harmonic contamination problem caused by the soft starter, Shinohara et al. have proposed a method of compensating harmonic currents using the shunt active filter [4]. However, a large capacity PWM converter is needed because the PWM converter

supplies the fundamental reactive power for the IM in the steady state. Yamada et al. have proposed the harmonic currents compensator for wind power generation systems with the soft starters using a small-rated hybrid acitve filter [5]. However, the inrush current for the passive filter and leading currents must suppress under the starting condition of the proposed harmonic current compensator.

This paper proposes a novel method of suppressing the inrush current of the IM using matrix converter (MC) in wind power generation systems. The input side of the MC is connected in pararrel to the receiving-end. The output side of the MC is connected in series through matching transformers between the receiving-end and IM. As the series-connected side of the MC behaves as a resistor to the IM current, no inrush phenomena occurs without a soft starter. Even though the active power flows into the MC during the inrush current suppression, the flow active power goes through the MC into the source side. The direct duty ratio pulse width modulation (DDPWM) method [6] is used to generate the switching pattern of the MC in this paper. In DDPWM, the output voltage is synthesized by updating the duty ratio values at each switching period directly. Moreover, the DDPWM uses all phases of the input voltage to synthesize the output voltage during the switching period. Thus, the input current is not distorted. Digital computer simulation is implemented to confirm the validity and practicability of the proposed suppression method on PSCAD/EMTDC. Simulation results demonstrate that the proposed inrush current suppressor can suppress the inrush current perfectly.

II. CONVENTIONAL WIND POWER GENERATION SYSTEM

Fig. 1 shows a wind power generation system with a conventional soft starter. The soft starter consists of anti-parallel thyristors. α is the control angle for the thyristors. The soft start works from the 0.95pu of the rotating speed to the synchronous speed for increasing the speed. The applied voltage of the IM is regulated by this soft starter. The soft starter is shorted when the rotating speed reaches at the synchronous speed.

978-1-4799-2706-7/14 $31.00 © 2014 IEEE

The 2014 International Power Electronics Conference

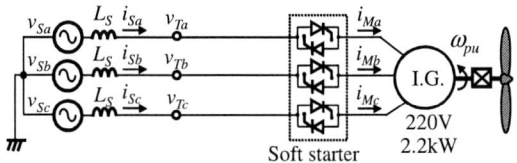

Fig. 1. A wind power generation system with a conventional soft starter.

TABLE I.   IM PARAMETERS.

Rated power		[kW]	2.2
Phase voltage	V_S	[V]	127
Angular frequency	ω_{IM}	[rad/s]	379.6
Magnetizing susceptance	b_0	[Ω]	65.3
Stator resistance	r_1	[Ω]	0.86
Stator leakage impedance	x_1	[Ω]	1.93
Rotor resistance	r_2'	[Ω]	1.30
Rotor leakage impedance	x_2'	[Ω]	1.93

Digital computer simulation is implemented to review the conventional starting methods of wind turbines with IMs. Table I indicates the IM parameters. A 220V, 8A, 2.2kW squirrel-cage IM is used in this paper. The wind turbine which is the radius r=2.2m, pitch angle β=10°, inertia J=0.44kgm^2, and output power P_w=2.2kW is used. The IM is connected to the source when the rotating speed reaches 0.95pu of the synchronous speed (1200rpm).

Fig. 2 shows the simulated waveforms when the IM is connected to the source directly. v_{Ta} is the a-phase voltage at the receiving terminals. i_{Sa} is the a-phase current, ω_{pu} is the IM speed, and v_w is the wind speed. v_w is increased from 2m/s to 3m/s gradually. And then, ω is also increased. The IM is connected to the source whne the ω reaches at 0.95pu. From this simulation result, the inrush current occur by connecting the IM to the source directly. The maximum value of the source current is about four times the rated current 8A of the IM. The voltage sag of the a-phase voltage v_{Ta} is caused by this inrush current.

Fig. 3 shows simulated waveforms of connecting IM to the source with the conventional soft starters. The control angle α is reduced from 180 to 90 degrees gradually. From this simulation result, it can be seen that the inrush current is suppressed by the soft starters. ω is increased gradually as compared to Fig. 2. However, a large amount of harmonic currents generated by the soft starters is included in the source current i_{Sa}.

III.   PROPOSED INRUSH CURRENT SUPPRESSOR

Fig. 4 shows the circuit diagram of the proposed inrush current suppressor with the MC. The IM is connected to the source through the switch SW1 when the IM speed reaches 0.95pu of the synchronous speed. The MC consists of 18 anti-pararrel switches, the input side of the MC is connected in pararrel and the output side of the MC is connected in series between the receiving-end and the IM [7]. The passive filters are connected to the both sides of the MC for reducing the switching ripple caused by switching action. When the IM speed reaches

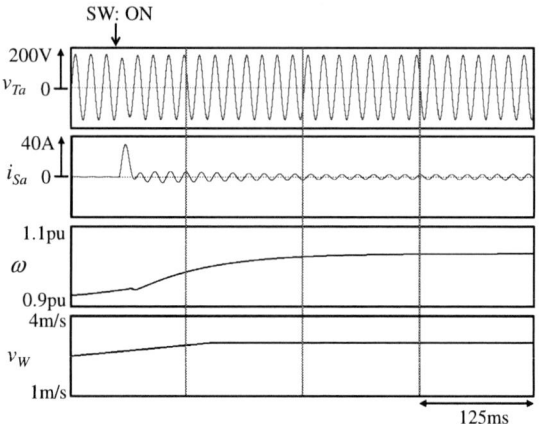

Fig. 2. Simulated waveforms when the IM is connected to the source directly.

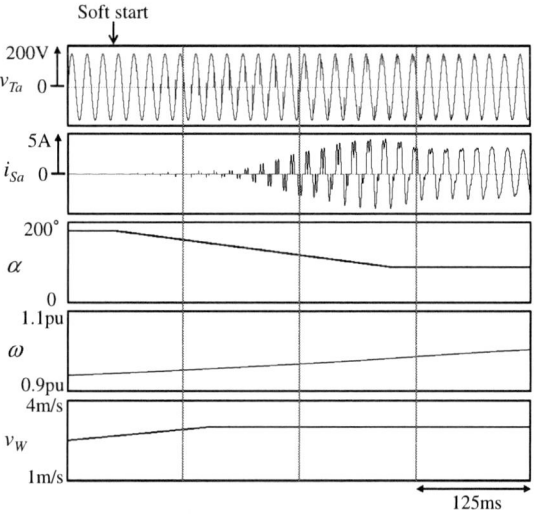

Fig. 3. Simulated waveforms with the conventional soft starter.

the synchronous speed, the SW2 is ON to improve the power factor.

In DDPWM, the output voltage is synthesized by updating the duty ratio values at each switching period directly. Moreover, the DDPWM uses all phases of the input voltage to synthesize the output voltage during the switching period. Thus, the input current is not distorted. Fig. 5 shows the receiving-end voltage and sectors. The MAX, MID and MIN denote the maximum, medium and minimum input voltage values respectively. From the receiving-end voltage in Fig. 5, the MAX, MID, MIN and sectors are divided. The DDPWM has two switching pattern, which the switching pattern I (P-I) is when MAX-MID is larger than MID-MIN and the switching pattern II (P-II) is when MAX-MID is smaller than MID-MIN.

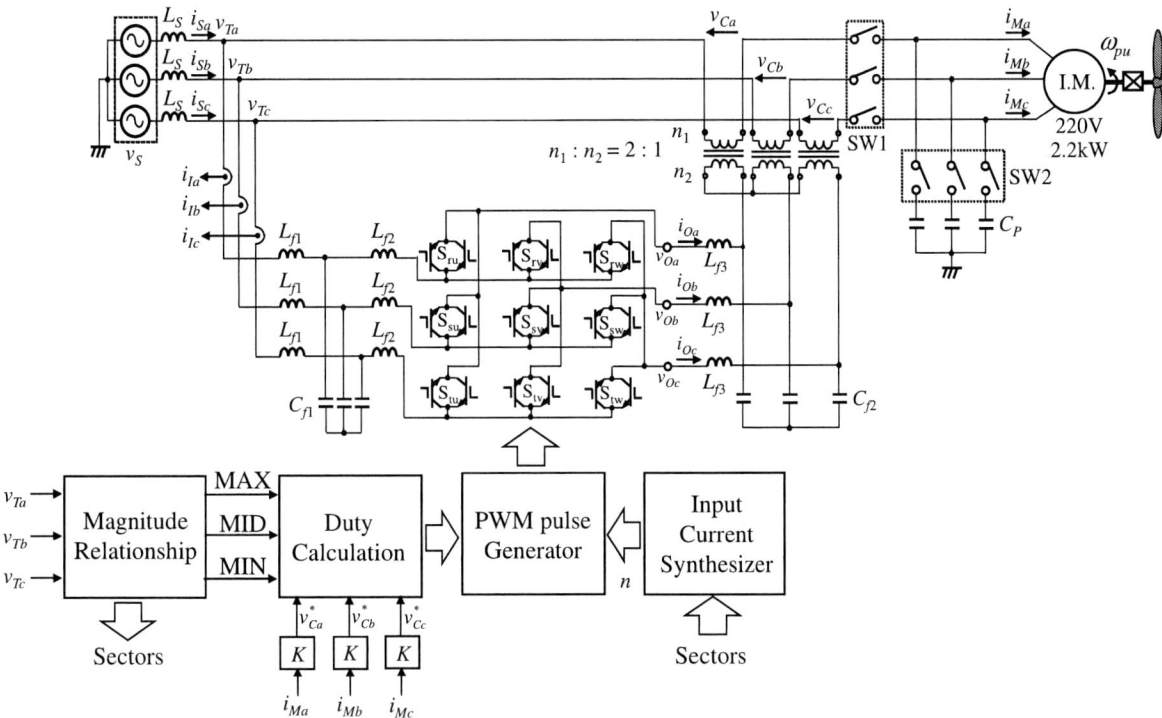

Fig. 4. A circuit diagram of the proposed inrush current suppressor with the MC based on DDPWM.

The duty ratio d_A for a-phase is determined by

$$d_A = \begin{cases} \dfrac{v_{Ca}^* - \mathrm{MAX}}{n\mathrm{MIN} - n\mathrm{MID} + \mathrm{MID} - \mathrm{MAX}} & \text{for P-I} \\[3mm] \dfrac{v_{Ca}^* - (n\mathrm{MAX} - n\mathrm{MID} + \mathrm{MID})}{\mathrm{MIN} - n\mathrm{MAX} - \mathrm{MID} + n\mathrm{MID}} & \text{for P-II,} \end{cases}$$

(1)

where n is monotonous value from 0.5 to 1.0 for synthesizing input current in each sectors and v_{Ca}^* is the output reference voltage. The duty ratio d_B and d_C are obtained in the same manner. Table II indicates the determination of n [6]. β_i is the input current angle which is defined as $\beta_i = \alpha_i + \phi$. α_i is the receiving-end voltage angle. ϕ is the displacement angle between the receiving-end voltage and current. In this paper, $\phi = 0$. Thus n is calculated by the sectors and receiving-end voltage angle.

Fig. 6(a) and Fig. 6(b) show the switching states for a-phase output in the P-I and P-II periods. The switching states generate by comparing the carrier and duty ratio. From this switching states, the duty ratio of the MIN, MID and MAX are decided. Fig. 7 shows the output voltage at the a-phase. $\bar{v}_{OA}$ is the average output voltage. $\bar{v}_{OA}$ is generated by MIN, MID and MAX.

The output reference voltage v_{Cn}^* is given as

$$v_{Cn}^* = K \cdot i_{Mn}, \qquad (2)$$

where K is the control gain and n is the phase a, b, or c. From (1) and (2), the duty ratios are calculated.

Fig. 8 shows the relationship between the speeds and control gain. The proposed inrush current supressor starts

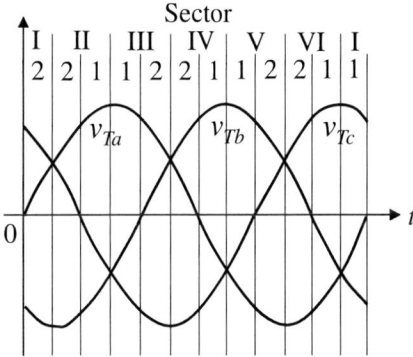

Fig. 5. Receiving-end voltage and sectors.

when the wind speed reaches at the cut-in speed. When the rotating speed of the IM reaches at the synchronous speed, the control gain of the proposed inrush current suppressor is reduced gradually to avoid the traisient phenomenon. t_2-t_1 is 50ms in this paper.

Fig. 9 is the equivalent circuit of the proposed inrush current suppressor in each phase. The MC acts as a resistor for the IM current i_M. The control gain is decided the IM impedance Z_m when the rotating speed is 0.95pu and the rating current of the IM. Therefore, $K > Z_m$.

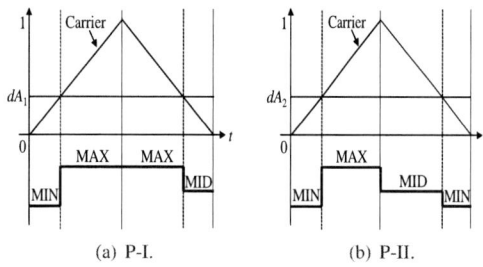

(a) P-I.   (b) P-II.

Fig. 6.   Switching states at a-phase output for one cycle of the carrier.

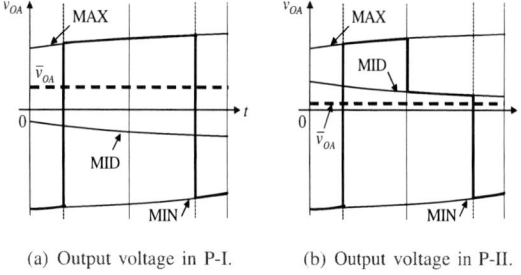

(a) Output voltage in P-I.   (b) Output voltage in P-II.

Fig. 7.   Output voltage at the a-phase for one cycle of the carrier.

IV.   SIMULATION RESULTS

Digital computer simulation is implemented to confirm the validity and practicability of the proposed inrush current suppressor. Table III shows the circuit parameters of Fig. 4. The control gain K is 30Ω.

Fig. 10 shows the simulation waveforms with the proposed inrush current suppressor. v_{Ca} is the output voltage of the MC. When ω_{pu} reaches 0.95pu of the synchronous speed of the IM, the switch SW1 is connected to the source. v_{Ca} and i_{Ma} are the same phase during the starting condition. This means that the series-connected side of the MC performs as a resistor for the IM current. Although the IM is connected to the source directly, no inrush phenomena occur. The peak IM current is about 6A. This means the proposed inrush current suppressor can suppress the inrush current drastically. And then, the voltage sag of the receiving-end voltage was avoided by this inrush current suppression. When ω reaches the synchronous speed 1200rpm, the control gain K is reduced to 0 gradually.

Fig. 11 the FFT analysis of the soft starter and proposed inrush current suppressor. The harmonic componets has been decreased compared between the soft starter and proposed inrush current suppressor. The THD of the soft starter is 64%. In contrast, the THD of the proposed inrush current suppressor is about 11%.

V.   CONCLUSION

This paper has proposed a novel method of suppressing inrush currents of IM using the MC. The output side of the MC which is connected in series between the source and IM behaves a resistor to the IM current. The switching pattern for the MC is generated by DDPWM.

TABLE II.   DETERMINATION OF n.

Sector	n	Monotonicity
I-2	$-\sin(\beta_i + 2\pi/3)/\sin(\beta_i - 2\pi/3)$	1-0.5 down
II-2	$-\sin(\beta_i)/\sin(\beta_i - 2\pi/3)$	0.5-1 up
II-1	$-\sin(\beta_i - 2\pi/3)/\sin(\beta_i)$	1-0.5 down
III-1	$-\sin(\beta_i + 2\pi/3)/\sin(\beta_i)$	0.5-1 up
III-2	$-\sin(\beta_i)/\sin(\beta_i + 2\pi/3)$	1-0.5 down
IV-2	$-\sin(\beta_i - 2\pi/3)/\sin(\beta_i + 2\pi/3)$	0.5-1 up
IV-1	$-\sin(\beta_i + 2\pi/3)/\sin(\beta_i - 2\pi/3)$	1-0.5 down
V-1	$-\sin(\beta_i)/\sin(\beta_i - 2\pi/3)$	0.5-1 up
V-2	$-\sin(\beta_i - 2\pi/3)/\sin(\beta_i)$	1-0.5 down
VI-2	$-\sin(\beta_i + 2\pi/3)/\sin(\beta_i)$	0.5-1 up
VI-1	$-\sin(\beta_i)/\sin(\beta_i + 2\pi/3)$	1-0.5 down
I-1	$-\sin(\beta_i - 2\pi/3)/\sin(\beta_i + 2\pi/3)$	0.5-1 up

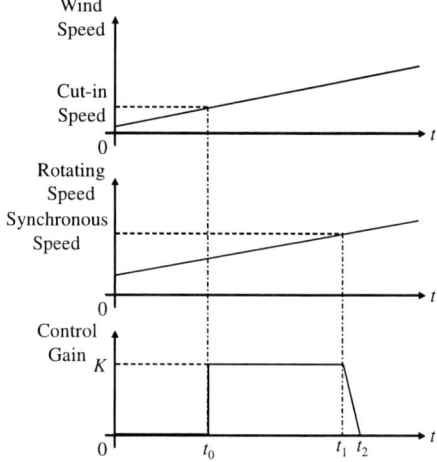

Fig. 8.   Relationship between speeds and control gain.

The basic principle of the proposed method of suppressing the inrush current have been discussed in detail. At first, we have described about the proposed inrush current suppressor. The output side of the MC performs as resistors for the IM current, the inrush current can suppress drastically. Therefore, the soft starters are unnecessary in wind power generation systems with squirrel-cage type induction machines. Digital computer simulation has been implemented to demonstrate the proposed suppressor using PSCAD/EMTDC software. From simulation results, the inrush current has been suppressed within the rated current of the IM. The THD of the source current is 11%. These results reveals that the proposed inrush current suppressor can suppress the inrush current without the soft starter.

REFERENCES

[1] F. D. Kanellos, and N. D. Hatziargyriou, "The Effect of Variable-Speed Wind Turbines on Operation of Weak Distribution Networks ", *IEEE Trans. Energy Conversion*, Vol. 17, No. 4, pp. 543–548, 2002

[2] Y. Sasaki, N. Harada, T. Kai, and T. Sato, A Countermeasure against the Voltage Sag due to a Inrush Current of Wind Power Generation System Interconnecting to a Distribution Line *IEEJ Trans. on Power and Energy*, Vol. 120, No. 2, pp. 180-186, 2000

[3] Torbjörn Thiringer, "Grid-Friendly Connecting of Constant-Speed Wind Turbines Using External Resistors " *IEEE Trans. Energy Conversion*, Vol. 17, pp. 537–5422002

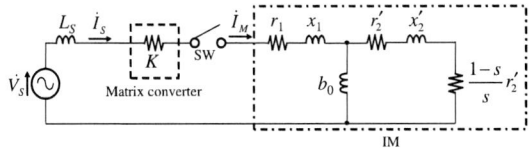

Fig. 9. Equivalent circuit of the proposed inrush current suppressor in each phase.

TABLE III. CIRCUIT CONSTANTS OF FIG. 4.

Source line-to-line voltage	V_S	[V]	200
Source frequency	f	[Hz]	60
Source side impedance	L_S	[mH]	2.9(5%)
Input filter inductance	L_{f1}	[mH]	10
Input filter capacitance	C_{f1}	[μF]	25
Input filter inductance	L_{f2}	[mH]	0.5
Output filter inductance	L_{f3}	[mH]	5
Output filter capacitance	C_{f2}	[μF]	20
Switching frequency	f_{sw}	[kHz]	10

[4] K. Shinohara, K. Yamamoto, K. Iimori, Y. Minari, O. Sakata, and M. Miyake Compensating for Magnetizing Inrush Currents in Transformers using PWM Inverter , *IEEJ Trans. on Industry Applications*, Vol. 124, No. 12, pp. 1173-1181, 2004

[5] H. Yamada, E. Hiraki, and T. TanakaA Starting Method of the Harmonic Current Compensator Using a Hybrid Active Filter for Wind Power Generation Systems with Soft Starters, IEEJ Trans. on Electrical and Electronic Engineering, Vol. 7, S1, pp.S139S145, 2012

[6] Y. Li, N. Choi, B. Han, Kyoung M. Kim, B. Lee, and J. Park, " Direct Duty Ratio Pulse Width Modulation Method for Matrix Converters", International Journal of Control, Automation, and Systems, Vol.6, No.5, pp.660-669, 2008

[7] J.Monteiro, J.Fernand Silva, S.F.Pinto, and J.Palma, Matrix Converter- Based Unified Power-Flow Controllers: Advanced Direct Power Control Method, *IEEE Trans on Power Delivery*, Vol.26, No.1, pp. 420-430, 2011

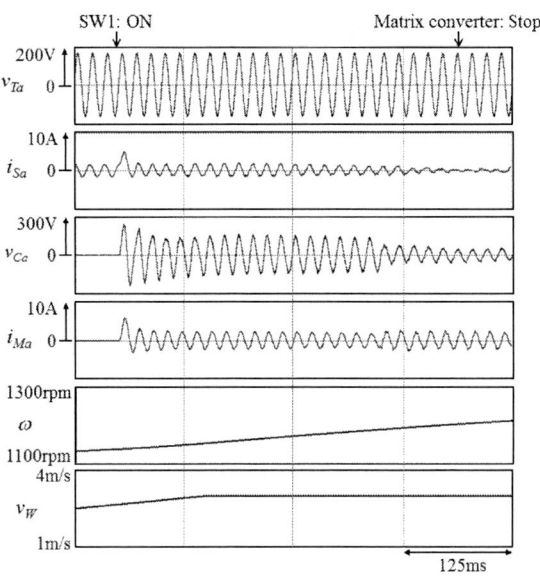

Fig. 10. Simulation waveform with the proposed inrush current suppressor.

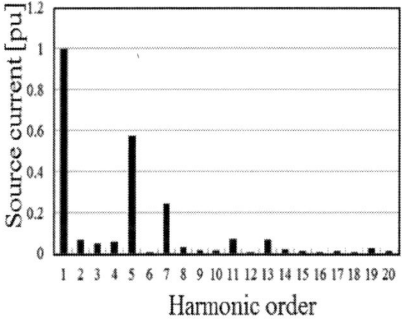

(a) Soft starter.

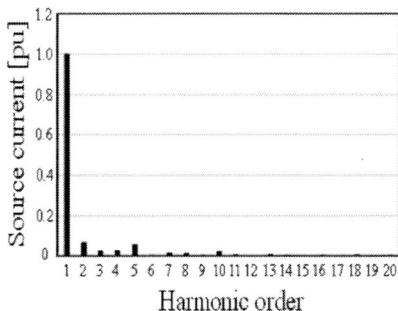

(b) Proposed inrush current suppressor.

Fig. 11. FFT analysis of the soft starter and proposed inrush current suppressor.

Nonlinear Pitch Control Design for Load Reduction on Wind Turbines

Shuai Xiao, Geng Yang, and Hua Geng

Tsinghua National Laboratory for Information Science and Technology
Department of Automation, Tsinghua University
Beijing 10084, China

Abstract—As the modern wind turbines (WTs) become larger and larger in size, the structural loads experienced by WTs increase dramatically. Such loads will shorten the working life of WTs and increase the maintenance cost. To mitigate such loads, this paper proposes a nonlinear pitch control strategy for WTs operating in the above-rated wind speed region. With the proposed strategy, nonlinear dynamic inversion method is employed to transform the non-affine nonlinear WT model into a pseudo linear one. Then the linear control theory can be applied for control design. The simulation results based on the aero-elastic model of Fatigue, Aerodynamics, Structures, and Turbulence (FAST) verify the superiority of the proposed nonlinear controller over the tranditional gain-scheduled proportional-integral (GSPI) controller. The proposed nonlinear controller can work well in the whole above-rated wind speed region, and easy to implement in the real application.

Keywords—load reduction, nonlinear control, pitch control, wind turbines

I. INTRODUCTION

With the increasing size of the modern large wind turbines (WTs), the structural loads experienced by the WTs are becoming dramatically larger, which shorten the working life of WTs and increase the cost of maintenance [1]. In order to reduce the cost of wind power generation, there is increasing interest in reducing loads for WTs through active load reduction control strategies in the past decades [2].

For WTs operating in the above-rated wind speed region, the main control task is to regulate rotor speed/output power by modifying pitch angles. Besides, additional control objective to reduce loads is usually taken into account, since such wind conditions account for most of the loads [3]. This study focuses on the collective pitch control strategies of WTs for load reduction in the above-rated wind speed region. For the WT system, the cost of tower occupies a large proportion of the total cost, especially for offshore WTs, it is up to about 40% [3]. Therefore, tower load is a very important factor for control design of WTs, and tower load reduction is included in the control objectives for load reduction in this study.

Many meaningful studies have been done on active load reduction control of WTs in the above-rated wind speed region. Most of them are linear control strategies designed based on the linearized model of WT around a specific operating point [1][3]-[8]. Such linear control strategies can work well around the specific operating point. However, to extend the operation region, several linear controllers have to be designed at different operating points, and proper gain scheduling method is required to schedule these controllers to ensure the stability of the whole system. Thus, it complicates the control design process for WTs. Other investigations focus on nonlinear control design directly based on the nonlinear model of WT. In this way, only one controller is required to work for the whole operation region. However, to the knowledge of the authors, such studies are rarely reported, and the existing methods [9][10] are usually complex and difficult to implement for real time application.

Since the WT system model is non-affine nonlinear, it is difficult to design a nonlinear controller directly based on the nonlinear model. Nonlinear dynamic inversion method [11][12], which can transform a non-affine nonlinear model into a pseudo linear one, is a powerful tool in solving non-affine nonlinear control problem. It is employed for control design in this study. The proposed nonlinear pitch controller is designed for WTs operating in the above-rated wind speed region, with the control objectives of both rotor speed regulation and tower load reduction. The simulation results with Fatigue, Aerodynamics, Structures, and Turbulence (FAST) software validate the effectiveness of the proposed control strategy by comparing it with the tranditional gain-scheduled proportional-integral (GSPI) controller. The proposed controller can work well in the whole above-rated wind speed region and easy to implement for application.

The remainder of the paper is organized as follows. In section II, the reduced nonlinear model of WT for control design is introduced. Then, the design of the proposed nonlinear controller is described in section III. Afterwards, the simulation results are shown in section IV, followed by the conclusion in section V.

II. THE REDUCED NONLINEAR MODEL OF WT

The full nonlinear model of WT is too complicated for control design, and a reduced nonlinear model [10] is used

---

This work is supported by National Natural Science Foundation (NNSF) of China under Grant (61104046 and 61273045), Tsinghua University Initiative Scientific Research Program and grant from Beijing Higher Education Young Elite Teacher Project.

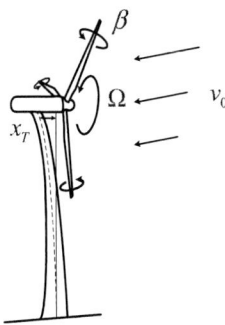

Fig. 1. Schematic diagram of the reduced nonlinear model

instead. The reduced nonlinear model includes three degrees-of-freedom (DOFs), as shown in Fig. 1. The control input is the collective pitch angle β, and the disturbance input is the rotor effective wind speed v_0. The generator torque M_g is kept constant in the above-rated wind speed region and not looked as control input here [6]. It is assumed that the measurable outputs are the rotor speed Ω and the tower top fore-aft velocity $\dot{x}_T$ [3].

The reduced nonlinear model of WT can be expressed as [10][6]

$$J\dot{\Omega} = M_a(\dot{x}_T, \Omega, \beta, v_0) - M_g \cdot i \qquad (1)$$

$$M_T\ddot{x}_T + C_T\dot{x}_T + K_Tx_T = F_a(\dot{x}_T, \Omega, \beta, v_0) \qquad (2)$$

$$T_{ac}\dot{\beta} + \beta = \beta^* \qquad (3)$$

The equation (1) is the drivetrain dynamic model, where J is the rotor equivalent moment of inertia, M_a is the aerodynamic torque, x_T is tower top fore-aft displacement, and i is the gear box ratio. The equation (2) models the tower fore-aft dynamics, where F_a is the aerodynamic thrust, and M_T, C_T, K_T are the tower equivalent modal mass, structural damping and bending stiffness, respectively. The model of the blade pitch actuator is represented by a 1st-order inertia system as (3) [6], where T_{ac} is the equivalent time constant of the actuator, and β^* is the control command from the collective pitch controller.

The nonlinearity of the model lies in the aerodynamic torque M_a and aerodynamic thrust F_a, which can be given by

$$\begin{aligned} M_a &= \tfrac{1}{2}\rho\pi R^3 C_q(\lambda, \beta) v_{rel}^2 \\ F_a &= \tfrac{1}{2}\rho\pi R^2 C_t(\lambda, \beta) v_{rel}^2 \end{aligned} \qquad (4)$$

where ρ is the air density, R is the rotor radius, $v_{rel} = v_0 - \dot{x}_T$ is the relative wind speed considering the effect of tower top speed, and λ is tip speed ratio defined as

$$\lambda = \frac{\Omega R}{v_{rel}} \qquad (5)$$

C_q and C_t are the effective torque and thrust coefficients, respectively, and they are non-affine nonlinear functions of λ and β . For the simulated 1.5 MW reference model provided by FAST, the curves of C_q and C_t are illustrated in Fig. 2. Each curve can be represented by a polynomial fit to two-dimensional table, which can be obtained by steady state

simulation or field test. The accuracy of the reduced nonlinear model for control design has been discussed in [10], and it is not addressed here.

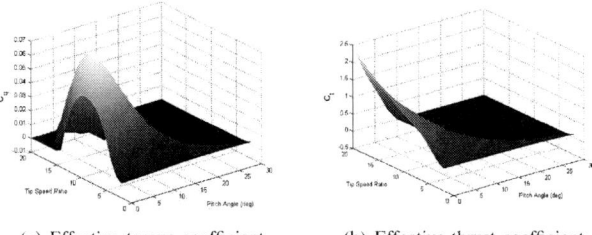

(a) Effective torque coefficient   (b) Effective thrust coefficient

Fig. 2. Non-affine nonlinear characteristics of WT

It is noted that the ignored dynamics and parameter uncertainty of the reduced model should be covered by the robustness of the proposed controller, and this point will be examined by simulation with full nonlinear aero-elastic model of WT in this study.

III. CONTROL DESIGN

Based on the reduced nonlinear model (1) - (3), nonlinear control strategy is designed in this section. As mentioned above, the control objectives of control design are to regulate rotor speed and reduce tower loads. Only tower fore-aft motion is considered here for tower load reduction, since it is the main driver for the tower fatigue [4].

The proposed nonlinear controller consists of a rotor speed regulator and a cascaded tower oscillation damper. Typically, the bandwidth of the rotor speed regulator is chosen to be around 1 rad/s, which is well below the tower fore-aft mode (usually above 2 rad/s). Thus, the rotor speed control loop and tower load control loop can be regarded as decoupled from the perspective of frequency domain [3], and the rotor speed control loop and tower load control loop can be designed independently [5]. The rotor speed regulator and tower oscillation damper are both designed using nonlinear dynamic inversion method. With nonlinear dynamic inversion method, the nonlinear systems (1) and (2) can be looked as two pseudo linear SISO systems, and then linear control theory can be applied for control design. The actuator model (3) is ignored in the control design, as the actuator can act fast to the control commands [13]. The block diagram of the control system is shown in Fig. 3, and the detailed description of each part of the control system is given as follows.

A. Rotor speed regulator

From the drivetrain dynamics (1), if the aerodynamic torque M_a is looked as the control input, then the drivetrain model becomes a simple pseudo linear 1st-order integral system. In order to eliminate the steady errors caused by errors of wind speed measurements and parameter perturbations of the reduced model, proportional integral (PI) control method

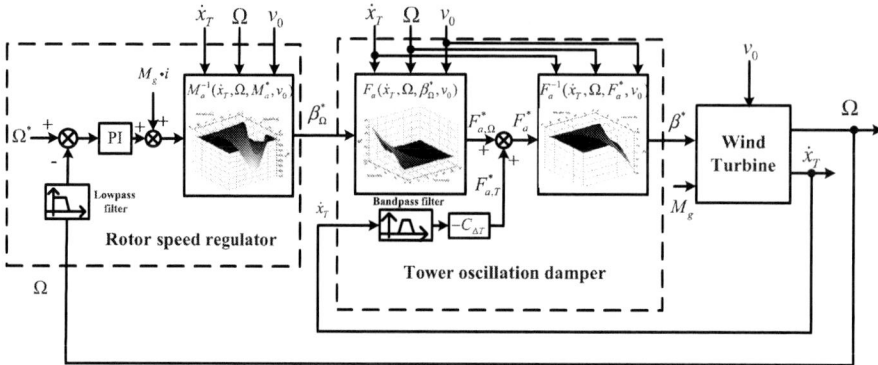

Fig. 3. The block diagram of the control system

is adopted here. With the feedforward of generator torque, the control command of the linear regulator can be given by

$$M_a^* = K_{\Omega,p}(\Omega^* - \Omega) + K_{\Omega,i} \int (\Omega^* - \Omega) dt + M_g \cdot i \quad (6)$$

where $K_{\Omega,p}$ and $K_{\Omega,i}$ are the proportional and integral parameters, respectively. On the right side of (6), the first two parts are the outputs of PI control, and the third part is the feedforward of generator torque.

With the aerodynamic torque command M_a^* of the linear regulator, the corresponding pitch angle command β_Ω^* can be obtained as

$$\beta_\Omega^* = M_a^{-1}(\dot{x}_T, \Omega, M_a^*, v_0) \quad (7)$$

where $M_a^{-1}(\dot{x}_T, \Omega, M_a, v_0)$ is the inversion of the nonlinear function $M_a(\dot{x}_T, \Omega, \beta, v_0)$, and it can be solved from

$$C_q(\lambda, \beta_\Omega^*) = \frac{2M_a^*}{\rho \pi R^3 v_{rel}^2} \quad (8)$$

It can be seen that, besides Ω and $\dot{x}_T$, v_0 is needed to solve the inverse function. In this study, v_0 is assumed to be known, as the modern wind measurement devices, such as Light Detection And Ranging (LIDAR), can provide accurate measurements of the incoming wind flow [14]. From Fig. 2(a), it can be known that, C_q has a parabola-like relation with β if λ is kept constant. This will result in one or two solutions for the inverse function. However, there is a unique stable solution for the inverse function. This point has already been explained in our previous work in [13], and it is omitted here for brevity. The curve of (7) can also be represented by a two-dimensional table, which can be obtained by the linear interpolation method from the aerodynamic torque data. Then solving the inverse function can be simply realized by looking up the table, which makes it easy to implement for application.

Besides, the rotor speed signal passes through a lowpass filter to filter out the 3p (1p is once-per-revolution) frequency caused by rotationally sampled turbulence and tower shadow, so as to mitigate high frequency action of the pitch actuator [5]. The control block diagram of rotor speed regulator is shown in Fig. 3.

B. Tower oscillation damper

Similar to the design of the rotor speed regulator, when the aerodynamic thrust F_a is looked as the control input, the tower dynamic model turns to be a pseudo linear 2nd-order system. Since the bandwidth of rotor speed regulator is well below the tower fore-aft mode, the aerodynamic thrust produced by the rotor speed regulator

$$F_{a,\Omega}^* = F_a(\dot{x}_T, \Omega, \beta_\Omega^*, v_0) \quad (9)$$

can be regarded as the steady state value for tower control approximately. On the basis of $F_{a,\Omega}^*$, a damping thrust $F_{a,T}^*$ is added in order to damp the tower fore-aft oscillations. Then, the tower model (2) can be rewritten as

$$M_T \ddot{x}_T + C_T \dot{x}_T + K_T x_T = F_{a,\Omega}^* + F_{a,T}^* \quad (10)$$

The small-signal model of (10) can be expressed as

$$M_T \Delta \ddot{x}_T + C_T \Delta \dot{x}_T + K_T \Delta x_T = F_{a,T}^* \quad (11)$$

To damp the tower fore-aft oscillations, let $F_{a,T}^* = -C_{\Delta T} \Delta \dot{x}_T$, and then (11) is equal to

$$M_T \Delta \ddot{x}_T + (C_T + C_{\Delta T}) \Delta \dot{x}_T + K_T \Delta x_T = 0 \quad (12)$$

It can be observed that the damping ratio of the tower fore-aft motion can be improved through the feedback control.

With the aerodynamic thrust command as

$$F_a^* = F_{a,\Omega}^* + F_{a,T}^* \quad (13)$$

the corresponding pitch angle command output of the tower oscillation damper can be obtained as

$$\beta^* = F_a^{-1}(\dot{x}_T, \Omega, F_a^*, v_0) \quad (14)$$

where $F_a^{-1}(\dot{x}_T, \Omega, F_a, v_0)$ is the inversion of the nonlinear function $F_a(\dot{x}_T, \Omega, \beta, v_0)$, and it can be solved from

$$C_t(\lambda, \beta^*) = \frac{2F_a^*}{\rho \pi R^2 v_{rel}^2} \quad (15)$$

Not the same as C_q, C_t has a monotonous relation with β if λ is kept constant, as shown in Fig. 2(b). Thus, only one solution exists for the inverse function. Similarly, both (9) and (14) are represented by two-dimensional tables, and the solutions can be found by looking up tables.

Moreover, a bandpass filter is used to extract the tower fore-aft mode from the measured tower signal, in order to avoid coupling of the tower control loop to rotor speed control loop and exciting other modes [3]. The control block diagram of the tower oscillation damper is also shown in Fig. 3.

IV. SIMULATION RESULTS

The 1.5 MW reference WT model provided by FAST is simulated to verify the proposed nonlinear control strategy.The main parameters of the 1.5 MW WT are given in Appendix A, and the detailed parameters can be found in the FAST software package. The time constant of pitch actuator is 0.2 s. During the simulation, FAST [15] is used to provide the nonlinear aero-elastic model of WT, and the controller design is implemented in MATLAB/Simulink. All related DOFs are enabled in FAST to make the simulation results more credible.

To evaluate the control performance, the traditional gain-scheduled proportional integral (GSPI) controller is used as a baseline controller, which is designed according to [16]. For the nonlinear controller, the parameters are tuned to be $K_{p,\Omega} = 5.74 \times 10^5$ and $K_{i,\Omega} = 4.10 \times 10^5$ in the rotor speed regulator, and $C_{\Delta T} = -5 \times 10^4$ in the tower oscillation damper.

Step wind condition is simulated to test the control performance of the controller. In this simulation, the wind speed is uniform all over the wind field, and it increases stepwise from 14 m/s to 22 m/s with a step of 2 m/s (the above-rated wind speed region is [12, 24] m/s), as shown in Fig. 4. It can be seen that, the proposed nonlinear controller can work well for the full nonlinear aero-elastic model of WT in the whole above-rated wind speed region. There is no steady state error of rotor speed, and the control performance is good for different wind speeds. Compared to the baseline controller, the overshot of the rotor speed dynamic response is well reduced, and the oscillation of the tower is also damped out more quickly. The blade mode is not excited and kept the same level as GSPI, since filter is used in the controller not to interact with the blade mode. Therefore, It can be concluded that the dynamic control performance is obviously improved with the proposed nonlinear controller. It can also be observed that, since the nonlinear controller acts to the wind disturbance more rapidly, the pitch rate is increased, but it is still within the pitch rate limit 0.26 rad/s.

V. CONCLUSION

This paper presents a nonlinear pitch controller for load reduction on WTs operating in the above-rated wind speed region. Nonlinear dynamic inversion method is used to cope with the non-affine nonlinearity of the WT model by transforming the non-affine nonlinear model into a linear one. Then the linear control theory is applied for control design. The simulation results with aero-elastic model of FAST indicate that, the proposed controller can improve the rotor speed regulation performance and reduce tower loads simultaneously

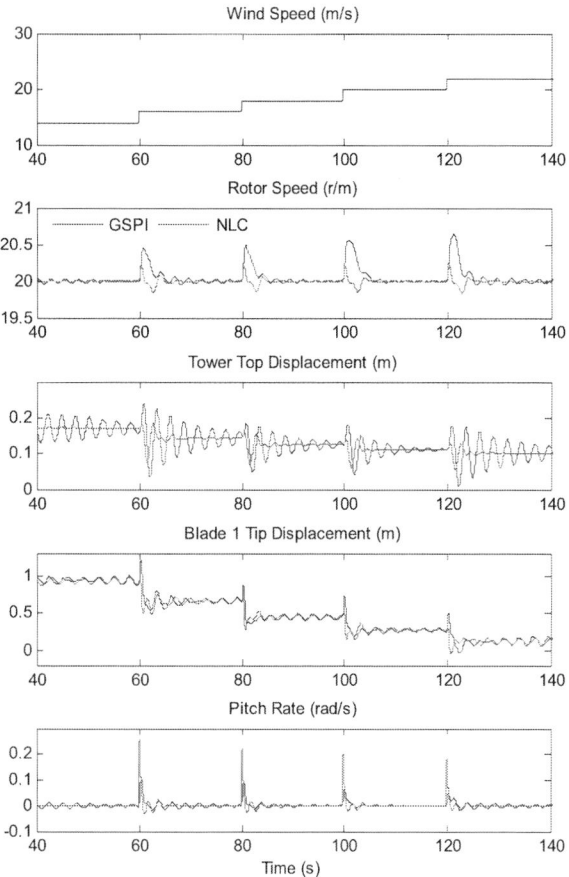

Fig. 4. The simulation results in step wind condition

compared with the traditional GSPI controller. Moreover, the proposed controller can work well in the whole above-rated wind speed region, and it is simple to implement for the real application.

APPENDIX A
THE MAIN PARAMETERS OF THE 1.5MW WT

Rated rotor speed: 20 rpm
Rated power: 1.5 MW
Cut-in wind speed: 3 m/s
Rated wind speed: 12 m/s
Cut-off wind speed: 25 m/s
Blade length: 35 m
Tower height: 82.39 m
Rotor inertia: 2962.44×10^3 kg$\times$ m^2
Generator inertia: 53.036 kg$\times$ m^2
Gearbox ratio: 87.965

REFERENCES

[1] E. A. Bossanyi, "Wind turbine control for load reduction," *Wind Energy*, vol. 6, pp. 229-244, 2003.

[2] J. H. Laks, L. Y. Pao and A. D. Wright, "Control of wind turbines: Past, present, and future," *American Control Conference*, 2009.

[3] W. E. Leithead and S. Dominguez, "Analysis of tower-blade interaction in the cancellation of the tower fore-aft mode via control," *European Wind Energy Conference*, 2004

[4] W. E. Leithead and S. Dominguez, "Coordinated control design for wind turbine control systems," *European Wind Energy Conference*, 2006.

[5] T. G. van Engelen, E. L. van der Hooft and P. Schaak, "Development of wind turbine control algorithms for industrial use," *European Wind Energy Conference*, 2003.

[6] A. D. Wright, "Modern control design for flexible wind turbines," Technical Report NREL/TP-500-35816, NREL, 2004.

[7] M. Geyler and P. Caselitz, "Robust multivariable pitch control design for load reduction on large wind turbines," *Journal of Solar Energy Engineering*, vol. 130, pp. 031014-12, 2008.

[8] M. Kristalny, D. Madjidian and T. Knudsen, "On using wind speed preview to reduce wind turbine tower oscillations," *IEEE Transaction on Control Systems Technology*, vol.21, iss.4, pp.1191-1198, July 2013.

[9] A. A. Kumar and K. A. Stol, "Simulating MIMO feedback linearization control of wind turbines using FAST," *46th AIAA Aerospace Sciences Meeting and Exhibit*, 2008.

[10] D. Schlipf, D. J. Schlipf and M. Khn, "Nonlinear model predictive control of wind turbines using LIDAR," *Wind Energy*, 2012.

[11] S. H. Lane and R. F. Stengel, "Flight control design using non-linear inverse dynamics," *Automatica*, vol. 24, pp. 471-483, 1988.

[12] S. A. Snell, D. F. Enns and W. L. Garrard, "Nonlinear inversion flight control for a supermaneuverable aircraft," *Journal of Guidance Control and Dynamics*, vol. 15, pp. 976-984, 1992.

[13] H. Geng, G. Yang, "Output power control for variable-speed variable-pitch wind generation systems," *IEEE Transactions on Energy Conversion*, 2010, 25(2):494-502.

[14] E. Simley, P. Lucy, F. Rod, K. Neil, J. Bonnie, and S. Eric, "Analysis of wind speed measurements using continuous wave LIDAR for wind turbine control," *49th AIAA Aerospace Sciences Meeting including the New Horizons Forum and Aerospace Exposition*, 2011.

[15] J. Jonkman, J. Buhl , FAST users guide, *Technical Report NREL/EL-500-38230, NREL*, 2005.

[16] J. Jonkman, S. Butterfield, W. Musial, et al, Definition of a 5MW reference wind turbine for offshore system development, *Technical Report NREL/TP-500-38060, NREL*, 2009.

Device Loading of Modular Multilevel Converter MMC in Wind Power Application

L. Popova, J. Pyrhönen
Department of Electrical Engineering
Lappeenranta University of Technology
Lappeenranta, Finland
liudmila.popova@lut.fi; juha.pyrhonen@lut.fi

K. Ma, F. Blaabjerg
Department of Energy Technology
Aalborg University
Aalborg, Denmark
kema@et.aau.dk, fbl@et.aau.dk

Abstract— **Modular multilevel converter (MMC) is a recently emerged multilevel topology for high-voltage high-power applications. However, in wind power application the performance of the MMC has not been deeply investigated. In this paper the application of MMC in wind energy systems is studied. The converters used to connect the wind turbine to the grid having the rated active powers of 2MW and 10MW are designed and investigated. Electrical losses and thermal loading of the power devices in the proposed converter solutions are analyzed. The efficiency of the MMC converter under different *P/Q* boundaries defined by grid codes is investigated and compared with two-level and three-level NPC converters. It is concluded that it is possible to use the MMC in wind power application and the losses are distributed evenly between the sub-modules of the MMC converter. However, inside a sub-module the losses of the power devices are not equal that can lead to the de-rating of the converter.**

Keywords— MMC Converters, renewable energy sources, thermal stresses, wind power generation.

I. INTRODUCTION

The amount of wind turbines installed worldwide has been growing constantly during the latest decades. As the power capacity of wind turbines has increased up to 10 MW the connection to grid using traditional two-level converters requires a large amount of series or parallel connected power devices in order to achieve the required power. The reliability of such a converter is decreased while the complexity is increased. Multilevel converters are known to be a promising solution in wind turbine applications due to their ability to work with higher output voltage levels and therefore obtain larger output powers using the power devices available nowadays [1].

Among various multilevel topologies such as Neutral Point Clamped (NPC), Flying Capacitor (FC), and Cascaded H-Bridge (CHB) the Modular Multilevel Converter (MMC) is a fairly new topology which has been introduced in 2002 [2]. The advantages of the MMC which make it an attractive solution for wind power generation system are the modular design, redundancy, simple voltage scaling, possibility to connect the converter directly to the grid without a transformer [3]. While the application of the MMC converter for High-Voltage DC (HVDC) transmission systems is well reported in

the literature [4], [5], not many papers are published about the application of this topology for wind turbines.

A thorough analysis of the MMC converter is required in order to investigate the suitability of the topology for wind power systems. Reliability and power density aspects of the converter are of great importance for high power wind turbines, which tend to be installed off-shore in a remote and harsh environment as the power is increasing. These aspects are affected by the loading and the thermal performance of the switching power devices, which will be studied in this paper.

II. MODULAR MULTILEVEL CONVERTER FOR CASE STUDY IN WIND APPLICATION

A simplified schematic of the MMC used in wind power application is presented in Fig. 1. The topology is based on series connection of sub-modules (SM). The number of SMs in each phase arm is equal. The circuit diagram of a half-bridge SM which contains a floating dc capacitor and two IGBTs with free-wheeling diodes is also presented in Fig. 1. The basic operation principles of the MMC converter are described in detail [3], [6].

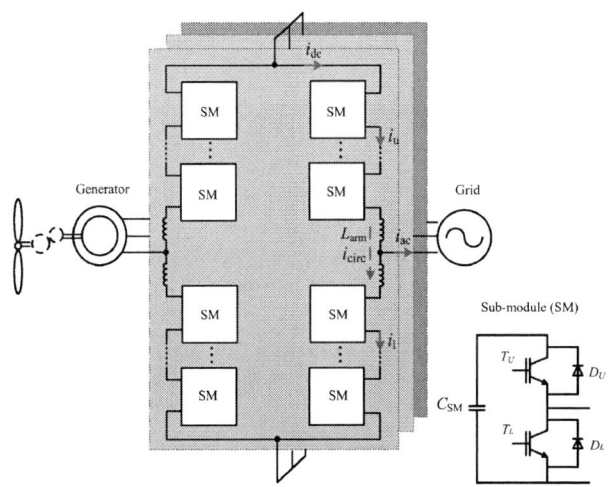

Fig. 1. Simplified schematic of the MMC for wind turbine application.

A. Selection of the components

Considering the operation principles of the MMC converter the IGBT modules and the capacitors of the SMs have to be selected with the rated voltage

$$U_{\text{SM}} = \frac{U_{\text{DC}}}{N} \tag{1}$$

where U_{DC} is a DC link voltage and N is a number of SM in an arm. A safety margin has to be taking into account in order to have kind of redundancy.

The selection of the SM capacitor is a compromise between the capacitor size, cost and the voltage requirements. The capacitance of the SM capacitor can be calculated based on desired voltage ripple factor ε having a value between 0 and 1 as it is presented in [7]

$$C_{\text{SM}} = \frac{\Delta W_{\text{SM}}}{2 \cdot \varepsilon \cdot U_{\text{SM}}^2} \tag{2}$$

where ΔW_{SM} is the energy change in one SM which is estimated by

$$\Delta W_{\text{SM}} = \frac{2}{3} \cdot \frac{S}{k \cdot \omega_{\text{n}} \cdot N} \left(1 - \left(\frac{k \cdot \cos \varphi}{2} \right)^2 \right)^{3/2} \tag{3}$$

where S is the apparent power of the converter, ω_{n} is the output angular frequency, k is the voltage modulation index, N is a number of SM in an arm, and $\cos\varphi$ is power factor (PF).

The arm inductors L_{arm} are used in the MMC converter for the suppressing the circulating currents and limiting the fault current during short circuit between the DC link terminals. The analytical expressions for the arm inductance calculation based on these two issues are given in [8]. In this paper the value of the arm inductance is selected to be 0.15 per unit which is a commonly used value for the MMC converter [9]. However, it should be checked that the selected SM capacitance and the arm inductance do not create a resonance as shown in [10].

B. Low voltage solution for wind power application

Converter with minimum amount of SMs in a phase leg is not presented in the literature, however, this solution may be beneficial due to reduced component counts in comparison with the MMC having more SMs. Thus, the first studied converter has a rated power of 2 MW and only 2 SMs in a phase leg. This converter produces 690 V line-to-line voltage and is used to connect a low-voltage generator to the grid (Fig. 2). It can be an alternative to the existing two-level wind power converter based on 690 V grid voltage. The parameters of the converter and ratings of the selected IGBT module

5SNA 1800E170100 from ABB [11] are summarized in Table I.

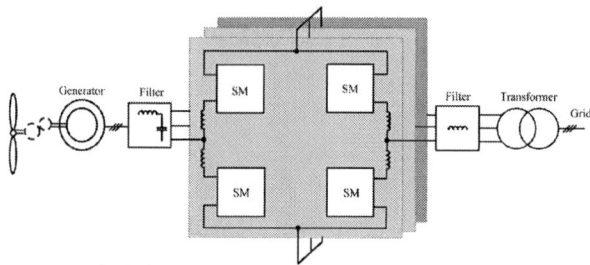

Fig. 2. 2 MW MMC converter used in a wind turbine.

TABLE I
PARAMETERS OF SIMULATED 2 MW MMC CONVERTER

Rated active power, MW	2
Number of SMs per arm	1
Rated line-to-line voltage, kV	0.69
DC voltage, kV	1.1
Rated current, A	1674
Arm inductance, µH	113.7 (0.15 p.u.)
SM capacitance, mF	20.8
Carrier frequency, Hz	1550
Filter inductance, mH	0.15 (0.2 p.u.)
IGBT module	ABB 5SNA 1800E170100
IGBT commutated voltage, kV	1.1
IGBT maximum current, kA	1.8
Total number of IGBT modules	12
Total MVA rating of all IGBT modules	23.76

The converter is simulated using Matlab/Simulink and PLECS Blockset [12]. Level-shifted pulse-width modulation (PWM) is applied to obtain the desired arm voltage. For simplicity the reference to the modulator is generated using a direct strategy presented in [13]. Simulated waveforms are shown in Fig. 3. The power factor of grid side converter is in Fig. 3 selected to be $PF = 1$.

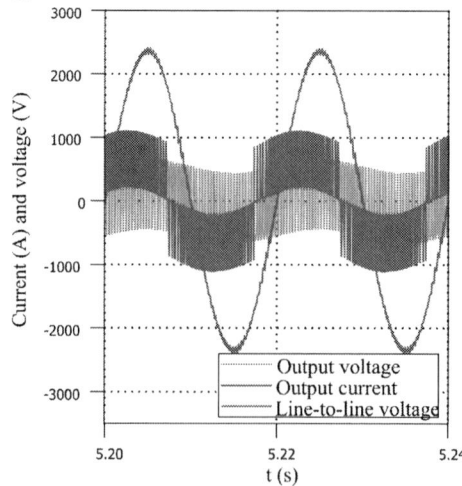

Fig. 3. Output current, output voltage, and line-to-line voltage of 2 MW MMC converter ($P = 2$ MW, $PF = 1$).

C.   High voltage solution for wind power application

The second solution of the MMC converter is a converter with 10 SMs in a phase leg. This converter is suitable for a high voltage application which is a preferred choice for the next generation wind turbines with 10 MW rated power as the power capacity grows quickly. This converter generates 10 kV line-to-line voltage using available on the market IGBT module 5SNA 0600G650100 from ABB [11]. An equal voltage sharing among the capacitors in each arm is assured by applying the selection mechanism described in [14]. The selection of the SM to be inserted or bypassed is made based on the measured arm current and relative capacitor voltage. The parameters of the 10 MW converter with 10 SMs in a phase leg are given in Table II. Simulated waveforms are presented in Fig. 4.

TABLE II
PARAMETERS OF SIMULATED 10 MW MMC CONVERTER

Rated active power, MW	10
Number of SMs per arm	5
Rated line-to-line voltage, kV	10
DC voltage, kV	16.3
Rated current, A	577
Arm inductance, mH	4.8 (0.15 p.u.)
SM capacitance, mF	3.2
Carrier frequency, Hz	1050
Filter inductance, mH	0.95 (0.03 p.u.)
IGBT module	ABB 5SNA0600G650100
IGBT commutated voltage, kV	3.6
IGBT maximum current, kA	0.6
Total number of IGBT modules	60
Total MVA rating of all IGBT modules	129.6

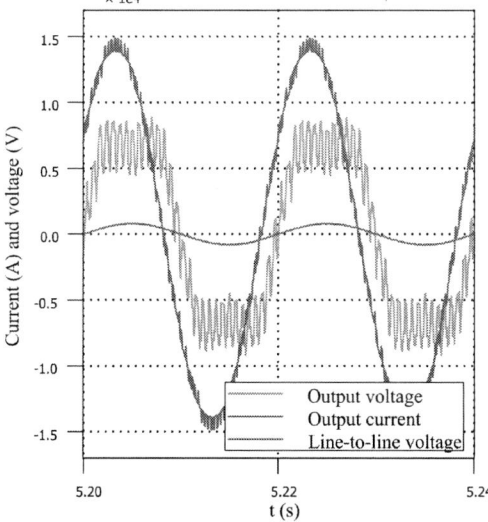

Fig. 4. Output current, output voltage, and line-to-line voltage of 10 MW MMC converter (P=10 MW, PF=1).

III.   LOADING OF POWER DEVICES

The current, loss and thermal distributions of the studied converters are considered in order to analyze the loading of the devices.

A.   Current distribution

Current distributions among the power devices in one SM of the 2 MW and the 10 MW converters operating with PF=1 are presented in Fig. 5 and Fig. 7.

It is shown in Fig. 5 and Fig. 7 that the current of the switching devices is about a half of a peak output current of the converter. Thus, even though the MMC converter with 2 SMs has two voltage levels as a traditional two-level converter, the switching devices with lower rated current can be selected for the MMC converter with the same rated power as a two-level converter.

Uneven current distribution between the power devices of the SM is observed. As shown in Fig. 5 the upper diode D_U and the lower IGBT T_L are heavily loaded when the converter operates with PF=1. When the converter operates with PF=-1 (Fig. 6) the upper IGBT T_U and the lower diode D_L are mostly loaded. Uneven current distribution among the power devices in SM is explained by the presence of a dc component i_{dc} in the upper and lower arm currents i_u and i_l that equals to one-third of the total dc current and provides the actual dc to ac power transfer (or ac to dc when the PF=-1) [6].

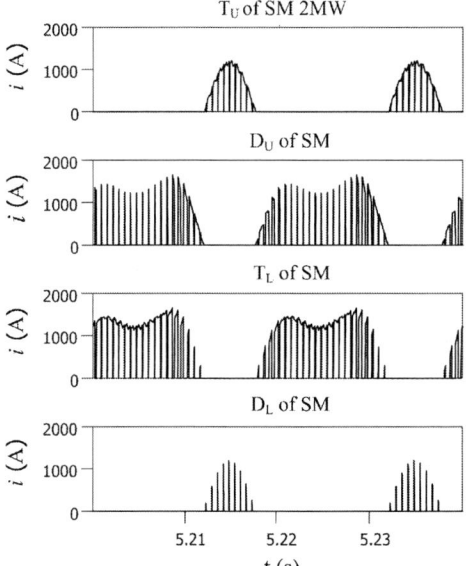

Fig. 5. Current distribution in the SM of the 2 MW MMC converter at PF=1 (U_{LL} = 690 V, f_s = 1550 Hz, I_{rated} = 1674 A).

The 2014 International Power Electronics Conference

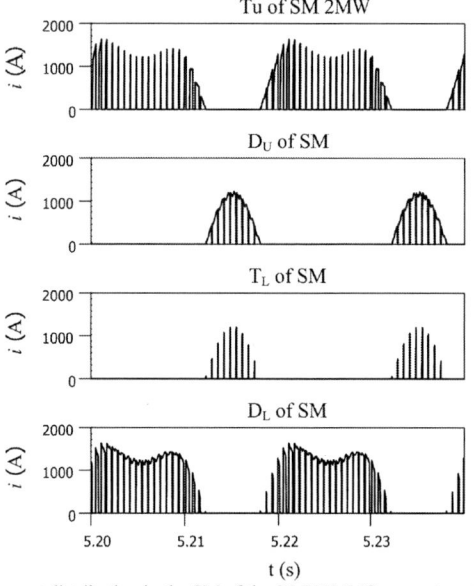

Fig. 6. Current distribution in the SM of the 2 MW MMC converter at PF=-1 (U_{LL} = 690 V, f_s = 1550 kHz, I_{rated} = 1674 A).

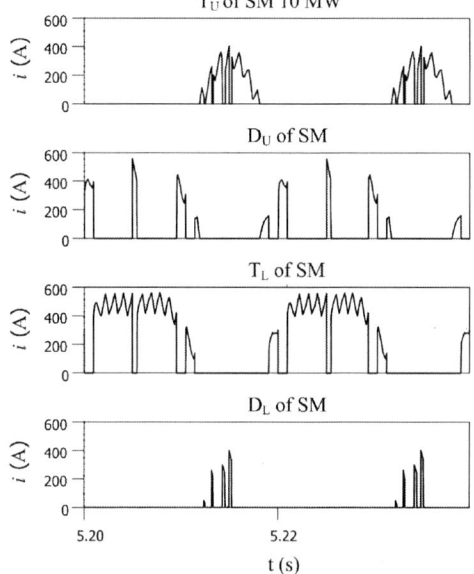

Fig. 7. Current distribution in the SM of the 10 MW MMC converter at PF=1 (U_{LL} = 10 kV, f_s = 1050 Hz, I_{rated} = 577 A).

Apart from the dc and ac components i_{dc} and i_{ac} the arm currents i_u and i_l (Fig. 1) contain also double-line frequency component which is called the circulating current i_{circ} as presented in [15]

$$i_u = i_{dc} + \frac{i_{ac}}{2} + i_{circ} \qquad (4)$$

$$i_l = i_{dc} - \frac{i_{ac}}{2} + i_{circ} \qquad (5)$$

$$i_{ac} = i_u - i_l \qquad (6)$$

The circulating current does not influence the output current but increase the rms value of the arm currents and consequently also increases the losses. The suppression of the circulating current has been under research in the last decade and several control methods have been proposed [16], [13], [17]. In Fig. 8 the current distribution among the devices of the SM of 2 MW MMC converter with the applied distributed control proposed in [16] are shown. The double-line frequency component in the arm current is suppressed but the unequal loading has become even more severe.

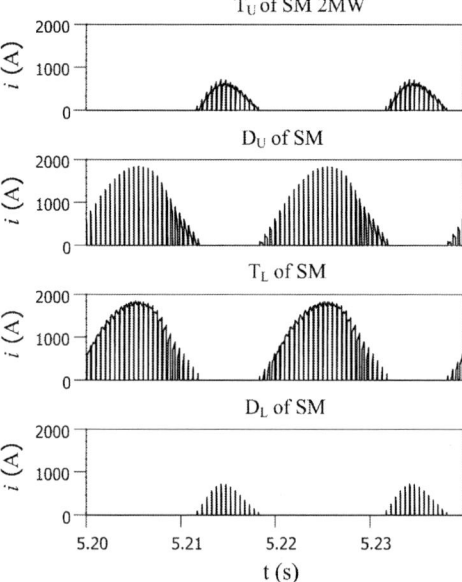

Fig. 8. Current distribution in the SM of the 2 MW MMC converter with circulating current control (U_{LL} = 690 V, f_s = 1550 kHz, I_{rated} = 1674 A, PF=1).

B. Loss distribution

The loss model presented in [18] is applied and simulation is carried out in PLECS Blockset in Simulink [12]. The loss distribution of the SM of the 2 MW converter without circulating current control is shown in Fig. 9. It is observed that the losses among the devices in the SM are unequally distributed. As shown in Fig. 10 the losses of the converter SM with applied circulating current control are lower but the unequal loading is more significant. Fig. 11 illustrates the loss distribution of the SMs of the 10 MW converter. While the losses in the SM are still unequally distributed, the loss distribution between five SMs of the arm is quite uniform. It is noted that equal loss distribution among different SMs is achieved by applying a selection mechanism for the voltage sharing between the SM capacitors, as shown in Fig. 11.

978-1-4799-2706-7/14 $31.00 © 2014 IEEE

The 2014 International Power Electronics Conference

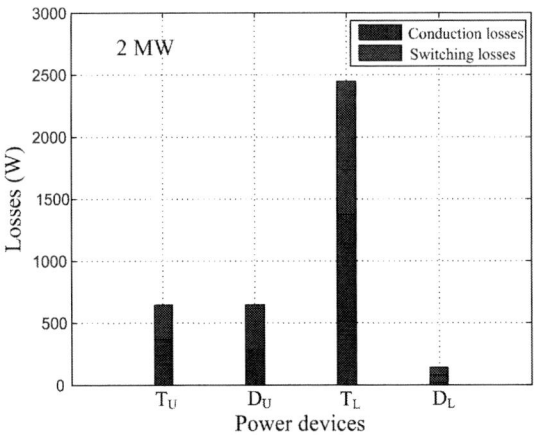

Fig. 9. Loss distribution in the SM of the 2 MW MMC converter (T_U – upper IGBT, D_U – upper diode, T_L – lower IGBT, D_L – lower diode).

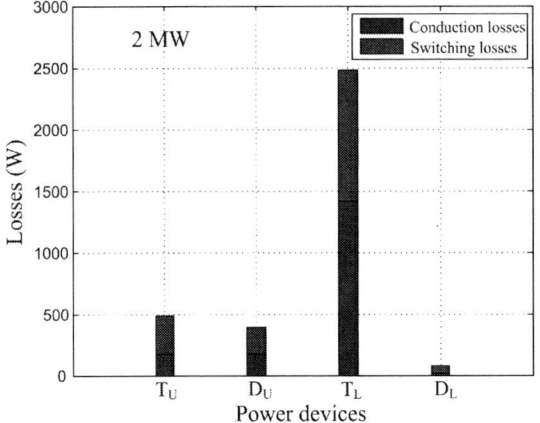

Fig. 10. Loss distribution in the SM of the 2 MW MMC converter with circulating current control (T_U – upper IGBT, D_U – upper diode, T_L – lower IGBT, D_L – lower diode).

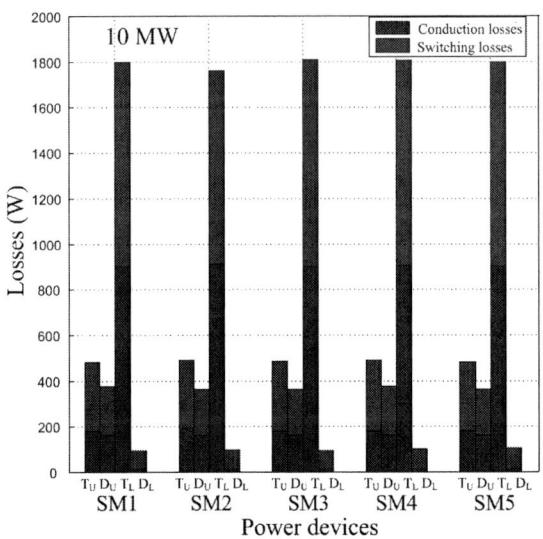

Fig. 11. Loss distribution in the SMs of the upper arm of the 10 MW MMC converter (T_U – upper IGBT, D_U – upper diode, T_L – lower IGBT, D_L – lower diode).

C. Thermal distribution

Thermal performance of the power devices is of great importance in high power wind turbine converters since it directly affects the converter reliability, power density and the cost of the whole system. Accurate estimation of the temperatures of the power devices is possible only with a good thermal model. In this paper the thermal model shown in Fig. 12 which is able to estimate both case and junction temperatures is employed. The description of the model is presented in [18]. The thermal parameters of multi-layer Foster model of the power device impedances are taken from the datasheets and summarized in Table III and IV [11]. Liquid cooling system is used to dissipate the thermal losses and fluid temperature is assumed to be 40 °C. The estimated temperatures of the power devices of the SM of the 2 MW and the 10 MW converters are presented in Fig. 13 and Fig. 14. It is clear that the temperatures of the devices are related to the losses presented in Fig. 9 and Fig. 11. Observed unequal temperatures of the power devices in the SM may lead to de-rating of the power converter and decreasing the converter power capability. They are also the weakness of the whole converter in respect to the reliability.

TABLE III
THERMAL IMPEDANCE OF THE IGBT MODULE 5SNA 1800E170100

Thermal impedance	$Z_{(j-c)}$				$Z_{(c-h)}$
	Layer 1	Layer 2	Layer 3	Layer 4	
R_{iIGBT}, K/kW	6.24	1.73	0.704	0.345	6
τ_{iIGTB}, s	192	20.4	1.97	0.52	
R_{iDiode}, K/kW	11.6	2.91	1.28	1.27	
τ_{iDiode}, s	204	29.3	6.96	1.5	

TABLE IV
THERMAL IMPEDANCE OF THE IGBT MODULE 5SNA 0600G650100

Thermal impedance	$Z_{(j-c)}$		$Z_{(c-h)}$
	Layer 1	Layer 2	
R_{iIGBT}, K/kW	8.5	2	6
τ_{iIGTB}, s	151	5.84	
R_{iDiode}, K/kW	17	4.2	
τ_{iDiode}, s	144	5.83	

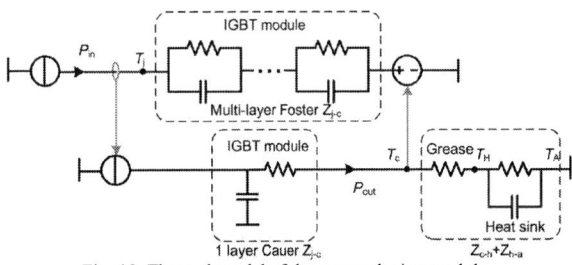

Fig. 12. Thermal model of the power device module.

978-1-4799-2706-7/14 $31.00 © 2014 IEEE 552

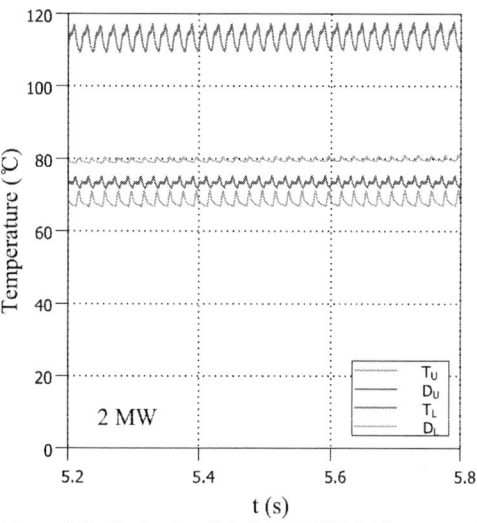

Fig. 13. Thermal distribution in a SM of the 2 MW MMC converter (T_U – upper IGBT, D_U – upper diode, T_L – lower IGBT, D_L – lower diode).

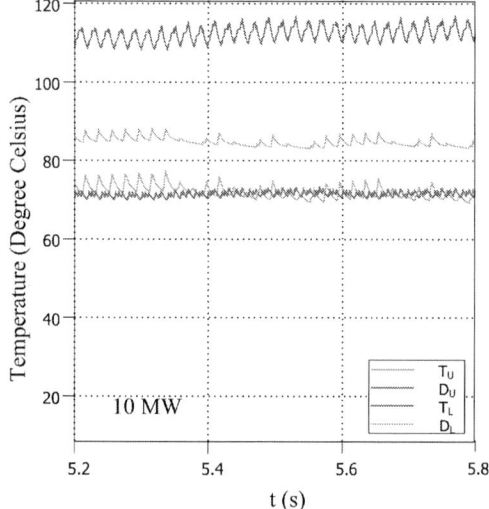

Fig. 14. Thermal distribution in a SM of the 10 MW MMC converter (T_U – upper IGBT, D_U – upper diode, T_L – lower IGBT, D_L – lower diode).

IV. CONVERTER EFFICIENCY UNDER DIFFERENT P/Q BOUNDARIES

Due to high penetration of wind turbine systems in many countries strict regulations concerning the performance of wind power converters under normal and fault operation have been introduced. Thus, the reactive power delivered by wind turbine has to be regulated in a certain range. The widest ranges are found in German grid code where three variants of the allowed boundaries of reactive power versus active power are defined as presented in Fig. 15 [19]. Usually the variant is chosen in agreement with the grid operator. As it can be seen in Variant 1 the underexcited reactive power should be less than 23 % of the rated active power P_{rated} and overexcited reactive power should be less than 48 % of P_{rated}.

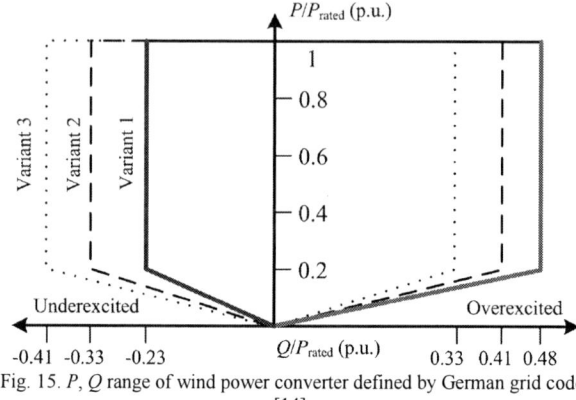

Fig. 15. P, Q range of wind power converter defined by German grid code [14].

In order to analyze the suitability of the MMC converter for wind application the efficiency of the converter under different active powers is studied. The MMC converter with 2 SMs in a phase leg (Table I) is compared with two level (2L) and three-level NPC (3L-NPC) converters which have the parameters presented in Table V. The efficiencies of three converters are presented in Fig. 16 where the cases when the converters operate at underexcited and overexcited reactive power boundaries of Variant 1 and without reactive power are indicated. It is observed that the efficiency of the converters at low active power is lower when the grid code requirements for extra reactive power are met. The 3L-NPC converter shows slightly better efficiency than 2L and MMC converter. It should be mentioned that by varying the switching frequency of the power devices the efficiency of the converters can be adjusted, and the filter size has to be also taken into account.

TABLE V
PARAMETERS OF SIMULATED 2L AND 3L-NPC CONVERTERS

	2L	3L-NPC
Rated active power, MW	2	2
Rated line-to-line voltage, kV	0.69	0.69
DC voltage, kV	1.1	1.1
Rated current, A	1674	1674
Carrier frequency, Hz	3450	1050
Filter inductance, mH	0.15 (0.2 p.u.)	0.15 (0.2 p.u.)
IGBT module	ABB 5SNA2400E170100	Infineon FZ2400R12HE4 B9
IGBT commutated voltage, kV	1.1	0.6
IGBT maximum current, kA	2.4	2.4
Total number of IGBT modules	6	18
Total MVA rating of all IGBT modules	15.84	25.92

The 2014 International Power Electronics Conference

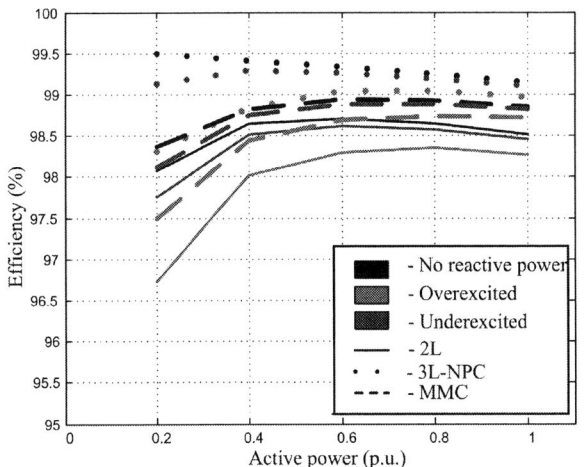

Fig. 16. Efficiency of 2L, 3L-NPC, and MMC converter under different *P/Q* boundaries defined by German grid code (Variant 1).

V.　CONCLUSIONS

In this paper two solutions of the wind turbine converters based on the MMC topology are evaluated. The first solution has a minimum number of SMs and requires only 12 IGBT modules (3x2x2*N*). This solution might be more feasible for wind application where the converter is installed in the nacelle and the size is critical, although in the HVDC application where the space is not an issue the MMC with a large number of the SMs is popular. The second solution has ten SMs in a phase leg and are using 60 IGBT modules.

The analysis of the converters is based on the study of the loading and thermal performances of the switching devices. The efficiency of the MMC converter under different *P/Q* boundaries defined by grid codes is presented.

It is observed that losses are uniformly distributed between the SMs of the MMC converter with ten SMs. However, severely unequal loss distribution between the devices in a SM is discovered which may lead to a de-rated converter power capacity when it is practically designed.

The requirement of extra reactive power modifies the loss distribution of the power devices and lower efficiency of the converter at low active power is observed.

REFERENCES

[1] F. Blaabjerg, F. Iov, T. Terekes, R. Teodorescu, K. Ma, "Power electronics - key technology for renewable energy systems," Power Electronics, Drive Systems and Technologies Conference (PEDSTC), 2011 2nd, pp.445-466, 16-17 Feb. 2011.

[2] R. Marquardt, "Current Rectification Circuit for Voltage Source Inverters with Separate Energy Stores Replaces Phase Blocks with Energy Storing Capacitors," German Patent (DE10103031A1), vol. 25, 2002.

[3] A. Lesnicar, R. Marquardt, "An innovative modular multilevel converter topology suitable for a wide power range," Power Tech Conference Proceedings, 2003 IEEE Bologna , vol.3, pp.6, 23-26 June 2003.

[4] G. Bergna, E. Berne, P. Egrot, P. Lefranc, A. Arzande, J.-C. Vannier, M. Molinas, "An Energy-Based Controller for HVDC Modular Multilevel Converter in Decoupled Double Synchronous Reference Frame for Voltage Oscillation Reduction," IEEE Transactions on Industrial Electronics, vol.60, no.6, pp.2360-2371, June 2013.

[5] H. Saad, J. Peralta, S. Dennetiere, J. Mahseredjian, J. Jatskevich, J.A. Martinez, A. Davoudi, M. Saeedifard, V. Sood, X. Wang, J. Cano, A. Mehrizi-Sani, "Dynamic Averaged and Simplified Models for MMC-Based HVDC Transmission Systems," IEEE Transactions on Power Delivery, vol.28, no.3, pp.1723-1730, July 2013.

[6] M. Glinka, R. Marquardt, "A new AC/AC multilevel converter family," IEEE Transactions on Industrial Electronics, vol.52, no.3, pp.662-669, June 2005.

[7] R. Marquardt and A. Lesnicar, "Modulares Stromrichterkonzept für Netzkupplungsanwendung bei hohen Spannungen," ETGFachtagung, Bad Nauheim, Germany, 2002.

[8] Q. Tu, Z. Xu, and H. Huang, "Parameter design principle of the arm inductor in modular multilevel converter based HVDC," Power System Technology, pp. 0-5, 2010.

[9] J. Peralta, H. Saad, S. Dennetiere, J. Mahseredjian, S. Nguefeu, "Detailed and Averaged Models for a 401-Level MMC–HVDC System," *IEEE Transactions on Power Delivery*, vol.27, no.3, pp.1501-1508, July 2012.

[10] M. Zygmanowski, B. Grzesik, R. Nalepa, "Capacitance and inductance selection of the modular multilevel converter," 15th European Conference on Power Electronics and Applications (EPE), 2013, pp.1-10, 2-6 Sept. 2013.

[11] Website of ABB semiconductors. (Available on :http://www.abb.com/product/us/9AAC910029.aspx)

[12] User manual of PLECS blockset version 3.4, October 2013. (Available: http://www.plexim.com/files/plecsmanual.pdf).

[13] A. Antonopoulos, Lennart Angquist, H-P. Nee, "On dynamics and voltage control of the Modular Multilevel Converter," 13th European Conference on Power Electronics and Applications, 2009. EPE '09., pp.1-10, 8-10 Sept. 2009.

[14] L. Angquist, A. Antonopoulos, D. Siemaszko, K. Ilves, M. Vasiladiotis, H-P. Nee, "Open-Loop Control of Modular Multilevel Converters Using Estimation of Stored Energy," IEEE Transactions on Industry Applications, vol.47, no.6, pp.2516-2524, Nov.-Dec. 2011.

[15] J-W. Moon, C-S. Kim, J-W. Park, D-W. Kang, J-M. Kim, "Circulating Current Control in MMC Under the Unbalanced Voltage," IEEE Transactions on Power Delivery, vol.28, no.3, pp.1952-1959, July 2013.

[16] M. Hagiwara, H. Akagi, "Control and Experiment of Pulsewidth-Modulated Modular Multilevel Converters," IEEE Transactions on Power Electronics, vol.24, no.7, pp.1737-1746, July 2009.

[17] Q. Tu, Z. Xu, L. Xu, "Reduced Switching-Frequency Modulation and Circulating Current Suppression for Modular Multilevel Converters," IEEE Transactions on Power Delivery, vol.26, no.3, pp.2009-2017, July 2011.

[18] K. Ma, F. Blaabjerg, M. Liserre, "Electro-thermal model of power semiconductors dedicated for both case and junction temperature estimation," Proceedings of PCIM Europe 2013, pp. 1042-1046, May 2013.

[19] M. Altın, O. Goksu, R. Teodorescu, P. Rodriguez, B.-B. Jensen, L. Helle, "Overview of recent grid codes for wind power integration," 12th International Conference on Optimization of Electrical and Electronic Equipment (OPTIM), pp.1152-1160, 20-22 May 2010.

978-1-4799-2706-7/14 $31.00 © 2014 IEEE

The 2014 International Power Electronics Conference

A Novel Optimal Design of DFIG Crowbar Resistor during Grid Faults

Sheng Hu
School of Electrical &Electronic Engineering
Huazhong University of Science and Technology
Wuhan, China
Yong Kang
School of Electrical &Electronic Engineering
Huazhong University of Science and Technology
Wuhan, China

XuDong Zou
Shool of Electrical &Electronic Engineering
Huazhong University of Science and Technology
Wuhan, China

Abstract- **In order to protect the converter of doubly fed induction generator(DFIG) and realize low voltage ride through(LVRT) during grid faults, the crowbar is usually used , since the resistor value of the crowbar effects DFIG's rotor current, reactive power output and electrical torque, it 's design is very crucial for LVRT.A novel optimal design of DFIG crowbar resistor for fault ride-through is proposed in this paper, simulations have been carried out to validate the proposed optimal design, the results demonstrate its effectiveness.**

Keywords—wind power, DFIG, low voltage ride through, crowbar

I. INTRODUCTION

Wind energy penetration is rapidly increasing over the past decades by the reason of many merits of wind power generation, such as cleanness, short construction cycle and low running cost. Nowadays, doubly fed induction generator (DFIG) is the most employed generator for wind turbines above 1MW, due to the advantages of variable-speed-constant-frequency based operation and decoupled control of active and reactive power, moreover, as the converter only manages the rotor power, it can be rated at around 25–35% of the generator rating [1]-[3], which can achieve advantages of small size, low cost, light weight, and small losses compared to wind power generation systems with a full power converter.

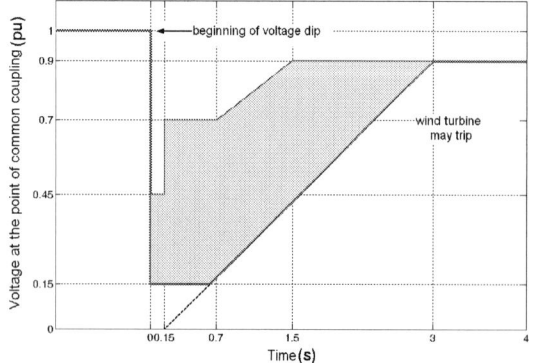

Fig. 1. Low voltage ride through requirement proposed by E.ON.

However, because the stator of DFIG is directly connected to the electrical grid, it is extremely sensitive to grid voltage disturbances. Voltage dips at stator due to grid faults induce overvoltages in the rotor windings, resulting in overcurrents of the rotor circuit, which may cause severe damage to vulnerable rotor-side power electronic converter and large fluctuation of dc-link voltage. Such large rotor inrush current, dc-link overvoltage and torque oscillations caused by grid faults are quite harmful for the DFIG-based wind turbines [1-5], and can lead to the destruction of converter and mechanical parts. Traditionally, once overcurrents occur in rotor windings, passive crowbar is used to protect the rotor converter by short-circuiting the rotor windings, at the same time, the DFIG was tripped from power grid.

Nevertheless, with the continuously increasing penetration level of wind energy, many system operators have revised the grid code for grid connection and fault ride-through capability of wind turbines. The E.ON standard is a notable grid code in the world, whose LVRT capability is specified in Fig.1, wind turbines must remain connected to the grid when the voltage at the point of common coupling (PCC) drops to 0.15pu for 0.625s. If the voltage does not fall below the minimum voltage and returns to 90% of the nominal voltage in Fig.1 within 3s after the beginning of the voltage drop, the wind turbine must operate continuously. In addition, active power output must resume immediately following fault clearing and the reactive power must be provided to support the power grid voltage within 20 ms after fault identification [9], [10]. Hence, in order to meet the LVRT requirement, active crowbar instead of passive crowbar should be used to protect power electronic converter,

Fig. 2 shows the block diagram of a DFIG equipped with active crowbar [1],[2],[5],[11]-[13]. As can be seen, the active crowbar consists of resistors, which are controlled by switches. Once overcurrents caused by grid faults occur in rotor windings, active crowbar is used to protect the rotor converter by short-circuiting the rotor windings and bypass the converter, the DFIG was still connected to power grid. When the rotor current was reduced to the permitted value by active crowbar, the converter can be restarted to realize active power and reactive power control of DFIG.

Since the resistor value of the crowbar effects DFIG's rotor current, reactive power output and electrical

978-1-4799-2706-7/14 $31.00 © 2014 IEEE

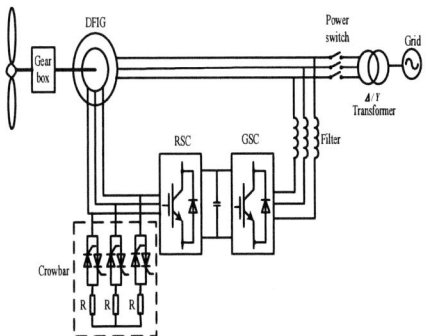

Fig. 2. DFIG based wind turbine equipped with crowbar.

torque, it's design is very crucial for LVRT. some authors [4],[5],[12] have proposed schemes for active crow-bar resistor design, but the derivation of crowbar resistor value is based on a serials of approximations, moreover, the derivation is also not based on the extreme conditions, therefore, such schemes for crowbar resistor design lead to large errors which can greatly degrade the LVRT capability of DFIG, differed from the scheme mentioned above, a novel optimal design of DFIG crowbar resistor for fault ride-through is proposed in this paper, simulations have been carried out to validate the proposed optimal design, the results demonstrate its effectiveness.

II. MODEL AND BEHAVIOR DURING GRID FAULTS

The Park model [1], [3] is usually used to analyze the behavior for DFIG, Fig. 3 shows the park model of DFIG, all of the variables are referred to as the rotor. According to this model, stator voltage vector and rotor voltage vector are expressed as

$$\begin{cases} v_s^r = R_s i_s^r + d\psi_s^r / dt + jw_r \psi_r^r \\ v_r^r = R_r i_r^r + d\psi_r^r / dt \end{cases} \quad (1)$$

Where superscript "r" denotes that the equations are expressed in rotor frame reference, subscripts "s" and "r" denote stator and rotor, respectively, v_s^r, v_r^r represent voltage, i_s^r, i_r^r represent current, R_s, R_r represent resistance, ω_r is the angular velocity of rotor, ψ_s^r, ψ_r^r represent fluxes.

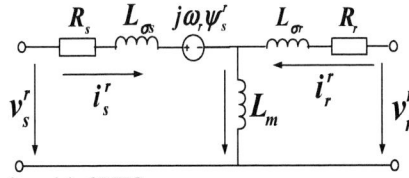

Fig. 3. Park model of DFIG.

The stator and rotor fluxes are express as

$$\begin{cases} \psi_s^r = L_s i_s^r + L_m i_r^r \\ \psi_r^r = L_m i_s^r + L_r i_r^r \end{cases} \quad (2)$$

L_s, L_r represent self-inductance, $L_{\sigma r}$, $L_{\sigma s}$ represent rotor and stator leakage inductance, L_m is mutual inductance between stator and rotor. Rotor flux can be obtained with (2) as

$$\psi_r^r = \frac{L_m}{L_s} \cdot \psi_s^r + \sigma L_r \cdot i_r^r \quad (3)$$

Where $\sigma = 1 - L_m^2 / L_s L_r$, by substituting (3) into (1), the rotor voltage is

$$v_r^r = \frac{L_m}{L_s} \cdot \frac{d}{dt} \psi_s^r + (R_r + \sigma L_r \frac{d}{dt}) \cdot i_r^r \quad (4)$$

From above expression, it can be seen that the first term in (4) is the electromotive force induced by stator flux, therefore, the equivalent DFIG model viewed from rotor side [2] can be obtained with (4) and shown in Fig. 4. where E is the electromotive force induced by stator flux, it is the first term in (4) and can be expressed as

$$E = \frac{L_m}{L_s} \cdot \frac{d}{dt} \psi_s^r \quad (5)$$

According to Fig. 4, the rotor current is determined by the injected rotor voltage and the induce electromotive

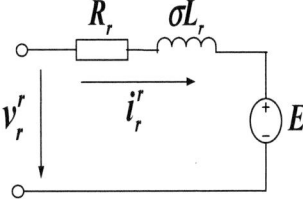

Fig. 4. Equivalent DFIG model viewed from rotor side.

force (EMF) which is the derivative of the stator-flux linkage. During severe grid faults, the induced EMF may exceed the maximum voltage of the rotor converter and cause rotor current is out of control, the dc-link voltage also dramatically increases [2], at the same time, large inrush currents appear in rotor circuit, if these currents exceed the limit that rotor converter can bear, rotor side converter can be severely damaged. In order to protect converter, crowbar can be used to short circuit the rotor and bypass the converter, in this manner, large rotor currents caused by grid faults flow into crowbar circuit and the converter is protected.

III. OPTIMIZATION OF CROWBAR RESISTANCE

Since the derivation of crowbar resistor value is based on a serials of arithmetic approximations and not based on the extreme conditions, therefore, such schemes for crowbar resistor design lead to large errors, furthermore, on the condition of guaranteeing the drop of crowbar resistor is lower than the maxim output voltage of rotor side converter, only the influence of crowbar resistor on rotor current is considered, so the LVRT capability of DFIG can be greatly degraded, in order to improved LVRT capability of DFIG, crowbar resistor can be optimized as following steps by using simulations: firstly, the influence of running conditions on rotor current at pre-faults is considered, then the extreme running condition to crowbar resistor design can be confirmed; secondly, at such extreme running condition, the influences of crowbar resistor value on rotor current, stator current, power and torque are totally considered, at the same time, the drop of crowbar resistor is lower than the maximum output voltage of rotor side converter.

978-1-4799-2706-7/14 $31.00 © 2014 IEEE

The 2014 International Power Electronics Conference

TABLE I
Simulative parameters of DFIG

Capacity	1.5WM
Stator voltage/Frequency	690V/50Hz
Stator resistance	0.023 PU
Rotor resistance	0.016 PU
Stator leakage inductance	0.18 PU
Rotor leakage inductance	0.16PU
Mutual inductance	2.9PU
Pole pairs	3
Turn ratios	0.4
Connection type	Y/Y

According to such optimal design scheme, we firstly confirm the extreme running conditions by simulations, the simulative parameters of a 1.5WM DFIG are shown on TABLE I. Assuming that the crowbar resistance is 0.48p.u., the rotated speed is 1.3p.u., the reactive power is 0p.u. and the stator voltage has an 80% dip depth at 1s, Fig. 5 shows the rotor current with different active power at pre-faults, as can be seen, in the cases that active power Ps is equal to 0.1pu,0.4pu,0.6pu and 1pu at pre-faults, respectively, rotor currents during grid faults are gradually increased, therefore, it is concluded that rotor current is increased with the increasing output active power of DFIG.

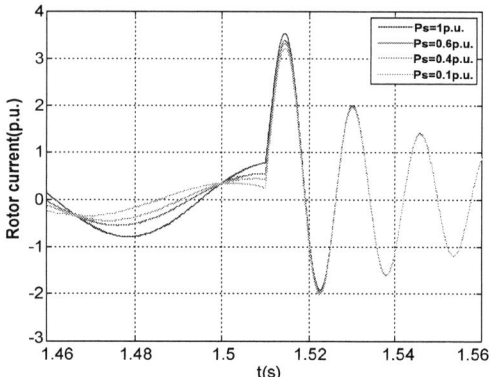

Fig. 5. Rotor current in the cases of different active power .

If slip rate of DFIG is between -0.3 and +0.3, the maximum rotated speed is 1.3p.u.. Assuming that the crowbar resistance is 0.48p.u., the rotated speed is 1.3p.u., the active power is 1p.u. and the reactive power is 0p.u., the stator voltage has an 80% dip depth at 1s, Fig. 6 shows the rotor current with different rotated speed at pre-faults, according to Fig.6, in the cases that rotated speed is equal to 0.8p.u.,0.9p.u.,1.1p.u., 1.2p.u. and 1.3p.u. at pre-faults, respectively, rotor currents during grid faults are also gradually increased, similarly, it is also concluded that rotor current is increased with the increase of DFIG rotated speed.

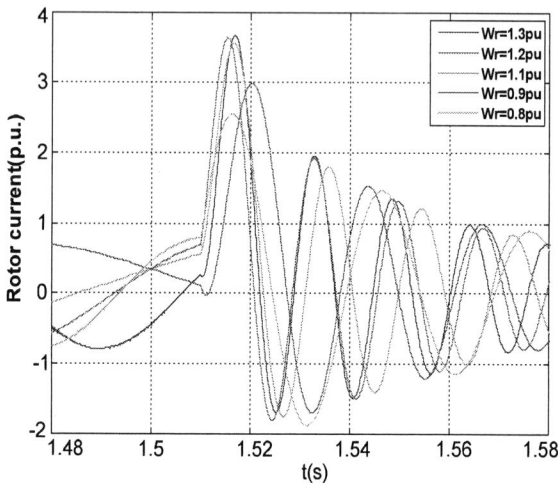

Fig. 6. Rotor current in the cases of different rotated speed .

As previous analysis, the extreme running condition at pre-fault is that the output active power of DFIG is full load, e.g. 1p.u. and rotated speed is 1.3p.u.in such extreme case, DFIG is under super-synchronous state. Under the extreme running condition, the crowbar resistor is designed by simulations based on maximal elimination on oscillation of rotor current, stator current, power and torque, at the same time, the drop of crowbar resistor is lower than the maximum output voltage of rotor side converter. Assuming that the stator voltage has an 85% dip depth at 0.7s, the fault lasts 250ms and stator voltage recovery at 0.95ms, when the faults occur, crowbar is activated to short circuit rotor windings, and it is cancelled at 100ms later than stator voltage recovery, e.g.1.05s,at the same time, the DFIG recovery normal vector control.

Fig. 7~ Fig. 11 show the rotor current, stator current, stator active power, stator reactive power and electromagnetic torque, during simulation, crowbar resistance varies from one time of rotor resistance to 30 times of rotor resistance. If crowbar resistance is larger than 30 times of rotor resistance, the drop of crowbar resistance exceeds the maximum output voltage of rotor side converter which leads the current flow into converter and makes the overcurrents of converter and larger fluctuation of DC link.

Fig.7 shows rotor current of phase a, according to the figure, when crowbar resistance is one time of rotor resistance, the peak value of rotor current is equal to 4.5 p.u. under stator voltage dip, while crowbar resistance is 30 times of rotor resistance, the peak value of rotor current is equal to 3.2 p.u. under stator voltage dip, furthermore, the rotor current with high value crowbar resistor is decayed much faster. The same conclusion can be drawn under stator voltage recovery from Fig.7, hence, whatever stator voltage dip or recovery, high value crowbar resistor is much more effective to eliminate the rotor overcurrents than low value crowbar resistor.

Fig.8 shows stator current of phase a, according to the figure, when crowbar resistance is one time of rotor resistance, the peak value of stator current is equal to 3.7

978-1-4799-2706-7/14 $31.00 © 2014 IEEE

The 2014 International Power Electronics Conference

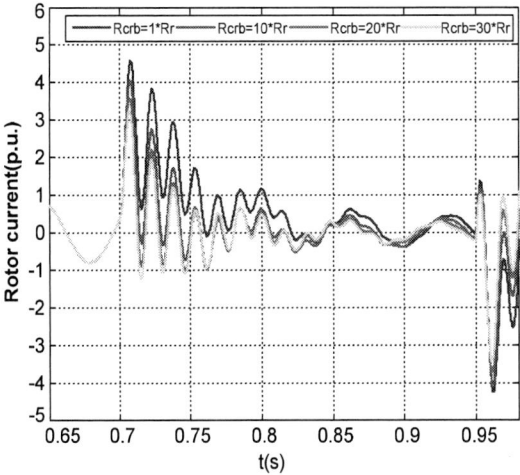

Fig. 7. Rotor current under different crowbar resistance value

p.u. under stator voltage dip, while crowbar resistance is 30 times of rotor resistance, the peak value of stator current is equal to 2.9 p.u. under stator voltage dip, similarly, the stator current with high value crowbar resistor is decayed much faster. The same conclusions can be also drawn under stator voltage recovery, therefore, high value crowbar resistor is much more effective to eliminate the stator current oscillations than low value crowbar resistor, which can do good at power system transient stability.

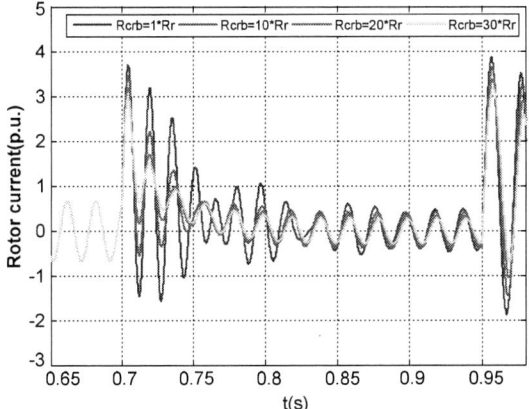

Fig. 8. Stator current under different crowbar resistance value

Fig.9 shows stator active power, according to the figure, during stator voltage dip, the higher crowbar resistance value, the smaller oscillation of active power and the stator active power also decays to stable state much more faster. During stator voltage recovery, active power to power grid under high crowbar resistance is larger than that under low crowbar resistance, so it benefits frequency stability of power grid.

Fig.10 shows stator reactive power, although active crowbar is able to protect rotor side converter, it simultaneously changes DFIG into a conventional squirrel-cage induction generator, which absorbs reactive power from power grid, with this behavior, it is adverse to grid voltage recovery, but according to Fig.8, whatever during grid voltage dip or during grid voltage recovery, the higher crowbar resistance value, the smaller reactive

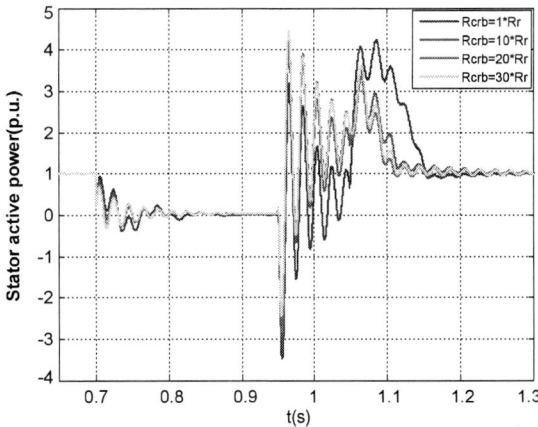

Fig. 9. Stator active power under different crowbar resistance value

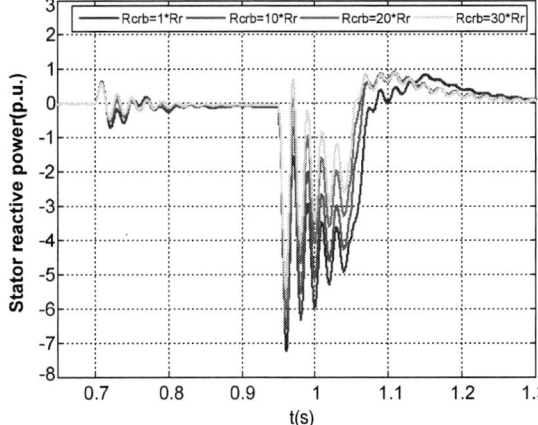

Fig. 10. Stator reactive power under different crowbar resistance value

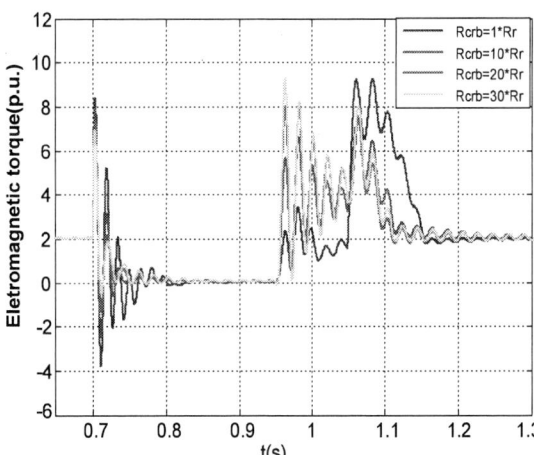

Fig.11. Electromagnetic torque under different crowbar resistance value

power of reactive power absorb from power grid, and the reactive power is also decayed to stable state much more faster, so it benefits voltage stability of power grid.

Fig.11 shows electromagnetic torque, according to the figure, the influence of crowbar resistance on electromagnetic torque is similar to its on active power, during stator voltage dip, the higher crowbar resistance value, the smaller oscillation of electromagnetic torque and the electromagnetic torque also decays to stable state much more faster. During stator voltage recovery,

978-1-4799-2706-7/14 $31.00 © 2014 IEEE   558

electromagnetic torque under high crowbar resistance is larger than that under low crowbar resistance. Similarly, the transit period of electromagnetic with high crowbar resistance is shorter than that with low crowbar resistance.

According to Fig.5~Fig.11, when the oscillation of rotor current is eliminated at most, simultaneously, the oscillation of stator current, active power, reactive power and electromagnetic torque are also eliminated at most. Therefore, according to Fig.7, when crowbar resistance is equal to 30 times of rotor resistance, e.g 0.48 p.u., the oscillation of rotor current is eliminated at most. In addition, if crowbar resistance is larger than 30 times of rotor resistance, the drop of crowbar resistance exceeds the maximum output voltage of rotor side converter which leads the current flow into converter and makes the overcurrents of converter and larger fluctuation of DC link. To sum up, the crowbar resistance is 0.48p.u..

IV. CONCLUSIONS

In this paper, a novel optimal design of DFIG crowbar resistor for fault ride-through is proposed, crowbar resistor can be optimized as follows: firstly, the influence of running conditions on rotor current at pre-faults is considered to confirm the extreme running conditions; secondly, at such extreme running condition, the influences of crowbar resistor value on rotor current, stator current, power and torque are totally considered, at the same time, the drop of crowbar resistor is lower than the maximum output voltage of rotor side converter. Simulations have been carried out to validate the proposed optimal design.

ACKNOWLEDGMENT

This work is financially supported by the Natural Postdoctoral Science Foundations of China under the Award Number 2013M540579 and by the National Basic Research Program of China under Award Number 2009CB219701

REFERENCES

[1] J. Lopez, E. Guba, P. Sanchis, E. Olea, J. Ruiz, L. Marroyo, "Ride through of wind turbines with doubly fed induction generator under symmetrical voltage dips," *IEEE Trans. on Industrial Electronics*, vol. 56, no. 10, pp. 4246–4254, Oct. 2009.

[2] P.S. Flannery, G. Venkataramanan., " Fault tolerant doubly fed induction generator wind turbine using a parallel grid side rectifier and series grid side converter," *IEEE Trans. on Power Electronics*, vol. 23, no. 3, pp. 1126-1135, May 2008.

[3] S. Hu, X. Lin, Y. Kang, X. Zou." An improved low voltage ride through control strategy of doubly-fed Induction generator during grid faults," *IEEE Trans. on Power Electronics*, vol.26, no. 12, pp.3653-3665, Dec.2011

[4] J. Morren and S. W. H. de Haan, "Short-circuit current of wind turbines with doubly fed induction generator," *IEEE Trans. Energy on Conversion*, vol. 22, no. 1, pp. 174–180, Mar. 2007.

[5] J. Morren and S. W. H. de Haan, "Ride through of wind turbines with doubly-fed induction generator during a voltage dip," *IEEE Trans. on Energy Conversion*, vol. 20, no. 1, pp. 435–441, Jun. 2005.

[6] J. Yang, J. E. Fletcher, and J. O'Reilly, "A Series dynamic-resistor-based converter protection scheme for doubly-fed induction generator during
various fault conditions," *IEEE Trans. on Energy Conversion,*,

vol. 25, no. 2, pp. 422–432, Jun. 2010.

[7] J. Lopez, P. Sanchis, X. Roboam, and L. Marroyo, "Dynamic behavior of the doubly-fed induction generator during three-phase voltage dips," *IEEE Transactions on Energy Conversion*, vol. 22, no. 3, pp. 709–717, Sep. 2007.

[8] J. Lopez, E. Guba, P. Sanchis, X. Roboam, and L. Marroyo, "Wind turbines based on doubly fed induction generator under asymmetrical voltage dips," *IEEE Transactions on Energy Conversion*, vol. 23, no. 1, pp. 321–330, Mar. 2008.

[9] E.ON. Netz Gmbh, Germany, "Grid code regulations for high and extra high voltage," status: 1st April 2006, Report ENENARHS2006, 46 pages, 2006, www.eon-netz.com.

[10] M. Tsili and S. Papathanassiou, "A review of grid code technical requirements for wind farms," Renewable Power Generation, IET: 308-332, 2009.

[11] S.M. Muyeenl, R. Takahashil, T. Muratal, et al, "Low voltage ride through capability enhancement of wind turbine generator system during network disturbance," IET Renew, Power Gener., vol.3, no.1, pp.65–74, 2009

[12] L. Zhang, X.Jin, Li.Zhan, "A novel LVRT control strategy of DFIG based rotor active crowbar," Proceedings 2011 Asia-Pacific Power and Energy Engineering Conference (APPEEC 2011), pp.1-6, Mar. 2011

[13] Y. Ren ,W. Zhang , "A novel control strategy of an active crowbar for DFIG-based wind turbine during grid faults," 2011 IEEE International Electric Machines&Drives Conference (IEMDC), pp. 1137 – 1142, May 2011

The 2014 International Power Electronics Conference

DC-Voltage Regulation of a Five Levels Neutral Point Clamped Cascaded Converter for Wind Energy Conversion System

Farid Merahi[1,2], Saad Mekhilef[2]

[1]Automatic Laboratory of Setif (LAS), Department of Electrical Engineering, F. Abbas University, Setif. Algeria
[2]Power Electronics and Renewable Energy Research Laboratory (PEARL), Department of Electrical Engineering, University of Malaya, 50603 Kuala Lumpur Malaysia.
merahif@gmail.com, saad@um.edu.my

El Madjid Berkouk[3]

[3]Process Control Laboratory (LCP), High Polytechnic National School. Hassan Badi Street, BP 182, Algiers, 16000-Algeria
emberkouk@yahoo.fr

Abstract— **Multilevel converters are widely recognized as a suitable solution for directly interfacing different types of power sources and energy storage systems to the medium voltage grids, due to their ability in high-voltage and high-power applications. The DC-voltage regulation of a five-level neutral-point clamped (NPC) in closed loop is presented. It consists to regulate the average value of the DC-voltage by using one loop instead of four loops. The modeling and the control of the different components of the wind energy conversion system are presented. The wind turbine is controlled using the maximum power point tracking algorithm (MPPT) based on the wind speed estimation. The vector control of active and reactive power is used to control the doubly fed induction generator (DFIG) through the rotor. The dynamic behavior of the global system is simulated in MATLAB/Simulink interface programming. The results are shown to validate the effectiveness of the proposed system.**

Keywords—NPC multilevel converters, DC voltage regulation; wind energy; DFIG.

I. INTRODUCTION

Multilevel converters are characterized by the ability to supply staircase-like voltage waveforms. Their applications become very interesting, especially with the insertion of the new renewable sources which is originally random into the electrical grid [1]. This new type of great power converters can contribute to the improvement of the quality of energy produced by attenuating the fluctuations which can occur. This feature enables converters with a high number of levels to reduce voltage and current total harmonic distortion (THD), and their architecture enables high-voltage high-current operation [2], [3], [4]. The fundamental principle of the multilevel conversion techniques is essentially based on an association series/parallel of the electronic power switches. They allowing bypassing the problem involved in maximum limit of the blocking voltage of the principal power semiconductors. These structures allow increasing in voltage beyond the limits of the semiconductors with a very good resolution of the output voltage [5], [6], [7]. The DC-voltages balancing in the DC-link is the main concern in the NPC

multilevel converter topology. The active NPC control approach [8], passive NP controls approach [9], dual current controller approach [10] has been proposed as solution to regulate the DC-voltage. However, these approaches suffer from constraints in high modulation, the system delivered unbalanced currents to the grid under unbalanced voltage dips and the systems required a linearization method for their implementation

This paper proposes the regulation of the DC-voltage of five-level NPC cascaded converters in closed loop. One closed loop is used instead of four loops, and the details of the control principal are theoretically explained. The wind energy conversion system description based on five-level NPC cascaded converters is shown in Fig. 1.

II. MODELING OF THE WIND GENERATOR

A. Modeling of the Wind Turbine and the Gearbox

The aerodynamic power P_{aer}, which is converted by a wind turbine, is dependent on the power coefficient C_p. It is given by [11]:

$$P_{aer} = \frac{1}{2} C_p(\lambda) \rho \pi R^2 v^3 \qquad (1)$$

Where ρ is the air density, R is the blade length and v the wind velocity.

The turbine torque is the ratio of the output power to the shaft speed. It is given by:

$$T_{aer} = \frac{P_{aer}}{\Omega_{aer}} \qquad (2)$$

The turbine is normally coupled to the generator shaft through a gearbox. Neglecting the transmission losses, the torque and shaft speed of the wind turbine, referred to the generator side of the gearbox, are given by [12]:

$$T_g = \frac{T_{aer}}{G} \qquad (3)$$

978-1-4799-2706-7/14 $31.00 © 2014 IEEE

The 2014 International Power Electronics Conference

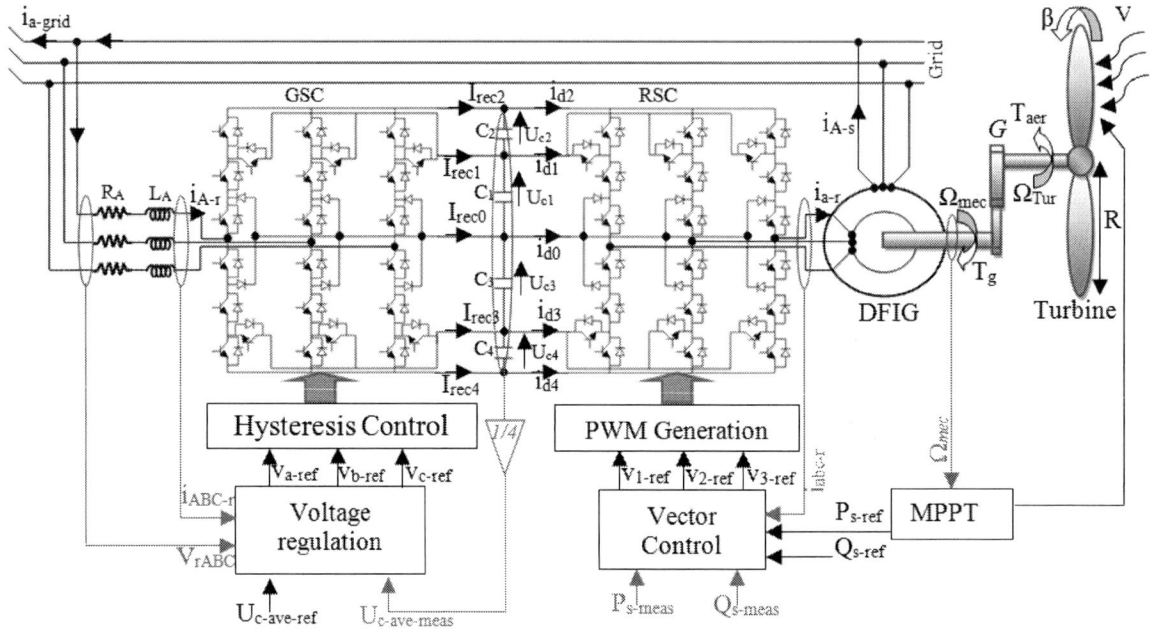

Fig. 1. Wind conversion system description

$$\Omega_{Tur} = \frac{\Omega_{mec}}{G} \tag{4}$$

Where T_g is the driving torque of the generator and Ω_{mec} is the generator shaft speed

A wind turbine can only convert just a certain percentage of the captured wind power. This percentage is represented by $C_p(\lambda)$ which is function of the wind speed, the turbine speed and the pitch angle of specific wind turbine blades β [12]. A C_p depends on the ratio λ between the turbines angular velocity Ω_{Tur} and the wind speed v. This ratio is called the tip speed ratio:

$$\lambda = \frac{\Omega_{Tur}.R}{v} \tag{5}$$

For a 1.5-MW wind turbine, the expression for the power coefficient is given as follows:

$$C_p(\lambda, \beta) = (0.5 - 0.167(\beta - 2)).\sin\left[\frac{\pi(\lambda + 0.1)}{10 - 0.3\beta}\right] -$$

$$0.00184(\lambda - 3).(\beta - 2) \tag{6}$$

We note that there is a value for which C_p and P_{mec} achieve their maximum. To maximize the generated power, therefore, the generator must have a power characteristic that follows the maximum $C_{p-\max}$.

The control must attenuate the action of the aerodynamic torque, which is a disturbance.

If the wind speed is measured and the mechanical characteristics of the wind turbine are known, it is possible to deduce in real-time the optimal mechanical power which can

be generated using the MPPT. The optimal mechanical power can be expressed as:

$$P_{mec-opt} = \frac{C_{p\_\max}}{\lambda^3_{C_{p\_\max}}} \frac{\rho\pi R^5}{2} \frac{\Omega^3_{mec}}{G^3} \tag{7}$$

B. Modeling of the DFIG

The electrical equations of the DFIG in the Park reference frame are expressed as follows [12], [13]:

$$\begin{cases} v_{ds} = R_s i_{ds} + \dfrac{d}{dt}\phi_{ds} - \omega_s \phi_{qs} \\[2mm] v_{qs} = R_s i_{qs} + \dfrac{d}{dt}\phi_{qs} + \omega_s \phi_{ds} \\[2mm] v_{dr} = R_r i_{dr} + \dfrac{d}{dt}\phi_{dr} - (\omega_s - \omega)\phi_{qr} \\[2mm] v_{qr} = R_r i_{qr} + \dfrac{d}{dt}\phi_{qr} + (\omega_s - \omega)\phi_{dr} \end{cases} \tag{8}$$

Where R_s and R_r are the stator and rotor phase resistance, respectively, $w = p.\Omega_{mec}$ is the electrical speed and p is the pair pole number.

The stator and the rotor flux can be expressed as:

$$\begin{cases} \phi_{ds} = L_s i_{ds} + L_m i_{dr} \\ \phi_{qs} = L_s i_{qs} + L_m i_{qr} \end{cases} \text{And} \begin{cases} \phi_{dr} = L_r i_{dr} + L_m i_{ds} \\ \phi_{qr} = L_r i_{qr} + L_m i_{qs} \end{cases} \tag{9}$$

Where i_{ds}, i_{qs}, i_{dr} and i_{qr} are the direct and quadrature stator and rotor currents, respectively.

The active and reactive power at the stator and the rotor are defined as:

$$\begin{cases} P_s = v_{ds}i_{ds} + v_{qs}i_{qs} \\ Q_s = v_{qs}i_{ds} - v_{ds}i_{qs} \end{cases} \text{ And } \begin{cases} P_r = v_{dr}i_{dr} + v_{qr}i_{qr} \\ Q_r = v_{qr}i_{dr} - v_{dr}i_{qr} \end{cases} \quad (10)$$

The electromagnetic torque is expressed as:

$$T_{em} = p(\phi_{ds}i_{qs} - \phi_{qs}i_{ds}) \quad (11)$$

III. CONVERTER MODELING

A. Five level converter model

The topology of a five level NPC converter is shown in fig. 2.

For the operation of this converter, seven configurations are possible [13], [14]. The table I gives the electric magnitude characterizing each one of these configurations (with M origin of potentials and V_{KM} the potential of node K of arm k).

The DC voltages U_{c1}, U_{c2}, U_{c3} and U_{c4} are provided by a rectifier.

Each switch TD_{ks} introduces a connection function F_{ks} that describes its state, such as:

$F_{ks} = 1$ If the switch is closed.

$F_{ks} = 0$ If the switch is open.

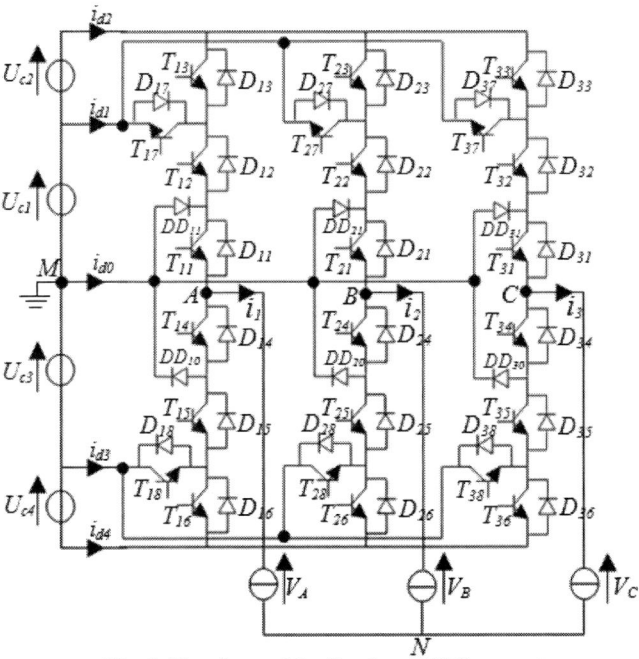

Fig. 2. Topology of the five-level NPC converter

B. Complementary Control

In purpose to avoid short circuit of the voltage sources and the converter become completely controlled, we adopt the complementary law [13], [15], [16], the connection functions of the switches on the arm are bound by the following expressions:

$$\begin{cases} B_{k4} = \overline{B_{k2}} \\ B_{k5} = \overline{B_{k1}} \\ B_{k6} = \overline{B_{k3}} \\ B_{k7} = B_{k1}B_{k2}\overline{B_{k3}} \\ B_{k8} = B_{k4}B_{k5}\overline{B_{k6}} \end{cases} \quad (12)$$

$$\begin{cases} F_{k4} = 1 - F_{k2} \\ F_{k5} = 1 - F_{k1} \\ F_{k6} = 1 - F_{k3} \\ F_{k7} = F_{k1}F_{k2}\overline{F_{k3}} \\ F_{k8} = F_{k4}F_{k5}\overline{F_{k6}} \end{cases} \quad (13)$$

Moreover, we define a connection function of the half-arm, noted with:

k : Number of the arm;

$m = 1$: For the high half-arm, and $m = 0$ for the low half-arm.

For an arm k, the connection functions of the half-arm are expressed by means of the switches connection functions as follows:

$$\begin{cases} F_{k1}^b = F_{k1}.F_{k2}.F_{k3} \\ F_{k0}^b = F_{k4}.F_{k5}.F_{k6} \end{cases} \quad (14)$$

TABLE I
ELECTRIC MAGNITUDES OF EACH CONFIGURATION OF ARM K

Configurations	Electric magnitudes
E_0	$i_k = 0$
E_1	$V_{kM} = U_{c1} + U_{c2}$
E_2	$V_{kM} = U_{c1}$
E_3	$V_{kM} = 0$
E_4	$V_{kM} = -U_{c3}$
E_5	$V_{kM} = -U_{c3} - U_{c4}$
E_6	$V_{kM} = 0$

F_{k1}^b : is associated to the high half-arm ($TD_{k1}, TD_{k2}, TD_{k3}$)

and F_{k0}^b to the low half-arm ($TD_{k4}, TD_{k5}, TD_{k6}$).

The simple voltages output of five level inverter are given in the following matrix form:

$$\begin{bmatrix} V_{AM} \\ V_{BM} \\ V_{CM} \end{bmatrix} = \begin{bmatrix} F_{17} + F_{11}^b \\ F_{27} + F_{21}^b \\ F_{37} + F_{31}^b \end{bmatrix}U_{c1} + \begin{bmatrix} F_{11}^b \\ F_{21}^b \\ F_{31}^b \end{bmatrix}U_{c2} - \begin{bmatrix} F_{18} + F_{10}^b \\ F_{28} + F_{20}^b \\ F_{38} + F_{30}^b \end{bmatrix}U_{c3} - \begin{bmatrix} F_{10}^b \\ F_{20}^b \\ F_{30}^b \end{bmatrix}U_{c4}$$

$$(15)$$

C. Connection Functions

By using the connection functions of the half-arm, the input inverter currents can be expressed according to the load currents i_1, i_2 and i_3 as follows:

$$\begin{cases} i_{d1} = F_{17}i_1 + F_{27}i_2 + F_{37}i_3 \\ i_{d2} = F_{11}^b i_1 + F_{21}^b i_2 + F_{31}^b i_3 \\ i_{d3} = F_{18}i_1 + F_{28}i_2 + F_{38}i_3 \\ i_{d4} = F_{10}^b i_1 + F_{20}^b i_2 + F_{30}^b i_3 \end{cases} \quad (16)$$

The current i_0 is expressed according to the load currents by the relation:

$$i_{d0} = (i_1 + i_2 + i_3) - (F_{17} + F_{18} + F_{11}^b + F_{10}^b)i_1 - (F_{27} + F_{28} + F_{21}^b + F_{20}^b)i_2$$
$$- (F_{37} + F_{38} + F_{31}^b + F_{30}^b)i_3 \quad (17)$$

D. PWM Control Strategy for the Five Level Inverter

Several control strategies are applicable for the five-level NPC voltage inverter [13], [14], [17], [18]. In this study, the triangular-sinusoidal pulse width modulation is employed to control the five-level NPC inverter.

1) Algorithm

After we have chosen the control strategy and the three voltage references, two stages characterize the control algorithm, where $U_c = U_{c1} = U_{c2} = U_{c3} = U_{c4}$.

a) Stage 1: intermediate voltage determination:

$$\begin{cases} U_{refk} \geq U_{car-4} \Rightarrow V_{k4} = 2U_c \\ U_{refk} < U_{car-4} \Rightarrow V_{k4} = U_c \end{cases} ; \begin{cases} U_{refk} \geq U_{car-3} \Rightarrow V_{k3} = U_c \\ U_{refk} < U_{car-3} \Rightarrow V_{k3} = 0 \end{cases}$$

$$\begin{cases} U_{refk} \geq U_{car-2} \Rightarrow V_{k2} = 0 \\ U_{refk} < U_{car-2} \Rightarrow V_{k2} = -U_c \end{cases} ; \begin{cases} U_{refk} \geq U_{car-1} \Rightarrow V_{k1} = -U_c \\ U_{refk} < U_{car-1} \Rightarrow V_{k1} = -2U_c \end{cases}$$

b) Stage 2: Determination of V_{km} and switch order control B_{ks} : $V_{km} = V_{k1} + V_{k2} + V_{k3} + V_{k4}$

$$\begin{cases} V_{km} = 2E \Rightarrow B_{k1} = 1; B_{k2} = 1; B_{k3} = 1 \\ V_{km} = E \Rightarrow B_{k1} = 1; B_{k2} = 1; B_{k3} = 0 \\ V_{km} = 0 \Rightarrow B_{k1} = 1; B_{k2} = 0; B_{k3} = 0 \\ V_{km} = -E \Rightarrow B_{k1} = 0; B_{k2} = 0; B_{k3} = 1 \\ V_{km} = -2E \Rightarrow B_{k1} = 0; B_{k2} = 0; B_{k3} = 0 \end{cases}$$

E. Control of Three-Phase Five Level Current Rectifier

The five-level NPC current rectifier provides four DC voltages to its output. These voltages must be kept constant, which requires their control in a closed loop. Instead of using four loops to control the four DC voltages, we propose to control the average value of the DC voltage using only one loop [13]. The control principle of the complete system is

shown in fig. 3. The voltage model is given by the equations (17) [13].

F. Control Strategy

The hysteresis current control limits the current variations in the hysteresis band. The measured and reference currents are compared in real time. If we indicate with ε_k the gap between the real current i_k and the reference current i_{ref-k} : $\varepsilon_k = i_{grid-k} - i_{ref-k}$.

The switch control orders B_{ks} of an arm k are determined by the algorithm below.

$$\begin{cases} C_1 \dfrac{dU_{c1}}{dt} = I_{rec1} + I_{rec2} - i_{d1} - i_{d2} \\ C_2 \dfrac{dU_{c2}}{dt} = I_{rec2} - i_{d2} \\ C_3 \dfrac{dU_{c3}}{dt} = -I_{rec3} - I_{rec4} + i_{d4} + i_{d3} \\ C_4 \dfrac{dU_{c4}}{dt} = -I_{rec4} + i_{d4} \end{cases} \quad (18)$$

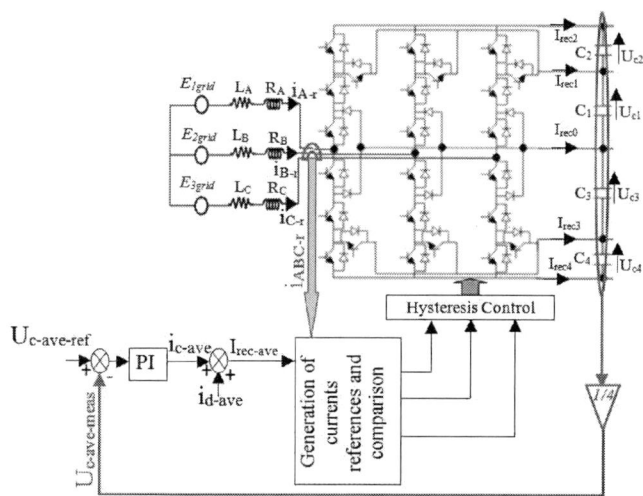

Fig. 3. Control principle scheme of the five level current rectifier

$$\begin{cases} \varepsilon_k > 2\Delta i \Rightarrow B_{k_1} = 0; B_{k_2} = 0; B_{k_3} = 0; \\ \Delta i < \varepsilon_k > 2\Delta i \Rightarrow B_{k_1} = 0; B_{k_2} = 0; B_{k_3} = 1; \\ -2\Delta i < \varepsilon_k < -\Delta i \Rightarrow B_{k_1} = 1; B_{k_2} = 1; B_{k_3} = 0; \\ \varepsilon_k < -2\Delta i \Rightarrow B_{k_1} = 1; B_{k_2} = 1; B_{k_3} = 1; \\ Else, \text{ the control remains unchanged} \end{cases}$$

IV. DFIG POWER CONTROL

The connection of the wind generator to the electrical grid must be established in three steps [12], [19]:

- Synchronization of the stator voltages with the grid voltages;
- Connection of the stator to the electrical grid;
- Power transit regulation between the DFIG and the grid.

A d–q reference-frame synchronized with the stator flux is employed. By setting the quadratic component of the stator to the null value as follows, [11], [20]:

$$\phi_{ds} = \phi_s \text{ And } \phi_{qs} = 0 \tag{19}$$

The torque relation becomes:

$$T_{em} = -p(\frac{L_m}{L_s})\phi_{ds}i_{qr} \tag{20}$$

The electromagnetic torque and the power active will be dependent only on the rotor current along the q-axis. By neglecting the stator resistance R_s, we will have:

$$v_{ds} = 0 \text{ And } v_{qs} = v_s \tag{21}$$

By choosing this reference, the stator voltages and flux can be rewritten as follows:

$$\begin{cases} v_{ds} = 0; v_{qs} = v_s = \omega_s\phi_{ds} \\ \phi_{ds} = \phi_s = L_s i_{ds} + L_m i_{dr}; \quad \phi_{dr} = L_r i_{dr} + L_m i_{ds} \\ \phi_{qs} = 0 = L_s i_{qs} + L_m i_{qr}; \quad \phi_{qr} = L_r i_{qr} + L_m i_{qs} \end{cases} \tag{22}$$

The active, reactive power and rotor voltages can be written according to the rotor currents:

$$\begin{cases} P_s = -\frac{v_s L_m}{L_s} i_{qr} \\ Q_s = \frac{v_s^2}{\omega_s L_s} - \frac{v_s L_m}{L_s} i_{dr} \end{cases} \tag{23}$$

$$\begin{cases} v_{dr} = R_r i_{dr} + \left(L_r - \frac{L_m^2}{L_s}\right)\frac{d}{dt}i_{dr} - s\omega_s\left(L_r - \frac{L_m^2}{L_s}\right)i_{qr} \\ v_{qr} = R_r i_{qr} + \left(L_r - \frac{L_m^2}{L_s}\right)\frac{d}{dt}i_{qr} - s\omega_s\left(L_r - \frac{L_m^2}{L_s}\right)i_{dr} + s\frac{L_m v_s}{L_s} \end{cases} \tag{24}$$

where $s = (\omega_s - \omega) / \omega_s$

The third terms are the cross-coupling terms, and they can be neglected because they have no significant contribution. The principle of the active and reactive power vector control of the DFIG is shown in Fig. 4.

V. SIMULATION RESULTS

The performances of the wind energy conversion system are illustrated in the figures bellow. The simulation results are carried out using the following parameters:

A. Simulation Parameters of the Grid Side Converter

RMS Value of the grid voltage $V_g = 690$ V, average DC voltage $U_{c-ave-ref} = 200$V, Capacitors value:

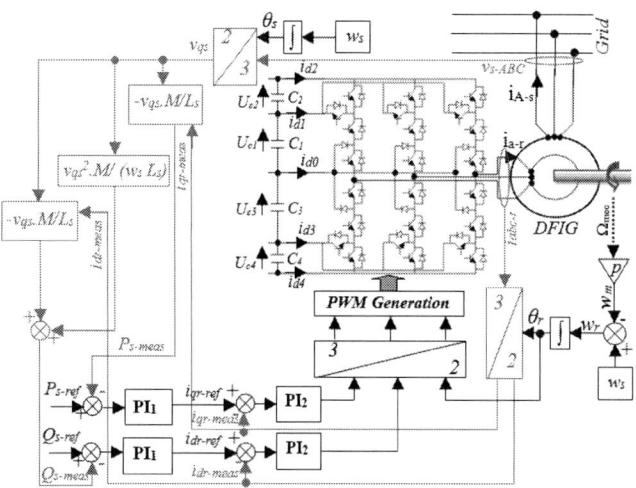

Fig. 4. The principle of the active and reactive power vector control of the DFIG

$C_1 = C_2 = C_3 = C_4 =$ 1100µf, and filter parameters (L =10mH, R =1Ω).

B. Simulation Parameters of the Rotor Side Converter

1) Turbine Parameters

Diameters = 70.5m, number of blades = 3, hubs height = 85m, gearbox = 90.

2) DFIG Parameters

Nominal power (P = 1.5 MW), nominal speed (N = 1.500 r/m), V_s = 690V, pairs of pole number (p = 2), f = 50Hz, Inertial (J (Turbine + DFIG) = 50kg.m^2), viscous coefficient (f (Turbine + DFIG) = 7.1e^{-2}Nm s/rad), stator resistance (R_s = 0.012Ω), stator inductance (L_s = 13.732mH), rotor resistance (R_r = 0.021Ω), rotor inductance (L_r = 13.703mH), mutual inductance (Lm = 13.528mH).

The average speed of the supposed wind profile is approximately 11m/s. Fig. 5 shows the DFIG mechanical speed which undergoes the variations as supposed on the wind profile. According to Fig. 5, the oscillation is around 140 rad/s. The generator operates in sub-synchronous mode.

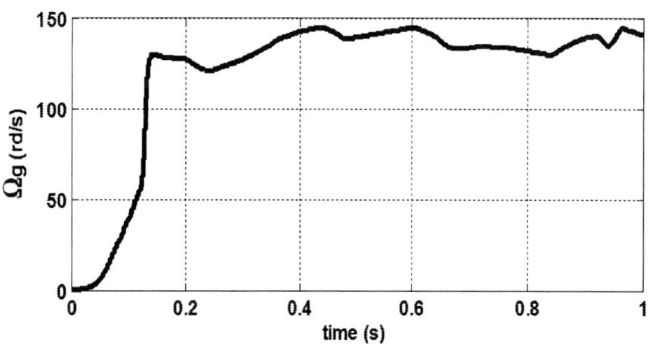

Fig. 5. Generator speed

The 2014 International Power Electronics Conference

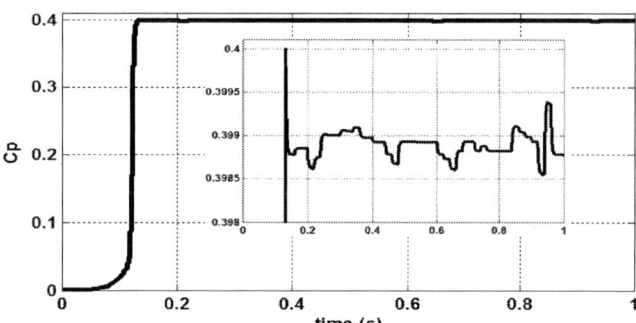

Fig. 6. Power coefficient

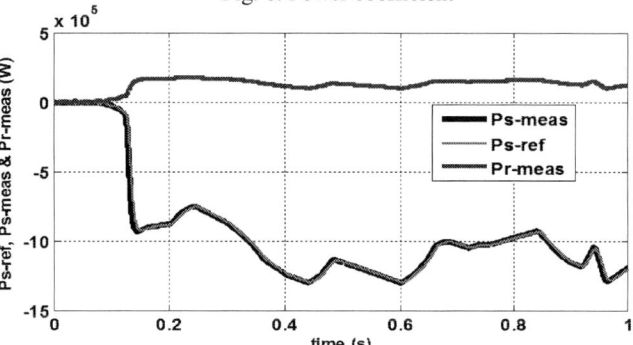

Fig. 7. DFIG active powers

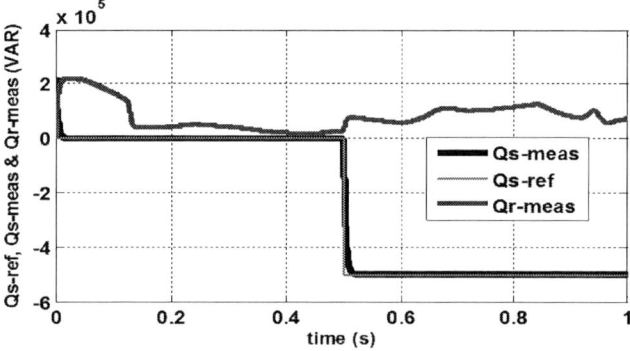

Fig. 8. DFIG reactive powers

Fig. 6 shows the power coefficient C_p obtained. It reaches its maximum (about 0.4). Fig. 7 shows the DFIG active powers.

The reference of the stator active power depends on the speed and the torque developed by the wind turbine. The power produced by the stator (power measured) agrees with active power reference. Also, we note that the produced power is maximized. On the same figure, we observe that the rotor active power is positive. It means that the rotor consumes the active power; therefore, the generator operates in hypo-synchrone mode.

The stator reactive power is kept at zero ($Q_{s-ref} = 0$) until t=0.5s, as shown in Fig. 8. A good follow-up of the reactive power reference by the measured reactive power Q_{s-meas} is observed after the compensation of the constant imposed by the grid $v_s^2 / w_s . L_s$. At t = 0.5s, a negative grade of reactive

power (-500kVAR) is imposed to the generator, we note that the DFIG is able to produce a reactive power when is necessary. The reactive power Q_r consumed by the rotor is almost negligible after the magnetization. Thus, the total reactive power Q_{dfig} consumed by the generator is equal to the reactive power Q_r consumed by the rotor.

Fig. 9 shows the average DC voltage. The measured DC voltage $U_{c-ave-meas}$ tests to reach its reference $U_{c-ave-ref}$ (200 V). A negligible error is observed between the reference and the measured DC voltage. Also, the appearance of the fluctuations identical to those appeared at the generator speed.

Fig. 10 shows the DC voltages (U_{c1}, U_{c2} , U_{c3} and U_{c4}). They reach the values that ensure power balance at a much shorter time than the low capacity value. Fig. 11 shows the normal and zoomed sinusoidal waveform of the stator currents. The frequency is 50 Hz and undergoes the same variations as those in the wind speed, which is random.

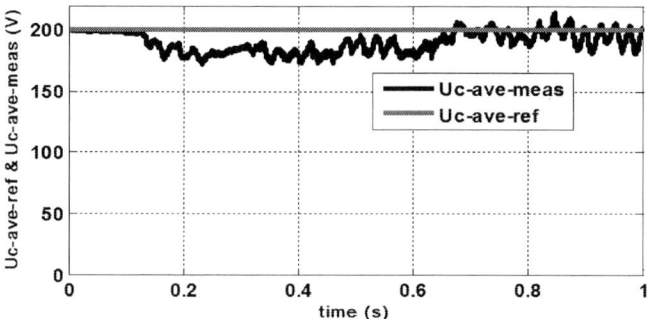

Fig. 9. Average DC voltages

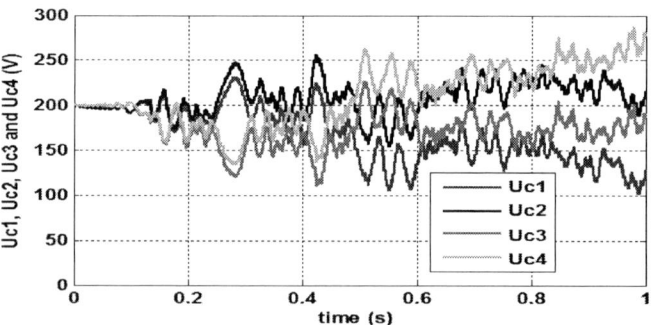

Fig. 10. DC voltages

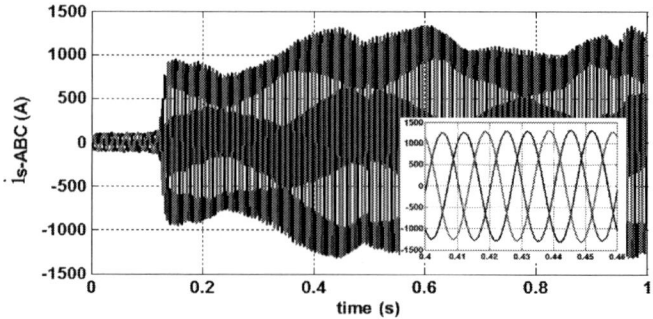

Fig. 11. Stator currents

The 2014 International Power Electronics Conference

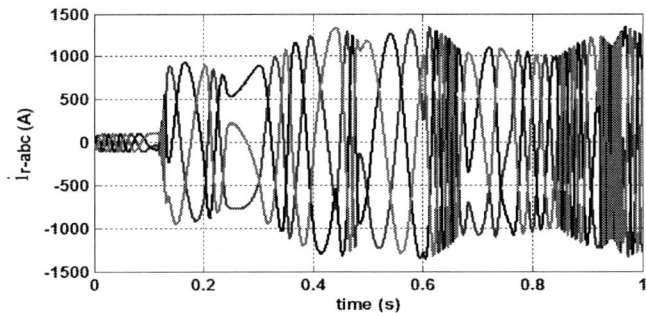

Fig. 12. Rotor currents

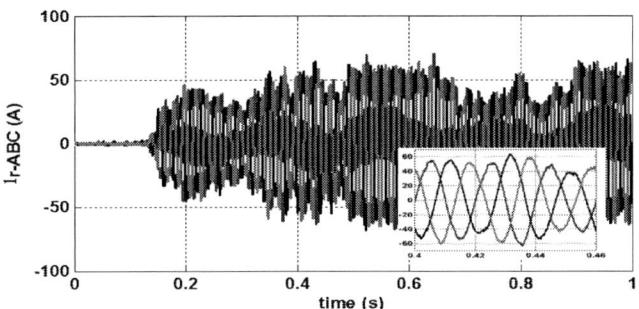

Fig. 13. Grid side converter currents

The rotor currents waveform is shown in Fig. 12. Its frequency varies because it depends on slip' s ' or on the generator speed.

The GSC currents are shown in Fig. 13. Their amplitudes vary and are affected by the wind profile fluctuations; however, the frequency remains at 50 Hz.

VI. CONCLUSION

This paper has presented a control method in closed loop of the DC voltage of the back-to-back NPC five-level converter. The converter cascade was applied in a wind energy conversion system to control the active and reactive power delivered to the grid using DFIG. The proposed control algorithm incorporated the control in closed loop of the average DC voltage. Modeling the different parts of the studied system has been presented. The proposed scheme can be extended and applied to high-power systems.

ACKNOWLEDGMENT

The authors would like to thank the Ministry of Higher Education of Algeria and the Ministry of Higher Education of Malaysia and University of Malaya for providing financial support under the research grant No. UM.C/HIR/MOHE/ENG/16001-00-D000024 and UMRG project RP015A-13AET.

REFERENCES

[1] T. Kompa, A. Samour, D. Schulz and R. Hanitsch, "Comparative Power Quality Investigations of Variable Speed Wind Energy Converters with Doubly-fed Induction and Synchronous Generator," PCIM Power Quality Conference Nuremberg, 39-44, May 14-16, 2002.

[2] J. D. Barros, J. Fernando A. Silva and É. G. A. Jesus "Fast-Predictive Optimal Control of NPC Multilevel Converters," IEEE Transactions on Industrial Electronics, volume 60,no. 2, pp 619–627, February, 2013.

[3] N. Yousefpoor, S. H. Fathi, N. Farokhnia, and H. A. Abyaneh, "THD minimization applied directly on the line-to-line voltage of multilevel inverters," IEEE Trans. Ind. Electron., vol. 59, no. 1, pp. 373–380, Jan. 2012.

[4] M. Marchesoni, M. Mazzucchelli, and S. Tenconi, "A non conventional power converter for plasma stabilization," in Proc. IEEE Power Electron. Spec. Conf., Apr. 1988, pp. 122–129.

[5] A. Talha, M.O. Mahmoudi, D. Berbier and E.M. Berkouk, "Study and control of two two-level PWM rectifiers-clamping bridge-two three-level NPC VSI cascade. Application to double stator induction machine," IEEE Power Electronics Specialists Conference. Aachen Germany 3894–3899, 2004.

[6] S. El Aimani, "Modelling and simulation of doubly fed induction generator for variable speed wind turbines integrated in a distribution network," In: 10th European conference on power electronics and application, Toulouse, France 2003.

[7] D. Aouzellag, K. Ghedamsi, E.M. Berkouk "Network power flux control of a wind generator," ScienceDirect, Renewable Energy 34 (2009) 615–622.

[8] J. Rodriguez, S. Bernet, P.K. Steimer and I.E. Lizama, "A survey on neutral-pointclamped inverters," IEEE Transactions on Industrial Electronics, vol. 57, no. 7, pp. 2219–2230, July 2010.

[9] Z. Mohzani, B.P. McGrath, D.G. Holmes,"The balancing properties of DC link compensation for 3-phase neutral point clamped (NPC) converter," IEEE 7th International Power Electronics and Motion Control Conference, ECCE Asia, pp. 574–579, June 2-5, 2012.

[10] J.B. Emilio, C. Santiago, F.J. Rodriguez, A. Hernandez and F. Espinosa,"Design of a backto-back NPC converter interface for wind turbines with squirrel-cage induction generator," IEEE Transactions on Energy Conversion, vol. 23, no. 3, pp. 932–945, September 2008.

[11] D.J Adams, F.Z. Peng and J.W. Mckeever, "Cascaded Multilevel Inverter for Utiling Applications," IEEE, 0-7803-3932-0, 2000.

[12] A. Tilli, S. Peresada and A. Tonielli, "Indirect Stator Flux-Oriented Output Feedback Control of a Doubly Fed Induction Machine," IEEE Transactions on Control Systems Technology, Vol.11, N°.6, pp. 25-33, November 2003.

[13] O. Motoyoshi and M. Yamamoto, "Active and reactive power control for doubly-fed wound rotor induction generator," IEEE Transactions on Power Electronics, Vol.6, N°.4, 624-629, October 1991.

[14] F. Poitiers, "Control of a doubly fed induction generator for windenergy conversion systems," GE44-LARGE. Saint Nazaire, France.

[15] A. Bendre and G. Venkataramanan, "Reciprocity-Transposition-Based Sinusoidal Plusewidth Modulation for Diode-Clamped," IEEE Transactions on Industrial Electronics, Vol.49, N°5, 2002.

[16] F. Merahi and E.M. Berkouk, "Commande d'une GADA par un onduleur de tension à cinq niveaux à structure NPC," International Conference on electronic and oil, from theory to applications (ICEO'11), Ouargla-Algiers, March 01-02, 2011.

[17] F.Z. Peng, J. Rodriguerz and J.S. Lai, "Multilevel Inverters: A Survey of Topologies, Controls, and Applications," IEEE Transactions on Industrial Electronics, Vol.49, N°.4, August 2002.

[18] A. Nabae, I. Takahashi and H. Akagi, "A new-neutral-point-clamped PWM inverter," IEEE Transactions on Industrial Applications IA–17, 518–523, September–October, 1981.

[19] P.N. Tekwani, R.S. Kanchan and K. Gopakumar, "Five-level inverter scheme for an induction motor drive with simultaneous elimination of common-mode voltage and DC-link capacitor voltage imbalance," IEE Proc.-Electr. Power Appl. Vol. 152, N°.6, November 2005.

[20] A. Tilli, S. Peresada and A. Tonielli, "Indirect Stator Flux-Oriented Output Feedback Control of a Doubly Fed Induction Machine," IEEE Transactions on Control Systems Technology, Vol.11, N°.6, pp. 25-33, November2003.

A Reactive Power Sharing Method Based on Virtual Capacitor in Islanding Microgrid

Haizhen Xu,Xing Zhang,Fang Liu,Rongliang Shi,
Changzhou Yu
School of Electrical Engineering and Automation
Hefei University of Technology
Hefei, China
xhzicy@sina.com, honglf@ustc.edu.cn

Wei Zhao, Yong Yu, Wei Cao
Sungrow Power Supply Co., Ltd.
Hefei, China
zhaow@sungrowpower.com , yuy@sungrowpower.com,
caow@sungrowpower.com

Abstract—**For the problem that paralleled inverters cannot share reactive power proportional to their power capacity in islanding microgrid. Originally, the effect of inverter output impedance and line impedance to reactive power sharing with conventional droop control is analyzed. Following, a method adjusting the output reactive power by paralleling capacitor at the inverter output is proposed, and the relationship between paralleled capacitors, line impedance and reactive power is analyzed. On the basis, virtual capacitor method is realized to improve reactive power balance. Finally, the proposed method and analysis are verified through simulation.**

Keywords— Reactive power sharing, paralleled inverters, microgrid, virtual capacitor.

I.   INTRODUCTION

With the power electronics technology development and awareness of environmental and energy requirements increase, conventional centralized power generation systems are turning into more intelligent distributed generation systems [1]. In order to integrate the advantages of distributed powers, improve the efficiency of energy management, microgrid is paid more and more attention [2]. For improvement of reliability and stability of power supply, microgrid should operate in islanding mode, in which reactive power sharing problem of distributed powers operating in parallel through inverter interface becomes an important issue.

Existing paralleled inverter control schemes are split into wired and wireless control. Wired control scheme is limited by the location of distributed power and complexity of communication protocol [3, 4]. Hence, the research is focused on wireless control and the mainly scheme is P-ω, Q-U droop control. Due to the difference of inverter output impedance and line impedance to PCC (Point of Common Coupling) between paralleled inverters, conventional droop control may result in reactive power sharing problem, which causes circulate current and overload of inverters furthermore. Therefore, many improved droop control methods have been proposed.

Paper [5, 6] propose methods of modifying reference voltage by changing droop curve or droop formulas which require accurate line impedance in grid-connected mode. Paper [7] uses the coupling of active and reactive power to adjust power sharing, but invariance of load should be guaranteed during the process, and this method may cause active power fluctuation.

In this paper, conventional droop control method is introduced firstly [8, 9], and then the effect of inverter output impedance and line impedance to reactive power sharing with tradition droop control is analyzed. Based on this, a self-adapting method to adjust reactive power balance by adding capacitor current feedback is proposed. Finally, the proposed method and analysis are verified by simulation.

II.   CONVENTIONAL DROOP CONTROL METHOD

Fig. 1 shows the model of two paralleled distributed power inverters system. DG1 and DG2 are connected to PCC by single phase voltage source inverter. U_{ref1}, U_{ref2} indicate reference voltage of inverter 1 and inverter 2. Z_{11}, Z_{21} represent inverter control output impedance. $U_{01}\angle\delta_1, U_{02}\angle\delta_2$ are inverter output voltage, $Z12$, $Z22=Ri2+jXi2$ are line impedance to PCC, $U_0\angle 0$ is PCC voltage, and Z_L is common load.

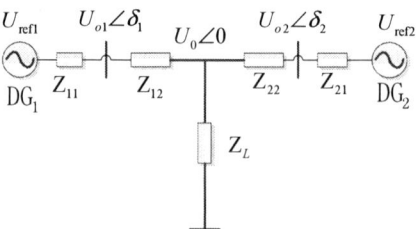

Fig. 1. Model of two paralleled distributed power inverters system.

The two inverters' output active power and reactive power can be calculated according to the system model:

$$\begin{cases} P_i = \dfrac{1}{R_{i2}{}^2 + X_{i2}{}^2}(R_{i2}U_{0i}{}^2 - R_{i2}U_{0i}U_0\cos\delta_i + X_{i2}U_{0i}U_0\sin\delta_i) \\ Q_i = \dfrac{1}{R_{i2}{}^2 + X_{i2}{}^2}(X_{i2}U_{0i}{}^2 - X_{i2}U_{0i}U_0\cos\delta_i - R_{i2}U_{0i}U_0\sin\delta_i) \end{cases} \quad (1)$$

The line impedance ratio varies with voltage levels. Line impedance is mainly inductive in high voltage system, while resistive in low voltage system. Analysis in this paper is based on inductive line impedance system ($X_{i2} \gg R_{i2}$).

Since δ_i is small enough, it can be approximated that $\sin \delta_i \approx \delta_i, \cos \delta_i \approx 1$. Applying it to (1), active and reactive power can be established as:

$$\begin{cases} P_i = \dfrac{1}{X_{i2}} U_{0i} U_0 \delta_i \\ Q_i = \dfrac{1}{X_{i2}} (U_{0i}{}^2 - U_{0i} U_0) \end{cases} \tag{2}$$

As can be seen from formula (2), active power is mainly related to voltage phase angle, while reactive power is related to voltage amplitude. Therefore, active power can be controlled by adjusting angular frequency and reactive power can be controlled by regulating voltage amplitude. P-ω, Q-U droop control formula is shown as:

$$\begin{cases} \omega_i = \omega^* - m_i P_i \\ E_i = E^* - n_i Q_i \end{cases} \tag{3}$$

Where ω*, U* are inverter output voltage angular frequency and amplitude with empty load, P_i, Q_i are No.i inverter active and reactive power, w_i, U_i are No.i reference voltage angular frequency and amplitude from power control loop.

III. THE EFFECT OF INVERTER OUTPUT IMPEDANCE AND LINE IMPEDANCE TO REACTIVE POWER SHARING

For inverter i the output power can be approximated as:

$$\begin{cases} S_i = P_i + jQ_i = u_{0i} i_{0i} \approx u_{0i} i_{0i} \\ i_{0i} = (u_{0i} - u_0) / jX_{i2} = (u_{refi} - u_0) / j(X_{i2} + X_{i1}) \end{cases} \tag{4}$$

Jointed with Formula (2), the output power of inverter i can be approximated as:

$$\begin{cases} P_i = \dfrac{1}{X_{i1} + X_{i2}} U_0{}^2 \delta_i \\ Q_i = \dfrac{1}{X_{i1} + X_{i2}} (U_{refi} U_0 - U_0{}^2) \end{cases} \tag{5}$$

Applying (3) to (5),

$$\begin{cases} P_i = \dfrac{U_0{}^2 (\omega^* - \omega_0)}{s(X_{i1} + X_{i2}) + m_i U_0{}^2} \\ Q_i = \dfrac{U^* U_0 - U_0{}^2}{X_{i1} + X_{i2} + n_i U_0} \end{cases} \tag{6}$$

Where, ω_0 is the angular frequency of PCC.

To realize power sharing of inverter 1 and inverter 2, $m_1 P_1 = m_2 P_2, n_1 Q_1 = n_2 Q_2$ should be satisfied. Solving (6) with it, it can be gained that active power sharing is not affected by output impedance and line impedance, while formula (7) should be obeyed to ensure reactive power

balance.

$$n_1 / n_2 = (X_{11} + X_{12}) / (X_{21} + X_{22}) \tag{7}$$

Inverter i output impedance can be gained according to control structure and control block shown in Fig. 2, 3[10]:

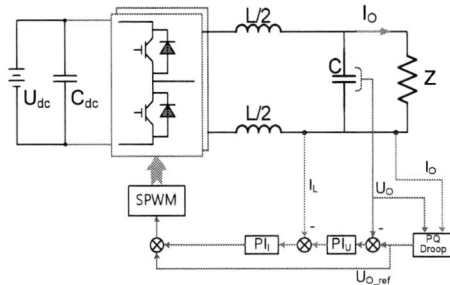

Fig. 2. Diagram of control structure of inverter i.

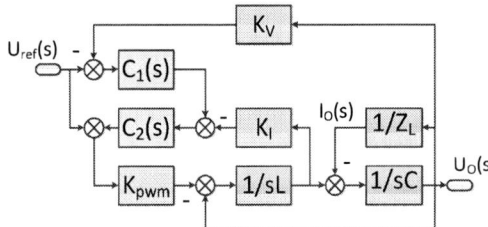

Fig. 3. Diagram of control block of inverter i.

Where, C1(s) is voltage loop PI regulator, C2(s) is current loop PI regulator, K_{pwm} is inverter gain, K_v, K_i are feedback gain of voltage and current loop, L,C are inductor and capacitor of inverter output filter.

From Fig. 3, transfer function of inverter output to input is [11]:

$$U_O(s) = \frac{K_{pwm}(1 + C_1(s)C_2(s))}{s^2 LC + sCK_i C_2(s) K_{pwm} + K_v C_1(s) C_2(s) K_{pwm} + 1} U_{ref}(s)$$
$$- \frac{sL + K_i C_2(s) K_{pwm}}{s^2 LC + sCK_i C_2(s) K_{pwm} + K_v C_1(s) C_2(s) K_{pwm} + 1} I_0(s) \tag{7}$$

And impedance of inverter i is:

$$Z_0 = \frac{L k_v s^2 + k_{p2} k_i s}{LC k_v s^3 + k_{p2} k_i C s^2 + (k_{p1} k_{p2} k_v + k_v)s + k_{i1} k_{p2} k_v} \tag{8}$$

Inverter output impedance is related to the regulator of dual-loop, the output filter, and feedback gains. When the same control structure is adapted, the output impedance is mainly inductive in a wide range of the fundamental frequency, and is inversely proportional to rated power capacity [12], Fig. 4 shows Bode diagram of inverter output impedances rated 3KW and 6KW.

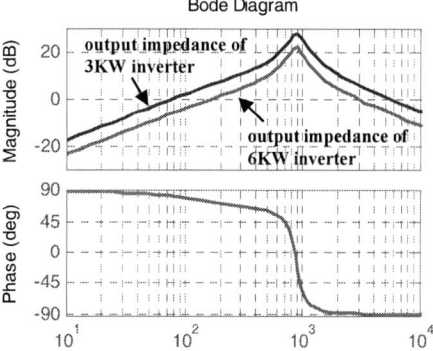

Fig. 4. Bode diagram of inverter output impedance.

While the distance of distributed powers to PCC are different, formula (7) cannot always meet, which lead to unequal sharing of reactive power. Therefore, a method should be found to solve reactive power sharing problem.

IV. REACTIVE POWER SHARING METHOD BASED ON VIRTUAL CAPACITOR

If a capacitor C_x is paralleled to the output of an inverter as Fig. 5:

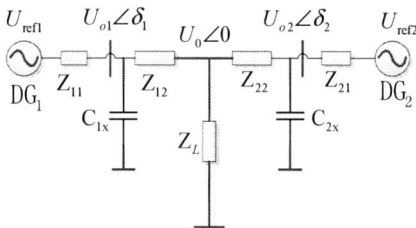

Fig. 5. Model of two paralleled distributed power inverters system with capacitor C_x.

According to formula (5), its output reactive power can be gained:

$$Q_{ix} = \frac{U_0(U_{refi} - U_0) + U_0^2 X_{i1} C_{ix} \omega_0}{X_{i1} + X_{i2} - X_{i1} X_{i2} C_{ix} \omega_0} \qquad (9)$$

Because $X_{i1} X_{i2} C_{ix} \omega_0 \ll X_{i1} + X_{i2}$, formula (9) can be simplified as:

$$Q_{ix} = \frac{U_0(U_{refi} - U_0) + U_0^2 X_{i1} C_{ix} \omega_0}{X_{i1} + X_{i2}} \qquad (10)$$

It can be known that reactive power has a linear relationship with C_x. If $C_x=0$, formula (10) is equal to formula (5).

Fig. 6 gives the relationship among C_x, Q_{ix} and Q_i, axis x represent C_x/C, axis y represent Q_{ix}/Q_i, axis z represent Q_i/S_L ((rated capacity of load).The output impedance is obtained from formula (8), and line impedance is chosen according to typical electrical parameters of high-voltage lines. It can be noticed that reactive power Q_{ix} raises with the increase of C_x, and the rise extent reduces with load power increases.

Fig. 7 is the relationship among C_x, Q_{ix} and X_{i2}, axis x

represent C_x/C, axis y represent Q_{ix}/Q_i, axis z is the value of $\log X_{i2}$. It can be seen that reactive power Q_{ix} rises with the increase of C_x, and the rise extent reduces with line impedance increases.

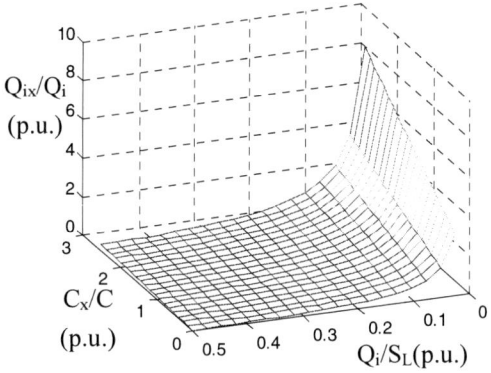

Fig. 6. Relationship among Cx, Qix and Qi.

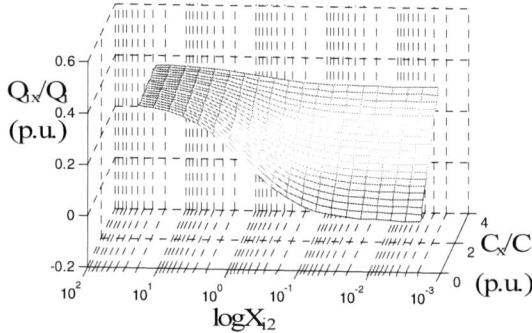

Fig. 7. Relationship among Cx, Qix and Xi2.

From the above analysis, it can be known that distributed power inverters cannot share reactive power proportional with their rated capacity because of unsatisfactory of formula (7), and output reactive power decreases if line impedance is larger. While the paralleled capacitor C_x can increase reactive power of inverter without changing reference voltage, and the larger C_x is, the larger reactive power is. Hence, the method to parallel a larger capacitor to the inverter with smaller reactive power and a smaller capacitor with larger reactive power is proposed.

Paralleling a real and continuously adjustable capacitor is hard and unrealistic. For this reason, a virtual capacitor with software algorithm by adding filter capacitor current feedback is adopted. The control block diagram of virtual capacitor is shown in Fig. 8 with dash line:

978-1-4799-2706-7/14 $31.00 © 2014 IEEE

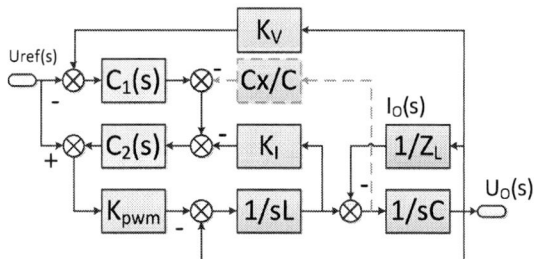

Fig. 8. Control block diagram of virtual capacitor.

From Fig. 8, transfer function of inverter output to input is:

$$U_O(s) = \frac{K_{pwm}\left(1+C_1(s)C_2(s)\right)}{s^2LC+s(C+C_x)K_iC_2(s)K_{pwm}+K_vC_1(s)C_2(s)K_{pwm}+1}U_{ref}(s)$$
$$-\frac{sL+K_iC_2(s)K_{pwm}}{s^2LC+s(C+C_x)K_iC_2(s)K_{pwm}+K_vC_1(s)C_2(s)K_{pwm}+1}I_0(s) \quad (11)$$

From (11) it can be seen that virtual capacitor control algorithm is equivalent to a capacitor paralleled to the real one C.

Virtual capacitor C_x is obtained according to $C_x = -k_{ci}Q_i$, k_{ci} is droop factor of C_x with Qi ,and should fulfill the requirement:

$$k_{ci} \le C_{ix\max} / Q_{i\max} \quad (12)$$

As k_{ci} is not calculated according to line impedance, reactive power sharing error still exists between paralleled inverters as a result. If line impedance is known and accurate value of k_{ci} is gained, reactive power can be shared precisely. Considering line impedance difference, k_{ci} is modified as:

$$k_{ci} = a_i n_i \quad (13)$$

Where $\square_i$ represent the weight of inverter i line impedance.

To get the value range of C_x, several factors should be thought:

A. *The steady of the system,since Cx is virtual and is in control loop instead of a real one,it may lead to system unsteable*

B. *The value of filter capacitor*

C. *The max reactive power of the inverter*

D. *The difference ratio of line impedance*

V.   SIMULATION AND EXPERIMENT RESULTS

In order to verify the correctness of above analysis, the model of two paralleled single-phase inverter system is built using Matlab/Simulink. The main parameters of the system are shown in Table I.

TABLE I
FUNDAMENTAL PHYSICAL CONSTANTS

Symbol	Meaning	Value
S	Inverter capacity	3KVA
L	Filter inductor	1mH
C	Filter capacitor	100uF

f_0	Microgrid frequency	50Hz
V_{PCC}	Voltage of pcc	230VAC
f_{sw}	Inverter switch frequency	10kHz
fc	Inverter sample	10kHz

The active power (AP) and reactive power (RP) waves of two 3KVA paralleled inverters are shown in Fig. 9. The power of common load is $S_L = 3000+1500\,jVA$, the line impedances are $Z_{12} = 0.06+0.191j\Omega, Z_{12} = 0.12+0.382j\Omega$ · The inverters are of the same active power, but different reactive power because of line impedance error.

Adding virtual capacitor algorithm to inverter current loop with other parameters unchanged, as shown in Fig.10. At steady state, active powers of two inverters are the same as Fig. 9, while reactive power difference reduces 50%.

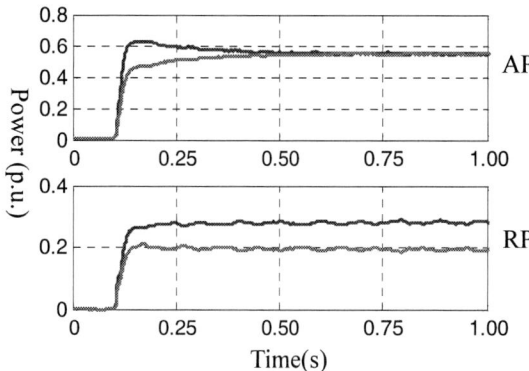

Fig. 9. Active and reactive power waves of two 3KVA paralleled inverters without virtual capacitor algorithm.

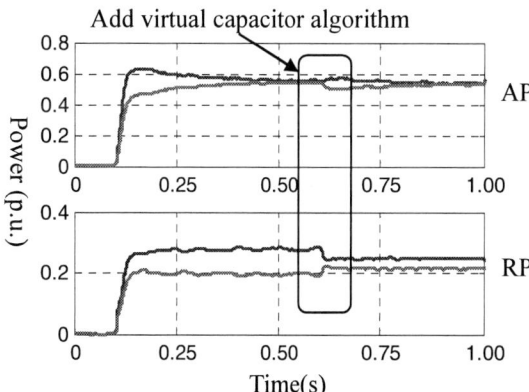

Fig. 10. Active and reactive power waves of two 3KVA paralleled inverters with virtual capacitor algorithm.

The active power of load increases 1000W as shown in Fig. 11 at 1.2s. It can be seen that the virtual capacitor algorithm does not affect the dynamic characteristic of the system, and still works in new steady state .The reactive power of both inverters increase slightly at 1.0s because of active and reactive power coupling.

978-1-4799-2706-7/14 $31.00 © 2014 IEEE

The 2014 International Power Electronics Conference

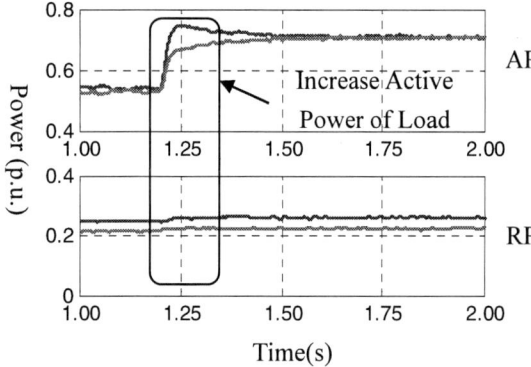

Fig. 11. Active and reactive power waves of sudden increase of load power with virtual capacitor algorithm.

In Fig. 12 are the active and reactive power waves of two paralleled inverters with the capacity ratio of 1:2. Due to the influence of line impedance, inverters cannot share reactive power proportional to their capacity. After adding the virtual capacitor algorithm, the line impedance effect is reduced, and the reactive power of two inverters is adjusted to 1:2.

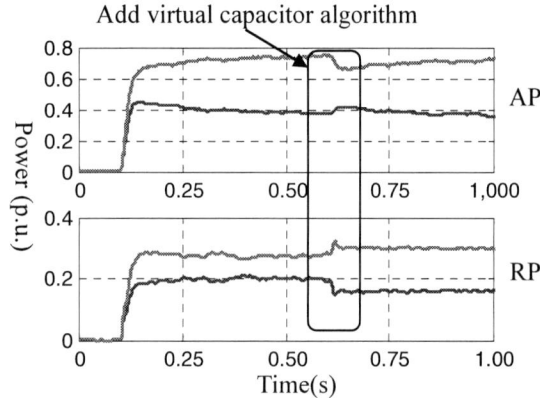

Fig. 12. Active and reactive power waves of two paralleled inverters with the capacity ratio of 1:2.

The experiment is done based on a two paralleled three phase inverters platform with 5KVA rated power capacity. Fig. 13 shows the experiment result of two inverters with a common load of 9KW+1.8KVA, for the reason that the parameters and line impedance are not exactly the same, the output power of inverter 1 is less than that of inverter 2, and current angle phase of inverter 1 lead current angle phase of inverter 2 .At 66ms a capacitor is paralleled to inverter 1 and the two inverters can share reactive load more accurately and circulating current is reduced.

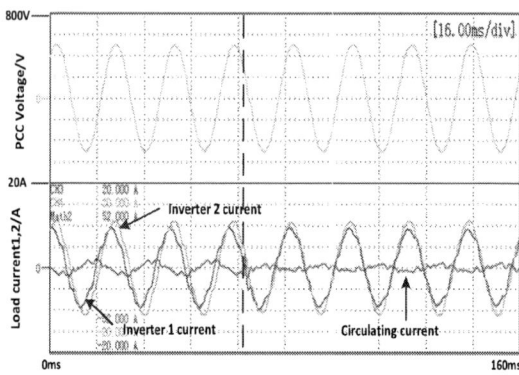

Fig. 13. Experiment results of two paralleled inverter system.

VI. CONCLUSIONS

In this paper, a reactive power sharing method based on inverter current loop control is proposed from the point of relationship of output power and current. It can be used to reduce impact of line impedance on power sharing. This method does not need the impedance of lines or its ratio, achieves reactive power sharing more quickly and easily, and the structure of the system requirement is lower.

REFERENCES

[1] Mohamed, Y. A -R I; El-Saadany, E.F., "Adaptive Decentralized Droop Controller to Preserve Power Sharing Stability of Paralleled Inverters in Distributed Generation Microgrids," *Power Electronics, IEEE Transactions on* , vol.23, no.6, pp.2806,2816, Nov. 2008.

[2] Xunwei Yu; Zhenhua Jiang; Yu Zhang, "Control of Parallel Inverter-Interfaced Distributed Energy Resources," *Energy 2030 Conference, 2008. ENERGY 2008. IEEE* , vol., no., pp.1,8, 17-18 Nov. 2008.

[3] Duan Shanxu; Meng Yu; Xiong Jian; Kang Yong; Chen Jian, "Parallel operation control technique of voltage source inverters in UPS," *Power Electronics and Drive Systems, 1999. PEDS '99. Proceedings of the IEEE 1999 International Conference on* , vol.2, no., pp.883,887 vol.2, 1999.

[4] Golestan, S.; Joorabian, M.; Rastegar, Hasan; Roshan, A.; Guerrero, J.M., "Droop based control of parallel-connected single-phase inverters in D-Q rotating frame,"*Industrial Technology, 2009. ICIT 2009. IEEE International Conference on* , vol., no., pp.1,6, 10-13 Feb. 2009.

[5] Yun Wei Li; Ching-Nan Kao, "An Accurate Power Control Strategy for Power-Electronics-Interfaced Distributed Generation Units Operating in a Low-Voltage Multibus Microgrid," *Power Electronics, IEEE Transactions on* , vol.24, no.12, pp.2977,2988, Dec. 2009.

[6] Aghasafari, M.A.; Lopes, L. A C; Williamson, S., "Frequency regulation and enhanced power sharing in microgrids including modified droop coefficients and virtual resistances," *Electrical Power & Energy Conference (EPEC), 2009 IEEE* , vol., no., pp.1,6, 22-23 Oct. 2009.

[7] Jinwei He; Yun Wei Li, "An Enhanced Microgrid Load Demand Sharing Strategy," *Power Electronics, IEEE Transactions on* , vol.27, no.9, pp.3984,3995, Sept. 2012.

[8] Guerrero, J.M.; Garcia de Vicuna, L.; Matas, J.; Castilla, M.; Miret, J., "A wireless controller to enhance dynamic performance of parallel inverters in distributed generation systems," *Power Electronics, IEEE Transactions on* , vol.19, no.5, pp.1205,1213, Sept. 2004.

978-1-4799-2706-7/14 $31.00 © 2014 IEEE

[9] Sao, C.K.; Lehn, P.W., "Autonomous load sharing of voltage source converters," *Power Delivery, IEEE Transactions on* , vol.20, no.2, pp.1009,1016, April 2005.

[10] [10] Shungang Xu; Jianping Xu, "Parallel control strategy of single-phase inverter based on virtual impedance," *Communications, Circuits and Systems (ICCCAS), 2010 International Conference on* , vol., no., pp.589,592, 28-30 July 2010.

[11] [11] WeiYu; Dehong Xu; Chaoyang Zhou, "Parallel control of the UPS inverters," *Applied Power Electronics Conference and Exposition, 2008. APEC 2008. Twenty-Third Annual IEEE* , vol., no., pp.939,944, 24-28 Feb. 2008.

[12] [12] Guerrero, J.M.; Garcia de Vicuna, L.; Matas, J.; Castilla, M.; Miret, J., "Output Impedance Design of Parallel-Connected UPS Inverters With Wireless Load-Sharing Control," *Industrial Electronics, IEEE Transactions on* , vol.52, no.4, pp.1126,1135, Aug. 2005.

Storage Capacity Performance for Hybrid PV/Diesel System in Sabah Malaysia

Nabil M Hidayat
Faculty. of Electrical Engineering
Universiti Teknologi MARA
Shah Alam, Selangor, Malaysia
mnabil@salam.uitm.edu.my

Mat Nasir Kari, Mohd Johari Mohd Arif
Dept. Of Electrical Engineering
Jabatan Kerja Raya
Malaysia
Nasir@jkr.gov.my

Abstract— **This paper presents the efficient limit setting for state of charge (SOC) to match with respect to the solar irradiance, load condition while special consideration based on analyzing the number of operational hours of diesel generator and lifetime of the battery. The hybrid PV/diesel standalone system is located in Ranau, Sabah, Malaysia.**

Keywords— Hybrid Photovoltaic System, Photovoltaic Monitoring, Performance ratio, Solar Photovoltaic System.

I. INTRODUCTION

Solar photovoltaic generated energy is one of the inexhaustible energy sources in effort to reduce dependence on fossil fuels. The technology enables the implementation of renewable energy system in remote location [1-2] Malaysia being gifted with abundance of solar irradiation, has a wide potential of solar energy applications to meet electricity demand for remote areas [3] Having a more stable system, the combination of solar system with diesel generator (hybrid PV/diesel system) has seen good potential application in remote located facilities. [4]

Malaysia, moving rapidly into a developed nation has a considerable number of under-developed rural areas that is located far away from utility grid. Out of more than 10,000 schools in Malaysia, 809 schools in Sabah and Sarawak still lack 24-hours electricity supply and is normally powered by standalone diesel generator to cope for the demand of electricity. In order to facilitate the energy demand and, to ensure these areas are not left behind as stated in the rural development strategy, the Ministry of Education (MOE) has initiated a large electrification program for rural school [5]. In this study, Ranau Sabah has been identified to be the sample site for evaluation and analysis process.

The several previous researches have been conducted on hybrid PV/ diesel system. Among the works that be highlighted are the work conducted by K.Y.Lau, M.F.M. Yousof, S.N.M. Arshad, M. Anwari, A.H.M Yatim in 2010 they did performance analysis of hybrid PV/diesel energy system under Malaysian Condition. This work is to analyze the potential use of hybrid PV/diesel system in remote locations [4]. The investigations demonstrated the impact of PV penetration, battery storage on energy production, cost of energy and number of operational hour of diesel generator. Besides that, it utilizes the hybrid PV/diesel over standalone diesel system based on different solar irradiance and diesel prices. The others previous researches have been conducted are the work by Abd El-Shafy A. Nafeh in 2010 through the study in proposed technique for optimally sizing a PV/diesel hybrid System. This work is to develop in optimizing the operation of diesel generator and to calculate the optimum number of PV module and batteries that would achieve minimum cost and desired depth of discharge for battery storage. In this work, to guarantee safe operation of battery used, the charging and discharging process have to be restricted within the design value of maximum and minimum SOC. [6]

II. SYSTEM CONFIGURATION

A. Overall System

The flow of power from PV panel for hybrid PV/ diesel system will be decided by the condition of solar irradiance and SOC battery. If the battery SOC is lower than allowable value, the system will cut off the usage of batteries and the diesel generator will kick start automatically to supply the load. The diesel generator keeps on running until battery reaches a certain preset value of SOC and then it will stop [7]. Due to this reason, the estimation value of battery SOC is one of the big issues for battery management [8]. Typically the desired value of SOC is kept within appropriate limits ranges, exp.: $20\% \leq SOC \leq 80\%$ [9]. In this study, the authors intended to come out with new ways of controlling limit setting of SOC level and have a less daily battery depth of discharge (DOD) value. The low level of DOD daily would help to increase battery lifetime. Besides taking into account of solar irradiance, the load demand need to be included in order to make sure system will choose the best SOC setting for various factors and conditions.

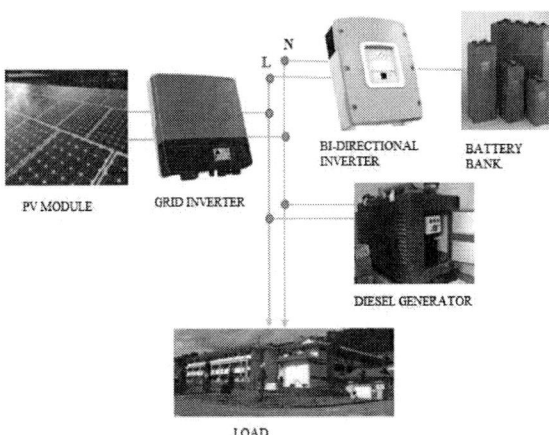

Fig. 1. Solar Hybrid System in Ranau Sabah.

B. Energy Storage

The hybrid PV/diesel system is an AC Coupled System as shown in Fig. 1. The 60kWp PV array is used to supply power to the load and excess energy charges the battery during daytime. Priority will be given to satisfy the day time load. Lead acid batteries are most widely used as energy storage and play the important part in hybrid PV/diesel System [10]. There are two different types of lead acid battery namely VRLA/OPzV (Valve Regulated Lead Acid) and FLA/OPzS (Flooded Lead Acid). The type of batteries used in this system is VRLA Model SMG 1500 by FIAMM manufacturer. This is sealed lead acid batteries with combination of gel electrolyte. In the first design, this system had the 48V DC/4500Ah energy storage installation but after a few months, additional 1500Ah battery bank upgrade to 48V DC/ 6000Ah is been installed to cater load for mostly during night time. The bidirectional inverter converts the DC-AC voltage and if there were insufficient power from PV to charge the battery, the 30 kVA diesel generator will turn on automatically. The advantage of using AC coupled system compared to DC coupled system is that the system can be expandable in the future and even if battery failed during daytime, the system is capable to supply power to the load. The DC coupled system will shut down if the battery occurs fault.

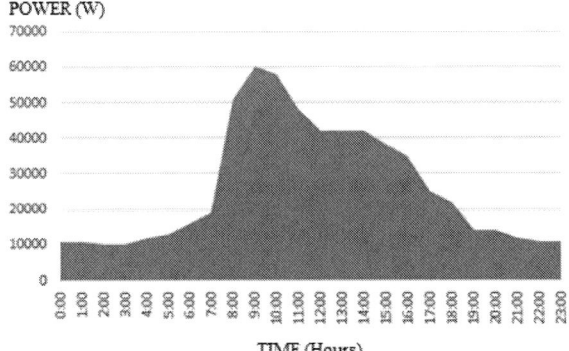

Fig. 2. System load profile .

C. Load Profile

The expected load profile distributed over 24hours is shown in Fig.2. The maximum power of the day is at 60kW while the minimum power is 11kW. Total energy consumption during daytime is much higher than night time. Daytime is considered normally from 06:00hours to 18:00 hours, which is the normal sunrise and sunset at this location. The system is basically designed to supply electricity for every building in the school exp.: classrooms, guard house, teacher's quarters and also student hostel.

III.   RESULT AND DISCUSSION

A. Evaluation on data for a year (July 2011-Jun2012)

To analyze the battery SOC, it is essential to know radiation data at intended site, Ranau Sabah. Often in PV system design, a more practical approach to determine solar irradiance is by the use of a concept called Peak Sun Hour (PSH). Technically PSH can be defined as the equivalent number of hours in a day when solar irradiance (G) is 1000Wm-2. [11].The preceding equation for PSH daily can be numerically calculated as;

$$H = \sum_{0}^{23} Gn\Delta t \qquad (1)$$

Where, G is Solar Irradiance (Wm-2) and t is time. From the equation of PSH, we can see the trend of solar irradiance in respect to data average rainfall (mm) at Ranau, Sabah from July 2011 until Jun 2012.

TABLE I
SOLAR IRRADIATION DATA

Month	Total PSH	Average Rainfall(mm)
July 2011	120	265.6
Aug 2011	127	245.1
Sept 2011	119	297.1
Oct 2011	110	411.3
Nov 2011	116	278.8
Dec 2011	112	210.9
Jan 2012	106	99.7
Feb 2012	139	79.3
Mar 2012	142	85.9
Apr 2012	148	75.2
May 2012	130	212.5

Table 1 tabulates that at Ranau area, more solar irradiance is to be expected from the month of February to May while less solar irradiance is expected from October to December. Apart from that, the annual PSH at Ranau area was estimated to be 125 hours in a month. This means in one day, an average of 4 hours of solar irradiance at the rate of 1000Wm-2.

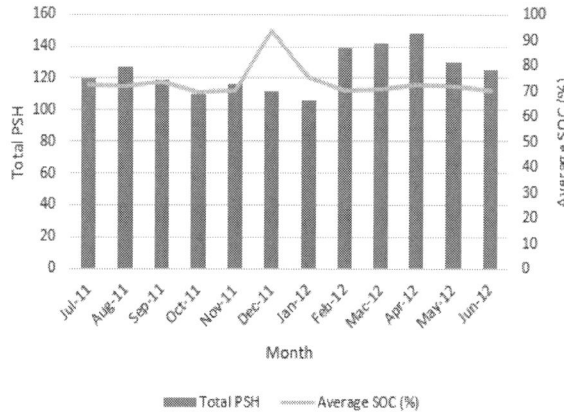

Fig. 3. Total PSH and Average SOC from Jul-11 to Jun-12.

Based on climate graph in Fig. 3, it is shown that there is slight difference between theory and actual measured value where that average SOC will increase if total PSH increases. But, from data, average SOC decrease when total PSH increase in December 2011 and January 2012. This is probably due to additional battery capacity installed on that month. The adding up of battery capacity gives effect to the average SOC battery especially when the graph grew rapidly in the first quarter and maintain around 70% SOC but reached a peak about 90% SOC in December 2011. Since then, it has quickly dropped around 70%. SOC on January 2012 and remains stable for the next month.

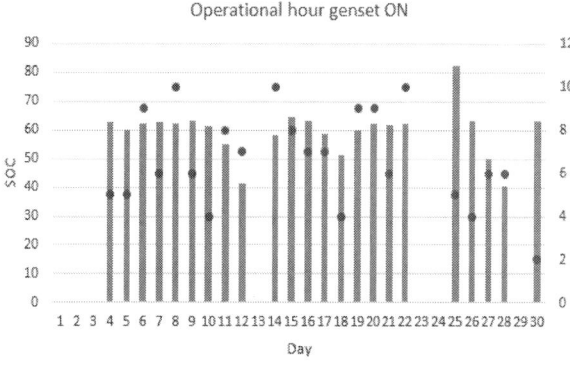

Fig. 4. Diesel generator operating hours

B. Evaluation on data for a month (April 2012)

The month of April 2012 with high solar irradiance is considered in order to evaluate SOC of battery. The evaluation is also based on daily DOD, load profile and number of operational hour of diesel generator. Average Solar Irradiance on April 2012 is around 220Wm-2 with average SOC 72.14%

Fig 4 shows the operational hour of the diesel generator in respect to the battery SOC value. The diesel generator operational time depends on the charging process of the battery, the load consume at that particular time and also the condition of solar irradiance. Overall

for this month, the diesel generator will operate when the DOD reach average 60%. The total number of daily operational hour for diesel generator is between 2-10hours. This allows lower quantity of diesel to be used and the cost burden compared to by using only diesel generator. In order to set the limit of SOC and DOD for each day, it will consider in analyzing small scope of time which is on the 1st week in April 2012. Fig.5 shows that in a random week in April, most of the day does receive average solar irradiance in exception of day 6. With the high solar irradiance, the system supplies the load and at the same time charges the battery until maximum storage. The load in each day as illustrated in Fig. 6 is almost the same where the much higher energy consumption is at night. Since then, the power withdrawn from the battery in every night will drop and suddenly reach the limit of DOD battery value. As much as SOC and DOD is concerned, the diesel generator will operate to supply the load and also charge the battery as shown in Fig. 7.

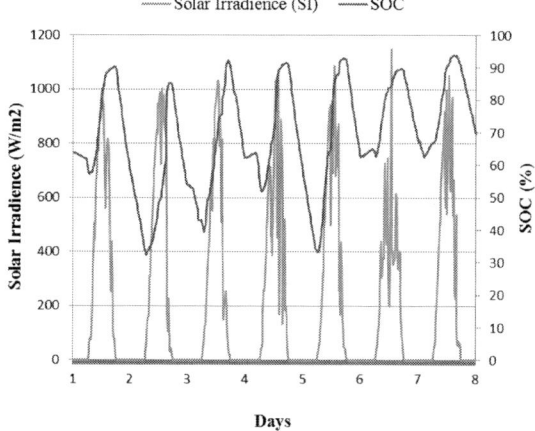

Fig. 5. Relationship of solar irradiance with battery SOC for a random week in April 2012

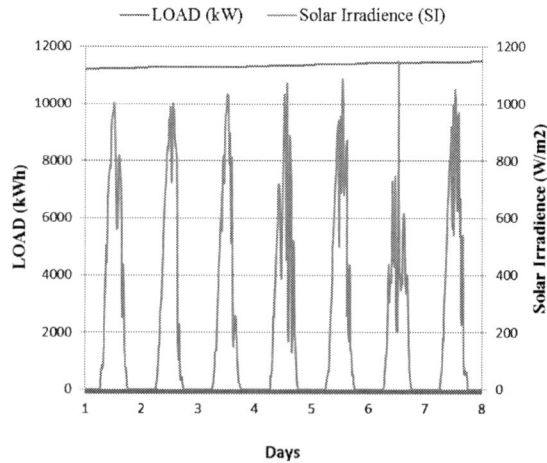

Fig. 6. Relationship of solar irradiance with load consumption for a random week in April 2012

The total operational time of diesel generator with 60% DOD setting for one random week is around 27hours. Compared to the manufacture's DOD limit recommendation of 80% for each day, the total operational time of the diesel generator is around 99hours in a week. From that, it is shown that with 60% setting DOD, the diesel generator will operate in a shorter time compared to 80% DOD as shown in Fig.8. This will affect the battery lifetime but will also reduce fuel consumption of the diesel generator.

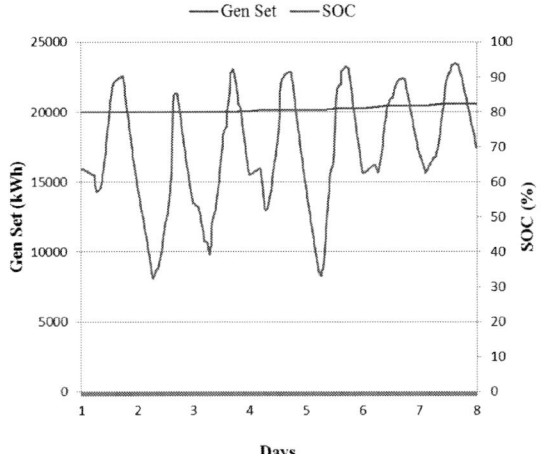

Fig.7. Diesel generator operating hours for random week in April 2012

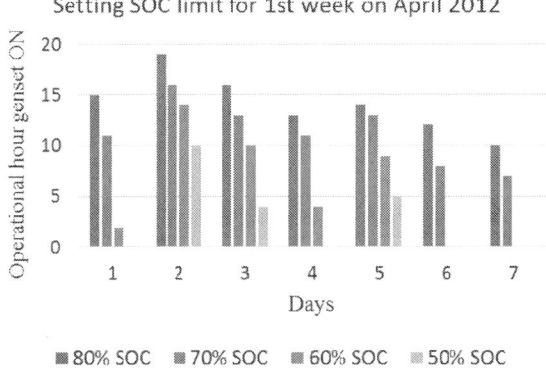

Fig. 8 SOC limit for random week in April 2012

IV. CONCLUSIONS

This paper has presented the evaluation energy storage capacity performance respect to the solar irradiance and load condition. In this work, an attempt is made by considering operational hours of the diesel generator that will meet the load continuously. This study shows the heart of the system is lies on the batteries as one of the key component in Hybrid PV/Diesel System. A more detailed analysis with hypothesis equation will be presented in the 8 page final manuscript.

ACKNOWLEDGMENT

Research was conducted under grant (600-RMI/RAGS 5/3 (44/2012) provided by Kementrian Pelajaran Malaysia and managed by Research Management Institute of Universiti Teknologi MARA (UiTM). Authors would also like to thank Bahagian Elektrik Jabatan Kerja Raya (JKR) Selangor for providing the data.

REFERENCES

[1] S.M Shaahid and M.A Elhadidy, "Economic Analysis Of Hybrid Photovoltaic-Diesel-Battery Power System For Residential Loads In Hot Regions- A Step To Clean Future", Renewable And Sustainable Energy Reviews, Vol.12 pp.488-503,2008

[2] Himri Y, Boudghene Stambouli A, Draoui B, Himri S. Techno-"Economical Study Of Hybrid Power System For Remote Village In Algeria" Energy 2008;33:1128-36

[3] Abatcha H. G, Ahmad M.Y.Jumba, Ladan Maijama'a. "Design And Simulation Of A Hybrid PV/Fuel Cell Energy System" Wilolud Journals, 2011

[4] K.Y. Lau, M.F.M. Yousof, S.N.M Arshad, M. Anwari ,H.M Yatim, " Performance Analysis of Hybrid PV/Diesel Energy System Under Malaysian Conditions", Energy 2010;3245-325

[5] National Green Energy Technology Policy, (2009), Ministry of Green, Green Technology and Water, Malaysia

[6] Abd El-Shafy A.Nafeh "Proposed Technique for Optimally Sizing a PV/Diesel Hybrid System", ICREPQ 2010

[7] W.X. Shen, "Optimally Sizing of Solar Array and Battery in a Standalone PV System in Malaysia" Renewable Energy, Vol. 34,pp 348-352, 2009

[8] David Linden and Thomas Reddy, Handbook of Batteries, 4th Ed. McGraw Hill, 2010

[9] J. Chiasson and B. Vairamohan, "Estimating the State of Charge of a Battery," IEEE Trans. on Control Systems Technology, Vol. 13, No. 3, pp. 465-470, (2008)

[10] "Design and Development of Batteries For Solar Photovoltaic Applications" collaboration between CSIR and MNRE New Delhi Jun2012

[11] Sulaiman Shaari, Ahmad Maliki Omar, Ahmad Hadri Haris, Shahril Irwan Sulaiman, Solar Photovoltaic Power: Fundamental, First Edition 2010.

[12] Daniel Aschwanden "Battery State of Charge Modelling" Swiss Federal Institute of Technology Zurich Sept 2011

The 2014 International Power Electronics Conference

New Techniques for Measuring Islanded Microgrid Impedance Characteristics Based on Current Injection

Lixiang Hou, Baoquan Liu, Hongtao Shi, Hao Yi* and Fang Zhuo

School of Electrical Engineering
Xi'an Jiaotong University
Xi'an, Shaanxi, China
*Corresponding author E-mail: hawkhou666@163.com

Abstract-**Impedance is an important parameter to analysis the stability of microgrid. Stability criteria have been developed in terms of source and load impedances for both dc and ac systems, and it is often helpful to have techniques for impedance measurement. First, the theoretical impedance of microgrid is calculated in this paper. Second, a new method of impedance measurement for microgrid is introduced. By injecting an unbalanced line-to-line square wave current between two lines of the ac system, all impedance information in the traditional synchronous reference frame d q model can be determined. The proposed measurement device also can be used for both ac and dc interfaces. The impedance identification method proposed in this paper greatly reduces online testing time by injecting an unbalanced line-to-line square wave current. Calculation, simulation and laboratory measurements verify the measurement results at last.**

I. INTRODUCTION

In recent years, distributed power generation won more and more attention and application and the dispersed small capacity of distributed power generation systems, energy storage system have become the research focus [1-4].

Compare to the power grid, the microgrid have smaller capacity, larger proportion of new energy, contains a lot of power electronic devices, such as all kinds of inverter and control system. So the design and operation of microgrid must consider the stability.

Many impedance measuring methods based on the variation of voltage and current at the point to be measured have already been proposed, and can be mainly classified into three kinds:(1) using harmonic current from existing non-linear loads;(2) with the help of switching an auxiliary element such as a shunt capacitor, reactor or transformer; (3) using a direct injection of harmonic currents. Literature In this paper, some stability criteria based on impedance parameters are introduced at first. Secondly the theoretical impedances of microgrid are calculated. In the third part we choose the harmonic current injection method to measure the harmonic impedance of the microgrid. Calculation, simulation and laboratory measurements verify the the measurement results at last.

II. IMPEDANCE MEASUREMENT FOR STABILITY ANALYSIS

Many research works in the area of power electronic system stability analysis are based on the small-signal

impedance, a variety of criteria have been introduced to stability issues in both dc and three-phase ac systems [3] – [7].

For symmetrical and balanced three-phase ac systems, an equilibrium point can be obtained from the d – q model in a reference frame synchronous to the fundamental frequency. When all quantities are transformed into the d – q reference frame, the small-signal impedance characteristics of a source or load describe the frequency-dependent relationship between small variations of voltages and currents. In the steady state, the source impedances are related to the source voltages and source currents by

$$\begin{bmatrix} \Delta V_q \\ \Delta V_d \end{bmatrix} = \begin{bmatrix} Z_{qq} & Z_{qd} \\ Z_{dq} & Z_{dd} \end{bmatrix} \begin{bmatrix} \Delta I_{q-source} \\ \Delta I_{d-source} \end{bmatrix} \tag{1}$$

On the load side, the admittance can be expressed in terms of the load voltages and load currents by

$$\begin{bmatrix} \Delta I_{q-load} \\ \Delta I_{d-load} \end{bmatrix} = \begin{bmatrix} Y_{qq} & Y_{qd} \\ Y_{dq} & Y_{dd} \end{bmatrix} \begin{bmatrix} \Delta V_q \\ \Delta V_d \end{bmatrix} \tag{2}$$

In (1) and (2), Δ denotes a small deviation of the respective variable from the equilibrium point.

Three criteria in an ac system have been defined in order to determine stability by the combination of source impedance and load admittance matrices [8]

$$ZY_{\infty 1} = \left\| Z_{dqS} \right\|_{\infty} \left\| Y_{dqL} \right\|_{1} \tag{3}$$

$$ZY_{GG} = \left\| Z_{dqS} \right\|_{G} \left\| Y_{dqL} \right\|_{G} \tag{4}$$

$$ZY_{\sigma\sigma} = \sigma \left(Z_{dqS} \right) \sigma \left(Y_{dqL} \right) \tag{5}$$

where σ is the largest singular of a matrix, and the infinity norm, the unity norm, and the G-norm are defined by

$$\left\| Z_{dqS} \right\|_{\infty} = \max_{i} \left[\sum_{j} \left| Z_{ij} \right| \right] \tag{6}$$

$$\left\| Y_{dqS} \right\|_{1} = \max_{j} \left[\sum_{i} \left| Y_{ij} \right| \right] \tag{7}$$

$$\left\| A \right\|_{G} = \max \left(\left| a_{qq} \right|, \left| a_{qd} \right|, \left| a_{dq} \right|, \left| a_{dd} \right| \right) \tag{8}$$

In (8), A represents a matrix with elements aij and can be impedance or admittance, and (3) – (5) are called unity-norm criterion, G-norm criterion, and singular criterion, respectively.

978-1-4799-2706-7/14 $31.00 © 2014 IEEE

According to the criteria, the system will be stable if $ZY_{\sigma\sigma1} < 0.5$ or $ZY_{GG} < 0.25$ or $ZY_{\sigma\sigma} < 1$. The most conservative criterion is the G-norm criterion and singular criterion is the least conservative one [8].

III. THE IMPEDENCE CALCULATE

As was shown in the previous section, knowledge of impedance is essential for determination of system stability. To verify the experiment results, the impedance parameters of microgrid are calculated. The equivalent circuit of microgrid as shown in Fig.1.In the equivalent circuit diagram, S1, S2 represents two micro source, they are all access microgrid through the inverter; R1, L1 and L2, R2, represents load. Rl1, Ll1, represents line impedances between the source .

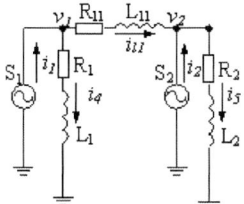

Fig.1.microgrid equivalent circuit

A. voltage source inverter harmonic impedance calculate

The voltage source inverter's control diagram is shown in Fig.2.

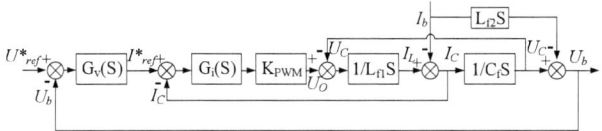

Figure.2.current source inverter Control diagram

From the Fig.2 the following formula can be got.

$$U_o = [(U^*_{ref} - U_b)G_v(S) - I_C]G_i(S)K_{PWM} \qquad (9)$$

$$U_C = [(U_o - U_C) \bullet \frac{1}{L_{f1}S} - I_b] \bullet \frac{1}{C_f S} \qquad (10)$$

$$U_b = U_C - I_b \bullet L_{f2}S \qquad (11)$$

$$U_b = G_{kv}(S)U^*_{ref} - Z(S)I_b \qquad (12)$$

U^*_{ref} is reference voltage, U_o is inverter output voltage, U_b is ac bus voltage, K_{PWM} is inverter transfer function, U_C is filter capacitor voltage, I_C is filter capacitor current, $G_i(S)$ is current regulator transfer function, $G_v(S)$ is voltage regulator transfer function.

$Z(S)$ is Inverter output impedance. G_{kv} is Current gain.

$$G_{kv}(S) = \frac{G_v(S)G_i(S)K_{PWM}}{L_{f1}C_f S^2 + C_f G_i(S)K_{PWM}S + G_v(S)G_i(S)K_{PWM} + 1} \qquad (13)$$

$$Z(S) = \frac{L_{f1}L_{f2}C_f S^3 + C_f L_{f2}G_i(S)K_{PWM}S^2 + (L_{f1} + L_{f2})S}{L_{f1}C_f S^2 + C_f G_i(S)K_{PWM}S + G_v(S)G_i(S)K_{PWM} + 1} \qquad (14)$$

If the current regulator voltage regulator 's transfer function is:

$$G_v(S) = K_{vp} + K_{vi}/S \qquad (15)$$

$$G_i(S) = K_{ip} \qquad (16)$$

$$Av\delta : K_{PWM} = \frac{U_{DC}}{2} \qquad (17)$$

The Inverter output impedance is:

$$Z(S) = \frac{L_{f1}L_{f2}C_f S^4 + C_f L_{f2}K_{ip}\frac{U_{DC}}{2}S^3 + (L_{f1} + L_{f2})S^2}{L_{f1}C_f S^3 + C_f K_{ip}\frac{U_{DC}}{2}S^2 + (K_{vp}K_{ip}\frac{U_{DC}}{2} + 1)S + K_{vi}K_{ip}\frac{U_{DC}}{2}} \qquad (18)$$

B. load harmonic impedance calculate

If the system load is Y connection and Three-phase symmetric. Load impedance of each phase is

$$Z = R + jX = R + j\omega L \qquad (19)$$

$$I_a + I_b + I_c = 0 \qquad (20)$$

$$I_a \bullet (R + jX) - I_b \bullet (R + jX) = U_{ab} \qquad (21)$$

$$I_b \bullet (R + jX) - I_c \bullet (R + jX) = U_{bc} \qquad (22)$$

The result from (29) ~ (31) is

$$\begin{bmatrix} I_a \\ I_b \\ I_c \end{bmatrix} = \frac{1}{3(R + jX)}\begin{bmatrix} 2 & -1 & -1 \\ -1 & 2 & -1 \\ -1 & -1 & 2 \end{bmatrix}\begin{bmatrix} V_a \\ V_b \\ V_c \end{bmatrix} \qquad (23)$$

And

$$Y_{abc} = \frac{1}{3(R + jX)}\begin{bmatrix} 2 & -1 & -1 \\ -1 & 2 & -1 \\ -1 & -1 & 2 \end{bmatrix} \qquad (24)$$

Thus the Positive sequence harmonic impedance of load is

$$Z_{012}(h) = (R + jL\omega h)\begin{bmatrix} \infty & 0 & 0 \\ 0 & 1 & 0 \\ 0 & 0 & 1 \end{bmatrix} \qquad (25)$$

C.microgrid system impedance calculate

According to the figure 1. If microgrid is running in the island state, each inverter is controlled to voltage source. $Y_{I1}(h)$, $Y_{I2}(h)$ is output harmonic admittance of the two inverters. From the equation (18), $Y_{I1}(h)$ is

$$Y_{I1} = \frac{1}{Z_{I1}(h)} = \frac{L_{f1}C_f(j\omega h)^3 + C_f K_{ip}\frac{U_{DC}}{2}(j\omega h)^2 + (K_{vp}K_{ip}\frac{U_{DC}}{2} + 1)(j\omega h) + K_{vi}K_{ip}\frac{U_{DC}}{2}}{L_{f1}L_{f2}C_f(j\omega h)^4 + C_f L_{f2}K_{ip}\frac{U_{DC}}{2}(j\omega h)^3 + (L_{f1} + L_{f2})(j\omega h)^2} \qquad (26)$$

Suppose two inverters have the same parameters and capacity.

$$Y_{I1} = Y_{I2} \qquad (27)$$

Suppose the two load is Y connection and Three-phase symmetric the. Harmonic impedances of load is

$$Y_{L1}(h) = Y_{L2}(h) = \frac{1}{Z_{L1}(h)} = \frac{1}{R_1 + jL_1\omega h} \qquad (28)$$

From equation (26) ~ (28), the microgrid system impedance can be calculated. The result is

The 2014 International Power Electronics Conference

$$Y_M(h)=\frac{-2L_{f1}C_f j(\omega h)^3-C_f K_{ip}U_{DC}(\omega h)^2+(K_{vp}K_{ip}U_{DC}+2)j\omega h+K_{vi}K_{ip}U_{DC}}{L_{f1}L_{f2}C_f(\omega h)^4-C_f L_{f2}K_{ip}(U_{DC}/2)j(\omega h)^3-(L_{f1}+L_{f2})(\omega h)^2}+\frac{3}{R_1+jL_1\omega h}+\frac{1}{R_2+jL_2\omega h}$$

(29)

IV. THE MEASUREMENT METHOD OF HARMONIC IMPEDANCE

The main idea of the microgrid impedance measurement is: inject harmonic current with certain size and frequency into the system, and voltage fluctuation will be caused in measuring point. So the harmonic impedance can be calculated. The diagram is shown in Fig.3.

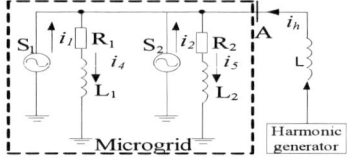

Figure 3.measurement principle diagram

For a three-phase system without zero-sequence current, the impedance matrix consists of four unknowns Zqq, Zqd, Zdq , and Zdd. Therefore, at least two sets of measurements are needed in order to solve the linear system of equations. To solve this problem, (1) can be extended by combining two or more measurement results in one set of equations with each measurement corresponding to a unique angle. A minimum of two measurements are necessary to obtain a solution.

The injection methods used for harmonic impedance measurement in AC systems mainly includes the switching capacitor method, the thyristor-controlled sub circuit switching method and the harmonic current injection method. The harmonic current injection method is shown in Fig.4.

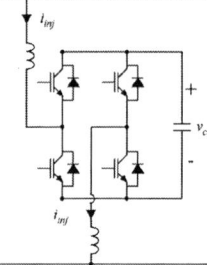

Figure 4.harmonic current injection method

Both the switching capacitor method and the thyristor-controlled sub circuit switching method have the same disadvantage. i.e. the current being generated is uncontrollable and does not contain all frequency current. The harmonic current injection method overcomes the problem mentioned above. Due to its controllability of harmonic frequencies and magnitude, the inaccuracy in some frequency ranges caused by little harmonic current can be avoided. This current injection is between only two phase lines, the proposed measurement device can be used for both ac and dc interfaces. The impedance identification method proposed in this paper greatly reduces online testing time by injecting an unbalanced line-to-line square wave current comparing with the traditional

sweep method. Because the square wave contains many harmonic current.

V. LABORATORY VALIDATION

A microgrid impedance measurement experimental platform is established, as shown in Fig.5, which consists of two inverters, two loads, and a current injection. Experimental platform Structure is shown in Fig.3.Meanwhile we have established a simulation model by PSCAD. Simulation model parameters are the same with experimental platform.

In order to reduce the measurement time we inject square wave harmonic currents into the system. Contrast figures of experiment result, simulation result and calculated result (predicted result) are shown below.

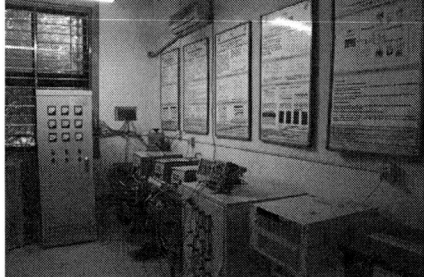

Fig.5 microgrid impedance measurement experimental platform

The injection current amplitude is inverter output current's 5% and frequency 50 Hz. The source impedances comparing results are shown in figure 6.

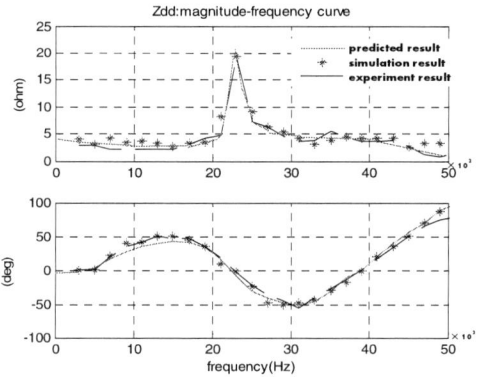

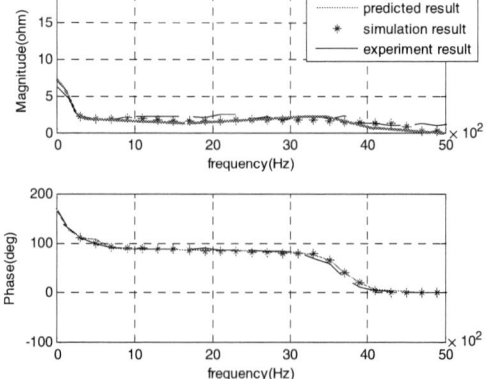

The 2014 International Power Electronics Conference

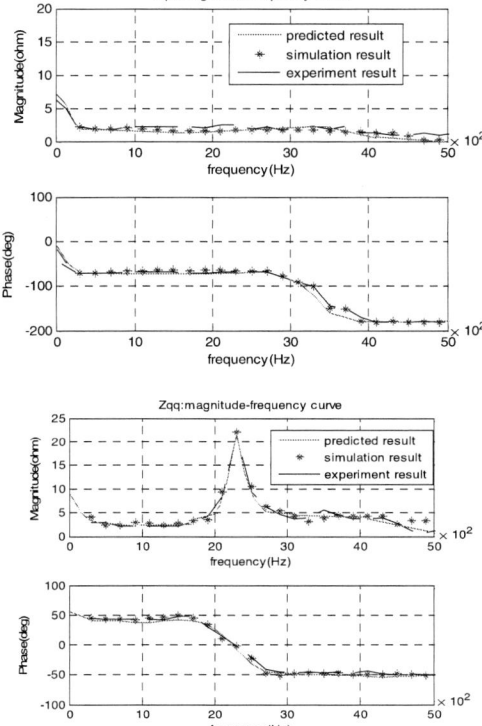

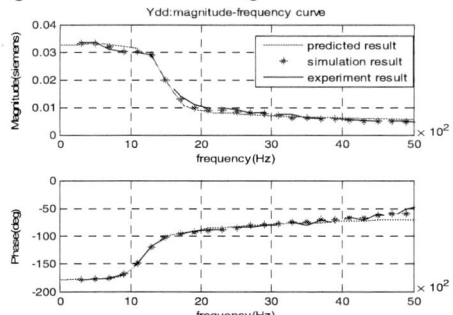

Fig.6 source impedances comparing results

In figure 6, there are some deviations among the measured, calculated and simulation curve in some frequency. Because in these frequencies system itself contains some background harmonic currents. Overall, the measured Amplitude-frequency curve and Phase-frequency curve is good agreement with the calculated and simulation curve. The load admittance comparing results are shown in figure7.

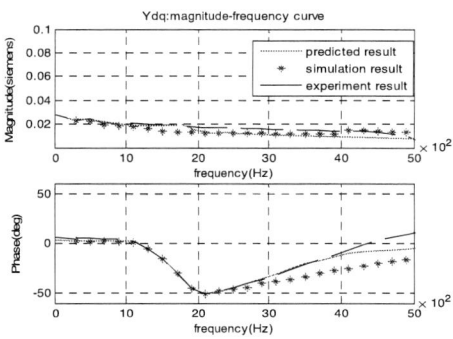

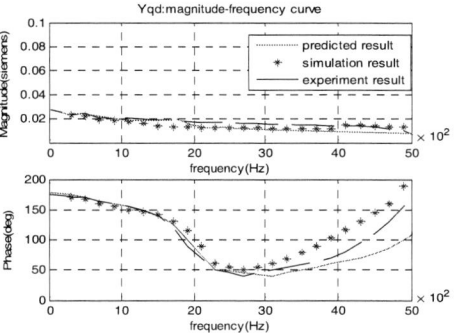

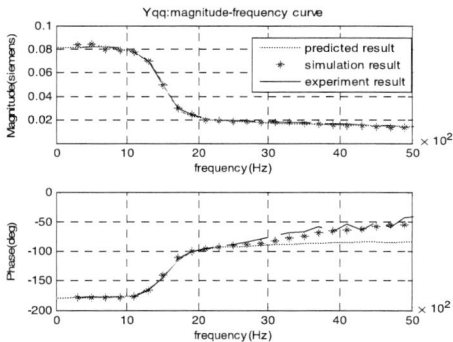

Figure7.load admittance comparing results

In figure 7, there are some deviations in Phase-frequency curve at high frequency. Because admittance is close to zero. Overall, the measured Amplitude-frequency curve and Phase-frequency curve is good agreement with the calculated and simulation curve.

VI. CONCLUSIONS

In this paper, the theoretical impedance parameters of microgrid were calculated .A novel impedance measuring method is proposed. Impedance parameters are also measured based on this new method. The theoretical, Simulation model is built through PSCAD. Simulation, measurement results and calculation results are compared and verify the measurement results at last.

ACKNOWLEDGMENT

This work was supported by2011 Natural Science Foundat-ion of China (No.51177130), and 2011 Application Form for

978-1-4799-2706-7/14 $31.00 © 2014 IEEE

Delta Science & Technology Educational Development Program. (No.DREK2011002).

REFERENCES

[1] S. V. Iyer, M. N. Belur, and M. C. Chandorkar, "A Generalized Computational Method to Determine Stability of a Multi-inverter Microgrid", IEEE Trans on Power Electronics, 2010, vol. 25, no. 9, pp. 2420- 2432.

[2] N. Pogaku, M. Prodanovic, T. C. Green, "Modeling, analysis and testing of autonomous operation of an inverter based MicroGrid", IEEE Trans on Power Electronics, 2007, vol. 22, no. 2, pp. 613 -625.

[3] T. K. Panagrahi, S. Chowdhury, S. P. Chodhury, N. Chakraborty, Y. H. Song, "Control & Reliability Issue of Efficient Microgrid Operation Using Hybrid Distributed Energy Resources", IEEE PES Power Systems Conference and Exposition, October 29-November 1, 2006, pp. 797-802.

[4] E. Barklund, N. Pogaku, M. Prodanovic, "Energy Management in Autonomous Microgrid Using Stability-Constrained Droop Control of Inverters", IEEE Trans on Power Electronics, 2008, 23(5): 2346-2352.

[5] R. D. Middlebrook, "Input filter consideration in design and application of switching regulators", IEEE Industry Applications Society Annual Meeting, October 11-14, 1976, Chicago, USA.

[6] Xiaogang Feng, Zhihong Ye, Kun Xing, F. C Lee, D. Borojevic, "Individual load impedance specification for a stable DC distributed power system", 14th Annual Applied Power Electronics Conference and Exposition, March 14-18, 1999, Dallas, USA.

[7] S. D. Sudhoff, S. F. Glover, P. T. Lamm, D. H. Schmucker, D. E. Delisle, "Admittance space stability analysis of power electronic systems", IEEE Transactions on Aerospace and Electronic Systems, vol. 36 no. 3, pp. 965 – 973, July 2000.

[8] M. Belkhayat, "Stability criteria for AC power systems with regulated loads" Ph.D. dissertation, Purdue Univ., West Lafayette,, IN, 1997.

A General Framework to Design Operation Modes of DC Microgrids without Communication Links

Miao Pan[1], Na Shen[2], Geng Yang[1], Kazunori Morita[3], Kazuya Ogura[3], Weiyang Wu[2]

[1]Tsinghua University
Beijing, China

[2]Yanshan University
Qinhuangdao, China

[3]Meidensha Corporation
Tokyo, Japan

Abstract— **This paper proposes a general framework to design operation modes of decentralized controlled DC microgrids without communication links. Under this framework, power units of the DC microgrid are firstly classified into several types according to their physical features. Then, a utilization priority of different types of units is determined by certain economic index, in order to achieve good economic performance. Autonomous transitions among these operation modes are realized based on DC bus voltage variation. The method is validated by MATLAB/Simulink simulations as a case study.**

Keywords— *Autonomous control, DC microgrid, mode design.*

I. INTRODUCTION

A lot of research work has been done on AC microgrids, owing to that the utility grid supplies customers with AC electricity [1]. However, DC microgrids are calling more of people's concern, because 1) power units, such as photovoltaic, fuel cell and storage, are delivering DC power, and using DC bus is more energy efficient. 2) DC microgrids are lower cost and easier to control. Lawrence Berkeley National Laboratory (LBNL) has demonstrated to the data center industry that DC delivery systems are viable, can be 20% or more efficient than current AC delivery systems, be more reliable, and potentially cost less in the long run [2]. In recent years, DC microgrid has been a hot topic. DC microgrid without communication links stands out above the rest types of microgrids, because of its good reliability and low cost [3]. However, it's difficult to coordinate the operations of all the power units in this DC system.

Due to the stochastic characteristics of renewable generation and load, some dispatchable sources, like batteries and diesel generators, are indispensable to maintain power balance. Because there is no communication link, unit controller can only use detected terminal quantity, that is to say the DC bus voltage, to work together. A conventional way of decentralized microgrid is the droop control [4][5]. In case of multiple converters, the power will be shared between the power units based on their droop coefficients. But a main drawback of droop control is that the power units are not prioritized. In most cases, it's not reasonable to use the non-renewable energy before we take full advantage of renewable energy. This is validated in [6] that it's possible to schedule the power units in a utilization priority, by using the DC bus voltage as an information carrier in a renewable "nanogrid". However, the proposed system in [6] is not connected to the utility grid and the types of transitions considered are limited. Based on the basic idea of [6], an improved four mode control strategy is implemented to a modular designed photovoltaic generation system in [7]. The strategy in [7] considers interaction with the utility grid and some extreme conditions such as the islanding operation with a fully charged battery. But four modes may be simplified to three modes. In [8], a three mode strategy is implemented to a grid-connected system, which contains solar and wind energy. Yet renewable generation curtailment has to be carried out to maintain the power balance. Some other operational transitions have been studied in [3], [9], [10] in terms of specific grid configurations. The common ground of the above strategies is that several operation modes are well defined in advance, and system works among these modes. However, designing operation modes depends on specific system configurations and might not apply to others. To the best knowledge of the author of this paper, the general method of designing operation modes for DC microgrid without communication links is rarely reported.

This paper introduces a general framework to design operation modes for DC microgrids without an extra communication link. This framework takes both physical feature and economic feature of power units in the DC microgrid into consideration. Under this framework, power units of the DC microgrid are firstly classified into several types according to their physical features. Then, a utilization priority of different types of units is determined by certain economic index, in order to achieve good economic performance. For instance, fuel price, maintenance charge, and emission cost are often taken into consideration.

This paper is organized as follows, section II is problem description and introduces the framework to design operation modes. This strategy is implemented to a typical system in Section III as case study. And Section IV is the conclusion.

This work is supported by National Natural Science Foundation (NNSF) of China under Grant (61104046 and 61273045), Tsinghua University Initiative Scientific Research Program and grant from Beijing Higher Education Young Elite Teacher Project.

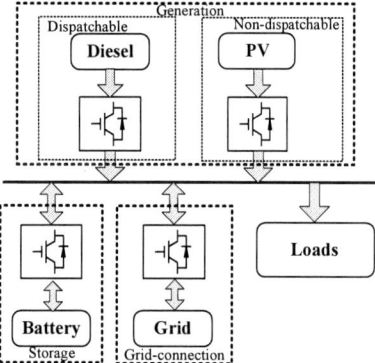

Fig 1. Typical DC system configuration

II. PROBLEM DESCRIPTION

The typical system configuration of a DC microgrid is shown in Fig.1. The power units within the system can mainly be categorized into five types: renewable generation, non-renewable generation, storage, grid-connected converter and load. Generally, renewable sources, such as solar and wind power, are non-dispatchable; while non-renewable sources, like diesel generators, are dispatchable, and they are usually used as backup generation in case of system power failure.

The most fundamental requirement of DC microgrids is to achieve generation load balance, maintaining the DC bus voltage within an acceptable range. After that, the secondary requirement for a DC microgrid is to achieve maximum profit with minimum cost.

Taking the system in Fig.1 as an example, the power flows of PV modules and loads are stochastic, and all units have power limitations. Generally, different types of power units have quite different usages, so power units are firstly classified into several types according to their physical features. In each operation mode, we choose one and only one type of units to work as voltage source, called the master unit, while other units work as current sources, because: 1) at least one of the system units should be selected to work to compensate for the stochastic output of other units, maintaining the DC bus voltage in an acceptable range; 2) only one voltage source operates in each mode, which means that the dc bus voltage can be regulated at constant in each mode. This characteristic ensures the system stability.

The most common operation mode for the DC microgrid is that the grid-connected converter is selected as the master unit, and it works to regulate the dc bus at a constant voltage. PV unit and WT unit are working at MPPT, and battery is operating with maximum charging current if it has not been fully charged. And the diesel generator is offline. The generated power is prior to feed the local load. If the generation is not enough, the DC bus would absorb power from the grid; otherwise the DC bus would send electricity to the grid.

However, some other working modes should be designed in case of grid fault. During grid fault, the grid-connected converter is disabled and the DC microgrid works in islanding mode. In some applications, we want

to maximize the usage of renewable energy, and the operation modes are designed as follows. If system is underload, renewable sources are selected as the master unit. They are no longer working at MPPT, because generation is more than load even when battery has reached its maximum charging current. Of course, the diesel generator is offline now. As load increases, the master unit increases its output power until maximum power point. Then the master unit becomes the battery. If load still increases when battery has reached its maximum discharging current limit, the diesel generator has to come to work. This time the diesel generator is selected as the master unit. And all modes are summarized in Table I.

TABLE I
OPERATION MODES

	Grid-connected	Grid fault		
		Underload <-------------> *Overload*		
Grid	Master	Disabled	Disabled	Disabled
Renewable energy	MPPT	Master	MPPT	MPPT
Storage	Max current Charging	Max current Charging	Master	Max current discharging
Non-renewable energy	Off	Off	Off	Master

The mentioned control strategy can be realized based on DC Bus Signaling [3], and each unit is assigned to a voltage threshold to trigger the point at which it begins discharging or charging. Autonomous transitions between operation modes are actualized by controlling the power units to exhibit three states of operation: off, constant voltage, and constant power. Load changes cause the power units to switch between constant voltage and constant power operation, changing the level of the dc bus. Thus, the primary requirement for a DC microgrid is completed.

Through Table I, it can be seen that power units join to work in order of a utilization priority, which is dependent on the discharge/charge thresholds. And a higher threshold means a higher priority. This helps a lot in achieving the secondary requirement for DC microgrids. Different types of power units have different economic performance. Here "economic performance" includes not only financial benefits, but also environmental effect and social efficiency. In order to achieve best economic performance with minimum cost, the utilization priority of power units should be determined according to certain economic index, and units with better economic performance are supposed to have higher priority.

In this case shown in Table I, the renewable sources have the highest priority, while the diesel generator is lowest. But the utilization priority should change as a result of environment change. For example, PV units ought to be highly prior in daytime, but at night it should be less prior. In this paper, only generation side management is considered. However, it's worth noting that this designing method could also be implemented in the demand side management. This characteristic plays an important role in the era of smart grid. One possible

TABLE II
CASE STUDY: AN APPLICATION OF AUTONOMOUS MASTER-SLAVE METHOD

	Grid-connected	Grid fault		
	Mode I (360V)	Mode II(370V)	Mode III(350V)	Mode IV(340V)
Master	Grid	PV	Battery	Diesel generator
Slaves	PV (MPPT)	Battery (Max charging if not full)	PV (MPPT)	PV (MPPT)
	Battery (Max charging if not full)	Diesel generator (Off)	Diesel generator (Off)	Battery (Max discharging or empty)
	Diesel generator (Off)	Grid (Disabled)	Grid (Disabled)	Grid (Disabled)

application is "Wind-EV complementation" [11]. At night, EVs charges to reduce the wind abandon. And in daytime, EVs discharges to support the grid during the peak load time. Since the application background of DC microgrids is manifold, this paper would give a case study as a guidance of this method in section III.

III. STRATEGY IMPLEMENTATION: CASE STUDY

The main components and devices of the previously cited system are shown as follows:

• a DC bus with a normal voltage of 360 V and connected DC capacitance of 9600 uF;

• a 13.5kW PV, whose maximum power point is 292.8V and 15.54A;

• a 15kW grid-connected converter, connecting the DC bus with the 200V single-phase ac grid;

• a diesel generator rated at 20kW, whose output is rectified first and then connected to a dc/dc converter;

• a 168V/20Ah battery connected through a bidirectional dc/dc converter rated at 30A;

• a resistive dc load, which can change from 0 to 20kW continuously.

Based on the autonomous master-slave method, a control scheme is proposed to maintain the DC bus voltage within an acceptable range, shown in Table II. The operations of the system are categorized into four modes relative to different master unit, and the master unit is cited in brackets: Mode I (grid-connected converter), Mode II (PV), Mode III (battery) and Mode IV (diesel generator). The basic idea of designing these four modes has been explained in section II.

Although the detailed control methods for different power units have a great many differences, the control principle is the same. For each unit, if it has been selected as the master unit, it works as a voltage source to regulate the DC bus voltage. Otherwise, for the renewable units, they work at MPPT. For the storage unit, it is charging or discharging with its maximum charging/discharging current. For the utility grid and the diesel generator, they are disabled if they are not working as the master unit. The control diagrams are shown as follows, in Fig 2, Fig 3, Fig 4, and Fig 5.

Simulations have been completed to validate the feasibility of this control method. And all possible routes of operation mode transition are shown in Fig 6.

Transition from Mode IV to Mode I would not happen, because we suppose that load would not change abruptly. But only the results of transitions cited in dashed line are shown in view of limited space.

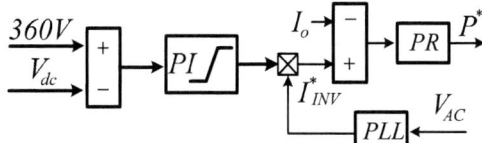

Fig 2. Control diagram for grid-connected converter

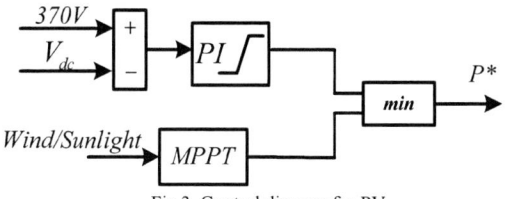

Fig 3. Control diagram for PV

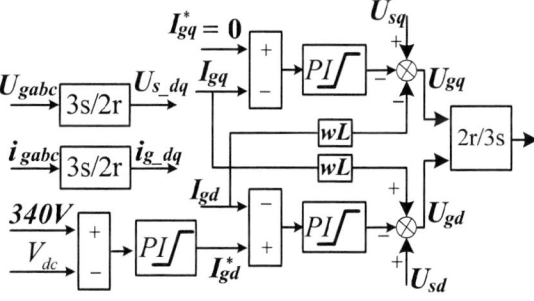

Fig 4. Control diagram for diesel generator

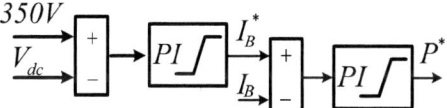

Fig 5. Control diagram for battery

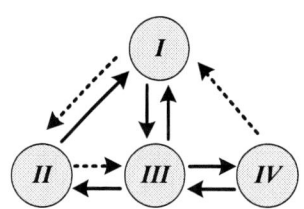

Fig 6. All possible mode transitions

The 2014 International Power Electronics Conference

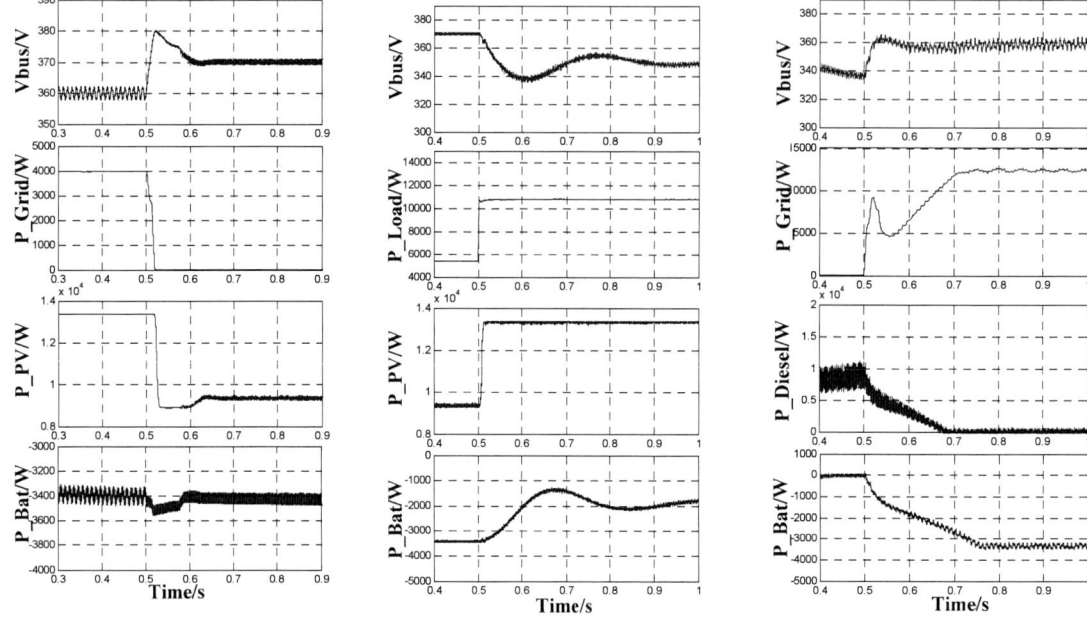

| Fig 7. Mode I to Mode II | Fig 8. Mode II to Mode III | Fig 9. Mode IV to Mode I |

A. Mode I to Mode II

Fig.7 is the simulation results of transition process from Mode I to Mode II. In Fig. 7, when fault occurs in the grid, the grid-connected converter is disabled. The DC bus voltage increases and the PV unit decreases its generation and begins to regulate the DC bus voltage at 370V. In this condition, the system is underload. Only the PV unit with high priority is working.

B. Mode II to Mode III

Fig.8 shows the transition process from Mode II to Mode III. In Fig 8, the load increases so that the PV generation increases as well. And the load still increases when the PV unit has reached its MPP. Then the battery unit with the second priority comes to work. It decreases its charging current and begins to regulate the DC bus voltage at 350V.

C. Mode IV to Mode I

Fig.9 shows the transition process from Mode IV to Mode I. In Fig. 9 the grid recovers and the grid-connected converter takes charge of the DC bus voltage again. The DC bus voltage increases to 360V and the diesel generator stops working. At the same time, the battery begins to charge with its maximum charging current, because the grid-connected converter has higher priority.

IV. CONCLUSIONS

Decentralized control for DC microgrid without communication links has drawn much attention for its good reliability and low cost. However, the existing strategies depend on specific system configurations and

the mode designing method might not apply to others. This paper proposes a general framework to design operation modes for DC microgrids without communication links. Under this framework, power units of the DC microgrid are firstly classified into several types according to their physical features. Then, a utilization priority of different types of units is determined by certain economic index, in order to achieve good economic performance. In this paper, the framework is only used in the generation side management, which could also be implemented in the demand side.

REFERENCES

[1] R.H. Lasseter, P. Paigi, "Microgrid: a conceptual solution," *Power Electronics Specialists Conference*, 2004. 2004 IEEE 35th Annual , vol.6, no., pp.4285,4290 Vol.6, 20-25 June 2004.

[2] B. Fortenbery, E.C. EPRI, W. Tschudi, "DC power for improved data center efficiency," 2008.

[3] T.L. Vandoorn, B. Meersman, L. Degroote, B. Renders, L. Vandevelde, "A Control Strategy for Islanded Microgrids With DC-Link Voltage Control," *IEEE Trans. on Power Delivery*, vol.26, no.2, pp.703,713, April 2011.

[4] H. Kakigano, Y. Miura, T. Ise, "Low-Voltage Bipolar-Type DC Microgrid for Super High Quality Distribution," *IEEE Trans. on Power Electronics*, vol.25, no.12, pp.3066,3075, Dec. 2010.

[5] E. Barklund, N. Pogaku, M. Prodanovic, C. Hernandez-Aramburo, T.C. Green, "Energy Management in Autonomous Microgrid Using Stability-Constrained Droop Control of Inverters," *IEEE Trans. on Power Electronics*, vol.23, no.5, pp.2346,2352, Sept. 2008.

[6] J. Schonberger, R. Duke, S.D. Round, "DC-Bus Signaling: A Distributed Control Strategy for a Hybrid Renewable Nanogrid," *IEEE Trans. on Industrial Electronics*, vol.53, no.5, pp.1453,1460, Oct. 2006.

[7] K. Sun; Z. Li, X. Yan, J.M. Guerrero, "A Distributed Control Strategy Based on DC Bus Signaling for Modular Photovoltaic Generation Systems With Battery Energy Storage," *IEEE Trans. on Power Electronics*, vol.26, no.10, pp.3032,3045, Oct. 2011.

978-1-4799-2706-7/14 $31.00 © 2014 IEEE

[8] Y. Ito, Z. Yang, H. Akagi, "DC microgrid based distribution power generation system," *Power Electronics and Motion Control Conference*, 2004. IPEMC 2004. The 4th International, vol.3, no., pp.1740,1745 Vol.3, 14-16 Aug. 2004.

[9] L. Xu, D. Chen, "Control and Operation of a DC Microgrid With Variable Generation and Energy Storage," *IEEE Trans. on Power Delivery*, vol.26, no.4, pp.2513,2522, Oct. 2011.

[10] D. Salomonsson, L. Soder, A. Sannino, "An Adaptive Control System for a Dc Microgrid for Data Centers," *Industry Applications Conference*, 2007. 42nd IAS Annual Meeting. Conference Record of the 2007 IEEE , vol., no., pp.2414,2421, 23-27 Sept. 2007.

[11] Z. Li, Q. Guo, H. Sun, Y. Wang, S. Xin, "Emission-Concerned Wind-EV Coordination on the Transmission Grid Side With Network Constraints: Concept and Case Study," *IEEE Trans. on Smart Grid*, vol.4, no.3, pp.1692,1704, Sept. 2013.

The 2014 International Power Electronics Conference

Implementation Design of the Converter-Based Galvanic Isolation for Low Voltage DC Distribution

A. Mattsson, V. Väisänen, P. Nuutinen, T. Kaipia, A. Lana, P. Peltoniemi, P. Silventoinen, J. Partanen.

LUT Energy

Lappeenranta University of Technology

Lappeenranta, Finland

Abstract-In this paper the implementation of the galvanic isolation in the low voltage direct current (LVDC) network is discussed. The galvanic isolation can be implemented either with a 50 Hz transformer located at the customer AC output of the customer-end inverter (CEI) or with an isolated DC-DC converter at the input DC network side of the CEI. However, the former solution results in a significant increase of the volume and mass of the system and has a high amount of no-load losses. Therefore, an isolated DC-DC converter is studied in this paper. Based on calorimetric measurements conducted for the passive 50 Hz transformer, the requirements for the DC-DC converter are set. Selection of the converter topology is carried out and the selected topology is studied by simulations and calculations. The results are compared with the 50 Hz transformer.

Keywords— Converter, DC-DC, DC distribution.

I. INTRODUCTION

The LVDC system is an emerging power distribution technology that has been under intense research and development globally. LVDC power distribution has been proposed for versatile use-cases from datacenters [1] to public utility grid distribution [2], [3], [4], [5]. The LVDC distribution research network [6] studied in this paper utilizes voltage levels of ±750 VDC. Customers are connected to either between +750 VDC and 0 VDC or between 0 VDC and -750 VDC. The customer-end 50 Hz AC voltage of 230/400 VAC$_{RMS}$ is supplied by the CEI. The power demand in the end-user installations, and thus the load of the CEI varies on a wide range [7]. This distinguishes the distribution system application of the converter systems from the common industry applications.

Even though the design and implementation of the CEI plays a significant role in the optimization of the LVDC system, one of the main contributors to increased efficiency and power density is the galvanic isolation between the DC network and the customer-end AC network. In this paper the galvanic isolation is addressed and a selection for the configuration of the isolation is carried out by utilizing simulated and calculated comparisons.

Two methods for the galvanic isolation are discussed. The simpler of the two methods is the utilization of a passive transformer on the AC output side of the CEI. The main drawbacks of this method are its poor power density, limited performance under unbalanced load and distorted currents and a significant increase in the volume and mass which restrict its use. The second method is the utilization of power electronics and a high frequency transformer on the DC input side of the CEI in the form of an isolated DC-DC converter. The advantage of this method is its high power density and a possibility for increased efficiency which result in a reduction of the volume and mass of the system. However, as a drawback the complexity of the system increases which may increase the cost. The DC-DC converter also introduces a possibility to regulate the input DC-voltage of the CEI.

II. GALVANIC ISOLATION

The system under study utilizes an earth isolated (IT) bipolar DC network with a positive +750 VDC pole, a negative -750 VDC pole and a middle pole which is represented as 0 VDC. However, as the network is floating the middle pole can have a potential difference of 750 VDC referenced to earth and therefore 0 VDC should be thought as 0 VDC only when referenced to the DC poles. In contrast the customer-end AC network is a functionally earthed (TN) network in which the neutral conductor is connected to earth due to protection and safety requirements and therefore, the two networks have to be galvanically isolated from each other. The scheme of the LVDC distribution network is depicted in Fig. 1.

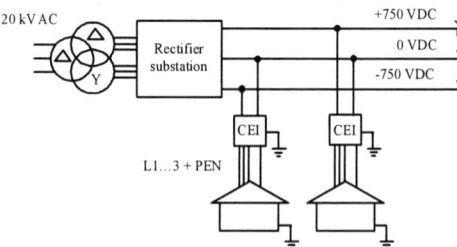

Fig. 1. Scheme of the LVDC distribution network. The rectifier substation is connected to the 20 kV distribution network by a double-tier transformer with Ddy5 connection. The DC network is an earth isolated (IT) network and the CEIs have 50 Hz isolation transformers at the AC output to allow the customer-end functionally earthed (TN) network earthing which is required due to protection and safety requirements.

The galvanic isolation can be implemented utilizing two different methods. Diagrams for the different isolated configurations are depicted in Fig. 2. In the six-pack inverter bridge configurations depicted in Fig. 2a and 2b, the minimum required DC-voltage level is 565 VDC. In

978-1-4799-2706-7/14 $31.00 © 2014 IEEE

587

the single phase converter-inverter configuration depicted in Fig. 2c, a minimum of 325 VDC is required as the converter outputs would be floating and therefore, the maximum achievable voltage difference between two phases is $2U_{DC,out}$.

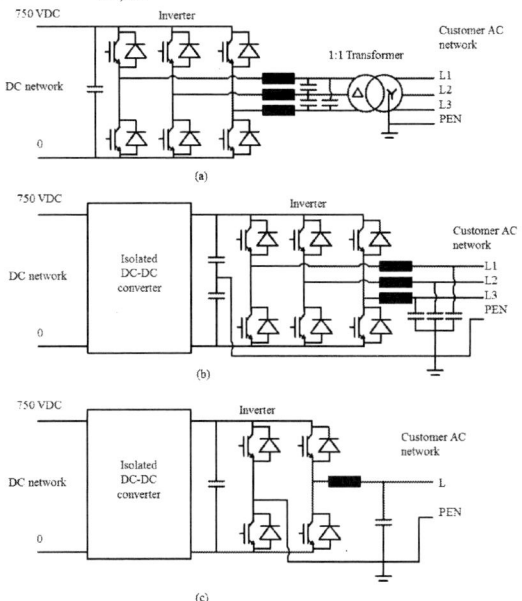

Fig. 2. Different configurations for the customer-end galvanic isolation. Three-phase 1:1 ratio 230/400 VAC_{RMS} 50 Hz transformer at the AC output of a three-phase inverter (a), with an isolated DC-DC converter at the DC input of a three-phase inverter (b) and with a DC-DC converter and a single-phase inverter configuration feeding single phases of a three-phase AC network (c).

The LVDC research site discussed in [6] has been implemented by utilizing the configuration in Fig. 2a. Even though the selected method is the simplest of the three, the 50 Hz isolation transformer adds a significant amount of volume and mass to the CEI resulting in a significant decrease in its power density. Also, a high percentage of the total losses of the CEI are generated by the transformer as depicted by calorimetric measurement results in Fig. 3.

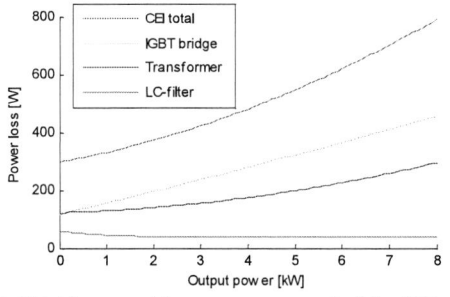

Fig. 3. Total losses and losses per component of the CEI measured with a calorimeter. The nominal power of the CEI is P_{nom} = 16 kVA and the switching frequency is f_{sw} = 16 kHz.

To achieve a higher power density and a higher power to weight ratio the configurations in Fig. 2b and Fig. 2c that utilize an isolated DC-DC converter are more feasible. Furthermore, the introduction of wide bandgap (WBG) switching devices utilizing Silicon Carbide (SiC)

and Gallium Nitride (GaN) has enabled the realization of high efficiency power converter systems [8], [9], [10], [11]. However, the new technology is still relatively expensive and the possibility of increased efficiency needs to be compared against the increase in the cost of the converter.

Utilization of DC-DC converters in direct customer-end DC distribution for datacenters and households have previously been studied for instance in [1], [12], [13]. Implementation of the galvanic isolation in the LVDC network with an isolated DC-DC converter has also been proposed in [14]. In this paper a comparison of the characteristics of an isolated DC-DC converter and a 50 Hz isolation transformer is carried out. A selection for the converter topology is presented and the topology is analyzed by simulations and calculations.

A. Volume and mass

The isolation transformer of the system in [6] is a 16 kVA, 400 VAC_{RMS}, 1:1 ratio unit with Dyn11 vector group. It weighs 93 kg and measures 42.5 cm x 37.0 cm x 18.5 cm. The power to weight ratio and power density of the transformer are thus very poor. An isolated 19.2 kW DC-DC converter with a power density of 10 W/cm^3 applying WBG devices was studied in [13]. The converter in [13] is designed for a voltage level of 384 VDC. However, the rating is achieved by connecting several 48 VDC modules in series and in parallel. Therefore, if the number of series connected modules is doubled and the number of paralleled modules is halved the converter could be configured for a 750 VDC system while sustaining the power density and nominal power ratings. The power density of the converter in [13] is 18 times higher in comparison to the 16 kVA isolation transformer. Therefore, if the transformer is replaced with a converter solution it would be possible to significantly reduce the volume and mass of the CEI.

B. Efficiency

As depicted in Fig. 3, the transformer contributes to a high percentage of the total losses of the CEI. However, Fig. 3 also shows that the losses of the IGBT bridge are even higher in comparison and therefore, a hard switched topology cannot be used in the isolated converter as a reduction of the losses is required. The measured transformer losses in combination with the transformer cost set the requirements for the efficiency and cost of the isolated converter.

III. DESIGN OF THE CONVERTER

A selection for the configuration in Fig. 2c was made due to the properties of the load. Therefore, the system will consist of a minimum of three converters, one per each of the phases. The converter is connected between the DC network and the CEI and therefore, the input voltage of the converter is U_{in} = 750 VDC. The converter was selected to be operated similarly to the isolation

978-1-4799-2706-7/14 $31.00 © 2014 IEEE

transformer and therefore, an output voltage value of $U_{out} = U_{in} = 750$ VDC was selected.

A. *Selection of the nominal power*

One of the main concerns in the converter design is the selection of the nominal power P_{nom}. The range of possible values of P_{nom} can be narrowed down based on the properties of the load [7]. The typical load type for the studied LVDC system is a detached house utilizing electrical heating. Therefore, the peak load is in the range of $P_{max} = 10$—16 kVA which represents the required sum of the nominal power of all the converters that are supplying the inverter. As the configuration depicted in Fig. 2c was selected the maximum nominal power of one converter becomes $P_{nom,max} = 5300$ W when $P_{max} = 16$ kVA. However, due to the highly varied load a case in which each of the phases utilizes a minimum of two parallel converter-inverter modules was included in the study. Therefore, the maximum nominal power value of one converter becomes $P_{nom,max} = 2700$ W. For illustrative purposes, a range of $P_{nom} = 200$—6000 W in which the difference between the consecutive values of P_{nom} is $P_{nom,step} = 200$ W while $P_{nom} = 200$—1000 W and $P_{nom,step} = 1000$ W while $P_{nom} - 1000$—6000 W was selected. The value of $P_{nom} = 6000$ W defines the case in which only a single converter per phase is operated and therefore, no converters are connected in parallel.

B. *Selection of the converter topology*

Different types of isolated converter topologies have previously been studied in [15]. Even though the primary effort in [15] was put on fuel cell applications utilizing a low input voltage and current-fed converter topologies, a comparison of voltage-fed topologies was also carried out. Some of the studied topologies are also suitable for high input voltage and low current applications like the LVDC system which utilizes an input voltage level of $U_{DC} = 750$ VDC in combination of low load currents.

Especially the phase-shifted full-bridge with a voltage doubler secondary side rectifier (PSFBVD) [15], [16] has favorable characteristics if utilized in the LVDC system. The PSFBVD is a soft-switching topology that utilizes a full-bridge on the primary side and a voltage doubler consisting of two semiconductors and two capacitors on the secondary side. The secondary side can also be implemented with diodes but then the possibility of bidirectional power flow is lost. The primary side lagging-leg switches are operated in phase-shift relative to the leading-leg switches which in combination with the magnetizing and leakage inductance of the isolation transformer is utilized to enable zero voltage switching (ZVS) for the primary side switches. The leakage inductance in combination with the secondary side voltage doubler capacitors form an LC-resonance circuit which allows the secondary and also the primary side to switch under zero current switching (ZCS). A more thorough explanation of the working principle and schematic of the topology can be found in [15], [16]. The voltage doubler also behaves as an LC-filter and therefore

a bulky output inductor is not required. As a result, the voltage stress on the rectifier semiconductors is clamped to the output voltage and RC snubbers that are typically required to suppress the voltage ringing can also be eliminated. The PSFBVD topology enables ZVS operation of the switches over a wide load range [16] which is required in the LVDC system. The main drawback of the PSFBVD topology is its poor output voltage regulation as a function of the duty cycle [16]. However, as the LVDC system utilizes an inverter which regulates the output voltage of the system to a value of 230/400 VACR$_{MS}$ and 50 Hz, a regulated DC-voltage is not a requirement. The lack of a sophisticated control of the DC-DC converter would also considerably simplify the system. The converter could be operated at a fixed duty cycle and the output voltage of the converter and therefore, the input voltage of the inverter would vary as a function of the network voltage similar to the configuration in Fig. 2a. If a duty cycle value near $D_{eff} = 0.5$ is selected the DC-voltage is still well regulated as a function of the load [16].

C. *Design criteria*

The configuration of the converter highly depends on the selected design criteria. The converter can for example be optimized for maximum efficiency, power density, minimization of the production cost, or total lifetime cost. Because the converters are to be utilized in a distribution network, the most significant criteria is the total lifetime cost of the converter which can be defined as

$$\min \int_0^T (C_{inv}(t) + C_{loss}(t) + C_{interr}(t) + C_{main}(t))dt$$
$$\approx \min \sum_{t=1}^T \left[C_{inv}(t) + C_{loss}(t) + C_{interr}(t) + C_{main}(t) \right] \tag{1}$$

in which C_{inv} is the investment cost, C_{loss} the loss cost, C_{interr} the interruption cost, and C_{main} the maintenance cost [17]. A utilization period of $t_u = 10a$ was selected for the study as this roughly estimates the expected lifetime of the converter. For the purpose of this paper only C_{inv} and C_{loss} are included in the calculations as C_{interr} and C_{main} represent the cost due to failure and replacement of the component and downtime of the system. The value of C_{inv} is selected to include the cost of the main power semiconductors, gate drivers, DC capacitors, transformers and heat sink. The converter is assumed to be operated at a fixed duty cycle and therefore, a sophisticated control system is not a requirement. The cost of the losses C_{loss} is calculated as present value of an annuity by

$$C_{loss} = P_{tot} C_e \frac{1 - (1 + p)^{-t_u}}{p} \tag{2}$$

in which p is the interest rate and C_e the market price of electricity [17]. A value of $p = 5\%$ and $C_e = 5.0$ cent/kWh was used in the calculations.

The losses are calculated by using a measured average load profile for detached households [7]. The annual energy consumption was selected to be $E = 20$ MWh as this represents an average value for a detached household with direct electrical heating. The load profile of the selected customer type is illustrated in Fig. 4.

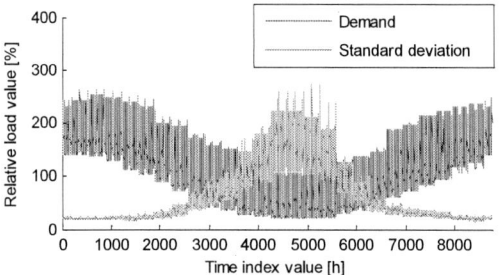

Fig. 4. Measured average load profile for the selected customer type which represents a percentage value of the average hourly power of the customer [7].

The parallel converters are operated in a way in which a minimum of three converters, one per phase, are in operation at all times and when the nominal power of the single converter is exceeded the required number of parallel converters are turned on to supply the increased load as defined by

$$N_{\text{modules, on}} \geq \frac{P_{\text{load}}}{P_{\text{nom}}}, N_{\text{modules, on}} \in [1,2,3...m]. \qquad (3)$$

By only operating the required number of converters $N_{\text{modules,on}}$ to supply the load, the no-load losses of the system can be minimized. The total number of the required converter modules for the customer is calculated by

$$N_{\text{modules}} \geq \frac{P_{\text{max}}}{P_{\text{nom}}}, N_{\text{modules}} \in [1,2,3...m] \qquad (4)$$

and is used to determine the investment cost of the system.

D. Transformer parameters

The effect of P_{nom} and f_{sw} on the transformer parameters was studied to determine if dividing the nominal power of the converter among several smaller converters would be feasible and how much the power density of the system could be increased by increasing the value of f_{sw}. The transformers were designed using ETD and PM cores [18]. The magnetic material of the cores is N87 and N97 ferrite [19], [20]. For the winding, copper foil and litz was selected. C_{inv} of the transformer was calculated using retail prices for 250 pcs quantities for the cores [21] and the price of copper per ton [22] for the winding. The losses of the transformer were also included in the design process and the cost of the losses was calculated by (2) and (3) and by utilizing the load curve in Fig. 4. Because the design process includes the

effect of the costs C_{inv} and C_{loss} the resulting parameters for the transformer do not necessarily result in the maximum efficiency of the system if the increased efficiency is shadowed by an increase in investment cost.

Fig. 5 shows the effect of P_{nom} and f_{sw} on the volume of the transformer core when the design goal is set for minimum lifetime cost and copper foil is used for the winding. The foil height is defined as 80% of the height of the winding window and the minimum thickness of the copper is 0.025 mm. The thickness of the insulation is the same in each case.

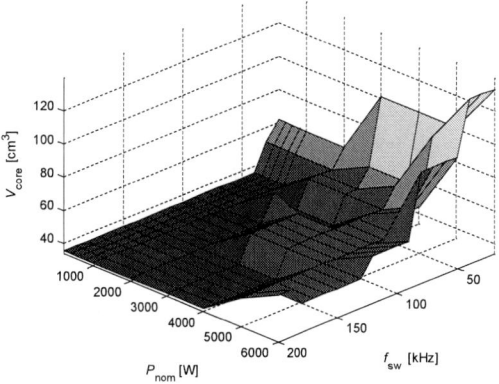

Fig. 5. Core volume of a single transformer as a function of P_{nom} and f_{sw} when the design goal is set for minimum lifetime cost. The number of transformers is defined by (4)

Fig. 5 shows that decreasing the value of P_{nom} does not have an effect on the core selection when $P_{\text{nom}} = 200$—4000 W while $f_{\text{sw}} = 140$—200 kHz. Similar behavior is observed when $P_{\text{nom}} = 200$—2000 W while $f_{\text{sw}} = 40$—140 kHz. The AC flux density B_{ac} of the transformer is defined by

$$B_{\text{ac}} = \frac{U_{\text{pri}} \cdot 10^4}{k N_{\text{pri}} A_{\text{c}} f_{sw}}, \qquad (5)$$

in which k is a waveform coefficient and A_{c} the core cross-sectional area [23]. As can be seen from (5) the value of U_{pri}, which is constant, limits the selection of the core. However, from (5) it can also be seen that by increasing the value of f_{sw} or N_{pri} results in the reduction of B_{ac} and the selection of a smaller core model becomes possible if the higher number of winding turns can be implemented or the design is not limited by the reduction in thermal performance of the smaller core. Because the dependency between f_{sw} and B_{ac} is linear but the dependency between f_{sw} and core loss density is exponential a significant increase in f_{sw} does not necessarily provide further reduction of the volume of the core as the core loss density to B_{ac} to f_{sw} ratio begins to saturate.

Fig. 6 illustrates the number of winding turns N_{pri} as a function of P_{nom} and f_{sw} when the design goal is set for minimum lifetime cost and copper foil is used for the winding. The number of secondary turns is $N_{\text{pri}}/2$ in each of the cases. The foil height is defined as 80% of the height of the winding window and the minimum thickness of the copper is 0.025 mm. The thickness of the insulation is the same in each case.

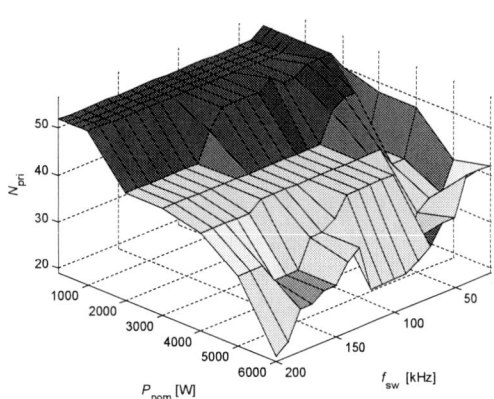

Fig. 6. Winding turns N_{pri} of a single transformer as a function of P_{nom} and f_{sw} when the design goal is set for minimum lifetime cost and copper foil is used for the winding.

By comparing Fig. 5 and Fig. 6 it can be seen that the design is limited by the maximum number of winding turns N_{pri} which results in the selection of an unnecessarily large core if the value of P_{nom} is decreased. Because the thickness of the copper foil cannot be decreased indefinitely the thickness and cross-sectional area of the foil limits further increasing of N_{pri}. Therefore, calculations were carried out for the same input parameters using litz wire for the winding. The litz wire calculation methodology is presented in [24]. Utilization of litz wire makes it possible to reduce the cross-sectional area of the winding significantly compared to the foil and therefore, a higher value of N_{pri} can be implemented when the available window area is the same. However, as a result the same value of N_{pri} can also be fitted to a smaller core and therefore, a smaller core can be selected if any other parameter is not limiting the design.

Fig. 7 illustrates the volume of the transformer core as a function of P_{nom} and f_{sw} when the design goal is set for minimum lifetime cost and litz wire is used for the winding. Fig. 8 depicts the number of winding turns N_{pri} as a function of P_{nom} and f_{sw} when the design goal is set for minimum lifetime cost and litz wire is used for the winding. The number of secondary turns is $N_{pri}/2$ in each of the cases. By comparing Figs 5—8 the effect of the possibility to reduce the winding cross-sectional area by using litz wire and therefore, increase the value of N_{pri} is obvious. N_{pri} is increased significantly when $P_{nom} < 4$ kW compared to the foil case and therefore, the value of B_{ac} decreases. As a result the value of A_c and therefore, the core volume can be decreased.

Fig. 9 illustrates the total core volume of the transformers when the number of transformers is calculated by (4). Fig. 9 shows that the minimum volume of the transformers is achieved when $P_{nom} = 800$ W. When P_{nom} is reduced further the total volume begins to increase exponentially as the number of transformers begins to increase rapidly as a function of P_{nom} and the

volume of the single core cannot be decreased further as the smallest ETD core is already selected in several cases.

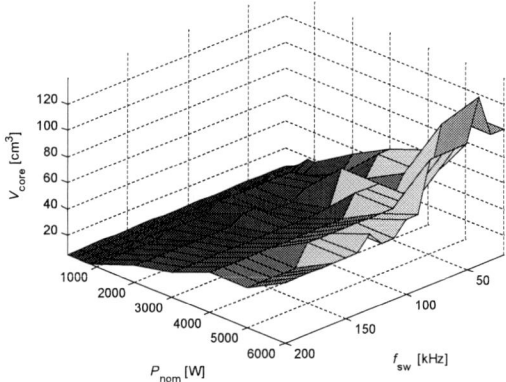

Fig. 7. Core volume of a single transformer as a function of P_{nom} and f_{sw} when the design goal is set for minimum lifetime cost and litz wire is used for the winding.

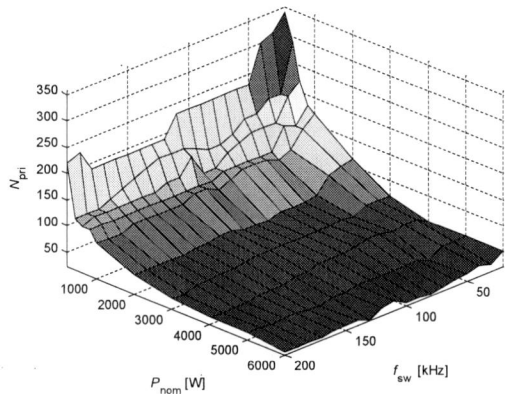

Fig. 8. Winding turns N_{pri} of a single transformer as a function of P_{nom} and f_{sw} when the design goal is set for minimum lifetime cost and litz wire is used for the winding.

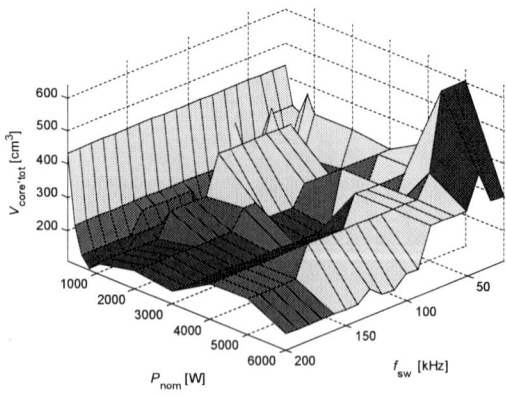

Fig. 9. Total core volume $V_{core,tot}$ of the transformers as a function of P_{nom} and f_{sw} when the design goal is set for minimum lifetime cost and litz wire is used for the winding. The number of transformers is defined by (4).

Fig. 10 illustrates the lifetime cost C_{tot} of the transformers as a function of P_{nom} and f_{sw} when litz wire is used as the winding and the cost of the cores, winding and losses are combined. In the loss calculation (2) and (3) are used, and (4) is used to calculate the required number of transformers for the investment cost.

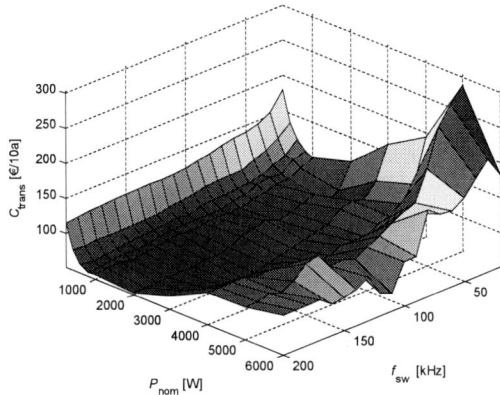

Fig. 10. Total lifetime cost C_{trans} of the transformers as a function of P_{nom} and f_{sw} when the design goal is set for minimum lifetime cost and litz wire is used for the winding. The number of transformers is as defined by (4).

Fig. 10 shows that the minimum value for C_{trans} is reached when P_{nom} = 600—3000 W while f_{sw} = 70—200 kHz. Based on the transformer cost C_{trans} we can observe that the selection of the value of P_{nom} could be narrowed down to values of P_{nom} = 800—3000 W. A preference toward values of f_{sw} = 70—200 kHz is also observed. Results for the foil are not discussed further as the litz wire resulted in reduced values of C_{trans} when P_{nom} < 4000 W. The differences were in the favor of the foil in some cases when $P_{nom} \geq$ 4000 W but the differences were negligible. Because the selection cannot be made based purely on the transformer cost, the semiconductor, DC capacitance, gate driver and heat sink cost was also evaluated.

E. Semiconductor selection

Because the studied nominal power range is so wide, semiconductors with two different current ratings were selected [25], [26]. The selection of the semiconductors for each combination of P_{nom} and f_{sw} was carried out comparing the safe operating area of the switching device against the pulse length and peak current of the switch which is defined as

$$I_{peak} = n\frac{\omega_r I_{out} T}{2} \qquad (6)$$

in which n is the transformer turn ratio, I_{out} the output current of the converter and ω_r the resonance frequency of the converter [15]. Because the converter is supplying an inverter generating a 50 Hz sine wave, the value of I_{out} is modulated as a function of the output current of the inverter and needs to be considered when selecting the semiconductor switches. The converter is operated near the maximum duty ratio and therefore $D_{eff} = 0.43$. Due to the voltage doubler configuration the current I_{peak} is doubled at the secondary side switches and therefore, becomes $2I_{peak}$ when compared to the value of I_{peak} on the

primary side. The primary side has four switches and the secondary side two switches. The selection of the semiconductors is illustrated in Table I.

TABLE I
SELECTION OF THE SEMICONDUCTOR SWITCHES

Secondary / Primary side	Transistor	Number of semiconductors per switch	P_{nom} [kW]
S	[23]	3	> 4.3
		2	2.2-4.3
		1	0.3-2.1
	[22]	1	< 0.3
P	[23]	2	> 4.3
		1	0.6-4.3
	[22]	1	< 0.6

Due to ZVS/ZCS switching and the utilization of silicon carbide semiconductors the switching loss is assumed negligible and only the conduction losses are included in the loss calculation. The RMS current of a single switch is calculated by

$$I_{RMS} = \sqrt{\frac{1}{T}\int_0^{\sqrt{L_{lk}C_r}\pi}\left(n\frac{\omega_r I_{out} T}{2}\sin(\omega_r t)\right)^2 dt} \qquad (7)$$

in which L_{lk} is the leakage inductance of the transformer and C_r the resonance capacitance [15]. For the secondary side semiconductors I_{RMS} is calculated with $n = 1$ and for the primary side with $n = N_{sec}/N_{pri}$. The value of I_{RMS} is divided between parallel semiconductors as defined by Table I and the value of I_{out} between parallel converters as defined by (3). Fig. 11 depicts the total lifetime cost of the main semiconductors as a function of P_{nom} and f_{sw}.

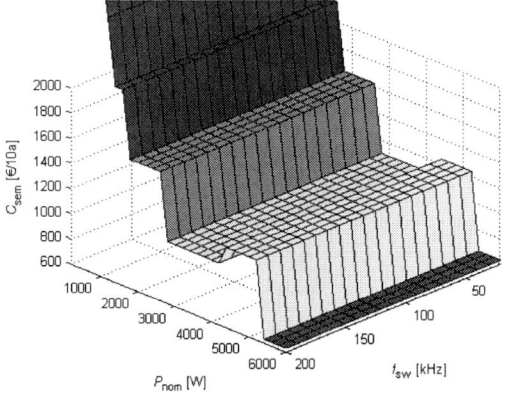

Fig. 11. Total lifetime cost C_{sem} of the main semiconductors as a function of P_{nom} and f_{sw}. The number of semiconductors is as defined by (4) and Table I. The values of P_{nom} = 1000—6000 W have been calculated with a step of $P_{nom,step}$ = 200 W to emphasize the points in which the semiconductor configuration is changed.

From Fig. 11 the points in which the number of parallel semiconductors or the number of parallel converters is increased as well as the points in which the semiconductor is changed to the more expensive alternative can clearly be observed as an increase in the value of the cost C_{sem}. Based solely on the semiconductor cost C_{sem} an obvious preference toward a single $P_{nom} > 5.2\,kW$ converter per phase is observed. If the results in Fig. 11 are compared against the transformer cost C_{trans} in Fig. 10 it can be seen that C_{sem} is clearly the more dominating of the two. Further, the dominating component in the value of C_{sem} is the investment cost of the semiconductor switches. From Fig. 11 it is also obvious that this type of configuration of the converter is not a feasible solution if $P_{nom} < 2.7\,kW$.

F. Total cost of the converter

In order to determine the most feasible configuration the cost of the heat sinks, DC-capacitance and gate drivers were also evaluated. For the gate drivers a similar configuration as in [27] was selected and prices for the main components in 250 pcs quantities was used in the calculation making the estimated cost 90 € for the required six drivers. The required thermal resistance of the heat sink for a 50°C rise above 50°C ambient was calculated from the semiconductor losses at P_{max} and a function for the relation between the price per K/W and P_{nom} was formulated using the price data from [28].

Essentially the total DC-capacitance is the same in each of the cases and therefore, it can be said that the total cost is very similar. In the multiple converter cases the capacitance is merely divided into smaller parts among the parallel converters. The output side capacitance of the converter is considered part of the inverter as the inverter includes a DC-capacitance also in the 50 Hz isolation transformer configuration and is therefore not included in the calculations. A similar value as in [6] was selected for the network side DC-capacitance of the converter, making the cost of the capacitor 4 €/kW if a similar 250 pcs price [29] is used as in the previous cases. Fig. 12 depicts the total cost of the converter as a function of P_{nom} and f_{sw}.

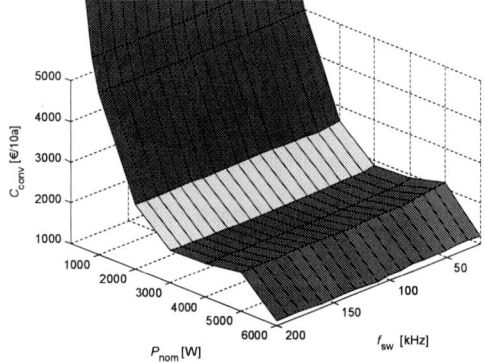

Fig. 12. Total lifetime cost C_{conv} of the converters as a function of P_{nom} and f_{sw}. C_{conv} includes the cost of the transformers, main semiconductors, heat sink, DC-capacitance, gate drivers and losses.

By comparing Fig. 11 and Fig. 12 it can be seen that the dominating component in the total lifetime cost C_{conv} of the converter is the cost C_{sem} of the main semiconductor switches. The total cost of the converter was then compared to the 50 Hz isolation transformer. The loss cost of the 50 Hz transformer was calculated by

(2) using the data in Fig. 3 and Fig. 4. Retail price of 900€ of the same transformer was used as the investment cost.

Fig. 13 illustrates the difference in lifetime cost between the passive 50 Hz transformer and the converter system as a function of P_{nom} and f_{sw} of the converters. The nominal power of the 50 Hz transformer is 16 kW in each case.

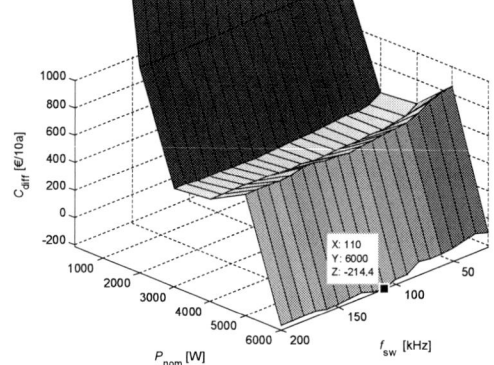

Fig. 13. Difference in total lifetime cost C_{diff} between the converters and the 16 kW 50 Hz isolation transformer as a function of P_{nom} and f_{sw} of a single converter.

Fig. 11 shows that despite the high cost of the converter, a clear advantage in losses still make it a feasible option if a single $P_{nom} = 6000\,W$ converter per phase is used. The main contributor to the difference in this case are the no-load losses, which are decreased significantly if the converter is used. However, in the parallel converter cases the advantage is shadowed by the high cost of the increasing number of semiconductors and the difference changes in the favor of the 50 Hz transformer. The effect of the paralleling should also be investigated on the inverter side to determine its effect on the inverter cost and this will be discussed in further publications. The transformer gained a clear advantage of the paralleling and therefore, if the semiconductor cost could be decreased the parallel converter system could prove to be a feasible option. The parallel converters should also be studied as a partly series connected system which would enable the use of semiconductors with lower voltage ratings and decrease the cost further.

IV. CONCLUSION

Galvanic isolation in the LVDC network is discussed and a comparison between two isolation methods is carried out. The passive isolation method utilizing a 50 Hz transformer is shown to have poor power density and high losses. The alternative method utilizing an isolated converter is discussed in more depth and a selection for a suitable converter topology is carried out. Due to the properties of the load a case in which the nominal power could be divided between several parallel converters was also studied. A comparison of the converter parameters as a function of the switching frequency and nominal power was presented. In the studied application a preference toward a single converter per phase is observed due to a significant increase in the semiconductor cost if several converters are paralleled. However, with the parallel system a further reduction of the losses is made possible. The conduction losses of the semiconductors and the transformer winding decrease as the load current is shared among parallel converters. Further, the no load losses could be decreased if the parallel converters are turned off during low loads. However, this requires the transformer designed for the lower nominal power to have lower core losses. The total lifetime cost of the two isolation methods was compared for a ten year utilization period. It was concluded that despite the relatively high cost of the converter system a significant decrease in the losses made the converter a feasible alternative over the 50 Hz transformer. However, it is to be noted that the estimated investment cost of the converter system does not reflect the actual production cost if the converters were to be manufactured and should be considered for illustrative purposes only.

REFERENCES

[1] Fukui, A.; Takeda, T.; Hirose, K.; Yamasaki, M.; , "HVDC power distribution systems for telecom sites and data centers", *Power Electronics Conference (IPEC), 2010 International* , vol., no., pp.874-880, 21-24 June 2010

[2] T. Kaipia, P. Salonen, J. Lassila, and J. Partanen, "Possibilities of the low voltage dc distribution systems", *In proc. of NORDAC 2006*, 20-21 August 2006.

[3] Kakigano, H.; Miura, Y.; Ise, T., "Configuration and control of a DC microgrid for residential houses", *Transmission & Distribution Conference & Exposition: Asia and Pacific, 2009* , vol., no., pp.1,4, 26-30 Oct. 2009

[4] Kakigano, H.; Miura, Y.; Ise, T., "Low-Voltage Bipolar-Type DC Microgrid for Super High Quality Distribution", *IEEE Trans. Power El.*, vol.25, no.12, pp.3066,3075, Dec. 2010

[5] Boroyevich, D.; Cvetkovic, I.; Dong Dong; Burgos, R.; Fei Wang; Lee, F., "Future electronic power distribution systems a contemplative view," *Optimization of Electrical and Electronic Equipment (OPTIM), 2010 12th International Conference on* , vol., no., pp.1369,1380, 20-22 May 2010

[6] T. Kaipia, P. Nuutinen, A. Pinomaa, A. Lana, J. Partanen, J. Lohjala, and M. Matikainen, "Field test environment for lvdc distribution – implementation experiences", in Proc. of CIRED Workshop 2012, 29-30 May 2012, Lisbon, Portugal.

[7] Suomen Sähkölaitosyhdistys ry. Sähkön käytön kuormitustutkimus 1992 (Finnish Electricity Association Sener, research on electricity consumption), in Finnish, SLY 7103, Helsinki 1992, ISSN 0786-7905.

[8] Liu J., Wong K. L., Kierstead P., "Increase Efficiency and Lower System Cost with 100KHz, 10kW Silicon Carbide (SiC) Interleaved Boost Circuit Design", *PCIM Europe 2013*, 14-16 May 2013

[9] Rabkowski, J.; Peftitsis, D.; Zdanowski, M.; Nee, H., "A 6kW, 200kHz boost converter with parallel-connected SiC bipolar transistors", *Applied Power Electronics Conference and Exposition (APEC), 2013 Twenty-Eighth Annual IEEE*, vol., no., pp.1991,1998, 17-21 March 2013

[10] Mitova R., Dentela A., Wang M., "Half bridge inverter with 600V GaN power switches", *PCIM Europe 2013*, 14-16 May 2013

[11] Shirabe, K.; Swamy, M.; Jun-Koo Kang; Hisatsune, M.; Yifeng Wu; Kebort, D.; Honea, J., "Advantages of high frequency PWM in AC motor drive applications", *Energy Conversion Congress and Exposition (ECCE), 2012 IEEE* , vol., no., pp.2977,2984, 15-20 Sept. 2012

[12] Simanjorang, R.; Yamaguchi, H.; Ohashi, Hiromichi; Takeda, T.; Yamazaki, M.; Murai, H., "A high output power density 400/400V isolated DC/DC converter with hybrid pair of SJ-MOSFET and SiC-SBD for power supply of data center", *Applied Power Electronics Conference and Exposition (APEC), 2010 Twenty-Fifth Annual IEEE* , vol., no., pp.648,653, 21-25 Feb. 2010

[13] Hayashi, Y., "Power density design of SiC and GaN DC-DC converters for 380 V DC distribution system based on series-parallel circuit topology", *Applied Power Electronics Conference and Exposition (APEC), 2013 Twenty-Eighth Annual IEEE* , vol., no., pp.1601,1606, 17-21 March 2013

[14] Mattsson A., "Modular Customer-end Inverter for an LVDC Distribution Network", *PCIM Europe 2013*, 14-16 May 2013

[15] Väisänen V., "Performance and Scalability of Isolated DC-DC Converter Topologies in Low Voltage, High Current Applications", doctoral dissertation, Lappeenranta: Lappeenranta University of Technology, 2012.

[16] Woo-Jin Lee; Chong-Eun Kim; Gun-Woo Moon; Sang-Kyoo Han, "A New Phase-Shifted Full-Bridge Converter With Voltage-Doubler-Type Rectifier for High-Efficiency PDP Sustaining Power Module", *IEEE Trans. Ind. El.*, vol.55, no.6, pp.2450,2458, June 2008

[17] E. Lakervi; J. Partanen, Sähkönjakelutekniikka (Electricity distribution technology), Helsinki, Finland, 2008 (in Finnish)

[18] EPCOS AG, Ferrites and accessories, ETD and PM Ferrite cores, www.epcos.com/ferrites

[19] EPCOS AG, Ferrites and accessories, SIFERRIT Material N87 , http://www.epcos.com/blob/528882/download/3/pdf-n87.pdf

[20] EPCOS AG, Ferrites and accessories, SIFERRIT Material N97, http://www.epcos.com/blob/528886/download/3/pdf-n97.pdf

[21] Retail prices for EPCOS ETD and PM cores, www.mouser.com

[22] Price of copper per ton, London Metal Exchange, www.lme.com/metals/non-ferrous/copper

[23] McLyman W.T. *Transformer and inductor design handbook*, 3rd ed., New York, 2004.

[24] Vaisanen, V.; Hiltunen, J.; Nerg, J.; Silventoinen, P., "AC resistance calculation methods and practical design considerations when using litz wire", *Industrial Electronics Society, IECON 2013 - 39th Annual Conference of the IEEE* , vol., no., pp.368,375, 10-13 Nov. 2013

[25] Cree Inc, C2M0160120D Silicon Carbide Power MOSFET Technical documentation, www.cree.com/power/products

[26] Cree Inc, C2M0080120D Silicon Carbide Power MOSFET Technical documentation, www.cree.com/power/products

[27] Cree Inc, CRD-001 gate driver, www.cree.com

[28] Fischer SK514/100 2.2 K/W heat sink http://www.tme.eu/en/details/sk514-100sa/radiators/fischer-elektronik/sk-514-100-sa/

[29] Cornell Dubilier 947C 1kV 470µF polypropylene capacitor http://www.digikey.fi/product-detail/en/947C471K102CD MS/338-1400-ND/191693

Peak Detection Method using Two-Delta Operation for Single Voltage Sag

Woo-Cheol Lee
Department of Electrical Electronic & Control Engineering
Hankyong National University
Anseong, Gyeonggi-do, Korea
woocheol@hknu.ac.kr

Taeck-Kie Lee
Department of Electrical Electronic & Control Engineering
Hankyong National University
Anseong, Gyeonggi-do, Korea
tklee@hknu.ac.kr

Abstract— In this paper, the proposed system is presented the detection method of sags/swells when a fault occurs in the grid-voltage. The conventional sag detector is from a single-phase digital phase lock loop (DPLL) based on a d-q transformation using an all-pass filter(APF). The d-q transformation is typically used in the 3-phase coordinate system. As present system using single-phase coordinate system, q-axis doesn't exist to convert stationary reference frame. To realize this transformation, the All-Pass Filter used to generate virtual q-axis with 90° through the grid voltage. However this method doesn't immediately recognize when it occurred the sag or momentary power failure that is variable to drop the power-quality in grid voltage. Due to delayed phenomenon, the signal of sag transmits to system late. In this paper, the rapid sag / peak detector is proposed using 2nd component of change of Vgds. To review the validity of proposed system, compare to conventional and proposed method through the simulation and experimental results.

Keywords— *APF, UPS, voltage sag, outages, compensation, sag detection method..*

I. Introduction

Because of the widespread automation of critical processes throughout industry, the importance of precise voltage control and voltage stability has increased dramatically. It is highly trusted power quality in domestic industrial. However it is not exist for a sure regulation of power factor of grid voltage and current. Therefore the equipment will cause malfunction and damage, because grid voltage is not compensated what power factor and harmonics in usual condition. Such equipment should be operated 24 hours a day, and it is required the grid voltage with high reliability. And there are voltage sag and outage for another type of the input power disturbance. It is occurred by failure on grid-connected. Due to these influences, it is occurred serious loss to equipment when to supply unstable grid voltage. [1][2][3]

Increasing the installed capacity of renewable power generation sites, it is concerned about the impact on the grid voltage. It is important that the system detects the phase and amplitude of grid voltage when the problem occurs in the utility. For these reason, sag voltage detector has been proposed. This detector is using the D-Q Synchronously Rotating Reference Frame Transformation of DPLL. Most conventional peak detector methods have time delay of detection. The RMS method has 2-9ms of time delay [4], the hybrid KF-RMS method has 0.5-4ms of time delay [5], and DVR has 2ms of time delay [6]. The sag detection method have been proposed using 1st component of change ΔV_{ds} for fast sag detection.[7] But the disadvantage of this method is not detected around zero-crossing. To reduce the delay time about these issues, the sag detector is proposed in this paper.

This paper was made up following three sections; System configuration, Simulation, Experimental result.

II. System Configration

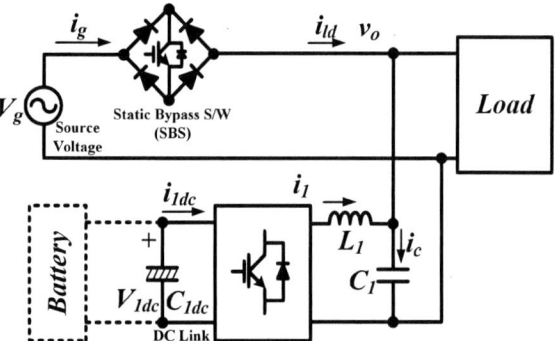

Fig. 1. Line-Interactive instantaneous voltage compensation system

Fig. 1 shows that the configuration of instantaneous voltage compensation system of Line-Interactive. This system operates to supply the voltage of constant amplitude, when sag occurred in grid voltage. The compensation system consists of 3-parts as follows; Bi-directional PWM converter of Single-phase, the LC Low-Pass filter for creating sinusoidal, the DC-Link consists of a battery or capacitor. And the compensation voltage of converter output to be added to grid voltage is applied to load. Unlike conventional voltage compensator, it is omitted the series transformer and it isn't required the extra charging equipment to charge the capacitor. An IGBT switch with diodes is used as a static bypass switch, meaning that the IGBT switch is turned on and off according to the current gate signal. As noted above, such system is composed of the single-phase grid voltage. The single-phase voltage can't always to keep the same amplitude and the sag / swell of grid voltage can be caused by the load. To figure out the fault of grid voltage, would need to the sag / peak value detection method. In this paper presented about this detection method as follows.

A. All-Pass Filter and Digital Phase-Locked-Loop

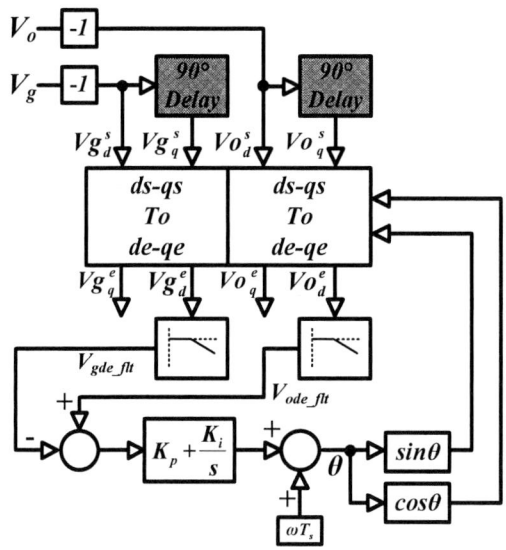

Fig. 2. Digital Phase-Locked-Loop and sag detection block diagram

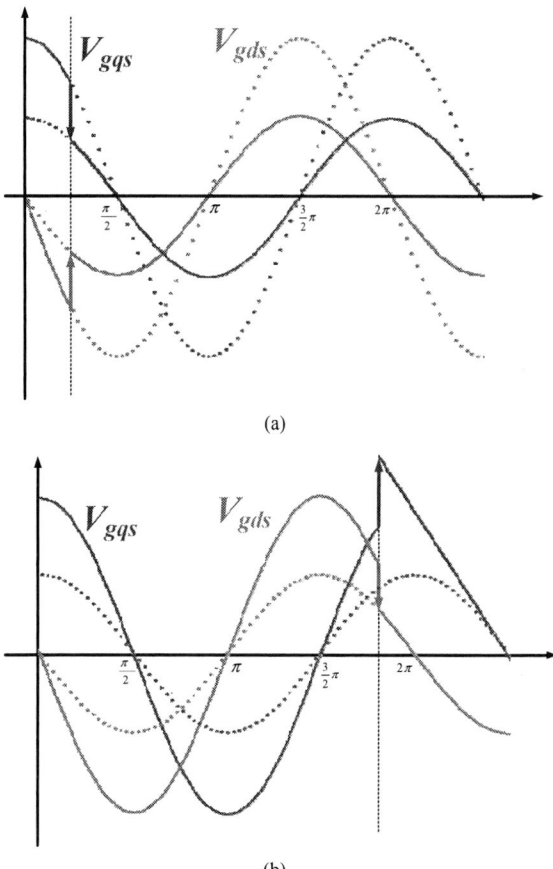

(a)

(b)

Fig. 3. The characteristic of All-Pass Filter
(a)Normal condition of All-Pass Filter(0~90°, 180~270°), (b)Abnormal condition of All-Pass Filter(90~180, 270~360)

The Fig. 2 shows the block diagram of digital PLL and sag detection block diagram. The virtual phase V_{gq}^{s} can be obtained from the measured source voltage $V_{gd}^{s} = -V_g$ by using the all-pass filter in the discrete-time domain as follows:

$$V_{gd}^{s}(t) = -V_g(t) \tag{1}$$

$$V_{gq}^{s}(t) = -kV_{gq}^{s}(t-1) + kV_{gd}^{s}(t) + V_{gd}^{s}(t-1) \tag{2}$$

The Fig. 3 shows characteristic of All-Pass-Filter that the sag occurred in entire section. The voltage of V_{gd}^{s} changes in real-time on occurred the sag. Because the grid voltage takes the negative value. The Fig. 3(a) shows feature in case of different sign between ds and qs voltage. At this point, the voltage of V_{gq}^{s} changes in real-time. But the fig. 3(b) shows feature in case of equal sign between ds and qs voltage, in this case can be found the qs voltage to increase the opposite direction. Because of this phenomenon is related to the voltage of ds to obtain the voltage of qs. Due to the non-ideal phenomena of qs voltage, the delay occurred to detect the peak value when converts rotating reference frame.

B. Conventional Sag Detection based on DPLL

The 2014 International Power Electronics Conference

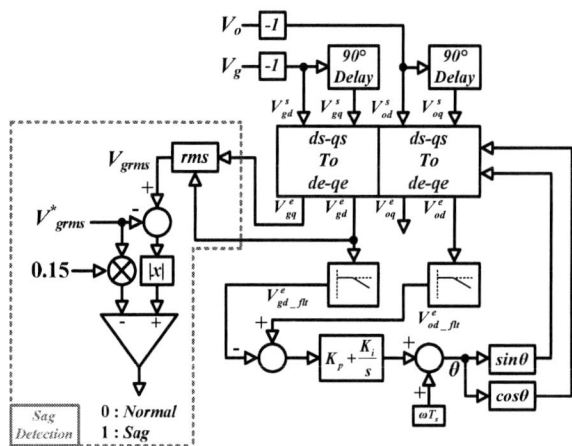

Fig. 4 Conventional sag detection based on DPLL

Fig. 4 shows the block diagram of conventional sag detection method.

By using a synchronous rotating reference frame, the rms value of the source voltage is obtained as follows;

$$V_{grms} = \sqrt{\frac{(V_{gd}^e)^2 + (V_{gq}^e)^2}{2}} \qquad (3)$$

$$V_{grms\_err} = V_{grms} - V_{grms}^* \qquad (4)$$

If the V_{grms_err} is beyond the 15% of absolute source voltage (reference; $V^*_{grms}*0.15$), the source voltage sag is detected. However, the delay for detecting the sag is occurred because of the non-ideal phenomenon of All-Pass Filter.

C. The proposed Sag/Peak Detection Method

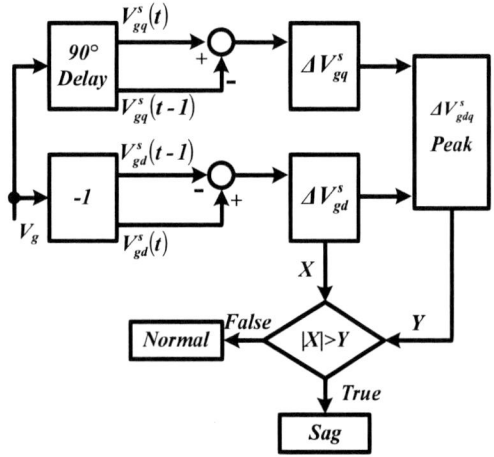

Fig. 5. Sag detection scheme using the difference between current and previous value of the voltage

Fig. 5 shows the block diagram of proposed detection method. The 1st-component of change was generated through ΔV_{gd}^s and ΔV_{gq}^s which based on the value of present / one step ahead for V_{gd}^s and V_{gq}^s.

$$\Delta V_{gd}^s = V_{gd}^s(t) - V_{gd}^s(t-1) = -A\cos(t) \qquad (5)$$

$$\Delta V_{gq}^s = V_{gq}^s(t) - V_{gq}^s(t-1) = -A\sin(t) \qquad (6)$$

$$\Delta\left(\Delta V_{gd}^s\right) = \Delta V_{gd}^s(t) - \Delta V_{gd}^s(t-1) = B\sin(t) \qquad (7)$$

$$\Delta\left(\Delta V_{gq}^s\right) = \Delta V_{gq}^s(t) - \Delta V_{gq}^s(t-1) = -B\cos(t) \qquad (8)$$

Where $V_{gd}^s(t-1), V_{gq}^s(t-1)$ are the values of one step ahead, respectively. In the same way, using (5) and (6) obtained the 2nd-component of change. And using these values, it will obtain the peak value as follows.

$$\Delta(\Delta V_{gdq}^S)peak = \sqrt{\left\{\Delta(\Delta V_{gd}^S)\right\}^2 + \left\{\Delta(\Delta V_{gq}^S)\right\}^2} \qquad (9)$$

By comparing $\Delta(\Delta V_{gdq}^s)_{peak}$ with $\Delta(\Delta V_{gd}^s)$, $\Delta(\Delta V_{gd}^s)$ existed within $\pm\Delta(\Delta V_{gdq}^s)_{peak}$ when the grid voltage was normal. But $\Delta(\Delta V_{gd}^s)$ became greater than $\pm\Delta(\Delta V_{gdq}^s)_{peak}$ when the failure occurred at the grid voltage. At this point, the system determined that the failure occurred in the grid voltage.

D. Simulation / Experimental Result

Simulations of the proposed algorithms have been carried out by using the simulation tool PSIM (Powersim Inc.), and system parameters which were used in the experiment and simulation are shown in Table I.

TABLE I
SYSTEM PARAMETERS

Parameters	Value
Source Voltage (V_g), Rating	208V, 60Hz, 5KVA
Sag Source Voltage(V_g)	104V, 60Hz, 5KVA
Switching Frequency	11kHz
DC-link Capacitor(C_{1dc})	6800μF
DC-link Voltage(V_{dc})	330V
L_1	0.4mH
C_1	100μF,
Load(50%,100%)	15Ω+31mH, 8Ω+15mH

The 2014 International Power Electronics Conference

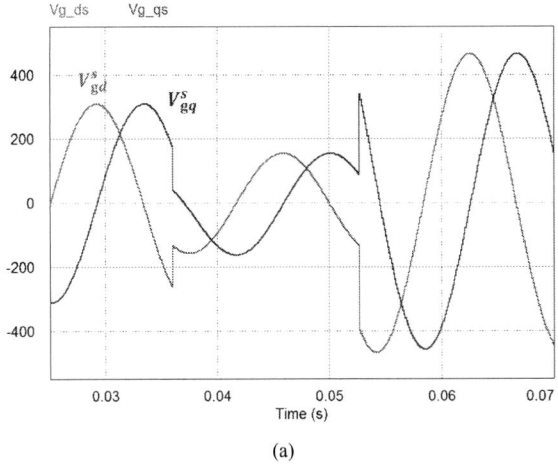

(a)

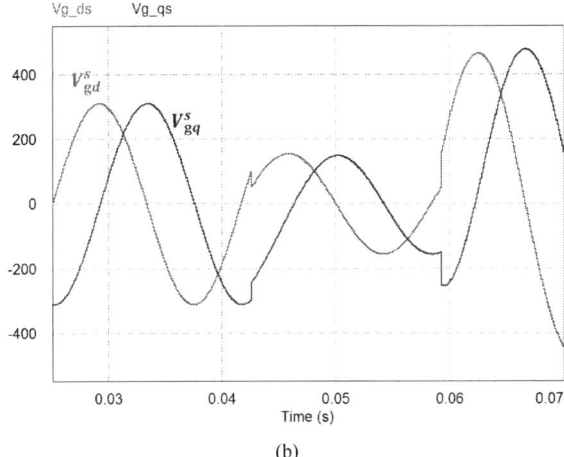

(b)

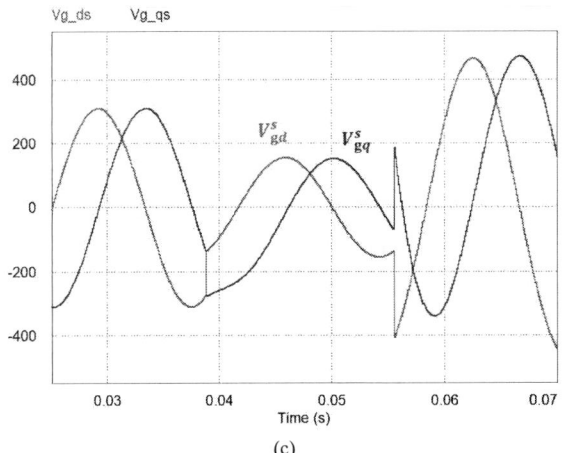

(c)

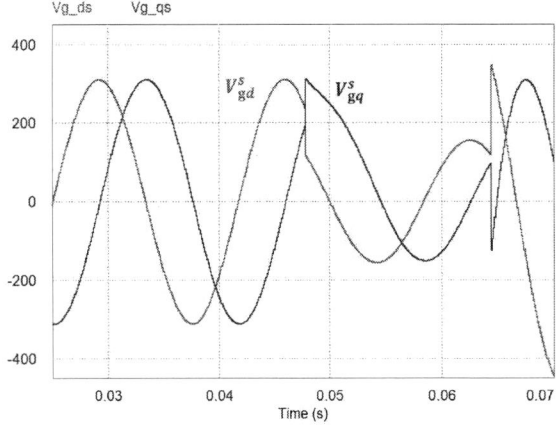

Fig. 6. 50% sagged/swelled grid voltage in the conventional All-Pass Filter. (a) sagged/swelled grid voltage in around 60° (b) sagged/swelled grid voltage in around 200° (c) sagged/swelled grid voltage in around 120° (d) sagged/swelled grid voltage in around 310°

The Fig. 6 shows the characteristic of All-Pass-Filter when the sag/swell occurred in specific phase. As shown in Fig. 6(a), 6(b) the All-Pass-Filter generates proper virtual phase when the sag/swell occurs during 0°~90°, and 180°~270°. In this case the sign of V_{gd}^s and V_{gq}^s are opposite to each other. However, Fig. 7(c),(d) show that the All-Pass-Filter generates opposite virtual phase when the sag/swell occurs during 90°~180°, and 270°~360°. In this case the sign of V_{gd}^s and V_{gq}^s are the same.

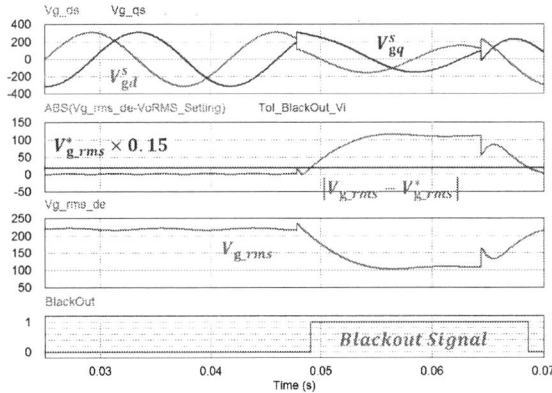

Fig. 7. 50% sagged/returned grid voltage in the conventional method around 310°. (a) stationery reference frame V_{gd}^s and V_{gq}^s (b) sag detection level (c) detected rms voltage V_{g_rms} (d) blackout signal

Fig. 7 shows the grid voltage in the conventional method around 310° when sagged by 50%. Fig. 8(a) shows V_{gd}^s from grid voltage and 90° phase lagged virtual waveform V_{gq}^s generated by All-Pass-Filter. When V_{gd}^s sags at 310°, the All-Pass-Filter generates opposite virtual phase, and it takes times to detect the sag voltage.

978-1-4799-2706-7/14 $31.00 © 2014 IEEE   598

sag event.

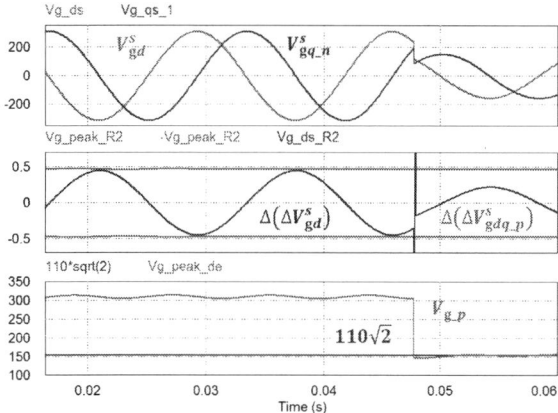

Fig. 8. 50% sagged grid voltage around 310°
(a) stationery reference frame V_{gd}^s and $V_{gq\_n}^s$ (b) difference voltage and $\Delta(\Delta V_{gd}^s)$ and peak voltage $\Delta(\Delta V_{dg}^s)_{peak}$ (c) peak grid voltage

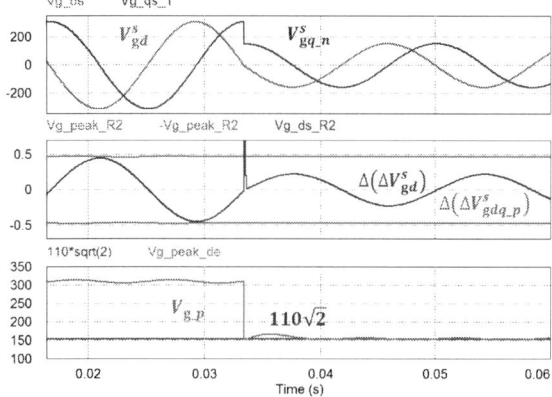

Fig. 9. 50% sagged grid voltage around 0°
(a) stationery reference frame V_{gd}^s and $V_{gq\_n}^s$ (b) difference voltage and $\Delta(\Delta V_{gd}^s)$ and peak voltage $\Delta(\Delta V_{dg}^s)_{peak}$ (c) peak grid voltage

Fig. 8 and 9 show the proposed sag/peak detection method. The value of V_{gpeak} is the peak of the grid voltage by using the synchronous d-q transformation. When the sag occurs during 0°~90°, 90°~180°, 180°~270°, and 270°~360° the sag is detected even though the sag is occurred at around 0°, and 180°. Owing to the sag detection, the peak grid voltage is acquired immediately.

In order to verify the proposed strategy, the algorithm has been implemented by using a digital signal processor (TMS320C33). A single 32-bit floating-point DSP with the single-cycle execution time of 13.3ns is used. The switching period of PWM is 91μs. The system parameters are shown in Table I. The nominal input voltage is 208V. Voltage sags are simulated with a tap changing transformer and switching devices, so that various voltage sag level can be generated. In this paper, the source voltage is dropped from 208V to 104V in the

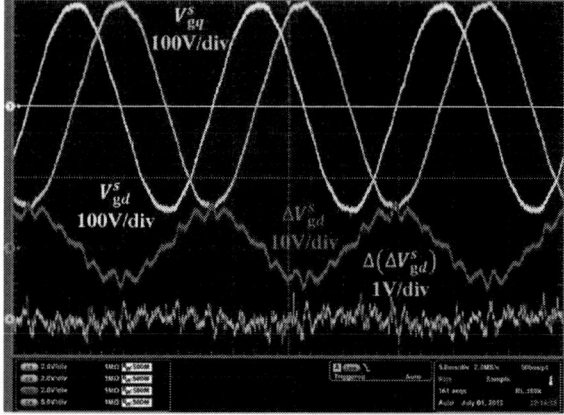

Fig. 10. Experimental results of the grid voltage
Ch 1. V_{gd}^s: grid voltage (100 V/div). Ch 2. V_{gq}^s: virtual source voltage (100V/div). Ch 3. $\Delta(\Delta V_{gd}^s)$: conventional difference voltage(10V/div). Ch 4. $\Delta(\Delta V_{dg}^s)_{peak}$: proposed difference voltage(1V/div).

Fig. 10 shows the derived components from the grid voltage. The waveforms of the d-axis voltage component (V_{gd}^s) and q-axis voltage component (V_{gq}^s) are a little bit distorted, and the difference voltages (ΔV_{gd}^s, $\Delta(\Delta V_{gd}^s)$) are distorted seriously.

Fig. 11 shows harmonic analysis of the grid voltage by power analyzer. It shows that the grid voltage includes harmonic components of 3th, 5th, and 7th. Therefore, the difference voltages (ΔV_{gd}^s, $\Delta(\Delta V_{gd}^s)$) are distorted seriously.

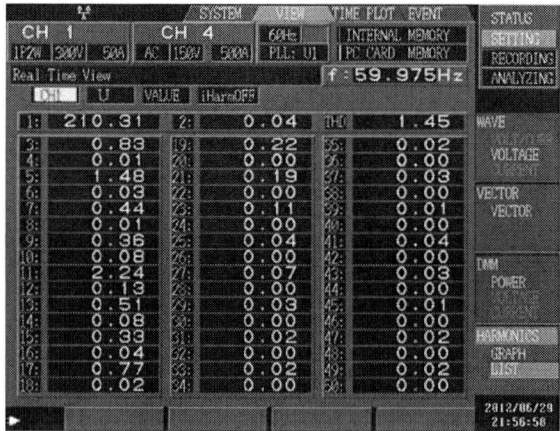

Fig. 11. Harmonic analysis of the grid voltage by power analyzer.

Fig. 12 shows the waveforms of comparing with conventional and proposed method. In case of Fig. 12(a) All-Pass filter operated in abnormal condition. As shown in Fig. 12(a), detecting time of the sag was delayed. The Fig. 12(b) shows that the voltage sag was detected by the proposed method immediately.

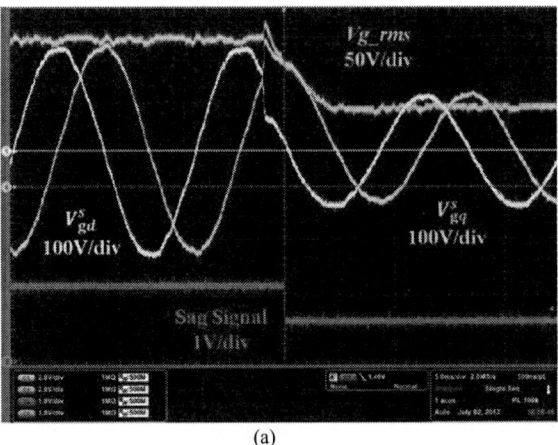

(a)

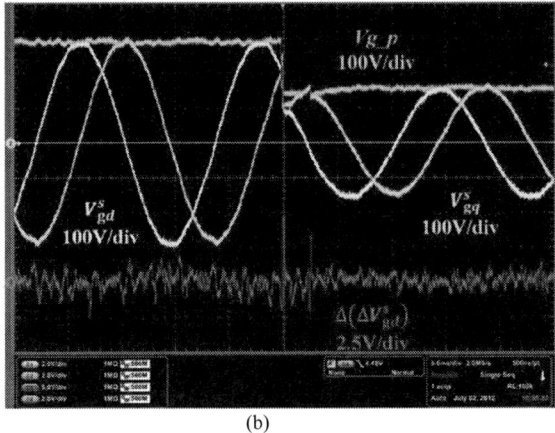

(b)

Fig. 12. The comparison waveforms of sag detections
(a) The occurrence of sag in 310° using conventional method
(b) The occurrence of sag in 310° using proposed method

III. Conclusions

In this paper, the proposed system is presented the detection method of sag / peak when a fault occurs in the grid-voltage. In the event of the failure in grid voltage, the delay was occurred to detect the sag using conventional method. So, it takes time to recognize the sag in the system. This is due to the characteristic of abnormal in All-Pass filter. To reduce these delays, the new method was proposed using the operation by the value of present and one step ahead. In case of 1st-component of change, the system can't recognize the sag around zero-crossing. So, the proposed method adapted the 2nd-component of change. According to the simulation and experimental result, it confirmed that the proposed detection method was accurate to detect the sag / peak immediately.

References

[1] N. Woodley, L. Morgan, A. Sundaram, "Experience with an inverter-based dynamic voltage restorer," *IEEE Trans. on Power Delivery*, Vol.14, No.3, pp.1181-1186, Jul. 1999.

[2] W. Lee, D. Lee, T. Lee, "New control scheme for a unified power quality compensator-Q with minimum active power injection," *IEEE Trans. on Power Delivery*, Vol.25, No.2, pp.1068-1076, Apr. 2010.

[3] W. Lee, W. Lee, T. Lee, "Integrated inverter system for compensation of Power Quality events", Power Electronics Specialists Conference, 15 – 19 June. 2008, pp. 3754 – 3759

[4] N. Tuaboylu, E. Collins, P. Chaney, "Voltage Disturbance Evaluation Using the Missing Voltage Technique", Harmonics And Quality of Power, vol.1, 1998, pp. 577-582

[5] M. Gonzalez, V.Cardenas, R. Alvarez, "A Fast Detection Algorithm for Sags, Swells, and Interruptions Based on Digital RMS Calculation and Kalman Filtering", in Proc. *Int. Power Electronics Congress* 2006, 16-18 Oct. 2006, pp.1-6

[6] J. Jeong, Ji. Lee, B. Han "Three-Phase Line-Interactive Dynamic Voltage Restorer with a New Sag Detection Algorithm", *Journal of Power Electronics*, vol. 10, no. 2, 2010, pp.210-217.

[7] H. Cha, S. Lee, T. Vu "A new fast peak detector for single or three-phase unsymmetrical voltage sags", Energy Conversion Congress and Exposition (ECCE), 12-16 Sept. 2010, pp.434 - 440.

Line Loss Minimization in Radial Distribution System using Multiple STATCOMs and Static Capacitors

Kensuke Miyazaki, and Takaharu Takeshita
Nagoya Institute of Technology
Gokiso, Showa, Nagoya, Japan
k.miyazaki.828@nitech.jp, take@nitech.ac.jp

Abstract—Recently, distribution networks based smart grid have gained much attention due to the availability of installing hybrid power generators in addition to the merits of installing FACTS (Flexible AC Transmission System) equipments in order to control its node voltages and the power flow. This paper presents a new technique for achieving line loss minimization in radial distribution system using multiple Static Synchronous Compensators (STATCOMs) and Static Capacitors (SCs). The proposed technique depends on injecting controlled reactive currents at certain nodes in the system in order to minimize the total line loss. The injected reactive currents compensate the reactive currents required by the inductive loads supplied from the system and hence reduce the feeders currents and the total line loss in the distribution system. The optimal number of the installed static capacitor should be considered in order to further reduce the total line loss. Therefore, the relation between the input number of static capacitors and the reactive current line loss in conjunction with the reactive current control by STATCOMs is derived. The effectiveness of the proposed control scheme has been verified experimentally using a laboratory prototype system.

Keywords—STATCOM, static capacitor, reactive current compensation, line loss minimization, shunt compensators.

I. INTRODUCTION

Concerns over global warming have led to an average increase in reducing the emissions of CO_2 by replacing the energy sources based fossil fuels with that based on natural renewable energies in addition to reducing the power loss in distribution networks since it has a high ratio of X/R compared with the transmission one [1]. Therefore, the operation and design standards of distribution networks have recently been changed to consider the energy saving by line loss minimization. However, the spread of distributed generation based renewable energies such as wind and solar in the distribution systems complicate the power flow calculations and affect the power quality. Its main disadvantages is that the node voltages may exceed the permissible limits during light load conditions [2]. Therefore, a new distribution system generation has been emerged named Smart Grid [3], which has the ability to monitor and control the power flow from the load side and supply side by using FACTS series and shunt power electronics devices in order to maintain the load voltages within the permissible voltage limit [4][5].

Distribution system has been classified as radial and loop systems. Although, the loop system has better voltage profile and loss, Its protection scheme is very expensive. Moreover, it is very difficult to decide the fault location in the loop distribution system [6]. Therefore, in Japan the radial distribution system has been used due to its simple and inexpensive protection scheme. Moreover, in the radial distribution system detection and recovery of the faulted part are very fast. However, the node voltage at the feeder far ends is usually low due to the line voltage drop along the feeders. Therefore, smart-grid have full monitor and control of the load voltage.

Much of recent research in distribution system has been focused on line loss minimization and load voltage regulation by using FACTS devices based power electronic components [7]. The authors have obtained the line loss minimum conditions in loop distribution system by eliminating the loop current from the loop system using UPFC [8]. Moreover, the authors have presented regulations of all node voltages simultaneously with minimizing the line loss of a loop distribution system using UPFC [9]. Back-to-Back converter has been used to achieve line loss minimization in distribution systems by controlling the power flow of two-adjacent feeders [10]. A comparison between shunt and series compensators in controlling the loop distribution system has been achieved. Also, the authors have already proposed line loss minimization control method to control reactive line current by the STATCOMs in radial distribution system considering multiple STATCOMs [11]-[14]. The main advantages of compensating the reactive currents in radial system is the availability of achieving unity power factor at the substation. Also, static capacitors have been considered as a simple method that improve the power factors and reduces the line loss [15]. However, the starting current in addition to its uncontrollable values and the life-time are considered the main drawbacks.

This paper presents an new control technique for controlling the existing radial distribution system in order to achieve line loss minimization using hybrid shunt compensators based STATCOMs and static capacitors. The insertion of these shunt compensator has the ability to compensate the reactive power in the radial distribution

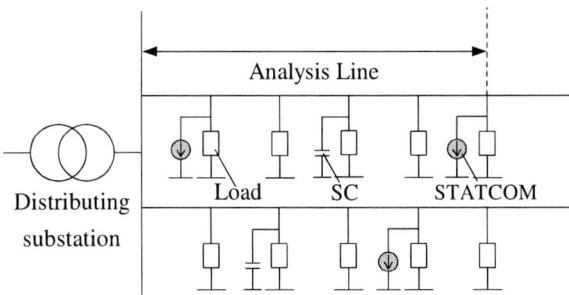

Fig. 1.   Model of Radial distribution system.

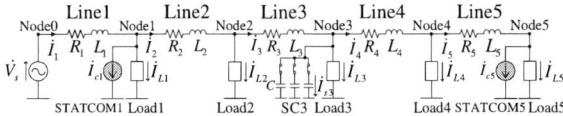

Fig. 2.   Analytical radial distribution model.

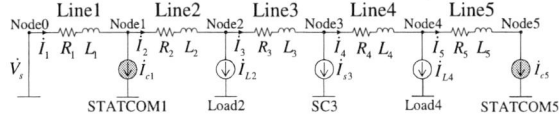

Fig. 3.   Approximate radial distribution model.

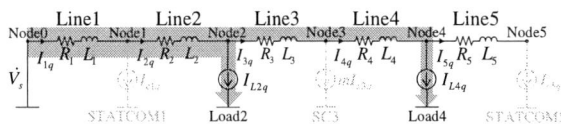

(a) Reactive current flows of the loads.

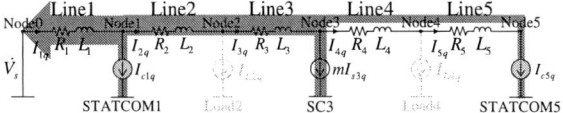

(b) Reactive current floes of STATCOMs and SCs.

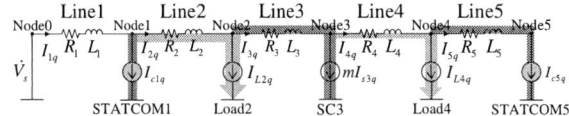

(c) Reactive current flows after the control.

Fig. 4.   Reactive current flows before and after the control.

system and hence minimize the total line loss by providing the inductive loads with the required reactive power, which in-turn will reduce the reactive current component along the distribution feeders. The selection method of the optimal installed capacitor number in conjunction with the reactive current control by STATCOMs is proposed. The effectiveness of the proposed control technique has been verified experimentally using a laboratory prototype system. Experimental results shows that the proposed control technique has the ability to achieve line loss minimization in radial distribution system. Moreover, compensator locations obtained by the proposed mathematical model are the optimal ones.

II. LINE LOSS IN RADIAL DISTRIBUTION SYSTEM

Fig.1 shows a schematic diagram of radial distribution model. Fig.2 shows one feeder of radial distribution system, shown in Fig.1. The feeder shown in Fig.2 consists of five-lines feeding five loads in addition to two STATCOMs and static capacitor bank unit. Each line has been model as a series resistance $R_1 \sim R_5$ and an inductance $L_1 \sim L_5$. The substation has been considered as a voltage source V_s. Each STATCOM has been considered as a current source I_c. The main function of the installed STATCOMs and static capacitors is to achieve line loss minimization in the radial feeder in addition to unity power factor at the substation.

A. Analytical Model

Fig.2 shows an analytical model of a radial distribution system that has five lines; Line $1 \sim 5$. In the distribution system, SCs and two STATCOMs, represented as current sources, are connected. In the analytical model

of Fig.2, Load $1 \sim 5$ are connected to Node $1 \sim 5$, STATCOM 1 and STATCOM 5 are connected to Node 1 and Node 5, respectively. At Node 3, three-parallel static capacitors are connected through three switches. The parallel capacitors have the same capacitance C. The compensation currents $\dot{I}_{c1}$ and $\dot{I}_{c5}$ of the STATCOMs are given as follows.

$$\dot{I}_{cn} = jI_{cnq} \quad (n=1,5) \tag{1}$$

Fig.3 shows the approximated distribution model. In Fig.3, the loads on the point of installation of the S-TATCOMs and SCs are considered as unity power factor. Therefore the loads at these nodes can be ignored for simplification. Load 2 and Load 4 are replaced with the current source $\dot{I}_{L2}$ and $\dot{I}_{L4}$, whereas the capacitor bank SC 3 is expressed by the current source $\dot{I}_{s3}$ for m capacitors. The shunt compensator currents $\dot{I}_{L2}, \dot{I}_{L4}$ and $\dot{I}_{s3}$ can be formulated as follows:

$$\dot{I}_{Ln} = I_{Lnp} + jI_{Lnq} \quad (n=2,4) \tag{2}$$
$$\dot{I}_{s3} = jmI_{s3q} \quad (m=0,1,2,3) \tag{3}$$

B. Before Installing the Shunt Compensators

Fig.4 shows the reactive currents that flow in the radial distribution system model before and after installation of the STATCOMs and the capacitor bank. Before the installation, the reactive currents of Load 2 and Load 4 I_{L2q} and I_{L4q} are supplied directly from the substation voltage $\dot{V}_s$, as shown in Fig.4 (a). In this case, the line current $\dot{I}_{01} \sim \dot{I}_{05}$ before installing of STATCOMs and static capacitors can be formulated as follows:

$$\left. \begin{aligned} \dot{I}_{01} &= I_{01p} + jI_{01q} = \dot{I}_{L2} + \dot{I}_{L4} \\ \dot{I}_{02} &= I_{02p} + jI_{02q} = \dot{I}_{L2} + \dot{I}_{L4} \\ \dot{I}_{03} &= I_{03p} + jI_{03q} = \dot{I}_{L4} \\ \dot{I}_{04} &= I_{04p} + jI_{04q} = \dot{I}_{L4} \\ \dot{I}_{05} &= I_{05p} + jI_{05q} = 0 \end{aligned} \right\} \tag{4}$$

The p-axis line loss and the q-axis line loss can be divided because the active and reactive current are orthogonal relation. Therefore, the line loss before the control P_0 is divided into the p-axis current line loss P_{0p} and the q-axis current line loss P_{0q}. Each line current before the control from (4), P_0, P_{0p} and P_{0q} are given as follows.

Based on the line currents and the line parameters, the total line loss can be formulated as follows:

$$P_0 = R_1 \left| \dot{I}_{01} \right|^2 + R_2 \left| \dot{I}_{02} \right|^2 + R_3 \left| \dot{I}_{03} \right|^2$$
$$+ R_4 \left| \dot{I}_{04} \right|^2 + R_5 \left| \dot{I}_{05} \right|^2$$
$$= (R_1 + R_2) \left| \dot{I}_{L2} + \dot{I}_{L4} \right|^2 + (R_3 + R_4) \left| \dot{I}_{L4} \right|^2 \quad (5)$$

Using The Park/Clark transformation, the line currents of the radial system can be divided into p-q components. Considering the p-q axis current components in line 2 and Line 4, the total power loss can be formulated as follows:

$$P_0 = P_{0p} + P_{0q} \quad (6)$$

$$P_{0p} = (R_1 + R_2)(I_{L2p} + I_{L4p})^2$$
$$+ (R_3 + R_4) I_{L4p}^2 \quad (7)$$

$$P_{0q} = (R_1 + R_2)(I_{L2q} + I_{L4q})^2$$
$$+ (R_3 + R_4) I_{L4q}^2 \quad (8)$$

C. After Installing the Shunt Compensators

Fig.4 (b) shows the compensating currents of STAT-COMs and capacitor bank that are injected to the radial distribution system model. It is cleared that these injected currents are leading currents in order to compensate the load lagging currents. Therefore, the summation of both currents, shown in Fig.4 (a) and (b), results in the line currents in radial system with the effect of STATCOMs and capacitor bank current as shown in Fig.4 (c). It is cleared that the total currents along the feeder will reduced after installing the shunt compensators. Also, the installed shunt compensator will compensate all the reactive current supplied by the substation, and hence unity power factor at the substation side will be achieved. Therefore, the line current $\dot{I}_1 \sim \dot{I}_5$ after installing the shunt compensators of STATCOMs and capacitors can be formulated as follows:

$$\left.\begin{aligned}
\dot{I}_1 &= I_{1p} + jI_{1q} = \dot{I}_{L2} + \dot{I}_{L4} + \dot{I}_{c1} + \dot{I}_{s3} + \dot{I}_{c5} \\
\dot{I}_2 &= I_{2p} + jI_{2q} = \dot{I}_{L2} + \dot{I}_{L4} + \dot{I}_{s3} + \dot{I}_{c5} \\
\dot{I}_3 &= I_{3p} + jI_{3q} = \dot{I}_{L4} + \dot{I}_{s3} + \dot{I}_{c5} \\
\dot{I}_4 &= I_{4p} + jI_{4q} = \dot{I}_{L4} + \dot{I}_{c5} \\
\dot{I}_5 &= I_{5p} + jI_{5q} = \dot{I}_{c5}
\end{aligned}\right\} \quad (9)$$

The active currents do not change before and after installing the shunt compensators because the compensating currents of STATCOMs and capacitors are only reactive currents. Therefore, the power loss based active power components P_p will not change, where as the power loss based reactive power component P_q will be changes according to the change in the q-axis line currents.

According to the line current given in (9), the line loss after installing the shunt compensators can be formulated based on the p-q axis line current components as follows:

$$P = R_1 \left| \dot{I}_1 \right|^2 + R_2 \left| \dot{I}_2 \right|^2 + R_3 \left| \dot{I}_3 \right|^2$$
$$+ R_4 \left| \dot{I}_4 \right|^2 + R_5 \left| \dot{I}_5 \right|^2$$
$$= P_p + P_q \quad (10)$$

$$P_p = P_{0p} \quad (11)$$

$$P_q = R_1 (I_{L2q} + I_{L4q} + I_{c1q} + mI_{s3q} + I_{c5q})^2$$
$$+ R_2 (I_{L2q} + I_{L4q} + mI_{s3q} + I_{c5q})^2$$
$$+ R_3 (I_{L4q} + mI_{s3q} + I_{c5q})^2$$
$$+ R_4 (I_{L4q} + I_{c5q})^2 + R_5 I_{c5q}^2 \quad (12)$$

III. ANALYSIS OF COMPENSATING CURRENTS

In this section, the compensating currents of STAT-COMs and static capacitors for minimize the line loss in radial distribution system are derived. The compensating current of the static capacitors is a stepwise value based on the number of the connected capacitors. Therefore, the static capacitor compensating current is only supporting the STATCOMs to achieve the main objective. Based on that, the controlled STATCOMs reactive current is used to optimize the power loss in the radial distribution system model. The capacity of the installed STATCOMs is also optimized by the installed static capacitors. Therefore, the derivation of the power loss minimization in radial distribution system will be based on the analysis of the STATCOMs injected currents.

A. Analysis of STATCOMs Compensating Current

Since the injected power of the STATCOMs and the static capacitors is only reactive power, the p-axis currents in each line of the radial distribution system in addition to the related power loss given in (11) will not change. The STATCOMs currents and the static capacitors currents are only q-axis current and will affect only the q-axis currents flowing along the feeder in addition to the related power loss given in (12). It is cleared that the q-axis related power loss P_q is a function of the STATCOMs currents I_{c1q} and I_{c5q}. Therefore, line loss minimization in the radial system can be formulated based on the STATCOMs currents as follows:

$$\left.\begin{aligned}
\frac{\partial P_q}{\partial I_{c1q}} &= 2R_1 (I_{L2q} + I_{L4q} + I_{c1q} + mI_{s3q} + I_{c5q}) = 0 \\[2mm]
\frac{\partial P_q}{\partial I_{c5q}} &= 2R_1 (I_{L2q} + I_{L4q} + I_{c1q} + mI_{s3q} + I_{c5q}) \\
&\quad + 2R_2 (I_{L2q} + I_{L4q} + mI_{s3q} + I_{c5q}) \\
&\quad + 2R_3 (I_{L4q} + mI_{s3q} + I_{c5q}) \\
&\quad + 2R_4 (I_{L4q} + I_{c5q}) + 2R_5 I_{c5q} = 0
\end{aligned}\right\} \quad (13)$$

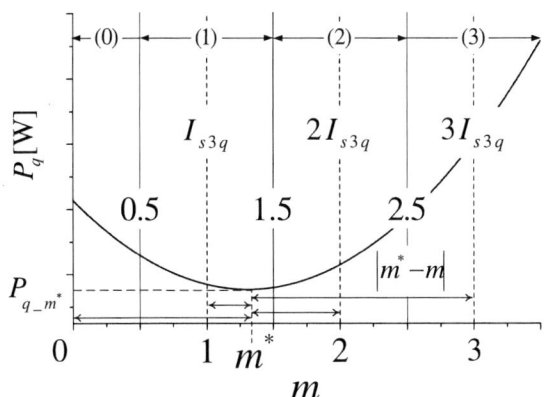

Fig. 5. Relation between the input number of the SC and the reactive current line loss.

According to (13), the injected STATCOMs currents I_{c1q} and I_{c5q} for minimizing the line loss are given as follows:

$$I_{c1q} = -\frac{\sum\limits_{i=3}^{5} R_i I_{L2q} + \sum\limits_{i=4}^{5} R_i m I_{s3q} + R_5 I_{L4q}}{\sum\limits_{i=2}^{5} R_i} \quad (14)$$

$$I_{c5q} = -\frac{R_2 I_{L2q} + \sum\limits_{i=2}^{3} R_i m I_{s3q} + \sum\limits_{i=2}^{4} R_i I_{L4q}}{\sum\limits_{i=2}^{5} R_i} \quad (15)$$

According to (14) and (15), all load reactive currents I_{L2q}, I_{s3q} and I_{L4q} can be compensated from the S-TATCOMs currents resulting in zero reactive current supplied from the substation. Based on the line parameters of the radial distribution system, STATCOMs share in compensating the load reactive currents. For example, the reactive current of Load 2 I_{L2q} is compensated from the STATCOM 1 side by the $R_3 + R_4 + R_5$ to summation of line resistances ratio, and from the STATCOM 2 side by R_2 to summation of line resistances ratio. This results in compensating load 2 reactive current by 100%.

B. Compensating Current and Installation number of SC

The line loss minimization can be achieved if the S-TATCOMs currents injected to the distribution system are as given in (14) and (15). Therefore, the reactive current line loss for line loss minimization can be obtained by substituting (14) and (15) into (12), as follows:

$$P_q = \frac{\sum\limits_{i=2}^{3} R_i \sum\limits_{i=4}^{5} R_i}{\sum\limits_{i=2}^{5} R_i} \left(m I_{s3q} + \frac{R_2}{\sum\limits_{i=2}^{3} R_i} I_{L2q} + \frac{R_5}{\sum\limits_{i=4}^{5} R_i} I_{L4q} \right)^2$$
$$+ \frac{R_2 R_3}{\sum\limits_{i=2}^{3} R_i} I_{L2q}^2 + \frac{R_4 R_5}{\sum\limits_{i=4}^{5} R_i} I_{L4q}^2 \quad (16)$$

where m is the number of the connected shunt capacitors.

The capacity of installed capacitors should be optimized, since the reactive current line loss given in (16) depends on the installed capacitor number. Therefore, the optimal capacitor number installed in the distribution system can be formulated as follows:

$$m^* = -\frac{\dfrac{R_2}{R_2 + R_3} I_{L2q} + \dfrac{R_5}{R_4 + R_5} I_{L4q}}{I_{s3q}} \quad (17)$$

According to (17), the compensating currents of STAT-COMs to realize the line loss minimization, given in (14) and (15), can be formulated as follows:

$$\left. \begin{array}{l} I_{c1q} = -\dfrac{R_3}{R_2 + R_3} I_{L2q} \\[3mm] I_{c5q} = -\dfrac{R_4}{R_4 + R_5} I_{L4q} \end{array} \right\} \quad (18)$$

From coefficient of the reactive current I_{L2q} and I_{L4q} in (17) and (18), the reactive current of the load is completely compensated by two compensators which exist closest to the left and right sides of the load. From (16), if the number of installed shunt capacitors is m^*, the reactive current line loss during line loss minimization $P_{q\_m^*}$ is given as follows:

$$P_{q\_m^*} = \frac{R_2 R_3}{R_2 + R_3} I_{L2q}^2 + \frac{R_4 R_5}{R_4 + R_5} I_{L4q}^2 \quad (19)$$

Substituting (17) and (19) into (16), the reactive current line loss P_q is given as follows:

$$P_q = (m^* - m)^2 \frac{\sum\limits_{i=2}^{3} R_i \sum\limits_{i=4}^{5} R_i}{\sum\limits_{i=2}^{5} R_i} I_{s3q}^2 + P_{q\_m^*} \quad (20)$$

According to (20), the reactive current line loss P_q is determined by the difference between the optimal number of installed static capacitors and the actual one $|m^* - m|$. Fig.5 shows the relation between the reactive current line loss P_q and the actual installed number of static capacitors m. The minimum point on the curve indicting the optimal number of the installed shunt capacitor. Therefore, the between the optimal and actual number of installed shunt capacitors $|m^* - m|$ should be minimized in order to achieve minimum line loss. For example, if the the range of $m^* \leq 0.5$, the actual input number of the installed static capacitor is $m = 0$ in order to achieve reactive current line loss minimization. In addition, if the optimal number of the installed capacitors equals $m^* = 0.5$, the actual installed capacitors can be $m = 0$ or $m = 1$ for obtaining line loss minimization. If the optimal number is in the range of $1 < m^* < 1.5$, the actual installed capacitors should be $m = 1$. Therefore, selection of the actual number of the installed shunt capacitors m is executed as follows:

$$
\begin{array}{llll}
0) & m^* \leq 0.5 & & m = 0 \\
1) & 0.5 \leq m^* \leq 1.5 & & m = 1 \\
2) & 1.5 \leq m^* \leq 2.5 & & m = 2 \\
3) & 2.5 \leq m^* & & m = 3
\end{array}
$$

The 2014 International Power Electronics Conference

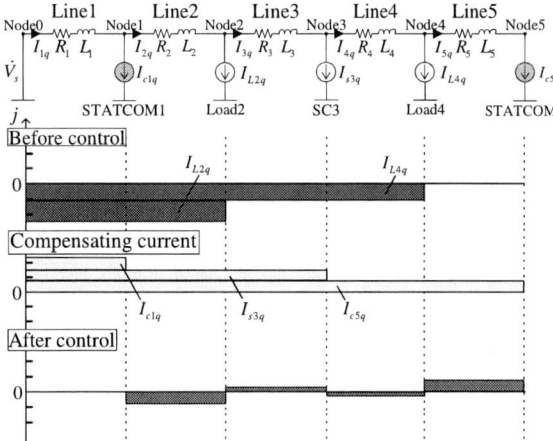

Fig. 6. Changes of the q-axis line current by compensation.

The optimal number of the installed shunt capacitors m^* is determined by the line parameters in addition to the reactive current of loads as given in (17). Line loss minimization is realized if the input number of the installed capacitors m has been selected according to the optimal number m^* as given in the above conditions.

C. Line Reactive Current Calculations

If the input number of installed shunt capacitors $m = m^*$, the resultant reactive currents in each line of the distribution system can be formulated by substituting the compensating currents given in (17) and (18) into the line currents given in (9) as follows:

$$\left.\begin{aligned} I_{1q} &= I_{01q} + I_{c1q} + mI_{s3q} + I_{c5q} = 0 \\ I_{2q} &= I_{02q} + mI_{s3q} + I_{c5q} = \frac{R_3}{R_2 + R_3} I_{L2q} \\ I_{3q} &= I_{03q} + mI_{s3q} + I_{c5q} = -\frac{R_2}{R_2 + R_3} I_{L2q} \\ I_{4q} &= I_{04q} + I_{c5q} = \frac{R_5}{R_4 + R_5} I_{L4q} \\ I_{5q} &= I_{05q} + I_{c5q} = -\frac{R_4}{R_4 + R_5} I_{L4q} \end{aligned}\right\} \quad (21)$$

Fig.6 shows the reactive currents flow before and after installing the shunt compensators of STATCOMs and static capacitors. Before installing the shunt compensators, the negative reactive currents, absorbed by the inductive loads, flow through the line 1, 2, 3 and 4 as shown in (4). Considering only shunt compensators installed in the distribution system, the injected positive reactive currents flow through line 1, 2, 3, 4 and 5 as shown in (17) and (18). After installing the shunt compensators, the reactive currents flow through the distribution system lines as given in (21). Fig.6 shows that the resultant reactive currents decrease as compared with the reactive currents flows in the distribution system before installing the shunt compensators. Also, it is cleared that the reactive current in line 1 I_{1q} is completely compensated, resulting in unity power factor at the substation side.

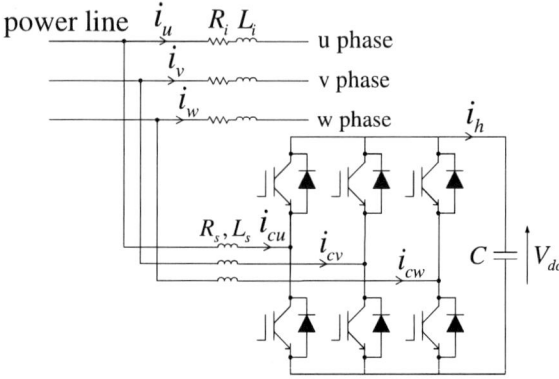

Fig. 7. Configuration of the STATCOM.

In this model, the line loss reduction effect increases if line resistance R_1 from the substation to STATCOM 1 increases. According to (20), the change in reactive current line loss ΔP_q due to the installation of the shunt compensators is given as follows:

$$\Delta P_q = P_{q\_m} - P_{q\_m^*}$$
$$= (m^* - m)^2 \frac{(R_2 + R_3)(R_4 + R_5)}{R_2 + R_3 + R_4 + R_5} I_{s3q}^2 \quad (22)$$

In the range of $-0.5 \leq m^* \leq 3.5$, the difference between the optimal and actual number of the installed static capacitors is always $|m^* - m| \leq 0.5$. Therefore, the range of change in reactive current line loss given in (22) after installing the shunt compensators is given as follows:

$$0 \leq \Delta P_q \leq \frac{1}{4} \frac{(R_2 + R_3)(R_4 + R_5)}{R_2 + R_3 + R_4 + R_5} I_{s3q}^2 \quad (23)$$

The change reactive current line loss ΔP_q due to the difference between the actual and optimal number of shunt static capacitors is small as shown in (23).

IV. CONTROL OF STATCOM

Fig.7 shows configuration of the STATCOM. The DC capacitor voltage is controlled by using the active current component, whereas the reactive current of the distribution system is controlled by using the reactive current component. The proposed control of STATCOMs is realized by a DSP (Digital Signal Processor, TM-S320C6713, TI) which calculates the new reference value in each control period based on the control algorithm using the detected voltages and line currents. Although, the reactive currents of Load 2 and 4 I_{L2q} and I_{L4q} are used to calculate the reference value in (17) and (18), the line currents of the distribution system model is used in the experimental system in order to avoid load variation effects on the controller. Therefore, the reference values of the shunt compensators given in (17) and (18) can be

978-1-4799-2706-7/14 $31.00 © 2014 IEEE 605

The 2014 International Power Electronics Conference

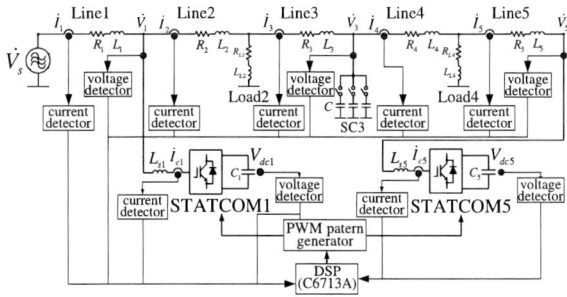

Fig. 8. Experimental system configuration.

formulated as follows:

$$
\left.
\begin{aligned}
I_{c1q}^* &= -\frac{R_3}{R_2 + R_3}\left(I_{2q} - I_{3q}\right) \\
m^* &= -\frac{\dfrac{R_2}{R_2+R_3}\left(I_{2q}-I_{3q}\right) + \dfrac{R_5}{R_4+R_5}\left(I_{4q}-I_{5q}\right)}{I_{3q}-I_{4q}} \\
I_{c5q}^* &= -\frac{R_4}{R_4 + R_5}\left(I_{4q} - I_{5q}\right)
\end{aligned}
\right\} \quad (24)
$$

Therefore, the controller is used to control the injected currents of both STATCOMs to follow the reference values given in (24).

V. EXPERIMENTAL RESULT

A. Experimental System configuration

Fig.8 shows the 3kVA-200V three-phase three-wire laboratory prototype system configuration that has five-lines Line 1 ∼ 5 supplied from the voltage source $\dot{V}_s$, which is same as the distribution system model shown in Fig.2. In this system, STATCOM 1 and STATCOM 5 are connected to Node 1 and 5, respectively, Load 2 and 4 are connected to Node 2 and 4, respectively, and three-shunt static capacitors are connected to Node 3. Load 1, 3 and 5 are ignored. TableI shows the parameters of the prototype system. The STATCOMs are connected to the distribution system lines through shunt transformers, which are ignored in the experimental system model since the focus of this research is only for the power loss in the distribution system lines. The loads are considered as inductive loads representing by constant resistance $R_{Ln}(n = 2, 4)$ and inductance $L_{Ln}(n = 2, 4)$, with a power factor of 0.85. The static capacitor bank consists of three capacitors with same capacitances C that is connected to the system through switches. The laboratory prototype has been carried out in three modes. First, the system has been carried out before installing any shunt compensator. Second, the experimental system has been carried out with the effects of the two STATCOMs only for line loss minimization. Finally, the system has been carried out with the effects of installing the shunt capacitors in addition to the the STATCOMs. TableII shows the theoretical values of the reactive currents and the optimal value of the connected shunt capacitors. From TableII it is cleared that minimization of the reactive current line loss is realized by selecting the actual number of the static capacitors $m = 1$, since the optimal value is $m^* = 1.22$.

TABLE I.   EXPERIMENTAL PARAMETERS.

Send-ending voltage V_s	200 V , 60 Hz
Load2($\dot{Z}_{load2}$) R_{load2}, L_{load2}	21.90 Ω , 36 mH
Load4($\dot{Z}_{load4}$) R_{load4}, L_{load4}	24.33 Ω , 40 mH
Static capacitor $3 \times C$	$3 \times 30\ \mu F$
Line1($\dot{Z}_1$) R_1, $L_1(X_{L1})$	0.66 Ω , 2.0 mH (0.75 Ω)
Line2($\dot{Z}_2$) R_2, $L_2(X_{L2})$	0.17 Ω , 0.5 mH (0.19 Ω)
Line3($\dot{Z}_3$) R_3, $L_3(X_{L3})$	0.33 Ω , 1.0 mH (0.38 Ω)
Line4($\dot{Z}_4$) R_4, $L_4(X_{L4})$	0.33 Ω , 1.0 mH (0.38 Ω)
Line5($\dot{Z}_5$) R_5, $L_5(X_{L5})$	0.17 Ω , 0.5 mH (0.19 Ω)
Input inductance L_{s1}, L_{s5}	8.0 mH , 8.0 mH
Electrolytic capacitor C_1, C_5	3000 μF, 3000 μF
DC link voltage V_{dc1}, V_{dc5}	300 V, 290 V
Control period T_s	160 μs

TABLE II.   THEORETICAL VALUES OF THE REACTIVE CURRENT AND OPTIMAL VALUES OF THE STATIC CAPACITOR.

Number (m)	Reactive current value mI_{s3q} [A]	Optimal number (m^*)	Optimal value $m^* I_{s3q}$ [A]
1	2.04		
2	4.09	1.22	2.50
3	6.13		

B. Consideration of Experimental Results

The laboratory prototype of the radial distribution system is carried out before and after installing the STATCOMs and the static capacitors. The line currents of each line is transformed from the three-phase to p-q axis using the Park/Clark transformation. The p-axis current represents the active current component, whereas the q-axis current represents the reactive current component. The substation voltage phase angle is considered as the reference for the whole system. The compensating current references of the two STATCOMs I_{c1q}^* and I_{c5q}^* in addition to the compensating current optimal value $m^* I_{s3q}$ of the shunt capacitors are calculated based on the line currents. Fig.9 shows the experimental waveforms of the laboratory prototype before and after installing the shunt compensators. The waveforms show the STATCOM 1 injected current I_{c1q}, STATCOM 5 injected current I_{c5q}, optimal and actual static capacitor currents I_{s3q}, and the p-q axis line current components of the five lines of the distribution system model, respectively. The DC capacitor voltages V_{dc1} and V_{dc5} are controlled to be constant by using the active currents I_{c1p} and I_{c5p}.

At the first 100 ms, from the starting time till T_0, the system has been carried out without the effect of the STATCOMs and the static capacitors. Therefore, the shunt compensating currents I_{c1q}, I_{c5q} and I_{s3q} are zero. The active and reactive line current components in the laboratory prototype are shown in the waveforms of $I_{1q} \sim I_{5q}$. Since the loads are inductive, the reactive current components flowing in the distribution system lines are negative values.

At the time from T_0 till T_1, only the reactive current control of the two STATCOMs starts. Therefore, the compensating current of the static capacitors I_{s3q} is zero. since the two STATCOMs currents I_{c1q} and I_{c5q} are injected to compensate the reactive current components absorbed by the inductive loads, the injected currents are positive values. Fig.9 shows that after installing the two STATCOMs, the reactive current components of line 1∼ 4, represented by $I_{1q} \sim I_{4q}$, are reduced.

978-1-4799-2706-7/14 $31.00 © 2014 IEEE   606

The 2014 International Power Electronics Conference

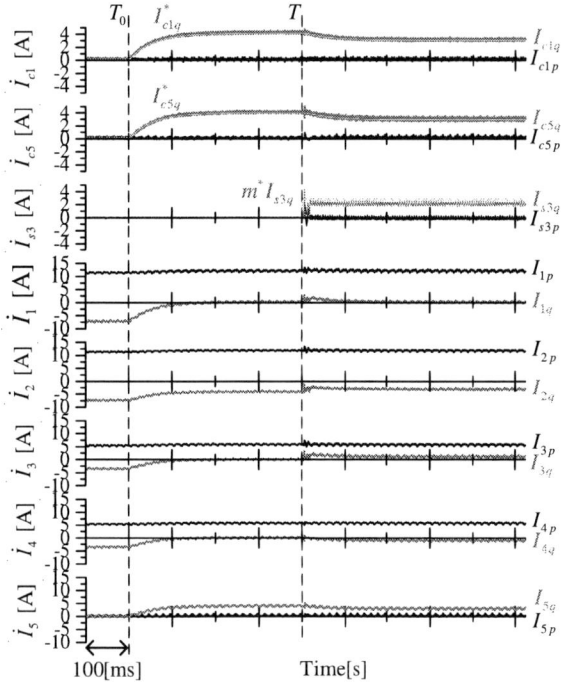

Fig. 9.   Experimental waveforms.

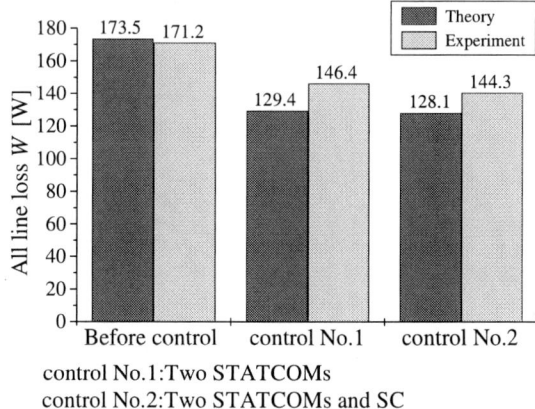

control No.1:Two STATCOMs
control No.2:Two STATCOMs and SC

Fig. 10.   Experimental results of all line loss using Power Meters.

TABLE III.   THEORETICAL AND DETECTED VALUES OF EACH
CURRENT BEFORE AND AFTER INSTALLING SC.

Theoretical Values			
	before control	Two STATCOMs	SC
STATCOM1 $\dot{I}_{c1\_V1}$	$0.00 + j0.00$A	$0.00 + j3.83$A	$0.00 + j2.80$A
STATCOM5 $\dot{I}_{c5\_V5}$	$0.00 + j0.00$A	$0.00 + j3.53$A	$0.00 + j2.50$A
SC3 $\dot{I}_{s3\_Vs}$	$0.00 + j0.00$A	$0.00 + j0.00$A	$0.06 + j2.04$A
Load2 $\dot{I}_{L2\_Vs}$	$5.93 - j3.90$A	$5.93 - j3.90$A	$5.93 - j3.90$A
Load4 $\dot{I}_{L4\_Vs}$	$5.17 - j3.45$A	$5.17 - j3.45$A	$5.17 - j3.45$A
Line1 $\dot{I}_{1\_Vs}$	$11.1 - j7.35$A	$11.3 - j0.00$A	$11.3 - j0.00$A
Line2 $\dot{I}_{2\_Vs}$	$11.1 - j7.35$A	$11.3 - j3.83$A	$11.3 - j2.81$A
Line3 $\dot{I}_{3\_Vs}$	$5.17 - j3.45$A	$5.29 + j0.07$A	$5.31 + j1.10$A
Line4 $\dot{I}_{4\_Vs}$	$5.17 - j3.45$A	$5.29 + j0.07$A	$5.25 - j0.95$A
Line5 $\dot{I}_{5\_Vs}$	$0.00 + j0.00$A	$0.12 + j3.53$A	$0.08 + j2.50$A
Experimental Values			
	before control	Two STATCOMs	SC
STATCOM1 $\dot{I}_{c1\_V1}$	$0.25 + j0.23$A	$0.16 + j4.28$A	$0.23 + j3.21$A
STATCOM5 $\dot{I}_{c5\_V5}$	$0.21 + j0.16$A	$0.14 + j4.12$A	$0.23 + j3.04$A
SC3 $\dot{I}_{s3\_Vs}$	$-0.01 + j0.01$A	$0.01 - j0.00$A	$-0.07 + j2.14$A
Load2 $\dot{I}_{L2\_Vs}$	$5.95 - j3.90$A	$6.04 - j4.19$A	$6.14 - j4.20$A
Load4 $\dot{I}_{L4\_Vs}$	$5.28 - j3.60$A	$5.40 - j3.97$A	$5.39 - j3.98$A
Line1 $\dot{I}_{1\_Vs}$	$11.5 - j7.15$A	$12.2 + j0.25$A	$12.2 + j0.19$A
Line2 $\dot{I}_{2\_Vs}$	$11.4 - j7.35$A	$11.9 - j4.05$A	$11.9 - j3.01$A
Line3 $\dot{I}_{3\_Vs}$	$5.45 - j3.45$A	$5.86 + j0.14$A	$5.76 + j1.19$A
Line4 $\dot{I}_{4\_Vs}$	$5.46 - j3.46$A	$5.86 + j0.15$A	$5.83 - j0.95$A
Line5 $\dot{I}_{5\_Vs}$	$0.18 + j0.14$A	$0.46 + j4.12$A	$0.44 + j3.03$A

At the time from T_1 till the end of experimental running time, the static capacitor is installed at node 3 in addition to the two STATCOMs installed at node 1 and 5. From the theoretical results given in TableII and the experimental waveform of the shunt capacitor in Fig.9, only one capacitor C is connected to the distribution system since the optimal value is $m^* = 1.22$. The compensating current of the static capacitors I_{s3q} in addition to its optimal values are nearly same. Fig.9 shows that after installing the two STATCOMs in addition to the static capacitor, the reactive current components of line 1 I_{1q} is completely compensated. Also, the reactive current components of line 2 and 5 represented by I_{2q} and I_{5q}, are reduced. The absolute values of the reactive currents on the Line 3 and 4 I_{3q} and I_{4q} are increased slightly. Moreover, the two STATCOMs currents I_{c1q} and I_{c5q} are reduced after installing the shunt capacitor.

The bar-chart shown in Fig.10 presents a comparison between the theoretical and experimental results of all the line loss in the distribution system before and after installing the shunt compensators. The experimental values of the line loss are obtained by the Digital Power Meters (Yokogawa Electric Corporation, WT1600). Before installing the shunt compensators, the total line loss in the distribution system model is theoretically 173.5 W and experimentally 171.2 W. After installing the two shunt STATCOMs, the injected reactive currents of them reduces the total line loss to be theoretically 129.4 W and experimentally 146.4 W. Installing the two STATCOMs reduces the total line loss of the distribution system model by 14.5 %. After installing the shunt capacitor in addition to the two STATCOMs, the total line loss in the distribution system model is theoretically 128.1 W

and experimentally 144.3 W. The total line loss after the control of STATCOMs and static capacitor is reduced by 15.7 % from its original value before installing the shunt compensators. The line loss in experimental results after the control increase. Although the load currents approximated to the current source does not change before and after the control in theory, the load currents increase in experiment because the line currents are reduced and the node voltages are increased by reactive current injection of STATCOMs and static capacitor to the distribution system. TableIII shows the theoretical and experimental results of the line currents before and after installing the shunt compensators. The theoretical values are derived considering the load current $\dot{I}_{L2}$ and $\dot{I}_{L4}$ to be constant. The line currents, the load currents and the compensating current of the shunt capacitor are converted the p-q axis on the basis of power supply voltage $\dot{V}_s$, whereas the compensating current of STATCOMs are converted to the p-q axis on the basis of each node voltages $\dot{V}_1$ and $\dot{V}_5$. From TableIII, the experimental results are almost coincident with the theoretical values.

978-1-4799-2706-7/14 $31.00 © 2014 IEEE

The 2014 International Power Electronics Conference

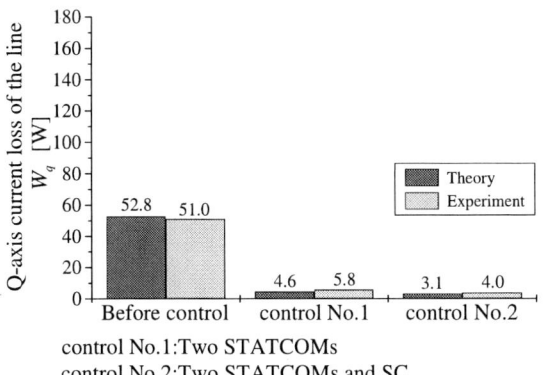

control No.1:Two STATCOMs
control No.2:Two STATCOMs and SC

Fig. 11. Experimental results of the q-axis current line loss using detection values.

The bar-chart shown in Fig.11 presents a comparison between the theoretical and experimental results of the reactive current line loss before and after the control. Before installing the shunt compensators, the total reactive current line loss is theoretically 52.8 W and experimentally 51.0 W. After installing the two STATCOMs, the total reactive current line loss is theoretically 4.6 W and experimentally 5.8 W. The reactive current line loss is reduced by 88.6 % after installing the two STATCOMs. After installing the static capacitors in addition to the two STATCOMs, the total reactive current line loss is theoretically 3.1 W and experimentally 4.0 W. Therefore, the reactive current line loss after the control of STATCOMs and static capacitor is reduced by 92.2 % compared with its original value before installing the shunt compensators. Experimental and theoretical results prove the effectiveness of the proposed control in reduction of all the line loss in the radial distribution system.

VI. CONCLUSION

This paper has presented the line loss minimization control of radial distribution system using hybrid shunt compensators of STATCOMs and static capacitors. The theoretical analysis and experimental validation of the proposed control are described. Line loss can be divided into the active and reactive current line loss. According to the proposed method, it is possible to reduce only the reactive current line loss. The reactive current reference values of STATCOMs and the optimal number of installed static capacitors in the radial distribution system in order to minimize the line loss can be determined by the coefficients of the load reactive currents. The relation between the input number of static capacitors and the reactive current line loss in conjunction with the reactive current control by STATCOMs is derived. The change in reactive current line loss between the optimal and actual number of the installed static capacitor is obtained by selecting the actual number closest to the optimal number. The effectiveness of the the proposed control method is verified experimentally and the experimental results match with the theoretical one.

VII. ACKNOWLEDGMENT

This research is partially supported by Grants-in-Aid for Scientific Research #24356187 in 2012-2013.

REFERENCES

[1] Hiromu Kobayashi, "Development Trend and Subjects of New Distribution Network Technology", *IEEJ Transactions on Power and Energy*, Vol. 124, No. 4, pp.517-520 (2004) (in Japanese).

[2] Mahmoud A. Sayed, and Takaharu Takeshita: "All Nodes Voltage Regulation and Line Loss Minimization in Loop Distribution Systems Using UPFC", *IEEE Trans. on PELS*, Vol.26, No.6, pp.1694-1703 (2011).

[3] Akihiko Yokoyama, "Development of Smarter Grid (II)", *The Journal of The Institute of Electrical Engineers of Japan*, Vol. 130, No. 3, pp.163-167 (2010) (in Japanese).

[4] Taro Kondo, Jumpei Baba, Akihiko Yokoyama, "Voltage Control of Distribution Network with a Large Penetration of Photovoltaic Generations using FACTS Devices", *IEEJ Transactions on Power and Energy*, Vol. 126, No. 3, pp.347-358 (2006) (in Japanese).

[5] Noriyuki Kimura, Tatsuhito Nakajima, Naoki Gibo, "Trend of Control Technique of Power Electronic Equipments for Power System",*IEEJ Transactions on Power and Energy*, Vol.128-B, No.9, pp.1067-1070 (2008) (in Japanese).

[6] Naotaka Okada, Kosuke Kurokawa, "Control Tests of 6.6kV-100kVA BTB Type Loop Power Flow Controller",*IEEJ Transactions on Power and Energy*, Vol. 125, No. 4, pp.390-398 (2005) (in Japanese).

[7] Mahmoud A. Sayed, and Takaharu Takeshita: "Line Loss Minimization in Isolated Substations and Multiple Loop Distribution Systems Using UPFC", *IEEE Trans. on PELS*, 10.1109/T-PEL.2014.2301833 (2014).

[8] Mahmoud A. Sayed, Nobuyuki Inayoshi, Takaharu Takeshita and Fukashi Ueda: "Line Loss Minimization of Loop Distribution systems Using UPFC", *IEEJ Trans. on IA*, Vol.128, No.4, pp.508-515 (2008).

[9] Mahmoud A. Sayed and Takaharu Takeshita: "All Nodes Voltage Regulation and Line Loss Minimization in Loop Distribution Systems Using UPFC", *IEEE Energy Conversion Congress and Exposition (ECCE · 2009)*, California-USA, 2009.

[10] R. Simanjorang,Y. Miura, T. Ise, and S. Sugimoto: "A method for effective utilization of power distribution apparatus in industrial park by looping distribution lines", *IEEJ Trans. Power Energy*, vol. 127-B, pp. 1127 · 136, 2007.

[11] Kazuhei Yamada, Ryuta Saino, Takaharu Takeshita, "Line Loss Minimization Control of Distribution System Using Static Synchronous Compensator",*Annual Meeting Record IEE of Japan*, 4-417, (2010) (in Japanese).

[12] Ryota Asakura, Takaharu Takeshita, "Line Loss Minimization Control of Distribution System Using Multiple Sources of Reactive Current",*2012 Annual Conference of I.E.E of Japan,Industry Applications Society*, R-1-5-2, 1-101, pp.433-436 (2012) (in Japanese).

[13] Kensuke Miyazaki, Ryota Asakura, Takaharu Takeshita, "Line Loss Minimization Control of Loop Distribution System Using Multiple Reactive Current Sources",*Annual Meeting Record IEE of Japan*, 4-156, pp.267-268 (2013) (in Japanese).

[14] Kensuke Miyazaki, Takaharu Takeshita, "Line Loss Minimization Control of Distribution System using Multiple Reactive Current Sources and Static Capacitors",*2013 Annual Conference of I.E.E of Japan,Industry Applications Society*, R1-7, 1-49 (2013) (in Japanese).

[15] Hiroaki Uejima, Taichi Naruse, Mutsumi Aoki, Hiroyuki Ukai, Hiroshi Kobayashi, "Method for Improving Power Quality in Distribution System by Static Capacitors based on Power Line Loss", *IEEJ Transactions on Power and Energy*, Vol. 132, No. 3, pp.226-234 (2012) (in Japanese).

978-1-4799-2706-7/14 $31.00 © 2014 IEEE

The 2014 International Power Electronics Conference

A Novel Control Method for Individual DC Voltage Balancing in H-bridge Cascaded STATCOM

Rong Xu, Yong Yu, Rongfeng Yang, Lizhi Qu, Wei Sun, Dianguo Xu

Dept. of Electrical Engineering
Harbin Institute of Technology
Harbin, China

Abstract—**H-bridge cascaded static synchronous compensator (STATCOM) is often used to compensate the harmonic and reactive current derived from non-linear load. However, with the special topology structure, the voltage balancing of DC capacitors of H-bridge cascaded STATCOM is a key problem. Many papers adjust active voltage to solve this problem with the conventional PI controller which needs to regulate the complex parameters of PI and introduces the delay. Hence, this paper proposed a novel balance control method for individual DC voltage balancing based on the popular carrier phase-shifted sinusoidal PWM (CPS-SPWM) strategy. With the upper and lower translations of modulation wave in CPS-SPWM, the proposed method could more simply and flexibly balance the DC-voltages during the charging and discharging processes. Finally, the simulation and experiment results show the effectiveness of the proposed method in controlling individual capacitor voltages.**

Keywords—*H-bridge cascaded; static synchronous compensator; DC capacitor voltage; voltage balancing control; modulation wave translation*

I. INTRODUCTION

In recent years, with the development of the power electronic technology, non-linear and impact loads have been applied widely in the power system which leading to large generation of reactive power and harmonics. Rapid local reactive power and harmonics compensation with power quality conditioner has important practical significance to improve power factor and the stability of the grid, to reduce the capacity of powered devices [1-2].

Then, in the high-voltage, high-power application field, multilevel technology has become an effective and practical solution. Among various multilevel STATCOMs, the type of H-bridge cascaded inverter with separated capacitors has been most widely accepted[3]. However, with an isolated DC capacitor without any power source, the individual DC voltage balancing problem has been an inherent problem of H-bridge cascaded STATCOM which restricts the development and application.

In order to solve the problem of individual DC voltage balancing, the various strategies have been presented. Akagi [4] proposed a common control strategy for DC voltage balancing by means of regulating active voltage to control the STATCOM reference output voltage and achieve the individual DC voltage balancing. This method is easily to understand. However, in the process of implementation, the equipment with much cascaded units needs abundant of PI controllers. It would bring the problem of delay and the difficulty of regulating the complex parameters of PI. The article [5] achieves a balance control of the DC capacitor voltage by adjusting the phase-shift angle of every single unit in the early days. This approach can achieve a good effect by eliminating phase difference which will result in pulse delay. However, to the high power converters, the tiny changing range of phase shift angle will gain a larger modulation scope at the cost of reducing its stability. In [6], with the addition of circuit, the energy flows among the units through the DC and AC bus. The individual capacitors can exchange active power. Because of the extra circuits, the size and the cost of STATCOM will be increased and the control method will be complex. The article [7] presents the method of adjusting the pulse sequence which achieving the DC voltage balancing from the view of energy. Firstly, the power units are sorted according DC capacitor voltage levels. Then, according to the flowing direction of power, the switching state of the highest DC voltage level capacitor or the lowest DC voltage level capacitor will be changed. This method has a good balancing effect and is suitable for all kinds of the modulation algorithms. But, because of the different loss characteristics of the power units and the different switching frequency of each unit, it could lead frequently action in certain units and influence the working life of STATCOM.

In this paper, a simple and flexible balance control method for individual DC voltage balancing which based on the popular CPS-SPWM strategy was proposed. It is different from the methods above. With the upper and lower translations of modulation wave in CPS-SPWM, the proposed method would balance the DC-voltages more simply and flexibly during the charging and discharging processes. The simulation and experiment results show the effectiveness and superiority of the proposed method in controlling individual capacitor voltages.

Project 51237002 supported by NSFC; Sponsored by grants from the Power Electronics Science and Education Development Program of Delta Environmental & Educational Foundation（DREM2012001）.

978-1-4799-2706-7/14 $31.00 © 2014 IEEE

II. VOLTAGE BALANCING CONTROL METHOD BASED ON MODULATION WAVE TRANSLATION

A. Basic Structure and Principle of STATCOM

Fig.1 shows the topology of H-bridge cascaded STATCOM which adopts N H-bridges connecting in series in each phase and it can be expanded according to requirement. Three phases are built in star configuration. STATCOM was accessed to the grid by reactor. Directly controlling the current of STATCOM, it can absorb or provide required quantity of harmonic and reactive current to achieve the purpose of dynamic harmonic and reactive current compensation. Finally, the quality of the grid was improved and the grid offers active current only.

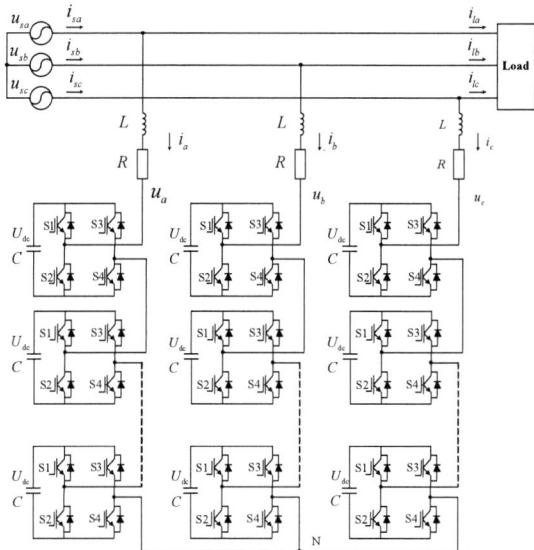

Fig.1. Topology of H-bridge cascaded STATCOM

B. H-bridge Converter and CPS-SPWM strategy

Fig.2 illustrates the basic unit of H-bridge. The voltage U_{dc} of DC capacitor is the input end. The main circuit is the single phase full bridge circuit which composed of four IGBTs. The AC is the output end. With changing the on-off compound mode of power switches, the amplitude and frequency of AC u is controllable. Due to each power switch having two working conditions, the output voltage has four conditions when the power switches in the same bridge arm are turning on at different times. When the S1 and S4 are turning on, the output voltage is Udc. When the S2 and S3 are turning on, the output voltage is -Udc. Otherwise, the output voltage is 0.

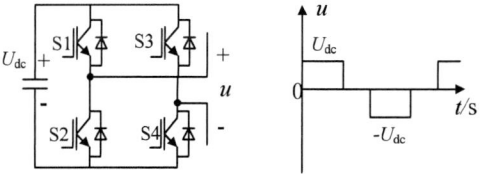

a) H-bridge unit   b) output voltage

Fig.2. H-bridge unit and output voltage

Fig. 3 shows the unipolar CPS-SPWM in the case of two H-bridge in series with inverted modulation wave. Each unit share the same modulation wave, and the triangular carrier of the second unit lags 1/2 period of the carrier cycle. Compared the modulation wave with every triangular carrier, two groups of PWM signals are derived to drive two H-bridges. Each unit outputs a three-level voltage. The output voltage of cascade STATCOM is multilevel step wave. This strategy can achieve a higher equivalent switching frequency, which results in relatively low switching loss and fast dynamic response.

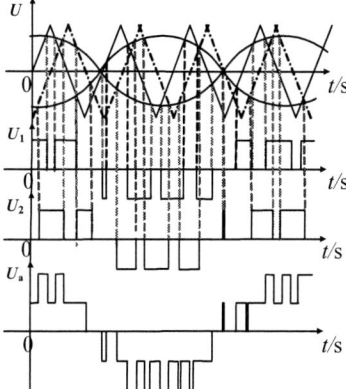

Fig.3. Unipolar CPS-SPWM

C. Reason for DC Capacitor Voltage Unbalance

In ideal conditions, if the parameters and the trigger signals of each unit are same, the DC capacitor voltage will keep balancing. However, in actual system, there are different losses among units. Although, the command of trigger pulse are the same. The signals that reaching the grid electrode need to pass many gate circuits and amplifying elements and have different delays. Thus, the unbalanced problem was leaded by the real parameters and the working state.

$$E_{DC}=\frac{1}{2}CU_{dc}^{2} \qquad (1)$$

Analyzing from the view of energy, the equation (1) can illustrate it. The primary cause of DC capacitor voltage unbalance is the difference of DC energy. The energy flow of capacitor can be divided into two types: The energy exchanged between capacitor and grid by H-bridge circuit. The flow direction of exchange energy was decided by switching element of H-bridge; The energy was lose in system working which includes the conduction loss of switching element, the off-state loss, switching loss, capacitor loss and reactor loss.

D. Principle of DC Capacitor Balancing Control

According the polarity of output voltage and current of unit AC side, the instantaneous of capacitor charging or discharging could be judged. Then, combining the relative size of unit DC voltage, the capacitor voltage would be adjusted, as is illustrated in Fig.4.

978-1-4799-2706-7/14 $31.00 © 2014 IEEE   610

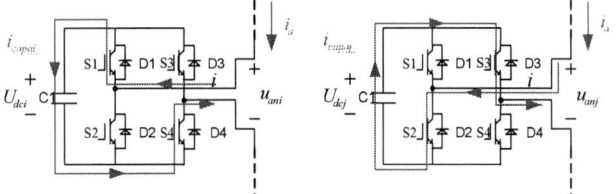

a) Charging process.   b) Discharging process.

Fig.4. Charging and discharging processes of one unit.

One moment in time, the current direction is from grid to system. If the S1 and S4 are opening, the output voltage of unit i is positive. The current flows into DC side along the direction which showed in Fig.4 (a) and charges the capacitor. Likewise, if the S2 and S3 are opening, the output voltage of unit j is negative. The current flows into DC side along the direction which showed in Fig.4 (b) and discharges the capacitor. Obviously, in order to make the capacitor voltage of each unit tend to be consistent, the turn-on time of unit with lower voltage should be extended and it with higher voltage should be decreased for the certain power units in the charging process. The control for the certain power units in the discharging process is on the contrary.

From what has been discussed above, the principle of DC capacitor balancing control can be summarized as follows: As Fig.4 depicted, i_a is output current of a phase. U_{ani} is AC output voltage of the i-th of a-phase (i =1,2 , …N). $U_{dc\_a\_i}$ and U_{dc_a} are capacitor voltage and average voltage of the i-th of a phase.

When $\left(i_a \times u_{ani} \right) > 0$, if $U_{dc\_a\_i} < U_{dc\_a}$, the duty cycle D was increased. If $U_{dc\_a\_i} > U_{dc\_a}$, the duty cycle D was reduced.

When $\left(i_a \times u_{ani} \right) < 0$, if $U_{dc\_a\_i} > U_{dc\_a}$, the duty cycle D was increased. If $U_{dc\_a\_i} < U_{dc\_a}$, the duty cycle D was reduced.

E. Voltage Balancing Control Method with Modulation Wave Translation

According to the above method judging the duty cycle and adjusting the direction and magnitude for power units. Assuming the need to reduce the unit charging and discharging time and the half cycle that the modulation wave is positive to analysis, the output electrical level of unit is 1 level or zero level. When the level is zero, the capacitor is not inserted into the main circuit and it does not charging and discharging. If want to reduce the unit charging time, it needs to reduce the action time of 1 level. And it can be achieved to reduce the turn-on time of S1and S4. Because of the state of left bridge arm was decided by normal phase modulation wave and positive triangular carrier. The state of right bridge arm was decided by opposition phase modulation wave and negative triangular carrier. Then, with x axis bounded, the duty cycle was reduced by shifting down the positive modulation wave and shifting up the negative modulation wave as the following equations (2) and (3).

$$u_i = u_{i0} - k * e_{Udc\_i} \tag{2}$$

$$u_i = u_{i0} + k * e_{Udc\_i} \tag{3}$$

When $e_{Udc\_i} = U_{dc\_i} - U_{dc\_avg}$, and k is regulation coefficient. u_{i0} is the previous modulation wave. u_i is the new modulation wave.

The definite process is illustrated in Fig.5.

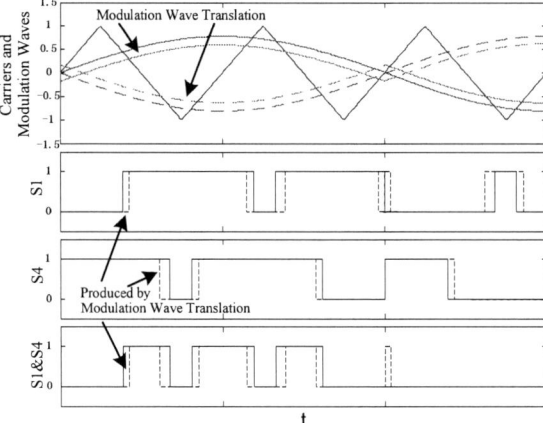

Fig.5. Process of the modulation wave translation

The above principle is suitable for prolonging the unit charge and discharge time. Summing up the two situations, the method of modulation wave translation could be illustrated as follows:

- If the adjustment requirement is to reduce the duty cycle, it needs to shift down the positive modulation wave and shift up the negative modulation wave.

- If the adjustment requirement is to enlarge the duty cycle, it needs to shift up the positive modulation wave and shift down the negative modulation wave.

The shift value was decided by the difference value of the capacitor voltage and average voltage.

The flow chart is shown in Fig.6.

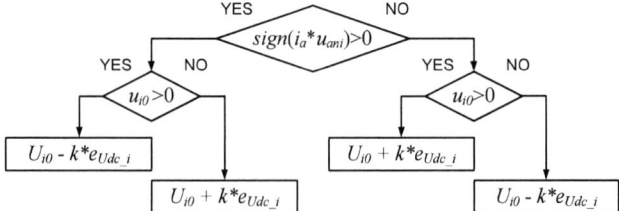

Fig.6. Flow chart of the modulation wave translation

The aboved method is the modulation strategy that was based on CPS-SPWM in this paper. However, it is not to say that this method must be used like this only. In order to regulating the duty cycle, as long as the pulse signal was achieved by the compare modulation wave and carrier, the modulation strategy can be used this method. The implementation block diagram of the individual DC voltage balancing control method was shown as Fig.7.

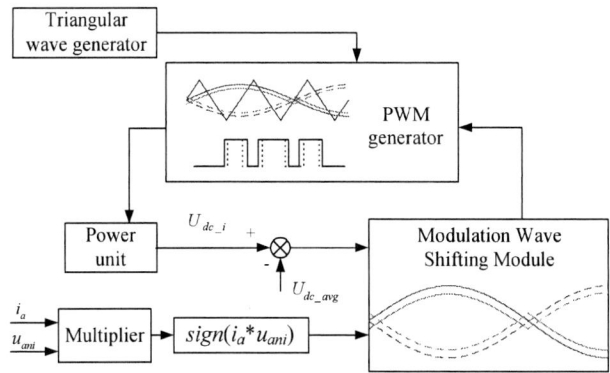

Fig.7. Implementation block diagram of the individual DC voltage balancing control method

III. SIMULATION AND EXPERIMENT RESULTS

To verify the performance of the H-Bridge cascaded STATCOM and the effectiveness of proposed DC voltage balancing control strategy, the simulation and experiment are conducted.

A. Simulation Results

According to the mathematical model of STATCOM, the simulation model with three phase star connection was established in MATLAB/Simulink program. The simulation parameters are listed in Table I. The simulated load is pure perceptual reactive load and three-phase thyristor rectifier bridge load.

TABLE I

SIMULATION PARAMETERS

Parameter	Symbol	Value
Grid voltage	u_s	1500V
Grid frequency	f	50Hz
Number of H-bridges	N	6
Inductance	L	10mH
DC capacitor	C	$5600\mu F$
DC voltage	U_{dc}	350V

In order to proving the superiority of this method simultaneously, the comparison simulation was conducted between this control method and the conventional PI control.

Fig.8 shows the simulation results of individual DC capacitor voltage in a phase. In Fig.8 (a), without any DC voltage balancing control strategy, it is easy to see that the phenomenon of DC voltage unbalance is serious and the voltage presents divergent tendency. Fig.8 (b) is the DC voltage value of each unit with the conventional PI control. The DC voltage value of each unit is still different after the voltage stability. Fig.8(c) is the result with the proposed DC voltage balancing control strategy in this paper. After the voltage stability, the DC voltage of each unit tends to be

uniform. It proves the effectiveness of proposed method above. In addition, the DC voltage ripple wave is smaller and the value is 0.35 percent of reference voltage. The result demonstrates that the DC capacitor voltage exhibits a favorable steady and dynamic performance with the proposed method above combining the CPS-SPWM strategy.

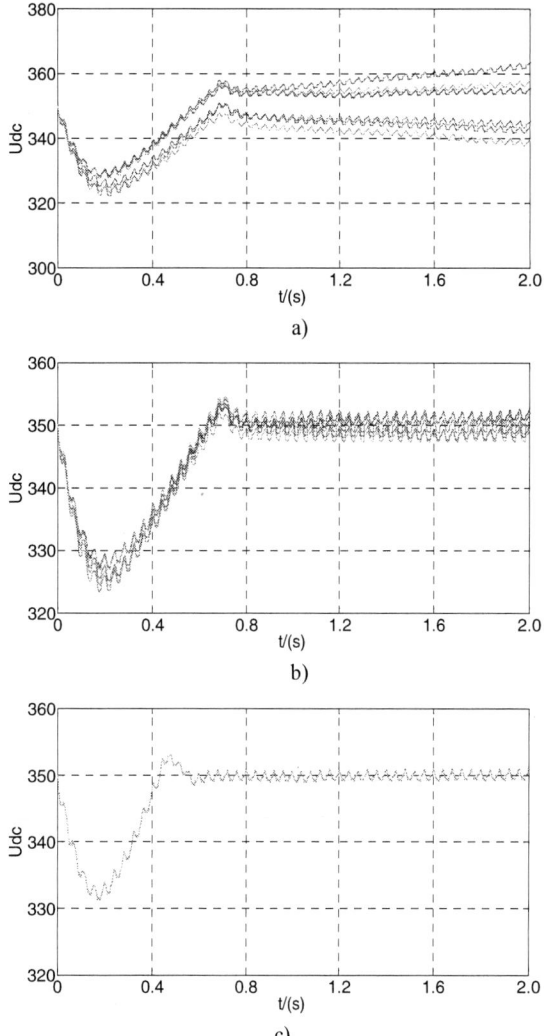

Fig.8. simulation results of individual DC capacitor voltage in a phase. a) without any control strategy. b) with conventional PI control strategy. c) with proposed control strategy

The magnitude of parameter k in equations (2) and (3) reflects the fierce degree of energy exchange in DC capacitor voltage balancing process. It affects the voltage balancing effect and the voltage balancing rate.

Fig.9 shows the simulation results of performance of voltage balancing with different k (k=10 in Fig.9(a) and k=300 in Fig.9(b)). Compared with Fig.8(c) (k=100), increasing parameter k can improve the voltage balancing rate. But, the oversize k makes energy exchange violently and leads DC voltage out of control. Hence, it needs to elect parameter k appropriately for achieving the best control performance.

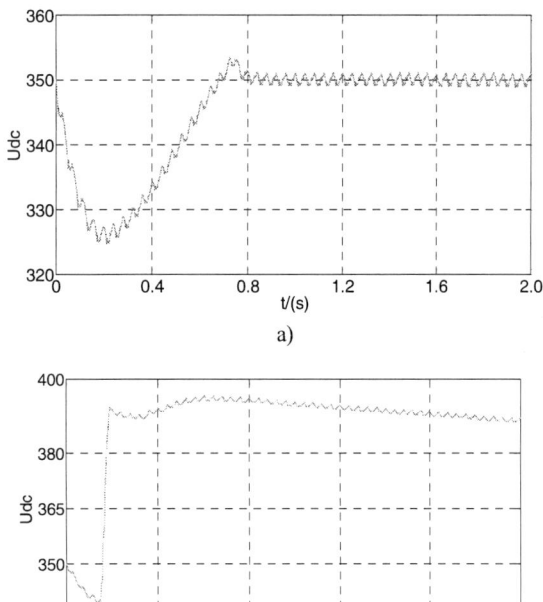

Fig.9. performance of voltage balancing with different k. a) k=10. b) k=300.

B. Experiment Result

To verify the DC capacitor voltage balancing control by experiment, the real STATCOM based on 25-level (n=12) cascaded multilevel converter experiment system was developed. The parameters of experiment system are list in Table II.

TABLE II

EXPERIMENT SYSTEM PARAMETERS

Parameter	Symbol	Value
Grid voltage	u_s	10000V
Grid frequency	f	50Hz
Number of H-bridges	N	12
Inductance	L	10mH
DC capacitor	C	$5600\mu F$
DC voltage	U_{dc}	800V

The main controller consists of TI-DSP and Altera-FPGA. The proposed control method is working in FPGA. The DC voltage values of 12 units in a phase was loaded into DSP. By using the CCS software, the experiment result could be observed and recorded. Then, the recorded data were plotted as experiment waveforms with MATLAB. The experiment waveforms were given as Fig.9.

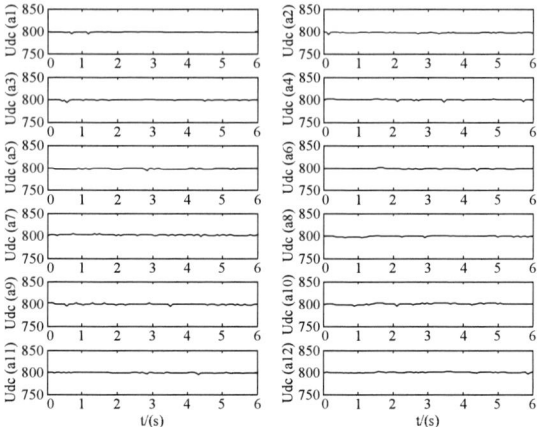

Fig.9. Experiment results of DC capacitor voltage balancing.

From the Fig.9, it is evidently to see that the DC voltage values of all units in a phase were kept 800V primely. With the proposed method, it achieves the satisfactory control performance.

IV. CONCLUSION

In this paper, a simple and flexible balance control method for individual DC voltage balancing which based on the popular CPS-SPWM strategy was proposed. With the upper and lower translations of modulation wave in CPS-SPWM, the proposed method could balance more simply and flexibly the DC-voltages during the charging and discharging processes. Compared with the conventional PI control, it omits the regulation the complex parameters of PI and eliminates the delay. The simulation and experiment results showed the effectiveness and superiority of the proposed method in controlling individual capacitor voltages.

REFERENCES

[1] Akagi H, "New trends in active filters for power conditioning," *IEEE Trans Industry Applications.* Vol. 32, pp.1312-1322,1996.

[2] Dixon J W, Venegas G, and Moran L A, "A series active power filter based on a sinusoidal current-controlled voltage-source inverter," *IEEE Trans Industrial Electronics.* Vol. 44, pp.612-620,1997.

[3] Rudnick H, Dixon J, and Moran L, "Delivering clean and pure power," *IEEE Power and Energy Magazine.* Vol. 1, pp.32-40,2003.

[4] Hirofumi Akagi, Shigenori Inoue, "Control and performance of a transformerless cascade PWM STATCOM with Star Con fi guration". *IEEE Transactions on Industry Applications,* vol.43(4), pp.1041-1049, July/August 2007.

[5] Wei Wenhui, Teng Letian, Liu Wenhua, Song Qiang and Geng Juncheng. Balancing Control for DC Capacitors Voltage of STATCOM Based on Cascade Multilevel Inverters.[J] *Automation of Electric Power Systems，* Apr. 2004，28(7)：4–9.

[6] Chen Yuanhua, Lin Wenhua, Yan Gangui and Song Qiang. Research of Capacitor Voltage Balance Control of H-bridge Inverter of Hybrid 9-level Inverter[J]. *Automation of Electric Power Systems，* Jan. 2004，28(1)：35–40 .

[7] H. Iman-Eini, J.-L. Schanen, S. Farhangi, J. Roudet. A Modular Strategy for Control and Voltage Balancing of Cascaded H-Bridge Rectifiers[J]. *IEEE Transactions on Power Electronics，* Sept. 2008，23(5): 2428-2442.

Research on the Control Strategy of STATCOM Based on Modular Multilevel Converter

Wei Zhang
Institute of Power Electronics and Electrical Driving
Harbin Institute of Technology
Harbin, China
zhangwei_0231@sina.cn

Qiang Gao
Institute of Power Electronics and Electrical Driving
Harbin Institute of Technology
Harbin, China
gq651@hit.edu.cn

Bonan Su
Institute of Power Electronics and Electrical Driving
Harbin Institute of Technology
Harbin, China
subonan1989@126.com

Miaoxin Jin
Institute of Power Electronics and Electrical Driving
Harbin Institute of Technology
Harbin, China
jinmiaoxin1986@163.com

Dianguo Xu
Institute of Power Electronics and Electrical Driving
Harbin Institute of Technology
Harbin, China
xudiang@hit.edu.cn

Jianyu Liu
School of Mechatronics Engineering
Harbin Institute of Technology
Harbin, China
jeromejy@163.com

Abstract- **The paper conducted a theoretical analysis and simulation research on the control strategy of STATCOM based on modular multilevel converter (MMC). Pre-charging control strategy, capacitor voltage balance control strategy as well as STATCOM control strategy are proposed. According to the simulation model of SVG, the strategy proposed is proposed. The control methods are tested with the help of PSCAD.**

Key words: MMC; STATCOM; control strategy; reactive power compensation.

I. INTRODUCTION

With the widely utilization of power switching devices, there are lots of non-liner loads in grids, which seriously decreases the quality of electronic energy. As a result, it has important significance in improving power factor, reducing the capacitor of power supply equipments as well as improving the stability of grid to apply reactive power compensation rapidly. As Static Synchronous Compensator (STATCOM) has advantages in faster response, lower loss, smaller volume of energy storage elements and lower harmonic waves in output current [1], it has become an important developing direction in the field of dynamic reactive power compensation [2].

In the high-voltage and high-power electric field, there are irreconcilable contradictions between power rating and switching frequency of the existing switching device. In recent years, with the advantages of its individual topology, multilevel converter is widely studied in the field above. Cascaded multilevel converter, taking cascaded H-bridge multilevel converter (CHMC) as a representation, has many applications in industrial

production because this type of topology adopts modularity designing, which can easily enlarging the capacitor of system and taking the control strategy of redundant scheme [3]. Nevertheless, when it comes to the field of energy processing, CHMC should be used with independent DC power sources for each power unit, which causes the system complex and the cost high [4].

As there are some disadvantages for traditional multilevel converters in high-voltage and high-power electric field, Marquardt R. and his partners put forward the topology of modular multilevel converter based on cascaded structure. Besides its modular characteristic, MMC has public DC bus. It can realize the AC-DC and DC-AC transform. Meanwhile, MMC can be used in the high-voltage and high-power field without any other equipments [5, 6]. In industrial applications, MMC is only used in High Voltage Direct Current (HVDC) [7, 8].

In this paper, the control strategy for STATCOM based on MMC is proposed. Pre-charging control strategy, capacitor voltage balance control strategy and STATCOM control strategy are mainly discussed.

Most literatures were on the study of MMC topology and its control strategy. Fewer literatures discussed its application in the field of reactive power compensation. [9] proposed a method for pre-charging, which is widely used. In this method, an extra DC power source was added. This method only fits the condition of low-voltage and low-power. For STATCOM, it cannot be selected. [10] proposed a strategy to control the capacitor voltage balance. An outer controlling circuit was build up, which was easily to achieve the purpose. Nevertheless, it increased the costs of the system. Meanwhile, the

978-1-4799-2706-7/14 $31.00 © 2014 IEEE

dependability and complexity of the system were required higher. This paper proposed other strategies to avoid these disadvantages.

II. THE WORKING PRINCIPLE OF MMC

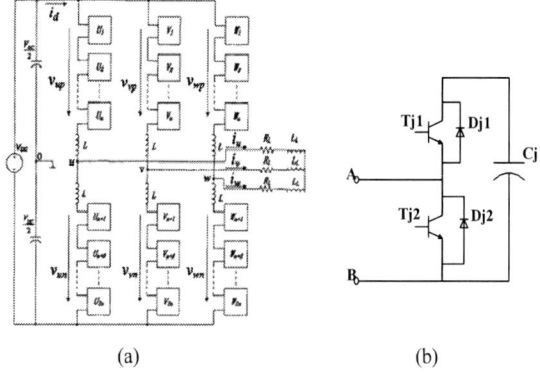

(a)   (b)

Fig. 1. (a)Topology of MMC (b)Topology of sub-model.

Fig. 1(a) shows the topology of MMC and Fig. 1(b) shows the topology of sub-model (SM). A half-bridge sub-model is composed of a half-bridge converter and a DC energy storage capacitor. Table 1 describes the 6 working modes of a SM. The current path of each working mode is shown in Fig. 2.

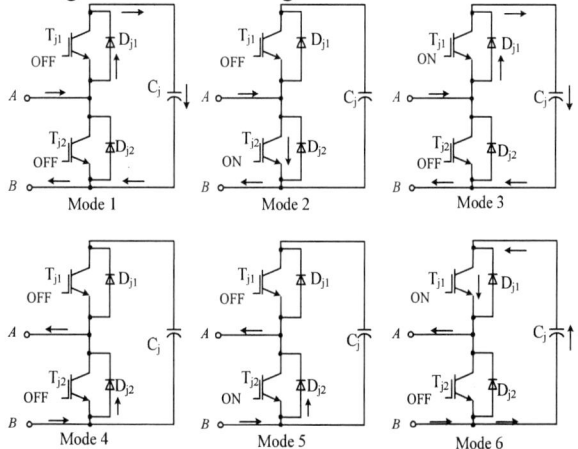

Fig. 2. The current path of each working mode.

TABLE I
WORKING MODES OF A SM

Mode	Tj1	Tj2	Dj1	Dj2	Current Direction	U_{AB}	Description
1	OFF	OFF	ON	OFF	A→B	Uc	Charging
2	OFF	ON	OFF	OFF	A→B	0	Bypass
3	ON	OFF	ON	OFF	A→B	Uc	Charging
4	OFF	OFF	OFF	ON	B→A	0	Bypass
5	OFF	ON	OFF	ON	B→A	0	Bypass
6	ON	OFF	OFF	OFF	B→A	Uc	discharging

Mode 1 and Mode 4 are used in starting statue and fault statue while other modes are used in power transmission statue. Each SM can be treated as a

controllable voltage source. The output voltage can be switched from 0 and V_c by controlling the switching statue of IGBTs. Thus the SMs are controlled on or off the system to get output multilevel step waves, which are used to fit sine waves.

III. PRE-CHARGING CONTROL STRATEGY

In the stage of decoupling control, instantaneous over-current appears in the bridge-arms of MMC, which may damage the power devices. In industrial applications, the sub-model controller (SMC) gets power by means of partial voltage from resistance. At the start of starting statue, the voltage on the capacitor is too low that SMC cannot get power to drive. Thus, pre-charging for MMC is of necessity.

First, all the SMs in MMC are pre-charged by AC grid. Take one bridge-arm in a-phase as an example. Break all the switches paralleled with the charge-resistances in order to connect the charge-resistances with the system (Shown in Fig.3). If $i_{pu}>0$, the current can flow over all the diodes named D1 in the SMs of the corresponding bridge-arm, which can charge the capacities. In this statue, the voltage of each capacitor can be charged to 1/N of line voltage. Though the capacities cannot be charged to the required voltage by this means of charging, the SMC can get power from the capacities to work regularly [11].

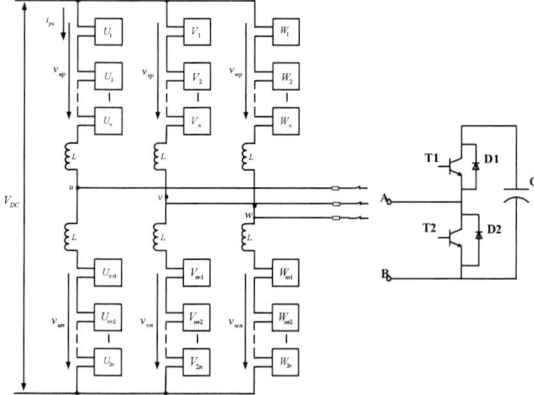

Fig. 3. General view of MMC charged by AC grid.

Then the SMs are pre-charged to the required voltage one by one by means of self-excitation. The specific method is as followed. Take one bridge-arm in a-phase as an example as before. At first, all the IGBTs named T1 in the SMs are turned off while all the IGBTs named T2 are turned on. If i_{pu} is detected to be higher than 0, the IGBT named T2 in the SM which is waited to be pre-charged should be turned on. Then the SM mentioned above is pre-charged. Otherwise, if i_{pu} is detected to be lower than 0, the IGBT named T2 is turned on to avoid the capacitor's discharging. Repeat the above-mentioned process till the voltage of the capacitor in charging gets the required voltage, and then the IGBT named T2 is turned off. In this way, all the SMs can be pre-charged to the required voltage. The flow chart of the process is shown in Fig. 4.

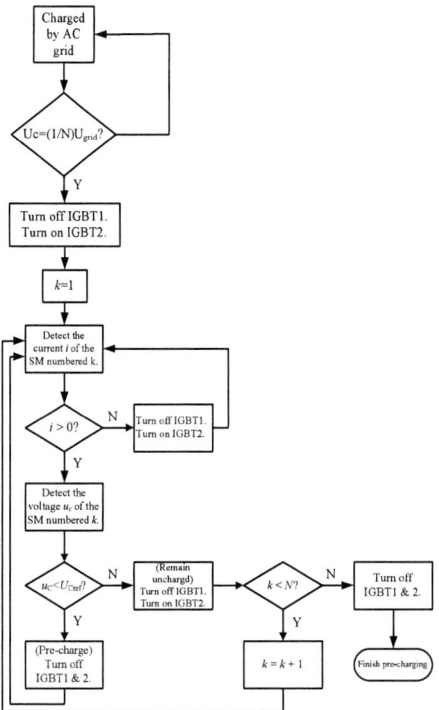

Fig. 4. Flow chart of pre-charging process.

IV. CAPACITOR VOLTAGE BALANCE CONTROL STRATEGY

To control the voltage balance of each capacitor, the instantaneous voltage of the capacitor is compared with the reference voltage in order to approximate rapidly and exactly. The flow chart of capacitor voltage balance control strategy is shown in Fig.5. In the chart, V_{cref} is the reference voltage of the capacitor; V^* is the reference signal of input voltage; V_{cj} is the instantaneous voltage of the capacitor; ΔV_{cj} is the error signal of instantaneous and reference voltage; ΔV_j is the voltage-steady signal generated by voltage-steady controlling; And V_j is the voltage command signal. There takes the upper bridge-arm as an example.

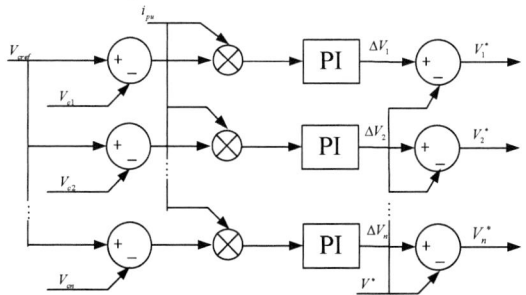

Fig. 5. Flow chart of capacitor voltage balance control strategy.

If $V_{cref} < V_{cj}$, the SM needs to be charged by taking in power from the AC grid till $V_{cref} = V_{cj}$. When $i_{pu} > 0$, ΔV_j is generated from ΔV_{cj} by the measure of PI, where $\Delta V_j > 0$. Added ΔV_j to the sine modulation waves, time that the SM working in mode 3 is lengthened, which means that time for SM to be charged gets longer. When $i_{pu} < 0$, ΔV_j^- generated from ΔV_{cj} by the measure of PI is less than 0.

Added ΔV_j to the sine modulation waves, time that the SM working in mode 6 is shortened, which means that time for SM to be discharged gets shorter.

If $V_{cref} > V_{cj}$, the SM needs to be discharged by giving out power to the loads till $V_{cref} = V_{cj}$. When $i_{pu} > 0$, ΔV_j is generated from ΔV_{cj} by the measure of PI, where $\Delta V_j < 0$. Added ΔV_j to the sine modulation waves, time that the SM working in mode 3 is shortened, which means that time for SM to be charged gets shorter. When $i_{pu} < 0$, ΔV_j^- generated from ΔV_{cj} by the measure of PI is more than 0. Added ΔV_j to the sine modulation waves, time that the SM working in mode 6 is lenthened, which means that time for SM to be discharged gets longer.

Compared the new reference wave with the triangular carrier wave, the pulse signal for controlling the IGBT is generated.

V. STATCOM CONTROL STRATEGY

As a result of the working efficiency, bridge circuits, which can be regarded as voltage source converters under regular working statues, are widely used in SVG. Considering the loss in reactance and the converter itself, the equivalent circuit of SVG is described in Fig. 6. The voltage of AC grid is described by U_s; the output voltage of SVG is described by U_l; and the voltage of reactance and equivalent resistance is described by U_L. Due to the equivalent resistance, there is a phase-difference described by δ between U_s and U_l, which means that the converter needs to take in active power. To regulate the reactive power taken from the AC grid, δ and U_l are changed, meanwhile the phase and amplitude of I are changed as well.

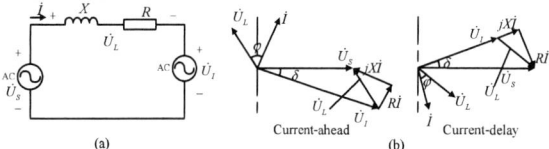

(a)   (b)

Fig. 6. (a) Equivalent circuit of SVG (b) Vector diagram.

According the circuit shown in Fig.6 and KCL, (1) can be obtained.

$$L\frac{d}{dt}\begin{bmatrix} i_a \\ i_b \\ i_c \end{bmatrix} = \begin{bmatrix} V_{sa} \\ V_{sb} \\ V_{sc} \end{bmatrix} - \begin{bmatrix} V_{ca} \\ V_{cb} \\ V_{cc} \end{bmatrix} - R\begin{bmatrix} i_a \\ i_b \\ i_c \end{bmatrix} \quad (1)$$

Applying dq transformation in a rotating coordinate, (1) can be changed into:

$$\frac{d}{dt}\begin{bmatrix} i_d \\ i_q \end{bmatrix} = \frac{d}{dt}\left\{ T_{abc-dq}\begin{bmatrix} i_a \\ i_b \\ i_c \end{bmatrix} \right\} = T_{abc-dq}\frac{d}{dt}\begin{bmatrix} i_a \\ i_b \\ i_c \end{bmatrix} + \frac{dT_{abc-dq}}{dt}\begin{bmatrix} i_a \\ i_b \\ i_c \end{bmatrix}$$

$$= \begin{bmatrix} -\dfrac{R}{L} & \omega \\ -\omega & -\dfrac{R}{L} \end{bmatrix}\begin{bmatrix} i_d \\ i_q \end{bmatrix} + \frac{1}{L}\begin{bmatrix} V_{sd} - V_{cd} \\ V_{sq} - V_{cq} \end{bmatrix} \quad (2)$$

Considering that the higher frequency of switching may lead to more loss of converting, indirect current controlling strategy is adopted in this paper. In this strategy, SVG is treated as a controllable VSI. Output current can be controlled indirectly by controlling output voltage.

In this paper, V_{cd} and V_{cq}, the components of SVG's output voltage on dq-axis, are selected as the control variables. As $V_c = \sqrt{V_{cd}^2 + V_{cq}^2}$ and $\delta = tg^{-1}\left(V_{cd}/V_{cq}\right)$, the power exchanging between SVG and AC grid can be controlled by controlling V_{cd} and V_{cq}. According to (2), V_{cd} and V_{cq} can be described as:

$$\begin{bmatrix} V_{cd} \\ V_{cq} \end{bmatrix} = \begin{bmatrix} V_{sd} \\ V_{sq} \end{bmatrix} + \begin{bmatrix} -R & \omega L \\ -\omega L & -R \end{bmatrix} \begin{bmatrix} i_d \\ i_q \end{bmatrix} - L\frac{d}{dt}\begin{bmatrix} i_d \\ i_q \end{bmatrix} \quad (3)$$

Obviously, SVG is a typical coupling system, which cannot be controlled simply. To make the control strategy simple, some measures must be carried out in order to make dq-axis decoupled. As the current is transformed into DC variable, non-static error regulation can be achieved by the traditional PI regulation.

Defining:

$$\begin{cases} x_1 = V_{sd} - V_{cd} + \omega L i_q \\ x_2 = V_{sq} - V_{cq} - \omega L i_d \end{cases} \quad (4)$$

Based on (2) and (3), (5) can be got.

$$\frac{d}{dt}\begin{bmatrix} i_d \\ i_q \end{bmatrix} = \begin{bmatrix} -R/L & 0 \\ 0 & -R/L \end{bmatrix}\begin{bmatrix} i_d \\ i_q \end{bmatrix} + \frac{1}{L}\begin{bmatrix} x_1 \\ x_2 \end{bmatrix} \quad (5)$$

Design the control strategy of x_1 and x_2:

$$\begin{bmatrix} x_1 \\ x_2 \end{bmatrix} = \begin{bmatrix} K_1(i_d^* - i_d) + \dfrac{K_1}{T_1}\int(i_d^* - i_d)dt \\ K_2(i_q^* - i_q) + \dfrac{K_2}{T_2}\int(i_q^* - i_q)dt \end{bmatrix} \quad (6)$$

Control strategy diagram is shown in Fig. 7.

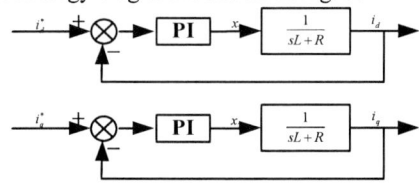

Fig. 7. Control strategy diagram.

The currents on dq-axis are designed into two PI regulations, whose outputs are x_1 and x_2. Thus, the currents on dq-axis decoupled. The feed-forward decouple controlling diagram of SVG is shown in Fig. 8.

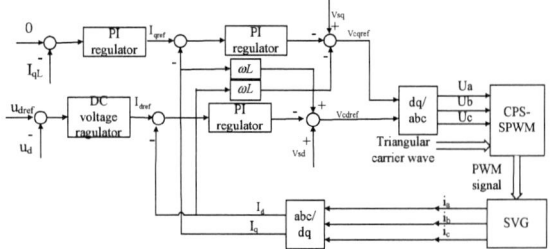

Fig. 8. Diagram for voltage and current's double close-loop control of feed-forward decoupled.

The controller in Fig. 8 is composed of voltage outer-loop and current inner-loop. The outer-loop is used to control the average voltage of the capacitors in DC side, whose output is the reference current variable of active power described by I_{dref}. Regulating reactive power current of the loads by PI, the output is the reference current variable of reactive power described by I_{qref}. The dq transformation of SVG's output current is treated as the feedback of inner-loop. In order to adopt PI regulation, the feedback current variable is compared with the reference variable. Based on (4), the required output voltages, which are described by V_{cdref} and V_{cqref}, are obtained. By using dq reverse transformation, the modulation waves for three-phase waves are generated. Compared the modulation waves with the triangular carrier waves, the driving signals for IGBTs are obtained.

VI. SIMULATION RESULT

In order to verify the above strategy, nine-level line voltage of SVG is simulated in PSCAD. The simulation model is shown in Fig. 9. Simulation parameters are as follows: $n = 4$, the voltage of capacitors are 250V.

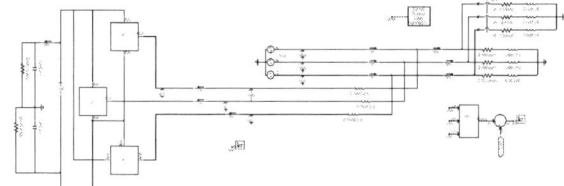

Fig. 9. Simulation model of SVG.

The capacitors' voltage waves in pre-charging statue is shown in Fig. 10. Based on it, the voltage of each capacitor increased stably from 0 ~ 3s. After 3s, the capacitors were charged one after another till the voltage reached 250V.

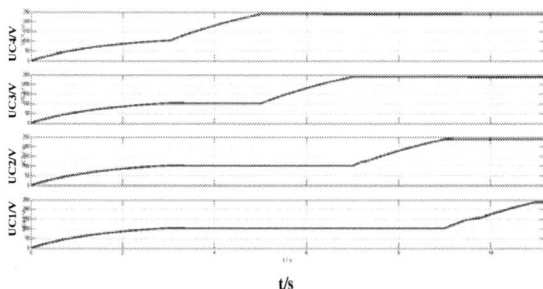

Fig. 10. Capacitors' voltage waves in pre-charging statue.

In the simulation model, 4 SMs are consisted in one bridge-arm. 5-level phase voltage and 9-level line voltage of U-phase are shown in Fig. 11. and Fig, 12.

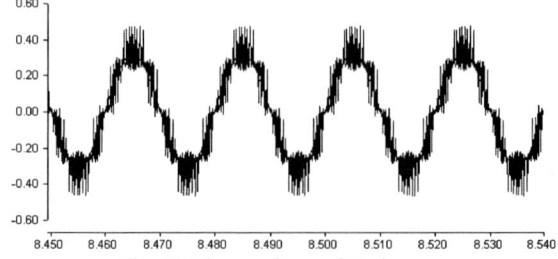

Fig. 11. Phase voltage of U-phase.

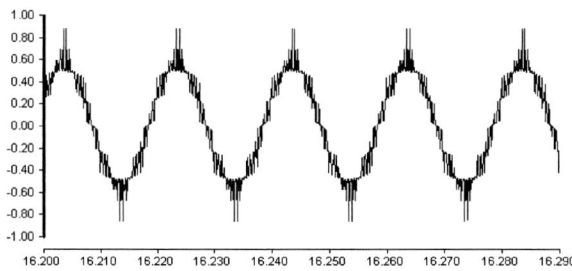

Fig. 12. Line voltage of U-phase.

The capacitors' voltage of U-phase upper bridge-arm is shown in Fig. 13. Based on it, the voltage was stabilized at the value of 250V.

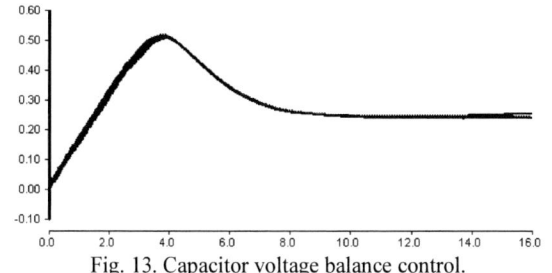

Fig. 13. Capacitor voltage balance control.

The voltage and current waves of grid is shown in Fig.14. ; And the power factor is shown in Fig. 15. In the figures, the voltage's phase of grid equaled to the current's. Thus, the reactive power of the loads was compensated. The function of reactive power compensation was achieved.

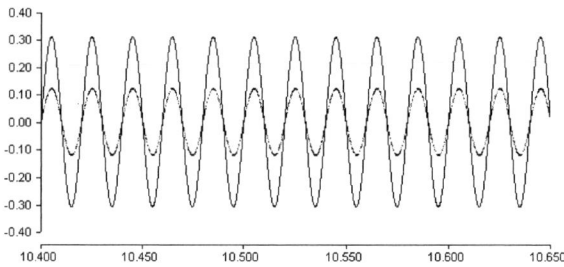

Fig. 14. Voltage & current of grid.

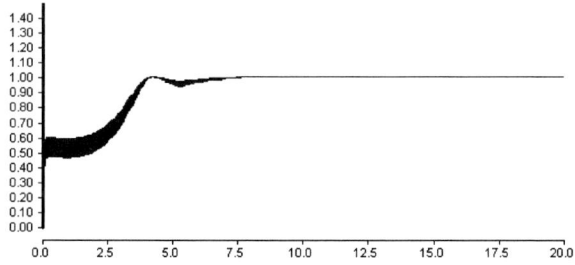

Fig. 15. Power factor of grid.

VII. CONCLUSION

In this paper, a series of strategies are proposed for the STATCOM based on MMC in order to achieve the function of reactive power compensation. In pre-charging

statue, free-charging and control-charging are adopted in order that the capacitor voltage achieve the required value. The proposed voltage balance control strategy, which is based on comparing of capacitor voltages and reference voltages, proves the validity in the results. By designing the PI regulator, SVG worked regular and the propose was achieved. The simulation result confirms the feasibility of the control strategy.

REFERENCES

[1] Shiping Li, Guiying Liu, "Static reactive power compensation technique", Beijing: China Electronic Express, 2006: 225-230.

[2] Yongkang Luo, Yong Ji, Daoyi Liu, "The principle and achieving of STATCOM", Zhenjiang: China Electronic Express, 2005: 5-7.

[3] Peng Fangzheng, Lai Jihsheng, McKeever J W, et al, "A multilevel voltage-source inverter with separate DC sources for static VAr generation", *IEEE Transactions on Industry Applications*, vol. 32, no. 5, pp. 1130-1138, 1996.

[4] Rodriguez J, Bernet S, Wu Bin, et al, "Multilevel voltage source-converter topologies for industrial medium-voltage drivers", *IEEE Transactions on Industrial Electronics*, vol. 54, no. 6, pp. 2930-2945, 2007.

[5] Makoto Hagiwara, Hirofumi Akagi, "Control and experiment of Pulsewith Modulated Modular Multilevel Converters", *IEEE Transactions on Power Electronics*, vol. 24, no. 7, pp. 1737-1740, 2009.

[6] Hirofumi Akagi, "Classification, Terminology, and Application of the Modular Multilevel Cascade Converter(MMC)", *IEEE Transactions on power electronics*, vol. 26, no. 11, pp. 3119-3124, 2011.

[7] Knaak H. J, "Modular multilevel converters and HVDC/FACTS: A success story". *Power Electronics and Applications(EPE 2011), Proceedings of the 2011-14th European Conference*, vol. 1 no. 1, pp. 1-6.

[8] Xiaofeng Yang, Zhiqin Lin, Qionglin Zheng, "Research summary on modular multilevel converter", *Proceedings of the CSEE*, vol. 33, no. 6, pp. 1-7, 2013.

[9] Guanjun Ding, Ming Ding, Guangfu Tang, "Strategy on voltage balance of sub-modles in a new type of VSC", *Proceedings of the CSEE*, vol. 29, no. 30, pp. 1-6, 2009.

[10] Zhao Liu, Bangyin Liu, Shanxu Duan, "Strategy on voltage balance control STSTCOM", *Proceedings of the CSEE*, vol. 29, no. 30, pp. 7-12, 2009.

[11] Zheng Xu, "Flexible DC transmission system", Beijing: China Mechanical Industry Press, 2013:121-124.

Fault Diagnosis in Large Format LiFePO$_4$ ESS Application through DWT-based MRA

Jonghoon Kim / Chosun University
Department of Electrical Engineering
Gwangju, Republic of Korea

Abstract - **An unexpected and instantaneous voltage sag in the energy storage system (ESS) application may cause irrespective breakdown and permanent internal damage of ESS. Therefore, this paper gives insight to the design and implementation of a novel methodology based on the application of discrete wavelet transform (DWT) to the fault diagnosis of the LiFePO$_4$ ESS application using discharging/ charging voltage signal (DCVS). This DCVS is applied as source data in the DWT-based analysis due to its ability to extract variable information from the non-stationary and transient phenomena in both time and frequency domain simultaneously. Specifically, DWT-based multi-resolution analysis (MRA) technique in processing voltage-sag having DCVS dealing with a high frequency component D_n enables us to diagnose fault effectively in the ESS. The experimental results show that DWT-based approach is clearly appropriate for the high-efficient fault diagnosis in large format LiFePO$_4$ ESS application.**

I. INTRODUCTION

Nowadays, with the increased concerns on environmental and energy cost, more renewable energy sources such as fuel cell, solar cell, wind are integrated into the power grid in the form of distributed generation (DG) [1]. Unfortunately, due to the variable characteristics of more renewable energy sources, the energy generated from these renewable have a constant fluctuation and may only be available during limited hours of the day, which makes it difficult to maintain a reliable and stable supply of power [2]. In order to overcome these shortcomings, an energy storage system (ESS) of multi-cell battery string in series and parallel connections shown in Fig. 1 can be employed. Efficient use of ESS enables us to smooth the energy generated from renewable sources and to reduce the power demand during peak hours. Thus, with the requirement of smart grid and smart load management, such ESS can be absolutely considered as a key role in stabilizing the smart grid and ensuring adequate supply of power to meet peak demand.

However, unexpected and instantaneous voltage sag in the ESS application may cause irrespective breakdown and permanent internal damage of ESS [3]. Specifically, considered in battery management system (BMS) including state-of-charge (SOC) estimation and state-of-health (SOH) prediction in order to guarantee the overall system performance for ESS, an accurate and reliable knowledge of fault diagnosis definitely need to be investigated.

This work deals with the discrete wavelet transform (DWT)-based approach for diagnosis of the fault generated by unexpected and instantaneous voltage sag of the ESS. Voltage sag sets by two parameters including magnitude and duration [4]. In this work, the discharging/charging

Fig. 1. Energy storage system (ESS) of Samsung SDI.

voltage signal (DCVS) of LiFePO$_4$ ESS is applied as source data in the DWT-based fault diagnosis due to its ability to extract information from the non-stationary and transient phenomena simultaneously in both the time and frequency domain [5][6]. By using the wavelet decomposition including the multi-resolution analysis (MRA) for filtering and down-sampling, several information on the electrochemical characteristics of LiFePO$_4$ ESS can be extracted from a signal over a wide frequency range [7]-[9]. The DWT decomposes the DCVS into time and frequency domains and focuses on short time intervals for high frequency component (detail; D_n) and on long time intervals for low frequency component (approximation; A_n). Specifically, DWT-based MRA technique in processing voltage-sag having DCVS dealing with a high frequency component D_n enables us to diagnose fault effectively in the ESS. The experimental results show that DWT-based approach is clearly appropriate for the high-efficient fault diagnosis in large format LiFePO$_4$ ESS. This approach has been validated by experimental results conducted series-cell configured battery pack of 70S1P including 70 prismatic LiFePO$_4$ cells that had a rated capacity of 50Ah produced by Samsung SDI [10] (Fig. 1).

II. DISCRETE WAVELET TRANSFORM (DWT)

A. Wavelet Analysis-Mathematical Representation.

For non-stationary signal, the Fourier analysis is not effective since it transforms the signal into the frequency domain the time information is lost [11]. This deficiency of the Fourier analysis can be removed to some extent by analyzing a small section of the signal at a time, called as windowing. This type of analysis, known as the short-time Fourier transform (STFT) [12], however, has the drawback in that the size of the time window is the same for all frequencies. On the other hand, the wavelet transform involves a varied time-frequency window and can provide good localization property in both the frequency and time domain. Window functions of two signal processing transform are shown in Fig. 2.

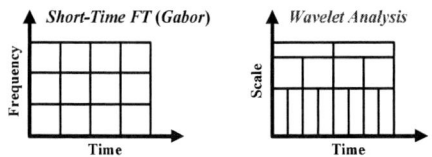

Fig. 2. Window functions of two signal processing transform.

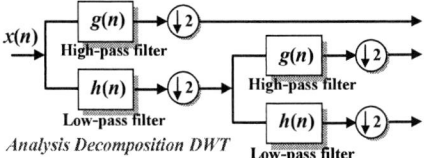

Fig. 3. DWT-based multi-resolution analysis (MRA) of the signal.

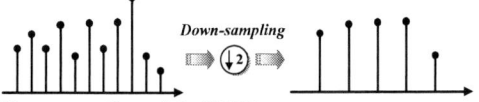

Fig. 4. Down-sampling of the DWT.

Basically, a wavelet is a function belonging to $L^2(R)$ with a zero average [11][12]. The continuous wavelet transform (CWT) of a signal $x(t)$ is defined as

$$\text{CWT}(a,b) = \frac{1}{\sqrt{a}} \int_{-\infty}^{\infty} x(t)\psi^*\left(\frac{t-b}{a}\right)dt \qquad (1)$$

where $\psi(t)$ is called the mother wavelet, the asterisk denotes complex conjugate, while a and b ($a,\ b \in R$) are scaling (dilation and translation) parameters, respectively. The scale parameter a determines the oscillatory frequency and the length of the wavelet, and the translation parameter b determines its shifting position. During CWT analysis, the wavelet is shifted smoothly over the full domain of the analyzed signal. It thus calculates the wavelet coeffcients at every scale, generating a huge amount of data. Due to the huge amount of data generated through CWT, training classifiers based on its coefficients at different scales can often become cumbersome.

The DWT is derived from CWT, and its advantage is that it does not shift and scale continuously, and can only be operated in discrete steps. It also provides sufficient information and offers high reduction in the computational time. The DWT can be defined as

$$\text{DWT}(a,b) = \frac{1}{\sqrt{2^j}} \int_{-\infty}^{\infty} x(t)\psi^*\left(\frac{t-k2^j}{2^j}\right)dt \qquad (2)$$

where a and b are replaced by 2^j and $2^j k$. This defines a dyadic-orthonormal wavelet transform are provides the basis for MRA.

B. DWT-based Multi-Resolution Analysis (MRA)

In MRA process, any time series $x(t)$ can be completely decomposed in terms of approximations, provided by scaling functions $\phi_{j,k}(t)$ and the details, provided by the wavelet function $\psi_{j,k}(t)$, and are defined as the following:

$$\phi_{j,k}(t) = 2^{-\frac{j}{2}}\phi(2^{-j}t-k) \qquad \psi_{j,k}(t) = 2^{-\frac{j}{2}}\psi(2^{-j}t-k) \qquad (3)$$

where j and k are integers. The scaling function is associated with the low-pass filters with filter coefficients of $h(n)$, and the wavelet function is associated with the high-pass filters with filter coefficients of $g(n)$ such that $g(L\text{-}1\text{-}n)=(-1)^n h(n)$ (filter length L). The approximations are the low-frequency components of the time series and the details are the high-frequency components. The MRA leads to a hierarchical

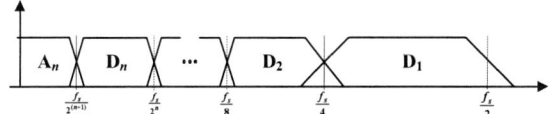

Fig. 5. Frequency bands corresponding to the DWT signal.

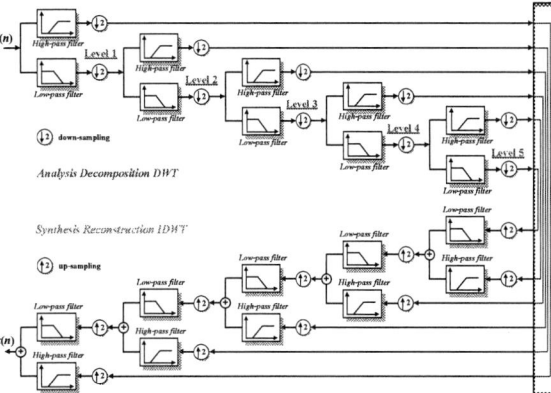

Fig. 6. Window functions of two signal processing transform.

and fast scheme. This can be implemented by a set of successive filter banks as shown in Fig. 3. Considering the filter bank implementation in Fig. 3, the relationship of the approximation coefficients and detail coefficients between two adjacent levels are given as

$$a_{j,k} = \sum_n h(n-2k)a_{j-1,n} \qquad d_{j,k} = \sum_n g(n-2k)a_{j-1,n} \qquad (4)$$

where $a_{j,k}$ and $d_{j,k}$ represent the approximation coefficients and detail coefficients of the signal at level j, respectively. Since $x(t)$ has a length of 2^N, then there is a maximum of N levels of decomposition, or a maximum $N(J{\leq}N)$ of applications of $h(n)$ and $g(n)$ [12]. Therefore, the J-level DWT representation of a signal $x(t)$ can be defined as (5).

$$x(t) = \sum_{k=0}^{2^{N-J}-1} a_{j,k} 2^{-\frac{j}{2}}\phi(2^{-j}t-k) + \sum_{j=1}^{J}\sum_{k=0}^{2^{N-j}-1} d_{j,k} 2^{-\frac{j}{2}}\psi(2^{-j}t-k) \qquad (5)$$

where $J(J{\leq}N)$ is the number of levels of decomposition. Each set of wavelet transform coefficients $a_{j,k}$ and $d_{j,k}$ for $1{\leq}j{\leq}J$, represents a band-pass filtered and down-sampled version (Fig. 4) of the original signal $x(t)$. The decomposition process can be iterated, with successive approximations being decomposed in turn, so that a signal $x(t)$ is broken down into many lower resolution components. This is called the wavelet decomposition tree. In practice, a suitable level number n is selected based on the nature of the signal, or on some other suitable criterion.

Fig. 5 shows a qualitative representation of the transfer function of the filters used in the transform. The resulting frequency bands are consecutive; the limits for these bands depend on the level of the signal and on the sampling rate. For the detail D_1, the upper limit is half the sampling rate ($f_s/2$), whereas the lower limit is $f_s/4$. For a generic detail D_n, the upper limit of its frequency band coincides with the lower limit of the band of the previous detail D_{n-1}, and its bandwidth is half the bandwidth of the previous detail. Therefore, a generic detail D_n contains the information concerning the signal components with frequencies included within the interval $[2^{-(n+1)}f_s, 2^{-n}f_s]$Hz. On the other hand, the approximation A_n contains the low frequency components of the original signal $x(t)$ included in the interval $[0, 2^{-(n+1)}f_s]$Hz [12].

The 2014 International Power Electronics Conference

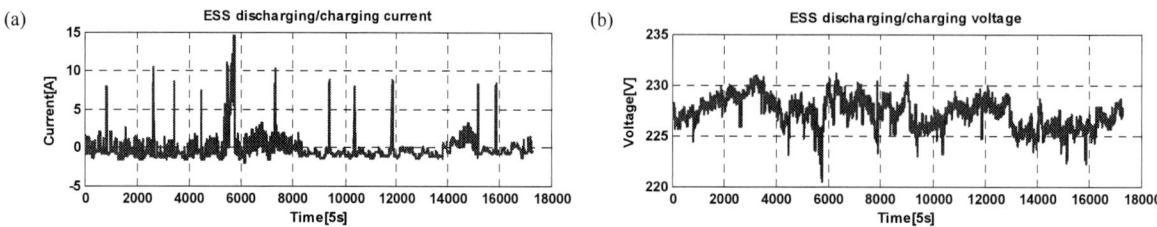

Fig. 7. Discharging/charging experimental data of the LiFePO$_4$ ESS (70S1P) including 70 prismatic cells that had a rated capacity 50Ah. (a) Discharging/charging current profile. (b) Discharging/charging voltage signal (DCVS).

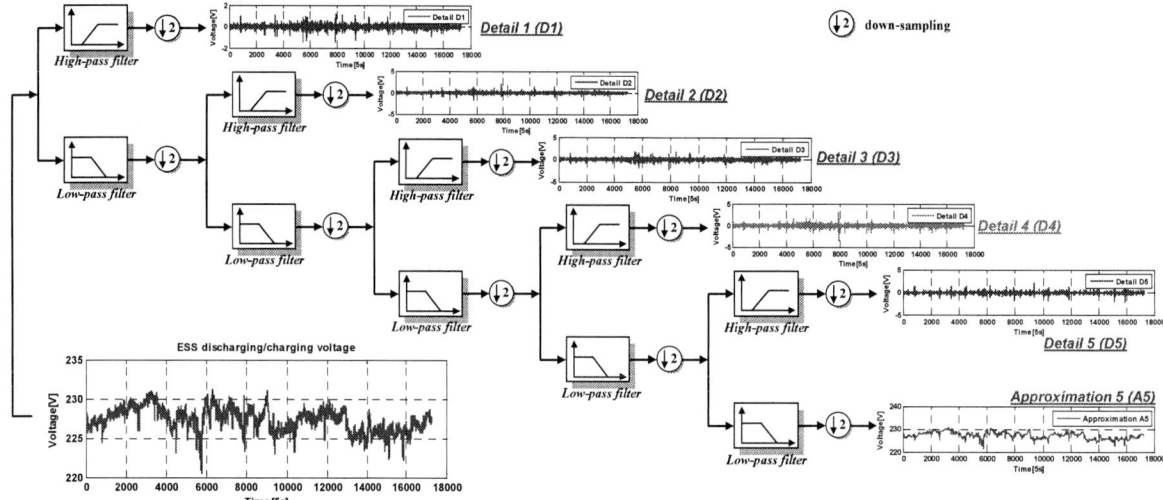

Fig. 8. Five-level decomposition process of the DWT-based MRA using the Daubechies wavelet as mother wavelet (dB3).

III. PROPOSED APPROACH

In this section, an approach of the DWT-based MRA for fault diagnosis of the unexpected and instantaneous voltage sag in the ESS is newly presented. For a discharging/charging current profile of the ESS (70S1P, LiFePO$_4$), the discharging/charging voltage signal (DCVS) is shown in Fig. 7. With only DCVS of Fig. 7(b), it is not easy to diagnose fault in the ESS. Therefore, in order to analyze and evaluate the DCVS with a non-stationary and transients, five-level decomposition of the multi-resolution analysis (MRA) is implemented for extracting information from these phenomena, as shown in Fig. 8. The original signal $x(t)$ is passed through high/low pass filters. Then, by showing down-sampling of the MRA, the outputs from both filters are decimated to obtain two components of detail and approximation at level 1 (A_1 and D_1). The approximation components are sent to the second state to repeat this procedure. Finally, the original signal $x(t)$ is decomposed at the expected level. After filtering, the two components contain redundancies and it is valid to down-sample each of the components by a factor of two, without losing any information. According to five-level decomposition shown in Fig. 8, A_5 with the lowest frequency band and $D_1 \sim D_5$ with high frequency bands can be obtained. The orthogonal wavelet filter 'dB3' of the Daubechies family that suitable for a wide of application is selected as the optimum wavelet function of the proposed work.

As aforementioned, high frequency resolution of decomposed signal can detect transient disturbance better

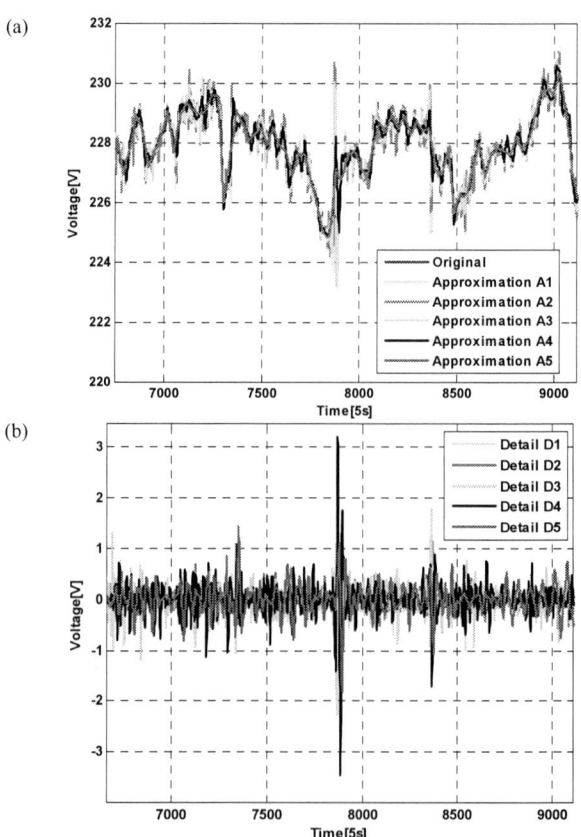

Fig. 9. Approximation and detail components by five-level decomposition. (a) Approximation $A_1 \sim A_5$. (b) Detail $D_1 \sim D_5$.

The 2014 International Power Electronics Conference

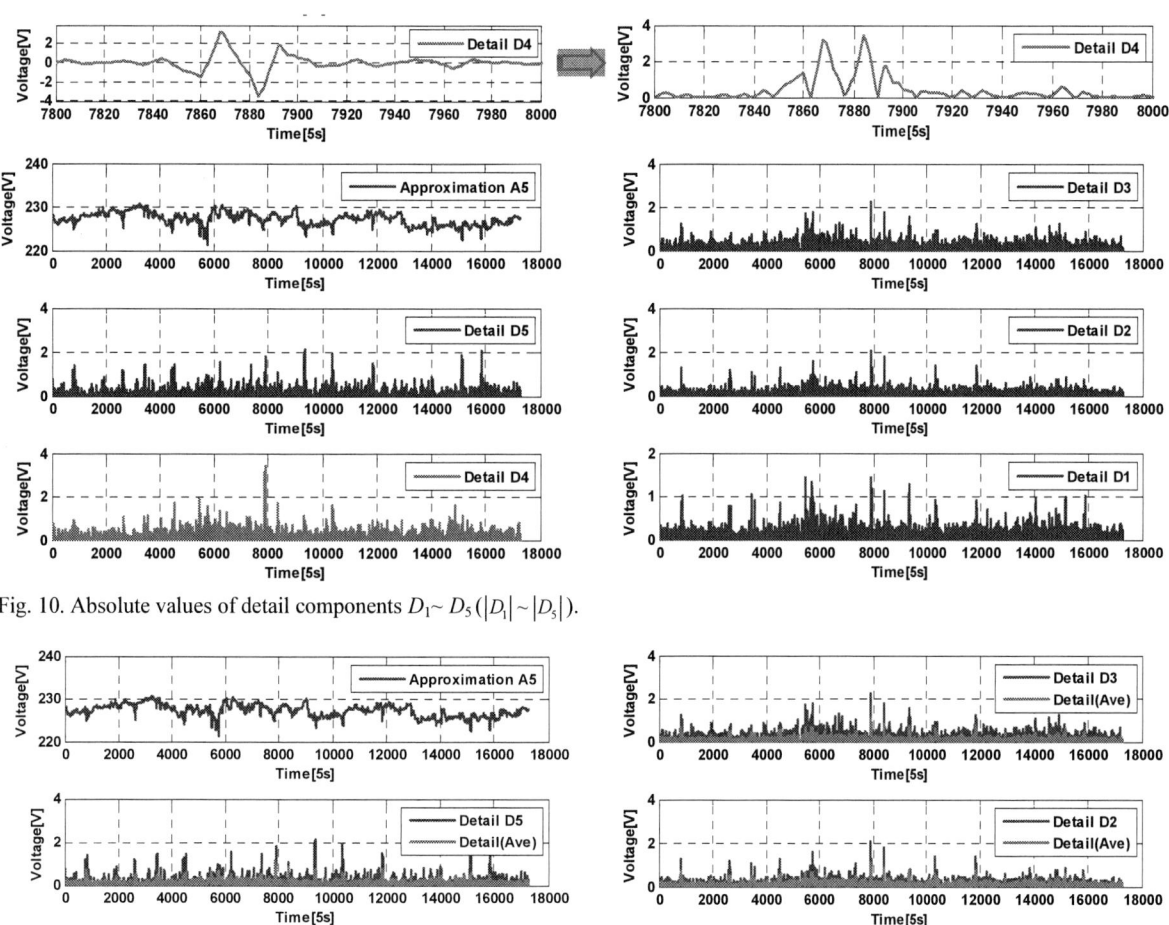

Fig. 10. Absolute values of detail components $D_1 \sim D_5$ ($|D_1| \sim |D_5|$).

Fig. 11. Plots for comparison between absolute values of detail components ($|D_1| \sim |D_5|$) and averaged value D_{avr} among detail $D_1 \sim D_5$ components.

than low resolutions. It can be seen that the fourth level (D_4) of the decomposed signal clearly shows a peak at the beginning and end of the voltage sag (Figs 8 and 9). In addition, as shown in Fig. 9, in comparison with approximation $A_1 \sim A_5$ components, a marked difference D_4 among detail $D_1 \sim D_5$ components is clearly shown. For reference, approximation components are generally similar with original signal DCVS. Therefore, detail D_4 component is considered for diagnosis of the interval fault caused by voltage sag in the ESS.

For verification of the selection D_4, as shown in Fig. 10, each absolute values of detail $D_1 \sim D_5$ components, $|D_1| \sim |D_5|$ are considered. In contrast to approximation A_n component, all detail D_n components have positive or negative values with zero basis. Fig. 11 shows the plots for comparison between absolute values of detail components $|D_1| \sim |D_5|$ and averaged value D_{avr} among detail $D_1 \sim D_5$ components (Eq. (6)).

$$\frac{(|D_1| + |D_2| + |D_3| + |D_4| + |D_5|)}{5} = D_{avr} \qquad (6)$$

Table I. Maximum value $|D_1|^{\max} \sim |D_5|^{\max}$ and ratio between $D_{avr}^{\max}$ and $|D_1|^{\max} \sim |D_5|^{\max}$

	$|D_1|^{\max}$	$|D_2|^{\max}$	$|D_3|^{\max}$	$|D_4|^{\max}$	$|D_5|^{\max}$	$D_{avr}^{\max}$
Mamimum value	1.4558	2.0627	2.2722	3.4434	2.1059	1.11958
Comparison with $D_{avr}^{\max}$	1.3003	1.8424	2.0295	3.0756	1.8810	-

It can be shown that a sharp difference between detail D_4 component and averaged value D_{avr}. Specifically, the magnitude of abrupt peak voltage is much larger than that of D_{avr}. For reference, the range of $D_{avr}^{\max}$ is considered as key factor for stable operation of the ESS. Using the absolute value of detail components $|D_1| \sim |D_5|$, each maximum value $|D_1|^{\max} \sim |D_5|^{\max}$ is obtained and listed in Table I. The comparison between $D_{avr}^{\max}$ and $|D_1|^{\max} \sim |D_5|^{\max}$ is additionally presented. In this approach, threshold value of 2.5 that decides fault occurrence is used. The ratio ($|D_n|^{\max} / D_{avr}^{\max}$) between $D_{avr}^{\max}$ and $|D_1|^{\max} \sim |D_5|^{\max}$ is larger than 2.5, it is determined that instantaneous fault is occurred in the ESS. As listed in Table I, the ratio between $|D_4|^{\max}$ and $D_{avr}^{\max}$ is only

978-1-4799-2706-7/14 $31.00 © 2014 IEEE

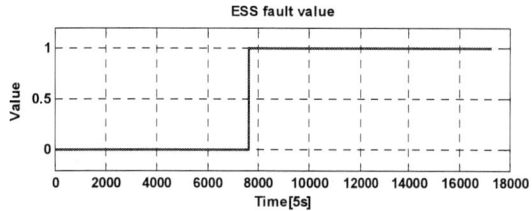

Fig. 12. Fault diagnosis in the ESS (sensing).
(0 = No fault, 1 = fault)

larger than 2.5 (Eq (7)). In contrast to $\left|D_4\right|^{max}$ and D_{avr}^{max}, four ratios between $\left|D_1\right|^{max}, \left|D_2\right|^{max}, \left|D_3\right|^{max}, \left|D_5\right|^{max}$ and D_{avr}^{max} are smaller than 2.5. Consequently, through information of detail component D_4, it is possible to correctly diagnose internal fault caused by the unexpected and instantaneous voltage sag in the ESS.

$$\frac{\left|D_4\right|^{max}}{D_{avr}^{max}} = 3.0756 > 2.5\,(threshold\ value) \qquad (7)$$

Fig. 12 shows the fault diagnosis using voltage/current sensors in the ESS. The value of '0' and '1' mean that there is no fault or fault in the ESS, respectively. As shown in Fig. 12, it can be known that fault was occurred at identical time (comparison with Figs 10 and 11). Thus, the proposed approach is very useful for fault diagnosis without specific voltage/current sensors (7900[5s]).

IV. CONCLUSION

This work focuses on the issue of fault diagnosis in the ESS based on DWT-based MRA. Using this technique in processing voltage-sag having DCVS dealing with a high frequency component D_n enables us to diagnose fault effectively in the ESS. The experimental results indicate the robustness of the proposed approach for the high-efficient fault diagnosis.

ACKNOWLEDGEMENT

This research was supported by Chosun University and Samsung SDI, Co. Ltd., republic of Korea.

REFERENCES

[1] H. Qian, J. Zhang, J.-S. Lai, and W. Yu, "A High-Efficiency Grid-Tie Battery Energy Storage System," *IEEE Transactions on Power Electronics*, vol. 26, no. 3, pp. 886-896, 2011.

[2] X. Fang, N. Kutkut, J. Shen, and I. Batarseh, "Analysis of generalized parallel-series ultracapacitor shift circuits for energy storage systems," *Renewable Energy*, vol. 36, pp. 2599-2604, 2011.

[3] C. Sen, Y. Usama, T. Carciumaru, X. Lu, and N. C. Kar, "Design of a Novel Wavelet Based Transient Detection Unit for In-Vehicle Fault Determination and Hybrid Energy Storage Utilization," *IEEE Transactions on Smart Grid*, vol. 3, no. 1, pp. 422-433, 2012.

[4] Ö. Gencer, S. Öztürk, and T. Erfidan, "A new approach to voltage sag detection based on wavelet transform," *Electrical Power and Energy Systems*, vol. 32, pp. 133-140, 2010.

[5] A. Y. Goharrizi and N. Sepehri, "A Wavelet-Based Approach to Internal Seal Damage Diagnosis in Hydraulic Actuators," *IEEE Trans. Ind. Electron.*, vol. 57, no. 5, pp. 1755–1763, May. 2010.

[6] M. R.-Guasp, J. A. A.-Daviu, M, P.-Sanchen, R. P.-Panadero, and J. P.-Cruz, "A General Approach for the Transient Detection of Slip-Dependent Fault Components Based on the Discrete Wavelet Transform," *IEEE Trans. Ind. Electron.*, vol. 55, no. 12, pp. 4167–4180, Dec. 2008.

[7] D. Giaouris, J. W. Finch, O. C. Ferreira, R. M. Kennel, and G. M,. El-Murr, "Wavelet Denoising for Electric Drives," *IEEE Trans. Ind. Electron.*, vol. 55, no. 2, pp. 543–550, Feb. 2008.

[8] M. Uzunoglu and M. S. Alam, "Modeling and Analysis of an FC/UC Hybrid Vehicular Power System Using a Novel-Wavelet-Based Load Sharing Algorithm," *IEEE Trans. Energy Convers.*, vol. 23, no. 1, pp. 263–272, Mar. 2008.

[9] A. Bouzida, O. Touhami, R. Ibtiouen, A. Belouchrani, M. Fadel, and A. Rezzoug, "Fault Diagnosis in Industrial Induction Machines Through Discrete Wavelet Transform," *IEEE Trans. Ind. Electron.*, vol. 58, no. 9, pp. 4385–4395, Sep. 2011.

[10] http://www.samsungsdi.com.

[11] A. Bouzida, O. Touhami, R. Ibtiouen, A. Belouchrani, M. Fadal, and A. Rezzoug, "Fault diagnosis in industrial machines through discrte wavelet transform," *IEEE Trans. Ind. Electron.*, vol. 58, no. 9, pp. 4385-4395, Sep. 2011.

[12] J. Kim, G.-S. Seo, B. H. Cho, W. Kim, J. Park, and T. Ishikawa, "Discrete Wavelet Transform-Based Characteristic Analysis and SOH Diagnosis for a Li-Ion Cell," in Proc. IEEE Int. Power Electronics and Motion Conference, Haerbin, China, Jun. 2012.

The 2014 International Power Electronics Conference

Comparison of Different IGBT based Designs of Power Electronic Transformer

Xinyu WANG, Shaodi Ouyang, Jinjun LIU, Fei MENG, Riffat JAVED
State Key Lab of Electric Insulation and Power Equipment and School of Electrical Engineering
Xi'an Jiaotong University
Xi'an, China
ericwsino@gmail.com

Abstract—Power electronic transformer (PET) has many merits as compared with conventional power transformer. For high-voltage high-power applications, silicon based IGBT and the cascaded modular structure are still the mainstream power semiconductor device and structure, respectively. In this paper, a basic design process of the power stage of an H-bridge based three-stage PET system is presented. Based on the design process, different PET designs based on different voltage rating IGBTs are done and compared. This is a reference for practical PET system design.

Keywords— *Power electronic transformer, three-stage, cascaded modular structure.*

I. INTRODUCTION

In last two decades, power electronic transformer (PET) or solid state transformer (SST) has attracted increasing attention from both academia and industry [1-9]. As compared with conventional copper-iron based power transformer, PET has many advantages in the aspects of size, weight, and functions (i.e. ability of power factor correction, renewable energy interfacing, etc.). It has been regarded as one of the most emerging technologies.

Most of the released literatures focus on the PET topologies and related control strategies for different applications [1-7]. All the PET topologies can be classified into three categories: the single-stage ones without DC link, the two-stage ones with one DC link at either high-voltage side or low-voltage side, and the three-stage ones with two DC links at both high-voltage side and low-voltage side. The more DC links are, the more potential functions can be developed. Therefore, the three-stage ones attract the most researching attention. However, to the best of the author's knowledge, only a few papers focus on the system design [8-9]. From the practical application point of view, the optimum design of the PET system is reasonable and important.

The power electronics related technologies have a very fast development in the past half century. However, according to the state-of-the-art power semiconductor device technology, commercial SiC devices with reasonable price are still not available. Therefore, for high-voltage high-power applications, the silicon based

IGBT and the cascaded modular structure are the popularly practical power semiconductor device and topological structure, respectively. In literature [7], comparison of different three-phase three-stage modular PET topologies was given. With the comprehensive comparison, the topology as shown in Fig. 1 is selected for high-voltage high-power applications which do not require a common low-voltage DC bus. In literature [6], the authors proposed the overall control strategy for the topology.

In this paper, authors focus on the system power stage design. A design process for the PET power stage is presented. Different designs based on different voltage rating IGBTs are done and compared. Based on the design process and comparison, an optimizing design method is proposed as a reference for the practical system design.

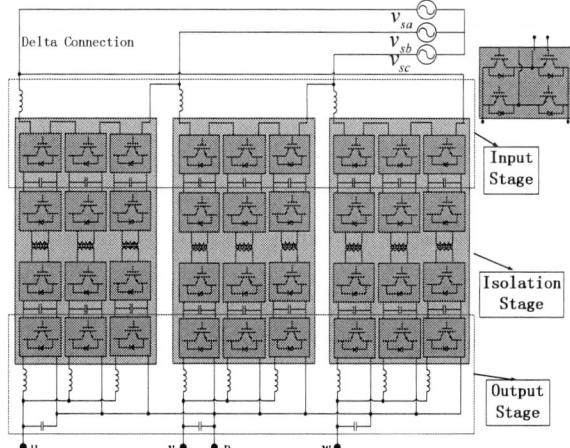

Fig. 1   Three-stage H-bridge based modular PET topology.

II. SYSTEM POWER STAGE DESIGN

The PET topology as depicted in Fig. 1 consists of three stages: a cascaded multilevel AC/DC rectifier, several individual dual active bridges (DAB), and a three-phase DC/AC inverter. All the converters of the three stages are H-bridges. The input stage is responsible for realizing unit power factor and input-side DC-link voltage balancing. Each DAB is responsible for maintaining output-side DC-link voltage and electrical isolation. The output stage is responsible for output

978-1-4799-2706-7/14 $31.00 © 2014 IEEE   624

voltage regulation and parallel-module current sharing control. The control block diagrams are given in [6].

In this paper, the PET system specification is set 11-kV/1-kV 5-MW. The designs are based on 6.5-kV, 3.3-kV and 1.7-kV IGBTs. In the following design process, the 3.3-kV IGBT based design will be demonstrated as an example. In this paper, the ratio that DC link voltage amplitude can be applied to IGBT is set 2/3 of the rated value; the ratio that current stress can be applied to IGBT is set 1/2 of the rated value. In practical design, these ratios can be adjusted according to different application requirements.

A. Input stage design

The PET input stage configuration is the same as the H-bridge cascaded STATic synchronous COMpensator (STATCOM). Therefore, the system design method of STATCOM [10] can be utilized. The input-side DC link is calculated by (1). The cascade number is calculated by (2). The modulation ratio is reevaluated by (3). The rated current stress is calculated by (4). The rated current stress can be adjusted according to additional requirements such as grid voltage sag. Herein, for the sake of simplicity, only the rated current stress is selected. Therefore, 400-A/3.3-kV IGBT is selected.

$$V_{HVDC} = 3.3\text{kV} \div 1.5 = 2.2\text{kV} \tag{1}$$

$$N = \frac{11\text{kV} \times 1.6}{2.2\text{kV}} = 8 \tag{2}$$

$$m_{in} = \frac{11\text{kV} \times \sqrt{2}}{8 \times 2.2\text{kV}} = 0.884 \tag{3}$$

$$I_{in} = \frac{5\text{MW} \times \sqrt{2}}{11\text{kV} \times 3} \approx 214.3\text{A} \tag{4}$$

where V_{HVDC} is the input-side DC link voltage; N is the input-stage cascade number; m_{in} is the input-stage modulation index; I_{in} is the input-stage current rated peak amplitude.

B. Output stage design

The output stage is comprised of parallel modules. The output line-to-line voltage is 1kV. The output-side DC link voltage can be evaluated by (5). Therefore, the output-side DC link voltage is set 950-V. The rated current stress can be calculated by (6). The modulation index is reevaluated by (7). Therefore, 1200-A/1.7-kV IGBT is selected for the output stage.

$$V_{LVDC} = 1000\text{V} \div \sqrt{3} \times 1.6 = 924\text{V} \tag{5}$$

$$I_{out} = \frac{5\text{MW} \times \sqrt{2}}{1000\text{V} \div \sqrt{3} \times 3 \times 8} \approx 510.3\text{A} \tag{6}$$

$$m_{out} = \frac{1000\text{V} \div \sqrt{3} \times \sqrt{2}}{950\text{V}} = 0.859 \tag{7}$$

where V_{LVDC} is the output-side DC link voltage; m_{out} is the output-stage modulation index; I_{out} is the output-stage current rated peak amplitude.

C. Isolation stage design

The isolation stage is comprised of individual DABs [11]. Phase shift modulation is applied to DAB which can realize zero voltage switching (ZVS) to improve the efficiency. Theoretically, when d is set 1, full ZVS range of φ can be achieved. Therefore, the transformer turns ratio can be set by (9). In this design, the switching frequency f_S is set 2 kHz, and the steady state phase shift angle φ is set $\pi/4$. The transformer leakage inductor can be calculated by (11). The rated current stresses of primary-side and secondary-side converters of DAB can be calculated by (12) and (13), respectively. Therefore, 400-A/3.3-kV IGBT is selected for the primary-side converter, and 600-A/1.7-kV IGBT is selected for the secondary-side converter.

$$d = \frac{V_{HVDC} \cdot n_S}{V_{LVDC} \cdot n_P} = 1 \tag{8}$$

$$n = n_p : n_s = 44 : 19 \tag{9}$$

$$P_{DAB} = \frac{5\text{MW}}{8 \times 3} \approx 208.3\text{kW} \tag{10}$$

$$L_S = \frac{n \cdot V_{HVDC} \cdot V_{LVDC}}{2\pi f_S \cdot P_{DAB}} \varphi \left(1 - \frac{|\varphi|}{\pi}\right) = 1.089\text{mH} \tag{11}$$

$$I_{P\max} = \frac{P_{DAB}}{V_{HVDC} \cdot (1 - \frac{|\varphi|}{\pi})} = 126\text{A} \tag{12}$$

$$I_{S\max} = \frac{P_{DAB}}{V_{LVDC} \cdot (1 - \frac{|\varphi|}{\pi})} = 292.4\text{A} \tag{13}$$

where n_p and n_s are the transformer's primary and secondary turns, respectively; n is the transformer turns ratio; P_{DAB} is the rated power transmitted by one DAB; L_S is the transformer leakage inductance; f_S is the switching frequency; φ is the phase shift angle; I_{Pmax} is the DAB primary-side converter current rated peak amplitude; I_{Smax} is the DAB secondary-side converter current rated peak amplitude.

D. Input-side interfacing inductor and output LC filter

For the inductor design, there are some requirements, such as ability of four-quadrant operation, fast current tracking response, suppressing high-frequency harmonic and reducing switching-frequency current ripple. According to these requirements, the inductor range can be fixed by (14). However, the range is normally very large. For the input-side interfacing inductor, it can be evaluated by the empirical equation (15), where the '10%' can be replaced by 5%-15% according to different papers. For the LC filter, many papers addressed different design methods. In this paper, the LC filter parameters are design by (16) and (17). In this system, $f_{har(min)}$ is around 10 kHz, therefore, f_L can be chosen as 1-kHz. The output line frequency is 50-Hz.

$$\begin{cases} \dfrac{E_m(nV_{dc}-E_m)}{\Delta i_{max} \cdot nV_{dc}} T_S \le L \le \dfrac{V_{dc}}{I_m \omega} \\ L \le \dfrac{(nV_{dc}-E_m)}{I_m \omega} \end{cases} \tag{14}$$

$$L_{in} = \frac{10\% \cdot U_n}{\omega_{in} I_n} = 23.109 \text{mH} \tag{15}$$

$$L_{out} = \sqrt{\frac{\left[\frac{\omega_o U_o^2}{\omega_L^2}+\frac{\omega_o^3 U_o^2}{\omega_L^4}\right]}{\omega_o I_o^2}} = 0.255 \text{mH} \tag{16}$$

$$C_{out} = \frac{1}{\omega_L^2 L} = 99.35 \mu\text{F} \tag{17}$$

where E_m and I_m are the AC-side voltage and current peak amplitudes, respectively; n is the H-bridge module number; V_{dc} is the H-bridge DC link voltage; TS is the equivalent switching period; Δi_{max} is the maximum current fluctuating peak-to-peak amplitude; ω is the AC-side line angular frequency; L_{in} is the input-stage interface reactor; L_{out} is the output-stage LC filter inductor; C_{out} is the output-stage LC filter capacitor; U_o and I_o are the output voltage and current, respectively; ω_o and ω_L are the output line angular frequency and LC filter resonant angular frequency.

E. DC link capacitors

All the converters in the PET topology are H-bridges. Therefore, the DC link capacitor is mainly to suppress the second-order harmonics. The required capacitance can be calculated by (18).

$$C_{dc} = \frac{P_m}{2\omega V_{dc} \Delta V_{dc}} \tag{18}$$

where P_m is the power amount transmitted through one module; ΔV_{dc} is the DC link voltage fluctuating peak-to-peak value; ω is the line angular frequency.

III. COMPARISON OF DIFFERENT DESIGNS

Based on the above design process, different designs based on 6.5kV, 3.3kV and 1.7kV IGBTs are conducted. Tab I demonstrates the IGBT parameter specification and amount of different designs. Tab II depicts the amounts of the module and other components, semiconductor cost of different designs.

TABLE I
IGBT AMOUNT AND REQUIREMENT OF DIFFERENT DESIGNS

Designs	IGBT in different stages		
	AC-DC	DC-DC	DC-AC
6.5kV	48×6.5kV/400A	48×6.5kV/400A +48×1.7kV/1200A	48×1.7kV/2400A
3.3kV	96×3.3kV/400A	96×3.3kV/400A +96×1.7kV/600A	96×1.7kV/1200A
1.7kV	192×1.7kV/450A	192×1.7kV/300A +192×1.7kV/300A	192×1.7kV/600A

TABLE II
MODULE AND OTHER COMPONENT AMOUNT, SEMICONDUCTOR COST OF DIFFERENT DESIGNS

Design	Module	L	Transformer	C	Semi Cost
6.5kV	48	3+12	12	24	$301,641.60
3.3kV	96	3+24	24	48	$212,217.60
1.7kV	192	3+72	48	96	$150,853.76

A. Efficiency evaluation

For high-power systems, the dominant power loss is the semiconductor device loss. Literature [12] addressed detailed power loss calculation criteria of IGBT. Equation (19) addresses the IGBT and diode conducting model, where u_{th} and r_{dson} can be obtained from datasheet. The conducting losses can be calculated by (20). IGBT switching loss and diode reverse-recovery loss can be calculated by (21) and (22), where coefficients A, B, and C can also be obtained from datasheet. The efficiency of input stage and output stage can be calculated by (19)-(22).

$$u_{on} = u_{th} + r_{dson} \cdot i_{con} \tag{19}$$

$$P_{con} = \frac{1}{2\pi} \int_{\alpha_1}^{\alpha_2} d_{con} \cdot u_{on} \cdot i_{con} d\omega t \tag{20}$$

$$E_{SW} = (A + B \cdot i_{sw} + C \cdot i_{sw}^2) \frac{V_{dc}}{V_{ref}} \tag{21}$$

$$P_{SW} = f_s \cdot \frac{1}{2\pi} \int_{\alpha_1}^{\alpha_2} E_{SW} d\omega t \tag{22}$$

where u_{th} is IGBT or diode threshold voltage; r_{dson} is IGBT or diode turning-on resistor; i_{con} is conducting current; d_{con} is conducting duty ratio; V_{ref} is testing voltage V_{ce} in the device datasheet; A,B and C are curve coefficients that can be obtained from datasheet; f_S is switching frequency; E_{SW} is switching energy that can be obtained from datasheet.

Since ZVS is applied to all switches in the DAB, the IGBT turning-on loss and diode reverse-recovery loss can be neglected. Therefore, the switching loss is just the turning-off loss. The efficiency of the isolation stage power can be calculated by (23) and (24).

$$P_{con} = f_S \cdot \int_0^{1/f_S} [u_{th} + r_{dson} \cdot i(t)] \cdot i(t) dt \tag{23}$$

$$P_{SW} = f_S \cdot E_{SW}(\tfrac{1}{2f_S}) \tag{24}$$

As the rated voltage of IGBT is higher, the highest switching frequency that is practical is lower. For higher rated voltage IGBT, the increasing switching frequency will rapidly decrease the efficiency. In order to obtain the same input stage terminal equivalent switching frequency 32 kHz of each design, the input stage switching frequency is set in the second column of Tab III. The DAB frequency is set as the third column. The switching frequency of 6.5kV IGBT based design is set only 1 kHz so that the efficiency will be not too lower than the others. All the output stage switching frequencies are set as the same. Based on the above analysis and theoretical calculation, the efficiency of each design is listed at the fifth column in Tab III.

TABLE III
SWITCHING FREQUENCY AND EFFICIENCY

Designs	AC-DC f_s (kHz)	DC-DC f_s (kHz)	DC-AC f_s (kHz)	Total Efficiency
6.5kV	4	1	5	89.72%
3.3kV	2	2	5	93.61%
1.7kV	1	5	5	94.31%

B. Size and weight evaluation

The size and weight of the whole system can be divided into three parts: the main circuit part (semiconductor devices, heatsinks and mechanical steadier), the passive component part (DC-link capacitors, inductors and transformers), and extra part (including driving circuit, power supply, etc.). Since the power rating and voltage level of the system is very high, the extra part makes up such a small proportion of the whole system that can be neglected. The passive component part as compared with the main circuit part also makes up a small proportion of the whole system.

According to the power rating that high-frequency transformer need to handle, different magnetic cores are needed. The core cross-sectional area A_e, core windows area A_w, and the product A_p of A_e and A_w are provided by the manufacturers and can represent the size and weight of the transformer. Equation (25) shows the single transformer power handling capability, which is reversely proportional to operating frequency. Each design has the same total power. Therefore, the 1.7 kV IGBT based design has the smallest size and weight of the transformer part among the three designs.

$$A_p = A_e A_w = \frac{(P_{in} + P_o)(10^4)}{B_{ac} f J K_f K_u}$$
(25)

The inductor can be designed by (26). Since the whole system power ratings are the same, the size and weight of this part of different designs are approximately the same.

$$n \cdot A_p = \frac{2P_{system}(10^4)}{B_{ac} f J K_f K_u}$$
(26)

Where P_{in} and P_o are the transformer input and output power, respectively; A_e is core cross-sectional area; A_w is windows area; B_{ac} is operating flux density, T; f is operating frequency, Hz; K_f is waveform factor, 4.0 for square wave; and 4.44 for sine wave; K_u is windows area utilization factor; J is current density, A/cm^2; n is the transformer number.

According to (18), the DC link capacitor parameters of each design are calculated. Tab IV lists the capacitor parameters of each design. These capacitor requirements should be met by unit capacitor with series-connection and parallel-connection. In this paper, the DC link capacitor size and cost is evaluated by (27) and (28). If these two ratios of different designs are the same, then, the total amount of unit capacitor of each design are the same.

TABLE IV
DC LINK CAPACITOR

Designs		6.5kV	3.3kV	1.7kV
Input side	Capacitance	342.5uF	685.1uF	1370.1uF
	Voltage Stress	4620V	2310V	1155V
Output side	Capacitance	7347.9uF	3673.9uF	1837.0uF
	Voltage Stress	997.5V	997.5V	997.5V

$$\eta_V = \frac{V_{dc}}{n \cdot V_{unit}} \times 100\%$$
(27)

$$\eta_C = \frac{C_{dc}}{m \cdot C_{unit}/n} \times 100\%$$
(28)

where V_{dc} is the capacitor voltage stress; V_{unit} is rated voltage stress of the selected unit capacitor; C_{dc} is the required capacitance; C_{unit} is the capacitance of the unit capacitor; n is the series-connected number; m is the parallel-connected number.

The main circuit's size and weight are the dominant one as compared with others. The main circuit's size and weight are mainly dependent on the heatsink since the semiconductor's size and weight are much smaller as compared with heatsink. The authors of [13] provide a method to estimate the size of heatsink. The ambient temperature and the semiconductor device junction temperature for different designs are set 30 °C and 100 °C, respectively. The ratio x of the required heatsink's size over the unit heatsink's size for different designs can be calculated by (30). Obviously, the most efficient design will have the smallest size and weight.

$$T_j = T_{ambient} + P(R_{thJC} + R_{heatsink})$$
(29)

$$x = \frac{R_{heatsink}}{\frac{T_j - T_{ambient}}{P_{swtich}} - R_{thJC}}$$
(30)

where $T_{ambient}$ is ambient temperature; T_j is the semiconductor device junction temperature; P is semiconductor device power loss; R_{thJC} is IGBT thermal resistance regarded to heatsink; $R_{hestsink}$ is heatsink thermal resistance.

Tab V shows the parameters of reference heatsink and the relationship between an arbitrary heatsink and the reference heatsink. According to the efficiencies and the reference heatsink, the size and weight of each design can be evaluated as demonstrated in Tab VI. It can be found that the 6.5kV IGBT based design has the biggest size and most heavy weight. The 3.3 kV and 1.7 kV IGBTs based designs have close performance on the size and weight.

TABLE V
REFERENCE HEATSINK AND ARBITRARY HEATSINK

Size	500W*200D*80H(mm)	x*100000mm^2*80mm
Weight	7.8kg	x*7.8kg
Thermal resistance	0.021°C/W	0.021/x °C/W

TABLE VI
SIZE AND WEIGHT OF EACH DESIGN

Schemes	Size[500W*200D *80H(mm)]	Weight		
	Amount	Heatsink	Device	Total
6.5kV	546.6	4263.5kg	204kg	4467.5kg
3.3kV	102.72	801.22kg	249.60kg	1050.80kg
1.7kV	89.76	700.13kg	242.40kg	942.53kg

According to the above analysis, the comprehensive performance on cost, efficiency, size and weight are demonstrated in Fig. 2. For the cost, size and weight, the 1.7 kV and 3.3 kV IGBT based designs are normalized to 6.5 kV IGBT based design. It can be easily found that the 1.7 kV and 3.3 kV IGBT based designs have very close performance on size and weight, however, the cost of the 3.3 kV IGBT based design is higher than the other one. However, the 1.7 kV IGBT based design has 192 modules. The large module amount will lead the control more complex. The 3.3 kV IGBT based design has 96 modules, which is only 50% of the module amount of 1.7 kV IGBT based design.

From a comprehensive point of view, the 3.3 kV IGBT based design is more suitable and practical.

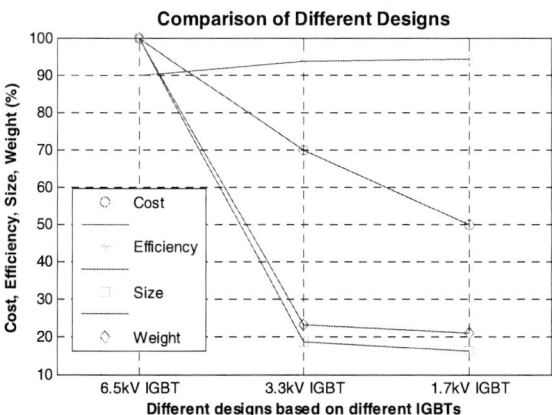

Fig. 2   Comparison of different designs

IV. CONCLUSIONS

In this paper, the PET design process is presented. Different designs based on different voltage rated IGBT

are compared. An optimizing method is proposed as a reference for the practical PET design.

REFERENCES

[1] M. Kang, P. N. Enjeti, and I. J. Pitel, "Analysis and design of electronic transformers for electric power distribution system," *IEEE Trans. Power Electron.*, vol.14, no. 6, pp. 1133–1141, Nov. 1999.

[2] E. R. Ronan, S. D. Sudhoff, S. F. Glover, et al, "A Power Electronic-Based Distribution Transformer," *IEEE Trans. Power Delivery*, vol. 17, pp. 537 - 543 , Apr. 2002.

[3] A. Q. Huang, M. L. Crow, G. T. Heydt, J. P. Zheng, and S. J. Dale, "The future renewable electric energy delivery and management system: The energy internet," *Proc. IEEE*, vol. 99, no. 1, pp. 133–148, Jan. 2011.

[4] D. Dujic, C. Zhao, A. Mester, J. K. Steinke, et al, " Power electronic traction transformer-low voltage prototype," *IEEE Trans. Power Electron.*, vol.28, no. 12, pp. 5534–5522, Dec. 2013.

[5] T. Zhao, G. Wang, S. Bhattacharya, and A.Q. Huang, "Voltage and power balance control for a cascaded H-bridge converter-based solid state transformer," *IEEE Trans. Power Electron.*, vol. 28, no. 4, pp. 523-1532, Apr. 2013.

[6] X. Wang, J. Liu, T. Xu, X. Wang, et al, "Control of a three-stage three-phase cascaded modular power electronic transformer," in *Proc. IEEE. Appl. Power Electron. Conf.*, Mar. 2013, pp. 1309-1315.

[7] X. Wang, J. Liu, T. Xu, et al, "Comparisons of different three-stage three-phase cascaded modular topologies for power electronic transformer," in *Proc. IEEE ECCE*, Sep. 2012, pp. 1420-1425.

[8] G. Wang, S. Baek, J. Elliott, A. Kadavelugu, et al, "Design and hardware implementation of gen-1 silicon based solid state transformer," in *Proc. IEEE. Appl. Power Electron. Conf.*, Mar. 2011, pp. 1344-1349.

[9] J. E. Huber, and J. W. Kolar, "Optimum number of cascaded cells for high-power medium-voltage multilevel converters," in *Proc. IEEE ECCE*, Sep. 2013, pp. 359-366.

[10] H. Akagi, S. Inoue, T. Yoshii, "Control and performance of a transformerless cascade PWM STATCOM with star configuration," *IEEE Trans. Ind. Appl.*, vol. 43, no. 4, pp. 1041-1049, Jul./Aug., 2007.

[11] R. W. De Doncker, D. M. Divan, and M. H. Kheraluwala, "A three phase soft-switched high-power density dc/dc converter for high-power applications," *IEEE Trans. Ind. Appl.*, vol. 27, no. 1, pp. 63–73, Jan./Feb. 1991.

[12] D. Graovac, M. Pürschel, "IGBT power losses calculation using the data-sheet parameters," Infineon, Application Note, V 1.1, Jan. 2009.

[13] T. Zhao, J. Wang, A.Q. Huang, A. Agarwal, "Comparisons of SiC MOSFET and Si IGBT based motor drive systems," in *Proc. IEEE 42nd Ann. Ind. Appl. Conf.*, 2007, pp. 331–335.

Semi-adaptive harmonic control for power balancing device for AC traction

Masataka AKAGI
Power Supply Laboratory
Railway Technical Research Institute (RTRI)
Tokyo, Japan
akagi@rtri.or.jp

Hironori TSURUTA
JRTT
Tokyo, Japan

Hiroshi OSO
Fuji Electric Co., Ltd.
Tokyo, Japan

"RPC: Railway Static Power Conditioner" is a proven static voltage compensator system with a pair of single phase voltage source inverters in the feeding substation to avoid a voltage fluctuation of 3-phase power grid under an single phase AC electric railway load. Several substations for Shinkansen or conventional rail substations have RPCs until now.

In Hokuriku Shinkansen section, there is another RPC with a function of harmonic compensation. There, in the case of emergency, length of the extended feeding circuit would exceed 100km. In such a case, stray capacitance between catenary and the earth would cause severe harmonic resonance. Thus RPC for this section introduce a semi-adaptive harmonic control algorithm to avoid a self-resonance. RPC watches active load current continuously and if load current is smaller than threshold level, RPC stops compensating harmonics. Also if compensating current for each harmonics order is too large, RPC stops the order. Authors have already conformed an appropriate operation for this semi-adaptive harmonic control algorithm with a simulation. RPC will test its ability in winter 2013 and will in service in spring 2015.

Keywords— RPC, Traction, Harmonics.

I. INTRODUCTION

Since 1964, JR groups' "Shinkansen" high speed train system is a backbone of Japanese transportation. Total length of Shinkansen exceeds 2,300km until 2013, and two new lines "Hokuriku" and "Hokkaido" are under construction toward 2015 and 2016, respectively.

Shinkansen is fed by single phase commercial frequency (50Hz in eastern part of Japan and 60Hz in western part) 25kV from catenary. The power of AC feeding electric railways should be received from a powerful and high-tension source. In the railway substation, the feeding transformer transfers three-phase electric power to a pair of single-phase feeding circuits. They are single-phase loads, changing independently of each other, sometimes including harmonics. A single phase load would cause a voltage unbalance in a three-phase electric power system and harmonics would

constitute an obstacle. In Japan, railways consuming large power are expected to run faster and faster, and they often pass an area of weak power supply.

Therefore JR groups should take countermeasures for a voltage unbalance and voltage fluctuations of a three-phase electric power system and harmonics. For such a situation, a kind of the configuration of compensator using power electronics called RPC (Railway static Power Conditioner) is commonly used in Japan.[1][2][3]

II. CONFIGURATION OF RPC

Railway substations receive electric power from 3-phase power grid. In Japan, 3-phase to 2-phase feeding transformer such as Scott-connected transformer is commonly used for AC fed traction. In this case, if the volume of feeding current for both feeding circuit - Main phase and Teaser - are equal and their power factors are 1.0, then 3-phase side is balanced.

3-phase voltage fluctuations ratio is calculated by the following equations. Where V_{RS}, V_{ST}, V_{TR} are 3-phase side voltages; I_M, I_T are load currents of Main phase (M-Phase) and Teaser; θ_M, θ_T are load phase angle of M-phase and Teaser; P_S is short circuit capacity of power source; V_0 is rated voltage of power source.

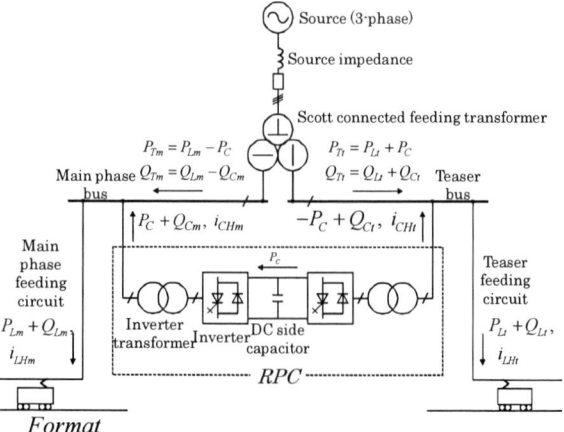

Fig. 1.   RPC (Railway Static Power Conditioner).

$$\Delta V_{RS} = \frac{V_0}{P_S}\{I_M\cos(\theta_M - \gamma + \pi/3) + \sqrt{3}I_T\cos(\theta_T - \gamma - \pi/6)\}$$

$$\Delta V_{ST} = \frac{V_0}{P_S}\{I_M\cos(\theta_M - \gamma - \pi/3) + \sqrt{3}I_T\cos(\theta_T - \gamma + \pi/6)\}$$

$$\Delta V_{TR} = \frac{2V_0}{P_S}I_M\cos\theta_M$$

$$- (1)$$

These equations show that when the value of 3-phase each voltage fluctuations ratio Δ_{RS}, Δ_{ST}, Δ_{TR} become minimum (0), which means the condition of $|I_{M|} = |I_{T|}$ and $\theta_M = \theta_T$; a pair of single phase load powers are equal and both power factors are equal. And voltage fluctuations, become minimum 0 when $\theta_M = \theta_T = 0$ (both power factors are equal to 1).

RPC is made up of a pair of self-commutated voltage source inverters, which connect both phases of single phase feeding circuit, coupled with DC side capacitor, shown as Fig. 1. PWM control circuit drives IGBTs (Insulated Gate Bipolar Transistor) of the inverters.

RPC can compensate a voltage unbalance and fluctuation in a three-phase electric power grid, with keeping a balance in a pair of feeding circuits. To compensate the voltage unbalance of a three-phase side, RPC controls the reactive power of the feeding circuit, and the active power also between two feeding circuits rapidly to railway substations. Therefore, the reactive power doesn't have to pass the feeding transformer.

Since 2013, all the Shinkansen rolling stocks are PWM controlled so that they produce less harmonic distortion current than former trains such as the thyristor phase control car. However, harmonic distortion current causes induced voltage on communication lines near the rail. So an electric railway usually has to control harmonic distortion. Self-commutated voltage source inverters can produce a voltage pulse from DC capacitor with optional timing. This way, RPC configured at the feeding substation, can act as an active filter to compensate harmonic distortion of feeding circuits.

Thus the RPC can handle all troubles caused by traction and it can be expected to be an ideal conditioner for the electric railway.

Fig. 2. Second section of Hokuriku-Shinkansen

III. HOKURIKU SHINKANSEN LINE

A. Second section of Hokuriku Shinkansen line

Hokuriku Shinkansen line was already opened first 117km section between Takasaki and Nagano since 1997. Next 229km section between Nagano and Kanazawa will go into operation in 2015 (Fig. 2).

Because of the historical reason, Japan uses two set of commercial frequency of power grid, 50Hz and 60Hz. The first section had passed the frequency border once,

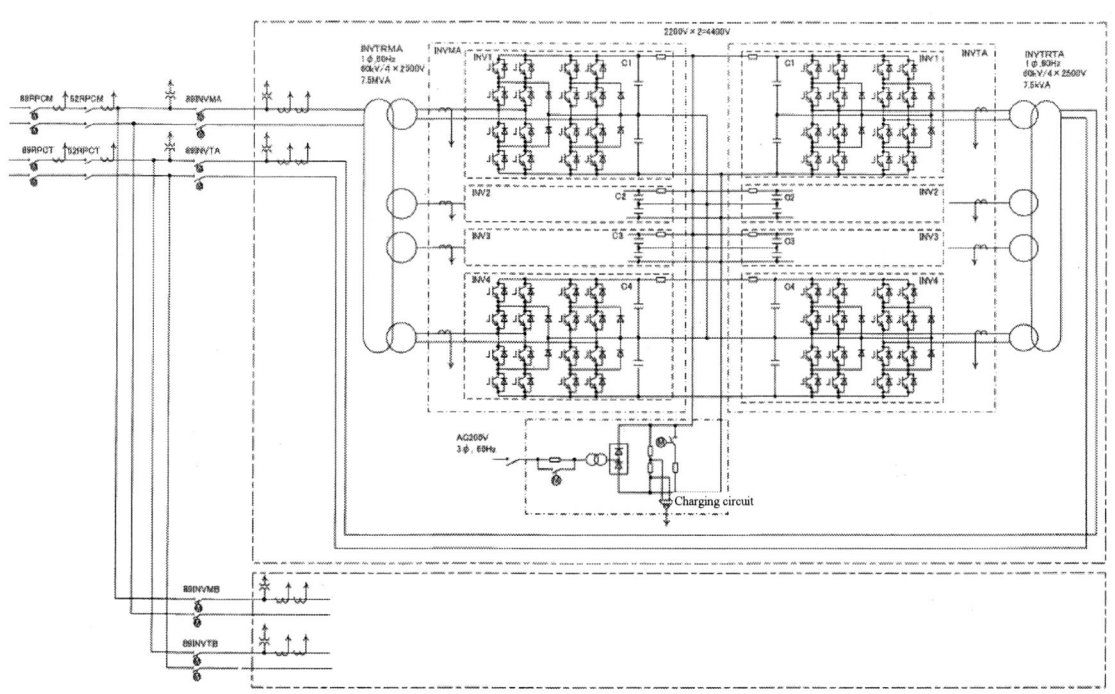

Fig.3 Main circuit of RPC for Shin-Kurobe substation.

978-1-4799-2706-7/14 $31.00 © 2014 IEEE   630

but, the new section has to across borders twice, shown as Fig. 2.

Since intermediate 50Hz section has only one substation – Shin-Joetsu –, in the case of emergency, 60Hz substations have to feed power into intermediate section. If Shin-Kurobe substation feed an extended feeding into intermediate section, the feeding length would reach about 100km.

B. RPC set for Shin-Kurobe substation

In the second section, Shin-Kurobe substation has a relatively smaller short circuit capacity than other substations. So that RPC would set in Shin-Kurobe substation to adjust voltage unbalance there. Rated capacity of RPC in Shin-Kurobe is 30MVA.

Fig.3 shows the main circuit of RPC using IGBTs. 30MVA RPC was divided into a pair of back-to-back converter sets to keep proper redundancy. Each converter set is composed with 4 multiple set of 3-level 2 parallel 4500V – 1200A IGBTs. Rated voltage of DC stage is 2200V * 2. Fig. 4 shows a quadrant multiple transformer and Fig.5 shows one of the inverter unit of Shin-Kurobe RPC, respectively.

Every IGBT is cooled by pure-water and driven with switching speed of 540Hz, and 4 multiple configurations realizes equivalent switching speed of 2160Hz for total.

Fig. 4. Multiple Transformer of RPC

Fig. 5. Inverter Unit of RPC

IV. ACTIVE POWER AND REACTIVE POWER COMPENSATION

An RPC has three independent function modes and can switch modes automatically. The first two functions (termed the RPC mode and SCV-Q mode) are for substations, and the third (SVC-V mode) is for sectioning posts.

A. Feeding substation functions

RPC mode: The RPC mode is the basic operating mode that is applied to the main phase and teaser feeding loads at a sub-station.

At first, the control circuit of RPC maintains the voltage of the DC capacitor. Next, RPC compares the active power (P) of the main phase with the one of teaser loads and the back-to-back converter supplies active power from the larger load circuit to the smaller one via a DC capacitor. If RPC has sufficient capacity to equalize the loads on the pair of feeding circuits, it will compensate for the active power P_C as shown in the following equation, where P_{LM} and P_{LT} are the active power of the main phase and teaser loads, respectively.

$$P_C = (P_{LM} - P_{LT})/2 \quad \text{..} \quad (2)$$

In this mode, the back-to-back converter also compensates independently for the reactive power of main phase and teaser loads (Q_{LM} and Q_{LT}) within their capacity as shown in the following equations. Fig. 6 shows this work with current and voltage vectors.

$$
\begin{aligned}
Q_{CM} &= Q_{LM} \\
Q_{CT} &= Q_{LT}
\end{aligned}
\quad \text{..} \quad (3)
$$

The back-to-back converter can likewise compensate independently for the harmonic distortion currents of feeding circuits within the capacity on both the main phase side and teaser side.

The rated capacity of the Shin-Kurobe RPC is not sufficient to provide the complete compensation required to cover all the RPC-mode functions to regulate active power P, and reactive power Q and harmonics. Therefore, Shin-Kurobe RPC will always keep a ratio of compensation of P and Q, for measured unbalance of P and measured Q in the feeding circuit. Compensating harmonic distortion is a second duty for RPC, so that if very high feeding load for P- and Q-compensation should expend the whole capacity, function of harmonic compensation function was to be neglected.

To control reactive power (Q) and harmonic distortion

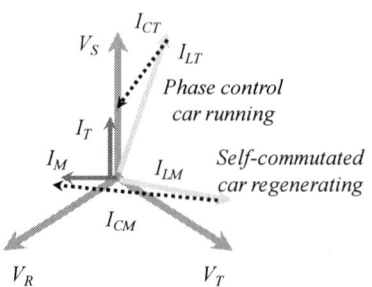

Fig. 6. Compensation in RPC mode

currents, inverters operate independently both at M-phase side and Teaser. All of the control circuits that control active power passing, DC voltage fluctuation, reactive power compensation and harmonic distortion compensation issue orders to their respective objects. And current control circuit integrates these orders and decides the current object of inverters. P

SVC-Q mode: The SVC-Q mode is one of the varieties of the RPC mode for single feeding phase substation operations. In this case, the RPC cannot handle active power so that it acts just like an SVC with both inverters works to control reactive power (Q) and harmonic distortion currents. Allocation strategy of compensating capacity between Q and H is as same as RPC mode: in the case of a much more feeding load, harmonics is to be neglected, either.

$$Q_C = Q_L \qquad (4)$$

B. Feeding sectioning post functions

If an RPC-equipped Shin-Kurobe substation is to be used as a sectioning post in an extended feeding situation, the RPC set at the sectioning post (former substation) automatically changes the function to SVC-V mode. In this mode, the RPC can produce reactive power with any scale at any time and can compensate for the feeding voltage drop at the sectioning post caused by the feeding load. In this way, the RPC operates as a single phase, self-commutated SVC. Fig. 7 and equation 5 show the principle of voltage compensation. Where P is the active power; Q is the reactive power of feeding load; R is resistance; X is reactance of the feeding circuit; and Q_C is a compensation output of reactive power from RPC for the deviation of voltage might become zero.

At sectioning posts, RPCs cannot deal with harmonics.

$$Q_C = Q + (R/X)P \qquad (5)$$

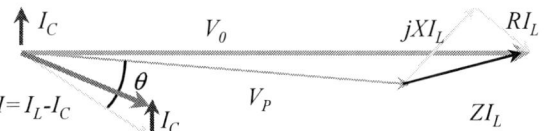

Fig. 7. Feeding voltage compensation in SVC-V mode

V. SEMI-ADAPTIVE CONTROL OF HARMONICS COMPENSATING

A. Very long extended feeding circuit

RPC for Shin-Kurobe substation has to prepare for this situation. In the fundamental frequency, length of the feeding circuit does not have much effect for active power and reactive power. Almost of active power is caused by rolling-stock load so that RPC can balance active power without any difficulty.

However, in such a very long feeding circuit, eigenvalue of resonant frequency will fall into 200-300Hz, because of the stray capacitance between catenary wires and the earth.

B. Harmonic Compensation

RPC of Shin-Kurobe substation drives IGBTs with equivalent switching speed of 2140Hz, which can handle harmonic distortion up to frequency of about 600Hz. So that RPC can compensates 3rd, 5th, 7th and 9th harmonics, in another expression 180, 300, 420, and 540Hz. If target frequency is so close to the resonant frequency of the very long feeding circuit, compensating action by RPC sometimes would cause terrible self resonance.

Thus authors developed a semi-adaptive control method of controlling harmonic compensation as follows.

At first, the control circuit evaluates the current of fundamental frequency (60Hz) and each target harmonics. Second, it calculates the ratio between fundamental current and each harmonic current to check the content.

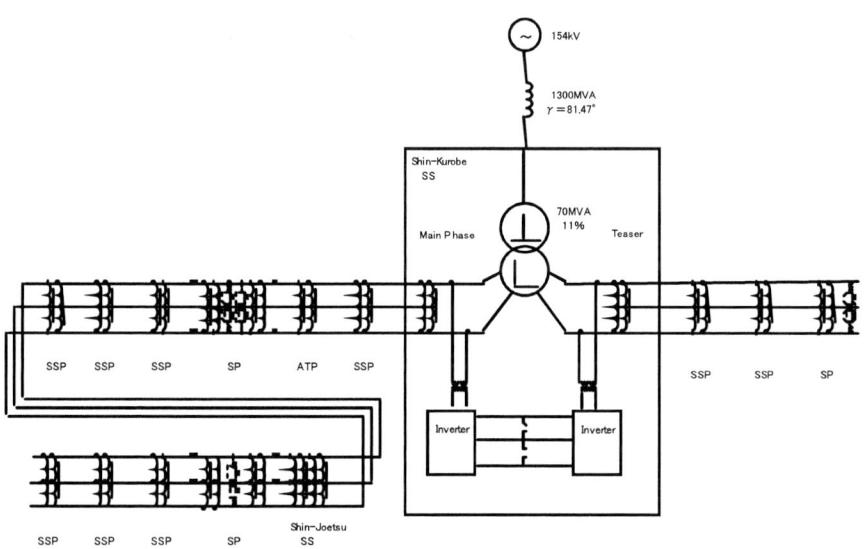

Fig.7. Target feeding circuit for simulation

If the fundamental current is smaller than proper threshold level, RPC does not have to suppress harmonics so that it stops compensation. Also if the ratio of harmonic is bigger than the proper level, there should be a possibility of harmonic resonance so that RPC stops compensating harmonics. Figure 6 shows a flowchart of this algorithm.

Anyway, the first object of RPC is to balance active power of feeding circuit. Harmonic compensating works in the margin of inverters.

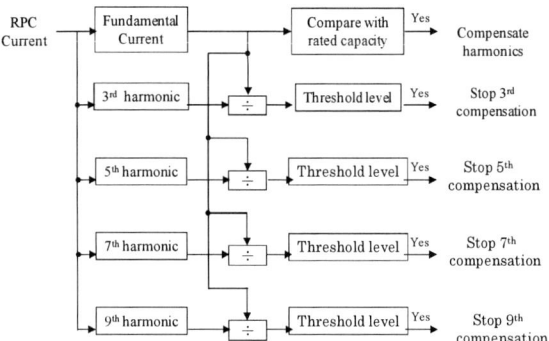

Fig.6. Semi-adaptive control algorithm for harmonic compensation

VI. SIMULATION RESULT OF SEMI-ADAPTIVE CONTROL ALGORITHM

Fig. 7 shows a simulation installation circuit to verify balancing power and compensating harmonics by RPC in Shin-Kurobe substation. In the longest feeding circuit reaches about 120km.

According to the series of simulation results, authors can expect that RPC balances the active power of feeding circuits properly in the case of normal condition and very long circuit condition. Also semi-adaptive control algorithm of compensating harmonics works properly to prevent self-resonance phenomena.

For example, RPC with the very long feeding circuit suppress the harmonic distortion caused by transformer inrush current within 5-6 cycles (Figure 8).

Fig. 9 shows another example. In this case, very large 3rd harmonic current was injected into the feeding circuit. Soon RPC stopped compensation of 3rd harmonic with the semi-adaptive control algorithm within about 160ms.

VII. CONCLUSION

Toward a special condition of Hokuriku Shinkansen, authors developed a semi-adaptive algorithm for compensating harmonics with RPC voltage compensator. 30MVA RPC for Shin-Kurobe substation has following features:

(1) RPC, as a basic configuration, will balance a pair of feeding side loads. Thus it can compensate voltage unbalance and voltage fluctuation at 3-phase side.

(2) RPC can compensate harmonic distortion in the feeding circuit. In the case of very long extended feeding mode, RPC watches a resonance there and stops compensation by comparing the ratio between

fundamental current and harmonic current.

(3) If Shin-Kurobe substation stops feeding power, RPC can act as an SVC for compensating voltage drop at the end of feeding circuit with a reactive power.

RPC will confirm its ability in winter 2013 and will in service in spring 2015. We are going to report successful result of the field.

REFERENCES

[1] T. Uzuka, et.al:"A Static Voltage Compensator for AC Electric Railway using Self-commutated Inverters", Proc. of IPEC-Yokohama, pp.1077-1082, 1995
[2] T. Uzuka, et.al:"Railway Static Power Conditioner Field Test", Quarterly Report of RTRI, Vol.45, No.2, 2004.
[3] K. Kunomura, "The introduction case of the power converters in Tokaido Shinkansen," Proc. of the 2007JIAS Conf., III, 3-S9-3, 2007 (in Japanese).

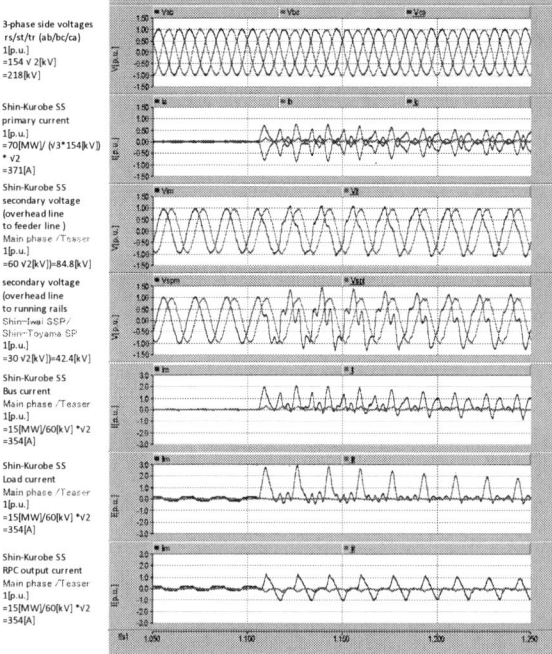

Fig.8. Simulation result of semi-adaptive control algorithm for harmonic compensation: compensation of inrush current

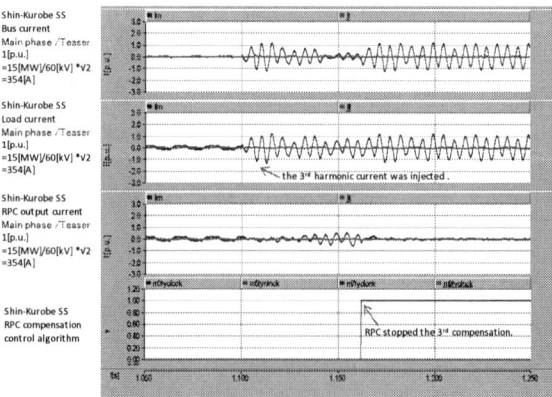

Fig.9. Simulation result of semi-adaptive control algorithm for harmonic compensation: against 3rd harmonic injection

Research of Efficient main power equipment using SiC power device

Mitsuo Shinbo
EAST JAPN RAILWAY COMPANY
Tokyo, Japan

Hiroshi Abiko
EAST JAPN RAILWAY COMPANY
Tokyo, Japan

Hideki Sonoda
EAST JAPN RAILWAY COMPANY
Tokyo, Japan

Kenichi Shibanuma
EAST JAPN RAILWAY COMPANY
Tokyo, Japan

Takahito Ishida
MITSUBISHI ELECTRIC
Hyogo, Japan

Yoshinori Chiba
MITSUBISHI ELECTRIC
Hyogo, Japan

In this report, energy saving characteristics of a main circuit system having the inverter which applied SiC elements and Motor whose v/f is reduced were reported.

As a main circuit system, an effective way of energy saving was examined, and evaluated. It was also put on the experimental vehicle, E995, and the electricity reduction effect against the existing system was evaluated.

Keywords— SiC,Energy saving,Regenerative brake,VVVF inverter,

I. INTRODUCTION

The railroad vehicle propulsion systems changed with the innovation of the main circuit semiconductor. A new semiconductor called SiC (silicon carbide) came up 20 years after the birth of IGBT, and its application to railroad vehicles has been carried out [1].

The withstanding pressure of SiC is approximately 10 times of Si as used for conventional power devices. The basic structure of Si-diode and SiC-Schottky Barrier Diodes (SBD) is shown in Fig. 1. The semiconductor layer of SiC-SBD is thinner than Si-diode. With this structure, the loss of the power device decreases. By the features of this low-loss, high frequency drive can be achieved.

The breakdown of energy loss in the railway system is shown in Fig.2. The lost energy from the traction motor and air brake is great. To save energy, it would be effective to reduce these losses.

In this research, the device which applied SiC power module was utilized, and, as a main circuit system, an effective way of energy saving was examined, and evaluated.

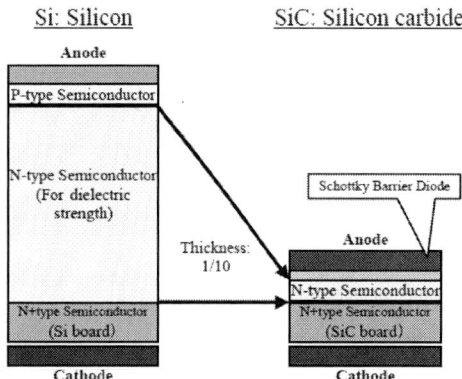

Fig.1 Basic structure of Si and SiC Power Devices

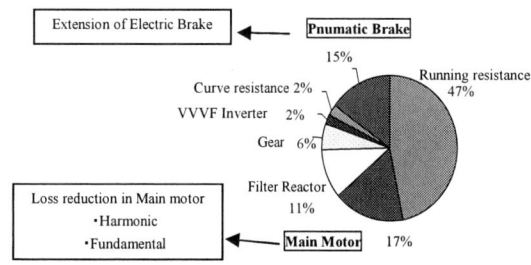

Fig.2 Energy loss of Train

II. ENERGY-SAVING EXAMINATION UTILIZING SiC

Using a characteristic of SiC methods reducing losses of traction motor and air brake were examined.

1) Reducing motor harmonic loss by improving switching frequency and reducing motor current destruction.

2) By the high frequency drive of VVVF and the low inductance main motor, stall torque is increased and regenerative braking region is expanded.

978-1-4799-2706-7/14 $31.00 © 2014 IEEE

III. SUMMARY OF THE TEST CONSTITUTION

A. Characteristics of an experimental vehicle used in the field examination.

The experimental vehicle, E995, was used in the field examination. This vehicle was developed as a "catenary and storage hybrid train" aimed to reduce burden on environment [2].

The main specifications and composition of this train are shown in Table1 and Figure 3, respectively. The main circuit of this train receives DC 1500V through the electrified line, converts it into DC 600V with DC/DC converter, and supplies the converted electricity to storage batteries, traction motors, and the auxiliary power supply unit. (Fig.4)

Table1. Main specifications of experimental car

Item	Details
Car dimensions (length × width × height)	19500mm× 2800mm× 4052mm
Car weight	39.9 t
Max. speed	100 km/h
Range on batteries	Approx. 20 km
Pantograph	Prototype mounted on roof that can collect large current from catenary while stopping
Power converter	Bi-directionally converts between DC 1,500V of catenary and 600V for storage batteries
Storage battery for main circuit	lithium-ion battery : 600V, 72kWh
Motor controller	VVVF inverter system (input: DC600V)
Motor system/output	Two 95 kW asynchronous motors (mounted on one bogie)

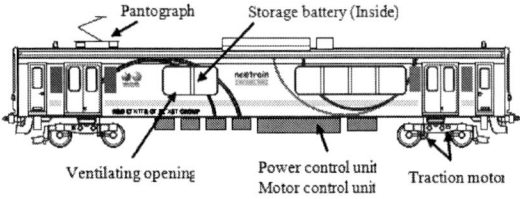

Fig.3 Composition of experimental car

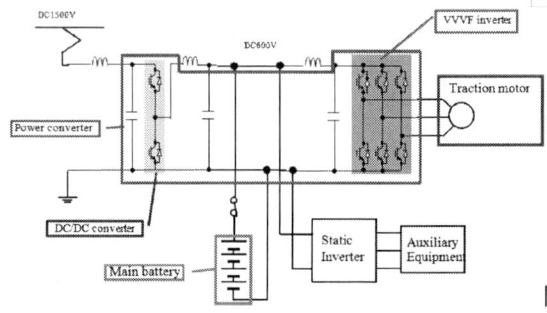

Fig.4 Hybrid system

B. Experimental model

Three steps consisting of different devices were evaluated. The constitutions of the main circuits used in each step are shown in Table 2. The main differences in each step are listed below.

STEP 0: Evaluation of the existing inverter（Si）
STEP 1: Evaluation of the inverter which applied SiC elements
STEP 2: Evaluation of the inverter which applied SiC elements and Motor whose V/f is reduced

1) The inverter which applied a SiC element

For VVVF inverter, we developed the power unit using SiC power device and added it to the current VVVF inverter. Si-system (STEP 0) must have cooling system using heat pipe. SiC-system (STEP 1) can adopt heat sink for the cooling system. We confirmed weight of VVVF inverter is reduced. (Fig.5)

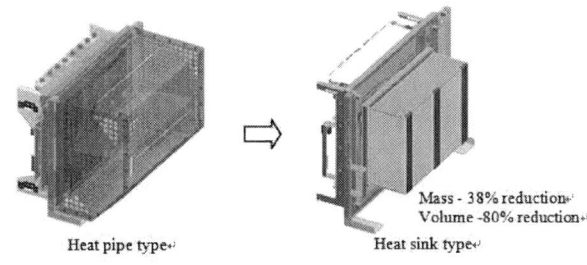

Fig.5 Comparison of Power Unit size

Fig.6 VVVF inverter outline

2) The traction motor whose V/f is reduced

Element loss is reduced with the above-mentioned SiC element application inverter. By increasing the domain of high-frequency switching with this characteristic, motor current is output in the form nearer to a sine wave, making it possible to reduce the harmonic loss on the side of motor.

In this case, motor voltage is lower due to restriction of pulse width modulation factor of VVVF inverter and motor current is larger. Therefore, motor should be designed to reduce coil turns for reducing coil resistance.

But if coil turns are reduced, magnetic flux is larger and size of motor must be enlarged. Magnetic flux is proportional to V/f. We designed the motor to reduce V/f and constructed it as same size as the current one.In

selecting the V / f, taking into account the constraints due to the allowable current of the VVVF and traction motor mass, were selected and 95Hz.

Fig.6 The traction motor

Table.2 Main circuit construction

		Step 0	Step 1	Step 2
electrical mode		DC1500V(catenary)/DC600V(storage battery)		
set		0.5M0.5T		
traction motor	type	three-phase induction motor		three-phase induction motor (The motor whose V/f is reduced)
	rated output	95kw×2(One hour rating)		115kW×2(One hour rating)
	rated current	150A(One hour rating)		245kW(One hour rating)
	Minimum frequency at maximum motor voltage	53Hz		95Hz
VVVF inverter	circuit structure	three-phase 2level PWM inverter		
	control function	torque control(vector control)		
	Main circuit element	1S1P2A×three-phase×1group standarad:3.3kV 1.2kA (1 in 1)	1S1P2A×three-phase×1group standarad:1.7kV 1.2kA (2 in 1)	
	cooling system	Heat pipe type	Heat sink type	
DC/DC converter	circuit structure	step-down and -up type chopper 600kW		
main battery	type	lithium-ion battery		
	system voltage	605V		
	system capacity	72kWh		

3) Expansion of the regenerative brake domain

As a reduction measure of an air brake loss, It is effective to expand a regeneration domain to the high speed side, and to prevent operation of air brake as much as possible.

For that purpose, it is necessary to expand the stalling torque of a motor. The traction motor whose V/f is reduced made as an experiment is low inductance type, and can make stalling torque increase. In STEP2, we made the regeneration domain expand to the high speed side as compared with STEP0, and decided to check an effect. (Fig. 7)

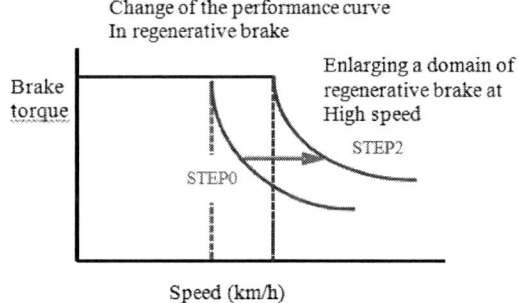

Fig. 7 Expansion of the regenerative brake domain

4) The modulation method of VVVF inverter

Modulation method of the VVVF inverter affects the power consumption and efficiency of the system. Generally, as VVVF inverter switching frequency is higher, motor current distortion, which has powerful effect on motor harmonics loss, decreases. Therefore, high frequency modulation method of VVVF inverter can decrease the power consumption of the system and improve the efficiency of the system.

However, high frequency modulation method increases VVVF inverter switching loss. We applied the following three modulation method to VVVF inverter and compared the efficiency of the system.

  a) *Ordinary modulation method*

  b) *Continuous asyncronous modulation method*

  c) *New modulation method, which changes from Asyncronous modulation to Synchronous modulation of some number of pulse in case of train exceeing particular speed*

We applied the new modulation method of VVVF inverter to the STEP2 system and confirmed the method kept the VVVF inverter loss lower and the motor harmonics loss lower in case of using SiC power module.

Applying the new modulation method to VVVF inverter using SiC module has advantage of being able to decrease the power consumption of the system and improve the efficiency of the system.

IV. TESTING CONDITION

By test composition shown in Chapter Ⅲ, the bench examination and the field examination were carried out and various evaluations were performed. In this paper, the following result is introduced as an example.

A. Bench test

The Propulsion Systems of the experimental vehicle E995 was removed and converted, and the bench test was carried out at the factory.

  1) *Electric energy loss comparison of Si and SiC*
      (comparison between STEP 0 and STEP 1)

  2) *Regenerative brake energy comparison*

B. Field test

The experimental model was carried in the experimental vehicle E995, and the field test was carried out.

  1) *Inductive obstruction examination*
      (comparison among three steps)

  2) *Running test on the real line*
      Evaluation of running energy and regenerative energy (comparison among three steps)

V. BENCH TEST

A. Electric energy loss comparison of Si and SiC

In powering up to 90 km/h, and the brake from 90 km/h, the electric energy in each main circuit apparatus was measured.

The electric energy loss in STEP0 (Si) and STEP1 (SiC) is shown in Fig. 8. SiC has checked that loss electric energy was reduced by no less than 50% compared with Si.

B. Regenerative brake energy comparison

In only regenerative braking from 90km/h to stop, the electric energy in each main circuit apparatus was measured.

The regenerative energy in STEP1 and STEP2 is shown in Fig.9. STEP2 has checked that regenerative brake energy was increased 44% compared with STEP1.

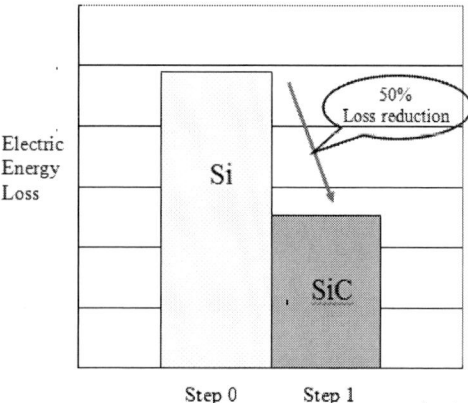

Fig.8 The loss electric energy
in STEP0 (Si) and STEP1 (SiC)

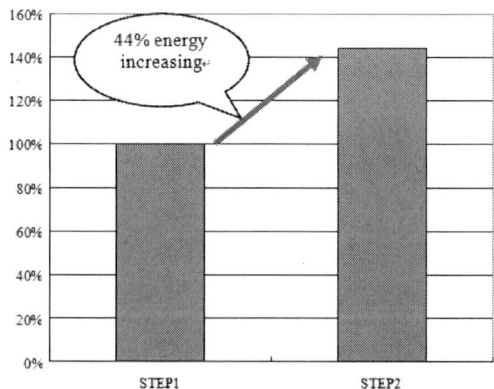

Fig.9 The regenerative braking energy
in STEP1 and STEP2

VI. FIELD TEST

A. Inductive obstruction examination (Running test at Omiya general rolling stock center)

We must confirm how the return current of the train affects to the track circuit of the protective device before commercial running of the train.

Harmonic current which mainly occurs by switching of an element generates the influence on track circuit by flowing into return current. Therefore we measured the harmonic current of the return current and confirmed effect of the return current in Omiya general rolling stock center.

1) Result in STEP1

In each operation mode of VVVF inverter (for example, Hybrid mode, no battery mode), the harmonic current of the return current is not affect to all track circuits of the protective device.

2) Result in STEP2

As same as STEP1, we confirmed the harmonic current of the return current is not affect to all track circuits of the protective device.

The above result showed not affecting signal equipment with this composition.

Fig.10 Inductive obstruction examination

B. Running test on the real line

After doing the above-mentioned examination, the running test was carried out in Utsunomiya Line. The electric energy of running (5N) and the electric energy of a regenerative brake (7N) of each step are shown in Table 5.

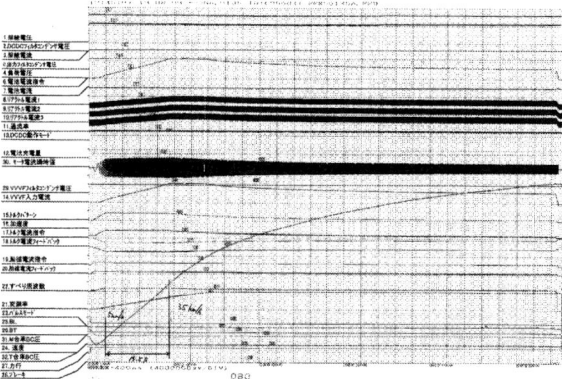

Fig.11 Powering examination result
from 0km/h to 90km/h

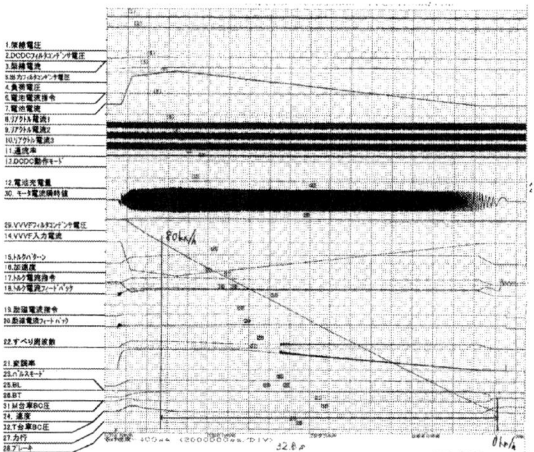

Fig.12 Braking examination result
from 90km/h to 0km/h

As shown in Fig.11 and Fig.12, we confirmed normal performance after applying SiC module and V/f reduction motor.

Table.5 The electric energy of running and regenerative brake

	STEP0	STEP1	STEP2
Running electric energy (kWh)	5.635 (100%)	5.624 (99.8%)	5.567 (98.8%)
Regenerative electric energy (kWh)	1.749 (100%)	1.779 (101.7%)	2.330 (133.2%)
Total consumption electric energy (kWh)	3.886 (100%)	3.845 (98.9%)	3.237 (83.3%)

978-1-4799-2706-7/14 $31.00 © 2014 IEEE

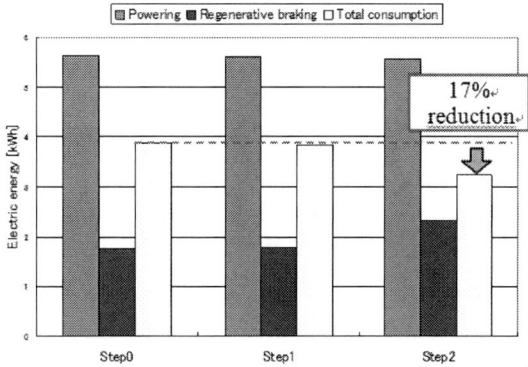

Fig. 13 The electric energy measurement result

As compared with STEP0, electric energy decreased 1% at STEP1, and it has checked no less than 17% of electric energy reduction at STEP2. In the STEP2, although there were few reduction effects of running energy, regeneration energy increased about 33% by expanding the domain of a regenerating brake to the high speed side as compared with STEP0.

VII. OPTIMIZATION OF SiC PROPULSION SYSTEM

The test result and effect in a SiC propulsion system have been described. This composition is only an example of a SiC propulsion system, and there are various ways, using SiC effectively. Although it made to increase regeneration energy by this time into the main purpose, probably the way aiming at reduction of running energy will also be possible, and the way aiming at a weight saving is also possible. The design concept with emphasis on what the conditions of the maximum spped, wire voltage, and vehicles design is important. Optimization of the system set by various kinds of conditions is important.

VIII. CONCLUSIONS

The following conclusions were obtained by this development.

1) Only by changing the power element of a VVVF inverter into a SiC element, loss reduction of an inverter is possible.

2) By the combination of the motor which reduced V/f, and a SiC element VVVF inverter device, expansion of a regeneration domain and the increase in the amount of regenerative electric power are possible.

3) The above-mentioned policy attains 17% of consumption energy reduction.

I would like to further consider the optimization as a propulsion system, such as improving career frequency, taking into consideration the balance of an inverter loss and a motor loss from now on.

REFERENCES

[1] Yoshinori Yamashita, Kenichi Kusano, Tomohiro Kobayashi, Yukio Nakashima, Kenta Kaneko (MitsubishiElectric) Nobuyuki Ubukata, Shingo Hukazawa, Hideaki Yamanoi(Tokyo Metro) ; Energy Saving Operation for Railway Inverter System with SiC Power Module, 2012 IEE Japan, No.5-078, pp.128-129, 2012.3,

[2] Kenichi Shibanuma, koji Yoshida (East Japan Railway) ; Development and evaluation of storage battery hybrid train system ; translog2011 , 2011.12.

A High Performance Control Strategy for Three-Level NPC EMU Converters

Song Kejian, Wu Mingli, Wang Hui
School of Electrical Engineering
Beijing Jiaotong University, BJTU
Beijing, China
12117360@bjtu.edu.cn; mlwu@bjtu.edu.cn

Vassilios Georgios Agelidis
School of Electrical Engineering and Telecommunications
The University of New South Wales, UNSW
Sydney, Australia
vassilios.agelidis@unsw.edu.au

Abstract—**A high performance control strategy for a three-level neutral-point-clamped (NPC) converter employed in electric multiple units traction drive system is proposed in this paper. The strategy is based on a multiple frequency tuned quasi-proportional resonant (PR) controller for the AC current loop. A notch filter is used in the DC bus voltage loop to filter out the ripple affecting the controller. A feed-forward loop is included in the inner current control loop to deal with AC power supply disturbances. The impact of the quasi-PR controller parameters on its performance is investigated through frequency-domain analysis. The generation of the low-order harmonics of the line current caused by the DC-link voltage fluctuations is analyzed from the control point of view. Simulation results presented in the paper validate that the proposed control strategy provides a better current tracking performance and a lower current harmonic content when compared with other more conventional control strategies.**

Keywords— Power electronics converters, quasi-PR controller, notch filter, harmonic analysis

I. INTRODUCTION

Modern high-speed railways utilize electric multiple units (EMU) based on AC drive technology [1]. EMUs utilize single-phase four-quadrant converters (4QCs) as rectifiers to either draw power from the traction network or inject power back into the network during electric braking.

The 4QC is typically controlled by a conventional double closed-loop control strategy to implement a constant DC-link voltage, a sinusoidal line current and unity power factor on the network side. This traditional control system consists of an AC current loop (inner loop) with a proportional (P) or proportional-integral (PI) controller and a DC voltage loop (outer loop) with a PI controller.

For the inner loop, the reference is a sinusoidal current that means both P and PI controller cannot ensure that the actual current tracks its reference without a steady-state error [2]. The error is often a phase-shift between the actual input current and the reference. This means that the power factor could be compromised. The P and PI controllers have another drawback, namely their poor disturbance rejection capability [3]. However, for the inner loop the AC supply voltage is a disturbance. Particularly, when the disturbance voltage contains

harmonics (practically it often contains a certain amount of 3rd, 5th and 7th harmonics), the system could not provide a strong rejection capability of such harmonics. This may cause an increase in the harmonic content. These harmonics injected into the traction network may lead to detrimental effects on the traction power supply system and other electric locomotives [4].

For the outer loop, the ripple (mainly at twice the fundamental frequency) of the DC-link voltage will cause relative low-order harmonics in the reference input current of the 4QC through the PI regulator, as explained in Section IV. This also means an increase in the harmonic content when the actual current tracks its reference which contains such harmonics.

Ref. [5] proposed a proportional resonant (PR) controller to eliminate the steady-state error of the inner current loop and then expanded the controller to be a multiple frequency tuned one to further deal with the impact of the supply voltage harmonics. Using the internal model principle, ref. [6] explained why such a resonant element in the current controller can implement zero steady-state error tracking for a sinusoidal signal. Ref. [7] explored the corresponding relationship between the PR controller in the stationary frame and the PI controller in the rotating frame and then proposed a modified PR controller, named as quasi-PR controller in some other papers ([3], [8], [9]), which is easier to implement in a practical system and has a better frequency tolerance when compared with a conventional PR controller. Ref. [3] used a multiple frequency tuned quasi-PR controller for a three-phase reversible PWM rectifier to eliminate the steady error and current distortion caused by the harmonics of the grid voltage. However, the harmonic compensation elements in multiple frequency tuned PR or quasi-PR controller could react when the reference current contains specific harmonics. A combination of a notch filter in the outer loop and a multiple frequency tuned PR controller in the inner loop for harmonic elimination of the line current of interlaced single-phase two-level EMU rectifiers was proposed in [4].

However, there is no technical paper that has documented the performance of a single-phase three-level (3L) NPC converter with a multiple frequency tuned quasi-PR controller in the input current controller and a multi-notch filter in the DC bus voltage controller.

978-1-4799-2706-7/14 $31.00 © 2014 IEEE

The objective of this paper is to eliminate the line current harmonics and to track a sinusoidal reference current with zero steady-state error for interlaced single-phase 3L NPC EMU converters by using a high performance control strategy based on multiple frequency tuned quasi-PR controller and multi-notch filter. The performance of the proposed control strategy is documented against conventional control strategies.

The rest of the paper is organized as follows. Section II presents briefly the circuit configuration of a typical EMUs and gives the circuit parameters as reference for the analysis and simulation discussion that follows. Section III builds a control system model of the input current loop and gives detailed analysis for the quasi-PR controller. Section IV explains the generation of low-order harmonics of the 4QC input current caused by the ripple of DC-link voltage, and then proposes a notch filter to eliminate such harmonics. Section V provides simulation results to validate the performance of the proposed control strategy.

II. A TYPICAL TRACTION DRIVE SYSTEM OF EMUs

Fig. 1 shows a typical configuration of a traction drive system used in EMUs, which is employed in CRH2, CHR380A and CRH380AL (three types of EMUs of China railway high-speed).

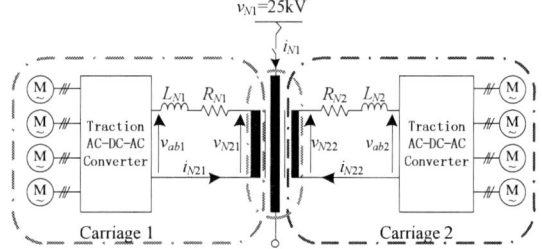

Fig. 1. Simplified diagram of typical configuration of traction drive system of EMUs

In Fig. 1, a 1Φ, 50Hz, 25kV AC supply powers the system through a traction transformer. Two traction converters are interlaced and fed from the same primary winding. Each traction converter supplies power through an AC-DC-AC configuration to four induction motors. The front-end AC-DC conversion stage is a single-phase 4QC. The 4QCs in those EMUs which have a traction drive system like the one shown in Fig. 1 are all single-phase three-level neutral-point-clamped (3L-NPC) 4QCs shown in Fig. 2.

The total impedance seen into the transformer secondary winding and referred to the secondary winding is represented by L_{N1} and R_{N1} where L_{N1} is the carriage 1 inductance and R_{N1} resistance respectively, and L_{N2} and R_{N2} are the carriage 2 counterparts. It is assumed that both circuits are identical therefore, $L_{N1}=L_{N2}=L_N$ and $R_{N1}=R_{N2}=R_N$. v_{N21} and v_{N22} are the supply voltages at the secondary side of the traction transformer feeding the converters of carriage 1 and 2 respectively. It can therefore also be assumed that $v_{N21}=v_{N22}=v_{N2}$.

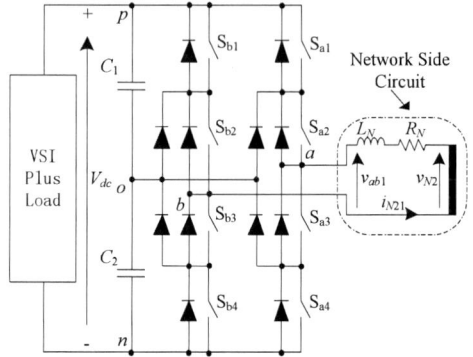

Fig. 2. Single-phase three-level neutral-point-clamped voltage-source four-quadrant rectifier

The main parameters of the proposed EMU traction drive system shown in Fig. 1 and 2 are listed in Table I.

TABLE I
SYSTEM PARAMETERS

Supply parameters		
@50Hz	V_{N1}=25kV	V_{N2}=1658V
L_N=2mH	R_N=0.05Ω	
Converter parameters		
V_{dc}=3000V	C_1=C_2=8mF	P_{out}=2.4MW

where P_{out} is the output power of one 4QC.

III. INPUT CURRENT CONTROLLER

A. System Model

The EMU 4QC has a number of suitable control strategies but almost all of them are based on the double closed-loop control which consists of a DC-link voltage loop (outer loop) and an input current loop (inner loop). The inner current loop is the core of the whole control system. Fig. 3 shows the control diagram of the input current loop with a feed-forward path incorporating the network voltage.

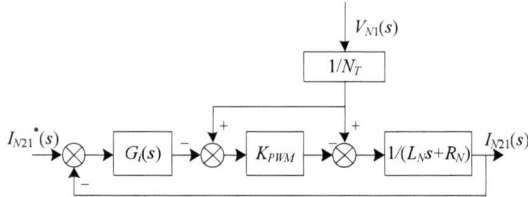

Fig. 3. Control diagram of the input current loop

In Fig. 3 $G_i(s)$ is the transfer function of the controller in this control system. K_{PWM} is the gain of the converter. $1/(L_N s+R_N)$ is the transfer function of the control object of this current loop. The network voltage $V_{N1}(s)$ is the disturbance of this control system, and N_T is the turns-ratio of the traction transformer. According to Table I N_T=25000/1658.

From the control diagram shown in Fig. 3 the input current of the 4QC can be expressed as:

$$I_{N21}(s) = G_I(s)I_{N21}^*(s) + G_V(s)V_{N1}(s) \qquad (1)$$

where $G_I(s)$ and $G_I(s)$ are the closed-loop transfer functions of $I_{N21}(s)$ refer to $I_{N21}^*(s)$ and $V_{N1}(s)$ respectively:

$$G_I(s) = \frac{K_{PWM}G_i(s)}{K_{PWM}G_{PR}(s)+L_N s+R_N} \qquad (2)$$

$$G_V(s) = \frac{1 - K_{PWM}}{N_T(K_{PWM}G_i(s) + L_N s + R_N)} \quad (3)$$

According to the system's model proposed above, the input current consists of two parts: one is related to the reference current; the other one is the disturbance current composition caused by the network voltage (both fundamental voltage and the harmonic voltages). The aim of the controller is to keep the current tracking its reference without magnitude and phase errors and to provide a rejection capability to the disturbance network voltage.

Commonly, the value of K_{PWM} is close to 1. Consequently (3) is close to 0. This means that the feed-forward loop is able to offset the effects of the network voltage disturbances in most cases satisfactorily but not necessarily in all cases.

When $G_i(s)$ is the transfer function of a P or PI controller, the control system is a conventional one. In this paper $G_i(s)$ is firstly considered as the transfer function of a PR controller as follows [7]:

$$G_{PR}(s) = K_p + \frac{2K_r s}{s^2 + \omega_o^2} \quad (4)$$

where K_p and K_r are the proportional and resonant gains respectively. Through providing an infinite gain at the resonant frequency ω_o, the PR controller can track a sinusoidal signal with frequency of ω_o with zero steady-state error. Fig. 4 gives the Bode diagram of (4) at the frequency of 50Hz ($\omega_o=2\pi*50$rad/s).

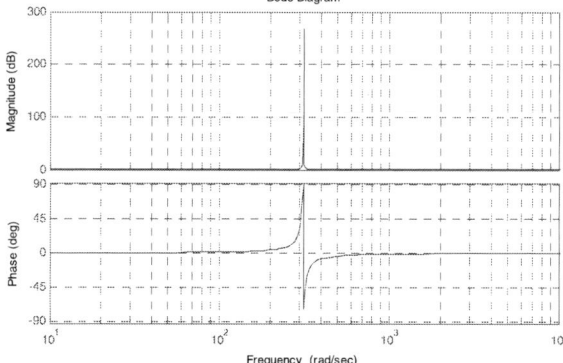

Fig. 4. Bode diagram of a PR controller

Unfortunately, the PR controller is not suitable for the application [8]. Firstly, its infinite gain leads to an infinite quality factor which cannot be achieved in either analog or digital implementation. Secondly, it provides very little gain out of the selected frequency due to its narrow bandwidth. In this paper a quasi-PR controller is chosen.

B. Quasi-PR Controller

Introducing a modification to (4), the transfer function of a quasi-PR controller can be obtained [10]:

$$G_{quasi-PR}(s) = K_p + \frac{2K_r \omega_{cut} s}{s^2 + 2\omega_{cut} s + \omega_o^2} \quad (5)$$

There are three parameters associated with the quasi-PR controller, namely, K_p, K_r and ω_{cut}.

The role of K_p in the quasi-PR controller is similar to the proportional gain of a PI controller [11]. K_p mainly

determines the stability and dynamics of the system. A higher value of K_p could bring a faster dynamic response but may lead to instability.

Bode diagrams of (5) with variations of K_r are plotted in Fig. 5, assuming $K_p=0$, $\omega_{cut}=1$rad/s. It can be seen that K_r determines the gain at the resonant frequency, but has no effect on the bandwidth. A higher value of K_r corresponds to a larger gain.

ω_{cut} is the lower breakpoint frequency of the DC transfer function from which (5) is derived [7]. Bode diagrams of (5) with variations of ω_{cut} are shown in Fig. 6, assuming $K_p=0$, $K_r=100$. It is seen that ω_{cut} determines the bandwidth of the controller, but has no effect on the gain at the resonant frequency. A higher value of ω_{cut} corresponds to a larger bandwidth. This means that when the network voltage is not exactly at its nominal frequency (50Hz in this paper), the controller can provide a better adaptability.

Therefore, by choosing proper values of K_r and ω_{cut}, the quasi-PR controller can provide a nearly infinite gain at the selected frequency and a suitable bandwidth.

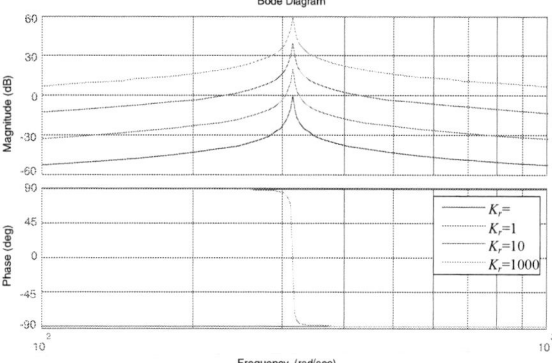

Fig. 5. Bode diagrams of a quasi-PR controller for varying K_r

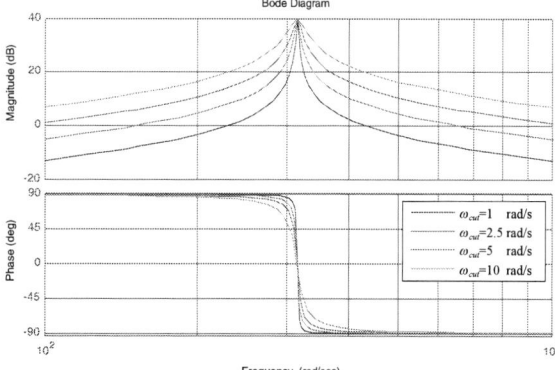

Fig. 6. Bode diagrams of a quasi-PR controller for varying ω_{cut}

Furthermore, the quasi-PR controller can be easily designed to offer selected harmonics (such as 3rd, 5th, 7th) compensation without influencing the controller dynamics [2] and the transfer function can be expressed as:

$$G_{multi-quasi-PR}(s) = K_p + \sum_{h=1,3,5,7} \frac{2K_{rh}\omega_{cuth}s}{s^2 + 2\omega_{cuth}s + (h\omega_o)^2} \quad (6)$$

where K_{rh} and ω_{cuth} determine the gain and the bandwidth for h order harmonic, respectively. This will allow a suitable gain and bandwidth for each harmonic frequency to be considered and designed independently. The controller with a transfer function of (6) is named as multiple frequency tuned quasi-PR controller in this paper.

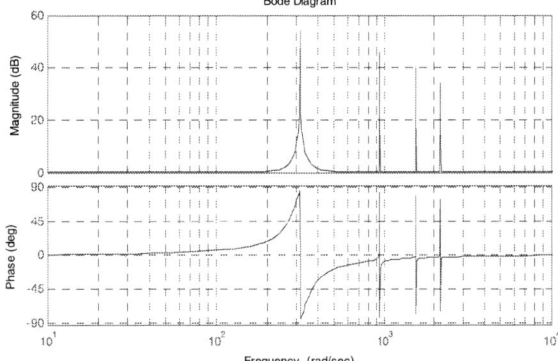

Fig. 7. Bode diagram of a quasi-PR controller with 3rd, 5th and 7th harmonics compensation

Fig. 7 is the Bode diagram of quasi-PR controller with selected harmonics compensation. It is seen that at selected harmonic frequencies the controller can bring also nearly infinite gains. This means that the controller can provide compensation for selected harmonics.

However, if the reference input current of the 4QC contains the selected harmonics, the actual input current will track these harmonics. This means a harmonic amplification. Unfortunately, the reference current indeed contains those harmonics caused by the ripple of the DC-link voltage of the 4QC. Consequently, there is a need of a notch filter to deal with this problem, which is described in detail in the next Section.

IV. DC-LINK VOLTAGE CONTROLLER

A. Low-order Harmonics of 4QC Input Current – Caused by DC-link Voltage Control

Generally, the 4QC input current contains inherited high-order harmonic components generated by 4QC's PWM process [12]. In practical applications, the carrier phase-shift control technique [13] is used to cancel some of these harmonics in the interlaced 4QCs of an EMU. Consequently the line current of an EMU contains few high-order harmonics.

However, from testing data, it is seen that there exist a certain amount of low-order harmonics in EMUs line current [14]. These harmonics are mainly caused by the fluctuations of 4QCs DC link voltage.

Assuming that the 4QC shown in Fig. 2 works under unity power factor, $v_{N2} = \sqrt{2}\,V_{N2}\cos\omega_o t$, and $i_{N21} = \sqrt{2}\,I_{N21}\cos\omega_o t$, the input instantaneous power p_{in} of the 4QC can be expressed as:

$$p_{in} = v_{N2} \times i_{N21} = V_{N2}I_{N21}[1 + \cos(2\omega_o t)] \quad (7)$$

Note that i_{N21} has just preserved its fundamental component and the high-order sideband harmonic components of i_{N21} have been neglected, because this analysis focuses on the low-order harmonics only. The output power p_{out} of the 4QC is

$$p_{out} \approx \overline{V}_{dc}C\frac{d\tilde{v}_{dc}}{dt} + \overline{V}_{dc}\overline{I}_L \quad (8)$$

where $\overline{V}_{dc}$ and $\tilde{v}_{dc}$ are the mean value and pulsating component of the DC-link voltage V_{dc}, $\overline{I}_L$ is the equivalent mean current of DC-link load, C is the equivalent capacitance of the series combination of C_1 and C_2. Assuming that both the converter conduction and switching losses and the power losses in L_N and R_N are negligible, then the input and output power of the 4QC are equal, i.e. $p_{in}=p_{out}$. Using (7) and (8), $\tilde{v}_{dc}$ can be calculated as:

$$\tilde{v}_{dc} = \frac{V_{N2}I_{N21}}{2\omega_o\overline{V}_{dc}C}\sin(2\omega_o t) \quad (9)$$

The voltage ripple has been derived analytically. The simplified traditional DC-link voltage control diagram is shown in Fig. 8(a).

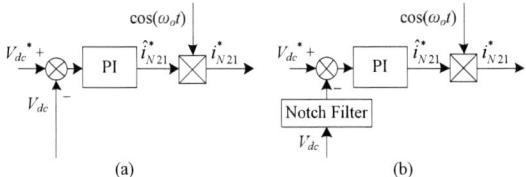

Fig. 8. Simplified DC-link voltage control diagram (a) without notch filter (b) with a notch filter

In the control system, a PI controller is used to regulate the deviation of the DC link voltage between the reference value and the actual one. The output of the PI controller is the amplitude reference $\hat{i}^*_{N21}$ of the input current of 4QC. A phase-locked loop (PLL) is used to obtain the phase of the supply voltage ($\cos(\omega_o t)$ shown in Fig. 8) and synchronize the various traction converters. The reference signal i^*_{N21} for the input current of the 4QC can be obtained:

$$i^*_{N21}(t) = K(V^*_{dc} - \overline{V}_{dc} - \tilde{v}_{dc})\cos(\omega_o t)$$
$$= K(V^*_{dc} - \overline{V}_{dc})\cos(\omega_o t) + \quad (10)$$
$$K\frac{V_{N2}I_{N21}}{4\omega_o\overline{V}_{dc}C}[\sin(3\omega_o t) + \sin(\omega_o t)]$$

In (10) the PI controller is replaced by a P controller with its parameter K for simplifying the calculation without affecting the spectrum.

From (7)-(10), it is found that the fundamentals of v_{N2} and i_{N21} cause a pulsation at twice the fundamental frequency in V_{dc} ($\overline{V}_{dc} + \tilde{v}_{dc}$) due to power processing. This pulsation is reflected in the control algorithm making the reference current i^*_{N21} to contain a 3rd harmonic. Therefore, during a control cycle the actual current i_{N21} traced with the reference signal i^*_{N21} will also contain a 3rd harmonic. Similarly, the new i_{N21} contained 3rd harmonic will cause a 5th harmonic during the next control cycle. Repeatedly, i_{N21} will contain more low-order odd harmonics such as 3rd, 5th, 7th, 9th, 11th etc. It should be noted that as the harmonic order increases the respective magnitude will decrease.

978-1-4799-2706-7/14 $31.00 © 2014 IEEE   643

Traditionally, an LC series branch with a resonant frequency at twice the fundamental frequency connected in parallel with the DC-link capacitor or capacitors could cancel most of the ripple of the DC-link voltage. However, some types of EMUs (such as CRH2, CRH380A and CRH380AL which have their traction drive system and 4QC topology like the ones shown in Fig. 1 and 2) leave out the LC series branch for saving space inside the carriages. So that a notch filter is needed to eliminate the DC-link voltage ripple to cancel the low-order harmonics of the reference input current without adding any extra hardware circuit.

B. Notch Filter

The simplified DC-link voltage control system with a notch filter is shown in Fig. 8(b). The typical transfer function of a second order notch filter can be expressed as:

$$G_{Notch}(s) = A_0 \frac{s^2 + \omega_n^2}{s^2 + \omega_n s / Q + \omega_n^2} \quad (11)$$

where A_0 is the gain of the filter; ω_n is the characteristic angular frequency; Q is the equivalent quality factor.

Choosing 100Hz for the characteristic frequency (i.e. $\omega_n=2\pi*100$rad/s) and assuming $A_0=1$, the Bode diagram of the notch filter with variations of Q is shown in Fig. 9. It is seen that the filter can eliminate the voltage at the frequency of ω_n without imposing an effect on other frequencies. The higher value of Q corresponds to a higher bandwidth but a smaller attenuation.

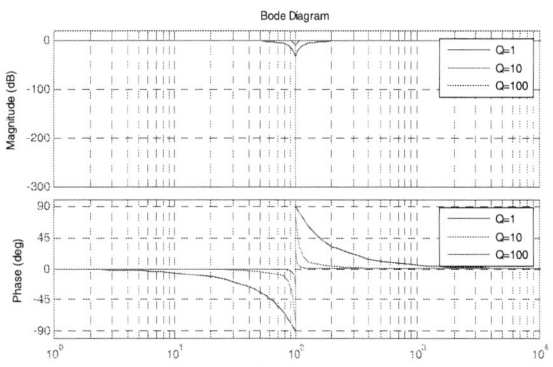

Fig. 9. Bode diagrams of notch filter for variation in Q

Similarly, the characteristic frequencies of notch filter can be designed at 100Hz, 200Hz and 300Hz for a further harmonics cancellation. The modified notch filter has a transfer function as:

$$G_{multi-Notch}(s) = \prod_{k=1,2,3} A_k \frac{s^2 + (k\omega_n)^2}{s^2 + k\omega_n s / Q_k + (k\omega_n)^2} \quad (12)$$

where $\omega_n=2\pi*100$rad/s yet. The filter with a transfer function of (12) is named as multi-notch filter in this paper.

According to the analysis of quasi-PR controller and notch filter above, a high performance control strategy for the single-phase 4QC can be presented. The simplified control diagram of the proposed control strategy is shown in Fig. 10.

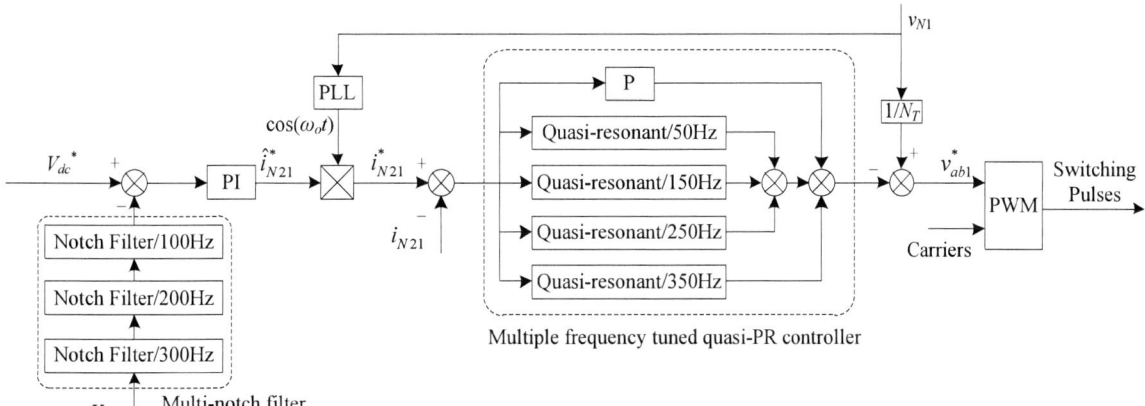

Fig. 10. Simplified control diagram of the proposed control strategy based on a multiple frequency tuned quasi-PR controller and a multi-notch filter

V. SIMULATION RESULTS

Simulation results obtained from Matlab/Simulink are provided to validate the proposed control strategy and its effect on input current tracking and harmonic suppressing. The simulation model is built according to Fig. 1 and 2 and Table I. The two interlaced 4QCs are operated by 3-level phase-opposition disposition (POD) PWM modulator with a $\pi/2$ carrier phase-shift for high-order harmonics eliminating [15]. The carrier frequency is 1250Hz. The control parameters are given in Table II below.

TABLE II

CONTROL PARAMETERS

Multiple frequency tuned quasi-PR controller		
$K_p=1$	$K_{r1}=200$	$K_{r3}=K_{r5}=K_{r7}=50$
$\omega_o=2\pi*50$rad/s	$\omega_{cut1}=1$	$\omega_{cut3}=\omega_{cut5}=\omega_{cut7}=0.5$
Multi-notch filter		
$\omega_n=2\pi*100$rad/s	$A_1=A_2=A_3=1$	$Q_1=Q_2=Q_3=10$

The input currents of the 4QC compared with their reference when using a multiple frequency tuned quasi-PR controller and a traditional P controller in the current loop are respectively shown in Fig. 11 (a) and (b).

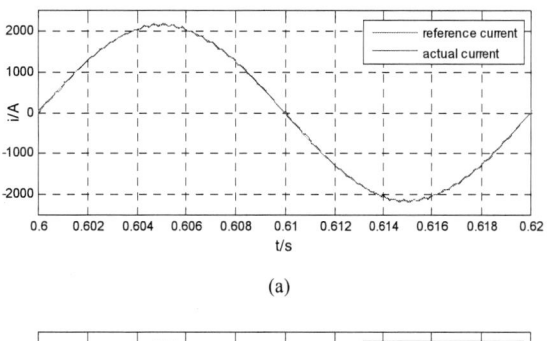

(a)

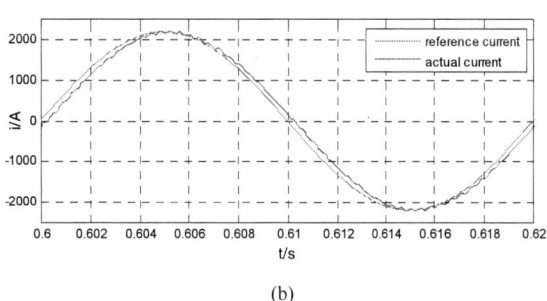

(b)

Fig. 11. Input current of the 4QC compared with its reference (a) using multiple frequency tuned quasi-PR controller (b) using P controller

For the results shown in Fig. 11(b), the proportional coefficient K of the P controller is set at 7. The actual current lags the reference current with a steady-state error. A bigger value of K (means a bigger gain) can decrease the error but may cause the system to become unstable. From Fig. 11(a) it can be deduced that a multiple frequency tuned quasi-PR controller can easily accomplish zero steady-state error tracking.

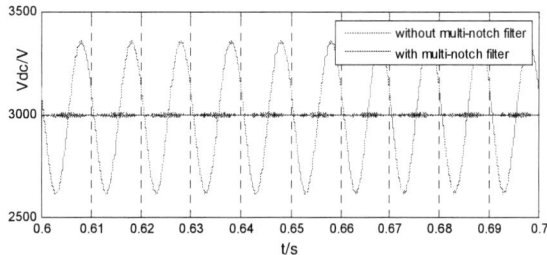

Fig. 12. DC-link voltages of the 4QC with multi-notch filter and without multi-notch filter

The DC-link voltage waveforms before being fed-back to the voltage controller with the notch filter and without one are shown in Fig. 12. So the contribution of the proposed multi-notch filter is validated.

The FFT analysis results of the line current of the primary winding of the traction transformer (i_{N1} in Fig. 1) when using conventional control strategy, multiple frequency tuned quasi-PR controller alone, and multi-notch filter alone strategies are shown in Figs. 13, 14 and 15, respectively.

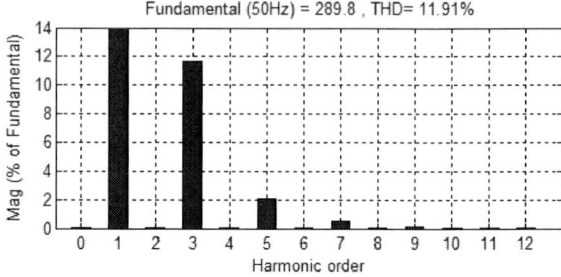

Fig. 13. FFT analysis of the i_{N1} for the conventional control strategy

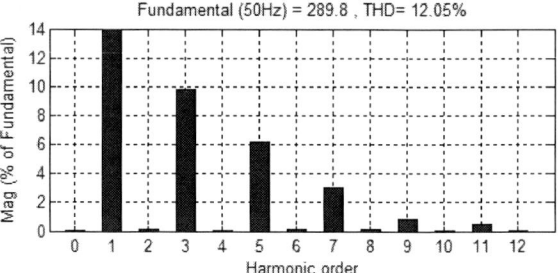

Fig. 14. FFT analysis of the i_{N1} for multi-frequency tuned PR controller alone

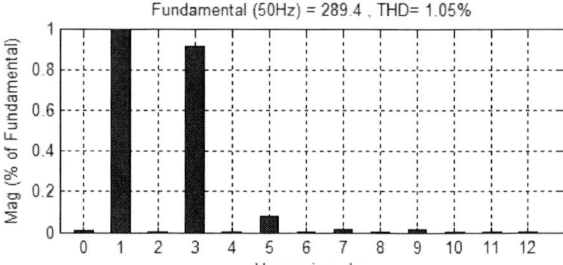

Fig. 15. FFT analysis of the i_{N1} for multi-notch filter alone

Comparing Figs. 13 and 14, it is found that when using only the multiple frequency tuned quasi-PR controller there is a small increase on the total harmonic distortion (THD) (from 11.91% to 12.05%) although the current tracking performance is better. The 5th and 7th harmonic currents increased rapidly, because of the resonant gains of the multiple frequency tuned quasi-PR controller at those frequencies.

Comparing Figs. 13 and 15, it is seen that a multi-notch filter can largely decrease the harmonic content. The THD of i_{N1} declined from 11.91% to 1.05%.

The spectrum and waveform of i_{N1} under the proposed control strategy (shown in Fig. 10) are shown in Fig. 16 with a reference waveform of the traction network voltage (v_{N1} in Fig. 1). It is found that the proposed control strategy combining multiple frequency tuned quasi-PR controller and multi-notch filter provides a further harmonic elimination. The THD is just 0.39%. Moreover the phase of i_{N1} is closely tracking that of the v_{N1}.

For quantitative comparison the characteristic harmonic ratios and THD of i_{N1} under the four control strategies are given in Table III.

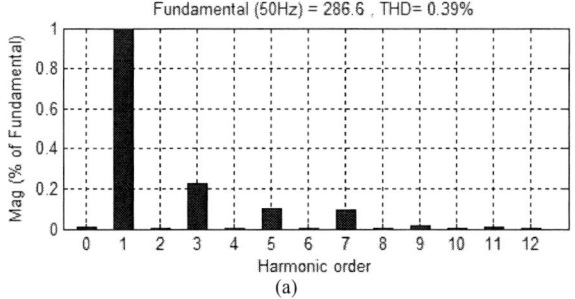

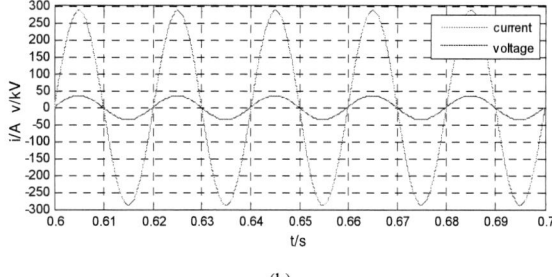

Fig. 16. FFT analysis and waveform of the i_{N1} for the proposed control strategy with a combination of multiple frequency quasi-PR controller and multi-notch filter (a) FFT analysis (b) current and voltage waveform

TABLE III
COMPARISON OF HARMONIC RATIOS

Harmonic Order	Conventional: P current controller& without DC voltage filtering	Multiple frequency tuned quasi-PR controller in current loop	Multi-notch filter for DC voltage filtering	Multiple frequency tuned quasi-PR & Multi-notch filter
3	11.71%	9.80%	0.91%	0.22%
5	2.05%	6.22%	0.08%	0.10%
7	0.49%	3.03%	0.02%	0.09%
THD	11.91%	12.05%	1.05%	0.39%

VI. CONCLUSION

A high performance control strategy based on a multiple frequency tuned quasi-PR controller for the current loop and a multi-notch filter for the voltage loop for EMU four-quadrant converters has been proposed in this paper. Simulation results obtained from Matlab/Simulink have validated the following (1) the proposed control strategy can eliminate the line current harmonics of the EMU and (2) provide a superior current tracking performance when compared with more conventional control strategies.

ACKNOWLEDGMENTS

Professor Agelidis wishes to acknowledge the support received from the Chinese Government through the High-End Experts Affairs Bureau for making the collaboration with Beijing Jiaotong University and his regular visits possible.

REFERENCES

[1] K. Sato, M. Yoshizawa, T. Fukushima, "Traction systems using power electronics for Shinkansen high-speed electric multiple units", in *Proc. of International Power Electronics Conference 2010*, pp. 2859- 2866.

[2] A. Timbus, M, Liserre, R, Teodorscu, P. Rodriguez, and F. Blaabjerg, "Evaluation of current controller for distributed power generation systems", in *IEEE Trans. Power Electron.*, Vol. 24, no. 3, pp. 654-664, Mar, 2009.

[3] Y. Huacheng, L. Hua, L. Yongcan, L. Yong, W. Xingwei, "A multi-resonant PR inner current controller design for reversible PWM rectifier", *2013 Twenty-Eighth Annual IEEE Applied Power Electronics Conference and Exposition (APEC)*, pp. 316- 320.

[4] H. Jin, L. Yang, Z. Bo, W. Na, S. Wensheng, "Harmonic current elimination for single-phase rectifiers based on PR controller with notch filter", *International Conference on Electrical Machines and Systems (ICEMS), 2012 15th*, pp. 1- 5

[5] Y. Sato, T. Ishizuka, K. Nezu, and T. Kataoka, "A new control strategy for voltage-type PWM rectifiers to realize zero steady-state control error in input current", in *IEEE Trans. Ind. Applicat.*, Vol. 34, pp. 480–486, May/June 1998.

[6] S. Fukuda and T. Yoda, "A novel current-tracking method for active filters based on a sinusoidal internal model for PWM invertors", in *IEEE Trans. Ind. Applicat.*, vol. 37, pp. 888–895, May/June 2001.

[7] D. N. Zmood & D.G. Holmes, "Stationary frame current regulation of PWM inverters with zero steady state" in *IEEE Trans. Power Electron.*, Vol. 18, no. 3, pp. 814-822, May, 2003.

[8] G. Xiaoqiang, Z. Qinglin, W. Weiyang, "A single-phase grid-connected inverter system with zero steady-state error", *IEEE 5th International Power Electronics and Motion Control Conference*, 2006. pp. 1- 5.

[9] Bo Liu, Xu Yang; Yan Zhang, Haizhong Ye, Fanlin Kong, "A new control strategy combing PI and Quasi-PR control under rotate frame for three phase grid-connected photovoltaic inverter", *8th International Conference on Power Electronics and ECCE Asia (ICPE & ECCE), 2011 IEEE*, pp. 882-888.

[10] D. N. Zmood & D.G. Holmes, "Stationary frame current regulation of PWM inverters with zero steady state" in *Proc. IEEE Power Electronics Specialists Conference.(PESC)*, 1999 , pp. 1185 – 1190.

[11] K. Jalili and S. Bernet, "Design of LCL filter of active-front-end two-level voltage-source converters" in *IEEE Trans. Ind. Electron.* Vol.56, No. 11, pp. 4483-4491, Nov. 2009.

[12] J. A. Taufiq, J. Shen, "Frequency domain modeling of a PWM converter traction system", in *Proc. of 5th European Conference on Power Electronics and Applications*, 1993, pp. 63- 67.

[13] J. Shen, J. Taufiq, A.D. Mansell, "Analytical solution to harmonic characteristics of traction PWM converters", in *IEE Proc. Electr. Power Appl.*, 1997, 144(2): 158-168.

[14] Y. Shaobing, W. Mingli, "Study on harmonic distribution characteristics and probability model of high speed EMU based on measured data", *Journal of the China Railway Society*, 2010, 32(3): 33-38 (in Chinese).

[15] D. G. Holmes and T. A. Lipo, *Pulse width modulation for power converters: principles and practice*, IEEE Press, 2003.

978-1-4799-2706-7/14 $31.00 © 2014 IEEE

A design of inrush current identification system for high-speed train's traction transformer

Weikai Yu, Xiankai Liu
National Engineering Research Center for High-speed EMU
CSR Qingdao Sifang Co., Ltd.
Qingdao, China

Yuzhuo Zhang, Yuan Cao
School of Electronic and Information Engineering
Beijing Jiaotong University
Beijing, China
08212055@bjtu.edu.cn

Weigang Ma, Xinhong Hei
School of Computer Science and Engineering
Xi'an University of Technology
Xi'an, China

Zhenhui Huang, Dawang Jiang
Project Management Department
CNR Tangshan Railway Vehicle Co., Ltd.
Tangshan, China

Abstract—**Distinguishing inrush current and short circuit current is a key issue for the longitudinal differential protection of high-speed train's traction transformer. This paper introduced virtual instrument technology to build an inrush current identification system, which utilized software to realize the identification algorithm and applied digital signal processing approach to the differential protection of the transformer. It developed the system software in virtual instrument tool LabWindows/CVI. The algorithm regards the inrush current and short circuit current as random signals, calculates the self-correlation functions of sampling data, then compares them with standard sine self-correlation coefficient vector table, and finally uses the magnitude of similarity coefficient to distinguish inrush current and short circuit current. Theoretical analysis and experiment result have proved that the system has high accuracy, fast decision rate and good expansibility.**

Keywords— inrush current; short circuit current; identification system; virtual instrument;.

I.   INTRODUCTION

The traction power supply of a high-speed train is commonly obtained from a 220kV/25kV single-phase distribution transformer energized from any 2 phases of the 3-phase supply. In order to make the high single-phase traction load appearing as an approximately balanced 3-phase load, the overhead contact wires are supplied from A-B, B-C and C-A phases repeatedly at regular intervals (about 20 Km). An un-energized neutral section (about 6 m) is provided between two consecutive sections of overhead wires energized from different phases to avoid short circuits [1].

When the train passes through the neutral section, the pantograph touches the next section contact wire, therefore the transformer is de-energized and then again re-energized. Accordingly, the inrush current is produced repeatedly. When a fault is cleared and the overhead wire is still energized, the inrush current can become very high,

6-8 times of the current rating, therefore leading to transformer protection misoperation. As the differential protection device reflects the sensitivity of internal short circuit of the transformer winding or internal insulation arc fault, the misoperation reasons must be find out once it occurs. How to distinguish the inrush current and short circuit current faster and more accurately has long been the focus of transformer differential protection research.

Many identification principles and methods have been put forward, which can be roughly divided into model-based methods and waveform-based methods. Model-based methods, such as Hidden Markov Model method [2], equivalent instantaneous inductance method [3], have some difficulties to apply in actual engineering. That is because some electrical parameters are needed, but cannot be accurately measured. These methods still need improvement in electrical measurement technology and theoretical study for field applying. Waveform-based methods, such as second harmonic restraint method, dead zone method and some improvements [4], have the advantage of simple principle and flexible functional configuration. These methods will be influenced by CT (Current Transformer) saturation and non-periodic components. There are also some new techniques [5] applying in this field, such as the artificial neural networks [6], the fuzzy logic [7], wavelet transformation.

As for the identification systems, many traditional ones are designed and implemented by hardware circuit to realize complex logic, which is quite expensive and limits its accuracy. With the development of computer technology, virtual instrument regards computer as a unified hardware platform, and utilize its powerful data processing ability to store, compute and analyze. It replaces the specialized function and panel of traditional instrument with software, to constitute a new instrument system that has the same appearance and function with traditional instrument, but virtual instrument owes the advantage of processing more complex test data with faster speed, lower price and quicker technological updating. Therefore, this technology has very promising future.

This work was supported by "the Fundamental Research Funds for the Central Universities" under the grant No. 2014JBM008.

This paper comprehensively analyzed the difference between inrush current and internal short circuit current of high-speed train's traction transformer, then put forward an inrush current identification method after a train passing through a phase separation point with self-correlation function principle, and realized the identification algorithm in virtual instrument tool LabWindows/CVI. Finally, the effectiveness of this approach was investigated.

II. PRINCIPLE OF IDENTIFICATION METHOD

The typical inrush current waveform and short circuit waveform can be seen in Fig. 1, where waveform A and waveform B are asymmetrical inrush current, waveform C is symmetrical inrush current short circuit current, while waveform D is short circuit current. It is clear that discontinuous angles exist in waveform A, B, C, and they are greater than $60°$ [4]. It is these discontinuous angles that cause the asymmetry between the preceding semi-cycle and the latter semi-cycle in the inrush current's waveform. Waveform D shows the short circuit waveform's preceding semi-cycle and latter semi-cycle are symmetrical. Therefore, this feature can be used to distinguish the inrush current and short circuit current of the transformer. In digital signal processing, as random signal cannot be precisely predicted, usually the similarity of two pieces of random signal or self-similarity of one piece in statistical significance can only be investigated, for the detection, identification and extraction of the signal. Hence, it is feasible to apply the principle of correlation function for the waveform similarity analysis to identify the feature.

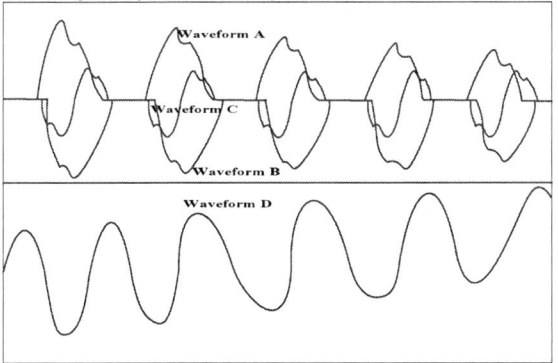

Fig.1. Waveform comparison of inrush current and short circuit current

Choose differential current between phases as the research object. In order to obtain it, first regard transformer secondary side's current magnitude and phase as the original side's ones, calculate the differential current of each phases. Then, calculate the differential current between phases and get 3 groups of current waveform, namely I_a, I_b, I_c. Take I_a as an example, the process can be formulated as:

$$\begin{cases} I_{ax} = I_{a1} - I_{a1}^{r} \\ I_{bx} = I_{b1} - I_{b1}^{r} \\ I_a = I_{ax} - I_{bx} \end{cases} \quad (1)$$

Where I_{a1}, I_{b1} are the original side current of phase a and b; I'_{a1}, I'_{b1} are the secondary side current which are regard as original side; I_a is the differential current expected [8].

Then the differential current needs smoothing and difference to remove high-frequency interference wave and DC component before processing. At the same time, assume 2N (the value of N is related to the system's sampling rate) points are sampled during every cycle, half cycle data window is used to calculate the projected area of data window on the time line. And choose the starting point corresponding to maximum area as the protection actuation point. The formula is as follows:

$$I(n) = I_a(n) - I_a(n-1) \quad (2)$$

$$S(k) = \sum_{k}^{k+N-1} abs(I(j)) \quad (3)$$

Where $I(n)$ is the differential current between phases after smoothing and difference; $S(k)$ is the projected area of half cycle data window on the time line.

This paper uses half cycle as data window to estimate self-correlation function of differential current. After 3/4 cycle delay, take the protection actuation point as the starting point of data window, move the window point by point, calculate N/2 sets of self-correlation function in order, and then conduct normalization and absolute value[9].

$$p(m) = \frac{1}{N} \sum_{n=0}^{N-1-abs(m)} I_N(n) I_N(n+m) \quad (4)$$

$$\bar{p}_i = \frac{abs(p_i)}{max[abs(p_i)]} \quad (i = 1,2,...,2N-1) \quad (5)$$

$$q(m) = \frac{1}{N} \sum_{n=0}^{N-1-abs(m)} I_{SN}(n) I_{SN}(n+m) \quad (6)$$

$$\bar{q}_i = \frac{abs(q_i)}{max[abs(q_i)]} \quad (i = 1,2,...,2N-1) \quad (7)$$

Where $I_N(n)$ and $I_{SN}(n)$ are the sampling values of differential current and fundamental wave current; $p(m)$ and $q(m)$ are the estimation value of differential current's and sinusoidal current's self-correlation function, their length are all 2N-1; $\bar{p}_i$ and $\bar{q}_i$ are the absolute values after normalization. Through N points self-correlation calculation of standard power frequency sine current, 2N-1 points self-correlation vector can be obtained, which is called standard sine self-correlation coefficient vector table. Take self-correlation value of $I(n)$ point by point with standard sine self-correlation coefficient vector table to conduct similarity calculation, and pick the minimum value of every similarity coefficient set. The calculation process is as follows:

$$\rho_i = \frac{1}{1 + [abs(\bar{p}_i - \bar{q}_i)]^2} \quad (i = 1,2,...,2N-1) \quad (8)$$

$$\rho = min(\rho_i) \quad (9)$$

Where ρ_i is the similarity of every sets, and ρ is the minimum value of ρ_i. Calculate the average value T of N/2 sets of minimum similarity coefficient, and set a decision value $T_{threshold}$. If $T < T_{threshold}$, it is judged as inrush current, otherwise it is judged as internal fault current.

III. DESIGN OF IDENTIFICATION SYSTEM

To realize the method above, an identification system is designed to prove its effectiveness. The system uses CT (Current Transformer) to acquire the current of the transformer's phase A, B, C respectively. We have chosen SCT264 type of CT, which inputs 5 A current and outputs 0.35A current. After transforming the output signal into voltage signal through A/V conversion, the signal needs gain matching. Hence, send the signal to

voltage amplifier circuit to make the actual voltage equals integral multiples of conditioning voltage. Then for the impedance matching, voltage follower circuit is needed to perform isolation function, as its input is high impedance while output is low impedance. The whole conditioning circuit is shown in Fig.2. Finally the acquired ±5 V voltage is sent to the data acquisition card and input into PC. The data acquisition card adopts PCI-MIO-16E-1 of NI because of its good real-time and high precision, which has the function of D/A, A/D, digital input/output, counter/timer.

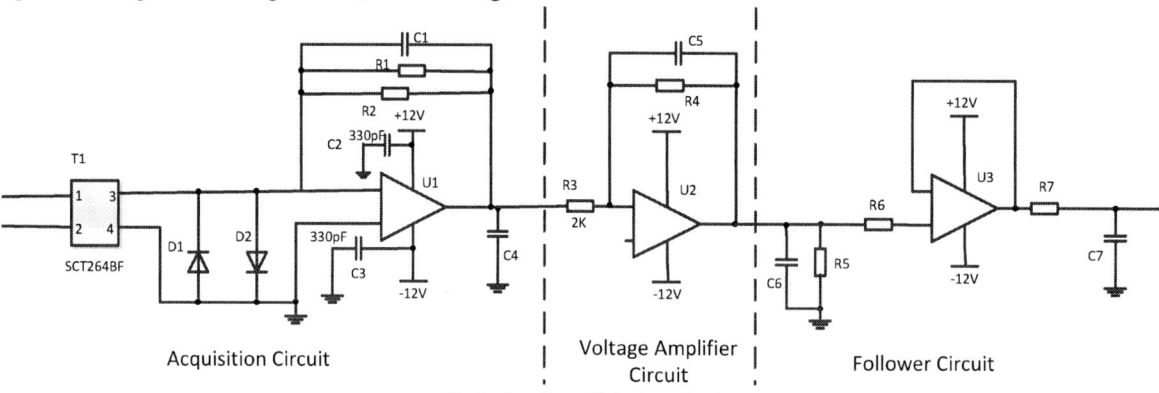

Fig.2. signal conditioning circuit

The system software plays an important role in the identification system and is developed in LabWindows/ CVI(C for Virtual Instruments), which has various function library, powerful function module and interface module, supplies an ideal platform for virtual instrument developers.

. The software includes two aspects, one is application software, and the other one is data acquisition drive software. The application software adopts the idea of design functional modules and contains the 5 modules, namely data acquisition, data storage module, display and recall module, data measurement and the main module. The main module is the core part of the software, which realizes the algorithm to distinguish waveform type according to standard sine self-correlation coefficient vector table, and the flow chart is shown as Fig.2.

In order to precisely calculate the average value of minimum self-correlation coefficient, the following measures should be taken:(1) guarantee the continuity of signal acquisition, especially for the periodic signal;(2) high-frequency interference and DC component should be removed by software smoothing and difference calculation; (3) boost sampling rate to guarantee analytical precision.

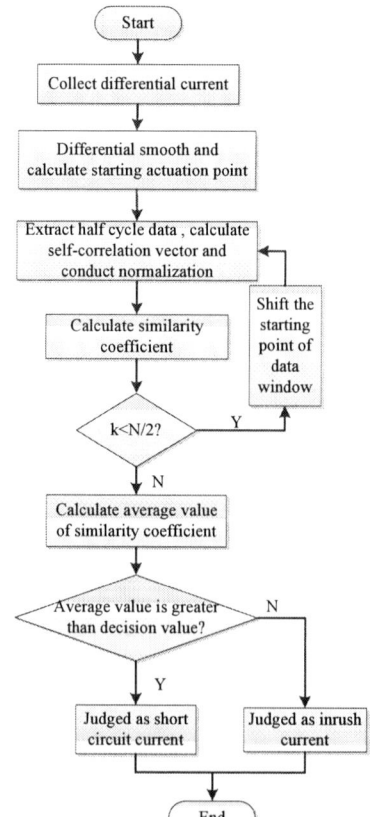

Fig.3. software design flow

IV. RESULTS AND ANALYSIS

Using signal generator in the laboratory, the inrush current and short circuit current are simulated and self-correlation function analysis is conducted. The obtained normalized self-correlation function of continuous N/2 sets of half cycle data window can be seen in Fig. 4, where waveform C is inrush current and waveform D is short circuit current. Intuitively, standard power frequency sine wave's self-correlation has poor similarity with inrush current's self-correlation function. However, it has good similarity with short circuit current's self-correlation function.

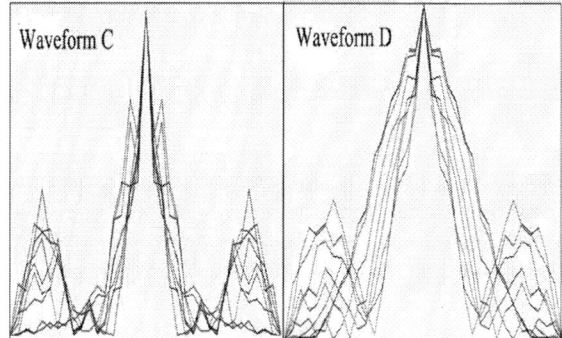

Fig.4. Self-correlation function comparison of inrush current and short circuit current

Run the software 50 times at different cases, each time we can get the minimum similarity coefficient of inrush current and short circuit current. After repeated experiment, the overall results are compared in Fig.4.

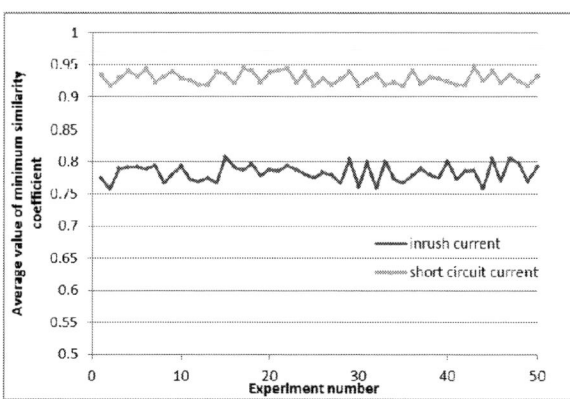

Fig. 5. Average value comparison of inrush current's and short circuit current's minimum similarity coefficient

From Fig.5, it is obvious that the minimum similarity coefficient decision value $T_{threhold}$ can be set to 0.86. If the average value of minimum similarity coefficient T is less than $T_{threshold}$, it can be identified that the secondary current at this moment is inrush current; otherwise, it is internal fault current. As long as the decision value is chosen properly, the method is effective, and this value can be adjusted to meet the requirements of various situations in field test.

V. CONCLUSIONS

This paper introduces virtual instrument into the measurement and protection of high-speed train transformer, and supply an approach based on correlation function in digital signal processing for the identification of inrush current and short circuit current. Theoretical analysis and experiment result have proved its correctness and feasibility. The future work lies in three aspects:

(1) improve the accuracy and decision rate of the identification algorithm ;

(2) supply measurement function extension to the system framework;

(3) establish experimental platform to verify the effectiveness of the suggested system in real environment.

REFERENCES

[1] K. P. Basu, S. Morris, "Reduction of magnetizing inrush current in traction transformer," *IEEE International Conference on Electric Utility Deregulation and Restructuring and Power Technologies,* pp.2302-2305, 2008.

[2] S. Jazebi, B. Vahidi, S. H. Hosseinian, "A new stochastic method based on hidden Markov models to transformer differential protection," *IEEE International Conference on Optimization of Electrical and Electronic Equipment,* pp. 179-184, 2008.

[3] C. Feng, D. Bi, B. Ge, "Equivalent instantaneous inductance based inrush current identification for three-phase V/V traction transformer," *IEEE International Conference on Power System Technology (POWERCON),* pp. 1-6, 2010.

[4] X. Tian, X. Li, Y. Li, "Using combined characteristics to identify inrush current and internal fault current of traction transformer," *IEEE Conference on Industrial Electronics and Applications (ICIEA),* pp.1948-1952, 2011.

[5] Y. Chu, B. Zhang, Z. Hao, "Application of new techniques in magnetizing inrush current identification for transformer differential protection," *Electric Power Automation Equipment,* vol.24, pp.87-91, 2004. In Chinese.

[6] A. Nosseir, A. S. Attia, F. Tahoon, et al, "A fast and reliable transformer protection system based on the transformer magnetizing characteristics and artificial neural networks," *IEEE International Middle-East Power System Conference,* pp. 511-516, 2008.

[7] A.Wiszniewski, and B. Kasztenny, "A multi-criteria differential transformer relay based on fuzzy logic, " *IEEE Transaction on Power Delivery,* Vol.10, pp.1786~1792,1995.

[8] X. Lin, P. Liu O. P. Malik, "Studies for identification of the inrush based on improved correlation algorithm," *IEEE Transaction on Power Delivery,* Vol. 17, pp.901~907, 2002.

[9] G. Shang, W. Li, "Study of transformer protection on Voltage current similarity theory," *2001 IEEE Power Engineering Society Summer Meeting,* pp.622~625, 2001.

Current Source Inverter Based Cascaded Solid State Transformer for AC to DC Power Conversion

Sudhin Roy[1], Ankan De[2]

Department of Electrical and Computer Engineering,
North Carolina State University, USA
[1]sudhin.roy@gmail.com; [2]ade@ncsu.edu

Subhashish Bhattacharya

Department of Electrical and Computer Engineering,
North Carolina State University, USA
sbhattacharya@ncsu.edu

Abstract— There are applications like railway traction system where power converters for HVAC to MVDC or MDAC to LVDC are required with a high power/weight ratio. The conventional way is to use heavy line frequency or medium frequency transformers followed by a controlled rectifier. The availability of high voltage SiC devices has made it possible to raise the switching frequency to higher value and hence the size and weight of the transformer can be reduced. This paper proposes a cascaded current source converter based topology with high frequency isolation as a replacement of the huge line frequency transformer. The paper also presents experimental results as a validation of the functionality of the converter.

Keywords— *Cascaded converter, Solid state transformer, and current source converter.*

I. INTRODUCTION

The conventional power converters for MVAC to LVDC applications require a line frequency transformer at the front end. With the invention of solid state transformer, the bulky and heavy line frequency transformer can be replaced with the compact high frequency one. The problem in selecting high switching frequency for the transformer is the unavailability of suitable devices (low switching loss) at that high voltage. For some particular application like traction system, the heavy and bulky transformer occupies considerable space of the train. The initial effort was to use the available devices in medium frequency range and bring the size of the transformer down to some extent. In today's world the availability of high voltage silicon carbide devices has made it possible to operate at high frequency even for high power converters. Now research is going on to replace the medium frequency transformer with the high frequency one so as to bring down the size and weight of the transformer further. [1] [2] and [3] have shown a cascaded converter topology with a medium frequency transformer. The converter topology with high frequency transformer is discussed in [4]. Here also cascaded converters are used to distribute the high voltage among the individual sections. The front end converters (HV side) are connected in series and the rear end converters (LV side) are connected in parallel. These rear end

converters produce the required DC bus voltage for a particular application. The topology is based on voltage source converters. [5, 6] discuss about basic power electronic based high frequency transformer concept which can be used to replace the bulky line frequency transformer.

In this paper, a cascaded converter based topology with high frequency transformer is proposed where current source converters are used. It is well known that in medium and high voltage applications current source converters are preferred compared to the voltage source converters. It is because of the fact that current source converters have better control on the short circuit current during fault conditions. [7] (Chapter 10) has shown a current source converter topology with high frequency fly-back transformer in the middle. A new topology for solid state transformer and its cascaded form for MV application is discussed in [8].

This paper proposes the cascaded CSI based converter for MVAC to LVDC application. A laboratory prototype of this converter has been build and experimental results are presented here to validate the operation of the converter.

II. CONVERTER TOPOLOGY

Fig.1 shows the proposed cascaded CSI based converter topology for the MVAC to LVDC application. Each of the converter section consists of two cascaded full-bridge inverters and a high frequency fly-back inductor. The fly-back inductor with two windings provides electrical isolation between the sources. It also allows to step-down the link voltage to the specified level. The input side full bridges are connected in series to distribute the source voltage among the sections. This allows the use of low voltage devices for the converters. The output sections of each cascaded converters are connected in parallel as shown in Fig.1. This allows the low voltage and low current devices to be used in the rear end converter of each cascaded sections. Filter capacitors are connected at the input as well as the output side. The input side AC capacitors are connected in series to distribute the source voltage to each of the cascaded sections. The switching scheme of the cascaded sections is same. Therefore, the current in the series connected front end converters may differ and hence the voltage distribution of the capacitors can be unequal. This unequal voltage sharing appears as an extra voltage burden to some of the switches on the high voltage side. To mitigate this problem, either external hardware circuits balancing the capacitor voltages can be implemented, or special control algorithm for the cascaded converters can be implemented so as to balance the series connected capacitor voltages. Moreover, the design and fabrication of the fly-back inductors should be identical to the extent it is possible so that the unequal voltage sharing issue is minimized at the root.

The output sections of the cascaded converters are connected in parallel to the DC bus capacitor. Therefore, any mismatch in the secondary side of the fly-back inductors current should not lead to any problem. As shown in Fig.1, suitable resistances are connected across each series connected capacitor at the input side to have proper distribution of the input voltage.

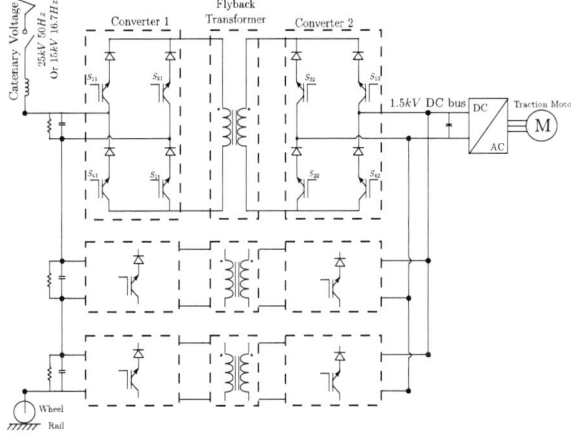

Figure 1: Proposed cascaded converter topology for the railway traction drive system

III. PRINCIPLE OF OPERATION

Fig.2 shows a section of the cascaded converter discussed above. It consists of three parts as (i) front end converter, (ii) fly-back inductor and (iii) rear end converter. The front end converter is used to store energy in the fly-back inductor depending on the front and rear end capacitor voltages. The direction of the current flow in the fly-back inductor remains always same. The converter switches to be turned ON are selected so as to make the inductor current always positive. Therefore, switches (IGBT in series with diode) which can carry unidirectional current are chosen for this application. For a particular instant switches corresponding to the positive current in the primary winding of the fly-back inductor are turned ON. This allows some energy to get stored in the air gap of

the fly-back inductor. During this charging time, the rear end converter remains in sleeping mode. The rear end converter switches are turned ON and the front end pairs are turned OFF when the inductor current reaches a specified value. The switches are so selected that the energy stored in the air gap is dumped into the output capacitor. During this mode the front end converter goes to sleeping mode like the rear end converter in inductor charging mode. Fig. 3(a) and (b) show the possible inductor charging and discharging current paths respectively. In practice, there should be an overlapping time between the two converters to avoid voltage spike due to high Ldi/dt. This charging and discharging of the inductor is performed within a short period compared to the fundamental period of the source voltage. The fly-back transformer size and weight depends on the selected frequency of this charging and discharging cycle.

The switching sequence of the cascaded converters remains same. This equalizes the voltage sharing of the series connected capacitors at the input side. However, there may be differences in the switching characteristics of the switches and the fly-back inductors may not be identical. This will lead to unequal voltage distribution in the input capacitors. The control oc converters should be able to take care of this unequal voltage distribution at steady state.

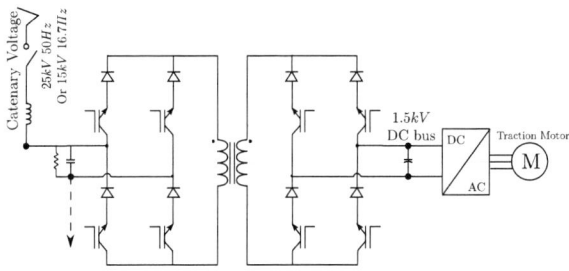

Figure 2: A section of the cascaded converter

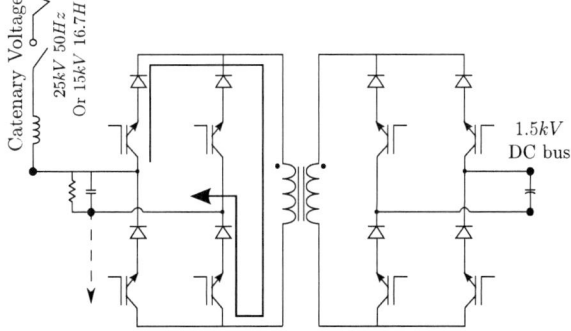

(a) During charging period of the fly-back inductor

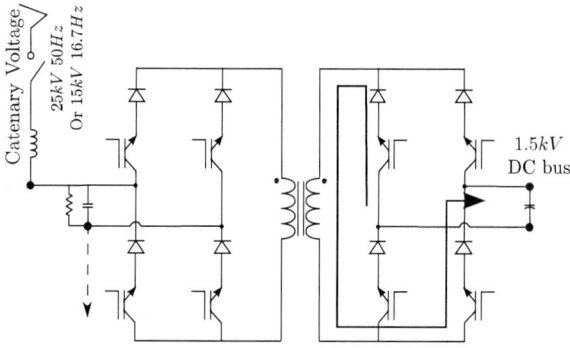

(b) During charging period of the fly-back inductor

Figure 3: Fly-back inductor charging and discharging of the cascaded converter

IV. CONTROL STRUCTURE

The cascaded converter sections have two-winding fly-back inductors which isolated the two electrical sides as well as works similar to a step-down transformer. If the inductors are fabricated identically, the converter switches can be gated in a synchronized manner. However, little bit of mismatch between the inductors is unavoidable practically. Therefore, synchronized gating of the semiconductor switches will lead to unbalance in the capacitor voltages. In order to mitigate this issue, the converters need to be controlled separately. The following paragraph discusses the control topology followed in this paper.

In this control topology, one of the converters is set as master which takes care of the output voltage regulation and the rest are controlled as slave. The reference to the slave converters is 1/N times the RMS/peak value of the main grid voltage where N is the number of cascaded sections. Fig. 4 shows the structure of the control system used to control the cascaded converter. Normal PI (proportional integral) controllers are used to track the references for the individual converters.

V. EXPERIMENTAL RESULTS

The functionality of the proposed cascaded converter has been experimentally verified on a laboratory prototype. The prototype is built with three series connected converter sections at the input side. The output sides are connected in parallel as mentioned in the above paragraphs. In practice, the number of cascaded sections can be decided based on the level of input voltage and the available IGBT voltage rating. The rear end converter devices are selected on the basis of available device current ratings and number of cascaded sections. The input of the prototype is fed from normal single phase utility supply (120V; 60Hz). A resistive load (40Ω) is connected at the output side.

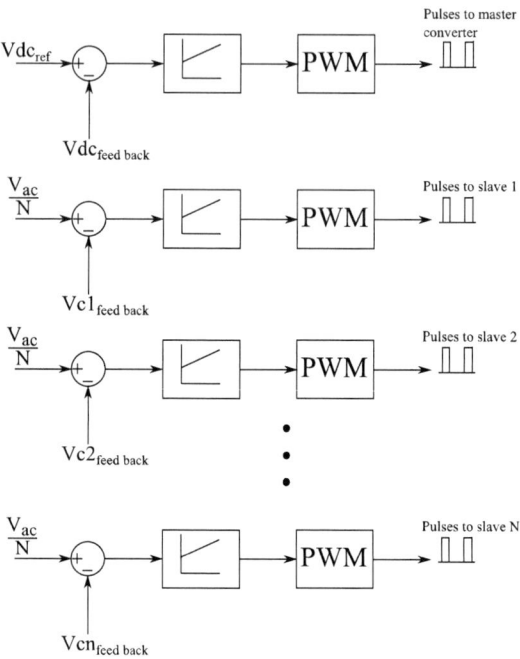

Figure 4: Control topology of the cascaded converter for AC to DC application.

It is discussed in the earlier sections that the use of similar fly-back inductors for the cascaded sections is needed in order to keep the voltages of the input side capacitors balanced under steady state. This ensures similar voltage stress on the devices of the front end converter. However, there can be some sort of mismatch in the fly-back inductors. This results in unequal voltage distribution on the input side capacitors and hence unequal voltage stress on the devices. In order to avoid this, the cascaded front-end converters are controlled separately so as to make the capacitor voltages balanced. In the following paragraphs the experimental results are shown to justify the functionality of the converter.

The inductances of the fly-back inductors referred to their primary windings are 256µH (top one), 324µH (middle one) and 280µH (bottom one). Therefore, the middle converter draws less current compared to the other two converters. Consequently, the middle capacitor voltage will be more compared to the other two voltages. Thus, the middle converter switches are stressed more electrically. This is true if the same PWM scheme is used for all the converters. During the experimentation, the converters are run for same PWM scheme as well as with the voltage balancing PWM topology. These waveforms will show the difference between the converter operations with the two control topology.

Fig.5 shows the fly-back transformer primary winding voltages (top and middle converter) with and without the input capacitor voltage balancing algorithm working respectively. The primary winding voltages of the top and bottom converter are shown in Fig.6. As per expectation, the middle inductor voltage is more compared to the other two with the simple PWM scheme. This voltage gets distributed equally between the converter sections with the modified control scheme working.

Fig. 7 shows the input side capacitor voltages (top and middle) with and without the voltage balancing algorithm working. Without the voltage balancing algorithm, the capacitor peak voltages are 51.1V and 67.3V respectively. These voltages become 58.6V and 58.1V respectively when the algorithm is working. Similarly, the top and bottom capacitor voltages are shown in Fig. 8.

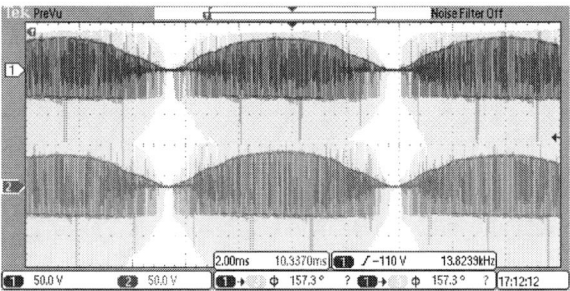

Fig. 5(a): With the voltage balancing algorithm of the input capacitors working.

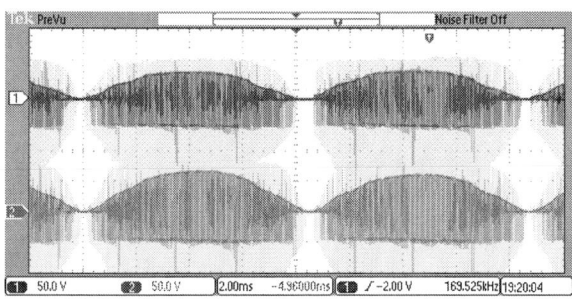

Figure 5(b): Without the voltage balancing algorithm of the input capacitors working.

Figure 5: Primary winding voltages of the top (ch1) and middle (ch2) fly-back transformers.

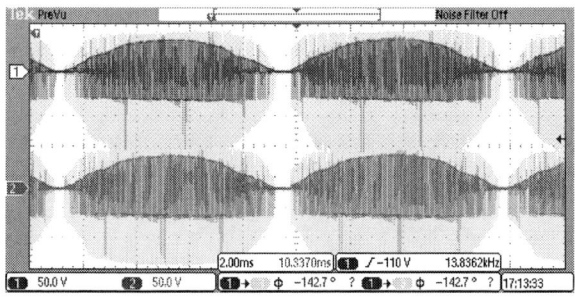

Figure 6(a): With the voltage balancing algorithm of the input capacitors working.

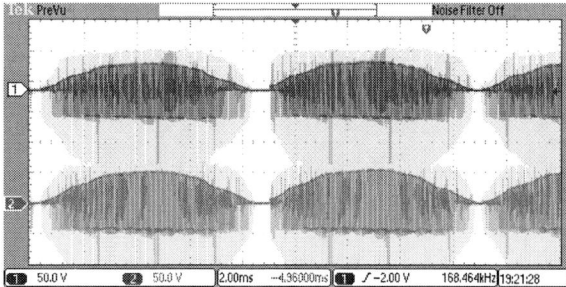

Figure 6(b): With the voltage balancing algorithm of the input capacitors working.

Figure 6: Primary winding voltages of the top (ch1) and bottom (ch2) fly-back transformers

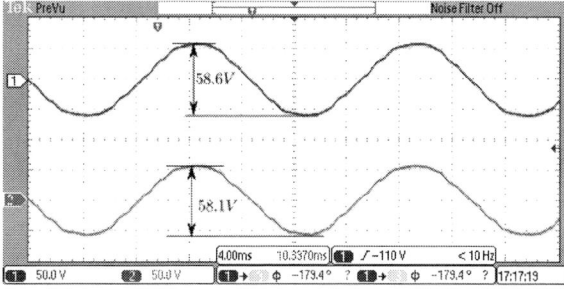

Figure 7(a): With the voltage balancing algorithm of the input capacitors working.

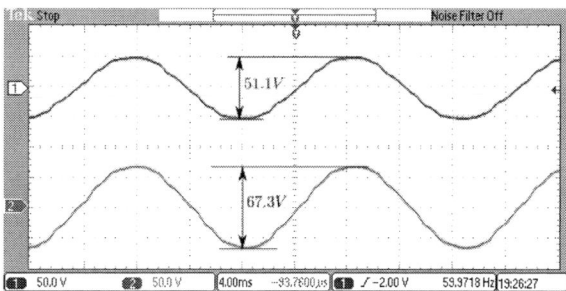

Figure 7(b): Without the voltage balancing algorithm of the input capacitors working.

Figure 7: Input side capacitor voltages of the top (ch1) and middle (ch2) converter

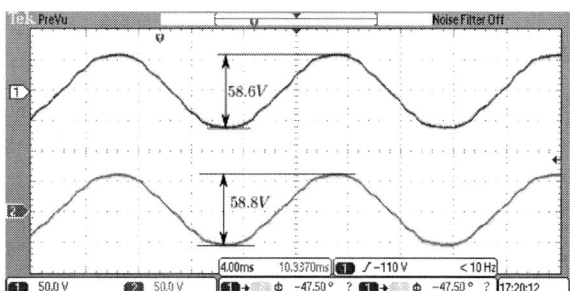

Figure 8(a): With the voltage balancing algorithm of the input capacitors working.

978-1-4799-2706-7/14 $31.00 © 2014 IEEE

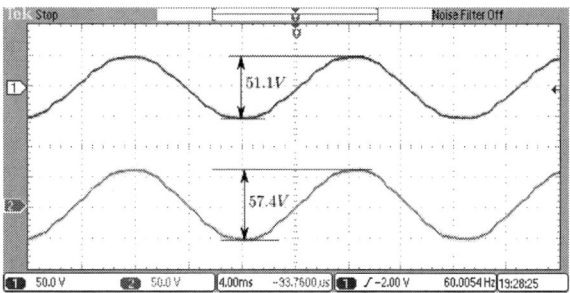

Figure 8(b): Without the voltage balancing algorithm of the input capacitors working.

Figure 8: Input side capacitor voltages of the top (ch1) and bottom (ch2) converter

The DC output voltage and the top fly-back transformer primary voltage is shown in Fig. 9.

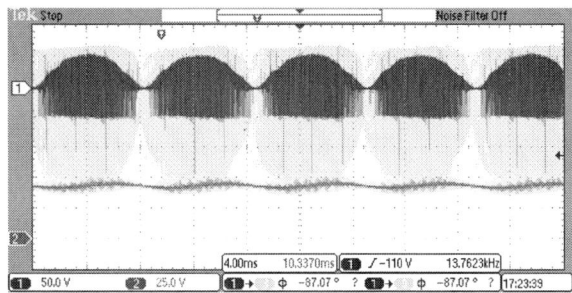

Figure 9: Top fly-back transformer primary voltage (ch 1) and the DC output voltage (ch 2) of the converter

Looking at the waveforms it is clear that without the voltage balancing algorithm working, the input capacitor voltages are unbalanced. This is because of the fact that the inductance of the top fly-back transformer is less than the other transformers. Therefore, converter 1 draws more current than the other two. In this situation, the voltage balancing algorithm comes into picture and it maintains the converter 1 average current same as the other converters. Thus the voltage profile of the input capacitors remains similar with this algorithm. The RMS value of the input capacitor voltages are (i) 42V (top capacitor), (ii) 41V (middle capacitor) and (iii) 40V (bottom capacitor) while the voltage balancing algorithm is working. These voltages become (i) 35V (top capacitor), (ii) 46V (middle capacitor) and (iii) 39V (bottom capacitor) without the voltage balancing algorithm added with the control system. Therefore, the voltage balancing control is a must for this converter.

It allows us to use switches with the same voltage rating. However, the peak currents become different for the switches. This voltage balancing control works fine under steady state condition. Special care has to be taken for transient voltage spike and unbalanced voltage of the series connected capacitors.

VI. CONCLUSIONS

The topology for replacing the line frequency transformer with the high frequency one in MVAC to LVDC

applications can be implemented easily with the available 6.5kV IGBTs. There are reverse blocking IGBTS available in the market which can be used in place of the series connected IGBT and diode. This will reduce the device count further compared to the conventional voltage source converter based topologies. In this paper the converter topology is implemented on a laboratory prototype with three cascaded sections. This can easily be extended for real application with more number of cascaded sections. The report also discusses about the voltage balancing of the input side series connected capacitors with non-identical inductors connected to the system. This method of achieving a balanced capacitor voltage works only at steady state. Extra measures are needed to consider for the transient condition.

VII. APPENDIX

The following are the design specifications of the laboratory prototype built in FREEDM system center.

- Input supply- 120V, 60Hz single phase supply.
- Fly-back inductance- 256µH (top one), 324µH (middle one) and 280µH (bottom one).
- Input side (ac side) capacitors: 10µF, 370 Volt AC capacitors.
- Output side capacitors: 470µF, 500V DC capacitors.
- Load resistance- 40Ω
- AC side filter inductor- 5mH

ACKNOWLEDGMENT

This work made use of FREEDM ERC shared facilities supported by National Science Foundation under Award Number EEC-0812121.

REFERENCES

[1] J. Martin, P. Ladoux, B. Chauchat, J. Casarin, and S. Nicolau, "Medium frequency transformer for railway traction: Soft switching converter with high voltage semiconductors," *International Symposium on Power Electronics, Electrical Drives, Automation and Motion*, SPEEDAM, 2008, pp. 1180–1185.

[2] B. Engel, "15kv/16.7 hz energy supply system with medium frequency transformer and 6.5kv igbts in resonant operation," *EPE03, Toulouse France, 2003.*

[3] H. Reinold and M. Steiner, "Medium frequency topology in railway applications," *European Power Electronics Conference* EPE, 2007.

[4] L. Heinemann, "An actively cooled high power high frequency transformer with high insulation capability," *Applied Power Electronics Conference, Dallas,* 2002.

[5] J. van der Merwe and H. du T. Mouton, "The solid-state transformer concept: A new era in power distribution," *AFRICON, 2009.* AFRICON '09., 2009, pp. 1–6.

[6] L. Heinemann and G. Mauthe, "The universal power electronics based distribution transformer, an unified approach," *Specialists Conference, 2001.* PESC, vol. 2, 2001, pp. 504–509.

[7] K. D. T. Ngo, "Topology and analysis in pwm inversion, rectification and cycloconversion," Ph.D. dissertation, California Institute of Technology, May 1984.
Patent Publication # US20130201733 A1, Aug 8, 2013. "Isolated dynamic current converters", by Deepakraj M. Divan, Anish Prasai, Hao Chen.

978-1-4799-2706-7/14 $31.00 © 2014 IEEE

Evaluation of High Voltage 15 kV SiC IGBT and 10 kV SiC MOSFET for ZVS and ZCS High Power DC -DC Converters

Shiva Moballegh, Sachin Madhusoodhanan, Subhashish Bhattacharya

FREEDM Systems Center, Department of Electrical and Computer Engineering, North Carolina State University
Raleigh, NC, USA

Abstract - **The advent of Silicon Carbide (SiC) devices has made possible high switching frequency operation of PWM power converters. In this paper, SiC devices are compared in detail with Si devices in a high power (1 MW) DC -DC converter application. The converter is designed as the building block for traction drives which requires it to operate at high power, high input voltage (11 kV) and low output voltage (800 V) levels. A dual active bridge (DAB) and a series resonant converter (SRC) topology are compared to achieve highly efficient operation. The performance and efficiency of these converters are compared by simulations using two different combinations of switches; the SiC combination consists of 10 kV/10 A SiC MOSFET at High Voltage (HV) side and 1200 V/100 A SiC MOSFET at Low Voltage side (LV), and the Silicon combination consists of 6.5 kV/10 A Si IGBT at HV side and 1200 V/100 A Silicon IGBT at LV side. For further understanding, efficiency analysis using the newly developed 15 kV/20 A SiC IGBT on the HV side is also carried out.**

Keywords— Dual Active Bridge, Series Resonant Converter, Silicon Carbide, ZVS and ZCS Operation

I. INTRODUCTION

The efficiency of high power converters have been always an important factor in their design suited for all applications. One of the main applications of such converters is in traction [1]. Choosing the right topology which only satisfies the basic electrical specifications of application is not enough. It is absolutely required to add

features such as soft switching techniques which will enhance the efficiency, especially as the converter operates at high switching frequency [2].

High operating frequencies result in higher switching losses of the semiconductor devices [3]. Therefore, operating at high frequency demands the reduction of the switching losses in high power applications. One typical solution in low power applications is the use of resonant converters with soft switching and hence high switching frequency operations [4]. However, these converters have been hardly able to find their way into high power applications as they have considerable conduction losses and also require high device ratings due to large voltage and current stresses on their component devices [4]. As an alternative, the pseudo-resonant converters have demonstrated lower component ratings in addition to low switching losses. Among such converters, the dual-active-bridge topology has become very popular [5].

Nonetheless, with considerable progress in developing SiC devices for high power and high frequency applications, the resonant converters are to be re-investigated in terms of overall efficiency. SiC devices have many advantages over Si devices such as higher critical field strength, lower intrinsic carrier concentration due to wide band gap, higher thermal conductivity and lower on-resistance [6, 7]. In this paper, the efficiency and performance of Dual Active Bridge (DAB) and

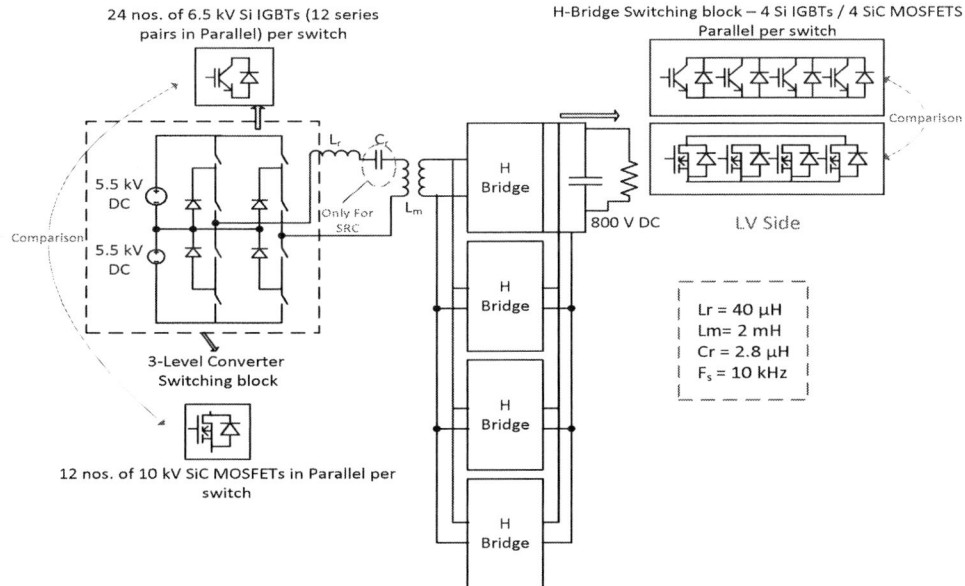

Fig.1. High Power DC-DC Converter Schematic

Series Resonant Converter (SRC) using 3-level poles on HV side are studied by their simulation models as they operate with high switching frequency (10 kHz) and use SiC or Si switches.

The topology is designed to use as the building block for traction drives or MVDC taps. The other applications are isolated MVDC systems for city in-fields and MVDC to LVDC collector systems for city in-fields. The specifications require 11 kV input level and 800 V output level, transferring 1 MW power from the high voltage side to the low voltage side. The high voltage side is implemented by a three level converter which decrease the voltage stress on these devices. Hence, 10 kV SiC MOSFET from Cree and 6.5 kV Si IGBT from ABB can be considered to equip the converter for the SiC and Silicon cases respectively [8] – [10].

As 10 kV SiC MOSFET is used at the HV input side, the output LV side is equipped with 1200 V/100 A SiC MOSFET from Cree [11]. In turn, 1200 V/100 A Si IGBTs from Semikron are used at the LV side when the HV side is equipped with 6.5 kV Si IGBTs (Fig. 1) [12]. For further understanding, efficiency analysis using the newly developed 15 kV/20 A SiC IGBT on the HV side is also carried out [7].

In *Section II*, the operation and design of both DAB and SRC are described. The simulation results are presented in *Section III* and the efficiency comparison of DAB/SRC using Si and SiC IGBTs and MOSFETs are compared in *Section IV*. Finally, conclusions are drawn in *Section V*.

II. SRC AND DAB OPERATION PRINCIPLE

A. Dual Bridge Series Resonant Converter

The topology of the three-level SRC is shown in Fig.2. The two input capacitors, C_{in1} and C_{in2} are connected in series to form the neutral-point. The eight high voltage switches, S_1-S_8, form the three-level full-bridge structure. The resonant tank connects the bridge's center-points (poles) by the resonant inductor L_r, the resonant capacitor C_r and the magnetizing inductor L_m. Four diodes, D_1-D_4, clamp the voltage stress on each device to half the DC bus voltage.

In order to perform simple analysis on the converter operation, it is assumed under steady state condition and

the input capacitors C_{in1}&C_{in2} acting as the voltage source. The output capacitor C_o is also considered large enough to represent the output voltage source V_{out}. The resonant operation of the converter can be divided into 10 intervals during each switching period.

Interval T_1: four switches, S_1, S_2, S_7 and S_8 are turned on. This provides a square-wave voltage source to the resonant tank, causing the primary resonant current I_r to increase in a sine-wave fashion. The magnetizing current I_m also increases yet in a linear fashion as the voltage of the secondary side of the transformer is clamped by output voltage. The secondary side current is proportionate to the difference between resonant current and magnetizing current.

Interval T_2: As the resonant current passes through its sinusoidal peak, it starts to decline and becomes equal to the linearly growing magnetizing current. At this moment, the secondary current reaches zero and the corresponding low voltage switches turnoff. Therefore, the secondary switches benefit from ZCS soft switching. In the primary circuit, the magnetizing inductor of the transformer L_m is now isolated from the secondary and participates in the resonance.

Interval T_3: As the current continues resonating, S_1 and S_8 are turned off with ZCS and their parasitic capacitance C_{S1} and C_{S8} obtain voltage charge by the circulating current. S_2 and S_7 still continue conducting and help discharge the parasitic capacitors of S_4 and S_5 correspondingly. These capacitors also participate in the resonance process.

Interval T4: As the voltage across the capacitors of C_{S1} and C_{S8} reach to $V_{in}/2$, the clamping diodes D_1 and D_4 start conducting, and the voltages will be clamped by input capacitors. The resonant current I_r flows through S_2, S_7, D_1 and D_4. At the same time, voltage across C_{S4} and C_{S5} are clamped to zero.

Interval T5: The voltage being clamped to zero leads to the soft switching turn off of S_2 and S_7. Therefore, C_{S2} and C_{S7} are charged and C_{S3} and C_{S6} are discharged by the resonant current. However, in this stage the anti-parallel diodes of S_4 and S_5 conduct and the discharge current is fed back to input, which causes reversal of the voltage applied on the resonant tank. Therefore the rectifying switches conduct and the magnetizing inductor L_m is clamped by the output voltage and exits from the

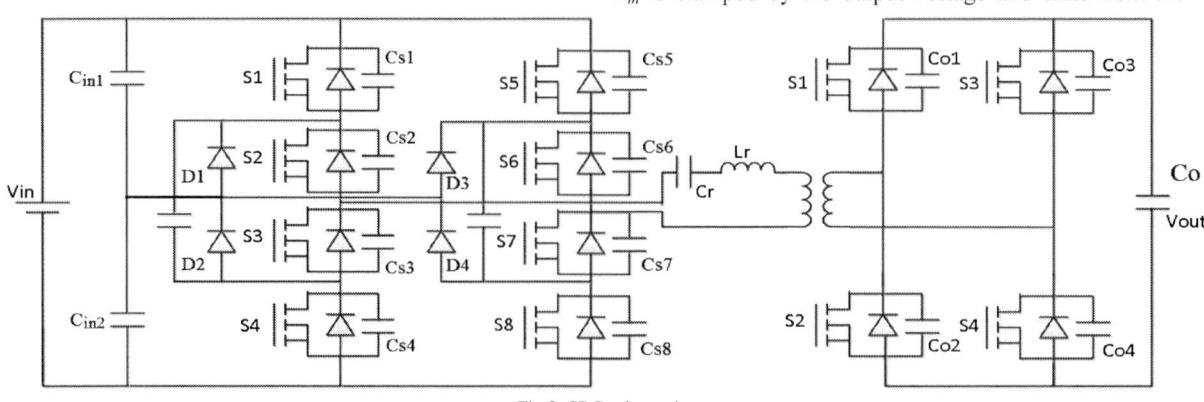

Fig.2. SRC schematic

resonance mode again.

At the end of this interval, the voltage of C_{S2} and C_{S7} are fully charged to $V_{in}/2$, C_{S3} and C_{S6} are also discharged to zero and the resonant current I_r flows through the anti-parallel diodes of S_3 and S_6. Therefore, S_3 and S_6 will turn on with ZVS. This ends the half of switching period discussed and begins the next similar half cycle. As discussed, the upper switches S_1 and S_2 and the lower switches S_7 and S_8 benefit from soft switching turn off while the other four switches exhibit ZVS turn on opportunity.

In SRC, when the switching frequency is below the resonant frequency, switches turn off at zero current condition, which is especially desirable for high power IGBTs. When the switching frequency is above the resonant frequency, the zero voltage turn on is achieved, ideal for high power MOSFETS [13].

The resonant parameters are designed based on the voltage gain, the load impedance and desired resonant frequency. For the sake of comparison between ZVS and ZCS capabilities, the resonant frequency of SRC is chosen above the switching frequency (10 kHz), being equal to 15 kHz. The resonant design parameters are shown in Fig. 1.

B. Dual Active Bridge Converter

The dual active bridge (DAB) converter is another prominent topology which is proven to be an alternative for high power applications [14]. It allows bidirectional power flow and soft switching operation upon having the same topology as SRC. However, it differs by its outstanding feature which is to take advantage of the transformer leakage inductance as the main energy transfer component. Therefore, a minimum number of resonant reactive components are required for realization. Using this converter has some benefits over SRC such as a better control of the power transfer while keeping the soft switching operation of the converter. Like SRC, it is possible to operate in both ZVS and ZCS modes of soft switching. In the case of operating in the triangular-current mode, DAB is able to exhibit ZCS capabilities similar to SRC when switched below the resonance frequency. Although, ZCS cannot be accomplished in the

LV side and the switches can only operate under ZVS soft switching mode. Several strategies have been proposed to maximize the DAB efficiency by enhancing the soft switching techniques. However, they may result in modified current shapes which have adverse impact on the converter performance. For the sake of comparison with SRC, this converter has been designed to meet the same specifications as described in previous section.

The DAB is operated at constant frequency with 50 percent duty cycle modulation, and the power which is transferred through the transformer leakage inductance is regulated by phase shift modulation technique. The three-level full bridge isolated converter is used to reduce the voltage across the switches.

The converter may operate in nine operation intervals, depending on the period between the turn off of the HV switches and also the value of the parallel capacitor (C_O). It should be noted that the converter can only operate under ZVS at the LV side and therefore for simplicity, the whole converter is designed to operate under ZVS. Through all different operation intervals, it maintains the ZVS operation as illustrated in the following:

The topology of three-level DAB is shown in Fig.3. An analysis of its operation under soft switching is presented in this section. The resonant mode occurs between L_s inductance and the parallel capacitors of the switches. In interval T1, as the switches S3, S4, S5 and S6 are conducting in the HV side, the other active switches are turned off and the power is transferred from the HV side to LV side. At interval T2, S4 and S5 are turned off at zero voltage level (ZVS) while S1 and S8 are kept conducting. The magnetizing inductance L_m and the parallel capacitors C_{s1} and C_{s4}, C_{s8} and C_{s5} will resonate together. Since, the resonant operation takes place during a small fraction of the switching period, this topology is called quasi-resonant ZVS converter. Now, the switch voltages of S4 and S5 are equal to the input capacitor voltages and the diodes parallel to S1 and S8 side conduct the load current. In the T3, the resonant inductor is charged and S1 and S8 can be turned on at any instant of time.

Because the anti-parallel diodes are conducting, the switches are ensured to have ZVS turn on. When the load

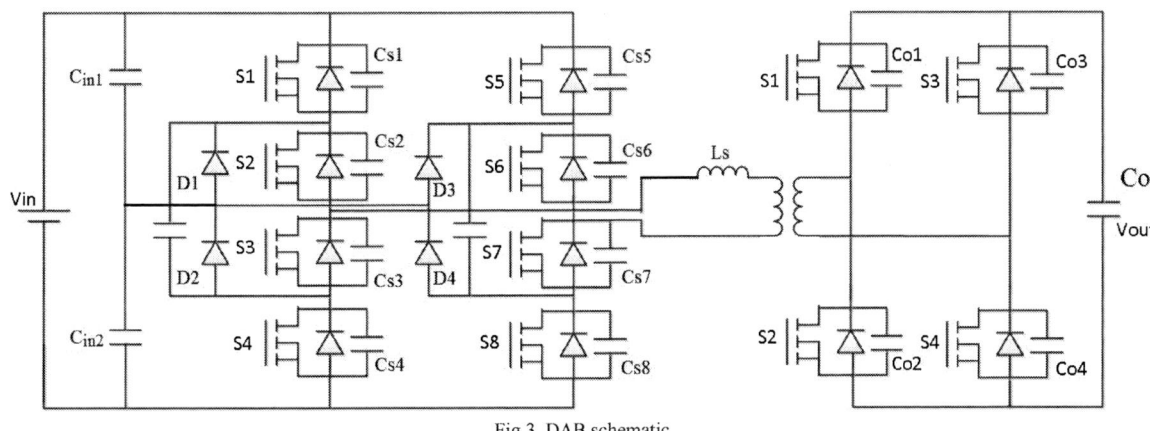

Fig.3. DAB schematic

current becomes positive, S1 and S8 start conducting and can be turned off by the control logic to start the next resonance operation [14]. The converter operates under soft switching for all switches as illustrated above. For DAB topology, the soft switching conditions are guaranteed only if the inductor current reaches zero between the zero crossings of transformer terminal voltages. Therefore, the phase shift between the voltages which determines the slope of the current will have a great impact on the range of ZVS operation.

III. SIMULATION STUDY OF CONVERTER SOFT SWITCHING

DAB and SRC converters are simulated in MATLAB – Plecs with the configuration as illustrated in Fig. 1. The 3-level NPC converter helps reduce the voltage stress on the HV switches. For the case with 10 kV SiC MOSFET, however, each switch is realized by twelve parallel MOSFETs to meet the current requirements. As there are eight switches in the 3-level topology, the HV switching block contains totally 96 switches. For the case with 6.5 kV Si IGBT, there are not only 6 switches in parallel but each switch is also realized by two switches in series. Hence, the total number of switches adds up to twice the number of MOSFETs, i.e. 192 switches. The output side consists of four H-bridges in parallel to carry the high load current, i.e. 1250A. Each switching block in one H-bridge contains four 1200V/100A SiC MOSFETs or 1200V Si IGBTs in parallel, resulting in a total of 64 switches.

A. Dual Active Bridge Converter

Soft switching in DAB occurs when the S1 and S2 in 3-level NPC are turned off and the inductor current discharges its capacitance. This process continues until either the body diode of lower switch starts to conduct or until the switch itself is turned on. Hence, S1 and S2, the upper switches are turned on with the forward voltage of diode that is very small, as shown in Fig. 4 (a).

At the time when the lower switches S3 and S4 are turned off, however, their body diodes continue to conduct due to the negative value of their current. The upper switches S1 and S2 are then turned on and their drain currents start to increase after the turn-on time delay expires. The current through the body diode of lower switches accordingly decreases and crosses zero, and becomes negative due to reverse recovery effects, and reaches the peak of its reverse recovery current. Then, the diode starts to block and the reverse current, which is still present in lower switch, decreases to zero. This switching operation is referred to as hard switching operation and typically generates high switching losses. This is shown in Fig. 4 (b). Fig. 4 (c) shows the voltage and current waveforms on the LV side switches.

B. Series Resonant Converter

Similarly in SRC, the HV upper switches S1 and S2 (S5 and S6) and the lower switches S3 and S4 (S7 and

S8) produce unequal losses. However, unlike DAB, the upper switch operates with soft switching at turn off, as shown Fig. 5 (a), and the lower switch benefits from the ZVS at turn on, as shown in Fig. 5 (b). The reason for such operation is the switching frequency being lower than the resonance frequency as discussed in Section II. Fig. 5 (c) shows the voltage and current waveforms on the LV side switches.

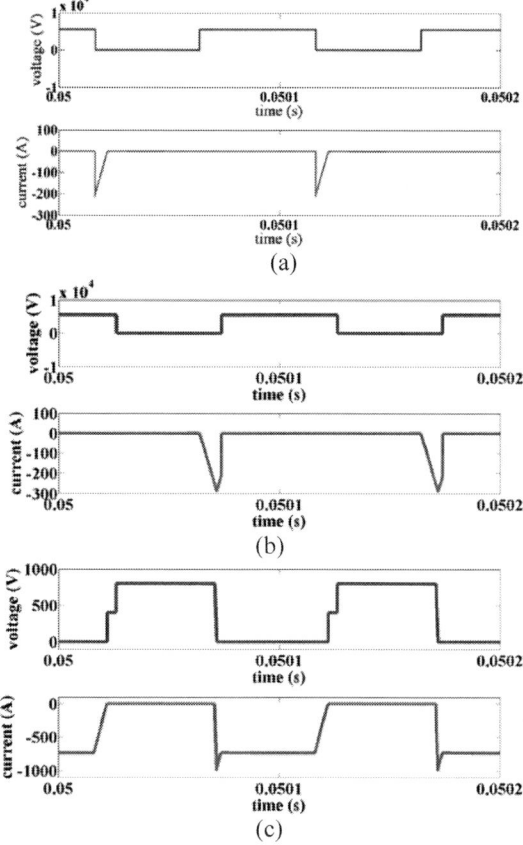

Fig.4. Current-voltage waveforms across DAB switches (a) HV side upper switch (b) HV side lower switch (c) LV side switch

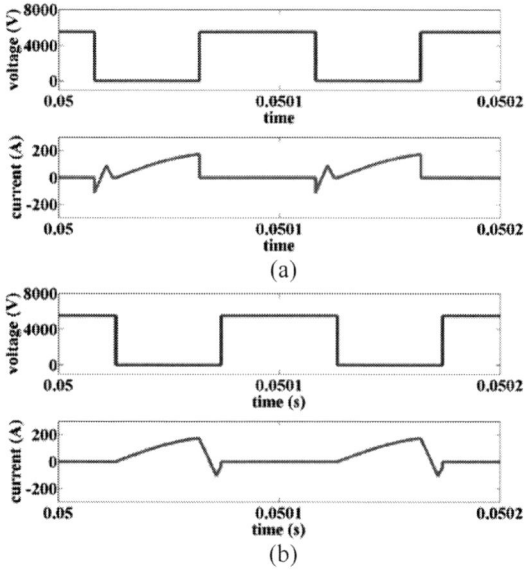

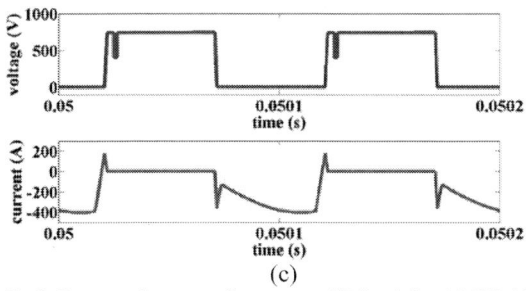

(c)

Fig.5. Current-voltage waveforms across SRC switches (a) HV side upper switch (b) HV side lower switch (c) LV side switch

TABLE 1 DAB LOSS DISTRIBUTION TABLE WITH SiC SWITCHES

Loss per switch / SiC switch	Switching loss (W)	Conduction loss (W)	Total loss (W)
10 kV MOSFET (upper)	0.23	4.27	4.5
10 kV MOSFET (lower)	10.5	51.5	62
1200 V MOSFET	0	183	183
Efficiency = 98.5 %			

TABLE 2 SRC LOSS DISTRIBUTION TABLE WITH SiC SWITCHES

Loss per switch / SiC switch	Switching loss (W)	Conduction loss (W)	Total loss (W)
10 kV MOSFET (upper)	9.2	17.8	27
10 kV MOSFET (lower)	7.2	19.8	27
1200 V MOSFET	6.95	56.05	63
Efficiency = 99.14 %			

TABLE 3 DAB LOSS DISTRIBUTION TABLE WITH Si SWITCHES

Loss per switch / Si switch	Switching loss (W)	Conduction loss (W)	Total loss (W)
6.5 kV IGBT (upper)	2.4	4.2	6.6
6.5 kV IGBT (lower)	940	40	980
1200 V IGBT	0	299	299
Efficiency = 88.6%			

TABLE 4 SRC LOSS DISTRIBUTION TABLE WITH Si SWITCHES

Loss per switch / Si switch	Switching loss (W)	Conduction loss (W)	Total loss (W)
6.5 kV IGBT (upper)	750	20	770
6.5 kV IGBT (lower)	485	15	500
1200 V IGBT	61	77.8	138.8
Efficiency = 86.9%			

TABLE 5 DAB LOSS DISTRIBUTION TABLE WITH 15 kV SiC IGBT ON HV SIDE

Loss per switch / SiC switch	Switching loss (W)	Conduction loss (W)	Total loss (W)
15 kV IGBT (upper)	1.04	0.01	1.05
15 kV M IGBT (lower)	248.5	51.5	300
1200 V MOSFET	0	183.5	183.5
Efficiency = 97.38 %			

TABLE 6 SRC LOSS DISTRIBUTION TABLE WITH 15 kV SiC IGBT ON HV SIDE

Loss per switch / SiC switch	Switching loss (W)	Conduction loss (W)	Total loss (W)
15 kV IGBT (upper)	200	18.5	218.5
15 kV IGBT (lower)	133.5	16.5	150
1200 V MOSFET	7	56	63
Efficiency = 97.83 %			

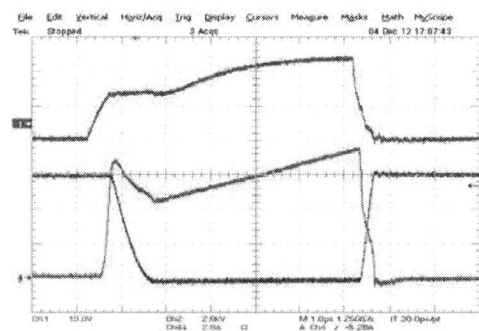

(a)

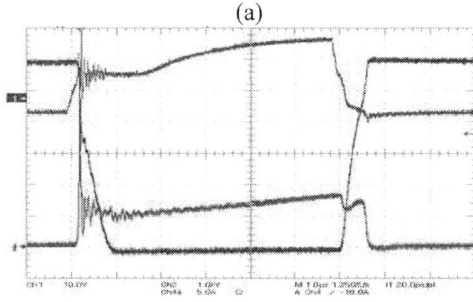

(b)

Fig.6. 10 kV SiC MOSFET switching characteristics at 6kV (b) 15 kV SiC IGBT switching characteristics at 6kV (Green: Current, Blue: Collector-Emitter Voltage, Brown: Gate Voltage)

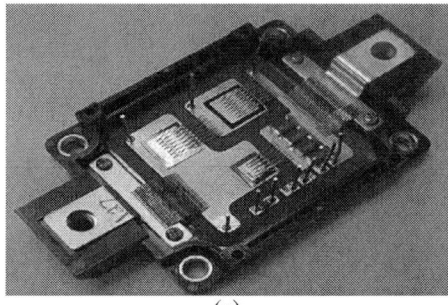

(a)

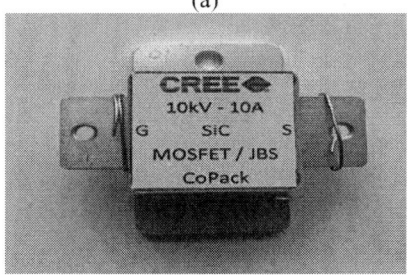

(b)

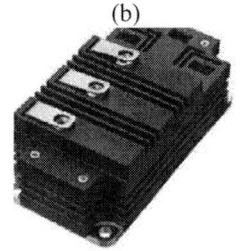

(c)

Fig.7.(a) 15 kV/ 20 A SiC IGBT co-pack module, (b) 10 kV/ 10 A SiC MOSFET co-pack module, (c) 6.5 kV/ 25 A Si IGBT dual module prototype

The 2014 International Power Electronics Conference

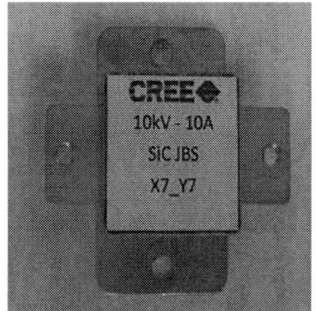

Fig.8. 10 kV/10 A SiC JBS Diode

Fig.9 1200 V/100 A SiC MOSFET Co-pack Module

IV. Loss Comparison Study and Discussions

The actual device loss data measured using tests conducted in the laboratory have been used in thermal library of Plecs simulation software for the loss calculation in the case of SiC devices. For the Si devices, the loss data have been taken from [10] and [12]. Fig. 6 shows the switching characteristics of 10 kV SiC MOSFET and 15 kV SiC IGBT measured at 6 kV blocking voltage [7, 8]. Each device on the HV side blocks 5.5 kV for the topology under consideration (Fig.1). Thus, these device characteristics up to 6 kV are considered in this paper. Fig. 7 shows the photos of co-pack modules of all HV active switches. Fig. 8 shows the 10 kV/10 A SiC JBS diode used as the clamping diode on the HV side. Fig. 9 shows the 1200 V/100 A SiC MOSFET co-pack module from Cree [11]. Fig. 10 shows the schematic of the standard Double Pulse Test (DPT) circuit. In all the cases, the inductance used in the test circuit is 8 mH. On-gate resistance of 100 Ω and off-gate resistance of 10 Ω are considered for these measurements till 6 kV. Fig. 11 shows the experimental setup used for testing these devices.

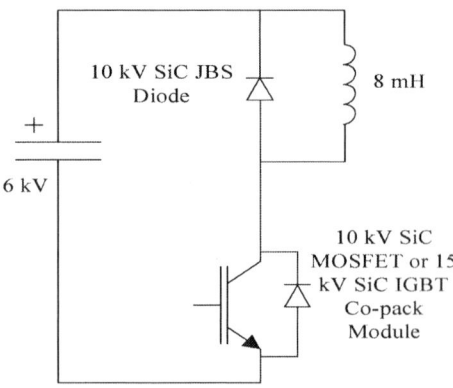

Fig.10 Double Pulse Test Circuit

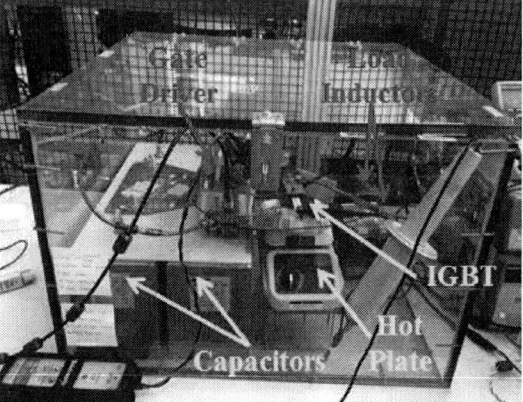

Fig. 10. Experimental setup of the Double Pulse Test circuit

6.5 kV Si IGBT Switching Loss data

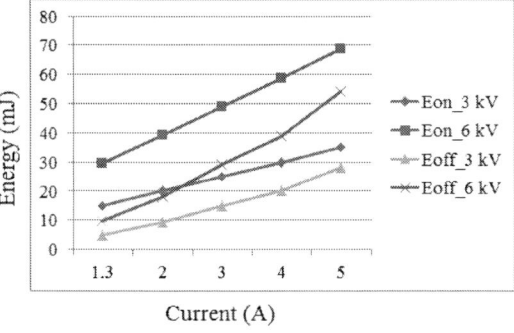

Fig. 12. Switching Loss data for the 6.5 kV Si IGBT

10 kV SiC MOSFET Switching Loss Data

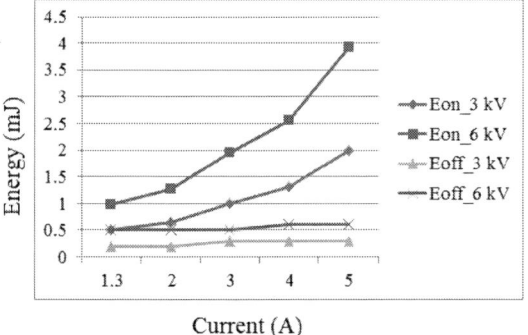

Fig. 13. Switching Loss data for the 10 kV SiC MOSFET

15 kV SiC IGBT Switching Loss Data

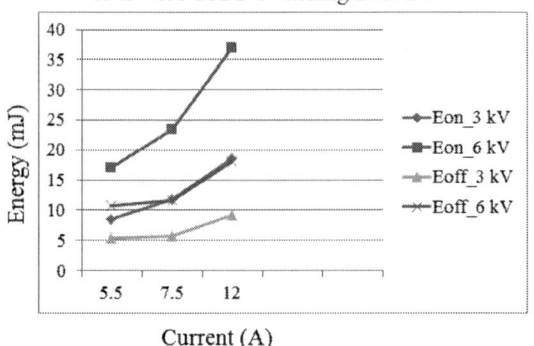

Fig. 14. Switching Loss data for the 15 kV SiC IGBT

978-1-4799-2706-7/14 $31.00 © 2014 IEEE   661

Figs. 12 to 14 give the switching loss data for the different devices considered on the HV side used for the loss calculation of the overall converter. The conduction loss data for the silicon devices are taken from [10] and [12]. The conduction loss data for the high voltage SiC devices are calculated from the forward characteristics of the devices as reported in [7, 8]. For the 15 kV SiC IGBT, the forward drop is 6 V at 20 A collector-emitter current with game-emitter voltage at 20 V and for 10 kV SiC MOSFET it is 5.2 V at 10 A. The switching and conduction loss data for the 1200 V/100 A SiC MOSFET are given in [11].

Tables.1 to 6 give the loss distribution for both DAB and SRC for the different combination of these devices. It can be observed that SiC devices have considerably lower loss compared to Si devices. DAB and SRC topologies are discussed separately in this section.

A. *Dual Active Bridge Converter*

DAB takes advantage of ZVS on the HV upper switches and ZCS on the HV lower switches, generating switching losses mainly due to reverse recovery effects. Tables.1 and 3 show the loss distribution for the DAB with SiC devices and Si devices respectively. The switching frequency considered is 10 kHz and the power transferred is 1 MW. The 10 kV SiC JBS diode has very small reverse recovery loss which has resulted in very small overall switching loss.

From Tables.1 and 3, the switching loss of the upper switch is very negligible while the lower switch produces considerable switching loss. The conduction loss of the upper switch is also lower than the conduction loss of the lower switch. From Tables.1 and 3, the conduction loss of 10 kV SiC MOSFET is slightly higher than the conduction loss of 6.5 kV Si IGBT. The higher conduction loss of MOSFET can be attributed to its unipolar nature. Also it has to be noted that the 10 kV SiC MOSFET drift region is designed to block 10 kV. This results in higher on-state resistance which contributes to this slight increase in conduction loss [3, 6]. However, this difference in conduction loss is very small compared to the drastic improvement in switching loss with SiC devices.

The switching loss is considerably lower in 10 kV SiC MOSFET. This counters the slightly higher conduction loss. Each pair of upper and lower MOSFETs results in 10.73 W switching loss while the same pair of Si IGBTs suffers from 942.4 W switching loss. The lower switching loss is partly because of the unipolar nature of MOSFET. Minority charge removal increases the turn-off times in the case of bipolar IGBTs [6]. Also, the smaller minority carrier life-time in SiC devices has resulted in faster switching time. The LV SiC MOSFET also presents better performance than its Si IGBT counterpart. Even though the 1200 V SiC MOSFET is unipolar, smaller life time and specific on-resistance of SiC compared to silicon resulted in lower conduction loss [3,

6]. Overall, the SiC devices based DAB has a very high efficiency of 98.5 % compared to 88.6 % for the Silicon based DAB.

The upper devices see very small switching loss. This fact can be used for cost optimization. For reducing the cost, using 6.5 kV Silicon IGBT for the upper switches and 10 kV SiC MOSFET or 15 kV SiC IGBT for the lower switches in the HV side 3-level poles will be a better alternative in the case of DAB.

B. *Series Resonant Converter*

Since SRC operates with switching frequency lower than its resonance frequency, the HV upper switch turns off with ZCS and the HV lower switch turns on with ZVS soft switching. Therefore, the upper switch generates more switching losses than the lower switch, mainly due to reverse recovery effects. This also contributes to higher overall switching losses in SRC as compared to DAB. Tables.2 and 4 show the loss distribution for the SRC with SiC devices and Si devices respectively. The switching frequency considered is 10 kHz and the power transferred is 1 MW. In fact, these tables confirm the previous discussion about Si and SiC HV devices in DAB. Both SiC and Si devices exhibit close values for conduction losses while the SiC MOSFET has considerably lower switching losses. This makes SRC and DAB significantly more efficient when equipped with SiC devices rather than with Si equivalent devices. Furthermore, 1200V SiC MOSFET exhibits lower switching losses and conduction losses than 1200V Si IGBT device.

Overall, the SiC devices based SRC has a very high efficiency of 99.14 % compared to 86.5 % for the Silicon based SRC. Another important point to note here is that with SiC devices, SRC is more efficient than DAB while with Si devices this reverses. This is because, SRC has higher switching loss compared to DAB, but lower conduction loss compared to DAB. With change from SiC devices to Si devices, the conduction loss do not change much, while the switching loss increases drastically. Thus, SRC shows lower efficiency with Si devices but better efficiency with SiC devices. This gives option for selecting the devices while going with different topologies for these high power applications.

C. *15 kV SiC IGBT on HV Side*

Tables.5 and 6 show the loss distribution of these topologies when the newly developed 15 kV SiC IGBT is used on the HV side [7]. Higher turn-on gate resistance of 100 Ω is used for measuring the switching loss of these devices. This has resulted in higher switching loss. It is expected that these devices will have improved switching performance when the gate resistance is reduced. The selection of the gate resistances in all the cases depends on the trade-off between the acceptable dv/dt and the switching loss. High dv/dt will cause EMI issues. Hence, it might not be possible to use small gate resistances in all the applications. At the same time, higher gate resistance

978-1-4799-2706-7/14 $31.00 © 2014 IEEE

will slow down the switching speed and thus, increase the switching loss. Therefore, an optimum choice of gate resistance has to be made based on both EMI and thermal considerations. The 15 kV SiC IGBT will be a suitable candidate for scaling up the power due to low conduction losses because of conductivity modulation of the drift region [6, 7].

V. CONCLUSION

In this paper DAB and SRC based 1 MW DC-DC power converter systems have been analyzed in detail. Same topology is studied for different combinations of SiC devices and Si devices. Both converters exhibit promising efficiencies when using SiC switches. Furthermore, it was found that SRC suffers from more HV switching loss than DAB, hence, it operates better with SiC devices and its efficiency is actually higher than DAB for that case. On the other hand, DAB generates more conduction losses than SRC and therefore it is most suited for Si devices. Its efficiency is indeed higher than SRC when both are equipped with Si devices.

ACKNOWLEDGEMENTS

This work made use of FREEDM ERC shared facilities supported by National Science Foundation under award no. EEC-0812121.

REFERENCES

[1] C. Zhao, D. Dujic, A. Mester, J. K. Steinke, M. Weiss, S. Lewdeni-Schmid, T. Chauduri, and P. Stefanutti, "Power Electronic Traction Transformer—Medium Voltage Prototype", *IEEE Transactions on Ind. Electron.*, vol. 61, no. 7, pp. 3257–3268, July 2014.

[2] N. Flourentzou, V. Agelidis, and G. Demetriades, "VSC-based HVDC power transmission systems: An overview," *IEEE Transactions on Power Electron.*, vol. 24, no. 3, pp. 592–602, Mar. 2009.

[3] S. Madhusoodhanan, K. Hatua, S. Bhattacharya, S. Leslie, S. H. Ryu, M. Das, A. Agarwal, D. Grider, "Comparison study of 12kV n-type SiC IGBT with 10kV SiC MOSFET and 6.5kV Si IGBT based on 3L-NPC VSC applications", in *proc. 2012 IEEE Energy Conversion Congress and Exposition*, Raleigh, NC, pp. 310 - 317.

[4] M. H. Kheraluwala, R. W. Gascoigne, and D. M. Divan, "Performance characterization of a high-power dual active bridge dc-to-dc converter," *IEEE Transactions on Power Electron.*, vol. 28, no. 6, pp. 1294-1301.

[5] A. K. Jain and R. Ayyanar, "PWM control of dual active bridge: Comprehensive analysis and experimental verification," in *Proc. IEEE Ind.Electron. Conf. (IECON 2008)*, Nov., pp. 909–915.

[6] B. Jayant Baliga, Fundamentals of Semiconductor Devices, New York: Springer, 2008, pp. 20-27.

[7] A. Kadavelugu, S. Bhattacharya, S. Ryu, E. V. Brunt, D. Grider, A. Agarwal, and S. Leslie, "Characterization of 15 kV SiC n-IGBT and its application considerations for high power converters", in *proc. 2013 IEEE Energy Conversion Congress and Exposition*, Denver, CO, pp.2528- 2535.

[8] J. Wang, T. Zhao, A.Q. Huang, R. Callanan, F. Husna, and A. Agarwal, "Characterization, modeling and application of 10 kV SiC MOSFET", *IEEE Transactions on Electron Devices*, vol. 55, no. 8, pp. 1798-1806, Aug. 2008.

[9] M. K. Das, R. Callanan, C. Capell, B. Hull, F. Husna, M. O'Loughlin, M. Paisley, A. Powell, J. Richmond, and Q. Zhang, "State of the art 10 kV NMOS transistors," in *proc. of 20th Annual ISPSD Conference*, Orlando, FL, pp. 253-255, 2008.

[10] A. Kopta, M. Rahimo, U. Schlabach, D. Schneider, E. Carroll, and S. Linder, "A 6.5 kV IGBT module with very high safe operating area", *IEEE IAS-05*, vol. 2, pp. 794-798, 2005.

[11] S. Hazra, S. Madhusoodhanan, G. Karimi-Moghaddam, K. Hatua, and S. Bhattacharya, "Design considerations and performance evaluation of 1200 V, 100 A SiC MOSFET based converter for high power density application", in *proc. 2013 IEEE Energy Conversion Congress and Exposition*, Denver, CO, pp.4278 - 4285.

[12] http://www.semikron.com/products/data/cur/assets/SK_100_GB_1 2T4_T_24914930.pdf.

[13] X. Li and A. K. S. Bhat, "Analysis and design of high-frequency isolated dual-bridge series resonant DC/DC converter," IEEE Trans. Power Electron., vol. 25, no. 4, pp. 850–862, Apr. 2010.

[14] A. Kadavelugu, S. Baek, S. Dutta, M. Das, A. Agarwal, J. Scofield, "High-frequency design considerations of dual active bridge 1200 V SiC MOSFET DC-DC converter," in *proc. Applied Power Electronics Conference and Exposition (APEC)*, 2011 Twenty-Sixth Annual IEEE, 2011, pp. 314-320.

The Direct Yaw-Moment Control to Follow the Neutral steering Path regardless of Velocity

Young-Jin Jang
Department of Electrical Engineering
POSTECH
Pohang, Republic of Korea
youngjin530@postech.ac.kr

Kwang-Hee Nam
Department of Electrical Engineering
POSTECH
Pohang, Republic of Korea
kwnam@postech.ac.kr

Abstract—**By using differential force between left and right wheel, lateral motion can be controlled known as Direct Yaw-moment Control (DYC). In previous researches, DYC control is proposed to increase the stability of the vehicle, but maneuverability has not been discussed sufficiently. The car handling condition which is called the index parameter of maneuverability is dependent on the vehicle velocity. To achieve the desired vehicle's cornering path, the car handling condition must be considered sufficiently. In this paper, the novel DYC method is proposed which gives the car handling condition regardless of the longitudinal speed. The proposed controller is based on the PI controller to feedback the curvature parameter. The controlled system shows the advantages of DYC regarding to the reference trajectory by the dual motor system. With respect to the uncontrolled model, the effectiveness of the proposed method is validated by numerical examples.**

Keywords—*DYC, Neutral steer, EV, maneuverability control.*

I. INTRODUCTION

Many papers about DYC have been proposed in [1]-[2]. These control methods design DYC to increase stability of the EV. The DYC was developed in [3] with the yaw moment observer which estimates the immeasurable yaw rate terms. Using the yaw moment observer, the controller compensates the disturbance applied to the EV to keep the reference trajectory. When the yaw rate is not the same as the desired yaw rate for cornering, yaw moment observer finds the differences between real value and desired value of yaw rate. The perspective of DYC control is feedback the difference of yaw rate for desired cornering regardless of disturbances.

However, there are few papers about maneuverability design. In [4] the feasibility of the steering condition is analyzed. But overall control system has not been discussed. In [5] synthesis problem of control system has been discussed. To achieve the desired path by driving, the paper proposed the DYC controller regardless of car handling condition. But the mechanical saturation of steering angle has not been discussed and changing input value of steering angle has not been discussed also.

In this paper, the DYC control algorithm for desired car handling condition is derived when the EV drives at the arbitrary velocity. This control structure allows us to set the same car handling condition whenever at any longitudinal speed. First, the dynamics of the EV and the

steering condition is introduced. Then, the reference law which is used to find the reference curvature of cornering is introduced. Thirdly, the DYC method is derived by using PI controller. Finally, the result of simulation is introduced to show the effectiveness of the proposed method.

II. VEHICLE DYNAMICS FOR DYC OF THE CORNERING CURVAUTURE

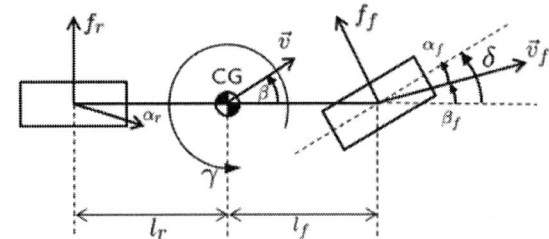

Fig. 1.   The bicycle model of the vehicle.

As shown in Fig.1, the lateral motion dynamics of vehicle is modeled as the bicycle model [6]. The motion equations of the lateral motion can be derived as:

$$mv(\frac{d\beta}{dt} + \gamma) = 2(f_f + f_r), \qquad (1)$$

$$I\frac{d\gamma}{dt} = N_z + N_t + N_d, \qquad (2)$$

where f_r and f_f are the lateral force of front and rear tires. β is the vehicle slip angle. γ is the yaw rate of the vehicle. The lateral force is proportional to the value of tire slip angle. Using above equations, the lateral dynamics of sideslip angle and yaw rate can be derived as following equation. Yaw rate and body slip angle can be varied by the input values of differential motor torque moment and steering angle.

$$
\begin{bmatrix} \dot{\beta} \\ \dot{\gamma} \end{bmatrix} =
\begin{bmatrix} -\frac{2(C_f+C_r)}{mv} & -1 - \frac{2(l_fC_f-l_rC_r)}{mv^2} \\ -\frac{2(l_fC_f-l_rC_r)}{I} & -\frac{2(l_f{}^2C_f+l_r{}^2C_r)}{Iv} \end{bmatrix}
\begin{bmatrix} \beta \\ \gamma \end{bmatrix}
+ \begin{bmatrix} \frac{2C_f}{mv} & 0 \\ \frac{2l_fC_f}{I} & \frac{1}{I} \end{bmatrix}
\begin{bmatrix} \delta \\ N_z \end{bmatrix}. \qquad (3)
$$

978-1-4799-2706-7/14 $31.00 © 2014 IEEE

$$\frac{\gamma(s)}{\delta(s)} = \frac{g_{f1}s + g_{f0}}{s^2 + a_1 s + a_2}, \frac{\gamma(s)}{N_z(s)} = \frac{g_{m1}s + g_{m0}}{s^2 + a_1 s + a_2}$$

$$\frac{\beta(s)}{\delta(s)} = \frac{g_{b0}}{s^2 + a_1 s + a_2}$$

$$g_{f1} = \frac{2l_f C_f}{I}, \ g_{f0} = \frac{4lC_f C_r}{mIV}$$

$$g_{m1} = \frac{1}{I}, \ g_{m0} = \frac{2(C_f + C_r)}{mIV}$$

$$g_{f0} = -\frac{2l_f C_f}{I} - \frac{2(l_f C_f - l_r C_r)}{mIV^2}$$

$$a_1 = \frac{2(C_f + C_r)}{mV} + \frac{2(l_f{}^2 C_r + l_r{}^2 C_r)}{IV}$$

$$a_2 = \frac{4l^2 C_f C_r}{mIV^2} - \frac{2(l_f C_r + l_r C_r)}{I}$$

The followings are the symbols of EV model.

α : tire slip angle

δ : steering angle

$\vec{v}$: velocity vector

M : mass of the EV

l_f, l_r : distances between the C.O.G. and wheel axle

l_p : wheelbase

C_f, C_r : stiffness at rear and front tire

f : differential motor torque moment

where the stiffness at each tire is set as a constant value. The differential motor torque moment is transformed from the differential motor force f (Fig.2). $N_z = 2l_p \times f$ presents the moment caused by f. So, the yaw rate of the vehicle can be controlled by not only steering angle but also by differential motor force f.

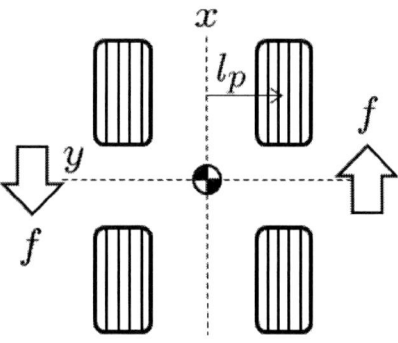

Fig. 2.   The vehicle model.

A. Dynamics for the curvature control system

From the definition of ankermann steering angle, we can define the cornering curvature by using longitudinal velocity and yaw rate when the vehicle's driving curvature keeps constant. In this case, the vehicle drives at constant velocity. So the curvature of the cornering can be expressed by simple ratio of velocity to yaw rate (Fig.3).

where the yaw rate γ and curvature R are given as

$$\gamma = \dot{\theta} = \frac{v_x}{R},$$

$$R = \frac{v_x}{\gamma}. \tag{4}$$

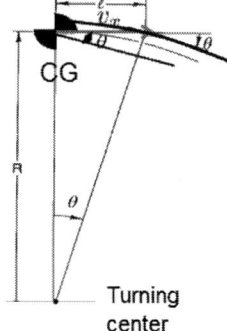

Fig. 3.   the vehicle model at constant curvature.

When we control the cornering curvature, we can control the curvature by controlling yaw rate only in cased of turning at constant curvature. However, yaw rate is not an only factor but also a differential vehicle slip angle $\dot{\beta}$ that affect the cornering curvature. The lateral acceleration is given as following equation in Fig. 4.

$$a_y = v_x \gamma + \dot{v}_y,$$

$$= v_x \gamma + v_x \dot{\beta}. \tag{5}$$

The centrifugal acceleration for keeping radius is proportional to the yaw rate.

$$a_{cent} = \frac{v_x^2}{R} = v_x \gamma. \tag{6}$$

Hence, the overall lateral accelerate equation regarding the centrifugal acceleration is obtained.

$$a_y = v_x \gamma + \dot{v}_y,$$

$$= a_{cent} + v_x \dot{\beta}. \tag{7}$$

The second term is a direct lateral acceleration term from the vehicle slip angle. So, we must consider the differential vehicle slip angle when we control the cornering curvature.

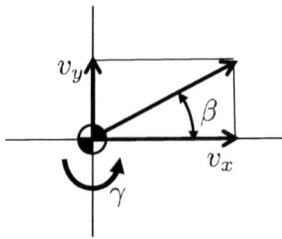

Fig. 4.   vehicle frame axis when the vehicle turns the corner.

B. Vehicle handling condition

We can classify the handling condition by varying the speed of the vehicle. Assuming the EV is accelerated with a constant steering angle. When the cornering's curvature becomes smaller, vehicle is on an oversteering action;If the vehicle draws the same circle, the vehicle is neutral steer; when the curvature of the vehicle becomes larger, the vehicle is understeer.

Usually the vehicle is understeer when the vehicle turns a corner. The vehicle's cornering curvature gets larger when it accelerates longitudinal velocity. The handling condition is affected by vehicle's parameters. The understeer vehicle satisfies the condition equation; $l_f C_f - l_r C_r < 0$.

In the understeering condition, the cornering curvature varies as the vehicle's longitudinal speed changes. As the vehicle's speed gets faster, the vehicle slip angle gets larger too. Which means that the direct lateral acceleration from vehicle slip angle affects the curvature to large. the vehicle slip angle in case of understeering is illustrated in Fig.5.

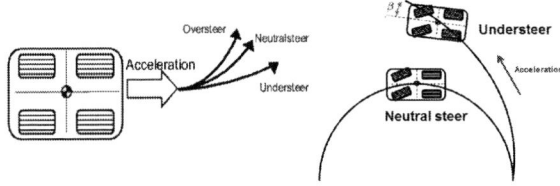

Fig. 5. The vehicle handling condition.

In this paper, we developed a control system that understeer vehicle drives like the neutralsteer vehicle. The controlled vehicle keep the radius when the vehicle accelerates the longitudinal speed.

1) Reference model of neutral steering: The vehicle state equation of neutral steering is express as

$$
\begin{aligned}
\dot{\tilde{x}} &= A(v_c)\tilde{x} + B(v_c)\tilde{u}, \\
\tilde{x} &= \begin{bmatrix} \dot{\tilde{\beta}} \\ \dot{\tilde{\gamma}} \end{bmatrix}, u(t) = \begin{bmatrix} \delta \\ N_z \end{bmatrix}, \\
A(v_c) &= \begin{bmatrix} -\frac{2(C_f+C_r)}{mv_c} & -1-\frac{2(l_fC_f-l_rC_r)}{mv^2} \\ -\frac{2(l_fC_f-l_rC_r)}{I} & -\frac{2(l_f{}^2C_f+l_r{}^2C_r)}{Iv_c} \end{bmatrix}, \\
\tilde{B}(v_c) &= \begin{bmatrix} \frac{2C_f}{mv_c} & 0 \\ \frac{2l_fC_f}{I} & \frac{1}{I} \end{bmatrix}.
\end{aligned}
\tag{8}
$$

v_c is constant longitudinal velocity.

2) Reference model of under steering: Recall to equation (3). If parameters satisfy $l_f C_f - l_r C_r < 0$, the vehicle

handling condition is understeer.

$$
\begin{aligned}
\dot{x} &= A(v_x)x(t) + B(v_x)u(t), \\
x(t) &= \begin{bmatrix} \beta \\ \gamma \end{bmatrix}, u(t) = \begin{bmatrix} \delta \\ N_z \end{bmatrix}, \\
A(v_x) &= \begin{bmatrix} -\frac{2(C_f+C_r)}{mv} & -1-\frac{2(l_fC_f-l_rC_r)}{mv^2} \\ -\frac{2(l_fC_f-l_rC_r)}{I} & -\frac{2(l_f{}^2C_f+l_r{}^2C_r)}{Iv_x} \end{bmatrix}, \\
B(v_x) &= \begin{bmatrix} \frac{2C_f}{mv} & 0 \\ \frac{2l_fC_f}{I} & \frac{1}{I} \end{bmatrix}.
\end{aligned}
\tag{9}
$$

III. DYC CONTROL SYSTEM

A. Review on the conventional DYC control method

The performance of the yaw moment control method based on the yaw moment observer was presented in[1]-[3](Fig.6). These papers focus on the vehicle's stability. The conventional DYC algorithm feedbacks the yaw rate only to stabilize the vehicle motion. Because the conventional system controls the yaw rate only, they don't regard maneuverability and curvature control.

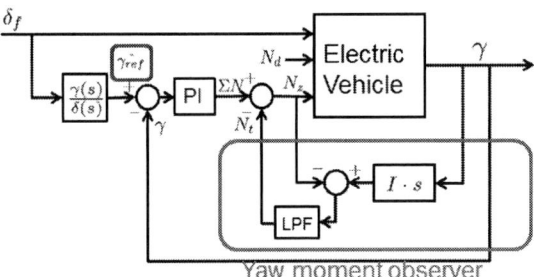

Fig. 6. The conventional DYC algorithm

B. The Proposed DYC control for neutral steering

The proposed DYC system controls not the yaw rate but the curvature radius. However, the curvature radius is hard to measure. If we want to measure the cornering radius of the vehicle, we would need an expensive experiment equipment like the three-dimensional laser instrument to measure the vehicle and center of the corner.

In this paper, a new parameter, k, to estimate curvature radius is presented. The new parameter can be measured by a cheep IMU sensor easily. The overall schematic of the proposed DYC system is illustrated in Fig.7.

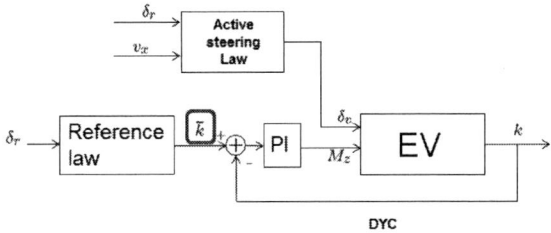

Fig. 7. The Schematic of the proposed system.

The functional schematic consists of three subsystems; Reference law; active steering law; PI controller. Using the reference law, we find the reference curvature parameter ($\tilde{k}$)that vehicle follows the reference curvature trajectory when the vehicle's velocity is a constant. In PI control system, the motor torque moment feedbacks the difference between reference parameter $\tilde{k}$ and measured parameter k. In active steering law part, the active steering angle is derived. With active steering angle, the EPS can help the vehicle to follow the reference path easily. The active steering angle can be treated as the the feedforward system.

C. Curvature expression by using the new parameter k

1) Definition of k: The curvature radius R can be denoted by k. We set $R = 1/|k|$. k can be expressed as follow.

$$k = \frac{\dot{\beta} + \gamma}{v}.$$

The parameter k can be regarded as a measure of the curve in a moment. As the definition of the curvature, the position of vehicle can be expressed by using k[13].

$$
\begin{aligned}
\theta(t) &= \int_0^t v(\tau)k(\tau)d\tau + \theta(0), \\
x(t) &= \int_0^t v(\tau)\cos\theta(\tau)d\tau + x(0), \\
y(t) &= \int_0^t v(\tau)\sin\theta(\tau)d\tau + y(0). \quad (10)
\end{aligned}
$$

k is a powerful measurement of curve because k is expressed by yawrate, body slip angle and steering angle. We made vehicle state equation for yaw rate and body angle. The state equation's input terms are steering angle and motor torque moment. It means that k is an output of the vehicle system. So we can determine the curvature by controlling the input variable steering angle and motor torque moment.

2) Measurability of k: The lateral acceleration can be obtained from the kinematical relationship in Fig.4.

$$
\begin{aligned}
a_y &= v_x\gamma + \dot{v}_y, \\
&= v_x\gamma + v_x\dot{\beta}. \quad (11)
\end{aligned}
$$

Let the A_{ymeas} the lateral acceleration output value from the IMU sensor. By using the IMU sensor, we can find the lateral acceleration, longitudinal acceleration and yaw rate.

$$A_{ymeas} = (\gamma + \dot{\beta})v_x.$$

From the definition of the parameter k, we rearrange the equation by using the output value of the IMU sensor.

$$k = \frac{(\gamma + \dot{\beta})}{v_x} = \frac{(\gamma + \frac{A_{ymeas}}{v_x} - \gamma)}{v_x} = \frac{A_{ymeas}}{v_x^2}. \quad (12)$$

Contrary to the curvature, R, which need the expensive instrument to measure, the new parameter k needs the

IMU sensor only to measure. The IMU sensor is a basic equipment for the stability control. Using the parameter k, we can control the curvature independently without additional equipments.

D. Reference law

We must set the information of reference curvature value before we make the PI controller. Reference curvature which make vehicle turn constant radius can be derived from the neutral steering model. Based on the reference model of neutral steer (8), we set the constant velocity v_c to define cornering radius. If the reference curvature parameter is denoted as $\tilde{k}$, reference parameter can be expressed as

$$
\begin{aligned}
\tilde{k} &= \frac{\dot{\tilde{\beta}} + \tilde{\gamma}}{v_c}, \\
&= \frac{1}{v_c}\left[-\frac{2(C_f + C_r)}{mv_c}\tilde{\beta} - \left(\frac{2(C_f l_f - C_r l_r)}{mv_c^2} - 1 \right)\tilde{\gamma} \right. \\
&\quad \left. + 2\frac{C_f}{mv_c}\delta + \tilde{\gamma} \right], \\
&= -\frac{2(C_f + C_r)}{mv_c^2}\tilde{\beta} - \frac{2(C_f l_f - C_r l_r)}{mv_c^3}\tilde{\gamma} \\
&\quad + 2\frac{C_f}{mv_c^2}\delta. \quad (13)
\end{aligned}
$$

The reference value of $\tilde{k}$ can be designed by setting the steering angle δ as the input. The state equation from (3), the transfer function of k can be obtained as

$$
\begin{aligned}
\frac{\tilde{k}(s)}{\delta(s)} &= -\frac{2(C_f + C_r)}{mv_c^2}\frac{\tilde{\beta}(s)}{\delta(s)} - \frac{2(C_f l_f - C_r l_r)}{mv_c^3}\frac{\tilde{\gamma}(s)}{\delta(s)} \\
&\quad + 2\frac{C_f}{mv_c^2}, \\
&= -\frac{2(C_f + C_r)}{mv_c^2}\frac{g_{f1}s + g_{f0}}{s^2 + a_1 s + a_2} \\
&\quad - \frac{2(C_f l_f - C_r l_r)}{mv_c^3}\frac{g_{b0}}{s^2 + a_1 s + a_2} \\
&\quad + 2\frac{C_f}{mv_c^2}. \quad (14)
\end{aligned}
$$

where $\frac{\beta(s)}{\delta(s)}$, $\frac{\gamma(s)}{\delta(s)}$ is defined in (3).

Equation (14) is a model-based reference model for neutral steering. It should be noted that the reference law in (14) has only input variance, δ. The kinematical relation about parameter $\tilde{k}$ in (12) has advantage of measuring values. Instead, The equation (14) is easy to use as the reference value because it has only one parameter.

E. PI controller

In the reference model, the reference curvature parameter($\tilde{k}$) given by the reference law. To make the curvature of driving path equal to that of the reference model, we apply the motor torque moment to real vehicle with the original steering angle δ_r.

The designed PI control law for curvature parameter is defined as:

$$\begin{cases} e = \frac{(\tilde{\gamma}+\dot{\tilde{\beta}})}{v_c} - \frac{(\gamma+\dot{\beta})}{v_x} = \tilde{k} - k, \\ N_z = K_p e + K_i \int_0^t e(\tau)d\tau. \end{cases} \quad (15)$$

The real value of curvature in a moment is not equal to the reference trajectory because the longitudinal velocity is not a constant value. The error terms of the PI controller is difference between reference $\tilde{k}$ and real k. By setting the output as the motor torque moment, we can control the curvature in a moment to equal to the reference signal.

F. Active steering law

When the vehicle accelerates the speed with constant steering angle, the curvature of the path becomes larger because of the characteristic of understeer. In the case of acceleration, we must control the steering angel regarding the speed adaptively to maintain the curvature of the circular path. Recalling (18), the relationship between steering angle and velocity at constant curvature can be derived [7].

$$\begin{aligned} F_{yf} + F_{yr} &= m\frac{v_x^2}{R}, \\ F_{yf}l_f - F_{yr}l_r &= 0. \end{aligned} \quad (16)$$

where F_{yf} is lateral force at front tire; F_{yr} is lateral force at rear tire. The sum of lateral forces is equal to centrifugal force at steady state cornering. Also the sum of torque moment of vehicle is zero. Steering angle (δ) is expressed tire slip angle and curvature

$$\delta = \frac{L}{R} + \alpha_f - \alpha_r. \quad (17)$$

where L is a distance between front and rear wheel axle. Subtracting (4)(5)(6), The curvature from steering angle can be derived as following equation.

$$\frac{1}{R} = \frac{1}{l_f + l_r + \left(\frac{ml_rC_r - ml_fC_f}{2C_fC_r(l_f+l_r)}\right)v_x^2}|\delta|. \quad (18)$$

Assuming $\frac{1}{R} = \frac{1}{R} = const$,

$$\delta_v = \frac{l_f + l_r + \left(\frac{ml_rC_r - ml_fC_f}{2C_fC_r(l_f+l_r)}\right)v_x^2}{l_f + l_r + \left(\frac{ml_rC_r - ml_fC_f}{2C_fC_r(l_f+l_r)}\right)v_c^2}\delta_r. \quad (19)$$

where δ_v is a active steering angle,

The active steering angle can be treated as the the feedforward system. Actually, without active steering law, we can control the vehicle to follow the neutral steer path. AT that case, the higher motor torque moment is needed to realize the control system. With active steering law, the EPS can help the vehicle to follow the reference path easily. It just act as the feedforward control system. If the system has error caused by the disturbance torque and parameter errors, motor torque feedback the disturbance by using the measured parameter 'k'.

IV. SIMULATION

The simulations are carried out to verify the performance of the proposed system compared to the system which doesn't have DYC law for the curvature control. We compared the trajectory of the controlled vehicle with that of the non-controlled vehicle.

TABLE I.   PARAMETERS OF DYC

Symbol	Meaning	Value
Cf	Front tire stiffness	50000
Cr	Rear tire stiffness	80000
m	Vehicle mass	400kg
l_f	Distance between COG and F.tire	0.689 m
l_r	Distance between COG and R.tire	0.591 m

A. Active steering law effectiveness without motor torque moment

The EV is accelerated from 18km/h to 30km/h for 20 seconds. The velocity of the reference model is chosen as $v_c = 18km/h$.(Fig. 8, 9, 10, 11).

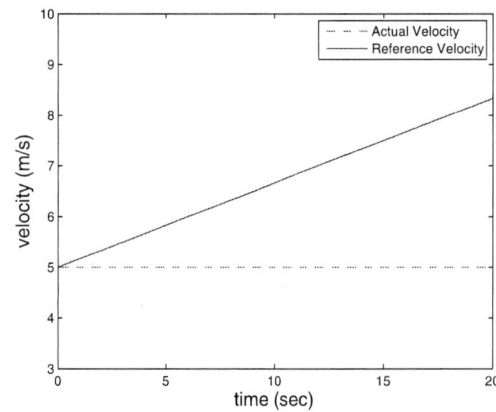

Fig. 8.   Velocities of the vehicle.

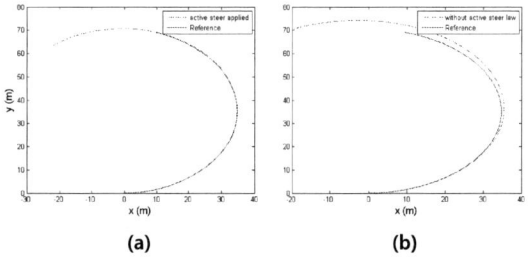

Fig. 9.   The trajectory with reference in case of active steer law applied and no control. (a) active steer law applied , (b) no control.

B. DYC when the velocity changes without active steering law

The EV is accelerated from 18km/h to 30km/h for 20 seconds (Fig. 12). The velocity of the reference model is chosen as $v_c = 18km/h$. By using the reference

The 2014 International Power Electronics Conference

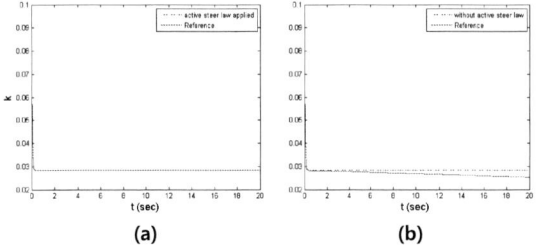

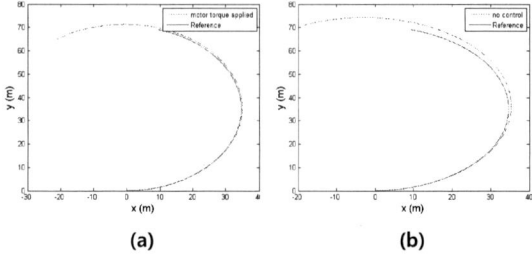

Fig. 10. The curvature parameter in case of active steer law applied and no control (a) active steer law applied , (b) no control.

Fig. 13. The trajectory with reference in case of DYC control applied and no control. (a) motor torque applied , (b) no control.

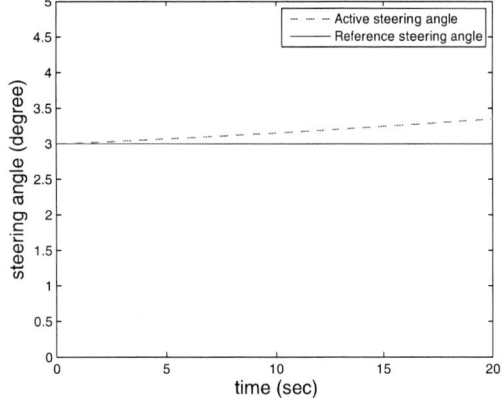

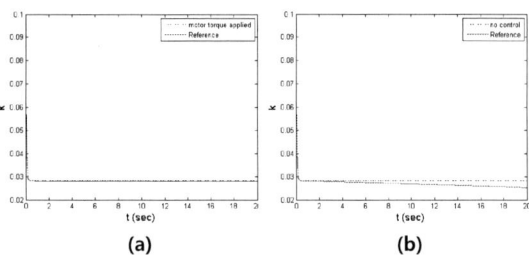

Fig. 11. The active steering angle and reference steering angle

Fig. 14. The curvature parameter in case of DYC control applied and no control (a) motor torque applied , (b) no control.

law(14), The reference curvature $\tilde{k}$ can be determined. that make the vehicle keep its trajectory with constant speed (Fig. 13). Initial position of the vehicle are (0,0). We can see that the vehicle's curvature with no control becomes larger as velocity becomes higher.(Fig. 14, 15). Kp=3; Ki=240.

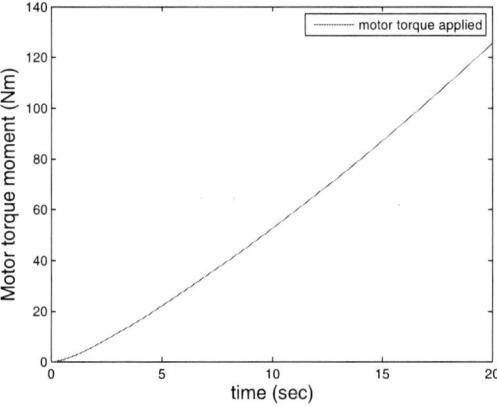

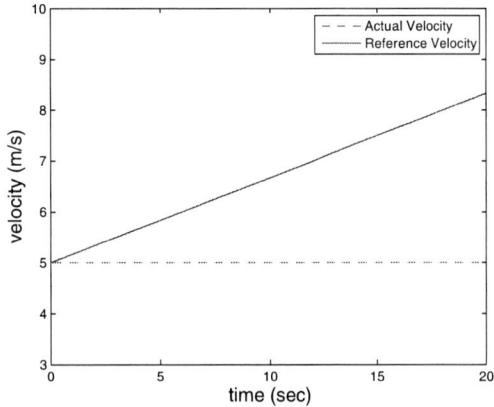

Fig. 15. The motor torque moment in case of DYC control applied and no control

Fig. 12. Velocities of the vehicle.

(0,0). When the vehicle turns the corner, the disturbance torque moment applies to the vehicle. Kp=3; Ki=240.(Fig. 16, 18, 19, 20, 17).

C. DYC when the disturbance torque moment applied

The EV is accelerated from 18km/h to 30km/h for 20 seconds (Fig. 16). The velocity of the reference model is chosen as $v_c = 18km/h$. Initial position of the vehicle are

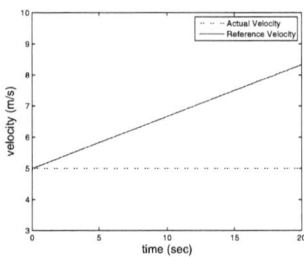

Fig. 16. Velocities of the vehicle.

978-1-4799-2706-7/14 $31.00 © 2014 IEEE   669

The 2014 International Power Electronics Conference

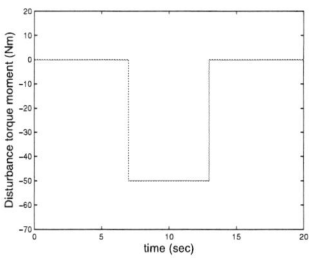

Fig. 17. The disturbance torque moment.

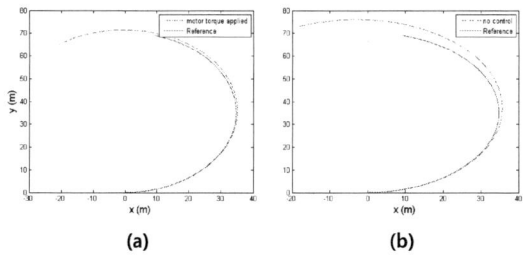

(a)   (b)

Fig. 18. The trajectory with reference in case of DYC control applied and no control. (a) motor torque applied , (b) no control.

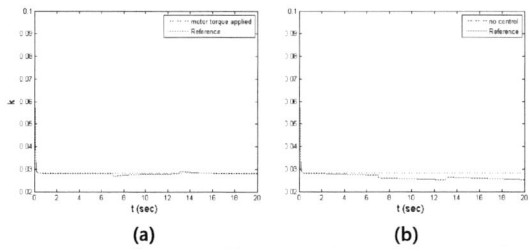

(a)   (b)

Fig. 19. The curvature parameter in case of DYC control applied and no control (a) motor torque applied , (b) no control.

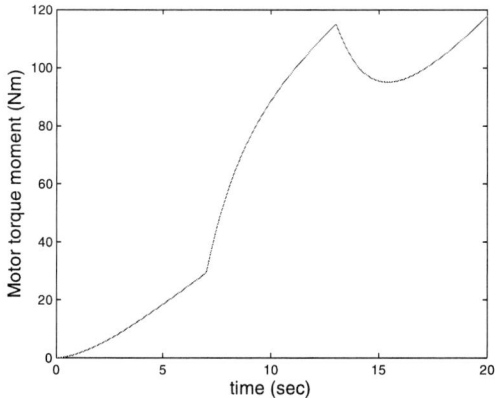

Fig. 20. The motor torque moment in case of DYC control applied

V. CONCLUSION

The DYC using the curvature parameter is proposed. The measurable parameter k is introduced to control the curvature. The PI control feedbacks the measured 'k' so that the vehicle follows the neutral steer path. The active steering law is applied to the EPS. The active steering law's role is like the feedforward control to the vehicle. It contributes the vehicle to follow the reference path with small motor torque moment.

We are going to do a experiment. By using a electric vehicle equipped with a IMU sensor, we can verify the DYC algorithm for neutral steering.

REFERENCES

[1] Y. Hori, "Vehicle Driven by Electricity and Control. REsearch on Four-Wheel-Motord 'UOT Electric March II'", in *IEEE Trans. on Industrial Electronics, vol.* 51, no. 5, pp. 954 - 962, 2004.

[2] S. Sakai, H. Sado, and Y. Hori, "MOtion control in an Electric Vehicle with 4-independently Driven In-Wheel Motors", in *IEEE Trans. on Mechatronics, vol.* 4, no. 1, pp. 9 - 16, 1999.

[3] H. Fujimoto, T. Saito, and T. Noguchi, "Motion Stabilization control of Electric Vehicle under Snowy conditions Based on Yaw-Moment Observer", in *Proc. 8th IEEE International Workshop on Advanced Motion Control(AMC'04)*, pp 35-40,2004.

[4] H. Okjima, S.Yonaha, N. Matunaga and S. Kawaji, "Direct Yaw-Moment Control Method for Electric Vehicles to Follow the Desired Path by Driver", in *Proceedings of SICE Annual conference*, pp 642 - 647, 2010.

[5] N. Fukushima, "DYC control for Turning Vehicle over Critical Range : 2nd Report, Application to Rear Ensrine Rear Wheel Drive Vehicle", in *Japan Society of Mechanical Engineers*, pp 1583 - 1590, 2005.

[6] R. Rajamani, "Vehicle Dynamics and Control", in *Springer*, 2012.

[7] J. M. Lee, "Riemannian Maniforlds : An Introduction to Curvature", in *Springer*, 1997.

The 2014 International Power Electronics Conference

Next-generation IGBT Module Structure for Hybrid Vehicle with High Cooling Performance and High Temperature Operation

Akira Morozumi, Hiromichi Gohara, Fumihiko Momose, Takashi Saito, Yoshitaka Nishimura, Eiji Mochizuki and Yoshikazu Takahashi

Electronic Device Laboratory, Fuji Electric Co., Ltd.
4-18-1 Tsukama, Matsumoto, Nagano 390-0821, Japan
morozumi-akira@fujielectric.co.jp

Abstract— We developed a new technology for the Fuji 4th generation power module for hybrid vehicle application. It achieved a 30-% volume reduction, and a 33-% weight reduction compared to the 3rd generation. Requirement trends for the power module for hybrid vehicles are high energy-efficient, downsizing, lightweight and higher reliability. To achieve these requirements, IGBT module needs low thermal resistance and high operating temperature. We developed new technologies, 1) New cooling structure with high cooling performance. 2) The Tjmax.=175°C continuous operation with automotive required quality.

Keywords— *IGBT Module, Direct Liquid Cooling, High Temperature Operation, High Reliability, Hybrid Vehicle.*

I. INTRODUCTION

Hybrid electric vehicle drive system consists of motors, batteries, and power modules to control them. The characteristics of the power modules are the key issues to make the vehicles more energy-efficient. In particular, volume and weight reduction of them are very important. Furthermore, power control unit will be integrated with electric motor or transmission.

Figure 1 shows the 3rd gen IGBT module [1,2] for hybrid vehicle which was released in 2012. This module is the world's first released module which has the direct liquid cooling structure with a solder jointed aluminum cooling fin with isolated substrate. In this paper, the newly developed Fuji 4th gen. IGBT module for hybrid vehicle is introduced (Fig. 2). It achieved a 30-% volume reduction, and 33-% by weight reduction compared to the 3rd gen..

Fig. 1. The 3rd gen. direct liquid cooling module.

Fig. 2. The 4th gen. direct liquid cooling module.

II. NEW COOLING STRUCTURE WITH HIGH COOLING PERFORMANCE

At first, we investigated which factor has the influence on the design of the cooling fin. Cooling capacity has relation to convective heat transfer coefficient of the fluid (α). The relationship between convective heat transfer coefficient and Nusselt number (Nu) is defined by Eq.(1)

$$\alpha = \frac{Nu\lambda}{L} \qquad (1)$$

Where λ is thermal conductivity of the fluid, L is the length. Nu is shown by following Eq.(2)

$$Nu = 0.664Re^{1/2}Pr^{1/3} \qquad (2)$$

where Re is Reynolds number, Pr is Prandtl number. Reynolds number and Prandtl number are shown by following Eq.(3) and (4)

$$Re = \frac{\rho v L}{\mu} \qquad (3)$$

where ρ is the density of the fluid, v is the mean velocity of the fluid, μ is the coefficient of viscosity of the fluid.

$$Pr = \frac{\mu Cp}{k} \qquad (4)$$

where Cp is the specific heat, k is the thermal conductivity. According to Eqs.(1), (2), (3) and (4), the relationship between the convective heat transfer

978-1-4799-2706-7/14 $31.00 © 2014 IEEE   671

coefficient of the fluid and the mean velocity of the fluid can be explained as follows;

$$\alpha \propto v \tag{5}$$

Therefore, the velocity of the fluid is important for the cooling system design.

Figure 3 shows the comparison of the cooling structure between the Fuji 3rd gen. and 4th gen.. It is to improve the cooling performance by using a cooler which is a unified cooling unit of the heat sink and water jacket. The 3rd gen. IGBT module has tightened up water jacket with gaskets by screws. This conventional structure has a gap between the cooling fin and the water jacket.

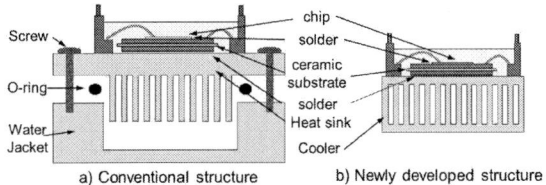

Fig. 3. Comparison of the cooling unit.

The relationship between this gap and thermal resistance (Rj-w) of IGBT module was investigated by simulation. Fig. 4 shows the simulation model. And Fig. 5 shows the result of the simulation. It is clear that the gap reduction results in the thermal resistance reduction.

The reason thermal resistance becomes large in proportion to the gap length is that the flow velocity of the coolant becomes slower because some of the coolant flows through the gap instead of fins area, and the cooling capability is reduced. The newly developed structure is integrated water jacket and cooling fin design change that top of the fin joint to water jacket (Fig. 3 (b)).

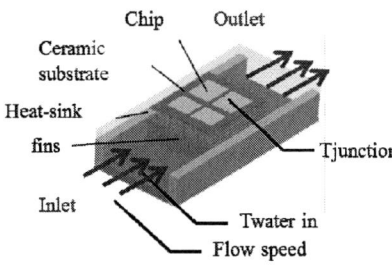

Fig. 4. Simulation model.

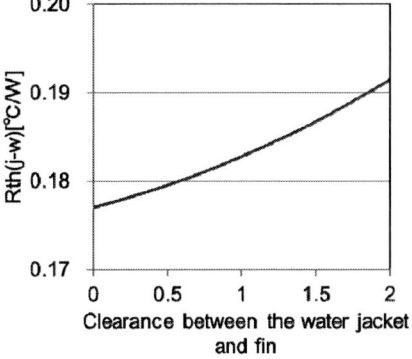

Fig. 5. Clearance dependence of the thermal resistance.

In the newly developed structure, the cooling unit has a higher stiffness than the conventional one. So we investigated this influence on reliability of solder joint layer. Fig. 6 shows the relationship between the top side thickness of cooling fin and solder strain during thermal cycling test (-40°C ~ 105°C). As the top side thickness of cooling fin became thinner, the solder strain decreased. Based on this relationship, the thickness of the top plate was optimized, and enough reliability characteristics of solder have been achieved.

When using this structure, thermal cycling lifetime is longer about 40-% compared with the conventional structure.

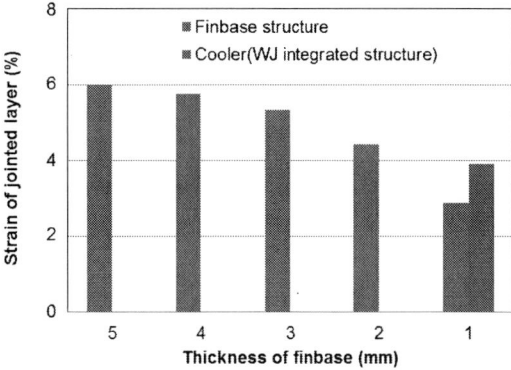

Fig. 6. Strain of solder joints in the cooling structure difference.

Finally, coolant flow speed and pressure loss were designed. Fig. 7 shows the flow speed and pressure loss in thermal fluid analysis. The coolant flow speed affects the thermal resistance of IGBT module and the pressure loss affects the coolant pump capacity. By these effects, 20-% of thermal resistance was improved (Fig. 8).

The 4th gen. was designed based on these technologies, which has unified cooling fin and water jacket structure, thinner thickness top plate and optimized form of a fin and a channel structure. These technologies realized the low thermal resistance and high reliability.

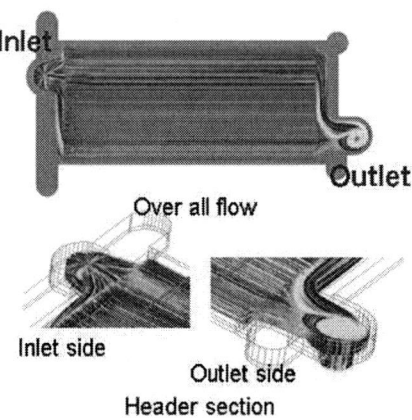

Fig. 7. State of the flow of coolant.

The 2014 International Power Electronics Conference

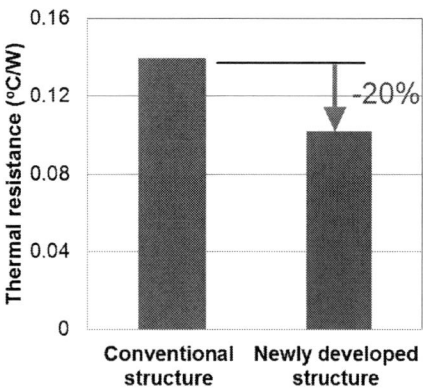

Fig. 8. Comparison of thermal resistance.

III. 175°C CONTINUOUS OPERATING GUARANTEE

For hybrid vehicle application, ambient temperature is higher than industrial application. This means that power module operating temperature range is severe than industrial application. We achieved the Tjmax increase to 175°C and widen the operating temperature range [3].

Continuous operation temperature of conventional IGBT module is 150°C and only limited time can be 175°C. This continuous temperature is limited from power cycling reliability.

Failure mode of IGBT module in power cycling test is caused by two factors [4]. One is the mismatch of CTE (coefficient of thermal expansion) between the structural components. Other is the influence of thermal fatigue. Influences of the two factors are expected to become larger at higher temperature. With the point of these views, the failure modes in power cycling test at Tjmax.=175°C operation was investigated.

The following three major failure modes were observed after power cycling test at Tjmax.=175°C:

  (a) Bond wire
  (b) Die to DCB solder joint
  (c) Top side die metallization

The observation results are shown in Fig. 9. In order to analyze these three failure modes in detail, the influence of ΔTj (junction temperature swings range) was investigated. Several samples with different strength material were tested to reveal the effect of these failure modes with the ΔTj on the lifetime of the module.

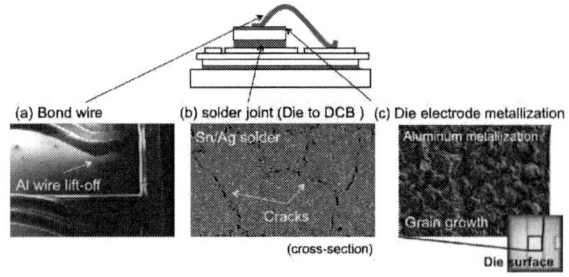

Fig. 9. Main failure modes in power cycling test at Tjmax. = 175℃.

Figure 10 shows the images of power cycling capability for each failure modes based on the experimental results. It is first time to clear that each jointing material lifetime is different depending on ΔTj region. These lifetime lines have a different gradient and these are some intersection points. When ΔTj range is increased, IGBT module failure modes are changed from (a) Al wire to (b) solder, and (c) die metallization. It means that, with improving these three materials with conventional technologies lifetime at each region, it possible to achieve keeping and higher reliability at Tjmax.=175°C.

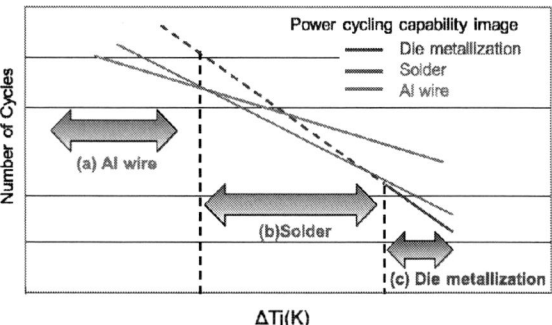

Fig. 10. Power cycling failure mode at each dTj region.

A. New Aluminum Wire

Figure 11 shows cross-section of Al wire bond after power cycling test. The failure of Al wires is caused by cracks propagating into the wire materials.

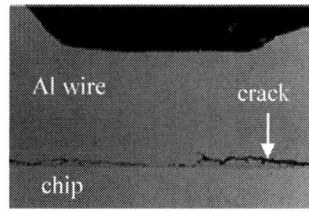

Fig. 11. Cross section of Al wire bond after power cycling.

Generally, the recrystallization temperature of metal material can be empirically assumed to be 0.4 times of melting point Kelvin for pure metal and 0.6 times of melting point Kelvin for alloy. The melting point of Al is 660°C and calculated recrystallization temperature is about 150°C to 200°C. These properties show Al grains are thought to grow under power cycling test at Tjmax.=175°C.

Relation between metal grain size and strength is presented in Eq. (6) by Hall-Petch [5]

$$\sigma_y = \sigma_o + kd^{-1/2} \qquad (6)$$

where σ_y is the yield strength; d is the average crystal grain size; and k and σ_o is a constant.

This equation shows that grain growth causes strength reduction. Thus, our development target of the new Al

978-1-4799-2706-7/14 $31.00 © 2014 IEEE   673

wire characteristics defined that recrystallization temperature is higher than 175°C. Figure 12 shows cross section observation and EBSD (Electron Back Scatter Diffraction) analysis of the conventional Al wire and the new Al wire before and after power cycling test at Tjmax.=175°C.

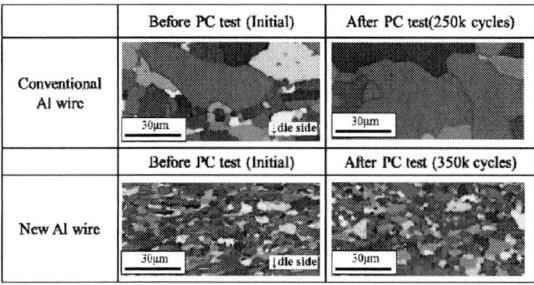

Fig. 12. Cross section and EBSD analysis of Al wire before and after power cycling test at Tjmax.=175℃.

The conventional Al wire grain size had grown after power cycling test, while the grain size of the new Al wire had not changed. And this new Al wire has twice the strength of conventional Al wire. As for the newly developed Al wire, heat-resistant improvement is achieved.

B. Sn-Sb Solder

Figure 13 shows the tensile strength of conventional Sn (Tin)-Ag (Silver) and Sn-Sb (Antimony) solder after high temperature storage test. Both solders have similar initial strength. However, the strength of Sn-Ag solder significantly decreased after 1,000 hours at 150°C and 175°C, while Sn-Sb solder strength was not changed.

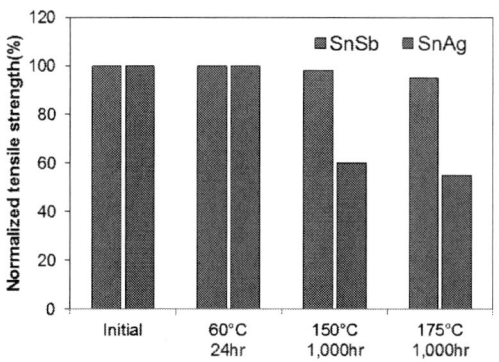

Fig. 13. Tensile strength results after high temperature storage test.

Solder microstructure influenced these solder strength. Figure 14 shows schematic diagram of solder microstructure at initial and after high temperature storage test. Sn-Ag solder is the representative of precipitation strengthening. This generates fine particles of Ag$_3$Sn intermetallic compound in grain boundaries, causing resistance against crack propagation. Reliability tests under high temperature cause grain growth of Sn

and Ag$_3$Sn aggregation, which affect cracks at grain boundaries. On the other hand, Indium or Sb is the representative of solid solution strengthening. These elements dissolve in Sn grains and restrain the grain growth of Sn under high temperature.

Precipitation Strengthening

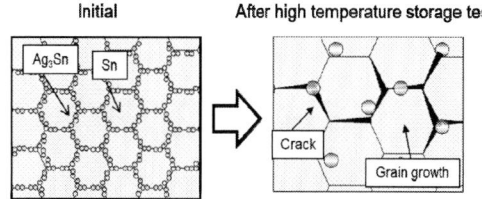

Solid Solution Strengthening

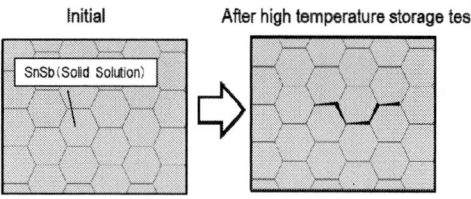

Fig.14. Schematic diagram of solder microstructure at initial and after high temperature storage test.

Next, power cycling tests were carried out for both samples using Sn-Ag, and Sn-Sb solders at Tjmax.=175°C. Sn-Sb solder sample has longer lifetime than Sn-Ag solder one. Figure 15 shows cross-section of solder joint after power cycling tests.

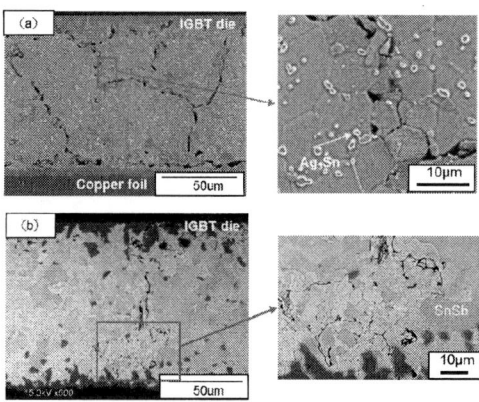

Fig. 15. Cross-section of solder joint after power cycling test at Tjmax.=175℃. (a)Sn-Ag solder, (b)Sn-Sb solder.

Sn-Ag solder showed thermal fatigue with Ag$_3$Sn aggregation and horizontal/vertical cracks along grain boundaries due to grain growth of Sn. On the other hand, Sn-Sb does not show thermal fatigue because Sn grain did not grow. Vertical cracks were caused by mechanical stress. These results indicate that the Sn-Sb solder can improve power cycling capability because this solder has higher temperature withstanding than Sn-Ag solder.

In addition, a new solder material which has both strengthening mechanisms of precipitation and solid solution has been developed. Figure 16 shows the tensile strength of Sn-Sb solder and new solder. This new solder material has a higher tensile strength than Sn-Sb solder. And this new solder was confirmed that it has high power cycling capability at Tjmax.=175°C.

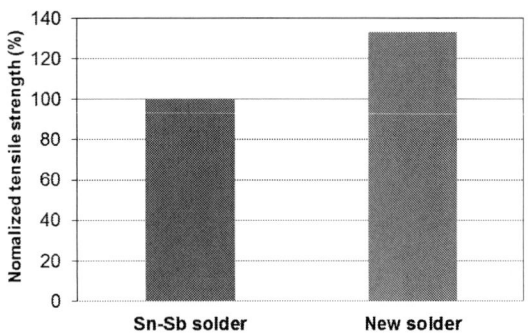

Fig. 16. Comparison of tensile strength of new solder.

C. Ni-layered metallization on the top side die

Conventionally, the top side of a die is metallized by pure Al or Al alloy with doping such as Si and /or Copper. Power cycling test at a larger ΔTj is known to cause cracks in the top side die metallization [6].

The reasons are as follows:

· Al grain growth and strength decrease
· Stress caused by the difference of CTE from Si die

A structure with a Ni (nickel) metallization layer on the top side die was developed to reduce the grain growth and cracks. Ni has a higher recrystallization temperature than Al and a closer CTE to Si. Figure 17 shows SEM (Scanning Electron Microscopy) observation of top side die metallization after power cycling test at Tjmax.=175°C.

 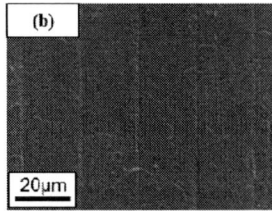

Fig. 17. SEM observation of top side die metallization after power cycling test at Tjmax.=175°C.
(a)Conventional Al metallization, (b)Ni-layered metallization

Al grain growth and cracks were observed in the conventional Al metallization, while Ni-layered metallization did not show grain growth or cracks. This result indicates that the Ni-layered metallization can improve power cycling capability.

D. Power Cycling Test Results

A power cycling test of the sample with newly developed technologies, the new Al wire bonding, the new solder joint under the chip, and the Ni layered electrode layer to confirm the performance. The power cycling test result with developed and conventional technologies are shown in Fig. 18. The result shows that the power cycling capability with developed technologies at Tjmax.=175°C is more than twice higher in all ΔTj regions than conventional technologies at Tjmax.=150°C.

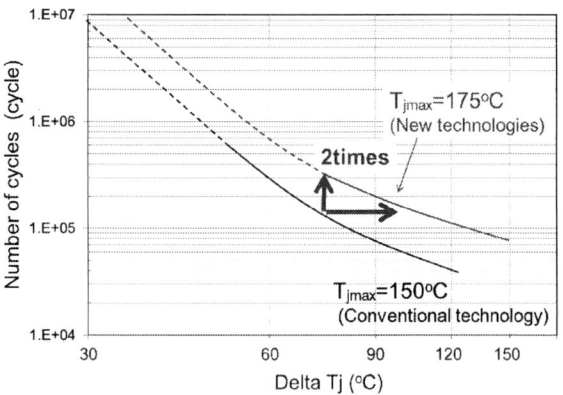

Fig. 18. Power cycling lifetime curve with new technologies at Tjmax.=175℃.

IV. CONCLUSION

To achieve minimization and high power density of IGBT module, the high cooling capacity which has the unified cooling fin and water jacket was developed.

It is cleared that velocity of the fluid is influence to cooling capacity of cooler. The velocity of the fluid has a relation to the gap distance of cooling fin and water jacket. The cooling fin for 4th gen. IGBT module is designed based on these phenomena.

Furthermore, the high temperature operation technologies which possible to guarantee Tj=175°C continuous operation was developed. The new Al wire has excellent characteristic in heat resistance, the new solder has excellent property in high temperature intensity, and an electrode protective metallization of low stress.

The Fuji 4th gen. IGBT module for automotive application has been developed using these new technologies. The new IGBT module achieved 30-% of volume reduction, 33-% of weight reduction in comparison with the 3rd gen.

REFERENCES

[1] H. Gohara, A. Morozumi, T. Ichimura, Y. Nishimura, K. Higuchi, P. Dietrich, A. Nishiura and Y. Takahashi, "Ultra Compact and Light Weight Intelligent Power Semiconductor Module for Hybrid System," Proceedings of the 27th EVS, Barcelona, Nov 17-20, 8C, 2013.

[2] A. Morozumi, H. Hokazono, Y. Nishimura, Y. Ikeda, Y. Nabetani and Y. Takahashi, "Direct Liquid Cooling Module with High Reliability Solder Joining Technology for Automotive

applications," *Proceedings of the 25th ISPSD & ICs,* Kanazawa, May, 26-30, 5-4, 2013.

[3] T. Saito, Y. Nishimura, K. Kido, F. Momose and E. Mochizuki, "New assembly technologies for Tjmax.=175°C continuous operation guaranty of IGBT module, " *Proceedings of PCIM Europe 2013*, Nuremberg, pp. 455-461, 2013.

[4] A. Morozumi, K. Yamada, T. Miyasaka, S. Sumi and Y. Seki, "Reliability of Power Cycling for IGBT Power Semiconductor Modules, " *IEEE Trans. on Industry Applications,* vol. 39, no. 3, pp. 665-671, 2003.

[5] N. J. Petch, *J. Iron Steel Inst.*, 174, Part I, 25-28, 1953.

[6] Y. Ikeda, et al, "A Study of the bonding wire reliability on the chip surface electrode in IGBT, " *Proceedings of The 22nd International Symposium on ISPSD, Hiroshima, June 6-10,* pp. 289-292,2010.

The 2014 International Power Electronics Conference

Integration of Plug-in Electric Vehicles in Power Systems Using Charging Mode Switching

Wen-Tai Li, Chao-Kai Wen
Institute of Communications Eng.,
National Sun Yat-sen University,
Kaohsiung 804, Taiwan

Jung-Chieh Chen
Dept. of Optoelectron. & Commun. Eng.,
National Kaohsiung Normal University,
Kaohsiung 802, Taiwan

Jen-Hao Teng
Dept. of Electrical Eng.,
National Sun Yat-sen University,
Kaohsiung 804, Taiwan

Pangan Ting
Ind. Tech. Res. Inst. (ITRI),
Hsinchu 310, Taiwan

Abstract—**In this paper, we develop an energy management algorithm for charging and discharging plug-in electric vehicles (PEVs). The algorithm considers both user convenience and electricity market factors. A performance metric, called "user convenience", is formulated to measure the effective coordination of PEV charging. As a response to the electricity market, we address an economic dispatch problem by considering power flows, grid losses, voltage profile patterns, and vehicle-to-grid costs. Both of the concerned factors can be optimized by simply switching a PEV during charging, discharging, or in off-mode. The mode switch allows easy integrate of PEVs into power systems. The PEV state-decision problem can be efficiently solved by using convex optimization tools. Moreover, the problem can be solved in a decentralized manner, wherein each sub-aggregator can determine to charge, discharge, and turn to off-mode *locally*. Simulation results indicate that the proposed PEV energy management algorithm can obtain the same final state-of-charge of the batteries as those obtained by unconstraint cases. Meanwhile, the effect of PEV charging on the power system is significantly mitigated.**

Keywords—*Energy management, plug-in electric vehicles, user convenience, economic dispatch.*

I. Introduction

Plug-in electric vehicles (PEVs) provide substantial environmental benefits and allow the delivery of electricity into the grid to alleviate the requirement for peaking generation sources. As such, several governments worldwide are attempting to accelerate the penetration of PEVs. For example, the present administration of the United States targets on reaching one million PEVs on the road by 2015 [1].

A PEV in power systems fulfills multiple roles such as a controllable load, a distributed energy storage, and a distributed generator. Given that most PEVs charge at the same time of day (e.g., during early evening after people come home), several reports noted that a high level of PEV usage will significantly affect existing power systems [2–7]. However, because of vehicle utilization behavior, most PEVs in domestic settings can be scheduled for nighttime charging without sacrificing the comfort level of the uses [8–12]. In this sense, PEVs serve

as controllable loads. Vehicle-to-Grid (V2G) techniques also allow PEVs to feed energy back into the power grid when the power demand is high. This feature leads to the classification of PEVs as a special class of distributed generators.

Large-scale penetration of EVs is expected to influence the behavior electricity markets strongly. Therefore, energy management by combining large-scale distributed PEVs has drawn significant attention in academic and industrial studies [2–7, 12]. The present study focuses on designing an energy management algorithm to control PEV charging and discharging. The algorithm considers both user convenience and electricity market factors.

From the perspective of user convenience, each user should aim to achieve the shortest charging time while maintaining the largest final state-of-charge (SOC). However, given the self-centered nature of most human beings, serious voltage drops and severe congestion resulting from PEV charging can cause inconvenient to all users. A global user convenience level will serve as an effective means to stimulate cooperation among users. To prioritize the needs of users, we quantize user convenience as a function of charging time, SOC, and allowable price. These factors will be specified in the latter part of this paper. The proposed energy management algorithm attempts to maximize the sum of user convenience while maintaining the power system within its acceptable operating limits.

From the perspective of electricity market, economic dispatch problems should be carefully addressed by considering power flows, grid losses, voltage profile patterns, and V2G costs. In response to electricity prices, a smart energy management algorithm should perform load shifting to be able to provide energy at a lower cost to PEVs. In addition, this algorithm can buy electricity at peak hours through a V2G technique to charge PEVs with urgent power needs. The proposed energy management algorithm also attempts to minimize energy prices without sacrificing user convenience.

The EV charging problem and the economic dispatch problem are investigated separately in most of the existing

978-1-4799-2706-7/14 $31.00 © 2014 IEEE

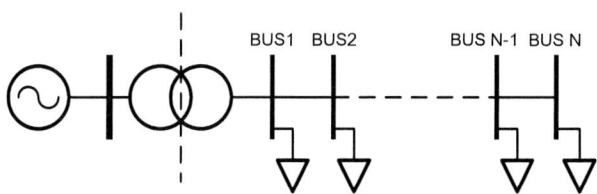

Fig. 1. A power network topology.

studies although these two design issues are closely related. In the present paper, a bi-objective optimization problem is formulated for the energy management algorithm by using a convex optimization framework. The algorithm aims to maximize total user convenience in the system and minimize energy prices simultaneously. The weighted sum method is applied to transform the bi-objective problem into a scalar objective problem.

Given the large number of PEVs, simply switching a PEV during charging and discharging or during the off-mode will be more practical than attempting to modulate the charging and discharging rates of each PEV [13]. Based on the switching strategy, our objective is to decide which optimal PEV states can maximize the scalar objective stated above. Moreover, an approach to solve the PEV state-decision problem in a decentralized manner is developed. In the decentralized algorithm, each sub-aggregator can decide the states of its associated PEVs *locally*. Finally, the performance advantage of the proposed algorithm is demonstrated through the simulation results. These results show that the proposed algorithm, which considers both user convenience and electricity market factors, can effectively mitigate the effects of charging PEVs on the power system.

II. SYSTEM MODELS AND PROBLEM FORMULATION

In our design for energy management algorithm the two performance utility factors, i.e., electricity prices and user convenience, are considered simultaneously when it optimizes PEV states. Toward this end, we briefly describe the classical economic dispatch problem in a distribution grid and then present the charging strategy to allow the simple integration of PEVs into the grid.

A. ECONOMIC DISPATCH

As shown in Fig. 1, we consider a power network with the main line consisting of N load buses. Let $\mathbf{v} = [V_1, \ldots, V_N]$ and $\mathbf{i} = [I_1, \ldots, I_N]$ with V_j and I_j being the voltage and current source in bus j, $\mathbf{Y} \in \mathbb{C}^{N \times N}$ be the bus admittance matrix. Based on Kirchhoff's law, we can write nodal equations for the power network by $\mathbf{i} = \mathbf{Y}\mathbf{v}$. Thus, the complex power delivered to bus j is

$$P_j + \mathrm{j}Q_j = V_j \left[\sum_{n=1}^{N} Y_{jn} V_n \right]^*. \tag{1}$$

If P_j's and Q_j's are specified as input data, V_j can be computed based on (1) by using a power-flow program [14]. If bus j is a load bus with no generation, P_j is negative.

Let $C(P_j^{\mathsf{G}})$ denote the generating cost versus real power output P_j^{G} for plant j. For example, the fuel-cost could be quadratic in the form [14]

$$C(P) = \alpha_2 P^2 + \alpha_1 P + \alpha_0 \tag{2}$$

with $\alpha_2, \alpha_1, \alpha_0 > 0$. The classical economic dispatch problem is to minimize the generator cost subject to the constraints that given load demands P_j^{D}'s must be met; that is,

$$P_j + \mathrm{j}Q_j = \left(P_j^{\mathsf{G}} - P_j^{\mathsf{D}}\right) + \mathrm{j}\left(Q_j^{\mathsf{G}} - Q_j^{\mathsf{D}}\right), \ \forall \, i, \tag{3}$$

satisfy some system operation constraints expressed as follows:

$$P_j^{\min} \leq P_j \leq P_j^{\max}, \ \forall j, \tag{4a}$$

$$V_{\min} \leq |V_j| \leq V_{\max}, \ \forall j. \tag{4b}$$

Here, (4a) describes the limited ability of the power injection on the buses and (4b) captures the voltage magnitude constraint. For transformers, these constraints can be regarded as normal operating constraints. In this case, the minimum limit $P_j^{\min}$ in (4a) can be removed.

Let $\mathbf{E}_j \in \mathbb{R}^{N \times N}$ with $E_{jj} = 1$ and all the other entries be equal to zero. Following the idea of [15], (1) can be written as

$$P_j + \mathrm{j}\, Q_j = \mathsf{tr}(\mathbf{A}_j \mathbf{U}) + \mathrm{j}\, \mathsf{tr}(\mathbf{B}_j \mathbf{U}), \tag{5}$$

where

$$\mathbf{A}_j \triangleq \frac{1}{2}(\mathbf{Y}^H \mathbf{E}_j + \mathbf{E}_j \mathbf{Y}), \tag{6a}$$

$$\mathbf{B}_j \triangleq \frac{1}{2\mathrm{j}}(\mathbf{Y}^H \mathbf{E}_j - \mathbf{E}_j \mathbf{Y}), \tag{6b}$$

$$\mathbf{U} \triangleq \mathbf{v}\mathbf{v}^H. \tag{6c}$$

The active/reactive power can be expressed as function of $\mathbf{U}$ (or, equivalently, $\mathbf{v}$). With (5), the constraints can be written in (4) as

$$P_j^{\min} \leq \mathsf{tr}(\mathbf{A}_j \mathbf{U}) \leq P_j^{\max}, \ \forall j, \tag{7a}$$

$$V_{\min}^2 \leq \mathsf{tr}(\mathbf{E}_j \mathbf{U}) \leq V_{\max}^2, \ \forall j. \tag{7b}$$

The classical economic dispatch problem is an optimization problem with variables of $\mathbf{U}$ or $\mathbf{v}$. However, this problem is *nonconvex* because of the constraint of $\mathbf{U} = \mathbf{v}\mathbf{v}^H$. In [15], the problem can be exactly solved without this rank one constraint if the power network has a tree topology. Consequently, the classical economic dispatch problem can be efficiently solved by using convex optimization tools.

B. CHARGING STRATEGY

For PEV batteries, we adopted the battery data provided in [5]. The battery characteristics used by a PEV can be described in terms of the battery storage capacity and the power needed to charge it. For ease in modeling, a discrete-time system is used to approximate the continuous system. The sampling rate of 15 mins was adopted, which is commonly used in a power system [3]. Let $\text{SOC}_{ji}(t)$ be the battery SOC of PEV i at bus j at the start of time slot t. Then, we have

$$\text{SOC}_{ji}(t+1) \approx \text{SOC}_{ji}(t) + \frac{15 \text{ mins.}}{60 \text{ mins.}} \gamma_{ji}, \qquad (8)$$

when the charging rate is at γ_{ji} level.

Let $p_{ji}(t)$ be the demanded power of the i-th PEV in the j-th bus during time slot t. If $p_{ji}(t) > 0$, PEV i charges (or discharging) at time slot t, whereas if $p_{ji}(t) < 0$, PEV i discharges at time slot t. When $p_{ji}(t)$ is zero, the charging and discharging of PEV i must be turned off at time slot t.

To formulate the demanded power of each PEV, it is convenient to have the following definitions. Let p_{ji} be the charging power of the i-th PEV at the j-th bus.[1] Thus, $\mu_{ji}p_{ji}$ is reasonably assumed as power send back to the grid by PEV i in vehicle-to-grid (V2G) mode, where $\mu_{ji} \leq 1$ represents the battery discharge efficiency. When a PEV is plugged into the power system, actual charging/discharging does not immediately start. The process will be controlled by our methodology as follows: Let $s_{ji}(t)$ denote a ternary PEV state during time slot t. Then, $s_{ji}(t) = 1$ if the i-th PEV is selected for charging, $s_{ji}(t) = -1$ if the i-th PEV is selected for discharging, and $s_{ji}(t) = 0$ if charging/discharging is turned off. Therefore, the power consumed by the i-th PEV during time slot t is given by

$$p_{ji}(t) = p_{ji}\langle s_{ji}(t) \rangle^+ + \mu_{ji}p_{ji}\langle s_{ji}(t) \rangle^-, \qquad (9)$$

where $\langle x \rangle^+ = \max\{x, 0\}$ and $\langle x \rangle^- = \min\{x, 0\}$. Let

$$\dot{P}_j(t) \triangleq \sum_{i=1}^{K_j(t)} p_{ji}\langle s_{ji}(t) \rangle^+, \qquad (10a)$$

$$\ddot{P}_j(t) \triangleq -\sum_{i=1}^{K_j(t)} \mu_{ji}p_{ji}\langle s_{ji}(t) \rangle^-, \qquad (10b)$$

where $K_j(t)$ is defined as the number of PEVs at the j-th bus at time slot t. Clearly, $\dot{P}_j(t)$ represents the total demand power for charging (i.e., G2V), whereas $\ddot{P}_j(t)$ indicates the total power for V2G at the j-th bus during time slot t. Let $P_j^{\text{B}}(t)$ be the residential load (in the absence of PEV charging) in the j-th bus, and $P_j^{\text{G}}(t)$ be the generating power of the DG in the j-th bus. Then, we have

$$P_j(t) = P_j^{\text{G}}(t) + \ddot{P}_j(t) - \left(P_j^{\text{B}}(t) + \dot{P}_j(t) \right). \qquad (11)$$

Note that $P_j(t)$ in (11) has the same rule as P_j in (4a) where we have removed the time slot index for convenience.

The problem that must be addressed is which PEV should be charging or discharging under the constraint of (4). This decision depends on our objective function. The objective function being considered in this project is the minimization of the generator cost and the maximization of the convenience levels

of PEV users. The generator cost subject to the operating limits of a power system has been discussed previously. Therefore, we will discuss the formulation of user convenience.

User's convenience will be indicated by the charging time and the final SOC of each PEV. If a vehicle has a lower initial SOC and lesser remaining charging time, this PEV should have a higher priority for charging. Using these concepts, a weight factor is given based on the importance of each user

$$w_{ji}(t) = \frac{T_{ji}^{\text{SOC}}(t)}{T_{ji}(t)}, \qquad (12)$$

where $T_{ji}(t)$ is the remaining time for the i-th PEV before it leaves the charging system and $T_{ji}^{\text{SOC}}(t)$ is the required time for the i-th PEV to achieve its full SOC state. According to (8), $T_{ji}^{\text{SOC}}(t)$ can be estimated by

$$T_{ji}^{\text{SOC}}(t) = \frac{1 - \text{SOC}_{ji}(t)}{\gamma_{ij}} \quad \text{(h)}. \qquad (13)$$

The weight factor $w_{ji}(t)$ is dynamic because $\text{SOC}_{ji}(t)$ and $T_{ji}(t)$ tend to vary from time to time. In particular, if the i-th PEV is charged (or discharged) during time slot t, its $T_{ji}^{\text{SOC}}(t)$ will decrease (or increase) at the start of time slot $t+1$. The corresponding $T_{ji}(t)$ will always monotonically decrease.

From the setting of the weight factor, a rule is established to determine the state of the PEVs. PEVs with large $w_{ji}(t)$ have to be considered for priority charging. On the other hand, PEVs with very small $w_{ji}(t)$ that allow long charging time and are at full SOC state can transfer power to the grid to alleviate the peak power demand. The battery discharge efficiency is lower and frequent charging/discharging switching affects the battery life, so the weight factor should be increased in the V2G mode, such as $\gamma_{ji}w_{ji}(t)$ with $\gamma_{ji} \geq 1$. If γ_{ji} is large, we avoid turning on the V2G mode for the PEV as much as possible; thus, the V2G mode is used only when the power demand is urgent.

C. PROBLEM FORMULATION

Our objective is the maximization of the sum of user convenience and minimization the generating cost. Clearly, the problem is a bi-objective optimization problem, which is usually difficult to solve. A common logic to remove this difficulty is to replace the original multiple objectives by their linear weighted sum. For ease of notation, we drop index (t) from all the succeeding notations because we can apply our description to each time slot. Let

$$\dot{\mathbf{s}}_j \triangleq \left[\langle s_{j1} \rangle^+, \ldots, \langle s_{jK_j} \rangle^+ \right]^T, \quad \ddot{\mathbf{s}}_j \triangleq \left[\langle s_{j1} \rangle^-, \ldots, \langle s_{jK_j} \rangle^- \right]^T,$$

$$\dot{\mathbf{w}} \triangleq \left[w_{j1}, \ldots, w_{jK_j} \right]^T, \quad \ddot{\mathbf{w}} \triangleq \left[\gamma_{j1}w_{j1}, \ldots, \gamma_{jK_j}w_{jK_j} \right]^T,$$

$$\dot{\mathbf{p}} \triangleq \left[p_{j1}, \ldots, p_{jK_j} \right]^T, \quad \ddot{\mathbf{p}} \triangleq \left[\mu_{j1}p_{j1}, \ldots, \mu_{jK_j}p_{jK_j} \right]^T.$$

---

[1] In contrast to p_{ji}, the value of $p_{ji}(t)$ is determined by the state (charging, discharging, or turning off) of the i-th PEV in the j-th bus during time slot t.

Then, the optimization problem can be formulated as

$$\max \quad \sum_{j=1}^{N} \left(\dot{\mathbf{w}}_j^T \dot{\mathbf{s}}_j + \ddot{\mathbf{w}}_j^T \ddot{\mathbf{s}}_j \right)$$

$$- \xi \left[\sum_{j=1}^{N} \left(C(P_j^{\mathsf{G}}(t)) + \ddot{c}_j \ddot{P}_j(t) \right) \right]$$

$$\text{s.t.} \quad \dot{P}_j = \dot{\mathbf{p}}_j^T \dot{\mathbf{s}}_j, \quad \ddot{P}_j = -\ddot{\mathbf{p}}_j^T \ddot{\mathbf{s}}_j, \quad \forall j, \tag{14a}$$

$$P_j = P_j^{\mathsf{G}} + \ddot{P}_j - \left(P_j^{\mathsf{B}} + \dot{P}_j \right), \quad \forall j, \tag{14b}$$

$$(7a), (7b), \tag{14c}$$

$$s_{ji} \in \{0, +1, -1\}, \quad \forall i, j, \tag{14d}$$

$$\text{over} \quad s_{ji}, P_j^{\mathsf{G}}, \dot{P}_j, \ddot{P}_j, \forall i, j, \mathbf{U}.$$

In (14), ξ is the linear weighted sum parameter, $C(P_j^{\mathsf{G}}(t))$ denotes the generating cost, $\ddot{c}_j \ddot{P}_j(t)$ represents the cost due to the V2G mode, constraints (14a) and (14b) correspond to (10) and (11) respectively, and constraint (14c) ensures the power system within its acceptable operating limits.

Considering the integer nature of $\{s_{ji}\}$'s, an exhaustive search is required to provide the optimal solution of (14). However, enumerating all possible combinations of $\{s_{ji}\}$'s is impractical, especially for a large number of PEVs. To address problem (14), we relax the constraint of $s_{ji} \in \{0, +1, -1\}$ and use the efficient convex optimization tool to obtain the relaxed results in the next section.

III. EFFECTIVE ALGORITHMS

We adopt a simplification by relaxing the PEV selection variables from ternary integer to real number. In particular, we change $\langle s_{ji} \rangle^+$ to $\dot{s}_{ji} \in [0, 1]$ and $\langle s_{ji} \rangle^-$ to $\ddot{s}_{ji} \in [-1, 0]$. Thus, constraint (14d) can be approximated by

$$\mathbf{0} \le \dot{\mathbf{s}}_j \le \mathbf{1}, \quad -\mathbf{1} \le \ddot{\mathbf{s}}_j \le \mathbf{0}, \quad \forall j. \tag{15}$$

The problem is now a *convex problem* that can be solved using semidefinite programming [16].

The problem with the relaxed constraint (15) is *not equal* to the original problem (14); the relaxed optimization results $\{\dot{s}_{ji}^\star, \ddot{s}_{ji}^\star\}$ can be fractional. $\dot{s}_{ji}^\star$ (or $\ddot{s}_{ji}^\star$) can be interpreted as a probability for charging (or discharging) PEV i. Each PEV should belong to one of the following three states: charge, discharge, and off. Thus, their sum $(\dot{s}_{ji}^\star + \ddot{s}_{ji}^\star)$ must be mapped into 0, 1, or -1 to obtain the desired results. This procedure can be performed as

$$\hat{s}_{ji}^\star := \begin{cases} 1, & \text{with probability } \max\{\dot{s}_{ji}^\star + \ddot{s}_{ji}^\star, 0\}, \\ -1, & \text{with probability } -\min\{\dot{s}_{ji}^\star + \ddot{s}_{ji}^\star, 0\}, \\ 0, & \text{with probability } 1 - |\dot{s}_{ji}^\star + \ddot{s}_{ji}^\star|. \end{cases}$$

$$\tag{16}$$

Solving problem (15) requires a control center. In the centralized approach, a very large dimensional matrix $\mathbf{U}$ defined in (6) has to be computed. As the power network topology becomes complex, the centralized approach becomes inconvenient in terms of computational complexity and flexibility.

TABLE I. THE STATISTICAL RESULTS OF THE SOC. HERE, Pr_x MEANS $\text{Pr}(\text{SOC} > x\%)$ AND $\overline{\text{SOC}}$ REPRESENTS THE MEAN OF THE FINAL SOC.

	Pr_{75}	Pr_{85}	$\overline{\text{SOC}}$	Cost
uncon. charging	98.70%	98.23%	99.20%	1.00 p.u.
20% pen. level	92.92%	89.60%	95.90%	0.86 p.u.
50% pen. level	88.43%	85.57%	93.34%	0.90 p.u.
100% pen. level	84.88%	82.08%	91.14%	0.97 p.u.

Because of the convexity nature of the formulated problem, we propose the use of alternating direction method of multipliers (ADMM) [17] for decentralized optimization on problem (15). Due to the space limitation, we do not presents the detailed algorithm. In the decentralized fashion, we only need to deal with a very small sub-matrix matrix of $\mathbf{U}$ (e.g., 2×2 or 3×3 matrices). Thus, computational complexity is distributed to each sub-aggregator, where the computing complexity of each sub-aggregator is very low. In addition, each sub-aggregator can decide the states of its associated PEVs locally.

IV. CASE STUDY AND DISCUSSIONS

In this section, simulations are conducted to verify the performance of the proposed algorithm. The test model is assumed to be a residential distribution circuit as developed in [18]. The usage pattern of the PEVs is same as that used in [12].

We consider the case when all the PEVs have V2G capabilities. In this case, the PEVs provide a method for "peak shaving" by sending power back to the grid so the available power for PEV charging can be increased. The battery discharge efficiency μ_{ji} is set to 0.75 and the increased weight factor γ_{ji} is 4 for all the PEVs. The parameters of the generating cost in (2) are $[\alpha_2, \alpha_1, \alpha_0] = [0, 8, 4]$ and the cost for V2G is $\ddot{c}_j = 24/0.75 \ \forall j$. By averaging over 10^4 independent trials, the final SOC statistical results are summarized in Table I, which shows that the final SOC results are as good as the uncontrolled charging system when the daily peak demand level $P_j^{\max}$ is similar to that without PEVs. We emphasize that the results of the uncontrolled charging system serve as the best-SOC benchmark. In fact, in the uncontrolled charging system, the grid has exceed it acceptable operating limits.

V. CONCLUSION

we develop an energy management algorithm for charging and discharging plug-in electric vehicles

We have developed an energy management algorithm by properly decide a PEV state to maximize the user convenience levels and simultaneously meet circuit-level demand limits. Our results indicated that, compared with the uncontrolled case, the proposed charging selection algorithm reduced the final SOC by 9% even at full PEV penetration level. In addition, we also developed a decentralized optimization algorithm that can solve the PEV state decision problem in a decentralized fashion, where charging, discharging, and off-mode are made *locally* by each sub-aggregator.

ACKNOWLEDGMENT

The authors gratefully acknowledge the technical and financial supports for this work from National Science Council of Taiwan and Taiwan Power Company under contract NSC 100-3113-P-110-004 and NSC 100-2221-E-017-006-MY3

REFERENCES

[1] *ANSI EVSP Standalization Roadmap for Electric Vehicles*, Apr. 2012.

[2] K. Schneider, C. Gerkensmeyer, M. Kintner-Meyer, and R. Fletcher, "Impact assessment of plug-in hybrid vehicles on pacific northwest distribution systems," in *Proc. Power Energy Soc. Gen. Meet.*, Jul. 2008, pp. 1–6.

[3] K. Clement-Nyns, E. Haesen, and J. Driesen, "The impact of charging plug-in hybrid electric vehicles on a residential distribution grid," *IEEE Trans. Power Syst.*, vol. 25, no. 1, pp. 371–380, Feb. 2010.

[4] K. Dyke, N. Schofield, and M. Barnes, "The impact of transport electrification on electrical networks," *IEEE Trans. Ind. Electron.*, vol. 57, no. 12, pp. 3917–3926, Dec. 2010.

[5] P. Fernández, T. G. S. Román, R. Cossent, C. M. Domingo, and P. Frías, "Assessment of the impact of plug-in electric vehicles on distribution networks," *IEEE Trans. Power Syst.*, vol. 26, no. 1, pp. 206–213, Feb. 2011.

[6] M. A. Kejun Qian, Chengke Zhou and Y. Yuan, "Modeling of load demand due to ev battery charging in distribution systems," *IEEE Trans. Power Syst.*, vol. 26, no. 2, pp. 802–810, May 2011.

[7] J. Lopes, F. Soares, and P. Almeida, "Integration of electric vehicles in the electric power system," *Proc. IEEE*, vol. 99, no. 1, pp. 168–183, Jan. 2011.

[8] W. Su and M.-Y. Chow, "Performance evaluation of an eda-based large-scale plug-in hybrid electric vehicle charging algorithm," *IEEE Trans. Smart Grid*, vol. 3, no. 1, pp. 308–315, Mar. 2012.

[9] W. Su and M.-Y. Chow, "Computational intelligence-based energy management for a large-scale phev/pev enabled municipal parking deck," *Applied Energy*, vol. 96, pp. 171–182, Aug. 2012.

[10] S. Shao, M. Pipattanasomporn, and S. Rahman, "Demand response as a load shaping tool in an intelligent grid with electric vehicles," *IEEE Trans. Smart Grid*, to appear, 2012.

[11] E. Sortomme, M. M. Hindi, S. D. MacPherson, and S. S. Venkata, "Coordinated charging of plug-in hybrid electric vehicles to minimize distribution system losses," *IEEE Trans. Smart Grid*, vol. 2, no. 1, pp. 198–205, Mar. 2011.

[12] C.-K. Wen, J.-C. Chen, J.-H. Teng, and P. Ting, "Decentralized plug-in electric vehicle charging selection algorithm in power systems," *IEEE Trans. Smart Grid*, vol. 3, no. 4, pp. 1779–1789, Dec. 2012.

[13] A. Brooks, E. Lu, D. Reicher, C. Spirakis, and B. Weihl, "Demand dispatch," *IEEE Power and Energy Mag.*, vol. 8, no. 3, pp. 20–29, Jun. 2010.

[14] J. D. Glover, M. S. Sarma, Thomas Overbye, *Power Systems Analysis and Design*. CL Engineering, 5 edition, 2011.

[15] J. Lavaei and S. H. Low, "Zero duality gap in optimal power flow problem," *IEEE Trans. Power Syst.*, vol. 27, pp. 92–107, Feb 2012.

[16] S. Boyd and L. Vandenberghe, *Convex Optimization*. Cambridge Univ. Press, 2004.

[17] S. Boyd, N. Parikh, E. Chu, B. Peleato, and J. Eckstein, *Distributed Optimization and Statistical Learning via the Alternating Direction Method of Multipliers*. Foundations and Trends in Machine Learning, 2011.

[18] S. Shao, T. Zhang, M. Pipattanasomporn, and S. Rahman, "Impact of TOU rates on distribution load shapes in a smart grid with PHEV penetration," in *Transmission and Distribution Conference and Exposition (2010 IEEE PES)*, 2010, pp. 1–6.

The 2014 International Power Electronics Conference

A Novel Compensation Method for a Motor Phase Current Sensor Offset Error Varied during a VSI-Motor Drive

Hiroshi Tamura
Hitachi Ltd., Hitachi Research Laboratory.
1-1 Omika-cho 7-chome, Hitachi city Ibaraki, Japan
Hiroshi.tamura.je@hitachi.com

Yasuo Noto
Hitachi Automotive Systems, Ltd.
2520 Takaba, Hitachinaka city Ibaraki, Japan
yasuo.noto.wh@hitachi.com

Toshiyuki Ajima
Hitachi Ltd., Hitachi Research Laboratory.
1-1 Omika-cho 7-chome, Hitachi city Ibaraki, Japan
toshiyuki.ajima.fz@hitachi.com

Jun-ichi Itoh
Nagaoka University of Technology
1603-1 Kamitomioka-cho Nagaoka city Niigata, Japan
itoh@vos.nagaoka.ac.jp

Abstract— This paper presents a method for reducing a torque ripple resulting from a motor phase current sensor offset error for motor drive applications such as a hybrid electric vehicle. The proposed method compensates for the motor phase current sensor offset error by subtracting a motor phase current sensor offset error value estimated by using a discrete Fourier transform from a detected motor phase current value. Its main feature is its ability to compensate for the motor phase current sensor offset error that varies due to temperature change during the inverter-motor drive. Simulation results reveal that the proposed method can estimate the motor phase current sensor offset error with less than or equal to +-2[A] difference and reduce the average value of the motor torque ripple to half (1/2) of that without the proposed method in a steady state.

Keywords— *Current sensor offset error compensation, Discrete Fourier transform, Motor torque ripple reduction, Automotive application*

I. INTRODUCTION

Hybrid electric vehicles (HEVs) are rapidly becoming more common due to increasing global environmental regulations. A HEV is an automobile that has an electric power train (EPT) added to its engine [1-3]. The EPT consists of a battery, a voltage source inverter (VSI), a UVW-phase current sensor (a UVW-phase CS), and a permanent magnet synchronous motor (PMSM). The HEV generates driving force by using the EPT. Therefore, to improve the ride quality of the HEV, the motor torque ripple of the PMSM must be reduced. The motor torque ripple is caused by spatial harmonics of a permanent magnet flux, a dead-time of the VSI, and a motor phase CS offset error.

Various motor torque ripple reduction methods have been developed on the basis of motor configuration and motor control. Motor configuration solutions include skew optimization [4] and new rotor design [5, 6]. However, these solutions complicate production and increase cost. Hence, motor control solutions have drawn attention. These include methods for generating a motor current reference value that cancels the harmonic distortion of a back electromotive force [7], reducing motor torque ripple on the basis of a motor torque

estimation equation [8-10], and compensating for dead-time using a disturbance observer [11, 12]. Though these motor control solutions can reduce the motor torque ripple resulting from spatial harmonics of the permanent magnet flux and the dead-time of the VSI, the motor torque ripple resulting from the motor phase CS offset error cannot be reduced. Accordingly, the motor phase CS offset error is usually compensated for by setting a detected motor phase current value to zero while the VSI-motor is stopped. However, a high accuracy motor phase CS must be applied to the EPT because the general compensation method for the motor phase CS offset error cannot compensate for the motor phase CS offset error value, which varies due to temperature change while the VSI-motor is driven. As a result, the EPT cost increases. A unique motor torque ripple reduction method has been developed for applying a low accuracy and low cost motor phase CS to the EPT [13]. The conventional method reduces the motor torque ripple by compensating for the motor phase CS offset error that varies during the VSI-motor drive by using the VSI output voltage reference value. However, the conventional method takes a long time to design the compensation algorithm for the motor phase CS offset error. This is because the optimal parameters required to estimate the motor phase CS offset error must be derived by repeating a simulation, and the response of the motor phase CS offset error estimation is slow. In addition, the conventional method can only estimate the motor phase CS offset error when the sum total value of the motor phase CS offset error is zero.

This paper presents a novel algorithm that can compensate for the motor phase current sensor offset error that varies due to temperature change during a VSI-motor drive with simple design, even when the sum total value of the motor phase CS offset error is not zero. The proposed method compensates for the motor phase CS offset error by subtracting a motor phase CS offset error value estimated by using a discrete Fourier transform from a detected motor phase current value.

In this paper, Section II describes the main problem of the low accuracy motor phase CS. Next, Section III explains the general compensation method for the motor

978-1-4799-2706-7/14 $31.00 © 2014 IEEE

682

phase CS offset error. After that, Section IV describes the proposed compensation algorithm for the motor phase CS offset error. Finally, Section V discusses the effectiveness of the proposed method for reducing the motor torque ripple with reference to the simulation results. These simulation results reveal that the proposed method can compensate for the motor phase CS offset error with less than or equal to +-2.0[A] difference and reduce the average value of the motor torque ripple to half (1/2) of that without the proposed method in a steady state in which the motor torque reference is from 10[Nm] to 80[Nm] and the motor rotation speed is 1000[rpm].

II. Main Problem of Low Accuracy Motor Phase Current Sensor

Fig. 1 shows the general control block diagram for a motor drive in a HEV. The detected UVW-phase current value (i_{us}, i_{vs}, i_{ws}) is obtained by adding the UVW-phase CS offset error value (i_{uoet}, i_{voet}, i_{woet}) to the true UVW-phase current value (i_{ut}, i_{vt}, i_{wt}) and is represented by Eq. (1).

$$i_{us} = i_{ut} + i_{uoet}$$
$$i_{vs} = i_{vt} + i_{voet} \qquad (1)$$
$$i_{ws} = i_{wt} + i_{woet}$$

Eq. (2) is obtained by transforming the detected UVW-phase current value (i_{us}, i_{vs}, i_{ws}) shown in Eq. (1) into the detected dq-axis current value (i_{ds}, i_{qs}).

$$i_{ds} = i_{dt} + i_{doet}$$
$$i_{qs} = i_{qt} + i_{qoet} \qquad (2)$$

where i_{dt} is the true d-axis current value, i_{qt} is the true q-axis current value, i_{doet} is the d-axis current offset error value, and i_{qoet} is the q-axis current offset error value. The true dq-axis current value (i_{dt}, i_{qt}) and the dq-axis current offset error value (i_{doet}, i_{qoet}) shown in Eq. (2) is calculated by Eq. (3) using the magnetic pole position θ_e.

$$i_{dt} = \sqrt{\frac{2}{3}}\left\{ i_{ut}\cos\theta_e + i_{vt}\cos\left(\theta_e - \frac{2\pi}{3}\right) + i_{wt}\cos\left(\theta_e + \frac{2\pi}{3}\right) \right\}$$

$$i_{qt} = -\sqrt{\frac{2}{3}}\left\{ i_{ut}\sin\theta_e + i_{vt}\sin\left(\theta_e - \frac{2\pi}{3}\right) + i_{wt}\sin\left(\theta_e + \frac{2\pi}{3}\right) \right\}$$

$$\qquad (3)$$

$$i_{doet} = \sqrt{\frac{2}{3}}\left\{ i_{uoet}\cos\theta_e + i_{voet}\cos\left(\theta_e - \frac{2\pi}{3}\right) + i_{woet}\cos\left(\theta_e + \frac{2\pi}{3}\right) \right\}$$

$$i_{qoet} = -\sqrt{\frac{2}{3}}\left\{ i_{uoet}\sin\theta_e + i_{voet}\sin\left(\theta_e - \frac{2\pi}{3}\right) + i_{woet}\sin\left(\theta_e + \frac{2\pi}{3}\right) \right\}$$

The motor drive system applying the auto current regulator (ACR) equalizes the detected dq-axis current value (i_{ds}, i_{qs}) shown in Eq. (2) and the dq-axis current reference value (i_d^*, i_q^*). For this reason, the true dq-axis current value (i_{dt}, i_{qt}) is represented by Eq. (4).

$$i_{dt} = i_{ds} - i_{doet} = i_d^* - i_{doet}$$
$$i_{qt} = i_{qs} - i_{qoet} = i_q^* - i_{qoet} \qquad (4)$$

TABLE I
PARAMETERS FOR MOTOR TOQUE CALCULATION

Number of pole pairs P_n [-]	4
Magnet flux linkage φ_a [Wb]	0.049
d-axis inductance L_d [mH]	0.21
q-axis inductance L_q [mH]	0.49
Magnetic pole position θ_e [rad]	0 to 2π
True U-phase current value i_{ut} [A]	$200\cos(\theta_e + 9\pi/12)$
True V-phase current value i_{vt} [A]	$200\cos(\theta_e - \pi/12)$
True W-phase current value i_{wt} [A]	$200\cos(\theta_e - 17\pi/12)$
U-phase CS offset error value i_{uoet} [A]	+20
V-phase CS offset error value i_{voet} [A]	+ 5
W-phase CS offset error value i_{woet} [A]	-25

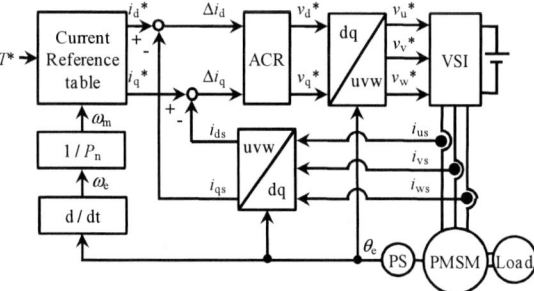

Fig. 1. General control block diagram for motor drive in HEV.

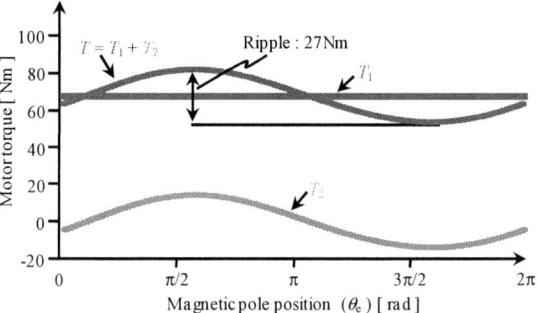

Fig. 2. Calculation results of motor torque.

$$T = T_1 + T_2$$
$$T_1 = P_n\left\{ \varphi_a i_q^* + (L_d - L_q) i_d^* i_q^* \right\} \qquad (5)$$
$$T_2 = P_n\left\{ (L_q - L_d)\left(i_d^* i_{qoet} + i_q^* i_{doet} - i_{doet} i_{qoet} \right) - \varphi_a i_{qoet} \right\}$$

The total torque value (T) of the motor is calculated by Eq. (5) using the true dq-axis current value (i_{dt}, i_{qt}) shown in Eq. (4), the number of pole pairs (P_n), the magnet flux linkage (φ_a), and the dq-axis inductance (L_d, L_q).

As is obvious from Eq. (5), the total torque (T) of the motor is the sum total value of the torque (T_1) corresponding to the dq-axis current reference value (i_d^*, i_q^*) and the torque (T_2) resulting from the dq-axis current offset error value (i_{doet}, i_{qoet}). As an example, Fig. 2 is the motor torque waveform calculated by using Eqs. (1) to (5) and the parameter shown in Table I. The total torque (T) of the motor is rippled on fundamental frequency by the influence of the torque (T_2) resulting from the motor phase CS offset error. The motor torque ripple deteriorates ride quality in a HEV. Therefore, the main problem of a low accuracy motor phase CS system is that it increases the motor torque ripple.

III. GENERAL COMPENSATION METHOD FOR MOTOR PHASE CURRENT SENSOR OFFSET ERROR

Fig. 3 shows the flow chart of the general motor phase CS offset error compensation, and Fig. 4 shows the motor phase current detection value (i_{cp}) compensated using the flow chart shown in Fig. 3. The general motor phase CS offset error compensation method compensates for the motor phase current detection value (i_{cp}) during the VSI-motor stop. First, Step 1 checks to see if the VSI-motor has stopped. If it has, then Step 2 detects the motor phase current. That is, the motor phase current is detected when the true motor phase current is zero. Next, Step 3 calculates the motor phase CS offset error value (i_{oet}). The calculated motor phase CS offset error value (i_{oet}) is equal to the motor phase current value (i_s) detected in Step 2. Finally, Step 4 compensates for the motor phase current detection value (i_{cp}) by subtracting the calculated motor phase CS offset error value (i_{oet}) from the detected motor phase current value (i_s). As a result, the compensated motor phase current detection value (i_{cp}) becomes zero. As described above, the true motor phase current value is zero because the VSI-motor has stopped. Therefore, the compensated motor phase current detection value (i_{cp}) coincides with the true motor phase current value. However, the general motor phase CS offset error compensation method cannot compensate for the motor phase CS offset error that varies due to temperature change while the VSI-motor is driven.

Fig. 5 shows the motor torque ripple with the general motor phase CS offset error compensation method when the U-phase CS offset error varies while the VSI-motor is driven. The worst case is assumed in the variation of the U-phase CS offset error value (i_{uoet}). Soon after the U-phase CS offset error value (i_{uoet}) is varied, the detected U-phase current value (i_{us}) and the true U-phase current value (i_{ut}) start to differ. Finally, the detected U-phase current value (i_{us}) and the true U-phase current value (i_{ut}) differ greatly. As a result, the motor torque ripple becomes 35Nm, which is 25Nm lager than that before the U-phase CS offset error varies. To solve this problem, a motor phase CS offset error compensation algorithm must be developed that can compensate for the motor phase CS offset error that varies while the VSI-motor is driven.

IV. PROPOSED COMPENSATION ALGORITHM FOR MOTOR PHASE CURRENT SENSOR OFFSET ERROR

Fig. 6 shows the proposed control block diagram for the motor drive in the HEV. The main feature of the proposed control block diagram is that the motor phase CS offset error compensator is added between the UVW-phase current sensor and the UVV/dq coordinate transformer. The motor phase CS offset error compensator compensates for the motor phase CS offset error by subtracting the motor phase CS offset error value (i_{udcx}, i_{vdcx}, i_{wdcx}) estimated by the motor phase CS offset error estimator from the motor phase current value (i_{us}, i_{vs}, i_{ws}) detected by the motor phase CS. It then obtains a compensated UVW-phase current value (i_{ucp}, i_{vcp}, i_{wcp}) almost equal to the true UVW-phase current value (i_{ut}, i_{vt},

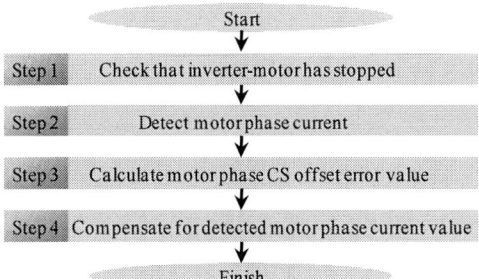

Fig. 3. Flow chart of general motor phase CS offset error compensation.

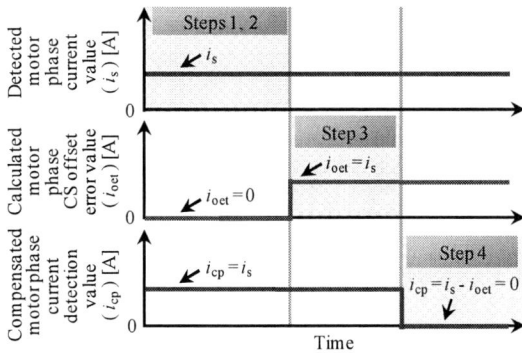

Fig. 4. Motor phase current detection value compensated for by general method.

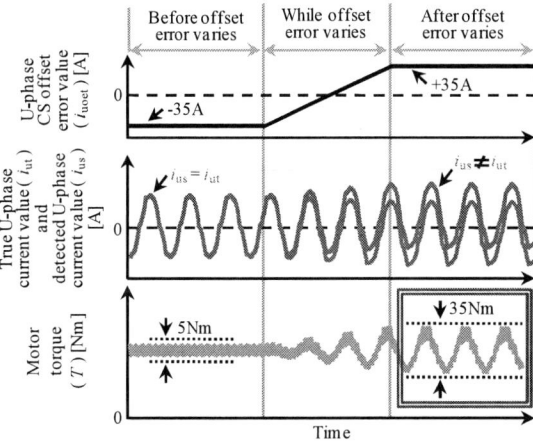

Fig. 5. Motor torque ripple with general motor phase CS offset error compensation method when U-phase CS offset error varies while VSI-motor is driven.

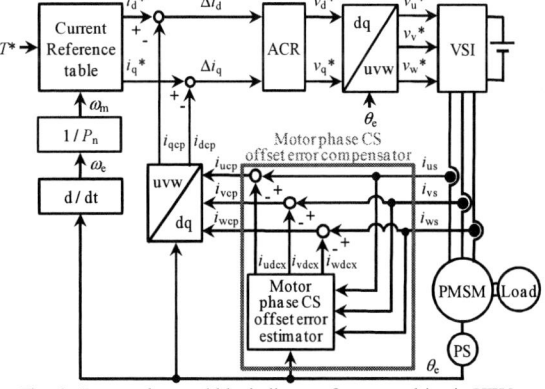

Fig. 6. Proposed control block diagram for motor drive in HEV.

i_{wt}) as shown in Eq. (6).

$$i_{ucp} = i_{us} - i_{udcx} \approx i_{ut}$$
$$i_{vcp} = i_{vs} - i_{vdcx} \approx i_{vt} \tag{6}$$
$$i_{wcp} = i_{ws} - i_{wdcx} \approx i_{wt}$$

Fig. 7 shows the block diagram inside the motor phase CS offset error estimator shown in Fig. 6. The motor phase CS offset error estimator consists of several direct current calculators (DCCs) per phase. The following describes the DCC operation for U-phase. The direct current calculator (DCC_1) calculates the U-phase direct current value (i_{udc1}) included in the detected U-phase current value (i_{us}) by using the discrete Fourier transform shown in Eq. (7).

$$i_{udc1} = \sum_{n=0}^{N-1} i_{us}(n) \Big/ N \tag{7}$$

where N is the number of the detected U-phase current value (i_{us}) sampled in 1-cycle of the magnetic pole position (θ_e) and the detected U-phase current value (i_{us}) is sampled by the DCC_1 whenever the magnetic pole position (θ_e) changes only the predetermined value ($\Delta\theta_e$) shown in Eq. (8).

$$\Delta\theta_e = 2\pi/m \quad (m = 2, 3, \cdots) \tag{8}$$

where m is the division number of 2π. In the proposed compensation method, to minimize the calculation amount of DCC_1, the division number (m) and the predetermined value ($\Delta\theta_e$) are set to 2 and π, respectively. Then, the U-phase direct current value (i_{udc1}) calculated by DCC_1 is represented by Eq. (9).

$$i_{udc1} = \{i_{us}(0) + i_{us}(1)\}/2 \tag{9}$$

where $i_{us}(0)$ is the U-phase current value sampled by DCC_1 before 1-operation-cycle of the DCC_1 and $i_{us}(1)$ is the newest U-phase current value sampled by DCC_1. The direct current value (i_{udc1}) is calculated by using DCC_1 whenever the magnetic pole position ($\Delta\theta_e$) changes only π.

Fig. 8 shows the detected U-phase current value (i_{us}) and the U-phase direct current value (i_{udc1}) calculated by DCC_1 when the amplitude of the detected U-phase current value (i_{us}) is constant. As shown in Fig. 8, the U-phase current value ($i_{us}(-2)$) sampled before 3-operation-cycle of the DCC_1 and the U-phase current value ($i_{us}(0)$) sampled before 1-operation-cycle of the DCC_1 have the same value when the amplitude of the detected U-phase current value (i_{us}) is constant. The U-phase current value ($i_{us}(-1)$) sampled before 2-operation-cycle of the DCC_1 and the newest U-phase current value ($i_{us}(1)$) sampled by the DCC_1 also have the same value. Here, Eq. (10) represents the differential values (i_{udifax}, i_{udifbx}) between the final U-phase direct current value (i_{udcx}) calculated by the number X of the DCCs and the true offset error value (i_{uoet}) included in the detected U-phase current value (i_{us}).

where X is the number of the used DCCs and Eq. (5) is the third order equation approximated on the basis of

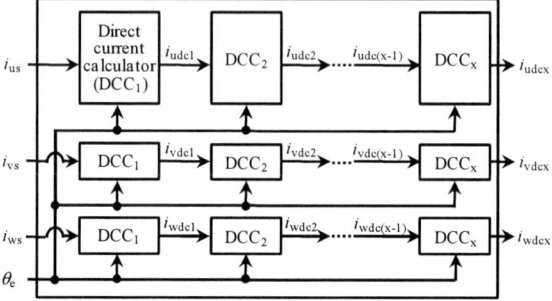

Fig. 7. Block diagram inside motor phase CS offset error estimator.

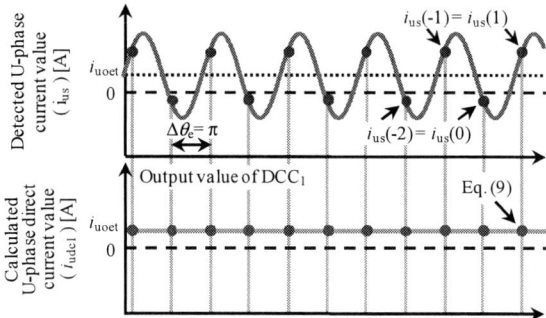

Fig. 8. Detected U-phase current value and U-phase direct current value calculated by DCC_1 when amplitude of detected U-phase current value is constant.

$$i_{udifax} \approx \frac{\{i_{us}(1) - i_{us}(-1)\}}{0.00258X^3 - 0.0931X^2 + 1.45X - 0.275}$$
$$i_{udifbx} \approx \frac{\{i_{us}(0) - i_{us}(-2)\}}{0.00258X^3 - 0.0931X^2 + 1.45X - 0.275} \tag{10}$$

a least-squares method and the output values of the 15 used DCCs obtained in the analysis conditions in which the detected U-phase current value (i_{us}) changes in step. Therefore, the U-phase direct current value (i_{udc1}) calculated by the DCC_1 coincides with the true offset error value (i_{uoet}) included in the detected U-phase current value (i_{us}). This is because the differential values (i_{udifa1}, i_{udifb1}) between the U-phase direct current value (i_{udc1}) calculated by DCC_1 and the true offset error value (i_{uoet}) included in the detected U-phase current value (i_{us}) are zero when the amplitude of the detected U-phase current value (i_{us}) is constant. However, if the amplitude of the detected U-phase current value (i_{us}) varies in 1-cycle of the magnetic pole position (θ_e), the direct current value (i_{udc1}) calculated by using only DCC_1 differs from the true offset error value (i_{uoet}) included in the detected U-phase current value (i_{us}) because the differential values (i_{udifa1}, i_{udifb1}) become large. Hence, the motor phase CS offset error estimator shown in Fig. 7 consists of several DCCs connected in series to make the differential values (i_{udifax}, i_{udifbx}) small. Furthermore, to calculate the direct current value (i_{udcx}) included in the detected U-phase current value (i_{us}) with high accuracy and a small number of the used DCCs, the motor phase CS offset error estimator sets the U-phase direct current value (i_{udcx}) calculated by the DCC_X as the final U-phase direct current value only when the values of numerators shown in Eq. (10) are small.

TABLE II
NUMBER OF DCCs REQUIRED TO MAKE DIFFERENTIAL VALUES LESS
THAN OR EQUAL TO PLUS-MINUS TWO AMPERE

Value of numerators shown in Eq. (10)	Number X of DCCs
+-2	1
+-4	2
+-6	3
+-8	4
+-10	6

TABLE III
SIMULATION PARAMETERS

Symbol	Meaning	Value
P_n	Number of pole pairs	4
φ_a	Magnet flux linkage	0.03Wb
L_d	d-axis inductance (at the point of 100 [A])	0.21mH
L_q	q-axis inductance (at the point of 100 [A])	0.37mH
$V_{vattery}$	Battery voltage	376V
$f_{carrier}$	Carrier frequency	10kHz
t_d	Dead time	4μs

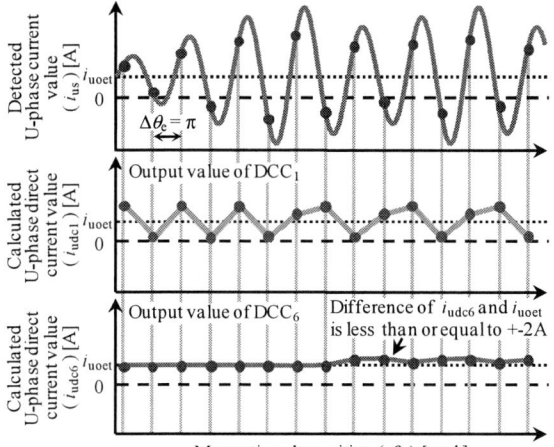

Fig. 9. Comparison of U-phase direct current value with one DCC
and U-phase direct current value with six DCCs.

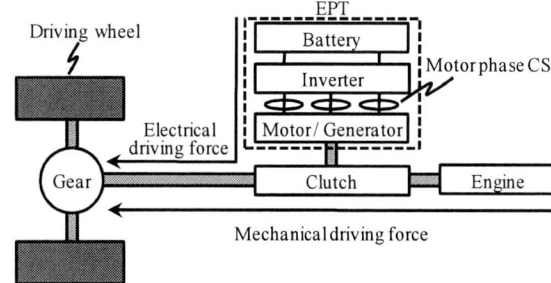

Fig. 10. Power train of HEV for simulation.

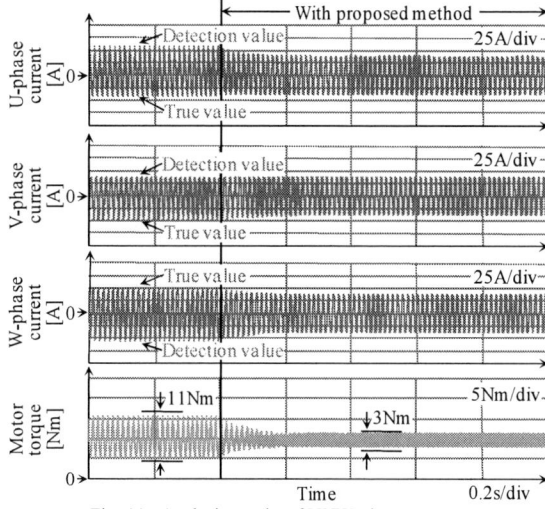

Fig. 11. Analysis results of UVW-phase current
and motor torque in steady state.

As an example, Table II shows the number X of the DCCs required to make the differential values (i_{udifax}, i_{udifbx}) less than or equal to +-2[A] corresponding to the value of numerators shown in Eq. (10). The number X of the DCCs is determined in consideration of the balance of the calculation amount, data amount memorized in a microcomputer's memory, and the calculation accuracy of the direct current value included in the detected motor phase current.

Fig. 9 compares the U-phase direct current value (i_{udc1}) with one DCC and the U-phase direct current value (i_{udc6}) with six DCCs while the amplitude of the detected U-phase current value (i_{us}) is varied at less than or equal to +-10[A]. The calculated U-phase direct current value (i_{udc1}) with one DCC includes the large ripple and greatly differs from the true offset error value (i_{uoet}) included in the detected U-phase current value (i_{us}). On the other hand, the ripple included in the calculated U-phase direct current value (i_{udc6}) with six DCCs is small, and the calculated U-phase direct current value (i_{udc6}) coincides with the true offset error value (i_{uoet}) included in the detected U-phase current value (i_{us}) with less than or equal to +-2[A] difference. Thereby, the proposed method can compensate for motor

phase CS offset error that varies due to temperature change during the VSI-motor drive and can reduce the motor torque ripple due to the motor phase CS offset error.

V. SIMULATION RESULTS

This section verifies the effectiveness of the proposed motor phase CS offset error compensation method for reducing the motor torque ripple. Fig. 10 shows the power train of the HEV for simulation, which is the parallel hybrid system. The EPT model consisted of the battery of 376[V], the general three-phase VSI, the interior permanent magnetic motor (IPMSM) of 18[kW], and the motor phase CS of the maximum offset error +-5[%] of 700A full scale. The UVW-phase CS offset error was set to 35[A], 30[A], and -5[A], and the sum total value was not zero. Moreover, the motor phase CS offset error estimator consisted of six DCCs per phase, and the motor phase CS offset error estimator set the direct current value (i_{udc6}) calculated by the DCC_6 as the final direct current value only when the values of numerators shown in Eq. (10) were less than or equal to +-10[A]. In addition, the DCCs calculate the direct current value included in the detected motor phase current value whenever the predetermined value ($\Delta\theta_e$) changes only π. The other parameters used in the simulation are listed in Table III.

Fig. 11 shows the analysis results of the true UVW-phase current value and the UVW-phase current detection

978-1-4799-2706-7/14 $31.00 © 2014 IEEE

value before and after compensation and the motor torque in steady state by which the motor rotation speed (ω_{m}) and the motor torque reference (T^{*}) were set as 1000[rpm] and 10[Nm], respectively. The compensated UVW-phase current detection value approaches the true UVW-phase value soon after the proposed compensation method is applied. After applying the proposed method, the compensated UVW-phase current detection value coincides with the true UVW-phase current value with less than or equal to +-2.0[A] difference in about 0.2 seconds. As a result, the motor torque ripple with the proposed compensation method is 3[Nm] and is reduced to about one-third (1/3) of that without the proposed method. However, the response of the motor phase CS offset error estimation changes with the motor rotation speed, and the response also becomes quick if the motor rotation is quick.

Fig. 12 compares the frequency analysis results of the motor torque from 0th-order to 15th-order shown in Fig. 11. The motor torque ripple of the fundamental frequency component is removed almost completely by using the proposed method. Remaining motor torque ripples are of 6th- and 12th-order-components resulting from the spatial harmonics of the permanent magnet flux and the dead-time of inverter. In this simulation case, the motor torque ripple of the fundamental component without the proposed method is 3.9[Nm], the motor torque ripple added with the 6th-order component and 12th-order component is 1.2[Nm], the percentage of the motor torque ripple resulting from the motor phase CS offset error is 76[%], and that resulting from the spatial harmonics of the permanent magnet flux and the dead-time of the inverter is 24[%].

Fig. 13 shows the motor torque ripple reduction results in steady state. Fig. 12 (A) compares the motor torque ripple with and without the proposed method when the motor rotation speed is 1000[rpm] and the motor torque reference is from 10[Nm] to 80[Nm]. In these conditions, the average value of the motor torque ripple with the proposed method is 10[Nm] and is reduced to half (1/2) of that without the proposed method. Fig. 13 (B) compares the motor torque ripple with and without the proposed method when the motor rotation speed is 2000[rpm] and the motor torque reference is from 10[Nm] to 80[Nm]. Also, in the same conditions, the average value of the motor torque ripple with the proposed method is 13[Nm] and is reduced to two-third (2/3) of that without the proposed method.

Fig. 14 compares the true UVW-phase current offset error value and the estimated UVW-phase current offset error value. The true UVW-phase current offset error value is varied from 35[A], 30[A], -5[A] to 25[A], 25[A], -10[A] in three seconds. The estimated UVW-phase current offset error value follows the true UVW-phase current offset error value. Finally, the estimated UVW-phase current offset error value coincides with the true UVW-phase current offset error value with less than or equal to +-2.0[A] difference. It also obtains the stable motor torque without the motor torque ripple of the

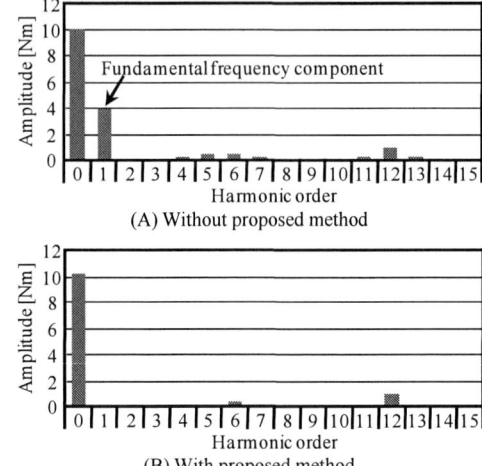

Fig. 12. Frequency analysis results of motor torque from 0th-order to 15th-order shown in Fig. 11.

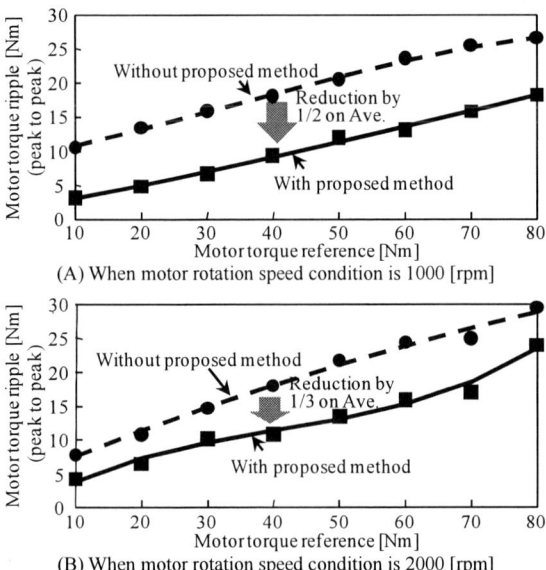

Fig. 13. Motor torque ripple reduction results in steady state.

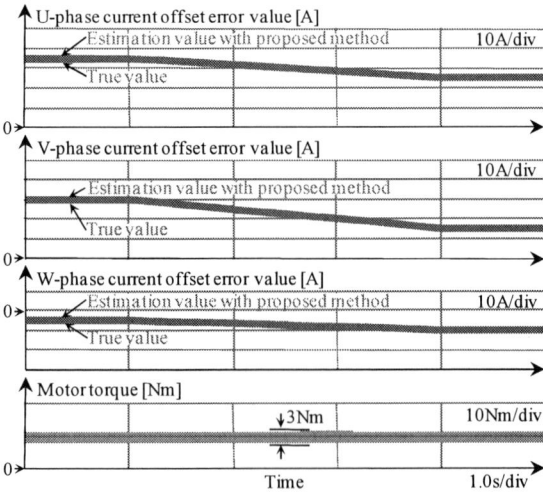

Fig. 14. Comparison of true UVW-phase current offset error value and estimated UVW-phase offset error value.

fundamental frequency component.

Figs. 15 and 16 show the torque ripple reduction results in the motor rotation speed variation and the motor torque reference variation, respectively. Also, in the conditions in which the motor rotation speed and the motor torque reference vary, the proposed method can compensate for the motor phase CS offset error normally and obtain the stable motor torque without the motor torque ripple of the fundamental frequency component.

These simulation results confirm that the proposed method reduces the motor torque ripple of the fundamental frequency component resulting from the motor phase CS offset error that varies due to temperature change while the VSI-motor is driven.

VI. CONCLUSION

This paper presented a novel compensation method for the motor phase current sensor offset error that varies due to temperature change while the VSI-motor is driven. The effectiveness of the proposed compensation method based on the simulation results is as follows.

- The motor phase current sensor offset error can be compensated for with less than or equal to +-2.0A difference even when the sum total value of the UVW-phase current sensor offset error is not zero.
- The motor torque ripple of the fundamental frequency component of the magnetic pole position (θ_e) resulting from the motor phase current sensor offset error is removed almost completely.
- In the steady condition of the motor torque reference from 10[Nm] to 80[Nm] and the motor rotation speed of 1000[rpm], the average value of the motor torque ripple with the proposed method is 10[Nm] and is reduced to half of that without the proposed method.

Consequently, the proposed method can effectively reduce the motor torque ripple resulting from the motor phase CS offset error.

REFERENCES

[1] G. Maggetto and J. Van Mierlo, "Electric and electric hybrid vehicle technology: a survey," *IEE Seminar Electric, Hybrid and Fuel Cell Vehicles*, pp.1-11, 2000.

[2] J. M. Miller, "Power electronics in hybrid electric vehicle applications," *Applied Power Electronics Conference and Exposition*, vol. 1, pp.23-29, 2003.

[3] G. Nanda and N. C. Kar, "A Survey and Comparison of Characteristics of Motor Drives Used in Electric Vehicles," *IEEE Canadian Conference on Electrical and Computer Engineering*, pp. 811-814, 2006.

[4] T. M. Jahans and W. L. Soong, "Pulsating Torque Minimization Techniques for Permanent magnet AC Motor Drives – A Review," *IEEE Trans. Industrial Electronics*, vol.43, no.2, pp.321-330, 1996.

[5] K. Hiramoto, Y. Takeda, M. Sanada, and S. Morimoto, "Torque Ripple Reduction of Reluctance Torque Assisted Motors Using Asymmetric Flux Barriers," *IEEJ Trans. Industry Applications*, vol. 124, no.2, pp. 208-214, 2004 (in Japanese).

[6] G. H. Kang, Y. D. Son, G. T. Kim, and J. Hur, "A Novel Cogging Torque Reduction Method for Interior-Type Permanent-Magnet Motor," *IEEE Trans. Industrial Electronics*, vol.45, no.1, pp. 161-167, 2009.

[7] Y. Kawano and A. Kawamura, "Analysis on Current Regulation for Torque Control for a Very Distorted CEMF Type IPMSM" *Power Conversion Conference*, pp.694-698, 2007.

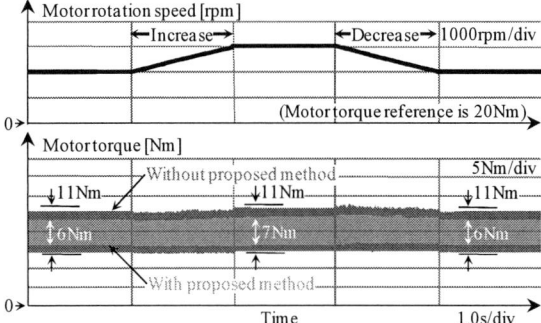

Fig. 15. Motor torque ripple reduction results in motor rotation speed variation.

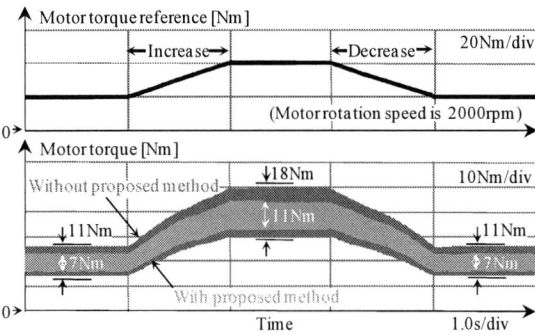

Fig. 16. Motor torque ripple reduction results in motor torque reference variation.

[8] H. Hida, Y. Tomigashi, K. Inoue, and S. Morimoto, "New Torque Estimation Method Considering Spatial Harmonics and Torque Ripple Reduction in Permanent Magnet Synchronous Motors," *IEEJ Trans. Industry Applications*, val. 130, no.9, pp.1051-1058, 2010 (in Japanese).

[9] N. Nakao and K. Akatsu, "A New Control Method for Torque Ripple Compensation of Permanent Magnet Motors," *International Power Electronics Conference*, pp.1421-1427, 2010.

[10] N. Nakao and K. Akatsu, "Torque Ripple Control for Synchronous Motors Using Instantaneous Torque Estimation," *IEEE Energy Conversion Congress and Exposition*, pp.2452-2459, 2011.

[11] H. S. Kim, H. T. Moon, and M. J. Youn, "On-line dead-time compensation method using disturbance observer," *IEEE Trans. on Power Electronics*, vol. 18, no. 6, pp. 1136-1345 (2003).

[12] N. Urasaki, T. Senjyu, K. Uezato, and T. Funabashi, "An Adaptive Dead-Time Compensation Strategy for Voltage Source Inverter Fed Motor Drives," *IEEE Trans. on Power Electronics*, vol. 20, no.5, pp. 1150-1160, 2005.

[13] H. Tamura, Y. Ajima, and Y. Noto, "A Torque Ripple Reduction Method by Current Sensor Offset Error Compensation," *European Conference on Power Electronics and Applications*, paper number 0185, pp.1-10. 2013.

The 2014 International Power Electronics Conference

Investigation of Calculation Method of Losses in PWM Inverter with Voltage Booster using both DC link Voltage Control and Flux Weakening Control

Akihiro Imakiire, Masayuki Hikita
Department of Electrical and Electronic Engineering,
Kyushu Institute of Technology
Kitakyushu, Japan

Kichiro Yamamoto, Ryo Yonemori
Department of Electrical and Electronics Engineering,
Kagoshima University
Kagoshima, Japan

Abstract— **In this paper, authors investigate calculation method of losses in system which consists of interior permanent magnet (IPM) motor and PWM inverter with voltage booster. This system has two driving methods for motor in high speed region. One is DC link voltage Control. The other is flux weakening control. However, it is not cleared how combination of these two control methods is suitable for this system. It is necessary to quantitatively estimate parts of system losses in voltage booster, inverter and IPM motor, because total efficiency of system is decided by parts of system losses. So, the simulation model, which is capable of calculating parts of system losses to find suitable driving method for IPM motor using combination of the two control methods, is introduced and simulation results are discussed.**

Keywords— interior permanent magnet motor, voltage booster, DC link voltage control, flux weakening control.

I. INTRODUCTION

Recently, permanent magnet (PM) motors form an increasingly important class of high performance ac motor drive and are used for not only industrial applications but also electric vehicles and other applications.

Authors have been investigated inverter motor drive system using low voltage supply such as lead-acid battery, fuel cell. Especially, they investigated the system limited by volume or weight to realize miniaturization and high efficiency [1] [2].

In this paper, simulation model, which is capable of calculating losses in voltage booster, inverter and PM motor to find operating condition of driving PM motor at high efficiency, is built and simulation results are discussed.

II. CALUCULATION METHOD OF PM MOTOR LOSS

Fig.1 shows the circuit configuration for IPM motor drive. This system consists of an IPM motor, a conventional PWM inverter, one bidirectional voltage booster and battery. This system has two driving methods for motor in high speed region. PM motor can be driven by using DC link voltage control with voltage booster. Alternatively, PM motor can be also driven by using flux weakening control with inverter.

In this section, we describe calculating method of IPM motor loss in using both DC link voltage control and flux weakening control.

Table I shows the parameters of an IPM motor. This IPM motor is concentrated winding motor (called D model) which is based on information of investigating R&D committee on practical performance evaluation techniques of rotating machines by electromagnetic field analysis [3].

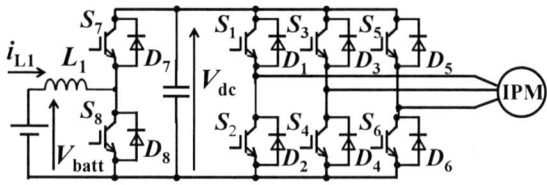

Fig. 1. Circuit configuration of inverter with voltage booster.

TABLE I
Parameters of IPM motor.

Interior Permanent Magnet motor		
Rated output power	-	940 W
Rated rotational speed	-	4800 r/min
Rated torque	-	1.87 N m
Rated voltage	-	165 V
Rated current	-	5 A
Pole number	P	4
Flux of field	Φ_a	0.11 Wb
Armature resistance	R_a	0.44 Ω
d-axis inductance	L_d	12 mH
q-axis inductance	L_q	20 mH

Fig.2 shows torque and output power versus rotational speed. First, it is assumed that the rated rotational speed is 4800 r/min as shown in the figure. Then, constant torque region was chosen from 0 r/min to 4800 r/min and constant power region was from 4800 r/min to 14400 r/min. As a result, the ratio of base speed and maximum speed was one to three. These are specification of the IPM motor to examine.

Next, the calculation method of the IPM motor loss is described.

At first, a calculation method of the iron loss P_i in the IPM motor is explained. As a results, the iron loss P_i can

978-1-4799-2706-7/14 $31.00 © 2014 IEEE

689

be calculated from equation (1) using hysteresis factor and eddy current factor as shown in Fig.3.

Fig.3 shows dependence of hysteresis factor and eddy current factor of the IPM motor on magnetic flux density B_m (35A300).

$$P_i = (K_h \cdot f \cdot B_m^2 + K_e \cdot f^2 \cdot B_m^2) \times M \qquad (1)$$

K_h hysteresis factor
K_e eddy current factor
f frequency of flux density
B_m flux density
M weight

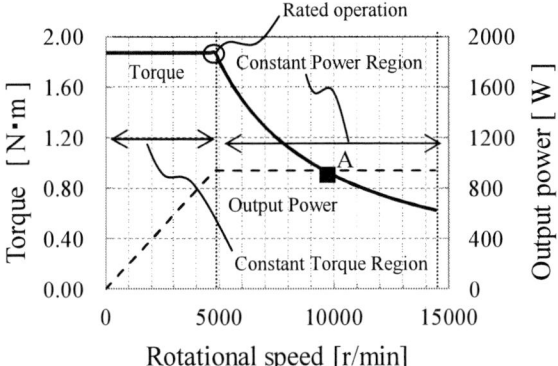

Fig. 2. Torque and output power vs. rotational speed characteristics.

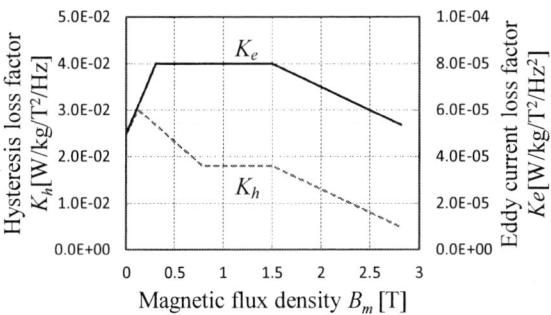

Fig. 3. Iron loss coefficients of 35A300 characteristic.

In fact, flux linkage of the IPM motor Φ_0 is calculated from equation (2), on the other hands, hysteresis factor and eddy current factor are function of flux density B_m. So, the flux linkage of the IPM motor Φ_0 on the rated operation is defined as 1.5 T (100% Φ_0). From this, hysteresis factor and eddy current factor change depending on the driving condition of the IPM motor.

Therefore, iron loss P_i changes depending on the driving condition of the IPM motor from equation (1).

$$\Phi_0 = \sqrt{(\varphi_a + L_d i_d)^2 + (L_q i_q)^2} \qquad (2)$$

φ_a flux linkage
L_d d-axis inductance
L_q q-axis inductance
i_d d-axis current
i_q q-axis current

Secondly, a calculation method of the copper loss P_c in the IPM motor is explained. Copper loss P_c can be calculated from equation (3) - (5), when maximum torque per ampere control (called MTPAC) is applied. Then, equation (3) is d-axis current reference for MTPAC, and, equation (4) is torque of the IPM motor.

$$i_d = \frac{\varphi_a}{2(L_q - L_d)} - \sqrt{\frac{\varphi_a^2}{4(L_q - L_d)^2} + i_q^2} \qquad (3)$$

$$T = \frac{p}{2}\left\{ \varphi_a i_q + (L_d - L_q) \, i_d i_q \right\} \qquad (4)$$

p pole

$$P_c = R_a (i_d^2 + i_q^2) \qquad (5)$$

Similarly, copper loss P_c can be calculated from equation (4) - (6), when flux weakening control (called FWC) is applied. Then, equation (6) is d-axis current reference for FWC.

$$i_d = \frac{-\varphi_a + \sqrt{\left(\dfrac{V_{om}}{\omega_e}\right)^2 + (L_q i_q)^2}}{L_d} \qquad (6)$$

V_{om} voltage limit of inverter
ω_e electrical angular frequency

Then, mechanical loss P_m in the IPM motor is assumed a constant value of 5.7 W at 1500 r/min based on D model IPM motor.

Finally, total loss of the IPM motor is obtained from equation (7).

$$P_{motor\_Loss} = P_i + P_c + P_m \qquad (7)$$

P_i iron loss
P_c copper loss
P_m mechanical loss

III. CALCULATION METHOD OF INVERTER AND VOLTAGE BOOSTER LOSS

First, the same kinds of power modules are assumed for inverter and voltage booster simply in simulation. Therefore, calculation method about inverter loss and voltage booster loss are fundamentally same as that for power module loss. But, in the voltage booster, copper loss of the DC reactor is added to the power module loss. However, iron loss of the reactor is not considered in this paper.

Then, power module loss is constructed by a switching loss and a conduction loss. Calculation method of IGBT (Insulated Gate Bipolar Transistor) loss and FWD (Free Wheeling Diode) loss are fundamentally same. IGBT loss can be calculated from equation (8) to (10) in the subsection of *A*. FWD loss can be calculated from equation (11) to (13) in the subsection of *B*.

A.   IGBT Loss Calculation

$$P_{SW\_Loss\_IGBT} = \left\{ E_{on}(I_c) + E_{off}(I_c) \right\} f_{SW} \, V_{dc} / V_{data} \qquad (8)$$

$$P_{CON\_Loss\_IGBT} = V_{ce}(I_c) \, I_c \, Duty_{IGBT} \qquad (9)$$

$$P_{IGBT\_Loss} = P_{SW\_Loss\_IGBT} + P_{CON\_Loss\_IGBT} \qquad (10)$$

$P_{sw\_Loss\_IGBT}$	switching loss
$P_{con\_Loss\_IGBT}$	conduction loss
P_{IGBT_Loss}	IGBT loss
$E_{on}(I_c)$	turn on loss
$E_{off}(I_c)$	turn off loss
f_{sw}	switching frequency
V_{dc}	DC link voltage
V_{data}	DC link voltage
	(Value of data sheet or switching test)
I_c	collector current
$V_{ce}(I_c)$	collector - emitter voltage
$Duty_{IGBT}$	ON duty

where $E_{on}(I_c)$, $E_{off}(I_c)$ and $V_{ce}(I_c)$ are function of the collector current I_c.

B.   FWD Loss Calculation

$$P_{SW\_Loss\_FWD} = E_{rr}(I_F) \, f_{SW} \, V_{dc} / V_{data} \qquad (11)$$

$$P_{CON\_Loss\_FWD} = V_F(I_F) \, I_F \, Duty_{FWD} \qquad (12)$$

$$P_{FWD\_Loss} = P_{SW\_Loss\_FWD} + P_{CON\_Loss\_FWD} \qquad (13)$$

$P_{sw\_Loss\_FWD}$	switching loss
$P_{con\_Loss\_FWD}$	conduction loss
P_{FWD_Loss}	FWD loss
$E_{rr}(I_F)$	reverse recovery loss
f_{sw}	switching frequency
V_{dc}	DC link voltage
V_{data}	DC link voltage
	(Value of data sheet or switching test)
I_F	forward current
$V_F(I_F)$	forward voltage
$Duty_{FWD}$	ON duty

where $E_{rr}(I_F)$ and $V_F(I_F)$ are function of the forward current I_F.

Finally, inverter loss and voltage booster loss can be calculated from equation (14) and (15).

$$P_{inv\_Loss} = \left(P_{IGBT\_Loss} + P_{FWD\_Loss} \right) \times 6 \qquad (14)$$

$$P_{chop\_Loss} = \left(P_{IGBT\_Loss} + P_{FWD\_Loss} \right) + P_{chop\_DC\_Loss} \qquad (15)$$

$P_{chop\_DC\_loss}$	copper loss of the DC reactor

Table Ⅱ lists the parameters of the power module made by Fairchild semiconductor, and the characteristics on data sheet is used in the simulation.

TABLE Ⅱ
Parameters of the power module used in the simulation.

FAIRCHILD SEMICONDUCTOR	
Rated voltage	1200 V
Rated current	30 A
Maximum power Dissipation	186 W
Appearance of Power module	
Length×Width 21.0×15.8 [mm]	

IV. SIMULATION MODEL IN CONSIDERATION OF PARTS OF SYSTEM LOSS

A.   Simulation model and Simulation condition

Fig.4 shows simulation model in consideration of the parts of system loss which is shown by equation (7), (14) and (15). These losses can be calculated, and, considered as the equivalent current sources connected parallel to the DC link capacitor as shown in the figure. These current sources can be calculated from equation (16), (17) and (18). The power module parameters are based on the data sheet of the power module listed the table Ⅱ.

In this way, simulation model, which is capable of calculating parts of system losses to find suitable driving method for IPM motor using combine two control methods, was created.

$$i_{pm} = P_{motor\_Loss} / V_{dc} \qquad (16)$$

$$i_{inv} = P_{inv\_Loss} / V_{dc} \qquad (17)$$

$$i_{chop} = P_{chop\_Loss} / V_{dc} \qquad (18)$$

Table Ⅲ shows parameters of the simulation model. The battery voltage is set 100 V to use both DC link voltage control and flux weakening control as shown in the table. Rated operation of the PM motor was assumed that rated speed is 4800 r/min and rated torque is 1.87 N m to investigate of the system loss and system efficiency using two control methods as shown in the Fig. 2. Switching frequencies of inverter and voltage booster in the table Ⅲ are used as the power module parameter as shown in Fig. 4.

Fig.5 shows the definition amount of the controlled voltage using two control methods. Vertical axis represents controlled voltage and horizontal axis represents %DVC. DC link voltage can be controlled within the range from power supply voltage to limit of DC link voltage as shown in the figure. So, operation point can be also controlled in this range using the two control methods. In short, operation point means DC link voltage.

%DVC is defined as the ratio of the controlled voltage by DVC versus standard controlled voltage. Controlled voltage by DVC is defined as the value from power supply voltage to operation point as shown in the figure. Standard controlled voltage is defined as the value from power supply voltage to required voltage V_{dc_ref} for PM

The 2014 International Power Electronics Conference

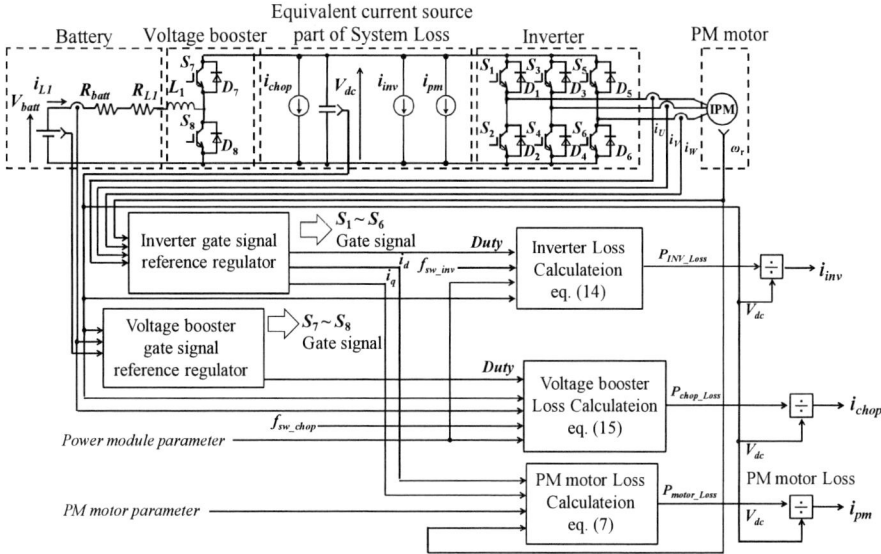

Fig. 4. Simulation circuit in consideration of parts of system loss.

TABLE III
Parameters of Simulation model.

Voltage booster and PWM inverter		
Battery voltage	V_{batt}	100V
Equivalent internal resistance of battery	R_{batt}	330mΩ
DC link Capacitor	C_{dc}	1500uF
Inductance of DC reactor	L_1	5mH
Resistance of DC reactor	R_{L1}	0.12Ω
Controller		
Switching frequency of inverter	f_{sw_inv}	5kHz
Switching frequency of voltage booster	f_{sw_chop}	8kHz

motor as shown in the figure. And, %FWC is also defined as the ratio of the controlled voltage by FWC versus standard controlled voltage. Controlled voltage by FWC is defined as the value from operation point to required voltage V_{dc_ref} for PM motor as shown in the figure.

Required voltage V_{dc_ref} for PM motor can be calculated from (19).

Therefore, %DVC, %FWC and standard controlled voltage relationships are represented as equation (20) and (21). As the result, %FWD decreases according to increase of the %DVC.

$$V_{dc\_ref} = \sqrt{2}\sqrt{V_d^2 + V_q^2} \qquad (19)$$

V_d d-axis voltage
V_q q-axis voltage

$$\%DVC = V_{OS}/V_{PS} \times 100 \qquad (20)$$

$$\%FWC = V_{PO}/V_{PS} \times 100 \qquad (21)$$

V_{PS} standard controlled voltage
V_{OS} value from power supply voltage to operation point
V_{PO} value from operation point to required voltage for PM motor

Fig.6 shows simulation condition. Vertical axis represents the ratio of %DVC and %FWC. In this case, %DVC increases from 29% to 73%, on the other hands, %FWC decreases from 71% to 27% according to changing the simulation condition from A-1 to A-4.

Then A-1, A-2, A-3 and A-4 are the simulation condition of the operation point A shown as Fig. 2. In operation point A, rotational speed is 9600 r/min, and torque is 0.94 N m. The operation point is on the constant power region. So, both DVC and FWC can be used. Then A-1, A-2, A-3 and A-4 are different control ratio of %DVC and %FWC as shown in Fig. 6. So, control ratio of %DVC and %FWC changed to realize two control methods. And value of the DC link voltage is represented for each simulation condition as shown in the figure. It should be note that on the operation point A-3 and A-4 PM motor cannot be driven because DC link voltage is over the limit value about 230 V.

B.   Simulation results and discussion

Fig.7 shows simulation results. It represents system efficiency and parts of system efficiency for voltage booster, inverter and PM motor on the simulation condition shown in Fig. 6. Vertical axis represents efficiency and horizontal axis represents DC link voltage. Dotted line indicates the limit of DC link voltage.

At first, system efficiency more increases when DC link voltage more increases from this result. Then, it reaches peak at the DC link voltage is around 260 V.

After that, it decreases according to increase of DC link voltage. These results can be explained about

978-1-4799-2706-7/14 $31.00 © 2014 IEEE

The 2014 International Power Electronics Conference

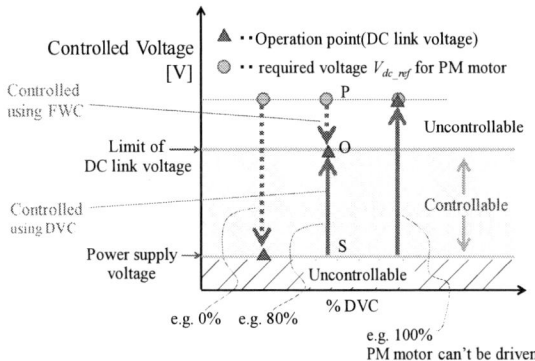

Fig. 5. Definition of control voltage using both DVC and FWC.

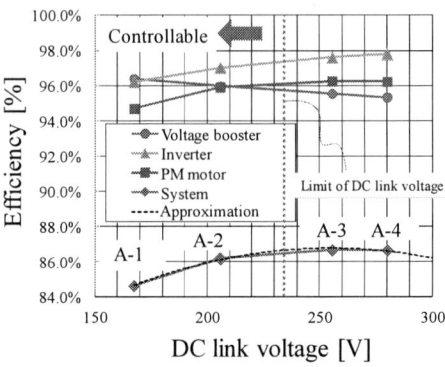

Fig. 7. Simulation results about system efficiency.

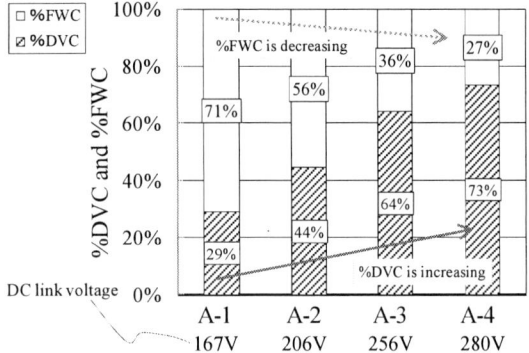

Fig. 6. Simulation condition.

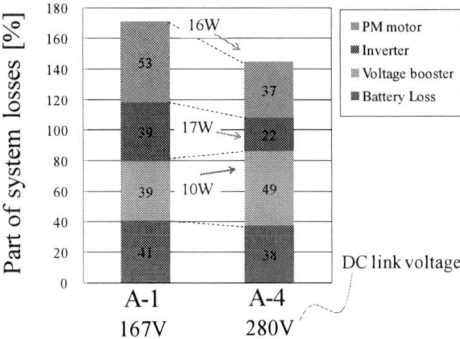

Fig. 8. Simulation results about system loss.

investigating the parts of system efficiency.

Then, focusing on the parts of system efficiency, the efficiencies of the PM motor and the inverter increase according to increase of DC link voltage. The efficiencies have a tendency to be saturated, and they will be near constant values. On the other hands, the efficiency of voltage booster decreases.

So, in the simulation case, the more high efficiency, this PM motor drive system has the more %DVC value is. However, DC link voltage is limited at about 230V as deadline shown in Fig. 7. Therefore, it is better that DC link voltage is controlled around the 230 V near limit value in this PM motor drive system using DVC and FWC. The value of %DVC will be better between 44% and 64% as shown in Fig. 6. These results can be explained from investigating of parts of system loss.

Fig.8 shows simulation results. It represents system loss and parts of it on the simulation condition A-1 and A-4 as shown in the figure. Where, battery loss is that loss of the interior resistance in the battery. It is clear that %DVC increase gives the great advantage for the PM motor drive system in this simulation case, because voltage booster loss increases but inverter and PM motor loss more decrease according to increase of DC link voltage.

V. CONCLUSION

In this paper, authors created simulation model capable

of calculating parts of system losses to find suitable driving method for IPM motor using combination DC link voltage control and flux weakening control, and the following conclusion was obtained.

Parts of system loss and PM motor drive system loss can be calculated using this simulation model. So, if PM motor or power module is changed for applying to the application. Operation points at the more high efficiency for the PM motor drive system can be found from investigating by this simulation model.

References

[1] K. Yamamoto, A. Imakiire, K. Iimori, "PWM inverter with voltage boosters with regenerating capability augmented by electric double-layer capacitor," *IEEJ Journal of Industry Applications*, vol. 131, No. 5, pp. 671-678, 2011.

[2] K. Yamamoto, K. Shinohara, A. Imakiire, "Steady state characteristics of PWM inverter with voltage boosters for permanent magnet synchronous motor drives," *Power Conversion Conference - Nagoya 2007*, pp. 296-301, 2007.

[3] Investigating R&D Committee on Practical Performance Evaluation Techniques of Rotating Machines by Electromagnetic Field Analysis, "Practical Performance Evaluation Techniques of Rotating Machines by Electromagnetic Field Analysis," *IEEJ* , vol. 1244, pp. 1-87, 2012.

Dynamic and Steady-State Behavior of a Paralleling Three-Phase AC-to-DC Converter with Reduced DC Bus Capacitor

Uthen Kamnarn
Department of Electrical Engineering,
Faculty of Engineering,
Rajamangala University of Technology
Lanna, RMUTL,
Chiang Mai, Thailand
uthen@rmutl.ac.th

Yutthana Kanthaphayao
Department of Electrical Engineering,
Faculty of Engineering and Architecture,
Rajamangala University of Technology
Suvarnabhumi, RMUTSB,
Nonthaburi, Thailand
yutthana_k44@yahoo.com

Viboon Chunkag
Department of Electrical Engineering,
Faculty of Engineering,
King Mongkut's University of Technology
North Bangkok, KMUTNB,
Bangkok, Thailand
vck@kmutnb.ac.th

Abstract — **Dynamic and steady-state behavior of a paralleling three-phase AC-to-DC converter with reduced dc bus capacitor and nearly unity power factor is investigated in this paper. The circuit is simulated and the simulation results are analyzed and verified by measurements. A 1.5 kW laboratory prototype with small dc bus capacitance 690 µF (0.46 µF /W) is implemented and tested to verify the feasibility of the proposed system. Steady-state and dynamic behavior is the other main advantage of the proposed system. The possibility of using, in the proposed system, a reduced ratio capacitance/watt lower than the typical values used in commercial applications, while maintaining the accurate input current and output voltage regulation, is also theoretically proved.**

Keywords— Small DC Bus Capacitor, Fast Response, Parallel Configuration, Power Factor Correction.

I. INTRODUCTION

DC bus in any Distributed Power System (DPS) is normally equipped with an electrolytic capacitor, which provides decoupling between the rectifier and the load converter. Nevertheless, the dc bus capacitor is a large, heavy, and expensive component [1].

The electrical performance requirements play a big part in determining the amount of capacitance required. The transient requirements of the proposed system are very important. The load transient amplitude, voltage deviation requirements, energy transfer, and capacitor impedance each affects capacitor selection. Other important issues to consider are minimizing PCB area, weight, size and capacitor cost. This problem is well recognized by the industry. Manufacturers of low cost and high volume systems are looking for ways to reduce the size and cost of this electrolytic capacitor [2-3].

This paper presents, an investigation of reduced dc bus capacitor in parallel operation of a three-phase ac-to-dc converter using single-phase isolated CUK rectifier module with nearly unity-power factor. The proposed system is implemented. It will result in the following advantages: small capacitor size, reduced converter volume, high power factor, good steady-state behavior and fast dynamic transient response.

This work was supported by the Commission on Higher Education (CHE), Rajamangala University of Technology Lanna (RMUTL) and the Thailand Research Fund (TRF), Under Contract MRG5480098.

II. THE PROPOSED SYSTEM

A. Power Circuit

The circuit diagram for the proposed three-phase configuration is shown in Fig. 1(a). L_{A1}, L_{B1} and L_{C1} represent the input inductors and L_{A2}, L_{B2} and L_{C2} are the output inductors for the respective three phases. C_{A1}, C_{A2}, C_{B1}, C_{B2}, C_{C1} and C_{A2} represent the transfer capacitors. v_{AN}, v_{BN} and v_{CN} and v_o represent input phase voltages for the three phases and output voltage, respectively. *IGBT* and *D* represent active switches and freewheeling diodes, respectively. A single capacitor C_o is connected at the output terminals for filtering the output voltage ripple.

The analytical model derivation is based on the power stage schematic shown in Fig. 1(a).

$$\sum_{i=A}^{C} \sum_{y=1}^{N} V_{gi} I_{L_{1yi}} = V_o I_o \qquad (1)$$

V_{gi} are the RMS value of rectified voltages and $I_{L_{1yi}}$ represent the RMS value of input inductor currents. V_o and I_o is average value of the output voltage and current, respectively.

The peak value of the inductor current is

$$\sum_{i=A}^{C} \sum_{y=1}^{N} V_{gi} I_{L_{1yi}} = k_s V_o I_{load} \qquad (2)$$

$$\hat{I}_{L_{1yi}} = \frac{k_{rms} V_o I_{load}}{3 N V_{gi}} \qquad (3)$$

Based on this theory, the reference value of the inductor current is

$$\hat{I}_{Lref_{1yi}} = \hat{I}_{L_{1yi}} + I_{VR} \qquad (4)$$

Here, $\hat{I}_{Lref_{1yi}}$ and I_{VR} represent the reference value for the peak value of input inductor currents and correcting signal from proportional-integral (PI) controller, respectively. A suitable PI controller is used which regulates the output voltage to the reference value. The dynamic equation of the output voltage is

$$\sum_{i=A}^{C} \sum_{y=1}^{N} n I_{oyi} = C_o n^2 \frac{d(V_o / n)}{dt} + n I_{load} \qquad (5)$$

Applying the perturbations in (1), (3), (4), and (5), and performing the small-signal approximation ($\tilde{v} \cdot \tilde{v} = 0$). Therefore, the output voltage can be expressed as

$$\tilde{v}_o = T_C \tilde{v}_{oref} + T_A \tilde{v}_{g_i} - Z_o \tilde{i}_{load} \qquad (6)$$

The transfer function of Fig. 1 can be obtained from the average small-signal model proposed (T_A and $Z_o = 0$):

$$\frac{\tilde{v}_o}{\tilde{v}_{oref}} = \frac{G_{VR}(s)\left(3Nk_{rms}\overline{V}_{gz}\Big/\overline{V}_o\right)(1/sC_o)}{1 + G_{VR}(s)\left(3Nk_{rms}\overline{V}_{gz}\Big/\overline{V}_o\right)(1/sC_o)k_{fb}} \qquad (7)$$

B. Control Circuit

This section presents the design of the proposed controller. The controller is comprised of six inner current controllers for waveshaping the input inductor current and a dc output voltage loop using PI controller to regulate the output voltage as shown in Fig. 1(b). A proportional integral compensator is selected for providing output voltage regulation. The output voltage (v_o) is sensed and compared with reference output voltage (v_{oref}) and the error is corrected by the PI controller to maintain the output voltage constant.

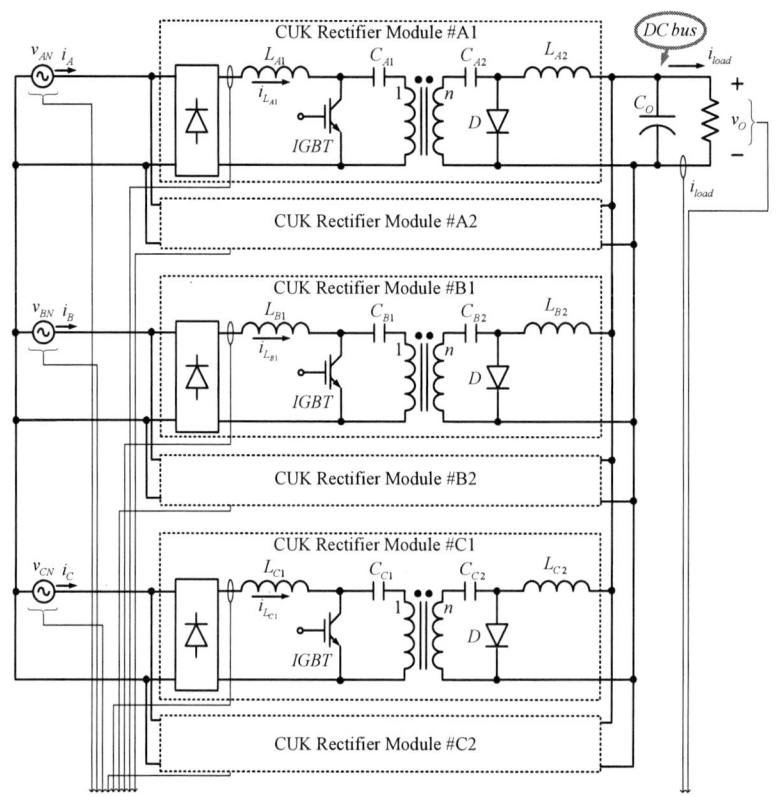

(a) Power circuit of the paralleling three phases AC to DC converter using CUK rectifier modules

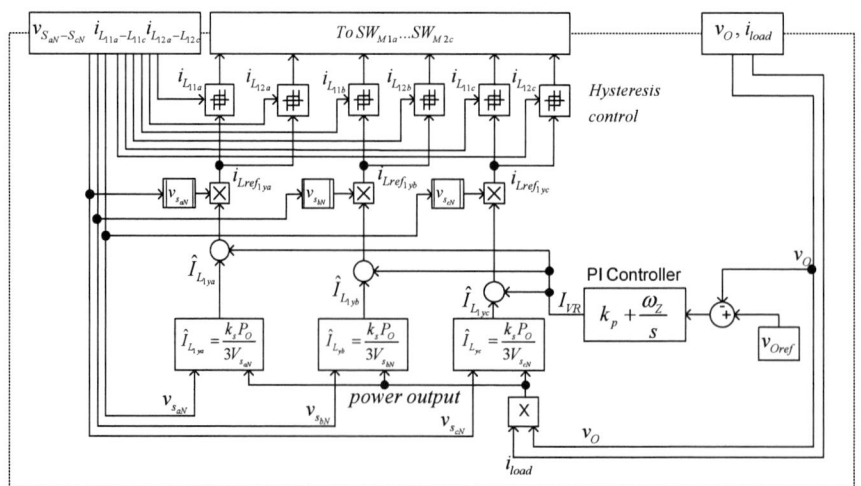

(b) Control circuit of the proposed system

Fig. 1 The proposed system: parallel operation of a modular three-phase ac-to-dc converter using six isolated CUK rectifier modules.

The values of K_p and ω_z are found by frequency response analysis and design method. In this method, the values for K_p and ω_z are found to be 1 and 150, respectively. Reference value for the peak value of input inductor currents are generated using the correcting signal from the PI controller, input phase voltages v_{AN}, v_{BN} and v_{CN}, load currents i_{load} and output voltage v_O. The instantaneous values of actual input inductor current and reference inductor current are compared using hysteresis control. The result of comparison is the PWM signals, which are used for driving the modular CUK converter switches SW_{M1a} to SW_{M2c}.

C. Analysis and Design of a DC Bus Capacitor

The value of energy-storage capacitor C_o is indicated in (8).

$$C_o \geq \frac{2P_o \Delta th}{V_o^2 - V_{o,\min}^2} \quad (8)$$

The partial is replaced by ordinary derivative equation as follows.

$$\frac{\partial^2 \Delta th}{\partial V_o^2} = \frac{d^2 \Delta th}{dV_o^2} = 0 \quad (9)$$

From equation (9) integrate twice will get the result as follows.

$$\Delta th = A V_o + B \quad (10)$$

From equation (10) A and B are the constant value of a second order differential equations. The hold-up time as shown in (10) is depended on the output voltage, the constant value (A and B). This is considered by the boundary conditions as follows: the tight output voltage, minimum overshoot and undershoot of output voltage and the stability of the whole system. The crossover frequency of the total open-loop transfer function should be well below the switching frequency. Next, the hold-up time is considered to minimize DC bus capacitor as follows.

$$\Delta th = \frac{1}{f_{sc}} \quad (11)$$

Where f_{sc} is the crossover frequency.

The parameter value for the small dc output capacitor calculation are as follows: $V_o = -48\text{V} \pm 5\%$, $P_o = 1.5\text{kW}$, $\Delta V_{droop} = \Delta V_{overshoot} = 5\%$ and $f_{sw} = 30\text{kHz}$.

We can calculate the value of the output capacitance as indicated in (8) as follows.

In case $\Delta th = \dfrac{4}{f_{sw}}$

$$C_o = \frac{8P_o}{f_{sw}(V_o^2 - V_{o,\min}^2)} = 1780.63\mu\text{F} \quad (12)$$

In case $\Delta th = \dfrac{4V_o}{f_{sw}\Delta V_{droop}}$

$$C_o = \frac{8P_o V_o}{f_{sw}V_{droop}(V_o^2 - V_{o,\min}^2)} = 1874.34\mu\text{F} \quad (13)$$

In case $\Delta th = \dfrac{4V_o}{f_{sw}\Delta V_{overshoot}}$

$$C_o = \frac{8P_o V_o}{f_{sw}V_{overshoot}(V_o^2 - V_{o,\min}^2)} = 1695.84\mu\text{F} \quad (14)$$

In practice, pick output capacitance C_o in difference values are $C_{o,min} = 470\mu\text{F}$ and $C_{o,max} = 28,270\mu\text{F}$ for approve and confirm the design procedure of the output DC capacitance. Then the calculated parameter values are listed in the Table I.

III. SIMULATION RESULTS

The configuration of a parallel connection of a single-stage three-phase ac to dc converter using six isolated CUK rectifier modules based on power balance control technique is simulated using MATLAB/SIMULINK program. The simulation results of the proposed scheme are shown in Fig. 2-5.

Fig. 2 shows the load current and dc bus voltage waveforms of the proposed converter at dynamic load condition that is obtained, respectively. It is shown that the reduced-capacitance based power balance controller is capable of offering good dynamical response to parallel operation of three-phase ac-to-dc converter using CUK rectifier module, with $2,470\mu\text{F} \leq C_o \leq 28,270\mu\text{F}$, little overshoots and droops, when the output power is alternated between 0.36 kW and 1.5 kW.

Fig. 3 shows the simulation results of the proposed system for individual input phase currents i_A, i_B, i_C, input inductor currents for six phases i_{LA11}, i_{LA12}, i_{LB11}, i_{LB12}, i_{LC11}, i_{LC12}, load current I_o and output voltage V_o for symmetric mains condition, module loss in phase A and B, unbalance phase voltage, voltage sage and over voltage condition.

TABLE I
DESIGN PARAMETERS FOR THE PROPOSED SYSTEM

Parameters	Design values
RMS value of input phase voltages ($V_{AN}=V_{BN}=V_{CN}$)	220V $\pm 15\%$
Line frequency (f_s)	50 Hz
Output voltage (V_o)	-48 V
Rated Power per module (P)	250 W/module
Output power (P_o)	1.5 kW
Input inductor (L_{A11}, L_{A21})	5.08 mH, 1.06 mH
Input inductor (L_{B11}, L_{B21})	5.06 mH, 1.08 mH
Input inductor (L_{C11}, L_{C21})	5.06 mH, 1.04 mH
Output inductor (L_{A12}, L_{A22})	5.05 mH, 1.05 mH
Output inductor (L_{B12}, L_{B22})	5.07 mH, 1.07 mH
Output inductor (L_{C12}, L_{C22})	5.05 mH, 1.05 mH
Transfer capacitors (C_{1a}, C_{1b})	0.68 μF
Output capacitor ($C_{o,min}$, $C_{o,max}$)	470 μF, 28,270 μF
Turn ratio of high-frequency transformer (n)	0.5
Gain of PI controller (k_p, ω_z)	1, 150
IGBT (G4PH50K)	1,200V UltraFast 4-20 kHz Discrete IGBT in a TO-247AC package
Freewheeling diode (HFA15TB60)	V_R=600V, V_F=1.7V, 125°C, Ultrafast, Very Low I_{RRM}, Soft Recovery Diode, TO-220AC package

The 2014 International Power Electronics Conference

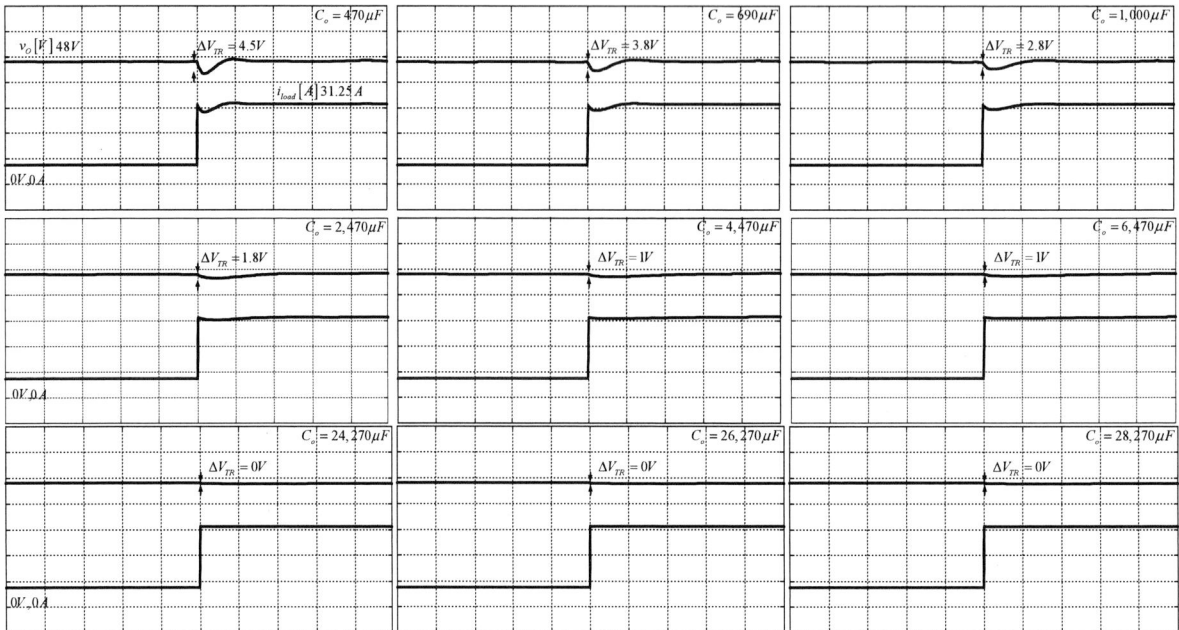

Fig. 2 Simulation results of different output voltage and load current waveforms of a paralleling three-phase ac-to-dc converter when ΔP_o is subject to a change from 0.36kW to 1.5kW, and different C_o values (C_o = 470 μ F, 690 μ F, 1000 μ F, 2470 μ F, 4470 μ F, 6470 μ F, 24270 μ F, 26270 μ F and 28270 μ F, respectively). (Ch1: v_o, 10V/div. Ch2: i_{load}, 10 A/div, Time scale is 1 ms/div).

Module loss occurs in phases A and B between 0.1-0.15 s and module loss occurs in phase A between 0.2-0.25 s. Hence, from Fig. 3, it is observed that, even under single or two module loss, unbalance supply voltage and voltage sage conditions, the system continues to work and maintain the output voltage nearer to the desired values. It can be seen from Fig.4 that nearly sinusoidal current operation and almost unity power factor is achieved.

The simulation result of transient response of the output voltage and current waveforms at load change from 1.5kW to 0.36kW and vice versa are shown in Fig. 4-5 and it is observed that the output is maintained constant for a step load change. The ripple output voltage at full load is 0.06V (0.125% when output capacitor C_o = 690μF). The output voltage is recovered to its steady state within 500μs.

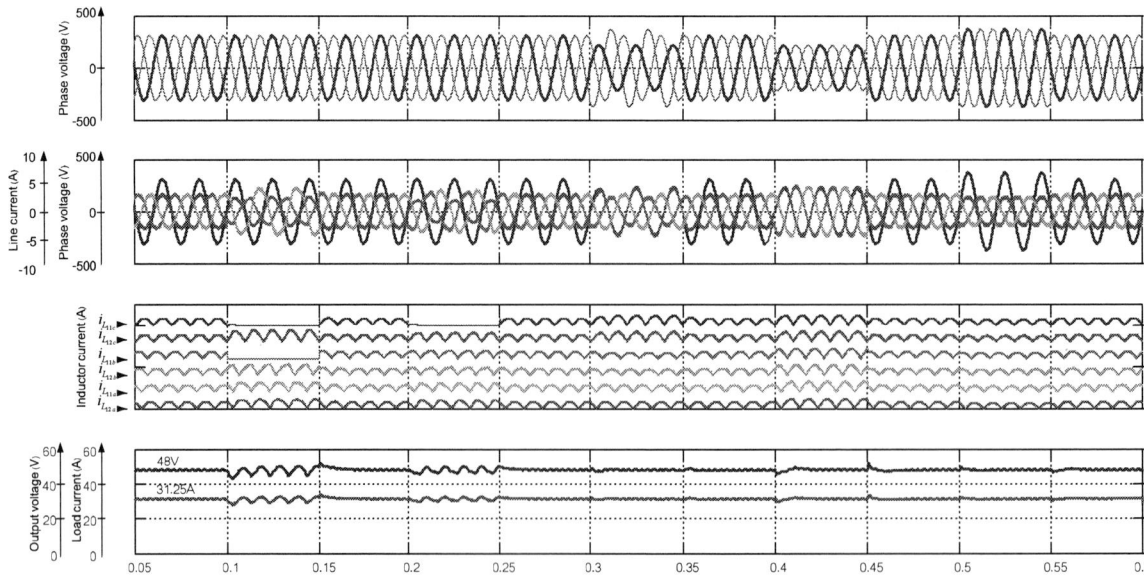

Fig. 3 Simulation results for the phase voltages, line currents, input inductor currents, output voltage and load current for symmetric mains condition, modules loss in phase A, module loss in phase A and B, unbalance in supply voltage and voltage sag condition for the proposed control system with small dc bus capacitor (C_o = 690F).

978-1-4799-2706-7/14 $31.00 © 2014 IEEE   697

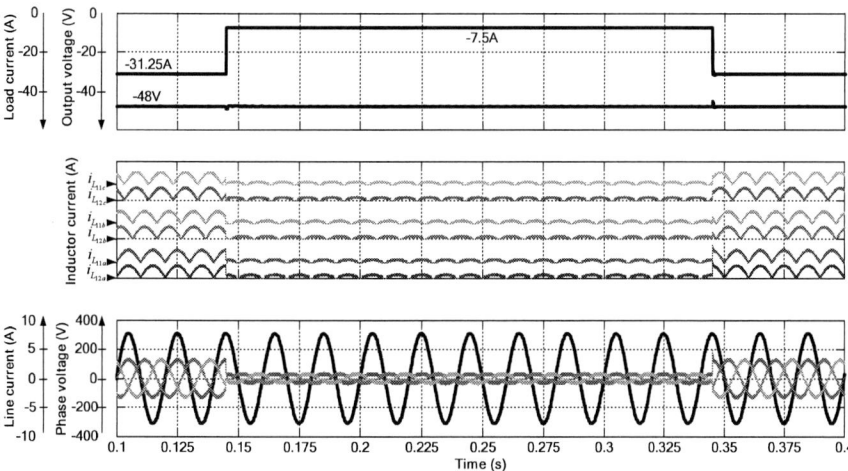

Fig. 4 Simulation results of the output voltage and load current waveforms of a paralleling three-phase ac-to-dc converter when ΔP_o is subject to a change from 0.36kW to 1.5kW and vice versa with small dc bus capacitor (C_o=690 μ F).

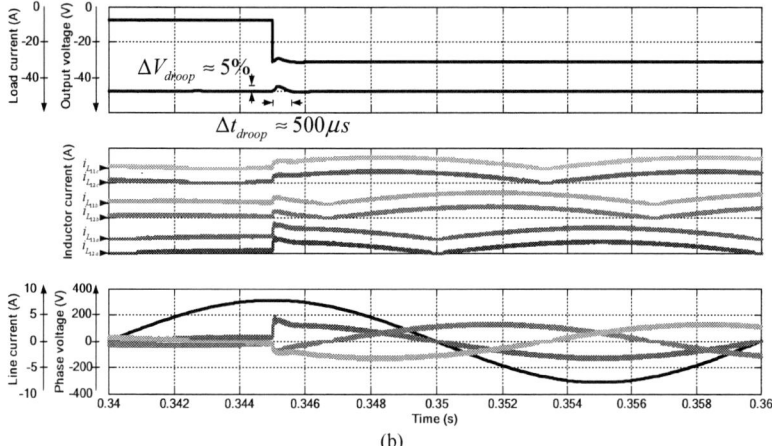

(b)

Fig. 5 Closed-up of the output voltage and load current waveforms of a paralleling three-phase ac-to-dc converter when ΔP_o is subject to a change from 0.36kW to 1.5kW and vice versa with small dc bus capacitor (C_o=690 μ F).

IV. EXPERIMENTAL RESULTS

Fig. 6-8 shows the experimental results of the proposed system. The performance features of the proposed system for apparent power (S), active power (P), reactive power (Q), input power-factor (PF) and input current total harmonic distortion (THD$_i$) are plotted in Figs. 6-7 as a function of the different DC capacitor size ($C_{o,max}$= 28,270µF, and $C_{o,min}$ = 470µF) at rated input and output voltage and output power (220V, -48V and 1.5kW). They are measured with the Digital power meter YOKOGAWA model WT1600. The results show high performance property for the dc bus capacitance range of 470µF $\leq C_o \leq$ 28,270µF. The influence of its value on ΔV_{droop}, $\Delta V_{overshoot}$ and settling time are shown in Fig. 8. It may then be concluded that three important cases can be identified: 1) Power factor is nearly unity for all difference values of output capacitor, 2) THD$_i$ remains low (lower than 3%) and 3) Transient response results of load current and bus voltage can be divided into three categories as shown in table II. Fig. 9 shows the load current and dc bus voltage waveforms of the proposed converter at dynamic load condition that is obtained, respectively, from experiment. It is shown that the

reduced-capacitance based power balance controller is capable of offering good dynamical response to parallel operation of three-phase ac-to-dc converter using CUK rectifier module, with 100µs $\leq \Delta t_{droop} \leq$ 1,000µs, little overshoots and droops, when the output power is alternated between 0.36 kW (quarter load) and 1.5 kW (full load).

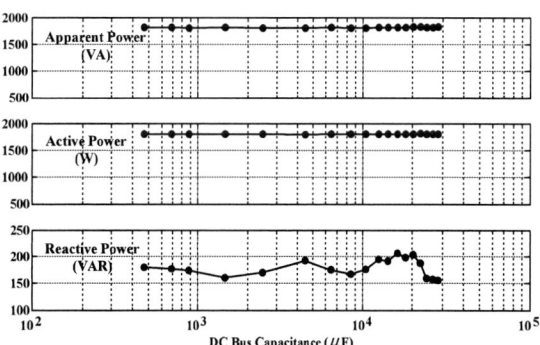

Fig. 6 Apparent power (S), active power (P) and reactive power (Q) at full-load condition in function of the dc bus capacitance C_o ($C_{o,max}$ = 28,270 µF and $C_{o,min}$ = 470µF).

The 2014 International Power Electronics Conference

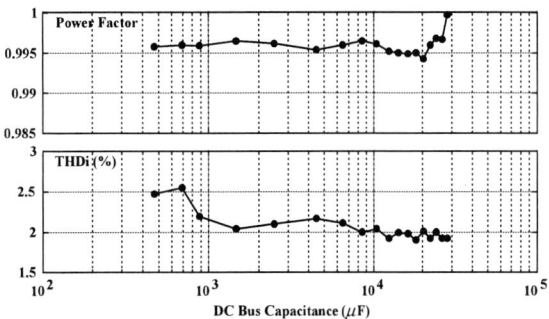

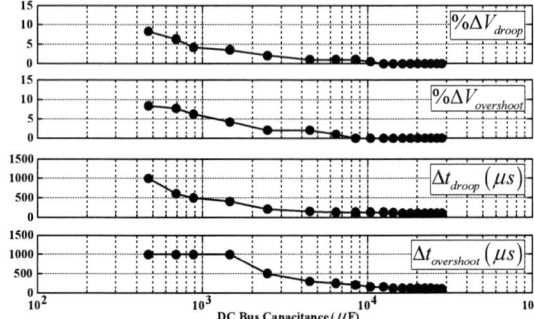

Fig. 7 Power Factor (*PF*) and THD$_i$ at nominal input voltage in function of dc bus capacitance C_o ($C_{o,max}$ = 28,270 µF and $C_{o,min}$ = 470µF).

Fig. 8 %ΔV_{droop}, %$\Delta V_{overshoot}$, Δt_{droop} and $\Delta t_{overshoot}$ in function of dc bus capacitance C_o ($C_{o,max}$ = 28,270 µF and $C_{o,min}$ = 470µF).

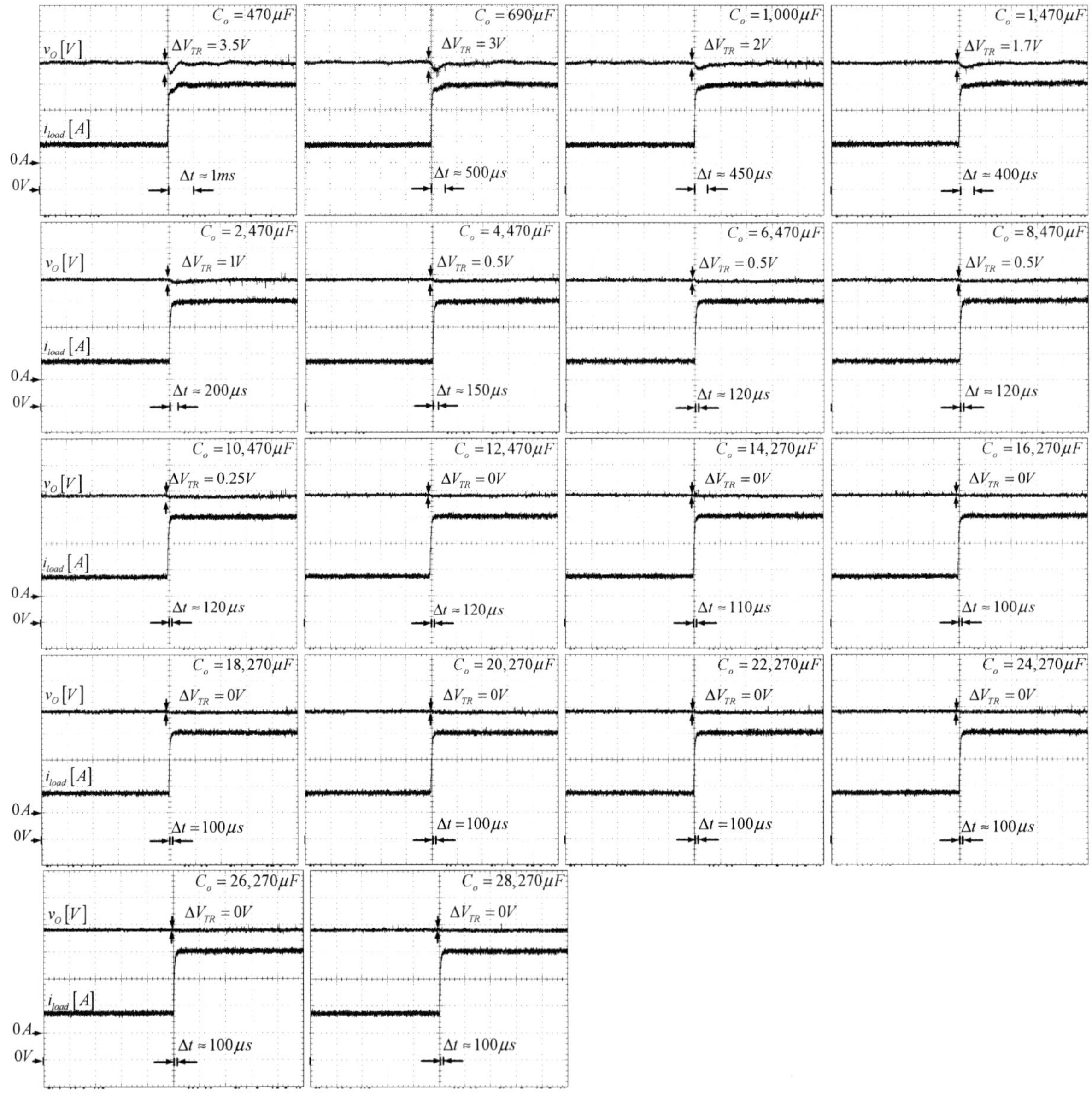

Fig. 9 Different output voltage and load current waveforms of a paralleling three-phase ac-to-dc converter when ΔP_o is subject to a change from 0.36kW to 1.5kW, and different C_o values (C_o = 470 µF, 690 µF, 1000 µF, 1470 µF, 2470 µF, 4470 µF, 6470 µF, 8470 µF, 10470 µF, 12470 µF, 14270 µF, 16270 µF,18270 µF, 20270 µF, 22270 µF, 24270 µF, 26270 µF and 28270 µF, respectively). (Ch1: v_o, 10V/div. Ch2: i_{load}, 10 A/div, Time scale is 1 ms/div).

978-1-4799-2706-7/14 $31.00 © 2014 IEEE

TABLE II EXPERIMENTAL RESULTS OF PARALLEL OPERATION OF THREE-PHASE AC-TO-DC CONVERTER WITH UNITY POWER FACTOR

Parameter / Categories	%ΔV_{droop}	%$\Delta V_{overshoot}$	Δt_{droop} (μs)	$\Delta t_{overshoot}$ (μs)	PF	THD$_i$	μF/W	$\Delta W_{Co}/W_{Co}$
470μF $\leq C_o \leq$ 690μF	6.25-7.29	7.70-8.33	500-1000	1000	> 0.99	< 3%	0.46	12%-14%
690μF $\leq C_o \leq$ 2,470μF	2.08-6.25	2.08-7.70	200-500	500-1000	> 0.99	< 3%	1.64	4%-12%
2,470μF $\leq C_o \leq$ 28,270μF	0.00–2.08	0.00–2.08	100-200	100- 500	> 0.99	< 3%	18.84	0%-4%

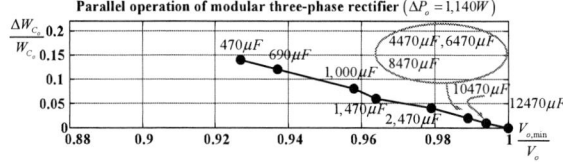

Fig. 10 Normalized delivered energy-storage capacitor and minimum energy-storage-capacitor voltage for the proposed system.

Moreover, the excellent consistency in the dynamical behaviour of the output voltage response between the load resistance step up and step down clearly demonstrates the property of the controller in large-signal operations.

The optimal sizing results are presented in Table II. It shows a key issue of a paralleling three-phase ac-to-dc converter with unity power factor. For the optimal range of dc bus capacitor with 690μF $\leq C_{o,opt} \leq$ 2,470μF, the total (%)ΔV_{droop} and (%) $\Delta V_{overshoot}$ is lower than 6.25%, 7.70% and Δt_{droop} of the settling time is very small (lower than 500μs). Hold-up time specifications vary depending on the application.

In this paper, requirement is 500μs at nominal input voltage, which is a more restrictive option. Thus, there is a capacitance range to comply with these specifications. The less restrictive option can be usually complied with a 0.46-μF/W capacitor. The most restrictive option can be complied with 0.46-μF/W or 1.64-μF/W. When, ΔW_{Co} is the amount of delivered energy to the output during hold-up time and W_{Co} is the total stored energy in C_o at V_{Co}. As seen from Fig. 10, only 12% of the stored energy is delivered to the load during the hold-up time if $V_{Co,min}$ is selected to be 0.958V_{Co} ($C_{o,minimized}$ = 690μF).

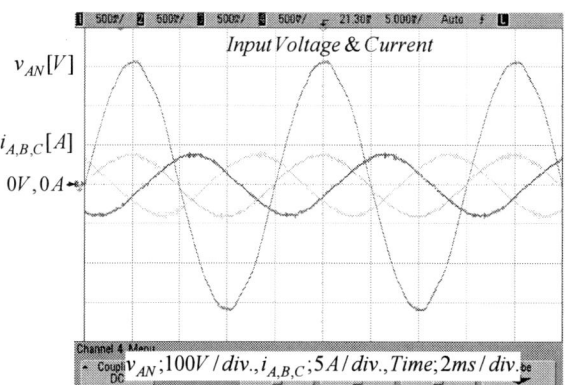

Fig. 11 Experimental results of the input voltage and phase currents for steady-state condition of a paralleling three-phase ac-to-dc converter with $C_{o,minimized}$ = 690μF (PI controller k_p = 1 and ω_z = 150).

Fig. 11 shows the experimental results of the proposed system. The closed-loop control of the proposed system forces the source current to follow the source voltage more effectively. The THD of the line currents were 2.75%, 2.74%, and 2.73%. They are nearly sinusoidal in shape. The power factor per phase was: 0.9962, 0.9961, and 0.9963, respectively.

Fig. 12-13 shows the operation of the proposed system under load variations (1.5 kW to 0.36 kW and vice versa). As a result, a fast response is obtained. The phase voltage, phase currents and inductor currents are illustrated under step load changes.

Fig. 14 shows that the converter gets back its reference voltage within 500μs for step load change from 24% (0.36kW) to 100% (1.5kW).

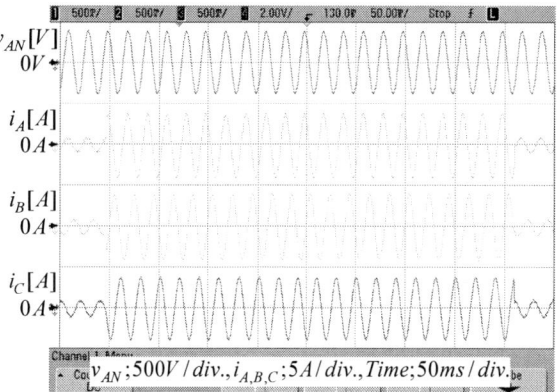

(a) step load change from 24% (0.36kW) to 100% (1.5kW) and vice versa

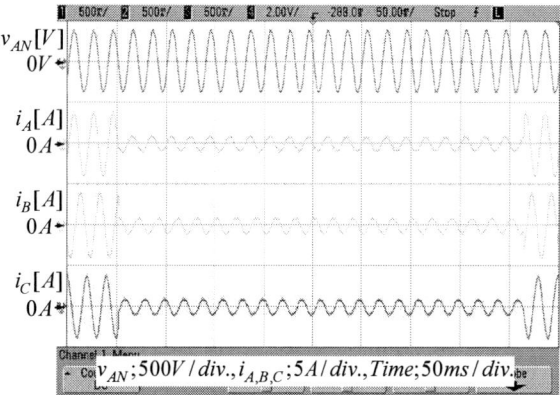

(b) step load change from 100% (1.5kW) to 24% (0.36kW) and vice versa

Fig. 12 Experimental waveforms for transient condition of a paralleling three-phase ac-to-dc converter with $C_{o,minimized}$ = 690μF (PI controller k_p = 1 and ω_z = 150). (a) and (b) input voltage and phase currents.

The 2014 International Power Electronics Conference

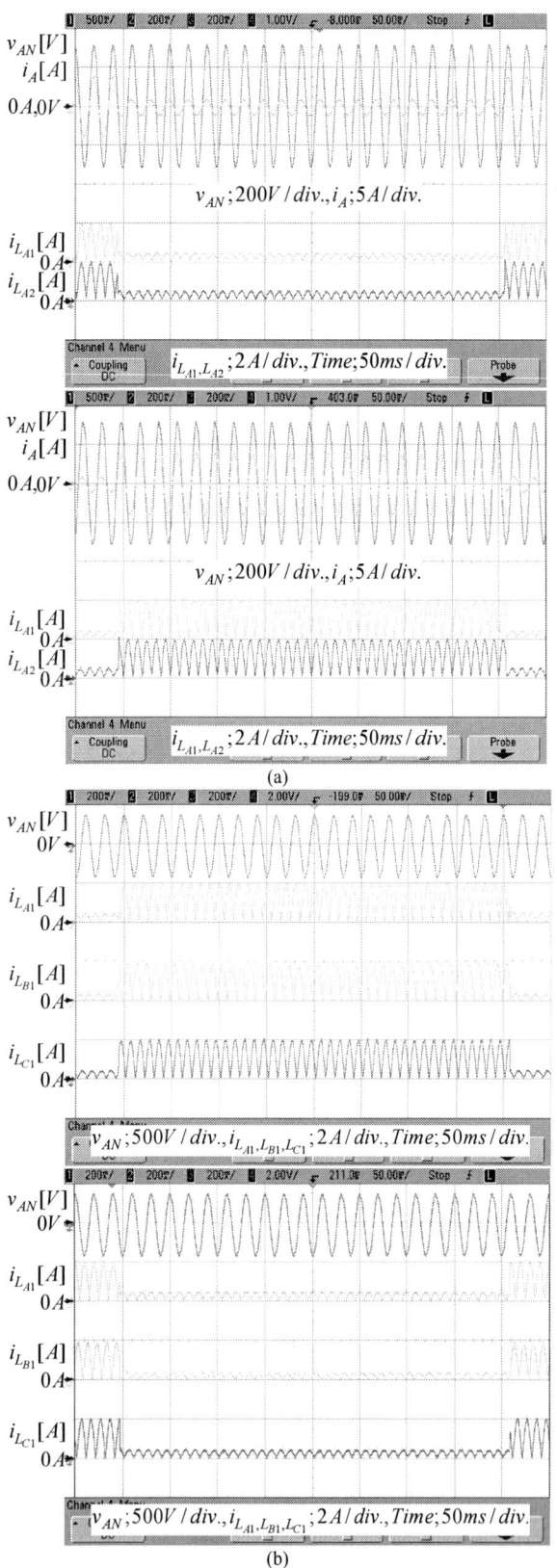

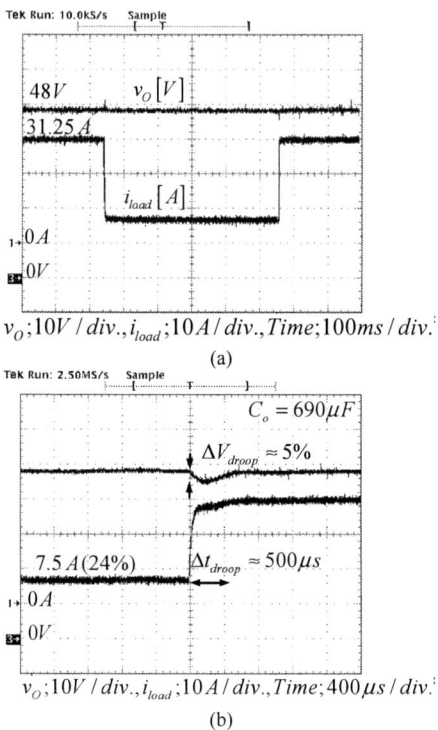

$v_O;10V/div.,i_{load};10A/div.,Time;100ms/div.$

(a)

$v_O;10V/div.,i_{load};10A/div.,Time;400\mu s/div.$

(b)

Fig. 14 Experimental waveforms: (a) dc bus voltage and load current at step load change from 24% (0.36kW) to 100% (1.5kW) and vice versa, (b) closed-up of step load change.

V. CONCLUSIONS

In this paper, an experimental methodology has been introduced in order to minimize dc bus capacitors. For the proposed system, the optimal range of dc bus capacitor with 690µF ≤ $C_{o,opt}$ ≤ 2,470µF, choose $C_{o,minimized}$ = 690µF (0.46µF/W) according to ΔV_{droop}, $\Delta V_{overshoot}$, Δt_{droop}, $\Delta t_{overshoot}$ and delivered energy. Maximum dc bus capacitor of the proposed system is 12470µF (8.31µF/W). The paper has presented that, with minimized dc bus capacitor, nearly unity power factor, good inductor current sharing and fast dynamic transient response can be achieved for parallel operation of three-phase ac-to-dc converter using six isolated CUK rectifier modules. The experimental results prove that proposed system is meet IEC 61000-3-2 Class A limits.

REFERENCES

[1] B. Singh, B.N. Singh, A. Chandra, K. Al-Haddad, A. Pandey, D.P. Kothari, "A review of three-phase improved power quality AC-DC converters," *IEEE Trans. Ind. Electron.,* vol.51, Issue 3, pp. 641 – 660, June 2004.

[2] Hava, A.M.; Ayhan, U.; Aban, V.V., "A DC bus capacitor design method for various inverter applications," Energy Conversion Congress and Exposition (ECCE), 2012 IEEE, vol., no., pp.4592,4599, 15-20 Sept. 2012.

[3] Ghias, A.M.Y.M.; Ciobotaru, M.; Pou, J.; Agelidis, V.G., "Performance evaluation of a five-level flying capacitor converter with reduced DC bus capacitance under two different modulation schemes," Power Electronics for Distributed Generation Systems (PEDG), 2012 3rd IEEE International Symposium on , vol., no., pp.857,864, 25-28 June 2012

Fig. 13 Experimental waveforms for transient condition of a paralleling three-phase ac-to-dc converter with $C_{o,minimized}$ = 690µF (PI controller k_p =1 and ω_z=150). (a) input voltage, phase current and inductor currents in phase A, (b) input voltage and inductor currents in phase A, B and C at step load change from 24% (0.36kW) to 100% (1.5kW) and vice versa.

The 2014 International Power Electronics Conference

Reactive Power Loss Optimization Method for Bi-directional Isolated DC-DC Converters

Huiqing Wen
Department of Electrical and Electronic Engineering
Xi'an Jiaotong-Liverpool University (XJTLU)
Suzhou, Jiangsu Province, China
Huiqing.Wen@xjtlu.edu.cn

Abstract- **In this paper, both the reactive power which delivered back to the source and from the load in the isolated bidirectional dual-active-bridge (DAB) dc-dc converter using the conventional phase-shift (CPS) control strategy are analyzed with mathematical expressions. An extended dual-phase-shift (EDPS) control is proposed and the reactive power losses are defined with regard to different operating cases. The experimental results reveal that the EDPS can effectively enhance the overall efficiency especially with the proposed method of reactive power losses minimization.**

I. INTRODUCTION

Recently distributed power system (DPS) has been gaining more attention. A DPS can be viewed as a group of generation sources that could be operated isolated from the main grid. Typically, DPS could be designed to be AC or DC and each type has its own challenges and benefits. One of the main advantages of DC DPS is the simplicity in controlling the power flow direction. Considering that the variable nature of the renewable energy including photovoltaic (PV) and wind energy [1], which strongly relies on natural phenomenon, energy storage systems (ESS) are required and Fig. 1 illustrates the configuration of distributed power systems with a dc bus.

In order to control the power flow and integrate the ESS with the dc link, bidirectional dc–dc converters are required to allocate dynamically the load current demand. Among various bidirectional DC–DC converters (BDCs), DAB has been proposed for interfacing hybrid electric vehicle and fuel cells. The key advantages of using DAB for distributed storage interface are bi-directional power flow capability, power controllability and its inherent soft switching [2].

The common modulation technique is the conventional phase-shift (CPS) modulation, where both bridges follow the same switching frequency and constant 50% duty cycles. However, the CPS cannot avoid the reactive power and circulating current flowing through the transformer. When the voltage of primary side is different from the reflected secondary side, the CPS allows significant current stresses and reactive power, which lead to conduction losses.

Considerably minimized converter power loss can be achieved with the use of alternative modulation strategies,

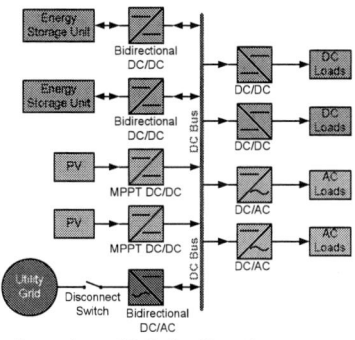

Fig. 1. Configuration of DC distributed power systems

which are basically active-power-oriented strategies by the variation the duty ratio of the H-bridge output voltage and the phase shift angle. A phase-shift plus pulse width modulation (PSPWM) control strategy is presented in [3] and one more degree of control freedom is added to extend the ZVS range. Similar methods also include Triangular modulation (TRM) [4], Trapezoidal modulation (TZM) [5] and hybrid modulation [6]. Besides, the method to modify the hardware components is also proposed. In [7], an adaptive inductor is proposed as the main energy transfer element of a DAB converter to reduce the circulating energy and maintain ZVS at light load. However, the design of adaptive inductor is complicated because the bias current needs to be sensed to control the inductance.

In this paper, both the reactive power which delivered back to the source and from the load in the isolated bidirectional DAB dc-dc converter using the proposed extend dual-phase-shift (EDPS) control strategy are analyzed with mathematical expressions. A prototype of DAB converter was built to verify the operating modes analysis.

II. REACTIVE POWER LOSS WITH CPS

Fig. 2 shows the topology and main waveforms of a DAB isolated bidirectional dc-dc converter with CPS control. The phase of current i_{B1}, which is the input current to phase leg B_1, is not always the same as the primary voltage V_{T1}. Similarly, the phase of current i_{B2}, which is the output current of phase leg B_2, is also not always the same as the secondary voltage V_{T2}. A considerable portion of the current, as represented in Fig.

978-1-4799-2706-7/14 $31.00 © 2014 IEEE   702

The 2014 International Power Electronics Conference

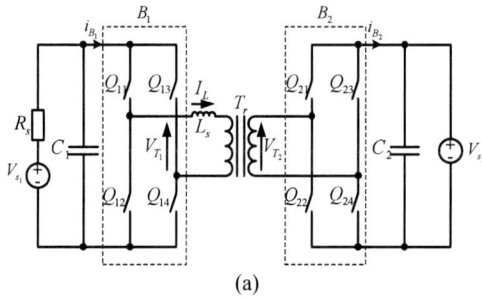
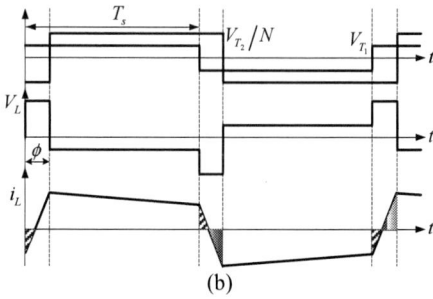

Fig. 2. Topology and main waveforms of a DAB isolated bidirectional dc-dc converter with CPS control. (a) Topology; (b) main waveforms.

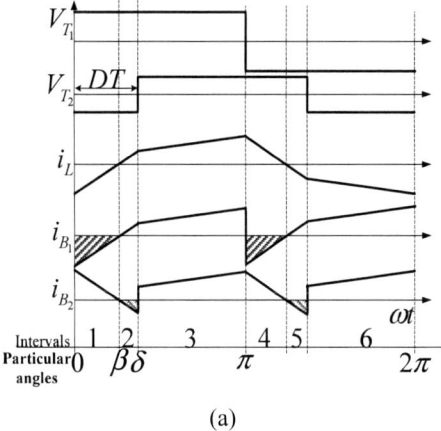
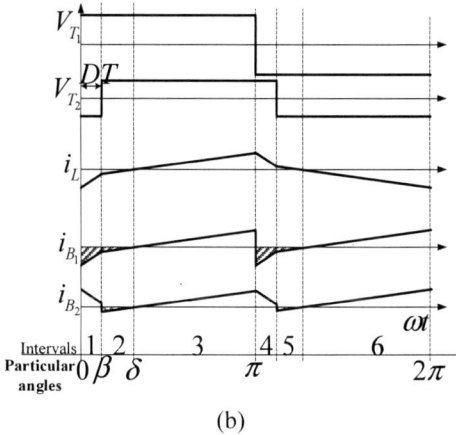

Fig. 3. Two operating cases of the DAB converter with CPS for buck operation: (a) case_I: heavy load conditions; (b) case_II: light load conditions.

2 (b) with blue shaded area for input side and red shaded area for output side, is contributing to the reactive power.

The operating modes of DAB converter with CPS can be distinguished according to the phase-shift angles, load conditions and output voltage. Under heavy load conditions, the inductor current i_L increases from an initial negative value at the beginning of the switching cycle. However, under light load conditions, the current i_L increases from a positive value at the beginning of the cycle and drops to a negative value at the end of the half switching cycle. Correspondingly, the distributions of reactive current and reactive power loss are different. Fig. 3 illustrates the steady-state voltage and current waveforms corresponding to two cases for buck operation. Where β corresponds to the zero crossing instant of the inductor current i_L, and δ represent the phase shift angles. The meaning of v_{T1}, v_{T2}, i_{B1} and i_{B2} are shown in Fig. 2 (a). The voltage conversion ratio d is defined as

$$d = V_{T2}/(NV_{T1}) \qquad (1)$$

The time duty-ratio of reactive current in the input and output side are defined as D_{B1} and D_{B2}, respectively. The expressions of the D_{B1}, D_{B2} and the reactive power losses for both cases P_{RL_B1} and P_{RL_B2} for buck operation are shown in TABLE I. Where V_{CE1} and V_{CE2} represent the voltage drop of the MOSFETS in bridge 1 and bridge 2 respectively. All quantities are normalized by the base values expressed as (2) in this paper.

$$I_b = V_{s1}/(4L_s f_s), P_b = dV_{s1}^2/(4L_s f_s) \qquad (2)$$

Fig. 4 illustrates the distribution of reactive current time duty-ratio and reactive power loss for buck operation with the same parameters shown as: V_{CE1}=1V, V_{CE2}=3V, V_{S1}=20V, N=6. The black line corresponds to the case_II and the magenta line represents the case_I. The phase-shift ratio D is assigned to within 0.5 considering that it covers the whole output power range and the reactive current in this range is smaller than the range of [0.5, 1]. Fig. 4 shows that the changing tendency of the reactive current time duty-ratio and reactive power losses for two cases are same and both their minimum values locate in the boundary of case_I and case_II. However, under buck operation, the time duty-ratio of D_{B1} is always higher than zero, as shown in Fig. 4 (a). With the 2% as reference, the range of the output power with P_{RL}/P_o less than 2% will vary with d. A wide range of from 0.13 to 0.4 of normalized output power exists to achieve P_{RL}/P_o less than 2% for d=0.8. No corresponding range existed for d=0.5, as shown in Fig. 4 (c).

The range of the ratio P_{RL}/P_o coincides with the high efficiency range irrespective of the voltage conversion ratio. Furthermore, as shown in Fig. 4, minimal reactive power loss happens in the boundaries, where ZVS Soft-switching is still achievable in both bridges B_1 and B_2 under the minimal reactive power loss conditions. This motivates to adopt the reactive power loss as the sole design index to improve the conversion efficiency of DAB converter over a wide operating range.

With CPS, the range with low reactive power loss varies with the voltage conversion ratio and output power

978-1-4799-2706-7/14 $31.00 © 2014 IEEE   703

The 2014 International Power Electronics Conference

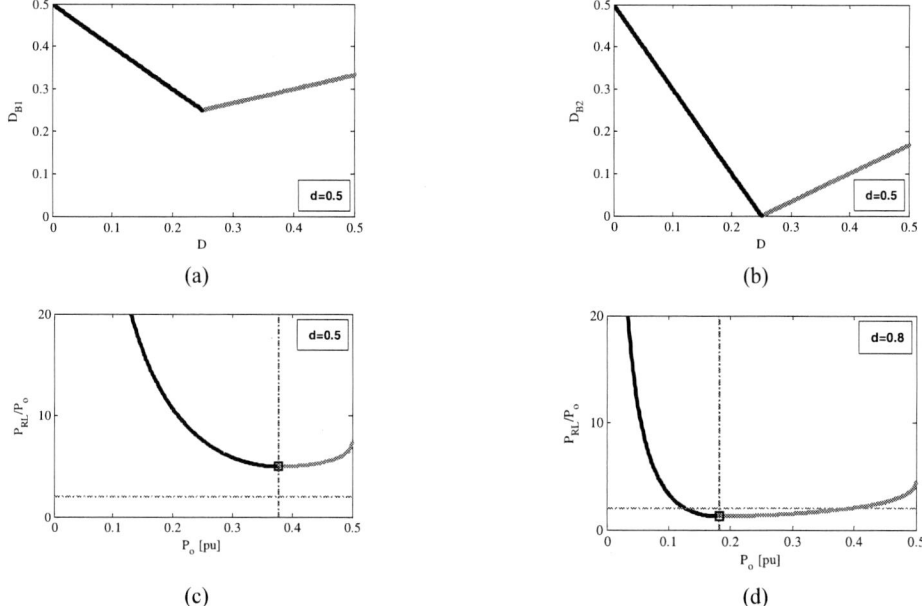

(a)   (b)

(c)   (d)

Fig. 4. The distribution of reactive current time duty-ratio and reactive power loss for buck operation; (a) D_{B1}; (b) D_{B2}; (c) P_{RL}/P_o versus P_o for d=0.5; (d) P_{RL}/P_o versus P_o for d=0.8.

TABLE I

EXPRESSIONS OF REACTIVE POWER LOSSES WITH CPS FOR BUCK OPERATION (PU)

Cases	D_{B1}	D_{B2}	P_{RL_B1}	P_{RL_B2}
I	$\dfrac{2dD-d+1}{2(d+1)}$	$\dfrac{2D+d-1}{2(d+1)}$	$\dfrac{V_{CE1}(2dD-d+1)^2}{2V_{S1}(d+1)d}$	$\dfrac{V_{CE2}(2D+d-1)^2}{2NV_{S1}(d+1)d}$
II	$\dfrac{1-2dD-d}{2(1-d)}$	$\dfrac{1-2D-d}{2(1-d)}$	$\dfrac{V_{CE1}(-4d^2D^2+8dD^2+4d^2D-4dD+d^2-2d+1)}{2V_{S1}(1-d)d}$	$\dfrac{V_{CE2}(2D+d-1)^2}{2NV_1(1-d)d}$

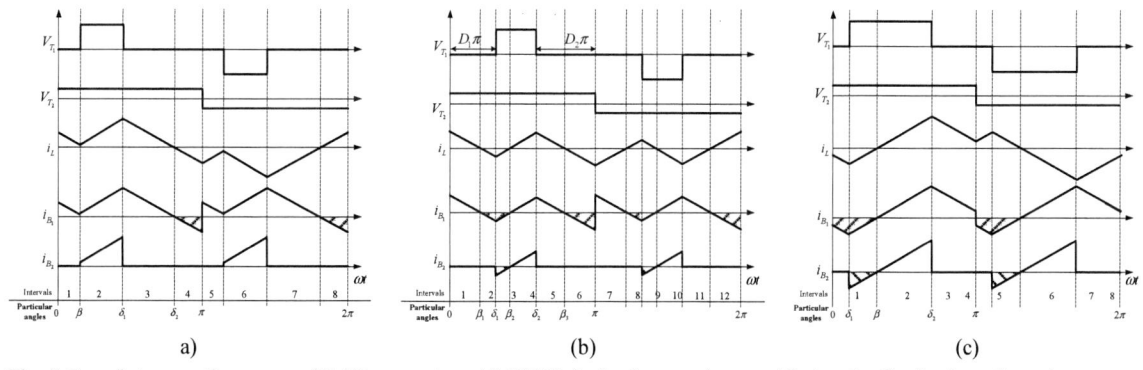

a)   b)   c)

Fig. 5 Complete operating cases of DAB converter with EDPS for buck operation considering the distribution of reactive current time duty-ratio and reactive power loss; (a) EDPS_I; (b) EDPS _II; (c) EDPS _III.

range, as shown in Fig. 4. Due to only one phase-shift variable in CPS control, it's impossible to tune the DAB converter with high efficiency for the whole operating range. However, the efficiency optimization process with other strategies is complex because the power losses cover all elements including the conducting losses, switching losses, transformer losses and other losses. In order to simplify the optimization process, the reactive power loss is defined in this paper and the method of minimization the reactive power loss is proposed to

improve the efficiency of DAB converter in the whole operating range by induce an extra phase-shift.

III. REACTIVE POWER LOSS WITH EDPS

Two phase shifts are adopted in the EDPS control and the corresponding phase-shift ratio are D_1 and D_2. D_1 is the phase shift between the diagonal control signals of the leg with the higher voltage, for instance, between the gate signals of Q_{11} and Q_{14} for the buck operation. D_2 is the phase shift between the primary control signal and the

978-1-4799-2706-7/14 $31.00 © 2014 IEEE

corresponding secondary gate signals, for instance, between the gate signals of Q_{11} and Q_{21}. Each device shown in Fig. 2 is driven by a 50% duty-cycle gate signals. Possible multiple zero crossing points of the inductor current corresponds to the half switching cycle with EDPS, which is different from CPS. The complete operating cases with EDPS for buck operation considering the distribution of reactive current time duty-ratio and reactive power los can be obtained and shown in Fig. 5. Where, EDPS _I and EDPS _III has one zero crossing point of the inductor current corresponds to the half switching cycle, and EDPS _II shows three zero crossing points.

Fig. 6 illustrates the comparison of the ratio of reactive power loss P_{RL} to the output power P_o using CPS and EDPS versus P_o under the condition of d=1.5. It shows that the ratio of reactive power loss P_{RL} to the output power P_o using EDPS is greatly reduced to less than 0.2% for majority range. Furthermore, the range is also extended with the 2% as reference, as shown the green line in Fig. 6. Among three operating cases, EDPS_II shows the best performance with the lowest P_{RL}/P_o and the widest effective range.

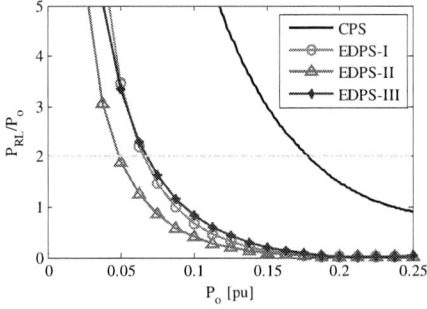

Fig. 6. Comparison of the ratio of reactive power loss P_{RL} to the output power P_o versus P_o under the condition of d=1.5.

The phase-shift pairs are determined with two methods including the minimum reactive power loss and the minimum rms current. Main parameters for the comparison are shown as follows: L_s=1.73uH, f_s=100 kHz, V_{S1} =20V. Fig.7 shows the phase-shift pair distribution and the rms current obtained by two different methods.

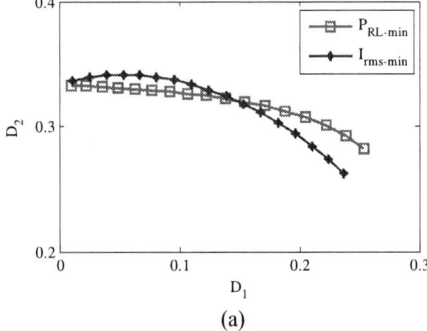

(a)

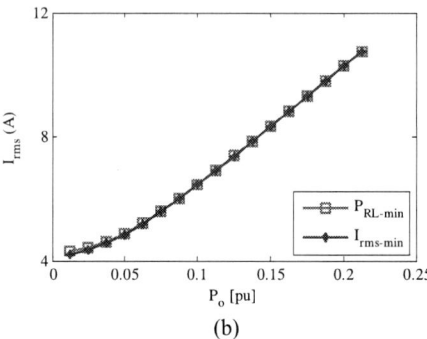

(b)

Fig. 7. Comparison of the phase-shift pair distribution and the rms current obtained for d=1.5; (a) the phase-shift pair distribution; (b) the rms current.

Considering that the conduction loss is the dominating part in the total losses distribution, the proposed method is effective in minimizing the losses because the inductor rms current obtained by the proposed method is coincide with the actual minimum rms current, as shown in Fig. 7 (b).

IV. EXPERIMENTAL EVALUATION

To verify the proposed design method of DAB converter in terms of the defined reactive power loss, a prototype of DAB converter was built. The digital control is implemented with TMS320F2808 DSP. Main parameters are: L_s=1.73uH, f_s=100 kHz, V_{S1} =20V, d=0.5~2, and deadtime T_d=0.2μs.

Fig. 8 shows the experimental waveform of v_{T1}, v_{T2}, and i_L using the DPS_II under the condition of d=0.5. The measured efficiency is improved from 67.5% of CPS to 90.1% of EDPS_II.

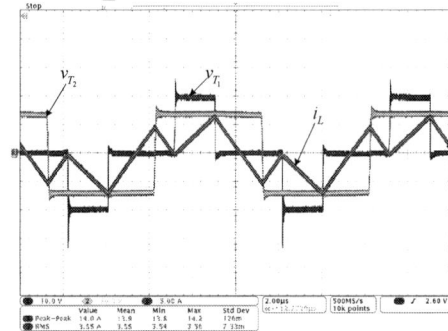

Fig. 8. Experimental waveform of v_{T1}, v_{T2}, and i_L using the EDPS_II to deliver the output power of 25W with d=0.6.

V. ACKNOWLEDGMENT

This research is supported in part by Jiangsu Province University Natural Science and Research Program (13KJB470013) and in part by Research development fund of XJTLU (REF-12-02-04).

REFERENCES

[1] A. Etxeberria, I. Vechiu, H. Camblong, and J. M. Vinassa, "Hybrid Energy Storage Systems for renewable Energy Sources Integration in microgrids: A review," in *IPEC, 2010 Conference Proceedings*, 2010, pp. 532-537.

[2] H. Wen, B. Su, and W. Xiao, "Design and performance evaluation of a bidirectional isolated dc-dc converter with extended dual-phaseshift scheme," *Power Electronics, IET,* vol. 6, 2013.

[3] G. G. Oggier, Garci, x, G. O. a, and A. R. Oliva, "Modulation Strategy to Operate the Dual Active Bridge DC-DC Converter Under Soft Switching in the Whole Operating Range," *Power Electronics, IEEE Transactions on,* vol. 26, pp. 1228-1236, 2011.

[4] F. Krismer, S. Round, and J. W. Kolar, "Performance Optimization of a High Current Dual Active Bridge with a Wide Operating Voltage Range," in *Power Electronics Specialists Conference, 2006. PESC '06. 37th IEEE,* 2006, pp. 1-7.

[5] Y. Wang, S. W. H. De Haan, and J. A. Ferreira, "Optimal operating ranges of three modulation methods in dual active bridge converters," in *Power Electronics and Motion Control Conference, 2009. IPEMC '09. IEEE 6th International,* 2009, pp. 1397-1401.

[6] Z. Haihua and A. M. Khambadkone, "Hybrid Modulation for Dual-Active-Bridge Bidirectional Converter With Extended Power Range for Ultracapacitor Application," *Industry Applications, IEEE Transactions on,* vol. 45, pp. 1434-1442, 2009.

[7] F. Haifeng and L. Hui, "High-Frequency Transformer Isolated Bidirectional DC-DC Converter Modules With High Efficiency Over Wide Load Range for 20 kVA Solid-State Transformer," *Power Electronics, IEEE Transactions on,* vol. 26, pp. 3599-3608, 2011.

Power Supply for a Wireless Sensor Network: Airliner Flight Test Case Study

P. Durand Estebe, V. Boitier, M. Bafleur, J-M. Dilhac
CNRS, LAAS
Toulouse, France
dilhac@laas.fr

S. Berhouet
AIRBUS Flight Test Instrumentation
Toulouse, France

Abstract— In this paper we present hands-on experience related to on-going implementation in aircraft of power supply for a wireless sensor network deployed for aerodynamic flight tests. This autonomous battery-free power supply is capturing, managing and storing primary energy from the environment, using solar light and photovoltaic cells. For practical purposes, it is also equipped with an auxiliary power input. The specifications are detailed, the general architecture is presented and justified, and test results are discussed.

Keywords— Flight Tests, Sensor Network, Solar Energy, Ultracapacitors.

I. INTRODUCTION

Air transport is now an essential pillar of modern society, and a major driver of global economic prosperity. Aviation forecasts show massive traffic increase, both for passengers and cargo, in the next decades. Therefore, air transport is facing various challenges such as fossil fuel consumption, pollutant and noise emission, air traffic congestion, business model, safety and security.

Technological innovations are therefore pursued in the area - among others - of engines, aircraft structural design, and structural new materials. Other innovations are taking place through the concept of More Electric Aircraft (MEA) that is an evolution starting with fly-by-wire (a full electronic interface replaces the classical electro-mechanical interface between pilots and aircraft systems), and evolving through drive-by-wire (no more physical connection – mechanical, pneumatic or hydraulic – between sensor and actuator) to power-by-wire (where pneumatic, mechanical and hydraulic power distributions are replaced by an electric power grid, except for propulsive power) [1]. It is worth to mention that MEA obviously calls for an increase in electrical power wiring size and complexity.

However, aviation's progress will also be carried out by an increased integration of cyber-physical systems into aircraft [2] for managing baggage, cargo, personal electronic devices, flight deck tasks and helping maintenance through Structural Health Monitoring

(SHM) [3]. There again the weight of wiring is likely to increase.

In that context, Wireless Sensor Networks (WSN) have been considered for various aeronautical applications, such as SHM [4] and flight tests. Each node of this network would perform sensing, data processing and wireless transmission of information. Such wireless networks will hopefully permit (often new) functions without the need of adding extra wiring to an already large burden.

However, these nodes need to be self-powered, many of the advantages of wireless sensor networking being obviously lost if external (i.e. wired) power sources were used. For this purpose, batteries offer a high energy density at low cost. However, there are critical drawbacks associated with the use of batteries: first environmental concerns and economical aspects associated with the replacement of primary batteries. Second, batteries (either primary or secondary) may be prohibited since the wireless nodes are often implemented in locations without temperature regulation that could result both in a drastically reduced yield together with safety issues. Fortunately, batteries can be avoided through the use of environmental energy capture allowing a solution for long term, deploy and forget, WSN. However, environmental energy transducers, unlike batteries, are not energy reservoirs, and energy buffer storage is mandatory whatever the context. We will explain why it can be achieved through electrostatic ultracapacitors (UC).

In this paper we present hands-on experience related to on-going implementation of energy harvesters in airliners for the purpose of flight tests. Even if aircraft manufacturers alone are involved in this application, this issue is critical as the implementation in the experimental aircraft of the extra wiring required for flight tests is complex and time-consuming.

In the second section we will describe the requirements of our application, and we will present the devised architecture in the third section. In a fourth section we will present preliminary results and identify directions for future work in the conclusion.

978-1-4799-2706-7/14 $31.00 © 2014 IEEE

II. SYSTEM GLOBAL REQUIREMENTS

In Fig. 1 the total electric cable lengths implemented in a car, a jet fighter and an airliner are compared. These lengths are also converted in kilometer per unit length of the car/aircraft as a measurement of complexity. From Fig. 1, it appears that fighter jets and airliners are very complex objects considering wiring, thus praising for wireless solutions when the deployment of new avionic is considered. As a consequence of such a choice, measurement data need of course to be transmitted wirelessly and the device has to be autonomous in energy. Specific components are therefore required to allow for this autonomy. Nevertheless, if stringent weight constraints (common in aeronautics) cannot afford these extra components, and if a power line runs close by, a connection to this line will be preferred.

Fig. 1 also gives the extra wiring related to flight tests, i.e. 60% of the regular wiring in the case of the A380. Generally speaking, this wiring is often difficult to implement and may be realized by bonding the cables using adhesive tapes for the instrumentation outside aircraft structure, or passing cables through multiple bulkheads by removing screws, bolts or rivets, or even by drilling holes when permitted (rarely allowed in composite materials) that is a cumbersome process, not to mention the multiplication of connectors. So, even if the application detailed in this paper is of interest to flight test departments of aircraft manufacturers only, it is a critical issue because flight-testing is essential to validate a design and obtain certifications from government agencies, while being expensive, complex and time-consuming.

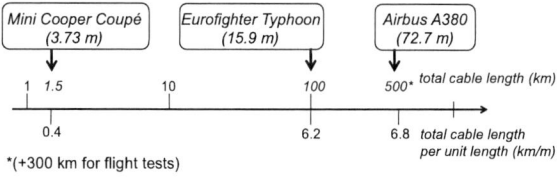

Fig. 1. Wiring complexity for various means of transportation (estimations).

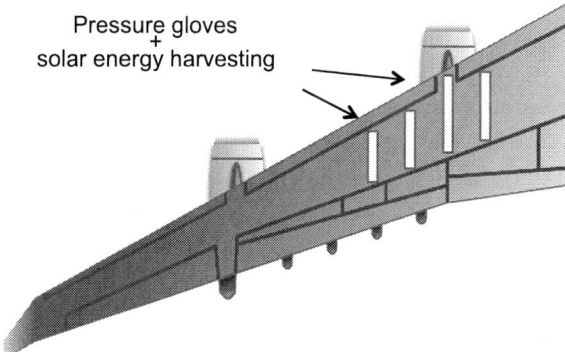

Fig. 2. Schematic of pressure gloves affixed on top of a wing. The system here is supposed wireless.

Among the typical measurements performed during flight-tests, are pressure sensing at tens of positions around wings, tail, rudder… For this purpose, pressure gloves, themselves accommodating tens of sensors are affixed to the aircraft outside skin together with cables connecting them to data loggers (Fig. 2). The replacement of this apparatus by WSN would reduce design load, installation and relocation time.

Considering that a wireless implementation is chosen, the simplest way of gaining autonomy in energy is to use an energy reservoir such as a primary electrochemical battery. Such batteries are products of a mature industry: they offer a high energy density – and even a high power density when in parallel with an ultracapacitor – together with a low self-discharge. But, as already mentioned, their performance is strongly altered by the extreme (low) temperature encountered outside aircraft at cruise altitude, not to mention safety issues raised by some technologies.

Energy capture from the environment is then mandatory and fortunately, for the considered application, solar energy can be harvested because i) the sensors are located outside the aircraft ii) for security and simplicity reasons most flight tests are done at daylight.

It is worth mentioning that energy storage is here still required for two reasons: first to accommodate for transient power surges from the load exceeding the mean electrical harvested power, and then to maintain the energetic autonomy in the case of an intermittent environmental source as will be illustrated later. UC performing electrostatic storage may do the job; while being affected by extreme positive ambient temperatures similarly to most batteries, they exhibit better performance at low temperature and provide an almost infinite lifetime. Consequently, they do not need any replacement. Because of their low energy density, they are safer than batteries, and supposed to be more environment-friendly. Unfortunately, they are impaired by strong self-discharge especially during the first hours [5] and efficient long-term storage is not feasible. Finally, they exhibit a charge-dependent output voltage asking for a voltage regulator when used for circuit biasing, and suffer from a low maximum voltage rating (as low as 2.2V for some models tested below) related to the dielectric properties of the electrolyte.

Solar energy is qualitatively one of the most abundant ambient energy and easily translates itself into electricity through mature and still progressing technology, while UC are an appealing alternative to batteries but, for supplying power to the above pressure sensors, specific requirements from the environment are to be fulfilled. They are two fold.

First, the power supply (comprising photovoltaic cells, maximum power point tracking circuit, energy management devices and buffer storage ultra capacitor) must be both very thin not to disturb the streamline flow, and globally flexible to adapt to curvy profiles. Considering areas, flexibility main impact is on photovoltaic (PV) cells, restricting the choice to less efficient cell technologies.

Then, all devices must operate at low temperature (down to -55°C) and reduced atmospheric pressure (200hPa); fortunately low temperatures have a favorable impact on PV cell yield. Conversely, references of energy buffer UC correctly performing at such extreme temperatures and ambient pressure must be identified.

Finally the amount of power delivered to the sensing and communicating devices has to be assessed, given the sampling frequency, number of sensors and wireless data rate.

This system development has to be carefully carried out, considering the environment (i.e. solar flux vs. altitude, time, season, location…) and the harvesting system itself (photovoltaic cells and associated electronics). However, the objective is not to install a permanent system aiming at monitoring a critical parameter during the airliner lifetime, but to deploy a temporary test rig using WSN. Long-term dependability is then not at stake.

In the next section we will now detail quantitative requirements and derive the general architecture for the power supply unit.

III. General Architecture

General requirements of our application translate into the following. The flexible power supply dedicated to be bonded to the extrados of an aircraft wing (see Fig. 2):
- uses solar cells as main source of energy,
- is less than a few millimeters in thickness,
- delivers a mean electrical power of at least 2W under 5V (later shifted to 48V, see below), under modest (cloudy) illumination,
- operates between -50°C (cruise altitude) and +85°C (parking under direct sunlight), and down to 200hPa (i.e. 35 000feet),
- incorporates an energy storage (UC) unit in order to deal with transient absence of sunlight, the targeted autonomy being 1 hour,
- is capable of autonomous start-up, with the help of solar cells alone, the storage unit being empty [6].

The typical use case is illustrated in Fig. 3.

Time	☽	🌥	☀	🌥	🌥	🌥	☀	☽
operation	park	taxi + take off	cruise	land + taxi	park	taxi + take off	cruise	land + taxi
power source	ground	storage	PV	storage	ground	storage	PV	storage
flight profile	ground	ground + up	FL 200 to 350	down + ground	ground	ground + up	FL 200 to 350	down + ground

Fig. 3. Power supply: example of a use case as given by the final user. FL stands for *Flight Level*, and roughly corresponds to an altitude in hundreds of feet.

Considering the above specifications, we devised the general architecture depicted in Fig. 4. The load corresponds to a datalogger used to manage the pressure measurement sensors; its typical characteristics are in the range 48V / 2W.

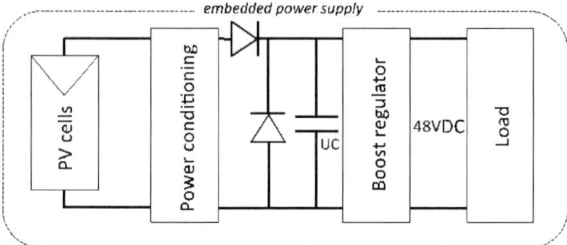

Fig. 4. Power supply: general architecture. Power conditioning includes MPPT (Maximum Power Point Tracking), UC overvoltage protection and buck-boost regulator command (not shown).

The PV cells are connected to a power conditioner whose task is to optimize the energy transfer through *Maximum Power Point Tracking* (MPPT) and to store the energy in the UC. It also protects the UC from overvoltage. A blocking diode is inserted between power conditioning and UC to prevent the latter from discharging itself when photovoltaic cells are not delivering any energy.

The UC is connected to the load through a boost regulator. In a start-up phase, as long as regulator input voltage is below min operating value, power conditioner does not bias the regulator, in order to speed-up UC charge.

IV. Experimental Results

For the solar cell testing, key points were a flexible photovoltaic panel with the best efficiency at low light, as the system has to power the load even in cloudy weather. We used thin film solar cells, reaching 9% efficiency at 72 W/m². Fig. 5 shows flexibility and main characteristics of the selected PV panel (SolbianFlex SP50-L).

DATASHEET

PHYSICAL CHARACTERISTICS	
Lenght	1110 mm
Height	285 mm
Thickness	1,5 mm
Weight	0,79 kg
ELECTRICAL CHARACTERISTICS	
Vmp (max power voltage)	9,20 V
Imp (Max power current)	5,51 A
Wp (peak power at stc)	50,70 W
Voc (Open circuit voltage)	10,88 V
Isc (short circuit current)	5,85 I
Efficiency (module)	17,14%
Max bending	25%

Fig. 5. Main characteristics of SolbianFlex SP50-L.

Considering energy storage, we tested 8 types of UC from six different manufacturers in climatic chambers to identify technologies compatible with our requirements. According to data sheets, the operating temperatures were ranging between [-20, +70°C] and [-40, +85°C] depending upon model. In other words, no model was initially rated for the extended range of [-50, +85°C].

With a SP-240 4A potentiostat from BioLogic, we first submitted the devices to constant current charges and discharges in a Temptronic Thermostream climatic chamber, at temperatures between -50°C and +100°C. Then, at -50°C only, we submitted them to an identical charge and discharge cycle, at 200hPa, in another climatic chamber (CLIMATS EX5423-TE).

During these two sets of experiments, we plotted the V(t) curves and the drift vs. temperature (see the example of Fig. 6).

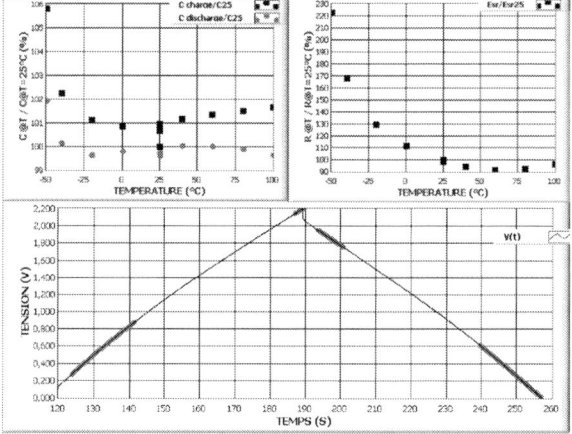

Fig. 6. Variations of capacitance and ESR with temperature and example of charge and discharge of a Maxwell PC10 ultracapacitor. The voltage drop after the peak is due to the equivalent serie resistor of the UC.

Finally, we measured the self-discharge performance in open circuit of all models by plotting, during more than three days, the output voltage vs. time after an initial charge at 2.2V.

From the above tests we made a trade-off and identified the best reference, also considering both thickness and capacitance per cm², in order to comply with the already mentioned thickness requirements and to minimize the area devoted to energy storage. Nevertheless, the targeted autonomy asked for the association of multiple UCs.

We then devised an electronic circuit (see Fig. 7) incorporating all devices shown in Fig. 4 except PV cells and the ground power supply. We carefully limited the thickness of the circuit, specifically that of the inductor.

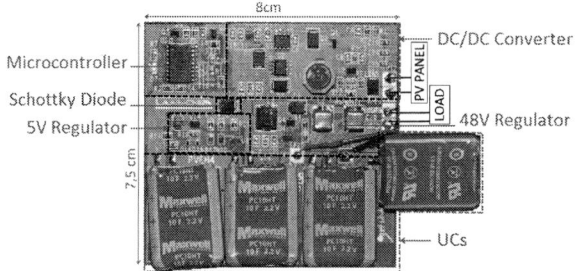

Fig. 7. Power management board.

Many electronic devices, such as the microcontroller and the converters, require a bias voltage. For this purpose, a 5 volts LDO voltage regulator TPS71550 (Vout = 5V , 6V ≤Vin ≤ 24V) is directly connected to solar panel. Consequently, if the photovoltaic voltage falls under 6V, the Power Conditioning device in between photovoltaic panel and UCs is cut off and, the load is powered by UCs only. This Power Conditioning is a DC/DC synchronous buck converter in series with a blocking Schottky diode.

To deal with various atmospheric conditions, a MPPT algorithm is implemented in a PIC18F1220 microcontroller based upon fractional Open-Circuit Voltage Method with the advantage of using one sensor only. The open circuit voltage of photovoltaic cells is measured every 10 seconds. When the UCs voltage reaches almost its maximum rated value, the algorithm will move from MPPT mode to UC voltage regulation mode.

The output stage was first designed to deliver a 5V output. Because of late changes in specifications requiring a 48V output, we simply add a 5V to 48V boost regulator due to a lack of time for another solution.

A boost regulator (LT3539) is placed between the UCs and the load in order to regulate the output voltage to 5V. The input voltage range of this component is 2.3-4.4V when it supplies 2W.

We used LT3539 Shutdown pin to keep the output disabled and to allow the photovoltaic panel to first charge UCs [6]. When the UCs terminal voltage reaches 4V (the higher the voltage across the UCs, the better the LT3539 efficiency is), the LT3539 is turned on by a command from the microcontroller. Then the solar panel will both power the load and charge the UCs whenever the sun is shining strongly enough. If not, the UCs alone will power the load.

The 5V to 48V conversion is realised by a MAX1523 boost controller, supplied by the LT3539 output, and thus is only activated when the LT3539 is on.

Finally, for the bonding onto the aircraft wing, the electronic power management board together with ultracapacitors was inserted into a thin plastic packaging whose shape was designed in order to minimize disorders induced to the airflow. Pictures of this arrangement are shown in Fig. 8. Paths for cables to the photovoltaic panel and the sensors are set up in the beveled edges of the packaging.

The overall system supplies the 2W load from a solar irradiance as low as 80W/m² using a 0.32m² SolbianFlex SP50L PV panel, rated with peak power of 50W.

Fig. 8. (a) power management module and ultracapacitors inserted into package (b) package as it will appear once bonded to the aircraft wing.

V.   CONCLUSIONS

We have reported the design details of a power supply for a wireless sensor network deployed for airliners in-flight tests. The system has been successfully tested on ground. Real flight-tests are scheduled to confront with the operational requirements.

REFERENCES

[1]   J. A. Rosero, J. A. Ortega, E. Aldabas, L. Romeral, "Moving Towards a More Electric Aircraft," *IEEE A&E Systems Magazine*, vol. 22, no. 3, pp. 3-9, 2007.

[2]   K. Sampigethaya, R. Poovendran, "Aviation Cyber-Physical Systems: Foundations for Future Aircraft and Air Transport," *Proceedings of the IEEE*, vol. 101, no. 8, pp. 1835-1855, 2013.

[3]   L. G. Dos Santos, "Embraer Perspective on the Introduction of SHM into Current and Future Commercial Aviation Programs," *Structural Health Monitoring 2011, Proceedings of 8th IWSHM, DEStech Publications*, pp. 19-29, 2011.

[4]   J-M. Dilhac, M. Bafleur, "Energy Thermo Generation in Aeronautics for Battery-free Wireless Sensor Networks," *Thermoelectrics Goes Automotive II, Daniel Jänsch ed., Expert Verlag*, pp. 135-143, 2013.

[5]   Y. Diab, P. Venet, H. Gualous, G. Rojat, "Self-Discharge Characterization and Modeling of Electrochemical Capacitor Used for Power Electronics Applications," *IEEE Trans. on Power Electronics*, vol. 24, no. 2, pp. 510-517, 2009.

[6]   V. Boitier, P. Durand Estebe, R. Montheard, M. Bafleur, J-M. Dilhac, "Under voltage lock-out design rules for proper start-up of energy autonomous systems powered by supercapacitors", *Proceedings of PowerMEMS International Conference, London, 2013*.

A Configurable Three-Phased Inverter for Teaching Power Electronics

Ansgar Kern

Department of Informatics, Electrical Engineering and
Mechatronics
University of Applied Sciences Giessen Friedberg
Friedberg, Germany

Abstract - This paper introduces a configurable three-phased inverter designed to demonstrate the basic operation and concepts of frequency inverters to electrical engineering and mechatronics students. The inverter consists of an FPGA-based space vector controller, a controller-based communication interface, and the actual power stage. Frequency inverters employing space vector controllers are usually complex and interdisciplinary devices, which are difficult to cover completely in one semester. They require an extensive knowledge of power electronics and control theory. In this paper, an approach is proposed based on the familiarity of students with modern communication methods like web browsers and information processing infrastructure derived from current information technology. The paper describes the application topology, methods used, and the results obtained.

Keywords – customised system, FPGA, power electronics, space vector

I. INTRODUCTION

Many frequency inverters use space vector-based algorithms to control their power stages to drive the attached induction or synchronous motors. They also include elaborate methods to identify the attached machines. In order to familiarise students with the workings of a modern frequency inverter, the complex system has to be stripped down to its essential components. These are the space vector controller and the power stage. Machine identification is an add-on which can be explained in sufficient detail separately. The closed loop control algorithm for the frequency inverter is of secondary interest only in this context.

The focus, therefore, is on an easy to understand and handle implementation of a space vector controller driving a power stage which will enable students to create simple pulse patterns. It would be especially important to be able to re-create standard textbook examples with regard to these pulse patterns [1]. This can usually not be done with off the shelf inverters since they are designed to interpolate a large number of points of the space vector control plane. They also can not be

coaxed to assume only a limited number of points. The power stage will be used to verify the shape of the produced voltage signal and to generate load currents under various load conditions.

The power stage has not been designed to drive large loads. A maximum DC-link voltage of 200 volts is sufficient. A maximum load current of 20A will also suffice. It would be, however, important to cover a wide input voltage range in order to use standard lab-power supplies.

The required control electronics including FPGA and driver electronics are treated as a black box from a didactical point of view. Its major purpose is to generate defined pulse patterns; an introduction into circuit design of methods of information processing or informatics is not the goal.

II. SYSTEM TOPOLOGY

Figure 1 shows the general components of the system. All components are connected to a computer, which serves as a general control centre. An oscilloscope and a spectrum analyser are used to observe the generated pulse patterns. An induction motor is attached as load. Its general behaviour is of no importance in this context except to demonstrate that there is a connection between pulse patterns and rotor speed.

The space vector controller has been implemented on an FPGA and communicates with the outside world via an I2C-interface. The actual implementation and the methods used are not explained in any way to the students. The actual space vector implementation can be treated as a black box.

The configurability of control signals, voltages, and currents generated by the power electronics are, however, the main focus of this work. Output voltages are measured by a spectrum analyser and displayed by an oscilloscope. All devices are connected to a computer, which serves as control centre. The programming and other software aspects are treated as a black box and only the use of application interfaces is explained to students.

978-1-4799-2706-7/14 $31.00 © 2014 IEEE

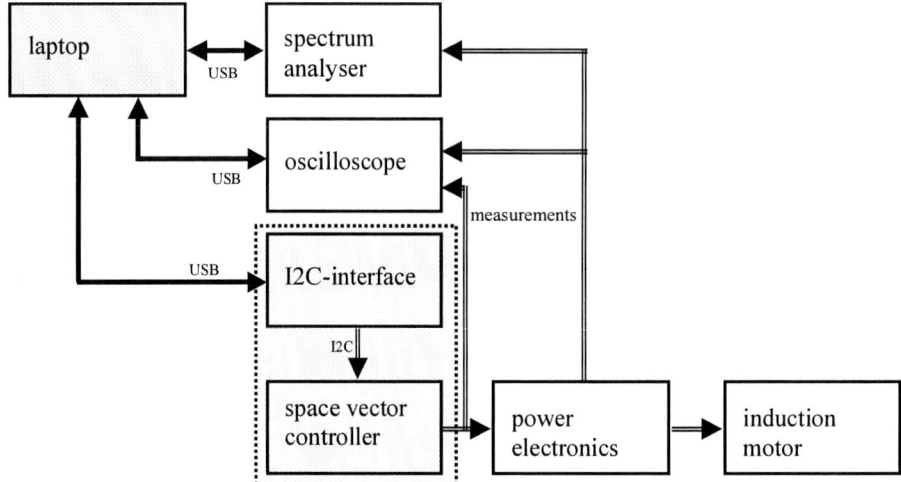

Fig. 1: Development environment

A. Communication Interface

Frequency inverters and other power electronics-related equipment are usually controlled via various bus systems like PROFIBUS [2] or industrial Ethernet. These bus systems require a protocol layer handled by the devices attached to the bus that add complexity detrimental to the purpose of teaching basic inverter behaviour. Also, the added security and redundancies provided by more complex protocols are not required in this case since the system will never operate unattended. A simple bus interface would have the advantage of enabling students to directly observe cause and effect while modifying parameters of the space vector controller.

In order to keep things simple and transparent, the I2C-bus [3] was chosen as a means to communicate with the controller. Since no protocol has to be implemented, parameters can be transmitted to and read from the controller directly. Likewise, a large number of USB to I2C converters are available for most operating systems. Students can configure the inverter and observe the effects of single bytes of information sent immediately.

An in-depth knowledge of the I2C-bus workings is not required. Bus traffic will not be observed by any means; only the bytes of information sent are displayed.

The disadvantage is that the I2C-interface has to be implemented additionally on the FPGA alongside the space vector controller. This makes the design process of the inverter controller more complex. Great care has been taken in order to keep I2C-bus-interface and space vector controller separate entities. Practical results have shown that the I2C-interface can be separated from the space vector controller quite neatly and be reused for other FPGA-based applications. The I2C-interface implementation can thus be treated as a black box within the black box of the inverter controller, which just receives or transmits parameters.

Figure 2 shows a simple transmission of three bytes of data via the I2C-bus. As can be seen, no protocol overhead is needed; only a simple acknowledge signal A is required.

B. General application topology

The principal workings of a space vector controller are

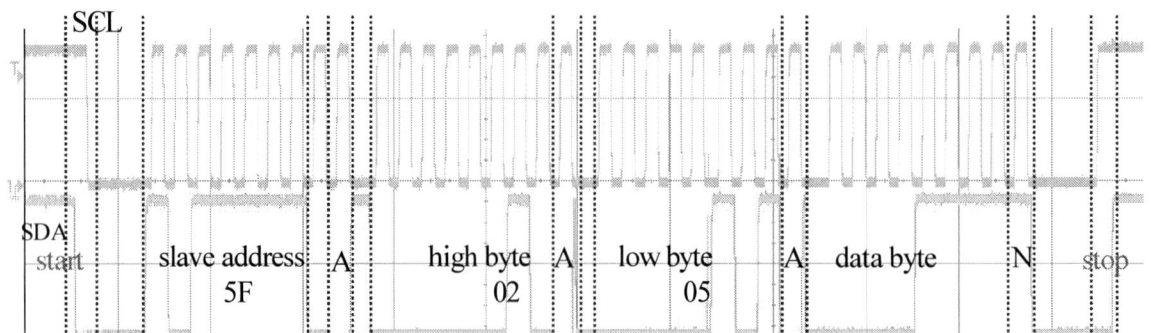

Fig. 2: I2C-interface operation

widely known and will therefore be touched only upon with regard to the implementation specific points. The theory is based on a continuously moving vector where vector length and speed can be adjusted in a quasi-continuous manner. For the implementation as a digital system, certain limitations apply depending on the platform used. With the FPGA based implementation, in this case, mathematically complex operations like divisions and sinus-functions have to be ruled out. They are not only too complex and time-consuming to implement, but they also would require too many logical connections used on the FPGA side. As a consequence, all mathematical operations are implemented based on integer arithmetic. Therefore, the vector can not be depicted as a quasi continuously moving entity but rather like an entity that can only assume a limited number of positions in the complex number plane (Fig. 3). The number of different positions is at a maximum for a space vector u_{1max} of maximum length, and reduced dramatically for a vector of minimum length.

The number of possible positions P_n depends on the binary representations of the controller values. In this case, 126 positions P_n have been sampled per 90 degrees at a resolution of eight bits. These 126 positions can be imagined as fence posts positioned at equal intervals along the 90 degree angular distance. They create 125

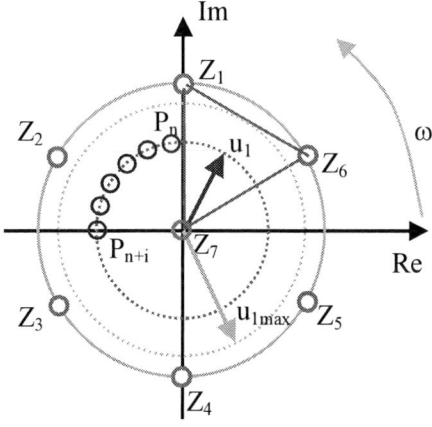

Fig. 3. Possible positions in the complex number plane

gaps. For a pointer to rotate from 0 degrees to 180 degrees, it will pass first from sample 1 to sample 126 and then back to sample 1. Thus a total of 252 samples will be used to generate the pulses required for 180 degrees.

This results in a total of 504 positions per 360 degrees. The space vector length can range from 1 to 255. In other words, theoretically 128520 different space vector positions are possible. The sample values are implemented via a look-up table which is initialised at start-up time.

The space vector algorithm has further been simplified by implementing a single-phase version and introducing a phase shift of 120 and 240 degrees to generate the required three phased system. An example of how pulses are obtained for a single phase system is shown in Fig. 4.

C.   Application implementation

The space vector controller is implemented on an FPGA board running at a base clock frequency f_b of 50MHz. The fastest space vector speed f_S at maximum resolution is 389Hz (Eq. 1).

$$f_S = \frac{50MHz}{504 * 255} = 389Hz \tag{1}$$

This is too fast for ordinary electrical drives. Therefore, a clock divider is used to slow down the base frequency f_B of the system by a factor of N. As a result, only vector speeds of 389Hz over N are possible. This results in significant gaps in the speed spectrum of the attached drive. To minimise these gaps, sample values can be skipped or repeated in order to influence the vector speed. This will result in distortions of the generated output voltages and additional losses in the form of higher harmonics. Figure 5 shows a simple example of the switching pulses generated in an operation mode called half-bridge.

In so called, half-bridge mode, two times three control signals are generated. Within each set, the signals are phase shifted 120 degrees and the two sets are phase shifted 180 degrees to each other. Of the six signals generated, Fig. 5 shows only three.

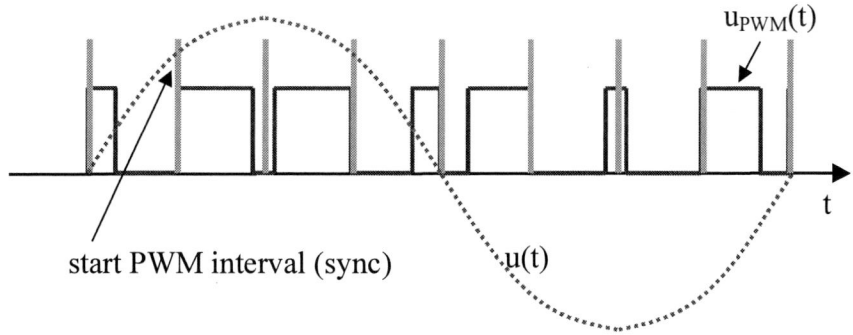

Fig. 4. Single phase voltage and digital equivalent

978-1-4799-2706-7/14 $31.00 © 2014 IEEE

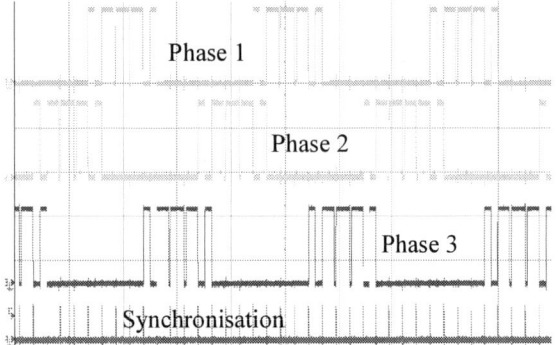

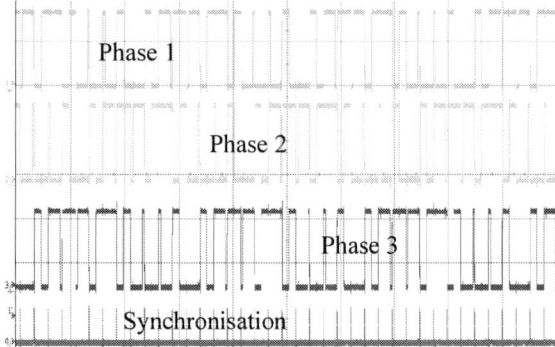

Fig. 5. Three phases in a half-bridge mode, angle increment set to 30 degrees

Fig. 6. Three phases in full-bridge mode

In contrast, only three control signals are generated in full-bridge mode. They are phase-shifted 120 degrees and can be used to control a three phased inverter. Figure 6 shows an example of these control signals. The major difference between half-bridge and full-bridge mode is the combining of two signals into one. The full-bridge control signal consists of a half-bridge signal and its inverted version.

D. Application hardware

The application hardware consists of an FPGA board and the power electronics. The FPGA board, as of now, is a standard development board and only used to implement space vector control for didactical purposes. Implementation issues are not considered here. As such, the FPGA board is treated as a black box.

What is of greater interest is the power electronics. The main aim in developing the power electronics is to create an easy to use and highly modular system. The power electronics consists of three half-bridges, each requiring only one control signal for operation. Two IGBTs connected in series (Fig. 7) are controlled in such a way that only one of the switches is active at any given time. An optional enable signal can be used to force both IGBTs into the off mode.

Each half-bridge is powered by an external voltage source to provide U_{DC}. An additional 5V power source is required for the driver electronics of the IGBTs. Figure 8

shows two half-bridges constituting a full-bridge with heat sink.

III. SWITCHING PULSES

Industrial space vector controllers are basically black box systems where pulse patterns and pulse generation are controlled by higher control structures. The user can not influence the pulse patterns or decide to generate pulse patterns associated with defined space vector positions.

The introduced system, on the other hand, provides full control with regard to the generated pulse patterns. The space vector can be moved around the complex plain in defined steps, generating patterns that are similar to the patterns depicted in standard literature. Figure 9 shows the line voltage produced by a low frequency pulse pattern representing a space vector rotating at a frequency of about 90Hz. The spectrum of the generated voltage can not be estimated except for the fact that there is a sinusoidal first harmonic present that enables

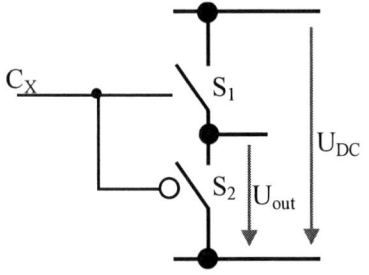

Fig. 7: Principal concept of the half bridge

Fig. 8: Principal concept of the half bridge

The 2014 International Power Electronics Conference

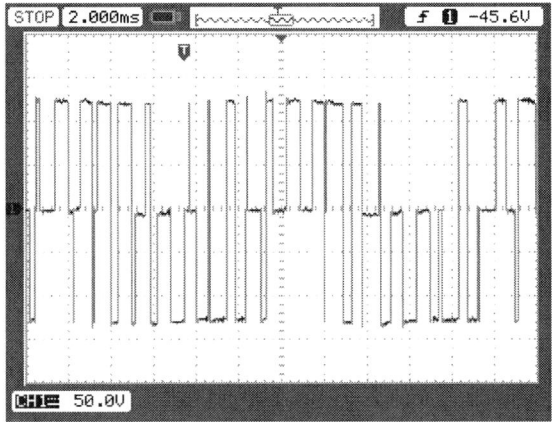

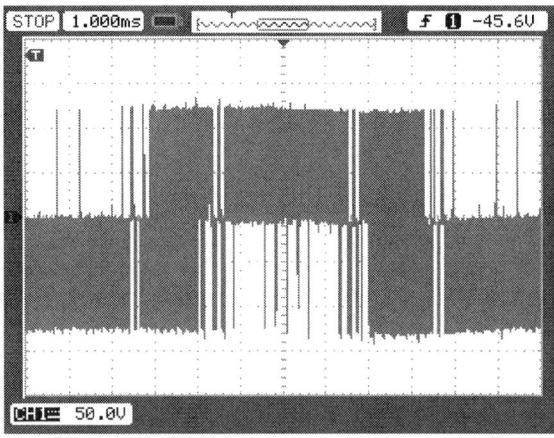

Fig. 9. Customised low frequency pattern

Fig. 11. Customised high frequency pattern

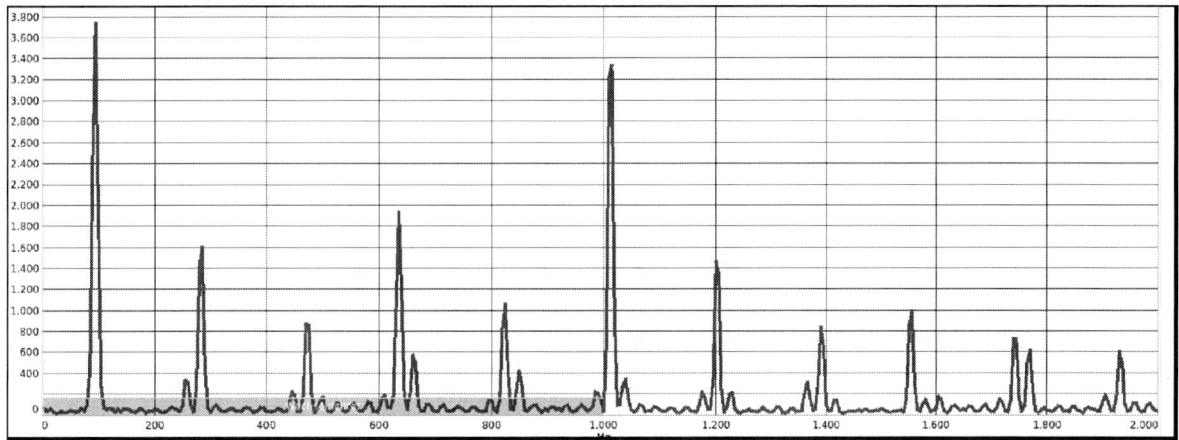

Fig. 10. Frequency spectrum of the voltage shown in Fig. 9

the rotor shaft to turn.

Very few discrete space vector positions are used, as can be seen by the low number of changes in the output voltage.

The frequency spectrum of the voltage is shown in Figure 10. The first harmonic is positioned at approximately 90Hz. Further harmonics are at approximately 270, 450, 630, 810Hz and so on. It can be noted that the 11th harmonic at around 1kHz is very pronounced. It can further be noticed that the harmonics are not as clearly defined and discrete as one would have expected from a mere analytical examination of the idealised pattern.

Figure 11 shows a pulse pattern of a space vector with 504 positions per 360 degree. The frequency of the space vector generated is slightly below 100Hz. The single pulses of the output voltage can no longer be displayed. The spectrum of the voltage (Fig. 12) shows

reduced higher harmonics with very discrete frequency lines.

These two examples demonstrate to students quite clearly the connection between space vector resolution and frequency content of the spectrum which immediately translates into varying losses with regard to the attached load.

IV. CONCLUSIONS AND OUTLOOK

The introduced concept was designed to familiarise students with the principal concept of space vector control. The approach taken was from a point of view that heavily relied on the concepts taught during informatics sessions. In this view, space vector control is just another application and not something specific to power electronics or electrical drives.

Designing an easy to use and robust power stage proved more difficult than expected but the resulting concept has so far provided easy to understand pulse patterns

978-1-4799-2706-7/14 $31.00 © 2014 IEEE

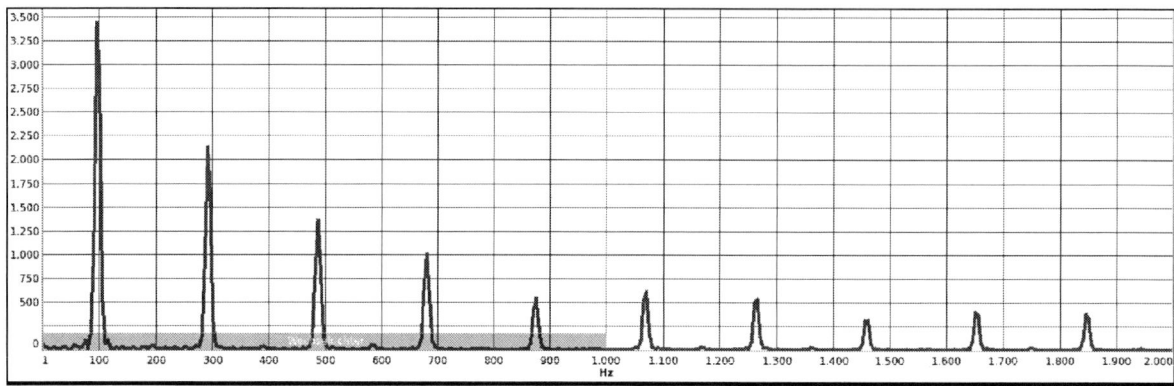

Fig. 12. Frequency spectrum of the voltage shown in Fig. 11

which can be compared with standard textbook examples. Students have the opportunity to build an understanding starting with basic configurations and expanding these configurations into relevant application examples.

Including a frequency analyser greatly enhanced the understanding of students with regard to the basic properties of different pulse patterns and their associated losses. The system, furthermore, induces students to experiment and adapt it to their own applications.

V. REFERENCES

[1] Nguyen Phung Quang, Jörg-Andreas Dittrich, Vector Control of Three-Phase AC Machines: System Development in the Practice, Springer-Verlag 2008, pp. 18-22.

[2] Max Felser, PROFIBUS Manual: A collection of information explaining PROFIBUS networks, epubli GmbH 2011, pp. 30-39.

[3] Philips Semiconductors, The I2C-bus and how to use it, Philips 1992.

A Bachelor-Student Project:
Buck-Boost Operation of an Integrated H-Bridge for Variable-Speed Energy Storage Systems Using Measurement Coils in the Stator of a DC-Machine

Frederik De Belie[1], Araz Darba and Jan Melkebeek

Dept. Electrical Energy, Systems & Automation, Ghent University, Sint-Pietersnieuwstraat 41, 9000 Gent, Belgium
[1] corresponding author, tel: +32 9 264 79 14, fax: +32 9 264 35 82, e-mail : Frederik.DeBelie@UGent.be.

Topic number: 13, Education in Power Electronics and Electrical Engineering

Abstract—**This paper discusses an interdisciplinary project to be performed by 3th year bachelor student groups during their last semester. The topic is related to the energy storage in micro dc-grids supplying several electrical drives. The energy storage in the dc-grid is based on the kinetic energy in variable-speed flywheels or in masses performing a linear acceleration-deceleration motion. Such devices have been designed and used at high power already such as in F1-vehicles and public transportation systems. For students, low-power components are more suitable to achieve a self-built setup. One of the challenges in this student project is the low budget of 50 Euro per student pushing them to reuse components found in broken devices or even scrap yards, to look for integrated components or to build low-budget sensors themselves.**

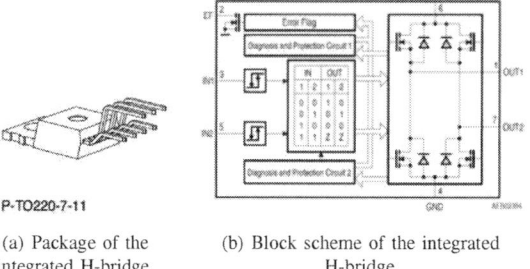

(a) Package of the integrated H-bridge.

(b) Block scheme of the integrated H-bridge

Fig. 1: Integrated H-bridge, used as either boost or buck converter for motor or generator operation respectively

I. INTRODUCTION

Students in Electromechanical Engineering at UGent finish their bachelor studies by performing an interdisciplinary project besides their regular courses. For this, some students participate in projects related to power electronics, control and dynamics of electrical drives as these projects are often related to the electrical systems in the growing industry of renewable energy sources. In particular, the storage system for electrical systems has received a lot of attention, a broad overview can be found in [1].

Most of the prerequisite courses and skills (mechanical construction, control theory, dc-machines, Matlab, electronics) for this project have been teached before the start of the project or during the project (power electronics). As most courses are mainly theoretical supported by computer exercises and simulations, practical and development skills are not fully developed yet at the project start. For this, the project tries to sharpen the students engineering skills.

In the student project, by means of an integrated H-bridge and reused dc-machine, power flows has to flow from the rectified ac-grid to a variable-speed storage system such as a flywheel, [2]. Such an energy flow illustrates to the students the behaviour of a converter in buck-operation.

Disconnecting the small dc-grid from the ac-grid, kinetic energy has to be transformed back to electric energy, slowing down the flywheel. To control this energy flow back to the micro dc-grid, the stator voltage induced in the dc-machine is boosted with the same integrated H-bridge as used during buck-operation. During this operation, the machine inductance is used as the coil of a boost converter. This means that proper dc-machines have to be found having a sufficient machine inductance for boost operation as well as having a small resistor to keep the voltage drop low during generator as well as motor operation. The small-scale setup allows students to study the use of buck and boost converters. Furthermore, the interaction with other converters is studied. For this, the dynamic behaviour of the entire system has firstly been evaluated by using SimPowerSystems, a toolbox within Simulink/Matlab. In a second step, the root-loci of such variable-speed system are studied by using Sisotool, available within Matlab as well.

Additionally to this study, a voltage-control loop has been developed by the students to achieve the voltage boost operation during generator mode. A voltage tracker is implemented, based on the back-emf of the dc-machine

The 2014 International Power Electronics Conference

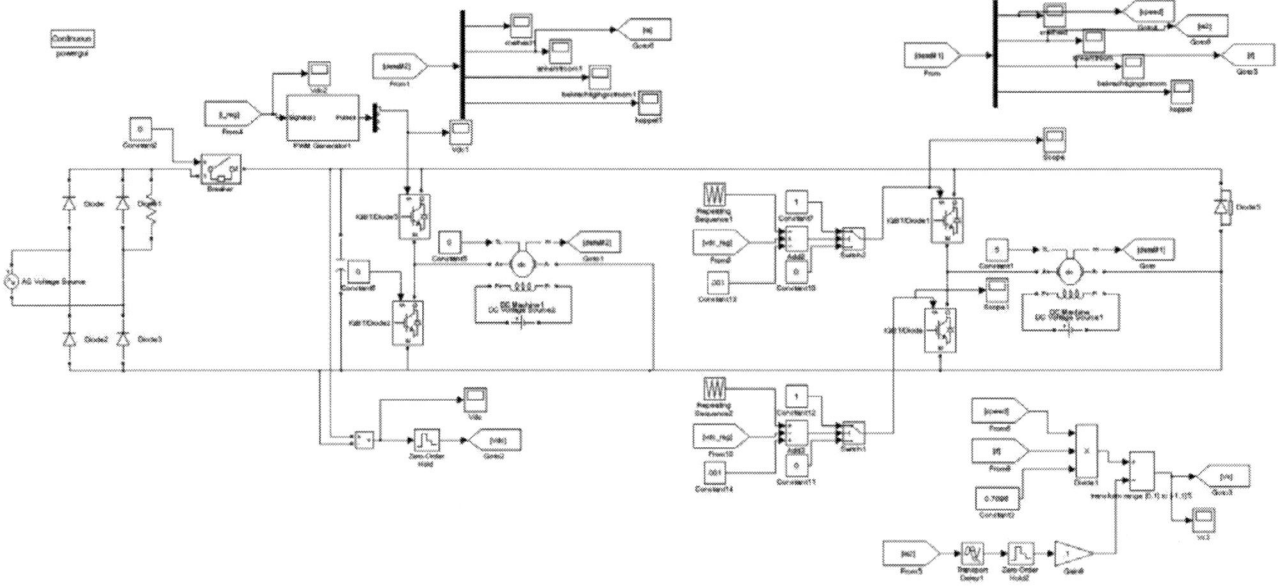

Fig. 2: Overall simulink model, from left to right, ac-grid, rectifier, dissipator, H-bridge, dc-motor as load, H-bridge, dc-generator.

and the voltage drop across the stator resistor. Instead of measuring the speed-induced voltage directly, an estimation is made using voltage measurements of a copper coil with windings placed in the slots of the stator. Despite the constant magnetic sources in dc-machines such as current and permanent magnets, a measurable voltage is induced in these coils during machine rotation due to the salient rotor configuration in low-power dc-machines as a result of the small number of deep rotor slots per pole.

II. Energy Storage System

The energy storage system, based on a variable-speed flywheel, together with the load and charging AC grid is given here by a Simulink model as shown in Fig. 2. From the right to the left, one can observe a single phase ac-grid, a passive diode rectifier, relay, dc-bus with capacitor, first part of an H-bridge inverter, load dc-machine, second part of the H-bridge inverter and finally the dc-machine connected with the flywheel (not drawn). By connecting the ac-grid, electric energy is transformed and stored in the rotating flywheel by using the H-bridge. Once a sufficient rotor speed is reached, the ac-grid is disconnected by opening the relay and kinetic energy is converted back to electric energy, supplying power to a second motor used as load of the micro dc-grid. As the ac-grid is no longer connected, the dc-voltage of the dc-grid can be boosted to a higher value more suitable for the second motor.

A. System Components

The main component is an integrated H-bridge TLE 5205-2 from Infineon Technologies, [3] shown in Fig. 1, making bach-

elor students able to swiftly implement a low-power converter in a buck or boost mode. The bridge is of interest as it includes diagnosis and protection circuits. Driving such a bridge is rapidly done by sending low-voltage pulses, generated by using a PWM-chip or waveform generator. Special care is taken on the cooling of the H-bridge, particular in the starting-up of the dc-motor connected with the flywheel. For this, losses are predicted by using the datasheet of the H-bridge, making students more familiar with using such documents.

Proper chooses have to be made regarding the electrical machines. Student have to search for used motors obtained from household appliances or toys, often ending up with low-efficient machines. Besides the nominal values for current and voltage, machine parameters have to be measured by using RLC-meters or by applying easy identification tests such as a voltage step response. Machines with low value resistors and high inductances are preferable to support the generator mode. In generator mode the H-bridge together with the inductance of the machine constitute a boost converter. The input voltage of this boost converter corresponds to the speed-dependent back-emf added with the voltage drop across the stator resistor. For bachelor students the voltage drop across the semiconductor devices is neglected or considered as a distortion. As the speed of the flywheel decreases during generator mode, boosting to a constant dc-bus voltage means the duty ratio of the lower transistor needs to decrease.

To control the dc-bus voltage by using this duty ratio, the generated voltage at the input of the boost converter needs to be measured. However, this voltage is unknown as the voltage measured at the output of the generator is one of the dc-bus voltages (this voltage includes the voltage drop across the machine inductance as well). As an alternative students are

Fig. 3: Rotor of dc-machine showing deep rotor slots, a measurement coil of 10 windings is visible around the stator yoke.

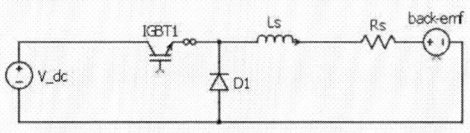

(a) Motor operation: kinetic energy is stored in flywheel through buck-mode operation of the H-bridge.

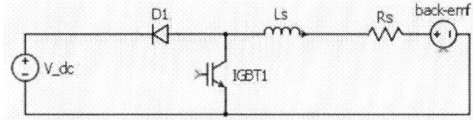

(b) Generator operation: kinetic energy is converted back to electric energy through the boost-mode operation of the H-bridge.

Fig. 4: Buck and boost operation mode of the integrated H-bridge

advised to estimate the input voltage of the boost converter from the current and the back-emf. As the latter is proportional to the speed, a speed measurement should be obtained instead. Instead of buying the more expensive resolver or encoder, students install a copper coil over the machine stator yoke themselves, generating a signal proportional to the speed. This variation is due to the rotor saliency and is large in low-power dc-machines as the rotor shows rotor slots relative deep compared to the rotor diameter, Fig. 3.

The mechanical design concerns the bearings, shaft connections and mass for the flywheel. The latter is often made of composite materials [7], however, for a student project, dumbbell disks are used instead taking into account the low budget boundary. For high masses the bearings of the dc-machine are computed to be insufficient to support the weight and hence additional bearings with low friction are required. Students have to choose for themselves a horizontal or vertical positioning of the flywheel shaft. Different bearings need to be chosen depending on a setup with vertical or horizontal shaft. Magnetic bearings, as described in [5], [4], [6], would result in low friction losses. However, adding magnetic levitation would increase the complexity of the system and is regarded as a follow-up project for next years.

B. Control System

Within the controller, two operation modes are implemented. One mode, referred to as motor-mode operation, accelerates the flywheel. For this mode, the micro dc-grid is coupled to the ac grid by means of a rectifier. The H-bridge together with the inductance of the machine are used as buck converter, Fig. 4(a). Electric power, coming from the ac-grid, is rectified and then transformed to and stored as kinetic energy. By measuring the speed of the dc-machine connected with the flywheel, the kinetic energy stored can be computed and controlled. Measuring the speed is done by using a speed measurement system described in more detail in the next section.

A second operation mode within the controller concerns the generator mode. For this mode, the ac-grid is disconnected from the micro dc-grid. By using the H-bridge together with the dc-machine as a boost converter, Fig. 4(b), the voltage

of the dc-bus is controlled to a set value appropriate to the operating point of the load, here a second dc-machine. Considering the duty ratio δ of the higher switch of the inverter leg supplying the dc-bus and considering complementary steering of upper and lower transistors, then the relation from input to output of the boost converter within the H-bridge is given as $1/\delta$. This means that to obtain a certain dc-voltage v_{wens} and by knowing the input voltage of the boost converter v_s, the duty ratio can be computed as

$$\delta = v_s / v_{wens}. \tag{1}$$

The resulting value is used in the feed-forward path of the controller. As the load current (to the dc-bus) can distort the dc-voltage, a PI-controller is added to support the dc-voltage control. The basic scheme of the controller is shown in Fig. 5, the set value of the dc-bus being 32.4V. Details are given in the Simulink model, Fig. 6. Analog circuits based on opamps are build, some of the electric schemes shown in Fig. 7.

C. Measurement System

The input voltage of the boost-converter is given as the back-emf added to the voltage drop across the stator resistor of the dc-machine connected with the flywheel. To estimate the back-emf of this dc-machine, a measurement coil with 10 windings is placed around the stator yoke, Fig. 3. Placing such a coil can be done with little effort as small dc-machines have many opening in the stator and a removable front lid. When closing the lid again, care has to be taken to place the brushes

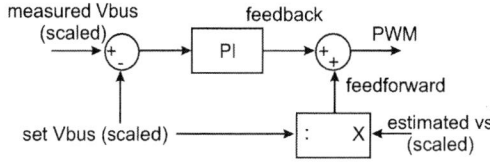

Fig. 5: Block diagram of the control loop in generator mode.

The 2014 International Power Electronics Conference

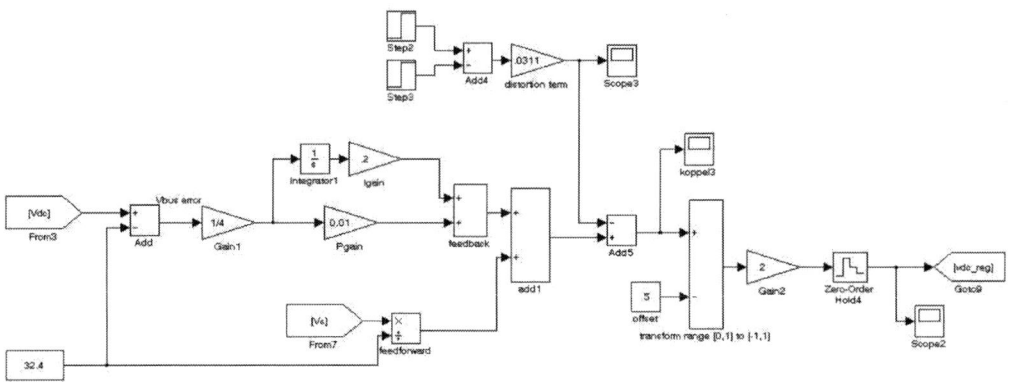

Fig. 6: Simulink blocks, control of the dc-voltage by using feedforward and PI-control on the dc-voltage error.

on the commutator in a good manner to avoid short circuiting the voltage supply. Due to the large reluctance variation along the air gap, a signal is measured within the measurement coil that depends on the speed. The measurement coil and rotor for two different rotor position is illustrated in Figs. 8(a)-8(b).

Processing the speed-dependent signal includes rectification, Fig. 9(a) and filtering, Fig. 9(b) and results for a 20V stator voltage in signals shown in Fig. 10. Loading and unloading of the capacitor in the rectifier can be observed clearly. To verify the speed-voltage relationship, the dc-machine is supplied with a controllable dc-source making it able to vary the speed of the machine. For different voltages, hence rotor speeds, the voltage of the measurement coil is given in Fig. 11. From this, it follows that the relationship is by approximation linear. To estimate the back-emf, this measurement signal is multiplied by a voltage constant. From this result and by measuring the stator current, the input voltage of the boost converter is reconstructed and used in the control algorithm previously described.

III. DYNAMICAL MODEL

The setup studied comprises two electrical machines, one connected to the flywheel and regarded as generator, the other machine operating as a motor, generating a desired torque.

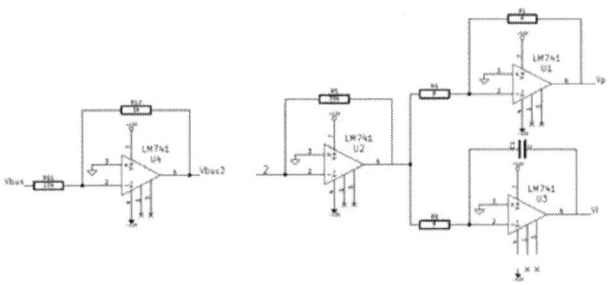

Fig. 7: Opamp-circuits concerning V_{bus} measurements (left) and PI-controller (right).

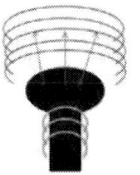

(a) Position of a rotor tooth under measurement coil, maximum inductance.

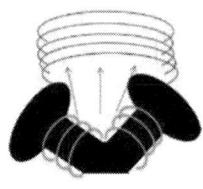

(b) Measurement coil axis located between two rotor teeth, minimum inductance.

Fig. 8: Different rotor positions with respect to axis of measurement coil

A first model is constructed in Simulink as shown in Fig. 2 for which design components from the toolbox Sim-PowerSystems has been included. The control strategy for dc-voltage control of the generator drive has been programmed in Simulink as well and are given in previous Fig. 6. Proper care has been taken to transform the boundaries [0,1] used by the output controllers to the boundaries [-1,1] defined by the PWM sawtooth carrier. Such a time-stepping model includes the switching harmonics and is often difficult to interpret by students. Therefore, a second model is obtained as a transfer function by which the influence of parameters on the dynamic behaviour can be studied directly by using the root-locus method.

Neglecting the discrete-time character of the controller and the high-frequency content of the signals due to the switching actions of the semiconductor devices, a dynamic model can be obtained in the Laplace-domain. This model is given by the transfer function from rotor speed variation to dc-voltage variation as follows

$$\frac{\Delta v_{dc}}{\Delta v_s} = \frac{R_L + sL_L}{R_L + sL_L + (R_s + sL_s)(1 + SC(R_L + sL_L))}, \tag{2}$$

for which subscripts L and s refer to the load machine and voltage source respectively. Using time constants $\tau_s = L_s/R_s$ and $\tau_L = L_L/R_L$ for source and load respectively the

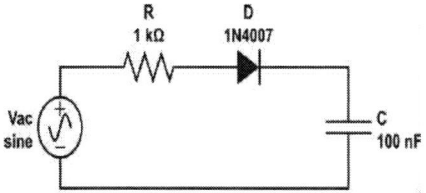

(a) peak detection of the output of the measurement coil

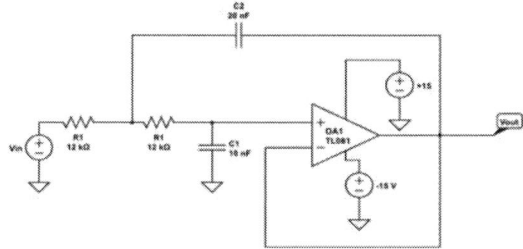

(b) second order filter applied to the output of the measurement coil.

Fig. 9: Electronic schemes related to the measurement coil

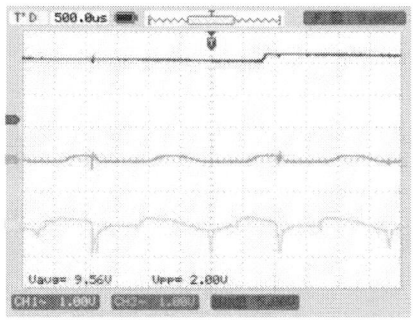

Fig. 10: Oscilloscope: output coil (yellow), filtered output (blue), rectified filtered output (pink).

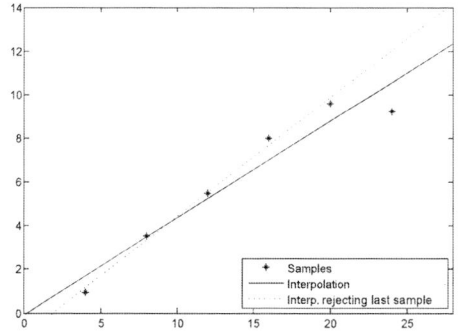

Fig. 11: Output voltage of the measurement coil for an unloaded dc-machine at different stator voltages (and rotor speed).

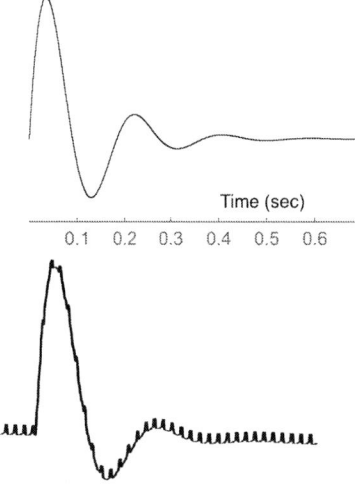

Fig. 12: dc-voltage response on impulse, computed by time simulations (below) or by using SISOTOOL (above).

characteristic equation is given by

$$1 + KF(s) = 0 \tag{3}$$

where the open-loop gain K is given as

$$K = \frac{L_s + L_L}{CL_sL_L} \tag{4}$$

depending on the inductances of source and load as well as the dc-bus capacitor; the function F(s) is given as

$$F(s) = \frac{s + \tau_t^{-1}}{s(s + \tau_s^{-1})(s + \tau_L^{-1})}, \tag{5}$$

for which a new time constant $\tau_t = (L_L + L_s)/(R_L + R_s)$ is defined. From this equation, students can examine the stability of the system by using SISOTOOL, a graphical environment in Matlab. In particular, the location of closed-loop poles can be studied depending on the open-loop gain K by using the root locus method.

Using either the transfer function (3) and SISOTOOL or the time simulation in Simulink, the response of the dc-voltage on variations in the speed of the fywheel can be studied. Fig. 12 shows the results for both methods in case an impuls is given in the speed. From this the good agreement between both models follows, meaning the high-frequency signals can be disregarded when studying the dynamic behaviour in a more qualitative manner.

IV. EXPERIMENTS

A test setup has been made by students with a limited budget of 50 Euro for each student. As each group was made out of 4 student, the total budget was 200 Euro. Most of this budget went to the electronics, PCB boards, additional bearings and materials to support the test ridge. The flywheel and motor were recycled from other devices. The measurement coil is a self-made coil placed over the yoke of the stator. Fig. 13

shows the overall test setup with self-made ac-dc power supply, integrated H-bridge and self-made PCBs for measurement and control purposes where Fig. 14 shows the flywheel and motor. The front panel includes switches in order to set the test ridge in different operation modes, e.g. motor mode, discharging capacitor, using the dissipator, generator mode, control of the second dc-motor as a load of the dc-grid, feedforward control or added feedback control of generator with flywheel. Students have limited help from experienced technicians to make some mechanical support components. The integrated H-bridge has been the main lead to design the setup. With limited resources and limited schemes to work from, students are pushed to search, ask and gather information themselves. Furthermore, close contact has been made with PhD-students around them who were familiar with (power) electronic design.

Fig. 13: Student test setup of a micro dc-grid with variable-speed flywheel en motor load.

Fig. 14: Closer view on flywheel, bearings and dc-generator.

V. ACKNOWLEDGEMENTS

This work was supported by the FWO project G.0083.13N. The research was performed as part of the Interuniversity Attraction Poles-Programme (IUAP) P7/02 financed by the Belgian government.

VI. CONCLUSIONS

This paper discusses an interdisciplinary project (for 6 credit points of a total of 60 credit points a year) which can be performed by multiple groups of 3th year bachelor students and without much guidance of university personnel. Each group consisted out of 4 students working during the period of the last semester for 20% of the available study time. The project concerns the design of a micro dc-grid for which energy is stored in a variable-speed flywheel. The idea is to firstly store energy coming from an ac-grid and to use this stored energy to drive an electrical machine. When re-using the stored energy, the ac-grid is disconnected and the dc-voltage is boosted to a more appropriate voltage level for the load and its operating point. For this project a limited fund is made available to the students forcing them to use integrated modules, to recycle many of the components such as machines and inertias or to make sensors themselves. In this paper, the latter is done by putting a copper wire around the stator of the dc-generator in order to estimate speed and from it the back-emf. By using simulation models in Simulink or transfer functions studied within the environment SISOTOOL of Matlab the dynamic behaviour of the micro dc-grid is studied, hence supporting the decision in system parameters and tuning of the controllers. Reaching the goals set forward in this project, students can then focus on the applications such as a micro dc-grid with energy storage for a transportation system in which many vehicles accelerate and decelerate (e.g. tram network). Each group of students has to design a A0-poster for a Project Presentation Day on which the setup is shown as well to fellow students and professors participating to the different projects. Additionally, each group has to write a final report.

REFERENCES

[1] Kusko, A. and J. DeDad, "Stored energy - Short-term and long-term energy storage methods", *Industry Applications Magazine, IEEE*, 13(4):66-72, 2007.

[2] Ruddel, Dr Alan, "Investigation on Storage Technologies for Intermittent Renewable Energies: Evaluation and recommended R & D Strategy", *Technisch Rapport, INVESTIRE Thematic Network, 2003.*

[3] Infineon technologies, "5-A H-Bridge for DC-Motor Applications", *datasheet of the TLE 5205-2, 2001.*

[4] Filatov, Alexei V. and Eric H. Maslen, "Passive magnetic bearing for ywheel energy storage systems", *IEEE Trans. on Industry Applications*, 37(6):3913-3924, 2001.

[5] Andrade, R. de, G.G. Sotelo, A.C. Ferreira, L. G B Rolim, J.L. da Silva Neto, R.M. Stephan, W.I. Suemitsu, and R. Nicolsky, "Flywheel Energy Storage System Description and Tests.", *Applied Superconductivity, IEEE Transactions on*, 17(2):2154-2157, 2007, ISSN 1051-8223.

[6] Lee, Jisung, Sangkwon Jeong, Young Hee Han, and Byung Jun Park, "Concept of Cold Energy Storage for Superconducting Flywheel Energy Storage System. Applied Superconductivity", *Superconductivity, IEEE Transactions on*, 21(3):2221-2224, 2011.

[7] Bitterly, J.G., "Flywheel technology: past, present, and 21st century projections", *Aerospace and Electronic Systems Magazine, IEEE*, 13(8):13-16, 1998.

Development of a Web-Based Remote Experiment System for Electrical Machinery Learners

Makoto Ishibashi*, Hisao Fukumoto*, Tatsuya Furukawa*, Hideaki Itoh* and Masashi Ohchi[†]
*Graduate School of Science and Engineering,
Saga University, Saga, Japan
[†]Department of Electrical, Electronics and Computer Engineering,
Chiba Institute of Technology, Narashino, Japan

Abstract—**Experiments have critical roles in engineering education. However, it is not easy to conduct experiments because of their high costs of equipment and unavailability. Remote experiment systems have coped with this issue. Using remote experiment systems, users can conduct experiments regardless of their locations or times. Remote experiment systems have been developed worldwide, but most of the systems are developed for personal computers, and therefore cannot be used with smartphones or tablets. This study aims to develop a remote experiment system which can be used with smartphones or tablets. The proposed system provides remote control of a DC motor by PWM control and remote measurement of a single phase alternator. The user interface of the system is designed without Web plug-ins so that users can use the systems with their smartphones or tablets as well as their personal computers.**

Keywords—electrical machinery, remote experiment system, smartphone, web

I. INTRODUCTION

Electrical machinery is used for various purposes, for example, generating electrical power, transforming voltage and controlling robots. Education of electrical machinery is important in order to improve electrical power engineering. For education of electrical machinery, it plays a very important role to carry out experiments on electrical machinery. However, experiments need high costs of their equipment and limit locations and times, which makes it difficult to conduct experiments. To cope with these issues, remote experiment systems have been developed. By sharing measurement instruments, the costs of experiments are lowered. By conducting experiments from a remote location, users can experience experiments regardless of their locations and times. Remote experiment systems have been developed worldwide [1]–[3], and the authors also have developed remote experiment systems for electrical engineering [4]–[6].

However, the conventional remote experiment systems are designed for personal computers, and most of the systems cannot be used with smartphones or tablets. The present study is aimed at developing educational support systems on electrical machinery, which can be used with smartphones or tablets as well as personal computers. The system's target is undergraduate students of electrical engineering. It is estimated that the user makes use of this system for confirmation of theoretical results, or for preparation of actual laboratory experiment.

The authors have developed a remote experiment system as a part of the educational support systems. The remote experiment system enables the user to control a DC motor by PWM (Pulse Width Modulation) control and measure a voltage generated by a single phase alternator from a remote location by using a Web browser. The user interface of the system is developed without Web plug-ins, e.g., Adobe Flash or Java applet, so that the system can be used with smartphones or tablets as well as a personal computers. In this paper, the remote experiment system of the educational support systems is described.

II. OVERVIEW OF THE SYSTEM

The authors are going to develop educational support systems for electrical machinery learners. Fig. 1 shows the configuration of the educational support systems. The system is going to integrate the following three systems.

1) Principles of Operation Learning System
2) Simulation System
3) Remote Experiment System

In the present study, the authors developed a 3) Remote Experiment System, as a part of the educational support systems.

The remote experiment system enables a user to conduct an experiment on a DC motor and a single phase

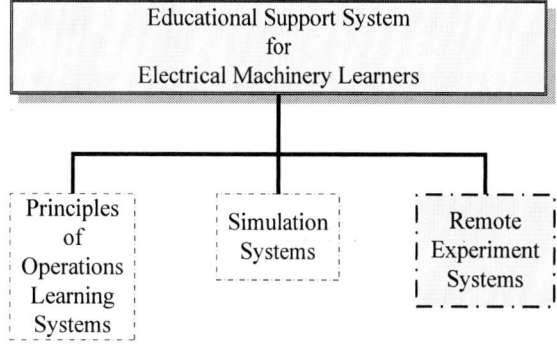

Fig. 1. Configuration of the educational support system

alternator from a remote location with a Web browser. Figure 2 shows an overview of the system. The user can operate a microcomputer board from a remote location to control the DC motor and measure a generated voltage of the alternator. The system needs only a Web browser, and the user can conduct the remote experiment without installing a special software. Furthermore, the system does not need Web plug-ins, e.g., Adobe Flash or Java applet, so that the user can use the system with a Web browser of his/her smartphone.

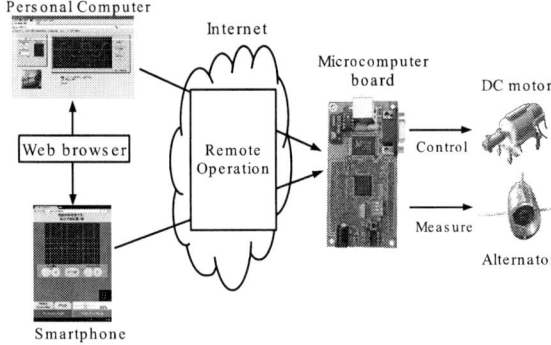

Fig. 2. Overview of remote experiment system

As compared with other remote experiment systems, this remote experiment system has some merits as follows.

- Can be used with smartphones or tablets as well as personal computers

- Can build a remote experiment system at a relatively low cost

As mentioned above, the system can be used with smartphones or tablets also. In the remote experiment system, a microcomputer board which is at a relatively low price, and easy electronic circuits are used to control the DC motor and measure the generated voltage of the alternator.

III. MEASUREMENT AND CONTROL

A. Instrument

The system provides a remote experiment on a DC motor and a single phase alternator. To control the DC motor and measure the voltage generated by the alternator, the authors adopted an H8/3069F microcomputer board (hereinafter referred to as H8/3069F) [7]. H8/3069F is a microcomputer board, which has a 16-bit microcomputer, Renesas H8/300. Typical features of the H8/3069F include the following.

- Relatively low price, about US$40.

- Embedded with 20 MHz CPU, 512 kB flash ROM, 16 kB RAM.

- Include various functions: eight channels 10-bit A/D converter, Ethernet controller, PWM signal output and so on.

The OS of the H8/3069F is MES (Micro Embedded System) [8], which is an embedded OS made by Yukio Mitsuiwa for microcomputers. MES enables multi tasking on the H8/3069F, and makes development of systems with the H8/3069F easy.

B. Handmade alternator

There are few educational reports of single phase alternators whose motions are visible. However, visibility of the motion is important for learning the essentials of electrical machinery. Therefore, the authors made a single phase alternator whose motion is visible. Figure 3 shows the single phase alternator made by the authors. The alternator is made by setting a pair of permanent magnets fixed with a pole inside a coil. When the pole rotates, the magnets are rotated and the flux linkage in the coil changes. Thereby the alternator generates electricity in synchronization with the rotation of the magnets.

Figure 4 shows voltage-rotor speed characteristics of the alternator. The generated voltage is proportional to the rotor speed. The alternator should be suitable for being used for remote experiment owing to its theoretical characteristic and visibility.

C. Control a DC motor

The DC motor is controlled by PWM control. PWM signals output from the H8/3069F are controlled by setting the H8/3069F's parameters: period and duty cycle. The H8/3069F outputs the PWM signals and controls the

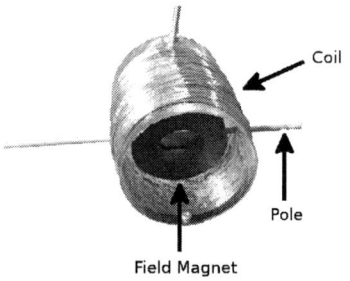

Fig. 3. Handmade single phase alternator

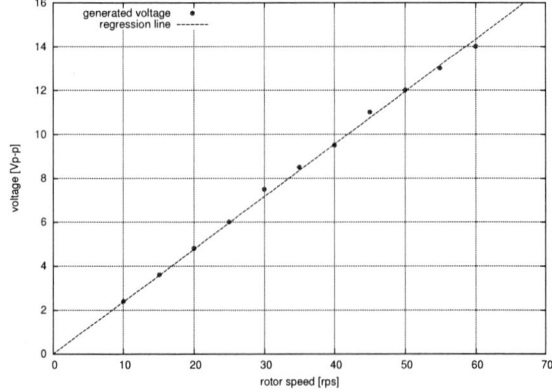

Fig. 4. Voltage-rotor speed characteristics of the alternator

DC motor through a TA7291P motor driver IC [9]. The DC motor and the alternator is connected to each other. Therefore, when the DC motor rotates, the alternator's rotor rotates in synchronization with the motor. Thus, the H8/3069F controls the rotor speed by PWM control.

D. Measure an alternator output

As mentioned above, the rotor speed is controlled by the H8/3069F, then the alternator generates electricity depending on the rotor speed. The voltage generated by the alternator is measured by an A/D converter of the H8/3069F. The sampling frequency of the A/D converter is approximately 37.4 kHz as the maximum. The A/D converter can read a 0 to +5 V analogue input. Thus, a voltage input into the A/D converter needs to be limited within this voltage range.

Figure 5 shows the circuit inserted between the alternator and the A/D converter. This circuit attenuates the generated voltage, applies a DC bias voltage to the generated voltage and removes noises. The alternator generates electricity at most 14 V peak-to-peak. Thereby, the generated voltage is attenuated by about one third. Besides, the circuit applies a DC bias voltage of +2.5 V to set a standard voltage to be the middle of the voltage range. To remove noises, the circuit include a low pass filter, whose cut-off frequency is approximately 160 Hz.

Fig. 5. The circuit to attenuate the voltage, apply a DC bias voltage and remove noises

IV. NETWORK OF THE SYSTEM

Figure 6 shows a network of the system. To establish communication between a Web browser and the H8/3069F, a Web server is introduced. Therefore, the remote experiment is conducted with a Web browser through the Web server. The communication between the Web server and the H8/3069F is made with TCP and UDP, and the communication between a Web browser and the Web server is made with the WebSocket protocol [10]. The WebSocket is a TCP-based protocol, and makes the communication between a Web browser and a Web server more interactive and real-time. Unlike the usual HTTP protocol, the WebSocket does not functions as a request-response protocol, so that the server push is enabled with the WebSocket. Furthermore, the size of the WebSocket header is much smaller than that of the HTTP header:

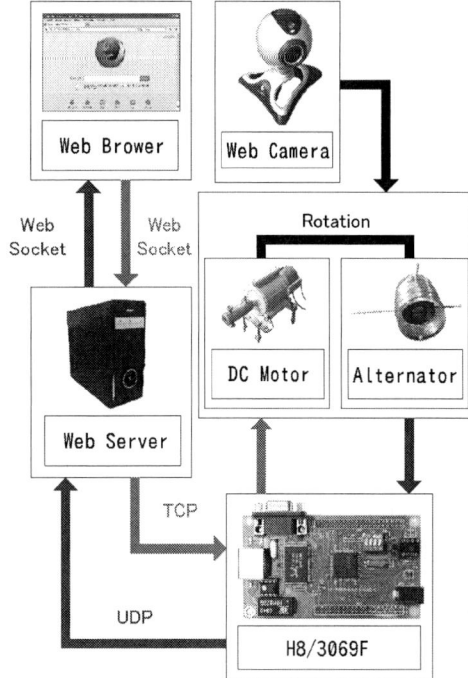

Fig. 6. Network of the system

the WebSocket header is about 10 Byte, and the HTTP header is about 300 Byte.

The authors adopted Node.js [11] as a Web server. Node.js is a server-side solution for JavaScript, for receiving and responding to HTTP request and for sending and receiving the WebSocket data. Node.js has the characteristics of lightness, tolerance to multiple connections and so on.

The value of a duty cycle of PWM is decided with a Web browser, and the value is sent to the H8/3069F from the Web browser through the Web server. The H8/3069F outputs PWM signals depending on the duty cycle decided in the Web browser, and controls the DC motor. Then, the rotor of the alternator rotates, and the alternator generates electricity. The generated voltage is converted to digital values by the A/D converter. Then, the values are sent to the Web browser through the Web server. Thereby, the remote control and the remote measurement with a Web browser are enabled.

A Web camera is connected to the Web server, and takes a video of the DC motor and the alternator. The video is converted into an image and sent to the Web browser every 33 ms (30 fps). Thus, learners can see the motions of the DC motor and the alternator from a remote location and make use of it for learning.

V. USER INTERFACE

The user interface of the system is developed with HTML, CSS and JavaScript so that the system can be used with smartphones or tablets as well as personal computers. Figure 7 shows a Website displayed on a Web

978-1-4799-2706-7/14 $31.00 © 2014 IEEE 726

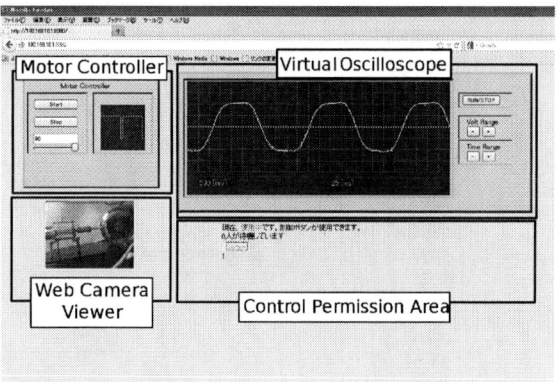

Fig. 7.   User interface

browser when we access the Web server. The Website consists of four components: the motor controller, the virtual oscilloscope, the Web camera viewer, and the control permission area. On this Website, it is possible to control the DC motor, measure the generated voltage and observe the video of the DC motor and the alternator, which is taken by the Web camera.

A. The motor controller

Figure 8 shows the magnified figure of the motor controller. The motor controller makes it possible to control the DC motor. Using the motor controller, the rotation of the DC motor can be controlled: start, stop and change the motor speed. An image of the PWM signal is displayed in order to make it easy to understand the relation between the duty cycle and the DC motor speed.

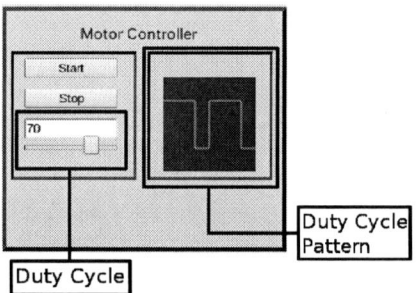

Fig. 8.   The motor controller

B. The virtual oscilloscope

The virtual oscilloscope in Fig. 7 makes it possible to measure the generated voltage. The virtual oscilloscope indicates voltage data as waveforms like a practical oscilloscope does. The user reads the scale of the virtual oscilloscope, then measures the voltage and period of the generated voltage. Also, the user can stop the waveform and change the vertical or horizontal scales just like a practical oscilloscope.

The authors carried out an experiment in order to confirm the validity of the virtual oscilloscope. The experiment was carried out by the following methods.

1) Output AC 2 V peak-to-peak, 20 Hz with a function generator.
2) Input the output of the function generator into the circuit shown in Fig. 7
3) Confirm if the displayed result on the virtual oscilloscope is correct or not

Figure 9 shows the waveform displayed on the virtual oscilloscope when an AC voltage of 2 V peak-to-peak and 20 Hz is input to the circuit shown in Fig. 9. By reading the scale, it turns out that the voltage is 2 V peak-to-peak and the frequency is 20 Hz. Therefore, the virtual oscilloscope displays the correct waveform. Thus, the validity of the virtual oscilloscope is confirmed.

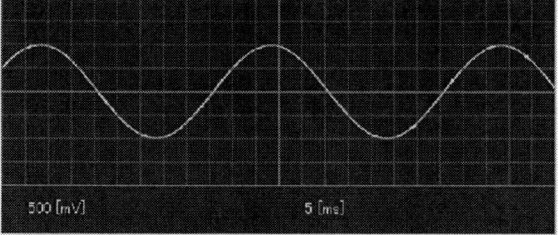

Fig. 9.   The virtual oscilloscope monitor when input AC 2 Vp-p, 20 Hz

C. The Web camera viewer

If the motions of the DC motor and the alternator are not displayed, for the user there is little difference between the experiment system and a simulation system. However, experiments with real machines are important for learning the essentials of electrical machinery. Therefore, the motion of them needs to be displayed in order for the user to see that they carry out experiments with real machines. The Web camera viewer makes it possible to observe the video of the DC motor and the alternator. By displaying the motion of them, the user can see the relation between the duty cycle of PWM and the DC motor speed, and can see the the relation between the generated voltage and the rotor speed. Thus, the Web camera viewer help the user to understand the engineering phenomena.

D. The control permission area

The remote experiment system has one DC motor, one alternator and one control/measurement device (H8/3069F). Therefore, the remote experiment system cannot be used by more than one people. If more than one people want to use the system, it is necessary to restrict who can use the system to only one.

Figure 10 shows a situation where there are more than one users in this system. To use the motor controller and the virtual oscilloscope, the user has to click (or tap) a button in the control permission area. After clicking the button, if no one is permitted to operate the motor controller and the virtual oscilloscope, the user who clicked the button is permitted to operate them, and only

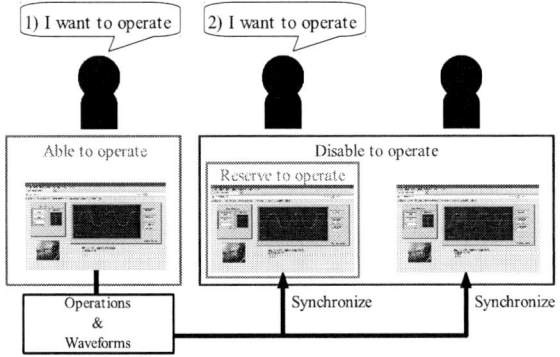

Fig. 10. Situation where there are more than one users

the user who is permitted can operate the motor controller and the virtual oscilloscope for 1 minute. When clicked the button, if someone is operating them, the user cannot operate them while he/she is operating them. Instead, the user can reserve operation of them, and operate them when the user's turn comes. Users who are not permitted can only observe the waveform in the virtual oscilloscope and the video in the Web camera viewer. Thus, the system copes with the situation when more than one users want to operate the system.

Screens displayed on every user are in synchronization with each other, and every user can see the relation between the duty cycle of PWM and the DC motor speed and between the generated voltage and the rotor speed, both when the user is not permitted to operate them and when the user is permitted to operate them.

In this method, the authors examined that three users could use the system simultaneously. The authors surmise that more than three user can use the system simultaneously owing to tolerance to multiple connections of Node.js.

VI. THE USER INTERFACE FOR A SMARTPHONE

The authors have been developing the remote experiment system which can be used with smartphones or tablets. However, the previous user interface of the system was not suitable for a smartphone which has a small display. Therefore, the authors made a new user interface for a smartphone. Figure 11 shows the user interface for a smartphone. The user interface is designed in order to make it easier to use the system with a smartphone. The user interface consists of four same component as the user interface in Fig. 7. The control permission area is displayed at the top of the user interface, and the motor controller is displayed at the bottom of it. The virtual oscilloscope and the Web camera viewer is displayed at the middle of the user interface, and the display of two components is switched by the switching tab at the bottom of the user interface. While the user operates the slide bar of the motor controller, the image of PWM control is displayed on the motor controller. Thus, the user interface for a smartphone has a same function as the user interface for a personal computer.

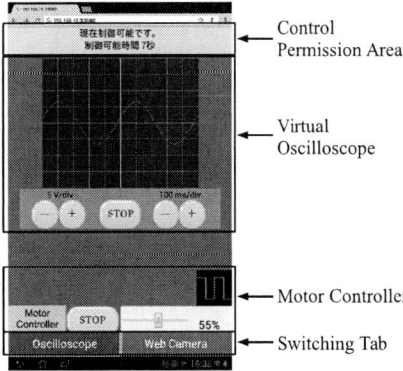

(a) Virtual oscilloscope

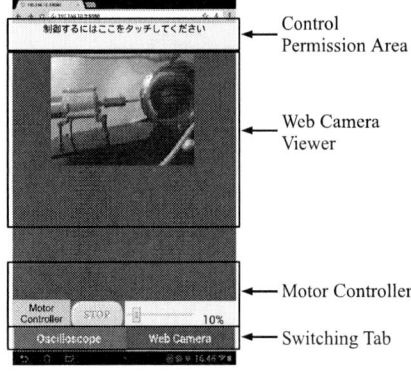

(b) Web camera viewer

Fig. 11. User interface for a smartphone

VII. GAME USING THE REMOTE EXPERIMENT SYSTEM

Figure 12 shows a game designed for the user to have fun studying the relation between the generated voltage and the rotor of alternator. The user can play the game with a Web browser. The game consists of a circle bar, a duty cycle meter and a graph.

When the user rotates the circle bar, the duty cycle is decided on the basis of the rotating speed of the circle bar. The duty cycle meter indicates the duty cycle visually. The faster the circle bar is rotated, the higher the duty cycle becomes. Then, the DC motor rotates faster, and the alternator generates a higher voltage. After the user

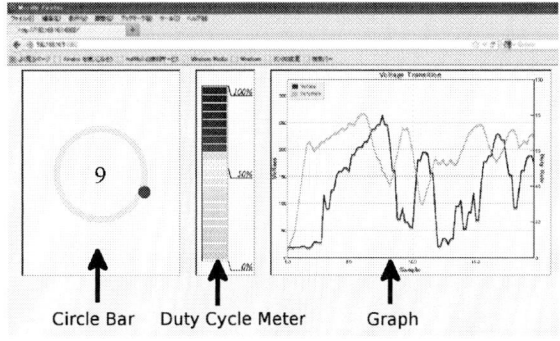

Fig. 12. Game Using the Remote Experiment System

rotates the circle bar for 20 seconds, the graph displays the transition of the duty cycle and the generated voltage.

Two lines in the graph indicate transitions of the generated voltage and the duty cycle. The dark blue line shows the transition of the generated voltage, and the light green line shows the transition of the duty cycle. The graph indicates that the transition of the generated voltage follows that of the duty cycle. Thus, the user can have fun studying the relation between the generated voltage and the rotor speed.

VIII. CONCLUSION

In the present study, the authors developed a remote experiment system which makes it possible to control a DC motor with PWM control and measure a single phase alternator from a remote location with a Web browser. Using the remote experiment system, the user can conduct an experiment regardless of his/her location or time. Compared with other remote experiment systems, the system has some features: the system can be used with smartphones or tablets, the system needs relatively low costs.

The user interface of the system is developed without Web plug-ins so that the system can be used with smartphones or tablets as well as with a personal computers. Furthermore, the authors developed a user interface for a smartphone in order to make it easy to use the system with a smartphone. For remote control and remote measurement, an H8/3069F microcomputer board, which is at a relatively low price, so that the system can be build at a low cost.

In further studies, the authors are going to tackle the following four works: (1) development of learning materials and simulation systems for electrical machinery, (2) addition of control/measurement targets, e.g., stepping motor and so on, (3) assessment of educational effects of the remote experiment system and (4) check of tolerance to many users.

REFERENCES

[1] B. Balamuralithara and P. C. Woods, "Virtual Laboratories in Engineering Education: The Simulation Lab and Remote Lab", Comput Appl Eng Educ, vol. 17, iss. 1, pp. 108-118, 2009.

[2] M. Tawfik, E. Sancristobal, M. Sergio, R. Gil, G. Diaz, A. Colmenar, K. Nilsson, J. Zackrisson, L. Hkansson and I. Gustafsson, "Virtual Instrument Systems in Reality (VISIR) for Remote Wiring and Measurement of Electronic Circuits on Breadboard," IEEE Trans. on Learning Technologies, vol. 6, iss. 1, pp. 60-72, 2013.

[3] B. Popovic, N. Popovic, D. Mijic, S. Stankovski and G. Ostojic, "Remote control of laboratory equipment for basic electronics courses: A LabVIEW-based implementation," Comput Appl Eng Educ, vol. 21, iss. s1, pp. E110-E120, 2013.

[4] M. Ohchi, T. Furukawa, S. Tanaka, "Development of Education Support System for Numerical Electromagnetic Analysis Based on Server-Client Model using Java," IEEJ Trans. FM, vol.125, no. 8, pp.621-628, 2005. (in Japanese)

[5] M. Ohchi, T. Yoshitomi, K. Akagi, S, Sasaki, T. Furukawa, "Design and Implementation of Study Support System for Electronic Circuits with Observable V-I Characteristics," IEEJ Trans. FM, vol. 128, no.4 pp. 275-282, 2008. (in Japanese)

[6] T. Yoshitomi, T. Furukawa, S. Sasaki, "Design and Implementation of Web based Measurement System for Beginner of Electric and Electronic Circuit," IEEJ Trans. FM, vol. 130, no. 1, pp. 67-73, 2010. (in Japanese)

[7] AKIZUKI DENSHI TUSHO (in Japanese) http://akizukidenshi. com/catalog/g/gK-01271/

[8] Micro Embeded System/MES(H8/OS 5.0) - SourceForge.J http: //mes.sourceforge.jp/mes/

[9] TA7291P — Products — TOSHIBA Semiconductor & Storage Products Company http://www.semicon.toshiba.co.jp/info/lookup. jsp?pid=TA7291P\&lang=en

[10] WebSocket http://ajf.me/websocket/

[11] node.js http://nodejs.org

The 2014 International Power Electronics Conference

Development of Power Measurement System in Simulated Micro Grid System for Education

Yuki Hira, Tatsuya Furukawa, Seichiro Yakabe, Hisao Fukumoto, Hideaki Itoh
Graduate School of Science and Engineering
Saga University
Saga, JAPAN

Masashi Ohchi
Faculty of Engineering
Chiba Institute of Technology
Chiba, JAPAN

Abstract—In recent years, "Smart Grid" obtains much attention. In the our laboratory, "Simulated Micro Grid System" for supporting learning of power engineering has been designed and implemented. The system helps young engineers to understand the principles of the Smart Grid technology. In this study, we have developed a power measurement system for our Simulated Micro Grid System. AD conversion is essential in the power measurement system to control components in the Simulated Micro Grid System and deliver measured data on the Web. In this paper, we have proposed two methods, AD conversion using ART–Linux and AD conversion using SH–3 microcomputer. The method by ART–Linux is used when real-time control is required. Moreover, the method by SH–3 microcomputer is used when there are many measurement targets.

I. INTRODUCTION

In recent years, the new power network called "Smart Grid" obtains much attention. The Smart Grid is a power network, where power supplies and demands will be automatically optimized by using IT technology.

Recently, studies on the smart grid are being carried out in many universities and companies. Research fields of the smart grid include a lot of fields such as measurment, communication and control. For young engineers who begin to study the smart grid, it is difficult to understand those fields. In the our laboratory, "Simulated Micro Grid System" for supporting the learning of power engineering has been designed and implemented.[1] The system helps young engineers to understand the principles of the Smart Grid technology using the technology of the CAI systems which have developed in our laboratory[2][3].

AD conversion is essential in the power measurement system to develop the system for delivery and display of measurement data on the Web. Thus, in this study, we have developed a power measurement system in the Simulated Micro Grid System which is a basis of the whole system. We will propose two methods, that is AD conversion using ART–Linux and AD conversion using SH–3 microcomputer. This micro grid system must measure large electric power and it is necessary to perform electric power control in real time. Thus, when measuring the target which needs real-time control, the method by ART–Linux is used. The method with the SH–3 microcomputer is used when there are a lot of measurement targets because the system is inexpensive.

II. OVERVIEW OF SIMULATED MICRO GRID SYSTEM

Fig. 1 shows the overview of the Simulated Micro Grid System which has been designed and implemented in our laboratory.

The main PC and the microcomputer will measure the power which has been generated by the simulated wind power station which is made in our laboratory and by solar panels. The main PC is equipped with an AD / DA converter board in order to perform the measurement and control of the power. The power which is generated by solar panels is charged in the battery. Because the battery is a DC power supply, it can not be used for electrical devices operating at commercial power supply. Therefore, the system uses an inverter in order to convert DC power to AC power. AC power supply which is converted by the inverter is intended to supply electrical devices such as the light bulb attached to the robot arm. The robot arm is used to simulate the solar power. In Addition, by using a network camera, our system can be monitored on the web, and hence the principles of the Smart Grid technology can be learned in an intuitive way.

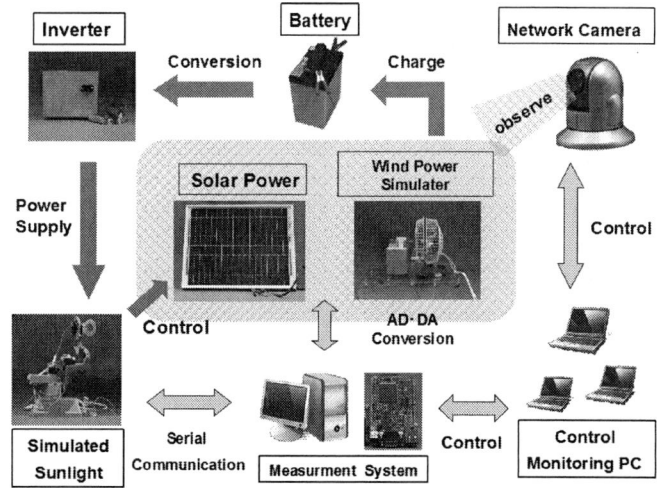

Fig. 1. Overview of the Simulated Micro Grid System.

III. PRINCIPLE OF POWER MEASUREMENT

We intend to develop the system which performs measurement, communication, and control simultaneously by using the

978-1-4799-2706-7/14 $31.00 © 2014 IEEE   730

main PC in which ART–Linux was installed. For example, we intend to develop a system which enables power measurement using AD conversion, real–time control of the actuators of both the simulated wind power station and the robot arm, and transmission of he measured data to other PCs using ART–Linux. In this study, we developed a power measurement system using AD conversion which is a basis of the whole system. AD conversion using ART–Linux has the feature that AD conversion can be performed in real–time, as compared with the usual AD conversion.

A. ART–Linux

In the main PC in the Simulated Micro Grid System, ART–Linux is installed as an operating system. The ART–Linux is a purely domestic Real Time Operating System which was developed by Yoich Ishiwata et al. at the National Institute of Advance Industrial Science and Technology[4].

The seven system calls is defined in ART–Linux.

- `art_enter`
- `art_exit`
- `art_wait`
- `art_wait_phase`
- `art_yield`
- `art_adjust`
- `art_shift`

The real–time process is not supported in usual Linux OS. However, using ART-Linux, it will be feasible using above system calls.

B. Power Measurement

We measured the power of the circuit that imitates the consumers using this system.

1) Circuit of the Consumers: Fig. 2 shows the simulated consumer circuit[5]. The circuit was built by connecting all the components in parallel, using an inverter–type fluorescent light of 32 [W] and four incandescent light bulb of 40 [W].

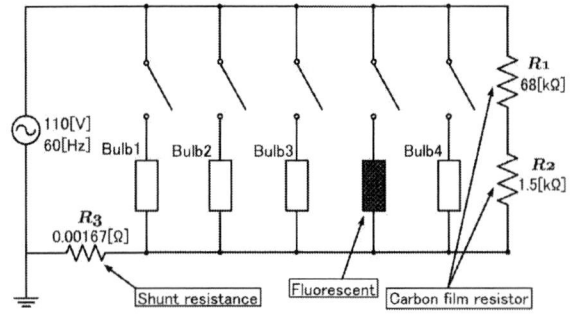

Fig. 2.   Circuit of the consumers.

2) Power Measurement Circuit: To measure the power of the simulated consumer circuit, we made an average power measurement circuit using an analog device[5]. Fig. 3 shows the circuit diagram. It is possible to measure in real time the average value of the signal by entering the signal to the circuit. The instantaneous electric power used as an input signal is generated by inputting current and voltage into the analog multiplication element (AD633) of ANALOG DEVICES company.

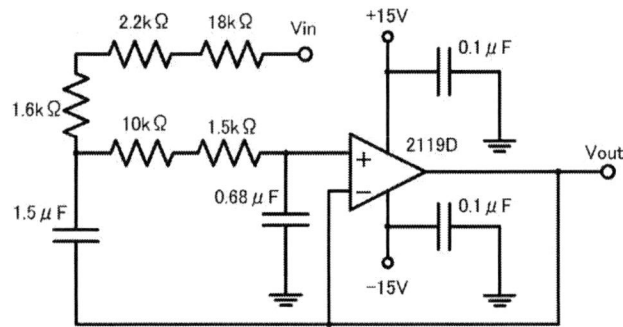

Fig. 3.   Output circuit of average power.

3) Measurement of the Power: The voltage of 110 [V] is applied to the circuit of the consumer by the slidac. Then, all the filament lamps connected to the circuit of the consumer were made turned on. Moreover, the sampling frequency of the AD convertion is set to 5 [kHz]. Using AD conversion using ART–Linux and a digital oscilloscope, we measured the electricity of the consumers circuit when an inverter-type fluorescent lamp is turned on and when it is turned off. When the inverter type fluorescent lamp is turned off, the result of the measurement with ART–Linux and a digital oscilloscope is shown in Figs.4 – 7 When the inverter type fluorescent lamp is turned on, the result of the measurement with ART–Linux and a digital oscilloscope is shown in Figs.8 – 11. The maximum, the minimum, average, and standard deviation of the execution cycle of the created program are shown in Tables I and II, when inverter–type fluorescent is on and off.

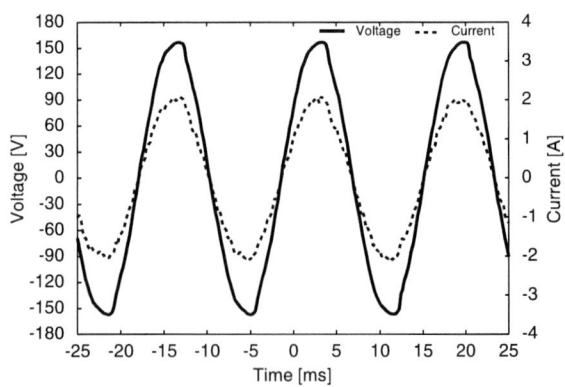

Fig. 4.   Voltage and current waves at the time of inverter–type fluorescent putting out lights(ART–Linux).

Comparing Fig.6,7,10 and 11, it can be seen that the current wave in Fig.10 and the instantaneous power wave in Fig.11 is

The 2014 International Power Electronics Conference

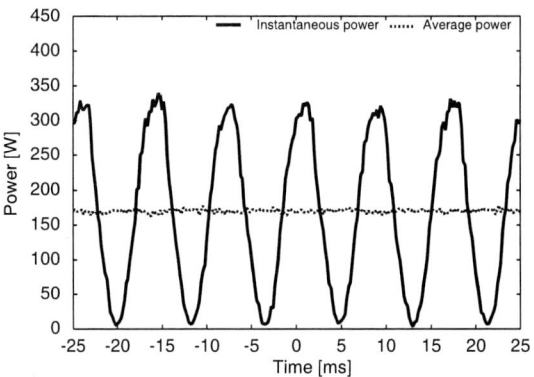

Fig. 5. Product and average waves at the time of inverter–type fluorescent putting out lights(ART–Linux).

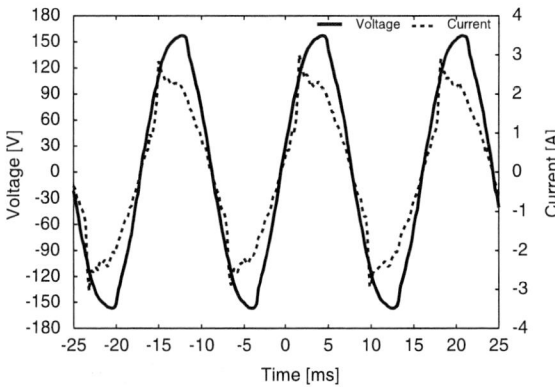

Fig. 8. Voltage and current waves at the time of inverter–type fluorescent lighting(ART–Linux).

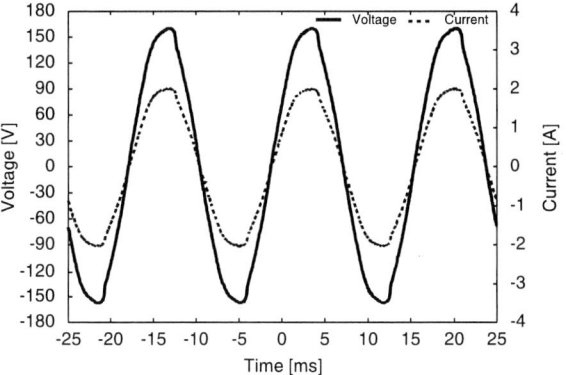

Fig. 6. Voltage and current waveform at the time of inverter–type fluorescent putting out lights(digital oscilloscope).

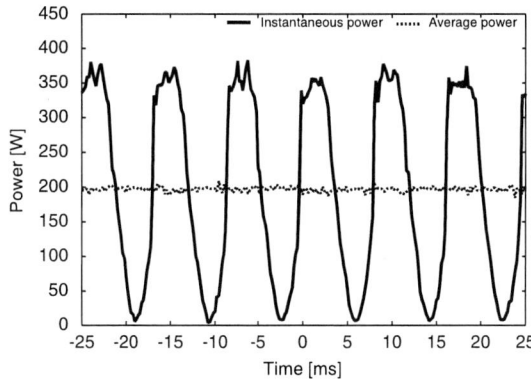

Fig. 9. Product and average waves at the time of inverter–type fluorescent lighting(ART–Linux).

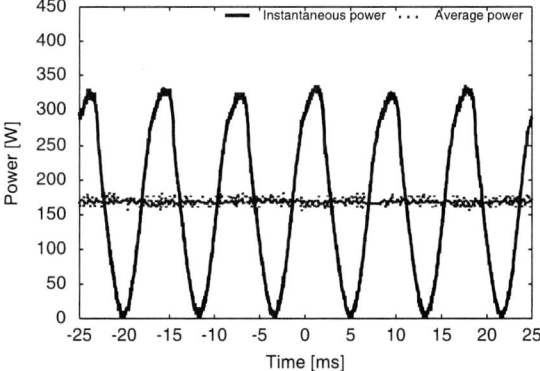

Fig. 7. Product and average waves at the time of inverter–type fluorescent putting out lights(digital oscilloscope).

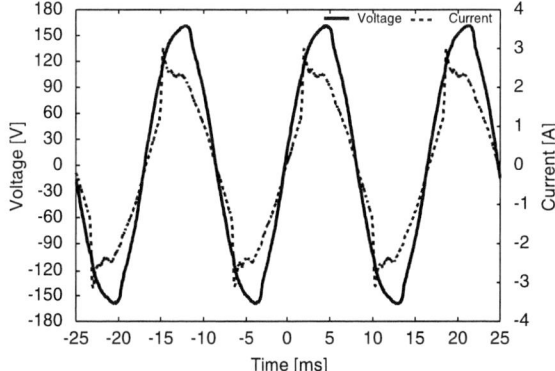

Fig. 10. Voltage and current waves at the time of inverter–type fluorescent lighting(digital oscilloscope).

TABLE I. EXECUTION CYCLE OF POWER MEASUREMENT PROGRAM(INVERTER–TYPE FLUORESCENT UNLIT).

maximum[μs]	minimum[μs]	average[μs]	standard deviation[μs]
314	125	201	18.040399

TABLE II. EXECUTION CYCLE OF POWER MEASUREMENT PROGRAM(INVERTER–TYPE FLUORESCENT LIT).

maximum[μs]	minimum[μs]	average[μs]	standard deviation[μs]
314	130	201	19.007867

978-1-4799-2706-7/14 $31.00 © 2014 IEEE

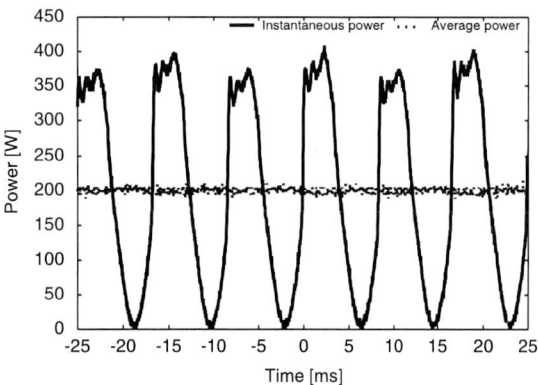

Fig. 11. Product and average waves at the time of inverter–type fluorescent lighting(digital oscilloscope).

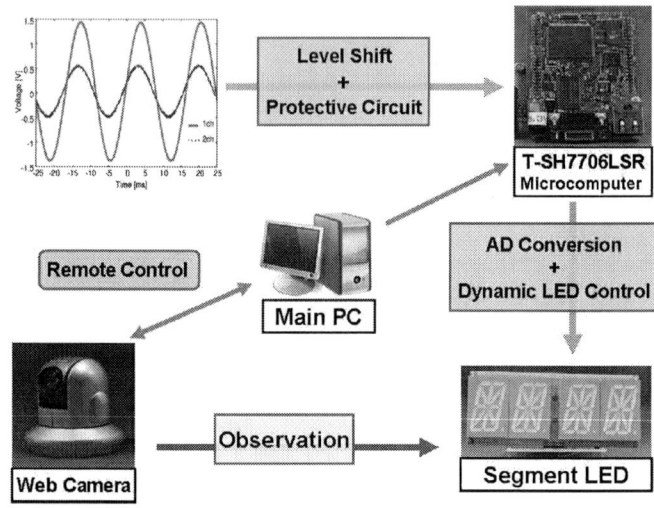

Fig. 12. Overview of the system.

distorted by the harmonics that the inverter type fluorescent lamp generates. Furthermore, it can be seen from Fig.4 to Fig.11 that the wave measured with the ART–Linux is similar to the wave measured with a digital oscilloscope. Thus, it could be confirmed that it is possible to perform measurement of harmonics by measurement using ART–Linux. In addition, there is a gap between the maximum value and the minimum value in Table I and Table II but the average value is 200[μs]. Namely, it is confirmed that AD conversion is performed at the sampling frequency 5[kHz].

IV. REMOTE MEASUREMENT SYSTEM WITH THE MICRO COMPUTER

We have developed another inexpensive remote measurement system using microcomputer. It uses an SH–3 microcomputer (T–SH7706LSR by TAC, 32bit CPU)[6] incorporating MES, which is an embedded OS for microcomputers. We added a system which visualizes a voltage digital value in order to promote understanding of the principle of the "Simulated Micro Grid System" using the microcomputer based system.

A. Remote Measurement System using SH–3 Microcomputer

The SH–3 microcomputer board has various functions such as the AD conversion function of a maximum of 4ch, DA conversion, and network communication through LAN. The MES, which operates on embedded microcomputers like H8 and SH series by Renesas has been adopted as the OS of the microcomputer. The features of MES is easy environmental construction, a multiple task system, etc.

The microcomputer is controlled by the main PC from a remote place, regarding the start and end of the measurement. The waveform to be measured is inputted to the AD converter port of the microcomputer board. The AD–converted values are displayed on a segment LED by dynamic lighting. The displayed values are monitored by a network camera and the information can be also checked with the main PC.

The present visualization of the value by segment LED was used so that the system user can see it not virtually but intuitively. The overview of the system is shown in Fig. 12.

B. Usefulness and Limit of Application

Suppose that the sensor output happens to deviate from the true value. When the real waveforms of the voltage and the current are expressed as $v = V \sin(\omega t - \theta)$ and $i = I \sin(\omega t - \theta - \varphi)$ respectively, where θ is an initial phase angle and φ is a power–factor angle, the active power P_a and the reactive power P_r are given by:

$$P_a = VI \cos\varphi + 2VI \sin(\varphi - \frac{\Delta\varphi}{2}) \sin\frac{\Delta\varphi}{2} \quad (1)$$

$$P_r = VI \sin\varphi - 2VI \cos(\varphi - \frac{\Delta\varphi}{2}) \sin\frac{\Delta\varphi}{2} \quad (2)$$

where $\Delta\varphi$, that is, the phase shift is a deviation from a true value of the real power factor angle.

The second terms in (1) and (2) are estimation errors due to the phase shift. Fig. 13 shows the relative error to the exact value in the case of a higher power factor $\cos\varphi \cong 90.6\%$ and a lower power factor $\cos\varphi = 50\%$, respectively. It is evident that the phase shift brings about a serious estimation error in both the higher and lower power factor loads. The errors of the estimated reactive and active power amount to about 10% at $\Delta\varphi = 3$ degree in the case of the higher and lower power factor loads respectively.

When measuring an electric power, we must measure a voltage value and a current value. In order to measure multiple channels, the function of the multiplexer on the microcomputer board is used. However, if this function is used, time delay will occur between channels. In electric power measurements, a small conversion delay is fatal. Thus, each channel's phase difference of the AD converter has been verified. For the experiment, a triangle wave of 0.5 to 1.5 [V] is simultaneously inputted into 1ch to 4ch of the AD conversion port. The frequency of the triangle wave was 1 [kHz]. The results are shown in Fig. 14. Fig. 15 is an expanded figure of Fig. 14.

The 2014 International Power Electronics Conference

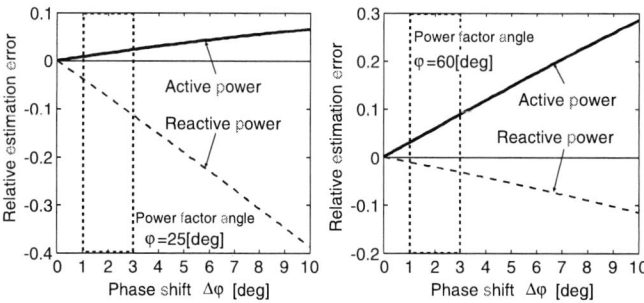

(a) the case of a higher power factor   (b) the case of a lower power factor

Fig. 13.   Relative estimation error due to a phase shift.

We have measured how much phase difference of the AD translation for the multiple channels there are. The phase difference between channels was 7.5 [μs] in 1ch and 4ch at 1 [kHz]. This time lag is 2.7 degree when expressed as an electric angle. The phase difference between channels was 0.67 degree in 1ch and 2ch. Since 1 degree or less is permissible, it turned out that reliable electric power measurement is possible by using two channels of an AD translation port.

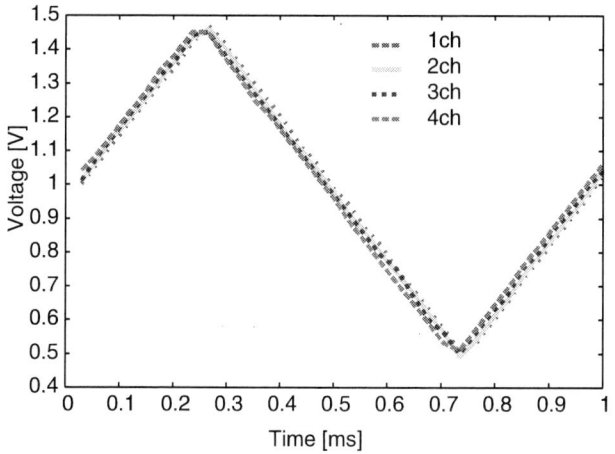

Fig. 14.   The result of the phase difference measurement between each conversion port.

C. Measurement Example

MES with a boot loader function starts T–SH7706LSR automatically by supplying a power. Boot and setup of a network device are performed automatically, and it is connected to a network by both an IP connection and a shell script file which starts the httpd daemon. After a connection is established, the remote control and measurement can be performed through a Web server. Then, an AD conversion program boots up by CGI that is written in the C language, after a measurement start button on the Web page is pushed.

In the previous section, we have showed that reliable electric power measurement is possible by this counting system. Thus, using two channels of a microcomputer, two kinds of voltage values are measured and they are actually displayed on

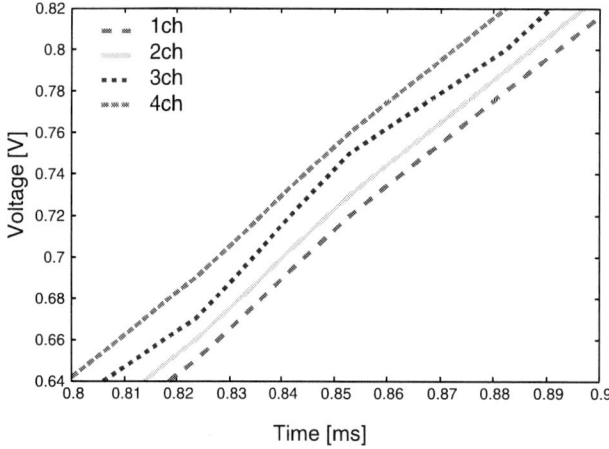

Fig. 15.   The graph of the expanded Fig. 14.

the segment LED. At the same time, the numerical value of the LED is monitored by a network camera from a web browser. The network camera is controllable by CGI, and hence the whole measurement system can be overviewed easily. Using the system, the voltages of effective value 50 [V] and 100 [V] are measured, as shown in Fig. 16.

Fig. 16.   The display of LED using Web camera.

V.   CONCLUSION

In this study, we developed a power measurement system in the Simulated Micro Grid System for education.

The electric power of the circuit which emulated the consumer was measured with ART–Linux. It was possible to measure harmonics also by ART–Linux, and therefore we have verified the usefulness of real–time measurement by ART–Linux.

We have verified the usefulness of the multiple channel AD conversion by the multiplexer of the SH microcomputer. It is confirmed that measurement of multiple AC is possible by an inexpensive SH–3 board. In addition, we succeeded in the digital display of the voltage value. We also have succeeded in monitoring it using a network camera and in remote controlling the AD conversion through the Web browser.

978-1-4799-2706-7/14 $31.00 © 2014 IEEE

In the future, we are going to integrate the two methods that we have proposed into the Simulated Micro Grid System for learning support.

REFERENCES

[1] Y. Hira, T. Furukawa, H. Fukumoto, H. Itoh, R. Nagakura and M. Outi, "Development of Simulated Micro Grid System for Learning Support" The 20th IEICE Kyushu Branch Student Conference Gakuseikaikoenkai A-06, 2012 (in Japanese)

[2] T. Yoshitomi, M. Ohchi, S. Sasaki and T. Furukawa, "Development of Remote Measurement System for Studying Electronic Circuits on Web Browser" IEEJ Trans. EIS, Vol. 129, No. 9, pp. 1635-1641, (2009) (in Japanese)

[3] T. Yoshitomi, S. Sasaki and T. Furukawa, "Design and Implementation of Web based Measurement System for Beginner of Electric and Electronic Circuit" IEEJ Trans. FM, Vol. 130, No. 1, pp. 67-73, (2010) (in Japanese)

[4] http://www.dh.aist.go.jp/en/research/assist/ART-Linux/

[5] S. Muta, T. Furukawa, H. Fukumoto, H. Itoh, T. Inazawa and R. Nagakura, "Study on Power Measurment System for Educational Simulated Microgrid" IN Conference of Instrumentation and Measurement, Electrical and Electronics Engineers, IM-12-047, pp.7-12, 2012 (in Japanese)

[6] http://www.tacinc.jp/T–SH7706/T–SH7706LSR.htm

Power Electronic Technologies for Flexible DC Distribution Grids

Rik W. De Doncker, *IEEE Fellow*

E.ON ERC at RWTH Aachen University

Institute for Power Generation and Storage Systems, Aachen, Germany

Abstract-**Market liberalization has significantly changed the energy supply system in Europe, i.e. from a top-down centralized power generation system towards a more decentralized system. In addition, partially due to incentive programs, vast amounts of renewable power generator systems (mostly wind and PV) have been installed.**

More flexible grid structures are needed to cope with this new landscape of distributed generation. This paper explores the role of state-of-the-art power electronics to realize the required infrastructure. The potential efficiency gains and cost savings that can be realized by using DC-to-DC converters in electronic substations are presented in detail.

I. INTRODUCTION

Issues like global climate change, local air quality and cost of fossil fuels have made many researchers and politicians aware that innovations towards a sustainable energy supply is mandatory to maintain or improve quality of life. Already in the mid 1970', during the first oil crisis, and spurred by the first report of the Club of Rome, many proposals were made to minimize dependency on fossil fuel and reduce energy consumption. These measures included, among others, increased use combined heat and power (CHP) plants, renewable power systems and energy savings. Actually, the use of converter fed variable speed drives to minimize energy consumption in pumps, compressors and air-conditioning units was initiated during this time and power electronic converter development was considerably accelerated. Although PV and wind generator demonstrators were built, many governments considered further investments in large-scale nuclear and coal power plants as the only viable solution to reduce the dependency on oil and other expensive fossil fuels. These large (above 1 GW) central power stations were located far away from load centers. Consequently, in many countries far distance heat distribution networks were considered too costly and CHP plants were not implemented.

Under the motto, "big is beautiful" power plants with generation capacities up to 1,3 GW were built. These units could be justified by the fact that (under continuous base load conditions) larger units are more efficient. In addition, economies of scale were the only target. Indeed, financing these systems and building the required high-voltage transmission grid was not a major political issue or investment risk, because in most countries utilities were public or national, government controlled entities. Indeed, economic growth was primordial and the benefits of central power stations outweighed strategic ecological considerations or economical risk assessments.

Ever since the first transformers were conceived and the first AC grid system was demonstrated by Westinghouse [20], power quality, efficiency and operational safety issues greatly determined the AC grid topology. Today, one can differentiate typically between three voltage layers, i.e. the high-voltage (above 100 kV) transmission grid, the medium-voltage distribution grid, and the low-voltage (below 1000 V) network (see Fig. 1). The transmission grid is interconnected, i.e. multi-terminal. This topology was needed to provide very high reliability and safety by use of redundancy. If one (large-scale) power plant unexpectedly goes off-line, other generating units can take over energy supply through the grid, for instance to keep all critical loads and auxiliary systems (for example, cooling systems of nuclear reactors) operating. The AC transmission grid is a tremendous engineering feat as its reliability (due to redundancy) is extremely high.

By contrast, the distribution grid and the low-voltage networks are designed radial. As short circuits and fault conditions occur more often at the load points, i.e. in distribution and low-voltage grids, it is extremely important to protect both the consumers and the grid and prevent fault propagation. Hence, coordination of fuses and/or circuit breakers, while limiting fault currents, is one of the main reasons why distribution AC systems are designed radial. To provide N+1 redundancy, AC distribution grids are equipped with either transfer switches (solid state MV transfer switches between multiple feeders have been demonstrated to allow voltage sag free transfers [7]) or the so-called open ring bus concept (as shown in Fig. 1). In both cases, the power is transferred to the healthy feeder when upstream a fault condition occurs. The consequence of this approach is the fact that the healthy feeder must be able to take over the additional load. It is not uncommon that the distribution cable network is therefor massively overdesigned for worst case scenarios.

Accidents with nuclear power plants, political considerations and free-market policies made by the turn of the century, changed the energy supply landscape in a significant way. For example, following the US, since 2000, Europe enforced market liberalization of the energy market (modeled similar to the communication market) successively in EU countries. By 2010, in just one decade, about 110 GW^{peak} of wind and PV was added. What went unnoticed to the public is the fact that in the same period more than 70 GW of CHP units were installed, mostly in factories and large buildings, of which 25 GW was based on biomass (see Fig. 2).

978-1-4799-2706-7/14 $31.00 © 2014 IEEE

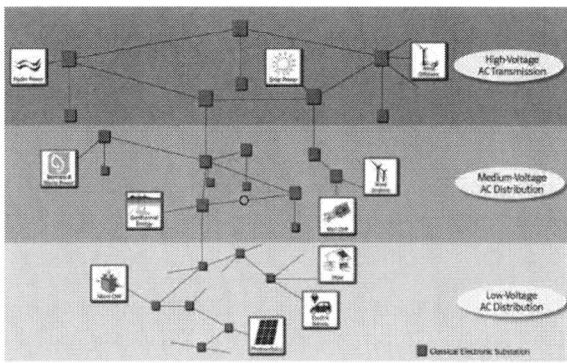

Fig. 1. Classical grid structure (here shown with decentralized power generation units) comprises a multi-terminal inter-connected transmission grid (top) to which large-scale central power stations are connected and a radial medium-voltage (middle) and the low-voltage (bottom) distribution grid.

Also shown in Fig. 2 is the increased linkage between the electrical grid and heat storage systems, as well as with the gas grid. The total system exerts a higher exergy, because with decentralized mini-CHP less heat is wasted compared to central power units. So far, the system has proven to remain stable. However, the amount of renewables and decentralized power units cannot be increased indefinitely without additional measures [1,4].

Based on the political desire to develop a so-called CO_2-neutral society, the question arises if the construction of a CO_2 free or a sustainable electrical power generation system is technically feasible. Taking the European power system as an example and extrapolating the 2011 new installation capacities of wind, PV and CHP [1-3] systems, a CO_2 neutral electrical power system could be built in approximately 15 years (see Fig. 3). The author considered innovations such as local (medium duration) heat storages in homes [5] and (strategic) long-term storage in gas, using electrolyzers. Local, short-term storage for grid support can be provided, for example, by charging and discharging intelligently (i.e. extending life) batteries of electric vehicles [17, 18].

II. FLEXIBLE ELECTRICAL GRID TOPOLOGY

To transmit power that is generated at all three voltage levels, the grid has to become more flexible to minimize expensive power quality devices and large (battery) storage systems. In the past, flexible AC Transmission systems (FACTS) [6] and Custom Power solutions for AC distribution systems [8] were proposed to guarantee power quality and stability of the AC grid. However, FACT systems assume that almost all power production is centralized at the transmission level. This may explain why few such installations have been commissioned so far. Apparently, in this changing energy market FACTS systems are deemed expensive and investment risks were too high. Also, as is illustrated in Fig. 3, from a power electronics viewpoint, the concept of FACTS seems flawed when power generation becomes equally distributed over all three voltage levels and when many decentralized generators are present. Indeed, with FACTS additional power electronics must be installed in the

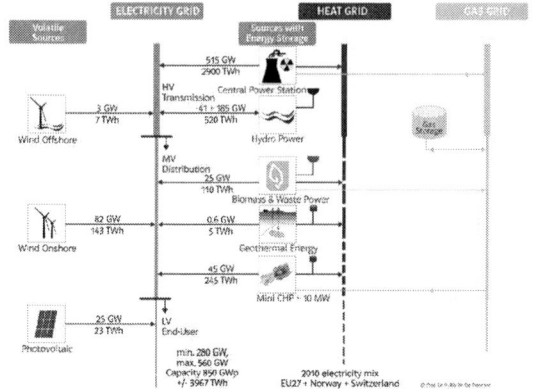

Fig. 2. Simple line diagram representation of the EU29 (= EU27, plus Norway and Switzerland) energy grids in 2010, ten years after market liberalization. The dramatic change towards more decentralized and renewable power sources is visible, totaling an installed 70 GW of CHP, biomass & waste power systems, 85 GW^{peak} of wind power and 25 GW^{peak} of PV systems|.

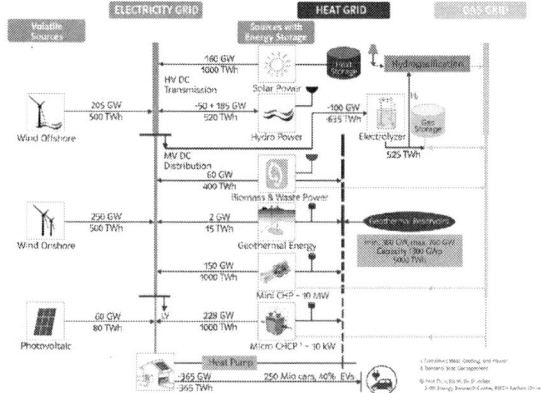

Fig. 3. Simple line diagram representing a possible future CO_2 neutral electrical energy production for EU29. About 1/3 of the installed energy production and power base is installed at the transmission level, 1/3 at the distribution level and 1/3 at the low-voltage networks in buildings and homes, providing balancing power at each voltage level.

transmission grid, when it is foreseeable that in the future most generating systems (PV, wind, high-speed generators) will have inherently power electronic converters that can control VARs and load flow in the grid. Furthermore, FACTS or Custom Power solutions do not solve the fundamental problem that the distribution grids are radial. Hence, all power generated locally at the distribution or low-voltage level that is not consumed locally has to be "sold" to the transmission grid, causing high peak power fluctuations and bottle-necks, i.e. overload condition, in the transmission grid which is supposed to transmit the energy of far distance wind and solar farms [11,12]. Oversizing the transmission grid for peak loads is an expensive proposition. In addition, in densely populated areas adding overhead transmission lines is not socially acceptable any more (not-in-my-backyard). To overcome these problems, this paper shows that modern medium-voltage DC technology can provide technical solutions that are less expensive and are socially

The 2014 International Power Electronics Conference

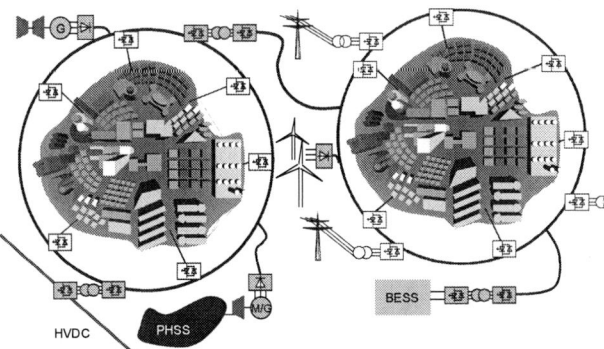

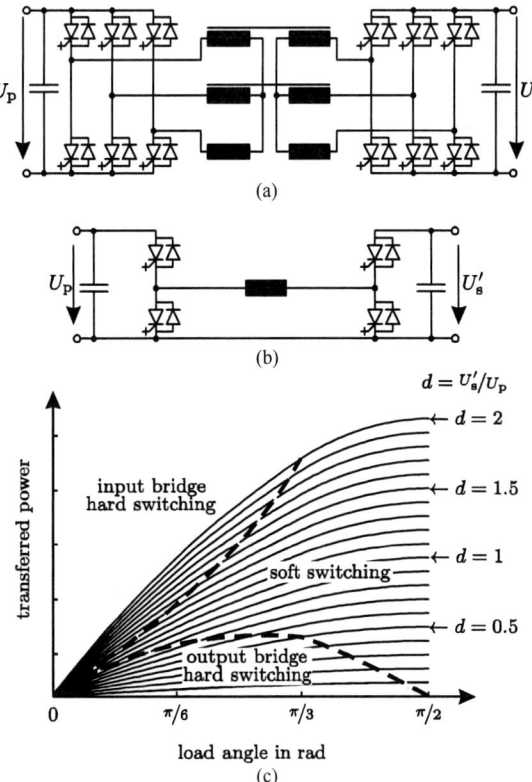

Fig. 4 The future energy grids will comprise multi-terminal HVDC networks that feed power, via electronic transformers, in a medium-voltage DC ring bus cable network. The DC-to-DC converters provide an interface to DC power sources, such as PV, electrolyzers and battery energy storage systems (BESS). Wind turbines, CHP units and pumped hydro storage systems (PHSS) are linked to the MVDC bus via simple rectifiers or inverters.

acceptable. To this end, next to HVDC, two innovation concepts have to be introduced in the distribution system, namely, MVDC distribution networks and multi-terminal DC inter-connections at all voltage levels (link to HVDC transmission and LVDC in city quarters and buildings), as illustrated in Fig. 4 [29].

III. HIGH-POWER MEDIUM-VOLTAGE ELECTRONIC TRANSFORMER – EDISON'S MISSING LINK

Key enabling innovations to realize a MVDC multi-terminal distribution network are (1) DC XLPE [30] and possibly HT-superconductive [31] cable technology, and (2) power electronics for protection devices and electrical conversion systems. Power electronics is critical, in particular to build efficient electronic, i.e. intelligent substations to create a multi-terminal distribution grid. Indeed, DC electronic substations mainly comprise electronic transformers (DC-to-DC converters) [10] and hybrid DC circuit breakers [9]. The fact that in the early days of electricity, power electronic conversion systems were not available, led to the development of the AC grid as we know it today. The only electrical conversion device available was the so-called secondary generator, nowadays called transformer, which enabled HV transmission and which was promoted primarily by Westinghouse. One can truly state that power electronic DC-to-DC converters were Edison's missing link to realize a multi-terminal DC grid that is more flexible, cheaper and efficient than the existing AC grid technology.

Over the past 25 years, aerospace and traction applications, as well as on-board power supply systems for automotive, HS trains, locomotives and ships have supported continued research and development of high-power medium-voltage DC-to-DC converters [32, 33]. Among these, the so-called (three-phase) Dual Active Bridge Converter (DABC) was proposed in 1988 at UW Madison by the author as an alternative DC or HF AC power supply systems for NASA space station [13, 33], and is illustrated in Fig. 5a. The entire topology can be

Fig. 5. Dual active bridge, three-phase DC-to-DC converter schematic (a). Single line diagram of DABC (b). Soft-switching area of DABC (c).

perceived as a three-phase bi-directional DC-to-DC converter (operating in block-mode, without PWM) or as a galvanically isolated bi-directional buck-boost converter (when looking at a single-phase diagram, see Fig. 5b.). The latter explains the fact that the DABC can provide buck-boost voltage control and zero-voltage soft-switching operation (see Fig. 5c.) over a relative wide operating range.

System levels studies on DC grids, in particular on DC collector grids for offshore wind farms, showed the great potential cost savings on investments and efficiency improvements as compared to wind farms with classical AC collector fields [35]. These results, spurred work to develop control algorithms for more efficient operation, better dynamic response [36, 37, 46] and to compensate asymmetric currents in classical three-phase transformers designs [38]. In addition, new power electronic devices, such as the MTO [39], ICT [40], IETO [41] and Dual ICT [42] were developed to improve robustness (short circuit protection), reduce cost and further reduce semiconductor device losses in DABCs. Detailed design tools were developed to select the proper topology (DABC versus Dual Active Bridge Series Resonant Converter (DABSRC) [43], single- or two-phase versus three-phase variants), lay-out, materials and components (magnetic transformer core, semiconductor devices), while optimizing overall efficiency over a defined operating range (power cycling, variable input-output voltage range) [15].

978-1-4799-2706-7/14 $31.00 © 2014 IEEE

The 2014 International Power Electronics Conference

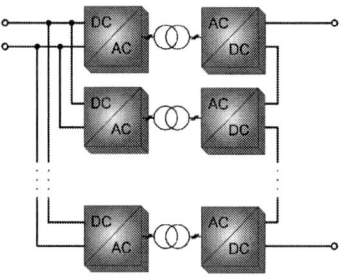

Fig. 6. Modular design allows adaptation to different voltages levels, linking medium-voltage (left) to high-voltage (right) terminals.

Using available state-of-the-art IGCT devices, materials and components, these studies led at RWTH Aachen University to the realization of a 5 MW, 5 kV DC-to-DC converter, that is optimized for utility applications. The selected topology is a three-phase dual active bridge bi-directional converter (see Figure 5). For maximum efficiency, this demonstrator operates at 1 kHz, uses a 180 μm Si-steel lamination core and standard IGCT power stacks [15]. By adding resonant capacitors, the IGCT stacks (three 3.3 kV devices in series for N+1 redundancy) can be modified to allow for full soft-switching operation. Detailed analysis and experiments show that at rated power and voltage, the DABC converter can achieve efficiencies of up to 99.2 % at 1 kHz [14,15]. Designed for the same hot-spot temperature, the transformer has a ten-fold power density increase as compared to 50 Hz transformers [14]. At ETHZ, two-phase DABCs were successfully demonstrated in traction applications [45]. At the FREEDM NSF Engineering Research Center at North Carolina State University, DABCs are used in AC-to-AC solid state transformers [46, 47]. 10 kV SiC MOSFET devices are tested in DABCs up to 12 kHz [48]. No doubt, high-voltage SiC devices, operating at higher frequencies than silicon devices, can further reduce size and material costs of DABCs. This illustrates that the developments of new materials for DABCs, both for improved semiconductor devices and magnetic cores, are still ongoing and have high potentials to make DABCs ever more competitive against classical AC transformers.

Due to its galvanic isolation, the DC-to-DC converter can be used as a building block to link medium-voltage systems to high-voltage systems or low-voltage DC systems (see Figure 6). The modularity offers high potential to define standard power electronic building blocks that can be mass-produced. Hence, the "plug-and-play" concept can be readily realized with DABCs for DC grids.

Furthermore, as the power flow can be fully controlled by the DC-to-DC converters (due to its buck-boost capability), intelligent substations can be built to control power flow in a multi-terminal DC distribution grid. Studies have shown that such interconnected multi-terminal system can maintain very high part-load efficiency [21] and can avoid overload conditions. Innovative concepts of linking DC systems with existing AC systems (hybrid grids) allows higher utilization of

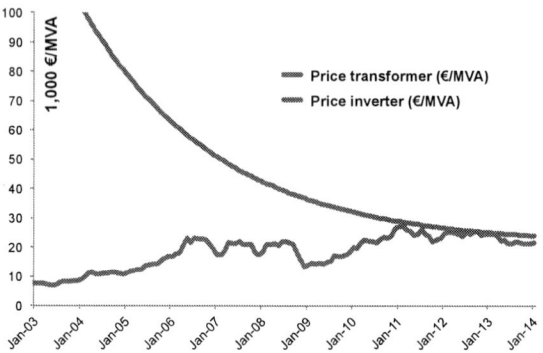

Fig. 7. Estimated price (€/MVA) evolution of 50Hz transformers (bottom curve) versus medium-voltage VSI PWM inverters. Price transformers (30% Cu, 70% Fe) based on evolution of copper and cold rolled steel materials [51], multiplied with factor (typically 1.4) to account for Cu-wire and lamination production and a factor (typically 3) to produce the whole transformer. Price converters based on sales prices of medium-voltage drive converters (having two inverters).

existing AC infrastructure [22, 23], while improving reliability. Indeed, DC back-to-back systems allow power flow control between multiple AC substations (closing the ring bus), which allows increased utilization of the infrastructure of existing AC distribution grid, while providing higher levels of redundancy.

IV. ECONOMICAL BENEFITS OF DC DISTRIBUTION NETWORKS IN WIND FARMS

Over the past decades, the price regression of medium-voltage power electronic converters has been spectacular, while the cost of transformers has steadily increased due to increased cost of copper wire and Si-steel lamination (see Fig. 7). One can state that, since 2013, the price per kVA of medium-voltage VSI power inverters (DC-to-AC) equals that of 50-60 Hz transformers. This power electronic development allows engineers to rethink the electrical grid infrastructure, in particular when more decentralized, renewable or small-scale, power generation is being installed.

Current off-shore wind turbine technology is based on medium-voltage full power converter concepts, using 5MW, 3 kV_{ac} generators, variable frequency AC-to-DC PWM voltage source inverters (VSI) for rectification, followed by fixed frequency DC-to-AC PWM inverters that feed power in the (medium-voltage) AC collector grid. Using 50 Hz transformers, the collected wind farm energy is than converted to high-voltage AC and subsequently rectified to transmit the energy in the HVDC undersea cable. On-shore, this energy is converted back from DC to AC using inverter substations and is fed into the AC transmission network. From the generator up to the HV AC on-shore substation, it can be easily calculated, that, assuming best case, i.e. full power (rated) efficiencies, the total wind farm conversion losses of all converters (97 to 99% efficiency) and transformers (98 to 99%) amount to a staggering 8,7%. At full power, a wind farm with a DC collector field feeding in an on-shore HVDC transmission systems would have only 3% losses. In addition, switching the entire wind farm and the on-

978-1-4799-2706-7/14 $31.00 © 2014 IEEE   739

shore transmission system to a multi-terminal DC system would reduce substantially the maintenance costs of the wind turbines, because the complex DC-to-AC PWM grid side VSI converter [50] is eliminated. Furthermore, with DABC switching at 1 kHz, the weight of the off-shore conversion platforms reduces down to 30%, which reduces construction costs significantly. Furthermore, it is estimated that with DC collector fields the infrastructure investment costs of the wind farm can be reduced by 8% [35].

Using a typical off-shore North Sea wind profiles and wind turbine performance data, the load cycle efficiency (LCE) of the wind farm can be computed [35]. The net output energy delivered of a 1 GW^{peak} (installed generator capacity) wind farm with DC collector field, simulated over a one-year period, is approximately 6% higher due to the higher (full and part load) efficiency of the DABCs as compared to the PWM inverters and low-frequency 50 Hz transformers. Hence, assuming an average feed-in tariff of 40 €/MWh, a 1.0 GW$_{peak}$ wind farm operator with AC collector field looses approximately 8.9 M€ on revenue each year due to conversion losses compared to the DC wind farm operator (who, on top of this, paid 8% less CAPEX and operates with less maintenance costs).

V. ECONOMICAL BENEFITS OF DC DISTRIBUTION NETWORKS

The impact of medium-frequency electronic DC transformers (DC-to-DC converters) on the cost of the power generation and voltage conversion infrastructure can be estimated. In Fig. 8a., the number of voltage transformation steps are shown for a typical AC power plant, transmission and distribution system. Note that in some countries (like Germany) an additional regional 110 kV HV-grid is used, which adds one more voltage transformation level. Typically, the active material of 50 Hz transformers weighs around 2.5 kg/kVA, while 50 Hz generators at 1,500 rpm weigh between 8 to 10 kg/kW. Hence, to transmit 1 GW of power (we assume unity power factor), approximately 17,500 ton up to 20,000 ton of copper and Si-steel are used.

The 5 MW, 5 kV DC-to-DC demonstrator, operating at 1 kHz, had a power density of 4 kVA/kg (0.25 kg/kVA) [15]. Using the future decentralized generation scenario as shown in Fig. 3 (wind farms, low- and high-speed CHP units) and assuming a full DC electrical grid, the weight of all medium-frequency transformers in the DC-to-DC converter stations can be estimated. As illustrated in Fig. 8b, the low-power conversion systems (from medium voltage to low voltage) are also assumed to have the same power density as the high-power HV units, i.e. operate at 1 kHz. In reality, low-power converter can switch at much higher frequencies and have higher power densities. Hence, the weight of the medium-frequency transformers per GW transmitted power is a worst-case estimate. On average, about 7,560 ton of copper and steel is required per GW. Assuming the cost (not price) of transformer materials, one can estimate that a DC grid system can save at least 10,000 ton/GW of transformer materials, leading to a material cost saving of ca. 29 M€/GVA on copper and steel, or a savings (on price, as

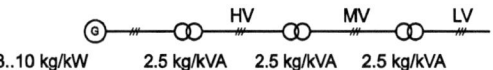

Fig. 8a. Material use in generators and transformers in classical 50 Hz systems.

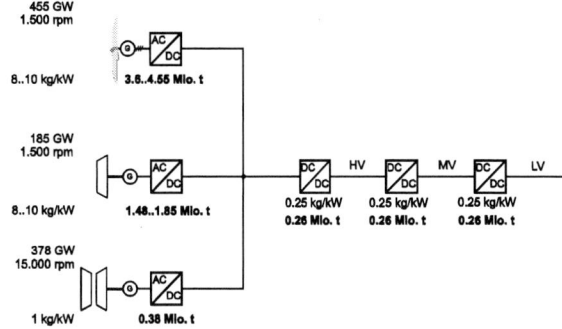

Fig. 8b. Material use in generators and transformers in the CO_2 sustainable scenario as illustrated in Fig. 3.

illustrated in Fig. 7) of ca. 87 M€/GVA. In comparison to AC transmission and distribution grid solutions, the same decentralized, CO_2 neutral generation base (as shown in Fig. 3), the added cost of power electronic converters (at today's prices of 20 €/kVA, to be expected to reduce to less than half in the next 10 years) can be estimated to be around 66 M€/GVA (estimated 33 M€/GVA by 2025). Note that no additional inverters are required in case of far distance wind farms, PV farms or high-speed generators that feed in HVDC grids. In addition, the AC grid requires additional FACTS converters to control power quality (VAR compensator, DVR, etc), which were not accounted for in this analysis.

VI. IMPACT OF DC DISTRIBUTION ON LOW-VOLTAGE NETWORKS

Higher efficiency (up to 60 % less losses) and lower cost can be realized when the low-voltage network in buildings and homes would also use a low-voltage DC bus. Newly built homes will have heat-pumps, PV and soon fast electric vehicle (EV) chargers. Today, many appliances, heat pumps and PV systems use power electronic converters with unity power factor corrector (UPFC) capability, i.e. PWM VSI with LC filters and EMI filters to connect to the AC grid. These interfaces tend to be complex and remain relatively expensive (estimated about 50 to 60% of the converter cost is the grid side UPFC). Major cost savings can be realized by avoiding the complex inverter systems that provide unity power factor and low-harmonic content.

Future smart homes will use DC-to-DC bi-directional converter modules to connect to the DC distribution grid, which would further reduce material costs of the distribution grid substantially. All large micro-generators and loads can be connected to this DC building- or domestic-grid, such as PV modules, heat pumps, washing machines and fast DC chargers for EVs. Studies show that using EV batteries as bi-directional storage devices in homes could actually extend the operating lifetime of the battery [17]. Hence, using bi-directional chargers [18] would be beneficial. If the vehicles use low-voltage

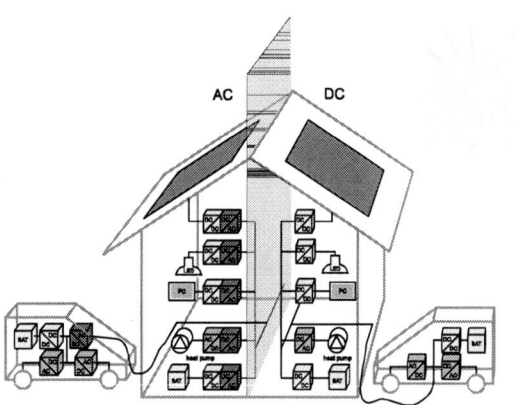

Fig. 9. Future "smart homes" with PV and heat pumps (right side of picture) will use DC bus systems, eliminating the need for multiple UPFCs and can provide fast charging to EVs.

batteries with DC-DC converters, then a direct connection to the DC grid is possible without any additional power electric infrastructure.

The proposed DC home grid configuration provides thru "prosumer" interfaces with high level of automation by distributed intelligence [28], demand side management at lower cost and higher flexibility. Again, DC-to-DC converters are essential components to realize this vision.

VII. SEMICONDUCTOR DEVICE DEVELOPMENTS FOR FUTURE FLEXIBLE GRIDS

To achieve high efficiency, proper selection of power semiconductor devices is essential. For medium-voltage DC converters further development of IGCTs would be beneficial [15] as the converter operates soft-switched and has negligible switching losses. Hence, high-voltage thyristor based devices with low conduction losses have an advantage [24, 25]. To reduce cost, in particular of gate drivers, the integrated commutated thyristor (ICT) was developed [40]. Further refinements, such as short-circuit turn-off capability and optimized switching and conduction losses were realized with the integrated emitter turn-off (IETO) [41] and the dual ICT [42]. The latter could further improve the performance of the dual active bridge DC-to-DC converter with 0,2%.

Low-voltage DC-to-DC converters operate at higher switching frequencies and IGBTs and or MOSFETs are commonly used [26]. The interface between AC and DC interconnected grids could be based on current source inverters (CSI) to protect against short circuit faults on the DC grid side. In this case, bi-direction blocking devices (IGBT, IGCT) are needed [27].

No doubt, future wide band gap power semiconductors, in particular SiC high-voltage unipolar devices (e.g. SiC 10 kV MOSFET [48]) can have a major impact on the design of high-power medium-voltage DC-to-DC converters, because higher switching frequencies are possible, which leads to further reduction of size and weight of the transformers and the DC-link capacitors. It is however, not clear if these higher frequencies (10 to 15 kHz) would lead to higher efficiencies when cost of the soft-switching DC-to-DC converter should remain the same as with the low-frequency (1 kHz) design that can

be realized with mature silicon devices. Indeed, contradictory to mobile applications, for grid applications the reduction of losses matters more than the potential space or weight savings one can realize with further increasing frequency. The wide band gap devices do provide substantial efficiency improvements in hard-switching PWM voltage source converters. One can conclude that the benefits of GaN and SiC will be more prevalent in AC systems and low-voltage DC-to-DC converters.

VIII. CONCLUSIONS

In summary, the main cost advantages of DC multi-terminal networks with large number of decentralized power generators, are:

- Higher efficiency in energy conversion systems, due to the fact that less conversion steps are needed and DC-to-DC converters can have less losses than 50 Hz transformers
- Higher power capacity in DC cables
- Lower infrastructure costs in the distribution grid and less material use (less copper and Si-steel) in energy conversion systems (electronic transformers)
- Higher reliability of the decentralized power generators that require otherwise (for AC) grid side PWM VSIs. Less maintenance cost.
- Potential of further cost reductions, especially with new existing materials (SiC, magnetic materials)

ACKNOWLEDGEMENT

The author wishes to thank the E.ON ERC gGmbH Foundation for supporting the work on the 5 MW demonstrator. Furthermore, the author is indebted to all PhD assistants at ISEA and E.ON ERC|PGS who contributed to the development of DC-to-DC converter technology. In particular, to Dr. Christoph Meyer, who made the first studies on DC hybrid circuit breakers and MVDC collector fields. Dr. Meyer passed away at too young age. This contribution is dedicated to him.

REFERENCES

[1] „European Energy and Transport – EU27 Energy Baseline Scenario to 2030- Update 2007", European Directorate-General for Energy and Transport, April 2008, ISBN 978-92-79-07620-6

[2] European Wind Energy Association, "Wind in Power – 2010 European Statistics," Feb. 2011, http://www.ewea.org

[3] European Photovoltaic Industry Association, "Market Outlook 2010," Feb. 2011, on-line: www.epia.org

[4] EUROSTAT „Energy, Transport and Environment Indicators," Pocketbooks 2010, ISSN 1725-4566, regular updates on-line: http://epp.eurostat.ec.europa.eu/

[5] Kristian Huchtemann, Dirk Müller, „Evaluation of Field Study Data on Domestic Heat Pump Systems," E.ON ERC Series ISSN 1868-7415, April 2011, Volume 3, Issue 2

[6] C. Schauder, E. Stacey, M. Lund, L. Gyugyi, L. Kovalsky, A. Keri, A. Mehraban, A. Edris, „AEP UPFC project: installation, commissioning and operation of the ±160 MVA STATCOM (phase I)," IEEE Transactions on Power Delivery, Oct. 1998, Vol. 13. Nr. 4, pp.1530-1535

[7] J. Schwarzenberg, R. W. De Doncker, „15 kV Medium Voltage Static Transfer Switch," IEEE IA Conference, 1995, Thirtieth Annual Meeting, Oct. 1995, conference records, Vol. 3, pp. 2515-2520

[8] N. Hingorani, „Introducing Custom Power," IEEE Spectrum, June 1995, Vol. 32, Issue 6, pp. 41-48

[9] C. Meyer, S. Schröder, e.a., „Solid-state circuit breakers and current limiters for medium-voltage systems having distributed power systems," IEEE Transactions on Power Electronics, Sept. 2004, Vol. 19, Issue 5, pp. 1333-1340

[10] R. De Doncker, C. Meyer, R. Lenke, F. Mura, "Power Electronics for Future Utility Applications," IEEE International Conference on Power Electronics and Drives Systems, Thailand, Bangkok, 2007

[11] "Clean Power from Deserts – the DESERTEC concept for energy, water and climate security," DESERTEC Foundation, Whitebook, 4th Edition, Protext Verlag, Bonn, Feb. 2009, on-line www.desertec.org

[12] "Wind Parks in the North Sea," DENA, German Energy Agency, www.offshore-wind.de

[13] R.W. De Doncker, D. Divan, M. Kheraluwala, "A Three-Phase Softswitched High-Power-Density dc/dc Converter for High-Power Applications," IEEE Transactions of IA, Vol. 27, No. 1, Jan/Feb 1991, pp. 63-73

[14] R. Lenke, S. Rhode, F. Mura, R. De Doncker, "Characterization of Amorphous Iron Distribution Transformer Core for use in High-Power Medium-frequency Applications," IEEE Energy Conversion Congress and Expo, Sept. 2009, pp. 1060-1066

[15] R. Lenke, "A Contribution to the Design of Isolated DC-DC Converters for Utility Applications," PhD thesis, RWTH Aachen University, E.ON ERC|PGS, 2012, ISBN 978-3-942789-05-9

[16] R. De Doncker, C. Meyer, R. Lenke, F. Mura, F, "Power Electronics for Future Utility Applications," PEDS '07, 7th Int. Conference, pp.K-1,K-8, 27-30 Nov. 2007

[17] André Hackbarth, Benedikt Lunz, Reinhard Madlener, Dirk Uwe Sauer, Rik W. De Doncker, „Plug.in Hybrid Electric Vehicles for CO2-Free Mobility and Active Storage Systems for the Grid , Part I," E.ON ERC Series Series ISSN 1868-7415 , Dec. 2010, Volume 2, Issue 3

[18] M. Rosekeit, R. De Doncker, " Galvanically Isolated Bidirectional Charger for Electric Vehicles – Analysis and Implementation," APEC 2011, Conference Proceedings

[19] R.W. De Doncker, "Towards a Sustainable Energy Supply - The New Landscape of Energy Technologies," Panasonic Technical Journal, Vol. 57, No. 4, Jan. 2012

[20] Jill Jonnes, "Empires of Light," Random House, 2004, ISBN 0-375-75884-4

[21] Florian Mura, Rik W. De Doncker, "Design Aspects of a Medium-Voltage Direct Current (MVDC) Grid for a University Campus," 8th International Conference on Power Electronics (ICPE 2011 - ECCE Asia), May 2011, Jeju, Korea

[22] De Graaff, Roald; Duarte, Jorge; Kling, W.; Vaessen, Peter, "Intelligent Nodes in distribution systems - Transition from radial to meshed operation," Electricity Distribution - Part 1, 2009. CIRED 2009. 20th International Conference and Exhibition, pp.1,4, June 2009

[23] European Spallation Source, Feb. 27, 2012, news on web, http://news.cision.com/european-spallation-source-ab/r/simulation-will-ensure-reliable-energy-systems-for-ess,c9223946

[24] Kollensperger, P.; Bragard, M.; Plum, T.; De Doncker, R.W., "The Dual GCT—A New High-Power Device Using Optimized GCT Technology," Industry Applications, IEEE Transactions, vol.45, no.5, pp.1754,1762, Sept.-Oct. 2009

[25] Van Brunt, E.; Huang, A.Q.; Butschen, T.; De Doncker, R.W., "Dual-GCT design criteria and voltage scaling," Energy Conversion Congress and Exposition (ECCE), 2012 IEEE , vol., no., pp.2596,2603, 15-20 Sept. 2012

[26] J. Hudgins, R. De Doncker, "Power Semiconductor Devices" IEEE Industry Applications Magazine, vol.18, no.4, pp.18,25, July-Aug. 2012

[27] Itoh, J.-I.; Sato, I.; Odaka, A.; Ohguchi, H.; Kodatchi, H.; Eguchi, N., "A novel approach to practical matrix converter motor drive system with reverse blocking IGBT," 35th Power Electronics Specialists Conference. IEEE PESC 04, Vol.3, pp.2380-2385 20-25 June 2004

[28] Monti, A.; Ponci, F., "Power Grids of the Future: Why Smart Means Complex," Complexity in Engineering, 2010. COMPENG '10. , vol., no., pp.7,11, 22-24 Feb. 2010

[29] J. von Bloh, "Multilevel-Umrichter zum Einsatz in Mittelspannungs-Gleichspannungs-Übertragung," Ph. D. dissertation, Institute for Power Electronics and Electrical Drives (ISEA), RWTH Aachen University, 2001

[30] Hanley, T.L.; Burford, R.P.; Fleming, R.J.; Barber, K.W., "A general review of polymeric insulation for use in HVDC cables," Electrical Insulation Magazine, IEEE, vol.19, no.1, pp.13,24, Jan.-Feb. 2003

[31] Brilinskiy, A. et al., "Application of high-temperature superconductive cable lines as components of DC transmission systems for enhancing the reliability of megalopolis electric power supply," Proceedings of the 3rd International Youth Conference on Energetics, July 2011

[32] Krismer, F.; Kolar, J.W., "Efficiency-Optimized High-Current Dual Active Bridge Converter for Automotive Applications," IEEE Transactions on Industrial Electronics, vol.59, no.7, pp.2745,2760, July 2012

[33] De Doncker R. W.; R. King, "Technique for decoupling the energy storage system voltage from the DC link voltage in AC electric drive systems,", US Patent US5373195 A, Dec. 1992/ Dec. 1994

[34] De Doncker R. W.; Divan, D. M.; Kheraluwala, M. H., "Power conversion apparatus for DC/DC conversion using dual active bridges," U.S. patent US5027264 A, Sep. 29, 1989/June 1991

[35] C. Meyer, "Key Components for Future Offshore DC Grids", Ph. D. dissertation, Institute for Power Electronics and Electrical Drives (ISEA), RWTH Aachen University, Germany, 2007

[36] Engel, S.P.; Soltau, N.; Stagge, H.; De Doncker, R.W., "Improved Instantaneous Current Control for High-Power Three-Phase Dual-Active Bridge DC-DC Converters," IEEE Transactions on Power Electronics, 2013

[37] Segaran, D.; Holmes, D.G.; McGrath, B.P., "Enhanced Load Step Response for a Bidirectional DC–DC Converter," Power Electronics, IEEE Transactions on , vol.28, no.1, pp.371,379, Jan. 2013

[38] Engel, S. P.; Soltau, N.; Stagge, H.; De Doncker, R. W., "Dynamic and Balanced Control of Three-Phase High-Power Dual-Active Bridge DC–DC Converters in DC-Grid Applications," IEEE Transactions on Power Electronics, vol.28, no.4, pp.1880,1889, April 2013

[39] Plum, T.; "Design and Realization of High-Power MOS Turn-Off Thyristors," Ph. D. dissertation, Institute for Power Electronics and Electrical Drives (ISEA), RWTH Aachen University, Germany, 2013

[40] Koellensperger, P.; "The Internally Commutated Thyristor - Concept, Design and Application," Ph. D. dissertation, Institute for Power Electronics and Electrical Drives (ISEA), RWTH Aachen University, Germany, 2011

[41] Bragard, M.; "The Integrated Emitter Turn-Off Thyristor," Ph. D. dissertation, Institute for Power Electronics and Electrical Drives (ISEA), RWTH Aachen University, Germany, 2012

[42] Butschen T., "Dual-ICT - A Clever Way to Unite Conduction and Switching Optimized Properties in a Single Wafer," Ph. D. dissertation, Institute for Power Generation and Storage Systems (E.ON ERC PGS), RWTH Aachen University, Germany, 2013

[43] Jacobs, J.; Averberg, A.; De Doncker, R., "A novel three-phase DC/DC converter for high-power applications," Power Electronics Specialists Conference PESC 04. IEEE 35th Annual , Vol.3, no., pp.1861,1867, June 2004.

[44] Soltau, N.; Apeldoorn, O.; Stagge, H.; De Doncker, R. W., „Development and Demonstration of a Medium-Voltage High-Power DC-DC Converter for DC Distribution Systems," IEEE 5th International Symposium on Power Electronics for Distributed Generation Systems, 2014 (submitted for publication)

[45] Peng Shuai; Biela, J., "Design and optimization of medium frequency, medium voltage transformers," Power Electronics and Applications (EPE), 2013 15th European Conference on , vol., no., pp.1,10, 2-6 Sept. 2013

[46] Zhao, T.; Wang, G.; Bhattacharya, S.; Huang, A. Q., "Voltage and Power Balance Control for a Cascaded H-Bridge Converter-Based Solid-State Transformer," Power Electronics, IEEE Transactions on , vol.28, no.4, pp.1523,1532, April 2013

[47] Gangyao Wang; Seunghun Baek; et. al. "Design and hardware implementation of Gen-1 silicon based solid state transformer," Twenty-Sixth Annual IEEE Applied Power Electronics Conference and Exposition, March, 2011

[48] Gangyao Wang; Xing Huang; Jun Wang; Tiefu Zhao; Bhattacharya, S.; Huang, A.Q., "Comparisons of 6.5kV 25A Si IGBT and 10-kV SiC MOSFET in Solid-State Transformer application," IEEE Energy Conversion Congress and Exposition (ECCE), pp.100,104, Sept. 2010.

[49] De, A.; Roy, S.; Bhattacharya, S., "Efficiency Comparison of AC-Link and TIPS (SST) Topologies based on accurate device models," Energy Conversion Congress and Exposition (ECCE), 2012 IEEE., pp.3862,3868, 15-20 Sept. 2012

[50] Faulstich S., Hahn, B.; Tavner, P. J., "Wind turbine downtime and its importance for offshore deployment," Jornal Wind Energy, Vol. 14, No. 3, pp. 327-337, 2011

[51] Historical data of copper and steel prices, e.g. www.infomin.com or www.metalprices.com or www.lme.com

The 2014 International Power Electronics Conference

2.5kV, 200kW Bi-Directional Isolated DC/DC Converter for Medium-Voltage Applications

Yuji Matsuoka
Power Electronics Department
Toshiba Mitsubishi-Electric Industrial Systems Corporation
Fuchu-City, Japan
MATSUOKA.yuji@tmeic.co.jp

Keiji Wada, Mizuki Nakahara
Tokyo Metropolitan University
Hachioji, Tokyo, Japan

Kazuto Takao
Corporate Research & Development Center, Toshiba Corporation
Kawasaki, Kanagawa, Japan

Kyungmin Sung
Ibaraki National College of Technology
Hitachinaka, Ibaraki, Japan

Hiromichi Ohashi, Shinichi Nishizawa
National Institute of Advanced Industrial Science and Technology
Tsukuba, Ibaraki, Japan

Abstract—A bi-directional isolated DC/DC converter for medium-voltage applications have been discussed for the next-generation electrical grid, such as smart girds. To realize the DC/DC converter for installing on a power distribution system, it should be achieved higher-efficiency and lower-volume. Higher switching frequency enables to reduce the volume of the transformer. However, it is difficult to realize the both higher-efficiency and lower-volume of the DC/DC converter, because it could not operate higher frequency with high power devices such as medium-voltage IGBTs. This paper discusses switching losses and operation limitations for the actual medium-voltage isolated DC/DC converter, in the case of using Si-IGBT.

In addition, the experimental system rated at 2.5 kV, 200 kW, and 5kH of the DC/DC converter with a medium-frequency transformer is designed, and the experimental results are shown.

Index Terms—Isolated DC/DC converter, Medium-frequency transformer, Si-IGBT

I. INTRODUCTION

Applicability of bi-directional power converters for medium-voltage applications has been discussed for a next-generation electrical grid, such as smart girds. One of the bi-directional circuits is an isolated DC/DC converter with a medium-frequency transformer. The isolated DC/DC converters have been proposed for low-voltage applications less than 400 V, and they are applied to power supply for data centers and industry applications. Many type of circuit topology for isolated DC/DC converters are proposed, but a dual active bridge DC/DC converter can be applied to the medium voltage applications.

Recently, the medium voltage isolated DC/DC converters have been proposed for railway applications

and power distribution systems. As a result, a line-frequency transformer can be replaced to the DC/DC converter, and the volume will be reduced to less than 1/10. Therefore, the DC/DC converter is expected as one of the most important converter circuit.

To realize the isolated DC/DC converter for installing on the market, it should be achieved higher-efficiency and lower-volume simultaneously. Higher switching frequency enables to reduce the volume of the transformer. However, it is difficult to realize the both higher-efficiency and lower-volume of the DC/DC converter because it could not operate higher frequency with high power devices such as medium-voltage IGBTs. Ref [8] shows a 100 kW DC/DC converter is presented, and SiC-MOSFETs are used as power devices and Nanocrystalline is used as core material of the transformer. Researches of the power devices for SiC-MOSFET and SiC-IGBT have been discussed [1], [2]. Moreover, Si-IGBTs for medium-voltage applications are widely used in several applications, and the researches for improving the Si-IGBT have also been presented [3].Furthermore, the medium-frequency transformers under non-sinusoidal excitation conditions have been discussed in details. Refs. [4] and [5] have presented core loss and winding structures of the medium-frequency transformers. Ref. [6] is also discussed analysis procedure for the medium-voltage transformer. However, they could not present experimental results of medium-voltage and -frequency transformers under non-sinusoidal excitation conditions.

This paper evaluates limitation of the switching frequency of a medium-voltage isolated DC/DC converter. The DC/DC converter utilizes 4.5 kV Si-IGBT modules. The operation condition with soft switching of the IGBT has been theoretically analyzed considering the influence of the dead time. The limitation of the switching frequency is estimated based on the losses of

978-1-4799-2706-7/14 $31.00 © 2014 IEEE

the IGBT and the junction temperature of it. In the analysis, both a conventional type IGBT and a fast switching type IGBT are considered. Core loss of a medium-frequency transformer is estimated by using improved generalized Steinmetz equation (iGSE) [6]. The analysis results are experimentally verified by using the prototype 2.5 kV, 200 kW DC/DC converter.

II. MEDIUM-VOLTAGE HIGH-POWER BI-DIRECTIONAL ISOLATED DC/DC CONVERTER

Fig. 1 shows the circuit of the bi-directional isolated DC/DC converter which is discussed in this paper. Primary- and secondary-full-bridge converters are connected through a medium-frequency transformer. Each switching device operates at 50% duty cycle, and the phase difference δ between the two bridges controls the power transmission from primary- to secondary-side.

In this study, 4.5 kV IGBT are applied as high-power Si switching devices for the DC/DC converter.

III. OPERATING FREQUENCY AND LOSSES OF THE DC/DC CONVERTER

Higher switching frequency enables to reduce the volume of the medium-frequency transformer. On the other hand it increases the switching loss of the IGBT. Therefore, the loss reduction of IGBT is required to realize both high-efficiency and lower-volume.

The relationships between switching frequency and junction temperature of IGBT are discussed. In addition, the relationship between switching frequency and losses of the DC/DC converter are also estimated. The switching losses using in this discussion can be calculated by the switching test results. Fig. 2 shows the waveforms of the turn-off switching test using the generally used existing IGBT rated 4.5 kV and 150 A. The turn-on loss does not occur because the zero-voltage turn-on is achieved in the DC/DC converter.

A. Relationships between Switching Frequency and Junction Temperature

Fig. 3 (a) shows the relationships between switching frequency and junction temperature of existing IGBT. Note that this calculation is based on under the condition of the direct liquid cooling of IGBT for medium-frequency operations [7]. As seen in Fig. 3 (a), when the limit of junction temperature rise ΔT is set to 85 K, the maximum allowed switching frequency of the DC/DC converter is approximately 6 kHz. However, it is assumed that the 4.5 kV IGBT used in this calculation is not designed for medium-frequency switching operations such as several kHz. We therefore focus on the trade-off between on-state voltage and turn-off loss of the IGBT.

Fig. 4 shows the trade-off curve between on-state voltage and turn-off loss of the 4.5 kV IGBT [3]. The point (a) is the characteristic of the existing 4.5 kV IGBT used in the calculation of Fig. 3 (a). As mentioned previously, because the existing 4.5 kV IGBT is not optimized for the medium-frequency switching, the switching loss accounts for a large proportion of IGBT losses in medium-

frequency switching operations. Therefore, if the characteristics of the IGBT can be adjusted to reduce the switching loss, the total IGBT losses in medium-frequency switching operation can be reduced and the maximum allowed operational frequency of the DC/DC converter will be increased.

Specifically, Fig. 3 (b) shows the relationship between switching frequency and junction temperature of an ideal IGBT whose trade-off characteristic is adjusted for the medium-frequency switching operation, as shown in Fig. 4 (b). As seen in Fig. 3 (b), the maximum allowed operational frequency of the DC/DC converter using the IGBT for medium-frequency switching operation reaches approximately 9 kHz.

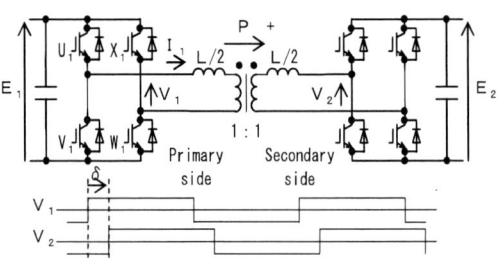

Fig. 1. Circuit of Bi-directional Isolated DC/DC Converter

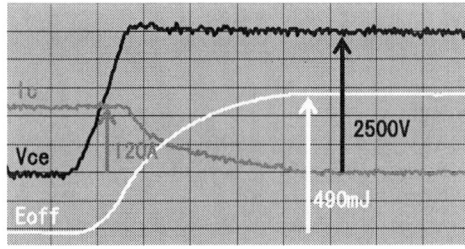

Fig. 2 Experimental Waveforms of Turn off

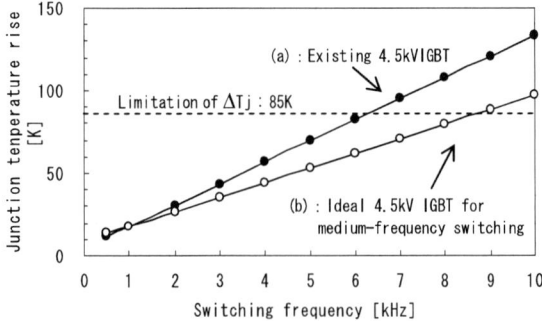

Fig. 3 Relationship between switching frequency and estimated junction temperature of IGBT

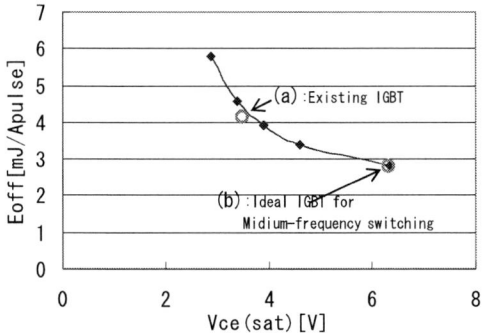

Fig. 4. Trade-off between Eoff and Vce(sat)

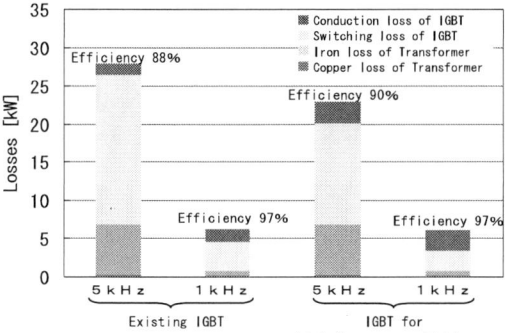

Fig. 5. Losses of Bi-directional isolated DC/DC converter

B. Estimated Losses of the DC/DC Converter

Fig. 5 shows the analysis results of loss estimation of the DC/DC converter. In this analysis, the switching loss and the conduction loss of the IGBT and the loss of the medium-frequency transformer are considered as the total losses of the DC/DC converter. In this case, it is analyzed by using the existing IGBT and the ideal IGBT adjusted to medium-frequency switching. In order to estimate the iron loss of the transformer, improved generalized Steinmetz equation (iGSE) [7] is used. The iGSE is expressed as Eq (1), where α, β, k_1 are Steinmetz constants. In this study, these constants were calculated from CS-125 sized core (121 mm x 63 mm x 35 mm).

$$P_{iGSE} = \frac{1}{T} \int_0^T k_i \left| \frac{dB}{dt} \right|^\alpha (\Delta B)^{\beta - \alpha} dt \qquad (1)$$

The switching frequencies of these estimations are set to 1 kHz or 5 kHz operations. The 5 kHz is a value close to maximum allowed switching frequency of existing 4.5 kV IGBT, and the 1 kHz is conventional switching frequency of existing 4.5 kV IGBT.

In the case of the 5 kHz operation, the efficiency of the DC/DC converter reaches 88% by using the existing IGBT, whereas it is to 90% by using the ideal IGBT adjusted to medium-frequency switching. In both cases, it

is clear that the switching loss accounts for a large proportion of the total IGBT losses.

In addition, although the silicon steel plates are used as the core material of the medium-frequency transformer for calculating the iron loss, if using the Nanocrystalline instead of the silicon steel plate, the iron loss can be reduced one twelfth. As a result, the efficiency of the DC/DC converter will be increased by 1 %.

On the other hand, in the case of the 1 kHz operation, the efficiencies are 97 % for both kinds of IGBTs.

IV. INFLUENCE OF THE DEAD TIME TO OPERATIONAL RANGE IN ZERO-VOLTAGE TURN-ON

The dead time of the medium-voltage inverter using 4.5 kV IGBT is longer than that of low-voltage switching devices. In this section, the influence of dead time to the bi-directional isolated DC/DC converter is discussed.

In ideal condition: the dead time T_d can be set to 0, the operation range of the DC/DC converter with soft-switching operation (zero-voltage turn-on) can be calculated by [8]:

$$\delta_1 > \frac{(E_2 - E_1)\pi}{2E_2} \qquad (2)$$

When the primary DC voltage E_1 is equal to the secondary DC voltage E_2, the IGBT can be driven as a soft-switching operation in the entire range of the phase-difference δ.

However, in actual condition: the dead time T_d can not be set to 0, the operation range of the DC/DC converter in soft-switching can be calculated by :

$$\delta_2 > \frac{2T_d(E_1 + E_2) - (E_1 - E_2)\pi}{2E_2} \qquad (3)$$

Eq. (3) shows that even if the primary DC voltage E_1 is equal to the secondary DC voltage E_2, the soft-switching operation range of the DC/DC converter is restricted. The right side of Eq. (3) indicates the boundary condition of the phase difference including the dead time. If the phase difference is set lower compared with the right side of Eq. (3), the current waveform become discontinuous conditions and the IGBT can not be operated in the soft switching conditions. Fig. 6 shows waveforms of non-zero-voltage turn-on when δ_2 it is not satisfied the condition of Eq. (3).

Furthermore, in actual condition, the discontinuous current shown in Fig. 6 diminishes the power transmission of the DC/DC converter. The transmitted power is expressed as Fig. 7 (a) in ideal condition and it is expressed as Fig. 7 (b) in actual condition.

Eq. (3) shows that longer dead time makes δ_2 larger. When the medium voltage IGBTs are applied to the DC/DC converter, Eq. (3) and these matter should be considered carefully.

The 2014 International Power Electronics Conference

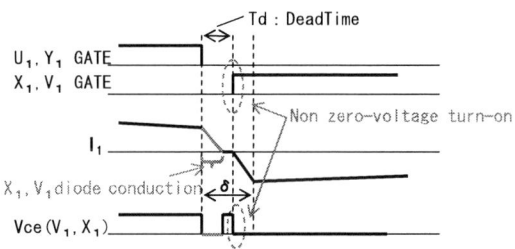

Fig. 6. Waveforms of non-zero-voltage switching

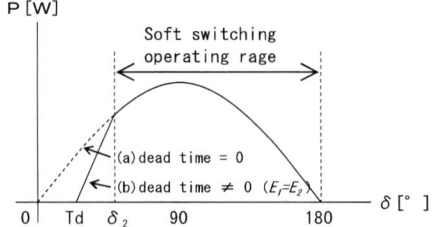

Fig. 7. Characteristic of the power transmission

V. EXPERIMENTAL RESULT

A. Experimental Circuit

Fig. 8 shows an overview of a full-bridge converter of the bi-directional isolated DC/DC converter. The specifications of the DC/DC converter are listed in Table 1. 4.5 kV Si-IGBT modules are employed as high-power devices. The rated output power of the DC/DC converter is 200 kW. The input and output voltages are set to 2.5 kV.

An experimental setup and the circuit configuration of the DC/DC converter are shown in Fig. 9. In order to re-generate the output power of the DC/DC converter, the output terminal of the secondary converter is connected to the input terminal of the primary converter. In this experimental setup, the utility line supplies only the power losses in the DC/DC converter. Therefore, the power losses are equal to the input power of the DC/DC converter.

Fig. 8. Overview of one bridge of prototype DC/DC converter
(medium -frequency transformer is not included)

TABLE 1
SPECIFICATIONS OF THE DC/DC CONVERTER

Rated power P	200 kW
DC voltage $E_1 = E_2$	2.5 kV
Switching frequency fsw	5 kHz
L	0.7 mH
IGBT rating	4.5 kV-150 A

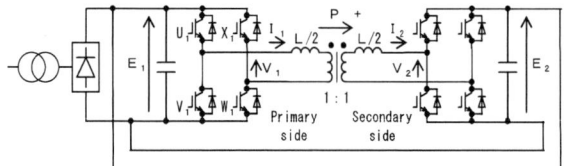

(a) Circuit configuration of experimental system

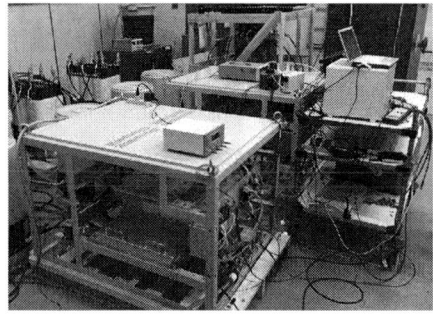

(b) Overview of the experimental system

Fig. 9. Overall structure of the experimental circuit

B. Experimental Waveforms and measured losses

Fig. 10 shows the operation waveforms of the DC/DC converter obtained at f_{sw} = 5 kHz, δ = 60°, and P = 200 kW. Around the rated power of 200kW, the experimental efficiency is 87%. The theoretical efficiency calculated in section III. B is 88%. The efficiency obtained in the experiment is 1% lower than that of the theoretical value. The difference may be caused by that the influences of the power losses of the wire and inductor are not taken into account the theoretical calculation.

It is difficult to distinguish the iron loss from the all loss of the transformer. However, the iron loss becomes a dominant component in the loss of the transformer under the condition of phase difference δ = 0 because the load current of the transformer is nearly zero. Therefore, the validity of the calculated iron loss is confirmed at phase difference δ = 0. The calculated iron loss is 11.6 kW at phase difference δ = 0, on the other hand, measured iron loss is 11.9 kW at phase difference δ = 0. Both losses are almost same, therefore it is assumed that calculated iron loss is validated.

978-1-4799-2706-7/14 $31.00 © 2014 IEEE   747

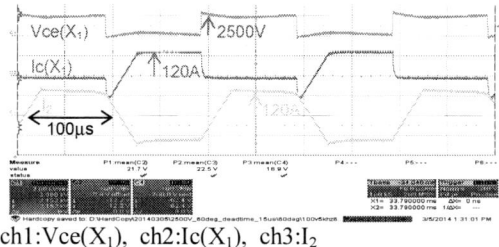

ch1:Vce(X_1), ch2:Ic(X_1), ch3:I_2

Fig. 10. Experimental waveforms

at δ=60 deg and fsw=5 kHz

C. Influence of The Dead Time to Operational Range in Zero-voltage Turn-On

Fig. 11 shows experimental waveforms of the non-zero-voltage turn-on conditions. In this experiment, the dead-time T_d is set to 15 µs, and the switching frequency f_{sw} is 5 kHz. In this condition, the boundary condition of Eq. (3) is calculated δ_2 = 54 deg.

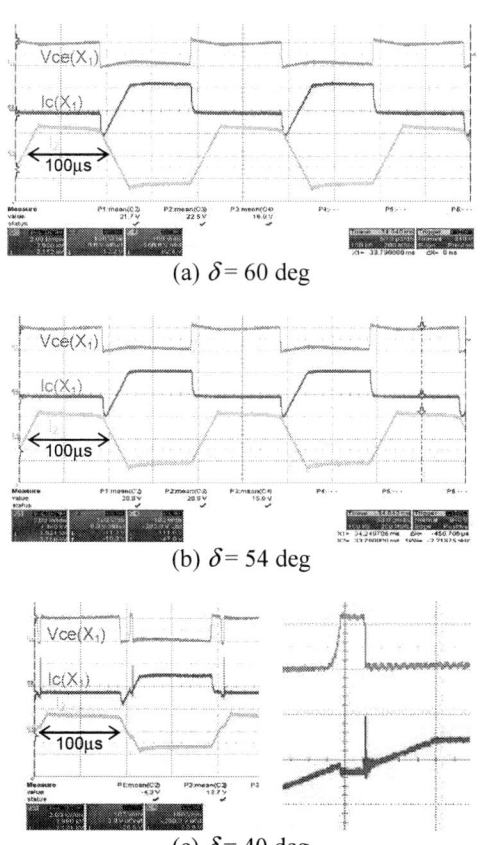

(a) δ = 60 deg

(b) δ = 54 deg

(c) δ = 40 deg

Fig. 11. Actual waveforms of Non-zero-voltage turn-on
(ch1:Vce(X_1), ch2:Ic(X_1), ch3:I_2)

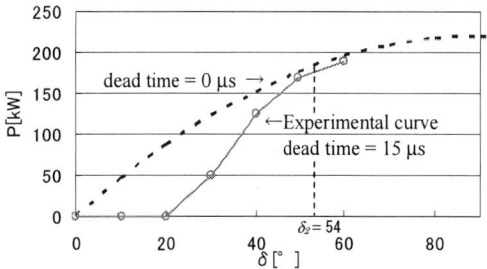

Fig. 12. Experimental characteristic of the power transmission

As seen in Fig. 11, the zero-voltage turn-on is achieved when the phase difference δ is set over 54 deg and the turn-on losses are a very few in these operation modes. On the other hand, when the phase difference δ is set less than 54 deg, the zero-voltage switching could not be archived as above mentioned theoretically, and the turn-on loss occurred.

Although this experimental DC/DC converter does not use any snubber capacitors around the IGBT, the surge current of IGBT which looks like discharging of the snubber capacitor is observed as shown in Fig. 11(c). This is due to discharge of the capacitance of the IGBT, and this can be inhibited by adjusting the gate resistor.

Fig. 12 shows experimental characteristic of power transmission of the DC/DC converter. Under 54 deg: the boundary condition between the continuous and discontinuous current, the transmitted power diminishes as Fig. 7.

VI. CONCLUSIONS

This paper discusses a medium-frequency DC/DC converter. Limitation of switching frequency of the medium-voltage isolated DC/DC converter using 4.5 kV IGBTs has been evaluated. The maximum switching frequency of a conventional 4.5 kV IGBT is estimated to 6 kHz. The 9 kHz operation could be achieved with an ideal IGBT. In addition, the boundary condition of the zero-voltage turn-on which is influenced by dead time is revealed.

A prototype medium-voltage isolated DC/DC converter rated 2.5 kV, 200 kW and 5kHz is designed and built by using a conventional type 4.5 kV IGBTs, the 5 kHz operation is confirmed. It is confirmed that the analysis results are experimentally verified.

REFERENCES

[1] A. Q. Huang, M. L. Crow, G. T. Heydt, J. P. Zheng, S. J. Dale, `` The Future Renewable Electric Energy Delivery and Management (FREEDM)System: The Energy Internet," Proceedings of the IEEE, Vol. 99 , no. 1, pp. 133 - 148, 2011

[2] G. Ortiz, J. Biela, D. Bortis, J. W. Kolar, `` 1 Megawatt, 20 kHz, isolated, bidirectional 12kV to 1.2kV DC-DC converter for renewable energy applications," IPEC 2010, pp. 3212-3219

[3] K.Nakamura, K.Sadamatsu, D.Okada, H.Shigeoka, K.Hadade: "Wide Cell Pitch LPT(II)-CSTBTTM(III) TechnologyRating up to 6500 V for Low Loss", Proc. ISPSD'10, pp. 387-390, 2010

[4] G. Ortiz, M. Leibl, J. W. Kolar, and O. Apeldoorn, ``Medium frequency transformers for solid-state-transformer applications -- Design and experimental verification," IEEE PEDS, pp. 1285-1290, 2013

[5] P. Shuai, J. Biela, ``Design and Optimization of Medium Frequency, Medium Voltage Transformers," EPE 2013

[6] K. Venkatachalam, C. R. Sullivan, T. Abdallah, and H. Tacca,"Accurate prediction of ferrite core loss with nonsinusoidal waveforms using only Steinmetz parameters''Proc. IEEE Workshop Comput. Power Electron., 2002, pp. 36-41.

[7] K.Wada, K.Jumpei, K.Takao, T.kanai, H.Oohashi: "Thermal Analysis for Hybrid Pair Module of Si-IEGT and SiC-PiN Diode", The 2010 International Power Electronics Conference, pp. 2135-2140

[8] S.Inoue, H.Akagi: "Operating Voltage and Loss Analysis of a Bi-Dir-ectional Isolated DC/DC Converter", IEEJ Trans. on IA, Vol. 127, 2007

Power-Loss Breakdown of a 750-V, 100-kW, 20-kHz Bidirectional Isolated DC-DC Converter Using SiC-MOSFET/SBD Dual Modules

Hirofumi Akagi and Tatsuya Yamagishi
Tokyo Institute of Technology, Japan

Nadia M. L. Tan
Universiti Tenaga Nasional, Malaysia

Shin-ichi Kinouchi, Yuji Miyazaki, and Masato Koyama
Mitsubishi Electric Corporation, Japan

Abstract—This paper describes the design, construction and testing of a 750-V, 100-kW, 20-kHz bidirectional isolated dual-active-bridge dc-dc converter using four 1.2-kV 400-A SiC-MOSFET/SBD dual modules. The maximum conversion efficiency from the dc-input to the dc-output terminals is accurately measured to be as high as 98.7% at 42-kW operation. The overall power loss at the rated-power (100 kW) operation, excluding the gate-drive and control circuit losses, is divided into conduction and switching losses produced by the SiC modules, iron and copper losses due to magnetic devices, and an unknown loss. The power-loss breakdown concludes that the summation of the conduction and switching losses is about 60% of the overall power loss, and that the conduction loss is nearly equal to the switching loss at the 20-kHz, 100-kW operation.

Keywords—Bidirectional isolated dc-dc converters, conversion efficiency, dual-active-bridge configuration, SiC-MOSFET.

I. INTRODUCTION

Bidirectional isolated dc-dc converters have been expected to play an important role in integrating battery energy storage systems and renewable energy sources such as solar and wind energy into medium-voltage and/or low-voltage dc power systems. Particular attention has been paid to a bidirectional isolated dc-dc converter combining two single-phase or three-phase full-bridge voltage-source converters with a single-phase or three-phase high-frequency transformer [1].

Fig. 1 shows the circuit configuration of the bidirectional isolated dc-dc converter based on two single-phase voltage-source converters using eight MOSFETs and diodes as the power semiconductor devices. This kind of dc-dc converter is often referred to just as the "dual-active-bridge converter," the naming of which results from circuit topology [2], while the original term "bidirectional isolated dc-dc converter" comes from functionality. Strictly speaking, this dc-dc converter should be referred to as the "bidirectional isolated dual-active-bridge dc-dc converter," to identify itself and to avoid confusion.

The dc-dc converter is symmetrical in circuit topology when the two voltage-source converters are seen upstream and downstream of the high-frequency transformer with a turns ratio of 1:1 ($N = 1$ in Fig. 1). The dc-dc converter has less device stress and reasonable ripple current levels in the dc capacitors because it can achieve zero-voltage switching (ZVS)

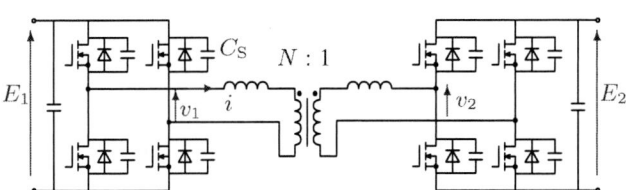

Fig. 1. Bidirectional isolated dual-active-bridge dc-dc converter.

with the help of a snubber capacitor C_S, connected across each MOSFET. This makes it possible to construct bidirectional isolated dc-dc converters in a range from a few tens to hundreds of kilowatts, depending on operating frequencies. However, one of the most major research targets for putting them into practical use is to improve conversion efficiency at high-frequency operation [3]–[14].

Recently, a significant progress has been made in SiC (Silicon Carbide) power devices as a result of continuous efforts by research scientists and engineers. This progress has made it possible to fabricate not only discrete SiC-SBDs (Schottky barrier diodes), SiC-JFETs and SiC-MOSFETs but also SiC-MOSFET/SBD modules [15]. Conventional 1.2-kV Si-IGBT/PND modules can be replaced with emerging 1.2-kV SiC-MOSFET/SBD modules, thus leading to higher conversion efficiencies and/or higher switching frequencies [16][17].

This paper presents a 750-V, 100-kW, 20-kHz bidirectional isolated dual-active-bridge dc-dc converter using four SiC-MOSFET/SBD dual modules. The dual module forming one leg by itself is also referred to as the "two-in-one" module. It integrates two pairs of SiC-MOSFETs and SiC-SBDs into a single module, where each pair of an SiC-MOSFET and an SiC-SBD is connected in anti-parallel. The SiC-MOSFET and the SiC-SBD have the same voltage and current ratings as 1.2 kV and 400 A.

One objective of this paper is to measure the overall power loss accurately, excluding the gate-drive and control circuit losses. The other is to divide the overall power loss at the rated-power (100 kW) operation into five power losses under reasonable and acceptable assumptions. The five components include conduction and switching losses produced by the SiC

978-1-4799-2706-7/14 $31.00 © 2014 IEEE

modules, iron and copper losses due to the magnetic devices, and an unknown loss. This power-loss breakdown gives an insight into which power-loss component is the most dominant among the five components. As a result, it would encourage power electronics research scientists and engineers to do further and effective improvements in conversion efficiency.

Experimental results obtained from the 750-V, 100-kW, 20-kHz dc-dc converter verify that the maximum conversion efficiency from the dc input terminals to the dc output terminals is 98.7% at 42 kW and that the conversion efficiency at 100 kW is 97.9% with a tolerance of 0.02%. The power-loss breakdown at the rated-power operation leads to the following conclusions:

- The summation of the conduction and switching losses produced by the SiC modules occupies about 60% of the overall power loss.
- The conduction loss is almost equal to the switching loss.
- The sum of the iron and copper losses due to the magnetic devices occupies about 30% of the overall power loss, and it is nearly equal to either conduction loss or switching loss.

II. THE DUAL-ACTIVE-BRIDGE DC-DC CONVERTER

The bidirectional isolated dual-active-bridge dc-dc converter shown in Fig. 1 is suitable for high-power applications [1]–[3]. Two single-phase voltage-source converters produce the so-called "180°-conducting" rectangular voltage with different amplitudes equal to dc-input and dc-output voltages. The dc-output power can be controlled by adjusting the phase-shift angle δ between v_1 and v_2 is given by

$$P = \frac{E_1 N E_2}{\omega L} \delta \left(1 - \frac{|\delta|}{\pi} \right). \tag{1}$$

Here, δ takes a positive value when v_1 leads v_2 by δ, N is a turns ratio of the transformer, f is the operating frequency ($\omega = 2\pi f$), and

$$L = L_A + l, \tag{2}$$

where L is the equivalent inductance given by the summation of the four auxiliary inductors and the leakage inductance of the transformer, in which both inductance values are referred to the primary side of the transformer. When δ takes a positive value, power flow occurs from left to right in Fig. 1. In this case, the right-side voltage-source converter (bridge 2) is operated as a "synchronous rectifier" in which an arm current flows through a MOSFET. Therefore, no current flows through the SBD connected in anti-parallel to the MOSFET under such an ideal condition that no dead time exists in each leg of the voltage-source converter.

III. EXPERIMENTAL SYSTEM

Fig. 2 shows the circuit configuration of a 750-V, 100-kW, 20-kHz dc-dc converter that was designed, built and tested to measure the overall power loss from the dc input terminals to the dc output terminals, as well as to achieve power-loss breakdown at the rated-power (100 kW) operation. Table I

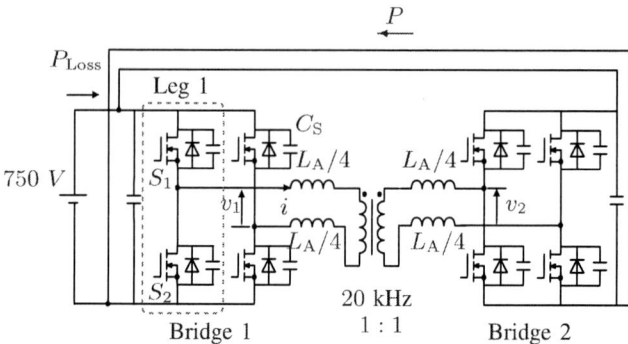

Fig. 2.   Experimental circuit system.

TABLE I.   CIRCUIT PARAMETERS OF THE BIDIRECTIONAL ISOLATED DC-DC CONVERTER.

Power rating		100 kW
Input/output voltage rating	E	750 V
Switching frequency	f	20 kHz
Auxiliary inductor	L_A	16.6 μH (37.1%)
Leakage inductance	l	0.7 μH (1.6%)
Phase shift angle	δ	0 − ±26.6°
Snubber capacitor	C_S	4.5 nF / 9 nF
Dead time	T_D	0.6 μs / 0.8 μs
Transformer core material		FINEMET*
Maximum flux density		0.48 T at 20 kHz
Transformer turns ratio		1 : 1

* Nano-crystalline soft-magnetic material

summarizes the circuit parameters used in Fig. 2. The dc-dc converter consists of four 1.2-kV, 400-A SiC-MOSFET/SBD dual modules, a single-phase 20-kHz transformer and four auxiliary inductors.

Generally, a university laboratory has the difficulty of providing the 750-V, 100-kW dc power supply and load necessary for this experiment. This issue can be solved by actively using the function of galvanic isolation between the dc input terminals and the dc output terminals. In other words, the dc output terminals of bridge 2 are directly connected to the dc input terminals of bridge 1, so as to regenerate the output power P to the dc input terminals [4]. As a result, the 750-V dc power supply feeds a small amount of electric power, which corresponds to the overall power loss P_{Loss}. This configuration allows the direct measurement of P_{Loss}, independent of the measurement of P. Since P_{Loss} is much smaller than P, the conversion efficiency from the dc input terminals to the dc output terminals, η can be calculated by

$$\eta = \frac{P}{P + P_{\text{Loss}}}. \tag{3}$$

A digital power analyzer (HIOKI 3390-10) was used to measure P and P_{Loss} simultaneously and individually. As a result, the accurate conversion efficiency obtained from (3) has a tolerance as small as 0.02% under such an accceptable assumption that P_{Loss} is much smaller than P. It would be impossible to attain such a small tolerance when the dc-input power and the dc-output power were measured simultaneously and individually even if a high-performance digital power analyzer were used.

As shown in Fig. 2, the auxiliary snubber capacitor C_S is connected across each SiC-MOSFET/SBD in order to reduce both switching loss and overvoltage. Like the equivalent inductor given by (2), the equivalent capacitance C is defined by

$$C = C_S + C_{ds}, \tag{4}$$

where C_S is the auxiliary snubber capacitor, and C_{ds} is the parasitic drain-to-source capacitance across each SiC-MOSFET/SBD. The series resonance between the equivalent inductance and the equivalent capacitance C allows the dc-dc converter to achieve zero-voltage switching (ZVS).

The authors of [18] have presented an optimal dead time T_{DOpt} for minimizing the snubber loss as follows:

$$T_{DOpt} = \frac{\pi\sqrt{LC}}{2}. \tag{5}$$

Equation (5) includes the equivalent capacitance C, so that the optimal dead time should be changed, according to the capacitance value of C_S. The following experiment sets the optimal dead time to 0.6 μs for $C_S = 4.5$ nF and to 0.8 μs for $C_S = 9$ nF. Under the ZVS operation, the duration of the current flowing through an SBD is shorter than the optimal dead time. Moreover, the duration of the current flowing through the equivalent capacitance C becomes longer as the output power P decreases.

IV. EXPERIMENTAL RESULTS

A. Experimental Waveforms, Overall Power Loss, and Conversion Efficiency

Fig. 3 shows experimental waveforms of ac terminal voltages appearing across bridges 1 and 2, v_1 and v_2, respectively, and the ac current flowing at the ac side of bridge 1, i, at $f = 20$ kHz when $C_S = 4.5$ nF and $T_{DOpt} = 0.6$ μs. Fig. 3 (a) presents those at the rated-power ($P = 100$ kW) operation, where the waveform of v_1 leads that of v_2 by 26.6°. This means that power flow occurs from bridge 1 to bridge 2. The peak voltage appearing in the waveforms of v_1 and v_2 is lower than 950 V, which is within an acceptable level because of using the 1.2-kV SiC modules in Fig. 2. Fig. 3 (b) and (c) are in the cases of $P = 30$ kW and $P = 10$ kW, respectively.

Fig. 4 plots overall conversion efficiencies η, calculated from (3), with operating frequencies of $f = 20$ kHz and 10 kHz when $C_S = 4.5$ nF and $T_{DOpt} = 0.6$ μs. The maximum conversion efficiency η_{max} is 98.7% at $P = 42.0$ kW, and the conversion efficiency at $P = 100$ kW is 97.9%.

Fig. 5 plots overall power losses P_{Loss} under the same conditions as those in Fig. 4. In the ZVS (zero-voltage-switching) region ranging from 20 to 100 kW, the power loss decreases as the output power decreases. In the incomplete-ZVS region ranging from 0 to 20 kW, however, the power loss increases as the output power decreases. This interesting

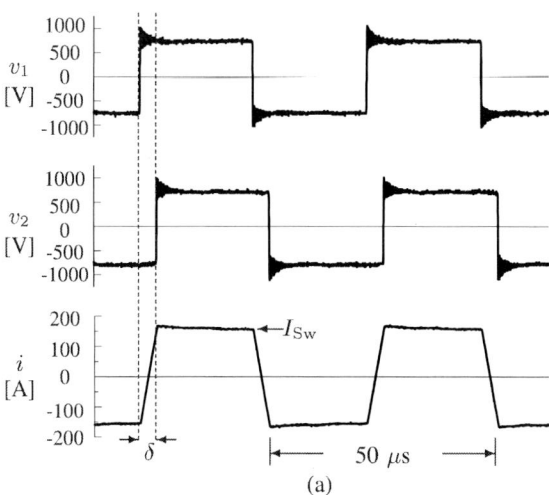

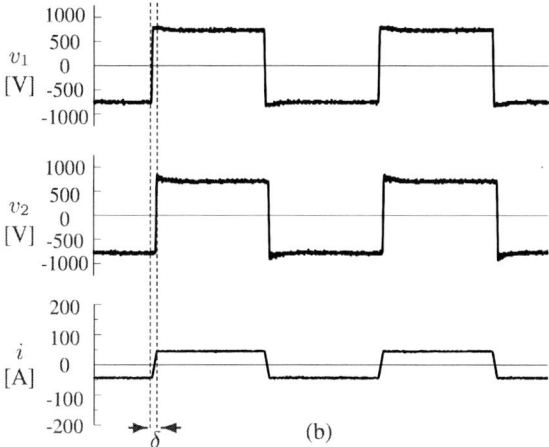

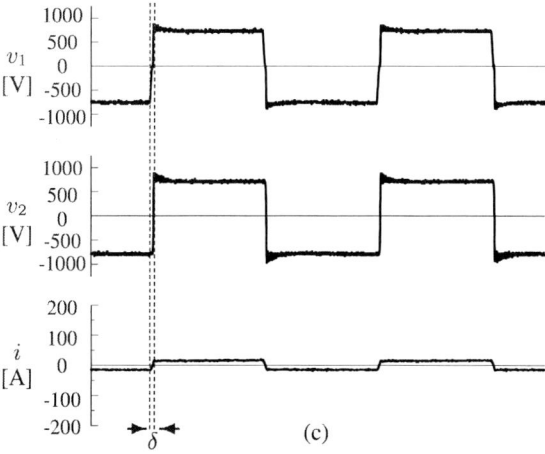

Fig. 3. Experimental waveforms when $f = 20$ kHz, $C_S = 4.5$ nF and $T_D = 0.6$ μs. (a) $P = 100$ kW and $\delta = 26.6°$. (b) $P = 30$ kW and $\delta = 7.0°$. (c) $P = 10$ kW and $\delta = 1.5°$.

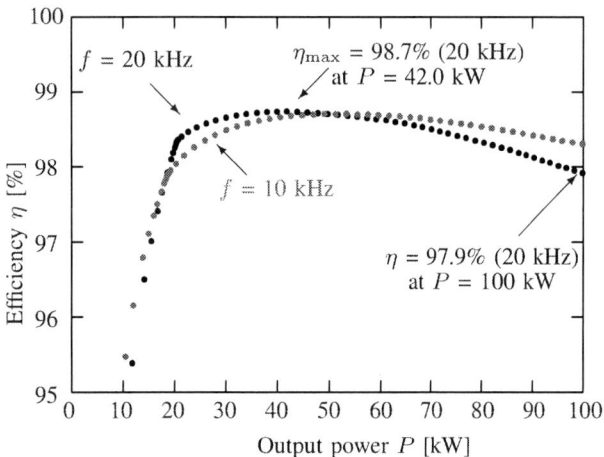

Fig. 4.   Conversion efficiency when $C_S = 4.5$ nF and $T_D = 0.6$ μs.

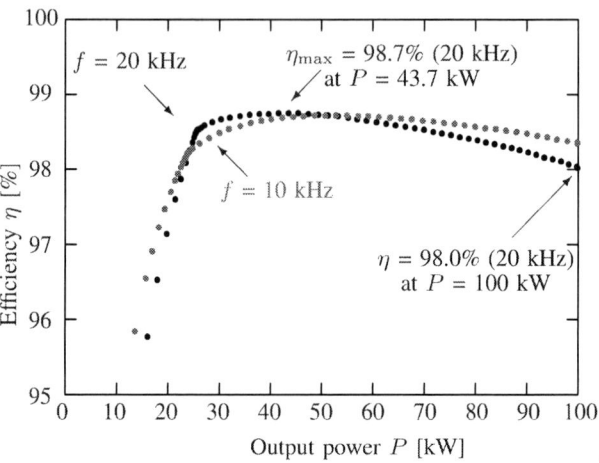

Fig. 6.   Conversion efficiency when $C_S = 9$ nF and $T_D = 0.8$ μs.

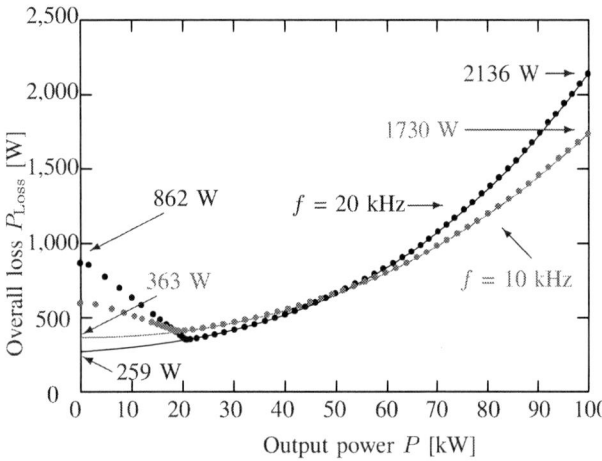

Fig. 5.   Measured overall power loss when $C_S = 4.5$ nF and $T_D = 0.6$ μs.

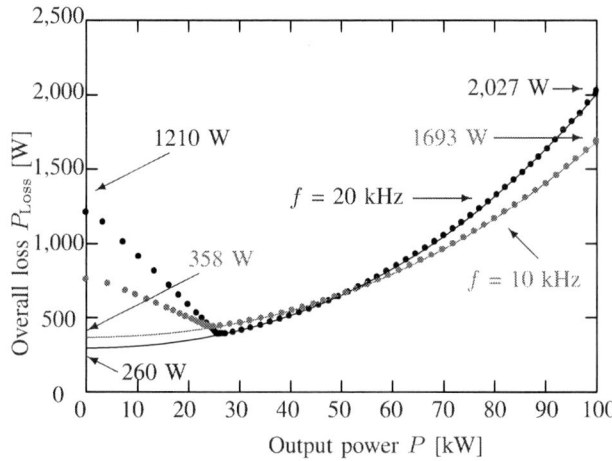

Fig. 7.   Measured overall power loss when $C_S = 9$ nF and $T_D = 0.8$ μs.

trend comes from an increase in the "subber loss" caused by incomplete ZVS.

Figs. 6 and 7 plot overall conversion efficiencies and overall power losses at $f = 20$ kHz and 10 kHz when $C_S = 9$ nF and $T_{DOpt} = 0.8$ μs. The incomplete ZVS region expands into a range from 0 to 26 kW because the snubber capacitor is larger in capacitance value than that in Fig. 5. The overall power loss at $P = 0$ increases from 862 W in Fig. 5 to 1210 W in Fig. 7 because the snubber loss gets higher as the snubber capacitor C_S becomes larger. On the other hand, the overall power loss at $P = 100$ kW decreases from 2136 W to 2027 W. This will be discussed in section VI.

B.   Comparison With a 750-V, 60-kW, 4-kHz DC-DC Converter Using Si-IGBT/PND Dual Modules

Table II summarizes comparisons in conversion efficiency between the 750-V, 100-kW, 20-kHz dc-dc converter using

the 1.2-kV 400-A SiC-MOSFET/SBD dual modules and the 750-V, 60-kW, 4-kHz dc-dc converter using 1.2-kV 300-A Si-IGBT/PND dual modules [19]. The power ratings are not the same because the current ratings of the SiC and Si modules are not the same. Moreover, the two operating frequencies are not the same because it would be impractical to operate the dc-dc converter using the Si modules at 20 kHz in terms of increased switching loss. However, note that the SiC module adopts the same packaging in size, shape, and terminal/pin arrangement as the Si module does. A main reason for having the same packaging in appearance is to make it easy for power electronics engineers to replace the Si module with the SiC module.

Table II concludes that the use of the SiC modules makes the maximum efficiency higher by 0.9% and the rated-power efficiency higher by 1.0% than the use of the Si modules, although the operating frequency is increased from 4 kHz to

TABLE II. COMPARISON BETWEEN THE TWO DC-DC CONVERTERS USING SiC-MOSFET/SBD AND Si-IGBT/PND MODULES.

Power module	1.2-kV, 400-A SiC-MOSFET/SBD	1.2-kV, 300-A Si-IGBT/PND
Power rating	100 kW	60 kW
DC voltage rating	750 V	750 V
Operating frequency	20 kHz	4 kHz
Snubber capacitor	4.5 nF	10 nF
Maximum efficiency	98.7%	97.8%
Rated power efficiency	97.9%	96.9%

20 kHz.

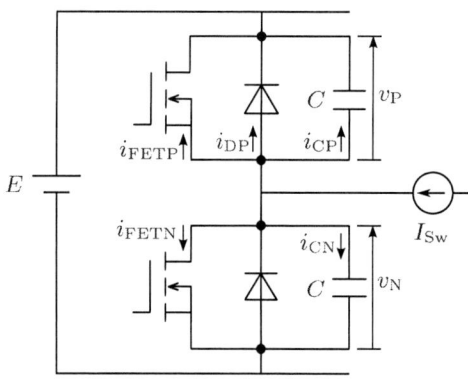

Fig. 8. Simplified circuit of a leg.

V. SWITCHING-LOSS ANALYSIS DURING ZERO-VOLTAGE-SWITCHING (ZVS) OPERATION

The switching loss caused by a switching-mode power converter can be divided into turn-on and turn-off losses. However, the dc-dc converter shown in Fig. 2 produces no turn-on loss as long as it is operated at ZVS (zero-voltage switching).

Fig. 8 shows the equivalent circuit used for the following analysis of the turn-off switching loss with focus on one leg in Fig. 2. Fig. 9 depicts voltage and current waveforms during the dead time T_D. From the period of $t = 0$ to the turn-off time $t = T_1$ on the horizontal axis in Fig. 9, the positive-side MOSFET is kept turned off, and $i_\mathrm{DP} = 0$. The following assumptions are made to analyze the turn-off loss.

1) The ac current during the turn-off switching is constant to be I_Sw, so that the turn-off-switching current is represented as a dc current source of I_Sw in Fig. 8. The reason is that the dead time of $T_\mathrm{D} = 0.6\ \mu\mathrm{s}$ ($C_\mathrm{S} = 4.5$ nF) is much shorter that the operating period of $1/f = 50\ \mu\mathrm{s}$.

2) The drain current of the negative-side MOSFET, i_FETN is falling down linearly with the passage of time from $t = 0$ to $t = T_\mathrm{f}$.

3) The drain current is being commutated equally to the positive-side and negative-side equivalent capacitors defined by (4) as $C\ (= C_\mathrm{S} + C_\mathrm{ds})$.

The first and second assumptions yield

$$
\begin{aligned}
i_\mathrm{FETN} &= I_\mathrm{Sw} - \frac{I_\mathrm{Sw}}{T_\mathrm{f}} t \\
&= I_\mathrm{Sw}\left(1 - \frac{1}{T_\mathrm{f}} t\right).
\end{aligned}
\tag{6}
$$

The combination of (6) and the second assumption gives the capacitor current $i_\mathrm{C}\ (= i_\mathrm{CP} = i_\mathrm{CN})$ as follows:

$$
\begin{aligned}
i_\mathrm{C} &= \frac{1}{2}(I_\mathrm{Sw} - i_\mathrm{FETN}) \\
&= \frac{I_\mathrm{Sw}}{2 T_\mathrm{f}} t.
\end{aligned}
\tag{7}
$$

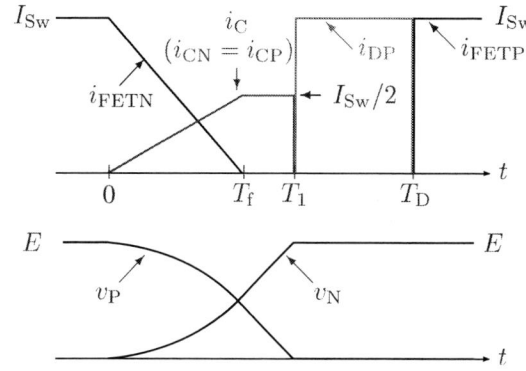

Fig. 9. Voltage and current waveforms during ZVS operation.

Integration of (7) with respect to time from 0 to t gives the following equation.

$$
\begin{aligned}
v_\mathrm{N} &= \frac{1}{C} \int_0^t i_\mathrm{C}\, dt \\
&= \frac{I_\mathrm{Sw}}{4 C T_\mathrm{f}} t^2
\end{aligned}
\tag{8}
$$

The instantaneous power formed by v_N and i_FETN is given as

$$
\begin{aligned}
p_\mathrm{Sw} &= i_\mathrm{FETN} \cdot v_\mathrm{N} \\
&= \frac{I_\mathrm{Sw}^2}{4 C T_\mathrm{f}}\left(t^2 - \frac{1}{T_\mathrm{f}} t^3\right)
\end{aligned}
\tag{9}
$$

Integration of (9) with respect to time from 0 to T_f gives the following energy loss of the negative-side MOSFET per turn-off.

$$
\begin{aligned}
E_\mathrm{Sw} &= \int_0^{T_\mathrm{f}} p_\mathrm{Sw}\, dt \\
&= \frac{T_\mathrm{f}^2}{48 C} I_\mathrm{Sw}^2.
\end{aligned}
\tag{10}
$$

Equation (10) suggests that the switching loss of the dc-dc converter during ZVS operation is proportional to a square of the turn-off-switching current I_{Sw} under the three reasonable assumptions mentioned above.

VI. POWER-LOSS BREAKDOWN FROM THE MEASURED OVERALL POWER LOSS

In general, it would be more difficult to achieve accurate measurement of switching and iron losses in a bidirectional isolated dual-active-bridge dc-dc converter as the converter efficiency becomes higher. To overcome the difficulty, measurement has often been replaced with "estimation" relying on data sheets from power-device and magnetic-core manufactures [5].

This paper proposes a unique method to break down the overall power loss P_{Loss} that was measured accurately from the circuit configuration shown in Fig. 2. This breakdown method is so simple and practical as not to require any equipment, instrument, or data sheet.

A. Assumptions for Achieving Power-Loss Breakdown

The following reasonable assumptions are made to achieve breakdown of the overall power loss:

1) The accurately-measured overall power loss P_{Loss} can be divided into the following five power-loss components; the switching loss P_{Sw} and the conduction loss P_{Cond}, both of which are produced by the SiC modules, the copper loss P_{Copp} and the iron loss P_{Iron}, both of which occur in the magnetic devices, and the other unknown loss P_{Unkn}.

2) The switching loss is proportional to the operating frequency f, and is proportional to a square of the turn-off-switching current I_{Sw}, as derived in the previous section. This results in neglecting a component proportional to the turn-off-switching current and another component independent of the turn-off-switching current from the switching loss.

Fig. 10 shows the relationship among the five power-loss components coming from the first assumption. This leads to the following equation.

$$
\begin{aligned}
P_{Loss} &= P_{Iron} + P_{Sw} + P_{Ohm} \\
&= P_{Iron} + P_{Sw} + (P_{Cond} + P_{Copp} + P_{Unkn})
\end{aligned}
$$
(11)

The ohmic loss produced in the power circuit, P_{Ohm}, which is proportional to a square of the rms value of the ac current i, is divided into the conduction loss P_{Cond}, the copper loss P_{Copp}, and the unknown loss P_{Unkn}. The unknown loss includes the following power-loss components:

*   an ohmic-loss component related to the laminated dc-bus bars,
*   temperature-dependent conduction-loss and copper-loss components related to the SiC modules and the magnetic devices, respectively,
*   a slightly-increased copper-loss component of Litz wires in the magnetic devices due to skin and proximity effects,

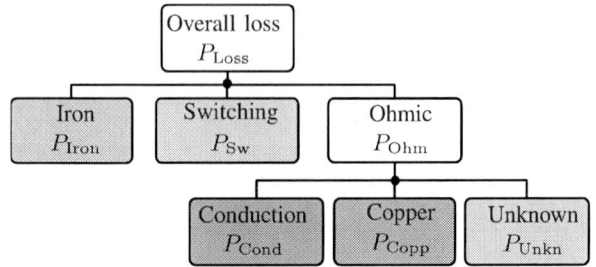

Fig. 10. Classification and relation of the five power-loss components separated from the measured overall power loss.

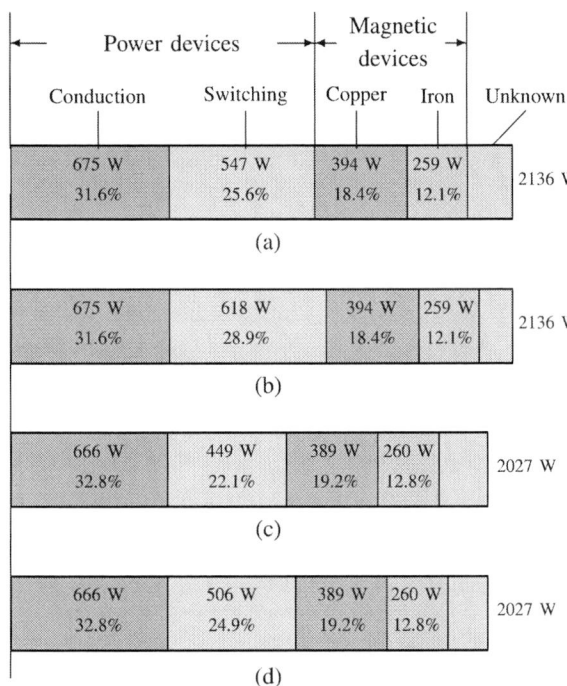

Fig. 11. Power-loss breakdown at 750-V, 100-kW, 20-kHz operation. (a) $C_S = 4.5$ nF and $P_{Sw} \propto I_{Sw}^2$. (b) $C_S = 4.5$ nF and $P_{Sw} \propto I_{Sw}$. (c) $C_S = 9$ nF and $P_{Sw} \propto I_{Sw}^2$. (d) $C_S = 9$ nF and $P_{Sw} \propto I_{Sw}$.

*   the power-loss component dissipated by the equivalent series resistance in the dc film capacitors in Fig. 2.

The temperature-dependent conduction-loss component related to the SiC modules would be the most dominant in the unknown loss.

B. Estimation of Iron Loss

The overall power-loss data plotted in the ZVS region from 20 to 100 kW can be fitted with a third-order polynominal as shown in Fig. 5. Generally, the power-loss data may be fitted with a second-order polynominal because most of the dominant power-loss components are exactly or nearly proportional to a square of the output power P in the ZVS region. However, Fig. 5 adopts the third-order polynominal, considering the effect of an increased phase-shift angle on the current waveform of i.

When the 20-kHz curve is extrapolated to zero power, an overall power loss of 259 W represents the iron loss at 20 kHz. The reason is that the dc-dc converter would produce it, even at $P = 0$ or $i = 0$, if ZVS operation is maintained across the power range of 0 to 100 kW. The same is true of the other 10-kHz curve, in which an overall power loss of 363 W represents the iron loss. Note that adjusting a phase-shift angle from 0 to $26.6°$ is assumed to produce no effect on the iron loss obtained from this extrapolation, and that the iron loss is assumed to be independent of the ac current i.

The same transformer with the parameters summarized in Table I was used for the experiments at 20 kHz and 10 kHz in Fig. 2. Therefore, the maximum magnetic flux density at 10 kHz is two times as high as that at 20 kHz under the same dc input voltage as 750 V. The experimental fact that the iron loss of 363 W at 10 kHz is higher than that of 259 W at 20 kHz coincides with the following fact estimated from magnetic-core data sheets; the iron loss at 10 kHz is higher than that at 20 kHz.

C. Extraction of Switching Loss and Ohmic Loss from Overall Loss

When $C_S = 4.5$ nF, the iron loss was estimated as 259 W out of the overall power loss of $P_{\text{Loss}} = 2136$ W at the 20-kHz, 100-kW operation, and as 363 W out of 1730 W at the 10-kHz, 100-kW operation in the previous subsection. Fig. 10 and (11) provide the following relation: The subtraction of P_{Iron} from P_{Loss} is equal to the summation of P_{Sw} and P_{Ohm}. Applying the second assumption in subsection VI-A to the above relation derives the following set of equations.

$$2136^{\text{W}} - 259^{\text{W}} = P_{\text{Sw}}^{20k} + P_{\text{Ohm}}^{20k} \tag{12}$$

$$1730^{\text{W}} - 363^{\text{W}} = \frac{1}{2}(\frac{I_{\text{Sw}}^{10k}}{I_{\text{Sw}}^{20k}})^2 P_{\text{Sw}}^{20k} + (\frac{I_{\text{rms}}^{10k}}{I_{\text{rms}}^{20k}})^2 P_{\text{Ohm}}^{20k} \tag{13}$$

Here, I_{Sw} represents such a turn-off-switching current as described in Fig. 3 (a) and I_{rms} represents the rms current of i. Note that superscripts "20k" and "10k" represent operating frequencies 20 kHz and 10 kHz, respectively. The current waveforms obtained from the experiments at the 20-kHz, 100-kW operation and the 10-kHz, 100-kW operation yield $I_{\text{Sw}}^{20k} = 152$ A, $I_{\text{Sw}}^{10k} = 132$ A, $I_{\text{rms}}^{20k} = 152$ A, and $I_{\text{rms}}^{10k} = 142$ A. Solving the set of equations leads to extracting P_{Sw}^{20k} and P_{Ohm}^{20k} from 2136 W. As a result, P_{Sw}^{20k} and P_{Ohm}^{20k} were estimated as 547 W (25.6%) and 1330 W (62.3%), respectively.

D. Estimation of Conduction Loss and Copper Loss

Fig. 10 and (11) can divide the ohmic loss P_{Ohm} into the following three components.

$$P_{\text{Ohm}} = P_{\text{Cond}} + P_{\text{Copp}} + P_{\text{Unkn}} \tag{14}$$

The subtraction of P_{Cond} and P_{Copp} from P_{Ohm} enables the estimation of P_{Unkn}. The use of the rms currents of i

at the 20-kHz and 100-kW operation as $I_{\text{rms}}^{20k} = 152$ A and $I_{\text{rms}}^{10k} = 142$ A results in estimating $P_{\text{Cond}} = 675$ W (31.6%) and $P_{\text{Copp}} = 394$ W (18.4%). Here, the 1.2-kV, 400-A SiC-MOSFET used in this experiment has an on-resistance value of 7.3 mΩ at room temperature. The "skin and proximity" effects at the 20-kHz operation would be negligible from the estimated value of P_{Copp} because Litz wires are used for both transformer and auxiliary inductors. As a result, P_{Unkn} includes temperature-dependent components present in P_{Cond} and P_{Copp}, a slightly-increased, but practically negligible, component contained in P_{Copp}, which is caused by the skin and proximity effects.

E. Tolerance of Power-Loss Breakdown

The above breakdown method may include a breakdown error of the iron loss. This breakdown error brings another breakdown error to the switching loss and the unknown loss because the conduction loss and the copper loss, as well as the overall power loss are independent of the iron loss. As a phase-shift angle is increasing from 0 to $26.6°$ gradually, the waveform of the magnetic flux density in the transformer is changing from a triangle to a trapezoid without dc bias. As a result, the peak value of the magnetic flux density is decreasing slightly. This makes an actual iron loss at the 100-kW operation slightly lower than the iron loss obtained by the extrapolation in Fig. 5. This change in the waveform of the magnetic flux density may bring a tolerance of around 8% to the iron loss, according to the magnetic-core date sheet. However, an aim of the breakdown achieved in this paper is to divide the accurately-measured overall power loss into the five power-loss components with an acceptable level of tolerance from a practical point of view. Another aim is to reveal which power-loss component is the most dominant among the five components, thus leading to further improvements in conversion efficiency.

F. Results of Power-Loss Breakdown

Fig. 11 shows four results of the power-loss breakdown at the 20-kHz and 100-kW operation under four different breakdown conditions. Fig. 11 (a) presents the breakdown result with $C_S = 4.5$ nF and the second assumption in subsection VI-A. Fig. 11 (b) is the result with $C_S = 4.5$ nF but with the following different assumption from that in (a): The switching loss P_{Sw} is proportional to the turn-off-switching current I_{Sw}. The aim of (b) is to look into how sensitive the second assumption is to the switching loss. Comparing (a) with (b) indicates that the second assumption is not so sensitive to the switching loss from a practical point of view. However, a ratio of P_{Sw} with respect to P_{Unkn} is changed slightly although no change happens in the summation of P_{Sw} and P_{Unkn}. Strictly speaking, the actual switching loss is given as the summation of a term proportional to a square of the turn-off-switching current, a term proportional to it, and a term independent of it. The first term seems to be the most dominant among the three. The same is true of Fig. 11 (c) and (d) with $C_S = 9$ nF.

Fig. 11 (a) indicates that the summation of P_{Sw} and P_{Cond} occupies about 60% of P_{Loss}, and that P_{Cond} is nearly equal to, but slightly higher than, P_{Sw}. Moreover, the summation of P_{Copp} and P_{Iron} occupies about 30% of P_{Loss}. It is reasonable to assume that bridge 1 has the conduction and switching losses as bridge 2 because bridge 2 is operated at synchronous rectification mode. This assumption allows one to estimate either ac-to-dc or dc-to-ac conversion efficiency as 99.4% ($= 100 - 0.622/100$) at the 20-kHz and 100-kW operation, excluding the copper and iron losses due to the magnetic devices. Besides, such an high efficiency does not take into account an increased conduction loss accompanying an increased temperature during continuous operation.

Comparing Fig. 11 (a) with Fig. 11 (c) suggests that the switching loss can be reduced by about 20% when the snubber capacitor C_S is changed from 4.5 nF to 9 nF. However, comparing Fig. 5 with Fig. 7 indicates that the overall power loss in the incomplete ZVS region of Fig. 7 is higher than that of Fig. 5 and that the incomplete ZVS region in Fig. 7 is extended slightly. In particular, the overall power loss at zero power ($P = 0$) increases from 862 W to 1210 W. This means that a trade-off or a compromise between power efficiency and energy efficiency exists in designing the snubber capacitor. For example, it is better to install a small snubber capacitor in terms of improving energy efficiency when low-power operation is longer in working time than high-power operation.

VII. CONCLUSION

This paper has designed, built, and tested a 750-V, 100-kW, 20-kHz bidirectional isolated dual-active-bridge dc-dc converter using four 1.2-kV, 400-A SiC-MOSFET/SBD dual modules. The maximum conversion efficiency from the dc-input terminals to the dc-output terminals is as high as 98.7% at 42 kW, which is calculated from an accurately-measured overall power loss excluding gate-drive and control circuit losses. Moreover, this paper has proposed a simple and practical method to divide the measured overall power loss into five power-loss components with an acceptable level of tolerance. This power-loss breakdown suggests that the sum of conduction and switching losses produced by the SiC modules occupies about 60% of the overall power loss, and that the switching loss is nearly equal to the conduction loss at the 20-kHz, 100-kW operation. In addition, the breakdown has revealed that it is important to reduce the conduction and switching losses produced by the SiC modules as well as the copper and iron losses due to the 20-kHz transformer and auxiliary inductors, aiming at further improvements in conversion efficiency.

VIII. ACKNOWLEDGEMENT

This work was supported by the Cabinet Office, Government of Japan and the Japan Society for the Promotion of Science (JSPS) through the Funding Program for World-Leading Innovative R&D on Science and Technology (FIRST Program).

REFERENCES

[1] R. W. De Doncker, D. M. Divan, and M. H. Kheraluwala, "A three-phase soft-switched high-power-density dc/dc converter for high power applications," *IEEE Trans. Ind. Appl.*, vol. 27, no. 1, pp. 63–73, Feb. 1991.

[2] M. H. Kheraluwala, R. W. Gascoigne, D. M. Divan, and E. D. Baumann, "Performance characterization of a high-power dual active bridge dc-to-dc converter," *IEEE Trans. Ind. Appl.*, vol. 28, no. 6, pp. 1294–1301, Nov./Dec. 1992.

[3] R. L Steigerwald, R. W. De Doncker, and M. H. Kheraluwala, "A comparison of high-power dc-dc soft-switched converter topologies," *IEEE Trans. Ind. Appl.*, vol. 32, no. 5, pp. 1139–1145, Sept./Oct. 1996.

[4] S. Inoue and H. Akagi, "A bidirectional isolated dc-dc converter as a core circuit of the next-generation medium-voltage power conversion system," *IEEE Trans. Power Electron.*, vol. 22, no. 2, pp. 535–542, Mar. 2007.

[5] S. Inoue and H. Akagi, "A bidirectional dc-dc converter for an energy storage system with galvanic isolation," *IEEE Trans. Power Electron.*, vol. 22, no. 6, pp. 2299–2306, Nov. 2007.

[6] N. M. L. Tan, S. Inoue, A. Kobayashi, and H. Akagi, "Voltage balancing of a 320-V, 12-F electric double-layer capacitor bank combined with a 10-kW bidirectional isolated dc-dc converter," *IEEE Trans. Power Electron.*, vol. 23, no. 6, pp. 2755–2765, Nov. 2008.

[7] C. Zhao, S. D. Round, and J. W. Kolar "An isolated three-port bidirectional dc-dc converter with decoupled power flow management," *IEEE Trans. Power Electron.*, vol. 23, no. 5, pp. 2443–2453, Sept. 2008.

[8] H. Bai and C. Mi, "Eliminate reactive power and increase system efficiency of isolated bidirectional dual-active-bridge dc-dc converters using novel dual-phase-shift control," *IEEE Trans. Power Electron.*, vol. 23, no. 6, pp. 2905–2914, Nov. 2008.

[9] G. G. Oggier, G. O. García, and A. R. Oliva, "Switching control strategy to minimize dual active bridge converter losses," *IEEE Trans. Power Electron.*, vol. 24, no. 7, pp. 1826–1838, July 2009.

[10] H. Zhou and A. M. Khambadkone, "Hybrid modulation for dual-active-bridge bidirectional converter with extended power range for ultracapacitor application," *IEEE Trans. Ind. Appl.*, vol. 45, no. 4, pp. 1434–1442, July/Aug. 2009.

[11] F. Krismer and J. W. Kolar, "Accurate power loss model derivation of a high-current dual active bridge converter for an automotive application," *IEEE Trans. Ind. Electron.*, vol. 57, no. 3, pp. 881-891, Mar. 2010.

[12] G. Guidi, M. Pavlovsky, A. Kawamura, T. Imakubo, and Y. Sasaki, "Improvement of light load efficiency of dual active bridge dc-dc converter by using dual leakage transformer and variable frequency," in *Proc. IEEE ECCE*, Sept. 2010, pp. 830–837.

[13] N. M. L. Tan, T. Abe, and H. Akagi, "Design and performance of a bidirectional isolated dc-dc converter for a battery energy storage system," *IEEE Trans. Power Electron.*, vol. 27, no. 3, pp. 1237–1248, Mar. 2012.

[14] R. T. Naayagi, A. J. Forsyth, and R. Shuttleworth, "High-power bidirectional dc-dc converter for aerospace applications," *IEEE Trans. Power Electron.*, vol. 27, no. 11, pp. 4366–4379, Nov. 2012.

[15] R. T. Wood and T. E. Salem, "Evaluation of a 1200-V, 800-A all-SiC dual module," *IEEE Trans. Power Electron.*, vol. 26, no. 9, pp. 4366–4379, Sept. 2011.

[16] K. Shirabe, M. Swamy, J. Kan, M. Hisatsune, M. Das, R. Callanan, and H. Lin, "Design of 400V class inverter drive using SiC 6-in-1 power module," in *Proc. IEEE ECCE*, Sep. 2013, pp. 2363–2370.

[17] G. Wang, F. Wang, G. Magai, Y. Lei, A. Huang, and M. Das, "Performance comparison of 1200V 100A SiC MOSFET and 1200V 100A Silicon IGBT," in *Proc. IEEE ECCE*, Sep. 2013, pp. 3230–3234.

[18] N. M. L. Tan, T. Abe, and H. Akagi, "Experimental discussions on operating frequencies of a bidirectional isolated dc-dc converter for a battery energy storage system," in *Proc. IEEE ECCE*, Sep. 2013, pp. 2333–2340.

[19] T. Chocktaweechock, K. Hasegawa, and H. Akagi: "System design and experimental verification of a 750-V, 60-kW, 4-kHz bidirectional isolated dc-dc converter," in *Proc. IEEJ IAS Annual Meeting*, 1-94, 2012.

Design Considerations of a 15kV SiC IGBT Enabled High-Frequency Isolated DC-DC Converter

Awneesh Tripathi, Krishna Mainali, Dhaval Patel,
Arun Kadavelugu, Samir Hazra, Subhashish Bhattacharya
Department of Electrical and Computer Engineering
North Carolina State University, Raleigh, NC 27606
{aktripat}, {kmainal}, {sbhatta4}@ncsu.edu

Kamalesh Hatua
Department of Electrical Engineering
IIT Madras, Chennai, INDIA
kamalesh@ee.iitm.ac.in

Abstract—The advent of the $15 kV\, SiC$ **IGBT device has made a single series stage medium-voltage (MV) and high-frequency (HF) DC-DC Dual Active Bridge (DAB) converter application viable. The $Y : Y/\Delta$ three-phase DAB is a high-power isolated DC-DC converter based on three-level neutral-point clamped (NPC) on the MV side. A MV/HF transformer used in the DAB, has significant parasitic capacitances, which cause ringing in the DAB current under high dV/dT switching. In addition, the converters need sufficient dead-time between complimentary switches to avoid possibility of any shoot-through. The length of the dead-time depends on switching characteristics. Both the dead-time and transformer parasitics affect zero voltage switching (ZVS) performance of the DAB. Thus, the DAB design has to be closely coupled with the switching characteristics of the devices and MV/HF transformer parasitics. For the ZVS mode, the current-vector needs to be between converter voltage vectors with a certain margins defined by dead-time, parasitics and desired duty ratio of three-level MV converter. This paper addresses these design challenges for the MV DAB application.**

I. INTRODUCTION

The Dual Active Bridge (DAB) is an isolated bidirectional DC-DC converter proposed in [1] and [2] for high-power applications. It is used as an intermediate DC-DC stage for applications like PV-Grid, Wind-Grid or Grid-Grid interface. The Transformer-less Intelligent Power Substation (TIPS) is a Grid-Grid interface [3] being developed at the FREEDM Systems Center Lab. It interfaces $13.8\,kV$ MV Grid to $480\,V$ low voltage (LV) Grid. It utilizes a three phase DAB in intermediate stage, switching at $10\,kHz$ to step down $22\,kV$ DC to $800\,V$ DC. The three-phase DAB offers advantage of a smaller DC capacitor, smaller parasitic ringing and higher magnetic utilization as compared to single-phase DAB of same rating [5].

The single series stage DAB is simple to control compared to multi-stage structure in terms of voltage and power balancing. But it poses additional challenges at the MV level as the devices need to be optimized for dV/dT and dead-time. Both these parameters also affect the transformer and overall DAB behavior. The MV/HF transformer has considerable parasitics which are ignored for a LV DAB. The HF transformer is used to save size and still provide isolation. Hence there is limit on the size and hence parasitics can not be kept very low. But the

presence of high dV/dT of the SiC IGBT/MOSFET switches which have been characterised in [6]–[8] at this voltage level causes undesired ringing in the current. The slowing down of devices will require more dead time of converters which affect the DAB with respect to the ideal operation. Also it cause more switching losses. This paper presents these challenges in the design of a medium voltage (MV) DAB converter system.

The design of three-phase $Y : Y/\Delta$ DAB topology shown in Fig 1 which has been introduced in [4], is briefly described in section II. The selected composite topology saves MV side device voltage rating as well as the current rating since the maximum step voltage applied across transformer MV winding is $V_{dc}/3$. The D-Q control of the $Y : Y/\Delta$ DAB has been presented in [4], [9] considering the parasitic effects. These issues are investigated in this paper from the design point of view. The MV DAB design must consider the non-idealities on the top of the basic DAB design [2]. Section III presents the effects of MV/HF transformer's parasitic, dead-time and duty-ratio on the $100\,kW$ three-phase DAB ZVS operation. Section IV presents an optimum design for the DAB while considering the device and transformer characteristics. The phase-angle and three-level converter duty-cycle control make it possible to set current-vector at optimum angle of voltage-vectors ensuring ZVS and minimum reactive power-flow [9]. Only an optimally designed DAB can be operated at an optimum point. The analysis have been validated with experimental results. Section V shows the experimental results for the MV $Y : Y/\Delta$ DAB followed by conclusion.

II. OPERATION AND DESIGN OF THREE-PHASE $Y : Y/\Delta$ DAB

The DAB topology comprises of a three-level NPC converter on the MV side which feeds the star configured primary-windings of three single-phase HF transformers. The star-connected NPC converter is operated in six-step with 41.7% duty ratio of top switch to generate minimum 5th, 7th harmonics in pole voltage. Fig. 2 shows the $100\,kW$ simulation waveforms for a phase converter voltages and currents for each of Y_{MV}, Δ_{LV} and Y_{LV} sides. Eq. 1 shows the nth harmonic primary phase voltage V_{prim} with $\beta = 15^{\circ}$ zero-period which is same as 41% duty-ratio. The two-level converters are in 30°

The 2014 International Power Electronics Conference

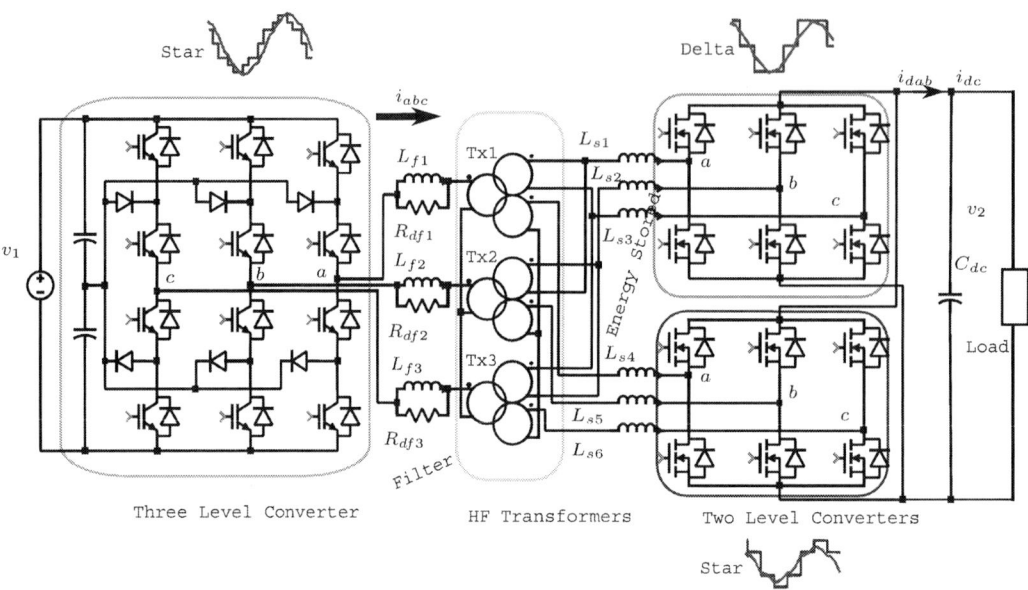

Fig. 1. Three Phase DAB Circuit-Diagram

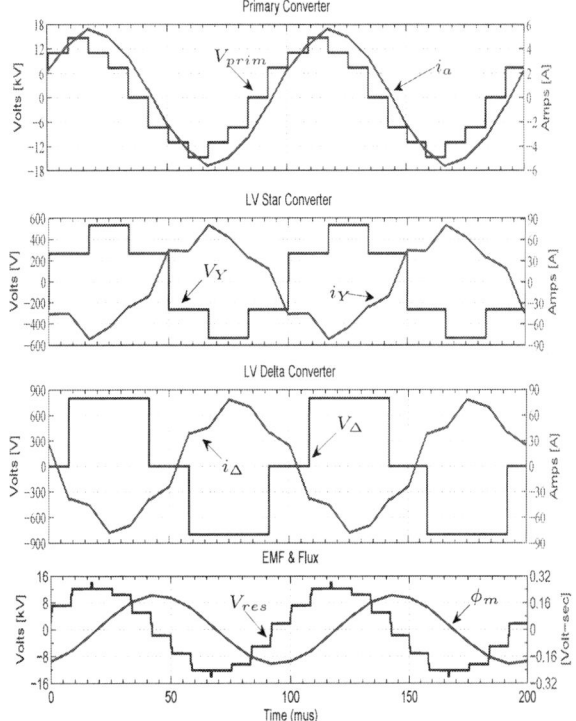

Fig. 2. Three Phase DAB Steady State Simulation $100\,kW$

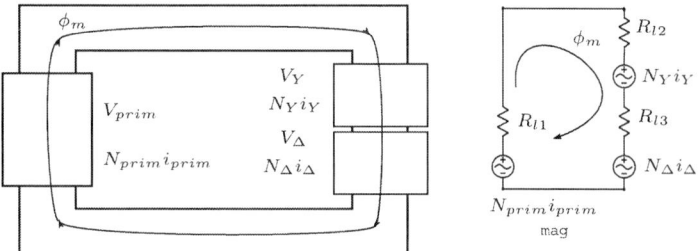

Fig. 3. Cancellation of Higher Order Harmonics

$$V_{n\Delta} = \sqrt{3}\frac{4V_{dc2}}{2n\pi}sin[n\omega t + \frac{\pi}{6}]\forall n = 1,3,5.. \qquad (3)$$

Equation 2 and 3 confirm that 5th and 7th harmonics of Y and Δ sides will add in phase-opposition. As shown in Fig. 3, the MMF in the core, contributed by secondary higher order odd-harmonics ($6m \pm 1$) tend to be in opposite direction due to opposite voltage phase ($30[6n \pm 1] \pm 30, \forall n = 1,3,5..$) [10]. Hence these harmonic voltages do not excite transformer EMF and there are harmonic currents flowing between secondary converters. The predominantly fundamental DAB voltages interact to generate a nearly sinusoidal current through primary winding. The rated parameters for the simulation-study are shown in Table I.

TABLE I.   RATED DAB PARAMETERS

Parameters	Value[Units]
Input DC Voltage	$22\,kV$
Output DC Voltage	$800\,V$
Input side Duty-Ratio	$41\%\,nominal$
Primary External Inductors	$440\,\mu H$
Secondary External Inductors	$80\,\mu H$
Real Power Transfer	$100\,kW$
Switching-frequency	$10\,kHz$

phase-shift so that they produce synchronous fundamental-flux in the transformer.

$$V_{n^{prim}} = \frac{4V_{dc1}}{2n\pi}cos[n\beta]sin[n\omega t] \qquad (1)$$

$$V_{n^Y} = \frac{4V_{dc2}}{2n\pi}sin[n\omega t] \qquad (2)$$

Maximum step-voltage applied on transformer primary is

978-1-4799-2706-7/14 $31.00 © 2014 IEEE

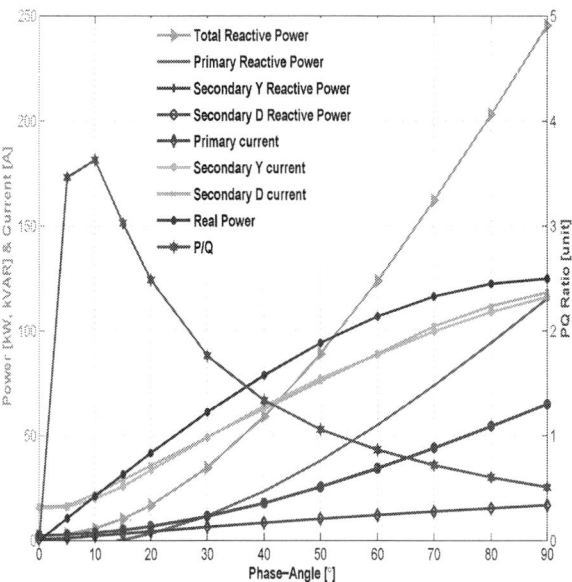

Fig. 4. $P, Q - \delta$ curve for $Y : Y/\Delta$

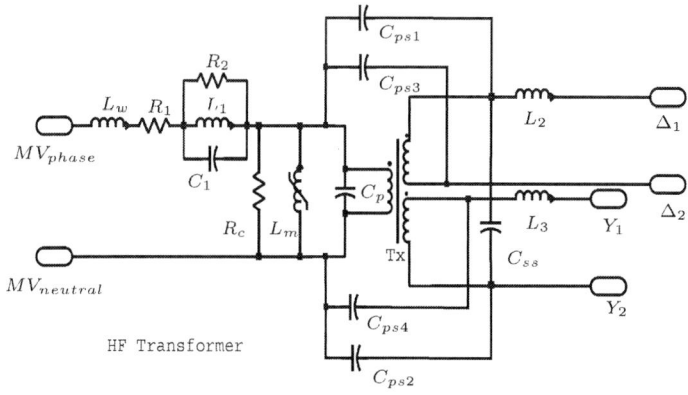

Fig. 5. Equivalent circuit of $Y : Y/\Delta$ DAB Transformer

capacitances as well. The primary-current would be high at the maximum step-voltage application instant. To limit the peak-noise current, non-capacitive external-inductors are also required at the MV-primary side, whose value depends on peak-current to be allowed through device at the resonance and the ringing-frequency [9]. Star-connection at the primary three-level converter side reduces the maximum step-voltage applied to individual transformer and makes it feasible for $15kV$ IGBTs to supply the peak current. To limit the device peak-current, non-capacitive external-inductors are required at the MV-primary side as well, whose value depends on the peak current to be allowed through the SiC device.

$$Z_p = \frac{1}{\frac{1}{sL_{f1}} + \frac{1}{R_{df1}}} + sL_w + R_1 + \frac{1}{\frac{1}{sL_1} + \frac{1}{R_2} + sC_1}$$
$$+ \frac{1}{\frac{1}{R_c} + \frac{1}{sL_m} + s[C_p + C_m]} \qquad (4)$$

$$Z_p(\omega_{peak}) = \sqrt{\frac{L_{f1} + L_1}{C_p + C_m}} \qquad (5)$$

$$\omega_p = \sqrt{\frac{1}{[L_{f1} + L_1][C_p + C_m]}} \qquad (6)$$

Eq. 4 shows the impedance of input-section of the equivalent-circuit. Peak current is given by the ratio of maximum step voltage to Z_p at the natural frequency. Eq. 5 shows the simplified impedance value at the resonance frequency calculated from Eq. 6.

III. ZVS Criteria for the $Y : Y/\Delta$ DAB

ZVS mode for both side converters is active as long as the DAB current-vector is between primary and secondary voltage-vectors. To meet ZVS for IGBTs for NPC converter at a given operating point, the DAB current lag can be increased by either reducing output DC voltage setting of the DAB or increasing the duty-cycle of the NPC converter. The smaller duty-ratio results in distorted and reactive current. The ZVS range is limited by DAB-angle. Lagging the current without increasing DAB-angle will cause ZVS to loose for

near $\frac{V_{dc}}{3}$, which would be $\frac{V_{dc}}{2}$ for single-phase DAB with three-level half-bridge converter and $\frac{2*V_{dc}}{3}$ for two-level Y three-phase converter. This topology has been compared with derivative three-phase DAB topologies in [11] and is more efficient with the given SiC device loss data. Fig. 4 shows the DAB $P, Q - \delta$ curve for the basic magnetic design. The full load DAB-angle is selected $45°$ keeping in view that large reactive power will result in inflated conduction loss. A small full-load DAB angle makes the controllability difficult. With ideal simulations, the HF transformer is designed. The details of the transformer design for this converter can be found in [12]. The estimated HF transformer leakage-inductance is insufficient for DAB operation and hence external-inductors are required. The six additional inductors can be put for Y and Δ coupling, since they help to reduce 5th and 7th harmonic-currents circulating between star and delta windings in this topology. Fig. 5 shows a simplified parasitic equivalent-circuit of the HF transformer similar to [13] based on measurements. The parameters are shown in the Table II.

TABLE II.   Parasitic Parameters of the MV/HF Transformer

Parameters	Absolute	p.u.	Parameters	Absolute	p.u.
L_1	$150\,\mu H$	0.0052	$C_{ps'i}$	$180\,pF$	(0.02)
R_1	$1.3\,\Omega$	0.0007	N	$26.7 : 1 : \sqrt{3}$	
R_2	$4000\,\Omega$	2.2	R_{df1}	$5000\,\Omega$	2.7
C_1	$100\,pF$	0.05	C_{ss}	$4000\,pF$	0.0005
L_m	$293\,mH$	10.14	L_2	$4\,\mu H$	0.05
L_w	$500\,nH$	10^{-5}	L_3	$7\,\mu H$	0.05
C_p	$50\,pF$	0.0057	R_c	$5\,M\Omega$	2700
L_{f1}	$400\,\mu H$	0.0152	L_{s1}	$80\,\mu H$	2.09

The differential capacitance measured from MV side is $250pF$ and includes the reflected secondary and inter-winding

The 2014 International Power Electronics Conference

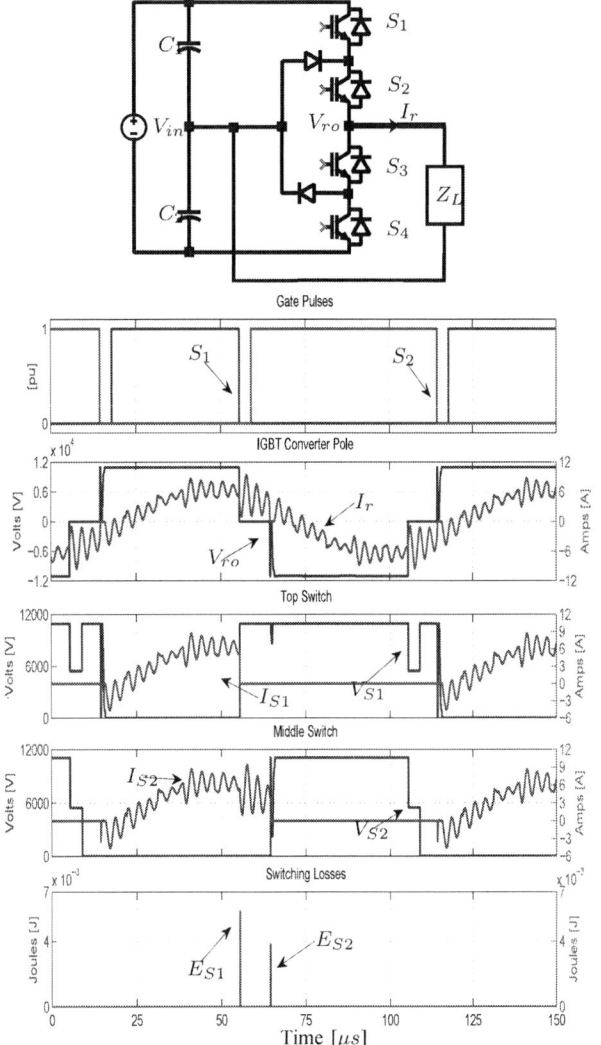

Fig. 6. IGBT NPC Waveforms with Dead Time in DAB ZVS Operation

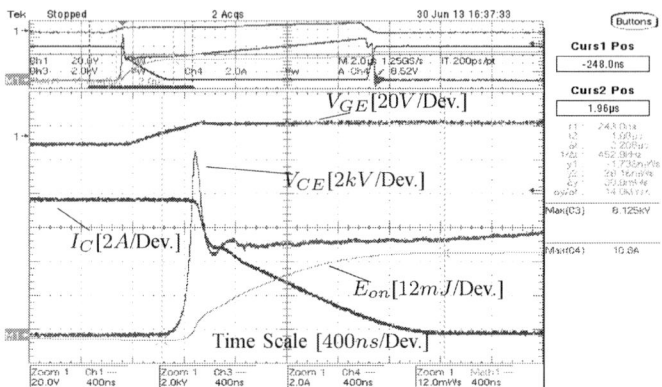

Fig. 7. IGBT Characterization Waveforms Hard-switching Turn-ON

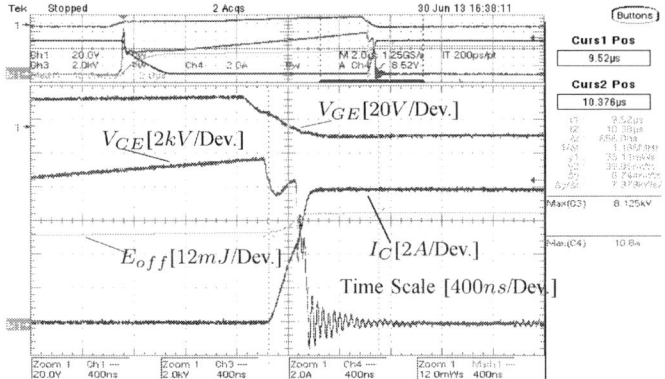

Fig. 8. IGBT Characterization Waveforms Hard-switching Turn-OFF

The ZVS of MOSFETs will be at boundary for higher current-lag.

B. Effect of MV Converter Dead Time

Fig. 7 and 8 show the SiC IGBT hard-switching characteristics at 8kV and 10A. The detailed characterization is provided in [7]. These devices can not be switched at a very high dV/dT due to EMI issues. The limited dV/dT also helps the transformer spike currents to be limited. Considering safety margin a dead-time of $3.5\mu s$ has been selected for MV converter. The dead-time is significant for $10kHz$ DAB since the time period is $100\mu s$. The dead-time will occur in both positive and negative half cycles. If DAB-current does not lag sufficiently, the dead-time can clip out considerable portion of pole-voltage (7% for $10kHz$). This may increase reactive power, result in non-ZVS operation and reduce maximum power-transfer capability significantly. A slight mismatch in the dead-times of top two and bottom two switches, can result in DC mid-point mismatch.

When NPC converter current is lagging as desired by ZVS, converter voltage will be same duty-cycle as desired by controls and there will be no dead-time effect. This is because of free-wheeling action by anti-parallel diodes. In Fig. 6 pole-voltage for 41.7% duty-cycle is exactly same as it rises to $+Vdc/2$ when bottom-middle switch turns-off (path through

the secondary converters. Increasing the DAB frequency for given magnetic design, will result in increased DAB-angle for a given load.

A. Effect of MV Converter Duty Cycle

The minimum phase-lag required by current can be calculated from waveforms for six-step shown in Fig. 6 for the MV side IGBT converter. With reduced duty-ratio (41.7%) gate signals for NPC converter, the fundamental voltage vector lags the turn-off instant of bottom-middle device and leads the turn-on instant of top device by $15°$. The DAB current must lag the voltage-vector by this minimum angle because of duty-ratio. Lagging the current vector for ZVS will cause turn-off to occur at higher current instant, now generating more turn-off loss. Thus it is an optimization problem keeping experimental behavior of the devices in view. The outer IGBTs of the NPC converter have narrow ZVS range compared to the inner ones. Lagging the current too much will also cause more reactive power-flow and hence more current for the same power level.

The 2014 International Power Electronics Conference

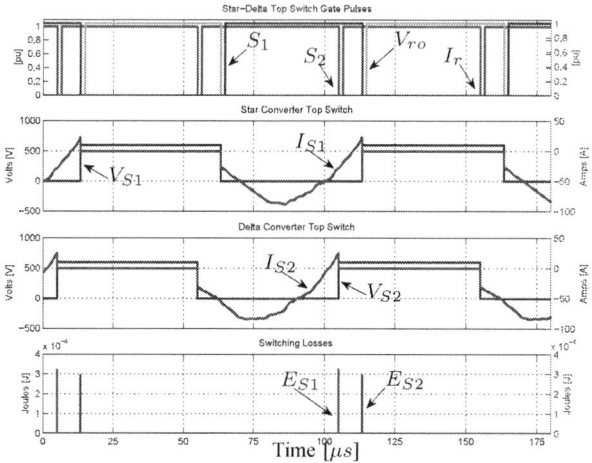

Fig. 9. MOSFET Waveforms with Dead Time in DAB ZVS Operation

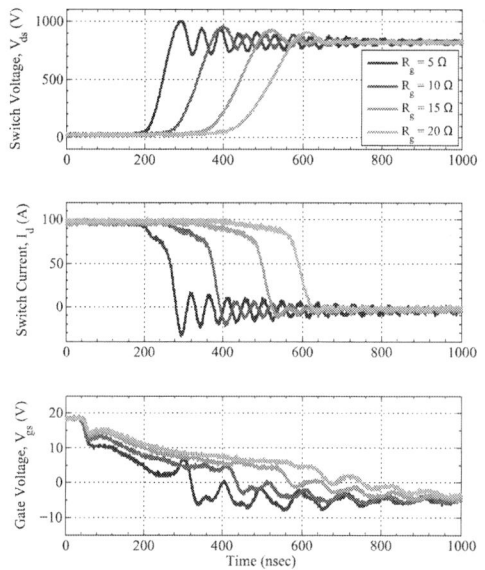

Fig. 10. MOSFET Characterization Waveforms Hard-switching Turn-OFF

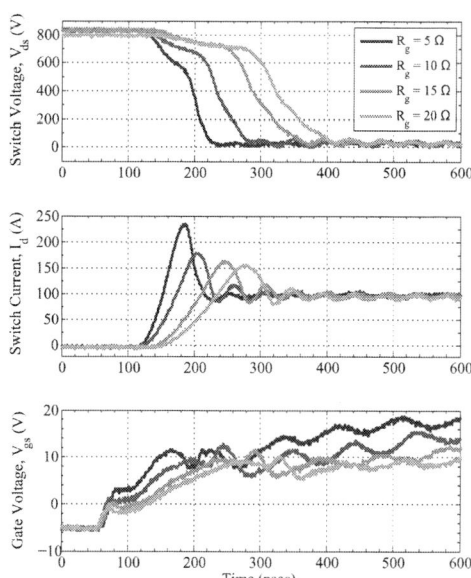

Fig. 11. MOSFET Characterization Waveforms Hard-switching Turn-ON

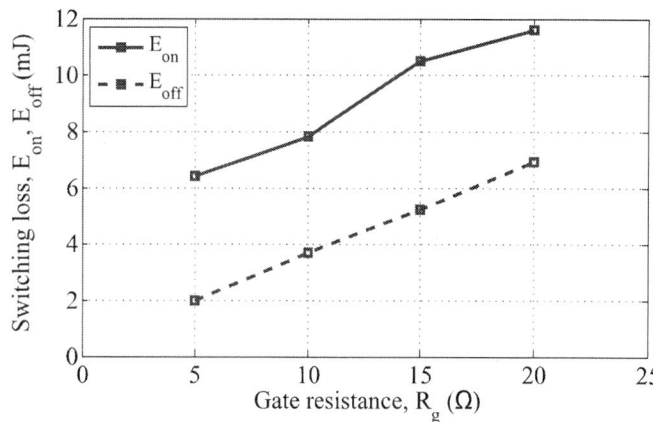

Fig. 12. MOSFET Characterization Effect of Gate Resistance

top diode in free-wheeling mode). The pole voltage falls to $Vdc/2$ when top device is turned-off. Importantly IGBT tun-off transient applies slower dV/dT compared to turn-on, hence this helps to reduce the DAB peak current ringing and does not cause false shoot-through in gate-drivers. An additional phase-lag of $12.6°$ needs to be added for dead-time.

C. Effect of LV Converter Dead Time

Fig. 9 shows the MOSFET switch waveforms of the LV converters in the DAB. Since the DAB is operating at 100kW in ZVS, the anti-parralel diodes freewheel and pole voltage changes polarity as soon the outgoing device turns OFF. There is a small range between hard and soft switching equivalent to the dead time where ZVS can not be ensured. Fig. 10, 11 and 12 show the SiC MOSFET hard-switching characteristics for turn ON, turn OFF and associated switching losses with different gate resistances. The detailed characterization is provided

in [8]. The LV side external inductors have high p.u. value and hence the effect of MOSFET dV/dT is expected to be limited provided they have small parasitics. In practice the device needs to be slowed such that losses do not increase much. A dead time of $1.5\mu s$ is taken for the MOSFET converters. With practical device data, the SiC IGBT based DAB has following loss and efficiency as shown in Table III.

TABLE III. 100kW DAB EXPECTED EFFICIENCY WITH PRACTICAL CONSIDERATIONS

Parameters	Value
Primary Switching Loss	601.8[W]
Secondary Switching Loss	86.4[W]
Primary Conduction Loss	84.1[W]
Secondary Conduction Loss	511.0[W]
Total Converter Loss	1283.3[W]
Transformer Coil losses	50[W]
Transformer Core losses	80[W]
Total Losses	1413.3[W]
Efficiency	98.47[%]

The 2014 International Power Electronics Conference

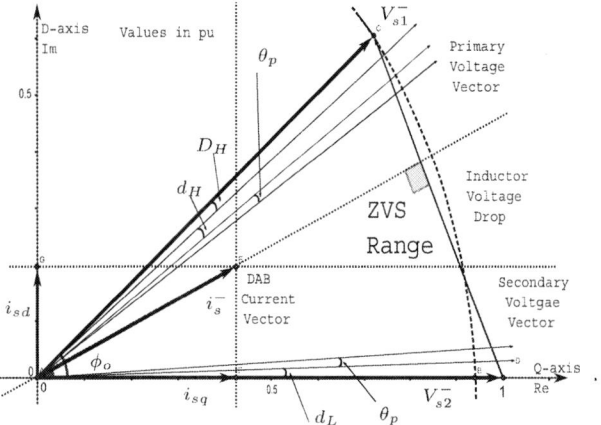

Fig. 13. Phasor Diagram showing DAB ZVS Constraints

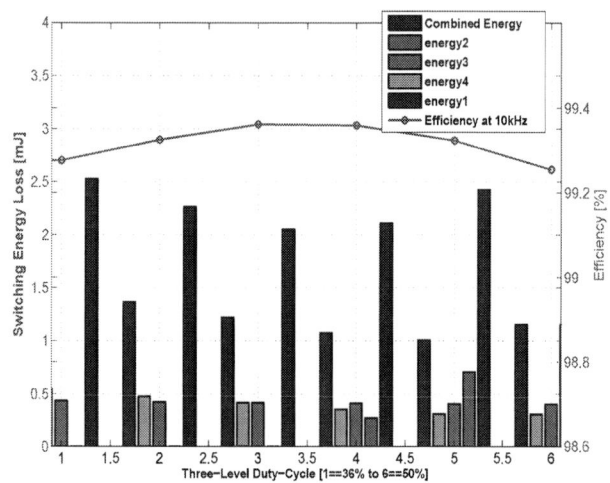

Fig. 14. Duty-Cycle Effect on Switching Loss at Nominal Turn-Ratio

D. Effect of HF Transformer Magnetizing Inductance and Parasitic Capacitance

The ringing current does not result in turn-on loss if the DAB current still lags the converter pole-voltage by a margin. The magnitude of the margin depends on the magnitude of final parasitic current flowing in primary. As transformer current has parasitic current-ripple, which occurs when bottom-middle switch is turned off and ZVS switching happens after the dead-time period passes. To ensure that due to ripple turn-on instant current does not go positive, two things can be done. Firstly, parasitic frequency can be adjusted with filter inductor design such that at turn-on instant it is in negative phase. Secondly, a margin equivalent to ringing ($5°$ here) can be added to the current-lag angle. With voltage phase-angle $45°$ for $100kW, 22kV - 800V$, ZVS can be ensured near full load as current lag over $32°$ is required. Smaller magnetizing-inductance L_m makes DAB current to lag behind and helps MV converter for ZVS but makes it easier for DAB MOSFET secondary converters to slip out of ZVS-mode. SiC MOSFET converter hard switching does not affect efficiency much.

IV. OPTIMUM DESIGN

The above analysis suggests that with practical considerations of the devices for dead time and parasitics, ZVS range is limited. There is a large operating area where the pole voltage shape and magnitude can vary as it is defined by parasitics and device dV/dT. The pole voltage can be defined only when ZVS occurs. The duty mismatch for voltage balancing is limited, hence in the hard switching modes the mid point voltage can be highly unbalanced though it does not saturate the transformer as it appears as common mode voltage for star connected transformers. Fig. 13 shows the space-vector phasor-diagram for the DAB. ZVS range shrinks for NPC converter smaller duty-ratio ($< 50\%D$), smaller magnetizing-inductance, dead-time of converters, current ringing and light-loading of the DAB. Table IV shows the factor for ZVS consideration. D-Q control can ensure ZVS operation of the DAB, as it can maintain D-axis current independent of Q-axis current by adjusting duty-cycle of primary converter and hence controlling primary voltage magnitude [14]. To calculate the optimum power-factor angle for DAB current, device switching-characteristics are taken in to consideration along with reactive power-flow. SiC IGBT turn-off losses are high and so current can not be let to lag by large angle from converter pole voltage. For minimum ZVS lag-angle, a margin for parasitic currents and dead-time are added to the converter zero-state angle.

Fig 14 shows the switching loss as a function of the MV converter duty cycle. The middle switch of the three-level converter marked as $energy2$ can be made to have zero switching losses that happens when the MV side converter operates at smaller phase-angle and near unity power-factor. As the switching and conduction losses in this application have opposite requirement for minimum value turn-ratio, total loss still finds nominal point most efficient for $10kHz$. Total converter loss which depends not only on reactive power-flow but also on used device characteristics, become smallest for around nominal turn-ratio. The D-current reference is generated from look-up table which is a function of Q-axis current for optimum-operation trajectory calculated analytically [15]. The optimum-operation ensures ZVS condition keeping minimum reactive power-flow in DAB magnetic components.

V. EXPERIMENTAL RESULTS

Fig 15 shows the effect of soft-switching on dV/dT slope for the IGBT MV converter. The lagging current causes free-wheeling action of the anti-parallel diodes which applies slower dV/dT due to natural commutation. The hard switch-

TABLE IV. NON-IDEAL FACTORS FOR ZVS LIMITATION

Symbol	Description
D_H	MV Converter Duty Ratio
d_H	MV Converter Dead Time
d_L	LV Converter Dead Time
θ_p	Parasitic Ringing Margin

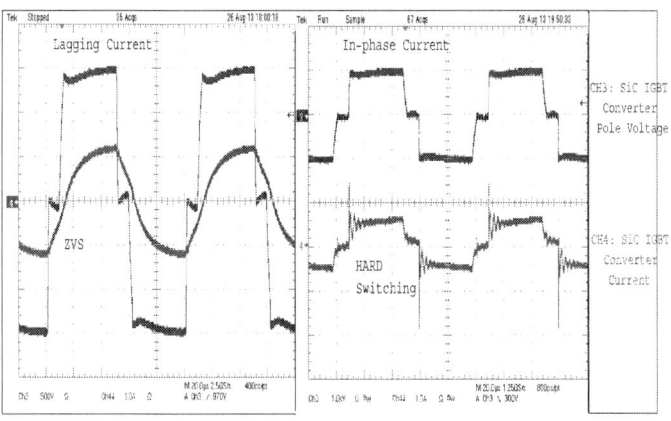

Fig. 15. Experimental Results showing ZVS effect on dV/dT slope

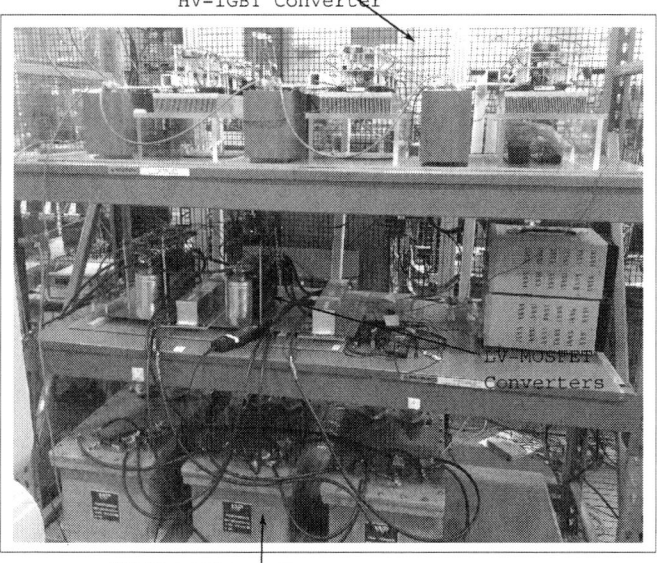

Fig. 16. MV Hardware Set-up for $Y : Y/\Delta$ DAB Topology

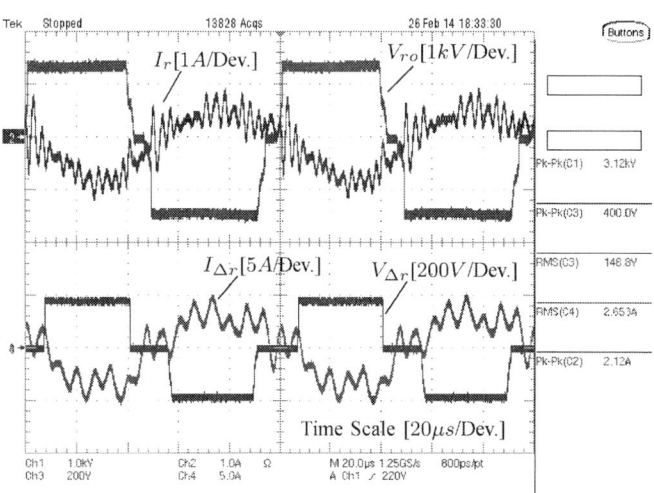

Fig. 17. $Y : Y/\Delta$ DAB Converter Waveforms at $3kV - 200V DC, 2kW$

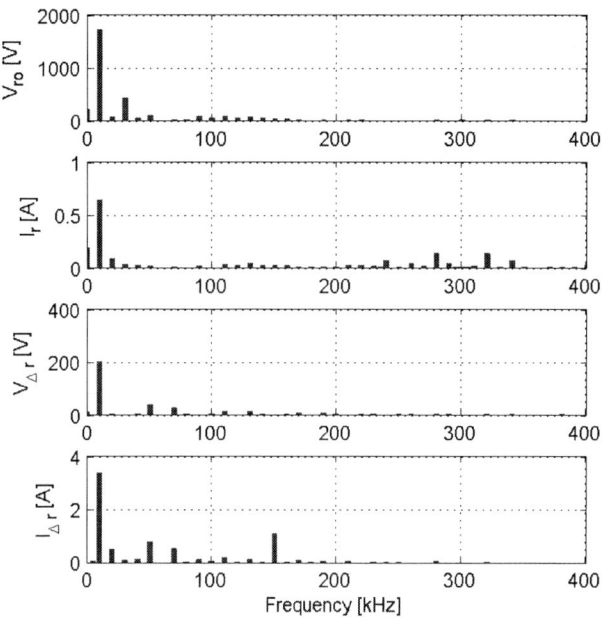

Fig. 18. $Y : Y/\Delta$ DAB FFT of the Waveforms at $3kV - 200V DC, 2kW$

ing dV/dT depends on the gate resonance only and hence generates spike in the device current.

Fig. 16 shows the MV-DAB setup for TIPS project working upto $4.2kV$. Fig. 17 shows converter waveforms at $3kV - 200V DC, 2kW$ three-phase DAB operation. As expected, the DAB currents have ringing due to parasitics. The frequency of the ringing matching with series inductance on each side and the transformer shunt capacitance. Fig. 18 shows the frequency spectra of the converter waveforms. It has $300kHz$ ringing on the MV side and an additional $150kHz$ component on the LV side current.

VI. CONCLUSION

From the ideal switching design aspect, the transformer turn-ratio is kept in ratio of DC voltages which ensures minimum reactive power flow and hence maximum efficiency. The leakage inductance for a given power-level can be designed optimum by keeping small full load DAB angle with high

P/Q ratio but still holding enough precision in the control. The device characteristics are important in determination of dead time and hence the DAB frequency. The dead time to DAB period ratio should not be high otherwise it will affect the ZVS mode, pole voltage, reactive power and optimum efficiency point. Since the LV converter has smaller dead time, the turn-ratio of the transformer and duty-cycle in three-phase DAB along can be selected such that the current vector lags the MV converter sufficiently. This may result in non-ZVS mode for LV converters at a smaller load. The MV converter is more important to maintain ZVS since the hard-switching will generate high dV/dT and hence more ringing. The duty-cycle control provides a little flexibility for optimum operation as it may result in non-ZVS mode and deteriorate current shape which is important for $D - Q$ control. Hence the ZVS can be maintained with $D - Q$ control if the design parameters such that transformer turn-ratio, duty ratio and DAB rated angle, take the dead-time and parasitics into consideration on top of the basic DAB design.

ACKNOWLEDGMENT

The authors would like to acknowledge ARPA-E for funding the TIPS Project. This work made use of ERC shared facilities supported by the National Science Foundation under Award Number $EEC - 0812121$.

REFERENCES

[1] R. W. A. A. DeDoneker, D. M. Divan, and M. H. Kheraluwala, "A three-phase soft-switched high-power-density dc/dc converter for high-power applications", *IEEE Trans. Ind. Appl.*, vol.27, no.1, pp.797-806, Jan./Feb. 1991.

[2] M. H. Kheraluwala, R. W. Gascoigne, D. M. Divan, and E. D. Baumann, "Performance characterization of a high-power dual active bridge dc-to-dc converter", *IEEE Trans. Ind. Appl.*, vol.28, no.6, pp.1294-1301, Nov./Dec. 1992.

[3] K. Hatua, S. Dutta, A.Tripathi, S. Baek , G. Karimi, S. Bhattacharya, "Transformer less Intelligent Power Substation design with 15kV SiC IGBT for grid interconnection", *IEEE Energy Conversion Congress and Exposition (ECCE)*, pp.4225-4232, Sept. 2011.

[4] A.K. Tripathi, K.Hatua, H. Mirzaee, S. Bhattacharya, "A three-phase three winding topology for Dual Active Bridge and its D-Q mode control", *Applied Power Electronics Conference and Exposition (APEC)*, pp.1368-1372, Feb. 2012.

[5] D. Segaran, D. G. Holmes, and B. P. McGrath, "Comparative analysis of single and three-phase dual active bridge bidirectional DC-DC converters," in *Australasian Universities Power Engineering Conference (AUPEC)*, pp. 1-6, 2008.

[6] A. Kadavelugu, S. Bhattacharya, S. Ryu, E. V. Brunt, D. Grider, A. Agarwal, S. Leslie, "Characterization of 15 kV SiC n-IGBT and its Application Considerations for High Power Converters", presented at 2013 IEEE Energy Conversion and Congress Exposition.

[7] A. Kadavelugu, S. Bhattacharya, "Design Considerations and Development of Gate Driver for 15 kV SiC IGBT," accepted for publication in the IEEE Applied Power Electronics Conference and Exposition (APEC) 2014.

[8] S. Hazra, S. Madhusoodhanan, S. Bhattacharya, G.K. Moghaddam, K. Hatua, "Design considerations and performance evaluation of 1200 V, 100 a SiC MOSFET based converter for high power density application", *Energy Conversion Congress and Exposition (ECCE)*, pp. 4278-4285, Sep. 2013.

[9] A. K. Tripathi, K. Mainali, D. Patel, K. Hatua, S. Bhattacharya, "Closed loop D-Q Control of High-Voltage High-Power Three-Phase Dual Active Bridge Converter in Presence of Real Transformer Parasitic Parameters", presented at 2013 IEEE Energy Conversion and Congress Exposition.

[10] H. A. Kojori, S. B. Dewan, and J. D. Lavers, "A two stage inverter large scale static var compensator with minimum filtering requirements," *IEEE Trans. on Magnetics*, vol. 26, no. 5, pp. 2247-2249, 1990.

[11] A.K. Tripathi, K.Hatua, S. Bhattacharya, "A comparative study of three-phase dual active bridge topologies and their suitability for D-Q mode control," *IEEE Energy Conversion Congress and Exposition (ECCE)*, pp.1719-1724, Sept. 2012.

[12] K. Mainali, A. Tripathi, D. C. Patel, S. Bhattacharya, T. Challita, "Design, Measurement and Equivalent Circuit Synthesis of High Power HF Transformer for Three-Phase Composite Dual Active Bridge Topology", accepted for publication in the IEEE Applied Power Electronics Conference and Exposition (APEC) 2014.

[13] L. Dalessandro, F. da S. Cavalcante, J. W. Kolar, "Self-Capacitance of High-Voltage Transformers",*IEEE Trans. on Power Electronics*, vol. 22, no. 5, pp.2081-2092, Sep. 2007.

[14] H. Bai and C. Mi, "Eliminate Reactive Power and Increase System Efficiency of Isolated Bidirectional Dual-Active-Bridge DC-DC Converters Using Novel Dual-Phase-Shift Control",*IEEE Trans. on Power Electronics*, vol. 23, no. 6, pp.2905-2914, Nov. 2008.

[15] G. G. Oggier, G. O. Garcia, A. R. Oliva, "Switching Control Strategy to Minimize Dual Active Bridge Converter Losses",*IEEE Trans. on Power Electronics*, vol. 24, no. 7, pp.1826-1838, July 2009.

Common-Mode Currents in Multi-Cell Solid-State Transformers

Jonas E. Huber and Johann W. Kolar
Power Electronic Systems Laboratory
ETH Zurich
8092 Zurich, Switzerland
huber@lem.ee.ethz.ch

Abstract—**Solid-State Transformers (SSTs) are considered important building blocks of a future Smart Grid, where they will replace conventional transformers and provide enhanced functionality and controllability. To deal with the medium-voltage levels at the input of an SST for distribution applications, commonly cascaded cells converter topologies are proposed. This paper presents a detailed analysis of the common-mode currents appearing in cascaded cells converter systems as a consequence of steep changes of the cells' potentials caused by the individual cells' switching actions. Common-mode chokes placed at the AC terminals of each individual cell are identified as the most feasible way of mitigating these common-mode current disturbances. Detailed, analytic design guidelines are provided and applied to a 1 MVA, 10 kV/400 V SST.**

Keywords—**Cascaded Cells Converters, Common-Mode Currents, Multilevel Converters, Solid-State Transformers.**

I. INTRODUCTION

With the increasing need of substituting conventional energy sources such as fossil fuel and nuclear fission by renewables, which often involves geographical dispersion of the power sources, completely new demands on the future power grid, especially on the distribution level, arise. These can be summarized in the requirement that the grid needs to transition towards a so-called "smart-grid", which allows for a precise distributed control of sources, loads, voltage levels and power flows. However, such functionality cannot be provided by passive components such as line-frequency transformers, which are ubiquitous in today's distribution grids. While power electronic equipment is used to, at least partly, overcome the limits of passive transformers, e. g., with FACTS or STATCOM devices, complete power electronic replacements could integrate higher functionality and therefore offer much more flexibility.

Solid-State Transformers (SSTs) are such power electronic systems that could replace line-frequency transformers at the interfaces between medium-voltage (MV) and low-voltage (LV) levels. An SST interfaces either grid by means of a combination of power electronic converter systems, provides galvanic isolation using medium-frequency (MF) transformers and, most prominently, enables control of the active power

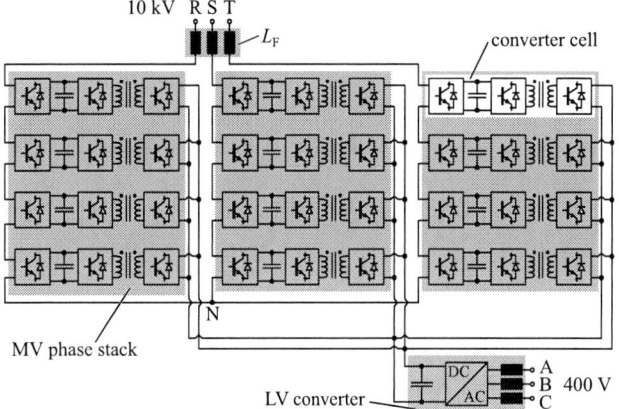

Fig. 1. Overview on the considered SST topology.

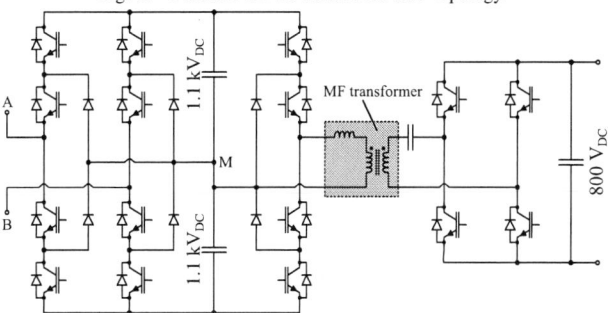

Fig. 2. Power circuit of a single MVAC-LVDC converter cell.

flows of the phases, independent reactive power compensation on both sides, active filtering of harmonics, and features a LV DC bus to interface energy storage systems or a local DC distribution infrastructure, etc. The first patent of an "electronic transformer" has been issued already in 1970 [1], however, only in recent years the concept has been scaled up to higher voltage and power ratings for grid [2]–[6] and traction applications [7]–[10].

Since the blocking voltage ratings of today's power semiconductors are limited, usually cascaded converter systems are used to interface the MV grid. The choice of the blocking voltage and hence the number of cascaded cells is an important design parameter. For the voltage and power ratings considered here (cf. Table I), it has been shown that systems based on low-voltage power devices, i. e., 1700 V IGBTs, are Pareto-optimal regarding efficiency and power density [11]. Fig. 1 shows a corresponding example SST system, which employs

TABLE I
SPECIFICATIONS OF THE CONSIDERED SST.

Rated power	1 MVA
MV grid voltage (line-line, rms)	10 kV
LV grid voltage (line-line, rms)	400 V
Cell rectifier stage switching frequency	1 kHz
Cell MV DC link voltage	2×1100 V
Cell LV DC link voltage	800 V

four cascaded cells per phase on its MV side and Fig. 2 shows the power circuit of a single cell, consisting of a single-phase neutral-point-clamped (NPC) rectifier stage and an isolated, resonant DC/DC converter [12].

While there are many publications dealing with the reduction of common-mode voltages at the cascaded cells converter's output, which is important for, e. g., drive systems, by means of special modulation schemes [13]–[15], so far only a single paper discusses another common-mode-related phenomenon inherent to cascaded cells converter systems [16]: the potential of a cascaded cell's MV side with respect to ground cannot be fixed, instead it changes during every switching transition of any subjacent cell in the same phase stack. Each cell, however, has a certain capacitance to ground arising from, e. g., transformer windings, heatsinks, etc., that needs to be charged or discharged for the cell's midpoint potential to change, which gives rise to common-mode currents. Furthermore, it has been shown that severe oscillations between these parasitic ground capacitances and parasitic inductances of the cell interconnections can occur [16].

This paper provides a more detailed analysis of the phenomenon in Sections II and III, and discusses mitigation means in Section IV, where also comprehensive design guidelines for the most promising solution, i. e., common-mode chokes at each cell's AC terminals, are given.

II. COMMON-MODE EQUIVALENT CIRCUIT

In this and the next section, an analytical description of the common-mode currents arising in the cascaded cells converter is presented, where the analysis confines to a single phase without loss of generality, since the star point, N, is assumed to be solidly grounded. To start, suitable equivalent circuits are derived.

A. Common-Mode Equivalent Circuit of a Converter Cell

A single converter cell's power circuit as shown in Fig. 2 can be simplified for common-mode analysis purposes, as for example suggested in [13]. To do so, the MV side switches and DC link capacitors are replaced by stepped voltage sources with amplitude $V_{dc} = 1100$ V (corresponding to one half of the total MV side DC link voltage), which results in the equivalent circuit shown in Fig. 3(a). The grayed-out capacitances, $C_{sc,I}$ and $C_{sc,D}$, represent capacitances between semiconductor dies and the heatsink, which is connected to the MV side midpoint, M. Any switching causes "local" common-mode currents through these capacitances. However, these currents flow only within a single cell and are no different from those present in any switching power converter. Therefore, these "local" common-mode currents are not considered further.

In contrast, any switching of the voltage source v_B, or of any other voltage source subjacent in the phase stack, will move the potential of the cell midpoint, M, with respect to ground. This will give rise to a current flow through the capacitance between the transformer winding (note that one end of the MV winding is connected to the midpoint) and ground, C_T, as well as through C_{HS}, the capacitance between the cell structures on

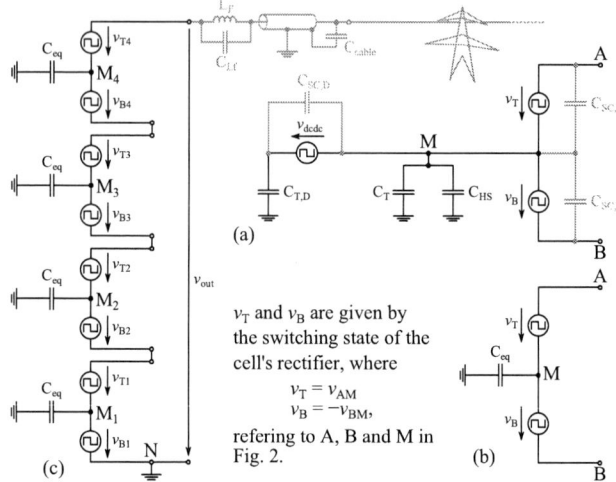

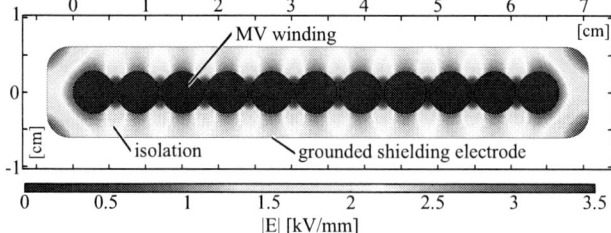

v_T and v_B are given by the switching state of the cell's rectifier, where
$$v_T = v_{AM}$$
$$v_B = -v_{BM},$$
refering to A, B and M in Fig. 2.

Fig. 3. Common-mode equivalent circuit of the converter cell shown in Fig. 2, including "local" common-mode effects and the DC/DC converter (a), reduced version suitable for the analysis of "global" common-mode effects (b), and corresponding common-mode equivalent circuit of a complete phase stack (c).

Fig. 4. Electrostatic FEM simulation of the transformer MV winding.

midpoint potential, such as the heatsink, and adjacent grounded parts such as cabinet walls. Additionally, also the switching of the DC/DC converter, represented by v_{dcdc}, causes common-mode currents through the capacitance between the transformer winding and ground, $C_{T,D}$. However, these are independent of the cell's position in the phase stack and therefore non-specific to cascaded cells systems; thus the DC/DC converter is not considered any further. Applying these simplifications yields the common-mode equivalent circuit of one converter cell as shown in Fig. 3(b), where C_{eq} summarizes C_T and C_{HS}.

Due to the close geometric proximity of the MV winding on cell potential and grounded parts such as the core, which is mandatory to achieve good magnetic and thermal performance of the MF transformer, C_T can be assumed to dominate C_{eq}. Finite element simulation has been used to determine the parasitic capacitance of a transformer designed for the system under consideration. Fig. 4 shows the corresponding electric field distribution, from which $C_{eq} \approx C_T = 650$ pF is obtained.

B. Common-Mode Equivalent Circuit of a Phase Stack

Fig. 3(c) finally shows the common-mode equivalent circuit of a complete SST phase stack. The neutral point, N, is assumed to be solidly grounded, or at least directly connected to the aforementioned structures such as cabinet walls or transformer cores. In this second case, no direct ground connection of the star point would be required, however, for safety reasons the voltage between N and ground would have to be kept small

The 2014 International Power Electronics Conference

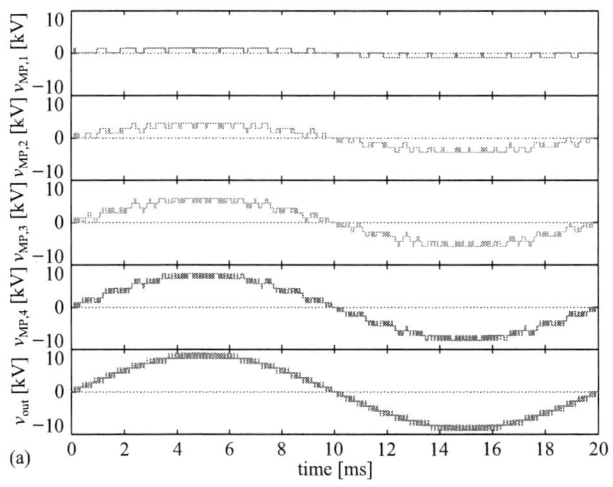 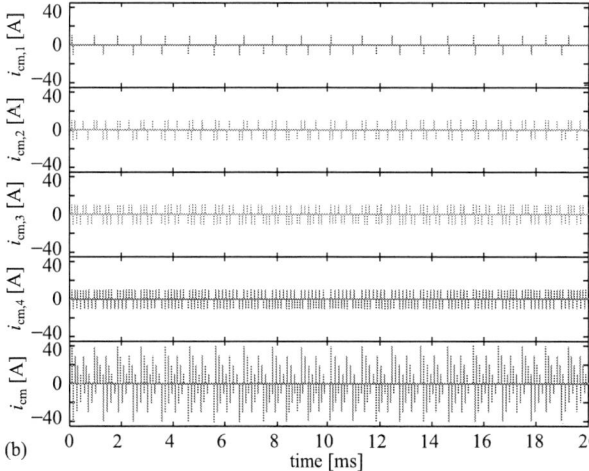

Fig. 5. Full-system simulation results for one SST phase. (a) shows the voltages between the cell midpoints and ground as well as the total phase stack output voltage, v_{out}, and (b) shows the corresponding common-mode currents flowing through the cell's parastitic capacitances to ground, and also the total common-mode current, i_{cm}, flowing back through the node N.

by suitable means.

Note that, in addition to C_{eq}, there is also a capacitive path to ground from the phase terminal via the parasitic capacitance of the filter inductor and the cable capacitance of a possibly connected MV cable. However, while these capacitances might be in the same order of magnitude as C_{eq} and need to be charged or discharged at every switching of any converter cell, the corresponding current spikes do not appear as common-mode current inside the converter structure, but only as a minor addition to the load current.

III. ANALYTICAL DESCRIPTION OF COMMON-MODE CURRENTS

The idealized equivalent circuit (not considering any parasitic inductances) depicted in Fig. 3(c) is now used to derive a basic analytic description of the common-mode currents flowing in the system.

The power semiconductors are impressing a high dv/dt during their switching transitions, $dv/dt = 15\,\text{kV}/\mu\text{s}$ being a typical value for the IGBTs considered here. Assuming a constant dv/dt, the corresponding charging current for a cell's parasitic capacitance is given as

$$\hat{I}_{\text{cm}} = C_{\text{eq}} \frac{dv}{dt}. \tag{1}$$

A constant current with this magnitude flows until the capacitor is charged to the new voltage level. The current spike is therefore rectangular with magnitude $\hat{I}_{\text{cm}}$ and a duration of

$$\Delta t = \frac{V_{\text{dc}}}{\frac{dv}{dt}}, \tag{2}$$

where $V_{\text{dc}} = 1100\,\text{V}$, which is the voltage change resulting from any switching transition.

Consider now a cell at position k in the stack, where $1 \leq k \leq 4$ in the example system. If the voltage source $v_{\text{B},k}$ or the voltage source $v_{\text{T},(k-1)}$ is changing its voltage value as a result of corresponding switching actions, the midpoint potentials of cell k and of all cells at positions $l > k$ will change, requiring a current flow through their respective common-mode capacitances as described above.

Using PWM with phase-shifted carriers creates the desired multilevel output voltage waveform but implies that all cells are continuously switching with the same frequency, f_{s}. Thus, the higher a cell is positioned in the stack, the more common-mode current pulses it experiences per time.

A. RMS Common-Mode Current of Cell k

More precisely, every voltage source shown in Fig. 3(c) switches twice during a switching period, $T_{\text{s}} = 1/f_{\text{s}}$. Therefore, the midpoint of cell k experiences $2 \cdot (2k-1)$ potential changes during T_{s}. The rms common-mode current flowing through the common-mode capacitance of cell k is thus given as

$$\tilde{i}_{\text{cm},k} = \sqrt{\frac{1}{T_{\text{s}}} \int_0^{T_{\text{s}}} i_{\text{cm},k}(t)^2\, dt} = \sqrt{\frac{1}{T_{\text{s}}} \sum_{j=1}^{(4k-2)} \Delta t \hat{I}_{\text{cm}}^2}$$

$$= C_{\text{eq}} \cdot \sqrt{f_{\text{DC}} V_{\text{DC}} \frac{dv}{dt}} \cdot \sqrt{4k-2}. \tag{3}$$

Note that therefore an effective capacitance for a cell at position k in the stack could be defined as $C_{\text{eq},k}^* = C_{\text{eq}} \cdot \sqrt{4k-2}$, which is increasing for cells at higher stack positions.

B. Total RMS Common-Mode Current

The common-mode currents of all cells flow back through the ground connection at node N in Fig. 3(c). The magnitude of these current pulses depends on which cell, denoted by its position in the stack, k, is switching, namely $\hat{i}_{\text{cm}\leftarrow k} = (N-k+1) \cdot \hat{I}_{\text{cm}}$. As mentioned above, there are two switching transitions of each voltage source in the equivalent circuit during one switching period. For $k > 1$ a switching transition of $v_{\text{B},k}$ has the same effect as one of $v_{\text{T},(k-1)}$, resulting in four corresponding current pulses per switching period. Thus, the total common-mode rms current becomes

$$\tilde{i}_{\text{cm}} = \sqrt{f_{\text{s}} V_{\text{DC}} C_{\text{eq}}^2 \frac{dv}{dt} \cdot \left(2N^2 + \sum_{k=2}^{N} 4 \cdot (N-k+1)^2 \right)}$$

$$= C_{\text{eq}} \cdot \sqrt{f_{\text{S}} V_{\text{DC}} \frac{dv}{dt} \cdot \left(\frac{4}{3} N^3 + \frac{2}{3} N \right)}. \tag{4}$$

978-1-4799-2706-7/14 $31.00 © 2014 IEEE   768

TABLE II
CALCULATED AND SIMULATED RMS COMMON-MODE CURRENTS.

	calc. [mA]	sim. [mA]	error [%]
cell 1	118.1	119.5	-1.17
cell 2	204.5	206.7	-1.06
cell 3	264.0	266.7	-1.01
cell 4	312.4	315.6	-1.01
total	783.2	790.9	-0.96

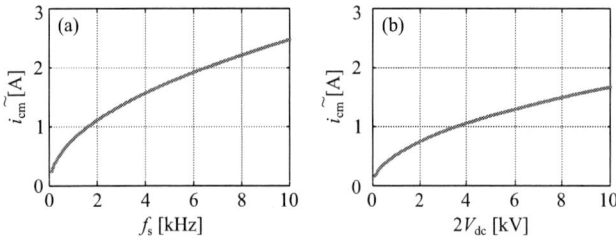

Fig. 6. Dependence of the total rms common-mode current, $\tilde{i}_{\mathrm{cm}}$, on the cells' switching frequency, f_{S}, at $V_{\mathrm{dc}} = 1.1\,\mathrm{kV}$ (a), and on the cells' DC link voltage, V_{dc}, at $f_{\mathrm{S}} = 1\,\mathrm{kHz}$ (b).

C. Verification

A complete simulation model of a single MV phase stack, featuring four cascaded cells including the DC/DC converter stages, has been implemented in GeckoCICRUITS [17] to verify the above calculations. Note that the fixed simulation time-step has to be chosen according to (2) in order to obtain a desired dv/dt. Fig. 5(a) shows the simulated cell midpoint voltages to ground and Fig. 5(b) shows the corresponding common-mode currents. The calculated and simulated rms common-mode currents are listed in Table II, confirming the accuracy of the analytic calculations.

D. Parameter Influences

The validated analytic expressions allow to investigate the influence of certain system parameters on these common-mode currents. From (4) it can directly be seen that $\tilde{i}_{\mathrm{cm}} \propto \sqrt{f_{\mathrm{s}}}$ and $\tilde{i}_{\mathrm{cm}} \propto \sqrt{V_{\mathrm{dc}}}$, which is illustrated by Fig. 6. Note that if instead of a constant dv/dt a constant switching time for the voltage to rise from $0\,\mathrm{V}$ to V_{dc} was considered, a linear dependency, $\tilde{i}_{\mathrm{cm}} \propto V_{\mathrm{dc}}$, would result.

Furthermore, there is a linear dependence on the parasitic capacitance, i.e., $\tilde{i}_{\mathrm{cm}} \propto C_{\mathrm{eq}}$ as shown in Fig. 7. There, also the fact that the rms common-mode current increases with higher position of a cell in the stack, corresponding to increasing k, is clearly visible.

E. Influence of the Number of Cascaded Converter Cells

The choice of the optimum number of cascaded converter cells, N, for a given grid voltage level has been discussed in [11]. It is thus interesting to assess the influence of N on the common-mode current stress in the system.

It has been shown that the following relations hold for a given output filter and constant output current ripple, i.e.,

$$V_{\mathrm{dc}} = V_{\mathrm{dc},0}\frac{N_0}{N} \quad \text{and} \quad f_{\mathrm{s}} = f_{\mathrm{s},0}\frac{N_0^2}{N^2}, \quad (5)$$

where $V_{\mathrm{dc},0}$, $f_{\mathrm{s},0}$ and N_0 are the parameters of a reference cascaded cells system. Coarsely approximated, the scaling behavior of the transformer stray capacitance is as follows:

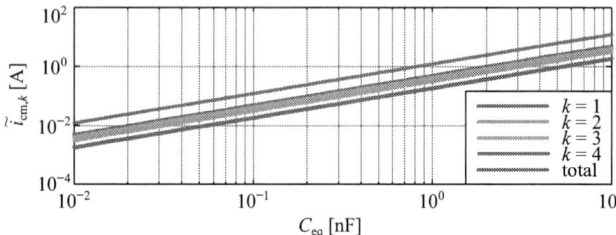

Fig. 7. Dependence of the individual cells' rms common-mode currents, $\tilde{i}_{\mathrm{cm},k}$, and of the total rms common-mode current, $\tilde{i}_{\mathrm{cm}}$, on C_{eq}.

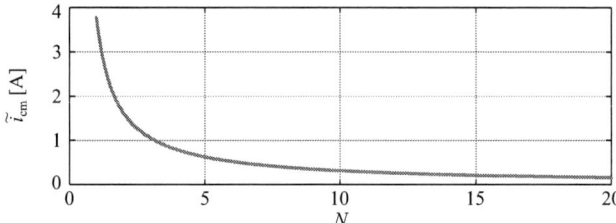

Fig. 8. Total rms common-mode current, $\tilde{i}_{\mathrm{cm}}$, as a function of the number of cascaded converter cells, N.

the cell power scales with $1/N$, but so does the cell DC voltage, resulting in the transformer current being independent of N, i.e., the required copper cross section remains constant and the same holds for the isolation distances, essentially leaving the cross section shown in Fig. 4 unchanged, assuming an equivalent number of turns. Assuming further a constant switching frequency of the DC/DC stage, the required core area scales with $1/N$ since the applied voltage-time area does so. Consequently, the third dimension of the transformer scales with $1/N$, corresponding to an according reduction of the stray capacitance, i.e.,

$$C_{\mathrm{eq}} = C_{\mathrm{eq},0}\frac{N_0}{N}. \quad (6)$$

Inserting these dependencies in (4) yields

$$\tilde{i}_{\mathrm{cm}}(N) = \sqrt{f_{\mathrm{S}}\frac{N_0^2}{N^2}V_{\mathrm{DC}}\frac{N_0}{N}C_{\mathrm{eq}}^2\frac{N_0^2}{N^2}\frac{dv}{dt}\cdot\left(\frac{4}{3}N^3 + \frac{2}{3}N\right)}, \quad (7)$$

which is plotted in Fig. 8. Even though the approximations are coarse and for example a scaling of the dv/dt with the switch voltage rating is not considered, the qualitative notion that an increasing number of cascaded cells results in lower total rms common-mode current stress is very pronounced, since $\tilde{i}_{\mathrm{cm}}(N) \propto \sqrt{N^3/N^5} \propto 1/N$.

F. Parasitic Inductances and Common-Mode Oscillations

So far, an idealized situation has been used to introduce the basic relations. However, additional parasitic elements, most prominently series inductances of the interconnections of the cells, are present in a real system. Fig. 9(a) shows a common-mode equivalent circuit extended accordingly, where the series inductances are summarized as $L_{\mathrm{eq}} = 100\,\mathrm{nH}$. Severe oscillations between the parasitic common-mode capacitances and these series inductances have been described in [16]. Here, a more detailed theoretical analysis is presented.

Considering a single switching transition of the source $v_{\mathrm{B},3}$ from $0\,\mathrm{V}$ to V_{dc}, the circuit from Fig. 9(a) can be redrawn

The 2014 International Power Electronics Conference

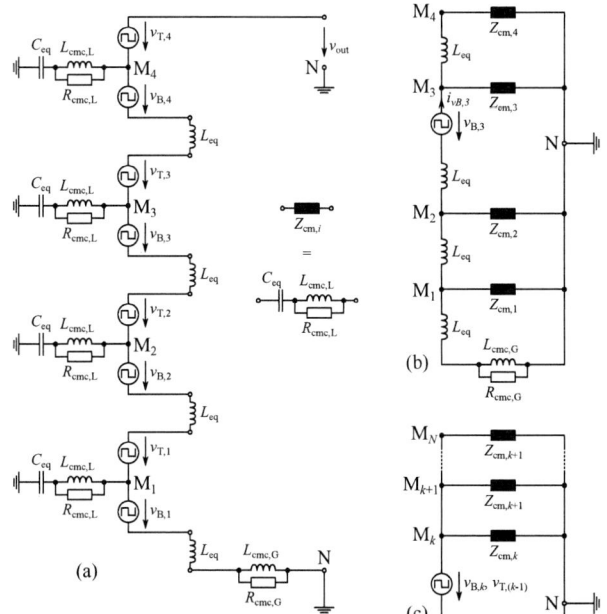

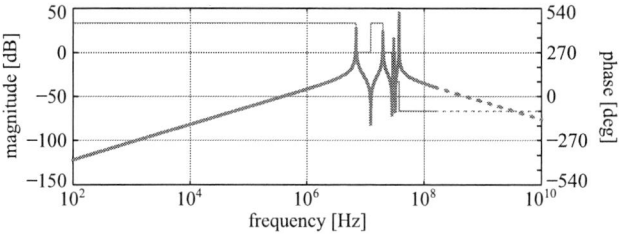

Fig. 9. (a) Common-mode equivalent circuit including parasitic inductances as well as local, i.e., per cell, common-mode chokes, $L_{\mathrm{cmc,L}}$, and a global, i.e., per phase, common-mode choke, $L_{\mathrm{cmc,G}}$. (b) Corresponding equivalent circuit for the case of a switching transition of the equivalent source $v_{\mathrm{B,3}}$. (c) Simplified and generalized equivalent cicuit suitable for designing the local common-mode chokes.

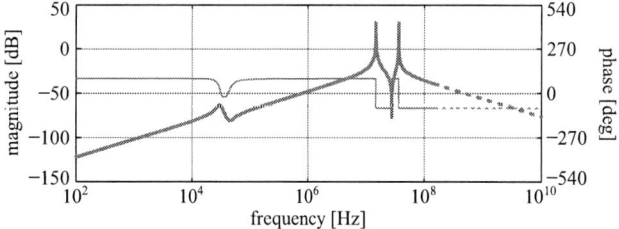

Fig. 10. Bode plot of the transfer function $G_3(s)$ for $L_{\mathrm{eq}} = 100\,\mathrm{nH}$ and without any common-mode chokes, i.e., $L_{\mathrm{cmc,G}} = L_{\mathrm{cmc,L}} = 0$.

to obtain Fig. 9(b), where for now $L_{\mathrm{cmc,L}} = L_{\mathrm{cmc,G}} = 0$ is assumed. Fig. 10 shows the Bode diagram of the transfer function

$$G_3(s) = \frac{I_{\mathrm{vB,3}}(s)}{V_{\mathrm{B,3}}(s)} = \frac{1}{Z_{\mathrm{T,3}}(s)}, \qquad (8)$$

i.e., the transfer function from the voltage step to the resulting current flow through the switching voltage source, where $Z_{\mathrm{T,3}}(s)$ is the total impedance seen by the source in the circuit of Fig. 9(b), including L_{eq}. As expected, severe resonances in the MHz-range are present. Although series resistances, e.g., semiconductor on-state resistances, etc., are not considered, these are required to be small from an efficiency point-of-view and therefore the provided damping is only minor.

IV. COMMON-MODE CURRENT REDUCTION STRATEGIES

The analysis presented so far provides the basis to evaluate means for reducing the common-mode currents circulating in the system and to suppress the severe common-mode oscillations indicated in Fig. 10.

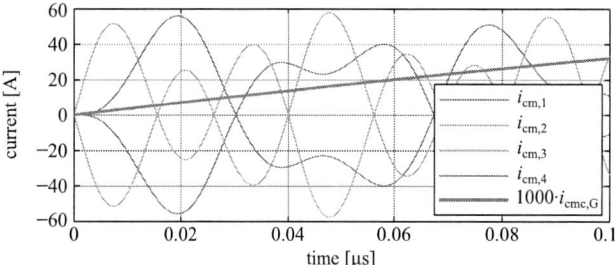

Fig. 11. Bode plot of the transfer function $G_3(s)$ for $L_{\mathrm{eq}} = 100\,\mathrm{nH}$, a global common-mode inductance of $L_{\mathrm{cmc,G}} = 10\,\mathrm{mH}$ and a parallel damping resistor of $R_{\mathrm{cmc,G}} = 10\,\mathrm{k\Omega}$.

Fig. 12. Step responses of the common-mode currents, $i_{\mathrm{CM,k}}$, and the current through the common-mode impedance, $i_{\mathrm{cmc,G}}$, for $L_{\mathrm{eq}} = 100\,\mathrm{nH}$, $L_{\mathrm{cmc,G}} = 10\,\mathrm{mH}$ and $R_{\mathrm{cmc,G}} = 10\,\mathrm{k\Omega}$.

Regarding the dependency of $\tilde{i}_{\mathrm{cm}}$ on the switching frequency, an obvious solution to mitigate common-mode currents would be to reduce the switching frequency, e.g., by using other modulation methods based on fundamental switching frequency at least for most of the cells [18]. However, such measures inevitably require increased passive filtering efforts to comply with harmonic standards.

Another approach would be to use dv/dt filters, as known from drive systems and described in, e.g., [19], [20], at the outputs of the individual converter cells. The downside of such filters is their differential-mode nature, which makes them potentially quite bulky.

Therefore, the most promising third option for reducing common-mode currents in the SST, common-mode choking, is investigated closer in the following.

A. Global Common-Mode Choke

A first approach could be to place a common-mode choke (CMC) at the output terminals of the complete phase stack. Such a CMC, including a parallel damping resistor, appears in the common-mode equivalent circuit from Fig. 9(b) as $L_{\mathrm{cmc,G}}$ and $R_{\mathrm{cmc,G}}$. Fig. 11 shows the Bode plot of the accordingly adjusted transfer function $G_3(s)$. The high common-mode inductance creates a second, damped resonance at a much lower frequency, and increases the common-mode impedance at the stack output, however, it does hardly affect some of the internal high-frequency resonances occurring between the common-mode capacitances, C_{eq}, and the parasitic inductances, L_{eq}, within the phase stack, which is in accordance with the findings reported in [16].

Similar to $G_3(s)$, transfer functions from the step voltage change to the common-mode currents flowing in the parasitic capacitances, C_{eq}, can be derived and Fig. 12 shows the

978-1-4799-2706-7/14 $31.00 © 2014 IEEE

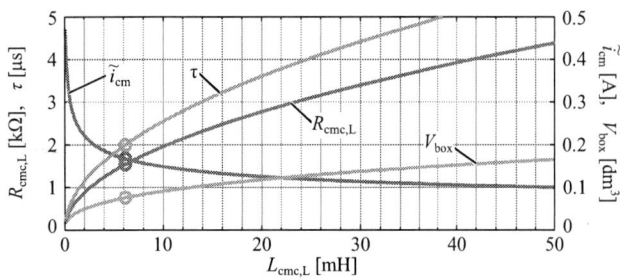

Fig. 13. Required damping resistor, $R_{\mathrm{CMC,L}}$, resulting time constant, τ, and rms common-mode current, $\tilde{i}_{\mathrm{cm}}$, as functions of $L_{\mathrm{cmc,L}}$. The estimated box volume of the CMC, V_{box}, is also shown.

corresponding step responses. As expected, nearly undamped oscillations occur, where it should be noted that the initial current flow in the common-mode capacitances of the cells residing at lower stack positions than the switching voltage source is in the negative direction, i.e., discharging these capacitances, which results in an undesired lowering of these cells' midpoint potentials. This is a result of $L_{\mathrm{cmc,G}}$ very effectively suppressing the common-mode current flowing back through node N; the only other path for the required charging current inevitably involves the common-mode impedances of those cells initially not affected by the switching.

All in all, the above shows that common-mode choking at the phase stack terminals is not a feasible solution for mitigating common-mode current flows inside an SST.

B. Local Common-Mode Chokes at Cell Inputs

Another option is to place "local" CMCs between the AC terminals of each individual cell, which appear as $L_{\mathrm{cmc,L}}$ and $R_{\mathrm{cmc,L}}$ in the equivalent circuit from Fig. 9(a). This solution has been proposed in [16], but without consideration of the components design. Therefore, a detailed design guideline is provided in the following.

1) Design Equations: The full transfer function, $G_3(s)$ given in (8), is a very complicated expression. However, since the parasitic inductances, L_{eq}, are much smaller than the common-mode inductances, $L_{\mathrm{cmc,L}}$, they can be neglected to reduce complexity and enable the derivation of straightforward design equations. In addition, no global CMC is present, which allows to reduce the equivalent circuit to that shown in Fig. 9(c), where a generalized case is shown. With these simplifications, the simplified total impedance, $Z_{\mathrm{T,s,k}}$, seen by the switching source $v_{\mathrm{B,k}}$ (or, equivalently, $v_{\mathrm{T,(k-1)}}$), is given as

$$Z_{\mathrm{T,s,k}} = \frac{Z_{\mathrm{CM}}}{N-k+1}, \quad \text{where}$$
$$Z_{\mathrm{CM}} = \frac{1}{sC_{\mathrm{eq}}} + \frac{R_{\mathrm{cmc,L}} \cdot sL_{\mathrm{cmc,L}}}{R_{\mathrm{cmc,L}} + sL_{\mathrm{cmc,L}}}. \quad (9)$$

The simplified transfer function from the voltage step to the total common-mode current flowing through the voltage source is thus

$$G_{\mathrm{s,k}}(s) = \frac{1}{Z_{\mathrm{T,s,k}}(s)}$$
$$= \frac{(N-k+1)(s^2C_{\mathrm{eq}}L_{\mathrm{cmc,L}} + sR_{\mathrm{cmc,L}}C_{\mathrm{eq}})}{s^2C_{\mathrm{eq}}L_{\mathrm{cmc,L}}R_{\mathrm{cmc,L}} + sL_{\mathrm{cmc,L}} + R_{\mathrm{cmc,L}}}, \quad (10)$$

which describes a second-order system.

With respect to reducing EMI, no oscillations are desired. On the other hand, while an overdamped response features a faster decay rate, this comes at the price of higher peak current values (and thus increased rms current), since the total amount of charge that needs to be transferred is given by C_{eq} and the applied voltage step. Therefore, critical damping should be aimed at for an optimum design, which corresponds to the imaginary part of the poles of (10) being zero, i.e.,

$$L_{\mathrm{cmc,L}} = 4C_{\mathrm{eq}}R_{\mathrm{cmc,L}}^2. \quad (11)$$

Applying this condition to $G_{\mathrm{s,k}}(s)$, the response to a voltage step of source $v_{\mathrm{B,k}}$ (or $v_{\mathrm{T,(k-1)}}$) with magnitude V_{dc} can be calculated as

$$i_{\mathrm{cm,T,k}}(t) = \frac{(N-k+1)V_{\mathrm{dc}}}{4R_{\mathrm{cmc,L}}^2 C_{\mathrm{eq}}} \left(4C_{\mathrm{eq}}R_{\mathrm{cmc,L}} - t\right) e^{-\frac{t}{2C_{\mathrm{eq}}R_{\mathrm{cmc,L}}}}. \quad (12)$$

Note that the time integral of such a current pulse directly corresponds to the amount of charge required to increase the voltage of all affected, i.e., $(N-k+1)$, common-mode capacitances by V_{dc},

$$\int_0^\infty i_{\mathrm{cm,T,k}}(t)dt = (N-k+1) \cdot C_{\mathrm{eq}}V_{\mathrm{dc}}. \quad (13)$$

The step response described by (12) features a time-constant

$$\tau = 2C_{\mathrm{eq}}R_{\mathrm{cmc,L}}, \quad (14)$$

and a peak value of

$$\hat{I}_{\mathrm{T,k}} = (N-k+1)\frac{V_{\mathrm{dc}}}{R_{\mathrm{cmc,L}}}. \quad (15)$$

Following the same procedure as in the idealized case (cf. (3) and (4)), the resulting rms common-mode current flowing through node N can be calculated by replacing the rectangular current blocks with such of shape $i_{\mathrm{cm,T,k}}(t)$. Defining

$$X(k) := \int_0^\infty i_{\mathrm{cm,T,k}}(t)^2 dt = \frac{5}{8}(N-k+1)^2\frac{C_{\mathrm{eq}}V_{\mathrm{dc}}^2}{R_{\mathrm{cmc,L}}}, \quad (16)$$

the desired total rms common-mode current becomes

$$\tilde{i}_{\mathrm{cm}} = \sqrt{f_{\mathrm{S}} \cdot \left[2X(1) + \sum_{k=2}^{N} 4X(k)\right]}$$
$$= \sqrt{f_{\mathrm{s}} \cdot \frac{5}{8}\frac{C_{\mathrm{eq}}V_{\mathrm{dc}}^2}{R_{\mathrm{cmc,L}}} \cdot \left[\frac{4}{3}N^3 + \frac{2}{3}N\right]}. \quad (17)$$

Similarly, the rms value of the common-mode current in cell l for $l \geq k$ is given by

$$\tilde{i}_{\mathrm{cm,l}} = \sqrt{f_{\mathrm{S}} \cdot (4l-2) \cdot \frac{5}{8}\frac{C_{\mathrm{eq}}V_{\mathrm{dc}}^2}{R_{\mathrm{cmc,L}}}}. \quad (18)$$

Using the current divider rule, the current flowing in $R_{\mathrm{cmc,L}}$ can be calculated as

$$\tilde{i}_{\mathrm{R,l}} = \sqrt{f_{\mathrm{s}} \cdot (4l-2)\frac{C_{\mathrm{eq}}V_{\mathrm{dc}}^2}{2R_{\mathrm{cmc,L}}}}. \quad (19)$$

Therefore, the power dissipated in the damping resistor becomes

$$P_{\mathrm{l,R}} = \tilde{i}_{\mathrm{l,R}}^2 \cdot R_{\mathrm{cmc,L}} = f_{\mathrm{s}} \cdot (2l-1)C_{\mathrm{eq}}V_{\mathrm{dc}}^2, \quad (20)$$

which is independent of the CMC and damping resistor, and amounts to negligible 5.5 W for $l = N = 4$ in the example system. The power dissipation in the CMC's series resistance, i.e., the windings, is anyway dominated by the differential-mode load current and will be addressed in the next section.

The 2014 International Power Electronics Conference

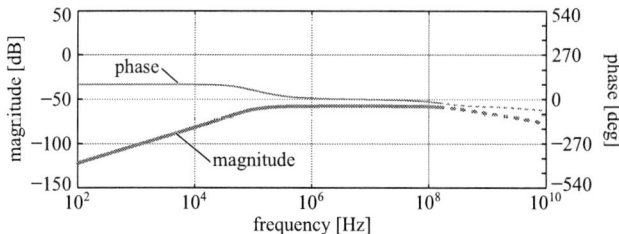

Fig. 14. Bode diagramm of $G_3(s)$ considering the designed local CMCs and $L_{eq} = 100\,\text{nH}$.

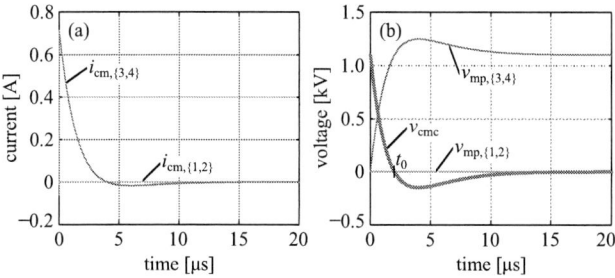

Fig. 15. Step responses of the common-mode currents (a) and the corresponding midpoint voltages (b), where also the voltage across a local CMC, v_{cmc}, is shown. Note that $L_{eq} = 100\,\text{nH}$ is considered in addition to the designed local CMCs.

2) Example Design: Eq. (17) serves together with (11) as design equation for the local CMCs and Fig. 13 illustrates the design trade-offs. However, to choose a specific value of $L_{cmc,L}$, an additional specification is required. In order for the common-mode current pulses to be short compared with the PWM pulses, the requirement $\tau \le 0.2\,\% \cdot T_s = 2\,\mu\text{s}$ could be introduced. This results in $L_{CMC,L} = 6.2\,\text{mH}$, $R_{CMC,L} = 1.539\,\text{k}\Omega$ and $\tilde{i}_{cm} = 167\,\text{mA}$, as is indicated also in Fig. 13. Note that the rms common-mode current is reduced by almost a factor of 5 when compared to the initial case (cf. Table II); a further reduction would come at the cost of very high $L_{cmc,L}$ values.

Fig. 14 shows the Bode plot of $G_3(s)$, i.e., the full transfer function including L_{eq}, where the above values for $L_{cmc,L}$ and $R_{cmc,L}$ are used. The high-frequency resonances are successfully removed by the critically damped CMCs. This is also visible in Fig. 15(a) and (b), where the corresponding responses of the common-mode currents and cell midpoint voltages to a step transition of the voltage source $v_{B,3}$ are shown.

3) Verification: Fig. 16 shows waveforms of the output voltage and total common-mode current (i.e., flowing through the ground connection at node N) obtained from a full-system simulation including the designed local CMCs. A massive reduction of the common-mode currents compared to those shown in Fig. 5(b) is obvious. Table III compares calculated and simulated rms common-mode current values, which confirms the accuracy of the proposed analytic modeling including the assumption $L_{eq} \approx 0$.

C. Common-Mode Choke Design

With the required values of the local CMCs now being calculable, this section discusses the resulting size and losses of possible actual realizations. A key design parameter is

TABLE III
CALCULATED AND SIMULATED RMS COMMON-MODE CURRENTS WITH LOCAL CMCs.

	calc. [mA]	sim. [mA]	error [%]
cell 1	25.3	25.2	0.352
cell 2	43.8	43.5	0.717
cell 3	56.5	56.2	0.571
cell 4	66.9	66.6	0.411
total	167.7	166.7	0.615

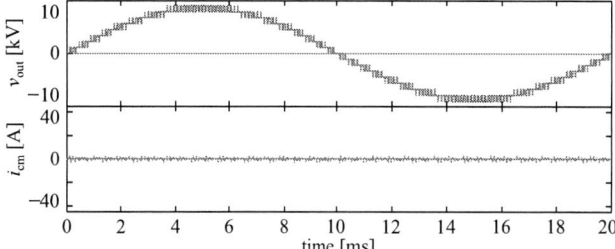

Fig. 16. Simulated phase stack output voltage, v_{out}, and total common-mode current, i_{cm}, for the case of optimally damped local CMCs.

the applied voltage-time integral, which follows from $v_{cmc}(t)$ shown in Fig. 15(b) as

$$(vt) = \int_0^{t_0} v_{cmc}(t)\,dt = \sqrt{L_{cmc,L}C_{eq}}\,V_{dc}e^{-1}, \tag{21}$$

where v_{cmc} is the difference between the applied voltage step and the voltage across the parasitic capacitance. Note that the simplified step response resulting from $G_{s,k}(s)$ has been used to obtain this analytic result. However, the deviation from numerical integration of the full step response shown in Fig. 15(b) is only about $0.35\,\%$ for the values used here.

The two well-known inductor design equations become thus

$$\hat{B} = \frac{\sqrt{L_{cmc,L}C_{eq}}\,V_{dc}e^{-1}}{N A_c} \quad \text{and} \tag{22}$$

$$J_{rms} = \frac{2N I_{rms}}{k_w A_w}, \tag{23}$$

where the factor 2 in (23) accounts for the second CMC winding. From these equations, the area product can be derived as

$$A_c A_w = \frac{2\sqrt{L_{cmc,L}C_{eq}}\,V_{dc}e^{-1}I_{rms}}{\hat{B}k_w J_{rms}} \propto \sqrt{L_{cmc,L}}. \tag{24}$$

Commonly, CMCs are realized using toroidal cores with an area product of (cf. Fig. 17(a))

$$A_c A_w = (h \cdot (r_a - r_i)) \cdot (\pi r_i^2)$$
$$= \pi s_h s_r^2 (1 - s_r) \cdot r_a^4, \tag{25}$$

where $s_r = r_i/r_a = 0.7$ and $s_h = h/r_a = 0.7$, as found from averaging a range of suitable toroidal cores, are introduced to establish a dependency between the area product and the box volume of the resulting CMC. Neglecting the winding, the box volume of a toroidal coil is thus given by

$$V_{box} = (2r_a)^2 \cdot h = (2r_a)^2 \cdot s_h r_a. \tag{26}$$

Solving (25) for r_a and inserting in (26) yields an estimate of the required CMC volume as a function of the area product,

$$V_{box} = 4s_h \left(\frac{A_c A_w}{\pi s_h s_r^2 (1 - s_r)} \right)^{3/4}, \tag{27}$$

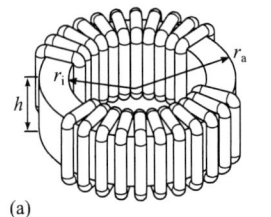
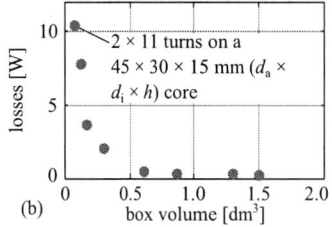

Fig. 17. (a) Geometric dimensions of the considered CMC and (b) thermally feasible CMC designs based on standard Vitroperm cores.

and substituting $A_\mathrm{c}A_\mathrm{w}$ with the expression from (24) leads to a direct dependence on the system parameters and the desired $L_\mathrm{cmc,L}$ value:

$$V_\mathrm{box} = 4s_\mathrm{h} \left(\frac{2\sqrt{L_\mathrm{cmc,L}C_\mathrm{eq}}V_\mathrm{dc}\mathrm{e}^{-1}I_\mathrm{rms}}{\hat{B}k_\mathrm{w}J_\mathrm{rms}\pi s_\mathrm{h}s_\mathrm{r}^2(1-s_\mathrm{r})} \right)^{3/4} \propto L_\mathrm{cmc,L}^{3/8}$$
(28)

Assuming $J_\mathrm{rms} = 5\,\mathrm{A/mm}^2$, $\hat{B} = 0.7\,\mathrm{T}$ for nanocrystalline core material such as Vitroperm [21], $k_\mathrm{w} = 0.1$ for a single-layer winding, and $I_\mathrm{rms} = 56.6\,\mathrm{A}$ corresponding to the phase current resulting from the specifications in Table I, yields the curve shown in Fig. 13 and a volume estimate of $V_\mathrm{box} = 0.076\,\mathrm{dm}^3$ for the designed CMC with $L_\mathrm{cmc,L} = 6.2\,\mathrm{mH}$.

To back-up these approximations, a number of CMCs for $L_\mathrm{cmc,L} = 6.2\,\mathrm{mH}$ have been designed based on standard toroidal Vitroperm cores from VAC [21]. For each core geometry, the required number of turns to meet the above stated peak flux density has been calculated. Together with the requirement for a minimum spacing between the two windings (isolation requirements), the maximum possible wire diameter for single-layer windings and hence the losses caused by the load current can be calculated. Core losses are neglected due to the moderate peak flux density and the comparably low switching frequencies. Finally, the box volume is calculated taking into account also the windings. To account for thermal limitations, the box surface area, a heat transfer coefficient of $\alpha = 20\,\mathrm{W/Km}$ for natural convection, a maximum surface temperature of $100\,^\circ\mathrm{C}$ and an ambient temperature of $50\,^\circ\mathrm{C}$ are used to calculate the maximum allowable power dissipation, and any design exceeding this value is discarded.

Fig. 17 shows the resulting designs in a loss-vs.-volume plot, which confirms that a design with a volume of roughly $0.1\,\mathrm{dm}^3$ and about $10\,\mathrm{W}$ power dissipation is possible (cf. example design details in Fig. 17) and the estimate of the CMC volume based on $L_\mathrm{cmc,L}$ presented above gives acceptable results.

Even though fifteen such CMCs are required for the complete three-phase SST, the additional volume and losses are negligible. Therefore, local CMCs designed as described above are a feasible means to suppress common-mode currents and associated resonant phenomena in cascaded cells converter systems.

V. CONCLUSION

Due to the high grid voltage levels, commonly multi-cell systems are employed in the medium-voltage interface of solid-state transformers. Without proper countermeasures, the steeply changing potentials within such cascaded cells converter

systems give rise to high common-mode currents flowing through parasitic capacitances and potentially resonating with parasitic inductances. An analytic description of these effects as well as of the mitigation thereof by means of common-mode choking has been presented. In accordance with literature, placement of common-mode chokes at the AC terminals of each individual converter cell has been identified as the most feasible solution and a detailed design guideline for such common-mode chokes has been presented. Furthermore, the impact of the additional volume and losses has been shown to be negligible considering the overall SST system.

REFERENCES

[1] W. McMurray, "Power converter circuits having a high frequency link," U.S. Patent 3,581,212, 1970.

[2] S. D. Sudhoff, "Solid state transformer," U.S. Patent 5,943,229, 1999.

[3] M. Kang, P. Enjeti, and I. Pitel, "Analysis and design of electronic transformers for electric power distribution system," *IEEE Trans. Power Electron.*, vol. 14, no. 6, pp. 1133–1141, 1999.

[4] M. Manjrekar, R. Kieferndorf, and G. Venkataramanan, "Power electronic transformers for utility applications," in *Conf. Rec. IEEE Industry Applications Conf.*, Rome, Italy, 2000, pp. 2496–2502.

[5] L. Heinemann and G. Mauthe, "The universal power electronics based distribution transformer, an unified approach," in *Proc. 32nd Annu. IEEE Power Electronics Specialists Conf. (PESC)*, Vancouver, Canada, 2001, pp. 504–509.

[6] E. Ronan, S. Sudhoff, S. Glover, and D. Galloway, "A power electronic-based distribution transformer," *IEEE Trans. Power Del.*, vol. 17, no. 2, pp. 537–543, Apr. 2002.

[7] S. Dieckerhoff, S. Bernet, and D. Krug, "Power loss-oriented evaluation of high voltage IGBTs and multilevel converters in transformerless traction applications," *IEEE Trans. Power Electron.*, vol. 20, no. 6, pp. 1328–1336, Nov. 2005.

[8] M. Steiner and H. Reinold, "Medium frequency topology in railway applications," in *Proc. 12th European Conf. Power Electronics and Applications (EPE)*, Aalborg, Denmark, 2007.

[9] C. Zhao and S. Lewdeni-Schmid, "Design, implementation and performance of a modular power electronic transformer (PET) for railway application," in *Proc. 14th European Conf. Power Electronics and Applications (EPE)*, Birmingham, UK, Sep. 2011.

[10] D. Dujic, F. Kieferndorf, F. Canales, and U. Drofenik, "Power electronic traction transformer technology," in *Proc. 7th Int. IEEE Power Electronics and Motion Control Conf. (IPEMC)*, Harbin, China, Jun. 2012, pp. 636–642.

[11] J. E. Huber and J. W. Kolar, "Optimum number of cascaded cells for high-power medium-voltage multilevel converters," in *Proc. Energy Conversion Congr. and Expo. (ECCE USA)*, Denver, CO, USA, Sep. 2013.

[12] J. Huber, G. Ortiz, F. Krismer, N. Widmer, and J. Kolar, "η-ρ Pareto optimization of bidirectional half-cycle discontinuous-conduction-mode series-resonant dc/dc converter with fixed voltage transfer ratio," in *Proc. IEEE Appl. Power Electronics Conf. (APEC)*, Long Beach, CA, USA, Mar. 2013.

[13] D. Rendusara, E. Cengelci, P. Enjeti, V. Stefanovic, and J. Gray, "Analysis of common mode voltage-'neutral shift' in medium voltage PWM adjustable speed drive (MV-ASD) systems," *IEEE Trans. Power Electron.*, vol. 15, no. 6, pp. 1124–1133, 2000.

[14] P. Loh and D. Holmes, "Reduced common-mode modulation strategies for cascaded multilevel inverters," *IEEE Trans. Ind. Appl.*, vol. 39, no. 5, pp. 1386–1395, 2003.

[15] P. Loh, D. Holmes, and T. Lipo, "Implementation and control of distributed PWM cascaded multilevel inverters with minimal harmonic distortion and common-mode voltage," *IEEE Trans. Power. Electron.*, vol. 20, no. 1, pp. 90–99, Jan. 2005.

[16] R. Lai, M. H. Todorovic, and J. Sabate, "Analysis and suppression of a common mode resonance in the cascaded H-bridge multilevel inverter," in *Energy Conversion Congr. and Expo. (ECCE)*, Atlanta, GA, USA, Sep. 2010, pp. 4564–4568.

[17] Gecko-Simulations AG, "GeckoCIRCUITS Power Electronics Circuit Simulator," www.gecko-simulations.com.

[18] A. Rufer, "An aid in the teaching of multilevel inverters for high power applications," in *Proc. 26th Annu. IEEE Power Electronic Specialists Conf.*, Atlanta, GA, USA, Jun. 1995, pp. 347–352.

[19] Y. Murai, S. Member, T. Kubota, Y. Kawase, and A. L. C. Loop, "Leakage current reduction for a high-frequency carrier inverter feeding an induction motor," *IEEE Trans. Ind. Appl.*, vol. 28, no. 4, pp. 858–863, 1992.

[20] T. Habetler, R. Naik, and T. Nondahl, "Design and implementation of an inverter output LC filter used for dv/dt reduction," *IEEE Trans. Power Electron.*, vol. 17, no. 3, pp. 327–331, May 2002.

[21] VAC Vacuumschmelze, "Nanocrystalline VITROPERM EMC products," 2010.

Single-Stage Reconfigurable DC/DC Converter for Wide Input Voltage Range Operation in HEVs

Sandra Zeljkovic, Tomas Reiter

Infineon Technologies AG
Neubiberg, Germany

Dieter Gerling

University of Federal Defense Munich
Neubiberg, Germany

Abstract— This paper proposes a single-stage reconfigurable topology [1] for an isolated DC/DC converter that supplies the 12V power net from a HV-battery pack in hybrid and electric vehicles. The proposed topology solves the problem of reduced converter efficiency in the upper range of HV-battery voltages. The converter is based on the ZVT phase shift full bridge DC/DC converter that is depending on the instantaneous value of battery voltage reconfigured into the push-pull converter. The ZVT phase shift configuration covers upper range of input voltages whereas it is reconfigured into the less efficient, hard-switching, push-pull configuration to cover the lower range. The point of reconfiguration is chosen to maximize the average efficiency according to the histogram of HV-battery voltage during typical driving cycle. Experimental validation of the proposed converter and the efficiency improvement is also presented.

Keywords—automotive, DC/DC converter, driving cycle, (H)EVs, phase shift converter, push-pull converter, reconfigurable, wide input voltage range, ZVT.

I. INTRODUCTION

In the application of the high voltage (HV) to low voltage (LV) DC/DC converter in hybrid and electric vehicles (HEVs), a zero voltage transition (ZVT) phase shift (PS) full bridge DC/DC converter (Fig. 1) is commonly used. This topology presents a good compromise between efficiency and power density. One of its advantages is a soft switching behavior during turn-on of the HV side switches, achieved by parasitic components in the circuit: the transformer leakage inductance and the output capacitance of switches [2], [3]. In Table I from [4], typical requirements given by the automotive industry for such a converter are listed.

When operating in the wide range of input voltages and load currents, the converter cannot maintain the high efficiency in all operating points. Since the transformer turns ratio is chosen according to the minimal specified input voltage $V_{in,min}$, the converter will have reduced efficiency at higher input voltages. This topology exhibits the highest efficiency when the converter is designed for a 'tight' input voltage range, when it operates with the phase shift of nearly 180° (close to the unity duty cycle).

Methods for efficiency improvement of isolated DC/DC converters in wide input voltage range application proposed in the literature can be roughly classified into two main groups:

1) two-stage converters (a non-isolated voltage regulator as input stage)

2) single-stage modifications of topology or control method in commonly used topologies.

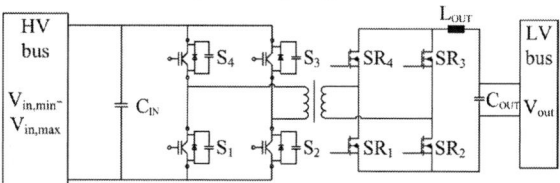

Fig. 1. Conventional ZVT phase shift full bridge DC/DC converter from [2].

TABLE I
CHARACTERISTICS OF HV TO LV DC/DC CONVERTER IN (H)EVS [4]

Parameter	Value
energy flow	unidirectional
$V_{out,min} ... V_{out,max}$	10.6...16V
$V_{out,ripple}$	< 0.2·Vpp
$V_{in,min}$	(PH)EV 240V / HEV 180V
$V_{in,max}$	(PH)EV 420V / HEV 310V
P_{out}	3kW
$I_{out,max}$	210A @ 14V

It has been shown that in certain carefully optimized designs the reduced amount of total losses and the reduction of the converter volume can be achieved with a two-stage topology in automotive applications with wide input and output voltages [5]. Still, an additional hard-switching converter in series that is required in such cases is usually undesirable due to additional components and power losses [6]. Furthermore, with a HV-battery pack at the input of the converter, there is no need for a power factor correction stage (PFC) typical for front-end converters, what is additional reason against the use of pre-regulating stage.

Single-stage topologies adapted for wide input voltage range in the literature often need multiple power transformers and/or rectifiers [7]–[11]. In addition, their control strategy may require modifications as well (e.g. [8], [12], [13]).

Another approach to increase the efficiency of wide input voltage range single-stage converters is the attempt to improve different resonant topologies (e.g. series-parallel, LLC described in [14]). Resonant converters, when designed for the narrow range of operating points, exhibit extremely low switching losses and consequently high efficiency at the price of an additional passive resonant tank. The design of the resonant tank becomes more complicated when the converter operates in a wide

978-1-4799-2706-7/14 $31.00 © 2014 IEEE

range of operating conditions. Several adaptations of resonant converters for improved operation at wide input voltages have been recently proposed in the literature [15]–[17]. Still, the applications in focus are intended for considerably lower load currents compared to the specification from Table I. This fact makes offered solutions less efficient when used as HV to LV DC/DC converters in (H)EVs.

Most of the methods reviewed here are either focused on fuel cells [10], [18], [19] or photovoltaic applications [12], [17], [20] as typical examples of applications with wide input voltage range, or optimized based on the hold-up requirements for front-end converters [8], [21]. They do not exactly match the application requirements in (H)EVs, where the distribution of input voltages is determined by the discharge behavior of a HV-battery pack, usually the Lithium Ion (Li-Ion) or the Nickel-Metal Hydride (NiMH) battery type. Methods to optimize electric machines [22] or power electronic components in (H)EVs (e.g. a main inverter [23] or a boost stage [24]) according to standard driving cycles have been proposed.

Operation of a DC/DC converter that supplies 12V system from the HV-battery is not directly affected by the vehicle's mission profile in the sense of output power that has to be delivered. It is affected rather indirectly, through the variations of the HV-battery voltage during its discharge cycles.

In this paper, a single-stage reconfigurable converter topology based on the ZVT PS converter is proposed for use in (H)EVs. It requires no additional passive elements, but applies only two additional switches that do not exhibit switching losses. The control strategy for both operation modes of the converter is unique, and the difference comes only at the level of gate driving patterns. The converter design procedure takes into account specifics of the discharge curve of automotive battery (Fig. 2) in order to optimize the converter's voltage-weighted average efficiency.

The remainder of the paper is organized as follows: Section II gives the background of the battery voltage behavior during a typical discharge cycle in EVs, Section III presents the proposed single-stage reconfigurable topology, Section IV deals with the proposed design procedure that maximize the voltage-weighted average efficiency of the reconfigurable converter according to the histogram of battery voltages. Section V gives the experimental validation of the proposed solution and discusses the efficiency improvement. Section VI concludes the work.

II. OPERATING CONDITIONS OF HV TO LV DC/DC CONVERTER

The voltage of the HV-battery pack at the input of a DC/DC converter V_{in} changes during a driving cycle, depending on the state of charge (SoC), but is also affected by the temperature and the discharge current. Since EVs operate in the charge-depletion mode, they are characterized by the deep discharge cycle with its SoC dropping down during the full cycle almost linearly. On the contrary, HEVs operate in the charge-sustaining

mode, where the battery SoC may increase or decrease over a driving profile, but on average remains at its initial level. According to this, wide ranges for V_{in} are specified in Table I for both HEVs and EVs. Furthermore, from the typical discharge curve of today's automotive batteries (Fig. 2(a), the example of the Li-Ion battery obtained using the simplified battery model from [25]), it can be concluded that for EVs, the battery voltage remains close to a certain value for most of the driving cycle, while in a significantly shorter part its voltage varies more notably (up to the maximum or down to the minimum specified battery voltage). Thus, the highest converter efficiency would be desirable in the range of mostly present battery voltages, and not, as usual, at $V_{in,min}$.

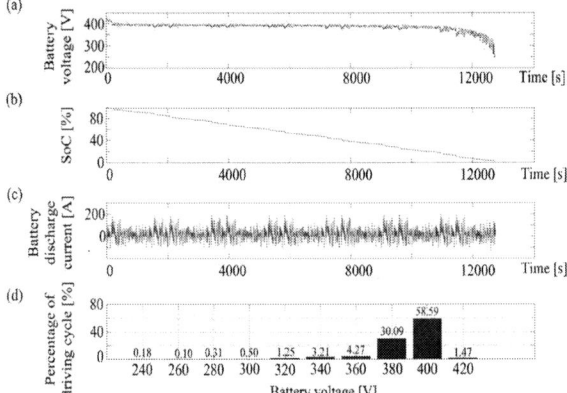

Fig. 2. Histogram of HV-battery voltages (d) during one discharge cycle of the Li-Ion battery in EV derived from (a) the battery voltage (b) the SoC and (c) the battery discharge current obtained using the FTP75 driving cycle data.

The histogram of HV-battery voltages (Fig. 2(d)), used in this paper for the design of the proposed reconfigurable converter, is derived from the example of the EV Li-Ion battery with the nominal voltage of 365V, operating range from 240 to 420V, ampacity of 66.2 Ah and the peak power of electric motor of 80kW. The data describing the battery voltage (Fig. 2(a)) and the discharge current (Fig. 2(c)) are used to obtain the histogram. The US city driving cycle FTP75 is assumed for the analysis. To quantify the relationship between the battery voltage histogram and the average converter efficiency, weighting factors k_{Vin}, can be defined from the histogram in Fig. 2(d) for different values of V_{in} in the range from $V_{in,min}$ to the maximal specified input voltage $V_{in,max}$. Coefficients k_{Vin} are time percentage of the presence of a certain battery voltage over the complete battery discharge cycle.

III. PROPOSED TOPOLOGY

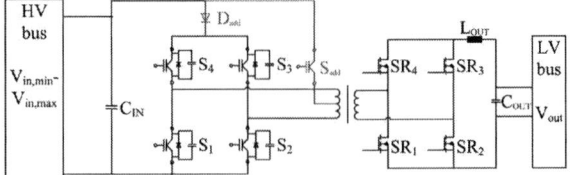

Fig. 3. The proposed reconfigurable topology based on the ZVT PS converter (S_{add} and D_{add} are additional components compared to the conventional ZVT PS topology that do not exhibit repetitive switching losses, but only conduction losses).

Fig. 3 presents the proposed reconfigurable topology [1] based on the ZVT PS converter. The configuration of the proposed converter, depending on the value of V_{in}, can be either a ZVT PS or a push-pull. In the upper range of V_{in}, the conventional ZVT PS converter is extended by addition of a diode D_{add} (see Fig. 4(a)). The diode D_{add} will conduct the input current in the upper range of V_{in}. In the lower range of V_{in}, the converter operates as a push-pull converter (see Fig. 4(b)).

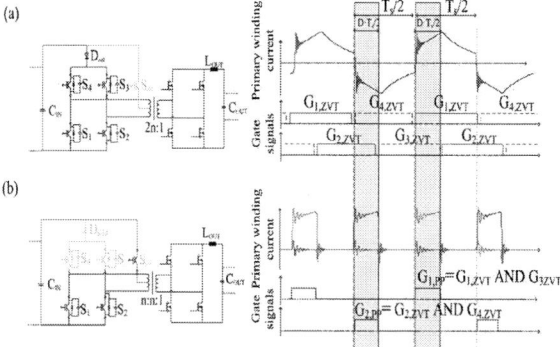

Fig. 4. Two different configurations of the proposed topology: (a) the ZVT PS configuration in the upper range of V_{in} (on the left) (b) the push-pull configuration in the lower range of V_{in} (on the left) with corresponding primary winding currents and gate signals on the right. Gate drive signals for the push-pull operating mode can be derived from those for the ZVT PS configuration using two AND logic gates.

In this configuration, the transformer operates with a center-tapped primary winding, and its middle point is connected to a voltage source (battery) through an additional active switch S_{add}. This switch is continuously on and no switching losses occur during the operation in the push-pull configuration. In the lower range of V_{in}, D_{add} needs to disable the conduction of the freewheeling diodes parallel to the IGBTs S_3 and S_4, not used in the push-pull configuration. The switch S_{add} in the ZVT PS configuration has to be able to block specified $V_{in,max}$. In both operating modes, the proposed converter operates with the duty cycle longer than 50%, allowing increased duty cycle utilization. The principles of the operation of a ZVT PS and a push-pull converter are explained in detail in [3] and [14] respectively.

The phase shift modulation strategy [3] is used to generate the gate driving signals for the ZVT PS converter. The modulation strategy required for the push-pull configuration is considerably simpler and can be easily derived from the driving signals for the ZVT PS configuration. In phase shift modulation, during the power transfer period, two diagonal switches in the HV side H-bridge (e.g S_1 and S_3, or S_2 and S_4) are always conducting simultaneously, so that they can be used to generate the gate driving signal for the corresponding push-pull switches S_1 and S_2. For the purpose, in e.g. analog implementation of the controller, two additional AND gates can be activated when the converter starts its operation in the push-pull mode (see Fig. 4(b)).

IV. DESIGN PROCEDURE FOR PROPOSED RECONFIGURABLE CONVERTER

The aim of the proposed procedure is to find the value of the input voltage V_{reconf} at which the converter will be reconfigured from one topology to another in order to achieve the optimized average converter efficiency η_{ave}, taking into account the typical battery voltage histogram (Fig. 2(d)). The procedure consists of following steps (Fig. 5):

Step 1 - Specification of operating conditions. The design procedure of the proposed converter takes into account requirements given in Table I. From the typical discharge curve of the Li-Ion battery in Fig. 2 it can be concluded that the voltage range around the voltage plateau is the most present at converter's input. In this range of input voltages, it is beneficial that converter operates with more efficient ZVT PS configuration. During the search for maximum η_{ave}, V_{reconf} is incremented by the voltage step ΔV_{in} until maximum input voltage is reached. In this specific case, an increment of 20V is used in the analysis.

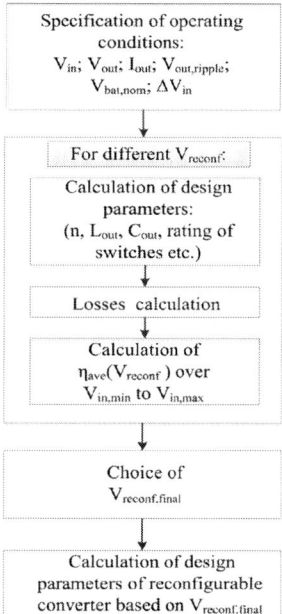

Fig. 5. Flowchart of the design procedure for proposed reconfigurable converter.

Step 2 - Calculation of η_{ave} for different values of V_{reconf}. The losses and the efficiency for both converter configuration are calculated in this step: in the range from $V_{in,min}$ to V_{reconf} for the push-pull configuration and in the range from V_{reconf} to $V_{in,max}$ for the ZVT PS configuration. V_{reconf} is being increased with ΔV_{in} defined in step 1. For that purpose, the transformer turns ratio and the size of the output filter inductor L_{out} are calculated for each considered value of V_{reconf}.

The parameter calculation based on V_{reconf} is addressed in Section IV-A in detail. Based on the calculated efficiency of the reconfigurable converter for the complete range from $V_{in,min}$ to $V_{in,max}$, the average converter efficiency (1) is calculated using voltage weighting factors $k_{Vin,m}$ defined in Section II.

Loss models for both converters are addressed in Section IV-B. Fig. 6 presents results of η_{ave} calculation over different values of V_{reconf}.

$$\eta_{ave} = k_{V_{in,1}} \cdot V_{in,1} + ... + k_{V_{in,m}} \cdot V_{in,m}. \quad (1)$$

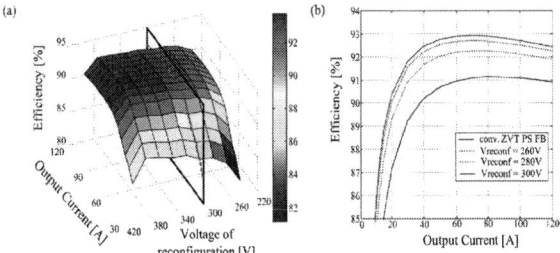

Fig. 6. (a) Voltage-weighted average efficiency of the proposed converter as a function of chosen V_{reconf} (b) Improvement of voltage-weighted average efficiency achieved by the proposed converter with different V_{reconf} up to 300V compared to the conventional ZVT PS converter.

Step 3 - Converter design based on chosen value of reconfiguration voltage $V_{reconf,final}$. Fig. 6(a) can be used to decide on $V_{reconf,final}$ and final design parameters can be calculated accordingly. Certain practical limitations in the choice of $V_{reconf,final}$ are discussed in Subsection IV-A together with results of the proposed design procedure.

A. Design Parameters and Results of Design Procedure

The transformer turns ratio used in the proposed design procedure n is chosen according to V_{reconf} considering the ZVT PS configuration as the main one, while the proper operation at $V_{in,min}$ of the converter operating in push-pull mode must be guaranteed (2). In push-pull mode, the same transformer will be used with the turns ratio $n/2 : n/2 : 1$.

$$n \leq \min \begin{cases} 2 \cdot \dfrac{(V_{in,min} - V_{drop,prim}) \cdot D_{max,pp}}{V_{out,max} + V_{drop,sec}}, \\[2mm] \dfrac{(V_{reconf} - V_{drop,prim}) \cdot D_{max,zvt}}{V_{out,max} + V_{drop,sec}} \end{cases} \quad (2)$$

The tighter range of V_{in} for both considered configurations will affect the size of L_{out} (3), leading to the lower value compared to the conventional ZVT PS converter when the same output current ripple, $I_{out,ripple}$, is specified, while the required value of output capacitor C_{out} (4) will not change.

$$L_{out} \geq \max \begin{cases} \dfrac{1}{2} \cdot \dfrac{V_{out} \cdot (1 - D_{min,pp}) \cdot T_s}{I_{out,ripple}} \\[3mm] \dfrac{1}{2} \cdot \dfrac{V_{out} \cdot (1 - D_{min,zvt}) \cdot T_s}{I_{out,ripple}} \end{cases} \quad (3)$$

$$C_{out} \geq \dfrac{I_{out,ripple} \cdot T_s}{4 \cdot 8 \cdot V_{out,ripple}} \quad (4)$$

Results from Fig. 6(a) show the increase of the voltage-weighted average efficiency when V_{reconf} increases, with the HV side switches assumed to have the same blocking voltage capability over the whole range of V_{in}. However, this analysis does not consider the constraint for the choice of blocking voltage of the HV side switches imposed by operation in the push-pull

configuration. In the push-pull configuration, the HV side switches must be able to block two times V_{reconf} (plus margin for the voltage overshoot caused by interaction of the transformer $L_{leak,prim}$ and C_{oss} of the HV side switches).

In the case of the IGBT switches (used for the study in this paper), this leads to the need for another class of switches regarding blocking voltage after a certain value of V_{in} chosen for V_{reconf}. Thus, the practical limitation for the efficiency improvement given by the blocking voltage capability of the IGBTs is marked at Fig. 6(a) at V_{reconf} of 300V. Furthermore, Fig. 6(b) shows the improvement of η_{ave} when V_{reconf} is increased from 260 to 300V as well as compared to the conventional ZVT PS converter designed for $V_{in,min} = 240$V.

The transformer design in the reconfigurable converter differ from the one of the conventional converter as follows: the transformer turns ratio, $n : 1$, whereas the size of transformer core is not considerably affected. Furthermore, thanks to the longer duty cycle that leads to lower maximum ripple of the output current, L_{out} in the reconfigurable converter is decreased.

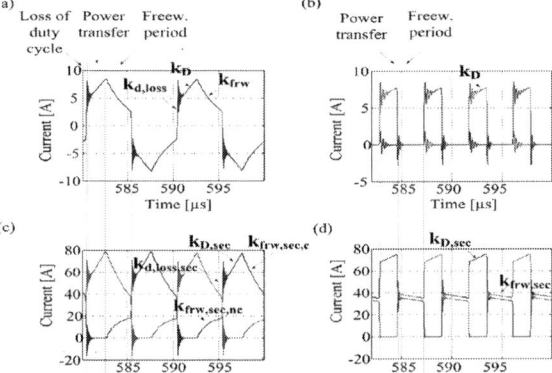

Fig. 7. Waveforms of transformer currents for the ZVT PS configuration in: (a) the primary winding and (b) the secondary winding and for the push-pull configuration in: (c) the primary winding and (d) the secondary winding. Slope coefficients used for modeling of current waveforms are marked here and defined in Appendix B.

The choice of the rectifier topology for the converter proposed in this paper is made according to the model given in [26], which also takes into account a voltage overshoot caused by the transformer $L_{leak,sec}$ and C_{oss} of switches. Thus, a bridge rectifier is adopted for remaining analysis, although other possible rectifier topologies can be used as well. It has to be assured at this point that blocking voltage rating of LV side switches are chosen according to the worst case of two possible converter configurations.

In the presented analysis, the LV side MOSFETs with the same blocking voltage capability are used in the conventional as well as in the reconfigurable converter. In case of reconfigurable converter, there is still a room for reduction of the LV side MOSFETs blocking voltage by approximately 20%. The benefit of MOSFETs with lower voltage class is reduced $r_{ds,on}$ and consequently the conduction loss reduction.

B. Loss Modeling for Design Procedure

An overview of segments of a switching period T_s that are significant for the modeling of primary winding currents and currents of LV side switches are given in Table II and III respectively, for both the ZVT PS and the push-pull configuration. At the HV side of the converter, one half of T_s can be divided into the following sub-periods: a loss of duty cycle ($d_{loss} \cdot T/2$), a power transfer (from $d_{loss} \cdot T/2$ to $D \cdot T/2$, duty cycle determined by controller) and a freewheeling period (Fig. 7(a)). Two transitions, in the leading (consisting of S_2 and S_3) and in the lagging leg (consisting of S_1 and S_4) of H bridge, occur during dead-times $t_{d,lead}$ and $t_{d,lagg}$. On the other hand, in the push-pull configuration mode, there is no loss of duty cycle and no current in the primary winding during the freewheeling period (Fig. 7(b)). Time-domain piece-wise defined waveforms of the primary $I_{tr,prim}$ as well as the secondary winding current $I_{tr,sec}$ are define for each of considered parts of T_s in Appendix B. Due to the significantly high magnetizing inductance of the transformer, the magnetizing current is neglected in this analysis.

The loss mechanisms considered in the presented design procedure are as follows:

- conduction losses of HV side switches (IGBTs and diodes):

$$P_{cond,IGBT} = I_{ave,IGBT} \cdot V_{ce\,0} + I_{RMS,IGBT}^2 \cdot r_{ce}, \quad (5)$$

$$P_{cond,diode} = I_{ave,diode} \cdot V_{d\,0} + I_{RMS,diode}^2 \cdot r_d, \quad (6)$$

where $V_{ce,0}$ and $V_{d,0}$ are threshold voltages of the HV side IGBT and diode, r_{ce} and r_d differential resistances and I_{ave} and I_{RMS} are average and RMS currents of each switch calculated using time-domain piece-wise defined switch currents over the switching period T_s;

TABLE II
CONDUCTION PERIODS AND SWITCHING TRANSITIONS OF HV-SIDE SWITCHES DURING T_s (ZVT PS FB VS. PUSH-PULL)

Part of T_s	Loss of Duty Cycle	Power Transfer	Leading Leg Transition	Freewheeling Period	Lagging Leg Trans.
Conducting switches/ Switching events in ZVT PS mode	1. half: S_1, S_3 2. half: S_2, S_4	1. half: S_1, S_3 2. half: S_2, S_4	1. half: S_3 off S_2 on 2. half: S_2 off S_3 on	1. half: S_1, S_2 2. half: S_3, S_4	1. half: S_1 off S_4 on 2. half: S_4 off S_1 on
Intervals in ZVT PS mode	0 to $d_{loss} \cdot T_s / 2$	$d_{loss} \cdot T_s / 2$ to $D \cdot T_s$	$t_{d,lead}$	$D \cdot T_s / 2$ to T_s	$t_{d,lagg}$
$I_{t,prim}$ in ZVT PS mode	Eq. 13	Eq. 14		Eq. 15	
Conducting switches/ Switching events in push-pull mode	/	1. half: S_1 2. half: S_2	1. half: S_1 off 2. half: S_2 off	/	1. half: S_2 on 2. half: S_1 on
Intervals in push-pull mode		0 to $D \cdot T_s / 2$		$D \cdot T_s / 2$ to $T_s / 2$	
$I_{t,prim}$ in push-pull mode		Eq. 14		0	

- conduction losses of the LV side synchronous rectifiers:

$$P_{cond,SR} = I_{RMS,SR}^2 \cdot r_{ds,on}, \quad (7)$$

where $r_{ds,on}$ is channel resistance of the LV side MOSFETs and $V_{bd,0}$ voltage drop of a conducting body diode;

- the body diode (BD) conduction losses during dead-times:

$$P_{cond,BD} = I_{ave,BD} \cdot V_{bd\,0}, \quad (8)$$

where $I_{ave,BD}$ is the current through the body diode of the LV side MOSFET during the dead-time period (that has to be estimated) that is averaged over complete T_s;

- copper losses of the power transformer:

$$P_{copp,prim} = I_{RMS,prim}^2 \cdot r_{prim}, \quad (9)$$

$$P_{cond,sec} = I_{RMS,sec}^2 \cdot r_{sec}, \quad (10)$$

where r_{prim} and r_{sec} are primary and secondary winding resistances, whereas $I_{RMS,prim}$ and $I_{RMS,sec}$ are RMS currents in primary and secondary transformer windings;

- copper losses of the output filter inductor (12):

$$P_{copp,L} = I_{RMS,Lout}^2 \cdot r_{Lout}, \quad (11)$$

where r_{Lout} is resistance of the output inductor;

- equivalent series resistance (ESR) losses of the output capacitors (13):

$$P_{ESR} = I_{RMS,ripple}^2 \cdot ESR, \quad (12)$$

where ESR is internal resistance of the output capacitor;

- losses due to trace resistances of a printed circuit board (PCB) on the LV side;
- switching losses on the HV side switches (IGBTs and diodes);
- core losses of the power transformer and
- reverse recovery of the diode (for push-pull operating mode).

TABLE III
CONDUCTION PERIODS AND SWITCHING TRANSITIONS OF LV-SIDE SWITCHES DURING T_s (ZVT PS FB VS. PUSH-PULL)

Part of T_s	Loss of Duty Cycle	Power Transfer	Dead Time	Free-wheeling Period	Dead Time
Conducting sw.	1.half: all SRs 2.half: all SRs	1. half: SR_1, SR_3 2. half: SR_2, SR_4	1. half: SR_1, SR_3 BD_2, BD_4 2. half: BD_1, BD_3 SR_2, SR_4	1. half: all SRs 2. half: all SRs	1. half: BD_1, BD_3 SR_2, SR_4 2. half: SR_1, SR_3 BD_2, BD_4
Intervals	0 to $d_{loss} \cdot T_s$	$d_{loss} \cdot T_s / 2$ to $D \cdot T_s$	$t_{d,lead,s}$	$D \cdot T_s / 2$ to T_s	$t_{d,lagg,s}$
I_{sw}	only ZVT PS mode 1.half: Eq.20 2.half: Eq.17	ZVT PS and push-pull mode: Eq. 16	/	ZVT: 1. half: Eq. 19 2. half: Eq. 18 Push-pull mode: Eq. 21	/

Switching losses at the HV side, transformer core losses and the diode reverse recovery are based on the characterization data of chosen components scaled to the operating point of interest. If dead times are chosen properly, the turn-off of the body diodes is always soft and the voltage rises after the body diode current has already fallen to zero. Thus, switching losses in the rectifier are neglected in this analysis. Also, the body diode turns on before the gate signal occurs and the MOSFET channel only takes over the diode's current causing no turn-on losses.

V. EXPERIMENTAL VALIDATION

(a)
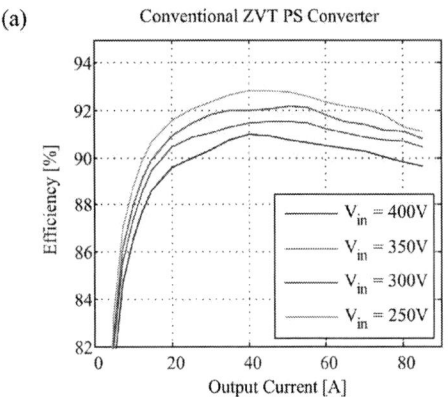

(b)
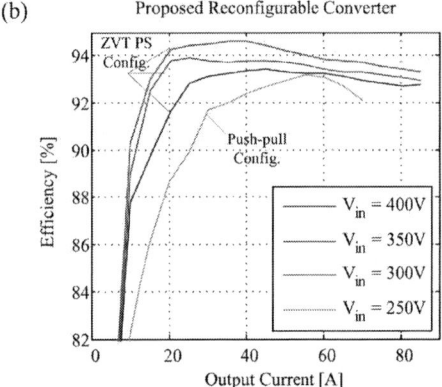

Fig. 8. Family of efficiency curves (prototype measurements) of (a) the conventional ZVT PS FB with the transformer turns ratio of 10:1 (b) the proposed reconfigurable converter with the transformer turns ratio of 10:10:1 (20:1) and $V_{reconf} = 300V$.

Details of the prototype used for experimental validation of the proposed converter and its design procedure can be found in Appendix A. The measured efficiency of the conventional ZVT PS converter for several values of V_{in} is compared in Fig. 8 to the efficiency of the converter proposed in this work. The power consumption of the auxiliary supply (for a controller, gate drivers etc.) is not included in presented efficiency results but does not differ in the reconfigurable converter operating in the ZVT PS configuration compared to the standard ZVT PS. The switching frequency in both converters is set to 100 kHz. It is confirmed that in the upper range of V_{in}, with values close to the nominal battery pack voltage, the proposed

reconfigurable topology operates with increased efficiency (curves from 300V to 400V). In the lower range of voltages, e.g. at 250V, the reconfigurable topology operating in the hard-switching push-pull configuration indeed cannot reach the efficiency of the conventional converter.

Loss distribution is roughly determined using the efficiency measured for both the proposed and the conventional converter from the example of $V_{in} = 350V$ in Fig. 9(b)-(d). At this value of V_{in}, the reconfigurable converter operates as well in the ZVT PS configuration with D_{add} conducting. It can be concluded that the main contributor to this efficiency improvement achieved with the proposed converter are reduced HV side currents and shorter freewheeling period. This leads to significant reduction of both switching and conduction losses at the HV side (Fig. 9(b)). Furthermore, in both reconfigurable and conventional converter prototypes, the same LV side switches are used, and thus no considerable difference in losses can be observed (Fig. 9(b)). Still, under the considered operating conditions the proposed reconfigurable converter allows use of LV side switches with 25% lower blocking voltage capability the additional loss reduction on the LV side of the proposed converter can be assured.

Nevertheless, the overall converter operation is barely affected by this fact knowing that this operating voltage is present at the converter input for a significantly shorter period of time compared to higher voltage values.

To quantify the efficiency improvement obtained with the proposed reconfigurable converter during one battery discharge cycle, the voltage-weighted average efficiency of both converters is calculated based on (1) and the data measured from Fig.8. Fig. 9(a) shows that when the converter efficiency is quantified in this way, the proposed reconfigurable converter offers improved voltage-weighted average efficiency compared to the conventional ZVT PS in the complete range of load currents. In this way, up to 50% of losses that occur at the HV side switches can be saved in the considered example, with further positive impact on the chip temperature swing and lifetime. From the converter cost optimization point of view, the increase of the effective converter's efficiency up to 2% presents the potential to use a smaller chip size.

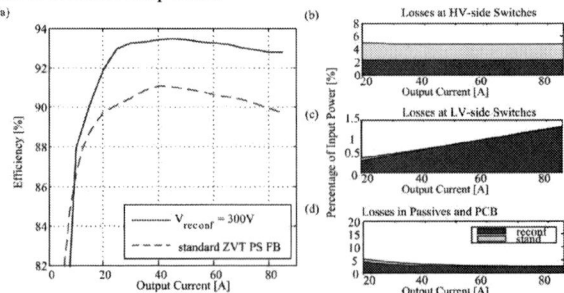

Fig. 9. (a) Voltage-weighted average efficiency of the proposed converter compared to the conventional one calculated based on prototype measurements from Fig. 8. On the right, the comparison of losses distribution in the reconfigurable and the standard ZVT PS converter at $V_{in} = 350V$ at (b) HV side switches (c) LV side switches and (d) passive elements (transformer, output filter and losses in PCB).

VI. CONCLUSIONS

This paper introduces an adaptation of the conventional ZVT PS converter to improve its operation over a wide range of input voltages. As typical operating voltages are far away from the worst case design values in the considered (H)EV application, the target is to increase the efficiency in the range of the most common (typical) operating voltages. The optimization is achieved in a way that the transformer turns ratio of the ZVT PS converter is designed for the higher, more common range of input voltages. As the ZVT PS configuration with a given turns ratio is designed for minimum input voltage higher than the one required in the specification, it is not able to transfer the complete required power in the lower range of input voltages. From that reason converter is reconfigured to push-pull, that will have no problem delivering required power to the output with the same, given turns ratio down to specified $V_{in,min}$. The reconfiguration happens at runtime, when HV-battery voltage falls below specified value of V_{reconf}. It is implemented using an additional diode and a switch, which add only insignificant conducting losses. Required gate driving patterns for the push-pull configuration are derived from the ZVT PS controller with two logic AND gates.

The concept of the proposed reconfigurable converter and the design methodology are validated using the prototype converter. Despite slightly lower efficiency under the less common input voltage conditions (lower voltage range), the effective converter efficiency (the histogram-weighted average over an application-based voltage profile) is increased by about 2%, which is a significant improvement knowing the fact that the conventional converter topology already has an efficiency greater than 90%. The achieved improvement leads to slower aging of switches due to the lower junction temperature transients and can be utilized in future work for further reduction of chip sizes resulting in more cost-effective and compact designs.

APPENDIX A: DESCRIPTION OF PROTOTYPE CONVERTER

Components of the prototype converter used for the experimental verification in Section IV are as follows:

Component	Type	Ratings
HV IGBTs/ diodes	F4-50R07W1H3 B11A	$V_{bl} = 650V$
LV MOSFETs	IPB019N08N3 G (x2 per switch)	$V_{bl} = 80V$
Transformer	CoilCraft NA6237-AL	$n = 10{:}10{:}1$
L_{out}	TDK T7921-51	1.7µH
C_{out}	3x TDK B41792B7158Q	3x 1500µF

APPENDIX B: MODELING OF SWITCH AND TRANSFORMER CURRENTS

$$i_{d_{loss}}(t) = k_{d_{loss}} \cdot t \tag{13}$$

$$i_D(t) = \frac{I_{min}}{n} + k_D \cdot (t - d_{loss} \cdot \frac{T_s}{2}) \tag{14}$$

$$i_{frw}(t) = \frac{I_{min}}{n} + k_D \cdot ((D - d_{loss}) \cdot \frac{T_s}{2}) - k_{frw} \cdot t \tag{15}$$

$$i_{D,sec}(t) = I_{min} + k_{D,sec} \cdot t \tag{16}$$

$$i_{d_{loss2,sec}}(t) = k_{frw,non-cond} \cdot (1 - D) \cdot \frac{T_s}{2} + \frac{k_{d_{loss}}}{n} \cdot t \tag{17}$$

$$i_{frw1,sec}(t) = k_{frw,non-cond} \cdot t \tag{18}$$

$$i_{frw2,sec}(t) = I_{min} + k_{D,sec}(D - d_{loss}) \cdot \frac{T_s}{2} - k_{frw,non-cond} \cdot t \tag{19}$$

$$i_{d_{loss},sec}(t) = I_{min} + k_{D,sec} \cdot (D - d_{loss}) \cdot \frac{T_s}{2} - k_{frw,cond} \cdot (1 - D) \cdot \frac{T_s}{2} - \frac{k_{d_{loss}}}{n} \cdot t \tag{20}$$

$$i_{frw,pp}(t) = i_{D,sec}(D \cdot \frac{T_s}{2}) - k_{frw,pp} \cdot t \tag{21}$$

Coefficients used in equations (13) - (21) are defined as follows:

$$I_{ripple} = \frac{V_{out} \cdot (1 - D) \cdot \frac{T_s}{2}}{L_{out}} \tag{22}$$

$$I_{min} = I_{out,ave} - \frac{I_{ripple}}{2} \tag{23}$$

$$d_{loss} = \frac{2 \cdot I_{min} \cdot L_{leak}}{n \cdot V_{in} \cdot T_s} \tag{24}$$

$$k_{d_{loss}} = \frac{V_{in}}{L_{leak,prim}} \tag{25}$$

$$k_D = \frac{\frac{V_{in}}{n} - V_{out}}{L_{out} \cdot n} \tag{26}$$

$$k_{frw} = \frac{V_{drop,prim}}{L_{leak,prim}} \tag{27}$$

$$k_{frw,nc} = \frac{\frac{n \cdot V_{drop,prim}}{L_{leak,prim}} + \frac{V_{out}}{L_{out}}}{2} \tag{28}$$

$$k_{frw,c} = \frac{\frac{n \cdot V_{drop,prim}}{L_{leak,prim}} - \frac{V_{out}}{L_{out}}}{2} \tag{29}$$

$$k_{D,sec} = \frac{\frac{V_{in}}{n} - V_{out}}{L_{out}} \tag{30}$$

$$k_{frw,PP} = \frac{2 \cdot V_{out}}{L_{out}} \tag{31}$$

where:

- $I_{out,ave}$ is average value of output inductor current;
- I_{ripple} is value of ripple of output inductor current;
- $L_{leak,prim}$ is leakage inductance of transformer primary winding;
- D is a duty cycle determined by the controller;
- $V_{drop,prim}$ is voltage drop over transformer primary winding during freewheeling period.

REFERENCES

[1] S. Zeljkovic and T. Reiter, "Reconfigurable ZVT phase shift converter topology," *Patent Pending*.

[2] J. Sabate, V. Vlatkovic, R. Ridley, F. Lee, and B. Cho, "Design considerations for high-voltage high-power full-bridge zero-voltage switched PWM converter," in *Proc. 5th IEEE Appl. Power Electron.*, 1990, pp. 275– 284.

[3] L. Mweene, C. Wright, and M. Schlecht, "A 1 kW 500 kHz front-end converter for a distributed power supply system," *IEEE Trans. Power Electron.*, vol. 6, no. 3, pp. 398–407, 1991.

[4] VDA EA Working Committee 4.5 Electric Drives, "VDA specification for power electronic generation 1 for electric power trains," Working Group 2 Power Electronics, Tech. Rep., 2011.

[5] F. Krismer, J. Biela, and J. Kolar, "A comparative evaluation of isolated bi-directional DC/DC converters with wide input and output voltage range," in *Proc. 40th IEEE Ind. Appl. Conf.*, 2005, pp. 599–606

[6] I. Barbi and R. Gules, "Isolated DC/DC converters with high-output voltage for TWTA telecommunication satellite applications," *IEEE Trans. Power Electron.*, vol. 18, no. 4, pp. 975–984, 2003.

[7] Z. Wang and H. Li, "A soft switching three-phase current-fed bidirectional DC/DC converter with high efficiency over a wide input voltage range," *IEEE Trans. Power Electron.*, vol. 27, no. 2, pp. 669–684, 2012.

[8] X. Wu, W. Lu, J. Zhang, and Z. Qian, "Extra wide input voltage range and high efficiency DC/DC converter using hybrid modulation," in *Proc. 41st IEEE Ind. Appl. Conf.*, vol. 2, 2006, pp. 588–594.

[9] J.-Y. Lee, "Single-stage AC/DC converter with input-current dead-zone control for wide input voltage ranges,*" IEEE Trans. Ind. Electron.*, vol. 54, no. 2, pp. 724–732, 2007.

[10] M. Harfman Todorovic, L. Palma, and P. Enjeti, "Design of a wide input range DC/DC converter with a robust power control scheme suitable for fuel cell power conversion," *IEEE Trans. Ind. Electron.*, vol. 55, no. 3, pp. 1247–1255, 2008.

[11] H. Wu, P. Xu, W. Liu, and Y. Xing, "Series-input interleaved forward converter with a shared switching leg for wide input voltage range applications," *IEEE Trans. Ind. Electron.*, vol. 60, no. 11, pp. 5029–5039, 2013.

[12] C. Yao, X. Ruan, X. Wang, and C. Tse, "Isolated buck-boost DC/DC converters suitable for wide input-voltage range," *IEEE Trans. Power Electron.*, vol. 26, no. 9, pp. 2599–2613, 2011.

[13] F. Krismer and J. Kolar, "Efficiency-optimized high-current dual active bridge converter for automotive applications," *IEEE Trans. Ind. Electron.*, vol. 59, no. 7, pp. 2745–2760, 2012.

[14] R. Erickson, and D. Maksimovic, *Fundamentals of Power Electronics*, 2nd ed., University of Colorado Boulder, 2001.

[15] I.-O. Lee and G.-W. Moon, "Analysis and design of a three-level LLC series resonant converter for high- and wide-input-voltage applications," *IEEE Trans. Power Electron*, vol. 27, no. 6, pp. 2966–2979, 2012.

[16] A. Coccia, F. Canales, P. Barbosa, and S. Ponnaluri, "Wide input voltage range compensation in DC/DC resonant architectures for on-board traction power supplies," in *Proc. European Conf. Power Electron. and Appl.*, 2007, pp. 1–10.

[17] H. Hu, X. Fang, F. Chen, Z. J. Shen, and I. Batarseh, "A modified high efficiency LLC converter with two transformers for wide input-voltage range applications," *IEEE Trans. Power Electron*, vol. 28, no. 4, pp. 1946–1960, 2013.

[18] L. Zubieta and G. Panza, "A wide input voltage and high efficiency DC/DC converter for fuel cell applications," in *Proc. 20th IEEE Appl. Power Electron. Conf.*, 2005, pp. 85–89.

[19] U. Prasanna and A. Rathore, "Extended range ZVS active-clamped current-fed full-bridge isolated DC/DC converter for fuel cell applications: Analysis, design, and experimental results," *IEEE Trans. Ind. Electron.*, vol. 60, no. 7, pp. 2661–2672, 2013.

[20] Z. Liang, R. Guo, G. Wang, and A. Huang, "A new wide input range high efficiency photovoltaic inverter," in *Proc. IEEE Energy Convers. Congr. and Expo.*, 2010, pp. 2937–2943.

[21] Y. Xing, L. Huang, X. Cai, and S. Sun, "A combined front end DC/DC converter," in *Proc. 18th IEEE Appl. Power Electron. Conf.*, 2003, pp. 1095–1099.

[22] L. Chen, J. Wang, P. Lazari, and X. Chen, "Optimizations of a permanent magnet machine targeting different driving cycles for electric vehicles," in *Proc. IEEE Intern. Electric Machines Drives Conf.*, 2013, pp. 855–862.

[23] E. Knischourek, K. Muehlbauer, and D. Gerling, "Power losses reduction in an electric traction drive at partial load operation," in *Proc. IEEE Intern. Electr. Vehicle Conf.*, 2012, pp. 1–5.

[24] J. Biela, S. Waffler, and J. Kolar, "Mission profile optimized modularization of hybrid vehicle DC/DC converter systems," in *Proc. 6th IEEE Power Electron. and Motion Control Conf.*, 2009, pp. 1390–1396.

[25] O. Tremblay, L.-A. Dessaint, and A.-I. Dekkiche, "A generic battery model for the dynamic simulation of hybrid electric vehicles," in *Proc. IEEE Vehicle Power and Propul. Conf.*, 2007, pp. 284–289.

[26] S. Zeljkovic, T. Reiter, and D. Gerling, "Analysis of rectifier topologies for automotive HV to LV phase shift ZVT DC/DC converter," in *Proc. 15th IEEE Power Electron. and Motion Control Conf.*, 2012, pp.DS1b.4-1–DS1b.4-7

A Two Stage DC/DC Converter with Wide Input Range for EV

Peng Wen, Changsheng Hu, Haitao Yang, Longlong Zhang, Cheng Deng, Yashun Li, Dehong Xu

Institute of Power Electronics

Zhejiang University

Hangzhou, China

xdh@cee.zju.edu.cn

Abstract—In electrical vehicles (EVs), the DC/DC converter is needed to interface high voltage power battery with 12 V electrical system. The converter is required to adapt the wide range input voltage as well as low-voltage high-current output. Moreover, the converter needs high efficiency and high power density. This paper investigates a two-stage converter with high efficiency including two-phase interleaved Boost converter and full-bridge LLC converter. Firstly the design of the two-stage converter is studied. Then the loss breakdown analysis is given. The conversion efficiency of the two-stage converter is evaluated with both theoretical analysis and experiment. A 1.5 kW prototype is built to verify the analysis.

Keywords—Boost-LLC converter, Electrical Vehicle (EV), Wide Input Range, Two-stage converter

I. INTRODUCTION

Hybrid electric vehicles (HEVs) and electric vehicles (EVs) have attracted a lot of attention as one of the potential solutions to ease the crisis of environmental pollution and energy shortage.

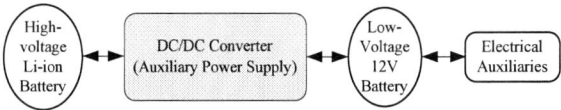

Fig. 1. Auxiliary power supply in EV

As shown in Fig. 1, the auxiliary power supply in EV is the interface between high voltage power battery and 12 V auxiliary battery in EVs. The 12 V auxiliary battery is to provide power for electrical auxiliaries such as audio accessories, driving control electronics, EMS electronics, horn, lamps, power steering units, power windows, windshield wipers etc [1]. The average power consumption of the electric auxiliaries in vehicles is in kW level. Since the voltage of the Lithium-ion power battery varies in the range of 200~400 VDC in general, the auxiliary power supply is required to adapt the wide range input voltage. At the same time, it should provide low-voltage high-current output. Generally, electrical isolation is needed for the sake of safety for the auxiliary power supply. For car application, it also needs high efficiency, high power density and high reliability.

There are many studies on the auxiliary power supply. DC/DC converters for this application can be roughly classified into two types, one type is single-stage converter, and the other is two-stage converter as shown in Fig. 2. In order to achieve high efficiency, soft switching single-stage DC/DC converters have been investigated such as phase-shifted full-bridge converter, series resonant converter, LLC converter and so on. However, these converters suffer from efficiency decrease in applications with wide input range. In order to adapt wide input range and maintain high conversion efficiency, some improvement measures have been proposed for these converters, such as adding auxiliary transformer, adding auxiliary circuit, asymmetrical winding half bridge and hybrid digital control method [2~5]. However, these improvement measures are at the expense of increasing the complexity of circuit or control method, which results in complex design procedure and less reliability [6].

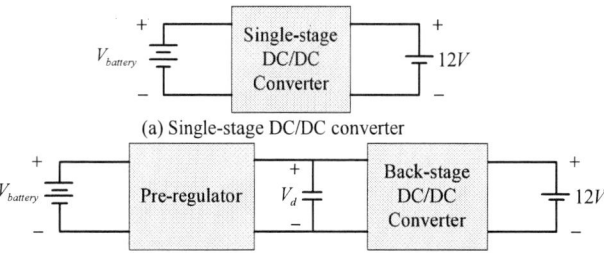

(a) Single-stage DC/DC converter

(b) Two-stage DC/DC converter

Fig. 2. Classification of the DC/DC converter for auxiliary power supply

In order to achieve high efficiency in wide input range, two-stage DC/DC converter attracts attention. It is composed of a pre-regulator and a back-stage DC/DC converter. The pre-regulator is used to output a stabilized DC voltage, so that the back-stage converter can be optimized with high efficiency based on fixed input voltage. A two-stage topology for 24V input and low voltage high current output DC/DC converter is proposed, which consists of a two-phase interleaved Boost converter and a fixed 50% duty cycle full-bridge converter [7]. A two-stage DC/DC converter is proposed for PV application, which consists of an interleaved Boost converter and a LLC resonant converter [8].

In this paper, a two-stage converter composed of a two-phase interleaved Boost converter and a full-bridge

978-1-4799-2706-7/14 $31.00 © 2014 IEEE

LLC converter is investigated. Detail design process and theoretical loss breakdown analysis of Boost-LLC converter are given. A 1.5 kW prototype is built to verify the analysis.

II. THE TWO-STAGE DC/DC CONVERTER

Generally speaking, LLC converter can achieve ZVS for its main transistors from zero load to full load conditions, it is an excellent converter in applications with the demand of high power density. However, the efficiency of LLC converter declines with large variation of input voltage. Therefore, Boost converter is used to stabilize the input voltage of LLC converter at 400 V, which provides a good operating condition for LLC converter. The two-stage DC/DC converter (Boost-LLC converter) which consists of two-phase interleaved Boost converter and full-bridge LLC converter is shown in Fig. 3.

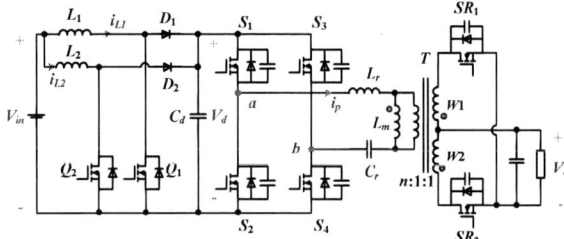

Fig. 3. The two-stage Boost-LLC converter

The Boost converter adopts interleaved structure. Since the ripple current flowing through the intermediate capacitor C_d is reduced, a smaller C_d can be chosen. In this way, the power density of the converter can be improved [9]. The interleaved Boost converter is designed to operate in CCM, and SiC diodes are adopted in order to minimize the reverse recovery loss.

The full-bridge LLC converter is used as the back-stage converter. The primary MOSFETs $S_1 \sim S_4$ operate in ZVS state in full load range. The resonant tank consists of the transformer magnetizing inductance L_m, resonant inductor L_r and the resonant capacitor C_r. The secondary side of the transformer is center tapped. Two or more synchronous rectifier MOSFETs can be paralleled to reduce the conduction power loss in the secondary side of the transformer.

III. DESIGN OF BOOST-LLC CONVERTER

Technical specifications of the target DC/DC converter for EV are shown as follows:
(1) Input voltage: $V_{in} = 200 \sim 375$ V
(2) Output voltage: $V_o = 12$ V
(3) Output power: $P_o = 1500$ W
(4) Ratio of output voltage ripple: $\Delta V_o \leq 1\ \%$
Since the control of interleaved Boost converter and full-bridge LLC converter is decoupled in this paper, the optimal design can be realized for each stage separately.

A. Desin of Interleaved Boost Converter

1) Switching Frequency f_{s_boost}

Switching frequency f_{s_boost} has significant influence on the performance of the converter. With higher switching frequency, the size of inductors can be reduced, and the dynamic response of the Boost converter can be improved. However, the switching loss of power devices and the loss of inductors will increase accordingly. Hence, there is a tradeoff between efficiency and power density to choose f_{s_boost}.

Efficiency of the Boost converter and volume of Boost inductor corresponding to different switching frequency are shown in Fig. 4 and Fig. 5, respectively. Under 50 kHz, efficiency of the converter at rated load of 1.5 kW is 99.2%, and the size of the inductor is 51.3 cm³. When the switching frequency increases to 150 kHz, the efficiency is reduced to be 98.7% at the same load. However, the size of the inductor can be reduced to be 24.2 cm³ accordingly. Since the size of the inductor under 100 kHz is only 1.9 cm³ larger than that under 150 kHz and its efficiency can reach 98.9%, 100 kHz is chosen as f_{s_boost}.

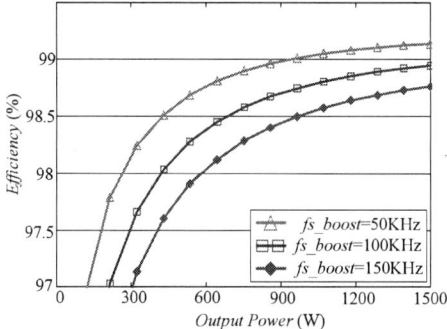

Fig. 4. Efficiency of the Boost converter correspond to different switching frequency

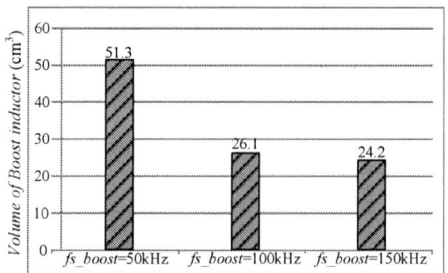

Fig. 5. Volume of Boost inductor correspond to different switching frequency

2) Ripple Ratio r_L of the Inductor Current

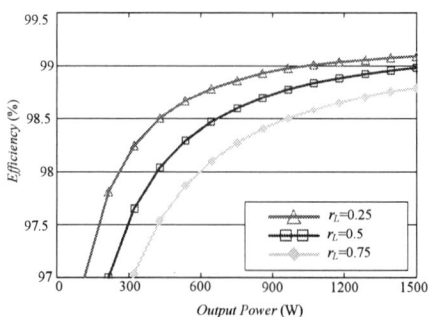

Fig. 6. Efficiency of the Boost converter under different ripple ratio r_L of the inductor current

The ripple ratio r_L of the inductor current also has an effect on the efficiency and power density of the converter. As shown in Fig. 6, under the condition of $r_L =$

0.25, the efficiency of the converter at rated load of 1.5 kW is 99.1%. When the current ripple ratio increases to 0.5, the efficiency will reduce to 99%, and the size of the inductor will be halved according to TABLE I. When r_L continues to increase, the efficiency of the converter is reduced further. But the size of the inductor is not significantly reduced further. Therefore, r_L is chosen to be 0.5.

TABLE I
COMPARISON ON PARAMETERS OF THE INDUCTORS IN BOOST
CONVERTER UNDER DIFFERENT CURRENT RIPPLE RATIO

Inductance (μH)	Current Ripple Ratio: r_L	Part Number of Magnetic Core	Loss (W)	Size:D×H (mm×mm)
630	0.25	77083	3.5	40.77×15.4
320	0.5	77071	5.94	33.66×11.5
210	0.75	77071	7.92	33.66×11.5

3) *Inductors L_1 and L_2* [10]

The inductance of L_1 and L_2 can be calculated by formula (1), and $L_1=L_2=320$ μH.

$$L_1 = L_2 = \frac{V_{in}^2}{r_L \cdot P_o \cdot f_{s\_boost}} \cdot (1 - \frac{V_{in}}{V_d}) \qquad (1)$$

4) *Main Power Switches Q_1 and Q_2*

The voltage stress and the current stress are two key parameters to choose main power switches. In our design, the input voltage is 200~375 V, so the breakdown voltage of Q_1 and Q_2 should not be lower than 500V considering enough voltage margin. The peak current of Q_1 and Q_2 occurs when V_{in}=200 V and P_o=1.5 kW, which is 5.3 A. Considering 1.5 times margin, the drain current of the main transistors is at least 8 A. There are four choices: IPP60R099CPA, IPW60R075CPA, STB27NM60ND and STB36NM60N. Since IPP60R099CPA has the minimum loss among the four MOSFETs as shown in Fig. 7, IPP60R099CPA is selected for Q_1 and Q_2.

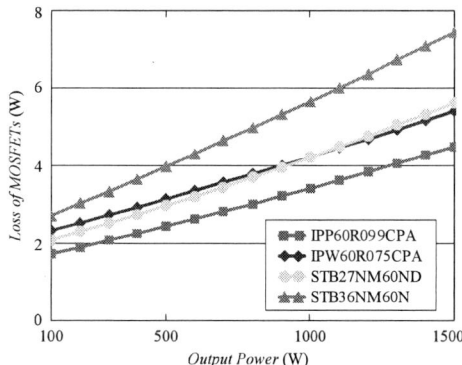

Fig. 7. Power loss of Q_1 and Q_2 with different MOSFETs

4) *Diodes D_1 and D_2*

As for the diodes, Silicon Carbide (SiC) diode of STPSC1206 is selected in order to reduce the reverse recovery loss.

B. *Design of LLC Converter* [11~13]

1) *Resonant Frequency f_r*

According to First Harmonic Approximation (FHA) [11], the AC equivalent circuit of full-bridge LLC converter is shown in Fig. 8.

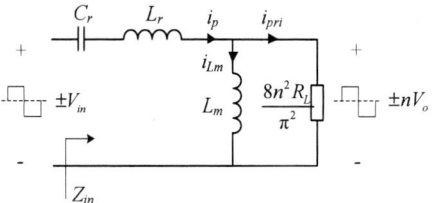

Fig. 8. AC equivalent model of full-bridge LLC converter

There are two resonant frequencies f_r and f_m in LLC converter. f_r is determined by L_r and C_r, and f_m is determined by C_r, L_r and L_m.

$$f_r = \frac{1}{2\pi\sqrt{L_r \cdot C_r}} \qquad (2)$$

$$f_m = \frac{1}{2\pi\sqrt{(L_r + L_m) \cdot C_r}} \qquad (3)$$

The normalized frequency f_n, the proportional coefficient k and the quality factor Q are defined as follows, respectively.

$$f_n = \frac{f_s}{f_r} \qquad (4)$$

$$k = \frac{L_m}{L_r} \qquad (5)$$

$$Q = \sqrt{\frac{L_r}{C_r}} \cdot \frac{\pi^2}{8n^2 R_L} \qquad (6)$$

where f_s is the switching frequency, R_L is the load resistance at rated output power.

According to the formula (A.1) for DC gain G_{dc} and (A.2) for input impedance Z_{in} of LLC converter in Appendix, the DC voltage gain characteristics can be drawn in Fig. 9.

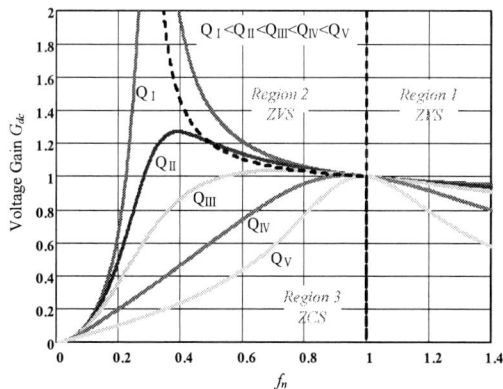

Fig. 9. DC voltage gain characteristics of LLC resonant converter

As shown in Fig. 9, the working plane of LLC converter can be divided into three regions by dotted lines. The following characteristics can be obtained in different regions [12].

- In Region 1, Z_{in} is inductive, resonant network current i_p is lagging input voltage, so ZVS turn-on for the primary switches S_1~S_4 can be achieved.
- In Region 2, Z_{in} is still inductive, and ZVS turn-on can be achieved for S_1~S_4. More than that, since $f_s<f_r$, the secondary currents flowing through synchronous rectifiers SR_1 and SR_2 will be discontinuous, then ZCS turn-off can be achieved for the parasitic diodes of SR_1 and SR_2.
- In Region 3, Z_{in} is capacitive, resonant network current i_p is leading input voltage, so ZCS turn-off can be achieved for S_1~S_4.

Through the above comparisons, the LLC converter is expected to operate in Region 2.

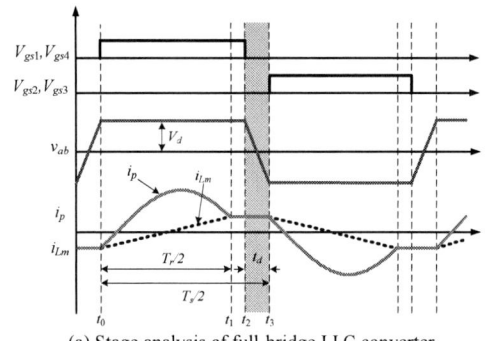

(a) Stage analysis of full-bridge LLC converter

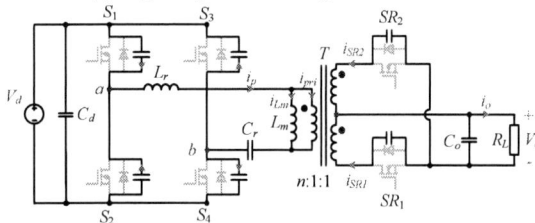

(b) Equivalent circuit of LLC converter in commutation process

Fig. 10. ZVS process of LLC converter

As shown in Fig.10, at the moment of t_2, S_1 and S_4 are switched off, then the primary current i_p begins to charge the output capacitors C_{oss} of S_1 & S_4 and discharge the output capacitors C_{oss} of S_2 & S_3. It takes some time for the voltage across S_2 and S_3 to drop from V_d to zero. So a dead-band time t_d is inserted to ensure ZVS switch on for S_2 and S_3. Then, there is

$$T_s \geq T_r + 2 \cdot t_d \quad (7)$$

During the time interval of t_1~t_3, transformer is not delivering power to the load, so this time interval should not be too long. In other words, f_s should not get far away from f_r, either.

$$f_r = (1.05 \sim 1.1) \cdot f_s \quad (8)$$

In order to obtain high conversion efficiency and ensure ZVS, f_r can be chosen according to (8). Since f_s is set to 100 kHz according to the power density requirements of DC/DC converter, f_r is chosen as 110 kHz.

2) Transformer Turns Ratio n

As shown in Fig. 9, the DC voltage gain of LLC converter is always unity when $f_s = f_r$.

$$G_{dc@\,fs=f_r} = \frac{nV_o}{V_{in}} = 1 \quad (9)$$

Since LLC converter is designed to operate near resonant frequency f_r, the transformer turns ratio n is designed to be 34 according to formula (9).

3) Magnetizing Inductance L_m

According to formulae (A.3) and (A.4) in Appendix , the root mean square (RMS) value of primary current I_{RMS_p} and that of secondary current I_{RMS_s} are entirely determined by L_m.

The total conduction power loss P_c of LLC converter is

$$P_c = I_{RMS\_p}^2 \cdot \left(2R_{ds\_S} + R_{Lr} + R_{p\_T}\right) + I_{RMS\_s}^2 \cdot \left(R_{s\_T} + R_{ds\_SR}\right) (10)$$

where R_{ds_S} is the conduction resistance of S_1~S_4, R_{Lr} is the AC equivalent resistance of L_r, R_{p_T} and R_{s_T} are the AC equivalent resistance of the transformer primary winding and DC equivalent resistance of the transformer secondary windings, respectively. R_{ds_SR} is the conduction resistance of SR_1 and SR_2.

Switching power loss P_{sw} of LLC converter mainly comes from the turn-off loss P_{off} of S_1~S_4, which is also determined by L_m according to (A.5) in Appendix.

The total power loss (except magnetic core loss) P_l of LLC converter is

$$P_l = P_c + P_{sw} \quad (11)$$

Then the curves of total power loss P_l versus L_m at different load conditions are plotted in Fig. 11. It shows that there exists an optimal L_m which can bring about the lowest power loss under constant output condition [13]. In order to achieve high efficiency at rated working condition, 600 μH is chosen as L_m.

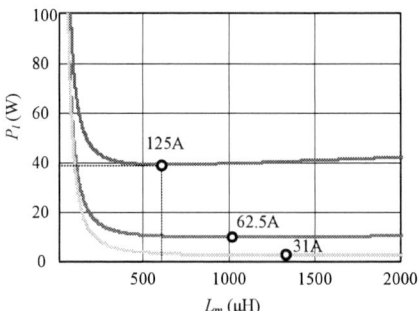

Fig. 11. P_l versus L_m with different load

4) Resonant Inductor L_r and Resonant Capacitor C_r

The input voltage of LLC converter V_d is allowed to range from 390~410 V. In order to prevent LLC converter from working in ZCS region, the peak voltage gain G_{pk} of LLC converter should meet constraints (12) on account of some margin.

$$G_{pk} \geq \left(1.1 \sim 1.2\right) \cdot \frac{V_{d\_\max}}{V_{d\_\min}} \quad (12)$$

where V_{d_min} and V_{d_max} are the minimum and the maximum intermediate voltage, respectively.

At rated output power, in response to a certain L_m, the product of k and Q is constant.

$$k \cdot Q = \frac{L_m}{L_r} \cdot \sqrt{\frac{L_r}{C_r}} \cdot \frac{\pi^2}{8 \cdot n^2 R_L} = \frac{\pi^2 \cdot \omega_r \cdot L_m}{8 \cdot n^2 \cdot R_L} \quad (13)$$

Different combinations of k and Q may correspond to different voltage gain [13]. According to the peak voltage gain requirement and formula (13), k and Q are selected as 10 and 0.386, respectively. According to formulae (5) and (6), L_r and C_r are calculated as 60 µH and 35 nF, respectively.

4) Primary Switches $S_1 \sim S_4$

Since the input voltage of LLC converter is DC 400 V, 600 V MOSFET is selected in order to maintain enough voltage margin. According to formula (A.6) in Appendix, the current stress of $S_1 \sim S_4$ is 6.2 A. Considering some current stress margin, IPP65R190CFDA is selected, and its major features are shown in TABLE II.

TABLE II
MAJOR FEATURES OF IPP65R190CFDA (INFINEON)

V_{DSS}	I_D (@T_c=100℃)	$R_{DS(on)}$
650 V	11 A	0.19 Ω

5) Synchronous Rectifiers SR_1 and SR_2

Since the secondary circuit realizes the full wave rectification, the voltage stress of SR_1 and SR_2 is twice of the output voltage V_o, i.e. 24 V. The synchronous rectifier MOSFET with breakdown voltage of 45 V is selected considering the ring caused by parasitic parameters. According to the following formula, the current stress of the synchronous rectifier is 217 A.

$$I_{s\_pk} = n \cdot I_{p\_pk} \quad (14)$$

In this paper, IRLB3036GPbF (IR) is selected, and its major features are shown in TABLE III. In order to reduce the conduction loss of the synchronous rectifier, three synchronous rectifiers are used in parallel.

TABLE III
MAJOR FEATURES OF IRLB3036GPBF (IR)

V_{DSS}	I_D (@T_c=100℃)	$R_{DS(on)}$
60 V	190 A	2.4 mΩ

6) Output Capacitor C_o

The capacitor with 16 V withstand voltage is selected since the output voltage is 12 V and a certain margin is kept. The RMS value of output ripple current can be calculated by formula (15), and its value is 67.5 A.

$$I_{Co} = \sqrt{\frac{T_s}{2} \cdot \int_0^{\frac{T_s}{2}} [i_s(t) - I_o]^2 \, dt} \quad (15)$$

TABLE IV
MAIN PARAMETERS OF NIPPON PSF CAPACITOR

Parameter	Value
Withstand voltage (V)	16
Rated ripple current (Arms @ 100 kHz)	5
Capacitance (µF)	270
ESR (mΩ @ 100 kHz)	10
Dimension D×L (mm²)	8×8

In order to meet the demand of the voltage ripple ratio, 16 capacitors are used.

7) Transformer T

According to formula (A.4) in Appendix, the RMS value of secondary current is 145.8 A. Based on the depth of the skin and the width of the ferrite core, copper foils with the height h_{cu} of 10mm and the thickness d_{cu} of 0.3mm are used.

The structure of the transformer is shown in Fig. 12, secondary windings W_1 and W_2 use the interleaving parallel method with six groups of two layer copper foils. These copper foils are perpendicular to the center leg of magnetic core. The primary windings with strand wire roil among the six group copper foils. Advantages of using this architecture are listed as follows:

- The cooling effect of both primary and secondary windings is enhanced.
- Symmetry of the two secondary windings W_1 and W_2 is improved, which will improve the current sharing situation of SR_1 and SR_2 and decrease the ripple of output voltage.

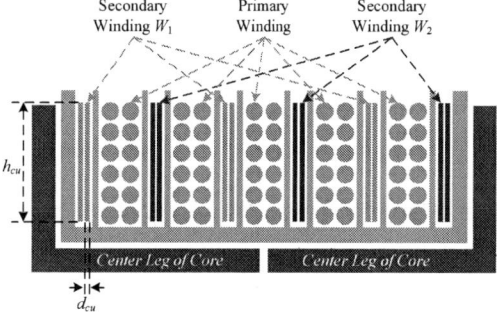

Fig. 12. Architecture of the transformer

C Control Scheme of Boost-LLC converter

The two-stage DC/DC converter as shown in Fig. 13 includes two-phase interleaved Boost converter and full-bridge LLC converter.

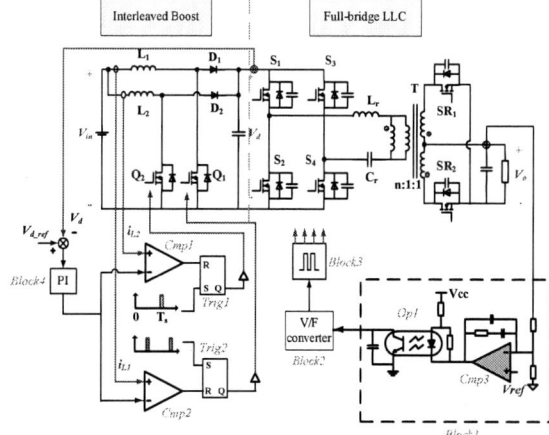

Fig. 13. System structure of Boost-LLC converter with controllers

The two-phase interleaved boost converter utilizes outer voltage loop and inner current loop control. The inner current loop includes inductor current sensor, peak current controller based on comparator $Cmp1$ ($Cmp2$), and PWM generator based on R-S flip-flop trigger $Trig1$ ($Trig2$) in each phase in order to achieve current sharing and fast dynamic control. The outer voltage loop utilizes PI controller ($Block4$) to stabilize the intermediate bus voltage V_d, which is the input of the full-bridge LLC converter.

The full-bridge LLC converter utilizes a single voltage loop including compensation network of PI controller ($Block1$) based on operational amplifier $Cmp3$ and opto-coupler $Op1$, a voltage to frequency converter ($Block2$), and PWM generator ($Block3$) to regulate the output voltage at different load. In order to achieve high efficiency, the output rectifier use synchronous rectifier MOSFETs instead of fast recovery diode or schottky diode, which are controlled by synchronous rectifier controllers.

IV. POWER LOSS ANALYSIS OF BOOST-LLC CONVERTER

A. Power Loss of Interleaved Boost Converter

1) Loss of Main Switches Q_1 and Q_2

The turn-on loss P_{on_Q} is:

$$P_{on\_Q} = \left(V_d I_{L\_\min} t_r + \frac{3}{2} V_d I_{L\_\min} t_{rr} + \frac{5}{3} Q_{rr} V_d \right) \cdot f_{s\_boost} \quad (16)$$

where $I_{L\_\min}$ is the minimum value of inductor current, Q_{rr} is the reverse recovery charge of diodes D_1 and D_2, t_r is the current rise time when Q_1 or Q_2 turns on, and t_{rr} is the reverse recovery time of D_1 and D_2.

The conduction loss P_{con_Q} is:

$$P_{con\_Q} = 2 \cdot D \cdot I_{Lrms}^2 \cdot R_{ds\_Q} \quad (17)$$

where R_{ds_Q} and D are the conduction resistance and duty cycle ratio of Q_1 and Q_2, respectively. I_{Lrms} is the RMS value of inductor current.

The turn-off loss P_{off_Q} is:

$$P_{off\_Q} = 2 \cdot \left[\frac{1}{2} \cdot V_d \cdot I_{L\_\max} \cdot (t_f + t_{sp}) \cdot f_{s\_boost} \right] \quad (18)$$

where $I_{L\_\max}$ is the peak value of inductor current, t_f and t_{sp} are the turn-off time and Miller platform time of Q_1 and Q_2, respectively.

2) Loss of Diodes D_1 and D_2

Conduction voltage drop V_D can be calculated by the following equation according to the volt-ampere characteristics curve of diodes.

$$V_F(i_F) = V_D + i_F \cdot R_D \quad (19)$$

where V_D and R_D are the parameters fitted from diodes datasheet.

Then the conduction loss P_{con_D} is:

$$P_{con\_D} = 2 \cdot \left[(1-D) \cdot V_F \cdot I_L + (1-D) \cdot I_{Lrms}^2 \cdot R_D \right] \quad (20)$$

Where I_{Lrms} is the RMS value of ripple current in inductors L_1 and L_2.

The reverse recovery loss P_{rr_D} is:

$$P_{rr\_D} = 2 \cdot \left(\frac{1}{6} \cdot V_d \cdot Q_{rr} \cdot f_{s\_boost} \right) \quad (21)$$

3) Loss of Inductors L_1 and L_2

Copper loss P_{cu_L} of one Boost inductor can be calculated as:

$$P_{Cu\_L} = I_L^2 \cdot R_{dc\_L} + I_{ac\_L}^2 \cdot R_{ac\_L} \quad (22)$$

where R_{dc_L} and R_{ac_L} is the DC equivalent resistance and AC equivalent resistance of L_1 and L_2, respectively. I_{ac_L} is the AC component of inductor current.

Core loss P_{fe_L} of one Boost inductor is determined by Steinmetz formula:

$$P_{fe\_L} = P_{v\_L} \cdot V_{core\_L} = C_m \cdot f_{s\_boost}^\alpha \cdot \Delta B_L^\beta \cdot V_{core\_L} \quad (23)$$

where V_{core_L} is the volume of inductor magnetic core, C_m, α and β are the parameters given by magnetic core manufacturers, and ΔB is the flux density ripple.

Then, the total loss of two Boost inductors P_L is:

$$P_L = 2 \cdot (P_{fe\_L} + P_{Cu\_L}) \quad (24)$$

B. Power Loss of LLC Converter [14]

1) Loss of Primary Switches $S_1 \sim S_4$

The conduction loss P_{con_S} is:

$$P_{con\_S} = I_{RMS\_p}^2 \cdot 2 \cdot R_{ds\_S} \quad (25)$$

The switching loss P_{sw_S} is:

$$P_{sw\_S} = 4 P_{off} \quad (26)$$

2) Loss of Synchronous Rectifiers SR_1 and SR_2

The power loss of SR_1 and SR_2 is mainly conduction loss P_{con_SR}:

$$P_{con\_SR} \approx I_{RMS\_s}^2 \cdot R_{ds\_SR} \quad (27)$$

3) Loss of Resonant Inductor L_r

Copper loss P_{Cu_Lr} and core loss P_{fe_Lr} of resonant inductor can be calculated according to formula (22) and (23) respectively. Therefore, the total loss P_{Lr} of resonant inductor is:

$$P_{Lr} = P_{fe\_Lr} + P_{Cu\_Lr} \quad (28)$$

4) Loss of Transformer T

The cooper loss P_{Cu_T} is:

$$P_{Cu\_T} = I_{RMS\_p}^2 \cdot R_{p\_T} + I_{RMS\_s}^2 \cdot R_{s\_T} \quad (29)$$

The core loss P_{fe_T} of the transformer can be calculated by (23). Then, the total loss P_T of the transformer is:

$$P_T = P_{Cu\_T} + P_{fe\_T} \quad (30)$$

5) Loss of Output Capacitor C_o

$$P_{Co} = I_{Co}^2 \cdot R_{esr\_co} \qquad (31)$$

where I_{Co} is the RMS value of AC current which flows through the output capacitor, and R_{esr_Co} is the equivalent series resistance of C_o.

C. Theoretical Efficiency of Boost-LLC Converter

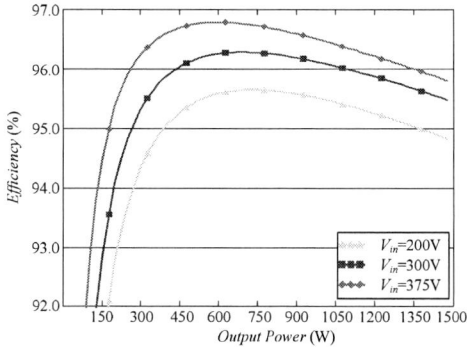

Fig. 14. Theoretical efficiency of Boost-LLC converter

Based on the above design, the conversion efficiency curves of Boost-LLC converter can be obtained and shown in Fig. 14. When the converter operates at 500 W with input voltage of 375 V, the maximum efficiency of Boost-LLC converter is obtained, which is 96.8%. When the input voltage V_{in} drops to 200 V, the maximum efficiency decreases to 95.7%. The efficiency declines slightly when the load becomes heavier. At rated power, the efficiency of the converter becomes 95.8% with the input voltage of 375 V and 94.82% with the input voltage of 200 V.

According to the theoretical efficiency as shown in Fig. 14, the conversion efficiency of Boost-LLC converter over the whole input voltage range of 200~375 V is pretty high, which is more than 95% over wide load range.

V. EXPERIMENT RESULT

A 1.5 kW prototype of Boost-LLC converter based on the above design is built and evaluated. Fig. 15 shows the inductor current i_{L1} of the interleaved Boost converter at rated output power when V_{in} equals 200 V, it can be seen that the interleaved Boost converter operates at CCM mode. The resonant tank current i_p of LLC converter is shown in Fig. 16 (a). From Fig. 16 (b), we can observe that the voltage V_{ds_S1} across the primary MOSFET S_1 has dropped down to zero before S_1 turns on, so S_1 is operating in ZVS mode. The driving signal of the SR_1 and its voltage stress are shown in Fig. 16 (c).

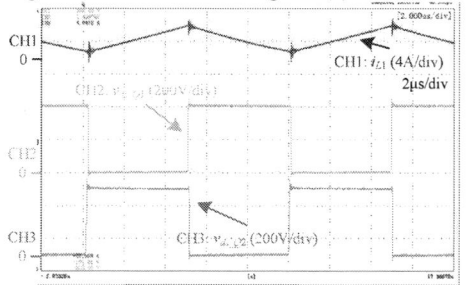

Fig. 15. Waveforms of interleaved Boost converter with rated output when V_{in} equals 200 V

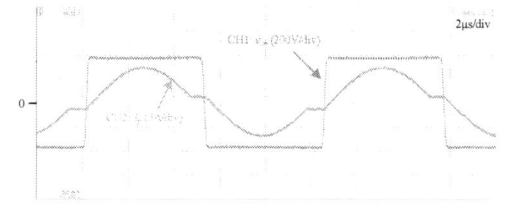

(a). Resonant tank current of LLC converter at rated output

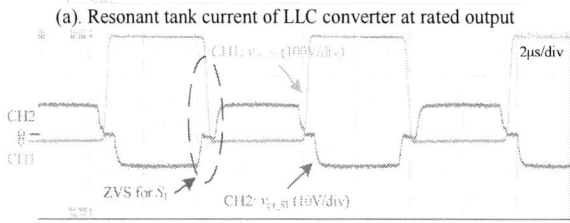

(b). ZVS of primary MOSFET at no-load state

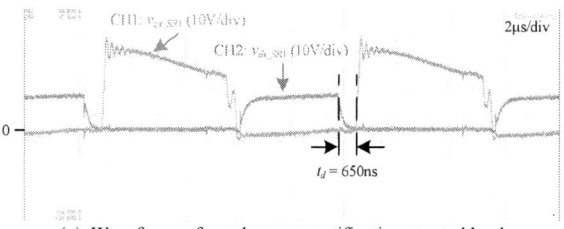

(c). Waveforms of synchronous rectification at rated load

Fig. 16. Waveforms of full-bridge LLC converter

The efficiency of the prototype is measured and depicted as shown in Fig. 17. With the input voltage range of 200~375 V, the efficiency is more than 95% in the load range of 20~60%. The highest efficiency (95.8%) is obtained when the input voltage is 375 V and output power is 600 W, which is 1% less than the theoretical analysis. With the load being increased, the efficiency declines, and it drops to 93.3% at rated output power with the input voltage of 375 V, which is lower than the theoretical efficiency by 2.4%.

Overall trend of the efficiency curves obtained in experiment are in accordance with that of theoretical analysis. The highest conversion efficiency has reached 95.8%, and there is still room to improve efficiency according to theoretical analysis.

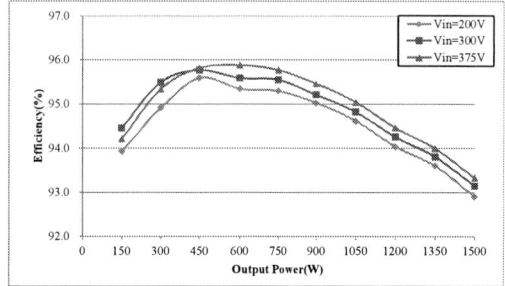

Fig. 17. Conversion efficiency of the prototype

VI. CONCLUSION

In order to meet the requirements of EV, this paper investigates a two-stage DC/DC converter composed of a two-phase interleaved Boost converter and full-bridge LLC converter. The detailed design process of a 1.5 kW prototype based on Boost-LLC converter is elaborated. Then the conversion efficiency of Boost-LLC converter is evaluated with both theoretical analysis and experiment. The highest efficiency of the prototype has reached 95.8%.

ACKNOWLEDGMENT

This work is supported by the National Natural Science Foundation of China (51277163), the Specialized Research Fund for the Doctoral Program of Higher Education of China (20120101130010).

APPENDIX

DC voltage gain of LLC converter based on FHA [11]

$$G_{dc} = \frac{nV_o}{V_{in}} = \frac{1}{\sqrt{\left[1 + \frac{1}{k}\left(1 - \frac{1}{f_n^2}\right)\right]^2 + \left[\left(f_n - \frac{1}{f_n}\right)Q\right]^2}} \quad (A.1)$$

Input impedance of LLC converter [11]

$$Z_{in} = \frac{(k \cdot f_n)^2 \cdot Q}{1 + (k \cdot f_n \cdot Q)^2} + j \cdot \left[\frac{k \cdot f_n}{1 + (k \cdot f_n \cdot Q)^2} + f_n - \frac{1}{f_n}\right] \quad (A.2)$$

RMS current in primary side of LLC converter [15]

$$I_{RMS\_p} = \sqrt{\frac{n^2 \cdot V_o^2 \cdot T_r^2}{32 L_m^2} \cdot \left(2 - \frac{f_s}{f_r}\right) + \frac{\pi^2 \cdot I_o^2 \cdot f_r}{8 \cdot n^2 \cdot f_s}} \quad (A.3)$$

RMS current in secondary side of LLC converter [15]

$$I_{RMS\_s} = \sqrt{\frac{\left(5\pi^2 - 48\right) \cdot n^4 \cdot V_o^2 \cdot f_s}{96 \cdot \pi^2 \cdot L_m^2 \cdot f_r^3} + \frac{\pi^2 \cdot I_o^2 \cdot f_r}{8 \cdot f_s}} \quad (A.4)$$

Turn-of loss of the main transistors in LLC converter [15]

$$P_{off} = \frac{n^2 \cdot V_o^2 \cdot t_f^2 \cdot f_s}{192 \cdot C_{oss} \cdot L_m^2 \cdot f_r^2} \quad (A.5)$$

Peak current in primary side of LLC converter [15]

$$I_{pk} = \sqrt{\left(\frac{\pi \cdot I_o \cdot f_r}{2 \cdot n \cdot f_s}\right)^2 + \left(\frac{n \cdot V_o}{4 \cdot f_r \cdot L_m}\right)^2} \quad (A.6)$$

REFERENCES

[1] C. C. Chan and K. T. Chau, Modern Electric Vehicle Technology. London: Oxford University Press, 2001

[2] Yang, B.; Peng Xu; Lee, F.C., "Range winding for wide input range front end DC/DC converter," Applied Power Electronics Conference and Exposition, 2001. APEC 2001. Sixteenth Annual IEEE , vol.1, no., pp.476,479 vol.1, 2001

[3] Xiangcheng Wang; Feng Tian; Yinxing Li; Batarseh, I., "High efficiency high power density DC/DC converter with wide input range," Industry Applications Conference, 2006. 41st IAS Annual Meeting. Conference Record of the 2006 IEEE , vol.5, no., pp.2115,2120, 8-12 Oct. 2006

[4] Weiyun Chen; Peng Xu; Lee, F.C., "The optimization of asymmetric half bridge converter," Applied Power Electronics Conference and Exposition, 2001. APEC 2001. Sixteenth Annual IEEE , vol.2, no., pp.703,707 vol.2, 2001

[5] Abu-Qahouq, J.A.; Shoubaki, E.; Yangyang Wen; Batarseh, I.; Potter, G., "Piecewise digital control method for DC-DC

converter," Applied Power Electronics Conference and Exposition, 2005. APEC 2005. Twentieth Annual IEEE , vol.1, no., pp.41-49 Vol. 1, 6-10 March 2005

[6] Wei Lu; Xinke Wu; Zhengyu, L.; Zhaoming Qian, "Wide input voltage isolated DC-DC converter with interleaving control,"Power Electronics and Motion Control Conference, 2006. EPE-PEMC 2006. 12th International , vol., no., pp.94-99, Aug. 30 2006-Sept. 1 2006

[7] Laili Wang, Weiming Zhang, Yunqing Pei, Xu Yang, Zhaoan Wang, "A two-stage topology for 24V input low voltage high current DC/DC converter,"Power Electronics and Motion Control Conference, 2009. IPEMC 2009. IEEE 6th international. pp. 804-809

[8] Eun-Soo Kim; Kwang-Ho Yoon; Phum, S.; Meas, S.; Jong-Seob Won; Seong-Jin Oh; Young-Sang Bae, "Operation characteristics of two-stage DC/DC converter for photovoltaic system," Applied Power Electronics Conference and Exposition (APEC), 2012 Twenty-Seventh Annual IEEE , vol., no., pp.681,685, 5-9 Feb. 2012

[9] Newlin D.J.S, Ramalakshimi R, Rajasekaran S., "A performance comparison of interleaved boost converter and conventional boost converter for renewable energy application," Green High Performance Computing(ICGHPC), 2013 IEEE International Conference, March 2013, pp. 1-6

[10] Robert W.Erickson and Dragan Maksimovic. Fundamentals of Power Electronics. New York: Kluwer Academic/Plenum Publishers, 2010

[11] S. DeSimone, C. Adragna,C. Spini and G. Gattavari, "Design-oriented steady state analysis of LLC resonant converters based on FHA," Power Electronics, Electrical Drives, Automation and Motion, 2006. pp.200,207.

[12] Yang B, Lee F.C, Zhang A.J., Guisong Huang, "LLC resonant converter for front end DC/DC conversion," Applied Power Electronics Conference and Exposition, 2002. APEC 2002. Seventeenth Annual IEEE. Vol.2, pp 1108-1112

[13] Lu Bing, Liu Wenduo, Liang Yan, et al. "Optimal design methodology for LLC resonant converter[C]," IEEE APEC 2006: 6.

[14] Zhang Yanjun, Dehong Xu, Mino Kazuaki, Sasagawa Kiyoaki, "1 MHz-1kW LLC resonant converter with integrated magnetic," Applied Power Electronics Conference, APEC 2007, Twenty Second Annual IEEE, pp 955-961

[15] Zhang Yanjun, Dehong Xu, Min Chen, Yu Han, "LLC resonant converter for 48 V to 0.9 V VRM," Power Electronics Specialists Conference, 2004. PESC 04. 2004 IEEE 35th Annual, Vol.3, pp 2848-1854

[16] Cheng Deng, Dehong Xu, Zhang Yanjun, Yi Chen, Okuma Y., Mino Kazuaki, "Impact of dielectric material on passive integration in LLC resonant cnverter", Power Electronics Specialists Conference, 2008. PESC 2008. pp. 269-272

Intermediate and Light Load Efficiency Improvement of a High-Power Density Bidirectional DC-DC Converter in Hybrid Electric Vehicles with MR Fluid Gap Inductor

Furqan Ahmed, Su-Han Kim, Honnyong Cha
School of Energy Engineering
Kyungpook National University
Daegu, Korea
furqanhmd164@gmail.com

Dong-Hun Kim and Heung-Geun Kim
Department of Electrical Engineering
Daegu, Korea

Abstract— **High power-density bidirectional DC-DC converters consisting of buck and boost converters are widely used in hybrid electric vehicles due to their smaller size and low cost. With a conventional constant-value inductor, the light load efficiency of this converter is greatly reduced because of the large constant peak-to-peak current swing and relatively constant conduction losses that result in the entire load range. This paper proposes a new type of nonlinear inductor using magneto-rheological (MR) fluid to improve the intermediate and light load efficiencies. The proposed nonlinear inductor has high current capability, and is gradually changeable. For comparison, a conventional constant-value inductor and the proposed nonlinear inductor are fabricated using the same EE-type ferrite cores with the same number of turns and gap length. The only difference is that in the proposed inductor, the gap is filled with MR fluid instead of air. A 3.7 kW prototype converter is built, and the efficiency from heavy to light loads is measured using both inductors separately, to validate the advantages of the proposed MR fluid-gap inductor.**

Keywords— *Bidirectional DC-DC converter, conduction losses, intermediate and light load efficiencies, magnetorheological (MR) fluid, nonlinear inductor, peak-to-peak current swing.*

I. INTRODUCTION

The use of bidirectional DC-DC converters with energy storage devices such as a battery packs or ultra-capacitors has been advanced in hybrid electric vehicles to enhance the performance of the vehicle and to reduce operating costs by decreasing fuel consumption [1-3]. A typical bidirectional DC-DC converter consists of a combination of buck and boost converters in a half-bridge configuration, as shown in Fig. 1(a). To achieve high power-density, a small inductor is used, and rather than operating in traditional discontinuous conduction mode (DCM), the inductor current is made to flow in the negative direction as well [4]. Fig. 1(b) shows the gating signals and inductor current waveform of the converter under buck-mode operation. During the interval when the

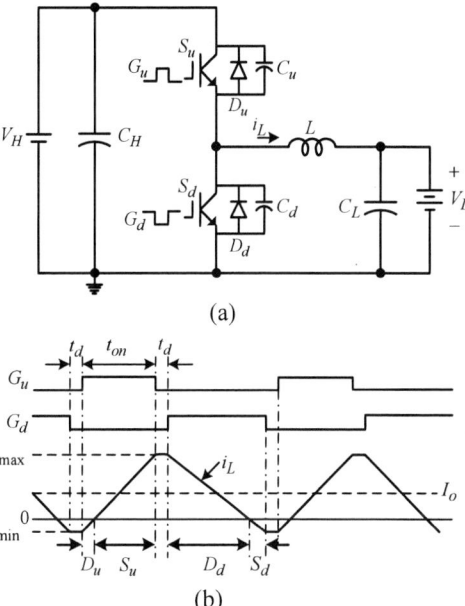

(a)

(b)

Fig. 1. Bidirectional DC-DC converter. (a) Circuit configuration. (b) Gating signals and inductor current waveforms under buck-mode operation.

gate signal G_u is high, switch S_u conducts, and current through the inductor starts increasing. During dead time t_d when both switches are off, the inductor current charges capacitor C_u and discharges C_d. After C_d is discharged, the inductor current starts freewheeling through diode D_d, and it keeps on decreasing until it becomes zero. Then, switch S_d starts conducting, and the inductor current flows in the negative direction. Since switch S_d turns on under zero voltage across it as its anti-parallel diode D_d is carrying the freewheeling current, it has zero voltage switching (ZVS). When switch S_d turns off during dead time t_d, inductor negative current charges capacitor C_d and discharges C_u

. After C_u is discharged during the dead time, diode D_u freewheels the negative inductor current, and when the current goes in the positive direction, upper switch S_u takes on the inductor current under ZVS. This converter operation has benefits of soft switching in a wide load range, as well as low diode reverse recovery loss, and compact design. Also, the parasitic ringing between the inductor and output capacitor associated with the traditional DCM mode, is fully avoided due to the continuity of the inductor current [5].

However, with constant value inductor when load is changed from heavy to intermediate and light, the peak-to-peak current swing of this converter remains constant with increasing negative peak. This results in large circulating energy and large conduction losses even at light load. Therefore, light load efficiency of this converter is greatly reduced as devices are conducting significant rms current at light load. To improve efficiency over wide load range of a buck converter, current ripple adoption is proposed in [6], but this method is too complicated. In [7]-[8] nonlinear inductors using commercially available cores and external dc bias windings are designed and used to improve efficiency of resonant converter and dual half bridge (DHB) converter. But, use of external dc bias increase volume of inductor and also winding losses and not suitable for high current applications. [9]-[10] adopted low-temperature cofired ceramics (LTCC) technology in which nonlinear inductor is designed by integrating ferrite tapes of different permeability. But each of these inductors have some limitations like low current capability, low varying range of inductance or very complicated inductor structure.

In order to overcome the drawbacks or limitations of aforementioned inductors and control techniques, this paper proposes design of a gradually changeable nonlinear inductor with high current handling capability by using commercial EE type ferrite core and filling the gap with MR fluid. The designed inductor due to its increasing inductance when load is changed from heavy load to light load, results in decreased peak to peak current swing and corresponding conduction losses at intermediate and light load. Proposed inductor has simple structure and small volume same as that of air gap inductor but has the benefits of improving both intermediate and light load efficiencies without adding any additional circuit or control complexity.

II. DESIGN OF MR FLUID GAP INDUCTOR

For years, magneto-rheological (MR) fluids have been widely used in rotary brakes, clutches and dampers due to their unique mechanical features such as magnetically controllable stress and viscosity [11]. But the most promising characteristic of MR fluids is that unlike ferromagnetic materials or air, the magnetic permeability of MR fluid is nonlinear with changes in the strength of the magnetic field applied. This magnetic field strength-dependent permeability feature of the MR fluids may bring significant improvements in nonlinear inductor design. For the first time, design of nonlinear inductor

using MR fluid and U-shaped ferrite core was presented in [12], but the work was mainly limited to the magnetic characteristics, measurement of BH-curve, losses considerations and usefulness of MR fluid inductor. While in this paper, application of MR fluid gap inductor in special type of high-power density bidirectional DC-DC converters for hybrid electric vehicles is presented.

A. B-H Curve of MR Fluid

MR fluid consists of magnetic particles suspended in a carrier oil. The size of magnetic particles and their concentration in oil are represented by the mass density (g/cm^3), which effects the permeability of the MR fluid. MR fluid with greater mass density will have higher permeability under a weak magnetic field, and shows more nonlinearity than one with lower mass density. MR fluid is available in various mass densities, including $1.7\,g/cm^3$, $2.4\,g/cm^3$ and $3\,g/cm^3$. In this paper, MR fluid with a mass density of $2.4\,g/cm^3$ is used to design a nonlinear inductor. Fig. 2 shows the B-H curve for the MR fluid and air. It can be seen that as compared to air having linear B-H curve, B-H curve for MR fluid is nonlinear. This is because unlike air having constant permeability irrespective of magnetic field, permeability of MR fluid changes depending upon strength of magnetic field applied. When magnetic field is weak, permeability of MR fluid is high and more nonlinear, i.e., decreasing rapidly with increase in magnetic field strength. As magnetic field gets stronger, permeability of MR fluid decreases and its characteristics changes from nonlinear towards linear. When magnetic field is strong, then permeability of MR is low and further increase in strength of magnetic field causes a small decrease in permeability. It is also worth noticing that even though permeability of MR fluid is high as compared to that of air, it is still very low than that of ferrite. Due to aforementioned properties of MR fluid, it is used to replace air gap in ferrite core, to realize nonlinear current vs. inductance characteristics.

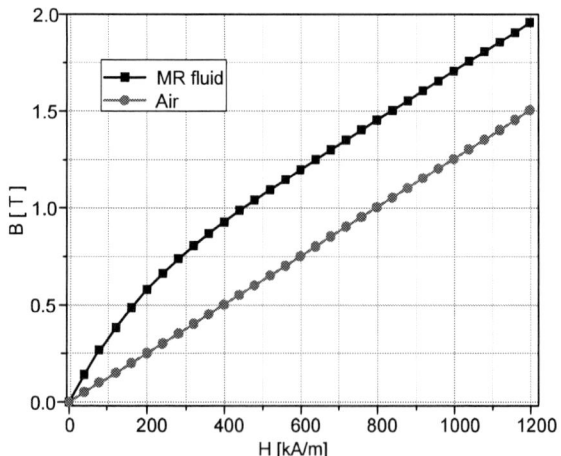

Fig. 2. Comparison of B-H curves for MR fluid and air.

B. Theoretical Calculation for the Inductance of MR Fluid-Gap Inductor

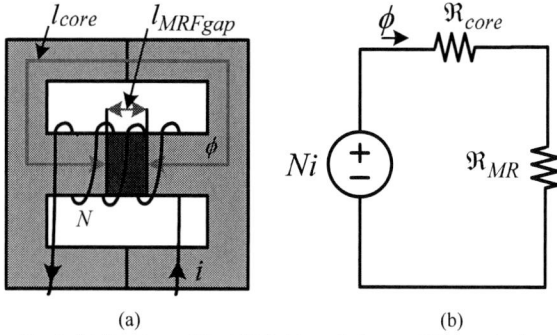

(a)   (b)

Fig. 3. (a) Structure of the MR fluid gap inductor. (b) Its equivalent reluctance model.

$$L_{(i)} = \lambda / I = N^2 (P_{core(i)} // P_{MR(i)}) \qquad (1)$$

$$P_{core(i)} = \frac{1}{\Re_{core(i)}} = \mu_{core(i)} \cdot \frac{S}{l_{core}} \qquad (2)$$

$$P_{MR(i)} = \frac{1}{\Re_{MR(i)}} = \mu_{MR(i)} \cdot \frac{S}{l_{MR}} \qquad (3)$$

Fig. 3 (a) shows the structure and various parameters of the MR fluid-gap inductor, where S is the area of the cross section of the core, l_{MR} and l_{core} are the lengths of the MR fluid gap and core, respectively, and N is the number of turns. The reluctance model of the proposed inductor is shown in Fig. 6(b), where $\Re_{MR}$ and $\Re_{core}$ are the reluctances of the MR fluid and ferrite core respectively, and i is the instantaneous current through the inductor. Equations (1)–(3) are derived from this reluctance model in which P_{MR} and P_{core} are the permeances of the MR fluid and core, respectively, ϕ is the magnetic flux and λ is the magnetic flux linkage for N turns of the winding. The inductance of the MR fluid-gap inductor at any arbitrary current flowing through it is given by (5). From (3), it is clear that for a fixed cross sectional area of the core, as well as a fixed gap length and number of turns, the permeance P_{MR} of the MR fluid is not constant unlike air, as μ_{MR} depends on magnetic field strength. At low current, the magnetic field is weak, and the high permeability of MR fluid results in a large effective inductance of the MR fluid-gap inductor. An increase in the current causes a stronger magnetic field, which decreases the permeability of the MR fluid, and consequently decreases the inductance of the MR fluid-gap inductor. In order to theoretically estimate the inductance of the MR fluid-gap inductor by using B-H curves for the MR fluid and ferrite for a specific cross sectional area of the core, a fixed gap length, and a fixed number of turns, (1) is used in this paper. By properly selecting the core, MR fluid gap length and number of turns, nonlinear inductor of desired current vs. inductance characteristics is obtained.

C. Inductance-vs.- Current Characteristics for Two Inductors

Fig. 7 demonstrates the inductance-vs.-current characteristics of the proposed MR fluid-gap inductor. It is clear that the inductance of the proposed inductor decreases with increases in inductor current. Due to the fact that the inductance of the proposed inductor varies with the current ripple irrespective of the constant output current in the same load conditions, there exists an effective average inductance experienced by the inductor current ripple corresponding to one load condition. With the change in load condition and corresponding peak-to-peak current ripple, the average inductance of the MR fluid-gap inductor also changes. At heavy load, the positive peak current is at its maximum, and the average inductance of the proposed inductor during one switching period is close to that of the air-gap inductor. But when the load is changed from heavy to light, the average inductance experienced by the inductor current becomes larger than that of the air-gap inductor then due to decrease in positive current ripple and increase in negative current ripple. This increase in the average inductance of the MR fluid-gap inductor with the decrease in output current can effectively decrease the peak-to-peak current ripple and corresponding conduction losses, which is not possible with the conventional constant-value inductor.

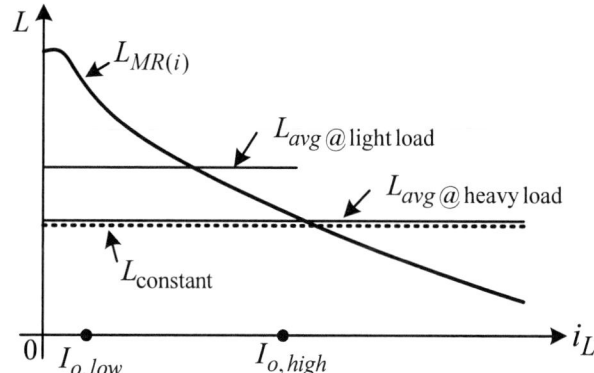

Fig. 4. Conceptual inductor current vs. inductances for two inductors.

D. Practical Inductor Design

To verify the efficiency improvement of the proposed converter with the MR fluid-gap inductor over the conventional converter with an air gap inductor, two inductors were fabricated using the same ferrite cores, the same number of turns, and the same gap length, but filling gap of the proposed inductor with MR fluid having mass density of $2.4 g / cm^3$. The practical design parameters of the two inductors are listed in Table I.

Figs. 5(a) and (b) show the MR fluid used in this paper and the proposed inductor with MR fluid filled in the center leg. In the prototype inductor, an adhesive tape is used to seal the MR fluid in the gap. Fig. 6 shows the proposed inductor and constant air-gap inductor. The proposed inductor has the same structure and volume as the air gap inductor.

TABLE I
DESIGN PARAMETERS FOR TWO INDUCTORS

Inductor	MR fluid gap inductor	Air gap inductor
Core type	EE 7066 Ferrite	EE 7066 Ferrite
N	9	9
Gap length	2.5 mm	2.5 mm

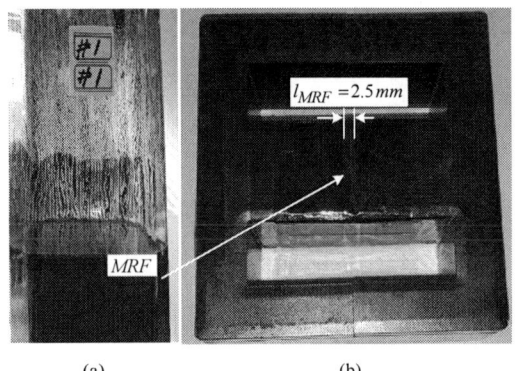

(a)   (b)
Fig. 5. Prototype inductor. (a) MR fluid in the glass tube. (b) Gap in the center leg of ferrite core filled with MR fluid.

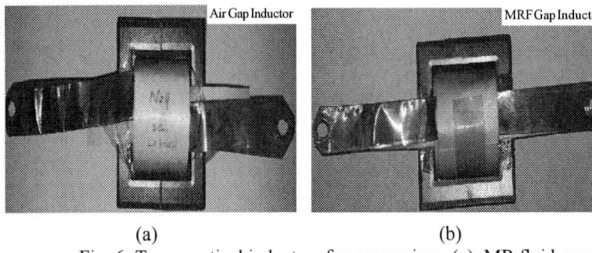

(a)   (b)
Fig. 6. Two practical inductors for comparison. (a) MR fluid-gap inductor. (b) Air-gap inductor.

From Fig. 7, it can be seen that inductance of the air-gap inductor is about 38 μH, and it is constant irrespective of the current flowing through it until saturation. But the proposed MR fluid-gap inductor has larger inductance at low current, and its inductance decreases with increases in current flowing through it. Due to the large initial inductance of the MR fluid-gap inductor, its saturation is more rapid compared to that of the air-gap inductor. A large inductance variation in the proposed inductor is realized by only replacing the air gap with the MR fluid with the same simple structure and same volume of the core. For the bidirectional DC-DC converters used in hybrid electric vehicles, the volume of the converter and efficiency over a wide load range are two critical design parameters, both of which can be achieved by using the proposed MR fluid-gap inductor.

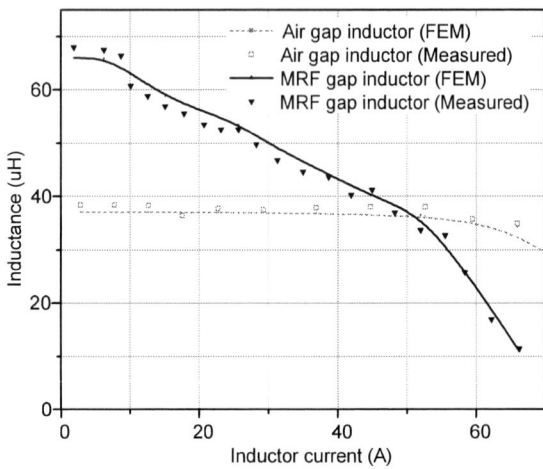

Fig. 7. Practical inductance-vs.-current of air gap inductor and proposed MR fluid gap inductor.

III. PRACTICAL RESULTS

TABLE II
TEST CONDITIONS FOR EXPERIMENT

Operational conditions	• V_{in} = 280 V • V_o = 120 V • Io = 33 A to 6 A • P_o = 3.7 kW to 600W • f_s = 25 kHz

To compare the performance of the proposed MR fluid-gap inductor and the air-gap inductor, a 3.7-kW prototype bidirectional DC-DC was designed. Applying a constant input voltage of 280 V and regulating the output voltage at 120 V, the output current of the converter was decreased from 33A to 6A, and the inductor current waveforms and efficiencies were measured for the two inductors. The experimental conditions are given in table. II.

Fig.8 shows the measured waveforms of the converter under buck-mode for the proposed inductor and the air-gap inductor at worst heavy load condition when the output current is 33 A and the other operating conditions are the same for both inductors. At this heavy load condition, the current ripple has only a positive peak, which is about double the output current. The inductance experienced by the current ripple flowing through the proposed nonlinear inductor is continuously changing compared to the air gap inductor. At low current, the inductance of the MR fluid-gap inductor is large and decreasing as current rises, and eventually becomes smaller than that of the air-gap inductor such that at this heavy load, the average inductance of the proposed inductor is about the same as that of the air-gap inductor. Due to this, the peak-to-peak current swing for both inductors is the same at almost 70 A. For the same heavy load output current, the saturation in the proposed MR fluid gap inductor is more rapid because of its larger initial inductance.

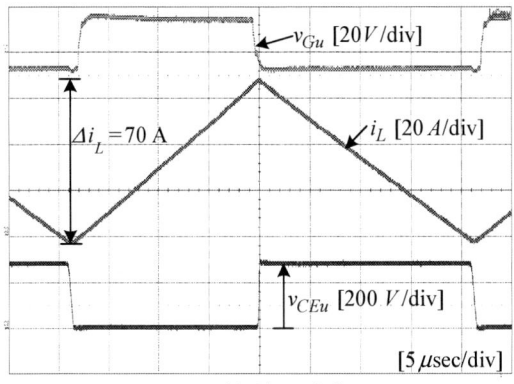

(a) Air gap inductor.

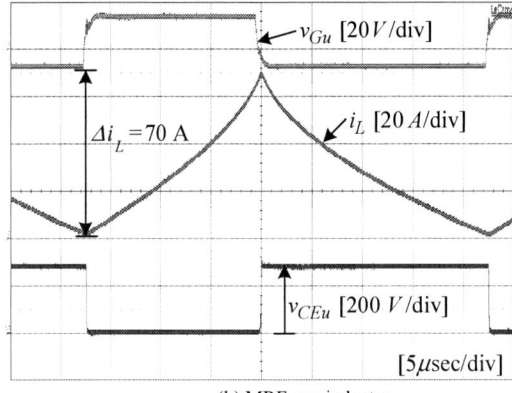

(b) MRF gap inductor.

Fig. 8. Measured waveforms for upper device gate voltage v_{Gu}, device voltage v_{CEu} and inductor current i_L at heavy load, I_o =32 A.

Fig. 9 shows the waveforms for two inductors at light load, where the output current is 6 A and all other operating conditions are the same as those at heavy load. At this relatively low output current, the positive peak of the current decreases, and the negative peak increases. Compared to the constant-value inductor with a constant peak-to-peak current ripple of 70 A (the same as that at heavy load), the peak-to-peak current ripple of the proposed inductor has been decreased to 47 A because of the increased average inductance of the proposed inductor at this relatively low output current. Also, due to the large inductance of the MR fluid gap inductor at low current, the negative current ripple is only 18 A, in contrast to the 30 A current ripple for the constant-value inductor. Due to the decreased circulating current and peak-to-peak current ripple, the conduction losses of the converter using the MR fluid gap inductor have decreased compared to that of the air-gap inductor.

Fig. 10 shows an efficiency comparison of the converter over a wide load range for the constant-value inductor and the MR fluid-gap inductor. At a small part of the worst heavy load, the efficiency of the converter with the proposed inductor is slightly less than that with the air-gap inductor. This is because of the relatively

higher core loss and absence of negative inductor current for soft- switching at some range of heavy load.

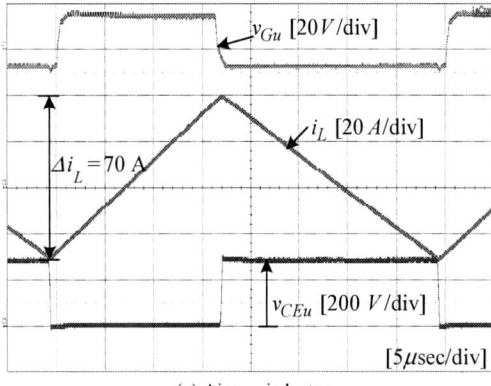

(a) Air gap inductor

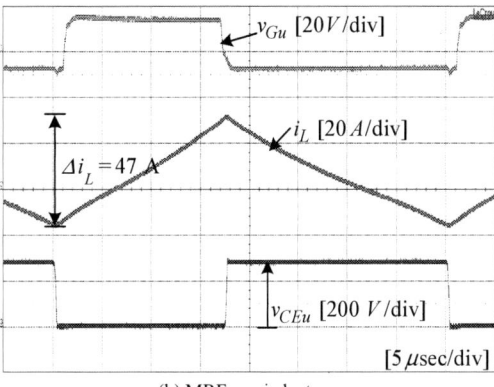

(b) MRF gap inductor.

Fig. 9. Measured waveforms for upper device gate voltage v_{Gu}, device voltage v_{CEu} and inductor current i_L at light load, I_o = 6 A.

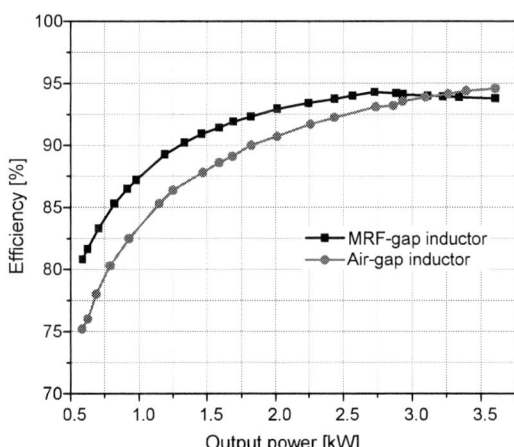

Fig. 10. Measured efficiencies of converter for two inductors.

It can be seen that as the load changes from heavy to light, the efficiency of the proposed converter using the MR fluid-gap inductor is significantly improved compared to that of the constant-value inductor.

In hybrid electric vehicles, where there is no specific rated load, and a bidirectional DC-DC converter operates most of the time at intermediate and light load, the use of the proposed inductor can significantly improve the performance of the vehicle and also increase the life time of the battery pack used.

IV. CONCLUSIONS

This paper proposed design of a new type of high current capability, gradually changeable nonlinear inductor to improve intermediate and light load efficiency of a high power density bidirectional dc-dc converter used in hybrid electric vehicles. Proposed inductor is designed using commercially available EE type ferrite core and has simple structure same as that of air gap inductor, only difference is that the gap of air is filled with MR fluid. Magnetic field strength dependent permeability feature of MR fluid causes the inductor to show inductance variation with change in inductor current. Gradually increasing inductance of MR fluid gap inductor with decrease in current flowing through it helps to decrease peak to peak current and corresponding conduction losses of components, which results in improved efficiency of converter at intermediate and light load. For comparison, two inductors were fabricated using same ferrite cores, same number of turns and same gap length with one conventional air gap and other filled with MR fluid. A 3.7 KW prototype bidirectional converter was built and performance of converter was tested at wide range of load from heavy load to light load, using both inductors separately. Experiment results show that use of proposed nonlinear inductor significantly improve intermediate and light load efficiencies of converter without increasing circuit complexity or volume of converter.

ACKNOWLEDGMENT

This research was supported by Basic Science Research Program through the National Research Foundation of Korea(NRF) funded by the Ministry of Science, ICT and future Planning(NRF-2013R1A2A2A01069038)

REFERENCES

[1] J. Moreno, M. E. Ortuzar and J. W. Dixon, "Energy-management system for a hybrid electric vehicle, using ultracapacitors and neural networks," *IEEE Trans. Ind. Electron.*, vol. 53, no. 2, pp. 614-623, April 2006.

[2] Y. Gao and M. Ehsani, "Parametric design of the traction motor and energy storage for series hybrid off-road and military vehicles," *IEEE Trans. Power Electron.*, vol. 21, no. 3, pp. 749-756, May 2006.

[3] J.-S. Lai and D. J. Nelson, "Energy management power converters in hybrid electric and fuel cell vehicles," *in Proc. IEEE Ind. Electron.*, April 2007, pp. 766 – 777.

[4] J. Zhang, J.-S. Lai, R.Y. Kim and W. Yu, "High-power density design of a soft-switching high-power bidirectional DC-DC converter," *IEEE Trans. Power Electron.*, vol. 22, no. 4, pp. 1145–1153, July 2007.

[5] X. Huang, X. Wang, T. Nergaard, J.-S. Lai, X. Xu, and L. Zhu, "Parasitic ringing and design issues of digitally controlled high power interleaved boost converters," *IEEE Trans. Power Electron.*, vol. 19, no. 5, pp. 1341–1352, Sep. 2004.

[6] S. Zhou and G.A. Rincon-Mora, "A high efficiency, soft switching dc-dc converter with adaptive current-ripple control for portable applications," *IEEE Trans. Circuits Syst.*, vol. 53, no. 4, pp. 319–323, Apr. 2006.

[7] D. Medini and S. Ben-Yaakov, "A current-controlled variable inductor for high frequency resonant power circuits," *in Proc. IEEE Appl. Power Electron.* Conf., Feb. 1994, pp. 219–225.

[8] H. Fan and H. Li, "High-Frequency transformer isolated bidirectional dc-dc converter modules with high efficiency over wide load range for 20 kVA solid-state transformer, " *IEEE Trans. Power Electron.*, vol. 26, no.12, pp. 3599–3608, Dec. 2011.

[9] M. H. F. Lim, J. D. van Wyk, and F. C. Lee, "Hybrid integration of a low voltage, high-current power supply buck converter with an LTCC substrate inductor," *IEEE Trans. Power Electron.*, vol. 25, no. 9, pp. 2287–2298, Sep. 2010.

[10] M. H. F. Lim, J. D. van Wyk, and Z. Liang, "Internal geometry variation of LTCC inductors to improve light-load efficiency of DC-DC converters," *IEEE Trans. Compon. Packag. Technol.*, vol. 32, no. 1, pp. 3–11, Mar. 2009.

[11] M. Lita, N. C. Popa, C. Velescu and L. N. Vekas, "Investigations of a magnetorheological fluid damper," *IEEE Trans. Magn.*, vol. 40, no. 2, pp. 469–472, Mar. 2004.

[12] D. Kim, H. Cha, S.-H. Lee, and D.-H. Kim, "Characteristic of a variable inductor using magnetorheological fluid for efficient power conversion," *IEEE Trans. Magn.*, vol. 49, no. 5, pp. 1901–1904, May. 2013.

Regenerative Control of Bi-directional DC-DC Converter Controlling Variable DC-link for FCEV

Il-Kuen Won[1], An-Yeol Ko[2], Do-Yun Kim[3], Chung-Yuen Won*

Dept. of Information and Communication Engineering
Sungkyunkwan University
Suwon, Republic of Korea
ganzipers@skku.edu

Young-Ryul Kim[4]

Electrical engineering
Anyang University
Anyang, Republic of Korea
yrkim@anyang.ac.kr

Abstract— FCEV (Fuel cell electric vehicle) is provided with energy for operating the motor from the fuel cell. But fuel cell is impossible to be charged at the regenerative mode and also, it has slow dynamic characteristics. So, additional battery is essential as an auxiliary power source in FCEV system. The output current of the VSI (Voltage Source Inverter) is distorted due to the dead-time and the delay time of switching devices [1].

The ripple of the output current is proportional to the DC-link voltage. So, variable DC-Link control is required to alleviate this problem. In this paper, the bi-directional DC-DC converter controlling variable DC-link is proposed. The proposed converter controls DC-link both motoring and regenerative mode. Also, it chooses the source from among the battery and fuel cell corresponding to the motor speed. The operating modes of converter at motoring and regeneration are presented. Finally, simulation results are presented to verify the reliability of proposed bi-directional converter.

Keywords— Regenerative braking control, Variable DC-link, Bi-directional converter, FCEV.

I. INTRODUCTION

Presently, fuel cell is one of the most promising renewable power sources because of its high power density and eco-friendly power generation system. Therefore, the research of FCEV is actively in progress at the automotive industry around the world. But the fuel cell is unable to be charged at the regenerative braking and it outputs lower voltage per unit volume and its response time is slower than that of the battery [2].

Electric vehicle needs instantaneous high power in order to output torque at the sudden acceleration and deceleration. Therefore, in the fuel cell electric vehicle system, the additional battery is essential to improve dynamic characteristic that outputs instantaneous torque and to charge the power generated from the motor at the regenerative braking. Fig. 1 shows the conventional battery and fuel cell hybrid system of fuel cell electric vehicle. In the conventional system, there are two converters connected to the battery and fuel cell. The output power of fuel cell charges directly to the battery.

But, it has disadvantages in terms of volume and cost of total system configuration. Also, the DC-link voltage is controlled always constantly in this system [3], [4].

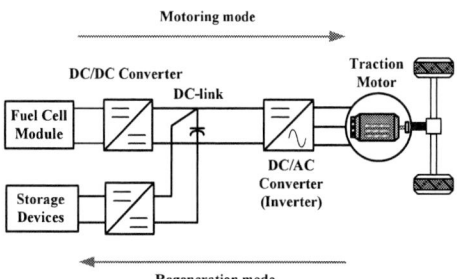

Fig. 1. The conventional hybrid system of fuel cell electric vehicle

But, the output current of 3-Phase VSI is distorted because of dead-time and delay time of switching device, and its distorted output is proportional to the value of DC-link voltage.

In order to alleviate these problems, this paper proposes the bi-directional converter that controls variable DC-link voltage in the hybrid FCEV system.

In this paper, converter controls the DC-link voltage low at the low speed region, reducing unnecessary input current ripple of 3-phase inverter and the torque ripple of the motor. Also, in the high speed region, the motor can be operated at high speed by controlling the DC-link voltage high that make the DC-link voltage not subjected to the influence of the back electromotive force. The battery can charged from fuel cell output through DC-link and also charged from motor at regenerative braking.

The input power for inverter is controlled by using the switches at the terminal of the battery and fuel cell and the converter which controls DC-link voltage is composed to half-bridge converter added additional switches. By using the proposed converter, it chooses the power source from among the battery and fuel cell and operates buck, boost and regenerative mode and outputs torque corresponding to the motor speed.

This paper is arranged as follows: Section 2 explains the distorted current and torque output of VSI related to the DC-link voltage; Section 3 introduces the proposed Bi-directional DC/DC converter and its control algorithm in the Hybrid FCEV System; Section 4 presents the operation modes of converter for operating motor and at the regenerative braking; Section 5 gives simulation results for evaluates the performance of variable DC-link control and regenerative braking torque control of proposed converter; and Section 6 draws some conclusion.

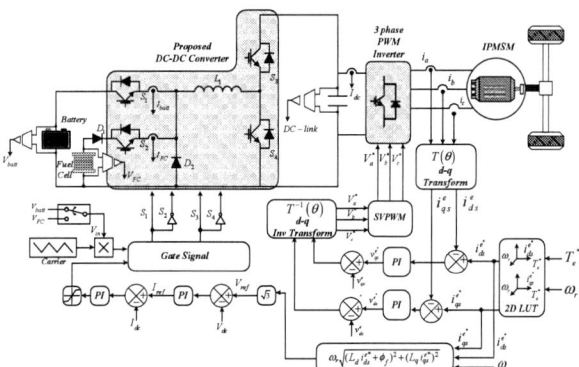

Fig. 2. The control algorithm of proposed FCEV system

II. DISTORTED VOLTAGE

The output of VSI is distorted by the dead-time and the delay time of the semiconductor switching devices in the inverter. Dead-time is applied in order to prevent the short circuit of the switch. But the switching time is delayed as long as the applied dead-time when operating Turn-on and Turn-off. Therefore the output of the terminal voltage is distorted and it affects to the phase current of the inverter. The equations of each distorted phase voltage are as follows:[5].

$$
V_{as\_err} = \frac{T_d + T_{on} - T_{off}}{2T_S} V_{dc} \left\{ \frac{2\sin n(i_{as}) - 2\sin n(i_{bs}) - 2\sin n(i_{cs})}{3} \right\}
$$
(1)

$$
V_{bs\_err} = \frac{T_d + T_{on} - T_{off}}{2T_S} V_{dc} \left\{ \frac{2\sin n(i_{bs}) - 2\sin n(i_{as}) - 2\sin n(i_{cs})}{3} \right\}
$$
(2)

$$
V_{cs\_err} = \frac{T_d + T_{on} - T_{off}}{2T_S} V_{dc} \left\{ \frac{2\sin n(i_{cs}) - 2\sin n(i_{as}) - 2\sin n(i_{bs})}{3} \right\}
$$
(3)

Where V_{as_err} is distortion phase voltage, T_d is the applied time of dead-time, T_{on} and T_{off} are the delay time of the switch at Turn-on and Turn-off operation, T_S is the switching period. Stationary reference frame of distorted Voltage V_{as_err}, V_{bs_err} and V_{cs_err} can be converted to Fourier series expansion. The Fourier series expansions of synchronous d-q axis current distortion considering load impedance and load impedance angle θ can be represented as follows:

$$
I_{ds\_err}^e = \frac{4}{\pi} \frac{\Delta V}{Z_L} \left(-\sin\theta + \frac{1}{5}\sin(6\omega_e t - 5\theta) + \frac{1}{7}\sin(6\omega_e t - 7\theta) \right)
$$
$$
+ \frac{1}{11}\sin(12\omega_e t - 11\theta) + \frac{1}{13}\sin(12\omega_e t - 13\theta)...)
$$
(4)

$$
I_{qs\_err}^e = \frac{4}{\pi} \frac{\Delta V}{Z_L} \left(-\cos\theta + \frac{1}{5}\cos(6\omega_e t - 5\theta) - \frac{1}{7}\cos(6\omega_e t - 7\theta) \right)
$$
$$
+ \frac{1}{11}\cos(12\omega_e t - 11\theta) - \frac{1}{13}\cos(12\omega_e t - 13\theta)...)
$$
(5)

Through this equation, 6 harmonics and 12 harmonics appear at the output d-q axis currents of synchronous rotating reference frame and these harmonics make the torque ripple of IPMSM as equation (6).

$$
T_{e\_err} = \frac{P}{2}\frac{3}{2}\left\{ \phi_f I_{qs\_err}^e + (L_{ds} - L_{qs})I_{ds\_err}^e I_{qs\_err}^e \right\}
$$
(6)

These distorted phase voltage and torque ripple are proportional to the value of the DC-link voltage based on the equation (1), (2) and (3).

Therefore, in order to reduce the current and torque ripple, the DC-link voltage should be controlled depending the each situation; required DC-link is changed considering motor speed.

Proposed converter controls the DC-link voltage low at the low speed and torque region of FCEV. It can improve the operation characteristics of IPMSM.

III. PROPOSED TOPOLOGY

Fig. 2 shows the control algorithm of proposed converter and hybrid FCEV system. Proposed algorithm determines main power considering the measured current using current sensors at the terminal of battery and the fuel cell.

At the case of transient character of fuel cell, whose output current is low, the battery is used for the main power. Also the converter output the DC-link voltage reference as much as back electromotive force originated from the IPMSM considering the measured speed from speed sensor.

In that case, in order to output minimum DC-link voltage reference, its DC-link reference voltage is equal as voltage limit circle by the back electromotive force. The equation of minimum DC-link voltage reference is as follows:

$$
V_{ref} = K \cdot \omega_r \sqrt{\left(L_d i_{ds}^{r*} + \phi_f \right)^2 + \left(L_q i_{qs}^{r*} \right)}
$$
(7)

K is determined considering to the maximum voltage, that SVPWM can output. So, the value of K is defined with $\sqrt{3}$. The left side of Fig. 2, proposed bi-directional DC/DC converter is presented. The switch S_1 and S_2 are used for the power control and S_3, S_4 are for the control of the DC-link voltage of inverter input. Diode D_1 is the blocking diode for preventing reverse current to the fuel cell, D_2 is operate as the freewheeling diode at the case of battery buck mode operation.

Through the complementary operation of S_1 and S_2, the power source of converter can be selected from among the battery and fuel cell. The torque control of inverter is controlled based on 2D Look-up table

978-1-4799-2706-7/14 $31.00 © 2014 IEEE

The 2014 International Power Electronics Conference

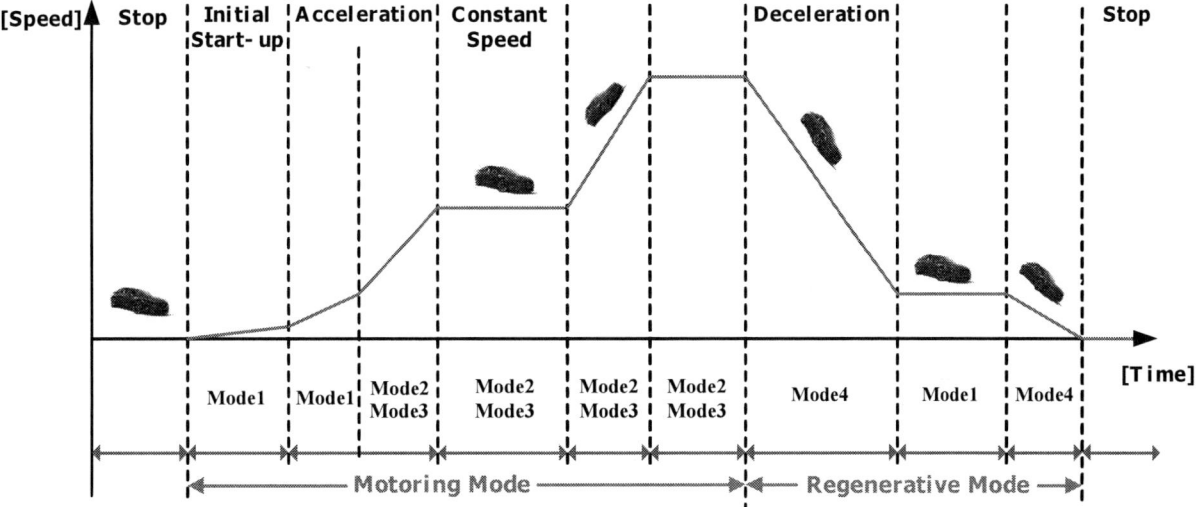

Fig. 3. Converter operation mode considering the speed

IV. OPERATION MODE

A. Motoring mode

Fig. 4 presents the operation mode of converter corresponding to the driving mode of FCEV. The driving mode can be classified to the initial start up, acceleration and deceleration. The operation mode of converter is changed depending on speed and driving mode of FCEV.

Mode 1 is the operation mode at the initial start-up of the FCEV representing battery buck mode. Due to the nature characteristic of the fuel cell, there is some delay time for initial operating fuel cell. So, fuel cell is unable to output the torque for initial start-up of FCEV at that time. Therefore, the battery should be used for power source at mode 1.

But, at mode 1, it has relatively low start-up speed. Also, after the initial start-up, the speed of FCEV gradually increases, little DC-link is needed at that speed region. Therefore, the converter operates buck mode using battery input.

Mode 2 is battery boost mode and mode 3 is fuel cell boost mode. At the case of acceleration, due to the speed of the vehicle is gradually increased, required DC-link voltage corresponding to the speed is greater than the battery input. So, at this time, required DC-link is followed by the boost operation of converter. At this case, depending on the SOC (State of charge) of the battery and the remaining amount of gas of fuel cell or required DC-link voltage, available power source is selected from among the battery and fuel cell.

When FCEV is driving at the constant speed driving, it is divided into case of constant speed in a flat road when the inverter does not require any torque reference and case of constant speed in uphill and downhill where the inverter requires torque reference depending on load torque to maintain constant speed. Also, available power source is selected from among the battery and fuel cell.

Each converter operating modes has been shown as following [6].

B. Regenerative mode

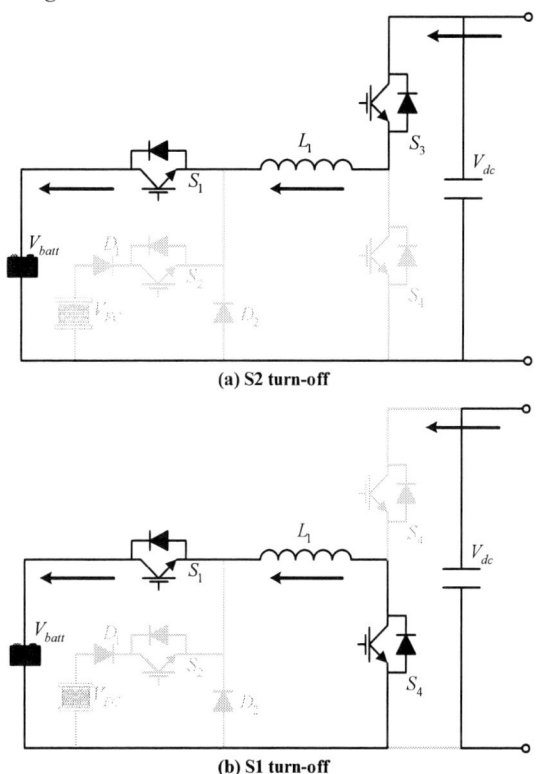

(a) S2 turn-off

(b) S1 turn-off

Fig .4. Mode 4 : Regenerative mode operation

At mode 4 in the Fig .3 describes as the deceleration of FCEV. As the speed of FCEV decreased, the required DC-link voltage is decreased and the converter operates regenerative mode in order to charge battery.

Fig. 4 represents the operation mode of proposed converter of regenerative mode. At regenerative mode, the battery is charged through buck operation of converter that DC-link voltage to the battery voltage.

978-1-4799-2706-7/14 $31.00 © 2014 IEEE

The 2014 International Power Electronics Conference

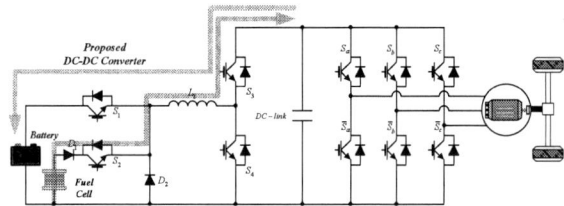

Fig. 5. The power flow of battery charging from fuel cell to battery

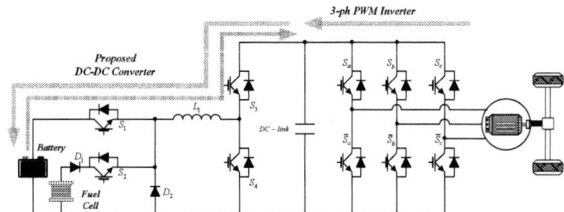

Fig. 6. The power flow of regenerative mode : case 1

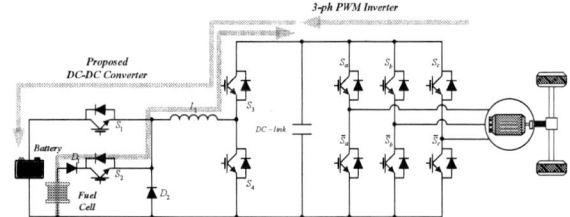

Fig. 7. The power flow of regenerative mode : case 2

At, the case of regenerative buck operation, the converter controls the DC-link voltage through duty-ratio control of switch S_3. The current path of from DC-link to battery is formed through S_3 and body diode of S_1. The buck operation is performed through operation of S_3 and S_4. The operation of S_3 and S_4 is complementary as seen in Fig. 4.

Fig. 5, 6 and 7 show the methods of charging battery. Conventional hybrid system of FCEV charges battery directly from fuel cell using the converters at the terminal of fuel cell and battery.

On the other hands, proposed converter charges battery indirectly through DC-link. Fig. 5 shows the power flow of battery charging from fuel cell to battery. The battery is charged from the DC-link which is charged from fuel cell before. It is possible when the FCEV is activated, but not driving. Fig. 6 represents the power flow of regenerative mode at the battery input. The required torque reference is followed from the battery. And the battery is charged at regenerative braking.

Fig. 7 shows the power flow of regenerative mode at the fuel cell input. In this case, the fuel cell should be in the range of speed that allows the inverter to output a torque reference. At regenerative braking torque is input, the back electromotive force generated by the motor charges the battery.

V. SIMULATION RESULT

TABLE I
PARAMETER OF CONVERTER FOR FCEV DRIVING SYSTEM

Meaning	Value
Battery	96[V]
Fuel Cell	48[V]
Inductor	1.014[mH]
Switching Frequency	20[kHz]

TABLE II
PARAMETER OF IPMSM

Meaning	Value
L_d	0.303[mH]
L_q	0.907[mH]
Pmflux	0.045501[Wb]
Pole	8[pole]
Rated Current	46[A]
Torque Max	14[Nm]
Stator resistance	0.072[Ohm]

In order to verify the reliability of proposed system, the simulation was performed controlling various torque. The waveforms of simulation results shows the variable speed, and DC-link voltage corresponding to the torque and load torque.

The simulation parameter of converter for FCEV driving system and IPMSM are shown in the table.

Fig. 8 and 9 show the simulation results representing variable DC-link control depending on torque control of FCEV which describes driving mode in the flat road and Fig .10 is the simulation waveforms representing FCEV driving in uphill and downhill which is describe by load torque.

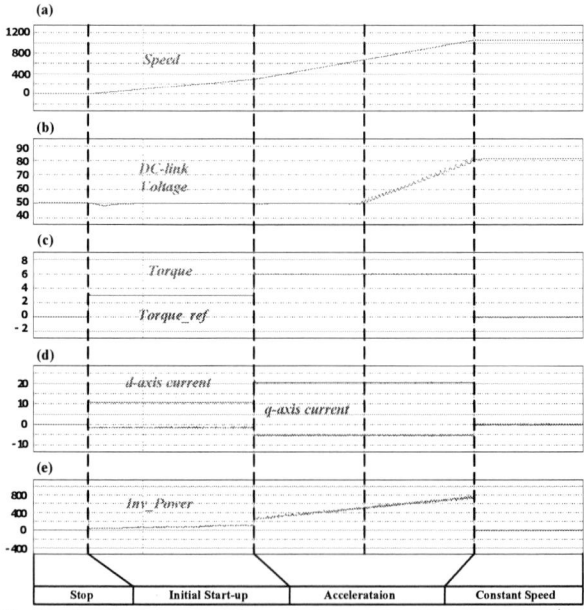

Fig .8. The simulation waveforms under variable torque at motoring mode (a) Motor speed, (b) DC-link voltage, (c) Torque, (d) d-axis, q-axis current, (e) Power of Inverter

The 2014 International Power Electronics Conference

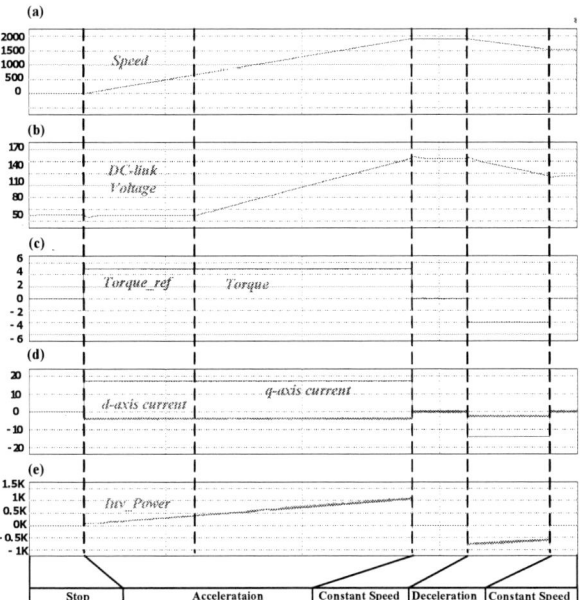

Fig .9. The simulation waveforms under variable torque including regenerative mode (a) Motor speed, (b) DC-link voltage, (c) Torque, (d) d-axis, q-axis current, (e) Power of inverter

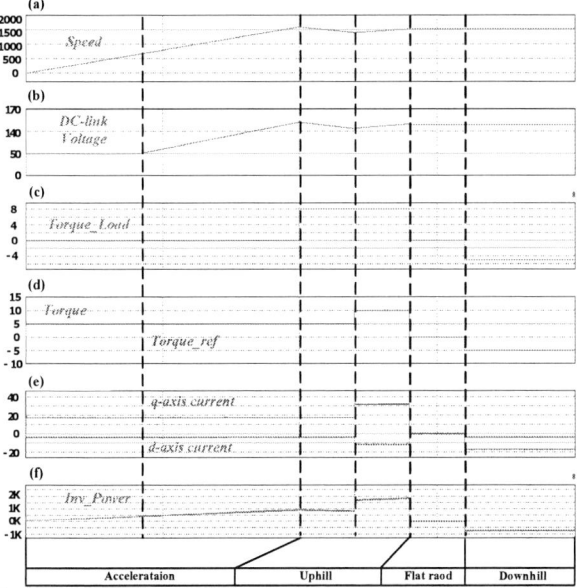

Fig .10. The simulation waveforms under variable torque under load torque (a) Motor speed, (b) DC-link voltage, (c) Load Torque, (d) Torque, (e) d-axis, q-axis current, (f) Power of inverter

As seen in Fig. 8, when the FCEV is initially started-up, dc-link voltage is constant 50[V] through buck operation. The output torque for initial start-up of FCEV is shown in Fig. 8(b). Due to the gradual increasing of motor speed after initial start-up, the DC-link voltage starts to increase when the motor speed reaches 700[rpm] in acceleration section. The torque references of initial start-up and acceleration section are 3[Nm] and 6[Nm]. Depending on the output of the torque reference, the power of inverter is positive value.

Fig. 9 is the waveforms including regenerative braking control.

Regenerative braking torque control is performed at 1900[rpm]. The torque for initial acceleration is 5[Nm]. Compared with Fig. 9 and Fig. 10, the inclination of increasing speed is different because of different torque reference. But it can be confirmed that the DC-link voltage changes at the same speed; 700[rpm].

The phase of the q-axis current is opposite and the power of inverter is negative value comparing to the acceleration because of breaking torque control.

Fig. 10 is the simulation waveforms representing FCEV driving in uphill and downhill which is described by load torque.

In the acceleration section, there is no load torque reference representing driving at flat load. After the load torque reference describing uphill is input, the speed is gradually decreased. The load torque representing uphill is 8[Nm].

But, the speed is increased again after input 10[Nm] torque reference. At the Downhill, negative load torque was input but the speed of FCEV is constant because of negative torque reference; -5[Nm]. In this section, the power of inverter has negative value due to the regenerative braking.

VI. CONCLUSIONS

In this paper, the bi-directional DC-DC converter controlling variable DC-link for FCEV is presented. By using this converter, DC-link voltage is controlled variable corresponding to the torque reference and speed.

The regenerative operation modes of converter and torque control simulation are given. Simulation results prove that the proposed converter is a promising approach for electric vehicle application.

ACKNOWLEDGMENT

This work was supported by Samsung Electronics Co. Ltd..(No. 1000871)

REFERENCES

[1] Blaabjerg, F., Pedersen, J.K., Thoegersen, P. "Improved modulation techniques for PWM-VSI drives" IEEE Trans Vol. 44, pp 87-95, 1997.

[2] Zhang Xuhui, Xuhui Wen, Zhao Feng, Guo Xinhua. "A new control strategy for bi-directional DC-DC converter in electric vehicle" Electrical Machines and Systems. pp 1-4, 2011.

[3] F.Laurencelle, R. Chahine, J. Hamelin, K. Agbossou, M. Fournier, T.K.Beso , A. Laperriere. "Characterization of a Ballard MK5-E Proton Exchange Membrane Fuel Cell Stack," Fuel Cells Journal, vol 1, no. 1,pp. 66-71, 2001.

[4] Haiping Xu, Gang Ma, Changfu Sun, Xuhui Wen, Li Kong. "Implementation of a bi-directional DC-DC converter in FCEV" Electrical Machines and Systems, pp 375-378, Nov, 2003.

[5] Seon-Hwan Hwang, Jang-Mok Kim "Dead Time Compensation Method for Voltage-Fed PWM Inverter" IEEE Trans. pp1-10, March, 2010.

[6] An-Yeol Ko; Do-Yun Kim; Jung-Pill Hwang; Il-Kuen Won; Chung-yuen Won, "Design of Bidierctional DC-DC Converter using Optimal Control DC-link for FCEV, pp102-103, July, 2013.

Large Driving Range Increase of Series Chopper Based Power Train Using Motor Test Bench

Yu Hosoyamada
Yokohama National
University,
Yokohama, Japan
Email:hosoyamada-yu-vm
@ynu.jp

Masashi Takeda
Yokohama National
University,
Yokohama, Japan
Email:takeda-masashi
@ynu.jp

Naoki Motoi
Kobe University,
Kobe, Japan
Email:motoi@maritime.
kobe-u.ac.jp

Atsuo Kawamura
Yokohama National
University,
Yokohama, Japan
Email:kawamura @ynu.ac.jp

Abstract—Authors proposed the series chopper based power train aimed for the extension of the driving range. In previous paper [1], under the condition of the constant DC-supply voltage, effectiveness of the optimization of the profile of the chopper output voltage was shown. However, the output voltage of the battery used in the actual EV varies depending on the state of charge (SOC). Because of this difference, the precise comparison of the power train was not performed. In this paper, by emulating the battery including the internal resistance and output voltage dependency on SOC, the driving range is compared in the motor test bench. It is shown that the chopper output voltage adjustment depending on the rotor speed and the weight of the chopper has large effect on the driving range. In comparison to non-chopper power train, the driving range of the series chopper power train increases by 2.1%. In addition, by decreasing the weight of the chopper by a half and zero, the driving range further extends by 1.4% and 3.0%.

I. INTRODUCTION

The use of the Electrical Vehicle (EV) is said to save total energy consumption [2]. The ultimate goal of the EV is to run a longer distance. Considering the present battery storage capability technology and cost, it is said that the EV is suitable for the urban use including short distance commuter [3][4]. From the view point of practical use of commuter EV's in a city area, the urban driving pattern such as JC08 mode in Japan or LA-4 mode in USA gives a prediction of actual driving range in a city area, thus the measurement regulation is strictly controlled by laws [5]. Under such a driving mode, the total driving distance of the EV can be treated as a function of multiple factors, such as the efficiency of each component, battery characteristics, and the total system design.

The power train system design has a large influence to the total driving distance. Therefore, many researchers developed and proposed the systems that have advantage over the conventional power train as shown in Fig. 1 [6]. In the paper, Fratta *et al.* proposed the system that the boost converter is inserted between the battery and the inverter. In this system, the EV can also run at high speed region even if the SOC (State of Charge) is low. However, the minimum voltage of the inverter input voltage is fixed to the battery voltage.

Therefore, authors proposed the series chopper based power train using the buck-boost chopper as shown in Fig. 2. This system has the equivalent characteristic and the efficiency

of the inverter can be increased by step-down conversion of the chopper at the low speed region. In the past research [8], authors were experimentally compared between with and without the chopper by the use of the actual EV. However, the condition of the experiment has some problems. For example, the cost of the experiment and the scale of the experimental facility are very large. From this viewpoint, authors constructed the motor test bench that had the same characteristic of the actual EV [7].

In the previous paper [1], the profile of the chopper output voltage was optimized for high efficiency. In addition, the decrease of the energy consumption was shown. However, in this previous paper, the DC-supply voltage was constant. Thus, the driving range per charge could not be compared. In the series chopper based power train, the increase of weight of the chopper is an important factor. However, in the same way, the effect of weight of the chopper to the driving range per charge could not be also compared.

In this paper, by emulating the Li-ion battery, the DC-supply voltage becomes variable. Therefore, when the profile of the chopper output voltage or weight of the chopper is varied, the variation of the driving range per charge can be compared. From the comparison, without making large-scale experiments of the actual EV, it is possible to obtain new knowledge regarding variations of the chopper weight and the profile of the chopper output voltage.

This paper is organized as follows. In the section II, the power train is described. Section III shows the equipment of the motor test bench and the comparison between the motor test bench and actual EV. Section IV shows the effects of profile of the chopper output voltage and weight of the chopper with the series chopper based power train. Finally, section V concludes this paper.

II. POWER TRAIN FOR EV

The typical EV power train is battery-inverter-motor combination as shown in Fig. 1. The DC voltage is converted to three phases by an inverter and it drives a motor. This kind of power train is called "non-chopper type" power train in this paper. This power train has advantages because of the simplicity and well-established technology. However, from the view point of energy saving, this power train has disadvantages. At the high speed region, the high AC voltage is required

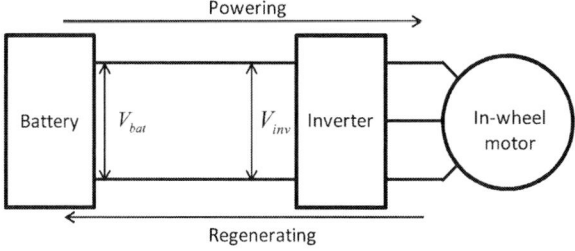

Fig. 1. Non-chopper type power train (Conventional power train)

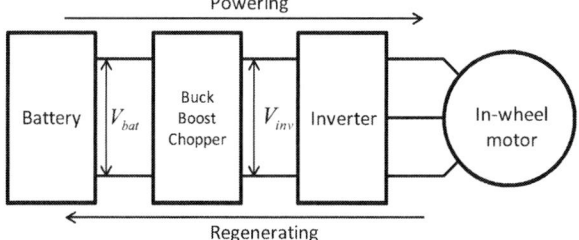

Fig. 2. Series chopper based power train

for a motor. However the inverter input voltage is fixed to the battery voltage. Therefore, the minimum voltage of the battery is designed based on the voltage required to drive at the high speed region. On the other hand, at the low speed region, the high AC voltage is not required for a motor. Therefore, the inverter is applied the extra voltage, and the efficiency of the inverter and the motor decreases. In addition, at the low SOC and high speed region, the lack of the voltage is compensated by the flux weakening control [9]. However, as the battery voltage is dropped, this compensation is not achieved. Thus, the maximum speed is dropped as shown in Fig. 3.

If the DC link voltage is variable, the motor and inverter design can be optimized. Then, the higher efficiency of the operation is expected at the both of the low and high speed region. From these considerations, the series chopper based power train was proposed as shown in Fig. 2.

In the system, the DC-DC chopper is required to have a function of the buck-boost, high efficiency, compactness, light weight, and high reliability. The SAZZ (Snubber Assisted Zero voltage Zero current transition chopper) is one of candidates [10]. The SAZZ has these characteristics. In the literature, 98% efficiency was reported as four quadrant operation and 25kW output power. Also a very high power density was reported in the previous paper [11], and about 30kW/l was achieved.

III. MOTOR TEST BENCH [7]

A. System of motor test bench

In this subsection, the system of the motor test bench is explained. The motor test bench consists of a chopper, a inverter, a motor, and DC supply as a battery. The experimental equipment and the diagram of this system are shown in Figs. 4, 5 and 6. And, the specification of each equipment is shown

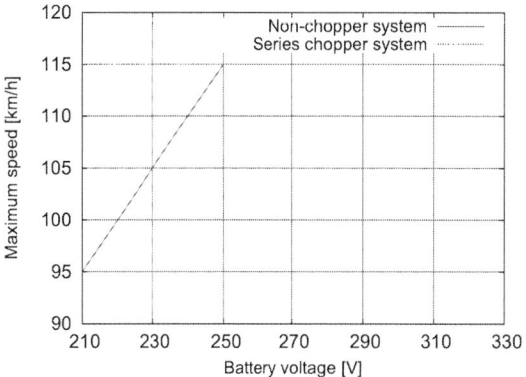

Fig. 3. Battery voltage and the maximum motor speed

in TABLE I, II and III.

The motor test bench is modeled based on EV "KANA" as shown in Fig. 7 [8]. The EV is composed of the battery, the chopper, the inverter, and the motor. On the other hand, the motor test bench has the DC-supply, the chopper, the inverter, and the motor. There is a difference of DC-supply and the battery between the motor test bench and the actual EV. For this reason, the DC-supply needs to be emulated the Li-ion battery.

In order to emulate the Li-ion battery, the system of DC-supply has the two characteristics. First, the voltage change corresponding to the discharge capacity is emulated. The discharged capacity of the DC-supply output is measured by the power analyzer and the data is sent to the computer. On the computer, the DC-supply output voltage is calculated by the look-up table. At that time, the double discharged capacity is used in the calculation. At the second, the internal resistance of the battery is emulated. There is a difference of the internal resistance between the DC-supply and the Li-ion battery. In order to eliminate the error of the internal resistance, the resistance (0.4Ω) is inserted in series between the DC-supply and the chopper as shown in Fig. 6.

In the motor test bench, the driving resistance such as the acceleration and deceleration force, the friction force and the aero-dynamic force does not exist. Therefore, the load motor is used in order to emulate the driving resistance. At this time, the total driving resistance torque is calculated from the function of the acceleration and deceleration force F_{acc}, the friction force F_{rol}, and the aero-dynamic force F_{air} from (1) to (3). The load torque is calculated by (4) and TABLE IV. In addition, the load torque is used as the command of the load motor. By emulating the load torque generated by the load motor, the variable load torque according to the change of the weight of the chopper and the EV can be achieved.

$$F_{acc} = M\frac{d(r\omega_m)}{dt} \tag{1}$$

$$F_{rol} = \mu M g \tag{2}$$

$$F_{air} = \frac{1}{2}\rho A C_d (r\omega_m)^2 \tag{3}$$

$$T_{load} = r \times F_{all} = r \times (F_{acc} + F_{rol} + F_{air}) \tag{4}$$

The 2014 International Power Electronics Conference

Fig. 4. Experimental equipment in motor test bench

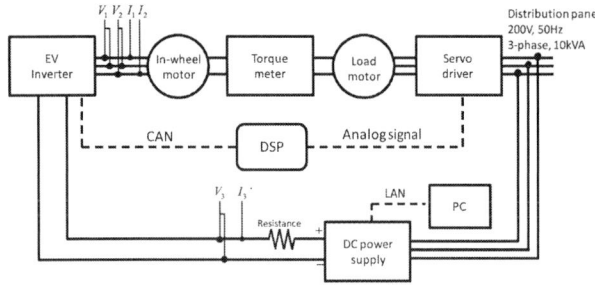

Fig. 5. Diagram of motor test bench without chopper

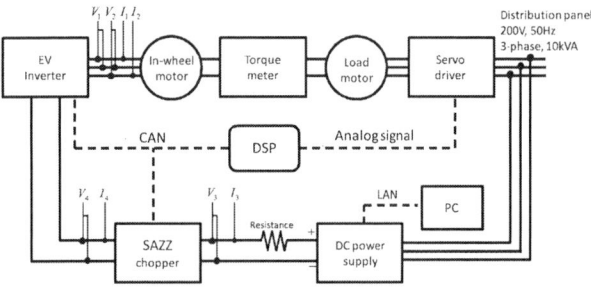

Fig. 6. Diagram of motor test bench with chopper

Fig. 7. EV "Kana"

where ω_m is angular velocity of the motor, T_{load} is the load torque, F_{all} is total driving resistance force, and other variables are shown in TABLE IV.

TABLE I. SPECIFICATIONS OF IN-WHEEL IPM MOTOR

Maximum output power	20kW(382rpm,20sec)
Maximum torque	500Nm(20sec)
Rated output power	6kW(521rpm)
Rated torque	110Nm
Rated speed	1113rpm(115km/h)

TABLE II. SPECIFICATIONS OF INVERTER

Control method	Vector control
Switching frequency	6kHz
Signal	CAN

TABLE III. SPECIFICATIONS OF SAZZ CHOPPER

Rated output power	8kW
Maximum output power	25kW(20sec)
Rated voltage(Battery side)	300VDC(240-370VDC)
Rated voltage(Inverter side)	400VDC(20-420VDC)
Switching frequency	25kHz
Dimensions	339mmW × 170mmH × 350mmD

TABLE IV. PATAMETER OF THE EV AND THE MOTOR TEST BENCH

Rolling resistance	μ	0.00944
Coefficient of air resistance	C_d	0.33
Frontal projected area	A [m^2]	1.648
Radius of tire	r [m]	0.275
Atmospheric density	ρ [kg/m^3]	1.23
Gravitational acceleration	g [kg/s^2]	9.8
Inertia	J [kgm^2]	0.45
Coefficient of viscous friction	D [kgm^2/s]	0.065
Weight of car	M_{car} [kg]	834
Weight of chopper	M_{chop} [kg]	38

B. Comparison methods between motor test bench and chassis dynamometer experiment

This part shows the comparison between the motor test bench and the actual EV by running the JC08 mode. JC08 mode is the driving pattern that is regulated by Ministry of Land, Infrastructure, Transport and Tourism in Japan. The speed command of the driving pattern is shown in Fig. 8.

In the previous paper [8], the experimental results by using the actual EV were shown. In this experiment, the EV ran JC08 mode on the chassis dynamometer. The chassis dynamometer is an instrument used to measure power by running on the roller. This experiment is divided into two cases as shown in TABLE V. Case 1 is the non-chopper type power train. Case 2(a) is the series chopper based power train. In the series chopper system, the motor test bench is repeated running of JC08 mode until SOC of the battery is reduced from 100% to 0% (from 334V to 208V). In non-chopper system, as shown in Fig. 3, the maximum speed of the specifications of the EV cannot be achieved if the battery voltage becomes below 250V. Thus, the motor test bench is repeated running of JC08 mode until the battery voltage is reduced from 334V to 250V. Under these conditions, these driving ranges per charge are compared between the results from the motor test bench and those from the actual EV.

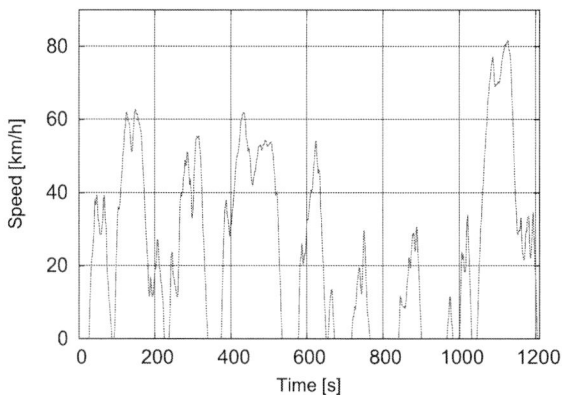

Fig. 8. Driving pattern of JC08 mode

TABLE V. COMPARISON METHODS OF THE DRIVE SYSTEM

	Power train	Total weight [kg]
Case 1	Non-chopper	M_{car}
Case 2(a)	Series chopper	$M_{car} + M_{chop}$

TABLE VI. RESULTS OF COMPARISON BETWEEN THE MOTOR TEST BENCH AND THE CHASSIS DYNAMOMETER EXPERIMENT

	Results from motor test bench		Results from actual EV	
	Distance	Ratio in Case 1	Distance	Ratio in Case 1
Case 1	86.3 km	100.0 %	85.6 km	100.0 %
Case 2(a)	86.8 km	100.5 %	85.8 km	100.2 %

C. Comparison results between motor test bench and chassis dynamometer experiment

The results from the motor test bench and the actual EV are shown in TABLE VI. From these results of the driving range per charge, the difference of Case 1 between the motor test bench and the actual EV was about 0.8%. And, the difference of Case 2(a) between the motor test bench and the actual EV was about 1.2%.

From these results of the ratio in driving range to one of Case 1, these values of Case 2(a) of the ratio in Case 1 are respectively 100.5% and 100.2% in the motor test bench and the actual EV. Thus, the difference of Case 2(a) was about 0.3%. Therefore, similar characteristics are confirmed.

Next, these results of the energy consumption are shown in Fig. 9 and 10. From these results, the difference becomes large. In non-chopper system, there is the small difference of about 2Wh between the actual EV and the motor test bench. On the other hand, in the series chopper system, there is the large difference of about 10Wh. The reasons of these errors are due to a little difference of some equipment. The inverter used in the motor test bench is different from the one used in the actual EV. In addition, the chopper efficiency in itself is also different. And, there is the difference of the chopper efficiency due to the variation of the output power between the right wheel and the left wheel. Except for the error by the difference of these devices, the shapes of the each energy consumption shows the same characteristic. Therefore, the motor test bench has the compatibility to the actual EV.

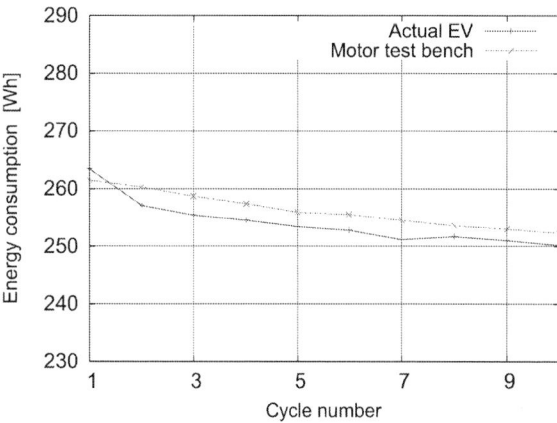

Fig. 9. Comparison of energy consumption without chopper between the motor test bench and the chassis dynamometer experiment

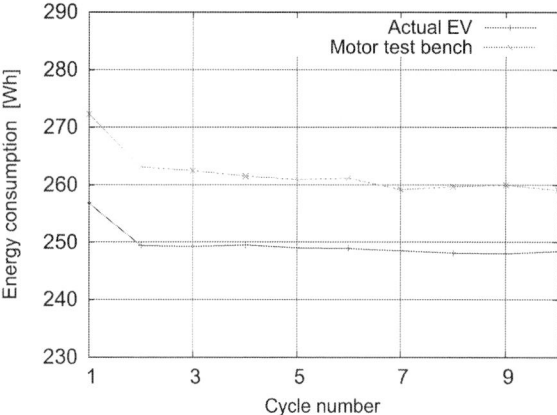

Fig. 10. Comparison of energy consumption with chopper between the motor test bench and the chassis dynamometer experiment

IV. VERIFICATION OF NEW KNOWLEDGE USING MOTOR TEST BENCH

This section shows the effects of the optimized DC voltage profile and the weight of the chopper.

A. Optimization of DC voltage profile

Fig. 11 (iii) shows the minimum voltage to achieve running the speed when the load torque is equal to zero. Fig. 11 (i) is the profile used in the actual EV. In the high speed range, the chopper output voltage is lower than the minimum voltage of the achieve running the speed. Therefore, the inverter was required to be compensated by the flux weakening control. Thus, the copper loss was increased, and the efficiency of the motor was decreased. In the previous paper [1], authors experimentally optimized the profile of the chopper output voltage by using the motor test bench. This optimization requires to the following two experiments.

978-1-4799-2706-7/14 $31.00 © 2014 IEEE

The 2014 International Power Electronics Conference

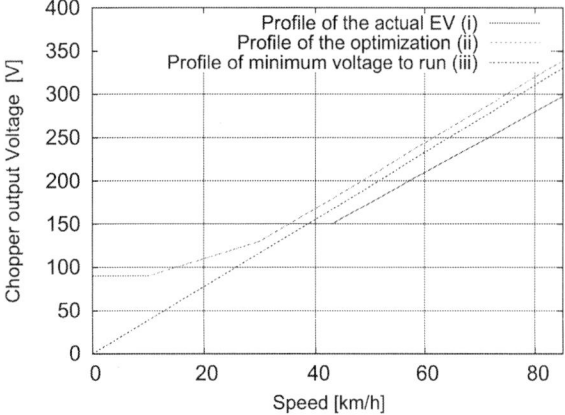

Fig. 11. Chopper output voltage profile

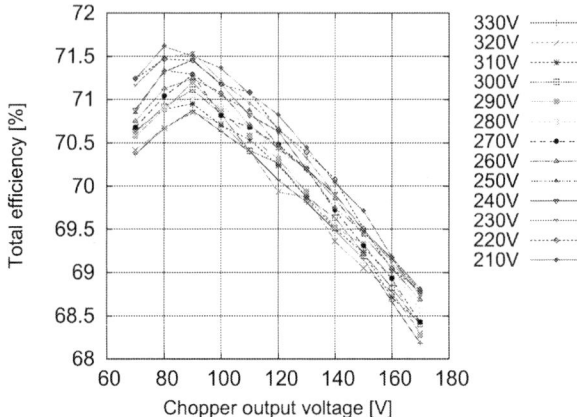

Fig. 12. Total efficiency of inverter and motor and chopper at Pm=1kW

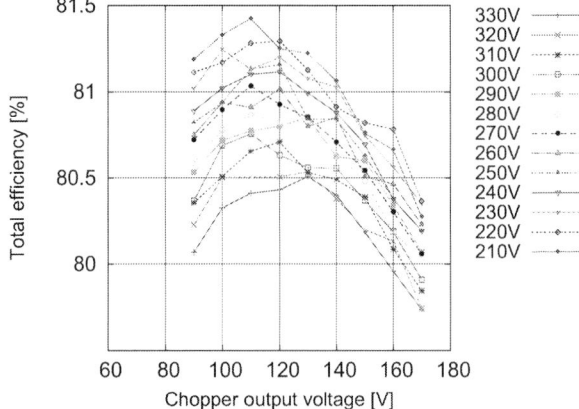

Fig. 13. Total efficiency of inverter and motor and chopper at Pm=2kW

(1) Measurement test of efficiency of inverter and motor

(2) Measurement test of efficiency of chopper

In the measurement test (1), the total efficiency of the inverter and the motor is measured. In that time the speed reference is increased from 10km/h to 70km/h and the torque is increased from 10Nm to 70Nm. And the inverter input voltage is changed from 10V to 330V. However, the inverter input voltage of the highest efficiency was the minimum voltage to achieve running the speed. So, in this paper, this optimization characteristic is used.

In the measurement test (2), if the stepping down ratio is too large, the efficiency of the chopper drops significantly. Therefore, only in the low speed range, the chopper efficiency is measured by changing the value of chopper input voltage and that of chopper output voltage. The chopper input voltage is changed from 210V to 330V by 10V steps. And the chopper output voltage is also changed in the range to achieve running the speed. From these results, the optimized DC voltage profile is obtained by connecting the highest efficiency voltage.

These experimental results of the total efficiency of the chopper, the inverter and the motor are shown in Figs. 12 and 13 when the speed is 10km/h and 20km/h. And from these figures, the optimized DC voltage profile is obtained and shown in Fig. 11 (ii).

B. Experimental comparison methods

TABLE VII shows the comparison candidates of the experiments of this section. In Case 2(a), Case 2(b) and Case 2(c), voltage profile (ii) as shown in Fig. 11 is used. And they differ in the chopper weight such as a half and zero. On the other hand, in Case 3(a), Case 3(b) and Case 3(c), voltage profile (i) as shown in Fig. 11 is used. And they also differ in the chopper weight.

By comparing to these results, the effects of these following effects are verified.

- The effects of the optimized DC voltage profile

- The effects of the weight of the chopper .

C. Experimental results

The results of the driving range per charge and the ratio in Case 1 are shown in TABLE VIII. And, the results of the driving range per charge are shown in Figs. 14 and 15.

From these results, by decreasing the weight of the chopper, it was confirmed that the driving range per charge became longer. From Figs. 14 and 15, the energy consumption was changing almost linearly with respect to the change of the weight of the chopper. And, from results of the ratio of driving range per charge in Case 3(a), the ratio of Case 3(b) and Case 3(c) are 1.4% and 3.0%, respectively. Therefore, the driving range per charge was also changing almost linearly with respect to the change of the weight of the chopper. However, there is some possibility that the characteristic of the JC08 mode using this experiment is affected these results. Therefore, the verification in the driving pattern of the other could be desirable.

By optimizing the profile of the chopper output voltage,

978-1-4799-2706-7/14 $31.00 © 2014 IEEE   805

TABLE VII.   COMPARISON METHODS OF THE DRIVE SYSTEM (2)

	Power train	Total weight [kg]	Profile
Case 2(b)	Series chopper	$M_{car} + M_{chop}/2$	(i)
Case 2(c)	Series chopper	M_{car}	(i)
Case 3(a)	Series chopper	$M_{car} + M_{chop}$	(ii)
Case 3(b)	Series chopper	$M_{car} + M_{chop}/2$	(ii)
Case 3(c)	Series chopper	M_{car}	(ii)

TABLE VIII.   COMPARISON OF MOTOR TEST BENCH RESULTS

	Distance	Ratio in Case 1
Case 2(b)	90.3 km	104.7 %
Case 2(c)	90.6 km	105.0 %
Case 3(a)	89.2 km	102.1 %
Case 3(b)	90.5 km	104.9 %
Case 3(c)	92.0 km	106.6 %

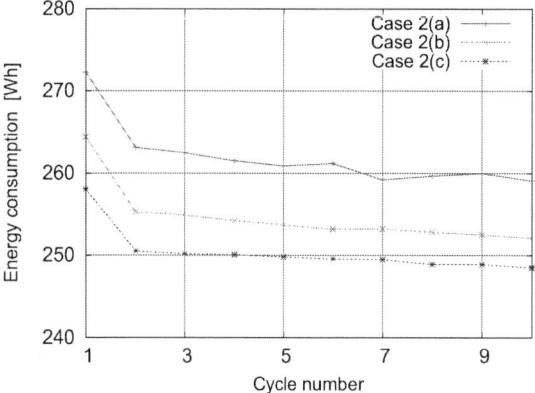

Fig. 14.   Comparison of energy consumption by change of the chopper weight in using profile (i)

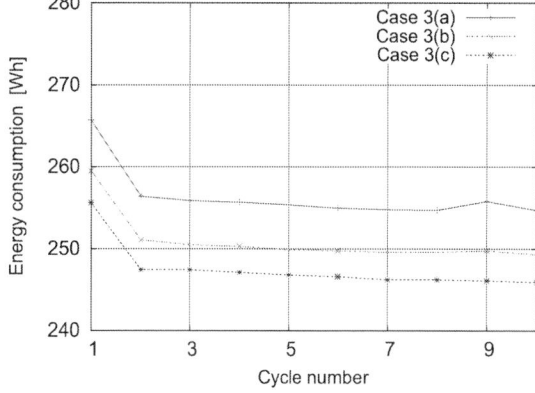

Fig. 15.   Comparison of energy consumption by change of the chopper weight in using profile (ii)

it was confirmed that the energy consumption per cycle was decreased by about 5Wh and the driving range per charge was increased by about 1.6%. And, the driving range per charge of the series chopper system was increased by 2.1% to one of the non-chopper system. If it is assumed that the weight of the chopper became sufficiently light to the weight of the electric vehicle by weight saving of the element, it is estimated that the driving range per charge of the series chopper system could be increased by 6.6% to that of the non-chopper system.

V.   CONCLUSIONS

In this paper, the driving range of the series chopper based power train was measured at the motor test bench, emulating the Li-ion battery. As the results, by optimizing the chopper output voltage profile depending on the rotor speed, the driving range of the series chopper based power train becomes 2.1% larger than that of the non-chopper based power train. In addition, by decreasing the weight of the chopper by a half and zero, the driving range further increased by 1.4% and 3.0%. This means 6.6% increase compared with the non-chopper power train. But, these data may include some error between the motor test bench and the actual EV measured at the chassis dynamometer two years ago.

REFERENCES

[1]   M. Takeda, N. Motoi, G. Guidi, Y. Tsuruta, A. Kawamura, "Driving Range Extension by Series Chopper Power Train of EV with Optimized dc Voltage Profile," *Proceedings of IEEE Industrial Electronics Society Annual Conference (IECON)*, pp. 2918-2923, 2012.

[2]   H. Shimizu, "To protect global warming," Random House Pub. Co. , 2007.

[3]   Y. Hori, "Application of Advanced Motion Control Techniques and the Prospect of Super Capacitors in Electric Vehicles", *Proceedings of IEEE International Conference on Industrial Technology*, 2006 (Plenary lecture).

[4]   T. Hosokawa, K. Tanihata, H. Miyamoto, "Development of i-MiEV Next-Generation Electric Vehicle(Second Report)," *Mitsubishi Motors Technical Review*, No.20, pp.53-60, 2008.

[5]   "TRIAS (Traffic Safety and Nuisance Research Institute's Automotive Type Approval Test Standards) 5-10-2008", *Ministry of Land, Infrastructure, Transport and Tourism*.

[6]   A. Fratta, P. Guglielmi, F. Villata, A. Vagati, "Efficiency and cost-effectiveness of AC drives for electric vehicles improved by a novel, boost DC-DC conversion structure," *IEEE Transactions on Industrial Electronics*, Vol.IE-56, NO.1, pp.361-371, 2009.

[7]   M. Takeda, Y. Hosoyamada, N. Motoi, A. Kawamura, "Development of the Experiment Platform for Electric Vehicles by using Motor Test Bench with the Same Environment as the Actual Vehicle", *The 13th International Workshop on Advanst Motion Control*, March, 2014.

[8]   A. Kawamura, G. Guidi, S. Tsutsuki, Y. Watanabe, Y. Tsuruta, N. Motoi, "Experimental data analysis on total driving performance of series chopper based EV power train," *Proceedings of IEEE Industrial Electronics Society Annual Conference (IECON)*, pp. 1348-1353, 2011.

[9]   S. Morimoto, M. Sanada, Y. Takeda, "Wide-speed operation of interior permanent magnet synchronous motors with high-performance current regulator," *IEEE Transactions on Industry Applications*, vol. 30, no. 4, pp. 920-926, 1994.

[10]   Y. Tsuruta, Y. Ito, A. Kawamura, "Snubber Assisted Zero Voltage and Zero Current Transition Bilateral Buck and Boost Chopper for EV Drive Application and Test Evaluation at 25kW," *IEEE Transactions on Industrial Electronics*, Vol.IE-56, NO.1, pp.361-371, 2009.

[11]   M. Pavlovsky, Y. Tsuruta, and A. Kawamura, "Recent Improvement of Efficiency and Power Density of DC-DC Converters for Automotive Applications," *Proceedings of International Power Electronics Conference*, Vol. 23, pp. 1866-1873, 2010.

The Power Electronics Program at Beijing Jiaotong University

Fei Lin, Zhongping Yang, and T.Q. Zheng
School of Electrical Engineering
Beijing Jiaotong University
Beijing, China

Abstract— **This paper introduces the undergraduate and graduate program in power electronics engineering at Beijing Jiaotong University (BJTU).The curriculum in this area has been designed and developed for specifically educational goals. Unique feature of the program is the application of power electronics in railway and renewable energy area. The curriculum is supported by case studies, design problems, plant visits. This paper also reviews the activities of the power electronics area graduate and research programs.**

Keywords— Power Electronics; Specialty Courses; Specialty Elective Courses.

I. INTRODUCTION

Since the establishment of "Advanced Electrical Engineering Class A" by the predecessor of Beijing Jiaotong University—Institute of Jiaotong Heritage and Learning in 1912, the discipline of Electrical Engineering has experienced the vicissitudes, and developed nearly a century. After 1949, Beijing Railway Institute established the Department of Electrification Engineering in accordance with the needs of the national economic construction in 1958. In 2000, the current School of Electrical Engineering was established at Beijing Jiaotong University. There are 132 staff members in the school, 24 professors and 43 associate professors. The school can culture professionals at all levels from undergraduate students to postgraduate students, from bachelor's degrees to doctorate degrees.

There are five majors in this school: power electronics, power systems, electrical machines, high voltage and Electrical theory and new technology. Traditionally, the power electronics is one of the key major, which has been named as the Provincial Key Disciplines. In 2005, the school set up a special discipline (direction) of Rail Transportation Electrification, and was honored as the National Academic Specialty in 2009. Actually, this major is almost the application of power electronics in railway area. Replying on education and research, the Research Center of the Electric Traction Engineering of the Ministry of Education, the Research Center of National Wind Power Engineering and Technology, National Demonstration Center of the Electro-techniques and Electronic Practice Teaching have been established in these years.

In china, the schools of Electrical Engineering in universities agreed to that is the growing importance of power electronics, specifically, the education of future professionals on undergraduate and graduate levels. The Beijing Jiaotong University (BJTU) has traditionally recognized the need of a program in Power Electronics and continued to provide courses in the area since 1980's, which is one of the first universities in China. The purpose of the program at BJTU is to train the future engineers in the field of Power Electronics. This includes analysis, design, and, to some extent, research. Particularly, there are some elective courses about application of power electronics in some areas related to rail transportation and renewable energy. These objectives are achieved by utilizing a curriculum which emphasizes the basic laws with principle background information and the application method with examines current engineering problems. In other words, we pay attention to combine the theory with practice. As the power electronics is the fast developing engineering, modern trends and the state of the art are also added continuously in these courses. Power Electronics Project Laboratory course ensure the practice opportunity for students in the area.

A group of faculty actively interested in power electronics is responsible for teaching and research in this area. One of the recent trends in engineering education in China is the reduction in the number of required core courses and the increase in the selection of elective courses. The curriculum at BJTU reflects this trend. The curriculum defines a combination of required core courses and a number of elective courses. This type of curriculum offers the possibility of foundation knowledge but also permits greater breadth for related applications. Thus, a program can meet specific employment objectives, such as railway traction and renewable energy systems, while ensuring that graduate student has the necessary broad fundamental background.

II. EDUCATIONAL PROGRAM

A. Required Course

Should any students in our school want a minor of similar discipline, they shall complete required specialized courses or restricted elective specialized courses as per the minor program. Students of other schools who pursue the minor of our specialty are required to complete the courses designated under this program on top of successful fulfillment of credit requirement of their major program.

There are nine required courses for all the electrical engineering students as table 1. Power electronics is one of the required courses proceeded in semester 6.

TABLE I
NINE REQUIRED COURSES IN ELECTRICAL ENGINEERING

Course Name	Credits	Hours Distribution		
		Total	Lecture	Practice
Electric Circuits	6	96	82	14
Analog Circuits	4	64	48	16
Digital Circuits	4	64	48	16
Engineering Electromagnetic Field	3.5	56	54	2
Automation Control	4	64	56	8
Microcomputer Principles and Interface Technology	3.5	56	40	16
Electric Machinery	5	80	72	8
Power Electronics	4	64	58	6
Power System Analyses	4	64	56	8

Power Electronics is one of the major courses of Electrical Engineering discipline. It is a branch of electronics associated with the use of semiconductor switch components to efficiently convert electric power of various inputs into high quality electrical output for industrial applications such as powering precise instruments, driving motors, performing process control. This course covers: power semiconductor devices of power diode, thyristor, MOSFET and IGBT in terms of electrical symbols, features and working principles; composition of power losses, driving characteristics and safety and driving norms; basic circuit topologies of DC/DC, DC/AC, AC/DC and AC/AC conversions and theorems of operation, features of performance and methods of control power electrical converters in terms of absorbing circuit, PWM, harmonic suppression and power factor improvement technologies.

B. Elective Course

In the semester 6 and 7, there are some elective courses for student who want to study further in a certain major. For power electronics area, seven elective courses are provided for students as table 2 shown. At least 10 credits should be completed for every students in power electronics major.

TABLE II
SPECIALTY ELECTIVE COURSES-POWER ELECTRONICS

Course Name	Credits	Hours Distribution		
		Total	Lecture	Practice
Applications of Power Electronics	2	32	32	
High Power Conversion	2	32	32	
Switching Power Supply	2	32	32	
AC Drives	2	32	28	4
Electric Drive and Automatic Control System	2	32	28	4
Power Electronics Simulation Technology	1	16	16	
Integrated Design for Electrical Engineering	2	Two weeks		Two weeks

We also offer other two direction courses, Rail

Traction Electrification and Renewable Energy and Its Power Generation. Most of these course are related to power electronics.

TABLE III
SPECIALTY ELECTIVE COURSES- RAIL TRACTION ELECTRIFICATION

Course Name	Credits	Hours Distribution		
		Total	Lecture	Practice
Power Traction Drive and Control	3	48	32	16
Rail Transit Traction Power Supply	3	48	40	8
Traction substation and its Automation	2	32	30	2
Electrical Equipment of Locomotive	2	32	32	
Catenary Principles and Technology	2	32	32	
Integrated Design for Electrical Engineering	2	Two weeks		Two weeks

TABLE IV
SPECIALTY ELECTIVE COURSES- RENEWABLE ENERGY AND ITS POWER GENERATION

Course Name	Credits	Hours Distribution		
		Total	Lecture	Practice
Wind Farm Construction, Lightning Protection and Grounding	2	32	32	
Principles and Control of Wind Turbine	2	32	30	2
Photovoltaic Generation	2	32	32	
Renewable Energy Intergration into Power Systems	2	32	28	4
Integrated Design for Electrical Engineering	2	Two weeks		Two weeks

C. Course for Graduate Students

A more specialized advanced course sequence limited to graduate students in power electronics is also available. This includes the advanced power electronics technology , digital control of power electronics, modern energy conversion, motor control, rail traction control systems, which deal with the more specialized knowledge and new applications of power electronics. Computer simulation software and experimental workbench are also established for graduated students.

III. EDUCATIONAL TECHNIQUES AND FACILITIES

Practical training is also available in addition to theoretical education, through laboratory courses, research and design assignments. The design of sequences makes extensive studies possible in areas that represent the more significant sectors in the power electronics.

To better prepare the electrical engineers of the future, all juniors are required to take a lecture and laboratory course in power electronics or power systems.

Two laboratories are opened 24 hours a day and 7 days a week. A total of six faculty members are responsible for these labs.

Design and case studies are used to provide an in-depth analysis of various power electronics engineering concepts.

Fig.1　One of the labs

IV. RESEARCH AND EDUCATION

A. Power Electronics in Railway and Urban Transit

As the international trend to the global friendly environment, the world is focusing on rail vehicles for the high energy efficiency and high capacity. In this situation, the development of railway and urban transit vehicles is fast in China recently. The demand for high-speed train, heavy-load locomotive and urban transit train is growing significantly. The power electronics equipment plays an important role in rail vehicles.

(1) *Traction control*

The professors and graduated students developed the traction control systems with NSR used in China's AC driven locomotive.

Fig.2　China's first AC motor driven AC4000 electric locomotive

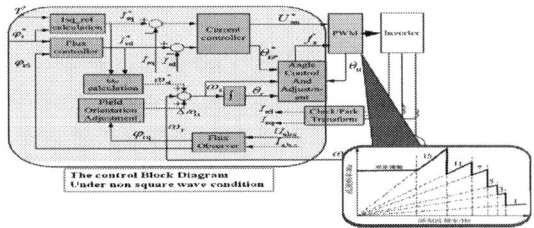

Fig.3　The control block diagram

(2) *Auxiliary converter*

Since 1990s, a group mainly composed of graduated students in power electronics researched and developed continuously for auxiliary converter. These converters have been used in DC and AC locomotive for many years.

 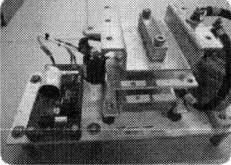

Fig.4　The auxiliary converter for HXD2

(3) *LIM for urban trasit*

Steel wheeled linear motor driven metro system was employed in Guangzhou Metro Line 5 operated in 2005 and Beijing Airport Express operated in 2008. The large air gap between the primary air space and secondary air space of LIM lowers the efficiency of the motors by 30%. The High Temperature Superconducting Linear Induction Motor (HTS LIM) has higher power density, higher efficiency, lower noise than regular LIM. To explore the potential for applying HTS LIM in rail transit and to accumulate more experience of HTS LIM design and control, a research into HTS LIM was carried out by Beijing Jiaotong University.

Fig.5　The prototype of HTS LIM

(4) *EDLC energy storage system for urban transit*

In recent years, China's railway sector pays more attention to the R&D of energy conservation technology, in which the energy storage devices used in urban rail transit system serves as a key point. A 200kW EDLC energy storage system was developed at BJTU for Beijing metro company.

Fig.6　The platform of EDCL energy storage system

978-1-4799-2706-7/14 $31.00 © 2014 IEEE

B. Power Electronics in Renewable Energy Systems

(1) Battery charger and EMS

The Charger & Battery Experimental Platform is the characteristic of our school. One of the contributions is that the Charging System for EV developed by our school has gained wide acceptance and applied in Olympic games, EXPO and Asian Games, successfully.

Fig.7　The platform of battery EMS

(2) High power wind turbine converter

1.5MW and 3MW wind turbine converters have been developed in BJTU for DFIG and PMSG.

Fig.8　Wind turbine

(3) PV Systems

The graduated students developed up to 600kW grid side inverter for PV systems.

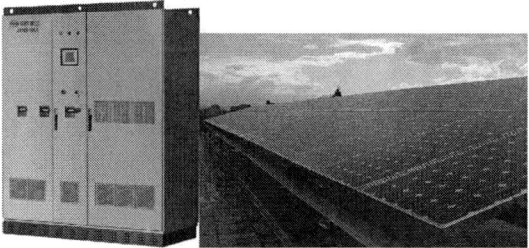

Fig.9　The platform for PV systems at BJTU

C. Integration of research and education

Both required and elective course topics have been developed with some innovative features and characteristics. The purpose as to promote the accumulation of new knowledge and to integrate research results into the curriculum while maintaining a high level of competency in power electronics engineering education.

V. CONCLUSION

In conclusion, the main objective of the power electronics program at BJTU is to provide industry and graduated school with engineers or graduated students technically competent in both theory and application.

Efforts for Power Electronics Education in A Start-Up Company

Fumiya Hattori
POWERELE ACADEMY CO., LTD.
Matue, Shimane, Japan
hattori@powerele-academy.co.jp

Jun Imaoka
Interdisciplinary Faculty of Science
and Engineering
Shimane University
Matue, Shimane, Japan
s139901@matsu.shimae-u.ac.jp

Manabu Ishitobi
Electrical Engineering
Nara National Collage of Technology
Yamatokoriyama, Nara, Japan
ishitobi@elec.nara-k.ac.jp

Shinichiroh Nagai
PONY ELECTRIC CO., LTD.
s-nagai@pony-e.jp
Yokohama, Kanagawa, Japan
s-nagai@pony-e.jp

Masayoshi Yamamoto
Interdisciplinary Faculty of Science and Engineering
Shimane University
Matue, Shimane, Japan
yamamoto@ecs.shimane-u.ac.jp

Abstract— In recent years, an energy issue is becoming serious or severe. Therefore, power electronics technology is one of the key technologies since this technology can deal with electric energy efficiently, and some company tries to introduce this technology. Our company offers the opportunity to study this technology such as a circuit topology, designs of inductor or transformer with practical experiments. In this paper, we present an effort of our company for power electronics education and introduce educational kits from an associated company.

Keywords— *a start-up company, power electronic, education*

I. INTRODUCTION

Development related to energy, especially, electrical energy, is currently world worldwide topic. Energy-saving of electrical appliance, hybrid vehicle, fuel cell vehicle, renewable energy, and other things are developed and researched[1]-[3]. We believe that its core technology is Power Electronics. Power electronics is leading-edge technology with power semiconductor devices, which needs high know-hows and experiments. Therefore, a job offer to power electronics engineers substantially increase even for a company selling power semiconductor devices and passive component such as a capacitor and inductor. In other words, power electronics education is required more and more. In the first step, beginners usually read a power electronics text book and analysis the state of circuitry with a pen and paper. Thereafter, in order to understand a power electronics circuitry deeply, a simulation soft-ware is necessary such as PSIM, SPICE, and GECKO CIRCUIT[6]-[10]. And also, there is a web-based introductory power electronics course [11]. As mentioned above, they use these kinds of the education tools for confirming and comprehending many circuitry and/or control strategy. However, we believe that practical experiments and/or visual

Fig.1. Picture representation of basic course of power semiconductor devises.

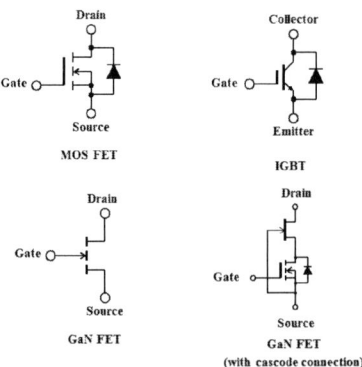

Fig.2. Circuit symbol of each switching device.

educational tools are certainly entailed. Operating a power electronics circuitry is not easier for beginners even after simulating the power electronics circuitry because there are real components; an inductor, a transformer, a capacitor, an IC and other components. We guess that a course or seminar including a practical experiment of power electronics circuitry not only for one day or two days but also for one week or one month is required. On the other hands, we think that setting a

978-1-4799-2706-7/14 $31.00 © 2014 IEEE

place for the technical exchange is also needed for experts. Some companies have much higher knowledge in their field such as power semiconductor devices, inductors, and capacitors. In case that they talk with each other, we believe that Japanese technology would be more developed, and new ideas will create from there.

In this paper, an approach and effort of our company are described. Furthermore, we introduce an educational tool from an associated company, which is useful for power switching semiconductor company.

II. POWER ELECTRONICS EDUCATION COURSE

Nowadays, there are many companies focusing on power electronics so that a company holds a power electronics seminar or course to people having a keen on learing it. A word, power electronics is simple, however, we have to learn and figure out a lot of subjects such as circuit topoloies, how to drive power semiconductor devices, designs of a Print Ciriuit Board(PCB), thermal diffusion technic, control technic, and other technics. We also need to experiment for aquiring skills. Generally speaking, practical experiments are necessary for several years to several decades to become an expert in power electronics[12], and the companies don't have a much time to educate the employees in their company. For that, our company offers an education service focused on a basic power electronics. Especially, this service aims at a begineer of power electronics, who would like to start to study power electronics. Trainees would learn basic technics of power electronics as follows:

· Basic course of power semiconductor devices

· Basic course of practical power electronics circuitry

· Basic cource of magnetics component

In the basic course of power semicondutor devices(Fig.1.), Trainees learn switching components such as MOS FETs, IGBTs and GaN FETs in Fig.2, and their drive methods. In addtion, we offers practical experiments to understand the switching behavior. As an example, buck-step converter with an inductive load is shown in Fig.3. This circuit is quite simple, however, by builting this circuit, the trainees would learn how the ringing happens on the switching devices and the circuitry, why reconvery phenomenon occurs, what decided the peak surge voltage of switching devices and so on. These are quite important to built the PCB circuit to understand which circuit roop we should shorten or not. This activity is even simple, but we think that studying with the basic experiment practically could be useful for beginners.

In the basic course of practical power electronics, first we introduce some simple RLC circuit to understand the circuit behavior. We give trainees the simple circuit which is important to catch the image of RLC transient phenomenon. Fig.4~Fig.6 are examples. We prepare

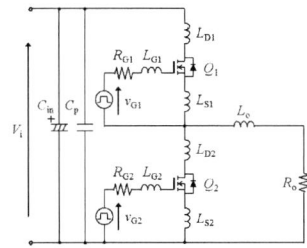

Fig.3. Step-down converter for comprehension of switching behavior.

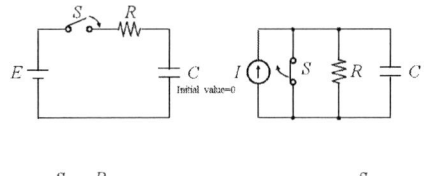

Fig.4 Circuits for understanding capacitor behavior.

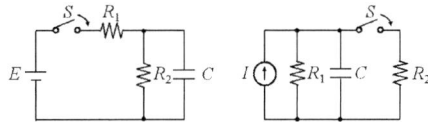

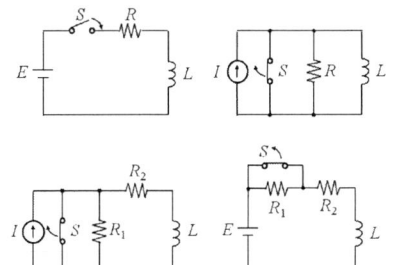

Fig.5 Circuits for understanding inductor behavior.

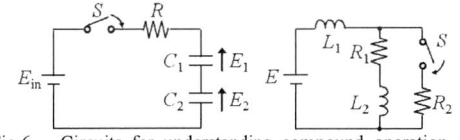

Fig.6 Circuits for understanding compound operation of capacitor and inductor.

over 41 different kind of circuits to understand a capacitor, an inductor, and their resonant phenomenon. These studies are helpful to understand the circuit behavior since the circuits are composed of a capacitor, an inductor and their resonant phenomenon. For example, the operation of each circuit topologies can be separated to the several stages. In case after learning the basic operation of RLC transient phenomenon, the trainees can reach the convergent point of each stage on their image. Thereafter, we teach how to analysis the circuit topologies and operation. In the next step, we think to invite some lecturer from the university and/or college of technology in accordance with the customer order (this is option).

In the basic course of magnetics devices (Fig.7), trainees learn designs of magnetics devices such as an

978-1-4799-2706-7/14 $31.00 © 2014 IEEE   812

inductor and a transformer.

Designs of magnetics devices are quite complicated generally, and substantially important for reducing the power electronics circuit volume. Moreover, they say that people who can design magnetic are few. Thus, we think that this course have much demands than any other courses. In this course, we teach the design of the magnetics devices for applications such as a step-up, a step-down, non-coupled interleaved converters. A design principle and image are basically given to trainees. Furthermore, the trainees would build a prototype of the inductor or transformer for DC-DC converter with the practical experiment. These courses are also corresponded to a customer's request. In case of the requirement from some trainees, the special inductor or transformer design such as for the coupled interleaved and three-phase coupled interleaved converter described in Fig.8 and Fig.9. The trainees can get many experiments through this course. In addition, we have strong support from prof. Masayoshi Yamamoto in Shimane University. He is famous for a coupled-inductor in Japan, so that our company can offer how to design of the magnetic components with the unique way [13]-[17].If you are interested in this course, please contact us.

III. POWER ELECTRONICS SEMINAR

We hold a power electronics seminar not only for power electronics engineers, but also for beginners of power electronics. Although the individual company has strong technic for its field, some company doesn't know its application. We believe that the contact among the different kind of companies each other can be good opportunity for facilitating the Japanese industry. Hence, we aim at exchanging and sharing their knowledge. When holding a seminar, we try to gather people from a different company, letting them know how the things provided by their company are used in the power electronics circuit. Accordingly, people from a different company would be able to learn the application of their company's product with the power electronics engineer or lecturer, and we hope the new ideas are given through our seminar between a power electronics company and a different field companies. For instant, the power electronics seminar (Fig.10.) programs of Prof. Yamamoto from Shimane University are below:

1. Power Electronics market

2. Technical trends of Power Electronics in ten years

3. Technology and future of Power Electronics regarding car, ship, and aerospace fields

4. Power semiconductor devices in future ten years

5. Control technology in future ten years

6. Magnetic components in future ten years

Fig.7. Picture representation of basic course of magnetic devices.

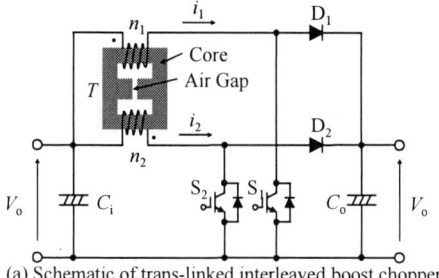

(a) Schematic of trans-linked interleaved boost chopper circuit.

(b) Core structure of trans-linked interleaved boost chopper circuit.

Fig.8. Coupled interleaved step-up circuit.

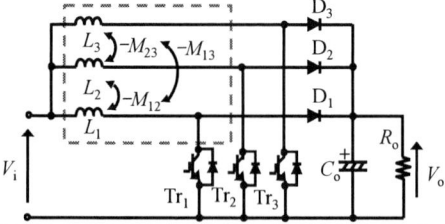

(a) Schematic of three phase interleaved dc-dc converter.

(b) Core structure of three-phase interleaved dc-dc converter.

Fig.9. Three phase interleaved dc-dc converter.

7. Passive components in future ten years

8. Circuit topology in future ten years

From above, prof. Yamamoto's topic is interesting for people from device and passive component companies, so that they could be come to the seminar. In the end of a seminar, we usually hold a place for technical exchange (Fig.3). We wish that they have the strong connection with the different field, and the evolution of new business and explore of the industry field are facilitated. We are glad that this triggers dynamizing the industry economy in Japan, and it will eventually bring us the increase of the employment for younger people, the profits, and the other benefit.

IV.   EDUCATIONAL TOOLS

We have considered regarding educational tools. In this section, we introduce an educational (development) kit of the associated company, and also educational tools in the progress. An educational kit of the associated company which is called Power Assist Technology Co., Ltd. is shown in Fig.11. This educational kit is composed of an inverter, a display, and Filter. The power switching devices in the inverter, and the filter corresponding to the switching frequency for 20 kHz and 50 kHz (Fig.12) are possible to replace. Si-MOS FET (Fig.13), SiC-MOS FET (Fig.14), and IGBT (Fig.15) are used for the switching of the inverter respectively, and this educational kit shows the performance of each switching device visually. For example, in case of comparing the heat sink volume of the Si MOS FET and SiC-MOS FET on the inverter, SiC-MOS FET can reduce the heat sink volume of inverter down to approximately one fifth (Fig.14). Furthermore, in case that the customers would like to use the own gate drive circuit by themselves, they can develop and replace the new gate drive circuit suitable for each of the switching device, so that we can study not only the performance of each switching device visually, but also the gate drive circuit for each switching device.

On the other hands, this educational kit also describes the performance of the switching frequency. As can be seen in Fig. 12. We can visually recognize that the filter volume is reduced by increasing the switching frequency. Additionally, because the filter on the educational kit can be replaced, the design of the filter is also possible to learn.

We introduce another educational tool for students in our progress, and however, this is still in the image step. As mentioned before, the practical experiments are quite important to understand the power electronics circuit. However, building the test circuit to understand the power electronics circuit is difficult and takes much time. Therefore, we are planning to make the educational kits to resolve this problem.

The concept of the educational tool is shown in Fig. 16. We think that this tool has a lot of components such as an inductor, a transformer, a capacitor, a power

Fig.10   Place for technical exchange.

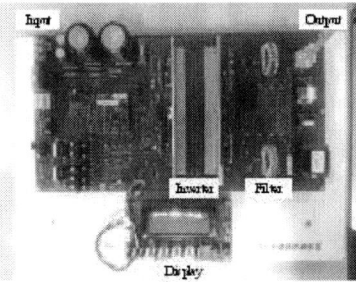

Fig.11   Device Evaluation kits.

Fig.12   Filter (left:20 kHz right:50kHz).

Fig13   Si-MOS FET inverter.

Fig.14 SiC-MOS FET inverter.

Fig.15   IGBT inverter.

semiconductor device, micro-computer and the other components for assembling the power electronics circuit in the block. The simple DC-DC converters can be assembled such as a step-up, step-down, fly-back converters and the other circuit topologies. We suppose that assembling the power electronics circuit is not so complicated with this educational kit. On the other hands, we think of putting the trap into the educational tool to learn dangerous parts of the power with the alert sound. This must reduce the mistake of users in the practical circuit.

Fig.17.   State of practical lecture.

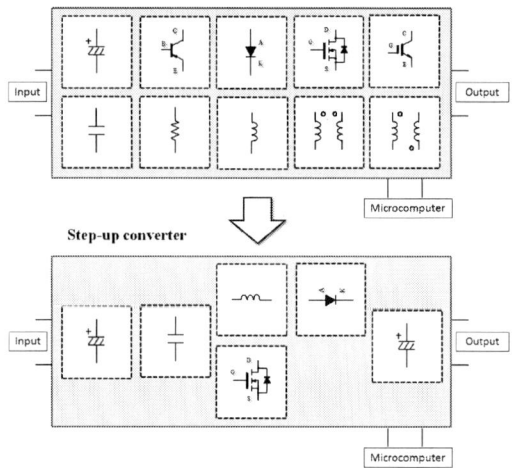

Fig.16.   Educational tool in progress.

V.   LECTURE TO LOCAL AREA

Last year, we joined a power electronics workshop in Shimane prefecture and sometimes educated basic power electronics to people from the different types of companies there. Its lecture is shown in Fig.17. In this lecture, we taught how to drive the gate drive circuit with a simple circuit shown in Fig.18. At the moment, they almost didn't know how to drive the switching devices and set the capacitor because of a beginner. Hence, in the first step, we taught switching devices with a water tap to understand easily. Then, the role of each device such as capacitor, gate resistor, and photo-coupler was explained.

After building circuit, the influent of the gate resistor is confirmed in term of the switching speed and loss with an oscilloscope by changing the gate resistance. Perhaps, this takes time, however, we believe that it was very good to understand because they touched the practical experiment and stay this experiment in their head. In this lecture, they stayed there over 1 hour even though the lecture finished. Thus, this lecture for beginners is succeeded.

We hope to expand this lecture to the elementary school, junior high school, high school students to interest. We believe that one of the key factors for education is interest. By keeping this lecture, we hope that people being interested in power electronics increase.

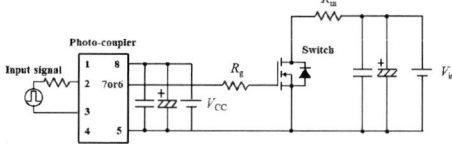

Fig.18.   Simple circuit to drive switching device.

If you are interested in this, please contact us.

VI.   CONCLUSION

Education for power electronics is quite important in the future of Japan because power electronics engineers have been being needed in terms of an environmental and energy issues. Therefore, we introduced the some effort to educate the power electronics beginners or develop the Japanese industry.

Our company tries to develop the human resources for basic power electronics, and prepare some courses:

· Basic course of power semiconductor devices

· Basic course of practical power electronic circuit

· Basic course of magnetic devices

These courses can be changed corresponding to the customer demand, and the day for the course can be determined by the customers.

We also introduced a seminar which is aimed at setting a place for the technical exchange. Actually, some company had the contact and corporation to another company. Now, the project is in the progress.

The educational (development) kit of the associated company was also introduced in this paper. This educational kit showed the performance of the new power semiconductor devices and high frequency operation. SiC-MOS FET can reduce the heat sink for the switching devices down to approximately one fifth. On the other hands, the high frequency operation achieves to decrease the volume of the filters down to one third. Next, the ideas of our company for the power electronics education were described. This tool is supposed to be used by students to understand power electronics easily, and students would understand by the trap deeply.

Finally, for one example, we picked up the state of the practical lecture for the power electronics beginners in a power electronics lecture of Shimane prefecture. This lecture went well since some people stayed even over 1 hour after finishing this lecture to complete the practical experiments.

We wish that not only the company specializing in power electronics but also the company specializing in the different types of field such as the power semiconductor device and passive components company cooperate with each other and even develop Japan more.

REFERENCES

[1] National Renewable Energy Laboratory, ''Energy innovations science & Technology at NREL'', available at http://www.nrel.gov/docs/fy10osti/46770.pdf, winter 2010.

[2] Wei Zhang, Dong Dong, Igor Cvetkovic, Fred C. Lee and Dushan Boroyevich, "Lithium-based Energy Storage Management for Dc Distributed Renewable Energy System", *Energy Conversion Congress and Exposition (ECCE)*, pp. 3270-3277, 2013.

[3] M. Nouni, S. Mullick, and T. Kandpal, "Providing electricity access toremote areas in India: An approach towards identifying potential areas for decentralized electricity supply," *Renewable & Sustainable Energy Reviews*, vol. 12, no. 5, pp. 1187–1220, Jun. 2008.

[4] S,Khader., A. Hadad, and A. A. Abu-Aisheh, "The application of PSIM & Matlab/ Simulink in power electronics courses", Global Engineering Education Conference pp. 118-121, 2011

[5] S,Khader., A. Hadad, and A. A. Abu-Aisheh, "Using SPICE for power system simulations", European Conference Power Electronics and Applications, pp. 1-10, 2011

[6] A. Cross, and D.Strickland, "Novel Online Simulator for Education of Power Electronics and Electrical Engineering", International Power Electronics Conference, pp. 1105–1111, 2010

[7] Müsing, A., Drofenik, U., and J. W. Kolar, "New circuit simulation applets for online education in power electronics", e-learning in Industrial Electronics International Conference, pp. 70-75, 2011

[8] J. Choi and Hyungsoo, Mok, "Simulation based Power Electronics Education in Korea", Power Conversion Conference, pp. 491-495, 2007.

[9] D. Beatty, A., Khan, and I., Batarenh, "Power electronics education:courses and laboratory", European Conference Power Electronics and Applications, pp. 1-10, 2009

[10] Canesin, C.A., Goncalves, F.A.S., and Sampaio, L.P. "Simulation tools of DC-DC converters for power electronics education", Southcon/95. Conference Record, pp. 240-245, 1995.

[11] Drofenik, U., and J. W. Kolar, "Interactive Power Electronics Seminar (iPES)-a web-based introductory power electronics course employing Java-applets", Power Electronics Specialists Conference, pp. 443-448, 2002

[12] Rikunabi, "Analog circuit is revived. The other side of the engineer needs of expanding", available at http://next.rikunabi.com/tech/docs/ct_s03600.jsp?p=002003. [in Japanese]

[13] Shota Kimura, Jun Imaoka, Masayoshi Yamamoto, "Potential Power Analysis and Evaluation for Interleaved Boost Converter with Close-Coupled Inductor", *Power Electronics and Drive Systems(PEDS)*, pp. 26–31, 2013

[14] Jun Imaoka, Masayoshi Yamamoto, "A Novel Integrated Magnetic Structure Suitable for Transformer-linked Interleaved Boost Chopper Circuit", *Energy Conversion Congress and Exposition (ECCE)*, pp. 3279-3284, 2012

[15] Yuta Nakamura, Masayoshi Yamamoto, "Optimal control design method for trans-linked type multi-phase boost converter ", *Power Electronics and Drive Systems (PEDS)*, pp. 991-994, 2011.

[16] Yasuki Kanazawa, Masayoshi Yamamoto, "A Novel Compact Single Phase Inverter Using Transformer-linked Interleaved Flyback Converter", *Renewable Energy Research and Applications (ICRERA)*, pp. 1-3, 2012.

[17] Masayuki Nakahama, Masayoshi Yamamoto, Yuki Satake "Trans-linked Multi-phase Boost Converter for Electric Vehicle", Energy Conversion Congress and Exposition (ECCE), pp. 2458-2463, 2010

Education for the engineers of traction power supply division in East Japan Railway Company

Toshiaki Takino,Tetsuro Iwakami/ East Japan Railway Company
Electrical & Signal Network system department
Tokyo, Japan

Abstract— In terms of run a railroad system, many techniques are required. I will introduce technician training program in the power sector in the East Japan Railway Company. We are positioned as one of the most important management issues human resource development. In recent years, time employees a large amount of leave has come, inheritance of technology and training of technical personnel are urgently needed. Therefore, education to maintain the technical capabilities of employees, further improve its technical capabilities are required. Current educational program is made from what is according to the level, such as mid-level employees and younger employees, aimed at a higher level. In this paper, I will introduce the three main development programs. I describe the contents and characteristics of the program.

Keywords—education, expert, railway, traction power supply

I. Introducton

Group of East Japan Railway Company (Hereinafter, JR-EAST) has developed the Group Management Vision "V" (Ever Onward) in October 2012. Here, as an important pillar to "pursuing unlimited potential" and "eternal mission", and has come up with the basic direction of six. One of them is that development employees and creating a corporate culture that maximizes human potential. Recognizing that there is no corporate growth and without growth of employees, are working positioned as one of the most important issues in corporate management on human resource development.

In this paper, among the technology sector, for technician training traction power sector, as the topic that is focusing in particular, we introduce three in particular. I will introduce each encourage the voluntary efforts of young engineers "Technical Project", to train engineers of each branch office as "Academy of Engineering" and "expert training".

II. Concept of engineer's education in JR-East

The traction power sector , there is a large number of fields interact with direct current of conventional lines , Shinkansen , substations , power distribution , and train lines . To operate the system and equipment like this, it is necessary to know the railway system a whole, including other departments not only one department. It is important to set the deepen how expertise in charge in that.

The University of Japan, to learn practical knowledge about the railway is small, and to learn to after joining the company in many cases. Therefore, from training after joining the company, through a variety of programs OJT, OFF-JT, etc., you are doing the training of employees. Need to learn quickly and effectively a variety of knowledge and skills in stages according to the skill also experience, and joined annual. As a program, there are three in order to improve the technical level, especially in "Technical Project", "Academy of Engineering", it is "expert training". Set for engineers who are the target respectively, make their own curriculum, we aim to improve the technical capabilities. I describe in detail below.

III. Technical Project

"Technical Project" is a program for young employees called potential adoption to be 2-4 years and I joined. Its purpose is to approach proactively research about solving technical problems in the workplace, and the skills of the next.

1) Deep understanding of the knowledge and know-how about the specialized field
2) Construction of logical thinking of as an engineer
3) Improvement of problem-solving ability and presentation skills

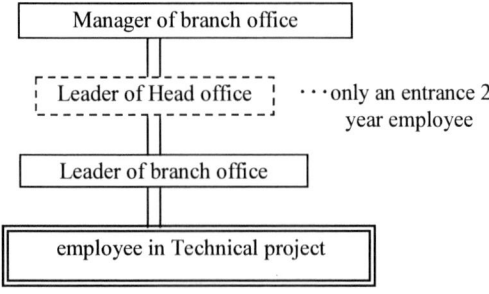

Fig.1 Instruction of Technical Project

How to proceed with the technical project is carried out by junior staff flow sets a theme related to the charge of their own business, and draw conclusions that objectivity in one year for that theme . He started two years at the company, with the setting of the research theme every year the next three years, We are conducting the research. But because it is the engineer years of experience less, employees of interest constitutes a support system for training. First , (10 years about joining) mid-level employees in the same work place becomes the guidance in charge of younger employees , every day , I have advice on how to proceed and progress of research . Manager of each branch office also becomes their leader, have advice research methods and significance, the direction of research on a regular basis. For (an employee of two years at the company that is) employees to be first year further , mid-level employees experienced head office work becomes leader , you are supporting them . (See figure 1).

In because experience is shallow young employees, knowledge of how to proceed and rail business is insufficient, there is a need to go to learn the skills to advance the research. Therefore, support systems of each leader is important. However, one feature of this program, to study themes related to the business, but it is to be subjected to time management yourself, is performed autonomously. That for young engineer, go to challenge the issues to consider themselves is important, it is carried out intensively in the period of 2-4 years at the company about it.

Examples of themes, some cause investigation of equipment trouble, the efficiency of equipment, introduction of new technology, maintenance techniques, and construction methods. To the theme such, is carried out technical approach for the resolution of the problem, conduct discussion and analysis and acquisition of the stator, put together logically.

A year later, they create a report of the study, we report the research results in the research report meeting of the headquarters. At this time, so that the presentation capabilities and ability to be described is nourished. On significant research technically, also reported in the study group, such as conferences and external further. For example, research on RPC that was presented at IPEC-Sapporo2010 is (Omi Shotaro, etc.), it is one of the research that was carried out in this effort.

As one of the training of young engineers, "technical project" is responsible for the end of the technical capabilities of improving young employees in this way.

IV. **Technical Academy**

A. *Outline of Technical Academy*

We were established in March 2010, "Technical Academy". The purpose of this academy is that the engineers of our inherit the values and skills of railway technology, they become an excellent technician. This program, has entered the 5th day now.

Leave the normal work in the field of branch offices and technical personnel to take this program has placed enrolled for a year at the headquarters during the period that is selected from each technology sector, and to attend. So they received a varied menu, so as to improve the technical capabilities. In-house training normally and

personnel selection in the company, but the appointment of this program will be determined by the selection of applicants. It is an employee of a level able to doing on your own to learn the business, employees of interest, often in their 30s to 15 years six years.

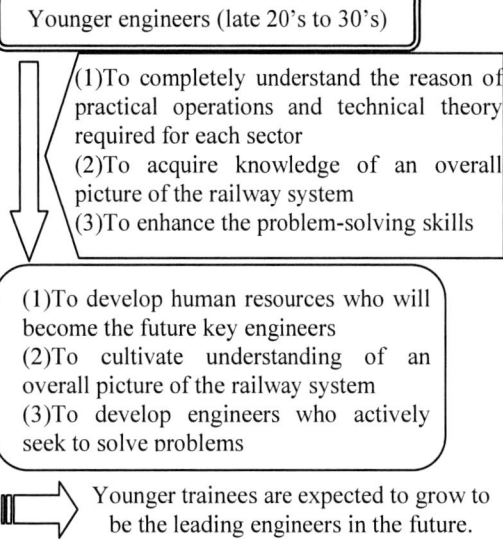

Fig.2 Image of Technical Academy

Course is divided vehicle, civil engineering, mechanical, signal, power, etc. In each course, there are those of all common system, the contents something professional that fall into each system curriculum. The goal in this academy, and that to further improve the expertise of their own, it is to understand a wide range of railway system as a whole. Therefore, "to learn the theory and basis of practice in the department that is in charge of their own", "to acquire knowledge of the railway system as a whole", the aim of this program is to acquire the skills necessary to "solve problems put "is that.

In a common program, the student thinks important that you have a sense of ownership and feel the feelings of customers who use the railway, to solve the problems and creative problem unknown is said to be themselves. Through this fact, as an engineer as a service industry, they are trying to acquire the necessary knowledge and skills.

The program, that they would solve the problem standing in broad perspective beyond the expertise of their own and mutual learning, teach the students of other departments training and lectures by outside experts, the contents of your area of expertise and the like groups studies aimed. There is also a lecture it is considered that the things you think of as an engineer, what is or will act in "Engineering Ethics".

In addition, position the coach Exclusive with each course, it is promoting and teaching a variety of research and planning, and management of student programs, in support system overall.

B. Course of traction power sector

I then describe the power sector course. How to proceed is a lecture by engineers and manufacturers lectures from each authority. The authority is a researcher of railway Technical Research Institute and OB of the company. Not only mere knowledge, you are learning a wide range of basis for technical provisions, such as the latest technology trends from the transition of the technology here. For example, I have learned lecture by the thought of design and manufacturers involved in each summary from the train line, electricity system, substation facilities. (See figure 3).

Because it is a position to service the train, we are advancing the business while emphasizing rules usually. However, when an attempt is made to resolve the technical problems various do not understand the rationale and technical rules, there may not be the fundamental resolution. For example, if the equipment accident occurs, and thus it is preferred that the first is allowed to recover. Then, why led to such accidents? , Mechanism of failure something? , I do pursue the cause that led to trouble such. Similar events may occur also, if there was not struck the effective measures.

What is important, is to understand the rationale and technical concepts of the various equipment and systems, to learn. It is possible as long as it can result, have found problems in the field, consider the measure. In this program, we are aiming to train engineers of problem-solving like this.

In addition to attending lectures, decided their theme, they are also working on technical research. And that to proceed with the research, it is supported by the adviser within the company, and is working under the guidance in the laboratory of the University of Tokyo. (Hidaka laboratory of the University of Tokyo, Iwafune laboratory, and Koseki laboratory). While learning how to parse necessary on not only a working knowledge of the company, to promote the research, We are conducting the research.

In addition, in order to look both internally and externally wearing a global perspective, overseas training are also incorporated. We also provide technical

Fig.3 Tour of the high-pressure equipment manufacturers

Fig.4 Technology exchange meeting in Southwest Jiaotong University

exchanges with China Southwest Jiaotong University with the exchange so far, I have done research report to each other. Through discussion in there, I'm working to further depth of research (See figure 4). Having a wide field of view by touching also railway overseas, to know the trend of overseas railway and the difference with Japan is expected.

And to report an in-house research is when you complete a one-year program, they are also reported in the outside of society, such as the results of research then. Technician of 24 people have power until the end of the course in March 2014, they have been working as an employee at the core of technology in each branch office then. Based on what you have learned in this program, you are responsible for the technical work will continue in order to become a technical leader in the future. Currently, finished the program of the 5th, we have started the program of the 6th.

Table.1 The main theme of study

1	Study on effective utilization of regenerative power in the Shinkansen using linear programming [1]
2	Consideration of Ground Fault Protection in AC Feeding System [2]
3	Numerical evaluation of the effectiveness between action of regenerative braking and voltage setting of DC interval substation. [3]

v. Expert Training

In the training program of the Department of power, "expert training" is training high technical level the most. This feature is to 3 years entry trainees selected from each branch. Target employees are experienced employees were selected at each branch office, it is an object of the present invention that it is through this program, to become a leader in the technology sector in the future.

Courses, some as "path" and "power distribution substation," course. The path course, the content-related train line, the distribution line mainly, of course, power distribution substation is in a content related to equipment and systems.

Trainees receive a set of training for a few days about a year. The curriculum of the training set, factory tour of the manufacturer of heavy electrical equipment and wires, lectures of experts railway manufacturers and Research Institute, are also included.

By Efforts three years the research described above, are summarized the results, in order to improve the problem-solving ability with technical capabilities. Trainees who have completed this training, it is possible to play an active part as a key player technical each branch office is expected.

VI. Conclusion

In the technical department of the Company, that the generation change progresses rapidly over the next 10 years is expected. Making a succession of technique is essential in order to continue to provide rail services of high standards and quality. Railway is a system that holds while combining various techniques, it is necessary to inherit reliably techniques that train specific. Training of technical personnel at the core of each workplace, human resources to lead the innovation and more are also required to do so. Younger employees has a high morale and motivation, we must so that we create the opportunities for many for the skill up. The discussions in the future, we will aim to further level up the training of engineers.

Further, in order to promote new development and innovation, we understand the theory underlying the field other techniques well railway field is also required. To do so, it is important for the eyes to the latest technology trends, such as which technology is what has attracted attention in various fields, or have been studied. Then sent to both internal and external, the results of research and the introduction of new technology in-house, we will aim to exchange technical.

In addition, of course, necessary qualifications to work is recommended for employees to acquire the technical qualifications. Examples of qualifications, other "professional engineer" and "electrical chief engineer" of, a "Professional Railway Design Engineer" and as the qualification relating to the railway. Because it leads up skills self, to qualify is promoted to obtain positively.

References

[1] K. Takagi. IEEJ. 2012 Industry Application Society Conference.R5-5-27.V-245

[2] M. Koyanagi. IEEJ. 2011 Industry Application Society Conference.R3-2.3-16. III-147

[3] K. Tanaka, IEEJ. 2011 Conference of Tokyo Branch.TER-11-048

The 2014 International Power Electronics Conference

Successful Online Education – GeckoCIRCUITS as Open-Source Simulation Platform

Andreas Müsing

Gecko-Simulations AG

Physikstrasse 3, 8092 Zurich, Switzerland

www.gecko-simulations.com

andreas.muesing@gecko-simulations.com

Johann W. Kolar

Power Electronic Systems Laboratory, ETH Zurich

Physikstrasse 3, 8092 Zurich, Switzerland

www.pes.ee.ethz.ch

kolar@lem.ee.ethz.ch

Abstract—**Modern lectures in power electronics should be complemented by interactive content, e.g. computer assisted learning via web-based applets or exercises employing a circuit simulator. The online e-learning system "iPES" was established in 2001; since then the iPES Java applets are frequently used in academia and industry to teach the basics of power electronics. This paper introduces new features of the open-source power electronics circuit simulator GeckoCIRCUITS, which enable a more flexible way to generate online content in comparison to iPES. In particular, the animation of current paths in a circuit model as well as a new language translation and internationalization scheme will be presented in detail. Finally, the popular iPES platform will be enhanced or replaced with Java applets based on GeckoCIRCUITS.**

Keywords—*Power electronics circuit simulation, online education, open-source, current path animation.*

I. INTRODUCTION

Many typical engineering problems are quite complex, show dynamic behavior and a large number of different operating states. It is difficult for a lecturer to describe complex interrelationships and parameter dependencies solely via verbal description and static pictures. In power electronics, current conduction paths depend on diode and switch conduction states, and PWM control adds another level of complexity. The student often finds it difficult to understand how a system works if he is only passively following a lecture in the classroom.

To address this issue, the Power Electronic Systems Laboratory at ETH Zurich introduced the online education platform iPES in 2001 [1]-[4]. An example of a typical iPES Java applet is shown in Fig. 1 [5]. Later on, the circuit simulator GeckoCIRCUITS was developed and used as an extension to the iPES platform [6]. Since 2013, GeckoCIRCUITS is available as open-source software published under the free GNU Public License [7]. Hence the software can now be used for free. All exercises in the basic and advanced power electronics courses at ETH Zurich include a GeckoCIRCUITS simulation model [8]. Additionally, specific simulation exercises motivate the students to use a modern simulation tool to understand complex lecture contents.

This paper introduces two new features of Gecko-CIRCUITS which are especially attractive for educational purposes: A dynamic current path visualization and a novel scheme for translating the graphical user interface into arbitrary languages. The main intention for implementing the new functionality was to reach a wider international audience in education for the simulation tool GeckoCIRCUITS.

This paper is structured in the following way: In the next section, the special requirements for a circuit simulation platform in education are discussed. The concept of the original iPES applets is reviewed in Section II. The new current path visualization and animation feature of GeckoCIRCUITS is then introduced in Section III, including some of the technical implementation aspects.

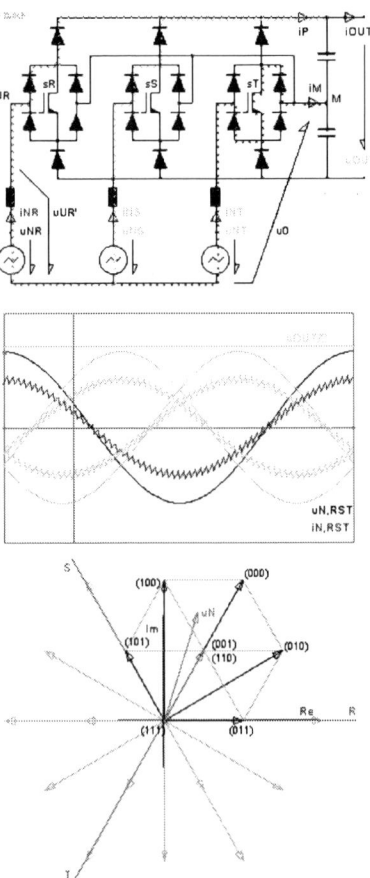

Fig. 1. A Vienna Rectifier iPES education applet including a current path animation and a space-vector visualization [5]. The red time slider in the waveform graph can be used to modify the animated circuit switching state and the space-vector display.

978-1-4799-2706-7/14 $31.00 © 2014 IEEE

821

Section III describes the new GUI translation and internationalization mechanism of GeckoCIRCUITS. Finally, a conclusion and an outlook to the future development of the education platform is given.

A. Requirements for a Successful Education Tool

Considering circuit simulators for power electronics, there are several commercial and few free tools available. However, a proper choice of a circuit simulator for education in power electronics depends on many requirements, namely the following criteria:

1. Students of a power electronics lecture are problem oriented, i.e. their primary goal is to understand the teaching content. Thus, the introduction of a new simulation tool should not lead to any additional challenges. Most important, the software has to be intuitive and easy-to-use.

2. For many academic institutions, the software licensing costs of commercial circuit solvers are too high, and it is often not possible to provide classroom licenses to groups of typically 10-50 individual students. Various software vendors offer academic institutions discounts on their software licenses. However, due to the large number of needed licenses, the annual cost still may remain too high. Another approach is that software vendors distribute a limited version of the software tools for free, e.g. the number of circuit components is constrained to a certain maximum. However, ideally the software used in combination with a course should be available for free without any limitations. A further benefit is that students may use the software at home or on their laptop computer outside the university classroom for solving exercises.

3. Circuit simulation in power electronics often suffers from stability and convergence issues. Therefore, a SPICE-based solver is not very well-suited for power electronics circuit simulation. Often, a piece-wise linearization of non-linear components such as switches and diodes is faster, more stable and is adequately accurate [15]. Additional desired features for a good power electronics circuit solver are thermal modeling and loss calculation, analog and digital control blocks, and a feature-rich toolbox for simulation result visualization.

4. The graphical user interface of the simulation software should be available in many different languages; proper internationalization is unfortunately not always available in commercial tools. It is often assumed that software users in industry and academia are sufficiently versed in the English language. For students not having English as their mother tongue, the missing translation is a great hurdle.

5. Other desired but not absolutely necessary features for educational purposes are platform independence, i.e. the tool has to run on the main operating systems Windows, Linux or OS X, as well as the capability to provide course content online.

B. GeckoCIRCUITS as Open-Source Simulation Platform

The power electronics simulator GeckoCIRCUITS was developed in the platform independent programming language Java, under consideration of all the above mentioned requirements. The software was published at the beginning of 2013 as free software under the GNU General Public License (GPL). This open-source software license does not only permit to use the software for free, but additionally the Java source code of GeckoCIRCUITS is available as download [7]. It can be inspected, modified and extended by anyone. The GPL ensures that the software is available also in future without any license cost. Hence, GeckoCIRCUITS evolved into a very powerful, easy-to-use, and easily accessible simulation tool for education in power electronics.

II. REVIEW OF THE iPES ONLINE EDUCATION PLATFORM

A. The Basic Concept of iPES

One of the problems in studying and teaching power electronics is the large number of possible reactions of a system to parameter changes. Another challenge in understanding circuit topologies is that a converter system typically goes through a sequence of different states. The traditional way of presenting such content is by using a sequence of slides showing a single system state displaying the behavior of the circuit for a particular set of parameters.

A main idea of iPES [4] is to provide Java-based animations where a "time slider" can be dragged by the student via the mouse; the slider position directly shows the related system state. Figure 2 shows an iPES Java-applet of a single phase PWM converter [9]. The time slider is the vertical red line in the waveform diagrams. According to the position of the time slider, the current flow of the associated path in the topology is animated employing moving dots.

Another basic concept of iPES applets is the possible manipulation of some predefined system parameters via the graphical user interface and interaction points, which are always indicated in red color. The user interaction allows to immediately investigate the resulting behavior of the system. In order to draw focus of the applet user to the most important circuit properties, the number of possible interactions is kept thereby intentionally low. Thus, the user is not distracted by irrelevant parameters and the learning effect is maximized. The variable system parameters allow the animation of interrelations within a converter system, for example the influence of switching frequency or the inductance on a current ripple.

Besides iPES, other projects of web-based interactive animations have been developed for power electronics in Java or Adobe Flash, e.g. [10] - [14]. Some of these projects used the basic principle and design of iPES as foundation.

978-1-4799-2706-7/14 $31.00 © 2014 IEEE

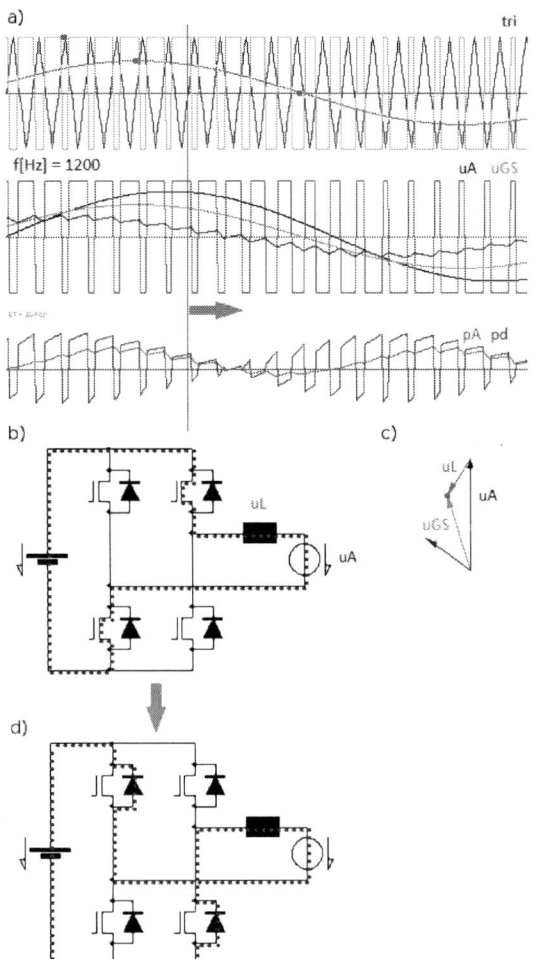

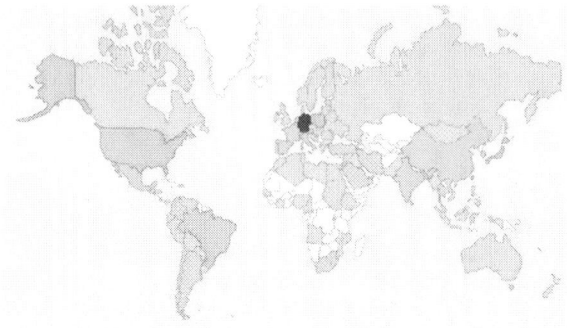

Fig. 3. The figure shows the global distribution of iPES-visitors based on 10'000 visits per month from more than 100 countries. The color intensity indicates the number of visitors per country. 30% of the visitors are from Germany (1st), 5% are from Japan (2nd).

Fig. 2. A single-phase PWM converter as iPES online Java applet [9]. a) The model parameters indicated in red color (PWM frequency, load phase shift, and time-slider) can be modified interactively by the user. The voltage vector diagram c) is updated according to the load current phase shift and amplitude. By moving the time-slider in direction of the green arrow, the animated current path changes its state from b) to d).

B. Acceptance and Users of iPES

The education platform iPES went online in August 2001. Up to now, it has been translated into 12 languages (in alphabetical order: Arabic, Chinese, Dutch, English, Farsi, French, German, Italian, Japanese, Korean, Spanish and Vietnamese). Currently, iPES has up to 500 visitors per day and on average more than 10'000 visits per month. The worldwide user distribution is shown in Fig. 3. The good acceptance of iPES in Asia can be attributed to the translation into many Asian languages. This indicates that a GUI translation of the circuit simulator GeckoCIRCUITS is desired, too, for generating a sufficient user acceptance. Therefore, a new software translation approach will be presented in Section IV.

C. Shortcomings of iPES Applets

Applets in iPES allow only a small number of parameters to be changed by the user. A further restriction is

that parameter changes have to be calculated numerically, but the user should experience an immediate effect. Accordingly, the underlying simulations, e.g. long-term transients in converter systems with low damping caused by a parameter change have to be calculated extremely fast.

The technical solution when implementing the iPES applets was to accept numerical inaccuracies by employing large numerical time steps and adding artificial damping resistors. The parameter variations then modify the shown waveforms. The changes are felt as immediate due to the fast but partly inaccurate simulation. The relationships and tendencies are all correct, but the values might show errors up to 10%. Numeric values are therefore not given in many iPES applets. After having gained experience and understanding in power electronics with the educational iPES-applets, many users have asked for real circuit simulation and for much more flexibility. This was the motivation to integrate online circuit simulation with the software GeckoCIRCUITS into the platform iPES [6].

Running Java applets requires the user to have a Java Runtime Environment (JRE) installed on the computer and enabled for the internet browser. At the moment, more than 90% of all computers have Java installed. However, running iPES or GeckoCIRCUITS on an Android based tablet computer is not yet possible. Recently, Java had some negative headlines in the media concerning security issues. Since then, many internet browsers allow to run Java applets only in case the most recent Java update is installed. This is an inconvenience which must be taken into account when using online education platforms based on Java. Nevertheless, computer security with Java is superior in comparison to online contents written in other internet technologies such as Adobe Flash.

The biggest shortcoming of iPES is the static property of its compiled Java binaries. The applet collection cannot cover all aspects of lectures at other courses outside of ETH Zurich and a modification or extension is difficult for electrical engineers without programming skills in Java. Another disadvantage of iPES is its large administration effort. Currently, more than 150 single Java applets are available in 12 different languages. Due to this large number, software maintenance and bug-fixing is quite challenging.

Hence, our goal is a replacement of most iPES applets with simulation models in GeckoCIRCUITS. The new features as introduced in this paper are a major step to achieve this effort. The main benefits of an easy-to-use but yet powerful circuit simulator can then be used by the power electronics engineer to generate educational applets tailored for a specific lecture.

III. CURRENT PATH VISUALIZATION AND ANIMATION IN GECKOCIRCUITS

A main feature of many iPES applets is the dynamic visualization and animation of current paths within the circuit schematic, see Figures 1 and 2, and Section II-A. The idea of including similar current path animations in GeckoCIRCUITS is obvious and adds educational value to the circuit solver. Besides its use in education, this feature is also useful for debugging complex simulation models. Missing or wrong gate signals, short circuits, etc. are thereby detected easily in a circuit model.

The concept of circuit simulation with animated currents is not new. Besides iPES, there is a similar functionality in other software packages. However, the tools available at [18] and [19] are just simple circuit applets comparable to iPES, which do not permit the user to build more complex circuits for power electronics. The power electronics circuit solver Caspoc [16], [17] includes a current animation feature, as well. However, the proprietary software Caspoc is only available as commercial tool. Furthermore, its current visualization does not show the current flow direction, connectors are here displayed with varying colors depending on the current flow.

At first glance, the implementation of a current animation seems to be straight forward. However, during the development and application of the new feature for GeckoCIRCUITS, some implementation issues have become apparent which had to be solved for a full functionality within the circuit solver. The are discussed in detailed in the following paragraphs.

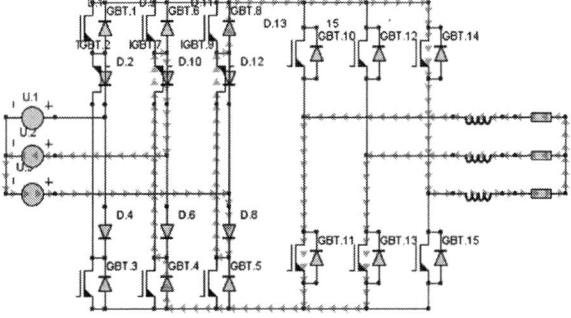

Fig. 4. Current path animation of a Sparse Matrix Converter in GeckoCIRCUITS. In the applet [20], the arrows in magenta move into current direction.

TABLE I. MEMORY CONSUMPTION OF SPARSE MATRIX CONVERTER CURRENT ANIMATION FROM FIGURE 4).

Approach	Memory consumption
Float-storage of all current values	13 MB
Single-byte representation of data	3.2 MB
Commutation events are considered only	80 kB
Repeated states saved as object reference	30 kB

A. Memory Consumption of the Current Path Animation

The simplest approach for the new animation feature would be to store all component current values during a simulation run. Then, the software can generate the animation based on the data stored in memory. However, this naive approach would lead to an excessive RAM memory consumption. Java applets are limited in memory, and thus this simple approach would only be applicable to very small simulation models. Therefore, the software takes advantage of the following properties:

- For the current animation, it is not necessary to store component currents as floating point values. All required information can be stored within three states: Forward conduction, backward conduction, and no current (off-state). Hence, the smallest usable data type has 8 bits (1 byte) in contrast to 32 bits of a float data type.

- Data storage during all simulation steps is not necessary. The actual animation state only changes its value after switching events and/or a current commutation. Hence, the frequency of state changes is in the order of the converter switching frequency and not the simulation stepwidth Δt.

- In power electronics, identical or similar conduction states are repeating several times during a simulation. Therefore, the software has to save a specific animation state only once. An additional list contains those points in time where an animation state needs to be changed. The list entries are then linked to the corresponding animation state.

Applying all above mentioned optimizations, the memory requirement of the animation feature was greatly reduced. Table I summarizes the memory consumption for the Sparse Matrix Converter example from Fig. 4. The circuit model consists of 42 components, and the simulation duration and step width were set to $T = 20\,\mathrm{msec}$ and $\Delta t = 250\,\mathrm{nsec}$. During the simulation, 1550 commutation events with 117 unique circuits states were detected by the software. Table I indicates that the presented approach allows to run even large simulation applet models without memory problems.

B. Current Animation of Complex Interconnections

Within GeckoCIRCUITS, the user can set up arbitrary component placements with complex wire interconnections. Thereby, wire layout properties have to be considered in addition to the ordinary circuit simulation. This is illustrated in Fig. 5. The current animation direction is immediately obvious for the example in Fig. 5a): The marked connector flow directions ① and ② at star point

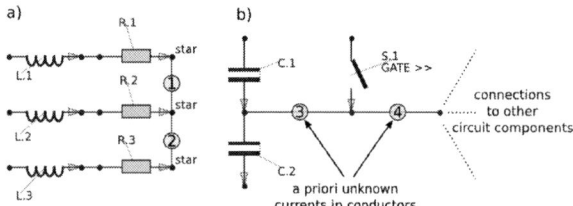

Fig. 5. Component configurations, where the current animation of wire segments has to be calculated as post-processing step based on Kirchhoff's current law.

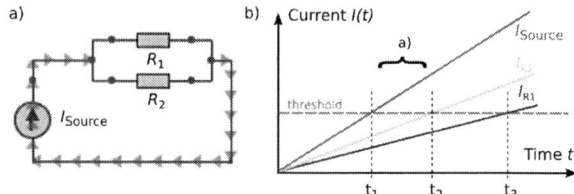

Fig. 6. Current path visualization without error correction: a) shows a loophole in the current animation. The problem can be recognized in b): The total current is larger than the threshold, whereas both resistor currents are below the threshold value.

potential are identical to the current values of resistors R.1 and R.3, respectively. The situation is more complicated for Fig. 5b). Here, the current flow direction is not immediately available at the wire segments ③ and ④. A circuit solver considers wire connections just as identical potentials of the connected circuit component terminals. Therefore, GeckoCIRCUITS had to be extended to take the connection topology into account. Basically, all simple wire connections without branching should be considered here with a new current measurement component. Finally, the missing current measurement is achieved by applying Kirchhoff's nodal rule, i.e. the sum of all incoming node currents has to be equal to zero. If several basic wire sections are connected to each other, Kirchhoff's current law has to be applied recursively until the system of equations is exclusively based on known circuit component currents. Once the network graph is analyzed, the additional effort for calculating all current directions in a postprocessing step is negligible during the circuit simulation. In the example of the Sparse Matrix Converter in Fig. 4, the calculation of the DC link currents is done by summation of 6 components at the converter output stage, respectively.

C. Current Commutation and Minimum Threshold Value

Based on a provided threshold value, the circuit solver determines if the animation of a conductor or component is in on- or off-state at a specific point in time. However, this is leading to the problem as depicted in Fig. 6. For $t = 0$, no current is visible at all, all components and conductors are in off-state. Then, the erroneous current animation state for the time interval between t_1 and t_2 is shown in the figure. The total current is larger than the threshold value. However, it is split into two resistor currents that are below the threshold value. This problem is inevitable when a fixed current threshold is used in combination with branching into more than two components. Hence, this issue will appear in most power electronic circuits.

The following approach helps to avoid the above mentioned visualization problem: The time intervals with wrong animations are typically quite short. Therefore it is possible to remove this time interval from the set of animation states. The problem can be detected quite easily: As soon as an activated current animation enters a node, at least one additional animation path with opposite sign has to exist. If this is not the case, the corresponding

state should be removed by merging nearby current states at the center of the flawed current state time interval.

IV. A NOVEL APPROACH FOR LANGUAGE TRANSLATION AND INTERNATIONALIZATION OF GeckoCIRCUITS APPLETS

A very important requirement for a successful online simulation and education tool is the software's capability for internationalization, i.e. the graphical user interface and the applet contents should be available in several or even all possible languages. Many commercial tools unfortunately do not offer any internationalization capability or only a translation into a few common languages. English is the main language in the scientific community. However, the amount of students having difficulties to work with education tools that do not allow internationalization should not be underrated. Therefore, the open-source software GeckoCIRCUITS was prepared to be available in all possible languages, e.g. Japanese, Chinese, Portugese, etc.

Often, commercial software manufacturers build language databases for all GUI component strings. This language database is then given to translation agencies for doing the internationalization as paid service. This approach has several difficulties: Professional translations are expensive, and software that evolves over time has to be updated repeatedly. The more languages that are involved, the more expensive it will be to keep all translations up-to-date. Furthermore, if the software contains highly technical terms, even professional translators will have difficulties in translating those terms correctly.

Hence, a new internationalization approach for Gecko-CIRCUITS, which reflects both the open-source character and the networking capability of the software, is presented in the following paragraph.

A. User Feedback for Language Translations

Allowing multilingual users to contribute directly to GeckoCIRCUITS's internationalization resolves the above mentioned problems when dealing with highly technical terms. Users who are well-versed in the technical jargon in their native language can provide the translations and upload their improved suggestion vocabulary to a centrally maintained server. In analogy to the various social/professional networks, GeckoCIRCUITS provides a virtual network of users who contribute to the translation database in a specific language.

Initially, the software provides a database of translations that was generated automatically, i.e. applying the translation API's of Google, Yandex and Microsoft. The initial database contains already more than 50 different languages, which covers most of the users' native languages. Obviously, the quality of the auto-generated database is rather bad.

Users who do not accept poor translations have the option to use English as fall-back language instead of database contents that were auto-generated. As second option, in case the user encounters a translation into his native language which is flawed, he or she has the opportunity to immediately improve the translation and contribute to the vocabulary database.

The following example clarifies the new translation procedure: Let's assume a software user whose native language is German. Most parts of the GUI are translated correctly. However, the string "Listening at Port:", was auto-translated to German as "Hören am Hafen", which does not make sense within its context, since a re-interpretation into English results in "hearing at the harbor". Then, the user can open an internationalization wizard by shift-clicking onto the component that contains the flawed string. A screenshot of the corresponding dialog window is shown in Fig. 7. Here, the user has the possibility to modify any GUI related string. Additionally, an optional comment can be inserted, preferably in English, which clarfies the modification. Comments are later useful for a database administrator without knowledge in German. Possibly, the administrator has to select from several different user suggestions. After pressing the "Confirm" button, the suggestion for improvement is sent via network to a database server and is, in principle, from now on available for the large user community of GeckoCIRCUITS.

Besides the input of a single text selection, the toolbox has a batch translation functionality accessible via the button "Open Translation Toolbox" in Fig. 7. The batch

translation can be used for a series of user data input, where all available GUI strings are presented sequentially. Currently, there are about 500 single entries in the language database. In batch mode, the strings are sorted by priority. Since many of the used strings are rarely used, they probably do not require an immediate translation. Details on the actual client-server architecture and the procedure of approving user contributions are given in the next paragraph.

B. Client-Server Architecture of the Translation Toolbox

The architecture of the new translation approach is built on a client-server model: The Java software runs as stand-alone tool on a personal computer or alternatively within an internet browser and has a network connection to the server that provides the translation database. If a translation is missing or incorrect, the user can directly insert or edit the string and send the new content to the server.

In order to prevent abuse, i.e. to filter undesired content, the contributors of translations are grouped into different categories: Anonymous online users without identification, trusted translators with login and password, and administrators. Translations that are sent to the server have to be peer-reviewed by trusted translators. Finally after a review-process, translations are integrated into the

Fig. 8. Browser screenshot of the MediaWiki translation feature which is very useful for database administration. Language administrators can be notified when new suggestions for improvements are available. Furthermore, convenient features as e.g. language statistics and the history are available within MediaWiki.

Fig. 7. Dialog window which pops up when the GeckoCIRCUITS user shift-clicks on a GUI component that contains an incorrect native language translation.

master database which is actually used by the software all around the world. Without such a review process, it may be possible that undesirable discriminatory text or misleading contributions could degrade the translation database.

A network connection is mandatory for Java applets which run inside an internet browser. Due to security and privacy reasons, Java applets are executed in a so-called "sandbox" within the web browser, preventing them from accessing local data and the file system. The available Java applet memory is limited to 30-120 Megabytes. In order not to waste the limited RAM memory, the applet exclusively loads the language database which was actually requested by the user. For the stand-alone version of GeckoCIRCUITS, a copy of all translation data is stored locally on the hard drive. Since an internet connection is not always available or a firewall might prevent the network communication, the locally stored files are necessary as a backup.

On the server side, quite a special and unique configuration was chosen as database solution. Instead of creating a database system from scratch, the MediaWiki engine runs on the server. This is the same software on which Wikipedia is based. MediaWiki is very easy to use and maintain, all data gets stored in Wiki pages. Tools for administering user accounts together with e-mail notifications and various other aspects of the system are included. All changes made by the user community and administrators are automatically tracked and the page history is stored. These are very convenient features to have in a database. Additionally, there is an extension available for MediaWiki which provides a user interface for translating text within the Wiki. The "Translate Extension" allows an easy administration of the numerous languages and progress statistics. A screenshot of the online MediaWiki interface is shown in Fig. 8. Hence, the effort for setting up the server system of the translation toolbox was negligible.

A schematic overview of the new translation toolbox is given in Fig. 9, and the GUI translation of Gecko-CIRCUITS into Japanese is shown in Fig. 10. Since the new software internationalization feature was published just recently, most languages are currently initialized with

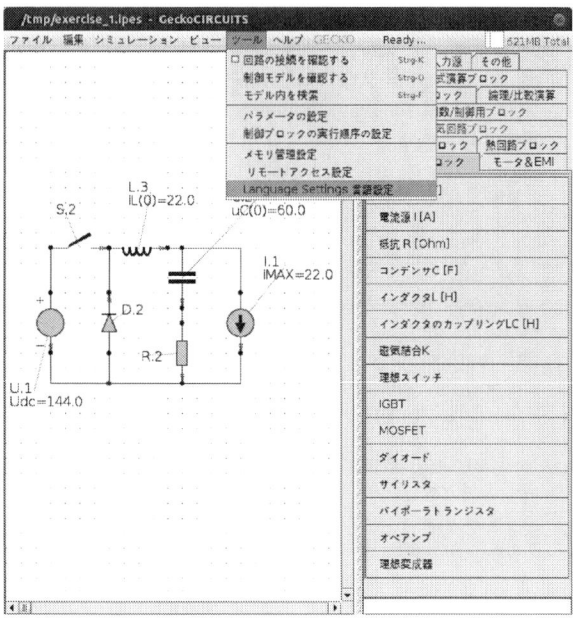

Fig. 10. Screenshot of the GeckoCIRCUITS graphical user interface translated into Japanese.

auto-generated contents which provides not the best data quality. However, with an active user community in the background, the database contents should evolve and improve over time. Every reader of this paper is invited to contribute to the project, for instance online using the Java applet in [20], or alternatively via the stand-alone open-source version of GeckoCIRCUITS, which can be downloaded from [7] for free. The only requirement for a user contribution is the installation of a Java runtime environment, good language skills and some patience.

V. CONCLUSIONS

In the first part of this paper, some requirements for interactive education platforms were briefly discussed. Our conclusions were that such a platform should be easy-to-use and available for free, preferably as open-source software. Then, the well-known iPES applet collection was reviewed, including a discussion of its advantages and shortcomings. iPES is very useful in demonstrating the behavior of power electronic circuits, but does not perform numerically accurate simulations. In order to allow unlimited simulations, the circuit simulation program GeckoCIRCUITS has been integrated into the iPES-platform as an online simulator.

Furthermore, two new features of the open-source circuit simulator GeckoCIRCUITS were introduced which improve the user experience and the educational value: A unique dynamic current path visualization now complements GeckoCIRCUITS as well as a novel language translation toolbox for its graphical user interface. The technical implementation of both features was explained in some detail.

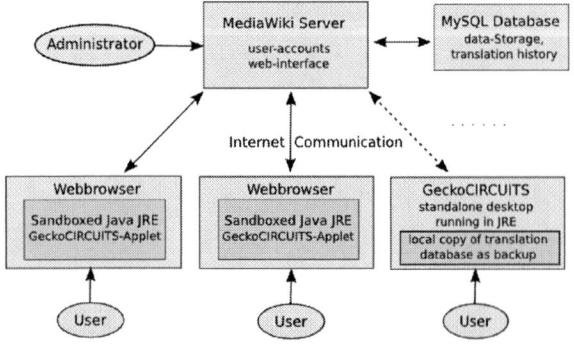

Fig. 9. Schematic overview of the GeckoCIRCUITS translation toolbox client-server configuration.

978-1-4799-2706-7/14 $31.00 © 2014 IEEE

VI. OUTLOOK

The possibilities of online education via Java applets are not yet fully exploited by iPES and GeckoCIRCUITS. Our intention on the long term is a replacement of traditional iPES applets with models implemented in Gecko-CIRCUITS. The necessary software adjustments and new features will allow users without in-depth programming skills to generate lecture-specific applets. The following list of features is yet missing in the circuit simulator to finally achieve this goal:

GeckoCIRCUITS has a built-in scripting capability, i.e. all component values, parameters and the simulation itself can be controlled via a simple source code. In order to include all iPES functionalities in the circuit solver, component values of resistances, inductances, etc. should be modifiable via mouse gestures.

The graph visualization of GeckoCIRCUITS should be more closely integrated into the circuit model. Then, for instance, it will possible to drag simulation curves for changing the amplitude of an sinusoidal input voltage or the duty ratio of a gate signal.

As an option, all irrelevant GUI components for the final education applet should be removed when the program is run in its applet mode. This would help the student to focus on the actual education content instead of the GUI of the program.

Finally, the software should provide compatibility with a vector graphics file format, for example the svg format (scalable vector graphics). The contents of many iPES applets is composed by static images and dynamic content. Then it is would be possible to easily draw the static images with an external program, e.g. Adobe Illustrator or Inkscape, whereas the dynamic contents would be implemented within GeckoCIRCUITS.

REFERENCES

[1] U. Drofenik, J. W. Kolar, P. J. Van Duijsen and P. Bauer, "New web-based interactive e-learning in power electronics and electrical machines," Proc. IEEE Ind. Appl. Conf. 36th IAS Annual Meeting, Chicago, vol. 3, pp. 1858-1865, October 2001.

[2] U. Drofenik and J. W. Kolar, Survey of modern approaches of education in Power Electronics, Proc. of the 17th Applied Power Electronics Conference (APEC) , vol. 2, pp. 749-755, Dallas, March 2002.

[3] U. Drofenik and J. W. Kolar, "Interactive power electronics seminar (iPES) - a web-based introductory power electronics course employing Java-applets," Proc. of the 33rd IEEE Power Electronics Spec. Conf. (PESC), Cairns, vol. 2, pp. 443-448, June 2002.

[4] Online: www.ipes.ethz.ch

[5] Online: www.ipes.ethz.ch/ipes/2002Vienna1/vr1.html

[6] U. Drofenik, A. Müsing and J. W. Kolar, "Novel online simulator for education of power electronics and electrical engineering," Proc. of the Int. Power Electronics Conf. (IPEC-ECCE Asia), Sapporo, Japan, June 2010.

[7] Online: www.gecko-simulations.com/geckocircuits.html

[8] Online: www.pes.ee.ethz.ch/en/our-range/education/courses.html

[9] Online: www.ipes.ethz.ch/ipes/Inverter/e_H_Bruecke.html

[10] V. Ramaswamy, "Interactive power electronics online text," online: http://services.eng.uts.edu.au/venkat/pe_html/peintro.htm

[11] S. Harb, K. Kalaldeh, A. Harb, I. Batarseh, "Interactive java applets for power electronics e-learning," Proc. of the IEEE Workshop on Power Electronics Education, pp. 26 -33, 2005.

[12] J. Hamar , H. Funato , S. Ogasawara , O. Dranga and C. K. Tse, "Multimedia based e-learning tools for dynamic modeling of DC-DC converters," Proc. of IEEE Int. Conf. on Industrial Technology (ICIT), Hong-Kong, December 2005.

[13] J. Hamar, I. Nagy, H. Funato, S. Ogasawara, O. Dranga and Y. Nishida, "Virtual power electronics: novel software tools for design, modeling and education", Proc. of the Power Conversion Conference (PCC), pp. 502 - 509, Nagoya, April 2007.

[14] U. Probst, "Object-oriented Toolbox for simple generation of Java-Applets for web-based teaching of electrical engineering applications," Interactive Conference on Computer Aided Learning (ICL), Villach, September 2007.

[15] J. H. Allmeling and W. P. Hammer, "PLECS – piece-wise linear electrical circuit simulation for Simulink," Proc. of the IEEE Int. Conf. on Power Electronics and Drive Systems (PEDS), vol.1, pp.355-360, 1999.

[16] P. J. van Duijsen, P. Bauer and B. Davat, "Simulation and animation of power electronics and drives, requirements for education," Taiwan Power Electronics Conf. and Exh., pp. 1048-1053, 2005.

[17] Online: www.caspoc.com

[18] Online: www.icircuitapp.com

[19] Online: www.falstad.com/circuit/e-index.html

[20] Online: www.gecko-simulations.com/IPEC/sparsematrix.html

An Electric Vehicle Project

for ECO-RUN Race

- As a student project in Ibaraki National College of Technology -

Shinichi YAMAGATA, Yoshinori ODA, Masanobu TANAI and Kyungmin SUNG
Engineering and Education Support Center &   Depart. of Electrical and Electronic System Engineering
Ibaraki National College of Technology
Hitachinaka-si, Ibaraki, Japan

Abstract— This paper presents the student's manufacture activity of a racing Electric Vehicle (EV) which is a very small size and single rider type for participate ECO-RUN Race in Sugo Japan. The racing EV requires a light body weight, small wind resistance, high electric energy efficient, etc. to achieve a good score in the ECO-RUN racing tournament in Sugo. Not only power electronics but also mechanical design and manufacture technique have been demanded. In this paper, firstly, the Ibaraki National College of Technology (INCT) of EV project team will be introduced. Next, the mechanical manufacture process and designed power circuit mainly bidirectional DC-DC converter in order to introduce Super Capacitor (EDLC; Electric Double Layer Capacitor) to EV will be presented. Furthermore, the new approach system which is a series or parallel changeover system using battery with EDLC for assist power in short term is proposed. Finally, the experimental result of the power circuit in Sugo racing circuit will be presented.

Keywords— *Electric Vehicle, ECO-RUN Race, DC-DC Converters, EDLC; Electric Double Layer Capacitor.*

I. INTRODUCTION

Ibaraki National College of Technology (INCT) is five years Course College after graduated middle school. After graduation of main five year, an advanced two years course is established to educate a special engineering study. The lecture of main major starts from two year degree. The study of power electronics is started from five years degree. In the five year, students have to study a graduation research of each major field. During the advanced two years course, students start more specialty research than main course. Also the academic conference presentation is attached to the duty of graduation.

In this paper, the INCT student's project of Electric Vehicle (EV) manufacturing is introduced. The four, five degree and the advanced course student of Electric and Electronic System Engineering Department join in this EV project. Since 2008, the members have been manufacturing a solar car and EV for ECO-RUN Racing tournament.

The INCT team of this project member has been participated the EV ECO-RUN race in SUGO in japan since 2009. The racing circuit of SUGO is popular with many up and downhill and high angle gradient.[1] Especially, the 400 meter with 10 % gradient uphill is difficult hurdle for small power EV. The charge and recharge technic of regenerative energy between a EDLC and driving motor of EV is the most important element technic to enter high level teams.

This paper presents the INCT student's manufacture activity of a racing EV which is a very small size and single rider type for participate ECO-RUN Race in SUGO Japan. The racing EV of light body weight, small wind resistance, high electric energy efficient, etc. are demanded to achieve a good score in the ECO-RUN racing in SUGO. Not only power electronics but also mechanical design and manufacture technique have been required. In this paper, firstly, the mechanical

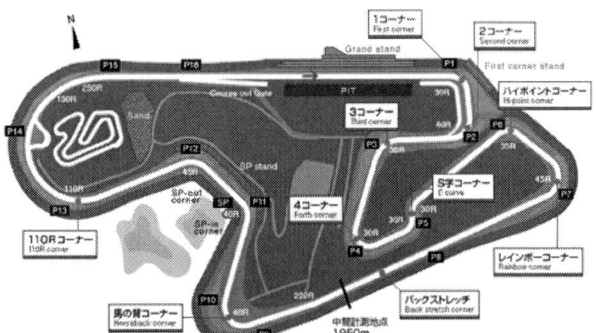

Fig. 1. Racing Circuit.

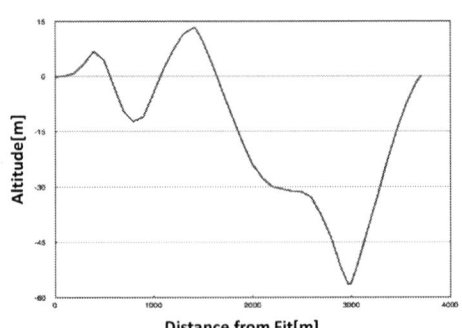

Fig.2. Altitude of racing circuit.

978-1-4799-2706-7/14 $31.00 © 2014 IEEE

Fig. 3. Body structure with three wheels.

TABLE I
SIZE OF VEHICLE

Overall Length	2500mm
Overall Width	750mm
Overall Height	600mm
Wheel Base	1300mm
Wheel Distance	700mm
Overall Body Weight	50kg

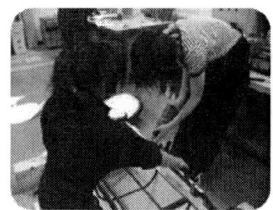

Fig. 4. Process of body frame manufacturing.

Fig. 5. Armature winding of BLDC motor.

manufacture process and designed power circuit mainly bidirectional DC-DC converter in order to introduce EDLC to EV be presented. Moreover, the new approach system which is a series or parallel changeover system using battery with EDLC for assist power in short term is proposed.[2] Finally, the experimental result of the power circuit in the SUGO racing circuit will be presented.

II. ECO-RUN RACING FOR EV

The EV ECO-RUN Racing is held in the Sport Land SUGO located in Sendai city in Japan. The rules of this racing compete the running distance after two hours run with a determined battery weight. Three or four wheel of EV machine can participate. There are no limit of machine weight except the length and width of machine. Only the battery weight is limited by racing regulation. So, in terms of the mechanical design, the light weight and sharp machine body to reduce wind resistance when running circuit has advantage.

The racing circuit is shown in Fig. 1. The full-length is 3705 meter. The best altitude between the peak and low place become approximately 70 meter. Fig. 2 shows the altitude on the circuit. From the Point of P7 in Fig. 1 to P9 is a straight dawn-hill course about 1000 meter. Normally, this course gives an opportunity to recharge super capacitors or batteries regenerative energy by drive motor. And after this dawn-hill course, EV dash up the 10% gradient about 500 meter up-hill course for pit's

straight. In the point of P12 to P15, EV has to output the maximum power by itself.

III. DESIGN & MANUFACTURE

A. Mechenical Manufacture

Students who participated in this project have a major of Electric and Electronics System Dept. So, the mechanical design and manufacture is first time for the member. First, we looking for many pictures and drawing related small EV from the internet. And we consider the enough mechanical strength that EV can be safe driving and technically easy structure for manufacturing.

The body structure of prototype is shown in Fig. 3. It is canoe type body that likes small ship with three wheels. The main body is manufactured by welding. The lateral semicircular bone of steel pipe is used by welding to increase mechanical strength.

Fig. 4 shows the process of body frame manufacturing. The welding of frame is carried out by Engineering and Education Support Center technical staff in INCT. Also, the mechanical drawing, manufacture steel and process step is supported by technical staff.

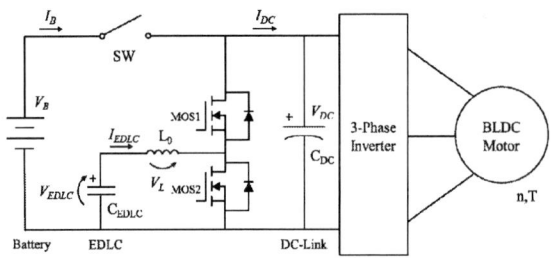

Fig. 6. Drive system of BLDC motor with bidirectional DC/DC converter.

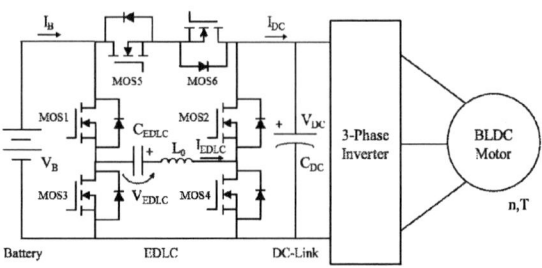

Fig. 8. Proposed new circuit.

TABLE II
ELECTRICAL SPECIFICATION

Motor	Brushless DC motor (1kW continuous, 2kW peak)
Battery Pack	48V 7.5Ah (4×12V, 7.5Ah cells in series)
EDLC Bank Capacitance C_{EDLC}	30V max 58.3F (12×2.5V max 350F cells in series and 2 in parallel)
DC link Capacitance C_{DC}	63V max 4980µF (6×63V max 820µF cells in parallel)
Inductance L_0	2mH
Switching Frequency	5kHz
Duty Ratio	0.5 constantly

TABLE III
MODE OF PROPOSED CIRCUIT.

	State of MOSFET	Energy Flow
Mode1	MOS5:ON	Battery → MOS5 → MOS6(D) → DC link
Mode2	MOS2:ON/OFF MOS3:ON	DC link → MOS2 → EDLC (→ MOS3)
Mode3	MOS4:ON/OFF MOS5:ON	Battery → MOS5 → MOS6(D) → DC link & EDLC → MOS2(D) → DC link (→ MOS3(D))
Mode4	MOS1:ON	Battery → MOS1 → EDLC → MOS2(D) → DC link

inner rotor type with outer frame, in which is field circuit by Neodymium magnet. A Commercial available 1kW rate three phase inverter drives this motor.[3]

IV.   BIDIRECTANALA DC/DC CONVERTERS

The configuration of drive system is shown in Fig. 6. The three inverter and bidirectional DC/DC converter are connected by DC link capacitor. In the EV, usually, a bidirectional DC/DC converter is required to interface an energy storage device such as an EDLC or a battery to the DC bus which is the DC link of a motor drive inverter [4-8]. The bidirectional DC/DC converter is designed and assembled by discrete electronics parts. In this chapter, the conventional system using a bidirectional DC/DC converter is presented to compare with our proposed system. A 16bit micro-processors used for this DC/DC converter. The photo of drive system of BLDC and bidirectional DC/DC converter is shown in Fig. 7.

The battery and the EDLC bank are connected to the DC link in parallel. The EDLC bank is interfaced to the DC link by the bidirectional DC/DC converter, which consists of MOS1, MOS2, and L_0. The vehicle specification and the electric specification are shown in Table I and Table II.

The operation mode of this system is basically divided into three modes such as the driving mode by the battery, the regenerative braking mode with the EDLC bank, and the driving mode by the EDLC bank. In the driving mode

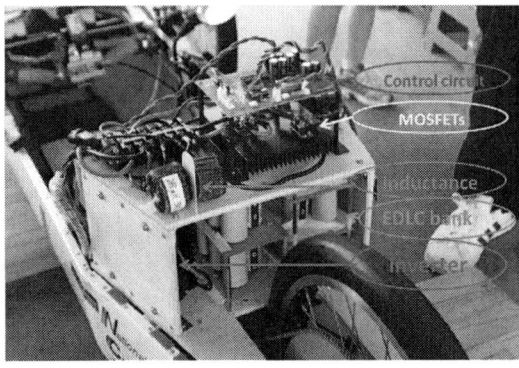

Fig. 7. Photo of the drive system of BLDC motor with real wheel.

B. Power Circuit

The prototype EV machine is adopted a Rear Wheel Drive method. The 1kW rate brushless DC motor (BLDC) motor is connected rear wheel by mechanical gear and chain. The front wheel is connected only steering. Due to climb the 10% gradient up-hill course, the gear ration is determined with rated torque and rotation speed of BLDC motor.

Fig. 5 shows the photo of BLDC motor's armature winding. A commercial assembling BLDC kit is made by a hand. It is three phase 6 poles revolving- armature and

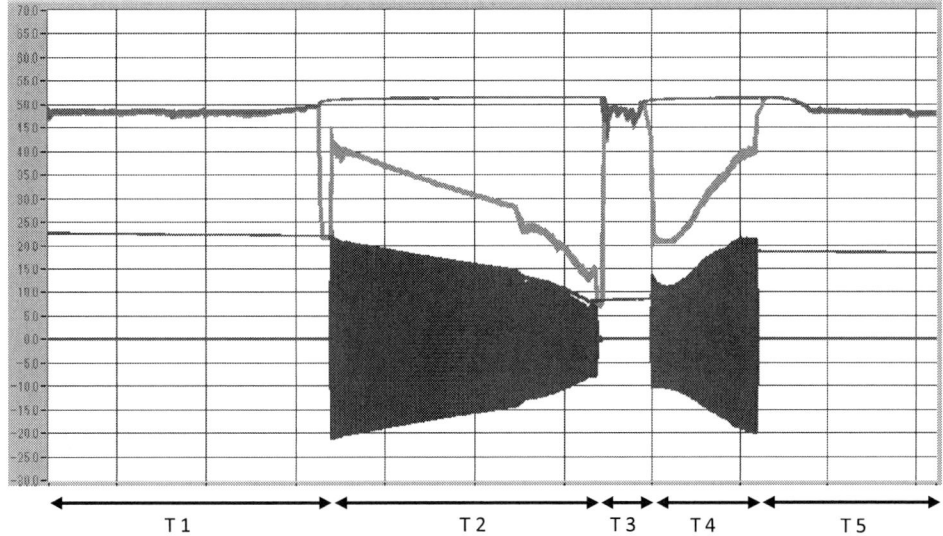

Fig. 9. Measured data in Sugo circuit. (From the top the battery voltage, DC link voltage, EDLC bank voltage and inductor voltage, 5[V/div]. 20[s/div]

by the battery, the mechanical switch (SW) is on state and the bidirectional DC/DC converter is not operated. At this time, only the battery provides the energy for the drive motor and the EDLC bank keeps the stored energy. When the EV changes the operation mode into the regenerative braking mode with the EDLC bank, the SW is turned off and the bidirectional DC/DC converter is operated as the step down converter by the switching operation of MOS1. As a result, the BLDC motor generates electric energy and it charges the EDLC bank.

In the driving mode by the EDLC bank, the SW is off state. And, the bidirectional DC/DC converter is operated as the boost converter by the switching of MOS2. When the DC/DC converter is a boost conversion mode, the EDLC bank can discharge the energy into the drive motor. On the other hand, in this boost mode, the operation voltage range of the DC/DC converter is limited by a voltage level of the EDLC bank. A high step-up voltage conversion over several times is difficult due to a saturation of L_0 or an excessive decrease in an EDLC terminal voltage under low voltage range.

V. PROPOSED NEW CIRCUIT FOR POWER ASSISTANCE

In order to improve the operating voltage range of the EDLC bank, the proposed system enables to connect the battery and the EDLC bank not only in parallel but also in series. When the bidirectional DC/DC converter becomes a situation under uncontrollable voltage level, the proposed system changes the connection of the battery and the EDLC bank from parallel to series. Therefore, even if V_{EDLC} is near zero, the proposed system can discharge the stored energy in the EDLC bank. The circuit configuration and the operation mode are presented in this chapter.

Fig. 8 shows the circuit configuration of the proposed system. The energy storage system is comprised of a battery pack, an EDLC bank, an inductor and six MOSFETs. The battery is interfaced to the DC link by the bidirectional switch, which replaces the conventional SW in Fig.1 with MOS5 and MOS6. The EDLC bank is interfaced to the DC link by the bidirectional DC/DC converter, which is composed of MOS2, MOS4, and L_0.

This bidirectional DC/DC converter is the same topology as the conventional configuration. The proposed system is introduced only two MOSFETs in comparison with the conventional system. MOS1 is used to connect the battery and the EDLC bank in series. And, MOS3 is used in the regenerative braking mode. Each mode is summarized in Table III.

Mode1: Driving Mode by Battery

During the driving mode by the battery, only MOS5 is on state. As a result, the battery is connected to the DC link through the bidirectional switch and the DC link voltage V_{DC} is equal to the battery voltage V_B.

Mode2: Regenerative Braking Mode with EDLC bank

In order to carry out the regenerative braking mode, after MOS3 is turned on, MOS2 is operated by the control signal. At this time, the bidirectional DC/DC converter is operated as the step down converter and the braking energy flows into the EDLC bank.

Mode3: Driving Mode by Parallel Connection

When the energy amount of the EDLC bank is sufficient, the driving mode by the parallel connection of the battery and the EDLC bank can be operated. In this mode, MOS5 is on state and the bidirectional DC/DC converter is operated as the boost converter by the switching of MOS4. As a result, this mode provides the EV with the power assist by the battery and the EDLC bank energy. But, for essentially the same reason as the conventional system, this mode is restricted by the EDLC bank voltage V_{EDLC}.

During this mode, V_{DC} is equal to V_B. And assuming that the duty ratio of MOS4 is D, the battery current is I_B and the EDLC bank current is I_{EDLC}, V_{DC} and the DC link current I_{DC} are shown in below equations.

$$V_{DC} = V_B = \frac{1}{1-D} \cdot V_{EDLC} \tag{1}$$

$$I_{DC} = I_B + (1-D) \cdot I_{EDLC} \tag{2}$$

Mode4: Driving Mode by Series Connection

The driving mode by the series connection of the battery and the EDLC bank is also used as the power assist. But this mode is only used when V_{EDLC} is lower than the limitation of Mode3. In this mode, only MOS1 is on state. Then the battery and the EDLC bank are connected in series. Therefore, V_{DC} is equal to the sum of V_B and V_{EDLC}. Additionally, I_{DC}, I_B and I_{EDLC} are equivalent. As a result, V_{DC} and I_{DC} are shown in (3),(4).

$$V_{DC} = V_B + V_{EDLC} \tag{3}$$

$$I_{DC} = I_B = I_{EDLC} \tag{4}$$

VI. EXPERIMENTAL RESUL

A. Conventional DC/DC converter

In order to confirm the behavior of the conventional configuration, the prototype small EV is experimented on a road. Fig.9 shows the experimental result with the conventional configuration. This data is measured while the EV drives up and down hill. From the top, waveforms show the battery voltage V_B, the DC link voltage V_{DC}, the EDLC bank voltage V_{EDLC} and the inductor voltage V_L. According to waveforms, this data can be divided into each term from T1 to T5.

During T1, T3 and T5, the EV is in the driving mode by the battery. Then V_{DC} is equal to V_B and V_{EDLC} is kept the same value. The bidirectional DC/DC converter is not operated. During T4 shows the regenerative braking mode with the EDLC bank. In this term, V_{EDLC} is increased by the braking energy. In this case, V_{EDLC} is about half of V_{DC} because the duty ratio has constant value of 0.5.

During T2 shows the driving mode by the EDLC bank. V_{DC} is about twice as large as V_{EDLC} because the duty ratio of MOS2 is 0.5 constantly. And V_{EDLC} is decreased with the discharge of the charged energy. As this result, when V_{EDLC} reaches to 15V, the bidirectional DC/DC converter becomes uncontrollable by a low terminal voltage of the EDLC bank. Therefore, this system can operate the EDLC bank only between 15V and the rated voltage 30V of terminal voltage.

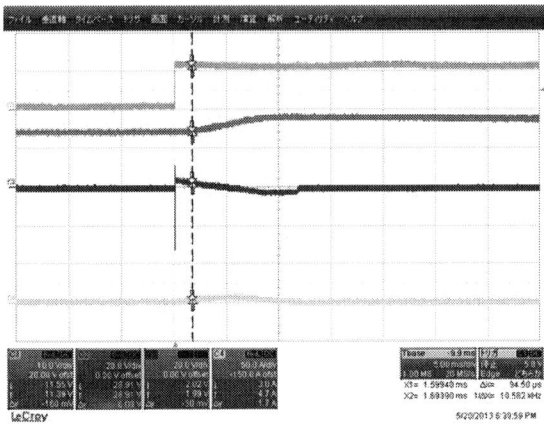

(a) Measured data by proposed circuit. (From the top, the Gate-Source voltage of MOS1 V_{GS1}, DC link voltage V_{DC}, inductor voltage V_L, and inductor current I_L)

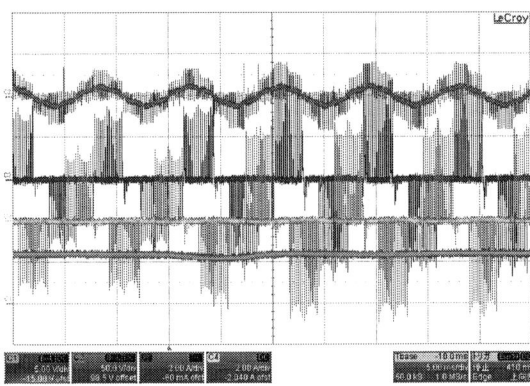

(b) Waveforms of BLDC Motor and EDLC Voltage
(From the top, U-V line voltage V_{UV}, U-phase current I_U, W-phase current I_W, and EDLC bank voltage V_{EDLC})
Fig. 10. Proposed system

B. Power Assistance Circuit

In order to verify that the proposed system can shift into the different mode in safety, the prototype is experimented. The mode change from Mode1 to Mode4 is presented. As the composition of the main circuit in this experiment, the battery is replaced with the DC power supply, which is fixed V_B, 30V. The EDLC bank consists of 6 cells in series (15V max 58.3F). The energy storage system is interfaced to the DC link with the inductor, which is adopted to prevent the inrush current due to the mode change. The inductance value is 2 mH.

The BLDC motor is the no-load condition. Fig.10 shows the experimental result. Fig.10 (a) are waveforms of the proposed energy storage system, which are the Gate signal of MOS1 V_{GS1}, the DC link voltage V_{DC}, the inductor voltage V_L, and the inductor current I_L. Fig.10 (b) shows terminal voltage and current waveforms of the BLDC motor with the EDLC bank voltage. From the top, waveforms are the U-V line voltage V_{UV}, the U-phase current I_U, the W-phase current I_W, and the EDLC bank voltage V_{EDLC}. In this experiment, the initial voltage of the EDLC bank is 6V. When MOS1 is turned on, the

978-1-4799-2706-7/14 $31.00 © 2014 IEEE

proposed system changes the operation mode from Mode1 to Mode4. At this time, V_{DC} is increased from V_B to the sum of V_B and V_{EDLC}. As a result, V_{UV}, I_U and I_W are increased in proportion to the increase in V_{DC}. Therefore, even if V_{EDLC} is lower, the BLDC motor is provided with the power assist by Mode4.

VII. CONCLUSIONS

In this paper, the INCT student's EV project, in which is a design and manufacturing process of the small EV car and about participation result of the ECO-RUN racing tournament is introduced. Through this project, the young student has understanding not only the power electronics technic by to implement bidirectional DC/DC converters also the electric machine like a BLDC motor. The archived result from ECO-RUN racing was presented in the 2013 15th European Conference.

The most important thing by this project is to give an opportunity which lead students to manufacturing field and making things. The INCT student's challenge for ECO-RUN Racing will be continued with a young passion.

REFERENCES

[1] Sport land Sugo, http://www.sportsland-sugo.co.jp/

[2] Katuhiro Hata, Naoya Watanabe, and Kyungmin Sung, "A Series or Parallel Changeover System using Battery with EDLC for EV", 2013 15th European Conference, pp.1-10.

[3] Mitsuba Corp. https://www.mitsuba.co.jp/

[4] Juan Dixon, Ian Nakashima, Eduardo F. Arcos, and Micah Ortuzar, " Electric Vehicle Using a Combination of Ultracapacitors and ZEBRA Battery", IEEE Trans. IE, Vol.57, No.3, March 2010, pp.943-949.

[5] Garcia, F.S. Ferreira, A.A. Pomilio, and J.A, "Cotrol strategy for battery-ultracapacitor hybrid energy storage system", Applied Power Electronics Conference and Exposition, 2009. APEC2009. Twenty-fourth Annual IEEE, pp.826-832.

[6] Jian Cao and Ali Emadi, "A New Battery/UltraCapacitor Hybrid Energy Storage System for Electric, Hybrid, and Plug-In Hybrid Electric Vehicles", IEEE, Tran. PE, vol27, NO.1, JAN 2012.

[7] Fangcheng. Liu, Jinjun. Liu, and Linyuan. Zhou, "A Novel Control Strategy for Hybrid Energy Storage System to Relieve Battery Stress", 2010 2nd IEEE International Symposium on Power Electronics for Distributed Generation Systems, pp.929-934.

[8] G.Guidi, T.M. Undeland, and Y. Hori, "Optimized Power Electronics Interface for Auxiliary Power Buffer Based on Supercapacitors", IEEE Vehicle Power and Propulsion Conference (VPPC), Sep. 3-5, 2008, Harbin, China.

[9] Luca Solero, Alessandro Lidozzi, and Jose Antenor Pomilio, "Design of Multiple-Input Power Converter for Hybrid Vehicles", IEEE Trans. PE, Vol.20, No.5, Sep. 2005, pp.1007-1016.

[10] Sung-Geun Yoon, Jae-Moon Lee, Jong-Hu Park, In-kyu Lee, and B.H. Cho, "A frequency controlled bi-directional synchronous rectifier converter for HEV using super-capacitor", 2004 34th Annual PESC, Aachen Germany 2004, pp.135-140.

[11] Dobri Cundev and Pavel Mindl, "European Driving Schedule of Hybrid Electric Vehicle with Electric Power Splitter and Super-capacitor as Electric Storage Unit", Proc. Of the 2008 International Conference on Electrical Machines, Paper ID 833, pp.1-5.

Multi-loop Controller Design for Diode-assisted Buck-boost Voltage Source Inverter

Yan Zhang, Student Member, IEEE, Jinjun Liu, Senior Member, IEEE, Xiaolong Ma and Junjie Feng

State Key Lab of Electrical Insulation and Power Equipment, Department of Electrical Engineering
Xi'an Jiao Tong University, Xi'an, China
Email: zhy831112@gmail.com

Abstract— Diode-assisted buck-boost voltage source inverter extends voltage regulation range and avoids the extreme boost duty ratio of switching devices in front boost circuit by introducing diode-capacitor network. It provides an attractive dc-ac power conversion with enhanced voltage buck-boost capability, low cost and high efficiency in high voltage gain application. However, due to the chopped intermediate dc-link voltage, the front boost circuit and inverter bridge have to be controlled coordinately, which increases the complexity of controller design. This paper presents model analysis and reveals that the converter can be equivalent to conventional dc-dc boosted voltage source inverter with modified modulation index. Therefore, the dc and ac-side regulator are designed separately. Like boost-type dc-dc converter, the intermediate dc-link voltage has a non-minimum-phase system transient response. This undesired influence may be transferred to the ac-side, resulting in initial voltage undershoot and oscillatory. Therefore, the close-loop control bandwidth of the inverter stage is relatively much larger than that of the front boost circuit to minimize such an influence on the output ac-side. Finally, all the theoretical findings and proposed control algorithms are verified by simulation and experiment results.

Keywords— *Buck-boost voltage source inverter, Controller design, Dynamic performance, Model analysis, Right half plane zero.*

I. Introduction

Renewable energy source such as solar and fuel cells are promising in the future. However, the obvious characteristic of them is low voltage supply with wide range voltage drop. In photovoltaic (PV) generation and fuel cells supplied motor drive system, power electronic interface has to regulate the amplitude and frequency to obtain required output ac voltage. These applications raise the stringent requirements for power converters such as low cost, high efficiency and wide range voltage regulation ability [1]. The conventional voltage source inverter (VSI) can only perform buck voltage regulation. Therefore, the various novel high step-up dc-dc, dc-ac topologies along with the related control methods have been proposed to solve this issue [2]-[17].

This work was supported in part by the National Basic Research Program (973 Program) of China under Project 2009CB219705 and in part by the State Key Laboratory of Electrical Insulation and Power Equipment under the Project EIPE09109.

The conventional two-stage VSI obtains required output voltage by introducing dc-dc boost circuit. In view of additional power stage increasing cost and lowing efficiency, Peng [2] proposed Z-source inverter. Z-source inverter introduces a unique impedance network between the source and the inverter bridge to obtain both voltage buck and boost properties. It boosts the dc source voltage by adopting shoot-through (ST) operation mode and provides a potential cheap and single-stage power conversion. However, the ST state accompanies the large ST current. Shen et al. [9] made comparison of two converters based on fuel cell supplied electric vehicle driver system. The results reveal that Z-source inverter demonstrates low cost and high efficiency when boost ratio is low (1~2). The authors in [10] and [11] conducted the same comparison based on the grid-tied renewable energy generation. It reveals that the conventional dc-dc boosted VSI bears smaller power rating of switching devices and has higher efficiency than that of Z-source inverter under the same input and output conditions.

Although both of them can boost output voltage to any desired value, the degradation of efficiency and increasing requirement of power devices are prominent under high voltage gain applications. Gao et al. [3] proposed diode-assisted buck-boost VSI shown in Fig.1 to further extend voltage gain and avoid the extreme boost duty ratio. The diodes are naturally conducting to perform parallel capacitive charging and series capacitive discharging to achieve high step-up ratio. In view of chopped intermediate dc-link voltage, the front boost circuit and inverter bridge have to be controlled coordinately. Therefore, various PWM strategies are proposed to maximize voltage gain as well as to minimize voltage stress of switches [4]-[6]. With the improved PWM strategies, it is a competitive topology for wide range voltage regulation [4]. However, taking into account of the chopped intermediate dc-link voltage, the controller design is more complicated than that of conventional VSI. Meantime, the transient analysis and controller design have not been discussed before. This paper makes a detailed model analysis and reveals that the converter can be equivalent to conventional dc-dc boosted voltage source inverter with modified modulation index. Like boost-type dc-dc converter, the transfer function of boost duty ratio to intermediate capacitor voltage contains right-half plane (RHP) zero. This non-minimum influence may propagate to the ac-

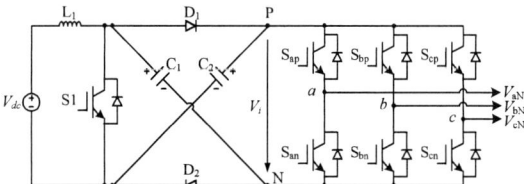

Fig. 1. Diode-assisted buck-boost VSI

side. Then, advanced multi-loop control algorithms for both dc-side boost circuit and ac-side inverter are designed, respectively [4]-[17].

II. Operation Principle and Modulation Strategies

The operation principles of aforementioned topology have been described in literature [3]. Fig.2 shows two operation modes according to the switching state of S1. When S1 is turned on, L_1 absorbs energy from dc source and both capacitors are connected in series to supply the inverter stage. When S1 is turned off, the energy accumulated in L_1 is transferred to C_1, C_2 and both capacitors are connected in parallel to supply the inverter stage. With parallel capacitive charging and series capacitive discharging, the relatively high intermediate dc-link voltage can be easily achieved under the same boost duty ratio and capacitor voltage rating.

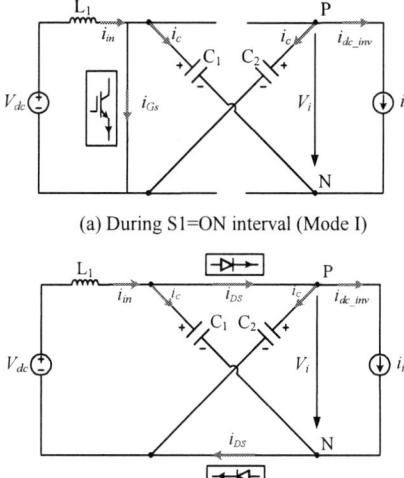

(a) During S1=ON interval (Mode I)

(b) During S1=OFF interval (Mode II)

Fig. 2. The equivalent circuit of diode-assisted buck-boost VSI viewed from the dc-link of the inverter bridge.

The standard two-level three-phase VSI can produce six active vectors and two zero vectors. Fig.3 shows all the available voltage vectors with regard to the variable intermediate dc-link voltage (V_i). Among which, the outer six large active vectors V_2, V_4, V_6, V_8, V_{10}, V_{12} are produced when V_i is twice the capacitor voltage $2V_C$ ($V_P=V_C$, $V_N=-V_C$) during S1=ON interval; the inner six small active vectors V_1, V_3, V_5, V_7, V_9, V_{11} are produced when V_i is the same as the capacitor voltage V_C ($V_P=V_C$, $V_N=0$) during S1=OFF interval; V_r is the reference voltage vector which represents the balanced three-phase

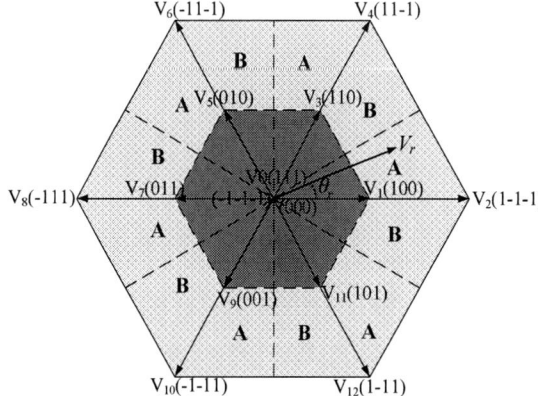

Fig. 3. Voltage vectors for diode-assisted buck-boost VSI.

output voltage [4][8]. With regard to the chopped voltage of V_i during one switching time period (T_s), the coordinate control of the front boost circuit and inverter stage has to be applied. Fig.4 shows the sequence of voltage vectors in the first sextant for improved PWM strategies introducing small active vectors to extend output voltage in [4]. Fig.5 shows the combination of a reference vector V_r. As is illustrated in Fig.5, reference vector V_r is divided into two voltage vectors OO' and $O'R$. In which, OO' is synthesized with small vectors V_1 and V_3; $O'R$ is synthesized with large vectors V_2 and V_4. With improved PWM, diode-assisted buck-boost VSI obtains the maximum linear voltage gain in theory, as well as minimize the voltage stress of switching devices. As is described in [4], the output ac voltage and corresponding voltage gain can be written as follows.

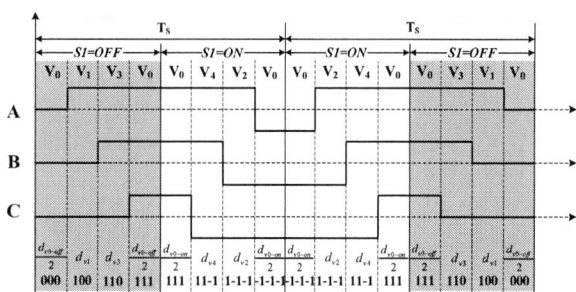

Fig. 4. Sequences of voltage vectors in the first sextant for improved PWM strategy in literature [4].

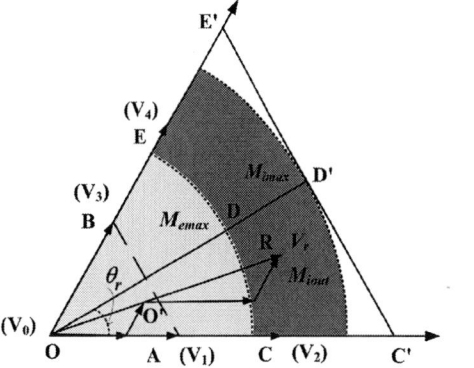

Fig. 5. Combination of reference vector V_r in the first sextant for improved PWM strategy in literature [4].

978-1-4799-2706-7/14 $31.00 © 2014 IEEE

$$\hat{v}_{ac} = \frac{1+k_{on}}{1-k_{on}} M_i \cdot \frac{V_{dc}}{2} \qquad (0 \le k_{on} \le 1, 0 \le M_i \le \frac{2}{\sqrt{3}}) \quad (1)$$

$$G = \frac{1+k_{on}}{1-k_{on}} M_i \qquad (0 \le k_{on} \le 1, 0 \le M_i \le \frac{2}{\sqrt{3}}) \quad (2)$$

Where: $\hat{v}_{ac}$ is the peak value of output phase voltage; G is defined as $\hat{v}_{ac}$ over the half of dc source voltage $V_{dc}/2$.

III. Model Analysis and Multi-loop Controller Design

A. Model and Controller Design for the Front Boost dc-dc Circuit

As for diode-assisted buck-boost VSI supplying three-phase balanced *RL* load, the three-phase load is equivalently transferred to the dc-side according to the power balance [14]. The inverter bridge can be simplified by a pair of controllable switching device (S2) and freewheeling diode (D$_x$) in buck converter. The equivalent dc-side circuit is shown in Fig.6. In which, S1 controls the boost duty ratio (k_{on}) and S2 controls the modulation index of inverter bridge (M_i) to regulate the output voltage. The active power of three-phase *RL* load and dc load shown in Fig.6 can be expressed as (3) and (4), respectively.

$$P_{ac} = \frac{3}{2} \cdot \hat{v}_{ac} \cdot \frac{\hat{v}_{ac}}{|Z_{ac}|} \cdot \cos\varphi \qquad (3)$$

$$P_{o\_dc} = \frac{\overline{v}_o^2}{R_e} \qquad (4)$$

Where: $\hat{v}_{ac}$ is the peak value of output phase voltage; $Z_{ac}=R_{ac}+j\omega L_{ac}$ is the impedance of ac load; $cos\varphi$ is load power factor; R_e is the equivalent resistor of dc load.

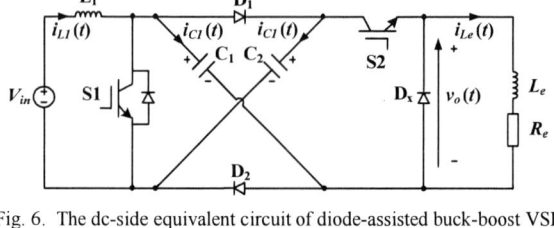

Fig. 6. The dc-side equivalent circuit of diode-assisted buck-boost VSI

Assuming the average value of output dc voltage $\overline{v}_o$ is twice the ac output voltage $\hat{v}_{ac}$, thus the duty ratio of S2=ON is M_i. According to the active power balance between dc load and previous ac load, and inductor with the same time constant, R_e and L_e can be expressed as:

$$R_e = \frac{8|Z|}{3\cos\varphi} \qquad L_e = R_e \cdot \frac{L_{ac}}{R_{ac}} \qquad (5)$$

According to switching state of S1 and S2, there are four operation modes for the dc-side equivalent circuit. As illustrated in Fig.7, during S1=ON and S2=ON interval, the inductor absorbs energy from dc source and both capacitors are connected in series to supply the load. This equivalent circuit represents that the inverter outputs large active vectors. During S1=OFF and S2=ON interval, (c) represents that the inverter outputs small active vectors. (b) and (d) represent the inverter output zero voltage vectors under the different intermediate dc-link voltage.

Taking i_{L1}, i_{Le} and V_C as state variables and using state-space averaging method in one switching time period (T_s), the converter average model can be obtained as (6)-(8).

$$L_1 \frac{di_{L_1}}{dt} = v_{in} - (1-k_{on}) \cdot v_c \qquad (6)$$

$$L_e \frac{di_{Le}}{dt} = \delta \cdot v_c - R_e \cdot i_{Le} \qquad (7)$$

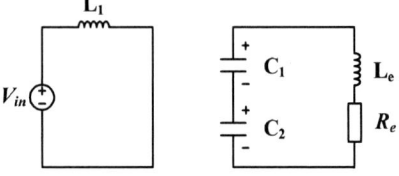

(a) During S1=ON and S2=ON interval.

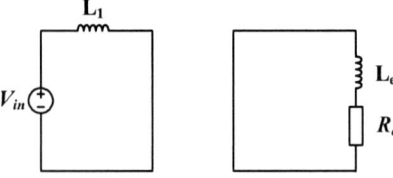

(b) During S1=ON and S2=OFF interval.

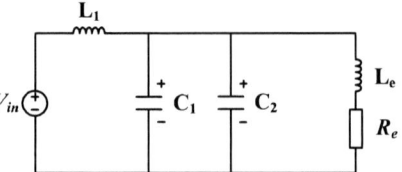

(c) During S1=OFF and S2=ON interval.

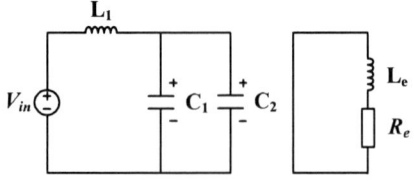

(d) During S1=OFF and S2=OFF interval.

Fig. 7. The different operation modes of dc-side equivalent circuit.

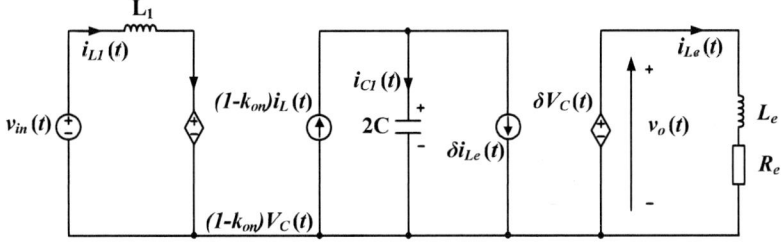

Fig. 8. The average model of dc-side equivalent circuit.

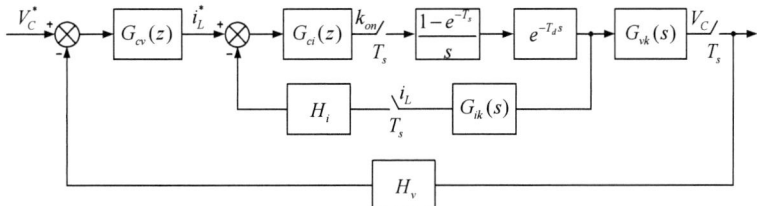

Fig. 9. The control diagram of dc-side boost circuit of diode-assisted buck-boost VSI.

$$2C\frac{dv_c}{dt} = (1-k_{on})\cdot i_{L_1} - \delta \cdot i_{Le} \qquad (8)$$

where $\delta=(1+k_{on})M_i$; k_{on} is on-state duty ratio of S1; M_i is the modulation index of the inverter stage; the control variable is k_{on}; the state variable to be controlled is v_C.

From the aforementioned state equations (6)-(8), the average model equivalent circuit can be obtained and shown in Fig.8. As illustrated in Fig.8, diode-assisted buck-boost VSI can be simplified into a conventional boost circuit connected with three-phase VSI if the v_C is seemed as the fixed dc link voltage of inverter bridge. And a new control variable δ is seemed as the equivalent modulation index of the inverter stage.

Performing small signal perturbation at the equilibrium operation point (I_{L1}, V_C, I_{Le}, k_{on}, δ), the transfer functions of k_{on} to i_L and v_c are as:

$$G_{ik}(s)=\frac{\hat{i}_{L1}(s)}{\hat{k}_{on}(s)}$$
$$=\frac{2CL_eV_C\cdot s^2+2(CR_eV_C+(1-k_{on})L_eI_{L1})\cdot s+\delta^2V_C+(1-k_{on})R_eI_{L1}}{2CL_1L_e\cdot s^3+2R_eCL_e\cdot s^2+\left((1-k_{on})^2L_e+\delta^2L_1\right)\cdot s+(1-k_{on})^2R_e}$$
$$(9)$$

$$G_{vk}(s)=\frac{\hat{v}_c(s)}{\hat{k}_{on}(s)}$$
$$=\frac{-L_1L_eI_{L1}\cdot s^2+\left((1-k_{on})L_eV_C-L_1R_eI_{L1}\right)\cdot s+(1-k_{on})R_eV_C}{2CL_1L_e\cdot s^3+2R_eCL_e\cdot s^2+\left((1-k_{on})^2L_e+\delta^2L_1\right)\cdot s+(1-k_{on})^2R_e}$$
$$(10)$$

It can be observed in (10) that the transfer function of $\hat{k}_{on}$ to $\hat{v}_C$ contains the right-half-plane (RHP) zero. With RHP zero, v_C falls before rising to the reference when the step up of control command is given [15] [16]. This non-minimum phase system exhibits a worse dynamic response. And the undesired performance may be transferred to the output ac-side.

In general, dual-loop controller is applied to deal with such an influence and obtain the good dynamic response. As is illustrated in (9) and (10), $G_{ik}(s)$ is the minimum phase system transfer function. Therefore, the inner current regulator is easily designed to meet the bandwidth requirement. As for the control of V_C, the RHP zero may limit the close-loop bandwidth. Thus, the voltage loop has to be designed with the relatively small bandwidth, which is beneficial to reduce the initial voltage undershoot and oscillatory. Fig.9 shows the digital control diagram of dc-side boost circuit, which contains the current and voltage regulators $G_{ci}(z)$ and $G_{cv}(z)$, the zero order hold ($(1-e^{-Ts})/s$), the computational delay (e^{-Tds}) and the transfer function of controlled objects $G_{id}(s)$, $G_{vd}(s)$. $G_{ci}(z)$ and $G_{cv}(z)$ are the typical PI regulators expressed as (11). H_i and H_v are the coefficients of current and voltage sampling units, respectively.

$$G_c(z)=K_p+K_iT_s\cdot\frac{z}{z-1} \qquad (11)$$

B. Model and Controller Design for the Equivalent Inverter Stage

Fig.10 shows diode-assisted buck-boost VSI supplying three-phase RL loads through LC filter in the ac side. It deserves to be noted that the modulation ratio of the equivalent inverter stage is δ from the average model shown in Fig.8, when the intermediate capacitor voltage (v_C) is seemed as the dc-link voltage of the inverter bridge (v_{PN}). By applying Kirchhoff's voltage and current law, the state equation of each phase can be expressed as:

$$\begin{cases} V_{sa}-V_{ao}-V_{No}=R_fi_a+L_f\dfrac{di_a}{dt} \\[2mm] V_{sb}-V_{bo}-V_{No}=R_fi_b+L_f\dfrac{di_b}{dt} \\[2mm] V_{sc}-V_{co}-V_{No}=R_fi_c+L_f\dfrac{di_c}{dt} \end{cases} \qquad (12)$$

978-1-4799-2706-7/14 $31.00 © 2014 IEEE

$$\begin{cases} i_a - i_{ao} = C_f \dfrac{dV_{ao}}{dt} \\[2mm] i_b - i_{bo} = C_f \dfrac{dV_{bo}}{dt} \\[2mm] i_c - i_{co} = C_f \dfrac{dV_{co}}{dt} \end{cases} \tag{13}$$

$$i_{dc} = d_{Sa} \cdot i_a + d_{Sb} \cdot i_b + d_{Sc} \cdot i_c \tag{14}$$

Where: $V_{si}=[V_{sa}\ V_{sb}\ V_{sc}]^T$ is each phase voltage; $d_{si}=[d_{sa}\ d_{sb}\ d_{sc}]^T$ is the on-state duty ratio of the upper switch in each phase; thus $V_{si}=d_{si}\times V_{PN}$. For three-phase balanced RL load, $\sum V_{io}=0$ and $\sum i_i=0$, thus $V_{No}=1/2v_{PN}$.

For three-phase system, the widely used method is to transfer the state equations in *abc* frame into *d-q* rotating frame to simplify the analysis. The transfer matrix is

$$T = \frac{2}{3}\begin{bmatrix} \cos(\omega t) & \cos(\omega t - \dfrac{2}{3}\pi) & \cos(\omega t + \dfrac{2}{3}\pi) \\[2mm] \sin(\omega t) & \sin(\omega t - \dfrac{2}{3}\pi) & \sin(\omega t + \dfrac{2}{3}\pi) \\[2mm] \dfrac{1}{2} & \dfrac{1}{2} & \dfrac{1}{2} \end{bmatrix} \tag{15}$$

Where: the factor 2/3 is included so that it means the peak value of the output phase voltage equals the amplitude of *d-axis* component.

The state equations in *d-q* rotating frame are:

$$\begin{bmatrix} \dfrac{di_d}{dt} \\[2mm] \dfrac{di_q}{dt} \end{bmatrix} = -\frac{1}{L_f}\begin{bmatrix} v_{do} \\ v_{qo} \end{bmatrix} - \frac{R_f}{L_f}\begin{bmatrix} i_d \\ i_q \end{bmatrix} + \begin{bmatrix} 0 & -\omega \\ \omega & 0 \end{bmatrix}\begin{bmatrix} i_d \\ i_q \end{bmatrix}$$
$$+ \frac{1}{L_f}\begin{bmatrix} d_d \\ d_q \end{bmatrix} v_{PN} \tag{16}$$

$$\begin{bmatrix} \dfrac{dv_{do}}{dt} \\[2mm] \dfrac{dv_{qo}}{dt} \end{bmatrix} = \frac{1}{C_f}\begin{bmatrix} i_d \\ i_q \end{bmatrix} - \frac{1}{C_f}\begin{bmatrix} i_{do} \\ i_{qo} \end{bmatrix} + \begin{bmatrix} 0 & -\omega \\ \omega & 0 \end{bmatrix}\begin{bmatrix} v_{do} \\ v_{qo} \end{bmatrix} \tag{17}$$

$$i_{dc} = \frac{3}{2}\begin{bmatrix} d_{sd} & d_{sq} \end{bmatrix}\begin{bmatrix} i_d \\ i_q \end{bmatrix} \tag{18}$$

Fig.11 shows the average model of the inverter stage of diode-assisted buck-boost VSI in *d-q* rotating frame. In which, the inductor current (i_d, i_q) and capacitor voltage (v_{do}, v_{qo}) are the fixed dc component in steady state. Therefore, the classical controller design approaches for conventional buck dc-dc converter can be applied. Evidently, the coupling terms exists between the *d-axis*

and *q-axis*, which leads to non-linear system. Thus, the decoupling methods are included into the inner current and outer voltage controllers.

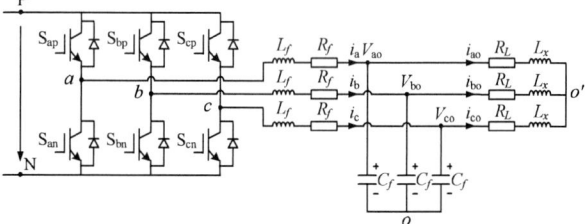

Fig. 10. The equivalent inverter stage supplying three-phase *RL* load with star connection.

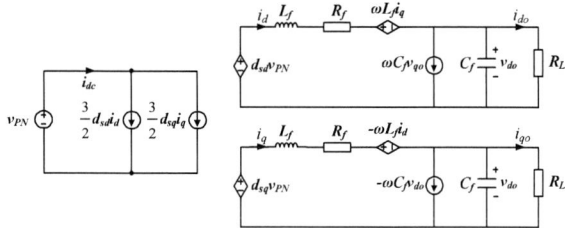

Fig. 11. Average model of the equivalent inverter stage in *d-q* rotating reference frame.

Similarly, performing small signal perturbation, the transfer functions of $\hat{d}_{Sdq}$ to $\hat{i}_{dq}$ and $\hat{v}_{dqo}$ without coupling terms can be obtained and written as (19) and (20). The control diagram of the inverter stage is shown in Fig.12. The corresponding inner current and outer voltage PI regulators H_{idq}, H_{vdq} as (11) can be designed.

$$\frac{\hat{i}_{dq}(s)}{\hat{d}_{Sdq}(s)} = \frac{v_{PN}(C_f R_L \cdot s + 1)}{L_f C_f R_L \cdot s^2 + (L_f + C_f R_f R_L)s + (R_f + R_L)} \tag{19}$$

$$\frac{\hat{v}_{dqo}(s)}{\hat{d}_{Sdq}(s)} = \frac{R_L v_{PN}}{L_f C_f R_L \cdot s^2 + (L_f + C_f R_f R_L)s + (R_f + R_L)} \tag{20}$$

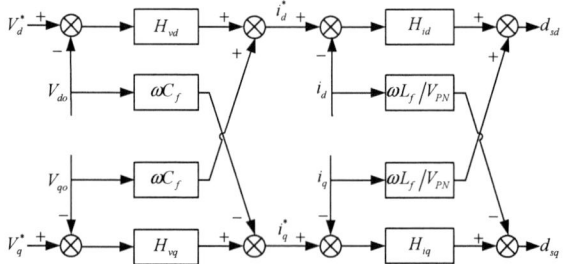

Fig. 12. The control diagram of the equivalent inverter stage of diode-assisted buck-boost VSI in *d-q* rotating reference frame.

IV. Simulation and Experiment Verification

Numerical simulations and experiments have been performed to verify above theoretical analysis and effectiveness of the multi-loop controllers. The main circuit specification and key parameters of controllers are listed as follow: $V_{in}=120V$; $L=2mH$; $C1=C2=150uF$; $R_L=40\Omega\sim100\Omega$; $R_f=0.25\Omega$; $L_f=2mH$; $C_f=50uF$; $T_s=100us$; $H_i=0.1$; $H_v=0.01$; $G_{ci}(z)=0.25+0.0156\cdot z/(z-1)$; $G_{cv}(z)=3.47+0.0913\cdot z/(z-1)$; $H_{idq}(z)=0.85+0.181\cdot z/(z-1)$; $H_{vdq}(z)=0.8+0.0191\cdot z/(z-1)$. The output of controllers in dc- and ac-side is k_{on}, d_{sd} and d_{sq}, respectively. According to the operation principle of improved PWM strategies, the PWM module generates drive signals of switches from control variable k_{on}, d_{sd} and d_{sq}.

Fig.13 and Fig.14 show the amplitude-frequency characteristic for the transfer function of boost duty ratio $\hat{k}_{on}$ to inductor current $\hat{i}_L$ and capacitor voltage $\hat{v}_C$ in frequency domain. The red line is the amplitude-frequency curve by simulation with PSIM. And the green line is the amplitude-frequency curve by calculation from (9) and (10). It is clear that the green curve is almost consistent with the red one in low frequency range, which verifies that the simplified model fully reflects the characteristic in the specialized frequency range. Fig.13(b)

and Fig.14(b) show the corresponding amplitude-frequency characteristic curve from experiment measurement. The location of zeros and poles of transfer function is almost close to the theoretical value.

Fig.15 shows the simulation waveforms of the converter supplying three-phase RL load under different step changes of the output voltage reference. Evidently, the intermediate capacitor voltage (V_C) responses slowly due to the RHP zero which is consistent with the aforementioned theoretical analysis. However, the control of ac output current and voltage are designed with relatively much large bandwidth to minimize the undesired influence of RHP zero. Therefore, the ac output voltage follows the reference value quickly. With the designed multi-loop controller, diode-assisted buck-boost VSI demonstrates good steady state and dynamic performance for output voltage.

Fig.16 shows the simulation waveforms of diode-assisted buck-boost VSI supplying three-phase RL load under different step changes of load resistor. As the load reference voltage rapidly. However, the response of dc-

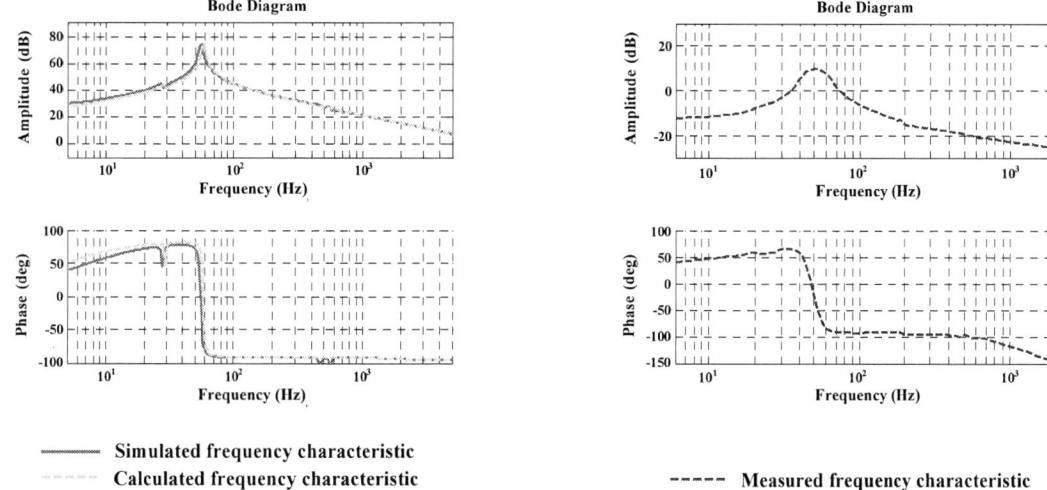

Fig. 13. The amplitude-frequency characteristic for the transfer function of boost duty ratio k_{on} to inductor current i_L.

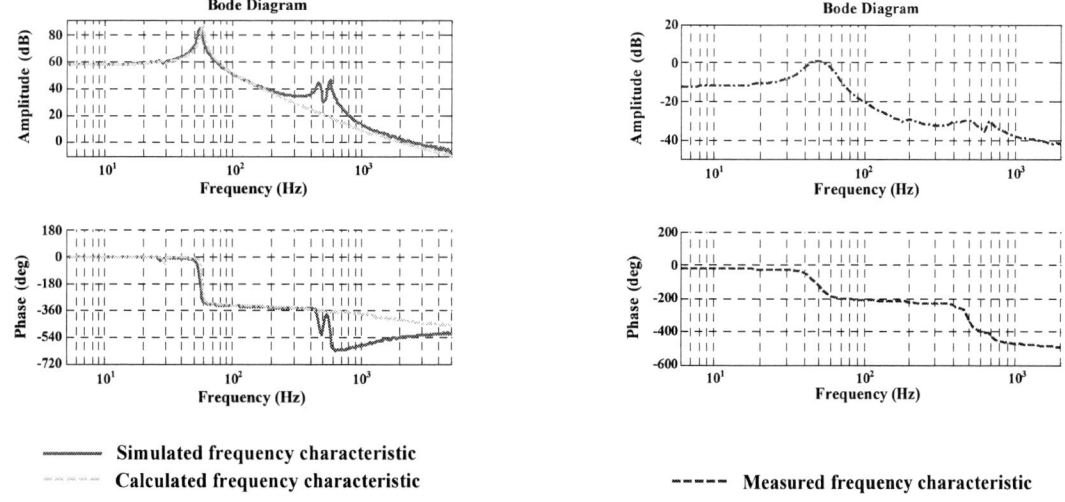

Fig. 14. The amplitude-frequency characteristic for the transfer function of boost duty ratio k_{on} to capacitor voltage v_C.

link capacitor voltage is relatively slow to balance the step change of output power. As illustrated in Fig.16, diode-assisted buck-boost VSI demonstrates good capacity of anti-disturbance of load change with the aforementioned multi-loop controller. Fig.17 and Fig.18 shows the corresponding experiment results under different step changes of the output voltage reference and load resistor, respectively.

v. Conclusions

The paper starts by model analysis of diode-assisted buck-boost VSI in wide range voltage regulation. From the average model, diode-assisted buck-boost VSI can be equivalent to the conventional dc-dc boost circuit supplying three-phase inverter bridge with the modified modulation ratio, if the intermediate capacitor voltage is seemed as the fixed dc-link voltage of the inverter bridge. The derived average model helps to reduce the complexity of controller design. for dc- and ac-side circuits. Furthermore, like boost-type dc-dc converter, the intermediate capacitor voltage has a non- minimum-phase transient response. Taking into consideration of this characteristic, the RHP zero influence is well overcome by appropriate close-loop bandwidth design of front boost dc-dc circuit and the inverter bridge. With the designed multi-loop controller, diode-assisted buck-boost

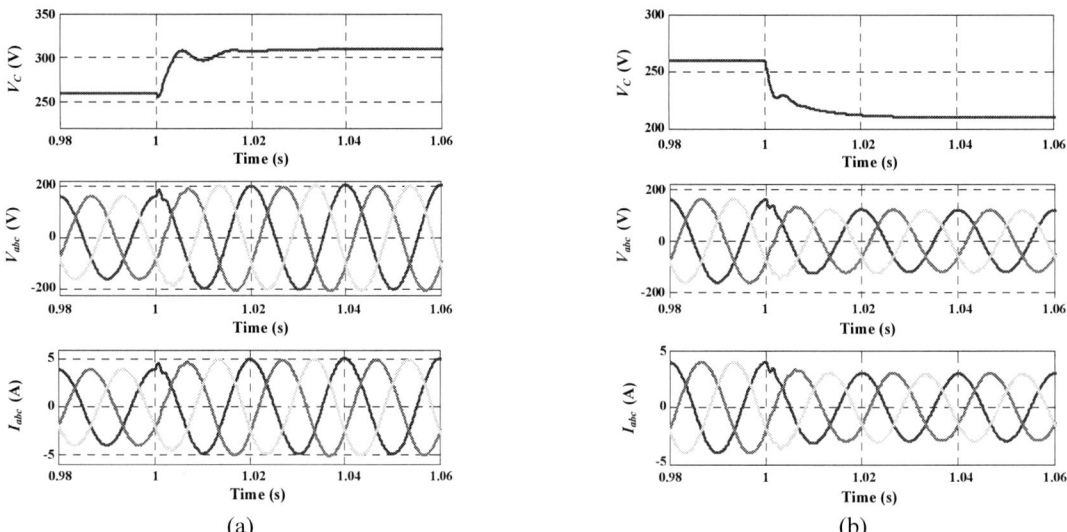

(a)   (b)

Fig. 15. Simulation results under step change of output voltage reference. (a): step-up 160V-200V; (b): step-down 160V-120V.

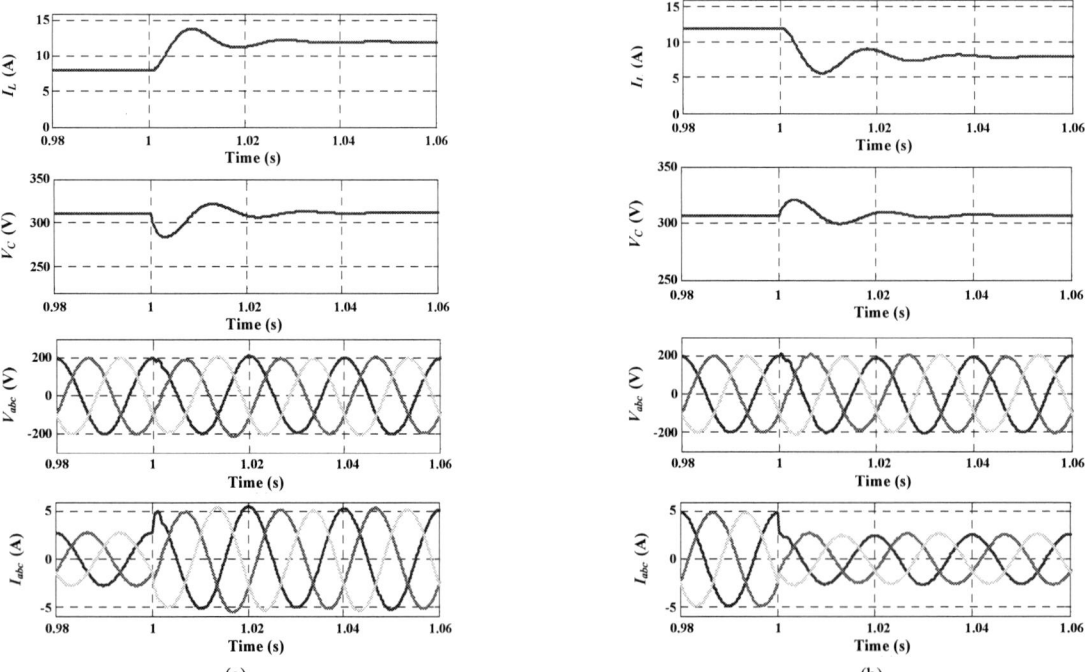

(a)   (b)

Fig. 16. Simulation results of diode-assisted buck-boost VSI under step change of load resistor. (a): step-down 80Ω-40Ω; (b): step-up 40Ω-80Ω.

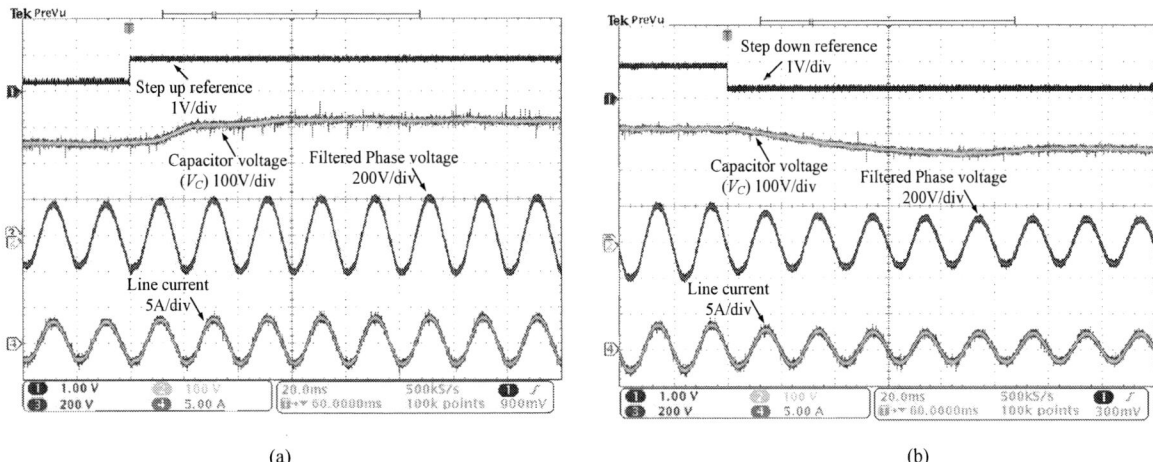

(a)   (b)

Fig. 17. Experiment results of diode-assisted buck-boost VSI under the step change of output voltage reference. (a): step-up 180V-200V; (b): step-down 200V-160V.

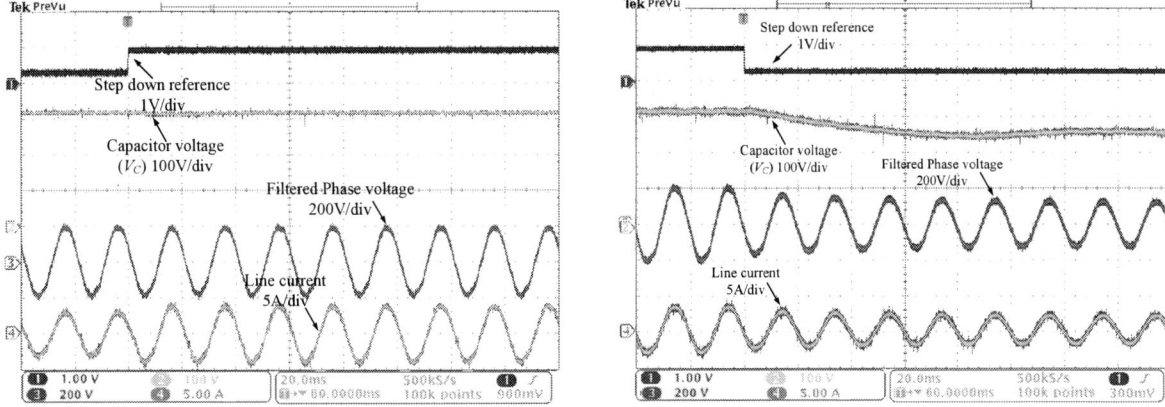

Fig. 18. Experiment results of diode-assisted buck-boost VSI under the step change of load resistor. (a): step-down 80Ω-50Ω; (b): step-up 60Ω-100Ω.

VSI demonstrates good dynamic performance of output voltage regulation and disturbance rejection.

References

[1] Wuhua Li; Xiangning He, "Review of Nonisolated High-Step-Up DC/DC Converters in Photovoltaic Grid-Connected Applications," IEEE Trans. Industrial Electronics, vol.58, pp. 1239–1250, May 2011.

[2] Fang Zheng Peng, "Z-Source Inverter," IEEE Trans. Industry Applications, vol.39, pp. 504–510, March 2003.

[3] Feng Gao, Poh Chiang Loh and Teodorescu R, "Diode-Assisted Buck–Boost Voltage-Source Inverters," IEEE Trans. Power Electronics, vol.24, pp. 2057–2064, September 2009.

[4] Yan Zhang, Jinjun Liu, "Improved Pulse -Width Modulation of Diode-Assisted Buck-Boost Voltage Source Inverter," IEEE Trans. Power Electron., vol. 28, no. 8, pp. 3675–3699, Aug. 2013.

[5] Antal, R, "Improved control of diode-assisted buck-boost voltage-source inverters," EPE/PEMC 2010, pp.93-99, Sept. 2010.

[6] Yan Zhang, Jinjun Liu and Xiaolong Ma, "Operation Modes Analysis and Limitation for Diode-Assisted Buck–Boost Voltage Source Inverter With Small Voltage Vector," IEEE Trans. Power Electron., vol. 29, no. 7, pp. 3525–3536, July. 2014.

[7] Improved Pulse -Width Modulation of Diode-Assisted Buck-Boost Voltage Source Inverter," IEEE Trans. Power Electron., vol. 28, no. 8, pp. 3675–3699, Aug. 2013.

[8] Frede Blaabjerg, Sigurdur Freysson, "A new optimized space-vector modulation strategy for a component-minimized voltage source inverter," IEEE Trans. Power Electron., vol.12, no.4, pp. 704–714, July 1997.

[9] Miaosen Shen, Jin Wang and Fang Zheng Peng, "Comparison of Traditional Inverters and Z-Source Inverter for Fuel Cell Vehicles," IEEE Trans. Power Electronics, vol.22, pp. 1453 – 1463, July 2007.

[10] Mohr, M. Franke, W.T and Wittig, B, "Converter Systems for Fuel Cells in the Medium Power Range—A Comparative Study," IEEE Trans. Power Electronics, vol.57, pp. 2024 – 2032, June 2010.

[11] Jin Li, Jinjun Liu and Zeng Liu, "Comparison of Z-source inverter and traditional two-stage boost-buck inverter in grid-tied renewable energy generation," IPEMC 2009, pp. 1493 – 1497, May 2009.

[12] Yan Zhang, Jinjun Liu, Chaoyi Zhang, "Comparison of traditional two-stage buck-boost voltage source inverter and diode-assisted buck-boost voltage source inverter," APEC 2012, pp. 141-148, Feb, 2012.

[13] R. D. Middlebrook and S. Cuk, "A general unified approach to modelling switching-converter power stages," in Proc. IEEE Power Electron. Spec. Conf., 1976, pp. 18–34.

[14] M. Shen, Q. Tang, and F. Z. Peng, "Modeling and controller design of the Z-source inverter with inductive load," in Proc. IEEE Power Electron. Spec. Conf., 2007, pp. 1804–1809.

[15] P. C. Loh, and D. M. Vilathgamuwa, "Transient modeling and analysis of pulse-width modulated Z-source inverter," IEEE Trans. Power Electron., vol. 22, no. 2, pp. 498–507, Mar. 2007.

[16] C. J. Gajanayake, D. M. Vilathgamuwa, and P. C. Loh, "Development of a Comprehensive Model and a Multiloop Controller for Z-Source Inverter DG Systems" IEEE Trans. Ind. Appl., vol. 54, no. 4, pp. 2352-2359, Aug. 2007.

[17] Omar Ellabbam and Joeri Van Mierlo, "A DSP-Based Dual-Loop Peak DC-link Voltage Control Strategy of the Z-Source Inverter," IEEE Trans. Power Electron., vol. 27, no. 9, pp. 4088–4096, Sept. 2012.

AUTHOR INDEX

Abe, Kodai 3153
Abe, Seiya 177, 1179, 2216, 2222, 3652
Abe, Shigeru 1109, 1115
Abe, T. 3007
Abe, Takashi.......................2183, 2189, 3024
Abe, Tomohiko 1575
Abiko, Hiroshi 634
Achara, Pichetjamroen.............................3687
Adachi, Mitsuo 92
Adhikari, Jeevan 1775
Agelidis, Vassilios G................. 1458, 3758, 3764, 3933
Agelidis, Vassilios Georgios.....................640
Aguglia, Davide 3371
Ahmed, Furqan 480, 790
Ahssanuzzaman, S. M. 3582
Aiso, Kohei..1141
Ajima, Toshiyuki............................383, 682
Akagi, Hirofumi 750, 1586, 1761, 2290, 2323, 3742
Akagi, Masataka 629
Akahane, Masashi...................................2302
Akatsu, Kan................ 1128, 1141, 1234, 2673, 3828
Aketa, M. 2074
Akira 3784
Akiyama, Satoru 2285
Alemi, Payam..1201
Alipoor, Jaber.......................................3298
Aljankawey, A. S. 2156
Allen, Scott...3447
Almer, Stefan...3563
Ama, Naji Rajai Nasri 2413, 2988, 3278
Amanci, Adrian Z. 1303
Amano, Yuki..1824
Amma, Ryosuke 2027
Anazawa, Yoshihisa 3801
Andersen, Michael A. E......... 78, 506, 2842, 3352, 3905
Ando, Itaru 1516
Ando, Masato 1317
Anthon, Alexander....................................78
An-Yeol Ko 796
Aoki, Mutsumi..2400
Aoyagi, Shigehisa 2451
Aoyama, Fumio 2644
Aoyama, Kohei 2266
Aoyama, Masahiro..................................1405
Aoyama, Tomohiro 3823
Ara, Takahiro...3044
Arai, Haruki 403
Arai, Manabu 3440
Araki, Jun..1728
Araki, Takahiro 1613
Arata, Masanori......................................1874
Arikawa, S. 3007
Arimatsu, Kenji.......................................415
Arita, Hideaki...2673
Asai, Inami...123

Asakimori, Koki 1567
Asama, Junichi 988
Asano, Katsunori 3440
Asano, Yoshinari 1997
Asano, Yuji 3872
Ashikaga, Tadashi 1886
Athab, Hussain S. 3695
Atsushi, Manabe 2745
Awaji, Sosuke..3194
Ayano, Hideki 2385
Azuma, M. 1892
Baba, Jumpei 1849
Babasaki, Tadatoshi 1567
Bac Xuan Nguyen 2722
Bafleur, M. 707
Bahman, A. S. 2862
Bahrani, Behrooz....................................1386
Bak, Claus Leth 3320
Bakran, Mark-M. 2113, 3255
Bang, Deok-Je 2427
Bani Shamseh, Mohammad 2794
Baoquan Liu 1155, 3546
Barater, Davide 433
Barrade, Philippe 1081
Barth, Henry 2881
Basari, Amat A. 3194
Basu, Kaushik...3061
Bauer, Florian 3898
Bauer, Pavol 1193, 3200
Baumgartner, Thomas 1707
Beczckowski, Szymon 2547
Beczkowski, S. M. 2862
Beczkowski, Szymon 2850
Belanger, Jean 2644
Ben Guo..3129
Ben, Hongqi 2318
Beres, Remus 3320
Berhouet, S. 707
Berkouk, El Madjid..................................560
Bessegato, Luca 1087
Besselmann, Thomas 3563
Bhat, Ashoka K. S. 1721
Bhattacharya, Subhashish 651, 656,
758, 1626, 2562, 3225, 3286, 3447, 3726
Bianda, Enea 3432
Biela, J. 868
Biela, Jurgen 1788
Bilal, Akin 230
Bin Wu 3482, 3695
Binbin Li 3680
Bizen, Yosio...2983
Blaabjerg, F. 548, 1912, 2862
Blaabjerg, Frede 216, 857, 1529,
1634, 1801, 2610, 3320
Blank, Frederic 264

AUTHOR INDEX

Bo Wen..944
Bocker, J. 2887
Bocker, Joachim.....................346, 1501, 1508
Boehm, Andreas.....................................283
Boillat, David O. 1073
Boitier, V. 707
Boroyevich, Dushan 944, 2626, 3850
Bortis, D. 1291
Bortis, Dominik 1309, 2079, 3864
Bosshard, R. 2167
Bosshard, Roman 1904
Boyu Wang 2893
Braz Cardoso, F. 3225
Burger, Niklaus 1386
Burgos, Rolando..................944, 2626, 3850
Burkart, Ralph M. 891, 3460
Buticchi, Giampaolo 433
Byoungchang Jung................................1185
Byung Moon Han...................................937
Byung-Geuk Cho...................................2802
Byung-Gyu Yu 3784
Cai, Zheng-Xiu 429
Canales, Francisco 1043, 3432
Cao, Guoen 2587
Cao, Wei 567
Cao, Yuan 647
Cardoso, Braz J.....................................3270
Caris, M. L. A.......................................2954
Carvalho, Eden Luiz..............................1276
Casolari, Ronaldo Pedro.........................1276
Castellazzi, A. 2503
Castellazzi, Alberto 433, 2920, 3718
Ceballos, Salvador 3758, 3764
Cha, Honnyong...................110, 480, 790
Chai Feng 3129
Chang, C.-H. 2050
Chang, Chien-Hsuan.....................2523, 3333
Chang, Hsiu-Feng 330
Chang, Kai-Chi 105
Chang, L. 2156
Chang, Yuan-Chih 330, 1832
Changsheng Hu 782
Changwoo Kim......................................1646
Chao Wang 2950
Chao-Fu Wang 2758
Chattopadhyay, Ritwik............................3225
Chen, Ching-Guo 1734
Chen, H. 1471
Chen, Hsin-Chih 1639
Chen, Hung-Chi 2580
Chen, Jiann-Jong 2910
Chen, Jung-Chieh 677
Chen, Min 485
Chen, Qianhong 1425
Chen, Shen-Li..236

Chen, Wei.......................................66, 72
Chen, Wenjie 2950
Chen, Yaow-Ming..................................3592
Chen, Ying-Zuo 351
Chen, Zhe 3538
Chen-Feng Chuang 3379
Cheng Deng 782
Cheng, Chun-An...........................2523, 3333
Cheng, Hung-Liang.......................2523, 3333
Cheng, Po-Tai 1261, 1639
Cheng, Shih-Jen 199, 2593
Cheng, Stone 3425
Cheng-Chieh Yu 2910
Cheng-Wei Chen 3592
Cheol-O Yeon 1738
Cheon, Jun P. 3358
Chia-Chi Chu 3379
Chiang, Hsin-Wei 2100
Chiba, Akira 982, 988, 3513
Chiba, Yoshinori 634
Chien-Yu Lin 2758
Chih Wei Chen 3938
Ching-Hsiang Yang 1639
Ching-Tasi Pan 3379
Ching-Wei Wang 1639
Chiu, Chian-Song 440
Chiu, Huang-Jen 172, 199, 2593, 3328
Chiu, Tse-Wei 440
Cho, Bo-Hyung 2272, 2575
Choi, Bo H. 2232, 3358
Choi, Byungcho 3638
Choi, Hangseok 2575
Choi, Seong-Chon 409
Choi, Sewan 1394, 2247
Choi, Su Y. 1103
Chokchai, Chuenwattanapraniti......................3789
Chou, Tzu-Han 208, 421
Chow, T. Paul 2208
Chu, Xi 1322
Chun, Chang Yoon 2272
Chung, Tsung-Yuan 2523
Chung-Chuan Hou 2821
Chung-Yi Lin 3185
Chung-Yuen Won 796, 3532
Chunkag, Viboon 694
Chu-Shen Chang 3928
Ciftci, Baris 3734
Coldevin, Grete H..................................1861
Colmenares, Juan 3712
Concari, Carlo 433
Cortes, Patricio 3864
Cortizio, Porfirio C. 3225
Cosovic, Mirsad 1148
Daesu Han 1185
Dahidah, Mohamed S. A. 1283

AUTHOR INDEX

Dahono, Pekik Argo 3893
Dai, Wei-Fu 330, 1832
Daikoku, A. 1892
Daikoku, Akihiro 2011, 2673
Dan Chen 3938
Darba, Araz 718
D'Arco, Salvatore 1544
Darus, Rosheila 3758, 3764
Daskalos, Mike 2330
Davoodnezhad, R. 1482
Dawson, Francis P. 1303
De Belie, Frederik 718
De Carvalho, Kelly Caroline
Mingorancia 2413, 2618, 2988, 3278
De Doncker, R. W. 3898
De Doncker, Rik W. 736, 2729, 3145
De Haan, Sjoerd 2787
De Mallac, Louis 3371
De Miranda, Rubens Domingos 1276
De Paula, Helder 3225
De S. Brito, Jose A. 3225
De Vega, Angel Ruiz 2547, 2850
De, Ankan 651, 2562, 3286, 3447
De, D. 2503
De, Dipankar 433
Deguchi, Tadayoshi 3440
Dehong Xu 782
Dekka, Apparao 3468
Demetriades, Georgios D. 1220
Deng, Lirong 465
Dianguo Xu 3174, 3680
Dianguo, Xu 341
Diduch, C. P. 2156
Dilhac, J-M. 707
Diniz, Rogerio Azevedo 3270
Doki, Shinji 907, 2445, 3079, 3823
Domoto, Kazuhide 3652
Dong Le 1837
Dong-Hee Lee 994, 2693
Dong-Jing Lee 1452
Dongkook Son 2914
Dongouk Kim 925
Dongwook Kim 2914
Dou, Qinyun 3604
Dowaki, Kiyoshi 1207
Do-Yun Kim 796
Drofenik, Uwe 1043
Du Yan 2668
Du, Yimian 1721
Duarte, J. L. 2954
Dujic, Drazen 3476
Durand Estebe, P. 707
Dutta, R. 2679
Endo, Takahisa 2541, 2977
Eni, E. 1912

Enomoto, Toshio 2421
Enomoto, Yuji 1997
Erturk, Feyzullah 3734
Eui-Cheol Nho 2763
Fang Zheng 1342
Fang Zhuo 1155, 3546
Fang, Xiaocun 335
Fassler, Lukas 3864
Fei Lin 807
Fei Meng 2815
Fei Zhang 3857
Fernandes, B. G. 2433
Ferrari, Bruno Augusto 3278
Ferreau, Joachim 3563
Ferreira, J. A. 1935
Ferreira, Jan A. 2787
Figueredo, Ricardo Souza 2413, 2618
Fletcher, J. 2679
Fletcher, John 2926, 2932
Foo, Gilbert 2722
Fosso, Olav B. 1861
Foureaux, Nicole C. 3225
Fournier-Bidoz, Sebastien 3496
Franca, Gleisson J. 3270
Franceschini, Giovanni 433
Franke, Toke 78
Fritz, Dominik 3476
Frohleke, N. 2887
Fujii, Junji 1654
Fujii, Kansuke 1748
Fujii, Toshiyuki 2663
Fujimoto, Hiroshi 1671, 2421
Fujimoto, Takafumi 3857
Fujimoto, Yasutaka 1685, 1968
Fujisaki, Keisuke 289, 2856, 2874
Fujisawa, Hiroyuki 3440
Fujita, Hideaki 1006, 1160, 1350, 2027, 2042
Fujitsuna, Masami 3079
Fuketa, Hiroshi 2228
Fukuda, Kenji 3440
Fukuhara, Shuhei 289
Fukumoto, Hisao 724, 730, 3067, 3249
Fukuoka, Hiroki 3341
Fukushima, Kentaro 2189
Fulin Zhou 1050
Funabiki, Shigeyuki 2470
Funaki, Tsuyoshi 1621
Funato, Hirohito 1728, 2517
Furukawa, Kimihisa 383
Furukawa, Tatsuya 724, 730, 3067, 3249
Furukawa, Yutaka 2252
Furuta, R. 2120
Gaing, Zwe-Lee 278
Gao Qiang 3050
Gao, Qiang 614

AUTHOR INDEX

Geng, Hua 543
Gerling, Dieter 774
Ghimire, Pramod 2547, 2850
Giaretta, Antonio Ricardo 1276
Goehler, Lutz 2554
Goh Teck Chiang 1028
Gohara, Hiromichi 671
Goto, Akira 130
Goto, Yasuyuki 1490
Goto, Yuichi 1671
Graus, Johannes 270
Grider, Dave 3726
Gruber, Wolfgang 1691, 1701
Gu, Beom W. 1103
Gueldner, Henry 2554
Guidi, Giuseppe 1544
Gunasekaran, Deepak 1342
Guo, Wei 160, 475
Gurpinar, Emre 433, 3718
Ha, Jung-Ik 3140
Hafner, Jurgen 3667
Haga, Hitoshi 415, 3153
Haghbin, Saeid 1373
Hagiwara, Makoto 1586, 1761, 2323, 3742
Hahn, Ingo 270, 283
Haining Wang 3702
Haitao Yang 782
Hak-Soo Kim 2763
Hakutou, Takuma 2297
Hama, Ryota 2470
Hamasaki, Shin-Ichi 2775, 3674
Hamasaki, Sin-Ichi 3093
Hamazaki, Yasuhiro 2126
Hanada, T. 2074
Hanamoto, Tsuyoshi 538, 1811
Han-Shin Youn 1743
Hao Huang 2967
Hao Yi 1155, 2960, 3546
Hao, Xiang 2950
Hao-Chien Cheng 3379
Hara, Hidenori 1654, 1898
Harada, Shingo 1671
Harada, Shinsuke 3440
Harakawa, Masaya 2638
Hariya, Akinori 3630
Hasegawa, Isamu 1365
Hasegawa, Kohei 3707
Hasegawa, Masaru 183, 907, 2445
Hasegawa, Masataka 2212
Hasegawa, Shinya 294, 299, 2972, 3055, 3159, 3162
Hasegawa, Tomonori 2126
Hashimoto, Kento 1974
Hashimoto, S 1471
Hashimoto, Seiji 3194
Hashimoto, Shizuka 3018

Hassanpoor, Arman 3667
Hatanaka, Ayumu 2285
Hattori, Fumiya 811
Hatua, Kamalesh 758, 1626
Hau-Chen Yen 3928
Hava, Ahmet M. 498, 2034, 3734
Hayase, Masanori 1207
Hayashi, Makoto 1950
Hayashi, Toshihiko 3440
Hayashi, Yusuke 1560
Hayashiya, Hitoshi 1062
Hazeyama, M. 1892
Hazra, Samir 758, 1626, 3447
He, Guofeng 485
Hee-Jun Lee 3532
Hei, Xinhong 647
Hella, Mona 2208
Hermansson, Willy 1220
Hernandez, Juan C. 3352
Heung-Geun Kim 790, 2763
Hibino, Shinya 2638
Hidaka, Akira 2216
Hidayat, Nabil M 573, 2529
Higuchi, Shinichi 1522
Higuchi, T. 3007
Higuchi, Tsuyoshi 3024
Hijikata, Hiroki 2673, 3828
Hikita, Masayuki 689
Hinata, Toshifumi 919
Hinkkanen, Marko 2489
Hino, Wataru 3525
Hintz, Andrew 2343
Hira, Yuki 730
Hirahara, Hideaki 3044
Hirai, Junji 1974
Hirakawa, Yuki 3024
Hiraki, Eiji 3292
Hirano, Yosei 1956
Hirao, Kuniaki 191, 1365
Hirase, Yuko 1552
Hirokado, K. 146
Hirose, Toshiro 2252
Hirota, Yukitsugu 1728
Hisada, Yoshihiro 3292
Hisato, Hosoyama 2745
Ho, Kung-Min 3942
Hoene, Eckart 2366
Hoffmann, Stefan 2366
Hofmann, Wilfried 2881
Hojo, Masahide 2152
Hojo, Toshiaki 1276
Hokazono, Hiroaki 2870
Holm, Toni 3432
Holmes, D. G. 1482, 2019, 2372, 3306
Homma, Hiroshi 1880

AUTHOR INDEX

Hong Li 3314
Hong, Ki-Nam 2598
Hong-Hee Lee 1013
Hongqi Ben 3213
Hori, Yoichi 2421
Horiguchi, Takeshi 2290
Horita, Yasuhisa 1317
Hosaka, Tatsuya 1350
Hoshi, Nobukazu 1242
Hosoyamada, Yu 801
Hou, Chih-Hao 1796
Hou, Jiaxin 526
Hou, Lixiang 577
Hredzak, Branislav 1458
Hsieh, Guan-Cyun 526
Hsieh, Hung-I 526, 2380
Hsieh, Min-Fu 278
Hsieh, Yao-Ching 429, 1796
Hsin-Chih Chen 1261
Hsin-Ping Su 2821
Hu, Jia-Sheng 278
Hu, Shang-Hung 2606
Hu, Sheng 555
Hu, Taiyuan 335
Huang, Hsin-Wei 421
Huang, Jia-Wei 3233
Huang, Lang 2950
Huang, Min 2610
Huang, T. D. 2195
Huang, Wen-Nan 1734
Huang, Zhenhui 647
Huang-Jen Chiu 2758, 2810, 3185, 3913
Huber, Jonas E. 766
Huber, Tobias 1508
Hui Liu 1634
Hui Zhang 1365, 3455
Huisman, H. 2954
Hull, Brett 3447
Huu-Nhan Nguyen 1013
Hwang, Seon-Hwan 2427
Hwang, Yuh-Shyan 2910
Hwu, K. I. 204, 2754, 3190, 3392
Hyoyol Yoo 1646
Ichihara, Junichi 2189
Ichiya, Takahiro 370
Ichiyanagi, Katsuhiro 1490
Ide, Kozo 933
Ieda, Jun 2663
Iga, Yuichi 3341
Igarashi, Kazunori 2983
Igarashi, S. 3702
IIda, Mikiya 2977
IIjima, Ryuji 117
IIjima, Yukihia 2095
Ikawa, O. 2569, 3702

Ikeda, Hidehiro 2476
Ikeda, Masahiro 2183
Ikeda, Tomohiko 1575
Ikeda, Y. 2569
Ikeda, Yoshinari 2870
Il-Kuen Won 796
Ilves, Kalle 1087
Imai, Jun 2470
Imakiire, Akihiro 689
Imamura, Yasutaka 863
Imanishi, Takao 2663
Imaoka, Jun 811, 883, 2497
Inamori, Mamiko 3509
In-Dong Kim 2763
Inomata, Kentaro 1654
Inoue, Kaoru 3872
Inoue, Keita 130
Inoue, M. 1892
Inoue, Tatsuki 363
Inoue, Y. 246, 258, 312, 390
Inoue, Yukinori 324, 356, 363, 370, 2183, 3018, 3519
Irokawa, Shoichi 1357
Ise, Tomofumi 1430
Ise, Toshifumi 1536, 1560, 2632, 3298, 3687
Ishibashi, Makoto 724
Ishida, Koichi 2228
Ishida, M. 146
Ishida, Masaaki 3707
Ishida, Masaki 3162
Ishida, Takahito 634
Ishigami, Takashi 1880
Ishigma, Satoru 403
Ishihara, Chio 1984
Ishihara, Yuji 1135
Ishii, Hirotaka 294
Ishikawa, Hiroki 1135, 2183, 2189
Ishikawa, Katsumi 2140, 2285
Ishikawa, Takeo 252, 1697
Ishimaru, Yusuke 92
Ishimori, Hitoshi 3440
Ishitobi, Manabu 811
Ishizuka, Tomotsugu 2644
Ishizuka, Yoichi 2222, 2252, 2737, 3630, 3652
Isida, Takashi 1950
Itako, Kazutaka 3244
Ito, Yasuhide 3823
Ito, Yoichi 403
Itoh, Hideaki 724, 730, 3067, 3249
Itoh, Jum-Ichi 1943
Itoh, Jun-Chi 1253
Itoh, Junichi 130
Itoh, Jun-Ichi 84, 138, 152, 191, 682,
        1021, 1028, 1095, 1613, 2277, 3659, 3815
Itoh, Tomomichi 850
Itoh, Youichi 415

AUTHOR INDEX

Itoh, Yuki 883, 2497
Iwaji, Yoshitaka 2451
Iwakami, Tetsuro 817
Iwasaki, Makoto 1665
Iwasaki, Shinya 2663
Iwata, Tetsuki 403
Iyer, Kartik V 3037, 3061
Iyer, Shivkumar 3482
Izumi, Toru 3440
Jacobson, Bjorn 3667
Jae-Bum Lee 1738
Jaeho Choi 2656
Jae-Hun Jung 2763
Jae-Hyun Kim 1738
Jaesig Kim 2656
Jang, Jinhaeng 3638
Jang, Young-Jin 664
Jang-Hwan Kim 925
Jardini, Jose Antonio 1276
Jauch, Felix 1788
Javed, Riffat 624
Jayoon Kang 1185
Jen-Hao Teng 1452
Jenn-Jong Shieh 3190
Jeon, Jin-Yong 166
Jeong, Seog Y 1103
Jeong, Seon-Yeong 2406
Jeongjoong Kim 1185
Jhen-Yu Jian 3928
Jia Liu 1536
Jia, Y. 1594
Jianfeng Li 3718
Jiang, Dawang 647
Jiang, Maoh-Chin 105
Jiang, W. 1471
Jiang, W. Z. 204, 3190
Jiang, Yongjie 458
Jianhui Su 2668
Jiann-Fuh Chen 2714
Jianwen Zhang 3124
Jih-Hua Yu 2910
Jin Miaoxin 3050
Jin, Miaoxin 614
Jin, Xu 341
Jin, Yasuhiro 3207
Jing Bian 3314
Jing-Hsiao Chen 3233
Jing-Yuan Lin 2758
Jinjun Liu 835
Jinno, Masahito 1781, 3333
Jin-Woo Ahn 994, 2693
Jinyong Zhang 3213
Ji-Shiang Lee 3346
Joebges, Philipp 2729
Jokipii, J. 514, 2240

Jokipii, Juha 1466
Jong Kyou Jeong 937
Jonghyung Park 1185
Jonishi, Akihiro 2302
Jou, Sung-Tak 224
Jung, Hochang 1990
Jung, Jae-Jung 1268
Jung, Sang-Yong 1990
Jung, Yong-Chae 166, 409
Junghum Lee 2656
Junjie Feng 835
Juntao Fei 3168
Juyoung Jang 2656
Kabasawa, Yuichiro 2175
Kabiri, R. 3306
Kadavelugu, Arun 758, 1626, 3726
Kai, Masahiko 1054
Kai-Hui Chen 2750, 3346
Kaipia, T. 587
Kajiwara, Kazuhiro 3950
Kakishima, Takeo 3513
Kalogera, Maria 1193, 3200
Kameshiro, Norifumi 2140
Kamikura, Mamoru 2064
Kamnarn, Uthen 694
Kanagawa, Kinji 2983
Kanai, Yasuyuki 1567
Kanamori, Masaki 2541, 2977
Kaneko, Junji 2745
Kaneko, Yasuyoshi 1109, 1115
Kanematsu, Masato 2421
Kanemoto, Daisuke 2737
Kang, Feel-Soon 2260
Kang, Yong 555
Kanno, Hiroshi 2302
Kano, Yoshiaki 2004, 2457
Kanoda, Akihiko 3920
Kanouda, Akihiko 2058
Kantar, Emre 2034
Kanthaphayao, Yutthana 694
Kari, Mat Nasir 573
Karki, Ujjwal 1342
Karvonen, Andreas 1373
Kasai, Makoto 3194, 3194
Kashihara, Yugo 1943
Kasper, Matthias 2079
Katade, Motohumi 130
Katakami, Shuji 3440
Kataoka, Yasuhiro 3801
Katayama, Noboru 1207, 1227
Kato, Hideaki 2972, 3162
Kato, Koji 403, 415
Kato, Shinji 2175
Kato, Takashi 3828
Kato, Taro 2972

AUTHOR INDEX

Kato, Tomohisa 3440
Kato, Toshiji...............................2183, 2189, 3872
Kato, Yutaka 2644
Katoh, Kaoru..2285
Katoh, Shuji..850
Katsuki, Akihiko 1575, 3624
Katsura, Seiichiro 1679
Kawachi, Konosuke..863
Kawaguchi, Shinichi.....................................3959
Kawahara, Keiji..1062
Kawakami, Noriko...2095
Kawamura, Atsuo 801, 2266, 2794, 3403
Kawamura, Mitsuhiro 3012
Kawamura, Wataru..3742
Kawano, Daisuke 1671
Kawano, Kenji..883
Kawazoe, Yosuke..2011
Kazuya, Ogura..452
Kempen, S...2887
Kenji, Matsumoto..3218
Kern, Ansgar...712
Khan, Ashraf Ali..110
Khan, Faisal H...2161
Khant, Hlaing Kyi Pyar....................................183
Khomfoi, Surin..2392
Kicin, Slavo...3432
Kihyun Lee..1646
Kikuchi, Takuya...1328
Kim, Bong C...3358
Kim, Chong-Eun..................................1738, 1743
Kim, Dong-Hun..790
Kim, Dong-Rak...409
Kim, Hee-Jun...2587
Kim, Heung-Geun................................110, 480
Kim, Hyejin..2575
Kim, Jae-Hyun...1743
Kim, Jang-Mok...2406
Kim, Ji H...2232
Kim, Ji-Won...2427
Kim, Jonghoon..619
Kim, Minjae...2247
Kim, Seonghye...2260
Kim, Su-Han..480
Kim, Sungmin..1268
Kim, Yong-Jae...1990
Kimoto, Tsunenobu..3440
Kimura, Hiroshi...1920
Kimura, Noriyuki...................1299, 1806, 2183, 3341
Kimura, Shota..883, 2497
Kinouchi, Shin-Ichi..............................750, 2290
Kish, Gregory J...951
Kitabayashi, Tatsuaki.....................................2517
Kitagawa, Wataru...............................2310, 3809
Kitajima, Jun...1247
Kitazawa, Satochi..1438

Kiyota, Kyohei..3513
Kleinecke, John...2330
Kluge, Andreas..2554
Knott, Arnold..506
Kobayashi, H...2569
Kobayashi, Hiroya..2517
Kobayashi, Ryota...1115
Kobayashi, Takenori......................................1868
Kobayashi, Y...2569
Kodama, Takashi...1365
Kogi, Ryosuke...2874
Kogoshi, Sumio.....................................1207, 1227
Kohama, Teruhiko.................................522, 2781
Kohno, Yusuke...2183
Koiwa, Kazuhiro.....................................84, 130, 1028
Kolar, J. W...1291, 2167
Kolar, Johann W. 766, 821, 891, 899, 975,
    1073, 1309, 1707, 1904, 2079, 2834, 3365,
    3460, 3864
Komada, Satoshi...1974
Komatsu, Wilson...................1276, 2413, 2988
Komeda, Shohei..1160
Komiya, Hiroshi...2421
Kon, Saytaro..3263
Kondo, Keiichiro...................................1438, 2126
Kondo, Seiji..415
Kondo, Takeshi..1365
Kondou, Masahiko...2421
Kono, Y...2120
Kono, Yasuhiko..2140
Konoto, Masaaki..2189
Konstantinou, Georgios 1458, 3758, 3764
Korner, Olaf..2113
Kosaka, T..2438
Kosaka, Takashi...................................1984, 1997
Koschik, Stefan....................................3145, 3898
Koseki, Takafumi...................................1334, 2126
Kotegawa, Ryo...1317
Kotera, Keito...3872
Kouki, Matsuse..3134
Kouno, Yusuke...2189
Kounoto, Masaaki..2175
Koyama, Masato...750
Krafft, Eberhard..2113
Krismer, Florian..2834
Kuan-Hsien Chou...3346
Kubo, H..395, 1594
Kubo, Hajime...1601
Kubo, Yuji...3134
Kubota, Hisao.....................919, 1929, 3119, 3134
Kubota, Yutaka..2183
Kudo, Takahiro..1109
Kuga, Shotaro..3955
Kukita, Akio..1444, 2351
Kumagai, Shunji..3194

AUTHOR INDEX

Kumakura, Yoshito.................................1715
Kume, Tsuneo.....................................1898
Kumsuwan, Yuttana..............................3417
Kun-Hung Chen....................................3592
Kunomura, Ken 1054
Kuo, Kuan-Yi.......................................278
Kuperman, A......................................2240
Kurabayashi, Toshiyuki 1962
Kuribayashi, H....................................2569
Kurihara, Takeshi.................................299
Kurihara, Yoshihiro 1874
Kurita, Nobuyuki...........................252, 1697
Kuroda, Y..1892
Kurokawa, Fujio 2108, 3611, 3950
Kusaka, Keisuke...................................191
Kusukawa, Jumpei...............................2904
Kusunoki, Hironobu.............................2330
Kutsuki, Tomohiro...............................2064
Kuwahara, Akinobu 3179
Kuzumaki, Atsuhiko.............................1929
Kwasinski, Alexis 2649
Kwasinski, Andres 2649
Kwon, Soon-Kurl................................2359
Kyungbae Lim....................................2656
Kyungmin Sung....................................744
Kyungsub Jung 1646
Lai, Yen-Shin.....................................3942
Lamantia, A.......................................2503
Lana, A..587
Lang, Klaus-Dieter..............................2366
Larsson, Tomas...................................1220
Laska, Bernd......................................2113
Law, Kah Haw....................................1283
Le Hoai Nam 3659
Lee, Chia-Tse.....................................1639
Lee, Dong-Choon 1201, 2406
Lee, Eun S.................... 1103, 2232, 3358
Lee, Hong-Hee...................................2826
Lee, Jae-Bum.....................................1743
Lee, June-Hee..............................493, 532
Lee, June-Seok.............................493, 532
Lee, Kyo-Beum.....................224, 493, 532
Lee, Min-Hua......................................236
Lee, Seong Ryong...............................3292
Lee, Shiu-Hui.....................................1734
Lee, Sung W.......................................1103
Lee, Taeck-Kie....................................595
Lee, Tzung-Lin...................................2606
Lee, Woo-Cheol...................................595
Lee, Ya-Ting.......................................440
Lee, Yuang-Shung...........................208, 421
Lehmann, Oliver.................................3085
Lehn, Peter W......................................951
Lei, Wanjun..................................160, 475
Leibl, Michael.....................................899

Lelie, Markus 2729
Leslie, Scott 3726
Leuenberger, D. 868
Leuer, Michael 346
Li Yan 2899
Li, Ding 341
Li, Haiqing 2095
Li, Hong 2893
Li, Ning 160, 475
Li, Qian 2161
Li, Yanxiang 3002
Lian, K. L. 2195
Liang Hao 3174
Liangyi Tang 3695
Liao, Jhen-Yu 2580
Lie Guo 3489
Lin Cheng 3447
Lin, Chiao-Chien 3072
Lin, Chia-Yu 1832
Lin, Chien-Yu 172
Lin, Chung-Yi 199, 2593
Lin, Fei 335, 1322, 2133
Lin, Jing-Yuan 172
Lin, L.-C 2050
Lin, Z. Y. 1471
Lindberg-Poulsen, Kristian 2842
Liping Zheng 1837
Liserre, Marco 857, 3320
Liu, Baoquan 577
Liu, Fang 567
Liu, Fangcheng 3604
Liu, Fuxin 458, 2768
Liu, Hanchao 967
Liu, Jianyu 614
Liu, Jilong 66, 72
Liu, Jinjun.........................624, 2815, 3604
Liu, Kangzhi 3568
Liu, Ning 2156
Liu, Rongqiang 3099
Liu, Tai-Chun 105
Liu, Xiankai 647
Liu, Xiaosheng 3002
Liu, Yi-Hua 3233
Liu, Yu-Chen 199, 2593
Liuchen Chang 1476, 2668, 3842
Lo, Yu-Kang 172, 199, 2593
Lobsiger, Yanick 1309
Loh, Poh Chiang.............216, 1529, 1634, 1801, 2610
Longlong Zhang 782
Lopez-Arevalo, Saul 3718
Lovatt, Howard 2679
Low, K. S. 446
Lu, Dylan D. C. 3553
Lu, Kao-Yi 105
Luo, Guomin 2145

AUTHOR INDEX

Luthardt, Sven 3029
Ma, K. 548, 2862
Ma, Weigang 647
Madawala, Udaya K. 2722
Madhusoodhanan, Sachin 656, 1626
Maekawa, Sari 919, 1929
Maemura, Akihiko..1898
Maeyama, Shigetaka......................1575, 3624
Maezono, Paulo Koiti.....................................1276
Mahdavikhah, Behzad.....................................3582
Mainali, Krishna 758, 1626
Makaino, Yuki...914
Makita, Shinji..3823
Mamun, Mostafa 97
Manias, Stefanos N...1606
Mannen, Tomoyuki..2042
Manolas, Iakovos 1606
Mao, Meiqin 2156
Maret, C...3239
Marrero Sosa, Juan Alberto 3476
Martinz, Fernando Ortiz...................2413, 2988, 3278
Maru, Naoki 2285
Marukawa, Yasuhiro.......................................1984
Marumori, Hiroki 3055
Maruta, Hidenori 3611
Marz, Andreas...2113
Marzouk, Ahmad Diab 3496
Masaki, Kenji 2663
Mashino, Masahiro 3162
Masic, Semsudin..1148
Maskell, D. L..3598
Masuda, Hiroyuki 92
Masui, Takeshi 1317
Masutomo, Kazufumi 3624
Masuzawa, Hiroshi 1054
Masuzawa, Takashi..2366
Matakas, Lourenco 2413, 2618, 2988, 3278
Matsubara, Masakatsu 1874
Matsuda, Katsuhiro...415
Matsuhashi, Daiki 1886
Matsuhashi, Masataka.....................................1516
Matsui, Hitoshi 1586
Matsui, Keiju 183
Matsui, Mikihiko 3489
Matsui, N. 2438
Matsui, Ryota 1128
Matsui, Yoshihiro...2385
Matsui, Yoshinobu 2745
Matsumoto, Akira 1560
Matsumoto, Atsushi 2445
Matsumoto, Kazushi 3440
Matsumoto, Satoshi...2216
Matsumoto, Shuhei..1929
Matsumoto, Yasushi 1920
Matsuo, Hirofumi..1781

Matsuo, Keisuke 1886
Matsuo, Yusuke 1671
Matsuoka, Kazumasa 3207
Matsuoka, Yuji 744
Matsushima, Yoshitarou 3801
Matsushita, Makoto 3012
Matsuura, Kei 1516
Matsuura, Ken 3630
Matsuzaki, Ryohei 1978
Mattavelli, Paolo 3850
Mattsson, A. 587
Mauerer, M. 1291
McGrath, B. P....................1482, 2019, 2372, 3306
McLean, Kenneth 3496
Meiqin Mao 1476, 2668, 3842
Mekhilef, Saad 560, 3574
Merahi, Farid 560
Meng, Fei 624
Meng, Tao 2318
Merahi, Farid 560
Messo, T. 514, 2240
Messo, Tuomas 1466
Mihara, Teruyoshi 1728
Mii, Kenji 2737
Miiura, Yushi 1430
Mikihiko 3784
Mills, Liam 3718
Ming Yang 3174
Mingfei Wu 3553
Mingyan Wang 3129
Mino, Kazuaki 1920
Minoshima, N. 2438
Minowa, Masanao 3828
Minsoo Jang 3933
Mira, Maria C. 506
Mishima, Tomoakzu 2533
Mishra, Santanu 2707
Mishra, Santanu Kumar 3587
Misu, Daisuke 1874, 3012
Mitterhofer, Hubert 1701
Miura, Yushi 1536, 3298
Miyajima, Hiroki 1054
Miyajima, Takayuki 2421
Miyakawa, Takayuki 2421
Miyama, Yoshihiro...2673
Miyashita, S. 3702
Miyawaki, Satoshi 84
Miyazaki, Hideki 383
Miyazaki, Kensuke 601
Miyazaki, Toshimasa 1956
Miyazaki, Yuji 750
Mizoguchi, Takahiro 1660
Mizukami, Makoto 3440
Mizuki, Tatsuya 1575, 3624
Mizuma, Takeshi 2126

AUTHOR INDEX

Mizuno, Takayuki 1886
Mizusaki, Hiroshi 3093
Moballegh, Shiva 656
Mochikawa, Hiroshi 1929
Mochizuki, Eiji 671, 2870
Mochizuki, K. 2569
Mohamed, Essam Ebaid 3877
Mohamed, Tarek Hassan 3877
Mohan, Ned 1036, 1412, 3037, 3061, 3750
Mohd Arif, Mohd Johari 573
Molinas, Marta 1861
Momose, Fumihiko 671
Moo, Chin-Sien 1796, 3796, 3928
Moon, Dongok 1394
Moon, Gun-Woo 1738, 1743
Moon, Sang-Ho 224
Moorthy, Radha Sree Krishna 2087, 3616
Moraes, Lenin 3225
Mori, Tomohiro 2983
Morikawa, R. 258
Morimoto, Masayuki 3509
Morimoto, S. 246, 258, 312, 390
Morimoto, Shigeo 324, 356, 363, 370, 1997, 3018, 3519
Morimoto, Shinya 1654
Morishita, Shin 130
Morita, Hiroshi 1490
Morita, Kazunori 191, 582
Morita, Kosuke 3624
Morita, M. 1892
Morizane, Toshimitsu 1299, 1806
Morizane, Tosimitsu 3341
Moroi, Takayuki 3134
Morozumi, Akira 671
Mory, David 2554
Motizuki, Shun 2745
Motoi, Naoki 801, 2266
Motomura, Masashi 3611
Mouri, Masayuki 1728
Mrak, Branimir 1701
Mukai, Ryosuke 2775
Mukunoki, Makoto 97, 1950
Munk-Nielsen, Stig 2547, 2850
Murai, Kensuke 1567
Murai, Toshiaki 1122
Murakami, Daichi 1728
Murakami, Kouhei 2385
Murakami, Toshiyuki 1962
Murata, Koji 2108
Murata, Munehiro 1173
Murata, Yuichiro 2064
Musing, Andreas 821
Mustapa, Rijalul Fahmi 2529
Muta, Shoichiro 3067
Nag, Soumya Shubhra 3587

Nagai, Shinichiroh 811
Nagano, Tetsuaki 2638
Nagano, Tsuyoshi 1253
Nagano, Y. 146
Nagashima, Tomohiro 2175
Nagata, Shun 2252
Nagatomo, Yoshinobu 2663
Nagel, Andreas 2113
Nagura, Hirokazu 2451
Nagy, Istvan 2700
Naitoh, Haruo 1135
Nakagawa, Hidehiko 1552
Nakagawa, Yuki 2533
Nakahara, Mizuki 744, 2511
Nakajima, Yoichiro 403
Nakamura, M. 376
Nakamura, Ritaka 92
Nakamura, Sota 2400
Nakamura, T. 2074
Nakamura, Tatsuya 1575
Nakamurame, Fuminori 2632
Nakanishi, Toshiki 1095
Nakano, Y. 2074
Nakao, Hiroshi 2745
Nakao, Noriya 1128, 1141
Nakaoka, Mutsuo 2359, 2533
Nakaoka, Mutuo 3341
Nakashima, Yoshiyasu 2745, 3386
Nakata, Yuki 138
Nakatsu, Kinya 2904
Nakatsugawa, Junnosuke 2451
Nakayama, Koji 3440
Nakayama, Naoyuki 3857
Nakayama, Yasushi 2290
Nakazawa, Yosuke 1357
Nam, Kwang-Hee 664
Narita, Takayoshi 294, 299, 3055, 3159
Nashida, N. 2569
Nayanasiri, D. R. 3598
Nee, Hans-Peter 3712
Neubert, Markus 3145
Nguyen, D. 2679
Nguyen, Quoc Khanh 318
Nguyen, Thanh Hai 2406
Nha, Quang Trong 3913
Nho Van Nguyen 2826
Nian Heng 843
Nicolae, Ileana-Diana 2996
Nicolae, Marian-Stefan 2996
Nicolae, Petre-Marian 2996
Niijima, Koji 1299, 1806
Nilssen, Robert 1412
Nimura, Tomohiro 3079
Ning Liu 2668

AUTHOR INDEX

Ninomiya, Tamotsu 177, 1179, 2216, 2222, 3630, 3652
Nishida, Katsumi 2359
Nishida, Yasuyuki 2189
Nishikata, Shoji 959
Nishimura, T. 3702
Nishimura, Tomohiro 2870
Nishimura, Yoshitaka 671
Nishio, Haruhiko 2302
Nishioka, Tomoya 2152
Nishisu, Koji 2285
Nishiyama, Noriyoshi 2011
Nishizawa, Shinichi 117, 744
Niu, Ruigen 160, 475
Noda, Koji 2541
Noda, Taku 2175
Noguchi, Kenji 2277
Noguchi, S. 2569
Noguchi, Toshihiko 1173, 1405
Noh, Yong-Su 166
Nomura, Naofumi 1522
Nomura, Shinichi 3218
Nonaka, Hirotaka 2737
Norigoe, Isami 117
Noro, Osamu 1552
Norrga, Staffan 1087
Noto, Yasuo 682
Nozaki, Takahiro 1660
Nozawa, Ryosuke 1115
Nussbaumer, Thomas 975, 3365
Nuutinen, P. 587
Oboe, Roberto 1679
O'Byrne, Sean 2926, 2932
Oda, Yoshinori 829
Odawara, Shunya 289, 2856, 2874
Ogasawara, Satoshi 1728, 2977, 3525
Ogashi, Yoshihiro 92
Ogawa, Kazutoshi 2140, 2285
Ogawa, Takashi 2285
Ogura, Kazuya 582, 3455
Ogura, Tsuneo 2068
Oh, Min-Seok 166
Ohara, Shinya 850
Ohashi, Hiromichi 117, 744
Ohashi, Shunsuke 3410
Ohchi, Masashi 724, 730, 3067, 3249
Ohishi, Kiyoshi 1247, 1516, 1956, 3153
Ohnishi, Kouhei 1660, 2483
Ohnuma, Takumi 914
Ohnuma, Yoshiya 84
Ohse, Naoyuki 3440
Ohtake, Asuka 3857
Oi, Kazunobu 452
Oi, Takeshi 2290
Oishi, K. 376

Oishi, Koji 3012
Oiwa, Takaaki 988
Ojika, Satoshi 1430
Oka, T. 376
Oka, Toshiaki 2330
Okamoto, Dai 3440
Okamoto, Masayuki 3292
Okamoto, Shoji 2290
Okamura, Kazuki 3674
Okazaki, Fumihiro 1728
Okazaki, Yuhei 1586
Okitsu, Takashi 1886
Okubo, Toshikazu 2058
Okuma, Jun 1978
Okuma, Yasuhiro 2834
Okumura, Hajime 3440
Okuyama, Yoshihiro 1811
Omata, Shinpei 2944
Omi, Masataro 1317
Omori, Hideki 1299, 1806, 3341
Omote, Kenichiro 1950
Omura, Mototsugu 1685
Ong, Andrew 2722
Onishi, Mitsuru 1054
Ooishi, Eiji 183
Ooshima, Masahide 1715
Orikawa, Koji 191, 1613, 2277, 3659
Ortiz, G. 1291
Ortiz, Gabriel 1309
Oshima, Ryo 1021
Oshinoya, Yasuo 294, 299, 2972, 3055, 3159, 3162
Oso, Hiroshi 629
Ota, Chiharu 3440
Ota, Satoru 2095
Otsuki, Midori 1054
Ouchi, Takayuki 3920
Ouyang, Shaodi 624
Ouyang, Ziwei 2842
Ozaki, Takayuki 2638
Ozkan, Ziya 498
Pala, Vipindas 2208
Palmour, John 3447
Pan, Miao 582
Panda, S K 1775
Panda, Sanjib Kumar 1580
Pansier, F. 1935
Papafotiou, Georgios 1606
Papastergiou, Konstantinos 1220
Park, Gyeong-Jae 1990
Park, Junsung 1394
Park, Yongsoon 2598
Parker, S. G. 2019, 2372
Partanen, J. 587
Patel, Dhaval 758, 1626
Pedersen, Kristian Bonderup 2547

AUTHOR INDEX

Peftitsis, Dimosthenis 3712
Peltoniemi, P. 587
Peng Gao 2926, 2932
Peng Wang 3124
Peng Wen 782
Peng, Han 2208
Peretti, L. 3111
Peters, A. 2887
Peters, Wilhelm 1508
Petersen, Lars P. 3352
Petersen, Lars Press 2842
Petrich, Matthias 318
Pettersson, Sami 3432
Pham Phu Hieu 3913
Pidaparthy, Syam Kumar 3638
Piepenbreier, Bernhard 1816, 3029
Ping-Heng Wu 1261
Pires, Igor A. 3225, 3270
Pittet, Serge 3371
Pittini, Riccardo 3905
Po-Chien Chou 3425
Poh Chiang Loh 857
Po-Jung Tseng 2810, 3328
Popa, Lucian-Dinut 2996
Popova, L. 548
Popovic, J. 1935
Poshtkouhi, Shahab 2336
Pou, Josep 3758, 3764
Prasanna, I. V. 1580
Prasanna, U. R. 230, 395, 1594
Prasanna, Udupi. R. 2343
Prodic, Aleksandar 3582
Pyrhonen, J. 548
Qi Zhang 3489
Qu, Lizhi 609
Qunzhan Li 1050
Rabkowski, Jacek 3712
Radic, Aleksandar 3582
Radman, Karlo 1691
Rae-Sung Yu 925
Rahman, F. 2679
Rahman, M. A. 982
Rahman, M. F. 2686
Rajashekara, K. 395, 1594
Rajashekara, Kaushik 230, 2343, 3134
Raju, Siddharth 1036
Ramadan, Husam A. 863
Rambetius, Alexander 270, 3029
Rannestad, Bjorn 2547
Rannested, Bjorn 2850
Rathore, Akshay K. 1775
Rathore, Akshay Kumar 2087, 3616
Ray, Olive 2707
Razik, H. 3239
Reiter, Tomas 774

Ren, Kangle 465
Riffat, Javid 2815
Rikitake, Jungo 2216
Rim, Chun T. 1103, 2232, 3358
Rivera, Marco 3574
Robbins, William P 3037, 3061
Rodriguez, Jose 3574
Rongfeng Yang 3680
Rosekeit, Martin 2729
Roth-Stielow, Jorg 264, 318, 3085
Roy, Sudhin 651, 2562, 3286
Ruan, Xinbo 458, 2768
Ruda, Harry E. 1303
Ruderman, Michael 1665
Rufer, Alfred 1081, 1386
Ryo, Mina 3440
Ryu, Sei-Hyung 3726
Saarakkala, Seppo E. 2489
Saga, Yasunao 1748
Sahoo, Ashish Kumar 3750
Saikusa, H. 3007
Saito, Eiichi 1679
Saito, Katsuhiko 2064
Saito, Ryo 3397
Saito, Ryoji 2189
Saito, Takashi 671
Saitoh, Ryoh 914
Sakaba, Kouichi 3159
Sakai, Kazuto 240
Sakai, Tomoyasu 1490
Sakai, Toshifumi 2451
Sakaino, Sho 1978
Sakimoto, Kenichi 1552
Sakurai, Naoki 2297
Sakurai, Takayasu 2228
Sampath, Prasad K. 2722
Sanada, M. 246, 258, 312, 390
Sanada, Masayuki 324, 356, 363, 370, 3018, 3519
Sand, Kjell 1861
Sariyildiz, Emre 2483
Sasaki, Tomotake 2745
Sasongko, Firman 1761
Sato, Daisuke 3815
Sato, Koji 1671
Satria, Andri 3893
Sauer, Dirk Uwe 2729
Sawada, Tadashi 1122
Sayed, Khairy 2359
Sayed, Mahmoud A. 3877
Schob, Reto. T. 1691
Schon, Andre 3255
Schrittwieser, L. 1291
Schupbach, Marcelo 3447
Schuster, Johannes 3085
Segaran, D. S. 2372

AUTHOR INDEX

Segsa, Karl-Heinz 2554
Seilmeier, Markus 1816
Sekiba, Yoichi 2175
Sekisue, Takayuki 2175, 2189
Seo, Gab-Su 2272
Seok-Jin Hong 3532
Seunghoo Song 1646
Seung-Ki Sul 925, 2802
Severson, Eric 1412
Shah, Shahil 843, 967
Shao Zhang 1342
Shaodi Ouyang 2815
Shaofeng Xie 1050
Shaohua Sun 3213
Shaohui Zhong 1155
Shen, Na 582
Shen, Zhiyu 3850
Shenghui Cui 1268
Sheng-Kai Kao 2714
Shi, Hongtao 577
Shi, Rongliang 567
Shibahara, Kohei 1575, 3624
Shibahara, Ryota 2222
Shibanuma, Kenichi 634
Shibata, Yuichiro 3950
Shieh, Hsin-Jang 351
Shieh, Jenn-Jong 204
Shigematsu, Koichi 2183, 2189
Shih, Bing-Jyun 105
Shih, Sheng-Fang 2380
Shih-Jen Cheng 3185
Shimada, Takae 2058
Shimamori, Hiroshi 2745
Shimao, Toshihiro 415
Shimatou, T. 2939
Shimizu, Kyohei 1968
Shimizu, Toshihisa ... 876, 1166, 2944, 2983, 3044, 3771
Shimizu, Toshimasa 1054
Shimode, Daisuke 1122
Shimomura, Junichi 2183
Shimono, Tomoyuki 1685
Shin Shiung Wang 3938
Shin, Hyunhak 110
Shin, Yesl 493
Shinagawa, Syuhei 252
Shinbo, Mitsuo 634
Shindo, Yuji 1552
Shinnaka, Shinji 1824
Shinohara, Atsushi 324
Shinozaki, Ikki 1728
Shinozuka, Yasuhiro 2228
Shioda, Masashi 130
Shirakawa, Kazuhiro 304, 1379
Shiraki, N. 2120
Shirasawa, Koki 3106

Shishida, Yasuhiro 2644
Shiting Weng.. 1476
Shixi Hou 3168
Shoeiby, B. 1482
Shoji, Hiroyuki 2058
Shoyama, Masahito 863, 3386
Shu-Hung Liao 1452
Shuitao Yang.. 1342
Shun-Chung Wang 3233, 3778
Shunke Sui 3680
Shuren Wang 3194
Shu-Wei Kuo 3185
Siemaszko, Daniel 3371
Silva, Marcelo 3864
Silva, Sidelmo M. 3225
Silventoinen, P. 587
Sin, Min-Ho 409
Singh, B. N. 3482
Sintamarean, C. 1912
Sitbon, M. 2240
Sivakumar K... 1400
Siwakoti, Yam P. 1801
Skuriat, Robert 2920
Smadi, Issam 1968
Smaka, Senad....................................... 1148
Smiththisomboon, Somrat 3885
Sogawa, Yuki 522, 2781
Solomon, Adane Kassa 2920
Sone, Kodai 3525
Song Kejian 640
Song, Z. Q. 2686
Sonoda, Hideki...................................... 634
Soo-Cheol Shin 3532
Specht, Andreas 1501
Srirattanawichaikul, Watcharin 3417
Steinert, Daniel 975
Steinke, Gina 1081
Stumpf, Peter 2700
Su, Bonan 614
Su, Hong-Wei 1781
Su, Jianhui 2156
Suetsugu, Tadashi................................ 3955
Sugao, Kazumi 403
Sugimoto, Hiroya 982
Sugiura, Makoto 1135
Suh, Yongsug 1185, 1646
Su-Han Kim 790
Sul, Seung-Ki 1268, 2598
Sumida, Hitoshi 2302
Sun, Jian 843, 967, 2202
Sun, Shaohua 2318
Sun, Wei 609
Sunaga, Keita 3162
Sung, Kyungmin 117, 829
Sunsoon Park 1646

AUTHOR INDEX

Suntio, T. 514, 2240
Suntio, Teuvo 1466
Suryadevara, Rohit 2433
Suto, Kenji 3194, 3194
Suul, Jon Are 1544
Suwankawin, Surapong 3885
Suzuki, Genri 1697
Suzuki, Hirokazu 3503
Suzuki, Katsumi 959
Suzuki, Michiaki 883
Suzuki, Nobuyuki 919, 2541
Suzuki, Ryosuke 2972
Suzuki, Shun 1166
Suzuki, Takashi 1062
Suzuki, Toshiki 907
Svensson, Jan R 1220
Tabira, K. 2939
Tadano, Yugo 1242
Tadokoro, D. 390
Taeck-Kie Lee 3532
Taekyun Kim 3933
Tae-Won Chun 2763
Taga, Hironori 1328
Tajima, G. 2438
Takada, Hiromu 1697
Takagi, Ryo 1328
Takahashi, Akiko 2470
Takahashi, Hiroki 152, 1021
Takahashi, Hirotaka 1068
Takahashi, Hisashi 3106
Takahashi, K. 2569
Takahashi, Naoya 3920
Takahashi, Nobuhiro 3207
Takahashi, Osamu 2285
Takahashi, Takehiro 3207
Takahashi, Yoshikazu 671, 2870
Takamiya, Makoto 2228
Takao, Kazuto 744, 3440, 3707
Takasaki, Mika 2252
Takashita, Haruomi 3386
Takasu, Shinji 3440
Takatsuka, Yushi 1898
Takayama, Masakazu 3801
Takayanagi, Atsushi 2794
Takeda, Kotaro 1654
Takeda, Masashi 801
Takeda, Takashi 1490
Takei, Manabu 3440
Takemoto, Masatsugu 1000, 3525
Takenaka, Kensuke 3440
Takenami, Fumiaki 2737
Takenoiri, S. 2939
Takeshita, Takaharu 123, 601, 2310, 3809, 3877
Takeuchi, Katsutoku 3012
Takeuchi, Shun 3646

Takezaki, Kenichi 3525
Taki, Hiroshi 876, 1379
Takino, Toshiaki 817
Tam Khanh Tu Nguyen 2826
Tamada, Shunsuke 1357
Tamura, Hiroshi 682
Tan, Nadia M. L. 750
Tanabe, Ryo 1234
Tanai, Masanobu 829
Tanaka, Daiki 3179
Tanaka, Junya 240
Tanaka, Kiminori 2222
Tanaka, Koutaro 1006
Tanaka, Seiyu 982
Tanaka, Takahide 2302
Tanaka, Toshihiko 3292
Tanaka, Yasunori 3440
Tanaka, Yuichiro 1880
Tanifuji, Hikaru 1115
Taniguchi, Shun 2465
Tao Meng 3213
Tatsuta, Fujio 959
Tauchi, Yuki 3119
Teng, Jen-Hao 677
Teodorescu, R. 1912
Tera, Takahiro 304, 876
Terabe, Ryosuke 2638
Terao, Yutaka 2644
Teshima, Masato 1068
Thiringer, Torbjorn 1373
Thogersen, Paul 2547
Thogersen, Paul Bach 2850
Tian, Yanjun 3538
Ting, Pangan 677
Ting, Yeh 2787
Tint Soe Win 3292
Toba, Akio 2011
Toda, Hiroaki 1984
Togashi, Ryo 356
Toi, Takahiro 1109
Tokiwa, Tsuyoshi 2977
Tokuda, Hirokazu 2175
Tokumasu, Akira 1379
Tokuyama, Takeshi 2904
Tominaga, Shinji 2290
Tomioka, Satoshi 3630
Tomita, Mutuwo 907
Tonogi, K. 2438
Toru 3784
Tosaka, Shuhei 1207, 1227
Town, Graham E. 1801
Toyoda, Hajime 1560
Tran, Q. V. 446
Trescases, Olivier 2336, 3496
Trillion Zheng 2893

AUTHOR INDEX

Trintis, Ionut 2547
Tripathi, Awneesh......................758, 1626
Trompa, Thomas..............................2554
Tsai, Jiung-Lin 1781
Tsai, Ming-Hsiao 278
Tsan Chen......................................2810
Tse, Chi K. 1425
Tseng, K. J. 2145
Tsorng-Juu Liang 2750, 3346
Tsubakidani, Takashi 3674
Tsuboi, Yoshiki 1160
Tsuboi, Yuichi 92
Tsuchida, Kazuo..............................3397
Tsuda, Junichi 1929
Tsuji, Mineo......................2775, 3093, 3674
Tsuji, Satoshi 522, 2781
Tsuji, Toshiaki 1978
Tsukakoshi, Masahiko 92, 2330
Tsuruma, Yoshinori 1054, 2644
Tsuruta, Hironori 629
Tsuruta, Ryoji 1350
Tsuruta, Yukinori 2266, 3403
Tsuyoshi, Hanamoto 2476
Tu, Yunwu......................................465
Tukiman, Rahayu 2529
Turpin, Santiago 1974
Tuysuz, Arda 1904
Tzou, Ying-Yu 3072
Uddin, Muslem 3574
Ueda, K. 312
Ueda, Tetsuya 403
Ueda, Tetsuzo 2075
Ueda, Yoshinobu 1855
Uemura, Hirofumi 891, 2834
Ukai, Hiroyuki.................................2400
Umeda, Nobuhiro..............................2183
Umeno, Masayoshi 183
Umesh B S 1400
Umetani, Kazuhiro..............................304
Undeland, Tore.................................1412
Uno, Masatoshi 1444, 2351
Urushibata, Hiroaki 2290
Urushibata, Shota 1365, 3455
Ushiro, Nobumasa.............................3106
Vaisanen, V. 587
Vajk, Istvan 2700
Van Brunt, Edward 3726
Van Wyk, J. D. 1935
Van-Long Tran 1214
Vasiladiotis, Michail 1386
Vasquez-Arnez, Ricardo Leon 1276
Veerasamy, Balaji 2310
Vieto, Ignacio 843
Viinamaki, J. 2240
Vilathgamuwa, D. M.....................2722, 3598

Vogt, T..2887
Wada, Keiji 744, 1379, 2511, 3646
Wahlstroem, Jonas 3476
Wajima, Kiyoshi 1984
Wallscheid, Oliver 1501
Wang Hui 640
Wang, Bin 2133
Wang, Chao-Fu 172
Wang, Fei 470
Wang, Fusheng 465
Wang, H. 1912
Wang, Hengli 72
Wang, Jun 944
Wang, Lingxiang 465
Wang, Lipeng...................................458
Wang, Xiaojian 2815
Wang, Xinyu 624, 2815
Wang, Xiongfei............216, 1529, 3320, 3538
Wang, Yanbo 3538
Wang, Yong 470
Wang, Yue 160, 475
Wang, Zhao'An 160, 475
Watanabe, Daisuke 988
Watanabe, Hiroki 84
Watanabe, S. 2939
Watanabe, Shoichiro 1334, 2126
Watashima, T...................................2939
Wei Jiang 3194
Wei Liu 1050
Wei Wang 3680
Wei Yan 3455
Wei, Guo 2318
Wei, Sun 341
Weili Dai.......................................3168
Weirong Chen 3695
Wei-Ting Hsu 2910
Wen, Bo 3850
Wen, Chao-Kai 677
Wen, Huiqing 702
Wen-Chien Hsu 2714
Wenjie Chen 2967
Wen-Tai Li 677
Wheeler, Pat 2920
Won, Chung-Yuen 166, 409
Wong, Siu-Chung 1425
Wonsuk Choi 2914
Woojin Choi 1214
Wu Mingli 640
Wu, Bin 3468
Wu, Chun-Wei 330, 1832
Wu, G. F. 1471
Wu, Gwo-Bin 3796
Wu, T.-F. 2050
Wu, Tsung-Hsi 1796
Wu, Weimin 2610

AUTHOR INDEX

Wu, Weiyang 582
Wu, Wenlong 470
Wu, Wen-Zhe 429
Wunsch, B. 2167
Xia, Huan 2133
Xiangdong Sun 3489
Xiang-Dong Sun 3784
Xiao, D. 2686
Xiao, Fei 66, 72
Xiao, Shuai 543
Xiaojie You 2893
Xiaojie Zhuang 2638
Xiaolong Ma 835
Xiaomei Song 2967
Xie, Ruiliang 2950
Xiong, Li 230
Xiuqin Wei 3955
Xu Cai 1842, 3124
Xu Dianguo 3050
Xu Yang 2967
Xu, David 3099
Xu, Dehong 485
Xu, Dianguo 609, 614, 3002
Xu, Haizhen 567
Xu, Rong 609
Xue, Danhong 3604
Xuling Chen 2768
Yablecki, Jessica 3496
Yachi, Toshiaki 3959
Yakabe, Seichiro 730
Yamada, Hiroaki 538, 1811
Yamada, Kenji 1898
Yamada, Ryuji 1920
Yamada, Takatoshi 2212
Yamada, Tatsuji 3263
Yamagata, Shinichi 829
Yamagishi, Tatsuya 750
Yamaguchi, Shota 3771
Yamaguchi, Takashi 1242
Yamaichi, Katsuya 2517
Yamaji, Masaharu 2302
Yamamoto, Eiji 1654
Yamamoto, Junichi 177, 1179
Yamamoto, Kenji 3106
Yamamoto, Kichiro 689
Yamamoto, Kohei 1438
Yamamoto, Masayoshi 811, 883, 2497
Yamamoto, Shu 3044
Yamamoto, Takashi 3707
Yamamoto, Yasuhiro 1601
Yamamura, N. 146
Yamanaka, Kenji 2152
Yamanaka, Tatsuya 1207, 1227
Yamanoi, Takashi 1062
Yamashita, Nobuyuki 2983

Yamashita, Shigeharu 2745
Yamazaki, Akira 933
Yan Li 3124
Yan Zhang 835
Yanagi, Hiroshige 3630
Yang Chuan 1962
Yang, Cs 199, 2593, 3185
Yang, Daeki 2247
Yang, Geng 543, 582
Yang, Guorun 66
Yang, Hong-Tzer 2100
Yang, Rongfeng 609
Yang, Shih-Sian 2606
Yang, Sihun 863
Yang, Xu 2950
Yang, Zhongping 335, 1322, 2133
Yanhong, Zhang 452
Yano, Yoshihiro 2775
Yanru Zhong 3489
Yaramasu, Venkata 3695
Yashiro, Daisuke 1974
Yashun Li 782
Yasubayashi, Mikio 183
Yasui, Kazuya 2465
Yasumura, Yuji 3568
Yasuno, Takashi 3179
Yau, Y. T. 2754, 3392
Yazdkhasti, Pegah 2156
Yi, Hao 577
Yi-Chun Lin 2714
Yi-Hsun Chiu 3778
Yi-Hua Liu 3778
Yixin Zhu 1155, 3546
Yizhanyi Tang 2977
Yoda, Kazuyuki 1748
Yokoi, Y. 3007
Yokoi, Yuichi 3024
Yokokura, Yuki 1956
Yokoyama, H. 2120
Yokoyama, Natsuki 2285
Yokoyama, Tomoki 3397, 3410
Yonemori, Ryo 689
Yonezawa, Hikaru 3055
Yonezawa, Yoshiyuki 3440
Yonezawa, Yu 2745, 3386
Yong Ding 1476, 3842
Yong, Yu 341
Yong-Cheol Kwon 925
Yongdong Tan 3538
Yongjae Lee 3140
Yoon, Sung Hyun 2272
Yoshida, Morito 3410
Yoshida, S. 2569
Yoshida, T. 2438
Yoshida, Yoshiaki 3503

AUTHOR INDEX

Yoshikawa, Yuichi 2011
Yoshimoto, Kantaro...................2421
Yoshimura, Eiji 1552
Yoshino, Teruo..................2644, 3834
Yoshino, Yukio..................2745
Yoshioka, S. 246
Yoshioka, Takashi..................1956
Yoshizawa, Daisuke 97, 1950
Young-Do Kim 1738, 1743
Youngjoon Choi..................1185
Young-Ryul Kim 796
Yu, Changzhou..................567
Yu, F. Y. 1471
Yu, Ling-Chia 208
Yu, Shuai 2318
Yu, Weikai 647
Yu, Yifan..................1458
Yu, Yong..................567, 609
Yu-Chen Liu..................2810, 3328
Yue Chen..................2768
Yue, Xiaolong 2960
Yu-Jen Wang..................1420
Yu-Kang Lo 2758, 2810, 3185, 3328, 3913
Yuki, Kazuaki 2465
Yukita, Kazuto 1490
Yukutake, Seigo 2140
Yunchang Kwak..................2693
Yun-Chu Chiu 3328
Yung-Ching Huang..................1452
Yunmei Fang 3168
Yunwei Li 3482
Yura, Masashi 2297
Yu-Shan Cheng 3233, 3778
Yuzurihara, Itsuo 2794
Zaitsu, Toshiyuki 177, 1179
Zanma, Tadanao 3568
Zargari, Navid R. 3468
Zehelein, Matthias..................3085
Zeliang Shu 1050
Zeljkovic, Sandra 774
Zhang Wei 3050
Zhang Yajing..................2899
Zhang, Haodong 3604
Zhang, Huiguo..................2202
Zhang, Tao..................485
Zhang, Wei 614, 1425
Zhang, Xing 465, 567
Zhang, Xuning..................2626
Zhang, Yuzhuo 647
Zhang, Zhe..................78
Zhao, Wei..................567
Zhao-Qin Guo 1580
Zhe Wang 3129
Zhe Zhang..................3905
Zheng Dong 3842

Zheng Li 1842
Zheng, T. Q. 807
Zheng, Trillion Q. 2899, 3314
Zhengzhi Han 3124
Zhenyao Xu 994
Zhongping Yang 807
Zhu, B. 395, 1594
Zhu, Honglin 3002
Zhuo, Fang 577, 2960
Zian Qin 857
Zingerli, Claudius M. 3365
Zitouni, Y. 3239
Zong-Zhen Yang..................3778
Zou, Xudong 555
Zwyssig, Christof 1707

9781479927067